DICTIONARY
OF
SCIENTIFIC BIOGRAPHY

PUBLISHED UNDER THE AUSPICES OF
THE AMERICAN COUNCIL OF LEARNED SOCIETIES

The American Council of Learned Societies, organized in 1919 for the purpose of advancing the study of the humanities and of the humanistic aspects of the social sciences, is a nonprofit federation comprising forty-five national scholarly groups. The Council represents the humanities in the United States in the International Union of Academies, provides fellowships and grants-in-aid, supports research-and-planning conferences and symposia, and sponsors special projects and scholarly publications.

MEMBER ORGANIZATIONS

AMERICAN PHILOSOPHICAL SOCIETY, 1743
AMERICAN ACADEMY OF ARTS AND SCIENCES, 1780
AMERICAN ANTIQUARIAN SOCIETY, 1812
AMERICAN ORIENTAL SOCIETY, 1842
AMERICAN NUMISMATIC SOCIETY, 1858
AMERICAN PHILOLOGICAL ASSOCIATION, 1869
ARCHAEOLOGICAL INSTITUTE OF AMERICA, 1879
SOCIETY OF BIBLICAL LITERATURE, 1880
MODERN LANGUAGE ASSOCIATION OF AMERICA, 1883
AMERICAN HISTORICAL ASSOCIATION, 1884
AMERICAN ECONOMIC ASSOCIATION, 1885
AMERICAN FOLKLORE SOCIETY, 1888
AMERICAN DIALECT SOCIETY, 1889
AMERICAN PSYCHOLOGICAL ASSOCIATION, 1892
ASSOCIATION OF AMERICAN LAW SCHOOLS, 1900
AMERICAN PHILOSOPHICAL ASSOCIATION, 1901
AMERICAN ANTHROPOLOGICAL ASSOCIATION, 1902
AMERICAN POLITICAL SCIENCE ASSOCIATION, 1903
BIBLIOGRAPHICAL SOCIETY OF AMERICA, 1904
ASSOCIATION OF AMERICAN GEOGRAPHERS, 1904
HISPANIC SOCIETY OF AMERICA, 1904
AMERICAN SOCIOLOGICAL ASSOCIATION, 1905
AMERICAN SOCIETY OF INTERNATIONAL LAW, 1906
ORGANIZATION OF AMERICAN HISTORIANS, 1907
AMERICAN ACADEMY OF RELIGION, 1909
COLLEGE ART ASSOCIATION OF AMERICA, 1912
HISTORY OF SCIENCE SOCIETY, 1924
LINGUISTIC SOCIETY OF AMERICA, 1924
MEDIAEVAL ACADEMY OF AMERICA, 1925
AMERICAN MUSICOLOGICAL SOCIETY, 1934
SOCIETY OF ARCHITECTURAL HISTORIANS, 1940
ECONOMIC HISTORY ASSOCIATION, 1940
ASSOCIATION FOR ASIAN STUDIES, 1941
AMERICAN SOCIETY FOR AESTHETICS, 1942
AMERICAN ASSOCIATION FOR THE ADVANCEMENT OF SLAVIC STUDIES, 1948
METAPHYSICAL SOCIETY OF AMERICA, 1950
AMERICAN STUDIES ASSOCIATION, 1950
RENAISSANCE SOCIETY OF AMERICA, 1954
SOCIETY FOR ETHNOMUSICOLOGY, 1955
AMERICAN SOCIETY FOR LEGAL HISTORY, 1956
AMERICAN SOCIETY FOR THEATRE RESEARCH, 1956
SOCIETY FOR THE HISTORY OF TECHNOLOGY, 1958
AMERICAN COMPARATIVE LITERATURE ASSOCIATION, 1960
AMERICAN SOCIETY FOR EIGHTEENTH-CENTURY STUDIES, 1969
ASSOCIATION FOR JEWISH STUDIES, 1969

DICTIONARY
OF
SCIENTIFIC BIOGRAPHY

CHARLES COULSTON GILLISPIE

Princeton University

EDITOR IN CHIEF

Volume 11

A. PITCAIRN – B. RUSH

CHARLES SCRIBNER'S SONS · NEW YORK

Copyright © 1970, 1971, 1972, 1973, 1974, 1975, 1976, 1978, 1980
American Council of Learned Societies.
First publication in an eight-volume edition 1981.

Library of Congress Cataloging in Publication Data

Main entry under title:

Dictionary of scientific biography.

  "Published under the auspices of the American Council
of Learned Societies."
  Includes bibliographies and index.
  1. Scientists—Biography. I. Gillispie, Charles
Coulston. II. American Council of Learned Societies
Devoted to Humanistic Studies.
  Q141.D5 1981   509'.2'2 [B]   80-27830
  ISBN 0-684-16962-2 (set)

ISBN 0-684-16963-0 Vols. 1 & 2   ISBN 0-684-16967-3 Vols. 9 & 10
ISBN 0-684-16964-9 Vols. 3 & 4   ISBN 0-684-16968-1 Vols. 11 & 12
ISBN 0-684-16965-7 Vols. 5 & 6   ISBN 0-684-16969-X Vols. 13 & 14
ISBN 0-684-16966-5 Vols. 7 & 8   ISBN 0-684-16970-3 Vols. 15 & 16

Published simultaneously in Canada
by Collier Macmillan Canada, Inc.
Copyright under the Berne Convention.

5 7 9 11 13 15 17 19   V/C   20 18 16 14 12 10 8 6

Printed in the United States of America

Editorial Board

Editorial Staff

Panel of Consultants

Contributors to Volume 11

The following are the contributors to Volume 11. Each author's name is followed by the institutional affiliation at the time of publication and the names of the articles written for this volume. The symbol † means that an author is deceased.

GIORGIO ABETTI
Osservatorio Astrofisico di Arcetri
RICCÒ

S. MAQBUL AHMAD
Aligarh Muslim University
AL-QAZWĪNĪ

D. J. ALLAN
University of Glasgow
PLATO

TORSTEN ALTHIN
Royal Institute of Technology, Stockholm
RINMAN

G. C. AMSTUTZ
University of Heidelberg
POSĚPNÝ; PUMPELLY; ROTH

TOBY A. APPEL
Johns Hopkins University
QUOY; RANVIER

WILBUR APPLEBAUM
Illinois Institute of Technology
ROOKE

ROGER ARNALDEZ
University of Paris
IBN RIDWĀN

JOZEF BABICZ
Polish Academy of Sciences
E. M. ROMER

HANS BAUMGARTEL
RULEIN

WILLIAM B. BEAN
University of Texas Medical Branch
REED

ROBERT P. BECKINSALE
University of Oxford
A. C. RAMSAY; RATZEL; RICHTHOFEN

HEINRICH BEHNKE
University of Münster/Westphalia
RADEMACHER

LUIGI BELLONI
University of Milan
REDI; RIMA; RIVA-ROCCI

RICHARD BERENDZEN
Boston University
POOR; RITCHEY

ALEX BERMAN
University of Cincinnati
POGGIALE; ROBIQUET

A. E. BEST
University of Edinburgh
POURFOUR DU PETIT; ROUGET

BRUNO A. BOLEY
Northwestern University
POLENI; PORRO

J. MORTON BRIGGS, JR.
University of Rhode Island
ROBINS

T. A. A. BROADBENT †
RAMSEY

W. H. BROCK
University of Leicester
PROUT

THEODORE M. BROWN
City College, City University of New York
PITCAIRN

VERN L. BULLOUGH
California State University, Northridge
RATHKE

WERNER BURAU
University of Hamburg
PLÜCKER; REYE; ROHN; ROSANES; J. G. ROSENHAIN

J. J. BURCKHARDT
University of Zurich
RUDIO

E. H. S. BURHOP
University College London
C. F. POWELL

JOHN G. BURKE
University of California, Los Angeles
P. PREVOST; QUENSTEDT

HAROLD L. BURSTYN
William Paterson College of New Jersey
POURTALÈS; REDFIELD

H. L. L. BUSARD
State University of Leiden
PITISCUS; ROOMEN

PERRY BYERLY
University of California, Berkeley
REID

G. V. BYKOV
Academy of Sciences of the U.S.S.R.
A. N. POPOV

JEROME J. BYLEBYL
University of Chicago
RIOLAN

WILLIAM F. BYNUM
University College London
B. W. RICHARDSON; RINGER

R. W. CAHN
University of Sussex
W. ROSENHAIN

LUIG CAMPEDELLI
University of Florence
M. RICCI; RICCIOLI

J. B. CARLES
Institut Catholique de Toulouse
RAULIN

ERIC T. CARLSON
Cornell University Medical Center
RUSH

ALBERT V. CAROZZI
University of Illinois at Urbana-Champaign
RASPE; RENARD

ETTORE CARRUCCIO
Universities of Bologna and Turin
P. RUFFINI

CARLO CASTELLANI
POLI; ROLANDO

PIERRE CHABBERT
Hôpitaux de Castres
PORTAL

JOHN CHALLINOR
University College of Wales
J. PLAYFAIR; PRESTWICH

SEYMOUR L. CHAPIN
California State University, Los Angeles
PONS

GEORGES CHAUDRON
University of Paris
PORTEVIN

FREDERICK B. CHURCHILL
Indiana University
W. ROUX

ARCHIBALD CLOW
ROEBUCK

N. H. CLULEE
Frostburg State College
RULAND

DAVID W. CORSON
University of Arizona
POLINIÈRE

PIERRE COSTABEL
École Pratique des Hautes Études
REYNEAU

J. K. CRELLIN
Wellcome Institute of the History of Medicine
POUCHET; J. REY; R. ROSS

F. A. E. CREW
PUNNETT

CONTRIBUTORS TO VOLUME 11

W. S. DALLAS
A. E. REUSS

J. M. A. DANBY
North Carolina State University
H. C. PLUMMER

GLYN DANIEL
University of Cambridge
PITT-RIVERS

KARL H. DANNENFELDT
Arizona State University
RAUWOLF

A. DELAUNAY
Institut Pasteur
RAMON; P. P. É. ROUX

SUZANNE DELORME
Centre Internationale de Synthèse
A. REY

Y. A. DEMIDOVICH
Academy of Sciences of the U.S.S.R.
POTANIN

JEAN DIEUDONNÉ
POINCARÉ

HÂMIT DILGAN
Istanbul University
QĀḌĪ ZĀDA AL-RŪMĪ

AUBREY DILLER
Indiana University
PYTHEAS OF MASSALIA

YVONNE DOLD-SAMPLONIUS
AL-QŪHĪ

J. G. DORFMAN †
A. S. POPOV; ROZHDESTVENSKY

HAROLD DORN
Stevens Institute of Technology
ROBISON

G. S. DUNBAR
University of California, Los Angeles
RECLUS

JOY B. EASTON
West Virginia University
RECORDE

DAVID E. EICHHOLZ
University of Bristol, England
PLINY

GUNNAR ERIKSSON
University of Umeå
PONTEDERA

VASILIY A. ESAKOV
Academy of Sciences of the U.S.S.R.
PRZHEVALSKY

JOSEPH EWAN
Tulane University
PURSH; RAFINESQUE

JOAN M. EYLES
RAMOND DE CARBONNIÈRES; RENNELL

W. V. FARRAR
University of Manchester
L. PLAYFAIR; K. L. REICHENBACH

MARTIN FICHMAN
Glendon College, York University
PRIVAT DE MOLIÈRES

BERNARD S. FINN
Smithsonian Institution
RÜHMKORFF

DONALD W. FISHER
New York State Museum and Science Service
RUEDEMANN

PAUL FORMAN
Smithsonian Institution
RITZ; C. D. T. RUNGE

ROBERT FOX
University of Lancaster
B. POWELL; REGNAULT

PIETRO FRANCESCHINI
University of Florence
A. RUFFINI

F. FRAUNBERGER
QUINCKE; REGENER

RICHARD D. FRENCH
Science Council of Canada
J. G. PRIESTLEY

HANS FREUDENTHAL
State University of Utrecht
A. PRINGSHEIM; QUETELET; RIEMANN

KURT VON FRITZ
University of Munich
PYTHAGORAS OF SAMOS

J. Z. FULLMER
Ohio State University
REMSEN

GERALD L. GEISON
Princeton University
N. PRINGSHEIM; ROLLESTON

A. GEUS
University of Marburg
PREYER; ROESEL VON ROSENHOF

OWEN GINGERICH
Smithsonian Astrophysical Observatory
REINHOLD

THOMAS F. GLICK
Boston University
RÍO-HORTEGA; RUIZ

STANLEY GOLDBERG
Hampshire College
RIECKE

J. B. GOUGH
Washington State University
RÉAUMUR

EDWARD GRANT
Indiana University, Bloomington
POMPONIUS MELA

I. GRATTAN-GUINNESS
Middlesex Polytechnic of Enfield
RIESZ

FRANK GREENAWAY
Science Museum, London
PLATTNER; POTT; ROBERTS-AUSTEN

A. T. GRIGORIAN
Academy of Sciences of the U.S.S.R.
PRIVALOV

M. D. GRMEK
Archives Internationales d'Histoire des Sciences
ROBIN; ROSTAN

JACOB W. GRUBER
Temple University
PUTNAM

JOHN HALLER
Harvard University
RENEVIER

SAMI K. HAMARNEH
Smithsonian Institution
IBN AL-QUFF

KOKITI HARA
ROBERVAL

RICHARD HART
Boston University
POOR

E. RUTH HARVEY
University of Toronto
QUSṬĀ IBN LŪQĀ

RALPH F. HAUPT
U.S. Naval Observatory
F. E. ROSS

JOHN L. HEILBRON
University of California, Berkeley
RICHMANN

BROOKE HINDLE
Smithsonian Institution
RITTENHOUSE

ERICH HINTZSCHE
Medicohistorical Library, University of Bern
REMAK

E. HLAWKA
University of Vienna
RADON

FREDERIC L. HOLMES
University of Western Ontario
RICHET

WILLIAM T. HOLSER
University of Oregon
RÍO

R. HOOYKAAS
State University of Utrecht
ROMÉ DE L'ISLE

MICHAEL A. HOSKIN
University of Cambridge
ROBERTS

PIERRE HUARD
René Descartes University
RABELAIS

JEAN ITARD
Lycée Henry IV
J. A. RICHARD; L. P. É. RICHARD; ROLLE

CONTRIBUTORS TO VOLUME 11

A. C. JERMY
British Museum (Natural History)
PRESL

DANIEL P. JONES
Oregon State University
F. F. RUNGE

PAUL JOVET
Centre National de Floristique
PLUMIER; POIVRE; RUEL

HANS KANGRO
University of Hamburg
PLANCK; E. PRINGSHEIM; RUBENS

ROBERT H. KARGON
Johns Hopkins University
REYNOLDS; ROSCOE

ALEX G. KELLER
University of Leicester
PLOT; RONDELET

MARTHA B. KENDALL
Vassar College
RIVIÈRE DE PRÉCOURT

HUBERT C. KENNEDY
Providence College
POST

DANIEL J. KEVLES
California Institute of Technology
ROOD; ROWLAND

PAUL A. KIRCHVOGEL
Landesmuseum, Kassel
REPSOLD FAMILY

MARC KLEIN
Louis Pasteur University
RASPAIL

FRIEDRICH KLEMM
Deutsches Museum
POGGENDORFF

OLE KNUDSEN
University of Aarhus
A. O. RANKINE

L. KOLDITZ
Humboldt University of Berlin
ROSENHEIM

ZDENĚK KOPAL
University of Manchester
RHEITA; O. C. RÖMER

SHELDON J. KOPPERL
Grand Valley State College
T. W. RICHARDS; ROWE

ELAINE KOPPELMAN
Goucher College
PLATEAU

HANS-GÜNTHER KÖRBER
Zentralbibliothek des Meteorologischen Dienstes der DDR
PULFRICH

CLAUDIA KREN
University of Missouri, Columbia
ROGER OF HEREFORD

A. D. KRIKORIAN
State University of New York at Stoney Brook
F. J. RICHARDS

VLADISLAV KRUTA
Purkyně University, Brno
PLENČIČ; PROCHÁSKA; PROWAZEK; PURKYNĚ; REICHERT; A. A. RETZIUS; RUDOLPHI

FRIDOLF KUDLIEN
University of Kiel
RUFUS OF EPHESUS

P. G. KULIKOVSKY
Academy of Sciences of the U.S.S.R.
RUMOVSKY

PAUL KUNITZSCH
IBN QUTAYBA

LOUIS I. KUSLAN
Southern Connecticut State College
RAOULT

GISELA KUTZBACH
University of Wisconsin
ROSSBY

P. S. LAURIE
Royal Greenwich Observatory
J. C. ROSS

JOHN E. LESCH
Princeton University
ROMANES

JACQUES R. LÉVY
Paris Observatory
RAYET; ROCHE

JOHN H. LIENHARD
University of Kentucky
PRANDTL

CAMILLE LIMOGES
University of Montreal
PLUCHE; QUATREFAGES DE BRÉAU

DAVID C. LINDBERG
University of Wisconsin
RISNER

STEN LINDROTH
University of Uppsala
RUDBECK

R. B. LINDSAY
Brown University
RICHTMYER

A. C. LLOYD
University of Liverpool
PLOTINUS

JAMES LONGRIGG
University of Newcastle Upon Tyne
PRAXAGORAS OF COS

J. M. LÓPEZ DE AZCONA
Comisión Nacional de Geología
RIBEIRO SANTOS

EDGAR R. LORCH
Columbia University
RITT

ROBERT M. McKEON
Babson College
PRONY

ROBERT J. McRAE
Eastern Montana College
RITTER

MICHAEL McVAUGH
University of North Carolina
RUFINUS

MICHAEL S. MAHONEY
Princeton University
RAMUS; J. K. F. ROSENBERGER

J. C. MALLET
Centre National de Floristique
PLUMIER; POIVRE; RUEL

JEROME H. MANHEIM
California State University, Long Beach
J. H. PRATT

FREDERICK GEORGE MANN
University Chemical Laboratory, Cambridge
POPE

ARNALDO MASOTTI
Polytechnic of Milan
M. RICCI; O. RICCI

OTAKAR MATOUŠEK
Charles University, Prague
RÁDL

SEYMOUR MAUSKOPF
Duke University
PROUST

ERNST MAYR
Museum of Comparative Zoology, Harvard University
RIDGWAY

OTTO MAYR
Smithsonian Institution
PITOT; REDTENBACHER; G. F. VON REICHENBACH; REULEAUX

A. J. MEADOWS
University of Leicester
PRITCHARD; RANYARD

JAGDISH MEHRA
Université Libre de Bruxelles
RAMAN

S. R. MIKULINSKY
Academy of Sciences of the U.S.S.R.
ROUILLIER

LORENZO MINIO-PALUELLO
University of Oxford
PLATO OF TIVOLI

A. M. MONNIER
University of Paris
PORTIER; ROSENBLUETH

GLENN R. MORROW †
PROCLUS

ANN MOZLEY
New South Wales Institute of Technology
RIVETT

xi

CONTRIBUTORS TO VOLUME 11

PIERCE C. MULLEN
Montana State University
RICKETTS

LETTIE S. MULTHAUF
ROTHMANN

ROBERT P. MULTHAUF
Smithsonian Institution
ROLFINCK

SEYYED HOSSEIN NASR
*University of Teheran and Arya Mehr
University of Technology*
QUṬB AL-DĪN

A. NATUCCI
University of Genoa
J. RICCATI; V. RICCATI

J. D. NORTH
University of Oxford
PLASKETT; PROCTOR; RICHARD OF
WALLINGFORD

GORDON W. O'BRIEN
University of Minnesota
H. POWER

HERBERT H. ODOM
*Sir George Williams University,
Montreal*
PRICHARD

OYSTEIN ORE †
RAMANUJAN

A. PABST
University of California, Berkeley
RAMMELSBERG; G. ROSE

JOHN PARASCANDOLA
University of Wisconsin
F. B. POWER

E. M. PARKINSON
Worcester Polytechnic Institute
W. J. M. RANKINE

KURT MØLLER PEDERSEN
University of Aarhus
POISEUILLE

STUART PIERSON
Memorial University of Newfoundland
H. ROSE

P. E. PILET
University of Lausanne
PLATTER; J.-L. PREVOST

MARTIN PINE
*Queens College, City University of New
York*
POMPONAZZI

SHLOMO PINES
The Hebrew University
AL-RĀZĪ

DAVID PINGREE
Brown University
PLANUDES; AL-QABĪṢĪ; RĀGHAVĀNANDA
ŚARMAN; RAṄGANĀTHA; PSELLUS

JAMES H. POTTER
Stevens Institute of Technology
RATEAU

EMMANUEL POULLE
École Nationale des Chartes, Paris
RAYMOND DE MARSEILLES

YVONNE POULLE-DRIEUX
RUFFO

LORIS PREMUDA
University of Padua
RUINI

HANS QUERNER
University of Heidelberg
ROSA

SAMUEL X. RADBILL
College of Physicians of Philadelphia
A. PLUMMER; PRINGLE

VARADARAJA V. RAMAN
Rochester Institute of Technology
P. C. RAY

PAUL RAMDOHR
University of Heidelberg
ROSENBUSCH

RHODA RAPPAPORT
Vassar College
G.-F. ROUELLE; H.-M. ROUELLE; J.
ROUELLE

M. HOWARD RIENSTRA
Calvin College
G. DELLA PORTA

GUGLIELMO RIGHINI
Osservatorio Astrofisico di Arcetri
RESPIGHI; ROSSETTI

GUENTER B. RISSE
University of Wisconsin-Madison
REIL

GLORIA ROBINSON
Yale University
PLATE; F. H. PRATT; I.-B. PRÉVOST;
PRUDDEN; RABL

JOHN RODGERS
Yale University
H. D. AND W. B. ROGERS

FRANCESCO RODOLICO
University of Florence
RISTORO D'AREZZO; ROVERETO

JACQUES ROGER
University of Paris
ROBINET

EDWARD ROSEN
City University of New York
REGIOMONTANUS; RHETICUS; RICHER

K. E. ROTHSCHUH
University of Münster/Westphalia
RUBNER

HUNTER ROUSE
University of Iowa
REECH

GERHARD RUDOLPH
University of Kiel
M. G. RETZIUS

M. J. S. RUDWICK
Free University, Amsterdam
L.-C. PRÉVOST

A. S. SAIDAN
University of Jordan
AL-QALAṢĀDĪ

A. T. SANDISON
University of Glasgow
RUFFER

HANS SCHADEWALDT
University of Düsseldorf
POLLENDER

ROBERT E. SCHOFIELD
Case Western Reserve University
J. PRIESTLEY; ROWNING

CYNTHIA A. SCHUSTER
University of Montana
H. REICHENBACH

JOHN A. SCHUSTER
University of Leeds
ROHAULT

CHRISTOPH J. SCRIBA
Technical University, Berlin
REIDEMEISTER

E. M. SENCHENKOVA
Academy of Sciences of the U.S.S.R.
PRYANISHNIKOV

DIANA M. SIMPKINS
Polytechnic of North London
RAFFLES

H. A. M. SNELDERS
State University of Utrecht
RICHTER; ROOZEBOOM

I. N. SNEDDON
University of Glasgow
ROUTH

PIERRE SPEZIALI
University of Geneva
RICCI-CURBASTRO

NILS SPJELDNAES
University of Aarhus
RENAULT

FRANS A. STAFLEU
State University of Utrecht
REDOUTÉ

WALLACE STEGNER
Stanford University
J. W. POWELL

F. STOCKMANS
Université Libre de Bruxelles
RAMES

D. J. STRUIK
Massachusetts Institute of Technology
RIBAUCOUR

CHARLES SÜSSKIND
University of California, Berkeley
PUPIN

CONTRIBUTORS TO VOLUME 11

LOYD S. SWENSON, JR.
University of Houston
O. W. RICHARDSON

FERENC SZABADVÁRY
Technical University, Budapest
PREGL

GIORGIO TABARRONI
Universities of Modena and Bologna
RIGHI

RENÉ TATON
École Pratique des Hautes Études
POINSOT; PONCELET; POUILLET; PUISEUX

DOUGLASS W. TAYLOR
University of Otago
RAMÓN Y CAJAL

HEINZ TOBIEN
University of Mainz
RECK; A. E. REUSS; F. A. REUSS;
F. ROEMER; F. A. ROEMER

G. J. TOOMER
Brown University
PTOLEMY

THADDEUS J. TRENN
University of Regensburg
W. RAMSAY

F. G. TRICOMI
Academy of Sciences of Turin
PLANA

HENRY S. TROPP
Humboldt State University
RADÓ

G. L'E. TURNER
University of Oxford
RÖNTGEN

JUAN VERNET
University of Barcelona
REY PASTOR

KURT VOGEL
University of Munich
RIES; RUDOLFF

E. H. WARMINGTON
University of London
POSIDONIUS

DIEDRICH WATTENBERG
Archenhold Observatory
POWALKY; ROSENBERG;
O. A. ROSENBERGER

CHARLES WEBSTER
University of Oxford
J. RAY

RODERICK S. WEBSTER
Adler Planetarium, Chicago
RAMSDEN

SIGFRID VON WEIHER
*Sammlung von Weiher zur Geschichte
der Technik*
RÜDENBERG

DORA B. WEINER
Manhattanville College
RENAUT; ROLLET

JOHN W. WELLS
Cornell University
RAYMOND

RONALD S. WILKINSON
Library of Congress
REYNA

A. E. WOODRUFF
Yeshiva University
POYNTING

DENISE WROTNOWSKA
Musée Pasteur
PRUNER BEY

A. P. YOUSCHKEVITCH
Academy of Sciences of the U.S.S.R.
PORETSKY; PRIVALOV; RAZMADZE

BRUNO ZANOBIO
University of Pavia
L. PORTA

DICTIONARY
OF
SCIENTIFIC BIOGRAPHY

DICTIONARY OF SCIENTIFIC BIOGRAPHY

PITCAIRN—RUSH

PITCAIRN, ARCHIBALD (*b.* Edinburgh, Scotland, 25 December 1652; *d.* Edinburgh, 20 October 1713), *medicine, physiology.*

Pitcairn was the son of Alexander Pitcairn, a merchant and magistrate. He was first urged by his father toward a clerical career and entered the University of Edinburgh in 1668 with that intention, graduating M.A. in 1671. By the time of his graduation, however, Pitcairn had already decided on a career in law. He began his legal studies in Edinburgh, then moved to Paris, where his interests widened to include medicine. While in Paris, Pitcairn spent much time with a group of medical students, even joining them on their hospital rounds. He returned to Edinburgh and was stimulated by his close friend David Gregory to take up mathematical studies, which he pursued with verve and some ability. He returned to Paris about 1675 and resumed his medical studies. On 13 August 1680 Pitcairn obtained an M.D. from Reims, then returned to Edinburgh, where in 1681 he joined in the founding of the Royal College of Physicians as probably the youngest of the original fellows. On 4 September 1685 he and another physician, James Halkett, were invited to join Robert Sibbald as the second and third medical professors at the University of Edinburgh. Although Pitcairn does not appear actually to have delivered any lectures, he was awarded the title of professor because of his "abilities and great qualifications."[1] Pitcairn's good fortune was altered with the Glorious Revolution, for his family had long been identified with the Stuart cause; and Pitcairn himself was known as a Jacobite and a scoffer, or at least a skeptic, about puritanical religion. He was also averse to suppressing his strong political views.

Pitcairn was an eager party to the correspondence between his intimate friend David Gregory and Newton, preparing manuscripts, comparing results, and copying letters for Gregory's side of the exchange.[2] In March 1692 Pitcairn visited Newton at Cambridge. Their conversation doubtless covered a wide range of topics, and Newton gave Pitcairn an autograph copy of "De natura acidorum."[3] The essay was not quite complete, and Pitcairn, after adding a few lines and two short remarks, gave the manuscript to Gregory.

Pitcairn's visit to Newton seems to have occurred on his journey to Leiden, where in November 1691 he had been invited to become professor of the practice of medicine at an annual salary of 1,000 guilders.[4] Although the post was a prestigious one, Pitcairn's interest in it probably derived equally from the opportunity it provided to leave Edinburgh honorably. He delivered his inaugural lecture on 26 April 1692. Later that year he was granted the right to lecture on the institutes of medicine and to participate in bedside teaching. All was not success and enthusiasm in Leiden, however. Pierre Bayle reported that Pitcairn's lectures were unpopular because of their abstruse and mathematical character.[5] In the summer of 1693 Pitcairn left Leiden for a holiday in Scotland, from which he never returned.

In 1693, after returning from Leiden, Pitcairn married Elizabeth Stevenson, eldest daughter of Sir Archibald Stevenson, a leading senior figure at the Royal College of Physicians. Pitcairn now became closely identified with Stevenson's group in the College. Fed by both professional rivalries and contemporary political and religious differences, bitter factional disputes continued until, in 1704, Pitcairn and those who sided with him were reinstated at the College and an Act of Oblivion was passed, deleting from the minutes all references to these extended quarrels. Although restored to his fellowship, Pitcairn subsequently attended few College meetings. He was granted a second M.D. by the University of Aberdeen in 1699 and in 1701 was elected to the Royal College of Surgeons (Edinburgh), a rare honor for a physician.

Pitcairn's principal intellectual campaign, for a "mathematical physick," was publicly launched in his inaugural lecture at Leiden. He called upon physicians to abandon all false philosophies and to rely instead in theory and in practice only upon such principles as cannot possibly be called in question by "Mathematicians and Persons who are the least entangled with

Prejudice." His model science was mathematical astronomy. The physician, Pitcairn asserted, must learn to avoid crude hypotheses and "physical causes," like the best mathematical astronomers, Newton and Gregory, who produced the undeniable mathematical laws that interrelate celestial phenomena.

The full meaning of Pitcairn's call for radical medical change became clearer in subsequent lectures at Leiden. He was really aiming at a theory closely similar to that recently published by the Italian iatromechanist Lorenzo Bellini in his *De urinis et pulsibus et missione sanguinis* (1683). Bellini claimed that the prime fact of life was the circulation of the blood and that illness was merely one form or other of impaired circulatory hydraulics. In fevers, for example, there was a fault in the blood, whether in "motion, quantity or quality." A typical symptom, excessive perspiration, resulted from "an augmented Velocity of the Blood . . . [which produces] a greater Number of Particles set at Liberty, and if they can perspire, and be perceived by the Touch, will appear to be greatly augmented in Efficacy and Quantity."[6]

Pitcairn relied heavily on Bellini, but he differed in that he consistently embedded his own theoretical statements in an aggressive context. He asserted that the hydraulic theory of sickness and health was the "true" and "mathematical" account of the animal economy; all others were "false," "hypothetical," and "philosophical." The iatrochemical and iatromechanical systems of such predecessors as Sylvius and Willis had to be abandoned utterly. Pitcairn's system had to be correct, he believed, because the terms with which he constructed his theory—fluids, velocities, dimensions of vessels—were all, at least in principle, observable entities. Pathological and etiological theories constructed from them must, therefore, have the same degree of mathematical certainty as the propositions of Newton's *Principia*.

Pitcairn's radical program also encompassed an assault on the institutes of medicine. Such classic conceptions as "sign," "symptom," and "indicant" had no meaning to him. The only basic pathological condition was impaired circulation, from which all symptoms could be logically and inexorably deduced. Similarly, therapeutic methods could be determined with deductive certainty. Remedies worked not by providing some imaginary "alkali" to neutralize a fictive "acid"; rather, successful therapy was hydraulic in nature, restoring the circulatory system to good order. The physician could either manage the blood by manipulating its secretory products (perspiration, urine) or, occasionally, by adding material of appropriate kind and quantity directly to the bloodstream.

Pitcairn's views provoked intense reactions from his medical colleagues. In Scotland, public reaction largely took the form of rival pamphlets bearing such titles as *A Short Answer to Two Lybels Lately Published . . . by Drs. Cheyne and Pitcairn* (1702). In England, the impact of "mathematical physick" was felt in a less personal, more philosophical way. William Cockburn and Richard Mead, two young physicians who had heard Pitcairn lecture at Leiden, led the way in establishing public support for "mathematical physick." They saw Pitcairn's system as part of the quickly rising Newtonian philosophy. Through their papers and books they helped the "Newtonian" medical system triumph over the "Cartesian" in the early eighteenth-century Royal Society, of which Newton was elected president in 1703. Arguments were advanced along "mathematical" and "corpuscular" lines, and the debate continued for several years. By the second decade of the century, Pitcairn's views were triumphant not only at the Royal Society but also at the Royal College of Physicians (London). While members of the Royal Society and the College of Physicians were being converted to the "mathematical" theory of the animal economy, Stephen Hales and James Keill applied sophisticated, quantitative experimental techniques to the verification and extension of the hydraulic system. Much of Hales's *Haemastaticks* (1733) was conceived in 1708 as an experimenter's response to Pitcairn's new theory.

On the Continent, Pitcairn also exerted a considerable influence. Bellini, from whom Pitcairn derived much of his general style and many specific ideas, was himself stimulated by Pitcairn's efforts for a "mathematical physick." In 1695 he dedicated *Opuscula aliquot* to Pitcairn. Boerhaave was also greatly impressed by Pitcairn, whom he had heard at Leiden. Later, in his lectures and published works he often referred to Pitcairn's ideas and medical program with respect. He thought mechanical and mathematical reasoning an essential part of medicine but differed from Pitcairn in his stress on the need for a studious empiricism in pathology and therapeutics. Albrecht von Haller showed two crucial traces of Pitcairn's influence. Many of his physiological ideas were based essentially on Pitcairn's hydraulic theories; and his methodology, important in his recognition and articulation of the phenomena of irritability, was explicitly Newtonian.

NOTES

1. Robert Peel Ritchie, *The Early Days of the Royall Colledge of Phisitians, Edinburgh*, p. 25.
2. H. W. Turnbull, ed., *The Correspondence of Isaac Newton*, III (Cambridge, 1961), 172.
3. *Ibid.*, 212–213.

4. G. A. Lindeboom, "Pitcairn's Leyden Interlude Described From the Documents," p. 274; further details about Pitcairn's Leiden professorship are also derived from Lindeboom.
5. Quoted in "Archibald Pitcairn," in *Dictionary of National Biography*, new ed., XV, 1221.
6. Lorenzo Bellini, *A Mechanical Account of Fevers* (London, 1720), 89.

BIBLIOGRAPHY

I. ORIGINAL WORKS. Pitcairn's two principal medical works are *Dissertationes medicae* and *Elementa medicinae physico-mathematica*. The first, a collection of such shorter pieces as his inaugural oration at Leiden, was published at Rotterdam in 1701 and at Edinburgh in 1713. It was also translated into English as *Works* (London, 1715) and *Whole Works* (London, 1727). The latter translation was done by George Sewell and J. T. Desaguliers. *Elementa medicinae* purports to present the substance of Pitcairn's lectures at Leiden as recorded in student notes. It was published in 1717 at London and appeared a year later in English trans. Several of Pitcairn's satirical poems, plays, and Latin verses were also published; there is a full listing in the British Museum *General Catalogue of Printed Books*, CXC, cols. 660–661, which also records many of the pamphlets he wrote or provoked. The fullest extant ed. of Pitcairn's various writings, including his medical works, poetry, and some satirical essays, is *Opera omnia medica* (Venice, 1733; Leiden, 1737).

II. SECONDARY LITERATURE. There is no full study of Pitcairn, although there are many useful shorter pieces. The entry in the *Dictionary of National Biography*, new ed., XV, 1221–1223, is quite rich, as is Robert Peel Ritchie's biographical account in *The Early Days of the Royall Colledge of Phisitians, Edinburgh* (Edinburgh, 1899), 159–187. There is an early appreciation by Charles Webster, *An Account of the Life and Writings of the Celebrated Dr. Archibald Pitcairne* (Edinburgh, 1781). G. A. Lindeboom, "Pitcairne's Leyden Interlude Described From the Documents," in *Annals of Science*, **19** (1963), 273–284, provides useful information.

Also of considerable help are accounts of early eighteenth-century iatromechanism and Newtonian philosophy that include references to Pitcairn and his followers. Particularly useful are Kurt Sprengel, *Histoire de la médecine*, A. J. L. Jourdan, trans., V (Paris, 1815), 131–195; Charles Daremberg, *Histoire des sciences médicales*, II (Paris, 1870), 849–887; Theodore M. Brown, "The Mechanical Philosophy and the 'Animal Oeconomy'" (Ph.D. diss., Princeton, 1968), chs. 4–6; and Robert Schofield, *Mechanism and Materialism: British Natural Philosophy in an Age of Reason* (Princeton, 1970), 49–62, 68–79.

THEODORE M. BROWN

PITISCUS, BARTHOLOMEO (*b.* Grünberg, Silesia [now Zielona Góra, Poland], 24 August 1561; *d.* Heidelberg, Germany, 2 July 1613), *mathematics*.

Very little is known of Pitiscus' life. He was court chaplain at Breslau, pursued theological studies in Heidelberg, and for more than a score of the last years of his life he was court chaplain and court preacher for Elector Frederick IV of the Palatinate. Although Pitiscus worked much in the theological field, his proper abilities concerned mathematics, and particularly trigonometry. His achievements in this field are important in two respects: he revised the tables of Rheticus to make them more exact, and he wrote an excellent systematic textbook on trigonometry, in which he used all six of the trigonometric functions.

The word "trigonometry" is due to Pitiscus and was first printed in his *Trigonometria: sive de solutione triangulorum tractatus brevis et perspicuus*, which was published as the final part of A. Scultetus' *Sphaericorum libri tres methodicé conscripti et utilibus scholiis expositi* (Heidelberg, 1595). A revised edition, *Trigonometriae sive de dimensione triangulorum libri quinque*, was published at Augsburg in 1600. It consists of three sections, the first of which comprises five books on plane and spherical trigonometry. The second section, "Canon triangulorum sive tabulae sinuum, tangentium et secantium ad partes radij 100000 et ad scrupula prima quadrantis," contains tables for all six of the trigonometric functions to five or six decimal places for an interval of a minute, and a third section, "Problemata varia," containing ten books, treats of problems in geodesy, measuring of heights, geography, gnomometry, and astronomy. The second enlarged edition of the first and third section was published at Augsburg in 1609. The largely expanded tables in "Canon triangulorum emendatissimus" are separately paged at the end of the volume and have their own title page, dated 1608. The same arrangement as in the first edition occurs in the third edition of Frankfurt (1612). In this edition the "Problemata varia" are enlarged with one book on architecture.

Soon after its appearance on the Continent, the *Trigonometria* of Pitiscus was translated into English by R. Handson (1614); the second edition of this translation was published in 1630; the third edition is undated. Together with these editions were also published English editions of the "Canon" of 1600: "A Canon of Triangles: or the Tables, of Sines, Tangents and Secants, the Radius Assumed to be 100000." There exists also a French translation of the "Canon" of 1600 published by D. Henrion at Paris in 1619. Von Braunmühl remarks in his "Vorlesungen" that in the Dresden library there is a copy of a lecture of M. Jöstel entitled "Lectiones in trigonometriam (Bartholomaei) Pitisci. Wittenbergae 1597," which indicates that the *Trigonometria* was one of the sources for the lectures in trigonometry that were given in the universities of Germany at the close of the sixteenth century.

The first book of the *Trigonometria* considers definitions and theorems from plane and spherical geometry. The names "tangent" and "secant" that Pitiscus used proceeded from the *Geometria rotundi* (Basel, 1583) by T. Finck; instead of "cosinus," Pitiscus wrote "sinus complementi." The second book is concerned with the things that must be known in order to solve triangles by means of the tables of sines, tangents, and secants. This book includes the definitions of the trigonometric functions, a method for constructing the trigonometric tables, and the fundamental trigonometric identities. From the "sinus primarii," that is, the sines of 45°, 30°, and 18°; Pitiscus derived the remaining sines, the "sinus secundarii." Book III is devoted to plane trigonometry, which he consolidated under six "Axiomata proportionum," the first three of which he combined into one in his editions of 1609 and 1612. What other authors designated propositions or theorems, Pitiscus called axioms. The spherical triangle is considered in Book IV, which he drew together in four axioms, the third of which is the sine law; the fourth is the cosine theorem for which Pitiscus was the first to give a real proof (for the theorem relative to angles). By means of these four axioms Pitiscus solved right and oblique spherical triangles. He did not study the polar triangle in this book on spherical triangles but treated it briefly in Book I in much the same way as P. Van Lansberge did. Book V contains such propositions as: "The difference of the sine of two arcs which differ from sixty degrees by the same amount is equal to the sine of this amount." Pitiscus referred to T. Finck and Van Lansberge as also giving this theorem; his proof is the same as the one given by Clavius. After publication in Leipzig of his "Canon doctrinae triangulorum" in 1551, and for at least a dozen years before his death in 1576, Rheticus and a corps of calculators carried on colossal computations in preparing the manuscript for his *Opus Palatinum de triangulis* (Neustadt, 1596). Shortly after the *Opus Palatinum* was published, it was found that the tangents and secants near the end of the quadrant were very inaccurate. Pitiscus was engaged to correct the tables. Because Rheticus seems to have realized that a sine or cosine table to more than ten decimal places would be necessary for such correction, Pitiscus sought the manuscript and finally after the death of V. Otho, a pupil of Rheticus, he found that it contained (1) the ten-second canon of sines to fifteen decimal places; (2) sines for every second of the first and last degree of the quadrant to fifteen decimal places; (3) the commencement of a canon for every ten seconds of tangents and secants, to fifteen decimal places; and (4) a complete minute canon of sines, tangents, and secants, to fifteen decimal places. With the canon (1) in hand Pitiscus recomputed to eleven decimal places all of the tangents and secants of the *Opus Palatinum* in the defective region from 83° to the end of the quadrant. Then eighty-six pages were reprinted and joined to the remaining pages of the great table. In 1607 the whole was issued with a special title page. After his discovery of the new Rheticus tables, Pitiscus started to prepare a second work, *Thesaurus Mathematicus*, which was finally published in 1613 and contained the following four parts: (1) (Rheticus) canon of sines for every 10″ to fifteen decimal places; (2) (Rheticus) sines for 0 (1″) 1°, 89° (1″) 90°, to fifteen decimal places; (3) (Pitiscus) the fundamental series from which the rest were calculated to twenty-two decimal places; and (4) (Pitiscus) the sines to twenty-two decimal places for every tenth, thirtieth, and fiftieth second in the first thirty-five minutes.

BIBLIOGRAPHY

For the full titles of the Pitiscus editions, see R. C. Archibald, "Bartholomäus Pitiscus (1561–1613)," in *Mathematical Tables and Other Aids to Computation*, **3** (1949), 390–397; and "Pitiscus' Revision of the Opus Palatinum Canon," *ibid.*, 556–561.

Secondary literature includes A. von Braunmühl, *Vorlesungen über Geschichte der Trigonometrie*, I (Leipzig, 1900), 221–226; G. J. Gerhardt, *Geschichte der Mathematik in Deutschland* (Munich, 1877), 93–99; N. L. W. A. Gravelaar, "Pitiscus' Trigonometria," in *Nieuw archief voor wiskunde*, V. 3, S. 2 (Amsterdam, 1898), 253–278; and M. C. Zeller, *The Development of Trigonometrie From Regiomontanus to Pitiscus* (Ann Arbor, Mich., 1944), 102–104.

H. L. L. BUSARD

PITOT, HENRI (*b.* Aramon, Languedoc, France, 3 May 1695; *d.* Aramon, 27 December 1771), *hydraulics.*

Pitot was the son of Antoine and Jeanne Julien Pitot. He was born into a patrician family of Aramon, a small town between Avignon and Nîmes. His scholarly career began inauspiciously. In his youth he felt such extreme repugnance against any form of study that his parents, in despair, let him begin a military career, which lasted only briefly. His chance discovery of a geometry text in a Grenoble bookstore led him to return home and spend some three years studying mathematics and astronomy.

In September 1718 Pitot set out for Paris. He was introduced to Réaumur, and won his goodwill. Réaumur advised him on his further studies and let him use his library, and through him Pitot began, at first informally, the association with the Académie des Sciences that was to last more than twenty years. In 1723 Pitot became Réaumur's assistant in the chemical laboratory of the Academy, a position that Pitot filled without neglecting his original interest in geometry. He was promoted *adjoint mécanicien* in 1724 and in 1733 became *pensionnaire géomètre*. In 1740 Pitot accepted an invitation from the Estates General of Languedoc to supervise the draining of swamps in the lower parts of the province. After the successful completion of that task, he became director of public works of one of the three districts of the province and superintendent of the Canal du Languedoc (now Canal du Midi). He lived in Montpellier until his retirement in 1756, when he returned to Aramon. In 1738 he married Marie-Léonine de Saballoua; their only surviving child, a son, was later attorney general in Montpellier.

Pitot's scientific career was both theoretical and practical. During his two decades at the Paris Academy he published a variety of papers dealing with astronomy, geometry, and mechanics, especially hydraulics. The papers offered competent solutions to minor problems, arrived at by elementary mathematics, and without lasting significance. From 1740 to 1756, he was active as a civil engineer. At least two significant engineering projects of his survive—the road bridge attached to the first level of the Pont du Gard, the famous Roman aqueduct near Nîmes, and the aqueduct that supplies drinking water to the city of Montpellier.

Pitot's only book, *La théorie de la manoeuvre des vaisseaux* (Paris, 1731), dealt with a subject of much contemporary interest (books on it had been published in 1689 by Bernard Renau, in 1714 by Johann I Bernoulli, and in 1749 by Leonhard Euler), and was well received; it was translated into English and earned him membership in the Royal Society of London.

Pitot is also remembered for his invention of a simple instrument for measuring the velocity of flowing fluids and of moving ships, which he reported in "Description d'une machine pour mesurer la vitesse des eaux courantes, et le sillage des vaisseaux," in *Mémoires de l'Académie royale des sciences*, **34** (1735), 363–373. The device consisted in principle of a simple glass tube, the end of which was bent at right angles. It was inserted into a river in such a way that the opening faced upstream. Since the water level in the tube rose considerably above that of the river, Pitot reasoned that the difference at the level was equal to the height from which water had to fall to acquire the velocity prevailing at the location of the tube's opening. In other words, the device permitted the determination of the local velocity of fluids directly; that is, without clock and yardstick, and even at locations deep below the surface. The instrument was quickly accepted, and it has remained, under the name "Pitot tube," one of the basic experimental tools of fluid dynamics.

BIBLIOGRAPHY

I. ORIGINAL WORKS. Apart from *La théorie de la manoeuvre des vaisseaux* (Paris, 1731), Pitot published (virtually all in the *Histoire et mémoires de l'Académie royale des sciences*) a considerable number of papers that are listed in Poggendorff, II (1863), 459.

II. SECONDARY LITERATURE. The basic source to all later biographical efforts is Grandjean de Fouchy, "Éloge de M. Pitot," in *Histoire de l'Académie royale des sciences*, **73** (Paris, 1774), 143–157. Discussions of Pitot's contributions to mathematics and hydraulics are Pierre Humbert, "L'oeuvre mathématique d'Henri Pitot," in *Revue d'histoire des sciences et de leurs applications*, **6** (1953), 322–328; and René Chevray, "A Man of Hydraulics: Henri de Pitot (1695–1771)," in *Journal of the Hydraulics Division, American Society of Civil Engineers*, **95** (1969), 1129–1138.

OTTO MAYR

PITT-RIVERS, AUGUSTUS HENRY LANE FOX (*b.* Yorkshire, England, 14 April 1827; *d.* Rushmore, England, 4 May 1900), *archaeology, anthropology, ethnology.*

Pitt-Rivers was known by his father's surname of Lane Fox until 1880, when he inherited the estates of his great-uncle George Pitt, second Baron Rivers, and took his names. He was trained as a soldier in Sandhurst Military College, saw service in the Crimea and India, and specialized in the development of firearms. His main work was on the use and improvement of the rifle, and he was commandant of the Hythe school of musketry. In 1845 he received a commission in the grenadier guards, and in 1882 he became a lieutenant-general. His wide travels and his special military interest in the evolution of firearms drove him to acquire all kinds of artifacts, and to an interest in the comparative study of material culture of primitive and prehistoric societies. His private collections—which soon outgrew his own house—were temporarily housed in the Bethnal Green and South Kensington museums; in 1883 they were moved to the specially created Pitt-Rivers Museum at Oxford. But by then he had started a fresh collection that was housed at Farnham in Dorset. The artifacts were arranged in typological sequences, and Pitt-Rivers was one of the chief nineteenth-century exponents of the very valuable typological method for the study of prehistoric artifacts.

When Pitt-Rivers succeeded to the Rivers estates, he took up residence in Dorset and set to work excavating sites—which included villages, farmsteads, forts, burial mounds, and linear earthworks—on Cranborne Chase. He excavated with meticulous care and, with Flinders Petrie, may be said to have invented the modern technique of British field archaeology. He stressed the importance of recording all finds, especially ordinary things. His accounts of his work and the models he made of his excavations have enabled modern archaeologists to follow his work from beginning to end and to reinterpret his finds in the light of modern knowledge. He insisted on rapid publication, and he had his five-volume *Excavations in Cranborne Chase* (1887–1898) privately printed and distributed free.

In 1876 Pitt-Rivers became a fellow of the Royal Society and in 1886 an Honorary D.C.L. at Oxford. When the Ancient Monuments Preservation Act was made law in 1882, largely at the instigation of his friend Sir John Lubbock (Lord Avebury), he became the first inspector of ancient monuments in Britain; but he soon gave up this appointment, finding the authorities in Whitehall insufficiently cooperative.

Pitt-Rivers had very clear ideas about the cultural process in prehistory and once declared, "History is evolution." His detailed study of British firearms, combined with his belief in the Darwinian concept of evolution, made him formulate the idea that all material objects developed in an evolutionary way and could be arranged in typological sequences. His work on firearms was the model for his thinking, as Henry Balfour, first keeper of the Pitt-Rivers Museum in Oxford said (J. L. Myres, ed., *The Evolution of Culture*, p. 78), and it was from firearms that "he was led to believe that the same principles must probably govern the development of the other arts, appliances, and ideas of mankind."

Pitt-Rivers began what may justly be called a sociological approach to artifacts, whether contemporary or prehistoric. He insisted that his collections were "not for the purpose of surprising anyone, either by the beauty or value of the objects exhibited, but solely with a view to instruction. For this purpose ordinary and typical specimens rather than rare objects have been selected and arranged in sequence." With Flinders Petrie he was a leader not only of new techniques in archaeological excavation but also of the

revolution that led archaeology away from the contemplation of objects of art to the contemplation of all objects.

BIBLIOGRAPHY

For Pitt-Rivers' lectures on the principles of classification, see "The Evolution of Culture," in J. L. Myres, ed., *The Evolution of Culture and Other Essays* (Oxford, 1906). See also the memoir by H. St. George Gray in the privately published *Excavations in Cranborne Chase*, IV.

GLYN DANIEL

PLANA, GIOVANNI (*b.* Voghera, Italy, 6 November 1781; *d.* Turin, Italy, 20 January 1864), *mathematics, astronomy.*

Plana was the son of Antonio Maria Plana and Giovanna Giacoboni. In 1796 his father sent him to complete his studies in Grenoble, where two of his uncles lived. Plana soon became noted for his scientific ability; in 1800 he was admitted to the École Polytechnique at Paris, where he remained for the next three years. This period was decisive in shaping his career, for one of his teachers was Lagrange. In Grenoble, Plana became a close friend of his famous contemporary Stendhal.

In 1803 Plana returned to Italy. Fourier, greatly impressed by Plana, had tried unsuccessfully to procure his nomination as professor of mathematics at the artillery school of Grenoble but managed to obtain a similar post for him at the artillery school in the Piedmont, which was then annexed to France; the school was located partly at Turin and partly at Alessandria. In 1811, on Lagrange's recommendation, Plana was named professor of astronomy at the University of Turin. He remained there until his death, teaching astronomy and infinitesimal analysis, as well as other subjects at the local military academy. For half a century he also directed and stimulated the development of the astronomical observatory of Turin.

Plana is generally considered one of the major Italian scientists of his age because, at a time when the quality of instruction at Italian universities had greatly deteriorated, his teaching was of the highest quality, quite comparable with that of the *grandes écoles* of Paris, at which he had studied.

Plana's scientific contributions cover a wide range: mathematical analysis (Eulerian integrals, elliptical functions), mathematical physics (the cooling of a sphere, electrostatic induction), geodesy (the extension of an arc of latitude from Austria to France), and astronomy (particularly the theory of lunar movement). His study of the moon was inspired by Barnaba Oriani, director of the Brera Observatory in Milan. Oriani had suggested that he and Francesco Carlini, who had done geodetic work with Plana, should attempt to compile reasonably precise lunar tables solely on the basis of the law of universal gravity—that is, using only the observational data essential to determine the arbitrary constants of the problem. Plana soon quarreled with Carlini, who withdrew in disgust; and Plana succeeded alone, after almost twenty years. The results were presented in the three-volume *Théorie du mouvement de la lune* (Turin, 1832). The work was not widely read and received criticism that was not always unfounded; but it is of notable scientific and philosophical value, and as such it was well received.

In 1827 Plana was named astronomer royal; in 1844 he became a hereditary baron; in 1848, a senator; and in 1860, *associé étranger* of the Paris Academy of Sciences.

BIBLIOGRAPHY

A list of about 100 works by Plana is in Poggendorff, II, cols. 460–463. Most of Plana's writings and his portraits are listed in Albert Maquet, "L'astronome royal de Turin, Giovanni Plana (1781–1864); un homme, une carrière, un destin," in *Mémoires de l'Académie royale de Belgique. Classe des sciences*, **36** (1965), fasc. 6.

The following articles were published on the centenary of Plana's death: G. Agostinelli, "Della vita e delle opere di Giovanni Plana," in *Atti dell'Accademia delle scienze*, **99** (1964–1965), 1177–1199; Jacopo Lanzi de Rho, ed., "Ultrapadum," in *Bollettino della Società di storia . . . dell'Oltrepò (pavese)*, **17** (Dec. 1964; pub. Dec. 1966), fasc. 27; and F. G. Tricomi, "Giovanni Plana (1781–1864). Cenni commemorativi," in *Atti dell'Accademia delle scienze*, **99** (1964–1965), 267–279.

F. G. TRICOMI

PLANCK, MAX KARL ERNST LUDWIG (*b.* Kiel, Germany, 23 April 1858; *d.* Göttingen, Germany, 4 October 1947), *theoretical physics, philosophy of physics.*

Planck was the fourth child of Johann Julius Wilhelm von Planck of Göttingen, professor of civil law at Kiel, and Emma Patzig of Greifswald; the family also included two children of Wilhelm von Planck by his first wife, Mathilde Voigt of Jena, who had died in Greifswald. Max Planck's ancestors on his father's side included clergymen and lawyers.

In the spring of 1867 the family moved from Kiel, where Max had completed the first classes of elementary school, to Munich. There he entered the classical Königliche Maximilian-Gymnasium in May 1867. His mathematical talents emerged early, and thus he gratefully recalled his teacher Hermann Müller, who also taught him astronomy and mechanics. Müller's explanation of the principle of conservation of energy made a strong impression on him, and this principle became one of the foundations of Max Planck's later work.

When Planck graduated from the Gymnasium in July 1874, he had not yet decided in what subject he wanted to continue his studies. As a child he had already displayed considerable ability in music; he was an excellent performer on the piano and organ. The encouragement of this talent was one result of an educational tradition that has largely been lost in Germany today: the many-sided stimulation provided by home and school, which was known as *Bildung* ("education," "cultivation," "culture"), and which encompassed religious training just as much as, say, mountain climbing with one's family.[1] From such a background Planck emerged prepared in principle to study the mathematical disciplines, or music, or even classical philology, which also attracted him because of its grammatical and moral harmony. It is not known why he did not choose philology. He probably gave up his musical career when, on seeking advice from a professional musician, he was told: "If you have to *ask*, you'd better study something else!" Nevertheless music remained a lifelong hobby for Planck.

Planck matriculated at the University of Munich on 21 October 1874 and initially decided to study mainly mathematics, influenced by the lectures of Gustav Bauer, who also taught the calculus of variations and probability theory, as well as other subjects.[2] He was soon attracted to physics, although Philipp von Jolly tried to persuade him that nothing essentially new remained to be discovered in this branch of learning.[3] But Planck stood by his rejection of pure mathematics because of his deep interest in questions concerning the nature of the universe[4] ("Weltanschauung"). Student notes[5] show that he attended Gustav Bauer's lectures on "Analytische Geometrie," Ludwig Seidel's course on "Höhere Algebra," Jolly's "Mechanische Wärmetheorie," and the physics lectures of Wilhelm Beetz. Such lectures were predominantly concerned with experimental physics, although lectures titled "Mathematical Physics" can be traced back to the beginning of the nineteenth century.[6] In any case this was the only time in Planck's life when he carried out experiments (for example, on the osmosis of gases).

Because of illness, he had to interrupt his studies during the summer term of 1875. He went to the University of Berlin for the winter semester of 1877–1878 and the summer of 1878. There he heard, in addition to the lectures of Weierstrass, those of Kirchhoff and Helmholtz, although he was not convinced that he learned very much from the latter two. By himself he studied Clausius' *Mechanische Wärmetheorie* in detail and later remarked that this private study was what had finally drawn him into physics.[7] His attempt to master thermodynamics as independently as possible he labeled "Nur nach eigener Überzeugung" ("Only when I have convinced myself"). These investigations led him to the preparation of his doctoral dissertation on the second law of thermodynamics, for which he was awarded the Ph.D. degree at the University of Munich on 28 July 1879. As customary then, he had already, in October 1878, passed the *Staatsexamen für das Höhere Lehramt* for a teaching certificate in mathematics and physics; and he taught these subjects for a few weeks at the Maximilian-Gymnasium.

On 14 June 1880 Planck was given the *venia legendi* at the University of Munich for his paper *Gleichgewichtszustände isotroper Körper in verschiedenen Temperaturen*. In this paper he extended the mechanical theory of heat, using the entropy concept, to treat elastic forces acting on bodies at different temperatures. It may be noted, however, that his habilitation lecture in the same year was "Über die Prinzipien der mechanischen Gastheorie," in accord with the lectures given by his colleagues at Munich. Later, when he was at Kiel and Berlin, he enjoyed a stimulating correspondence on the current problems of thermodynamics with his friend Leo Graetz, then *Privatdozent* at Munich. At Munich he also made friends with Carl Runge,[8] who in later years gave him valuable mathematical assistance.

An appointment as professor extraordinarius at the University of Kiel on 2 May 1885 gave Planck greater scientific independence. Positions of this kind were then rather new in Germany and were restricted primarily to theoretical physics, which did not have a very high status compared to experimental physics.[9] It seems that, as a result, Planck had relatively few students and so had correspondingly more time available for research in his new subject. Yet it is remarkable that in the winter semester of 1887–1888 he announced simultaneously four lecture courses: "Vorträge und Übungen aus der Electricitätslehre," "Theoretische Optik," "Mechanische Wärmetheorie," and "Besprechung wichtiger Literaturerscheinungen auf dem Gebiete der Wärmelehre." These topics, combined with the report of Hertz's recently performed experiments on and simplification of Maxwell's electromagnetic theory, point toward Planck's combination of these fields in his radiation theory in the 1890's.

The appointment at Kiel also gave him some personal security, with an adequate annual salary (2,000 marks), so that he was able to marry his fiancée from Munich, Marie Merck, and establish a household.

In his publications during this period, Planck still concentrated, as he had done in Munich, on applications of his ideas to physical (or "general"[10]) chemistry. After completing his prize essay on *Das Princip der Erhaltung der Energie* (1887), which included a ninety-one-page historical introduction, he turned again to the "second principle" and in three papers tried to generalize it to cover the theory of dilute solutions and thermoelectricity. These studies later culminated in his monograph *Grundriss der allgemeinen Thermochemie* (1893), which had a thirty-one-page historical introduction, and in his *Vorlesungen über Thermodynamik* (1897).

On the basis of these successful researches in thermodynamics (or *Thermomechanik* as it was then called), Planck was appointed on 29 November 1888 to be the successor of Kirchhoff, as assistant professor at the University of Berlin and director of the Institute for Theoretical Physics (newly founded for him). He served as professor ordinarius in Berlin from 23 May 1892 to 1 October 1926. He quickly attained professional recognition; he was at once made a member of the Physikalische Gesellschaft zu Berlin and was elected to the Königlich-Preussische Akademie der Wissenschaften zu Berlin on 11 June 1894. Before 1900 he participated demonstrably in the meetings of the Gesellschaft Deutscher Naturforscher und Ärzte, namely in 1891, 1898, and 1899, on which occasions he took the opportunity to engage in scientific exchanges with Boltzmann. His circle of colleagues included such men as Emil du Bois-Reymond, Hermann von Helmholtz, Ernst Pringsheim, Wilhelm Wien, Max B. Weinstein (who in 1883 had edited a translation of Maxwell's *Treatise on Electricity and Magnetism*), the physicist Carl A. Paalzow of the Technische Hochschule in Berlin-Charlottenburg, August Kundt, Werner von Siemens, and also the theologian Adolph von Harnack, the historian Theodor Mommsen, and the Germanic philologist Wilhelm Scherer. He was, in particular, closely connected with the experimental physicists at the Physikalisch-Technische Reichsanstalt (founded in 1887)—Otto Lummer, W. Wien, Ludwig Holborn, Ferdinand Kurlbaum, and others. An enormous correspondence began to develop with scientists outside of Berlin—H. Hertz, Ernst Lecher, Leo Koenigsberger, A. Sommerfeld, P. Ehrenfest, Albert Schweitzer, and others.

As an admirer of Helmholtz it was appropriate for Planck to combine his physics with music, but in contrast to Helmholtz, tempered scale he preferred the natural scale, and commissioned the construction of a harmonium with 104 tones in each octave. Such interests went hand in hand with private home concerts, in which the violinist Joseph Joachim and Maria Scherer participated.

In his scientific work at Berlin, Planck endeavored to give an independent character to "mathematical physics." An indication of this was the lecture "System der gesammten Physik," in which he followed an approach somewhat similar to that of Kirchhoff.[11] Planck moreover was drawn into discussions on more general ideas of his time. This was a consequence of his striving for generalization. Thus, in 1895, he defended Clausius' form of the second law of thermodynamics against the gross oversimplifications of the "new energetics" of G. Helm, W. Ostwald, and later E. Mach. Planck also was an indefatigable advocate of the absolute validity of laws of nature, a position that places him in the mainstream of the search for absolute constants in the second half of the nineteenth century. His views on both these points already hint at his later interests in the philosophical foundations of science.

The bulk of his systematic work may be divided into thermodynamics, radiation theory, relativity, and philosophy of science. The first culminated in his *Vorlesungen über Thermodynamik*, already mentioned. By 1895 he was fully occupied with irreversible processes, especially in electrodynamics. He connected his early studies on thermodynamic irreversibility with Maxwell's electromagnetic theory of light, in the form given it by Helmholtz, O. Heaviside, and H. Hertz. The theory was at that time the subject of fundamental experimental and theoretical investigations by Berlin scientists, who showed in many lectures on the history of the exact sciences that they considered this new theory of light to be in keeping with the most up-to-date physics. Planck combined Clausius' phenomenological method with Kirchhoff's theorem that light and heat radiation in thermal equilibrium are independent of the nature of the substance, a theorem that filled Planck with enthusiasm. This combination, along with the statistical methods of calculation, was what would lead him in 1900 to the energy elements of the new radiation law. His third major interest, relativity, arose in the winter of 1905–1906 from the publications of Lorentz and Einstein, but not without knowledge of the work

of Poincaré. It is characteristic of Planck that, in 1907, he connected the "principle of relativity" with his quantum of action h.

Planck had always been inclined toward generalization. Encouraged by finding himself in the spotlight of publicity, he now attacked even more general questions in some twenty published popular lectures (as well as in unpublished letters) devoted to "developing and explaining" his "scientific views." This pursuit of general ideas going back to the 1890's really started with his lecture "Einheit des physikalischen Weltbildes" given at Leiden in 1908, and continued in following years with numerous reflections on the relations of science to philosophy, religion, and human nature.

Planck received the high distinction of the 1918 Nobel Prize in physics, but his personal life was clouded by misfortune. His wife died on 17 October 1909, his son Karl during World War I (1916), and his two daughters Margarete and Emma during childbirth (1917 and 1919). His older son from this marriage was executed in 1944 on suspicion of conspiracy to assassinate Adolf Hitler. On 14 March 1911 Planck married the niece of his first wife, Marga von Hoesslin; they had one son, Hermann.

Planck lived through two world wars, and his correspondence with Lorentz, Schweitzer, and others shows that he maintained an uncorrupted independent viewpoint and a positive attitude toward life. In 1944 almost all his manuscripts and books in Berlin were destroyed during an air raid. From 1943 to 1945 he lived in Rogätz, near Magdeburg, and then for the last two and a half years of his life he was in Göttingen, where he witnessed the founding of the Max Planck Gesellschaft zur Förderung der Wissenschaften, successor to the Kaiser Wilhelm Gesellschaft founded in 1911. (He had been its president from 1930 to 1937.)

Thermodynamics. Several principal features characterize Planck's treatment of this, his earliest field of research. Aside from his extremely consistent procedure for extending the theory of cyclic processes to an arbitrary number of arbitrary bodies in 1879, Planck at that time preferred to base his analysis on the distinction between neutral and natural (irreversible) processes. Although he was trying to continue the work of Clausius, he deviated from his predecessor's calculation of "equivalence values," which he eliminated by including all heat reservoirs, and simplified the statement of the second law by requiring the compensation of all losses such as those by friction and heat conduction during a complete change of state, so as to make the process reversible. Planck was still implicitly using the concept of "state" to mean a function of condition depending on temperature, pressure, and volume; in the twentieth century, under the influence of J. Willard Gibbs, this concept would acquire another meaning arising from the theory of elementary regions in phase space. In 1882 Planck did cite Gibbs's first paper of 1873, but he later admitted in his autobiography that Gibbs had anticipated most of his results on this topic.

In 1880, and also at the Versammlung Deutscher Naturforscher und Ärzte at Halle in 1891, Planck conformed closely to Clausius' phenomenological method. He even argued that problems could be solved "without the help of special assumptions about the molecular constitution of bodies," then a basis of his project to determine the effect of temperature in the theory of elasticity. This attitude brought him into protracted internal conflict over the corpuscular hypothesis and the statistical interpretation of nature, which was not yet really established by experiment in the nineteenth century, a conflict that only occasionally came to the surface in his discussions with Boltzmann.

Finally, in 1897, Planck abandoned also the "second" view[12] of the nature of heat—that heat consists of some kind of motion of particles, the precise nature of which was not specified (cf. Clausius 1850)—and turned to the "third" method, in which one "abstains completely from any definite assumption about the nature of heat." He retained that viewpoint even in the 1905 edition of his *Vorlesungen über Thermodynamik*, although he noted that the "principle of increase of entropy" has "no independent significance" but comes rather from "known theorems of the probability calculus." Even after incorporating the Nernst heat theorem (1906), he stated in 1910 that he was leaving the atomic theory "completely out of the picture." In 1912 he still saw both the quantum hypothesis[13] and the Nernst theorem simply as "recent thermodynamic theories."

Planck's *Vorlesungen* was effective for more than thirty years as an exceptionally clear, systematic, and skillful presentation of thermodynamics.[14]

Heat Radiation and Electrodynamics. Planck's contribution to the theory of heat radiation comprised the adroit combination of his studies on irreversibility with the new electrodynamics. A recently discovered manuscript reveals that his interest in heat radiation may have been stimulated by John Tyndall's book *Heat Considered as a Mode of Motion*,[15] on which he made critical notes in 1878.[5] In place of the older concepts "accord" and "discord" for heat absorption, adopted by Tyndall, Planck introduced, particularly in the case of gases, the principle of energy conservation to explain the equilibrium between ether motion and the heated body (later to become his "resonator"). This exchange process was mathematically formulated in the 1890's with the inclusion of radiation damping. In particular, Tyndall's "calorescence"—the opposite of fluorescence—perplexed Planck because of the old question of whether an individual atom has one or more vibration frequencies. He reduced light absorption by solid bodies to the conduction of heat from atom to atom by oscillations. Thus, for Planck in the following years, heat conduction problems were closely connected with radiation phenomena as they had been in the first half of the nineteenth century. He put a question mark against Tyndall's remark that the period of vibration of an atom that corresponds to its maximum amplitude is the same for all bodies; this involved the question of the displacement of maximum wavelength with temperature, about which Planck also made a note on page 569 of Kirchhoff's *Gesammelte Abhandlungen*, which had appeared in 1882.

Soon afterward Planck was influenced by the increasing German interest in Maxwell's theory. It is understandable that, as a theoretical physicist, he would seek to unite his earlier thermodynamic studies with the new theory, an attempt in the spirit of his lecture "System der gesammten Physik." Thus, in his inaugural lecture at the Berlin Academy in 1894, he clearly expressed the hope "that we can also explain those processes which are directly dependent on temperature, as manifested especially in heat radiation, without first having to make a laborious detour through the mechanical interpretation of electricity." Indeed, he had already asserted more clearly in 1891 that "the principle of entropy increase must extend to all forces of nature . . . not only thermal and chemical, but also electrical and other processes."[16] Consequently, in 1895, Planck began just in electrodynamics to prepare for this undertaking by treating the irreversibility of heat radiation, before he introduced entropy into his equations in 1898. Planck's ultimate goal was the investigation of irreversible processes through the study of conservative effects (that is, conservative or radiation damping).[17]

Starting from a concrete Hertzian "secondary conductor" (receiver), he confirmed his first mathematical steps by Vilhelm Bjerknes' experiments with oscillators, also called resonators.[18] Planck then gradually eliminated all their special properties in his calculation, since he knew nothing about their actual nature. Consequently, as soon as the Berlin experimenters succeeded in constructing Kirchhoff's second blackbody, the *bloss Hohlraum*, Planck equated the effect of certain real radiating bodies to that of the "radiation state of the vacuum,"[19] but continued to use the expression "resonator" in his calculations.

He also referred explicitly to W. Wien's cavity radiation of 1894, which Wien partly abandoned in 1896. Like Wien, Planck felt compelled to include certain centers of radiation in addition to the radiation itself, following the approach of W. A. Michelson, who had published the first application of statistical-molecular theory to radiation in 1887. According to Planck's intention in 1897 the resonator was to produce by its vibrations an energy exchange between absorbed and emitted radiation. L. Boltzmann, who had been debating with Loschmidt and others the interpretation of irreversibility, now attacked the equilibrating role of the resonator and so made it central to Planck's considerations.[20] Consequently in 1898 Planck replaced it by a "spectral-analyzing resonator."

He also introduced the concept of natural radiation, the analogue of Boltzmann's assumption of molecular chaos in the kinetic theory of gases. This assumption of randomness in the radiation allowed Planck to establish a causal relation between the energy U of a resonator of frequency ν_0 and the intensity of the surrounding radiation field of the same frequency. Planck derived the equation

$$U = I_0(t) - \frac{1}{2\sigma\nu_0} \frac{dI_0(t)}{dt}.$$

Here σ is the damping constant (logarithmic decrement) of the resonator and $I_0(t)$ measures the energy of the radiation field at the frequency ν_0. Planck had not introduced any molecular statistics at this stage of his work.

In the following year, 1899, Planck completed a derivation of the spectral distribution law for heat radiation, obtaining the form first given by Wien in 1896. For his purpose Planck had to introduce the entropy S of the resonator by means of a definition, writing

$$S = -\left(\frac{U}{a\nu}\right) \ln \left(\frac{U}{eb\nu}\right). \tag{1}$$

U and ν are the energy and frequency of the resonator as before, e is the base of the natural logarithms, and a and b are "two universal positive constants" which appeared here for the first time. The resulting expression for the energy U of the resonator had the form

$$U = b\nu \exp \left(-\frac{a\nu}{\theta}\right), \tag{2}$$

where θ is the absolute temperature. Making use of the relationship between the resonator energy and the spectral intensity of the radiation at equilibrium, a

relationship he derived in this same article, Planck obtained Wien's form of the distribution law,

$$E_\lambda = \frac{2c^2 b}{\lambda^5} \exp \left(-\frac{ac}{\lambda\theta}\right),$$

where c is the velocity of light, λ is the wavelength, and E_λ is the radiant energy per unit volume in a unit interval of wavelength.

Planck evaluated the constants a and b numerically, obtaining for b the value 6.885×10^{-27} erg sec, but he made no attempt to give either constant a physical interpretation.

In this regard there has arisen the myth to which Planck himself in 1901 gave voice: comparing the constants a and b of equation (2) with those in his 1900 energy equation, which is only slightly different from (1):

$$U = h\nu \left\{\exp \left(\frac{h\nu}{kT}\right) - 1\right\}^{-1}, \tag{3}$$

he set h equal to b, k equal to b/a. But these equalities are theoretically (and experimentally) wrong.[21] It is characteristic that until the end of his life Planck continued in this error—apparently because of his bias in favor of theories of a supposedly timeless nature, in contrast to those that come and go in the course of historical change. Thus he made the experiments responsible for the differences between the two sets of constants, writing that "the divergence of the figures corresponds to the deviations in the measurements of the various observers. . . ."

With equation (2), Planck had established Wien's distribution law, which was confirmed by experiment over a wide range of frequencies. By arguing backward from (2), he could even conclude that his definition of the entropy (1) was confirmed. He was clearly not satisfied with this result.

Planck obtained a little-known additional result in 1899: the difference between the absorbed and the emitted energy is given by the equation

$$dU_0 = Z \frac{df}{dt} dt - \frac{8\pi^2\nu_0^2}{3c^3} \left(\frac{df}{dt}\right)^2 dt$$

where Z is the intensity of the exciting wave, and f is the electric moment. Planck concluded that "the absorbed energy would in some circumstances be negative. . . . In this case the 'exciting wave' [Z] would extract energy from the resonator."[22] Einstein treated this situation on the basis of the quantum theory in 1916, when he gave a new derivation of Planck's radiation law. In 1921 Planck gave it the name "negative Einstrahlung." This effect is now known as stimulated emission and is the basis of the "MASER" (microwave amplification by stimulated emission of radiation), invented in 1954.

It is also of interest to note that Planck, like Wien before him, treated the temperature of radiation.[23]

By the end of 1899 Planck noted that the experimental results published by Rubens, Lummer, and E. Pringsheim in September 1899 showed deviations from Wien's law and thus from the predictions of his own theory of oscillators, which he still connected with ponderable atoms.[24] He attempted to save the phenomena (as Rubens had done in 1898) and in March 1900 introduced the second derivation with respect to the resonator energy U:

$$\frac{d^2S}{dU^2} = -\frac{\alpha}{U}.$$

By integrating this expression and applying the definition

$$\frac{dS}{dU} \equiv \frac{1}{T},$$

he arrived again at Wien's law.

We can only offer external reasons as to why in October 1900 Planck made the modification

$$\frac{d^2S}{dU^2} = \frac{\alpha}{U(\beta + U)} \tag{4}$$

and thereby arrived at the new radiation formula. Planck's inference from the behavior of an individual oscillator to the collective behavior of n oscillators was criticized by Lummer and Wien at the Congrès International de Physique at Paris in August 1900, and by E. Pringsheim at the Versammlung Deutscher Naturforscher und Ärzte at Aachen in September 1900. (Unfortunately nothing is known about Planck's conversation with Boltzmann at Munich in September 1899, where he learned from the experimentalists about more significant experimental deviations.) The decisive proof for *curved* "isochromatics" (lines of the temperature function for constant wavelength) against those of Wien's law (straight lines) encouraged the experimenters, who reported it orally in February 1900, although only at the end of September did they publish their results. Stimulated by a visit of Rubens on 7 October, Planck on the very same day wrote down equation (4) together with his new radiation equation. He was already predisposed in October to associate a logarithmic function of U with a probability calculation, presumably influenced by Boltzmann. In any case, Planck admitted in his first paper on the new theory, in October 1900, that the

so-called n-resonator problem disturbed him. He was not able to resolve part of this problem until the end of 1906. In the last instance he was guided by the old principle of greatest simplicity.[25]

Planck, in December 1900, relied on Boltzmann for the statistical basis of his formula for the resonator energy S, and proposed

$$S = k \ln R_0,$$

where R_0 is the maximum number of his "complexions" of a group of resonators with definite frequency. Boltzmann in turn had used the device of approximating a continuum by finite intervals, a tradition going back to the seventeenth century.[26] For the energy U_N of each group of N resonators, Planck introduced a finite number of equal energy elements, $\epsilon = h\nu$, and as his accompanying table shows, these elements were associated with each individual resonator.[27] Or, as he wrote three weeks later, let U_N be conceived as a "discrete quantity [Grösse], compounded of a whole number of equal finite parts." With the help of a combinatorial argument he computed R as

$$\frac{(N + P - 1)!}{(N - 1)! \, P!},$$

where $U_N = NU = P\epsilon$. After simplification by Stirling's formula, and with

$$\frac{dS}{dU} = \frac{1}{T}$$

he obtained U as the function of frequency and temperature already given above in equation (3). The corresponding equation for the spatial density of the spectral distribution of the radiation, Planck's distribution law, had the form

$$u = \frac{8\pi h\nu^3}{c^3} \left\{ \exp\left(\frac{h\nu}{kT}\right) - 1 \right\}^{-1},$$

where u is the energy per unit volume in a unit frequency interval.

At least in 1905 Planck felt that the finite "energy quantum" was "a new hypothesis alien to the resonator theory [of classical electrodynamics]." Thus, in 1910, he abandoned the hypothesis of discrete absorption, and in 1914 he even gave up discontinuous emission. From the calculation in 1910 arose Planck's concept of zero point energy, $h\nu/2$.

In his book *Vorlesungen über die Theorie der Wärmestrahlung*, published in 1906, Planck introduced a new interpretation of his constant, h. He examined the resonator's states with the help of its phase plane, whose axes represent coordinate and momentum. The locus of phase points corresponding to a fixed energy U for a resonator of frequency ν is an ellipse enclosing an area equal to (U/ν). Planck considered a series of concentric ellipses, each having an area exceeding that of its predecessor by the amount h. The energy difference ΔU of successive ellipses would then be given by the equation

$$\frac{\Delta U}{\nu} = h.$$

The total area enclosed by successive ellipses would be h, $2h$, $3h$, $\cdots$. The number of resonators having a definite amount of energy would now become, in the new language, resonators "falling in a *definite* energy region," the size of which depends on h. Within an elementary region h (elliptic ring surface) of the state space, oscillators of different frequencies ν are distributed according to the assumption of elementary disorder, that is, "an almost uniform" distribution, prevails.[28] Henceforth Planck preferred this "quantity of action" to Einstein's "energy quantum." One difficulty was that Planck, in 1906, no longer used the maximum energy but, rather, only the average energy in his calculation.[29]

A supplemental result that should be mentioned is the first proof of the applicability of Maxwell's theory in the infrared region, furnished in 1903 by Planck and his friend the experimentalist Rubens.[30] The attempt at assimilation into the classical theory led Planck in 1911–1912 to the application of a method of identification of parameters in the quantum and classical theories, which Bohr cited in 1913 and called "correspondence" in 1920.

When, in the course of time, more serious consideration was given to rotating dipoles, Planck turned to Adriaan D. Fokker's generalization of the Einstein fluctuation theory and in 1917 proved the basic equation of Fokker's theory.

Relativity Theory. In 1906 Planck was one of the first scientists to take up what he called "the principle of relativity introduced by H. A. Lorentz and stated more generally by A. Einstein," and to extend this theory (to which H. Poincaré had also contributed) from electrodynamics to mechanics.

Thus he showed that one could write for the x component, X, of the force acting on a particle of mass m,

$$X = \frac{d}{dt}\left\{ m\dot{x} \Big/ \sqrt{1 - \frac{q^2}{c^2}} \right\},$$

where $\dot{x}$ is the x component of the particle's velocity and $q = \sqrt{\dot{x}^2 + \dot{y}^2 + \dot{z}^2}$ is the magnitude of that velocity. Planck also showed how these relativistic equations of motion of a particle could be put into Lagrangian and Hamiltonian form by a proper choice of the Lagrangian function, H ("kinetische Potential").

In 1907 Planck clearly stated that the classical separation of the energy into an internal energy of state, independent of the velocity of the body, and an external part that depends only on velocity, could no longer be maintained. He made a connection with his radiation researches by investigating the dynamics of moving blackbody radiation and its relation to the quantity of action, W. He found that this quantity,

$$W = \int_1^2 H\, dt,$$

remains invariant under Lorentz transformations: "to each change in nature there corresponds a definite number of elements of action, independent of the choice of coordinate system."

Planck felt that the relativity principle was experimentally confirmed only by the negative results of the experiment of Michelson and Morley. On the other hand, the measurements of simultaneous magnetic and electric deviations of electron beams by Walter Kaufmann, which involved the dependence of the mass of the electron on its velocity, gave some difficulties. Kaufmann, in January 1906, asserted that his "results are not compatible with the Lorentz-Einstein basic assumption. The equations of Abraham [for a rigid spherical electron, 1903] and of Bucherer [deformable electron of fixed volume, 1904] represent the results of observations equally well." Nine months later, Planck recognized that because of the disagreement of Kaufmann's values with the new Lorentz-Einstein theory, "there is still an essential gap in the theoretical interpretation of the measured quantities," especially since the calculated electron velocity was higher than the speed of light. Toward the middle of the year 1907 Planck changed his calculation of the "apparatus constants" and, with the help of the new value of Adolf Bestelmeyer for the charge-to-mass ratio of the electron, succeeded in arriving at the conclusion that "the chances of relativity theory are somewhat better." This historic episode is yet another demonstration of Planck's close attachment to experiment—in this case, the only positive one available at the time to test relativity.

Philosophy, Religion. Planck's writings on general subjects, published between 1908 and 1937, have received scarcely any historical appreciation.[31] These writings emerged from his occupation with the basis of physical theories. Just as he had found a generalizing synthesis of electrodynamic and thermodynamic principles in the theory of heat radiation, so he was now concerned to comprehend the character of physics as a whole. Having in 1891 ascribed to an ideal process "the role of a pathfinder whose statements have very great generality" even though they lack "probative force"—a role just like that assigned by W. Wien to the thought experiment[32]—Planck suggested in 1894 that "the time is past when one person can deal with both, specialized knowledge [physics] and general knowledge [theory]." He took every opportunity to exhibit the "role of the theoretician in scientific progress." At that time Planck believed that in principle all natural phenomena can be reduced to mechanics, yet he conceded that thermal phenomena could be described by only two nonmechanical laws, and that the connection between electrodynamics and optics, and perhaps also heat radiation, did not depend on mechanics. He postulated "the attainment of a permanently inalterable goal, which rests on the establishment of a single grand connection among all forces of nature"—a foreshadowing of what he called in his 1908 Leiden lecture "the unity of the physical picture of the world" (Einheit des physikalischen Weltbildes).

Along the same line was his search for "natural units" independent of particular bodies, which would "retain their meaning for all times and for all cultures, including extraterrestrial and nonhuman ones."[33] Consequently he turned against Mach's positivism in 1908, and, in his lectures at Columbia University in 1909 ("Das gegenwärtige System der theoretischen Physik"), Planck stressed the path away from observation and anthropomorphism in physics toward a "constant world picture" (Weltbild). He rejected pure subjectivity on the grounds that it would allow any two physicists to maintain two equally valid but different interpretations of a phenomenon, from the standpoint of their different world views. It is remarkable that, by referring to historical examples, Planck supported his ideas on definitions and theorems (1908) and, in 1909, on the unification of empirical knowledge and practice by theory. The characteristic of the theoretician, and especially of one with conservative attitudes like Planck, was frequent reference to history.[34]

On the other hand, Planck warned against overestimating the value of physical theories by applying them to the life of the spirit. In a letter to the theologian Adolph von Harnack in 1914, he clearly separated Weltanschauung, that is, "to grasp the whole in its totality," from Wissenschaft. While "philosophical systems succeed one another, the later one being not necessarily the better . . . ," he wrote, "there

is only one unique science, and this is binding on all mankind . . . it marches forward though it never will and never can attain its ideal goal."

In the rewarding article "Die Stellung der neueren Physik zur mechanischen Naturanschauung," on which he lectured in 1910 at Königsberg, Planck said that theories have tottered under the impact of new experimental techniques and therefore one needs a "working hypothesis" that can "be generated only from an appropriate world view." Since the mechanical world view is no longer acceptable in all areas, for example, in the case of the "aether" (Planck mentioned Nernst's neutrons), one must look for a new world view. Having renounced the requirement of intuitive clarity [Anschaulichkeit], Planck saw that the new physical system of the world would have to be based on the constants of nature as cornerstones. Of course Planck also thought that "if a hypothesis has once proved to be fruitful, one becomes accustomed to it and then little by little it acquires a certain intuitive clarity quite on its own." In 1913 Planck added the ever valid physical principles as invariants in nature, although he admitted that, for example, the principle of immutability of atoms had not remained valid. He now equated the world view to an unprovable hypothesis and recommended that physics also should adopt "faith, at least the faith in a certain reality outside us." This was to be the kernel of Planck's subsequent philosophy, in contrast to his positivistic attitude during the Kiel period, which he confessed had been the basis for his earlier phenomenological methods.

Planck had the whole of the human condition in his purview. Thus, in 1922, shortly after the revolution, he emphasized clearly that in science disputed questions "can not be settled by joint manifestos or even by majority votes"; "the whole of science . . . is an inseparable unity." He drew a contrast between science in itself and the discussion of controversial scientific issues. In the last years of life he wrote again on scientific controversies. He approved of them in principle but warned against personal interest in "dogmatically attempting to defend one's own opinion," to which he attributed "the great majority of scientific controversies."

Closely linked to Planck's conception of an external world are his statements concerning Kausalgesetz and Willensfreiheit (1923). He argued that the contradiction between the two concepts is only apparent. Causality is not subordinate to logic but, rather, is a category of reason (Vernunft). In agreement with Kant, Planck associated causality with metaphysics[35] and assumed that it is valid in nature as well as in mental life; moreover, that it is unprovable. As in statistics, it does not even need to be recognized unequivocally; indeed one cannot get along without the products of the "power of imagination" (Einbildungskraft), which cannot be reduced to causality (for example, concepts such as shortest light-path, virtual motions). Causality itself must be given its appropriate meaning in each individual field of intellectual interest; thus philosophy cannot be placed above the special sciences. Planck's assumption of the lawfulness of nature is presumed, namely, that accuracy and simplicity dominate natural law. In history and psychology Planck attributed to causality the "motive of action"—excepting the "I" since one cannot predict one's own actions on the basis of causality. Within this gap reigns "freedom of the will," including belief in miracles. God alone has insight into man's own causality. Planck supported the moral law, ethical obligation, and the categorical imperative. Such causality demands that men remain responsible to their consciences, even those "whose excessive involvement in immature social theories has disturbed their impartiality and removed their natural inhibitions." Thus each religion is compatible with a rigorous scientific point of view, if it neither comes into contradiction with itself nor with the law of causal dependence of all external processes. Each complements the other. Science also brings to light ethical values, it teaches us veracity (Wahrhaftigkeit) and reverence (Ehrfurcht)—by the "glance at the divine secret in one's own breast."

In 1930 Planck declared that youthful yearning for a comprehensive world view need not decay into the extremes of mysticism and superstition. A science that is not conceived merely rationally invites a faith in the future upon entering into it. Planck elaborated this theme in Die Physik im Kampf um die Weltanschauung (1935), where he placed "abstraction" alongside such faith or working hypothesis and emphasized the utility as well as the ideal character of thought experiments. It was precisely in the inseparability of knowledge from the scholar that Planck saw the favorable influence of science. He addressed himself to "science, religion, and art [including music]" as a whole. Given the abstractions required, he declared that neither science nor ethics can be considered ideally complete.

In his 1937 lecture "Religion und Naturwissenschaft," Planck expressed the view that God is omnipresent and held that "the holiness of the unintelligible Godhead is conveyed by the holiness of intelligible symbols." Atheists attach too much importance merely to the symbols. On the other hand, "understanding without symbols would be impossible." Planck, who from 1920 until his death was a churchwarden at Berlin-Grunewald, professed his belief in an almighty, omniscient, beneficent God, although he did not personify him. The Godhead is "identical in character with the power of natural law." Both science and religion, although starting from different standpoints, wage a "tireless battle against skepticism and dogmatism, against unbelief and superstition," with the goal: "toward God!"

In his last lecture (1946), "Scheinprobleme der Wissenschaft," Planck held that there are more pseudoproblems [Scheinprobleme] "than one commonly assumes." They arise when assumptions are wrong (as in the problem of perpetual motion) or unclear (nature of the electron) or when there is no connection at all between things (such as body and soul). At the end Planck returned to the confusion of viewpoints (for example, pain and wound), his concern of the preceding four decades; but he denied that everything is just a matter of different viewpoints. To such "shallow relativism" he contrasted absolute values: in the exact sciences the absolute constants, in the religious domain truth (Wahrhaftigkeit). Striving toward them was for Planck the task of practical life, the value of which should be recognized by the fruits.

NOTES

PAV = Abhandlungen und Vorträge
PVW = Vorlesungen über die Theorie der Wärmestrahlung

1. Planck participated in this favorite pastime with his family throughout his entire life. It is not surprising, therefore, that his son Karl was coauthor of Führer durch die Mieminger Berge, with R. Burmeister (Munich, 1920).
2. Curriculum vitae, handwritten by Planck on 22 July 1922. I thank Dr. G. Ross of Hildesheim, who made the MS available.
3. Jolly may have remembered the advice once given to him by Ettinghausen in Vienna, who dissuaded him from puzzling his brain over a problem other people had failed to solve; see G. Böhm, Philipp von Jolly (Munich, 1886), 12.
4. A. Hermann, Max Planck (Hamburg, 1973), 11.
5. In the office of the Physikalische Gesellschaft, Magnus Haus, Am Kupfergraben, Berlin, D.D.R.
6. Lectures on mathematical physics had been given long before this at the University of Berlin. In early 1875 G. R. Kirchhoff was appointed professor of this subject there, and at the University of Innsbruck, Ferdinand Peche already had a similar position in 1868.
7. Planck tried without success to contact Clausius; he did not appear to derive much benefit from his correspondence with the mathematician Carl Neumann at Leipzig (son of Franz Neumann).
8. See article in Dictionary of Scientific Biography by Paul Forman.
9. The famous chemist Adolph von Baeyer let Planck know, in his 1879 examination, that he did not think much of theoretical physics (Planck, 1949, 4).
10. See J. R. Partington, A History of Chemistry, IV (London, 1964), 595–596.

11. G. R. Kirchhoff, *Untersuchungen über das Sonnenspectrum und die Spectren der chemischen Elemente*, H. Kangro ed., (Osnabrück, 1972), II, 9.
12. The *first* view was the *statistical*. Cf. Planck, *Vorlesungen über Thermodynamik* (Leipzig, 1897), forewords.
13. Although Planck first created this concept only after 1906, he retained it—among other expressions for it—at least until 1915; but in 1909 he also used "quantum theory," in accord with his philosophical views.
14. Translations : Russian (1900), English (1903), French (1913), Spanish (1922), Japanese (1932); the 11th German ed. appeared in 1966.
15. Planck followed the German translation, Helmholtz and Wiedemann, eds. (Brunswick, 1867), based on the second English ed. (London, 1865).
16. PAV III, 3, and PAV I, 382, respectively.
17. See L. Rosenfeld, "La première phase de l'évolution de la théorie des Quanta," in *Osiris*, **2** (1936), 149–196; M. J. Klein, "Planck, Entropy and Quanta 1901–1906," in *The Natural Philosopher*, **1** (1963), 87–99; and "Thermodynamics and Quanta in Planck's work," in *Physics Today*, **19** (Nov. 1966), 23–27.
18. PAV I, 452 and 484.
19. G. R. Kirchhoff (1972), [1], 42 (300), [2], 16.
20. On "The Development of Boltzmann's Statistical Ideas," see M. J. Klein, in *Acta physica austriaca*, supp. **10** (1973), 53–106; Planck did not support Zermelo's views in all respects.
21. H. Kangro (1970), 144–148. (Cited in Bibliography, below.)
22. PAV I, 562; PVW (1906), 113 and 186; (1913), 156; (1921), 174.
23. PAV I, 594 and 684; II, 756–757; PVW (1906), 127 and 167–168; (1913), 170; (1921), 186; see H. Kangro (1970), 100, 143.
24. H. Kangro (1970), 180.
25. A systematic treatment of the principle was given by Joachim Jungius in the seventeenth century; see H. Kangro, "Organon Joachimi Jungii ad demonstrationem Copernici hypotheseos Keppleri conclusionibus suppositae," in *Organon*, **9** (1973), 169–183.
26. Stephen G. Brush, *Kinetic Theory*, II (Oxford, 1966), 119; *Planck's Original Papers in Quantum Physics*, annotated by H. Kangro, translated by D. ter Haar and S. G. Brush, edited by H. Kangro (London, 1972), n. 32.
27. In 1973 T. S. Kuhn announced to me a different interpretation and his plan to write a book on it, which I have not yet seen. Our discussion stimulates me to cite for the reader's consideration the following passages of sources in the original language: PAV 700–701, 720–721; PVW (1906) 153. PAV II, 244–247 and 452–454. A. E. Haas, J. W. Nicholson, N. Bohr, and W. Wien, among others, followed Planck's view of discrete resonator energies from 1900 on: see *Sitzungsberichte der Kaiserlichen Akademie der Wissenschaften*, Mathematisch-naturwissenschaftliche, Abt. IIa, **119** (1910), 125; *Monthly Notices of the Royal Astronomical Society*, **72** (1912), 677; *Philosophical Magazine*, 6th ser., **26** (1913), 4, and letters of Planck to Wien. Einstein supported this view, which Planck held up to 1909, in *Physikalische Zeitschrift*, **10** (1909), 822, although Einstein himself tried to shift the accent to independent light quanta.
28. PVW (1906), 151; (1913), 136–139.
29. On the question of constructing the entropy maximum, see H. Kangro (1970), chs. 8, 9.
30. H. Kangro, "Ultrarotstrahlung bis zur Grenze elektrisch erzeugter Wellen, das Lebenswerk von Heinrich Rubens," in *Annals of Science*, **26** (1970), 235–259.
31. Pertinent information will be found in H. Hartmann, *Max Planck als Mensch und Denker* (Basel–Thun–Düsseldorf, 1953).
32. *Gedankenexperiment* presumably coined by H. C. Ørsted: see *Berichte über die 9. Versammlung Deutscher Naturforscher und Ärzte* (Hamburg, 1830), 18.
33. M. Planck, in *Sitzungsberichte der Königlich Preussischen Akademie der Wissenschaften* (1899), 479–480; M. J. Klein, in *Physics Today*, **19** (November 1966), 26.
34. M. Planck, in *Kultur der Gegenwart*, I (Leipzig–Berlin, 1915), pt. 3, sec. 3, pp. 692–702, also 714–731; see also Planck's addresses from 1919 to 1930 commemorating the founding of the Academy by Leibniz, in *Max Planck in seinen Akademie-Ansprachen*, Deutsche Akademie der Wissenschaften zu Berlin, ed. (Berlin, 1948). Instructive examples are also in Planck's "Das Wesen des Lichts" (1919) and "Theoretische Physik" (1930), in PAV III, 108–120, 209–218.
35. In 1929 Planck added *axiomatists* to positivists and metaphysicians as "workers on the physical world-picture" but deplored their tendency toward formalism without content; see PAV III, 183.

BIBLIOGRAPHY

I. ORIGINAL WORKS. Poggendorff, III–VIIa, gives a fairly complete bibliography, although it is not free of errors: for example, two titles, "Theorie der hyperkomplexen Grössen" and "Die Formen der Landoberfläche und Verschiebungen des Klimagürtels," are not by Planck. The list in *Max Planck in seinen Akademie-Ansprachen* (Berlin, 1948) is almost complete, lacking only the full number of editions of several works.

Planck's *Abhandlungen und Vorträge*, 3 vols. (Brunswick, 1958), is a selection of commemorative addresses and papers, mainly on physical topics, although vol. III contains important general ones. Other important works are *Vorlesungen über die Theorie der Wärmestrahlung* (Leipzig, 1906; 6th ed., 1966), also in English trans. (Philadelphia, 1914); and *Einführung in die theoretische Physik*, 5 vols. (Leipzig, 1916–1930), also in English (London, 1932–1933), vol. I in Spanish (Cuenza, 1930), and all 5 vols. in Japanese under the general title *Riron butsurigaku hanron* ("Survey of the Theory of Physics"; 1926–1932; 6th ed., 1945—the author is indebted to Tetu Hirosige for this information).

Planck's lectures on general topics are collected in *Acht Vorlesungen über theoretische Physik, gehalten an der Columbia University in the City of New York im Frühjahr 1909* (Leipzig, 1910), also in English trans. (New York, 1915); and *Physikalische Rundblicke. Gesammelte Reden und Aufsätze von Max Planck* (Leipzig, 1922), also in English trans. (London–New York, 1925), enl. as *Wege zur physikalischen Erkenntnis* (Leipzig, 1933), 5th ed., enl., entitled *Vorträge und Erinnerungen* (Stuttgart, 1949), 8th ed. (Stuttgart, 1970), English ed., *The Philosophy of Physics* (New York, n.d. [1936]). Planck's Nobel Prize lecture is *Die Entstehung und bisherige Entwicklung der Quantentheorie* (Leipzig, 1920), English eds. (New York, 1920; London, 1922), see also *Abhandlungen und Vorträge*, III, 121–136.

Planck's orations are listed in a MS prefixed to a collection of his publications in *Sitzungsberichte der Deutschen Akademie der Wissenschaften zu Berlin* and are preserved at the Academy's archives. Selected extracts are in *Max Planck in seinen Akademie-Ansprachen* (Berlin, 1948).

Material on Planck's life includes his curricula vitae of

14 Feb. 1879, given by A. Hermann, in *Max Planck in Selbstzeugnissen und Bilddokumenten* (Hamburg, 1973), 15—the MS is presumably in the archives of the University of Munich; of 1920, in *Abhandlungen und Vorträge*, III, 135–136; and of 22 July 1922, "An die Akademie der Wissenschaften in Wien," in the possession of G. Roos. Other autobiographical source materials are "Persönliche Erinnerungen" (1935), in *Abhandlungen und Vorträge*, III, 358–363; and "Zur Geschichte der Auffindung des physikalischen Wirkungsquantums" (1943), *ibid.*, 255–267, also in *Vorträge und Erinnerungen* (Stuttgart, 1949), 15–27. Historically valuable sources are "Persönliche Erinnerungen aus alten Zeiten," in *Naturwissenschaften*, **33** (1946), 230–235, also in *Vorträge und Erinnerungen* (Stuttgart, 1949), 1–14; *Erinnerungen*. 1 (Berlin, 1948), published as a MS; and *Wissenschaftliche Selbstbiographie* (Leipzig, 1948), also in *Abhandlungen und Vorträge*, III, 374–401, English eds. (New York, 1949; London, 1950), French ed. (Paris, 1960), and published under the title *Scientific Autobiography and Other Papers* (New York, 1968).

A phonograph record, *Über die exakte Wissenschaft*, in the series Stimme der Wissenschaft, contains, in addition to speeches made on the occasion of Planck's eightieth birthday (1938), an intro. by W. Gerlach, Planck's comments, and his lecture "Über die exakte Wissenschaft" (Mar. 1947).

None of the existing lists of Planck's manuscripts is even nearly complete. Some are included in T. Kuhn, J. L. Heilbron, P. Forman, and L. Allen, eds., *Sources for the History of Quantum Physics* (Philadelphia, 1967), which is not free of errors. See also A. Hermann, *Frühgeschichte der Quantentheorie* (Mosbach, 1969), 36–37; and *Max Planck in Selbstzeugnissen und Bilddokumenten* (Hamburg, 1973), 130–134; and H. Kangro, *Vorgeschichte des Planckschen Strahlungsgesetzes. Messungen und Theorien der spektralen Energieverteilung bis zur Begründung der Quantenhypothese* (Wiesbaden, 1970), 251.

A substantial number of copies of Planck's letters are at the Library of the American Philosophical Society, Philadelphia; the Center for History of Physics, American Institute of Physics, New York; the Bibliothek des Deutschen Museums, Munich; and the Staatsbibliothek Preussischer Kulturbesitz, MS Div., West Berlin, which has 149 recently acquired letters from Planck to Wien (1900–1928), typewritten copies of three letters from Wien to Planck (12 June 1914, 1 May 1915, and 12 Feb. 1916); to H. A. Lorentz (22 Jan. 1914); and from Wilhelm Hallwachs to Wien (10 June 1914). Letters, cards, and other documents are at the Max Planck Gesellschaft zur Förderung der Wissenschaften e. V., Munich. A useful guide for locating MSS, still being compiled, is *Zentralkartei der Autographen*, at the Staatsbibliothek Preussischer Kulturbesitz.

Photographs of paintings, medals, and a 30-pfennig postage stamp are listed in the mimeographed typewritten catalog *Max Planck (1858–1947), Gedächtnisausstellung zum 20. Todesjahr in der Staats- und Universitäts-Bibliothek Hamburg vom 5.4. bis 13.5. 1967 und in der Universitäts-*

bibliothek Kiel vom 22.5 bis 10.6. 1967, V. Wehefritz, ed. (n.p., n.d. [Hamburg, 1967]). A considerable number of photographs are at the Max Planck Gesellschaft, Munich. Books from Planck's own library, some with his marginal notes (partially in *Gabelsberger* stenography), are at the Physikalische Gesellschaft, Berlin, D.D.R.

On Planck's activity as an editor, see *Max Planck, Gedächtnisausstellung. . .*, V. Wehefritz, ed. (see above), 4–5; *Max Planck in seinen Akademie-Ansprachen* (Berlin, 1948), 199—which is incomplete; and Poggendorff, VI–VIIa.

II. SECONDARY LITERATURE. Poggendorff, VI–VIIa, gives an uncritical bibliography, mostly of biographical works. See also M. Whitrow, ed., *Isis Cumulative Bibliography*, II (London, 1971), which covers the period 1913–1965, and the annual *Isis* bibliographies from 1966 on. The works by A. Hermann and H. Kangro, cited above, also include lists of recent secondary material. On genealogy, see Otto Kommerell, "Die Planck in Untergruppenbach," in *Südwestdeutsche Blätter für Familien- und Wappenkunde*, **11** (1960), 77–85; "Südwestdeutsche Ahnentafeln in Listenform"; Max Planck, in *Blätter für Württembergische Familienkunde*, nos. 1–9 (Stuttgart, 1921–1942), *passim*; and Lothar Seuffert, "Planck, Johann Julius Wilhelm von," in A. Bettelheim, ed., *Biographisches Jahrbuch und deutscher Nekrolog*, V (Berlin, 1903), 14–18.

Less well-known biographical studies are G. Grassmann, "Max Planck," in *V. [ereins] Z. [eitung] des Akademischen Gesangvereins München i. S.V.*, spec. no. "Max Planck" (Munich, n.d.); and Bernhard Winterstetter, "Zum 100. Geburtstag von Max Planck," in *Stimmen aus dem Maxgymnasium*, **6** (1958), 1–6. A. Hermann's recent biography, *Max Planck in Selbstzeugnissen und Bilddokumenten* (Hamburg, 1973) includes anecdotes and forty portraits of Planck. Also worthwhile is the personal recollection of Lise Meitner, "Max Planck als Mensch," in *Naturwissenschaften*, **45** (1948), 406–408; cf. "Lise Meitner Looks Back," in *Advancement of Science*, **20** (1964), 39–46.

On Planck's introduction of the quantum of action, see L. Rosenfeld, *Max Planck Festschrift* (Berlin, 1959), 203–211; M. J. Klein, "Max Planck and the Beginnings of Quantum Theory," in *Archive for History of Exact Sciences*, **1** (1962), 459–479; A. Hermann, *The Genesis of Quantum Theory (1899–1913)* (Cambridge, Mass., 1971), translated from *Frühgeschichte der Quantentheorie* (Mosbach, 1969); and H. Kangro, *Vorgeschichte des Planckschen Strahlungsgesetzes. Messungen und Theorien der spektralen Energieverteilung bis zur Begründung der Quantenhypothese* (Wiesbaden, 1970).

On other aspects of Planck's work, see E. N. Hiebert, "The Concept of Thermodynamics in the Scientific Thought of Mach and Planck," in *Wissenschaftlicher Bericht, Ernst Mach Institut* (Freiburg im Breisgau), **5** (1968); and T. Hirosige and S. Nisio, "The Genesis of the Bohr Atom Model and Planck's Theory of Radiation," in *Japanese Studies in the History of Science*, **9** (1970), 35–47.

HANS KANGRO

PLANUDES, MAXIMUS (*b.* Nicomedia, Bithynia, *ca.* 1255; *d.* Constantinople, 1305), *polymathy.*

Manuel Planudes, as he was named by his well-to-do family, was probably brought to Constantinople shortly after Michael VIII recovered the city from the Latins in 1261. When he became a monk, in or shortly before 1280, he took the name Maximus, by which he is now known. He was a leading intellectual and teacher in Byzantium in the early Palaeologan period, first at the Chora monastery and, after *ca.* 1300, at the Akataleptes monastery.

His knowledge of Latin was sufficient for him to translate many theological and classical works into Greek, including á translation of Macrobius' *Commentum in somnium Scipionis.* It has also been claimed that he is the translator of the Aristotelian *De plantis* from the Latin of Alfred Anglicus. But his primary scholarly interest was the editing of texts and the training of younger scholars, such as Demetrius Triclinius and Manuel Meschopulus, to carry on this work. His school was concerned mainly with editing and commenting on the Greek poets and dramatists, although it did not neglect authors of prose. His editions include Aratus' *Phaenomena,* with scholia; Theodosius' *Sphaerica;* Euclid's *Elements;* Ptolemy's *Geography;* pseudo-Iamblichus' *Theologumena arithmeticae;* and Diophantus' *Arithmetica,* with scholia on the first two books.

Another scholarly activity in which Planudes delighted was the compilation of florilegia. His anthology of Greek poetry, the *Anthologia Planudea,* is well known; and his *Very Useful Collection Gathered From Various Books* preserves much of what we possess of John Lydus' *De mensibus.*

For historians of science his most important original work is *Calculation According to the Indians, Which Is Called Great,* which deals with addition, subtraction, multiplication, division, sexagesimals, and squares and square roots. Although it contains little original material, it is significant for its use of the eastern Arabic form of the Indian numerals. Planudes' source may have been his contemporary, Gregory Chioniades, who had studied astronomy at Tabriz in the early 1290's and was using the same forms of the Indian numerals in Byzantium in 1298–1302. The medical works sometimes attributed to Planudes— that on uroscopy, for example—were probably written by Nicophorus Blemmydes.

BIBLIOGRAPHY

The fundamental articles remain C. Wendel, "Planudea," in *Byzantinische Zeitschrift,* **40** (1940), 406–445; and his article in Pauly-Wissowa, XX, 2202–2253.

DAVID PINGREE

PLASKETT, JOHN STANLEY (*b.* Hickson, near Woodstock, Ontario, 17 November 1865; *d.* Esquimalt, near Victoria, British Columbia, 17 October 1941), *astronomy.*

Plaskett's chief contributions to astronomy were in instrumental design and the supervision of a lengthy program of observation, especially of spectroscopic binary stars.

Plaskett was the son of Joseph and Annie Plaskett. He was educated at the local village school and the Woodstock High School. He left school to work on the family farm but soon was employed by a mechanic in Woodstock. He then went to the Edison Electric Company, first in Schenectady, New York, and later in Sherbrooke, Quebec. In 1889 Plaskett was a mechanic in the department of physics at the University of Toronto. He took the opportunity first to matriculate and then to take up undergraduate studies, which he followed concurrently with his other work. In 1892 he married Rebecca Hope Hemley, and with her encouragement he eventually graduated in physics and mathematics in 1899. He remained a mechanic until 1903, but in his final years at Toronto he was engaged in research in color photography.

From 1903 Plaskett was on the staff of the new Dominion Observatory in Ottawa. He worked with the spectrograph and measured radial velocities with the fifteen-inch reflector; and after making a careful study of the mechanical problems involved and discussing the matter with the staffs of many American observatories, he designed a new spectrograph for the reflector. It is claimed that his design so increased the speed of this instrument that for radial velocity work it became the equal of the Yerkes refractor.

After persistently recommending that the Canadian Parliament build a seventy-two-inch reflector, Plaskett prevailed and the contracts were signed in 1913. Much of the design was Plaskett's own. In 1917 he was appointed director of the Dominion Astrophysical Observatory in Victoria, British Columbia, which opened in 1918. Again he and his staff concentrated on the observation of radial velocities. New binaries were discovered and their orbits measured, these binaries including "Plaskett's twins." This star system, which Plaskett was the first to realize was not a single star (B.D. + 6° 1309), was long the most massive known (1922). He gave particular attention to early type O and B stars, and he studied the motion and distribution of interstellar calcium. But perhaps his most notable achievement was his application of spectroscopic evidence to the problem of galactic rotation and the distance and direction of the center of gravity of the galaxy, those being determined from the motions of stars of spectral types O5 to B7 (1930).

Plaskett retired in 1935. He was elected a fellow of the Royal Society, was president of the Royal Astronomical Society of Canada, and received medals from six societies in North America and Great Britain. After his retirement he supervised the grinding and polishing of the eighty-two-inch mirror for the McDonald Observatory at the University of Texas. One of his two sons, Harry Hemley Plaskett, became Savilian professor of astronomy at Oxford.

BIBLIOGRAPHY

I. ORIGINAL WORKS. C. S. Beals *et al.*, "Bibliography of the Published Papers of J. S. Plaskett," in *Journal of the Royal Astronomical Society of Canada*, **35** (1941), 408–411, lists 85 works. Those cited in the text are "A Very Massive Star," in *Monthly Notices of the Royal Astronomical Society*, **82** (1922), 447–500; "The Radial Velocities of 523 O- and B-Type Stars Obtained at Victoria," in *Publications of the Dominion Astrophysical Observatory, Victoria, B.C.*, **5** (1931), 1–98, written with J. A. Pearce; and "A Catalogue of the Radial Velocities of O- and B-Type Stars," *ibid.*, 99–165, written with J. A. Pearce. Plaskett's most substantial memoirs appeared in vols. **1–5** (1920–1934) of this journal.

II. SECONDARY LITERATURE. See A. C. D. Crommelin, "Address . . . on the Award of the Gold Medal of the Society . . .," in *Monthly Notices of the Royal Astronomical Society*, **90** (1930), 466–477. Obituaries are F. S. Hogg, *ibid.*, **102** (1942), 70–73; R. O. Redman, in *Observatory*, **64** (1942), 207–211; and R. F. Sanford, in *Astrophysical Journal*, **98** (1943), 137–141, with portrait. See also the article by R. O. Redman in *Dictionary of National Biography*, supp. 6 (1941–1950), 675–676.

J. D. NORTH

PLATE, LUDWIG HERMANN (*b.* Bremen, Germany, 16 August 1862; *d.* Jena, Germany, 16 November 1937), *zoology.*

Plate was a proponent and defender of Darwin's theory of evolution who allowed the inheritance of acquired characters and certain other factors to supplement natural selection.

Plate's parents were Heinrich Plate, a language teacher, and the former Phoebe Hind. As a boy, Plate collected plants and animals and attended lectures of the local group of natural scientists. He studied mathematics and natural science at the University of Jena, where he was fascinated by Haeckel's evolutionary theories, although he considered the concepts too dogmatic. At Bonn and Munich, Plate studied invertebrate zoology under Richard Hertwig and

completed his dissertation in 1885, returning to take the examinations for his doctorate at Jena. He qualified as lecturer at Marburg in 1888 after further study and became *Privatdozent* there.

From 1898 he was titular professor at Berlin, where he was named curator of the Museum für Meereskunde in 1901; he was also ordinary professor at the Landwirtschaftliche Hochschule. Plate's museum experience especially impressed Haeckel, whom he succeeded in 1909 as professor of zoology and director of the Phyletische Museum at Jena. Haeckel, who had strongly favored Plate's appointment, expected to have his wishes followed; but bitter dissension soon broke out between the two. No doubt there were contributing factors in both personalities, but the former pupil and protégé did his best to derogate Haeckel and to make his position untenable.

Under Plate's administration the facilities of the museum and the Zoological Institute were expanded and the collections augmented, in part with material gathered in the course of his travels. He made expeditions to South America (1893–1896), Greece and the Red Sea area (1901–1902), and the West Indies (1904–1905). His trip to Ceylon and India (1913–1914) yielded data and specimens for research and museum use at Berlin and Jena.

Plate was a founder and the coeditor for biology of the *Archiv für Rassen- und Gesellschaftsbiologie* from its establishment in 1904. The publication was dedicated to the study of the laws of variability and inheritance, of the development of races, and the relation of the family, society, and the state to "racial and social hygiene," a goal perhaps initially eugenic but later having increasingly political application. Plate's views on zoological matters became entwined with his political opinions; and for years he digressed freely during his biological lectures to express his convictions as an avowed member of the Right, as a Pan-Germanist, and as an anti-Semite. He expounded his ideas on the labor movement and was a declared pro-Fascist long before 1933.

Plate married Hedwig von Zglinicki in 1902, and after her death he married Klara König (1933). He became professor emeritus in 1934.

Plate's most influential work was in the exposition and defense of Darwin's theory of natural selection; he also worked in genetics and studied problems of Mendelian heredity. In the wake of the enthusiasm engendered by Darwin's *Origin of Species*, new uncertainties had arisen when many observers had experienced difficulties in applying the single all-encompassing theory. The mechanism of evolution was still unclear, and the post-Darwinians of the late nineteenth and early twentieth centuries devised various syntheses. Some biologists thought that Darwin's theory of natural selection had been superseded, while among those who considered themselves faithful Darwinists two factions developed. The strict selectionists, such as Weismann, argued that natural selection alone could account for the evolutionary process; others maintained, when they supported natural selection with auxiliary theories, that they were actually following in Darwin's footsteps—for had not Darwin himself admitted some influence of external conditions upon heredity and a certain degree of Lamarckism?

Plate, competent and painstaking in his zoological investigations, was a member of the latter group although he was a staunch Darwinist. He examined the theory of natural selection as it might apply to numerous specific instances in which evolutionary changes were discernible. Plate experimented and evaluated the various theories of heredity and evolution of his colleagues before coming to his own neo-Darwinist conclusions. His studies were widely cited by his contemporaries; and his *Über Bedeutung und Tragweite des Darwin'schen Selectionsprincips* (Leipzig, 1900), first published the previous year in *Verhandlungen der Deutschen zoologischen Gesellschaft* was enlarged and, with change of title, in its fourth edition by 1913. Plate also wrote numerous articles on evolution, genetics, the Mollusca (*Chiton*), and on more general zoological subjects. His texts appeared in various editions into the 1930's. He was a member of scientific academies in Germany, Hungary, and Sweden.

BIBLIOGRAPHY

I. ORIGINAL WORKS. The work on evolution cited in text was retitled *Selectionsprinzips und Probleme der Artbildung. Ein Handbuch des Darwinismus*, 3rd ed. (Leipzig–Berlin, 1908; 4th ed., 1913). "Vererbungslehre und Descenztheorie," in *Festschrift zum sechzigsten Geburtstag Richard Hertwigs*, II (Jena, 1910), 535–610, was Plate's inaugural lecture as professor at Jena. *Allgemeine Zoologie und Abstammungslehre*, 2 vols. (Jena, 1922–1924), is a more general text. Plate gave a brief account of his life and outlook in "Kurze Selbstbiographie," in ·*Archiv für Rassen- und Gesellschaftsbiologie*, **29** (1935), 84–88.

II. SECONDARY LITERATURE. Vernon L. Kellogg, in *Darwinism Today* (New York, 1908), 145, 147, 166, 188 ff., 203, 247, 264, 272–273, 278–279, discusses Plate's ideas and studies, and outlines some of the problems that concerned him in relation to the climate of post-Darwinian evolutionary theory in the early 1900's. The best account of Plate's life and work is in Georg Uschmann's *Geschichte der Zoologie und der zoologischen Anstalten in Jena 1779–1919* (Jena, 1959), 141, 172–174, 176, 196, 201 ff., which describes Plate's relations with Haeckel and his activities at the educational institutions of Jena and contains an invaluable bibliography with comprehensive listings of MS material. Certain key letters pertaining to Plate's career and differences with Haeckel are reproduced in Georg Uschmann, ed., *Ernst Haeckel, Forscher, Kunstler, Mensch* (Leipzig–Jena–Berlin, 1961). References to Plate's role at the University at Jena, especially in regard to politics, appear in *Geschichte der Universität Jena 1848/58–1948*, 2 vols. (Jena, 1958), I, 510, 540, 543, 545, 552, 564, 584, 601, 605; II, 160, 447 ff., 603 ff., 608.

GLORIA ROBINSON

PLATEAU, JOSEPH ANTOINE FERDINAND (*b.* Brussels, Belgium, 14 October 1801; *d.* Ghent, Belgium, 15 September 1883), *physics, visual perception.*

Plateau was one of the best-known Belgian scientists of the nineteenth century. The son of an artist, he received his early education at schools in Brussels. In 1822 he entered the University of Liège as a student in the law faculty. He became interested in science; and in 1824, after he received a diploma in law, he enrolled as a candidate for an advanced degree in the physical sciences and mathematics.

Since Plateau had been orphaned at the age of fourteen, he had to support himself during his studies by serving as professor of elementary mathematics at the *athénée* in Liège. He received his *docteur ès sciences* in 1829 and the following year returned to Brussels, where he became professor of physics at the Institut Gaggia, then one of the most important teaching institutions in Belgium. In 1835 he was called to the State University of Ghent, as professor of experimental physics. He accepted the offer, and in 1844 became *professeur ordinaire*—a post that he held until his retirement in 1872. He was a successful teacher and was also active in organizing the physics laboratory at the university.

In 1834 Plateau was elected a corresponding member of the Royal Academy of Belgium and in 1836 a full member. He was also a member of a large number of foreign scientific organizations, including the Institut de France, the royal academies of Berlin and Amsterdam, and the Royal Society. In Belgium his honors included the office of *chevalier de l'ordre de Léopold* (1841), and in 1872 he rose to the rank of commander. In 1854 and in 1869 he also won the Prix Quinquennal des Sciences Physiques et Mathématiques of the Royal Academy of Belgium.

Plateau's long (he continued to do research even after his retirement) and productive career is especially remarkable because he was totally blinded in 1843.

This was apparently the result of an 1829 experiment in physiological optics, during which he stared into the sun for twenty-five seconds. At that time he was blinded for several days, but his sight returned partially. In 1841 he showed signs of serious inflammation of the cornea, which became steadily worse and ended in blindness. During his blindness he was aided in his work by colleagues—particularly, E. Lamarle, F. Duprez, his son Félix Plateau (a noted naturalist), and his son-in-law G. Van der Mensbrugghe.

Plateau's early work was in the field of physiological optics. The basis of much of this work was his observation that an image takes an appreciable time to form on, and to disappear from, the retina. In his dissertation (1829) Plateau showed, among other things, that the total length of an impression, from the time it acquires all its force until it is scarcely sensible, is approximately a third of a second. He applied his results to the study of the principles of the color mixture produced by the rapid succession of colors. This led to the formulation of the law (now known as the Talbot-Plateau law) that the effect of a color briefly presented to the eye is proportional both to the intensity of the light and the time of presentation. Plateau also studied various optical illusions that result from the persistence of the image on the retina. In 1832 he invented one of the earliest stroboscopes, which he called a "phénakistiscope." Plateau's device consisted of pictures of a dancer that were placed around a wheel. When the wheel was turned, the dancer was seen to execute a turn. Plateau sent his stroboscope to Michael Faraday.

Plateau studied in great detail the phenomena of accidental colors and irradiation, both of which he considered as arising from a similar cause related to the persistence of the image on the retina. Accidental colors are those that appear after staring for some time at a colored object and then at a black surface, or closing one's eyes and pressing one's hands over them. An image of the object appears, usually in complementary color and slightly diminished in size. Plateau's results include his discovery that accidental colors combine both with each other and with real colors according to the usual laws of color mixture. In irradiation luminous objects on a dark background appear enlarged, a factor clearly of interest to astronomers, among whom the question of the extent of the enlargement was causing controversy. Plateau showed that enlargement occurs regardless of the distance from the object and—explaining the varied experiences of the controversialists—that the mean amount of enlargement from the same source varied considerably from one individual to another.

Plateau was one of the first to attempt to measure sense distance. He used the method of bisection, presenting artists with white and black papers and asking them to produce a color midway between the two. Throughout his career, Plateau was interested in visual perception, and between 1877 and 1882 he published a critical bibliography of what he called "subjective phenomena of vision." He analyzed works from antiquity to the end of the eighteenth century, and listed, with short summaries, nineteenth-century works. Plateau's optical work has been neglected, perhaps because it contained theoretical errors, but his experiments were imaginative and interesting and earn him a name as a pioneer in physiological psychology.

In the 1840's Plateau turned his major energies to the study of molecular forces, through the consideration of a weightless mass of liquid. By immersing a quantity of oil in a mixture of water and alcohol, the density of which was equal to that of the oil, Plateau effectively annulled the action of gravity and showed that under these conditions the oily mass formed a perfect sphere. He then introduced centrifugal force and found that the sphere flattened at the poles and bulged at the equator. By controlling the velocity, he transformed the sphere into a ring, then a ring with a sphere at the center. He also formed a system of small spheres, which rotated about a central axis, each rotating around its own axis; this corresponded strikingly with the image of the formation of the rings of Saturn, and with that of the formation of the planets in Laplace's nebular hypothesis. (There are, of course, essential differences between conditions of the experiment and the astronomical situation, as Plateau himself indicated.)

Plateau also varied the conditions of his experiment by introducing metal wires to which the oil could adhere. He studied the forms of equilibrium that occurred, particularly the cylinder. Based on the assumption that the action was due to a very thin layer at the surface, he concluded that these forms should have surfaces of constant mean curvature. He obtained five different forms and showed geometrically that these were the only possible ones. He was, despite his blindness, a superb geometer, with a gift both for visualizing physical results and for physically interpreting geometric results.

Another way in which Plateau studied the effects of molecular forces—not influenced by the force of gravity—was by using thin films. In these studies he employed a treated mixture of soapy water and glycerin that he himself had developed. This liquid had the property that, with proper precautions, a bubble or film would last up to eighteen hours. Among other things Plateau studied the films that formed within wire contours dipped into the solution. His theoretical work led him to conclude that the surfaces formed were always minimal surfaces, and his experimental results confirmed this. But because his mathematical analysis was not rigorous, other mathematicians were led to formulate what is known as the problem of Plateau—to show that across any Jordan space curve there may be stretched a minimal surface. The question led to the study of functions of a complex variable and attracted the attention of Riemann, Weierstrass, and Schwarz. In 1931 Jesse Douglas gave the first mathematical solution.

Plateau's work on molecular forces was published in a series of memoirs between 1843 and 1868, and again, with some revision, as a book in 1873. In the work on thin films Plateau was drawn to the question of surface tension. He concluded that molecular forces alone were not sufficient to account for it. This probably indicates why Plateau is not as well-known as he was in his own time. He was nonetheless an able and ingenious experimenter and his work on thin films is remarkable for the results he obtained with the simplest of apparatus. His theoretical explanations, both in his optical investigations and in the study of molecular forces, are not, however, generally accepted.

Plateau also did interesting work in magnetism, proving that it is impossible to suspend something in the air using magnetic forces alone. His mathematical writings include papers on the theory of numbers. In addition he was the coauthor, with Adolphe Quetelet, of a long article on physics in the *Encyclopédie populaire*.

BIBLIOGRAPHY

I. ORIGINAL WORKS. Plateau's works on optics include *Dissertation sur quelques propriétés des impressions produites par la lumière sur l'organe de la vue* (Liège, 1829); "Sur quelques phénomènes de vision," in *Correspondance mathématique et physique*, **7** (1832), 288–294; "Sur un nouveau genre d'illusions optique," *ibid.*, 365–369, in which the stroboscope is described; "Essai d'une théorie générale comprenant l'ensemble des apparences visuelles qui succèdent à la contemplation des objets colorés et de celles qui accompagnent cette contemplation," in *Annales de chimie*, **58** (1835), 337–407; "Sur la mesure des sensations physiques et sur la loi qui lie l'intensité de ces sensations à l'intensité de la cause excitante," in *Bulletin de l'Académie royale de Belgique*, **33** (1872), 376–388; and "Bibliographie analytique des principaux phénomènes subjectifs de la vision depuis les temps anciens jusqu'à la fin du XVIII siècle, suivie d'une bibliographie simple pour la partie écoulée du XIX siècle," in *Mémoires de l'Académie royale de Belgique* (1877–1883), **42, 43, 45**; the latter part also lists Plateau's own works.

The memoirs on the weightless masses of liquids and thin films are in *Statique expérimentale et théorique des liquides soumis aux seules forces moléculaires*, 2 vols. (Ghent–Paris, 1873). The first six memoirs were translated into English, and appear with commentary by Joseph Henry in *Annual Report of the Board of Regents of the Smithsonian Institution* (1863), 207–285; (1864), 285–369; (1865), 411–435; (1866), 255–289.

The miscellaneous writings include "Sur un problème curieux de magnétisme," in *Mémoires de l'Académie royale de Belgique* **34** (1864), sect. 4; "Note sur une récréation arithmétique," in *Les mondes*, **3** (1864), 536–538; "Sur une récréation arithmétique," *ibid.*, **36** (1875), 189–193; and "Physique" in M. Jamar, ed., *Encyclopédie populaire* (Brussels, 1851–1855), written with A. Quetelet.

A bibliography of Plateau's work is in Poggendorff, II (1863), 466–467; III (1898), 1048; IV (1904), 1173; and in Royal Society, *Catalogue of Printed Papers*, IV (1870), 936–938; XI (1896), 33; XII (1902), 579.

II. SECONDARY LITERATURE. Accounts of Plateau's life are scarce, but his work is discussed at length in Charles Bergmans, "Joseph Plateau," in *Biographie nationale*, **17** (1903), 768–788; J. Delsaulx, "Les travaux scientifiques de Joseph Plateau," in *Revue des questions scientifiques*, **15** (1884), 114–158, 518–577, and **16**, 383–437; and C. V. Overbergh, ed., *Le mouvement scientifique en Belgique 1830–1905*, I (Brussels, 1907–1908), 368–396. For a history of the problem of Plateau, see Jesse Douglas, "Solution of the Problem of Plateau," in *Transactions of the American Mathematical Society*, **33** (1931), 263–321.

ELAINE KOPPELMAN

PLATER, FÉLIX. See **Platter, Félix.**

PLATO (*b.* Athens [?], 427 B.C.; *d.* Athens, 348/347 B.C.), *theory of knowledge; advocacy, in theory and practice, of education based on mathematics; organization of research.*

Plato's enthusiasm for mathematics, astronomy, and musical theory appears everywhere in his writings, and he also displays a far from superficial knowledge of the medicine and physiology of his day. In ancient times competent judges held that he had promoted the advance of mathematics, especially geometry, in his lifetime.[1] Theodore of Cyrene and Archytas of Tarentum were his friends, and Eudoxus of Cnidus, Theaetetus, and Menaechmus his colleagues or pupils. His critics assert that his theory of knowledge rules out any empirical science and that, owing to his idealism, he had a radically false idea of the procedure and value of the mathematics that he admired. Even so, it can be said that the Academy, founded by him at Athens at a date not exactly known (380 B.C.?), became a center where specialists—not all of them sympathizers with his philosophy and epistemology—could meet and profit by discussion with him and with one another.

Our object here must be to trace Plato's intellectual development and, incidentally, to submit part of the material upon which an estimate of his services or disservices to science must be based. It must be considered how far he is likely to have carried out in his school the project, sketched in the *Republic*, of a mathematical training preparatory for and subordinate to dialectic, and whether, in the writings believed to belong to the last twenty years of his life, he took note of recent scientific discoveries or was influenced by them in matters belonging to philosophy.

As for sources of information, the account of Plato's life and doctrine by Diogenes Laertius (probably early third century A.D.) is based on previous authorities of unequal value. He reports some evidently reliable statements by men who were in a position to know the facts and who were neither fanatical devotees nor detractors, and he has preserved the text of Plato's will. Aristotle gives us a few details, and Cicero a few more.

The Epistles, ascribed to Plato and printed in the Herrmann and Burnet editions of the Greek text of his works, would, if genuine, furnish us with a personal account of his conduct at important crises in his life; and, what is more, they would tell far more about his ideals of education and the work of the Academy than can be gathered from the dialogues. Unfortunately, opinion regarding authenticity of the *Epistles* is so divided that caution is essential. In the ensuing account, where reference is made to this source, the fact has been indicated.

Plato's writings have been preserved entire. But the double fact that they are dialogues and that the scene is usually laid in the past leads to difficulties of interpretation which are sufficiently obvious. Moreover, nothing definite is known about either the manner of their first publication or their relative order, still less the dates. Some hypothesis about the latter is a presupposition of fruitful discussion of Plato's development.

A statistical study of the style of Plato's works, in which the pioneer was the Reverend Lewis Campbell (1869), has led to results which have met with wide approval: many scholars hold that *Parmenides* and *Theaetetus* were written later than the *Republic*, and that a group of dialogues having close stylistic affinity to one another and to the *Laws* (which is plainly a work of old age) came still later. This is credible from a philosophical point of view, and its correctness is assumed here; but such a method can only yield a probable result.

Several members of Plato's family are mentioned, or appear as characters, in his dialogues. He himself was the son of Ariston[2] and Perictione, and was born either at Athens or Aegina, where his father may have gone as a settler when the Athenians occupied the island. Nothing reliable is known of his father's ancestors, but those on his mother's side were men of distinction. Perictione was descended from Dropides, a close friend (some say brother) of Solon, the famous poet-statesman of the sixth century B.C. She was a cousin of Critias, son of Callaeschrus, an intellectual daring in both speculative thought and action. It was Critias who in 404 B.C. led the extremists among the Thirty Tyrants and put to death the moderate Theramenes. He became guardian of Perictione's brother Charmides and drew him into public affairs. Both perished in the battle which put an end to the Thirty's six months of power.

Plato was one of four children. His brothers Adeimantus and Glaucon take a leading part in the *Republic*, where they are depicted with admiration and a clear impression of their personality is left. They appear once more briefly in the *Parmenides*, and Xenophon presents Socrates proving to Glaucon the folly of his trying to address the Assembly when he is not yet twenty.[3] The brothers were considerably older than Plato, and his sister Potone (mother of Speusippus, who followed him as head of the Academy) was doubtless born in the interval.

Plato's father, Ariston, appears to have died young. Perictione then married Pyrilampes, son of Antiphon, who had been prominent in state affairs as a close associate of Pericles; he was probably her uncle. Another son, called Antiphon after his grandfather, was born; this half brother of Plato's has a part in the *Parmenides*. Most of these persons are mentioned either in *Charmides* (155–158), or in *Timaeus* (20E), or at the opening of *Parmenides*.[4]

Plato's social position was such that he might well have aspired to an active part in public affairs, but it could not have been easy for him to decide what role to assume. Some scholars take it for granted that the example and writings of Critias left a deep impression upon him; others point out that it was only in the concluding phase of the Athenian struggle against the Peloponnesians that Plato's maternal relatives emerged as reactionary extremists, and that in his stepfather's home he would have been imbued with liberal opinions and respect for the memory of Pericles.

No one can say with certainty what the complexion of his views was at the age of twenty-four, except that he was obviously no friend to egalitarianism and full democracy. The story which he tells, or is made to tell, in Epistle VII seems probable. His friends and relatives among the Thirty at once called upon him to join them, but instead he determined to wait and see what they would do. They soon made the former regime seem highly desirable by comparison. Socrates was commanded to help in the arrest of a man who was to be put to death illegally under a general sentence, so that he would either be involved in their impious actions or refuse and thus expose himself to punishment. When the Thirty were overthrown, Plato again thought of public affairs—but with less eagerness than before. The democratic leaders restored to power showed moderation at a time when ruthless acts of revenge might have been expected. Nevertheless Socrates was brought to trial on the pretext of impiety and found guilty. As Plato grew older and the politicians, laws, and customs of the day displeased him more and more, he was thrown back on a theorist's study of ways of reform.[5]

None of this is inconsistent with what is otherwise known. Socrates' disobedience to the illegal command of the Thirty was a fact widely spoken of. Plato would probably have been impressed to an equal degree by Socrates' courageous independence in such matters and by his faith in argument (argument with himself when he could not find a respondent). He became conspicuous among Socrates' habitual companions, as distinct from the occasional listeners to his conversation. With Adeimantus he heard Socrates' provocative defense in court against the charge of impiety;[6] when a majority had found him guilty, Plato was one of those who induced Socrates to offer to pay a substantial fine, for which he would be a guarantor.[7]

Owing to illness, Plato was absent from the last meeting of Socrates with his friends.[8] After the tragedy, he retired, with other Socratics, to Megara, the home of Euclid.[9] The attack on Socrates was personal, and perhaps the prosecutors did not desire his death. His Athenian friends can hardly have been in danger. But there are some hints in the *Phaedo* that he advised those present, among whom were the Megarians Euclid and Terpsion, to pursue the search for truth in common and not lose heart when plausible reasoning led them nowhere; rather, they must make it their business to master the "art of argument," *logōn technē*.[10]

Probably his wish was piously carried out by his followers, and a few years elapsed before the different direction of their interests became clear. As a metaphysician Euclid was a follower of Parmenides, and accepted the Socratic thesis that there is a single human excellence, not a plurality of "virtues." His thought had no religious coloring, nor was he an educational reformer. His younger disciples turned in earnest to the hoped-for *logōn technē*, and not without result; they prepared the way for the propositional

logic of the Stoic school.[11] This might be supposed to go together with an interest in the sciences, but this is not recorded of the Megarians. Plato, on the other hand, began to turn in that direction; his first dialogues and the *Apology* must have been written during the years 399–388 B.C. He felt it his duty to defend the memory of Socrates, especially since controversy about his aims had been revived by hostile publications. As the chance of political action remained remote, he gradually developed the idea of a training of the young not in rhetoric but in mathematics—and in Socratic interrogation only after the mathematical foundation had been laid. Part of his diagnosis of the ills of Athens was that young men had bewildered themselves and others by engaging too soon in philosophical controversy; these ideas probably found little sympathy among the Megarians. How long he remained among them is not recorded, but he was liable to Athenian military service, probably as a cavalryman. A statement has indeed come down to us[12] that he went on expeditions to Tanagra (in Boeotia) and Corinth. This is credible in itself, and in the latter case the reference could be to an engagement in 394 B.C. outside Corinth, in which the Spartans and their allies defeated the Athenians and Thebans. But neither does it seem inconsistent with Plato's regarding Megara for a time as his home. About 390 he resolved to visit the West, where Archytas of Tarentum survived as a maintainer of the Pythagorean system of education and was also active in research.

Plato's views at the time of departure on his journey to the West are well seen in the *Gorgias*. It is his first major constructive effort as a moralist, but there is as yet no positive doctrine of knowledge and reality. When Callicles spurns conventional justice, as a means of defrauding the strong and energetic of what naturally belongs to them, and declares that temperance is not a virtue (why should a clear-sighted man choose to curb his own desires?), Socrates confidently develops an answering thesis: the supervision of the soul must be supposed comparable in its operation to the arts, which impose form and design (*eidos, taxis*) and preserve the natural subordination of one part of a subject to another (*kosmos*). Human good does not consist in the ceaseless satisfaction of desires, irrespective of their quality (if it did, man would stand apart from the general world order), and self-discipline is the basis of happiness. But the statesmen of Athens, the dramatists and musicians, the teachers and learners of rhetorical persuasion, have all alike failed to understand this and have flattered rather than guided the public.

In his use of the varied senses of *kosmos* (which, according to the context, means world or world order, moral discipline, or adornment), Socrates is here on Pythagorean ground; and ideas are already present which Plato expanded only in his later writings and his oral instruction.[13] The *Gorgias* passage is also an emphatic answer to the friends who had sought to draw Plato into Athenian politics.[14]

Concerning the journey itself, in Epistle VII Plato says, or is made to say, that he was then forty years old (324A) and that in Italy and Sicily he was appalled by the sensuous indulgence which he found taken for granted there. On crossing to Syracuse he made the acquaintance of Dion, the young brother-in-law of the tyrant Dionysius the Elder, who listened attentively to his discourses and aroused his admiration by his intelligence and preference for a sober life. In the tyrant's entourage this was viewed as an affectation of singularity and led to Dion's becoming unpopular.[15]

If this evidence is set aside as suspect, the next best source is Cicero.[16] He says that Plato visited Egypt before proceeding to Italy; that he spent a considerable time with Archytas and with Timaeus of Locri; and that the object of the voyage was to gain acquaintance with Pythagorean studies and institutions. To this some reservations must be made. First, it can hardly be true—if Cicero means this—that when he boarded the ship Plato was altogether ignorant of mathematics. In his own dialogues there is clear evidence that the sciences were to some extent taught to boys at Athens and that there was an opportunity of learning from specialists in mathematics and astronomy, no less than from those in music, meter, and grammar. About Pythagoreanism also Plato already had some information, judging from the *Gorgias* passage mentioned above; he could have obtained this (as Wilamowitz suggests) from the Thebans Simmias and Cebes, pupils of Socrates who are said to have met Philolaus.[17]

Secondly, it does not seem likely that Timaeus of Locri was still alive at the time of Plato's journey. In *Timaeus* 20A he is described as a man of intellectual distinction who has already held high office, and this is at a time certainly previous to 415 B.C. (It is possible that at this time Plato met Philistion of Locri, and derived from him the interest in the physiology of the Sicilian Empedocles, which is visible in both *Meno* and *Phaedo*.) Cicero's report may be wrong in some of its detail, but it seems true in spirit. Plato's purpose in visiting the West was to see for himself how the Pythagoreans conducted their science-based educational system, and he did at this time establish a connection with Archytas.

Plato returned to Athens, after two years' absence, in 388 B.C. (Ancient biographers related, with some circumstantial detail, that at Syracuse he had exasperated the tyrant Dionysius the Elder by open criticism of his rule and had been handed over as a prisoner to a Spartan envoy. But such insolence is hardly in character for Plato, and probably his voyage home was of a less sensational kind.) He might at this time have visited the Pythagoreans at Phlius, in the Peloponnesus. The setting of the *Phaedo* suggests personal acquaintance with their leader Echecrates, and Cicero confirms this.[18]

Nothing definite is recorded about Plato's personal life during the ensuing twenty-two years. But the Academy was founded, or gradually grew up, during this time, and he composed further dialogues in Socratic style. The *Meno* and *Euthyphro*, *Euthydemus*, *Phaedo*, *Symposium*, and *Republic* must all be assigned to these years. In them he puts forward the distinctive account of knowledge which has taken shape in his mind; explains his purpose and method in education and shows the continuity of his aims with those of Socrates; and differentiates himself, where necessary, from the Italian Pythagoreans. It is natural to place the *Republic* at the end of this series, and to regard it as either a prospectus for a proposed school or as a statement to the Athenian public of what was already being carried out among them.

Aristotle gives a clear analysis of the factors which produced Plato's doctrine of Forms.[19] Plato was acquainted from youth with an Athenian named Cratylus, who declared with Heraclitus that there is no stable substance, or hold for human knowledge, in the sense world. Plato did not deny this then or later but, wishing to take over and continue the Socratic search for universals, in the sphere of morals, which do remain permanent, he necessarily separated the universals from sensible particulars. It was he who termed them Ideas and Forms. In his view particulars (that is, things and states of things, actions and qualities) derive reality from Forms by "participation"; and when we name or speak of these particulars, we in effect name Forms.

In the dialogues Plato often starts from a contrast between knowledge and opinion. To live in a state of opinion is to accept assertions, either of fact or of principle, on authority or from mere habit. The opinion may be true and right; but since it is held without a rational ground, it may be driven from the mind by emotion and is less proof against forgetfulness than knowledge is. The holder of it may also be deceived in an unfamiliar instance. Based as it is on habit, an opinion cannot easily be transmitted to another; or, if the transmission takes place, this is not teaching. In terms of the theory of Forms, the holder of knowledge knows the Forms and can relate particular instances to them (although Plato did not successfully explain how this occurs), whereas the contented Holder of opinions moves about among half-real particulars.

In middle life, then, Plato had advanced from his Socratic beginnings toward beliefs, held with assurance, from which many practical consequences flowed. The chief elements were the knowledge-opinion contrast; the belief in a realm of immutable Forms, with which human minds can make intermittent contact and which on such occasions the minds recognize as "their own" or as akin to them;[20] given this, the soul, or its intellectual part, is seen to be likewise eternal; and the belief that the Forms, each of which infuses reality into corresponding particulars, in turn derive their existence, intelligibility, and truth from one supreme Form, the Good.

The advance from the plurality of Forms to their source is in consequence regarded as the ultimate stage in human study, *megiston mathēma*;[21] it is a step which will be taken by only a few, but for the welfare of mankind it is important that a few should take it. Within the dialogue it is described but cannot be accomplished. There are hints of a methodical derivation of the other Forms from the Good; but for the present the image, whereby the Good is shown to have the same relation to other objects of intellection as the sun has to other visible things, takes its place. In reading the *Republic* and later dialogues, one must therefore reckon with the possibility that in the school Plato amplified or corrected the exposition which he chose to commit to writing.

The Athenians thought it suitable that young men should exercise themselves in argument on abstract themes before turning to serious business, and were prepared to tolerate "philosophy" on these terms. But Plato, as has been said, speaks out against this practice and holds that it has brought philosophy into discredit. Indeed, according to him, the order of procedure should be reversed. Argument, or its theory, is the hardest branch of philosophy and should come later. Men and women to whom legislation and administration are ultimately to be entrusted should undergo discipline in the sciences (including reflection on their interrelation) before they embark, say, at the age of thirty, on dialectical treatment of matters which have to be grasped by the intellect without the help of images. Such a discipline will single out those who have capacity for dialectic. To them will fall the task of making good laws, if these are not found in existence, and of interpreting and applying them if they are. For this purpose knowledge must be reinforced by experience.[22] Lawless government is the common fault of despotism and democracy.

Plato holds that ignorance of mathematical truths which are in no way recondite, for example, the wrong belief that all magnitudes are commensurable, is a disgrace to human nature.[23] It is not, however, this that is emphasized in his educational plan in the *Republic*. He explains that it is characteristic of mathematical studies that they gently disengage the intellect from sensible appearances and turn it toward reality; no other discipline does this. They induce a state of mind (which Plato terms *dianoia*, discursive thought) clearer than "opinion" and naïve trust in the senses but dimmer than knowledge and reason. In geometry, for instance, the learner is enabled or compelled, with the aid of figures, to fix his attention on intelligible objects. Also, mathematicians "lay down as hypotheses the odd and even, various figures, and the three kinds of angles and the like"[24] but leave them unexamined and go on to prove that the problem that gave rise to their investigation has been solved. In this respect mathematical procedure tends to divert the mind from reality and can provide only conditional truth. But such studies, pursued steadily and without continual talk of their practical use, are a good preparation for methodical treatment of such relations among Forms as cannot be visibly depicted.

Arithmetic and plane geometry will be the basis of an education which is to end in knowledge; the geometry of three-dimensional figures must also be studied. When, in the dialogue, Glaucon observes that this hardly yet exists as a science, Socrates says that there are two reasons for this: first, no state at present honors the study and encourages men to devote themselves to it; and second, a director is needed in order to coordinate the research.[25] Such a man will be hard to find; and at present even if he existed the researchers are too self-confident to defer to him. Even without these conditions, and even when the researchers do not succeed in explaining what they are striving to achieve, the intrinsic charm of the study of three-dimensional figures is carrying it forward. This is one of the passages in which speakers in Plato's dialogues refer prophetically, but in veiled terms, to circumstances at the time of writing. It is somewhat enigmatic for us. The intention is perhaps to compliment Theaetetus, who had discovered constructions for inscribing in a sphere the regular octahedron and icosahedron. Either he or Plato himself is cast for the role of a director, and there is a plea for public support of the Academy so that research can continue.

There is a similar personal reference in the treatment of the sciences of astronomy and musical theory.[26] Socrates dissociates himself from the Pythagoreans while approving of their statement that the two sciences are closely akin. Their theory of harmony is restricted to a numerical account of audible concords; and the aim of their astronomy is to discover the proportion between month, year, and the period of revolution of the planets. Instead of this, heard harmonies should be studied as a special case of the harmonies between numbers; and the proportion between month, year, and so forth (which is doubtless not unvarying) as an application of some wider theory dealing with the spatial relations of any given number of bodies of any shape, moving at any regular speeds at any distances. The visible universe will be to the "true astronomer" what a beautifully contrived diagram might be to the geometer, that is, an aid to the science, not the object of contemplation. Here Plato is indicating that he has not simply established on Athenian soil a replica of the Pythagorean schools. Attachment to the sense world must be loosened and the sciences taught with emphasis on their affinity to one another; for it is the power of synopsis, the perception of common features, that is to be strengthened. Those few who excel in this are to be set apart and trained to pursue another method which treats hypotheses as provisional until they have been linked to the unconditionally real and so established.[27]

There is no question here, as some have thought, of a fusion of the existing special sciences. The ideal held out is probably this: The One is an aspect of the Good (this was common ground to Euclid of Megara and Plato); from the concept of unity, the number series can be made to emerge by deduction; and from the Good, by a process which makes more use of Socratic questioning, the system of moral and aesthetic concepts can likewise be evolved.[28]

Boys will have some introduction to all the sciences that have been mentioned. Compulsion is to be avoided in the training of the mind, especially since what is learnt under duress leaves no lasting impression. At the age of twenty when state-enforced physical exercise is to cease, those judged capable of further progress must begin to bring together the knowledge that they have hitherto acquired randomly and to consider in what way the sciences are akin to one another and to reality. Until they are thirty this will be their occupation. After this there can follow five years of that dialectical exercise that Athenian custom regards as a fitting occupation for mere adolescents.[29]

The program here outlined is our safest guide to the actual institutions of the Academy. But it is an ideal that is sketched, and one need not insist that it was carried out in detail. For instance, Plato did not admit women, and it is highly improbable that he gave no philosophical instruction to those under thirty; it would doubtless occur to him that since he could not prevent them from obtaining such instruction, it would be better to make sure that they were well taught.

"Academy" was an area to the northeast of the city which had been laid out as a park, including a public gymnasium. According to Lysias, the Spartans encamped there during the troubled year 403 B.C. Plato may have commenced teaching in the gymnasium itself; but he soon purchased an adjoining garden and erected buildings there, and from this moment he may be said to have instituted a school. Hitherto he could not exclude chance listeners. The buildings may have included lodgings for students or visitors, and Plato himself presumably lived in the neighborhood. Common meals had been a feature of Pythagorean life, and this precedent was followed. Legal recognition was secured by making the Academy a religious fraternity devoted to the Muses.

Plato nominated his nephew Speusippus to be his successor, but later the head seems to have been elected. Presumably the power of admission rested with the head, and those accepted contributed to the maintenance of the school according to their means. The story that the words "Let no man unversed in geometry enter" were inscribed over the door cannot be traced back further than John Philoponus (sixth century). In the first century B.C., "academic" teaching was being given in the gymnasium of Ptolemy near the agora of Athens.[30]

In the teaching, Plato will doubtless have been assisted from an early stage by Theaetetus, who, as the dialogue named after him shows, was a boy of exceptional promise in 399 B.C. He was an Athenian. The dialogue records his death from wounds received in battle at Corinth, complicated by illness, and was obviously written soon afterward (ca. 368–367 B.C.). Hence his collaboration with Plato may have lasted about fifteen years.

Somewhat later, Eudoxus of Cnidus resided occasionally at the Academy. He is known to have died at the age of fifty-three, and the dates for his life now usually accepted are ca. 400–ca. 347 B.C. In mathematics he was a pupil of Archytas, but it is not clear when this instruction was received. During a first visit to Athens at the age of twenty-three, he heard lectures from Plato. Later he established a school at Cyzicus in northwest Asia Minor, one of those Greek cities which had been abandoned to Persian control under the Peace of Antalcidas in 387 B.C.; Eudoxus probably felt insecure there and was glad to maintain a connection with the Platonists.

Eudoxus was at Athens when Aristotle first arrived as a student in 367 B.C., and it is plain from several Aristotelian passages[31] that he deeply influenced his juniors and played an important part in the life of the school. In Plato's *Philebus* there is probably some concealed allusion to him. Mathematics and astronomy were his main business, but he took part as a friendly outsider in philosophical debate arising from Plato's writings. He was not one to pledge himself to accept philosophical dogmas, nor was this expected of him, and his work exhibits the close attention to phenomena which the Socrates of the *Republic* deprecates.

But Eudoxus recognized that philosophy has a legitimate role in criticizing the procedure of specialists. His explanation of the celestial movements in terms of homocentric spheres, rotating about a stationary spherical earth, was put forward in answer to a problem posed by Plato: "What are the uniform and ordered movements by the assumption of which the apparent movements of the planets can be accounted for?" Eudoxus did not abandon his own school and merge it with that of Plato; and Proclus' statement that he "became a companion of Plato's disciples" does not mean this, for we know that Aristotle later carried on the connection with Cyzicene mathematicians which Plato had established. Epistle XIII, ascribed to Plato, mentions Helicon as a pupil of Eudoxus and implies the presence of other Cyzicenes at Athens in the period 365–362 B.C.

If we consider Plato's relation to Theaetetus and Eudoxus and also bring in the evidence of Aristotle, his personal role in the Academy begins to appear. He probably committed all the specialist instruction to them. He took note of their research and sometimes criticized their methods, speaking as a person with authority; he guided the juniors in that reflection about first principles and about the interrelation of sciences that in *Republic* VII is designated as suitable for them;[32] and he confided to some an ethico-mathematical philosophy in which two ultimate principles, from which Form-Numbers were derived, were found by analysis.

The last item is outside the scope of this article, and concerning the other two a few examples must suffice. Theaetetus added to previously known constructions of regular solid figures those for the octahedron and icosahedron, and gave a proof[33] that there can be no more than five regular solids. He also tried to classify incommensurable relations, connecting them with the three kinds of means. Plato's inspiration may be seen in this effort to systematize. Plato is credited by Proclus with beginning the study of conic sections, which his followers developed, and with either discovering or clearly formulating the method of analysis, which proved fruitful in geometry in the hands of Leodamas of Thasos.[34] According to Plutarch, Plato criticized Eudoxus, Archytas, and Menaechmus for trying to effect the duplication of the cube by mechanical means and so losing the benefit of geometry.[35] Aristotle says that Plato (in oral instruction) "used to object vigorously to the point, as a mathematical dogma, and on the other hand often posited his indivisible lines."[36]

The view that the sciences contemplate objects situated halfway between Forms and sensibles and having some of the characteristics of both, which is not found in Plato's dialogues, was probably at home in the same discussion. Observation of the heavens went on, perhaps under the direction of Eudoxus or his companions. Aristotle says that he had seen an occultation of Mars by the moon (then at half-moon); this must have been in April or May 357 B.C., a calculation first made by Kepler.[37]

It was maintained by H. Usener in 1884 that the Academy was the first known institute for scientific research, a statement which initiated a debate not yet closed. It has been opposed from one point of view by those who compare the Academy to a modern school of political science (and perhaps jurisprudence) thoroughly practical in its orientation; and from another especially by Jaeger, who thinks that it was no part of Plato's intention to teach science in encyclopedic fashion and promote its general advance.[38] The Academy was not a place in which *all* science was studied for its own sake but one in which selected sciences were taught and their foundations examined as a mental discipline, the aim being practical wisdom and legislative skill, which in Plato's opinion are inseparable from contemplative philosophy.

Evidently the crux of this matter is whether empirical sciences, which had no place in the curriculum projected for the guardians in the *Republic*, were, in fact, pursued under Plato's auspices. Jaeger seems to be right in his skepticism about the apparent evidence of such activity. It may be added that proofs that Plato was personally interested in (for instance) medicine and physiology, which *Timaeus* affords, are not quite what is wanted. One might argue that he was indeed an empirical scientist manqué, pointing to his interest in the manufacturing arts in the *Statesman*, his marvelous sketch of the geology of Attica in the *Critias*, and his attention to legislative detail in the *Laws*. It was an imperfectly suppressed love of the concrete and visible, rather than any retreat from his avowed opinions, that led him to pose the problem concerning celestial phenomena which Eudoxus later solved. But it is a long step from these admissions to the pronouncement that the Academy became an institute of scientific research.

Plato's activity at Athens was interrupted by two more visits to Sicily. When Dionysius the Elder died in the spring of 367 B.C., two problems in particular demanded solution: what the future form of government Syracuse itself should be; and when and how several Greek cities, whose populations had been transferred elsewhere under Dionysius' policy, should be refounded. Plato's admirer Dion planned to make his nephew, Dionysius the Younger, a constitutional monarch, and appealed for Plato's aid in educating him for his responsibility. His proposal was that Plato should come to Syracuse and take charge of a group of earnest students, which Dionysius might unobtrusively join.

Plato yielded to pressure and arrived in the spring of 366 B.C. But, according to Epistle VII, he found a situation of intrigue. Some Syracusans believed that Dion's aim was to occupy his nephew with interminable study while he himself wielded effective power. From another side there was pressure for the restoration of full democracy. A war with Carthage (or against the Lucanians) was in progress. Three months after Plato's arrival the young ruler charged Dion with attempting to negotiate with the Carthaginians and expelled him. Plato left the following year (365 B.C.), after obtaining what he took to be a promise that Dion would be recalled.

In 362 B.C. he once more left the Academy (appointing, it is said, Heraclides Ponticus as his deputy) and returned to Syracuse. There were reports that Dionysius was now genuinely interested in philosophy; but in consenting to go Plato was more influenced, according to Epistle VII, by a promise that a favorable settlement of Dion's affairs could be reached on condition that he return. Dion had spent his years of exile in Athens and had become a friend of Plato's nephew Speusippus.

Plato's mission ended in failure. Dion's agents were forbidden to send him the revenue from his estates; and Plato made his escape with some difficulty, returning home in 361 B.C. Dion thereupon took steps to effect his return to Syracuse by force, urging Plato to aid him and to punish Dionysius as a violator of hospitality. But Plato answered (still according to the Epistle) that he was too old, that after all Dionysius had spared his life, and that he wished to be available if necessary as a mediator. Other members of the Academy, however, joined the expedition. Dion succeeded in his enterprise but failed to reconcile the warring parties; and after a brief period of power he was murdered by the Athenian Callippus (354 B.C.). Epistles VII and VIII, addressed to Dion's partisans, belong to this time or were fabricated as belonging to it. They contain constructive advice which, in language and spirit, closely resembles *Laws* III. Plato died in 348/347 B.C., at the age of eighty.

Was Plato at all influenced in his later years by progress in the special sciences? It may be replied that later modifications in his general theory of knowledge seem to have been the product of his own reflection rather than of any remarkable discoveries and were not such as to affect his educational ideals. He did, however, absorb new scientific ideas and make use of them for his purposes as a moralist. This is seen in one way in the *Timaeus*, where the world is shown to be a product of beneficent rational design, and in another in the *Laws*, where theological consequences are drawn from the perfect regularity of the planetary movements.

In the *Timaeus*, Plato has not abandoned his theory of Forms, for Timaeus declares that the perpetually changing, visible cosmos can at best be an object of "right opinion." The view earlier attributed to Socrates in *Phaedo*—that the only satisfactory explanation of physical facts is a teleological one—is assumed from the start and carried out in detail. Timaeus opposes mechanistic views with a superior atomism, in which Theaetetus' construction of the mathematical solids plays a part. Sicilian influence is especially strong. From Empedocles' system the doctor Philistion of Locri, with whom Plato was personally acquainted,[39] developed a theory whereby the heart is the center of life and consciousness, and the veins and arteries carry pneuma along with blood. Timaeus adopts this physiology and the explanation of disease which went with it, except that in making the brain the organ of consciousness he follows Alcmaeon. (It appears then to have been the Academy, with its Sicilian connections, which brought the knowledge of Empedoclean medicine to the mainland.) All this leads up to the conclusion that man can learn to regulate his life by study of the cosmos, which is a divine artifact but also an intelligent being.

The Athenian Stranger in the *Laws* says that he was no longer a young man when he became persuaded that each of the planets moves in a single path, and that we malign them when we call them "wanderers." He evidently speaks for Plato, and the remark is a formal withdrawal of what was said in *Republic* VII of the erratic celestial movements; but in favor of what new system? Surely that of Eudoxus. This may not have lasted long, and was open to an objection that was soon seen, but it was the first scientific astronomy. From it Plato could, and did, argue that all movement stems from a soul (matter is inactive); perfectly regular movement stems from a wise and beneficent soul; and the rotation of the stars, sun, and planets is perfectly regular.

In these later developments Plato may appear not as a lover of science but as a biased user of it. But "an intense belief that a knowledge of mathematical relations would prove the key to unlock the mysteries of the relatedness within Nature was ever at the back of Plato's cosmological speculations" (Whitehead, *Adventures of Ideas*, p. 194).[40]

NOTES

1. Proclus in his commentary on Euclid's *Elements* gives a summary account of the development of geometry from the time of Thales. He claims that Plato brought about a great advance, and names some of those who worked with his encouragement in the Academy. Closely similar language is used in a fragmentary papyrus found at Herculaneum, "Academicorum philosophorum index Herculanensis," where Plato's role is said to be that of a supervisor who propounded problems for investigation by mathematicians. Simplicius in his commentary on Aristotle's *De Caelo* informs us that Eudoxus' astronomical system was a solution of a carefully formulated problem set by Plato. The common source of the first two reports is probably the history of mathematics by Eudemus of Rhodes, and Simplicius says that his statement is derived from Book 2 of Eudemus' *History of Astronomy*. Greek texts are in K. Gaiser, "Testimonia Platonica" (Appendix to *Platons Ungeschriebene Lehre*), nos. 15–21, 460–479. An English translation of the latter part of the Proclus passage is in Heath, *A Manual of Greek Mathematics*, 184–185.
2. *Apology* 34A; *Republic* II, 368A.
3. *Memorabilia* III, 6.
4. For a family tree, see John Burnet, *Greek Philosophy From Thales to Plato* (London, 1924), appendix; or C. Ritter, *Platon*, I, 13.
5. *Epistles* VII, 324–326.
6. *Apology* 34A.
7. *Ibid.*, 38B.
8. *Phaedo* 59B, C.
9. Diogenes Laërtius III, 6, and II, 106, quoting Hermodorus, a pupil of Plato.
10. *Phaedo* 78A, 107B.
11. W. and M. Kneale, *The Development of Logic* (Oxford, 1962), ch. 3; B. Mates, *Stoic Logic* (Berkeley–Los Angeles, 1961), intro.
12. Aelian, V, 16; VII, 14; and Diogenes Laërtius, III, 8, where there is confusion between Plato and Socrates.
13. Robin, *Platon*, pp. 101–103.
14. Cornford, *Republic of Plato*, Intro., p. xvii.
15. *Epistles* VII, 326–327.
16. *De Republica* I, 10.
17. *Phaedo* 61D.
18. *De finibus* V, 29.87.
19. *Metaphysics*, 987A32.
20. *Phaedo* 75E, 76E.
21. *Republic* VI, 503–505.
22. *Republic* VI, 484C, D.
23. *Laws* VII, 819–820.
24. *Republic* VI, 510C.
25. *Ibid.*, VII, 528B.
26. *Ibid.*, 528E ff.
27. *Ibid.*, 533.
28. Cornford, who gives this exposition, relies upon *Parmenides* 142D ff. See his trans. of the *Republic*, p. 245, and his "Mathematics and Dialectic in the *Republic* VI–VII."
29. *Republic* VII, 535–540.
30. See Cicero, *De finibus* V, 1, for an eloquent description of a walk to the gymnasium's original site, now deserted.
31. Especially *Nicomachean Ethics* X, ch. 1.
32. *Republic* VII, 535–540.
33. Euclid XIII, 18A.
34. See Cornford, "Mathematics and Dialectic in the *Republic* VI–VII," pp. 68–73.
35. *Quaestiones convivales* 718F.
36. *Metaphysics A*, 992a20; see Ross *ad. loc.*
37. See Stocks's note in Oxford trans. of Aristotle, vol. II; and Guthrie's ed. of Aristotle's *On the Heavens* (Loeb), p. 205.
38. Aristotle, p. 18 and elsewhere.
39. *Epistles* II, 314D.
40. That the system presented through Timaeus is formulated in antithesis to mechanist atomism—whether or not Plato knew of Democritus as its chief author—is clearly shown by Frank, *Platon u. die sogenannten Pythagoreer*, pp. 97–108. On the meaning of *Timaeus* 40B, C, concerning the earth's rotation, see Cornford, *Plato's Cosmology*, pp. 120–134. On Philistion, see Jaeger, *Diokles von Karystos*, pp. 7, 211. For the argument that *Laws* 822A–E can refer only to the system of Eudoxus, and for Plato's astronomy in general, see Burkert, *Weisheit und Wissenschaft*, pp. 302–311.

BIBLIOGRAPHY

I. ORIGINAL WORKS. For present purposes the most important dialogues are *Meno, Phaedo, Republic, Parmenides, Theaetetus, Timaeus,* and *Laws.* The *Epistles* and *Epinomis,* even if they are not authentic, are informative. There are translations with introductions and notes in many modern languages.

Some useful English versions are *Phaedo,* R. Hackforth, trans. (Cambridge, 1955); *Republic,* F. M. Cornford, trans. (Oxford, 1941); *Theaetetus, Parmenides,* and *Timaeus,* F. M. Cornford, trans., in vols. entitled, respectively, *Plato's Theory of Knowledge* (London, 1935), *Plato and Parmenides* (London–New York, 1939), and *Plato's Cosmology* (London, 1937); *Timaeus* and *Critias,* A. E. Taylor, trans. (London, 1929); *Philebus and Epinomis,* A. E. Taylor, trans., R. Klibansky *et al.,* eds. (Edinburgh–London, 1956), with intro. to the latter by A. C. Lloyd; *Epinomis,* J. Harward, trans. (Oxford, 1928); *Epistles,* J. Harward, trans. (Cambridge, 1932); *Epistles,* Glenn R. Morrow, trans. (Urbana, Ill., 1935; repr. New York, 1961); *Laws,* A. E. Taylor, trans. (London, 1934; repr. 1960).

The appropriate vols. in the French trans., *Oeuvres complètes,* 13 vols. in 25 pts. (Paris, 1920–1956), published under the patronage of the Association Guillaume Budé, should also be consulted.

II. SECONDARY LITERATURE. On Plato's life, see H. Leisegang, *Platon,* in Pauly-Wissowa, *Realenzyklopädie,* 1st ser., XX. The following books and articles are important for his views concerning method and for an estimate of his understanding of science.

J. Adam, *The Republic of Plato,* II (Cambridge, 1902); W. Burkert, *Weisheit und Wissenschaft* (Nuremberg, 1962), containing a useful bibliography; F. M. Cornford, "Mathematics and Dialectic in the *Republic* VI–VII," in *Mind,* 41 (1932), 37–52, 173–190, repr. in *Studies in Plato's Metaphysics,* R. E. Allen, ed. (London, 1965), 61–95; and E. Frank, *Platon und die sogenannten Pythagoreer* (Halle, 1923).

See also K. Gaiser, *Platons ungeschriebene Lehre,* 2nd ed. (Stuttgart, 1968), which contains in an app. the *Testimonia Platonica,* ancient evidence relative to the organization of the Academy and to Plato's esoteric teaching; and "Platons *Menon* und die Akademie," in *Archiv für Geschichte der Philosophie*, **46** (1964), 241–292; T. L. Heath, *History of Greek Mathematics* (Oxford, 1921), and *Aristarchus of Samos* (Oxford, 1913), esp. chs. XV and XVI; G. E. R. Lloyd, "Plato as Natural Scientist," in *Journal of Hellenic Studies*, **88** (1968), 78–92; C. Ritter, *Platon*, 2 vols. (Munich, 1910–1922), containing a chapter on Plato's doctrine of nature and his attitude toward the problems of science; L. Robin, *Platon* (Paris, 1935); F. Solmsen, "Platonic Influences in the Formation of Aristotle's Physical System," in I. Düring and G. Owen, eds., *Aristotle and Plato in the Mid-Fourth Century* (Göteborg, 1960); and A. E. Taylor, *Plato, the Man and His Work*, 4th ed. (London, 1937).

There is a review of literature relating to Plato, 1950–1957, by H. Cherniss in *Lustrum*, **4** (1959), 5–308, and **5** (1960), 323–648; sec. II.B deals with the Academy, sec. V.E with mathematics and the sciences.

On the Academy, the comprehensive account in E. Zeller, *Philosophie der Griechen*, 6th ed. (Hildesheim, 1963), has not yet been superseded; and H. Usener, "Organisation der wissenschaftlichen Arbeit," in *Preussische Jahrbücher*, **53** (1884), repr. in *Vorträge und Aufsätze* (1907), 67 ff., has been a starting point for later discussion. See also E. Berti, *La filosofia del primo Aristotele* (Padua, 1962), pp. 138–159; H. Cherniss, *The Riddle of the Academy* (Berkeley, 1945); H. Herter, *Platons Akademie*, 2nd ed. (Bonn, 1952); W. Jaeger, *Aristotle*, 2nd ed. (Oxford, 1948), *Diokles von Karystos* (Berlin, 1938), and *Paideia* (Oxford, 1944), II and III; G. Ryle, *Plato's Progress* (Cambridge, 1966); P. M. Schuhl, "Une école de sciences politiques," in *Revue philosophique de la France et de l'étranger*, **84** (1959), 101–103; J. Stenzel, *Platon der Erzieher* (Leipzig, 1928); and U. von Wilamowitz, *Platon* (Berlin, 1919; 5th ed. 1959).

On Eudoxus, see K. von Fritz, "Die Lebenszeit des Eudoxos v. Knidos," in *Philologus*, **85** (1929–1930), 478–481; H. Karpp, *Die Philosophie des Eudoxos v. Knidos* (Würzburg, 1933); and E. Frank, "Die Begründung der mathematischen Naturwissenschaft durch Eudoxus?" in L. Edelstein, ed., *Wissen, Wollen, Glauben* (Zurich, 1955), 134–157.

On the mathematical intermediates in Plato's philosophy, see W. D. Ross, *Metaphysics of Aristotle* (Oxford, 1924), I, 166 ff.; J. A. Brentlinger, "The Divided Line and Plato's Theory of Intermediates," in *Phronesis*, **7**, no. 2 (1963) 146–166; S. Mansion, "L'objet des mathématiques et de la dialectique selon Platon," in *Revue philosophique de Louvain*, **67** (1969), 365–388.

On the number in *Republic* VIII, 546, see *The Republic of Plato*, Adam, ed., II, 201–203, 264–312; A. Diès, "Le Nombre de Platon," in *Mémoires de l'Académie des inscriptions*, **14** (1940); Gaiser, *Platons ungeschriebene Lehre*, pp. 271–273, 409–414. On the shape of the earth in the *Phaedo*, see Burkert, *Weisheit und Wissenschaft*, pp. 282–284, and earlier literature there mentioned. On Plato as geographer and as physicist, see P. Friedländer, *Platon*, 2nd ed. (Berlin, 1954), I, chs. 14, 15. On the

Epinomis and the history of science, see E. des Places, "Notice" to the *Epinomis*, in *Platon, Oeuvres complètes*, Collection Budé, XII, 118–128.

D. J. ALLAN

PLATO OF TIVOLI (*fl.* Barcelona, first half of twelfth century), *mathematics, astronomy, astrology, medicine* [?].

Presumably an Italian, Plato is known only through his work, at least part of which was produced at Barcelona between 1132 and 1146. He was one of the first scientist-scholars active in the Iberian Peninsula to provide the Latin West with some of the works of Greek authors as transmitted or elaborated in Arabic and Hebrew and with works originally written in those languages; he was the first to edit Ptolemy in Latin and to help in the translation of the most important Hebrew treatise on geometry. His name appears only as a translator from the Arabic and Hebrew—or, better, as an editor of translations made in collaboration with the Jewish mathematician and polymath Abraham bar Ḥiyya ha-Nasi (Savasorda). It is impossible to determine to what extent the translations ascribed to Plato alone or to both men reflect Plato's linguistic and scientific knowledge; the editor of the most difficult of the translations finds the Latin rendering exact and clear. In the introductions to two translations Plato says that he was prompted by selective interests: he preferred al-Battānī to Ptolemy because al-Battānī was less verbose, and Ibn al-Ṣaffār to many other authors on the astrolabe because he was more reliable and scientific. He was a friend of at least one other translator of Arabic scientific literature, John, son of David, to whom he dedicated his translation of the work on the astrolabe. In the middle of the thirteenth century Plato was mentioned as an eminent Christian mathematician by the author of the *Summa philosophie* wrongly ascribed to Robert Grosseteste. His influence as a translator and editor is shown by the use of his versions by Leonardo Fibonacci and Albertus Magnus, by the relatively large number of manuscripts of his texts still surviving, and by the number of printed editions of some of them produced in the late fifteenth and sixteenth centuries.

The four truly scientific works—two mathematical and two astronomical—that in manuscripts or printed editions carry the name of Plato of Tivoli, with or without the name of Abraham bar Ḥiyya, as a translator are the following:

1. The *Liber Embadorum* ("Book of Areas," or "Practical Geometry") of Savasorda, identified by Steinschneider as the *Ḥibbur ha-meshiḥah we-ha-tish-*

boret of Abraham bar Ḥiyya, translated from the Hebrew in 1145 [the manuscripts give the right astronomical equivalent of this date but the wrong Hegira year, 510/1116 instead of 540/1145] and preserved in at least five manuscripts.

2. The *Spherica* by Theodosius of Bithynia, translated from the Greco-Arabic version of Ḥunayn ibn Isḥāq or of Qusṭā ibn Lūqā and extant in eleven manuscripts and four early editions.

3. Al-Battānī's *al-Zīj* ("Astronomical Treatise"), or *De motu stellarum*, in ten manuscripts and two editions.

4. The *De usu astrolabii* of Abu'l-Qāsim Maslama (Ibn al-Ṣaffār), in three manuscripts—quite convincing linguistic reasons have been given by M. Clagett to confirm a suggestion (based on the place of this text in the three existing manuscripts) that Plato is the author of the Latin translation from the Arabic of Archimedes' *In quadratum* [for *Quadratura?*] *circuli*, or *De mensura circuli*.

The translations from the Arabic of seven other works (five astrological, one geomantical, and one medical [now lost]) are ascribed to Plato, with or without Abraham bar Ḥiyya, in the manuscripts.

1. Ptolemy's *Quadripartitum*, translated in 1138 and preserved in nine manuscripts and five editions.

2. The *Iudicia Almansoris*, which is the *Centum* (or *Centumquinquaginta*) *propositiones* or *Capitula* (*Stellarum*) by, or dedicated to, one al-Manṣur (or by ar-Rāzī [?]), translated in 1136 and preserved in over forty manuscripts and a dozen printed editions.

3. The *De electionibus horarum* of 'Ali ibn Aḥmad ál-Imrani, translated in 1133 or 1134 and extant in eighteen manuscripts.

4. The *De nativitatibus* or *De iudiciis nativitatum* of Abu 'Ali al-Khayaṭ, translated in 1136 and preserved in nine manuscripts.

5. The *De revolutionibus nativitatum* by Abū Bakr al-Ḥasan (Albubather), preserved in one manuscript.

6. The *Questiones geomantice* or *Liber Arenalis scientie* by "Alfakini, son of Abizarch" or "son of Abraham," probably translated in 1135 and preserved in two printed editions and one manuscript (copied from the earlier edition [?]).

7. A *De pulsibus et urinis* by "Aeneas" (Ḥunayn ibn Isḥāq [?]), now lost.

A translation of an anonymous *Commentary on the Tabula Smaragdica*, extant in one manuscript, has been tentatively attributed to Plato by its recent editors.

The translation of Abraham bar Ḥiyya's *Geometry* (*Liber embadorum*) contributed much to both the structure and the contents of Leonardo Fibonacci's *Practica geometriae*. In particular it was the first work to introduce into the Latin study of mathematics the solution of quadratic equations and, with Theodosius' *Spherica*, it furthered the development of trigonometry. As late as 1819, Plato of Tivoli's translation of al-Battānī's *al-Zīj* was the basis for Delambre's detailed study of that work.

BIBLIOGRAPHY

I. ORIGINAL WORKS. Printed eds. of the works cited in text are Abraham bar Ḥiyya, *Liber embadorum*, M. Curtze, ed., in *Urkunden zur Geschichte der Mathematik im Mittelalter und der Renaissance*, I, which is vol. XII of *Abhandlungen zur Geschichte der Mathematischen Wissenschaften* (Leipzig, 1902), 10–182, with German trans. on facing pages; Theodosius of Bithynia, *Spherica* (Venice, 1518; Vienna, 1529; Messina, 1558); al-Battānī, *De motu stellarum* (Nuremberg, 1537; Bologna, 1645), both eds. with additions and notes by Regiomontanus; Archimedes, *In quadratum circuli*, Marshall Clagett, ed., in "The *De mensura circuli*," in *Osiris*, **10** (1952), 599–605, and in Clagett's *Archimedes in the Middle Ages*, I (Madison, Wis., 1964), 20–26, with English trans. on opposite pages— see also 691–708 for a selective index of geometrical terms that includes those used in this trans.; Ptolemy, *Quadripartitum* (Venice, 1484, 1493; Basel, 1551); *Iudicia Almansoris* (Venice, 1484, 1492; Basel, 1533; Ulm, 1641, 1647); al-Imrani, *De electionibus horarum*, partly edited by J. M. Millás-Vallicrosa in *Las traducciones orientales . . .* (see below), 328–339; *Questiones geomantice* (Verona, 1687, 1705); and *Commentary on the Tabula Smaragdica*, R. Steele and D. W. Singer, eds., in *Proceedings of the Royal Society of Medicine*, Hist. sec., **21** (1928), 41–57.

No printed eds. are known to exist of *De usu astrolabii*, *De iudiciis nativitatum*, *De revolutionibus nativitatum*, and *De pulsibus et urinis*.

No complete list of MSS containing the translations ascribed to Plato of Tivoli has been published, but many are mentioned in L. Thorndike and P. Kibre, *A Catalog of Incipits of Mediaeval Scientific Works in Latin*, rev. ed. (Cambridge, Mass., 1963), see index, col. 1889. See also the works of Boncompagni, Carmody, Haskins, Millás-Vallicrosa, Sarton, and Steinschneider cited below.

II. SECONDARY LITERATURE. The four essential works dealing with Plato of Tivoli are F. J. Carmody, *Arabic Astronomical and Astrological Sciences in Latin Translation: A Critical Bibliography* (Berkeley–Los Angeles, 1956), see index, 192; C. H. Haskins, *Studies in the History of Mediaeval Science* (Cambridge, Mass., 1927), 9–11; G. Sarton, *Introduction to the History of Science*, II, pt. 1 (Baltimore, 1931), 177–179; and M. Steinschneider, *Die Europäischen Übersetzungen aus dem Arabischen bis Mitte des 17ten Jahrhunderts* (1904–1905; repr. Graz, 1956), 62–66.

Further information is in the introductions to the modern eds. of Plato's works and in B. Boncompagni, "Delle versioni fatte da Platone Tiburtino," in *Atti dell' Accademia pontificia dei Nuovi Lincei*, **4** (1851), 249–286;

J. A. von Braunmühl, *Vorlesungen über die Geschichte der Trigonometrie*, I (Leipzig, 1900), 48–50; C. H. Haskins, "The Translations of Hugo Sanctallensis," in *Romanic Review*, **2** (1911), 2, a note on the date of the *Liber embadorum*; J. M. Millás-Vallicrosa, *Assaig d'història de les idees físiques i matemàtiques a la Catalunya medieval*, I (Barcelona, 1931), 29 ff., on the *De usu astrolabii*; *Las traducciones orientales en los manuscritos de la Biblioteca catedral de Toledo* (Madrid, 1942); and *Estudios sobre historia de la ciencia española* (Barcelona, 1949), 219 ff., "La obra enciclopédica de Abraham bar Ḥiyya"; C. A. Nallino's ed. of al-Battānī, *Opus astronomicum* (*De motu stellarum*), Arabic text and new Latin trans., I (Milan, 1903), l–lvi; M. Steinschneider, "Abraham Judaeus: Savasorda und Ibn Esra . . .," in *Zeitschrift für Mathematik*, **12** (1867), 1–44, important for the relationship between Plato and Abraham, for the authorship of the *Iudicia Almansoris*, and for al-Imrani's *De electionibus horarum*; L. Thorndike, *A History of Magic and Experimental Science*, II (London, 1923), esp. 82–83; and P. Ver Eecke's introduction to his French trans. of Theodosius of Bithynia's *Spherica* (Bruges, 1926; repr. Paris, 1959), xxxv–xli, on the Latin printed eds., on the use made of them, and on Regiomontanus.

LORENZO MINIO-PALUELLO

PLATTER, FÉLIX (*b.* Basel, Switzerland, October 1536; *d.* Basel, 28 July 1614), *botany, medicine, psychiatry.*

Platter was the son of Thomas Platter, a well-known printer, who sent him to study medicine at Montpellier. At the age of twenty-one he returned to Basel to present his medical thesis, for which he was awarded the doctorate on 20 September 1557. A few years later he was teaching applied medicine at the University of Basel, and in 1571 the city council named him chief physician. His interest in natural history led him to assemble a remarkable herbarium, which was highly admired by scholars of the time.

Platter is known today for his medical activity and his works on human pathology, especially *De corporis humani structura*, which made him famous. He was a faithful disciple of Eustachi, Falloppio, and above all, of Vesalius, from whose *De humani corporis fabrica* much of his own writing on anatomy was derived. From his books and especially from public autopsies, which he performed in Basel, he soon acquired a reputation as an important anatomist.

In *Praxeos seu de cognoscendis* . . . (1602), an outstanding work on pathology, Platter proposed a classification of diseases, which differed greatly from that followed by practitioners. In *Praxis medica* he presented precise descriptions of illnesses and an analysis of their symptoms. *Observationum in hominis affectibus* (1614) contains rigorous descriptions of human

ailments and the search for their causes, as well as accounts of gynecological diseases and investigations of the infectious nature of illnesses.

One of the first to study certain mental disturbances scientifically, Platter refused to consider them the work of a demon—unlike most of his contemporaries—but sought their physiological causes. Insanity, he believed, must be attributed to natural causes, whether they resulted from the influence of overeating or from dissolute living.

Platter's statistical studies of, and memoirs on, the plague contain an abundance of useful data. As a practicing pediatrician he was ahead of his time, and his works were authoritative until the beginning of the eighteenth century.

BIBLIOGRAPHY

I. ORIGINAL WORKS. Platter's chief works are *De corporis humani structura et usu libri III* (Basel, 1583); *Praxeos seu de cognoscendis, praedicendis, praecavendis curandisque affectibus homini incommodantibus tractatus tres* (Basel, 1602); *Praxis medica* (Basel, 1603); and *Observationum in hominis affectibus* (Basel, 1614).

II. SECONDARY LITERATURE. On Platter and his work, see D. A. Fechter, *Thomas Platter und Félix Platter, zwei Autobiographien* (Basel, 1840); A. Bruckner, *Briefe an F. Platter von seine Mutter* (Basel, 1932); H. Buess, "Gynäkologie und Geburtshilfe bei Félix Platter," in *Journal suisse de médecine*, **22** (1941), 1244; F. Ernst, *Die beiden Platter* (Zurich, 1927); J. A. Häfliger, "F. Platter sogenannte Hausapotheke," in *Pharmaceutica acta Helvetiae*, **11** (1936), 351–360; R. Hunziker, "Platter als Arzt und Stadtarzt in Basel" (thesis, Univ. of Basel, 1939); and H. Karcher, "Félix Platter," in *Journal suisse de médecine*, **24** (1943), 1232–1238.

P. E. PILET

PLATTNER, KARL FRIEDRICH (*b.* Klein-Waltersdorf bei Freiberg, Germany, 2 January 1800; *d.* Freiberg, Germany, 22 January 1858), *metallurgy.*

From 1817 to 1820 Plattner studied at the Königliche Bergakademie, Freiberg. He then entered the metallurgical industry, where he gained rapid promotion and acquired a great reputation as an assayer and analyst. He specialized in the use of the blowpipe, which had become a highly versatile and qualitative instrument in many hands, from Gahn to Berzelius. Plattner perfected the quantitative use of the blowpipe and in 1835 published a treatise (translated into several languages) on its use.

Plattner's blowpipe techniques were grounded in an understanding of chemical reactions, which he utilized in his later work (1856) on smelting. At the age of thirty-eight he left Freiberg for Berlin in order to work for a year with Heinrich Rose. Upon his return to Freiberg in 1840, he was appointed to a senior assaying post and then to a chair of metallurgy. Pursuing the chemistry of smelting, he sought to solve the problem (as much a pollution problem as a technical one) of the efficient conversion to sulfuric acid of the sulfur dioxide produced from sulfide ores. Conversion by the chamber process was not difficult in principle, but the necessary, massive structures were so difficult to accommodate in a smelting plant that a simpler process was desirable. Contact catalysis was known (Peregrine Phillips patented his method in 1831) but not perfected. Plattner carried out many experiments, but his work was impeded by failing health and eventually cut short by his death in 1858. He had laid the foundation of the success achieved in 1875 by his colleague at Freiberg, Clemens Winkler.

BIBLIOGRAPHY

On Plattner's life and work, see I. Westermann, "Aus Plattners Leben und Werken," in *Metall und Erz*, **30** (1933), 101–103, with portrait.

His works include *Die Probirkunst mit dem Lötrohre* (Leipzig, 1835); *The Use of the Blowpipe in the Examination of Minerals, Ores, Furnace Products and Other Metallic Combinations*, J. S. Muspratt, trans. (London, 1845), with a preface by Liebig and notes; and *Die metallurgischen Röstprozesse* (Freiberg, 1856). See also Poggendorff, II (Leipzig, 1863), col. 469.

FRANK GREENAWAY

PLAYFAIR, JOHN (*b.* Benvie, near Dundee, Scotland, 10 March 1748; *d.* Edinburgh, Scotland, 20 July 1819), *mathematics, physics, geology.*

Playfair was the eldest son of the Reverend James Playfair. At the age of fourteen he went to the University of St. Andrews, primarily to qualify for the ministry, but he also showed remarkable mathematical ability. In 1769 he left St. Andrews for Edinburgh. On the death of his father in 1772, he succeeded him in the living of Benvie, which he resigned in 1782 to become tutor for a private family. From 1785 to 1805 Playfair held a professorship of mathematics at the University of Edinburgh. He edited the *Transactions of the Royal Society of Edinburgh* for many years, and most of his papers appeared in that periodical. They were concerned almost entirely with mathematics, physics, and biographies. His *Elements of Geometry* was published in 1795. Following the death in 1797 of his

friend James Hutton, Playfair proceeded to make a careful analysis, clarification, and amplification of Hutton's *Theory of the Earth*, which had originally been presented as a paper read in 1785 to the Royal Society of Edinburgh and was expanded into the two-volume work of 1795. Playfair's efforts resulted in *Illustrations of the Huttonian Theory of the Earth* (1802). In 1805 he became professor of natural philosophy at Edinburgh, where his lectures now embraced physics and astronomy. His *Outlines of Natural Philosophy* was published in 1814.

Playfair's professional work was thus in mathematics and physics. His book on geometry is a full presentation of the first six books of Euclid, with much additional material. The formal treatment of linear parallelism requires axioms. Finding Euclid's axioms on this matter to be unsatisfactory, Playfair proposed "that two straight lines, which intersect one another, cannot be both parallel to the same straight line." This is what became known as "Playfair's axiom," as it is given in his *Elements of Geometry*.

Playfair's fame as a scientist, however, rests almost entirely on his work in geology—hardly a "professional" study at the time—in presenting Hutton's momentous theory in a clear and palatable form (which Hutton himself had failed to do), and in adding materially to the geological knowledge of the time. The precision and elegance of the style of his mathematical exposition is here applied to a descriptive, inductive science. As Archibald Geikie remarked (1905): "How different would geological literature be to-day if men had tried to think and write like Playfair!" The publication of the *Illustrations* is indeed one of the most conspicuous landmarks in the progress of British geology. It ended the early period in the history of that science, a termination that happened to coincide with the end of the eighteenth century. There was a pause in the advance of geology during the early years of the nineteenth century; but powerful forces were gathering an impetus that was released in the second decade with the publication of important works by William Smith, John Farey, and Thomas Webster, which were summarized in the next landmark of the literature, Conybeare and Phillips' *Outlines of the Geology of England and Wales* (1822). Playfair lived well into this second period of activity but did not take any part in it. A project that was very much in his mind was the preparation of a comprehensive work on geology, which was to have been a greatly amplified edition of his *Illustrations*. The peace of 1815 enabled Playfair to make an extensive tour of France, Switzerland, and Italy, in order to extend his observations for this purpose; but although we have details of the journey, nothing of the projected work was composed.

By the end of the eighteenth century the rocks of Britain had been classified into two main groups, the Primary (at first called "Primitive") and the Secondary. This division was based on observed superposition, particular attention being paid to unconformities and, as the names show, on the consequently inferred relative ages. The grouping is in fact a natural twofold occurrence in many regions, but the stratigraphical (time) gap is not everywhere at the same part of the general succession; for instance, over much of Scotland it is below the Old Red Sandstone, in northwest England below the Carboniferous, and in south Wales and southwest England below the Permo-Triassic or the Lias. This discrepancy was not known at the time; and long after the end of the eighteenth century the Devonian and Carboniferous rocks of Devon and Cornwall were thought to be equivalent in age to rocks below the Old Red Sandstone elsewhere. As for the igneous rocks, the large granite masses were generally believed to be among the most ancient; but since their intrusive nature had been demonstrated, this assumption was found to rest on no very secure basis. They had not, however, yet been discovered as being intrusive into any but Primary rocks. The smaller intrusions, dikes and sills, found among Secondary rocks, were necessarily accepted as being comparatively young. In fact, the logical position had been reached of a preliminary classification of rocks, regardless of age, into the two lithological groups: igneous and sedimentary.

It is not surprising that no classification of the Primary rocks had been attempted. Particular rocks were simply described by such lithological or mining terms as schistus, slate, killas, and clay-slate, which had no very precise meaning. Within the Secondary group a very rough succession from the Coal Measures to the Chalk had been given by John Strachey (1725), and John Michell (1788) had offered a more accurate succession of the same range of strata. In those regions where the Carboniferous succession had been observed —in Scotland (by John Williams, 1789) and, particularly, in Derbyshire (by John Whitehurst, 1778)—the Limestone was found to underlie the Coal Measures, with Millstone Grit (if present) intermediate. It does not seem that the stratigraphic relation of the Old Red Sandstone (the equivalent in age to the marine Devonian) to the Carboniferous rocks had been observed. The question arises of the extent to which a geological map of Britain could have been constructed from the observations recorded up to 1802. The map would have been very sketchy; and no one made any serious attempt at such a compilation, although William George Maton drew a very inaccurate "mineralogical map" of southwest England in 1797.

The geological researches of William Smith had begun about 1790. In 1801 he colored geologically a small map of England and Wales, and in 1815 his great map of England and Wales was published.

Such was the position when Playfair wrote his book, which divided two eras in the history of geological investigation. His own observations, inferences, and expressions were the final contributions to the first era and can be classed under three heads.

First, Playfair realized the importance of unconformity in the manifestation of the geological cycle; and he searched throughout Britain for signs of this kind of structural relation, to add to the instances already recorded by Hutton. (Unconformity implies the operation of the "geological cycle"—deposition, deformation, emergence, erosion, submergence, and deposition. The concept of the geological cycle is the essence of Hutton's theory.) Thus Playfair observed the unconformity between the Permo-Triassic and Devonian to be seen at places on the coasts of both north and south Devonshire, and that between the Old Red Sandstone and Dalradian ("Primary schistus") on the east and west coasts of Scotland. He graphically described the unconformity between the Carboniferous Limestone and pre-Devonian rocks in the Ingleborough district of Yorkshire (the British region that shows this phenomenon most clearly) and gave a glimpse of the structure of the English Lake District with its rim of unconformable Carboniferous rocks.

Second, Playfair made miscellaneous observations, of which the more significant were the fossiliferous nature of the Primary Devonian limestone at Plymouth; the fractures, curiously plane without shattering, in the Old Red Sandstone conglomerate at Oban in Scotland; the general constancy of the east-northeast/west-southwest trend in the structure of the older rocks of Britain; the form of the intrusive sill at Salisbury Craigs and the metamorphism at the volcanic neck of Arthur's Seat, in the neighborhood of Edinburgh; the small-scale folding in the Dalradian schist of Ben Lawers, which he noticed resembled that in the Alpine region; intrusive veins in Ayrshire and Arran and the contact metamorphism produced by them; the flint-gravels of southern England as the residue of dissolved flinty chalk; and the submerged forest of the Lincolnshire coast.

Third, Playfair's book used many more-or-less ordinary words (arenaceous, consolidated, petrifaction) in modern geological senses, most of them probably for the first time, and introduced several highly significant terms into geological literature (geological cycle, igneous origin). His name is attached to a geomorphological "law," "Playfair's law of accordant junctions," which, as given in the *Illustra-*

tions, states that "Every river appears to consist of a main trunk, fed from a variety of branches, each running in a valley proportioned to its size, and all of them together forming a system of vallies, communicating with one another, and having such a nice adjustment of their declivities that none of them join the principal valley on too high or too low a level,—a circumstance which would be infinitely improbable if each of these vallies were not the work of the stream that flows in it."

BIBLIOGRAPHY

I. ORIGINAL WORKS. Playfair's chief writings are *Elements of Geometry* (Edinburgh, 1795); *Illustrations of the Huttonian Theory of the Earth* (Edinburgh, 1802); *Outlines of Natural Philosophy* (Edinburgh, 1814); and *A General View of the Progress of Mathematical and Physical Science Since the Revival of Letters in Europe*, vols. II and IV of *Encyclopaedia Britannica*, supp. (1816). In *The Works of John Playfair*, James G. Playfair, ed., 4 vols. (Edinburgh, 1822), "are contained all the publications to which Mr. Playfair affixed his name, with the exception of the *Elements of Geometry*, and of the *Outlines of Natural Philosophy*, which were intended only for the use of students, and although excellently adapted to their object, would possess but little interest for the general reader." Vol. I contains a biographical memoir (see below) and *Illustrations of the Huttonian Theory*; vol. II, the *Progress of Mathematical and Physical Science*; vol. III, various papers on mathematics and physics, plus a "lithological survey of Schehallien" (in Scotland), all of which had appeared either in *Transactions of the Royal Society* or *Transactions of the Royal Society of Edinburgh*; and vol. IV, various biographical accounts and reviews, a biography of James Hutton being particularly important. His works are listed in the British Museum *General Catalogue of Printed Books*, CXCI, cols. 378–379.

II. SECONDARY LITERATURE. The source for details of the life of Playfair is F. Jeffrey, "Biographical Account of the Late Professor Playfair," prefixed to J. G. Playfair's ed. of the *Works* (see above). See also B. B. Woodward, in *Dictionary of National Biography*, XLV (1896), 413–414. In reviewing Hutton's theory, Playfair's *Illustrations* is nearly always referred to, since it is an essential part of the authoritative exposition of Hutton's principles. See particularly A. Geikie, *The Founders of Geology* (London, 1905), 280–316; and "Lamarck and Playfair," in *Geological Magazine*, **43** (1906), 145–153, 193–202; C. C. Gillispie, *Genesis and Geology* (New York, 1951; repr. 1959), *passim*; and R. J. Chorley *et al.*, *The History of the Study of Landforms* (London, 1964), 57–68. Playfair's original contributions to geological knowledge are reviewed in J. Challinor, "The Early Progress of British Geology—III," in *Annals of Science*, **10** (1954), 107–148, see 137–143.

JOHN CHALLINOR

PLAYFAIR, LYON (*b.* Chunar, India, 21 May 1818; *d.* London, England, 29 May 1898), *chemistry.*

The son of a medical officer in the East India Company, Playfair was brought up by relatives in Scotland. In 1835 he became a medical student at the Andersonian Institution, Glasgow, where Thomas Graham awakened his interest in chemistry. Playfair finished his medical studies at Edinburgh but had to give up a medical career because of eczema. A mercantile career in India was also abandoned after a brief trial, and he returned to London to become Graham's assistant.

Playfair spent 1839–1841 in Liebig's laboratory at Giessen, where he did outstanding work on myristic acid and caryophyllene; he increased his reputation by translating Liebig's *Die organische Chemie in ihre Anwendung auf Agricultur und Physiologie* into English. After leaving Germany, Playfair became chemical manager at the Primrose Mill, near Clitheroe, Lancashire; but, sensing a decline in trade, he left in the autumn of 1842 to become professor of chemistry at the Royal Institution, Manchester. During his three years there, Playfair collaborated with Joule on laborious but ultimately fruitless work on the atomic volumes of substances in solution; he also worked with Bunsen on the analysis of blast-furnace gases.

In 1842 Playfair's application for a chair in chemistry at Toronto had elicited the unprecedented intervention of the prime minister (Sir Robert Peel), who offered Playfair an important post if he would stay in Britain. This offer could not be implemented until 1845, when he became chemist to the Geological Survey and professor of chemistry at the new School of Mines. Meanwhile, however, he was much in governmental favor, and served on important Royal Commissions on the Health of Towns and on the Irish Potato Famine. In 1848 Playfair was elected a fellow of the Royal Society and was subsequently Prince Albert's chief adviser on the Great Exhibition of 1851. While holding the post of inspector of schools of science in London (1850–1858) he made his most important discovery, the nitroprussides.

In 1858 Playfair accepted the chair of chemistry at Edinburgh with high hopes, although in his eleven years there he achieved little. In 1868, as a Liberal party candidate, he was elected member of Parliament for the universities of Edinburgh and St. Andrews, and the rest of his life was spent in politics. He was active in the cause of universal compulsory education (1870) and took a prominent part in the reform of the Civil Service; he was briefly postmaster general (1873–1874) under Gladstone and deputy speaker of the House of Commons (1880–1883). He was knighted on retirement from this office, and in 1892 he was raised to the peerage as Baron Playfair of St. Andrews.

Short in stature and with more than his share of Victorian pomposity and wordiness, Playfair openly sought the company of the exalted and successful, and was consequently disliked by many. He played an important role, however, as one of the first government scientists in Britain.

BIBLIOGRAPHY

I. ORIGINAL WORKS. Apart from the usual scientific papers, of which the two most important are "Report on the Gases Evolved from Iron Furnaces," in *British Association Report for 1845*, 142–186, written with R. W. Bunsen, and "On the Nitroprussides, a New Class of Salts," in *Philosophical Transactions of the Royal Society*, **139** (1849), 477–578, Playfair's publications are mainly in the form of government reports, of which *Report on the State of Large Towns in Lancashire* (London, 1845) is the most important.

II. SECONDARY LITERATURE. Shortly after Playfair's death, Wemyss Reid published *Memoirs and Correspondence of Lyon Playfair* (London, 1899), consisting largely of extracts from his journal and letters. There is a chapter on Playfair in J. G. Crowther, *Statesmen of Science* (London, 1965), 105–174. There is a notice on him in supp. III of *Dictionary of National Biography* (1900), 1142–1144. His obituary notices were very numerous, and only that by H. E. Roscoe need be mentioned: *Nature*, **58** (1898), 128–129. See also R. G. W. Norrish, in *Journal of the Royal Society of Arts*, **99** (1951), 537–548.

W. V. FARRAR

PLENČIČ (or **PLENCIZ**), **MARCUS ANTONIUS** (*b.* Solkan, Austria [now Yugoslavia], 28 April 1705; *d.* Vienna, Austria, 25 November 1786), *medicine*.

Plenčič received his early education at Gorizia, near his birthplace, then studied medicine at Vienna and at Padua, where Morgagni was one of his teachers. Having taken the degree at the latter university, Plenčič returned to Vienna in 1735. He established a successful practice and remained there for the rest of his life. In 1770 Maria Theresa ennobled him in recognition of his achievements; he was also accorded the freedom of the city and the province of Gorizia.

As a practitioner, Plenčič made careful observations of contagious diseases; applying a rigorous logic to these observations, he reached a remarkable theory of the nature of contagion. He was one of the first to recognize the etiological significance of Leeuwenhoek's animalcules, and his own views were developed a century later by Pasteur, Koch, and their followers.

Plenčič's theory is set out in his *Opera medico-physica* of 1762. The work consists of four separate treatises, of which the first contains Plenčič's general theory of disease, while the second and third deal with the specific diseases of smallpox and scarlet fever, presented in illustration of that theory. (The fourth treatise is a digression on the great earthquake of 1755, which destroyed Lisbon.)

Plenčič believed contagious diseases to be caused by microorganisms, which he called "*animalcula minima*," or "*animalcula insensibilia*." These microorganisms, he stated, are both specific and constant—a given animalcule always causes the same disease, and attacks a specific host. They are carried by the air; Plenčič's descriptive terms for the means of infection—"*materia animata*," "*miasma animatum*," "*miasma verminosum*," "*seminia animata*," and "*principium aliquod seminale verminosum*"—emphasizes his conviction concerning the animal nature of the microorganisms. The microorganisms are, in addition, very small (Leeuwenhoek estimated that a drop of water could contain between two and three million) and reproduce extremely rapidly in an appropriate medium, even outside the infected body. The speed of reproduction, Plenčič stated, accounts for why a minute amount of an inoculum (as of smallpox) can cause disease. He also noted the incubation period in infectious diseases and discussed the possibility that microorganisms might have periods of latency, after which, conditions having become more suitable, they might resume their pathogenic activity. He was aware of disposition toward a specific disease, immunity (including that resulting from a previous attack), mixed infection, antibiosis, and chemotherapy.

In recommending his theory, Plenčič pointed out that it permitted a simple explanation for the propagation of diseases and offered an occasion for their rational prevention and treatment. He suggested the use of remedies acting directly upon the microorganisms, among them anthelmintics and antiseptics (mostly compounds of heavy metals). A further advantage of his etiological theory was that it allowed all the contagious diseases of man (including smallpox, plague, and scarlet fever), animals (including cattle plague), and even plants (for example, wheat rust) to be considered on a common basis.

Although Plenčič's work was preceded by that of Fracastoro, Kircher, and Lancisi, his theory is the most comprehensive and consistent. His book presents it clearly, and the examples he gives are convincing. Nonetheless, his theory attracted little attention. As late as 1828 K. Sprengel wrote, in the fifth volume of his history of medicine, that "His rich experience with scarlet fever did not prevent Plenčič from proposing a flighty hypothesis of *seminiis animatis*." It was only in 1840 that the germ theory of contagious diseases was again advanced, against considerable opposition, by F. G. J. Henle.

BIBLIOGRAPHY

I. ORIGINAL WORKS. Plenčič's principal work was published as *Opera medico-physica in IV tractatus digesta . . .* (Vienna, 1762). The third treatise was translated into German by J. P. G. Pflug as *Abhandlung vom Scharlachfieber* (Copenhagen, 1779); it was emended by Plenčič and published separately as *Tractatus de scarlatina . . .* (Vienna, 1780). Another work is *Dissertatio physicoeconomica sive nova ratio frumenta aliaque legumina quam plurimis annis integra salvaque conservandi* (Vienna, 1764), also published in French and German.

II. SECONDARY LITERATURE. See I. Fischer, "M. A. Plenčičz, ein Wiener Vorläufer der modernen Bakteriologen," in *Wiener klinische Wochenschrift*, **26** (1913), 1804–1807; and "Marc Anton Plenciz," in *Wiener medizinische Wochenschrift*, **77** (1927), 735–736. A more recent treatment is A. Berg, "Marc Anton Plencicz und seine Lehre von den belebten Krankheitserregern (1762)," in *Medizinische Welt* (1962), 2425–2428, 2480–2483. See also Branko and Ivan Marušič, *Solkanski rojak dr. Marko Anton pl. Plenčič (1705–1786)* (Solkan, 1967); S. Peller, "Marc Anton von Plenciz," in *Pirquet Bulletin of Clinical Medicine*, **10** (1963), 2, 11–12; and I. Pintar, "Marko Anton Plenčič in ujegov nauk o 'Contagium vivum,' " in *Zdravstveni vestnik*, **16** (1947), 64–71.

VLADISLAV KRUTA

PLINY (GAIUS PLINIUS SECUNDUS) (*b.* Como, Italy, *ca.* A.D. 23; *d.* near Pompeii, Italy, 25 August A.D. 79), *natural history.*

Pliny's parents are not known to have been distinguished, but his father had means and was no doubt a respected member of the community. By the age of twelve Pliny was in Rome, where he must have received a thorough education in literature, oratory, and law, as well as some military training. His attitude to religion was a typically Roman blend of credulity and skepticism. There is no evidence that he married. He adopted as his son and heir, perhaps in his will, his nephew Pliny the Younger. Even in later life, when he suffered from respiratory trouble, his mental stamina was exceptional. He never needed much sleep, his motto being "To live is to be awake" (*Natural History,* preface, sec. 18).

Promotion to the senatorial order and the highest offices was achieved by Pliny's heir; but thanks to his father's position and his own education, Pliny himself was able at the age of about twenty-three to begin the official career open to members of the second great Roman order, the equestrian. The early stages of such a career were military. While in command of a cavalry squadron stationed on the Rhine frontier, Pliny began his career as a writer with a monograph concerning the use of javelins by cavalry. It was soon followed by a history of the Roman campaigns in Germany and by a biography of Pomponius Secundus, who may have been his superior officer in Germany and who had become his close friend. Pomponius was a poet and dramatist as well as an administrator and a general. This range of activities foreshadowed and perhaps encouraged Pliny's own versatility.

By A.D. 57 or 58 Pliny had completed his military duties and had returned to Italy, where for the next ten years he wrote works on oratory and grammar, and possibly practiced as a lawyer. One reason for this change of course may have been his dislike of Nero's regime; another may have been Nero's disfavor or Pliny's fear of it. When Vespasian became emperor in A.D. 69, Pliny was able to resume his official career. He had probably served with Vespasian's son Titus in Germany, and this connection may have helped to bring about the series of appointments as financial overseer of a province that he is said to have held with the utmost integrity. One of these posts took him to a Spanish province, another perhaps to the province of Africa, which is now Tunisia and Tripolitania. Rather than impede Pliny's literary activities, these duties seem to have spurred him to almost feverish efforts. At this time he was working on a history, published posthumously, of the period covering roughly A.D. 44–71.

Roman historians were usually men of affairs, and in this respect Pliny was exceptional only because he was not a senator. While working on his history, he was drafting his only extant work, the thirty-seven books of the *Natural History*, although the collecting of notes for it must have started earlier. The work was dedicated to Titus in 77, but the final redaction may have been carried out after Pliny's death by his nephew. Toward the end of his life Pliny became a recognized counselor or "friend" of Vespasian and then of Titus. His last official post was that of commander of the fleet based at Misenum, at the northwest extremity of the Bay of Naples; and it was from Misenum that he started on the voyage that led him to his death near Pompeii, where he was overcome by the fumes from the eruption of Vesuvius.

Pliny was a savant who was also a man of affairs, and the latter must not be overlooked at the expense of the former. It is misleading to approach the *Natural History* as if it were self-contained. Pliny's preoccupation with his public status is evident in the digression on the history of the equestrian order in book 33 (secs. 29–36) and in frequent references to members of the order. His campaigns in Germany bear fruit at the beginning of book 16 in his vivid description of North Sea fisherfolk living precariously on their artificial mounds, as well as in other eyewitness reports. Pliny's interest in history, not least the history of the period that was the subject of his own historical writing, is reflected throughout the work. Oratory too reverberates through it, notably in numerous denunciations of greed, extravagance, and moral decadence, and in glowing panegyrics of nature, the Roman Empire and Italy, and of such statesmen as Pompey, Cicero, and Titus. Although much of the *Natural History* consists of dry facts, both style and observations continually reveal the personality of the man who wrote it.

In his preface to the *Natural History*, Pliny claims—rightly—that the enterprise is a novel one. There had been other encyclopedias—for example, of the liberal arts—but, as he says (preface, sec. 14), no Greek by himself had compiled an encyclopedia of the whole of nature; and no Roman had done so by himself or with others. The novelty of the task was one of its attractions. Among others were Pliny's inexhaustible curiosity, his conviction that he must be of service—"It is godlike," he writes (bk. 2, sec. 18), "for man to help man"—his anxiety to save the science of past ages from the forgetful indifference of the present, and his desire to make his reputation secure. The result was aptly described by Pliny the Younger as "a diffuse and learned work, no less rich in variety than nature itself" (III.5.6).

The preface addressed to Titus is followed by a novelty, in that book 1 consists of an index of topics and authorities for each of the succeeding thirty-six books. The general plan of the treatise itself is conventional, proceeding from the world to the earth, and from the earth to its products—animal, vegetable, and mineral. But this simple outline is blurred. Book 2 duly surveys the universe, ending with the earth conceived as its center and with terrestrial phenomena. It is followed by books 3–6 (geography), 7 (man), 8–11 (other animals), 12–19 (botany), and 20–27 (materia medica from botanical sources). These last eight books are complemented by 28–32 (materia medica from animal sources); books 33–37 concern metals and stones, including their uses in medicine, architecture, and especially art.

Pliny states in the preface (sec. 17) that 100 principal authors have provided him with 20,000 important facts for his work. The incomplete figures in his index (book 1), however, add up to far more authors and

"remedies, researches and observations"—473 and 34,707, respectively. These statistics indicate the predominantly practical aim and factual character of the treatise. Book 2 necessarily contains the most theory, and in it Pliny generally adopts Stoic doctrines, directly or indirectly derived from Posidonius, a distinguished philosopher with special scientific interests. Otherwise theorizing is spasmodic—for example, in book 37 there are scattered traces of a theory concerning the formation of stones that also may have originated with Posidonius.

Yet merely as a compilation of facts the *Natural History* is unique. Comprehensiveness is all: "Things must be recorded because they have been recorded," remarks Pliny (bk 2, sec. 85); and criticism will not deter him. In book 37 (secs. 30–46), through his own knowledge and observation Pliny gives an almost entirely correct account of the nature and provenance of amber, but not before he has related all the myths and speculations about it that have come to his notice. Still, this uncritical and all-inclusive method has its advantages. A nonsensical reference to Indian amber may be an indication that shellac was known. Pliny would have felt that knowledge preserved even in this way justified the means. Although such diffuseness interfered with the practical aims of the work, Pliny's influence in the succeeding centuries was nevertheless great and abridgments were made, especially of his medical and geographical material.

Less apparent in antiquity was Pliny's lack of reliability in quoting or using his sources, a result of the speed with which he worked. This fault, coupled with his eccentric style, has tended to lower his standing with modern classical scholars, although he might at least have been credited with an eye for landscape and a bizarre imagination that are by no means to be expected in a classical author. The phrase "Mountains and ridges soaring away into the clouds" (bk. 27, sec. 3) illustrates the former. The latter is exemplified by a strange vision of the city of Rome: "If the whole were massed together and thrown on to one great heap, the grandeur that would tower above us would be as if some other world were being described, all concentrated in one single place" (bk. 36, sec. 101).

Students in other fields are rather more appreciative. One of Pliny's happiest thoughts—the series of digressions on painting (bk. 35, secs. 50–148), sculpture in bronze (bk. 34, secs. 15–93), and sculpture in marble (bk. 36, secs. 9–43)—has provided art historians with what amounts to the earliest surviving history of art. Historians of science find that his descriptions often make the identification of species uncertain, if not impossible, but gratefully acknowledge that he is indispensable. Without him their material would be much depleted; and even when he is grossly inaccurate, he can be illuminating. As a purveyor of information both scientific and nonscientific, Pliny holds a place of exceptional importance in the tradition and diffusion of Western culture.

BIBLIOGRAPHY

The *Natural History* is available in Latin text with English trans. by H. Rackham *et al.* in the Loeb Classical Library, 10 vols. (London–Cambridge, Mass., 1942–1963), and as *Histoire naturelle*, Latin text with French trans. and commentary by A. Ernout *et al.*, in the series Belles Lettres (Paris, 1947–). The modern numbering by short sections is now generally used in references.

Biographical sources include two letters of Pliny the Younger, III.5 (listing and describing his uncle's writings, with details of his habits and references to his career) and VI.16 (describing the eruption of Vesuvius and his death), in Pliny, *Letters*, Latin text with English trans., rev. by W. M. L. Hutchinson, in Loeb Classical Library, 2 vols. (London–Cambridge, Mass., 1927). For discussions and commentaries see A. N. Sherwin-White, *The Letters of Pliny, a Historical and Social Commentary* (Oxford, 1966), 215–225, 371–375.

Pliny as historian is presented in R. Syme, *Tacitus*, 2 vols. (Oxford, 1958); and B. H. Warmington, *Nero: Reality and Legend* (London, 1969). Special commentaries are K. C. Bailey, *The Elder Pliny's Chapters on Chemical Subjects*, 2 vols. (London, 1929–1932); D. J. Campbell, *C. Plini Secundi Naturalis historiae liber secundus* (Aberdeen, 1936), on the cosmology of bk. 2; K. Jex-Blake and E. Sellers, *The Elder Pliny's Chapters on the History of Art* (London, 1896); and L. Urlichs, *Chrestomathia Pliniana* (Berlin, 1857). A general reference is W. Kroll *et al.*, "Plinius (5)," in Pauly-Wissowa, *Real-Encyclopädie der classischen Altertumswissenschaft*, XXI, pt. 1 (Stuttgart, 1951), 271–439.

DAVID E. EICHHOLZ

PLOT, ROBERT (*b.* Borden, Kent, England, 13 December 1640; *d.* Borden, 30 April 1696), *natural history, archaeology, chemistry.*

Of all the British naturalists of the late seventeenth century, few represent the omnivorous curiosity of the Baconian tradition and its passion for collecting specimens and observations for their own sake so well as Robert Plot. The son of Robert Plot and Elisabeth Patenden, he was educated at the Wye Free School and at Oxford, which he entered in 1658, graduating B.A. in 1661, M.A. in 1664, and LL.D. in 1671. For many years he served as a college tutor. About 1674 he drew up an itinerary patterned on those of earlier English

antiquaries; but whereas they had been concerned with books and buildings to the exclusion of natural history and technology, Plot intended to tour England and Wales in search of "all curiosities both of Art and Nature such . . . as transcend the ordinary performances of the one and are out of the ordinary Road of the other." He began with the county in which he was then living, starting work on his *Natural History of Oxfordshire* in June 1674; by November 1675 he had a fine collection of minerals to exhibit to the Royal Society, and the book appeared in 1677. On the strength of the *Natural History*, Plot was elected fellow of the Royal Society in 1677. He was secretary in 1682–1684 and thus joint editor of the *Philosophical Transactions*, most of which were printed at Oxford during his term of office; he was elected secretary again in 1692. His success as a collector of rarities must also have helped when, in March 1683, the University of Oxford appointed him first keeper of the newly acquired Ashmolean Museum.

At the same time there was to be an Ashmolean professor of chemistry, with a laboratory in the museum, equipped at great expense; Plot was chosen for this post too. In this field his researches were dominated by the hunt for a wonderful menstruum, to be extracted from spirits of wine, which could act as a general solvent. Since no adequate funds had been set aside to pay salaries, the professor and his assistant supplemented their incomes by making up iatro-chemical drugs. Although Plot did not believe in transmutation in the old sense—it was "but a kind of dying" and it would be sufficient to achieve "a fixt penetrating colour"—he held that the methods of alchemy were essential for true medicine. The medicinal springs of Oxfordshire were subjected to analysis by methods taken from the work of Boyle and Thomas Willis.

With his new dual position at Oxford, Plot had to give up his offices in the Royal Society, although he continued to supply it with specimens from his travels. In 1684 he helped to revive the Oxford Philosophical Society and became its director of experiments. Meanwhile, although he had abandoned his plan to survey the whole country, Plot was invited to resume his explorations, this time in Staffordshire, the natural history of which he published in 1686. He subsequently concentrated more on archaeology, searching in particular for Roman remains. In 1690 he resigned all his Oxford posts, married Rebecca Burman of London, and settled on his ancestral estate, Sutton Baron. Two more natural histories, of Kent and Middlesex, were never completed. In the summer of 1695 he began to suffer severely from calculi; by September he was sufficiently recovered to go on an archaeological tour of East Anglia. The disease returned, however, and he died of it in April 1696.

Plot's stress on the unusual and anomalous, and his expectation that more can be learned from exceptions than from the general rule, apparently stemmed from his interpretation of the Baconian inheritance; this approach gives his natural histories a rather bizarre and curious flavor—his zoology tends to be teratology. He started with the heavens—curious meteorological phenomena observed in the county—then its airs (acoustic researches into sites famous for their echoes), waters—especially mineral and medicinal—and earths. The phenomena of erosion, which he called "deterration," are discussed. He had some notion of stratigraphy, observing that "the Earth is here [Shotover Hill], as at most other places, I think I may say of a bulbous nature, several folds of diverse colour and consistencies still including one another," and listed those observed at localities where their disposition was of economic interest.

Plot also made an extensive study of "formed stones" or fossils, without appreciating that they could be used to identify strata. The controversy on the origin of fossils was then at its height. Plot argued, from the differences between fossil shells and any known specimens of the living shellfish they were thought to represent, that fossil shells were crystallizations of mineral salts; their zoomorphic appearance was as coincidental as the regular shapes of stalactites or snowflakes. Large quadruped fossils he considered the remains of giants, except for one identified as that of an elephant through comparison with an elephant skull in the Ashmolean Museum.

In 1684 Plot produced a treatise on the origin of springs; the bulk of it was reprinted in the natural history of Staffordshire. Here he marshaled evidence, from his own observations and from contemporary geographers, purporting to show that it is quantitatively impossible for all springs to be supplied entirely from rainfall, so that most must come through underground channels from the sea. In investigating such wonders as fairy rings or rains of frogs he tried to offer naturalistic explanations.

In archaeology Plot failed to allow enough for pre-Roman structures: he usually ascribed barrows and henges to Saxons or Danes. One of his main objectives was to describe local crafts and farming techniques, in the hope of diffusing successful practices or new inventions throughout the country. Thus technological information is scattered through both his works on natural history, providing useful evidence on contemporary agriculture, mines, and such industries as the Staffordshire potteries. Whatever the shortcomings of Plot's works, in their time they fulfilled a need. Plot had many imitators, for the books proved to be a useful means of assembling data on the distribution of rocks, fossils, minerals, flora, and fauna.

BIBLIOGRAPHY

I. ORIGINAL WORKS. Plot's books are *The Natural History of Oxfordshire* (Oxford, 1677); *De origine fontium* (Oxford, 1684); and *The Natural History of Staffordshire* (Oxford, 1686). Papers on various curiosities of nature were published in the *Philosophical Transactions of the Royal Society*, esp. **12-20** (1682–1686).

II. SECONDARY LITERATURE. R. T. Gunther, *Early Science in Oxford*, XII, *Dr. Plot and the Correspondence of the Philosophical Society of Oxford* (Oxford, 1939); E. Lhwyd, "A Short Account of the Author," prefixed to *Natural History of Oxfordshire*, 2nd ed. (London, 1705); Anthony à Wood, *Athenae Oxonienses*, P. Bliss, ed., IV (London, 1820), 772–776; and see also F. Sherwood Taylor, "Alchemical Papers of Robert Plot," in *Ambix*, **4** (1949), 67–76.

A. G. KELLER

PLOTINUS (*b. ca.* A.D. 204; *d.* southern Italy, A.D. 270), *philosophy*.

Plotinus' family is unknown, but his education and culture were entirely Greek. He may have been born in Egypt; he studied philosophy at Alexandria and later taught it at Rome, where he settled about 243. A biography by his disciple Porphyry is informative about his manner of teaching and contains a reliable chronological list of his writings. His works were published shortly after his death in the form of six books, called *Enneads* because each contained nine tracts.

The contribution of the *Enneads* to scientific thought is minimal. They represent a combination, which is for the most part original and which is the broad meaning of Neoplatonism, of personal mysticism and a special interpretation of Plato's metaphysics. In this interpretation there is an ineffable One, which is known only by mystical union with it, and two lower levels of reality ("hypostases"): Intellect, which is also the realm of the Platonic Ideas; and then Soul, which also contains nature. Each hypostasis is a reflection or, as it is often called, an "emanation," of the one before it. Until modern times Neoplatonism was taken as the accepted philosophical meaning of Plato. But Plato's interest, as it is shown by the *Timaeus*, in natural science for its own sake, and above all in the application of mathematics to it, is not shared by Plotinus. And his Neoplatonic successors in Athens and

Alexandria, whose learning was transmitted to the Arabs in the sixth century, turned to Aristotle rather than to Plotinus for their physics.

By historical standards—or the standards of Greek science—Plotinus' model of scientific explanation was backward-looking. The reason was his metaphysical theory of nature, which was in effect a kind of panpsychism. Matter as such was unreal, so that the properties of nature fell under the same type of explanation as those of conscious behavior; there was no discontinuity in the chain of being. Such a type of explanation is represented by the concept of sympathies, which Plotinus used not merely to explain psychophysical interaction, such as one's awareness of a pain in his foot, but also to explain external sense perception: there is no need, he argued, for a medium such as air or light between sense organ and object (*Enneads* IV, 5). Sympathy was a Stoic concept; but according to Stoics it operated by mechanical means —exactly the explanatory mechanism that Plotinus rejected.

In a more general and more indirect way Plotinus' metaphysics may have helped later scientists to recognize the importance of mathematics. For Neoplatonists, light was the visible manifestation of goodness and of power; effects of causes could be described as their "reflections" and knowledge as "illumination." If this belief is combined with the fact, well-known since Euclid, that the behavior of light seems to follow the laws of geometry, it can be thought both to prompt further study of optics and astronomy and to justify the place of geometry in nature. Thus for Robert Grosseteste light was not only an analogy of the divine light but also the first form possessed by a physical solid. But it must be borne in mind that Plotinus' writings were unknown in the Middle Ages, and the chief intermediary would have been St. Augustine, who was not interested in mathematical science.

The *Enneads* were first printed in a Latin translation by Marsilio Ficino in 1492 and the Greek text in 1580. They were then of interest to humanists and philosophers, although through Ficino the tract on astrology (*Enneads* II, 13) aroused wider controversy among savants. In it Plotinus argued that the relative positions of stars were not causes but signs of future events; this significance, which was in any case limited, depended on the mutual sympathy of the parts of the universe.

BIBLIOGRAPHY

The standard ed. is *Plotini Opera*, P. Henry and H. R. Schwyzer, eds. (Paris–Brussels, 1951–1973). English translations are by S. MacKenna, 3rd ed., rev. by B. S. Page (London, 1962), and, with Greek text, by A. H. Armstrong, in Loeb Classical Library (London–Cambridge, Mass., 1966–   ); a German translation with explanatory notes is by R. Harder, R. Beutler, and W. Theiler (Hamburg, 1956–1967).

Secondary sources are Porphyry's "Life of Plotinus," included in all above eds.; *Cambridge History of Later Greek and Early Medieval Philosophy*, A. H. Armstrong, ed. (Cambridge, 1967), chs. 12–16; and A. C. Crombie, *Robert Grosseteste and the Origins of Experimental Science 1100–1700* (Oxford, 1953), chs. 5, 6.

A. C. LLOYD

PLUCHE, NOËL-ANTOINE (*b.* Reims, France, 13 November 1688; *d.* La Varenne-Saint-Maur, near Paris, France, 19 November 1761), *scientific popularization*.

Pluche, whose father died when he was twelve years old, came from a modest family. His mother intended that he should become a priest, and he began theological studies at an early age. He became a subdeacon at the age of fifteen and was ordained in 1712. In 1710 Pluche began teaching humanities at the seminary in Reims, where, upon his ordination, he obtained the chair of rhetoric. He was interested in ancient languages, science, and the fine arts, and he also played music, tried his hand at poetry, and wrote plays in which his pupils acted. About 1713 he wrote *Abrégé de l'histoire de Reims*, which was published under a pseudonym in 1721.

Meanwhile, Pluche had become identified with the Jansenists and was suspected of opposing the positions pronounced in the papal bull *Unigenitus* (1713). As a result, he had to give up his post and leave Reims in 1717. He was appointed principal at the *collège* of Laon but had to relinquish the post the following year when he refused to sign any retraction. Fearing that the king would grant his superiors the *lettre de cachet* that they were seeking against him, Pluche, under the assumed name of the Abbé Noël, took refuge with the intendant of Normandy, Monsieur de Gasville, who placed him in charge of his son's education. Having become acquainted at Rouen with a rich Englishman, Lord Stafford, Pluche learned English and gave what were then called physics lessons—actually natural history—to the latter's son.

Henceforth, Pluche gained his livelihood from teaching and from land rentals. He continued to support himself in the same manner when, soon after, he left Normandy for Paris, where he gave lessons in history and geography.

Pluche gave up teaching entirely for several years in order to write *Le spectacle de la nature*, which appeared in eight volumes between 1732 and 1750. His efforts were well rewarded, for the work enjoyed an immediate and immense success. According to the estimate of C. V. Doan, the work, along with its abridgments and adaptations, went through at least fifty-seven editions in France, seventeen in England, and several more in other European countries. In the inventory Daniel Mornet made of the books in five hundred private libraries of the period, he found *Le spectacle* in 206 of them. Well known by the educated public, the work played an important role in the education of children of wealthy families and was sometimes even used as a textbook of natural science. *Le spectacle* is explicitly didactic, and for a time Pluche had even thought of calling it "La physique des enfants." Composed mainly in the form of dialogues between a young nobleman, his parents, and a prior, it is an idealization of Pluche's activities as tutor to the Stafford family.

From the publication of the first volume of *Le spectacle*, contemporaries pointed out that Pluche had borrowed extensively from Derham. Pluche's familiarity with the English literature of the period is obvious, and he often cited Hales, Burnet, and Woodward, as well as Derham. Indeed, Pluche's work lay outside the mainstream of eighteenth-century French natural science, since he rejected most aspects of Enlightenment thought. He found his vein in the current of natural theology that had been developing in England since John Ray, and it was this way of thinking that he introduced with astonishing success to the French public.

Like Linnaeus, Pluche thought that nature was created for man to admire, a spectacle that cannot be conceived without the spectator. Man is necessary to the creation in a more fundamental way; in him, all beings find, directly or indirectly, their reason for existence, which is nothing other than to be useful and edifying to man.

The relationship of spectacle to spectator guarantees man's importance but at the same time sets a limit to knowledge. A spectacle is something that presents itself to the eye; it consists of the exterior of things and the relationships that can be perceived between them. Pluche's epistemology was completely utilitarian and pragmatic. "To claim to penetrate the very heart of nature; to wish to relate effects to their particular causes; to wish to understand the artifice and the play of the springs (*ressorts*) and the smallest elements of which these springs are composed: this is a risky enterprise, in which success is too uncertain" (*Spectacle*, I, ix). "It is not always the most brilliant speculations nor the choice of the most exotic materials that is most profitable. I prefer Monsieur de Réaumur busy exterminating moths by means of an oily fleece; or increasing fowl production by making them hatch without the help of their mothers, than Monsieur Bernoulli absorbed in algebra, or Monsieur Leibniz calculating the various advantages and disadvantages of the possible worlds" (*Spectacle*, I, 475). Always reluctant to adopt any theoretical stance, even one of limited generality, Pluche was never really a Newtonian; he inclined to accept the Abbé de Molières's compromise between Newtonian physics and Cartesian vortices.

Pluche combined this pragmatism and lack of concern for theoretical coherence with a devout utilitarianism. In his words: "We are here only to be virtuous" (*Spectacle*, I, 523). He contended that reason is escorted by arms and feet, with which man is endowed not in order to contemplate but to work; too much knowledge derived from the pleasure of contemplation would lead us to distraction. Man must be restored to nature and "reason [must be] brought back to earth." Then unity will reign everywhere and man, in his true place, will see that everything can be of service to him. The spectacle of nature provides him with a guide for action.

Through what might appear to be a paradoxical decision on the part of an author writing a work entitled *Le spectacle de la nature*, Pluche devoted as much attention to the arts and trades as he did to natural science; because, for him, the arts formed part of nature. Man is made not so much to understand the earth as to cultivate it, to exchange its products, and to transform them for his benefit. Reason, while admiring nature, does not have to penetrate her deepest mysteries but at most should imitate nature through the arts. The arts are already present in nature, as can be seen from the web of the spider and from the art of government displayed by bees, ants, and other animals. In Pluche's view, therefore, theoretical knowledge is relatively useless. There is no reason to know except in order to act. This notion accorded with his fundamentally conservative view of society, according to which the psychological constitution of each person defines his intellectual capacities in such a way that one is exactly suited to the position one occupies in the social order and is to be satisfied with one's place and function.

In a certain sense Pluche's work constituted a reassertion of the rights of feeling and of a global view of nature against the rational and analytical outlook of the philosophers. Pluche, moreover, deplored the whole development of rationalist science that had occurred in Europe since the end of the Middle Ages, an aberration that he attributed to the pernicious influence of the Arabs. To a sterile, theoretical science, he opposed an account based as much on emotion as on reason. The spectacle of nature, like a mirror, reflects something other than itself and speaks to us in the language by which Providence teaches us virtue as well as its own glory and perfection. *Le spectacle de la nature* thus provided its readers with a popular and pragmatic theology. Because of its distrust of theoretical ambitions, the analytic method, and abstract demonstration, the science it contained scarcely reflects the concerns of the scientific research of its time. As a popularizer, Pluche was no Fontenelle. His approach, rather than making science truly intelligible to the public through the medium of everyday language, actually tended to dissolve the scientific content into this language to such an extent that its specific nature is destroyed.

The first edition of Pluche's *Histoire du ciel* appeared in 1739; it enjoyed a certain success, but not on the scale of the success of the *Spectacle*. In the *Histoire* Pluche restated his opposition to the cosmologies proposed by the physicists—which he termed "romans philosophiques"—and sought to display the excellence of the physics of Moses, which supposedly conforms to the teachings of both "history and experimental physics." He also attempted to demonstrate that monotheism preceded polytheism.

Assured of a comfortable income from the contracts that he skillfully negotiated with his publisher, Pluche, who had gone deaf, left Paris in 1749 and retired to La Varenne-Saint-Maur, where he devoted himself to scholarship and contemplation. Until his death from apoplexy in 1761, he published books on linguistics, geography, and the Bible.

BIBLIOGRAPHY

The most important of Pluche's works are *Le spectacle de la nature*, 8 vols. (Paris, 1732–1750); *Histoire du ciel*, 2 vols. (Paris, 1739); *Révision de l'histoire du ciel* (Paris, 1749), which was recast in the new ed. of *Histoire du ciel* (Paris, 1759); *La mécanique des langues et l'art de les enseigner* (Paris, 1751); and *Concorde de la géographie des différents âges* (Paris, 1765).

An exhaustive bibliography of the various eds. can be found in Caroline V. Doan, "Un succès littéraire du XVIIIe siècle: le spectacle de la nature de l'abbé Pluche" (thesis, Sorbonne, 1957). Most accounts of eighteenth-century thought mention Pluche, but the only detailed study is Doan, who also cites secondary literature.

On Pluche's influence, see D. Mornet, "Les enseignements des bibliothèques privées, 1750–1780," in *Revue d'histoire littéraire de la France*, **17** (1910), 449–496.

CAMILLE LIMOGES

PLÜCKER, JULIUS (*b.* Elberfeld, Germany, 16 June 1801; *d.* Bonn, Germany, 22 May 1868), *mathematics, physics.*

Plücker was descended from a Rhenish merchant family of Aix-la-Chapelle (Aachen). After graduating from the Gymnasium in Düsseldorf, he studied at the universities of Bonn, Heidelberg, Berlin, and Paris until 1824, when he earned his doctorate *in absentia* from the University of Marburg. In 1825 he became *Privatdozent* at the University of Bonn, where in 1828 he was promoted to extraordinary professor. In 1833 he served in Berlin simultaneously as extraordinary professor at the university and as teacher at the Friedrich Wilhelm Gymnasium. In 1834 he became ordinary professor at the University of Halle. He then served as full professor of mathematics (1836–1847) and physics (1847–1868) at Bonn, where he succeeded Karl von Münchow. In 1837 he married a Miss Altstätten; his wife and one son survived him.

Although Plücker was educated primarily in Germany, throughout his life he drew much on French and English science. He was essentially a geometer but dedicated many years of his life to physical science. When Plücker began his work in mathematics, the only German mathematician of international repute was Gauss. In 1826, however, Crelle founded, in Berlin, his *Journal für die reine und angewandte Mathematik*; and the work of Plücker, Steiner, and others soon became well known. Their field of research was not the differential geometry of Monge and Gauss, but rather the analytic and projective geometry of Poncelet and Gergonne. But differences between the synthetic school in geometry, of which Steiner was the head in Berlin, and Plücker's analytical school—together with a conflict of personality between the two men—resulted in Plücker's being resident at Berlin for only a year.

In 1828 Plücker published his first book, volume I of *Analytisch-geometrische Entwicklungen*, which was followed in 1831 by volume II. In each volume he discussed the plane analytic geometry of the line, circle, and conic sections; and many facts and theorems—either discovered or known by Plücker—were demonstrated in a more elegant manner. The point coordinates used in both volumes are nonhomogeneous affine; in volume II the homogeneous line coordinates in a plane, formerly known as Plücker's coordinates, are used and conic sections are treated as envelopes of lines. The characteristic features of Plücker's analytic geometry were already present in this work, namely, the elegant operations with algebraic symbols occurring in the equations of conic sections and their pencils. His understanding of the so-called reading in the formulas enabled him to achieve geometric results while avoiding processes of elimination, and his algebraic elegance was surpassed in some matters only by Hesse. Plücker's careful treatment, in the first book, of conic sections that osculate with one another in different degrees is still noteworthy.

In 1829 Plücker introduced the so-called triangular coordinates as three values that are proportional to the distances of a point from three given lines. Simultaneously, Möbius introduced his barycentric coordinates, another type of homogeneous point coordinates. In his *Analytisch-geometrische Entwicklungen*, however, Plücker used only nonhomogeneous point coordinates. At the end of volume II he presented a detailed explanation of the principle of reciprocity, now called the principle of duality. Plücker, who stood in the middle of the Poncelet-Gergonne controversy, was inclined to support Poncelet's position: Plücker introduced duality by means of a correlation polarity and not in the more modern sense (as in Gergonne) of a general principle. Thus Plücker's work may be regarded as a transitional stage preceding the pure projective geometry founded by Staudt.

After 1832 Plücker took an interest in a general treatment of plane curves of a higher degree than the second. Although his next book, *System der analytischen Geometrie, insbesondere eine ausführliche Theorie der Kurven 3. Ordnung enthaltend* (1835), discussed general (or projective) point and line coordinates for treating conic sections, the greater part of the book covered plane cubic curves. Plücker's consideration of these curves began with the following theorem by Poncelet. The three finite points where the three asymptotes of a cubic intersect the curve lie on a straight line. Analytically this theorem is equivalent to the possibility of writing the curve equation in the form $pqr + \lambda s^3 = 0$ (p, q, r, s linear forms). A cubic curve is determined by the 4 lines with equations $p = 0$, $q = 0$, $r = 0$, $s = 0$, and one point on the curve. Plücker gave constructions for the cubics thus determined. A real affine classification based upon these constructions leads to 219 different types.

Plücker devoted the greater part of *Theorie der algebraischen Kurven* (1839) to the properties of algebraic curves in the neighborhood of their infinite points. He considered not only the asymptotic lines, but also asymptotic conic sections and other curves osculating the given cubic in a certain degree. For the asymptotic lines he corrected some false results given by Euler in *Introductio in analysin infinitorum* (1748).

Although the increasing predominance of projective and birational geometry abated interest in these particulars about the behavior of curves at infinity,

the second part of *Theorie der algebraischen Kurven* was of more permanent value. It contained a new treatment of singular points in the plane, a subject previously discussed in Cramer's work (1750) on the theory of curves. Plücker's work also resolved several doubts concerning the relation between the order and class of curves in the work of Poncelet and Gergonne.

In his 1839 publication Plücker proved the following celebrated formulas, known as "Plücker's equations":

$$k = n(n - 1) - 2d - 3s, \quad n = k(k - 1) - 2\delta - 3\sigma,$$

connecting the order n and class k of a curve, which contains as singular points d double points and s cuspidal points, and as dual singularities δ double tangents and σ inflexions. The same assumptions validate Plücker's formulas of the second group:

$$\sigma = 3n(n - 2) - 6d - 8s, s = 3k(k - 2) - 6\delta - 8\sigma.$$

A cubic without singular points therefore contains nine inflexional points, and Plücker discovered that no more than three inflexional points can be real. He thus prepared the foundation for the results later obtained by Hesse. In the last chapter of *Geometrie der algebraischen Kurven* Plücker dealt with plane quartic curves and developed a full classification of their possible singular points. A nonsingular quartic curve that possesses twenty-eight double tangents is the central fact in his theory of these curves.

Although Plücker's treatment of quartic curves and his theorems on their configuration were all wrong (Hesse later corrected his errors), Plücker had a clear insight, which eluded his predecessors, into the meaning of the so-called Cramer paradox and its generalizations. The crux of the paradox is that $1/2\, n(n + 3) - 1$ common points of two curves of degree $n \geqslant 3$ determine another set of $1/2(n - 1)(n - 2)$ common points. Severi, in his conferences and books on enumerative geometry, underlined the so-called Plücker-Clebsch principle in the following form: If a system of algebraic equations, depending on certain constants, generally has no common solution—except when the constants fill certain conditions—then the system in this latter case has not only one, but also infinitely many solutions.

In 1829, independent of Bobillier, Plücker extended the notion of polars (previously known only for conic sections) to all plane algebraic curves. He also studied the problem of focal points of algebraic curves, the osculation of two surfaces, and wave surface, and thus became concerned with algebraic and analytic space geometry. This field was also discussed in *System der Geometrie des Raumes in neuer analytischer Behandlungsweise* (1846), in which he treated in an elegant manner the known facts of analytic geometry. His own contributions in this work, however, were not as significant as those in his earlier books.

After 1846 Plücker abandoned his mathematical researches and conducted physical experiments until 1864, when he returned to his work in geometry. His mathematical accomplishments during this second period were published in *Neue Geometrie des Raumes, gegründet auf die Betrachtung der Geraden als Raumelement*, which appeared in 1868. Plücker's death prevented him from completing the second part of this work, but Felix Klein, who had served as Plücker's physical assistant from 1866 to 1868, undertook the task. Plücker had indicated his plans to Klein in numerous conversations. These conversations served also as a source for Plücker's ideas in *Neue Geometrie*, in which he attempted to base space geometry upon the self-dual straight line as element, rather than upon the point or in dual manner upon the plane as element. He thus created the field of line geometry, which until the twentieth century was the subject of numerous researches (see Zindler, "Algebraische Liniengeometrie," in *Encyklopädie der mathematischen Wissenschaften*, III, 2 [1921], 973–1228).

Plücker's work in line geometry can be related to several earlier developments: the notion of six line coordinates in space as well as a complex of lines intersecting a rational norm curve had already been discussed by Cayley; the researches of Poinsot and Möbius on systems of forces were closely related to line geometry; and researches on systems of normals to a surface were made by Monge and later generalized by W. R. Hamilton to a differential geometry of ∞^2 rays.

Notwithstanding these developments, Plücker's systematic treatment of line geometry created a new field in geometry. He introduced in a dual manner the six homogeneous line coordinates p_{ij}, now called "Plücker's coordinates," among which a quadratic relation $Q_4(p_{ij}) = 0$ exists. Subsequent work by Klein and Segre interpreted line geometry of R_3 as a geometry of points on a quadric Q_4 of P_5. But this development, as well as further generalizations of an S_k geometry in S_n to be interpreted as a point geometry on a Grassmannian variety $G_{n,k}$, was not anticipated by Plücker, who restricted his work to the domain of ordinary space and conceived therein a four-dimensional geometry with the line as element.

Plücker's algebraic line geometry was distinct, however, from the differential line geometry created by Hamilton. Plücker introduced the notions (still used today) of complexes; congruences; and ruled surfaces for subsets of lines of three, two, or one

dimension. He also classified linear complexes and congruences and initiated the study of quadratic complexes, which were defined by quadratic relations among Plücker's coordinates. (Complex surfaces are surfaces of fourth order and class and are generated by the totality of lines belonging to a quadratic complex that intersects a given line.) These complexes were the subject of numerous researches in later years, beginning with Klein's doctoral thesis in 1868. In *Neue Geometrie* Plücker again adopted a metrical point of view, which effected extended calculations and studies of special cases. His interest in geometric shapes and details during this period is evident in the many models he had manufactured.

In assessing Plücker's later geometric work it must be remembered that during the years in which he was conducting physical research he did not keep up with the mathematical literature. He was not aware, for example, of Grassmann's *Die Wissenschaft der extensiven Grösslehre oder die Ausdehnungslehre* (1844), which was unintelligible to almost all contemporary mathematicians.

Plücker was professor of mathematics and physics at the University of Bonn; he is said to have always been willing to remind other physicists that he was competent in both fields. It is particularly noteworthy that Plücker chose to investigate experimental rather than theoretical physics; Clebsch, in his celebrated obituary on Plücker, identified several relations between Plücker's mathematical and his physical preoccupations. In geometry he wished to describe the different shapes of cubic curves and other figures, and in physics he endeavored to describe the various physical phenomena more qualitatively. But in both cases he was far from pursuing science in a modern axiomatic, deductive style.

Plücker's guide in physics was Faraday, with whom he corresponded. His papers of 1839 on wave surface and of 1847 on the reflection of light at quadric surfaces concerned both theoretical physics and mathematics, though often counted among his forty-one mathematical papers. Plücker also wrote fifty-nine papers on pure physics, published primarily in *Annalen der Physik und Chemie* and *Philosophical Transactions of the Royal Society*. He investigated the magnetic properties of gases and crystals and later studied the phenomena of electrical discharge in evacuated gases. He and his collaborators described these phenomena as precisely as the technical means of his time permitted. He also made use of an electromagnetic motor constructed by Fessel and later collaborated with Geissler at Bonn in constructing a standard thermometer. Plücker further drew upon the chemical experience of his pupil J. W. Hittorf in his study of the spectra of gaseous substances, and his examination of the different spectra of these substances indicates that he realized their future significance for chemical analysis.

In 1847 Plücker discovered the magnetic phenomena of tourmaline crystal; and in his studies of electrical discharges in rarefied gases he anticipated Hittorf's discovery of cathodic rays. His discovery of the first three hydrogen lines preceded the celebrated experiments of R. Bunsen and G. Kirchhoff in Heidelberg. Although Plücker's accomplishments were unacknowledged in Germany, English scientists did appreciate his work more than his compatriots did, and in 1868 he was awarded the Copley Medal.

BIBLIOGRAPHY

I. ORIGINAL WORKS. Plücker's major works are *Analytisch-geometrische Entwicklungen*, 2 vols. (Essen, 1828–1831); *System der analytischen Geometrie* (Berlin, 1835); *Theorie der algebraischen Kurven* (Bonn, 1839); *System der Geometrie des Raumes in neuer analytischer Behandlungsweise* (Düsseldorf, 1846); *Neue Geometrie des Raumes, gegründet auf der geraden Linie als Raumelement* (Leipzig, 1868–1869); and *Gesammelte wissenschaftlichen Abhandlungen*, 1 vol. (Leipzig, 1895–1896).

II. SECONDARY LITERATURE. On Plücker and his work, see A. Clebsch, "Zum Gedächtnis an Julius Plücker," in *Abhandlungen der K. Gesellschaft der Wissenschaften zu Göttingen*, **15** (1872), 1–40; Dronke, *Julius Plücker* (Bonn, 1871); and Wilhelm Ernst, *Julius Plücker* (Bonn, 1933).

WERNER BURAU

PLUMIER, CHARLES (*b.* Marseilles, France, 20 April 1646; *d.* El Puerto de Santa María, near Cádiz, Spain, 20 November 1704), *natural history, botany.*

Plumier's father was a thrower, and he taught his son the craft. At the age of sixteen Plumier entered the Order of Minims and studied physics, mathematics, and drawing. Subsequently he was sent to Rome, where, influenced by Paolo Boccone, he abandoned mathematics for botany. Later he was sent back to Provence and lived at the convent in Bormes, near Hyères, where he met Tournefort, with whom he botanized.

In 1689, at the request of Louis XIV, Plumier agreed to accompany Joseph Surian, a physician from Marseilles, on a voyage to the Caribbean. With a view toward medical application of all the gathered plants, Surian made chemical analyses, while Plumier's task was to form natural history collections. The two men quarreled, and on a second voyage (1693) Plumier traveled alone as *botaniste du roi*. Upon his return he published *Description des plantes de l'Amérique*, which included 107 plates engraved at royal expense.

In 1695 Plumier set out on a third expedition. He visited Guadeloupe, Martinique, Santo Domingo, and also the southern coast of Brazil. After the voyage he composed two important works: *Nova plantarum americanarum genera* (1703), which contains forty plates and descriptions of 106 new genera, each of which was dedicated to a botanist; and *Traité des fougères de l'Amérique* (1705), with 172 plates.

In 1704 Plumier—inspired by Gui Crescent Fagon—decided to sail to Peru in search of the cinchona tree. While waiting for the departure of the ship he died, probably from pneumonia.

Plumier was one of the first naturalists interested in the Antilles. He is known for his excellent descriptions and drawings of a great number of species. Although Plumier's herbarium from the Antilles was lost in a shipwreck, his drawings and Surian's herbarium on which Plumier collaborated are extant. Plumeria, an American tree or shrub of the family Apocynaceae, was named in honor of Plumier.

BIBLIOGRAPHY

I. ORIGINAL WORKS. Plumier's works include *Description des plantes de l'Amérique, avec leurs figures, par le R. P. Charles Plumier* (Paris, 1693); "Réponse du Père Plumier à M. Pommet, marchand droguiste à Paris, sur la cochenille," in *Journal des sçavans* (1694); *L'art de tourner ou de faire en perfection toutes sortes d'ouvrages au tour . . . ouvrage très curieux et très nécessaire à ceux qui s'exercent au tour; composé en français et en latin, en faveur des étrangers* (Lyon, 1701; 1749); "Réponse du P. C. Plumier à une lettre de Mr. Baulot écrite de la Rochelle," in *Mémoires pour servir à l'histoire des sciences et des beaux-arts* (1702), 112–113; and "Lettre sur une espèce de moucherons bleus observés dans le montagne de Lure. Lettre (2de) sur la cochenille," *ibid.* (1703).

See also *Nova plantarum americanarum genera . . . catalogus plantarum americanarum* (Paris, 1703); *Filicetum americanarum, seu filicum, polypodiorum, adiantorum, etc. . . . in America nascentium, icones; auctore P. Car. Plumier* (Paris, 1703); *Réponse du P. Plumier à diverses questions d'un curieux sur le crocodile, sur le colibri et sur la tortue* (Paris, 1704); *Traité des fougères de l'Amérique* (Paris, 1705); and *Plantarum americanarum fasciculus primus (-decimus) continens plantas quas olim Carolus Plumierius . . . detexit eruitque atque in insulis Antillis ipse depinxit. Has primum in lucem edidit . . . aeneisque tabulis illustravit Joannes Burmannus* (Amsterdam, 1755–1760).

II. SECONDARY LITERATURE. Works on Plumier are Michaud, *Biographie universelle*, new ed., XXXIII, 536–539; A. Saverien, "Histoire des philosophes modernes avec leur portrait gravé dans le goût du crayon, d'après les desseins des plus grands peintres par M. Saverien," in *Histoire des naturalistes*, **8** (1773), 39–44; and Ignaz Urban, "Plumier, Leben und Schriften nebst einen Schlüssel zu einen Blüten-Pflanzen," in *Beihefte zum Repertorium specierum novarum regni vegetabilis*, **5** (1920). 1–196.

PAUL JOVET

J. C. MALLET

PLUMMER, ANDREW (*b.* Scotland, *ca.* 1698; *d.* Edinburgh, Scotland, 16 April 1756), *medicine*.

Plummer was a member of the first organized medical faculty at the University of Edinburgh and one of the founders of a society of physicians that published the journal *Medical Essays and Observations*. The journal first appeared in 1733 and contained meteorological observations; accounts of diseases known in Edinburgh; essays on the history of medicine, drugs, chemical operations; and reports on experiments. This material was collected and revised by the society according to rules laid down in the preface to volume I.

After completing his early education in Edinburgh, Plummer entered the medical school at Leiden on 5 September 1720. He received the M.D. on 23 July 1722, and on 4 February 1724 he petitioned the College of Physicians of Edinburgh "for tryall"; he was examined and then licensed to practice on 25 February. Plummer was elected a fellow of the College in September and a fellow of the Royal College of Physicians on 3 November 1724. A week later he and three other young physicians purchased a house next to the "physic gardens" for a chemical "elaboratory" in which to instruct students. Although a medical faculty had been appointed at Edinburgh in 1685, in 1720 Alexander Monro (Primus), who had recently returned from Leiden, was appointed professor of anatomy and thus became the "Father" of the Edinburgh Medical School.

For two years Plummer and his three colleagues lectured privately; but in February 1726 they were appointed to the faculty, with full powers "to profess and teach medicine in all its branches, examine candidates, and do every thing requisite to the graduations of doctors of medicine." The medical school thus became a formal part of the university; rooms were assigned in the college, and the administration of the examination for the M.D. degree was transferred from the Royal College of Physicians to the medical faculty.

Plummer taught medical chemistry, which he had studied under Boerhaave at Leiden. John Fothergill, who had been a student of Plummer's in 1734, stated that Plummer knew chemistry well and because of his broad knowledge of science was dubbed "a living library" although diffidence obscured his talents as a

lecturer. Plummer avoided the mysticism of Helmont and the alchemists and thus established a modern approach to teaching chemistry in the British Isles; his method was then more fully developed by his successors, William Cullen and Joseph Black. From 10 March 1756 Cullen was joint professor of chemistry with Plummer, but succeeded to the chair when Plummer died. Black received the chair ten years later and developed it into a famous seat of chemical learning.

Plummer made the first chemical analysis of the waters of Moffat Spa; and in 1745, in a letter to Cullen, Black alluded to experiments by Plummer on the analysis of pit coal. Plummer's personal wealth enabled him to leave his practice; and after suffering a stroke in 1755, he sold his laboratory to Cullen for £120—a great bargain since it was near the infirmary, where, from 1741, students had attended clinical lectures.

Plummer's name was familiar to the medical profession for nearly two centuries because of his popular "Plummer's pills," which were composed of calomel and golden sulfuret of antimony and were used to treat venereal diseases, cutaneous eruptions, and other complaints. His pill received great acclaim, especially after its popularization in Germany by Paul Werlhof and its introduction into European pharmacopoeias. In his formulation of the pill Plummer substituted calomel for the ethiops mineral (which had been condemned by Boerhaave) in the combination of mercury and antimony used in older ethiopic pills; thus his pill was sometimes called Plummer's ethiops. Antimony, which had been used as a cosmetic in antiquity and rhapsodized about in Basil Valentine's translation of Thölde, *The Triumphant Chariot of Antimony*, was present in kermes mineral, which Glauber produced about 1651. Golden sulfuret of antimony is a by-product in the preparation of this mineral. The famous *Bravo* case of 1876, in which a young barrister died of antimony poisoning, may have sounded the death knell of antimony; and antibiotics in the past thirty years have all but eliminated the medical use of mercury.

BIBLIOGRAPHY

I. ORIGINAL WORKS. Plummer's works are "De phthisi pulmoni" (M.D. diss., Univ. of Leiden, 1722); "An Alterative Mercurial Medicine," in *Medical Essays and Observations Revised and Published by a Society of Physicians in Edinburgh*, **1**, art. 6 (1733), 46–62, contains his pill formula and reports his "trial of a medicinal compound of *sulphur auratum antimonii* and *calomelas . . .* two Herculian medicines"; a Latin trans. is "Actorum medicorum Edinburgensium . . . De medicamento alterante ex mercurio . . . ex anglico sermone Latine reddidit 1735," in Paul Werlhof, *Opera medica collegit et auxit J. E. Wichmann*, III (Hannover, 1776), 641–652; Werlhof indicates that A. Hugo introduced Plummer's formula into the German pharmacopoeia as *pulveris alterantis Edinburgensium*.

Subsequent works include "Experiments on the Medicinal Waters of Moffat," in *Medical Essays and Observations* . . ., **1**, art. 8 (1733), 82–94; "A History of the Rabies Canina," *ibid.*, 3rd ed., **5**, pt. 2 (1747), 97; "Remarks on Chemical Solutions and Precipitations," in *Essays and Observations, Physical and Literary, Read Before a Society in Edinburgh*, **1** (1754), 284; "Experiments on Neutral Salts, Compounds of Different Acid Liquors, and Alcaline Salts, Fixt and Volatile," *ibid.*, 315; and "History of a Cure Performed by Large Doses of an Alterative Mercurial Medicine Communicated to Dr. Plummer by Mr. George Dennistown, Surgeon in Falkirk," *ibid.*, 390. This last journal was the continuation of the *Medical Essays and Observations*; Plummer contributed to every issue that appeared during his life.

II. SECONDARY LITERATURE. On Plummer and his work, see J. Comrie, *History of Scottish Medicine*, I (London, 1932), 266, 299; *Complete Collection of the Medical and Philosophical Works of John Fothergill*, J. Elliot, ed. (London, 1781), 643; R. Fox, *Dr. John Fothergill and His Friends* (London, 1919), 13, 14, 38, 45; *The Works of Alexander Monro, M.D., by his Son*, A. Monro (Secundus), ed. (Edinburgh, 1781), xii; R. Ritchie, *Early Days of the Royal College of Physicians* (Edinburgh, 1899), 116, 132, 135; J. Thomson, *Account of the Life, Lectures and Writings of William Cullen, M.D.*, I (Edinburgh, 1859), 39, 59, 82, 86, 89; and A. Wootton, *Chronicles of Pharmacy*, I (London, 1910), 351; II, 153.

SAMUEL X. RADBILL

PLUMMER, HENRY CROZIER (*b.* Oxford, England, 24 October 1875; *d.* Oxford, 30 September 1946), *astronomy*.

The son of an astronomer, Plummer graduated from Hertford College, Oxford, with first class honors and in 1901 became an assistant at the Oxford University Observatory. He is known primarily for his book *Dynamical Astronomy* (1918), which is still used as a text and reference in theoretical and practical celestial mechanics and is valued for both its content and its style. In addition to dynamics, Plummer was interested in the accuracy of measurements, the theory of errors, and in mathematical methods applied to astronomical computations. During his lifetime he worked on many topics, including the *Astrographic Chart and Catalogue*, cometary motion (including nongravitational forces), stellar motions, spectroscopic binaries, and the dynamics of globular star clusters.

In 1912 Plummer was appointed astronomer royal of Ireland, a post once occupied by William Hamilton. It is said that Plummer wanted to concentrate on observational astronomy but that the Irish climate frustrated him and enforced a preoccupation with theoretical topics. His isolation and the quality of instruments at the Dunsink Observatory (there had been no new equipment since the 1860's) increased his dissatisfaction. In 1921 he resigned from this post and became professor of mathematics at the Royal Artillery College, where he remained until his retirement in 1940. Plummer was elected a fellow of the Royal Society (1920) and was president of the Royal Astronomical Society (1939).

In addition to his many research papers and *Dynamical Astronomy*, Plummer wrote two other books: *Principles of Mechanics* (1929) and *Probability and Frequency* (1939). He was preparing an edition of Newton's works at the time of his death.

Although his contemporaries made greater innovations, Plummer did not lack originality and he was a superb expositor.

BIBLIOGRAPHY

Plummer wrote more than 100 scientific papers, most of which appeared in *Monthly Notices of the Royal Astronomical Society*. His textbooks are *Introductory Treatise on Dynamical Astronomy* (Cambridge, 1918); *Principles of Mechanics* (London, 1929); and *Probability and Frequency* (London, 1939).

On Plummer and his work, see W. M. H. Greaves's notice in *Dictionary of National Biography, 1941–1950* (1959), 667–668; *Obituary Notices of Fellows of the Royal Society*, no. 16 (May 1948); W. M. Smart, in *Monthly Notices of the Royal Astronomical Society*, **56** (1947), 107; and Edmund Whittaker, in *Observatory*, **66** (1946), 394.

J. M. A. DANBY

POGGENDORFF, JOHANN CHRISTIAN (*b.* Hamburg, Germany, 29 December 1796; *d.* Berlin, Germany, 24 January 1877), *physics, biography, bibliography*.

Poggendorff worked in three areas of physics. He was an excellent experimenter and devised various measuring devices that were appropriate to his research; he skillfully edited the *Annalen der Physik und Chemie* for more than half a century; and he took an intense interest in the history of physics. Most notably, he produced an indispensable biographical-bibliographical reference work in the history of the exact sciences that is now called simply "Poggendorff."

Poggendorff was born into a well-to-do family, but his father lost nearly the entire fortune during the French occupation of Hamburg (1806–1811). He was educated first at the Johanneum in Hamburg and then, from 1807, at a boarding school in Schiffbeck, near Hamburg. At age fifteen he became an apprentice in a Hamburg apothecary shop. Later he worked as an apothecary's assistant in Itzehoe. Because of his poor financial situation, he had scarcely any prospect of ever owning his own apothecary shop. Moreover, he was strongly drawn to the study of chemistry and physics. Consequently, in 1820 he followed the advice of a former schoolmate at Schiffbeck, F. F. Runge, to study science in Berlin. He shared lodgings with Runge, who later became a distinguished chemist. Poggendorff eagerly devoted himself to science and carried out many experiments. Already by 1820 he had invented, independently of Schweigger, a galvanoscope (multiplier), the sensitivity of which he increased in 1826 through an arrangement that enabled him to take readings by reflection. This device was used by Gauss in his observations on magnetism. In 1823 Poggendorff was commissioned for a small salary by the Berlin Academy to make meteorological observations.

In 1824, at age twenty-eight, Poggendorff took over the editorship of the renowned *Annalen der Physik und Chemie*. In 1830 he received, as a *Privatgelehrter*, the title *Königlicher Professor*. He married in 1831 and by 1838 had three children from his happy marriage. In 1834 he became extraordinary professor at the University of Berlin and in 1839 was named a member of the Prussian Academy of Sciences. He received offers of full professorships from other universities but rejected them; for he considered Berlin to be the right place for his activity as editor of the *Annalen* and for his historical, biographical, and bibliographical research. Until 1875, he gave lectures at the university, primarily on the history of physics and on physical geography. Despite his very full work schedule, Poggendorff was sociable and a generous host. In the spring of 1876 he began to suffer from a painful neuralgia, and he died the following year at the age of eighty.

Poggendorff's achievements in the field of experimental physics—in addition to the invention of the multiplier—included the compensating circuit for determining the electromotive force of constant and inconstant circuits; research on induction; experiments with A. W. Holtz's electrical influence machine (electric power transmission); determination of the electromotive series in dilute sulfuric acid and in potassium cyanide; construction of a silver voltameter, of a rheocord (for measuring resistances), of an electric balance (for the study of polarization phenomena), of an improved sine galvanometer, of a thermopile, of a mercurial air pump, and of an Atwood's machine; and, finally, the demonstration of the existence of the Peltier effect in magnetoelectric currents.

It might seem astonishing that Poggendorff, who was only twenty-seven at the time, was chosen to edit the *Annalen der Physik und Chemie* (founded in 1790) upon the death, in 1824, of the previous editor, L. W. Gilbert. Yet we know that Poggendorff was very resolute in his dealings with the publisher, J. A. Barth, and even hinted that if turned down he would found his own journal, especially as he had already obtained the support of many prominent scientists. In the course of his fifty-two years of editorial activity, Poggendorff brought out 160 volumes of the *Annalen*, for which he procured first-rate original contributions and in which he also presented translations of important foreign papers. In addition, the journal published supplementary volumes and pamphlets. Eventually Poggendorff gave up personal supervision of the articles on chemistry and confined himself to physics. Thoroughly imbued with the values of empiricism, he rejected manuscripts that were speculative in nature and placed the greatest stress on articles with an experimental basis. Following Poggendorff's death he was succeeded as editor by Gustav Wiedemann.

Poggendorff's historical interests, expressed in his lectures and in a book on the history of physics, issued in a project to which he wholeheartedly devoted himself: the *Biographisch-Literarische Handwörterbuch zur Geschichte der exakten Wissenschaften*. The original two-volume work first appeared in 1863. He included dates and bibliographical references for 8,400 researchers in the exact sciences of all periods and countries up to the year 1858. This useful publication attests its compiler's immense capacity for work. The work has continued to be published and by 1974 comprised some eighteen individual volumes.

BIBLIOGRAPHY

I. ORIGINAL WORKS. Poggendorff's works include *Lebenslinien zur Geschichte der exakten Wissenschaften* (Berlin, 1853); *Biographisch-Literarisches Handwörterbuch zur Geschichte der exakten Wissenschaften*, 2 vols. (Leipzig, 1863); and W. Barentin, ed., *Geschichte der Physik* (Berlin, 1879). Articles by Poggendorff are listed in Poggendorff, *Biographisch-Literarisches Handwörterbuch*, II (1863), 480–482; III (1898), 1052–1053.

Poggendorff edited the following works: *Annalen der Physik und Chemie*, **1** (**77** of the entire sers.) (1824) through **160** (**236** of the entire sers.) (1877); and *Handwörterbuch* *der reinen und angewandten Chemie*, I (Brunswick, 1842), edited with J. Liebig. Further vols. of this last work were published, but Poggendorff withdrew from the project after the first vol. appeared.

II. SECONDARY LITERATURE. On Poggendorff and his work, see W. Baretin, "Ein Rückblick," in W. Baretin, ed., *Jubelband der Annalen der Physik und Chemie dem Herausgeber J. C. Poggendorff zur Feier fünfzigjährigen Wirkens gewidmet* (Leipzig, 1874), ix–xiv; W. Baretin, "J. C. Poggendorff," in *Annalen der Physik*, **160** (1877), v–xxiv; E. Frommel, *J. C. Poggendorff, Leichenrede, nebst eigenhändigen Lebensnachrichten, Reden und Briefen* (Berlin, 1877); *Allgemeine Deutsche Biographie*, **26** (1888), 364–366; J. Volhard, *Justus Von Liebig*, I (Leipzig, 1909), 371–373, on the *Handwörterbuch der Chemie*; H. Salié, "Ein Standardwerk zur Geschichte der Naturwissenschaften. Hundert Jahre 'Poggendorff,' " in *Forschungen und Fortschritte*, **37** (1963), 202–205; and H. Salié, "Poggendorff and Poggendorff," in *Isis*, **57**, pt. 3, no. 189 (1966), 389–392.

FRIEDRICH KLEMM

POGGIALE, ANTOINE-BAUDOIN (or **Baudouin**) (*b.* Valle di Mezzana, near Ajaccio, Corsica, 9 February 1808; *d.* Bellevue, Seine-et-Oise, France, 26 August 1879), *chemistry, military hygiene, public health, military pharmacy.*

The son of a Corsican country doctor, Poggiale was educated in Ajaccio and Marseilles. In 1828 he entered the military teaching hospital in Strasbourg in order to train for a career as an army pharmacist. Upon graduation in 1828 he was assigned to several posts before obtaining a coveted position at the Val-de-Grâce in Paris. This position lasted from 1831 to 1837 and provided the opportunity to earn an M.D. from the Faculty of Medicine in 1833 and to pursue his scientific interests.

Poggiale served as professor of chemistry at the military teaching hospital in Lille (1837–1847) and as chief pharmacist and professor of chemistry at the Val-de-Grâce (1847–1858). In 1858 he became the top-ranking pharmacist (*pharmacien inspecteur*) in the French army and a member of the Army Health Council. In 1856 he was elected to the Paris Academy of Medicine and in 1860 to the Council of Public Hygiene and Health of the Seine Department.

Poggiale belonged to a scientific elite of career military pharmacists in France whose activities spanned more than a century and included such distinguished members as Bayen, Parmentier, Serullas, Fée, Millon, and F. Z. Roussin. Among Poggiale's most important investigations were his analyses of drinking and mineral waters, notably the sources of drinking water for the barracks and fortifications of the Paris region and the water from the Seine and Dhuis rivers. In 1853 he published a painstaking comparative analysis of the bread supplied to the troops in the French and European armies and in Paris custodial institutions; he also compared the quality of commercial flour with that found in military provisions.

Poggiale's other studies included the chemical composition of blood in humans and animals, the physiological chemistry of sugar and glycogen, and the analysis of milk. The publication of his *Traité d'analyse chimique par la méthode des volumes* (Paris, 1858) further enhanced his reputation as an outstanding analytical chemist. In 1860 Poggiale brilliantly defended the application of chemistry to pathology and therapeutics, against the attacks of the eminent clinician Armand Trousseau (1801–1867).

BIBLIOGRAPHY

I. ORIGINAL WORKS. For listings of Poggiale's publications, see A. Balland, *Travaux scientifiques des pharmaciens militaires français* (Paris, 1882), 90–95; and Royal Society *Catalogue of Scientific Papers*, IV (1870), 956–957; VIII (1879), 640.

Poggiale's exchange with Trousseau took place in a lengthy debate in the Paris Academy of Medicine. For Trousseau's views, see *Bulletin de l'Académie impériale de médecine*, **25** (1859–1860), 720–723, 733–746. Poggiale's rebuttal is to be found in the same vol. on pp. 760–786, 957–987.

II. SECONDARY LITERATURE. See A. Balland, *Les pharmaciens militaires français* (Paris, 1913), 66–70; P. P. H. Blondeau, "Discours prononcé par M. Blondeau, président de la Société de Pharmacie, aux obsèques de M. Poggiale," in *Journal de pharmacie et de chimie*, 4th ser., **30** (1879), 383–385; E. A. Bourgoin, "Discours prononcé par M. Bourgoin sur la tombe de Poggiale," in *Bulletin de l'Académie de médecine*, 2nd ser., **8** (1879), 921–924; P. J. Coulier, "Notice nécrologique sur M. Poggiale," in *Recueil de mémoires de médecine, de chirurgie et de pharmacie militaires*, 3rd ser., **35** (1879), 556–560; and A. Mattei, *Notice biographique sur Poggiale* (Clermont, Oise, 1879).

ALEX BERMAN

POINCARÉ, JULES HENRI (*b.* Nancy, France, 29 April 1854; *d.* Paris, France, 17 July 1912), *mathematics, mathematical physics, celestial mechanics.*

The development of mathematics in the nineteenth century began under the shadow of a giant, Carl Friedrich Gauss; it ended with the domination by a genius of similar magnitude, Henri Poincaré. Both were universal mathematicians in the supreme sense, and both made important contributions to astronomy and mathematical physics. If Poincaré's discoveries in number theory do not equal those of Gauss, his achievements in the theory of functions are at least on the same level—even when one takes into account the theory of elliptic and modular functions, which must be credited to Gauss and which represents in that field his most important discovery, although it was not published during his lifetime. If Gauss was the initiator in the theory of differentiable manifolds, Poincaré played the same role in algebraic topology. Finally, Poincaré remains the most important figure in the theory of differential equations and the mathematician who after Newton did the most remarkable work in celestial mechanics. Both Gauss and Poincaré had very few students and liked to work alone; but the similarity ends there. Where Gauss was very reluctant to publish his discoveries, Poincaré's list of papers approaches five hundred, which does not include the many books and lecture notes he published as a result of his teaching at the Sorbonne.

Poincaré's parents both belonged to the upper middle class, and both their families had lived in Lorraine for several generations. His paternal grandfather had two sons: Léon, Henri's father, was a physician and a professor of medicine at the University of Nancy; Antoine had studied at the École Polytechnique and rose to high rank in the engineering corps. One of Antoine's sons, Raymond, was several times prime minister and was president of the French Republic during World War I; the other son, Lucien, occupied high administrative functions in the university. Poincaré's mathematical ability became apparent while he was still a student in the *lycée*. He won first prizes in the *concours général* (a competition between students from all French *lycées*) and in 1873 entered the École Polytechnique at the top of his class; his professor at Nancy is said to have referred to him as a "monster of mathematics." After graduation he followed courses in engineering at the École des Mines and worked briefly as an engineer while writing his thesis for the doctorate in mathematics which he obtained in 1879. Shortly afterward he started teaching at the University of Caen, and in 1881 he became a professor at the University of Paris, where he taught until his untimely death in 1912. At the early age of thirty-three he was elected to the Académie des Sciences and in 1908 to the Académie Française. He was also the recipient of innumerable prizes and honors both in France and abroad.

Function Theory. Before he was thirty years of age, Poincaré became world famous with his epoch-making discovery of the "automorphic functions" of one complex variable (or, as he called them, the

"fuchsian" and "kleinean" functions). The study of the modular function and of the solutions of the hypergeometric equation had given examples of analytic functions defined in an open connected subset D of the complex plane, and "invariant" under a group G of transformations of D onto itself, of the form

$$T: z \rightarrow (az + b)/(cz + d)$$
$$(a, b, c, d \text{ constants}, ad - bc \neq 0) \qquad (1)$$

G being "properly discontinuous," that is, such that no point z of D is the limit of an infinite sequence of transforms (distinct from z) of a point $z' \in D$ by a sequence of elements $T_n \in G$. For instance, the modular group consists of transformations (1), where a, b, c, d are integers and $ad - bc = 1$; D is the upper half plane $\mathscr{I}z > 0$, and it can be covered, without overlapping, by all transforms of the fundamental domain defined by $|z| \geqslant 1$, $|\mathscr{R}z| \leqslant 1/2$. Using non-Euclidean geometry in a very ingenious way, Poincaré was able to show that for any properly discontinuous group G of transformations of type (1), there exists similarly a fundamental domain, bounded by portions of straight lines or circles, and whose transforms by the elements of G cover D without overlapping. Conversely, given any such "circular polygon" satisfying some explicit conditions concerning its angles and its sides, it is the fundamental domain of a properly discontinuous group of transformations of type (1). The open set D may be the half plane $\mathscr{I}z > 0$, or the interior or the exterior of a circle; when it is not of this type, its boundary may be a perfect non-dense set, or a curve that has either no tangent at any point or no curvature at any point.

Poincaré next showed—by analogy with the Weierstrass series in the theory of elliptic functions—that for a given group G, and a rational function H having no poles on the boundary of D, the series

$$\Theta(z) = \sum_k H(T_k \cdot z)\,(c_k z + d_k)^{-2m} \qquad (2)$$

where the transformations

$$T_k: z \rightarrow (a_k z + b_k)/(c_k z + d_k)$$

are an enumeration of the transformations of G, and m is a large enough integer, converges except at the transforms of the poles of H by G; the meromorphic function Θ thus defined in D, obviously satisfies the relation

$$\Theta(T \cdot z) = (cz + d)^{2m}\,\Theta(z)$$

for any transformation (1) of the group G. The quotient of two such functions, which Poincaré called thetafuchsian or thetakleinean, corresponding to the same integer m, gives an automorphic function (meromorphic in D). It is easy to show that any two automorphic functions X, Y (meromorphic in D and corresponding to the same group G) satisfy an "algebraic" relation $P(X, Y) = 0$, where the genus of the curve $P(x, y) = 0$ is equal to the topological genus of the homogeneous space D/G and can be explicitly computed (as Poincaré showed) from the fundamental domain of G. Furthermore, if $v_1 = (dX/dz)^{1/2}$, $v_2 = zv_1$, v_1, and v_2 are solutions of a linear differential equation of order 2:

$$d^2v/dX^2 = \varphi(X, Y)\, v,$$

where φ is rational in X and Y, so that the automorphic function X is obtained by "inverting" the relation $z = v_1(X)/v_2(X)$. This property was the starting point of Poincaré's researches, following a paper by I. L. Fuchs investigating second-order equations $y'' + P(x)\, y' + Q(x)\, y = 0$, with rational coefficients P, Q, in which the inversion of the quotient of two solutions would give a meromorphic function; hence the name he chose for his automorphic functions.

But Poincaré did not stop there. Observing that his construction of fuchsian functions introduced many parameters susceptible of continuous variation, he conceived that by a suitable choice of these parameters, one could obtain for an "arbitrary" algebraic curve $P(x, y) = 0$, a parametric representation by fuchsian functions, and also that for an arbitrary homogeneous linear differential equation of any order

$$y^{(n)} + P_1(x)\, y^{(n-1)} + \cdots + P_n(x)\, y = 0,$$

where the P_j are algebraic functions of x, one could express the solutions of that equation by "zeta-fuchsian" functions (such a function $\mathbf{F}$ takes its value in a space $\mathbf{C}^p$; in other words, it is a system of p scalar meromorphic functions and is such that, for any transformation (1) of the fuchsian group G to which it corresponds, one has $\mathbf{F}(T \cdot z) = \rho(T) \cdot \mathbf{F}(z)$, where ρ is a linear representation of G into $\mathbf{C}^p$). The "continuity method" by which he sought to prove these results could not at that time be made rigorous, due to the lack of proper topological concepts and results in the early 1880's; but after Brouwer's fundamental theorems in topology, correct proofs could be given using somewhat different methods.

Much has been written on the "competition" between C. F. Klein and Poincaré in the discovery of automorphic functions. Actually there never was any real competition, and Klein was miles behind from the start. In 1879 Klein certainly knew everything that had been written on special automorphic functions, a theory to which he had contributed by several beautiful papers on the transformation of elliptic functions. He could not have failed in particular to notice the connection between the fundamental domains of these functions, and non-Euclidean geometry, since it was he who, after Cayley and Beltrami, had clarified the concept of Euclidean "models" for the various non-Euclidean geometries, of which the "Poincaré half plane" was a special example.

On the other hand, Poincaré's ignorance of the mathematical literature, when he started his researches, is almost unbelievable. He hardly knew anything on the subject beyond Hermite's work on the modular functions; he certainly had never read Riemann, and by his own account had not even heard of the "Dirichlet principle," which he was to use in such imaginative fashion a few years later. Nevertheless, Poincaré's idea of associating a fundamental domain to any fuchsian group does not seem to have occurred to Klein, nor did the idea of "using" non-Euclidean geometry, which is never mentioned in his papers on modular functions up to 1880. One of the questions Klein asked Poincaré in his letters was how he had proved the convergence of the "theta" series. It is only after realizing that Poincaré was looking for a theorem that would give a parametric representation by meromorphic functions of all algebraic curves that Klein set out to prove this by himself and succeeded in sketching a proof independently of Poincaré. He used similar methods (suffering from the same lack of rigor).

The general theory of automorphic functions of one complex variable is one of the few branches of mathematics where Poincaré left little for his successors to do. There is no "natural" generalization of automorphic functions to several complex variables. Present knowledge suggests that the general theory should be linked to the theory of symmetric spaces G/K of E. Cartan (G semisimple real Lie group, K maximal compact subgroup of G), and to the discrete subgroups Γ of G operating on G/K and such that G/Γ has finite measure (C. L. Siegel). But from that point of view, the group $G = \mathbf{SL}(2, \mathbf{R})$, which is at the basis of Poincaré's theory, appears as very exceptional, being the only simple Lie group where the conjugacy classes of discrete subgroups Γ depend on continuous parameters (A. Weil's rigidity theorem). The "continuity" methods dear to Poincaré are therefore ruled out; in fact the known discrete groups $\Gamma \subset G$ for which G/Γ has finite measure are defined by arithmetical considerations, and the automorphic functions of several variables are thus much closer to number theory than for one variable (where Poincaré very early had noticed the particular "fuchsian groups" deriving from the arithmetic theory of ternary quadratic forms, and the special properties of the corresponding automorphic functions).

The theory of automorphic functions is only one of the many contributions of Poincaré to the theory of analytic functions, each of which was the starting point of extensive theories. In a short paper of 1883 he was the first to investigate the links between the genus of an entire function (defined by properties of its Weierstrass decomposition in primary factors) and the coefficients of its Taylor development or the rate of growth of the absolute value of the function; together with the Picard theorem, this was to lead, through the results of Hadamard and E. Borel, to the vast theory of entire and meromorphic functions that is not yet exhausted after eighty years.

Automorphic functions had provided the first examples of analytic functions having singular points that formed a perfect non-dense set, as well as functions having curves of singular points. Poincaré gave another general method to form functions of this type by means of series $\sum_n \dfrac{A_n}{z - b_n}$ of rational functions, leading to the theory of monogenic functions later developed by E. Borel and A. Denjoy.

It was also a result from the theory of automorphic functions, namely the parametrization theorem of algebraic curves, that in 1883 led Poincaré to the general "uniformization theorem," which is equivalent to the existence of a conformal mapping of an arbitrary simply connected noncompact Riemann surface on the plane or on an open disc. This time he saw that the problem was a generalization of Dirichlet's problem, and Poincaré was the first to introduce the idea of "exhausting" the Riemann surface by an increasing sequence of compact regions and of obtaining the conformal mapping by a limiting process. Here again it was difficult at that time to build a completely satisfactory proof, and Poincaré himself and Koebe had to return to the question in 1907 before it could be considered as settled.

Poincaré was even more an initiator in the theory of analytic functions of several complex variables— which was practically nonexistent before him. His first result was the theorem that a meromorphic function F of two complex variables is a quotient of two entire functions, which in 1883 he proved by a very ingenious use of the Dirichlet principle applied to the function $\log |F|$; in a later paper (1898) he deepened the study of such "pluriharmonic" functions for any number of complex variables and used it in the theory of Abelian functions. Still later (1907), after the publication of F. M. Hartogs' theorems, he

pointed out the completely new problems to which led the extension of the concept of "conformal mapping" for functions of two complex variables. These were the germs of the imposing "analytic geometry" (or theory of analytic manifolds and analytic spaces) which we know today, following the pioneering works of Cousin, Hartogs, and E. E. Levi before 1914; H. Cartan, K. Oka, H. Behnke, and P. Lelong in the 1930's; and the tremendous impulse given to the theory by cohomological ideas after 1945.

Finally, Poincaré was the first to give a satisfactory generalization of the concept of "residue" for multiple integrals of functions of several complex variables, after earlier attempts by other mathematicians had brought to light serious difficulties in this problem. Only quite recently have his ideas come to full fruition in the work of J. Leray, again using the resources of algebraic topology.

Abelian Functions and Algebraic Geometry. As soon as he came into contact with the work of Riemann and Weierstrass on Abelian functions and algebraic geometry, Poincaré was very much attracted by those fields. His papers on these subjects occupy in his complete works as much space as those on automorphic functions, their dates ranging from 1881 to 1911. One of the main ideas in these papers is that of "reduction" of Abelian functions. Generalizing particular cases studied by Jacobi, Weierstrass, and Picard, Poincaré proved the general "complete reducibility" theorem, which is now expressed by saying that if A is an Abelian variety and B an Abelian subvariety of A, then there exists an Abelian subvariety C of A such that $A = B + C$ and $B \cap C$ is a finite group. Abelian varieties can thus be decomposed in sums of "simple" Abelian varieties having finite intersection. Poincaré noted further that Abelian functions corresponding to reducible varieties (and even to products of elliptic curves, that is, Abelian varieties of dimension 1) are "dense" among all Abelian functions—a result that enabled him to extend and generalize many of Riemann's results on theta functions, and to investigate the special properties of the theta functions corresponding to the Jacobian varieties of algebraic curves.

The most remarkable contribution of Poincaré to algebraic geometry is in his papers of 1910–1911 on algebraic curves contained in an algebraic surface $F(x, y, z) = 0$. Following the general method of Picard, Poincaré considers the sections of the surface by planes $y = $ const.; the genus p of such a curve C_y is constant except for isolated values of y.

It is possible to define p Abelian integrals of the first kind on C_y, $u_1, \ldots, u_p$, which are analytic functions on the surface (or rather, on its universal covering). Now, to each algebraic curve Γ on the surface, meeting a generic C_y in m points, Poincaré associates p functions $v_1, \ldots, v_p$ of y, $v_j(y)$ being the sum of the values of the integral u_j at the m points of intersection of C_y and Γ; furthermore, he is able to characterize these "normal functions" by properties where the curve Γ does not appear anymore, and thus he obtains a kind of analytical "substitute" for the algebraic curve. This remarkable method enabled him to obtain simple proofs of deep results of Picard and Severi, as well as the first correct proof of a famous theorem stated by Castelnuovo, Enriques, and Severi, showing that the irregularity $q = p_g - p_a$ of the surface (p_g and p_a being the geometric and the arithmetic genus) is exactly the maximum dimension of the "continuous nonlinear systems" of curves on the surface. The method of proof suggested by the Italian geometers was later found to be defective, and no proof other than Poincaré's was obtained until 1965. His method has also shown its value in other recent questions (Igusa, Griffiths), and it is very likely that its effectiveness is far from exhausted.

Number Theory. Poincaré was a student of Hermite, and some of his early work deals with Hermite's method of "continuous reduction" in the arithmetic theory of forms, and in particular the finiteness theorem for the classes of such forms (with nonvanishing discriminant) that had just been proved by C. Jordan. These papers bring some complements and precisions to the results of Hermite and Jordan, without introducing any new idea. In connection with them Poincaré gave the first general definition of the genus of a form with integral coefficients, generalizing those of Gauss and Eisenstein; Minkowski had arrived independently at that definition at the same time.

Poincaré's last paper on number theory (1901) was most influential and was the first paper on what we now call "algebraic geometry over the field of rationals" (or a field of algebraic numbers). The subject matter of the paper is the Diophantine problem of finding the points with rational coordinates on a curve $f(x, y) = 0$, where the coefficients of f are rational numbers. Poincaré observed immediately that the problem is invariant under birational transformations, provided the latter have rational coefficients. Thus he is naturally led to consider the genus of the curve $f(x, y) = 0$, and his main concern is with the case of genus 1; using the parametric representation of the curve by elliptic functions (or, as we now say, the Jacobian of the curve), he observes that the rational points correspond on the Jacobian to a subgroup, and he defines the "rank" of the curve as the rank of that subgroup. It is likely that Poincaré conjectured that the rank is always finite; this funda-

mental fact was proved by L. J. Mordell in 1922 and generalized to curves of arbitrary genus by A. Weil in 1929. These authors used a method of "infinite descent" based upon the bisection of elliptic (or Abelian) functions; Poincaré had developed in his paper similar computations related to the trisection of elliptic functions, and it is likely that these ideas were at the origin of Mordell's proof. The Mordell-Weil theorem has become fundamental in the theory of Diophantine equations, but many questions regarding the concept of rank introduced by Poincaré remain unanswered, and it is possible that a deeper study of his paper may lead to new results.

Algebra. It is not certain that Poincaré knew Kronecker's dictum that algebra is only the handmaiden of mathematics, and has no right to independent existence. At any rate Poincaré never studied algebra for its own sake, but only when he needed algebraic results in problems of arithmetic or analysis. For instance, his work on the arithmetic theory of forms led him to the study of forms of degree $\geqslant 3$, which admit continuous groups of automorphisms. It seems that it is in connection with this problem that his attention was drawn to the relation between hypercomplex systems (over $\mathbf{R}$ or $\mathbf{C}$) and the continuous group defined by multiplication of invertible elements of the system; the short note he published on the subject in 1884 inspired later work of Study and E. Cartan on hypercomplex systems. A little-known fact is that Poincaré returned to noncommutative algebra in a 1903 paper on algebraic integrals of linear differential equations. His method led him to introduce the group algebra of the group of the equation (which then is, finite), and to split it (according to H. Maschke's theorem, which apparently he did not know but proved by referring to a theorem of Frobenius) into simple algebras over $\mathbf{C}$ (that is, matrix algebras). He then introduced for the first time the concepts of left and right ideals in an algebra, and proved that any left ideal in a matrix algebra is a direct sum of minimal left ideals (a result usually credited to Wedderburn or Artin).

Poincaré was one of the few mathematicians of his time who understood and admired the work of Lie and his continuators on "continuous groups," and in particular the only mathematician who in the early 1900's realized the depth and scope of E. Cartan's papers. In 1899 Poincaré became interested in a new way to prove Lie's third fundamental theorem and in what is now called the Campbell-Hausdorff formula; in his work Poincaré substantially defined for the first time what we now call the "enveloping algebra" of a Lie algebra (over the complex field) and gave a description of a "natural" basis of that algebra deduced from a given basis of the Lie algebra; this theorem (rediscovered much later by G. Birkhoff and E. Witt, and now called the "Poincaré-Birkhoff-Witt theorem") has become fundamental in the modern theory of Lie algebras.

Differential Equations and Celestial Mechanics. The theory of differential equations and its applications to dynamics was clearly at the center of Poincaré's mathematical thought; from his first (1878) to his last (1912) paper, he attacked the theory from all possible angles and very seldom let a year pass without publishing a paper on the subject. We have seen already that the whole theory of automorphic functions was from the start guided by the idea of integrating linear differential equations with algebraic coefficients. Poincaré simultaneously investigated the local problem of a linear differential equation in the neighborhood of an "irregular" singular point, showing for the first time how asymptotic developments could be obtained for the integrals. A little later (1884) he took up the question, also started by I. L. Fuchs, of the determination of all differential equations of the first order (in the complex domain) algebraic in y and y' and having fixed singular points; his researches were to be extended by Picard for equations of the second order, and to lead to the spectacular results of Painlevé and his school at the beginning of the twentieth century.

The most extraordinary production of Poincaré, also dating from his prodigious period of creativity (1880-1883) (reminding us of Gauss's *Tagebuch* of 1797-1801), is the qualitative theory of differential equations. It is one of the few examples of a mathematical theory that sprang apparently from nowhere and that almost immediately reached perfection in the hands of its creator. Everything was new in the first two of the four big papers that Poincaré published on the subject between 1880 and 1886.

The Problems. Until 1880, outside of the elementary types of differential equations (integrable by "quadratures") and the local "existence theorems," global general studies had been confined to linear equations, and (with the exception of the Sturm-Liouville theory) chiefly in the complex domain. Poincaré started with general equations $dx/X = dy/Y$, where X and Y are "arbitrary" polynomials in x, y, everything being real, and did not hesitate to consider the most general problem possible, namely a qualitative description of all solutions of the equation. In order to handle the infinite branches of the integral curves, he had the happy idea to project the (x, y) plane on a sphere from the center of the sphere (the center not lying in the plane), thus dealing for the first time with the integral curves of a vector field on a compact manifold.

The Methods. The starting point was the consideration of the "critical points" of the equation, satisfying $X = Y = 0$. Poincaré used the classification of these points due to Cauchy and Briot-Bouquet (modified to take care of the restriction to real coordinates) in the well-known categories of "nodes," "saddles," "spiral points," and "centers." In order to investigate the shape of an integral curve, Poincaré introduced the fundamental notion of "transversal" arcs, which are not tangent to the vector field at any of their points. Functions $F(x, y)$ such that $F(x, y) = C$ is a transversal for certain values of C also play an important part (their introduction is a forerunner of the method later used by Liapunov for stability problems).

The Results. The example of the "classical" differential equations had led one to believe that" general" integral curves would be given by an equation $\Phi(x, y) = C$, where Φ is analytic, and the constant C takes arbitrary values. Poincaré showed that on the contrary this kind of situation prevails only in "exceptional" cases, when there are no nodes nor spiral points among the critical points. In general, there are no centers—only a finite number of nodes, saddles, or spiral points; there is a finite number of closed integral curves, and the other curves either join two critical points or are "asymptotic" to these closed curves. Finally, he showed how his methods could be applied in explicit cases to determine a subdivision of the sphere into regions containing no closed integral or exactly one such curve.

In the third paper of that series Poincaré attacked the more general case of equations of the first order $F(x, y, y') = 0$, where F is a polynomial. By the consideration of the surface $F(x, y, z) = 0$, he showed that the problem is a special case of the determination of the integral curves of a vector field on a compact algebraic surface S. This immediately led him to introduce the genus p of S as the fundamental invariant of the problem, and to discover the relation

$$N + F - C = 2 - 2p \qquad (3)$$

where N, F, and C are the numbers of nodes, spiral points, and saddles. He then proceeded to show how his previous results for the sphere partly extend to the general case, and then made a detailed and beautiful study of the case when S is a torus ($p = 1$), so that there may be no critical point; in that case, he is confronted with a new situation—the appearance of the "ergodic hypothesis" for the integral curves. He was not able to prove that the hypothesis holds in general (under the smoothness conditions imposed on the vector field), but later work of Denjoy showed that this is in fact the case.

In the fourth paper Poincaré finally inaugurated the qualitative theory for equations of higher order, or equivalently, the study of integral curves on manifolds of dimension ≥ 3. The number of types of critical points increases with the dimension, but Poincaré saw how his relation (3) for dimension 2 can be generalized, by introducing the "Kronecker index" of a critical point, and showing that the sum of the indices of the critical points contained in a bounded domain limited by a transversal hypersurface $\sum$ depends only on the Betti numbers of $\sum$. It seems hopeless to obtain in general a description of all integral curves as precise as the one obtained for dimension 2. Probably inspired by his first results on the three-body problem (dating from 1883), Poincaré limited himself to the integral curves that are "near" a closed integral curve C_o. He considered a point M on C_o and a small portion $\sum$ of the hypersurface normal to C_o at M. If a point P of $\sum$ is close enough to M, the integral curve passing through P will cut $\sum$ again for the first time at a point $T(P)$, and one thus defines a transformation T of $\sum$ into itself, leaving M invariant, which can be proved to be continuously differentiable (and even analytic if one starts with analytic data). Poincaré then showed how the behavior of integral curves "near C_o" depends on the eigenvalues of the linear transformation tangent to T at M, and the classification of the various types is therefore closely similar to the classification of critical points.

After 1885 most of Poincaré's papers on differential equations were concerned with celestial mechanics, and more particularly the three-body problem. It seems that his interest in the subject was first aroused by his teaching at the Sorbonne; then, in 1885, King Oscar II of Sweden set up a competition among mathematicians of all countries on the n-body problem. Poincaré contributed a long paper, which was awarded first prize, and which ranks with his papers on the qualitative theory of differential equations as one of his masterpieces. Its central theme is the study of the periodic solutions of the three-body problem when the masses of two of the bodies are very small in relation to the mass of the third (which is what happens in the solar system). In 1878 G. W. Hill had given an example of such solutions; in 1883 Poincaré proved—by a beautiful application of the Kronecker index—the existence of a whole continuum of such solutions. Then in his prize memoir he gave another proof for the "restricted" three-body problem, when one of the small masses is neglected, and the other μ is introduced as a parameter in the Hamiltonian of the system. Starting from the trivial existence of periodic solutions for $\mu = 0$, Poincaré proved the existence of "neighboring" periodic solutions for small enough μ, by an application of Cauchy's method of majorants. He then showed that there exist solutions that are asymptotic to a periodic solution for values of the time tending to $+\infty$ or $-\infty$, or even for both ("doubly asymptotic" solutions). It should be stressed that in order to arrive at these results, Poincaré first had to invent the necessary general tools: the "variational equation" giving the derivative of a vector solution $\mathbf{f}$ of a system of differential equations, with respect to a parameter, as a solution of a linear differential equation; the "characteristic exponents," corresponding to the case in which $\mathbf{f}$ is periodic; and the "integral invariants" of a vector field, generalizing the particular case of an invariant volume used by Liouville and Boltzmann.

Celestial Mechanics. The works of Poincaré on celestial mechanics contrasted sharply with those of his predecessors. Since Lagrange, the mathematical and numerical study of the solar system had been carried out by developing the coordinates of the planets in series of powers of the masses of the planets or satellites (very small compared with that of the sun); the coefficients of these series would then be computed, as functions of the time t, by various processes of approximation, from the equations obtained by identifying in the equations of motion the coefficients of the powers of the masses. At first the functions of t defined in this manner contained not only trigonometric functions such as $\sin(at + b)$ (a, b constants) but also terms such as $t \cdot \cos(at + b)$, and so forth, which for large t were likely to contradict the observed movements, and showed that the approximations made were unsatisfactory. Later in the nineteenth century these earlier approximations were replaced by more sophisticated ones, which were series containing only trigonometric functions of variables of type $a_n t + b_n$; but nobody had ever proved that these series were convergent, although most astronomers believed they were. One of Poincaré's results was that these series cannot be uniformly convergent, but may be used to provide asymptotic developments of the coordinates.

Thus Poincaré inaugurated the rigorous treatment of celestial mechanics, in opposition to the semiempirical computations that had been prevalent before him. However, he was also keenly interested in the "classical" computations and published close to a hundred papers concerning various aspects of the theory of the solar system, in which he suggested innumerable improvements and new techniques. Most of his results were developed in his famous three-volume *Les méthodes nouvelles de la mécanique céleste* and later in his *Leçons de mécanique céleste*.

From the theoretical point of view, one should mention his proof that in the "restricted" three-body problem, where the Hamiltonian depends on four variables (x_1, x_2, y_1, y_2) and the parameter μ, and where it is analytic in these five variables and periodic of period 2π in y_1 and y_2, then there is no "first integral" of the equations of motion, except the Hamiltonian, which has similar properties. Poincaré also started the study of "stability" of dynamical systems, although not in the various more precise senses that have been given to this notion by later writers (starting with Liapunov). The most remarkable result that he proved is now known as "Poincaré's recurrence theorem": for "almost all" orbits (for a dynamical system admitting a "positive" integral invariant), the orbit intersects an arbitrary nonempty open set for a sequence of values of the time tending to $+\infty$. What is particularly interesting in that theorem is the introduction, probably for the first time, of null sets in a question of analysis (Poincaré, of course, did not speak of measure, but of "probability").

Another famous paper of Poincaré in celestial mechanics is the one he wrote in 1885 on the shape of a rotating fluid mass submitted only to the forces of gravitation. Maclaurin had found as possible shapes some ellipsoids of revolution to which Jacobi had added other types of ellipsoids with unequal axes, and P. G. Tait and W. Thomson some annular shapes. By a penetrating analysis of the problem, Poincaré showed that still other "pyriform" shapes existed. One of the features of his interesting argument is that, apparently for the first time, he was confronted with the problem of minimizing a quadratic form in "infinitely" many variables.

Finally, in one of his later papers (1905), Poincaré attacked for the first time the difficult problem of the existence of closed geodesics on a convex smooth surface (which he supposed analytic). The method by which he tried to prove the existence of such geodesics is derived from his ideas on periodic orbits in the three-body problem. Later work showed that this method is not conclusive, but it has inspired the numerous workers who finally succeeded in obtaining a complete proof of the theorem and extensive generalizations.

Partial Differential Equations and Mathematical Physics. For more than twenty years Poincaré lectured at the Sorbonne on mathematical physics; he gave himself to that task with his characteristic thoroughness and energy, with the result that he became an expert in practically all parts of theoretical physics, and published more than seventy papers and books on the most varied subjects, with a predilection

for the theories of light and of electromagnetic waves. On two occasions he played an important part in the development of the new ideas and discoveries that revolutionized physics at the end of the nineteenth century. His remark on the possible connection between X rays and the phenomena of phosphorescence was the starting point of H. Becquerel's experiments which led him to the discovery of radioactivity. On the other hand, Poincaré was active in the discussions concerning Lorentz' theory of the electron from 1899 on; Poincaré was the first to observe that the Lorentz transformations form a group, isomorphic to the group leaving invariant the quadratic form $x^2 + y^2 + z^2 - t^2$; and many physicists consider that Poincaré shares with Lorentz and Einstein the credit for the invention of the special theory of relativity.

This persistent interest in physical problems was bound to lead Poincaré into the mathematical problems raised by the partial differential equations of mathematical physics, most of which were still in a very rudimentary state around 1880. It is typical that in all the papers he wrote on this subject, he never lost sight of the possible physical meanings (often drawn from very different physical theories) of the methods he used and the results he obtained. This is particularly apparent in the first big paper (1890) that he wrote on the Dirichlet problem. At that time the existence of a solution inside a bounded domain D limited by a surface S was established (for an arbitrary given continuous function on S) only under rather restrictive conditions on S, by two methods due to C. Neumann and H. A. Schwarz. Poincaré invented a third method, the "sweeping out process": the problem is classically equivalent to the existence of positive masses on S whose potential V is equal to 1 in D and continuous in the whole space. Poincaré started with masses on a large sphere Σ containing D and giving potential 1 inside Σ. He then observed that the classical Poisson formula allows one to replace masses inside a sphere C by masses on the surface of the sphere in such a way that the potential is the same outside C and has decreased inside C. By covering the exterior of D by a sequence (C_n) of spheres and applying repeatedly to each C_n (in a suitable order) the preceding remark, he showed that the limit of the potentials thus obtained is the solution V of the problem, the masses initially on Σ having been ultimately "swept out" on S. Of course he had to prove the continuity of V at the points of S, which he did under the only assumption that at each of these points there is a half-cone (with opening $2\alpha > 0$) having the point as vertex and such that the intersection of that half-cone and of a neighborhood of the vertex does not meet D (later examples of Lebesgue showed that such a restriction cannot be eliminated). This very original method was later to play an important part in the renewal of potential theory that took place in the 1920's and 1930's, before the advent of modern Hilbert space methods.

In the same 1890 paper Poincaré began the long, and only partly successful, struggle with what we now call the problem of the eigenvalues of the Laplacian. In several problems of physics (vibrations of membranes, cooling of a solid, theory of the tides, and so forth), one meets the problem of finding a function u satisfying in a bounded domain D an equation of the form

$$\Delta u + \lambda u = 0 \qquad (4)$$

and on the boundary S of D the condition

$$u + k(du/dn) = 0 \qquad (5)$$

where du/dn is the normal derivative and λ and k are constants. Heuristic variational arguments (generalizing the method of Riemann for the Dirichlet principle) and the analogy with the Sturm-Liouville problem (which is the corresponding problem for functions of a single variable) lead to the conjecture that for a given k there exists an increasing sequence of real numbers ("eigenvalues"),

$$\lambda_1 < \lambda_2 < \cdots < \lambda_n < \cdots$$

such that the problem is only solvable when λ is equal to one of the λ_n, and then has only one solution u_n such that $\int_D u_n^2 \, d\tau = 1$, the "eigenfunctions" u_n forming an orthonormal system. In the case of the vibrating membrane, the u_n corresponds to the experimentally detectable "harmonics." But a rigorous proof of the existence of the λ_n and the u_n had not been found before Poincaré; for the case $k = 0$, Schwarz had proved the existence of λ_1 by the following method: the analogy with the Sturm-Liouville problem suggested that for any smooth function f, the equation

$$\Delta u + \lambda u + f = 0 \qquad (6)$$

would have for λ distinct from the λ_n a unique solution $u(\lambda, \mathbf{x})$ satisfying (5), and which would be a meromorphic function of λ, having the λ_n as simple poles. Schwarz had shown that, as a function of λ, the solution $u(\lambda, \mathbf{x})$ was equal to a power series with a finite radius of convergence. Picard had been able to prove also the existence of λ_2. In 1894 Poincaré (always in the case $k = 0$) succeeded in proving the above property

of $u(\lambda, \mathbf{x})$, by an ingenious adaptation of Schwarz's method, using in addition an inequality of the type

$$\iiint_D V^2 dx \, dy \, dz \leqslant$$

$$C \iiint_D \left[\left(\frac{\partial V}{\partial x} \right)^2 + \left(\frac{\partial V}{\partial y} \right)^2 + \left(\frac{\partial V}{\partial z} \right)^2 \right] dx \, dy \, dz \qquad (7)$$

(C constant depending only on D)

valid for all smooth functions V such that $\iiint_D V \, dx \, dy \, dz = 0$ (the forerunner of numerous similar inequalities that play a fundamental part in the modern theory of partial differential equations). But he could not extend his method for $k \neq 0$ on account of the difficulty of finding a solution of (6) having a normal derivative on S (he could only obtain what we now would call a "weak" derivative, or derivative in the sense of distribution theory).

Two years later he met similar difficulties when he tried to extend Neumann's method for the solution of the Dirichlet problem (which was valid only for convex domains D). Through a penetrating discussion of that method (based on so-called "double layer" potentials), Poincaré linked it to the Schwarz process mentioned above, and was thus led to a new "boundary problem" containing a parameter λ: find a "single layer" potential φ defined by masses on S, such that $(d\varphi/dn)_i = -\lambda(d\varphi/dn)_e$, where the suffixes i and e mean normal derivatives taken toward the interior and toward the exterior of S. Here again, heuristic variational arguments convinced Poincaré that there should be a sequence of "eigenvalues" and corresponding "eigenfunctions" for this problem, but for the same reasons he was not able to prove their existence. A few years later, Fredholm's theory of integral equations enabled him to solve all these problems; it is likely that Poincaré's papers had a decisive influence on the development of Fredholm's method, in particular the idea of introducing a variable complex parameter in the integral equation. It should also be mentioned that Fredholm's determinants were directly inspired by the theory of "infinite determinants" of H. von Koch, which itself was a development of much earlier results of Poincaré in connection with the solution of linear differential equations.

Algebraic Topology. The main leitmotiv of Poincaré's mathematical work is clearly the idea of "continuity": whenever he attacks a problem in analysis, we almost immediately see him investigating what happens when the conditions of the problem are allowed to vary continuously. He was therefore bound to encounter at every turn what we now call topological problems. He himself said in 1901, "Every problem I had attacked led me to *Analysis situs*," particularly the researches on differential equations and on the periods of multiple integrals. Starting in 1894 he inaugurated in a remarkable series of six papers—written during a period of ten years—the modern methods of algebraic topology. Until then the only significant step had been the generalizations of the concept of "order of connection" of a surface, defined independently by Riemann and Betti, and which Poincaré called "Betti numbers" (they are the numbers $1 + h_j$, where the h_j are the present-day "Betti numbers"); but practically nothing had been done beyond this definition. The machinery of what we now call simplicial homology is entirely a creation of Poincaré: concepts of triangulation of a manifold, of a simplicial complex, of barycentric subdivision, and of the dual complex, of the matrix of incidence coefficients of a complex, and the computation of Betti numbers from that matrix. With the help of these tools, Poincaré discovered the generalization of the Euler theorem for polyhedra (now known as the Euler-Poincaré formula) and the famous duality theorem for the homology of a manifold; a little later he introduced the concept of torsion. Furthermore, in his first paper he had defined the fundamental group of a manifold (or first homotopy group) and shown its relations to the first Betti number. In the last paper of the series he was able to give an example of two manifolds having the same homology but different fundamental groups. In the first paper he had also linked the Betti numbers to the periods of integrals of differential forms (with which he was familiar through his work on multiple integrals and on invariant integrals), and stated the theorem which G. de Rham first proved in 1931. It has been rightly said that until the discovery of the higher homotopy groups in 1933, the development of algebraic topology was entirely based on Poincaré's ideas and techniques.

In addition, Poincaré also showed how to apply these new tools to some of the problems for which he had invented them. In two of the papers of the series on analysis situs, he determined the Betti numbers of an algebraic (complex) surface, and the fundamental group of surfaces defined by an equation of type $z^2 = F(x, y)$ (F polynomial), thus paving the way for the later generalizations of Lefschetz and Hodge. In his last paper on differential equations (1912), Poincaré reduced the problem of the existence of periodic solutions of the restricted three-body problem (but with no restriction on the parameter μ) to a theorem on the existence of fixed points for a continuous transformation of the plane subject to certain conditions, which was probably the first example of an existence proof in analysis based on algebraic topology. He did not succeed in proving

that fixed point theorem, which was obtained by G. D. Birkhoff a few months after Poincaré's death.

Foundations of Mathematics. With the growth of his international reputation, Poincaré was more and more called upon to speak or write on various topics of mathematics and science for a wider audience, a chore for which he does not seem to have shown great reluctance. (In 1910 he even was asked to comment on the influence of comets on the weather!) His vivid style and clarity of mind enhanced his reputation in his time as the best expositor of mathematics for the layman. His well-known description of the process of mathematical discovery remains unsurpassed and has been on the whole corroborated by many mathematicians, despite the fact that Poincaré's imagination was completely atypical; and the pages he devoted to the axioms of geometry and their relation to experimental science are classical. Whether this is enough to dub him a "philosopher," as has often been asserted, is a question which is best left for professional philosophers to decide, and we may limit ourselves to the influence of his writings on the problem of the foundations of mathematics.

Whereas Poincaré has been accused of being too conservative in physics, he certainly was very open-minded regarding new mathematical ideas. The quotations in his papers show that he read extensively, if not systematically, and was aware of all the latest developments in practically every branch of mathematics. He was probably the first mathematician to use Cantor's theory of sets in analysis; he had met concepts such as perfect non-dense sets in his work on automorphic functions or on differential equations in the early 1880's. Up to a certain point, he also looked with favor on the axiomatic trend in mathematics, as it was developing toward the end of the nineteenth century, and he praised Hilbert's *Grundlagen der Geometrie*. However, Poincaré's position during the polemics of the early 1900's about the "paradoxes" of set theory and the foundations of mathematics has made him a precursor of ·the Intuitionist School. He never stated his ideas on these questions very clearly and mostly confined himself to criticizing the schools of Russell, Peano, and Hilbert. Although accepting the "arithmetization" of mathematics, Poincaré did not agree to the reduction of arithmetic to the theory of sets nor to the Peano axiomatic definition of natural numbers. For Poincaré (as later for L. E. J. Brouwer) the natural numbers constituted a fundamental intuitive notion, apparently to be taken for granted without further analysis; he several times explicitly repudiated the concept of an infinite set in favor of the "potential infinite," but he never developed this idea systematically. He obviously had a blind spot regarding the formalization of mathematics, and poked fun repeatedly at the efforts of the disciples of Peano and Russell in that direction; but, somewhat paradoxically, his criticism of the early attempts of Hilbert was probably the starting point of some of the most fruitful of the later developments of metamathematics. Poincaré stressed that Hilbert's point of view of defining objects by a system of axioms was only admissible if one could prove a priori that such a system did not imply contradiction, and it is well known that the proof of noncontradiction was the main goal of the theory which Hilbert founded after 1920. Poincaré seems to have been convinced that such attempts were hopeless, and K. Gödel's theorem proved him right; what Poincaré failed to grasp is that all the work spent on metamathematics would greatly improve our understanding of the nature of mathematical reasoning.

BIBLIOGRAPHY

See *Oeuvres de Henri Poincaré,* 11 vols. (Paris, 1916–1954); *Les méthodes nouvelles de la mécanique céleste,* 3 vols. (Paris, 1892–1899); *La science et l'hypothèse* (Paris, 1906); *Science et méthode* (Paris, 1908); and *La valeur de la science* (Paris, 1913).

On Poincaré and his work, see Gaston Darboux, "Éloge historique d'Henri Poincaré," in *Mémoires de l'Académie des sciences,* **52** (1914), lxxxi–cxlviii; and Poggendorff, III, 1053–1054; IV, 1178–1180; V, 990; and VI, 2038. See also references in G. Sarton, *The Study of the History of Mathematics* (Cambridge, Mass., 1936), 93–94.

JEAN DIEUDONNÉ

POINSOT, LOUIS (*b.* Paris, France, 3 January 1777; *d.* Paris, 5 December 1859), *mathematics, mechanics.*

At the end of October 1794 Poinsot, who was a student in his last year at the Collège Louis-le-Grand in Paris, presented himself as a candidate in the first competitive entrance examination to the future École Polytechnique. Admitted despite an insufficient knowledge of algebra, he left in 1797 in order to enter the École des Ponts et Chaussées, where he remained for three years. Neglecting his technical studies—which held little attraction for him—in favor of mathematics, he eventually gave up the idea of becoming an engineer. From 1804 to 1809 he taught mathematics at the Lycée Bonaparte in Paris; he was then appointed inspector general of the Imperial University.

Despite the frequent travels to the provinces necessitated by this new post, on 1 November 1809 Poinsot was named assistant professor of analysis and mechanics at the École Polytechnique, substituting for Labey. Although he held this position until the school was reorganized in September 1816, he actually taught there for only three years, after which time he arranged for A. A. L. Reynaud and later for Cauchy to substitute for him. He owed his appointment at the Polytechnique to the favorable reception given to his *Éléments de statique* (1803) and to three subsequent memoirs that dealt with the composition of momenta and the composition of areas (1806), the general theory of equilibrium and of movement in systems (1806), and polygons and polyhedra (1809). His reputation also resulted in his election on 31 May 1813 to the mathematics section of the Académie des Sciences, replacing Lagrange.

From 1816 to 1826 Poinsot served as admissions examiner at the École Polytechnique, and on several occasions after 1830 he worked with the school's Conseil de Perfectionnement. Although in 1824 he gave up his duties as inspector general, his nomination in 1840 to the Conseil Royal de l'Instruction Publique kept him informed of university problems. Meanwhile he continued research on number theory and on mechanics, publishing a small number of original and carefully executed memoirs. Named to the Bureau des Longitudes in 1843, he displayed a certain interest in celestial mechanics. Moderately liberal in his political opinions, he protested against the clericalism of the Restoration but later accepted nomination to the Chambre des Pairs (1846) and to the Senate (1852).

Poinsot was determined to publish only fully developed results and to present them with clarity and elegance. Consequently he left a rather limited body of work, which was devoted mainly to mechanics, geometry, and number theory. He showed almost no interest in algebra except for his early investigations concerning the fifth-degree equation and his remarkable analysis of Lagrange's *Traité de la résolution des équations numériques de tous les degrés* (1808). Similarly, the infinitesimal calculus appears in his work only in the form of extracts (published in 1815) from his course in analysis at the École Polytechnique.

Poinsot's contributions to number theory (1818–1849) have been analyzed by L. E. Dickson. They deal primarily with primitive roots, certain Diophantine equations, and the expression of a number as a difference of two squares.

A fervent disciple of Monge, Poinsot was one of the principal leaders of the revival of geometry in France during the first half of the nineteenth century. In particular, he was responsible for the creation in 1846 of a chair of advanced geometry at the Sorbonne, which was intended for Chasles. Poinsot, who had a presentiment of the importance of the geometry of position, established the theory of regular star polygons and discovered several types of regular star polyhedra (1809), a general study of which was carried out by Cauchy shortly afterward.

Yet it was in mechanics that Poinsot most effectively displayed his gift for geometry. Although *Éléments de statique* (1803) was merely a manual designed for candidates to the École Polytechnique, the work possessed the great merit of applying geometric methods to the study of elementary problems of mechanics and of introducing the concept of the couple. The latter notion, moreover, held a central place in two more highly developed memoirs that Poinsot presented to the Académie des Sciences in 1804 and published in the *Journal de l'École polytechnique* in 1806. The second of these memoirs inspired an interesting debate between Poinsot and Lagrange concerning the principles of mechanics.

Among Poinsot's other writings on mechanics, the most important is *Théorie nouvelle de la rotation des corps* (1834). Pursuing the theoretical study undertaken in the eighteenth century by Euler, d'Alembert, and Lagrange, Poinsot established in a purely geometric fashion the existence of the axes of permanent rotation and worked out a very elegant representation of rotary motion by the rolling of the ellipsoid of inertia of a body on a fixed plane (Poinsot motion). This theory was developed by Sylvester and was applied by Foucault to the discovery of the gyroscope. Poinsot's remarkable geometric intuition also enabled him to elaborate a purely geometric theory of the precession of the equinoxes (1858).

In frequent opposition to the French analytic school of the first half of the nineteenth century, Poinsot produced an original body of work by successfully submitting to geometric treatment a certain number of fundamental questions in the mechanics of solids.

BIBLIOGRAPHY

I. ORIGINAL WORKS. Poinsot's best known work, *Éléments de statique* (Paris, 1803), went through many editions: the 9th ed. (1848) was the last to appear in his lifetime; 12th ed. (1877). This work was accompanied, progressively, by some of his most important papers on statics; the 8th through 10th eds. contain four memoirs on the composition of momenta and of areas, on the unchanging plan of the system of the world, on the general theory of equilibrium and of the movement of systems and on a new theory of the rotation of bodies. There is an English trans. of this work by T. Sutton (Cambridge, 1847).

Most of Poinsot's other publications are listed in Royal Society *Catalogue of Scientific Papers*, IV, 960–961, which includes 31 articles and memoirs on mechanics, algebra, number theory, infinitesimal calculus, geometry, infinitesi-

mal geometry, and celestial mechanics, all of which were published in various academic collections or in mathematical or astronomical reviews. Many of these papers were also published as offprints and enjoyed a fairly broad distribution.

An important analysis of the 2nd ed. of Lagrange's *Traité de la résolution des équations numériques de tous les degrés* is in *Magasin encyclopédique*, **4** (1808), 343–375 (repr. in 1826 in the 3rd ed. of this work, pp. v–xx), as well as two pamphlets that do not appear to be offprints: *Recherches sur l'analyse des sections angulaires* (Paris, 1825) and *Théorie nouvelle de la rotation des corps présentée à l'Institut le 19 mai 1834* (Paris, 1834), with English trans. by C. Whitley (Cambridge, 1834).

The bulk of Poinsot's MSS have been preserved in Paris at the Bibliothèque de l'Institut de France (MSS 948–965, 4738); MS 4738, which is an offprint of one of Poinsot's first memoirs, "Théorie générale de l'équilibre et du mouvement des systèmes," in *Journal de l'École polytechnique*, **6** (1806), 206–241, contains in the margin MS criticisms by Lagrange and Poinsot's responses.

II. SECONDARY LITERATURE. Poinsot's life and work have not yet been subjected to the detailed study they merit. His geometric writings have been analyzed in detail by M. Chasles, in *Aperçu historique* . . . (Brussels–Paris, 1837), see index; and in *Rapport sur les progrès de la géométrie* (Paris, 1870), 13–17; those on number theory have been analyzed by L. E. Dickson, in *History of the Theory of Numbers*, 3 vols. (Washington, 1919), see index. Other aspects of his work remain to be examined.

The most complete study of Poinsot's work is J. Bertrand, "Notice sur Louis Poinsot," in *Journal des savants* (July 1872), 405–420; and in Poinsot, *Éléments de statique*, 11th ed. (Paris, 1872), ix–xxviii. See also J. Bertrand, "Éloge historique de Louis Poinsot, lu le 29 décembre 1890," in *Revue générale des sciences pures et appliquées*, **1** (1890), 753–762, repr. in *Mémoires de l'Académie des sciences de Paris*, **45** (1899), lxxiii–xcv, and in J. Bertrand, *Éloges académiques*, n.s. (Paris, 1902), 1–27.

Other articles (listed chronologically) include G. Vapereau, in *Dictionnaire universel des contemporains*, 2nd ed. (Paris, 1861), 1408; E. Merlieux, in F. Hoefer, ed., *Nouvelle biographie générale*, XL (Paris, 1862), 562–563; P. Mansion, in *Résumé du cours d'analyse infinitésimale* (Paris, 1887), 289–291; M. d'Ocagne, in *Histoire abrégée des sciences mathématiques* (Paris, 1955), 200–202; and P. Bailhache, "La théorie générale de l'équilibre et du mouvement des systèmes de Louis Poinsot et sa signification critique" (thesis, Paris, 1974).

RENÉ TATON

POISEUILLE, JEAN LÉONARD MARIE (*b.* Paris, France, 22 April 1797; *d.* Paris, 26 December 1869), *physiology*, *physics*.

Poiseuille was the son of Jean Baptiste Poiseuille, a carpenter, and Anne Victoire Caumont. From 1815 to 1816 he studied at the École Polytechnique in Paris.

In 1828 he became doctor of science, but we do not know what kind of positions he held until 1860, when he was elected inspector of the primary schools in Paris. In 1829 he married a daughter of M. Panay de la Lorette, *ingénieur en chef des ponts et chaussées*. In 1842 Poiseuille was elected to the Académie de Médecine in Paris and to the Société Philomathique in Paris. He was also a member of several foreign societies, which included the societies of medicine in Stockholm, Berlin, and Breslau. He received the Montyon Medal in 1829, 1831, 1835, and 1843 for his researches in physiology.

Poiseuille's name is permanently associated with the physiology of the circulation of blood through the arteries. Hales was the first to measure the blood pressure by allowing the blood to rise into a vertical glass tube. Poiseuille improved the experiment by using a mercury manometer instead of the long tube and by filling potassium carbonate into the connection to the artery in order to prevent coagulation. With this instrument, a hemodynamometer, he showed in his 1828 dissertation, "Recherches sur la force du coeur aortique," that the blood pressure rises and falls on expiration and inspiration. He also found that the dilatation of an artery at each heartbeat was about 1/23 of normal. Ludwig improved the instrument by adding a float, which he caused to write on a rotating drum.

Poiseuille's interest in blood circulation led him to experiment on the flow and outflow of distilled water in capillary tubes with diameters ranging from 0.03 mm. to 0.14 mm. Such experiments had been carried out before, especially by Franz Joseph von Gerstner and Pierre-Simon Girard; but since they used tubes with larger diameters, their experiments were disturbed by turbulence. In his 1840 paper, "Recherches expérimentales sur le mouvement des liquides dans les tubes de très-petits diamètres," Poiseuille announced the law $Q = k(D^4p/L)$, where Q is the volume discharged in unit time, k is a constant, p is the pressure difference in mm. of mercury at the two ends of the tube, D is the diameter, and L is the length. He also measured the variation of Q with the temperature T (from 0° C. to 45° C.) and found $Q = 1836.724$

$$\times \ (1 + 0.0336793T + 0.0002209936T^2)(D^4p/L),$$
which agrees within 0.5 percent with modern values. Poiseuille also found that the law was not valid if the length L (as a function of the diameter) was below a certain limit.

Poiseuille's paper was reviewed by a committee consisting of Arago, Piobert, and Regnault. They persuaded him to make further experiments with ether and mercury, and these investigations were published in 1847. He found that ether yielded the same law as distilled water, whereas mercury obeyed a different law. In 1870 Emil Gabriel Warburg found that mercury obeys the Poiseuille law, except for certain anomalies caused by amalgamation in metal tubes.

In 1839 G. H. L. Hagen had already found the same law as Poiseuille, using brass tubes with diameters from 2.5 mm. to 6 mm., the temperature varying from 1° C. to 15° C. Poiseuille and the committee reviewing his paper indicated at no point any knowledge of Hagen's researches, and it seems that Hagen's work was not appreciated at this time, probably because he used—besides the correct experimental law—a wrong velocity profile (a wedge profile) in his theoretical investigations. In 1860 Jacob Eduard Hagenbach named the law after Poiseuille, and it was not until 1925 that Wilhelm Ostwald argued that the law should be renamed the Hagen-Poiseuille law.

A correct analytical derivation of the Hagen-Poiseuille law was given independently by Franz Neumann and Hagenbach in 1860, both of whom derived the parabolic expression for the velocity distribution and identified the constant k as an expression for the viscosity of the fluid. In 1845 Stokes calculated the discharge of long straight circular pipes and rectangular canals. Since he compared his formulas with the experiments of Bossut and Du Buat, which were complicated by turbulence, he did not evaluate the constant but obtained the parabolic velocity profile.

BIBLIOGRAPHY

I. ORIGINAL WORKS. For Poiseuille's dissertation on blood pressure, see "Recherches sur la force du coeur aortique," in *Archives générales de médecine*, **8** (1828), 550–554. The outflow law was published in "Recherches expérimentales sur le mouvement des liquides dans les tubes de très-petits diamètres," in *Comptes rendus hebdomadaires des séances de l'Académie des sciences*, **11** (1840), 961–967, 1041–1048; **12** (1841), 112–115; also in *Annalen der Physik und Chemie*, **58** (1843), 424–447. His experiments on the flow of ether and mercury were published in "Recherches expérimentales sur le mouvement des liquides de nature différente dans les tubes de très-petits diamètres," in *Comptes rendus hebdomadaires des séances de l'Académie des sciences*, **24** (1847), 1074–1079, and in *Justus Liebigs Annalen der Chimie*, **21** (1847), 76–110.

II. SECONDARY LITERATURE. René Taton and Mlle Cazenave have collected some of Poiseuille's biographical dates in a MS at the Centre de Recherches Alexandre Koyré in Paris. Some biographies give 1799 or 1800 as his year of birth, but Taton has found Poiseuille's birth certificate, his date of birth being 22 April 1797. C. Sachaile,

Les médecins de Paris jugés par leurs oeuvres (Paris, 1845) mentions that Poiseuille was professor of experimental physics at the Institut de France but this is not corroborated by other evidence. Short biographies can be found in G. Vapereau, *Dictionnaire universel des contemporains* (Paris, 1861), 1408–1409; August Hirsch, ed., *Biographisches Lexikon der hervorragende Ärzte*, IV (Vienna, 1886), 599; Dechambre and Lereboullet, in *Dictionnaire encyclopédique des sciences médicales*, **26** (1888), 425; and C. P. Callisen, *Medicinisches Schriftsteller-Lexikon*, XXXI (Copenhagen, 1843), 265. The Archives de l'Académie de Médecine in Paris has a short notice on the works of Poiseuille in *Mélanges scientifiques recueil de mémoires, discours, rapports*, . . . **1**, no. 22; **40**, nos. 17, 18.

In "Ueber die Bestimmung der Zähigkeit einer Flüssigkeit durch den Ausfluss der Röhren," in *Annalen der Physik und Chemie*, **109** (1860), 385–426, E. Hagenbach named the outflow law "Poiseuille's law"; whereas W. Ostwald, in *Kolloidzeitschrift*, **36** (1925), 99, argued in favor of naming it the Hagen-Poiseuille law. Much useful information can be found in L. Schiller, ed., *Drei Klassiker der Strömungs Lehre: Hagen, Poiseuille, Hagenbach*, Ostwalds Klassiker der exakten Wissenschaften no. 237 (Leipzig, 1933). See also Hunter Rouse and Simon Ince, *History of Hydraulics* (New York, 1963). C. Truesdell has commented on G. G. Stokes's derivation of the Hagen-Poiseuille law in the intro. to the reprint of Stokes, *Mathematical and Physical Papers* (New York, 1966), pp. ivF., ivG. For the history of viscosity of liquids, see E. N. da C. Andrade, "The Viscosity of Liquids," in *Endeavour*, **13** (1954), 117–127. For Poiseuille's contribution to physiology, see Fielding H. Garrison, *An Introduction to the History of Medicine*, 4th ed. (Philadelphia–London, 1929).

Kurt Møller Pedersen

POISSON, SIMÉON-DENIS (*b.* Pithiviers, Loiret, France, 21 June 1781; *d.* Paris, France, 25 April 1840), *mathematical physics.*

For a complete study of his life and works, see Supplement.

POIVRE, PIERRE (*b.* Lyons, France, 23 August 1719; *d.* La Freta, near Lyons, France, 6 January 1786), *botany.*

Poivre was the son of a respected silk merchant. Brought up in a missionary college where he received a Catholic education, he intended to become a priest. After finishing the course in theology, and before his ordination, he studied natural history, drawing, and painting for four years in preparation for his "missionary job." In 1749 he went to China and Cochin China in order to learn the languages of the countries, but he was also instructed to bring back anything curious or useful for his own country. Upon his return to France in 1745, the ships were encountered by the English, and Poivre was taken a prisoner to Batavia

(during the battle he lost his right arm, the result of which was his giving up of the priesthood). During his five-month imprisonment, he inquired about the culture of the spice plants that he wanted to introduce into the French possessions. Until then, spices were sold solely by the Dutch, but Poivre realized that the situation was due to lack of enterprise on the part of other nations. When he arrived at the Île-de-France (Mauritius) in 1746, he noted that the few spice plants introduced from the Moluccas seemed to grow quite well. He then decided to try to cultivate all the spice plants he had seen in the Moluccas. Poivre wanted to create a new center for spices in Cochin China, and to purchase spice plantlets for cultivation.

Returning to France in June 1748, Poivre convinced Mr. David, the director of the French East India Company, and Rouillé and de Montaran, the king's superintendents, to charge him to go to the Indies to execute his projects. In spite of some difficulties with one of the directors of the Company, Duvelaer, who did not seem quite truthful, Poivre returned to the East; but nothing was signed between him and the Company, and he doubted the promise would be thoroughly respected. Between 1749 and 1755, in spite of many problems, he introduced twenty plantlets of nutmeg. Instead of cooperating with him, the directors of the Company and several personalities of the Île-de-France—including the director of the Botanic Garden in Reduit, Fusée-Aublet—did everything to ruin the efforts and work of Poivre.

In 1757 Poivre returned to a peaceful life with his family in France. But in 1766 he was asked by the royal government to accept the charge of *Commissaire général ordonnateur* (general intendant) of the Île-de-France and Bourbon (Mauritius and Réunion). As intendant, Poivre had so many difficulties that as early as 1770 he wished to be released from his office. In October 1772 he returned to his home, near Lyon, France.

The clove and nutmeg plants that Poivre had introduced to the Grapefruit Garden ("Jardin des Pamplemousse") of the Île-de-France were not productive until 1775 and 1778 respectively. Consequently, he never saw the result of his efforts. Poivre himself published nothing, but he did leave many manuscripts, a great number of which are in the main library of the Muséum National d'Histoire Naturelle, Paris.

BIBLIOGRAPHY

I. Original Works. For the complete works of Poivre, see *Oeuvres complètes de Pierre Poivre . . . précédées de sa vie (par R. S. Dupont de Nemours) et accompagnées de notes [et d'une préface signée L. L. (Louis-Mathieu*

Langlès)] (Paris, 1797), which includes "Voyages d'un philosophe ou Observations sur les moeurs et les arts des peuples de l'Afrique, de l'Asie et de l'Amérique"; "Discours prononcé par Pierre Poivre à son arrivée à l'Isle de France"; "Discours prononcé à la première assemblée du nouveau Conseil supérieur de l'Isle de France, le 3.8.1767"; "Extrait du voyage fait en 1769 et 1770 aux Isles Philippines & Moluques par les vaisseaux . . . le Vigilant et . . . l'Étoile du Matin, sous le commandement de M. Evrard de Trémigon . . . présenté par le sieur d'Etcheverry . . . d'après les vues de Pierre Poivre pour la recherche des arbres à épiceries"; "Mission faite aux îles Moluques par le sieur d'Etcheverry, depuis le 10.3.1770 . . . jusqu'au 25 juin suivant"; "Lettre de M. Poivre au Père Coeurdoux, lettre du Père Coeurdoux"; "Rapport fait à l'Académie des Sciences sur le transport des plans de Cannelliers et de girofliers à L'Isle de France"; and "Extrait du voyage aux Indes, à la Chine par Sonnerat," vol. I, p. 81. A certain number of separate editions of every one of those works have been done in Europe, and can be found in big libraries.

II. SECONDARY LITERATURE. On Poivre and his work, see Henri Cordier, *Voyages de Pierre Poivre de 1748 jusqu'à 1757* (Paris, 1918); Marthe De Fels, *Pierre Poivre ou l'amour des épices* (Paris, 1968); Pierre Samuel Du Pont de Nemours, *Notice sur la vie de M. Poivre, chevalier de l'ordre du roi, ancien intendant des isles de France et de Bourbon* (Philadelphia, 1786); Yves Laissus, "Note sur les manuscrits de Pierre Poivre (1719–1786) conservés à la bibliothèque centrale du Museum national d'Histoire naturelle," in *Proceedings of the Royal Society of Arts and Sciences of Mauritius*, **4**, pt. 2 (1973); Madeleine Ly-Tio-Fane, "Mauritius and the Spice Trade. The Odyssey of Pierre Poivre," in *Bulletin of the Mauritius Institute*, **4** (1958); and Louis Malleret, "Pierre Poivre," in *Publications de l'École française d'Extrême Orient*, **92** (1974).

PAUL JOVET
J. C. MALLET

POLENI, GIOVANNI (*b.* Venice, Italy, 23 August 1683; *d.* Padua, Italy, 15 November 1761), *mathematics, physics, engineering, ancient history, archaeology.*

Poleni was the son of Jacopo Poleni and carried his title of marquis of the Holy Roman Empire, conferred by Emperor Leopold I and confirmed in 1686 by the Republic of Venice. In his early life he followed a variety of studies, and his intellectual endowment was soon known to be extraordinary. After completing his studies, first in philosophy and then in theology, at the school of the Padri Somaschi in Venice, he began, with his parents' encouragement, a judicial career. At the same time his father introduced him to mathematics and physics, and it became clear that the natural sciences were going to be his most prominent field of activity.

At the age of twenty-six he married Orsola Roberti of a noble family of Bassano del Grappa and accepted the chair of astronomy at the University of Padua; six years later he became professor of physics as well. The Venetian Senate invited him to investigate problems of hydraulics pertinent to the irrigation of lower Lombardy, Poleni soon acquired such proficiency in this field that he became the accepted arbiter of all disputes between states bordering on rivers. In 1719 he assumed the chair of mathematics at the University of Padua left vacant by Nikolaus I Bernoulli, upon the latter's return to Basel. His noteworthy opening lecture was published in 1720 as *De mathesis in rebus physicis utilitate.*

In 1738 Poleni established within a few months an up-to-date laboratory of experimental physics and began to lecture on that subject. He simultaneously conducted meteorological observations, corresponded with French, English, German, and Italian savants (particularly Euler, Maupertuis, the Bernoullis, and Cassini III), published memoirs on various subjects, and participated in the study of calendar reform that had been sponsored by Pope Clement XI. In 1733 Poleni received a prize from the Royal Academy of Sciences of Paris for a paper on a method of calculating—independently of astronomical observations—the distance traveled by a ship; and in 1736 he was awarded a prize for a study of ships' anchors. In 1739 he became a foreign member of the Academy, and in 1741 he received a prize for a study of cranes and windlasses.

Poleni's scientific activities were paralleled by classical researches, which were described in treatises on the temple of Ephesus, on ancient theaters and amphitheaters, on French archaeological findings, on an Augustan obelisk, and on several architectural topics. In 1748 he was called to Rome by Pope Benedict XIV to examine the cupola of St. Peter's basilica and to propose means of preventing its further movement, but he was soon recalled to Padua to assume judicial duties. Excessive work gradually affected his health, although not his enthusiasm, until his death at the age of seventy-eight. His remains were laid in the Church of St. Giacomo in Padua, where his sons placed a monument in his honor. The citizens of Padua subsequently decreed that a statue (one of the earliest works of Antonio Canova) of Poleni be placed among those of illustrious men in the Prato della Valle. A medal in his honor was struck by the Republic of Venice.

BIBLIOGRAPHY

I. ORIGINAL WORKS. Poleni's earliest paper was *Miscellanea: de barometris et thermometris; de machina quadam arithmetica; de sectionibus conicis in horologiis solaribus describendis* (Venice, 1709). A treatise on assorted topics, it includes a dissertation on barometers, which was followed by a second dissertation on this instrument, in *Giornale letterario d'Italia* (1711), and on thermometers, in which several improvements are proposed; also included is the design of an arithmetic machine based on reports that Poleni had received of those of Pascal and of Leibniz. Poleni actually built this machine, which was reportedly very simple and easy to operate; but when he heard of another machine presented to the emperor by the Viennese mechanician Brauer, he destroyed his own and never rebuilt it. A planned 2nd ed. of *Dialogus de vorticibus coelestibus* (Padua, 1712) was unrealized. His lecture, *De physices in rebus mathematicis utilitate oratio* (Padua, 1716), was reprinted in 1720, with some observations by J. Erhard Kapp, in *Clarissimorum virorum orationes selectae* (Leipzig, 1722).

Poleni's works on hydraulics and hydrodynamics include *De motu aquae mixto libri duo* (Padua, 1717), which contains information on estuaries, ports, and rivers; and *De castellis per quae derivantur aquae fluviorum habentibus latera convergentia liber* (Padua, 1718), with reports on experiments on water flow and on the force exerted by an impacting fluid. On the same subject are his corrections on Frontinus' treatise, *L. Julii Frontini de aquaeductibus urbis Romae commentarius restitutus atque explicatus* (Padua, 1722), which were in large part incorporated in Rondelet's trans. (1820) of Frontinus' commentary. A paper combining astronomical and anatomical subjects followed in 1723: "Ad abbatem Grandum epistolae duae de telluribus forma; observatio exlipsis lunaris Patavii anno 1723; et de causa motus musculorum." The memoir on the solar eclipse of 1724, *Ad Johan. Jacob. Marinonum epistola in quo agitur de solis defectu anno 1724 Patavii observato* (Vienna, 1725), was reprinted in *Acta eruditorum Lipsensium* (Leipzig, 1725). A collection of Poleni's letters is in *Epistolarum mathematicarum fasciculus* (Padua, 1728), to which Poleni appended a now very rare short treatise by Giovanni Buteo, *Misura delle acque*.

Some studies of the ancient world followed: a collection of ancient writings, *Utriusque thesauri antiquitatum Romanarum Graecarumque supplementa* (Venice, 1735); a study of Vitruvius' work, *Exercitationes Vitruvianae, seu commentarius criticus de Vitruvii architectura* (Venice, 1739); and an architectural criticism, *Dissertazione sopra il tempio di Diana in Efeso* (Rome, 1742). A reply to an anonymous critic of the latter work appeared in *Giornale dei Dotti* (July 1748). A description of the means employed by Poleni for the restoration of the cupola of St. Peter's Church is found in *Memorie istoriche della gran cupola del tempio Vaticano* (Padua, 1748). Other papers, too numerous to list, may be found in *Acta Lipsiensia*, in the memoirs of the Imperial Academy of Sciences at St. Petersburg, and in *Transazioni filosofiche*.

II. SECONDARY LITERATURE. Particulars about Poleni's life are in *Memorie per la vita, gli studi e i costumi del signor Giovanni Poleni* (Padua, 1762) and in his eulogy, which was inserted by Grandjean de Fouchy in the collec-

tion of Poleni's writing issued by the Academy of Sciences (1763) and by Fabroni in *Vitae Italorum*, XII, no. 2 (1763). A good biography of Poleni is in *Biografia universale antica e moderna*, XLV (1828), and in the 2nd French ed. of the same work (Paris, 1854). A more detailed account of his life appeared in E. de Tipaldo, ed., *Biografia degli Italiani illustri* (Venice, 1834).

BRUNO A. BOLEY

POLI, GIUSEPPE SAVERIO (*b.* Molfetta, Italy, 28 October 1746; *d.* Naples, Italy, 7 April 1825), *physics, natural sciences.*

Poli was the son of Vitangelo Poli and Eleonora Corlé. He completed his early education in Molfetta and then entered the University of Padua, where he studied medicine and natural sciences; among his teachers were Morgagni and Caldani. After obtaining his degree, Poli returned to Molfetta and established a medical practice, which he subsequently abandoned to dedicate himself to the natural sciences. To broaden his knowledge in this field, he traveled for several years throughout Italy, staying at the major universities.

Poli finally went to the Kingdom of the Two Sicilies and took up residence in Naples, where, in 1776, he received the chair of physics at the Royal Military Academy. Shortly thereafter he was sent to observe the teaching methods in France, England, and Germany and to purchase equipment for the physics course at the academy. Upon his return, Poli was offered a teaching post in physics at the Ospedale degli Incurabili, which was then a school of higher learning independent of the University of Naples. Ferdinand IV later summoned him to court to tutor his son, the future Francis I. Poli also held the chair of experimental physics at the university and was director of the military school.

Despite these commitments, Poli continued his scientific studies, for which he won an international reputation as well as membership in the principal academies of Italy and Europe, including the Royal Society of London. Poli gained distinction also as a patron of the arts and sciences, and at his insistence the library of the royal family in Naples was made available to scholars. On his initiative, too, the botanical garden of Naples was founded; and he supplied it with many rare plants from his own garden. Poli's scientific collections were donated to the Museo di Storia Naturale.

Poli's interest in electrical and magnetic phenomena led him, in 1772, to study the causes and effects of lightning and thunder; and he concluded that electricity, lightning, and magnetism were all caused by the same "fluid." His interest in this type of investigation was so passionate that, on his journey from France to England, he made observations in the middle of a thunderstorm, during which the ship itself was struck by lightning.

Poli was also interested in many other topics. He was an expert in numismatics, for example, and his collection of coins and medals became famous. He produced a series of lectures on military history and wrote a work on an earthquake that had hit Naples as well as an elaborate poetic work on astronomy, which failed to arouse enthusiasm for either its artistic or scientific merits.

In the field of natural sciences Poli did extensive research, particularly in marine biology. His proximity to the sea and financial support from the court, where he enjoyed an influential position, enabled him for twelve years to study Mediterranean mollusks and crustaceans. He was helped in this work by the anatomist Michele Troja. Poli's investigations appeared in two folio volumes, in which he outlined a systematic classification of the two animal groups and presented detailed anatomical and physiological descriptions. The work was illustrated with a series of beautiful engravings by contemporary artists. Because of this work Poli is recognized as the founder of the study of mollusks.

BIBLIOGRAPHY

I. ORIGINAL WORKS. Poli's works include *La formazione del tuono, della folgore e di varie altre meteore, spiegata giusta le idee del signor Franklin. Diretta al Sig. D. Daniello Avelloni* (Naples, 1772); *Riflessioni intorno agli effetti di alcuni fulmini* (Naples, 1773); *Continuazione delle Riflessioni intorno agli effetti di alcuni fulmini, ove si esamina la dissertazione del P. G. del Muscio relativa alle riflessioni medesime* (Naples, 1774); "Lettera sulla formazione delle meteore," in *Nuova raccolta di opuscoli scientifici e filologici*, **41** (1785), and **42** (1786–1787); *Lezioni di storia militare* (Naples, 1777); *Ragionamento intorno allo studio della natura* (Naples, 1781); *Elementi di fisica sperimentale*, 5 vols. (Naples, 1787); *Testacea utriusque Siciliae eorumque anathome tabulis aeneis*, 2 vols. (Parma, 1790–1795); *Memoria sul terremoto* (Naples, 1805); and *Viaggio celeste*, 2 vols. (Naples, 1805).

II. SECONDARY LITERATURE. On Poli and his work, see S. delle Chiaie, *De praestantissimi equitis Poli vita* (Naples, 1825); S. Gatti, *Elogio del commendatore Giuseppe Saverio Poli* (Naples, 1825); P. Giampaolo, *Elogio del commendatore Giuseppe Saverio Poli* (Naples, 1825); N. Morelli, "Il Cavalier G. Saverio Poli," in *Biografie degli uomini illustri del regno di Napoli* (Naples, 1826); and M. Tridenti, "G. Saverio Poli, antesignano della moderna biologia," in *Archivio di storia pugliese*, **3–4** (1950), 228–246, and "G. Saverio Poli e gli inizi della genetica sperimentale," *ibid.* (1951), 169–174.

CARLO CASTELLANI

POLINIÈRE, PIERRE (*b.* Coulonces, near Vire, Normandy, France, 8 September 1671; *d.* Coulonces, 9 February 1734), *physics, experimental natural philosophy.*

One of the first in France to present public lectures on experimental natural philosophy, Polinière enjoyed a considerable reputation as a popular demonstrator and as such was an important precursor of Nollet. He made independent, although unappreciated, discoveries in electroluminescence and was one of the earliest on the Continent to advocate Newton's theory of color. His name appears in several eighteenth-century variants: Poliniere, Polinier, and Polynier.

Polinière studied humanities at the University of Caen and, later, philosophy at the University of Paris, where, according to Michaud (*Biographie universelle*, new ed., XXXIII, 637), he studied mathematics under Varignon. In 1704 Polinière published *Éléments de mathématiques*, a work of little significance that was drawn from his experiences as a teacher of mathematics.

In the 1690's Polinière's interests turned to medicine and natural philosophy. Although he apparently received a medical degree, he subsequently devoted his energies to the pursuit and popularization of experimental natural philosophy. Drawing his instruments and techniques from a variety of sources, he worked diligently to perfect his early experiments and to put them in a form suitable for public presentation. Sometime around the turn of the century he began—at the request of members of the Faculty of Philosophy—to present demonstrations before the students at the Collège d'Harcourt and at other colleges of the University of Paris. From these lecture series, some of which lasted for more than two months, Polinière drew the materials for his *Expériences de physique*.

Published in Paris in 1709, this duodecimo volume contained 100 carefully detailed experiments, approximately half of which dealt with the weight and elasticity of air. The remaining experiments were concerned with chemistry, hydrostatics, sound, magnetism, light and colors, and selected aspects of physiology. A second, revised and considerably enlarged, edition was published in 1718; and a fifth and final edition appeared posthumously in 1741.

Although distinguished more for clarity of presentation and popularity than for originality, these volumes contained several experiments of particular interest, notably in the section on light and colors. In each of the editions Polinière included an experiment that elaborated his discovery of the luminescence produced by rubbing partially evacuated glass containers. This discovery is usually attributed to Francis Hauksbee "the elder" but was announced simultaneously and independently by Polinière. In the second edition Polinière abandoned the modification theory of color and—specifically citing Samuel Clarke's 1706 Latin translation—incorporated from Book I of Newton's *Opticks* a series of experiments demonstrating the heterogeneity of white light.

Polinière made his most significant contribution, however, as a popularizer of experimental natural philosophy. His courses were received with such enthusiasm that he continued to offer them annually until his death; and their great success, both with students and the educated public, brought him to the attention of both the court and scientific circles. In 1722 he presented a series of experiments before the young Louis XV and, according to Michaud, Fontenelle was himself a vocal supporter of Polinière and entrusted to him the education of his nephew. Despite the public esteem he enjoyed, Polinière was apparently a retiring and rather shy man, who preferred the satisfaction of his books and instruments to the fame and honors he might have attained. He was a member of the Louis de Bourbon-Condé, count of Clermont's Société des Arts, but he was neither a member nor a correspondent of the French Academy of Sciences.

BIBLIOGRAPHY

I. ORIGINAL WORKS. Polinière's works include *Éléments de mathématiques* (Paris, 1704); and *Expériences de physique* (Paris, 1709; rev. and enl., Paris, 1718, 1728, 1734, and 1741; with each of the latter 2 eds. in 2 vols.). J. C. Heilbronner, in *Historia matheseos universae* (Leipzig, 1742), also attributed to Polinière *Euclides alio ordine digestus et novis demonstrationibus munitus* (Paris, 1704). Five unpublished papers on electroluminescence, which Polinière read before the French Academy of Sciences, are in *Procès-Verbaux de l'Académie des Sciences* (Nov.-Dec. 1706).

II. SECONDARY LITERATURE. There is no general account of Polinière's life or accomplishments. The primary biographical sources are the anonymous "Abrégé de la vie de M. Polinière," in the 4th and 5th eds. of *Expériences de physique* (Paris, 1734, 1741), and Polinière's prefaces to earlier eds. of the same work. Polinière's discoveries in electroluminescence are discussed in David W. Corson, "Pierre Polinière, Francis Hauksbee, and Electroluminescence: A Case of Simultaneous Discovery," in *Isis*, **59** (1968), 402–413.

DAVID W. CORSON

POLLENDER, ALOYS (*b.* Barmen, Germany, 25 May 1800; *d.* Barmen, 15 August 1879), *medicine.*

Pollender's family originally came from Neuss on the Rhine. His father returned there in 1802 as secretary to the district prefect in the French-controlled administration; but he was soon transferred to Kleve, where Pollender attended the Gymnasium for three years. His father was next assigned to Monschau and Pollender prepared there for university study at the Collegium Monjoinse. He also received private instruction in French from his father. The latter was a heavy drinker; and his habit served as a warning to his son, who later became a teetotaler.

When the Prussians took over the Rhineland, his father lost his post, and Pollender had to leave school. In 1815 he became an apprentice to an apothecary in Neuss, but he was clearly unhappy with this choice of career and thus took private lessons to prepare for the university. He passed the entrance examination at Bonn in 1820, and began to study medicine there at the new university. He took the doctoral examination on 2 April 1824 and earned his medical degree for a dissertation entitled "De sanguine coccineo in venis."

The impoverished Pollender became interested in microscopy while working as a laboratory assistant in physics and chemistry courses; and he based his dissertation on a study of so-called scarlet blood in which he carried out physiological experiments on rabbits. Both his skill with the microscope—an instrument that would be greatly improved in the following decades—and his interest in experiments in pathological physiology, played a role in his discovery in 1849 of the anthrax bacillus.

After completing his studies, Pollender attempted to establish a medical practice in the small community of Lindlar, near Cologne. But there were already two doctors in the town; and after a short, unsuccessful stay, he moved to Wipperfürth. The exact year of this move is not known, but his name appears in the town records in 1830. In Wipperfürth, Pollender quickly became a popular and respected figure, but he preferred solitude and remained a bachelor until his seventieth year. He loved horses and riding and was a great amateur musician.

Unlike his colleagues, Pollender spent long hours at his microscope. He took a keen interest in agriculture and cattle raising, studied the microscopic structure of flax fibers, and made a microchemical analysis of the composition of the pollen. For his article on this subject, "Anatomischen Untersuchungen des Flachses, besonders der Bastfaser desselben," the Berlin Academy of Sciences awarded him its Cothenius prize, worth about 300 gold marks. His most important scientific research was on the etiology of anthrax, a disease he often confronted in his medical practice. But even after the publication in 1855 of his important studies on this subject, he was never given the opportunity to continue his research in a university laboratory. In recognition of his services to medicine, the Prussian government awarded him the title *Sanitätsrat*.

At age sixty-nine, Pollender fell in love with Therese Baumann, a beautiful working girl forty-two years his junior. The ensuing scandal obliged him to house his fiancée, along with his mother and sister, in Düsseldorf. He then moved to the Schaerbeeck district of Brussels, where he had another, wealthy sister. He married Therese there in May 1870. He did not succeed, however, in obtaining a license to practice medicine in Belgium; and thus, in 1872, he moved to his native Barmen with his wife and ten-month-old son, Max. Despite an inheritance of 80,000 francs from his brother's estate, he could no longer earn a sufficient income from his practice to maintain his household. They were eventually destitute. When he died of apoplexy, his wife and child had to seek refuge with his sister in Brussels, where the child died several years later. In 1929 a commemorative plaque was affixed to the house in Wipperfürth in which Pollender had practiced medicine. It states simply: "In this house Dr. Aloys Pollender (1800–1879) discovered the anthrax bacillus in 1849."

Pollender's outstanding and original achievement was his description of the causative agent of anthrax, which at the time was causing many losses among cattle in Germany and France and which not infrequently infected human beings, especially flayers. In the older literature one sometimes finds the statement that Pollender was a veterinarian—an idea falsely deduced from the fact that he studied an infectious disease that appears primarily in animals. According to the oral tradition drawn upon by one of Pollender's biographers, Reiner Müller, Pollender first encountered anthrax in 1841, when a flayer he was treating died in great pain a few days after contracting an anthrax carbuncle. In the fall of 1849 a large number of cows again died of anthrax; and Pollender decided to make a thorough investigation of the cause of the disease, which was frequently assumed to be the result of a general corruption of the blood. Shortly before the outbreak of 1849, Pollender had acquired a modern microscope from Simon Plössl of Vienna; and he used it to examine microscopically and chemically the blood of infected animals. It was not until 1855 that he published the results of his study: "Mikroskopische und mikrochemische Untersuchungen des Milz-

brandblutes, so wie über Wesen und Kur des Milzbrandes."

In this paper Pollender stated that "the wish to find out the morphological, physical, and chemical changes that the blood undergoes in anthrax, one of the most terrible contagious diseases, inspired me not to waste the opportunity offered in the fall of 1849 by the death of a number of cows in the area from this disease." Pollender examined blood taken from five cows eighteen to twenty-four hours after their death. He discovered certain pathological alterations in the blood corpuscles and the so-called chylus corpuscles. He noted particularly "a countless number of extremely fine, apparently solid, and not fully transparent rodlike corpuscles, of equal thickness throughout their length and [which are] neither coiled, nor wavy, nor constricted, but entirely straight and flat, with no branching along their course."

Chemical analyses of these bodies, which always appeared in the blood as well as in the spleen and carbuncle fluid of victims, led Pollender to the idea that they must be a type of plant life. He found that "their behavior in the presence of caustic potash, sulfuric acid, hydrochloric acid, and nitric acid is thoroughly characteristic [that is, of plants] and points to a plant-like nature." Pollender thought that this behavior proved that the corpuscles could be neither corrupted albuminous blood nor crystallization products. He observed that "Iodine solution colors them a pale yellow and thereby makes them visible." Thus he was the first to make use of what is actually a type of staining method. (It was not until 1869 that Hermann Hoffmann, Weigert, Salomonsen, and, particularly, Robert Koch introduced bacterial staining by means of specific dyes.)

Pollender was not yet certain whether the "corpuscles" he had observed were the cause of the disease or its so-called corruption products that had emerged following death. He admitted that he could report "nothing concerning the origin and emergence of these remarkable and mysterious corpuscles." Nor could he answer a number of questions about them: "Whether [they] exist in the blood of live animals infected by anthrax, or whether they appear only after death, as a result of fermentation or corruption? Whether they are entozoa or entophytes? Whether they are the infectious material itself or simply the bearers of it, or, perhaps, have no connection with it at all?"

Pollender restricted his claims to what he could ascertain with the means at his disposal. Undoubtedly he was not aware that in 1850 the French physician Pierre François Rayer had detected corpuscles similar to those he had seen. Rayer mentioned them in a report on the transmission of anthrax that he delivered to the Société de Biologie of Paris, but he did not attempt to assess their possible significance and devoted only three lines to them: "In the blood there were, in addition, little thread-like bodies whose length is approximately twice the size of a blood corpuscle. These little bodies display no spontaneous movements at all." Thirteen years later, Rayer's collaborator, Davaine, claimed this observation as his own, thereby asserting priority in the discovery of the cause of anthrax. In addition, the German veterinarian Christian Joseph Fuchs pointed out—although not until 1859—that as early as 1842 he had discovered "granulated threads" (*granulierte Fäden*) in the blood of animals that had died of anthrax; but he renounced any claim to priority. At the end of the nineteenth century and even into the twentieth, German and French scholars were quarreling over the question of priority.

From this dispute the following facts have been established. In 1842 Fuchs, a professor at the school of veterinary medicine at Karlsruhe, saw granulated fibers in the blood of animals attacked by anthrax; and apparently he was the first to do so. But he did not publish his findings until seventeen years later, after he had learned of Pollender's work of 1855. In the fall of 1849 Pollender saw through the microscope the body that he called a "corpuscle," and he subjected it to chemical analysis. But for reasons that are still unknown he delayed publication of his findings until 1855. Nevertheless, Pollender was the first to raise the question of whether these forms could be the cause of anthrax.

In 1850, seeking to study experimentally the transmission of anthrax, Rayer examined the spleen taken from an animal killed by the disease. (He received the organ on 26 June 1850.) Only in this connection did Rayer mention his co-worker Davaine, who did not publish his first independent work on anthrax until 1863. Davaine stressed that his research was inspired by Pasteur's work on butyric acid fermentation. He first challenged Pollender's priority in 1875.

Prior to Davaine's publication, two other authors published articles on the microscopic bodies associated with anthrax. In 1857 Friedrich Brauell published the first account of an observation of anthrax bacilli in blood taken from a live animal, but he obscured the issue by identifying them with vibrios. In 1860, Onésime Delafond published his observation of these bodies in the blood of sheep. (He had first seen them in 1856, before Brauell, but did not publish his findings until after the latter.) Delafond also noted their diagnostic value and their growth. Despite these earlier contributions, Davaine's works are of special importance since he was the first to designate unambiguously the anthrax bacillus as the cause of the disease. On the basis of transmission experiments, research on cultures, and the discovery of anthrax spores, Koch confirmed (1876) Davaine's findings. But Pollender was the first to recognize these spores as plant organisms, and he discussed the theory that these corpuscles could be the cause of anthrax.

BIBLIOGRAPHY

I. ORIGINAL WORKS. Pollender's major works are *De sanguine coccineo in venis* (Bonn, 1824), his M.D. diss.; "Anatomische Untersuchungen des Flachses, besonders der Bastfaser desselben," in *Abhandlungen der Königlichen Akademie der Wissenschaften zu Berlin*, (1847), 2–3; "Mikroskopische und mikrochemische Untersuchung des Milzbrandblutes, so wie über Wesen und Kur des Milzbrandes," in *J. L. Caspers Vierteljahrsschrift für gerichtliche und öffentliche Medicin*, 7 (1855), 103–114; "Chromsäure, ein Lösungsmittel für Pollenin und Cutin, nebst einer neuen Untersuchung über das chemische Verhalten dieser beiden Stoffe," in *Botanische Zeitung*, 20 (1862), 385–389, 397–405; and *Über das Entstehen und die Bildung der kreisrunden Oeffnungen in der äusseren Haut des Blütenstaubes, nachgewiesen an dem Baue des Blütenstaubes der Cucurbitaceen und Onagrarien* (Bonn, 1867).

Subsequent works include *Neue Untersuchungen über das Entstehen, die Entwichlung, den Bau und das chemische Verhalten des Blütenstaubes* (Bonn, 1868); and *Wem gebührt die Priorität in der Anatomie der Pflanzen, dem Grew oder dem Malpighi? Ein Vortrag gehalten in der Section für Botanik und Pflanzen-physiologie bei der 41. Versammlung Deutscher Naturforscher und Ärzte in Frankfurt a.M. im September 1867* (Bonn, 1868).

II. SECONDARY LITERATURE. On Pollender and his work, see O. Bollinger, "Historisches über den Milzbrand und die stäbchenförmigen Körperchen," in *Beiträge zur vergleichenden Pathologie und pathologischen Anatomie*, 2 (1872), 1–22; "Zur Pathologie des Milzbrandes. Vorläufige Mitteilung," in *Centralblatt für die medizinischen Wissenschaften*, 10 (1872), 417–420; *Zur Pathologie des Milzbrandes* (Munich, 1872); and "Milzbrand," in H. von Ziemssen, ed., *Handbuch der speciellen Pathologie und Therapie*, III (Leipzig, 1874), 447–500; F. A. Brauell, "Versuche und Untersuchungen betreffend den Milzbrand des Menschen und der Thiere," in *Archiv fur pathologische Anatomie und Physiologie und für klinische Medizin*, 11, n.s. 1 (1857), 132–144; C. Davaine, "Recherches sur les infusoires du sang dans la maladie connue sous le nom de sang de rate," in *Comptes rendus hebdomadaires des séances de l'Académie des sciences*, 57 (1863), 220–223; O. Delafond, "Communication sur la maladie régnante." in *Récord de médecine vétérinaire*, 37 (1860), 574, 726–748; C. J. Fuchs, "Über das Blut beim Milzbrande der Thiere," in *Magazin für die gesamte Thierheilkunde*, 25 (1859),

314–315; and H. Hiddenmann, "Aloys Pollender, ein Wegbereiter Robert Kochs," in *Naturwissenschaftliche Rundschau*, **3** (1950), 483.

Additional works are Robert Koch, "Die Aetiologie der Milzbrand-Krankheit, begründet auf die Entwicklungsgeschichte des Bacillus Anthracis," in *Beiträge zur Biologie der Pflanzen*, **2** (1876), 277–310, repr. in *Gesammelte Werke von Robert Koch*, J. Schwalbe, ed., I (Leipzig, 1912), 5–25; R. Müller, "Aloys Pollender 1800–1879," in *Zeitschrift des Bergischen Geschichtsvereins*, **53** (1922), 17–25; "80 Jahre Seuchenbakteriologie. Die Seuchenbakteriologie vor Robert Koch: Pollender 1849, Brauell 1856, Delafond 1856, Davaine 1863," in *Zentralblatt für Bakteriologie, Parasitenkunde, Infektionskrankheiten und Hygiene. Abt. 1*, **115** (1930), 1–17; "Aloys Pollender (1800–1879)," in *Münchener Medizinische Wochenschrift*, **77** (1930), 114; and "Aloys Pollender," in *Biographisches Lexikon der hervorragenden Ärzte aller Zeiten und Völker*, 3rd ed., IV (Munich-Berlin, 1962), 647.

See also O. Malm, "Die Entdeckung des Milzbrandbazillus. Eine historische Kritik," in *Zeitschrift für Infektionskrankheiten, parasitäre Krankheiten und Hygiene der Haustiere*, **15** (1914), 195–201; G. Olpp, *Hervorragende Tropenärzte in Wort und Bild* (Munich, 1932), 324–327; P. F. O. Rayer, "Inoculation du sang de rate," in *Comptes rendus des séances de la Société de biologie de Paris*, **2** (1850), 141–144; C. Schnee, "Das Lebenswerk Aloys Pollenders unter besonderer Berücksichtigung des Milzbrand-bazillus" (M.D. diss., Munich, 1955); G. Seidel, "Historische Betrachtungen über den Milzbrand," in *Zeitschrift für Geschichte der Naturwissenschaften, Technik und Medizin*, **1** (1960), 72–93; and J. Theodorides, "Casimir Davaine (1812–1882): A Precursor of Pasteur," in *Medical History*, **10** (1966), 155–165; and "Un grand médecin et biologiste Casimir-Joseph Davaine (1812–1882)," in *Analecta Medico-Historica*, IV (1968), esp. 72–77.

HANS SCHADEWALDT

POMPONAZZI, PIETRO (*b.* Mantua, Italy, 16 September 1462; *d.* Bologna, Italy, 18 May 1525), *natural philosophy.*

Pomponazzi was a professional philosopher in the Aristotelian tradition, who was associated with the universities of Padua, Ferrara, and Bologna. Educated at Padua under Nicholas Vernia, he learned the Scholastic analysis of Aristotle and his commentators, and the traditional separation of philosophy and faith. In 1488 he began his teaching career at Padua as extraordinary professor. Promoted to ordinary professor in 1495, he left the following year for Ferrara. With the death of Vernia in 1499, Pomponazzi was recalled to Padua to succeed his former teacher. When that university was closed because of war in 1509, he returned to Ferrara. After about a year at Ferrara he

was called to Bologna, where he remained until his death.

Pomponazzi's chief works concern immortality, miracles, and free will. In his treatises on immortality, *De immortalitate* (1516), *Apologia* (1518), and *Defensorium* (1519), Pomponazzi developed a theory of mortality based primarily on Aristotle's view of the knowing process. All knowing, he argued, depends on sensation, which in turn depends on the continuous functioning of bodily powers. When sensation ceases at death, knowing too must cease; the mind corrupts with matter. In the Peripatetic tradition only the mind or intellect can be considered immortal. Therefore this argument, indissolubly connecting the intellect to the corruptible powers, proves mortality. Pomponazzi insisted further that Aristotle's definition of the soul as the act of the body supports this analysis.

Following the interpretation of the late Greek commentator Alexander of Aphrodisias, Pomponazzi held that Aristotle's definition is univocal—that the entire soul, intellect included, is the act of the body. Inseparable from the bodily powers it actualizes, the intellect perishes with the decay of organic life. Mortality, then, is the conclusion of natural philosophy. To prevent a conflict with faith, Pomponazzi drew on the distinction traditional among professional Aristotelians since the thirteenth century: he sharply separated philosophy and faith, reserving truth for the doctrine of faith. The conclusions of philosophy, he said, are not true; they can be described as neutral, rational, or probable. As a doctrine of faith, immortality alone is true; its truth is guaranteed by revelation, the principles of which are above rational apprehension.

There is evidence that this disclaimer was a device that at once protected Pomponazzi from the censure of the Church and allowed him to prove a view quite the opposite of that dictated by faith. It is contradicted by another position in the philosophical discussion, in which Pomponazzi ascribed a human origin to immortality and asserted indirectly the truth of mortality. The doctrine of immortality, he said, is an invention of religious lawmakers who "do not care for truth." The aim of these lawmakers is to make men good rather than learned. Relying on the fear of punishment and the hope of reward, the doctrine of immortality aims at producing decent behavior in the masses. The truth these lawmakers "do not care for" is apparently mortality—the doctrine of philosophy.

In the *De incantationibus* (written 1520, published 1556), Pomponazzi investigated seemingly miraculous events reported by contemporary witnesses, pagan literature, and Christian doctrine. He developed naturalistic explanations for all these occurrences except, at first, the Christian miracles. The first event Pomponazzi considered was a recent series of cures supposedly produced by demons invoked by a magician. These cures, said Pomponazzi, are actually produced by the magician himself, who must have possessed occult properties that could cure these diseases. He based his argument on the observation that curative occult powers existing in plants and animals must also be found in man, whose essence unites all natural qualities.

The second "miracle" attracting Pomponazzi's attention was the recent appearance of St. Celestine to the people of Aquileia. During a long rainstorm the people of Aquileia prayed to their patron, St. Celestine, for relief. Soon their prayers were answered, for the storm ended and St. Celestine appeared in the sky. Pomponazzi believed that the people actually saw this vision; he doubted only that it was produced by the miraculous intervention of angels, demons, or God. Offering natural explanations for this occurrence, he suggested that the vision might have been produced by the people themselves in one of two ways. The sheer power of their collective wills might have projected the image from their internal imaginations into the heavens. Or perhaps the vapors emitted by the praying crowd were imprinted with the image that later appeared. Finally, this vision could have also been produced by the intervention of Intelligences who were favorably disposed at that moment to the cult of St. Celestine.

The third category of strange events derived from pagan literature and included animals talking and prophesying, spirits of the dead communing with the living, and statues moving, sweating, and bleeding. These events, explained Pomponazzi, are produced by the Intelligences as movers of the heavenly bodies. The intervention of the Intelligences is simply an aspect of the eternal movement of the heavens rather than a violation of natural processes. Guided by divine providence, the heavens can therefore produce these events within natural limits.

Pomponazzi realized that the application of natural explanations to Christian miracles would destroy their very nature. He noted that his position, if generalized, would mean that "there are no miracles." And if there are no miracles, the religions of Moses and Christ are imperiled, for their fundamental doctrines rest on miraculous events. To avoid this position Pomponazzi again exempted faith from the conclusions of philosophy. Biblical miracles, he stated, are indeed suspensions and reversals of the normal operations of nature. They are produced directly by God or by demons or angels. Knowledge of them is guaranteed by the truth of revelation. Thus there are no natural explanations for the raising of Lazarus, the gushing of the fountain at the word of Moses, the stopping of the sun by Joshua, or the exceptionally long eclipse at the crucifixion of Jesus.

This resolution of the conflict between the natural and the religious explanation in favor of religion is reversed at a certain point in the philosophical discussion. The Christian religion, which at first appeared as a final truth, above time and corruption, is now depicted as having a perfectly natural origin. All religions, said Pomponazzi, are born, flourish, and die. Their birth is produced not by the fiat of a personal deity but by the eternal movements of the heavens, guided by the Intelligences. Religious miracles, apparently interruptions of natural laws, are simply rare and wondrous events produced by the Intelligences. The Intelligences endow religious leaders with the power to perform all kinds of wonders, including curing the ill and raising the dead. The purpose of these "miracles" is easily explained. Occurring mostly at the birth of a new religion, they dispose men favorably toward a new cult. As religions decline, their "miracles" grow less frequent and finally cease. This is why, Pomponazzi held, there were so few miracles in his time. Christianity was dying, having run its life cycle.

If religions arise naturally, it follows that their doctrines may also have a natural basis, a human rather than a divine origin. In fact, the doctrines about angels and demons are not eternal truths but simply fictions. Like the doctrine of immortality, they were invented to lead men to the good life, "although their inventors knew very well that their existence was impossible."

In the *De fato* (written 1520, published 1567), Pomponazzi shifted his approach. Instead of a purely philosophical analysis the conclusions of which were apparently suspended by faith, he presented a double investigation of religious and philosophical determinism, examining each of these traditions in relation to free will.

Pomponazzi held that philosophical determinism, as derived from Stoic-Aristotelian formulas, eliminates free will. Every natural object, he noted, is a moved mover the motion of which could ultimately be traced to the Prime Mover, the source of all motion. All objects are thus subject to external motions that determine their course. As part of nature, man is also subject to movement; his actions, seemingly caused by internal decision, are actually controlled externally; his will, which propels him to action, is itself determined by external movement. Whenever man acts, therefore, he has chosen a course determined for him by the pressure of external motion on the will. We need not deny, said Pomponazzi, the experience of the internal power of choice or the actual deliberation about alternative courses of action. But, he added, these internal activities, attested to by experience, can in no way change the determining effect of external stimuli.

Where philosophical doctrine eliminates free will, Christian doctrine apparently can preserve it. The great problem in Christian doctrine is to reconcile God's omniscience and omnipotence with free will. Pomponazzi attempted this by qualifying God's eternal knowledge and power through the temporal sequences of past, present, and future. Included in God's omniscience is the certain knowledge of future events, even those within our willpower. Hence man appears unable to act other than as foreknown by God, and consequently his freedom disappears. Human freedom can nevertheless be preserved, argued Pomponazzi, if we hold that God foreknows the future in two ways. In His eternal vision of time, God sees the future both as present and as future. As present, God knows the future event as a settled occurrence, as it is "beyond its causes," reduced from potentiality to actuality. As future, God sees the future event as a possibility, a potentiality that can or cannot occur. The future event, foreknown as purely future, is not really determined by God. God is here limited by the nature of time; He cannot know as determined a future event that is by nature undetermined. Thus free will is not destroyed by God's foreknowledge. Apparently Pomponazzi meant that God knows both the future as present and the future as future in a single eternal intuition. Yet He is somehow able to set aside his knowledge of the future as present when he knows the future as future. It is this gap in God's foreknowledge that allows for human freedom.

Yet God not only foreknows the future but also caused it, for through His omnipotence He is the cause of all He knows. Furthermore, He has caused the future in the special sense that He has predestined it. Clearly He cannot be ignorant at any moment of what He Himself has predetermined. Pomponazzi solved this paradox by limiting predestination through the temporal concept of the future. In order to accord with the nature of future time, God's determination of the future as future is contingent. Just as He foreknows the future as future as a mere possibility, so He predestines the future as future as a mere contingency. This contingent determination, allowing freedom to the will created with liberty, saves human freedom.

It seems, then, that Christian doctrine rather than natural reason can save free will. Yet Pomponazzi still had doubts. He wondered if Christian doctrine could have saved the will from the determination by external stimuli to which the Stoic-Aristotelian view condemned it. Setting aside the earlier Stoic argument

which granted deliberation while denying its impact on choice, Pomponazzi was now prepared to argue that thought does lead to free choice. By a complex combination of the doctrines of Aquinas and Scotus, he tried to prove that the will always retains some freedom in choosing among the objects presented by the intellect. It is therefore not blindly led to choice but freely makes its own decisions. Although this Scholastic defense of free will was not as fully developed as the earlier Stoic determinism, it nevertheless reflected several key elements in Pomponazzi's thought.

The Scholastic doctrines, which Pomponazzi now appeared to accept, depend on the view that the psychological foundations of action are the real determinants of human conduct. It is the will, guided by the intellect, which seizes on external stimuli and directs them to its own end. Yet this view was completely opposed to the Stoic-Aristotelian tradition which he had hitherto defended. In spite of this he devoted a good part of the *De fato* to a defense of the religious doctrine of freedom. In the light of his emphasis in previous works that religious doctrines cannot have a rational foundation, this defense appears at first glance as a surprising reversal of his fundamental method of inquiry. But he was driven to this defense of free will by a deep conflict in his own thought.

Pomponazzi's vision of the universe was deeply split. On one side, he defended a view of nature which, by making the highest powers of intellect and will dependent on organic forces, eliminates autonomous activity. On the other side, his ethical doctrine always championed a view of man which places human dignity precisely in the subjection of and escape from the forces of nature. This conflict between a naturalism eliminating freedom and a freedom independent of material determinations remained unresolved. The irresolution of the *De fato*, as Cassirer has noted, was one of the fundamental themes of Renaissance thought. The Renaissance sense of nature which demanded an orderly universe to which man himself is subject conflicted with the Renaissance sense of life which demanded freedom for the individual. From the view of nature, freedom can only be miraculous, a force escaping the natural agents which determine all organic life. From the view of morality, nature itself must be subjected to and commanded by man's reason. Man may be part of nature but he cannot be absorbed by it. He must be able to turn against its blind forces and direct them to human ends.

Pomponazzi's doctrines on miracles and free will were not widely circulated in his lifetime since they remained in manuscript and were published posthumously. But his doctrines on immortality, first set forth in the *De immortalitate*, occasioned cries of outrage from prominent philosophers and theologians and produced the immortality controversy, one of the most important debates prior to the Reformation. In this battle, nine men wrote against Pomponazzi including such distinguished figures as Gaspar Contarini, Agostino Nifo, and the prominent Dominican theologian Bartolomeo de Spina. It was Spina who charged that Pomponazzi's views were heretical and demanded a trial for heresy. The other opponents contented themselves with arguments for the immortality of the soul derived from the classical and medieval traditions. The controversy reached such proportions that Pope Leo X demanded a retraction from Pomponazzi in 1518, and Pomponazzi's final work on immortality, the *Defensorium* (1519), was allowed to be published only with an appended list of orthodox conclusions supporting the immortality of the soul. Despite this furor, Pomponazzi was never formally charged with heresy nor was he forced to retract his views. Undoubtedly, he was protected by his close friends in the Church hierarchy. The fact that the height of the immortality controversy coincided with the Church's preoccupation with the more serious Lutheran problem may also have indirectly helped him: the Church had more serious matters to attend to.

Although Pomponazzi never recanted formally, he certainly became more circumspect about expressing his views after he was attacked. In his later works, the doctrine of mortality is treated less as an independent philosophical position and more as a purely exegetical view. This doctrine, he came to say, was simply the position of Aristotle, which he was professionally charged to expound and develop. He also became more restrained in his assertion that religious belief could be reduced to doctrines of great social utility but of dubious truth.

The general tendency of Pomponazzi's philosophy was a pervading naturalism. By binding the soul, will, and religion of man to the forces of nature, he tended to undermine those medieval categories that separated man from the natural universe. That universe, governed by eternal laws and freed from miraculous interruptions, was a necessary stage in the development of the later mathematical view of nature. Through his astrological determinism, Pomponazzi developed a strict causality for all occurrences within nature so that "once this astrological concept is replaced by that of mathematics and physics, the development of the new concept will find no inner obstacle to resist it. In this completely mediate sense, even Pomponazzi's strange and abstruse work [*De incantationibus*] helped pave the way for the new, exact scientific conception of

natural occurrences" (Cassirer, *The Individual and the Cosmos in the Renaissance*, trans. by M. Domandi [Oxford, 1963], 106).

BIBLIOGRAPHY

I. ORIGINAL WORKS. For full bibliographical references (up to 1954) on original works, modern eds., and secondary literature, see G. Morra, ed., *Tractatus de immortalitate animae* (Bologna, 1954), 17–31.

Collected eds. are *Tractatus acutissimi, utillimi et mere peripatetici, de intensione et remissione formarum ac de parvitate et magnitudine, de reactione, de modo agendi primarum qualitatum, de immortalitate animae, apologie libri tres, contradictoris tractatus doctissimus, defensorium auctoris, approbationes rationum defensorii, per fratrem chrysostomum theologum ordinis predicaturum divinum, de nutritione et augmentatione* (Venice, 1525); and *Opera, de naturalium effectuum admirandorum causis, seu de incantationibus liber: Item de fato, libero arbitrio, praedestinatione, providentia dei, libri V* (Basel, 1567). *Dubitationes in quartum Meteorologicorum Aristotelis librum* (Venice, 1563) and *De naturalium effectuum causis, sive de incantationibus* (Basel, 1556) were published separately.

Previously unpublished texts are L. Ferri, *La psicologia di Pietro Pomponazzi secondo un manoscritto della Biblioteca Angelica di Roma* (Rome, 1877); P. O. Kristeller, "Two Unpublished Questions on the Soul of Pietro Pomponazzi," in *Mediaevala et humanistica*, **9** (1955), 76–101; B. Nardi, *Studi su Pietro Pomponazzi* (Florence, 1965); and P. Pomponazzi, *Corsi inediti dell'insegnamento padovano*, A. Poppi, ed., I (Padua, 1966); II (Padua, 1970).

II. SECONDARY LITERATURE. See G. Di Napoli, *L'immortalità dell'anima nel Rinascimento* (Turin, 1963), 227–338; A. H. Douglas, *The Philosophy and Psychology of Pietro Pomponazzi* (Cambridge, 1910); F. Fiorentino, *Pietro Pomponazzi, Studi storici sulla scuola bolognese e padovana del secolo XVI* (Florence, 1868); E. Gilson, "Autour de Pomponazzi, problématique de l'immortalité de l'âme en Italie au début du XVIe siècle," in *Archives d'histoire doctrinale et littéraire du moyen âge*, **36** (1961), 163–279; A. Poppi, *Saggi sul pensiero inedito di Pietro Pomponazzi* (Padua, 1970); and M. Pine, "Pomponazzi and 'Double Truth,' " in *Journal of the History of Ideas*, **29** (1968), 163–176, and "Pietro Pomponazzi and the Scholastic Doctrine of Free Will," in *Rivista critica di storia della filosofia* (1973), 3–27.

MARTIN PINE

POMPONIUS MELA (*fl.* A.D. 44), *geography*.

The little that is known of Pomponius Mela is gleaned from his only known work, *De chorographia*, occasionally called *De situ orbis*. He was a Roman from Tingentera, a place (otherwise unidentified) in southern Spain, near Gibraltar, that was inhabited by Phoenicians brought over from Africa.[1] From an apparent reference to the triumph of Emperor Claudius (A.D. 41–54) in Britain in A.D. 43, it has been inferred that the *De chorographia* was written during that year.[2] A conjectured sojourn in Rome virtually completes the biographical data on Pomponius.[3]

Although the *De chorographia* is a compendious, and largely derivative, work on geography, it is the first extant geographical work in Latin and the only Roman treatise of the classical period devoted exclusively to that subject.[4] Unscientific in approach and devoid of mathematical or quantitative content (distances and measurements are wholly omitted), it offers only a general descriptive survey of the world as it was then known. To this Pomponius added uncritical and fantastic accounts of the customs, habits, and idiosyncrasies of peoples and nations, as well as spectacular phenomena drawn largely from earlier writers, including Herodotus.

In this three-book treatise, Pomponius assumed a spherical earth that lay in the middle of the world and was divided into two hemispheres and five climatic zones. Three continents—Europe, Africa, and Asia[5]—made up the habitable world, which was completely surrounded by a great ocean that intruded into the continents by pouring its waters into four seas or gulfs: the Caspian, which received its waters along a narrow strait directly from the Scythian, or northern, part of the surrounding ocean;[6] the Arabian and Persian gulfs, which drew their waters directly from the Indian, or southern, part of the ocean; and the Mediterranean Sea[7] (including the Black Sea), the source of which was the Atlantic, or western, part of the great ocean. Below the equator, south of the Indian Ocean in the south temperate zone, lay another world inhabited by the Antichthones; the region was inaccessible because of the heat of the torrid zone.

In describing the habitable world, Pomponius followed tradition traceable to the fourth century B.C.[8] Only the coastal regions were considered, thus ignoring the interiors of countries (Germany, Spain, and Gaul) and omitting many altogether (Dacia, Media, Bactria).[9] He began, in book I, with the Straits of Gibraltar, the area he knew best, and moved eastward along the southern coast of the Mediterranean, turning northward along the east coast of that sea until he reached the Tanaïs, or Don, River, which separated Asia from Europe. The European coast of the Mediterranean was described next, in book II, from the Tanaïs west to the Straits of Gibraltar (included in this survey were Scythia, Thrace, Macedonia, Greece, Italy, southern Gaul, and southern Spain); this segment concluded with descriptions of the major and minor Mediterranean islands.

The coastal survey shifted, in book III, to the outer fringes of the three continents bordering on the great ocean. Here Pomponius began with the Atlantic coast of Spain and then moved eastward along the northern coasts of Europe (Gaul, Germany, and Sarmatia) and Asia (Scythia), turning south along the farthest coast of Asia and then westward again along the southern coastlines of India, Arabia (following the Persian and Arabian gulfs), and Africa; after turning northward, he terminated at the Straits of Gibraltar, the starting point.

As a consequence of his major concern with the ocean, Pomponius found occasion to mention the tides. He believed that they rise and fall simultaneously all over the world. As for the cause of this phenomenon:

> There is no definite decision whether this is the action of the universe through its own heaving breath, attracting and repelling the waters everywhere (on the assumption of savants that the world is a single animate being); or whether there exist some cavernous depressions for the ebb-tides to sink into, thence to well out and rise anew; or whether the moon is responsible for currents so extensive.[10]

Despite his general inferiority as a geographer, Pomponius knew more than Strabo about the positions of Britain, Ireland, and the coasts of Gaul and north Germany; he was also the first to mention the Orkney Islands.[11] Pomponius exerted a considerable influence on early medieval authors, both on his own account and because Pliny used and cited his work.

NOTES

1. *De chorographia*, II.6.96.
2. "Closed for ages, Britain is now being opened up by the greatest of emperors, victorious over tribes not only unconquered before his own day, but actually unknown. The evidence of its peculiarities which he arrived at in his campaign, he brings home to render clear by his triumph." *De chorographia*, III.5.49.; trans. by J. W. Duff, *A Literary History of Rome in the Silver Age*, p. 130.
3. F. Gisinger, "Pomponius Mela," col. 2361.
4. Pliny's subsequent four books on geography, in which Pomponius is cited a number of times, formed only a part of the thirty-seven books of his *Natural History*.
5. The Tanaïs (Don) River was held to separate Europe and Asia; the Nile, Asia and Africa; and the Straits of Gibraltar, Europe and Africa.
6. A common confusion in the ancient world, derived perhaps from a vague hint of the Volga River. See E. H. Bunbury, *A History of Ancient Geography*, II, 363.
7. Pomponius calls it *nostrum mare* ("our sea;" I.1.7). "*Mediterraneum* as an adjective was first used by Solinus in the third century and, as a proper name, by Isidorus in the seventh century." Harry E. Burton, *The Discovery of the Ancient World*, p. 78.
8. See Bunbury, *op. cit.*, I, 384–385.
9. Pomponius' procedure was contrary to the practice of contemporaries, who usually considered the successive countries of a continent.

10. III.1.1–2; trans. by Duff, *loc. cit.*
11. See Bunbury, *op. cit.*, II, 358–361; and Burton, *op. cit.*, p. 79.

BIBLIOGRAPHY

Since the first printed ed. (Milan, 1471), more than 100 eds. of the *De chorographia* have appeared. Some are cited by F. Gisinger, "Pomponius Mela," in Pauly-Wissowa, *Real-Encyclopädie der classischen Altertumswissenchaft*, XXI (Stuttgart, 1952), cols. 2409–2410. Until recently the standard critical ed. was Karl Frick, *Pomponii Melae, De chorographia libri tres* (Leipzig, 1880). A reissue of this ed. (Stuttgart, 1968) contains an extensive bibliography (pp. 109–120) by Wiebke Schaub of the literature on Pomponius' work since 1880. A new ed. has been published by Gunnar Ranstrand, *Pomponii Melae, De chorographia libri tres una cum indice verborum*, Studia Graeca et Latina Gothoburgensia no. 28 (Goteborg, 1971).

Translations include the (apparently) only English version by Arthur Golding, *The Worke of Pomponius Mela the Cosmographer Concerning the Situation of the World* (London, 1585); a French version by Jean-Jacques-Nicolas Huot in *Macrobe* [*Oeuvres complètes*]; *Varron* [*De la langue latine*]; *Pomponius Méla Oeuvres complètes, avec la traduction en française* . . . (Paris, 1845); and two versions included in Schaub's bibliography, one in Italian by Domenico Pavone (Siena, 1893) and the other in German by Hans Philipp (Leipzig, 1912).

Perhaps the most extensive and detailed study of Mela and the *De chorographia* is F. Gisinger, "Pomponius Mela," in Pauly-Wissowa, *Real-Encyclopädie der classischen Altertumswissenschaft*, XXI (Stuttgart, 1952), cols. 2360–2411, which contains biographical data, an intensive analysis of the work, Pomponius' sources, subsequent influence, and bibliography. A useful analysis and evaluation are given by E. H. Bunbury, *A History of Ancient Geography Among the Greeks and Romans From the Earliest Ages Till the Fall of the Roman Empire*, 2nd ed., 2 vols. (London, 1883), II, 352–368, with a foldout map of Pomponius' view of the world inserted at 368. Briefer accounts appear in J. Wight Duff, *A Literary History of Rome in the Silver Age From Tiberius to Hadrian* (New York, 1927), 125–131; and Harry E. Burton, *The Discovery of the Ancient World* (Cambridge, Mass.: Harvard University Press, 1932), 76–81.

EDWARD GRANT

PONCELET, JEAN VICTOR (*b.* Metz, France, 1 July 1788; *d.* Paris, France, 22 December 1867), *geometry, theory of machines, industrial mechanics.*

Poncelet was the natural son (later legitimated) of Claude Poncelet, a rich landowner and advocate at the Parlement of Metz, and Anne-Marie Perrein. At a very early age Poncelet was entrusted to a family of the little city of Saint-Avold, and they saw to his earliest education. In 1804 he returned to Metz. After brief and highly successful studies at the city's *lycée*, he entered the École Polytechnique in October 1807, but fell behind a year on account of poor health. During his three years at the Polytechnique, his teachers included Monge, S. F. Lacroix, Ampère, Poinsot, and Hachette.

In September 1810 Poncelet was admitted to the corps of military engineers. Upon graduating from the École d'Application of Metz in February 1812, he was assigned to work on fortification on the Dutch island of Walcheren. Beginning in June 1812 he participated in the Russian campaign as a lieutenant attached to the engineering general staff. In the course of the subsequent retreat, on 18 November 1812, he was taken prisoner at the Battle of Krasnoï and brought to a camp on the Volga River at Saratov where he was held until June 1814. He profited from his enforced leisure by resuming his study of mathematics. Since he had no books at his disposal he was obliged to reconstruct the elements of pure and analytic geometry before undertaking the original research on the projective properties of conics and systems of conics that established the basis for his later important work in this domain. The notes from this period, which he later designated by the name "cahiers de Saratov," were published in 1862 in volume I of his *Applications d'analyse et de géométrie.*

Upon his return to France in September 1814, he was appointed captain in the engineering corps at Metz, where until May 1824 he worked on various projects in topography and fortification, and also on the organization of an engineering arsenal. Consequently, he was able to acquire a firm knowledge of the problems of fortification and industrial mechanics. Among the innovations to his credit is the development, in 1820, of a new model of the variable counterweight drawbridge, the description of which he published in 1822. His position as a military engineer left him sufficient time to pursue the research on projective geometry that he had commenced in Saratov.

After several preliminary works, which were not published until 1864 (in *Applications d'analyse et de géométrie* II), Poncelet published several articles in Gergonne's *Annales de mathématiques pures et appliquées* (starting in 1817). On 1 May 1820 he presented to the Académie des Sciences an important "Essai sur les propriétés projectives des sections coniques," which contained the essence of the new ideas he wished to introduce into geometry. The original version of this fundamental memoir also was not published until 1864 (in *Applications d'analyse . . .* II). Poncelet sought to show—taking the example of the conics—that the language and concepts of geometry could be generalized by the systematic employment of elements at infinity and of imaginary elements. This goal was within reach, he contended, thanks to the introduction of the concept of "ideal chord" and the use of the method of central projections and of an extension procedure called the "principle of continuity." Here Poncelet made explicit some of the ideas and methods underlying Monge's work. Thus in this remarkable paper, the fruit of investigations begun in 1813, Poncelet resolutely opened the way to the development of the subject of complex projective geometry. Cauchy, the Academy's referee, was little disposed to accept Poncelet's high estimation of the value of geometric methods. Cauchy criticized one of the paper's fundamental components, the principle of continuity, characterizing it as a "bold induction," "capable of leading to manifest errors" (report of 5 June 1820).

Poncelet was deeply affected by this serious criticism for in effect it made the principle of extension, which he considered to be an axiom, dependent on algebraic identities. Rejecting this recourse to analysis, he intransigently maintained his position against Cauchy, despite the latter's great authority. Next he reworked his 1820 "Essai" in order to make it the first section of a large *Traité des propriétés projectives des figures* (Metz–Paris, 1822). This work (discussed below) contained many extremely important innovations, ideas, methods, and original results, and it played a decisive role in the development of projective geometry in the nineteenth century.

Pursuing his geometric studies along with his professional duties, Poncelet then undertook the preparation of four important memoirs. The first two (on centers of harmonic means and on the theory of reciprocal polars) were presented to the Academy in March 1824, but the other two memoirs remained unfinished for several years following an important change in Poncelet's career.

On 1 May 1824, after long hesitation, Poncelet agreed, at the urging of Arago, to become professor of "mechanics applied to machines" at the École d'Application de l'Artillerie et du Génie at Metz. In preparing the courses he was scheduled to give starting in January 1825, Poncelet soon displayed an increasingly lively interest in the study of the new discipline he was to teach and in various aspects of applied mechanics. As a result he interrupted the elaboration of his geometric work, returning to it only occasionally to defend certain of his ideas or to complete unfinished papers.

In 1828 Poncelet published in Crelle's *Journal für*

die reine und angewandte Mathematik his first memoir of 1824, on the centers of harmonic means. A second 1824 memoir, on the theory of reciprocal polars, was not published until the beginning of 1829 in Crelle's *Journal*, because Cauchy had delayed his report on it until February 1828. On account of the long delay in publication, Poncelet had the misfortune of seeing certain of his ideas and a portion of his results attributed to other authors. Thus various articles appearing in 1826 and 1827 in Gergonne's *Annales de mathématiques* and Férussac's *Bulletin des sciences mathématiques* accorded priority to Gergonne and Plücker concerning the principle of duality and its chief applications, conceding merely vague anticipations to Poncelet. In December 1826 the latter vigorously protested to Gergonne and sent him an analysis of his memoir, which was published in March 1827 in the *Annales* accompanied by Gergonne's critical remarks. A regrettable priority dispute began; it lasted until May 1829 and opposed Poncelet first to Gergonne and then to Plücker. This distressing affair greatly upset Poncelet, who was already very hurt by Cauchy's criticisms and by the unenthusiastic welcome given his work by a group of the mathematicians at the Académie des Sciences.

Accordingly, Poncelet abandoned to his rivals the task of continuing the important work he had accomplished in projective geometry. He confined himself to completing, in the winter of 1830–1831, the two memoirs left unfinished since 1824. Even so, he published only the first memoir in the winter of 1832; in it he took up the application of the theory of transversals to curves and geometric surfaces. The second memoir he withheld until 1866, when it was included in volume II of the second edition of the *Traité des propriétés projectives des figures*.

Beginning in 1824 Poncelet had essentially shifted his attention from geometry to applied mechanics. Although he had previously studied certain machines and ways of improving them, it was during the summer of 1824 that he achieved his first important innovation: the design and realization of an undershot waterwheel with curved paddles, which possessed a much increased efficiency. The paper he wrote on this subject gained him a prize in mechanics from the Académie des Sciences in 1825. After new trials conducted on full-scale models, he presented a revised version of his study in 1827. Yet, most of his activity at this period was devoted to elaborating and continually updating the course in mechanics applied to machines that he gave from 1825 to 1834. In addition, he prepared a parallel course in industrial mechanics on a more elementary level which he gave to artisans and workers of Metz from November 1827 to March 1830.

In preparing these courses Poncelet drew on many earlier works concerning the theory of machines and the various branches of mechanics: theoretical, experimental, and applied. In addition, he kept himself informed about all the innovations in these fields. He also profited from his practical professional experience, which he rounded out during a trip taken in 1825 for the purpose of studying the machines in the principal factories of France, Belgium, and Germany.

Poncelet's course on mechanics applied to machines was lithographed for the students at Metz. Three main versions of the work were prepared under his personal supervision: the first (partial), in 1826; the second, in 1832, with the assistance of A. T. Morin; the third (definitive), in 1836. But many other editions, both lithographed and printed, were produced without his knowledge, thus giving his work a broad distribution. The first authentic printed edition, brought out by his disciple X. Kretz, reproduced—with some notes added to bring the work up to date—the text of the lithographed version of 1836 (*Cours de mécanique appliquée aux machines* [Paris, 1874–1876]).

As for the course on industrial mechanics, a first draft, written up by F. T. Gosselin, was lithographed in three parts between 1828 and 1830. Poncelet himself wrote up a more complete version of the first part, dealing with fundamental principles and applications; it was printed in 1829 and rapidly sold out. The following year the work was reprinted as the opening section of a general treatise. The printing of the latter, interrupted on four occasions, dragged on until 1841, when it finally appeared as *Introduction à la mécanique industrielle, physique ou expérimentale*, with the peculiar feature of containing successive passages printed at different dates. In addition to numerous pirated editions, this work was brought out in a third edition by Kretz in 1870.

All this writing and correcting obliged Poncelet to rethink the presentation of the principles of mechanics with a view to applying the theoretical results to the various machines employed in industry. His chief goal was to understand and to explain the functioning of real machines in order to attempt to improve them and to increase their efficiency. For him theory was only a tool in the service of practice. Along with this sort of work, Poncelet carried out several series of experiments on the laws of the discharge of water from large orifices (in 1827 and 1828, with J.-A. Lesbros) and on the formation of ripples on the surface of water (1829); the results were published a short time later.

In 1830 Poncelet became a member of the municipal council of Metz and secretary of the Conseil Général of the Moselle. Hence, it appears that he wished to remain permanently in his native city. In 1834 his nomination as scientific *rapporteur* for the Committee of Fortifications and as editor of the *Mémorial de l'officier du génie*, together with his election to the mechanics section of the Académie des Sciences, led him to move to Paris. He seems then to have forgotten his initial vocation for geometry, to which he made no more than brief references, in 1843 and in 1857. Contrariwise, he drew extensively on his technical knowledge in the memoirs he wrote for the *Mémorial* and in the reports he prepared at the request of the Académie des Sciences and of various official commissions.

Meanwhile he appears to have wanted to do more teaching, and at the end of 1837 he happily agreed to create a course on physical and experimental mechanics at the Faculty of Sciences of Paris. His assignment was of a different kind from what he had at Metz, and he sought to develop for his students a concrete conception of mechanics, midway between abstract theory and industrial application. Although his courses, which he gave until April 1848, were not published, they provided the inspiration, in part, for several books, including the last part of his *Introduction à la mécanique industrielle, physique ou expérimentale* (Metz–Paris, 1841) and H. Résal's *Éléments de mécanique* (Paris, 1851). In this manner a new current was introduced into the teaching of mechanics in France, which until then had an exclusively theoretical orientation.

In 1842 Poncelet married Louise Palmyre Gaudin. By this date he was no longer conducting truly original research. Dividing his time between teaching at the Sorbonne, his duties as an academician and military officer, and technical investigations for *Mémorial de l'officier du génie*, he waited for 1848, when he could retire as a colonel in the engineering corps.

But the Revolution of February 1848 abruptly changed this rather calm existence. Following his successive nominations as member of a commission for curriculum reform and then as professor of mechanics at an ephemeral school of administration created under the control of the Collège de France, he gave up his chair of physical and experimental mechanics at the Faculty of Sciences (16 April). A few days later he was entrusted with new responsibilities that soon monopolized his time. Having been elected to the Constituent Assembly at the end of April as deputy of the Moselle, he regularly participated as a moderate republican in the work of that body during its brief existence. At the same time, Arago, the new minister of war, appointed him brigadier general and then designated him for the post of commandant of the École Polytechnique, which he filled until October 1850. Entrusting the responsibility for current administration to his assistant, Poncelet devoted himself mainly to projects for reforming the school's curriculum. He also had to intervene directly to contain and to direct the activity of the school's students during the riots of May and June 1848.

Having retired at the end of October 1850, Poncelet gladly accepted the chairmanship of the division of industrial machines and tools at the Universal Expositions of London (1851) and Paris (1855). Profoundly interested in the rapid progress of mechanization, he transformed the simple report he was supposed to make on the London Exposition into a vast inquiry into the advances made since the beginning of the century in the conception and use of the different types of machines and tools employed in industry (textile and others). This work, to which he devoted nearly seven years, was published in 1857 in two large volumes that constitute a precious documentary source.

Several years later Poncelet undertook to regroup and edit the whole of his published and unpublished work. Of this ambitious project he was able to bring out only the four volumes of his writings on geometry, the two volumes of the *Applications d'analyse et de géométrie* (1862–1864), and the two volumes of the second edition of the *Traité des propriétés projectives des figures* (1865–1866). Certainly this was not a disinterested enterprise; the polemical character of the commentaries and introductions is obvious. Nevertheless, the compilation of these texts—to which were added a number of commentaries and notes intended to bring them up to date—does allow us to follow closely the evolution of Poncelet's thought, especially since the presentation is chronological: the unpublished notebooks from Saratov (1813–1814) in volume I of the *Applications;* published and unpublished geometric works of the period from 1815 to 1821 in volume II; republication of the original version of the *Traité des propriétés projectives des figures* (1822) in volume I of the *Traité;* and geometric works of the period 1823 to 1831 (including the writings from the polemic with Gergonne and Plücker) in volume II of the *Traité.*

Poncelet began similar work on his writings on the theory of machines and on industrial, physical, and experimental mechanics; but his death in 1867 interrupted this project. At the request of Mme Poncelet the project was resumed by Kretz, who republished the courses on industrial mechanics and on mechanics applied to machines that Poncelet had given at Metz. Kretz did not publish his course on physical and experimental mechanics given at the Faculty of Sciences, despite the many documents at his disposition. Unfortunately these documents disappeared along with the bulk of Poncelet's manuscripts during World War I.

Poncelet's scientific and technical work was concentrated in two very different areas, corresponding to two successive stages in his career: projective geometry and applied mechanics. In geometry, his work, conceived for the most part between 1813 and 1824, was published between 1817 and 1832, except for some writings that appeared too late to influence the development of this field. The most significant portion of this work is the *Traité des propriétés projectives des figures*, which was the first book wholly devoted to projective geometry, a new discipline that was to experience wide success during the nineteenth century. In this domain Poncelet considered himself the successor to Desargues, Blaise Pascal, and Maclaurin and the continuator of the work of Monge and his disciples. Concerned to endow pure geometry with the generality it lacked and to assure its independence *vis-à-vis* algebraic analysis, Poncelet systematically introduced elements at infinity and imaginary elements, thus constructing the space employed in complex projective geometry. Basing his efforts on the principle of continuity and the notion of ideal chords, he also made extensive use of central projections and profitably utilized other types of transformations (homology and transformation by reciprocal polars in two or three dimensions, birational transformation, and so forth).

The distinction Poncelet made between projective and metric properties prefigured the appearance of the modern concept of structure. Among the many original results presented in the *Traité* are those stating that in complex projective space two nondegenerate conics are of the same nature and have four common points (a finding that led to the discovery of cyclic points, imaginary points at infinity common to all the circles of a plane), and that all quadrics possess (real or imaginary) systems of generatrices. The decisive influence that *Traité des propriétés projectives des figures* exercised on the development of projective geometry—an influence underestimated by Chasles, Poncelet's direct rival—is brought to light by most commentators, particularly by E. Kötter, who made the most complete analysis of it, but also by A. Schoenflies and A. Tresse, J. L. Coolidge, C. B. Boyer, N. Bourbaki, and others (see Bibliography). Of the later memoirs, the most striking is devoted to the theory of reciprocal polars, which in Poncelet's hands became an extremely fruitful instrument of discovery, although he did not perceive the more general character of the principle of duality, which was pointed out shortly afterward by Gergonne, Plücker, Möbius, and Chasles. Although it was prematurely interrupted, Poncelet's geometric work marks the first major step toward the elaboration of the fundamental theories of modern geometry.

The bulk of Poncelet's work in applied mechanics and technology was conceived between 1825 and 1840. With regard to technological innovation, his principal contributions concern hydraulic engines (such as Poncelet's waterwheel), regulators and dynamometers, and various improvements in the techniques of fortification (a new type of drawbridge, resistance of vaults, stability of revetments). In applied mechanics Poncelet made important contributions to three broad, interrelated fields: experimental mechanics, the theory of machines, and industrial mechanics. He achieved remarkable successes by virtue of his training at the École Polytechnique, his experience as a military engineer, and his vast knowledge—all of which allowed him to utilize and combine the resources of mathematics and theoretical science, the results of systematic experiments, and the teaching of industrial and craft practice.

Poncelet, whose goal was always real applications, placed experimental and practical findings above theories and hypotheses. In the theory of machines, for example, he refused to make any theoretical classification such as Monge had, relying instead on the more concrete principles set forth by G. A. Borgnis in 1819. Instead of vast syntheses he preferred precise and limited studies, informed by a profound knowledge of the technical imperatives involved. Consequently his original work is to be found to a much greater degree in the realms of organization and improvements than in that of brilliant innovations. The influence of his thinking—a mixture of the theoretical and the concrete—on the creation of the field of applied mechanics is indicated by the success of his two lecture courses. From the two chief aspects of his work, the mathematical and the technological, it appears that Poncelet wished to contribute to the concurrent development of both pure science and its applications.

BIBLIOGRAPHY

I. ORIGINAL WORKS. Nearly complete—although occasionally imprecise—lists of Poncelet's various publications are in I. Didion, *Mémoires de l'Académie de Metz*, **50** (1870), 149–159; and in H. Tribout, *Un grand savant: Le général Jean-Victor Poncelet* (Paris, 1936), esp. 204–220. Accessible, but much less complete, are the bibliographies in Poggendorff, **2** (1863), cols. 496–497; **3** (1898), 1057–1058; and in Royal Society *Catalogue of Scientific Papers*,

IV (London, 1870), 479–481, which cites 41 articles. A partial supp. is in *Catalogue général des livres imprimés de la Bibliothèque nationale*, CXL (Paris, 1936), cols. 490–495.

As none of these bibliographies is fully satisfactory, we shall present here a selective bibliography based on the list of articles in Royal Society *Catalogue of Scientific Papers* and divided into 3 sections: geometry, other mathematical works, and applied mechanics (industrial and experimental).

Geometry. The list of the Royal Society *Catalogue of Scientific Papers*—from which the first entry, by an homonymous author, must be eliminated—contains 16 memoirs on geometry: nos. 2–9, 40 (published between 1817 and 1825), nos. 13–15 (1827–1829), nos. 18–20 (1832), and no. 31 (1843). Several works should be added to this list: "Problèmes de géométrie," in *Correspondance sur l'École polytechnique*, **2**, sec. 3 (1811), 271–274; various nn. on the projective properties of figures, on the principle of continuity, and on an instrument for tracing conics, are in *Mémoires de l'Académie de Metz*, **1–4** (Metz, 1820–1823), details of which can be found in the bibliographies of Didion and Tribout cited above; and certain items stemming from the polemic Poncelet commenced against Gergonne and Plücker: "Analyse d'un mémoire présenté à l'Académie royale des sciences," in *Annales de mathématiques pures et appliquées*, **17**, no. 9 (1827), 265–272, which is a partial text of a letter by Poncelet dated 10 December 1826; it is followed by Gergonne's "Réflexions," *ibid.*, 272–276; "Note sur divers articles . . .," in Férussac, ed., *Bulletin des sciences mathématiques*, **8** (1827), 109–117 (this note is cited as no. 13 in the list of Royal Society *Catalogue of Scientific Papers* in accord with its republication in the *Annales de mathématiques pures et appliquées*, **18**, no. 5 [1827], 125–149, with Gergonne's critical remarks and the corresponding passages from Poncelet's letter of 10 December 1826).

See also "Sur la dualité de situation et sur la théorie des polaires réciproques," *ibid.*, **9** (May 1828), 292–302; "Réponse de M. Poncelet aux réclamations de M. Plücker," *ibid.*, **10** (May 1829), 330–333; and "Sur la transformation des propriétés métriques des figures au moyen de la théorie des polaires réciproques," in *Comptes rendus hebdomadaires des séances de l'Académie des sciences*, **45** (1857), 553–554.

Three of Poncelet's important memoirs on geometry first became known through the critical reports on them that Cauchy presented to the Académie des Sciences. See Cauchy's reports on "Mémoire relatif aux propriétés projectives des coniques," in *Annales de mathématiques pures et appliquées*, **11**, no. 3 (1820), 69–83; on "Mémoire relatif aux propriétés des centres de moyennes harmoniques," *ibid.*, **16**, no. 11 (1826), 349–360; and on "Mémoire relatif à la théorie des polaires réciproques," in *Bulletin des sciences mathématiques*, **9** (1828), 225–229.

Poncelet's geometric *oeuvre* also includes two important books. The famous *Traité des propriétés projectives des figures* (Metz–Paris, 1822), the 2nd ed. of which (2 vols., Paris, 1865–1866) reproduces the text of the 1st ed., with annotations (in vol. I) and Poncelet's geometric articles posterior to 1822 and also the principal items from his polemic with Gergonne and Plücker (1825–1829) (in vol. II).

The 2nd book is *Applications d'analyse et de géométrie* . . ., 2 vols. (Paris, 1862–1864). Its first vol. contains, with some nn. and additions, the text of the 7 notebooks in analytic and pure geometry dating from his imprisonment at Saratov (1813–June 1814); the last notebook is the first, incomplete sketch of his future treatise on projective properties. The 2nd vol. contains the text of various geometric works dating from the period 1815–1817 (notebooks 1 to 3), an unpublished memoir from 1818–1819 on the principle of continuity (no. 4), the "Essai sur les propriétés projectives des figures," presented to the Académie des Sciences on 1 May 1820 (no. 5), the text of the articles published in *Annales* between 1817 and 1822 (no. 6), and a few letters and polemical writings (no. 7).

Other Mathematical Works. This relatively secondary portion of Poncelet's *oeuvre* is virtually limited to the 4 articles and memoirs cited as nos. 21–23 (1835) and no. 37 (1848) in Royal Society *Catalogue of Scientific Papers*.

Applied Mechanics (Industrial and Experimental). The Royal Society *Catalogue of Scientific Papers* mentions 20 articles from 1825 to 1852, dealing with this fundamental aspect of Poncelet's *oeuvre*: nos. 10–12, 16, 17, 24–30, 32–36, 38, 39, and 41.

To this very incomplete list should be added the reports and communications presented to the Académie de Metz between 1820 and 1834 and a portion of the reports presented to the Académie des Sciences between 1834 and 1857 (the bibliographies of Didion and Tribout cited above give references to the corresponding publications).

Eight important technical memoirs published between 1822 and 1844 are in different fascs. of the *Mémorial de l'officier du génie*: "Mémoire sur un nouveau pont-levis à contrepoids variable," no. 5 (1822); "Notice sur un pont-levis à bascule mouvante," no. 10 (1829); "Solution graphique des principales questions sur la stabilité des voûtes," no. 12 (1835); "Mémoire sur la stabilité des revêtements et de leurs fondations," no. 13 (1840); "Note additionnelle sur les relations analytiques qui tiennent entre elles la poussée et la butée des terres," *ibid.*; "Rapport et mémoire sur la construction et le prix des couvertures en zinc," *ibid.*; "Observation sur le mode d'exécution et de restauration du pont-levis à contrepoids variable," *ibid.*; and "Résumé historique de la question du défilement des tranchées," no. 14 (1844). See, in addition, the two short notes, "Extrait d'un mémoire sur les projets d'usines et de machines pour l'Arsenal du génie de Metz," in *Bulletin de la Société d'encouragement pour l'industrie nationale*, **23** (1824), 66–68; and "Réclamation de M. Poncelet," in *Bulletin de Férussac*, **13** (1830), 7–9.

Poncelet was also the author of two lecture courses, one on "mechanics applied to machines" and another on industrial mechanics. In addition, he wrote an important two-volume study on the different types of industrial machines and tools.

The course, *Cours de mécanique appliquée aux machines*, which grew out of lectures Poncelet had given from 1825 to 1834 to students at the École d'Application de l'Artillerie et du Génie de Metz, went through a large number of editions. The only ones among these that Poncelet himself supervised are a lithographed ed. of 1826, "Première partie du cours de mécanique appliquée aux machines . . ."; a lithographed reedition, in 1832, of secs. 1 and 3 and an ed. in the same year, with the assistance of A. T. Morin, of two supplementary sets of notes: "Leçons préparatoires au lever d'usines," which became secs. 6 and 7 of the course, and "Leçons sur les ponts-levis" (sec. 8); and a lithographed reedition, in 1836, of the whole of the course (the 4th sec. now including the 5th).

These eds.—each containing original or additional material—of the course went through many other printings, certain of which were done for the use of the students at the École de Metz; other printings were done without Poncelet's knowledge. Printed eds. were also brought out without the author's authorization (e.g., *Traité de mécanique appliquée aux machines*, 2 vols. [Liège, 1845]); these contained, in addition to the 6 secs. of the lithographed edition of 1836, a brief 4th sec. on friction in gears. See also definitive posthumous ed., *Cours de mécanique appliquée aux machines*, 3rd. ed. in 2 vols. (Paris, 1874–1876).

The course, *Cours de mécanique industrielle*, stemmed from lectures Poncelet gave between 1827 and 1830 to artisans and workers of Metz. It went through several eds., which are sometimes difficult to identify.

A lithographed version, prepared with Poncelet's permission by his colleague F. T. Gosselin, was published in 3 pts. as *Cours de mécanique industrielle fait aux artistes et ouvriers messins* . . .: Première Partie: "Résumé des leçons du Cours de mécanique industrielle . . ." (Metz, 1828), of which there also exists a more detailed version; Deuxième Partie: "Cours de mécanique industrielle professé de 1828 à 1829" (Metz, 1829; 2nd ed., 1831); and Troisième Partie: "Cours de mécanique industrielle professé du mois de janvier au mois d'avril 1830."

Several other unauthorized lithographed versions of this course were also produced. Poncelet published a partial printed version of the course, *Première partie du Cours de mécanique industrielle, fait aux ouvriers messins* (Metz, 1829). An ed. of the entire course, begun in 1830 and continued in 1835, 1838, and 1839, was completed in 1841: *Introduction à la mécanique industrielle, physique ou expérimentale*, 2nd ed. (Metz–Paris, 1841), which was republished under the same title (Paris, 1870).

Poncelet's report, *Inventaire des machines et outils. Travaux de la Commission française* [at the Universal Exposition of 1851], III: *Rapport fait au jury international . . . sur les machines et les outils employés dans les manufactures* (Paris, 1857), appeared in 2 vols.: I, *Machines et outils des arts divers*; II, *Machines et outils appropriés aux arts textiles*.

II. SECONDARY LITERATURE. Major accounts of Poncelet's life and work are I. Didion, "Notice sur la vie et les ouvrages du général J.-V. Poncelet," in *Mémoires de l'Académie de Metz*, **50** (1870), 101–159; J. Bertrand, "Éloge historique de Jean-Victor Poncelet," in *Mémoires de l'Académie des sciences de l'Institut de France*, **41**, pt. 2 (1879), i–xxv, repr. in J. Bertrand, *Éloges académiques* (Paris, 1890), 105–129; and H. Tribout, *Un grand savant: Le général Jean-Victor Poncelet* (Paris, 1936).

Among the briefer notices, see F. Hoefer, in *Nouvelle biographie générale*, **40** (Paris, 1862), cols. 735–736; J.-B. Dumas, C. Dupin, and M. Rolland, *Discours prononcés aux funérailles de M. le général Poncelet* (Paris, 1867), E. T. Bell, in *Les grands mathématiciens* (Paris, 1939), 226–238; and M. D'Ocagne, in *Histoire abrégée des sciences mathématiques* (Paris, 1955), 205–209.

Poncelet provided an analysis of his own work in geometry and applied mechanics before 1834, in *Notice analytique sur les travaux de M. Poncelet* (Paris, 1834).

Many studies have been devoted to Poncelet's geometric work. The principal studies are M. Chasles, *Aperçu historique sur le développement des méthodes en géométrie . . .* (Brussels, 1837), see index; and *Rapport sur les progrès de la géométrie* (Paris, 1870), 38–45; E. B. Holst, *Om Poncelet's betydning for Geometrien* (Christiania [Oslo], 1878); E. Kötter, "Die Entwicklung der synthetischen Geometrie von Monge bis auf Staudt (1847)," in *Jahresbericht der Deutschen Mathematiker-Vereinigung*, **5**, pt. 2 (Leipzig, 1901), see index; A. Schoenflies and A. Tresse, in *Encyclopédie des sciences mathématiques*, III, vol. 2, fasc. 1 (Leipzig–Paris), esp. pp. 6–15; J. L. Coolidge, *A History of Geometrical Methods* (Oxford, 1940), see index; and *A History of the Conic Sections and Quadric Surfaces* (Oxford, 1945), see index; C. B. Boyer, *A History of Analytic Geometry* (New York, 1956), see index; and N. Bourbaki, *Éléments d'histoire des mathématiques,* (Paris, 1969), 165–168.

No comprehensive study has as yet been devoted to Poncelet's technical work. Its essential aspects are briefly outlined in M. Daumas, ed., *Histoire générale des techniques*, III (Paris, 1968), see index.

On Poncelet's work in applied mechanics, see C. Comberousse, in *Nouvelles annales de mathématiques*, 2nd ser., **13** (1874), 174–185; and an anonymous author, "J. V. Poncelet. Son rôle en mécanique," in *Revue scientifique*, 2nd ser., **11** (1876), 256–258.

RENÉ TATON

PONS, JEAN-LOUIS (*b.* Peyre, Dauphiné, France, 24 December 1761; *d.* Florence, Italy, 14 October 1831), *astronomy.*

Although it started at an even lower level and remained even more narrowly restricted in its devotion to cometary observations, Pons's career was quite similar to that of Charles Messier.

Born to a poor family, Pons received a very incomplete education that did nothing to prepare him to become the most successful discoverer of comets in the history of astronomy. That preparation began only when, after entering the service of the Marseilles observatory in 1789 as its concierge, he was instructed in astronomy by the directors of that establishment. Pons progressed rapidly, devoting himself primarily to practical observation; an increasing familiarity with the normal aspect of the heavens enabled him quickly to perceive slight changes. Combined with excellent eyesight and great patience, this ability equipped him as a new "ferret of comets"; his first discovery of such a celestial object—which he shared with Messier— occurred in July 1801. Curiously, this discovery was also Messier's last. From then until August 1827 Pons rarely failed to discover at least one comet a year and found a total of thirty-seven during that twenty-six-year period.

These observations did not go unrewarded. In 1813 Pons was named *astronome adjoint* at the Marseilles observatory, and by 1818 he had become its assistant director. In the latter year he discovered three very small and tailless comets, for which feat the Académie des Sciences in Paris—aware that two of them would probably have gone unnoticed but for Pons—awarded him its Lalande Prize. One of these objects proved to be of extraordinary interest, for, pursuing Pons's suggestion that this comet was the one he (Encke) had discovered in 1805, Encke calculated its elements, announced that it had a period of only 1,208 days, and predicted its return in 1822. That return, visible only in the southern hemisphere, was seen by Karl Rümker in Australia. Since this was only the second instance of a recognized return of a comet, its importance was acknowledged in 1823, when the Astronomical Society of London awarded one of its first gold medals to Encke—after whom the comet has come to be named, despite his own insistence that it should be called Pons's Comet. The Society granted silver medals to Rümker and Pons, although technically Pons received his for the discovery of two new comets in 1822.

The latter two comets were not discovered at Marseilles, for, on the recommendation of F. X. von Zach, Pons had been called to Lucca in 1819 to become director of the new observatory created at Marlia by the Spanish infanta Maria Luisa, widow of the former King Louis of Etruria. There Pons added several comets to his list, for one of which—"la 26ᵉ ou la 27ᵉ qu'il a vue"—he shared another Lalande Prize with Joseph Nicolas Nicollet in 1821. This observatory was closed at the death of the Duchess Maria Luisa in 1824, and Pons, again due largely to Zach, was invited by Grand Duke Leopold II of Tuscany to become director of the Florence observatory in 1825. There he found seven more comets, three of which were acknowledged in 1827 in another joint Lalande Prize, shared with J. F. A. Gambart. Failing eyesight caused Pons to lose his position as the first announcer of comets after 1827 and, ultimately, to relinquish his observational duties a few months before his death.

BIBLIOGRAPHY

I. ORIGINAL WORKS. Exclusively an observer, Pons communicated his discoveries in informal letters to leading astronomers or learned societies rather than in formal papers. F. X. von Zach, especially, frequently quoted from these letters in his own communications to the *Zeitschrift für Astronomie und verwandte Wissenschaften.* He also published Pons's "Observation d'une tache fort singulière qui a paru sur le soleil, le 23 déc. 1823" in his *Correspondance astronomique, géographique, hydrographique et statistique,* **9** (1823), 603.

II. SECONDARY LITERATURE. The only treatments of any extent are those by René Alby in J. F. Michaud, ed., *Biographie universelle,* new ed., XXXIV, 62–63; and by Elizabeth Roemer, *Jean Louis Pons, Discoverer of Comets,* Astronomical Society of the Pacific Leaflet no. 371 (May 1960). Useful information may also be gleaned from the prize-awarding speeches and comments: Henry Thomas Colebrooke, "On Presenting the Silver Medal to M. Pons," in *Memoirs of the Royal Astronomical Society,* **1** (1822), 513–514; and *Procès-verbaux des séances de l'Académie tenues depuis la fondation de l'Institut, jusqu'au mois d'août 1835,* 10 vols. (Hendaye, 1910–1922), VI, 433; VII, 166–167; VIII, 549. Finally, since he had become an associate of the Royal Astronomical Society, that body's publications acknowledged his death with brief biographical notices: *Memoirs of the Royal Astronomical Society,* **5** (1833), 410–411; and *Monthly Notices of the Royal Astronomical Society,* **2,** 68.

SEYMOUR L. CHAPIN

PONTEDERA, GIULIO (*b.* Vicenza, Italy, 7 May 1688; *d.* Lonigo, Italy, 3 September 1757), *botany.*

Pontedera studied philosophy and medicine at the University of Padua, from which he graduated in 1715. He early became interested in botany and in 1718 produced his first botanical tract, *Compendium tabularum botanicarum,* which treated new or otherwise interesting plants of the province of Venetia. With few exceptions his plant system mirrored that of Tournefort. In 1719 Pontedera was appointed prefect of the botanical garden of the University of Padua and lecturer in botany. He held this combined post until his death.

Pontedera's chief work in botany was *Anthologia sive de floris natura* (1720). Its main subject was the negation of the sexuality of plants, which had been reported by Nehemiah Grew, Rudolf Camerarius, and Sébastien Vaillant. To prove his thesis Pontedera discussed in great detail the morphology of different flower types and their organs, including stamens and pistils. The many figures illustrating the text were not original but stemmed from Tournefort's *Éléments de botanique.* Pontedera ascribed to the pollen a function quite different from the sexual: in the pollen grains there was, according to him, a liquor essential to the growth and development of the embryo. But the liquor was not dispersed with the pollen grains. Instead, before the grains loosened from the anther, the liquor moved out of the grains, through the filaments, down to the receptacle, and from there to the embryos.

With a series of examples Pontedera tried to demonstrate the nonsexuality of plants. He mainly used dioecious species, of which he claimed to know female individuals isolated at considerable distances from male representatives of the same species and still bearing fully matured fruit. Linnaeus, who in the 1730's based his systematics upon the sexuality of plants, wrote polemics against Pontedera. Yet Pontedera's detailed studies of the flower and its parts and his insight into contemporary theories stimulated knowledge of the function of the flower in plant reproduction.

From the 1720's Pontedera became absorbed in classical studies and prepared new versions of Latin texts.

BIBLIOGRAPHY

A complete list of Pontedera's writings is in A. Béguinot, "Giulio Pontedera," in A. Mieli, ed., *Gli scienziati italiani,* I (Rome, 1921), 93–94. His botanical works are also listed in G. A. Pritzel, *Thesaurus literaturae botanicae* (Leipzig, 1872). Among them are *Compendium tabularum botanicarum* (Padua, 1718) and *Anthologia sive de floris natura* (Padua, 1720). An MS in 7 vols. concerning the history of the botanical garden of Padua is in the archives of the garden.

There is a biographical account in Béguinot (see above), pp. 90–94.

GUNNAR ERIKSSON

POOR, CHARLES LANE (*b.* Hackensack, New Jersey, 18 January 1866; *d.* New York, N.Y., 27 September 1951), *astronomy.*

The son of Edward Eri Lane and Mary Wellington, Lane obtained his undergraduate and part of his graduate education at the City College of New York, receiving a B.S. in 1886 and an M.S. in 1890. He completed his education at the Johns Hopkins University, where he was awarded a Ph.D. in 1892. At Johns Hopkins he studied under Simon Newcomb, who was then chairman of the departments of astronomy and mathematics. Newcomb's direction led Poor to general studies of comets and to a dissertation on the difficult problem of the orbit and motion of Comet 1889V.

Poor became known not only for his excellent work on comets but also for his polemical stand against relativity theory. He later published many papers and books vehemently criticizing Einstein and his work. The quality of his work on comets won Poor a post on the faculty of Johns Hopkins in 1892. He eventually became head of the department of astronomy and held the position until 1899, when he resigned to take over his father's cotton factoring business in South Carolina.

In 1903 Poor returned to science as a professor of astronomy at Columbia University, where he remained for the rest of his life, becoming professor emeritus in 1944.

Poor was elected a fellow of the American Academy of Arts and Sciences and a fellow of the Royal Astronomical Society. From 1901 to 1906 he was editor of the *Annals of the New York Academy of Sciences.* Many of his personal interests were related to his professional work in astronomy and celestial mechanics. An avid yachtsman, he wrote books on yachting and navigation and also invented numerous navigational instruments.

BIBLIOGRAPHY

I. ORIGINAL WORKS. Poor's works on comets and planetary studies include "Preliminary Note on the Comet 1889V," in *Astronomical Journal,* **14** (1894), 63; *The Solar System* (New York, 1908); and "Secular Perturbations of the Inner Planets," in *Science,* n.s. **54** (1921), 30–34. Some of his works on relativity are *Gravitation versus Relativity* (New York, 1922); "Relativity: An Approximation," in *Popular Astronomy,* **31** (1923), 661; "Relativity and the Motion of Mercury," in *Annals of the New York Academy of Sciences,* **29** (1925), 285–319; and "Relativity and the Law of Gravitation," in *Astronomische Nachrichten,* **238** (1930), 165–170. Among his books on navigation is *Simplified Navigation for Ships and Aircraft* (New York, 1918).

II. SECONDARY LITERATURE. Brief obituaries of Poor are S. A. Mitchell, in *Monthly Notices of the Royal Astronomical Society,* **112** (1952), 279–280; and the unsigned obituaries in *Publications of the Astronomical Society of the Pacific,* **64** (1952), 48; and New York *Herald Tribune* (28 Sept. 1951), 22. Poor's work is mentioned in A. M. Clerke, *A Popular History of Astronomy During the Nineteenth Century* (London, 1902); and H. S. Williams, *The Great Astronomers* (New York, 1930).

RICHARD BERENDZEN
RICHARD HART

POPE, WILLIAM JACKSON (*b.* London, England, 31 October 1870; *d.* Cambridge, England, 17 October 1939), *organic chemistry.*

Pope's parents, William Pope and Alice Hall, were staunch and active Wesleyans who had eight children, of whom William was the eldest. In 1878 he entered the Central Foundation School, in London, where his ability to learn rapidly gave him leisure at the age of twelve to carry out simple chemical experiments in his bedroom. While at school he also developed great skill as a photographer—many of his early photographs were in perfect condition fifty years later. Pope readily learned foreign languages and became proficient in French and German in his teens and in Italian somewhat later. He was known, on the eve of his departure to deliver an important lecture in Paris, to sit down at his typewriter, think deeply for a few minutes, and then rapidly type the complete lecture in French. This ready acquisition of foreign languages was not limited to Pope; his brother Thomas translated into English E. Molinari's *Treatise on General and Industrial Inorganic Chemistry* (2nd ed., 1920) and its two-volume companion, *Industrial Organic Chemistry* (2nd ed., 1921–1923).

Pope left school in 1885 with full marks on his final examinations in theoretical and practical chemistry and in theory of music, and obtained entrance scholarships to Finsbury Technical College and the City and Guilds of London Institute, South Kensington. At the latter he worked under H. E. Armstrong. A firm believer in the heuristic method of teaching, Armstrong forbade his students to take any examinations, so they ultimately departed without degrees. In 1897 Pope became head of the chemical department of the Institute of the Goldsmiths' Company at New Cross.

In 1901 Pope became head of the chemical department of the Municipal School of Technology and professor of chemistry at Manchester, and in 1908 he was appointed professor of chemistry at the University of Cambridge, a post he held until his death. At the time of his election, only the major departments were headed by a professor; and the appointment of Pope at the age of thirty-eight caused some surprise in nonchemical circles.

In the following years Pope's advancement of chemistry in various directions was recognized by the conferment of the freedom and livery of the Goldsmiths' Company by special grant in 1919, and he served as prime warden for 1928–1929.

During World War I, Pope was a consultant to the Board of Invention and Research; and his work, particularly on the manufacture of photographic sensitizers and of mustard gas, was recognized after the war by the award of a knighthood of the Order of the British Empire.

During 1922–1936 Pope presided at the chemical conferences of the Solvay Foundation in Brussels, and he worked for several years to promote the formation of the International Union of Pure and Applied Chemistry, of which he became the first president. He was president of the Chemical Society for 1918 and 1919 and received its Longstaff Medal in 1903 and the Davy Medal of the Royal Society in 1914.

Pope's first investigation, in collaboration with H. E. Armstrong, was an attempt to obtain a crystalline derivative of pinene. Following some very old work of Ascanio Sobrero, they exposed the terpene fraction of oil of turpentine to moist oxygen in sunlight. The solid deposit, which when purified had the composition $C_{10}H_{18}O_2$, was called sobrerol.[1] This compound was optically active; and the dextro and levo forms were isolated from turpentines of various origins and were studied in detail. For this purpose Pope's knowledge of crystallography and his skill with the goniometer were of great assistance to Armstrong.

Pope was then joined by F. S. Kipping, with whom he worked for several years. They showed that camphor was sulfonated by fuming sulfuric acid and by chlorosulfonic acid, which gave the corresponding sulfochlorides, compounds of exceptional crystallizing power. Furthermore, these sulfochlorides and sulfobromides lost sulfur dioxide when heated, forming the corresponding halogenocamphors, which they termed π-derivatives because of their pyrogenic formation.[2] Furthermore, Kipping showed that the sulfonation had occurred on one of the *gem*-dimethyl groups and (on the basis of J. Bredt's camphor formula) on the 8-methyl group—that is, on the methyl group furthest from the carbonyl group. Consequently the acid known as α-bromo-π-camphorsulfonic acid (Formula 1) should be termed 3-bromocamphor-8-sulfonic acid.[3]

(1)

A further investigation produced several new halogenocamphors. The study of these compounds, their sulfonic acids, and the conditions of their racemization revealed a number of points of crystallographic and theoretical interest and of the particular relationships that Kipping and Pope termed pseudoracemism.[4]

Pope and Kipping also investigated the crystallization of sodium chlorate from aqueous solution, whereby dextro (*d*) and levo (*l*) crystals are deposited in virtually equal quantities. The activity here must be due to the arrangement of the molecules in the crystal, since the molecule is symmetric. Each crystal, when dissolved in water, therefore gives an inactive solution. In an asymmetric environment, such as an aqueous solution of glucose, the weights of the deposited dextro and levo crystals of the chlorate are no longer equal. The investigation of this subject and later of the similar behavior of ammonium sodium tartrate in considerable detail was greatly aided by Pope's crystallographic skill.[5] This fruitful partnership ended when Kipping became professor of chemistry at University College, Nottingham, and Pope became head of the chemical department of the Goldsmiths' Institute at New Cross.

At the Goldsmiths' Institute, Pope and S. J. Peachey investigated "tetrahydropapaverine," which hitherto had resisted optical resolution. Pope recalled his earlier experience with bromocamphorsulfonic acid; and with the salt of this acid he readily resolved the base, which was later found by F. L. Pyman to be dihydropapaverine.[6] This was one of the earliest resolutions using bromocamphorsulfonic acid, and a number of resolutions of basic compounds or dissymmetric cations using salts of camphorsulfonic and bromocamphorsulfonic acids followed. These acids were of outstanding value at this stage of stereochemical elucidation.

Pope then started a new line of investigation, which was conspicuously successful and immensely increased his reputation. Compounds that showed optical activity in solution had hitherto all contained one or more asymmetric carbon atoms. Le Bel had claimed that by growing microorganisms in an aqueous solution of methylethylpropylisobutylammonium chloride (see Formula 2a), he obtained that salt in an optically active condition.[7] This work was carefully repeated by W. Marckwald and A. von Droste-Huelshoff, who were unable to confirm his results.[8] E. Wedekind similarly failed to resolve benzylphenylallylmethylammonium iodide (Formula 2b).[9]

$$[CH_3C_2H_5C_3H_7C_4H_9N]\,Cl \qquad (2a)$$

$$[C_6H_5CH_2C_6H_5C_3H_5CH_3N]\,I \qquad (2b)$$

Pope and J. Read, after a very carefully controlled repetition of Le Bel's work, decided that he had never obtained the chloride.[10] They then prepared Wedekind's iodide, converted it into the *d*-camphorsulfonate, and by fractional crystallization obtained the diastereoisomerides, from which they isolated the optically active iodides and bromides. This proved "that quaternary ammonium derivatives in which the five substituting groups are different, contain an asymmetric nitrogen atom which gives rise to antipodal relationships of the same kind as those correlated with an asymmetric carbon atom."[11] The chemical and optical properties of Wedekind's iodide and salts with other anions were examined in considerable detail.[12]

This type of work was extended by combining ethyl methyl sulfide with ethyl bromoacetate to yield methylethylthetin bromide (Formula 3a), which was converted into the *d*-bromocamphorsulfonate; recrystallization similarly gave the *d*- and *l*-forms

$$[CO_2HCH_2SCH_3C_2H_5]Br \qquad (3a)$$

$$[CO_2HCH_2SeC_6H_5CH_3]Br \qquad (3b)$$

of methylethylthetine bromide.[13] The analogous phenylmethylselenetine bromide (Formula 3b) was similarly resolved into optically active forms, again via the *d*-bromocamphorsulfonate.[14] The appalling and persistent stench of selenides of type R_2Se very seriously delayed this work.

Compounds of tin were next investigated. These compounds required careful manipulation, because they were highly toxic in both the liquid and the vapor states. Pope and Peachey were able, however, to prepare methyl ethyl propyltin iodide (Formula 4a).[15] This liquid was volatile without decomposition and was soluble in nonpolar solvents; thus it was a

$$CH_3C_2H_5C_3H_7SnI \qquad (4a)$$

$$CH_3C_2H_5C_3H_7Sn \ C_{10}H_{15}SO_3 \qquad (4b)$$

covalent compound. It reacted with silver *d*-camphorsulfonate to give the corresponding sulfonate (Formula 4b), which was soluble in water and was clearly a salt. Evaporation of an aqueous solution of the sulfonate gave solely the *d*-tin *d*-sulfonate (Formula 4b), indicating a ready racemization of the cation of the salt and a marked difference in the solubilities of the two diastereoisomerides. Treatment of the active sulfonate in aqueous solution with potassium iodide precipitated the active iodide, which readily underwent racemization.

The iodide was clearly of a type different from the well-recognized quaternary ammonium and the sulfonium and selenonium salts. It has been suggested that the sulfonate may have formed a tetrahedral

cation (Formula 5a) that readily gave the tetrahedral covalent iodide (Formula 5b).

$$[R_3Sn \leftarrow OH_2] + I^- \rightleftarrows ISnR_3 + H_2O \qquad (5)$$
$$(a) \qquad\qquad\qquad (b)$$

In a brief excursion into true organometallic chemistry, Pope and Peachey showed that methylmagnesium iodide reacted with platinum tetrachloride to give compounds of type $(CH_3)_3PtX$, where X is OH, Cl, I, or NO_3.[16]

Similarly, Pope and C. S. Gibson found that ethylmagnesium bromide reacted with auric tribromide to give two compounds considered to be $(C_2H_5)_2AuBr$ and $(C_2H_5)AuBr_3$.[17] Many years later Gibson and his collaborators reinvestigated these gold compounds and found that they were both dimeric bridged compounds of structures shown in Formulas 6a and 6b, respectively.

In 1906–1910 Pope devoted a great amount of time to attempting, with W. Barlow, to correlate chemical constitution and crystal structure. The "valency-volume theory of crystal structure," which they developed, states, in brief, that the space that an atom occupies in a crystal is proportional to its valency.[18] The labors of Barlow and Pope revealed many factors that have helped the evolution of modern crystallography, but the main theory has been discarded.

While in Manchester, Pope initiated an outstanding investigation, which was completed at

Cambridge. It was reasonably certain that an allene hydrocarbon carrying substituents as in Formula 7a would have pairs of bonds alternately in the horizontal plane (thick lines) and in a vertical plane running through the central carbon atoms (thin lines). This compound would possess molecular dissymmetry and should be resolvable into optically active forms. However, attempts by various chemists to synthesize such a compound, with a and/or b carrying acidic or basic groups for resolution purposes, had failed. (The realization of this synthesis and resolution did not occur until 1935.[19])

Pope, in collaboration with W. H. Perkin, Jr., therefore attempted to synthesize 1-methylcyclohexylidene-4-acetic acid (Formula 7b), a compound similar to the allene hydrocarbon (Formula 7a) but with one double bond of the latter expanded to a six-member ring. This synthesis was achieved, but by a laborious method giving a low overall yield.[20] By coincidence, W. Marckwald and R. Meth had been investigating the synthesis of the acid (Formula 7b);[21] their product, certainly different from Perkin and Pope's acid, was later identified as the isomeric acid (Formula 7c). Shortly afterward O. Wallach devised a greatly improved synthesis of the true acid (Formula 7b) and, without divulging the method, offered to supply Perkin and Pope with sufficient acid for the resolution if his name was subsequently included in the published results. The resolution was finally accomplished using the brucine salt and the active acid isolated had $[\alpha]_D \pm 81°$. By a very exceptional arrangement, the paper was published simultaneously in English and in German.[22]

Another gap in the stereochemistry of carbon was subsequently filled. Before 1914 no one had synthesized and resolved a compound having only one carbon atom, which must necessarily be asymmetric. After considerable work Pope and J. Read synthesized chloroiodomethane sulfonic acid (Formula 8a) and resolved it, using initially hydroxyhydrindamine (Formula 8b) and later brucine for this purpose.[23] The acid had $[M]_{5461} + 43°$ and considerable optical stability: its aqueous solution could be boiled for two hours without loss of activity.

$$(8)$$

(a) (b)

At the outbreak of World War I, Pope was in Australia presiding over Section B (chemistry) of the British Association for the Advancement of

Science. He immediately returned to England and was soon involved in chemical problems, one of which was the preparation of photographic sensitizers. Photographic plates at that time were prepared with a silver bromide-silver iodide emulsion, which was sensitive only in the ultraviolet, violet, and blue regions; aerial photographs taken at dawn in a predominantly red light were almost useless. The enemy had plates on which the emulsion was incorporated with a sensitizer extending the sensitivity well into the red region. Pope, with much help from W. H. Mills and a small group of research students, elucidated the structure and the preparation of the main sensitizer, Pinacyanol, and of many other such compounds, and clarified their chemistry. Pope and Mills, writing in 1920 about these sensitizers, stated: ". . . we commenced the study of the more necessary of these [latter] substances and devised methods for their preparation on a sufficiently ample scale. Throughout the war practically all the sensitizing dyestuffs used by the Allies in the manufacture of panchromatic plates were produced in this [the Cambridge] Laboratory."[24]

A second main project, carried out with C. S. Gibson, was the preparation of mustard gas, or 2, 2'-dichlorodiethyl sulfide. This was prepared by the interaction of ethylene and sulfur chloride:

$$2C_2H_4 + S_2Cl_2 \rightarrow (ClCH_2CH_2)_2S + S.$$

This method apparently was quicker than the German method.[25] In the latter ethylene chlorohydrin ($HOCH_2CH_2Cl$) was converted into 2,2'-dihydroxydiethyl sulfide ($[HOCH_2CH_2]S$), which, when treated with hydrogen chloride, gave the required product. The large-scale preparation of ethylene chlorohydrin was apparently the slow stage that restricted the output.

The chemistry of mustard gas was later investigated in considerable detail, as was that of the β-chlorovinylarsines.[26] In the course of this work, the action of "Chloramine-T," or p-toluenesulfonchlorosodioamide ($CH_3C_6H_4SO_2NNaCl$), on many other sulfides to give compounds of type $CH_3C_6H_4SO_2N \leftarrow SR_2$ was examined.[27] These compounds, named sulfilimines and later sulfonilimines, could be resolved into optically active forms if the sulfur atom carried two unlike groups; hence a coordinate link joined the S and N atoms.

Two main lines of research occupied Pope for several years: the metallic coordination compounds formed by aliphatic polyamines and the stereochemistry of certain spirocyclic compounds. Much earlier A. Werner had used ethylenediamine (Formula 9a) to coordinate with various metallic halides and had obtained some noteworthy examples of isomeric compounds, several of which were

$$H_2NCH_2CH_2NH_2 \qquad (9a)$$

$$H_2NCH_2CH(NH_2)CH_2NH_2 \qquad (9b)$$

resolved into optically active forms. Pope and Mann considered that 1, 2, 3-triaminopropane (Formula 9b) might offer an even wider field.

This triamine was synthesized in good yield and was found to give very stable octahedral complexes of type [trp$_2$M]X$_3$, where M = Co(III), Rh (III), and [trp$_2$M]X$_2$, where M = Ni (II), Cd (II), Zn(II), and "trp" indicates one molecule of the triamine; no examples of isomerism were, however, detected.[28] The stability of these complexes is well shown by [trp$_2$Zn]I$_2$ and [trp$_2$Pt(II)]I$_2$, which crystallize unchanged after their aqueous solutions have been boiled for one to two hours.

Chloroplatinic acid (H$_2$PtCl$_6$), however, reacted in boiling aqueous solution with the triamine to give a crystalline compound of composition trpPtCl$_4$HCl, in which only two amino groups had coordinated, the third forming a hydrochloride salt. This could occur in either the 1,2-diamino complex (Formula 10a) and the 1,3-diamino complex (Formula 10b). The compound was resolved as the d-camphorsulfonate into optically active forms and was therefore the 1,2-diamino form, which has an asymmetric carbon atom. This was the first example of a carbon atom becoming asymmetric by the operation of a coordinate link.[29]

(10a)

(10b)

$\beta\beta'\beta''$-triaminotriethylamine (H$_2$NCH$_2$CH$_2$)$_3$N, "tren," was found to coordinate as a tetramine; and with Ni (II) salts it gave two types of complexes, such as [Nitren]SO$_4$ and [Ni$_2$tren$_3$]I$_4$. The sulfate was shown to be monomolecular, and consideration of strain factors indicates that the nickel atom must have the tetrahedral configuration. Similar considerations apply to the compound PttrenI$_2$.[30] $\gamma\gamma'\gamma''$-triaminotripropylamine ([H$_2$NCH$_2$CH$_2$CH$_2$]$_3$N) gave similar compounds, such as

$$([H_2NCH_2CH_2CH_2]_3NNi)(SCN)_2 .^{[31]}$$

The work on complex metallic compounds initiated by Mann and Pope continued long after Pope's

death and in 1934 led to the first systematic use of tertiary phosphines and arsines, thus opening a very wide and important field of coordination chemistry.[32] Two centroasymmetric spirocyclo compounds attracted Pope's interest. Spiro-5,5-dihydantoin (Formula 11a), a known compound, is a member of this class; and

(11a)

(11b)

(11c)

Pope considered that the possibilities of tautomeric shift—to the structure of Formula 11b or 11c—would make the compound sufficiently acidic for stable salt formation with bases. The resolution was readily accomplished, for a hot ethanolic solution of the dibrucine salt deposited the l-form as the monobrucine salt, and the mother liquor slowly deposited the d-form as the dibrucine salt. The stereochemical and crystallographic relationships were examined in detail.[33]

A simpler compound of this class was 2,6-diaminospiro [3.3] heptane (Formula 12), which S. E. Janson

(12)

and Pope were able to prepare from the corresponding 2,6-dicarboxylic acid.[34] It was resolved as the di-d-camphor-β-sulfonate and converted into the dihydrochloride, having [M]$_{4358}$ ± 30° in aqueous solution. This diamine, by virtue of its simplicity and its rigid rings, was admirably suitable for the application of mathematical theories of optical rotatory power and chemical constitution; and it was utilized for this purpose by M. Born.[35]

NOTES

1. "Terpenes and Allied Compounds. Sobrerol . . ." (1891).
2. "Studies of the Terpenes and Allied Compounds; the Sulphonic Derivatives of Camphor" (1893, 1895).
3. F. S. Kipping, "Derivatives of Camphoric Acid," in *Journal of the Chemical Society*, **69** (1896), 913–971.

4. Kipping and Pope, "Optical Inversion of Camphor" (1897) and "Racemism and Pseudo-Racemism" (1897).

5. Kipping and Pope, "Enantiomorphism" (1898) and "The Crystallisation of Externally Compensated Mixtures" (1909).

6. Pope and S. J. Peachey, "The Resolution of Tetrahydro-papaverine Into Its Optically Active Components . . ." (1898) and "The Non-resolution of Racemic Tetrahydro-papaverine by Tartaric Acid" (1898); F. L. Pyman, "Iso-quiniline Derivatives. Part 2," in *Journal of the Chemical Society*, **95** (1909), 1610–1623.

7. J. A. Le Bel, "Sur La dyssymétrie et la création du pouvoir rotatoire dans les dérivés alcooliques du chlorure d'ammonium," in *Comptes rendus . . . de l'Académie des sciences*, **112** (1891), 724–726.

8. W. Marckwald and A. von Droste-Huelshoff, "Ueber die Methyl-äthyl-propyl-isobutyl-ammoniumbase," in *Berichte der Deutschen chemischen Gesellschaft*, **32** (1899), 560–564.

9. E. Wedekind, "Ueber das fünfwerthige asymmetrische stickstoffatom," *ibid.*, 517–529.

10. Pope and J. Read, "Asymmetric Quinquevalent Nitrogen Compounds of Simple Molecular Constitution" (1912).

11. Pope and Peachey, "Asymmetric Optically Active Nitrogen Compounds; *d*- and *l*-benzylphenylallylmethylammonium iodides and bromides" (1899).

12. Pope and A. W. Harvey, "Optically Active Nitrogen Compounds and Their Bearing on the Valency of Nitrogen. . ." (1901).

13. Pope and Peachey, "Asymmetric Optically Active Sulphur Compounds . . ." (1900).

14. Pope and A. Neville, "Asymmetric Optically Active Selenium Compounds and the Sexavalency of Selenium and Sulphur . . ." (1902).

15. Pope and Peachey, "Asymmetric Optically Active Tin Compounds . . ." (1900) and "The Racemisation of Optically Active Tin Compounds" (1900).

16. Pope and Peachey, "The Alkyl Compounds of Platinum" (1909).

17. Pope and C. S. Gibson, "The Alkyl Compounds of Gold" (1907).

18. W. Barlow and Pope, "A Development of the Atomic Theory . . ." (1906); "The Relation Between the Crystalline Form . . ." (1907); "On Polymorphism . . ." (1908); "The Relation Between the Crystal Structure . . ." (1910).

19. P. Maitland and W. H. Mills, "Experimental Demonstration of the Allene Asymmetry," in *Nature*, **135** (1935), 994; and "Resolution of an Allene Hydrocarbon Into Antipodes by Asymmetric Catalysis," in *Journal of the Chemical Society* (1936), 987–998. See also "Mills, W. H.," in *Dictionary of Scientific Biography*, IX, 402–404.

20. W. H. Perkin, Jr., and Pope, "Experiments on the Synthesis of 1-Methylcyclohexylidene-4-Acetic Acid" (1908).

21. W. Marckwald and R. Meth, "Ueber optisch-active ver-bindungen, die rein asymmetrisches Atom enthalten," in *Berichte der Deutschen chemischen Gesellschaft*, **39** (1906), 1171–1177; and "Ueber die 1-Methylcyclohexyliden-4-essigsäure," *ibid.*, 2404–2405.

22. Perkin, Pope, and O. Wallach (1909), also in *Justus Liebigs Annalen der Chemie*, **371** (1909), 180–200. For later work on this acid, see Perkin and Pope, "Optically Active Derivatives of 1-Methylcyclohexylidene-4-Acetic Acid," in *Journal of the Chemical Society*, **99** (1911), 1510–1529.

23. Pope and Read, "The Optical Activity of Compounds of Simple Molecular Constitution . . ." (1914).

24. W. H. Mills and Pope, "Studies on Photographic Sensitisers . . .," 2 pts. (1920). For further details see F. G. Mann, "William Hobson Mills," in *Biographical Memoirs of Fellows of the Royal Society*, **6** (1960), 201–225.

25. Pope, Gibson, and H. F. Thuillier, British patent 142875 (1918); Gibson and Pope, "β, β'-Dichlorodiethyl Sulphide" (1920).

26. F. G. Mann, Pope, and R. H. Vernon, "The Interaction of Ethylene and Sulphur Monochloride" (1921); Mann and Pope, "The β-Chlorovinylarsines" (1922).

27. Mann and Pope, "The Sulphilimines . . ." (1922).

28. Mann and Pope, "1:2:3-Triaminopropane and Its Complex Metallic Compounds" (1925) and "The Configuration of the Bistriaminopropane Metallic Complexes" (1926).

29. Mann and Pope, "A Novel Type of Optically Active Complex Metallic Salt," in *Nature*, **119** (1927), 351; Mann, "Tetrachloro Platinum, an Optically Active Complex Salt," in *Journal of the Chemical Society* (1927), 1224–1232.

30. Mann and Pope, "ββ'β"-Triaminotriethylamine and Its Complex Metallic Compounds" (1925); "The Complex Salts of ββ'β"-Triaminotriethylamine With Nickel and Palladium" (1926).

31. Mann and Pope, "γγ'γ"-Triaminotripropylamine and Its Complex Compounds With Nickel" (1926).

32. See Mann, "The Development of Co-ordination Chemistry at the University of Cambridge, 1925–1965," Advances in Chemistry series, no. 62 (Washington, D.C., 1966), 120–146. See also Mann and D. Purdie, "The Constitution of Complex Metallic Salts. Part 3," in *Journal of the Chemical Society* (1935), 1549–1563; (1936), 873–890.

33. Pope and J. B. Whitworth, "Optically Active Di- and Tetramethylspiro-5:5-Dihydantoins" (1936).

34. S. E. Janson and Pope, "Optically Active Amines Containing No Asymmetric Atom," in *Chemistry and Industry* (1932), 316.

35. M. Born, "On the Theory of Optical Activity," in *Proceedings of the Royal Society*, **150A** (1935), 84–105.

BIBLIOGRAPHY

Pope wrote the following articles while at the City and Guilds of London Institute (1891–1897): "Terpenes and Allied Compounds. Sobrerol, a Product of the Oxidation of Terebenthene (Oil of Turpentine) in Sunlight," in *Journal of the Chemical Society*, **59** (1891), 315, written with H. E. Armstrong; "The Crystalline Forms of the Sodium Salts of the Substituted Anilic Acids," *ibid.*, **61** (1892), 581; "*o*-Benzoic sulphinide," *ibid.*, **67** (1895), 985; "The Crystalline Form of the Isomeric Dimethyl-pimelic Acids," in *Proceedings of the Chemical Society*, **11** (1895), 8; "Substances Exhibiting Circular Polarisation Both in the Amorphous and Crystalline States," in *Journal of the Chemical Society*, **69** (1896), 971; "The Refraction Constants of Crystalline Salts," *ibid.*, p. 1530; "A Compound of Camphoric Acid With Acetone," *ibid.*, p. 1696; "A Method of Studying Polymorphism, and on Polymorphism as the Cause of Some Thermochemical Peculiarities of Chloral Hydrate," in *Proceedings of the Chemical Society*, **12** (1896), 142, 249; and in *Journal of the Chemical Society*, **75** (1899), 455; "The Localisation of Deliquescence in Chloral Hydrate Crystals," in *Proceedings of the Chemical Society*, **12** (1896), 249; "Enantio-morphism," *ibid.*, 249–251, written with F. S. Kipping; and "Crystalline Form of Iodoform," *ibid.*, **14** (1898), 219; and in *Journal of the Chemical Society*, **75** (1899), 46.

While head of the chemical department of the Institute of the Goldsmiths' Company at New Cross, London, Pope published (1897–1901) "A Composite Sodium Chlorate Crystal in Which the Twin Law Is Not Followed," in *Journal of the Chemical Society*, **74** (1898), 949; "The Application of Powerful Optically Active

Acids to the Resolution of Externally Compensated Basic Substances: Resolution of Racemic Camphoroxime," *ibid.*, **75** (1899), 1105; "*d-ac*-Tetrahydro-β-naphthylamine," in *Proceedings of the Chemical Society*, **15** (1899), 170; "Homogeneity of *dl*-α-phenethylamine *d*-camphorsulphonate," in *Journal of the Chemical Society*, **75** (1899), 1110, written with A. W. Harvey; "Racemisation Occurring During the Formation of Benzylidene, Benzoyl, and Acetyl Derivatives of *d-ac*-tetrahydro-β-naphthylamine," in *Proceedings of the Chemical Society*, **16** (1900), 74, written with A. W. Harvey; "The Inversion of the Optically Active *ac*-tetrahydro-β-naphthylamines Prepared by the Aid of *d*- and *l*-bromocamphorsulphonic Acids," *ibid.* (1900), 206, written with A. W. Harvey; and in *Journal of the Chemical Society*, **79** (1901), 74; "Optically Active Nitrogen Compounds and Their Bearing on the Valency of Nitrogen. *d*- and *l*-α-Benzylphenylallylmethylammonium Salts," *ibid.*, p. 828, written with A. W. Harvey; "The Characterisation of 'Racemic' Liquids," *ibid.*, **75** (1899), 1119, written with F. S. Kipping; and "Method of Discriminating Between 'Non-racemic' and 'Racemic' Liquids," *ibid.*, p. 1111, written with S. J. Peachey.

While working at the Municipal School of Technology and at Victoria University of Manchester (1901–1908), Pope published "Asymmetric Optically Active Selenium Compounds and the Sexavalency of Selenium and Sulphur. *d*- and *l*-Phenylmethylselenetine salts," in *Journal of the Chemical Society*, **81** (1902), 1552, written with A. Neville. Works written with S. J. Peachey include "The Resolution of Tetrahydropapaverine Into Its Optically Active Components; Constitution of Papaverine," *ibid.*, **73** (1898), 893; "The Non-resolution of Racemic Tetrahydropapaverine by Tartaric Acid," *ibid.*, p. 902; "The Application of Powerful Optically Active Acids to the Resolution of Externally Compensated Basic Substances: Resolution of Tetrahydroquinaldine," *ibid.*, **75** (1899), 1066; "Asymmetric Optically Active Nitrogen Compounds; *d*- and *l*-benzylphenylallylmethylammonium Iodides and Bromides," *ibid.*, p. 1127; "Asymmetric Optically Active Sulphur Compounds; *d*-methylethylthetine Platinichloride," *ibid.*, **77** (1900), 1072; "Asymmetric Optically Active Tin Compounds: *d*-methylethyl-*n*-propyltin Iodide," in *Proceedings of the Chemical Society*, **16** (1900), 42; "The Racemisation of Optically Active Tin Compounds: *d*-methylpropyl Tin *d*-bromocamphorsulphonate," *ibid.* (1900), 116; "Preparation of the Tetra-alkyl Derivatives of Stanni-methane," *ibid.*, **19** (1903), 290; "A New Class of Organo-tin Compounds Containing Halogens," in *Proceedings of the Royal Society*, 72A, no. 7 (1903); and "A New Class of Organo-metallic Compounds; Preliminary Notice: Trimethylplatinimethyl Hydroxide and Its Salts," in *Proceedings of the Chemical Society*, **23** (1907), 86. See also "The Application of Powerful Optically Active Acids to the Resolution of Externally Compensated Basic Substances; Resolution of Tetrahydro-*p*-toluquinaldine," in *Journal of the Chemical Society*, **75** (1899), 1093, written with E. M. Rich; and the following works written with F. S. Kipping: "Genesis of

New Derivatives of Camphor Containing Halogens by the Action of Heat on Sulphonic Chlorides," in *Proceedings of the Chemical Society*, **9** (1893), 130; (1894), 212; "Studies of the Terpenes and Allied Compounds; the Sulphonic Derivatives of Camphor," in *Journal of the Chemical Society*, **63** (1893), 548; **67** (1895), 354; and in *Proceedings of the Chemical Society*, **10** (1894), 163, 211; "Dextrorotatory Camphorsulphonic Chloride," *ibid.* (1894), 164; "π-Halogen Derivatives of Camphor," in *Journal of the Chemical Society*, **67** (1895), 371; "The Melting Points of Racemic Modifications and of Optically Active Isomerides," in *Proceedings of the Chemical Society*, **11** (1895), 39; "π-Chlorocamphoric acid," *ibid.*, **11** (1895), 213; "Optical Inversion of Camphor," in *Journal of the Chemical Society*, **71** (1897), 956; "Derivatives of Camphoric Acid. Part II. Optically Inactive Derivatives," *ibid.*, p. 962; "Racemism and Pseudoracemism," *ibid.*, p. 989; "Enantiomorphism," *ibid.*, **73** (1898), 606; "The Separation of Optical Isomerides," in *Proceedings of the Chemical Society*, (1898), 113; and "Characterisation of Racemic Compounds," *ibid.*, p. 219; and in *Journal of the Chemical Society*, **75** (1899), 36.

Other works include "Resolution of Tetrahydro-*p*-toluquinaldine Into Its Optically Active Components," in *Journal of the Chemical Society*, **91** (1907), 458, written with T. C. Beck; "The Resolution of Externally Compensated Dihydro-α-methylindole," *ibid.*, **85** (1904), 1330, written with G. Clarke, Jr.; "The Alkyl Compounds of Gold," *ibid.*, **91** (1907), 2061; "*l*-Methylcyclohexylidene-4-acetic acid," in *Proceedings of the Chemical Society*, **22** (1906), 107, written with W. H. Perkin, Jr.; "Relation Between Crystalline Form and Chemical Constitution of the Picryl Derivatives," in *Proceedings of the Royal Society*, 80A (1908), 557, written with G. Jerusalem; and the following works, written with W. Barlow: "A Development of the Atomic Theory Which Correlates Chemical and Crystalline Structure and Leads to a Demonstration of the Nature of Valency," in *Journal of the Chemical Society*, **89** (1906), 1675; "The Relation Between the Crystalline Form and the Chemical Constitution of Simple Inorganic Substances," *ibid.*, **91** (1907), 1150; and "Note on the Theory of Valency," in *Proceedings of the Chemical Society*, **23** (1907), 15.

Pope's works at the University Chemical Laboratory, Cambridge (1908–1939), include "The Optical Activity of Compounds Having Simple Molecular Structure," in *Journal of the Chemical Society*, **93** (1908), 796, written with J. Read; "Experiments on the Synthesis of *l*-Methylcyclohexylidene-4-acetic Acid," *ibid.*, p. 1075, written with W. J. Perkin, Jr.; "On Polymorphism, With Especial Reference to Sodium Nitrate and Calcium Carbonate," *ibid.*, p. 1528, written with W. Barlow; "Optically Active Substances Which Contain No Asymmetric Atom" (Preliminary Note), in *Proceedings of the Chemical Society*, **25** (1909), 83, written with W. H. Perkin and O. Wallach; "The Crystallisation of Externally Compensated Mixtures," in *Journal of the Chemical Society*, **95** (1909), 103, written with F. S. Kipping; "The Condensation of Oxymethylenecamphor With Primary and Secondary Amino Compounds," *ibid.*, p. 171, written with J. Read; "The Alkyl Compounds of Platinum," *ibid.*, p. 571, written with S. J. Peachey; and "Ueber optisch active Substanzen, die kein asymmetrisches Atom enthalten," in *Justus Liebigs Annalen der Chemie*, **371** (1909), 180, written with W. H. Perkin and O. Wallach.

See also "The Resolution of Externally Compensated Acids and Bases," in *Journal of the Chemical Society*, **97** (1910), 987; "Externally Compensated Tetrahydroquinaldine (Tetrahydro-2-methylquinoline) and Its Optically Active Components," *ibid.*, p. 2199, written with J. Read; "The Resolution of Externally Compensated Pavine and α-Bromocamphor-ω-Sulphonic Acid," *ibid.*, p. 2207, written with C. S. Gibson; "The Rotatory Powers of the Salts of *d*- and *l*-Camphor-β-Sulphonic Acid With *d*- and *l*-Pavine," *ibid.*, p. 2211, written with C. S. Gibson; "The Relation Between the Crystal Structure and the Chemical Composition, Constitution, and Configuration of Organic Substances," *ibid.*, p. 2308, written with W. Barlow; "Optically Active Derivatives of *l*-Methyl*cyclo*hexylidene-4-acetic Acid," *ibid.*, **99** (1911), 1510, written with W. H. Perkin, Jr.; "Dihydroxydihydrindamine and Its Resolution Into Optically Active Components," *ibid.*, p. 2071, written with J. Read; "Asymmetric Quinquevalent Nitrogen Compounds of Simple Molecular Constitution," *ibid.*, **101** (1912), 519, written with J. Read; "Some Mixed Phosphonium Derivatives," *ibid.*, p. 735, written with C. S. Gibson; "The Alkaloidal Salts of Phenylmethylphosphinic Acid," *ibid.*, p. 740, written with C. S. Gibson; "The Resolution of Benzoylalanine Into Its Optically Active Components," *ibid.*, p. 939, written with C. S. Gibson; "The Externally Compensated and Optically Active Hydroxyhydrindamines, Their Salts and Derivatives," *ibid.*, p. 758, written with J. Read; "The Resolution of *sec*-Butylamine Into Optically Active Components," *ibid.*, p. 1702, written with C. S. Gibson; "The Relation Between Constitution and Rotatory Power Amongst Derivatives of Tetrahydroquinaldine," *ibid.*, p. 2309, written with T. F. Winmill; "The Absence of Optical Activity in the α- and β-2 : 5-Dimethylpiperazines," *ibid.*, p. 2325, written with J. Read; "A Novel Method for Resolving Externally Compensated Amines: Derivatives of *d*- and *l*-Oxymethylencamphor," *ibid.*, **103** (1913), 444, written with J. Read; "The Ten Stereoisomeric Tetrahydroquinaldinomethylenecamphors," *ibid.*, p. 1515, written with J. Read; and "The Resolution of 2 : 3-Diphenyl-2 : 3-Dihydro-1 : 3 : 4-Naphthaisotriazine Into Optically Active Components," *ibid.*, p. 1763, written with Clara M. Taylor.

Other works include "Ueber das Quecksilber-dibenzyl," in *Berichte der Deutschen chemischen Gesellschaft*, **46** (1913), 352; "The Relation Between Optical Activity and Molecular Complexity," in *Transactions of the Faraday Society*, **10** (1914), 118, written with J. Read; "Optically Active Substances of Simple Molecular Constitution," in *Proceedings of the Cambridge Philosophical Society, Mathematical and Physical Sciences* (1914), written with J. Read; *British Association. Section B. (Australia, 1914)*

Address to the Chemical Section; "The Chemical Significance of Crystalline Form," in *Journal of the American Chemical Society*, **36** (1914), 1675, written with W. Barlow; "The Identity of the Supposed β-2 : 5-Dimethylpiperazine," in *Journal of the Chemical Society*, **105** (1914), 219, written with J. Read; "The Variable Rotatory Powers of the *d*-α-Bromocamphorsulphonates," *ibid.*, p. 800, written with J. Read; "The Optical Activity of Compounds of Simple Molecular Constitution. Ammonium *d*- and *l*-Chloroiodomethanesulphonates," *ibid.*, p. 811, written with J. Read; "Enantiomorphism of Molecular and Crystal Structure," *ibid.*, **107** (1915), 700, written with W. Barlow; "On Topic Parameters and Morphotopic Relationships," in *Philosophical Magazine*, **29** (1915), 745, written with W. Barlow; "The Future of Pure and Applied Chemistry. Presidential Address," in *Journal of the Chemical Society*, **113** (1918), 289; "Chemistry in the National Service. Presidential Address," *ibid.*, **115** (1919), 397; "ββ'-Dichlorodiethyl Sulphide," *ibid.*, **117** (1920), 271, written with C. S. Gibson; "The Preparation and Physical Properties of Carbonyl Chloride," *ibid.*, p. 1410, written with R. H. Atkinson and C. T. Heycock; "Triphenylarsine and Diphenylarsenious Salts," *ibid.*, p. 1447, written with E. E. Turner; "Studies on Photographic Sensitisers. Part I. The Isocyanine Dyestuffs," in *Photographic Journal* (May 1920), written with W. H. Mills; and "Studies on Photographic Sensitisers. Part II. The Carbocyanines," *ibid.* (Nov. 1920), written with W. H. Mills.

See also "The Interaction of Sulphur Monochloride and Substituted Ethylenes," in *Journal of the Chemical Society*, **119** (1921), 396, written with J. L. B. Smith; "The Interaction of Ethylene and Sulphur Monochloride," *ibid.*, p. 634, written with F. G. Mann and R. H. Vernon; "Production and Reactions of ββ'-Dichlorodiethyl Sulphide," *ibid.*, **121** (1922), 594, written with F. G. Mann; "*iso*Quinoline and the *iso*Quinoline-reds," *ibid.*, p. 1029, written with J. E. G. Harris; "The Sulphilimines, a New Class of Organic Compounds Containing Quadrivalent Sulphur," *ibid.*, p. 1052, written with F. G. Mann; "The Chlorinated Dialkyl Sulphides," *ibid.*, p. 1166, written with J. L. B. Smith; "The β-Chlorovinylarsines," *ibid.*, p. 1754, written with F. G. Mann; "The αα'-Dichlorodialkyl Sulphides," *ibid.*, **123** (1923), 1172, written with F. G. Mann; "The Isomeric Trithioacetaldehydes," *ibid.*, p. 1178, written with F. G. Mann; "The Preparation of Sulphuryl Chloride," in *Recueil des travaux chimiques des Pays-Bas et de la Belgique*, **42** (1923), 939; "The Optically Active Sulphilimines," in *Journal of the Chemical Society*, **125** (1924), 911, written with F. G. Mann; and "The Resolution of *dl*-Diphenylpropylenediamine and *dl*-1 : 4-Diphenyl-2-methylpiperazine," *ibid.*, p. 2396, written with F. S. Kipping.

See also "1 : 2 : 3-Triaminopropane and Its Complex Metallic Compounds," in *Proceedings of the Royal Society*, **107A** (1925), 80, written with F. G. Mann; "ββ'β"-Triaminotriethylamine and Its Complex Metallic Compounds," *ibid.*, **109A** (1925), 444, written with F. G. Mann; "Dissymmetry and Asymmetry of Molecular Configuration," in *Journal of the Society of Chemical Industry*, **44**

(1925), 833, written with F. G. Mann; "The Complex Salts of $\beta\beta'\beta''$-Triaminotriethylamine With Nickel and Palladium," in *Journal of the Chemical Society*, **129** (1926), 482, written with F. G. Mann; "$\gamma\gamma'\gamma''$-Triaminotripropylamine and Its Complex Compounds With Nickel," *ibid.*, p. 489; "The Resolution of *dl*-Alanine and the Formation of *trans*-2 : 5-Dimethylpiperazine," *ibid.*, p. 494, written with F. S. Kipping; "Preparation and Resolution of *dl-cis*-2 : 5-Dimethylpiperazine," *ibid.*, p. 1076, written with F. S. Kipping; "The Configuration of the Bistriaminopropane Metallic Complexes," *ibid.*, p. 2675, written with F. G. Mann; "The Optically Active *spiro*-5 : 5-Dihydantoins," in *Proceedings of the Royal Society*, **134A** (1931), 357, written with J. B. Whitworth; "The Symmetrical *spiro*-Heptanediamine and Its Resolution Into Optically Active Components," in *Proceedings of the Royal Society*, **154A** (1936), 53, written with S. E. Janson; "Optically Active Di- and tetra-methyl*spiro*-5 : 5-dihydantoins," *ibid.*, **155A** (1936), 1, written with J. B. Whitworth; "Preparation of $\beta\beta'$-dichlorodiethyl sulphide" (Brit. Pat. 142875), in *Journal of the Chemical Society*, **118**, pt. 1 (1920), 523, written with C. S. Gibson and H. F. Thuillier; "Production of Aromatic Arsenic Compounds (mono- and di-aryl Arsenious Haloids)" (Brit. Pat. 142880), *ibid.*, p. 578; and "Production and Utilisation of Sulphur Dichloride" (Brit. Pat. 142879), *ibid.*, pt. 2, p. 484, written with C. T. Heycock.

The following papers on crystallography were published from the City and Guilds of London Institute (1896–1899): "Angular Measurement of Optic Axial Emergences," in *Proceedings of the Royal Society*, **60** (1896), 7; "Die Krystallformen der Natriumsalze der substituirten Anilsauren," *ibid.*, **24** (1895), 529; "Die Krystallformen der stereoisomeren $\alpha\alpha'$-Dimethylpimelinsauren," *ibid.*, 533; "Die Krystallform einiger neuer Halogenderivate des Camphers," in *Zeitschrift für Kristallographie und Mineralogie*, **25** (1895), 437, written with F. S. Kipping; "Ueber die Krystallform einiger organischer Verbindungen," *ibid.*, p. 450; "Ein bemerkenswerther Fall von Phosphorescenz," *ibid.*, **25** (1895), 567; "Ueber die Messung Winkels der optischen Axen," *ibid.*, **26** (1896), 589; "Ueber optisches Drehungsvermögen," *ibid.*, **27** (1896), 406; "Die Refractionsconstanten krystallisirter Salze," *ibid.*, **28** (1897), 113; "Eine Acetonverbindung der Camphersaure," *ibid.*, **28** (1897), 128; "Ueber Racemie und Pseudoracemie," *ibid.*, **30** (1898), 443, written with F. S. Kipping; "Ueber Enantiomorphismus," *ibid.*, p. 472, written with F. S. Kipping; "Eine nicht zwillingsartige Verwachsung von Natriumchloratkrystallen," *ibid.*, **31** (1899), 15; "Eine neue, partiell racemische Verbindung," *ibid.*, p. 11, written with S. J. Peachey; and "Ueber die Krystallformen einiger organischer Verbindungen," *ibid.*, p. 115.

FREDERICK GEORGE MANN

POPOV, ALEKSANDR NIKIFORIVICH (*b.* Vitebsk region [oblast], Russia, 1840 [?]; *d.* Warsaw, Russia [now Poland], 18 August 1881), *chemistry.*

Popov studied at Kazan University from 1861 to 1865, taught chemistry at that university until 1869, and then was professor of chemistry at Warsaw University. During the winter of 1871-1872 he worked in Kekulé's laboratory at Bonn.

Popov's first publication, which appeared while he was working in Butlerov's laboratory at Kazan (1865), was devoted to refuting Kolbe's views on the isomers of ketones and to proving the identity of the "units of affinity" (valencies) of the atom of carbon. His basic contribution lies, however, in the development of destructive oxidation, in which a "chromium mixture" ($K_2Cr_2O_7-H_2SO_4$) is used as an oxidizing agent for the study of the chemical structure of organic compounds.

Popov's best-known achievement is his rule relating to ketones. The results of his work in this area were generalized in his master's thesis (1869). He found that in the oxidation of ketones the radicals form the following order in their affinity for the former carbonyl group: C_6H_5 and $R_3C > CH_3 > RCH_2 > R_2CH > C_6H_5CH_2$ (R = alkyl)—and that in the oxidization of ketones the carbonyl group remains joined to the most stable radical. G. G. Wagner made Popov's rule more precise by showing that it deals with the main direction of the reaction. Popov's rule was used by himself and other chemists to determine the structures of acids, alcohols, and metallo-organic compounds used for the synthesis of ketones.

Popov also pointed out the possibility of using "chromium mixture" to determine the structures of organic acids and alcohols. In the oxidation of alcohols he suggested $>C=(OH)_2$ as an intermediate state.

Popov and Theodor Zincke showed that in the oxidization of alkylbenzenes, the splitting occurs so as to form the $C_6H_5CH_2$ and the alkyl radicals, which are further oxidized independently. This method permitted the determination of the structure of lateral chains and consequently of compounds used for the synthesis of alkylbenzenes.

Popov's letters to Butlerov show that Popov suggested to Kekulé an important rule concerning the splitting of olefins during oxidization according to the site of the double bond. Kekulé first stated it in print, so that it is frequently called Kekulé's rule.

BIBLIOGRAPHY

Popov's writings include "Ueber die Isomerie der Ketone," in *Zeitschrift für Chemie*, n.s. **1** (1865), 577–580; and in *Annalen der Chemie und Pharmacie*, **145** (1868), 283–292; *Ob okislenii ketonov odnoatomnykh* ("On the Oxidation of Monoatomic Ketones"; Kazan, 1869); "Bestimmung der Constitution von Alkoholradikalen durch Oxydation aromatischer Kohlenwasserstoffe," in *Berichte der Deutschen chemischen Gesellschaft*, **5** (1872), 384–387, written with T. Zincke; "Die Oxydation der Ketone als Mittel zur Bestimmung der Constitution der Säuren und Alkohole," in *Annalen der Chemie und Pharmacie*, **162** (1872), 151–160; *O zakonnosti okislenia ketonov i o primenenii ee k opredeleniyu stroenia alkogoley i kislot* ("On the Regularity of Oxidation of Ketones and Its Application to the Determination of the Structure of Alcohols and Acids"; Warsaw, 1872); and "Pisma k A. M. Butlerovu" ("Letters to A. M. Butlerov"), in *Pisma russkikh khimikov k A. M. Butlerovu. Nauchnoe nasledstvo* ("Letters of Russian Chemists to A. M. Butlerov. Scientific Legacy"), IV (Moscow, 1961), 305–359. Popov's last chemical works are in *Sbornik rabot khimicheskoy laboratorii Varshavskogo universiteta* ("Collection of Works of the Chemical Laboratory of Warsaw University"; Warsaw, 1876).

A secondary source is G. V. Bykov, "Ocherk zhizni i deyatelnosti Aleksandra Nikiforovicha Popova" ("A Sketch of the Life and Work of . . . Popov"), in *Trudy Instituta istorii estestvoznaniya i tekhniki . . .*, **12** (1956), 200–241, with a bibliography of Popov's writings on 241–245.

G. V. Bykov

POPOV, ALEKSANDR STEPANOVICH (*b.* Turinsk mining village, at the Bogoslov works [now Krasnoturinsk, Sverdlovsk oblast], Russia, 16 March 1859; *d.* St. Petersburg, Russia, 13 January 1906), *physics, technology.*

The son of a priest, Popov received a free seminary education to encourage him to follow his father's profession. After graduating from the seminary in Perm [now Molotov], he did not continue his clerical education, for he had become interested instead in physics, mathematics, and engineering. After preparing privately for the entrance examinations, he was admitted to the Faculty of Physics and Mathematics of St. Petersburg University in 1877. While still a student Popov began in 1881 to work at the Elektrotekhnik artel, which ran the first small electric power stations in Russia and the first electric lighting installations using arc lamps.

After finishing his studies and defending his dissertation (1882), Popov declined an offer to remain at the university in order to prepare for an academic career, because there was no opportunity there to conduct experimental research in electrical engineering. In 1883 he became an assistant at the Torpedo School in Kronstadt, which trained naval specialists in all branches of electrical engineering. An instructor at the school from 1888, Popov lectured and conducted laboratory sessions in electricity and magnetism, as well as on electrical machines and motors. He stayed at Kronstadt until 1900.

Popov became interested in electromagnetic waves following their discovery by Hertz in 1888. In 1890 E. Branly discovered the decrease of electrical resistance in metallic powders under the influence of electrical discharges. In 1894 Oliver Lodge used this discovery to construct an indicator of electromagnetic waves, which he called a coherer. Lodge's first indicator had a serious flaw: under the action of electromagnetic waves, the grains of powder stuck together and the sensitivity of the apparatus declined sharply. Lodge, and later Popov, improved this indicator, equipping it with an electric bell-like apparatus that automatically tapped the powder tube when the impulse of current was produced and thereby restored its sensitivity for receiving the original signal. By constructing the first continuously operating indicator, Popov made it possible, as he wrote in 1895, "to note separate, successive discharges of an oscillatory character."

Having screened the receiver from outside variable fields and having equipped it with a wire antenna, Popov demonstrated the possibility of receiving signals sent by Hertz's oscillator, at a distance of up to eighty meters.

In a public lecture, "Ob otnoshenii metallicheskikh poroshkov k elektricheskim kolebaniam" ("On the Relation of Metallic Powders to Electrical Oscillations"), presented on 7 May 1895 to the Physical Section of the Russian Physicochemical Society in St. Petersburg, Popov demonstrated the reception of electromagnetic signals for the first time. In January 1896 he published a more extensive account in "Pribor dlya obnaruzhenia i registratsii elektricheskikh kolebany" ("An Apparatus for Detecting and Recording Electrical Oscillations"), in which he introduced a detailed circuit diagram. The article concluded with "the hope that my apparatus, when perfected, may be used for the transmission of signals over a distance with the help of rapid electrical oscillations, as soon as a source of such vibrations with sufficient energy is discovered."

In the summer of 1895 Popov had adapted his instrument for the automatic registration of atmospheric oscillatory discharges; it was later called a storm indicator. Experiments with it led Popov to study the possible influence of atmospheric obstacles to the transmission of signals. By the beginning of 1896 Popov had substantially improved his receiver and had obtained important results in transmitting and receiving signals. Before the summer of 1896 the improved apparatus had been publicly demonstrated three times, in Kronstadt and St. Petersburg.

In the fall of 1896 the first published notice of Marconi's invention of a wireless telegraph appeared; his claim dated from June 1896. When he was issued a patent in 1897 and the diagram of his apparatus was published, it appeared to coincide almost completely with the description published in January 1896 by Popov. A commission of competence, established in 1908 by the Physical Section of the Russian Physicochemical Society to investigate the question of priority, concluded that Popov "was justified as being recognized as the inventor of the wireless telegraph."

Popov's improvements (1897-1900) in his radiotelegraph led to its practical use by the Russian navy and its introduction into the army, but the development of radio in the tsarist armed forces proceeded very slowly, compared with those of other nations.

In 1901 Popov became professor at the St. Petersburg Institute of Electrical Engineering, and in 1905 he was elected its director. In December 1905 he was ordered by the governor of St. Petersburg to take repressive measures against student political disturbances. Popov refused, and this event severely affected his health. He died soon afterward.

BIBLIOGRAPHY

I. ORIGINAL WORKS. The claim of Popov's priority as the inventor of radio rests on a one-paragraph summary of the report he made at a meeting on 7 May 1895, in *Zhurnal Russkago fiziko-khimicheskago obshchestva* . . ., **27** (1895), 259–260; and on a more extensive and updated account, *ibid.*, **28** (1896), 1–4, describing his recorder of electrical disturbances, which expresses his hopes for the signaling possibilities of his apparatus "when perfected." An English trans. of this article appeared in *Electrical Review* (London), **47** (1900), 845–846, and 882–883. A bibliography of Popov's writings, lectures, and patents was compiled by A. M. Lukomskaya in *A. S. Popov* (Moscow, 1951). A more extensive, annotated bibliography of the field is in A. I. Berg, ed., *Izobretenie radio. A. S. Popov. Dokumenty i materialy* ("The Invention of Radio. A. S. Popov. Documents and Materials"; Moscow, 1966).

II. SECONDARY LITERATURE. Of the extensive scientific and popular literature about Popov, three book-length studies are noteworthy: A. I. Berg and M. I. Radovsky, *Izobretatel radio A. S. Popov* ("A. S. Popov: Inventor of Radio"; Moscow, 1945); I. V. Brenev, *Izobretenie radio A. S. Popovym* ("The Invention of Radio by A. S. Popov"; Moscow, 1965); and *A. S. Popov v kharakteristikakh i vospominaniakh sovremennikov* (". . . Popov in the Characterizations and Recollections of His Contemporaries"; Moscow, 1959). The first, an account by a surviving collaborator, appeared in English in the Men of Russian Science series as M. Radovsky, *Alexander Popov: Inventor*

of Radio (Moscow, 1957). For a critical review of Popov's contribution, see C. Süsskind, *Popov and the Beginnings of Radiotelegraphy* (San Francisco, 1962, 1973).

J. G. DORFMAN

PORETSKY, PLATON SERGEEVICH (*b.* Elisavetgrad [now Kirovograd], Russia, 15 October 1846; *d.* Joved, Grodno district, Chernigov guberniya, Russia, 22 August 1907), *mathematics, astronomy.*

The son of a military physician, Poretsky graduated from the Poltava Gymnasium and from the Physical-Mathematical Faculty of Kharkov University; in 1870 he was attached to the chair of astronomy to prepare for a professorship. For several years Poretsky worked as an astronomer-observer at the Kharkov observatory and, from 1876, at Kazan University, where he conducted observations of stars in the Kazan zone according to the program of the International Astronomical Society. In 1886 he defended a thesis for his master's degree, the theoretical portion of which dealt with reducing the number of unknowns and equations for certain systems of cyclic equations that occur in practical astronomy. For this work he was awarded a doctorate in astronomy. In the same year Poretsky became *Privatdozent* at Kazan University and in 1887-1888, for the first time in Russia, he lectured on mathematical logic, in which he had become interested soon after going to Kazan through the influence of A. V. Vasiliev.

From 1882 to 1888 Poretsky was secretary and treasurer of the Physical-Mathematical Section of the Kazan Society of Natural Science, supervising the publication of its *Proceedings*; for several years he edited a liberal newspaper, *Kazansky telegraf*, sometimes publishing in it his translations of Pierre Béranger's poems. At the beginning of 1889 poor health forced Poretsky to retire, but he continued his research in mathematical logic for the rest of his life.

Poretsky's main achievement was the elaboration of the Boolean algebra of logic; he considerably augmented and generalized the results obtained by Boole, Jevons, and E. Schröder. In papers published from 1880 to 1908, Poretsky systematically studied and solved many problems of the logic of classes and of propositions. He developed an original system of axioms of logical calculus and proposed a very convenient mode of determining all the conclusions that are deducible from a given logical premise, and of determining all possible logical hypotheses from which given conclusions may be deduced. He also applied the logical calculus to the theory of probability. Poretsky was the first eminent Russian scholar in mathematical logic. His research was continued by E. Bunitsky, Couturat, Archie Blake, and N. Styazhkin.

BIBLIOGRAPHY

A nearly complete list of Poretsky's writings is in the work by Styazhkin (see below), 291–292.

Secondary literature includes A. Blake, *Canonical Expressions in Boolean Algebra* (Chicago, 1938); L. Couturat, *L'algèbre de la logique* (Paris, 1905); D. Dubyago, "P. S. Poretsky," in *Izvestiya Fiziko-matematicheskogo obshchestva pri (Imperatorskom) kazanskom universitete*, 2nd ser., **16** (1908), 3–7; and N. I. Styazhkin, *Stanovlenie idei matematicheskoy logiki* (Moscow, 1964), ch. 6, sec. 2, trans. into English as *History of Mathematical Logic From Leibniz to Peano* (Cambridge, Mass., 1969).

A. P. YOUSCHKEVITCH

PORRO, IGNAZIO (*b.* Pinerolo, Italy, 25 November 1801; *d.* Milan, Italy, 8 October 1875), *topography, optics, geodesy.*

After receiving his education in Turin, Porro entered the Piedmontese Corps of Engineers, in which he remained until 1842, attaining the rank of major. In that capacity he conceived and constructed several optical surveying instruments that revolutionized topographical and geodetic practice, and brought it essentially to its modern status. The science of rapid surveying (tachymetry; in Italian, *celerimensura*) was then new, and Porro states in his text on the subject that it originated in Italy in 1823. His first invention was the stereogonic telescope, which permitted the optical measurement of the distance from the objective to a graduated stadia rod; attaching this telescope to a theodolite, an instrument then becoming popular, Porro constructed the telemeter in 1835. It thus became possible to measure distance, elevation, and orientation with one instrument. The remarkable saving in time and the improvement in accuracy that it afforded led prominent French engineers to write impressive testimonials to Porro's innovations in the introduction to the French edition of his *Tachéométrie.*

The practical test of his instruments had been made in the construction of roads, railroads, canals, and military fortifications by the time Porro left the army. From 1842 to 1861, in Turin and then in Paris, he improved existing instruments and developed new ones; he also wrote articles and textbooks popularizing them. The results of this activity, despite its acknowledged scientific merit, were not satisfactory, partly because of Porro's apparent lack of organization and administrative skill and partly because of a tendency to modify his work before it had been fully tested.

After returning to Italy in 1861, Porro taught tachymetry in Florence and then in Milan, where he founded the journal *Tecnomasio italiano* and the Società Filotecnica (1865). He continued his scientific activities there until his death.

In 1848 Porro invented a telescopic objective (*obiettivo anallattico*) that furnished several views of the same subject at different scales. The forerunner of range finders and of modern objectives for telemetry and telephotography, it was adapted by him to topography. In the last years of his life Porro, who in 1855 had divided a circle with a 35-millimeter diameter into 4,000 parts, constructed the cleps, an improved tachymeter that he hoped would become the universal instrument, easy to transport without impairing accuracy and facility of use. Porro's other contributions include the introduction of inverting prisms in binoculars in 1851 and the subsequent construction of the modern prismatic instrument; the application of photography to topography, showing how the measurement of both distances and angles could be performed on the basis of photographs; and some of the earliest astronomical photographs of the eclipse of 1857.

BIBLIOGRAPHY

I. ORIGINAL WORKS. Porro's principal work is the text *La tacheometria* (Turin, 1854), best known in its rev. French ed., *La tachéométrie, ou l'art de lever des plans et de faire les nivellements avec beaucoup de précision et une économie de temps considérable* (Paris, 1858). The French ed. was an expanded rev. of a memoir in the *Annales des ponts et chaussées*, sec. 5, **4** (1852), 273–390, and appeared with the first-name initial J. He also wrote *Applicazioni della celerimensura alla misura generale porcellaria ed altimetria in Italia* (Florence, 1862).

II. SECONDARY LITERATURE. A review of the development of optical instruments, which places some of Porro's inventions in historical perspective, may be found in *Encyclopaedia Britannica*, 11th ed. (1910), articles "Binocular," "Photography," "Range-Finder," and "Stereoscope." For a comprehensive historical treatment of tachymetry, see N. Jadanza, "Per la storia della celerimensura," in *Rivista di topografia e catasto*, supp. to *Giornale dei lavori pubblici e delle strade ferrate* (1894).

BRUNO A. BOLEY

PORTA, GIAMBATTISTA DELLA (*b.* Vico Equense, Italy, between 3 October and 15 November 1535; *d.* Naples, Italy, 4 February 1615), *natural philosophy, mathematics.*

The modest fortunes of the Porta family, who belonged to the ancient nobility of Salerno, were improved when Nardo Antonio della Porta, father of Giambattista, entered the service of Emperor Charles V in 1541. From that year the family residence alternated between a villa in Vico Equense and a house in Naples. Giambattista was the second of three sons. Only two of his teachers are known: Antonio Pisano, a royal physician in Naples, and Domenico Pizzimenti, a translator of Democritus. The nature of his formal education is unknown, but early accounts of his life suggest that he was self-taught. His informal education, however, was clearly the convivial, and sometimes profound, discussion of scientific and pseudoscientific topics.

Porta was examined by the Inquisition some time prior to 1580, and in 1592 all further publication of his works was prohibited. This ban was not lifted until 1598. In the same period his religious activity is first mentioned. By 1585 he had become a lay brother of the Jesuits, and his participation in the charitable works of both the Jesuits and the Theatines in Naples demonstrates his devotion to the ideals of the Catholic Reformation. The relationship of this overt piety to his difficulties with the Inquisition and to his personal relations with Fra Paolo Sarpi in Venice after 1579 and with Campanella in Naples in 1590 cannot easily be determined.

Nothing is known of Porta's marriage except that it produced his only child, a daughter, about 1579. He suffered from various psychosomatic ailments, which by his own account were cured when his anxiety was relieved. His most frequent illness was a persistent fever, which on occasion confined him to bed for several months.

Porta's relationship to the academies of late Renaissance Italy is of great importance. The academies of Naples were closed under suspicion of political intrigue in 1547 and began to reopen only after 1552. In the following decade the Altomare was the outstanding literary academy of Naples, and several of Porta's close friends were members. Porta himself established the Accademia dei Segreti (Academia Secretorum Naturae) some time prior to 1580. It met in Porta's house in Naples; was almost certainly founded on the model of the earlier literary academies; and was devoted to discussion and study of the secrets of nature. It seems to have been closed by order of the Inquisition and may have been the cause for Porta's original process by the Inquisition.

This early academy was but a vague anticipation of the Accademia dei Lincei, founded at Rome in 1603 by Federico Cesi and three friends. The relationship of Porta to the Lincei is difficult to establish. In 1604 Cesi traveled to Naples and often visited Porta. In the same year Porta wrote a compend of the history of the Cesi family. Cesi, who wrote the constitution of the Lincei, known as the *Lynceographum*, acknowledged that the idea of such an academy preceded the fact; and he seems to have known of Porta and his academy. The documented meeting of Cesi and Porta in 1604 was followed by a respectful correspondence which culminated in the enrollment of Porta among the Lincei on 6 July 1610. Porta's reputation among his contemporaries was second only to that of Cesi, but the enrollment of Galileo on 25 April 1611 soon overshadowed Porta and gave a new direction to the academy. It is significant, however, that the choice of the lynx with the motto "Auspicit et Inspicit" for the Lincei was derived from Porta's *Phytognomonica* (Naples, 1588). In 1611 Porta was enrolled among the Oziosi in Naples, then the most renowned literary academy.

Porta's first book, published in 1558 as *Magiae naturalis*, was a treatise on the secrets of nature, which he began collecting when he was fifteen. The secrets are arranged in four books, and the conception implied in the title is that natural magic is the perfection of natural philosophy and the highest science. This small collection of secrets constituted the basis of a twenty-book edition of the *Magiae naturalis* published in 1589, which is Porta's best-known work and the basis of his reputation. It is an extraordinary hodgepodge of material representing that unique combination of curiosity and credulity common in the late Renaissance. But combined with the author's insatiable desire for the marvelous and apparently miraculous is a serious attempt to define and describe natural magic and some refined application of both mathematical and experimental techniques in science. Book XVII, on refraction, is the basis of the attribution of priority to Porta in inventing the telescope and demonstrates his involvement with both theory and practice.

Natural magic is no longer quite so pretentiously conceived as in the first edition. It presumes an orderly and rational universe into which the magician-scientist has insights that are revealed to him because of his virtue and his study. Natural magic entails a survey of the whole of nature, but with a modicum of modesty Porta acknowledges that it may merely be the practical part of natural philosophy. The 1589 edition represents in part the work, discussions, and experiments that took place at Porta's academy—hence the emphasis on experimentation and application in his definition of natural magic.

Behind Porta's conception of natural magic lie the Hermetic and Neoplatonic traditions given new life in Renaissance philosophy, and these traditions present Porta with the possibility of an intellectual synthesis founded on the conviction that rational orderliness exists behind all the marvels and prodigies of nature that he has collected. In this belief he is a philosophical, if not a religious, mystic. But natural magic is the art and practice of such mysticism. Natural magic is not simply philosophy or religion; it is both of these brought into practice and subjected to experiment. And experiment is merely refined experience. The contribution of Porta's conception and practice of natural magic to the emerging idea of science is not merely rational or theoretical or contemplative. Rather, science must represent theory and contemplation coming to practical and experimental expression. Such a conception of natural magic as science is ideally represented in his work on concave and convex lenses. His theoretical and experimental work prepared the way for the invention of the telescope.

The range of Porta's scientific and literary interests is easily demonstrated by his works. The first published after the *Magiae* was a treatise on cryptography, *De furtivis literarum notis*, in 1563. It was followed in 1566 by *Arte del ricordare*, a book on the art of memory and mnemonic devices. Both reveal his fascination with hidden and marvelous things. In the 1570's Porta composed his first plays and wrote a treatise on the physiognomy of hands. The latter, based on his observations in the prisons of Naples, is often cited as a precursor of criminal physiognomy. The plays were not published until much later, and the treatise on the physiognomy of hands did not appear until after his death.

The early biographies of Porta suggest that his writing of drama was occasioned by the Inquisition's scrutiny of the activity of his academy, and Porta himself says that he turned to comedy as a diversion from his more serious studies. In 1584 and in 1585 he published treatises on horticulture and agriculture that were based on careful study and practice. In 1586 he published a treatise on human physiognomy, *De humana physiognomonia*, in which he clearly established the doctrine of the correspondence between the external form of the body and the internal character of the person. The doctrine of signatures—that the external form of a plant indicates its medicinal properties—is worked out in Porta's treatise on the physiognomy of plants, *Phytognomonica* (1588), in which he established the claim that physiognomy of plants is the theoretical part of agriculture. His work on physiognomy attracted the attention of the Inquisition, and a proposed Italian edition of his treatise on human physiognomy was prohibited in 1593.

In the same year, Porta published *De refractione optices*, an expansion of book XVII of the *Magiae* of 1589 on the properties of refracting lenses. In 1601 he brought out a curious treatise on celestial physiognomy, in which, after a prefatory denunciation of astrology, he proceeded to develop a theory of astral signatures that he had confirmed by experience and observation. Also in 1601 he published a small book on the mechanics of water and steam and another on the elements of curved lines in which he addressed, with some finesse, the ancient topic of squaring the circle. In 1605 he issued a translation of book I of Ptolemy's *Almagest* together with the commentary of Theon of Alexandria. In 1608 there appeared his short treatise on military fortification and a longer study on the alchemical technique of distillation (*De distillatione*). The last book published during Porta's lifetime was a treatise on meteorology (1610). Among his many special studies, those on agriculture and refraction have received the highest praise from both his contemporaries and posterity. The others are modest, even though careful study usually reveals that they are not devoid of merit.

Porta's contribution to the theory and practice of Renaissance optics is found in book XVII of the *Magiae* of 1589 and in the *De refractione* of 1593. He did not invent the camera obscura, but he is the first to report adding a concave lens to the aperture. He also juxtaposed concave and convex lenses and reports various experiments with them. But in both the 1589 and 1593 treatises he limits his purposes to clarifying the image and to a geometrical explanation of the refracting properties of such lenses. Despite his claim to priority he did not invent the telescope. His comparison of the lens in the camera obscura to the pupil in the human eye did provide an easily understandable demonstration that the source of visual images lay outside the eye as well as outside the darkened room. He thus ended on a popular level an age-old controversy. Porta's work lies conceptually and chronologically between Risner's *Opticae thesaurus* of 1572 and Kepler's *Ad Vitellionem paralipomena* of 1604. He was thoroughly familiar with the former and did not attain the geometrical certainty of the latter.

Both Porta and his position in the history of late Renaissance science are tragically revealed in two of his unpublished works. The first is a treatise known as "De telescopiis." Among the Lincei, Porta was thought to have priority in the invention of the telescope, but this treatise reveals his secondary position both in theory and in practice. He acknowledges that Galileo brought his early (1589) theory to fruition, but he fails to go beyond Galileo in any way.

He returns to the elusive quest for a parabolic mirror that will permit him to see to infinity. The second treatise, known as the "Taumatologia," is an unfinished work that exists only in manuscript; but when trying to get it published, Porta claimed that it would be the consummation of his lifework. His correspondence about it reveals his return to his youthful enthusiasm for the secrets of nature and for the arcane and marvelous. It was to be another expanded version of his *Magiae*. Thus Porta devoted his last years more to discovering the philosophers' stone and the quintessence of nature than to the disciplined mathematical and experimental work of his younger contemporary Galileo. His devotion to experiment and his study of mathematics brought him in the 1580's to the verge of greatness, but he was soon overwhelmed again by the lure of the occult and the marvelous. Perhaps Porta's most compelling virtue and weakness was this youthful enthusiasm for the things of nature. There is a joy in his studies that not even the fatigue of working on the telescope and parabolic mirrors could diminish.

BIBLIOGRAPHY

I. ORIGINAL WORKS. Porta's first published work is *Magiae naturalis libri iiii* (Naples, 1558), repr. several times before publication of the expanded *Magiae naturalis libri xx* (Naples, 1589). There were Italian and other translations of both eds. There is an English trans., *Natural Magick* (London, 1658; repr. New York, 1957). There is a convenient list of the 1st eds. of all of Porta's scientific and literary works in Louise George Clubb, *Giambattista Della Porta, Dramatist* (Princeton, 1965), 316–342. Fuller bibliographical information for all eds. and translations is in Giuseppe Gabrieli, "Bibliografia Lincei I: G. B. della Porta," in *Atti dell'Accademia nazionale dei Lincei. Rendiconti*, Cl. Sc. Mr., 6th ser., **8** (1932), 206 ff.

The most important MSS of Porta are in the Library of the Accademia dei Lincei in Rome. MS Archivio Linceo IX contains an autograph copy of the index of the "Taumatologia." MS Archivio Linceo X is the MS base for Porta's book on distillation. MS Archivio Linceo XIV contains primarily the text for Porta's proposed book on the telescope. This portion of the MS has been edited and published by Vasco Ronchi and Maria Amalia Naldoni as *De telescopio* (Florence, 1962). MS Archivio Linceo XV contains the MS base for Porta's book on meteorology published in 1610. The library of the Faculty of Medicine of the University of Montpellier, MS H 169, contains parts of the "Taumatologia" as well as treatises on the magnet and the physiognomy of the hand. MS portions of the treatise on the physiognomy of the hand are also in Naples, Paris, and Toronto but are of minor importance.

II. SECONDARY LITERATURE. In addition to the book by Louise G. Clubb, three articles by Giocchino Paparelli are

important: "La Taumatologia di Giovambattista Della Porta," in *Filologia romanza*, **2** (1955), 418–429; "La data di nascita di G. B. Della Porta," *ibid.*, **3** (1956), 87–89; and "Giambattista Della Porta: Della Taumatologia e liber medicus," in *Rivista di storia delle scienze mediche e naturali*, **47** (1956), 1–47.

M. HOWARD RIENSTRA

PORTA, LUIGI (*b.* Pavia, Italy, 4 January 1800; *d.* Pavia, 9 September 1875), *surgery*.

Porta's parents were of modest means and his father died when he was young. He therefore received his schooling only at considerable sacrifice. He was awarded a degree in surgery from the University of Pavia in 1822, and was immediately sent by his teacher Antonio Scarpa to Vienna for further study. He returned to Pavia in 1826 and took the degree in medicine in the same year. In 1832 he was appointed professor of clinical surgery, a position that he held for more than forty years.

Porta's work lay in many surgical fields; that in which he was preeminent was experimental pathological surgery. By his research on the pathological changes caused to arteries by ligation and torsion he contributed to establishing the foundations of modern vascular surgery. Beginning in 1835 he made an extensive series of animal experiments on more than 270 animals of various species, including dogs, sheep, goats, horses, asses, oxen, and rabbits. He continued these investigations for nine years, and his results, coupled with his clinical observations, led him to significant discoveries concerning experimentally induced pathological changes of the arteries. Among these, his findings on the manner in which collateral circulation is established following the obliteration of parts of the arteries are of particular interest.

In his work Porta distinguished between direct collateral circulation, which occurs by anastomosis from the ends of the trunk of an obliterated artery, and indirect collateral circulation, which arises from the anastomosis of the secondary vessels of the limb to the periphery of the main trunk of the obliterated artery. He also made a distinction between two types of anastomosis produced by ligation of the artery: primitive, or preformed anastomoses originate through dilation of vasa vasorum, which are enlarged by the action of the ligature, while newly formed anastomoses arise from the vasa vasorum through hyperplasia, rather than dilation. He further pointed out that direct collateral circulation through anastomoses, in which anatomical continuity was maintained, could follow simple ligature of an artery, and that it was of great functional importance in cases in which the vessel was poor in collateral branches.

Indirect collateral circulation, Porta concluded, occurs through anastomoses of deep muscular vessels and superficial subcutaneous ones, although intramuscular, intermuscular, intranervous, periosteal, and subcutaneous vessels may all play a part. The establishment of an indirect collateral circulation was, he added, substantially the result of the conversion of the lateral anastomotic system to the chief channel of circulation; in the process the lateral anastomotic system is modified, since its branches, initially numerous and slender, decrease in number and increase in diameter. He found that these alterations were not constant, however, but varied in the form and length of time that they took according to animal species. Nor was the time required for the establishment of collateral equilibrium constant, since according to Porta's results it varied from case to case.

In addition Porta did work in anesthesiology, thyroid pathology, urology, traumatology, and autoplasty. He made a significant contribution, too, to lithotripsy, for which procedure he developed a special instrument, a combination of drill and pincers. He made a number of anatomical and pathological collections, which he gathered in the museum that he founded and, in 1860, gave the university; upon his death, the university received all his property.

Porta received many honors from both Italian and foreign academies. He was a senator of the Kingdom of Italy, head of the Medical and Surgical Faculty at Pavia, and for a time the rector of the university. He never married, and cared for his mentally deranged sister throughout her lifetime.

BIBLIOGRAPHY

I. ORIGINAL WORKS. Porta wrote some fifty scientific works, listed in Alfonso Corradi, in "Porta Comm. Luigi Senatore del Regno," in *Annuario della Reale Università di Pavia* (1875–1876), 34–60. The most important of these are *Delle alterazioni patologiche delle arterie per la legatura e la torsione. Esperienze ed osservazioni . . .* (Milan, 1845); and *Della litotrizia di Luigi Porta* (Milan, 1859). A number of his manuscripts, as well as numerous other mementos and original preparations (including some concerning collateral circulation), are preserved in the historical museum at the University of Pavia.

II. SECONDARY LITERATURE. In addition to Corradi, cited above, see Angelo Scarenzio, "Commemorazione di Luigi Porta," in *Rendiconti dell'Istituto lombardo di scienze e lettere*, 2nd ser., **20** (1887); and Bruno Zanobio,

"The Research of Luigi Porta on Morbid Changes of the Arteries Induced by Ligation and Torsion," in *Clio medica*, **8** (1973), 305–313.

BRUNO ZANOBIO

PORTAL, ANTOINE (*b.* Gaillac, France, 5 January 1742; *d.* Paris, France, 23 July 1832), *medicine, history of medicine.*

Portal was the eldest son of Antoine Portal, an apothecary, and Marianne Journès. After attending the Jesuit college in Albi, he went to Montpellier, where he studied medicine from 1762 to 1765. Then, with letters of recommendation to Sénac, Lieutaud, and Buffon, he traveled to Paris, where he remained until his death. In 1774 he married Anne Barafort; they had two daughters.

In 1766 Portal was appointed anatomy teacher to the twelve-year-old dauphin (the future Louis XVI). In Paris, as in Montpellier, he organized private courses in anatomy for students who were dissatisfied with the antiquated instruction given at the Faculté de Médecine. He also established a medical practice and rapidly built up a large clientele. Because of his knowledge of anatomy and medicine Portal was one of the most respected physicians in Paris, especially for the diagnosis, through physical examination, of abdominal disorders; to win such repute, however, he employed methods bordering on charlatanism.

Portal was elected an adjunct member of the Académie des Sciences in 1769, an associate member in 1774, and a pensioner in 1784. He served as professor of anatomy at the Collège de France from 1769; and in 1778, with Buffon's support, he was assigned the chair of anatomy at the Jardin du Roi that Vicq d'Azyr had hoped to win. Portal's lectures, which were very popular at both institutions, were supplemented by experiments in physiology and experimental pathology. No important anatomical discovery, however, can be attributed to Portal; and on certain points he was even unaware of contemporary knowledge: he mistakenly viewed the urachus as the suspensory ligament of the bladder rather than the remains of an embryonal canal; and he believed that anencephaly was caused by a difficult childbirth.

The French Revolution had little effect on Portal's career; he continued to treat a distinguished clientele and to teach at the Collège de France and the Muséum d'Histoire Naturelle (formerly the Jardin du Roi). He was named to the Institut de France when that body was created in 1795 in place of the old academies.

Portal's works appeared between 1764 and 1827, when he published *Observations sur la nature et le traitement de l'épilepsie* at the age of eighty-five.

Although he demonstrated, even in his earliest works, an interest in pathological anatomy, Portal is known chiefly for his *Histoire de l'anatomie et de la chirurgie*, in which he recounted (in the form of a biobibliographical dictionary) the evolution of anatomy and its surgical applications from "the Flood and the Trojan War" to 1755. Although the notices on contemporary authors led to several polemics, the work is still a valuable reference. Portal also published new editions of his patrons' works: for Lieutaud, *Historia anatomica medica* (1767) and *Anatomie historique et pratique* (1776), and for Sénac, *Traité de la structure du coeur* (1774).

Portal published several *Instructions*, commissioned by the government and addressed to physicians and the public concerning problems of public health (including asphyxiation in "mephitic vapors" and measures of first aid for the drowning and for persons bitten by rabid animals). In an *Instruction* published in 1775 he advocated the method of mouth-to-mouth resuscitation.

Portal's *Observations sur la nature et sur le traitement de la phthisie pulmonaire* (1792), which was translated into Italian (1801) and German (1802), was a synthesis of currently accepted knowledge, in which he affirmed the noncontagiousness of phthisis. In his work on liver ailments Portal cited the hepatic congestion that accompanies deficient cardiac action, the cardiac cirrhosis that can result from it, and the types of hepatitis induced by overexertion.

In 1818, an old man still attired in the style of the Ancien Régime, he was named first physician to the king, Louis XVIII, an honor of which he had dreamed all his life. He retained this office under Charles X, who named him a baron. In 1820, after repeated requests from Portal, the Académie Royal de Médecine was created, and Portal was named its permanent honorary president.

BIBLIOGRAPHY

I. ORIGINAL WORKS. Portal's major works are *Histoire de l'anatomie et de la chirurgie*, 7 vols. (Paris, 1770–1773); *Instruction sur les traitemens des asphixiés par le méphitisme, des noyés, . . .* (Paris, 1795–1796); *Mémoires sur la nature et le traitement de plusieurs maladies*, 5 vols. (Paris, 1800–1825); *Cours d'anatomie médicale ou élémens de l'anatomie de l'homme*, 5 vols. (Paris, 1803–1804); and *Observations sur la nature et le traitement des maladies du foie* (Paris, 1813).

II. SECONDARY LITERATURE. There is no complete biography of Portal. See Paul Busquet, "Antoine Portal," in *Biographies médicales*, **1** (1928), no. 17, 261–272; no. 18, 277–288; François Granel, "La prestigieuse carrière d'Antoine Portal," in *Monspeliensis Hippocrates*, no. 2 (1961), 9–21; and Étienne Pariset, "Discours aux obsèques du baron Portal," in *Revue médicale française et étrangère*, **3** (1832), 152–155.

PIERRE CHABBERT

PORTEVIN, ALBERT MARCEL GERMAIN RENÉ (*b*. Paris, France, 1 November 1880; *d*. Abano Terme, Italy, 12 April 1962), *metallurgy*.

Portevin was brought up by his mother after his father's early death. Trained as an engineer at the École Centrale des Arts et Manufactures, he was named professor at that school in 1925. He also taught at the École Supérieure de Fonderie and at the École Supérieure de Soudure. In 1907 he became editor in chief of the *Revue de métallurgie* and, with Henry Le Chatelier and Léon Guillet, was principally responsible for the great success of the journal. In 1942 Portevin was elected unanimously to the Académie des Sciences. He was also a grand officer of the Legion of Honor.

Portevin was a precursor in many aspects of the science of metallurgy. In 1905, while working in the metallurgical laboratory of the Établissements de Dion-Bouton, he conducted a micrographic study of chrome steels and was struck by their resistance to ordinary reagents when the chrome content exceeded 9 to 10 percent. He systematically studied the ways in which these steels could be corroded, and in his report to the Carnegie Foundation (1909) he gave precise figures on the chemical resistance of tempered chrome steels (in tempering, the chrome was put into solid solution) to such oxidizing reagents as nitric acid and picric acid.

In order to use these steels, however, it was necessary to make them amenable to standard milling processes. Through a judicious interpretation of the equilibrium diagrams, Portevin determined the appropriate thermal treatment for softening them. This achievement was even greater than his initial discovery, for his opinion was then in contradiction with that of metallurgists who were working from incomplete or erroneous diagrams. Today stainless steels are essential in the chemical industry, and their immense development since the 1940's is intimately linked with the metallurgical progress made in their production.

Portevin's interest in chemical resistance never ceased, and he set forth the general principles for obtaining it: the rules of homogeneity, of self-screening, and of concentration limit—which together enabled him to achieve constant progress in the field.

Portevin's discoveries and studies extended to the major areas of metallurgy. He was a pioneer in the scientific tempering of steels, in the structural hardening of light alloys, and in the introduction of scientific methods in the casting and welding of metals. Albert Sauveur paid tribute to Portevin in 1937 in his treatise on metallography: "The French metallurgist Albert Portevin is able to explain the most subtle subjects in metallurgy; his thought has penetrated all the darkest corners of this science and in them he has cast a bright light."

BIBLIOGRAPHY

I. Original Works. Among the more important of his papers are "L'équilibre du système nickel-bismuth," in *Comptes rendus ... de l'Académie des sciences,* **145** (1907), 1168–1170; "Contribution to the Study of Special Ternary Steels," in *Carnegie Scholarship Memoirs,* **1** (1909), 230–364; a patent, "Procédés pour rendre usinables des aciers de très grande dureté," B.F. 430,362 of 31 May 1911; "Sur les aciers au chrome," in *Comptes rendus ... de l'Académie des sciences,* **153** (1911), 64–66; "Sur les propriétés thermoélectriques du système fer-nickel-carbone," *ibid.,* **155** (1912), 1082–1085, written with E. L. Dupuy; "Le coefficient d'écoulement et son importance dans la coulée des lingotières métalliques," in *Revue de métallurgie,* **10** (1913), 948–951; and "Influence de la rapidité du refroidissement sur la trempe des aciers au carbone," in *Bulletin de la Société d'encouragement pour l'industrie nationale,* **132** (1920), 198–226, 297–346.

Portevin also wrote many important memoirs in the *Comptes rendus ... de l'Académie des sciences* (1920–1940), particularly in the fields of steel hardening, metallographic structures and transformations of metals and alloys, corrosion, and protection of metallic materials.

II. Secondary Literature. There is no full-scale biography, but see G. Chaudron, "Albert Portevin," in *Centenaire de la Société chimique de France* (Paris, 1957), 224; "Discours prononcés aux obsèques d'Albert Portevin, le 17 avril 1962," in *Revue de métallurgie,* **59** (May 1962); and obituaries by G. Chaudron in *Comptes rendus ... de l'Académie des sciences,* **254** (1962), 4109–4112, and P. Bastien, in *Journal of the Iron and Steel Institute,* **200** (1962), 1070–1071.

Georges Chaudron

PORTIER, PAUL (*b.* Bar-sur-Seine, France, 22 May 1866; *d.* Bourg-la-Reine, France, 26 January 1962), *biology, physiology.*

Portier came from a family that traditionally served in the national bureaucracy. When he was in his early twenties, he succeeded in a competitive examination, and was offered a career in the Ministry of Finance. He had, however, been deeply interested in biology since he was a child, and his family decided to allow him to follow his own inclination, provided that he

first study medicine. Portier received the M.D. and the degree of *docteur ès sciences* from the University of Paris, having worked there under Albert Dastre, who held the chair of general physiology. His own chief work was directed to comparative physiology, and a chair of that subject was established for him in 1923. He was also the director of the Institut Océanographique, and as such supervised more than 100 doctoral dissertations.

Portier's chief discovery grew out of his interest in marine biology. He was a regular member of the scientific staff that accompanied Albert I of Monaco on his summer oceanographic voyages on the *Princesse Alice II.* In July 1901 Portier and a fellow physiologist, Charles Richet, professor at the Paris Faculty of Medicine, undertook the study of the nature of the contact toxin secreted by coelenterates. An abundant catch of *Physalia* was brought on board off the Cape Verde Islands, and Portier and Richet ground up some of the specimens with sand and seawater. They filtered the mixture, then injected the resulting fluid into pigeons and guinea pigs, which soon fell into a state of profound anesthesia. Portier and Richet then determined to investigate whether an immunity might be produced by the injection of repeated doses of the toxin at long intervals; since such long-term experimentation could not be performed at sea, they decided to resume their work in Paris.

In their second series of experiments they injected a toxin derived from *Actinia* into dogs. Rather than becoming immune, the animals became sensitized to the toxin with repeated injections, and the toxic symptoms increased. In the most dramatic instance, a dog named Neptune received a weak injection that produced no effect; three days later the dose was repeated, again with no visible result. Three weeks later, however, Neptune had become hypersensitized to the degree that a third injection of the weak toxin caused him to collapse and die in a state of severe shock. Portier and Richet named this negative immunization "anaphylaxis," and by their experiment opened the new field of allergy studies. The scientific community only gradually realized that Portier had had a most significant role in this major discovery. Almost sixty years later, in 1958, Professor Portier received a special homage from the Third International Congress of Allergology in his old laboratory.

Portier also made numerous contributions to marine biology. He was the first to show that the spout of a blowing marine animal is visible because of the condensation of water vapor contained in the expelled air, the cooling brought about by the expansion of that air being always sufficient to produce condensation. From 1909 he investigated the physiological role of surface tension in aquatic insects. As early as 1922 he showed, in a collaborative work, that the *milieu intérieur* of fishes—which he had made the subject of extensive study—varies with exterior water pressure and salinity. His interest extended to marine birds, and in 1934 Portier demonstrated that the death of birds that have become covered with oil from spillage on the surface of the sea is the result of the loss of heat due to their oil-impregnated feathers.

In studies related to his practice of medicine, Portier showed that artificial respiration must be accompanied by warming the bulbar region, so as to reactivate the respiratory center. In biochemical studies, he insisted that carbon dioxide was necessary to synthetic and regenerative processes, a conclusion that was demonstrated experimentally thirty years later. Old age did not diminish his scientific activities, and when he was in his eighties he published a large treatise on butterflies, bringing together the results of lifelong observations on one of his favorite subjects. He was for many years an active member of the Academy of Medicine and of the Academy of Sciences.

BIBLIOGRAPHY

I. ORIGINAL WORKS. Portier's earlier writings include "Sur les effets physiologiques du poison des filaments pêcheurs et des tentacules des coelentérés (hypnotoxine)," in *Comptes rendus . . . de l' Académie des sciences,* **134** (1902), 247–248, written with C. Richet, also in *Travaux du Laboratoire de Charles Richet,* **5** (1902), 506; "De l'action anaphylactique de certains venins," in *Comptes rendus des séances de la Société de biologie,* **54** (1902), 170-172, written with Richet, also in *Travaux du laboratoire de Charles Richet, loc. cit.;* "Nouveaux faits d'anaphylaxie ou sensibilisation aux venins par doses réitérées," in *Comtes rendus des séances de la Société de biologie,* **54** (1902), 548–551, written with Richet; also in *Travaux du laboratoire de Charles Richet,* **5** (1902), 510; "Études sur la respiration. Mécanisme qui s'oppose à la pénétration de l'eau dans le système trachéen," in *Comptes rendus des séances de la Société de biologie,* **66** (1909), 422–424; "Généralité du mécanisme de fermeture de l'appareil trachéen," *ibid.,* 452–454; "Action des corps gras sur l'appareil stigmatique. Mécanisme de la lutte des larves aquatiques contre les phénomènes d'asphyxie," *ibid.,* 496–499; "Sort des corps gras introduits dans les trachées. Conséquences touchant le mode d'infection des insectes aquatiques et les procédés de destruction de ces animaux," *ibid.,* 580–582; and *Recherches physiologiques sur les insectes aquatiques* (Paris, 1911).

Later works are "Variation de la pression osmotique du sang des poissons téléostéens d'eau douce sous l'influence de l'accroissement de salinité de l'eau ambiante," in *Comptes rendus . . . de l' Académie des sciences,* **174** (1922), 1366–1368, written with Marcel Duval; "Variation de la pression osmotique du sang des sélaciens sous l'influence de la modification de salinité de l'eau environnante," *ibid.,* 1493–1495, written with Duval; "Variation de la pression osmotique du sang de l'anguille en fonction des modifications de la salinité du milieu extérieur," *ibid.,* **175** (1922), 324–326, written with Duval; "Pression osmotique du sang de l'anguille essuyée en fonction des modifications de la salinité du milieu extérieur," *ibid.,* 1105, written with Duval; "Role physiologique du gaz carbonique. Son intervention dans les phénomènes de synthèse et de régénération," in *Bulletin de l' Académie de médecine,* **100** (1928), 1274; and "La biologie des lepidoptères," in *Encyclopédie biologique* (Paris, 1949).

II. SECONDARY LITERATURE. For discussion of Portier and his work see M. Fontaine's obituary in *Bulletin de la Société scientifique d'hygiène alimentaire,* **50** (1962), 75–76; M. Fontaine, "La découverte de l'anaphylaxie," in *Bulletin de l'Institut océanographique,* **997** (1951), 1–9; and Léon Binet, "Mon laboratoire de la faculté et ses souvenirs," in *Biologie médicale,* **58** (1962), 443–480.

A. M. MONNIER

POŠEPNÝ, FRANZ (*b.* Starkenbach, Bohemia, 30 March 1836; *d.* Döbling, near Vienna, Austria, 27 March 1895), *geology, economic geology.*

Pošepný studied natural science at Prague and, in 1857–1859, mining geology at the School of Mines in Příbram, Bohemia. The lectures on ore deposits given by its director, Grimm, appear to have influenced Pošepný, especially toward his theory that ore deposits are characteristically confined to decomposed rocks. From 1859 to 1879 Pošepny held various positions that enabled him to become thoroughly acquainted with the mines on which he published monographs; some of them are, partly through his writings, still famous examples of type deposits.

From 1873 to 1879 Pošepný was a geologist at the Royal Imperial Ministry of Agriculture in Vienna. In 1876 he visited the United States, where he established close contacts with the American Institute of Mining and Metallurgy, especially with Rossiter W. Raymond, who was vital in the translation and publication of Pošepný's major work, "The Genesis of Ore-Deposits."

From 1879 to 1888 Pošepný taught mining geology at the School of Mines in Příbram. In 1882 he became associate professor and in 1887 full professor of special mining geology and analytical chemistry. In 1888 he retired in order to devote himself fully to the study of ore genesis. He visited ore deposits in Transylvania, Germany, Switzerland, the Ural Mountains, France, England, Sweden, Norway, Italy (including Sardinia), Greece, and Palestine.

The results of Pošepný's scientific work were published in more than 100 papers on ore deposits and other geological topics. Much of his work appeared in the *Archiv für praktische Geologie* (Vienna), which he founded in 1880. The summary of his system of ore genesis is his "Über die Genesis der Erzlagerstätten" (1895).

Pošepný's scientific achievement and influence can be attributed to two main properties of his work: (1) his extensive and systematic observations, which led to the introduction of mining geology into the curricula of the schools of mines at Příbram and Leoben and into the investigations of ore deposits, both old and new; and (2) his genetic views, which remained controversial and which he shared largely with his friend and colleague Alfred Stelzner. These views are today essentially termed panepigenetic and do not differ in essence from those held by Pošepný's teacher, Grimm. Both insisted on using genetic (interpretative) classifications, whereas their contemporary opponent, Groddeck, was the first to offer a geometric, observational classification of ore deposits in 1879, far ahead of his time, this being the modern approach since about 1860.

Pošepný's relationship with the American Institute of Mining and Metallurgy led to publication of his major work in the United States in 1895, together with numerous comments and criticisms by American authorities, whereas the textbook by Groddeck was little known in the English-speaking world.

BIBLIOGRAPHY

I. ORIGINAL WORKS. Pošepný's writings include *Cu-Vorkommen der Perm-Formation in Böhmen* (Ziva, Czechoslovakia, 1861); "Blei- und Galmei-Erzlagerstätte von Raibl, Kärnten," in *Geologisches Jahrbuch* (Vienna), **23** (1873), 317–424; *Geologisch-Montanistische Studie der Erzlagerstätten von Rézbánya, Ungarn* (Budapest, 1874); "Lateral-Secretionstheorie zur Erklärung der Erzgangfüllung," in *Österreichische Zeitschrift für Bergwesen*, **30** (1882), 607–609, 619–622; "Über die Adinolen von Příbram in Böhmen," in *Mineralogische und petrographische Mittheilungen*, **10** (1889), 175–202; "Über die Entstehung der Blei- und Zinklagerstätten in auflöslichen Gesteinen," in *Berg- und Hüttenmännische Jahrbuch* (Vienna), **42** (1894), 77–130; and "Über die Genesis der Erzlagerstätten," *ibid.*, **43** (1895), 1–226, trans. into English as "The Genesis of Ore-Deposits," in *Transactions of the American Institute of Mining Engineers*, **23** (1894), 197–369, 587–608, and **24** (1895), 942–1006.

II. SECONDARY LITERATURE. See R. W. Raymond, "Biographical Notice of Franz Pošepný," in *Transactions of the American Institute of Mining Engineers*, **25** (1896), 434–446; and the unsigned obituaries in *Österreichische Zeitschrift für Bergwesen*, supp. **43** (1895), 40–42; and *Zeitschrift für praktische Geologie*, **3** (1895), 261–262.

G. C. AMSTUTZ

POSIDONIUS (*b.* Apameia, Syria, *ca.* 135 B.C.; *d. ca.* 51 B.C.), *philosophy, science, history.*

Of Greek parentage and upbringing, Posidonius studied at Athens under the Stoic Panaetius of Rhodes and devoted himself to philosophy and learning. On travels in the western Mediterranean region, especially at Gades (Cádiz), he observed natural phenomena. Between 100 and 95 B.C. he became head of the Stoic school at Rhodes, where he at least once held some political office. In 87–86 as ambassador of Rhodes he reached Rome, visited the dying Marius, and was befriended by such conservatives as Publius Rutilius Rufus (a former fellow student), Pompey (Gnaeus Pompeius Magnus), and Cicero, who had heard him lecture at Rhodes and hoped for a historical memoir from him. An admired friend, Pompey also heard Posidonius at Rhodes in 67 and 62 B.C., when Posidonius was crippled but unconquered by gout. He died at the age of about eighty-five. His works have been lost but he was used or mentioned by authors whose writings are extant.[1]

For Posidonius, fundamental principles depended on philosophers and individual problems on scientists; and he believed that, among early men, the philosophically wise managed everything and discovered all crafts and industry. He stressed the Stoic ordering of philosophy—physical, ethical, logical—as a connected entity. For true judgment the standard is right reasoning; but precepts, persuasion, consolation, and exhortation are necessary; and enquiry into causes, especially as opposed to matter, is important.[2]

In scientific philosophy,[3] inspired partly by Aristotle, Posidonius tried to shape the achievements of others into coherent doctrine. He postulated three causing powers: everlasting God, supreme, having forethought or providence and mind or reason, a fiery breath, thinking, penetrating everything, taking all shapes; Nature; and Fate. God, artificer of everything, ordained and manages the Universe, which is His substance pervaded by reason in varying intensity. Of two Stoic principles (unborn, undestroyed, incorporeal), the passive is substance without quality, or (what we can envisage in thought only) matter, and the active is reason, equivalent to God, in matter. Every substance is material. Posidonius alone distinguished three bodily causes: matter, through which something secondary exists; soul, the prime active power; and reason, the principle of activity.[4]

Posidonius described the one spherical universe, set finite within eternal time and indefinite void, as a living, sentient organism endowed with a soul and having "sympathy" throughout; it includes a spherical revolving heaven, which plays a "leading" part, and the minute, spherical, motionless earth. The universe, which as a whole is the "being" of God, developed from pure "fiery nature" into moisture, which condensed into earth, air, and fire. Mixture of these elements—which have always existed, the real first origin—produced all else. He denied the real existence of qualified matter as such and of creation of elements from it. He denied that in the Stoic periodic destruction of the universe—if it occurs—substance (matter) is annihilated.[5]

All heavenly bodies are divine, ether-made, animate, moving, and nourished by the earth. Posidonius made a portable, spherical orrery illustrating the motion of the sun, moon, and five planets round the earth. The spherical sun, a star of pure fire, is about 3 million stades in diameter; the moon about 2 million stades from the earth, which is smaller than the moon and sun, and the sun is 500 million stades beyond the moon. If we assume 8.75 stades as equivalent to the English mile, or ten stades to one geographical mile, these are remarkable estimates, however conjectural, if we can rely on Pliny's figures.

In *On Ocean* (astronomical, geographical, geological, historical), based on Eratosthenes and Hipparchus and supplemented from his own observations, Posidonius dealt with the entire globe. Disagreeing with Eratosthenes' excellent calculation of 252,000 stades as the meridian circumference of the earth, he apparently first calculated it at 240,000 from the behavior of the star Canopus; later he preferred 180,000—a figure far too small.[6] It was a disastrous error, which nevertheless encouraged Columbus from the time he began planning his voyage. Posidonius believed that one deep ocean surrounds the globe and, as indicated by voyages and uniform behavior of tides, its known sling-shaped landmass (Europe, Asia, Africa) and possibly unknown continents. Oceanic transgressions and regressions have occurred, as have terrestrial sinkings and uprisings, of both seismic and volcanic origin. In his theory of tides Posidonius improved on his predecessors by observation. But, gravitation being then unknown, he said that not the sun but the moon only caused tides by its different positions and phases and by stirring up winds. He criticized the conception of five latitudinal zones projected onto the earth from heaven and favored two additional earthly ones. It would be sensible to divide the known landmass into narrow latitudinal belts, each having uniform characteristics. Posidonius' belief that

longitude affects life was wrong, and he overstressed the influence of climate.[7] He speculated fancifully on the effects of the sun and moon on the products of the earth.

In meteorology Posidonius relied greatly on Aristotle. Winds, mists, and clouds reach upward at least four miles from the earth; then all is clear brightness. He discussed winds (believing them to be produced mainly by the moon), rain, hail, and frost. A rainbow, he thought—not knowing it to be a dioptric and not a catoptric effect—is a continuous image of a segment of the sun or moon on a dewy cloud acting as a concave mirror. Lightning is nourished by dry, smoky exhalations from the earth which cause thunder (produced by moving air) if they disrupt clouds. Earthquakes are caused by enclosed air, which produces trembling, lateral tilt, or vertical upjolt, resulting in displacements or chasms. He described an earthquake that nearly destroyed Sidon and was felt over a vast area. Posidonius was interested in volcanic activity and described how a new island appeared in the Aegean. He also studied comets and meteors.[8]

In moral philosophy,[9] like most Stoics, Posidonius arranged ethics into topics: impulse; good and evil; emotions; virtue; the aim of life; primary values and actions; average duties; and inducements and dissuasions. His ethic, confined to mankind, was both psychological and moral. Man's highest good is to promote the true order of the universe, refusing leadership by the irrational, animal faculties of the soul; man's first "art" is virtue within his fleeting flesh —for thither Nature leads. Virtue is teachable and not self-sufficing; one needs health, strength, and means of living. There are various virtues, and animals other than man have some besides emotions. But there is no justice, or right, between men and animals. Evil is rooted in man; not all comes from outside. Average duties, not being part of morals, but indifferent, should be simply concomitants to life's object.

Every man's soul is a fragment of the universe's warm animating breath, a "form" holding body together as real surface holds a solid. It has three faculties, one being rational, one emotional, and one appetitive; the soul strives not for redemption but for knowledge, the one logical virtue. How far Posidonius believed in the human soul's immortality is uncertain.[10] Unlike other Stoics he did not compare the diseases of the soul with those of the body. His approach to emotions was psychological: their comprehension is the basis of ethics and is closely concerned with the understanding of virtues and vices and the object of life. Like reason, they are real. Posidonius, favoring older views, rejected the Stoic Chrysippus' opinion that emotions are errors of judgment. Not confined to mankind, as Stoics think, they are movements of illogical faculties; uncontrolled, they produce unhappy disharmony through man's inconsistency with his inner "daimon" (Latin *genius*). Men who progress morally feel only appropriate emotions. Their intensity of emotions and their characters can be indicated and even caused by bodily features and are affected by bodily condition, country, and education.[11]

Posidonius was no more "mystic" than other Stoics but, unlike Panaetius, regarded divination by man's clairvoyant soul, especially when death is near, as proved by fulfilled oracles and omens. The act of divination manifests Fate (a causing power with God and Nature) in action in an endless chain of causation of future by past and mediates (as dreams do) between gods and men. He also believed, if we can rightly so judge from a passage in St. Augustine and from more doubtful hints, that configurations of heavenly bodies could affect the futures of children conceived or born under them; but we ought not to conclude that Posidonius encouraged astrology.[12]

Posidonius' great *Histories* described, with much lively detail, events from 146 B.C. to perhaps 63: the subjugation of the Hellenistic monarchies by Rome, the rise of Parthia, the menace of Mithridates VI (Eupator), completion of Roman control throughout Mediterranean areas, the earlier civil wars of Rome, and a new growth of Greco-Roman contacts with backward "barbarians." Critically appreciative of Roman peace and order and desiring to reconcile other peoples to the Romans, Posidonius produced, as part of moral philosophy, contemporary history (Greek-Roman-"barbarian") based on written records and personal contacts. He took special interest in the peoples and products of Spain and Gaul and in wars against slaves and pirates. He made important contributions to the ethnology of the Germans (Cimbri and Teutones), Celts, and others, and to geography, sociology, anthropology, folklore, customs, and resources. Biased more toward "conservative" than to "popular" politics, he criticized and praised all classes and races.

Posidonius' narrative became more directly contemporary as it progressed and more personal, perhaps reaching a climax with Pompey. He stressed ethical and psychological motives and other processes as reasons for events, believing in a causal connection between physical environment and national character. His central feeling was that old Roman virtues had languished—hence perfidious and grasping behavior toward other peoples, and civil war. Cruelty begets cruelty. Men should be "decent" and lovers of men. Rule by the bigger and stronger is a habit of other animals, whereas free men are equals.[13]

Strabo, Seneca, Galen, and others testify to Posidonius' merits.[14] As philosopher or philosopher-scientist he was not comparable with Plato or Aristotle. It is wrong to regard him as the chief influence on thought and practice of two centuries; as the source of Neoplatonism; as a deep religious thinker; as a fuser of Greek and Oriental thought; or as an exponent of a philosophy based on sciences. Some of his beliefs were refuted in his own time, and his scientific skill is doubtful. But in following up results of others' demonstrations and research, and his own, he was better than most Stoics; and without being very original or deeply critical, he was a good thinker, investigator, observer, and recorder. Posidonius upheld the Stoics' moral dignity but modified their doctrines. In "psychology" (theory of the soul) and ethics he diverged widely from them, his chief differences leading him to a partial return from Chrysippus and even Panaetius to early philosophy.[15] He had a following; but even in his lifetime the influence of the old "Academics" and the Epicureans was greater than the Stoics', and it was the Old Stoa that became dominant in the first century of the Christian era. His works were neglected and by the fourth century were forgotten; he ended a Greek era and began no new one.

NOTES

1. These authors are Athenaeus, Cicero, Cleomedes, Diogenes Laërtius, Galen, Pliny the Elder, Plutarch, Priscianus Lydus, Proclus, Seneca the Younger, Sextus Empiricus, Stobaeus, Strabo, and a number of other writers, in varying degree. Our knowledge of Posidonius is incomplete, and (since right attribution, extent, interpretation, and correlation of the materials are sometimes hard) inexact here and there.
2. Posidonius' attitudes are revealed especially in Diogenes Laërtius, VII. 39, 87, 103, 124, 129; Seneca, *Epistulae*, 83.9–11; 87.31–40; 88.21-28; 90.5,20,30; 92.10; 95.65–67; Sextus Empiricus, *Adversus logicos*, I.19.
3. For the scientific fragments as a whole see L. Edelstein and I. G. Kidd, *Posidonius*, I, *The Fragments* (Cambridge, 1972), F92–149, pp. 98–137; F195–251, pp. 176–220; and F4–27, pp. 39–48.
4. Diogenes Laërtius, VII. 134, 138, 148; *Scholia in Lucani Bellum Civile*, H. Usener, ed., pars 1, 9.578; Stobaeus, *Eclogae*, I.1.29b; I.5.15; Cicero, *De divinatione*, I.125; Joannes Lydus, *De mensibus*, IV.71.48 and 81.53.
5. Diogenes Laërtius, VII. 134, 138, 139, 140, 142, 143, 148; Stobaeus, *Eclogae*, I.8.42; 11.5c; 18.4b; 20.7; Simplicius, *In Aristotelis de caelo*, IV.3.310b.
6. Sun and the moon: Cleomedes, *De motu circulari corporum caelestium*, Ziegler, ed., I.11.65; II.1.68,79–80; II.4.105; Pliny, *Natural History*, II. 85; Diogenes Laërtius, VII. 144, 145. Planetarium or orrery: Cicero, *De natura deorum*, II.88. Earth: Cleomedes, *op. cit.*, I.10.50–52; Strabo, *Geography*, II.2.2, C95; J. O. Thomson, *History of Ancient Geography* (Cambridge, 1948), 212, 213, 407.
7. Posidonius' *On Ocean*: Strabo, *Geography*, II.2.1–3.8. See also Priscianus Lydus, *Solutiones ad Chosroem*, I. Bywater, ed., 69–76; Cleomedes, *op. cit.*, I.6.31–33; F. Schülein, *Untersuchungen über die Posidonische Schrift*

Περὶ Ὠκεανοῦ (Erlangen, 1901); L. Edelstein and I. G. Kidd, *op. cit.*, F49, pp. 65–77; F214–221, pp. 191–201.

8. Seneca, *Naturales Quaestiones*, I.5.10,13; II.26.4; 54.1; IV.3.2; VI.17.3; 21.2; 24.6; VII.20.2; 20.4; Diogenes Laërtius, VII. 144, 145, 152–154.

9. For the fragments, see L. Edelstein and I. G. Kidd, *op. cit.*, F29–41,pp.49–58; F150a–186, pp. 137–172. Cf. L. Edelstein, "The Philosophical System of Posidonius," in *American Journal of Philology*, 57 (1936), 305–316; Marie Laffranque, *op. cit.*, 449–514.

10. Each human soul, as a part of that of the universe, would simply be conscious in the human body during the body's lifetime.

11. Most of our knowledge of Posidonius' psychological and ethical thinking comes from a score of passages in Galen, *De Placitis Hippocratis et Platonis*, I. Müller, ed. (Leipzig, 1874), supplemented from some other sources. They are all in L. Edelstein and I. G. Kidd, *op. cit.*, 137–172; see also p. xxiv.

12. Cicero, *De divinatione*, I.64 and 129–130; II.33–35 and 47 (there may be much more from Posidonius in this work and in *De natura deorum*); *De fato*, 5–7; Diogenes Laërtius, VII.149; Nonnus Abbas, in Migne, *Patrologia Graeca*, XXXVI, 1024; Suidas ("The Suda"), *s.v.* Διαίρεσις οἰωνιστικῆς; Boethius, *De diis et praesensionibus*, 20,77; Augustine, *De civitate Dei*, V. 2–5. In classical times "astrology" usually meant astronomy.

13. Historical fragments are in L. Edelstein and I. G. Kidd, *op. cit.*, F51–79 (or 81), pp. 77–90, and F252–283, pp. 220–252 (see also pp. xxi–xxii and 335–336); and in Jacoby, *op. cit.*, IIA, 225–252 (cp. 252–267) and IIC. The latest datable fragment relates to 83 B.C. See also Marie Laffranque, *op. cit.*, 109–151 and H. Strasburger, "Poseidonios on Problems of the Roman Empire," in *Journal of Roman Studies*, 55 (1965), 40–53. There is little doubt that the historian Diodorus Siculus used Posidonius, particularly in books V and XXXIV–XXXV, but one cannot extract Posidonian fragments from him.

14. Strabo, *Geography*, II.3.5, C102; Seneca, *Epistulae*, 104.22 and 90.20; Galen, *De placitis*, I. Müller, ed., IV. 402–403, p. 376; and *Scripta Minora*, II.77–78, p. 819.

15. K. Reinhardt, "Poseidonios," in Pauly-Wissowa, *Real-Encyclopädie der classischen Altertumswissenschaft*, XXII.i, cols. 570–624; L. Edelstein, *op. cit.*, *American Journal of Philology*, 57 (1936), 286–288, 321–325; Marie Laffranque, *op. cit.*, 1–44, 515–518; A. D. Nock, "Posidonius," in *Journal of Roman Studies*, 49 (1959), 1–15; V. Cilento, "Per una ricostruzione di Posidonio," in *Annali della Facoltà di Lettere e Filosofia. Bari, Università*, 9 (1964), 52–75.

BIBLIOGRAPHY

None of Posidonius' writings is extant; apparently known titles of them, or their contents, are *Against Zenon of Sidon*, on geometry; *On Average Duties* (or fitting actions); *On Divination* (and prophecy); *On Emotions*; *Exhortation to Philosophy; On Fate; On Gods; On Heroes and Spirits; Histories*, of which *History of Pompey's Campaigns in the East* may have formed a detailed part and *Tactics* a small one; *Introduction to Diction*; posthumously edited lectures; pamphlets, commentaries on Plato, and historical monographs; *On Meteorology* (or *Elements of Meteorology*); *On Ocean; On the Soul; On the Standard* (of truth or judgment); *Treatise on Ethic(s); Treatise on Physics; Treatise on Virtues; On the Universe;* and (doubtful) *On Void*. Titles of other works are recorded, but they probably indicate part of the works here listed.

The chief secondary literature on Posidonius includes studies by L. Edelstein and I. G. Kidd, *Posidonius*, I, *The Fragments* (Cambridge, 1972), esp. T1a–115, pp. 3–35; F. Jacoby, *Die Fragmente der griechischen Historiker*, (Berlin, 1926), IIA, 220–317, and IIC, 154–220; K. Reinhardt, "Poseidonios," in Pauly-Wissowa, *Real-Encyclopädie der classischen Altertumswissenschaft*, XXII.i, cols. 558–826, repr. as *Poseidonios von Apameia* (Stuttgart, 1954); M. Pohlenz, *Die Stoa*, 2 vols. (Göttingen, 1948, 1953); and Marie Laffranque, *Poseidonios d'Apamée* (Paris, 1964).

See also V. Cilento, "Per una ricostruzione di Posidonio," in *Annali della Facoltà di lettere e filosofia. Bari, Università*, 9 (1964), 52–75; H. Cherniss, "Galen and Posidonius' Theory of Vision," in *American Journal of Philology*, 54 (1933), 154–161; J. F. Dobson, "The Posidonius Myth," in *Classical Quarterly*, 12 (1918), 179–195; L. Edelstein, "The Philosophical System of Posidonius," in *American Journal of Philology*, 57 (1936), 286–325; R. M. Jones, "Posidonius and Cicero's *Tusculan Disputations* I. 17–81," in *Classical Philology*, 18 (1923), 202–228; "Posidonius and the Flight of the Mind Through the Universe," *ibid.*, 21 (1926), 97–119; and "Posidonius and Solar Eschatology," *ibid.*, 27 (1932), 113–135; M. P. Nilsson, *Geschichte der griechischen Religion*, II (Munich, 1950), 250 ff.; A. D. Nock, "Posidonius," in *Journal of Roman Studies*, 49 (1959), 1–15; and K. Reinhardt, *Kosmos und Sympathie* (Munich, 1926).

E. H. WARMINGTON

POST, EMIL LEON (*b.* Augustów, Poland, 11 February 1897; *d.* northern New York, 21 April 1954), *mathematics, logic.*

Post was the son of Arnold J. and Pearl D. Post. In May 1904 he arrived in America, where his father and his uncle, J. L. Post, were in the fur and clothing business in New York. As a child Post's first love was astronomy, but the loss of his left arm when he was about twelve ruled that out as a profession. He early showed mathematical ability, however; and his important paper on generalized differentiation, although not published until 1930, was essentially completed by the time he received the B.S. from the College of the City of New York in 1917. Post was a graduate student, and later lecturer, in mathematics at Columbia University from 1917 to 1920, receiving the A.M. in 1918 and the Ph.D. in 1920.

After receiving the doctorate, Post was a Proctor fellow at Princeton University for a year and then returned to Columbia as instructor, but after a year he suffered the first of the recurrent periods of illness that partially curtailed his scientific work. In the spring of 1924 he taught at Cornell University but again became ill. He resumed his teaching in the New York City high schools in 1927. Appointed to City College in

1932, he stayed there only briefly, returning in 1935 to remain for nineteen years. Post's family was Jewish; while not orthodox in his adult years, he was a religious man and proud of his heritage. He married Gertrude Singer on 25 December 1929 and they had one daughter.

Post was a member of the American Mathematical Society from 1918 and a member of the Association for Symbolic Logic from its founding in 1936. His extrascientific interests included sketching, poetry, and stargazing.

Post was the first to obtain decisive results in finitistic metamathematics when, in his Ph.D. dissertation of 1920 (published in 1921), he proved the consistency as well as the completeness of the propositional calculus as developed in Whitehead and Russell's *Principia mathematica*. This marked the beginning, in important respects, of modern proof theory. In this paper Post systematically applied the truth-table method, which had been introduced into symbolic logic by C. S. Peirce and Ernst Schröder. (Post gave credit for his method to Cassius J. Keyser when he dedicated his *Two-Valued Iterative Systems* to Keyser, "in one of whose pedagogical devices the author belatedly recognizes the true source of his truth-table method.") From this paper came general notions of completeness and consistency: A system is said to be complete in Post's sense if every well-formed formula becomes provable if we add to the axioms any well-formed formula that is not provable. A system is said to be consistent in Post's sense if no well-formed formula consisting of only a propositional variable is provable. In this paper Post also showed how to set up multivalued systems of propositional logic and introduced multivalued truth tables in analyzing them. Jan Łukasiewicz was studying three-valued logic at the same time; but while his interest was philosophical, Post's was mathematical. Post compared these multivalued systems to geometry, noting that they seem "to have the same relation to ordinary logic that geometry in a space of an arbitrary number of dimensions has to the geometry of Euclid."

Post began a scientific diary in 1916 and so was able to show, in a paper written in 1941 (but rejected by a mathematics journal and not published until 1965), that he had attained results in the 1920's similar to those published in the 1930's by Kurt Gödel, Alonzo Church, and A. M. Turing. In particular, he had planned in 1925 to show through a special analysis that *Principia mathematica* was inadequate but later decided in favor of working for a more general result, of which the incompleteness of the logic of *Principia* would be a corollary. This plan, as Post remarked, "did not count on the appearance of a Gödel!"

If Post's interest in 1920 in multivalued logics was mathematical, he also wrote in his diary about that time: "I study Mathematics as a product of the human mind and not as absolute." Indeed, he showed an increasing interest in the creative process and noted in 1941 that "perhaps the greatest service the present account could render would stem from its stressing of its final conclusion that *mathematical thinking is, and must be, essentially creative.*" But this is a creativity with limitations, and he saw symbolic logic as "the indisputable means for revealing and developing these limitations."

On the occasion of Post's death in 1954, W. V. Quine wrote:

> Modern proof theory, and likewise the modern theory of machine computation, hinge on the concept of a recursive function. This important number-theoretic concept, a precise mathematical substitute for the vague idea of "effectiveness" or "computability," was discovered independently and in very disparate but equivalent forms by four mathematicians, and one of these was Post. Subsequent work by Post was instrumental to the further progress of the theory of recursive functions.

If other mathematicians failed to recognize the power of this theory, it was forcefully shown to them in 1947, when Post demonstrated the recursive unsolvability of the word problem for semigroups, thus solving a problem proposed by A. Thue in 1914. (An equivalent result had been obtained by A. A. Markov.) When reminded of his earlier statement, Quine in 1972 confirmed his opinion, adding: "The theory of recursive functions, of which Post was a co-founder, is now nearly twice as old as it was when I wrote that letter. What a fertile field it has proved to be!"

BIBLIOGRAPHY

I. Original Works. Except for abstracts of papers read at scientific meetings, the following is believed to be a complete list of Post's scientific publications: "The Generalized Gamma Functions," in *Annals of Mathematics*, **20** (1919), 202–217; "Introduction to a General Theory of Elementary Propositions," in *American Journal of Mathematics*, **43** (1921), 163–185, his Ph.D. dissertation, repr. in Jean van Heijenoort, ed., *From Frege to Gödel. A Source Book in Mathematical Logic, 1879–1931* (Cambridge, Mass., 1967), 264–283; "Generalized Differentiation," in *Transactions of the American Mathematical Society*, **32** (1930), 723–781; "Finite Combinatory Processes. Formulation I," in *Journal of Symbolic Logic*, **1** (1936), 103–105, repr. in *The Undecidable* (see below), 289–291; "Polyadic Groups," in *Transactions of the American Mathematical Society*, **48** (1940), 208–350; *The*

Two-Valued Iterative Systems of Mathematical Logic, Annals of Mathematics Studies no. 5 (Princeton, 1941); "Formal Reductions of the General Combinatorial Decision Problem," in *American Journal of Mathematics*, **65** (1943), 197–215; "Recursively Enumerable Sets of Positive Integers and Their Decision Problems," in *Bulletin of the American Mathematical Society*, **50** (1944), 284–316, repr. in *The Undecidable* (see below), 305–337, Spanish trans. by J. R. Fuentes in *Revista matemática hispano-americana*, 4th ser., **7** (1947), 187–229; "A Variant of a Recursively Unsolvable Problem," *ibid.*, **52** (1946), 264–268; "Note on a Conjecture of Skolem," in *Journal of Symbolic Logic*, **11** (1946), 73–74; "Recursive Unsolvability of a Problem of Thue," *ibid.*, **12** (1947), 1–11, repr. in *The Undecidable* (see below), 293–303; "The Upper Semi-Lattice of Degrees of Recursive Unsolvability," in *Annals of Mathematics*, 2nd ser., **59** (1954), 379–407, written with S. C. Kleene; and "Absolutely Unsolvable Problems and Relatively Undecidable Propositions—Account of an Anticipation," in Martin Davis, ed., *The Undecidable* (Hewlett, N.Y., 1965), 338–433.

II. SECONDARY LITERATURE. There is an obituary of E. L. Post in *The Campus* (City College), 27 April 1954. Part of Post's work was carried to a conclusion by S. V. Yablonsky and his students. See S. V. Yablonsky, G. P. Gavrilov, V. B. Kudryavtsev, *Funktsii algebry logiki i klassy Posta* (Moscow, 1966); translated into German as *Boolesche Funktionen und Postsche Klassen* (Brunswick, 1970).

HUBERT C. KENNEDY

POTANIN, GRIGORY NIKOLAEVICH (*b.* Semiyarsky, Russia, 4 October 1835; *d.* Tomsk, U.S.S.R., 30 June 1920), *geography, biogeography, ethnography.*

The son of a Cossack officer, Potanin studied at the military school in Omsk from 1846 to 1852 and then served as junior cavalry officer in the Cossack regiment in Semipalatinsk. From 1859 to 1862 he studied at the University of St. Petersburg. His scientific research was connected chiefly with the activities of the Russian Geographical Society, of which he became a member in 1862. In 1862–1864 he participated in the society's astronomical-geodesical and geographical expedition led by K. V. Struve, with whom he published a detailed geographical and ethnographical description of Lake Zaysan and the Tarbagatay mountain range. On a commission from the society, Potanin made five major expeditions through Inner Asia: into northwestern Mongolia through the Mongolian Altay (1876–1877); through the Tannu-Ola mountain range to China and Tibet (1879–1880); through the eastern border area of Tibet and Nan Shan (1884–1886), his most extensive journey, through Szechwan and the eastern regions of Tibet to Bhutan (1892–1893); and to the eastern border area of Mongolia and the Great Hsingan (1899). As a result of these journeys Potanin gave the first extensive and detailed information on the geography, geology, and economy of little-known regions of Mongolia, and his research helped to create a basic geographical framework for Inner Asia.

Potanin also gathered extensive zoogeographical data, published on the flora and fauna of Inner Asia, and assembled carefully collected herbariums. He discovered many plants, including three species of phanerogams, one of which is named for him. The ethnographical material on the life, economy, and culture of the peoples of Asia that he published are of great importance; the most valuable deal with Turkic and Mongolian tribes (Dungan, Tangut, Chinese, and Tibetans). In addition to organizing a series of expeditions to Siberia, Potanin was also the initiator and founder of the Society for the Study of Siberia, in Tomsk, and a number of museums and expeditions. A mountain in the Nan Shan and a glacier in the Mongolian Altay are named for him.

BIBLIOGRAPHY

I. ORIGINAL WORKS. Potanin's writings include "Puteshestvie na ozero Zaysan i v rechnuyu oblast Chernogo Irtysha do ozera Marka-Kul i gory Sar-Tau, letom 1863 god a" ("Travels to Lake Zaysan and the River Area of the Black Irtysh to Lake Marka-Kol and the Sar-Tau Mountains in the Summer of 1863"), *Zapiski Imperatorskago russkago geograficheskago obshchestva, po obshchei geografii*, **1** (1867), written with K. V. Struve; "Poezdka po Vostochnomu Tarbagatayu letom 1864 goda" ("Trip Through the Eastern Tarbagatay in the Summer of 1864"), *ibid.*, written with K. V. Struve; *Ocherki Severo-Zapadnoy Mongolii* ("Sketches of Northwestern Mongolia"), 4 pts. (St. Petersburg, 1881–1883); *Tangutsko-Tibetskaya okraina Kitaya i Tsentralnaya Mongolia* ("The Tangut-Tibetan Border of China and Central Mongolia"), 2 vols. (St. Petersburg, 1893); and "Poezdka v srednyuyu chast Bolshogo Khingana, letom 1899 goda" ("Trip to the Middle Part of the Great Hsingan in the Summer of 1899"), in *Zapiski Imperatorskago russkago geograficheskago obshchestva, po obshchei geografii*, **37**, no. 5 (1901).

II. SECONDARY LITERATURE. On Potanin and his work, see L. S. Berg, *Vsesoyuznoe Geograficheskoe obshchestvo za sto let* ("The All-Union Geographical Society for One Hundred Years"; Moscow–Leningrad, 1946); Y. N. Bessonov and V. Y. Yakubovich, *Po Vnutrenney Azii* ("Through Inner Asia"; Moscow, 1947); M. A. Lyalina, *Puteshestvie po Kitayu, Tibetu i Mongolii* ("Travels Through China, Tibet, and Mongolia"; St. Petersburg, 1898); and V. A. Obruchev, *Grigory Nikolaevich Potanin. Zhizn i deyatelnost* (". . . Life and Work"; Moscow–Leningrad, 1947); and *Puteshestvia Potanina* ("Potanin's Travels"; Moscow, 1953).

Y. A. DEMIDOVICH

POTT, JOHANN HEINRICH (*b*. Halberstadt, Saxony, 1692; *d*. Berlin, Prussia, 29 March 1777), *technical chemistry.*

Intended by his parents to study theology, Pott entered the University of Halle but soon turned to science, working under Friedrich Hoffmann and Georg Ernest Stahl. Both were professors of medicine who, in their teaching, separated chemistry from medical theory and practice. Pott was thus able to study chemistry as a subject in itself. He became professor of theoretical chemistry at the Collegium Medico-Chirurgicum in Berlin, where he succeeded Caspar Neumann in 1737 as professor of practical chemistry and director of the royal pharmacy. His career was characterized by great industry, wide reading, and fondness for personal controversy.

Pott's principal contribution to chemistry was in the systematic examination of mineral substances. He extended knowledge of several metals, at a time when the traditional notion of a fixed number of metals was changing. Zinc, which had long been used as a constituent of brass, had only recently been isolated in sufficient quantity for detailed study. Pott characterized it as a distinct metal and prepared a number of its compounds, including zinc sulfate, showing that white vitriol was a compound of zinc and sulfuric acid. He described bismuth fully and added to knowledge of its compounds and those of borax, alkalies, and alkaline earths. Although useful and meritorious, this work did not raise Pott above the level of many competent contemporaries; it was the current interest in porcelain that brought out his true talent.

The introduction of porcelain to Europe in the seventeenth century stimulated attempts to imitate its hardness and translucence. The secret of its manufacture, relying on the use of a flux (alabaster, marble, or feldspar) with the otherwise infusible clay, was worked out chiefly by J. F. Böttger (1682-1719) at Meissen. The processes were kept so strictly secret by Böttger's master, Frederick Augustus I of Saxony, that Meissen proved as impossible to duplicate as Chinese porcelain. Instructed by the king of Prussia to study the problem, Pott embarked on an extremely elaborate systematic investigation.

Pott reportedly made over 30,000 experiments on all manner of substances and mixtures subjected to heat in the furnace, and the results were published in his *Lithogeognosia* (1746). The elaborate tables of reactions in this work, quite apart from their relevance to manufacture, were a notable contribution to chemical analysis "in the dry way." Mineral chemistry was just entering a period of rapid development made possible through analysis by the blowpipe and other related means. Moreover, Pott's enormous repertoire of reactions was a model of comprehensiveness in chemical study and showed how relative reactivity could be found from planned, interrelated analysis. Although Pott did not speculate about affinity, his tables contributed greatly to that field of chemical theory.

Pott did not finally succeed in making porcelain, and that perhaps is the reason why Frederick the Great gave Marggraf (a younger man) preferment over Pott for the Berlin Academy. In 1754 Pott gave up chemical work, broke with the Berlin Academy, and burned his papers.

Pott held that phlogiston was the matter of fire but not of light. In his early practical writings he told his readers that he expected them to become familiar with current theories before reading his new work. His later writings were controversial and unproductive. By the time he reached old age, the tradition in which he had been brought up was about to be swept away by the systematic, experimental approach to chemistry that he had helped to create.

BIBLIOGRAPHY

I. ORIGINAL WORKS. Pott's main work is *Lithogeognosia: Chymische Untersuchungen* (Potsdam, 1746; Berlin, 1757), also trans. into French as *Lithogéognosie* (Paris, 1753). Other of his writings are collected in *Exercitationes chymicae* (Berlin, 1738), French trans., with additions, by J. F. Demachy (Paris, 1759); and *Observationum et animadversionum chymicarum*, 2 vols. (Berlin, 1739–1741).

II. SECONDARY LITERATURE. See J. H. S. Formey, *Histoire de l' Académie royale des sciences et belles lettres de Berlin* (Berlin, 1777), 55; J. R. Partington, *History of Chemistry*, II (London, 1961), 717–722, with bibliography and extensive notes; and J. R. Spielmann, *Instituts de chymie*, II (Paris, 1770), 409–417, with bibliography.

FRANK GREENAWAY

POUCHET, FÉLIX-ARCHIMÈDE (*b*. Rouen, France, 26 August 1800; *d*. Rouen, 6 December 1872), *biology, natural history.*

The son of an industrialist, Pouchet qualified in medicine in 1827 after studying at Rouen and at Paris. Almost immediately he became director of the Muséum d'Histoire Naturelle at Rouen, an institution with which he was associated throughout the remainder of his life. In addition Pouchet held teaching posts in Rouen, notably at the École Supérieure des Sciences et Lettres and the École de Médecine. A member of many learned societies and corresponding member of the Académie des Sciences, he became *chevalier* of the Legion of Honor in 1843.

A prolific author, Pouchet covered many areas of botany, zoology, physiology, and microbiology. He was also history-minded, writing, for instance, *Histoire des sciences naturelles au moyen age* (Paris, 1853). Widely read and, on many topics, of independent thought, Pouchet was also an excellent popularizer of science. Notable was his profusely illustrated general biology book, *L'univers* (Paris, 1865). "My sole object in writing this," Pouchet commented in the preface, "was to inspire and to extend to the utmost of my power a taste for natural science." In this, so far as can be judged, he was successful; certainly the English edition was very popular. A great deal of Pouchet's more specialized biological writing awaits detailed assessment, but it undoubtedly contains much of value: for instance, his clear recognition that human ovulation occurs within a limited period in the menstrual cycle.

In view of Pouchet's wide-ranging contributions to biology, it is unfortunate that he is often remembered only as the defeated adversary of Pasteur over the question of whether microorganisms could be spontaneously generated, although this was the most fundamental problem with which Pouchet dealt. Many other workers were involved in the fierce, often bitter, spontaneous generation controversies which, at least in France, reached a crescendo during 1858–1864. Pouchet wrote much on the subject, but it was his *Hétérogénie ou traité de la génération spontanée basé sur de nouvelles expériences* (Paris, 1859) that did much to arouse widespread interest.

Pouchet held that three factors—putrescent organic matter, air, and water—were the absolutely necessary conditions for spontaneous generation, which could be aided by, for instance, electricity and sunlight. Specifying the effects of light, Pouchet said that red light promoted the formation of animal "proto-organisms" and green light, vegetable "proto-organisms."

Pouchet naturally took great pains over the basic issue—whether contamination by existing organisms could account for apparent instances of spontaneous generation. He paid particular attention to airborne particles ("atmospheric micrography," as he called it), collecting them by such means as passing air through distilled water and obtaining deposits. From subsequent microscopic examination of the particles he recorded the presence of starch, cloth fibers, and carbon and mineral particles. He also concluded that the air contained only an occasional fungal spore or encysted infusorian, thus making airborne contamination highly unlikely. The fact that spontaneous generation apparently followed, under appropriate conditions, the use of heat-"sterilized" air also supported this view. Pouchet reported a great many experiments of his own and of others in which microorganisms appeared in nutrient media although rigorous attempts were made to avoid contamination.

Pasteur soon became the leading opponent of spontaneous generation, publishing his "Mémoire sur les corpuscules organisés qui existent dans l'atmosphère" in 1861. In it he claimed to have demonstrated that deposits from filtered air contained microorganisms or, at least, "organized corpuscles," and that their removal by heat or filtration meant that no growth occurred in sterilized nutrient media. This formed an important part of Pasteur's evidence for his outright repudiation of spontaneous generation. Although a commission of the Académie des Sciences accepted Pasteur's results in 1864, the controversies lasted until the late 1870's, largely because of the conflicting experimental evidence that continued to appear. The lasting value of the work of Pouchet and his supporters, now no longer held to be valid, was that it gave a great impetus to improving experimental technique in microbiology, which contributed to the rapid development of the subject during the last decades of the nineteenth century.

BIBLIOGRAPHY

I. ORIGINAL WORKS. The majority of Pouchet's many publications are listed in a useful article on Pouchet in J. Roger, *Les médecins normands* (Paris, 1890), 221–229. Publications up to 1862 are also listed in a shorter notice in Hoefer, *Nouvelle biographie générale*, XL (1866), 911. Full references to the spontaneous generation debate can be found in G. Pennetier, *Un débat scientifique. Pouchet et Pasteur* (Rouen, 1907). Pouchet's most significant work not cited in text is *Nouvelles expériences de génération spontanée et la résistance vitale* (Paris, 1864), a sequel to the *Hétérogénie* of 1859. Prior to these two books his most celebrated study was *Théorie positive de l'ovulation spontanée et la fécondation des mammifères et de l'espèce humaine* (Paris, 1847), which won a physiology prize of the Académie des Sciences in 1845.

II. SECONDARY LITERATURE. Pouchet's ovulation studies are noted by J. Rostand in "Félix-Archimède Pouchet et les méthodes contraceptives," in *Revue d'histoire des sciences et de leurs applications*, **22** (1969), 257–258. For some perspective on Pouchet's studies on airborne organisms, see J. K. Crellin, "Airborne Particles and the Germ Theory: 1860–1880," in *Annals of Science*, **22** (1966), 49–60.

J. K. CRELLIN

POUILLET, CLAUDE-SERVAIS-MATHIAS (*b.* Cusance, Doubs, France, 16 February 1790; *d.* Paris, France, 13 June 1868), *physics, technology.*

After attending the *lycée* of Besançon, Pouillet taught for two years at the *collège* of Tonnerre. From

1811 to 1813 he was a student at the École Normale Supérieure, to which he returned from 1815 until 1822 as *maître de conférences* in physics. During this phase of his career he also taught physics at the Athénée in Paris and at the Collège Royal de Bourbon, now the Lycée Condorcet (1819–1829). In 1826 he became assistant professor of physics at the Faculty of Sciences in Paris, first under Gay-Lussac and then under Dulong. Upon the latter's death in 1838, Pouillet assumed the chair of physics, which he held until 1852, when he was removed for refusing to swear an oath of allegiance to the imperial government.

Pouillet's lectures—which were partially collected in his *Éléments de physique expérimentale et de météorologie* (1827) and in the *Leçons de physique de la Faculté des sciences* (1828)—were widely read. (Pouillet published a popular account of *Éléments* in 1850.) Although offering no spectacular novelties, they presented, in clear language, a survey of the state of the various branches of physics and of recent developments in them. Simultaneously, Pouillet held important posts at the Conservatoire des Arts et Métiers. Appointed assistant director and demonstrator of machines there in 1829, he became professor of "physics applied to the arts" and administrator—in effect, director—in 1831. A few years later he published the two-volume *Portefeuille industriel du Conservatoire . . .* (1834–1836), a collection of annotated technical drawings.

Pouillet, who had taught several sons of Louis Philippe, was a confirmed supporter of that monarch, whose programs he had faithfully supported as deputy from the Jura (1837–1848). Although he abandoned political activity during the Revolution of 1848, Pouillet suffered the repercussions of that event. On 15 June 1849 he was dismissed from his post as administrator following an attempted revolt organized by Alexandre Ledru-Rollin on the premises of the Conservatoire des Arts et Métiers and on 12 November 1852 he lost his professorship for his refusal to take the oath of allegiance. Moreover, saddened by the death of his two children (a son of seventeen in 1849 and a daughter of twenty in 1850), he wished to retire. He therefore devoted his last years to the experimental research in which he had been engaged since the start of his career and to the work of the Académie des Sciences, to which he had been elected in 1837.

A very active member of the Société d'Encouragement pour l'Industrie Nationale, Pouillet served as *rapporteur* at several industrial expositions. His most important work, however, is contained in the forty memoirs he published between 1816 and 1868. After devoting his first studies, under the direction of Biot, to diffraction and interference phenomena, he turned to electrical and thermal phenomena. He experimented, before Regnault, on expansion and compressibility of gases. He also improved methods for measuring high temperatures and undertook to measure solar heat and atmospheric absorption (1837). In addition, he refined techniques for measuring weak currents through the introduction of the tangent galvanometer and sine galvanometer; hence he was able in 1839 to verify, with a very high degree of precision, Ohm's law of resistance (1827). In fact, some authors have wrongly attributed the law itself to Pouillet. It is true, however, that he helped considerably to make it more widely known. Pouillet's last research dealt mainly with atmospheric electricity and the construction of lightning rods, as well as with terrestrial magnetism. In sum, his works contained many individual advances but presented no major innovations.

BIBLIOGRAPHY

I. Original Works. Pouillet's principal publications are listed in Poggendorff, II, cols. 512–513, and III, 1063; in *Catalogue général des ouvrages imprimés de la Bibliothèque nationale*, CXLI (1937), cols. 828–833; and in the Royal Society *Catalogue of Scientific Papers*, IV, 998–999; VIII, 652.

Pouillet published three important books: *Éléments de physique expérimentale et de météorologie*, 2 vols. in 4 pts. (Paris, 1827; 7th ed., 3 vols., Paris, 1856), German trans. by J. H. J. Müller (Brunswick, 1842); *Portefeuille industriel du Conservatoire des arts et métiers, ou atlas et description des machines, appareils . . .*, 2 vols. (Paris, 1834–1836), written with V. Leblanc; and *Notions générales de physique et de météorologie à l'usage de la jeunesse* (Paris, 1850).

His physics course was published in A. Grosselin, ed., *Leçons de physique de la Faculté des sciences . . .*, II (Paris, 1828).

II. Secondary Literature. Only a few articles and studies have been devoted to Pouillet's life and work: G. Vapereau, in *Dictionnaire universel des contemporains*, 2nd ed. (Paris, 1861), 1420; E. Becquerel and H. Sainte-Claire-Deville, *Discours prononcé aux funérailles de M. Pouillet* (Paris, 1868); P. A. Bertin-Mourot, in *Caisse de secours mutuels des anciens élèves de l'École normale, 24e réunion générale annuelle* (Paris, 1869), 10–12; and the unsigned account in *Cent cinquante ans de haut enseignement technique au Conservatoire national des arts et métiers* (Paris, 1970), 3, 20, 25, 27, 33, 40, 78–80.

René Taton

POURFOUR DU PETIT, FRANÇOIS (*b.* Paris, France, 24 June 1664; *d.* Paris, 18 June 1741), *physiology, surgery.*

Although he is said to have shown a flair for science, it would appear that the early, conventionally classical education of Pourfour du Petit was something of a failure. On leaving school, therefore, he traveled and undertook private study before enrolling at the University of Montpellier, from which he received his medical degree in 1690. Before practicing, however, Pourfour du Petit continued his medical and scientific studies in Paris, and completed his surgical training at the Charité hospital. For extended periods until the Treaty of Utrecht (1713), he served as physician-surgeon in the armies of Louis XIV, and it was during this war service that he carried out some of his most important physiological investigations. On leaving the army, Pourfour du Petit returned to Paris and established himself as an eye specialist, although he was also esteemed for his more generally scientific pursuits; and in 1722 he was elected as a member of the Académie des Sciences.

Apart from skillfully removing cataracts and designing ophthalmic instruments, Pourfour du Petit is known for a number of important anatomical discoveries, including that of the canal between the anterior and posterior suspensory ligaments of the lens of the eye.

The physiological experiments with which Pourfour du Petit is especially associated were carried out at Namur between 1710 and 1712, and at Paris during the mid-1720's. In *Trois lettres d'un médecin* (1710) he described the head wounds and symptoms of paralysis of soldiers brought to him as patients. After many observations and postmortem dissections, the results of which he confirmed on dogs, he concluded that the movements of a limb are effected by animal spirits supplied by the side of the brain opposite the limb and that paralysis is complete only after the destruction of the contralateral corpus striatum. Pourfour du Petit's other experiments were concerned with the origin of the sympathetic—or, as it was called then, intercostal—nerve. Through an excusable misunderstanding, seventeenth-century anatomists, including Thomas Willis, the leading English authority on the anatomy of the brain and nervous system, conceived this nerve chain as an outflow of the fifth and sixth cranial nerves. Through brilliant experiments on dogs, Pourfour du Petit showed that whatever the site of superficial origin of the sympathetic chain, it was not in the cranium. It is remarkable that, although his results were definitive, they were largely ignored until the nineteenth century.

The experiments on the sympathetic nerve were carried out in Namur at the Hôpitaux du Roi in 1712 and at the Académie des Sciences in 1725. Pourfour du Petit's method of investigation, which originated at the time of Galen, consisted of dividing or ligating the trunk of a nerve and observing the syn-

dromes. In this case the nerve selected was the extension of the sympathetic nerve in the neck, and the observed syndromes were those of the eye and surrounding tissue.

In the Namur experiments Pourfour du Petit observed, after cutting the cervical extension of the sympathetic nerve, that the affected eye became filled with tears and that the nictitating membrane encroached upon the cornea. This effect showed that the supply of "animal spirits" normally controlling these phenomena passes upward into the head through the nerve in the neck. The second series of experiments revealed other syndromes that had escaped his notice: inflammation of the conjunctiva and contraction of the pupil. Hence Pourfour du Petit concluded that the control of the nictitating membrane, of the lacrimal points (which normally drain off liquid from the eye), of the expansion of the pupil, and of the supply of blood to the eyeball originates in that nerve in the neck now recognized as the uppermost trunk of the orthosympathetic system. Thus he demolished the erroneous view that the sympathetic system was an outflow of one of the cranial nerves.

BIBLIOGRAPHY

I. ORIGINAL WORKS. The primary sources for Pourfour du Petit's original research are: *Trois lettres d'un médecin sur un nouveau système du cerveau* (Namur, 1710); "Mémoire sur les yeux gelés, dans lequel on détermine la grandeur des chambres qui renferment l'humeur aqueuse" (1723), in *Mémoires de l'Académie Royale des Sciences* (1753), 38–54; "Mémoires sur plusieurs découvertes faites dans les yeux de l'homme, des animaux à quatre pieds, des oiseaux et des poissons" (1726), *ibid.* (1753), 69–83; "Mémoire dans lequel on détermine l'endroit où il faut piquer l'oeil dans l'opération de la cataracte" (1726), *ibid.*, 262–272; "Mémoire dans lequel il est démontré que les nerfs intercostaux fournissent des rameaux qui portent des esprits dans les yeux" (1727), *ibid.* (1739), 1–19 (this describes the series of experiments of 1712 and 1725 respectively); "Pourquoi les enfants ne voyent pas clair en venant au monde, et quelque temps après qu'ils sont nés" (1727), *ibid.*, 246–257; "Démontrer que l'uvée est plane dans l'homme" (1728), *ibid.* (1753), 206–224; and "Différentes manières de connoître la grandeur des chambres de l'humeur aqueuse dans les yeux de l'homme" (1728), *ibid.*, 289–300 (this describes Pourfour du Petit's ophthalmometer).

II. SECONDARY LITERATURE. Biographies of Pourfour du Petit are in *Biographie Universelle*, L. G. Michaud, ed., XXXIII (Paris, 1823), 500–501; Poggendorff, II, 415; *Biographisches Lexikon der hervorragenden Ärzte*, August Hirsch, ed., IV (Berlin–Vienna, 1932), 567–568. For a longer biography see J. J. Dortous de Mairan, *Éloges des académiciens de l'Académie royale des sciences morts dans les années 1714, etc.* (Paris, 1747). Most general histories of medicine make passing reference to Pourfour du Petit's neurological research. For a fuller study of his research see A. E. Best, "Pourfour du Petit's Experiments on the Origin of the Sympathetic Nerve," in *Medical History*, **13** (1969), 154–174.

A. E. BEST

POURTALÈS, LOUIS FRANÇOIS DE (*b.* Neuchâtel, Switzerland, 4 March 1823 or 1824; *d.* Beverly Farms, near Salem, Massachusetts, 17 or 18 July 1880), *oceanography.*

One of Louis Agassiz's favorite students, Pourtalès joined his master's "scientific factory" in Neuchâtel about the age of fifteen. He came to the United States in 1847 when Agassiz resettled there, and he prepared most of the illustrations for Agassiz and A. A. Gould's *Principles of Zoology* (Boston, 1848). Pourtalès joined the U.S. Coast Survey in 1848, and he served as assistant under three superintendents, making use of his education as an engineer by surveying and by heading the tidal division after 1854. The death of his father about 1870 brought Pourtalès the title of count and financial independence. He resigned from the Coast Survey and returned to Massachusetts in 1873.

In his choice of a successor to direct the Museum of Comparative Zoology that he had founded at Harvard, Agassiz vacillated between Pourtalès and his son Alexander. After his death, Pourtalès became "keeper" and Alexander Agassiz, "curator." Between them they presided over one of the world's greatest museums of natural history, with Pourtalès responsible for much of the administrative work from 1873 until his brief, final illness in 1880. On his frequent trips home to Switzerland, Pourtalès often visited England, where he was well known among naturalists for his deep-sea work. Pourtalès was exceptionally modest and unassuming. He married Elise Bachmann of Boston; the couple had one daughter.

Pourtalès was a pioneer in two fields where America was abreast of Europe: marine biology and submarine geology. Upon taking over the Coast Survey in 1844, Bache turned it from hydrography to oceanography by insisting on the preservation of the specimens brought up by the sounding lead. Because the lead brought up only small samples of the bottom, the first fruits of Bache's new policy came in the study of sediments. Using the microscope, J. W. Bailey of West Point characterized the sediments obtained from Coast Survey vessels. When Bailey died in 1857, Pourtalès, who had earlier studied the foraminifera, inherited the entire work. The chart that he prepared in 1870, from nine thousand samples, showed the distribution of bottom sediments along the coast between Cape Cod and Florida. Pourtalès demonstrated that glaciers had extended offshore as far south as New Jersey, and his results have not been greatly improved upon by the research of the 1960's.

In addition to studying sediments, Pourtalès collected whatever marine fauna he could find, especially corals. At Bache's invitation, Louis Agassiz had begun collecting in 1847 from Coast Survey vessels in shallow water off Massachusetts. In 1851 Agassiz accompanied Pourtalès on a Coast Survey party to the Florida reefs, where Pourtalès collected sipunculids and holothurians, on which he later reported. With this background in shallow-water forms, Pourtalès was ready to extend his researches into deeper water when after the Civil War the Coast Survey resumed operations under Louis Agassiz's close friend Benjamin Peirce. In the steamers *Corwin* (1867) and *Bibb* (1868, 1869), Pourtalès extended the technology of dredging, hitherto confined to shallow water, to depths as great as 850 fathoms off the east coast of Florida. These pioneering dredgings—slightly earlier in time although not as deep as those of W. B. Carpenter and C. W. Thomson in Britain—led to the "opening of a new era in zoological and geological research" (L. Agassiz, *Report of the Superintendent of the U. S. Coast Survey for 1869,* Appendix 10 [Washington, D.C., 1872], 208). Accompanied by Louis Agassiz on the last of these three cruises, Pourtalès based his most distinguished work, *Deep-Sea Corals* (1871), on the collections he had made. He also specialized among the coelenterates in halcyonarians and among the echinoderms in crinoids and holothurians.

Like his colleagues on both sides of the Atlantic, Pourtalès was astonished by the abundance of living forms at depths previously believed to be largely uninhabited. He eagerly joined the planning for what was to be the great American dredging expedition: the cruise of the new Coast Survey steamer *Hassler* westward around South America. Louis Agassiz led the 1871–1872 expedition, with Pourtalès in charge of dredging. Although *Hassler*'s cruise failed—because of faulty equipment—to achieve its hoped-for results, it inspired Carpenter to promote in Great Britain the far more successful *Challenger* Expedition (1872–1876). Pourtalès' researches, left unfinished at his death, are commemorated in the Pourtalès Plateau, an area off southeast Florida rich in corals, and the sea urchin *Pourtalesia*, named in 1869 by Alexander Agassiz and found by H.M.S. *Challenger* to be one of nature's most widely distributed genera. Pourtalès was a member of the National Academy of Sciences and the American Academy of Arts and Sciences.

BIBLIOGRAPHY

I. ORIGINAL WORKS. A list of Pourtalès' major publications appears in the longer version of the obituary notice by Alexander Agassiz, cited below. The two most important works are *Deep-Sea Corals* (Cambridge, 1871) and "Der Boden des Golfstroms und der Atlantischen Kuste Nord Amerika's," in *A. Petermanns Mitteilungen aus J. Perthes geographische Anstalt*, **11** (1870), of which an English version, without the essential colored charts, was published as *Report of the Superintendent of the U. S. Coast Survey for 1869*, app. 11 (Washington, D.C., 1872). Publications other than those listed by Agassiz may be found in *Coast Survey Reports* (1853–1872) and in *Bulletin of the Museum of Comparative Zoology at Harvard College*. Pourtalès' correspondence is among the Coast Survey records in the U. S. National Archives and in the Museum of Comparative Zoology.

II. SECONDARY LITERATURE. Contemporary obituaries are Alexander Agassiz, *Proceedings of the American Academy of Arts and Sciences*, **16** (1881), 435–443; and National Academy of Sciences, *Biographical Memoirs*, V (Washington, D.C., 1905), 79–89, of which a shorter version is in *American Journal of Science*, 3rd ser., **20** (1880), 253–255; and *Nature*, **22** (1880), 371–372. Agassiz gives Pourtalès' birth date as 1824; his assistant H. L. Clark, in *Dictionary of American Biography*, gives 1823. See also Theodore Lyman, in *Proceedings of the Boston Society of Natural History*, **21** (1880), 47–48; Henry N. Moseley, in *Nature*, **22** (1880), 322–323; and P. Martin Duncan, in *Nature*, **22** (1880), 337.

Modern evaluations of Pourtalès' achievements are Thomas J. M. Schopf, *Atlantic Continental Shelf and Slope of the United States—Nineteenth Century*, U.S. Geological Survey Professional Paper 529-F (Washington, D.C., 1968); and Rudolf S. and Amelie H. Scheltema, "Deep-Sea Biological Studies in America, 1846 to 1872— Their Contribution to the *Challenger* Expedition," in *Proceedings of the Royal Society of Edinburgh*, B ser., **72** (1971–1972), 133–144; these achievements are set in a wider historical context in Susan B. Schlee, *The Edge of an Unfamiliar World. A History of Oceanography* (New York, 1973), chs. 1–3.

HAROLD L. BURSTYN

POWALKY, KARL RUDOLPH (*b.* Neudietendorf, near Gotha, Germany, 19 June 1817; *d.* Washington, D.C., 11 July 1881), *astronomy.*

The details of Powalky's scientific education are unknown. He received his practical training as assistant, from 1842 to 1847, at the Hamburg observatory, then directed by K. L. C. Rümker, whose student he later called himself. From 1850 to 1856 he was at the duke of Mecklenburg's private observatory at Seeberg, near Gotha, where he had an important part in the computation of P. A. Hansen's lunar and solar tables.

In the summer of 1856 he moved to Berlin, where he won deserved acclaim with his calculations for the *Berliner astronomisches Jahrbuch.* Powalky received his doctorate in 1864, at the age of forty-seven; his dissertation, submitted to the University of Kiel, was entitled "De transitu stellae Veneris ante discum solis anno 1769 peracto ad solis parallaxin accuratius determinandam." In this work he derived a solar parallax between 8.832" and 8.86", which agreed very closely with the value of 8.86" previously obtained by Foucault. This finding led to a controversy between Powalky and Le Verrier. Powalky left Berlin in 1873 for Washington, where he worked at the Bureau of the Census and at the Naval Observatory. He was, however, never able to find a post in the United States that fully matched his desires.

After taking part at Hamburg in observations of stellar occultations from 1845 to 1847, Powalky turned exclusively to theoretical work. He reduced Rümker's observations of Halley's comet (1835III) and computed the elements of the comets 1846III (Brorsen), 1846V (Hind), 1846VIII, 1847V (Brorsen-Metcalf), and 1847VI. Later, at Berlin, he made corresponding calculations for the comets 1858VIII (Encke) and 1860III. It was through Powalky's ephemerides that W. J. Foerster sighted Encke's comet on 7 August 1858 at Berlin.

Also at Berlin, Powalky engaged in extensive theoretical research. In 1863 the *Berliner astronomisches Jahrbuch* presented solar ephemerides computed on the basis of the sun charts of Hansen and Olufsen. Considering such data important for determining orbits of the minor planets, Powalky published corrections of the solar positions for 1845–1862. In particular he calculated the following ephemerides for the Berlin yearbook: the moon (using Hansen's tables, 1863–1875), the sun (1871–1875), Saturn and Uranus (1860), Mercury to Uranus (1861), Mercury to Neptune (1862–1868), Jupiter to Neptune (1871–1872), Mars to Neptune (1873–1875), the satellites of Jupiter (1870, 1873–1875), Saturn's rings (1868–1870, 1873–1875), eclipses (1871–1872), the transit of Venus (1874), star positions (employing J. P. Wolfer's *Tabulae reductionum*, 1862–1868), and stellar occultations (1868–1876).

Especially interested in the minor planets, Powalky based his orbital computations on formulas taken from Hansen's "Theorie der absoluten und speziellen Störungen der kleinen Planeten" (1853–1859). Between 1856 and 1881 Powalky published computations of the orbital elements and ephemerides of eleven minor planets. In 1859 he identified a star missing in Bessel's zone with the planetoid Amphitrite. For the *Berliner astronomisches Jahrbuch* he calculated the annual ephemerides of Juno (1860–1869, 1873) and computed those of Ceres (1873), Pallas (1873), Vesta (1868, 1873), and Astraea (1871, 1873). At various times between 1861 and 1883 he worked out the ephemerides for Fortuna (1860–1875), Harmonia (1860–1868, 1872, 1873), Nysa, Aglaia, Doris, and Pales (1861–1875), as well as those for thirty-two other minor planets. Even while in Washington, Powalky continued to contribute these computations to the *Berliner astronomisches Jahrbuch* until his death.

At Berlin, Powalky made a critical examination of the Greenwich observatory's 1852 lunar observations and compared them with Hansen's moon charts. He also published a reduction of the observations of the fixed stars made by Rümker between 1826 and 1828 at Paramatta, Australia. In connection with his dissertation, he made several studies of phenomena pertaining to the transits of Venus and to the determination of solar parallax.

BIBLIOGRAPHY

I. ORIGINAL WORKS. Powalky's writings are listed in *Generalregister der Astronomische Nachrichten, Bände 41 bis 80* (Kiel, 1938), 78–79; and . . . *Bände 81 bis 120* (Kiel, 1891), 91; and in Poggendorff, III, 1063. Among them is "Neue Untersuchung des Venusdurchgangs von 1769 zur Bestimmung der Sonnenparallaxe," in *Schriften der Universität Kiel*, **11**, no. 1 (1865). While in Washington he published "The Combination of the Different Results of Various Series of Observations," in *Monthly Notices of the Royal Astronomical Society*, **34** (1874), 476–479; "Die Elemente der Sonnenbahn und die Massen der Planeten Venus und Mars," in *Astronomische Nachrichten*, **88** (1876), 257–276; "Reduction von Lacaille's AR-Bestimmungen südlicher Sterne durch correspondierende Höhen (1751–1752)," *ibid.*, **89** (1877), 183–190, 193–204; **90** (1877), 21–28; and "Comparison of the Observations of the Sun Made at Washington, 1866–1875, With Hansen's Tables," in *Monthly Notices of the Royal Astronomical Society*, **41** (1881), 1–17.

II. SECONDARY LITERATURE. A short biography is Günther, "Karl Rudolf Powalky," in *Allgemeine deutsche Biographie*, XXVI (1888), 494. An obituary is A. Hall, "Carl Rudolph Powalky," in *Astronomische Nachrichten*, **100** (1881), 159.

DIEDRICH WATTENBERG

POWELL, BADEN (*b.* Stamford Hill, England, 22 August 1796; *d.* London, England, 11 June 1860), *physics.*

Powell was the eldest son of Baden Powell of Langton, Kent. After private education he entered Oriel College, Oxford, in 1814 and graduated in 1817 with a first-class degree in mathematics. After ordination in 1820 he was curate at Midhurst, Sussex, and, from 1821 to 1827, vicar of Plumstead, Kent. The quality of his early researches in optics and radiant heat, which he performed at Plumstead, was recognized by his election as a fellow of the Royal Society in 1824 and by his appointment as Savilian professor of geometry at Oxford in 1827. He remained a prominent and respected figure at Oxford until 1854, when, despite the fact that he was to retain his chair until his death, he took up residence in London. He was married three times—in 1821, 1837, and 1846—and had fourteen children. One of his sons by his last marriage was Robert Baden-Powell, later Lord Baden-Powell of Gilwell, the founder of the scouting movement.

Most of Powell's scientific research dates from the 1820's and 1830's. His best-known experiments were those on the heating effect produced beyond the red end of the solar spectrum (described in *Philosophical Transactions of the Royal Society* in 1825 and 1826). Powell used his experiments on the transmission of radiant heat and light through glass screens as the basis for some unusual speculations on the nature of radiation. According to Powell, all sources, whether luminous or not, emitted radiant heat of a kind that could be intercepted by a glass screen; but once a body became luminous, it began to exercise an additional and quite separate "heating power," characterized by the ability to pass through glass.

Although in this way he associated at least a part of the heating effect with luminosity, Powell left his conjecture unexplained and certainly did not commit himself to John Leslie's view that radiant heat and light had a common cause. Powell's theory won little support and, as evidence of the identity of radiant heat and light accumulated in the 1830's and 1840's, it was ignored. Indeed, once J. D. Forbes had demonstrated the polarization of radiant heat in 1835 (an observation that Powell had failed to make in his experiments on polarization some five years earlier), Powell's own conviction weakened and he became increasingly sympathetic to the view that radiant heat, like light, consisted of vibrations in an all-pervading ether.

Of Powell's other experimental work, the most notable was the investigation of the dispersion of light (described in the *Philosophical Transactions of the Royal Society* between 1835 and 1838), which he used to support the wave theory as treated by Cauchy.

Powell's scientific reputation was not founded solely on his skill in original research. After 1830 he became even better known as a commentator on the work of others—for example, in the reports on the state of the study of radiant heat that he read to the British Association for the Advancement of Science in 1832, 1840, and 1854. As a public speaker he was noted for his lucidity and sedateness, and he was much in demand both as a lecturer on science and as a preacher. Powell wrote a number of elementary textbooks of mathematics, a popular *History of Natural Philosophy* (1834) for the Cabinet Cyclopaedia, and an exposition of the wave theory of light in 1841.

Throughout his life, but especially after 1850, Powell was involved in religious controversy. Opposed equally to the strict Sabbatarianism of the Evangelicals and to the Tractarians, he adopted a position on matters of doctrine similar to that later associated with the Broad Church movement. Like the Broad Churchmen, he maintained that the searching re-examination of traditional beliefs in the light of scientific knowledge did not endanger faith. Repeatedly he affirmed the irrelevance to Christian belief of the Old Testament (which he saw as little more than a Jewish parable) and argued, with Thomas Chalmers and against William Whewell, for the possibility of life on other planets. An admirer of Charles Lyell, Powell enthusiastically accepted the principle of the uniformity of nature and strongly criticized forms of natural theology that incorporated geological catastrophes and supposed suspensions of the normal laws of nature as evidence of divine intervention. Miracles, being deviations from uniformity, were inconceivable to Powell and merely reflected either the unreliability of the witness or man's ignorance of the true laws of nature.

Although Powell presented his views on miracles most fully in his contribution to *Essays and Reviews* (1860), in which he elaborately distinguished the "external accessories" of Christianity (like miracles) from the "essential doctrines," he had been no less committed to the principle of uniformity in the 1830's, when such a position was unorthodox among theologians. It was entirely characteristic of Powell's independence in Anglican circles (as also of his zeal for the principle of uniformity) that he was one of the first major theologians to treat Robert Chambers' *Vestiges of the Natural History of Creation* (1844) sympathetically, as he did in his *Essays on the Spirit of the Inductive Philosophy* (1855). Like Chambers, he maintained that belief in the uniformity of nature could be the basis for a new natural theology that would yield a far grander conception of the Creator than the old.

Powell was also active in educational reform, and throughout his tenure of the Savilian chair he was a forthright spokesman for science at Oxford. As a member of the Royal Commission of Enquiry into Oxford University between 1850 and 1852, he urged the extension of science teaching at the university, against strong opposition; and he must take much credit for the modest reforms of the 1850's.

BIBLIOGRAPHY

I. Original Works. Powell's most important researches are described in the following papers: "An Experimental Inquiry Into the Nature of Radiant Heating Effects From Terrestrial Sources," in *Philosophical Transactions of the Royal Society*, **116** (1825), 187–202; "An Account of Some Experiments Relating to the Passage of Radiant Heat Through Glass Screens," *ibid.*, **117** (1826), 372–382; "Researches Towards Establishing a Theory of the Dispersion of Light," *ibid.*, **126** (1835), 249–254; **127** (1836), 17–20; **128** (1837), 19–24; **129** (1838), 64–72; "Remarks on the Theory of the Dispersion of Light as Connected With Polarization," *ibid.*, **129** (1838), 253–264, with supplement, **130** (1840), 157–160. He was also a frequent contributor to the *Philosophical Magazine*. His scientific and mathematical books, which were generally pitched at an elementary level, include: *The Elements of Curves* (Oxford, 1828); *A Short Treatise on the Principles of the Differential and Integral Calculus*, 2 vols. (Oxford, 1829–1830); *An Elementary Treatise on the Geometry of Curves and Curved Surfaces* (Oxford, 1830); *A General and Elementary View of the Undulatory Theory, As Applied to the Dispersion of Light and Some Other Subjects* (London, 1841). His *History of Natural Philosophy From the Earliest Periods to the Present Time* (London, 1830) is popularization of a rather different kind. The most interesting of his publications on educational matters is *The Present State and Future Prospects of Mathematical and Physical Studies in the University of Oxford* (Oxford, 1832). Many of his theological views are expounded in *The Connexion of Natural and Divine Truth* (London, 1838), but the fullest treatments are his contribution to *Essays and Reviews* (London, 1860) and three sets of essays: *Essays on the Spirit of the Inductive Philosophy* (London, 1855); *Christianity Without Judaism* (London, 1857); and *The Order of Nature Considered in Reference to the Claims of Revelation* (London, 1859).

II. Secondary Literature. The standard obituary of Powell is in *Monthly Notices of the Royal Astronomical Society*, **21** (1861), 103–105, repr. in *Proceedings of the Royal Society*, **11** (1860–1862), xxvi–xxix. A detailed if uncritical account of Powell's life and work appears in W. Tuckwell, *Pre-Tractarian Oxford* (London, 1909), 165–225; a thorough study has yet to be written. The critical unsigned article "Recent Latitudinarian Theology," in *Christian Remembrancer*, **38** (1860), 388–427, is useful for an understanding of Powell as a theologian.

Robert Fox

POWELL, CECIL FRANK (*b.* Tonbridge, Kent, England, 5 December 1903; *d.* near Bellano, Lake Como, Italy, 9 August 1969), *physics.*

The son of Frank Powell, a gunsmith, and Elizabeth Caroline Bisacre, Powell was educated at Judd School, Tonbridge, and Sidney Sussex College, Cambridge. After graduating in 1925 with first-class honors in the natural science tripos, he entered the Cavendish Laboratory, Cambridge, where he was influenced by Lord Rutherford and especially by C. T. R. Wilson, under whose supervision he worked for the Ph.D.; it was awarded in 1927 for a dissertation on condensation phenomena important in the operation of the Wilson cloud chamber. Powell went to Bristol University in 1927 as research assistant to A. M. Tyndall and remained there for the rest of his career. He was appointed Melville Wills professor of physics in 1948 and Henry Overton Wills professor and director of the H. H. Wills Physics Laboratory in 1964, a post from which he retired shortly before his death. He received the Nobel Prize in physics for 1950; was fellow and Hughes medalist (1949) and Royal medalist (1961) of the Royal Society, and Lomonosov gold medalist of the Soviet Academy of Sciences (1967); held honorary doctorates from six universities; and was honorary member of four academies of science.

Following some important work with Tyndall on ionic mobility in gases, Powell began in the late 1930's to investigate the possible use of photographic emulsions to record the tracks of fast-moving electrically charged particles. After development of the emulsion, blackened grains of silver appear in the positions of atoms that have been ionized by the fast particles, so that the particle trajectories are visible when the processed plate is examined under a suitable microscope. Powell and his collaborators transformed this method for estimating the masses, charges, and energies of the particles producing the tracks.

The power of the method was fully revealed when Powell exposed stacks of specially designed emulsion pellicles to cosmic radiation at mountain altitude. A new charged particle, the pi-meson or pion, of mass 273 times the mass of the electron, was discovered in 1947.

The existence of a particle of about this mass had been predicted by the Japanese physicist Hideki Yukawa in 1935, to explain the short-range attractive forces between neutrons and protons that hold atomic nuclei together. Physicists searching for this particle with Geiger counters or cloud chamber detectors had found a charged particle, the muon of mass 206 times the electron mass. By 1946, however, experiments had shown that the muon interacted only weakly with atomic nuclei and could not be the particle predicted by Yukawa. The pion discovered by Powell and his colleagues had the required properties, however. They showed also that it was unstable, with a lifetime of about 2×10^{-8} second, decaying into a muon and a neutrino.

The pion was the first of a large number of unstable elementary particles found since then, many by means of the photographic emulsion method. Their study dominates high-energy physics and has led to the discovery of important symmetry properties in nature: nonconservation of parity and charge conjugation in weak interactions like beta radioactivity, and unitary symmetry in strong interactions like those concerned with nuclear forces. Powell's work marks the beginning of elementary particle physics as it is known today.

Powell established a large international collaboration of many laboratories to study elementary particles in the cosmic radiation by means of large stacks of emulsion in free balloons, and continued to sponsor large-scale international collaboration in studying particle physics by use of accelerators. He played an important part in CERN, the laboratory of the European Organization for Nuclear Research, serving as chairman for three years of its Scientific Policy Committee. His passionate belief in science as the great transforming force of society and his deep commitment to the social responsibility of scientists led to his key role in the Pugwash Movement for Science and World Affairs, of which he was a founder, and also in the World Federation of Scientific Workers, of which he was president.

The research school that Powell built up at Bristol occupied a dominant position in particle physics during 1946–1956, prior to the building of the large accelerator laboratories, and has exerted a profound influence on the development of the field.

BIBLIOGRAPHY

Many of Powell's writings have been brought together in *Selected Papers of Cecil Frank Powell,* E. H. S. Burhop, W. O. Lock, and M. G. K. Menon, eds. (Amsterdam–London, 1972). His works include *The Cosmic Radiation* (Stockholm, 1951), his Nobel Prize address; *The Study of Elementary Particles by the Photographic Method* (London, 1959), written with P. H. Fowler and D. H. Perkins; "Cosmic Radiation," in *Proceedings of the Institute of Electrical Engineers,* **107B** (1960), 389–394, the Kelvin lecture; "Priorities in Science and Technology in Developing Countries," in *The Science of Science* (London, 1964), 71–92, speech given at Pugwash Conference, Udaipur, India; "The Role of Pure Science in European Civilization," in *Physics Today,* **18** (1965), 56–64, address to the Council of CERN; "Promise and Problems of Modern

Science," in *Nature*, **216** (1967), 543–546; and "The Nature of the Primary Cosmic Radiation," in *Scientific World*, **13**, (1969), 5–13, the Walther Bothe memorial lecture.

E. H. S. BURHOP

POWELL, JOHN WESLEY (*b.* Mount Morris, New York, 24 March 1834; *d.* Haven, Maine, 23 September 1902), *geology, ethnology.*

Powell's parents, Joseph and Mary Dean Powell, were Methodist immigrants from England bent upon carrying the Gospel to border cabins. As a consequence the family led a wandering life, moving always westward toward newer frontiers: in 1838 to Jackson, Ohio, near Chillicothe; in 1846 to South Grove, Wisconsin; in 1851 to Bonus Prairie, Illinois; the next year to Wheaton, Illinois; and finally, after John Wesley had broken away to follow his scientific inclinations rather than devote himself to the ministry for which his father had intended him, to Emporia, Kansas. The clarity with which Powell later understood the problems of western settlement derived to some extent from his boyhood frontier experience.

His education was often interrupted and in good part homemade; but a frontier man of learning, George Crookham of Jackson, early initiated Powell into natural history and into the habit of collecting trips in the field. Later, when he was working his father's Wisconsin farm, teaching in a country school, or snatching short periods of instruction at Illinois Institute (now Wheaton College), Illinois College, and Oberlin, Powell continued collecting. A persistent and omnivorous amateur at that stage, he made long solitary trips through Wisconsin, down the Mississippi to New Orleans, down the Ohio from Pittsburgh to Cairo, through the Iron Mountain country of Missouri, down the Illinois, up the Des Moines. In 1858 he was made secretary of the Illinois Society of Natural History. Two years later, as principal of schools in Hennepin, Illinois, he won a prize with his mollusk collection at the fair of the Illinois State Agricultural Society and seemed on his way to becoming a rural savant on the model of his instructor, Crookham.

The Civil War eased Powell into a wider world. Enlisting as a private in May 1861, he was a captain within six months and a member of Grant's staff, considered something of an authority on fortifications. Despite the loss of his right arm, shattered by a Minié ball at Shiloh, he remained in the army throughout the war, rising to the command of the artillery of the 17th Army Corps. In January 1865 he resigned with the rank of brevet lieutenant colonel. He had married a cousin, Emma Dean, in 1862.

Shortly after returning to civilian life, Powell accepted a professorship of geology at Illinois Wesleyan University, moving a year later to its sister institution, Illinois State Normal University. From there in the summer of 1867 he led a party of students to the Rocky Mountains under the sponsorship of the Illinois State Natural History Society. In 1868 he repeated the expedition, exploring west of the continental divide and wintering on the White River in western Colorado. There he conceived and prepared for his 1869 boat exploration down the unknown canyons of the Green and Colorado rivers from Green River, Wyoming, to the mouth of the Virgin. From that exploration, the last major one within the continental United States, he emerged a national hero; and when, on 12 July 1870, Congress created an early, informal version of what would become the Geographical and Geological Survey of the Rocky Mountain Region, parallel to and in competition with the King, Hayden, and Wheeler surveys already in the field, Powell was placed in charge of it.

He remained in charge of it throughout its nine years of existence; and out of its continuing investigation of the lands, water, and people of the plateau country of Utah, western Colorado, and northern Arizona he developed most of the broad principles upon which he built his later career. The career was multiple, involving a constant interaction among Powell's personal knowledge of the west, his active involvement in the nascent sciences of geology and American ethnology, and his increasing influence as a government scientist.

Powell's pioneering trip down the Green and Colorado was supplemented by a second, and scientifically more productive one, in 1871–1872. The geologists he enlisted in the later years of the survey were men of real stature; and in collaboration with them, especially with Grove Karl Gilbert, Clarence E. Dutton, and W. H. Holmes, he did much to formulate the basic principles of structural geology. His own publications, notably *The Exploration of the Colorado River of the West* (1875) and *The Geology of the Eastern Portion of the Uinta Mountains* (1876), as well as the reports and monographs of his collaborators, are of lasting scientific importance. Powell on antecedent and subsequent streams; Gilbert on stream erosion, recession of cliffs, and laccolithic uplifts; Dutton on isostasy and volcanism; and all of them together on the large problems of orogeny are still basic and indispensable after nearly a century. Their contributions were less individual than mutual. Both Gilbert and Dutton testified to the provocative fecundity of Powell's ideas and the impossibility of separating out individual contributions.

Knowing the West from years of field experience, Powell early came to feel that the land laws under which it was being settled were both destructive to the land and ruinous to the individual settlers. He believed in the role of government as a source of unbiased scientific information for the use of both citizens and lawmakers; and when President Hayes came into office in 1877 on a platform of reform, Powell seized the opportunity to present to Secretary of the Interior Carl Schurz a program for land laws and settlement policies appropriate to the West. His extremely important *Report on the Lands of the Arid Region* (1878) made use of the staff and the findings of his survey to state the conditions of the West and the necessary institutional and legal changes if it was to be settled without great individual hardship and disaster to the land. It proposed for the lands beyond the 100th meridian a revision of the near-sacred 160-acre formula of the Homestead Act in favor of smaller irrigated farms and very much larger grazing farms; the stopping of the rectangular surveys and the practice of contract surveying; the institution of surveys based upon drainage divides and the location of perennial water; and—in the interest of governmental efficiency—the transfer of the cadastral surveys to the Coast and Geodetic Survey from the General Land Office, and the consolidation of the four competing western surveys into a single bureau under the Interior Department.

In the ensuing struggle between the reformers and the forces engaged in western promotion, Congress put none of that program into effect except the consolidation of the western surveys into the U.S. Geological Survey. Having supported Clarence King for the directorship of the new bureau, Powell moved over to the Smithsonian Institution to direct the newly created Bureau of Ethnology. His *Report on the Lands of the Arid Region*, years ahead of the public and governmental acceptances of the times, did not become truly influential until the Dust Bowl years of the 1930's, when its principles became the basis for the attempt to heal conditions that might have been largely prevented if the *Report* had been acted on when it was presented.

As head of the Bureau of Ethnology, renamed the Bureau of American Ethnology in 1894, Powell put into practice the same collaborative and directive effort that had marked his administration of the Powell Survey. Even before he undertook to direct the governmental study of the Indians, he published his seminal *Introduction to the Study of the Indian Languages* (1877), building upon and greatly enlarging the pioneer studies of Albert Gallatin. His effort was to create first the alphabet for ethnological study, and then the systematic classification of the Indian tribes. Personally and through his collaborators James Pilling, Garrick Mallery, Cyrus Thomas, and others, he moved from classification of the language stocks to other sorts of classification and study. Pilling's bibliographies of the various language stocks grew out of an assignment from Powell. The reports and monographs of the Bureau of Ethnology, beautifully printed and carefully made, mark not only the systematization of the study of the Indian tribes but also its first notable achievements. And Frederick Webb Hodge's *Handbook of the American Indian* (1907–1910) was the summation of the work of many, all of them working under Powell's direction and guided by his powerfully synthesizing and organizing mind.

When Clarence King resigned as director of the U.S. Geological Survey in 1881, Powell succeeded him, without relinquishing his position as head of the Bureau of Ethnology. Thereafter for a dozen years, as head of two major bureaus, he was perhaps the most powerfully placed scientist in the United States.

The aridity which Powell saw as the compelling fact of western life—the fact which should enforce changes in laws and institutions as well as in patterns of agriculture and land use—was impressed upon even the most extreme western boosters by the ten-year drought of the 1880's. In consequence Powell's proposals, made in the *Arid Region* report, had a second chance to gain political support. Put in charge of a program of western irrigation surveys in 1888, Powell made use of the opportunity to throw all the resources of the new survey, as well as much of the effort of the U.S. Geological Survey, behind topographical mapping and land classification as the preliminary for the spotting of dam and canal sites. The bill authorizing the irrigation survey was loosely and ambiguously phrased. Upon interpretation by the attorney general, it proved to have withdrawn from settlement the entire public domain until Powell could complete his designation of reclamation sites and could certify the lands under them as irrigable and hence open to filing.

Possessed of sudden, unexpected power, Powell labored to finish his mapping, hoping to forestall in much of the West the unhappy consequences of dryland homesteading. But the so-called irrigation clique, headed by Senator William Stewart of Nevada, had in mind only a quick spotting of irrigation sites upon existing maps, with all the speculation such a program would give rise to. The cries from the West grew louder as Powell's work dragged on. In 1890 Stewart and his colleagues sharply cut the Irrigation Survey budget, and in the following year they pursued their feud with Powell by cutting the budget of the U.S. Geological

Survey as well. In 1894, defeated for the second time in his program for scientific planning of western settlement, Powell resigned from the Geological Survey. He devoted the rest of his life to the Bureau of American Ethnology and to the writing of philosophical treatises. Of these last the most important, although never influential upon the history of thought, was *Truth and Error, or the Science of Intellection* (1898).

As a thinker on the origins and patterns of society, Powell should be associated with the social Darwinism of Lester Ward, one of his employees and close collaborators, and with the systematic anthropological theories of Lewis Morgan, another of his close intellectual companions. His synthetic formulations have proved less durable than his contributions in geology and his sound organization of the science of man. Powell was one of the founders and systematizers of government science in the United States, and his bureaus proved to be models for many subsequent ones covering other fields of science. Above all else, his perception of the abiding aridity of much of the West, and of the importance of lack of water upon all the institutions of the men who must live there, entitles him to be called one of the prophetic pioneers.

BIBLIOGRAPHY

I. ORIGINAL WORKS. Powell's principal geological contributions were made in *Report on the Exploration of the Colorado River of the West and Its Tributaries* (Washington, D.C., 1875), and in *Report on the Geology of the Eastern Portion of the Uinta Mountains and a Region of Country Adjacent Thereto* (Washington, D.C., 1876). *Exploration of the Colorado River* has been repr. several times, most recently with an intro. by Wallace Stegner (Chicago, 1957). The important *Report on the Lands of the Arid Region of the United States*, 45th Congress, 2nd Session, HR Executive Document 73 (Washington, D.C., 1878), has been repr., also with an intro. by Wallace Stegner (Cambridge, Mass., 1962). Later developments of the ideas in the *Arid Region* report may be found in three articles: "The Irrigable Lands of the Arid Region," in *Century*, **39** (1890), 766–776; "The Non-Irrigable Lands of the Arid Region," *ibid.*, 915–922; and "Institutions for the Arid Region," *ibid.*, **40** (1890), 111–116.

Powell's ethnological publications are extensive, beginning with the seminal *Introduction to the Study of the Indian Languages* (Washington, D.C., 1877). Representative later studies are "Human Evolution," in *Transactions of the Anthropological Society of Washington*, **2** (1883), 176–208; "From Savagery to Barbarism," *ibid.*, **3** (1885), 173–196; and "From Barbarism to Civilization," in *American Anthropologist*, **1** (1888), 97–123. *Truth and Error, or the Science of Intellection* (Chicago, 1898) is the expression of Powell's philosophical system.

Further bibliographical information is in Darrah and in Stegner (see below). Powell's papers, including the extensive and valuable letter files of his two bureaus, are in the National Archives, Washington, D.C.

II. SECONDARY LITERATURE. For biographical information as well as estimates of Powell's scientific contributions, see Grove Karl Gilbert *et al.*, *John Wesley Powell, a Memorial* (Chicago, 1904); W. M. Davis, "Biographical Memoir of John Wesley Powell," in *Biographical Memoirs. National Academy of Sciences*, **8** (1915), 11–83; William Culp Darrah, *Powell of the Colorado* (Princeton, 1951); and Wallace Stegner, *Beyond the Hundredth Meridian: John Wesley Powell and the Second Opening of the West* (Boston, 1954).

The reports and monographs of the U.S. Geographical and Geological Survey of the Rocky Mountain Region (Powell, 1870–1879), as well as those of the U.S. Geological Survey (1881–1892) and the Bureau of [American] Ethnology (1879–1902), are important as revealing not only Powell's general interests but also, often, his specific ideas, passed on to collaborators.

WALLACE STEGNER

POWER, FREDERICK BELDING (*b*. Hudson, New York, 4 March 1853; *d*. Washington, D.C., 26 March 1927), *chemistry, pharmacy*.

The son of Thomas Power and the former Caroline Belding, Power received his early education at a private school and at the Hudson Academy. He worked for several years in a local drugstore and briefly in a Chicago pharmacy before obtaining a position in the establishment of the noted Philadelphia pharmacist Edward Parrish. He also attended the Philadelphia College of Pharmacy, from which he graduated in 1874, in the same class as his lifelong friend Henry Wellcome. After spending two more years with the Parrish firm, Power studied chemistry and the pharmaceutical sciences at the University of Strasbourg. Upon receiving his doctorate in 1880, he returned to America to teach analytical chemistry at the Philadelphia College of Pharmacy.

In 1883 Power went to the University of Wisconsin to serve as the first director of the newly created department of pharmacy. After guiding the Wisconsin pharmacy program through its infant years and placing it on a scientific footing that was to make it a leader in American pharmaceutical education, he left in 1892 to become scientific director of the chemical laboratories of Fritzsche Brothers in New Jersey. In 1896 Henry Wellcome established the Wellcome Chemical Research Laboratories in London and appointed Power to be director, a position he held until 1914. From 1916 until his death, Power was head of the phytochemical laboratory of the Bureau of Chemistry, U.S. Department of Agriculture.

Power married Mary Van Loan Meigs in 1883; they had two children. Among his honors were the Hanbury gold medal (1913), the Flückiger gold medal (1922), and election to the National Academy of Sciences.

Power's research was concentrated entirely in phytochemistry. He added significantly to the knowledge of plant chemistry by isolating and purifying numerous constituents from a host of different plant substances, and by determining the chemical structures of many of these constituents. Probably his best-known work involved the investigation of chaulmoogra oil, a traditional remedy for leprosy in the Orient, which entered Western medicine in the nineteenth century. Sulfones have since largely replaced chaulmoogra oil in the treatment of leprosy, and some investigators have questioned whether the oil is really of any therapeutic value. Power and his co-workers at the Wellcome Laboratories isolated from the oil two new fatty acids, chaulmoogric and hydnocarpic acids, which were considered to be the active ingredients. They also determined the structures of these two acids, although the formulas assigned have been modified. Other important studies include his researches on the constituents of the essential oils of nutmeg, of the cotton plant, and of certain fruits.

BIBLIOGRAPHY

I. Original Works. For a bibliography containing most of Power's works, see Ivor Griffith, "A Half Century of Research in Plant Chemistry: A Chronological Record of the Scientific Contributions of Frederick Belding Power," in *American Journal of Pharmacy*, 96 (1924), 605–613. Many of his publications are listed in Poggendorff, V, 1001, and VI, 2067–2068.

The work on the isolation and determination of the structure of chaulmoogric and hydnocarpic acids was reported in F. B. Power and F. H. Gornall, "The Constituents of Chaulmoogra Seeds," in *Journal of the Chemical Society*, 85 (1904), 838–851; F. B. Power and F. H. Gornall, "The Constitution of Chaulmoogric Acid. Part I," *ibid.*, 851–861; F. B. Power and M. Barrowcliff, "The Constituents of the Seeds of *Hydnocarpus Wightiana* and of *Hydnocarpus anthelmintica*. Isolation of a Homologue of Chaulmoogric Acid," *ibid.*, 87 (1905), 884–896; and F. B. Power and M. Barrowcliff, "The Constitution of Chaulmoogric and Hydnocarpic Acids," *ibid.*, 91 (1907), 557–578.

The Kremers Reference Files of the University of Wisconsin School of Pharmacy contain a significant amount of MS materials related to Power: correspondence, drafts of papers, and an unpublished biographical memoir by his daughter.

II. Secondary Literature. The most detailed discussion of Power's scientific work appears in Max Phillips, "Frederick Belding Power, Most Distinguished American Phytochemist," in *Journal of Chemical Education*, 31 (1954), 258–261. Other biographical sketches include C. A. Browne, "Frederick Belding Power," in *Journal of the Association of Official Agricultural Chemists*, 11 (1928), iii–vi; E. G. Eberle, "Frederick Belding Power, Ph.D., LL.D., F.C.S.," in *Journal of the American Pharmaceutical Association*, 11 (1922), 403–405; and Lyman Newell, "Frederick Belding Power," in *Dictionary of American Biography*, XV (1935), 154–155. His early years at Wisconsin are described in Edward Kremers, "Dr. Power at Wisconsin," in *Badger Pharmacist* (published by the Wisconsin chapter of Rho Chi), no. 13 (Dec. 1936), 1–13.

John Parascandola

POWER, HENRY (*b.* Halifax, Yorkshire, England, 1623; *d.* New Hall, Yorkshire, England, 23 December 1668), *microscopy, physics, medicine.*

Power was the son of John Power, a merchant. He took his B.A. (1644) and M.A. (1648) at Cambridge and, with the encouragement of Sir Thomas Browne (a friend of Power's father), the M.D. degree there in 1655. He practiced medicine in Halifax and later in the neighboring community of New Hall. Between 1646 and 1659 he corresponded extensively with Browne, largely on medical and scientific matters. He was admitted to the Royal Society in 1661.

Aside from the letters to Browne and a mercifully short poem in heroic couplets on the microscope, Power's only published work is *Experimental Philosophy, in Three Books* (on the microscope, atmospheric pressure, and magnetism), completed in 1661 and published in 1664. It was the first book in English on microscopy and the first in any language to describe (along with flora and fauna) the nature of various metals as seen through a microscope. Power's test of Boyle's "spring of the air" hypothesis shows that he understood the need for precise instruments and that he could conduct meticulously controlled experiments. Although his work on microscopy was shortly eclipsed by that of Hooke and Swammerdam, Power remains important as one who helped materially to realize the principles and set the standards of inquiry and exposition formulated by the progenitors and charter members of the Royal Society.

BIBLIOGRAPHY

I. Original Works. Power's unpublished papers are in the Sloane Collection of the British Museum and are listed by Thompson Cooper in his life of Power in the *Dictionary of National Biography*, XVI (1967–1968), 256. Most of them appear to be early drafts of matter that later appeared in *Experimental Philosophy*. Among those which do not so appear are "Experiments Recommended to Him by the Royal Society," Sloane MS 1326, art. 10—reviewed by Cowles (see below); "The Motion of the Earth Discovered by Spotts of the Sun," Sloane MS 4022, art. 3—reviewed by Thorndike (see below); and "Poem in Commendation of the Microscope," Sloane MS 1380, art. 16—published as "Dr. Henry Power's Poem on the Microscope," in *Isis*, **21** (1934), 71–80, with annotations by Thomas Cowles. The correspondence with Browne is repr. in *The Works of Sir Thomas Browne*, G. Keynes, ed., IV (Chicago, 1964), 254–270; two of these letters also appear in *A Collection of Letters Illustrative of the Progress of Science in England*, J. O. Halliwell, ed. (London, 1841), 91–92. *Experimental Philosophy* (London, 1664) has been repr. with a new biographical and critical intro. by Marie Boas Hall (New York, 1966), which provides the most illuminating commentary on Power's place in the history of science.

II. SECONDARY LITERATURE. Other commentary is in L. Thorndike, *A History of Magic and Experimental Science*, VIII (New York, 1958), 211–216; and T. Cowles, "Dr. Henry Power, Disciple of Sir Thomas Browne," in *Isis*, **20** (1934), 344–366. For Power's relations with the Royal Society see Thomas Birch, *The History of the Royal Society*, I (London, 1756), 22 and *passim*.

GORDON W. O'BRIEN

POYNTING, JOHN HENRY (*b.* Monton, near Manchester, England, 9 September 1852; *d.* Birmingham, England, 30 March 1914), *physics*.

The youngest son of a Unitarian minister, Poynting attended his father's school and then Owens College, Manchester (1867–1872). In 1872 he received the B.Sc. from London University. He studied at Trinity College, Cambridge, from 1872 to 1876 and was third wrangler on the mathematical tripos in 1876. He returned to Owens College as demonstrator in physics under Balfour Stewart. Made a fellow of Trinity in 1878, Poynting returned to Cambridge and did research in the Cavendish Laboratory, under the direction of Maxwell. He was appointed professor of physics at Mason College, Birmingham, on its founding in 1880 and remained there for the rest of his life. When Mason College became the University of Birmingham in 1900, Poynting became dean of the Faculty of Science, a position he held for twelve years. He received the Sc.D. from Cambridge in 1887 and became a fellow of the Royal Society the following year. His best-known work, the derivation of the expression for the flow of energy in an electromagnetic field (Poynting flux), appeared in 1884.

In 1893 Poynting was awarded the Adams Prize of Cambridge for his essay *The Mean Density of the Earth*. He was president of the Physical Society in 1905, a member of the Council of the Royal Society in 1909–1910, and vice-president of the Royal Society in 1910–1911. His honors included the Hopkins Prize of the Cambridge Philosophical Society (1903) and the Royal Society's Royal Medal for 1905. Poynting married in 1880 and had three children. He loved the countryside where he lived, and was chairman of the Birmingham Horticultural Society; he also served as a Justice of the Peace. An excellent and beloved teacher, he died from a diabetic attack brought on by a bout of influenza.

From the time of his stay at the Cavendish Laboratory, Poynting performed painstaking experiments to measure the mean density of the earth or, equivalently, the constant of universal gravitation. Instead of using a torsion balance, as had Cavendish a century before, Poynting employed a beam balance. His best result, reported in 1891, differs from the presently accepted value by about four parts in a thousand. He admitted, however, that the use of the quartz-fiber torsion balance by C. V. Boys for the same purpose had proved to be inherently more accurate; and he employed that instrument in his later experiments on radiation pressure. In research performed with assistants at Birmingham, Poynting placed small upper limits on any dependence of the gravitational force between crystals on their orientation, and on any effect of temperature on gravitation.

The Poynting flux was derived in his paper of 1884 from Maxwell's electromagnetic field theory. The term flux appears in an equation representing the energy balance in a closed region ("Poynting's theorem"), and the concept was applied to trace the flow of energy around a conducting wire, a discharging capacitor, and a voltaic cell, and in an electromagnetic wave. Other work in electricity included the design of electrical instruments and discussion of Lodge's models for representing the electromagnetic field. Poynting also did experiments on radiation pressure. At the turn of the century, P. N. Lebedev, and E. F. Nichols and G. F. Hull demonstrated the pressure exerted normal to a material surface. In 1904, with the aid of his colleague Guy Barlow, Poynting measured the tangential stress when a beam of light was reflected at an angle from a partially absorbing surface. This effect of the momentum carried by radiation had the experimental advantage of being less masked than the normal pressure by forces exerted by the unevenly heated residual gas in the apparatus. Poynting also demonstrated the existence of a torque on a prism so arranged that a beam of light emerged parallel to, but shifted from, the line of incidence. Perhaps of greater importance was his work with Barlow on the recoil of a heated, radiating body as a result of its own radiation, which work formed the subject of the Bakerian lecture for 1910.

The effect of radiation pressure on dust in the solar system, and the use of the law for the intensity of radiation from a blackbody to estimate the temperature of the planets, also interested Poynting. Other theoretical work involved more thorough discussion of the phase transition between the solid and liquid states (1881) and of osmotic pressure (1896). He developed instruments for research and for lecture demonstration, and performed research confirming his own predictions concerning the behavior of loaded wires under torsion (1905, 1909). Among Poynting's earliest works were statistical studies on drunkenness in England (1877, 1878) and on fluctuation of commodity prices (1884). Besides his Adams Prize essay he wrote *The Pressure of Light* and *The Earth*, and was coauthor with J. J. Thomson of a series of physics textbooks.

BIBLIOGRAPHY

Poynting's *Collected Scientific Papers*, G. A. Shakespear and G. Barlow, eds. (Cambridge, 1920), contains a bibliography and lists his books, of which the Adams Prize essay, *The Mean Density of the Earth* (London, 1894), and *The Earth: Its Shape, Size, Weight and Spin* (Cambridge, 1913) were most significant.

The *Collected Scientific Papers* contains several biographical notices, particularly one by J. J. Thomson from *Proceedings of the Royal Society*, **92A** (1915–1916), i–ix.

A. E. WOODRUFF

PRANDTL, LUDWIG (*b.* Freising, Germany, 4 February 1875; *d.* Göttingen, Germany, 15 August 1953), *fluid mechanics.*

Prandtl was the founder of boundary layer theory and the originator of the German school of aerodynamics. His own work and that of his many students over half a century made Göttingen University the source of most of the elements of modern fluid mechanics. The notion of the boundary layer, the role of bound vortices in airfoil theory, the foundations of supersonic aerodynamics, explanations of drag coefficients and friction factors for a host of situations, the concept and applications of the turbulent mixing length, and the theory of wakes were results of his work.

Prandtl was the only child of Alexander Prandtl, an engineering professor at the agricultural college at Weihenstephan. Owing to the protracted illness of his mother, the former Magdalene Ostermann, he was particularly close to his father, who led him early in life to an interest in natural phenomena.

In 1894 Prandtl entered the Technische Hochschule at Munich to study engineering. He graduated in 1898 and two years later completed his doctorate under the famous mechanics professor, August Föppl, who had been one of his undergraduate teachers. His thesis dealt with the lateral instability of beams in bending. Although the mechanics of solids was to become a secondary interest, it was an area in which Prandtl continued to contribute significantly. He discovered the well-known soap film analogy for the Saint Venant torsion problem in 1904 and did fundamental work in plastic deformation during the early 1920's.

Prandtl's interest in fluid flow began immediately after his graduation from Munich. He went to work in the Maschinenfabrik Augsburg-Nürnberg, where he was asked to improve a suction device for the removal of shavings. In the process of greatly improving the device, he came to recognize some basic weaknesses in the current understanding of fluid mechanics.

One weakness was the inability of existing fluid mechanics to explain why the moving fluid in a pipe would separate from the wall in a sharply divergent section instead of expanding to fill the pipe. In the course of the next three years Prandtl attacked this problem in a way that was to be characteristic of his later work. After accepting a professorship at the Hannover Technische Hochschule in 1901, he proceeded to ask what the missing element in the existing analyses of pipe flows might have been. By 1904 he had developed his celebrated paper on the flow of fluids with small viscosity. In it he showed that no matter how small the viscosity was, the fluid had to be stationary on the walls of the pipe. Thus the classical theory of inviscid fluids could never be employed without taking cognizance of the way in which a thin viscous region near the wall shaped the flow. An understanding of this region, the boundary layer, was to facilitate the subsequent explanation of lift and drag on airfoils, and aspects of almost all other aerodynamic behavior. Prandtl's insight thus led him to the large problem that lay behind a small one, and he set in motion a program of theoretical and experimental research that is still being worked out today.

By this time a new wind had begun to blow through the German system of higher education. The mathematician Felix Klein, feeling that the gulf between mathematics and technology was too wide, had established several technical institutes at Göttingen University. On Föppl's recommendation Klein gave Prandtl a chair at Göttingen and placed him in charge of the Institute for Technical Physics. Just after this move Prandtl's presentation of the boundary layer paper at the Third International Congress of Mathematicians at Heidelberg won high praise from Klein, who was quick to understand what his new man had accomplished.

At Göttingen, Prandtl had very good graduate students and access to the resources and interest of industry. He also was encouraged to pursue more theoretical lines of research than he had been at Hannover. During his first years there he made lasting contributions to the theory of supersonic flow. He combined Riemann's theories with Mach's *Schlieren* flow visualization apparatus to obtain the first explanation of the behavior of supersonic nozzle efflux. Later he was responsible for the first mathematical description of supersonic flow around slender bodies. He also was instrumental in developing the first German wind tunnel, which was completed at Göttingen in 1909.

Meanwhile, in the autumn of 1906, Prandtl's most notable student arrived in Göttingen. Theodore von Kármán sought Prandtl out to direct his research on the theory of nonelastic buckling. He was only six years Prandtl's junior; and their lives and pursuits touched many times between 1906 and their last meeting in 1945, when Kármán, a Hungarian Jew who had become an American citizen, returned to Germany with an army interrogation team.

In his autobiography Kármán gives a picture of Prandtl that clearly mingles affection with annoyance. Prandtl's life was, he tells us, "particularly full of overtones of naïveté." In 1909, for example, Prandtl decided that he really ought to marry; but he didn't know how to proceed. Finally he wrote to Mrs. Föppl, asking for the hand of one of her daughters. But which one? Prandtl had not specified. At a family conference the Föppls made the practical decision that he should marry their eldest daughter, Gertrude. He did and the marriage was apparently a happy one. Two daughters were born in 1914 and 1917.

Another facet of his naïveté lay in Prandtl's inclination to absorb himself totally in what interested him. The details of toys and magic tricks fascinated him to the exclusion of his surroundings. His greatness as a teacher did not include greatness as a lecturer because "he could not make a statement without qualifying it" and consequently was tedious. Nevertheless Prandtl was personable, gracious, and unassuming. He was an accomplished pianist with a good musical sense. His importance as a researcher was matched by an extraordinary ability to work fruitfully with his individual students.

Prandtl's aerodynamic work developed steadily from

1909 to the end of World War I. Between 1909 and 1912 he helped to establish test codes for fans. In 1914 he explained a puzzling, sudden drop of the drag coefficient for a sphere that occurred with increasing velocity. He did this by again recognizing the important phenomenon underlying a minor problem: the laminar-to-turbulent transition of the boundary layer on the body.

Prandtl's most significant contribution during this period was his work on airfoil theory. The lift force on wings was fairly well understood, and he turned to the explanation of drag. The boundary layer gave rise directly to skin friction which was much too small to account for wing drag. In 1911 and 1912 Kármán, stimulated by the experimental work going on in Prandtl's laboratory, did much to explain another component, profile drag. He described the so-called Kármán street of alternating vortices, parallel with an aerodynamic body, that must be pulled along behind it. Prandtl continued to work on a third contribution, induced drag. He had recognized that the presence of lift causes a trailing vortex to be induced in the shape of a long distorted horseshoe with its base at the airport where the flight began and its ends at the wing tips, which continually generate the vortex.

Prandtl's attempts to analyze this source of drag subsequently fell under the veil of wartime secrecy, and his descriptions of the effect finally emerged in restricted Göttingen publications in 1918 and 1919. By 1920 the idea reached a larger public and wrought sharp changes in the wing design and streamlining of airplanes. It led in general to more cleanly shaped wings, to higher aspect ratios, to today's swept-back wings, and to the use of streamlining fillets.

The elusive problem of describing turbulent flow yielded to Prandtl's and Kármán's competitive efforts in the mid-1920's. Kármán, who then was teaching at Aachen, discussed the structure of turbulent flows in 1924 but failed to produce a strategy for analyzing it. In 1926 Prandtl provided the conceptual device needed to make an analysis. This was the "mixing length," or average distance that a swirling fluid element would travel before it dissipated its motion. The idea resembled the notion of a mean free path in the kinetic theory of gases and was used in a somewhat similar way. Subsequent experiments by Prandtl and theoretical work by Kármán made it possible for Prandtl to present a summary paper on the subject in 1933 that is the basis for chapters on turbulence in today's textbooks.

During the 1930's and 1940's Prandtl, now an elder statesman of fluid mechanics, continued to contribute to the basic literature and technology, and reaped honors for his accomplishments. Nevertheless, his interrogation by the U.S. Army in 1945 found that the mainstream of wartime technology had more or less bypassed Prandtl and Göttingen. Hitler had shown greater interest in the rocketry at Peenemunde and the laboratory at Brunswick. The team, arriving in Göttingen from the slave labor camp at Nordhausen, found Prandtl complaining peevishly about bomb damage to his roof and asking how the Americans planned to support his ongoing work.

While the picture is one of a technician who felt little obligation to anyone's politics, it forms a consistent view of Prandtl, who, after all, made enormous contributions to his world as a technologist. He was then seventy-one and just turning his attention to meteorology, on which he published material as late as 1950.

BIBLIOGRAPHY

I. ORIGINAL WORKS. Poggendorff, V, 1002; VI, 2069–2070; VIIa, pt. 3, 620, lists some 80 articles and books written between 1901 and 1950. Among the most important are "Zur Torsion von prismatischen Stäben," in *Physikalische Zeitschrift*, **4** (1903), 758–767; "Über die stationären Wellen in einem Gasstrahl," *ibid.*, **5** (1904), 599–601; "Über Flüssigkeits-Bewegung bei sehr kleiner Riebung," in *Verhandlungen der III Internationaler Mathematiker-Kongress* (Leipzig, 1905); "Die Luftwiderstand von Kugeln," in *Nachrichten von der Gesellschaft der Wissenschaften zu Göttingen* (1914), 177–189; "Tragflügel Theorie, 1 und 2, Mitteilungen," *ibid.*, (1918), 451–477, and (1919), 107–137; "Über die ausgebildete Turbulenz," in *Proceedings of the Second International Congress for Applied Mechanics* (Zurich–Leipzig, 1927), 62; and *Führer durch die Strömungslehre*, 3rd ed. (Brunswick, 1949).

II. SECONDARY LITERATURE. Poggendorff, VIIa, pt. 3, 619–620, provides many references to writings about Prandtl and his work. Hermann Schlichting, *Boundary Layer Theory*, 4th ed. (New York, 1960), which includes 61 references to Prandtl and far more than that number to his students, provides the best overview of his impact on viscous flow theory. H. W. Liepmann and A. Roshko, *Elements of Gasdynamics* (New York, 1957), gives a comparable picture of his role in compressible flow theory. Additional biographical material appears in Kármán's autobiography, *The Wind and Beyond* . . . (New York, 1967).

JOHN H. LIENHARD

PRATT, FREDERICK HAVEN (*b.* Worcester, Massachusetts, 19 July 1873; *d.* Wellesley Hills, Massachusetts, 11 July 1958), *physiology.*

During a long teaching career, Pratt's main researches were in heart and muscle physiology; his investigations into the phenomena of muscle fiber contraction are classics of their time.

The son of Frederick Sumner Pratt, a merchant descended from early Massachusetts settlers, and Sarah McKean Hilliard, Pratt entered Harvard after attending preparatory school in his native Worcester, receiving the A.B. in 1896 and the A.M. in 1898. In the latter year he reported on experiments he had been conducting for some time to determine the role of the veins of Thebesius and of the coronary veins in the nutrition of the heart, a problem to which he returned years later.

His training enabled Pratt to combine a keen interest in the theoretical aspects of physiological research, which he examined in relation to their historical antecedents, with an expertise in the use of apparatus in experimentation. After studying at the University of Göttingen and a stay at the Worcester Polytechnic Institute, he became assistant in physiology at Harvard in 1901. He graduated from Harvard Medical School in 1906. He taught physiology in Wellesley College's department of hygiene until 1912, when he was appointed professor of physiology at the University of Buffalo. In that year he married Margery Wilerd Davis; they had five children.

The first of Pratt's researches into skeletal muscle fiber contraction applying the "all-or-none" principle was published while he was head of the department of physiology in the medical school at Buffalo. From 1919 to 1920 he was an honorary fellow in biology at Clark University, and the next year he served as teaching fellow in physiology at Harvard. Named professor of physiology at Boston University in 1921, he remained there until he retired emeritus in 1942. Later he headed a firm supplying apparatus for physiological research. A member of the American Physiological Society and other scientific societies, he was interested in the work of the Marine Biological Laboratory at Woods Hole, Massachusetts, and the Bermuda Biology Station. He wrote several biographical studies and was especially intrigued by the biological concepts of Emanuel Swedenborg, finding them in many ways modern and in some respects anticipatory of certain of his own regarding cardiac nutrition.

The gradation of muscle activity had engaged investigators before Pratt, with his undergraduate assistant John P. Eisenberger, set out to devise a method of determining by direct observation the response of single muscle fibers to stimuli. After Henry Pickering Bowditch in 1871, studying cardiac muscle, had set forth the "all-or-none" principle, others had explored the further applicability of the rule that independent of its strength, a stimulus, if it elicited a response, evoked one that was maximal. Francis Gotch had studied nerve fibers; Keith Lucas had maintained that the contraction of individual skeletal muscle fibers in response to stimuli was an all-or-none phenomenon. In 1917 Pratt published his description of a method of stimulating a single muscle fiber, or a few at most. He utilized a capillary pore electrode having a pore of 8μ, smaller in diameter than the contractile element (a single muscle fiber of the sartorius of the frog), but noted that still smaller pores had been made. The technique provided for direct microscopic observation and for photomicrographic tracings. Also in 1917 Pratt reported studies in which the movements of a mercury globule lying on the surface of a muscle preparation were recorded, and he followed fatigue and staircase gradients as well as graded responses.

Pratt saw the problem of muscle contraction as essentially one of energy transformations and of the changes involved between stimulus and response. He demonstrated discontinuities in muscle fiber response that were interpreted as direct confirmation of the all-or-none effect; thus the graded response of skeletal muscle was explained as a summation of maximal individual fiber contractions. He drew an analogy with quantum hypotheses in describing the all-or-none response in muscle fibers.

Pratt's researches in muscle physiology, which he carried further at Boston University, were a stimulus to other researchers. The results of investigators who later found gradation in muscle fiber response to minute excitation too small to be propagated throughout the fiber led to continuing efforts to define conducting and contractile mechanisms, as well as the recognition that certain conditions were the normal and others quite unusual.

Pratt's experiments mark an era in the history of muscle physiology; his special interest lay in the use of apparatus, often of his own design, to study mechanical responses in muscle. He pursued research on salientian lymph hearts and on problems of nerve-muscle and cardiac physiology, and developed electrical recording and timing devices. As an educator he was interested in making medical history accessible to the student of medicine.

BIBLIOGRAPHY

I. ORIGINAL WORKS. The early papers on all-or-none response are "The Excitation of Microscopic Areas: A Non-Polarizable Capillary Electrode," in *American Journal of Physiology*, **43** (1917), 159–168; "The All-or-None Principle in Graded Response of Skeletal Muscle," *ibid.*, **44** (1917), 517–542; and "The Quantal Phenomena in Muscle: Methods, With Further Evidence of the All-or-None Principle for the Skeletal Fiber," *ibid.*, **49** (1919), 1–54, written with John P. Eisenberger. Later, when physiological investigations had demonstrated certain localized graded responses within skeletal fibers, he wrote "Localized Response of the Muscle Fiber to Excitation Traversing a Quiescent Area," *ibid.*, **122** (1938), 27–33, with S. E. Steiman. His articles also included "Scientific Apparatus and Laboratory Methods," in *Science*, **72** (1930), 431–433; and "Homolateral Synchronism of Lymphatic Hearts," in *Proceedings of the Society for Experimental Biology and Medicine*, **29** (1932), 1019–1022, written with Marion A. Reid.

Some correspondence is in the John F. Fulton Papers at the Yale Medical Library.

II. SECONDARY LITERATURE. The significance of Pratt's work is discussed in John F. Fulton, *Muscular Contraction and the Control of Reflex Movement* (Baltimore, 1926), 48–52, 127–128; and in John F. Fulton, ed., *Selected Readings in the History of Physiology* (Springfield, Ill.–Baltimore, 1930), 221–222, and 2nd ed., compiled by Fulton and completed by Leonard G. Wilson (Springfield, Ill., 1966), 237–238.

Biographical sources include *National Cyclopaedia of American Biography*, XLVI (New York, 1963), 556; and Jaques Cattell, ed., *American Men of Science*, 9th ed., II (Lancaster, Pa.–New York, 1955), 897.

GLORIA ROBINSON

PRATT, JOHN HENRY (*b*. London, England, 4 June 1809; *d*. Ghazipur, India, 28 December 1871), *mathematics*.

Pratt was the son of Rev. Josiah Pratt, secretary of the Church Missionary Society. He received the B.A. at Gonville and Caius College, Cambridge, in 1833 and the M.A. at Christ's and Sidney Sussex colleges in 1836. His missionary zeal, coupled with his exceptional scientific aptitude, propelled him into both arenas, sometimes separately, sometimes simultaneously.

Through the influence of Bishop Daniel Wilson, Pratt obtained a chaplaincy appointment with the East India Company in 1838 and, in 1844, became chaplain to the bishop of Calcutta. He was appointed archdeacon of Calcutta in 1850, a post he held until he died. In 1866 he became a fellow of the Royal Society.

Of Pratt's several books only his first, *The Mathematical Principles of Mechanical Philosophy* (Cambridge, 1836; revised 1842; expanded and republished in 1860 as *A Treatise on Attractions, LaPlace's Functions, and the Figure of the Earth*), is exclusively concerned with science. The focal point of this book is the shape of the earth.

The theory of fluids suggests that the earth is essentially spheroidal. Pratt began by calculating the

Newtonian attractive force exerted by a homogeneous sphere on a point, then successively relaxed the constraints on composition and shape. He ended the first part of the treatise by considering local gravitational effects, those due to irregularities in the earth's crust.

In the second part Pratt turned his attention to the fact, first demonstrated by Newton, that the earth is not a sphere, showing that the fluid hypothesis leads to an oblate spheroidal shape. He next produced a sequence of arguments intended to show that the lower bound of the thickness of the earth's crust is 1,000 miles (current estimates range from five to eight miles). Pratt concluded by showing that the precession of the equinoxes, the period differential for pendulums as a function of latitude, and geodetic data all support the conclusion that the earth is an oblate spheroid. He gave the difference between the equatorial and polar axes as 26.9 miles, a figure that compares favorably with current measurements.

For most of his professional life Pratt was concerned primarily with the propagation of the faith. He was a not undistinguished member of that informal fraternity committed to the proposition that revelation and science are complementary avenues to the acquisition of knowledge.

BIBLIOGRAPHY

With the exception of *A Treatise on Attractions*, Pratt's works do not appear to have survived in the major collections; nor have his publishers, through a complex of unfortunate circumstances, been able to preserve copies of his works, the more scientifically oriented of which include *Scripture and Science Not at Variance; With Remarks on the Historical Character, Plenary Inspiration, and Surpassing Importance of the Earlier Chapters of Genesis Unaffected by the Discoveries of Science* (London, 1856; 7th ed. 1872); *The Descent of Man, in Connexion With the Hypothesis of Development* (London, 1871); and *Difficulties in Receiving the Bible as a Divine Revelation Arising From the Progress of Human Knowledge* (Calcutta, 1864).

JEROME H. MANHEIM

PRAXAGORAS OF COS (*b.* Cos, *ca.* 340 B.C.), *anatomy, physiology.*

Little is known about the life of Praxagoras. He was born into a medical family. His father, Nicarchus, was an eminent physician and it is possible that his grandfather, too, was a doctor. It has been claimed that he was the father of the poet Theocritus, but it is more likely that the latter's father was a contemporary namesake. Galen clearly regarded Praxagoras as an influential figure in the history of medicine and lists him as a member of the so-called "logical" or "dogmatic" school of medicine. There is evidence that Praxagoras wrote several treatises, notably a work entitled Φυσικά in at least two books, presumably dealing with natural sciences; another on anatomy, which probably comprised more than one book; a work on diseases in foreign countries comprising at least two books; a comprehensive work on treatments (Θεραπεῖαι) containing at least four books; a work with at least three books on diseases; three books on symptoms; a treatise on acute diseases; and a work or collection of works on causes, diseases, and treatments. Although none of these works has survived, it is likely that at least some of them were extant in Galen's time.

Praxagoras subscribed to a variation of the humoral theory. According to Galen, he distinguished eleven humors (including the blood) and we learn elsewhere that he thought that health and diseases were ultimately dependent upon them. He believed that when heat was present in the organism in due proportion, nourishment was transformed naturally by the process of digestion into blood; but an excess or deficiency of heat gave rise to the other humors, which he regarded as morbid and ultimately productive of various pathological conditions (including epilepsy). Like Empedocles and Diocles of Carystus before him, Praxagoras considered the digestive process to be a kind of putrefaction. The blood produced as the end product of this process he restricted to the veins; the arteries, however, served as vessels for carrying the psychic pneuma, which issues from the heart. Praxagoras is generally credited with this distinction between veins and arteries and with the theory that πνεῦμα moves through the latter and blood through the former. But a passage in Galen also attributes this differentiation to Nicarchus; and Diocles, too, must have approximated to these doctrines. For the explanation of paralysis as the result of a gathering of thick, cold phlegm in the arteries is attributed to both Diocles and Praxagoras, and the same source tells us that both men regarded the arteries as the channels "through which voluntary motion is imparted to the body." (The pneuma is the cause or agent of this motion.) Furthermore, both men are agreed upon the immediate cause of epilepsy, which they ascribe to the blocking of the passage of the psychic pneuma from the heart through the aorta by an accumulation of phlegm. Like Philistion, Aristotle, and Diocles, Praxagoras thought that the heart was the seat of the intelligence and the central organ of thought. He differed from them, however, in his views upon the purpose of respiration, believing that its function was to provide nourishment for the psychic pneuma, rather than to cool the innate heat.

Whatever the degree of their originality, Praxagoras' views upon the nature of the arteries were greatly influential in the development of physiology, since he not only prescribed channels specifically for the transmission of the pneuma but also took an important step toward the discovery and theory of the nerves. He conjectured that some arteries become progressively thinner until their walls ultimately fall together and their lumen ($\kappa o\iota\lambda\acute{o}\tau\eta s$) disappears. To this attenuated part of the artery he applied the term $\nu\epsilon\hat{v}\rho o\nu$, and he explained the movement of fingers and other parts of the hands by the operation of these $\nu\epsilon\hat{v}\rho a$—operations we now associate with the nerves. Although Praxagoras did not himself discover the nerves, as Solmsen points out, he evidently wondered about the nature of the organ to which the bodily extremities owe their movement, identified this organ to his own satisfaction, and discussed its connection with the center of vitality and energy. It is assuredly no accident that Herophilus, who actually discovered both sensory and motor nerves, was a pupil of Praxagoras.

Praxagoras' influence on Herophilus may also be seen in the latter's interest in pulsation. Praxagoras was apparently the first to direct attention to the diagnostic importance of the arterial pulse. He defined pulsation as every perceptible (that is, natural) movement of the arteries as distinct from morbid tremors ($\pi a\lambda\mu\acute{o}s$, $\tau\rho\acute{o}\mu o s$, and $\sigma\pi a\sigma\mu\acute{o}s$). He also maintained that the arteries pulsed by themselves independently of the heart.[1] Although he was followed in this particular belief by Phylotimus, Herophilus was not convinced of its validity and he sought at the beginning of his book "On Pulses" to refute this doctrine.

According to Galen, Praxagoras occasionally displayed too little care in anatomy. He criticized him for his contention, contrary to evidence ($\pi a\rho\grave{a}\ \tau\grave{o}\ \varphi a\iota\nu\acute{o}\mu\epsilon\nu o\nu$), that the heart is the starting point of the nerves. Praxagoras' belief that the heart is the central organ of the intelligence and the seat of the soul undoubtedly led him to adopt not only this belief but also the view that the brain is a kind of overgrowth and excrescence ($\dot{v}\pi\epsilon\rho a\acute{v}\xi\eta\mu a\ \tau\iota\ \kappa a\grave{\iota}\ \beta\lambda\acute{a}\sigma\tau\eta\mu a$) of the spinal cord. Galen's criticism suggests that Praxagoras did not arrive at this theory on the basis of dissection.

Praxagoras exercised considerable influence upon the development of Greek medicine. His belief, for example, that the arteries contain not blood but pneuma was dominant for four-and-a-half centuries.

Several of his pupils achieved eminence, notably Phylotimus, Plistonicus, and Xenophon, all of Cos. But his most famous pupil was Herophilus, who not only developed his master's teaching, but also vigorously attacked certain of his views. Praxagoras' influence was not limited solely to medicine, but was also apparent in philosophy. We find the Stoic Chrysippus, some fifty years later, appealing to the authority of Praxagoras in support of his contention that the $\dot{\eta}\gamma\epsilon\mu o\nu\iota\kappa\acute{o}\nu$ is located in the heart. Previously, scholars have regarded Praxagoras as a slavish follower of Diocles but, although Praxagoras shared certain views with the so-called Sicilian school of medicine, in many other respects he departed radically from this body of doctrine and proposed important and influential theories of his own, thereby forming a transition between the old medicine and the new medicine soon to arise in Alexandria.

NOTE

1. Steckerl has suggested that Praxagoras conjectured that this pulsation of the arteries was due to bubbles derived from the blood in the veins (*The Fragments of Praxagoras*, pp. 19–36, 61–68). This view, together with Steckerl's belief that the psychic pneuma is (partly) derived from these bubbles, produced as a result of the digestive process, has recently been uncritically accepted by Phillips.

BIBLIOGRAPHY

None of Praxagoras' works has survived; for his importance in the history of medicine, see T. C. Allbutt, *Greek Medicine in Rome* (London, 1921); K. Bardong, "Praxagoras," in Pauly-Wissowa, *Real-Encyclopädie der klassischen Altertumswissenschaft*, XXII, pt. 2 (Stuttgart, 1954), 1735–1743; E. D. Baumann, "Praxagoras of Kos," in *Janus*, **41** (1937), 167–185; C. G. Kühn, in *Opuscula academica medica et philologica*, **2** (1828), 128; E. D. Phillips, *Greek Medicine* (London, 1973), 135–138; F. Solmsen, "Greek Philosophy and the Discovery of the Nerves," in *Museum Helveticum*, **18** (1961), 169–197; F. Steckerl, *The Fragments of Praxagoras of Cos and His School* (Leiden, 1958); K. Sudhoff, "Zur operativen Ileusbehandlung des Praxagoras," in *Quellen und Studien zur Geschichte der Naturwissenschaften und der Medizin*, **3** (1932–1934), 151–154; and C. R. S. Harris, *The Heart and the Vascular System in Ancient Greek Medicine* (Oxford, 1973), which was published after this article was written.

JAMES LONGRIGG

PREGL, FRITZ (*b.* Laibach, Austria [now Ljubljana, Yugoslavia], 3 September 1869; *d.* Graz, Austria, 13 December 1930), *analytical chemistry*.

Pregl's father, Raimund Pregl, a bank treasurer in Carniola, died when his son was still young. His mother was Friderike Schlacker. In 1887, after graduating from the Gymnasium in Laibach, he entered the University of Graz to study medicine. He spent almost his entire career at this university. His mentor in physiology quickly recognized Pregl's abilities and made him an assistant in his laboratory before he had completed his studies. Following his graduation, he remained at the laboratory but also practiced medicine, specializing in ophthalmology.

Although Pregl gained a thorough knowledge of physiology, his interest soon turned to physiological chemistry. He became an outstanding experimenter and investigated the reactions of cholic acid and the causes of the high values of the carbon : nitrogen ratio in human urine. The latter research qualified him in 1899 for the post of university lecturer in physiology. To broaden his scientific knowledge, Pregl spent a year (1904) in Germany, where his preference for chemistry became stronger. He studied physiological chemistry with Carl Hüfner at Tübingen, physical chemistry with Wilhelm Ostwald at Leipzig, and organic chemistry with Emil Fischer. At Berlin he also investigated the hydrolysis products of egg albumin with E. Abderhalden, who became a lifelong friend.

After his return to Graz, Pregl became an assistant at the medical-chemical laboratory of the university; he studied bile acids and also conducted research in protein chemistry. From these investigations he repeatedly found that the analytical methods of organic chemistry were too complicated, lengthy, and inexact for the determination of the composition of materials of biochemical origin. Such methods also required large test samples, which were often difficult to obtain with organic substances.

Pregl thus set himself the task of making the classical methods of elementary analysis feasible on a microscale. He approached this problem as a physiological chemist, hoping to create a suitable method for his investigations. The problem, however, proved to be more complicated than expected, and it gradually claimed his full attention. Organic microanalysis increasingly became the focus of his research, and he found that the methods of analytic chemistry were essential.

Pregl's methods did not deviate in principle from those of Liebig and Dumas; but now only one to three milligrams were sufficient for a determination, which could also be executed much more quickly and exactly. To create the necessary apparatus, this transformation of methods required a great deal of effort and technical skill, and it was of such importance that

Pregl was awarded the Nobel Prize in chemistry.

In 1910 Pregl was appointed professor of medical chemistry at the University of Innsbruck, where he was able to realize his plans. His first task was to find a balance more sensitive than that used in analytical chemistry. W. H. Kuhlmann had recently constructed balances for the determination of noble metals that measured with an exactness of 0.01 to 0.02 milligrams. Pregl found that with smaller quantities and under suitable circumstances measurements could be made with an exactness of $\pm$ 0.001 milligrams. He soon established methods for the microdetermination of carbon and hydrogen; these methods were followed by others for determining nitrogen, halogen sulfur, carboxyl, and other substances. Pregl extended his research to include a broad range of organic substances; and in 1917 he published *Die quantitative organische Mikroanalyse*, which was preceded by only a short publication in Abderhalden's *Handbuch der biochemischen Arbeitsmethoden*. Pregl's book underwent numerous and frequently revised editions and was translated into several languages.

Pregl's development of microanalysis was an immeasurable advance in both science and industry, and much of the groundwork in biochemistry evolved from his developments. In 1923, only six years after the appearance of his book, he was awarded the Nobel Prize—the first to be awarded for accomplishments in the field of analytical chemistry.

In 1913 Pregl had been recalled to Graz; he was named professor of medical chemistry and was active in research until his death. (His laboratory had become a world-renowned center of organic microanalysis.) Pregl was an earnest, energetic man, often given to sarcastic humor. He frequently engaged in mountain climbing and bicycling and was an enthusiastic automobile driver. He never married.

BIBLIOGRAPHY

Pregl's major work is *Die quantitative organische Mikroanalyse* (Berlin, 1917; 2nd ed., 1923; 3rd ed., 1930), 4th ed., revised with supp., von Hubert Roth, ed. (Berlin, 1935); later eds. were published in Vienna.

On Pregl and his work, see H. Lieb, "Fritz Pregl," in *Berichte der Deutschen chemischen Gesellschaft*, **64A** (1931), 113, and "Fritz Pregl," in *Mikrochemie*, 3 (1931), 105. See also R. Strebinger, "Fritz Pregl," in *Österreichische Chemikerzeitung*, 34 (1931), 10; F. Szabadváry, *History of Analytical Chemistry* (Oxford–New York, 1966), 302–303, and *Geschichte der analytischen Chemie* (Brunswick, 1966), 304–305.

FERENC SZABADVÁRY

PRESL

PRESL, KAREL BORIWOJ (*b.* Prague, Czechoslovakia, 17 February 1794; *d.* Prague, 2 October 1852), *botany.*

As a youth Presl explored the countryside in Bohemia with his brother Jan and developed an interest in natural history. Under the guidance of W. Seidl, they collected a herbarium of cryptogamous plants and in 1812 published a work on the cryptogamous flora of Bohemia. Karel subsequently went to Charles University to study medicine and during that time traveled to Italy and Sicily, where he met the well-known botanists Michele Tenore and Giovanni Gussoni. Following this visit he published a work on the grasses and sedges of Sicily and then, in 1826, his first major work, *Flora sicula*, which dealt with the vascular plants (both cultured and wild) of that country.

In 1818 Presl received the doctorate in medicine and later in obstetrics but made only brief use of these qualifications; in 1822 he was appointed custodian at the National Museum in Prague at the instigation of its director, Kaspar von Sternberg. His first task was to sort Thaddaeus Haenke's large collection of plants from South America and the Mariana and Philippine islands. Presl also edited an account of Haenke's specimens in *Reliquiae Haenkeanae*, a two-volume folio published between 1825 and 1835. Many of the specimens were of previously unknown species, and Presl chose a complete set of these for his own herbarium, which he took with him when, in 1836, he accepted the chair of natural history and technology at Charles University.

As the result of his study of Haenke's ferns, Presl himself began anatomical and morphological studies of ferns; and he established an entirely new classification, which he described in *Tentamen Pteridographiae* (1836). This work was followed in 1844 by an account of Hymenophyllaceae and in 1846 by his studies of Marattiaceae and more primitive ferns (*Supplementum Tentaminis Pteridographiae*, 1846). In his last major work, *Epimeliae botanicae* (1851), which he completed shortly before his death, he described many new species and several new genera that were noted in material from various sources, particularly Hugh Cuming's research in the Philippines and southeast Asia.

Presl bequeathed his herbarium to the Botanical Institute of Charles University, where it is still preserved.

BIBLIOGRAPHY

I. ORIGINAL WORKS. Presl's works include *Reliquiae Haenkeanae, seu descriptiones et icones plantarum, quas in America meridionali et boreali, in insulis Philippinis et Marianis collegit Thadaeus Haenke*, 2 vols. (Prague, 1825–1835); *Flora sicula, exhibens plantas vasculosas in Sicilia aut sponte crescentes aut frequentissime cultas, secundum systema naturale digestas* (Prague, 1826); *Symbolae botanicae, sive descriptiones et icones plantarum novarum aut minus cognitarum*, 2 vols. (Prague, 1830–1858); *Tentamen pteridographiae, seu genera filcacearum praesertim juxta vena rum decursum et distributionem exposita* (Prague, 836), also in *Abhandlungen der Böhmischen Gesellschaft der Wissenschaften*, 4th ser., **5** (1837), 1–290; *Hymenophyllaceae. Eine botanische Abhandlungen* (Prague, 1844), also *ibid.*, 5th ser., **3** (1845), 93–162; *Supplementum tentaminis pteridographiae, continens genera et species ordinum dictorum Marattiaceae, Ophioglossaceae, Osmundaceae, Schizaeaceae et Lygodiaceae* (Prague, 1846), also *ibid.*, 5th ser., **4** (1847), 261–380; and *Epimeliae botanicae* (Prague, 1851), also *ibid.*, 5th ser., **6** (1851), 361–624.

II. SECONDARY LITERATURE. See R. E. Holttum, "A Commentary on Some Type Specimens of Ferns in the Herbarium of K. B. Presl," in *Novitates botanicae Institutus botanici Universitatis Carolinae* (1968), 3–57; and Frans A. Stafleu, "Taxonomic Literature. A Selective Guide to Botanical Publications With Dates, Commentaries and Types," in *Regnum vegetabile*, **52** (1967), 365–367.

A. C. JERMY

PRESTWICH

PRESTWICH, JOSEPH (*b.* Clapham, London, England, 12 March 1812; *d.* Shoreham, Kent, England, 23 June 1896), *geology.*

Prestwich was descended from an old Lancashire family. He joined his father's wine-merchant business and remained in that trade for nearly forty years; but his scientific energy and enthusiasm impelled him to accomplish, in his spare time, work that placed him among the foremost English geologists. He was awarded the Wollaston Medal of the Geological Society of London (1849) and was president of that society in 1870–1872. He was also elected a vice-president of the Royal Society (1870), having received a Royal Medal in 1865. In 1874, after retiring from business, he was appointed professor of geology at Oxford. He was knighted in 1896, shortly before his death. By his will he established the Prestwich Medal of the Geological Society.

Prestwich's most important geological investigations were those into the older Tertiary formations in southeastern England and those into the Quaternary deposits of England and France containing the traces of early man.

In a series of papers from 1846 to 1857 (with a supplement in 1888) Prestwich elucidated the previously unexplored stratigraphy of the beds lying between the Chalk and the main part of the London

Clay in England, and he correlated these with the corresponding beds that had been described in France and Belgium. He called these beds the Lower London Tertiaries and divided them into the Thanet Sands, the Woolwich and Reading series, and the basement beds of the London Clay (later called the Blackheath and Oldhaven beds). He thus established the classification of this important part of the stratigraphical succession and became the acknowledged leader in the Tertiary geology of Europe.

In 1859 Prestwich began investigating claims made in France concerning the occurrence and age of shaped flints and bones of extinct mammals found by Boucher de Perthes in the valley-gravels of the Somme. Prestwich confirmed these claims and showed conclusively that the flints were man-made implements; that they were made when now-extinct mammals were living; and that the implements and bones occurred in, and were contemporaneous with, deposits laid down at an early stage in the development of the valley and were thus of an age to be measured in tens of thousands, if not hundreds of thousands, of years. Prestwich was thus the first authority to confirm Boucher de Perthes's evidence for the remote antiquity of man. In connection with this work Prestwich eventually demonstrated that rivers produce the valleys in which they flow, showing, in particular, that river terraces are normal features of valley development.

Prestwich's earliest work was an important monograph (1840) on the small coalfield of Coalbrookdale in Shropshire; later he assisted the commission inquiring into the potential coal supplies of Britain. He became a leading authority on questions of water supply and also discussed the conditions affecting the construction of a tunnel beneath the English Channel. Oceanography and changes in sea level were among the other subjects on which he wrote important papers. The broad range of his research and thought was incorporated in his well-known comprehensive textbook, *Geology—Chemical, Physical, and Stratigraphical* (1886–1888).

BIBLIOGRAPHY

I. ORIGINAL WORKS. A complete bibliography is given in the *Life and Letters* (see below). His major works include "On the Geology of Coalbrook Dale," in *Transactions of the Geological Society of London*, 2nd ser., **5** (1840), 413–495; *A Geological Inquiry Respecting the Water-bearing Strata of the Country Around London* (London, 1851); "On the Structure of the Strata Between the London Clay and the Chalk in the London and Hampshire Tertiary Systems—The Woolwich and Reading Series," in *Quarterly Journal of the Geological Society of London*, **10** (1854), 75–170; "Theoretical Considerations on the Conditions Under Which the Drift Deposits Containing the Remains of Extinct Mammalia and Flint Implements Were Accumulated, and on Their Geological Age," in *Philosophical Transactions of the Royal Society*, **154** (1864), 247–309; "Report on the Quantities of Coal in the Coal-Fields of Somersetshire" and "Report on the Probabilities of Finding Coal in the South of England," in *Report of the Commissioners Appointed to Inquire Into . . . Coal in the United Kingdom* (London, 1871); "On the Geological Conditions Affecting the Construction of a Tunnel Between England and France," in *Proceedings of the Institute of Civil Engineers*, **37** (1874), 110–145; "Tables of Temperatures of the Sea at Different Depths Beneath the Surface," in *Philosophical Transactions of the Royal Society*, **165** (1876), 587–674; *Geology—Chemical, Physical, and Stratigraphical*, 2 vols. (Oxford, 1886–1888); and "The Raised Beaches . . . of the South of England," in *Quarterly Journal of the Geological Society of London*, **48** (1892), 263–343.

II. SECONDARY LITERATURE. On Prestwich and his work, see H. W[oodward], "Eminent Living Geologists: Professor Joseph Prestwich, D.C.L., F.R.S., . . .," in *Geological Magazine*, **30** (1893), 241–246; [Sir] J. E[vans], obituary notice in *Proceedings of the Royal Society of London*, **60** (1897), xii–xvi; H. Hicks, obituary notice in *Proceedings of the Geological Society*, **53** (1897), xlix–lii; G. A. (Lady) Prestwich, *Life and Letters of Sir Joseph Prestwich*, with a chapter summarizing his scientific work by Sir Archibald Geikie (Edinburgh–London, 1899); and T. G. Bonney, *Dictionary of National Biography*, supp. III (1901), 284–287.

JOHN CHALLINOR

PRÉVOST, ISAAC-BÉNÉDICT (*b.* Geneva, Switzerland, 7 August 1755; *d.* Montauban, France, 10 June 1819), *natural philosophy, mathematics, physics, chemistry.*

Prévost was an astute observer whose knowledge of the sciences was largely self-acquired. His memoir of 1807 on the cause of the smut or bunt of wheat was remarkable in its time and its validity has endured. Although its implications in identifying a plant parasite as the agent of the disease were not fully appreciated then, the work later became influential and contributed to the background against which developed an understanding of contagious disease.

The son of Jean-Jacques Prévost and Marie-Élisabeth Henri, young Isaac-Bénédict was sent to a nearby village boarding school but had neither the disposition nor the opportunity to obtain more than the essentials for his later scholarship. He worked briefly as an engraver's apprentice and then as an apprentice in a grocery house, where he mulled over questions of weight and force, odors, and other subjects that were later to concern him. Deciding against a commercial career, he resolved instead to devote himself to the sciences. He became tutor to the sons of M. Delmas of Montauban; Montauban was his home and the family were his friends until his death. Encouraged by the family, Prévost decided to give his full energies to his studies, and he set out to remedy his education. At first, he was most attracted to mathematics but later turned to physics and natural history, which he was able to study when the family spent part of each year in the country. Although the necessary books were sometimes hard to obtain, he persevered in reading, observing, and experimenting.

A founding member of the Société des Sciences et des Arts du Département du Lot in Montauban, Prévost belonged to a number of learned societies and corresponded with many colleagues, including his cousin Pierre Prevost. He communicated papers on various subjects to the scientific journals and established himself as a savant. Although he had declined in 1784 an offer to succeed Pierre Prevost at the Berlin academy, he later accepted an appointment (1810) as professor of philosophy at the new Protestant academy in Montauban.

La carie or *charbon* (the bunt, smut, or stinking smut of wheat) was vastly destructive of crops when Prévost began a ten-year study to find the direct cause of the disease. Within the affected kernels, forming a dark powder, were the particles Tillet had described but considered a contaminating virus (1755). Prévost examined these brown-black, spherical granules beneath the microscope and estimated that millions were contained in a bunted kernel. He noted their likeness to the globules seen in certain uredos and recognized them specifically as the seeds of a cryptogam. By placing the spores in water, he was able to cultivate microscopic plants; and with the use of controls, he experimented in the fields to observe the destruction of young wheat plants. Although he could not discern the mycelia, he inferred that the penetrating parasite ultimately reached the kernel, there forming its gemmae. Prévost also discovered that copper solutions (particularly copper sulfate) can prevent the disease.

In 1807 Prévost finally identified a fungus parasite as the causative agent. Since at that time the spontaneous generation of lower plants was still an accepted hypothesis, Prévost's discovery was extraordinarily important. His work gained wider recognition in 1847, when the Tulasnes cited his memoir in their own studies.

BIBLIOGRAPHY

I. ORIGINAL WORKS. Prévost's most important work was *Mémoire sur la cause immédiate de la carie où charbon des blés, et de plusieurs autres maladies des plantes, et sur les préservatifs de la carie* (Paris, 1807), trans. by George Wannamaker Keitt as *Memoir on the Immediate Cause of Bunt or Smut of Wheat, and of Several Other Diseases of Plants and on Preventives of Bunt*, Phytopathological Classics, no. 6 (Menasha, Wis., 1939).

II. SECONDARY LITERATURE. The main source on Prévost's life and character is Pierre Prevost, *Notice de la vie et des écrits d'Isaac-Bénédict Prévost* (Geneva–Paris, 1820), which describes published and unpublished works, journals, notes, and correspondence. A shorter biography, P. P. [Pierre Prevost], "Prévost, (Isaac-Bénédict)," is in the *Biographie Universelle, Ancienne et Moderne*, XXXVI (1823), 59–60. Keitt (see above) includes a biography and an evaluation of the *Mémoire*; see also G. W. Keitt, "Isaac-Bénédict Prévost, 1755–1819," in *Phytopathology*, **46** (1956), 2–5; E. C. Large, in *The Advance of the Fungi* (New York, 1940; 2nd ed., 1962), 76–79, also assesses Prévost's work.

GLORIA ROBINSON

PRÉVOST, JEAN-LOUIS (*b.* Geneva, Switzerland, 1 September 1790; *d.* Geneva, 14 March 1850), *physiology, embryology, medicine.*

While still a child Prevost knew the outstanding naturalists of Geneva: François and Jean-Pierre Huber, Jean Senebier, Jean-Pierre Vaucher, and Augustin-Pyramus de Candolle. Like many of them, he began theological studies and later turned to medicine. In 1814 he went to Paris, where he remained for two years, and then to Edinburgh, where from 1816 to 1818 he studied for his doctorate in medicine. Having obtained it, he went to Dublin, doubtless to familiarize himself with the details of medical practice.

Prevost then returned to Geneva, where he set up a practice; but research soon seemed more attractive than the sick, and he turned to microscopy and the study of modern chemistry. He assembled a group of collaborators with whom he wrote important memoirs: A. Le Royer, H. Lebert, Antoine Morin, and Jean-Baptiste Dumas. Prevost was concerned with the care of the indigent sick; and assisted by a few friends, including Louis Gosse, he founded a hospital where the poor could be cared for without charge—undoubtedly the first outpatient hospital in Europe.

In "Sur les animalcules spermatiques de divers animaux" (1821), written with Dumas, Prevost made a histological examination of spermatozoa and proved, for the first time, that these cells originate in certain tissues of the male sex glands. His observations were the culmination of a series of experimental researches, including those of Spallanzani, which prepared the way for modern discoveries in fertilization.

In 1824, again in collaboration with Dumas, Prevost published three memoirs on generation in the *Annales des sciences naturelles* that are considered the foundation of experimental embryology. Through their analysis of segmentation of the frog's egg, Prevost and Dumas confirmed the research of Swammerdam and of K. E. von Baer, and set forth the classic laws governing the development of the fertilized egg. These achievements brought Prevost and Dumas the Prix Montyon of the Paris Academy of Sciences.

Prevost next turned to new fields and, with Le Royer, published "Observations sur les contenus du canal digestif" (1825), a six-page memoir in which they expounded an advanced theory of digestion. For Prevost physiology was inseparable from chemistry, and thus he became one of the first biochemists. A work published in 1828 was the first of a series of investigations on the circulation of the blood, and a note on the circulation in the ruminant fetus also showed original observations.

With Morin, Prevost published "Recherches physiologiques et chimiques sur la nutrition du foetus" (1841), followed by two works written with Lebert, "Sur la formation des organes de la circulation et du sang chez les batraciens" (1844), and "Observations sur le développement du coeur chez le poulet" (1847).

Prevost should be considered a pioneer in hematology. Through his analysis of the composition of the blood and the nature of its circulation, as well as in his studies of the origin and evolution of the heart and blood vessels, he proved to be a remarkable physiologist and embryologist. He was one of the first to suggest the possibility of blood transfusions between individuals. Prevost was also interested in microscopy, conducted experiments on muscle contraction, and was the first to use a "galvanic current" in his experiments. He published observations on the reproduction of mollusks and reported findings on neuromuscular relations.

BIBLIOGRAPHY

I. ORIGINAL WORKS. The Royal Society *Catalogue of Scientific Papers*, V, 14–16, lists about 50 memoirs written by Prevost or in collaboration with others. Those cited in the text are "Sur les animalcules spermatiques de divers animaux," in *Mémoires de la Société de physique et d'histoire naturelle de Genève*, 1 (1821), 180–207, written with J.-B. Dumas; "Sur la génération," in *Annales des sciences naturelles*, 1 (1824), 1–29, 167–187, 274–293; 2 (1824), 100–120, 129–149; 3 (1824), 113–138, written with J.-B. Dumas; "Observations sur les contenus du canal digestif chez les foetus des vertébrés," in *Bibliothèque universelle*, 1st ser., 29 (1825), 133–139, written with A. Le Royer; "Note sur la circulation du foetus chez les ruminants," in *Mémoires de la Société de physique et d'histoire naturelle de Genève*, 4 (1828), 60–66; "Recherches physiologiques et chimiques sur la nutrition du foetus," *ibid.*, 9 (1841), 235–260, written with A. Morin; "Sur la formation des organes de la circulation et du sang chez les batraciens," in *Annales des sciences naturelles* (Zoologie), 1 (1844), 193–229, written with H. Lebert; and "Observations sur le développement du coeur chez le poulet," in *Comptes rendus . . . de l'Académie des sciences*, 24 (1847), 291–292, written with H. Lebert.

II. SECONDARY LITERATURE. Writings on Prevost include H. Lebert, "Éloge du Dr. Prevost," in *Mémoires de la Société de biologie*, 2 (1850), 60–65; P. E. Schazmann, *Un ami genevois de Stendahl: le Dr. J. L. Prevost* (Geneva, 1936); and L. A. Gosse and T. Herpin, "Notice biographique sur le Dr. J. L. Prevost," in *Bibliothèque universelle de Genève*, 4th ser., 15 (1850), 265–300.

P. E. PILET

PRÉVOST, LOUIS-CONSTANT (*b.* Paris, France, 4 June 1787; *d.* Paris, 16 August 1856), *geology*.

After deciding against a career in law, his stepfather's profession, Prévost studied medicine but under the influence of Cuvier and Brongniart turned to geology. During a residence in Austria he studied the Tertiary strata of the Vienna basin; and in 1820, following his return to a teaching post at the Paris Athenée, he showed that these strata were much younger than those of the Paris basin and were thus more closely related to the present condition of the earth's surface. In 1823-1824 he collaborated with Lyell in comparing the Tertiary and "Secondary" strata on both sides of the English Channel. As early as 1809 Prévost had been skeptical of Cuvier's and Brongniart's interpretation of the Parisian strata. In 1827 he argued that the strata with marine and freshwater fossils did not alternate in the succession but were lateral equivalents and therefore did not prove that periodic incursions of the sea had occurred over the continent. Prévost argued instead that all the strata had been formed within a single saltwater gulf analogous to the present English Channel, the freshwater fossils being present because of the outflow of large rivers similar to the present Seine. This work emphasized the importance of interpretations based on analogy with present conditions and thus made Prévost a supporter of Lyell's, whose *Principles* he later intended to translate.

In 1830 Prévost was one of the founders of the Société Géologique de France (serving as president in 1834, 1839, and 1851); and in 1831 he was appointed, with Cuvier's support, to a new chair of geology at the Sorbonne. In the same year he accompanied an official expedition to study a new volcanic island off

Sicily. Returning through Italy, he became convinced, like Lyell, that volcanoes are formed solely by the accumulation of ejected material and not, as Leopold von Buch and his followers maintained, by elevation from below. His views on both sedimentation and vulcanism placed him in an unorthodox position, especially within French geology. He opposed the paroxysmal dynamics of Cuvier and later of Élie de Beaumont, emphasizing instead the efficacy of "actual causes." But he also questioned the validity of d'Orbigny's use of fossils simply as indices of geological age, stressing instead their importance as indicators of ecological conditions that might have recurred at many periods. Yet despite this Lyellian approach, Prévost was explicitly influenced at the theoretical level by Deluc. Seeing geological history as a process of gradual withdrawal of the sea from the present continents, Prévost rejected the idea that powerful forces within the earth were capable of producing substantial elevation. Prévost was elected to the Académie des Sciences in 1848, having been an unsuccessful candidate in 1835 for the place vacated at Cuvier's death.

BIBLIOGRAPHY

I. ORIGINAL WORKS. Prévost's principal publications are "Sur la constitution physique et géognostique du bassin à l'ouverture duquel est située la ville de Vienne en Autriche," in *Journal de physique, de chimie, et d'histoire naturelle*, **91** (1820), 347–364, 460–473; "De l'importance de l'étude des corps organisés vivants pour la géologie positive . . . ," in *Mémoires de la Société d'histoire naturelle de Paris*, **1** (1824), 259–268; "Observations sur les schistes oolithiques de Stonesfield en Angleterre, et sur les ossements de mammifère qu'ils renferment," in *Annales des sciences naturelles*, **4** (1825), 389–417; "Les continents actuels, ont-ils été, à plusieurs reprises, submergés par la mer?" in *Mémoires de la Société d'histoire naturelle de Paris*, **4** (1828), 249–346; "Notes sur l'île Julia, pour servir à l'histoire de la formation des montagnes volcaniques," in *Mémoires de la Société géologique de France*, **2** (1835), 91–124; "Essai sur la formation des terrains des environs de Paris," in *Académie des sciences. Section de géologie et minéralogie, Candidature de M. Constant Prévost* (Paris, 1835), 93–124, which includes "Les continents actuels" and later notes and a valuable autobiographical summary of his work; "Sur la théorie des soulèvements," in *Bulletin de la Société géologique de France*, **11** (1840), 183–203; and "Sur la chronologie des terrains et le synchronisme des formations," *ibid.*, 2nd ser., **2** (1845), 366–373.

II. SECONDARY LITERATURE. A brief account of Prévost's life and work is given in Hoefer, ed., *Nouvelle biographie générale*, XLI (Paris, 1862), 15–17.

M. J. S. RUDWICK

PREVOST, PIERRE (*b.* Geneva, Switzerland, 3 March 1751; *d.* Geneva, 8 April 1839), *physics, philosophy, literature.*

Prevost's principal contribution to physics was his theory of exchanges, enunciated in 1791, in which he stated that all bodies regardless of temperature are constantly radiating heat and that an equilibrium of heat between two bodies consisted in an equality of exchange. W. C. Wells's theory of the formation of dew was a consequence of Prevost's hypothesis, and Fourier took Prevost's work as the basis for his mathematical analysis of heat radiation.

Prevost's father, a Calvinist minister and principal of the college of Geneva, made every effort to assure the excellence of his children's education. Consequently, Prevost studied not only the classical languages and literatures but also the sciences under H. B. de Saussure, Lesage, and Mallet. In accordance with his father's wishes he pursued theology for several years but then turned to law and received his doctorate in 1773.

From 1773 to 1780 Prevost worked as a teacher and tutor in Holland, in Lyons, and finally in Paris. Simultaneously, he was engaged in a translation of the works of Euripides; in 1778 he published *Orestes*, which gained him fame among classical scholars. As a result, he was invited to Berlin in 1780 by Frederick the Great as a member of the Academy of Sciences and Belles-Lettres. While in Berlin, Prevost contributed several memoirs on moral philosophy and poetry, and stimulated by Lagrange, published articles on scientific subjects. Upon the death of his father in 1784, he returned to Geneva and filled the chair of literature there for a year before going to Paris to work on an edition of Greek drama. Late in 1786 he returned to Geneva and became active in politics, serving as a member of the Council of Two Hundred, as attorney general, and as a member of the national assembly created to change the constitution of Geneva.

In 1788 Prevost published *De l'origine des forces magnétiques*, which made him known among physicists, and thereafter he became principally interested in heat phenomena owing to the publication in 1790 of *Essai sur le feu*, the work of a fellow Genevan, Marc Auguste Pictet. Pictet, following Deluc, believed that heat consisted in a continuous material fluid and that the radiation of heat between two objects at unequal temperatures was accomplished by the progressive expansions and contractions of this fluid, drawing an analogy between this process and the transmission of sound. Rejecting this view in "Sur l'équilibre du feu" (1791), Prevost conceived of heat as a "discrete fluid" or medium composed of particles the intervals of which are very great in comparison to their dimensions; during the process of radiation these particles, in the form of rays, stream continuously between the two radiating bodies. When the equilibrium of heat is upset, it is gradually reestablished by the unequal exchange of particles. In later writings Prevost clarified his ideas while retaining his conviction in the materiality of heat.

In 1793 Prevost was named to the chair of philosophy and general physics at Geneva, and he retained this post until his retirement in 1823. He wrote extensively on political economy, psychology, public education, probability theory, electricity, and meteorology. He translated Adam Smith's *Wealth of Nations*, Malthus's *Essay on Population*, the moral philosophy of Dugald Stewart, and Hugh Blair's works on rhetoric. In addition, Prevost maintained an active correspondence with scientists, philosophers, and classical scholars throughout Europe. He was named a corresponding member of the Royal Society of Edinburgh in 1796, of the Royal Society of London in 1801, and of the Institute of France in 1801.

Early in the French Revolution Prevost argued for the continued independence of Geneva and pleaded for moderation in the social and political reform of the city-state, which led to his detention for three weeks in 1794 by the revolutionary party. In 1798 he was named to the commission that regulated the union of Geneva with France; and in 1814, when Geneva was restored to the status of a republic, he was elected a member of the representative council.

Prevost remained mentally active until his death. In his last years he turned his attention to the study of aging in human beings, writing in detail about the progressive physical infirmities that he observed in himself.

BIBLIOGRAPHY

I. ORIGINAL WORKS. Prevost's principal works are *Euripide* (Paris, 1782); *De l'économie des anciens gouvernemens comparée à celle des gouvernemens modernes* (Berlin, 1783); *De l'origine des forces magnétiques* (Geneva, 1788); "Sur l'équilibre du feu," in *Journal de physique*, **38** (1791), 314–332; *Recherches physico-mécaniques sur la chaleur* (Geneva, 1792); *Genève, égalité, indépendance, liberté* (Geneva, 1793); *Essais philosophiques par feu Adam Smith* (Geneva, 1797); *Des signes envisagés relativement à leur influence sur la formation des idées* (Paris, 1798); *Essais de philosophie*, 2 vols. (Geneva, 1804); *Dugald Stewart: Éléments de la philosophie de l'esprit humain*, 2 vols. (Geneva, 1808); *Malthus: L'essai sur le principe de population* (Geneva, 1809); *Du calorique rayonnant* (Paris, 1809); *Deux traités de physique mécanique* (Geneva, 1818); and *Exposition élémentaire des principes qui servent de base à la théories de la chaleur rayonnante* (Geneva, 1832). An

extensive collection of Prevost's MSS is in the archives of the Bibliothèque Publique et Universitaire de Genève.

II. SECONDARY LITERATURE. On Prevost's life and work, see C. Bartholmerz, "Pierre Prevost," in *Dictionnaire des sciences philosophiques* (Paris, 1851), 208–211; A. P. de Candolle, "Notice sur Pierre Prevost," in *Archives des sciences physiques et naturelles*, **19** (1839), 1–10; and A. Cherbuliez, *Discours sur la vie et les travaux de feu Pierre Prevost* (Geneva, 1839).

JOHN G. BURKE

PREYER, THIERRY WILLIAM (*b.* Moss Side, near Manchester, England, 4 July 1841; *d.* Wiesbaden, Germany, 15 July 1897), *physiology*.

Preyer was educated first in London and then in Duisburg, following his immigration to Germany in 1857. He began to study science at Heidelberg in 1859, concentrating on physiology and chemistry, and received the doctorate in 1862. While still a student he traveled to Iceland with the geologist Ferdinand Zirkel. The results of their journey were published in 1862. Turning from natural science to medicine, Preyer studied in Paris with Claude Bernard and Charles-Adolphe Wurtz, and later in Berlin, Vienna, and Bonn. In 1865 Preyer qualified at Bonn as a lecturer in zoophysics and zoochemistry in the Faculty of Philosophy. The following year he received his medical degree at Bonn. In 1867 he qualified as a teacher of physiology at the Faculty of Medicine of the University of Jena, where in 1869 he succeeded Johann Nepomuk Czermak as professor of physiology and was appointed director of the Physiology Institute. Poor health obliged him to retire in 1888, and he spent the rest of his life as a private scholar in Wiesbaden.

Preyer's initial scientific work dealt with physiological chemistry (hemoglobin, gases in the blood, curare) and sense physiology (myophysical law). His lactic acid theory of sleep (1876) became widely known. His most significant contributions, however, are in his writings on psychology, especially *Die Seele des Kindes* (1882). With this work Preyer established himself as one of the founders of modern developmental psychology—an approach by which the descriptive natural sciences could exhibit the attributes of the determinate and empirically oriented disciplines. Its prime method was the chronological arrangement of observations made throughout the psychological life of the child.

Preyer frequently tackled more general scientific problems. In *Naturwissenschaftliche Thatsachen und Probleme* (1880) he discussed the possibility that inorganic matter had emerged from living systems.

The sole difference between them, he asserted, was that the former can come into being in various ways, whereas all living creatures derive from organic bodies and therefore must undergo a process of development. This development, which cannot be interrupted, goes as far back as the period when the earth was in a fluid, incandescent state; and Preyer interpreted movements within it as themselves signs of life. All inorganic compounds were at first absent from the earth, which in Preyer's view was teeming with life. Only with increasing cooling did compounds condense, and these became steadily more like protoplasm, the material basis for the evolution of the organisms.

Preyer's definition of protoplasm, in *Elemente der allgemeinen Physiologie* (1883), directly challenged the view of the older biologists, who considered protoplasm to be a homogeneous substance. For Preyer it was a "mixture of solids and liquids, of very complex chemical compounds that in the present stage of life are going through rapid and uninterrupted decomposition and re-formation." In his genetic system of the elements (1893) Preyer took into account Mendeleev's periodic system as well as the periodic laws formulated by Julius Lothar Meyer and the seven divisions first recognized by John A. R. Newlands.

BIBLIOGRAPHY

I. ORIGINAL WORKS. Preyer's earlier writings include *Reise nach Island* (Leipzig, 1862), written with F. Zirkel; *Rétablissement de l'irritabilité des muscles roides* (Paris, 1865); *De haemoglobino observationes et experimenta* (Bonn, 1866), his dissertation; *Empfindungen* (Berlin, 1867); *Über einige Eigenschaften des Haemoglobins und Methaemoglobins* (Bonn, 1868); *Grenzen des Empfindungsvermögens und des Willens* (Bonn, 1868); *Die Blausäure*, 2 vols. (Bonn, 1868–1879); *Kampf ums Dasein* (Bonn, 1869); *Die fünf Sinne des Menschen* (Leipzig, 1870); *Die Blutkrystalle* (Jena, 1871); *Die Erforschung des Lebens* (Jena, 1871); *Das myophysische Gesetz* (Jena, 1874); *Aufgaben der Naturwissenschaften* (Jena, 1876); *Über die Grenzen der Tonwahrnehmung* (Jena, 1876); *Über die Ursachen des Schlafes* (Stuttgart, 1877); *Kataplexie und Hypnotismus* (Stuttgart, 1878); and *Naturwissenschaftliche Thatsachen und Probleme* (Berlin, 1880).

Later works include *Die Entdeckung des Hypnotismus* (Berlin, 1881); *Farben- und Temperatursinn* (Jena, 1881); *Concurrenz in der Natur* (Breslau, 1882); *Die Seele des Kindes* (Leipzig, 1882; 8th ed., 1912); *Elemente der allgemeinen Physiologie* (Leipzig, 1883); *Ein neues Verfahren zur Herabsetzung der Körpertemperatur* (Jena, 1884); *Aus Natur- und Menschenleben* (Stuttgart, 1885); *Spezielle Physiologie des Embryo* (Leipzig, 1885); *Naturforschung und Schule* (Stuttgart, 1887); *Biologische Zeitfragen* (Berlin, 1889); *Über die Erhaltung der Energie. Briefe*

Robert von Mayers an Wilhelm Griesinger nebst dessen Antwortschreiben aus den Jahren 1842 bis 45 (Berlin, 1889), which Preyer edited; *Der Hypnotismus* (Vienna–Leipzig, 1890); *Die geistige Entwicklung in der ersten Kindheit, nebst Anweisung für Eltern, dieselbe zu beobachten* (Stuttgart, 1893); *Das genetische System der chemischen Elemente* (Stuttgart, 1893); *Der Prozess Czynski. Thatbestand desselben und Gutachten über Willensbeschränkung durch hypnotisch-suggestiven Einfluss* (Stuttgart, 1895); and *Zur Psychologie des Schreibens* (Hamburg, 1895; 2nd ed., Leipzig, 1912). See also Poggendorff, III, 1069–1070.

II. SECONDARY LITERATURE. Two short biographies are in *Biographisches Lexikon der hervorragenden Ärzte der letzten fünfzig Jahre* (Munich–Berlin, 1962), 1246; and *Biographisches Lexikon hervorragender Ärzte des neunzehnten Jahrhunderts* (Berlin–Vienna, 1901), 1323–1325, with portrait.

A. GEUS

PRICHARD, JAMES COWLES (*b.* Ross, Herefordshire, England, 11 February 1786; *d.* London, England, 22 December 1848), *anthropology, natural history, linguistics.*

Prichard was the eldest of four children born to Thomas and Mary Prichard. His early schooling was conducted at home by his father, and by tutors. He displayed a precocity in languages that was further stimulated when the family moved to Bristol, a cosmopolitan port. Prichard later dated his interest in anthropology to this period.

In 1802 Prichard began his apprenticeship in medicine (the only profession readily accessible to Dissenters) under Thomas Pole of Bristol and then under Robert Pope and William Tothill of Staines, all prominent Quakers. In 1805 he attended lectures at Saint Thomas's Hospital, London, and in the summer of 1806 entered the Medical Faculty of the University of Edinburgh, then at the peak of its renown. At Edinburgh he met weekly with members of a private debating society, the Azygotic, in which anthropological topics were often discussed. In addition to medical courses, he attended Dugald Stewart's lectures on moral philosophy; and it was a remark in one of these that stimulated Prichard, in 1808, to devote his medical dissertation to human races. This short monograph was later expanded into his *Researches Into the Physical History of Mankind*, a monumental work which, in its third edition, comprised five lengthy volumes and remained a major reference work in anthropology into the 1870's. Having received his M.D., Prichard studied at Trinity College, Cambridge, and, after converting to the Church of England (from Quakerism), at Saint John's and Trinity colleges, Oxford. In 1811 he became physician to Saint Peter's Hospital, Bristol, and in 1814 to the Bristol Infirmary; he also developed a substantial private practice. Prichard was a pioneer in the "moral" treatment of insanity, a subject on which he wrote several influential books.

In 1811 Prichard married Anne Maria Estlin, whose father was a friend of Coleridge, Priestley, and Southey. Prichard's home was a center of intellectual life in Bristol and attracted many eminent visitors. In 1835 he received an honorary doctorate from Oxford and in 1845 was appointed a commissioner of lunacy in London. He was also elected a fellow of the Royal Society (1827) and served as president of the Ethnological Society of London (1847–1848).

Prichard wrote that his anthropological interests were first aroused by hearing a challenge to the Scripture regarding the single origin of human races. Although he was convinced of single origin, he was unable to believe that racial differences were caused by the direct action of environmental factors, as maintained by Buffon, Blumenbach, and other authorities. In his medical dissertation he argued that changes due to external factors affect only the individual and are not transmitted to the next generation. The origin of races, and of animal and plant varieties in general, he attributed instead to the accumulation of what he called "connate" variations, which appeared for unknown reasons in the ovum or germ of the parents and which were invariably transmitted to the offspring. He supported this claim by cases from the medical literature (including albinism and other hereditary skin conditions) and by the accomplishments of animal breeders. Prichard had thus sketched a position similar to that upon which Darwin later based his theory of natural selection.

In successive expansions of his dissertation, Prichard elaborated his theory of connate variations, although he equivocated increasingly about the role of external conditions. In 1813 he introduced John Hunter's view that civilization (or domestication, in the case of animals) conduced to the appearance of lighter coloration and proposed that racial differences were due to varying progress from an originally dark, uncivilized stock. In 1826 he abandoned this view but was pressed, because of evidence of racial adaptation, to acknowledge a factor in the environment that produced variants, which were able to survive in specific regions. Prichard recognized the inconsistency in his position but was unable to resolve it; and in the final version of his work on races, published from 1836 to 1847, he omitted the discussion of heredity, variation, and race formation. Since this was the edition read by most scientists, including Darwin, Prichard's direct influence on thinking about these topics was probably minor. Indirectly, however, he did have an effect; William Lawrence drew extensively upon the 1808 dissertation in his much-reprinted and influential *Lectures on Physiology, Zoology and the Natural History of Man* (London, 1819).

In the study of geographical distribution, Prichard's influence was more immediate. He addressed this topic, from 1813, in order to establish that if man is a single species, in the sense of having no "constant and perpetual" racial differences, then he is likely to be descended from a single original stock. Taking this point for granted was a fallacy for which Buffon had been criticized. Prichard assembled masses of evidence to show that every species is either isolated in the region of its origin or is spread only over regions across which plausible courses of migration (either now or in the past) can be charted. Finding such localization to be the rule, rather than the exception, he concluded that man, if a single species, must have originated in a single place and, therefore, most likely from a single stock. Contemporary scientists, including Lyell and Swainson, cited Prichard's evidence together with Candolle's and gave his work the highest praise.

Prichard's greatest influence was as an anthropologist. In his effort to show mankind to be a single species, he compiled evidence in four different fields. First, he examined physiological and psychological characters of races; indeed, he was one of the first to conceive the possibility of a comparative psychology. Second, he sought examples of stable populations formed by racial hybridization. Third, he followed Blumenbach in comparing racial anatomy, endeavoring to show racial variation comparable to variation within accepted animal species. Fourth, he conducted what he called his "ethnographical investigation," in which he surveyed the entire world, country by country, assembling not only physical descriptions but also linguistic and cultural evidence of connections among races. The first three of these investigations were fairly rapidly superseded by the work of others. The fourth, however, which was quite original in the context of race theory, had a more lasting usefulness. Prichard was increasingly convinced, as his work proceeded, that cultural and linguistic "artifacts" were the surest index to the history of races. He therefore devoted four volumes to such material in the third edition of *Researches*, as well as two more specialized treatises, *An Analysis of the Egyptian Mythology* (London, 1819; second edition, London, 1838) and *Eastern Origin of the Celtic Nations* (London, 1831). (In the latter work, he anticipated Adolphe Pictet in arguing for the Indo-European character of the Celtic languages.) These studies, which had their roots in the British antiquarian tradition, were given serious although somewhat grudging attention by German historical linguists and students of mythology, and were later praised by E. B. Tylor.

Prichard failed to convince his successors of the single origin of mankind. Shortly after his death, the majority view of anthropologists swung definitely to the contrary position; and, somewhat later, the entire question was put on a different footing by the theory of evolution. Prichard's methodology, which had no element of fieldwork, was also rapidly superseded. Yet, in assembling an enormous store of organized data on human populations, which even in the 1870's Paul Topinard called the anthropologist's *vade mecum*, Prichard laid important groundwork for later research.

BIBLIOGRAPHY

I. ORIGINAL WORKS. Prichard's ideas are best illustrated in "Disputatio inauguralis de generis humani varietate" (M.D. diss., Univ. of Edinburgh, 1808) and the three eds. of *Researches into the Physical History of Mankind* (London, 1813; 2nd ed., 1826; 3rd ed., 1836–1847); the first ed. has the word *Man* rather than *Mankind*. The first vol. of the third ed. was reissued with the legend, "Fourth Edition," but the actual fourth ed. (1851), edited by W. Norris, had nothing new by Prichard himself. See also his more popular *Natural History of Man* (London, 1843) and his Anniversary Addresses to the Ethnological Society of London in *Journal of the Ethnological Society*, **1** (1846–1848), 301–329, and **2** (1848–1850), 119–149. In another context, his *Review of the Doctrine of a Vital Principle* (London, 1829) has some significance. Prichard also wrote several books on medical topics, mostly on insanity, and numerous scattered articles and reviews. Note also the references in the text.

II. SECONDARY LITERATURE. The major source of information is an obituary by Thomas Hodgkin, in *Journal of the Ethnological Society*, **2** (1848–1850), 182–207, to which a few details are added by John Addington Symonds' memoir of March 1849, reprinted in his *Miscellanies* (London, 1871), 116–144. Other obituaries (e.g., that of the Royal Society) add nothing. See also Isabel Southal, *Memorials of the Prichards of Almeley and their Descendants*, 2nd ed. (Birmingham, 1901). The major treatment of his biological ideas has long been E. B. Poulton, "A Remarkable Anticipation of Modern Views on Evolution," in *Science Progress*, n.s. **1** (1897), 278–296; this is now largely superseded by George Stocking's introductory essay to the reprint of *Researches into the Physical History of Man* (Chicago, 1973), which also gives an extended treatment of Prichard's ethnological work. For Prichard's influence on Lawrence, see Kentwood D. Wells, "Sir William Lawrence (1783–1867). A Study of Pre-Darwinian Ideas on Heredity and Variation," in *Journal of the History of Biology*, **4** (1971), 319–361.

HERBERT H. ODOM

PRIESTLEY, JOHN GILLIES (*b.* Cottingley Hall, Bingley, Yorkshire, England, 10 December 1879; *d.* Oxford, England, 9 February 1941), *physiology, medicine*.

Priestley was the son of Charles Henry Priestley and Ann Ford Gillies. His family possessed private means, derived from the textile industry. In 1908 he married Elizabeth Stewart. Priestley was educated at Eton and Christchurch, Oxford, where he took first-class honors in physiology soon after the turn of the century. He subsequently achieved medical qualifications at St. Bartholomew's Hospital, and following an additional year of training in Vienna he was appointed director of the chemical pathology laboratory at St. Bartholomew's. The emergence in 1912 of a tubercular condition (from which he suffered for the rest of his life) necessitated his move from London to Oxford in that year. With the outbreak of World War I, Priestley joined the Royal Army Medical Corps; he served with distinction in France and Belgium and was awarded the Military Cross. After the war Priestley returned to Oxford, where he was appointed reader in clinical physiology. There he spent the remainder of his life, teaching and doing research in physiology and medicine.

Priestley's scientific reputation rests primarily upon his collaboration with his Oxford mentor, John Scott Haldane, which resulted in their classic paper on the regulation of lung ventilation (1905). In the words of G. E. Allen, this paper "offered for the first time a completely workable system for the regulation of the respiratory rate both in terms of the changes in the external environment, and in terms of the regulatory mechanisms within the body" ("J. S. Haldane . . .," p. 406). Haldane and Priestley were able to demonstrate that the carbon dioxide content of the blood—which depends upon the concentration of carbon dioxide in alveolar air—is principally responsible for changes in both depth and frequency of respiration. Haldane and Priestley built upon Haldane's previous work and that of F. Miescher-Rüsch and Léon Fredericq; but most earlier research on respiratory regulation had failed to clarify the relative significance of neural, as opposed to blood-chemical, mechanisms and of carbon dioxide, as opposed to oxygen, levels in the blood.

Priestley's share in this work is difficult to assess. The 1905 paper was merely the most notable of Haldane's many technical and conceptual contributions to the field of respiratory regulation, contributions which had begun in the early 1890's before Priestley's arrival at Oxford as an undergraduate and which continued after he went to London to study medicine at St. Bartholomew's. Priestley's later work, principally on regulatory mechanisms in respiration, renal physiology, and blood chemistry, were marked by the technical precision and rigor that he had learned from Haldane. Late in World War I, Priestley joined Haldane and some of the latter's other students in an investigation of the aftereffects of gas poisoning.

An expert bibliographer, Priestley was responsible for the index to the first sixty volumes of the *Journal of Physiology* and for the editing for some years of *Physiological Abstracts*. He was joint editor of *Human Physiology. A Practical Course* (Oxford, 1924) and assisted Haldane in the preparation of the 1935 edition of his *Respiration,* based on Haldane's Silliman lectures at Yale and originally published in 1922.

BIBLIOGRAPHY

I. ORIGINAL WORKS. Priestley's papers and books are listed in *Index medicus.* His major paper with J. S. Haldane is "The Regulation of the Lung-Ventilation," in *Journal of Physiology*, **32** (1905), 224–266. MS material relating to his work on the aftereffects of gas poisoning is in the Sinclair Collection of the Woodward Biomedical Library, University of British Columbia.

II. SECONDARY LITERATURE. There are obituaries of Priestley in *British Medical Journal* (1941), **1**, 299–300; *Nature*, **147** (1941), 319–320; and *St. Bartholomew's Hospital Journal* (wartime ed.), **2** (1941), 167. The principal historical source is Garland E. Allen, "J. S. Haldane: The Development of the Idea of Control Mechanisms in Respiration," in *Journal of the History of Medicine and Allied Sciences*, **22** (1967), 392–412. For the wider context within which the Priestley-Haldane work emerged, see J. F. Perkins, Jr., "Historical Development of Respiratory Physiology," in W. O. Fenn and H. Rahn, eds., *Handbook of Physiology. Section 3: Respiration*, I (Washington, D.C., 1964), 1–62.

RICHARD D. FRENCH

PRIESTLEY, JOSEPH (*b.* Birstal Fieldhead, Yorkshire, England, 13 March 1733; *d.* Northumberland, Pennsylvania, 6 February 1804), *chemistry, electricity, natural philosophy, theology.*

Educated for the dissenting ministry and employed most of his adult life as a teacher or preacher, Priestley wrote books, pamphlets, and articles on theology, history, education, metaphysics, language, aesthetics, and politics, as well as on scientific subjects. In his own day he was as well known for his religious and political views as for his science, although he is chiefly remembered as the discoverer of oxygen.

Priestley was the eldest son of a Yorkshire cloth dresser, Jonas Priestley, and his first wife, Mary Swift. He spent much of his infancy with his maternal grandparents. Shortly after the death of his mother in 1739, he was sent to live with an aunt, Sarah Priestley Keighley, who was left a propertied, childless widow in 1745. Priestley lived with his aunt until he was nineteen. Perhaps some of his lifelong independence from authority can be credited to this continual separation from his immediate family. Certainly, by the age of eighteen he had rejected the sterner Calvinism of his family's Independency for the Arminianism of eighteenth-century English Presbyterians, whose clergy were sometimes entertained in his aunt's house.

Having early demonstrated an affinity for learning, Priestley was encouraged to study for the ministry. He attended local parish schools and supplemented his formal training by private lessons and self-directed studies in ancient languages, mathematics, and introductory natural philosophy, reading on his own 's Gravesande's elementary text, *Mathematical Elements of Natural Philosophy Confirmed by Experiments.* When ill health prompted a change in plans, Priestley learned modern languages in preparation for employment as a clerk, but his health returned and at the age of nineteen he entered the dissenting academy at Daventry, the successor to the famous Northampton Academy of Philip Doddridge, which was continued under two of Doddridge's former students, Caleb Ashworth and Samuel Clark. Because of his assiduous self-education, he was excused the first year of the course of studies and part of the second, but subsequent years were to disclose some weaknesses in his preparation. Although he claimed to have read Latin, Greek, Hebrew, some Syriac and Arabic, French, Italian, and High Dutch (German) and to have learned various branches of theoretical and practical mathematics (all prior to his admission to Daventry), he later admitted having to relearn German in order to read scientific works in that language, and his mathematics was always weak.

At Daventry, Priestley was sufficiently grounded in Latin and Greek to hold his own in subsequent disputes with university-trained scholars. He was more generally introduced to a range of subjects in natural philosophy, but more significantly, he was there formally instructed in logic and metaphysics. Doddridge had favored an educational method of free inquiry on controversial subjects, based on a study and comparison of conflicting texts. This method was continued at Daventry through the use of Doddridge's lectures, which were later published as *A Course of Lectures on the Principal Subjects in Pneumatology, Ethics, and Divinity* (1763). From these lectures Priestley acquired a knowledge of the empirical philosophy of Locke, the theology of Richard Baxter, the speculations of the Cambridge Neoplatonists, and the Newtonian natural philosophy and theology of the early Boyle lecturers. At Daventry he also read Rowning's *Compendious System of Natural Philosophy*, Hartley's *Observations on Man, His Frame, His Duty, and His Expectations*, and one of the English translations of Boerhaave's *Elementa chemiae*. By the time he left Daventry in 1755, Priestley was tending toward an Arian position in theology and had begun compounding a metaphysics of Locke, Newtonian corpuscular philosophy, and the determinism and associationist psychology of Hartley. In all of his subsequent work (theological, educational, political, and scientific) Priestley illustrated the utilitarian, reductionist approach based on this system.

Completing his studies, Priestley went to preach, first at Needham Market, Suffolk, and then at Nantwich, Cheshire. He was successful in neither of these posts. Difficulties arose in the former because of his heterodox position on the Trinity, and in both pulpits he was handicapped by an inherited speech impediment. At Nantwich he opened a school that proved so successful that he was invited to become tutor of languages and belles lettres at the recently founded dissenting academy at Warrington, to which he moved in 1761. There he seems to have taught languages, modern history, law, oratory and criticism, and even anatomy—nearly everything, that is, but theology and experimental sciences, which, he said, he should most like to have taught. While at Warrington he published *The Rudiments of English Grammar* (1761) and *A Course of Lectures on the Theory of Language* (1762), both cited by modern scholars for their early recognition of recently rediscovered linguistic principles and for their singular utilitarian insistence on usage as the only proper guide to correct English. He also prepared *A Course of Lectures on Oratory and Criticism*, published several years later (1777). It emphasizes, on Hartleian grounds, practical rules of oratory and an associationist aesthetic criticism, which—as further developed by Archibald Alison— was greatly to influence the poetry of Coleridge and Wordsworth.

Priestley's general philosophy of education was described in an *Essay on a Course of Liberal Education for Civil and Active Life* (1765) and *Miscellaneous Observations Relating to Education* (1778), and his teaching in history and law was outlined in *Lectures on History and General Policy*, prepared at Warrington but not published until 1788. The *Essay on the First Principles of Government* (1768), from which Jeremy Bentham declared that he had derived his Utilitarian formula: "The greatest happiness of the greatest number," was, therefore, but one of many protoutilitarian works written by Priestley.

While at Warrington, Priestley was ordained and obtained an LL.D. from the University of Edinburgh (1764) in recognition of his work in education. There he also began his scientific career, with the writing of his *History of Electricity* for which he enlisted the support of Benjamin Franklin, John Canton, Richard Price, and William Watson, whom he met in London late in 1765. At their suggestion—before the *History* was published but after some of his experiments were known privately to his sponsors—he was nominated and elected F.R.S. in 1766.

In 1762 Priestley married Mary Wilkinson, the daughter of Isaac and the sister of John and William Wilkinson, three of the great ironmasters of eighteenth-century England. William had been a pupil of Priestley at Nantwich and Warrington. Early in 1767, because of growing family responsibilities and the perennial financial and sectarian problems of Warrington, Priestley resigned his teaching position to become minister of Mill-Hill Chapel, a major Presbyterian congregation in Leeds. The *History of Electricity* (1767) and the *History of Optics* (1772) were published while he was at Leeds, and he there began his most famous scientific researches, those into the nature and properties of gases. He also completed his itinerary from orthodoxy by adopting Unitarian (then called Socinian) principles and commenced his writing of controversial theological works.

In 1773 William Petty, Earl of Shelburne and later Marquis of Lansdowne, prevailed on Priestley to enter his service. Priestley became Shelburne's resident intellectual, although officially he served as librarian and adviser to the household tutor. He probably also was useful as unofficial liaison between Shelburne and the politically active dissenting interest. Although Priestley regularly disclaimed any particular concern with politics, his disclaimer must be viewed against the Test and Corporation Acts, which virtually forced all eighteenth-century nonconformists to become involved in politics. Priestley had already written some politico-religious pamphlets (for example, *Remarks on . . . Dr. Blackstone's Commentaries on the Laws of England* [1769], regarding the legality of dissent); he was to continue with others, particularly at the suggestion of his friend Franklin, in support of the American colonies.

Shelburne, a disciple of Lord Chatham and sometime associate of the Rockingham Whigs, excited the animosity and distrust of most professional politicians; but toward Priestley his actions were always honorable and friendly. Together they toured the continent in the summer and fall of 1774—Priestley's only visit to Europe, during which he met many scientists in Paris. Priestley and his family had a house in Calne, near Bowood, the Shelburne estate in Wiltshire, and during the season lived at Shelburne House in London. During this period Priestley did most of his scientific work, preparing five of the six major volumes of experiments on gases. He also wrote his major philosophical works, including the only books that come close to relating explicitly his theological and scientific philosophies, the *Disquisitions Relating to Matter and Spirit* (1777) and the *Doctrine of Philosophical Necessity, Illustrated* (1777). During 1779 the relationship between Priestley and Shelburne cooled (possibly as a result of Shelburne's second marriage in that year) and Priestley chose to leave Shelburne's service in 1780, retaining till his death the annuity promised him should such a separation occur.

At the suggestion of his brother-in-law, John Wilkinson, Priestley settled with his family in Birmingham for what were to be the happiest years of his life. He became preacher at New Meeting House, one of the most liberal congregations in England, and was soon associated with the Lunar Society, an informal collection of provincial intellectuals, scientists, and industrialists. The Lunar Society—comprised during Priestley's years of Matthew Boulton, Erasmus Darwin, Richard Lovell Edgeworth, Samuel Galton, Jr., Robert Augustus Johnson, James Keir, Jonathan Stokes, James Watt, Josiah Wedgwood, and William Withering—combined widely ranging curiosity about nature with pragmatic concerns that would most appeal to Priestley. The members supported his researches intellectually and financially. He began there his opposition to the new chemical system of Lavoisier and devoted much of his experimentation to the application of scientific phenomena to practical pursuits.

Priestley's major preoccupations, however, were increasingly theological. He became the chief propagandist and protagonist for Unitarian beliefs in England, writing annual defenses against attack, and developing in various historical and polemical works (for example, *An History of the Corruptions of Christianity* [1782] and *An History of Early Opinions Concerning Jesus Christ* [1786]) a rationalist theology that suggests, in some measure, the ideas of textual and "higher criticism" of the New Testament. In the eyes of the church establishment, he came to represent the intolerable encroachments of dissent, and on him was focused their theological and political animus.

The opportunity to silence Priestley came in 1791, as a result of his support of early phases of the French Revolution and his related criticism of continued political discrimination against theological dissent in England. A "Church-and-King" mob in Birmingham destroyed the New Meeting House and Priestley's house and laboratory and threatened his personal safety. He removed to London and was briefly associated with the dissenting academy at Hackney, where he taught natural philosophy and preached; but increasing signs of political persecution, including economic sanctions against his sons, prompted him to emigrate to the United States in 1794.

Although he was warmly received on his arrival and offered a position as professor of chemistry at the University of Pennsylvania, he chose to settle in Northumberland, Pennsylvania, near what was intended to become a settlement for British émigrés fleeing political repression. The settlement failed to develop, but Priestley and his family remained. Increasingly remote from current developments, he continued his theological writings and his scientific work. Taking the side of Jeffersonian opposition to the administration of John Adams, Priestley was vilified in the Federalist press; but the election of Jefferson to the presidency in 1800 changed his situation. Priestley declared that for the first time in his life he was living in a country in which the political authorities were friendly to him. Jefferson personally befriended the aging Priestley, who was ill and lonely; his youngest and favorite son had died in 1795 and his wife in 1796. Priestley died in 1804, at the age of seventy-one, leaving two sons in the United States and a married daughter in England. He is buried at Northumberland, where his house has been turned into a museum.

Scientific Work. Priestley's scientific work was begun as a logical extension of his interests in education. *The History and Present State of Electricity, With Original Experiments* was conceived as a methodized account of previous discoveries and an assessment of contemporary electrical studies, to encourage further work on the subject. That is, the work was to be a "history" in the Baconian sense; and as a chronicle of near-contemporary and contemporary electrical researches, lucidly and simply described, it was very successful. During Priestley's lifetime the work went through five English editions and was translated into Dutch, French, and German.

The first edition was marred by Priestley's slight access to the work of German and Scandinavian electricians (a deficiency corrected in later editions through reference to the historical accounts by Daniel Gralath in the *Versuche und Abhandlungen der Naturforschenden Gesellschaft in Danzig*), but he made a serious attempt to consult primary sources and his account is fairly impartial, although it understandably favors the currently popular one-fluid theory of electricity as developed and demonstrated by his friend Benjamin Franklin and the Franklinian school.

His advisers, with whom he regularly exchanged correspondence and to whom he sent sections of the *History* for criticism, supplied Priestley with books and references and gave him hints for experiments, which he initially performed to verify or elucidate the work of others. From these experiments he was soon drawn into original investigations, which he described in the *Philosophical Transactions of the Royal Society* and included in successive editions of the *History*. His experiments relate primarily to conductivities of different substances, although he also examined other modes of the motion of the electrical fluid. He discovered the conductivities of charcoal and of metallic salts, ranged the metals in a table of comparative conductivities, first noted the distinctive marks left by spark discharges on metallic surfaces—now known as "Priestley's rings"—and examined the phenomena of "electric wind" and sideflash. His most remarkable electrical discovery came as an interpretation of an experiment by Franklin. From the observation that pith balls lowered within an electrified metallic cup were not influenced by electricity, Priestley deduced, on Newtonian grounds, the inverse-square form of the force law between electrical charges. The publication of this deduction in the *History* passed nearly unnoticed (as had that of Daniel Bernoulli in 1760), but it probably inspired Cavendish's subsequent experimental determination of the force law.

The success of the *History of Electricity* involved Priestley in a scientific career in which his original experiments had, initially, little place. Almost at once he commenced an international correspondence with scientists such as Bergman and Volta, who were anxious either to correct misunderstandings in the *History* or to win a place in a continuation of that work beyond 1766, which Priestley had promised. This continuation never appeared; in subsequent revised editions Priestley merely corrected and improved accounts contained in the first edition and added additional experiments of his own.

The educational function of the *History* was extended by *A Familiar Introduction to the Study of Electricity* (1768), which was intended for beginners; and this was followed by *A Familiar Introduction to the Theory and Practise of Perspective* (1770), written from Priestley's own experience in drawing illustrations of apparatus for the *History*. The reception given the *History* encouraged Priestley to undertake a multivolumed history of all the experimental sciences. After some hesitation he determined that the next volume should be on optics and in 1772 published his *History and Present State of Discoveries Relating to Vision, Light, and Colours* (usually referred to as the *History of Optics*).

Although the *History of Optics* contains much useful information, it was considerably less successful than the *History of Electricity*. Until recently optics had not attracted substantial historical interest, and although Priestley's *History of Optics* had but one English edition and a translation into German, it remained the only English work on the subject for a hundred and fifty years and the only one in any language for over fifty. Eighteenth-century optical concerns were primarily ontological and mathematical; even Priestley's own experimental speculations on the variation of optical parameters with electrification and on the indices of refraction of different gases, relate to his electrical or pneumatic studies and do not appear in the *History of Optics*. Yet Priestley was not mathematically minded and avowed his intention of presenting the material so as to make it perfectly accessible to readers with little or no knowledge of mathematics. His Baconian design of an exhaustive chronicle was precluded by the masses of material requiring condensation, and he lacked the discrimination to select wisely a ruling principle around which the work might be organized.

Priestley permitted himself one theoretical judgment in expressing strong reservations on the reality of Newton's optical ether, but the most interesting and original sections of the *History of Optics* were derived from the work and suggestions of John Michell, rector of the neighboring parish of Thornhill. Michell, a geologist, astronomer, and student of magnetism, assisted Priestley in the preparation of the *History of Optics*, advised him in the interpretation of optical experiments and phenomena, and provided him with an account of an experiment to measure the momentum of light, which appeared to confirm its particulate nature. Apparently, it was also Michell who called Priestley's attention to Bošković's theory of matter, as contained in the *Philosophiae naturalis theoria* (1763 edition); however, this was not Priestley's first introduction to a theory involving alternating, concentric spheres of repulsion and attraction surrounding fundamental particles of matter. Rowning's *Compendious System of Natural Philosophy* (1735–1743), which he had read at Daventry and from which he borrowed plates to illustrate the *History of Optics*, contains a discussion of just such a theory. Boškovič described the details and implications of the theory of matter with far more completeness than anyone had earlier done and he extended the theory to the dissolution of any material substratum for the particles, contracting them into geometrical points.

Here was the ultimate in reduction of phenomena to simple elements; the prospect fascinated Priestley. He described Boškovič's theory in detail in the *History of Optics* and used a variation of the theory—conceived by Michell—in explanation of the phenomenon of Newton's rings, for which Newton had introduced the reciprocal interaction of vibrations in an ether and in refracting bodies. Priestley also used the theory in his metaphysics and theology.

From the publication of the *History of Optics*, Priestley continued into his major metaphysical writings. His *Examination of Dr. Reid's Inquiry Into the Human Mind . . .* (1774) criticized the Scottish philosophy of common sense for its multiplication of entities (in this case independent instincts or affections of the mind) in contradiction to Newton's "Rules of Reasoning" and to the reductionist principles of Hartley, whose work, minus the mechanistic physiology, Priestley edited as *Hartley's Theory of the Human Mind . . .* in 1775. His edition of Hartley emphasized the deterministic and associationist aspects of Hartley's psychology and led him to the *Disquisitions Relating to Matter and Spirit* (1777).

The *Disquisitions* is primarily a theological work, but in the first part Priestley borrowed from Boškovič and also from Hartley in order to construct a theory of matter and of mind and soul as material substance. As the "material" is that of Boškovič's geometrical points surrounded by alternating concentric spheres of attracting and repelling force, it would appear that Priestley had really dissolved all matter in a matrix of forces. These forces were, at least to the Cambridge Neoplatonists and early Newtonian natural philosophers, identifiable with the omnipresent will of God. As Priestley further insisted (on the grounds of Revelation) that body and soul were to be reconstituted on a resurrection of man at the Second Coming of Christ, the storm of criticism that stigmatized Priestley as an atheistic materialist is not easy to understand. Although he defended himself vigorously —for example, in a friendly controversy with Richard Price (*A Free Discussion of the Doctrines of Materialism and Philosophical Necessity* [1778]) in which he declared that matter is not self-existent but has meaning only in terms of its properties—Priestley seldom again ventured into an extended discussion of his theory of matter.

This was particularly unfortunate, for an understanding of his theoretical position might help to clarify what appear to be anomalies in Priestley's work on gases ("different kinds of air"), which he had begun concurrently with continued researches on electricity and the writing of the *History of Optics*. The failure of that *History* to pay the costs of the books that were collected to write it, and the evident problems involved in condensing the profusion of materials for use in any other subject, prompted Priestley to drop the idea of a history of all the experimental sciences. Besides, he had already embarked on those researches on gases, which were more than sufficient to occupy all the time he had to spare for his scientific work.

Priestley's concentration on pneumatic studies began comparatively late in his career (at the age of thirty-seven). This interest lasted for the remainder of his life; and during those thirty-odd years, and chiefly in the first ten of them, he was to establish himself as one of the world's foremost pneumatic chemists. His discoveries of new gases and new processes were to make the chemistry of his day seem untenable; but Priestley never developed a new system to encompass his discoveries, and he refused to adopt the system developed by Lavoisier, which did so.

Attempts to explain this apparent failure of creative synthesis have generated a legend for which Priestley is nearly as responsible as his critics and biographers. Accounts of his scientific work (left by Priestley in his *Memoirs*), supported by memorable quotations out of the context of his books and papers, have made it appear that he worked in ignorance of the chemical writings of others, that he used the crudest of apparatus, and that the discoveries of this most prolific of eighteenth-century British chemists were made haphazardly and accidentally. Detailed reading of all of Priestley's scientific writings and of his correspondence suggests that the germs of truth in this legend were exaggerated by Priestley in order to emphasize the contrast between his career and that of his great rival, Lavoisier. Subsequent writers enhanced the exaggeration in order to explain anomalies in his work as a chemist, but Priestley as a scientist is not be understood so easily.

As early as 1755 Priestley had read Boerhaave's textbook on chemistry. In 1762 he participated in planning, and then assisted in, a course of chemical lectures given at Warrington Academy by Matthew Turner, a physician-chemist of Manchester. It appears that chemical apparatus was acquired by the Academy and that the course of lectures was repeated in subsequent years. In 1766, in his first letter on electricity to Canton, Priestley refers to his reading of William Lewis' translation of Caspar Neumann's lectures, while further studies on electrical conductivity of gases and carbon in 1770 cite the opinion of "Macquer and other chemists." Finally, the "Catalogue of Books of Which Dr. Priestley Is Already Possessed or to Which He Has Access, for Compiling the History of Experimental Philosophy," appended to the *History of Optics* and published the year of his first paper on gases, included over fifty titles (exclusive of papers in the listed scientific journals), which relate primarily to chemistry. Clearly Priestley was aware, at least, of contemporary chemical literature as he commenced his own studies.

Yet in a sense most of this literature was irrelevant to Priestley's enterprise, for it relates primarily to that activity of separating substances into their constituents and recombining them, which defined the nature of eighteenth-century chemistry. Priestley showed so little interest for this activity that he repeatedly denied any particular knowledge of or concern with chemistry. Priestley was interested in the nature of gases; this was true as early as 1766 when he began his study of common, mephitic, and inflammable airs, and the relationships of the airs to one another, During 1769 a fourth edition of Hales's *Vegetable Staticks* was published, and early in 1770 Priestley wrote his friend Theophilus Lindsey, "I am now taking up some of Dr. Hales' inquiries concerning air." From this time Priestley's experimental interests were almost exclusively related to the study of "different kinds of air."

"Dr. Hales' inquiries concerning air" were those described in a long chapter, "A Specimen of an Attempt to Analyse the Air," in *Vegetable Staticks*, which had already been the inspiration of British pneumatic chemists before Priestley: Brownrigg, Black, Cavendish, and Macbride. Of Brownrigg's work, Priestley can have known very little prior to his own first paper on "airs," as the most significant parts were unpublished before 1774. Black's work was available in the *Essays and Observations, Physical and Literary. Read Before a Society in Edinburgh*, which Priestley cited in the appendix of the *History of Optics*. Although he knew of Black's work, Priestley seems never seriously to have studied it. Cavendish is cited in Priestley's correspondence by 1771, while Priestley knew of Macbride at least as early as 1767, when a laudatory review of the *History of Electricity* referred him to Macbride's studies on fixed air as an antiscorbutic. This, and the accident of Priestley's first settling in Leeds next to a brewery, with its ample supply of carbon dioxide, may explain why his first "chemical" publication was a pamphlet: *Directions for Impregnating Water With Fixed Air, in Order to Communicate to it the Peculiar Spirit and Virtue of Pyrmont Water, . . .* (1772).

The pamphlet was widely and favorably noticed; within the year it was translated into French, and it was an important factor in the awarding to Priestley of the Copley Medal of the Royal Society for 1773. The incongruity of a science honor awarded in part for the discovery of artificially carbonated water is more than balanced by the appearance of Priestley's magisterial "Observations on Different Kinds of Air," read to the Society in March 1772, with a supplement in November, and printed in the *Philosophical Transactions of the Royal Society* for that year. This paper reports Priestley's pneumatic researches since 1770, including the isolation and identification of nitric oxide and anhydrous hydrochloric acid gases; the beginnings of the discovery of photosynthesis; and a scarcely noted reference to an "air extracted from nitre," which appeared extraordinary and important to him, but on which he was not to experiment further for several years. Both the range and quality of the research described in this paper illustrate the major influence on Priestley of Hales's work.

Without in any way detracting from the personal characteristics of Priestley's work as a scientist—the ingenuity with which he diversified his experiments, the increasingly skilled manipulation of simple apparatus, and the tenacity with which he followed minor variations in results—it is necessary to emphasize the great difference in his mode of working before and after his reading of Hales. In the paper of 1772 and thereafter, the experiments performed, the instruments used, and the way of using them—but particularly the thinking that informed the experiments and guided their interpretation—are all developed from the chapter on airs of the *Vegetable Staticks*.

Hales had interpreted gases in terms of a Newtonian mechanical model in which a single elastic substance, air, could be fixed and made inelastic in other substances, and then could be released and made elastic again by processes of distillation (heating) or fermentation (mixing with acids, alkalies, or other fluids). He had thus released a variety of gases, but in spite of the title of his chapter, he had not chemically analyzed any of them; for he regarded them all as a single air made various by differing amounts of differing impurities. Black and Cavendish had each identified a particular species of air, but no one had examined the natures of all of the airs that might be released from substances.

This examination was the task Priestley set for himself. First he examined the airs as generated by Hales; and then, as his confidence grew, he set about generating new varieties—by heating substances, mixing them, or taking the residues from containers in which such processes as calcination, vegetation, or electrical discharge had taken place. He adapted the pneumatic trough with inverted receiver, pedestal apparatus, and supporting rack that had been used and improved by Hales, Brownrigg, and Cavendish. The occasional advantage of substituting mercury for water in the pneumatic trough was, for example, first noted by Cavendish. For his early experiments Priestley transformed household utensils (a laundry tub, beer and wine glasses, clay tobacco pipes) into chemical apparatus; but soon he was designing equipment to meet his particular requirements. Josiah Wedgwood freely supplied him with ceramic tubes, dishes, crucibles, and mortars; and the London firm of William Parker and Sons was his supplier of glassware, including the burning lenses that he frequently used to heat substances within the receivers of the pneumatic trough. An inventory of his apparatus destroyed in the Birmingham riots of 1791 (published as an appendix to H. C. Bolton's edition of Priestley's correspondence) reveals a well-stocked laboratory of sophisticated apparatus and a variety of reagents.

The tests Priestley initially used in distinguishing the airs that he produced were quite simple: Did they turn lime water turbid? Would they burn or support combustion? What was their appearance and taste? How long would a mouse live in a container filled with one of them? As he gained in knowledge his tests became more comprehensive. He developed techniques of eudiometry using nitric oxide, noted flame size and color in gases that burned or supported combustion, and even recorded the different colors of electric sparks through different gases. Most of his experiments were qualitative; when he did quantitative work it was generally volumetric and not gravimetric. For however skilled in experimental manipulation Priestley became, he never lost the conviction that the important pneumatic parameters were physical and mechanical rather than substantive and chemical.

Priestley's emphasis on mechanical considerations provided the rationale for an experimental program, which, from a chemist's viewpoint, appears chaotic. Of course, any successful set of experiments generates a kind of momentum in which one operation suggests another; this can frequently be seen in Priestley's career, as when the surprising reactivity of marine acid air (anhydrous hydrochloric acid) led to attempts to produce other highly active anhydrous acids and alkalies, and thus to the discovery of sulfur dioxide and ammonia. Moreover, any systematic investigation of the differences between airs must almost of necessity lead to the discovery of new airs. Priestley's well-reported "accidental" discovery of oxygen in 1774–1775 was not an accident of producing an air. Priestley had expected the discovery, having deliberately created

the conditions for it when he placed a piece of mercuric oxide within a receiver inverted in a pneumatic trough and heated it with his newly acquired burning lens. The surprise was in the unexpected nature of the gas released, which he had expected to be the same as the carbon dioxide he had found in heating impure red lead. Routine examination of substances could produce new airs but did not permit predictions as to their natures.

Internal momentum of experimentation and routine investigation of substances aside, it was the mechanistic mode of interpretation that provided continuity for Priestley's pneumatic investigations. This mode of investigation could be, and ultimately was, a weakness. All of his life Priestley persisted in believing that physical operations on gases (compression or rarefaction, agitation in water, electrification) would somehow transform one kind of air into another kind of air; and his search for mechanical-force explanations in a kind of premature physical chemistry blinded him to the prior necessities of classifying elements and compounds. Yet the same ideas were also involved in leading him in 1766 from a generalized concern to know the nature of the changes by which combustion or respiration made common air mephitic, to the development of eudiometry, the differentiation of oxygen from nitrous oxide, the work on photosynthesis, and his experiments that led Cavendish and Watt to discover the compound nature of water.

Priestley's experiments were carried on at such a prolific rate, that following the paper of 1772, it was decided that he should publish his accounts of them in book form. The first volume of *Experiments and Observations on Different Kinds of Air* appeared in 1774, the second in 1775, and the third in 1777. In 1779 Priestley began a new series, *Experiments and Observations Relating to Various Branches of Natural Philosophy*, continued with a second volume in 1781 and a third in 1786. (These six volumes are generally cited as forming a single series; in 1790 they were combined and edited in three volumes as *Experiments and Observations on Different Kinds of Air, and Other Branches of Natural Philosophy*.) These works were supplemented by an occasional paper in the *Philosophical Transactions* (including the "Account of Further Observations on Air" [1775], in which he announced his discovery of "dephlogisticated air," later to be defined as oxygen), and an extensive correspondence with other scientists in Britain and on the Continent.

During this period—in addition to his discovery of oxygen—Priestley described the isolation and identification of ammonia, sulfur dioxide, nitrous oxide and nitrogen dioxide, and silicon tetrafluoride.

He discussed the properties of mineral acids; further extended the knowledge of photosynthesis; defined the role of the blood in respiration; and noted, unknowingly, the differential diffusion of gases through porous containers. More than any other person, he established the experimental techniques of pneumatic chemistry. For over a decade Priestley dominated the scientific scene in Britain and attracted the attention of scientists throughout Europe. In 1784 he was elected one of the eight foreign associates of the Royal Academy of Sciences in Paris, and he was similarly honored by nearly a score of memberships in other scientific societies from Boston and Philadelphia to Stockholm and St. Petersburgh. His reign came to an end with the development of Lavoisier's new chemistry.

Priestley had met Lavoisier during his visit to Paris in 1774, when he exhibited the gas (as yet unidentified) that he extracted from mercuric oxide. Their first clash occurred the following year when Priestley asserted his claim to the discovery of a new gas and set Lavoisier right as to its essential properties. Subsequently, there were minor disagreements, but the confrontation did not become a major one until after the discovery in 1783, by Cavendish and Watt, of the compound nature of water. It was these two discoveries, of oxygen and the composition of water, that formed the experimental basis of Lavoisier's new, oxidation, chemistry; yet Priestley refused to accept Lavoisier's interpretation of either of them. First in papers in the *Philosophical Transactions*, then in privately printed pamphlets, and, finally, from the United States in papers in the *Transactions of the American Philosophical Society* and the *New York Medical Repository* (papers frequently republished in Nicholson's *Journal of Natural Philosophy* and the *Monthly Magazine and British Register*), Priestley presented experimental arguments to counter those of Lavoisier and his growing school of disciples. Many of these papers continue and repeat confusions that had dogged Priestley from his earliest experiments on gases: impurities in his reagents, for example, or difficulties produced by gaseous diffusions. Some of the objections that he raised did have merit. Not all acids do contain oxygen, as he demonstrated in the case of hydrochloric acid. Some of Priestley's experiments required the definition of yet another new gas, carbon monoxide, to which he has some claim of discovery.

Priestley's opposition to Lavoisier is frequently described as the conservatism of an old man unable to give up the doctrine of phlogiston, the principle of combustion whose use he had learned early in his career. There is no doubt but that Priestley employed both the language and the concepts of phlogiston chemistry in his arguments; but he regularly insisted that he was ready to abandon phlogiston should the advantage of doing so be demonstrated to him, and his entire career is that of a man not bound to tradition or convention.

Yet to penetrate behind the obvious to a deeper understanding of the differences between Priestley and Lavoisier remains a conjectural operation. Priestley always concealed the theoretical considerations that prompted his experimental investigations in a "Baconian" conviction that only "facts" were important, and he therefore adopted a mode of argument in which Lavoisier's experiments were countered by his own. But experiments do not stand independent of interpretation, and in the implications of their variant interpretation it is possible to see a fundamental ontological difference between Priestley and Lavoisier.

In his *Heads of Lectures on a Course of Experimental Philosophy, Particularly Including Chemistry* (1794), Priestley declared that changes in the properties of bodies may result from the addition of substances, from a change in the texture of the substance itself, or from the addition of something not a substance. It was the first of these methods of interpretation that had introduced the imponderable fluids of electricity, heat, and phlogiston and it was in this mode of explanation that Lavoisier's chemistry achieved its revolution, through its emphasis on mass as a parameter and on gravimetrics as a technique for defining the elements that entered into chemical composition. Priestley's training and instincts led him to prefer the second method of explanation as he twice declared—in the *Experiments on the Generation of Air From Water* (1793) and again in "Miscellaneous Observations Relating to the Doctrine of Air," in *New York Medical Repository* (1802), when he emphasized that the principle and mode of arrangement of elements in substances was the object of his investigations.

From the beginning of his scientific labors, Priestley was concerned to elucidate the dynamic, corpuscular view of matter outlined by Newton in the queries to the first edition and to the Latin edition of the *Opticks*. Although he adopted the fluid theory of electricity as the one that was most successful currently, his own experiments led him (at least in correspondence) to doubt the existence of an electric fluid, *sui generis*. In the same manner he was ambivalent about heat as a fluid, admitting the concept into his *Heads of Lectures* but earlier suggesting that heat was the vibratory motion of the parts of bodies. In the preface and in a concluding section to the *History of Electricity* he explicitly noted that electricity, optics, and chemistry combine to give information on the internal structure of bodies, on which their sensible properties depend.

In 1777 Priestley described his reason for experimenting with gases as the exhibiting of substances in the form of air, thus advancing nearer to their primitive elements. These are the elements, which, in the *Heads of Lectures*, he was to define as combinations of shared properties of extension and powers of attraction and repulsion. Except in the *History of Optics* and the *Disquisitions Relating to Matter and Spirit*, there is no indication that Priestley had adopted the theory of matter outlined by Bošković, but there are many suggestions that he had retained the idea—learned at school and reinforced by much of his reading—that a final scientific explanation required the reduction of phenomena into terms of the sizes, shapes, and motions of the fundamental particles of matter and the forces of attraction and repulsion between them. He could accept phlogiston, as he had tentatively accepted the fluids of electricity and heat, prior to the ultimate achievement of true, mechanical explanations. But he could not have accepted Lavoisier's antiphlogistic chemistry, for this was based upon a ratification of a multiplicity of substances. Priestley's work in science is thus consistent with that in his other endeavors, where his publications on language, aesthetics, psychology, politics, and particularly on theology, all reveal an attempt to reduce these subjects to a few basic elements interacting according to determinant laws.

BIBLIOGRAPHY

I. ORIGINAL WORKS. The major source for Priestley's personal life is the *Memoirs of Dr. Joseph Priestley, to the Year 1795, Written by Himself, With a Continuation to the Time of His Decease, by His Son, Joseph Priestley: and Observations on His Writings by Thomas Cooper* (Northumberland, Pa., 1805; London, 1806), which has been reprinted many times. No collected ed. of all of Priestley's writings exists. His *Theological and Miscellaneous Works*, John Towill Rutt, ed., 25 vols. (London 1817–1831), includes prefaces to the scientific works; Rutt annotated the various works included in his ed. but used no consistent policy for the eds. included and sometimes modified their form. A reissue of Rutt's ed. has been published.

Single-volume collections of extracts of Priestley's writings have been edited by Ira V. Brown, *Joseph Priestley, Selections From His Writings* (University Park, Penn., 1962); John A. Passmore, *Priestley's Writings on Philosophy, Science and Politics* (New York, 1965); and P. Kovaly, *Joseph Priestley Vybrané Splsy* (Prague, 1960), a Czech trans. of selections. A complete listing of all of Priestley's publications, even in 1st eds. alone, would require more space than can reasonably be devoted to it. An essentially complete bibliography of titles can be found in Ronald E. Crook, *A Bibliography of Joseph Priestley, 1733–1804* (London, 1966). Recent eds. of the works cited in the text above are *Lectures on Oratory and Criticism* (Carbondale, Ill., 1965); *Directions for Impregnating Water With Fixed Air* (Washington, D.C., 1945); and the *History and Present State of Electricity*, 3rd. ed. (London, 1775), with electrical papers from the *Philosophical Transactions of the Royal Society* (New York, 1966). Extracts from *Experiments and Observations on Different Kinds of Air*, II (1775), describing the discovery of oxygen, reprinted in *Alembic Club Reprints*, no. 7 (1901). One of Priestley's pamphlets, *Considerations on the Doctrine of Phlogiston and the Decomposition of Water* (1796), with the contemporary response of John Maclean, was reprinted (Princeton, 1929). The published articles are listed by Crook, cited above, and in an appendix to R. E. Schofield, ed., *Scientific Autobiography*, cited below.

The major collection of scientific MSS (all published) is that of the Royal Society, the archives of which also contain a few MS letters, nearly all published in Bolton or Schofield (cited below). Manchester College, Oxford, possesses some MS sermons, notes in the shorthand of Peter Annet, and a few letters. The Pennsylvania State University Library holds 2 copies of the *Memoirs*, in the hand of an amanuensis, as are many other "Priestley" MSS. Priestley letters are in various public and private collections around the world. The primary collection of nonscientific correspondence is in the Dr. Williams Library, London, from which J. T. Rutt took the extracts included in his ed. of Priestley's *Memoirs*, vol. I, pts. 1, 2, of the *Works*. Rutt's editing of the letters is even worse than that of the printed materials, and his versions cannot be depended upon. A substantially complete ed. of Priestley's scientific correspondence—drawn from major collections of the Royal Society, the American Philosophical Society, the Bodleian Library, and other archives listed in an appendix—is R. E. Schofield, ed., *A Scientific Autobiography of Joseph Priestley, 1733–1804* (Cambridge, Mass., 1963). An earlier, smaller ed. of letters, including some not in the *Scientific Autobiography*, was edited by H. C. Bolton and privately printed as *Scientific Correspondence of Joseph Priestley* (New York, 1892).

II. SECONDARY LITERATURE. Books and articles about Priestley are almost as profuse as those by Priestley. Probably the best full-length biography is F. W. Gibbs, *Joseph Priestley: Adventurer in Science and Champion of Truth*, in the British Men of Science Series (London, 1965). Anne D. Holt, *A Life of Joseph Priestley* (London, 1931), is perhaps the most knowledgeable treatment of Priestley's theology. The most detailed examination of his scientific philosophy is in the commentary in Robert E. Schofield, ed., *A Scientific Autobiography of Joseph Priestley, 1733–1804* (Cambridge, Mass., 1963).

ROBERT E. SCHOFIELD

PRINGLE, JOHN (*b.* Roxburgh, Scotland, 10 April 1707; *d.* London, England, 18 January 1782), *medicine*.

Pringle, the son of Sir John and Magdalen Eliot Pringle, received a classical education at St. Andrews and Edinburgh universities and his M.D. at Leiden in 1730. After studying at Paris, he returned to practice medicine in Edinburgh, where in 1734 he was appointed professor of pneumatics (metaphysics) and moral philosophy. In 1742 he embarked upon a military career as head of the hospital of the British army in Flanders, and the following year he initiated the idea that military hospitals should be sanctuaries from enemy action, thus anticipating the "Red Cross" concept. Successively physician general and physician to the royal hospitals, he settled in London in 1749 to practice medicine, obtaining medical appointments to Queen Charlotte, King George III, and other royalty and being created baronet. He was elected to the Royal Society in 1745, and was president from 1772 to 1778, having achieved renown in a wide range of scientific subjects. A fellow of the College of Physicians of London *speciali gratia*, he was a member of scientific societies in Haarlem, Amsterdam, Göttingen, Kassel, Hanau, Madrid, Paris, St. Petersburg, Naples, and Edinburgh.

Pringle's enduring contribution to medicine was his *Diseases of the Army*, a classic that entitles him to be regarded as a founder of modern military medicine. In it he was the first to publish a text that presented the results of extensive experience and research, clarifying knowledge about typhus, malaria, epidemic meningitis, dysentery, and other army scourges, and laying down practical rules of military hygiene. Pringle mentioned the "ague cake" spleen of malaria, described ulcers now called Peyer's patches in typhoid, and called attention to a theory of contagion caused by animalcules but, surprisingly, did not mention tetanus.

BIBLIOGRAPHY

I. ORIGINAL WORKS. Pringle's inaugural dissertation was *De marcore senili* (Leiden, 1730). Three (of seven) papers entitled "Experiments on Substances Resisting Putrefaction," which introduced the words "septic" and "antiseptic" into medical terminology and won for him the Copley Medal of the Royal Society, were published in *Philosophical Transactions of the Royal Society*, **46**, no. 495 (1750), 480, and no. 496, 525, 550. *Observations on Diseases of the Army* (London, 1752), went through numerous eds. during his lifetime (1753, 1761, 1764, 1768, 1778) and one posthumously (1783), each revised and improved by the author. The final London ed. was printed in 1810. The 2nd ed. was translated into French as *Observations sur les maladies des armées dans les camps et dans les garnisons . . .* (Paris, 1755, 1771, 1793), also into Italian (Naples, 1757; Venice, 1762, 1781), German (Altenburg, 1754, 1772), Dutch (Middelburg, 1763; Amsterdam, 1785–1788), and Spanish (Madrid, 1775). There is also an American ed. with notes by Benjamin

Rush and a biography of Pringle extracted from Benjamin Hutchison's *Biographica medica* (Philadelphia, 1810; repr. 1812).

Observations on the Nature and Cure of Jayl-Fevers in a Letter to Dr. Mead (London, 1750), dealing with typhus, as well as Pringle's papers on septics and antiseptics, were revised and added as appendixes to *Diseases of the Army. Six Discourses Delivered by Sir John Pringle, Bart. When President of the Royal Society . . .* (London, 1783), collected and arranged by Pringle, is prefaced by a biography of Pringle by his friend Andrew Kippis. Just before his death Pringle presented 10 MS vols. entitled "Medical and Philosophical Observations," to the Royal College of Physicians of Edinburgh with the proviso that they should not be published.

II. SECONDARY LITERATURE. The basic biography is that by Kippis in *Six Discourses*, cited above. William Mac-Michael, *Lives of British Physicians* (London, 1830), 172–182, adds a few facts. Thomas Joseph Pettigrew, *Medical Portrait Gallery With Biographical Memoirs of the Most Celebrated Physicians*, II, no. 14 (London, n.d. [1840]), includes the portrait by Sir Joshua Reynolds for the Royal Society. See also William Munk, *The Roll of the Royal College of Physicans of London*, II (1878), 252; and J. F. Payne, in *Dictionary of National Biography* (London–New York, 1896), XLVI, 386.

SAMUEL X. RADBILL

PRINGSHEIM, ALFRED (*b.* Ohlau, Silesia, Germany, 2 September 1850; *d.* Zurich, Switzerland, 25 June 1941), *mathematics*.

Pringsheim studied at Berlin and Heidelberg in 1868–1869, received the Ph.D. at Heidelberg in 1872, and qualified as *Privatdozent* at Munich in 1877. He was appointed extraordinary professor at Munich in 1886 but did not become full professor until 1901. He retired in 1922. Pringsheim was a member of the Bavarian Academy of Sciences.

Pringsheim, who came from a rich family, was a lover and promoter of music and fine arts. In his youth he had been a friend of Richard Wagner; and with his wife, Hedwig Dohm, he made his home into a center of Munich's social and cultural life. The novelist Thomas Mann, who was his son-in-law, wrote a novel based on the Pringsheim family. Pringsheim's refined wit was famous. His sprightly *Bierrede* was the acme of the yearly meeting of the Deutsche Mathematiker-Vereinigung and was mentioned by mathematicians throughout the year. His puns were famous: Once when he was asked about his son, who at that time worked as a physicist under Nernst, he answered "Peter ist in Berlin und lernt da den Nernst der Lebens kennen."

After 1933 he was subjected to persecution as a "non-Aryan"; Pringsheim was forced to sell his house to the Nazi party, which tore it down to erect a party building. Having been forced to give up his library and to move several times, he was finally allowed to sell his celebrated majolica collection to a London dealer, although he had to surrender the greater part of the proceeds. In 1939 he moved to Zurich, where he died two years later.

In mathematics Pringsheim was the most consequent follower of Weierstrass. His field was pre-Lebesgue real functions and complex functions; his work is characterized by meticulous rigor rather than by great ideas. His best-known discovery concerns power series with positive coefficients: they have a singularity in the intersection of the positive axis and the circle of convergence. His elaboration of the theory of integral transcendental functions was exemplary and influential, and his extremely simple proof of Cauchy's integral theorem has been generally accepted.

Pringsheim was a brilliant lecturer and conversationalist, but his writings do not reflect this brilliance. This is even true of his celebrated *Festrede*, a paragon of stylistic and oratorical splendor only for those who had heard him speak. The voluminous edition of his courses is one of the dreariest specimens of epsilontics.

BIBLIOGRAPHY

Pringsheim's writings include *Die Grundlagen der modernen Wertlehre: Daniel Bernoulli, Versuch einer neuen Theorie der Wertbestimmung von Glücksfällen*, Sammlung Älterer und Neuerer Staatswissenschaftlicher Schriften, no. 9 (Leipzig, 1896); "Grundlagen der allgemeinen Funktionentheorie," in *Encyklopädie der mathematischen Wissenschaften*, II, pt. 1, fasc. 1 (1899), 1–52; "Über den Goursat'schen Beweis des Cauchy'schen Integralsatzes," in *Transactions of the American Mathematical Society*, **2** (1901), 413–421; "Elementare Theorie der ganzen transcendenten Funktionen von endlicher Ordnung," in *Mathematische Annalen*, **58** (1904), 257–342; "Über Wert und angeblichen Unwert der Mathematik," in *Jahresbericht der Deutschen Mathematiker-Vereinigung*, **13** (1904), 357–382; "Algebraische Analysis," in *Encyklopädie der mathematischen Wissenschaften*, II, pt. 3, fasc. 1 (1908), 1–46; "Table générale," in *Acta mathematica* (1913), 164, with portrait; and *Vorlesungen über Zahlen- und Funktionenlehre*, 2 vols. in 5 pts. (Leipzig–Berlin, 1916–1932).

An obituary is O. Perron, "Alfred Pringsheim," in *Jahresbericht der Deutschen Mathematiker-Vereinigung*, **56** (1952), 1–6.

HANS FREUDENTHAL

PRINGSHEIM, ERNST (*b.* Breslau, Germany [now Wrocław, Poland], 11 July 1859; *d.* Breslau, 28 June 1917), *theoretical and experimental physics.*

Pringsheim was the son of Siegmund Pringsheim, a merchant and lord of a manor, and Anna Guradze. After attending the Magdalenengymnasium and the Johannesgymnasium, leaving the latter at Easter 1877, he studied mathematics for three semesters at Heidelberg and from 6 November 1878 to 9 August 1879 at Breslau, then physics and mathematics at Berlin from the autumn of 1879. He received the Ph.D. under Helmholtz on 3 July 1882 and qualified as lecturer by habilitation in physics at the University of Berlin on 5 January 1886. He was given the title of professor on 30 October 1896 and on 28 August 1905 was appointed full professor of theoretical physics at the University of Breslau, where his close collaborator Otto Lummer had preceded him six months earlier.

Pringsheim's first lecture at Berlin (1886) was "Mechanische Wärmetheorie und kinetische Theorie der Gase." In the winter of 1886–1887 he discussed "Thermodynamik elektrischer Vorgänge," a topic connected with R. Clausius' earlier studies as well as with Planck's investigations nine years later. From 1889–1890 to 1905, he delivered nearly every year a one-hour experimental lecture on "Physik der Sonne"; a book on the subject, containing twelve lectures, appeared in 1910. Pringsheim often connected solar physics with a lecture entitled "Einführung in die physikalische Chemie." His lectures from 1897–1898 to 1904 covered alternately "Interferenz und Polarisation des Lichtes" and astrophysics.

Despite his later appointment in theoretical physics at Breslau, Pringsheim's scientific production was largely experimental. At Berlin it was characterized by a period of research done alone and in experimental cooperation with Lummer from 1896. Pringsheim's doctoral dissertation of 1882 ultimately determined the direction of his research, heat and light radiation.

In 1881 he had replaced the lenses of the spectrometer with hollow specula and thus had made more accurate measurements of wavelengths in the infrared with the diffraction grating. He was the first to develop the radiometer into a useful instrument for measuring infrared radiation. In the following years Pringsheim had not yet come to specialize in radiation but dealt, among other things, with chemical effects of light on hydrochloric acid gas. He also cooperated with the philologist E. Schwan of Jena to investigate the French accent phonometrically and with the lawyer Otto Gradenwitz of Königsberg to reconstruct old palimpsests by photography.

In physics, at the beginning of the 1890's Pringsheim studied the limits of the validity of Kirchhoff's law. He argued that it should apply only in the case of pure temperature radiation, as Kirchhoff had stated, and not, for example, to the radiation in Geissler's tube or in a flame (because of chemical reactions). This view led to a dispute with Friedrich Paschen, who stated that gases may also radiate when stimulated by temperature alone. The entire question, in which Robert von Helmholtz, the son of Hermann von Helmholtz, and Willem Henri Julius of Utrecht had earlier been interested, to a certain degree contributed to recognition of the limits of the law in connection with the role of line spectra. Pringsheim denied that line spectra of a flame could be included in Kirchhoff's law, because they would be caused by chemical influences, but it is known today that he went too far. Kirchhoff's law again interested Pringsheim in 1900 and 1901, when he gave a new simple theoretical proof of it that did not presume the complete blackbody, completely diathermanous substances, and completely reflecting walls. He was attacked for that proof by David Hilbert in 1912.

Pringsheim's period of cooperation with Lummer began toward the end of the nineteenth century. First they treated the experimental determination of the ratio (κ) of the specific heats for various gases, researches that had begun in the 1880's. In 1896 they turned to investigations of heat radiation. Continuing Wilhelm Wien's work at the Physikalisch-Technische Reichsanstalt at Berlin, Pringsheim assisted Lummer in implementing Kirchhoff's concept of the blackbody. By this new means they began to verify the law of Joseph Stefan and Ludwig Boltzmann for the temperature dependence of total radiated energy. In 1900 they recognized that the small systematic deviations of their observations from the theoretical law were due to an insufficient connection of the thermoelectric temperature scale with the scale of the gas thermometer (in 1907 and in the 1960's the scale had to be corrected again); Stefan's law was thus verified. Pringsheim and Lummer then measured the spectral distribution of the radiation energy with the aid of the cylindrical blackbody.

Although in 1899 they stated a variability of the exponential constant of Wien's equation with the wavelength and although they discovered even a slight curvature of the isochromatics in contradiction to that equation, it was not until September 1900 that Pringsheim and Lummer published a paper stating the "invalidity of the Wien-Planck spectral equation" on these grounds. Such negative statements were the main stimuli for Planck to seek a new radiation expression. Eugen Jahnke's equation

$$E = C'T^5(\lambda T)^{-\mu}\, e^{-c'/(\lambda T)^{\nu}},$$

in which Lummer and Pringsheim proposed $\mu = 4$ and $\nu = 1.3$, was soon superseded by Planck's law,

although the equation for $\mu = 4.5$ and $\nu = 1$ (Max Thiesen) remained experimentally in competition with Planck's for about ten years. In 1899 Pringsheim and Lummer—reversing the question—utilized Planck's new law, specifically the so-called Wien's limit of it (with the new constants), to develop three methods for measuring high temperatures (up to 2,300° K). They did so by applying Stefan's law, Wien's T^5 law, and spectrophotometry (founded by Paschen and H. Wanner) of the "black isochromatics," as they called them. The displacement law, first called that by Lummer and Pringsheim in 1899, proved experimentally impracticable for measuring high temperatures. In 1903 Lummer and Pringsheim defined a "strahlungstheoretische Temperaturskala," that is an absolute temperature scale of radiation theory. Attacking Heinrich Rubens' determination of the temperature of Welsbach light in 1905, they provided a discrimination between "true" and "black" temperature.

At Breslau, Pringsheim established a six-term course in theoretical physics in 1906. His delivery was clear and animated by humor. He had sought to return to his native town, Breslau, since 1895, and for much of his life he was a member of the Schlesische Gesellschaft für vaterländische Cultur. As early as 1879 he read a paper on his geomagnetic measurements before this society, and later he was its secretary and also its president.

BIBLIOGRAPHY

I. ORIGINAL WORKS. The Staatsbibliothek Preussischer Kulturbesitz, Berlin-Dahlem, has the following letters by Pringsheim: one to Kultusminister Julius Robert Bosse (15 June 1895); 4 to Heinrich Kayser, Bonn (1 June 1894, 29 June 1913, 26 July 1913, 19 Aug. 1913 [written from England]); 2 to Max Iklé (27 July 1905, 11 Dec. 1906); one postcard to Iklé (4 Feb. 1915); and 2 to Ludwig Darmstaedter (7 June 1903, 13 June 1903). One letter dated 24 Oct. 1900, signed by Pringsheim and Lummer, is in the Bibliothek des Deutschen Museums, Munich. The University of Breslau has personal documents from 1878–1879, a portrait, a curriculum vitae written by Pringsheim on 3 Nov. 1905, and notes in the minutes of the university senate from 1903.

Published works by Pringsheim include "Über das Radiometer," in *Annalen der Physik*, **254** (1883), 1–32, his doctoral dissertation; "Eine Wellenlängenmessung im ultrarothen Sonnenspectrum," *ibid.*, 32–44; "Über die chemische Wirkung des Lichts auf Chlorknallgas," in *Verhandlungen der Physikalischen Gesellschaft zu Berlin*, **4** (1886), 64–65, and **6** (1888), 23, and *Annalen der Physik*, **268** (1887), 384–428; "Eine neue Anwendung des Telephons zur Messung elektrischer Widerstände," in *Verhandlungen der Physikalischen Gesellschaft zu Berlin*, **5**

(1887), 80–82; "Das labile Gleichgewicht der Atome," in *Zeitschrift für physikalische Chemie*, **3** (1889), 145–158; "Der französische Accent (eine phonometrische Untersuchung)," in *Archiv für das Studium der neueren Sprachen und Litteraturen*, **85** (1890), 203–268, written with E. Schwan; "Argandlampe für Spectralbeobachtungen," in *Annalen der Physik*, **281** (1892), 426–427; "Das Kirchhoff'sche Gesetz und die Strahlung der Gase," *ibid.*, 428–458, and **285** (1893), 347–365; "Bemerkungen zu Hrn. Paschen's Abhandlung 'Über die Emission erhitzter Gase,'" *ibid.*, **287** (1894), 441–447; "Über die Leitung der Elektricität durch heisse Gase," *ibid.*, **291** (1895), 507–512; "Nachruf an Franz Schulz-Berge," in *Verhandlungen der Physikalischen Gesellschaft zu Berlin*, **13** (1895), 53–55; and "Photographische Reconstruction von Palimpsesten," *ibid.*, 58–60, written with O. Gradenwitz.

Later works are "Die Strahlungsgesetze und ihre Anwendungen," in *Naturwissenschaftliche Rundschau*, **15** (1900), 1–2, 17–19; "Sur l'émission des gaz," trans. by E. Rothé, in *Rapports présentés au Congrès international de physique réuni à Paris en 1900* (Paris, 1900), II, 100–132; "Einfache Herleitung des Kirchhoff'schen Gesetzes," in *Verhandlungen der Deutschen Physikalischen Gesellschaft*, **3** (1901), 81–84; "Über die Gesetze der schwarzen Strahlung nach gemeinschaftlich mit O. Lummer ausgeführten Versuchen (Vorträge und Diskussionen von der 72. Naturforscherversammlung zu Aachen 16.–22.9.1900)," in *Physikalische Zeitschrift*, **2** (1901), 154–155, also in *Verhandlungen der Gesellschaft Deutscher Naturforscher und Ärzte*, *73. Versammlung in Hamburg* (1901), Abt. Physik, 27–30; "Über Temperaturbestimmungen mit Hülfe der Strahlungsgesetze (nach gemeinsamen Untersuchungen mit Herrn O. Lummer)," in *Verhandlungen der Gesellschaft Deutscher Naturforscher und Ärzte*, *73. Versammlung in Hamburg* (1901), pt. 2, 31–36; "Künstliche Chromosphäre," in *Verhandlungen der Deutschen Physikalischen Gesellschaft*, **7** (1905), 14–15; *Vorlesungen über die Physik der Sonne* (Leipzig–Berlin, 1910); "Zur Theorie der Lumineszenz," in *Physikalische Zeitschrift*, **14** (1913), 129–131; "Bemerkungen zur der Abhandlung des Herrn Hilbert: 'Begründung der elementaren Strahlungstheorie,'" *ibid.*, 589–591; and "Über Herrn Hilberts axiomatische Darstellung der elementaren Strahlungstheorie," *ibid.*, 847–850.

Other works are listed in *Poggendorff*, IV (1904), 1194–1195; V (1926), 1006. The British Museum *Catalogue of Printed Books* credits Ernst Pringsheim with *Studien zur heliotropen Stimmung und Präsentationszeit* (Breslau–Halle, 1909); this work is, however, by the botanist Ernst Georg Pringsheim.

Papers published with Otto Lummer include "Neue Bestimmung des Verhältnisses der beiden specifischen Wärmen," in *Verhandlungen der Physikalischen Gesellschaft zu Berlin*, **6** (1888), 136–140; "Die Strahlung eines 'schwarzen Körpers' zwischen 100 und 1300°C," in *Annalen der Physik*, **299** (1897), 395–410; "Bestimmung des Verhältnisses (κ) der specifischen Wärmen einiger Gase," in *Annalen der Physik*, **300** (1898), 555–583, trans. as "A Determination of the Ratio (κ) of the Specific Heats at

Constant Pressure and at Constant Volume for Air, Oxygen, Carbon-Dioxide, and Hydrogen," in *Smithsonian Contributions to Knowledge*, **29**, no. 1126 (1903), 1–29; "Vertheilung der Energie im Spektrum des schwarzen Körpers," in *Zeitschrift für Instrumentenkunde*, **19** (1899), 214–215; "Die Vertheilung der Energie im Spectrum des schwarzen Körpers," in *Verhandlungen der Deutschen Physikalischen Gesellschaft*, **1** (1899), 23–41; "Notiz zu unserer Arbeit: Über die Strahlung eines 'schwarzen' Körpers zwischen 100°C und 1300°C," in *Annalen der Physik*, **308** (1900), 159–160; "Energieverteilung im Spektrum des schwarzen Körpers," in *Zeitschrift für Instrumentenkunde*, **20** (1900), 148–149; and "Über die Strahlung des schwarzen Körpers für lange Wellen," in *Verhandlungen der Deutschen Physikalischen Gesellschaft*, **2** (1900), 163–180.

See also "Kritisches zur schwarzen Strahlung," in *Annalen der Physik*, **311** (1901), 192–210; "Temperaturbestimmung hocherhitzter Körper (Glühlampe etc.) auf bolometrischem und photometrischem Wege," in *Verhandlungen der Deutschen Physikalischen Gesellschaft*, **3** (1901), 36–46; "Temperaturbestimmung mit Hilfe der Strahlungsgesetze," in *Physikalischen Zeitschrift*, **3** (1902), 97–100; "Zur Temperaturbestimmung der Flammen," *ibid.*, 233–235; "Die strahlungstheoretische Temperaturskala und ihre Verwirklichung bis 2300°abs.," in *Verhandlungen der Deutschen Physikalischen Gesellschaft*, **5** (1903), 3–13; "Zur anomalen Dispersion der Gase," in *Physikalische Zeitschrift*, **4** (1903), 430; "Über das Emissionsvermögen des Auerstrumpfes," *ibid.*, **7** (1906), 89–92; "Bemerkungen zu der Abhandlung von H. Rubens: 'Über die Temperatur des Auerstrumpfes,'" *ibid.*, 189–190; and "Über die Jeans-Lorentzsche Strahlungsformel," *ibid.*, **9** (1908), 449–450.

With Eugen Jahnke and Otto Lummer, Pringsheim published "Kritisches zur Herleitung der Wien'schen Spectralgleichung," in *Annalen der Physik*, **309** (1901), 225–230.

II. SECONDARY LITERATURE. The most reliable—and almost the only—biography is the obituary by Clemens Schaefer, in *Fünfundneunzigster Jahres-Bericht der Schlesischen Gesellschaft für vaterländische Cultur. 1917*, I (Breslau, 1918), 32–36. On Pringsheim's scientific work see H. Kangro, *Vorgeschichte des Planckschen Strahlungsgesetzes. Messungen und Theorien der spektralen Energieverteilung bis zur Begründung der Quantenhypothese* (Wiesbaden, 1970), *passim*. See also H. Kangro, "Ultrarotstrahlung bis zur Grenze elektrisch erzeugter Wellen, Das Lebenswerk von Heinrich Rubens, I (experimenteller Beweis der elektromagnetischen Lichttheorie für das Ultrarot)," in *Annals of Science*, **26** (1970), 235–259.

HANS KANGRO

PRINGSHEIM, NATHANAEL (*b.* Wziesko, Silesia, 30 November 1823; *d.* Berlin, Germany, 6 October 1894), *botany, plant physiology.*

Pringsheim belonged to that group of young German botanists—including Ferdinand Cohn, Hofmeister, and Mohl—who revolutionized the science during the middle years of the nineteenth century by shifting attention from collection and taxonomy to the dynamics of cell development and life history. This movement, inspired in large part by Schleiden, created a new appreciation for the degree of unity among all plants—and indeed between plants and animals as well. The new insights coincided with, and depended importantly upon, a reexamination of previously neglected organisms. The larger and more familiar flowering plants (phanerogams) increasingly yielded pride of place to the relatively obscure cryptogams, a name long applied to the nonflowering plants because their mode of reproduction seemed hidden.

Hofmeister's discovery that the higher cryptogams and phanerogams share a common reproductive pattern, in which sexual generation alternates with asexual reproduction, gave special force to the new botanical movement. Pringsheim's chief contribution to this movement was to identify in the lower cryptogams, and especially in the algae, those basic reproductive modes (notably sexual union and the alternation of generations) that Hofmeister and others had already established for the higher cryptogams.

Pringsheim's father, the director of an industrial concern, had foreseen a quite different career for his son. During Nathanael's early education at home and at the Gymnasiums in Oppeln (now Opole) and Breslau (now Wrocław), his father nourished the hope that he might become an industrialist or merchant. In the face of Pringsheim's attraction to science, his father urged that he at least direct those interests toward a career in the secure and practical field of medicine. In the winter term of 1843–1844, to meet his father's wishes, Pringsheim transferred to the medical faculty at the University of Breslau, having already spent several months as a student in the philosophical faculty there. In the spring term of 1844 he moved to the University of Leipzig, also as a medical student. But Pringsheim had never been more than a nominal medical student; and when he moved to the University of Berlin, it was as a declared student of natural science.

At Berlin the professor of botany was Karl Kunth; but Pringsheim's orientation and approach were more deeply influenced by Purkyně, the exacting physiologist who taught him at Breslau, and especially by Schleiden's famous *Grundzüge der wissenschaftliche Botanik* (1842–1843). In March 1848 Pringsheim very briefly joined the revolutionary movement in Berlin but then withdrew permanently from overt political activity. He took his Ph.D. at Berlin in April 1848. His doctoral thesis defended the controversial view

that the cell wall was built up by apposition from the interior and not by adhesion from without. Following additional study in Paris and London, he returned in September 1849 to Berlin, where he became *Privatdozent* in 1851. Although his *Habilitationsschrift*, "Die Entwickelungsgeschichte der *Achlya prolifera*," wrongly classified an organism that Pringsheim himself later identified as *Saprolegnia ferax* (1857), it represented a pioneering attempt to follow the complete developmental history of a lower cryptogam. On 20 May 1851 Pringsheim married Henriette Guradze, the daughter of a leading merchant in Oppeln. Three daughters were born into this happy union, which lasted more than forty years. Pringsheim's final months were darkened by his wife's death in February 1893.

Almost ironically, Pringsheim's career in pure research benefited immensely from the financial success of his pragmatic father. Upon the latter's death in 1868, Nathanael inherited an estate in Silesia and sufficient means to pursue his research freely. In fact, except for his years as *Privatdozent* at Berlin and a brief period as professor of botany at the University of Jena, Pringsheim held no teaching posts. When called to Jena in 1864 (as Schleiden's successor), he accepted only after the collapse of preliminary negotiations to secure him a post at Berlin and only on the condition (quickly met) that a new botanical institute be constructed for him at Jena. There Pringsheim lectured in the summer on general botany and in the winter on cryptogams, offering microscopical instruction both semesters.

Despite the gratification of seeing eager students (notably Strasburger) hard at work in his new and well-designed institute, Pringsheim never achieved real happiness at Jena, perhaps because of friction with authorities there, perhaps because he did not really enjoy teaching, but certainly because of increasingly severe attacks of a lung ailment (reportedly asthma) that had troubled him for years and that ultimately caused his death. Pringsheim resigned the post at Jena in 1868 and settled in a house near the botanical gardens in Berlin, where he thereafter conducted his research in a private laboratory attached to his home. Periodic trips to the Riviera and to the northern coast of France brought temporary benefits to his health and permanent additions to his collections of specimens and to his circle of foreign botanical friends.

Pringsheim's earliest works bore less directly on plant reproduction than on the controversy over Schleiden's original version of the cell theory. From observations of the vegetative and reproductive cells of many algae, and also of the pollen mother cells of the higher cryptogams and flowering plants, Pringsheim concluded that cell division (and not free-cell formation in a structureless blastema) is the basic mode of cell multiplication. In this conclusion he joined the swelling attack against Schleiden's "watch-crystal" conception of cell formation. On another front, however, Pringsheim refused to endorse the reformist cause, at least insofar as it was embodied in Mohl's conception of the "primordial utricle" *(Primordial-schlauch)*. Mohl claimed that he had seen this peculiar nitrogenous structure surrounding the cell contents but within the cell wall. Insisting that it was a true membrane, he assigned it an important role in cell formation and function and made it part of a wider effort to direct attention away from the cellulose cell wall (so prominent in Schleiden's scheme) toward the protoplasmic cell contents.

In a monograph of 1854, Pringsheim attacked Mohl's conception. He argued that Mohl had mistaken a mere layer of the cell contents for a true membrane and emphatically denied Mohl's claim that cell division took place by the infolding of the primordial utricle, followed by the secretion of a cellulose wall on its outer surface. Instead, Pringsheim insisted that the cell wall arose by direct transformation of the primordial utricle, and that the latter played no active role in cell division, being merely dragged along as a passive lining on the true septa, which were always cellulose. The issues surrounding this debate are extremely complex; and although Mohl's views more nearly approach the later notion of the cell as a "naked clump of protoplasm with a nucleus," neither his nor Pringsheim's position has survived intact. What gives their controversy special interest is the fact that they were observing almost precisely the same cells under almost identical conditions. That they could disagree so fundamentally about what they had seen suggests that their observations were "theory-loaded," being shaped especially by differing conceptions of cell division.

Pringsheim's interest in this debate receded somewhat as he made his immediately famous contributions to the discovery of sexuality in the algae. Like most important discoveries in science, this one cannot be assigned to a single worker or to a single moment; it evolved gradually. After Unger, Carl Naegeli, and Hofmeister had provided evidence of sexuality in the higher cryptogams, attention shifted to the previously neglected algae and fungi. Between 1849 and 1851 Thuret, Alexander Braun, and others drew attention to the presence in numerous algae of three different kinds of spores—the familiar resting spores and two types of motile spores that Braun called macrogonidia and microgonidia. But no one had yet fully perceived that these spores are sexually differentiated and that the resting spores arise from the sexual union of the two kinds of motile spores. Pringsheim himself had described the germination of resting spores in *Achlya [sic Saprolegnia]* (1851) and in *Spirogyra* (1852) without recognizing them as the products of sexual union. In 1854, however, Thuret showed that the resting spores in the brown marine alga *Fucus* germinate only after the microgonidia (or spermatozoids) attach themselves to the macrogonidia (or egg cell). Moreover, by mixing the spermatozoids from one species with the egg cells from another, he managed to produce hybrids.

After confirming Thuret's results in *Fucus* and other higher marine algae, Pringsheim turned to lower freshwater species. In 1855 he discovered spermatozoids in the unicellular alga *Vaucheria terrestris* and carefully described their role in the production of the resting spore: after they attach themselves to the egg mass, a true cell membrane forms on the latter, converting it into a resting spore. Although he also observed a colorless corpuscle inside the egg cell and clearly believed that this corpuscle was a sperm cell, it is an exaggeration to claim (as many do) that he had already observed the actual penetration of the spermatozoids into the egg mass. Until he observed this penetration, many botanists (including Cohn) resisted the startling discovery of so advanced a mode of reproduction in a unicellular plant.

But in this same paper of 1855 Pringsheim had also drawn attention to the wide distribution of differentiated spores in other species, and he now set out to establish his belief that sexual reproduction is a general phenomenon among the algae. Between 1856 and 1858, and in a manner that convinced even the previously skeptical, Pringsheim extended sexuality to the freshwater algae *Oedogonium* and *Coleochaete*. In the *Oedogonium*, he definitively observed the actual sexual act, following the spermatozoid as it forced its way through the outer layer of the egg cell and penetrated into its protoplasmic mass, where it dissolved. On this basis he emphasized the crucial point that fertilization involves a real material fusion of the two sexual cells and not (as Thuret had believed) a mere dynamical reaction between them.

Pringsheim's attempt, during the same period, to establish sexuality in the *Saprolegnia* (later placed among the fungi) gave rise to an inconclusive debate with Anton de Bary lasting into the 1880's. De Bary insisted that the family was apogamic and that any sex organs found in it were nonfunctional; Pringsheim maintained that fertilization took place via "spermamoebo" in *Saprolegnia* with developed anthers. The debate drew interest because of a wider controversy over the existence and origin of sexuality in fungi.

From 1861 to 1863 Pringsheim produced valuable memoirs on the "Water-net" *(Hydrodictyon)*, on *Chara* (arguing, on the basis of the striking correspondence between its prothallium and the protonema of mosses, that it belonged among the mosses and not the algae, as commonly supposed), and on *Salvinia* (his most important contribution to our knowledge of the higher cryptogams). His inaugural lecture at Jena (1864) emphasized the important place that the cryptogams had come to occupy in the history of botany and general biology. While at Jena, Pringsheim published no papers; but in 1869 he reported his discovery of the conjugation of the swarm spores in *Pandorina*, a process that he viewed as the most primitive form of sexual reproduction. Comparing the anterior ciliated ends where the conjugating spores first united with the "receptive spot" of the ovum in higher algae, the canal cells of archegoniate plants, and the synergidae of angiosperms, Pringsheim insisted upon the embryological unity of the entire plant kingdom.

In 1873 Pringsheim returned to the *Sphacelaria*, a group of brown marine algae on which he had written a decade before. His main concern now was the bearing of Darwinian evolutionary theory on this family, which displays a progressive complexity from a filamentous to a bulblike structure. Acknowledging that this series supported the doctrine of descent, Pringsheim nevertheless assigned natural selection a minor role. Like many German botanists (including Braun, Cohn, and especially Naegeli), he insisted that the progressive accumulation of structural complexity was a "purely morphological" phenomenon, conferring no survival value, taking place independently of the struggle for existence, and having as its cause an "inherent directing force" *(inneren richtenden Kräfte)*.

Pringsheim's attempts to extend the law of the alternation of generations to the lower cryptogams excited much controversy. As early as 1856–1858, in his famous papers on alga sexuality, he had introduced the notion that the law might apply in modified form to some species of algae. He showed in particular that the fertilized egg cell in *Oedogonium* and *Coleochaete* does not develop directly into a new plant but gives rise instead to a cluster of four asexual spores, each of which becomes a new plant; and he perceived an especially close analogy between the life histories of the *Coleochaete* and the mosses. Following William Farlow's observation of 1874 that the asexual phase of the fern could arise directly (that is, vegetatively) from the sexual phase (apogamy), Pringsheim sought to achieve the converse production of the sexual phase directly from the asexual without the intervention of spores (a phenomenon later called apospory). After

many experiments he succeeded in producing protonemata (and hence gametes) directly from the divided seta of a moss capsule. This achievement suggested to Pringsheim that plant sexuality was a fluid property, which could be lost or recovered according to circumstances, and that the gametophyte and sporophyte stages should be considered "homologous" since one could give rise to the other vegetatively. On this basis he claimed that the difference between the alternation of generations in the lower cryptogams and that in the mosses and liverworts was one only of degree, with the series of recurrent asexual (sporophyte) generations in the former being reduced to a single such generation in the latter. Put another way, an irregular but homologous alternation of generations in the lower cryptogams gives way to a regular but equally homologous alternation in the mosses and liverworts. In opposition to Pringsheim's "homologous" theory of the alternation of generations, Ladislav Celakovsky proposed his "antithetic" theory, according to which a regular alternation of generations occurs in the higher cryptogams only through the interpolation of a new, nonhomologous generation (the sporophyte) arising from the division and progressive sterilization of the zygote of a primitive, sexually reproducing plant. Although Pringsheim's theory had certain advantages from an evolutionary point of view, Celakovsky's theory was more consonant with some of the histological differences between the lower and higher cryptogams. By the time of Pringsheim's death, Celakovsky's theory was the more widely accepted, especially in England, where Bower developed it to a new level of completeness.

Beginning about 1874 Pringsheim turned his attention increasingly to plant physiology, and especially to the function of chlorophyll. In twelve papers extending into the late 1880's, he developed and defended the remarkable theory that chlorophyll played no direct role in photosynthesis, but rather acted as a protective screen for the plant protoplasm and as a regulator of plant respiration. These functions, he supposed, depended on the capacity of chlorophyll to absorb deleterious and high-energy light rays. His views derived in part from an experimental observation of great significance—namely, that highly concentrated light destroys chlorophyll within a few minutes and plant protoplasm upon prolonged exposure. But he minimized a wide range of alternative explanations for these phenomena, and his startling hypothesis never enjoyed much favor. Nonetheless, it stimulated a flood of work on the subject between 1880 and 1885, including some that established chlorophyll's role in the absorption of the effective rays in photosynthesis. As part of his theory Pringsheim claimed to have discovered a new substance ("hypochlorin") that he supposed to be the first product of assimilation and the source of starch and other compounds. His critics argued effectively that hypochlorin, if it existed at all, was not an assimilation product but a degradation product of chlorophyll itself. In general Pringsheim's work in plant physiology did little to enhance, and may have damaged, the reputation built upon his earlier algological work.

In addition to his contributions to research, Pringsheim served botany through his talents for organization and leadership. An advocate of the agricultural benefits of botanical research, he was appointed in 1862 to the Central Commission for Agricultural Experimentation by the Prussian Ministry of Agriculture. In this capacity he submitted four reports over the next decade—on potato growth and the potato blight—based on the work of the Prussian agricultural stations. Pringsheim's three most important organizational contributions survived him: the *Jahrbücher für wissenschaftliche Botanik*, which he founded in 1857 and edited until his death; the German Botanical Society, of which he was chief founder in 1882 and president from then until his death; and a biological station that he helped to establish on Helgoland, an algae-rich island in the North Sea off the German coast. After his death, a museum named for him was erected on Helgoland with funds supplied by his children. Pringsheim was elected to the Berlin Academy of Sciences in 1860 and awarded the title of *Geheimen Regierungsrat* by the Prussian government in 1888.

BIBLIOGRAPHY

I. Original Works. Pringsheim's only book or monograph was *Untersuchungen über den Bau und die Bildung der Pflanzenzelle. Erste Abtheilung. Grundlinien einer Theorie der Pflanzenzelle* (Berlin, 1854). A projected second part never appeared. Fifty-seven reports, addresses, and papers (including the monograph above) are collected in *Gesammelte Abhandlungen*, edited by his children, 4 vols. (Jena, 1895–1896). His major contributions to the discovery of sexuality in the algae appeared initially in three papers in *Monatsberichte der Königlicher Akademie der Wissenschaften* (1855), 133–165; (1856), 225–237; and (1857), 315–330. These papers were subsequently expanded into a four-part treatise published serially between 1857 and 1859 in his own *Jahrbücher für wissenschaftliche Botanik*, **1**, 1–81, 284–306; **2**, 1–38, 205–236. An epilogue, which seeks to give a critical history of the discovery of algae sexuality, appeared in 1860 (*ibid.*, **2**, 470–481). The fact that the bound volumes of the *Jahrbücher* sometimes include papers published two or more years before the date of binding, together with the fact that the four-part treatise was an expanded version of three earlier papers, helps to account for the variation in dating some of Pringsheim's contributions.

Among the other papers discussed at some length in the text above are "Über Paarung von Schwärmsporen, die morphologische Grundform der Zeugung im Pflanzenreiche," in *Monatsberichte der Königlicher Akademie der Wissenschaften* (1869), 721–738; "Über den Gang der morphologischen Differenzung in der Sphacelarien-Reihe," in *Abhandlungen der Königlichen Akademie der Wissenschaften* (1873), 137–191; and "Über Sprossung der Moosfrüchte und den Generationwechsel der Thallophyten," in *Jahrbücher für wissenschaftliche Botanik*, **11** (1877), 1–46.

For a chronological bibliography of Pringsheim's works, see Karl Schumann, *Verhandlungen der Botanischen Vereins der Provinz Brandenburg*, **36** (1894), xl–xlviii. The Royal Society *Catalogue of Scientific Papers*, V, VIII, XI, XII, and XVII, lists fifty-two papers by Pringsheim (sometimes erroneously paginated) and gives references to English and French abstracts or translations of some of his works. Upon his death, according to Cohn (see below), Pringsheim's library went to the Berlin botanical gardens, with duplicates sent to the biological station on Helgoland.

II. Secondary Literature. Of the obituary sketches of Pringsheim, the most valuable are Ferdinand Cohn, in *Bericht der Deutschen botanischen Gesellschaft*, **13** (1895), (10)–(33), repr. in *Jahrbuch für wissenschaftliche Botanik*, **28** (1895), i–xxxii (following p. 321); and D. H. Scott, in *Nature*, **51** (1895), 399–402. See also G. Wunschmann in *Allgemeine deutsche Biographie*, LIII, 120–124. For an additional list of obituary notices, see the Royal Society *Catalogue of Scientific Papers*, XVII, 1020.

On Pringsheim's place in the history of botany, see Julius von Sachs, *Geschichte der Botanik vom 16. Jahrhundert bis 1860* (Munich, 1875), trans. as *History of Botany, 1530–1860* by Henry E. F. Garnsey, rev. by Isaac Bayler Balfour (Oxford, 1890), esp. 203, 209–213, 318, 372, 442–443; J. Reynolds Green, *A History of Botany, 1860–1900* (Oxford, 1909), esp. pp. 47, 51–52, 227, 233, 237, 291–294, 302, 314, 451; and R. J. Harvey-Gibson, *Outlines of the History of Botany* (London, 1919), esp. 141–144, 161, 170–171, 206–208. For a summary of the controversy over Schleiden's cell theory, but without reference to Pringsheim's role, see G. L. Geison, "The Protoplasmic Theory of Life and the Vitalist-Mechanist Debate," in *Isis*, **60** (1969), 273–292, esp. 273–278. The controversy is examined at greater length in G. L. Geison, "Toward a Substance of Life: Concepts of Protoplasm, 1835–1870" (M.A. thesis, Yale University, 1967), where Pringsheim's work receives fleeting and somewhat misleading attention.

Gerald L. Geison

PRITCHARD, CHARLES (*b.* Alberbury, Shropshire, England, 29 February 1808; *d.* Oxford, England, 28 May 1893), *astronomy, astrophysics.*

Various dates have been given for Pritchard's birth; 29 February 1808 seems the most likely. Pritchard graduated from Cambridge in 1830 and soon afterward moved to London, where he helped found Clapham Grammar School. He remained in charge of the school until 1862, when he retired to the Isle of Wight. While in London, Pritchard developed an interest in astronomy and erected an observatory at Clapham Grammar School. But although he became an influential member of the Royal Astronomical Society, serving as president in 1866–1868, he had no major research to his credit when, in 1870, he was appointed Savilian professor of astronomy at Oxford. Pritchard persuaded the university to provide funds for building an observatory on the edge of the University Parks. When the observatory was completed, it was partly equipped by Warren de la Rue, who donated his own instruments.

The most important aspect of Pritchard's work at Oxford was his role in convincing the astronomical community that accurate measurements of position could be obtained from photographic plates. One of the first programs he undertook—at de la Rue's suggestion—was an attempt to determine the lunar librations from photographs of the moon. The final results apparently were never published, but a paper on the moon's diameter did appear. Pritchard next stressed the possibility of using photography for the determination of stellar parallaxes. He made detailed parallax observations of a few stars, including 61 Cygni, but also employed the method in a more wholesale form to derive an average parallax for all stars of the second magnitude visible at Oxford. The latter measurements received some criticism, but J. Kapteyn subsequently followed a quite similar approach in his much more extensive and important investigations.

Pritchard's other major project at Oxford was photometric. He devised a program for measuring the magnitudes of all naked-eye stars up to 100° from the North Pole using a wedge photometer. The results, published in 1886 as *Uranometria nova Oxoniensis*, paralleled work carried out not long before at the Harvard College Observatory. Agreement between the two sets of results was quite good, thus providing a generally acceptable magnitude sequence for the brighter stars.

During his time at Oxford, Pritchard was also involved in a number of researches of lesser importance such as the determination of a few double star orbits. Although dependent in all these investigations on research assistants for help, he personally participated in the work as well. He remains one of the very few scientists who carried out all his important research work after the age of sixty.

BIBLIOGRAPHY

There is a detailed obituary of Pritchard in *Proceedings of the Royal Society*, **54** (1893), iii–xii, which contains a representative list of his more important papers. His daughter, Ada Pritchard, subsequently published a biography, *The Life and Work of Charles Pritchard* (London, 1897).

A. J. MEADOWS

PRIVALOV, IVAN IVANOVICH (*b*. Nizhniy Lomov, Penza guberniya [now oblast], Russia, 11 February 1891; *d*. Moscow, U.S.S.R., 13 July 1941), *mathematics*.

Privalov, son of Ivan Andreevich Privalov, a merchant, and Eudokia Lvovna Privalova, graduated from the Gymnasium in Nizhniy Novgorod (now Gor'kiy) in 1909 and in the same year entered the department of physics and mathematics of Moscow University. He graduated in 1913 and remained at the university to prepare for an academic career. His scientific supervisor was D. F. Egorov, and his work was greatly influenced by N. N. Lusin. In 1916 he passed the examinations for the master's degree and began teaching at Moscow University as lecturer; in 1917 he became professor at Saratov University. Privalov returned to Moscow in 1922 and for the rest of his life was professor of the theory of functions of a complex variable; from 1923 he also taught at the Air Force Academy. He received his doctorate in physics and mathematics in 1935 without defending a dissertation. He was an active member of the Moscow Mathematical Society, of which he was vice-president from 1936. He was elected a corresponding member of the Soviet Academy of Sciences in 1939.

Privalov's first works, dealing with orthogonal series and integral equations, appeared in 1914; he then turned to the study of properties of Fourier series. His principal interests soon concentrated, however, upon boundary properties of analytic functions, that is, their properties in the vicinity of the set of their singular points; a considerable part of his seventy-nine published works is concerned with these problems. Privalov was closely preceded in this field by V. V. Golubev, another Moscow mathematician who taught at Saratov University and in 1916 published his master's degree thesis on analytic functions with a perfect set of singular points.

In 1917 Privalov and Lusin established a wide-ranging program of studies on the theory of analytical functions by means of the theory of measure and Lebesgue integrals, and began to put it into effect at once. In "Cauchy Integral" (1918), which continued the works of Pierre Fatou (1906) and Golubev (1916) and was initially intended as a master's degree thesis, Privalov described many new discoveries in the theory of boundary properties of analytic functions defined in the domain bounded by one rectifiable curve. Thus it was proved that under conformal mapping of such domains the angles are preserved on the boundary almost everywhere. Privalov and Lusin established the invariance of a point set with a measure equal to zero on the boundary; and Privalov solved many problems on the unicity of analytical functions, proved the existence almost everywhere of the Cauchy type of integral, established its boundary properties, and investigated in detail the problem of determining the analytical function with its values on the boundary by means of the Cauchy type of integral. Because "Cauchy Integral" appeared at a time when scientific contacts between Russia and other countries were almost nonexistent, it did not attract attention abroad. In 1924–1925 some of the results obtained in that work were reported by Privalov in two articles in French, the second of which was written with Lusin. These results were considerably supplemented here by the solution of a number of new and difficult problems of unicity of analytic functions determined by the set of their values on the boundary.

In 1934 Privalov began to study subharmonic functions, which had been introduced as early as 1906 and became the subject of Riesz's works in 1925–1930. In *Subgarmonicheskie funktsii* Privalov presented an original systematic construction of the general theory of this class of functions in close connection with the theory of harmonic functions. He also elaborated the ideas of his work on Cauchy's integral. Shortly before his death Privalov summarized many studies in *Granichnye svoystva odnoznachnykh analiticheskikh funktsy.*

Some of Privalov's manuals, especially university-level courses on the theory of functions of complex variables and a manual of analytical geometry for technological colleges, became very widely used in the Soviet Union.

BIBLIOGRAPHY

I. Original Works. Privalov's writings include "Cauchy Integral," in *Izvestiya Saratovskogo gosudarstvennogo universiteta*, **11** (1918), 1–94; "Sur certaines propriétés métriques des fonctions analytiques," in *Journal de l'École polytechnique*, **24** (1924), 77–112; "Sur l'unicité et la multiplicité des fonctions analytiques," in *Annales scientifiques de l'École normale supérieure*, **42** (1925), 143–191, written with N. N. Lusin; *Analiticheskaya geometria* (Moscow, 1927; 30th ed., 1966); *Vvedenie v teoriyu funktsy kompleksnogo peremennogo* ("Introduction to the Theory of Functions of a Complex Variable"; Moscow, 1927;

11th ed., 1967); *Subgarmonicheskie funktsii* (Moscow–Leningrad, 1937); and *Granichnye svoystva odnoznachnykh analiticheskikh funktsy* ("The Boundary Properties of the Single-Valued Analytic Functions"; Moscow, 1941).

II. Secondary Literature. For bibliographies of Privalov's works see *Matematika v SSSR za tridtsat let* ("Mathematics in the U.S.S.R. During the [Last] Thirty Years"; Moscow–Leningrad, 1948); and V. Stepanov, "Ivan Privalov. 1891–1941," in *Izvestiya Akademii nauk SSSR*, Ser. math., **6** (1941), 389–394.

A. P. Youschkevitch
A. T. Grigorian

PRIVAT DE MOLIÈRES, JOSEPH (*b.* Tarascon, Bouches-du-Rhône, France, 1677; *d.* Paris, France, 12 May 1742), *physics, mathematics.*

The son of Charles Privat de Molières and Martine de Robins de Barbantane, Privat de Molières was born into a prominent Provençal family. He showed an early aptitude for philosophical and scientific studies and received an excellent education at Oratorian schools at Aix, Marseilles, Arles, and, finally, Angers, where he studied under the mathematician Charles-René Reyneau during 1698–1699. Against his parents' wishes, Privat de Molières chose an ecclesiastical life and entered the Congregation of the Oratory. He taught at the order's colleges at Saumur, Juilly, and Soissons but left in 1704 to pursue a more active scientific career in Paris. There he became an intimate of Malebranche, studying mathematics and metaphysics with him until the latter's death in 1715. Elected to the Académie Royale des Sciences as *adjoint mécanicien* in 1721, Privat de Molières succeeded to the chair of philosophy at the Collège Royal in 1723, following the death of Varignon. He was raised to the rank of *associé* in the Academy in 1729 and became fellow of the Royal Society of London in the same year.

A major figure in the protracted struggle against the importation of Newtonian science into France, Privat de Molières devoted his career to developing and improving Cartesian physics. Cognizant of the superiority of Newtonian precision in comparison with Cartesian vagueness in the explication of natural phenomena, he was nonetheless convinced of the rectitude of Descartes's ideal of a purely mechanical science. In a series of memoirs read to the Academy, in articles in the *Journal de Trévoux*, and in the published version of his lectures at the Collège Royal, the four-volume *Leçons de physique, contenant les éléments de la physique déterminés par les seules lois des mécaniques* (1734–1739), Privat de Molières offered an emended Cartesian program which, by incorporating Newton's calculations and mathematical techniques, would accord with exact experimental and observational data. Central to his system was the existence of small vortices (*petits tourbillons*), an idea borrowed from Malebranche to replace the discredited Cartesian theory of matter. Unlike Descartes's elements, the *petits tourbillons* were elastic rather than hard particles and constituted the basic structural units of the universe.

The hypothesis of *petits tourbillons* was adopted to establish the superiority of the concepts of the plenum and impulsion over the rival ideas of the void and attraction. Privat de Molières sought to answer Newton's refutation of the vortex theory (propositions LII and LIII of book II of the *Principia*) by obviating the objection that planetary vortices were incompatible with Kepler's laws. He offered an elaborate mathematical demonstration showing that the subtle movements of *petits tourbillons* within the larger planetary vortices could produce the motion of the planets required by astronomical data. His system, extended to include electrical and chemical phenomena, was influential in France and was cited by Fontenelle, secretary of the Academy, as one of the most effective rehabilitations of Cartesian science. Privat de Molières's ingenious use of the vortex hypothesis, intended as a reconciliation between Cartesian and Newtonian ideas, succumbed, however, to the cogent attacks by French Newtonians, notably Pierre Sigorgne.

BIBLIOGRAPHY

I. Original Works. Privat de Molières's major books include *Leçons de mathématiques, nécessaires pour l'intelligence des principes de physique qui s'enseignent actuellement au Collège royal* (Paris, 1725), trans. into English by T. Haselden as *Mathematic Lessons . . . Delivered at the College Royal of Paris . . .* (London, 1730); *Leçons de physique, contenant les éléments de la physique déterminés par les seules lois des mécaniques*, 4 vols. (Paris, 1734–1739); and *Traité synthétique des lignes du premier et du second genre, ou éléments de géométrie dans l'ordre de leur génération* (Paris, 1740). Among his more important memoirs presented to the Academy are "Loix générales du mouvement dans le tourbillon sphérique," in *Mémoires de l'Académie royale des sciences* (1728), 245–267; and "Problème physico-mathématique, dont la solution tend à servir de réponse à une des objections de M. Newton contre la possibilité des tourbillons calesses," *ibid.* (1729), 235–244.

II. Secondary Literature. For biographical details consult Jean-Jacques Dortous de Mairan, *Éloges des académiciens de l'Académie royale des sciences, morts dans les années 1741, 1742, 1743* (Paris, 1747), 201–234; F. Hoefer, ed., *Nouvelle biographie générale*, XXXV (Paris, 1861), 887–889; and Alexandre Savérien, *Histoire des philosophes modernes*, VI (Paris, 1773), 217–248. The best recent study of Privat de Molières's physics is Pierre Brunet, *L'introduction des théories de Newton en France au XVIIIe siècle. I: Avant 1738* (Paris, 1931), 157–165, 240–262, 327–338. Also useful is E. J. Aiton, *The Vortex Theory of Planetary Motions* (London–New York, 1972), 209 ff. A brief discussion of the use of *petits tourbillons* in chemistry is Hélène Metzger, *Les doctrines chimiques en France du début du XVIIe à la fin du XVIIIe siècle*, I (Paris, 1923), 462–467.

Martin Fichman

PROCHÁSKA, GEORGIUS (JIŘÍ) (*b.* Blížkovice, Moravia, 10 April 1749; *d.* Vienna, Austria, 17 July 1820), *anatomy, physiology, embryology, ophthalmology.*

Procháska was the son of a smith and smallholder. Two of his older brothers also became smiths, but Procháska, being thought too weak physically to work at the forge, was sent to the Jesuit Gymnasium at Znojmo. When Procháska's father died in 1763, the family was left with little money to support his further education, but a nearly fatal accident (in which he was poisoned by fumes from a leaking stove in his room) brought Procháska to the attention of the wealthy parents of his schoolmates. He was thereby offered the opportunity to become a private, paid tutor, which enabled him to remain at the Gymnasium. Procháska studied philosophy at Olomouc from 1765 until 1767, then went to Vienna, where he studied medicine from 1770 to 1776. In Vienna he won the patronage of Anton de Haen, to whose clinic he had been taken, critically ill and unable to pay for treatment; de Haen recognized his skill in anatomical drawing, and engaged him as an assistant in his department of clinical teaching.

De Haen died before Procháska had finished his studies, but Procháska found another supporter in Joseph Barth, the cultivated but eccentric professor of anatomy, who gave him an assistantship and taught him his methods of blood-vessel injection and his operation for cataract. In 1778 Procháska was appointed professor of anatomy and ophthalmology at the University of Prague, where from 1786 he also taught "higher anatomy," physiology, and ophthalmology. In 1791 he returned to Vienna as Barth's successor as professor of anatomy and ophthalmology. He remained in that post until his retirement in 1819.

Procháska began his own anatomical and physiological research while he was still a student. In 1778 he published his first treatise, on certain questions concerning the strength of the heart and the movement of blood through the blood vessels, entitled *Controversae quaestiones* . . .. In this work Procháska demonstrated, through physical models, that the velocity of the blood diminishes as it passes from larger to smaller vessels, in proportion to the increase in the sum of the cross-sectional area of the branches. The treatise provoked a violent reaction from Spallanzani, who was annoyed by the relatively unknown Procháska's largely justified criticism of his own position. Spallanzani's attack on Procháska was published under the pseudonym "A. Castiglioni"; it was shortly answered with a defense signed "Antonius Slawik," a name that does not appear in any contemporary records or among learned men of the time. Procháska published two other anatomical treatises soon thereafter. These, *De carne musculari* (1778) and *De structura nervorum* (1779), were illustrated and contained a number of new observations. In 1780 Procháska published a work on the abrasion of human teeth with age, which his pupil J. Pešina later used to work out a system for estimating the age of horses.

Procháska's 1781 treatise on the generation and origin of monsters set out his views on embryology. He opposed the prevailing notion of preformation, and, based upon his own observations of monsters, argued against the ideas of Bonnet, Haller, Spallanzani, and their supporters. He chose, rather, to champion the view offered by C. F. Wolff, whereby the fetus develops progressively by differentiation from uniform tissues with the emergence of organs and parts that had not existed previously. Procháska went on to point out that such epigenesis offered the best explanation for the development of monsters, an idea that was taken up and elaborated only forty years later.

Procháska's principal work, *Commentatio de functionibus systematis nervosi*, was published in 1784. In analyzing the function of the nervous system, Procháska was careful to differentiate between facts and theories, endeavoring to explain the workings of the nerves on the basis of observation. He attempted to avoid philosophical considerations and to refute unfounded hypotheses, which he thought to be the main obstacle to a proper understanding of nerve functions. His formulation of reflex action, as an example of nerve activity, is of particular interest. It is based upon the notion of a nerve force (*vis nervosa*, an analog of Newton's *vis attractiva*) and of a *sensorium commune*, which coordinates all impressions passing to the individual nerve centers. The *vis nervosa* is divisible and can exist in even severed nerves; a portion of the nervous system can retain its activity although separated from the rest, and it may therefore be seen that the *vis nervosa* does not proceed from the brain. The neural force is, rather, latent in the nerves, and it remains latent until it is activated by a stimulus. The stimulus may be either internal or external, and the activity of the nerve is dependent upon its intensity. Procháska's *vis nervosa* is thus not unlike the modern notion of the nerve impulse.

Procháska postulated the existence of two kinds of nerve fibers. One of these conducts sensory impression from the periphery of the body to the *sensorium commune* (which extends from the spinal cord through the medulla oblongata and crura cerebri to the thalamus, corresponding approximately to the central gray matter), from which they are reflected as motor impressions. The other conducts "reflected" impulses from the nerve centers to the muscles and other effector organs. Reflection is thus automatic and independent of both the will and the soul; it is not subject to physical laws, and Procháska further rejected any mechanistic explanation. Reflex movements, he believed, could be produced either consciously or unconsciously, without cerebral control, and could even occur in decapitated animals or in anencephalous monsters. He drew a clear distinction between the *sensorium commune* and the seats of the intellect and the will, and thereby figures in the development of the idea of brain localization. His systematic account and rational synthesis of data were influential on Marshall Hall, F. A. Longet, and Pflüger, among others, largely between 1830 and 1860.

On his return to Vienna, Procháska published several editions of a textbook on physiology as well as some interesting work on the relation of the circulation of the blood to the nutrition of body tissues (such as the growth and simultaneous destruction of bone) and several observations on pathology. From about 1910, he attempted to interpret the phenomena of life according to the romantic tenets of *Naturphilosophie*.

BIBLIOGRAPHY

I. Original Works. Most of Procháska's important early works were published or reprinted in two collections: *Adnotationum academicaeum*, 3 vols. (Prague, 1780–1784); and *Operum minorum anatomici, physiologici et pathologici argumenti*, 2 vols. (Vienna, 1800). His principal treatise, *Commentatio de functionibus systematis nervosi* (Prague, 1784), was repr. with a Czech translation, *Úvaha o funkcích nervové soustavy* (Prague 1954). There is also an English trans. by T. Laycock: *A Dissertation on the Functions of the Nervous System* (London, 1851). Extracts are in R. J. Herrnstein and E. G. Boring, *A Source Book in the History of Psychology* (Cambridge, Mass., 1965), 289–299. Later works include *Bemerkungen über den Organismus des menschlichen Körpers und über die den-*

selben betreffende arteriösen und venösen Haargefässe, nebst den darauf gegründeten Theorie von der Ernährung (Vienna, 1810); and *Disquisitio anatomico-physiologica organismi corporis humani eiusque procesus vitalis* (Vienna, 1812). His last views are presented in *Physiologie oder Lehre über die Natur des Menschen* (Vienna, 1820).

II. Secondary Literature. A comprehensive biography of Procháska with a bibliography of his writings is V. Kruta, *Med. Dr. Jiří Procháska. Život, dílo, doba* (Prague, 1956). Philosophical aspects of his work were analyzed by J. Černý; *Jiří Procháska a dialektika v německé přírodní filosofii* (Prague, 1960). Short accounts are V. Kruta, "G. Procháska and the Reflex Theory," in *Scripta medica*, **34** (1961), 297–314; and *Epilepsia*, **3** (1962), 446–456; V. Kruta and Z. Franc, "Un ouvrage de physiologie romantique de G. Procháska," in *Castalia*, **22** (1966), 3–12; and M. Neuburger, "Der Physiologe Georg Procháska," in *Wiener medizinische Wochenschrift* (1937), 1155–1157. Procháska's place in the history of neurophysiology is discussed in G. Canguilhem, *La formation du concept de réflexe aux XVIIe et XVIIIe siècles* (Paris, 1955); F. F. Fearing, *Reflex Action* (Baltimore, 1930); M. Neuburger, *Die historische Entwicklung der experimentellen Gehirn- und Rückenmarksphysiologie vor Flourens* (Stuttgart, 1897); and J. Soury, *Le système nerveux central. Structure et fonctions* (Paris, 1899).

Vladislav Kruta

PROCLUS (*b.* Byzantium, 410 [412?]; *d.* Athens, 485), *philosophy, mathematics, astronomy.*

Proclus' parents, Patricius and Marcella, were wellborn citizens of Lycia; and his father had attained eminence as an advocate in the courts of Byzantium. Proclus received his early education at the grammar school of Xanthus, a city on the southern coast of Lycia. He was later sent to Alexandria, where he began the study of rhetoric and Latin in preparation for following his father's career. But on a visit to Byzantium during these years he experienced a "divine call," as his biographer Marinus tells us (chs. 6, 9, 10), to devote himself to philosophy. Returning to Alexandria, Proclus studied Aristotle with Olympiodorus the Elder and mathematics with a certain Heron, otherwise unknown. But these teachers did not satisfy him; and before he was twenty, he moved to Athens, where the Platonic Academy had recently undergone a notable revival under the headship of Plutarch of Athens. From this time until his death in 485 Proclus was a member of the Academy, first as student, then as a teacher, and finally as its head— whence the title Diadochus (Successor) which is usually attached to his name.

Proclus was the last great representative of the philosophical movement now called Neoplatonism.

The first notable exponent of this Hellenistic form of Platonism was Plotinus, from whom the doctrine had been transmitted, through Porphyry and Iamblichus, to Plutarch and Syrianus, Proclus' teachers at Athens. During these two centuries Neoplatonism had taken on a more pronounced religious coloration and had acquired a tincture of the Eastern predilection for magic, or "theurgy," as Iamblichus called it; on the other hand, its logical structure had become more precise and systematic, and its exponents had turned increasingly to scholarly examination and exposition of the writings of Plato and Aristotle.

Proclus had an extraordinarily acute and orderly mind. Because of his religious temperament he enthusiastically espoused Neoplatonism and devoted his talents and energies to perfecting it by systematizing and extending the views of his predecessors, strengthening their logical structure, and showing in detail their derivation from the teaching of Plato, who was taken as the source and final authority. But Proclus was more than a systematic metaphysician. He had a broad interest in all products of Greek culture, in religion, literature, science, and philosophy. His literary production was tremendous. Many of his writings have been lost, but those remaining constitute a priceless source of information regarding this last stage of Greek culture; and because of their underlying philosophy they embody an impressive restatement of Greek rationalism in its last confrontation with Christian thought.

The goal of philosophy, according to the Neoplatonists, was to attain a vision of and contact with the transcendent and ineffable One, the principle from which all things proceed and to which they all, according to their several natures and capacities, endeavor to return. But this synthesizing insight was to be attained only by the hard labor of thought. Proclus believed that a prerequisite to the study of philosophy was a thorough grounding in logic, mathematics, and natural science. One of the most important of his extant writings is the *Commentary on the First Book of Euclid's Elements*, in part certainly a product of his lectures at the Academy. Proclus was not a creative mathematician; but he was an acute expositor and critic, with a thorough grasp of mathematical method and a detailed knowledge of the thousand years of Greek mathematics from Thales to his own time. Because of his interest in the principles underlying mathematical thought and their relation to ultimate philosophical principles, Proclus' commentary is a notable—and also the earliest— contribution to the philosophy of mathematics. Its numerous references to the views of Euclid's predecessors and successors, many of them otherwise unknown to us, render it an invaluable source for the history of the science.

In the same vein but of more limited interest today is Proclus' *Hypotyposis* [Outline] *of the Hypotheses of the Astronomers*, an elaborate exposition of the system of eccentrics and epicycles assumed in Ptolemy's astronomy. This Ptolemaic system had arisen out of an effort to provide a mathematical explanation of the anomalies in the motions of the heavenly bodies as observed from the earth. Proclus approved the motives that led to its construction, and thought a knowledge of it desirable for his students; but he was understandably critical of its complexity as a whole and of the ad hoc character of its individual hypotheses. Several other writings on astronomy are attributed to Proclus: an elementary treatise entitled *Sphaera*, which appeared in more than seventy editions or translations during the early Renaissance; a paraphrase of Ptolemy's astrological *Tetrabiblos*; and another astrological essay entitled *Eclipses*, extant in two different Latin translations. Finally, his *Elements of Physics* offered a summary of books VI and VII of Aristotle's *Physics* and the first book of *De caelo*, arranged in geometrical form with propositions and proofs.

Proclus' most systematic philosophical work is his *Elements of Theology*, which presented in geometrical form, in a series of propositions, each supported by its proof, the successive grades of being that proceed from the superexistent and ineffable One downward to the levels of life and soul. The treatise *Platonic Theology*, probably a later work, presented this hierarchy of divine principles as they were revealed in Plato's dialogues, particularly in the *Parmenides*. Planned on a grandiose scale, this work either was not completed or has been imperfectly transmitted to us.

Better known is the impressive series of commentaries on Platonic dialogues: a lengthy commentary on the *Timaeus*—which, Marinus tells us (ch. 38), was Proclus' favorite—another equally long commentary on the *Parmenides*, another on the *First Alcibiades*, and still another on the *Republic*. The texts of all these have been preserved. His commentary on the *Cratylus* survives only in fragments, and those on the *Philebus*, the *Theaetetus*, the *Sophist*, and the *Phaedo* have been completely lost, as have those he is reported to have written on Aristotle. We possess only a fragment of Proclus' commentary on the *Enneads* of Plotinus, and we know of his *Eighteen Arguments for the Eternity of the World*, a tract against the Christians, only because it is extensively quoted in Philoponus' book written to refute it. His treatises *On Providence and Fate* and *On the Subsistence of Evils* were long known only in Latin translations, but

large portions of the Greek text have recently been recovered and edited.

Of Proclus' numerous works on religion, none survives except in fragments. Like Aristotle, he believed that ancient traditions often contain truth expressed in mythical form. Orphic and Chaldean theology engaged his attention from his earliest years in Athens. He studied Syrianus' commentary on the Orphic writings along with the works of Porphyry and Iamblichus on the Chaldean Oracles; and he undertook a commentary of his own on this collection, which, Marinus states (ch. 26), took five years to write. Proclus himself was a devout adherent of the ancient faiths, scrupulously observing the holy days of both the Egyptian and Greek calendars—for, he said, it behooves the philosopher to be the hierophant of all mankind, not of one people only (Marinus, ch. 19).

Proclus never married and made liberal use of his apparently ample means for the benefit of his relatives and friends. His diet was abstemious but not ascetic, although he customarily refrained, in Pythagorean fashion, from eating meat. He composed many hymns to the gods, of which seven survive, written in Homeric language and marked by literary quality as well as religious feeling. We are told that he lived in constant communication with the divine world, addressing his adoration and aspiration in prayers and ritual observances and receiving messages from the gods in dreams. His pious biographer also presents him as something of a wonder-worker who, having been initiated into the secrets of the hieratic art, practiced necromancy and other forms of divination and who was able by his arts to produce rain and to heal disease. Such beliefs, like the belief in astrology which both he and Ptolemy held, were almost universal in that age.

Proclus deserves to be remembered, however, not for these beliefs that he shared with almost all his contemporaries, but for the qualities he possessed that are exceedingly rare in any age and were almost unique in his: the logical clarity and firmness of his thought, the acuteness of his analyses, his eagerness to understand and his readiness to present the views of his predecessors on controversial issues, the sustained coherence of his lengthy expositions, and the large horizon, as broad as the whole of being, within which his thinking moved.

Proclus' thought indirectly exercised considerable influence in the early Middle Ages through the writings of the so-called Dionysius the Areopagite, whose teachings were a thinly disguised version of Proclus' doctrines. With the revival of learning in the fifteenth century and the desire of Renaissance thinkers to throw off the yoke of medieval Aristotelianism, Proclus' Platonism had a great vogue in the Florentine Academy and strongly influenced Nicholas of Cusa and Johannes Kepler. Modern criticism has tended, rather hastily, to discredit his interpretation of Plato; and with the decline of interest in speculative philosophy, his writings have fallen into neglect. But it is fair to say that the wealth of learning and insight in his works does not deserve to be neglected, and that the constructive philosophy they contain still awaits adequate appraisal and appreciation by modern philosophers.

BIBLIOGRAPHY

I. ORIGINAL WORKS. *Procli philosophi Platonici opera inedita*, Victor Cousin, ed., 2nd ed. (Paris, 1864), contains text of *Commentary on the Parmenides*, French trans. by A. Chaignet, 3 vols. (Paris, 1901–1903).

Separate eds. of other texts are: *Commentary on the First Alcibiades of Plato*, L. G. Westerink, ed. (Amsterdam, 1959), English trans. by William O'Neill (The Hague, 1965); *Commentary on the Timaeus*, Ernst Diehl, ed., 3 vols. (Leipzig, 1903–1906), French trans. by A. J. Festugière, 5 vols. (Paris, 1966–1968); *Commentary on the Republic*, Wilhelm Kroll, ed., 2 vols. (Leipzig, 1899–1901), French trans., by A. J. Festugière, 3 vols. (Paris, 1970); *Elements of Physics*, Albert Ritzenfeld, ed., with a German trans. (Leipzig, 1911); *Hypotyposis*, Charles Manitius, ed., with a German trans. (Leipzig, 1909); *Commentary on the First Book of Euclid's Elements*, Gottfried Friedlein, ed. (Leipzig, 1873), German trans. by Leander Schönberger (Halle, 1945), French trans. by Paul ver Eecke (Bruges, 1949), English trans. by Thomas Taylor, 2 vols. (London, 1792), and Glenn R. Morrow (Princeton, 1970); *Commentary on the Cratylus*, G. Pasquali, ed. (Leipzig, 1908); *Elements of Theology*, E. R. Dodds, ed., with an English trans., 2nd ed. (Oxford, 1963); *Platonic Theology*, H. D. Saffrey and L. G. Westerink, eds., with a French trans. (Paris, 1968), English trans. by Thomas Taylor (London, 1816); *Providence and Fate* and *Subsistence of Evils*, Helmut Boese, ed. (Berlin, 1960); and *Hymns*, E. Vogt, ed. (Wiesbaden, 1957), Thomas Taylor, ed., with an English trans. (London, 1793).

For a complete list of Proclus' writings, with a bibliography of eds. and translations of individual items during the modern period to 1940, see Rosán (below), 245–254.

II. SECONDARY LITERATURE. An ancient biography is *Vita Procli*, by his pupil and successor Marinus, J. F. Boissonade, ed. (Leipzig, 1814), repr. in Cousin, *Procli ... Platonici opera inedita.* See W. Beierwaltes, *Proklos* (Frankfurt am Main, 1965); Rudolf Beutler, "Proklos (4)," in Pauly-Wissowa, *Real-Encyclopädie der classischen Altertumswissenschaft*, XXIII (1957), 186–247; Raymond Klibansky, "Ein Proklus-fund und seine Bedeutung," which is *Sitzungsberichte der Heidelberger Akademie der Wissenschaften*, **19**, no. 5 (1929); Laurence J. Rosán, *The Philosophy of Proclus* (New York, 1949); A. E. Taylor, "The Philosophy of Proclus," in *Proceedings of the Aristotelian Society*, **18** (1918), 600–635; Friedrich Ueberweg, *Grundriss der Geschichte der Philosophie*, I, *Die Philosophie des Altertums*, 12th ed., Karl Praechter, ed. (Berlin, 1926), 621–631; Thomas Whittaker, *The Neo-Platonists*, 2nd ed. (Cambridge, 1928), 155–184, 231–314; and E. Zeller, *Die Philosophie der Griechen*, 5th ed., III, sec. 2 (Leipzig, 1921), 834–890.

GLENN R. MORROW

PROCTOR, RICHARD ANTHONY (*b.* Chelsea, London, England, 23 March 1837; *d.* New York, N.Y., 12 September 1888), *astronomy.*

Proctor acquired an immense popular reputation by virtue of his lucid presentation, both in books and in lectures. He was familiar not only with what was visible through the telescopes of the time but also with astronomy in a literary context. As an astronomer he was prone to speculation, and frequently he was wildly mistaken.

Proctor was the youngest son of William Proctor, a well-to-do solicitor. His father's death in 1850, followed by a protracted lawsuit, left the family in difficult circumstances. Proctor was a bank clerk (1854) until he was able to attend London University (1855) and then St. John's College, Cambridge (1856), where he studied theology and mathematics. He married before his degree examination in 1860, in which his results were disappointing (twenty-third wrangler); he then began to study for the bar, but the death in 1863 of the first of his many children turned him toward astronomy and mathematics as distractions. His first article was on colors in double stars (*Cornhill Magazine* for 1865); and his first book, privately published, was *Saturn and His System* (1865).

With the failure in 1866 of a New Zealand bank in which his money was invested, Proctor turned to a precarious and badly paid literary career. His popular articles were frequently rejected, and his early books barely met their printing costs. His best-selling *Half-Hours With the Telescope* (London, 1868), written for a fee of £25, reached twenty editions by the time of his death. One of his most widely read works was *Myths and Marvels of Astronomy* (1877), an entertaining collection of astronomical lore.

Proctor taught mathematics for some time at a private military school in Woolwich. After 1873 he went on several lecture tours of America and Australasia. His first wife died in 1879, and in 1881 he married an American widow, Mrs. Robert J. Crawley, and settled at St. Joseph, Missouri, her hometown. He moved to Orange Lake, Florida, in 1887 but died the following year in New York City, perhaps of yellow fever, on his way to England on business.

Proctor long served as an officer of the Royal

Astronomical Society, contributing many articles to its *Monthly Notices*. He founded (1881) the London scientific weekly *Knowledge and Illustrated Scientific News*, to which he contributed articles under a number of pseudonyms. In all he published nearly sixty books, none of them of much astronomical moment but many of them acknowledged by later astronomers as having introduced them to the subject.

Proctor's most original scientific work related to Venus and Mars, of which he drew the best maps then available, and to the Galaxy. As a means of determining the size of the astronomical unit, the transit of Venus of 8 December 1874 was to be used. The astronomer royal, Sir George Airy, had announced approximate data in 1857; and Proctor, after calculating more precise circumstances of the transit and urging the use of a method originated by Halley, found himself severely—and for the most part wrongly—criticized by Airy.

Proctor wrote a series of articles on the rotation of Mars and deduced an extremely accurate period of rotation ($24^h37^m22.735^s$), utilizing a drawing of the planet made by Robert Hooke in 1666. This is 0.1 second higher than the estimate subsequently made by F. Kaiser from observations made during seventeen nights of the opposition of 1862, and from their comparison with the observations of many earlier astronomers.

In 1870 Proctor charted the directions and proper motions of about 1,600 stars, as determined by E. J. Stone and Rev. R. Main, and found the phenomenon that he called "star drift," whereby large groups of stars share a proper motion vector in space. The Taurus stream, between Aldebaran and the Pleiades, was the most important group. Had Proctor continued these researches, he might well have reached important conclusions concerning the structure of the Galaxy and associated clusters. (The next significant study of the problem was made by L. Boss.) Proctor was helped by his abilities as a draftsman, and his charts of the distribution of nebulae indicated clearly to him the tendency of all but the gaseous nebulae to avoid the plane of the Milky Way. He wrongly concluded that since it is unlikely that such an arrangement is accidental, all must be part of a single system (1869). A paper by Cleveland Abbe published two years earlier used the same evidence as grounds for the opposite conclusion, but not for sixty years did something akin to Abbe's ideas prevail.

BIBLIOGRAPHY

I. ORIGINAL WORKS. Proctor's principal memoirs on the rotation of Mars appeared in *Monthly Notices of the Royal*

Astronomical Society, **28** (1868), 37–39; **29** (1869), 229–232; and **33** (1873), 552–558. See "On the Distribution of the Nebulae in Space," *ibid.*, **29** (1869), 337–344; and "On Certain Drifting Motions of the Stars," in *Proceedings of the Royal Society*, **18** (1870), 169–171; and in *Philosophical Magazine*, 4th ser., **39** (1870), 381–383.

The titles of more than half of Proctor's books are listed in the *Dictionary of National Biography* article by E. [sic] M. Clerke. An unfinished book, *New and Old Astronomy* (London, 1892), was completed by A. C. Ranyard. Most of Proctor's articles are listed in the Royal Society *Catalogue of Scientific Papers*, VIII, 666–668. He contributed 83 papers to *Monthly Notices of the Royal Astronomical Society* and was a regular contributor to *Intellectual Observer, Chambers's Journal, Popular Science Review*, and of course *Knowledge*, for which he wrote not only on astronomy but also on chess and whist.

II. SECONDARY LITERATURE. There are obituaries of Proctor in *The Times* (London) (14 Sept. 1888), 5; *Knowledge* (Oct. 1888), 265–266; *Monthly Notices of the Royal Astronomical Society*, **49** (1889), 164–168; and *Observatory*, **11** (1889), 366–368. For the somewhat futile work done by all concerned on the 1874 transit of Venus, see Agnes M. Clerke, *History of Astronomy in the Nineteenth Century*, 3rd ed. (London, 1893), 287–292. For early work of note on the dynamics of moving clusters, see A. S. Eddington, *Stellar Movements and the Structure of the Universe* (London, 1914), esp. ch. 4. See also Cleveland Abbe, "On the Distribution of the Nebulae in Space," in *Monthly Notices of the Royal Astronomical Society*, **27** (1867), 257–264.

J. D. NORTH

PROFATIUS TIBBON. See **Ibn Tibbon, Jacob ben Machir.**

PRONY, GASPARD-FRANÇOIS-CLAIR-MARIE RICHE DE (*b.* Chamelet, France, 22 July 1755; *d.* Asnières, France, 29 July 1839), *engineering*.

Prony's father, a prominent lawyer, wished his son to follow him into a legal career and had him trained in the classics. But after convincing his father and after spending more than a year studying mathematics, in 1776 Prony entered the École des Ponts et Chaussées, from which he graduated in 1780. He won the admiration of its director Perronet, who in 1783 had Prony brought from the provinces to Paris to assist him. Prony's defense of Perronet's bridge at Neuilly led to his first memoir, on the thrust of arches (1783), and to friendship with Monge, who personally initiated him in advanced analysis and descriptive geometry. In this memoir using Phillipe de La Hire's formula for the conditions of rupture of masonry arches, he argued that the Pont de Neuilly was structurally sound in spite of the thinness of its pillars and the breadth of its arches because it had sufficient strength to resist tipping caused by the flow of the river, bursting caused by the thrust of the arches, and crumbling caused by the jolting of heavily loaded wagons.

In 1790 Prony published the first volume of *Nouvelle architecture hydraulique*, which applies rational mechanics to engineering practice at an elementary level. It was a textbook for beginners, specifically for the students of the École des Ponts et Chaussées, of which he had just been appointed inspector of studies. He used Lagrangian analysis to prove theorems of statics, which he then represented graphically by use of descriptive geometry and for which he gave numerical methods of calculation that permitted the reader to solve problems without recourse to theory. In dynamics and hydrodynamics Prony, who used the calculus liberally, followed the same procedure. This textbook presented statics as the science of hydraulic machines, which Prony viewed as lifters of fluids. The second volume (1796) reverted to a nonanalytical and descriptive level in the vein of Bélidor's *Architecture hydraulique* (1737–1739).

During the Revolutionary period, which he survived thanks to Lazare Carnot's protection in 1793, Prony had an active career. In 1791 he was appointed director of the cadastral survey of France and in 1792 he supervised the drawing up of new logarithmic and trigonometric tables incorporating decimal divisions. Using the concept of the division of labor which he got from reading Adam Smith's *An Inquiry Into the Nature and Causes of the Wealth of Nations* (1776), he allocated the work of calculating the tables among several hundred men who knew only the elementary rules of arithmetic. In 1794 at the creation of the École Centrale des Travaux Publics (École Polytechnique) Prony became professor of analysis, a position that he held until 1815, when he resigned, reducing his commitment to that of *examinateur de sortie*. This position led him to write during this period several textbooks on analysis and mechanics, for instance "Cours d'analyse appliquée à la mécanique" (*Journal de l'École Polytechnique*, 1st cahier, 1794, 92–119, and 2nd cahier, 1795, 1–23) and *Leçons de mécanique analytique données à l'École . . . Polytechnique . . . seconde partie* (Paris, 1815). In the manual *Mécanique philosophique* (1799) he proposed to do for mechanics what Fourcroy had done for chemistry in *Philosophie chimique* (1792), namely to give a taxonomy of the theorems of mechanics. Although his textbooks on analytical mechanics drew largely on Lagrange's *Mécanique analytique* (1788), they had a practical orientation and employed geometrical representations freely. In them, Prony reasoned on the manipulative

level rather than on the conceptual one. Prony was named to the Institut de France on its founding in 1795. At the death of Antoine de Chezy in 1798, Prony succeeded him as the director of the École des Ponts et Chaussées and held this position until his death. He guided the school through great changes and modernized its curriculum for an industrial age. Drafting yielded to applied analytical mechanics; professional engineering scientists took the place of student teachers.

The Napoleonic period, during which Prony made several field trips to Italy to study river control and to draw up a project for the drainage of the Pontine Marshes, formed a highly productive part of his theoretical engineering career. To instruct engineers, he published treatises on earth thrust (*Recherches sur la poussée des terres*, 1802), on the measure of the flow of liquids through orifices (*Le jaugeage des eaux courantes*, 1802), and on their flow in pipes and in canals (*Recherches physico-mathématiques sur la théorie des eaux courantes*, 1804). In his treatise on earth thrust, Prony simplified Coulomb's analysis by assuming that the curve of maximum thrust is a straight line rather than some unknown curve that could only be found by using the calculus of variations. This simplification led him to conclude that the line of maximum thrust bisects the complement of the angle of repose. In his treatise on the gauging of water, Prony gave practical advice on how one should go about measuring water flow in such a way that the measurements are independent of hypotheses formulated concerning the laws of fluid flow. To measure, the engineer should use a pitot tube to find the speed of flow of the water through an orifice cut into a weir around which is built a system of channels to assure stagnant water behind the weir. To find all that is known about the flow of fluids through orifices, Prony referred the reader to Giovanni Battista Venturi, *Recherches expérimentales sur le principe de la communication latérale du mouvement dans les fluides, appliqué à l'explication de différens phénomènes hydrauliques* (Paris, 1797). In his study of flowing water Prony had a triple objective: build on solid principles of mechanics, respect experiment, and present results in a way that does not require difficult calculations on the part of the reader. He limited his study to cases where fluids flow along sufficient distances to assure constant speed and equality between the accelerating force of gravity and the resisting force of cohesion and friction. Drawing on memoirs by Charles Augustin Coulomb (1800) and Pierre-Simon Girard (1803 and 1804), Prony derived $U = a + \sqrt{b + cRI}$ where U is the average speed, R the mean radius (cross-sectional area/wetted perimeter), *l* the slope, and *a*, *b*, *c* constants. He presented the results of his study in graphical and numerical form for the use of the field engineer.

In 1805 when Prony became *inspecteur général des ponts et chaussées*, he obtained a seat on the general council of the Ponts et Chaussées, which gave technical and administrative direction to the Corps of the Ponts et Chaussées—no project could be undertaken without its prior approval. In this post, which he held for life, he greatly influenced the course of civil engineering in France and encouraged the construction of numerous bridges.

Under the Bourbon and Orleans monarchs Prony continued an active engineering career, which became gradually one of a highly placed technical consultant rather than a field engineer. He did not abandon all engineering research, however. In 1821 he developed the Prony brake while conducting tests on steam engines. He discovered that the work delivered by a motor, to the shaft of which was applied a heavy frictional drag by means of a brake, was measured by the moment of the weight that kept the brake from rotating with the shaft. He showed that the equation for the work delivered is independent of the nature of friction and of the equation relating friction and force.

France's various governments recognized Prony's useful service to the state. Even though he refused to accompany Napoleon to Egypt, Napoleon made him a member of the Legion of Honor when he founded it. Charles X made him baron, and Louis Philippe raised him to the peerage. Prony was the true successor of Jean-Rodolphe Perronet, becoming France's leading engineer and engineering educator during the period 1800–1840.

BIBLIOGRAPHY

I. ORIGINAL WORKS. A complete bibliography of Prony's writings includes over 100 titles; fortunately the library of the École des Ponts et Chaussées possesses Prony's library and many of his MSS; the *Catalogue des livres composant la bibliothèque de l'École des ponts et chaussées* (Paris, 1872) and the *Catalogue des manuscrits de la bibliothèque de l'École des ponts et chaussées* (Paris, 1886) give the necessary references; this library has bound together most of Prony's articles under the title *Ses opuscules, ou collection de toutes les brochures publiées par Riche de Prony*, 4 vols. Parisot's notice on Prony (cited below) lists Prony's major works. With A.-A. Parmentier, J.-P.-F. Guillot-Duhamel, J.-G. Garnier, *et al.*, Prony edited *Mémoires des sociétés savantes et littéraires de la République*, 2 vols. (Paris, 1801–1802). L. B. Francoeur, *Traité élémentaire de mécanique* (Paris, 1801), consists of Prony's notes. Prony briefly describes his method for measuring the work of a machine in "Note

sur un moyen de mesurer l'effet dynamique des machines de rotation," in *Annales de chimie et de physique*, **19** (1822), 165–173, and more fully in "Rapport . . . sur la nouvelle et l'ancienne machines à vapeur établies, à Paris, au Gros-Caillou . . .," in *Annales des mines*, **22** (1826), 3–100, plus 3 plates. At the Archives Nationales, Paris, the following references are of interest: F^{14} 2304², Prony's dossier, briefly details his career; F^{17} 1084¹³ contains documents concerning the printing and distribution of Prony's *Mécanique philosophique*; F^{14} 1023 contains a report by Prony on the Pontine Marshes; F^{14} 1031² contains correspondence with Prony on his mission in Italy; F^{14} 11055 has a few letters by Prony; F^{14} 11057 has *Sommaires des leçons du cours de mécanique de M. De Prony* (1811) and a series of documents by Prony, Navier, and Coriolis concerning an 1830 reform of the École des Ponts et Chaussées; F^{14} 11138 has a report by Prony dated 1810 concerning the Pont d'Iéna; F^{14} 13859² has documents by Prony concerning the 1830 reform of the École des Ponts et Chaussées and measurement of the effect of steam engines; F^{17} 1393 has documents concerning Prony and the cadastral survey of France. Some letters by Prony are at the Bibliothèque Nationale, Paris, n.a. fr. 2479, 2762, 2769, 15778, 22431. The archives of the École Polytechnique, Paris, carton 1829, has a report by Prony that remarks on Cauchy's teaching. The Académie des Sciences, Paris, has MSS by Prony and a MS report by Monge on Prony's memoir on arch thrust in the carton for 1783. This memoir, "Sur l'application de la mécanique des voûtes au Pont de Neuilly," is at the library of the École des Ponts et Chaussées, MS no. 2215.

II. SECONDARY LITERATURE. Parisot, "Prony," in Michaud's *Biographie universelle* LXXVIII (Paris, 1846), 79–91, gives an excellent survey of Prony's life. *Bulletin de la Société d'encouragement pour l'industrie nationale*, **139**, nos. 3 and 4 (Mar.–Apr. 1940), 66–98, commemorates Prony's death, and **139/140** (July 1940–June 1941), p. 166, details Prony's descendants. Other obituaries are J.-B. Tarbé de Vauxclairs, in *Moniteur universel*, no. 219 (7 Aug. 1839), 1605, and in *Annales des ponts et chaussées . . ., mémoires et documents*, **18** (2nd sem., 1839), 394–400; C. Dupin, ". . . Éloge de M. le Bon de Prony . . .," in *Chambre des pairs, Impressions diverses*, session 1840, no. 48 (Paris, 1840), sess. of 2 Apr.; and F. Arago, "Discours funèbre de Prony," in *Oeuvres complètes . . .*, III (Paris, 1855), 584–592. Archives Nationales, Paris, F^{14} 10906¹–11041¹⁸⁰ is the collection of the registers of the deliberations of the Conseil Général des Ponts et Chaussées. At the École des Ponts et Chaussées, Paris, office of the secretary, is "Registre des procès verbaux des séances du conseil de l'École des ponts et chaussées. . . ." See also A. Lorion, "Une mésaventure de Prony à Venise sous l'Empire (juillet 1805)," in *Revue de l'Institut Napoléon*, no. 93 (Oct. 1964), 164–168; and M. d'Ocagne, *Hommes et choses de science*, 3rd sers. (Paris, 1936), 174–179. C.-A. Vieilh de Boisjoslin, *Biographie universelle et portative des contemporains*, 4 (Paris, 1830), 1024–1027; and M. Edgeworth, *Lettres intimes . . .* (Paris, 1896), letters dated 18 Nov. 1802, 1 Dec. 1802, and 3 May 1820, give details concerning Prony. J. V. Poncelet, "Examen critique et historique des principales théories ou solutions concernant l'équilibre des voûtes," in *Comptes rendus hebdomadaires des séances de l'Académie des sciences*, **35** (1852), 494–502, gives the context of Prony's work on arches. C. S. Gillmor, *Coulomb and The Evolution of Physics and Engineering in Eighteenth-Century France* (Princeton, 1971), 100–115, 189–190; and H. Rouse and S. Ince, *History of Hydraulics* (New York, 1963), 139–143, discuss Prony briefly.

See also Jacques Payen, "La pratique des machines à vapeur au temps de Carnot," in *Actes du colloque "Sadi Carnot et l'essor de la thermodynamique"* (Paris, in press), which discusses tests that Prony made on steam engines and during which he first used the Prony brake.

ROBERT M. McKEON

PROUST, JOSEPH LOUIS (*b*. Angers, France, 26 September 1754; *d*. Angers, 5 July 1826), *chemistry*.

The second son of Joseph Proust, an apothecary, and Rosalie Sartre, Proust received his early education under the supervision of his godparents and continued it at the local Oratorian *collège*. He was then apprenticed to his father, to study pharmacy and to succeed him. Around 1774, despite parental opposition, he went to Paris to continue his training.

In Paris, Proust attached himself to Clérambourg, the apothecary, and studied chemistry with Hilaire-Martin Rouelle (not his elder brother, Guillaume-François Rouelle, as is often reported), with whom he became friends, ever after referring to him as his "master." In 1776 Proust secured the position of *pharmacien en chef* at the Salpêtrière. He published his first papers while at this hospital.

At the end of 1778, Proust took his first position in Spain, the country in which he spent the major part of his professional life. The post was a professorship in chemistry at the recently established Real Seminario Patriótico Vascongado at Vergara. This school was the creation of the first and most important of the "enlightened" societies which came into existence in the second half of the eighteenth century to bring modern learning and culture to Spain: the Real Sociedad Económica Vascongada de Amigos del País, established in 1765. Royal authorization had been obtained for professorships at the school in experimental physics, chemistry, mineralogy, and metallurgy. Proust stayed in Vergara only a short time, returning to France in June 1780.

During the next five years Proust's most important enterprises were those connected with Pilatre de Rozier. Until 1784 Proust taught chemistry at the Musée, founded by Pilatre in 1781. Also during this period Proust was involved with Pilatre, as well as the physicist Jacques Charles, in aerostatic experiments that culminated in the ascent of Proust and Pilatre in a balloon at Versailles on 23 June 1784, in the presence of the king and queen of France, the king of Sweden, and the French court.

In 1785 Proust again contracted to take a teaching position in Spain, this time at the invitation of the Spanish government and through the intermediary of Lavoisier. The offer was for a financially remunerative post, and after some hesitation Proust accepted and went to Spain in 1786. He first taught in Madrid; but in 1788 he moved to Segovia as professor of chemistry at the Royal Artillery College, where chemistry had been made a required subject. Proust taught and experimented there and also conducted geological and mineralogical surveys and analyses for the government. Assessments of his success as a teacher are contradictory; and there is some evidence that toward the end of this period he wished to return to France.

On 30 June 1798 Proust married a French resident of Segovia, Anne Rose Châtelain Daubigné. They had no children. In April 1799 Proust was brought from Segovia to Madrid to head a newly organized chemical laboratory that amalgamated the previously existing facilities in Madrid and Segovia, and to offer public courses in chemistry. The opulence of his laboratory became legendary. While in Segovia, Proust began to publish his papers on definite proportions and it was while he was in Madrid that the controversy with Berthollet over this issue took place.

Proust returned to France toward the end of 1806, for reasons not altogether clear (one account suggests that it was to settle his patrimony with his brother Joachim), and remained there. In 1808 Napoleon's forces invaded Spain; Carlos IV abdicated; and during the ensuing political dislocation and resistance, Proust's laboratory establishment was dispersed. Left in reduced circumstances by this sequence of events, Proust settled in Craon but spent considerable time at a family farm that he had inherited in the Loire valley. One of the few bright features of his return to France was his cordial reception by Berthollet. Both Proust and his wife were ill after 1810; she died in 1817, and Proust then moved to Angers, where in 1820 he took over the pharmacy of his brother Joachim, who was in poor health.

Awards came to Proust, although he remained aloof from the French scientific establishment. On 12 February 1816 he was elected to succeed Guyton de Morveau at the Institut de France. In 1819 he became a *chevalier* of the Legion of Honor, and in 1820 he was granted a pension by Louis XVIII.

Proust's published papers are notable for their clear style and exposition. Testimony differs over his abilities as a lecturer and teacher, and there is reason to believe that he was at times negligent of his duties. Proust left Pilâtre de Rozier's Musée in 1784 after criticism of his teaching, and there is evidence that his effectiveness as a teacher in Spain was less than his early French biographers suggested it had been.

Proust's historical importance derives primarily from his abilities as an analyst: his development of the use of hydrogen sulfide as an analytical reagent; his use of quantitative methods—consistently giving the results of his analyses in terms of percentage weight composition and sometimes (for oxides and sulfides of the same metal) the weight of oxygen or sulfur in comparison with a constant weight of the metal (without, however, drawing any conclusion from the results)—and most significant, his enunciation of the law of definite proportions.

There can be little question that the general tendency in late eighteenth-century chemistry toward quantitative analysis, associated particularly with the expression of weight composition which became common especially in the 1780's, served as the background to the formulation of the law of definite proportions. Yet the exact relationship between this development and the enunciation of the law remains obscure. The assumptions behind gravimetric analysis are not necessarily the same as the fundamental tenet of the law of definite proportions: that combining substances can combine in only a small number of fixed proportions.

Thus, to speak, as some have done, of Proust's law as a commonplace of eighteenth-century chemical thought is unwarranted, at least on the basis of the limited study that this question has received. There were, however, precursors of Proust's law in the literature of the late eighteenth century, notably Robert Dossie (1759), G. F. Venel (1765), L. B. Guyton de Morveau (1786), and Thomas Thomson (1801).[1]

These authors shared a common viewpoint and stated what I would call the "principle of constant saturation proportions." Their argument ran roughly as follows: Reacting substances had a unique proportion of combination, the saturation proportion, at which—and only at which—chemical combination took place. Observationally, the saturation concept derived from neutralization reactions and from saturation of solvent by solute. The satiation at constant proportions was due to the constant intensity of the affinity forces between the two reagents. Dossie's statement is a good summary and illustration of this principle.

In bodies that will commensurate strongly with each other, the specific attractions are in many cases limited only to certain respective proportions. For in some kinds, after they are combined together in a certain proportion, the compound becomes neutral, or indifferent with regard to further quantities of any of its constituents; in the same manner, as if those specific attractions, by which it was formed, had been wholly wanting.[2]

In addition to the chemical formulation of the principle of constant saturation proportions, the doctrine of fixed mineral species, enunciated by Romé de l'Isle (1784), Haüy (1793, 1801), and Dolomieu (1801),[3] presented defining characteristics of constant chemical composition and fixed crystal form. Although the mineralogical doctrine of fixed species seems to have developed more or less independently of the chemical principle of constant saturation proportions or of Proust's law, it was to have indirect relevance to Proust, since he was led to make his most comprehensive statement of the law of definite proportions in answer to a challenge actually directed at the mineralogists.

Proust's own formulation of the law of definite proportions was published rather suddenly in a paper on iron oxides, "Recherches sur le bleu de Prusse" (1794). Until further research is done on Proust's lecture notes and other materials from the Segovia period just prior to and encompassing this paper, it will be difficult to form an impression of the origins of his law.

Proust introduced his problem at the start of the 1794 paper:

If iron were, as is thought, susceptible of uniting with oxygen in all proportions between 27/100 and 48/100, which seem to be the two extreme terms of its union with this principle, ought it not to give as many diverse combinations with the same acid as it can produce different oxides?

He drew the following conclusion:

A great number of facts prove, on the contrary, that iron does not at all stabilize indifferently at all the degrees of oxidation intermediary between the two terms which we have just cited; and despite the different degrees of oxygenation through which one believes iron can pass when its sulfate is exposed to the air, only two sulfates of this metal are known.

At the end of the paper, Proust generalized his conclusion for other oxidizable substances:

I terminate . . . by concluding from these experiments the principle which I established at the beginning of this memoir; namely, that iron is, like several other metals, subject by that law of nature which presides over all true combinations, to two constant proportions of oxygen. It does not at all differ in this regard from tin, mercury, lead etc. and finally from virtually all of the known combustibles.[4]

Over the next thirteen years, and especially after 1797, Proust elaborated this conclusion in a series of papers in which he tried to show that most metals formed two distinct oxides at constant proportions— which he called the minimum and maximum—and that these two metal oxides were capable of forming two separate series of compounds, lead being recognized as an exception in forming three. Proust also asserted that there was only one sulfide per metal, with the exception of iron, which he came to recognize had two.

Although Proust's papers on metallic oxides appeared quite abruptly in the 1790's, interest in the different saturation proportions of metallic oxides had been growing in the previous decade with the recognition of the pneumatic chemistry and its gravimetric methods. Lavoisier had published two papers on the analyses of iron oxides (Proust cited data from the second). Proust's interest in metallic oxides and sulfides undoubtedly was also stimulated by his professional concern in Spain with metallurgical and mineral analysis.

Proust's strategy for proving his assertions about metallic oxides was well developed in his paper on Prussian blue and changed but little in general outline thereafter. He tried to show that there were two oxides of each metal, each oxide having a set of well-defined physical and chemical characteristics. Any metallic salts formed in chemical reactions had to have one of the oxides as its base. In oxidation or reduction reactions, Proust attempted to show that the oxide was raised (or lowered) directly from minimum to maximum (or vice versa) with no intermediate oxide produced.

As for oxides at intermediate proportions, Proust's strategy was to identify them as mixtures of the maximum and minimum oxides, mixtures of one of the oxides with uncombined metal, or a compound that really was not an oxide. Similarly, in the case of sulfides, Proust tried to show that reports of sulfides in variable proportions (such as those later offered by Berthollet) really concerned various kinds of mixtures or solutions, not true chemical compounds. This distinction between "true" compounds and mixtures or solutions became the crucial—and weakest— part of his strategy.

To separate what he took to be mixtures of the two oxides, Proust employed the diversity in chemical properties of the maximum and minimum compounds—for example, solubility in alcohol or some other solvent. Separating oxide from excess metal could also be accomplished by this means, as well as by more strictly physical ones, such as washing the oxide or heating a sulfide to sublimate the excess sulfur without changing the essential "physionomie" of the compound.

Proust consistently adopted a posture of empiricism, particularly in his controversy with Berthollet; indeed, there is little evidence of concern on his part with establishing any detailed theoretical underpinnings for his assertions of constant proportions. He chided Berthollet for elaborating an all-encompassing theoretical system, claiming that chemistry was not yet "mature" enough for such an approach but required more experimental data. Yet one can detect in Proust's justification of definite proportions more than the mere generalization from laboratory data that he claimed for his method. There were occasionally echoes of chemical affinity theory: "Election and proportion are the two poles around which the whole system of true combination rotates invariably, as much in Nature as in the hands of the chemist." And in one of his rare formulations of chemical combination in something like molecular terms, he wrote:

> For example, when a glass of potash is exposed to free air, every molecule of carbonic acid which approaches it is seized instantly by the number of alkali molecules which are needed to transform it into the carbonate. The attraction is there, as one knows; it keeps watch, it presides over this number. This reaction thus introduces into the potash new portions of carbonate, but of a complete carbonate.[5]

But even more general and fundamental to Proust's assertion was his belief in a natural principle of order, which he saw as regulating all chemical unions in nature as well as in the laboratory. This *pondus naturae*, as he called it, assured the invariability of chemical reaction and the constancy of the proportions of the product, regardless of the circumstances under which the reaction took place.[6] In 1788, six years before his first paper on definite proportions, Proust had already expressed this idea in a comment on the presence of phosphoric acid in minerals as well as in laboratory-produced compounds: "Presided over by the same laws, the ones and the others [minerals and artificially produced substances] are always alliances of choice and proportion."[7]

This notion of an overarching principle of proportion in nature and in the laboratory permeated Proust's writings on definite proportions. It remained disconcertingly empty of content; at times there is, indeed. more than a hint of rhetorical flourish in his invocation of it. Yet in one aspect it was endowed with some specificity: Proust held tenaciously to the idea that each metal could form only two oxides (with rare exceptions) and one sulfide (also with rare exceptions). Two oxides and one sulfide seem to have had an almost numerological significance for Proust; and he later defended this view despite Berzelius' analyses that contradicted it.

By 1801, through his refinements of the theory of chemical affinity, Berthollet had arrived at the proposition that chemical combination was, in principle, not necessarily restricted to definite proportions; combining substances could unite in a continuum of weight proportions between the minimum and the maximum. Berthollet attributed those combinations that seemed to occur in fixed proportions to physical properties—such as volatility, or tendency toward precipitation or crystallization—which tended to halt chemical reactivity and, hence, to stabilize the combinations. But there was nothing intrinsically "natural" about such compounds; and where reagents did not tend to produce compounds strongly endowed with such physical characteristics, a continuum of proportions might be produced.

Berthollet published a comprehensive exposition of his chemical theory, including his ideas on combining proportions, in *Essai de statique chimique* (1803). Despite his obvious disagreement with Berthollet's ideas, Proust was not the main object of Berthollet's attack on this issue in the *Essai*. Rather, in the theoretical sections of the first volume Berthollet devoted his most elaborate criticisms to Haüy's and Dolomieu's doctrine of fixed mineral species. But in the second volume, Proust did come under attack for his position on metallic oxides and sulfides—an attack strong enough to provoke his response in 1804.

The controversy through 1807 consisted of a paper or two every year from both parties to the dispute. Proust's most elaborate rebuttal of Berthollet's views, and defense of his own, was presented in "Sur les oxidations métalliques" (1804). Its arguments are typical. His strategy against Berthollet was twofold: to demonstrate inconsistencies and even contradictions between Berthollet's physicalistic and theoretical explanations for the appearance of some oxides at fixed proportions; and to show experimentally that instances of what Berthollet took to be oxides at intermediate proportions between minimum and maximum were really mixtures either of the two oxides or of the metal and one of its oxides.

To demonstrate such inconsistencies and contradictions, Proust noted that Berthollet had suggested that when metallic oxides were produced having stable proportions, those proportions were endowed with a greater degree of condensation than the metal itself. But Proust pointed out that two of Berthollet's examples—arsenic and antimony—had oxides at minimum that were more volatile than the metal. Berthollet also had argued that volatilization of the metal favored its rapid oxidation to maximum, because the metal could mix more thoroughly with the oxygen in the air; on the other hand, the solid state was a hindrance to complete oxidation because of the reciprocal attraction of the metallic particles. Proust pointed out that arsenic, which volatilized easily in the metallic state, formed the oxide at minimum readily, whereas tin, lead, antimony, copper, and bismuth, all of which volatilized with much greater difficulty than did arsenic, formed oxides at maximum by simple calcination.

Proust demonstrated that there were no oxides at intermediate proportions, either by showing that some of the unoxidized metal was mixed with the oxide by separating it mechanically (as by washing) or by separating mixtures of the two oxides through use of their different chemical properties. He also corroborated his results with mineral analyses: in the case of iron ores, for example, he claimed always to obtain oxides at either the minimum or maximum proportion, never at intermediate ones. Proust did express some uneasiness over his methodology and reasoning in the separation of the oxides in the latter case:

> Does one say that it is the means of analysis which changes the state of the oxide for another which did not exist in the mineral? I reply . . . that insofar as this has not been demonstrated, it would [best] conform to the true principles of science to admit nothing provisionally, or beyond that which the facts set forth at present.[8]

More difficult than the oxides for Proust to defend—and more openly attacked by Berthollet—were his views on metallic sulfides, particularly pyrite ores and the various antimonial compounds and complexes (glasses, livers, and others). Proust contended that all metals and semimetals, except iron, formed only one sulfide each; Berthollet had called the sulfides that occurred in variable proportions and the antimonial sulfide-oxide complexes "solutions"—and he considered them to be undifferentiable from other types of compounds. Proust seized on this use of "solution" to distinguish sharply these as well as other types of solutions (including salts in water) from true chemical compounds at fixed proportions. It was over this issue of sulfides that Proust was led to make this important distinction. But when challenged by Berthollet to provide criteria for distinguishing between "true" compounds and what Berthollet called "solutions," Proust was repeatedly unable to oblige.

Proust's most detailed consideration of the nature of chemical combination was given in answer not to Berthollet, but to a critic of Haüy's fixed mineral species concept. Proust's own view was that minerals were complexes of metallic oxides, sulfides, and other simple binary compounds. The binary compounds were the true combinations; the mineral complexes were merely "secondary assemblages of mineralizations, united and dissolved in all sorts of proportions."

This argument raised the question of what was a true combination. The heart of Proust's answer was that real compounds "have been given to us only under the rigorous condition of one proportion or two at most," whereas other types of intermixtures, such as solutions, can be obtained "in a latitude of proportions the extremes of which are infinitely separated."[9]

The circularity of Proust's definition of true "combination" and his failure to provide criteria for his distinction between compounds at definite proportions and other homogeneous systems have been noted in the literature on the Berthollet-Proust controversy. Proust was doubtless at a disadvantage here vis-à-vis Berthollet; on the other hand, his intuitive sensitivity to these practical distinctions was surer than Berthollet's.

The immediate impact of Proust's work can be considered under two related questions: Did Proust triumph over Berthollet and convince contemporary chemists that he was right? What was the relation of Proust's work to the development and implementation of the chemical atomic theory?

With regard to the first question, former historians of chemistry, perhaps seduced by the neat historical and logical relationship between Proust's law and Daltonian chemical atomism, were inclined to credit Proust with a more decisive victory over Berthollet and more general success in imposing his conviction on the chemical community of his time than the evidence warrants. The series of exchanges with Berthollet ended in 1807, unmarked by any obvious triumph on either side; each protagonist continued to maintain his position thereafter. Proust had scored notable successes against Berthollet's physicalistic explanation for the appearance of compounds at fixed proportions, but Proust remained unable to answer Berthollet's challenge to produce criteria, other than definite proportions, for "true" chemical combination.

The evidence in Proust's favor in the chemical community during the controversy is not much clearer, at least concerning the particulars of his position. Proust's analyses and the ensuing controversy with Berthollet had aroused interest in the numbers of existent metallic oxides and sulfides and their weight proportions, but it is difficult to find any kind of agreement between the various published analyses— much less agreement with Proust's particular position. An index to the reception of Proust's work and views is found in the early editions of Thomas Thomson's *A System of Chemistry* (1802; and ed., 1804). In these editions Thomson, who a few years earlier had proposed a version of the constant saturation principle and who, in the third edition of the textbook, was to give the first published account of Dalton's atomic theory, in fact adopted a compromise position between Proust and Berthollet (before their controversy had even erupted!). Recognizing that only a limited number of metallic oxides and their derivative compounds could be formed, Thomson cited Proust but disregarded his limit of two oxides per metal; and in later theoretical parts of the text, he explained these stable proportions in Bertholletian physicalist terms.[10]

The evidence of Proust's influence on Dalton is disappointing. While Dalton was certainly in a position to know of Proust's work and of the controversy with Berthollet, there is no evidence that Proust's work played any role in the genesis of Dalton's chemical atomic theory—even though Dalton was engaged in controversy with Berthollet over the issue of mixture versus true chemical compound (the air) in the early years of the Berthollet-Proust controversy, and Dalton's theory was to provide the theoretical foundation for Proust's position.

Indeed, the first published attempt to incorporate Proust's findings systematically into the atomic theory was not by Dalton but by Thomson. With suitable caution, not to say reluctance, Thomson introduced into the third (and Daltonian) edition of his textbook (1807) a short table exhibiting what he took to be multiple proportions for ten metals and semimetals.[11]

In the table, even though he derived much of his data from Proust's analyses, Thomson gave Proust no particular recognition. Moreover, one can understand Thomson's reluctance to have much confidence in the results of his table, given some of the lengths to which he had to go in order to square his data with multiple proportions. For example, in the case of the three oxides of lead, Thomson had to take the mean of his and Proust's weight proportions for two of them in order to arrive at reasonably simple proportions.

Dalton gave Proust very scant notice in the first two volumes of *A New System of Chemical Philosophy* (1808, 1810). It was apparently left to Berzelius

(1811) to give Proust his due by establishing the logical relationship between Proust's work (and the controversy with Berthollet) and Daltonian atomism, as well as to give Proust proper credit for the law of definite proportions.[12]

When Proust was confronted with Berzelius' post-Daltonian stoichiometry in Thenard's *Traité de chimie élémentaire, théorique et pratique*, I and II (1813–1814), he reacted adversely—partly out of pique at Thenard's attribution of the law of definite proportions to Berzelius—and passed over the brief exposition of the law of multiple proportions in this textbook, describing it merely as "another relationship" between sulfides.[13]

Second to his mineral and inorganic analyses associated with the law of definite proportions, Proust's most important work lay in organic chemistry and chemical technology. In 1799 he succeeded in isolating grape sugar, and he suggested that it could be manufactured and used to supplement the relatively expensive and uncertain supplies of cane sugar from the West Indies. After Proust's return to France and publication of his researches there, this idea attracted the attention of the Napoleonic government, which in 1810 offered him 100,000 francs to help establish a factory for grape sugar production. In 1818 he announced the discovery of what he called *oxide caséeux* (leucine) in cheese, which shortly afterward was independently discovered by Braconnot. He also devoted considerable time to attempts to prepare nutriments that could be cheaply made and conveniently stored.

NOTES

1. R. Dossie, *Institutes of Experimental Chemistry*, I (London, 1759), 11–13; [G. Venel], "Mixte et mixtion," in *Encyclopédie ou dictionnaire raisonné des sciences, des arts et des métiers*, X (Paris, 1765), 587; L. B. Guyton de Morveau, "Affinité," in *Encyclopédie méthodique. Chymie, pharmacie et métallurgie*, I (Paris, 1786–1789), 560–563; T. Thomson, "Chemistry," in *Encyclopaedia Britannica*, 3rd ed., supp., I (Edinburgh, 1801), 342–344.
2. Dossie, *op. cit.*, 11.
3. See S. Mauskopf, "Minerals, Molecules and Species," in *Archives internationales d'histoire des sciences*, **23** (1970), 185–206.
4. "Recherches sur le bleu de Prusse," in *Journal de physique*, **45** (1794), 334–335, 341.
5. "Sur les oxidations métalliques," *ibid.*, **59** (1804), 329.
6. "Recherches sur le cuivre," in *Annales de chimie*, **32** (1799), 31.
7. "Lettre de M. Proust à M. d'Arcet sur un sel phosphorique calcaire naturel," in *Journal de physique*, **32** (1788), 241.
8. *Ibid.*, **59** (1804), 332.
9. "Sur les mines de cobalt, nickel et autres," *ibid.*, **63** (1806), 369–370.
10. T. Thomson, *A System of Chemistry* (Edinburgh, 1802), I, 230–231, III, 196–203.
11. *A System of Chemistry*, 3rd ed. (Edinburgh, 1807), III, 520–522.
12. J. J. Berzelius, "Essai sur les proportions déterminées dans lesquelles se trouvent réunis les éléments de la nature inorganique," in *Annales de chimie*, **78** (1811), 5.
13. "Sixième lettre sur l'incertitude de quelques oxidations," in *Journal de physique*, **81** (1815), 258.

BIBLIOGRAPHY

I. Original Works. MSS of chemical lectures by Proust, which L. Silván dates from Proust's tenure at Segovia, are in the archives of the Diputación Provincial de Guipúzcoa in San Sebastián, Spain. A set of nine letters to Proust (including some by Berthollet) is in the archives of the Académie des Sciences; all are from 1810 or later. See also H. David, "Une correspondance inédite du grand chimiste, Joseph Louis Proust," in *Revue d'histoire de la pharmacie* (1938), 266–279. Unfortunately, David only paraphrases the letters, some of which date from the 1790's. Proust also edited *Anales del Real laboratorio de química de Segovia* (**1** and **2**, 1791 and 1795) and was an editor of *Anales de historia natural* **1** and **2** (1799–1800); title changed to *Anales de ciencias naturales*, **3–7** (1801–1804), contributing to both. Proust published many papers, most of them in either *Annales de chimie* or *Journal de physique*. Some were first published in Spain. Adequate lists of them (although not complete or without errors) are in Poggendorff, II, cols. 536–538; and Royal Society *Catalogue of Scientific Papers*, V, 31–33.

II. Secondary Literature. There is still no adequate general study of Proust and his place in the history of chemistry. With regard to biography, there are studies by both French and Spanish authors. The French tend to be ignorant about his Spanish career; the Spanish tend to be sketchy about his life in France. The following are useful but often erroneous: H. David, "Une correspondance inédite . . ."; and M. Godard-Faultrier, *Notice biographique sur le chimiste J. L. Proust* (Angers, 1852). Two opposing appraisals of Proust's career in Spain are J. Rodríquez Carracido, *Estudios histórico-críticos de la ciencia española* (Madrid, 1897), 153–166 (negative); and J. Rodriquez Mourelo, "L'oeuvre de Proust en Espagne," in *Revue scientifique*, **54** (1916), 257–266 (positive). More recent studies on Proust by Leandro Silván are "Proust en Vergara," in *Boletín de la Real sociedad vascongada de amigos del país*, **1** (1945), 237–247; *Los estudios científicos en Vergara a fines del siglo XVIII*, Monografía Vascongada no. 12 (San Sebastián, 1953); and *El químico Luis José Proust* (Vitoria, 1964), particularly rich in information on Proust's Spanish career. Background studies on the law of definite proportions and the Berthollet-Proust controversy are H. Guerlac, "Quantification in Chemistry," in *Quantification: A History of the Meaning of Measurement in the Natural and Social Sciences* (Indianapolis, 1961), 64–84; and S. Mauskopf, "Thomson Before Dalton," in *Annals of Science*, **25** (1969), 229–242, on the principle of constant saturation proportions.

The best and most recent study on the law of definite proportions and the controversy is S. C. Kapoor, "Berthollet, Proust and Propositions," in *Chymia*, **10** (1965), 53–110, although Kapoor's focus is on Berthollet, not Proust. R. Hooykaas, "The Concept of 'Individual' and 'Species' in Chemistry," in *Centaurus*, **5** (1958), 307–322, is oriented more toward the doctrine of fixed mineralogical species. J. R. Partington, *A History of Chemistry*, III, 640–650, has much useful information.

The following older studies are still very valuable: I. Freund, *The Study of Chemical Composition* (Cambridge, 1904), 127–143; and A. N. Meldrum, "The Development of the Atomic Theory: (1) Berthollet's Doctrine of Variable Proportions," in *Memoirs and Proceedings of the Manchester Literary and Philosophical Society*, **54**, no. 7 (1910), 1–16.

SEYMOUR MAUSKOPF

PROUT, WILLIAM (*b*. Horton, Gloucestershire, England, 15 January 1785; *d*. London, England, 9 April 1850), *chemistry, biochemistry*.

Prout was the eldest of three sons of John Prout, a tenant farmer whose fortunes had increased through the inheritance of land, and his wife Hannah Limbrick[?]. Educated at local charity schools until the age of thirteen, Prout worked on his father's farm until about 1802, when he attended the private classical academies of Rev. John Turner at Sherston, Wiltshire, and Rev. Thomas Jones at Bristol. In 1808, on Jones's recommendation, he entered Edinburgh University to study medicine; he graduated in 1811 with an unoriginal dissertation on fevers. He completed his medical training at St. Thomas's and Guy's hospitals in London, where he set up practice after gaining the licentiate of the Royal College of Physicians on 22 December 1812. During 1814 Prout gave a successful course of public lectures on animal chemistry in his London home and met Alexander Marcet, who praised him in letters to Berzelius. There is some evidence that from 1816 until 1817 he edited *Annals of Medicine and Surgery* with his friend John Elliotson.

When his father died in 1820, Prout passed the Horton estate, which he inherited, to his surviving brother. Little is known of Prout's personal life in London. He became a very successful, but not wealthy, physician who specialized in digestive and urinary complaints. His reputation in medicine and chemistry in Great Britain and on the Continent was considerable, both as an experimentalist and as a theorist. Unfortunately, deafness made him avoid scientific contacts after 1830. He subsequently made little effort to keep abreast of the rapid developments that took place in biochemistry and chemistry between 1830 and 1850; and although much of his biochemical research had foreshadowed that of Liebig and his school, Prout found himself eclipsed by their achievements during this period.

An extremely religious man, Prout was invited to write one of the eight Bridgewater treatises, which had the general title *On the Power, Wisdom and Goodness of God, as Manifested in the Creation*. An accomplished organist who composed music for his family, he also possessed artistic talents. He married Agnes Adam (1793-1863), the daughter of Alexander Adam, the Edinburgh educator; they had seven children, one of whom became a military surgeon.

Prout received the M.D. from Edinburgh (1811), and was a fellow of the Royal Society (1819) and of the Royal College of Physicians (1829). He was a Copley medalist of the Royal Society (1827) and a Gulstonian lecturer at the Royal College of Physicians (1831). He served on the Council of the Royal Society, and on several of its committees and on those of the British Association for the Advancement of Science. For a time he was an active member of the Medico-Chirurgical Society of London.

Prout's contribution to the concept of the unity of matter, which he adopted as a student, played a dominant role in the development of the theory of the elements and the fortunes of the atomic theory. Prout was much influenced by Humphry Davy's speculations on "undecompounded bodies" (1812) and by Dalton's atomic theory as modified by Berzelius. He hoped to develop a "mathematical" chemistry analogous, perhaps, to the scheme expressed tentatively by Thomas Thomson in his *First Principles* (1825). The inspiration to improve Gay-Lussac's and Berzelius' methods of organic analysis came largely from his attempt to find the mathematical laws that govern the formation of organic compounds from the elements carbon, hydrogen, oxygen, and nitrogen. In 1817, 1820, and 1827 Prout published accounts of elaborate and expensive analytical methods. His organic analyses were renowned for their accuracy, and he remained skeptical of Liebig's simple and successful technique (1830).

Between 1815 and 1827 Prout published a series of important papers on urine and digestion that opened up the areas of purine and metabolic chemistry. He found a boa constrictor's excrement to be 90 percent uric acid; he also extracted extremely pure urea from urine and attempted to synthesize it in 1818, ten years before Wöhler's accidental success. In 1821 Prout published a concise textbook on urine; but a similar work on digestion, partly printed in 1822, was withdrawn from publication. In 1840, however, Prout published a long and successful practical textbook of urinary and digestive pathology.

The brilliant demonstration in 1824 that the gastric juices of animals contain hydrochloric acid appeared incredible to many of Prout's contemporaries. Yet in 1827 they readily adopted his classification of foodstuffs into water, saccharinous (carbohydrates), oleaginous (fats), and albuminous (proteins). Although Prout promised detailed analyses of the three organic aliments, only those of the saccharinous class were published by him. As a vitalist, Prout maintained that organized bodies (which were composed from organic substances) contained "independent existing vital principles." Under the influence of these teleologic agents, the four aliments were transformed into blood and tissues. Prout termed the processes of digestion and blood formation "primary assimilation." "Secondary assimilation" (Liebig's "metamorphosis of tissues") included both the process of tissue formation from blood and the destruction and removal of unwanted parts from the animal system. The absorption and removal of water from processed aliments were the principal chemical features of chylification and sanguification, respectively. Organization of processed aliments could not occur, however, without the presence and admixture of minute amounts of water or of elements other than carbon, hydrogen, oxygen, and nitrogen. In 1827 Prout coined the word "merorganized" to denote the isomerism and vitalization of organic substances by the presence of these incidental materials. For some years this concept was a serious alternative to the structural interpretation of isomerism.

Much of this metabolic theory was speculative, as was the corpuscular theory upon which it was based. In unpublished lectures (1814) Prout supposed that hydrogen might be converted by electricity into other elements and the imponderable fluids: caloric, light, and magnetism. In 1815 he published an unsigned article in which he reconciled information on the combining weights of substances with their combining volumes when in real or imaginary gaseous states. After several dubious assumptions and adjustments, Prout calculated that the atomic weights of all elements were integral when the atomic weight of hydrogen was taken as unity. In a correction (1816) he added that hydrogen might be the primary matter from which all other "elements" were formed.

Prout quickly identified himself as the author of these papers; and the two hypotheses, of integral atomic weights and of the unity of matter, became known ambiguously and singularly as Prout's hypothesis. It was a continuous source of inspiration to chemists and physicists until the work of F. W. Aston on isotopes in the 1920's. T. Thomson, J.-B. Dumas,

J. C. Marignac, and L. Meyer supported Prout, and Berzelius, E. Turner, J. S. Stas, Mendeleev, and T. W. Richards opposed him. But whatever the attitude of individual scientists toward the hypotheses or their modifications, they stimulated the improvement of analysis and enforced interest in atomic weights and, therefore, in the atomic theory. They also gave impetus to the search for a system of classification of the elements, and, when the periodic law was achieved, they encouraged speculations about the evolution of the elements and structural theories of the atom. Few hypotheses have been so persistently fruitful.

Prout left it to others, notably T. Thomson, to work out the consequences of his suggestions; but in 1831 he added that there was no reason why the material unit of condensation might not be smaller than hydrogen (perhaps half or one-quarter of hydrogen). This offered a convenient explanation of such anomalous, nonintegral atomic weights as chlorine and copper, and it was used by Marignac and Dumas.

In his Bridgewater treatise (1834; 3rd ed., 1845) Prout revealed that his hypotheses were only part of an elaborate corpuscular philosophy in which spherical particles were imagined to revolve spontaneously, with mutually repulsive forces and velocities that were inversely proportional to their masses. In addition, there were attractive forces that were directly proportional to the masses of the rotating particles. Like Aepinus and Mossotti, Prout supposed that the force of gravitation was the difference between the attractive and repulsive forces. The model remained a speculation, because there was no way of determining angular velocities independently of atomic weights.

This polarity theory also involved Prout in conclusions similar to those drawn by Avogadro in 1811: that at the same temperature and pressure, equal volumes of all gases contain equal numbers of molecules, and that all molecules of elementary gases contain at least two submolecules, or atoms. Like Avogadro, Prout deduced from this that the ratio of the weights of two equal volumes of different gases, at the same temperature and pressure, was equal to the ratio of their molecular weights. Prout was attacked for these views by the chemist William Charles Henry; but Prout's support for Avogadro was without much influence, largely because he compromised with equivalent weights and because he was opposed to the use of chemical formulas.

Prout's other significant contributions included unpublished thoughts on the unity of sensations (1810), on the distinction between taste and flavor (1812), on elaborate self-experiments regarding carbon dioxide output (1813, 1814), on a study of the chemical changes in an incubating egg (1822), on the neologism "convection" (1834), and on the design of the Royal Society's standard barometer (1831–1836).

BIBLIOGRAPHY

I. ORIGINAL WORKS. A list of 34 papers by Prout is recorded in the Royal Society *Catalogue of Scientific Papers*, V, 34–35. To them should be added W[illiam] P[rout], "The Sensations of Taste and Smell," in *London Medical and Physical Journal*, **28** (1812), 457–461; and the unsigned "On the Relation Between Specific Gravities of Bodies in the Gaseous State and the Weights of Their Atoms," in *Annals of Philosophy*, **6** (1815), 321–330, and **7** (1816), 111–113. The latter is repr. with an unsigned intro. in L. Dobbin and J. Kendall, *Prout's Hypothesis*, Alembic Club Reprint no. 20 (Edinburgh, 1932), and in facs. in D. M. Knight, *Classical Scientific Papers. Second Series* (London, 1970). Note also Prout's Gulstonian lectures, "On the Application of Chemistry to Physiology, Pathology and Practice," in *Medical Gazette*, **8** (1831), 257–265, 321–327, 385–391; and his letter to Daubeny (1831), in C. Daubeny, *An Introduction to the Atomic Theory* (Oxford, 1831), 129–133; 2nd ed. (Oxford, 1850), 470–474.

Prout's books were *De febribus intermittentibus*, (Edinburgh, 1811), his M.D. thesis; *An Inquiry Into the Nature and Treatment of Gravel, Calculus, and Other Diseases* (London, 1821), 2nd ed. retitled *Inquiry . . . Treatment of Diabetes, Calculus and Other Affections* (London, 1825), 3rd ed. retitled *On the Nature and Treatment of Stomach and Urinary Diseases* (London, 1840), 4th ed. retitled *On the Nature . . . Stomach and Renal Diseases* (London, 1843; 5th ed., London, 1848). See also *Chemistry, Meteorology and the Function of Digestion*, eighth Bridgewater Treatise (London, 1834; 2nd ed., 1834; 3rd ed., 1845; 4th ed., 1855).

There are letters and papers of Prout's at the Royal Society, the Royal College of Physicians, and the Wellcome Institute for the History of Medicine, London.

II. SECONDARY LITERATURE. There were three principal obituaries: an unsigned one in *Medical Times*, **1** (1850), 15–17; an unsigned one in *Edinburgh Medical and Surgical Journal*, **76** (1851), 126–183; and C. Daubeny, "On the Great Principles Either Suggested or Worked out by Dr. William Prout," in *Edinburgh New Philosophical Journal*, **53** (1852), 98–102, repr. in Daubeny's *Miscellanies*, II (London, 1867), 123–127. The only comprehensive study is W. H. Brock, "The Chemical Career of William Prout," Ph.D. thesis (Leicester, 1966).

Important articles on Prout are O. T. Benfey, "Prout's Hypothesis," in *Journal of Chemical Education*, **29** (1952), 78–81; W. H. Brock, "The Life and Work of William Prout," in *Medical History*, **9** (1965), 101–126; "The Selection of the Authors of the Bridgewater Treatises," in *Notes and Records. Royal Society of London*, **21** (1966), 162–179; "Dalton Versus Prout: The Problem of Prout's Hypotheses," in D. S. L. Cardwell, ed., *John Dalton and the Progress of Science* (Manchester, 1968), 240–258; "Studies in the History of Prout's Hypotheses," in *Annals of Science*, **25** (1969), 49–80, 127–137, which reprints Prout's essay "De facultate sentiendi" and other notes; and "William Prout and Barometry," in *Notes and Records. Royal Society of London*, **24** (1970), 281–294; W. V. Farrar, "Nineteenth-Century Speculations on the Complexity of the Chemical Elements," in *British Journal for the History of Science*, **2** (1965), 297–323; A. M. Kasich, "Prout and the Discovery of Hydrochloric Acid in Gastric Juice," in *Bulletin of the History of Medicine*, **20** (1946), 340–348; D. F. Larder, "Prout's Hypothesis, a Reconsideration" in *Centaurus*, **15** (1970), 44–50; and J. R. Partington, *A History of Chemistry*, III (London, 1962), 713–714, and IV (London, 1964), *passim*.

Two portraits of Prout are reproduced in G. Wolstenholme, *The Royal College of Physicians Portraits* (London, 1964), 346–348.

W. H. BROCK

PROWAZEK (PROVÁZEK), STANISLAUS VON LANOV (*b.* Jindřichův Hradec, Bohemia, 12 November 1875; *d.* Cottbus, Germany, 17 February 1915), *microbiology, parasitology.*

Prowazek was descended from Czech peasant stock. His father, an officer in the Austro-Hungarian army, was ennobled in 1893; at about the same time Prowazek, then a student in the Plzeň Gymnasium, altered the spelling of his name from the original "Provázek." In 1895 Prowazek began to study natural science at the University in Prague, where he came under the influence of the zoologist Berthold Hatschek and of the physicist and philosopher Ernst Mach. After two years' study, he followed Hatschek to the University of Vienna, from which he received the Ph.D. in 1899, with a dissertation entitled *Protozoen Studien.* He continued his zoological work with Hatschek both in Vienna and at the zoological station in Trieste until 1901, when he went to Paul Ehrlich's Institute for Experimental Therapy at Frankfurt. The following year Prowazek worked in Richard Hertwig's department at Munich, where he continued his earlier cytological investigations of flagellates and investigated the reproductive modes of infusorians. In 1903 he accepted an invitation from Fritz Schaudinn, whom he had met in Trieste in 1901, to work as his assistant at the zoological section of the University of Berlin at Rovigno.

The friendship between Prowazek and Schaudinn was decisive in both their careers, and the brief time in which they worked together—less than two years— was of considerable importance to the development of protistology. In 1905 Schaudinn—following his discovery, with Hoffmann, of the spirochete that causes syphilis—was appointed director of the zoological section of the Institut für Schiffs- und Tropenkrankheiten in Hamburg; upon his premature death the following year (at the age of thirty-five), Prowazek was named his successor.

Prowazek had taken part in Neisser's 1906 expedition to Java; at Batavia he and Ludwig Halberstädter discovered the inclusions in the epithelial cells from the conjunctivas of trachomatous eyes that are now called Prowazek's bodies or Halberstädter-Prowazek bodies. At the same time, he began his studies of vaccinia. After a short visit to Japan, Prowazek returned to Hamburg, from which he made several other research expeditions. In 1908 he went to Rio de Janiero to study the etiology of vaccinia and variola at the Instituto Oswaldo Cruz; two years later he visited the German colonies of Western Samoa, Yap, and Saipan to explore the causes of a number of infectious diseases, among them trachoma, fowl pox, Newcastle disease, silkworm jaundice, epitheliosis desquamativa conjunctivae, and molluscum contagiosum.

In 1913 and 1914 Prowazek traveled to Serbia and Constantinople, where typhus was raging. He made observations on the etiology, mode of transmission, and life cycle of the parasite causing the disease, and devoted the last two years of his life to studying it. He himself died of typhus when, in 1915, he and Henrique da Rocha-Lima were sent to investigate an epidemic that had broken out among Russian prisoners confined in a camp near Cottbus. Da Rocha-Lima contracted the disease at the same time, but recovered to isolate the causative microorganism, which he called *Rickettsia prowazekii*, in honor of both Prowazek and H. T. Ricketts, who had also died of typhus while investigating it.

During his rather brief scientific career, Prowazek dealt with a wide variety of topics, including the origin of the axial filament of flagella, merotomy, and the mode of transmission of microbial diseases. He was the first to demonstrate that the parasite *Trypanosoma lewisi* passes through a special stage in the body of its host, the rat louse. He made a number of transplantation experiments on protista, and contributed widely to protistology and to its medical applications. Although he lacked formal medical education, Prowazek had a considerable grasp of medical problems, to which he applied his wide biological knowledge and his skill in chemistry and physics. He was an acute observer, and was able to master subtle techniques and find appropriate subjects through which to approach general biological problems. He considered general aspects of biology in even his specialized studies. His work led him from morphology and developmental studies to an investigation of unicellular organisms, which he examined by means of physicochemical methods, in an attempt to understand the underlying principles of life. He was not, however, a systematic theoretician, but rather developed the ideas of others by systematic research.

Prowazek was a cultivated man who wrote and worked with facility. His interests were wide, and his publications include works of fiction, which he signed "P. Laner." During his expeditions to the South Pacific he became enchanted by the tropics and fond of the primitive peoples; his book on the Mariana Islands is concerned with their history, flora, fauna, and ethnography.

BIBLIOGRAPHY

I. ORIGINAL WORKS. A bibliography of Prowazek's works by M. Hartmann, in *Archiv für Protistenkunde*, **36** (1916), xiii–xix, contains 209 titles but a few items seem not to have been included. The more important and more general are "Die pathogenen Protozoen (mit Ausnahme der Hämosporidien)," in W. Kolle and A. Wassermann, eds., *Handbuch der pathogenen Microorganismen*, I (Jena, 1903), 865–1006, written with F. Doflein; *Taschenbuch der mikroskopischen Technik der Protistenuntersuchung* (Leipzig, 1907; 3rd ed., 1922); and *Einführung in die Physiologie der Einzelligen* (Leipzig–Berlin, 1910). Prowazek also edited and contributed several articles to *Handbuch der pathogenen Protozoen*, 3 vols. (I–II, Leipzig, 1912–1915; III, 1931). Prowazek also wrote "Über Zelleinschlüsse parasitären Natur beim Trachom," in *Arbeiten des K. Gesundheitsamte*, **26** (1907), 44–47, written with L. Halberstädter; *Die deutschen Marianen, ihre Natur und Geschichte* (Leipzig, 1913); and "Ätiologische Untersuchungen über den Flecktyphus in Serbien 1913 und in Hamburg 1914," in *Beiträge zur Klinik der Infektionskrankheiten . . .*, **4** (1915), 5–31.

II. SECONDARY LITERATURE. A very good short account of Prowazek's life and achievements, with citations of about 10 obituaries, is by G. Olpp in *Hervorragende Tropenärzte in Wort und Bild* (Munich, 1932), 330–334. It is based mainly on the best single appreciation: Max Hartmann, in *Archiv für Protistenkunde*, **36** (1915), i–xii. See also F. K. St[udnička], in *Biologické listy*, **4** (1915), 95–96; J. Muk and F. Miller, *Prof. Dr. St. Prowazek rodák z J. Hradce a bojovník s epidemiemi* (Jindřichův Hradec, 1942); and R. B. Goldschmidt, *Portraits From Memory. Recollections of a Zoologist* (Seattle, 1956), 138–142.

VLADISLAV KRUTA

PRUDDEN, THEOPHIL MITCHELL (*b.* Middlebury, Connecticut, 7 July 1849; *d.* New York, N.Y., 10 April 1924), *pathology, bacteriology, public health, archaeology.*

A founder of pathology as a field of research and teaching in the United States, Prudden was a son of George Peter Prudden, a Congregational minister, and Eliza Anne Johnson Prudden. Ill health frequently interrupted his early schooling, and he was sent for long periods to his grandfather's farm. When the family moved to New Haven, where the father headed a boys' school, Prudden was enrolled as a student.

At seventeen Prudden entered the employ of a furniture manufacturing firm, but he soon realized that he must continue his education. To remedy his irregular schooling, he studied first at home, then at the Wilbraham Academy in Massachusetts; and at twenty, with a Connecticut state scholarship for his tuition, he began the course at the Sheffield Scientific School at Yale. Here he determined to enter medicine; and a new program of premedical courses, including zoology, botany, and chemistry, was organized for Prudden and his friend Thomas H. Russell. The two students received the close attention of Addison Verrill and Sidney Smith, and their scientific enthusiasms were broad. They avidly collected specimens of botanical, zoological, and mineralogical interest; and during the spring vacation of 1872 they dredged in Long Island and Vineyard sounds from a chartered yacht. After graduating Ph.B. that year, Prudden joined Verrill's dredging expedition based at Eastport, Maine, for the summer and explored the waters of the Bay of Fundy, collecting crustaceans and studying the algae with Daniel C. Eaton while the latter was with the group.

As an undergraduate at Sheffield, Prudden had begun to teach the freshman chemistry course as a substitute; his continued teaching of chemistry helped him to meet expenses when he entered the Yale School of Medicine. During the summer of 1873 he took part in the fossil-hunting expedition in the West headed by Othniel C. Marsh.

In the spring of 1875, while in his last year at medical school, Prudden attended lectures in New York at the College of Physicians and Surgeons, and worked in the laboratory of the pathologist Francis Delafield. He received the M.D. degree from Yale. While an intern at New Haven Hospital he was given an excellent microscope—one he could not have afforded—and this may have influenced his interest in microscopic investigation.

Prudden left for two years' study abroad in 1876. At Heidelberg, under Julius Arnold and Richard Thoma, he investigated the effects of various injurious substances upon the living cartilage cells of the frog, using their methods for *in vitro* studies. He visited laboratories in Vienna and Berlin; but upon returning to America, he found his training in histology and pathology in little demand by the medical schools and therefore established a practice in New Haven. William H. Welch, who pioneered in research and teaching of pathology, recommended Prudden for the post of first assistant in the new laboratory of pathology and histology that Delafield was organizing in New York in 1878. The laboratory received most of its funds from the alumni of the College of Physicians and Surgeons and for several years made up its annual deficits by reducing the salary of the director. Nevertheless the narrow room, crowded between a harness shop and an ice cream shop, represented a stride forward. Prudden became director of the laboratory in 1882 and drew a number of students and others interested in his investigations. The facility later had new quarters, and Prudden was professor of pathology at Columbia from 1892 until his retirement; but it was to this first cramped, all-purpose laboratory that Prudden brought the new concepts of Pasteur and Koch, applying them in his teaching on the etiology and spread of disease, and in dealing with the health problems of the city.

Prudden and Welch were among the first Americans taught by Koch at Berlin in 1885; Prudden carried a request from the Connecticut State Board of Health to report on Koch's method of studying the germs that cause diseases, particularly cholera. When he returned, Prudden partitioned a section of his laboratory for research in bacteriology.

Prudden worked tirelessly: from 1880 to 1886 he commuted to teach normal histology at the Yale School of Medicine; he carried out numerous routine investigations in the laboratory; he was consulting bacteriologist for the health department of New York City; and he served in many capacities on local and state bodies dedicated to improved hygiene. His advice was often sought: he was called during an outbreak of typhoid in Providence in 1888; he examined the water of Minneapolis that year when winter cholera struck; in 1891 he testified in the suit to obtain an injunction preventing the city of Passaic from emptying sewage into the Passaic River in New Jersey; and he assisted when cholera threatened the port of New York in 1892.

Prudden's activities drew many honors. He received the LL.D. from Yale in 1897 and was a member of the National Academy of Sciences. At the founding of the Rockefeller Institute for Medical Research in 1901 Prudden served as vice-president of its board of scientific directors; and his retirement from teaching responsibilities in 1909 gave him more time to devote to its administration, policy, and other matters.

With Delafield, Prudden wrote a landmark textbook of pathology, long considered the foremost English text in the field. He also wrote a manual of histology and numerous scholarly articles. Through popular works he endeavored to broaden the understanding of the infectious organisms that cause disease; he stressed proper sanitation and scored such common practices as the cutting of ice from the polluted Hudson River; he campaigned for clean ice as well as clean drinking water, and for assured purity of the milk supply. During summer vacations he made many trips to the West, and he wrote on the archaeology and Indians of the Southwest. He visited the prehistoric cliff dwellings of the region and contributed the materials gathered on his expeditions to museums; a collection is in the Peabody Museum at Yale.

BIBLIOGRAPHY

I. ORIGINAL WORKS. There is a collection of Prudden's papers, including correspondence, notes, and various other items, in the Manuscript Room of the Yale University Library. The notebooks of the 1872 trip to Eastport, Maine, and the 1873 trip to the West with O. C. Marsh are in the Beinecke Rare Book and Manuscript Library at Yale. Sets of Prudden's articles and reprints are in the libraries of the Rockefeller University, the New York Academy of Medicine, and the Yale University School of Medicine. With Francis Delafield he published *A Handbook of Pathological Anatomy and Histology* (New York, 1885; 7th ed., 1904); this influential text was first written as an extension and revision of Delafield's *A Handbook of Post-Mortem Examinations and of Morbid Anatomy* (1872); Francis Carter Wood was later coauthor with Prudden and subsequently made the revisions. Prudden wrote *A Manual of Practical Normal Histology* (New York, 1882); *The Story of the Bacteria and Their Relations to Health and Disease* (New York, 1889); and *Dust and Its Dangers* (New York, 1890). His articles include "Beobachtungen am lebenden Knorpel," in *Virchows Archiv für pathologische Anatomie und Physiologie* . . . 75 (1879), 1–14; and "Studies on the Action of Dead Bacteria in the Living Body," in *Medical Record* (New York), 53 (1891), 637–640, 697–704, written with Eugene Hodenpyl. "Pathology and the Department of Pathology," in *Columbia University Bulletin,* 19 (1898), 103–119; and "Progress and Drift in Pathology," in *Medical Record* (New York), 57 (1900), 397–405, provide reviews of the changes in pathology during Prudden's career. Among his archaeological contributions were "A Summer Among Cliff Dwellings," in *Harper's New Monthly Magazine,* 93 (1896), 545–561; and *On the Great American Plateau* (New York, 1907).

II. SECONDARY LITERATURE. Lillian E. Prudden, ed., *Biographical Sketches and Letters of T. Mitchell Prudden, M.D.* (New Haven, 1927), contains autobiographical and other material on Prudden's life and contributions, and a bibliography (pp. 293–300). See also Simon Flexner, "T. Mitchell Prudden, 1849–1924," in *Science,* 60 (1924), 415–419; Ludvig Hektoen, "Biographical Memoir of Theophil Mitchell Prudden 1849–1924," in *Biographical Memoirs. National Academy of Sciences,* 12 (1929), 73–93; and Francis Carter Wood, "Prudden, Theophil Mitchell," in *Dictionary of American Biography,* XXV (New York, 1935), 252–253; and *Obituary Record, Yale University, 1923–24* (New Haven, 1924), 1154–1155.

GLORIA ROBINSON

PRUNER BEY, FRANZ IGNACE (*b.* Pfreimd, Germany, 8 March 1808; *d.* Pisa, Italy, 29 September 1882), *medicine, ethnology, anthropology.*

Pruner was the son of Ignace Brunner, a civil servant, and Catherine Hochler, the daughter of a municipal councillor. He entered the University of Munich in 1826, and began his medical studies there the next year, becoming assistant to Ernest von Grossi, a specialist in experimental medicine. Pruner received the M.D. in 1830, then, with a fellow student, Sebastian Fischer, began to prepare an edition of the manuscripts that had been left unpublished at Grossi's death the preceding year. In 1831 Pruner went to Paris to continue his medical studies; he there met Étienne Pariset, who aroused his interest in traveling in the East. The same year, Pruner accompanied K. M. von Hügel on a voyage to Greece, Syria, Palestine, India, and Egypt, where they observed cholera and plague epidemics and, in Jerusalem, studied the treatment of lepers. At the same time, Pruner laid the basis for his anthropological work, making careful observations of various populations (particularly the Druses of Sidon) and relating their characteristics to their native soils and climates.

In September 1831 Pruner was in Alexandria, where the pasha offered him the chair of anatomy and physiology at the medical school at Abu Za'bal, which had been founded in 1825 by A. B. Clot. Pruner accepted, but returned in 1832 to Munich, where he resumed his work in publishing Grossi's papers. In 1833 Pruner went to Malta, Sicily, and Italy, where he studied ophthalmology at Pavia. In the same year he went back to Egypt to become director of a military hospital near Cairo; he then practiced ophthalmology at Hejaz, and, in 1835, went to Mecca to help combat a cholera epidemic. In 1836 he was given the rank of captain and appointed director of military hospitals in Cairo itself; he was particularly concerned with typhus and trachoma, which he treated with a mixture of Luxor water and a saturated solution of zinc and alum. He was subsequently named professor of ophthalmology at Cairo.

During his years in Egypt, Pruner studied the Arab peoples, Arab literature, and ancient Arab medicine. He observed the coastal regions and formed zoological and ethnographic collections, which he subsequently gave to the Bavarian government. In 1846, during a stay in Munich, he presented to the Royal Academy a report on the ancient races of Egypt and published a study of the medical topography of Cairo, the first work in the field of geographical pathology.

In 1849 Abbas Pasha named Pruner his personal physician, with the title "Bey." Ill health made it necessary for Pruner to go to Europe the following year, and although he returned to Egypt in 1852, he found that he could no longer tolerate the climate. He spent some time in Bavaria, Baden, and Geneva, but his health did not improve, and Abbas Pasha accepted his resignation in 1860. Pruner then moved to Paris. He continued his anthropological studies, and was elected an associate member of the Société d'Anthropologie, serving as its president in 1865.

Pruner considered anthropology the "science of sciences." He did a considerable amount of original work in the field, including investigations of comparative nosology, based upon the observation of the relation between individuals and their environments. Among his many communications to the Société d'Anthropologie, the most important concern craniology, including a series of three remarkable tables, recording more than 15,000 measurements of 507 craniums. His craniological work led Pruner into a dispute with Paul Broca, since Pruner held that the Basques (the origin of whose language he had also studied) were brachycephalic, while Broca maintained that they were dolichocephalic.

Pruner was a participant in a number of international scientific conferences and received honors and decorations from several countries. In 1872 he settled in Pisa, where he died ten years later, following a brief illness.

BIBLIOGRAPHY

I. ORIGINAL WORKS. Pruner Bey's books include *Tentamen de morborum transitionibus* (Munich, 1830), his doctoral thesis; his ed. of Grossi's *Historia morbi et descriptio sectionis cadaveris* (Munich, 1830); Grossi's *Opera medica posthuma*, 3 vols. (Stuttgart–Tübingen–Munich, 1831–1832), edited with S. Fischer; *Ist dem die Pest wirklich ein austeckendes Uebel?* (Munich, 1839); *Ueber die Ueberbleibsel der altenägyptischen Menscherace* (Munich, 1846); *Die Krankheiten des Orients vom Standpunkte der vergleichenden Nosologie betrachtet* (Erlangen, 1847); *Topographie médicale du Caire avec le plan de la ville et des environs* (Munich, 1847); *Die weltseuche Cholera oder die Polizei der Natur* (Erlangen, 1851); *Der Mensch im Raume und in der Zeit . . . Eine ethnographische Skizze* (Munich, 1859); and *Les carthaginois en France. La colonie lybio-phénicienne du Liby, canton de Bourg-Saint-Andéol* (Montpellier, 1870), written with J. Ollier de Marichard.

Among Pruner's many articles are "Communication sur les Druses," in *Bulletin de la Société d'anthropologie de Paris*, **1** (1859–1860), 454–456; "Sur la perfectibilité des races," *ibid.*, 479–492; "Mémoire sur les nègres," in *Mémoires de la Société d'anthropologie de Paris*, **1** (1860–1863), 293–336; "Recherches sur l'origine de l'ancienne race égyptienne," *ibid.*, 399–434; "Parallèle crâniométrique des races humaines," in *Bulletin de la Société d'anthropologie de Paris*, **3** (1862), 238; "Sur le climat de l'Égypte," *ibid.*, **4** (1863), 17–24; "De la chevelure comme caractéristique des races humaines, d'après des recherches microscopiques," in *Mémoires de la Société d'anthropologie de Paris*, **2** (1863–1865), 1–36, trans. as "On Human Hair as a Race Character, Examined by the Aid of Microscope," in *Anthropological Review* (London), **2** (1864), 1–23; "Resultats de crâniométrie," in *Mémoires de la Société d'anthropologie de Paris*, **2** (1863–1865), 417–432; "Questions relatives à l'anthropologie générale," in *Bulletin de la Société d'anthropologie de Paris*, **5** (1864), 64–154; "Sur l'origine asiatique des européens," *ibid.*, 223–242; "Importance de l'anthropologie," *ibid.*, **6** (1865), 1–10, his inaugural lecture; "De l'anthropologie en Espagne," *ibid.*, 361–370; "Sur la chevelure comme caractéristique des races humaines," *ibid.*, 376; and "L'homme et l'animal," *ibid.*, 522–562.

Other articles are "Os crâniens provenant des palafittes de la Suisse," in *Bulletin de la Société d'anthropologie de Paris*, 2nd ser., **1** (1866), 674–683; "Sur les caractères du crâne basque," *ibid.*, **2** (1867), 10–18, 21–28; "Sur la langue euskuara, parlée par les basques," *ibid.*, 39–71; "Description sommaire de restes humains découverts dans les grottes de Cro Magnon, près de la station des Eyzies, arrondissement de Sarlat (Dordogne)," in *Annales des sciences naturelles* (Zoologie), 5th ser., **10** (1868), 145–155; "Deuxième série d'observations microscopiques sur la chevelure," in *Mémoires de la Société d'anthropologie de Paris*, **3** (1868), 77–97; and "Sur le transformisme," in *Bulletin de la Société d'anthropologie de Paris*, 2nd ser., **4** (1869), 647–682.

II. SECONDARY LITERATURE. See P. Broca, "Histoire des travaux de la Société d'anthropologie," in *Mémoires de la Société d'anthropologie de Paris*, **2** (1863–1865), xix; *Archivio per l'anthropologia e la etnologia*, **12** (1882), 120; V. Gietl, "Pruner Bey," in *Allgemeine Zeitung* (1883), 339; N. Hirsch, "Pruner Bey," in *Allgemeine deutsche Biographie*, XXVI (Leipzig, 1888), 675; P. Lhimosof, *Livre d'or de la géographie. Essai de biographie-géographie* (Paris, 1902), 181; A. Proust, obituary, in *Bulletin de la Société d'anthropologie de Paris*, 3rd ser., **5** (1882), 547; A. Schäfer, "Leben und Wirken des Arztes Franz Pruner-Bey," in *Janus*, **35** (1931), 248–277, 297–311, 335–343, 360–375; and **36** (1932), 59–70, 114–127; G. S., "Pruner Bey," in *Enciclopedia italiana di scienzi, letteri ed* *arti*, XIV (Rome, 1935), 4231, col. 1; and C. Voit, "Pruner Bey," in *Sitzungsberichte der Bayerischen Akademie der Wissenschaften zu München*, **13** (1883), 241–246.

DENISE WROTNOWSKA

PRYANISHNIKOV, DMITRY NIKOLAEVICH (*b.* Kyakhta, Transbaikalia, Siberia, 7 November 1865; *d.* Moscow, U.S.S.R., 30 April 1948), *agricultural chemistry, plant physiology.*

Pryanishnikov's father, a native Siberian, was a bookkeeper; his mother was the daughter of exiles. After the father's death in 1868, the family moved to Irkutsk, where Pryanishnikov graduated from the Gymnasium with a gold medal in 1882. In the same year he entered Moscow University in the natural sciences section of the department of physics and mathematics. After his graduation in 1887, he decided to study agronomy and entered the third-year course at the Petrovskaya Agricultural Academy (now the Timiryazev Agricultural Academy); following his graduation in 1889, he remained there to prepare for an academic career. From then on, all of Pryanishnikov's work was connected with that academy, where in 1895 he received the chair that he occupied for the rest of his life, and began to give courses on fertilization and special agriculture. From 1892 Pryanishnikov lectured on agronomical chemistry at Moscow University and later presented the first course in Russia on the chemistry of plants; the latter formed the basis for the university's department of plant biochemistry. In 1896 he defended his master's thesis, "O raspadenii belkovykh veshchestv pri prorastanii" ("On the Breakdown of Albuminous Substances During Germination"), and in 1900 his doctoral dissertation, *Belkovye veshchestva i ikh prevrashchenie v rastenii v svyazi s dykhaniem i assimilyatsiey* ("Albuminous Substances and Their Transformation in Plants in Connection With Respiration and Assimilation").

Pryanishnikov devoted much time to the improvement of agricultural education. In 1896 he introduced into practical courses an arrangement for experimentation with plants, and did much to improve the teaching at the Petrovskaya Agricultural Academy (which later became the Moscow Agricultural Institute), where from 1907 to 1913 he was deputy director of the teaching division, then its director and dean. From 1907 he took part in the organization of women's courses in agriculture, directed them for ten years, and lectured on plant physiology and agronomical chemistry. Pryanishnikov wrote a number of textbooks: *Chastnoe zemledelie* ("Special Agriculture"; 1898; 10th ed., 1938), *Uchenie ob udobrenii* ("Studies

in Fertilization"; 1900; 5th ed., 1922), *Khimia rasteny* ("Plant Chemistry"; 1907; 2nd ed., 1925), and *Agrokhimia* ("Agricultural Chemistry"; 1934; 3rd ed., 1940). Almost all were translated into several languages. He is also considered the founder of the largest school of Soviet agricultural chemists.

In 1892–1894 Pryanishnikov worked in Göttingen with Alfred Koch, in Paris with Émile Duclaux, and in Zurich with F. E. Schulze. His travels abroad aided his scientific work and teaching considerably, as did his frequent participation in scientific congresses and his familiarity with the organization of higher education, experimental institutions, and the agriculture of various countries.

Pryanishnikov's basic scientific research was devoted to the study of plant nutrition and the application of artificial fertilizers to agriculture. His works on nitrogen nutrition and the replacement of nitrogen substances in the plant organism are especially well known. He provided a general scheme for the transformation of nitrogen substances in plants and noted the exclusive role of ammonia as the original and final product in this process. Having explained the role of asparagine in the plant organism, Pryanishnikov refuted the idea that this substance was the primary product of the breakdown of albumins, showing that it is synthesized from ammonia, which forms during the breakdown of proteins or enters it from without. Introducing an analogy between the role of asparagine in the plant and urea in the animal, Pryanishnikov discovered the general features of the exchange of nitrogen substances in plants and animals. This research provided a scientific foundation for the use of ammonia salts in agriculture and for their extensive production. Pryanishnikov also participated in the development of methods of evaluating the natural phosphorites as a source of phosphorus for plants and as raw material for industrial production of superphosphates. He described the physiological characteristics of natural potassium salts, studied various forms of nitrogen and phosphorus fertilizers, and examined the liming of acid soils and the use of gypsum on solonetz.

Pryanishnikov's works and his school of agricultural chemistry aided the wide introduction of mineral fertilizers into agricultural practice and the creation of a high-capacity chemical fertilizer industry in the U.S.S.R. For his work *Azot v zhizni rasteny i v zemledelii SSSR* ("Nitrogen in the Life of Plants and in the Agriculture of the U.S.S.R."; Moscow–Leningrad, 1945), the Soviet Academy of Sciences awarded him the K. A. Timiryazev Prize in 1946.

Pryanishnikov was also concerned with "green fertilizer" (enriching land by planting a legume crop), and with the use of peat, manure, and other organic fertilizers. He provided a scientific foundation for methods of plant feeding and for the application of various types of fertilizer. He also developed ways to study plant nutrition: isolation feeding, sterile cultures, and fluid solutions, as well as methods of analyzing soils and plants.

For his scientific services Pryanishnikov was elected corresponding member (1913) and full member (1929) of the Soviet Academy of Sciences, and from 1935 he was an active member of the V. I. Lenin All-Union Academy of Agricultural Sciences. He received the Lenin Prize in 1926 and the State Prize in 1941.

BIBLIOGRAPHY

I. ORIGINAL WORKS. Pryanishnikov's works were collected in *Izbrannye sochinenia* ("Selected Works"), V. A. Maksimov, ed., 4 vols. (Moscow, 1951–1955), and in a shorter collection with the same title, A. V. Peterburgsky, ed., 3 vols. (Moscow, 1965), with a bibliography of his writings in III, 577–608. His doctoral dissertation is *Belkovye veshchestva i ikh prevrashchenie v rastenii v svyazi s dykhaniem i assimilyatsiey* ("Albuminous Substances and Their Transformation in Plants in Connection With Respiration and Assimilation"; Moscow, 1899). There is also the autobiographical *Moi vospominania* ("My Recollections"; Moscow, 1957).

II. SECONDARY LITERATURE. See S. I. Vavilov, ed., *Dmitry Nikolaevich Pryanishnikov (1865–1948)* (Moscow–Leningrad, 1948), with a bibliography of his works and of secondary literature, 21–75; V. S. Nemchinov, ed., *Akademik D. N. Pryanishnikov* (Moscow, 1948), with a bibliography of his writings and of secondary literature; *Materialy nauchnoy konferentsii, posviashchennoy 100-lety so dnya rozhdenia akademika D. N. Pryanishnikova* ("Material of the Scientific Conference on the Occasion of the 100th Anniversary of the Birth of Academician D. N. Pryanishnikov"), B. A. Baranov, ed. (Perm, 1968); and S. I. Volfkovich, ed., *Dmitry Nicolaevich Pryanishnikov. Zhizn i deyatelnost* ("D. N. Pryanishnikov: Life and Work"; Moscow, 1972).

E. M. SENCHENKOVA

PRZHEVALSKY, NIKOLAY MIKHAYLOVICH (*b.* Kimbarovo, Smolensk guberniya, Russia, 12 April 1839; *d.* Karakol [now Przhevalsk], Russia, 1 November 1888), *geography, natural science.*

Przhevalsky studied at the Smolensk Gymnasium from 1849 to 1855 before entering the army as a cadet. From 1861 to 1863 he studied at the General Staff Academy in St. Petersburg; his graduation thesis, *Voenno-statisticheskoe obozrenie Priamurskogo kraya* ("A Military-Statistical Survey of the Amur Region," 1862) attracted the attention of the Russian Geographical Society. Commissioned a lieutenant upon graduation, he was appointed in 1864 to the Warsaw Military School, where he taught history and geography. While in Warsaw he also gave public lectures on the history of geographical discoveries and published a textbook on general geography (1867). In it he divided geography into physical and political geography; the former he considered as the science of the phenomena and processes of the three spheres of the earth's surface: lithosphere, hydrosphere, and atmosphere. To prepare himself for travel he studied the works of Humboldt and Karl Ritter on Asia, mastered avian taxidermy, and acquired a sound knowledge of plants.

At the end of 1866 Przhevalsky was assigned to eastern Siberia; and the following May he was sent by the Russian Geographical Society to the Ussuri region, where he remained until 1869 studying the area and correcting existing maps. In exploring the southern part of the Maritime Territory, he traveled by foot and boat along the Ussuri River and its tributaries, visited Lake Khanka, reached the Sea of Japan, and explored the Poseta and Amur bays, from which he went on foot along the shore to the Gulf of Olga and the lower reaches of the Tadusha River. Returning to the Ussuri he crossed the Sikhote-Alin mountain range. In his account of the expedition (1870) Przhevalsky described in detail the natural history, climate, and population of the area. He amassed a collection of great scientific interest, comprising 310 bird specimens, about 2,000 plants, 552 eggs of 42 bird species, and seeds of 83 plant species.

In 1870 the Russian Geographical Society sent Przhevalsky to Mongolia and northern China on a three-year expedition that covered about 12,000 kilometers. Setting out from Irkutsk, he passed through Kyakhta, Urga (now Ulan Bator), and Kalgan (Changkiakow) and visited Peking. From there he traveled to Lake Dalai Nor (Hulun Nor) in the north. He then explored the Ordos, the valley of the Yellow River, and the desert and mountains of Ala Shan, stopping at Lake Kuku Nor before reaching the upper Yangtze-T'ien-shui. On the return journey from Ala Shan to Urga the expedition crossed the eastern Gobi. The published results of the expedition (1875–1876) brought Przhevalsky an international reputation as an authority on Asia and attracted worldwide scientific attention.

Przhevalsky's second journey to central Asia (1876–1877)—to Lob Nor—began from Kuldja. Traveling along the valleys of the Ili and Tarima rivers to Lake Lob Nor and the Astin Tagh Mountains, the expedition returned to Kuldja after having explored Dzungaria. Przhevalsky intended to travel to Lhasa across Khama and Tsaidam, but illness prevented him from continuing and he returned to Zaisan. After having recovered he was ordered to postpone the expedition because of deteriorating relations with China. Przhevalsky wrote that although the ultimate purpose of the expedition, to reach Lhasa, was not achieved, the trip yielded rich results in both physical geography and natural history. One of the most important was the discovery and description of Lake Lob Nor, which, it later became clear, had changed position because of the migration of the channels that fed it. Przhevalsky described in detail his studies and discoveries (1877) and was subsequently elected an honorary member of the St. Petersburg Academy of Sciences.

In the spring of 1879 Przhevalsky embarked on his first Tibetan expedition, which lasted until the late autumn of 1880. He went on foot across Dzungaria, the western part of the Nan Shan (Kunlun) system, the eastern Tsaidam, the upper reaches of the Yangtze-T'ien-shui, and a series of mountain ranges to central Tibet. On the return journey the expedition again explored the upper Yangtze-T'ien-shui, the Yellow River, and Lake Kuku Nor, before returning by caravan road across Urga into Kyakhta. Przhevalsky's skilled and original description of the trip (1883) confirmed his reputation as one of the most distinguished explorers of central Asia.

As leader of the second Tibetan expedition (1883–1885) Przhevalsky paid special attention to the sources of the Yellow River on the borders of northern Tibet and Kozhgaria. The expedition discovered an entire mountain area that included the Tsaidam, Marco-Polo, Zagadochnaya (now Przhevalsky), Russian, and Moscovite ranges. After his return, Przhevalsky immediately began to prepare the results of the expedition for publication (1888).

In October 1888 Przhevalsky was in the foothills of the Tien Shan, at Lake Issyk Kul; Karakola (now Przhevalsk) had been chosen as the starting point for the expedition. But the indefatigable explorer was soon dead of typhus, contracted by drinking water from a river. Following his request, he was buried on the banks of Lake Issyk Kul, and the exploration was continued under the leadership of M. V. Pevtsov.

Przhevalsky was influenced by Humboldt and shared his ideas on the universal interrelationship of natural phenomena and process. Considering his main task to be the study of nature, he made many original contributions to the study of the orography and hydrography, climate, vegetation, and fauna of central Asia. By providing scientific analysis that explained many natural phenomena of the region, his expeditions made it possible to solve problems that were confronting scientists throughout the world. His research was based on compiling maps and describing characteristic landscapes and entire physicogeographical areas, such as the Gobi, Ordos, Ala Shan, and northern Tibet. He asserted that the Gobi Desert is concave—not convex, as scientists had believed. As a result of his research the northern border of the Tibetan upland was shifted 300 kilometers to the north.

Przhevalsky showed the error of the widely held belief that the mountain systems of central Asia had a lattice structure and ascertained that the mountain ranges lay in a primarily east-west direction. He was the first to visit and describe the mountainous regions of Nan Shan (Kunlun), and he discovered the Burkhan-buda, Humboldt, Ritter, Marco-Polo, Columbus, Zagadochnaya, Moscovite, and Tsaidam ranges. His orographical and hydrographical diagrams provided the foundation for present ideas on the directions of the main mountain ranges and river networks and on their interrelationships.

Przhevalsky amassed an impressive herbarium of more than 15,000 plants (1,700 species), including 218 new species and seven new genera. His zoological collection consisted of 702 specimens of small mammals, 5,010 birds, 1,200 reptiles and amphibians, and 643 specimens of fish. He discovered and described a wild camel and a wild horse, now named after him, and studied the growth conditions of plants and the habitats and habits of animals.

His meteorological observations provided the first basis for a climatology of central Asia. He obtained empirical data on maxima and minima and amplitudes of variation of temperature and winds in the Gobi Desert and northern Tibet; and by describing certain regularities of atmospheric processes he increased knowledge of the atmospheric circulation patterns. He also pointed out the influence of Indian and Chinese monsoons and western winds on the climate of central Asia.

Although he did not conduct systematic geological research, Przhevalsky collected rock samples, described the composition of mountain-forming rocks, and studied the topography and the activity of the exogenous factors that alter the contours of the earth's surface. His conclusions on the aeolian origin of the Kuzupchi Hills in Ordos are especially interesting.

The results of Przhevalsky's scientific expeditions were prepared for publication after his death by members of the St. Petersburg Academy of Sciences and the Russian Geographical Society. Six volumes on the zoology, botany, and meteorology of central Asia, based on the accounts of his travels, appeared between 1888 and 1912.

Przhevalsky was the founder of a school of Russian explorer-researchers of Central Asia through whose work the nature of the region was relatively well known by the end of the nineteenth century. He received gold and silver medals from many Russian and foreign scientific societies and academies. In 1891 the Russian Geographical Society established a silver medal and a prize in his honor, and in 1949 the Przhevalsky Gold Medal was established to be awarded by the Geographical Society of the U.S.S.R.

BIBLIOGRAPHY

I. ORIGINAL WORKS. Przhevalsky's writings include *Zapiski vseobshchey geografii* ("Notes on General Geography"; Warsaw, 1867; 2nd ed., 1870); *Puteshestvie v Ussuryskom krae 1867–1869* ("Journey to the Ussuri Region"; St. Petersburg, 1870; 2nd ed., Moscow, 1947; 2nd ed. repr. Vladivostok, 1949); *Mongolia i strana tangutov* ("Mongolia and the Land of the Tanguts"), 2 vols. (St. Petersburg, 1875–1876; 2nd ed., vol. I, Moscow, 1946); "Ot Kuldzhi za Tyan-Shan i na Lobnor" ("From Kuldja Past the Tien Shan to Lob Nor"), *Izvestiya Russkogo geograficheskogo obshchestva*, **13**, no. 5 (1877), also repr. separately (Moscow, 1947); *Iz Zaysana cherez Khami i Tibet i na verkhovya Zheltoy reki* ("From Zaisan Across the Khama to Tibet and the Upper Reaches of the Yellow River"; St. Petersburg, 1883; 2nd ed., Moscow, 1948); "Avtobiograficheskie zapiski N. M. Przhevalskogo" ("Autobiographical Notes . . ."), *Russkaya starina* (1888), no. 11; *Ot Kyakhty na istoki Zheltoy reki. Issledovania severnoy okrainy Tibeta i put cherez Lobnor po basseynu Tarima* ("From Kyakhta to the Sources of the Yellow River. Research on the Northern Border Regions of Tibet and the Route Across Lob Nor Along the Basin of the Tarim"; St. Petersburg, 1888; 2nd ed., Moscow, 1948).

Additional works are *Nauchnye rezultaty puteshestvy N. M. Przhevalskogo po Tsentralnoy Azii* ("Scientific Results of N. M. Przhevalsky's Travels Through Central Asia"), zoological sec., 3 vols. (St. Petersburg, 1888–1912), edited by E. Vikhner, V. Zelensky, *et al.*; botanical sec., 2 vols. (St. Petersburg, 1889–1895): edited by K. Maksimovich; meteorological sec. edited by A. I. Voeykov; and "Stati, dnevniki, pisma" ("Articles, Diaries, Letters"), in *Izvestiya Vsesoyuznogo geograficheskogo obshchestva*, **72**, nos. 4–5 (1940), 469–640.

II. SECONDARY LITERATURE. See D. N. Anuchin, "N. M. Przhevalsky," in *O lyudyakh russkoy nauki i kultury* ("People of Russian Science and Culture"; Moscow, 1950), 65–89; N. F. Dubravin, *Nikolay Mikhaylovich Przhevalsky* (St. Petersburg, 1890); V. A. Esakov, "Iz perepiski M. I. Venyukova s N. M. Przhevalskim" ("From the Correspondence of . . . Venyukov and . . . Przhevalsky"), in *Voprosy istorii estestvoznaniya i tekhniki*

1 (1956), 207–212; M. G. Kadek, "N. M. Przhevalsky," in *Lyudi russkoy nauki* ("People of Russian Science"), I (Moscow–Leningrad, 1948), 569–578; N. M. Karataev, *N. M. Przhevalsky—pervy issledovatel prirody Tsentralnoy Azii* (". . . First Investigator of the Nature of Central Asia"; Moscow–Leningrad, 1948); S. I. Khmelnitsky, *N. M. Przhevalsky. 1839–1888* (Leningrad, 1950); P. K. Kozlov, *Veliky russky puteshestvennik N. M. Przhevalsky* ("The Great Russian Traveler . . ."; Leningrad, 1929); and E. M. Murzaev, *N. M. Przhevelasky* (Moscow, 1952; 2nd ed., Moscow, 1953).

See also V. A. Obruchev, "N. M. Przhevalsky kak puteshestvennik i issledovatel Tsentralnoy Azii" (". . . Przhevalsky as Traveler and Investigator of Central Asia"), in *Vestnik Akademii nauk SSSR* (1939), no. 6, 71–77; *Pamyati Nikolaya Mikhaylovicha Przhevalskogo* ("Memories of . . . Przhevalsky"; St. Petersburg, 1890); P. P. Semenov, *Istoria poluvekovoy deyatelnosti Russkogo geograficheskogo obshchestva* ("History of a Half-Century of Activity of the Russian Geographical Society"), pt. 2 (St. Petersburg, 1896); P. P. Semenov *et al.*, "Rechi, proiznesennye na chrzevychaynom sobranii imp. Russkogo geografischeskogo obshchestva 9 noyabrya 1888 g., posvyashchennye pamyati N. M. Przhevalskogo" ("Speeches Given at the Extraordinary Meeting of the Imperial Russian Geographical Society, 9 November 1888, Dedicated to the Memory of N. M. Przhevalsky"), in *Izvestiya Russkogo geografischeskogo obshchestva*, **24**, no. 4 (1888), 233–272; M. Shokalsky, L. S. Berg, *et al.*, "Pamyati N. M. Przhevalskogo" (Recollections of Przhevalsky"), *ibid.*, **72**, no. 4–5 (1940); V. V. Potemkin, ed., *Veliky russky geograf Przhevalsky. K stoletiyu so dnya rozhdenia 1839–1939 gg.* ("The Great Russian Geographer Przhevalsky. On the Centenary of His Birth . . ."; Mosow, 1939), a collection of articles with a bibliography; and A. V. Zelenin, *Puteshestvia N. M. Przhevalskogo* ("The Travels of N. M. Przhevalsky"), 2 vols. (St. Petersburg, 1899–1900).

VASILIY A. ESAKOV

PSELLUS, MICHAEL (baptized **CONSTANTINE**) (*b.* near Constantinople, 1018; *d.* Constantinople [?], April/May 1078), *philosophy, transmission of knowledge.*

Born in a western suburb of Constantinople near the monastery of Narses,[1] Psellus belonged to an aristocratic but impoverished family originally from Bithynian Nicomedia. He studied under Nicetas of Byzantium and John Mauropus, but at the age of sixteen was forced by financial need to enter the Byzantine provincial administration. On the death of his sister, he returned to Constantinople and entered the legal profession. He became associated with the court of Michael IV Paphlagon (1034–1041) and was soon appointed judge in Philadelphia, under the protection of his former schoolmate Constantine Leichudes. During the brief reign of Michael V Calaphates (1041–1042), Leichudes became prime minister and Psellus the head of the imperial secretariat. During these years of developing his political career Psellus had continued his studies independently; and when Constantine IX Monomachus (1042–1055) reorganized higher education in Constantinople, Psellus became the head of the philosophical faculty (1045–1054) while retaining his high position at the court.

His teaching of pagan, and particularly Neoplatonic, philosophy, however, combined with the jealousy of his political rivals, opened the way for attacks upon Psellus. In 1053 he was forced to reaffirm his religious orthodoxy in a confession of faith.[2] His troubles were compounded by the death of his only child, his daughter Styliane.[3] Finding his relations with the emperor worsening, Psellus pleaded illness and retired to the monastery of Olympus in Bithynia in 1054; upon becoming a monk, he took the name Michael. On the death of Constantine in 1055, however, he was recalled to a position of influence in Constantinople by Empress Theodora (1055–1056) and participated in the intrigues that resulted in the elevation of Isaac I Comnenus (1057–1059) to the throne.

Although he had previously composed a group of works on alchemy, divination, and the like[4] for Patriarch Michael Caerularius (1043–1058), Psellus now attacked him for his interest in these nefarious subjects.[5] But, with the accession of a new emperor, Constantine X Ducas (1059–1067), Psellus was forced by his former protector, Leichudes, who had succeeded Caerularius as patriarch (1059–1063), to honor his monastic vows; he was forced to stay at the monastery of Narses, near his birthplace, from 1059 to 1063. In the latter year his old friend John Xiphilinus became patriarch and allowed Psellus, upon delivering a eulogy of Caerularius, to leave the monastery and return to his political activities. His influence at court continued, with various vicissitudes, until his death in April or May of 1078.[6]

During his very active political life Psellus found time to write an astonishing number of works: letters, history, poems, essays, orations, and treatises on grammar, rhetoric, law, philosophy, and science. An autodidact in philosophy and science, he was instrumental in reviving an interest in those subjects in Byzantium. In particular, he was concerned to promote later Neoplatonism and its theurgical component. He accepted the idea that a knowledge of the divine is acquired in stages through a study of the sensible objects in this world, followed by a thorough investigation of mathematics, which leads directly to

theology.[7] In this spirit he wrote compendia of excerpts[8] concerning the sciences and theology derived from various sources; since some of these sources are otherwise lost to us, Psellus' excerpts are frequently of great historical value. He also was instrumental in preserving for us various works of Iamblichus and Proclus[9], and one version of the *Corpus Hermeticum*.[10] He was not an original thinker but did his best to explain what he had learned; his best in technical subjects, such as chronology,[11] is not very impressive. Through his interpretation we are able to gain access to the theurgy, demonology, and alchemy of the third, fourth, and fifth centuries, an access extremely valuable to intellectual historians. Also, Psellus' activities stimulated in others an interest in philosophy and science. These two contributions more than compensate for the derivative character of the material he so assiduously compiled.

A list of Psellus' works on philosophy and science is given below. For unpublished treatises, see G. Weiss, "Untersuchungen zu den unedierten Schriften des Michael Psellos," in *Βυζαντινά*, **2** (1970), 335–378, reprinted with corrections, *ibid.*, **4** (1972), 1–52. Collected works are Bidez: J. Bidez, *Catalogue des manuscrits alchimiques grecs*, VI (Brussels, 1928); Boissonade: J. F. Boissonade, *Michael Psellus De operatione daemonum* (Nuremberg, 1838; repr. Amsterdam, 1964); Boissonade, *Anecdota*: J. F. Boissonade, *Anecdota Graeca*, I (Paris, 1829), 175–247; Boissonade, *Tzetzes*: J. F. Boissonade, *Tzetzae Allegoriae Iliados. Accedunt Pselli Allegoriae quarum una inedita* (Paris, 1851; repr. Hildesheim, 1967), 341–371; Ideler: I. L. Ideler, *Physici et medici Graeci minores*, 2 vols. (Berlin, 1841–1842; repr. Amsterdam, 1963); Kurtz and Drexl: E. Kurtz and F. Drexl, *Michaelis Pselli Scripta minora*, 2 vols. (Milan, 1936–1941); *PG* 122: J.-P. Migne, *Patrologia Graeca*, vol. 122 (Paris, 1889), cols. 476–1186; Sathas: C. N. Sathas, *Bibliotheca Graeca Medii Aevi*, IV and V (Paris, 1874–1876); Weinstock: S. Weinstock, *Catalogus codicum astrologorum Graecorum*, IX, pt. 1 (Brussels, 1951), 101–128.

I. Philosophy.

Ia. In general. "That the Movements of the Soul Resemble the Motions of the Heavenly Bodies," in Boissonade, *Michael Psellus*, 56–57; and *PG* 122, cols. 1075–1076; "On Philosophy," in Kurtz and Drexl, I, 428–432; "To Those Who Ask How Many Are the Kinds of Philosophical Discourses," *ibid.*, 441–450; "That Substance Is a Self-Subsistent," *ibid.*, 451–458; and "Greek Arrangements Concerning the Divine Creation," in Weinstock, 111–114. See also M. Sicherl, "Platonism und Textüberlieferung," in *Jahrbuch der*

Österreichischen Byzantinischen Gesellschaft, **15** (1966), 201–229, esp. 206–212.

Ib. Plato. "On Plato's Psychogony," edited by A. J. H. Vincent, "Notice sur divers manuscrits grecs relatifs à la musique," in *Notices et extraits des manuscrits*, XVI, pt. 2 (Paris, 1858), 1–600, esp. 316–337; C. W. Linder, editor, *M. Pselli in Platonis De animae procreatione praecepta commentarius* (Uppsala, 1854); and *PG* 122, cols. 1077–1114. That this work is based on Proclus' commentary on the *Timaeus* is shown by J. Bidez, "Psellus et le Commentaire du *Timée* de Proclus," in *Revue de Philologie*, new series **29** (1905), 321–327; "Commentary on the Platonic Chariot-Driving of the Souls and the Army of the Gods in the *Phaedrus*," in Kurtz and Drexl, I, 437–440, based on Hermeias' commentary on the *Phaedrus*; and "On the Ideas Which Plato Mentions," *ibid.*, 433–436, based on Plotinus. Psellus also excerpted passages from Proclus' commentary on Plotinus' *Enneads*; see L. G. Westerink, "Exzerpte aus Proklos' Enneaden-kommentar bei Psellos," in *Byzantinische Zeitschrift*, **52** (1959), 1–10.

Ic. Aristotle. In Physicen Aristotelis Commentarii, Latin translation by I. B. Camotius (Venice, 1554). Part of the beginning of the Greek text was edited by C. A. Brandis in *Aristotelis Opera*, IV (Berlin, 1836; repr. Berlin, 1961), 322b–324a; and Bidez, 211–212. A complete edition is promised by L. Benakis; for now see P. Joannou, *Christliche Metaphysik in Byzanz I. Die Illuminationslehre des Michael Psellos und Joannes Italos* (Ettal, 1956); L. Benakis, "Studien zu den Aristoteles-Kommentaren des Michael Psellos," in *Archiv für Geschichte der Philosophie*, **43** (1961), 215–238, and **44** (1962), 33–61; "Michael Psellos' Kritik an Aristoteles und seine eigene Lehre zur 'Physis'- und 'Materie-Form'-Problematik," in *Byzantinische Zeit-schrift*, **56** (1963), 213–227; and "Doxographische Angaben über die Vorsokratiker im unedierten Kom-mentar zur 'Physik' des Aristoteles von Michael Psellos," in *Χάρις Κωνσταντίνῳ 'Ι. Βουρβέρη* (Athens, 1964), 345–354. "Opinions Concerning the Soul," *PG* 122, cols. 1029–1076, is based on John Philoponus' commentary on Aristotle's *De anima;* see also E. A. Leemans, "Michel Psellos et les *Δόξαι περὶ ψυχῆς*," in *L'antiquité classique*, **1** (1932), 203–211; and a collection of four meteorological tracts, in Bidez, 49–70, based on Olympiodorus' commentary on Aristotle's *Meteorologica*. C. Zervos in *Un philosophe néoplatonicien . . .* mentions two other works, but I have not been able to check their existence and the reliability of their attribution to Psellus: *Paraphrase of the De interpretatione* (Venice, 1503), Latin translation by S. Boetius (Venice, 1541) and by C. Gesner (Basel, 1542); and *Synopsis of*

Aristotle's Logic, edited by E. Ehingerus (Vienna, 1597).

Id. Theurgy. "Commentary on the *Chaldaean Oracles*," edited by J. Opsopoeus, in *Oracula magica Zoroastris* (Paris, 1607), 52–113; *PG* 122, cols. 1123–1150. See also N. Terzaghi, "Parergon de quibusdam *Oraculis chaldaicis*," in *Studi italiani di filologia classica*, **16** (1908), 433–440; "Brief Exposition of the Doctrines of the Chaldaeans," in Opsopoeus, *op. cit.*, 112–121; *PG* 122, cols. 1149–1154; and D. Bassi, "Notizie di codici greci nelle bibliotheche italiane III; Michele Psello," in *Rivista di filologia e d'istruzione classica*, **26** (1898), 122–123; and "Summary Descrip-tion of the Ancient Doctrines of the Chaldaeans," in W. Kroll, *De Oraculis Chaldaicis* (Breslau, 1894; repr. Hildesheim, 1962), 73–76. Psellus made two excerpts from Proclus' commentary on the *Chaldaean Oracles*: *Proclus From the Chaldaean Philosophy*, edited by A. Jahn (Halle, 1891), and "Proclus on the Hieratic Art According to the Greeks," in Bidez, 137–151.

That Psellus' knowledge of the *Chaldaean Oracles* is derived from Proclus is demonstrated by Kroll, *op. cit.*; J. Bidez, "Proclus *Περὶ τῆς ἱερατικῆς τέχνης*," in *Annuaire de l'Institut de philologie et d'histoire orien-tales et slaves*, **4** (1936), 85–100; and H. Lewy, *Chaldaean Oracles and Theurgy* (Le Caire, 1956), 473–479. See also E. des Places, "Le renouveau platonicien du XIe siècle: Michel Psellus et les *Oracles chal-daïques*," in *Comptes rendus de l'Académie des Inscriptions et Belles-lettres* (1966), 313–324; "On Sacrifice," in Bidez, 155–158; and "Different Greek Opinions About the Soul," in Weinstock, 106–111. M. Sicherl, "Michael Psellos und Iamblichus *De mysteriis*," in *Byzantinische Zeitschrift*, **53** (1960), 8–19, has shown that Psellus wrote the scholium that appears at the beginning of the *De mysteriis* in all extant manuscripts, in which it is stated that Proclus attributed the work to Iamblichus, and that he also wrote the lemmata. Various of Psellus' excerpts from the *De mysteriis* are in Weinstock, 114, and in M. Sicherl, *Die Handschriften, Ausgaben und Über-setzungen von Iamblichos De mysteriis* (Berlin, 1957), 134–137.

Ie. Corpus Hermeticum. R. Reitzenstein, *Poiman-dres* (Leipzig, 1904), 319, surmised that Psellus' manuscript of the *Corpus* was the ancestor of the extant ones and that he made some corrections and additions to the text (p. 326). Indeed, one manuscript of the fourteenth century (Vat. Gr. 951) contains a scholium by Psellus on *Corpus Hermeticum* I, 18, which is in Boissonade, *Michael Psellus*, 153–154; *PG* 122, cols. 1153–1156; and Reitzenstein, *op. cit.*, 333–334. There are other indications that Psellus had read the *Corpus*: for example, Bidez, 214–215, 218. But it seems most probable that only the text of Vat. Gr. 951 represents the recension of Psellus, and that the other manuscripts descend from an independent archetype—see A. D. Nock and A.-J. Festugière, *Corpus Hermeticum*, I (Paris, 1945), xlix–li.

If. Demonology. "Timotheus, or On the Operation of Demons," in Boissonade, *Michael Psellus*, 1–36; *PG* 122, cols. 819–976. See also M. Wellnhofer, "Die thrakischen Euchiten und ihr Satanskult im Dialoge des Psellos 'Τιμόθεος ἢ περὶ τῶν δαιμόνων'," in *Byzantinische Zeitschrift*, 30 (1929–1930), 477–484; "What the Greeks Thought About Demons," in Boissonade, *Michael Psellus*, 36–43; *PG* 122, cols. 875–882; "On Demons," in Bidez, 113–131; "Greek Classifications of Demons," in Weinstock, 114–120. See also A. Delatte and C. Jesserand, "Contribution à l'étude de la démonologie byzantine," in *Annuaire de l'Institut de philologie et d'histoire orientales*, 2 (1934), 207–232; K. Svoboda, *La démonologie de M. Psellos* (Brno, 1927); and P. Joannou, *Démonologie populaire—démonologie critique au XIe siècle. La vie inédite de S. Auxence par M. Psellos* (Wiesbaden, 1971).

Ig. Marvels. The following four works apparently were all addressed to the Patriarch Michael Caerularius: *On the Powers of Stones*, edited by P. J. Maussacus and J. S. Bernard (Leiden, 1745); Ideler, I, 244–247; and *PG* 122, cols. 887–900; "On Marvelous Readings," edited by A. Westermann, in *Scriptores rerum mirabilium Graeci* (Brunswick–London, 1839), 143–148; "On Looking at Shoulder-Blades and Birds [as Omens]," edited by R. Hercher, in *Philologus*, 8 (1843), 166–168; and "On How to Make Gold," in Bidez, 1–47.

Ih. Astrology. "Solutions of Astrological Questions in Thirty Chapters"—what appears to be the unique manuscript, Monacensis Gr. 537, fols. 1–8v, ends in the middle of the sixth chapter; "On Seven-Month, Eight-Month, and Nine-Month Embryos," in Weinstock, 101–103; and "Is Astrology True?" edited by M. A. Šangin, in *Catalogus codicum astrologorum Graecorum*, XII (Brussels, 1936), 167.

II. Science.

IIa. Encyclopedias. "Instruction of All Sorts," edited by L. G. Westerink, appeared as *Michael Psellus. De omnifaria doctrina* (Utrecht, 1948). Four different redactions were composed by Psellus: "Brief Solutions of Physical Problems," in 137 sections; "Brief Solutions and Explanations of Physical Problems," in 193 sections, dedicated to Michael VII Ducas (1071–1078; co-emperor from about 1060); "Synoptic Answers and Explanations to Different Questions and Problems," in 201 sections, also dedicated to Michael; and "Instruction of All Sorts and Entirely Necessary," in 193 sections, addressed to Michael after he had become sole emperor in 1071. Westerink follows the order of "Synoptic Answers . . .," the longest redaction. See also P. Tannery, "Psellus sur la grande année," in *Revue des études grecques*, 5 (1892), 206–211, reprinted in his *Mémoires scientifiques*, IV (Toulouse–Paris, 1920), 261–268; and F. Boll, "Psellus und das 'grosse Jahr,'" in *Byzantinische Zeitschrift*, 7 (1898), 599–602. "Solutions to Various Problems" is in Boissonade, *Michael Psellus*, 63–69. Not by Psellus is *Brief Solutions to Physical Problems*, in two books, edited by G. Seebode (I, Gotha, 1840; II, Wiesbaden, 1857); and *PG* 122, cols. 783–810. In reality this is an incomplete text of Symeon Seth's "Synopsis of Physical Problems," edited by A. Delatte, in *Anecdota Atheniensia et alia*, II (Paris, 1939), 1–89. On this work see also C. Giannelli, "Di alcune versioni e rielaborazioni serbe delle 'Solutiones breves quaestionum naturalium' attribuite a Michele Psello," in *Studi byzantini e neoellenici*, 5 (1939), 445–468, reprinted in *ibid.*, 10 (1963), 1–25.

IIb. Numbers. "On the Properties of Numbers," in Weinstock, 103–106; "On Numbers," in P. Tannery, *Revue des études grecques*, 5 (1892), 343–347, reprinted in his *Mémoires scientifiques*, IV (Toulouse–Paris, 1920), 269–274; "From Diophantus' Arithmetic," edited by P. Tannery, in *Zeitschrift für Mathematik und Physik*, Historisch-literarische Abtheilung, 37 (1892), 41–45, reprinted in his *Mémoires scientifiques*, IV, 275–282, also in *Diophanti Alexandrini Opera omnia*, II (Leipzig, 1895), 37–42—these three works were apparently extracted from Iamblichus' *Collection of Pythagorean Dogmas*; and "On the Purpose of Learning Geometry," in Boissonade, *Michael Psellus*, 159–163.

IIc. Music. "Introduction to the Science of Rhythm," edited by J. Cäsar, in *Rheinisches Museum für Philologie*, new series 1 (1842), 620–633—see also R. Westphal, *Griechische Rhythmik und Harmonik*, second edition (Leipzig, 1867), supplement, 18–21. Three fragments on music are published as Psellus' by A. J. H. Vincent, *op. cit.*, 338–343; they belong to an *Anonymus De musica* (Paris, 1545), which Vincent claims is entirely by Psellus. But this work seems to be a part of the *Anonymi logica et quadrivium*, edited by J. L. Heiberg (Copenhagen, 1929), which was once incorrectly ascribed to Psellus; see also L. Richter, "'Des Psellos kurzen Inbegriff der Musik' bei L. Chr. Mizler," in *Studia Byzantina* (Halle, 1966), 149–157.

IId. Chronology. Psellus' work on chronology was partially edited by A. Mentz in *Beiträge zur Osterfestberechnung bei den Byzantinern* (Königsberg, 1906),

102–108; the complete text was edited by G. Redl, "La chronologie appliquée de Michel Psellos," in *Byzantion*, **4** (1927–1928), 197–236, and **5** (1929), 229–286. See also G. Redl, "Untersuchungen zur technischen Chronologie des Michael Psellos," in *Byzantinische Zeitschrift*, **29** (1929–1930), 168–187, and "Studien zur technischen Chronologie des Michael Psellos," in *Byzantinisch-neugriechische Jahrbücher*, **7** (1930), 305–351.

IIe. Medicine. "Medical Work in Iambs," in Boissonade, *Anecdota*, 175–232; and Ideler, I, 203–243; "On Common Terms in Illnesses," in Boissonade, *Anecdota*, 233–241; "On Bathing," in Ideler, II, 193; and "On Foods and Drinks," *ibid.*, 257–281—see also E. Renauld, "Quelques termes médicaux de Psellos," in *Revue des études grecques*, **24** (1909), 251–266.

IIf. Minor scientific works. "On Athenian Places and Names," in Boissonade, *Michael Psellus*, 44–48; and *PG* 122, cols. 1155–1160; and "On Agriculture," in Boissonade, *Anecdota*, 242–247.

NOTES

1. P. Joannou, "Psellos et le Monastère Τὰ Ναρσοῦ," in *Byzantinische Zeitschrift*, **44** (1951), 283–290.
2. A. Garzya, "On Michael Psellus' Admission of Faith," in Ἐπετηρὶς Ἑταιρείας Βυζαντινῶν Σπουδῶν, **35** (1966–1967), 41–46.
3. A. Leroy-Molinghen, "Styliané," in *Byzantion*, **39** (1969), 155–163. On Psellos' adopted daughter see R. Guilland, "À propos d'un texte de Psellos," in *Byzantinoslavica*, **20** (1959), 205–230, and **21** (1960), 1–37, repr. in his *Recherches sur les institutions byzantines*, I (Berlin–Amsterdam, 1967), 84–143; and A. Leroy-Molinghen, "La déscendance adoptive de Psellos," in *Byzantion*, **39** (1969), 284–317; and "À propos d'un jugement rendu contre Psellos," *ibid.*, **40** (1970), 238–239.
4. See section *Ig.* in text.
5. L. Bréhier, "Un discours inédit de Psellos. Accusation du patriarche Michel Cérulaire devant la Synode (1059)," in *Revue des études grecques*, **16** (1903), 375–416, and **17** (1904), 35–76.
6. P. Gautier, "Monodie inédite de Michel Psellos sur le Basileus Andronic Doucas," in *Revue des études byzantines*, **24** (1966), 153–170.
7. S. A. Sofroniou, "Michael Psellos' Theory of Science," in Ἀθηνᾶ, **69** (1966–1967), 78–90.
8. See section *IIa.* in text.
9. *Ibid. Id.*
10. *Ibid. Ie.*
11. *Ibid. IId.*

BIBLIOGRAPHY

Much of our information about Psellus comes from his *Chronography*, which is a history of Byzantium from 976 to 1077; the best ed. is by E. Renauld, 2 vols. (Paris, 1926–1928); see also R. Anastasi, *Studi sulla Chronographia di Michele Psello* (Catania, 1969). Additional information is in his orations and letters, some of which are in Boissonade, *Michael Psellus*, 170–188; *PG* 122,

cols. 1161–1186; Sathas, V; and Kurtz and Drexl; see also J. Darrouzès, "Les lettres inédites de Psellos," in *Revue des études byzantines*, **12** (1954), 177–180. Still useful is C. Zervos, *Un philosophe néoplatonicien du XIe siècle. Michel Psellos* (Paris, 1920). The most recent, but not entirely satisfactory, study is the article by E. Kriaras in Pauly-Wissowa, *Real-Encyclopädie der classischen Altertumswissenschaft*, supp. XI (Stuttgart, 1968), cols. 1124–1182, trans. into Greek in Βυζαντινά, **4** (1972), 53–128.

Dᴀᴠɪᴅ Pɪɴɢʀᴇᴇ

PTOLEMY (or **Claudius Ptolemaeus**) (*b. ca.* A.D. 100; *d. ca.* A.D. 170), *mathematical sciences, especially astronomy.*

Our meager knowledge of Ptolemy's life is based mostly on deductions from his surviving works, supplemented by some dubious information from authors of late antiquity and Byzantine times. The best evidence for his dates is the series of his observations reported in his major astronomical work, the *Almagest*: these are all from the reigns of the Roman emperors Hadrian and Antoninus, the earliest 26 March 127 and the latest 2 February 141.[1] Since he wrote several major works after the *Almagest*, this evidence fits well with the statement of a scholiast attached to works of Ptolemy in several late manuscripts—that he flourished under Hadrian and lived until the reign of Marcus Aurelius (161–180).[2] The only other explicit date is that of the "Canobic Inscription": this is found in manuscripts of Ptolemy's astronomical works and purports to be a copy of an inscription dedicated by Ptolemy to the "Savior God" at Canopus, a town at the western mouth of the Nile, in the tenth year of Antoninus (A.D. 147/148).[3] It consists mostly of lists of astronomical parameters determined by Ptolemy; although most of its contents are extracted from the *Almagest* and other genuine works of Ptolemy, I doubt its authenticity. A statement by the sixth-century philosophical commentator Olympiodorus that Ptolemy practiced astronomy for forty years in "the so-called wings at Canopus," and hence set up there the inscription commemorating his astronomical discoveries,[4] is probably a fictional elaboration on the "Canobic Inscription." In fact the only place mentioned in any of Ptolemy's observations is Alexandria, and there is no reason to suppose that he ever lived anywhere else. The statement by Theodore Meliteniotes that he was born in Ptolemais Hermiou (in Upper Egypt) could be correct,[5] but it is late (*ca.* 1360) and unsupported. The belief that he came from Pelusium is a Renaissance misinterpretation of the title "Phelud(i)ensis" attached to his name in medieval Latin texts, which in turn comes from a corruption of the Arabic "qalūdī," a misunderstanding of Κλαύδιος.[6] His name "Ptolemaeus" indicates that he was an inhabitant of Egypt, descended from Greek or hellenized forebears, while "Claudius" shows that he possessed Roman citizenship, probably as a result of a grant to an ancestor by the emperor Claudius or Nero.

It is possible to deduce something about the order of composition of Ptolemy's surviving works from internal evidence. The *Almagest* is certainly the earliest of the major works: it is mentioned in the introductions to the *Tetrabiblos, Handy Tables*, and *Planetary Hypotheses*, and in book VIII, 2, of the *Geography*; a passage of the *Almagest* looks forward to the publication of the *Geography*.[7] One can occasionally trace development: in the *Handy Tables* many tables are presented in a form more convenient for practical use than the corresponding sections of the *Almagest*, and some parameters are slightly changed. The *Planetary Hypotheses* exhibits considerably more change in parameters, and introduces a notable improvement in the theory of planetary latitudes and an entirely new system for calculating the absolute sizes and distances of the planets. The prime meridian of the *Geography* is not Alexandria, as is promised in the *Almagest*, but a meridian through the "Blessed Isles" (the Canaries) at the extreme west of the ancient known world (which had the advantage that all longitudes were counted in the same direction). The phenomenon of the apparent enlargement of heavenly bodies when they are close to the horizon is explained in the *Almagest* as due to physical causes (the dampness of the atmosphere of the earth),[8] whereas in the *Optics* Ptolemy gives a purely psychological explanation.[9] Presumably in the interval between the composition of the two works he had discovered that there is no measurable enlargement. Similarly the *Optics* discusses the problem of astronomical refraction,[10] which is never considered in the *Almagest* despite its possible effect on observation. It is hardly possible, however, to trace Ptolemy's scientific development, except in his astronomical work; and even there the later works contribute only minor modifications to the masterly synthesis of the *Almagest*.

We know nothing of Ptolemy's teachers or associates, although it is a plausible conjecture that the Theon who is said in the *Almagest* to have "given" Ptolemy observations of planets from between 127 and 132 was his teacher.[11] Many of the works are addressed to an otherwise unknown Syrus. It would be hasty, but not absurd, to conclude from the fact that the *Geography* and the *Harmonica* are not addressed to Syrus that they are later than all the works so

addressed (that is, all the other extant works except for the dubious Περὶ κριτηρίου and possibly the *Optics* and the *Phaseis*, the relevant sections of which are missing). Living in Alexandria must have been a great advantage to Ptolemy in his work (and perhaps his education). Although much declined from its former greatness as a center of learning, the city still maintained a scholarly tradition and must at the least have provided him with essential reference material from its libraries.

Ptolemy's chief work in astronomy, and the book on which his later reputation mainly rests, is the *Almagest*, in thirteen books. The Greek title is μαθηματικὴ σύνταξις, which means "mathematical [that is, astronomical] compilation." In later antiquity it came to be known informally as ἡ μεγάλη σύνταξις or ἡ μεγίστη σύνταξις ("the great [or greatest] compilation"), perhaps in contrast with a collection of earlier Greek works on elementary astronomy called ὁ μικρὸς ἀστρονομούμενος ("the small astronomical collection").[12] The translators into Arabic transformed ἡ μεγίστη into "al-majisti," and this became "almagesti" or "almagestum" in the medieval Latin translations. It is a manual covering the whole of mathematical astronomy as the ancients conceived it. Ptolemy assumes in the reader nothing beyond a knowledge of Euclidean geometry and an understanding of common astronomical terms; starting from first principles, he guides him through the prerequisite cosmological and mathematical apparatus to an exposition of the theory of the motion of those heavenly bodies which the ancients knew (sun, moon, Mercury, Venus, Mars, Jupiter, Saturn, and the fixed stars, the latter being considered to lie on a single sphere concentric with the earth) and of various phenomena associated with them, such as eclipses. For each body in turn Ptolemy describes the type of phenomena that have to be accounted for, proposes an appropriate geometric model, derives the numerical parameters from selected observations, and finally constructs tables enabling one to determine the motion or phenomenon in question for a given date.

In order to appreciate Ptolemy's achievement in the *Almagest*, we ought to know how far Greek astronomy had advanced before his time. Unfortunately the most significant works of his predecessors have not survived, and the earlier history has to be reconstructed almost entirely from secondary sources (chiefly the *Almagest* itself). Much remains uncertain, and the following sketch of that history is merely provisional.

The first serious attempt by a Greek to describe the motions of the heavenly bodies by a mathematical model was the system of "homocentric spheres" of Eudoxus (early fourth century B.C.). Although mathematically ingenious, this model was ill-suited to represent even the crude data on which it was based; and the approach proved abortive (it would have vanished from history had it not been adopted by Aristotle).[13] Equally insignificant for the development of astronomy were works on "spherics" that treat phenomena such as the risings and settings of stars in terms of spherical geometry (these appear from the fourth century B.C. on). The heliocentric theory of Aristarchus of Samos (early third century B.C.), perhaps developing ideas of Heraclides Ponticus (ca. 360 B.C.), was purely descriptive and also without consequence. After Eudoxus, however, the epicyclic and eccentric models of planetary motion were developed; and the equivalence of the two was proved by Apollonius of Perga (ca. 200 B.C.), if not earlier. Apollonius made an elegant application of these models to the problem of determining the stationary points of a planet.[14]

Meanwhile astronomical observations were being made in the Greek world from the late fifth century B.C. The earliest were mostly of the dates of solstices, but by the early third century Aristyllus and Timocharis in Alexandria were attempting to determine the positions of fixed stars and observing occultations. These observations, however, were few and unsystematic; no firmly based theory was possible until records of the observers in Babylon and other places in Mesopotamia, reaching back to the eighth century, became available to Greeks. It seems likely that period relations derived by the Babylonian astronomers from these observations were known as early as Eudoxus, but the first Greek who certainly used the observations themselves was Hipparchus; and it is no accident that Greek astronomy was established as a quantitative science with his work. Hipparchus (active from ca. 150 to 127 B.C.) used Babylonian eclipse records and his own systematic observations to construct an epicyclic theory of the sun and moon that produced reasonably accurate predictions of their positions. Hence he was able to predict eclipses. He measured the lunar parallax and evolved the first practical method for determining the distances of sun and moon. By comparing his own observations of the position of the star Spica with those of Timocharis 160 years earlier, he discovered the precession of the equinoxes. He employed plane trigonometry and stereographic projection. In the latter techniques and in the observational instruments he used he may have been a pioneer, but we know too little of his predecessors to be sure. Ptolemy expressly informs us that Hipparchus did not construct a theory of the five planets but contented himself with showing that existing theories did not satisfy the observations.[15] Between Hipparchus and Ptolemy the only advance was the work of Menelaus (ca. A.D. 100) on spherical trigonometry.

Greek astronomy, then, as Ptolemy found it, had evolved a geometric kinematic model of solar and lunar motion that successfully represented the phenomena, at least as far as the calculation of eclipses was concerned, but had produced only unsatisfactory planetary models. It had developed both plane and spherical trigonometry and had adopted the Babylonian sexagesimal place-value system not only for the expression of angles but also (although not systematically) for calculation. As a mathematical science it was already sophisticated. From the point of view of physics it was not a science at all: such physical theories as were enunciated were mere speculation. But there was available a fairly large body of astronomical observations, of which the most important, both for completeness and for the length of time it covered, was the series of eclipses observed in Mesopotamia. Ptolemy made no radical changes in the system he took over; but by intelligent use of available observations and ingenious modification of the basic principle of all existing kinematic models (uniform circular motion), he extended that system to include the five planets and significantly improved the lunar model.

Books I and II of the *Almagest* are devoted to preliminaries. Ptolemy begins by stating and attempting to justify his overall world picture, which is that generally (although not universally) accepted from Aristotle on: around a central, stationary, spherical earth the sphere of the fixed stars (situated at a distance so great that the earth's diameter is negligible in comparison) revolves from east to west, making one revolution per day and carrying with it the spheres of sun, moon, and planets; the latter have another, slower motion in the opposite sense, in or near a plane (the ecliptic) inclined to the plane of the first motion. He then develops the trigonometry that will be used throughout the work. The basic function is the chord (which we denote Crd) subtended by the angle at the center of a circle of radius 60. This is related to the modern sine function by

$$\sin \alpha = \frac{1}{2 \cdot 60} \operatorname{Crd} 2\alpha.$$

The values of some chords (for instance, Crd 60°) are immediately obtainable by elementary geometry. By geometrically developing formulas for Crd $(\alpha + \beta)$, Crd $(\alpha - \beta)$, Crd $\frac{1}{2}\alpha$, where Crd α and Crd β are known, and then finding Crd 1° by an approximation procedure, Ptolemy produces a table of chords, at intervals of 1/2° and to three sexagesimal places, which serves for all trigonometric calculations. There was probably nothing new in his procedure (Hipparchus had constructed a similar table).[16]

Since the sine function is sufficient for the solution of all plane triangles, the chord function too was sufficient, although the lack of anything corresponding to a tangent table often made the solution laborious, involving the extraction of square roots. For trigonometry on the surface of the sphere Ptolemy uses a figure that we call a Menelaus configuration. It is depicted in Figure 1, where all the arcs *AB*, *BE*, and

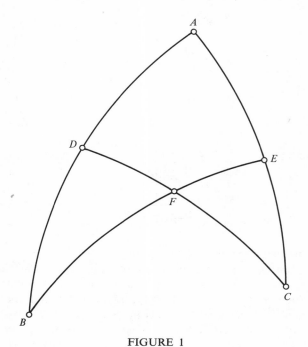

FIGURE 1

so on are segments of a great circle less than 90°. Ptolemy proves (following Menelaus) that

$$(1) \quad \frac{\text{Crd } 2EC}{\text{Crd } 2EA} = \frac{\text{Crd } 2CF}{\text{Crd } 2FD} \cdot \frac{\text{Crd } 2DB}{\text{Crd } 2BA},$$

$$(2) \quad \frac{\text{Crd } 2AC}{\text{Crd } 2AE} = \frac{\text{Crd } 2CD}{\text{Crd } 2DF} \cdot \frac{\text{Crd } 2FB}{\text{Crd } 2BE}.$$

It can be shown that the four basic formulas used in modern trigonometry for the solution of right spherical triangles are directly derivable from (1) and (2). Hence all problems soluble by means of the former can also be solved by the use of Menelaus configurations. Thus Ptolemaic trigonometry, although sometimes cumbersome, is completely adequate.

This trigonometry is applied in books I and II to various phenomena connected with the annual variation in solar declination. The sole numerical parameter used is the inclination of the ecliptic (ϵ).

This is also the amount of the greatest declination of the sun from the equator, and that is the basis of the two simple instruments for measuring ϵ which Ptolemy describes (I, 12). He reports that he measured 2ϵ as between 47; 40° and 47; 45°. Since the estimation of Eratosthenes and Hipparchus, that 2ϵ is 11/83 of a circle, also falls between these limits, Ptolemy too adopts the latter, taking ϵ as 23; 51, 20°. His failure to find a more accurate result, and hence to discover the slow decrease in the inclination, is explained by the crudity of his instruments. He can now construct a table of the declination of the sun as a function of its longitude, which is a prerequisite for solving problems concerning rising times. The rising time of an arc of the ecliptic is the time taken by that arc to cross the horizon at a given terrestrial latitude. Most of book II is devoted to calculating tables of rising times for various latitudes. Such tables are useful astronomically, for instance, for computing the length of daylight for a given date and latitude (important in ancient astronomy, since the time of day or night was reckoned in "civil hours," one civil hour being 1/12 of the varying length of day or night); but the space devoted to the topic is disproportionate to its use in the *Almagest*. It is essential in astrology, however (for example, in casting horoscopes); and this is one of the places where astrological requirements may have influenced the *Almagest* discussion (although they are never explicitly mentioned).

Book III treats the solar theory. By comparing his own observations of the dates of equinoxes with those of Hipparchus, and his observation of a solstice with one made by Meton and Euctemon in 432 B.C., Ptolemy confirms Hipparchus' estimate of the length of the tropical year as 365 1/4 − 1/300 days. This estimate is notoriously too long (the last fraction should be about 1/128), and the error was to have multiple consequences for Ptolemaic astronomy. It is derivable from the data only because Ptolemy made an error of about one day in the time of each of his observations. This is a gross error even by ancient standards and is the strongest ground of those modern commentators (such as Delambre) who maintain that Ptolemy slavishly copied Hipparchus, to the point of forging observations to obtain agreement with Hipparchus' results. The conclusion is implausible, but it is likely that in this case Ptolemy was influenced by his knowledge of Hipparchus' value to select such of his own observations as best agreed with it. For Hipparchus' solar and lunar theory represented the known facts (that is, eclipse records) very well, and Ptolemy would be reluctant to tamper with those elements of it that would seriously affect the circumstances of eclipses. Using the above year length, Ptolemy constructs a table of the mean motion of the sun that is the pattern for all other mean motion tables: the basis is the Egyptian calendar, in which the year has an unvarying length of 365 days (twelve thirty-day months plus five epagomenal days). The motion is tabulated to six sexagesimal places, for hours, days, months, years, and eighteen-year periods.

The main problem in dealing with all planets is to account for their "anomaly" (variation in velocity). In the case of the sun this variation is apparent from the fact that the seasons are of unequal length—for instance, in Ptolemy's day the time from spring equinox to summer solstice was longer than that from summer solstice to autumn equinox. Ptolemy proposes a general model for representing anomalistic motion. The "eccentric" version is depicted in Figure 2, where

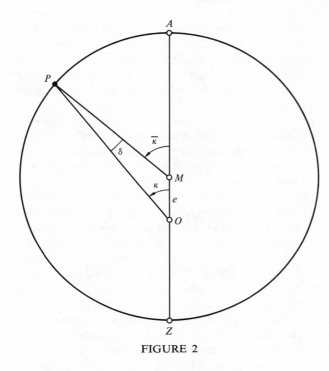

FIGURE 2

O represents the central earth. The body P moves with uniform angular velocity on a circle whose center M is distant from O by an amount e (the eccentricity). It is clear that if the motion of P appears uniform from M, it will appear nonuniform from O, being slowest at its greatest distance, the apogee A, and fastest at its least distance, the perigee Z. The angle κ which P has traveled from the apogee is derivable from its mean motion $\bar{\kappa}$ by the formula

$$\kappa = \bar{\kappa} \pm \delta.$$

δ is called by Ptolemy the προσθαφαίρεσις and by us, following medieval usage, the "equation." The same

motion can also be represented by an epicyclic model. In Figure 3 an epicycle, center C, moves with uniform

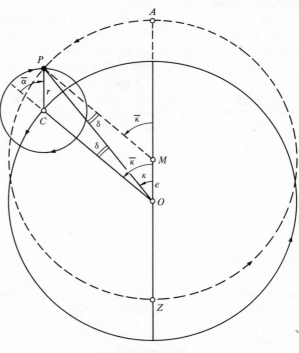

FIGURE 3

angular velocity on the circle (known as the deferent) about O, while the body P moves uniformly about C in the opposite sense. Provided that the angular velocities of body and epicycle are equal and the radius of the epicycle equals the eccentricity, the two models are completely equivalent, as can be seen from Figure 3, where $\bar{\alpha} = \bar{\kappa}$ and $r = e$. Such is the case of Ptolemy's solar model. But the angular velocity of P may be different from that of C, as in the lunar theory (this is equivalent to a rotation of the apogee in the eccentric model); or the rotation of P may be in the same sense as C, as in Ptolemy's planetary theory. The general model is, therefore, extremely versatile.

From his observations of solstices and equinoxes Ptolemy found the same length of seasons as Hipparchus. He therefore concluded that the apogee of the sun is tropically fixed and that its motion can be represented by the simple eccentric of Figure 2. He determined its parameters, as Hipparchus had, from the observed length of the seasons. In Figure 4 the sun is in T at spring equinox, in X at summer solstice, and in Y at autumn equinox. Ptolemy states that it moves from T to X in 94 1/2 days and from X to Y in 92 1/2 days. Thus the angles at M (the center of uniform motion), TMX and XMY, can be calculated from the mean motion of the sun; and the angles at O, the earth, are right. Hence one can determine the

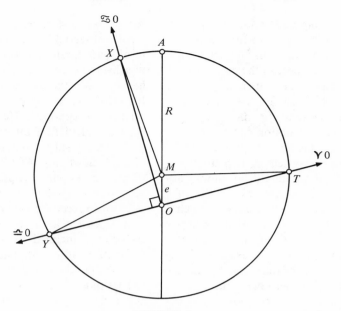

FIGURE 4

eccentricity *OM* and the longitude of the apogee, <*TOA*. Since Ptolemy uses the same data and method as Hipparchus, he gets the same results: an eccentricity of 2;30 (where *R*, the standard radius, is 60) and an apogee longitude of 65;30°. In fact the eccentricity had decreased and the apogee longitude increased since Hipparchus' time, but this could not be detected through equinox and solstice observations made with crude ancient instruments.

Using the parameters determined, Ptolemy shows how the equation δ can be calculated trigonometrically for a given mean anomaly κ̄ and sets up a table giving δ as a function of κ̄. The only thing now lacking in order to calculate the position of the sun at any date is its mean position at some given date. From his equinox observation of A.D. 132 Ptolemy calculates its position at the date he has chosen as epoch, Thoth 1 (first day of the Egyptian year), year 1 of the Babylonian king Nabonassar (26 February 747 B.C.). To complete the solar theory he discusses the equation of time. Since the sun travels in a path inclined to the equator and with varying velocity, the interval between two successive meridian transits of the sun (the true solar day) will not be uniform but will vary slightly throughout the year. Since astronomical calculations employ uniform units (that is, mean solar days), whereas local time, in an age when the sundial is the main chronometer, is reckoned according to the true solar day, one must be able to convert one to the other. This conversion is done by means of the equation of time, the calculation of which Ptolemy explains.

The lunar theory is the subject of books IV and V.

According to Ptolemy one must distinguish three periods connected with the moon: the time in which it returns to the same longitude, the time in which it returns to the same velocity (period of anomaly), and the time in which it returns to the same latitude. In addition one must consider the synodic month, the time between successive conjunctions or oppositions of the sun and moon. He quotes a number of previous attempts to find a period containing an integer number of each of the above (such a period would, clearly, be an eclipse cycle); in particular, from Hipparchus:

(1) In 126,007 days, 1 hour, there occur 4,267 synodic months, 4,573 returns in anomaly, and 4,612 sidereal revolutions less 7 1/2° (hence the length of a mean synodic month is 29; 31, 50, 8, 20 days).
(2) In 5,458 synodic months there occur 5,923 returns in latitude.

Ptolemy says that Hipparchus established these "from Babylonian and his own observations." We now know from cuneiform documents that these parameters and period relations had all been established by Babylonian astronomers.[17] At best Hipparchus could have "confirmed" them from his own observations. Such confirmation could be carried out only by comparison of the circumstances of eclipses separated by a long interval. Ptolemy gives an acute analysis of the conditions that must then obtain in order for one to detect an exact period of return in anomaly. On the basis of his own comparison of eclipses he accepts the above relations, with very slight modifications to the parameters for returns in anomaly and latitude (justified later). Thus he is able to construct tables of the lunar mean motion in longitude, anomaly, argument of latitude (motion with respect to the nodes in which the orbit of the moon intersects the ecliptic), and elongation (motion with respect to the mean sun).

The next task is to determine the numerical parameters of the lunar model. Ptolemy first assumes (although he knows better) that the moon has a single anomaly, that is, that its motion can be represented by a simple epicycle model (or eccenter with rotating apogee); this was the system of Hipparchus. To determine the size of the epicycle he adopts a method invented by Hipparchus. He takes a set of three lunar eclipses: the time of the middle of each eclipse can be calculated from the observed circumstances. Hence the true longitude of the moon at eclipse middle is known, since it is exactly 180° different from the true longitude of the sun (calculated from the solar theory). Furthermore, the time intervals between the three eclipses are known; hence one can calculate the travel in mean longitude and mean anomaly between the three points. Thus one has the situation of Figure 5: P_1, P_2, P_3 represent the positions of the moon on the epicycle at the three eclipses. The angles δ_1, δ_2 (the equational differences as seen from the earth O) are found by comparing the intervals in true longitude with the intervals in mean longitude; the angles θ_1, θ_2 are found by taking the travel in mean anomaly modulo 360°. From these one can calculate trigonometrically the size of the epicycle radius r in terms of the deferent radius $R = OC$, and the angle ACP_1 (which gives an epoch value for the anomaly). Ptolemy makes calculations

for two sets of eclipses—the first early Babylonian, the second observed by himself—and gets almost identical results. He finally adopts the value $r = 5;15$, and on this basis he constructs an equation table. Although he borrowed the above procedure from Hipparchus, Ptolemy's result seems to be a distinct improvement on his predecessor's. He tells us that through small slips in calculating intervals Hipparchus found two discrepant results from two eclipse triples—namely, 327 2/3 : 3144 and 247 1/2 : 3122 1/2. Independent evidence shows that Hipparchus adopted the latter eccentricity, although it is a good deal too small.[18]

In the preceding calculations the moon has been treated as if it lay in the plane of the ecliptic. In fact its orbit is inclined to that plane, but the angle of inclination is so small that one is justified in neglecting it in longitude calculations. Accurate knowledge of the latitude is, however, essential for eclipse calculations. The inclination of 5° which Ptolemy accepts was probably an established value. But his procedure for finding the epoch value and mean motion in argument of latitude from two carefully selected eclipses is both original and a great improvement over Hipparchus' method, which involved estimating the apparent diameter of the moon and of the shadow of the earth at eclipse, both of which are difficult to measure accurately.

The simple lunar model of book IV is essentially that of Hipparchus. When Ptolemy compared observed positions of the moon with those calculated from the model, he found good agreement at conjunction and opposition (when elongation of the moon

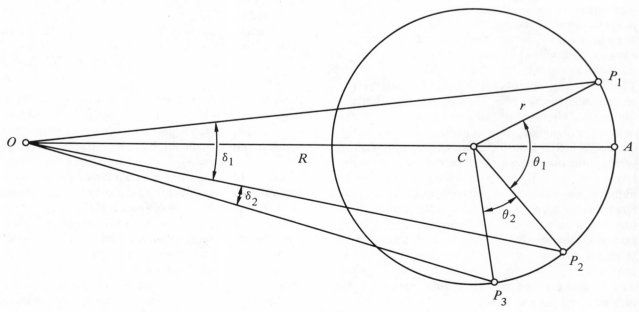

FIGURE 5

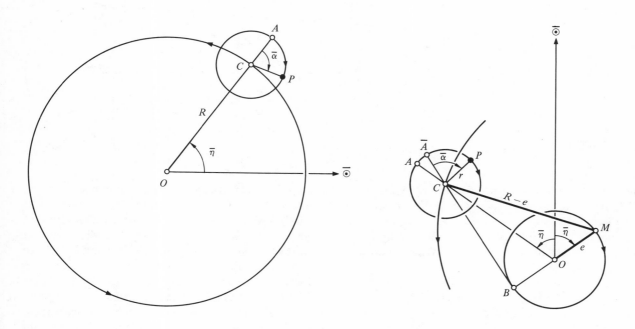

FIGURES 6A AND 6B

from the sun is 0° or 180°); this agreement was to be expected, since the model was derived from eclipse observations. But he found serious discrepancies at intermediate elongations. Hipparchus had already mentioned such discrepancies but failed to account for them. In book V Ptolemy develops his own lunar theory. Analyzing his observations, he finds that they seem to indicate an increase in the size of the epicycle between opposition and conjunction that reaches a maximum at quadrature (90° elongation). This increase he represents by incorporating in the model a "crank" mechanism that "pulls in" the epicycle as it approaches quadrature, thus making it appear larger. Compare Figures 6A and 6B, which depict the same situation according to the simple and refined models, respectively. In the latter the epicycle center C continues to move uniformly about the earth O, but it now moves on a circle the center of which is not O but M. M moves about O in the opposite sense to C, so that its elongation $\bar{\eta}$ from the mean sun is equal to that of C. Since $OM + MC$ of Figure 6B is equal to OC of Figure 6A, it is clear that the two models are identical when $\bar{\eta} = 0°$ or $180°$, that is, at mean conjunction and opposition. At intermediate elongations, however, the refined model pulls the epicycle closer to O, thus increasing the effect of the anomaly. This increase is greatest at quadrature ($\bar{\eta} = 90°$). From two observations of the moon near quadrature by himself and Hipparchus, Ptolemy finds that the maximum equation increases from about 5° at conjunction

to 7;40° at quadrature, and hence e in Figure 6B is 10;19 (where $R = 60$). As a further refinement he shows that one can obtain better agreement with observation if one reckons the anomaly $\bar{\alpha}$ not from the true epicycle apogee A but from a mean apogee $\bar{A}$ opposite the point B (B in turn being opposite M on the small circle about O). Thus a third inequality is introduced, also varying with the elongation but reaching its maximum near the octants ($\bar{\eta} = 45°$ and 135°). Ptolemy can now construct a table to compute the position of the moon. In contrast with previous tables, the tabulated function depends on two variables ($\bar{\alpha}$ and $2\bar{\eta}$). Ptolemy's solution to tabulating such a function (which may have been his own invention) became standard: he computes the equation at extreme points (in this case at conjunction and quadrature) and introduces an interpolation function (here varying with $2\bar{\eta}$) to be used as a coefficient for intermediate positions.

Ptolemy's refined lunar model represents the longitudes of the moon excellently. It is a major improvement on Hipparchus' model, and yet it does not disturb it at the points where it was successful—namely, where eclipses occur. But one effect of the crank mechanism is to increase greatly the variation in the distance of the moon from the earth, so that its minimum distance is little more than half its maximum. If correct, this should be reflected by a similar variation in the apparent size of the moon, whereas the observable variation is much smaller. This objection

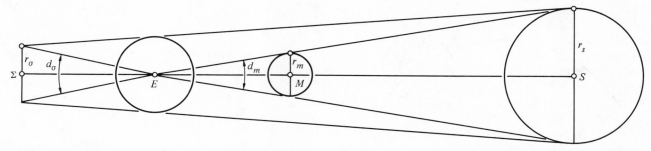

FIGURE 7

seems not to have occurred to Ptolemy: he treats the crank mechanism not merely as a convenient device for predicting the longitude but also as a real feature of the model (for instance, in his parallax computations). Fortunately the apparent size of the moon is of consequence only at eclipses, when the crank mechanism has no effect.

Having established a complete theory of the motion of sun and moon, Ptolemy can proceed to eclipse theory. But first he must deal with parallax, the angular difference between the true and apparent positions of a body that stems from the fact that we observe it not from the center of the earth but from a point on the surface. In practice only the parallax of the moon is significant, but the misconceptions of the ancients about the distance of the sun led them to estimate solar parallax as well. By comparing a suitable observed position of the moon with its computed position, Ptolemy obtains a value for the parallax. Since this value is equivalent to the angle under which the radius of the earth is seen from the moon in that position, he can immediately calculate the distance of the moon in earth radii, first at that position and then (from his model) at mean distance. His final figure, fifty-nine earth radii, is close to the truth but is reached by a combination of multiple errors in his data and model, which, by pure chance, cancel each other.

Ptolemy knew that solar parallax is too small to measure directly; but he calculates the distance of the sun, and hence its parallax, by a method invented by Hipparchus. The data required are the apparent diameters of moon and sun at distances such that their apparent diameters are equal, that distance for the moon, and the apparent diameter of the shadow of the earth at that distance of the moon. Ptolemy assumes that moon and sun have exactly the same apparent diameter when the moon is at greatest distance (which implies that annular eclipses cannot occur); he determines the apparent diameter of moon and shadow from two pairs of lunar eclipses by an ingenious method of his own (greatly improving on

the figures which Hipparchus had obtained by direct measurement). Then, in Figure 7 he knows EM ($= E\Sigma$), the distance of the moon; d_m, the apparent diameter of moon and sun; and d_σ, the apparent diameter of the shadow. From these data it is simple to calculate r_m and r_s, the radii of moon and sun, and ES, the distance of the sun. His value for the latter is 1,210 earth radii, too small by a factor of twenty. In fact, the method requires much too accurate a measurement of the apparent diameter of the sun to produce a reliable solar distance, but it continued to be used up to and beyond Copernicus. Ptolemy now constructs a table of solar and lunar parallaxes and explains how to compute the parallax for a given situation. This computation is both laborious and mathematically unsatisfactory. Ptolemaic parallax theory is perhaps the most faulty part of the *Almagest*. But it was not significantly improved until the late sixteenth century.

Eclipse theory, the topic of book VI, is easily derived from what precedes. Ptolemy sets up a table for calculating mean syzygies (conjunctions and oppositions), with the corresponding lunar anomaly and argument of latitude. He then determines (from the apparent sizes of the bodies) the eclipse limits, that is, how far from the node the moon can be at mean syzygy for an eclipse still to take place. The eclipse tables proper give the size in digits and duration of eclipses as a function of the distance of the moon from the node. Ptolemy explains minutely how to compute the size, duration, and other circumstances of both lunar and solar eclipses for any given place. But his method does not allow one to compute the path of a solar eclipse (a development of the late seventeenth century).[19]

Books VII and VIII deal with the fixed stars. The order of treatment is a logical one, since it is necessary to establish the coordinates of ecliptic stars to observe planetary positions. Ptolemy compares his own observations with those of Hipparchus and earlier Greeks to show that the relative positions of the fixed stars have not changed and that the sphere of the fixed stars moves about the pole of the ecliptic from east to west 1° in 100 years with respect to the tropical points. He ascribes the discovery of the latter motion (the precession of the equinoxes) to Hipparchus, who had estimated it as *not less than* 1° in 100 years. This figure is too low (1° in seventy years would be more accurate); the error is mostly due to Ptolemy's wrong figure for the mean motion of the sun.[20] The bulk of these two books is composed of the "Star Catalog," a list of 1,022 stars, arranged under forty-eight constellations, with the longitude, latitude, and magnitude (from 1 to 6) of each. To compile this entirely from personal observation would be a gigantic task, and Ptolemy has often been denied the credit. Delambre, for instance, maintained that Ptolemy merely added 2;40° to the longitudes of "Hipparchus' catalog."[21] This particular hypothesis has been disproved.[22] In fact, the evidence suggests that no star catalog in this form had been composed by Hipparchus or anyone else before Ptolemy (the quotations from Hipparchus in *Almagest* VII, 1, show that Ptolemy had before him not a catalog but a description of the constellations with some numerical data concerning distances between stars). Modern computations have revealed numerous errors in Ptolemy's coordinates.[23] In general, the longitudes tend to be too small. This too is explained by the error in his solar mean motion, which is embedded in the lunar theory: the moon was used to fix the position of principal stars (the only practical method for an ancient astronomer).[24] Book VIII ends with a discussion of certain traditional Greek astronomical problems, such as the heliacal risings and settings of stars.

The last five books are devoted to planetary theory. Here, in contrast with the moon and sun, Ptolemy had no solid theoretical foundation to build upon and much less in the way of a body of observations. The most striking phenomenon of planetary motion is the frequent occurrence of retrogradation, which had been explained at least as early as Apollonius by a simple epicyclic model (see Figure 3) in which the sense of rotation of planet and epicycle is the same. Such a model, however, would produce a retrogradation arc of unvarying length and occurring at regular intervals, whereas observation soon shows that both arc and time of retrogradation vary. No geometric model had been proposed that would satisfactorily account for this phenomenon. Certain planetary periods, however, were well established, and so was the law for outer planets that

$$Y = L + A,$$

where (in integer numbers) Y stands for years, L for returns to the same longitude, and A for returns in anomaly (Venus and Mercury have the same period of return in longitude as the sun, hence for them $Y = L$). Ptolemy quotes from Hipparchus such a period relation for each planet—for instance, for Saturn: "In 59 years occur 57 returns in anomaly and 2 returns in longitude." We now know that all the period relations quoted are in fact Babylonian in origin. From these Ptolemy constructs tables of mean motion in longitude and anomaly, first applying small corrections; it turns out, however, that the latter are in part based on the models he is going to develop, so he must have used the uncorrected period relations in the original development.

Analysis of observations revealed that each planet has two anomalies; the first varying according to the planet's elongation from the sun and a second varying according to its position in the ecliptic. Ptolemy isolated the first by comparing different planet-sun configurations in the same part of the ecliptic and the second by comparing the same planet-sun configurations in different parts of the ecliptic. He thus found that the first could be represented by an epicycle model in which the sense of rotation of planet and epicycle are the same, while the second was best represented by an eccenter (to avoid a double epicycle). The model he finally evolved is depicted in Figure 8. The planet P moves on an

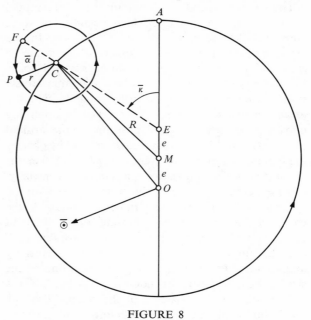

FIGURE 8

epicycle, center C. C moves in the same sense on a circle the center of which, M, is distant from the earth O by the eccentricity e. The uniform motion of C, however, takes place not about M but about another point E on the opposite side of M from O

and distant from M by the same amount e. The mean motion in anomaly is counted from a point F on the epicycle opposite to E. Figure 8 depicts the situation for an outer planet, in which the line CP always remains parallel to the line from O to the mean sun (thus preserving the law $Y = L + A$). Except for this feature the model for Venus is identical. The Mercury model has an additional mechanism to produce a varying eccentricity but is otherwise the same as that for Venus.

The most original element in this model is the introduction of the point E (known as the equant from medieval times). This element disturbed later theoreticians (including Copernicus), since it violated the philosophical principle that uniform motion should take place about the center of the circle of revolution. But it enabled Ptolemy to construct the first satisfactory planetary model. In heliocentric theory a Ptolemaic equant model with eccentricity e produces planetary longitudes differing from those of a Kepler ellipse with the same eccentricity by less than 10′ (even for the comparatively large eccentricity of Mars).[25] Ptolemy deduces the existence of the eccenter directly from observations only in the case of Venus. For the other planets he merely assumes it and regards this as justified by the resulting agreement between observed and computed positions.

The determination of the parameters of the model for the individual planets, which occupies most of books IX–XI, reveals Ptolemy's brilliance. In his mastery of the choice and analysis of observations in conjunction with theory he has no peer until Kepler. For Venus and Mercury the problem is comparatively simple: the center of their epicycles coincides with the mean sun, and so one "sees" the epicycle by observing maximum elongations. The position of the apsidal line ($OMEA$ in Figure 8) can be determined by observing symmetrical positions of the epicycle, and then it is simple to calculate the size of epicycle and eccentricity. For the outer planets such "direct observation" of the epicycle is not possible. So Ptolemy eliminates the effect of the epicycle by choosing three observed oppositions; this choice gives him three positions of C as seen from O. The problem of finding the eccentricity and apogee then resembles the problem of finding the size of the epicycle of the moon in book IV, but it is complicated by the existence of the equant. Ptolemy meets this complication with an ingenious iterative process: he first assumes that the equant coincides with the center of the eccentric; this produces an approximate apsidal line and eccentricity, which are used to compute corrections to the initial data; then the whole process is repeated as many times as is necessary for the results to converge. Such a procedure

is alien to classical Greek mathematics, although not unique in the *Almagest*.

Book XII is devoted to establishing tables for computing the arcs and times of retrogradation and (for Venus and Mercury) the greatest elongation. In connection with the former, a theorem of Apollonius on stationary points is adapted to fit Ptolemy's refined planetary model. In book XIII he deals with planetary latitudes. All five planets have orbits slightly inclined to the ecliptic. From a heliocentric point of view the situation is simple, but a geocentric theory encounters considerable difficulties. The simplest approximation in Ptolemaic terms would be as follows: For an outer planet set the plane of the deferent at an inclination to the ecliptic equal to that of the heliocentric orbit of the planet, and keep the plane of the epicycle parallel to the ecliptic; for an inner planet keep the plane of the deferent in the ecliptic, and set the inclination of the epicycle to the deferent equal to the inclination of the heliocentric orbit of the planet. Ptolemy eventually reaches that solution in the *Planetary Hypotheses* (except that the deferents of Venus and Mercury are inclined to the ecliptic at a very small angle); but in the *Almagest*, misled by faulty observations and the eccentricity of the orbit of the earth, he devises a much more complicated theory in which the epicycles of outer planets and both epicycles and deferents of inner planets undergo varying inclinations as the epicycle moves round the deferent. The resultant tables do, however, represent the actual changes in latitude fairly well and are no mean achievement. The work ends with a discussion of the traditional problem of the heliacal risings and settings of the planets.

As a didactic work the *Almagest* is a masterpiece of clarity and method, superior to any ancient scientific textbook and with few peers from any period. But it is much more than that. Far from being a mere "systematization" of earlier Greek astronomy, as it is sometimes described, it is in many respects an original work. Without minimizing Ptolemy's debt to Hipparchus (which Ptolemy himself admiringly acknowledges), we may say confidently that Hipparchus' pioneering work would have had very little effect had it not found its completion in the *Almagest*. The Ptolemaic system is indeed named after the right man.

The *Almagest* contains all the tables necessary for astronomical computations, but they are scattered throughout the work. At a later date Ptolemy published the tables separately, together with an introduction explaining their use, under the title Πρόχειροι κανόνες (*Handy Tables*). The tables themselves are extant only in the revised version of Theon of Alexandria (*fl.* 360), but from Ptolemy's introduction it is clear that Theon changed nothing essential. Ptolemy himself is responsible for the numerous differences from the *Almagest*. There are additions (such as a table of the longitudes and latitudes of principal cities, to enable one to convert from the meridian and latitude of Alexandria), changes in layout to facilitate computation, and even occasional improvements in basic parameters. The epoch is changed from era Nabonassar to era Philip (Thoth 1 = 12 November 324 B.C.). The tables became the standard manual in the ancient and Byzantine worlds, and their form persisted beyond the Middle Ages.

Later still Ptolemy published a "popular" résumé of the results of the *Almagest* under the title Ὑποθέσεις τῶν πλανωμένων (*Planetary Hypotheses*), in two books. Only the first part of book I survives in Greek, but the whole work is available in Arabic translation. It goes beyond the *Almagest* in several respects. First, it introduces changes in some parameters and even in the models, notably in the theory of planetary latitude already mentioned. Second, in accordance with Ptolemy's declaration in the introduction that one purpose of the work is to help those who aim to represent the heavenly motions mechanically (that is, with a planetarium), the models are made "physical," whereas in the *Almagest* they had been purely geometric. Ptolemy describes these physical models in detail in book II (most of book I is devoted to listing the numerical parameters). He argues that instead of assigning a whole sphere to each planet, it is sufficient to suppose that the mechanism is contained in a segment of a sphere consisting of a drum-shaped band extending either side of its equator. The most portentous innovation, however, is the system proposed at the end of book I for determining the absolute distances of the planets.

In the *Almagest* Ptolemy had adopted the (traditional) ascending order: moon, Mercury, Venus, sun, Mars, Jupiter, Saturn; but he admitted that this order was arbitrary as far as the planets are concerned, since they have no discernible parallax.[26] This order has no consequences, since the parameters of each planet are determined independently in terms of a conventional deferent radius of 60. In the *Planetary Hypotheses* Ptolemy proposes a system whereby the greatest distance from the earth attained by each body is exactly equal to the least distance attained by the body next in order outward (that is, the planetary spheres are touching, and there is no space wasted in the universe; this system conforms to Aristotelian thinking). He takes the distance of the moon in earth radii derived in *Almagest* V: its greatest distance is equal to the least distance of Mercury (if one assumes the above order of the planets). Using the previously determined parameters of the model

for Mercury, he now computes the greatest distance of Mercury, which is equal to the least distance of Venus, and so on. By an extraordinary coincidence the greatest distance of Venus derived by this procedure comes out very close to the least distance of the sun derived by an independent procedure in *Almagest* V. Ptolemy takes this finding as a striking proof of the correctness of his system (and incidentally of the assumption that Mercury and Venus lie below the sun). He goes on to compute the exact distances of all the bodies right out to the sphere of the fixed stars in earth radii and stades (assuming the circumference of the earth to be 180,000 stades). Furthermore, taking some "observations" by Hipparchus of the apparent diameters of planets and first-magnitude stars, he computes their true diameters and volumes. This method of determining the exact dimensions of the universe became one of the most popular features of the Ptolemaic system in later times.

A work in two books named *Phases of the Fixed Stars* (Φάσεις ἀπλανῶν ἀστέρων) dealt in detail with a topic not fully elaborated in the *Almagest*, the heliacal risings and settings of bright stars. Only book II survives; and the greater part of this book consists of a "calendar," listing for every day of the year the heliacal risings and settings, as well as the weather prognostications associated with them by various authorities. Predicting the weather from the "phases" of well-known stars long predates scientific astronomy in Greece, and calendars like this were among the earliest astronomical publications (Ptolemy quotes from authorities as early as Meton and Euctemon). The chief value of the Φάσεις today is the information it contains on the history of this kind of literature.

Much greater scientific interest attaches to two small works applying mathematics to astronomical problems. The first is the *Analemma* (Περὶ ἀναλήμματος), surviving, apart from a few palimpsest fragments, only in William of Moerbeke's Latin translation from the Greek. It is an explanation of a method for finding angles used in the construction of sundials, involving projection onto the plane of the meridian and swinging other planes into that plane. The actual determination of the angles is achieved not by trigonometry (although Ptolemy shows how that is theoretically possible) but by an ingenious graphical technique which in modern terms would be classified as nomographic. Although the basic idea was not new (Ptolemy criticizes his predecessors, and a similar procedure is described by Vitruvius *ca.* 30 B.C.),[27] the sophisticated development is probably Ptolemy's. The other treatise is the *Planisphaerium* (the Greek title was probably Ἅπλωσις ἐπιφανείας σφαίρας).[28] This treatise survives only in Arabic translation; a revision of this translation was made by the Spanish Islamic astronomer Maslama al-Majrīṭī (d. 1007/1008) and was in turn translated into Latin by Hermann of Carinthia in 1143. It treats the problem of mapping circles on the celestial sphere onto a plane. Ptolemy projects them from the south celestial pole onto the plane of the equator. This projection is the mathematical basis of the plane astrolabe, the most popular of medieval astronomical instruments. Since the work explains how to use the mapping to calculate rising times, one of the main uses of the astrolabe, it is highly likely that the instrument itself goes back to Ptolemy (independent evidence suggests that it goes back to Hipparchus).[29] These two treatises are an important demonstration that Greek mathematics consisted of more than "classical" geometry.

To modern eyes it may seem strange that the same man who wrote a textbook of astronomy on strictly scientific principles should also compose a textbook of astrology ('Αποτελεσματικά, meaning "astrological influences," or Τετράβιβλος, from its four books). Ptolemy, however, regards the Tetrabiblos as the natural complement to the Almagest: as the latter enables one to predict the positions of the heavenly bodies, so the former expounds the theory of their influences on terrestrial things. The introductory chapters are devoted to a defense of astrology against charges that it cannot achieve what it claims and that even if it can, it is useless. Ptolemy regards the influence of heavenly bodies as purely physical. From the obvious terrestrial physical effects of the sun and moon, he infers that all heavenly bodies must produce physical effects (that such an argument could be seriously advanced reflects the poverty of ancient physical science). By careful observation of the terrestrial manifestations accompanying the various recurring combinations of celestial bodies, he believes it possible to erect a system which, although not mathematically certain, will enable one to make useful predictions. Ptolemy is not a fatalist: at least he regards the influence of the heavenly bodies as only one of the determinants of terrestrial events. But, plausible as this introduction might appear to an ancient philosopher, the rest of the treatise shows it to be a specious "scientific" justification for crude superstition. It is difficult to see how most of the astrological doctrines propounded could be explained "physically" even in ancient terms, and Ptolemy's occasional attempts to do so are ludicrous. Astrology was almost universally accepted in the Roman empire, and even superior intellects like Ptolemy and Galen could not escape its dominance.

Book I explains the technical concepts of astrology,

book II deals with influences on the earth in general ("astrological geography" and weather prediction), and books III and IV with influences on human life. Although dependent on earlier authorities, Ptolemy often develops his own dogma. The discussion in books III and IV is confined to what can be deduced from a man's horoscope: Ptolemy ignores altogether the branch of astrology known as catarchic, which answers questions about the outcome of events or the right time to do something by consulting the aspect of the heavens at the time of the question. This omission helps to explain why the Tetrabiblos never achieved an authority in its field comparable with that of the Almagest in astronomy.

The Geography (Γεωγραφικὴ ὑφήγησις), in eight books, is an attempt to map the known world. The bulk of it consists of lists of places with longitude and latitude, accompanied by very brief descriptions of the chief topographical features of the larger land areas. It was undoubtedly accompanied in Ptolemy's own publication by maps like those found in several of the manuscripts. But knowing how easily maps are corrupted in copying, Ptolemy takes pains to ensure that the reader will be able to reconstruct the maps on the basis of the text alone: he describes in book I how to draw a map of the inhabited world and lists longitudes and latitudes of principal cities and geographical features in books II–VII. Book VIII describes the breakdown of the world map into twenty-six individual maps of smaller areas. Ptolemy tells us that the Geography is based, for its factual content, on a similar recent work by Marinus of Tyre. But it seems to have improved on Marinus' work (for which the Geography is the sole source of our knowledge) in several ways. From I, 7–17 (in which various factual errors of Marinus are corrected), it appears that the bulk of Marinus' text was topographical description (giving, for instance, distances and directions between places), and that this was supplemented by lists of places with the same longest daylight and of places the same distance (in hours) from some standard meridian (book VIII of the Geography, which looks as if it is a remnant of an earlier version, uses a system similar to the latter). Ptolemy was probably the first to employ systematically listings by latitude and longitude. Here, as always, he shows a sound sense of what would be of most practical use to the reader.

Ptolemy also criticizes Marinus' map projection, a system of rectangular coordinates in which the ratio of the unit of longitude to that of latitude was 4:5. Ptolemy objects that this system distorts distances except near the latitude of Rhodes (36°). While accepting such a system for maps covering a small

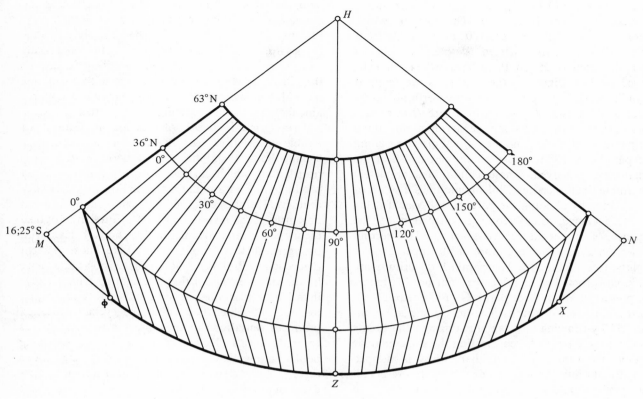

FIGURE 9

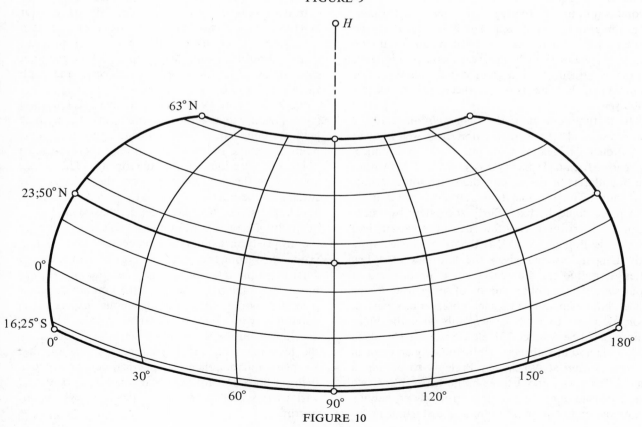

FIGURE 10

area, for the world map he proposes two alternative projection systems (I, 21–24). The known world, according to Ptolemy, covers 180° in longitude from his zero meridian (the Blessed Isles) and in latitude stretches from 16;25° south to 63° north. In his first projection (see Figure 9) the meridians are mapped as radii meeting in a point *H* (*not* the north pole), the parallels of latitude as circular arcs with *H* as center. Distances are preserved along the meridians and along the parallel of Rhodes, and the ratio of distances along the parallel of 63° to those along the parallel of the equator is preserved. These conditions completely determine the projection. Ptolemy modifies it south of the equator by dividing the parallel *MZN* as if it lay at 16;25° *north*, thus avoiding distortion at the expense of mathematical consistency. The second projection aims to achieve more of the appearance of a globe (see Figure 10). The parallels of latitude are again constructed as circular arcs, but now distances are preserved along three parallels: 63° north, 23;50° north, and 16;25° south. The meridians are constructed by drawing circular arcs through the points on these three parallels representing the same angular distance from the central meridian, which is mapped as a straight line along which distances are preserved. The first projection is (except for the modification south of the equator) a true conic projection; the second is not, but for the segment covered by the map is a remarkably good approximation to the true conic projection that was later developed from it (the Bonne projection, which preserves distances along all parallels). Ptolemy took a giant step in the science of mapmaking, but he had no successor for nearly 1,400 years.

The factual content of the *Geography* naturally is inaccurate. Only the Roman empire was well known; and Ptolemy's idea of the outline of, for instance, southern Africa or India is grossly wrong. Even within the boundaries of the empire there are some serious distortions. This was inevitable: although the latitude of a place could be fixed fairly accurately by astronomical observation, very few such observations had been made. Ptolemy says (I, 4) that of his predecessors only Hipparchus had given the latitudes of a few cities. Longitudinal differences could be determined from simultaneous observations of an eclipse at two places, but again almost no such observations existed. Ptolemy seems to have known only one, the lunar eclipse of 20 September 331 B.C., observed at Arbela in Assyria and at Carthage. Unfortunately an error in the observation at Arbela led Ptolemy to assume a time difference of about three hours between the two places instead of about two; this error was probably a major factor in the most notorious distortion in his map, the excessive length of the Mediterranean. In the absence of anything resembling a modern survey, Ptolemy, like his predecessors, had to rely on "itineraries" derived from milestones along the main roads and the reports of merchants and soldiers. In view of the inaccuracies in lengths, and especially directions, inevitable in such works, the *Geography* is a remarkable factual as well as scientific achievement.

The *Optics*, in five books, is lost in Greek. An Arabic translation was made from a manuscript lacking book I and the end of book V; from this translation, which is also lost, Eugenius of Sicily produced the extant Latin translation in the twelfth century. Despite the incompleteness and frequent obscurity of the text, the outlines of Ptolemy's optical theory are clear enough. The lost book I dealt with the general theory of vision. Like most ancient theoreticians Ptolemy believed that vision takes place by means of a "visual flux" emanating from the eye in the form of a cone-shaped bundle of "visual rays," the apex of which lies within the eyeball; this flux produces sensations in the observer when it strikes colored objects. References back in the surviving books show that besides enunciating the above theory, Ptolemy demonstrated that vision is propagated in a straight line and determined the size of the visual field (both probably by experiments).

In book II he deals with the role of light and color in vision (he believes that the presence of light is a "necessary condition" and that color is an inherent quality of objects); with the perception of the position, size, shape, and movement of objects; and with various types of optical illusion.

Books III and IV treat the theory of reflection ("catoptrics," to use the ancient term). First, three laws are enunciated: (1) the image appears at some point along the (infinite) line joining the eye to the point of reflection on the mirror. (2) The image appears on the perpendicular from the object to the surface of the mirror. (3) Visual rays are reflected at equal angles. The laws are then demonstrated experimentally. There follows a remarkable discussion on the propriety of assimilating binocular to monocular vision in geometric proofs. Ptolemy incidentally determines the relationships between the images seen by the left and right eyes and the composite image seen by both, using an ingenious experimental apparatus with lines of different colors. He then develops from the three laws a series of theorems on the location, size, and appearance of images, first for plane mirrors, then for spherical convex mirrors, then (in book IV) for spherical concave mirrors and various types of "composite" (such as cylindrical) mirrors.

Book V deals with refraction. The phenomenon is demonstrated by the experiment (at least as old as Archimedes) of the coin in the vessel that appears when water is poured in. There follows a very interesting experiment to determine the magnitude of refraction from air to glass, from water to glass, and from air to water. Ptolemy does this experiment by means of a disk with a graduated circumference. For each pair of media he tabulates the angles of refraction corresponding to angles of incidence of 10°, 20°, · · ·, 80°. The results cannot be raw observations, since the second differences are constant in all cases; but they are remarkably close to what one can derive from Snell's law $\frac{\sin i}{\sin \rho} =$ const., with a suitable index of refraction. The rest of the surviving part of the book is devoted to discussion of astronomical refraction, the relationship between the amount of refraction and the density of the media, and the appearance of refracted images.

It is difficult to evaluate Ptolemy's achievement in the *Optics* because so little remains of his predecessors' work. In "pure optics" we have only the work of Euclid (*ca.* 300 B.C.), consisting of some elementary geometrical theorems derived from a few postulates that are a crude simplification of the facts of vision. In catoptrics we have a corrupted Latin version of the work of Hero (*ca.* A.D. 60) and a treatise compiled in late antiquity from authors of various dates, falsely attributed to Euclid. From these we get only an occasional glimpse of Archimedes' catoptrics, which was probably highly original, particularly in its use of experiment. The establishment of theory by experiment, frequently by constructing special apparatus, is the most striking feature of Ptolemy's *Optics*. Whether the subject matter is largely derived or original, the *Optics* is an impressive example of the development of a mathematical science with due regard to the physical data, and is worthy of the author of the *Almagest*.

A work on music theory ('Αρμονικά), in three books, deals with the mathematical intervals (on a stretched string) between notes and their classification according to various traditional Greek systems. It seeks a middle ground between the two schools of the Pythagoreans and the followers of Aristoxenus, of whom the former, according to Ptolemy (I, 2), stressed mathematical theory at the expense of the ear's evidence, while the latter did the reverse. Again we see Ptolemy's anxiety to erect a theory that is mathematically satisfactory but also takes due account of the phenomena. According to the commentary of Porphyry (late third century), the *Harmonica* is mostly derivative, especially from the work of one Didymus

(first century).[30] We have no means of checking this statement.

Ptolemy's philosophical standpoint is Aristotelian, as is immediately clear from the preface to the *Almagest*. But it is clear from his astronomy alone that he does not regard Aristotle as holy writ; and influences from later philosophy, notably Stoicism, have been detected. An insignificant philosophical work entitled Περὶ κριτηρίου καὶ ἡγεμονικοῦ ("On the Faculties of Judgment and Command") goes under his name. There is nothing in its contents conflicting with Ptolemy's general philosophical position, but the style bears little resemblance to his other works; and the ascription, while generally accepted, seems dubious. Ptolemy's high reputation in later times caused a number of spurious works, mostly astrological, to be foisted on him. An example is the Καρπός (*Centiloquium*, in its Latin translation), a collection of 100 astrological aphorisms.

Ptolemy was active in almost every mathematical science practiced in antiquity, but several of his works are known to us only from references in ancient authors. These are a work on mechanics (Περὶ ῥοπῶν), in three books; a work in which he attempted to prove Euclid's parallel postulate; *On Dimension* (Περὶ διαστάσεως), in one book, in which he "proved" that there can be no more than three dimensions; and *On the Elements* (Περὶ τῶν στοιχείων).

In estimating Ptolemy's stature and achievement as a scientist, it is unfortunately still necessary to react against the general tendency of nineteenth-century scholarship to denigrate him as a mere compiler of the scientific work of his predecessors. This extreme view, exemplified in the writings of Delambre,[31] is no longer held by anyone competent but still persists in handbooks. It need not be refuted in detail. With a candor unusual in ancient authors Ptolemy freely acknowledges what parts of his theory he owes to Hipparchus. To say that he is lying when he claims other parts as his own work is a gratuitous slander, which, when it can be tested by independent evidence (which is rare), has proved false. It is certain that a great part of the theory in the *Almagest* is his personal contribution, and it is unlikely that the situation was radically different in all his other scientific work. On the other hand, his was not an original genius: his method was to take existing theory and to modify and extend it so as to get good agreement with observed facts. In this method, however, Ptolemy was no different from the vast majority of scientists of all periods; and he in no way deserves the reputation of a hack. His work is remarkable for its blend of knowledge, ingenuity, judgment, and clarity. The authority that it achieved in several fields is not surprising.

The *Almagest* was the dominant influence in theoretical astronomy until the end of the sixteenth century. In antiquity it became the standard textbook almost immediately. Commentaries to it were composed by Pappus (*fl.* 320) and Theon of Alexandria, but neither they nor any other Greek advanced the science beyond it. It was translated into Arabic about 800; and improved translations were made during the ninth century, notably in connection with the astronomical activity patronized by Caliph al-Ma'mūn. The Islamic astronomers soon recognized its superiority to what they had derived from Persian and Hindu sources; but since they practiced observation, they also recognized the deficiencies in its solar theory. An example of the influence of Ptolemy on Islamic astronomy and its improvements on his theory is the *Zīj* of al-Battānī (*ca.* 880). The first part of this work is closely modeled on the *Almagest*, the second on the *Handy Tables* (also translated earlier). Al-Battānī greatly improves on Ptolemy's values for the obliquity of the ecliptic, the solar mean motion (and hence precession), the eccentricity of the sun, and the longitude of its apogee. He substitutes the sine function (derived from India) for the chord function. Otherwise his work is mostly a restatement of Ptolemy's. This is typical of Islamic astronomy: the solar theory was refined (so that even the proper motion of the apogee of the sun was enunciated by al-Zarqāl *ca.* 1080), but Ptolemy's lunar and planetary theories were accepted as they stood. Such attempts as were made to revise them were based not on observation but on philosophical objections, principally to the equant. Alternative systems, preserving uniform circular rotation, were devised by Naṣīr al-Dīn al-Ṭūsī (*fl.* 1250) and his followers at Maragha, and also by Ibn al-Shāṭir (*fl.* 1350). The latter's lunar model is in one sense a real improvement over Ptolemy's, since it avoids the exaggerated variation in the lunar distance. But the influence of these reformers was very small (except for a hypothetical transmission from Ibn ash-Shāṭir to Copernicus, who adopted almost identical models).

The *Almagest* became known in western Europe through Gerard of Cremona's Latin translation from the Arabic in 1175 (a version made in Sicily from the Greek *ca.* 1160 seems to have been little known). The arrival from Islamic sources of this and other works based on Ptolemy led to a rise in the level of Western astronomy in the thirteenth century, but until the late fifteenth serious attempts to make independent progress were sporadic and insignificant. The first major blow at the Ptolemaic system was Copernicus' *De revolutionibus* (1543). Yet even this work betrays in its form and in much of its content the overwhelming influence of the *Almagest*. However

great the cosmological implication of the change to a heliocentric system, Copernican astronomy is cast in a firm Ptolemaic mold. This influence was not broken until Tycho Brahe realized that the "reform of astronomy" must be based on systematic accurate observation. So he compiled the first star catalog since Ptolemy based entirely on independent observation (earlier star catalogs, such as the famous one of al-Ṣūfī, epoch 964, had been mostly repetitions of Ptolemy's with the addition of a constant of precession; star coordinates independent of Ptolemy, such as those in the list of the *Zīj al-Mumtaḥan* compiled at the order of al-Ma'mūn, are very few in number); Brahe was also able to make the first real improvements in lunar theory since Ptolemy. Most important, he provided Kepler with the essential material for his treatise on Mars, justifiably entitled *Astronomia nova*. Kepler's work finally made the *Almagest* obsolete except as a source of ancient observations.

Ptolemy's system for establishing the exact distances and sizes of the heavenly bodies was also enthusiastically adopted in Islamic astronomy. One of the numerous adaptations of it found its way to the West in the twelfth-century Latin translation of the epitome of the *Almagest* by al-Farghānī (*fl.* 850). In this version it became part of the standard world picture of the later Middle Ages (for instance, in Dante's *Divina commedia*) and persisted into the sixteenth century, although its basis was removed by Copernicus.

The *Tetrabiblos* went through a process of transmission to the Islamic world and then to western Europe similar to that of the *Almagest* and also enjoyed an immense reputation in both, although never so unrivaled. The *Geography* was also translated into Arabic early in the ninth century. Its inadequacy for much of the territory subject to Islam was immediately obvious, and so it was soon replaced by revisions more accurate for those parts. The earliest is the *Kitāb Ṣūrat al-Arḍ* of al-Khwārizmī (*ca.* 820). The original Arabic translation of the *Geography* is lost, but its direct and indirect influence on Islamic geographical works is considerable. The *Geography* did not reach western Europe until the fifteenth century (in a Latin translation from the Greek by Jacobus Angelus, *ca.* 1406), when exploration was already making it obsolete. Nevertheless it became very popular, since it was still the best guide to much of the known world; most cartographic publications of the fifteenth and sixteenth centuries are based on it. Ptolemy's two map projections were an important stimulus to the development of cartography in the sixteenth century. The first is reflected in Mercator's map of Europe (1554), the second in various maps beginning with one by Bernardus Sylvanus in his 1511 edition of the *Geography*, which uses an equivalent of the Bonne projection.[32]

The *Optics* had little direct influence on western Europe. Eugenius' version was known to Roger Bacon and probably to Witelo in the later thirteenth century, but they are exceptions. It did however, inspire the great optical work of Ibn al-Haytham (*d.* 1039). The latter was an original scientist who made some notable advances (of which the most important was his correct explanation of the role of light in vision), but the form and much of the content of his *Kitāb al-Manāẓir* are taken from Ptolemy's work; and the inspiration for his remarkable experiments with light must surely also be attributed to Ptolemy. His treatise was translated into Latin and is the basis of the *Perspectiva* of Witelo (*ca.* 1270), which became the standard optical treatise of the later Middle Ages. The indirect influence of the *Optics* persisted until the early seventeenth century.

Ptolemy's treatise on musical theory never attained the authority of his other major works, since rival theories continued to flourish. But it was extensively used by Boethius, whose work was the main source of knowledge of the subject in the Latin West; and hence some of Ptolemy's musical doctrine was always known. When it became available to western Europe in Greek, it was no more than a historical curiosity. But it had a strange appeal for one great scientist. The last three chapters of book III of the *Harmonica* are missing; they dealt with the relationships between the planetary spheres and musical intervals. Kepler intended to publish a translation of book III, with a "restoration" of the last chapters and a comparison with his own kindred speculations, as an appendix to his *Harmonice mundi*. The appendix never appeared, but the whole work is a tribute to his predecessor.

NOTES

1. *Almagest* XI, 5; IX, 7 (Manitius, II, 228, 131).
2. Boll, "Studien," p. 53.
3. *Opera astronomica minora*, p. 155. An alternative MS reading is "the fifteenth year" (152/153).
4. Olympiodorus, . . . *In Platonis Phaedonem* . . ., Norwin ed., p. 59, l. 9.
5. Boll, *op. cit.*, pp. 54–55.
6. Buttmann, "Ueber den Ptolemäus in der Anthologie . . .," pp. 483 ff.
7. II, 13 (Manitius, I, 129).
8. I, 3 (Manitius, I, 9).
9. III, 59; Lejeune ed., p. 116.
10. V, 23–30; Lejeune ed., pp. 237–242.
11. E.g., X, 1 (Manitius, II, 156).
12. See Pappus, *Collectio*, VI, intro.; Hultsch ed., p. 474, with Hultsch's note *ad. loc.*
13. Aristotle, *Metaphysica*, *Λ*, 1073b17 ff.; Simplicius, *In Aristotelis De caelo*, Heiberg ed., pp. 491 ff.

14. *Almagest* XII, 1 (Manitius, II, 267 ff.).
15. *Almagest* IX, 2 (Manitius, II, 96–97).
16. Theon, *Commentary on the Almagest*, I, 10; Rome ed., p. 451. See Toomer, "The Chord Table . . .," pp. 6–16, 19–20.
17. Kugler, *Die Babylonische Mondrechnung*, pp. 4 ff.
18. See Toomer, "The Size of the Lunar Epicycle . . .," pp. 145 ff.
19. The first to compute and draw the path of an eclipse seems to have been Cassini I in 1664; see Lalande, *Astronomie*, II, 358.
20. Shown by A. Ricius, *De motu octavae sphaerae*, f. 39 (following Levi ben Gerson); see also Laplace, *Exposition*, p. 383.
21. Delambre, *Histoire de l'astronomie ancienne*, II, 250 ff.
22. By Vogt in "Versuch einer Wiederherstellung von Hipparchs Fixsternverzeichnis."
23. E.g., Peters and Knobel, *Ptolemy's Catalogue of Stars*, *passim*.
24. See *Almagest* VII, 4 (Manitius, II, 30).
25. See Caspar's intro. to his trans. of Kepler's *Neue Astronomie*, pp. 60*–61*.
26. IX, 1 (Manitius, II, 93–94).
27. Vitruvius, *De architectura* IX, 7.
28. Suidas, "Πτολεμαῖος ὁ Κλαύδιος"; Adler ed., IV, 254.
29. Synesius, *Opuscula*, Terzaghi ed., p. 138.
30. Porphyry, *Commentary on the Harmonica*; Düring ed., p. 5.
31. E.g., *Histoire de l'astronomie ancienne*, II, *passim*.
32. See Hopfner in Mžik, *Des Klaudios Ptolemaios Einführung*, p. 105.

BIBLIOGRAPHY

I. ORIGINAL WORKS. *Almagest*: The standard text is that of J. L. Heiberg, *Claudii Ptolemaei Opera quae exstant omnia*, I, *Syntaxis mathematica*, 2 pts. (Leipzig, 1898–1903). The only reliable trans. is the German one by K. Manitius: *Ptolemäus, Handbuch der Astronomie*, 2nd ed. (with corrections by O. Neugebauer), 2 vols. (Leipzig, 1963). For detailed discussion of the contents the best guide is still J. B. J. Delambre, *Histoire de l'astronomie ancienne*, II (Paris, 1817; repr. New York–London, 1965). Delambre's notes in the otherwise antiquated edition by N. Halma, *Composition mathématique de Claude Ptolémée*, 2 vols. (Paris, 1813–1816), are also useful. Pappus' lemmas to the Μικρὸς ἀστρονομούμενος are in Bk. VI (474–633) of his Συναγωγή, *Pappi Alexandrini Collectionis quae supersunt*, F. Hultsch, ed., 3 vols. (Berlin, 1875–1878; repr. Amsterdam, 1965). No satisfactory account exists of the early history of Greek astronomy; the best available is P. Tannery, *Recherches sur l'histoire de l'astronomie ancienne* (Paris, 1893). On the astronomical system of Eudoxus see G. V. Schiaparelli, "Le sfere omocentriche di Eudosso, di Calippo e di Aristotele," in *Memorie dell'Istituto lombardo di scienze e lettere*, Classe di Scienze Matematiche e Naturali, **13**, no. 4 (1877), repr. in his *Scritti sulla storia della astronomia antica*, II (Bologna, 1926), 5–112. The account by Simplicius is in his commentary on Aristotle's *De Caelo*, *Commentaria in Aristotelem Graeca*, J. L. Heiberg, ed., VII (Berlin, 1894), 491–510. Examples of early works on "spherics" are those by Autolycus (late fourth century B.C.), edited by J. Mogenet, *Autolycus de Pitane*, which is *Université de Louvain, Recueil de travaux* *d'histoire et de philologie*, 3rd ser., fasc. 37 (Louvain, 1950); and Menelaus' *Spherics* (available only in Arabic translation), M. Krause, ed., in *Abhandlungen der Akademie der Wissenschaften zu Göttingen*, Phil.-hist. Kl., 3rd ser., **17** (1936).

On the trigonometry of Ptolemy and Hipparchus see G. J. Toomer, "The Chord Table of Hipparchus and the Early History of Greek Trigonometry," in *Centaurus*, **18** (1973), 6–28. The cuneiform evidence for the Babylonian origin of Hipparchus' lunar periods is given in F. X. Kugler, *Die Babylonische Mondrechnung* (Freiburg im Breisgau, 1900), pp. 4–46. On the difference between the maximum lunar equations of Ptolemy and Hipparchus see G. J. Toomer, "The Size of the Lunar Epicycle According to Hipparchus," in *Centaurus*, **12** (1967), 145–150. On Hipparchus' determination of the lunar and solar distances see N. Swerdlow, "Hipparchus on the Distance of the Sun," in *Centaurus*, **14** (1969), 287–305. Information on the earliest mappings of the path of a solar eclipse is in Jérôme le Français (Lalande), *Astronomie*, 3rd ed., II (Paris, 1792), 357 ff. Augustinus Ricius gave the correct explanation of the error in Ptolemy's determination of the amount of precession in his *De motu octavae sphaerae*, 2nd ed. (Paris, 1521), f. 39. The same explanation was given independently by P. S. Laplace in his *Exposition du système du monde*, 4th ed. (Paris, 1813), p. 383.

A critical trans. of Ptolemy's star catalog containing detailed comparisons with modern computed positions was made by C. H. F. Peters and E. B. Knobel, *Ptolemy's Catalogue of Stars, a Revision of the Almagest* (Washington, D.C., 1915). This adopts Delambre's erroneous conclusion about Ptolemy's dependence on the (hypothetical) star catalog of Hipparchus, which was refuted by H. Vogt in "Versuch einer Wiederherstellung von Hipparchs Fixsternverzeichnis," in *Astronomische Nachrichten*, **224** (1925), 17–54. The star catalog of al-Ṣūfī was published by the Osmania Oriental Publications Bureau: Abu'l-Ḥusayn 'Abd al-Raḥmān al-Ṣūfī, *Ṣuwaru' l-Kawākib* (Hyderabad-Deccan, 1954). A French trans. was made by H. C. F. C. Schjellerup, *Description des étoiles fixes* (St. Petersburg, 1874). The star list of the *Zīj al-Mumtaḥan*, containing 24 items, was published in J. Vernet, "Las 'Tabulae probatae,' " in *Homenaje a Millás-Vallicrosa*, II (Barcelona, 1956), 519. For derivatives from this list see P. Kunitzsch, "Die arabische Herkunft von zwei Sternverzeichnissen," in *Zeitschrift der Deutschen morgenländischen Gesellschaft*, **120** (1970), 281–287.

A description of the Ptolemaic system is given by O. Neugebauer, *The Exact Sciences in Antiquity*, 2nd ed. (Providence, R.I., 1957), pp. 191–206. This should be consulted for such omissions from my account as the details of the Mercury model. Literature comparing Ptolemaic with modern theory will be found *ibid.*, pp. 182–183. A good mathematical presentation of Ptolemy's theory of celestial motion is given by N. Herz, *Geschichte der Bahnbestimmung von Planeten und Kometen*, I (Leipzig, 1887), 86–169. A mathematical comparison between the Kepler ellipse and a Ptolemaic equant model for Mars was made by M. Caspar in the intro. to his trans. of Kepler's

Astronomia nova: Johannes Kepler, *Neue Astronomie* (Munich–Berlin, 1929). For an analysis of the problem of determining the eccentricity from three observations in an equant model see G. W. Hill, "Ptolemy's Problem," in *Astronomical Journal*, **21** (1900), 33–35. An account of the astronomical instruments described by Ptolemy in the *Almagest*, and also of the astrolabe, is given by D. J. Price in *A History of Technology*, C. Singer and E. J. Holmyard, eds., III (Oxford, 1957), 582–609.

For details of the Arabic translations of the *Almagest* and other works of Ptolemy see M. Steinschneider, *Die arabischen Übersetzungen aus dem Griechischen* (Graz, 1960), (191)–(211). An extensive work by P. Kunitzsch on the history of the *Almagest* in Arabic and Latin translation is *Der Almagest—Die Syntaxis Mathematica des Claudius Ptolemäus in arabische-lateinischer Überlieferung* (Wiesbaden, 1974). The astronomical treatise of al-Battānī was published in a masterly ed. by C. A. Nallino, "Al-Battani sive Albatenii Opus astronomicum," *Pubblicazioni del R. Osservatorio astronomico di Brera in Milano*, **40**, 3 pts. (1899–1907). An account of the "improvements" to the Ptolemaic system made by al-Ṭūsī and other late Islamic astronomers, with references to further literature, is given by E. S. Kennedy in *The Cambridge History of Iran*, V (Cambridge, 1968), 668–670. The earliest Latin version of the *Almagest* is discussed in C. H. Haskins, *Studies in the History of Mediaeval Science* (Cambridge, Mass., 1924; repr. New York, 1960), pp. 157–165. Other versions are mentioned *ibid.*, pp. 103–112. An ed. of Copernicus' major work was published by F. and C. Zeller, *Nikolaus Kopernikus Gesamtausgabe*, 2 vols. (Munich, 1944–1949)—vol. I is a facsimile of the MS of the *De revolutionibus* in Copernicus' own hand; vol. II an unreliable "critical" ed. of the text. Brahe's star catalog of 1598 (published only after his death) is in *Tychonis Brahe Opera omnia*, J. L. E. Dreyer, ed., III (Copenhagen, 1916), 335–389. For details of Brahe's other star catalogs refer to the index, *ibid.*, XV (1929). His improvements to the lunar theory are discussed in V. E. Thoren, "Tycho Brahe's Discovery of the Variation," in *Centaurus*, **12** (1968), 151–166. For Kepler's *Astronomia nova* see the ed. by M. Caspar, pub. as vol. III of Johannes Kepler, *Gesammelte Werke*, W. von Dyck and Caspar, eds. (Munich, 1937).

The remains of Pappus' commentary on the *Almagest* were published by A. Rome, *Commentaires de Pappus et de Théon d'Alexandrie sur l'Almageste*, I (Rome, 1931), Studi e Testi, no. 54. He also published Theon's commentary to Bks. I and II, and that to Bks. III and IV, in the same series: no. 72 (1936) and no. 106 (1943). For the remainder of Theon's commentary one must still consult the *editio princeps* of the Greek text of the *Almagest: Claudii Ptolemaei Magnae constructionis . . . lib. xiii, Theonis Alexandrini in eosdem commentariorum lib. xi* (Basel, 1538).

Handy Tables: The only printed version—except for Ptolemy's intro., which is on pp. 159–185 of Heiberg's ed. of *Claudii Ptolemaei Opera quae exstant omnia*, II, *Opera astronomica minora* (Leipzig, 1907)—is the execrable ed. by N. Halma, *Tables manuelles astronomiques de Ptolémée et de Théon*, 3 pts. (Paris, 1822–1825). Pt. 1 contains Theon's smaller commentary on the tables (a larger one exists only in MS). B. L. van der Waerden makes some useful remarks on the parameters and use of the tables in "Bemerkungen zu den Handlichen Tafeln des Ptolemaios," in *Sitzungsberichte der Bayerischen Akademie der Wissenschaften zu München*, Math.-nat. Kl., **23** (1953), 261–272; and "Die Handlichen Tafeln des Ptolemaios," in *Osiris*, **13** (1958), 54–78.

Planetary Hypotheses: The Greek text was published by Heiberg in Ptolemy's *Opera astronomica minora*, pp. 70–106, together with a German trans. of the Arabic by L. Nix, *ibid.*, pp. 71–145. By a strange oversight this omits the end of Bk. I (missing from the Greek), which contains Ptolemy's theory of the absolute distances of the heavenly bodies; and thus Ptolemy's authorship of the latter theory remained unknown in modern times until the complete Arabic text (with English trans. of the part omitted by Nix) was published by B. Goldstein, "The Arabic Version of Ptolemy's Planetary Hypotheses," in *Transactions of the American Philosophical Society*, n.s. **57**, no. 4 (1967), 3–55. For an account of that theory and its medieval developments see N. Swerdlow, "Ptolemy's Theory of the Distances and Sizes of the Planets," Ph.D. thesis (Yale University, 1968). The text through which it reached the West was published, in its most widespread version, by F. J. Carmody, "Alfragani differentie in quibusdam collectis scientie astrorum" (Berkeley, 1943), multigraphed.

Phaseis, Analemma, Planisphaerium: The texts are printed in *Opera astronomica minora*, pp. 3–67, 189–223, and 227–259, respectively. On the *Phaseis* see H. Vogt, "Der Kalender des Claudius Ptolemäus (Griechische Kalender, ed. F. Boll, V)," in *SB Heidelberg. Akad. d. Wiss.*, in Phil.-hist. Kl. (1920), 15. Because of its excellent commentary, the *editio princeps* of the *Analemma* is still worth consulting: *Claudii Ptolemaei liber de Analemmate a Federico Commandino Urbinate instauratus* (Rome, 1562). See especially P. Luckey, "Das Analemma von Ptolemäus," in *Astronomische Nachrichten*, **230** (1927), 17–46. Vitruvius' description of an *Analemma* construction is in his *De architectura* IX, 7; V. Rose, ed. (Leipzig, 1899), pp. 230–233. The Arabic text of the *Planisphaerium* has never been printed (the *Opera astronomica minora* contains only the medieval Latin trans.); but it exists, in a version prior to Maslama's revision, in MS Istanbul, Aya Sofya 2671₃. There are related texts in MS Paris, Bibliothèque Nationale Ar. 4821, 69v ff.; see G. Vajda, "Quelques notes sur le fonds des manuscrits arabes de la Bibliothèque nationale," in *Rivista degli studi orientali*, **25** (1950), 7–9. On the relevance of the *Planisphaerium* to the astrolabe see O. Neugebauer, "The Early History of the Astrolabe," in *Isis*, **40** (1949), 240–256. The work of Synesius referring to Hipparchus and the astrolabe is in *Synesii Cyrenensis Opuscula*, N. Terzaghi, ed. (Rome, 1944), pp. 132–142.

Tetrabiblos: Critical text by F. Boll and A. Boer, *Claudii Ptolemaei Opera quae exstant omnia*, III, 1, *ΑΠΟΤΕΛΕΣΜΑΤΙΚΑ* (Leipzig, 1957). There is a text with English trans. by F. E. Robbins in Loeb Classical Library (Cambridge, Mass., 1940). On various aspects of the *Tetrabiblos*, Boll, "Studien," pp. 111–218, is useful. For

the content A. Bouché-Leclercq, *L'astrologie grecque* (Paris, 1899; repr. Brussels, 1963), remains unsurpassed.

Geography: There is no complete modern ed. For a complete text one must still use C. F. A. Nobbe, *Claudii Ptolemaei Geographia*, 2 vols. (Leipzig, 1843–1845; repr. in 1 vol., Hildesheim, 1966); but more satisfactory for the parts they contain are the eds. by F. W. Wilberg and C. H. F. Grashof, 6 fascs. (Essen, 1838–1845), Bks. I–VI; and by C. Müller, 2 vols. (Paris, 1883–1901), Bks. I–V. For other partial eds. see the article by E. Polaschek in Pauly–Wissowa (see below) and W. H. Stahl, *Ptolemy's Geography: A Select Bibliography* (New York, 1953). A German trans. of Bk. I by H. von Mžik, *Des Klaudios Ptolemaios Einführung in die darstellende Erdkunde*, Klotho 5 (Vienna, 1938), contains a good study of the projections by F. Hopfner, pp. 87–105. The geographical work of al-Khwārizmī was edited by H. von Mžik, *Das Kitāb Ṣūrat al-Arḍ des abū Ğaʿfar Muḥammad ibn Mūsā al-Ḫuwārizmī* (Leipzig, 1926). C. A. Nallino gave a classic account of its relationship to Ptolemy in "Al-Khuwārizmī e il suo rifacimento della Geografia di Tolomeo," in *Atti dell' Accademia nazionale dei Lincei. Memorie*, Classe di Scienze Morali, Storiche e Filologiche, 5th ser., **2**, no. 1 (1894), 3–53, repr. in his *Raccolta di scritti editi e inediti*, V (Rome, 1944), 458–532. Excellent reproductions of maps found in MSS of the *Geography* were published in *Claudii Ptolemaei Geographiae codex Urbinas graecus 82 phototypice depictus*, pts. I, 2, and II (Leiden–Leipzig, 1932). Pt. I, 1, by J. Fischer, contains (pp. 290–415) an account of early European cartographic publications influenced by the *Geography*. Bernardus Sylvanus' map is at the end of his *Claudii Ptholemaei Alexandrini liber geographiae* (Venice, 1511).

Optics: This was first printed by G. Govi, *L'Ottica di Claudio Tolomeo* (Turin, 1885). A. Lejeune provided an excellent critical ed.: *L'Optique de Claude Ptolémée, Université de Louvain, Receuil de travaux d'histoire et de philologie*, 4th ser., fasc. 8 (Louvain, 1956). The same author gave a good analysis of most parts of the work in two publications: *Euclide et Ptolémée, deux stades de l'optique géométrique grecque, ibid.*, 3rd ser., fasc. 31 (Louvain, 1948); and "Recherches sur la catoptrique grecque," in *Mémoires de l'Académie r. de Belgique. Classe des sciences*, **52**, no. 2 (1957). Euclid's *Optics* and the pseudo-Euclidean *Catoptrics* are in *Euclidis opera omnia*, J. L. Heiberg, ed., VII (Leipzig, 1895). The Latin trans. of Hero's *Catoptrics* was edited by W. Schmidt in *Heronis Alexandrini Opera quae supersunt omnia*, II, 1 (Leipzig, 1900). The Arabic text of Ibn al-Haytham's *Optics* has never been printed (an English trans. and ed. are being prepared by A. Sabra). The Latin trans. is in *Opticae thesaurus*, F. Risner, ed. (Basel, 1572), which also contains Witelo's optical work. For Ibn al-Haytham's experiments see M. Schramm, *Ibn al-Haythams Weg zur Physik*, which is *Boethius*, I (Wiesbaden, 1963). References to Ptolemy's *Optics* by Roger Bacon are found in several parts of his *Opus maius*, J. H. Bridges, ed., 2 vols. (Oxford, 1900), see index under "Ptolemaeus."

Harmonica: Text published by I. Düring, "Die Harmonielehre des Klaudios Ptolemaios," which is *Göteborgs högskolas årsskrift*, **36**, no. 1 (1930). The same author published Porphyry's commentary, *ibid.*, **38**, no. 2 (1932); and a German trans. of Ptolemy's work, with commentary on both works, "Ptolemaios und Porphyrios über die Musik," *ibid.*, **40**, no. 1 (1934). Boethius' *De institutione musica* was edited by G. Friedlein together with his *De institutione arithmetica* (Leipzig, 1877; repr. Frankfurt, 1966). Kepler's *Harmonice mundi* was published as vol. VI of his *Gesammelte Werke* by M. Caspar (Munich, 1940).

Doubtful, spurious, and lost works: Περὶ κριτηρίου καὶ ἡγεμονικοῦ was published by F. Lammert in *Claudii Ptolemaei Opera quae exstant omnia*, III, 2 (Leipzig, 1961). The same vol. contains an ed. of the Καρπός by A. Boer. On Ptolemy's philosophical position see Boll, "Studien," pp. 66–111. Fragments and testimonia to the lost works are printed in Heiberg's ed. of the *Opera astronomica minora*, pp. 263–270.

II. SECONDARY LITERATURE. *General*: B. L. van der Waerden *et al.*, "Ptolemaios 66," in Pauly-Wissowa, XXIII, 2 (Stuttgart, 1959), 1788–1859, is a good guide. The supplementary article on the *Geography* by E. Polaschek, *ibid.*, supp. X (1965), 680–833, is useful only for its bibliography.

Life: The evidence is assembled and discussed by F. Boll, "Studien über Claudius Ptolemäus," in *Jahrbücher für classische Philologie*, supp. **21** (1894), 53–66. This takes some note of the Arabic sources, which I omitted since they add nothing credible to the Greek evidence. The only formal biographical notice (wretchedly incomplete) is in the tenth-century Byzantine lexicon of Suidas ("the Suda"), *Suidae Lexicon*, Ada Adler, ed., IV (Leipzig, 1935), 254, no. 3033. The "Canobic Inscription" is printed in Heiberg's ed. of the *Opera astronomica minora*, pp. 149–155. The work of Olympiodorus was published by W. Norvin, *Olympiodori philosophi in Platonis Phaedonem commentaria* (Leipzig, 1913). The origin of the appellation "Phelud(i)-ensis" was first correctly explained by J. J. Reiske in the German trans. of B. d'Herbelot's *Bibliothèque orientale: Orientalische Bibliothek*, II (Halle, 1787), 375. It was thence repeated by Philip Buttmann, "Ueber den Ptolemäus in der Anthologie und den Klaudius Ptolemäus," in *Museum der Alterthums-Wissenschaft*, **2** (1810), 455–506. An exhaustive discussion is given by P. Kunitzsch, *Der Almagest* (above).

G. J. TOOMER

PUISEUX, VICTOR (*b.* Argenteuil, Val-d'Oise, France, 16 April 1820; *d.* Frontenay, Jura, France, 9 September 1883), *mathematics, mechanics, celestial mechanics*.

Puiseux spent his youth in Lorraine, where his father, a tax collector, was posted in 1823. He was educated at the Collège de Pont-à-Mousson and, from 1834, at the Collège Rollin in Paris, where he attended C. Sturm's course in special mathematics. After winning the grand prize in physics (1836) and mathe-

matics (1837) in the *concours général*, he was admitted in 1837 to the École Normale Supérieure. There he became friends with his future colleagues Briot and Bouquet. In 1840 Puiseux placed first in the *agrégation* in mathematics and then spent an additional year at the École Normale Supérieure as *chargé de conférences*, completing his training and preparing a dissertation in astronomy and mechanics, which he defended 21 August 1841.

Puiseux was professor of mathematics at the royal college of Rennes (1841–1844) and at the Faculty of Sciences of Besançon (1844–1849). During this period he published about ten articles on infinitesimal geometry and mechanics in Liouville's *Journal de mathématiques pures et appliquées*. In 1849 he was called to Paris as *maître de conférences* of mathematics at the École Normale Supérieure, a post he held until 1855 and again from 1862 to 1868.

In addition to his teaching duties, for several years Puiseux attended Cauchy's courses and became one of his closest followers. Under this fruitful influence, Puiseux wrote several important memoirs on the theory of functions of a complex variable before turning to celestial mechanics. In 1857, having substituted for various professors, including the astronomer Jacques Binet at the Collège de France and Sturm and Le Verrier at the Faculty of Sciences, Puiseux succeeded Cauchy in the chair of mathematical astronomy at the latter institution. He retained this post until 1882, publishing several important memoirs. Brief tenures as director of the Bureau de Calculs at the Paris observatory (1855–1859) and at the Bureau des Longitudes (1868–1872) permitted him to display his mastery of the techniques of astronomical computation. In 1871 he became a member of the mathematics section of the Académie des Sciences, succeeding Lamé.

In 1849 Puiseux married Laure Jeannet; of their six children only Pierre and André survived childhood; both became astronomers. An austere teacher and tireless worker, Puiseux devoted himself to the education of his children, was active in various catholic organizations, and took a passionate interest in botany and alpinism. He was, in fact, a pioneer in the latter sport and in 1848 was the first to scale one of the peaks (now bearing his name) of Mount Pelvoux.

Puiseux's scientific work encompassed infinitesimal geometry, mechanics, mathematical analysis, celestial mechanics, and observational astronomy. His first publication (1841), his doctoral dissertation, dealt with the invariability of the major axes of the planetary orbits and with the integration of the equations of motion of a system of material points. Although well-executed, the work lacked great originality. Similarly, his papers on infinitesimal geometry, most of which were published at the beginning of his career, attested his analytic virtuosity but constituted a rather limited contribution to the subject—notwithstanding his discovery of new properties of evolutes and involutes. The most interesting among these papers pertain to questions related to mechanics: the motion of the conical pendulum, tautochrones, a generalization of the top problem, and the study of the apparent movements of the surface of the earth.

In 1850 and 1851, however, Puiseux accomplished much more original work, developing, correcting, and completing major aspects of the theory of functions of a complex variable that had been elaborated by Cauchy. Examining functions of a complex variable z defined by an algebraic equation of the form $f(u,z) = 0$, Puiseux succeeded in separating the various branches and in formulating the expansions in corresponding series. He clearly distinguished, for the first time, the different types of singular points (poles, essential points, and branch points); determined the integrals of algebraic differentials over the paths of integration; specified the "mode of existence of non-uniform functions" (C. Hermite); and pointed out the applications of series containing fractional powers of the variable. Despite its intrinsic interest, Puiseux's theory was surpassed in 1857 when Riemann, in his *Theorie der Abelschen Funktionen*, approached the topic from a topological point of view and introduced the famous "Riemann surfaces." Puiseux subsequently turned to the study of celestial mechanics and astronomy and virtually never returned to his theory.

Following Cauchy, Puiseux sought to apply the most recent mathematical methods to the fundamental problems of celestial mechanics. His papers on the series expansions of the perturbation function, on long-term inequalities in planetary motions, and on related questions constitute an elaboration and refinement of earlier work by Cauchy. After presenting the lucid exposition "Sur les principales inégalités du mouvement de la lune" (*Annales scientifiques de l'École Normale Supérieure*, **1** [1864], 39–80), Puiseux took up the difficult problem of the acceleration of the mean motion of the moon. Although Laplace (1787) thought he could explain this phenomenon by the secular decrease in the eccentricity of the orbit of the earth, J. Adams showed in 1853 that Laplace's theory accounted for only half of the observed effect. After extensive calculations, Puiseux established (*Journal de mathématiques pures et appliquées*, 2nd ser., **15** [1870], 9–116) that the secular displacement of the ecliptic had no significant influence on the acceleration. Although a purely negative conclusion, Puiseux's finding led to a better delimitation of the problem, which was investigated by G. Hill in 1877.

Puiseux was also concerned with improving the computational methods employed in basic astronomy. At the Bureau de Calculs, he directed the reduction of both the lunar observations made at Paris from 1801 to 1829 and the meridional observations of 1837–1838. After comparing the different methods available for deducing the solar parallax from the observation of the transits of Venus, Puiseux participated in the preparations carried out for the observation of the 1874 and 1882 transits; he also worked on the observations made in 1874 by French astronomers. During his brief tenure at the Bureau des Longitudes, he served as principal editor of the *Connaissance des temps ou des mouvements célestes*.

BIBLIOGRAPHY

I. ORIGINAL WORKS. Puiseux's only separately printed publication was his dissertations, *Sur l'invariabilité des grands axes des orbites des planètes, thèse d'astronomie* . . . *Sur l'intégration des équations du mouvement d'un système de points matériels, thèse de mécanique* (Paris, 1841).

Forty-one articles and memoirs published between 1842 and 1880 are cited in the Royal Society *Catalogue of Scientific Papers*, V, 39–40; VIII, 672–673; IX, 77; and XII, 592. Most of these are also listed in Poggendorff, II, 542; and III, 1076. A summary of Puiseux's first publications is given in his *Notice sur les travaux scientifiques de M. Victor Puiseux* (Paris, 1856) and in two later, undated eds., the latter of which appeared in 1871.

II. SECONDARY LITERATURE. The chief biographical accounts of Puiseux are E. Glaeser, in *Biographie nationale des contemporains* (Paris, 1878), 620; C. Vapereau, in *Dictionnaire universel des contemporains*, 5th ed. (Paris, 1880), 1485; P. Gilbert, in *Revue des questions scientifiques*, **15** (1884), 5–37; J. Bertrand, "Notice lue à l'Académie des Sciences le 5 mai 1884," in *Bulletin des sciences mathématiques*, 2nd ser., **8**, pt. 1 (1884), pp. 227–234, repr. in *Mémoires de l'Académie des sciences de l'Institut de France*, 2nd ser., **44** (1888), lxvii–lxxviii; and in J. Bertrand, *Éloges académiques* (Paris, 1890), 275–285; J. Tisserand, in *Bulletin des sciences mathématiques*, 2nd ser., **8**, pt. 1 (1884), 234–245; and M. d'Ocagne, in *Histoire abrégée des sciences mathématiques* (Paris, n.d. [1955]), 283–284.

Information on certain aspects of Puiseux's career and work can be found in M. Chasles, *Rapport sur les progrès de la géométrie* (Paris, 1870), 180–182; and in J. Tannery, "L'enseignement des mathématiques à l'École," in *Le centenaire de l'École normale (1795–1895)* (Paris, 1895), 391–392.

RENÉ TATON

PULFRICH, CARL (*b.* Strässchen [near Burscheid], Solingen, Germany, 24 September 1858; *d.* Baltic Sea, near Timmendorferstrand, Germany, 12 August 1927), *physics.*

Pulfrich was an outstanding representative of the school of physics created by Ernst Abbe and a prominent expert in the field of photometry and refractometry. He is considered the father of stereophotogrammetry.

The son of a schoolteacher, Pulfrich attended the Realschule in Mülheim an der Ruhr and then studied physics, mathematics, and mineralogy at the University of Bonn. In 1881 he passed an examination enabling him to teach in secondary schools and in 1882 earned the doctorate under Clausius for his dissertation, "Photometrische Untersuchungen über die Absorption des Lichtes in isotropen und anisotropen Medien." From 1883 to 1889 he was an assistant to Clausius and then, after the latter's death, to Hertz. In 1891 he married Hertz's sister-in-law.

At Bonn, Pulfrich investigated the refraction of light in crystals, glasses, and fluids and constructed a total reflectometer and a refractometer. On the basis of this research he qualified in 1888 as a university lecturer; and in 1890 he met Abbe, who was then employed at the Zeiss Works in Jena. In the same year (1890) Pulfrich accepted an offer to work as a physicist, under Abbe's direction, for the Zeiss company, where he was made head of the department of optical measuring instruments, a post he held for many years.

Pulfrich's first task was to improve some of the instruments that Abbe had devised, including the micrometer, dilatometer, and Abbe refractometer for liquids. In 1895 he designed, for use by chemists, an improved model of the latter device, which became known as "the Pulfrich." He also constructed, in collaboration with R. Wollny, a butterfat refractometer; and he was instrumental in the development of heatproof glasses, which were manufactured at the Jena glassworks of Schott and Associates.

In 1899, at a scientific congress in Munich, Pulfrich presented the first model of a stereoscopic rangefinder with a staggered scale. He subsequently devoted himself completely to stereophotogrammetry. He introduced the so-called floating mark and constructed new surveying equipment and auxiliary devices, including the stereo comparator and stereo copying machine. (For these devices he employed the ideas of E. von Orel.) In 1925 Pulfrich devised a step photometer adapted to the range of sensitivity of the human eye. This photometer also finds application as a colorimeter and turbidimeter.

BIBLIOGRAPHY

I. Original Works. Pulfrich wrote approximately 100 papers and books. His major works include "Ein neues Totalreflectometer (I. Mitteilung)," in *Annalen der Physik und Chemie*, n.s. **30** (1887), 193–208, and *Zeitschrift für Instrumentenkunde*, **7** (1887), 16–27; "Das Totalreflectometer und seine Verwendbarkeit für weisses Licht (II. Mitteilung)," *ibid.*, **30** (1887), 487–502, and *ibid.*, **7** (1887), 55–65; "Das Totalreflectometer (III. Mitteilung)," *ibid.*, **31** (1887), 724–736, and *ibid.*, **7** (1887), 392–396; "Ein neues Refraktometer, besonders zum Gebrauch für Chemiker eingerichtet," in *Zeitschrift für Instrumentenkunde*, **8** (1888), 47–53; *Über das Totalreflektometer und das Refraktometer für Chemiker, ihre Verwendung in der Krystalloptik und zur Untersuchung der Lichtbrechung von Flüssigkeiten* (Habilitationsschrift, Univ. of Bonn, 1888–1890; published in book form by W. Engelmann, Leipzig, 1890); and "Universalapparat für refraktometrische und spektrometrische Untersuchungen," in *Zeitschrift für Instrumentenkunde*, **15** (1895), 389–394.

Subsequent works are "Über ein neues Refraktometer mit veränderlichem brechenden Winkel," *ibid.*, **19** (1899), 335–339; "Über den von der Firma Carl Zeiss in Jena hergestellten stereoskopischen Entfernungsmesser. Vortrag, gehalten auf der Naturforscherversammlung in München am 19. September 1899," in *Physikalische Zeitschrift*, **1** (1899), 98; "Über eine Prüfungstafel für stereoskopisches Sehen," in *Zeitschrift für Instrumentenkunde*, **21** (1901), 249–260; "Ueber neuere Anwendungen der Stereoskopie und über einen hierfür bestimmten Stereo-Komparator," *ibid.*, **22** (1902), 65–81, 133–141, 178–192, 229–246; "The Stereoscope," in *Encyclopaedia Britannica*, 11th ed. (1910); "Die 'drehbare wandernde Marke,' eine Neueinrichtung am Stereo-Komparator," in *Zeitschrift für Instrumentenkunde*, **34** (1914), 221–223.

Pulfrich's books on stereoscopy are *Neue stereoskopische Methoden und Apparate für die Zwecke der Astronomie, Topographie und Metronomie*, pt. 1 (Berlin, 1903); *Neue stereoskopische Methoden und Apparate* (Berlin, 1909); *Stereoskopisches Sehen und Messen, mit einem Literaturverzeichnis der Arbeiten über Stereoskopie etc. seit 1900* (Jena, 1911); *Über die Photogrammetrie aus Luftfahrzeugen und die ihr dienenden Instrumente* (Jena, 1919); and *Die Stereoskopie im Dienste der Photometrie und Pyrometrie* (Berlin, 1923).

Pulfrich's photometer was described in "Über ein den Empfindungsstufen des Auges tunlichst angepasstes Photometer, Stufenphotometer genannt, und über seine Verwendung als Farbmesser, Trübungsmesser, Kolloidometer, Kolorimeter und Vergleichsmikroskop," in *Zeitschrift für Instrumentenkunde*, **45** (1925), 35–44, 61–70, 109–120; and "Über einen Zusatzapparat zum Stufenphotometer, der zur bequemeren Ermittelung des Farbtones einer Farbvorlage dient, und daran anschliessend Forderung eines gleichabständigen Farbenkreises," *ibid.*, **45** (1925), 521–530.

Pulfrich also wrote on "Methoden zur Bestimmung von Brechungsindizes," in *Handbuch der Physik*, ed. by A. Winkelmann, **2** (1893), 302–344 (1st ed.); and **2** (1906), 583–625 (2nd ed.). A bibliography is given in Poggendorff, IV, 1199–1200; V, 1010–1011; VI, 2092; and VIIa, supp. 517, in obituaries below.

II. Secondary Literature. On Pulfrich and his work, see O. Lacmann, "Prof. Dr. Pulfrich zum Gedächtnis," in

Internationales Archiv für Photogrammetrie, **7**, pt. 2 (1932), 1–7, with a selected bibliography on Pulfrich's stereoscopic papers; F. Löwe, "Carl Pulfrich," in *Zeitschrift für Instrumentenkunde*, **47** (1927), 561–567, with a complete bibliography, "Carl Pulfrich," in *Vierteljahrsschrift der Astronomischen Gesellschaft*, **63** (1928), 7–12, "Prof. Dr. Carl Pulfrich. Ein Gedächtnisblatt zur 25. Wiederkehr seines Todestages (12. August 1927)," in *Monatsschrift für Feinmechanik und Optik*, **69** (1952), 147–148; F. Manek, "Pulfrich und der erste Stereoautograph Mod. 1908," in *Jenaer Jahrbuch* (1958), pt. 2, 7–18, "Pulfrich und der Stereoautograph Modell 1909," *ibid.* (1959), pt. 2, 7–23; "Pulfrich und die Stereoautographen Modell 1911 und Modell 1914," *ibid.* (1960), pt. 2, 327–340; and F. Schneider, "Carl Pulfrich. Prof. Dr. Carl Pulfrich zu seinem 100. Geburtstag am 24. September 1958," in *Jenaer Rundschau*, **3** (1958), 127–128.

HANS-GÜNTHER KÖRBER

PUMPELLY, RAPHAEL (b. Owego, New York, 8 September 1837; d. Newport, Rhode Island, 10 August 1923), *geology, geography, ethnology, archaeology.*

Pumpelly received his early education in private schools and in the Owego Academy, following which he attended the New Haven Collegiate and Commercial Academy for two years. He had thought to enter Yale, but in 1854 went instead to Europe with his mother. In the course of this tour, which lasted for several years, his interest in geology started; in his memoirs (*My Reminiscences*, published in 1928) he gave an account of a journey in 1855 through the Rhine Valley to the Drachenfels and the Laacher See, adding "Truly I was entering, though gropingly, into geology through the gate of romance." He met von Roemer, and, subsequently, visited Switzerland, France, Italy, and Corsica and observed the grandeur of their mountains, volcanoes, and solfataras. In Vienna he by chance met Emil Nöggerath, who advised him to enroll at the Freiberg Bergakademie.

Pumpelly entered the Bergakademie in 1856, and studied mining engineering and geology with Weisbach, Breithaupt, and von Cotta, who had particular influence upon him. He remained at Freiberg until 1860, a period that he interrupted to return to Corsica for several months. Pumpelly's memorialist and friend, Bailey Willis, suggests that it was during these years that Pumpelly formulated his chief investigative technique, that of drawing hypotheses from the two vantage points of sequential and multiple analysis.

Upon his return to the United States in 1860, Pumpelly received his first professional assignment when he was put in charge of the development of the Santa Rita silver mine in southern Arizona. The area was under constant siege by Apache Indians, and the smelting operations were performed hurriedly; Pumpelly nevertheless remained for a year, and conducted an independent geological study of the area, the results of which were presented before a meeting of the California Academy of Sciences in August 1861.

While in California Pumpelly received an appointment from the government of Japan to undertake a survey of the mineral resources of the Japanese empire and immediately sailed for that country, where he remained for about one year. In 1862 he went on to China, where he studied coal deposits and visited Peking. In 1865 he returned to West China, traveling overland through Mongolia, Siberia, and St. Petersburg; he summed up his experiences of almost five years of travel in the Orient in two works, "Geological Researches in China, Mongolia, and Japan" (1867) and "Across America and Asia" (1870). His adventurous spirit is apparent in both, particularly in his accounts of meeting the resistance of foreign officials and of the physical hardship of travels to the edge of the Tibetan desert.

Several months after his return to the United States, Pumpelly was offered the chair of mining at Harvard University, which he never occupied formally because the salary was too low. Since he was more interested in exploration and fieldwork, he turned instead to a study of the copper and iron deposits of the region surrounding Lake Superior, which he continued, with interruptions, from 1866 until 1877. In 1869 he settled in Cambridge, Massachusetts, and married Eliza Frances Shepard there in the same year.

From 1871 until 1872 Pumpelly served as geologist to the state of Missouri but resigned because of ill health. He then went to live in Balmville, New York, where he began work on the problem of the origin of metal-bearing ores. He returned to Michigan at intervals, and his major report on the copper-bearing rocks of that region was published in 1873. This paper represented pioneering work in two ways; in it Pumpelly both employed thin sections in petrographic examinations (according to B. Willis he was probably the first to do so in the United States) and recognized for the first time that copper was indigenous to the series of rocks in which it occurred. He did, however, continue to believe that copper was accumulated from the host rocks by the action of outside factors, most notably that of seawater, since the indigenous aqueous magmatic phase, that is, the hydrothermal or deuteric water, was not given its full importance at that time.

Pumpelly's contribution toward solving the problem of the origin of the Chinese loess—and of loess deposits in general—is also especially noteworthy. His "Relation of Secular Rock-Disintegration to Loess, Glacial Drift, and Rock Basins" was published in the *American Journal of Science* in 1879. In his pioneering study, F.P.W. von Richthofen had recognized the transport and deposition of loess by wind; to this Pumpelly added the notion that the deep decay of rocks in moist climates supplied the original material of loess. Von Richthofen himself acknowledged the value of Pumpelly's results.

From 1879 until about 1890 Pumpelly was concerned with organizing and directing a series of official geological surveys. He was first engaged in a survey, carried out in connection with the census of the mineral resources of the United States, exclusive of precious metals and petroleum. This investigation lasted until 1881, when Pumpelly undertook an economic survey of the resources and properties along the route of a projected northwestern railway, called the Northern Transcontinental Survey. His biographer Willis served on his staff.

In 1884 Pumpelly was placed in charge of the New England section of the United States Geological Survey. He investigated the Precambrian terrains of the region, especially the structure of the Green Mountains, largely in cooperation with C. R. Van Hise and R. D. Irving. His report, "Geology of the Green Mountains in Massachusetts," was published in 1894 and represented the application to this subject of all of Pumpelly's vast experience with ongoing geological processes. In it he discussed, among other things, the changes in facies that had previously been explained as faulting, tracing the fossil record of metamorphosed rocks. His great knowledge of the foreign literature is exemplified by his application of Albert Heim's principles of orogenic mechanisms to the New England areas. This was his last work in the public service.

From 1893 until 1895 Pumpelly traveled with his family in Europe, visiting mainly Italy, France, and Switzerland. Returning to the United States, he then worked on various consulting tasks (including some in the Lake Superior area) until 1903, when he embarked on the great exploration project that took him twice to Central Asia, in that year and the following. He journeyed to Turkestan with his son, Raphael W. Pumpelly, and with W. M. Davis and Ellsworth Huntington, under the auspices of the Carnegie Institution of Washington, D.C. The chief purpose of these expeditions was to determine "the physical basis of the human history," and to this end they combined geology, archaeology, and ethnology. In connection with this research Pumpelly also visited major museums in Europe and Russia, and made a side trip to Egypt to study the rate of growth of village mounds.

Elected president of the Geological Society of America in 1905, Pumpelly chose for his address (delivered in Ottawa in 1906) a major topic arising from his Central Asian researches, "Interdependent Evolution of Oases and Civilizations." He then spent three years completing the reports on this work, which appeared as the general and geological account "Archeological and Physico-Geographical Reconnaissance in Turkestan," Carnegie Institution Publication No. 26, *Explorations in Turkestan* . . . 1903 (1905), and the ethnologically oriented "Ancient Anau and the Oasis World," Carnegie Institution Publication No. 73, *Explorations in Turkestan* . . . 1904 (1908).

The results of the work of Pumpelly and his team included their proof that in several glacial periods High Asia had been covered by the ice that also covered most of Russia. The ice then withdrew, leaving an inland sea, of which the shores are still visible, as Pumpelly reported in 1908. The earliest men in the region, he went on, must have settled on the edges of the post-glacial lakes that remain only as the Transcaspian oases; he pointed out evidence of communities, agriculture, and domestic animals. The desiccation of the region then forced the migration of its early inhabitants, and they spread out in various directions, into Europe among others. Although Pumpelly had long been interested in the origin of the Aryans, and although he wrote a section on this question, he chose to suppress it, since he thought that the observations that he and his colleagues had made on this point were too meager to justify any conclusion. (This restraint in publishing is typical of his career, as was his willingness in the course of these and other studies to encourage his co-workers to publish their own work over their own names.)

Pumpelly's work in general—in geology, geography, and ethnology—may best be understood as a reflection of his personal approach to research. He insisted on the separation of observation and interpretation; he was firm in his belief that observation must be objective. He made a number of original investigations and utilized a number of new techniques (as, for example, his use of the microscope to examine thin sections of rocks in his petrographic studies of the Lake Superior region), but was often content to leave the description of details to others. He chose to reach a working hypothesis from two sides, through sequential analyses and multiple analyses, a method derived from his student years under von Cotta and Breithaupt. In his fairness and generosity in his dealings with his colleagues, he was an example to others. He was particularly skilled in organizing and conducting interdisciplinary research, perhaps impelled by his inner need to study man and mankind, rather than specific scientific problems divorced from the totality of life.

Pumpelly spent the last twelve years of his life at his winter and summer residences in Newport, Rhode Island, and Dublin, New Hampshire. His wife died in 1915; he survived her for eight years, and was himself survived by one son and two daughters. In 1925 a mineral from the Lake Superior copper district was named "pumpellyite" in his honor.

BIBLIOGRAPHY

I. ORIGINAL WORKS. Pumpelly's major works include "Geological Researches in China, Mongolia, and Japan in 1862–1865," which is in *Smithsonian Contributions to Knowledge*, **15** (1867); "The Paragenesis and Derivation of Copper and Its Associates on Lake Superior," in *American Journal of Science*, 3rd ser., **2** (1874), 188–198, 243–258, 347–355; "Copper District," in *Geological Survey of Michigan; Upper Peninsula, 1869–1873*, I, pt. 2 (New York, 1873); "Metasomatic Development of the Copper-bearing Rocks of Lake Superior," in *Proceedings of the American Academy of Arts and Sciences*, (1878), 253–310; "The Relation of Secular Rock Disintegration to Loess, Glacial Drift, and Rock Basins," in *American Journal of Science*, 3rd ser., **17** (1879), 133–144; "The Relation of Secular Rock Disintegration to Certain Transitional Crystalline Schists," in *Bulletin of the Geological Society of America*, **2** (1891), 209–223; and *My Reminiscences*, 2 vols. (New York, 1918).

II. SECONDARY LITERATURE. The best work on Pumpelly and his career is Bailey Willis's memorial in *Bulletin of the Geological Society of America*, **36** (1925), 45–84, with portrait and bibliography, and his shorter art. in *American Journal of Science*, ser. 5, **6** (1923), 375–376.

G. C. AMSTUTZ

PUNNETT, REGINALD CRUNDALL (*b*. Tonbridge, England, 20 June 1875; *d*. Bilbrook, Somerset, England, 3 January 1967), *morphology, genetics*.

Punnett was the elder son of George Punnett, the head of a Tonbridge building firm, and Emily Crundall. The family was comfortably bourgeois, conservative, and churchgoing. In 1913 he married Eveline Maude Froude, widow of Sidney Nutcombe-Quicke and daughter of John Froude Bellew; they had no children.

Having graduated from a preparatory school in Tonbridge, Punnett, in 1889, went to Clifton, where his chief interest was the classics. He then obtained a scholarship at Gonville and Caius College, Cambridge,

and registered as a medical student. For the first part of the natural science tripos he studied zoology, human anatomy, and physiology, but found the first of these so attractive that he decided to become a zoologist. For the second part of the tripos he offered zoology, with human physiology as his subsidiary subject, and gained a first in the examination and the Walsingham Medal in 1898. With the aid of the Shuttleworth studentship at Caius, he spent six months at the Zoological Station in Naples and a short time at Gegenbaur's laboratory in Heidelberg. In 1899 he went to St. Andrews University as demonstrator and part-time lecturer in the natural history department. In 1901 he was elected a fellow of Caius and in the following year returned to Cambridge as demonstrator in morphology in the department of zoology. He was Balfour student from 1904 to 1908, was awarded the Thurston Medal, and in 1909 was appointed superintendent of the Museum of Zoology. In 1910 he succeeded Bateson as professor of biology at the University of Cambridge. When this impermanent chair became adequately endowed and was transformed into the Arthur Balfour chair of genetics (the first such chair in the United Kingdom), Punnett became its first occupant. He held this post until his retirement in 1940. Punnett was elected a fellow of the Royal Society in 1912 and was awarded its Darwin Medal in 1922. He was a founding member of the (British) Genetical Society and was one of its secretaries from 1919 to 1930, when he became president. He was also an honorary member of the Genetical Society of Japan and of the Poultry Science Association of America.

While still an undergraduate, Punnett became keenly interested in nemertines; and he pursued this interest at the Naples Zoological Station, at the Marine Biological Station in Plymouth, at the Marine Laboratory at St. Andrews University, and at Nordgaard's laboratory in Bergen. Punnett wrote about a dozen papers on the morphology of these worms and his name was given to *Cerebratulus punnetti* and *Punnettia splendida* Stiasny-Wijnhoff (Keferstein).

In 1901, while convalescing from an appendectomy, Punnett became interested in a report that sex ratios can be influenced by diet; and he decided to test this hypothesis experimentally by using mice. On his return to Cambridge in 1902, he wrote to Bateson, who was conducting Mendelian experimentation in Grantchester, and asked if his proposed nutritional experiment might be tested to yield information concerning the inheritance of coat color. Although he learned that such investigations were already well advanced, this contact with Bateson became the turning point in Punnett's career. In 1903 an anonymous friend offered Bateson £150 a year, for two years, to continue his work. He invited Punnett to join him. Their collaboration was to last six years and produced many notable and enduring contributions to the nascent science of genetics. During the period 1904–1910 they confirmed several basic discoveries of classical Mendelian genetics, including the Mendelian explanation of sex determination, sex linkage, complementary factors and factor interaction, and the first example of autosomal linkage. For these experiments Bateson and Punnett used the domestic fowl and the sweet pea. Punnett also used the rabbit, and in addition to his hybridization work he was greatly interested in mimicry in the butterfly. In 1911 Bateson and Punnett launched the *Journal of Genetics*, which they edited jointly until Bateson's death in 1926. Thereafter Punnett became sole editor until 1946, when he was succeeded by J. B. S. Haldane.

During World War I, Punnett served in the Food Production Department of the Board of Agriculture (he was an authority on the scientific aspects of poultry breeding), and it was here that he conceived using sex-linked plumage-color factors in the fowl to reduce wastage. Feed stuffs were in short supply and the vast majority of male chicks were unwanted. If the sexes could be distinguished among day-old chicks the unwanted males could be destroyed and considerable economy achieved. He suggested various crosses involving such characters as silver and gold, barred and nonbarred; and his method was soon practiced on a huge scale, involving the production of millions of sex-linked chicks every year.

After the war the National Poultry Institute was founded and the Genetical Institute at Cambridge was selected as the center for research in poultry breeding. Michael Pease joined Punnett as research assistant; and the most dramatic production of this partnership was the Cambar, the first auto-sexing breed of poultry. This breed was synthesized by transferring the X-borne gene for barring (*B*) from the Barred Rock to the Golden Campine. Ten years later he produced the Legbar, which resulted from the introduction of the *B* gene into the genotype of the Brown Leghorn. The production of these autosexing breeds was a fine piece of biological engineering. But in 1930, when the authorities decided that this work should be greatly expanded, Punnett withdrew; and a separate center for poultry research was established with Pease as its director. Thereafter Punnett worked alone. He subsequently left Cambridge and went to the village of Bilbrook, where he continued his experimental work with fowl until 1955, when his incubator house was destroyed by fire. Thus, more than half a century of quiet, method-

ical, and highly intelligent Mendelian experimentation ended.

During the period 1899–1904 Punnett worked alone on various problems. But he was principally concerned with meristic variation in elasmobranchs; the morphology and systematics of nemertines; and, toward the end of this period, sex determination. He became both an authority on nemertines and an expert morphologist. Then, encouraged by Bateson, he abruptly shifted his interest to experimental biology and applied his previous experience to hybridization experiments, which confirmed and extended the findings of Mendel. Punnett defended Mendelism against the criticisms of various biometricians and zoologists, and he was instrumental in introducing genetics to the general public, and especially to fanciers and commercial breeders of livestock. He also made significant contributions to knowledge of the genetics of the fowl, the duck, the rabbit, the sweet pea, and of man; and his first book, *Mendelism* (1905), did much to make the science known. His *Heredity in Poultry* (1923) remained the standard work on the subject for nearly thirty years, and our present knowledge of poultry genetics has evolved largely from Punnett's early investigations. In 1940 he produced a map of the X chromosome of the fowl showing the tentative arrangement of seven genes. His last paper, which was concerned with recessive, black plumage color, appeared in 1957. Data from nearly thirty years of experimentation with the sweet pea were consolidated in his paper on linkage; and in it he showed that eighteen known recessive mutations fell into seven groups, seven being the haploid number of chromosomes in the sweet pea.

Punnett was a gentle, quiet, cultured man who never strove for prominence or priority. The slow rate of reproduction of the fowl and of the sweet pea suited his temperament. He was a Mendelian throughout his career and remained largely unaffected by the development of the theory of the gene, of cytogenetics, and of biometrical genetics. He belonged to the first group of scientists who carried forward the revolution in biological thought that was inspired by the rediscovery of Mendel's work.

BIBLIOGRAPHY

A complete bibliography of Punnett's writings is given in *Biographical Memoirs of Fellows of the Royal Society*, **13** (1967), 323–326. His scientific interests can be clearly discerned in the long series of papers on poultry genetics that appeared from 1911 to 1967 in the *Journal of Genetics*.

F. A. E. CREW

PUPIN, MICHAEL IDVORSKY (*b*. Idvor, Banat [now Yugoslavia], 4 October 1858; *d*. New York, N.Y., 12 March 1935), *applied physics*.

Pupin was born to a family of unlettered Serbian settlers in the Banat, a military buffer zone between the Ottoman and Austro-Hungarian empires. Because of his obvious gifts, his parents were encouraged to let him complete his secondary studies at a larger center. He was thus sent to Prague, where he stayed for more than a year; but before he was sixteen, he went alone to America, arriving in New York in 1874. During the next five years he worked at odd jobs on farms and in factories, studying at night to prepare himself for admission to Columbia University on a scholarship, an ambition he fulfilled in 1879.

Pupin graduated with distinction in 1883 and after additional study at the University of Cambridge went to Berlin, where he worked under Helmholtz and G. Kirchhoff, receiving the doctorate in 1889 with a dissertation on osmotic pressure. He then returned to Columbia to teach mathematical physics in the newly formed department of electrical engineering. He advanced rapidly and in 1901 was made professor of electromechanics, a post he occupied until his retirement in 1931.

During his studies of the distortions that arise when iron is magnetized by an alternating current, Pupin developed electrical resonators (by analogy with resonators used to study complex sound waves) that proved to be applicable to problems in telegraphy and telephony. His most important contribution grew out of a study of the electrical analogue of a vibrating string "loaded" at regular intervals. This work not only confirmed that the periodic insertion of inductance coils in telephone lines would improve their performance by reducing attenuation and distortion, but it also allowed him to calculate optimum coil size and spacing, an invention of considerable practical and commercial value. For a time such lines were called "pupinized."

Pupin also made many other contributions of an applied nature, for instance, in X-ray fluoroscopy, design of early radio transmitters, and electrical network theory. He was a popular and outstanding teacher. Among his pupils were several of the pioneers of radio communications, the most notable of whom was E. H. Armstrong. Pupin also became prominent in public affairs and was an adviser to the Yugoslav delegation to the Paris Peace Conference in 1919. He was an accomplished writer; and his best-selling autobiography, *From Immigrant to Inventor*, received the Pulitzer Prize in 1924. He received many honors, including eighteen honorary degrees, and was elected to the National Academy of Sciences. The physics laboratory at Columbia University is named in his honor.

In 1888 Pupin married a young widow, Sarah Katherine Jackson; she died in 1896. The couple had one daughter, Varvara.

BIBLIOGRAPHY

Pupin's publications follow his entries in Poggendorff, VI, 2094; and *Biographical Memoirs. National Academy of Sciences*, **19** (1938), 307–323. The latter (by Bergen Davis) also lists his many honors and thirty-four patents. Other biographical entries include *National Cyclopaedia of American Biography*, XXVI (1937), 5–6; and *Dictionary of American Biography*, XXI, supp. 1 (1944), 611–615.

CHARLES SÜSSKIND

PURKYNĚ (PURKINJE), JAN EVANGELISTA (*b*. Libochovice, Bohemia [now Czechoslovakia], 17 December 1787; *d*. Prague, Bohemia, 28 July 1869), *physiology, histology, embryology, education*.

Purkyně's name (usually spelled Purkinje, a form he adopted so as to have it pronounced correctly by German speakers) is known today in the eponyms Purkyně cells (in the cerebellum), Purkyně fibers (of the heart), Purkyně (or Purkyně-Sanson) images, Purkyně's phenomenon (shift in the relative apparent brightness of red and blue in dim light), and Purkyně's tree (the shadows of the retinal vessels). He was a versatile scholar with wide-ranging interests and an exceptional ability to observe, mainly subjective sensory phenomena and minute morphological structures. After 1850 Purkyně was concerned mainly with the role that knowledge and science should play in the life of his nation.

Purkyně's father was manager of an estate of Prince Dietrichstein in northern Bohemia. He stimulated interest in and knowledge of nature in his eldest son, although he died when Jan was only six. The local schoolteacher and parson helped the talented boy, who at the age of ten was admitted as a choirboy to a Piarist monastery on another of the Dietrichstein estates, at Mikulov (Nikolsburg) in southern Moravia, near the Austrian border. Initially handicapped because he knew only Czech, Purkyně soon learned both languages of instruction, German and Latin, and became one of the best students.

When he had completed his secondary education, Purkyně took orders and, after a year of novitiate, began teaching in a Piarist school at Strážnice, Moravia. In 1806 he was sent to Litomyšl in eastern Bohemia to continue his education at the Piarist Philosophical Institute, the obligatory preparation for "higher" university studies (theology, law, medicine). The writings of contemporary philosophers, however, mainly Fichte's *Über die Bestimmung des Gelehrten*, led him to abandon an ecclesiastical career (1807) and earn a meager living by tutoring while he completed his philosophical studies in Prague. At that time he attempted his first research, in physics: an analysis of "acoustic waves," ingeniously fixed on small vibrating glass plates. Lack of guidance, however, prevented him from achieving any significant results; but he did gain a good grounding in physics that was very valuable for his later work in biology. His most influential teacher was the philosopher and mathematician Bernard Bolzano.

After three years at the estate of Blatná (south of Prague) as tutor to the son of the owner, Baron Hildprandt, Purkyně began to study medicine at Prague, planning a career in science rather than in the practice of medicine. Before completing these studies, Purkyně, inspired by the pedagogical work and ideas of J. H. Pestalozzi and P. E. von Fellenberg, as well as by Novalis' *Lehrlinge zu Sais*, entertained the idea of founding an institution for education of future scientists.

Purkyně's main interest, however, was physiology; and a physiological topic—the subjective visual phenomena—was the subject of his inaugural dissertation (1818). He began his academic career as prosector and assistant in anatomy at Prague; but his liberal, nonconformist thinking and affiliations doomed to failure his attempts to obtain a permanent appointment. With the help of the Prussian surgeon general, J. N. Rust, and on the recommendation of the influential Berlin professor K. A. Rudolphi, who recognized his abilities, in 1823 he was appointed professor of physiology at Breslau, against the will of the Faculty of Medicine. He soon overcame the initial hostility, won the respect and friendship of his colleagues, and became one of the best-known teachers at the university. In 1827 he married Rudolphi's daughter Julia, whose death in 1835 left him with two young sons. He did not remarry. In 1850 he returned to Prague, where he remained as professor of physiology until his death.

Purkyně's research—which included experimental pharmacology, experimental psychology, phonetics, histology, embryology, and physical anthropology—falls between the Romantic period, which in central Europe was largely influenced by Schelling, and the period of empirical physiology. He considered physiology to be a natural science based on observation and experiment.

Between 1818 and 1825 Purkyně concentrated on the subjective sensory phenomena, studying them by observation and by experiments on himself because he lacked facilities for other experimental work. He began self-observation of unusual visual sensations as an amusement in his early years but later realized that these phenomena—errors in perception, sensations with no adequate external cause, discrepancies between physical cause and evoked sensation—are not chance but have a relationship to features in the structure or function of the eye and its nerve connections with the brain, or to some abnormal influence of certain stimulations.

Purkyně observed and studied the puzzling visual sensations produced by strong intermittent illumination (the "light-shadow figure"), by pressure on the eyeball, or by galvanic stimulation. He also showed the possibility of seeing the shadows of one's own retinal vessels (Purkyně's figure or tree) when these shadows fall on the neighboring sensitive elements (for instance, when light is concentrated on a spot of the sclera). In 1855 Heinrich Müller confirmed, through the geometrical relation of the movement of these shadows to that of the light source, that the light-sensitive layer of the retina was not on the inner surface of the eyeball but deeper, at the level of the apexes of the rods and cones.

Much attention has been paid to the "Purkyně phenomenon" or "Purkyně shift" (1825), a change in the apparent relative luminosity of colors in a dim light (scotopic vision) compared with that in full daylight (photopic vision) that is due—as became known later—to different visual sensory mechanisms (the rods and the cones, respectively). He also discovered the physiological inability of peripheral parts of the retina to distinguish colors, overlooked by all previous specialists. Owing to his exceptional ability to observe himself and to concentrate on the details of sensations, he detected many phenomena that other observers went to great pains to confirm.

In contrast to his contemporaries (mainly Goethe) who made similar observations, Purkyně was aware that the subjective sensory phenomena were neither exceptions to the otherwise clear laws of nature nor a matter of chance, but that they had a physiological basis, their determinism: "The sensory organs are the finest indicators and analyzers for exploring the pertinent qualities and material relations making it possible to study the laws of the material world." He postulated that to each subjective sensation there corresponds an objective physiological process in the relevant sensory organ, and that subjective phenomena, such as the visual errors or illusions, are an appropriate means of investigating the objective truth—that is, the physiological processes in the eye and its connections with the brain. Purkyně could not explain most of these observations, but his descriptions drew attention to them and stimulated further study. Some of them are not yet understood. He also followed these ideas in his studies of other subjective phenomena, mainly the effects of drugs and the phenomena of vertigo.

During the same period Purkyně investigated the possibilities of determining the physical properties of the sensory organs. This was very original, and some of the methods he recommended in 1823—determination of the limits of the visual field (perimetry), examination of the anterior segment of the eye in oblique illumination at the focus of a converging lens and with a microscope (developed later by Gullstrand), the usage of the reflected images, and illumination of the fundus of the eye—were developed later and are used still in routine clinical examinations.

The reflex image arising from the outer surface of the cornea had long been known; but the other three images described by Purkyně are extremely faint and thus not easy to detect, so that their discovery was a brilliant achievement. Images due to reflection from the anterior and posterior surfaces of the crystalline lens were independently rediscovered by Louis Sanson fourteen years later, but that of the posterior surface of the cornea was not confirmed until fifty years later—and then only by use of a special device. Purkyně also realized the importance of these reflex images in ophthalmology, for appraising the transparency of the optic media of the eye and for determining the curvature of each reflecting (and refracting) surface. He recommended measuring the sizes of the images by use of a microscope fitted with a micrometer and comparing them with similar images of glass balls of different sizes. Later Helmholtz and others developed special instruments for this purpose. The change in the size of the second image, measured by Langenbeck in 1849, elucidated the nature of the change in the accommodation.

Most important, but overlooked and unrecognized in its time, was Purkyně's recommendation that the interior of the eye be examined in light reflected into it by a concave lens, a principle later used by Helmholtz in his ophthalmoscope (1851).

His 1823 Breslau dissertation also contained his renowned classification of the fingerprints.

From 1820 to 1827 Purkyně studied vertigo and the physiological phenomena of the maintenance of posture and equilibrium. He was attracted by the observation of Erasmus Darwin that when one stops, after rotating for a period round the body axis, the apparent motion of the surroundings changes from horizontal to vertical when the head is inclined. He investigated this observation systematically and

found that there is also an exact determinism in these subjective sensory phenomena: the direction of the apparent motion is determined by the position of the head during the rotation (Purkyně's law of vertigo). The involuntary muscular reactions of the limbs and of the eyes (nystagmus) also depend on the position of the head during the primary rotation. He pointed out that these reactions have a compensatory character, that their purpose is to oppose the apparent motion, and that they follow the law of vertigo. Purkyně also studied "galvanic vertigo," the sensation of an apparent movement toward the anode that is compensated for by a real deviation in the opposite direction. His experiments on animals, made with his pupil C. H. W. Krauss, showed the importance of the cerebellum in these reactions. Purkyně's observations and experiments complemented those of Flourens; but neither man fully grasped the implications of the other's results, and the phenomena they described remained puzzling for the next fifty years. In 1873–1874 Breuer, Mach, and Crum Brown reported almost simultaneously on the role of the vestibular receptors in the maintenance of equilibrium and orientation. In the 1820's, however, the idea that the inner ear is the organ of hearing was so firmly fixed that no one could conceive that it is also the seat of special organs for sensing and transmitting to the brain the position and movements of the head.

In 1825–1832 Purkyně studied the early development of the avian egg in the body of the female. His discovery and isolation of a minute structure, the germinal vesicle ("Purkyně's vesicle"), on the spot of the yolk where the embryo develops—later identified with the cell nucleus—formed a bridge between the large avian egg and the small ova of other animals. It also stimulated the work of K. E. von Baer that led in 1827 to the discovery of the ovum in mammals and man. Purkyně's pupil A. Bernhardt contributed to the final elucidation of Baer's interpretation in 1835.

In this period Purkyně concluded his studies of vertigo (1827) and of the effects of drugs (1829). Noteworthy are his description of visual sensations produced by toxic doses of digitalis and belladona, and his conception of physiological pharmacology. He then began research on what he called the physiology of the human language (phonetics), again mainly by observations on himself. His work greatly influenced the further studies of J. N. Czermak and E. W. Brücke. His main report, however, was lost at the Berlin Academy of Sciences and was not published until the 1970's.

Stimulated by his colleague A. W. Henschel, Purkyně studied plant structures, mainly the elastic fibrous cells of the anthers and the form of the pollen and spores in relation to the mechanism of their dispersion. In his extensive comparative study, combining anatomy with physiology, he distinguished structural types, and drew attention to mechanical factors and the role of cells in the differentiation of plant tissues. His dynamic concept was recognized mainly by French botanists.

During his first years of teaching, Purkyně thought often of physiology, both as a science and as a medical discipline. He was opposed to the speculative treatment of physiology that prevailed in central Europe at that time. He repeatedly stated that physiology is a science based on observation and experiment and, like physics and chemistry, is an experimental science. This conviction entailed practical instruction, which he began in 1824. But as his practical course and his experimental research met with difficulties and obstacles, Purkyně realized the need for an independent department of physiology; and from 1831 he fought for its establishment against indifference, lack of understanding, and hostile egotism.

In 1832 Purkyně acquired a "great, modern" achromatic microscope made by S. Plössl, one of the best instruments at that time. This was the beginning of a new period in his research (1832–1845), a patient and systematic investigation of structure as the material basis of life phenomena. He wrote to Rudolph Wagner in 1841: "With boundless eagerness I investigated within the shortest time all areas of plant and animal histology, and concluded that this new field was inexhaustible. Nearly every day brought new discoveries, and soon I felt the necessity to make others share my enhanced vision, and to take pleasure in their discoveries."

Purkyně saw in the microscope many structures that had escaped the attention of other observers; but, once described, they seemed so obvious that younger specialists could hardly believe they had remained unnoticed for so long. He also paid great attention to the preparation of tissues for examination under the microscope—fixing, sectioning, staining, and other means of making visible structures that are not seen in fresh, untreated specimens (acetic acid, for instance, makes the cell nuclei visible). Purkyně constructed a compressorium for a finely graded squeezing of tissue specimens. Later, his assistant A. Oschatz constructed the first plate microtome for cutting thin sections. For the study of bone and teeth Purkyně developed a technique of decalcification prior to sectioning, and one of grinding to thin, transparent layers. He also used amber, copal varnishes, and Canada balsam for embedding.

Purkyně's systematic and detailed studies, in which his students participated, contributed to the knowledge of the microscopic structure of the skin and its glands (sweat glands and their spiral ducts, the "granular" structure of the basal layer of the epidermis), bone (bone cells, canaliculi with concentric lamellar structure of the matrix), teeth (structure of dentin) and their development (an investigation soon pursued by several other biologists), cartilage (cells, "ground substance"), and arteries and veins. His discovery of the gastric glands (described independently by Sprott Boyd) and of the cellular structure of all other glands led Purkyně to study the digestive action of extracts of gastric and intestinal mucosae and of the pancreas, and to discover several new factors in the digestive process.

The eponym "Purkyně cells" for the large, pear-shaped bodies in the cerebellum commemorates Purkyně's investigation of the structure of the nervous system. He was the first to describe cells as ubiquitous formations in the central nervous system of vertebrates and in their ganglia—as structures that play an important role in nervous activity, ". . . elementary centres of collection, production and distribution of the force within the nervous system." He showed that nerve fibers are not hollow tubes; the sheath envelops an axis cylinder formed of an albuminous matter. With his pupil D. Rosenthal, Purkyně examined the number of nerve fibers and their distribution according to their diameters in various roots of spinal nerves and in the cerebral nerves of several animal species, in the belief that both the total number and the relative proportion of thin and thick fibers have a functional significance. This was one of the earliest endeavors in quantitative neurobiology, inspired by the idea that "nature acts according to an eternal law even in the sphere of its most delicate microscopic structures."

Histology was then regarded as a branch of physiology rather than of morphology; and Purkyně hoped that the microscope would aid in the understanding of life phenomena, a hope that began to materialize more than a century later with electron microscopes of much greater resolving power. Investigating living objects, Purkyně and Gabriel Valentin discovered the ciliary motion in higher animals and thoroughly studied its physiology and pharmacology. Purkyně was also interested in the functional structure of such muscular organs as the heart and the uterus (both gravid and nongravid). In the heart he described a special type of fibers (Purkyně fibers, 1839), flat gray threads with transverse markings (and therefore muscular) under the lining of the cavity. In sheep they can be seen with the naked eye; some continue into the muscle columns and others form bridges between them. These fibers, forming "Purkyně's network," were later shown to have the important specific function of conducting the contraction to all parts of the heart.

The chief advance in Purkyně's time was Schwann's formulation of the cell theory. Purkyně was Schwann's immediate and most important predecessor. His finding of nucleated "granules" (cells) in many animal tissues made the analogy between the basic structural elements of plants and animals more evident, and thus speculations of earlier investigators became a well-founded and sound scientific theory that had a great impact on biological work and thought.

In 1839 Purkyně opened a modest independent physiological institute in Breslau, the first of its kind. The legal act of establishing such an institute with statutes and an appropriation was a breakthrough highly appreciated by such younger contemporaries as Rudolph Wagner, Rudolf Heidenhain, and Emil du Bois-Reymond. Purkyně's perseverance in pursuing his goal and restating his arguments led to his success and drew greater attention to the essential condition for the steady advancement of life sciences. Physiological institutes were very rare until the middle of the nineteenth century, but after that their number grew until they became a regular part of medical schools.

After 1850, as professor at Prague, Purkyně devoted his energy mainly to organizing and expanding science, especially to promoting education among his Czech countrymen. The return to the vernacular in the eighteenth century led to the adoption of German at the central European universities; thus for many nationalities, including the Czechs, knowledge was the privilege of the few who learned German or another more widely spoken language. Purkyně had had to learn German and Latin in secondary school, but he believed that each nation must have easier access to knowledge. In 1853, therefore, he began publishing a Czech scientific review, *Živa*, and sought to secure the conditions for encouraging science and learning in the Czech language. Purkyně struggled with his adamant German colleagues for the acceptance of Czech as a teaching language at the University of Prague and worked out a detailed plan for a national academy. He was interested in facilitating communication, exchange of knowledge and ideas, and freedom of the press; he also recommended unification of writing in the Slav languages by the general acceptance of the Roman alphabet. Purkyně repeatedly stressed the importance of science and knowledge in practical life. His work in the last period of his life was of great importance in the Czech national revival and exerted a lasting effect on the subsequent development of science in his country.

BIBLIOGRAPHY

I. ORIGINAL WORKS. Purkyně's writings have been collected in *Opera omnia*, 12 vols. (Prague, 1919–1973); a thirteenth and last vol., containing his autobiographical works, is in preparation. His *Opera selecta* (Prague, 1848) contains several important works.

The most complete bibliography, which includes translations and multiple editions, is that by V. Kruta, *J. E. Purkyně (1787–1869) Physiologist. A Short Account of His Contributions to the Progress of Physiology With a Bibliography of His Works* (Prague, 1969). Some of his correspondence has been collected in J. Jedlička, ed., *Jana Ev. Purkyně Korespondence*, 2 vols. (Prague, 1920–1925); and in V. Kruta, *Beginnings of the Scientific Career of J. E. Purkyně. Letters With His Friends From the Prague Years 1815–1823* (Brno, 1964). Some separate publications are his correspondence with A. Retzius, in *Lychnos* (1956 and 1959); with K. E. von Baer, *ibid.* (1971–1972); with Johannes Müller, in *Nova acta Leopoldina*, **22** (1961), 213–228; and, in connection with the Berlin meeting of German scientists held in 1828, in *Sudhoffs Archiv für Geschichte der Medizin und der Naturwissenschaften*, **57** (1973), 152–170.

II. SECONDARY LITERATURE. The first biography of Purkyně was published during his lifetime: F. J. Nowakowski, *Życie i prace naukowe Jana Purkiniego* (Warsaw, 1862). An authoritative short biography is by Heidenhain, in *Allgemeine deutsche Biographie* (1888). More recent are the biography by V. Kruta and M. Teich, *Jan Evangelista Purkyně* (Prague, 1962), in Czech, with English, French, German, Russian, and Spanish trans.; and the enthusiastic account by H. J. John, "Jan Evangelista Purkyně, Czech Scientist and Patriot, 1787–1869," in *Memoirs of the American Philosophical Society*, **49** (1949), which should not be considered altogether reliable. See also the biographical and bibliographical work by V. Kruta, listed above.

Two valuable collections of essays were published to commemorate the 150th anniversary of Purkyně's birth: *Jan Ev. Purkyně 1787–1937. Sborník statí* (Prague, 1937) and *In memoriam Joh. E. Purkyně* (Prague, 1937). Other symposia are V. Kruta, ed., *Jan Evangelista Purkyně 1787–1869 Centenary Symposium* (Brno, 1971); Bohumil Němec and Otakar Matoušek, eds., *Jan Ev. Purkyně, Badatel národní buditel* (Prague, 1955); and Rudolph Zaunick, "Purkyně Symposium der deutschen Akademie der Naturforscher Leopoldina in Gemeinschaft mit der tschechoslovakischen Akademie der Wissenschaften . . .," in *Nova acta Leopoldina*, **24**, no. 151 (1961). See also E. Lesky, *Purkyněs Weg. Wissenschaft, Bildung und Nation* (Vienna–Cologne–Graz, 1970).

VLADISLAV KRUTA

PURSH, FREDERICK (*b.* Grossenhain, Saxony [now German Democratic Republic], 4 February 1774; *d.* Montreal, Canada, 11 July 1820), *botany*.

Pursh was the first botanist to describe plants of the Pacific Coast in a flora of North America. The increase in knowledge of the North American flora is exemplified by the ferns: in 1753 Linnaeus accounted for only twenty-three species; fifty years later André Michaux, fifty-five species; and in only another ten years Pursh listed ninety-eight species.

Friedrich Traugott Pursch (his original surname) attended public schools in his birthplace. According to his brother, Carl August Pursch, Friedrich lacked the financial means to pursue the scientific education that he wished. After studying horticulture under the court gardener, Johann Heinrich Seidel, Pursh joined the staff of the Royal Botanic Gardens at Dresden. In January 1799 he sailed for the United States, and, according to his own statement, he was first employed at a garden near Baltimore. In 1803 he succeeded John Lyon at "The Woodlands," the estate near Philadelphia of William Hamilton (1745–1813). In Thomas Jefferson's opinion the estate was "the only rival in America to what may be seen in England" (Betts [1944], 323). Among Hamilton's visitors was Benjamin Smith Barton, who, among his many projects, planned a flora of North America to include the discoveries of Lewis and Clark. Barton employed Pursh on the first extended botanical exploration of North America sponsored by an American. In 1806 Pursh botanized as far south as the North Carolina line. The following year he journeyed to Niagara Falls and east to Rutland, Vermont. Upon his return, but not before reading proof and checking synonymy for some of Barton's publications, he left his patron to lodge with the Philadelphia nurseryman Bernard M'Mahon, who with Hamilton had been entrusted with the living novelties brought back from the Pacific Northwest by Lewis and Clark. Barton was to prepare the natural history account while Pursh was to assist with the plant descriptions and drawings, but Barton was overcommitted and made too little progress. Discouraged, Pursh in April 1809 took employment with the physician David Hosack, then developing his Elgin Botanic Garden near New York City.

Hosack, like Barton, envisioned an "American Botany, or a Flora of the United States," but that, too, did not materialize. Hosack wrote of Pursh in *Hortus Elginensis* (2nd ed., 1811): "I shall have a very industrious and skillful botanist to collect from different parts of the Union such plants as have not yet been assembled at the Botanic Garden." During 1810–1811, while awaiting hoped-for financial support from the state, Pursh visited five islands of the West Indies for his health. He returned to Wiscasset, Maine, and visited William Dandridge Peck of Harvard. Pursh sailed from New York for England, since Hosack had been unable to raise support. He took with him notes, drawings, and selected specimens, some of which were scissored from the Lewis and Clark gatherings in Barton's care.

In England Pursh came under the patronage of Aylmer Bourke Lambert, a wealthy cabinet-naturalist, and, reputedly fortified with quantities of spirits, completed his *Flora Americae Septentrionalis* (1814). The collections of Lambert and Joseph Banks were utilized, and the Sherardian Herbarium at Oxford was searched. Altogether the records of forty-one collectors were cited, a notable achievement. Among the genera he named is *Lewisia* named for Meriwether Lewis. Although Pursh's *Flora* was sometimes disparaged for its inadequacies, it spurred the publication of Nuttall's *Genera* (1818). Darlington praised the *Flora* in 1827 as a "valuable work, and the spirit of botanical research which it has excited amongst us."

Following his involvement with the publication of catalogs of the gardens of Cambridge, England, and of Count Orlov in St. Petersburg, Pursh was offered the curatorship of the newly launched botanic garden at Yale but declined. He was invited to accompany the exploration of the Red River by Thomas Douglas, fifth earl of Selkirk, but the expedition was abandoned after the murder of the leader Robert Semple. From 1816 Pursh lived in Montreal working desultorily on a flora of Canada. He botanized on Anticosti Island in 1818, and assisted the Scot John Goldie in his collecting, but that winter specimens not already shipped to Lambert (and probably notes accompanying) were destroyed by fire. Discouraged, destitute, and dependent during those years on the charity of friends, Pursh died at the age of forty-six. His marriage to a barmaid has been alluded to but the details are wanting. Buried in a potter's field, he was removed subsequently by the intercession of Montreal naturalists to Mount Royal Cemetery, where he rests beside Sir John William Dawson. Benjamin Silliman, who met Pursh in 1819, wrote that "his conversation was full of fire, point, and energy; and, although not polished, he was good humored, frank, and generous."

BIBLIOGRAPHY

I. ORIGINAL WORKS. For details on Pursh's *Flora* and minor publications, see J. Ewan, "Frederick Pursh, 1774–1820, and His Botanical Associates," in *Proceedings of the American Philosophical Society*, **96** (1952), 599–628. An extended account of the backgrounds, type collections, and locations of the collections in the *Flora* will accompany a facsimile, which is to appear in the series Classica Botanica Americana.

II. SECONDARY LITERATURE. For Pursh's relation to the Lewis and Clark expedition, see Edwin Morris Betts,

"Thomas Jefferson's Garden Book, 1766–1824," in *Memoirs of the American Philosophical Society*, **22** (1944), 1–704; Donald Jackson, *Letters of the Lewis and Clark Expedition With Related Documents, 1783–1854* (Urbana, 1962); and Paul R. Cutright, *Lewis and Clark. Pioneering Naturalists* (Urbana, 1969). Benjamin Silliman's comment appeared in his *Remarks Made on a Short Tour Between Hartford and Quebec in the Autumn of 1819* (New Haven, 1820), 323–325. For critical remarks on Pursh, favoring Nuttall, see Jeannette E. Graustein, *Thomas Nuttall, Naturalist* (Cambridge, Mass., 1967). Joseph Ewan and Nesta Ewan, "John Lyon, Nurseryman, and Plant Hunter, and His Journal, 1799–1814," in *Transactions of the American Philosophical Society*, **53**, pt. 2 (1963), 1–69, summarizes our sparse information on Hamilton's "Woodlands."

JOSEPH EWAN

PUTNAM, FREDERIC WARD (*b*. Salem, Massachusetts, 16 April 1839; *d*. Cambridge, Massachusetts, 14 August 1915), *archaeology, anthropology*.

Putnam was the son of Ebenezer Putnam and Elizabeth Appleton. In 1856, after early education at home and in local private schools, he entered Harvard College, where he became an assistant to Louis Agassiz. Although Putnam later left this assistantship to free himself from Agassiz' dominance, the latter's rigid empiricism, together with a general and popular approach to the natural sciences, provided the foundations for Putnam's later activities.

With Jeffries Wyman, first curator of the Peabody Museum of American Archaeology and Ethnology at Harvard, Putnam excavated shell heaps in New England. This work provided an important empirical base for American archaeology. In 1874, following Wyman's death, Putnam was named curator of the museum, which he established as the leading center for anthropological research in America. From 1887 he served as Peabody professor at Harvard. In 1891 he was appointed chief of the ethnology and archaeology department of the World's Columbian Exposition in Chicago (1893). Against great public opposition and ridicule, he gathered a comprehensive exhibit—based upon commissioned research and collections—that in itself defined the broad range of the field of anthropology and brought the term into common and accepted usage. His chief assistant in Chicago was Franz Boas, whom Putnam later employed when, in 1894, he became curator of the American Museum of Natural History. There he initiated a center for anthropological activity rivaling that which he had organized at Harvard.

Putnam served as permanent secretary of the American Association for the Advancement of Science from 1873 to 1898, when he became its president. He thus played a significant role in defining the association as the focal point for professional science and scientific communication.

Putnam's particular contributions to an investigative science were modest. His most important work was done as an educator and museum administrator. He sought not only to enlarge the activities and responsibilities of science and its practitioners but also to relate both to society.

BIBLIOGRAPHY

The *Putnam Anniversary Volume* (New York, 1909), a collection of anthropological essays by Putnam's friends and associates, was presented to him in honor of his seventieth birthday and contains a bibliography of his writings. See also Alfred M. Tozzer's notice in *Biographical Memoirs. National Academy of Sciences*, **16** (1936), 125–153, with portrait, comprehensive bibliography of Putnam's works, and brief list of secondary literature; and the Royal Society *Catalogue of Scientific Papers*, V, 47; VIII, 675; XI, 80; and XVII, 1051, which lists 29 memoirs by Putnam and 1 of which he was coauthor.

JACOB W. GRUBER

PYTHAGORAS OF SAMOS (*b*. Samos, *ca*. 560 B.C.; *d*. Metapontum, *ca*. 480 B.C.), *mathematics, theory of music, astronomy*.

Most of the sources concerning Pythagoras' life, activities, and doctrines date from the third and fourth centuries A.D., while the few more nearly contemporary (fourth and fifth centuries B.C.) records of him are often contradictory, due in large part to the split that developed among his followers soon after his death. Contemporary references, moreover, scarcely touch upon the points of Pythagoras' career that are of interest to the historian of science, although a number of facts can be ascertained or surmised with a reasonable degree of certainty.

It is, for example, known that in his earlier years Pythagoras traveled widely in Egypt and Babylonia, where he is said to have become acquainted with Egyptian and Babylonian mathematics. In 530 B.C. (or, according to another tradition, 520 B.C.) he left Samos to settle in Croton, in southern Italy, perhaps because of his opposition to the tyrant Polycrates. At Croton he founded a religious and philosophical society that soon came to exert considerable political influence throughout the Greek cities of southern Italy. Pythagoras' hierarchical views at first pleased the local aristocracies, which found in them a support against the rising tide of democracy, but he later met strong opposition from the same quarter. He was forced to leave Croton about 500 B.C., and retired to Metapontum, where he died. During the violent democratic revolution that occurred in Magna Graecia in about 450 B.C., Pythagoras' disciples were set upon, and Pythagorean meetinghouses were destroyed. Many Pythagoreans thereupon fled to the Greek mainland, where they found a new center for their activities at Phleius; others went to Tarentum, where they continued as a political power until the middle of the fourth century B.C.

The political vicissitudes of Pythagoras and his followers are significant in the reconstruction of their scientific activities. True to his hierarchical principles, Pythagoras seems to have divided his adherents into two groups, the ἀκουσματικοί, or "listeners," who were enjoined to silence, in which they memorized the master's words, and the μαθηματικοί, who, after a long period of training, were allowed to ask questions and express opinions of their own. (The term μαθηματικοί originally meant merely those who had attained a somewhat advanced degree of knowledge, although it later came to imply "scientist" or "mathematician.")

A few decades after Pythagoras' death, these two groups evolved into sharp factions and began a controversy over which of them was most truly Pythagorean. The ἀκουσματικοί based their claim on their literal adherence to Pythagoras' own words (αὐτὸς ἔφα, "he himself has spoken"); the μαθηματικοί, on the other hand, seem to have developed Pythagoras' ideas to such an extent that they were no longer in complete agreement with their originals. The matter was further complicated because, according to ancient tradition, Pythagoras chose to reveal his teachings clearly and completely to only his most advanced disciples, so that the ἀκουσματικοί received only cryptic, or even mysterious, hints. The later Pythagorean tradition thus includes a number of strange prescriptions and doctrines, which the ἀκουσματικοί interpreted with absolute literalness; the more rationalistic group (led at one time by Aristoxenus, who was also a disciple of Plato and Aristotle) preferred a symbolic and allegorical interpretation.

This obscurity concerning Pythagoras' intent has led historians of science into differences of opinion as to whether Pythagoras could really be considered a scientist or even an initiator of scientific ideas. It is further debatable whether those ancient authors who made real contributions to mathematics, astronomy, and the theory of music can be considered to have been true Pythagoreans, or even to have been influenced by authentically Pythagorean ideas. None-

theless, apart from the theory of metempsychosis (which is mentioned by his contemporaries), ancient tradition assigns one doctrine to Pythagoras and the early Pythagoreans that can hardly have failed to influence the development of mathematics. This is the broad generalization, based on rather restricted observation (a procedure common in early Greek science), that all things are numbers.

Pythagoras' number theory was based on three observations. The first of these was the mathematical relationships of musical harmonies—that is, that when the ratio of lengths of sound-producing instruments (such as strings or flutes) is extended to other instruments in which one-dimensional relations are involved, the same musical harmonies result. Secondly, the Pythagoreans noted that any triangle formed of three sticks in the ratio 3:4:5 is always a right triangle, whatever the length of its segments. Their third important observation derived from the fixed numerical relations of the movements of heavenly bodies. It was thereby apparent to them that since the same musical harmonies and geometric shapes can be produced in different media and sizes by the same combination of numbers, the numbers themselves must express the harmonies and shapes and even the things having those harmonies and shapes. It could thus be said that these things—or, as they were later called, the essences (οὐσίαι) of these things—actually were numbers. The groups of numbers that embodied the essence of a thing, and by which it might be reproduced, were called λόγοι ("words"), a term that later came to mean "ratio."

The translation of philosophical speculation into mathematics is thus clear. This speculation about numbers as essences was extended in several directions; as late as the end of the fifth century B.C., philosophers and mathematicians were still seeking the number of justice, or marriage, or even of a specific man or horse. (Attempts were made to discover the number of, for example, a horse, by determining the number of small stones necessary to produce something like the outline of it.) By this time, however, the Pythagoreans had split into a set of groups holding highly differing viewpoints, so it would be inaccurate to assume that all Pythagorean speculations about numbers were of this primitive, unscientific kind.

The theory of special types of numbers, which lay somewhere between these mystical speculations and true science, was developed by the Pythagoreans during the fifth century B.C. The two aspects of the theory are apparent in that the Pythagoreans distinguished between two types of "perfect" numbers. The number ten was the only example of the first group, and its perfection derived from its fundamental role in the decimal system and in its being composed of the sum of the first four numbers, $1 + 2 + 3 + 4 = 10$. Because of this second quality it was called the tetractys, and represented by the figure $\therefore$; it was considered holy, and the Pythagoreans swore by it. The second type of perfect numbers consisted of those equal to the sum of their factors, as, for example, six $(1 + 2 + 3)$ or twenty-eight $(1 + 2 + 4 + 7 + 14)$. Euclid, in the *Elements* (IX. 36) gave the general theory for this numerical phenomenon, stating that if $2^n - 1$ is a prime number, then $(2^n - 1)2^{n-1}$ is a perfect number.

Similar speculations prompted the search for "amicable" numbers—that is, numbers of which each equals the sum of the factors of the other—and for integers satisfying the Pythagorean formula $a^2 + b^2 = c^2$ (as, for example, $3^2 + 4^2 = 5^2$, or $5^2 + 12^2 = 13^2$). Only one pair of amicable numbers, 284 and 220, was known by the end of antiquity, and its discovery is attributed by Iamblichus to Pythagoras himself, who is said to have derived it from the saying that a friend is an alter ego. Proclus (in Friedlein's edition of Euclid's *Elements*, p. 426) also attributes to Pythagoras himself the general formula by which any number of integers satisfying the equation $a^2 + b^2 = c^2$ may be found,

$$n^2 + \left(\frac{n^2 - 1}{2}\right)^2 = \left(\frac{n^2 - 1}{2} + 1\right)^2,$$

where n is an odd number. If this tradition is correct (and it is doubtful), Pythagoras must have learned the formula in Babylonia, where it was known, according to O. Neugebauer and A. Sachs (in *Mathematical Cuneiform Texts* [New Haven, 1945], p. 38).

Figured numbers were of particular significance in Pythagorean arithmetic. These included triangular numbers, square numbers, and pentagonal numbers, as well as *heteromeke* numbers (numbers forming rectangles with unequal sides), stereometric numbers (pyramidal numbers forming pyramids with triangular or square bases), cubic numbers, and altar numbers (stereometric numbers corresponding to *heteromeke* numbers). These numbers were represented by points, with ., $\therefore$, $\therefore$, $\therefore$, $\therefore$, for example, for the triangular numbers, and $\boxed{\cdot \cdot}$ for the square numbers. The triangular numbers thus occur in the series $1, 3, 6, 10, 15, \cdots$, which can be expressed by the formula $n(n + 1)/2$, while square numbers have the value n^2 and pentagonal numbers may be given the value $n(3n - 1)/2$. *Heteromeke* numbers may be expressed as $n(n + 1)$, $n(n + 2)$, and so on; pyramidal numbers with triangular bases are formed by the successive sums of the triangular numbers $1, 4, 10, 20, 35, \cdots$.

Pythagorean authors gave only examples of the various kinds of figured numbers until the second century A.D., and it was only in the third century A.D. that Diophantus developed a systematic mathematical theory based upon Pythagorean speculations.

The theory of $\mu\epsilon\sigma\acute{o}\tau\eta\tau\epsilon\varsigma$, or "means," is also undoubtedly Pythagorean and probably of considerable antiquity. Iamblichus asserts (in Pistelli's edition of Iamblichus' commentary on Nicomachus' *Introductio arithmetica*, p. 118) that Pythagoras learned of arithmetic means during his travels in Babylonia, but this cannot be definitely proved. The theory was at first concerned with three means; the arithmetic, of the form $a - b = b - c$; the geometric, of the form $a:b = b:c$; and the harmonic, of the form $(a - b):a = (b - c):c$. Other means were added at later dates, particularly by the Pythagorean Archytas of Tarentum, in the first half of the fourth century B.C.

It would seem likely, as O. Becker (in *Quellen und Studien zur Geschichte der Mathematik*, III B, pp. 534 ff.) and B. L. van der Waerden have pointed out, that Euclid took the whole complex of theorems and proofs that are based upon the distinction between odd and even numbers from the Pythagoreans, and that these reflect the Pythagorean interest in perfect numbers. This adaptation would seem to be particularly apparent in the *Elements*, IX. 30, and IX. 34, which lay the groundwork for the proof of IX. 36, the general Euclidean formula for perfect numbers. The proofs given in Euclid are strictly deductive and scientific, however, and would indicate that one group of Pythagoreans had quite early progressed from mysticism to true scientific method.

Although the contributions made by Pythagoras and his early successors to arithmetic and number theory can be determined with some accuracy, their contributions to geometry remain problematic. O. Neugebauer has shown (in *Mathematiker Keilschrifttexte*, I, 180, and II, 53) that the so-called Pythagorean theorem had been known in Babylonia at the time of Hammurabi, and it is possible that Pythagoras had learned it there. It is not known whether the theorem was proved during Pythagoras' lifetime, or shortly thereafter. The pentagram, which played an important role in Pythagorean circles in the early fifth century B.C., was also known in Babylonia, and may have been imported from there. This figure, a regular pentagon with its sides extended to intersect in the form of a five-pointed star, has the interesting property that its sides and diagonals intersect everywhere according to the golden section; the Pythagoreans used it as a symbol by which they recognized each other.

Of the mathematical discoveries attributed by ancient tradition to the Pythagoreans, the most important remains that of incommensurability. According to Plato's *Theaetetus*, this discovery cannot have been made later than the third quarter of the fifth century B.C., and although there has been some scholarly debate concerning the accuracy of this assertion, there is no reason to believe that it is not accurate. It is certain that the Pythagorean doctrine that all things are numbers would have been a strong incentive for the investigation of the hidden numbers that constitute the essences of the isosceles right-angled triangle or of the regular pentagon; if, as the Pythagoreans knew, it was always possible to construct a right triangle given sides in the ratio 3 : 4 : 5, then it should by analogy be possible to determine the numbers by which a right-angled isosceles triangle could be constructed.

The Babylonians had known approximations to the ratio of the side of a square to its diagonal, but the early Greek philosophers characteristically wished to know it exactly. This ratio cannot be expressed precisely in integers, as the early Pythagoreans discovered. They chose to approach the problem by seeking the greatest common measure—and hence the numerical ratio of two lengths—through mutual subtraction. In the case of the regular pentagon, it may easily be shown that the mutual subtraction of its diagonals and sides can be continued through an infinite number of operations, and that its ratio is therefore incommensurable. Ancient tradition credits this discovery to Hippasus of Metapontum, who was living at the period in which the discovery must have been made, and who could easily have made it by the method described. (Hippasus, one of the $\mu\alpha\theta\eta\mu\alpha\tau\iota\kappa o\iota$, who is said to have been set apart from the other Pythagoreans by his liberal political views, is also supposed to have been concerned with the "sphere composed by regular pentagons," that is, the regular dodecahedron.)

An appendix to book X of the *Elements* incorporates a proof of the incommensurability of the diagonal of a square with its side. This proof appears to be out of its proper order, and is apparently much older than the rest of the theorems contained in book X; it is based upon the distinction between odd and even numbers and closely resembles the theorems and proofs of book IX. Like the proofs of book IX, the proof offered in book X is related to the Pythagorean theory of perfect numbers; an ancient tradition states that it is Pythagorean in origin, and it would be gratuitous to reject this attribution simply because other members of the same sect were involved in nonscientific speculations about numbers. It is further possible that the ingenious process by which the theory of proportions (which had been conceived in the form of ratios of integers) had been applied to

incommensurables—that is, by making the process of mutual subtraction itself the criterion of proportionality—was also a Pythagorean invention. But it is clear that the later elaboration of the theory of incommensurability and irrationality was the work of mathematicians who no longer had any close ties to the Pythagorean sect.

Pythagoras (or, according to another tradition, Hippasus) is also credited with knowing how to construct three of the five regular solids, specifically the pyramid, the cube, and the dodecahedron. Although these constructions can hardly have been identical to the ones in book XIII of the *Elements*, which a credible tradition attributes to Theaetetus, it is altogether likely that these forms, particularly the dodecahedron, had been of interest not only to Hippasus (as has been noted) but also to even earlier Pythagoreans. Their curiosity must have been aroused by both its geometrical properties (since it is made up of regular pentagons) and its occurrence in nature, since iron pyrite crystals of this form are found in Italy. An artifact in the form of a carved stone dodecahedron, moreover, dates from the tenth century B.C., and would seem to have played some part in an Etruscan cult.

The notion that all things are numbers is also fundamental to Pythagorean music theory. Early Pythagorean music theory would seem to have initially been of the same speculative sort as early Pythagorean mathematical theory. It was based upon observations drawn from the lyre and the flute, the most widely used instruments; from these observations it was concluded that the most beautiful musical harmonies corresponded to the most beautiful (because simplest) ratios or combinations of numbers, namely the octave (2:1), the fifth (3:2), and the fourth (4:3). It was thus possible to assign the numbers 6, 8, 9, and 12 to the four fixed strings of the lyre, and to determine the intervals of the diatonic scale as 9:8, 9:8, and 256:243. From these observations and speculations the Pythagoreans built up, as van der Waerden has pointed out, a deductive system of musical theory based on postulates or "axioms" (a term that has a function similar to its use in mathematics). The dependence of musical intervals on mathematical ratios was thus established.

The early music theory was later tested and extended in a number of ways. Hippasus, perhaps continuing work begun by the musician Lasos of Hermione, is said to have experimented with empty and partially filled glass vessels and with metal discs of varying thicknesses to determine whether the same ratios would produce the same harmonies with these instruments. (Contrary to ancient tradition, it would have been impossible to achieve sufficient accuracy by these means for him to have been altogether successful in this effort.) The systematic deductive theory was later enlarged to encompass the major and minor third (5:4 and 6:5), as well as the diminished minor third (7:6) and the augmented whole tone (8:7). The foundation for the enharmonic and chromatic scales was thus laid, which led to the more complex theory of music developed by Archytas of Tarentum in the first half of the fourth century B.C.

In addition to its specifically Pythagorean elements, Pythagorean astronomy would seem to have comprised both Babylonian observations and theories (presumably brought back by Pythagoras from his travels) and certain theories developed by Anaximander of Miletus, whose disciple Pythagoras is said to have been. It is not known precisely when Babylonian astronomy had begun, or what state it had reached at the time of Pythagoras, although ancient documents indicate that regular observations of the appearances of the planet Venus had been made as early as the reign of King Amisadaqa (about 1975 B.C.). The *mul apin* texts of about 700 B.C. give a summary of Babylonian astronomy up to that time, moreover, and contain divisions of the heavens into "roads of the fixed stars" (similar to the divisions of the zodiac), statements on the courses of the planets, and data on the risings and settings of stars that are obviously based on observations carried out over a considerable period of time. In addition, Ptolemy stated that regular observations of eclipses had been recorded since the time of King Nabonassar, about 747 B.C. The Babylonians of this time also knew that lunar eclipses occur only at full moon and solar eclipses at new moon, and that lunar eclipses occur at intervals of approximately six months; they knew seven "planets" (including the sun and moon), and therefore must have known the morning and evening star to be identical.

The Babylonians also knew that the independent motions of the planets occur in a plane that intersects the equator of the heavenly sphere at an angle. Greek tradition attributes the determination of this angle as 24° to Pythagoras, although the computation was actually made by Oenopides of Chios, in the second half of the fifth century B.C. Oenopides was not a Pythagorean, but he obviously drew upon Pythagorean mathematics and astronomy, just as the Pythagoreans drew upon the body of Babylonian knowledge.

Anaximander's contributions to Pythagorean astronomy were less direct. The Pythagoreans rejected his chief theory, whereby the stars were in fact rings of fire that encircled the entire universe; according to Anaximander these fiery rings were obscured by "dark air," so that they were visible only through the holes

through which they breathed. Pythagoras and his adherents, on the other hand, accepted the Babylonian notion of the stars as heavenly bodies of divine origin. They did, however, make use of Anaximander's assumption that the planets (or, rather, the rings in which they appear) are at different distances from the earth, or at any rate are nearer to the earth than are the fixed stars. This idea became an important part of Pythagorean astronomy (see Heiberg's edition, Eudemus of Rhodes in Simplicius' *Commentary* on Aristotle's *De caelo*, p. 471, and Diels and Kranz's edition of *Die Fragmente der Vorsokratiker*, sec. 12, 19).

Their knowledge of the periodicity of the movements of the stars undoubtedly strengthened the Pythagoreans in their belief that all things are numbers. They attempted to develop astronomical theory by combining it with this general principle, among others (including the principle of beauty that had figured in their axiomatic foundations of the theory of music). Their concern with musical intervals led them to try to determine the sequence of the planets in relation to the position of the earth (compare Eudemus, in the work cited, and Ptolemy, *Syntaxis*, IX, 1). According to their theory, probably the earliest of its kind, the order of the planets, in regard to their increasing distance from the earth, was the moon, Mercury, Venus, the sun, Mars, Jupiter, and Saturn—a sequence that was later refined by placing Mercury and Venus above the sun, since no solar transits of these bodies had been observed.

Further theories by which the distances and periods of revolution of the heavenly bodies are correlated with musical intervals are greatly various, if not actually contradictory. Indeed, according to van der Waerden (in "Die Astronomie der Pythagoreer," pp. 34 ff.), a number of them make very little sense in the context of musical theory. It is almost impossible to tell what the original astronomical-musical theory on which these variants are based actually was, although it was almost certainly of considerable antiquity. It may be assumed, however, that in any original theory the celestial spheres were likened to the seven strings of a lyre, and were thought to produce a celestial harmony called the music of the spheres. Ordinary mortals could not hear this music (Aristotle suggested that this was because they had been exposed to it continuously since the moment of their birth), but later Pythagoreans said that it was audible to Pythagoras himself.

Another mystical notion, this one adopted from the Babylonians, was that of the great year. This concept, which was used by the Pythagoreans and probably by Pythagoras, held that since the periods of revolution of all heavenly bodies were in integral ratio, a least common multiple must exist, so that exactly the same constellation of all stars must recur after some definite period of time (the "great year" itself). It thereupon followed that all things that have occurred will recur in precisely the same way; Eudemus is reported to have said in a lecture (not without irony) that "then I shall sit here again with this pointer in my hand and tell you such strange things."

Pythagorean ideas of beauty required that the stars move in the simplest curves. This principle thus demanded that all celestial bodies move in circles, the circle being the most beautiful curve, a notion that held the utmost importance for the development of ancient astronomy. If van der Waerden's ingenious interpretation of the difficult ancient texts on this subject is correct (in "Die Astronomie der Pythagoreer," pp. 42 ff.), there may have been—even before Plato asked the non-Pythagorean mathematician Eudoxus to create a model showing the circular movements of all celestial bodies—a Pythagorean theory that explained the movements of Mercury and Venus as epicycles around the sun, and thus represented the first step toward a heliocentric system.

Ancient tradition also refers to an entirely different celestial system, in which the earth does not rest in the center of the universe (as in the theories of Anaximander, the Babylonians of the fifth century B.C., and the other Pythagoreans), but rather revolves around a central fire. This fire is invisible to men, because the inhabited side of the earth is always turned away from it. According to this theory, there is also a counter-earth on the opposite side of the fire. Pythagorean principles of beauty and of a hierarchical order in nature are here fundamental; fire, being more noble than earth, must therefore occupy a more noble position in the universe, its center (compare Aristotle, *De caelo*, II, 13). This theory is sometimes attributed to Philolaus, a Pythagorean of the late fifth century B.C., and he may have derived the epicyclic theory from it, although the surviving fragments of Philolaus' work indicate him to have been a man of only modest intellectual capacities, and unlikely to have been the inventor of such an ingenious system. Other ancient sources name Hicetas of Syracuse, a Pythagorean of whom almost nothing else is known, as its author.

The decisive influence of this theory in the history of astronomy lies in its explanation of the chief movements of the celestial bodies as being merely apparent. The assumption that the solid earth, on which man lives, does not stand still but moves with great speed (since some Pythagoreans according to Aristotle explained the phenomenon of day and night by the movement of the earth around the central fire) was a bold one, although the paucity and vagueness of ancient records make it impossible to determine with any certainty how far this notion was applied to other celestial phenomena. Further details on this theory are also difficult to ascertain; van der Waerden (in "Die Astronomie der Pythagoreer," pp. 49 ff.) discusses the problem at length. It is nevertheless clear that this daring, and somewhat unscientific, speculation was a giant step toward the development of a heliocentric system. Once the idea of an unmoving earth at the center of the universe had been overcome, the Pythagorean Ecphantus and Plato's disciple Heraclides were, in about 350 B.C., able to teach that the earth revolves about its own axis (*Aetius*, III, 13). A fully heliocentric system was then presented by Aristarchus of Samos, in about 260 B.C., although it was later abandoned by Ptolemy because its circular orbits did not sufficiently agree with his careful observations.

It is thus apparent that the tendency of some modern scholars to reject the unanimous and plausible ancient tradition concerning the Pythagoreans and their discoveries—and to attribute these accomplishments instead to a number of unknown, cautious, and pedestrian observers and calculators—obscures one of the most interesting aspects of the early development of Greek science.

BIBLIOGRAPHY

I. ORIGINAL WORKS. The most complete collection of ancient Pythagorean texts, with critical apparatus, Italian trans., and commentary, is M. Timpanaro-Cardini, *Pitagorici. Testimonianze e frammenti*, 3 vols. (Florence, 1958–1964). See also H. Diels and W. Kranz, eds., *Die Fragmente der Vorsokratiker*, 7th ed. (Berlin, 1954), sec. 14 (Pythagoras) and secs. 37–58 (early Pythagoreans); K. Freeman, *Ancilla to the Presocratic Philosophers, a Complete Translation of the Fragments in Diels, Fragmente der Vorsokratiker* (Cambridge, Mass., 1948), which translates only the fragments but not the very important testimonies; and *The Pre-Socratic Philosophers, a Companion to Diels* (Oxford, 1946). A good selection of texts and testimonies is Cornelia J. de Vogel, *Greek Philosophy. A Collection of Texts, Selected and Supplied With Notes.* Vol. I, *From Thales to Plato* (Leiden, 1950).

II. SECONDARY LITERATURE. The modern literature on Pythagoras and the early Pythagoreans is enormous. A more complete selection than the following is found in Timpanaro-Cardini (chs. 11–19). The most important recent contributions to the history of Pythagorean science are B. L. van der Waerden, "Die Arithmetik der Pythagoreer," in *Mathematische Annalen*, **120** (1947–1949), 127–153, 676–700; "Die Harmonielehre der Pythagoreer," in *Hermes*, **78** (1943), 163–199; "Die Astronomie der Pythagoreer," in *Verhandelingen der*

K. akademie van wetenschappen, Afdeeling Natuurkunde, eerste Reeks, pt. 20 (1954), art. 1; "Das Grosse Jahr und die Ewige Wiederkehr," in *Hermes,* **80** (1952), 129–155; *Ontwakende Wetenschap* (Groningen, 1954), trans. by A. Dresden as *Science Awakening* (New York, 1963); *Erwachende Wissenschaft,* II, *Die Anfänge der Astronomie* (Groningen, 1967), for the Babylonian antecedents of Pythagorean astronomy.

See also the extensive articles on Pythagoras and the Pythagoreans by B. L. van der Waerden, K. von Fritz, and H. Dörrie in Pauly-Wissowa, *Real-Encyclopädie der classischen Altertumswissenschaft,* XXIV, pt. 1 (Stuttgart, 1963), cols. 171–300. Other important recent works are: Cornelia J. de Vogel, *Pythagoras and Early Pythagoreanism* (Assen, 1966); W. Burkert, *Weisheit und Wissenschaft, Erlanger Beiträge zur Sprach- und Kunstwissenschaft* (Nuremberg, 1962); W. K. C. Guthrie, *A History of Greek Philosophy,* I (Cambridge, 1962), 146–340; and E. Zeller and R. Mondolfo, *La filosofia dei Greci nel suo sviluppo storico,* II (Florence, 1938), 288–685.

For important special problems see R. Mondolfo, "Sui frammenti di Filolao," in *Rivista di filologia e d'istruzione classica,* n. s. **15** (1937), 225–245; O. Neugebauer, *The Exact Sciences in Antiquity* (Copenhagen, 1951), for the oriental antecedents of Pythagorean astronomy; and K. von Fritz, "Mathematiker und Akusmatiker bei den alten Pythagoreern," in *Sitzungsberichte der Bayerischen Akademie der Wissenschaften zu München,* Phil.-hist. Kl., no. 11 (1960).

See further Paul-Henri Michel, *De Pythagore à Euclide, Contribution à l'histoire des mathématiques préeuclidiennes* (Paris, 1950); A. Rostagni, *Il verbo di Pitagora* (Turin, 1924); J. A. Philip, *Pythagoras and Early Pythagoreanism* (Toronto, 1966); Erich Frank, *Platon und die sogenannten Pythagoreer* (Halle, 1923); and G. Sarton, *A History of Science* (Cambridge, Mass., 1952), 199–217, 275–297.

KURT VON FRITZ

PYTHEAS OF MASSALIA (*fl.* Massalia [now Marseilles, France], 330 B.C.), *geography.*

The tradition of Pytheas' work is defective and controversial. He seems to have been inspired by the new knowledge of the earth as a sphere, probably through the writings of Eudoxus of Cnidus. He was close in his measurement of the latitude of his native Massalia and corrected Eudoxus' position of the north celestial pole. His voyage to see the midnight sun led him to the exploration of Britain and the North Sea and to the discovery of the island that he called Thule, located on the Arctic Circle a six-day journey north of Britain. He described the sea near Thule as "neither land nor sea nor air but a mixture of all, like a sea-lung, in which sea and land and everything swing and which may be the bounds of all, impassable by foot or boat"; this description challenges explanation.

It is not known how Pytheas arrived at Thule or where it was located. The usual view is that Thule was Iceland, although some have argued vigorously for Norway. Iceland seems to fit the evidence better. Again, the usual view is that Pytheas navigated his own ship through the Strait of Gibraltar, but the alternative view that he went overland through France and traveled as a passenger on native vessels seems, on the whole, more feasible. The commercial situation, with Carthage in control of Gibraltar, and the Greek influence of Massalia radiating through France, seems to favor the latter view.

Pytheas had a good deal to tell about the nearer parts of northern Europe. He claimed to have traveled throughout Britain and along the entire coast of the Continent from Cádiz to the Don River (that is, to the boundary of Europe and Asia—wherever he supposed that was). But the preserved data are so meager, erratic, and corrupt as to be of little use. He certainly investigated the source of tin in south Britain and probably that of amber in the North Frisian Islands. His explorations were probably extended by inquiry and hearsay, and it is doubtful that he reached the Baltic Sea. Although his description of the shape of Britain is correct the dimensions he gave are much too large. The tradition is silent about Ireland, but he must have known of it.

Pytheas' report of his explorations elicited controversy from the start, apparently because it conflicted with the established theory of uninhabitable torrid and frigid zones. Polybius was altogether hostile to it, and Strabo followed suit with abusive language. They classified Pytheas with the writers of fiction Euhemerus and Antiphanes. According to Strabo, Polybius said that even Dicaearchus of Messina did not believe him; this attack is the earliest known reference to Pytheas' work. Polybius' attitude may have been part of his polemic against his predecessor, the historian Timaeus, who had used Pytheas' report in his digression on geography. The scientific geographers Eratosthenes and Hipparchus accepted Pytheas' work at least in great part. There is no doubt that Pytheas' opponents were mistaken, and he must rank as one of the greatest explorers of antiquity. The recorded history of Britain begins with him; and the name of Thule, which no other man of the ancient classical world reported on, is the hallmark of his influence.

BIBLIOGRAPHY

On Pytheas and his work, see F. Gisinger, "Pytheas I," in Pauly-Wissowa, *Real-Encyclopädie der classischen Altertumswissenschaft,* XXIV (Stuttgart, 1963), 314–366; R. Hennig, *Terrae incognitae I, Altertum bis Ptolemäus,*

2nd. rev. ed. (Leiden, 1944), 155–182; H. J. Mette, *Pytheas von Massalia*, Kleine Texte für Vorlesungen und Übungen no. 173 (Berlin, 1952); and J. O. Thomson, *History of Ancient Geography* (Cambridge, 1948), 143–151.

AUBREY DILLER

AL-QABAJĀQĪ. See Baylak al-Qibjāqī.

AL-QABĪṢĪ, ABŪ AL-ṢAQR ʿABD AL-ʿAZĪZ IBN ʿUTHMĀN IBN ʿALĪ (*fl. ca.* 950 in Aleppo, Syria), *astrology.*

Al-Qabīṣī, who came from either the Qabīṣa near Al-Mawṣil (Mosul) or that near Sāmarrāʾ (both are in Iraq), studied under ʿAlī ibn Aḥmad al-ʿImrānī in Al-Mawṣil and became a recognized authority on Ptolemy's *Almagest* (according to Ibn al-Qifṭī, in 980/981). Al-ʿImrānī, who died in 955/956, does in fact refer to al-Qabīṣī in his *De electionibus horarum* (J. M. Millás-Vallicrosa, *Las traducciones orientales en los manuscritos de la Biblioteca Catedral de Toledo* [Madrid, 1942], 338).

Although al-Qabīṣī's education was primarily in geometry and astronomy, his principal surviving treatise, *Al-madkhal ilā ṣināʿat aḥkām al-nujūm* ("Introduction to the Art of Astrology") in five sections, which he dedicated to Sayf al-Dawla, the Ḥamdānid ruler of Aleppo from 944/945 to 966/967, is on astrology. The date of this work is fixed by his use of the year 948/949 as an example in the fourth section. The book, as its title indicates, is an introductory exposition of some of the fundamental principles of genethlialogy; its present usefulness lies primarily in its quotations from the Sassanian Andarzghar literature and from al-Kindī, the Indians, Ptolemy, Dorotheus of Sidon, Māshāʾallāh, Hermes Trismegistus, and Valens. Although completely lacking in originality, it was highly valued as a textbook. There are many Arabic manuscripts (including some in Hebrew script), although it was never found to need a commentary, and it was translated into Latin by Joannes Hispalensis in 1144, and into French (presumably from the Latin) by Pelerin de Pousse in 1362. Joannes' Latin version was commented on by Joannes de Saxonia at Paris in 1331 and by V. Nabod in 1560, and was also the text commented on by Francesco degli Stabili, called Cecco d'Ascoli, who lived between 1269 and 1327.

A manuscript in Istanbul, MS 4832 of the Ayasofya Library, contains three short treatises of al-Qabīṣī, of which the first two are dedicated to Sayf al-Dawla. These are a *Risāla fī anwāʿ al-aʿdād* ("Epistle on the Kinds of Numbers"), a *Risāla fī al-abʿād wa al-ajrām* ("Epistle on Distances and Volumes"), and a commentary on the astronomical handbook written by al-Farghānī in the middle of the ninth century. Further, al-Qabīṣī himself refers in the introduction to his *Madkhal* to his now lost *Kitāb fī ithbāt ṣināʿat aḥkām al-nujūm* ("On Confirming the Art of Astrology"), composed as a response to an attack on the art by one ʿĪsā ibn ʿAlī, and in *Madkhal* IV 1 to a treatise on the namūdār (a significant point in a horoscope).

More doubtful are two other works sometimes attributed to our author. A poem describing the rainbow is at times said to have been written by Sayf al-Dawla, at times by al-Qabīṣī. And a *De planetarum coniunctionibus* of Alchabitius, which was translated into Latin by Joannes Hispalensis and commented on by Joannes de Saxonia, is not sections four and five of the *Madkhal* as has been suggested and was not known to either of our chief Arabic sources of information about al-Qabīṣī, al-Bayhāqī, and Ḥājjī Khalīfa. The Latin text was translated into French by Oronce Finé in 1551.

BIBLIOGRAPHY

The most complete article on al-Qabīṣī, in which all of the relevant literature is cited, is that by D. Pingree in the new ed. of the *Encyclopaedia of Islam*. The *Madkhal* in its Latin version (entitled *Isagoge*) was published many times in the fifteenth and sixteenth centuries. The eds. that can be confirmed are as follows: by Matheus Moretus de Brixia (Bologna, 1473); by E. Ratdolt (Venice, 1482), and with the commentary of Joannes de Saxonia (Venice, 1485); by I. and G. de Forlivio (Venice, 1491); by I. and G. de Gregoriis (Venice, 1502, 1503); by M. Sessa (Venice, 1512); by B. Trot, with the commentary of Joannes de Saxonia and the notes of Petrus Turrellus (Leyden, 1520?); by M. Sessa and P. de Ravanis, with the commentary of Joannes de Saxonia (Venice, 1521); by P. Liechtenstein, with the commentary of Joannes de Saxonia (Venice, 1521); and by Simon Colinaeus with the same commentary (Paris, 1521). V. Nabod's commentary was published as *Enarratio elementorum astrologiae* (Cologne, 1560), and Cecco d'Ascoli's *Commento all'Alcabizzo* was edited by P. G. Boffitto (Florence, 1905).

The possibly spurious *De planetarum coniunctionibus* was also often published by E. Ratdolt, with the commentary of Joannes de Saxonia (Venice, 1485); by I. and G. de Forlivio (Venice, 1491); by M. Sessa and P. de Ravanis (Venice, 1521); and by P. Liechtenstein (Venice, 1521). The French trans. by Oronce Finé was published as an appendix to his *Les canons et documents très amples touchant l'usage et practique des communs almanachz* (Paris, 1551, 1557).

DAVID PINGREE

QĀḌĪ ZĀDA AL-RŪMĪ (more properly **Salah al-Dīn Mūsā Pasha**) (*b.* Bursa, Turkey, *ca.* 1364; *d.* Samarkand, Uzbekistan, *ca.* 1436), *mathematics, astronomy.*

Most historians of science have erred concerning Qāḍī Zāda. A. Sédillot, for example, called him Hassan Tchélebī; and Montucla, in his history of mathematics, said that he was a Greek convert to Islam. Montucla may have been deceived by the surname Rūmī, for the peoples who lived in Asia Minor were called Rūm, meaning Roman (not Greek), because Asia Minor was once Roman. Qāḍī Zāda means "son of the judge."

After completing his secondary education at Bursa, Qāḍī Zāda became a student of the theologian and encyclopedist Mullā Shams al-Dīn Muḥammad al-Fanārī (1350–1431), who taught him geometry and astronomy. Sensing the great talent of his student, al-Fanārī advised him to go to Transoxiana, then a great cultural center, to continue his training in mathematics and astronomy. According to several historians, al-Fanārī gave Qāḍī Zāda letters of recommendation and one of his works (*Enmuzeğ al-ulum,* "Types of Sciences") to present to the scholars of Khurasan and Transoxiana.

The year of Qāḍī Zāda's departure from Bursa is not known. It must have been after 1383, however, for in that year, still at Bursa, he had composed a work on arithmetic, *Risāla fī'l-ḥisāb.* When he presented himself to Ulugh Beg in Samarkand (*ca.* 1410),[1] he had spent some time in Iran, Jurjan, and Khorasan. At Jurjan he had met the philosopher-theologian Seyyid al-Sharīf al-Jurjānī. Therefore, the year of departure from Bursa probably falls between 1405 and 1408.

In 1421 Ulugh Beg ordered the construction of a university at Samarkand and named Qāḍī Zāda as its rector. Serving in addition as professor of mathematics and astronomy, Qāḍī Zāda frequently had Ulugh Beg as a student in his classes. Also in 1421, under the direction of the young Persian astronomer and mathematician al-Kāshī, the construction of the observatory at Samarkand was completed. Astronomical observations had been made and the composing of astronomical tables had been begun. These tables were composed to correct and complete the tables of Naṣir al-Dīn al-Ṭūsī. Ulugh Beg made al-Kāshī director of the observatory; and after his death in 1429, Qāḍī Zāda took his place. Qāḍī Zāda died before the astronomical tables were completed. ʿAlī Kūshjī succeeded him, and was director when the Jurjanian Tables were published.

Qāḍī Zāda married in Samarkand and had a son named Shams al-Dīn Muḥammad, who married a daughter of ʿAlī Kūshjī; from this marriage was born Quṭb al-Dīn, father of the Turkish mathematician Mīram Chelebi.

One of Qāḍī Zāda's calculations is presented below as an example of his approach to a geometrical-algebraic problem. It concerns determining the value of sin 1°. The example is drawn from the *Dastūr al-ʿamal wa-taṣḥīḥ al-jadwal* ("Practical Formula and Correction of the Table") of Mīram Chelebi, who states that Qāḍī Zāda wrote it.

Qāḍī Zāda, finding al-Kāshī's work on the approximate determination of the value of sin 1° to be very precise, commented upon it and gave further explanation in his *Risāla fī'l-jayb.*[2]

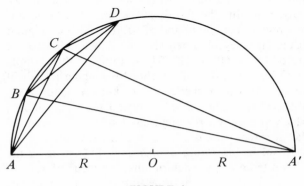

FIGURE 1

Like al-Kāshī, Qāḍī Zāda supposed that an arc *ABCD* taken on a circle with center *O* and diameter $\overline{AA'}$ is divided by the points *C* and *B* into three equal parts. Also like al-Kāshī, he was well aware of the impossibility of geometrically dividing an arc into three equal parts in order to obtain the chords $\overline{AB}$, $\overline{BC}$, $\overline{CD}$, $\overline{AC}$, $\overline{BD}$, and $\overline{AD}$. Applying Ptolemy's theorem to the inscribed quadrilateral *ABCD* thus constructed, he wrote the following equation:

$$(1) \qquad \overline{AB} \cdot \overline{CD} + \overline{BC} \cdot \overline{AD} = \overline{AC} \cdot \overline{BD}.$$

Considering the equalities

$$\overline{AB} = \overline{BC} = \overline{CD} \qquad \text{and} \qquad \overline{AC} = \overline{BD}$$

according to the hypothesis, equation (1) becomes

$$(2) \qquad \overline{AB}^2 + \overline{AB} \cdot \overline{AD} = \overline{AC}^2.$$

Qāḍī Zāda also supposed that arc *ABCD* is equal to 6°; therefore the chords $\overline{AB}$, $\overline{BC}$, and $\overline{CD}$ will belong to arcs of 2°.

Qāḍī Zāda next applied the iterative method of al-Kāshī to the determination of the chord belonging to the arc of 2°. That is, he algebraically divided the arc of 6°, the chord of which was known, into three equal parts. In taking as unknown (as a function of the parts of the radius) the chord belonging to the arc of 2° (that is, $\overline{AB} = x$) he obtained the equation

(3) $$x^2 + x \cdot \overline{AD} = \overline{AC}^2,$$

which is the equivalent of (2).

Since the chord $\overline{AD}$ belongs to the arc of 6°, Qāḍī Zāda (like al-Kāshī) obtained the following value of $\overline{AD}$ (in the sexagesimal system):

$$6^p.16.49.07.59.08.56.29.40$$

(where $\overline{OA} = R = 60^p$). This value of $\overline{AD}$ (in both authors) was determined by means of the arc of 72° and 60°, the chords of which were known geometrically. Knowing the chords of the arcs of 72° (the side of a regular pentagon) and of 60° (the side of a regular hexagon), they obtained the chord belonging to the arc of 72° − 60° = 12°. Then, applying the formula that gives the value of the chord belonging to half of an arc of which the chord is known, Qāḍī Zāda obtained the value of $\overline{AD}$. To find the value of $\overline{AC}$, which appears on the right-hand side of (1), he used the theorem ·discussed below (already utilized by al-Kāshī in his *Risāla al-muḥīṭiyya* ["Treatise on the Circumference"] for determining the value of π).

The theorem is that the ratio of the difference between the diameter of a circle and the chord belonging to any arc (when the arc is taken on this circle and one of its extremities passes through the extremity of the diameter) to the chord belonging to half of the supplement of this arc is equal to the ratio of this same chord to the radius of the circle.

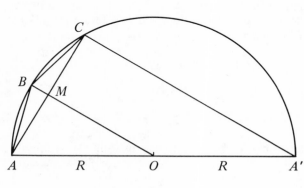

FIGURE 2

Taking $A'C$ as the arc, the theorem can be written in the following manner:

(4) $$\frac{\overline{AA'} - \overline{CA'}}{\overline{AB}} = \frac{\overline{AB}}{\overline{OA}}.$$

In equation (4)[3] one first separates $\overline{CA'}$, then for the value of $\overline{CA'}^2$ obtains (replacing $\overline{OA}$ with R and $\overline{AA'}$ with $2R$):

(5)
$$\overline{CA'}^2 = 4R^2 - 4\overline{AB}^2 + \frac{\overline{AB}^4}{R^2} = 4R^2 - 4x^2 + \frac{x^4}{R^2}.$$

From the right triangle ACA' one obtains

(6) $$\overline{CA'}^2 = 4R^2 - \overline{AC}^2;$$

and in equating the two values of $\overline{CA'}^2$ given by (5) and (6), one arrives at the equation

$$\overline{AC}^2 = 4x^2 - \frac{x^4}{R^2}.$$

Placing this value of $\overline{AC}^2$ in equation (3), one obtains (after simplifications) the equation

(7) $$3x = \overline{AD} + \frac{x^3}{R^2}.$$

The value of $\overline{AD}$ having been determined, the above equation can be written in the sexagesimal system as

$$3x = (6^p.16.49.07.59.08.56.29.40) + x^3.$$

Qāḍī Zāda, applying al-Kāshī's iterative method to this last equation, found for the unknown x (the value of the chord relative to 2°) the value

$$x = 2^p.05.39.26.22.19.28.32.52.33.$$

Half of this value would yield approximately the sine of 1°:

$$1^p.02.49.43.11.14.44.16.26.16.30.$$

This value is equal to that found by al-Kāshī. The value of sin 1° in the decimal system thus would be 0.017452406437, which is exact to within 10^{-12}, a degree of precision also achieved by al-Kāshī.[4]

NOTES

1. At this time Ulugh Beg was seventeen years old and was governor of Samarkand. He had appointed Qāḍī Zāda as private professor of mathematics and astronomy. *Qāmūs al-aʿlam*, 2, 1023.
2. The full title of this work by Qāḍī Zāda is *Risāla fī istikhrāj jayb daraja wāḥida bi'l-aʿmāl al-muʾassasa ʿalā qawāʿid ḥisābiyya wa-handasiyya* ("Treatise on the Determination of sin 1°, With the Aid of the Rules of Computation and of Geometry"). It is in the *Dastūr al-ʿamal wa-taṣḥīḥ al-jadwal* of Mīram Chelebi: Süleymaniyye Library, Istanbul, Hassan Hüsnü 1284; and in the same library, Tchorloulou Ali Pasha 342. Also in the Bibliothèque Nationale, in Paris, MSS Thévenol, Anc. Fonds 171.
3. Although the equation (4) that expresses the theorem is not in Euclid's *Elements*, it can readily be verified. From the obvious equalities $\overline{A'C} = 2\overline{OM}$ and $\overline{OM} = R - \overline{MB}$, one

immediately derives the relations $\overline{A'C} = 2R - 2\overline{MB}$ and $\overline{AA'} - \overline{A'C} = 2\overline{MB}$. Demonstrating the theorem thus reduces to verifying

$$\frac{2\overline{MB}}{\overline{AB}} = \frac{\overline{AB}}{R}, \text{ or } (4') \ \overline{AB}^2 = 2R \cdot \overline{MB}.$$

On the other hand, from the triangles (see Figure 2) AMB and OAM one can write the equations

$$\overline{AM}^2 = \overline{AB}^2 - \overline{MB}^2; \ \overline{AM}^2 = R^2 - \overline{OM}^2.$$

Then, from the fact that $\overline{OM} = R - \overline{MB}$, one obtains

$$AB^2 - MB^2 = R^2 - (R - \overline{MB})^2,$$

which, after simplification, gives equation $(4')$.

4. Abu'l Wafā' had found sin $1° \cong 0.017452414587$, which is exact to within 10^{-7}. Bibliothèque Nationale, Paris, MS 1138.

BIBLIOGRAPHY

I. ORIGINAL WORKS. *Risāla fi'l-ḥisāb* ("Treatise on Arithmetic"), which covers arithmetic, algebra, and measures, was written at Bursa in 1383. It is now in Šehid Ali Paša, Šehzada Ğami, Istanbul.

Sharḥ al-Mulaḥḥaṣ fi'l-hay'a, a commentary on the *Mulaḥḥaṣ* ("Compendium") of the astronomer 'Umar al-Jaghmīnī (*d.* 1444/1445), was written at Samarkand in 1412–1413 for Ulugh Beg. Several copies of the work are at Istanbul and in various European libraries. It has been printed at Delhi, Lucknow, and Teheran.

Sharḥ Ashkāl al-ta'sīs, a commentary on the *Ashkāl al-ta'sīs* ("Fundamental Theorems") of Shams al-Dīn al-Samarqandī, was written at Samarkand in 1412. It contains 35 propositions from Euclid's *Elements*. It is now in the Ayasofya (Süleymaniyya) Library, Istanbul, MS 2712. An author's autograph copy is at Bursa, Haradji-oğlou Library, no. 21. A Turkish trans. of *Ashkāl al-ta'sīs* by Muftu Zāda 'Abd al-Raḥīm Effendi (1795) is in the library of the Technical University of Istanbul, No. (4316–5075).

Risāla fi'l-hay'a wa 'l-handasa ("Treatise on Astronomy and Geometry"), in an author's autograph copy, is in Iné Bey Library, Bursa, MS 25.

Risāla fī samt al-qibla ("Treatise on the Azimuth of Qibla"), dealing with facing Mecca during prayer, is in Bursa, Iné Bey Library, MS 12. It also contains a short work by Mīram Chelebi, *Risāla fī taḥqīq samt al-qibla wa-barāhīniha* ("Treatise on the Verification and Proof of the Azimuth of Qibla").

Risālat al-jayb ("Treatise on the Sine"), Qāḍī Zāda's most original work, was written at Samarkand at the time of al-Kāshī. See Mīram Chelebi, *Dastūr al-'amal wa-taṣḥīḥ al-jadwal*, Süleymaniyya Library, Istanbul, Hassan Hüsnü 1284 and Tchorloulou Ali Pasha 342.

II. SECONDARY LITERATURE. See Adnan Adivar, *Osmanli Türklarinde Ilim* (Istanbul, 1943), 4; B. A. Rozenfeld and A. P. Youschkevitch, "Primechania k traktatu Kazi-Zade Ar-Rumi" ("Notes on a Treatise of Qāḍī

Zāda"), in *Istoriko-matematicheskie issledovania*, **13** (1960), 552–556; Süheyl Ünver, *Ali Kuschdju* (Istanbul, 1948), 73; and Sälih Zeki, *Assār-i Bāqiyya*, I (Istanbul, 1913), 186.

HÂMİT DİLGAN

AL-QALAṢĀDĪ (or AL-QALṢĀDĪ), ABU 'L-ḤASAN 'ALĪ IBN MUḤAMMAD IBN 'ALĪ (*b.* Basṭa [now Baza], Spain, 1412; *d.* Béja, Tunisia, December 1486), *arithmetic, algebra, Islamic law.*

Al-Qalaṣādī, the last known Spanish-Muslim mathematician, is also known by the epithets al-Qurashī and al-Basṭi, the latter referring to his place of birth. He remained in Basṭa until the city was taken by the Christians, at which time he began his journey through the Islamic world and studied with learned men.

Several books on arithmetic and one on algebra are attributed to al-Qalaṣādī. The one on algebra is a commentary on the *al-urjūzā al-Yāsmīnīya* of Ibn al-Yāsmīnī (*d.* 1204), which gave algebraic rules in verse. This *urjūzā* (poem) had been commented upon by several Western Muslims before al-Qalaṣādī.

One of his arithmetical works is a commentary on the *Talkhīṣ a'māl al-ḥisāb* ("Summary of Arithmetical Operations") of Ibn al-Bannā'; several summaries and extracts of this commentary have reached us. Al-Qalaṣādī's original work began with *al-Tabṣira fī 'lm al-ḥisāb* ("Clarification of the Science of Arithmetic"); it proved to contain difficult material, and he therefore simplified it in *Kashf al-jilbāb 'an 'ilm al-ḥisāb* ("Unveiling the Science of Arithmetic"). A shorter version of the same work is *Kashf al-asrār 'an waḍ 'ḥurūf al-ghubār* ("Unfolding the Secrets of the Use of Dust Letters [Hindu Numerals]"). The last two are said to have been used in some schools of North Africa for several generations, and the last is the work studied by F. Woepcke in his "Traduction du traité d'arithmétique d'Aboûl-Haçan Alî Ben Mohammed Alklçadî" (*Atti dell'Accademia pontificia de' nuovi lincei*, **12** [1858-1859], 399-438).

Since the 1850's al-Qalaṣādī has been credited with the following:

Dealing with the sequences $\sum n^2$ and $\sum n^3$.

Using the method of successive approximations for obtaining the roots of imperfect squares.

Using symbols in algebraic equations.

In the light of our present knowledge, the following can be stated:

First, al-Qalaṣādī has no claim to priority in the field of sequences. Those that he treated and more advanced ones, such as those of polygonal and pyramidal numbers, had been treated by Abū Manṣūr al-Baghdādī (d. 1037) and al-Umawī al-Andalusī (fl. fourteenth century).

Second, the method of finding square roots by successive approximation had been known to the Greeks and, probably, the Babylonians. In principle, it states that if r_1 is an approximation of $\sqrt{n}$, then let $r_2 = n/r_1$ and a better approximation is $r_3 = \frac{1}{2}(r_1 + r_2)$.

This method must have been known to the arithmeticians of eastern Islam, but they seem to have preferred to find roots expressed in sexagesimal fractions in almost the same way we use to find them to any desired decimal place. Al-Qalaṣādī, however, is the first mathematician known to have stressed it.

Third, al-Qalaṣādī used both short Arabic words and letters as symbols. The short words are

> wa (and) for addition
> illā (less) for subtraction
> fī (times) for multiplication
> ʿalā (over) for division.

Letters were also used to designate certain terms; these correspond to:

> j for jadhr (root)
> sh (from shayʾ, "thing"), or ∴ (the diacritical points from shayʾ) for x.
> m for māl (x^2)
> k for kaʿb (x^3)
> mm for mal mal (x^4)
> l (from the verb yaʿdilu) for equality.

The letters, more than the words, indicate a sense of symbolism. But al-Qalaṣādī has no claim to priority here, either; the same symbols were used in the same way by Ibn Qunfudh of Algiers (d. 1407/1408) and Yaʿqūb Ibn Ayyūb of Morocco (fl. ca. 1350), and many earlier writers in the East.

Like similar works from the thirteenth century on, al-Qalaṣādī's writings show Arabic arithmetic and algebra when their constituents—ancient manipulational tradition, Hindu techniques, and Greek number theory—are combined to form one entity. But they also reflect a civilization on the wane, for most of them are commentaries, summaries, or summaries of summaries of works by al-Qalaṣādī himself or by others.

BIBLIOGRAPHY

See C. Brockelmann, *Geschichte der arabischen literatur*, II, pt. 2 (Leiden, 1949), 343–344, and supp. II (Leiden, 1938), 363–369; al-Maqqarī, *Nafḥ al-ṭib*, Iḥsan ʿAbbās, ed., II (Beirut, 1968), 692–694; and H. Suter, *Die Mathematiker und Astronomen der Araber und ihre Werke* (Leipzig, 1900), no. 444, pp. 180–182. The best Arabic biography of al-Qalaṣādī is perhaps that written by M. Souissi of the University of Tunisia in the University periodical, **9** (1972), 33–49.

A. S. SAIDAN

AL-QARASHĪ. See **Ibn al-Nafīs.**

AL-QAZWĪNĪ, ZAKARIYĀ IBN MUḤAMMAD IBN MAḤMŪD, ABŪ YAḤYĀ (b. Qazwin [now Kasvin], Persia, ca. 1203; d. 1283), *cosmography, geography.*

An Arab by descent, al-Qazwīnī belonged to a family of jurists who had long before settled in Qazwīn. He seems to have left his native town at an early age, for in 1233 he was in Damascus, where he came under the influence of the mystic Ibn al-ʿArabī (d. 1240). He was a jurist by training and held the office of qāḍī (judge) in Wāsiṭ and Ḥilla (Iraq) under the last ʿAbbāsid caliph, al-Muʿtaṣim (1241–1258).[1]

Al-Qazwīnī wrote two works: one on cosmography, ʿAjāʾib al-makhlūqāt wa gharāʾib al-mawjūdāt ("Wonders of the Creation and Unique [phenomena] of the Existence"),[2] dedicated to ʿAṭā Malik-i Juwaynī (d. 1283); and one on geography, in two recensions, ʿAjāʾib al-buldān ("Wonders of the Lands"), written in 1262, and Āthār al-Bilād wa akhbār al-ʿibād ("Monuments of the Lands and Histories of the Peoples"), written in 1275.[3]

Muslim cosmography had its origins in the Greek cosmographical literature and was especially influenced by the works of Aristotle, but the first Muslim cosmographical works were not written until the twelfth century.[4] The world view of the Muslim writers on the subject, however, was mainly mystic-Islamic; they believed that there was an organic or spiritual relationship between everything that existed in the universe (including the angels) and that nothing, not even a single atom, was created by God without a purpose. Thus, the entire creation, including the wonderful and the unique phenomena, was a manifestation of divine wisdom and intelligence. It was for man to perceive and appreciate it and thereby achieve happiness in the world hereafter. This theme dominates the writings of al-Qazwīnī, who undoubtedly seems to have been influenced by Ṣūfī thought. Quoting a verse from the Koran, "Do they not look at the firmament above them? How We have made it and adorned it, and there are no flaws in it?" (S.L: v.6), al-Qazwīnī remarked that "looking" does not simply mean "turning the pupil of the eye towards it." It means reflection on the intelligible *(maʿqūlāt)* and sensible *(maḥsūsāt)* things and investigation of the wisdom underlying them and behind the changes *(taṣārīf)* so that one discovers the truth that leads to mundane pleasures and happiness in the next world. But contemplation of the intelligible could be accomplished only by one who, being righteous in character and pure of soul, had in addition the knowledge of the sciences. It is then that he acquires insight and perceives the wonderful in everything.[5] Hence, al-Qazwīnī accumulated in his work "the scattered material" and pieced together "the dissipated data" identifying the causes (of things) that an "unaware imbecile" would reject but that an "intelligent and reasonable" person would not refuse, even though it might be remote from "the normal practice."[6]

For his treatise on cosmography al-Qazwīnī cited more than a hundred written and oral sources, including Aristotle;[7] Ptolemy;[8] Dioscorides;[9] Balīnās;[10] al-Jāḥiẓ;[11] al-Rāzī;[12] Ibn Sīnā;[13] al-Bīrūnī;[14] Abū Ḥāmid al-Gharnāṭī (d. 1170);[15] the Koran and the Ḥadīth; the Torah;[16] *al-Filāḥa* by Ibn Waḥshiyya;[17] the *Tuḥfat al-gharāʾib*;[18] and a number of accounts by historians, geographers, and travelers. He also obtained oral information from his friends, who included jurists, judges, and explorers. There is great resemblance between the first part of *ʿAjāʾib al-makhlūqāt wa gharāʾib al-mawjūdāt*, which deals with the celestial world, and the contemporary Christian epistles on the same subject, and especially with the anonymous Syriac work *ʿEllath kull ʿEllān*.[19] These works have several common sources.

Al-Qazwīnī's vast knowledge of the sciences is reflected throughout his work; and he was well-versed in other subjects as well, for example, the Islamic sciences, history, and literature. To present systematically the various scientific concepts and theories that he so ably did, he must have studied the sciences as developed by the Muslims; and he must have collected and classified the data under appropriate headings. But he generally lacked critical ability and originality of thought. His aim was to present facts as they were available to him and to highlight, wherever necessary, the wonderful and the unique aspects of the universe. Throughout his work he also described the medicinal values, based on earlier authorities, of minerals, plants, and animals; and he stressed the influence of the heavenly bodies on human life. These aspects must have been an important reason for the popularity of his work in later centuries. Whatever the demerits of the work may be, he was undoubtedly the greatest Muslim cosmographer of medieval times.

Al-Qazwīnī's work on cosmography begins with an introduction in which he explains the meaning of the words *al-ʿajab* ("a perplexing phenomenon whose cause or mode of influence a man is unable to grasp") and *al-gharīb* ("every perplexing phenomenon, rare in occurrence, and contrary to the normal practice and usual observation"). He also describes the different categories of the created beings. The work is then divided into two parts, the first dealing with the celestial world, and the second dealing with the terrestrial world—the demarcation between the two being the sublunar region. The subdivisions of each part are given the name *al-Naẓar* ("contemplation").

In the first part al-Qazwīnī surveyed the astronomical knowledge of the Muslims and the astrological beliefs of the Arabs, described the inhabitants of the firmament (for example, the angels), and discussed in detail the calendars of the Arabs, the Romans, and the Persians, including their festivals and customs.

The spiritual approach of al-Qazwīnī to various astronomical and philosophical problems may be judged from the following examples. On the views of the philosophers that the sphere *(al-falak)* was limited and that beyond it space was neither empty *(khalāʾ)* nor full *(malāʾ)*, he remarks that Muḥammad ibn ʿUmar al-Rāzī (d. 1210), after having exposed the falsehood of this belief, said, "Any one who attempts to determine the (extent) of the kingdom of God with the yardstick of reason faces utter misguidance."[20] Again, after stating that Aristotle and "his companions" believed that time was the measure of the movement of the sphere, while others considered it the passage of days and nights, al-Qazwīnī said, "Time is the most precious capital by which all happiness is earned and it dwindles away gradually. Your Time is your life and its quantity is known to God even though it may not be known to you."[21]

On the angels al-Qazwīnī said that they are formed of a simple substance, possessing life and intelligence, but are free from sexual desire and anger. They are obedient to God and execute His will[22] and are meant for the welfare of the world and the perfection of being.[23] They also assist in the metabolic process of digestion, which, if unaided, men could not perform. Hence men are the apparent workers while angels are the imperceptible workers.[24]

The second part of *ʿAjāʾib al-makhlūqāt wa gharāʾib al-mawjūdāt* deals with the elements and their spheres —the sphere of fire, meteors, and thunderbolts; the sphere of air, clouds, rains, winds, thunder and lightning, and halo and rainbow; the sphere of water, seas, islands, and the fishes and animals found in them; the sphere of earth, shape, size, circumference and motion of the earth, mountains, rivers, and springs; as well as minerals, varieties of stones, quicksilver, sulfur, and ambergris; plants, animals, and man; birds, insects, reptiles, and hybrid animals; and shapes, dresses, and colors of angels. It is in this part of the book that al-Qazwīnī revealed his extensive knowledge and grasp of the sciences.

The element (*'unṣur*) forms the original substance and hence the bodies (*ajsām*), namely, fire, air, water, and earth, are called elements or pillars (*arkān*) because it is from them that minerals, plants, and animals originate. Each of the elements has a sphere of its own (although some overlapping others, as the sphere of air with those of fire and earth), dual temperaments and qualities, and a center wherein it rests naturally unless restrained by a hindering object (*māni'*). When this object is removed the element either pulls itself toward the center of the universe and in the process becomes heavy, or it moves toward the circumference of the universe, where it becomes light. The elements possess the property of interchangeability.[25]

All bodies originating from their sources (*umm*) either have the quality of growth or they do not have it. To the latter category belong the minerals. Those bodies that possess the quality of growth may or may not have the faculty of sensibility and motion. To the latter category belong the plants; animals belong to the former.

Of the material (*mādda*) out of which minerals, plants, and animals are produced, al-Qazwīnī stated that the philosophers assert that the first things the elements change into are vapors (*al-bukhār*) and extracts (*al-'aṣīr*). The finer parts of the sea and river waters rise into the air in the form of vapors because of the action of solar heat; the extracts of the rain waters imported inside the earth mix with the earth particles and become thick. The innate heat of the earth then "cooks it thoroughly" and turns the extract into the material needed for the production of plants, minerals, and animals. Each of these bodies is connected with the other by a wonderful arrangement and extraordinary system. Thus, the first (lowest) body in the order of the universe is the earth and the last (highest) is the pure angelic soul. The first (lowest) part of the minerals is joined to earth or water and the last (highest) part is joined to the plants; then the first of the plants is joined to minerals and the last to the animals; the first of the animals is joined to the plants and the last to man; and the first of human souls is joined to animals and the last to angelic souls.[26] Minerals are produced by the vapors and smokes inside the earth that become mingled in different types of mixtures in different quantities and modes. Hence al-Qazwīnī discussed in detail a variety of minerals with different qualities and properties.[27]

Plants occupy the middle position between minerals and animals. The "defect" of being in a solid state (like the minerals) is absent from them, yet they cannot attain the stage of complete sensibility and motion that is peculiar to animals. Yet plants do have a few qualities in common with the animals. God provides every genus with certain organs to ensure its protection, and any additional provision would be burdensome. Moreover, plants, unlike animals, do not need sensibility and motion. It is one of the wonders of divine workmanship that the seed and the stone, when planted in the earth, draw their nourishment from the sun and from the fine particles of the earth and water. These particles then "heap upon each other" by means of the faculties (*quwā*) created by God in them until they mature and grow into plants and trees with trunks, leaves, and fruits.[28]

Al-Qazwīnī also described the different stages of evolution in the birth of man, beginning with the change of food into semen in the body, conception, and the fetus.[29]

Al-Qazwīnī placed animals in the third stage of the creation, furthest removed from their sources. Although minerals retain their solid state, they belong to the first stage because of their proximity to the simple objects (*basā'iṭ*). Occupying the middle position between minerals and animals, plants belong to the second stage because they acquire the quality of growth and development. Animals occupy the third stage because they combine the qualities of growth, sensibility, and motion, which are present in every "animal," including flies and mosquitos.[30] Man is a combination of soul and body and is the noblest of the animals and the choicest of the creation. He is a small world in himself. He possesses the qualities of speech, intelligence, feeling, and vigor and has a brain; each of these performs its duty in protecting him. He is a plant by virtue of growth, an animal by virtue of sensibility and motion, and an angel by virtue of his knowledge of the reality of things.[31]

Like his work on cosmography, al-Qazwīnī wrote *Āthār al-bilād* to collect information on the fine objects created by God and on the wonders of his wisdom bestowed upon the countries and the peoples.[32] It is a geographical dictionary in which each town and country is described in alphabetical order within each of the seven Ptolemaic climes running parallel to the equator from east to west in the Northern Hemisphere. After describing the geographical position and some physical features of a location, al-Qazwīnī mentioned personalities connected with it and recorded interesting aspects of their lives; but such descriptions were confined mainly to the Islamic world. The Ṣūfīs and religious men, jurists and *imāms* found an honored place in his accounts. He then related the unique features, wonderful objects, talismans, and interesting habits and customs of the inhabitants of the place. The work includes three introductions: the first deals with the sociological need for the foundation of cities and villages; the second discusses the special properties of lands and is divided into two sections, one covering the influence of lands on its inhabitants and the other examining minerals, plants, and animals; the third introduction deals with the climes (*aqālīm*) of the earth.

Al-Qazwīnī utilized a number of sources for his book, including literary works, histories and legends, geographical works, and travel accounts; but in drawing information from them he kept in view his main objective, namely, selection of the wonderful and the unique aspects of the lands and their inhabitants. Much of the information and sources are identical with those of his work on cosmography. As a geographer his main contribution lies in the fields of human and physical geography.

NOTES

1. On the life of al-Qazwīnī, see Krachkovsky, *Arabskaya geograficheskaya literatura* (Moscow-Leningrad, 1957), with Arabic trans. by Ṣalāḥ al-Dīn 'Uthmān Hāshim as *Ta'rīkh al-adab al-jughrāfī al-'Arabī*, I (Cairo, 1963), 360–361. Cf. M. Streck, *Encyclopaedia of Islam*, 1st ed., II, 841–844.
2. The oldest MS is dated 1280; but Wüstenfeld's ed. (Göttingen, 1849) is based on the third recension of the work, hence it is considered arbitrary and unauthentic (Krachkovsky, *op. cit.*, 362; Streck, *op. cit.*, 842). The ed. used for the present article was printed on the margin of al-Damīrī's *Kitāb al-Ḥayawān* (Cairo, A.H. 1309), repr. by Dār al-Qāmūs (Beirut, n.d.). It was based on the second recension to which also belonged the oldest MS; see Streck, *op. cit.*, 841, 842, 844.
3. For a full description of the extant MSS of both *'Ajā'ib al-makhlūqāt* and *Āthār al-bilād* and on their Persian, Turkish, and Chaghatāy trans., see Krachkovsky, *op. cit.*, 362–365; C. Brockelmann, *Geschichte der Arabischen litteratur*, I (Leiden, 1943), 481 and supp., I (Leiden, 1937), 882–883; C. A. Storey, *Persian Literature*, II, pt. I (London, 1958), 124–128; E. G. Browne, *A Literary History of Persia*, II (Cambridge, 1951–1953), 482–483.
4. See Krachkovsky, *op. cit.*, 324.
5. *'Ajā'ib*, I, 3–4.
6. *Ibid.*, 6.
7. Aristotle was an important source of al-Qazwīnī's thought; on his *Kitāb al-Ḥayawān*, see Ḥājjī Khalīfa, *Kashf al-ẓunūn*, I (Istanbul, 1941), 696. The so-called *Petrology of Aristotle*, wrongly attributed to Aristotle, was also a source (see Streck, *op. cit.*, 843). Cf. R. Walzer, *Encyclopaedia of Islam*, 2nd ed., I, 632.
8. Ptolemy's *Almagest* formed an important source of al-Qazwīnī's knowledge of astronomy.
9. Dioscorides is quoted on stones and plants, probably from his *Kitāb al-Ḥashā'ish*; see Ibn Juljul, *Ṭabaqāt al-aṭibbā' wa 'l-ḥukamā'*, Fu'ād Sayyid, ed. (Cairo 1955), intro. and 21–23.

10. Balīnās of the Arabs was the sage whose personality was based on the tradition about Appolonius of Tyana in Cappadocia (1st century); his work *Kitāb al-Khawāṣṣ* referred to by al-Qazwīnī is a fiction in the opinion of Steinschneider and has not yet been traced; see M. Plessner, *Encyclopaedia of Islam*, 2nd ed., II, 994–995.
11. Al-Qazwīnī used his *al-Ḥayawān* extensively.
12. Al-Qazwīnī quotes al-Rāzī on subjects of medicinal value and even on geographical information that al-Rāzī borrowed from the original geographical sources, including al-Jayhānī.
13. Al-Qazwīnī used the *al-Shifā'* of Ibn Sīnā, who formed an important source of al-Qazwīnī's knowledge. He relates experience about the rainbow (*'Ajā'ib*, I, 162–163) on a mountain between Abīward and Ṭūs while the clouds were below him; as he descended the circle of the rainbow narrowed, and when he reached the cloud it disappeared; cf. *al-Shifā'*, *al-Ṭabī'iyyāt*, section 5. *al-Ma'ādin wa 'l-āthār al-'ulwiyya*, Ibrahim Madkour *et al.*, eds. (Cairo, 1965), 52–53.
14. Al-Qazwīnī used al-Bīrūnī's *al-Athār al-bāqiya 'an al-qurūn al-khāliya* and other works.
15. Al-Qazwīnī seems to have consulted his *Tuḥfat al-albāb wa nukhbat al-a'jāb*; but he also mentions a *Kitāb al-'Ajā'ib* (*'Ajā'ib*, I, 198, 205), which is probably the same as *Tuḥfat al-albāb*, Dubler, ed., with a Spanish trans. (Madrid, 1953).
16. The first book of the Tōrāh is quoted on genesis, although indirectly, from Arabic biographical works. On the various recensions and trans. of the Tōrāh into Syriac and Arabic, see Ḥājjī Khalīfa, *op. cit.*, 504–506.
17. All the references to *al-Filāḥa*, except the one on locusts (*'Ajā'ib*, II, 269), pertain to plants and fruits. The full title of Ibn Waḥshiyya's work is *al-Filāḥa al-nabaṭiyya* ("Nabataean Agriculture"); al-Qazwīnī also probably used Ibn Waḥshiyya's *Kitāb Asrār al-shams wa 'l-Qamar*; see al-Dimashqī, *Nukhbat al-dahr fī 'ajā'ib al-barr wa 'l-baḥr*, Mehren, ed. (Leipzig, 1923), 78. Cf. Brockelmann, *op. cit.*, supp. I, 430–431, on Ibn Waḥshiyya's works.
18. The work used by al-Dimashqī is also identified with *Tuḥfat al-'ajā'ib wa ṭurfat al-gharā'ib* by Mehren, who attributes it to Majd al-Dīn Abu 'l-Sa'āda Ibn al-Athīr al-Jazarī (*d.* 1210); see Mehren, al-Dimashqī, *op. cit.*, lxxxviii. Cf. Brockelmann, *op. cit.*, I, 609, for another name of the work: *al-Durr al-muḍī'a fī 'ajā'ib al-barriyya*. The Arabic biographical accounts, however, differ about the authorship of the work.
19. Kayser, ed. (Leipzig, 1889).
20. *'Ajā'ib*, I, 92.
21. *Ibid.*, 108.
22. *Ibid.*, 93.
23. *Ibid.*, 94.
24. *Ibid.*, 107–108.
25. *Ibid.*, 144–145.
26. *Ibid.*, 281–282.
27. *Ibid.*, 282 ff.
28. *Ibid.*, II, 2.
29. *Ibid.*, 93–97.
30. *Ibid.*, 78.
31. *Ibid.*, 80.
32. *Āthār al-bilād*. The ed. used for the present article was published in Beirut (1960).

S. MAQBUL AHMAD

QUATREFAGES DE BRÉAU, JEAN-LOUIS-ARMAND DE (*b.* Valleraugues, France, 10 February 1810; *d.* Paris, France, 12 January 1892), *medicine, zoology, anthropology.*

The son of Jean-François de Quatrefages and Marguerite-Henriette-Camille de Cabanes, Quatrefages received a Protestant education. From 1822 to 1826 he studied at the Collège Royal of Tournon, where he showed great enthusiasm for mathematics and the exact sciences. He was sent to study medicine at the University of Strasbourg, where he first defended the two theses required for the *doctorat ès sciences:* "Théorie d'un coup de canon" (1829) and "Du mouvement des aérolithes considérés comme des masses disséminées dans l'espace par l'impulsion des volcans lunaires" (1830). He obtained an assistantship in chemistry and physics at the Strasbourg Faculty of Medicine in 1830. After defending his medical thesis, "L'extroversion de la vessie" (1832), he established a medical practice in Toulouse (1833). In 1836 he founded the *Journal de médecine et de chirurgie de Toulouse,* in which he published many articles, especially a series of zoological observations that gained him a provisional appointment to the chair of zoology at the Toulouse Faculty of Sciences. He was, however, disappointed in his academic ambitions and left for Paris in 1840.

In the same year, at Paris, Quatrefages defended two theses for his third doctorate, this time in the natural sciences: "Sur les caractères zoologiques des rongeurs et sur leur dentition en particulier" and "Sur les rongeurs fossiles." During this period he earned a meager living by writing popular articles and—thanks to his talent for drawing—by preparing the plates for the new edition of Cuvier's *Le règne animal* then being prepared by Henri Milne-Edwards.

From 1840 to 1855 Quatrefages, with the support and advice of Milne-Edwards, studied the invertebrates, publishing more than eighty memoirs on them during this period. He also undertook a series of voyages in order to study animal life along the Atlantic coast of France, and in 1844 he participated in a zoological mission off the Sicilian coast with Milne-Edwards and Émile Blanchard. These expeditions are recorded in his two-volume *Souvenirs d'un naturaliste* (1854). In 1850 Quatrefages was awarded the chair of natural history at the Lycée Henri IV in Paris; in April 1852 he replaced Savigny as a member of the anatomy and zoology section of the Academy of Sciences; and in 1855, through the support of Milne-Edwards, he obtained the vacant chair in the natural history of man at the Muséum d'Histoire Naturelle. Thus began his career as anthropologist. He managed nonetheless to complete and publish *Histoire naturelle des annelés marins et d'eau douce* (1865). Moreover, twice between 1857 and 1859 he was sent on missions by the Academy of Sciences to study the devastating diseases attacking the silkworm and to find a means of controlling them.

Quatrefages's zoological views were basically those of his teacher Milne-Edwards. Preoccupied with "organic complication," Quatrefages wanted to study the degeneration (*dégradation*) of organisms and to ascertain the mechanism by which it took place. Marine invertebrates and especially annelids, because of the multiplicity and diversity of their typical structures, provided him with an abundance of material for this study: "Nowhere else does degeneration appear so extensively and in such variety." He viewed this organizational degeneration as a function of a decrease in the physiological division of labor. He accompanied his ideas with a critique of the Cuvierist position, which linked the importance of the functions to that of the organs.

It was on these theoretical positions that Quatrefages established his incorrect and highly controversial theory of *phlébentérisme* (see, for instance, "Mémoires sur les gastéropodes phlébentérés, ordre nouveau de la classe des gastéropodes," in *Annales des sciences naturelles,* 3rd ser., **1** [1844], 129–183). In accord with the principle of the physiological division of labor, which implies that when a group of organs with a specific function disappears, the accompanying function continues, Quatrefages assumed the disappearance of the circulatory system in certain creatures. He first hypothesized its disappearance in the gastropod mollusks, which he called *phlébentérés,* and later supposed that "almost all types of invertebrates have their derivative *phlébentérés.*" According to Quatrefages, the circulatory system was replaced by the digestive tube, substituted for circulating not the blood but chymified alimentary substances. At first this conception generated much interest and excited a rather hot controversy; but with time, it was quietly abandoned even by Quatrefages's friends like Milne-Edwards.

Quatrefages often extended his analysis to the tissues. For example, from the work done on his Sicilian expedition he produced a study on the histology of the amphioxus; and he was one of the first to work in comparative histology, although he never accepted the cell theory as applied to animals. Convinced, like Milne-Edwards, of the importance of embryological characteristics for the classification of organisms, he believed that he had surpassed Cuvier and had rediscovered the spirit of A.-L. Jussieu. Moreover, unlike Cuvier, Quatrefages believed that there were transitional forms between certain "embranchements"; and he created the class Gephyrea (a class now comprising Priapulida, Sipunculida, and Echiurida) as a link between the Articulata and the Radiata (see "Mémoire sur l'échiure de Gaertner," in *Annales des Sciences Naturelles,* 3rd ser., **7** [1847], 307–344).

It was also as a zoologist that Quatrefages approached anthropology when he began to teach that subject at the Muséum d'Histoire Naturelle in 1856. For him the fundamental problem of this discipline remained the unity of the human species. Nonetheless, he conceived of anthropology in a very broad sense, in which the natural history of man would be associated with a general survey of all human populations, both living and fossil. For the Paris Exposition of 1867, Quatrefages wrote *Rapport sur les progrès de l'anthropologie*, in which he defended his monogenetic ideas. In collaboration with his disciple E.-T. Hamy, who did most of the work, he published *Crania ethnica* (1882), surveying what was then known of the comparative craniology of living and fossil human races, for which he was trying to achieve a classification. Convinced by Lartet of the existence of fossil man, Quatrefages in 1863 supported Boucher de Perthes after the discovery at Abbeville of a human jaw accompanied by flint axes. In 1871, at the Fifth International Congress of Anthropology and Prehistoric Archaeology held at Bologna, he came out in favor of the view that man existed as far back as the Tertiary.

Nevertheless, Quatrefages remained a lifelong opponent of the Darwinian theory and of the simian origin of man; he also defended the conception of a human kingdom distinct from the animal one (*L'unité de l'espèce humaine* [1861]). In 1870 he published a collection of his antitransformist articles as *Darwin et ses précurseurs français*; two more volumes were published in 1894 as *Les émules de Darwin*.

Quatrefages was a member of many scholarly societies, both French and foreign, and was a founder of the Association Française pour l'Avancement des Sciences.

BIBLIOGRAPHY

I. ORIGINAL WORKS. An almost complete bibliography of Quatrefages's works is in Godefroy Malloizel, *Armand de Quatrefages de Bréau. Liste chronologique de ses travaux* (Autun, 1893). Also useful is *Notice sur les travaux zoologiques et anatomiques de M. A. Quatrefages* (Paris, 1850; 2nd rev. ed., 1852), prepared by Quatrefages at the time of his candidacy for the Academy of Sciences; it includes commentaries following the title of each memoir.

II. SECONDARY LITERATURE. In addition to several obituary notices listed by Malloizel, another valuable reference is Edmond Perrier's 95–page preface to vol. I of Quatrefages's *Les émules de Darwin.* See also Georges Hervé and L. de Quatrefages, "Armand de Quatrefages de Bréau, médecin, zoologiste, anthropologue (1810–1892)," in

Bulletin de la Société française d'histoire de la médecine, **20** (1926), and **21** (1927).

CAMILLE LIMOGES

QUENSTEDT, FRIEDRICH (*b.* Eisleben, Germany, 9 July 1809; *d.* Tübingen, Germany, 21 December 1889), *paleontology, mineralogy, geology.*

In crystallography Quenstedt extended the work of Christian Weiss, and in geology that of Leopold von Buch. His most significant work concerned the paleontology and the stratigraphy of the Jurassic series of Swabia.

Quenstedt received his early education at Eisleben and studied mineralogy and geology at the University of Berlin under Weiss and von Buch. He became associate professor of mineralogy and geology at Tübingen in 1837 and was named full professor in 1842. He held this post until his death. Because of his superior teaching ability Quenstedt attracted many students, and his dedication to paleontology aroused the interest in fossils of even the farmers in the areas surrounding Tübingen.

In his crystallographic writings Quenstedt extended the application of spherical geometry to crystallography, a technique introduced by Weiss. Von Buch had provided the basis for the geology of the Swabian Jura, but Quenstedt made important additions. He subdivided each of the three principal divisions of the Jura into six zones on the basis of petrographical development and paleontological evidence. He did not, however, compare the succession of Jurassic rocks found in Swabia with those in other countries. Quenstedt's *Petrefaktenkunde Deutschlands*, published in seven parts over thirty-eight years, contained 218 plates and was one of the best reference works on vertebrate fossils of the period.

For years before the publication of the Darwinian theory, Quenstedt taught that species were not sharply defined—that not only could variations be found in the same stratigraphic horizon, but also that the same species could be seen in several horizons. This approach was thus opposed to catastrophism, and his thinking was distinctly phylogenetic. His methods and views often brought him into sharp conflict with many of his contemporaries.

BIBLIOGRAPHY

Quenstedt's chief publications were *Methode der Krystallographie* (Tübingen, 1840); *Das Flözgebirge Württembergs* (Tübingen, 1843); *Petrefaktenkunde Deutschlands*, 7 pts.

(Tübingen, 1846–1884); *Über Lepidotus im Lias und Württembergs* (Tübingen, 1847); *Beiträge zur rechnenden Krystallographie* (Tübingen, 1848); *Handbuch der Petrefaktenkunde* (Tübingen, 1852); *Handbuch der Mineralogie* (Tübingen, 1854); *Der Jura* (Tübingen, 1856); *Sonst und Jetzt: Populäre Vorträge über Geologie* (Tübingen, 1856); *Epochen der Natur* (Tübingen, 1861); *Geologische Ausfluge in Schwaben* (Tübingen, 1864); *Schwabens Medusenhaupt* (Tübingen, 1868); *Klar und Wahr* (Tübingen, 1872); *Neue Reihe populärer Vorträge über Geologie* (Tübingen, 1872); *Grundriss der bestimmenden und rechnenden Krystallographie* (Tübingen, 1873); and *Die Ammoniten des schwäbischen Jura*, 3 vols. and atlas (Stuttgart, 1882–1889). Quenstedt also published approximately 40 articles in various scientific journals; see Royal Society *Catalogue of Scientific Papers*, V, 54–55; VIII, 678; XI, 84.

A brief biography is A. Rothpletz, "Friedrich Quenstedt," in *Allgemeine deutsche Biographie*, LIII, 179–180.

JOHN G. BURKE

QUERCETANUS, JOSEPHUS. See **Duchesne, Joseph.**

QUETELET, LAMBERT-ADOLPHE-JACQUES (*b.* Ghent, Belgium, 22 February 1796; *d.* Brussels, Belgium, 17 February 1874), *statistics*.

Adolphe Quetelet was the son of François-Augustin-Jacques-Henri Quetelet and Anne-Françoise Vandervelde. After graduating from the *lycée* in Ghent he spent a year as a teacher in Oudenaarde. In 1815 he was appointed professor of mathematics at the Collège of Ghent. He wrote an opera, together with his friend G. P. Dandelin (better known for a theorem on conics); he also published poems and essays. Quetelet was the first to receive a doctorate (1819) from the newly established University of Ghent, with a dissertation on geometry. The same year he was appointed professor of *mathématiques élémentaires* at the Athénée of Brussels. In 1820 he was elected a member of the Académie Royale des Sciences et Belles-Lettres of Brussels. During the next years he worked in geometry. His papers were published by the Academy and in the periodical *Correspondance mathématique et physique*, which he founded and coedited with J. G. Garnier, a professor at Ghent who had guided Quetelet's first steps in higher mathematics. From 1824 Quetelet taught higher mathematics at the Athénée and physics and astronomy at the Musée, which later became the Université Libre. His wife, whom he married in 1825, was a daughter of the French physician Curtet and a niece of the chemist van Mons; she bore him a son and a daughter. In 1826 he published popular books on astronomy and on probability.

From 1820 Quetelet had proposed founding an observatory, and in 1823 the government sent him to Paris to gain experience in practical astronomy. Here he met famous scientists. His increasing interest in probability was possibly due to the influence of Laplace and Fourier. In 1827 he went to England to buy astronomical instruments and to visit universities and observatories. The following year he was appointed astronomer at the Brussels Royal Observatory, which was not completed until 1833. Meanwhile he traveled extensively. In 1834 he was elected permanent secretary of the Brussels Academy.

From 1832 Quetelet lived at the observatory. His research there was more meteorological and geophysical than astronomical, with an emphasis on statistics. He had turned to statistics as early as 1825, and until 1835 he wrote a considerable number of papers on social statistics. In that year he published *Sur l'homme et le développement de ses facultés, essai d'une physique sociale*, which made him famous throughout Europe. Subsequently a great part of his activity consisted in organizing international cooperation in astronomy, meteorology, geophysics, and statistics. His work after 1855 was impaired by the consequences of a stroke he had suffered in that year.

Quetelet was an honorary member of a great many learned societies and received many decorations. His funeral was a gathering of princes and famous scientists, and his memory was honored by a monument, unveiled in Brussels in 1880.

By his contemporaries his personality has been described as gay, charming, enthusiastic, and gifted with wide intellectual interests. Though he exerted a tremendous influence in his lifetime, his fame hardly survived him. His work has not been republished since his death.

The word "*Statistik*," first printed in 1672, meant *Staatswissenschaft* or, rather, a science concerning the states. It was cultivated at the German universities, where it consisted of more or less systematically collecting "state curiosities" rather than quantitative material. The actual predecessor of modern statistics was the English school of political arithmetic; the first effort to describe society numerically was made by Graunt in 1661. This school, however, which included Malthus, suffered from a lack of statistical material. In 1700 Napoleon, influenced by Laplace and fond of numerical data, established the Bureau de Statistique. In 1801 the first general censuses were held in France and England. Statistics became a fashionable subject, but nobody knew what kind of data to collect or how to organize the material. Nothing was done to justify Fourier's plea:

QUETELET

"Statistics will not make any progress until it is trusted to those who have created profound mathematical theories."

With Quetelet's work of 1835 a new era in statistics began. It presented a new technique of statistics or, rather, the first technique at all. The material was thoughtfully elaborated, arranged according to certain preestablished principles, and made comparable. There were not very many statistical figures in the book, but each figure reported made sense. For every number, Quetelet tried to find the determining influences, its natural causes, and the perturbations caused by man. The work gave a description of the average man as both a static and a dynamic phenomenon.

This work was a tremendous achievement, but Quetelet had aimed at a much higher goal: social physics, as the subtitle of his work said; the same title under which, since 1825, Comte had taught what he later called sociology. Terms and analogies borrowed from mechanics played a great part in Quetelet's theoretical exposition. To find the laws that govern the social body, said Quetelet, one has to do what one does in physics: to observe a large number of cases and then take averages. Quetelet's average man became a slogan in nineteenth-century discussions on social science. The use of mathematics and physics in social sciences was praised, although none of the parties to the discussions knew what it should really mean.

The above statement also applied to Quetelet himself. There is not much more mathematics contained in his work than the vague idea that the reliability of an average increases with the size of the population—and even this idea was not understood by many of his contemporaries. It is evident that Quetelet knew more about mathematical statistics, but he never thought to apply it in his social statistics. Neither did he make significance tests, although as early as 1840 they came into use in medical statistics. He often urged that one should consider not only the average but also the deviation in order to know whether the latter is accidental or not, but he never followed up this suggestion. He always judged intuitively whether a statistical figure was constant or variable under different conditions.

In more theoretical work about 1845, Quetelet approached mathematical statistics more closely. For the first time he mentioned the normal distribution, or, rather, a binomial distribution of a high degree. As an example, he explained the error distribution by the theory of elementary errors. Possibly he made this discovery independently of Thomas Young, G. Hagen, and Bessel. In any case it was clearly

Quetelet's own achievement to unveil the normal distribution of the heights of a population of soldiers. The normal distribution, not only as a law of observation errors but also as a genuine natural law, was indeed an important discovery, although Quetelet's examples were not convincing.

Quetelet's impact on nineteenth-century thinking can in a certain sense be compared with Descartes's in the seventeenth century. He certainly gave science new aims and tools, although his philosophy was rather pedestrian and his thinking in somewhat sophisticated matters was rather confused. There was a strong emotional component in Quetelet's influence. In fact, he became famous for a passage, quoted again and again from *Sur l'homme*, in which he draws his conclusions from the statistics of the French criminal courts from 1826 to 1831:

> The constancy with which the same crimes repeat themselves every year with the same frequency and provoke the same punishment in the same ratios, is one of the most curious facts we learn from the statistics of the courts; I have stressed it in several papers; I have repeated every year: *There is an account paid with a terrifying regularity; that of the prisons, the galleys, and the scaffolds. This one must be reduced.* And every year the numbers have confirmed my prevision in a way that I can even say: there is a tribute man pays more regularly than those owed to nature or to the Treasury; the tribute paid to crime! Sad condition of human race! We can tell beforehand how many will stain their hands with the blood of their fellow-creatures, how many will be forgers, how many poisoners, almost as one can foretell the number of births and deaths.
>
> Society contains the germs of all the crimes that will be committed, as well as the conditions under which they can develop. It is society that, in a sense, prepares the ground for them, and the criminal is the instrument . . . .
>
> This observation, which seems discouraging at first sight, is comforting at closer view, since it shows the possibility of improving people by modifying their institutions, their habits, their education, and all that influences their behaviour. This is in principle nothing but an extension of the law well-known to philosophers: as long as the causes are unchanged, one has to expect the same effects.

H. T. Buckle, in England, and Adolph Wagner, in Germany, were Quetelet's most fervent supporters in social science. Florence Nightingale considered his work a new Bible.

BIBLIOGRAPHY

I. ORIGINAL WORKS. Quetelet's writings include *Sur l'homme et le développment de ses facultés, essai d'une physique sociale* (Paris, 1835); "Sur l'appréciation des documents statistiques . . .," in *Bulletin de la Commission de Statistique* (de Belgique) (1845), 205–286; *Lettres à S. A. R. le duc régnant de Saxe-Cobourg et de Gotha* (Brussels, 1846); and *Du système social et des lois qui le régissent* (Paris, 1848).

II. SECONDARY LITERATURE. On Quetelet and his works, see Hans Freudenthal, "De eerste ontmoeting tussen de wiskunde en de sociale wetenschappen," in *Verhandelingen van de K. vlaamse academie voor wetenschappen, letteren en schone kunsten van België*, **28**, no. 88 (1966); Maurice Halbwachs, *La théorie de l'homme moyen* (Paris, 1912); F. H. Hankins, *Adolphe Quetelet as Statistician* (New York, 1908); G. F. Knapp, "Bericht über die Schriften Quetelets zur Socialstatistik und Anthropologie," in *Jahrbücher für Nationalökonomie und Statistik*, **17** (1871), 106–174, 342, 358; J. Lottin, *Quetelet statisticien et sociologue* (Paris, 1912); and E. Mailly, *Essai sur la vie et les ouvrages de L.-A.-J. Quetelet* (Brussels, 1875).

HANS FREUDENTHAL

IBN AL-QUFF, AMĪN AL-DAWLAH ABŪ AL-FARAJ IBN MUWAFFAQ AL-DĪN YAʿQŪB IBN ISHĀQ AL-MASĪHĪ AL-KARAKĪ (*b.* Karak, Jordan, 22 August 1233; *d.* Damascus, Syria, 1286), *medicine, physiology, natural sciences, philosophy.*

Ibn al-Quff's father, Muwaffaq al-Dīn Yaʿqūb, was a Christian Arab (an adherent of the Imperial Orthodox Church), who held an important governmental position under the Ayyūbids in Karak. These facts are reflected in his cognomens al-Masīhī (the Christian) and al-Karakī. When Muwaffaq al-Dīn was promoted to the position of a secretary-scribe of the high court, the family moved to Sarkhad in Syria. There Muwaffaq al-Dīn met and formed a close friendship with the physician-historian Ibn Abī Usaybiʿah (1203–1270), who spoke of him as "a learned scholar, unequaled scribe in the elegance and perfection of his handwriting, a man of letters, a competent historian, and a pleasant companion, witty and respectable."

Upon the father's request, Ibn Abī Usaybiʿah agreed to teach young Ibn al-Quff the healing art. The tutor was soon impressed by the brilliance and aptitude for learning of his new student. He also found him fond of reading biographies of illustrious sages, and inclined to quiet, thoughtful meditations. Ibn Abī Usaybiʿah began to teach young Ibn al-Quff with the assistance of preliminary and fundamental texts on the healing art, such as the *Masāʾil* (an introduction to medicine) of Hunayn ibn Ishāq, and the *Aphorisms* and the *Prognosis* of the Hippocratic corpus in the Arabic version as rendered also by Hunayn ibn Ishāq. Through the study of leading manuals, such as those by al-Rāzī, Ibn al-Quff was instructed by Usaybiʿah in the classification and treatment of diseases, and their causes and symptoms.

Later, Ibn al-Quff's father was transferred to the high court in Damascus, and the family moved to the Syrian capital. Here, Ibn al-Quff studied metaphysics, philosophy, medicine, natural sciences, and mathematics. He was then appointed an army physician-surgeon at the citadel of Ajlun in Jordan, where he stayed for several years. After his fame had spread, he was transferred to Damascus, where until his death at the age of fifty-two, he taught medicine and performed his professional duties among the soldiers stationed at the citadel.

Despite his absorbing responsibilities as physician-surgeon for the Mamluk army, Ibn al-Quff was the best-known medical educator of his time in Syria and a prolific author. He wrote a philosophic commentary on the *Ishārāt* of Ibn Sīnā and *Al-Mabāhith* on natural sciences, but neither was completed nor published and presumably both have been lost. He also wrote ten books and commentaries on medical topics, at least seven of which are extant either whole or in part. His only edited work, thus far, is his *Kitāb al-ʿUmdah* on surgery, theory and practice, in twenty treatises. This is the largest Arabic text devoted to surgery written during the entire medieval period; and it superseded the surgical treatise in the *Al-Tasrīf* of al-Zahrāwī. In *Kitāb al-ʿUmdah*, Ibn al-Quff described the vital connection between the arteries and veins and the passage of life-giving blood and pneuma from the former to the latter. This reference to the capillaries was made nearly four centuries before the work of Malpighi, who benefited from the use of the microscope. Ibn al-Quff also explained the function of the cardiac valves, their number, and the direction in which they open and close. He also appealed for all the Arab lands to standardize the weights and measures used in pharmacy and medicine. His pleas were scarcely heeded on account of the intellectual decline that soon after gripped the Arab world.

In his commentary on the *Aphorisms* of Hippocrates, Ibn al-Quff included sayings and annotations made earlier by the Muslim theologian al-Rāzī. Ibn al-Quff's elaborate discussions show philosophical and metaphysical tendencies, but they have no new, independent medical concepts.

His *Jāmiʿ al-Gharad* on embryology, child growth, diet and drug therapy, the preservation of health, and physiognomy contains original approaches and ideas. For example, he theorized on the genesis of the embryo and the stages it passes through in its growth, especially the appearance of a foamlike cluster after the sixth day of fertilization, and on the early formation of the embryo after the twelfth day. He spoke of how "the head distinctly emerges as separate from the shoulders . . . and that the brain is the first major organ to develop." Also, his instructions on what should be done to the infant at birth and thereafter are of great historical interest.

Wars with the Crusaders during the twelfth and thirteenth centuries and internal upheavals in Syria, Iraq, and Egypt created new challenges and gave enduring vigor to practitioners of the healing art. Several eminent doctors, pharmacists, and educators appeared on the scene and contributed materially to activating or maintaining the high status of the health professions. Medical schools and hospitals were founded.

At the climax of this period, as a teacher, author, and practitioner, Ibn al-Quff played an important role. His contributions to surgery and physiology, as well as his personal observations on embryology, human environment, and health preservation, put him on a par with al-Majūsī, al-Zahrāwī, and Ibn al-Nafīs as one of the greatest physician-surgeons in medieval Islam.

BIBLIOGRAPHY

I. ORIGINAL WORKS. Ibn al-Quff's only edited work is his surgical manual *Kitāb al-'Umdah fī Ṣinā'at al-Jirāhah* (preferably 'Umdat al-Iṣlaḥ fī 'Amal Sinā'at al-Jarrāḥ), Osmania Oriental Publications Bureau, 2 vols. (Hyderabad, India, 1937), which is also extant in MS form in several libraries, including the British Museum Library, American University of Beirut Library, National Library and Archives, Cairo, and the Bibliothèque Nationale of Paris. Ibn al-Quff's work on hygiene, *Kitāb Jāmi' al-Gharaḍ fī Ḥifẓ al-Ṣiḥḥah wa-daf' al-Maraḍ* in 60 chs. is extant (complete or in part) in the Wellcome Institute of the History of Medicine (WMS.OR.116), London, and the British Museum Library. In Latin the work was rendered as *Corpus optatorum de servanda sanitate et depellendo morbo*.

His commentary on the Hippocratic *Aphorisms* (*Kitāb Al-Uṣūl fī Sharḥ al-Fuṣūl*) and on the *Al-Qānūn* of Ibn Sīnā are extant in several MSS (see, for example, National Library and Archives of Cairo, nos. 4 and 1732 Ṭibb; and Alexandria, 3352J). His two books on the art of healing, *Al-Shafī fī al-Ṭibb* and *Zubdat al-Ṭibb*, are reported in two incomplete MSS in the Vatican Apostolic Library and in the Rampur state library, India, respectively. His 2 treatises on the utilities of the organs of the human body, *Risālah fī Manāfi' al-A'ḍā' al-Insānīyah*, and on the preservation of health, *Fī Ḥifẓ al-Ṣiḥḥah*, are also reported in two unique MSS.

II. SECONDARY LITERATURE. The most reliable biography we have on the life of Ibn al-Quff is Ibn Abī Uṣaybi'ah, *'Uyūn al-Anbā'*, Būlāq ed., II (Cairo, 1882), 273–274. He was also mentioned by Quṭb al-Dīn al-Yunīnī, *Dhayl Mir'āt al-Zamān*, IV (Hyderabad, 1960), 312–314; and

Ḥājjī Khalīfah, *Kashf al-Ẓunūn*, II (Cairo, 1893), 88, 132.

In the modern period, Ibn al-Quff's biography was included in the works of Antoine Barthélémy Clot, *Note sur la fréquence des calculs vésicaux en Égypte et sur la méthode employée par les chirurgiens arabes pour en faire l'extraction* (Marseilles, 1830); F. Wüstenfeld, *Geschichte der arabische Aerzte und Naturforscher* (Göttingen, 1840), 146; Lucien Leclerc, *Histoire de la médecine arabe*, II (Paris, 1876), 203–204; and Ernst J. Gurlt, *Geschichte der Chirurgie und Ihrer Ausübung*, I (Berlin, 1898), 662–663.

More attention has been paid to Ibn al-Quff in the present century. See Carl Brockelmann, *Geschichte der arabische Litteratur*, II (Weimar, 1902; rev. with a supp., Leiden–Brill, 1937–1949), 649, 899–900 respectively; E. Wiedemann, "Beschreibung von Schlangen bei Ibn Quff," Beiträge 50, in *Sitzungsberichte der Physikalisch-medizinischen Sozietät in Erlangen*, **48** (1918), 61–64; G. Sarton, *Introduction to the History of Science*, II, pt. 2 (Baltimore, 1931), 1098–1099; Otto Spies, "Beiträge Zur Geschichte der arabische Zahnheikunde," in *Sudhoffs Archiv für Geschichte der Medizin und der Naturwissenschaften*, **46** (1962), 161–177; O. Spies and H. Müller-Bütow, "Drei urologische Kapital aus der arabische Medizin," in *Sudhoffs Archiv für Geschichte der Medizin und der Naturwissenschaften*, **48** (1964), 248–259; A. Z. Iskandar, *A Catalogue of Arabic Manuscripts on Medicine and Science*, Wellcome Institute of the History of Medicine Library, (London, 1967), 34, 45–47, 113–114; and S. Hamarneh, "Surgical Development in Medieval Arabic Medicine," in *Viewpoints*, 4 (1965), 17; "Arabic Texts Available to Practitioners in Medieval Islam," in *Bulletin de l'Institut d'Égypte*, **49** (1969), 69; "Medical Education and Practice in Medieval Islam," in C. D. O'Malley, ed., *The History of Medical Education* (Berkeley, Calif., 1970), 62; and *The Physician, Therapist and Surgeon Ibn al-Quff* (Cairo, 1974).

See also Otto Spies and Horst Mueller-Buetow, *Anatomie und Chirurgie des Schaedels, insbesondere der Hals-Nasen-und Ohrenkrankeiten nach Ibn al-Quff* (Berlin, 1971). For a more general discussion, see Hamarneh, "Thirteenth Century Physician Interprets Connection Between Arteries and Veins," in *Sudhoffs Archiv für Geschichte der Medizin und der Naturwissenschaften*, **46** (1962), 17–26; *Index of MSS on Medicine and Pharmacy in the Ẓāhirīyah Library*, Arab Academy (Damascus, 1968–1969), 325–329 (Arabic) and 20–21 in the English text; and "The First Recorded Appeal for Unification of Weight and Measure Standards in Arabic Medicine," in *Physis*, 5 (1963), 230–248; and "The Physician and the Health Professions in Medieval Islam," in *Bulletin of the New York Academy of Medicine*, **47** (1971), 1088–1110.

SAMI K. HAMARNEH

AL-QŪHĪ (or **AL-KŪHĪ**), **ABŪ SAHL WAYJAN IBN RUSTAM** (*fl.* Baghdad, *ca.* 970–1000), *mathematics, astronomy.*

Al-Qūhī's names indicate his Persian origin: al-Qūhī means "from Quh," a village in Tabaristan;

and Rustam is the name of a legendary Persian hero. At the peak of his scientific activity he worked in Baghdad under the Buwayhid caliphs 'Aḍud al-Dawla and his son and successor Sharaf al-Dawla.

In 969/970 al-Qūhī assisted at the observations of the winter and summer solstices in Shiraz. These observations, ordered by 'Aḍud al-Dawla, were directed by Abū'l Ḥusayn 'Abd al Raḥmān ibn 'Umar al-Ṣūfī; Aḥmad ibn Muḥammad ibn 'Abd al Jalīl al Sijzī and other scientists were also present. In 988 Sharaf al-Dawla instructed al-Qūhī to observe the seven planets, and al-Qūhī constructed a building in the palace garden to house instruments of his own design. The first observation was made in June 988 in the presence of al-Qūhī, who was director of the observatory; several magistrates (quḍāt); and the scientists Abū'l Wafā', Aḥmad ibn Muḥammad al-Ṣāghānī, Abū'l Ḥasan Muḥammad al-Sāmarrī, Abū'l Ḥasan al-Maghribī, and Abū Isḥāq Ibrāhīm ibn Hilāl ibn Ibrāhīm ibn Zahrūn al Ṣābī. Correspondence between Abū Isḥāq and al-Qūhī still exists. They very accurately observed the entry of the sun into the sign of Cancer and, about three months later, its entry into the sign of Libra. Al-Bīrūnī related that activity at al-Qūhī's observatory ceased with the death of Sharaf al-Dawla in 989.

Al-Qūhī, whom al-Khayyāmī considered to be an excellent mathematician, worked chiefly in geometry. In the writings known to us he mainly solved geometrical problems that would have led to equations of higher than the second degree. Naṣīr al Dīn al Ṭūsī adds to his edition of Archimedes' *Sphere and Cylinder* the following note by al-Qūhī: "To construct a sphere segment equal in volume to a given sphere segment, and equal in surface area to a second sphere segment —a problem similar to but more difficult than related problems solved by Archimedes—Al-Qūhī constructed the two unknown lengths by intersecting an equilateral hyperbola with a parabola and rigorously discussed the conditions under which the problem is solvable."

The same precision is found in *Risāla fī istikhrāj ḍilʿ al-musabbaʿ al-mutasāwī'l-aḍlāʿ fī'd-dāʾira* ("Construction of the Regular Heptagon"), a construction more complete than the one attributed to Archimedes. Al-Qūhī's solution is based on finding a triangle with an angle ratio of 1:2:4. He constructed the ratio of the sides by intersecting a parabola and a hyperbola, with all parameters equal. Al-Sijzī, who claimed to follow the method of his contemporary Abū Saʿd al-ʿAlā ibn Sahl, used the same principle. The latter, however, knew al-Qūhī's work, having written a commentary on the treatise *Kitāb ṣanʿat al-asṭurlāb* ("On the Astrolabe"). Another method used by

al-Qūhī is found in al-Sijzī's treatise *Risāla fī qismat al-zāwiya* ("On Trisecting an Angle").

Again, in *Risāla fī istikhrāj misāḥat al-mujassam al-mukāfī* ("Measuring the Parabolic Body"), al-Qūhī gave a somewhat simpler and clearer solution than Archimedes had done. He said that he knew only Thābit ibn Qurra's treatise on this subject, and in three propositions showed a shorter and more elegant method. Neither computed the paraboloids originating from the rotation of the parabola around an ordinate. That was first done by Ibn al-Haytham, who was inspired by Thābit's and al-Qūhī's writings. Although he found al-Qūhī's treatment incomplete, Ibn al-Haytham was nevertheless influenced by his trend of thought.

Analyzing the equation $x^3 + a = cx^2$, al-Qūhī concluded that it had a (positive) root if $a \leqslant 4c^3/27$. This result, already known to Archimedes, apparently was not known to al-Khayyāmī, whose solution is less accurate. Al-Khayyāmī also stated that al-Qūhī could not solve the equation $x^3 + 13.5x + 5 = 10x^2$ while Abū'l Jūd was able to do so. (Abū'l Jūd, a contemporary of al-Bīrūnī, worked on geometric problems leading to cubic equations; his main work is not extant.)

In connection with Archimedean mathematics, Steinschneider stated that al-Qūhī also wrote a commentary to Archimedes' *Lemmata*. In I. A. Borelli's seventeenth-century Latin edition of the *Lemmata* (or *Liber assumptorum*), there is a reference to al-Qūhī.

Al-Qūhī was the first to describe the so-called conic compass, a compass with one leg of variable length for drawing conic sections. In this clear and rather general work, *Risāla fī'l birkar al-tāmm* ("On the Perfect Compass"), he first described the method of constructing straight lines, circles, and conic sections with this compass, and then treated the theory. He concluded that one could now easily construct astrolabes, sundials, and similar instruments. Al-Bīrūnī asked his teacher Abū Naṣr Manṣūr ibn 'Irāq for a copy of the work; and in al-Bīrūnī, Ibn al-Ḥusayn found a reference to al-Qūhī's treatise. Having tried in vain to obtain a copy, Ibn al-Ḥusayn wrote a somewhat inferior work on the subject (H. Suter, *Die Mathematiker und Astronomen der Araber und ihre Werke* [Leipzig, 1900], p. 139).

Al-Qūhī also produced works on astronomy (Brockelmann lists a few without titles), and the treatise on the astrolabe mentioned above. Abū Naṣr Manṣūr ibn 'Irāq, who highly esteemed al-Qūhī, gave proofs for constructions of azimuth circles by al-Qūhī in his *Risāla fī dawāʾir as-sumūt fī al-asṭurlāb* ("Azimuth Circles on the Astrolabe").

BIBLIOGRAPHY

I. ORIGINAL WORKS. C. Brockelmann, *Geschichte der arabischen Literatur*, 2nd ed., I (Leiden, 1943), 254 and Supp. I (Leiden, 1937), 399, list most of the available MSS of al-Qūhī. See also G. Vajda, "Quelques notes sur le fonds de manuscrits arabes de la bibliothèque nationale de Paris," in *Rivista degli studi orientali*, **25** (1950), 1–10.

Translations or discussions of al-Qūhī's work are in A. Sayili, "A Short Article of Abū Sahl Waijan ibn Rustam al-Qūhī on the Possibility of Infinite Motion in Finite Time," in *Actes du VIII Congrès international d'histoire des sciences* (Florence–Milan, 1956), 248–249; and "The Trisection of the Angle by Abū Sahl Wayjan ibn Rustam al Kūhī," in *Proceedings of the Tenth International Congress of History of Science* (Ithaca, 1962), 545–546; Y. Dold-Samplonius, "Die Konstruktion des regelmässigen Siebenecks nach Abū Sahl al-Qūhī Waiğan ibn Rustam," in *Janus*, **50** (1963), 227–249; H. Suter, "Die Abhandlungen Thābit ben Kurras und Abū Sahl al-Kūhīs über die Ausmessung der Paraboloide," in *Sitzungsberichte der Physikalisch-medizinischen Sozietät in Erlangen*, **49** (1917), 186–227; and F. Woepcke, *L'algèbre d'Omar Alkhayyāmī* (Paris, 1851), 96–114, 118, 122, 127; and "Trois traités arabes sur le compas parfait," in *Notices et extraits de la Bibliothèque nationale*, **22**, p. 1 (1874), 1–21, 68–111, 145–175.

Edited by the Osmania Oriental Publications Bureau are *Risāla fī misāḥat al mujassam al mukāfī* ("On Measuring the Parabolic Body") (Hyderabad, 1948) and *Min kalāmi Abī Sahl fī mā zāda min al ashkāl fī amr al maqālat al ṣānīyati* ("Abū Sahl's Discussion on What Extends the Propositions in the Instruction of the Second Book") (Hyderabad, 1948).

II. SECONDARY LITERATURE. Ibn al-Qifṭī, *Ta'rīkh al-ḥukamā'*, J. Lippert, ed. (Leipzig, 1903), 351–354. Information on al-Qūhī the mathematician is also in Woepcke, *L'algèbre . . .*, 54–56; and A. P. Youschkevitch, *Geschichte der Mathematik im Mittelalter* (Basel, 1964), 258–259, 292. On the observations in Shirāz, see al-Bīrūnī, *Taḥdīd nihāyāt al-amākin li-taṣḥīḥ masāfāt al-masākin* (Cairo, 1962), 99–100; on the observations at Baghdad, A. Sayili, *The Observatory in Islam* (Ankara, 1960), 112–117; M. Steinschneider, "Die mittleren Bücher der Araber und ihre Bearbeiter," in *Zeitschrift für Mathematik und Physik*, **10** (1865), 480.

YVONNE DOLD-SAMPLONIUS

QUINCKE, GEORG HERMANN (*b.* Frankfurt-an-der-Oder, Germany, 19 November 1834; *d.* Heidelberg, Germany, 13 January 1924), *physics*.

Quincke's father was a physician; his mother, Marie Gabain, came from a Huguenot family. In 1843 the family moved to Berlin, where the father was promoted to a medical council. After graduating from the Werder Gymnasium, Quincke began to study physics at the University of Berlin at the age of eighteen. He continued his studies at Königsberg under Franz Neumann and then under Gustav Kirchhoff at Heidelberg. At the same time Quincke worked in the laboratory of Robert Bunsen. He then returned to Berlin, where he received his doctorate in 1858 with a dissertation on the capillary constant of mercury. Only a year later he obtained the *venia legendi*, for which he did not have to fulfill the usual requirement of presenting special *Habilitationsschrift*. He began teaching physics in 1859 at the Berlin Gewerbeakademie (the predecessor of the Technische Hochschule), and in 1865 he became extraordinary professor at the University of Berlin.

In 1872 Quincke was appointed full professor at the University of Würzburg, and in 1875 he succeeded Kirchhoff at Heidelberg. Following his retirement at the age of seventy-three, Quincke worked in his private laboratory at his country house.

During his lifetime Quincke was held in high regard by his peers, especially in England, where his friends included William Thomson, J. W. Strutt, Stokes, Tyndall, and Tait. He was a member of the academies of Göttingen, Berlin, Munich, Uppsala, and Halle and of the Royal Societies of London and Edinburgh. He also received honorary doctorates from the universities of Würzburg, Heidelberg, Oxford, Cambridge, and Glasgow.

Quincke's views on physics were rooted in the thought of the first half of the nineteenth century. He admired Faraday's method of working and ideas, but he never understood Maxwell's elaboration and mathematical reformulation of Faraday's discoveries. In 1859 Quincke discovered the so-called diaphragm currents. Above all, however, he was passionately interested in making measurements; and the bulk of his work consisted of the determination and collection of data concerning the properties and constants of materials. He frequently returned to capillarity, the subject of his doctoral dissertation. He found countless opportunities for research in extending the capillary-tube and angle-of-contact methods of measuring from simple liquids to solutions and fusions and in measuring interfacial tension between two liquids.

Quincke devoted a group of sixteen studies to problems in optics, basing his work on theories that viewed light as elastic vibrations in a mechanical medium. Between 1880 and 1897 he published the fifteen installments of his "Elektrische Untersuchungen," dealing with the behavior of materials in electrostatic and magnetic fields. In this connection Quincke developed his elegant meniscus-displacement method of determining diamagnetic and paramagnetic susceptibilities of liquids and gases. During the final years of his life he was concerned primarily with foams and their structures.

Quincke introduced the first practical laboratory work to be given in a physics course at a German university. The restricted means at his disposal obliged him to get along with little, and the students spoke jokingly of "Quincke's Cork-wax-penny System." But he took pride in this.

BIBLIOGRAPHY

I. ORIGINAL WORKS. Quincke's writings were published only in periodicals; the most important (until 1900) appeared in *Annalen der Physik und Chemie*. For a list of his memoirs, see Royal Society *Catalogue of Scientific Papers*, v, 64–65; VIII, 681–682; XI, 86; and XVIII, 8—which lists more than 90 works published to 1900—and Poggendorff, IV, 1203–1204; V, 1015; and VI, 2101–2102, for later writings.

Quincke's memoirs on capillarity include "Ueber die Capillaritätsconstanten des Quecksilbers," in *Annalen der Physik*, 4th ser., **105** (1858), 1–48; "Ueber die Capillaritätsconstanten fester Körper," *ibid.*, 5th ser., **134** (1868), 356–367; "Ueber die Capillaritätsconstanten geschmolzener Körper," *ibid.*, **135** (1868), 621–646; "Ueber die Entfernung, in welcher die Molecularkräfte der Capillarität noch wirksam sind," *ibid.*, **137** (1869), 402–414; "Ueber die Capillaritätsconstanten geschmolzener chemischer Verbindungen," *ibid.*, **138** (1869), 141–155; and "Ueber Capillaritäts-Erscheinungen an der gemeinschaftlichen Oberfläche von Flüssigkeiten," *ibid.*, **139** (1870), 1–89.

For his optical studies, see *Annalen* for 1862–1873; research on electricity was published in 1880–1897. See also "Eine physikalische Werkstätte," in *Zeitschrift für den physikalischen und chemischen Unterricht*, **5** (1892), 113–118; and **7** (1894), 57–72.

II. SECONDARY LITERATURE. On Quincke and his work, see F. Braun, "Hermann G. Quincke zum 70. Geburtstag," in *Annalen der Physik*, 4th ser., **15** (1904), i–iv, with portrait; and E. H. Stevens, "The Heidelberg Physical Laboratory," in *Nature*, **65** (1902), 587. Obituaries include A. Kalähne, in *Physikalische Zeitschrift*, **25** (1924), 649–659, with portrait; W. König, in *Naturwissenschaften*, **12** (1924), 621–627, with portrait; and A. Schuster and G. E. Allen, in *Nature*, **113** (1924), 280–281.

F. FRAUNBERGER

QUOY, JEAN-RENÉ-CONSTANT (*b.* Maillé, Vendée, France, 10 November 1790; *d.* St.-Jean-de-Liversay, France, 4 July 1869), *zoology*.

Quoy is best known to science for the zoological collections he brought back to France in his capacity as naturalist-voyager in the French navy. The eldest child in a large family, he was the son of Jean Quoy, a surgeon, and Louise Arsonneau. For several generations there had been surgeons in the family— even his grandmother had become a surgeon in 1760. To escape the dangers of the civil war in the Vendée, Quoy was sent to his aunt and grandfather at Luché. In 1806, at age sixteen, he entered the School of Naval Medicine at nearby Rochefort. Beginning in 1807, he served as a surgeon on vessels engaged in wartime missions. On the mission of the *Loire* to regain the Île de Bourbon (now Réunion), he collected and described natural history specimens at the request of the Rochefort Council of Health. This mission awakened his interest in natural history. In 1814 Quoy defended a Latin dissertation, *Epistolae dominae de nonnullis pavoribus effectibus*, and received the M.D. at Montpellier, although he had never studied there.

At the end of the Napoleonic Wars, Quoy was named surgeon-major of the corvette *Uranie*, under the command of Louis-Claude de Freycinet. The *Uranie* was to make a scientific voyage around the world, the main object of which was to determine the form of the terrestrial globe in the southern hemisphere. Also to be studied were magnetic and meteorological phenomena, natural history, ethnology, and geography. Freycinet established an important precedent for scientific voyages by insisting that all scientific work be carried out by members of the navy rather than professional scholars. This stipulation produced a group of French naturalist-voyagers, members of the Naval Health Service, who brought back natural history materials to the scientists in Paris.

On the *Uranie*, Quoy was placed in charge of zoology. His colleagues were J. P. Gaimard, second surgeon, who worked with Quoy on zoology, and Charles Gaudichaud-Beaupré, pharmacist, who handled botany. The corvette left France in September 1817 and visited Rio de Janeiro, the Cape of Good Hope, the Marianas, Hawaii, and the south of Australia. Returning to France via Cape Horn, the *Uranie* was shipwrecked in the Falkland Islands. All the collections were submerged, but parts were later recovered. A large portion of the bird and insect collections was lost. After the voyagers returned to France in November 1820, Quoy and Gaimard were called to Paris to publish the zoology of the voyage. Quoy then met Cuvier and Blainville, as well as the other scholars of the capital; and he and Gaimard presented several papers at the Academy of Sciences.

According to Cuvier, the zoology of the voyage, published in 1824, contained 254 animal sketches or anatomical studies, including 227 new species, as well as a textual description of eighty additional new species. The work was arranged according to Cuvier's classification, beginning with the cranium and a description of the Papuans. For each area Quoy and Gaimard discussed the geographical distribution and the habits of its animals. On the whole they confined themselves to detailed descriptions of their findings and avoided theorizing as much as possible.

After completing the zoology of the voyage, Quoy returned to Rochefort, where he won a competition for the professorship of anatomy at the School of Naval Medicine (1824). In 1826, despite poor health and an exemption from sea duty by virtue of his title of professor, Quoy sought the position of surgeon-major on the scientific voyage of the *Astrolabe* under the command of J.-S.-C Dumont d'Urville. Since Gaimard had already been appointed surgeon-major, Quoy was named zoologist. Pierre-Adolphe Lesson was adjoint for botany. The *Astrolabe* left Toulon in April 1826 and visited Australia, New Zealand, New Guinea, and several other nearby islands. Hoping to avoid losses like those incurred when the *Uranie* was shipwrecked, Quoy and Gaimard sent materials and sketches to Paris from each major stopover. Quoy carefully sketched and colored each object twice. Now aware that the mollusks and zoophytes most attracted the attention of the Paris zoologists, he concentrated on finding new species of these animals. Instead of merely collecting their shells, he sketched the entire animals as soon as possible in their natural colors. Quoy also particularly sought new and colorful fish for Cuvier's *Histoire des poissons*.

After a three-year voyage, the *Astrolabe* returned to Marseilles in March 1829 with two living babirusas, the first in Europe. Again Quoy went to Paris to prepare the zoology of the voyage for publication. He had brought back over 4,000 sketches, many his own, depicting 1,200 different species. His descriptive work was admired by Cuvier for its patient analysis and accuracy of detail. The materials were deposited at the Muséum d'Histoire Naturelle, where naturalists made exact catalogs. Quoy worked with Cuvier several hours a day on their classification.

The zoology of the *Astrolabe* voyage was published in four volumes plus an atlas from 1830 to 1832. In keeping with the expanded knowledge of comparative anatomy, the descriptions and sketches contained many more anatomical details than those of the *Uranie* voyage. For each animal Quoy gave all its dimensions and a long list of nomenclature. He gave special attention to the age, sex, and form of the sternum of each bird. The mollusks were presented with detailed anatomies, many made by Blainville. In the first section of the zoology, which dealt with anthropology, Quoy and Gaimard divided the natives of Polynesia into two races, the yellow and the black. Their findings, they said, supported Gall's ideas on the influence of climate on the physical constitution of man.

Quoy was elected correspondent of the Academy of Sciences in May 1830. In 1832, when Cuvier died, Blainville took his chair of comparative anatomy at the Muséum d'Histoire Naturelle, hoping that Quoy would succeed him as professor of mollusks and zoophytes at the museum. But in the wake of a series of complex intrigues, Quoy, who was the nominee of the museum, lost to Achille Valenciennes, the candidate supported by the Academy of Sciences. Quoy, who hated intrigue, never recovered from having lost this opportunity. He left natural history and henceforth devoted all his energies to the Naval Health Service. As first physician-in-chief, he served at Toulon and Brest until 1848, when he was named inspector general, the highest officer in the Health Service. He retired in 1858.

Quoy never married. He was known as a religious, modest, studious, somewhat austere man, with a strong sense of duty and of hierarchy.

BIBLIOGRAPHY

I. ORIGINAL WORKS. Quoy and Gaimard published the zoology of their two voyages: *Voyage autour du monde . . . exécuté sur les corvettes de S.M. l'Uranie et la Physicienne, pendant les années 1817, 1818, 1819 et 1820, publié par M. Louis de Freycinet, . . . Zoologie,* 2 vols. (Paris, 1824); and *Voyage de découvertes de l'Astrolabe exécuté par ordre du roi, pendant les années 1826, 1827, 1828, 1829, sous le commandant de M. J. Dumont d'Urville. Zoologie,* 4 vols. plus 2-vol. atlas (Paris, 1830–1832). Most of Quoy's memoirs were written with Gaimard. A list is in Royal Society *Catalogue of Scientific Papers,* IV, 66–67; as well as in Charles Berger and Henri Rey, *Répertoire bibliographique des travaux des médecins et des pharmaciens de la marine française, 1698–1873* (Paris, 1874), published as an app. to *Archives de médecine navale,* **31** (1874).

A very complete list of MSS is in the Noël thesis (see below), 201–209. The most important MS collection is "Dossier Quoy" at the Bibliothèque Municipale de La Rochelle. For an important letter from Quoy to J. Desjardins de Maurice describing his relations with Cuvier and other scholars, see E. T. Hamy, "Notes intimes sur Georges Cuvier, rédigées en 1836, par Dr. Quoy pour son ami J. Desjardins de Maurice," in *Archives de médecine navale,* **86** (1906), 450–475.

II. SECONDARY LITERATURE. Yvan Delteil collected all the MS sources for a life of Quoy but died (1957) before completing his work; he left his materials to Jean-Pierre Noël, who wrote an M.D. dissertation on Quoy: *J. R. C. Quoy (1790–1869). Inspecteur général du Service de santé de la marine. Médecin-naturaliste-navigateur. Sa vie—son milieu—son oeuvre* (Bordeaux, 1960). It has very complete information on Quoy's life but lacks a satisfactory analysis of his scientific achievements. There is a good bibliography of secondary sources relating to Quoy and his milieu, 197–200. Particularly to be noted are C. Maher, *Éloge de J. R. C. Quoy* (Rochefort, 1869), also in *Archives de médecine navale*, **12** (1869), 402–422; and Yvan Delteil, "L'enfance de J.-R.-C. Quoy," in *Histoire de la médecine*, **6** (1956), no. 11, 21–34.

TOBY A. APPEL

IBN QURRA. See **Thābit ibn Qurra.**

QUSṬĀ IBN LŪQĀ AL-BAʿLABAKKĪ (*fl.* Baghdad and Armenia, 860–900), *medicine, philosophy, translation of scientific literature.*

Qusṭā ibn Lūqā, a doctor from Baalbek (Heliopolis), was praised by the ancient Arab biographers as one of the famous authors and translators who provided Arabic and Syriac versions of Greek scientific works during the ʿAbbāsid period (749–1258). Indeed, Ibn al-Nadīm declares that Qusṭā was even greater than Ḥunayn ibn Isḥāq, especially in medicine. Qusṭā was a Christian of Greek origin; he had visited Byzantine regions and returned to Syria with Greek books which he then had translated by others or translated himself. He also revised earlier translations, and he was praised for his excellent literary style. Qusṭā was summoned to Baghdad, where he worked for Caliph al-Mustaʿīn (862–866); and there he probably knew al-Kindī and Thābit ibn Qurra. Al-Kindī revised Qusṭā's translations of Hypsicles' *Liber . . . de ascensionibus* and Autolycus' *De ortu et occasu,* and Thābit completed Qusṭā's version of Theodosius of Bithynia's *De sphaeris.* Probably before 865 Qusṭā was summoned to Armenia by the ruler Sanhārīb, and it was there that he wrote a number of works for the Patriarch Abū 'l-Ghiṭrīf, and composed his reply to Abū ʿĪsā ibn al-Munajjim on the prophetic mission of Muḥammad. Qusṭā remained in Armenia, greatly honored, until his death, and a noble tomb was erected for him there.

The biographers claim that Qusṭā ibn Lūqā was skilled in philosophy, geometry, arithmetic, music, astronomy, logic, and especially in medicine. They list more than sixty titles of works ascribed to him and mention some seventeen translations made by him. Qusṭā was quoted as a medical authority by al-Rāzī and Ibn al-Jazzār, and the majority of his writings seem to have been medical. He wrote treatises on various organs and diseases, diet, bathing, and bloodletting, and an introduction to medicine. Among his extant medical works are *Kitāb fī'l-sahar* ("On Insomnia"), a viaticum, a treatise on the four

humors, and a work on the origin of hair. A Latin translation of his book on poisons seems to have been available in sixteenth-century Italy.

Qusṭā's arithmetical works included a treatise on numerical questions in the third book of Euclid's *Elements* and a commentary on Diophantus, whose work he also translated into Arabic. He wrote an introduction to geometry, a work on Euclid, and a treatise entitled *Kitāb fī shakl al-kura wa 'l-usṭuwāna* ("Shape of the Sphere and the Cylinder"), possibly a version of Archimedes' work. None of these appears to be extant.

Qusṭā wrote an introduction to astronomy and several works on the use of astronomical instruments. The most widely known of these, *Kitāb fī 'l-ʿamal bi 'l-kura al-nujūmiyya* ("On the Use of the Celestial Globe"), in sixty-five chapters, exists in Arabic in two recensions; it was also translated into Latin (by Stephanus Arnaldus, as *De sphaera solida*), Hebrew, Spanish, and Italian.

Other works, no longer extant, dealt with logic, politics, and natural science; there were treatises on winds, mirrors, and on the atom. Some interesting works on psychophysical relations are extant; in them Qusṭā sought to define the effects within man of matter on form, and of form on matter. Of these, the treatise on the efficacy of amulets is extant only in the Latin translation of Arnald of Villanova, entitled both *De physicis ligaturis* and *De incantatione*; and the treatise *Risāla ilā Abī ʿAlī ibn Bunān . . .* ("On the Diversity of the Characters of Men") is extant only in Arabic. Most widely known and influential was the short treatise *Kitāb fī 'l-farq baina 'l-nafs wa 'l-rūḥ* ("On the Difference Between the Spirit and the Soul"), which, in the Latin translation of John of Seville, was used as an authority by Alfred of Sarashel, Albertus Magnus, Roger Bacon, and many others. It was frequently copied and commented upon with Aristotle's works, as if it were a clarification of or supplement to the *De anima.* Although Latin forms of Qusṭā's name (Costa ben Luca, or Constabulus) were current in the later Middle Ages, his works frequently were ascribed to other authors, especially to Constantine the African.

Qusṭā's greatest importance to his contemporaries, however, was as a translator of Greek scientific and philosophical works: in this respect he was the rival and associate of the better-known Ḥunayn ibn Isḥāq and Thābit ibn Qurra. Qusṭā's Arabic version of Hero of Alexandria's *Mechanics* provides the only text of the work extant today. Qusṭā is credited with an Arabic version of Aristotle's *Physics* with the commentary of Alexander of Aphrodisias on books IV, V, and VII; the first four books of the commentary by

John Philoponus on the same works; and also with part of Aristotle's *De generatione et corruptione* commented upon by Alexander. Treatises entitled *Maqāla fī ṭūl al-ʿumr wa qaṣrih* ("On Length and Shortness of Life"), and *Kitāb fī 'l-nawm wa 'l-ruʾyā* ("On Sleep and Dream") are ascribed to Qusṭa, but the relationship of these to Aristotle's works is not known.

Qusṭā's versions of works of Diophantus and Plutarch are no longer extant; but manuscripts of his translations of Aristarchus, Autolycus, Hypsicles, and Theodosius Tripolitanus survive. The Arabic versions of Galen's commentary on Hippocrates' *Aphorisms* and the *De horoscopo* of Asclepius are also ascribed to Qusṭā by the scribes of the manuscripts of these works, which are now in Florence. Gerard of Cremona's Latin translations of Theodosius of Bithynia's *De sphaeris* and *De habitationibus* were made from Qusṭā's Arabic versions. Although Qusṭā evidently played an important part in the transmission of Greek science to the West, he is a neglected figure and very few of his works have been published.

BIBLIOGRAPHY

I. ORIGINAL WORKS. Full details on what is known about Qusṭā's works are in Gabrieli (see below). The work of al-Munajjim on the prophetic mission of Muḥammad and Qusṭā's reply are extant in MS at Bibliothèque orientale, Université St.-Joseph, Beirut; see L. Cheikho, "Catalogue raisonné des manuscrits de la Bibliothèque orientale, VI: Controverses," in *Mélanges de l'Université St.-Joseph* (Beirut), **14**, fasc. 3 (1929), 44 (MS 664).

Gabrieli does not mention the *Kitāb al-faṣd thamaniyya ʿashara bāb* ("Phlebotomy in 18 Chapters"), listed by Ibn al-Nadīm, *Fihrist* (see below), 295. Among the extant works are *Kitāb fī 'l-sahar* ("On Insomnia"), Staatsbibliothek, Berlin, Arabic MS 6357; *Kitāb fī tadbīr al-abdān fī safar al-ḥajj* ("Viaticum"); and *Kitāb ʿilal al-shaʿr* ("On the Origin of Hair"), both in British Museum Add. 7527/3; and a treatise on the four humors, in Staatsbibliothek, Munich, Arabic MS 805. For the existence of *Costa ben Luca de venenis* in the sixteenth century, see M. Steinschneider, "Die toxicologischen Schriften der Araber bis Ende XII. Jahrhunderts. Ein bibliographischer Versuch, grossentheils aus handschriftlichen Quellen," in *Virchows Archiv für pathologische Anatomie und Physiologie und für klinische Medizin*, **52** (1871), 371–372.

The only extant arithmetical treatise of Qusṭā seems to be *Kitāb fī 'l-burhān ʿalā ʿamal ḥisāb al-khaṭaʾayn;* a German trans. was published by Heinrich Suter: "Die Abhandlung Qosṭā ben Lūqās und zwei andere anonyme über die Rechnung mit zwei Fehlern und mit der angenommen Zahl," in *Bibliotheca mathematica*, 3rd ser., **9**, no. 2 (1908), 111–122.

There has been some confusion over the works on the use of astronomical instruments. There appear to be three different works in question: (1) a doubtful work in Leiden University library (MS 1053) entitled *Kitāb al-ʿamal bi 'l-asṭurlāb al-kurī*, which is discussed by Hugo Seemann and T. Mittelberger, "Das kugelförmige Astrolab nach den Mitteilungen von Alfons X. von Kastilien und den vorhandenen arabischen Quellen," in *Abhandlungen zur Geschichte der Naturwissenschaften und der Medizin*, **8** (1925), 46–49; (2) a well-authenticated work in Bodleian Library MS Arabic 879, entitled *Hayʾat al-aflāk* ("On the Shape of the Celestial Spheres"); and (3) a very famous work, *Kitāb fī 'l-ʿamal bi 'l-kura al-nujūmiyya* ("On the Use of the Celestial Globe"), which is summarized and discussed by W. H. Worrell, "Qusta ibn Luqa on the Use of the Celestial Globe," in *Isis*, **35** (1944), 285–293. For the Latin version of Stephanus Arnaldus, see M. Steinschneider, "Der europäischen Übersetzungen aus dem arabischen bis Mitte des 17. Jahrhunderts," in *Sitzungsberichte der K. Akademie der Wissenschaften in Wien*, Phil.-hist. Kl., **149** (1904), 77. Steinschneider states that there are MSS of this trans. in the Bodleian Library (Coxe 693), in Vienna (MSS 5415 and 5273), and in the cloister of San Marco, Florence.

Other works include *De physicis ligaturis* (*De incantatione*), in *Constantini Africani, Opera omnia* (Basel, 1536), 317 f; Arnald of Villanova, *Opera* (Basel, 1585), cols. 619–624; and Heinrich Cornelius Agrippa, *Opera omnia*, 2 vols. (Lyons, n.d.), I, 741–745. The treatise ("Diversity of the Characters of Men") has been edited with a French trans. by Paul Sbath, "Le livre des caractères de Qosṭâ ibn Loûqâ: Grand savant et célèbre médecin au IXᵉ siècle," in *Bulletin de l'Institut d'Égypte*, **23** (1941), 103–169. The Arabic text of ("Difference Between the Spirit and the Soul") has been edited twice, from two different MSS; by G. Gabrieli, "La Risālah di Qusṭā b. Lūqā 'Sulla differenza tra lo spirito e l'anima,' " in *Atti della R. Accademia dei Lincei. Rendiconti*, cl. di scienze morali, storiche, e filologiche, 5th ser., **19** (1910), 622–655, which includes an Italian trans.; and by L. Cheikho in *al-Mashriq*, **14** (1911), 94–109. The Latin version was printed in *Constantini Africani Opera omnia* (Basel, 1536), 308–317; and it has been edited by C. S. Barach, *Costa-ben-Lucae: De differentia animae et spiritus liber*, vol. 3 of Bibliotheca Philosophorum Mediae Aetatis (Innsbruck, 1878). The Latin version is copied with the works of Aristotle in Balliol MSS 232A and 232B; in Corpus Christi College, Oxford MS CXI; and in Bodleian MS Auct. F.5.25.

Qusṭā's Arabic version of Hero of Alexandria, *Mechanica*, has been published by B. Carra de Vaux in *Journal asiatique*, 9th ser., **1** (1893), 386–472 and **2** (1893), 152–192, 193–267, 420–514; and by L. Nix, ed., *Heronis Alexandrini Opera quae supersunt omnia*, II, fasc. 1 (Leipzig, 1901), which includes a German trans. The treatise ("Length and Shortness of Life") is in a Staatsbibliothek, Berlin MS Arabic no. 6232; the incipit does not sound like a work by Aristotle. In the Bodleian Library are the Arabic texts of Aristarchus of Samos, *De magnitudine et distantia solis et lunae* (MS 875); Autolycus, *De ortu et occasione* *siderum inerrantium* (MS 895, there ascribed to Thābit ibn Qurra); Hypsicles, *De ascensionibus* (MS 875) and additions to Euclid's *Elements* (books 14 and 15), (MS 279; see Nicoll, p. 257); Theodosius Tripolitanus, *De sphaeris* and *De habitationibus* (MS 875). The Arabic text of Theodosius, *De diebus et noctibus*, is in MSS 271 and 286 at the Biblioteca Medicea Laurenziana, Florence. Galen's commentary on Hippocrates' *Aphorisms* and the *De horoscopo* of Asclepius are in MSS 271 and 260 at the Palatine Library.

For further information on the location of other MSS containing Qusṭā's translations, see the references in Carl Brockelmann, *Geschichte der arabischen Literatur*, 2 vols. and 3 supps. (Leiden, 1937–1942), I, 204, 512; supp. I, 365, 374. For Gerard of Cremona's translations of Theodosius, see M. Steinschneider, *Die arabischen Übersetzungen aus dem Griechischen* (Graz, 1960); 219. For MSS in the Bodleian Library, see J. Uri and A. Nicoll, *Bibliothecae Bodleianae codicum manuscriptorum orientalium*, 2 parts (Oxford, 1787–1821). For the Palatine Library, see Stephanus Evodius Assemanus, *Bibliothecae Mediceae Laurentianae et Palatinae codicum mss. orientalium* (Florence, 1742), 375, 381–383, 392. For Munich, see J. Aumer, *Catalogus codicum mss. Bibliothecae Regiae Monacensis* (Arabic), I, pt. 2 (Munich, 1866), 353–354. Fuat Sezgin, *Geschichte des arabischen Schrifttums*, III (Leiden, 1970), 270–274, contains a list of Qusṭā's medical works; IV (Leiden, 1971), mentions a work by Qusṭā on wine: *Kitāb al-nabidh wa shurbihi fī 'l-walā'im.*

II. SECONDARY LITERATURE. Biographical material is in Ibn al-Nadīm, *Kitāb al-fihrist*, G. Flügel, ed., I (Leipzig, 1871), esp. 295; Ibn al-Qifṭī, *Ta'rīkh al-ḥukamā'*, J. Lippert, ed. (Leipzig, 1903), 292; Ibn Abī Uṣaybi'a, *'Uyūn al-anbā' fī ṭabaqāt al-aṭibbā'*, A. Müller, ed., I (Cairo–Königsberg, 1882), 244; G. Gabrieli, "Nota biobibliografica su Qusṭā ibn Lūqā," in *Atti della R. Accademia dei Lincei. Rendiconti*, cl. di scienze morali, storiche, e filologiche, 5th ser., **21** (1912), 341–382; and Georg Graf, *Geschichte der Christlichen arabischen Literatur*, II (Vatican City, 1947), 30.

E. RUTH HARVEY

IBN QUTAYBA, ABŪ MUḤAMMAD 'ABDALLĀH IBN MUSLIM AL-DĪNAWARĪ AL-JABALĪ (*b.* Baghdad or Kufa, Iraq, 828; *d.* Baghdad, 884 or 889), *transmission of knowledge.*

Little is known of Ibn Qutayba's life. His family was from Merv (now Bairam-Ali), Transoxania, which leads to the conclusion that they might have been of Persian or Turkish stock. In some of his works, however, he speaks strongly in favor of the Arabs and points out their superiority to the Persians. Ibn Qutayba spent some years as *qāḍī*, or judge, in the town of Dīnawar, in northern Persia, then taught in

Baghdad, where he died. He was more a philologist and a lexicographer than a scientist in the proper sense of the word.

Among his several historical and philological works is *Kitāb al-anwā'* ("Book of the *anwā'*"), which is of special importance for the history of astronomical knowledge. In the great number of monographs and specialized treatises by the ancient Arabic philologists and lexicographers, the books of the *anwā'* formed a category of their own. (*Anwā'* is the Arabic plural of *naw'*—a verbal noun from *nā'a*—and signifies, in this context, the acronychal setting of a certain constellation, or lunar mansion, while another one just opposite is heliacally rising. This system was used for determining the dates of seasons, fixing certain agricultural activities, events, etc., in the ancient "prescientific" epoch.)

Two sorts of these books can be distinguished. The main group comprises compilations of all available information on the native Arabic knowledge of celestial and meteorological phenomena as found in the ancient sources—in folklore, poetry, and literature; strictly excluded from this group is the scientific knowledge taken from other civilizations through translations (such as the *Almagest* and Indian or Persian sources). More than twenty authors of such books are known, but almost none of their texts have survived. Only a few, or less important, portions of these texts are available in print: the *Kitāb al-azmina* ("Book of Seasons, or Times") by Quṭrub (*d.* 821/822), partially edited in *Revue de l'Académie de Damas*, **2**, pt. 1 (January 1922), 33–46; another *Kitāb al-azmina* by the physician Ibn Māsawayh (*d.* 857), edited by Paul Sbath in *Bulletin de l'Institut d'Égypte*, **15** (1932–1933), 235–257; French translation, by G. Troupeau, in *Arabica*, **15** (1968), 113–142; *Kitāb al-azmina wa 'l-amkina* ("Book of Times and Places") by al-Marzūqī (*d.* 1030), published in two volumes (Hyderabad, 1914); and *Kitāb al-azmina wa 'l-anwā'* ("Book of the Seasons and the *anwā'*") by Ibn al-Ajdābī (thirteenth century), edited by 'Izzat Ḥasan, number 9 in the series Iḥyā' al-Turāth al-Qadīm (Damascus, 1964). The most complete of the *anwā'* books is credited by Arabic sources to the historian and philologist Abū Ḥanīfa al-Dīnawarī (*d.* 895); only excerpts are preserved in some lexicographic works, such as Ibn Sīda's (*d.* 1066) *Mukhaṣṣaṣ*, IX (Cairo, 1901), and al-Marzūqī's *Kitāb al-azmina wa 'l-amkina*. Ibn Qutayba's *Kitāb al-anwā'* is significant primarily because the full text has survived and is available in print (edited by M. Hamidullah and C. Pellat, with a long introduction in Arabic, for the Osmania Oriental Publications Bureau, Hyderabad, 1956).

Anwāʾ books of the second type are arranged in the form of a calendar enumerating natural events of importance to peasants and herdsmen. One such book, for 961, also exists in a medieval Latin translation; a later edition is R. Dozy, *Le calendrier de Cordoue*, new edition by C. Pellat (Leiden, 1961). In form, this sort of *anwāʾ* books resembles the calendaric texts of antiquity, such as the Babylonian *mul apin* texts or Ptolemy's *Phainomena*.

Because Ibn Qutayba was a contemporary of Abū Ḥanīfa, whose *Kitāb al-anwāʾ* was highly renowned, it is difficult, if not impossible, to decide upon the charge that has sometimes been raised against him: that he largely or completely excerpted or copied the work of Abū Ḥanīfa. As can be seen from comparison with excerpts and quotations from other *anwāʾ* books spread over a great number of lexicographic and scientific texts (for example, *Kitāb ṣuwar al-kawākib* ["Book of the Constellations of the Fixed Stars"] of al-Ṣūfī), the contents of Ibn Qutayba's *anwāʾ* book seem to have been similar to all the others of this type. This book, the text of which has never been translated, is available only in Arabic; in summarizing its contents reference is always made to the Hyderabad edition. After a short introduction, there is a detailed description of the twenty-eight lunar mansions (pp. 4–88), which contains much information on other adjacent stars and constellations. It is followed by meteorological traditions concerning them (pp. 88–94). Other astrometeorological lore follows (pp. 94–120), mainly on seasons and events in the Bedouins' life. Then astronomical information is given on the twelve zodiacal signs (p. 120), on the poles (p. 122), on the Milky Way (p. 123), on the heavenly spheres (p. 124), on the planets (p. 126), on the sun and moon (p. 128), on risings, settings, and the dawns (pp. 141, 142, 143), and on famous fixed stars (those not included in the passage on the lunar mansions, pp. 145–158). Ibn Qutayba then treats meteorological subjects: winds, rain, clouds, lightning and thunder, and the prediction of rain. Finally he deals with different ways of explaining some constellations (p. 182) and the use of the stars for orientation (*al-ihtidāʾ*, p. 186).

None of this information represents the result of scientific research, nor were scientific methods employed to gather it. Moreover, the book is merely a collection of what the Arabs—not influenced by the "scientific" techniques of foreign astronomy and astrology, which came to be known through translations not earlier than the second half of the eighth century—possessed in terms of popular lore about the sky and the stars, and all the phenomena connected or supposedly connected with them. It was taken

mostly from the existing poetical literature as well as from earlier philological compilations which, in turn, were based on similar sources and perhaps, in some cases, also on popular traditions of desert tribes. The popular astronomical knowledge of the Arabs in the "prescientific" epoch contained many elements of older, non-Arabic origin. The *anwāʾ* books, therefore, form a source of great interest not only for the history of Arabic lore and literature but for the history of dissemination of scientific knowledge and the development of astronomical observations and activity among the Arabs and their predecessors in several civilizations.

BIBLIOGRAPHY

See Carl Brockelmann, *Geschichte der arabischen Litteratur*, 2nd ed., I (Leiden, 1943), 124 ff., and supp., I (Leiden, 1937), 184 ff.; I. M. Huseini, *The Life and Works of Ibn Qutayba*, Publications of the Faculty of Arts and Sciences, American University, Oriental Series, no. 21 (Beirut, 1950), originally his dissertation (London, 1934); G. Lecomte, *Ibn Qutayba, l'homme, son oeuvre, ses idées* (Damascus, 1965); and C. Pellat, "Le traité d'astronomie pratique et de météorologie populaire d'Ibn Qutayba," in *Arabica*, 1 (1954), 84–88; and "Dictons rimés, *anwāʾ* et mansions lunaires chez les arabes," in *Arabica*, 2 (1955), 17–41; intro. to his ed. of Ibn Qutayba's *Kitāb al-anwāʾ* (Hyderabad, 1956), in Arabic; and "Anwāʾ," in *Encyclopaedia of Islam*, new ed. I (Leiden, 1960), 523–524. The chapter on celestial phenomena in Ibn Qutayba's *Kitāb adab al-kātib* ("Handbook for Authors"; the full Arabic text was ed. by M. Grünert, Leiden, 1900) was separately ed., trans., and discussed by A. Sprenger in *Journal of the Asiatic Society of Bengal*, 17, pt. 2 (1848), 659 ff.

PAUL KUNITZSCH

QUṬB AL-DĪN AL-SHĪRĀZĪ (*b.* Shīrāz, Persia, Ṣafar 1236; *d.* Tabrīz, Persia, 17 Ramaḍān 1311), *optics, astronomy, medicine, philosophy.*

Quṭb al-Dīn Maḥmūd was born into a well-known family of physicians and Ṣūfis. His father, Ḍiāʾ al-Dīn Masʿūd, was both a Ṣūfī master attached to Shihāb al-Dīn al-Suhrawardī and a famous physician; and under his guidance Quṭb al-Dīn received his early training in both medicine and Ṣūfism. At the time of his father's death Quṭb al-Dīn was but fourteen years old, yet he was entrusted with his father's duties as physician and ophthalmologist at the Muẓaffarī hospital in Shīrāz, where he remained for ten years.

At the age of twenty-four his love of learning led Quṭb al-Dīn to leave his position at the hospital in order to devote himself fully to his studies, especially in medicine. He studied Ibn Sīnā's *Canon* with several of the best-known masters of his day, but he could not find a teacher who satisfied him completely. He therefore traveled from city to city, seeking masters who could instruct him in both the medicine and the philosophy of Ibn Sīnā, a figure who attracted him greatly. In his journeys Quṭb al-Dīn met many Ṣūfī masters, whose gatherings he frequented. He traveled in Khurasan, Iraq, and Anatolia, meeting most of the medical authorities of the day. Also during these journeys he was initiated formally into Ṣūfism at the age of thirty by Muḥyi'l-Dīn Aḥmad ibn 'Alī, a disciple of Najm al-Dīn Kubrā.

Around 1262 Quṭb al-Dīn became associated with his most famous teacher, Naṣīr al-Dīn al-Ṭūsī, at Marāgha; his superior intelligence soon made him al-Ṭūsī's foremost student. With al-Ṭūsī he studied both astronomy and the philosophy of Ibn Sīnā, particularly *Al-Ishārāt wa'l-tanbīhāt* ("Book of Directives and Remarks"). After a long period during which he was closely connected with the circle of Naṣīr al-Dīn, Quṭb al-Dīn left Marāgha for Khurasan to study with another well-known philosopher, Najm al-Dīn Dabīrān Kātibī al-Qazwīnī. His studies later took him to Qazvīn and Baghdad, where he stayed at the Niẓāmiyya school. From there he set out for Konya and became a follower of the celebrated Ṣūfī and disciple of Ibn 'Arabī, Ṣadr al-Dīn al-Qunyawī, with whom he studied the religious sciences such as Quranic commentary and Ḥadīth. After the death of Ṣadr al-Dīn, Quṭb al-Dīn left Konya to become judge in Sivas and Molatya, starting the period during which some of his major works appeared.

When he later moved to Tabrīz, Quṭb al-Dīn attracted the attention of the son of Hulāgu Khan, Aḥmad Takūdār, who was then ruling Persia. The latter sent him as ambassador to the court of the Mameluke ruler of Egypt, Sayf al-Dīn Qalā'ūn. This journey was of major scientific importance for him, for during this period he gained access to some of the important commentaries upon Ibn Sīnā's *Canon* which he had long sought and which were to serve him in the preparation of his major commentary upon this work. In 1283 he finally began to write this commentary, which occupied him for most of the rest of his life.

From Egypt, Quṭb al-Dīn returned to Tabrīz, where he met the important scholarly figures of his day, such as the learned vizier and historian Rashīd al-Dīn Faḍlallāh. It was in this capital of the Īl-Khanids that he died, after nearly fourteen years spent mostly in seclusion and devoted to writing. His love of learning became proverbial in Persia; he was given the honorific title 'Allāma, rare in medieval times, and the historian Abu'l-Fidā' gave him the title al-Muta-

fannin, "master in many sciences." He was also called "the scholar of the Persians." He was known as a master chess player and an excellent player of the lute, and he spent much of his time on these two pastimes.

Although Quṭb al-Dīn was among the foremost thinkers and scholars of Islam, only two of his works have been printed: the *Durrat al-tāj* and the *Sharḥ Ḥikmat al-ishrāq*, the latter only in a lithographed edition. The rest of his writings remain in manuscript. The entire body of his thought cannot be known until these works are edited and made accessible for study.

Quṭb al-Dīn's geometrical works are the following:

1. The Persian translation of Naṣīr al-Dīn al-Ṭūsī's *Taḥrir Uṣūl Uqlīdus* ("Recension of the *Elements* of Euclid").

2. *Risāla fī ḥarakat al-daḥraja wa'l-nisba bayn al-mustawī wa'l-munḥanī* ("Treatise on the Motion of Rolling and the Relation Between the Straight and the Curved").[1]

Those on astronomy and geography are the following:

3. *Nihāyat al-idrāk fī dirāyat al-aflāk* ("The Limit of Understanding of the Knowledge of the Heavens"). Quṭb al-Dīn's major astronomical work, it consists of four books: introduction, the heavens, the earth, and the "quantity" of the heavens. There are sections on cosmography, geography, geodesy, meteorology, mechanics, and optics, reflecting both the older scientific views of Ibn al-Haytham and al-Bīrūnī and new scientific theories in optics and planetary motion. This work was completed around 1281 and has been commented upon by Sinān Pāshā.

4. *Ikhtiyārāt-i muẓaffarī* ("Muẓaffarī Selections"). This work, one of Quṭb al-Dīn's masterpieces, contains his own views on astronomy and is perhaps the best work on astronomy in Persian. It is a synopsis of the *Nihāya*; is composed, like that work, of four sections; and was written sometime before 1304.

5. *al-Tuḥfat al-shāhiyya fi'l-hay'a* ("The Royal Gift on Astronomy." Composed shortly after the *Nihāya* (in 1284), to solve more completely problems begun in the earlier work, it constitutes, along with the *Nihāya* Quṭb al-Dīn's masterpiece in mathematical astronomy. About these two works Wiedemann wrote, "Ḳuṭb al-Dīn has in my opinion given the best Arabic account of astronomy (cosmography) with mathematical aids."[2] This work, like the *Nihāya*, was celebrated in later Islamic history and has been commented upon by Sayyid Sharīf and 'Alī Qūshchī.

6. *Kitāb fa'altu fa-lā talum fi'l-hay'a* ("A Book I Have Composed, But Do Not Blame [Me for It], on Astronomy").

7. *Kitāb al-tabṣira fi'l-hay'a* ("The Tabṣira on Astronomy").

248

8. *Sharḥ al-tadhkira al-naṣīriyya* ("Commentary Upon the Tadhkira of Naṣīr al-Dīn"). Commentary upon the famous *Tadhkira* of Naṣīr al-Dīn al-Ṭūsī and also on the *Bayān maqāṣid al-tadhkira* of Muḥammad ibn ʿAlī al-Himādhī.

9. *Kharīdat al-ʿajāʾib* ("The Wonderful Pearl").

10. *Khulāṣat Iṣlāḥ al-majisṭī li-Jābir ibn Aflaḥ* ("Extracts of *Correction of the Almagest* of Jābir ibn Aflaḥ").

11. *Ḥall mushkilāt al-majisṭī* ("Solution of the Difficulties of the Almagest"). A work that is apparently lost.

12. *Taḥrīr al-zīj al-jadīd al-riḍwānī* ("Recension of the New Riḍwānī Astronomical Tables").

13. *al-Zīj al-sulṭānī* ("The Sulṭānī Astronomical Tables"). These tables have been attributed to both Quṭb al-Dīn and Muḥammad ibn Mubārak Shams al-Dīn Mīrak al-Bukhārī.

Medical works by Quṭb al-Dīn include the following:

14. *Kitāb nuzhat al-ḥukamāʾ wa rawḍat al-aṭibbāʾ* ("Delight of the Wise and Garden of the Physicians"), also known as *al-Tuḥfat al-saʿdiyya* ("The Presentation to Saʿd") and *Sharḥ kullīyyāt al-qānūn* ("Commentary Upon the Principles of the *Canon* of Ibn Sīnā"). This is the largest work by Quṭb al-Dīn, in five volumes. He worked on it throughout his life and dedicated it to Muḥammad Saʿd al-Dīn, the vizier of Arghūn and the Īl-Khanid ruler of Persia.

15. *Risāla fiʾl-baraṣ* ("Treatise on Leprosy").

16. *Sharḥ al-Urjūza* ("Commentary Upon Ibn Sīnā's *Canticum*").

17. *Risāla fī bayān al-ḥāja ilaʾl-ṭibb wa-ādāb al-aṭibbāʾ wa-waṣāyāhum* ("Treatise on the Explanation of the Necessity of Medicine and of the Manners and Duties of Physicians").

Theosophical, philosophical, and encyclopedic works are the following:

18. *Durrat al-tāj li ghurrat al-dībāj fiʾl-ḥikma* ("Pearls of the Crown, the Best Introduction to Wisdom"). This encyclopedic philosophical and scientific work in Persian comprises an introduction on knowledge and the classification of the sciences; five books *(jumla)* dealing with logic, metaphysics, natural philosophy, mathematics, and theodicy; and a four-part conclusion on religion and mysticism. The introduction, and the books on logic, metaphysics, and theodicy, were published by S. M. Mishkāt (Teheran, 1938–1941), and book 4 on mathematics, excluding certain portions on geometry, by S. H. Ṭabasī (Teheran, 1938–1944).

The philosophical sections of *Durrat al-tāj* were greatly influenced by the writings of Ibn Sīnā and Suhrawardī; the geometry is mostly a Persian translation of Euclid's *Elements* with the paraphrases and commentaries of al-Ḥajjāj and Thābit ibn Qurra. The astronomy is a translation of the *Summary of the Almagest* of ʿAbd al-Malik ibn Muḥammad al-Shīrāzī, and the music is taken from al-Fārābī, Ibn Sīnā, and ʿAbd al-Muʾmin. In the sections on religion and ethics, Quṭb al-Dīn made use of the writings of Ibn Sīnā and Fakhr al-Dīn al-Rāzī; and in Ṣūfism or mysticism, of the *Manāhij al-ʿibād ilaʾl-maʿād* of Saʿd al-Dīn al-Farghānī, a disciple of Mawlānā Jalāl al-Dīn Rūmī and Ṣadr al-Dīn al-Qunyawī.

19. *Sharḥ Ḥikmat al-ishrāq* ("Commentary Upon the Theosophy of the Orient of Light"). The best-known commentary upon Suhrawardī's *Ḥikmat al-ishrāq*, it was published in a lithographed edition (Teheran, 1897).

20. *Sharḥ Kitāb rawḍat al-nāẓir* ("Commentary Upon the *Rawḍat al-nāẓir*"). A commentary upon Naṣīr al-Dīn al-Ṭūsī's *Rawḍat al-nāẓir* on questions of ontology.

21. *Sharḥ al-najāt* ("Commentary Upon the *Najāt*"). Commentary upon Ibn Sīnā's *Kitāb al-najāt*.

22. *al-Sharḥ waʾl-ḥāshiya ʿalaʾl-Ishārāt waʾl-tanbīhāt* ("Commentary and Glosses Upon the *Ishārāt*"). Commentary upon Ibn Sīnā's last philosophical masterpiece, the *Ishārāt*.

23. *Ḥāshiya ʿalā Ḥikmat al-ʿayn* ("Glosses Upon the *Ḥikmat al-ʿayn*"). The first commentary upon Najm al-Dīn Dabīrān al-Kātibī's well-known *Ḥikmat al-ʿayn*, upon which many commentaries appeared later.

24. *Unmūzaj al-ʿulūm* ("A Compendium of the Sciences").

25. *Wajīza fiʾl-taṣawwur waʾl-taṣdīq* ("A Short Treatise on Concept and Judgment").

26. *Risāla dar ʿilm-i akhlāq* ("Treatise on Ethics"). A treatise in Persian which is apparently lost.

The rest of the works by Quṭb al-Dīn treat the sciences of language and strictly religious questions, and there is no need to deal with them here. He also left a few poems of some literary quality.

Philosophy and Theology. Quṭb al-Dīn belonged to that group of Muslim philosophers between Suhrawardī and Mullā Ṣadrā who revived the philosophy of Ibn Sīnā after the attacks of al-Ghazālī, giving it at the same time an illuminationist quality drawn from the teaching of Suhrawardī. After his teacher Naṣīr al-Dīn al-Ṭūsī, Quṭb al-Dīn must be considered the foremost philosophical figure during the four centuries which separated Suhrawardī from Mullā Ṣadrā. Quṭb al-Dīn was also a leading example of the Muslim sage or *ḥakīm*, who was the master of many disciplines and wrote definitive works in each of them. His *Durrat al-tāj* is the outstanding Persian encyclopedia of Peripatetic philosophy. Written on the model of Ibn Sīnā's *al-Shifāʾ*, it has additional sections devoted to Ṣūfism and strictly religious matters not found in earlier Peripatetic works. His commentary upon the *Ḥikmat al-ishrāq*, although based mostly upon that of Shahrazūrī, rapidly replaced the latter as the most famous such work. Later generations saw Suhrawardī mostly through the eyes of Quṭb al-Dīn. His theological and religious writings also commanded great respect. The thirteenth and fourteenth centuries were marked in Persia by the gradual rapprochement of the four intellectual schools of theology *(kalām)*, Peripatetic philosophy *(mashshāʾī)*, illuminationist theosophy *(ishrāq)*, and gnosis *(ʿirfān)*. Quṭb al-Dīn was one of the key figures who brought this about and prepared the way for the synthesis of the Safavid period. He was at once a fervent disciple of Ibn Sīnā, the master of Peripatetics; a commentator on Suhrawardī, the founder of the *ishrāqī* school; and a student of Ṣadr al-Dīn al-Qunyawī, the closest disciple of the greatest expositor of the gnostic teachings of Islam, Ibn ʿArabī. Furthermore he was a theologian and religious scholar of note. To all of these he added his remarkable acumen in mathematics, astronomy, physics, and medicine, for which he has become known as much as a scientist as a philosopher.

Mathematics. Quṭb al-Dīn attached a metaphysical significance to the study of mathematics, which he viewed more in the Pythagorean than in the Aristotelian manner. He saw it as the means to discipline the soul for the study of metaphysics and theosophy. His greatest contributions came in astronomy and optics, which were then part of the mathematical sciences, rather than in pure mathematics in the modern sense.

Optics. After Ibn al-Haytham there was a relative lack of interest in optics among Muslims; the optical writings even of Naṣīr al-Dīn al-Ṭūsī show a definite decline in comparison. Probably mostly because of the spread of Suhrawardī's newly founded school of illumination, which made light synonymous with being and the basis of all reality, a definite renewal of interest in optics occurred in the thirteenth century, for which Quṭb al-Dīn was largely responsible. Although he did not write separate treatises on optics, his *Nihāyat al-idrāk* contains sections devoted to the subject. He was especially interested in the phenomena of the rainbow and must be considered the first to have explained it correctly. He concluded that the rainbow was the result of the passage of light through a transparent sphere (the raindrop). The ray of light is refracted twice and reflected once to cause the observable colors of the primary bow. The special attention paid by Quṭb al-Dīn and his students was in fact responsible for the creation in Islam of a separate science of the rainbow *(qaws qazaḥ)*, which

first appeared in the classification of the sciences at this time. The significance of Quṭb al-Dīn in optics also lies in his transmission of the optical teachings of Ibn al-Haytham to al-Fārisī, who then composed the most important commentary upon Ibn al-Haytham's *Optics*, the *Tanqīḥ al-Manāẓir*.

Also of interest in this field is Quṭb al-Dīn's theory of vision in his *Sharḥ Ḥikmat al-ishrāq*, in which he rejected both the Euclidean and the Aristotelian theories and confirmed the *ishrāqī* theory, according to which vision occurs when there is no obstacle between the eye and the object. When the obstacle is removed, the soul of the observer receives an illumination through which the whole of the object is perceived as a single reality.

Astronomy. Quṭb al-Dīn wrote at the beginning of his *Ikhtiyārāt* that the principles of astronomy fall under three headings: religion, natural philosophy, and geometry. Those who study this science become dear to God, and the student of astronomy becomes prepared for the understanding of the divine sciences because his mind is trained to study immaterial objects. Moreover, through the study of astronomy the soul gains such virtues as perseverance and temperance, and aspires to resemble the heavenly spheres. He definitely believed that the study of astronomy possessed a religious value and he himself studied it religiously and with reverence.

Quṭb al-Dīn played a major role in the observations made at Marāgha which led to the composition of the *Īlkhānī zīj*, although his name is not mentioned in its introduction. In his *Nihāya* he suggested that the values listed in the *Īlkhānī zīj* for the motion of the apogee were not based on calculation from the successive equinoxes but were dependent upon repeated observations. He asserted that the shift in the solar apogee could be confirmed by comparing the values found in Ptolemy and the later astronomical tables preceding the *Īlkhānī zīj*, which implies recourse to frequent observations. Quṭb al-Dīn was keenly interested in scientific observation, but this in no way reduced his viewpoint to empiricism or detracted from his theoretical interests or philosophical vision.

Quṭb al-Dīn emphasized the relation between the movement of the sun and the planets in the way that is found later in the writings of Regiomontanus, and which prepared the way for Copernicus. In fact, through the research of E. S. Kennedy and his associates, it has been discovered that new planetary models came out of Marāgha which represent the most important departure from the Ptolemaic model in medieval times and are essentially the same as those of Copernicus, provided one ignores the heliostatic hypothesis.

The Marāgha school sought to remove a basic flaw from the Ptolemaic model for planetary motion, namely, the failure of certain Ptolemaic configurations to conform to the principle that celestial motion must be uniform and circular. To remedy this situation Nasīr al-Dīn al-Ṭūsī proposed in his *Tadhkira* a rolling device consisting of two vectors (to use the modern terminology) of equal length, the second moving with a constant velocity twice that of the first but in the opposite direction. This device Kennedy has named the "Ṭūsī couple."

Quṭb al-Dīn in his *Nihāya* and *al-Tuḥfat al-shāhiyya*, both of which, like the *Tadhkira*, are divided into four parts, sought to work out this model for the different planets but apparently never did so to his full satisfaction, for he kept modifying it. In fact, he produced the two above-mentioned works within four years, in an attempt to achieve the final answer. In the several manuscripts of each, the two works contain successive endeavors to reach a completely satisfactory solution to what is definitely Quṭb al-Dīn's most important achievement in astronomy.

The planetary model which Quṭb al-Dīn used for all the planets except Mercury can be summarized as shown below (see Figure 1).[3]

As the figure shows, a vector of length 60 which is in the direction of the mean longitude is drawn from a point midway between the equant center and the deferent center. Another vector, with a length equal to half the eccentricity, rotates at the end of this vector.

Because of the great eccentricity of Mercury, its model requires special conditions. Figure 2 demonstrates how Quṭb al-Dīn was finally able to create a model which fulfilled the conditions for this planet. As E. S. Kennedy—to whom we owe this analysis and figure—states:

> The first vector r_1 is of length 60, it issues from the deferent center, and it has at all times the direction of the mean planet. The next four vectors, each of length $c/2$ [where $c = 3$, because the eccentricity of Mercury is 6], make up two Ṭūsī couples. The last vector r_6 has length c. The initial positions and rates of rotation of all the vectors are as shown on the drawing, where k is the mean longitude measured from apogee.[4]

This model represents the height of the techniques developed at Marāgha to solve the problems of planetary motion. Quṭb al-Dīn also applied these techniques to the solution of the problem of the moon, trying to remove some of the obvious flaws in the Ptolemaic model. But in this matter another Muslim astronomer who adopted these techniques, Ibn al-Shāṭir, was more successful. He produced a model that was greatly superior to that of Ptolemy, the same that was produced later by Copernicus.

Geography. The interest of Quṭb al-Dīn in observation is also evident in geography. Not only did he write on geography in his *Nihāya*, drawing from earlier Muslim geographers, especially al-Bīrūnī, but he traveled throughout Asia Minor, examining the route to be followed by the Genoese ambassador of the Mongol ruler Arghūn to the Pope, Buscarello di Ghizalfi. In 1290 he presented a map of the Mediterranean to Arghūn based on observations made of the coastal areas of Asia Minor.

Physics. In his Peripatetic works Quṭb al-Dīn generally followed the physics of Ibn Sīnā, but in the *Sharḥ Ḥikmat al-ishrāq* he developed a physics of light which is of particular interest. In it he considered light as the source of all motion, both sublunar and celestial. In the case of the heavenly spheres, motion is a result of the illumination of the souls of the spheres by divine light. He divided bodies into simple and compound, and these in turn into transparent and opaque, so that light and darkness, rather than the Aristotelian hylomorphism, dominate his physics. He also reinterpreted meteorological phenomena in terms of light and light phenomena.

Medicine. Quṭb al-Dīn's major contribution to medicine was his commentary upon Ibn Sīnā's *Canon*, which was celebrated in later centuries in the Islamic world but has not been analyzed thoroughly in modern times. This work seeks to explain all the difficulties in the *Canon* relating to general principles of medicine. Quṭb al-Dīn based it not only on his own lifelong study of the text and what he had learned from his masters in Shīrāz, Marāgha, and other cities, but also on all the important commentaries he found in Egypt, especially the *Mūjiz al-Qānūn* of Ibn al-Nafīs, the *Sharḥ al-kulliyyāt min kitāb al-Qānūn* of Muwaffaq al-Dīn Ya'qūb al-Sāmarrī, and the *Kitāb al-shāfī fī'l-ṭibb* of Abu'l-Faraj ibn al-Quff. In medicine, as in philosophy, Quṭb al-Dīn did much to revive the teachings of Ibn Sīnā and had an important role in the propagation of Avicennan medicine, especially from the fifteenth century onward in the Indian subcontinent.

Influence. The most famous students of Quṭb al-Dīn were al-Fārisī, the outstanding commentator on Ibn al-Haytham; Quṭb al-Dīn al-Rāzī, the author of many famous works, including the *Muḥākamāt*, a "trial" of the relative merits of the commentaries of Naṣīr al-Dīn al-Ṭūsī and Fakhr al-Dīn al-Rāzī upon the *Ishārāt* of Ibn Sīnā; and Niẓām al-Dīn al-Naishāpūrī the author of *Tafsīr al-Taḥrīr*, on Naṣīr al-Dīn al-Ṭūsī's *Recension of the Almagest*. Quṭb al-Dīn's influence continued through these and other students,

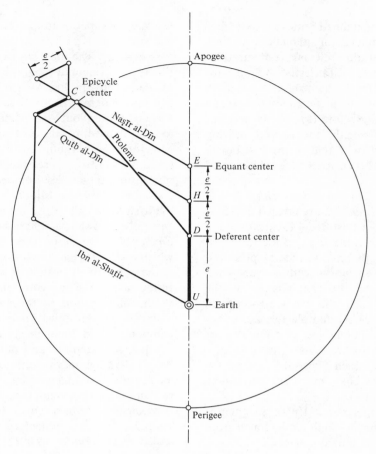

FIGURE 1

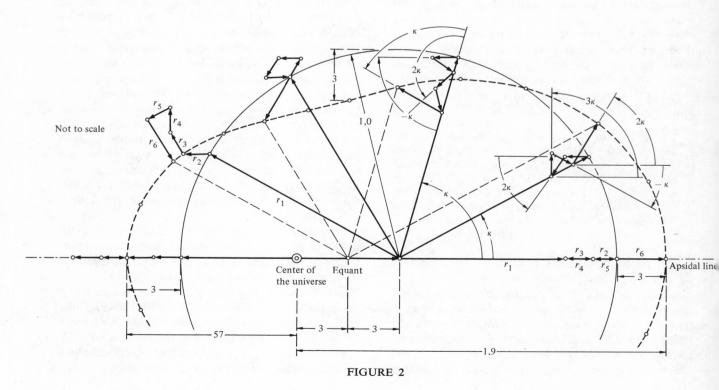

FIGURE 2

252

and also through his writings, especially the *al-Tuḥfat al-saʿdiyya* in medicine, *Nihāyat al-idrāk* in astronomy, and *Sharḥ Ḥikmat al-ishrāq* in philosophy, the last having become a standard text of Islamic philosophy in the traditional schools of Persia. His writings were also one of the influential intellectual elements that made possible the Safavid renaissance in philosophy and the sciences in Persia, and his name continued to be respected and his works studied in the Ottoman and the Mogul empires.

NOTES

1. One of the few treatises of Quṭb al-Dīn analyzed thoroughly in a European language, is E. Wiedemann, "Ueber eine Schrift ueber die Bewegung des Rollens und die Beziehung zwischen dem Geraden und dem Gekruemmten von Quṭb al Dīn Maḥmûd b. Masʿûd al Schîrâzî," in *Sitzungsberichte der Physikalisch-medizinischen Sozietät in Erlangen*, **58–59** (1926–1927), 219–224.
2. Article on Kuṭb al-Dīn al-Shīrāzī, in *Encyclopaedia of Islam*, 1st ed., II, 1167.
3. E. S. Kennedy, "Late Medieval Planetary Theory," in *Isis*, **57**, pt. 3 (1966), 367, 373.
4. *Ibid.*, 373–374.

BIBLIOGRAPHY

I. ORIGINAL WORKS. Quṭb al-Dīn's published works are *Durrat al-tāj*, pt. I, 5 vols., S. M. Mishkāt, ed. (Teheran, 1938–1941); pt. II, 5 vols. S. H. Ṭabasī, ed. (Teheran, 1938–1944); and *Sharḥ Ḥikmat al-ishrāq* (Teheran, 1897).

II. SECONDARY LITERATURE. See E. S. Kennedy, "Late Medieval Planetary Theory," in *Isis*, **57**, no. 3 (1966), 365–378; M. Krause, "Stambuler Handschriften islamischer Mathematiker," in *Quellen und Studien zur Geschichte der Mathematik, Astronomie und Physik*, Abt. B, Studien, **3** (1936), 437–532; M. Minovi, "Mullā Quṭb Shīrāzī," in *Yād-nāma-ye īrāni-ye Minorsky* (Teheran, 1969), 165–205; M. T. Mīr, *Pizishkān-i nāmī-ye pārs* (Shiraz, 1969), 110–117; S. H. Nasr, *Science and Civilization in Islam* (Cambridge, Mass., 1968), 56 and *passim*; G. Sarton, *An Introduction to the History of Science*, II (Baltimore, 1941), 1017–1020; A. Sayili, *The Observatory in Islam* (Ankara, 1960), *passim*; H. Suter, "Die Mathematiker und Astronomen der Araber," in *Abhandlungen zur Geschichte der mathematischen Wissenschaften* (1900), 158; Qadrī Ḥāfiẓ Ṭuqān, *Turāth al-ʿarab al-ʿilmī fi'l-riyāḍiyyāt wa'l-falak* (Cairo, 1963), 425–427; and E. Wiedemann, "Zu den optischen Kenntnissen von Quṭb al Dīn al Schîrâzî," in *Archiv für die Geschichte der Naturwissenschaften und der Technik*, **3** (1912), 187–193; "Ueber dei Gestalt, Lage und Bewegung der Erde sowie philosophisch-astronomische Betrachtungen von Quṭb al Dīn al Schîrâzî," *ibid.*, 395–422; and "Ueber eine Schrift ueber die Bewegung des Rollens und die Beziehung zwischen dem Geraden und den Gekruemmten, von Quṭb al Dīn Maḥmûd b. Masʿûd al Schîrâzî," in *Sitzungsberichte der Physikalisch-medizinischen Sozietät in Erlangen*, **58–59** (1926–1927), 219–224.

SEYYED HOSSEIN NASR

RABELAIS, FRANÇOIS (*b.* "La Devinière," near Chinon, France, 1494 [?]; *d.* Meudon, France, 9 April 1553 [?]), *medicine*.

Rabelais entered monastic life at the age of sixteen. A Franciscan for about fifteen years, he was ordained at the convent of Fontenay-le-Comte. About 1525 he joined the more intellectually liberal order of the Benedictines and began to visit university cities, acquiring a complete humanist and Hellenistic education. He subsequently abandoned regular orders and obtained a canonry. He became a physician at the age of thirty-six but continued his literary activity and his political, diplomatic, and administrative duties in the service of the king and of his patrons, the du Bellay family, notably Guillaume du Bellay, Seigneur de Langey, and Bishop Geoffroy d'Estissac. This article, however, will concern solely his medical activity.

Rabelais probably began his medical studies at Paris in 1525 (Marichal) but subsequently enrolled at the Faculty of Medicine of Montpellier, where he received his bachelor's degree on 1 November 1530. He spent the following year studying the *Aphorisms* of Hippocrates and Galen's *Ars parva*. He then went to Lyons, where he made many friends, including Symphorien Champier (1471/1472–1538/1539), Jean Canape, Étienne Dolet (1509–1546), and Clément Marot (*ca.* 1496–1544). In 1532 he was appointed physician at the Hôtel Dieu with a yearly salary of forty *livres*. Rabelais called Lyons, more tolerant and welcoming than Paris, "sedes studium meorum"; and his intense medical activity there found expression in three books published by Sébastien Gryphe (on Hippocrates and Galen, on Manardi, and on the *Aphorisms* of Hippocrates) and in four letters.

Rabelais's stay in Lyons soon ended. Hated by the *sorbonnards*, he twice left his hospital duties on very short notice—after having arranged for someone to replace him. During his first absence he was with Jean du Bellay in Italy from February to May 1534. It is not known what he did during his second absence, during which he lost his post at the Hôtel Dieu, despite the intervention of his friends (1535).

Rabelais returned to Montpellier, where he earned his *licence* and, on 22 May 1537, his doctorate in medicine. Five months later he gave a course on Hippocrates' *Prognostic*; and at the end of that year, at Lyons, he conducted a public dissection of the corpse of a hanged criminal, an event celebrated in a poem by Étienne Dolet. In August 1539 he again taught at Montpellier. From 1540 to 1543 he was in the service of Guillaume du Bellay (1491–1543). After holding the post of municipal physician of Metz from March 1546 to April 1547, he rejoined Cardinal Jean du Bellay until 1550. The latter obtained the rectorship of Meudon for him, and Rabelais spent his final years carrying out the duties of this position and writing his last books.

Apart from the brief interlude at Metz, Rabelais actually practiced medicine only from 1530 to 1539 (Marichal). He probably began so late in life because he was remarkably well prepared not only to study but also to teach science, which was then essentially philological and oriented toward the examination of authenticated ancient texts. Yet he was not content to expound an "ancient and worn-out medicine" and to play an important role in the translation of Latin and Greek medical terms into French. He was also an all-round practitioner, interested in his patients' well-being and diet. He described, following Galen, a *glottocomion* for treating thigh fractures by continuous extension and a syringotome with a concealed blade for incising strangulated hernias. These devices are mentioned in the writings of Ambroise Paré.

At Montpellier, as at Lyons and in Italy, Rabelais was considered an excellent physician. He never departed very far from traditional doxology, however, nor did he claim to be an innovator. The *Quart livre* (chs. XXX, XXXII) contains the anatomy and physiology of Quaresméprenant. Each organ is described by comparison with a familiar object; the comparisons are quite often accurate, but they do not reveal the anatomical discoveries that Ledouble claims for Rabelais. The many allusions to natural history (the itch mite and more general zoological and botanical subjects) also derive more from his reading than from direct and objective knowledge of nature.

Rabelais's literary works helped to introduce into the language of the time a great number of contemporary scientific terms as well as the anatomical and physiological concepts that they designated.

BIBLIOGRAPHY

For the most important works on Rabelais's works on medicine, see L. Bourilly, "Sa vie et son oeuvre d'après des travaux récents," in *Revue d'histoire moderne et contemporaine,* **7** (1905–1906), 588–609; Félix Bremond, *Rabelais médecin avec notes et commentaires* (Paris, 1879); C. G. Custom, "Reflections on the Obstetrical Science of Maître François Rabelais," in *American Journal of Obstetrics and Diseases of Women and Children,* **65** (1912), 1006–1020, "A Short Outline of the Medical Career of Maître François Rabelais," in *New York Medical Journal,* **95** (1912), 873–875; Paul Delaunay, *Oeuvres de François Rabelais,* V (Paris, 1931, 1953), *Humanisme et encyclopédisme in Histoire Générale des Sciences,* II (Paris, 1958), 3–10; C. F. Gillard, "Rabelais médecin. Les études et les opinions médicales de l'auteur du Gargantua et du Pantagruel," in *Thèse médicale Paris* (1920); and R. Gordon, *F. Rabelais à la Faculté de médecine de Montpellier* (Paris, 1876).

See also A. Heulhard, *Rabelais, chirurgien. Applications de son glossocomion dans les fractures du fémur, et de son syringotome dans le traitement des plaies pénétrantes de l'abdomen décrites à Arthur Heulhard* (Paris, 1885), *Rabelais, ses voyages en Italie, son exil à Metz* (Paris, 1891); H. Joly, *Rabelais à Lyon in: Le livre à Lyon des origines jusqu'à nos jours* (Lyon, 1933), 23–29; A. F. Le Double, "Rabelais anatomiste et physiologiste, quelques contenances de Quaresméprenant," in *Gazette médicale du Centre,* **11** (1906), 273–278; J. Margarot, "François Rabelais—médicin. Influence de la médecine sur son oeuvre," in *Biologie médicale,* **43** (1954); C. Mentre, "Rabelais et l'anatomie," in *Thèse médicale Nancy* (Nancy, 1970); and J. Plattard, *L'oeuvre de Rabelais* (Paris, 1910).

PIERRE HUARD
MARIE-JOSÉ IMBAULT-HUART

RABL, CARL (*b.* Wels, Oberösterreich, Austria, 2 May 1853; *d.* Leipzig, Germany, 24 December 1917), *anatomy, comparative embryology, cytology.*

Rabl's contemporaries saw him primarily as a zoologist whose anatomical studies encompassed many facets of morphology and whose researches in comparative embryology clarified various aspects of developmental history; but his role in the development of the chromosome theory of inheritance remains outstanding. It was Rabl who first clearly expressed the concept of the continuity of the chromosomes throughout cellular division. The idea of the individuality of the chromosomes, suggested by Edouard van Beneden and given its fullest definition in the basic work of Boveri, was fundamental to the understanding of the mechanism of heredity within the cell.

The son of Carl Rabl, a physician, Rabl entered the Gymnasium at Kremsmünster, intending to study medicine. He early showed a preference for natural history and was especially attracted to zoology and comparative anatomy. Books on biological aspects of man's place in nature became his favorites. But, most of all, Ernst Haeckel's *Natürliche Schöpfungsgeschichte* fired his enthusiasm, for in Germany that period was marked by fascination with Darwin's theory of evolution. Long before he finished his Gymnasium course in 1871, Rabl had resolved to study under Haeckel at Jena; but instead he studied medicine for two years at Vienna. Disappointed by the lack of stimulus or encouragement, he attended lectures and dissected both in the anatomical rooms and at home, meanwhile reading the works of Haeckel and Darwin.

In the fall of 1873 Rabl transferred to the University of Leipzig to work under the zoologist Rudolf

Leuckart, a happier experience, and managed to visit Jena and meet Haeckel. During the Easter vacation of 1874 he was already engaged in research on the early development of gastropods, which he planned to continue during the summer at Jena. He studied there under Haeckel in the summers of 1874 and 1875 and in the latter year published the first of several papers on his investigations in developmental biology. The general direction of his work was deeply influenced by Haeckel, who communicated his scientific enthusiasms and in later years remained his friend and correspondent. But Haeckel tended to broad theorizing, as Rabl recognized; returning to Vienna in 1875, he came under the tempering influence of another great teacher, the physiologist Ernst Brücke, who insisted on extremely careful observation in his histological studies and firmly placed fact before theory.

Rabl's zoological interests had caused him to take longer than usual to complete his medical course; finally, in 1882, he received his degree at Vienna, then became a prosector at the anatomical institute there, assisting Karl Langer. In addition to lecturing to medical students and carrying out special studies of vertebrate developmental anatomy, Rabl published a remarkable and carefully illustrated monograph on cell division, "Über Zelltheilung," in *Morphologisches Jahrbuch* (1885). He continued this work when he was appointed in 1885 to teach anatomy at the Ferdinand University in Prague, where in 1886 he attained the rank of ordinary professor.

While at Prague, Rabl served a term as the dean of his faculty, and he was rector of the Ferdinand University in 1903–1904. In 1891 he had married Marie Virchow, daughter of the pathologist Rudolf Virchow; but although they had many friends and Rabl was esteemed by his colleagues at Prague, he felt that he would be happier at a German university and accepted the chair at Leipzig as successor to Wilhelm His in 1904. At Leipzig, Rabl reorganized and improved the teaching of anatomy, expanding the facilities and research collections, and directed the anatomical institute until his death.

As Haeckel's student Rabl had begun investigations into the formation of the germ layers in the young embryo. Soon he was going back to early cleavage and to the structure of the egg cell itself. Over the years he returned to the formation of the germ layers and especially of the mesoderm and its derivatives. The various organs of which he traced the origins were not, however, followed merely to chronicle in detail their separate morphological histories; rather, they were examples that would lead, through induction, to general laws. Rabl was convinced that the events of cell division were precisely determined and that

embryological development was a mechanism in which the final position of each cell in the body had been predetermined, a conclusion he reached independently of His, but one similarly opposed to epigenetic explanations. The cell was to Rabl a complex and bilaterally differentiated and symmetrical organism; he stressed its polarities, and he concluded that the unsegmented egg already contained differentiated protoplasmic particles.

During the period when the new science of cytology was making known the course of cell division and was revealing that the chromosomes were halved lengthwise, to be distributed equally to the daughter cells, Rabl published his detailed description of the process, beautifully illustrated with his own drawings. In this study (1885) he maintained that the chromatin-staining filaments in the nucleus, later called the chromosomes, must actually persist through interphase even though, for a time, they seemed to disappear. "It is inconceivable," he wrote, "that in the resting cell no trace of this arrangement should exist any more." He had concluded, following his determination in studies of salamander larvae, that the organization of the cell must remain through division; that there was a constancy in the number of chromosomal filaments characteristic of a given tissue; and that a numerical law applied to each kind of cell. The significance of the continuity of the chromosomes for the understanding of the process of cell division soon became evident, and the important individuality of the chromosomes was shown in 1887 and afterward in the work of Boveri. Although Boveri acknowledged Rabl's contribution, embryologists in later years took sides in an unfortunate priority dispute—for, as one colleague put it, Rabl regarded the theory of the continuity of the chromosome as "his exclusive intellectual property."

Besides his work on the cell, Rabl's numerous special investigations included the development of the heart in amphibians, the formation of the lens of the vertebrate eye, and cranial segmentation, skeletal derivation, and the origin of the paired extremities. His contributions to morphology were widely cited by other comparative anatomists. Still, Rabl was no systematist, despite his careful working out of detail; rather, he sought to make a general application of his findings in a theory of development.

BIBLIOGRAPHY

I. Original Works. A curious but invaluable assessment of Rabl's ideas and the aims of his research is provided by his "critical analysis" of the work of his colleague Edouard van Beneden, a critique requested in the latter's will. It was to have been written by Rabl and Walther Flemming, who predeceased van Beneden, and became an entire volume that even included some new work by Rabl: "Edouard van Beneden und der gegenwärtige Stand der wichtigsten um ihm behandelten Probleme," in *Archiv für mikroskopische Anatomie . . .*, **88** (1915), 1–470. Rabl's paper on the cell, "Über Zelltheilung," in *Morphologisches Jahrbuch*, **10** (1885), 214–330, was followed by an essay on the same subject addressed to Kölliker, "Über Zelltheilung," in *Anatomischer Anzeiger*, **4** (1889), 21–30. Rabl's inaugural lecture at the University of Leipzig, *Über "Organbildende Substanzen" und ihre Bedeutung für die Vererbung* (Leipzig, 1906), outlines his researches and views, and defines problems of the day in regard to the cell, heredity, and embryonic development. Developmental studies include "Die Ontogenie der Süsswasser-Pulmonaten," in *Jenaische Zeitschrift für Naturwissenschaft*, **9** (1875), 195–240; "Ueber die Entwicklungsgeschichte der Malermuschel," *ibid.*, **10** (1876), 310–394; "Über den Bau und die Entwicklung der Linse," in *Zeitschrift für wissenschaftliche Zoologie*, **63** (1898), 496–572; **65** (1899), 257–367; **67** (1900), 1–138; and others listed in the obituaries cited below.

II. SECONDARY LITERATURE. See A. Fischel, "Carl Rabl," in *Anatomischer Anzeiger*, **51** (1918), 54–79; Hans Held, "Nekrolog auf Carl Rabl," in *Berichte über die Verhandlungen der Sächsischen Gesellschaft der Wissenschaften zu Leipzig*, math.-phys. Kl., **60** (1918), 363–380; and F. Hochstetter, "Carl Rabl," in *Wiener klinische Wochenschrift*, **31** (1918), 196–200; and his obituary notice in *Almanach der Akademie der Wissenschaften in Wien*, **69** (1918), 260–274. Georg Uschmann, *Geschichte der Zoologie und der zoologischen Anstalten in Jena 1779–1919* (Jena, 1959), 130–133, 167, and plate 41, describes Haeckel's influence on Rabl's work and their continuing relationship and correspondence.

GLORIA ROBINSON

RADEMACHER, HANS (*b.* Wandsbeck, Schleswig-Holstein, Germany, 3 April 1892; *d.* Haverford, Pennsylvania, 7 February 1969), *mathematics*.

Rademacher attended the University of Göttingen, where he studied real functions and the calculus of variations with Constantin Carathéodory and number theory with Edmund Landau. He received the doctorate in 1916 with a dissertation on single-valued mappings and mensurability. After teaching at a school in Thuringia run by teachers with modern ideas, Rademacher became *Privatdozent* at the University of Berlin in December 1916. There he was influenced by Erhard Schmidt, Issai Schur, and Hans Hamburger. In 1922 he was appointed associate professor with tenure at Hamburg. Under the influence

of Erich Hecke he turned to number theory, writing at first on the method devised by Viggo Brun and later on the additive prime-number theory of algebraic numbers.

Rademacher's chief field of interest for forty years was analytic number theory, particularly additive problems. At Easter 1925, after long hesitation, he went to Breslau. (Hecke had vainly attempted to procure a corresponding position for him at Hamburg.) There he was concerned in particular with the behavior of the logarithm of the function

$$\eta(\tau) = e\,\frac{2\pi i\tau}{24} \prod_n (1 - e^{2\pi i\tau n}), \qquad \text{Im}(\tau) > 0$$

with respect to modulus substitutions. The function $\eta(\tau)$ had appeared, in its essential aspects, in the writings of Euler. Dedekind had later treated it in the course of his comments on a fragment of Riemann's (see Riemann, *Gesammelte . . . Werke*, R. Dedekind and H. Weber, eds. [1876], 438–447). Rademacher devised a new proof of the results obtained with the function, utilizing the connection between modular functions and Dirichlet series over the Mellin integral (Hecke's method).

During this period Rademacher wrote, in collaboration with Otto Toeplitz, *Von Zahlen und Figuren*, addressed to a broad, nonprofessional audience. In many of his lectures Rademacher discussed fundamental problems in extremely diverse fields and the ways in which they had been treated. Forced to flee Germany in 1933 because of his pacifist views, he went to Swarthmore College and later to the University of Pennsylvania, all the while continuing his research on analytic number theory. His most outstanding achievement was the proof of his asymptotic formula for the growth of the function $p(n)$, which yields the number of representations of a natural number n as a sum of natural numbers. The question had been raised much earlier by Leibniz (see his *Mathematische Schriften*, C. I. Gerhardt, ed., III). Then Euler found that

$$\prod_{n=1}^{\infty} (1 - x^n)^{-1} = \sum_{\kappa=0}^{\infty} p(\kappa)\, x^\kappa.$$

The first asymptotic formula for $p(n)$ originated with Hardy and Ramanujan (*Proceedings of the London Mathematical Society*, 2nd ser., **17** [1918], 75–115); Rademacher proposed a simpler formula for $p(n)$ that led to remarkable results. As an example he calculated $p(599)$, a number that consists of twenty-four digits in the decimal system. Rademacher's asymptotic formula deviated from this by only about 0.5.

BIBLIOGRAPHY

A detailed bibliography of Rademacher's works is in *Jahresberichte der Deutschen Mathematiker-Vereinigung*, **71** (1969), 204–205. His writings include "Eindeutige Abbildung und Messbarkeit," in *Monatshefte für Mathematik und Physik*, **27** (1916), 183–290; "Über streckentreue und winkeltreue Abbildungen," in *Mathematische Zeitschrift*, **4** (1919), 131–138; "Über partielle und totale Differenzierbarkeit von Funktionen mehrerer Variablen," in *Mathematische Annalen*, **79** (1919), 340–359, and **81** (1920), 52–63, his *Habilitationsschrift*; "Beiträge zur Viggo Brunschen Methode in der Zahlentheorie," in *Abhandlungen aus dem Mathematischen Seminar, Universität Hamburg*, **3** (1924), 12–30; "Zur additiven Primzahltheorie algebraischer Zahlkörper," *ibid.*, 109–163, 331–337, and *Mathematische Zeitschrift*, **27** (1927), 324–426; "Zur Theorie der Modulfunktionen," in *Journal für die reine und angewandte Mathematik*, **167** (1932), 312–336; *Von Zahlen und Figuren* (Berlin, 1933; repr. 1968), written with Otto Toeplitz, also trans. into English as *The Enjoyment of Mathematics* (Princeton, 1957); "On the Partition Function $p(n)$," in *Proceedings of the London Mathematical Society*, 2nd ser., **43** (1937), 241–254; "On the Expansion of the Partition Function in a Series," in *Annals of Mathematics*, 2nd ser., **44** (1943), 416–422; "Zur Theorie der Dedekindschen Summen," in *Mathematische Zeitschrift*, **63** (1956), 445–463; and "A Proof of a Theorem on Modular Functions," in *American Journal of Mathematics*, **82** (1960), 338–340.

A short account of Rademacher's life appears in Max Pinl, "Kollegen in einer dunkeln Zeit," in *Jahresberichte der Deutschen Mathematiker-Vereinigung*, **71** (1969), 205 ff.

HEINRICH BEHNKE

RÁDL, EMANUEL (*b.* Pyšely, Bohemia [now Czechoslovakia], 21 December 1873; *d.* Prague, Czechoslovakia, 12 May 1942), *philosophy, history of biological sciences.*

Rádl was one of the eleven children of František Rádl, a poor peddler, and his wife, Barbara. Of six children who survived infancy, one brother became professor of mathematics at Prague Technical University, and Rádl himself was mathematically gifted. With the help of an uncle, who was an abbot, Rádl graduated from the Gymnasium in Domažlice, then entered a seminary to study theology. He gave up this study after some time, however, over the objections of his parents, and went to Prague, where he worked his way through the university, supporting himself by tutoring. He was a brilliant student and took the Ph.D. in natural sciences. While at the university Rádl was influenced by F. Vejdovský, a leading zoologist, and, especially, T. G. Masaryk,

who was then a professor of philosophy. During his last years as a student Rádl began to publish scientific papers; the first, in 1897, was a petrological study, while later works were concerned with biology.

Since no teaching posts were available at the University of Prague, Rádl was obliged to take the state education examinations and to accept a post at a secondary school in Pardubice, in eastern Bohemia. With the limited means at his disposal, Rádl continued to work enthusiastically in biology, the history of biology, comparative psychology, and philosophy. In 1904 he qualified as *Privatdozent* at the University of Prague, assigned to teach the biology of invertebrates "with special regard to its historical development." The post was unsalaried, and Rádl simultaneously taught at a secondary school in Prague; his connection with the university, however, allowed him access to its extensive library, and he continued his own studies, working day and night.

Rádl published steadily, in both Czech and German, on a wide variety of topics. Most of his papers written during this period were summarized in *Untersuchungen über den Phototropismus der Tiere* (1903) and *Neue Lehre vom zentralen Nervensystem* (1913). Even more important than his biological articles were his critical essays on such topics as the philosophy of naturalists, Czech *Naturphilosophen*, Goethe, Leibniz and Stahl, vitalism, and the mechanics of evolution. Chief among these was his study (1900) of the histological work of the Czech biologist Purkyně, in which Rádl attempted to analyze Purkyně's method and genius.

With the publication of *Geschichte der biologischen Theorien* (1905–1909) Rádl's name became widely known. The work is by no means an ordinary textbook, and Rádl's conception of the history of biology is still thought-provoking. His reputation, gained largely from this book, earned him a place on the editorial board of *Isis*, at its founding in 1913, and the first issue contains his article on Paracelsus.

World War I ended this international cooperation and increased the difficulty of Rádl's life in Prague, since he was known as a disciple of Masaryk, who was then in political exile. With the help of Masaryk, who became the first president of the Czechoslovak Republic in 1918, Rádl was appointed professor of philosophy and history of natural sciences at Prague. The war had changed Rádl's outlook, however, and he began to work toward a synthesis of philosophy and socially responsible religion. He traveled widely, disseminating his ideas, and took part in several international conferences. His chief concerns were reflected in *Západ a Východ* ("West and East," 1925), in which he proposed a nonformalized western Christianity as a unifying philosophy.

In 1907 Rádl, again with the encouragement of his uncle, married Marie Ptáčníková, the convent-educated daughter of a rich notary in Domažlice. His financial condition improved, but his domestic life was not always harmonious; his wife was a strongly conservative Catholic, while Rádl had left that church to become a liberal Protestant. When Rádl was invited to join the faculty of an American university during his travels around the world in 1921–1922, his wife refused to accompany him and he returned to Prague, where he devoted himself to his philosophical work.

Rádl worked to propagate Masaryk's social and philosophical ideals; he was himself, in fact, the more profound philosopher. During this time he also wrote a Czech textbook for his students, *Moderní věda* ("Modern Science," 1926), and applied his general erudition to the organization of the new Masaryk encyclopedia. Rádl's two-volume *Dějiny filosofie* ("History of Philosophy," 1932–1933) considered philosophers of all periods and reflected Rádl's view that they should accept personal responsibility to work for the good of mankind. In addition he wrote books on the political problems that had begun to concern not only Czechoslovakia but also the world; these included *Válka Čechů s Němci* ("The War of the Czechs Against the Germans," 1928), also published in German as *Der Kampf zwischen Tschechen und Deutschen*, and *O německé revoluci* ("On the German Revolution," 1933).

Rádl wrote on the contemporary mission of philosophy in such books as *Náboženství a politika* ("Religion and Politics," 1921); this was also the chief concern of the Eighth International Congress of Philosophy held at Prague in 1934, of which Rádl was organizer and president. These activities, together with increasing political tensions, limited the time and attention that Rádl was able to devote to the history of science; he initiated a collaboration with Otakar Matoušek, who later succeeded him at the university, to write a major work on this subject.

This project was not realized; in 1935 Rádl suffered a light stroke, which was followed by another in the next year. He recovered from these, but was soon afterward ordered to stay at home, where he spent the last five years of his life as an invalid, increasingly removed—through his wife's unhappy anxiety—from the world, from his friends, and often even from his own children. He continued to write, and his study "La philosophie de T. G. Masaryk" appeared in 1938; in 1939 he published *Věda a víra u Komenského* ("On the Science and Faith of Comenius," 1939), the last work that he himself was to see in print. The Nazi occupation of Czechoslovakia made further publication impossible. He died unexpectedly of pneumonia in the room to which he had been confined for five years. He left a penciled manuscript, *Útěcha z filosofie* ("Consolation from Philosophy"), which, although unpublishable during World War II, went through three editions after the war ended, prior to 1948. A fourth edition was published in 1968.

The predominant factor of Rádl's philosophy was idealism. Although he was influenced by Marx and by the Russian Revolution of 1917, he opposed materialism and asked for a deeper understanding of the human soul; he thus earned enemies on both sides. The "Consolation" is valuable chiefly as an illustration of his own spirit rather than as a general philosophical program, and as such it is an impressive legacy. He wished to halt the disintegration that he saw in human society, and he saw in philosophy an effective means for doing so; finding most philosophy to be without creative force, he criticized it, embracing a strong practical Christianity.

No complete study of Rádl has yet been made. Journals and newspapers printed articles on some aspects of his life and work in honor of his sixtieth birthday, but not even a eulogy was permitted at the time of his death—indeed, none was given at his funeral. His library was dispersed, and no really complete bibliography of his work survives. Only the first half of Rádl's career is known outside Czechoslovakia; his work awaits the reediting and translation that will establish him as what he was—one of the most original historians of biological ideas and one of the most original philosophers of his time.

BIBLIOGRAPHY

I. ORIGINAL WORKS. Rádl's writings, which dealt with scientific, political, social, and moral problems, were often written in both Czech and German. Many of his works appeared in *Sitzungsberichte der K. Böhmischen Gesellschaft der Wissenschaften*, also known as *Věstník K. České společnosti nauk* and hereafter cited as *Sitzungsberichte*. The following is a representative list of his published writings: "Gabbro ze Studeného v okolí Jílovském" ("Gabbro from Studená in the Vicinity of Jílové"), in *Sitzungsberichte* (1897), no. 24; "Sur quelques éléments des ganglions optiques chez les décapodes," in *Archives d'anatomie microscopique*, **2** (1898), 373–418; "Über den Bau und die Bedeutung der Nervekreuzungen im tractus opticus der Arthropoden," in *Sitzungsberichte* (1899), no. 23; "Über die Krümmung der zusammengesetzen Arthropodenaugen," in *Zoologischer Anzeiger*, **23** (1900), 372–379; "J. E. Purkyně práce histologické" ("Purkyně's Histological Works"), in *Sitzungsberichte* (1900), no. 15; "O dnešní filosofii přírodní" ("On Contemporary Naturalistic Philosophy"), in *Česká mysl*, **2** (1901), 206–211, 281–285, 362–366, 437–439; "Untersuchungen über die Lichtreactionen der Arthropoden," in *Archiv für die* *gesamte Physiologie,* **87** (1901), 418–466; *O morfologickém významu dvojitých očí u členovců* ("On the Morphological Character of the Double Eyes of Arthropoda"; Prague, 1901); "Über den Phototropismus einiger Arthropoden," in *Biologisches Zentralblatt,* **21** (1901), 75–86; and "Über die Bedeutung des Prinzips von der Korrelation in der Biologie," *ibid.,* 401–416, 490–496, 530–560, 605–621.

Subsequent works include "Über specifische Strukturen der nervösen Centralorgane," in *Zeitschrift für wissenschaftliche Zoologie,* **72** (1902), 31–99; "Bemerkungen zu den Vorschlägen von R. Pick, die wissenschaftliche Sprachverwirrung betreffend," in *Anatomischer Anzeiger,* **21** (1902), 27–29; "Nová pozorování o fototropismu zvířat" ("New Observations on Phototropism in Animals"), in *Sitzungsberichte* (1902), no. 55; *Untersuchungen über den Phototropismus der Tiere* (Leipzig, 1903); "O vitalismu" ("On Vitalism"), in *Česká mysl,* **5** (1904), 133–137, 206–210; "Über die Anziehung der Organismen durch das Licht," in *Flora, oder allgemeine botanische Zeitung,* **93** (1904), 167–178; *Geschichte der biologischen Theorien seit dem Ende des siebzehnten Jahrhunderts,* 2 pts. (Leipzig, 1905–1909), 2nd ed., extensively revised, *Geschichte der biologischen Theorien in der Neuzeit,* pt. 1 (Leipzig–Berlin, 1913); trans. of pt. 2 by E. J. Hatfield, as *The History of Biological Theories* (London, 1930); "Über das Gehör des Insekten," in *Biologisches Zentralblatt,* **25** (1905), 1–5; "Über ein neues Sinnesorgan auf dem Kopfe der Corethralarve," in *Zoologischer Anzeiger,* **30** (1906), 169–170; "Einige Bemerkungen und Beobachtungen über den Phototropismus der Tiere," in *Biologisches Zentralblatt,* **26** (1906), 677–690; *Dějiny vývojových theorii v biologii XIX. století* ("History of Biological Theories in the Nineteenth Century"; Prague, 1909); "Über spezifisch differenzierte Leitungsbahnen," in *Anatomischer Anzeiger,* **36** (1910), 385–401; *Neue Lehre vom zentralen Nervensystem* (Leipzig, 1913); "Paracelsus. Eine Skizze seines Lebens," in *Isis,* **1** (1913), 62–94; and *Zur Geschichte der Biologie von Linné bis Darwin* (Leipzig, 1915).

Rádl resumed publishing after World War I with *Romantická věda* ("Romantic Science"; Prague, 1918), followed by *Demokracie a věda* ("Democracy and Science"; Prague, 1919); *T. G. Masaryk* (Prague, 1919), with many reprints and translations; *Národ a stát* ("The Nation and the State"; Prague, 1921); *Dějiny filosofie* ("History of Philosophy"), 2 vols. (Prague, 1932–1933); and *Actes du huitième congrès international de philosophie à Prague . . . 1934* (Prague, 1936), of which Rádl was editor.

II. SECONDARY LITERATURE. On Rádl's life and work, see B. Koutník, *Emanuel Rádl* (Prague, 1933); and J. L. Hromádka, *Don Quijote české filosofie: E. Rádl, 1873–1942* ("The Don Quixote of Czech Philosophy . . . "; New York, 1943).

OTAKAR MATOUŠEK

RADÓ, TIBOR (*b.* Budapest, Hungary, 2 June 1895; *d.* New Smyrna Beach, Florida, 12 December 1965), *mathematics.*

The son of Alexander Radó and Gizella Knappe, Radó began his university studies in civil engineering at the Technical University in Budapest. In 1915 he enlisted in the Royal Hungarian Army, was trained, and then commissioned a second lieutenant in the infantry. He took part in two major battles on the Russian front before being captured on a scouting mission. Of his capture, Radó recounted: "I had spent six months traveling back and forth through the Russian lines, picking up information, cutting telephone wires and holding up supply trains. Then one day I was surrounded by Russians—I wasn't surprised."

His four years in prison camps read like a dramatic scenario. As an officer he found the camp in Tobolsk, Siberia, relatively comfortable in the period preceding the Revolution. Food was plentiful and cheap, but reading material was not readily available. The only books he could obtain happened to be on mathematics.

After the Revolution, the prisoners' life changed drastically. They were packed into boxcars and transported thousands of miles under harrowing conditions, in order to get them out of the fighting zone. During the confusion he and three fellow officers traded names with four private soldiers. As far as his family knew, Radó was dead. He spent the next year working as a laborer in railroad yards. He and a group of prisoners escaped by hijacking a train. Finally, in 1920 he returned to Budapest on an American financed boat which was assisting in the return of war prisoners. Back at the University of Szeged, he re-enrolled, this time as a mathematics major, and in 1922 he received his Ph.D. under Frigyes Riesz.

From 1922 to 1929 Radó was *Privatdozent* at the University of Szeged and also adjunct at the Mathematical Institute in Budapest. He was awarded an international research fellowship of the Rockefeller Foundation to study at Munich during 1928–1929. In 1929 Radó went to the United States as a visiting lecturer, first at Harvard (fall semester, 1929–1930) and then at Rice Institute (spring semester, 1930). In 1930 he moved to Ohio State University as full professor of mathematics.

In 1944–1945 Radó was a fellow at the Institute for Advanced Studies at Princeton. At the end of World War II, he went to Europe as a scientific consultant with the Army Air Force to recruit German scientists needed by the United States. He returned to Ohio State as chairman of the mathematics department in 1946. He resigned this post in 1948, when he was appointed the first Ohio State University research professor, a position created to enable distinguished faculty members to pursue creative activity.

Radó's research interests and contributions span a wide range of topics: conformal mapping, real variables, calculus of variations, partial differential equations, measure and integration theory, point-set and algebraic topology, rigid surfaces (very thin shells), logic, recursive functions, and what he called "Turing programs."

Radó's first major original contribution concerned Plateau's problem, finding the surface of minimal area bounded by a given closed contour in space. The problem, which originated in the initial phases of the calculus of variations, is named for Joseph Plateau, who conducted experiments on certain shapes with soap bubbles. The existence and uniqueness of solutions in the general case remained to be solved independently by Radó and Jesse Douglas in the early 1930's. Radó's interest in problems relating to surface measure dated from his work under Riesz's guidance on problems raised by Zoard de Geöcze. It was on the basis of the theory of functions of real variables of Lebesgue and Riesz that Radó was able to simplify and generalize Geöcze's results and help to create a modern theory of surface area measure.

Radó was active in mathematical societies. He was invited to give the American Mathematical Society Colloquim Lectures in 1945, and in 1952 he gave the first Mathematical Association of America Hedrick Memorial Lecture. He was also an editor of the *American Journal of Mathematics* and served as vice-president of the American Association for the Advancement of Science in 1953.

BIBLIOGRAPHY

I. ORIGINAL WORKS. Radó's major works include "On the Problem of Plateau," in *Ergebnisse der Mathematik und Ihrer Grenzgebiete*, **2** (Leipzig, 1933; repr. New York, 1951); *Sub-Harmonic Functions, Ergebnisse der Mathematik under Ihrer Grenzgebiete*, V (Leipzig, 1937; repr. New York, 1949); *Length and Area*, American Mathematical Society Colloquium Publication, XXX (New York, 1948); and *The Mathematical Theory of Rigid Surfaces: An Application of Modern Analysis* (Chapel Hill, N.C., 1954), a collection of lectures presented at a summer conference at the University of North Carolina in 1954.

II. SECONDARY LITERATURE. Antonio Mambriana, "Una visione dell'opera scientifica di Tibor Radó," in *Rivista di matematica della Università di Parma*, **1** (1950), 239–273, with portrait, covers in considerable detail Radó's contributions in conformal mapping, the Plateau problem, harmonic functions, and work related to surface area and the problems of Geöcze. It also contains an 83-item bibliography of Radó's work through 1949. Alice Holton, "Professor Recalls Siberian Prison Camp," in Columbus (Ohio) *Dispatch* (4 Dec. 1939), and Gwendolyn Riggle, "Years in Russian Concentration Camp Led Him to Professor's Post at O.S.U.," *ibid.*,

(2 Mar. 1941), are descriptive interviews with Radó that cover his early life in some detail. The Ohio State University News and Information Service has prepared a short biography as well as a more complete bibliography of 102 items that was drawn up by Radó in 1960. At his death he had written more than 140 papers and books, for which no single complete bibliography exists.

HENRY S. TROPP

RADON, JOHANN (*b*. Tetschen, Bohemia [now Decin, Czechoslovakia], 16 December 1887; *d*. Vienna, Austria, 25 May 1956), *mathematics*.

Radon entered the Gymnasium at Leitmeritz (now Litomerice), Bohemia, in 1897 and soon showed a talent for mathematics and physics. In 1905 he enrolled at the University of Vienna to study those subjects and was subsequently influenced by Gustav von Escherich, who introduced him to the theory of real functions and the calculus of variations. His doctoral dissertation (1910), on the latter subject, was also his first published paper. Radon spent the winter semester of 1911 at the University of Göttingen, served for a year as assistant professor at the University of Brünn (now Brno), then went to the Technische Hochschule of Vienna in the same capacity. He became *Privatdozent* at the University of Vienna in 1914 and achieved the same rank a year later at the Technische Hochschule.

In 1919 Radon was appointed associate professor at the University of Hamburg; he became full professor at Greifswald in 1922, at Erlangen in 1925, and at Breslau in 1928. He left Breslau in 1945 and in 1947 obtained a full professorship at Vienna, where he spent the rest of his life. In the same year he became a full member of the Austrian Academy of Sciences.

The calculus of variations remained Radon's favorite field because of its close connections with so many areas of analysis, geometry, and physics. His most important paper in this field (1927) greatly influenced its further development, especially of the difficult Lagrange problem. In 1928, in lectures at Hamburg, Radon presented his results in an expanded form. He was deeply interested in the applications of the calculus of variations to differential geometry and discovered the so-called Radon curves, which have found applications in number theory. In addition to his work in affine differential geometry (1918–1919), in conformal differential geometry (1926), and in Riemannian geometry, Radon treated mathematical problems of relativity theory. His important paper on algebra, "Lineare Scharen orthogonaler Matrizen," was inspired by his work on the calculus of variations and proved to have many applications.

Radon's best-known work, "Theorie und Anwendungen der absolut additiven Mengenfunktionen," which exerted a great influence, essentially combined the integration theories of Lebesgue and Stieltjes. Led to his research by physical considerations, he studied the most general distributions of masses in space and developed the concept of the integral, now known as the Radon integral. A continuation of this work is the paper "Über lineare Funktionaltransformationen und Funktionalgleichungen."

An important theorem in the calculus of variations, later generalized by Otton Nikodym, is the Radon-Nikodym theorem. Radon himself applied this theory to the Dirichlet problem of the logarithmic potential. He also developed a technique now known as the Radon transformation (1917), which has many applications. Radon's interest in the philosophy of mathematics was reflected in his paper "Mathematik und Wirklichkeit."

BIBLIOGRAPHY

Radon published 45 papers which still are of great importance. Among them are "Über das Minimum des Integrales $\int_{s_0}^{s_1} \mathfrak{F} \ (x,y,\theta,\chi) \ ds$," in *Sitzungsberichte der Akademie der Wissenschaften in Wien*, **119** (1910), 1257–1326; "Theorie und Anwendungen der absolut additiven Mengenfunktionen," *ibid.*, **122** (1913), 1295–1438, his *Habilitationsschrift*; "Über die Bestimmung von Funktionen durch ihre Integralwerte längs gewisser Mannigfaltigkeiten," in *Berichte über die Verhandlungen der Königlich Sächsischen Gesellschaft der Wissenschaften in Leipzig*, **69** (1917), 262–277; "Über lineare Funktionaltransformationen und Funktionalgleichungen," in *Sitzungsberichte der Akademie der Wissenschaften in Wien*, **128** (1919), 1083–1121; "Über die Randwertaufgaben beim logarithmischen Potential," *ibid.*, 1123–1167; "Lineare Scharen orthogonaler Matrizen," in *Abhandlungen aus dem Mathematischen Seminar, Universität Hamburg*, **1** (1921), 1–14; "Über statische Gravitationsfelder," *ibid.* (1922), 268–288; "Mathematik und Wirklichkeit," in *Sitzungsberichte der Physikalisch-medizinischen Sozietät in Erlangen*, **58–59** (1926–1927), 181–190; "Oszillationstheoreme der Konjugierten Punkte beim Problem von Lagrange," in *Sitzungsberichte der Bayerischen Akademie der Wissenschaften zu München*, **7** (1927), 243–257; and "Zum Problem von Lagrange," in *Abhandlungen aus dem Mathematischen Seminar, Universität Hamburg*, **6** (1928), 273–299, his lectures.

A monograph on the Radon transformation as well as applications to differential equations can be found in the book of F. John, *Plane Waves and Spherical Means* (New York, 1955). Obituaries can be found amongst others in *Monatshefte für Mathematik*, **62** (1958), 189–199, and in *Almanach. Österreichische Akademie der Wissenschaften*, **107** (1958), 363–368.

E. HLAWKA

RAFFLES, THOMAS STAMFORD BINGLEY

(*b.* at sea off Port Morant, Jamaica, 6 July 1781; *d.* London, England, 5 July 1826), *natural history.*

Raffles was born on board the ship *Ann*, of which his father, Benjamin Raffles, was the master, on a voyage from Jamaica to England. He grew up in London; but when the family lost its money, he went to work in 1795 as a clerk for the East India Company, studying languages and natural history in his spare time. In 1805 Raffles was sent to Penang as assistant to the chief secretary of the company; and before leaving he married a widow, Olivia Fancourt. During the voyage he learned Malay, an accomplishment rare among the company's staff, and was so competent an administrator that he was promoted to chief secretary in 1807. In 1807 and 1810 he visited Calcutta to meet Lord Minto, governor-general of India, and was appointed agent to the governor-general of India in Malaya, based at Malacca. From there Raffles planned the invasion of Java and was appointed its lieutenant governor, based at Batavia; this was politically expedient but not entirely welcomed by the East India Company, for new territories were expensive to administer. His government was both efficient and humane, but he suffered from overwork and tropical diseases. In 1814 his wife died and accusations of corruption were brought against him.

Raffles was not able to return to England until 1816; but once in London, he had great success. He was friendly with the prince regent, Princess Charlotte, the duke and duchess of Somerset (to whom some of his most important letters were written), and also scientists including Sir Joseph Banks. Raffles cleared himself of the charges laid with the court of directors of the East India Company, traveled in Holland, wrote his *History of Java*, was elected fellow of the Royal Society, and was knighted. He married his second wife, Sophia Hull, and returned to Bengkulu, Sumatra, in 1817, taking with him the botanist Joseph Arnold.

Back in the East, Raffles planned and executed the taking of Singapore and supervised the setting up of the new community there. He returned home in 1824, his health seriously damaged and four of his five children dead of tropical diseases. He was active in founding the Zoological Society of London and was its first president until his death.

Raffles' importance in the history of science is as a facilitator rather than as a discoverer. He went on long trips of exploration wherever he was based, surveying the country, customs, and natural history. In a letter of 1818 to the duchess of Somerset he described the parasitic plant later named *Rafflesia* as "perhaps the largest and most magnificent flower in the world" and sent home specimens from which Robert Brown published descriptions. He wrote on the little-known tapir and dugong, and sent specimens of these and many other animals and plants to the Indian Museum in Calcutta and to Banks in 1818, 1820, and 1824.

The collecting was organized at Raffles' own expense, first informally and later by scientists he brought out, including Arnold, the American Thomas Horsfield, and two Frenchmen, Pierre-Médard Diard and Alfred Duvaucel. He was highly regarded as a naturalist by scientists in England and abroad, including Wallich and Hardwick in India. The botanic gardens at Malacca and Bengkulu were planned and largely tended by Raffles himself; and he commissioned many drawings, some of which survive in the India Office library. Most of the specimens passed to the Zoological Society and then to the British Museum (Natural History).

BIBLIOGRAPHY

I. ORIGINAL WORKS. Raffles' most important work is his *History of Java* (London, 1817; 2nd ed., 2 vols., 1830; new ed., 1965); chs. 1 and 3 cover the geography, flora and fauna, and agriculture. Raffles acknowledges: "For all that relates to the natural history of Java I am indebted to the communications of Dr. Thomas Horsfield." His scientific papers are not easy to trace and are not listed in the Royal Society *Catalogue of Scientific Papers*. They include "Discourse to the Batavia Society of Arts and Sciences in 1813 on the State of Science in Java," in *Verhandelingen van het Bataviaasch genootschap van kunsten en wetenschappen*, **7** (1814), 1–35; "Some Account of the Dugong," in *Philosophical Transactions of the Royal Society*, **110**, pt. 1 (1820), 174–182, a detailed description, including measurements, food, habitat, and dissection; and "Descriptive Catalogue of a Zoological Collection, Made . . . in the Island of Sumatra and Its Vicinity . . . With Additional Notices Illustrative of the Natural History of Those Countries," in *Transactions of the Linnean Society of London*, **13** (1821), 239–340.

II. SECONDARY LITERATURE. The description of *Rafflesia* was first published by R. Brown, "An Account of a New Genus of Plants Named *Rafflesia*," in *Transactions of the Linnean Society*, **13** (1821), 201–234, and 8 plates; he quoted Raffles' own description and a letter from William Jack, and used the drawing of a specimen brought back by Horsfield to give the formal Latin description. Brown later published "Description of the Female Flower and Fruit of *Rafflesia Arnoldi*, With Remarks on Its Affinities," in *Transactions of the Linnean Society*, **19** (1834), 221–247 and 8 plates, with a synopsis of the Rafflesiaceae.

There is abundant material, both MS and published, mainly on Raffles' work as a colonial administrator. The first biography was Sophia Raffles, *Memoir of the Life and Public Services of Sir Thomas Stamford Raffles . . . and Selections From His Correspondence*, 2 vols. (London, 1830; new ed., abr., 1835), the best published source of letters. The definitive biography is C. E. Wurtzburg, *Raffles of the Eastern Isles* (London, 1954); M. S. Collis, *Raffles* (London, 1966), is short and readable. Both have good bibliographies, including sources of MS material, of which the India Office is the most important. A Malay clerk in Raffles' office, Abdullah, wrote a personal account of Raffles' scientific work, first published in 1874; the best translation is by A. H. Hill, repr. as Abdulla bin Abdul Kadir, *The Hikayat Abdullah* (Kuala Lumpur, 1970). There is a scientific memoir by Sir William Jardine in his *The Natural History of Game Birds* (Edinburgh, 1834), 17–88. Raffles' role in the early Zoological Society may be found in P. Chalmers Mitchell, *Centenary History of the Zoological Society of London* (London, 1929), 1, 7, 30, 61; and in John Bastin, "Dr. Joseph Arnold and the Discovery of *Rafflesia Arnoldi* in West Sumatra in 1818," in *Journal of the Society for the Bibliography of Natural History*, **6** (1973), 305–372. Bastin has also published "Raffles the Naturalist," in *Straits Times Annual* (Singapore, 1971), 59–63, and is editing his papers.

DIANA M. SIMPKINS

RAFINESQUE, CONSTANTINE SAMUEL (*b.* Galata, near Constantinople, 22 October 1783; *d.* Philadelphia, Pennsylvania, 18 September 1840), *natural history, archaeology.*

Rafinesque missed greatness by embracing too many fields of knowledge, yet the rule of priority in systematic biology, which requires the earliest validly published description to be honored, has forced the recognition of this rejected naturalist. His contemporaries gave scant heed to his voluminous, erratic writings. John Bohn's London sales catalog of natural history books, issued in 1835, listed 2,178 titles—but not one of Rafinesque's. Yet in Rafinesque's espousal of the emerging schemes of natural plant classification over the artificial "sexual system" of Linnaeus, he was ahead of his time. Rafinesque proposed a multiplicity of forms—over 6,700 binomials—consistent with the theory of organic evolution; his germinal idea was acknowledged by Charles Darwin in *On the Origin of Species*. Rafinesque saw life ruled by great laws:

> *Symmetry*, that gives the bodily forms to Genera, casting the mould of typical frames—*Perpetuity*, that by reproduction perpetuates these original primitive forms—*Diversity*, that bids and compels all living bodies to assume gradually a variety of slight changes when reproduced, and never evolves two individuals perfectly alike, nor two leaves quite similar in all points

on the very same tree. Lastly *Instability*, that does not allow any forms nor frames to be perpetual nor ever the same, giving to plants and animals birth, growth, decay and death! in succession, within a term of a few hours, a day, a month, a year, or 1000 years [*Flora telluriana*, I (Philadelphia, 1837), 99–100].

Rafinesque's father was Georges F. Rafinesque, a French merchant who had settled in the Levant but during the Napoleonic Wars took refuge in Philadelphia and died there of yellow fever. His mother, Madeleine Schmaltz, was a native Greek of German extraction who fled France during the Reign of Terror, taking Constantine, his brother, and his sister, to Leghorn, Italy, where he lived from 1792 to 1796 and was taught by private tutors. Rafinesque declared that he had read a thousand volumes on almost every subject by the age of twelve and had studied fifty languages, including Chinese, Hebrew, and Sanskrit, by the age of sixteen.

Rafinesque lived successively in Pisa, Genoa, and Marseilles until 1800, when he was apprenticed to a merchant friend of his father. After two years in Leghorn he sailed for Philadelphia, where he met Benjamin Rush (whose pupil he declined to become), the horticulturist Thomas Forrest, Moses Marshall, and William Bartram. He collected reptiles for François Daudin; and when Michaux's *Flora* appeared in 1803, he determined to prepare a supplement to it. In a letter to Thomas Jefferson in November 1804, Rafinesque inquired why the American government had not sent a botanist with Dunbar's Red River party. Jefferson's offer of that post reached Rafinesque after he had sailed for Italy.

Rafinesque considered his ten years in Sicily "the best epoch of [his] life." His exploration of Etna, during which he made hundreds of sketches and collected *naturalia*, was supported first by his serving as secretary to the United States consul and later, profitably, as manufacturer of squill for export. Various discouragements—his unhappy marriage in 1809 to Josephine Vaccaro, his failure to publish his Sicilian portfolios, and his being refused the chair of botany at the University of Palermo—caused Rafinesque to return to America in 1815. After a voyage of over a hundred days he was shipwrecked off Long Island. He lost his books, manuscripts, drawings —"everything"—the labors of twenty years. He was subsequently befriended in New York by Samuel Latham Mitchill, who introduced him to naturalists, as did Zaccheus Collins in Philadelphia.

Although Rafinesque's historical writings may be set aside as superficial, his contacts with so many important contemporaries are significant. He advocated the construction of a Panama canal, the culture of pearls in mussels, houses and ships of fireproof construction, and steam plowing; and he developed and marketed a vegetable remedy for tuberculosis, although he did not patent it. Rafinesque organized and managed a savings bank in order to provide funds for his later publications, but he was eventually impoverished by his frenzy to print more and more. He was first to suggest from a meager published account of 1822 that the Mayan system of ideographs was partly syllabic. An advocate of "rigid sobriety" and walking, he was frugal and overly sensitive but severely critical of others, and was described in 1837 as "a little dried up, muffy-looking old man, resembling an antiquated Frenchman" (Pennell, p. 61). Two verified portraits exist, and Rafinesque drew several of his Kentucky friends.

Beginning in 1818, with his 2,000-mile tour to the west of the Alleghenies, Rafinesque made important botanical explorations, mostly on foot, as far as Kentucky and Illinois. From 1819 to 1826 he was professor at Transylvania College; but instead of a summation of his firsthand acquaintance with Kentucky's flora and fauna, he left merely miscellaneous fragments. The misfortune of his sketchy descriptions was aggravated by damage to his collections by vermin and contemptuous curators, such as Élie Durand, who discarded most of Rafinesque's herbarium, once reputed to contain 50,000 specimens. Consequently modern interpretation of many of Rafinesque's organisms, notable and often overlooked, has been hopelessly unrewarding. "In spite of Rafinesque's idiosyncrasies," wrote Merrill (p. 296), "in spite of his careless work, in spite of his constant and often caustic criticism of his associates, much that he accomplished was distinctly worth while."

BIBLIOGRAPHY

I. ORIGINAL WORKS. *A Life of Travels* (Philadelphia, 1836), a resumé of Rafinesque's writings, was repr. in *Chronica botanica*, **8** (1944), 291–360, with an intro. by E. D. Merrill, portraits, and an annotated index by F. W. Pennell. T. J. Fitzpatrick, *Rafinesque, a Sketch of His Life With Bibliography* (Des Moines, Iowa, 1911), lists 938 titles of Rafinesque's writings. Rafinesque's *Medical Flora, or Manual of the Medical Botany of the United States of North America*, 2 vols. (Philadelphia, 1828–1830), contains 100 woodcut plates by Rafinesque printed in green—an early example of this process. Rafinesque's translation of C. C. Robin's *Florula ludoviciana* (New York, 1817) includes notes culled from William Bartram's *Travels* (1791). This and many other works by Rafinesque have been reprinted.

II. SECONDARY LITERATURE. The most important guide to Rafinesque's life and work is E. D. Merrill, *Index Rafinesquianus, the Plant Names Published by Rafinesque*

With Reductions, and a Consideration of His Methods, Objectives and Attainments (Jamaica Plain, Mass., 1949). An overview of the man and his position in science by Raymond L. Taylor, in *William and Mary Quarterly*, 3rd ser., **2** (1945), 213–221, has useful references. F. W. Pennell, "Life and Work of Rafinesque," in *Transylvania College Bulletin*, **15**, no. 7 (1942), 10–70, contains previously unpublished material; summaries of Rafinesque's contributions to archaeology by W. D. Funkhouser (*ibid.*, 78–80), to ichthyology by William E. Ricker (81–83), to herpetology by William M. Clay (84–90), and to materia medica by H. B. Haag (91–96) are included in this symposium sponsored by Transylvania College.

See also H. A. Pilsbry, "Rafinesque's Genera of Freshwater Snails," in *Nautilus*, **30** (1917), 109–114; and S. N. Rhoads, "Rafinesque as an Ornithologist," in *Cassinia*, **15** (1912), 1–12. A notice of the "enthusiastic and persevering Rafinesque" by William Swainson, who knew him in Sicily, appeared in Swainson's *Taxidermy With the Biography of Zoologists* (London, 1840), 300–301. The often quoted classification of twelve species of thunder and lightning, mentioned by Frederick Brendel in his "Historical Sketch of the Science of Botany in North America From 1635 to 1840," in *American Naturalist*, **13** (1879), 754–771, is an erroneous misinterpretation of Rafinesque. Recent commentaries include F. A. Stafleu, "Rafinesque's *Caratteri* and *Florula ludoviciana*," in *Taxon*, **17** (1968), 296–299; and Ronald L. Stuckey, "C. S. Rafinesque's North American Vascular Plants at the Academy of Natural Sciences of Philadelphia," in *Brittonia*, **23** (1971), 191–208.

JOSEPH EWAN

RĀGHAVĀNANDA ŚARMAN

RĀGHAVĀNANDA ŚARMAN (*fl.* Bengal, India, 1591–1599), *astronomy.*

A Bengālī Brāhmaṇa, Rāghavānanda wrote a series of astrological and astronomical works toward the end of the sixteenth century. In astronomy, in which he was a follower of the Saurapakṣa (see essay in Supplement), he wrote a commentary, *Sūryasiddhāntarahasya*, on the *Sūryasiddhānta* (1591), which remains unpublished, and two sets of astronomical tables. The *Viśvahita*, also written in 1591, was published with an incorrect attribution to Mathurānātha Śarman by Viśvambhara Jyotiṣārṇava and Śrīścandra Jyotiratna (Calcutta, 1913); and the *Dinacandrikā*, composed in 1599, was edited with his own commentary and Bengālī translation by Bhagavatīcaraṇa Smṛtitīrtha (Calcutta, 1913). Rāghavānanda's astrological works include a *Spaṣṭajātakapaddhati*, sometimes called *Vidagdhatoṣiṇī.*

No analysis has yet been made of any of Rāghavānanda's works.

DAVID PINGREE

RAMAN

RAMAN, CHANDRASEKHARA VENKATA (*b.* Tiruchirapalli [Trichinopoly], India, 7 November 1888; *d.* Bangalore, India, 21 November 1970), *physics, physical theory, physiology of vision.*

Raman was the son of Chandrasekhara Aiyar, professor of mathematics and physics at the A. V. N. College in Vizagapatam, and Parvati Ammal, who belonged to a family known for Sanskrit scholarship. From his father Raman gained an early interest in science and a love of music and musical instruments; his mother contributed to his sense of self-reliance and his strong personality. Raman was educated at the A.V.N. College and at the Presidency College of the University of Madras, from which he received the B.A. in 1904, when he was sixteen. He ranked first among the students and was awarded the gold medal for physics. He then read for the M.A. degree, which he was granted with highest honors in January 1907; during this time he also, in 1906, published his first paper, on the unsymmetrical diffraction bands produced by a rectangular aperture. Among the works that he studied, Rayleigh's *Theory of Sound* and Helmholtz' *Sensations of Tones* influenced him profoundly.

Ill health prevented Raman from pursuing higher studies in physics at one of the great British universities and India offered him no possibility of a further career in science. He therefore decided to enter the civil service. The Indian audit and accounts service was at this time the only department that did not require a training period in Great Britain, and Raman took the examination for a position in the Indian finance department. He placed first, and in June 1907 was posted to Calcutta as assistant accountant general. In the same year he married Lokasundari, an accomplished artist who shared his interest in musical instruments. Raman served in the finance department for ten years, in posts of increasing responsibility. At the same time he continued to perform research, particularly on vibrations and sound (including experimental and theoretical studies of the oscillations of strings) and on the theory of musical instruments, particularly the violin family and Indian drums. Most of this work was accomplished at the laboratory of the Indian Association for the Cultivation of Science, which had been founded in Calcutta by Mahendralal Sircar in 1876.

Raman's independent research, and the many publications that resulted from it, led to his being invited, in 1917, to fill the newly established Palit professorship of physics at the University of Calcutta. He gave up his lucrative government career to accept it. He held the chair for sixteen years, during which he continued to do research on acoustics and optics at the Association. He was assisted in his efforts by a growing number of collaborators and visitors; during this period such men as Meghnad Saha and S. N. Bose, inspired by Raman's example and success, rose to prominence as Calcutta became a center for scientific research. In the summer of 1921 Raman represented the University of Calcutta at the British Empire Universities Congress held at Oxford. While in England he also lectured on the theory of stringed instruments before the Royal Society.

Raman returned to India by way of a voyage through the Mediterranean. He was struck by the deep opalescent blue of the sea, and, back in Calcutta, undertook to discover its cause. Rayleigh had explained the azure of the sky as being caused by the scattering of sunlight by molecules in the gaseous atmosphere, but had gone on to state that the color of the sea had nothing to do with the color of the water, but rather merely reflected the blue of the sky. He had then further speculated that the apparent blueness of the sea might be caused by the absorption of light by the water, since "if a liquid is not absolutely clear, but contains in suspension very minute particles, it will disperse light of a blue character."

Raman did not find Rayleigh's explanation convincing, and in a paper published in the *Proceedings of the Royal Society* in 1922 he demonstrated that the scattering of light by water molecules could account for the color of the sea in precisely the same manner that the scattering of light by air molecules could explain the color of the sky. Raman applied the Einstein-Smoluchowski theory of fluctuations in his study and further obtained experimental confirmation that water under normal conditions scatters light into an angle of 30°—about 150 times greater than scattering of sunlight by air free of dust particles. His co-workers K. Seshagin and K. S. Krishnan then tested and generalized the Einstein-Smoluchowski theory itself.

Although he had begun to concentrate upon optics, Raman continued his work in acoustics as well. His researches in this field were admired by Rayleigh, and it was largely because of them that Raman was elected a fellow of the Royal Society in 1924. His definitive article, "Musikinstrumente und ihre Klänge," appeared in the third volume of H. Geiger and K. Scheel's *Handbuch der Physik* in 1927. Only the linear theory of vibrations of ideal strings had been developed before Raman began his investigations, and he obtained new results on the excitation of string vibrations, the motion of the bowed point, and on the effect of the bridge in coupling the motion of the string to the body of the violin. He also made quantitative studies of the vibration phenomena of the piano, the *sitar*, and the *veena*, and showed that the *mridangam*

and the *tabla*—Indian musical drums—possess harmonic overtones specific to their drumheads that do not normally occur in circular membranes.

Raman also traveled extensively during this time. In 1924 he spoke on the scattering of light before the meeting of the British Association for the Advancement of Science, held in Toronto in 1924, then journeyed across Canada and the United States. He represented India at the centennial celebrations of the Franklin Institute in Philadelphia, then, at the invitation of Robert A. Millikan, spent five months at the California Institute of Technology in Pasadena. Following his return to India early in 1925 Raman went to the U.S.S.R. to attend the bicentennial ceremonies of the Academy of Science; on his return trip he visited scientific institutions in Germany, Switzerland, and Italy. He was also active in organizing learned institutions in his own country; he was among the founders of the Indian Science Congress in 1924, and served as its secretary for several years, later becoming president of its Madras convention. In 1926 he established the *Indian Journal of Physics*.

Optical studies remained his chief concern, however. With his associates Raman studied the scattering of light of available frequencies by a number of substances, particularly fluids. In April 1923 Raman's associate K. R. Ramanathan observed a weak secondary radiation, shifted in wavelength along with normally scattered light, which was attributed to "fluorescence." S. Venkateswaran then noticed that highly purified glycerin does not appear blue under sunlight, but rather radiates a strongly polarized, brilliant green light.

Raman and K. S. Krishnan then undertook to isolate the effect under impeccable experimental conditions. They employed complementary light filters placed in the paths of the incident and scattered light, respectively, and observed a "new type of secondary radiation" from the scattering of focused beams of sunlight in both carefully purified liquid and dust-free air. They reported this discovery in a letter to *Nature* in February 1928. Raman then refined the experiment by using a mercury arc as the source of light; the effect was thus clearly seen for the first time on 28 February 1928 and was reported to the Science Congress at Bangalore the following month. The secondary radiation showed several lines shifted toward longer wavelengths, the shifts being characteristic of the substances being examined, and indicated the absorption of energy by the scattering molecule—the precise effect that had been predicted by A. Smekal in 1923. G. Landsberg and L. Mandelshtam, in the U.S.S.R., independently observed the same phenomenon in quartz, shortly after Raman and Krishnan made their discovery, but Raman's account of the effect reflected a much more detailed investigation. In 1929 Raman was knighted in recognition of his work, and the following year he was awarded the Nobel Prize for physics.

The changes of the wavelength of incident light in the Raman effect are caused by the internal motions of molecules of the scattering substance. Additional lines that are not present in the spectrum of the incident beam are visible in the spectrum of the scattered light. While the shifts of wavelengths attributable merely to translational molecular movements are usually small, irregular, and unobservable, the rotation of molecules in gases gives rise to closely spaced Raman lines on either side of the incident line. In the case of a dense fluid, the collision of molecules hinders molecular rotation, so that the Raman lines develop as a continuous band. Since the internal vibrations of molecules lead to shifts of great wavelength, the Raman lines attributed to them appear well separated from the parent line.

The quantum theory offers a satisfactory explanation for the Raman effect by which it may be seen to arise from the exchange of energy between the light quanta and the molecules of the substance on which they impinge. The incident photon of energy, $h\nu_o$, is absorbed by the molecule, which is then translated into the intermediate state m, from which it returns to the final state f, emitting a photon in the process, so that $h\nu_o - h\nu_m = h\nu_f$. If the scattering molecule imparts energy $h\nu_m$ to the photon, the total energy of the photon becomes $h\nu_o + h\nu_m$, which appears as scattered radiation of increased frequency $\nu_o + \nu_m$. Energy is conserved only between the initial and final stages of this system, so that virtual transitions to a whole gamut of intermediate states are possible. The intensity ratio of lines with frequencies $\nu_o + \nu_m$ and $\nu_o - \nu_m$ may be expressed as a function of temperature, T, and characteristic frequency of the molecule ν_m as $e^{-h\nu_m/kT}$. R. W. Wood called the Raman effect "one of the most convincing proofs of the quantum theory of light."

In 1933 Raman left Calcutta and went to Bangalore, where he served the Indian Institute of Science as both its president (1933–1937) and head of the physics department (1933–1948). In 1934 he founded the Indian Academy of Science, whose *Proceedings* have appeared continuously and regularly since that date. During this period Raman also continued his investigations of the Raman effect (as did others—almost 2,000 papers on the subject were published by other workers in the twelve years following its discovery). In 1935 and 1936, with N. S. Nagendra Nath, Raman published two important papers in the *Proceedings of the Indian*

Academy of Science on the scattering of radiation by ultrasonic waves in a liquid. Earlier workers, including P. Debye and F. W. Sears in the United States and R. Lucas and P. Biguard in France, had utilized L. Brillouin's theory whereby the sound waves create a grating because of the rarefaction and condensation of the fluid. Raman and Nath, however, as a conclusion of their researches proposed instead a new theory of pure phase grating that could also account for the intensities of higher-order Bragg reflections.

In 1940 Raman began to look for a new approach to the dynamics of crystal lattices. In a series of articles published in 1941 and 1942, he assumed a crystal to be a geometrical array with a coherent structure of similarly placed atoms (and molecules and ions) capable of exerting mutual influence; he was thereby able to explain a finite number of discrete monochromatic frequencies that he had observed in the Raman spectra. In this he drew upon Rayleigh's definition of a normal mode as being one in which all particles in a normal vibration oscillate at any instant with equal amplitude in the same or the opposite phase. Thus all particles in a crystal unit cell may be replaced by equivalent particles in adjacent cells, and the vibrations proceed along Bravais axes; the modes in a crystal therefore reduce to the modes in a supercell of twice the linear dimensions of a unit cell. The theory does not account for the continuum that is observed to be superimposed on the line spectra, a consideration that stimulated M. Born and his collaborators to reexamine the Born-Kármán theory of 1912, which was subsequently experimentally proved to be accurate.

In 1943 the government of Mysore gave the Indian Academy of Sciences eleven acres of land in Hebbal, a suburb of Bangalore. The Raman Research Institute was built on this site and, upon its completion in 1948, Raman became its first director. The government of the newly independent India appointed him national professor in the same year. He continued his research on optics and crystal structure—to which he added investigations of the physiology of vision—and his training of graduate students, many of whom he sent out into important positions throughout India.

From 1950 until 1958 Raman investigated optical effects in gems and minerals. He was particularly fascinated by diamonds—indeed, he had in the early 1930's studied their Raman spectra, and in 1944 and 1946 he had participated in two symposia on the structure of diamonds, held at Bangalore, in which he and his co-workers reported on the fluorescence, luminescence, absorption spectra, magnetic susceptibility, and second-order Raman effects of that gem. In the 1950's Raman returned to the subject and examined the specific heat, X-ray diffraction, and infrared spectrum of diamonds; in addition, he devised a new experimental technique in which he used a pencil beam of sunlight to demonstrate the reflections from gems and minerals as diffusion halos against a white screen. In particular, he studied iridescent feldspars, such as labradorite, which changes color with the angle of observation from peacock blue to green to golden yellow. He interpreted this effect as being caused by the mixture of chemical components within the lamellar structure of the mineral. He added that the Schiller effect of the moonstone is similarly caused, both of its components being birefringent. He also examined opals and pearls.

In the 1960's Raman turned to investigating colors and their perception. He was perhaps the first to study flowers spectroscopically, and in a series of papers published in 1963 he set out the results of his researches on the colors of the petals and the spectra of various floral species, including the rose, aster, and hibiscus. In 1964 he began to publish a series of papers in which he attempted to establish a new theory of color vision, in opposition to the trichromacy theory of human color perception; in particular he rejected Maxwell's work on the subject. These papers—there were eventually forty-three of them—were collected as *The Physiology of Vision*, published in 1968 in honor of Raman's eightieth birthday.

Raman was a man of great personal authority and presence; he was proud, even arrogant, and could be contemptuous of the power and authority of others. He was a great teacher—he once stated that "the principal function of the older generation of scientific men is to discover talent and genius in the younger generation and to provide ample opportunities for its expression and expansion"—and was generous in encouraging his students. He delighted in the public exposition of science, and gave annual lectures on general scientific topics at the Raman Research Institute on 2 October, Gandhi's birthday. He gave his last Gandhi Memorial Lecture, on the "Theory of Hearing," a few weeks before his death. As a scientist, Raman was a pioneer. Educated entirely in India, he did outstanding work at a time when the small Indian scientific community worked almost entirely in isolation and few made science a career; in fostering Indian science, Raman emerged as one of the heroes of the Indian political and cultural renaissance, along with Tagore, Gandhi, and Nehru.

Raman's private universe was one of light and color, of sound and tonality. The colors of flowers and of precious stones and the sound of strings and drums were for him the elements of eternal experience, which he could order with the aid of science. He loved roses more than anything else, and maintained a large rose garden. He was cremated there following his death at the age of eighty-two.

BIBLIOGRAPHY

Raman published more than 500 articles in *Philosophical Magazine, Nature, Physical Review, Proceedings of the Royal ·Society, Astrophysical Journal, Journal of the Optical Society of America, Transactions of the Optical Society of America, Zeitschrift für Physik, Bulletin. Indian Association for the Cultivation of Science, Proceedings of the Indian Association for the Cultivation of Science,* and *Indian Journal of Physics;* from 1934 on his papers appeared almost exclusively in *Proceedings of the Indian Academy of Science* and *Current Science.* Poggendorff gives an extensive bibliography of individual items.

Among Raman's most important works are "Dynamical Theory of the Motion of Bowed Strings," in *Bulletin. Indian Association for the Cultivation of Science,* **11** (1914); "On the Mechanical Theory of the Vibration of Bowed Strings and of Musical Instruments of the Violin Family with Experimental Verification of the Results," *ibid.,* **15** (1918); "On the Molecular Scattering of Light in Water and the Colour of the Sea," in *Proceedings of the Royal Society,* **64** (1922); "Musikinstrumente und ihre Klänge," in H. Geiger and K. Scheel, eds., *Handbuch der Physik,* VIII (Berlin, 1927); "A New Type of Secondary Radiation," in *Nature,* **121** (1928), 501, written with K. S. Krishnan; "A New Radiation," in *Indian Journal of Physics,* **2** (1928), 387; "The Diffraction of Light by High Frequency Sound Waves," in *Proceedings of the Indian Academy of Science,* **2A** (1935), 406, and 413, **3A** (1936), 35, 75, 119, 495, written with N. S. Nagendra Nath; "Crystals and Photons," *ibid.,* **13A** (1941), 1; "The Thermal Energy of Crystalline Solids; Basic Theory," *ibid.,* **14A** (1941), 459; "New Concepts of the Solid State," *ibid.,* **15A** (1942), 65; "Floral Colours," *ibid.,* **58A** (1963), 57 ff.; and "The New Physiology of Vision," *ibid.,* **60A** (1964), **61A** (1965), **62A** (1965), **63A** (1966), collected as *The Physiology of Vision* (Bangalore, 1968).

JAGDISH MEHRA

RAMANUJAN, SRINIVASA AAIYANGAR (*b.* Erode, near Kumbakonam, Tanjore district, Madras province, India, 22 December 1887; *d.* Chetput, near Madras, India, 26 April 1920), *mathematics.*

Ramanujan belonged to a Brahman family, but his father was poor and served as a bookkeeper in the firm of a cloth merchant in Kumbakonam. At the age of seven, after two years in elementary school, he transferred to the high school at Kumbakonam. In 1897 he placed first in the Tanjore primary examination. Ramanujan early studied some trigonometry on his own; but his real enthusiasm for the subject arose in 1903, when he was able to borrow an English text, Carr's *Synopsis of Pure Mathematics*. From then on, mathematics was nearly his only interest. He jotted down his results in a notebook which he carried with him and showed to people who were interested.

Quiet and meditative, Ramanujan was very fond of numerical calculations and had an unusual memory for numbers. In 1904 he won a fellowship at Government College, Kumbakonam. His excessive devotion to mathematics and neglect of English, a fundamental subject, led to his failure to be promoted. He returned to the college after traveling but could not graduate. He briefly attended a college in Madras, then returned to Kumbakonam, where he again failed. For several years he had no definite occupation but continued to record his mathematical results.

In 1909 Ramanujan married and was compelled to earn a living. He went to Ramaswami Aiyar, the founder of the Indian Mathematical Society, then deputy collector in the little town of Tirukkoyilur, to ask for a minor clerical job. He was sent on to Seshu Aayar, one of Ramanujan's former teachers, and obtained a substitute office job for a few months. He was then recommended to Ramachaudra Rao, collector of Nellore, eighty miles north of Madras, who was interested in mathematics; he later described Ramanujan at the time of the interview:

> A short uncouth figure, stout, unshaved, not overclean, with one conspicuous feature—shining eyes—walked in with a frayed notebook under his arm. He was miserably poor.—He opened his book and began to explain some of his discoveries. I saw quite at once that there was something out of the way; but my knowledge did not permit me to judge whether he talked sense or nonsense. Suspending judgment, I asked him to come over again, and he did. And then he had gauged my ignorance and shewed me some of his simpler results. These transcended existing books, and I had no doubt he was a remarkable man. Then, step by step, he led me to elliptic integrals and hypergeometric series and at last his theory of divergent series not yet announced to the world converted me. I asked him what he wanted. He said he wanted a pittance to live on so that he might pursue his researches.

Rao was convinced that a job as a clerk was not the answer to Ramanujan's troubles and sent him back to Madras, where Rao supported him for a while and tried unsuccessfully to get a fellowship for him. When this did not succeed, Ramanujan in 1912 found a job in the office of the Madras Port Trust. At this time he began his mathematical publication in the *Journal of the Indian Mathematical Society*. His first paper, "Some Properties of Bernoulli's Numbers," was followed by a number of brief communications

on series and infinite products and a geometric approximate construction of π.

Encouraged by influential friends interested in his mathematical work, Ramanujan began a correspondence with G. H. Hardy of Cambridge, one of the world's foremost specialists in analytic number theory. In his first letter Ramanujan mentioned his investigations on the distribution of primes and then added more than 100 theorems he had found in various parts of mathematics. Hardy was duly impressed and invited Ramanujan to come to England but, being a Brahman, he had scruples about leaving India. Instead a two-year fellowship was arranged for him at the University of Madras.

Hardy was disappointed and continued to attempt to persuade Ramanujan to come to Cambridge. When his colleague E. H. Neville lectured in Madras, he approached the young man and this time obtained his consent. Very favorable fellowship arrangements were made, and Ramanujan was admitted to Trinity College in 1914. He developed rapidly under the guidance of Hardy and Littlewood, who also helped him to publish his papers in English periodicals. Aside from the dozen papers in the *Journal of the Indian Mathematical Society*, questions proposed, and notes from the proceedings of meetings, Ramanujan published twenty-one papers during his five-year stay in Europe, several of them in collaboration with Hardy.

It was inevitable that a large portion of Ramanujan's results from his notebooks consisted of rediscoveries; he had never had systematic training in mathematics or access to a good library. To quote Hardy: "What was to be done in the way of teaching him modern mathematics? The limitations of his knowledge were as startling as its profundity." He worked with modular equations and theorems of complex multiplication, yet had no notion of doubly periodic functions; he worked with analytic number theory and had only the vaguest idea of what a function of a complex variable was. Most of the theorems in the notebooks were not proved in the standard sense but were only made plausible: "His ideas as to what constituted a mathematical proof were of the most shadowy description. All his results, new or old, right or wrong, had been arrived at by a process of mingled argument, intuition and induction, of which he was entirely unable to give any coherent account." Ramanujan's investigations on the distribution of primes suffered particularly from these weaknesses and therefore contributed little to the development of the general theory.

Ramanujan's first paper published in Europe, "Modular Equations and Approximations to π," contained a number of peculiar and very good approximations to π, many of them by means of square root expressions. Of greater systematic interest is his long memoir "Highly Composite Numbers," in which he derived some important properties of these numbers. Ramanujan returned to the question of the average number of prime divisors of a number in a paper written with Hardy, "The Normal Number of Prime Factors of a Number n."

The part of Ramanujan's work that stimulated most contributions from later mathematicians is probably his study of the partition of numbers into summands. From MacMahon's extensive numerical calculations of the number $p(n)$ of partitions of a number n, Ramanujan conjectured that these integers must have simple congruence properties with respect to small primes and their powers. He was able to prove some of these results by means of formulas from the theory of elliptic functions; others have been derived later. These papers by Ramanujan on partitions were followed by a joint paper with Hardy on the asymptotic value of $p(n)$. Their result was remarkable, since not only did their formula give good approximations to the values already calculated but also seemed to give an exact expression for $p(n)$; Rademacher later established that this was correct.

Ramanujan's other work comprised a variety of topics, mainly of a combinatorial nature. He wrote on the representation of integers as the sum of squares and on the lattice points inside a circle; in function theory he produced several papers on definite integrals, as well as on elliptic, hypergeometric, and modular functions.

In 1917 Ramanujan fell ill, possibly with tuberculosis; and the remainder of his stay in England was spent in several sanatoriums. In 1918 he was elected a fellow of the Royal Society and in the same year a fellow of Trinity College. These high distinctions seemed to improve his health and stimulate his mathematical production. Nevertheless, since the English climate did not seem to be beneficial, it was decided to send him back to India. An annual allowance of £250 for five years was awarded him by the University of Madras, with prospects of a later professorship.

Ramanujan returned to Madras in April 1919 in a precarious state of health. A difficult patient who refused medical aid, he went for a while to his home district but was prevailed upon to return to Madras for treatment. Until his last days he continued his mathematical research.

BIBLIOGRAPHY

Ramanujan's works have been brought together in *Collected Papers of Srinivasa Ramanujan*, G. H. Hardy,

P. V. Seshu Aiyar, and B. M. Wilson, eds. (Cambridge, 1927), and *Notebooks of Srinivasa Ramanujan*, 2 vols. (Bombay, 1957).

See also G. H. Hardy, *Ramanujan. Twelve Lectures on Subjects Suggested by His Life and Work* (Cambridge, 1940).

OYSTEIN ORE

RAMES, JEAN BAPTISTE (*b.* Aurillac, France, 26 December 1832; *d.* Aurillac, 22 August 1894), *botany, geology.*

Rames was the son of a pharmacist, who introduced him to the natural sciences at an early age. He earned his degree in pharmacy at Toulouse, where he also spent considerable time at the Faculté des Sciences and the Jardin des Plantes. Appointed an assistant at the university, he carried out research in the region of Toulouse that led him to discover fossil remains, notably of tortoises and crocodiles. This work made possible the precise dating of the Tertiary formation of Toulouse. Rames subsequently collaborated with Félix Garrigou and Henri Filhol in the publication of a work on the human fossil remains found in the caves at Lombrive and Lherm.

Rames then returned to Aurillac to take over his father's pharmacy. Despite the limits on his time imposed by this responsibility, he undertook an exhaustive study of the Cantal region, with the goal of producing a comprehensive description of its flora and geology. It was in the study of the volcanoes of the Cantal that he made his outstanding contributions. Following the work of Nicolas Desmarest, he showed that they were not caused by elevation (the *soulèvement* of Buch) but followed by an ejection of material which fell back around the crater. The sharp *pics* typical of the Cantal are the result of degradation by atmospheric agencies of flows of phonolite lava. Rames made an altogether larger number of distinctions within the stratigraphy of the Cantal than had his predecessors, and he described numerous petrographical varieties among the volcanic products. His important discovery, that the supposedly inclined basalts were Miocene, made it possible to establish the synchronism of the uplift of the Alpine chain and the earliest volcanic phenomena in central France. This volcanic action had continued until well into the Quaternary period; the huge massifs of the Cantal and of the Mont-Dore were erected during the Lower Pliocene. Rames presented a theory according to which the excavation of valleys started from the sides of extinct volcanoes and was accomplished by the action of the torrential erosion that succeeded the glaciers. He believed that the Cantal had been formed out of a single vast crater of which the only vestiges are the "puys." The general view today is that it is the result of a number of volcanoes.

Rames also pointed out in the Miocene alluvial formations the presence of flint that appeared to have been deliberately fashioned and Gabriel de Mortillet did not hesitate to baptize the hypothetical creature who supposedly shaped it *Anthropithecus ramesii*. This theory is no longer given much credence. Rames discovered the "flore des cinérites" of the Cantal, which the paleobotanist Gaston de Saporta considered among his most important findings.

Géogénie du Cantal (1873), Rames's major work, was preceded by *La création, d'après la géologie et la philosophie naturelle* (1869), in which he defended evolutionary concepts. A change in his philosophical opinions, probably inspired by his friend Msgr. Gérard, seems to have prevented him from completing a second edition of this work. The Musée Rames, founded in 1902 in Aurillac, contains his collections together with other specimens from the Cantal.

BIBLIOGRAPHY

I. ORIGINAL WORKS. Rames's writings include *L'homme fossile des cavernes de Lombrive et de Lherm* (Toulouse, 1862), written with Félix Garrigou and Henri Filhol; *Étude sur les volcans* (Paris, 1866); *La création, d'après la géologie et la philosophie naturelle* (Paris, 1869); *Géogénie du Cantal avec une étude historique et critique sur les progrès de la géologie dans ce département* (Paris, 1873); and *Topographie raisonnée du Cantal* (Aurillac, 1879).

II. SECONDARY LITERATURE. See M. Boule, "Notice sur Jean-Baptiste Rames," in *Bulletin de la Société géologique de France*, 3rd ser., **23** (1895), 192–202; J. Jung, "Les pionniers de la découverte des volcans d'Auvergne," part of "Hommage à J. B. Rames et inauguration du Musée," in *Revue de la Haute Auvergne*, **31** (1946), 251–259; L. Le Peletier d'Aunay, *Jean-Baptiste Rames (1832–1894). Sa vie—ses oeuvres—sa correspondance* (Aurillac, 1946), with 2 maps, a list of his works, and portrait; and G. de Saporta, "Les caractères propres à la végétation pliocène; à propos des découvertes de M. J. Rames dans le Cantal," in *Bulletin de la Société géologique de France*, 3rd ser., **1** (1873), 212–232; and "Forêts ensevelies sous les cendres éruptives de l'ancien volcan du Cantal, observées par M. J. Rames et conséquences de cette découverte pour la connaissance de la végétation dans le centre de la France à l'époque pliocène," in *Comptes rendus . . . de l'Académie des sciences*, **76** (1873), 290–294.

F. STOCKMANS

RAMMELSBERG, KARL (or **CARL**) **FRIEDRICH** (*b.* Berlin, Germany, 1 April 1813; *d.* Gross-Lichterfelde, near Berlin, 28 December 1899), *chemistry.*

Rammelsberg was the son of a merchant. He was an apprentice in a pharmacy, first in Berlin and then in Dardesheim, near Halberstadt, but soon turned to the study of chemistry at the University of Berlin under Mitscherlich and Heinrich Rose. He also devoted himself to related sciences, especially crystallography and mineralogy, under C. S. Weiss and Gustav Rose. His Latin doctoral dissertation, dealing with cyanides, was presented in 1837. In 1841 Rammelsberg became a *Privatdozent* in chemistry at the University of Berlin, in 1846 an associate professor, and in 1874 a full professor. He was also professor of chemistry at the Gewerbeakademie, the forerunner of the present Technical University of Berlin, but left that position in 1883 to become the director of the "second chemical laboratory" at the university.

During the early part of his career, Rammelsberg was a close friend of Heinrich and Gustav Rose. Both had been pupils of Berzelius, with whom Rammelsberg also formed a close association, translating one of Berzelius' last papers at his request and publishing *Berzelius' neues chemisches Mineralsystem* in 1847. In an address on the occasion of his election to the Berlin Academy of Sciences (1855), Rammelsberg referred to the need for specialization within the broad field of chemistry, saying he had selected mineral chemistry. He emphasized that the methods of chemistry and mineralogy should be the same and that substance and form and the resulting physical properties should be studied in both.

Rammelsberg's publications, including some twenty books, total about 430. The peak of his activity was reached in 1870, with the publication of twenty papers within a year. About 60 percent of Rammelsberg's papers were concerned with the chemical compositions, crystal forms, and physical properties of minerals. He made and published a staggering number of mineral analyses, worked on rock and meteorite analyses, improved analytical methods, synthesized new compounds, and wrote many papers discussing the analytical results of other scientists.

Rammelsberg was greatly interested in making known in Germany the results of foreign work in mineral chemistry and related fields. To this end he published many discussions of foreign work and translations of French, Italian, and Swedish papers. Thus he helped especially to spread the influence of St. Claire-Deville and others of the French school of chemical mineralogy of the middle of the century. One of his earliest works was a translation of Jean-Baptiste Dumas's *Leçons sur la philosophie chimique*.

Rammelsberg established the first university laboratory for the teaching of chemistry in Prussia and produced textbooks on stoichiometry, inorganic chemistry, qualitative and quantitative analysis, and crystallography for chemists. Most of these went through several editions, and one was translated into English as *Guide to a Course of Quantitative Chemical Analysis* . . . (Geneva, N.Y., 1871).

Rammelsberg's most important works were comprehensive compilations on mineral chemistry and on chemical crystallography. Each went through two editions, and at the end of the nineteenth century the new editions were the most important works in their respective fields. The *Handbuch der Mineralchemie*, a single-handed effort, was the forerunner of the compilation with the same title edited by C. A. Doelter with many co-workers (1912–1931), a fact acknowledged by Doelter in his introduction. Similarly, the *Handbuch der krystallographisch-physikalischen Chemie* may be considered a forerunner of Groth's five-volume *Chemische Krystallographie* (1906–1919), also the work of one man. Groth, however, made no reference to Rammelsberg in his introduction. This may possibly be ascribed to the fact that Rammelsberg continued to use the antiquated symbolism and classification of Weiss, whose great contributions belong in the first two decades of the nineteenth century. Although he produced crystallographic data throughout his career and made the most comprehensive compilations of such data, Rammelsberg appears to have ignored most of the progress of crystallography during the last sixty years of his life.

BIBLIOGRAPHY

I. Original Works. Some of Rammelsberg's most important works are *J. J. Berzelius' neues chemisches Mineralsystem nebst einer Zusammenstellung seiner älteren hierauf bezüglichen Arbeiten* (Nuremberg, 1847); "Antrittsrede," in *Monatsbericht der Deutschen Akademie der Wissenschaften zu Berlin* (1856), 373–375, a short address given at the time of Rammelsberg's election to the Academy; "Die Fortschritte der Mineralchemie, wie sie seit fünfzig Jahren aus *Poggendorff's Annalen* sich ergeben," in *Poggendorff's Annalen*, Jubelband (1874), 381–407, a summary of the progress of mineral chemistry as shown in the first 150 vols. of the *Annalen; Handbuch der Mineralchemie*, 2nd ed. (Leipzig, 1875), plus 2 supp. vols. (Leipzig, 1886, 1895); and *Handbuch der krystallographisch-physikalischen Chemie*, 2 vols. (Leipzig, 1881–1882).

II. Secondary Literature. Max Bauer, "Karl Friedrich Rammelsberg," in *Centralblatt für Mineralogie* (1900), 221–233, 319–329, 342–357, with portrait, a biography by one of his more distinguished pupils, is the most comprehensive account of Rammelsberg's life and work and includes a bibliography of 308 items deemed by Bauer to be of mineralogical interest. G. Wyrouboff, "Notice nécrologique de M. Rammelsberg," in *Bulletin de la Société française de minéralogie*, **24** (1901), 280–306, is a brief note

followed by the most nearly complete list, 419 items, of Rammelsberg's works. Strangely, among the omissions is the translation of Dumas's 1837 *Leçons* as *Die Philosophie der Chemie* (Berlin, 1839).

A. Pabst

RAMON, GASTON (*b*. Bellechaume, France, 30 September 1886; *d*. Paris, France, 8 June 1963), *immunology*.

Ramon completed his secondary studies at the *lycée* in Sens, then entered the veterinary school at Alfort, where he became interested in research. In 1911 he joined the Pasteur Institute at its Garches Annex, of which he was named director in 1926. In 1933 Ramon was appointed assistant director of the Paris Pasteur Institute, and, for several months in 1940, served as the sole director of the whole institute. From 1949 until his retirement in 1958 he was head of the Bureau of Epizootic Diseases in Paris.

Ramon made his most important discoveries during four years of intense research activity between 1922 and 1925. The first of these, the flocculation reaction, revolutionized the method of titrating anatoxins and microbial toxins. Ramon showed that when a zone of flocculation appears in a mixture of antitoxin and toxoid, the mixture is completely neutralized; the reaction therefore presented a simple way in which antitoxic serums might be standardized, and simplified the use of these substances.

Ramon also demonstrated a method, based upon what he called "the principle of anatoxins," by which a potentially strong toxin could be subjected to the combined action of Formol and heat to yield an anatoxin—a new substance that is itself harmless, although it retains the ability to stimulate the formation of antitoxins. This simple procedure had a number of practical consequences, of which the most immediate was the preparation of effective vaccines, new in both their composition and their mode of action, against diphtheria and tetanus. The method was later applied in the production of antiviral vaccines, including those effective against aphtha and poliomyelitis.

Ramon also discovered the substances that he called "adjuvantes et stimulantes de l'immunité," including calcium chloride, alum, and tapioca, which have the property of increasing the activity of antigens. These agents were first utilized in serotherapy laboratories to attain serums rich in antibodies, and they now play an important role in immunological research. His investigations of these substances led Ramon to the idea of "associated vaccinations," and he developed the method of combining several antigenic substances into one vaccine. This technique has been successfully applied in man, and offers several immunities with a single injection.

BIBLIOGRAPHY

I. ORIGINAL WORKS. Ramon's principal writings are *Le principe des anatoxines et ses applications* (Paris, 1950); *La lutte préventive contre les maladies infectieuses de l'homme et des animaux domestiques au moyen des vaccins* (Paris, 1955); and *Quarante années de recherches et de travaux* (Toulouse, 1960).

II. SECONDARY LITERATURE. See R. Debré, A. Blaizot, R. Richou, *et al.*, "Allocutions et discours," in *Revue d'épidémiologie, médecine sociale et santé publique*, **15**, no. 8 (1967), 707–728; A. Delaunay, in *Presse médicale*, **71**, no. 40 (28 Sept. 1963), 1891; and *Hommage à Gaston Ramon* (Paris, 1970), issued by the Pasteur Institute.

A. DELAUNAY

RAMOND DE CARBONNIÈRES, LOUIS FRANÇOIS ÉLISABETH (*b.* Strasbourg, France, 4 January 1755; *d.* Paris, France, 14 May 1827), *geology, botany.*

Ramond, the son of a state official in Alsace, took courses in law and medicine at the University of Strasbourg. In 1777 he visited Switzerland, where he met Voltaire, Haller, and Johann Lavater. He also traveled across the Alps in order to study their natural history. In 1776 the Englishman William Coxe had made similar journeys, and his observations were later published as *Sketches of the Natural, Civil and Political State of Swisserland in a Series of Letters to William Melmoth* (London, 1779). Ramond translated this work into French, adding notes based on his own observations in Switzerland, in particular giving details about the glaciers.

The success of this book, which was published in 1781, attracted the attention of Cardinal Louis de Rohan, and he engaged Ramond as his confidential secretary. In this capacity Ramond acted as go-between in Rohan's dealings with the notorious charlatan Cagliostro.

In 1787 he accompanied Rohan to Barèges, a small spa in the foothills of the Pyrenees. The mineralogical and botanical observations he made in the neighboring mountains were published in 1789 as *Observations faites dans les Pyrénées*, a work which aroused great interest. Ramond described glaciers in the Pyrenees which were quite unknown to the scientific world. He also gave an account of the fauna and flora, and described the changes that took place in the vegetation with increasing altitude.

During the early years of the Revolution Ramond was in Paris and in 1791 was elected to the Legislative Assembly as a deputy. In debates he supported Lafayette and after the events of 10 August 1792 found it necessary to leave the city immediately. He returned to Barèges, but in 1794 was imprisoned in Tarbes (Hautes-Pyrénées) for ten months because of his political views. After his release he was appointed professor of natural history at Tarbes.

In the summer of 1797 Ramond attempted to reach the summit of Mont-Perdu (now Monte Perdido, in Spain) in the central Pyrenees, which he erroneously believed to be the highest peak of the range. He was accompanied by Philippe Picot de la Peyrouse, botanist and inspector of mines, from Toulouse, and several students. The summit was not reached, but the party made the unexpected discovery of abundant fossil remains of marine shells in the limestone strata at an altitude of about 10,000 feet. Ramond and Picot de la Peyrouse, each anxious to claim credit for this discovery, sent separate accounts to the Institut National des Sciences et des Arts (the former Académie des Sciences) in Paris and these were published in the *Journal des Mines*, **8** (1798), 35–66.

Ramond continued his researches, botanical as well as geological, in the Pyrenees. In 1800 he returned to Paris and was elected to the Corps Législatif. He published a new account of his several journeys in the Pyrenees in *Voyages au Mont-Perdu et dans la partie adjacente des Hautes-Pyrénées* (Paris, 1801). He continued to visit the district and on 10 August 1802 at last succeeded in reaching the summit of Monte Perdido, an altitude of 10,997 feet.

Ramond was elected a member of the Institut National in 1802, and in 1806 Napoleon appointed him prefect of Puy-de-Dôme. In Auvergne he continued to pursue his botanical and geological researches, and barometric measurements also engaged his attention. During the invasion of France in 1814 his house in Paris was ransacked by Cossacks, and manuscripts that he was preparing for publication were destroyed.

Ramond's researches in the Pyrenees have received little notice in the histories of geology and botany, but his discovery of abundant fossils in calcareous sediments at a great altitude was undoubtedly a momentous one. At that time it was widely thought that the highest mountains were composed of granite, "the oldest work of the sea," and other "primitive" rocks; against them lay steeply inclined non-fossiliferous bedded rocks, chemically deposited or derived by erosion from the primitive mountains. Fossiliferous sediments, horizontally bedded or gently inclined, were believed to be confined to a lower level. These ideas, first clearly stated by Pallas, had been accepted and taught by Werner in Freiberg. Thus, Ramond's discovery was a revolutionary one, which required new explanations of geological structures. In his *Voyages* (1801) he also described granites, some of which he thought were less ancient than others, although he did not accept an igneous origin for them.

His name is commemorated by the genus *Ramonda*, beautiful little rock plants named after him.

BIBLIOGRAPHY

I. ORIGINAL WORKS. Ramond's first scientific work was *Lettres de M. William Coxe à M. W. Melmoth sur l'état politique civil et naturel de la Suisse et augmentées des Observations faites dans le même pays par le traducteur* (Paris, 1781); two more eds. (1782 and 1787) were almost unchanged. The *Observations* were reprinted and issued separately, H. Beraldi, ed. (Toulouse, 1929). An English translation was added to W. Coxe's *Travels in Switzerland* (London, 1802). *Observations faites dans les Pyrénées pour servir de suite des observations sur les Alpes . . .* (Paris, 1789) was translated into German (Strasbourg, 1789). An English trans. by F. Gold, *Travels in the Pyrenees* (London, 1813), omits the more interesting geological material found in the second part. *Voyages au Mont-Perdu* (Paris, 1801) was followed by *Voyage au sommet du Mont-Perdu* (Paris, 1803); the latter appeared in English in John Pinkerton's *General Collection of Voyages and Travels*, IV (London, 1809), and in an abbreviated form in *Nicholson's Journal of Natural Philosophy*, **6** (1803), 250–252. It was reprinted (Pau, 1914) and also reproduced in facs. by the Société Ramond (Bagnères-de-Bigorre, 1925). Some correspondence between Ramond and Picot de la Peyrouse was edited by C. Roumeguère and published in *Bulletin de la Société agricole, scientifique et littéraire des Pyrénées-Orientales*, **20** (1873). The Musée Pyrénéen du Chateau-fort in Lourdes has many of Ramond's personal possessions, his medals, certificates, scientific instruments and maps, as well as a suite of rocks he brought back from Monte Perdido in 1802.

II. SECONDARY LITERATURE. An *éloge* by Cuvier is in *Mémoires de l'Académie des Sciences de l'Institut de France*, **9** (1830), clxix–cxcv; there is a longer notice in Michaud's *Biographie universelle*, XXXV (Paris, 1843), 150–154. The most recent account of Ramond is C. M. Girdlestone, *Poésie, politique, Pyrénées: Louis-François Ramond 1755–1827, sa vie, son oeuvre littéraire et politique* (Paris, 1968).

JOAN M. EYLES

RAMÓN Y CAJAL, SANTIAGO (*b.* Petilla de Aragón, Spain, 1 May 1852; *d.* Madrid, Spain, 18 October 1934), *neuroanatomy, neurohistology.*

Santiago Ramón y Cajal was born in a poverty-stricken and isolated village in Navarre, the son of Justo Ramón y Casasús, a barber-surgeon who some years later—by hard work and considerable sacrifice—acquired a medical degree, and of his wife Antonia. Ramón y Cajal has left us a very full autobiographical record. His early educational experiences were troubled. An interest in art displeased his authoritarian father, who decreed that his son study medicine. The son, predictably, became totally unamenable to any sort of discipline and showed contempt for his teachers and for the whole educational process. Eventually, and possibly aided by enforced apprenticeship to a barber and then to a shoemaker, he acquired sufficient formal learning to enable him to begin the study of medicine at the University of Zaragoza, from which he graduated in 1873. He then joined the army medical service and in the following year was sent to Cuba. There he contracted malaria and within twelve months had to be discharged from the service and sent back to Spain.

Ramón y Cajal determined on an academic career—anatomy was the only subject of his medical course in which he showed any real interest or ability—and spent a further two years at Zaragoza studying for his doctorate. In 1883 he was appointed to the chair of anatomy at Valencia, having in the meanwhile made himself, virtually without aid, a highly competent microscopist and histologist. He had also, while convalescing from tuberculosis, become a skilled photographer. In 1887 Ramón y Cajal was appointed to the chair of histology at Barcelona and, in 1892, to the chair of histology and pathological anatomy at Madrid, which he held until his retirement in 1922.

Cajal was the recipient of numerous prizes, honorary degrees, and distinctions, both Spanish and foreign. In 1894 he was invited to give the Croonian lecture to the Royal Society, and in 1899 he was special lecturer at Clark University, Worcester, Massachusetts. He was elected a foreign member of the Royal Society in 1909. In 1906 he shared the Nobel prize for physiology or medicine with Golgi. He married Silvería Fanañás García in 1880; they had four sons and four daughters.

The picture of Ramón y Cajal that emerges from his own writings is full and candid. Interested in things rather than people, dedicated to neurohistology to the point of obsession, and prepared to submit his wife and family, at least in the earlier years, to considerable hardship while he financed his own laboratory and publications, Ramón y Cajal appears as proud, ashamed of his country's administrative inefficiency, corruption, and scientific backwardness, ambivalent in that he recognized the need to publish in one of the major scientific languages of Europe, but resented foreign ignorance of the language of Cervantes, and intensely patriotic and determined that Spain should have a place on the scientific and intellectual stage. He succeeded in founding a Spanish school of histology, and his many distinguished pupils included P. del Río-Hortega, F. de Castro, and R. Lorente de Nó.

In the course of more than half a century from 1880, Ramón y Cajal published numerous scientific papers and an imposing number of books. In the twenty years of his most intense activity, 1886–1906, he may be said to have laid the histological foundations of our present knowledge of the nervous system. He came to study the subject partly because he was systematically teaching himself the whole of histology, but partly also because he saw in the fine structure of the nervous system the material basis of thought and in the elucidation of that structure the answer to many of the problems of physiology and psychology.

Ramón y Cajal found that there was no clear notion of something so fundamental as how a sensory impulse was conducted to a motor fiber, since contemporary histological technique apparently was incapable of defining the course of nerve-cell processes in the gray matter of the central nervous system and, hence, the relationship of one nerve cell to another. He solved this problem by adopting Golgi's then largely unknown potassium dichromate–silver nitrate technique and applying it to thick sections of embryonic, as opposed to adult, material. The majority of neurologists at this time believed in the reticular theory of nervous interconnection, the only prominent dissentients being His and Forel. Schäfer's work on *Medusa*, published in 1878, seems to have been completely ignored.

Ramón y Cajal established first that axons end in the gray matter of the central nervous system in a number of different ways, but always independently and never so as to form a network with other axon terminals. He showed next that although these terminals were in close contact with the dendrites and cell bodies of other nerve cells, there was no physical continuity between one such cell and another. He thus confirmed what had been tentatively suggested by His and by Forel: that the nervous system was an agglomeration of discrete and definable units. The implications for theories of nervous function of such a structural scheme—the neuron doctrine, as it came to be known—are of course profound. It becomes possible to imagine much more clearly the existence of distinct functional pathways, in that a group of axons may be shown to terminate around one group of nerve cells and not another, instead of losing their identity in a reticulum. On the other hand, it poses acutely the problem of how "information" is passed across anatomical "gaps"—synaptic transmission in other

words. Ramón y Cajal's studies at this time, mainly on the cerebellum, spinal cord, retina, and olfactory mucosa, also convinced him of the truth of what he called the "theory of dynamic polarization": that the transmission of the nerve impulse is always from dendrites and cell body to axon.

Ramón y Cajal's success in delineating nerve cells all the way to the termination of their finest processes had already enabled him—for example, in the cerebellum and spinal cord—to classify neurons according to the form and direction taken by those terminal fibers. In 1897–1900, having adopted Ehrlich's methylene blue stain in addition to Golgi's, he extended his studies to the human cerebral cortex, where he was able to demonstrate the terminal arborizations of the afferent sensory fibers. He again described and classified the various types of neurons in such a way, he believed, as to permit the ascribing of specific structural patterns to different areas of the cortex; hence he was able to place the concept of cerebral localization on firm histological foundations. His descriptions of the cerebral cortex are still the most authoritative.[1] They led to the cytoarchitectonics of W. Campbell, K. Brodman, the Vogts, and later workers. Ample tribute has also been paid to the continuing value of his work on the cerebellum.[2]

If the cell body itself was concerned with conduction rather than, or as well as, mere nutrition, then a knowledge of its fine structure was obviously of importance. Neurofibrils had been described, but their staining was a highly uncertain business. In his autobiography Ramón y Cajal describes how in 1903 he discovered the reduced silver nitrate method for displaying these structures. Although he does not say so, his photographic expertise may well have been a subconscious factor.

In 1904 Ramón y Cajal published *Textura del sistema nervioso del hombre y de los vertebrados*, in which he brought together the results of the previous fifteen years and which must rank as a classic of medical science. This massive work, more than any other, contains the cytological and histological foundations of modern neurology, yet structural detail is seen never as an end in itself but only as a preliminary to the answering of three questions: What is the functional meaning of this pattern? How does it work? By what physicochemical processes has it reached its present state across the paths of phylogenetic and ontogenetic history?

Ramón y Cajal next turned his attention to the problem of traumatic degeneration and regeneration of nervous structures. He did this in response to what he considered a dangerous revival of the reticularist theory. The main facts had not been in dispute since the work of Waller, Ranvier, and others nearly half a century earlier; but there were two schools of thought about precisely how the degenerated peripheral end of the cut axon was restored to structural and functional continuity with its nerve cell. The polygenesists, who earlier had included E. F. A. Vulpian and C. E. Brown-Séquard and whose leader at the time was A. T. J. Bethe, maintained that the regenerated peripheral fibers were the result of progressive transformation and eventual fusion of the Schwann cells which had sheathed the degenerated fibers. The monogenesists, to whom Ramón y Cajal belonged, said that the regenerated fibers were the result of sprouting from the cylinders of the central stump, still in continuity with their nerve cells, and saw their opponents as reviving the reticular theory of nerve continuity in thinly disguised form. Ramón y Cajal, using his reduced silver nitrate method of staining, fully confirmed the monogenesist theory. The results of these researches were collected and published in 1913–1914 as *Estudios sobre la degeneración y regeneración del sistema nervioso*, still the fullest account of the subject.

Ramón y Cajal had always felt isolated from the mainstream of science, living in Spain and publishing almost exclusively in Spanish; and his isolation was increased by World War I. Nevertheless he continued to publish papers. The most important work of his later years centered on his discovery in 1913 of the gold sublimate method which he applied to the staining of neuroglia, first described by Virchow[3] and until recently believed to be merely a supporting skeleton for the nervous elements. This work did much to lay the foundation of current knowledge of the pathology of tumors of the central nervous system.

After formal retirement Ramón y Cajal remained director of the institute which the government had erected and named for him; he also continued to work with the tirelessness and patience which had characterized his adult life.

NOTES

1. See Edwin Clarke and C. D. O'Malley, *The Human Brain and Spinal Cord*, 446.
2. John C. Eccles, Masao Ito, and János Szentágothai, *The Cerebellum as a Neuronal Machine* (Berlin–Heidelberg–New York, 1967), 2.
3. See Clarke and O'Malley, *op. cit.*, 84.

BIBLIOGRAPHY

I. ORIGINAL WORKS. The foreign student of Ramón y Cajal's original work faces certain difficulties. He wrote some 20 books and 250 scientific papers. Many of his earlier papers were published in *Boletín médico valenciano*, *Gaceta médica catalana*, and *Gaceta sanitaria de Barcelona*, which outside Spain are likely to be found only in the largest and best-equipped medical libraries. Of the early numbers of *Revista trimestral de histología normal y patológica*, financed and largely written by Ramón y Cajal himself, only 60 copies were published and have long ranked as rarities. He himself reckoned that less than one third of his output had been read by foreign scientists. Only 800 copies of his magnum opus, *Textura del sistema nervioso del hombre y de los vertebrados*, 3 vols. (Madrid, 1894–1904), were printed. Most workers must use the French trans., altered and brought up to date by L. Azoulay, *Histologie du système nerveux de l'homme et des vertébrés*, 2 vols. (Paris, 1909). His *Estudios sobre la degeneración y regeneración del sistema nervioso*, 2 vols. (Madrid, 1913–1914), was translated into English and edited by Raoul M. May as *Degeneration and Regeneration of the Nervous System*, 2 vols. (London, 1928).

Of great value and interest for the light they shed on Ramón y Cajal's personality are *Reglas y consejos sobre investigación scientífica*, 7th ed. (Madrid, 1935), based on Ramón y Cajal's inaugural address following his election to the Royal Academy of Sciences in Madrid, translated by J. M. Sánchez-Pérez and edited and annotated by Cyril B. Courville, as *Precepts and Counsels on Scientific Investigation* (Mountain View, Calif., 1951); and *Recuerdos de mi vida*, 2 vols. (Madrid, 1901–1907), translated (with some abridgment) by E. Horne Craigie with the assistance of Juan Cano, as *Recollections of My Life* in *Memoirs of the American Philosophical Society*, 8 (1937), repr. as a book (Cambridge, Mass.–London, 1966). These books contain much good advice; they also exhibit a characteristically late nineteenth-century attitude to science, and a worship of "hard facts" which many no longer find congenial, together with a moralizing on science and scientists which reads less well when one bears in mind the polemical tone of some of Ramón y Cajal's scientific papers. For his general outlook on life see *Charlas de café; pensiamentos anécdotas y confidencias, por S. R. Cajal* (Madrid, 1920), parts of which are trans. in *The World of Ramón y Cajal with selections from his nonscientific writings*, E. Horne Craigie and William C. Gibson, eds. (Springfield, Ill., 1968), and *El mundo visto a las ochenta años. Impresiones de un arteriosclerótico*, 2nd ed. (Madrid, 1934).

Ramón y Cajal's Croonian Lecture, "La fine structure des centres nerveux," is in *Proceedings of the Royal Society of London*, 53 (1894), 444–468. A number of his most important papers have been translated into English: "Estructura del asta de Ammon y fascia dentada," in *Anales de la Sociedad española de historia natural*, 22 (1893), 53–114, translated by Lisbeth M. Kraft as *The Structure of Ammon's Horn* (Springfield, Ill., 1968); and four papers on the limbic cortex in *Trabajos del Laboratorio de investigaciones biológicas de la Universidad de Madrid*, 1 (1901–1902), 1, 141, 159, 189, translated by Lisbeth M. Kraft as *Studies on the Cerebral Cortex [with Limbic Structures]* (London,

1955). His work on the development of various nervous structures, published intermittently over a long period, was collected and translated into French as *Études sur la neurogenèse de quelques vertébrés* (Madrid, 1929); this French text was translated into English by Lloyd Guth as *Studies on Vertebrate Neurogenesis* (Springfield, Ill., 1960). *Studies on the Diencephalon*, compiled and translated by Enrique Ramón-Moliner (Springfield, Ill., 1966), is an anthology of papers and chapters, including some from the *Histologie du système nerveux*. Similarly, *The Structure of the Retina*, Sylvia H. Thorpe and Mitchell Gluckstein, trans. and eds. (Springfield, Ill., 1972), is based on three texts: "La rétine des vertébrés," in *Cellule*, **9** (1892) 121–246; the German trans. by R. Greeff, *Die Retina der Wirbeltiere* (Wiesbaden, 1894); and Ramón y Cajal's revision of his original article in *Travaux du laboratoire des recherches biologiques de l'université de Madrid*, **28** (1933).

Ramón y Cajal's address on receipt of the Nobel prize and useful biographical information are in *Nobel Lectures Including Presentation Speeches and Laureates' Biographies. Physiology or Medicine, 1901–1921* (Amsterdam–London–New York, 1967), 220–267. Not long before he died, Ramón y Cajal published "¿Neuronismo o reticularismo?" in *Archivos de neurobiología*, **13** (1933), 217–291, 579–646, translated by M. Ubeda Purkiss and Clement A. Fox as *Neuron Theory or Reticular Theory? Objective Evidence of the Anatomical Unity of Nerve Cells* (Madrid, 1954). Translated excerpts from his writings in historical context are Edwin Clarke and C. D. O'Malley, *The Human Brain and Spinal Cord* (Berkeley–Los Angeles, 1968).

II. SECONDARY LITERATURE. A. D. Loewy, "Ramón y Cajal and Methods of Neuroanatomical Research," in *Perspectives in Biology and Medicine*, **15** (1971), 7–36; F. H. Garrison, "Ramón y Cajal," in *Bulletin of the New York Academy of Medicine*, **5** (1929), 483–508; W. C. Gibson, "Santiago Ramón y Cajal (1852–1934)," in *Annals of Medical History*, n.s. **8** (1936), 385–394; and C. S. Sherrington, in *Obituary Notices of Fellows of the Royal Society of London*, **1**, no. 4 (1935), 425–441.

DOUGLASS W. TAYLOR

RAMSAUER, CARL WILHELM (*b.* Osternburg, Oldenburg, Germany, 6 February 1879; *d.* Berlin, Germany, 24 December 1955), *physics*.

For a detailed study of his life and work, see Supplement.

RAMSAY, ANDREW CROMBIE (*b.* Glasgow, Scotland, 31 January 1814; *d.* Beaumaris, Wales, 9 December 1891), *geology*.

Ramsay was the third child of William Ramsay, who owned a dye business. The family was well educated and French was spoken at the breakfast table on one morning a week and passages from famous authors read on other mornings. In 1827 the father died, and Andrew had to leave Glasgow

Grammar School to become a clerk in a cotton broker's office. Here he found consolation in literature and in composing humorous poems. Later he worked for a firm of linen merchants and entered into partnership in a cloth and calico business, which soon failed.

The turning point in Ramsay's life came in 1840, when the British Association for the Advancement of Science met at Glasgow, and he played a major part in collecting geological specimens and in making a geological model and map of the Isle of Arran. A paper he read on the subject was well received, and Ramsay was asked to guide a section of the Association to the island. Unfortunately, he worked so late into the night on notes for this journey that he overslept the following morning and missed the boat. The ability he had shown, however, led indirectly to his appointment to the Geological Survey of Great Britain in 1841. He served this institution for the next forty years and rose from assistant surveyor to local director (1845), senior director for England and Wales (1862), and finally to director general in 1871, a post that he held for ten years. To offset the poor pay, he held the professorship of geology at University College, London, from 1847 to 1852 and a lectureship at the Government School of Mines (1852–1871).

In 1852 Ramsay married a daughter of the Rev. Chancellor Williams of Llanfairynghornwy, Anglesey; they had four daughters and one son. He wrote nearly fifty scientific papers and books, and shared responsibility for numerous official maps and memoirs. He was elected a fellow of the Royal Society in 1862 and presided over the Geological Society of London (1862–1864) and the British Association (1880). On his retirement in 1881 he was knighted, and three years later went to live at Beaumaris, Anglesey.

Owing to overwork Ramsay suffered a few periods of nervous strain, one of which culminated in the removal of his left eye in 1878, but he is rightly held up as a delightful geologist, exuberant, a devotee and frequent quoter of the major English poets, eloquent, humorous, musical, and athletic.

Ramsay was essentially a stratigrapher and geomorphologist, and showed little regard for paleontology and petrology. Apart from detailed local surveys, his main contributions to geology concern general denudation, the development of river systems, and glaciation. In these his bold theories had a great and lasting influence. His early professional studies were summarized in "On the Denudation of South Wales . . ." (1846), which is of outstanding importance for three reasons.

First, it demonstrated that the Paleozoic sedimentary strata were composed of fragments derived from older rocks, thus proving that the district had experienced immense denudation in early geological periods.

Second, it postulated the power of the sea to cut wide plains (of marine planation) on a stationary or subsiding landmass, as distinct from the theories of Charles Lyell, who applied marine action to a rising landmass and to dissection rather than planation.

Third, it emphasized for the first time the visual reconstruction of former folded surfaces and the statistical assessment of the amount of material removed by denudation.

Ramsay's contributions to the knowledge of the physical development of the Rhine and of certain rivers in England and Wales proved of less importance and at best remain controversial, but they aided the cause of fluvialism. On the other hand, his ideas on glacial action, which caused tremendous contemporary disagreement, have for the most part been vindicated. Although at first skeptical of glaciation, after working in northern Wales he became, by 1850, the leading British advocate of glacial action. He went further and suggested that certain Permian conglomerates in the English Midlands had been transported by glacial action. These today are thought to have been deposited by flash floods in a semiarid environment, but Ramsay's idea led to the universal acceptance of extensive glaciation in Gondwanaland during Permo-carboniferous times. In 1858 he revisited the Alps with John Tyndall and soon added important suggestions on the power of glacial erosion.

Ramsay postulated that Alpine valleys were created by rivers, then deepened and widened by ice, and finally modified by river action. His most exciting contention was that many large lake basins in areas formerly occupied by land ice had been scooped out by glacial action. This principle of glacial over-deepening did not command general assent until at least sixty years after its announcement. His contemporaries thought it of little significance, and Sir Archibald Geikie considered Ramsay's elucidation of the Snowdonia region in northern Wales "a masterpiece of geological work and his own fittest and most enduring monument" ("Obituary . . .," 42).

BIBLIOGRAPHY

I. Original Works. Ramsay's writings include *The Geology of the Isle of Arran* (Glasgow, 1841); "On the Denudation of South Wales and the Adjacent Counties of England," in *Memoirs of the Geological Survey of Great Britain*, I (London, 1846), 297–335; "On the Superficial Accumulations and Surface Markings of North Wales," in *Quarterly Journal of the Geological Society*, 8 (1852), 371–376; "On the Occurrence of Angular . . . Fragments and Boulders in the Permian Breccia of Shropshire, Worcestershire . . .," ibid., 11 (1855), 185–205; "The Excavation of the Valleys of the Alps," in *Philosophical Magazine*, 4th ser., 24 (1862), 377–380; "On the Glacial Origin of Certain Lakes in Switzerland, the Black Forest, . . . and Elsewhere," in *Quarterly Journal of the Geological Society of London*, 18 (1862), 185–204; *The Physical Geology and Geography of Great Britain* (London, 1863; 6th ed., enl., 1894); "On the Erosion of Valleys and Lakes," in *Philosophical Magazine*, 4th ser., 28 (1864), 293–311; "The Geology of North Wales," which is *Memoirs of the Geological Survey of Great Britain*, III (London, 1866; 2nd ed., 1881); "On the River-Courses of England and Wales," in *Quarterly Journal of the Geological Society of London*, 28 (1872), 148–160; "The Physical History of the Valley of the Rhine," ibid., 30 (1874), 81–95; and "On the Physical History of the Dee, Wales," ibid., 32 (1876), 219–229.

II. Secondary Literature. See R. J. Chorley, A. J. Dunn, and R. P. Beckinsale, *The History of the Study of Landforms*, I (London, 1964), 301–313 and *passim*; A. Geikie, "Obituary of Sir A. C. Ramsay," in *Quarterly Journal of the Geological Society of London*, 48 (1892), 38–47; and *Memoir of Sir A. C. Ramsay* (London, 1895); and E. Walford, ed., *Andrew Crombie Ramsay* (London, 1866), a biographical memoir with photograph.

Robert P. Beckinsale

RAMSAY, WILLIAM (b. Glasgow, Scotland, 2 October 1852; d. Hazlemere [near High Wycombe], Buckinghamshire, England, 23 July 1916), *physical chemistry*.

Ramsay is best known for his discovery and isolation of the family of inert gases of the atmosphere. For this experimental work, along with the theoretical work that situated these elements in the periodic system, he was awarded the 1904 Nobel Prize in chemistry.

Ramsay was the only child of the civil engineer and businessman William Ramsay, whose forebears were chemist-dyers, and Catherine Robertson. He was raised in the Calvinist tradition. After completing his secondary education at the Glasgow Academy, Ramsay matriculated in November 1866 at the University of Glasgow, where he read the standard course in classics. Although he was originally intended for the ministry, his latent interests in science gradually developed. He attended the chemistry lectures of John Ferguson in 1869–1870 and the physics lectures of William Thomson in 1870. From 1869 he also worked for eighteen months as chemist apprentice for the local analyst, Robert Tatlock, further developing his interest and ability in chemistry.

In April 1871 Ramsay went to the organic chemistry laboratory of Fittig, who, in 1870, had accepted the chair of chemistry at the University of Tübingen. Under Fittig's guidance, Ramsay did research on nitrotoluic acids and in August 1872 received the Ph.D. at the age of nineteen.

Returning to Glasgow, Ramsay was appointed assistant under Georg Bischof at Anderson's College. In 1874 he became tutorial assistant to Ferguson at the university and published his first independent scientific paper. At that time he considered himself "a promising youth who will be most persistent and stick . . . to his work."[1]

In 1880 Ramsay was appointed professor of chemistry at the provincial University College, Bristol. The principalship of the college was entrusted to him in 1881, and he retained this post until 1887. In August 1881 he married Margaret Buchanan. They had two children.

In 1887 Ramsay succeeded Alexander Williamson in the chair of chemistry at University College, London. He set up a private laboratory at the college and worked there for twenty-five years until his retirement in 1912. At London, although no longer an educational administrator, Ramsay continued to struggle for educational excellence and for the independence of provincial institutions begun at Bristol.[2] In 1892 he led a successful attempt to improve the significance of both teaching and research at University College; he sought to make research an integral part of the basis on which degrees were awarded by the University of London.

Ramsay, the eternal optimist, whose motto was said to be "be kind," was above all a highly cultured gentleman. He was well traveled, fluent in several languages, and able to converse with colleagues, students, and acquaintances alike. He was admired and beloved by almost all who knew him; and his boyish vigor and simple charm, unaffected by the many honors showered upon him, remained with him throughout his life. His active inquiring mind, enthusiasm, and total involvement never allowed him to postpone anything. He was a good and patient teacher and was extremely interested in educational matters, particularly those concerning chemistry. He continually improved his own skills with laboratory techniques and apparatus and was always willing to acknowledge his mistakes. His tendency to persist in a given line of research even when no results were forthcoming was offset by a continuous line of faithful collaborators. Ramsay was "a great general who wanted an able chief-of-staff. All his best work was done with a colleague."[3] Not a great theorist, his best work was also inevitably along experimental lines.

The Royal Society elected Ramsay a fellow in 1888

and awarded him the Davy Medal in 1895. In addition to the Nobel Prize he received the Hodgkins Prize from the Smithsonian Institution (1895), the Longstaff Medal from the London Chemical Society (1897), and the Hofmann Medal from the German Chemical Society (1903), and many honorary degrees. In 1902 Ramsay was made a K.C.B.

His career in physical chemistry can be divided into four periods: the Glasgow period (1874–1880), during which he dealt with matters mostly pertaining to organic chemistry; the Bristol–London period (1880–1894), during which he dealt with the critical states of liquids and vapors; the early London period (1894–1900), during which he concentrated on the inert gases; and the final London period (1901–1916), during which he became increasingly interested in radioactivity.

At Glasgow, Ramsay inherited from Ferguson's predecessor, Thomas Anderson, preparations of pyridine bases of about 1854 vintage. From them Ramsay produced, by oxidation, a variety of pyridinic acids. He also succeeded in synthesizing pyridine itself. Collaborating with his senior student, James Dobbie, Ramsay studied the relationship between the acids formed from the oxidation of the alkaloids of both quinine and cinchonine and the acids he had earlier produced from the pyridine bases. These observations showed the important connection between pyridine and its derivative alkaloids. Ramsay was one of the first scientists to offer a plausible explanation for Brownian movement.[4]

At Bristol, Ramsay investigated critical states. This topic arose at Glasgow from a controversy with James Hannay; and with the appointment of Sydney Young as Ramsay's assistant,[5] it became his main line of research. From 1882 the "firm of Ramsay and Young" published[6] more than thirty papers concerning research on vapor pressure and critical states of liquids. "The question whether Kopp's quantitative laws hold at all pressures . . . remained in an unsatisfactory state until Ramsay and Young published their exhaustive researches on the vapour pressures of liquids."[7] Their series of investigations on evaporation and dissociation was continued even after Ramsay's 1887 appointment in London. During this important period of collaboration, Ramsay improved his laboratory techniques, learned the art of glass blowing, and laid the foundation for his later experimental research on vapors and gases.

While in London, Ramsay continued, until 1894, his researches into critical states with John Shields and others. He determined the molecular weights of associated (or aggregated) liquids and verified Eötvös's determination (1886) of a linear relationship between surface tension and temperature. With Shields, he developed an experimental method for determining the molecular weight of a substance in the liquid state. Ramsay also investigated the molecular complexity of liquids and distinguished between molecular associating liquids and nonassociating liquids. He became interested in "pseudo-solutions" and assisted his students Harold Picton and S. Ernest Linder.

Ramsay was a close friend of Wilhelm Ostwald and became a strong advocate of the ionic theory of Arrhenius and its relation to colligative properties (for example, osmotic pressure) of solutions.

In the period 1880–1894 Ramsay had already investigated various aspects of gas analysis. After the announcement from Lord Rayleigh (7 September 1892) reconsidering the discrepancy between atmospheric and chemical nitrogen (first noted by Cavendish in 1785), Ramsay had speculated upon its cause. He followed with particular interest Rayleigh's presentation to the Royal Society on 19 April 1894. Concerning this confirmation of the discrepancy, Ramsay recalled, in a letter of August 1894, that Rayleigh gave "numbers about which there could be no reasonable doubt. I asked him then if he minded my trying to solve the mystery. He thought that the cause of the discrepancy was a light gas in non-atmospheric nitrogen; I thought that the cause was a heavy gas in atmospheric nitrogen. He spent the summer in looking for the light gas; I spent July in hunting for the heavy one. And I have succeeded in isolating it."[8]

Ramsay had removed the oxygen from atmospheric air by sparking and had removed the nitrogen by combining it with heated magnesium to form magnesium nitride. On 7 August 1894 he wrote that the residue was definitely "a gas of density between 19 and 20, which is not absorbed by magnesium . . . and [which] appears to be a most astonishingly indifferent body. . . . Further experiments will show whether it displays such inertness towards all other elements. . . ."[9] Ramsay, having written Rayleigh on 4 August about his successful isolation, received on 7 August a reply from him that he had also been partially successful. By mutual agreement, Rayleigh (who had first established the discrepancy) and Ramsay (who had isolated its cause and partially determined its nature) joined in a preliminary announcement of their results on 13 August to the British Association at Oxford. During a further period of combined research, they named the gas argon, determined by the speed of sound in the gas that it is monatomic, and found that it has an atomic weight of about 40. With supporting evidence in January from Crookes and Olszewski concerning the unique-

ness of its spectrum and its well-defined critical points, Ramsay and Rayleigh formally presented their results on argon in a joint paper to the Royal Society on 31 January 1895.

On 1 February 1895 H. A. Miers wrote to Ramsay suggesting that the 1890 "nitrogen"[10] of William Hillebrand might also be argon. Ramsay then prepared a quantity of this gas by boiling cleveite in weak sulfuric acid and gave it to Crookes for spectrum analysis. In 1868 Janssen made solar spectroscopic observations that yielded a new spectral line, which suggested to Lockyer the presence of an unknown element in the sun. Crookes's spectroscopic analysis of the gas prepared by Ramsay confirmed the presence of this same element. On 27 March 1895 Ramsay announced the existence of terrestrial helium; and this was independently confirmed in Cleve's laboratory at Uppsala and reported 8 April 1895.[11] In August, Kayser announced the presence of atmospheric helium as well.

From April 1895 Ramsay was joined by Travers, and together they confirmed the inert nature of helium and identified its physical characteristics. Their discovery that helium arises from thorium as well as uranium minerals was a fact requiring developments in radioactivity (1902) for an explanation. With the confirmation in 1898 by Edward Baly of the presence of atmospheric helium in unexpectedly large quantities with respect to the less volatile gases, the existence of helium in thorium and uranium minerals was considered an explanation for the replenishing of atmospheric helium, but the connection with radioactivity had not yet been drawn.

In 1897, Ramsay, as President of the Chemistry Section of the British Association meeting in Toronto, delivered an address with the title "An Undiscovered Gas." He showed that in accordance with the Periodic Law, using a method of analysis[12] which avoided a difficulty arising from the fact that the position of argon in the Periodic Table seemed to be abnormal, there was a very high probability that there must exist a gas having properties intermediate between those of helium and argon.[13]

Olszewski's work on the liquefaction of argon had indicated that its residue might indeed contain another constituent. Thus, after improving their techniques of gas manipulation (1895–1897), Ramsay and Travers began to search for a third inert gas. In May 1898, shortly before receiving fifteen liters of liquid argon residue, they tested the procedure with a liter of liquid air. They collected the least volatile gaseous fraction as the liquid air evaporated and, after removing the oxygen and nitrogen, spectroscopically examined the inert residue. On 31 May 1898 they observed the yellow and green lines characteristic of krypton and on 6 June first announced their results. As soon as the liquid argon residue arrived, they separated (11 June) the most volatile gaseous fraction. After further preparation, they spectroscopically examined the residue and on 12 June 1898 observed the crimson presence of neon, which they announced on 16 June. In their excitement they also announced their discovery of "metargon," but this proved to be a mixture of impurities in the gas.

After confirming the presence of atmospheric krypton and neon, Ramsay and Travers continued to develop techniques to obtain quantities of these gases sufficient for the research on their chemical and physical properties. The krypton residue collected from preliminary attempts to obtain such quantities was found (July 1898) to contain about twenty percent of an even less volatile constituent, which added several new blue lines to the krypton residue but which totaled only 6×10^{-9} percent of the atmosphere. They announced this gas, xenon, on 8 September 1898.

By April 1899 Ramsay and Travers had secured their own liquid-air apparatus and were thus able to obtain enough krypton and xenon for physical analysis. Neon, however, being more volatile, resisted liquefaction by means of liquid-air techniques. Travers therefore rigged up a liquid hydrogen apparatus based on Dewar's demonstration in 1898 of the feasibility of obtaining adequate quantities of liquid hydrogen. They were thus able to solidify neon; and by July 1900 they had determined its properties.

Having identified, isolated, and determined the properties of five of the inert gases, Ramsay traveled to India and was engaged to advise on the organization of an Institute of Science to train Indian graduates.

By late 1902 Ramsay had become interested in the new gas, "emanation," that Rutherford had linked (1899) to thorium and F. E. Dorn had linked (1900) to radium. Rutherford was joined in 1901 by Soddy, who established the inert character of "emanation" by passing it, unchanged, over Ramsay's most extreme reagents (including red-hot magnesium powder and red-hot palladium black). By 1902 they had liquefied "emanation" and had obtained preliminary results on its well-defined points of volatilization and condensation. At Ramsay's invitation, Soddy joined Ramsay in March 1903 to initiate research on radioactivity at the latter's laboratory. Ramsay was particularly interested in determining whether radioactivity was a general characteristic of inert gases. Comparing the behavior of "emanation" with the other known inert gases, Soddy convinced Ramsay by May 1903 that it was unique in this respect.

In June 1903 Soddy obtained twenty milligrams of radium bromide, which had been produced commercially by Giesel. The emanation from a solution preparation of radium bromide was collected in a spectrum tube, and after several days they observed the spectrum of helium. In 1902 Rutherford and Soddy, on the basis of their disintegration theory, suggested that the well-known excess of atmospheric and terrestrial helium might be connected with radioactivity. The experimental proof by Ramsay and Soddy that helium is produced directly from radium emanation was a strong confirmation of their disintegration theory.

Using the emanation from about fifty[14] milligrams of radium bromide, Ramsay and Soddy attempted in 1903–1904 to map the spectrum of radium emanation; but the quantity proved to be too small to maintain the necessary spark discharge. During this same period Ramsay and J. N. Collie, using more than one hundred milligrams of radium bromide, were able to obtain sufficient emanation for a preliminary spectrum determination. By 1908 Ramsay and A. T. Cameron had experimentally confirmed the spectrum of radium emanation. In later experiments Ramsay and Whytlaw-Gray determined (1910) the density of this gas. In 1918 C. Schmidt gave the name radon to the emanation from radium to distinguish it from the emanations from thorium (thoron) and actinium (actinon). In 1904 Ramsay had suggested "exradio," "exthorio," and "exactino" for these emanations; he later suggested the term "niton" for the emanation from radium. They found that the quantity of radon in equilibrium from one gram of radium is 0.6 cubic millimeters, about one-half the value estimated in 1904 by Ramsay and Soddy. Weighing a small fraction of this quantity with an extremely delicate microbalance designed for the purpose, Ramsay and Gray determined the density of radon—assuming its monatomicity, they determined the atomic weight to be 223; that is, about four units lighter than radium. This determination further confirmed the theory that the transition from radium to radon involves the expulsion of helium.

With Richard Moore, Ramsay investigated (1907–1908) the residue from twenty tons of liquid air for possible nonradioactive inert gases lighter than helium or heavier than xenon; but they achieved only negative results. Stimulated by "the probable presence of nebulium in the nebulae and coronium in the sun,"[15] they repeated this investigation using the residue from more than one hundred tons of liquid air. Although these investigations were also negative, they did yield about 250 cubic centimeters of xenon and krypton and permitted a redetermination of their densities.

From 1903 Ramsay became increasingly interested in radioactivity. By 1905 he was convinced of the severable nature of the atoms of elements. Interpreting atomic disintegration as just one type of dissociation, involving highly endothermic substances, he considered it possible to dissociate and synthesize elements. Hahn's discovery of radiothorium in 1905 gave additional authority to the results on radioactivity coming from Ramsay's laboratory. Ramsay interpreted James Spencer's results (1906) concerning the dissociation of metal surfaces under ultraviolet radiation in the same way he had interpreted radioactive disintegration, except that the nonradioactive case required a slight input of energy. From 1907 Ramsay became involved in attempts to transmute elements, using radon as the external source of energy. Working with Cameron, by 1908 he had dissociated carbon monoxide into its components, had decomposed a variety of materials, and had allegedly obtained lithium from copper. But the latter result, which seemed a case of true transmutation, was not confirmed.

Ramsay adopted the view that the atom consisted of a vast number of electrons, and he considered some electrons more constitutive than others. "Ramsay's theory was that when an alpha particle struck a non-radio-active atom a glancing blow near the surface the atom was ionized; if it struck the atom squarely in the centre, the latter was broken up with the formation of new elements."[16] In 1912 Ramsay thought he had evidence to prove that neon was a product of radioactive change. (Inert helium had been so suggested by 1903.) In 1913 Rutherford's concept of the nuclear atom led Ramsay to refine his own atomic model. Rutherford also bombarded various substances with alpha particles and by 1919, using the scintillation method instead of chemical techniques, had detected the release of hydrogen from disintegrating nitrogen. This discovery was the beginning of the important and fruitful line of investigation by Rutherford concerning nuclear disintegration.[17]

Ramsay officially retired in 1912 but remained at his London post until succeeded by Donnan in March 1913. During World War I Ramsay continued to conduct scientific research at his home in Hazlemere but became deeply involved in the violent anti-German feeling of the time; in spite of this, he was mourned internationally at his death at the height of the conflict.

NOTES

1. Travers, *Life of Ramsay*, 30, 265.
2. His efforts were rewarded in 1899 when both Bristol and Owens colleges received university status; Bristol College became a full university in 1909 (Travers, *op. cit.*, 86).
3. Travers to Frederick Soddy, 10 October 1949, referring to the former's *Life of Ramsay*, esp. p. 61. This letter exists in the Bodleian Library, Oxford, Soddy-Howorth Collection, folder 43.
4. F. Cajori, *A History of Physics* (New York, 1962), 352; M. J. Nye, *Molecular Reality: A Perspective on the Scientific Work of Jean Perrin* (London, 1972), 26–27.
5. In 1887 Young succeeded Ramsay as professor of chemistry at Bristol.
6. Ostwald referred to their collaboration in this manner, inquiring of Ramsay if it would continue beyond 1887; cf. Travers, "Ramsay and University College," p. 7. Young continued independently this same line of research; their only extensive joint publication after 1888–1889 was on the properties of water and steam in 1891–1892.
7. S. Smiles, *The Relations Between Chemical Constitution and Some Physical Properties* (London, 1910), 234.
8. Ramsay to his aunt, the wife of the geologist Andrew Ramsay, August 1894. Part of the original letter is reproduced in Travers, *Life of Ramsay*, 103–104. For a well-balanced historical account of the discovery of argon, see Hiebert, "The Discovery of Argon," in H. H. Hyman.
9. Ramsay, "On a New Gas Contained in Air," unpublished MS (7 August 1894). The original exists at University College London, Library, Ramsay Papers, vol. 7/1, pp. 87–108. Travers, *Life*, 118 and 119, has the first and last pages reproduced. The quotation is from the last two pages of the manuscript, pp. 20–21.
10. Hillebrand, "On the Occurrence of Nitrogen in Uraninite and on the Composition of Uraninite in General," in *Bulletin of the United States Geological Survey*, **78** (1890), 43–78.
11. P. F. Cleve, "Sur la présence de l'hélium dans la clèvèite," in *Comptes rendus hebdomadaires des séances de l'Académie des sciences*, **120** (1895), 834.
12. This method was not an uncommon approach. By a similar analysis and comparison of the groups within the periodic table, Stoney had theoretically predicted in 1888 an unoccupied series of positions in which the entire family of inert gases was later placed; cf. Lord Rayleigh (J. W. Strutt), "On Dr Johnstone Stoney's Logarithmic Law of Atomic Weights," in *Proceedings of the Royal Society*, **85** (1911), 471–473.
13. Travers, "Ramsay and University College," p. 18.
14. The additional thirty milligrams had been lent to Soddy by Rutherford.
15. R. B. Moore, "Ramsay" (1918), 39.
16. *Ibid.*, 42. Moore was present during the beginning of Ramsay's transmutation investigations. This statement is unconditioned by Rutherford's successful results in 1919.
17. Cf. T. J. Trenn, "The Justification of Transmutation: Speculations of Ramsay and Experiments of Rutherford," in *Ambix*, **21** (1974), 53–77.

BIBLIOGRAPHY

I. Original Works. No complete bibliography of Ramsay's nearly 300 papers exists. Tilden included a brief list of the Ramsay-Young papers in his *Ramsay Memorials*, 99–101. More than half of his papers were listed by Richard B. Moore in his article "Sir William Ramsay," in *Journal of the Franklin Institute*, **186** (1918), 29–55.

The following list is intended only to supplement and correct Moore. In cases where only the abstract was listed by Moore, this supplement lists the main paper.

Obvious errors, including names of collaborators, titles of journals, and the inclusion of several articles by another William Ramsay, are not even noticed or included in this supplement. This supplement is furthermore also selective and does not normally cite multiple publication of articles. The lists by Tilden and Moore together with this supplement cover about 80 percent of his scientific contributions. Most of Ramsay's papers appeared in *Berichte der Deutschen Chemischen Gesellschaft, Chemical News and Journal of Physical (Industrial) Science*, and *Zeitschrift für physikalische Chemie*, but few of these were original publications.

Ramsay's supplementary papers are *Investigations on the Toluic and Nitrotoluic Acids* (Tübingen, 1872), his inaugural diss.; "On the Influence of Various Substances in Accelerating the Precipitation of Clay Suspended in Water," in *Quarterly Journal of the Geological Society of London*, **32** (1876), 129–132; "On Smell," in *Proceedings of the Bristol Naturalists' Society*, **3** (1882), 280–293; "On Brownian or Pedetic Motion," *ibid.*, **3** (1882), 299–302; "Some Thermodynamical Relations," in *Proceedings of the Physical Society of London*, **7** (1885), 289–306, 307–326, 327–334; **8** (1886), 56–61, 61–65; "The Estimation of Free Oxygen in Water," in *Journal of the Chemical Society*, **49** (1886), 751–761, written with K. I. Williams; "Note as to the Existence of an Allotropic Modification of Nitrogen," in *Proceedings of the Chemical Society*, **2** (1886), 223–225, written with K. I. Williams; and "Researches on Evaporation and Dissociation," in *Proceedings of the Bristol Naturalists' Society*, **5** (1888), 298–328.

See also "On the Destructive Distillation of Wood," in *Journal of the Society of Chemical Industry*, **11** (1892), 395–403, 872–874, written with John Chorley; "The Molecular Surface-Energy of Mixtures of Non-associating Liquids," in *Proceedings of the Royal Society*, **56** (1894), 182–191, written with Emily Aston; "Argon, a New Constituent of the Atmosphere," *ibid.*, **57** (1895), 265–287, written with Strutt; "L'argon," in *Revue scientifique*, 4th ser., **4** (1895), 545–547; "On the Discovery of Helium in Cleveite," in *Journal of the Chemical Society*, **67** (1895), 1107–1108; "On the Occlusion of Oxygen and Hydrogen by Platinum Black," in *Philosophical Transactions of the Royal Society*, **186** (1895), 657–693; **190** (1897), 129–153, written with L. Mond and John Shields; "On the Determination of High Temperatures With the Meldometer," in *Proceedings of the Physical Society of London*, **14** (1896), 105–113, written with N. Eumorfopoulos; "An Attempt to Determine the Adiabatic Relations of Ethylic Oxide," in *Philosophical Transactions of the Royal Society*, **189** (1897), 167–188, written with E. P. Perman and John Rose-Innes; and "Sur un nouvel élément constituant de l'air atmosphérique," in *Comptes rendus hebdomadaires des séances de l'Académie des sciences*, **126** (1898), 1610–1613, written with Travers, with trans. as "On a New Constituent of Atmospheric Air," in *Proceedings of the Royal Society*, **63** (1898), 405–408.

See also "On the Companions of Argon," *ibid.*, **63** (1898), 437–440, written with Travers; "On the Extraction From Air of the Companions of Argon and on Neon," in *Report of the British Association for the Advancement of Science* (1898), 828–830, written with Travers; "On the Occlusion of Hydrogen and Oxygen by Palladium," in *Philosophical Transactions of the Royal Society*, **191** (1898), 105–126, written with L. Mond and J. Shields; "On the Companions of Argon," *ibid.*, **197** (1901), 47–89, written with Travers; "The Vapour-Densities of Some Carbon Compounds: An Attempt to Determine Their Correct Molecular Weights," in *Proceedings of the Physical Society of London*, **18** (1902), 543–572, written with Bertram D. Steele; "Gases Occluded by Radium Bromide," in *Nature*, **68** (1903), 246, written with Soddy; "Further Experiments on the Production of Helium From Radium," in *Proceedings of the Royal Society*, **73** (1904), 346–358, written with Soddy; "Sur la dégradation des éléments," in *Journal de chimie physique et de physico-chimie biologique*, **5** (1907), 647–652; "Les gas inertes de l'atmosphère et leur dérivation de l'émanation des corps radioactifs," in *Archives des sciences physiques et naturelles*, **26** (1908), 237–262; and "Experiments With Kathode Rays," in *Nature*, **89** (1912), 502.

The following is a supplementary list of some of Ramsay's most important lectures and addresses. "Liquids and Gases" (8 May 1891), in *Proceedings of the Royal Institution of Great Britain*, **13** (1890–1892), 365–374, published 1893; "On Argon and Helium," Graham Lecture (8 January 1896), in *Proceedings. Philosophical Society of Glasgow*, **27** (1896), 92–116; "The Position of Argon and Helium Among the Elements," Fifth Boyle Lecture (2 June 1896), in *Transactions of the Oxford University Junior Scientific Club*, **2**, no. 41 (1896); "An Undiscovered Gas," presidential address—chemistry section, in *Rep. Brit. Ass.*, Toronto (1897), 593–601; "L'hélium," read to the Chemical Society of France, in *Annales de chimie et de physique*, 7th ser., **13** (1898), 433–480, with summary trans. in Tilden, *Ramsay Memorials*, 142–145; "The Newly Discovered Elements; and Their Relation to the Kinetic Theory of Gases," Wilde Lecture (28 March 1899), in *Memoirs and Proceedings of the Manchester Literary and Philosophical Society*, **43**, no. 4 (1899); "The Aurora Borealis," Watt Lecture (9 January 1903), in *Papers of the Greenock Philosophical Society* (1903); "Present Problems of Inorganic Chemistry," address to the International Congress of Arts and Sciences, St. Louis (Sept., 1904), in *Report of the Board of Regents of the Smithsonian Institution*, no. 1604 (1905), 207–220; "The Sequence of Events," Nobel Lecture (December 1904), published as *Les Prix Nobel en 1904* (Stockholm, 1906); and "The Electron as an Element," presidential address—London Chemical Society (26 Mar. 1908), in *Journal of the Chemical Society*, **93** (1908), 774–788.

See also *Die edlen und die radioaktiven Gase* (Leipzig, 1908); "Elements and Electrons," presidential address—London Chemical Society (25 Mar. 1909), in *Journal of the Chemical Society*, **95** (1909), 624–637; "Les mesures de quantités infinitésimales de matières" (20 April 1911), in *Journal de physique théorique et appliquée*, 7th ser., **1** (1911), 429–442; and "Le rôle de l'hélium dans la nature," read to the Chemical Society of Italy (2 June 1913), *Revue scientifique*, **51** (1913), 545–551.

Ramsay wrote and lectured extensively on educational matters and on general topics; these works include "Education in Science in Britain and in Germany," read 30 September 1896 (Bangor, 1896); "The Functions of a University," read 6 June 1901 (London, 1901), reprinted in *Essays Biographical and Chemical* (1908), 227–247; "Progress in Chemistry in the Nineteenth Century," in *Report of the Board of Regents of the Smithsonian Institution*, no. 1272 (1901), 233–257; "Les gaz de l'atmosphère," in *Revue générale des sciences pures et appliquées*, **13** (1902), 804–810; "The Inert Constituents of the Atmosphere," in *Popular Science Monthly*, **59** (1901), 581–595; "Über die Erziehung der Chemiker," in *Annalen der Naturphilosophie*, **4** (1905), 153–170; and "La Société Royale de Londres," in *Journal des savants*, **5** (1907), 61–70. Additional general works, including some of his biographical articles and obituary notices, are in his *Essays* (1908).

Ramsay's books are *Elementary Systematic Chemistry for the Use of Schools and Colleges* (London, 1891); *A System of Inorganic Chemistry* (London, 1891); *Kurzes Lehrbuch der Chemie: nach den neuesten Forschungen der Wissenschaft* (Anklam, 1893), prepared by G. C. Schmidt; and *Gases of the Atmosphere: The History of Their Discovery* (London, 1896; 2nd ed., 1900; 3rd ed., 1905; 4th ed., 1915). Each ed. was revised to include interim developments; a trans. of the third ed. by M. Huth was *Die Gase der Atmosphäre und die Geschichte ihrer Entdeckung* (Halle, 1907). *Modern Chemistry: Theoretical* (London, 1900) and *Modern Chemistry: Systematic* (London, 1900) were translated by M. Huth as *Moderne Chemie, Teil I: Theoretische Chemie* (Halle, 1905; 2nd ed., 1908) and *Moderne Chemie, Teil II: Systematische Chemie* (Halle, 1906; 2nd ed., 1914). This work was also translated into Russian by L. A. Tchougaeff (Moscow, 1909).

See also *Introduction to the Study of Physical Chemistry* (London, 1904); *Essays Biographical and Chemical* (London, 1908), with several of Ramsay's lectures and general essays; a trans. by Wilhelm Ostwald appeared as *Vergangenes und Künftiges aus der Chemie: Biographische und Chemische Essays* (Leipzig, 1909; 2nd ed., 1913), with an introductory thirty-five-page autobiography by Ramsay. Subsequent writings are *Elements and Electrons* (London, 1912); *Die Edelgase*, written with George Rudorf, which is vol. II in W. Ostwald's *Handbuch der allgemeinen Chemie* (1918); and *Life and Letters of Joseph Black, M.D.* (London, 1918), with an eleven-page intro. by Donnan dealing with the life and work of Ramsay.

Some of Ramsay's correspondence is included by Tilden in *Ramsay Memorials* and by Travers in *Life of Ramsay*. Other correspondence was published in R. J. Strutt, *Life of John William Strutt, Third Baron Rayleigh* (London, 1924), revised (Madison, 1968); and in E. E. Fournier D'Albe, *The Life of Sir William Crookes* (London, 1923).

Over seventy letters from Ramsay to Rayleigh, mostly from 1894 to 1898, are listed in John N. Howard, ed.,

"The Rayleigh Archives Dedication," special rep., 63 (Cambridge, Mass., 1967), B8-B11. Correspondence between Ramsay and Rutherford, mostly dated about November 1907, exists at Cambridge University Library Add. MSS 7653/R2–R15. By far the largest collection of Ramsay materials is preserved at the Library of University College London, largely through the efforts of M. W. Travers and the kindness of the Ramsay family. This collection consists of sixteen bound volumes of correspondence, mostly to Ramsay, his laboratory notebooks, lecture notes, off prints of many of his publications, and many other items. A twenty-three-page handlist entitled "Ramsay Papers" was prepared by the Library in 1969. This library also has correspondence from Ramsay in the Oliver J. Lodge Collection. A two-page autobiographical letter dated 15 April 1886 exists in the Krause *Album* held in the Sondersammlungen, Bibliothek, Deutsches Museum, Munich.

II. SECONDARY LITERATURE. The best biography is Morris W. Travers, *The Discovery of the Rare Gases* (London, 1928), expanded and revised, with a detailed biographical account, as *A Life of Sir William Ramsay, K.C.B., F.R.S.* (London, 1956). See also H. Pettersson, "Minnen av Sir William Ramsay," in *Götheborgske Spionen*, **4** (1940), 13–16. Frederick G. Donnan wrote the biographical account introducing Ramsay's *Life of Black* (1918) and the notice for the *Dictionary of National Biography 1912–1921* (London, 1927), 444–446.

See also Luigi Balbiano, "L'opera sperimentale di Guglielmo Ramsay," in *Atti dell'Accademia della scienze di Torino*, **52** (1916–1917), 29–38; J. Norman Collie's obituary notice "Sir William Ramsay, 1852–1916," in *Proceedings of the Royal Society*, **93A** (1917), xlii–liv; Eduard Farber, *Nobel Prize Winners in Chemistry 1901–1961* (London, 1963), 15–18; Philippe-A. Guye, "Sir William Ramsay," in *Journal de chimie physique*, **16** (1918), 377–387; W. Marckwald, "Sir William Ramsay," in *Zeitschrift für Elektrochemie*, **22** (1916), 325–327; Richard B. Moore, "Sir William Ramsay," in *Journal of the Franklin Institute*, **186** (1918), 29–55, with extensive bibliography; and Wilhelm Ostwald, "Scientific Worthies XXXVII: Sir William Ramsay, K.C.B., F.R.S.," in *Nature*, **88** (1912), 339–342, with portrait. Charles Moureu's notice in *Revue scientifique*, **10** (Oct. 1919), 609–618, was translated in *Annual Report of the Board of Regents of the Smithsonian Institution for 1919* (1921), 531–546; and reissued in Eduard Farber, ed., *Great Chemists* (New York, 1961), 997–1012. See also Edmond Perrier, "Eloge," in *Comptes rendus hebdomadaires des séances de l'Académie des sciences*, **163** (1916), 113–116.

Paul Sabatier, "Sir William Ramsay et son oeuvre," in *Revue scientifiques*, **54** (Oct. 1916), 609–616, gives a scholarly analysis of many of Ramsay's papers. See H. G. Söderbaum, "The Winner of the Nobel Prize in Chemistry for This Year," in *Svensk kemisk tidskrift*, **8** (1904), 183–187; and "Professor William Ramsay," in *Proceedings of the Bristol Naturalist's Society*, **8** (1895–1896), 1–5; Frederick Soddy's obituary "Sir William Ramsay, K.C.B., F.R.S.," in *Nature*, **97** (10 Aug.

1916), 482–484, with excerpts included in Travers, *Life of Ramsay*, 291–293; William A. Tilden, *Sir William Ramsay, K.C.B., F.R.S., Memorials of His Life and Work* (London, 1918), containing much otherwise unavailable documentation; *Famous Chemists: The Men and Their Work* (London, 1921), 273–287; and "Sir William Ramsay, K.C.B.," in *Journal of the Society of Chemical Industry*, **35** (31 Aug. 1916), 877–880, with repr. in *Journal of the Chemical Society*, **111** (1917), 369–376; P. Walden, "Lothar Meyer, Mendelejeff, Ramsay und das periodische System der Elemente," in Günther Bugge, *Das Buch der Grossen Chemiker*, 2nd ed., II (Weinheim-Bergstr., 1955), 229–287, with a biographical sketch of Ramsay on pp. 250–263; T. I. Williams, in *A Biographical Dictionary of Scientists* (London, 1969), 432–433; and A. M. Worthington, "Sir William Ramsay K.C.B., F.R.S.," in *Nature*, **97** (1916), 484–485. Additional works are listed in N. O. Ireland, *Index to Scientists of the World* (Boston, 1962); and M. Whitrow, ed., *Isis Cumulative Bibliography*, II (1971), 381.

Articles written for the 1952 centenary include E. Andrade, "William Ramsay: Great Discover and Leader of Chemical Research," in the *Times* (London), 2 Oct. 1952; Sir Harold Hartley, "Ramsay and the Inert Gases," in *Observer* (28 Sept. 1952); "The Ramsay Centenary," in *Notes and Records of the Royal Society*, **10** (1953), 71–80; J. R. Partington, "Sir William Ramsay Discoverer of Five Elements," in *Manchester Guardian Weekly* (2 Oct. 1952); "Sir William Ramsay 1852–1916," in *Nature*, **170** (1952), 554–555; M. W. Travers, "Sir William Ramsay (1852–1916)," in *Endeavour*, **11** (1952), 126–131, "The Scientific Work of Sir William Ramsay," in *Science Progress* (1952), 232–244; and *William Ramsay and University College London* (London, 1952), privately issued for the Ramsay centennial.

Probably the best account of Ramsay's work on argon is Erwin N. Hiebert, "Historical Remarks on the Discovery of Argon: The First Noble Gas," in H. H. Hyman, ed., *Noble-Gas Compounds* (Chicago, 1963), 3–20. David M. Knight, in R. Harré, ed., *Some Nineteenth-Century British Scientists* (Oxford, 1969), 232–259, discusses the determination of the monatomic character of argon and the analysis of the periodic law as a predictor of new elements. R. J. Havlik's article in Gerhard A. Cook, ed., *Argon, Helium and the Rare Gases*, I (New York, 1961), 17–34, and J. R. Partington, *A History of Chemistry*, IV (London, 1964), 915–918, give useful historical accounts of Ramsay's work. See also R. B. Moore, "Helium: Its History, Properties, and Commercial Development," in *Journal of the Franklin Institute*, **191** (1921), 145–197. Zdzisław Wojtaszek, "On the Scientific Contacts of Karlo Olszewski with William Ramsay," in *Actes du XIe Congrès International Histoire Sciences*, **4** (1965), 113–116, examines the significance of the liquefaction of gases. See also J. W. van Spronsen, *The Periodic System of Chemical Elements* (Amsterdam, 1969), *passim*.

Ramsay as seen by his contemporaries is partially revealed in Lawrence Badash, ed., *Rutherford and Boltwood: Letters on Radioactivity* (New Haven, 1969).

A nearly complete list of honors, awards, degrees, memberships, prizes, and other achievements is included in Tilden, *Ramsay Memorials*, pp. 307–308.

THADDEUS J. TRENN

RAMSDEN, JESSE (*b.* Halifax, England, October 1735; *d.* Brighton, England, 5 November 1800), *mathematics, optics, physics.*

Ramsden was acknowledged to be the most skillful and capable instrument maker of the eighteenth century. He developed techniques that permitted him to increase greatly the precision of astronomical, surveying, and navigational equipment.

He was the son of a Halifax innkeeper and the great-nephew of Abraham Sharp, mathematician, instrument maker, and assistant to the Astronomer Royal, John Flamsteed. Ramsden's education included three years at the Halifax free school and four years' study of mathematics under the Reverend Mr. Hall. He was apprenticed at sixteen to a clothworker in Halifax. In 1755, upon finishing this service, he moved to London, where he worked as a clerk in a cloth warehouse. In 1758 he apprenticed himself to the mathematical instrument maker Burton. Ramsden opened his own shop in the Haymarket in 1762, moving to larger quarters at 199 Piccadilly in 1775. His great skill brought him commissions from the foremost practitioners of the period, including J. Sisson, J. Adams, J. Dollond, and E. Nairne. In 1765 he married Dollond's youngest daughter and for her dowry received a share of Dollond's patent on achromatic lenses.

In 1774 Ramsden published *Description of a New Universal Equatorial Instrument*, which improved on the design of Short's portable telescope mounting. Ramsden's mounting was quickly accepted and served to enhance his growing reputation. A passion for precision was the motivating force behind Ramsden's development of the dividing engine, his greatest contribution to the technology of the era. His first machine, built around 1766, produced only moderate improvement in the accuracy he sought, but the machine constructed in 1775 reduced the error to less than one-half second of arc as compared to the three seconds of arc of the first machine. This achievement earned him a grant from the Commissioners of the Board of Longitude in 1777.

Ramsden's shop grew until the staff numbered about sixty workmen. At no time, however, did he permit the quality of his output to diminish. By 1789 he had produced some one thousand sextants, as well as theodolites, micrometers, balances, barometers, and the many philosophical instruments required by the physicists of the period. It appears that the demand far exceeded his capacity for production. In 1784 William Roy, who was conducting the trigonometric survey that linked England with the Continent, ordered a three-foot-diameter theodolite from Ramsden, who took three years to complete it.

Ramsden also supplied many of the observatories of Europe with new achromatic telescopes equipped with accurately divided circles. The altazimuth instrument he built for Piazzi for the observatory at Palermo was equipped with a five-foot-diameter vertical circle. Its graduations were read by means of the micrometer microscope that Ramsden had developed. This instrument was completed in 1789, only a year late, probably because Piazzi personally expedited the project.

Ramsden was elected fellow of the Royal Society on 12 January 1786, and in 1794 he was made a member of the Imperial Academy of St. Petersburg. The Royal Society awarded him the Copley Medal (1795) for "various inventions and improvements in philosophical instruments."

Many of Ramsden's instruments have been preserved and may be seen at the Smithsonian Institution, the Science Museum and the National Maritime Museum in London, the Museum of the History of Science in Florence, Teyler's Museum in Haarlem, the Conservatoire National des Arts et Métiers in Paris, and in many other museums. His portrait is in the Science Museum in London.

BIBLIOGRAPHY

I. ORIGINAL WORKS. Ramsden's works include *Description of a New Universal Equatorial Instrument* (London, 1774; 2nd ed., 1791); *Description of an Engine for Dividing Mathematical Instruments* (London, 1777); *Description of an Engine for Dividing Straight Lines on Mathematical Instruments* (London, 1779); "Description of Two New Micrometers," in *Philosophical Transactions of the Royal Society*, **69** (1779), 419–431; and "A New Construction of Eyeglasses for Such Telescopes as May Be Applied to Mathematical Instruments," *ibid.*, **73** (1782), 94–99.

II. SECONDARY LITERATURE. On Ramsden and his work, see Maurice Daumas, *Les instruments scientifiques au XVII et XVIII siècles* (Paris, 1953), 318; Nicholas Goodison, *English Barometers, 1680–1860* (New York, 1968), 202–204; Henry C. King, *The History of the Telescope* (London, 1955), 162–172, 230–231; W. E. Knowles Middleton, *The History of the Barometer* (Baltimore, 1968), 196–198, 452–453; J. A. Repsold, *Zur Geschichte der astronomischen Messwerkzeuge, 1450–1830* (Leipzig, 1908), 82–87; Leslie Stephen and Sidney Lee, eds., *Dictionary of National Biography*, XVI (London, 1917), 708–710; and E. G. R. Taylor, *The Mathematical Practitioners of Hanoverian England* (London, 1966), 57–59, 244–245.

RODERICK S. WEBSTER

RAMSEY, FRANK PLUMPTON (*b.* Cambridge, England, 22 February 1903; *d.* Cambridge, 19 January 1930), *mathematical logic.*

Ramsey, the elder son of A. S. Ramsey, president of Magdalene College, Cambridge, was educated at Winchester and Trinity colleges, Cambridge. After graduation in 1923 he was elected a fellow of King's College, Cambridge, where he spent the rest of his short life. His lectures on the foundations of mathematics impressed young students by their remarkable clarity and enthusiasm, and his untimely death deprived Cambridge of one of its most brilliant thinkers.

Whitehead and Russell, in their system of mathematical logic, *Principia Mathematica* (1910–1913), had avoided the antinomies (paradoxes) by creating both a theory of types, which dealt with the nature of propositional functions, and an axiom of reducibility. Ramsey accepted the Whitehead-Russell view of mathematics as part of logic but said of the axiom of reducibility that it is "certainly not self-evident and there is no reason to suppose it true." In his first paper (1926) he took up Wittgenstein's work on tautologies and truth functions, reinterpreting the concept of propositional functions and thus obviating the need for the axiom of reducibility. He also drew an important distinction between the logical antinomies, for example, that concerning the class of all classes not members of themselves, and those antinomies that cannot be stated in logical terms alone, for example, "I am lying." For the first class, he accepted Russell's solution; for the second, which are not formal but involve meaning, he applied his reinterpretation of the *Principia Mathematica.*

In a paper "On a Problem of Formal Logic" (1930) Ramsey discussed the *Entscheidungsproblem*—the search for a general method for determining the consistency of a logical formula—using some ingenious combinatorial theorems. To the Oxford meeting of the British Association in 1926, he described the development of mathematical logic subsequent to the publication of *Principia Mathematica*; this address was printed in the *Mathematical Gazette* (1926).

Two papers on the mathematical theory of economics appeared in the *Economic Journal*. In a biographical essay on Ramsey, J. M. Keynes described Ramsey's interest in economics, the importance of his two papers, and the way in which Cambridge economists made use of his critical powers to test their theories. An interesting, if not altogether convincing, study of the bases of probability theory was published posthumously in a collection of Ramsey's essays.

BIBLIOGRAPHY

I. Original Works. Ramsey's works include "The Foundations of Mathematics," in *Proceedings of the London Mathematical Society*, 2nd ser., **25** (1926), 338–384; "Mathematical Logic," in *Mathematical Gazette*, **13** (1926), 185–194; "A Contribution to the Theory of Taxation," in *Economic Journal* (Mar. 1927); "A Mathematical Theory of Saving," *ibid.* (Dec. 1928); and "On the Problem of Formal Logic," in *Proceedings of the London Mathematical Society*, 2nd ser., **30** (1930), 264–286.

A convenient source is the posthumous volume *Foundations of Mathematics and Other Essays*, R. B. Braithwaite, ed. (London, 1931); this work contains Ramsey's published papers, excluding the two on economics, and a number of items from his unpublished manuscripts.

II. Secondary Literature. An obituary notice by Braithwaite is in *Journal of the London Mathematical Society*, **6** (1931), 75–78. The essay on Ramsey by J. M. Keynes, in *Essays in Biography* (London, 1933), is valuable for Ramsey's work in economics and in philosophy.

T. A. A. Broadbent

RAMUS, PETER, also known as **Pierre de La Ramée** (*b.* Cuts, Vermandois, France, 1515; *d.* Paris, France, 26 August 1572), *logic and method, pedagogy, mathematics, astronomy, optics, mechanics.*

Born into a family that had lost its wealth but not its title of nobility with the sack of Liège in 1468, Ramus was the son of Jacques de La Ramée, a laborer, and Jeanne Charpentier. After a primary education at home, in 1527 he entered the University of Paris (Collège de Navarre), where he met his costs by working as a manservant. Apparently an outstanding student, he first drew widespread attention in 1536 with his defense of an M.A. thesis, "Quaecumque ab Aristotele dicta essent, commentitia esse," in which he attacked not only the accuracy but also the authenticity of traditional Aristotelian philosophy. The precise meaning of the thesis, of which there is no extant text, hinges on the term *commentitia*. Translated by some as "false," the word connotes, rather, something made up as opposed to factual. Ong[1] has analyzed the question closely and has argued for a meaning close to "badly organized, unmethodical."

Ramus' teaching career began at the Collège du Mans, from which he soon moved, together with Omer Talon and Bartholomew Alexandre, to the Collège de l'Ave Maria. Attracted by Johannes Sturm to the rhetorical logic and pedagogical ideas of Rudolf Agricola, Ramus undertook a program of critical reeducation that in 1543 culminated in a broad-scale attack on Aristotelian logic, *Aristotelicae animadversiones*, and plans for a new arts curriculum. A counterattack led by Antoine de Govéa soon succeeded in obtaining a royal edict forbidding Ramus to teach or write on philosophical topics. Consequently Ramus turned to rhetoric and mathematics, in part for their inherent importance but also as guises for his logical theories.

Ramus' fortunes began to improve in 1545 when, as a result of staff shortages caused by the plague, he was called to the Collège de Presles. Shortly thereafter he became principal of the college, a position he held, with some interruptions, until his death. Through the intercession of his patron, Charles Cardinal de Guise (later Cardinal de Lorraine), Ramus was released from the 1544 teaching ban upon the accession of Henry II in 1547. The release did not, however, still the controversy Ramus had aroused and was continuing to enflame through the popularity of his lectures at Presles. Moreover, the position of royal lecturer, to which Ramus was appointed in 1551, gave him even greater freedom to attack his scholastic opponents and to espouse his often radical ideas.

Beginning in 1562 Ramus' intellectual positions became increasingly fused with religious and political issues. A defense of the Roman church by the Cardinal de Lorraine at Poissy in 1561 had the unintended consequence of leading Ramus to embrace Calvinism, which he then pursued with his usual enthusiasm. In 1562 Ramus published a plan of reform for the University of Paris. This plan grew out of the work of a commission appointed by Henry II in 1557, to which Ramus had been recommended by a vote of the university faculty. Although the text appeared anonymously, internal evidence[2] makes clear Ramus' authorship but not whether the commission was defunct and, therefore, whether Ramus was acting largely on his own. He suggested a reduction of the teaching staff, the abolition of student fees, and the financing of the institution with income from monasteries and bishoprics. He also proposed a chair of mathematics, which he later endowed from his own estate; a year of physics in the arts curriculum; the teaching of civil law in the law faculty; chairs of botany, anatomy, and pharmacy, and a year of clinical practice in the medical faculty; and the study of the Old Testament in Hebrew and the New Testament in Greek in the theological faculty.[3] The plan hardly endeared him to some of his academic colleagues, who were quick to suggest a link between it and Ramus' religious persuasion. Hence, late in 1562, when Calvinists were ordered out of Paris, Ramus fled to Fontainebleau, where he found refuge for a time with the Queen Mother, Catherine de Medicis.

On his return to Paris under the Peace of Amboise in 1563, Ramus resolved to avoid controversy; but by 1565 he was leading opposition to the naming of Jacques Charpentier (no relation), a long-time adversary, to the royal chair of mathematics. Charpentier, who had by then succeeded Ramus as the Cardinal de Lorraine's protégé and who enjoyed Jesuit support, kept his chair; and Ramus, ever more threatened, in 1567 again fled Paris, taking refuge with the Prince de Condé.

Sensitive to the worsening political situation, in 1568 Ramus returned to Paris, where he found his library ransacked. He stayed just long enough to ask leave of the king to travel in Germany. From 1568 to 1570 he toured the Protestant centers of Switzerland and Germany, where he encountered an enthusiastic welcome strangely coupled with opposition to his permanent settlement in a teaching post because of his non-Aristotelian doctrines. Lured back to Paris in 1570 by promises of tolerance, Ramus soon found himself with titles and salaries, but banned from teaching. In the midst of a vast publication project, he was caught by the St. Bartholomew's Day Massacre and, despite explicit royal protection, was cruelly murdered, apparently by hired assassins.[4]

Ramus' general intellectual stance, from which his thoughts on the sciences derived, was the complex result of two distinct educations and of a life spent entirely within an academic setting. As Ong has emphasized,[5] Ramus was primarily a pedagogue, whose views on the content of philosophy were shaped by the exigencies of teaching in the arts faculty. Having received first a traditional scholastic education, with its emphasis on the Aristotelian corpus, he then immersed himself in the humanist teaching of Rudolf Agricola, who focused on Ciceronian rhetoric and dialectic and on the revival of the seven liberal arts of classical antiquity. The tensions brought about by Ramus' attempt to reconcile and combine these two traditions is best reflected in his attitude toward Aristotle. Like many "anti-Aristotelians" of his day, he aimed his criticism not so much at Aristotle himself, for whom he had genuine respect, but at contemporary Aristotelians. To concentrate solely on Aristotle's works was to ignore or to fail to appreciate a whole body of equally classical material that was often better adapted to the purposes of education. Aristotelians, Ramus argued, had lost sight of the proper goal of teaching and had become entangled in a sterile web of logical subtleties. In concentrating on forms of the syllogism, for example, scholastics forsook the main purpose of logic, to wit, the finding of arguments and their presentation in a manner designed to convince an audience.[6] By illustrating precisely this use of logic, the works of rhetoricians and dialecticians both before and after Aristotle (most notably, Cicero) provided a more effective means of teaching the subject.

Ramus' attitude reflected a basic epistemology quite close to Aristotle's, as Ramus himself realized. Reason was a natural faculty of man which, like all natural faculties, revealed itself in its actual exercise.[7] Just as general physical principles were the product of induction from particular phenomena of nature, so too the principles of logic should be derived from examples of its effective use by orators, rhetoricians, and dialecticians. Indeed, Ramus maintained, all teaching should be rooted in examples of the use of the subject, from which students could move more easily and naturally to the general precepts underlying that use. It is a mark of Ramus' continuing commitment to Aristotle that he sought the theoretical underpinnings of this method of teaching in the *Posterior Analytics*, and his attacks on Aristotle and his followers were generally based on supposed violations of the precepts contained in that text. Ramus borrowed from the *Posterior Analytics* his three "laws of method"—*kata pantos*, *kat' auto*, and *kath' holou prōton*—which required that all material taught should be in the form of propositions that are universally true, demonstrable within the strict confines of the subject, and as general as possible. Although trivial in content, the "laws" became a touchstone for Ramists.[8]

Thus "method" was for Ramus primarily a pedagogical concept; accordingly, his contributions to the sciences were essentially pedagogical and propagandistic in nature. In seeking a return to the curriculum of the seven liberal arts, he sought in particular to retrieve arithmetic, geometry, astronomy, and physics (the quadrivium[9]) from the neglect into which they had fallen. As taught (when they were taught at all) they suffered from a form of intellectual detachment that made them appear more abstruse, and hence less important, than they were. Ramus' solution to this problem was twofold: first, to make clear in a series of commentaries *(scholae)* where the teaching of the sciences had gone astray and, second, to reorganize the subjects according to his own method. The result was a series of textbooks which, together with his texts on grammar, rhetoric, and dialectic, circulated widely for the next hundred years.

Ramus' twofold approach emerges most clearly from his *Scholae mathematicae* (1569) and his texts on arithmetic (1555) and geometry (1569). In the first three books of the *Scholae*, which appeared separately in 1567 under the title *Prooemium mathematicum*, he sought first to defend mathematics against charges of its lack of utility and its obscurity. Surveying the history of Greek mathematics (largely on the basis of Proclus' summary), Ramus insisted on the practical origins of the subject and on the use to which the ancients had put it, both as a theoretical foundation for natural philosophy and as a practical tool in areas like astronomy and mechanics. A mere look at the contemporary scene, he argued, revealed the continuing utility of mathematics in commerce and industry; moreover, recent developments in astronomy and mechanics showed by contrast the sterility of a scholastic natural philosophy devoid of mathematics. The blame for the neglect of mathematics lay first with Plato for having shunned its practical application (a fault Archimedes shared for not having written about his engineering feats and mechanical inventions) and then with Euclid for having severed the precepts of geometry from their use and for having written the *Elements* in an obscure syllogistic form, ostensibly following Aristotle's precepts. The remaining books of the *Scholae* are devoted to analyzing in exhaustive detail the methodological faults of the *Elements*.

The cure for obscurity lay in a return to teaching mathematics on the basis of its application to practical problems. Arithmetic should deal with computational problems occurring in the market place and in the law courts; geometry should be concerned with measurement of distances, areas, volumes, and angles, and with the types of mechanical problems to which Aristotle had applied the properties of the circle in his treatise on mechanics; the theory of proportion should be rooted in pricing and exchange problems and in applications of the law of the lever. Ramus' textbooks on arithmetic and geometry sought to effect this cure by rearranging the content of traditional arithmetical texts and of Euclid's *Elements* (together with scraps from Archimedes, Apollonius, and Pappus) in terms of the bodies of related problems that the theorems helped to solve. Apparently Ramus was perplexed about the proper role of algebra, and a text attributed to him was published only some years after his death. At one point in the *Scholae mathematicae*, however, he did suggest a link between algebra and Greek geometrical analysis, a notion that was picked up and developed by Viète and Descartes.[10]

The same separation of theory and practice led Ramus to discard completely Aristotle's *Physics* as a suitable text for natural philosophy. In terms that Bacon would later echo, Ramus argued that the *Physics* dealt not with natural phenomena but with logical analysis addressed to concepts rooted in the mind alone. Far more revealing of Aristotle's philosophy of nature were his *Mechanical Problems*, his *Meteorologica*, and his biological texts. Beyond Aristotle, Hippocrates, Plato, Theophrastus, Virgil, Pliny, Witelo, Copernicus, and Georgius Agricola all belonged in the physics curriculum; in particular, despite Aristotle's strictures, astronomy, optics, and mechanics formed an integral part of physics, even if it was more convenient to teach them separately or as subtopics of geometry. Ramus' broad view of this subject remained largely programmatic. His *Scholae physicae* appeared in 1565; but he never did write a textbook, and his lectures suggest that he lacked the technical command necessary to do so.[11] As presented to his students, Ramus' physics consisted primarily of agricultural maxims and natural history culled from Virgil and Pliny.

Ramus turned to astronomy late in his career, and apparently the subject perplexed him. Filled with admiration for this most obviously useful and practical application of mathematics, he nonetheless felt that both Ptolemy and Copernicus had succumbed to the lure of Aristotelian metaphysics in their reliance on such "hypotheses" as the principle of uniform motion on circles. In a letter written to Rheticus in 1563[12] Ramus urged a return to the observational astronomy of the Babylonians and Egyptians in an attempt to determine the nonhypothetical, directly observable regularities of the heavens and to build astronomy on them. It is unclear from his letter and from other statements whether Ramus would have accepted as "nonhypothetical" a system based on sun-centered measurements (that is, the Copernican system), although Kepler did later claim to have met Ramus' demands.[13]

Although the problem of Ramus' influence, especially in the sciences, still requires much study, it is clear that he and his works enjoyed widespread popularity both during his lifetime and in the century following his death. If that popularity was concentrated in the Protestant areas of the Rhineland, the Low Countries, England, and New England, it also filtered back to France, particularly after the accession of Henry IV. The Latin and French editions of Ramus' *Dialectics* went through a hundred printings in as many years, and his other texts seem to have been only slightly less well known. For example, through Rudolph Snellius and his son Willebrord, Ramus' mathematical works became part of the Dutch curriculum by the early 1600's, and Ramist texts in mathematics and physics spread rapidly.[14]

In particular, however, Ramus and Ramism became almost synonymous with the term "method," and all writers who dealt with the subject in the early seventeenth century, including Bacon and Descartes, felt it necessary to come to terms with Ramus' ideas. Indeed, as Ong[15] points out, the lack of reference to Ramus in the seventeenth century often means not that he had been forgotten but, rather, that the content of his thought was so well known as to obviate the need of naming the source. By emphasizing the central importance of mathematics and by insisting on the application of scientific theory to practical problem-solving, Ramus helped to formulate the quest for operational knowledge of nature that marks the Scientific Revolution.

NOTES

1. Ong, *Ramus, Method, and the Decay of Dialogue*, 45–47.
2. Cf. Waddington, *Ramus*, 141.
3. *Ibid.*, 144 ff.
4. Waddington, in *Ramus*, ch. 10, lays the blame squarely on Charpentier; but Ong (*Ramus*, 29) feels the evidence is insufficient.
5. Ong, *Ramus*, passim but esp. ch. VII, emphasizes as a main theme the continuity of pedagogical concerns within the scholastic tradition and sees many of Ramus' ideas as new solutions to old problems.
6. Here Ramus contributed decisively to a Renaissance concept that largely erased Aristotle's careful distinction between scientific logic and rhetorical dialectic. For a careful analysis of the concept, see Ong, *Ramus*, ch. IV, esp. 59–63.
7. Cf. Hooykaas, *Humanisme, science et réforme*, ch. 5.
8. Cf. Ong, *Ramus*, 258–262.
9. The traditional quadrivium made music the fourth subject, but Ramus believed music, like astronomy and optics, belonged to the wider subject of physics.
10. Cf. M. S. Mahoney, "Die Anfänge der algebraischen Denkweise im 17. Jahrhundert," in *Rete*, **1** (1971), 15–30.
11. Apparently Ramus relied heavily on the work of his students, notably Henri de Monantheuil and Risner.
12. First published in the preface to *Professio regia* (1576).
13. Cf. Hooykaas, *op cit.*, ch. 9.
14. Viète clearly knew Ramus' works, and Descartes almost certainly learned of them through Beeckman, who had studied with Rudolph Snellius.
15. Ong, *Ramus*, 9.

BIBLIOGRAPHY

I. ORIGINAL WORKS. Ramus published extensively. Waddington (see below) provides an initial survey, which has been extensively supplemented by Walter J. Ong, *Ramus and Talon Inventory. A Short-Title Inventory of the Published Works of Peter Ramus (1515–1572) and of Omer Talon (ca. 1510–1562) in Their Original and in Their Variously Altered Forms* (Cambridge, Mass., 1958). There is no modern edition of Ramus' works, although recently some have been reprinted photostatically from the originals. Ramus' most important writings include *Dialecticae partitiones sive institutiones* (Paris, 1543), later replaced by *Dialectique de Pierre de la Ramée* (Paris, 1555) and *Dialecticae libri duo, Audomari Talaei praelectionibus illustrati* (Paris, 1556); *Aristotelicae animadversiones* (Paris, 1543); *Oratio de studiis philosophiae et eloquentiae conjungendis, Lutetiae habita anno 1546* (Paris, 1547); *Arithmeticae libri duo* (Paris, 1555); *Grammaticae libri quatuor* (Paris, 1559); *Scholae grammaticae* (Paris, 1559); and *Prooemium reformandae Parisiensis Academiae, ad regem* (Paris, 1562).

Subsequent writings are *Scholarum physicarum libri octo, in totidem acroamaticos libros Aristotelis* (Paris, 1565); *Scholarum metaphysicarum libri quatuordecim, in totidem metaphysicos libros Aristotelis* (Paris, 1566); *Actiones duae habitae in senatu, pro regia mathematicae professionis cathedra* (Paris, 1566); *Prooemium mathematicum* (Paris, 1567), which is bks. I–III of *Scholae mathematicae; Geometriae libri septem et viginti* (Basel, 1569); and *Scholarum mathematicarum libri unus et triginta* (Basel, 1569).

Three important writings appeared posthumously: *Testamentum* (Paris, 1576), which endowed a chair of mathematics at the Collège Royal; *Professio regia. Hoc est, Septem artes liberales, in Regia cathedra, per [Ramum] Parisiis apodictico docendi genere propositae . . .* (Basel, 1576); and *Collectaneae Praefationes, Epistolae, Orationes* (Paris, 1577). A work on optics is also attributed to Ramus, both by its title and by references in his letters, although his precise role in it is not clear: *Opticae libri quatuor ex voto Petri Rami novissimo per Fridericum Risnerum ejusdem in mathematicis adjutorem olim conscripti* (Cassel, 1606). Similarly, Lazarus Schoner published in Frankfurt in 1586 an *Algebrae libri duo*, which he attributed to Ramus. Ramus also appears to have had some hand in Henri de Monantheuil's edition of Aristotle's *Mechanical Problems* (Paris, 1557).

II. SECONDARY LITERATURE. Two major nineteenth-century studies, Charles Desmazes' *P. Ramus: Sa vie, ses écrits, sa mort (1515–1572)* (Paris, 1864) and Charles Waddington's *Ramus (Pierre de la Ramée), sa vie, ses écrits et ses opinions* (Paris, 1855), have been updated, but not entirely superseded, by Walter J. Ong's *Ramus, Method, and the Decay of Dialogue: From the Art of Discourse to the Art of Reason* (Cambridge, Mass., 1958), which contains the best scholarly account of Ramus' theories of logic and method.

For Ramus' influence in England, see Wilbur S. Howell, *Logic and Rhetoric in England, 1500–1700* (Princeton, 1956); for his influence in New England, see Perry Miller, *The New England Mind: The Seventeenth Century* (New York, 1939); for his influence in the Low Countries, see Paul Dibon, "L'influence de Ramus aux universités néerlandaises du XVII siècle," *Actes du XIᵉ Congrès international de philosophie*, XIV (Louvain, 1953), 307–311.

For a general survey of Ramus' scientific thought, see R. Hooykaas, *Humanisme, science et réforme. Pierre de la Ramée (1515–1572)* (Leiden, 1958); for his mathematics, see J. J. Verdonk, *Petrus Ramus en de wiskunde* (Assen, 1966), with extensive bibliography of his mathematical writings. Other, more specialized studies include P. A.

DuHamel, "The Logic and Rhetoric of P. Ramus," in *Modern Philology*, **46** (1949), 163–171; N. W. Gilbert, *Renaissance Concepts of Method* (New York, 1960); F. P. Graves, *P. Ramus and the Educational Reform of the 16th Century* (London 1912); Henri Lebesgue, "Les professeurs de mathématique du collège de France: Humbert et Jordan; Roberval et Ramus," in *Revue scientifique*, **59** (1922), 249–262; and N. E. Nelson, *Peter Ramus and the Confusion of Logic, Rhetoric and Poetry* (Ann Arbor, 1947).

See also W. J. Ong, "P. Ramus and the Naming of Methodism," in *Journal of the History of Ideas,* **14** (1953), 235–248; Edward Rosen, "The Ramus-Rheticus Correspondence," *ibid.,* **1** (1940), 363–368; Paolo Rossi, *Francesco Bacone. Dalla magia alla scienza* (Bari, 1957); "Ramismo, logica e retorica nei secoli XVI e XVII," in *Rivista critica di storia della filosofia*, **12** (1957), 357–365, with extensive critical bibliography; *Clavis universalis. Arti mnemoniche e logica combinatoria da Lullo a Leibniz* (Milan–Naples, 1960); L. A. Sédillot, "Les professeurs de mathématiques et de physique générale au Collège de France," in *Bollettino Boncompagni*, **2** (1869), 389–418; and J. A. Vollgraff, "Pierre de la Ramée (1515–1572) et Willebrord Snel van Royen (1580–1626)," in *Janus,* **18** (1913), 595–625.

MICHAEL S. MAHONEY

RAŇGANĀTHA (*fl.* 1603 at Benares, India) *astronomy.*

Rañganātha was the son of Ballāla, scion of a learned Brāhmaṇa family who traced his descent through Trimalla and Rāma to Cintāmaṇi of the Devarātagotra, a resident of Dadhigrāma on the Payoṣṇī River in central India. Among Rañganātha's brothers was the celebrated astrologer Kṛṣṇa (see D. Pingree, *Census of the Exact Sciences in Sanskrit,* series A, II [Philadelphia, 1971], 53a–55b). He should not be confused with the Benares astronomer Rañganātha who wrote a commentary, the *Mitabhāṣiṇī,* on the *Līlāvatī* of Bhāskara II (*b.* 1115) in 1630; a *Bhaṅgīvibhaṅgīkaraṇa* based on the work of Muniśvara Viśvarūpa (*b.* 17 March 1603), the son of our Rañganātha (the *Bhaṅgīvibhaṅgīkaraṇa* was edited by M. H. Ojhā [Varanasi, 1958]); *Lohagolakhaṇḍana* (edited by M. H. Ojhā [Varanasi, 1961]); and a *Palabhākhaṇḍana.* This second Rañganātha was a son of Nṛsiṃha and a brother of Muniśvara's principal rival, Kamalākara (*fl.* 1658; see D. Pingree, *op. cit.,* pp. 21a–23a). The only known work of Rañganātha, the son of Ballāla, is the *Gūḍhārthaprakāśikā,* a commentary on the *Sūryasiddhānta* that he wrote at Benares in 1603 (see essay in Supplement). Its importance for the history of Indian astronomy lies in the fact that Rañganātha's recension and interpretation of the *Sūryasiddhānta* were followed by E. Burgess in his

English translation of that text (*Journal of the American Oriental Society,* **6**, pt. 2[1860], 141–498; reprinted, New Haven, 1860, and Calcutta, 1935).

BIBLIOGRAPHY

The *Gūḍhārthaprakāśikā* accompanied the eds. of the *Sūryasiddhānta* by Fitz Edward Hall and Bāpū Deva Śāstrī (Calcutta, 1859), by Jīvānanda Vidyāsāgara (Calcutta, 1871; 2nd ed. Calcutta, 1891), at the Kāśisaṃskṛta Press (Benares, 1880), and by B. P. Miśra (Bombay, 1956). Short notices regarding Rañganātha are Sudhākara Dvivedin, *Gaṇakataraṅgiṇi* (*The Pandit,* n.s. **14** [1892], repr. Benares, 1933), 79–81 and Ś. B. Dīkṣita, *Bhāratīya Jyotiḥśāstra* (Poona, 1896; repr. Poona, 1931), 285.

DAVID PINGREE

RANKINE, ALEXANDER OLIVER (*b.* Guildford, England, 1881; *d.* Hampton, Middlesex, England, 20 January 1956), *physics, geophysics.*

Rankine's father, the Reverend John Rankine, was a Baptist minister; and Rankine himself remained a member of the Baptist church. He received his professional education at University College, London, where he studied under Trouton, obtaining the D.Sc. in 1910. In 1907 he married Ruby Irene Short, the daughter of Samuel Short of Reading.

Rankine was an assistant in the physics department of University College from 1904 to 1919 and professor of physics at the Imperial College of Science and Technology, South Kensington, from 1919 to 1937, when he resigned his chair to become chief physicist with the Anglo-Iranian Oil Company. He retired from this post in 1947 but continued to act as geophysical adviser to the company until 1954. He held military research posts in both world wars and received the Order of the British Empire in 1919 for his work during World War I. He was elected a fellow of the Royal Society (1934), was an active member of several professional societies, and held various honorary positions throughout his life.

Although Rankine's career involved him in much work of an administrative or advisory character, he also produced valuable research. His major contribution was the invention and use of an apparatus for measuring the viscosity of a gas. This apparatus, known as the Rankine viscosimeter, consisted of a vertical glass tube in which a pellet of mercury could slide downward, thereby forcing the air below it through a capillary tube connecting the lower end of the glass tube with the upper end. By enclosing the viscosimeter in a heat bath and measuring the viscosity of various gases as a function of temperature, Rankine was able to determine the so-called Sutherland constant. This constant entered into a theoretical formula —relating viscosity with temperature—that Sutherland had derived from kinetic theory and mean-free-path considerations; and the determination of its numerical value allowed Rankine to estimate the diameter of gas molecules and to arrive at some notions of their shape and atomic arrangement. This work was published in a series of articles during the period 1910–1926.

Rankine later became interested in geophysics and improved the Eötvös gravimeter; this work led him to construct a magnetometer of great sensitivity. This later instrument was used for measuring distortions, and thus susceptibilities, in weak magnetic fields.

BIBLIOGRAPHY

A complete list of Rankine's scientific papers is given in G. P. Thomson's biography, "Alexander Oliver Rankine," in *Biographical Memoirs of Fellows of the Royal Society*, **2** (1956), 249–255.

OLE KNUDSEN

RANKINE, WILLIAM JOHN MACQUORN (*b.* Edinburgh, Scotland, 5 July 1820; *d.* Glasgow, Scotland, 24 December 1872), *engineering, engineering education, physical science.*

Rankine was the son of David Rankine, an army lieutenant, and Barbara Grahame. Another son died in childhood. Because of poor health, Rankine received most of his early education at home, being taught at first by his father and later by private tutors. In 1836 he entered the University of Edinburgh, where he studied natural philosophy under J. D. Forbes. The following year he was introduced to railroad engineering while assisting his father, who had become a superintendent for the Edinburgh and Dalkeith Railway. Although a successful student, Rankine left Edinburgh in 1838 without a degree. He then went to Ireland and worked for four years on railroad, hydraulic, and various other projects as an apprentice to John MacNeill, a prominent civil engineer who had been Thomas Telford's chief assistant. After finishing his apprenticeship, Rankine returned to Scotland and practiced civil engineering.

By 1850, when he was elected a fellow of the Royal Society of Edinburgh, Rankine was beginning to work in both engineering and science. He was elected a fellow of the Royal Society of London (1853) and received the Keith Medal (1854) from the Royal Society of Edinburgh for his published researches into the mechanical theory of heat. Receiving this medal was an unusual distinction for a professional consulting engineer. Thus when he lectured at the University of Glasgow early in 1855, substituting for Lewis Gordon, regius professor of civil engineering and mechanics, Rankine was already well-known for his contributions to both engineering and science. When Gordon relinquished his professorship later that year, Rankine was appointed to the chair, a post he held until his death. In 1857 Trinity College, Dublin, conferred on him the degree of LL.D.

Although much of his time as regius professor was spent writing engineering textbooks and teaching, Rankine continued to act as a consultant. He also remained active in engineering and scientific societies, especially in Glasgow, where he became an extremely influential figure. In 1857 he became the first president of the Institution of Engineers in Scotland, a society of which he was the principal founder; and from 1862 he was an associate member of the Institution of Naval Architects. He enjoyed music and took delight in singing and composing his own songs. He never married.

Rankine's earliest investigations were directly related to the construction and operation of the rapidly growing network of British railroads. In 1841, while surveying on the Dublin and Drogheda Railway, he invented a technique, later known as Rankine's method, for laying out circular curves. This technique, one of the first to be based on the use of the theodolite, was more accurate and faster to use than any other then available. Occasionally he collaborated with his father, and together they performed experiments to determine the advantages of cylindrical over tapered wheels on railroad cars. They published their findings in 1842, but the latter continued to be used. Rankine also examined the serious problem of unexpected breaks in railroad axles. His conclusions, published in 1843, were generally adopted in later construction.

In 1849 Rankine began to publish papers in physical science. Initially the central theme of his theoretical researches, which he pursued while still working as an engineer, was a comprehensive theory of matter, which he called "the hypothesis of molecular vortices." In this, matter was composed of atoms, each comprising an atmosphere surrounding a comparatively small nucleus. The atmospheres consisted of innumerable vortices, or circulating streams, of matter that were elastic and automatically tended to increase their volume. The absolute temperature of an atom was proportional to the square of the vortical velocity as defined by Rankine. Although the nature of the nucleus was left vague, its function was clearly stated: the nuclei formed the luminiferous medium. Thus light and radiant heat were the vibrations of the nuclei, which exerted forces on each other at a distance. Rankine proposed this unorthodox medium largely because he could not conceive of the luminiferous ether as imagined by Fresnel, that surrounds ponderable matter and that is elastic enough to transmit transverse vibrations at the speed of light while offering no perceptible resistance to the motion of macroscopic bodies.

Rankine suggested in 1850 that double refraction could be explained by the aeolotropy of inertia of the atomic atmospheres that dampened slightly the vibratory motion of the nuclei. But in 1872 Stokes showed experimentally that double refraction could not depend on aeolotropy of inertia, a conclusion that later led to the general abandonment of this explanation. In his 1850 hypothesis Rankine also utilized his atmospheres and nuclei to analyze the elasticity of solid bodies. His analysis was based on the assumption that the elasticity could be divided into two parts, one arising from the mutual forces of the nuclei and the other arising solely from changes of volume of the atmospheres. But by 1855 he had concluded that elasticity could not be divided in this manner. Having rejected this hypothesis, he concentrated on determining the mutual relations and physical significance of the elastic constants of solids.

It was to the theory of heat that Rankine applied his hypothesis of molecular vortices most extensively, and most successfully. Probably his interest in heat and heat engines developed from his early work with railroads—in 1842 he had already attempted to reduce the theory of thermal phenomena to a branch of mechanics. He had had to delay this research for about seven years, however, because of the lack of suitable experimental data with which to compare his results. In 1849 he published preliminary formulas relating the pressure of saturation of a vapor to its temperature, and soon afterward, in 1850, published his theory based on molecular vortices. Among the various formulas he derived were relations for the pressure, density, and temperature of gases and vapors and the latent heat of evaporation of a liquid. He also obtained theoretical expressions for the real and apparent specific heats of gases and vapors. Among his conclusions was the prediction that the apparent specific heat of saturated steam must have a negative value. This conclusion was later confirmed experimentally.

Rankine then began to generalize his equations to include solids and liquids. Suggesting that his previous equations were probably valid only for perfect gases, he removed any restrictions on the shape of the atoms and vortices, except to demand that the matter in the vortices move in closed paths. By clever approximations, Rankine obtained the same equations as before and immediately concluded that they therefore applied to all substances, whether in the solid, gaseous, or liquid state.

At this time Rankine turned his attention to the problem of calculating the efficiency of heat engines. Like Clausius and William Thomson, who had been working independently on the theory of heat, Rankine attempted to derive Carnot's law. Clausius had derived the law from the axiom that heat could not, by itself, pass from a cold to a hot body. Thomson had based his proof on a similar axiom; but Rankine knew only of Thomson's conclusion, not the details of his derivation. Dissatisfied with Clausius' proof, Rankine set out in 1851 to deduce the law directly from his hypothesis of molecular vortices, without the aid of extraneous axioms. He showed that the efficiency of an engine when operating on a Carnot cycle between two temperatures depends only on those temperatures. Implicit in his proof was the assumption, equivalent to the axioms that Thomson and Clausius had used, that that cycle yields the maximum efficiency possible for any engine operating between those temperatures. He obtained the efficiency as a function of the temperatures and a universal constant. Rankine's theoretical expression differed from those of Clausius and Thomson, but numerically the differences were insignificant, as Rankine quickly pointed out. It was not until 1853 that he modified his expression for the efficiency, thus making it coincident with Clausius' and Thomson's.

Yet Rankine's examination of the behavior of heat engines was merely an initial step in the evolution of his theory of heat. He developed a general theory of energy that was independent of all mechanical hypotheses and then immediately recast his theory of heat in its mold. In 1853 he distinguished between two types of energy, "actual (or sensible) energy," which caused substances to change their state, and "potential (or latent) energy," which replaced the actual energy lost in any change. Rankine's examples of actual energy were thermometric heat and the *vis viva* of moving matter; and of potential energy, the mechanical powers of gravitation and static electricity. The sum of the actual and potential energies of the universe was assumed constant—a law already familiar as the conservation of energy. Rankine then proposed a law that determined the amount of energy transformed during any change of state of a substance. Apparently the actual form of the law was dictated by Rankine's desire to make it yield Carnot's law in the particular case of a heat engine operating on a

Carnot cycle. From 1854 he began to make frequent use of his "thermodynamic function," which he later identified with the entropy of Clausius.

In 1855 the law of the transformation of energy was assimilated by Rankine into a general theory of energy called "the science of energetics." Rankine was the founder of this science, which assumed major importance in the 1890's. Energetics is a striking illustration of Rankine's boldly speculative thinking as well as of his penchant for introducing terminology. (An inveterate nomenclator, he freely invented his own vocabulary, for example, "potential energy," to denote novel concepts or to replace terms he considered unsuitable.) Energy and its transformations, rather than force and motion, formed the basis of this science. All physical phenomena were described by changes in actual and potential energy. The rate of transformation of energy was represented by the "metamorphic function" and the equality of energy in the equilibrium state by the "metabatic function." In the theory of heat the metabatic function was the absolute temperature.

From the mid-1850's energetics replaced molecular vortices as the basis of Rankine's thermodynamics. He did not abandon the use of mechanical hypotheses completely, however, and often found them useful when expounding the theory of heat to engineers. As late as 1864 Rankine maintained that any mechanical hypothesis explaining the nature of heat must resemble his vortex theory of rotational motion rather than that of Clausius, Waterston, Maxwell, and others, in which heat was the result of the linear motion of the elementary bodies of a substance. It was not until 1869 that he acknowledged the possibility of treating heat as the result of linear motion. This acknowledgment appeared in a paper in which Rankine discarded his theory of nuclei and defended the hypothesis of a gas consisting of matter without any nuclei around which to congregate. His model of the atom had become almost identical with that of Thomson. One of Rankine's final applications of thermodynamics appeared in 1869, when he examined the propagation of finite longitudinal disturbances, later commonly called shock waves. The equations that Rankine derived for propagation in a perfect gas took into account the effects of heat conduction. These equations were later generalized by Hugoniot in 1887 and became known as the Rankine-Hugoniot relations.

Despite the acknowledged practical importance of Rankine's thermodynamics, most nineteenth-century physicists remained opposed to its molecular foundations. In his Baltimore lectures Thomson playfully but revealingly remarked that the most important aspect of Rankine's hypothesis of molecular vortices

was its name. Maxwell, too, criticized Rankine, suggesting facetiously that his definition of the second law and much else in his thermodynamics was inscrutable, although he was careful to place Rankine alongside Thomson and Clausius as one of the founders of theoretical thermodynamics. Part of the immense success of Rankine undoubtedly derived from his ability to solve complex problems by elementary methods, to simplify the theory of heat, and to present it systematically so that engineers could understand and use it. Moreover, his numerical computations on a wide variety of heat engines, his analysis (later called the Rankine cycle) of the operation of an ideal engine employing a condensable vapor such as steam, and the calculations of the properties of steam and other gases and vapors that he continued throughout his career enhanced his reputation as a major contributor to the understanding of thermal phenomena.

Another of Rankine's major interests was naval architecture. He turned his attention to this field about 1855, after he had settled permanently in Glasgow. With friend J. R. Napier, a shipbuilder, and others, Rankine wrote *Shipbuilding, Theoretical and Practical* (1866), which was intended to bring precision and theory to a British industry that was largely empirical. Rankine was the editor of and principal contributor to this treatise. He also played an active role in the Institution of Naval Architects, where he presented papers frequently and became a major contributor to the theory of the design and motion of ships. Indeed, he pursued his hydrodynamical researches, not with purely scientific intent, but primarily to improve naval design. Beginning in 1862 with an independent derivation of the trochoidal shape of waves in deep water, a result already published in 1804 by F. Gerstner, Rankine examined the rolling, dipping, and heaving motion of ships in waves. Similarly, his two-dimensional analysis of the flow of water around circular and oval bodies enabled him to determine the waterlines of a ship that would create a minimum of friction as it moved through the sea; he also calculated the efficiency of propellers. A number of his papers were devoted to the exposition of elementary ways of solving hydrodynamical problems. He devised a simple method for obtaining a graphical representation of streamlines to demonstrate propositions in hydrodynamics.

Rankine conducted most of his theoretical investigations with an eye to practical applications. To facilitate the calculation of the elasticities of various substances from an experimental knowledge of their vibratory behavior, he carried out a mathematical analysis of sound waves in solids and liquids of finite extent. He obtained diagrams of forces for such structures as arches and roofs by introducing his theorem on parallel projections of polygonal frames, which he extended in 1864 to polyhedral frames. The theorem was subsequently generalized by Maxwell in his geometrical study of reciprocal figures, and then simplified by R. H. Bow for engineering use. Rankine's papers on soil mechanics also served practical ends, as did his modification of Gordon's version of Hodgkinson's formula for the strength of iron pillars.

Rankine's teaching at Glasgow combined theory with practice. Indeed the chair had been founded in 1840 to promote instruction in the application of science to engineering, and it was in this spirit that Rankine taught. He insisted that the application of pure science to engineering constituted a separate subject in its own right and campaigned vigorously for the award of diplomas and degrees in engineering studies. In 1863 the university awarded its first Certificate of Proficiency in Engineering Science but did not confer degrees in the subject until a few years after Rankine's death.

Rankine published a highly successful set of engineering textbooks. These works were extremely comprehensive and combined practical knowledge with theory, often demanding from the reader a considerable background in mathematics; they also illustrate Rankine's concern for terminological precision. (For instance, he incorporated into the manuals his earlier definition of "strain" as the relative displacement of the particles of a body, a usage that became standard.) His first textbook, *A Manual of Applied Mechanics*, appeared in 1858 and was immediately hailed as a classic. *A Manual of Civil Engineering*, published in 1862, was unrivaled. Both books ran through numerous editions, became paradigmatic texts for later authors, and largely established the methods of engineering education in Britain and elsewhere. *A Manual of the Steam Engine and Other Prime Movers* (1859) and *A Manual of Machinery and Millwork* (1869) also saw many editions. The range and content of these manuals reflect both Rankine's versatility and the multifarious facets of his career.

BIBLIOGRAPHY

I. ORIGINAL WORKS. A comprehensive list of Rankine's papers appears in the Royal Society *Catalogue of Scientific Papers*, V, 93–96; VI, 747; VIII, 696–698, although this source omits his contributions to *Engineer* and *Transactions of the Institution of Engineers in Scotland*. Most of his major papers on thermodynamics, light, elasticity of solids, energetics, and hydrodynamics are included in

W. J. Millar, ed., *Miscellaneous Scientific Papers; by W. J. M. Rankine. From the Transactions and Proceedings of the Royal and Other Scientific and Philosophical Societies, and the Scientific Journals. With a Memoir of the Author by P. G. Tait . . .* (London, 1881).

Rankine's texts are *A Manual of Applied Mechanics* (London–Glasgow, 1858; 21st ed., 1921), with 1st ed. reprinted in *Encyclopaedia Metropolitana*, XXXIX (London, 1848–1858) and with trans. of 7th ed. by A. Vialay as *Manuel de mécanique appliquée* (Paris, 1876); *A Manual of the Steam Engine and Other Prime Movers* (London–Glasgow, 1859; 17th ed., 1908), with trans. of 8th ed. by G. Richard as *Manuel de la machine à vapeur et des autres moteurs* (Paris, 1878); *A Manual of Civil Engineering* (London, 1862; 24th ed., 1911), with trans. of 12th ed. by F. Kreuter as *Handbuch der Bauingenieurkunst* (Vienna, 1880); and *A Manual of Machinery and Millwork* (London, 1869; 7th ed., 1893). The posthumous eds. of these textbooks were edited and revised initially by E. F. Bamber and later by W. J. Millar. *A Mechanical Text-Book; or, Introduction to the Study of Mechanics and Engineering* (London, 1873; 3rd ed., 1884), written with Bamber, was intended to serve as an introduction to the manuals.

Rankine also published *Useful Rules and Tables Relating to Mensuration, Engineering, Structures, and Machines* (London, 1866; 7th ed., 1889), with later eds. being edited and revised at first by Bamber and later by Millar. Rankine was the corresponding and general editor of, as well as the main contributor to, *Shipbuilding, Theoretical and Practical* (London, 1866), written with I. Watts *et al.* He contributed a number of articles, mainly on heat, to J. P. Nichol's *Cyclopaedia of the Physical Sciences* (London–Glasgow, 1857; 2nd ed., 1860), and an essay on the strength and qualities of wood and metals to the *Cyclopaedia of Machine and Hand Tools* (London, 1869). His inaugural lecture as regius professor, "De Concordia inter Scientiarum Machinalium Contemplationem et Usum," was published as *Introductory Lecture on the Harmony of Theory and Practice in Mechanics* (London–Glasgow, 1856), with repr. in *A Manual of Applied Mechanics*. This lecture lucidly and eloquently presents Rankine's views on the relationship of theory to practice in engineering.

His main nontechnical writings are *A Memoir of John Elder, Engineer and Shipbuilder* (Edinburgh, 1871; 2nd ed., Glasgow, 1883); and the posthumous *Songs and Fables* (Glasgow–London, 1874).

II. SECONDARY LITERATURE. Of the few obituaries and memoirs sketching Rankine's life, the best informed and most enlightening are P. G. Tait, in *Glasgow Herald* (28 December 1872), with a modified version subsequently appearing in *Miscellaneous Scientific Papers*, xix–xxxvi; and L. D. B. Gordon, in *Proceedings of the Royal Society of Edinburgh*, **8** (1875), 296–306. These accounts form the main sources for A. Barr's eulogy "W. J. Macquorn Rankine, a Centenary Address," in *Proceedings of the Royal Philosophical Society of Glasgow*, **51** (1923), 167–187; J. Henderson, *Macquorn Rankine: An Oration Delivered in* the *University at the Commemoration of Benefactors on 15th June, 1932* (Glasgow, 1932), with repr. in *Engineer*, **153** (1932), 689–690; and J. Small, "The Institution's First President William John Macquorn Rankine," in *Transactions of the Institution of Engineers and Shipbuilders in Scotland*, **100** (1956–1957), 687–697. Other obituaries are listed in Royal Society *Catalogue of Scientific Papers*. Although these obituaries and eulogies contain brief impressions of Rankine's science and engineering, no detailed assessment of his work has ever been written.

See also G. Helm's interpretation of Rankine's thermodynamics and energetics in *Die Energetik nach ihrer geschichtlichen Entwickelung* (Leipzig, 1898), 110–120; J. C. Maxwell's witty and facetious, but very valuable, comments in his review "Tait's *Thermodynamics*," in *Nature*, **17** (1878), 257–259, 278–280; and the summaries of Rankine's writing on the elasticity of solids in I. Todhunter and K. Pearson's *History of the Theory of Elasticity and of the Strength of Materials*, 2 vols. (Cambridge, 1886–1893; repr. New York, 1960).

The only recent important accounts of any of Rankine's work are Sir Richard Southwell's selective survey "W. J. M. Rankine: A Commemorative Lecture Delivered on 12 December, 1955, in Glasgow," in *Proceedings of the Institution of Civil Engineers*, **5** (1956), pt. 1, 177–193; and E. E. Daub's article, "Atomism and Thermodynamics," in *Isis*, **58** (1967), 293–303, in which the author suggests a possible influence of Rankine's thermodynamics on Clausius.

E. M. PARKINSON

RANVIER, LOUIS-ANTOINE (*b.* Lyons, France, 2 October 1835; *d.* Vendranges, Loire, France, 22 March 1922), *histology.*

Ranvier was the foremost histologist in France in the latter part of the nineteenth century. A student of Claude Bernard, he combined the German histological tradition with the French physiological tradition. Ranvier's father, Jean-François-Victor Ranvier, was a businessman who had retired early to devote himself to public administration.

After studying medicine at Lyons, Ranvier went to Paris, and passing the examination to become an intern, entered the Paris hospitals in 1860. He received his medical degree in 1865. Shortly thereafter he and a friend, Victor Cornil, founded a small private laboratory where they offered a course in histology to medical students. Continuing the work of the Paris clinical school on the microscopical level, Cornil taught pathological anatomy while Ranvier taught normal anatomy. From this collaboration resulted the *Manuel d'histologie pathologique*, a unique work in France, where the leading pathological anatomists still disdained the microscope.

Ranvier's collaboration with Cornil ended in 1867 when Ranvier became *préparateur* for Bernard at the Collège de France. Ranvier made his lodgings there into a small histology laboratory, which in 1872 was annexed to Bernard's chair of experimental medicine and given official recognition as the Laboratoire d'Histologie of the École des Hautes Études. In 1875, through Bernard's influence, Portal's defunct chair of anatomy at the Collège de France was recreated for Ranvier as the chair of general anatomy; and for a time the histology laboratories at the École des Hautes Études and at the Collège de France were combined. But eventually the laboratories were separated, with Ranvier's student Louis Malassez becoming director of the laboratory at the École des Hautes Études. In 1887 Ranvier became a member of the Academy of Sciences.

Ranvier's teaching, which he conducted mainly in the laboratory, tended to be dryly technical and focused on his own research. He therefore attracted only a small audience, but his printed *leçons* were widely read. For several decades his *Traité technique d'histologie* (1875–1882) was a leading textbook in the field. At the beginning of his career, Ramón y Cajal took Ranvier's text as his scientific "Bible." From 1875 until about 1890 Ranvier's laboratory was a center of activity for a large number of students, both French and foreign, including Malassez, Louis de Sinéty, Maurice Debove, Renaut, William Nicati, and E. Suchard. After 1890 biological research moved away from morphology to chemistry, physical chemistry, and microbiology. For several years thereafter (until 1895) Ranvier continued his work, producing particularly important studies on cicatrization and on the development of the lymphatic vessels. In 1897 he and Balbiani founded the *Archives d'anatomie microscopique*, the first journal in France devoted exclusively to microscopical studies. By 1900 Ranvier felt isolated from the scientific community in France and retired to his estate in Thélys, where he spent the next twenty-two years almost totally removed from the scientific scene. He never married.

When Ranvier began his career, histology was well established in Germany but little pursued in France. Because he inherited the physiological tradition established by Bernard, Ranvier's work was not as strictly morphological as much of the work in Germany. He supplemented histological techniques with those of physiological experimentation, namely, ligation, excitation of nerves and muscles by electricity, nervous section, and graphical registration of movements. His biographers have looked upon his work as an extension into histology of Bernard's method.

Ranvier's work was noted for its precision, thoroughness, and simple but effective techniques. He preferred disassociations (or fine dissections) to sectional cuts; and whenever possible he worked with thin membranes that were naturally disassociated. His few sectional cuts were usually made by hand rather than with the microtome. Osmic acid, alcohol, and bichromates were his usual fixating agents. The remainder of his reagents included a few coloring agents for injections and solutions of gold and silver for impregnations. Most often Ranvier worked with adult tissues; he took little interest in histogenesis. Like Bernard, he had only contempt for statistics. Although he recognized the necessity of forming hypotheses in his experimental work, he disliked theorizing.

Ranvier's work embraced all organic systems, but he is best known for his researches on the peripheral nervous system. His earliest and most celebrated achievement was his discovery in 1871 of the annular constrictions of medullated nerves, now known as the nodes of Ranvier. Ranvier showed that the medullated nerves are not regularly cylindrical—at approximately equal distances there are constrictions in the form of rings where the myelin sheath, but not the cylinder axis, is interrupted. The nodes divide the nervous fiber into interannulary segments possessing a nucleus and protoplasm (neurilemma nucleus). Ranvier prepared his specimens either by the osmic acid method first employed by Schultze or by impregnation with silver nitrate.

Ranvier also investigated the degeneration and regeneration of sectioned nerves. In spite of Waller's earlier studies (1852), many scientists continued to believe that the cylinder axis persisted in the peripheral segment after section and therefore had an independent life. Ranvier's work, confirming Wallerian degeneration, showed that the cylinder axis of the peripheral segment does fragment and disappear, while the cylinder axis of the central segment hypertrophies and emits buds that are the point of departure for new nervous fibers. Ranvier believed that regeneration of nerves was a particular case of the general law of growth from the center to the periphery.

Using an improved method of impregnation by gold, Ranvier did extensive research into the nervous terminations in the skin, the muscles, the cornea, and the sensorial organs. His other notable work on the nervous system includes a description of the "laminous sheath" (perineurium) connecting bundles of nervous fibers and his discovery that the apparently unipolar cells of the spinal ganglia of mammals bifurcate in a T-branch. This research tended to support the theory that the cylinder axis is a prolongation of the nervous cell. Much of Ranvier's work on the nervous system was later used to support the neuron theory, but Ranvier himself did not become involved in the famous neuron-reticulum controversy. He did, however, support the controversial doctrine of the fibrillary structure of nerve cells.

Ranvier also studied the differences between the nervous terminations of voluntary and involuntary muscles. To study muscle contraction he devised a method for utilizing a spectrum of diffraction set up by an extended muscle, which he then subjected to a tetanizing current.

Ranvier studied the secretion of the salivary glands in the dog on a microscopical level. He obtained secretions by electric stimulation of the tympanic cord. He also investigated the nature of connective tissue and refuted Virchow's theory by the novel method of disassociation by an edematous papule *(boule d'oedème)*. He studied the problem of the origin of the lymphatics and invented a cardiograph for measuring the movements of lymphatic hearts.

Because Ranvier rarely discussed the theoretical implications of his works, he is almost unknown to historians today. Independent, unsociable, often rude, and seemingly insensitive, he was admired and feared more often than loved by his students and colleagues.

BIBLIOGRAPHY

I. ORIGINAL WORKS. Ranvier's works include *Manuel d'histologie pathologique* (Paris, 1869; 2nd ed. 1881–1884), written with Victor André Cornil; and *Traité technique d'histologie* (Paris, 1875–1882; 2nd ed., 1889). Several of Ranvier's courses at the Collège de France were edited and published: *Leçons d'anatomie générale faites au Collège de France, par L. Ranvier: Année 1877–1878. Appareils nerveux terminaux des muscles de la vie organique: coeur sanguin, coeurs lymphatiques, oesophage, muscles lisses, leçons recueillies par MM. Weber et Lataste; Année 1878–1879. Terminaisons nerveuses sensitives; cornée, leçons recueillies par M. Weber,* 2 vols. (Paris, 1880–1881); *Leçons sur l'histologie du système nerveux, par M. L. Ranvier,. . . recueillies par M. Ed. Weber,* 2 vols. (Paris, 1878); *Leçons d'anatomie generale sur le système musculaire, par L. Ranvier, . . . recueillies par M. J. Renaut* (Paris, 1880). *Exposé des titres et des travaux de M. L. Ranvier* (Paris, 1885) was written by Ranvier to support his candidacy for the Académie des Sciences and is a useful source of information on Ranvier's work.

The Royal Society *Catalogue of Scientific Papers,* V, 96; VIII, 699–700; XI, 104–105; and XVIII, 51–52, lists Ranvier's memoirs through 1900. A more complete list can be found in Jolly. (See below.)

II. SECONDARY LITERATURE. The most extensive biography of Ranvier is J. Jolly, "Louis Ranvier (1835–1922): Notice biographique," in *Archives d'anatomie microscopique,* **19** (1922), i–lxxii, with bibliography. Jolly's analysis of Ranvier's work is largely based on Ranvier's

Exposé . . .. Jolly, a student of Ranvier's, also wrote "Ranvier et la methode expérimentale," in *Le Collège de France (1530–1930)* (Paris, 1932), 209–219. See also J. Nageotte, "Louis-Antoine Ranvier," in *Comptes rendus des séances de la Société de biologie,* **36** (1922), 1144–1152. The *Isis Cumulative Bibliography* contains a list of obituary notices on Ranvier.

TOBY A. APPEL

RANYARD, ARTHUR COWPER (*b*. Swanscombe, Kent, England, 21 June 1845; *d*. London, England, 14 December 1894), *astrophysics.*

Ranyard was educated at University College, London, and then at Pembroke College, Cambridge. He eventually returned to London, where, in 1871, he was called to the bar. While at University College, he and George De Morgan began a student mathematical society which subsequently became the London Mathematical Society.

Ranyard was interested throughout his life in problems of photography. His main research field, however, was astronomy; and he was concerned especially with the observation of solar eclipses. He assisted Lockyer in organizing the British eclipse expedition in 1870, and he himself observed the eclipses of 1878 and 1882. His major scientific contribution resulted from this work. When the Royal Astronomical Society decided to publish a memoir devoted to solar eclipse observations, Ranyard undertook the task but gradually extended its scope. The volume finally appeared in 1879, and it covered all nineteenth-century eclipse observations up to that time. The work remained a standard source of information on eclipses for many years.

Ranyard found himself involved in some dispute with the Royal Astronomical Society over the length of time it took him to prepare this eclipse volume; and as a member of the council of the society during the 1870's and 1880's, he was, in fact, one of the central figures in several of the controversies that affected the society at that time.

In the years preceding his death, Ranyard became interested in the possibility of observing the sun with a spectroheliograph. He had such an instrument constructed but died before it could be brought into operation. This and other of his instruments were bequeathed to Evershed, whom he helped introduce to solar research.

BIBLIOGRAPHY

Ranyard's important work is "Observations Made During Total Solar Eclipses," in *Memoirs of the Royal Astronomical Society,* **41** (1879), 1–792. The Royal Society *Catalogue of Scientific Papers,* VIII, 700; XI, 105–106; and XVIII, 52, gives a bibliography of his works.

There is a short obituary of Ranyard in the *Monthly Notices of the Royal Astronomical Society,* **55** (1895), 198–201. Most of his research papers appeared in the publications of the Royal Astronomical Society.

A. J. MEADOWS

RAOULT, FRANÇOIS MARIE (*b*. Fournes, France, 10 May 1830; *d*. Grenoble, France, 1 April 1901), *chemistry.*

Raoult early chose a scientific career. Despite family encouragement, he lacked the financial resources to complete his work at the University of Paris, although he presented a brief note on electrolytic transport and on electrical endosmosis to the Académie des Sciences. This note, which appeared in *Comptes rendus . . . de l'Académie des sciences* in 1853, was the first of more than 100 published papers.

In 1853 Raoult was employed as a teacher at the *lycée* in Reims, then moved to the College of St. Dié as *régent de physique.* At St. Dié he received the *baccalauréat ès lettres* and *baccalauréat ès sciences,* passed the *licencié* examination, and was appointed an *agrégé de l'enseignement secondaire spécial.* In 1862 he moved to the *lycée* at Sens, where, on his own, he carried out research on the electromotive force of voltaic cells; this research led in 1863 to the *docteur ès sciences physiques* from the University of Paris.

In 1867 Raoult was called to the Faculté des Sciences of Grenoble as *chargé du cours de chimie* and in 1870 was elevated to the chair of chemistry. Raoult taught and carried out extensive research for thirty-one years at Grenoble, almost until the day he died. Beginning in 1899 he was elected dean of the Grenoble Faculté des Sciences for five consecutive terms and was instrumental in bringing about a major reorganization of the University of Grenoble in 1896.

In 1865 Raoult was appointed an *officier d'Académie.* In 1872 he became an *officier de l'Instruction publique,* and was awarded the Médaille des Sociétés Savantes. In 1883 he was presented the Prix International de Chimie La Caze, ten thousand francs, and was named *correspondant de l'Institut de France,* which awarded him its biennial prize in 1895. In 1892 he was awarded the Davy Medal of the Royal Society. He also became a foreign fellow of the Chemical Society of London in 1898, and of the Academy of St. Petersburg in 1899. In 1890 he was chosen chevalier of the Legion of Honor, elevated to officer in 1895, and honored by the high distinction of commander in 1900.

Raoult's work may be divided into three stages: physical, chemical, and physicochemical. During the first stage, which lasted until Raoult moved to Grenoble, he discovered that the chemical heat of reaction of galvanic cells of. the Daniell type was generally different from the heat equivalent of the electrical work done in these cells. He also used a voltameter, an instrument that measures the quantity of electricity by the amount of electrolysis, to study the heat evolved in voltaic cells. Raoult was one of the first to recognize that the electrical work done by voltaic cells was not equal to the heat evolved by the chemical reaction driving these cells. He showed that whenever the electrical work done in voltaic cells was less than the heat of all reactions, heat was evolved. Moreover, he stated that changes in concentration, oxidation, and in acid-base relationships, but not changes in aggregation such as dissolving, melting, or solidifying, were the sources of the electromotive force of voltaic cells.

These experiments did not excite much attention despite the fact that by 1870 Raoult had published twelve papers on his thermochemical and electrochemical work, almost all of which appeared in *Comptes rendus* and *Annales de chimie*. The theories through which his work could have been fruitful had not yet been evolved, and the appreciation of his pioneering work had to wait until the theoretical studies of Gibbs, Helmholtz, van't Hoff, Arrhenius, and Nernst made its value apparent to all.

From 1870 to about 1882 Raoult turned his attention to studies on the effect of carbon dioxide on animal respiration, on the absorption of ammonia by ammonium nitrate, and on the rate of inversion of cane sugar under the influence of sunlight. Fourteen papers appeared during this stage, mainly in *Comptes rendus*. The papers show excellent and careful work but were relatively trivial and far less interesting than the pioneering investigations of his earlier years. He had begun, however, to study the vapor pressure and freezing points of salt solutions, and the freezing point of alcohol solutions. This interest was to lead him to the physicochemical third phase in which he made his most significant contributions to chemistry.

Richard Watson in 1771 and Charles Blagden in 1788 had shown that the freezing point of salt solutions is lowered in proportion to the weight of salt dissolved. This work was extended in 1871 by Louis de Coppet, who calculated what he termed "atomic depressions" in compounds lowering the freezing point of water. In 1878 Raoult experimentally examined the lowering of the freezing point and of the vapor pressure of water caused by eighteen different salts. He found that the lowering of the vapor pressure

caused by each of these salts was approximately proportional to the lowering of the freezing point. This work confirmed Guldberg's theoretical, thermodynamic proof in 1870 of the proportionality of the lowering of the freezing point, the rise of the boiling point, and the lowering of vapor pressure that resulted from the dissolving of a homogeneous substance in water. In 1882 Raoult published a significant paper on his investigation of the lowering of the freezing point of a large number of aqueous solutions of organic compounds.

In a series of papers appearing in *Comptes rendus* from 1882 to 1884, Raoult described a method for finding the molecular weight of an organic compound by the determination of the lowering of the freezing point of water that resulted from dissolving that compound in water. He first found by experimentation the simplest formula of the compound and the coefficient of lowering of the freezing point. Then he calculated the "molecular depression" from the sum of the "atomic depressions" that he had previously elucidated. From the formula below, the molecular weight could be directly ascertained: $M = T/a$, where M is the molecular weight; T is the molecular depression; and a is the coefficient of lowering of the freezing point.

This method for determining molecular weight soon received wide application, although most chemists turned to the somewhat simpler but completely analogous elevation of the boiling point with the development of Beckmann's thermometer for this purpose in 1888. Victor Meyer and Emanuele Paternò publicized Raoult's method because chemists could ascertain with relative ease the molecular weight of organic compounds.

Raoult soon turned to the anomalous results with salts in water, which had puzzled previous investigators. He classified the salts he used according to the valence of the radicals and found that the lowering of the freezing point could be accounted for by assigning certain numbers to these radicals. He demonstrated that the freezing point lowering obtained with these salts was consistent with the hypothesis that the salt radicals themselves acted as if they existed independently in the solution, and that certain radicals were more effective than others in lowering the freezing point of water. With the statement that "the neutral salts of mono and di-basic salts . . . act as if the electropositive and electronegative radicals of these salts when dissolved in water solution do not combine, but remain as simple mixtures" (*Annales de Chimie*, **20** [1890], 355), Raoult showed that he had come to accept much of Arrhenius' work on ionization.

In 1892, using extremely dilute solutions, Raoult

found that the molecular constant for organic compounds dissolved in water was 18.7, whereas for sodium and potassium chloride, it was 37.4 and 36.4 respectively. Raoult argued that this was a strong argument for Arrhenius' hypothesis of electrolytic dissociation, which had been published in 1887.

During this time Raoult experimented with vapor pressure lowering in solutions. He confirmed the work of Adolph Wüllner and Lambert H. J. von Babo that the relative lowering of the vapor pressure, $(f-f')/f$ is independent of the temperature, where $f-f'$ is the lowering of the vapor pressure, and f is the vapor pressure in dilute solution. He also showed that the lowering of the vapor pressure of water by a nonvolatile solute is proportional to the concentration of the solute.

In a brilliant paper in 1886, Raoult derived an expression for the relative lowering of the vapor pressure of solutions in ether, which is still in use:

$$\left(\frac{f-f'}{f}\right)\frac{M}{P} = K$$

where P is the number of grams of solute in 100 grams of solvent; M is the molecular weight of solute; and K is the constant.

In 1887 Raoult made an extended study of a large number of solvents other than ether. He formulated the law: "One molecule of a non-saline substance dissolved in 100 molecules of any volatile liquid decreases the vapor pressure of this liquid by a nearly constant fraction, 0.0105" (*Comptes rendus*, **104** [1887], 1433).

Raoult's law precisely coincided with an equation that van't Hoff had derived thermodynamically in 1886. Raoult was deeply moved by this harmony in their work, and he declared that the "agreement between experiment and theory is therefore, on all points, as complete as one could desire in the case of these substances" (*Comptes rendus*, **105** [1887], 859).

Raoult was a pioneer in demonstrating that the lowering of the vapor pressure of solvents by solutes, the rise in the boiling point of the solvent by adding a solute, and the lowering of the freezing point of the solvent, depend only on the ratio between the number of solute molecules and solvent molecules.

Raoult was a careful and accurate experimenter. He built most of his own apparatus himself because he felt that he could rely only on his own craftsmanship to attain the precision he sought. He was not a theoretical chemist, and he rarely resorted to the kind of mathematical analysis introduced by the thermodynamic chemists of his time. He gloried instead in laborious and painstaking experiments

through which he accumulated a mass of useful information on the freezing point and vapor pressure depressions of dozens of solvents and solutes under a variety of conditions. These data were heavily drawn on by van't Hoff for his study of osmotic pressures and by Arrhenius, who explained Raoult's anomalous results with salts that ionize in solution.

Without question, Raoult was a leading French experimental physical chemist in the nineteenth century. As Victor Meyer said in 1890,

> What a change our conceptions will have to undergo if we have to accustom ourselves to regard a dilute solution of sodium chloride as one containing not undecomposed molecules of this salt, but separated atoms of sodium and chlorine.

We owe these revolutionizing innovations to the investigations of van't Hoff, Arrhenius, Ostwald, Planck, and de Vries, but in regard to experimental research especially to the splendid work of Raoult, which during recent years has effected this mighty theoretical progress [Victor Meyer, "The Chemical Problems of Today," trans. by L. H. Friedburg in *Annual Report of the Board of Regents of the Smithsonian Institution, 1890* (Washington, D.C., 1891), p. 371].

BIBLIOGRAPHY

I. ORIGINAL WORKS. Raoult wrote more than 100 papers, the majority of which are listed in Poggendorff, III (1898), 1089; IV (1904), 1212; and in Royal Society *Catalogue of Scientific Papers*, V, 97; VIII, 700; XI, 106; XVIII, 52–53. Among the most important are "Recherches sur les forces électromotrices et les quantités de chaleur dégagées dans les combinaisons chimiques," in *Annales de chimie*, 4th ser., **2** (1864), 317–372; "Loi de congélation des solutions aqueuses des matières organiques," in *Comptes rendus hebdomadaires des séances de l'Académie des sciences*, **94** (1882), 1517–1519; "Méthode universelle pour la détermination des poids moléculaires," in *Annales de chimie*, 6th ser., **8** (1886), 317–319; and "Sur les tensions de vapeur des dissolutions," in *Annales de chimie*, **20** (1890), 297–371. He summarized his work in two small volumes: *La tonométrie* (Paris, 1900); and *La cryoscopie* (Paris, 1901).

II. SECONDARY LITERATURE. There is no detailed study of Raoult's life and work. The major sources are J. H. van't Hoff, "Raoult Memorial Lecture," in *Journal of the Chemical Society*, **81** (1902), 969–981; William Ramsay, "Prof. François Marie Raoult," in *Nature*, **64** (1901), 17–18; Harry C. Jones, "François Marie Raoult," in *Science*, n.s. **13** (1901), 881–883; and Frederick H. Getman, "François-Marie Raoult—Master Cryoscopist," in *Journal of Chemical Education*, **13** (1936), 153–155. Excellent brief accounts of his scientific work are in M.M. Pattison Muir, *A History of Chemical Theories and Laws* (New York,

1907), 145–170; and J. R. Partington, *A History of Chemistry*, IV (London, 1964), 645–650.

Louis I. Kuslan

RASPAIL, FRANÇOIS-VINCENT (*b.* Carpentras, France, 29 January 1794; *d.* Arcueil, near Paris, France, 7 January 1878), *biology, medicine, politics*.

Raspail, who came from a poor family, prepared for the priesthood and thus received a thorough classical education at the seminary in Avignon. He was a brilliant student and an excellent speaker, and in 1811 he was assigned to teach philosophy and theology. He refused to take his religious vows and instead found a modest position at the *collège* in Carpentras. Having become involved in the political troubles that accompanied the change in regime at the end of the First Empire, Raspail went to Paris. He openly expressed his republican views during the Restoration and played an active role in the secret societies of the day. For a time he studied law but then turned his attention to the natural and physical sciences, supporting himself and his young family by coaching candidates for examinations. In science he was self-taught, a fact that later found expression in bitter priority claims and in incessant attacks against established scientific institutions. As early as 1824 he attracted attention, especially in certain German periodicals, by an investigation into the classification of the Gramineae that he presented to the Academy of Sciences.

After 1830 Raspail published the results of an extensive series of studies covering a broad range of disciplines. Simultaneously he continued his political activities, participating actively in the revolutions of 1830 and 1848. Raspail incarnated the citizen struggling against the political power of the State. He constantly asserted his highly independent political views, incurring persecution by the many regimes in France during his life. The resulting innumerable political trials, which led to long prison terms and to exile, increased Raspail's popularity with the public. Raspail often expressed pride in being the sole master of his works. His scientific publications are always enlivened by a polemical streak and are particularly interesting because of the many autobiographical details they contain, although these intrusions frequently prejudiced the judgment of his contemporaries. In later years Raspail often boasted of having held the pen in one hand and the sword in the other, symbolizing in a vivid manner his dual vocation of politician and scientist.

Raspail held a prominent place in the development of science in the nineteenth century. In organic chemistry he specified the properties of numerous substances, and he wrote pedagogical works that enjoyed a broad success and went through many editions.

Raspail belonged to the group of biologists who prepared the way for the rise of the cell theory. Although it would be too strong to call him the creator of the modern concept of the cell, the definitions and descriptions he gave of the cell are truly remarkable. On the basis of precise observation he described the general characteristics of the plant cell long before Mohl, who was unaware of Raspail's existence. One of the most original of Raspail's statements in this regard is the following:

> The plant cell, like the animal cell, is a type of laboratory of cellular tissues that organize themselves and develop within its innermost substance; its imperforate walls, to judge from our strongest magnifying instruments, have the property of drawing out by aspiration from the ambiant liquid the elements necessary for this elaboration. They thus have the property of acting as a sorter, of admitting certain substances and preventing the passage of others, and consequently of separating the elements of certain combinations in order to admit only a portion of them.

This functional definition can still serve as the guiding thread through the maze of contemporary cytology. It preceded by several years the formulation of the cell theory by Schwann and Schleiden, who knew Raspail's work and expressed the deepest contempt for it.

An expert microscopist, Raspail not only set forth theoretical considerations of great importance but also made many significant observations. He was skillful at handling the microscope and constructed a simplified model possessing a high magnification. Scientists now agree that he was one of the founders of cytochemistry. As he himself put it, he brought chemical analysis under the microscope. Raspail applied the iodine-starch color reaction to the cell and its products. He also discovered histochemical reactions specific to protides and glucides, although their importance was not fully recognized until the development of biochemistry. The praise accorded to him by Pearse in his classic treatise on histochemistry has substantially contributed to the recent increase in Raspail's reputation in the English-speaking countries.

Was Raspail the founder of cellular pathology, as certain of his hagiographers claim? In this field he was, along with other contemporary workers, a precursor of Virchow. Among his statements on this topic is the following: "Since disease originates in the elementary cell, the organization and microscopic functions of which reproduce the general organization exactly and in all its relationships, nothing is more suited to simplifying the work of classification and of systematic division than to take the elementary cell as the basis of division."

Raspail thus constructed a system of general pathology, which he set forth in his voluminous work on the history of health and illness. In this field too it would be unjust to consider Raspail as merely a theoretician, for he provided valuable new data on the causes of various diseases. For example, he determined the agent of scabies, the itch mite, at a time when medical doctrines, particularly homeopathy, contained exuberant speculations regarding that malady. Raspail is therefore rightly considered one of the founders of modern parasitology.

Although Raspail never acquired a medical degree, he was concerned with the art of healing throughout his life. He established his own pharmacopoeia, which proved to be a success. On several occasions he was prosecuted for practicing medicine illegally; but this did not stop him, especially since his convictions increased his popularity. He gave free consultations in dispensaries that he himself set up, and sought to spread medical and therapeutic knowledge throughout the population. Accordingly, in 1845 Raspail began to publish an annual health manual. This almanac enjoyed a huge success, the last edition appearing in 1945, when its annual printing was still 10,000 copies. After Raspail's death his son Xavier continued its publication, updating the successive editions with new prefaces. These prefaces, a collection of which appeared in 1916, contained extraordinarily harsh invectives against official medicine in all its forms.

Raspail's almanacs constituted veritable formularies with practical directions for the preparation and application of the medications. Camphor, given in various forms, was the best known; along with Raspail's sedative water and liquor it perpetuated his memory. In addition to the almanac Raspail published, when finances allowed, supplementary journals in monthly installments.

Raspail's social activities went far beyond the limits of applied science and of politics. Having survived prison and exile, he recorded his experiences and reflections in numerous newspaper articles, which he twice collected and published in book form (1839, 1872). Drawing on autobiographical material, he forcefully presented his criticisms and proposals for reform of economics and society. He was particularly concerned with improving the judiciary and the prisons.

In many respects Raspail was representative of the age in which he lived. His projects for reform—some of which were later carried out—allow him to be compared with the first "utopian" socialists: Saint-Simon, Charles Fourier, and especially Étienne Cabet, whom he knew personally. Raspail had an advantage over all the others through his scientific and medical experience, which brought him substantial influence with all groups of the population. Daniel Stern, a contemporary witness, wrote in his history of the Revolution of 1848, an event in which Raspail played an important role:

> He exerted a very great influence on the population of the *faubourgs*. His medical knowledge enabled him to . . . relieve, at any moment, the ills and sufferings that the orators of the clubs were happy to depict and that the ambitious knew how to exploit; but this was an isolated moral activity, secretly envied and opposed by the heads of the parties, and it never made a decisive impact on the revolutionary movement [*Histoire de la Révolution de 1848*, p. 9].

This judgment accurately evokes Raspail's dual vocation of political activist and biologist. He enjoyed not only the allegiance of the masses but also the friendship of such literary figures as Sainte-Beuve and Gustave Flaubert. At the same time he was exposed to the hatred of government officials and was the target of vindictiveness of scientists and members of the university community. Of all the polemics he conducted, the most famous is one that opposed him to Orfila, a much honored official scientist, during the sensational trial of Mme Laffarge, an affair not wholly forgotten even today.

Raspail was to a large degree the author of his own legend. After the Revolution of 1830 he refused the Legion of Honor and, in the same period, declined an important position offered him at the Muséum National d'Histoire Naturelle. He did, however, have devoted friends and students who were in a position to help him. For example, the fourteenth baron Vilain, president of the Belgian Chamber of Representatives, enabled Raspail to live and work in Belgium during his exile under the Second Empire.

Raspail's popularity was a constant source of support to him in the course of long periods of imprisonment, during which he never lost courage or ceased to write and publish. His wife unfailingly provided him with moral strength. When she died their daughter took her place—a display of familial devotion that in turn evoked further popular admiration. His sons gave him valuable assistance in the publication and illustration of his works, and in his later years Raspail had the satisfaction of witnessing their professional and political success. His youngest son, Xavier, continued to propagate and defend his father's ideas. Raspail's effigy was often sketched, engraved, and sculpted from his youth until far into his old age, and his portrait was known by many people for nearly half a century.

Very soon after Raspail's death enthusiastic scientific articles, written sympathetically, gave deserved recognition to his achievements. During his lifetime a detailed biography was published in Vapereau's *Dictionnaire des contemporains;* and in 1886 a biographical dictionary of famous physicians included an article on Raspail even though he was not a medical doctor. As early as 1889 a monument to him was erected in Paris at Place Denfert-Rochereau; at about the same date a marble plaque was placed on a house where Raspail had treated the ill without charge. In 1903 Blanchard, an informed and fervent biographer, made a complete list of Raspail's scientific works and gave his name to a laboratory room at the Faculty of Medicine of Paris. Since 1913 one of the capital's most important boulevards has borne Raspail's name. His name has also been given to streets and squares in many cities and even villages of France.

Raspail's life is a moving combination of original and fruitful scientific work and reforming, frequently prophetic political activity. Among Frenchmen his is one of the best known of names, and his fame has become worldwide, particularly since the upsurge in interest in the history of science among scientists and historians.

BIBLIOGRAPHY

I. ORIGINAL WORKS. Complete lists of Raspail's scientific writings are in the works of Blanchard and Weiner (see below).

His most important writings are "Observations et expériences propres à démontrer que les granules . . . ne sont pas des êtres organisés," in *Mémoires de la Société d'histoire naturelle de Paris,* **4** (1828), 347–361; "Premier mémoire sur la structure intime des tissus de nature animale," in *Répertoire général d'anatomie et de physiologie pathologiques,* **4,** no. 2 (1828), 148–161; *Essai de chimie microscopique appliquée à la physiologie . . .* (Paris, 1830); *Nouveau système de chimie organique* (Paris, 1833), 2nd ed., rev., 3 vols. (Paris, 1838); *Nouveau système de physiologie végétale et de botanique,* 2 vols. and atlas (Paris, 1837); *Réforme pénitentiaire. Lettres sur les prisons de Paris,* 2 vols. (Paris, 1839); and *Mémoire à consulter à l'appui du pourvoi en cassation de dame Marie Capelle, veuve Laffarge, sur les moyens de nullité que présente l'expertise chimique* (Paris, 1840).

Manuel annuaire de la santé ou médecine et pharmacie domestique . . . (Paris, 1845), 75th ed., rev. by Xavier Raspail (Paris, 1930), 77th ed. (Paris, 1945); *Revue élémentaire de médecine et pharmacie domestiques . . . ,* 2

vols. (Paris, 1847–1849), issued in monthly installments; *Revue complémentaire des sciences appliquées à la médecine et pharmacie, à l'agriculture, aux arts et à l'industrie,* 6 vols. (Brussels, 1854–1860), issued in monthly installments; *Histoire de la santé et de la maladie,* 3rd ed., 3 vols. (Paris–Brussels, 1860); *Réformes sociales* (Paris–Brussels, 1872); and *François-Vincent Raspail ou le bon usage de la prison,* selected writings with preface and notes by Daniel Ligou (Paris, 1968).

II. SECONDARY LITERATURE. See *Biographisches Lexikon der hervorragenden Aerzte aller Zeiten und Völker,* IV (Leipzig, 1886), 673; R. Blanchard, "Notices biographiques sur François-Vincent Raspail," in *Archives de parasitologie,* **8,** no. 1 (1903), 5–87; G. Duveau, *Raspail* (Paris, 1948); Marc Klein, "Histoire des origines de la théorie cellulaire," in *Actualités scientifiques et industrielles,* no. 328 (1936), 34 ff., 47; Eugène Jacquot de Mirecourt, *Raspail,* 3rd ed. (Paris, 1869); A. G. E. Pearse, *Histochemistry, Theoretical and Applied,* 2nd ed. (London, 1960), 1–12; X. Raspail, *Raspail et Pasteur: 30 ans de critique médicale et scientifique, 1884–1914* (Paris, 1916); Marie de Flavigny, comtesse d'Agoult [Daniel Stern], *Histoire de la Révolution de 1848,* 2nd ed., II (Paris, 1862), 9; G. Vapereau, *Dictionnaire des contemporains,* 4th ed. (Paris, 1870), 1501–1503, and 6th ed. (1893), 1299; and D. B. Weiner, *Raspail, Scientist and Reformer* (New York, 1968).

MARC KLEIN

RASPE, RUDOLF ERICH (*b.* Hannover, Germany, 1737; *d.* Muckross, Ireland, 1794), *literature, geology.*

Raspe was the son of an accountant in the department of mines and forests in Hannover. His interest in geology began early when he came in contact, in the Harz, with some of the oldest mining communities of Europe. But at age eighteen, he was studying law at Göttingen in order to qualify for a position in the bureaucracy. A year later Raspe left Göttingen for Leipzig where his interests ranged from science to literature. Following a three-year stay in Leipzig, he took a master's degree at Göttingen; and in 1760 he became a clerk in the Royal Library of Hannover, where he remained until 1767.

The impact of the great Lisbon earthquake of 1 November 1755 sent Raspe in search of an explanation for this catastrophic event. In the library he "discovered" the almost completely forgotten "Lectures and Discourses of Earthquakes and Subterraneous Eruptions" of Hooke, written in 1668 and published posthumously in 1705.

Instantly Raspe became convinced that Hooke's system was the best explanation for volcanoes and earthquakes and also for the origin of islands and continents in general. Thus he decided to write a theory of the earth that would defend Hooke's ideas and also present historically proven data on new islands and mountains. These documents were supposed to demonstrate that nonvolcanic islands and continents together with their fossiliferous beds could be explained, as Hooke maintained, by the uplifting of the sea bottom that resulted from the forces of earthquakes and subterranean fires. This theory of the earth, entitled *Specimen Historiae Naturalis Globi Terraquei . . .*, was printed in 1763 and met with great success. Raspe, in spite of his lack of formal geological training, had been able to discuss cleverly the major geological issues of his time and even to present some original theoretical ideas in structural geology. He ended his book with a plea to naturalists to investigate the islands that rose in 1707 in the Grecian Archipelago and in 1720 in the Azores.

As suggested by its title, the *Specimen* was presented as the introductory part of a supposedly complete system of the earth that Raspe was preparing. He dreamed of this major work for his entire life and repeatedly stated its impending completion, modifying its emphasis, in his typical opportunistic fashion, according to his everchanging needs. This great work was never published, and apparently never written. An analysis of Raspe's later writings shows that he actually considered the *Specimen* as a completed system, having realized his own limitations. Indeed, he had successfully taken advantage of Hooke's system and of the current popular interest in earthquakes and could not possibly have proceeded scientifically any further.

His *Specimen* was pompously dedicated to the Royal Society of London. Raspe thus hoped that the society would subsidize an expedition under his direction to newly born islands to prove Hooke's theory. This hope was not realized, but the *Specimen* eventually led to Raspe's election as a fellow in 1769.

Had his complete theory of the earth been more than a dream, Raspe would have continued his geological activity. Instead, having discovered in the library several of Leibniz' unpublished manuscripts, in 1765 he undertook their publication. This work was also received with great enthusiasm and had a strong influence on Kant. Meanwhile Raspe had written a poetic comedy (or *Lustspiel*), *Die Verlohrene Bäuerin*, followed by *Hermin und Gunilde, eine Geschichte aus den Ritterzeiten*, the first German ballad of high romance.

By 1767 Raspe had obtained success in the unusual combination of fields of science, philosophy, and literature. He was also corresponding with some of the outstanding men of his age, including Franklin. In the same year (1767) he accepted the curatorship of the collections owned by the Landgrave of Hesse-Kassel, the chair of antiquity at the Collegium Carolinum in Kassel, and a seat on the Hessian Privy Council. His academic post brought him a rather high salary, which temporarily eased his chronic financial difficulties. It coincided with a renewal of his interest in geology, particularly in the origin of basalt. Indeed, Raspe was among the few who read Desmarest's first written statement (1768) demonstrating the volcanic origin of columnar basalt in Auvergne. Raspe immediately understood the significance of the numerous occurrences of basalt around Kassel, which until that time had defied interpretation. On 24 October 1769 he wrote a paper entitled "Nachricht von einigen niederhessischen Basalten . . ." (published in 1771) in which he described the basalt occurrences in the vicinity of Kassel and interpreted the Habichtswald as the remains of an ancient volcano in which massive basalt could be seen overlying prismatic basalt.

Goethe considered this work epochal because it introduced in Germany the volcanic origin of basalt. On 31 October of the same year Raspe wrote an important letter to Sir William Hamilton, British ambassador to Naples and a keen observer of many eruptions of Vesuvius, asking him if any of the cooled lava flows of that volcano displayed structures similar to prismatic basalt. Hamilton answered that they did not, and Raspe hastily concluded that Desmarest's example of Auvergne was a unique and accidental occurrence from which generalizations could not be drawn. Indeed, Raspe felt that since the subaerial lava flows of both Vesuvius and Etna did not show any prismatic structure, columnar basalt (although volcanic) could not represent subaerial flows as interpreted by Desmarest.

Two years of fieldwork (1767–1769) had convinced Raspe that he could demonstrate, in the vicinity of Kassel, that the outcrops of prismatic basalt, such as those of the lower part of the Habichtswald, were always located below what he considered an ancient sea level; while those of massive basalt (for example, at the top of the Habichtswald) were located above that level. He concluded that prismatic basalt represented rapidly cooled submarine lava flows and that massive basalt represented slowly cooled subaerial flows. This erroneous interpretation, which Raspe later called his "bold hypothesis," was published in the fall of 1774 in his *Beytrag zur allerältesten und natürlichen Historie von Hessen . . ..*

This volume appeared just before Raspe, newly married, was accused of having pawned for several years medals and coins from the landgrave's collection in order to pay his creditors. The uncovering of the scandal led to Raspe's arrest. He soon escaped. By mid-April 1775 he had left Germany forever, going first to Holland and then to Great Britain, while his wife and two children took refuge in Berlin pending a divorce. The Royal Society of London expelled him in December 1775.

A fugitive, Raspe managed to survive and maintain his self-respect. His resourcefulness may even deserve admiration. He submitted to Lockyer Davis, the printer for the Royal Society, a proposal to give British scientists the benefit of new German works in geology in a series of annotated translations. The project was accepted and provided Raspe with a modest income. The first volume of the series, a new version of Raspe's own natural history of Hesse, appeared as *An Account of Some German Volcanoes and Their Productions With a New Hypothesis of the Prismatical Basaltes Established Upon Facts* . . . (1776). This work is of great interest when compared with the *Specimen* because it shows Raspe's conviction that he had not only found the physical evidence of the volcanoes postulated by Hooke for the uplifting of the sea bottom but also the prismatic basalts of their submarine eruptions. Although he still recommended investigating the newly born islands of the Grecian Archipelago, this time it was not to prove his idea, expressed in the *Specimen*, that they consisted of fossiliferous beds, but to demonstrate that they were of volcanic origin. Having postulated his own original concept, Raspe was no longer interested in defending Hooke's ideas. He sought to establish a new reputation in England, where he planned to stay and publicize his new theory of the origin of prismatic basalt in a country where the controversial origin of that rock had not as yet generated much interest.

Raspe's first translation was well received and his style praised. He next translated the Swedish geologist Johann Jacob Ferber's letters to Baron Inigo von Born, written during Ferber's travels through Italy in 1771 and 1772. This translation became important in the history of geology because it introduced Arduino's classification of Primary, Secondary, Tertiary, and Volcanic rocks to English-speaking naturalists. Raspe wrote a long preface in which, thirteen years after the publication of the *Specimen*, he again described an improved edition of his theory of the earth, allegedly being prepared for publication. This new version, which he called his "System of the Earth" and his "Natural History of the Earth," would emphasize volcanism as one of the most important geological processes. According to Raspe's extravagant ideas, volcanoes were responsible not only for the formation of islands and continents but also for the deposition of fossiliferous shales and limestones—interpreted as subaqueous volcanic productions—and even for the salty character of seawater.

Toward the end of 1776 Raspe began to realize that only mining and prospecting rather than translating scientific works would allow his financial survival. Thus the lengthy preface to his translation of Born's travels through the Banat of Timisoara, Transylvania, and Hungary (1777) contains an interesting discussion of the history of mining since antiquity, as well as a detailed presentation of the theoretical and practical aspects of mineralogy and mining at the end of the eighteenth century. The last mention of his elusive dream occurs in the preface as a discussion of a new version of the *Specimen*, now called "System of the Earth and Mountains." It was supposed to emphasize the origin of mountains and metallic veins; to present a general discussion of mining; and to advance a new system of mineralogy for miners, one combining geology, chemistry, and metallurgy.

The last eighteen years of Raspe's life, although eventful, were a period of scientific decline. He was still occasionally admitted to the best circles of London society, mainly under the sponsorship of Horace Walpole, who encouraged him to write an essay on oil painting (1781). He also became acquainted with Matthew Boulton. The latter realized that Raspe's technical knowledge could be useful in his expanding tin mines of North Cornwall and consequently offered him the post of assay-master at the Dolcoath mine; he held this post from 1783 to 1786. Raspe was also an industrial spy and in many other ways actively served Boulton's interests. It was apparently during this period that Raspe wrote *Baron Münchausen's Narrative of His Marvellous Travels and Campaigns in Russia*, his most famous work, which however brought him little remuneration. It was published anonymously in 1786. He wanted to be remembered for his serious work, and thought fiction unworthy of a scholar.

In 1787, when the mining industry in Cornwall deteriorated, Raspe moved to Edinburgh, where Sir John Sinclair of Ulbster, who was greatly interested in mineralogy, introduced him to high society, which included James Hutton and Joseph Black.

Between 1789 and 1792 the Highland Society of Scotland and a number of English patrons hired Raspe to undertake a mineralogical survey of the Highlands. Although he hinted that he had found quicksilver deposits, no commercial ores were ever discovered. During this period Raspe published (1791) an impressive catalog of the 15,833 ancient and modern gems from famous collections throughout Europe, of which his friend James Tassie had produced replicas. He also wrote his last major work, a translation of Baron Born's treatise on amalgamation of gold and silver ores. Yet Raspe remained penniless; and in

304

1793 he transferred his prospecting to Ireland, where, at the copper mine of Muckross, in a remote part of Kerry, he died of scarlet fever.

BIBLIOGRAPHY

I. ORIGINAL WORKS. Raspe's writings include *Specimen Historiae Naturalis Globi Terraquei, praecipue de novis e mari natis insulis, et ex his exactius descriptis et observatis ulterius confirmanda Hookiana telluris hypothesi, de origine montium et corporum petrefactorum* (Amsterdam–Leipzig, 1763), trans. by Audrey N. Iversen and Albert V. Carozzi as *An Introduction to the Natural History of the Terrestrial Sphere* . . . (New York, 1970); *Die Verlohrene Bäuerin, ein Lustspiel in einem Aufzuge, auf das Geburtsfest der Königin Sophia Charlotta von Grossbritt. auf dem deutschen Theater der Ackermannische Gesellschaft . . . vorgestellt* (Hannover, 1764), published anonymously; *Oeuvres philosophiques latines et françoises du feu Mr. de Leibniz, tirées de ses manuscrits qui se conservent dans la Bibliothèque Royale à Hanovre et publiées par Mr. Rud. Eric Raspe, Avec une Préface de Mr. Kaestner* . . . (Amsterdam–Leipzig, 1765); and *Hermin und Gunilde, eine Geschichte aus den Ritterzeiten, die sich zwischen Adelepsen und Usslar am Schäferberg zugetragen, nebst einem Vorbericht über die Ritterzeiten in einer Allegorie* (Leipzig, 1766).

Subsequent works are "Nachricht von einigen niederhessischen Basalten, besonders aber einem Säulenbasaltstein Gebürge bei Felsberg und den Spuren eines verlöschten brennenden Berges am Habichtswalde über Weissenstein nahe bei Cassel," in *Deutschen Schriften der Kgl. Societät der Wissenschaften in Göttingen,* **1** (1771), 72–83; "Anhang eines Schreibens an den Königl. Grosbritt. Gesandten Herrn William Hamilton, zu Neapolis," ibid., 84–89; *Beytrag zur alterältesten und natürlichen Historie von Hessen oder Beschreibung des Habichswaldes und verschiedner andern Niederhessischen alten Vulcane in der Nachbarschaft von Cassel* (Kassel, 1774); and *An Account of Some German Volcanos and Their Productions With a New Hypothesis of the Prismatical Basaltes Established Upon Facts; Being an Essay of Physical Geography for Philosophers and Miners, Published as Supplementary to Sir William Hamilton's Observations on the Italian Volcanos* (London, 1776).

Raspe's trans. are *Travels Through Italy in the Years 1771 and 1772 Described in a Series of Letters to Baron Born on the Natural History, Particularly of the Mountains and Volcanos of That Country, by John James Ferber* (London, 1776), from the German; and *Travels Through the Bannat of Temeswar, Transylvania and Hungary in the Year 1770; Described in a Series of Letters to Professor Ferber, on the Mines and Mountains of These Different Countries by Baron Inigo Born; to Which Is Added John James Ferber's Mineralogical History of Bohemia* (London, 1777).

See also *A Critical Essay on Oil Painting, Proving That the Art of Painting in Oil Was Known Before the Pretended Discovery of John and Hubert van Eyck; to Which Are Added Theophilus de Arte Pingendi, Eraclius de Artibus Romanorum, and a Review of Farinator's Lumen Animae* (London, 1781); *Baron Münchausen's Narrative of his Marvellous Travels and Campaigns in Russia* (Oxford, 1786), published anonymously, with a recent ed. by John Carswell, *Singular Travels, Campaigns and Adventures of Baron Münchausen* (London, 1948); *A Descriptive Catalogue of a General Collection of Ancient and Modern Engraved Gems . . . Taken From the Most Celebrated Cabinets in Europe and Cast . . . by J. Tassie . . . Arranged and Described by R. E. Raspe . . . to Which Is Prefixed an Introduction on the . . . Use of the Collection, the Origin of the Art of Engraving on Hard Stones, and the Progress of Pastes,* 2 vols. (London, 1791); and *Baron Inigo Born's New Process of Amalgamation of Gold and Silver Ores, and Other Metallic Mixtures . . . From the Baron's Own Account in German; to Which Are Added, a Supplement on a Comparative View of the Former Method of Melting and Refining; and an Address to the Subscribers, Giving an Account of Its Latest Improvements, and of the Quicksilver Trade* (London, 1791).

II. SECONDARY LITERATURE. On Raspe and his work, see J. Carswell, *The Prospector, Being the Life and Times of Rudolf Erich Raspe (1737–1794)* (London, 1950), with complete bibliography; R. Hallo, *Rudolf Erich Raspe, ein Wegbereiter von deutscher Art und Kunst,* Göttinger Forschungen no. 5 (Stuttgart–Berlin, 1934); and Thomas Seccombe's notice in *Dictionary of National Biography,* XVI, 744–746.

ALBERT V. CAROZZI

RATEAU, AUGUSTE CAMILLE EDMOND (*b.* Royan, Charente-Maritime, France, 13 October 1863; *d.* Paris, France, 13 January 1930), *fluid mechanics, turbomachinery.*

Rateau's father, also named Auguste, listed his occupation as "entrepreneur des travaux publics"; his mother was Lucie Chardavoine. He graduated at the top of his class at the École Polytechnique in 1883 and chose service in the Corps des Mines. During 1885 he made an extensive tour of technical installations in Belgium, Germany, Austria-Hungary, Italy, and Russia. He was promoted through the grades of third-, second-, and first-class student engineer, then third-, second-, and first-class *ingénieur ordinaire,* attaining the last rank in 1898.

In 1888 Rateau was named professor at the École des Mines at St. Étienne, a post he held for the next decade. He combined teaching duties with mining and railroad assignments in central and southern France. While on leave from the Corps des Mines during 1899, he served as secretary to the Commission on Construction Materials and as consultant to the firm of Houillères de St. Chamond. While at St. Étienne he became interested in mine ventilation, and this led to his lifelong interest in turbomachinery. He designed centrifugal pumps and multistage centrifugal gas compressors.

Rateau was appointed professor at the École Supérieure des Mines at Paris in 1902. The following year he formed his own company, Société Rateau, to manufacture his turbomachines. Sautter-Harlé et Cie. also manufactured apparatus and equipment designed by Rateau. During World War I he operated a munitions factory while continuing his research in other areas.

In the power industry Rateau is remembered primarily for his steam turbine. Before this, C. G. P de Laval had developed the high-speed single-stage steam turbine, which required reduction gearing for most applications and was limited in capacity. In England, Charles Parsons had developed a multistage steam turbine in which relatively small pressure drops were used in the nozzles and the moving blades. Rateau is credited with the invention of the pressure-stage impulse turbine. He restricted the expansion of steam to the nozzles but used larger pressure drops per stage. In his design the rotors were separated by diaphragms which contained the nozzles. This reduced the number of stages for a given power output and shortened the distance between bearings. The Rateau turbine has been dated at 1901; the U.S. patent was issued in 1903. His turbine found wide acceptance in both marine and central station applications. The use of a single Curtis wheel (velocity staging), followed by a number of Rateau stages (pressure staging), characterized General Electric Company practice in the United States over the succeeding half-century. The combination is still widely used today.

In the analysis of heat-power cycles, Rateau appreciated the advantages of thermal storage and was one of the early advocates of the steam accumulator. In mines and mills he applied low-pressure condensing steam turbines, which operated on steam exhausted from reciprocating engines.

Rateau performed many experiments on ejectors and proposed theoretical explanations for their performance. His investigations on the flow of water through nozzles and on the flow of steam through nozzles were widely acknowledged. His work in machine design led to such diverse results as a new deflection equation for Belleville springs, minimum friction gearing, a muzzle brake for artillery, stereo-optics for motion pictures, and servo systems for governing steam turbines.

As early as 1900 Rateau was interested in the problems associated with aviation. He studied aircraft performance, wing sections, and aircraft propellers, confirming his calculations with wind tunnel tests.

Although Rateau did not invent the turbosupercharger, he was perhaps the first to apply it to aircraft engines (1916–1917). He investigated the effect of the supercharged engine on the range and altitude capabilities of airplanes.

The Académie des Sciences awarded Rateau the Prix Laplace (1883), the Prix Fourneyron (1899), and the Prix Poncelet (1911). He was elected to the Institut de France in 1911 and to the Académie des Sciences in 1918. He was an honorary member of the American Society of Mechanical Engineers and of the Institution of Mechanical Engineers (London). In 1922 the latter organization awarded him the Hawkesley Medal. He was made a commander of the Legion of Honor in 1925. Honorary doctorates were conferred upon him by the universities of Louvain, Birmingham, and Wisconsin, and the Berlin Technische Hochschule.

BIBLIOGRAPHY

I. ORIGINAL WORKS. Rateau's books are *Études sur les appareils Picard pour la vaporisation des dissolutions salées et sur l'emploi du travail pour obtenir de la chaleur* (Paris, 1888); *Considérations sur turbo-machines, et particulièrement sur les ventilateurs* (St. Étienne, 1892); *Hypothèse des cloches sous-continentales* (Paris, 1893); *Recherches expérimentales sur l'écoulement de la vapeur d'eau par les tuyères et les orifices* (Paris, 1902); *Rapport sur les freins de bouche: théorie générale—Calculs pratiques—Résultats expérimentaux* (Paris, 1911); *Expériences d'Allevard sur les coups de bélier* (St. Cloud, 1914); *Étude théorique et expérimentale sur les coups de bélier dans les conduites forcées* (Paris, 1917), written with Jouquet and de Sparre; *Travaux scientifiques et techniques* (Paris, 1917); *Théorie des hélices propulsives marines et aérienne et des avions en vol rectiligne* (Paris, 1920); and *Turbines hydrauliques* (Paris, 1926), written with D. Eydoux and M. Garril.

On turbomachinery see also "Sur les turbo-machines," in *Comptes rendus hebdomadaires des séances de l'Académie des sciences*, **113** (1891); 463–465; "Considérations sur les turbo-machines," in *Bulletin de la Société de l'industrie minérale*, 3rd ser., **6** (1892), 47–228; "Ventilateurs et pompes centrifuges pour haute pressions," *ibid.*, **1** (1902), 73–141; "Steam Turbines," in *Engineering*, **76** (1903), 32, 105–106; "Steam Turbine Propulsion for Marine Purposes" in *Mechanical Engineering*, **26** (1904), 487–493; and "Turbo-compresseurs à haute pression et utilisation des vapeurs d'échappement," in *Bulletin de la Société de l'industrie minérale*, **9** (1908), 569–595. On steam accumulators and machinery see "Recuperateur de vapeur," in *Revue de Mécanique*, **8** (1901), 485–486; "Recuperateur de vapeur Rateau," *ibid.*, **12** (1903), 385–391; "Essais fait au mois d'Avril 1902 sur la turbine Rateau à basse pression des mines de Bruay," *ibid.*, **12** (1903), 400–402; "Développement des turbines à vapeur d'échappement," *ibid.*, **21** (1907), 356–387; "Perfectionner en quelque point la théorie des trompes" in *Comptes rendus hebdo-*

madaires des séances de l'Académie des sciences, **129** (1899), 1077–1078; "Nouvelle théorie des trompes," in *Revue de Mécanique*, **7** (1900), 265–314; "Théorie des ejecto-condenseurs," *ibid.*, **10** (1902), 300–301, 533–559; "Sur la théorie générale du mouvement varié de l'eau les tuyaux de conduite," *ibid.*, **14** (1904), 1–9; "États successifs d'un gaz à haute pression dans un récipient qui se vide par une tuyère," in *Comptes rendus hebdomadaires des séances de l'Académie des sciences*, **168** (1919), 435–439; "Quantité de mouvement totale et vitesse moyenne du jet de gaz sortant d'un réservoir qui se vide par une tuyère," *ibid.*, 581–587; "Formule pratique pour le calcul des Rondelles Belleville," *ibid.*, **104** (1887), 1690; "Sur les engrenages sans frottement," *ibid.*, **114** (1892), 580–582; "Adresse quelques indications sur un projet d'appareil qui permettrait d'obtenir la vision stereoscopique en cinématographie," *ibid.*, **127** (1898), 139; and "Étude théorique et expérimentale sur les coups de bélier dans les conduites forcées," *ibid.*, **174** (1922), 1598–1604.

For his work on aircraft see "Théorie des hélices propulsives," in *Comptes rendus hebdomadaires des séances de l'Académie des sciences*, **130** (1900), 486–489, 702–705; "Théorie des hélices et des surfaces sustentrices," in *Revue de Mécanique*, **25** (1909), 117–130; "Méthode d'expériences pour recherches aérodynamique," in *Comptes rendus . . . de l'Académie des sciences*, **148** (1909), 1662–1664; "Étude de la poussée de l'air sur un surface," *ibid.*, **149** (1909), 260–263; "Théorie du vol des aéroplanes aux diverses altitudes," *ibid.*, **168** (1919), 1142–1147; "Suite de la théorie des aéroplanes," *ibid.*, 1246–1251; and "Théorie de la montée rectiligne des aéroplanes," *ibid.*, 1295–1301. On superchargers, see "Théorie générale du turbo-compresseur pour moteurs d'avions," *ibid.*, **174** (1922), 1511–1516; and "Calcul des variations du plafond d'un aéroplane dues à une variation de son poids ou à l'emploi d'un turbo-compresseur," *ibid.*, 1669–1674.

II. SECONDARY LITERATURE. On Rateau's work see C. Dornier, *Beitrag zur Berechnung der Luftschrauben* (Berlin, 1912); A. J. Büchi, *Exhaust Turbocharging of Internal Combustion Engines* (Philadelphia, 1953); and E. T. Vincent, *Supercharging the Internal Combustion Engine* (New York, 1948). An obituary is H. H. Supplée, "Auguste C. E. Rateau," in *Mechanical Engineering*, **52** (1930), 571.

JAMES H. POTTER

RATHKE, MARTIN HEINRICH (*b.* Danzig, Prussia [now Gdańsk, Poland], 25 August 1793; *d.* Königsberg, Prussia [now Kaliningrad, U.S.S.R.], 15 September 1860), *embryology, anatomy*.

Rathke was born into a wealthy burgher family. His father was a master shipbuilder. In 1814, after studying at the local Gymnasium in Danzig, he went to the University of Göttingen to study natural history and medicine. In 1817 he left Göttingen for Berlin, where he received the M.D. degree in 1818.

He then returned to Danzig, where he practiced medicine and for three years served as assistant master in the city Gymnasium.

In 1825 Rathke became chief physician at the municipal hospital in Danzig, and in 1826 he was named district physician. During this time, he engaged in research and writing. He discovered branchial clefts, plates, and vascular arches in the embryos of the higher abranchiate animals and also published his study of the respiratory organs in birds and animals. In 1829 he was named professor of physiology and pathology at the University of Dorpat. While at Dorpat he traveled extensively in the Baltic states and Finland and visited St. Petersburg and Moscow.

During his travels, Rathke collected considerable information on animal and marine life, but his interests focused increasingly on embryology and comparative anatomy. He had established contact with Baer, who was then a professor at Königsberg; and he had been a student at Göttingen with Pander. These three men are recognized as the founders of modern embryology. In 1834, when Baer left Königsberg for St. Petersburg, Rathke succeeded him as professor of zoology and anatomy. He joined the faculty at Königsberg in 1835 and remained there until his death. He was an extremely influential force on the development of the university. Rathke was greatly admired by his students and colleagues and, shortly before his death, was given a twenty-fifth anniversary celebration for his years at Königsberg. Rathke continued his travels while at Königsberg; in 1839, for example, he visited Norway and Sweden.

Probably Rathke is best known for his discovery of gill slits and gill arches in embryo birds and animals. He followed the embryological history of these structures and found that the gill slits eventually disappear and that the blood vessels adapt themselves to the lungs, which develop from an expansion of the front part of the digestive canal. He also described and compared the development of the air sacs of birds and the larynx of birds and mammals. Rathke's pocket—a small pit on the dorsal side of the oral cavity that marks the point of invagination of the hypophysis in the development of vertebrates—is named after him.

Rathke was also interested in metamorphosis and tried to account for the regressive development of certain organs, including the gills, the tails of the tadpole, and the so-called primordial Wolffian bodies discovered by him. He characterized these bodies as "head kidneys" (pronephros) and described how, during early stages of embryonic development, they perform the function of excretal organs; they disappear as true kidneys develop. (In certain animals the efferent ducts of the Wolffian bodies serve as part of the sex organs.)

Rathke thought that regressive organs were either dissolved and reabsorbed by the body or knocked off and eliminated. In the former case they possess blood vessels through which their substance can be absorbed and used by the body; in the latter case they are horny and lack blood vessels. He thought also that the disappearance of one organ is always succeeded by the development of another to take its place—only an altered mode of living during more advanced stages of development can effect the total loss of previously existing organs.

Rathke was interested in marine research and was the first to describe the lancet fish, which previously had been considered the larva of a mollusk. He also wrote several monographs on crustaceans (both independent and parasitic), mollusks, and worms, as well as on a number of vertebrates, including the lemming and various reptiles. He was also interested in the embryonic development of sex organs. Rathke carried some of the embryological investigations launched by his friend Pander to their logical conclusions, but his work in embryology is less well known than that of Baer—perhaps because his generalizations were more embryological than transcendental. Nonetheless his work was as solidly grounded as that of the more influential Baer.

BIBLIOGRAPHY

I. ORIGINAL WORKS. Rathke wrote more than 125 articles, monographs, and books. A helpful bibliog. is L. Stieda, in *Allgemeine deutsche biographie*, XXVII, 352–355. His major works include *De Salamandrarum corporibus adiposis, ovariis et oviductibus, eorum evolutione* (Berlin, 1818), his M.D. diss.; *Beiträge zur Geschichte der Thierwelt* (1820–1824); *Bemerkungen über den inneren Bau der Prick* (Danzig, 1826); *Untersuchungen über die Bildung und Entwickelung das Flusskrebes* (Leipzig, 1829); and *Abhandlungen zur Bildungs- und Entwicklungsgeschichte der Menschen und der Thiere*, 2 vols. (Leipzig, 1832–1833).

Later works are *Entwickelungsgeschichte der Nätter* (Königsberg, 1839); *Bemerkungen über den Bau des Amphioxus lanceolatus, eines Fishes aus der Ordnung der Cyclostomas* (Königsberg, 1841); *Beiträge zur vergleichenden Anatomie und Physiologie, Reisebemerkungen aus Skandinavien, nebst einem Anhange über die rückschreitende Metamorphose der Thiere* (Danzig, 1842); *Über die Entwickelungen der Krokodile* (Brunswick, 1866); and *Entwickelungsgeschichte der Wirbeltier* (Leipzig, 1861).

II. SECONDARY LITERATURE. For a discussion of Rathke's influence, see Jane M. Oppenheimer, "The Non-Specificity of the German Layers," in *Essays in the History of Embryol-*

ogy and Biology (Cambridge, Mass., 1967), a repr. with updated bibliog. of an article that originally appeared in *Quarterly Review of Biology*, **15** (1940), 1–27. A brief sketch can also be found in Eric Nordenskiöld, *History of Biology* (New York, 1928).

VERN L. BULLOUGH

RATZEL, FRIEDRICH (*b*. Karlsruhe, Germany, 30 August 1844; *d*. Ammerland, Germany, 9 August 1904), *geography, ethnography.*

Ratzel was one of the four children of the manager of household staff of the grand duke of Baden. He spent six years at La Fontaine Gymnasium in Karlsruhe, before being apprenticed at the age of fifteen, to an apothecary at Eichtersheim, a village between Karlsruhe and Heidelberg. After six years as an apprentice there, at Rapperswil, Switzerland, and at Moers, in the Ruhr, he decided to follow an academic career. After a short time at the Karlsruhe Technische Hochschule he studied zoology at Heidelberg and then at Jena, where he produced a thesis for a degree in 1868. In the following year he published a popular history of creation which showed an uncritical acceptance of Darwin's main concepts and an obvious reliance on the views of Haeckel.

Ratzel subsequently traveled in the Mediterranean countries and assisted the French naturalist Charles Martin at Sète and Montpellier. After his recovery from wounds suffered in the Franco-Prussian war, he studied for a short while at Munich, where he met the geologist Karl von Zittel and Moritz Wagner, the curator of the university ethnographical museum, whose ideas on the importance of the migration of species made a lasting impression on him.

In 1871 Ratzel traveled widely in the Danube countries, the Alps, and Italy. In 1874 and 1875 he made a long tour of North America, where he was especially interested in the black population, the dwindling habitation areas of the American Indians, and the inflow of the Chinese into California. He wrote six large volumes on North America, treated it in parts of other works, and produced a shorter treatise entitled *Das Meer als Quelle der Völkergrosse* (1900).

On his return to Europe, Ratzel began a full-time academic career, lecturing on geography at the Technische Hochschule in Munich. He began to criticize Darwin's theory of natural selection and Haeckel's views on evolution but always remained convinced of the application of concepts of organic evolution to human societies. His lectures covered a wide range of geographical topics; and although he was mainly interested in human aspects, he also wrote extensively on physical aspects such as earth-pillars, limestone surfaces, and the snow line. During his eleven years at Munich, Ratzel produced the first volume of *Anthropo-geographie* (1882) and the first two volumes of his *Völkerkunde* (1885–1886), as well as at least 160 shorter items.

In 1886 Ratzel accepted the chair of geography at Leipzig, succeeding Ferdinand von Richthofen. During his years at Leipzig (1886–1904), he wrote thirteen books and about 350 sizable articles. His shorter articles range from useful observations on physical phenomena, such as the measurement of the density of snow (1889), to numerous obituaries and polemics. They also include several important analyses of his concepts of *Lebensraum* and "scientific political geography."

An accurate assessment of the scientific importance of Ratzel's work is difficult because of his changes of view, his enormous output, and the emotions aroused by the disastrous political results that stemmed from perverted versions of his concepts.

Besides the patient amassing of knowledge on terrestrial objects and phenomena and their spatial distribution, Ratzel's main scientific accomplishments were in ethnography and human geography. His chief contribution to ethnography was to popularize the importance of the diffusion of culture by migration and by borrowing. He assumed, often correctly, that agreement of form which was independent of the nature, material, and function of an object or artifact indicated a historical connection or borrowing.

The bridge between Ratzel's ethnography and geography appeared clearly in his *Anthropo-geographie*, in which the central themes are spatial distributions of a culture and their dependence on the physical environment and on migrations and borrowings over the centuries. The first volume deals mainly with the causes or dynamic aspects of human distributions, and the second emphasizes the static aspects. The development of a society or a people is considered to be governed largely by its situation relative to other geographical phenomena, by the space in which it moves, and by the frame that limits those movements and is itself related to the development of adjacent peoples.

In his later years Ratzel took a keen interest in contemporary philosophy, and in his *Politische Geographie* (1897) he attempted to combine practical politics with physical-philosophical material. He began to favor the idea of a "sense of space" that influences the collective psychology of its inhabitants, and believed that great space confers great rights and virtues. This important book was probably the first political geography to tackle each problem methodically and to treat the subject scientifically.

Ratzel's deep concern with spatial relations and especially with the covariants of cultural distributions were truly fruitful concepts. When carried to extremes their political aspects could be perverted into crude master-race theories, which had a disastrous effect during the rise of Hitler. Ratzel did occasionally criticize such extreme racial views, but his criticisms were drowned amid the mass of his writings. Apart from his major works he published many important articles on political geography and the spatial relationships of geographical phenomena. The most famous was the article "Der Lebensraum. Eine biogeographische Studie" (1901), which summarized his earlier efforts to associate organic activities with the physical environment and to link biology and biogeography with human geography.

Ratzel's ideas on man-milieu relationships could, if advertised more clearly, have developed into modern notions on the cultural landscape and on the conservation of the environment. Similarly, his ideas on spatial economic and cultural distributions were among the most fruitful concepts ever devised in scientific human geography. The *Lebensraum* concept, as he showed, was applicable to units of different size. His essays on the sites and development of cities and on the hinterlands of seaports (in which he defined the areal extent and overlap of the limits in regard to nature, politics, delivery, supply or production, and traffic) were the precursors of studies that appeared after 1930. Despite his misfortunes at the hands of extremists, Ratzel must be regarded as the true founder and greatest single contributor to the development of modern human geography.

BIBLIOGRAPHY

I. ORIGINAL WORKS. The most complete bibliography of Ratzel's publications is that by Viktor Hantzsch in Hans Helmolt, ed., *Kleine Schriften von Friedrich Ratzel*, II (Munich–Berlin, 1906), 1–lix. Harriet Wanklyn gives a selective bibliography in *Friedrich Ratzel*, 57–94.

Ratzel's books include *Sein und Werden der organischen Welt. Eine populäre Schöpfungsgeschichte* (Leipzig, 1869); *Die Vorgeschichte des europäischen Menschen* (Munich, 1874); *Die Vereinigten Staaten von Nordamerika*, 3 vols. (Munich, 1878–1893); *Die Erde* . . . (Stuttgart, 1881), 24 popular lectures on general geography; *Anthropogeographie oder Grundzüge der Anwendung der Erdkunde auf die Geschichte*, 2 vols. (Stuttgart, 1882–1891; I, 2nd ed., 1899); *Völkerkunde*, 3 vols. (Leipzig, 1885–1888; 2nd ed., 1894–1895), translated by A. J. Butler as *The History of Mankind*, 3 vols. (London, 1896–1898); *Die Schneedecke, besonders in deutschen Gebirgen* (Stuttgart, 1889); *Anthropogeographische Beiträge. Zur Gebirgskunde, vorzüglich Beobachtungen über Höhengrenzen und Höhengürtel*

(Leipzig, 1895); *Politische Geographie* (Munich–Leipzig, 1897), 2nd ed. enl. as *Politische Geographie oder die Geographie der Staaten, des Verkehrs und des Krieges* (Munich–Berlin, 1903), 3rd ed. edited by Eugen Oberhummet (Vienna–Berlin, 1925); *Beiträge zur Geographie des Mittleren Deutschlands* (Leipzig, 1899); and *Das Meer als Quelle der Völkergrosse. Eine politisch-geographische Studie* (Munich, 1900); *Die Erde und das Leben. Eine vergleichende Erdkunde*, 2 vols. (Leipzig, 1901–1902).

Among Ratzel's papers are "Zur Entwicklungsgeschichte des Regenwurms," in *Zeitschrift für wissenschaftliche Zoologie*, **18** (1868), 547–562; "Über die Entstehung der Erdpyramiden," in *Jahresbericht der Geographischen Gesellschaft in München* for 1877–1879 (1880), 77–88; "Über geographische Bedingungen und ethnographische Folgen der Völkerwanderungen," in *Verhandlungen der Gesellschaft für Erdkunde zu Berlin*, **7** (1880), 295–324; "Die Bestimmung der Schneegrenze," in *Naturforscher*, 19th annual special issue (1886), 245–248; "Über Messung der Dichtigkeit des Schnees," in *Meteorologische Zeitschrift*, **6** (1889), 433–435; "Über Karrenfelder im Jura und Verwandtes," in *Leipziger Dekanatsprogramm* (1892), 3–26; "Über die geographische Lage. Eine politisch-geographische Betrachtung," in *Feestbundel aan Dr. P. J. Veth aangeboden* (Leiden, 1894), 257–261; "Studien über politische Räume," in *Geographische Zeitschrift*, **1** (1895), 163–182, 286–302; "Die Gesetze des räumlichen Wachstums der Staaten . . ," in *Petermanns, A., Mitteilungen aus J. Perthes Geographischer Anstalt*, **42** (1896), 97–107; "Der Staat und sein Boden geographisch betrachtet," in *Abhandlungen der philologischen Klasse der Königlichen Sächsischen Gesellschaft der Wissenschaften zu Leipzig*, philolog. Kl., **17**, no. 4 (1896); "Der geographische Methode in der Ethnographie," in *Geographische Zeitschrift*, **3** (1897), 268–278; "Der Lebensraum. Eine biogeographische Studie," in *Festgaben für Albert Schaeffle* (Tübingen, 1901), 103–189; and "Die geographische Lage der grossen Städte," in *Grosstadt, Gehestiftung zu Dresden*, **9** (1902–1903), 33–72.

II. SECONDARY LITERATURE. See P. Vidal de la Blache, "Friedrich Ratzel," in *Annales de géographie*, **13** (1904), 466–467; J. O. M. Broek, "Friedrich Ratzel in Retrospect," in *Annals of the Association of American Geographers*, **44** (1954), 207; Jean Brunhes, "F. Ratzel (1844–1904)," in *Géographie*, **10** (1904), 103–108; W. J. Cahnmann, "The Concept of Raum and the Theory of Regionalism," in *American Sociological Review*, **9** (1944), 455–562; Paul Claval, *Essai sur l'évolution de la géographie humaine* (Paris, 1964), 18, 37–45; R. E. Dickinson, *The Makers of Modern Geography* (London, 1969), 62–76; Y. M. Goblet, *Political Geography and the World Map* (London, 1955), 8–12; Kurt Hassert, "Friedrich Ratzel, Sein Leben und Wirken," in *Geographische Zeitschrift*, **11** (1905), 305–325, 361–380; R. H. Lowie, *The History of Ethnographical Theory* (New York, 1937), 120 and *passim*; O. Marinelli, "Frederigo Ratzel e la sua opera geografica," in *Rivista geografica italiana*, **12** (1905), 8–18, 102–126; Hermann Overbeck, "Ritter, Riehl, Ratzel . . .," in *Erde*, **3** (1951–1952), 197–210; and "Das politischgeographische Lehre-

gebaude von Friedrich Ratzel . . .," *ibid.*, **9** (1957), 169–192 —articles reproduced in Overbeck's "Kulturlandschaftsforschung und Landeskunde," in *Heidelberger geographische Arbeiten*, no. 14 (1965), 60–103; Louis Raveneau, "L'élément humain dans la géographie: L'anthropogéographie de M. Ratzel," in *Annales de géographie*, **1** (1891–1892), 331–347; E. G. Ravenstein, "Friedrich Ratzel," in *Geographical Journal*, **24** (1904), 485–487; Johannes Steinmetzler, "Die Anthropogeographie Friedrich Ratzels und ihre Ideengeschichtlichen Würzeln," in *Bonner geographische Abhandlungen*, no. 19 (1956), an indispensable account of Ratzel's cultural anthropology; and Harriet G. Wanklyn, *Friedrich Ratzel: A Biographical Memoir and Bibliography* (Cambridge, 1961), a small but valuable book.

R. P. BECKINSALE

RAULIN, JULES (*b*. Mézières, France, 6 September 1836; *d*. Lyons, France, 26 May 1896), *plant physiology*.

Raulin was a pupil of Pasteur at the École Normale Supérieure, where he first began to study the mineral nutrition of plants. He continued this work at the *lycées* of Brest and Caen, then returned to the École Normale Supérieure as assistant director of Pasteur's laboratory.

Pasteur had grown yeast in a medium containing saccharose, ammonium tartrate, and ashes of yeast, which supplied all the mineral needs of the growing plant. Raulin therefore determined to nourish a plant experimentally with a carefully controlled mixture of specific minerals. He chose to work with the common mold *Sterigmatocystis nigra* (or *Aspergillus niger*), which thrives on saccharose. Having ascertained the ideal environment for its growth—a large dish containing 3 centimeters of water at 35° C., adequately aerated, in a room kept at 70 percent humidity—he set out to establish empirically the combination of mineral nutrients that would produce the best yield, measured in dry weight, in a given period of time.

After a number of attempts, Raulin determined the most efficient mixture of nutrients to be 1,500 grams water, 70 grams saccharose, 4 grams tartaric acid, 4 grams ammonium nitrate, 0.6 gram ammonium phosphate, 0.6 gram potassium carbonate, 0.4 gram magnesium carbonate, 0.25 gram ammonium sulfate, 0.07 gram zinc sulfate, 0.07 gram ferrous sulfate, and 0.07 gram potassium silicate. This mixture is now called Raulin's medium, or Raulin's fluid.

The chief value of Raulin's experiment lay in his establishment, by trial-and-error methods, of the importance of each of the nutrients in his mixture, especially the minerals. For example, he eliminated potassium from the mixture, and the weight of the yield dropped from 24.4 grams to 0.92 gram; he was therefore able to demonstrate that the presence of potassium in the medium resulted in a yield of 26.6 times greater weight—or, in other terms, that the specific utility of potassium was 87, since 0.271 gram of potassium increased the yield by 23.48 grams. Raulin studied each element in the same way, and found that nitrogen produced 153 times the yield, phosphorus 182, magnesium 91, sulfur 11, iron approximately 2, and zinc approximately 2.4.

In adding a few milligrams of iron and zinc to the medium, Raulin introduced the problem of trace elements. The purity of the chemical products that he used was at best doubtful, and there was no way for him to analyze how much iron or zinc might be contained in the four grams of ammonium nitrate or tartaric acid that he added to the medium. He was nonetheless able to isolate zinc as a trace element; its role in plant nutrition had not previously been recognized.

Raulin published the results of his experiments as "Études chimiques sur la végétation," for which he received the *doctorat ès sciences*. Pasteur himself was enthusiastic about Raulin's work, and in 1868 wrote to him that his researches had opened completely new horizons in plant research. In 1876 Raulin was appointed professor of chemistry at the Faculté des Sciences of Lyons, of which he later became dean. He also established a school of industrial chemistry, made a number of agronomic maps, and published a large number of works on a variety of subjects, of which the greatest part concern the production of silk.

BIBLIOGRAPHY

Raulin's chief work is "Études chimiques sur la végétation," in *Annales des sciences naturelles, botanique*, 5th ser., **51** (1870), 93–299; it was separately published as *Études chimiques sur la végétation* (Paris, 1905).

J. B. CARLES

RAUWOLF, LEONHARD (*b.* Augsburg, Bavaria, Germany, 21 June 1535; *d.* Waitzen [now Vac], Hungary, 15 September 1596), *botany.*

Rauwolf's father was probably of the merchant class in industrial Augsburg. The family was Lutheran and on 6 November 1556, "Leonhardus Rauwolff" matriculated at the University of Wittenberg. On 22 October 1560 he entered the renowned medical school at Montpellier, where Antoine Saporta became his chief adviser. The medical curriculum there was still within the Arabic tradition and included the works of Ibn Sīnā (Avicenna) and al-Rāzī (Rhazes). This training was useful for his later botanical studies in the Near East. He also studied the work of Dioscorides under the famous botanist Rondelet. Rauwolf explored the rich flora of Provence and Languedoc and collected more than 400 specimens, which were the beginning of his outstanding herbarium. On some of these field trips the future botanist Jean Bauhin of Basel was his companion.

In 1562 Rauwolf transferred to the University of Valence, where he received his M.D. degree. In 1563 he left Valence for Augsburg; en route he collected 200 plants in northern Italy and in Switzerland. At Zurich he met the famous naturalist Gesner. Rauwolf eventually settled in Augsburg, maintained a botanical garden, and in 1565 was visited by the botanist L'Écluse. Among his fellow botanists, Rauwolf was known as "Dasylycus" or "Shaggy Wolf." On 26 February 1565 he married Regina Jung, the daughter of a patrician Augsburg family. He later practiced medicine at Aich and Kempten; but in 1570 he returned to Augsburg and secured the post of "city physician," which offered an annual salary of 100 gulden.

On 18 May 1573 Rauwolf left Augsburg for Marseilles and the Near East, thus beginning the fulfillment of his dream "to discover and to learn to know the beautiful plants and herbs described by Theophrastus, Dioscorides, Avicenna, Serapion, etc. in the location and places where they grow." He also wished to encourage the apothecaries to stock such plants for drugs. The opportunity to make this field trip was given to him when he was appointed physician to the Near East factors of the Melchior Manlich merchant firm, which hoped to profit from his discovery of new drugs. Rauwolf's brother, Georg, a factor of the firm, resided in Tripoli, Syria.

The Manlich ship took twenty-nine days to reach Tripoli, where Rauwolf remained for several weeks before departing for Aleppo. At both of these trading cities, he wrote descriptions of the local plants and collected specimens for his herbarium. After a stay of nine months in Aleppo, he traveled to Bir, on the upper Euphrates, where he boarded a flat-bottomed boat for the trip to Baghdad. The river trip to Felugia, the Euphrates port of Baghdad, took fifty-five days; but it enabled Rauwolf to observe the riparian vegetation. (He was the first modern European botanist to do so.) Although he intended to go from Baghdad to India (the fabled land of spices and drugs), his plans were changed when he heard that the Manlich firm was bankrupt. He returned overland to Tripoli by way of Mosul, Nisibin (Nusaybin), Urfa, and

Aleppo, again a pioneering and dangerous trip. From Tripoli he made excursions to the famous cedars on the Lebanon mountains and, as a devout Lutheran pilgrim, to Jerusalem. On 7 November 1575 he sailed from Tripoli for Venice; he arrived in Augsburg on 12 February 1576. His entire trip lasted thirty-three months.

At Augsburg, Rauwolf resumed his medical practice, maintained a botanical garden, and exchanged plants with several botanists, including L'Écluse and Joachim Camerarius. By 1582, however, the ruling circles of Augsburg had become Catholic. Rauwolf, the city physician and a leader of the Protestant opposition to the introduction of Catholic festivals and the Gregorian calendar, was forced to leave Augsburg in the summer of 1588. He served as city physician at Linz for eight years but in 1596 joined the imperial troops fighting the Turks in Hungary, where he died of a severe case of dysentery.

In 1582, while at Augsburg and on the urgings of his friends, Rauwolf published his travel book. Written in Swabian German and based on the extensive notes taken on his trip, the work gives detailed information on peoples, places, trade, religions, customs, and political life, as well as descriptions of numerous plants. It illustrates very well the difficulties and dangers encountered on early scientific field trips.

Among the classical and Arabic authorities that Rauwolf cited are Dioscorides, Theophrastus, Pliny, Galen, Paul of Aegina, Strabo, Aristotle, Plutarch, Ptolemy, Josephus, Ibn Sīnā, al-Rāzī, and Serapion the Younger. Contemporary botanists cited include L'Écluse, da Orta, Dodoens, and L'Obel. The third edition of the book (1583) included forty-two woodcuts of exotic plants.

Rauwolf's fame was increased with the publications of others. Daléchamps's *Historia generalis plantarum* (1586–1587) contained a special section on exotic plants, probably written by Rauwolf himself, and was illustrated with thirty-five woodcuts from Rauwolf's travel book. Camerarius' *Hortus medicus et philosophicus* (1588) often cited Rauwolf and described him as "a most learned man, indefatigable in the study and investigation of nature." Bauhin cited the discoveries of Rauwolf in his posthumously published *Historia plantarum universalis . . .* (1650–1651). Rauwolf was also cited by Leonard Plukenet, John Ray, and Robert Morison.

Upon his return from the Near East, Rauwolf prepared four folio volumes of his herbarium of 834 European and Near Eastern plants. The herbarium was sold to Duke William of Bavaria and placed in the Kunstkammer in Munich, but was taken to Sweden during the Thirty Years' War. About 1650

Queen Christina presented the herbarium to Isaac Vossius, who later brought the collection to London, where it was studied by a number of botanists. In 1680 the volumes were purchased by the University of Leiden, where they remain a prized possession.

Rauwolf was the first modern botanist to collect and describe the flora of the regions east of the Levantine coast. Kurt Sprengel (*Historia rei herbariae*, I [Amsterdam, 1809]; 377–379) credits him with thirty-three new discoveries. His descriptions of the busy and cosmopolitan city life of the provincial capitals of Tripoli, Aleppo, and Baghdad are informative and interesting historically. His description of the preparation and drinking of coffee in Aleppo is the first such report by a European. In 1703, in recognition of Rauwolf's many achievements, Plumier dedicated to him a genus of tropical plants, *Rauwolfia serpentina*. Today *Rauwolfia* alkaloids are widely used in medical therapy for their calming and hypotensive effects.

BIBLIOGRAPHY

I. ORIGINAL WORKS. Rauwolf's travel book appeared as *Aigentliche beschreibung der Raisz so er vor diser zeit gegen Auffgang inn die Morgenländer fürnemlich Syriam, Iudaeam, Arabiam, Mesopotamiam, Babyloniam, Assyriam, Armeniam, usw. nicht ohne geringe mühe unnd grosse gefahr selbs volbracht; neben vermeldung vil anderer seltzamer und denckwürdiger sachen die alle er auff solcher erkundiget gesehen und obseruiert hat* (Lauingen, 1582). Later editions were published in Frankfurt am Main (1582) and Lauingen (1583) with 42 woodcuts in pt. 4.

Reyssbuch des heiligen Landes, Sigismund Feyerabend, ed. (Frankfurt am Main, 1584, 1609, 1629; Nuremberg, 1659), contains only pts. 1–3. An English trans. by Nicolas Staphorst was published in *A Collection of Curious Travels and Voyages*, John Ray, ed., I (London, 1693, 1705) and in *Travels Through the Low-Countries, Germany, Italy, and France*, John Ray, ed., II (London, 1738). Dutch ed. by Pieter van der Aa were published in Leiden in 1707, 1710, and 1727. Excerpts can be found in Max Pannwitz, *Deutsche Pfadfinder des 16. Jahrhunderts in Afrika, Asien und Südamerika* (Stuttgart, 1911), 122–138.

II. SECONDARY LITERATURE. Bibliographical data can be found in Karl H. Dannenfeldt, *Leonhard Rauwolf: Sixteenth-Century Physician, Botanist, and Traveler* (Cambridge, Mass., 1968). The most important works are Franz Babinger, "Leonhard Rauwolf, ein Augsberger Botaniker und Orientreisender des sechzehnten Jahrhunderts," in *Archiv für die Geschichte der Naturwissenschaften und der Technik*, **4** (1913), 148–161; Johann Fredericus Gronovius, *Flora orientalis sive Recensio Plantarum quas Botanicorum Coryphaeus Leonhardus Rauwolffus . . . observavit, et collegit* (Leiden, 1755); and Ludovic Legré, *La botanique en Provence au XVIe siècle: Leonard Rauwolff, Jacques Raynaudet* (Marseilles, 1900).

KARL H. DANNENFELDT

RAY, JOHN (*b*. Black Notley, Essex, England, 29 November 1627; *d*. Black Notley, 17 January 1705), *natural history*.

Ray may have acquired his interest in science during his early years at Black Notley, where his father, Roger Ray, was blacksmith and his mother, Elizabeth, attained local eminence for her skills as an amateur herbalist and medical practitioner. He acknowledged that he had been devoted to the study of botany since his earliest years (*Catalogus . . . Cantabrigiam . . .*, preface). After attending the grammar school at Braintree, he entered Trinity College, Cambridge, in 1644, graduating B.A. in 1648 and M.A. in 1651. An academic career unfolded smoothly during the Commonwealth and Protectorate; he was elected fellow of Trinity in 1649, and during the next decade he held college teaching positions in Greek, mathematics, and humanities, as well as other minor offices. This course was suddenly interrupted in 1662 with the Act of Uniformity. Ray refused to take the oath required by the Act and thus elected to sacrifice his fellowship and leave Cambridge.

Subsequently Ray's work was supported by the generous patronage of his younger Cambridge contemporary Francis Willughby. For more than a decade, Willughby's estates at Middleton Hall, Warwickshire, and Wollaton Hall, Nottingham, were the bases for Ray's expeditions throughout Britain. Ray and Willughby became close collaborators, their most ambitious journey (1663–1666) being through the Low Countries, Germany, and Italy, with short visits by Ray to Sicily and Malta. The return journey, through France and Switzerland, included a long stay in Montpellier, where Ray formed one of his main scientific friendships, with Martin Lister, who also was making a Continental tour.

Upon his return to England, Ray was elected fellow of the Royal Society on 7 November 1667; he very seldom attended the meetings, however, although some of his letters to Oldenburg were published in the *Philosophical Transactions*. Upon Oldenburg's death Ray was offered the secretaryship of the Society but refused it, probably for a mixture of conscientious and temperamental reasons; he found an obscure existence most convenient for his work as a naturalist. The close partnership with Willughby ended with the latter's death in 1672. In the same year Ray married Margaret Oakeley, a member of the household at Middleton Hall. With their four daughters they retired to Black Notley, where Ray spent the rest of his life engaged in prolific writing and correspondence.

Despite serious interest in natural history in Britain and aspirations to compose comprehensive floras and faunas, little had been achieved by Ray's time. The partnership of Ray and Willughby attempted to meet these ambitions by composing a *systema naturae* based on firsthand observation, collaboration, and the critical use of authorities. This ambitious enterprise emerged early in their partnership, Ray undertaking to compose a *historia plantarum*, while Willughby agreed to study "birds, beasts, fishes and insects" (Derham, *Memorials*, p. 33). Their close association prevented strict compartmentalization, making it possible for Ray to take over the entire work after Willughby's death.

Ray's scientific apprenticeship occurred during the early phase of enthusiasm for experimental philosophy, which was reflected in an outburst of scientific activity at both Oxford and Cambridge. At Cambridge overriding influence was exerted by the Platonists, who provided the most important formative influence on Ray's natural philosophy. He absorbed their deep religious motivation for the study of nature as a means to reveal the workings of God in His creation. Among the young Cambridge scholars Ray found many collaborators, including the anatomist Walter Needham. Their initial activities embraced the whole range of experimental natural philosophy, but comparative anatomy and botany emerged as Ray's central preoccupations.

Ray's interest in botany began after an illness in 1650. During country walks undertaken for recreation, he became interested in the precise study of the local flora. Soon he and his friends embarked on a systematic investigation of the flora of Cambridgeshire, establishing small botanical gardens at their colleges. The joint nature of this venture was underlined by the anonymous publication of the *Catalogus plantarum circa Cantabrigiam nascentium* (1660). But there is no doubt that Ray was the prime contributor. His first publication created entirely new standards for the composition of British floras. The preface surveyed the defective English botanical literature, which, if comprehensive, was a derivative compilation; where it was more original, it was sketchy and incomplete. Ray looked to the Continent for more exalted standards, finding particular satisfaction in the work of Jean and Gaspard Bauhin, who provided him with the immediate model for the Cambridge catalog and a more general guide to the composition of the *historia plantarum*.

The main section of the Cambridge catalog followed Gaspard Bauhin's *Catalogus plantarum circa Basileam sponte nascentium . . .* (Basel, 1622). Ray gave an alphabetical list of species, including descriptions of little-known species or problematical groups, as well as detailed consideration of nomenclature and localities. At many points he experienced difficulty in reconciling his observations with the descriptions and nomenclature of the standard authorities. On the whole he displayed great caution, resisting the temptation to multiply species, although some cases of variable flower color were subdivided unnecessarily (Raven, *John Ray*, pp. 93–94). Like other authors he was weak on nonflowering plants, trees, shrubs, and grasses; but whereas Bauhin had dealt only with wild species, Ray gave valuable information on cultivated varieties. The potato was introduced in the brief appendix which he published in 1663.

So successful was the Cambridge catalog that it was not supplanted until Babington's *Flora of Cambridgeshire* (1860), which acknowledged that Ray had recognized 558 species. A broader dimension was added to the catalog by brief concluding essays on morphology and classification. Ray had obtained from Hartlib the manuscript of Joachim Jungius' then unpublished *Isagoge phytoscopica*, which was recognized as sufficiently important for Ray to include a digest of its contents, giving definitions of the main plant parts. He also presented a brief system of classification derived from Jean Bauhin's *Historia plantarum universalis* (1650–1651). Thus over a wide area, from etymology to classification, Ray's first work gave indications of his later preoccupations.

No sooner was the Cambridge catalog completed than Ray began work on a *Phytologia Britannica* to replace the feeble British plant list of that title by William Howe (1650). Preparation for this work involved expeditions to most parts of Britain, during which he also collected materials for other works. The *Catalogus plantarum Angliae* . . . (1670) followed the same plan as his Cambridge catalog, with the addition of sections on pharmacology, a subject which had interested Ray since childhood. This material was gathered from a large number of sources and was assessed with Ray's characteristic caution. The alphabetical list of species and localities demonstrated his extensive acquaintance with the British flora, and gave more attention to trees, mountain plants, and grasses than in his previous catalog.

The two catalogs mark the first phase of Ray's botanical work. He then turned his attention to the wider issues of physiology, morphology, and taxonomy in a series of brief communications. An important step was the fulfillment of a request from John Wilkins to compose the tables of plants for the *Essay Towards a Real Character* (1668). This confronted Ray with the basic problems of classification, as well as imposing an arbitrary condition that taxonomic divisions be tripartite. Ray's tables indicate a search for consistent new taxonomic principles. Although the results were

not entirely satisfactory, his work scarcely deserved the strong censures of Robert Morison, the irascible Oxford botanist.

In response to a questionnaire from the Royal Society, Ray submitted four botanical papers, on the motion of sap, on germination, on specific differences, and on the number of species. In his work on germination he slightly antedated Malpighi in making the monocotyledon-dicotyledon distinction, which became fundamental to his ideas on taxonomy. Two papers were concerned with the crucial issue of evolving satisfactory criteria for the identification of species, in order to replace the inconsistent and ad hoc methods used by previous writers. Ray appealed for the use of a small range of invariable morphological characteristics, instead of features liable to wide variation, which would lead to unnecessary multiplication of species. Ray realized that many of the "species" then recognized had little permanence, and he believed on both Biblical and empirical grounds that the true species in nature were "fixed and determinate." The essays presented to the Royal Society were published as a brief Latin work, *Methodus plantarum* (1682), with the intention of providing general principles for botanists attempting to define species and to arrive at a sound classification of the seemingly endless profusion of nature.

Ray came to agree with Cesalpino and Morison that the seed vessel was sufficiently invariable to provide the soundest basis for natural classification. Nevertheless he claimed that the petals, calyx, and leaf arrangement must also be taken into account. According to these priorities Ray opened his book with three sections expanding his paper on seeds, reemphasizing the monocotyledon-dicotyledon distinction as the guide to major taxonomic categories. The germination of various seeds was described in detail, using illustrations drawn from Malpighi's *Anatomia plantarum* (1675–1679). The *Methodus* ended with two new sections. Part IV reverted to Ray's early interest in Jungius, analyzing the flower into its constituent parts, with reference to both structure and function. The distinction between simple and composite inflorescences was clearly recognized. Having clarified the structure of fruits, seeds, and flowers, in part V Ray described his own ideas on classification, which had been evolving since the publication of his Cambridge catalog. The influence of Bauhin was still apparent, but many novel features were introduced. Herbs were given a tripartite classification into imperfect flowers (cryptogams), dicotyledons, and monocotyledons. This basic division was sound, as were many of the thirty-six "family" groupings of dicotyledons. For example, emphasis on seed vessels brought the varied flower and inflorescence types of the Ranunculaceae into a single natural group. An unsatisfactory aspect of this classification was the treatment of trees and shrubs. Although it was recognized that these were arbitrary divisions, they were retained in deference to popular usage.

Sufficient time had now passed for Ray to feel it necessary to bring his work on the English flora up to date. A brief supplement, *Fasciculus stirpium Britannicarum* (1688), was a prelude to the full revision of the English catalog, *Synopsis stirpium Britannicarum* (1690), which included numerous additions, particularly of cryptogams and grasses, many collected by the young botanists in Ray's circle. Medical notes, greatly shortened, were placed in an appendix. Like Ray's other catalogs, this text was designed for use as a field guide. For larger flowering plants the *Synopsis* reached the scope of modern floras. Also, for the first time Ray dispensed with alphabetical order in favor of the classification outlined in his *Methodus*, another important step toward the form of later floras. The treatment of trees and shrubs remained anomalous and imperfect.

Other plant lists illustrate Ray's continuing interest in taxonomic principles. *Sylloge Europeanarum* (1694) consisted primarily of European plant lists, but it was less systematic and detailed than complementary English lists. In a long preface Ray gave an extensive critique of the principles of classification adopted by A. Q. Bachmann (Rivinus), defending the use of a range of morphological criteria against Bachmann's artificial system based on corolla types. Bachmann's reply and a further rejoinder by Ray were published as an appendix to the second edition of *Synopsis Britannicarum* (1696). Again Ray defended the *Methodus* and gave a summary of the classification used in his *Historia plantarum* as an illustration of its practical applicability. A postscript was added to the *Synopsis* in reply to J. P. de Tournefort, whose *Élémens de botanique* (1694) had criticized Ray's classification. Tournefort had such authority that Ray was prompted to compose a separate tract against him, *Dissertatio de methodus* (1696). He granted that a single obvious criterion for classification was desirable; but although writers from Cesalpino to Tournefort had shown the superiority of fruit characteristics, this method resulted in certain crucial anomalies. To reinforce his views still further, Ray completely revised the *Methodus*, which became *Methodus emendata* (1703), the final, clearest, and most satisfactory statement of his principles and system of classification.

Although Ray's catalogs are his best-known writings, a series of works indicates his continual concern with the principles of taxonomy, his gradually evolving views being vigorously defended against the more influential followers of Cesalpino. This work provided the foundation for his magnum opus, the *Historia plantarum* (1686–1704), a 3,000-folio-page work designed to supplant the encyclopedic herbals of the brothers Bauhin. Its aim was to classify and list all known plants, with the goal of being comprehensive for Europe. Its more difficult function was to integrate information about exotic floras, which had been subject to considerable attention during the century. Such histories were often begun but rarely completed, as witnessed by the work of Ray's antagonist Morison. Furthermore, the rapid rate of discovery quickly rendered such works obsolete. Fully aware of these difficulties, Ray began the *Historia* with great confidence. It began with an extensive general botanical treatise covering certain subjects not previously considered in his writings. Ray displayed wide and critical reading over the whole field, and on many topics he presented extensive firsthand knowledge. This is well illustrated by the opening section, in which Jungius' definition of a plant is adopted, including "lacking sensation" among its criteria. In order to retain this definition, it was necessary for Ray to overcome the anomaly of the sensitive plant, *Mimosa pudica*, which had led to an increasingly popular doctrine of plant sensitivity. After a careful study of plant movement, Ray adopted a mechanism which could explain all movement mechanically, without recourse to the doctrine of sensitivity.

Ray derived his anatomy and morphology of the vegetative organs primarily from Grew and Malpighi, and gave considerable details of the floral parts, fruits, seeds, and germination in terms of their relevance to classification. He took a generally favorable attitude toward Grew's theory that the stamens were the male sex organs. Considering the nature of species, Ray reemphasized the view that breeding true from seed was a necessary test of a natural species. His previously strong conviction about the fixity of species was slightly modified. Examination of many case histories convinced him that limited transmutation was possible. For classification the system developed in his *Methodus* was followed, as it had been in the more limited *Synopsis* of the British flora.

As in other writings, Ray embraced practical aspects of the subject, considering the techniques of propagation, seed sowing, and curing plant diseases. After this comprehensive introduction Ray commenced his systematic survey of the flora according to his taxonomic principles. The first, relatively short sections dealt with lower plants, which had become familiar to Ray only at a late stage in his career. He relied primarily on reports of other botanists for this section. In the books on flowering plants his surveys were most successful when dealing with clearly delimited families, or where generic differences were obvious. Herbaceous families formed the core of the two-volume work, trees appearing at the end of the second volume. This group was no longer neglected, and Ray even considered the numerous cultivated varieties of apple and pear.

Whatever the limitations, Ray attempted to carry out his scheme rigorously, taking no shortcuts. For each species, besides nomenclature and morphological description, he gave details of habitat, distribution, and medicinal uses. Illustration was completely abandoned. In previous herbals illustrations had been so defective and derivative that they were the source of considerable confusion. Ray probably recognized that adequate illustration of such a comprehensive treatise was beyond his technical and financial resources. The third, supplementary volume of the *Historia*, published shortly before Ray's death, attempted to synthesize the considerable botanical literature which had appeared since the publication of the original volumes. Ray compiled material from a wide range of sources, from descriptions of exotic floras to botanical garden catalogs from Amsterdam and Nuremberg. By this time he was entirely a compiler, although his experience as a field botanist greatly improved the quality of this work. The enormous size of the supplement indicates the difficulties involved in single-handed attempts to deal with the growing avalanche of botanical literature.

Apart from the tables on animals prepared for John Wilkins, Willughby died without publishing anything of his share of the *systema naturae*. He bequeathed to Ray miscellaneous notes which required considerable elaboration before publication was possible. Ray recognized an obligation to complete Willughby's survey of the animal kingdom in addition to his own enormous botanical undertakings. He had already served an apprenticeship in zoology at Cambridge and had assisted Willughby with the table of animals used by Wilkins. More important were the lists of English birds and fishes included in Ray's *Collection of English Words* (1673). Although he published works on fishes and birds under Willughby's name, Raven's researches leave no doubt that Ray was primarily responsible for their composition.

Ornithology was the first topic completed by Ray (*Ornithologiae*, 1676). English ornithology had been given an auspicious inception by William Turner in the sixteenth century; but little more was achieved, as witnessed by the inadequate lists compiled by Merret

and Charleton, Ray's contemporaries. Throughout his travels Ray had made ornithological notes; subsequently such naturalists as Sir Thomas Browne sent him notes on the birds of their localities. Ray and Willughby accounted for at least 230 descriptions in *Ornithologia*; but in order to increase comprehensiveness, it was necessary to draw upon Continental authorities for both text and illustrations. Hence the survey was very uneven in quality. An important feature of this work was Ray's pioneering attempt to classify birds according to habitat and anatomy. As in plant taxonomy, he was suspicious of divisions based on single criteria. The basic groups were ecological—land and water birds. Land birds were divided according to beak characteristics, while water birds were placed in two groups, waders and swimmers. Lower categories were decided by diet or beak or foot characteristics. Ray's descriptions concentrated on plumage, with notes on variation according to age or sex; occasionally behavior patterns were described.

Ray's next project was completion of the *Historia piscium* (1686), a task analogous to the ornithology. It is probable that Willughby left relatively little material; but Ray's travels had furnished considerable information, including descriptions of Mediterranean fishes. Appended to his *Collection of English Words* were extensive lists of English freshwater and marine fish. He had also contributed two relevant papers to the *Philosophical Transactions*: an account of the dissection of a porpoise (1671) and considerations of the function and anatomy of swim bladders (1675). One of Ray's main problems in completing the *Historia piscium* was incorporating the considerable body of ichthyological literature which had been accumulating since the Renaissance. As in other treatises, Ray began with introductory chapters on the definition, anatomy, physiology, and classification of fish. Cetaceans were included despite recognition of their mammalian affinities, but invertebrate aquatic creatures were omitted. At the highest level Aristotle's classification was adopted, fin structure providing the main criterion for designating the minor divisions. An appealing feature of this book is the illustrations, which were more numerous than in any of his other works, because of the generous financial assistance from members of the Royal Society.

With the good progress of this zoological work, Ray was encouraged by Tancred Robinson to draw his material into a comprehensive *systema naturae*, summarizing the contents of longer treatises along the lines of the botanical *Synopsis*. The *systema naturae* was undertaken somewhat reluctantly by Ray, the manuscripts on fishes and birds lying idle until published posthumously by Derham as *Synopsis avium et piscium* (1713). Not only were descriptions condensed and slight emendations of the classification made, but a considerable number of new species, primarily of birds, were included. This supports the general impression that the work on fishes was undertaken out of duty rather than enthusiasm.

The *Synopsis animalium quadrupedum et serpentini* (1693) was published promptly, for no general histories of these groups were undertaken by Ray. Here the debt to Willughby was minimal, thus vindicating Ray's abilities as an anatomist and zoologist. Introductory essays dealt with general issues; the Cartesians were attacked for accepting animal automatism; Redi was supported for his opposition to spontaneous generation; the Platonist doctrine of Plastic Spirit was adopted and the treatment of embryology inclined toward ovism. Classification was the outstanding part of the introduction. Aristotle's division into "blooded" and "bloodless" was maintained; the former were divided according to respiratory mechanism, cetaceans being firmly placed with the mammals. For the minor divisions foot types were basic, but reference was also made to internal anatomy and general morphology. Classification of the invertebrates was unsatisfactory, but at least Ray attempted to avoid large amorphous groups.

The culmination of Ray's zoological work was the study of invertebrates, which resulted in his posthumous *Historia insectorum* (1710). On this topic Ray was dealing with virtually virgin territory. The only previous substantial work on the subject was produced a century before by the English naturalist Thomas Moffett. But the potentialities of entomology in particular had been displayed in Ray's generation by the microscopists and the detailed study of spiders made by Ray's friend Lister (1678). Ray and Willughby had made parallel investigations into gall insects that had been reported in a series of letters to the *Philosophical Transactions*. However, botanical work prevented Ray from beginning serious work on the history of lower animals until 1690, but the labor was pleasant and the helpers numerous. Despite great progress the text was incomplete at his death; and in the absence of any willing editor, it was published as Ray left it, with the addition of a scheme of classification previously published as *Methodus insectorum* (1705). Although it was ostensibly concerned with all "bloodless" creatures, only insects were considered in detail. More specifically, only moths, butterflies, bees, wasps, and ichneumon flies were described in the comprehensive form intended by Ray. On each of these groups Ray considerably improved on Moffett, whose bias lay in a similar direction.

As indicated above, most of Ray's publications were concerned directly with taxonomy. He was, however, responsible for some more general works, many of which had considerable celebrity. Most had an obvious relationship to his taxonomic labors. *Observations Topographical and Physiological* (1673), an account of his Continental journeys, was the vehicle for the "Catalogue of Foreign Plants." Likewise, Ray's *Collection of English Words* (1673), of lasting interest for its dialect records, contained lists of English fishes and birds, as well as accounts of mining and industrial practices. The long travel diary of Rauwolf, and a number of other brief accounts of the Middle East, were published as *A Collection of Curious Travels and Voyages* (1693), in order to draw attention to the botanical observations contained in these writings.

None of these works compared in influence or significance with the two theologically oriented books, *Physico-Theological Discourses* and *The Wisdom of God*. The intensity of Ray's conviction that science served a religious end is apparent from the preface to the Cambridge catalog, in which he declared, "There is for a free man no occupation more worthy and delightful than to contemplate the beauteous works of nature and honour the infinite wisdom and goodness of God." Accordingly, his students were admonished to cultivate natural history as a means to mental satisfaction and bodily exercise. In the preface to the ornithology the potential religious value of the study of birds was proffered as his reason for taking up the incomplete work of Willughby. For Ray, following the Platonists' lead, it was necessary to evolve a system of nature which would reinforce man's conviction of the power of providence and guarantee against the incursions of materialism.

Miscellaneous Discourses Concerning the Dissolution and Changes of the World (1692), revised and entitled *Three Physico-Theological Discourses* within a year, exposed Ray's considerable knowledge of paleontology and geology. On the much-debated issue of "formed stones," he supported the view that they were generally organic remains, one of the results of divine intervention at the Deluge. In an attempt to explain the placing of fossils, John Woodward and William Whiston had evolved speculative cosmogonies giving natural explanations for the Deluge. Consideration of the Deluge and Creation, in an attempt to arrive at scientific theories reconcilable with the Scriptures, was the dominant feature of the *Discourses*.

Even more popular and general was *The Wisdom of God* (1691), a phrase that often appears in Ray's prefaces. This work went through four editions, greatly expanded and revised by Ray himself. After his death its popularity continued, and it formed the model for a genre of natural theology literature. Ray's own text was derived from notes used at Cambridge that were heavily indebted to the Cambridge Platonists, especially More and Cudworth in general conception and philosophical outlook. But in the systematic survey of nature, no author was better equipped than Ray. The text, beginning with the solar system, passing through the theory of matter, geology, and the vegetable and animal kingdoms, and ending with a detailed study of human anatomy and physiology, gave Ray an ideal opportunity to display his scientific virtuosity. The great bulk of the material was drawn from his own experience, which extended from geology to anatomy, a spectrum which could be matched by few scientists and none of his imitators. *The Wisdom of God* provided an epitome of his scientific achievement in the religious perspective, which furnished Ray's basic motivation. For Ray, to be a naturalist was to acknowledge that "Divinity is my profession" (*Further Correspondence*, p. 163).

BIBLIOGRAPHY

I. ORIGINAL WORKS. For a complete list of Ray's works, see G. L. Keynes, *John Ray, a Bibliography* (London, 1951). Ray's major works are *Catalogus plantarum circa Cantabrigiam nascentium* (Cambridge, 1660); *Appendix ad catalogum plantarum . . .* (Cambridge, 1663); *Catalogus plantarum Angliae et insularum adjacentium* (London, 1670; 2nd ed., 1677); *Francisci Willughbeii . . . ornithologiae* (London, 1676; English ed., 1678); *Methodus plantarum* (London, 1682), rev. ed., *Methodus emendata* (London, 1703); *Historia piscium* (Oxford, 1686), written with Willughby; *Historia plantarum*, 3 vols. (London, 1686–1704); *Synopsis methodica stirpium Britannicarum* (London, 1690; 2nd ed., 1696; 3rd ed., 1724, with facs. repr., London, 1973); *The Wisdom of God* (London, 1691; enl. eds., 1692, 1701, 1704); *Miscellaneous Discourses Concerning the Dissolution and Changes of the World* (London, 1692), rev. as *Three Physico-Theological Discourses* (1693, 1713); *Synopsis animalium quadrupedum et serpentini* (London, 1693); *Historia insectorum* (London, 1710); and *Synopsis avium et piscium* (London, 1713). Contributions to the Royal Society are listed by Keynes.

Various editors have assembled poor eds. of Ray's correspondence: *Philosophical Letters*, W. Derham, ed. (London, 1718); *Correspondence of John Ray*, E. Lankester, ed. (London, 1848); and *Further Correspondence*, R. T. Gunther, ed. (London, 1928).

II. SECONDARY LITERATURE. The main older source is W. Derham, *Select Remains* (London, 1760), repr. as *Memorials of John Ray* (London, 1846). Also valuable is R. Pulteney, *Sketches of the Progress of Botany*, I (London, 1790), 189–281. All other sources are surpassed by C. E. Raven, *John Ray His Life and Works* (Cambridge, 1942; 2nd ed., with minor additions, 1950). A useful study of

Ray's work on plant taxonomy is D. C. Gunawardena, "Studies in the Biological Works of John Ray" (unpub. M.Sc. diss., Univ. of London, 1932). On this subject see also W. T. Stearn, "Ray, Dillenius, Linnaeus and the *Synopsis methodica stirpium Britannicarum*," introduction to the facs. ed. of the 3rd. ed. of *Synopsis methodica . . .* (1973).

CHARLES WEBSTER

RAY, PRAFULLA CHANDRA (*b.* Raruli, Khulna, India [now Bangladesh], 2 August 1861; *d.* Calcutta, India, 16 June 1944), *chemistry, history of science.*

Ray was the third of seven children born to Harishchandra and Bhuvamohini Ray. After attending the village school he was sent to Calcutta in 1871, where he studied at Hare School. In 1879, soon after receiving his F.A. degree (awarded after two years' study) from Metropolitan College of the University of Calcutta, Ray went to the University of Edinburgh on the Gilchrist scholarship. Here he obtained the B.Sc. in 1885 and the D.Sc. in 1887, both in inorganic chemistry. He received the Hope Prize after the D.Sc. and thus was able to remain in Edinburgh for another year.

Returning to India in 1888, Ray abandoned Western dress and manners, and reassumed his native traditions. After some disappointments and waiting, even though he had letters of recommendation, in 1889 he secured a modest, specially created position as chemistry lecturer at the Presidency College in Calcutta. Here he distinguished himself as a devoted scholar, inspiring teacher, and tireless researcher. He soon became professor of chemistry and managed to receive funds for establishing a new chemistry laboratory in 1894. He began work on problems related to food adulteration, especially the purity of ghee (clarified butter) and mustard oil, which are used extensively in Bengali foods. His researches were published in *Journal of the Asiatic Society of Bengal* in 1894.

Ray also initiated research to discover some of the elements missing from the then incomplete periodic table. While analyzing certain rare minerals he found hitherto unobserved yellow crystalline deposits. Careful analyses revealed these to be mercurous nitrate, a compound that until then had been considered unstable. This discovery impressed Ray greatly, and he spent the next several years exploring this salt and its many derivatives. He also developed methods for preparing ammonium nitrite, alkylammonium nitrite, and other compounds. He wrote more than 100 papers, some in collaboration with his students, on mercury salts and related compounds.

Ray was not merely an academic chemist; he wanted to apply his scientific knowledge and skill for the benefit of his countrymen. To this end, in 1901, he founded Bengal Chemical & Pharmaceutical Works, Ltd., which began with a handful of workers and modest capital and grew into one of India's major chemical industries, employing more than 5,000 people in the 1960's. Ray also established Bengal Pottery Works, Calcutta Soap Works, and Bengal Canning and Condiment Works.

Ray was also an ardent scholar and investigator of scientific history. Stimulated by Berthelot's *Les origines de l'alchimie*, he decided to write a history of chemistry in ancient India and was encouraged in this project by Berthelot himself, who sent Ray his *Collection des anciens alchemistes grecs* and wrote on aspects of ancient Hindu chemistry after reading an essay on the *Rasendrasara Samgraha* that Ray had written and sent to him. The first volume of *The History of Hindu Chemistry* appeared in 1902 and the second in 1908. The work won high acclaim from the scholarly world, as is shown in Berthelot's review in *Journal des savants* (January 1903). Besides its historical value this work not only revealed ancient Hindu scientific lore to the Western world but also instilled in Ray's countrymen a sense of pride in their scientific heritage.

In 1916 Ray became professor of chemistry at the newly founded University College of Science and Technology, Calcutta, where he continued research and teaching for another two decades. Much of the money that his industrial ventures earned was given by Ray to workers, to students as stipends and scholarships, to laboratories, and to scientific organizations. He was also involved in social action and politics, and often urged his students to join India's struggle for independence.

Ray received many honors, including honorary doctorates from the universities of Benares, Calcutta, Dacca, and Durham; Companion of the Order of the Indian Empire (1911); knighthood (1919); the presidency of the Indian Chemical Society (1924); and honorary fellowship of the Chemical Society (London). He also came to be known as Acharya Ray; "Acharya," a title of respect that could be roughly translated as "erudite master," is conferred by the people without any ceremony on a distinguished scholar.

BIBLIOGRAPHY

I. ORIGINAL WORKS. A complete bibliography of Ray's works is in Gupta (see below); Poggendorff, V, 1025, and VI, 2130–2131, lists some of the more important papers. Ray's autobiography, *Life and Experiences of a Bengali Chemist*, 2 vols. (Calcutta, 1932–1935), reveals the many facets of his thought.

II. SECONDARY LITERATURE. For more details on Ray's life and works, see Monoranjan Gupta, *Profulla Chandra Ray: A Biography* (Bombay, 1966). The English version of a Bengali study by an associate of Ray's, it contains numerous quotations, eulogies, and appendixes, and presents many items of interesting information; but it does not deserve the description of "definitive biography" which its author claims for it; *Acharya Ray Commemorative Volume* (Calcutta, 1932), edited by leading Bengali thinkers on the occasion of Ray's seventieth birthday, which contains many articles of biographical and scientific interest; and the anonymous "Sir Prafulla Chandra Ray," in *Journal of the Indian Chemical Society*, **21** (1944), 253–260, which includes a complete bibliography. Shorter obituaries are in *Twentieth Century* (Allahabad), **10** (1944), 531–535; *Nature*, **154** (1944), 76; *Journal of the Chemical Society* (1946), 216–218; and J. L. Farmer, "Sir P. C. Ray," in *Journal of Chemical Education*, **22** (1945), 324.

VARADARAJA V. RAMAN

RAYET, GEORGES ANTOINE PONS (*b.* Bordeaux, France, 12 December 1839; *d.* Floirac, near Bordeaux, 14 June 1906), *astronomy.*

Rayet's family belonged to the Bordeaux upper middle class. His father, a magistrate removed from office in 1830 because of his Legitimist views, worked in the wine industry in Bordeaux and then, beginning in 1853, in Paris. It was not until the latter year that Rayet went to school, but his studies did not suffer from this delay: he was admitted to the École Normale Supérieure in 1859 and graduated *agrégé* in physics three years later. After teaching for a year at the *lycée* in Orléans, he obtained a post as physicist, in the weather forecasting service newly created by Le Verrier, at the Paris observatory.

In addition to his regular work Rayet engaged in astronomical studies, participating in the research of Charles Wolf, an astronomer at the observatory. Their first joint success came in 1865, when they photographed the penumbra of the moon during an eclipse. In the meantime they had improved spectroscopic techniques and had sought to obtain the spectra of bright comets. A nova appeared on 4 May 1866. Observing it on 20 May, when its brilliance was considerably diminished, Rayet and Wolf detected bright bands in the spectrum, a phenomenon that had never been noticed in star spectra. The bands were the result of a phase that can occur in the evolution of a nova after the explosion and is now known as the Wolf-Rayet stage.

The two astronomers next attempted to determine, through systematic investigation, whether permanently bright stars exhibit the same phenomenon, and in 1867 they discovered three such stars in the constellation Cygnus. "Wolf-Rayet stars," the spectra of which contain broad, intense emission lines, are rare—barely more than 100 are known. They are very hot stars the mechanism of which has not yet been explained. The expansion of the shell is not sufficient to account for the width of the lines, and there is a great disparity between the energy produced in the interior and the radiated energy. These stars permit the study of interstellar matter, the lines of which stand out against the bright background of the broad emission lines of the stars.

In 1868 Rayet participated in the mission sent by the Paris observatory to the Malay Peninsula to observe a solar eclipse; he was responsible for the spectroscopic work. His observations on the solar prominences provided valuable information, and in conjunction with other observations made during the same eclipse, notably those of J. Janssen, they contributed to the establishment of the first precise data on the sun. Rayet then concentrated on spectroscopic observations of the solar atmosphere and of solar prominences. He identified the brilliant lines of their spectrum with the spectral lines of the elements and developed a theory of the physical constitution of the sun that, at least in its descriptive portion, contains material found in modern theories. This work formed the basis of his doctoral dissertation, which he defended in 1871.

Rayet had no serious disagreement with Le Verrier until the latter entrusted him with running the meteorological service in 1873. Within less than a year Rayet's opposition to the practical organization of the forecasting of storms—which he had good reason for judging to be premature—led to his dismissal. Rayet then became lecturer in physics at the Faculty of Sciences of Marseilles and in 1876 was appointed professor of astronomy at Bordeaux.

In 1871 the French government proposed to construct several new observatories. Rayet organized a collective survey of the history and equipment of the principal observatories in the world. With collaborators, he wrote three of the five volumes of the study. Rayet took a great interest in the history of science and wrote short, authoritative studies on Greek sundials (1875) and the history of astronomical photography (1887).

Along with his appointment to Bordeaux, Rayet was also designated director of the observatory to be built at Floirac in 1878. He held the post from 1879 and equipped the observatory with the most modern astrometric instruments. Rayet was one of the first supporters of the *Carte international photographique du ciel*. In 1900 he made an important contribution to the problem of the reduction of photographic plates. Rayet had the satisfaction of being able to publish, a year before his death, the first volume of the *Catalogue photographique* of the Bordeaux observatory.

Rayet became a corresponding member of the Académie des Sciences in 1892 and of the Bureau des Longitudes in 1904. He had a sophisticated conception of scientific research and appreciated the advantages of teamwork. He worked thus himself but gave his subordinates great freedom to take initiatives. He died from the effects of a serious lung disease that troubled him during his last years. Possessed of a great firmness of character, Rayet, despite the state of his health, took part in the expedition to Spain to observe the eclipse of August 1905.

BIBLIOGRAPHY

I. ORIGINAL WORKS. Rayet's most notable early works are a note on the spectroscopic observation of the nova of 1866, inserted in U. J. J. Le Verrier, "Note sur deux étoiles . . .," in *Comptes rendus . . . de l'Académie des Sciences*, **62** (1866), 1108–1109, written then with C. Wolf; the announcement of the discovery of the Wolf-Rayet stars, in "Spectroscopie stellaire," *ibid.*, **65** (1867), 292–296, written with C. Wolf; and "Sur les raies brillantes du spectre de l'atmosphère solaire et sur la constitution physique du soleil," in *Annales de chimie et de physique*, 4th ser., **24** (1871), 5–80.

Rayet's meteorological observations and astronomical observations (spectroscopy, astrometry, eclipses) are contained in about thirty notes in the *Comptes rendus . . . de l'Académie des sciences* between 1866 and 1905. Rayet presented an important analysis of the evolution of the magnetic declination and inclination at Paris between 1667 and 1872 in "Recherches sur les observations magnétiques . . .," in *Annales de l'Observatoire de Paris. Mémoires*, **13** (1876), A*1–A*140. On photographic astrometry, see "Instructions pour la réduction des clichés photographiques . . .," in *Annales de l'Observatoire de Bordeaux*, **9** (1900), 1–87; and *Catalogue photographique du ciel, Observatoire de Bordeaux*, I (Paris, 1905).

Rayet created the *Annales de l'Observatoire de Bordeaux* in 1885 and edited the first 12 vols.; his studies on the observatory's longitude, on its latitude, and on practical astronomy are contained, respectively, in vols. **1**, **2**, and **3**; his climatological studies, in vols. **6** and **10**; and vol. **13** contains a posthumous article on the eclipse of 1905.

His works on the history of science are "Cadrans solaires coniques," in *Annales de chimie et de physique*, 5th ser., **6** (1875), 52–85; *L'astronomie et les observatoires*. I, *Angleterre*, and II, *Écosse, Irlande et colonies anglaises*, written with C. André (Paris, 1874); and V, *Italie* (Paris, 1878); "Notice historique sur la fondation de l'Observatoire de Bordeaux," in *Annales de l'Observatoire de Bordeaux*, **1** (1885), 1–62; and "Histoire de la photographie astronomique," *ibid.*, **3** (1889), 41–99, previously pub. in 5 pts. in *Bulletin astronomique*, **4** (1887), 165–176, 262–272, 307–320, 344–360, 449–456.

II. SECONDARY LITERATURE. A biography and a list of Rayet's publications are given by E. Stephan in *Annales de l'Observatoire de Bordeaux*, **13** (1907), A1–A31; it is followed by "Discours prononcés aux obsèques," B1–B31. See also "Obsèques de M. Rayet," in *Bulletin astronomique*, **23** (1906), 273–285; E. Esclangon, "Nécrologie," in *Astronomische Nachrichten*, **172** (1906), 111; and W. E. P., "Obituary," in *Nature*, **74** (1906), 382.

JACQUES R. LÉVY

RAYLEIGH. See **Strutt, John William.**

RAYMOND, PERCY EDWARD (*b.* New Canaan, Connecticut, 30 May 1879; *d.* Cambridge, Massachusetts, 17 May 1952), *paleontology, geology.*

Raymond entered Cornell University in 1897, intending to study engineering, but came under the influence of G. D. Harris; his interests then centered on paleontology, which soon became his lifework. He received the Ph.D. at Yale in 1904 and was appointed assistant curator of invertebrate paleontology at the Carnegie Museum in Pittsburgh. In 1910 he became paleontologist on the Geological Survey of Canada and in 1912 went to Harvard as assistant professor of paleontology at the Museum of Comparative Zoology, rising to professor in 1929 and becoming emeritus in 1945.

Throughout his scientific career Raymond published extensively in paleontology, stratigraphy, and sedimentation. His interests in these fields ranged widely, and he described many new Paleozoic fossils belonging to nearly every major animal group. Although he made many important contributions to knowledge of the stratigraphy of the Canadian Rockies and the Appalachians, his lifelong specialty was the trilobites; he made detailed studies of their anatomy, ontogeny, and classification, which culminated in his monograph *The Appendages, Anatomy and Relationships of Trilobites* (1920). While much has since been learned of the morphology, life history, and ecology of these long-extinct marine arthropods, Raymond's work remains a paleontological classic. In 1939 he published *Prehistoric Life*, an admirable and widely read summation of his concepts of organic evolution as exemplified by the fossil record, based on his years of teaching and research.

BIBLIOGRAPHY

Raymond wrote more than 75 monographs and papers, the most important being *The Appendages, Anatomy and Relationships of Trilobites*, Connecticut Academy of Arts and Sciences Memoir no. 7 (New Haven, 1920); and *Prehistoric Life* (Cambridge, Mass., 1939).

H. C. Stetson, "Memorial to Percy Edward Raymond," in *Proceedings. Geological Society of America*, for 1952 (1953), 121–126, includes a portrait and a bibliography of published works.

JOHN W. WELLS

RAYMOND OF MARSEILLES (*fl.* France, first half of twelfth century), *astronomy.*

Nothing is known of Raymond's life except that he wrote in Marseilles before and in 1141. His name appears on only one manuscript and, since Duhem was not acquainted with it, the first analysis of his work refers to him simply as an anonymous "astronomer of Marseilles." Raymond wrote three works, which contain references to each other, either because he had already written them or because he planned to do so.

The treatise on the astrolabe, "Vite presentis indutias silentio . . .," is preserved in only one manuscript (Paris, lat. 10266), which is unsigned. This treatise, which includes a section on construction and a section on use, belongs to a series of translations and original writings that between 1140 and 1150 afforded the Latin West the definitive mastery of the astrolabe. Earlier texts on this instrument date from the end of the tenth century and from the eleventh century, but they had not been able to assure such mastery due to their lack of scientific accuracy. From the technical point of view, the instrument described by Raymond is the astrolabe that became standard in university instruction from the thirteenth to the sixteenth centuries, even though he graduated the zodiac of the rete incorrectly by joining the center of the instrument to equal divisions of the equator. The astronomical data are borrowed from al-Zarqālī, whose figures Raymond preferred to those of Ptolemy: an obliquity of the ecliptic of 23°33′30″ and apsis of the sun at 17°50′ of Gemini. Raymond presented two star tables side by side. One is of twenty-seven stars, the ecliptic coordinates of which he affirmed; he attributed this table to Ptolemy but in fact it comes from the treatises on the astrolabe of Llobet of Barcelona and of Hermann the Lame.[1] The other is of forty stars in ecliptic coordinates *secundum sententiam modernorum*, which is the slightly incorrect transcription of al-Zarqālī's table.[2] Raymond demonstrated a sincere enthusiasm for astronomy and for al-Zarqālī; he had an opportunity to strengthen it a little later while writing his work on planetary movements.

This work, "Ad honorem et laudem nominis Dei . . .," dates from 1141 and contains three sections: a long introduction, canons of astronomical tables, and the tables themselves. The introduction, well known through the analyses of Duhem and R. Lemay, presents a cautious but sincere defense of astronomy and astrology that attests to some knowledge of the texts of Abū Ma'shar and of al-Zarqālī. The astronomical tables and their canons are an adaptation, for the Christian calendar and the latitude of Marseilles, of al-Zarqālī's tables. They entailed a remarkable effort to endow astronomy with a tool for calculating the planetary positions for every date. The only similar enterprise in the Latin West had been the translation by Adelard of Bath of al-Khwārizmī's tables.[3] The calculation of the planetary longitudes in al-Khwārizmī's work, however, is based on an approximation that transfers to the mean center a part of the correction that was later termed the *equatio argumenti secundo examinata*. This resulted in the presence of a coordinate, which Adelard translated by the word *sublimatio*, which depends on the apsis and the equation of the argument.[4]

Despite the facility that the use of the *sublimatio* brought to the determination of planetary positions, the Latin Middle Ages did not employ this method of calculation but retained that originating with Ptolemy and developed by al-Zarqālī for the Toledan Tables and by the authors of the Alphonsine Tables. The latter method consisted in correcting the equation of the argument by means of the proportional part. Raymond's application of al-Zarqālī's tables to the latitude of Marseilles, nearly thirty years before Gerard of Cremona's translation of the Toledan Tables and before that of the *Almagest*, thus appears to have given Latin astronomers one of the first contacts with what would be, for four centuries, the standard method for determining the positions of the planets.[5] It is still not known, however, how Raymond gained access to al-Zarqālī's work; nor do we know which were the tables the mediocrity of which he and two other astronomers of Marseilles confirmed in November 1140.[6]

In both the work on the astrolabe and that on the courses of the planets, Raymond mentioned his intention of writing a *Liber judiciorum*. This text appeared to be lost; but it has recently been identified with "A philosophis astronomiam sic difinitam . . .," a treatise on astrology formerly listed, on the strength of information furnished by one of its manuscripts, under the name of John of Seville.

It should be noted that a fourteenth-century manuscript from Cusa attributes to a certain "Ramundus civis Masiliensis" the translation of a treatise on alchemy, *Theorica occultorum*,[7] but it is difficult to link it to the work of Raymond of Marseilles.

NOTES

1. These tables were published by P. Kunitzsch, *Typen von Sternverzeichnissen in astronomischen Handschriften des 10. bis 14. Jahrhunderts* (Wiesbaden, 1966), 23–30. On the nature of their coordinates, see E. Poulle, "Peut-on dater les astrolabes médiévaux?" in *Revue d'histoire des sciences . . .*, 9 (1956), 301–322, esp. 316–320.
2. Published by Kunitzsch in *op. cit.*, 73–82.
3. A. Bjornbo, R. Besthorn, and H. Suter, eds., *Die astronomischen Tafeln des Muḥammed ibn Mūsā al-Khwārizmī . . .* (Copenhagen, 1914), K. Danske Videnskabernes Selskabs Skrifter, 7th ser., Hist.-phil. sec., III, no. 1.
4. O. Neugebauer, *The Astronomical Tables of al-Khwārizmī . . .* (Copenhagen, 1962), K. Danske Videnskabernes Selskabs Skrifter, Hist.-fil. Skrifter, IV, no. 2, 23–29.
5. The canons of the Toledan Tables apparently exist in two versions, one by Gerard of Cremona and one of unknown origin and date. See J. M. Millás Vallicrosa, *Estudios sobre Azarquiel* (Madrid–Granada, 1943–1950), 36; and F. J. Carmody, *Arabic Astronomical and Astrological Sciences in Latin Translation* (Berkeley, 1956), 157–161; comparative study and criticism of the two versions remain to be done.
6. The discussion of 1140 is among the passages translated by Duhem (see bibliography), who incorrectly dates it 1139, but he corrects almost all the figures given, which he misread and did not understand.
7. L. Thorndike, *A History of Magic and Experimental Science*, IV (New York, 1934), 17–18.

BIBLIOGRAPHY

I. Original Works. "Le traité d'astrolabe de Raymond de Marseille," E. Poulle, ed., is in *Studi medievali*, 3rd ser., 5 (1964), 866–909. M.-T. d'Alverny and E. Poulle are preparing an ed. that will contain the work on the movements of the planets, based on the three MSS known—Cambridge, Fitzwilliam Museum, McClean 165, fols. 44–47 and 51–66v (incomplete); Oxford, Corpus Christi College 243, fols. 53–62, without the tables; and Bibliothèque Nationale, Paris, lat. 14704, fols. 110–135v— and the *Liber judiciorum*, of which about ten MSS are available.

II. Secondary Literature. See P. Duhem, *Le système du monde*, III (Paris, 1915), 201–216; C. H. Haskins, *Studies in the History of Mediaeval Science* (Cambridge, 1924), 96–98; and R. Lemay, *Abū Ma'shar and Latin Aristotelianism in the Twelfth Century* (Beirut, 1962), 141–157.

Three articles by E. Chabanier, who is interested only in the list of geographic longitudes included in Raymond's tables, are inaccurate and present risky conclusions: "Un cosmographe du douzième siècle capable de mesures exactes de longitude," in *Académie des inscriptions et belles-lettres, comptes rendus des séances* (1932), 344–352; "La géographie mathématique dans les manuscrits de Ptolémée," in *Bulletin de la Section de géographie du Comité des travaux historiques et scientifiques*, **49** (1934), 6–7; and "La filiation de la table de Raymond de Marseille et les tables dites ptoléméennes du moyen âge," in *Comptes rendus du 15ᵉ Congrès international de géographie, Amsterdam, 1938*, II (Amsterdam, 1939), 101–117.

EMMANUEL POULLE

AL-RĀZĪ, ABŪ BAKR MUḤAMMAD IBN ZAKARIYYĀ, known in the Latin West as **Rhazes** (*b.* Rayy, Persia [now Iran], *ca.* 854; *d.* Rayy, 925 or 935), *medicine, alchemy, philosophy, religious criticism.*

We possess very little authentic information about Rāzī's life. He was born around 854 in Rayy, and directed a hospital in that town and later in Baghdad. Because of changes in the political situation and his personal standing at court, he returned on several occasions to Rayy. He died there in 925 or 935. Most of Rāzī's philosophical and antireligious works are lost. Two treatises on ethics, *Kitāb al-Ṭibb al-Rūḥānī* (*The Book of Spiritual Physick* in Arberry's translation) and *Sīrat al-Faylasūf* (*The Philosopher's Way of Life*), have been published. Several manuscripts exist of the work entitled *Doubts Concerning Galen,* which deals with philosophical as well as with medical questions. Abundant information about Rāzī's philosophical teaching can be gleaned from the quotations and references to him made by his critics, many of whom belonged to the Ismāʿīlī sect and defended the strict hierarchical principle maintained by this sect against the equalitarian views propounded by Rāzī.

Rāzī rejected the notion that men can be stratified according to their innate capacities. All of them have, according to Rāzī, their share of reason, which enables them not only to deal with practical matters, but also to reach correct views on theoretical questions. With regard to some of these questions the judgment of simple unsophisticated people may be more valuable than that of the learned, who are befogged by erudite quibblings and subtleties.

Rāzī's rejection of the hierarchical principle formed a part of his attack against religion. According to him, men, being naturally equal, did not need, in order to manage their affairs, the discipline imposed by religious leaders, who deceived them. The miracles supposed to have been worked by the prophets of the three monotheistic religions as well as by Mani were tricks. (*The Tricks of the Prophets* is the title of a lost treatise attributed to Rāzī.) Men of science like Euclid and Hippocrates were much more useful than the prophets. In fact, religion was definitely harmful, for fanaticism engendered hatred and religious wars.

Holding these views, Rāzī evidently could not subscribe to the doctrine, maintained by many Islamic authors, according to which the viable human societies were instituted by prophets. He may have followed a suggestion found in Plato in stating that society originated because of the need for a division of labor.

Rāzī's refusal to accept the principle of absolute authority is in evidence not only in his antireligious polemics, but also in his attitude toward the traditional verities of science and philosophy and the eminent authors who had established these verities. Thus, in justifying his *Doubts Concerning Galen*, he observes: "Medicine is a philosophy, and this is not compatible with renouncement of criticism with regard to the leading [authors]." In this context he points to the example of disciples of Aristotle who criticized the latter and to that of Galen himself. This attitude is closely connected with Rāzī's belief in the continuing progress of the sciences (which contrasts with the view of the Aristotelians that knowledge of the various sciences either has already reached its point of perfection or will necessarily reach it at some time in the future). According to Rāzī, a man of science, who knows the work of his predecessors, has because of this an advantage over them (however eminent they may have been) and can proceed to new discoveries. Rāzī's distrust of established scientific dogmas is evident also in his readiness to give the benefit of doubt to reports concerning various phenomena that appeared to have no theoretical explanation. Thus he wrote a treatise (still unpublished) on properties, which contains a jumble of miscellaneous information concerning bizarre phenomena, some of them of magical nature. At the beginning of his introduction, Rāzī voices his conviction that he will be blamed for composing his treatise, the critics referred to being people who hasten to deny the statements they cannot prove. In fact, they are constantly observing phenomena similar to those the truth of which they deny. For instance—and this is but an example of many cited by Rāzī—they often see a magnet attracting iron. Yet, if someone claimed that there exists a stone attracting copper or glass, they would be quick to give him the lie. This open-minded attitude is somewhat reminiscent of that shown by Francis Bacon when discussing magic. Both of them appear to feel that all recorded facts, however strange and inexplicable, should be taken into consideration, because all may be of scientific interest.

This position may account for Rāzī's interest in alchemy. In his writings on this science, he avoids the symbolism and occultism that are characteristic of Jābir ibn Ḥayyān, whom, as far as is known, he does not mention. The writings in question contain precise classification of various substances and precise accounts of the methods he follows in this science. Rāzī's view of human reason determined to a considerable extent his physical doctrine; for his conceptions of space and time were based on the presupposition that the immediate a priori certainties (put down by the Aristotelians to the effects of the imaginative faculty and discounted in consequence) are a conclusive proof of truth. Thus the fact that everyone (all men having a share of reason) has (unless his judgment has been warped by the discussions of Schoolmen) an immediate certainty that a tridimensional space would exist even if all the bodies were to disappear, and that this space has no boundaries, demonstrates by itself the truth of these conceptions and suffices to refute the Aristotelian theories and arguments.

Basing himself in the main on these a priori certainties, Rāzī affirmed the existence contested by the Aristotelians of absolute space, which is pure extension, independent of the bodies that it contains; parts of it may be empty. This space extends beyond the limits of the world, is infinite. Rāzī also posited the existence of relative or partial space, equated with the extension of each body.

Rāzī had a similar approach to the problem of time. In attempting to disprove the thesis of the Aristotelians, who define as the number of the motion (of the highest sphere) and thus according to prior and posterior posit the dependence of its existence on that of the world, Rāzī, as in the case of space, appealed to the a priori certainties of untutored people, who, when called upon to imagine that the world has ceased to exist, have no doubt whatever that even in that case there would be a flow of time; for time is a substance that flows.

As in the case of space, Rāzī distinguished between two species of time, absolute time and limited time. The Aristotelian definition is, according to him, applicable to the latter, but not to the former, which is unmeasured and which existed before the creation of the world and will continue to exist after the latter's dissolution. It is eternity (aiōn, dahr). These conceptions are in some respects reminiscent of certain Zoroastrian conceptions of time (zurvān). Analogies to them may also be found in Greek philosophy, in a passage attributed to Cicero (De natura deorum, I.9.21), to the Epicurean Velleius, and, what may be even more to the point, given Rāzī's view of himself

as a disciple of Plato (see below), in a quotation purporting to give the opinion of the school of the Platonist Attikos (Proclus in Timaeum, Diehl, ed., III, 37; cf. L. Tarán, "The Creation Myth in Plato's Timaeus," in J. P. Anton and G. L. Kustas, eds., Essays in Ancient Greek Philosophy [New York, 1971], 379, 397, n. 56).

Rāzī was an atomist, and his theory has a certain, although not very close, resemblance to that of Democritus; it differs profoundly from the teachings of Kalām. According to Rāzī, absolute matter, before the creation of the world, consisted of indivisible atoms having extension. For material, in contradistinction to geometrical, bodies are not divisible ad infinitum. On the other hand, if the visible material bodies were not composed of atoms, it would be necessary to suppose that the world was not created in time.

Having been mixed according to various proportions with particles of vacuum, the atoms produced the five elements, namely, earth, water, air, elemental fire, and the heavenly element. All the qualities of the elements, such as lightness and heaviness, opacity and transparence, and so forth, are determined by the quantity of matter, as compared with the particles of vacuum, which is found in them. The dense elements, namely earth and water, tend to move toward the center of the earth, while air and fire, in which the particles of vacuum are predominant, move upward. The heavenly element, in which there is an equilibrium between matter and the particles of vacuum, has circular motion.

In several passages, Rāzī states that he does not accept the philosophy of Aristotle, but that he is a disciple of Plato. The latter claim probably refers in the first place to some Hellenistic interpretation of the Timaeus. In Doubts Concerning Galen, Rāzī refers to two refutations made by the Greek physician of theories based on the doctrine of the Timaeus concerning the creation of physical bodies out of geometrical figures. In this connection Rāzī puts forward arguments for his personal brand of atomism. It may be noted that several of the physical conceptions of Rāzī are attributed in various Arabic philosophical and doxographical works to Plato. On the other hand, Rāzī's equalitarianism is at the antipode of Plato's political theory, which was adapted to a considerable extent by Arabic so-called Aristotelian philosophers.

As against the Aristotelians, Rāzī believed in the temporal creation of the world. In his cosmogony, which appears in his work Al-ʿilm al-ilāhī (The Divine Science), of which only fragments have been preserved, he appears to have adopted a gnostic myth. In this connection, it may be relevant to note that in the work that has just been mentioned he cites Mani's writings. Rāzī himself is reported to have asserted that this view of the creation of the world was held by Socrates. (See Nāṣir-i Khusraw, *Kitāb Jāmiʿal-Ḥikmatayn*, H. Corbin, ed. [Teheran, 1953], 211–213. This account agrees in all the main points with the one found in another work of Nāṣir-i-Khusraw entitled *Kitāb Zād al-Musāfirīn*. The second account may be found in P. Kraus's edition of Rāzī's *Opera philosophica*, 282, and in S. Pines, *Beiträge zur Islamischen Atomenlehre*, 59.) According to this cosmogony, there exist five preeternal principles: creator, soul, matter, time, and space; for the doctrine of temporal creation can be maintained only if it is supposed that several preeternal principles exist. The existence of one such principle only—which would be immutable—entails the eternity of the world.

The soul, which had life but no knowledge, conceived the desire to be conjoined with matter and to produce in the latter forms that procure pleasures of the body. Matter, however, was refractory. Thereupon, the creator in his mercy created this world, which contained forms in which the soul could take pleasure, and through the intermediary of which it could engender men. But the creator also sent the intellect, which is part of his substance, in order to waken the soul; for the latter is asleep in its temple, which is man. It is the task of the intellect to teach the soul that this created world is not its veritable home and that he cannot achieve in it happiness and tranquillity. Man can be freed from bondage to matter only through the study of philosophy. When all human soul will have achieved liberation, the world will be dissolved, and matter, being no longer provided with forms, will revert to its primeval state— that is, that of dispersed atoms.

A very similar cosmogony is attributed by the thirteenth-century author al-Kātibī to members of the Hellenistic pagan community of the town of Ḥarrān. Rāzī's cosmogony translated into German by H.H. Schaeder (see bibliography) was quoted and half-seriously adopted by Thomas Mann in *Josef und Seine Brüder*. Mann does not name his source, but refers to an exposé *(Referat)* which he used.

The cosmogony in which the conjunction of the soul with the matter is considered as the origin of suffering and Rāzī's pessimistic view (indicated in another fragment of the *Divine Science* [quoted by Maimonides]) that in our world evil is predominate over good, seem to call for an ascetic ethical doctrine. The two ethical treatises of Rāzī that are extant, however, are moderate in this respect. *The Book of Spiritual Physick* sets forth Plato's doctrine con-cerning the three souls of man: the rational, the spirited, and the concupiscent. Only the first is regarded as surviving the death of the body. It is for the sake of the rational soul that the two others are generated. All three souls should avoid both excess and deficiency in their activities and functions. In the case of the rational soul deficiency means failure to attempt to investigate both the things of this world and in particular the body in which the soul is lodged, and the destiny of the soul after death. Excess means such total preoccupation with this enquiry that the needs of the concupiscent soul are not met.

In his other ethical treatise, *The Philosopher's Way of Life*, Rāzī mentions that people point out that Socrates' way of life was incompatible with the workings of orderly society. According to Rāzī, this cynical tradition concerning Socrates is correct only as regards the beginning of his career (as a philosopher). Afterward, he took part in social activities and did not lead an excessively ascetic life. Like *The Book of Spiritual Physick*, this treatise, too, strikes the note of moderation. Rāzī's attitude toward animals is part of his ethics. Only carnivores and noxious animals such as snakes may be killed. The killing of the others is lawful for only one reason. Souls lodged in the bodies of animals cannot be liberated; only souls subsisting in human bodies achieve liberation. Hence, given the doctrine of transmigration, according to which a soul may pass from an animal to a human body, the killing of an animal may set it on the path of liberation.

Rāzī's medical works, some of which were translated in the Middle Ages into Latin, include monographs, such as a treatise on smallpox and measles, and comprehensive manuals, such as *Al-Ḥāwī (Continens of the Latin)*, a voluminous work containing extracts from Greek and Arabic physicians concerning the various medical problems and accounts of Rāzī's personal experiences and conclusions.

Rāzī's clinical observations have been edited by M. Meyerhof (see bibliography). His critical attitude toward the traditional, that is, Galenic, medical theories and observations comes out very clearly in his *Doubts Concerning Galen*. Thus, he observes, speaking of Galen's descriptions of fevers, that in the hospitals of Baghdad and of Rayy he noted throughout many years both the names of those whose illness followed the course laid down in "those books" and of those whose illness did not conform to these descriptions, and found that the latter were as numerous as the former.

Rāzī also indicated that on some points his medical experience was much more abundant than that of Galen. Thus he observed—in connection with a remark by Galen that he had noted only two cases of a certain urinary disease—that the disease may have been rare in Galen's country. Rāzī himself encountered in Iraq and al-Jebel more than a hundred cases of the disease.

One of the very numerous medical points on which Rāzī took issue with Galen was the rule formulated by Galen that a thing that has the property of cooling or warming is always colder or warmer than things cooled or warmed by it. According to Rāzī, this a priori rule is not valid in medicine. Experience shows that in cases of illness a beverage that is only moderately warm can heat to a degree that is much greater than its own heat. In such cases the beverage triggers in the human body a passage from potentiality into actuality.

In his critique of Galen's theory of vision, Rāzī indicates that some of the errors of his predecessor were caused by the latter's excessive recourse to mathematics. His own theory, which by and large is reminiscent of Aristotle's doctrine, has one peculiarity: Rāzī considered that the air carrying the visual images passes through the hollow optic nerve and reaches the ventricles of the brain that contain the animal spirit. A major point on which Rāzī disagreed with Galen has both medical and philosophical aspects. It concerns the nature of the soul, which Rāzī regarded as a substance, whereas Galen considered it as a mixture. Rāzī appeared to think that the brain is an instrument of the soul.

Rāzī's antireligious attitude and his interest in alchemy provoked attacks in which even his medical competence was questioned. Bīrūnī, who compiled a list of Rāzī's works, cites a dictum in which Rāzī was accused of destroying the wealth of people (by means of alchemy), destroying their bodies (by means of medicine), and corrupting their souls (by means of his defamation of the prophets). Bīrūnī had a great admiration for Rāzī as a physician (and may have, moreover, been influenced by his refusal to accept without question the authority of Aristotle), but he did not defend him, or not wholeheartedly, on the two other counts. To some extent this attitude was characteristic. But Bīrūnī's criticism was mild compared with that of many other authors. In medicine Rāzī had a great name; but in other areas, philosophy for example, his reputation was dubious.

There is a certain affinity between his attitude (notably his acceptance of what he regarded as a priori certainties) and that of the unorthodox philosopher Abu'l-Barakāt, and their conceptions of space are similar. On the other hand some of the leading so-called Aristotelian philosophers manifested their disdain for him.

Ibn Sīnā suggested that he should have confined himself to dealing with boils, urine, and excrement and should not have dabbled in matters beyond the range of his capacity. In doing so he exposed himself to contempt. Maimonides asserted that he was only a physician, that is, not a philosopher. It may be noted that Ibn Sīnā and Maimonides were themselves physicians. But they were also to some extent Aristotelians, and hence intolerant of Rāzī's reliance on immediate certainties—and perhaps also of his readiness to accept empirical evidence that was apt to upset established doctrines.

BIBLIOGRAPHY

See P. Kraus, ed., *Bīrūnī, Risāla fī Fihrist Kutub Muḥammad Ibn Zakariyā al-Rāzī* (Paris, 1936); J. Ruska, "Al-Bīrūnī als Quelle für das Leben und die Schriften al-Rāzī's," in *Isis*, **5** (1922), 26–50; *Al Ḥāwī*, the most voluminous of Rāzī's medical works (Hyderabad, 1955–1968); M. Meyerhof, "Thirty-Three Clinical Observations by Rhazes," in *Isis*, **23** (1933), 322 ff; *A Treatise on the Smallpox and Measles*, W. A. Greenhill, trans. (London, 1847); P. de Koning, *Traité sur le Calcul, les Reins et la Vessie* (Leiden, 1896); J. Ruska, "Al-Rāzī als Chemiker," in *Zeitschrift für angewandte Chemie* (1922), 719 ff; "Die Alchemie al-Rāzī's," in *Der Islam*, 22, 719 ff; and "Übersetzung und Bearbeitungen von al-Rāzī's Buch Geheimnis der Geheimnisse," in *Quellen und Studien zur Geschichte der Naturwissenschaften und der Medizin*, 4 (1935). *Kitāb al-Asrār wa-sirr al-Asrār* (Teheran, 1964). P. Kraus, *Abi Bakr Mohammadi Zachariae Ragensis (Razis) Opera philosophica, Fragmenta que quae supersunt*, Pars Prior (Cairo, 1939), contains the Arabic text of Rāzī's two ethical treatises and practically all the known philosophical fragments. Persian quotations are given both in Persian and in Arabic trans.

See also P. Kraus, "Raziana," in *Orientalia*, n.s. **4** (1935), 300 ff; **5** (1936), 35 ff; H. H. Schaeder, in *Zeitschrift der Deutschen Morgenländischen Gesellschaft*, **59**, 228 ff; S. Pines, *Beiträge zur Islamischen Atomenlehre* (Berlin, 1936); "Rāzī Critique de Galien," in *Actes du 7e Congrès international d'histoire des sciences* (Jerusalem, 1955), 480–487; and "What Was Original in Arabic Science," in *Scientific Change, Symposium on the History of Science* (Oxford, 1961), 181 ff; M. Mohaghegh, *Filsūf-Rayy Muḥammad Ibn-i-Zakariyā-i-Rāzī* (Teheran, 1970), in Persian, a comprehensive work on Rāzī as a philosopher.

A detailed bibliography, to 1924, is found in G. Sarton, *Introduction to the History of Science*, I, 609–610.

SHLOMO PINES

RAZMADZE, ANDREI MIKHAILOVICH (*b.* Chkhenisi, Russia [now Samtredia district, Georgian S.S.R.], 11 August 1889; *d.* Tbilisi, U.S.S.R., 2 October 1929), *mathematics*.

Razmadze was a son of a railway employee, Mikhail Gavrilovich Razmadze, and of the former Nino Georgievna Nodia. In 1906 he graduated from high school in Kutaisi and in 1910 from the mathematics department of the University of Moscow, then taught mathematics in high schools for several years. In 1917 he passed the examinations for the master's degree and for a short time was a lecturer at Moscow University. At the end of 1917 Razmadze returned to Georgia and became one of the most prominent organizers of higher education and scientific research there. He was one of the founders of Tbilisi University, opened in 1918, particularly of the Physics and Mathematics Faculty; and he spent the rest of his life as a professor there. Razmadze also had a major role in the elaboration of Georgian mathematical terminology and textbooks. He wrote texts in Georgian on infinitesimal calculus, on the introduction to analysis (1920), and on the theory of indefinite integrals (1922). These works were to have been parts of a complete course of analysis, which Razmadze was unable to complete.

Razmadze's investigations covered the calculus of variations. In this field he followed the classical direction of Weierstrass, and partly that of David Hilbert. In his first paper, published in 1914, he considered the problem of determining the plane curve minimizing the integral $\int_{t_1}^{t_2} F\left(x, y, \frac{dx}{dt}, \frac{dy}{dt}\right) dt$, when one end point of the curve is fixed and the other is absolutely free; he established the necessary and sufficient conditions of the existence of a minimum. Razmadze next largely generalized the so-called fundamental lemma of the calculus of variations; from Razmadze's lemma, Euler's differential equation for extremals of the integral $\int_a^b F\left(x, y, \frac{dy}{dx}\right) dx$ is deduced very simply and without partial integration.

Razmadze obtained his most important results when investigating discontinuous solutions. The first problems of the calculus of variations to be studied were those in which solutions are represented by smooth curves with continuously changing tangents. But there are problems which do not have such solutions and may be solved, for instance, by means of continuous curves with corners, where a slope of the tangent line has a jump. Such solutions, systematically studied by G. Erdmann and later by Carathéodory, were called discontinuous; but Razmadze designated them more properly as angular solutions. Proceeding further, Razmadze developed a comprehensive theory of the solutions represented by curves with a finite number of finite jumps. He presented a report on his research to the International Congress of Mathe-

maticians at Toronto in 1924, and for that paper he received the doctorate in mathematics from the Sorbonne in 1925.

BIBLIOGRAPHY

I. ORIGINAL WORKS. Razmadze's writings) include "Über Lösungen mit einem variablen Endpunkt in der Variationsrechnung," in *Mathematische Annalen*, **75** (1914), 380–401; "Deux propositions du calcul des variations," in *Bulletin de l'Université de Tiflis*, **1** (1919–1920), 157–172; *Introduction to Analysis* (Tbilisi, 1920), in Georgian; "Über das Fundamentallemma der Variationsrechnung," in *Mathematische Annalen*, **84** (1921), 115–116; *Theory of Indefinite Integrals* (Tbilisi, 1922), in Georgian; "Über unstetige Lösungen mit einem Unstetigkeitspunkt in der Variationsrechnung," in *Bulletin de l'Université de Tiflis*, **2** (1922–1923), 282–312; "Sur une condition de minimum nécessaire pour solutions anguleuses dans le calcul des variations," in *Bulletin de la Société mathématique de France*, **51** (1923), 223–235; and "Sur les solutions discontinues dans le calcul des variations," in *Mathematische Annalen*, **94** (1925), 1–52.

II. SECONDARY LITERATURE. See L. P. Gokieli, "A. M. Razmadze," in *Trudy Tbilisskogo matematicheskogo instituta*, **1** (1937), 6–10, with bibliography of Razmadze's works on 10; and L. Tonelli, "Andrea Razmadzé," *ibid.*, 11–13.

A. P. YOUSCHKEVITCH

RÉAUMUR, RENÉ-ANTOINE FERCHAULT DE (*b.* La Rochelle, France, 28 February 1683; *d.* near St.-Julien-du-Terroux, France, 18 October 1757), *mathematics, technology, natural history, biology, experimental physics.*

Réaumur was of an illustrious Vendée family, the Ferchaults, who prospered in trade and purchased the ancient Réaumur estate in the early seventeenth century. Through a protracted lawsuit Réaumur's grandfather, Jean Ferchault, obtained half the seignorial rights over his newly acquired fiefdom and thus entered into the ranks of the lesser French nobility. René Ferchault, Réamur's father, was a *conseiller au présidial* at La Rochelle, a position corresponding to an appellate judge in an intermediate provincial court. He married Geneviève Bouchel, the youngest daughter of a municipal magistrate from Calais, in April 1682; René-Antoine was born the following February. Réaumur's father died nineteen months later. A second son, Jean-Honoré, was born posthumously. Réamur and his brother were reared by their mother with the aid of several aunts and uncles.

Concerning Réaumur's early education, nothing is known with certainty. Probably he studied with either the Oratorians or the Jesuits at La Rochelle. Then, in accordance with established custom among the bourgeoisie and lesser nobility of the region, he would most likely have been sent to study with the Jesuits at Poitiers. In any case, Réaumur's early education probably included very little physics or mathematics. In 1699 his uncle, Gabriel Bouchel, summoned him to Bourges to study law. He went there with his younger brother and stayed for three years.

In 1703 Réaumur went to live in Paris, where he met a cousin on his mother's side, Jean-François Hénault, the future *président*.[1] Hénault was studying mathematics with a certain M. Guisnée, an obscure "student geometer" of the Academy of Sciences.[2] Réaumur decided to take lessons from Guisnée and, according to Hénault, after only three sessions knew more than his cousin and as much as his instructor. It was probably through Guisnée that Réaumur became acquainted with the great mathematician Pierre Varignon. Varignon became Réaumur's friend, teacher, and guide; and in March 1708 he nominated him to be his "student geometer" at the Academy of Sciences.

Réaumur's first three communications to the Academy, on geometrical subjects, were presented in 1708 and 1709, and demonstrate a degree of mathematical sophistication worthy of a student of Varignon. Had Réaumur decided to remain a mathematician, he might well have been one of the greatest geometers of his age. In November 1709, however, he quite suddenly changed the course of his scientific career by reading a paper on the growth of animal shells. From then on, Réaumur's work would be characterized by its extraordinary richness and diversity, but never again would he devote himself to the pure mathematical researches that had so fascinated him in his youth.

Technology and Instrumentation. Shortly after the formation of the Paris Academy of Sciences, Louis XIV's finance minister, Colbert, charged it with the task of collecting a description of all the arts, industries, and professions. This work was intended to be a sort of industrial encyclopedia which was to present the secret processes of industrial technology so that they might be better examined and improved. The French Academy, unlike the English Royal Society, was an integral part of the French bureaucratic system. This governmental role of the Academy became more and more pronounced throughout the eighteenth century as academicians assumed administrative control of French technology as consultants, inspectors, and even directors of industry.[3] Réaumur was one of the earliest and most enthusiastic supporters of this technocratic function of the Academy, and perhaps it was for this reason that he was given charge of

writing the vast industrial encyclopedia that Colbert had projected.

Réaumur began this enormous task in 1713 with his "Description de l'art du tireur d'or," a paper on the art of drawing gold into thread and wire. Two years later he published his investigations of the arts dealing with precious stones, in which he showed that certain turquoise stones were actually fossilized animal teeth. His most significant and original contribution to industrial technology was unquestionably his investigation of the iron and steel industry, the results of which he presented in a series of memoirs read before the Academy in 1720, 1721, and 1722.[4] Réaumur was not a trained engineer or metallurgist and knew only a little chemistry; but he did bring to these researches a profound mathematical ability, an extraordinarily keen power of observation, a lively experimental imagination, and a fine rational intellect.

The French government, especially the regent, Philippe II, duke of Orléans, took great interest in Réamur's work. It was primarily through the duke's good offices that Réaumur was able to obtain documentation concerning the iron and steel industries of foreign countries. The regent also subsidized Réaumur's researches by granting him a pension of 12,000 *livres* on the postal farm. All this generosity was based on the mercantilistic policy of the French government, which encouraged and subsidized native industries in hopes of improving the balance of trade. The French ferrous metals industry was technologically backward, and it was hoped that Réaumur's study would help to remedy the situation.

The first part of Réaumur's investigation concerns the production of steel. This was usually accomplished in the eighteenth century by the lengthy and expensive process of cementation. Small pieces of wrought iron were mixed with charcoal and heated for two or three weeks, until the iron was carburized or case-hardened into blister steel. In the first part of his study, Réaumur was concerned to find the best possible cement, that is, the best combination of substances to mix with the iron. His procedure was experimental and crudely empirical, although of course he knew from the experience of generations of ironmasters what kinds of substances to try. He made dozens of tiny, earthen crucibles of equal size and shape and capable of holding about half a pound of iron. This way he could make about forty trials at once in the ovens that were available to him. The experiments were rigidly controlled in such a way as to insure that the only variable would be the cement. After innumerable experiments of this nature, Réaumur concluded that the best mixture was a specific combination of chimney soot, charcoal, ashes, and common salt.

This result, however, is less significant than Réaumur's conclusion concerning the nature of steel. He recognized that steel, far from being a more refined form of iron, as most people thought, was in fact impure iron the small parts of which were interpenetrated with "sulfurous and saline particles." Réaumur is sometimes taken to task for not having recognized that steel is iron combined with a small quantity of carbon; but given the relatively primitive state of early eighteenth-century chemistry, he could hardly have reached such a conclusion. Also, it should be remembered that the "sulfurous particles" he believed to be one of the constituents of steel were not particles of common sulfur. In the parlance of eighteenth-century chemistry, the term "sulfurous" usually referred to an inflammable or oily principle that was present in combustible bodies, such as charcoal, chimney soot, and other carboniferous substances. Thus Réaumur was closer to realizing the truth about the nature of steel than many people give him credit for.

Réaumur also investigated the treatment of steel after it was manufactured. He was especially interested in the tempering process, for which he attempted, without much success, to give a scientific explanation. He also noted that by rupturing a steel sample and examining the texture of the grain at the point of breakage, one could determine the quality of the steel. The better to determine the relative hardness of metals, Réaumur set up a scale of seven substances not unlike the hardness scale contrived by Mohs a century later and still in use by mineralogists. He also invented an apparatus for measuring the flexibility of tempered steel wire.

The second part of Réaumur's study concerns his attempts to produce a malleable (nonbrittle) cast iron. In the eighteenth century cast iron was made by heating iron ore in a blast furnace at temperatures high enough to produce the metal in a molten state. The resulting iron always had a high carbon content, which had the advantage of lowering its melting point and thus making it suitable for casting, but the disadvantage of making it brittle and thus unsuitable for objects that had to withstand severe strain. Guns were often cast of iron instead of the preferred bronze; but they were liable to fracture and explode, causing more deaths among the friends at the firing end than among the enemies at the receiving end. It was of the greatest importance, not only to the art of war but also to the arts of peace, that a cast iron be produced that had the strength of steel and the resilience of bronze. This was the ultimate goal of Réaumur's entire study.

Although the composition of cast iron is fairly complex and the chemical tools available to Réaumur were quite primitive, he nonetheless came surprisingly close to identifying its true nature. Just as steel is made harder and more brittle than iron by the penetration between its parts of sulfurous and saline particles, so, Réaumur thought, cast iron is made harder and more brittle than steel by the interpenetration of its parts by still more sulfurous and saline particles. If one substitutes "carbon" for "sulfurous and saline particles," then one has the modern notion of what constitutes the basic differences between wrought iron, steel, and cast iron.

To remove the brittleness of cast iron, Réaumur reasoned, it was necessary to remove at least some of the sulfurous and saline particles. Heat would open the pores of the metal and force out many of the offending particles; but it was necessary to find a substance that would combine with them, thus preventing their reentry into the body of the metal. Again Réaumur's method for finding the best substance to achieve this end was largely inspired guesswork. At first he tried bone ash and powdered chalk, but the results were not entirely satisfactory.

He then turned his attention to an interesting substance that had appeared as a by-product of some of his experiments. He had noticed that when cast iron plaques were heated in his oven for several days, a thick layer of reddish powder formed on them. This substance was known to chemists as "saffron of Mars," and it was believed to be a calcined or burned iron. Réaumur reasoned that iron in this state was divested by fire of all its oily, sulfurous, and saline particles, and that therefore it would be a substance most fit to reabsorb those same particles, like a chemical sponge. He entirely surrounded pieces of cast iron with this powder, placed the mixture in a crucible, and heated it to bright red. After several days Réaumur discovered that the iron had become soft, resilient, and malleable, rather like wrought iron. Unwittingly, he had found the process for making European malleable castings or "whiteheart." Unfortunately for the French iron industry, Réaumur did not fully realize the importance of this discovery, and did not emphasize it sufficiently. The reason seems to be that he thought that the saffron of Mars (the red oxide of iron) was a particular substance obtained only through the firing of manufactured iron and that it was, therefore, quite uncommon, relatively expensive to produce, and thus unsuitable for large-scale enterprise. In fact it is, as we now know, the same substance as the common and inexpensive red ore of iron. Of so little importance did Réaumur consider this process that when he returned to the subject in his *Nouvel art d'adoucir le fer fondu* (1726), he neglected altogether to mention it. Only in the nineteenth century was the "Réaumur process" exploited on a large commercial scale.

Réaumur also studied and worked in the tinplate industry. Once encouraged and subsidized by the government, the French tinplate industry had fallen into a state of utter ruin during the early part of the eighteenth century. At the public assembly of the Academy of Sciences on 11 April 1725, Réaumur delivered a paper in which he revealed the basic industrial secrets of tin-plating. It was believed that once the essential processes of the industry became known, anyone possessing the necessary capital could set up a tinplate factory in France. Indeed, Réaumur helped to found such an operation at Cosne-sur-Loire; but the costs of production proved too great to meet competition from imported German tinplate.

From about 1717 Réaumur undertook a lengthy and intensive investigation of the porcelain industry. China porcelain was in great demand in Europe because of its delicate beauty and its soft, appealing translucence. European workmen had attempted to imitate it, but they were deceived by its vitrified texture into believing that common glass entered into its composition. In France factories were set up at Rouen (1673) and St. Cloud (1677) to manufacture an artificial porcelain made with a previously fired glassy mixture or frit of the type known as soft paste (*pâte tendre*) to distinguish it from hard paste, from which true porcelain is made. In Germany a technique for making true, hard-paste porcelain was discovered by Johann Friederich Böttger in association with Tschirnhausen; and as early as 1708 their factory at Dresden began to turn out a hard, red stoneware in almost every way comparable with a type of true china.

What seems to have initiated Réaumur's interest in porcelain was a letter published in 1717, from a Jesuit missionary to China, Father François-Xavier Entrecolles, in which he described in detail the processes used to manufacture porcelain in the famous Chinese factory at Ching-te-chen (Fou-liang). Along with his letter Father Entrecolles sent samples of the two substances used in the making of china. One of these substances, petuntse, was a feldspar; the other, kaolin, was a clay. Réaumur obtained the samples from Entrecolles's correspondent and began a two-year effort to identify them. The regent ordered intendants all over France to send specimens of all manner of sands, stones, earths, and other mineral substances found in their districts. Although large numbers of samples were collected, Réaumur was not able to identify the two Chinese specimens or to find their French equivalents.

Réaumur did, however, discover and make public, in a series of memoirs delivered before the Academy of Sciences, the secret processes for making soft-paste porcelain from glass. He confirmed by experiment that, despite their similar appearances, true China porcelain and the French imitation were quite differently composed. During the course of his experiments Réaumur invented a new type of crystalline ceramic that has proved useful in the twentieth century in protecting rocket nose cones from overheating.

Although Réaumur was unsuccessful in his attempts to discover the secret of making hard-paste porcelain, he did prepare the way for later investigators. His pupil Jean-Étienne Guettard discovered French sources of the two substances, kaolin and petuntse, necessary for the manufacture of porcelain. Later chemists and academicians, such as Pierre-Joseph Macquer, Jean Hellot, and the count of Milly, built on his pioneer researches a complete and detailed technological structure that was to make French porcelain among the finest in the world.

Réaumur was perhaps best known for the thermometer scale that he invented and that bears his name. Thermometers had been in use for about a century when he became interested in them, but there were not yet any universally accepted scales that would allow scientists who were not in the same place to compare their thermometric findings. Fahrenheit's scale was beginning to be adopted both in England and in Holland; but with its two fixed points and its scale divided linearly into a given number of degrees, it was accurate only if the inside diameter of the hollow thermometer tube was perfectly regular.

Réaumur sought to avoid this difficulty by constructing a thermometer with a single fixed point and a degree defined volumetrically (instead of linearly) in terms of some fraction of the total volume of liquid in the thermometer bulb. To fashion his thermometer according to these principles, he made a series of pipettes, the smallest equal to the volume of a single degree and the others equal to 25, 50, or 100 times the volume of the smallest. Then, using these pipettes, he filled a thermometer with 1,000 measures of liquid. Since it was not necessary to use the thermometric liquid itself when graduating the thermometer, Réaumur first used water and then switched to mercury, which he found more convenient. The place on the thermometer tube reached by 1,000 measures of liquid was marked 0°, and each degree above and below that single and arbitrary fixed point was equal to 1/1,000 the volume of the liquid at 0°. Then the graduated thermometer was emptied and refilled to a point just below the 0° mark with the thermometric liquid—in this case alcohol. The thermometer was then placed in ice water and alcohol was carefully added until it reached the 0° mark. Then the tube was hermetically sealed.

The one serious drawback to Réaumur's thermometer was that different strengths of alcohol have different coefficients of dilation, so that while one type of alcohol might expand one degree after the application of a certain amount of heat, another might expand two degrees under the same conditions. It was vital that all thermometers scaled according to his system have the same grade of alcohol. Réaumur suggested that the alcohol used in his thermometers be of a type that would dilate 80 degrees—that is, 8 parts in 100—between the temperature of ice and the temperature at which the alcohol began to boil in an open thermometer tube. Owing to an unfortunate confusion of language in his article on the thermometer, however, nearly everyone believed that 80° on his scale was the temperature of boiling water; and as a result, when so-called Réaumur thermometers began to be made by the artisans of Paris, they were nearly all scaled linearly with respect to two fiducial points, 0° for ice and 80° for boiling water. Scientists using mercury for their thermometers and basing their degree on the same value as Réaumur (1/1,000 the volume of the liquid at 0°) found the boiling point somewhere between 100° and 110°. In short, while there were many types of thermometers named for Réaumur, few were constructed in accordance with his specific instructions. As a result, it is often impossible to tell from the text of an eighteenth-century author claiming to use a Réaumur thermometer exactly what scale he is referring to, unless he happens to mention a universal fixed point, such as the temperature of boiling water.

Natural History. Réaumur was among the greatest naturalists of his or any age. In the breadth and range of his researches, in the patient detail of his observations, and in the brilliant ingenuity of his experiments, it would be difficult to name his equal. Thomas Henry Huxley has compared him favorably with Darwin.[5]

Réaumur's motives in pursuing natural history were a strange mixture of hardheaded practicality and frivolous delight in the curiosities of nature. In 1715 his investigation of artificial pearls led him to study the substance that gave luster to the scales of fishes. These researches were in turn linked to inquiries that he had undertaken since 1709 into the formation of mollusk shells, which he showed to grow by the addition of successive layers rather than by the incorporation of new matter into an already existing structure. In 1717 he attempted artificially to stimulate pearl formation in bivalves, and in 1711 he rediscovered the secret of making the purple dye of the ancient

Romans from the substance produced by a particular species of mollusk. Réaumur also investigated the means by which mollusks, starfish, and various other invertebrates move about. He was the first to describe ambulacral feet. In 1710 he published a memoir on the fabric he had made from spiders' silk. He presented the duke of Noailles with a pair of spiders' silk stockings, and he gave Jean-Paul Bignon some spiders' silk gloves. His memoir on spiders' silk was translated into Italian, English (it was inserted into the *Philosophical Transactions of the Royal Society*), and Chinese (by request of the Manchu emperor).

Réaumur's greatest work in natural history was his *Mémoires pour servir à l'histoire des insectes*, published in six volumes between 1734 and 1742. He had originally intended to publish ten volumes on insects; but after the six published during his lifetime, nothing remained of the project but fragments in manuscript, some of which were not published until the twentieth century.[6] It is not altogether clear why Réaumur stopped writing his great work at such an early date. There is some evidence, however, that he may have been discouraged by the jealous rivalry with his younger and more popular contemporary Buffon.[7]

The term "insect" was used in the early eighteenth century to designate almost any small invertebrate —not just hexapods or six-legged arthropods. The word refers primarily to creatures possessing segmented bodies; thus spiders, myriopods, and worms were usually included in the class. Réaumur's concept of the insects was even broader; he included polyps, mollusks, crustaceans, and even reptiles and amphibians. "The crocodile is certainly a fierce insect," he once proclaimed, "but I am not in the least disturbed about calling it one."[8]

Because of their large numbers and great diversity, insects were difficult to classify. Taxonomical schemata had been formulated by Aldrovandi and Vallisnieri which relied on superficial physiological and ethological, as well as morphological, characteristics. Swammerdam made things even more complicated by taking into consideration developmental characteristics. It is typical of Réaumur's approach that he tended to neglect morphology (which has since been used as the basis for insect classification) and concentrated instead on ethological characteristics. As a result, it is difficult at times to determine exactly what insect he was discussing. He seemed fascinated above all with insect behavior, obviously admiring the industry, diligence, ingenuity, organization, and skill of those small creatures. There is something almost of the eighteenth-century *bon bourgeois* in his description of the bees and the ants; and just as middle-class humans are categorized according to their professions, so, Réaumur thought, the insects might be classified according to their industries and occupations:

> . . . the portion of the history of insects to which I am most sensitive is that which concerns their ingenuity [*génie*],[9] their industries; also their industries will often decide the order in which I shall treat them. I have thought, for example, that one would prefer to see together all the insects that know how to clothe themselves, and which are above all remarkable for that, than to find them dispersed in different classes as they necessarily would be according to the methods of Swammerdam and Valisnieri.[10]

Another characteristic of Réaumur's approach to the natural history of insects is his persistent utilitarianism. There is in him little of the gimcrack virtuosity which seeks to know petty and useless details merely for the sake of knowing them. He doubtless admired the cunning practicality of the insects because it reflected his own turn of mind. He sought always to justify his researches by emphasizing their usefulness. Near the beginning of the first volume of his monumental study on insects he discussed at some length the economic value of entomological research. Silk, wax, honey, lacquer, and cochineal, to name just a few, are products of economic importance derived from the "industries" of insects.[11] Réaumur apparently believed that it might be possible to derive useful technological procedures from the observation of insect activities. He sought, for example, to mimic the industry of the bees by attempting to make wax from pollen.[12] He seemed to imply that perhaps the caterpillars and the spiders have something to teach us about weaving, that perhaps the useful resins manufactured by the ants could be made artificially.

The study of insects is profitable from another point of view as well—pest control. As early as 1728, Réaumur had investigated the life cycle of clothes moths in order to determine the best means of eradicating them. Again, at the beginning of his first volume on the natural history of insects he stressed the utility of this aspect of his research.

> An infinity of these tiny animals defoliate our plants, our trees, our fruits. . . . they attack our houses, our fabrics, our furniture, our clothing, our furs. . . . He who in studying all the different species of insects that are injurious to us, would seek means of preventing them from harming us, would seek to cause them to perish, to cause their eggs to perish, proposes for his goal important tasks indeed.[13]

Réaumur, in short, was a pioneer in applied entomological research.

The most widely read portion of Réaumur's natural history of insects is probably the nine memoirs of volume V on the history of the bees. Réaumur lavished an enormous amount of time and observational and experimental skill on these productive social insects. His descriptions were minute and exacting in every detail, and his experiments were among the most ingenious he ever contrived. Réaumur was one of the first to undertake extensive quantitative research on insects. He discovered that by immersing a hive in cold water he could, in effect, anesthetize the insects for a time, thus enabling him to separate the members of the bee community into their various classes and to count them. He carefully kept track of the number of bees leaving the hive in the course of a day, measured the average amount of pollen brought back by each, and from this estimated the weight of a single day's harvest. Réaumur also counted cells and larvae in a hive relative to its adult population; estimated the prodigious number of eggs laid by a single queen; weighed a swarm of bees in order to determine their number; measured the temperature of the hive and kept track of its variations in relation to both the season and the number and density of its inhabitants; and even made very careful investigations of the geometric form of the honey cells.

With regard to the position and function of the queen in bee society, Réaumur's researches were extensive and original. He discovered that all hives, even those very close to swarming, have only one queen; if others are introduced, they will be rejected or even killed. He found that if a colony is deprived of its queen, it must make a new one (by feeding a special substance called royal jelly to a developing larva) or it will die. He discovered that a hive without a queen will (under certain conditions) accept an exogenous ruler.

Réaumur kept track of individual bees by tinting them with various dyes. He dissected bees and their larvae and had detailed plates made to accompany his treatise. He made some of the first tentative studies of communication among the bees. In short, there was no aspect of the life cycle or behavior of bees too minute or too unimportant to escape his attention. He took every pain, every precaution to make his study as complete and exhaustive as possible. And so it was with the other insects he studied.

Biology and Genetics. Réaumur's biological and genetical notions were dominated by the ideas of the preformationists. Given the prevalence of the mechanical philosophy at the beginning of the eighteenth century, it was very difficult to imagine ways in which new biological forms could arise from undifferentiated matter. To deal with this problem, the preformationists went so far as to deny that there had ever been generation of any kind, whether of biological individuals or of members or parts. Thus when Leeuwenhoek peered through his microscope at semen, he fancied that he observed a tiny fetus encased in the head of each sperm. When properly "planted" in the female womb, these tiny creatures would simply enlarge into infants. Swammerdam believed that each structure of the adult butterfly was present in the infant caterpillar. All that was needed to change the latter into the former was that the caterpillar slough off its skin and allow the preexisting butterfly parts to unfold and grow. There was neither metamorphosis nor generation, only unfolding and growth. It follows, then, that every living creature that ever will exist actually exists now, with all its mature parts, in seeds or in seeds of seeds. The unborn are indescribably small, but they are nonetheless there, even though we may not be able to see them. In the biology of the preformationists there was literally nothing new under the sun and Adam was, in the realest of senses, the father of us all, or perhaps (because eighteenth-century opinion was divided on the matter) Eve was the universal mother.

Réaumur gave his cautious support to some of the ideas of the preformationists, largely because the alternative—the outmoded Aristotelian notions of epigenesis—seemed, in an era insistent upon mechanism, absurd and without foundation. For example, in his discussion of caterpillars and butterflies, which takes up the first two volumes of his natural history of the insects, Réaumur repeated Swammerdam's observations on the development of the chrysalis and came to the same conclusion—the parts of the adult butterfly are simply enlargements of structures preexistent beneath the skin of the caterpillar. It is indicative of the power of these preformationist conceptions that when he was unable to find certain preexistent structures in the caterpillar, he asserted that they were there nonetheless, although invisible.

Others of Réaumur's observations lent support to the ideas of preformation. For example, he discovered that different degrees of heat could retard or accelerate the growth of the butterfly chrysalis. This experiment showed that conditions in the environment played an important role in the rate of biological development. The implication was that preformed biological structures could, like the chrysalis, remain in a state of suspended animation until the conditions became suitable for their development.

Another observation that seemed to lend credence to preformationism was the discovery of parthenogenesis among the aphids. It had been noticed long before that female aphids seemed capable of producing abundant offspring even when there were no males around to fertilize them. Furthermore, no one had

ever observed aphids in the act of mating. Réaumur suggested that one could discover whether it was possible for aphids to reproduce without sexual contact by raising them in isolation from the time of their birth. He attempted the experiment himself, but his aphids died before reaching sexual maturity. His student, the young Genevan naturalist Charles Bonnet, repeated the experiment at Réaumur's suggestion and succeeded in producing aphids parthenogenetically. To Bonnet the experiment seemed to show that the preformed fetus was in the female egg rather than in the male sperm, and he became a leader of the "ovist" school of preformationism.

Réaumur never accepted all the ideas of preformationism, for he was too aware of the difficulties some aspects of the theory entrained. How can it be, for example, that offspring resemble both of their parents, since the preformed infant would have to originate in the seed of only one of them? In the second edition of his study of the technique of artificial incubation, Réaumur included an account of the inheritance of human polydactyly showing that the trait is passed to the offspring through male and female parents alike. It was precisely this kind of observation of biparental inheritance that led many biologists in the middle of the eighteenth century to abandon the notions of preformationism altogether and adopt pangenetic theories instead.

As early as 1712, Réaumur had apparently rejected the idea of *emboîtement* (the "encasement" of germs within germs). In that year he published a paper in which he demonstrated that certain crustaceans had the power to regenerate missing legs or parts of legs lost through misadventure. A preformationist account of this type of regeneration posed serious difficulties. Presumably any given place on the leg would have to possess a germ containing a preformed leg or a preformed section of leg similar to the section below the place where the germ was located. Furthermore, to account for secondary regenerations, one would have to assume that each of these tiny preformed legs contained multitudes of yet tinier preformed legs and sections of legs, and so on ad infinitum. Yet if the germs were not preformed, where did they come from? Were they, then, truly generated from undifferentiated matter? Réaumur was at a loss to explain the phenomenon. The preformationist account of regeneration was manifestly absurd; but then, so it seemed, were the alternative explanations.

If the discovery of regenerated animal members posed serious difficulties for preformationism, the discovery of regeneration in the fresh-water hydra nearly devastated the theory and many others as well. It was a cousin of Charles Bonnet, Abraham Trembley, who collected some of these tiny creatures from a stagnant pool in the summer of 1740 and observed that they had animallike powers of locomotion, extension, and contraction. Uncertain whether to classify them as animals or plants, he communicated his findings to Réaumur and asked for his opinion. After observing the creatures, Réaumur was convinced that they were indeed animals. Trembley had also cut one of his "polyps," as they were to be called, transversely in two and noted, to his great astonishment, that each part continued to manifest signs of life and that at the end of about nine days each had regenerated its other half. Later it was found that one could divide hydras into almost as many pieces as one pleased and each part would live independently and, after a time, regenerate the whole. It was Réaumur who announced these remarkable facts to a startled and somewhat incredulous scientific community in March and April 1741. The following summer his student Guettard went to the coast to test the regenerative powers of several marine animals.

The discovery of the hydra and its peculiar ability to sustain life even when divided into a number of small parts caused a profound shock among European naturalists. The observation of analogous regenerative powers in starfish, sea anemones, and worms was likewise highly disturbing to the commonly received notions of biology. Many of the materialist biological theories that arose in the middle of the eighteenth century were founded upon observations such as these.[14]

Réaumur also made a significant contribution to physiology in his brilliantly conceived experimental investigation of the process of digestion in birds. He demonstrated the enormous power of the gizzard by forcing several grain-eating birds to swallow tubes of glass and tin. When the animals were opened after two days, the glass tubes were found shattered, the pieces of glass smoothed and polished by the action of the gizzard; the tin tubes were crushed and flattened. In carnivorous birds he showed that digestion was more chemical than mechanical. He enclosed pieces of meat in a tube, the ends of which were closed with fine gratings, and forced a kite to swallow it. When the bird later regurgitated the tube, Reaumur found that the meat had been partially digested. There was no sign of putrefaction. Vegetable matter introduced into the bird's stomach in the same manner underwent little change.

Academic Activities and Death of Réaumur. Réaumur was perhaps the most prestigious member of the Academy of Sciences during the first half of the eighteenth century. Through his incessant labors and voluminous publications, through his extensive cor-respondence with scientists both in France and abroad, and through the reflected brilliance of his students, he acquired enormous authority and renown in the European scientific community. Many of the important discoveries of the day were announced by him—for instance, the discoveries of the parthenogenesis of aphids and of the regenerative powers of the hydra. It was also he who announced Musschenbroeck's discovery of the Leyden jar in 1746.

Réaumur rose quickly in the ranks of the Academy. He was elected *pensionnaire mécanicien* in May 1711; he was also made director of the Academy twelve times and subdirector nine times. He was elected to the Royal Society of London, to the Academies of Science of Prussia, Russia, and Sweden, and to the Institute of Bologna.

Réaumur never married, but devoted all his time to his scientific and academic career. From a needy relative he bought the title of commander and intendant of the Royal Military Order of Saint Louis, an honorific office possessing the dignity of a count. Two years before his death he inherited the castle and lordship of La Bermondière in Maine. There in October 1757 he suffered a fall from his horse while returning from Mass and died. He was buried in the parish church of the nearby village, St.-Julien-du-Terroux. An inscription dedicated to his memory was placed in the church at the time of its restoration in 1879.

NOTES

1. A *président* is a chief justice. Hénault was *président* of the most prestigious court in France, the Parlement of Paris.
2. No *éloge* was ever published for Guisnée; thus neither his first names nor the date and place of his birth are known.
3. For a discussion of the French Academy of Science's role in the bureaucracy of the *ancien régime*, see Roger Hahn, *The Anatomy of a Scientific Institution: The Paris Academy of Sciences, 1666–1803* (Berkeley, Calif., 1971).
4. They were collected and published under the title *L'Art de convertir le fer forgé en acier, et l'art d'adoucir le fer fondu, ou de faire des ouvrages de fer fondu aussi finis que de fer forgé* (Paris, 1722).
5. "From the time of Aristotle to the present day I know of but one man who has shown himself Mr. Darwin's equal in one field of research—and that is Réaumur." Quoted in Leonard Huxley, *Life and Letters of Thomas Henry Huxley*, I (New York, 1901), 515.
6. *The Natural History of Ants*, William Morton Wheeler, trans. and ed. (New York, 1926), French text included; *Histoire des fourmis*, Charles Perez, ed., with intro. by E.L. Bouvier; *Encyclopédie entomologique*, ser. A, XI (Paris, 1928); *Histoire des scarabés*, P. Lesne and F. Picard, eds., XXXII (Paris, 1955); *Les papiers laissés par de Réaumur et le tome VII des Mémoires pour servir à l'histoire des insectes . . . Introduction du tome VII des Mémoires pour servir à l'histoire des insectes*, Maurice Caullery, ed., *Encyclopédie entomologique*, ser. A, XXXIIa (Paris, 1929).
7. Jean Torlais, "Une rivalité célèbre: Réaumur et Buffon," in *Presse médicale*, **65** (11 June 1958), 1057–1058.

8. Quoted by Wheeler in *Natural History of Ants*, 29.

9. It is difficult to give an exact English equivalent for the word *génie* in this context. Réaumur seems to have had in mind their characteristic skills. The older English word "ingeniosity" would be etymologically nicer.

10. Réaumur, *Mémoires pour servir à l'histoire des insectes*, I, 42.

11. *Ibid.*, 4–5.

12. Réaumur did not know that the wax is made from special secretions and not directly from the pollen collected by the bee.

13. Réaumur, *Mémoires* . . ., I, 8.

14. See Aram Vartanian, "Trembley's Polyp, La Mettrie, and Eighteenth-Century French Materialism," in *Journal of the History of Ideas*, **11** (1950), 259–286.

BIBLIOGRAPHY

I. ORIGINAL WORKS. Complete lists of Réaumur's works are in William Morton Wheeler's trans. of Réaumur's *The Natural History of Ants* (New York, 1926), 263–274; and Jean Torlais, "Chronologie de la vie et des oeuvres de René-Antoine Ferchault de Réaumur," in *Revue d'histoire des sciences et de leurs applications*, **11** (1958), 1–12, with portrait.

Réaumur's major works are *L'art de convertir le fer forgé en acier, et l'art d'adoucir le fer fondu, ou de faire des ouvrages de fer fondu aussi finis que de fer forgé* (Paris, 1722), English trans. by Annelie Grünhaldt Sisco, with notes and intro. by Cyril Stanley Smith, *Réamur's Memoirs on Steel and Iron* (Chicago, 1956); *Mémoires pour servir à l'histoire des insectes*, 6 vols. (Paris, 1734–1742); *Art de faire éclorre et d'élever en toute saison des oiseaux domestiques de toutes espèces, soit par le moyen de la chaleur du fumier, soit par le moyen de celle du feu ordinaire*, 2 vols. (Paris, 1749), 2nd ed. entitled *Practique de l'art de faire éclore et d'élever en toute saison des oiseaux domestiques . . .* (Paris, 1751), also an English trans. of the 1st ed., *The Art of Hatching and Bringing up Domestic Fowls . . .* (London, 1750); and 3 vols. in the Paris Academy of Sciences' series Description des Arts et Métiers: *Art de l'épinglier . . . avec des additions de M. Duhamel du Monceau . . .* (Paris, 1761); *Fabrique des ancres . . . avec des notes et additions de M. Duhamel* (Paris, 1761); and *Nouvel art d'adoucir le fer fondu aussi finis que de fer forgé . . .*, with intro. by Duhamel du Monceau (Paris, 1762). Also see the works cited in note 6.

In addition see *Abhandlungen über Thermometrie, von . . . Réaumur . . .*, A. J. von Oettingen, trans., in *Ostwalds Klassiker der exakten Wissenschaften*, **57** (1894); and *Correspondance inédite entre Réaumur et Abraham Trembley*, Émile Guyenot, ed. (Geneva, 1943).

II. SECONDARY LITERATURE. For general biographical information see Jean-Paul Grandjean de Fouchy's *éloge* in *Histoire de l'Académie royale des sciences . . .* for 1757 (1762), 201–216; Frédéric Lusson, *Étude sur Réaumur* (La Rochelle, 1875); and Jean Torlais, *Réaumur et Ch. Bonnet d'après leur correspondance inédite* (Bordeaux, 1932); *Réaumur et sa société* (Bordeaux, 1932); and *Un esprit encyclopédique en dehors de "l'Encyclopédie": Réaumur, d'après des documents inédits* (Paris, 1936; 2nd ed., rev. and enl., 1961).

On Réaumur's mathematical papers, see René Taton, "Réaumur mathématicien," in *Revue d'histoire des sciences et de leurs applications*, **11** (1958), 130–133. On Réaumur's work on iron and steel technology, see A. Birembaut, "Réaumur et l'élaboration des produits ferreux," in *Revue d'histoire des sciences . . .*, **11** (1958), 138–166; and A. Portevin, "Réaumur, métallurgiste et chimiste," in *Archives internationales d'histoire des sciences*, **13** (1960), 99–103. On Réaumur's thermometer, A. Birembaut, "La contribution de Réaumur à la thermométrie," in *Revue d'histoire des sciences . . .*, **11** (1958), 302–329.

On Réaumur as a naturalist, see Pierre Grasse, *Réaumur et l'analyse des phénomènes instinctives*, Conférences du Palais de la Découverte, ser. D, no. 48 (Paris, 1957); Jean Torlais, "Réaumur et l'histoire des abeilles," in *Revue d'histoire des sciences . . .*, **11** (1958), 51–67; A. Davy de Virville, "Réaumur botaniste," *ibid.*, 134–137; Jean Rostand, "Réaumur et les premiers essais de léthargie artificielle," *ibid.*, **15** (1962), 69–71; and "Réaumur et la résistance des insectes à la congélation," *ibid.*, 71–72; and Jean Théodoridès, "Réaumur (1683–1757) et les insectes sociaux," in *Janus*, **48** (1959), 62–76.

On Réaumur's biology and genetics, see Jean Torlais, "Réaumur philosophe," in *Revue d'histoire des sciences . . .*, **11** (1958), 13–33; Jean Rostand, "Réaumur embryologiste et généticien," *ibid.*, 33–50. For general background on the problem of preformation, see Jacques Roger, *Les sciences de la vie dans la pensée française du XVIIIᵉ siècle* (Paris, 1971), 326–453, and Réaumur and preformation, 380–384. On Réaumur and polydactyly, see Bentley Glass, "Maupertuis, Pioneer of Genetics and Evolution," in Bentley Glass, ed., *Forerunners of Darwin: 1745–1859* (Baltimore, 1959), 63–66.

On Réaumur's correspondence, see Pierre Speziali, "Réaumur et les savants genevois. Lettres inédites," in *Revue d'histoire des sciences . . .*, **11** (1958), 68–80 (which contains eight letters to Gabriel Cramer, Théodore Tronchin, and André Roger); Jean Chaia, "Sur une correspondance inédite de Réaumur avec Arthur, premier médecin du roy à Cayenne," in *Episstème*, **2** (1968), 37–57, (which contains twelve letters); and Peeter Müürsepp, "Rapports entre le célèbre savant français Réaumur et le Tzar Pierre I," in *Actes du XIIᵉ Congrès internationale d'histoire des sciences*, XI (Paris, 1971), 95–101.

On Réaumur's last years, see A. Davy de Virville, "Réaumur dans la Mayenne," in *Revue d'histoire des sciences . . .*, **11** (1958), 81–82. The articles on Réaumur in *Revue d'histoire des sciences et de leurs applications*, **11** (1958), have been collected and published as *La vie et l'oeuvre de Réaumur (1683–1757)* (Paris, 1962).

J. B. GOUGH

RECK, HANS (*b.* Würzburg, Germany, 24 January 1886; *d.* Lourenço Marques, Mozambique, 4 August 1937), *volcanology, paleontology, paleoanthropology.*

Reck, who came from a family of officers, attended the universities of Würzburg, London, and Berlin, receiving his doctorate from Berlin in 1910 with the dissertation—supervised by Wilhelm von Branca—"Isländische Masseneruptionen." He then became Branca's assistant at the Geological and Paleontological Institute of the University of Berlin. In this capacity he led the expedition sent by the Institute in 1912–1913 to Tendaguru (now in Tanzania) to excavate dinosaurs of the Upper Jurassic. In 1913 Reck headed an expedition sent by the Prussian Academy of Sciences to investigate the geological structure of the Germany colonies then in West Africa (now Togo and Cameroun). In 1913–1914, on behalf of the universities of Berlin and Munich, he made the first systematic excavations in the Olduvai Gorge (now in Tanzania) in the hope of finding Pleistocene human and animal remains. From 1915 to 1919 he served as a government geologist in German East Africa. After World War I, Reck was a private scholar and extraordinary professor at the Geological and Paleontological Institute, Berlin. During these years he made several more expeditions through East and South Africa. In 1931–1932 he participated in Louis Leakey's expedition in the Olduvai Gorge.

By marshaling exceptional energy Reck overcame a congenital heart defect and was able to endure hardship and exertion over long periods, even surviving his experience as a soldier in the East African campaign and as a prisoner of war of the English in Egypt. He died of heart failure during a South African expedition. Shortly before his death the University of Athens awarded him an honorary doctorate for his volcanologic research on Santorini (Thira) Island in the Aegean Sea. Reck was married to I. von Grumbkow; they had no children.

Reck's first geological works (1911–1912) dealt with the geomorphology of southern Germany: the earliest course of the Danube, the development of Triassic-Jurassic escarpments and recent crust movements. The African expeditions led to investigation of Quarternary sediments in East Africa, of the age of the great East African graben system and of the Jurassic-Cretaceous boundary in Ethiopia. His first paleontologic studies (1911) were concerned with Neogene sea urchins of Java and—in collaboration with his friend Hans von Staff—the mode of life of the trilobites, as well as with the bivalves found in the malm of Solnhofen (Bavaria).

In later publications Reck confined his attention to the vertebrates, especially the Pleistocene mammals of Africa. Gathering material in the course of his expeditions, he reported on the excavations of dinosaurs and pterodactyls in the Tendaguru and of reptiles in the Karroo series of Natal.

Reck's excavations in the Olduvai Gorge became widely known. Its Pleistocene layers yielded not only numerous mammals but also a human skeleton that he reported in 1914. From then until 1937, he contributed to the *Wissenschaftliche Ergebnisse der Oldoway-Expedition*. In it he and other experts described the geology and fossil remains of the site. Following Leakey's Olduvai expedition, Reck again examined—with Leakey and other English colleagues—the human skeleton discovered in the Olduvai Gorge in 1913. Today it is known that the skeleton is of a post-Ice Age man of the Capsian culture, and that it was embedded in layers dating from the early Pleistocene. In 1933 Reck wrote a popular book on his work in the Olduvai Gorge and in 1936, with Kohl-Larsen, he published the first survey of the Pleistocene human and animal remains discovered by the latter in the Njarasa graben in East Africa.

Volcanologic studies constituted a large portion of Reck's scientific work. Beginning with his dissertation in 1910, he studied volcanism for more than twenty-five years. His publications in this field were devoted to volcanic regions or to specific volcanoes, both Tertiary and Recent. He wrote on the late Tertiary volcanoes of the Hegau (southwestern Germany), on volcanoes in Iceland, on African volcanoes, and on the eruptions and activity of Etna, Vesuvius, Krakatoa, and of Asian and South American volcanoes. Much of this material appeared in *Zeitschrift für Vulkanologie*, of which Reck was coeditor from 1923. He was particularly interested in the volcanic island group of the Santorini in the Aegean Sea. He devoted many brief studies and a three-volume work (1936) to the emergence and eruptions of the volcano there in the years 1925 to 1928.

Along with these regional descriptions, Reck was interested in general manifestations and problems of volcanism. Among the topics he treated were linear eruptions, craters of elevation, volcanic horsts, collapse calders, volcanic bombs, and the relationships between volcanism and tectonics. Hence his colleagues were indebted to him for having introduced many new ideas as well as for having expanded knowledge of recent and fossil volcanoes.

BIBLIOGRAPHY

I. ORIGINAL WORKS. Reck's writings include "Isländische Masseneruptionen," in *Geologische und paläontologische Abhandlungen*, n.s. **9** (1910), 83–184; "Die morphologische Entwicklung der süddeutschen Stufenlandschaft im Lichte der Davis'schen Cyklustheorie," in *Zeitschrift der Deutschen geologischen Gesellschaft*, **64** (1912), 81–232; "Physiographische Studie über vulkanische

Bomben," *Zeitschrift für Vulkanologie*, supp. 1 (1915); "Über vulkanische Horst-Gebirge," *ibid.*, **6** (1922), 155–182; *Die Hegau-Vulkane* (Berlin, 1923); "Die Kräfte-Gruppen des Vulkanismus und der Tektonik und ihre gegenseitigen Beziehungen," in *Zeitschrift der Deutschen geologischen Gesellschaft*, **76** (1924), 115–137; *Grabungen auf fossile Wirbeltiere in Deutsch-Ostafrika*, Geologische Charakterbilder, no. 31 (Berlin, 1925); "Über die Tätigkeit von Etna und Vesuv im Herbst 1928," in *Zeitschrift der Deutschen geologischen Gesellschaft*, **80** (1928), 345–352; "The Oldoway Human Skeleton," in *Nature*, **131** (1933), 397–398, written with L. B. Leakey *et al.*; *Oldoway. Die Schlucht des Urmenschen* (Leipzig, 1933); *Santorin, der Werdegang eines Inselvulkans und sein Ausbruch*, 3 vols. (Berlin, 1936); "Erster Überblick über die jungdiluvialen Tier- und Menschenfunde Dr. Kohl-Larsen's im nordöstlichen Teil des Njarasa-Grabens (Ostafrika)," in *Geologische Rundschau*, **27** (1936), 401–441, written with L. Kohl-Larsen; and *Wissenschaftliche Ergebnisse der Oldoway-Expedition*, n.s. 4 (1937).

II. SECONDARY LITERATURE. See A. T. Hopwood, "Prof. Hans Reck," in *Nature*, **140** (1937), 351; Poggendorff, VI (1938), 2136–2137, with partial bibliography; P. Range, A. Diehm & G. A. Schmidt, "Prof. Dr. Hans Reck," in *Tropenpflanzer*, **40** (1937), 365; and K. Sapper, "Hans Reck †," in *Zeitschrift für Vulkanologie*, **17** (1936–1938), 225–232, with bibliography and portrait.

HEINZ TOBIEN

RECLUS, ÉLISÉE (*b.* Sainte-Foy-la-Grande, France, 15 March 1830; *d.* Thourout, Belgium, 4 July 1905), *geography*.

Reclus was the son of a Protestant pastor, Jacques Reclus, and his wife, the former Zéline Trigant. After studying at Neuwied, Sainte-Foy, and Montauban, he entered the University of Berlin in February 1851 to continue his study of theology but also attended the popular lectures of Carl Ritter in geography. Reclus subsequently rejected religion, but he remained a lifelong admirer of Ritter; the geographical writings of the two are quite similar, except that Reclus altered Ritter's teleological explanations. Late in 1851 Reclus became embroiled in political events at home and again went abroad. He spent 1852–1857 in the British Isles, Louisiana, and Colombia, drawn to the last through admiration for Humboldt. Living in Paris after his return to France in 1857, Reclus occupied himself mostly with his geographical writings in the *Revue des deux mondes* and Guides Joanne and especially in the preparation of *La terre*, the great geographical synthesis that established his fame. *La terre* shows the influence of Humboldt and Ritter and also bears similarity to the contemporary works of Oscar Peschel and George Perkins Marsh.

Banished in 1872 for his part in the Paris Commune, Reclus chose Switzerland as his land of exile and plunged immediately into his most grandiose project, a world geography. Originally planned as a five- or six-volume work, it eventually appeared in nineteen volumes. Reminiscent of Ritter's *Erdkunde*, it was better organized and carried to completion. Also unlike Ritter, Reclus traveled widely, even returning to the Americas to improve his books with personal observations. Upon completion of the *Nouvelle géographie universelle* in 1894, Reclus removed to Brussels, where he became professor of geography at the newly established Université Nouvelle. He then began writing his last great work, *L'homme et la terre*, an ambitious historical-geographical synthesis, most of which was published posthumously. Also during his last years Reclus conceived great cartographic schemes, especially huge relief globes, which he hoped might be symbols of world unity and brotherhood. Of special interest was his proposal in 1895 to build a globe on the scale of 1:100,000 (diameter about 420 feet), a project which was supported by A. R. Wallace and Patrick Geddes but never came to fruition.

BIBLIOGRAPHY

I. ORIGINAL WORKS. Reclus's most important geographical works are those which he referred to as his "trilogy": *La terre*, 2 vols. (Paris, 1868–1869); *Nouvelle géographie universelle*, 19 vols. (Paris, 1876–1894); and *L'homme et la terre*, 6 vols. (Paris, 1905–1908). Good guides to his voluminous writings on geography and anarchism are the published catalogs of the Bibliothèque Nationale (Paris) and the Internationaal Institut voor Sociale Geschiedenis (Amsterdam). Also useful are Hem Day, *Essai de bibliographie de Élisée Reclus* (Paris–Brussels, 1956); and the bibliography accompanying Jean Maitron's *Histoire du mouvement anarchiste en France (1880–1914)* (Paris, 1951). The best bibliography is a typescript, "Élisée Reclus, géographie et sociologie" (n.p., n.d.), presumably prepared by Reclus's sister, Louise Dumesnil, in the department of printed books at the Bibliothèque Nationale.

The most significant collections of Reclus MSS are in Paris at the Institut Français d'Histoire Sociale, Fonds Élisée Reclus; and especially at the Bibliothèque Nationale, dépt. des manuscrits, N.A.F. 22909–22919, correspondance et papiers d'Élisée Reclus. Of fundamental importance is his *Correspondance*, 3 vols. (Paris, 1911–1925), arranged and edited by Louise Dumesnil, who left the collection of MSS to the Bibliothèque Nationale after her death in 1917.

II. SECONDARY LITERATURE. The most important biographical studies are Joseph Ishill, ed., *Élisée and Élie Reclus: In Memoriam* (Berkeley Heights, N.J., 1927); Max Nettlau, *Élisée Reclus: Anarchist und Gelehrter*

(1830–1905) (Berlin, 1928), enl. Spanish ed., *Eliseo Reclus, la vida de un sabio justo y rebelde*, 2 vols. (Barcelona, 1929); and Paul Reclus, *Les frères Élie et Élisée Reclus; ou, du protestantisme à l'anarchisme* (Paris, 1964). Perhaps the most important obituaries in geographical journals are by Patrick Geddes, in *Scottish Geographical Magazine*, **21** (1905), 490–496, 548–555; by Paul Girardin and Jean Brunhes, in *Geographische Zeitschrift*, **12** (1906), 65–79; by Peter Kropotkin, in *Geographical Journal*, **26** (1905), 337–343; and by Franz Schrader (Reclus's cousin), in *Géographie*, **12** (1905), 81–86.

G. S. DUNBAR

RECORDE, ROBERT (*b*. Tenby, Pembrokeshire, Wales, *ca*. 1510; *d*. London, England, 1558), *mathematics.*

Recorde was the second son of Thomas Recorde, whose father had come to Wales from Kent, and of Rose Johns of Montgomeryshire. He graduated B.A. from Oxford in 1531 and was elected a fellow of All Souls College in the same year. All Souls was a chantry and graduate foundation for the study and training of clerks in theology, civil and canon law, and medicine. At some time he removed to Cambridge and there received the M.D. degree in 1545. According to the Cambridge records he had been licensed in medicine at Oxford some twelve years earlier, and the B.M. usually went with the Oxford license. Tradition has it that he lectured on mathematics at Oxford and Cambridge; but details of his university career, and of any degrees other than the B.A. and M.D., are lacking.

Recorde was in London by 1547, probably practicing medicine. There is no evidence that he acted as physician to any of the Tudors, although he served the government in other capacities. In January 1549 he was appointed comptroller of the Bristol mint. In October 1549, at the time of Somerset's first fall, he sided with the protector, refusing to divert money intended for King Edward to the armies of the west under Lord John Russell and Sir William Herbert. He was accused of treason by Herbert (later earl of Pembroke) and was confined at court for sixty days while the mint ceased production. This was the beginning of a permanent quarrel with Herbert which had serious consequences for Recorde's later career.

From 1551 to 1553 Recorde was surveyor of the mines and monies in Ireland, in charge of the abortive silver mines at Wexford and, technically, supervisor of the Dublin mint. The venture was unsuccessful from the start. In addition to differences with the German miners over technology, and the personal animosity of Pembroke, the treasury was not able to bear the great expenses of the mines and their lack of profits. The work stopped in 1553 and Recorde was recalled. Not until 1570 was his estate compensated for some £ 1,000 due him for his services there.

In 1556 Recorde attempted to regain a position at court and laid charges of malfeasance as commissioner of the mints against Pembroke. Regardless of the merits of the case, it was a serious error in judgment on his part; Pembroke had the complete confidence of Queen Mary and King Philip, and it was impossible for a minor civil servant, whose last post had ended in failure, to survive the clash with the "politic old earl." Pembroke sued for libel in a bill of 16 October 1556. The hearing was held in January 1557, with a judgment of £ 1,000 damages against Recorde awarded 10 February. Presumably Recorde was imprisoned for failure to pay this sum. His will was written in King's Bench prison and was admitted to probate 18 June 1558.

Recorde has been justly called the founder of the English school of mathematical writers. He envisioned a course of instruction in elementary mathematics and its applications for mathematical practitioners. Deliberately choosing the vernacular, he wrote simple, clear English prose of a higher quality than his scientific contemporaries or immediate successors. Recorde made a special effort to find English equivalents for Latin and Greek technical terms, but very few of his innovations were adopted by later writers. His books indicate great skill as a teacher. His use of dialogue enabled him to carry a student step by step through the mastery of techniques, and to emphasize the proper order and method of instruction. Difficult questions were deferred until an understanding of fundamentals was achieved. Recorde took a rational view of his sources and was refreshingly critical of unquestioning acceptance of established authority. The mathematical books were written in the order in which he intended them to be studied: arithmetic, plane geometry, practical geometry, astronomy, and theoretical arithmetic and algebra. Projected works on advanced astronomy, navigation, and a translation of Euclid's *Elements* probably were never completed.

The arithmetic, *The Ground of Artes* (1543, enlarged in 1552), was the most popular of all Recorde's works. The first edition dealt only with whole numbers, covering the fundamental operations, reduction, progression, golden rule, and counter reckoning. In 1552 it was enlarged to include the same operations with fractions, and false position and alligation. There are three editions of the first version: 1543,

1549, and 1550[?]. The third has been dated formerly 1542[?] or 1545[?]; but bibliographers, on the basis of the state of the title-page border, now place it between the editions of 1549 and 1552.

The Pathway to Knowledge (1551) is a translation and rearrangement of the first four books of Euclid's *Elements*. Like Proclus before him, and Ramus later, Recorde separated the constructions ("things to be done") from the theorems ("things to be proved"). Proofs are not given, but explanations and examples are provided. Pedagogically, Recorde felt that it is not easy for a student to understand at the beginning both the thing that is taught and the reason why it is so.

The Gate of Knowledge dealt with measurement and the use of the quadrant. It has been lost and possibly was never published, although in the *Castle* it is referred to as complete.

The Castle of Knowledge (1556), on the construction and use of the sphere, is an elementary Ptolemaic astronomy with a brief, favorable reference to the Copernican theory. The often-cited edition of 1551 is a "ghost." The *Castle* is based chiefly on Ptolemy, Proclus, Sacrobosco, and Oronce Fine, but is much more than a synthesis of earlier writers. More than in any other of his books, Recorde was concerned here with sources. He devoted considerable space to a critical examination of the standard authorities, offering corrections of textual errors in the Greek authors and suggesting that the mistakes of Sacrobosco and others were caused by their lack of knowledge of Greek.

The Whetstone of Witte (1557) was the only one of Recorde's book not to have seen at least two editions, no doubt because it was less immediately useful to the London craftsmen than were his other works. It contains the "second part of arithmetic" promised in *The Ground of Artes* (from the arithmetic books of Euclid) and elementary algebra through quadratic equations. It is based on German sources, especially Johann Scheubel and Michael Stifel, and the algebra uses the German cossic notation. With Recorde's addition of the "equal" sign this algebra became completely symbolic. Although it is derivative, there are several noteworthy features: the use of zero coefficients in algebraic long division; the use of arbitrary numbers to check algebraic operations rather than the check by inverse operations; and the treatment of quadratics. Recorde did not admit negative roots but did use negative coefficients in equations. All quadratics are written with the square term equal to roots plus or minus numbers, or numbers minus roots. He still had to give the three usual rules of solution; but, in the case of an equation with two positive roots, $x^2 = px - q$, he stressed the solution using the relation between the roots and coefficients: $r_1 + r_2 = p$, $r_1 r_2 = q$.

In addition to his mathematical works, Recorde published *The Urinal of Physick* (1547), dedicated to the Company of Surgeons. A promised anatomy has not survived. A traditional medical work on the judgment of urines, full of sensible nursing practice, the *Urinal* is less modern than his mathematical works and less critical of authority.

Recorde was not only an able teacher and a skillful textbook writer but was also one of the outstanding scholars of mid-sixteenth-century England. He was well trained in mathematics, and was familiar with Greek and medieval texts as well as contemporary developments. His intelligent attitude toward authority, and his appeals to reason and observation, anticipated in a more moderate manner the anti-Aristotelianism of Petrus Ramus. Recorde was learned in medicine and the law. He was an able Greek scholar who stressed the importance of a knowledge of that language for an accurate understanding of sources. He had a wide range of learning in various fields: he was a historian interested in the antiquities of Britain, a collector of manuscripts, and one of the first students of the Anglo-Saxon language.

Recorde had no international reputation because all of his works were in English and on an elementary level. In England, however, his books remained the standard texts throughout the Elizabethan period. A generation of English scientists, especially the non-university men, stated that Recorde's books had been their first tutors in the mathematical sciences. The excellence of the English school of mathematical practitioners, fostered by growing geographical interests, has been attributed to the high quality of the vernacular movement in applied science begun by Recorde.

BIBLIOGRAPHY

I. ORIGINAL WORKS. Recorde's published works are *Ground of Artes* (London, 1543, 1549, 1550[?]; enl. ed., 1552, 1558, and many later eds.; last ed., 1699); *Urinal of Physick* (London, 1547, 1548, and others; last ed., 1665); *Pathway to Knowledge* (London, 1551, 1574, 1602); *Castle of Knowledge* (London, 1556, 1596); and *Whetstone of Witte* (London, 1557).

John Bale's contemporary autograph notebook lists other, unpublished, works by Recorde as well as MSS from his library. Bale's work has been published as *Index Britanniae scriptorum*, R. L. Poole, ed. (Oxford 1902).

II. SECONDARY LITERATURE. W. G. Thomas of Tenby, Pembrokeshire, who discovered the King's Bench records concerning Recorde and the Earl of Pembroke, is publishing a full account of the case.

For details on specific books see J. B. Easton, "The Early Editions of Robert Recorde's *Ground of Artes*," in *Isis*, **58** (1967), 515–532; and "A Tudor Euclid," in *Scripta mathematica*, **27** (1966), 339–355; L. D. Patterson, "Recorde's Cosmography, 1556," in *Isis*, **42** (1951), 208–218 (which must be read with caution); F. R. Johnson, "Astronomical Textbooks in the Sixteenth Century," in E. A. Underwood, ed., *Science, Medicine and History* (Oxford, 1953), 285–302; and L. C. Karpinski, "The Whetstone of Witte," in *Bibliotheca mathematica*, 3rd ser., **13** (1912–1913), 223–228.

Recorde's place in Renaissance science is discussed in F. R. Johnson and S. V. Larkey, "Robert Recorde's Mathematical Teaching and the Anti-Aristotelian Movement," in *Huntington Library Bulletin*, no. 7 (1935), 59–87; F. R. Johnson, *Astronomical Thought in Renaissance England* (Baltimore, 1937); and Edward Kaplan, "Robert Recorde (c. 1510–1558): Studies in the Life and Works of a Tudor Scientist," Ph.D. diss. (New York University, 1960), available on University Microfilms (Ann Arbor, Mich., 1967).

JOY B. EASTON

REDFIELD, WILLIAM C. (*b.* Middletown, Connecticut, 26 March 1789; *d.* New York, N.Y., 12 February 1857), *meteorology, paleontology.*

The eldest of six children of Peleg Redfield and Elizabeth Pratt, Redfield was apprenticed to a saddle- and harnessmaker in 1803, soon after his seafarer father's death. Completing his apprenticeship in 1810, he set out to visit his mother, who had remarried and moved to Ohio. The journey of 700 miles took Redfield and his companions twenty-seven days. Returning to Connecticut in 1811, Redfield made saddles and kept a store for a decade. He married Abigail Wilcox on 15 October 1814; the couple had three sons. After her death on 12 May 1819, he married her cousin Lucy on 23 November 1820. She died on 14 September 1821, soon after the birth of a son who did not long survive her. In 1824 Redfield moved to New York, where he worked in steam transportation until 1856. On 9 December 1828 he married Jane Wallace, who, together with two sons by his first wife, survived him.

A diligent student of science from his days as an apprentice, Redfield came to the insight that a storm was a "progressive whirlwind." He drew this inference after observing the fall of trees on a trip from Connecticut to Massachusetts following the hurricane of 3 September 1821, and he confirmed it after two violent storms that struck New York in August 1830. A chance meeting with Denison Olmsted of Yale led Redfield to publish in the *American Journal of Science and Arts* (1831) his theory that storm winds blow counterclockwise around a center that moves in the direction of the prevailing winds. This paper brought Redfield instant recognition on both sides of the Atlantic. From the information supplied him by sea captains sailing to and from New York, Redfield continued for twenty-five years to develop his ideas on the rotary nature of storms and to publish his ideas in American and British scientific journals. The practical consequences he drew from his theory were widely disseminated through E. and G. W. Blunt's *American Coast Pilot*. Neither Redfield nor his rival, James P. Espy, who proposed a theory of storms that emphasized convection and condensation in the vertical as the source of a storm's energy, perceived that both principles are necessary for a complete theory; the controversy between them and their respective supporters was a lively one.

When his son John published a short paper (1836) about the fossil fish he had found in the sandstone quarries of Durham, Connecticut, the elder Redfield began to examine sandstones not only in Connecticut but also in Massachusetts, New Jersey, and Virginia. In a series of seven short papers that appeared between 1838 and 1856 in *American Journal of Science and Arts* and *Proceedings of the American Association for the Advancement of Science*, Redfield established himself as the first American specialist on fossil fish. He named one genus and seven species of Triassic fish, and his elucidation of their taxonomic and stratigraphic relationships still stands virtually without revision.

The arduous journey to Ohio in 1810–1811, on which he encountered Fulton's steamboat as he passed through Albany, stimulated Redfield's interest in transportation. In 1822 he began his career as marine engineer and transportation promoter with a steamboat on the Connecticut River. By 1824 he and his associates' vessels were plying the Hudson. The frequent explosions of steam boilers led Redfield to the idea of passenger-carrying "safety barges" towed by steam vessels from New York to Albany. By 1826, when the passenger barge was no longer popular, the Steam Navigation Company, which Redfield served as superintendent, had inaugurated a towboat service for freight barges. Railroads, too, claimed his attention. In 1829 Redfield published a plan for a rail link between the Hudson and the Mississippi river valleys, and in the next decade he laid out the Harlem and the Hartford–New Haven railroads. He also served on the board of the Hudson Railroad.

Redfield's prominence among scientists and his organizational ability made him the leader of the transformation of the Association of American Geologists and Naturalists into the American Association for the Advancement of Science. He served as president of the new organization at its first meeting in Philadelphia in September 1848. Yale University awarded Redfield an honorary degree in 1839; his name is commemorated by one genus and several species of fish and by Mount Redfield in the Adirondacks.

BIBLIOGRAPHY

I. ORIGINAL WORKS. A list of sixty-two published works of Redfield is given in *American Journal of Science*, 2nd ser., **24** (1857), 370–373; among them the most significant scientific paper is "Remarks on the Prevailing Storms of the Atlantic Coast of the North American States," *ibid.*, **20** (1831), 17–51. A bound volume of these works is in the possession of Alfred C. Redfield, Woods Hole, Massachusetts; it contains other brief items by Redfield, chiefly from newspapers. There are three letter books in the Yale University Library, and the holograph diary of the journey to Ohio in 1810 belongs to Alfred C. Redfield.

II. SECONDARY LITERATURE. There are two principal biographical sources: his son's fragmentary autobiography, *Recollections of John Howard Redfield* (privately printed, 1900); and Denison Olmsted, "Biographical Memoir of William C. Redfield," in *American Journal of Science*, 2nd ser., **24** (1857), 355–373, which includes the bibliography cited above and all but a few paragraphs of "An Address in Commemoration of William C. Redfield," in *Proceedings of the American Association for the Advancement of Science*, **11** (1857), 9–34. W. J. Humphreys of the U.S. Weather Bureau wrote a modern biographical sketch in *Dictionary of American Biography*, VIII, 441–442.

A contemporary evaluation of Redfield's meteorology is Charles H. Davis, "Redfield, Reid, Espy and Loomis on the Theory of Storms," in *North American Review*, **58** (1844), 335–371. There is a modern treatment, more concerned with method but exceptionally useful for social background, in George Daniels, *American Science in the Age of Jackson* (New York, 1968), esp. chap. four and the biographical sketch in app. I.

Redfield's paleontology is authoritatively treated in George G. Simpson, "The Beginnings of Vertebrate Paleontology in North America," in *Proceedings of the American Philosophical Society*, **86** (1942), 130–188, esp. 167.

HAROLD L. BURSTYN

REDI, FRANCESCO (*b.* Arezzo, Italy, 18 February 1626; *d.* Pisa, Italy, 1 March 1697 or 1698), *entomology, parasitology, toxicology.*

Redi was the son of Gregorio Redi, a renowned Florentine physician who also worked at the Medici court, and Cecilia de' Ghinci. He graduated in philosophy and medicine from the University of Pisa on 1 May 1647. On 26 April 1648 he registered at the Collegio Medico in Florence. He served at the Medici court as head physician and superintendent of the ducal pharmacy and foundry. Friend, counselor, and virtual secretary to his employers, he was also a member of the small Accademia del Cimento, which flourished actively, although intermittently, at the Medici court from 1657 to 1667. This decade at the Academy coincided with the period in which Redi produced his most important works.

In 1664 there appeared the *Osservazioni intorno alle vipere*, which was closely related to the doctrine of the circulation of the blood. Redi was by then superintendent of the ducal pharmacy, where snakes were widely used in the preparation of theriaca. Contrary to prevailing belief, Redi held that snake venom was completely unrelated to its bile. It was rather the yellow humor produced by "two glands, which I have found in all vipers." The humor stagnates in the "two sheaths in which the viper conceals its fangs" and, "when the viper bares its fangs and strikes, it is of necessity spurted on the wound." He also exonerated the viper's animal spirits, which had been considered responsible for the effects of the bite inflicted by the enraged and vindictive serpent.

Sucking the tissues bitten by the viper was harmless, Redi discovered, since the poison was ineffective if swallowed: to be effective it had to be inoculated into the animal's tissues and enter the bloodstream. Hence he considered it advantageous to make "a tight ligature not far above the wound so that the poison is not carried to the heart by the circular movement of the blood and all the blood infected."

To achieve these results, which mark the first stirrings of experimental toxicology, Redi performed countless experiments on the effects of snakebite. With the venom obtained from living as well as dead snakes, he poisoned other animals. He either sprinkled liquid or powdered venom on wounds, or inserted into the flesh "sharp slivers of broom" soaked with the poison. The animal was not rendered poisonous to eat; and Redi examined the viscera in vain in an attempt to discover the mechanism of action of the fatal toxin.

Redi's masterpiece is considered to be *Esperienze intorno alla generazione degli insetti* (1668), in which he disproved the doctrine of spontaneous generation in insects, inherited from Aristotle and still considered dogma. The microscope revealed in insects an organization as marvelous as it was unsuspected. Redi prepared and observed the egg-producing apparatus in insects, and he also used the microscope to good advantage in observing the morphological elements characteristic of the eggs of each species.

Redi then set out to attack the doctrine of spontaneous generation in the lower animals. Even if decaying animals or plants appeared to "give birth to an infinity of worms [larvae]," the reality was quite different, he held. It must be assumed "that flesh and plants and other things whether putrefied or putrefiable play no other part, nor have any other function in the generation of insects, than to prepare a suitable place or nest into which, at the time of procreation, the worms or eggs or other seed of worms are brought and hatched by the animals; and in this nest the worms, as soon as they are born, find sufficient food on which to nourish themselves excellently." These organic bodies "never become verminous if they are kept in a place where flies and gnats cannot enter." Redi demonstrated this in experiments of almost unique simplicity, using wide-mouthed flasks containing such organic substances as meat or cheese.

Contagion by insects was therefore necessary before decaying substances could develop worms. Redi supposed it was necessary in living plants and animals as well—the worms and insects found in fruit, in the galls of plants, and inside the bodies of animals, for instance, the liver fluke *(Fasciola hepatica)* and the larva of the gadfly. After proposing these highly significant examples of entozoa, Redi ended *Esperienze* with a long review of ectozoa, observed in various animals, which he also felt "more inclined to believe . . . are born from the eggs laid by their mothers, fertilized by coitus." Despite Redi's research, the question remained open. Led astray by various observations—especially by the "many fibers and threads" that from the wall of the "gall run to the egg, almost like so many veins and arteries which convey the material suitable for formation of the egg and worm and for the nutriment of which they have need"—Redi confined himself, provisionally, to attributing a zoogenic property to vegetable organisms and even to living animals. The insects in galls were, like those in the cherry, "generated by the same spirit and the same natural power that give birth to the fruits of plants." It was left to Malpighi (1679) to deny spontaneous generation also for the insects of galls.

Redi's major parasitological treatise, *Osservazioni intorno agli animali viventi, che si trovano negli animali viventi* (1684), was an abundant compilation of endoparasitic helminths found in the organs of different classes of animals, including mollusks and crustaceans. Jules Guiart, who in 1898 attempted to identify the parasites described by Redi, compiled a list of 108 items, two-thirds of which were endoparasitic helminths, and one-third ectoparasitic insects and acarids. In the 1684 *Osservazioni* Redi also formulated the idea of an evolutionary cycle of parasitic worms.

Redi's studies on the generation of insects and on parasitism, which culminated in the acarian etiology of scabies, were extended by the physician Giovanni Cosimo Bonomo and the apothecary Giacinto Cestoni. Their results were presented in a full and concise formulation that had been subjected to Redi's corrections and embellishments: the *Osservazioni intorno a' pellicelli del corpo umano* (1687).

Like Galileo, Redi used the Italian language with mastery, both in his scientific works and in the "consultations" he composed as a practicing physician. His great knowledge of his mother tongue, including rare and obsolete terms, was a great help in his contributions to a revision of the dictionary of the Accademia della Crusca. But his literary contributions, unlike his scientific, were marked by fabrications; and he took pleasure in defending the correctness of terms he had invented himself. Even the history of science was not spared by these fabrications, since his apocryphal version of the invention of spectacles was accepted for over two centuries.

BIBLIOGRAPHY

I. Original Works. Redi's works include *Osservazioni intorno alle vipere scritte in una lettera a Lorenzo Magalotti* (Florence, 1664); *Esperienze intorno alla generazione degli insetti scritte in una lettera a Carlo Dati* (Florence, 1668); *Lettera sopra alcune opposizioni fatte alle sue Osservazioni intorno alle vipere scritta alli Ab. Bourdelot Sig. di Condé e S. Leger e Alessandro Moro* (Florence, 1670); *Esperienze intorno a diverse cose naturali, e particolarmente a quelle che ci sono portate dall'Indie, scritte in una lettera al P. Atanasio Chircher* (Florence, 1671); *Lettera intorno all'invenzione degli occhiali scritta a Paolo Falconieri* (Florence, 1678); *Osservazioni intorno agli animali viventi che si trovano negli animali viventi* (Florence, 1684).

For other editions, works, and collections of works, see Dino Prandi, *Bibliografia delle opere di Francesco Redi* (Reggio Emilia, 1941).

II. Secondary Literature. On Redi and his work, see Luigi Belloni, "Francesco Redi, biologo," in *Celebrazione dell'Accademia del Cimento nel tricentenario della fondazione (Domus Galilaeana, 1957)* (Pisa, 1958), 53–70; "Francesco Redi als Vertreter der italienischen Biologie des XVII. Jahrhunderts," in *Münchener Medizinische Wochenschrift,* **101** (1959), 1617–1624; and "L'influence exercée sur la médecine clinique par les sciences de base développées par l'école Galiléienne (génération spontanée et 'contagium vivum' de la Gale)," in *Clio medica,* **8** (1973), 143–149; Andrea Corsini, "Sulla vita di Francesco Redi. Nuovo contributo di notizie," in *Rivista di storia critica delle scienze mediche e naturali,* **13** (1922), 86–93; Jules Guiart, "Francesco Redi," in *Archives de parasitologie,* **1** (1898), 420–441; Dino Prandi, *Bibliografia delle opere di Francesco Redi* (Reggio Emilia, 1941); Ugo Viviani, *Vita ed opere inedite di Francesco Redi,* 3 vols. (Arezzo, 1928–1931);

and "L'autopsia e la data della morte di Francesco Redi," in *Archeion*, **16** (1934), 181–185; and Nella Volterra, "Sulla malattia e sulla morte di Francesco Redi," *ibid.*, **15** (1933), 73–77.

A further bibliography on Redi is contained in Viviani (I, 103–117) and in Prandi (pp. v–vi).

On Redi's literary falsifications and their influence on the history of the invention of spectacles, see Giuseppe Albertotti, "Note critiche e bibliografiche riguardanti la storia degli occhiali," in *Annali di ottalmologia e clinica oculistica*, **43** (1914), 328–356; and "Lettera intorno alla invenzione degli occhiali scritta da Giuseppe Albertotti all'Onorev.mo Senatore Isidoro del Lungo," *ibid.*, **50** (1922), 85–104; Isidoro del Lungo, "Le vicende d'un'impostura erudita (Salvino degli Armati)," in *Archivio storico italiano*, **78** (1920), 5–53; and *Chi l'inventore degli occhiali?* (Bologna, 1921); and Guglielmo Volpi, "Le falsificazioni di Francesco Redi nel vocabolario della Crusca," in *Atti della R. Accademia della Crusca* (1915–1916), 33–136.

LUIGI BELLONI

REDOUTÉ, PIERRE-JOSEPH (*b.* St.-Hubert, Belgium, 10 July 1759; *d.* Paris, France, 20 June 1840), *botanical illustration.*

Redouté was the son of a modest interior decorator and painter of church pieces, Charles-Joseph Redouté. In 1782 he moved to Paris to find scope for his artistic talent. While sketching flowers for pleasure in the Jardin du Roi, Redouté was discovered by the botanist C. L. L'Héritier de Brutelle, who trained him as a botanical artist and employed him to illustrate his publications. Through L'Héritier, Redouté's artistic development came under the guidance of Gerardus van Spaendonck (1746–1822), professor of flower painting at the Jardin du Roi. Through the patronage especially of Joséphine Bonaparte and later royal protectors, Redouté was able to illustrate many of the great French botanical works of his period.

Redouté developed for botany the technique of color printing by means of stipple engravings *à la poupée*, with which he produced some of the finest colored botanical illustrations ever published. The combination of scientific botanical precision and innate artistry made Redouté one of the greatest flower painters. Many of the paintings from his *Roses* are still reproduced on a large scale. The scientific value of his paintings and books lies especially in the documentation of the original specimens of plants new to science and described in his time.

BIBLIOGRAPHY

I. ORIGINAL WORKS. Main works to which Redouté contributed or which he published under his own name are the following (listed chronologically): C. L. L'Héritier de Brutelle, *Stirpes novae aut minus cognitae* . . . (Paris, 1785–1805); and *Sertum anglicum* . . . (Paris, 1789–1792); A. P. de Candolle, *Plantarum historia succulentarum* (Paris, 1799–1832); R. L. Desfontaines, *Flora atlantica* (Paris, 1800); E. P. Ventenat, *Description des plantes nouvelles et peu connues, cultivées dans le jardin de J. M. Cels* (Paris, 1800); H. L. Duhamel du Monceau, *Traité des arbres et arbustes que l'on cultive en pleine terre en Europe, et particulièrement en France par Duhamel*, 2nd ed., enl., 7 vols. (Paris, 1800–1819); A. Michaux, *Histoire des chênes de l'Amérique* (Paris, 1801); A. P. de Candolle, *Astragalogia* (Paris, 1802); P. J. Redouté, *Les liliacées*, 8 vols. (Paris, 1802–1808), text by Candolle (I–IV), Delaroche (V–VI), and Raffeneau Delile (VII–VIII); E. P. Ventenat, *Jardin de la Malmaison*, 2 vols. (Paris, 1803); J. J. Rousseau, *La botanique de J. J. Rousseau* (Paris, 1805); A. J. G. Bonpland, *Description des plantes rares cultivées à Malmaison et à Navarre* (Paris, 1813); and P. J. Redouté, *Les roses*, 3 vols. (Paris, 1817–1824); and *Choix des plus belles fleurs* (Paris, 1827–1833).

II. SECONDARY LITERATURE. See F. Léger, *Redouté et son temps* (Paris, 1945); F. A. Stafleu, "Redouté, peintre de fleurs," in G. H. M. Lawrence, ed., *A Catalogue of Redouteana* (Pittsburgh, 1963), 1–32; and *Taxonomic Literature* (Utrecht, 1967), 376–380, full information on studies, commentaries, bibliographies, and biographical publications; and Lotte Günthart, Andre Lawalrée, and Günther Buchheim, *P. J. Redouté* (Pittsburgh, 1972).

FRANS A. STAFLEU

REDTENBACHER, FERDINAND JAKOB (*b.* Steyr, Austria, 25 July 1809; *d.* Karlsruhe, Germany, 16 April 1863), *mechanics, mechanical engineering.*

Redtenbacher grew up in Steyr, a city noted for its metal industry and sometimes called the "Birmingham of Austria." His father, Alois Redtenbacher, a prosperous merchant and a man of considerable culture, did not favor the "humanist" education then conventional in the German upper classes and apprenticed his son, at age eleven, to a grocer (1820–1824). This experience developed in young Redtenbacher a sense of practicality and a reliance upon self-education. His decision to become an engineer was made in 1825, when he spent nine months as an assistant draftsman and surveyor in the Imperial Construction Department at Linz. Later in the same year he entered the Polytechnikum of Vienna, where he studied mechanical engineering while also attending lectures at the university. His chief teachers were the engineer Johann Arzberger and the mathematician Andreas von Ettingshausen. After his graduation in 1829 he served for four years as Arzberger's assistant in the chair of mechanics and theory of machines. In 1834 he became teacher, and later professor, of applied mathematics at the Obere Industrieschule of Zurich. Redtenbacher found his ultimate position when the Polytechnische Schule of Karlsruhe invited him to occupy the chair of mechanical engineering. He moved to Karlsruhe in 1841 and served there until his death in 1863.

Redtenbacher's significance lies in his role as engineering educator. It was he more than anyone else who gave the German *Technische Hochschule* its characteristic structure, and it was he who conceived and propagated the particular blend of theory and practice that constituted the success of German engineering for the following century. His influence was equally great through lecturing, writing, and administration.

Redtenbacher's literary production consisted mostly of textbooks. His own education had been theoretical, in the spirit of the École Polytechnique, and based on the study of Euler, Laplace, Poisson, Navier, and Poncelet. He soon realized that the theoretical insights contained in such works were difficult to apply in practice and that the practical engineers of the English tradition denied the relevance of theory. In all his activities Redtenbacher strove to bridge this gulf. In Zurich, during frequent visits to the waterwheel factory of his friend Caspar Escher, he recognized that the newly introduced water turbines presented an opportunity to demonstrate the utility of well-presented theory; a book on water turbines was his first work (1844). His desire to convince practical engineers of the value of theory comes out even more clearly in his *Resultate für den Maschinenbau* (1848), his best-known work. This is a handbook with tables and formulas for the solution of all common mechanical engineering problems, presented without mathematical derivation. This work was supplemented by his *Principien der Mechanik* (1852), which provided the theoretical background, and by *Der Maschinenbau* (1862–1865), which concentrated on the art of engineering design. Besides a number of specialized books on steam engines, locomotives, and caloric engines, Redtenbacher also wrote a work of philosophical intent, *Das Dynamidensystem* (1857), a kind of mechanistic theory of atomism based upon ideas of Dalton, Poisson, and Cauchy.

As Redtenbacher's administrative influence increased—he was soon the most famous professor, and after 1857 he was director, of the Karlsruhe Polytechnic—he was able to put his own conception of engineering education into effect. His students were required to serve apprenticeships in industry and were instructed in advanced mechanics and mathematics in a manner different from that of the École Polytechnique in its practical motivation and its

irreverence toward formal rigor. For Redtenbacher's model was not the tightly regimented École Polytechnique but the liberal German university with its system of lecturing, its separate faculties, and its academic freedom. The example of Karlsruhe had great influence; in the planning of the Swiss Federal Institute of Technology (founded 1855), for instance, it was Karlsruhe, not the École Polytechnique, that served as the exemplar, and Redtenbacher as the principal adviser.

BIBLIOGRAPHY

I. Original Works. Virtually all of Redtenbacher's literary output consisted of books: *Theorie und Bau der Turbinen und Ventilatoren* (Mannheim, 1844; 2nd ed., 1860); *Theorie und Bau der Wasserräder* (Mannheim, 1846; 2nd ed., 1858); *Resultate für den Maschinenbau* (Mannheim, 1848; 6th ed., 1875; 2nd French ed., 1873); *Principien der Mechanik und des Maschinenbaus* (Mannheim, 1852; 2nd ed., 1859); *Die calorische Maschine* (Mannheim, 1853); *Die Gesetze des Locomotivbaues* (Mannheim, 1855); *Die Bewegungsmechanismen* (Mannheim, 1857; new ed., 1861); *Das Dynamidensystem: Grundzüge einer mechanischen Physik* (Mannheim, 1857); *Der anfängliche und der gegenwärtige Erwärmungs-Zustand der Weltkörper* (Mannheim, 1861); and *Der Maschinenbau*, 3 vols. (Mannheim, 1862–1865; French ed., 1872).

II. Secondary Literature. There is no definitive biography of Redtenbacher. The most useful shorter works are the following (listed chronologically): Emil Kretzschmann, "Ferdinand Redtenbacher," in *Verein deutscher Ingenieure, Zeitschrift*, **9** (1865), 246–262; Franz Grashof, *Redtenbachers Wirken zur wissenschaftlichen Ausbildung des Maschinenbaus* (Heidelberg, 1866); Rudolf Redtenbacher, ed., *Erinnerungsschrift zur 70-jährigen Geburtstagsfeier Ferdinand Redtenbachers* (Munich, 1879); K. Keller, "Zum Gedächtnisse Redtenbachers," in *Bayerisches Industrie- und Gewerbeblatt* (1910), 351–355, 363–367; Franz Schnabel, "Die Anfänge des technischen Hochschulwesens," in *Festschrift Technische Hochschule Karlsruhe* (Karlsruhe, 1925), 1–44; and "Ferdinand Redtenbacher," in *Blätter für Geschichte der Technik*, **4** (1938), 66–71; Heinrich Ebeling, *Ferdinand Redtenbacher, Leben und Werk* (Karlsruhe, 1943); Otto Kraemer, "Ferdinand Redtenbacher," in *Die Technische Hochschule Fridericiana Karlsruhe: Festschrift zur 125-Jahrfeier* (Karlsruhe, 1950), 79–84; Karl Lindner, *Ferdinand Jakob Redtenbacher, der Begründer des wissenschaftlichen Maschinenbaues* (Graz, 1959); and Johannes Körting, "Ferdinand Redtenbacher," in *Verein deutscher Ingenieure, Zeitschrift*, **105** (1963), 449–451.

OTTO MAYR

REECH, FERDINAND (*b.* Lampertsloch, Alsace, France, 9 September 1805; *d.* Lorient, France, 6 May 1884), *marine engineering.*

After graduating sixth in a class of ninety-five from the École Polytechnique, Reech entered the École du Génie Maritime at Brest in 1825. Upon completion of his naval training he worked at the port of Brest and was promoted to the rank of lieutenant in 1829. After brief service at the arsenal of Cherbourg, he was assigned to Lorient, where the École du Génie had just been transferred. He soon devoted himself exclusively to teaching and research in geometry, mechanics, thermodynamics, and hydrodynamics. From 1831, Reech was director of the school for thirty-nine years, arranging in 1854 to have it transferred back to Paris, where it had been established in 1765. He also became a professor at the Sorbonne.

Reech published ten memoirs, ranging from a brochure supporting his unsuccessful candidacy for membership in the Académie des Sciences, through lecture notes on mechanics, to tomes on applied thermodynamics. The contribution for which he is still known was the first formulation of the hydraulic-model law of gravitational similitude: the necessary proportionality between the velocity of a ship and the square root of its length (more generally known as Froude's law of similarity). This is said to have been included in his lectures as early as 1831, but it is first mentioned in his memoir of 1844 and then discussed in detail in his course notes of 1852.

In 1870 Reech retired from active duty and returned to Lorient in Brittany rather than to Alsace because of the German occupation. During the period from 1875 to 1883 he was a member of the Association Française pour l'Avancement des Sciences, which possesses the most complete record of his life. At his request he was buried in Soultz-sous-Forêts, close to the village of his birth.

BIBLIOGRAPHY

A major work by Reech is *Mémoire sur les machines à vapeur et leur application à la navigation présenté a l'Académie royale des sciences* (Paris, 1844). A listing of his works is in *Catalogue générale des livres imprimés de la Bibliothèque nationale*, CXLVII (1938), cols. 840–841.

A discussion of some of Reech's work is in H. Rouse and S. Ince, *History of Hydraulics* (New York, 1963).

HUNTER ROUSE

REED, WALTER (*b*. Belroi, near Gloucester, Virginia, 13 September 1851; *d*. Washington, D.C., 23 November 1902), *medicine*.

Reed was the youngest of the five children of Lemuel Sutton Reed and Pharaba White, both North Carolinians of English descent. His father was a Methodist minister who was called to a new parish every year or two; Reed therefore grew up in a number of communities in southeastern Virginia and northeastern North Carolina. His schooling was haphazard until late in 1865, when the family moved to Charlottesville, Virginia, where Reed began two years' tutelage under the Confederate Lieutenant William R. Abbot, who later became headmaster of Bellevue School, near Bedford, Virginia. Reed studied the traditional subjects of Latin and Greek, English composition, grammar and rhetoric, and history and the humanities. He entered the University of Virginia at the age of fifteen, and after a year was permitted to take the medical course, a nine-month graded curriculum, twice as long as that offered anywhere else in the country. He received the M.D. in 1869, standing third in his class of fifty. Reed then went to New York to study at the medical school of Bellevue Hospital; he earned a second M.D. in 1870, when he was eighteen, but the degree was not awarded him officially until he was twenty-one.

Reed then became assistant physician at the New York Infants' Hospital. In 1871 he was resident physician at Kings County Hospital, Brooklyn, and in 1871–1872 at Brooklyn City Hospital. He simultaneously served as district physician for the New York Department of Public Charities. In June 1873 he was appointed sanitary inspector of the Brooklyn Board of Health. In June of the following year he received a commission as assistant surgeon, with the rank of first lieutenant in the Army Medical Corps and was ordered to a post in Arizona. Before he left for the west he married Emilie Lawrence, whom he had met and become engaged to while on a series of visits to his father in Murfreesboro, North Carolina. For the next eleven years he worked at bases in Arizona (including Fort Apache), Nebraska, Minnesota, and Alabama. His son, Lawrence, was born in Arizona in 1877 and a daughter, Blossom, was born in Omaha, Nebraska, in 1883.

As he neared forty, Reed became restless. He was aware of new discoveries in pathology and bacteriology, but garrison life afforded him no opportunities for further study. He therefore applied for a leave of absence to do advanced work, but was instead ordered to Baltimore, Maryland, as attending surgeon. He was authorized to study at Johns Hopkins Hospital where, after completing a brief clinical course, he undertook the study of pathology under William H. Welch, who early recognized his abilities. Reed did autopsies, learned new pathological techniques, and performed experiments. He was then sent for another two years to another isolated and dreary army outpost at Fort Snelling, Minnesota, but in 1893 was promoted major and, upon the appointment of George Sternberg as surgeon general of the United States, brought to Washington as curator of the Army Medical Museum and professor of bacteriology and clinical microscopy at the new Army Medical College. He worked closely with John Shaw Billings in the old Surgeon General's Library and Army Pathology Museum.

Reed's early work in Washington is buried in the *Reports of the Surgeon General of the Public Health Service of the United States*. It was not until 1896 that he was first able to demonstrate his skill as a medical investigator. In that year malaria was rampant in the Washington Barracks and at Fort Myer, in Virginia; it was generally believed to be caused by bad drinking water. Reed showed that there were no outbreaks of the disease in other areas of Washington that, like the military bases, obtained drinking water from the swampy Potomac. Moreover, although both officers and enlisted men shared much the same food, water, and housing, malaria was rife among only the enlisted class. The enlisted men would, however, often surreptitiously leave camp at night and go to the city along a trail through the swamps, and Reed was therefore able to state that not the drinking water, but rather the bad air—"malaria" in the ancient and literal sense—was responsible for the disease.

When the Spanish-American War broke out in 1898 Reed volunteered unsuccessfully for service in Cuba. He was instead appointed chairman of the board that had been hurriedly formed to investigate the outbreak of typhoid that had reached epidemic proportions in most large army camps. Hundreds of new cases of typhoid were found every day and many were of fatal outcome; bad water was again thought to be the mode of transmission of the bacillus, which had recently been identified. The board found that water was of little importance as a vector of the disease, which, they reported, was rather spread by flies and contact with fecal material. In addition, the Typhoid Board contributed to the control of the disease by discovering that it might be transmitted by carriers—persons who harbored the infectious organisms without being ill, or, indeed, without ever having had typhoid fever. The work of the board made it possible to end the epidemic which had killed between fifty and one hundred times as many soldiers as had died in combat or of wounds in Cuba; their massive two-volume report is a model for epidemiologists.

In 1900 Reed was appointed head of an army board to investigate the causes of yellow fever, which had broken out among American troops in Cuba. In the United States yellow fever was an annual scourge; the disease swept up the eastern seaboard and often spread to the Gulf states and up the Mississippi River valley. The places in which it occurred, the season of its occurrence, and the temperature coordinates of its spread were predictable and well known but not understood. No one knew what caused it, and the terror that accompanied it was equaled only by the misapprehension of its nature. A few lucky guesses had been made before Reed took up his work; as early as 1848 the Alabama physician Josiah Nott had suggested that an insect—perhaps the mosquito—caused the disease, although he had only tenuous evidence, and in 1881 Carlos Finlay, working closely with the United States Yellow Fever Commission in Havana, suggested that yellow fever was transmitted by *Culex fasciatus* (now called *Aëdes aegypti*).

The question of the cause of yellow fever was nonetheless still a vexed one when Reed took it up. In 1897 the Italian physician Giuseppe Sanarelli had stated that the organism *Bacillus icteroides* was the specific causative agent, and Sternberg, who also believed yellow fever to be caused by a bacterium, assigned Reed and James Carroll the task of investigating Sanarelli's claim. Reed and Carroll soon demonstrated that Sanarelli's organism bore no relationship to yellow fever. It was shortly thereafter that the disease broke out in Havana, invading even the headquarters of the governor, General Leonard Wood, and Reed was sent to Cuba as the head of the Yellow Fever Board, consisting of himself, Carroll, Jesse W. Lazear, and Aristides Agramonte.

Upon arriving in Havana, Reed and his co-workers pursued further investigations of Sanarelli's *Bacillus icteroides* and concluded that it was the hog-cholera virus. They then turned to the theory of the mosquito as the vector of yellow fever. Finlay had been conducting experiments along these lines, and provided them with mosquitoes, mosquito eggs, and instructions for raising the insects in the laboratory. Since there was no laboratory test for yellow fever, Finlay also aided them in the clinical verification of the disease. Reed then designed and conducted the experiments that elucidated the means of transmission of the disease once and for all. He used human subjects of necessity, and Lazear, who was accidentally bitten by an infected mosquito, contracted yellow fever and died; Carroll also became ill with the first experimental case of the disease but recovered. Using soldier volunteers, Reed and his team produced twenty-two other experimental instances of yellow fever, of which none was fatal. These experiments, coupled with the notes that Lazear had left, led Reed and his co-workers to the discoveries about the disease that resulted in its eradication.

Reed and his colleagues discovered that the female *Aëdes aegypti* mosquito can become infected by biting a victim of yellow fever only during the first three days of the course of the illness; she does not become infectious for two weeks thereafter, but may then remain infectious for up to two months in a warm season. The mosquito does not herself become ill, nor do her eggs, which remain infertile until she has fed on blood, harbor an infection that will affect the brood. The period of incubation of yellow fever is three to five days after the victim has been bitten by an infected mosquito, and having had the disease provides excellent immunity against subsequent attacks. Through experimentation Reed and his group established that whole blood taken from a patient early in the course of the disease will, upon injection into a susceptible person, cause yellow fever. They also determined that blood from an infected person could be passed through a Pasteur filter and still remain infectious; this was thus the first known filterable virus causing a human infection.

Within a year, as a direct result of the work of the Yellow Fever Board, Havana was freed of its age-old plague. Reed lived to see this triumph; he died late in 1902, following surgery for a ruptured appendix and abscess of the cecum which perhaps followed from an amoebic infection that he might easily have contracted in Cuba. Shortly before his death Harvard University awarded him an honorary M.A., citing him as "Walter Reed, medical graduate of Virginia, the Army Surgeon who planned and directed in Cuba the experiments that have given man control over that fearful scourge, yellow fever."

BIBLIOGRAPHY

I. ORIGINAL WORKS. Reed's papers on yellow fever are contained in *Yellow Fever. A Compilation of Various Publications, Results of the Work of Major Walter Reed, Medical Corps, United States Army, and the Yellow Fever Commission*, Senate document no. 822 from the Third Session of the 61st Congress (Washington, D. C., 1911). This compilation includes the major contributions of Reed's Yellow Fever Board and includes three notes by Reed on the etiology of yellow fever (Oct. 1900, Feb. 1901, and Jan. 1902), a note on *Bacillus icteroides* and *Bacillus cholerae suis* (1 Apr. 1899), "Experimental Yellow Fever" (May 1901), and "The Prevention of Yellow Fever" (Sept. 1901). Reed also wrote "Report of the Sanitary Inspection District Health Department, City of Brooklyn," in *Report of the Board of Health of the City of Brooklyn from May 8, 1873 to January 1, 1875* (New York, 1876); and *Report on the Origin and Spread of Typhoid Fever in US Military Camps During the Spanish War of 1898* (Washington, D.C., 1904), written with Victor C. Vaughn and Edwin O. Shakespeare.

II. SECONDARY LITERATURE. Biographies of Reed are Howard A. Kelly, *Walter Reed and Yellow Fever* (Baltimore, 1923); Laura N. Wood, *Walter Reed, Doctor in Uniform* (New York, 1943); Albert E. Truby, *Memoir of Walter Reed: The Yellow Fever Episode* (New York, 1943); and William B. Bean, "Walter Reed: 'He Gave Man Control of That Dreadful Scourge — Yellow Fever,'" in *Archives of Internal Medicine*, **89** (1952), 171–187. See also two memoirs of Reed by Walter D. McCaw and Jefferson R. Kean in Senate document no. 822 (above).

WILLIAM B. BEAN

REGENER, ERICH RUDOLPH ALEXANDER (*b.* Schleussenau, near Bromberg, Germany [now Bydgoszcz, Poland], 12 November 1881; *d.* Stuttgart, Germany, 27 February 1955), *physics.*

Regener studied physics at the University of Berlin. His doctoral dissertation, begun under Emil Warburg and concluded under Paul Drude, concerned the influence of shortwave light on the oxygen-ozone equilibrium, a subject to which he returned thirty years later in his research on stratospheric physics. Yet he considered his real teacher to be Heinrich Rubens, whom he assisted in Berlin. Regener turned to the new field of nuclear physics at Berlin and qualified as a university lecturer in 1909 with an outstanding *Habilitationsschrift* on the determination of the elementary electric charge by means of scintillation and charge measurements.

In 1913 he became professor of physics at the agricultural college in Berlin, and in 1920 he accepted a position at the Technische Hochschule in Stuttgart, from which he was dismissed in 1937 because his wife was Jewish. She then emigrated along with both their children. A son, Victor, was also a physicist, and a daughter was married to a physicist. About the time of his dismissal, the Kaiser-Wilhelm-Gesellschaft built for him a research station for stratospheric physics at Friedrichshafen on Lake Constance. In 1944 the facility was transferred to Weissenau, and in 1953 it was elevated by the Max-Planck-Gesellschaft to the status of an institute for stratospheric physics. Regener served as its director until his death. Also, after the war he again held a post at the Technische Hochschule in Stuttgart.

Regener furnished the first convincing proof of his skill as an experimentalist at Berlin in 1909, when he determined, through the use of α radiation, the elementary electric charge. He based his work on the procedure developed by Rutherford and Geiger—a combination of measuring the charge of a number of α particles and counting these particles by means of the number of points of light (scintillations) they produced as they struck a zinc sulfide screen, which consisted of pulverized zinc sulfide on a base-plate spread with an adhesive material. But this arrangement allowed some α particles to pass between the granules of zinc sulfide and thus escape being counted. Accordingly, Regener sought to find scintillating materials not possessing a granular structure; he settled upon thin layers of yellow diamond. These had the additional advantage of permitting the investigator to count points of light on the reverse side. He obtained the result $e = 4.79 \cdot 10^{-10}$ electrostatic units, which differs from the presently accepted figure of 4.802 by only two parts per thousand. At the beginning of his stay in Stuttgart, Regener and his co-workers attempted to find out why, when Millikan's oil-drop method was used to determine the elementary charge, there also appeared smaller charges, the so-called subelectrons. They were able to show that the reason was that the masses of the drops had been incorrectly determined through a failure to take into account the rate of vaporization.

Regener's contributions to cosmic radiation were of special importance. He recognized that the intensity of the rays had to be measured at the highest possible altitudes and at the greatest possible depths (in his case, in Lake Constance). For this purpose balloons equipped with measurement chambers were sent aloft and high-pressure chambers outfitted with ionization chambers, Geiger counters, and spectroscopes were lowered into Lake Constance. Regener developed appropriate recording devices in his own house. The balloon coverings were at first thin sheets of rubber glued together and later cellophane. The balloons reached altitudes as great as thirty kilometers. The measurements revealed that the intensity of the cosmic rays decreased at a height of twenty kilometers, which suggested that the greater ionic density in the layer below the incoming radiation must originate from secondary effects. (Today it is known that the effect is due to nuclear processes.) Regener also made a sensational finding from data gathered during an ascent conducted in March 1933. On this occasion he was able to link an unusually strong ionization with an eruption on the solar surface, thereby demonstrating that events occurring on the fixed stars are a source of cosmic radiation.

In his later years Regener investigated the formation of ice directly from the vapor state at low temperatures. He was especially interested in research on the nuclei around which freezing commences.

BIBLIOGRAPHY

I. ORIGINAL WORKS. An early work is "Über Zählung der α-Teilchen durch die Szintillation und über die Grösse des elektrischen Elementarquantums," in *Sitzungsberichte der Preussischen Akademie der Wissenschaften zu Berlin*,

(1909), 948–965. Memoirs on cosmic rays published in *Physikalische Zeitschrift* include "Über die durchdringende Komponente der Ultrastrahlung," **31** (1930), 1018–1019; "Die Absorptionskurve der Ultrastrahlung und ihre Deutung," **34** (1933), 306–323; "Weitere Messungen der Ultrastrahlung in der Stratosphäre," *ibid.*, 820–823, 880; and the series of short articles in **35** (1934), 779–793. See also the following, in *Zeitschrift fur Physik*: "Über das Spektrum der Ultrastrahlung," **74** (1932), 433–454; "Die Energiestrom der Ultrastrahlung," **80** (1933), 666–669; "Über Ultrastrahlungsmessungen in grossen Wassertiefen und über die Radioaktivität von Trockenbatterien," **100** (1936), 286; and "Über die Schauer der kosmischen Ultrastrahlung in der Stratosphäre," **111** (1938–1939), 501–507, written with A. Ehmert; a series of brief communications in *Naturwissenschaften*, **19–33** (1931–1946); and "Erforschungen und Ergebnisse mit Registrierballonen und Registrierapparaten in der Stratosphäre," in *Beiträge zur Physik der freien Atmosphäre*, **22** (1935), 249–260.

Regener's writings are listed in Poggendorff, V, I, 1031; VI, 2138–2139; and VIIa, pt. 3, 696–697.

II. SECONDARY LITERATURE. See Regener's Festschrift: W. Bothe and O. Hahn, "Zum 70. Geburtstag," in *Zeitschrift für Naturforschung*, **6a** (1951), 565 ff., with portrait. Obituaries include A. Ehmert and E. Schopper, in *Naturwissenschaften*, **43** (1956), 69–71; W. Gerlach, in *Jahrbuch der bayerischen Akademie der Wissenschaften* (1956), 222–229, with portrait; O. Hahn, in *Reden und Aufsätze der Technische Hochschule Stuttgart*, **21** (1956), 37; and L. Weikmann, in *Meteorologische Rundschau*, **8** (1955), 41.

F. FRAUNBERGER

REGIOMONTANUS, JOHANNES (*b.* Königsberg, Franconia, Germany, 6 June 1436; *d.* Rome, Italy, *ca.* 8 July 1476), *astronomy, mathematics.*

Nothing is known of Regiomontanus before he enrolled in the University of Vienna on 14 April 1450 as "Johannes Molitoris de Künigsperg."[1] Since the name of his birthplace means "King's Mountain," he sometimes Latinized his name as "Joannes de Regio monte," from which the standard designation Regiomontanus was later derived. He was awarded the bachelor's degree on 16 January 1452 at the age of fifteen; but because of the regulations of the university, he could not receive the master's degree until he was twenty-one. On 11 November 1457 he was appointed to the faculty, thereby becoming a colleague of Peuerbach, with whom he had studied astronomy. The two men became fast friends and worked closely together as observers of the heavens.

The course of their lives was deeply affected by the arrival in Vienna on 5 May 1460 of Cardinal Bessarion (1403–1472), the papal legate to the Holy Roman

Empire.[2] Bessarion's native tongue was Greek (he was born in Trebizond), and as part of his ardent campaign to bring ancient Greek authors to the attention of intellectuals in the Latin West, he persuaded Peuerbach to undertake a "briefer and more comprehensible" condensation, in Latin, of the *Mathematical Syntaxis* of Ptolemy, whose Greek style was formidable and whose ideas were far from simple. In those days Greek was not taught at the University of Vienna,[3] and Peuerbach did not know it. He had, however, made his own copy of Gerard of Cremona's Latin translation of Ptolemy's *Syntaxis*. Using this twelfth-century version, Peuerbach reached the end of book VI just before he died on 8 April 1461. On his deathbed he pledged Regiomontanus to complete the project.

Complying with Peuerbach's last wish, Regiomontanus accompanied Bessarion on his return trip to Rome, where they arrived on 20 November 1461.[4] When Regiomontanus finished the *Epitome*, as he entitled the translation by Peuerbach and himself, he dedicated it to Bessarion. In the parchment manuscript, which still survives, he did not address Bessarion as titular Patriarch of Constantinople, an honor bestowed on him on 28 April 1463,[5] a decade after the capital of the Byzantine Empire had been captured by the Turks. Thus, sometime before that date the Peuerbach-Regiomontanus *Epitome* was ready to go to press; but it was not actually printed until 31 August 1496, twenty years after the death of Regiomontanus.

At the end of the fifteenth century, Ptolemy's achievement remained at the pinnacle of astronomical thought; and by providing easier access to Ptolemy's complex masterpiece, the Peuerbach-Regiomontanus *Epitome* contributed to current scientific research rather than to improved understanding of the past. Moreover, the *Epitome* was no mere compressed translation of the *Syntaxis*, to which it added later observations, revised computations, and critical reflections—one of which revealed that Ptolemy's lunar theory required the apparent diameter of the moon to vary in length much more than it really does. This passage (book V, proposition 22) in the *Epitome*, which was printed in Venice, attracted the attention of Copernicus, then a student at the University of Bologna. Struck by this error in Ptolemy's astronomical system, which had prevailed for over 1,300 years, Copernicus went on to lay the foundations of modern astronomy and thus overthrow the Ptolemaic system.

Ptolemy was not only the foremost astronomer of antiquity but also its leading geographer; and Jacopo Angeli's widely used Latin translation (1406–1410)[6] of Ptolemy's *Geography* was condemned by Regiomontanus because the translator "had an inadequate knowledge of the Greek language and of mathematics."[7] Many of the obscure passages in Angeli's translation could not be explained by Peuerbach, who, as noted above, had not learned Greek. Hence Regiomontanus determined to master the language of Ptolemy. He acquired a remarkable fluency in Greek from his close association with Bessarion, and armed with a thorough comprehension of Ptolemy's language, he announced his intention to print an attack on Angeli's translation. But he died before completing this work. Nevertheless, "Johannes Regiomontanus' Notes on the Errors Committed by Jacopo Angeli in His Translation" formed the appendix (sig. P1r–Q8r) to a new version of Ptolemy's *Geography* (Strasbourg, 1525) by a scholar who had access to Regiomontanus' literary remains.

In a letter written not long after 11 February 1464 to the Italian mathematician Giovanni Bianchini, Regiomontanus reported that he had found an incomplete manuscript of Diophantus and, if he had the whole work, he would undertake to translate it into Latin—"since for this purpose the Greek I have learned in the home of my most revered master would be adequate."[8] Regiomontanus never translated Diophantus nor did he ever find a complete manuscript; nor did anyone else. Nevertheless, the recovery of Diophantus in modern times began with Regiomontanus' discovery of the incomplete manuscript.

When Bessarion was designated papal legate to the Venetian Republic, Regiomontanus left Rome with him on 5 July 1463.[9] In the spring of 1464[10] at the University of Padua, then under Venetian control, Regiomontanus lectured on the ninth-century Muslim scientist al-Farghānī. Although the main body of these lectures has not survived, "Johannes Regiomontanus' Introductory Discourse on All the Mathematical Disciplines, Delivered at Padua When He was Publicly Expounding al-Farghānī" was later published in *Continentur in hoc libro Rudimenta astronomica Alfragani . . .*, whose first item was John of Seville's twelfth-century Latin translation of al-Farghānī's *Elements of Astronomy* (Nuremberg, 1537).

Also included in this volume was Plato of Tivoli's twelfth-century Latin version, "together with geometrical proofs and additions by Johannes Regiomontanus," of al-Battānī's *The Motions of the Stars*. One such addition (to al-Battānī's chapter 11, although the printed edition misplaced it in the middle of chapter 12) may have been the germ from which Regiomontanus subsequently developed the earliest statement of the cosine law for spherical triangles. Although he employed the versed sine (1 − cos) rather than the cosine itself and used the law only once, he was the first to formulate this fundamental proposition of spherical trigonometry. He enunciated it as theorem 2 in book V of his treatise *On All Classes of Triangles* (*De triangulis omnimodis*).

The urgent need for a compact and systematic treatment of the rules governing the ratios of the sides and angles in both plane and spherical triangles had become apparent to Peuerbach and Regiomontanus while they were working on the *Epitome*. At the close of the dedication of that work Regiomontanus stated that he would write a treatise on trigonometry. The manuscript of the last four books contains many blank spaces, which, despite Regiomontanus' intentions, were never completed. Part of the volume had been written before he left Rome on 5 July 1463. At the end of that year or at the beginning of 1464 he told a correspondent: "I do not have with me the books which I have written about triangles, but they will soon be brought from Rome."[11] It may have been in Rome that Regiomontanus propounded, in theorem 1 of book II, the proportionality of the sides of a plane triangle to the sines of the opposite angles (or, in modern notation $a/\sin A = b/\sin B = c/\sin C$, the sine law). The corresponding proposition for spherical triangles appears in book IV, theorem 17. Theorem 23 in book II solves, for the first time in the Latin West, a trigonometric problem by means of algebra (here called the *ars rei et census*). Regiomontanus' monumental work on *Triangles*, the first publication of which was delayed until 12 August 1533, attracted many important readers and thereby exerted an enormous influence on the later development of trigonometry because it was the first printed systematization of that subject as a branch of mathematics independent of astronomy.

Regiomontanus dedicated his *Triangles* to Bessarion, whom Pius II, in 1463, had named titular Patriarch of Constantinople. When the pope died, Bessarion returned to Rome in August 1464 to take part in the election of a successor. Regiomontanus accompanied him and while in Rome composed a dialogue between a Viennese named Johannes (evidently himself) and an unnamed scholar from Cracow. The subject of their conversation was a thirteenth-century planetary theory that was still very popular. Some of its defects were discussed in the dialogue, which was printed by Regiomontanus when he later acquired his own press. Although he published the dialogue without a title, it was often reprinted under some such heading as *Johannes Regiomontanus' Attack on the Absurdities in the Planetary Theory of Gerard of Cremona* (Gerard's

pupils did not list this *Theorica planetarum* in the catalog of their teacher's productions).[12]

After an observation on 19 June 1465,[13] presumably in Viterbo, a favorite resort of Bessarion's, Regiomontanus' activities during the next two years are not known. In 1467, however, he was firmly established in Hungary, where the post of astronomer royal was held by Martin Bylica of Olkusz (1433–1493), who was also present in Rome during the papal election and in all likelihood is the unnamed interlocutor in Regiomontanus' dialogue on planetary theory.

In 1467, with Bylica's assistance, Regiomontanus computed his *Tables of Directions*, which consisted of the longitudes of the celestial bodies in relation to the apparent daily rotation of the heavens. These *Tables*, computed for observers as far north of the equator as 60°, were first published in 1490 and very frequently thereafter.[14] Regiomontanus wrote accompanying problems and in problem 10 he indicated the desirability of abandoning the sexagesimal character of the table of sines by putting $\sin 90° = 100,000$ (10^5) instead of 60,000 (6×10^4), the base he had used in *Triangles* (book IV, theorem 25). In that work he had not employed the tangent function; but in *Tables of Directions* he included a table of tangents (although he did not use this term) for angles up to 90°, the interval being 1° and $\tan 45° = 100,000$, thereby providing the model for our modern tables.

In 1468 in Buda, then the capital of the kingdom of Hungary, Regiomontanus computed a table of sines with $\sin 90° = 10,000,000$ (10^7). But before he realized the advantage of the decimal base, he had prepared a sexagesimal sine table, to which he had referred in the dedication of his *Triangles* and which he had used in computing his *Tables of Directions*, with $\sin 90° = 6,000,000$ (6×10^6), the interval being 1′ and the seconds being found by an auxiliary table of proportional parts. Both of Regiomontanus' major sine tables, the sexagesimal and the decimal, were first published at Nuremberg in 1541, together with his essay on the *Construction of Sine Tables*.

While still in Italy, Regiomontanus began to compute his *Table of the First Movable* [*Sphere*], or of the apparent daily rotation of the heavens. He completed this work, together with an explanation of its use, in Hungary and dedicated it to his friend King Matthias I Corvinus. He also expounded the geometrical basis of this *Table*. These three related works constituted an item in the list of his own writings that Regiomontanus intended to print on his own press, an intention he could not carry out. Of these three works, the first two were published in Vienna in 1514, and the third in Neuburg in 1557.

Regiomontanus wrote each of these works for the purpose of facilitating astronomical computations. But whatever use was made of them ended with the advent of logarithms.

In 1471 Regiomontanus left Hungary. "Quite recently I have made [observations] in the city of Nuremberg . . . for I have chosen it as my permanent home," he informed a correspondent on 4 July 1471, "not only on account of the availability of instruments, particularly the astronomical instruments on which the entire science of the heavens is based, but also on account of the very great ease of all sorts of communication with learned men living everywhere, since this place is regarded as the center of Europe because of the journeys of the merchants."[15] On 29 November 1471 the City Council of Nuremberg granted Regiomontanus residence in the city until Christmas of the following year. He installed a printing press in his own house in order to publish scientific writings, a class of books in which the existing establishments were reluctant to invest their capital, partly because the necessary diagrams required special craftsmen and additional expense.

Regiomontanus was the first publisher of astronomical and mathematical literature, and he sought to advance the work of scientists by providing them with texts free of scribal and typographical errors, unlike the publications then in circulation. His emphasis on correct texts was aided by his introduction into Nuremberg printing of the Latin alphabet and, for writings in the German language, rounded and simplified letters that approached the Latin alphabet in legibility.

Regiomontanus' first publication, a mark of his deep affection for his former teacher, colleague, and collaborator, was Peuerbach's New Theory of the Planets. This work was the first item in the catalog which Regiomontanus sent out in the form of a broadside, listing his publications, issued or projected, written by himself or others. The second item in the list of his own publications was the Ephemerides, which he issued in 1474 and which was the first such work to be printed. It gave the positions of the heavenly bodies for every day from 1475 to 1506. Of all the books written and published by Regiomontanus, this is perhaps the most interesting from the standpoint of general history: Columbus took a copy on his fourth voyage and used its prediction of the lunar eclipse of 29 February 1504 to frighten the hostile Indians in Jamaica into submission.[16]

The geographer Martin Behaim "boasted that he was a pupil of Regiomontanus"[17] in Nuremberg. More credit is given to the statement that Regio-

montanus attracted Bernhard Walther as a pupil. Walther, who was born in Memmingen, in 1467 became a citizen of Nuremberg, where he helped Regiomontanus with his observations and continued them after his teacher left for Rome in the summer of 1475. Regiomontanus' last observation in Nuremberg is dated 28 July 1475 and Walther's observations begin five days later.[18]

According to a Nuremberg chronicler, Regiomontanus went to Rome in response to a papal invitation to emend the notoriously incorrect ecclesiastical calendar. If this report is true, nothing positive resulted from his trip, for he died in less than a year.

In all probability Regiomontanus fell victim to the plague that spread through Rome after the Tiber overflowed its banks in January 1476. But a more sensational rumor concerning the cause of his death surfaced in a laudatory poem that served as the title page of a posthumous edition of his Latin Calendar (Venice, 1482). The rumor gained currency by being repeated in 1549 in Reinhold's commemorative eulogy of Regiomontanus and again in 1654[19] in Gassendi's biography of the astronomer. In his catalog Regiomontanus had announced his intention to publish an extensive polemic against George of Trebizond, whose "commentary on the Syntaxis he will show with the utmost clarity to be worthless and his translation of Ptolemy's work not to be free of faults." Although Regiomontanus never actually published his attack, which still remains in manuscript in Leningrad, George's sons poisoned him, according to the rumor. Yet Bessarion died unmolested on 18 November 1472, three years after his own devastating attack on George of Trebizond as a Calumniator of Plato was published in Rome (1469).

"The motion of the stars must vary a tiny bit on account of the motion of the earth." This portentous statement in the handwriting of Regiomontanus was excerpted from one of his letters by Georg Hartmann, the discoverer of the vertical dip of the magnetic needle and an early supporter of the Copernican cosmology. Hartmann regarded the excerpt as a treasure, undoubtedly because to his mind it provided clear proof that Regiomontanus, the greatest astronomer of the fifteenth century, had accepted the concept of the moving earth and realized one of its numerous implications; Regiomontanus was therefore a Copernican before Copernicus.

The letter from which Hartmann took this excerpt has not survived, nor has the excerpt itself. But it was copied by a professor onto the margin of his unpublished lecture in 1613 on Copernicus' planetary theory, with the explanation that Hartmann "recognized Regiomontanus' handwriting because he was also familiar with his features." Yet Hartmann was not even born until 1489, thirteen years after the death of Regiomontanus.

Nevertheless, it has been suggested that the letter in question may have been sent by Regiomontanus to Novara, who, in an unpublished essay on the duration of pregnancy, called Regiomontanus his teacher. Novara in turn became the teacher of Copernicus. Thus it can be inferred that the concept of the revolutionary geokinetic doctrine was first conceived by Regiomontanus and communicated to Novara, who then passed it to Copernicus. Nevertheless, in the voluminous published and unpublished writings of Regiomontanus, no other reference to the earth in motion has ever been found.

NOTES

1. *Die Matrikel der Universität Wien*, I (Graz–Cologne, 1954), 275. The Johannes Molitoris who entered the University of Leipzig on 15 October 1447 has been identified with Regiomontanus by Zinner, *Leben und Wirken des . . . Regiomontanus*, 13. The Leipzig rector, however, did not associate this namesake with any particular place and Molitoris, as a Latinized form of the surname Müller, was extremely common.
2. Ludwig Mohler, *Kardinal Bessarion als Theologe, Humanist und Staatsman*, Quellen und Forschungen aus dem Gebiete der Geschichte, no. 20 (Paderborn, 1923), 298.
3. Joseph Aschbach, *Geschichte der Wiener Universität im ersten Jahrhunderte ihres Bestehens* (Farnborough, 1967; repr. of Vienna, 1865), 539.
4. Mohler, *op. cit.*, 303.
5. Conrad Eubel, *Hierarchia catholica medii aevi*, II (Padua, 1960; repr. of 2nd ed., Münster, 1913–1923), 150. Bessarion's elevation is dated in April 1463, but the exact day is marked as unknown. However, Bessarion's predecessor, Isidore of Kiev, died on 27 April 1463 (Eubel, II, 36, n. 199).
6. Robert Weiss, "Jacopo Angeli da Scarperia," in *Medioevo e rinascimento, studi in onore di Bruno Nardi*, Pubblicazioni dell'Istituto di filosofia dell'Università di Roma (Florence, 1955), 824.
7. Regiomontanus' catalog of the books to be printed on his press; reproduced by Zinner, "Die wissenschaftlichen Bestrebungen Regiomontans," in *Beiträge zur Inkunabelkunde*, **2** (1938), 92.
8. Silvio Magrini, "Joannes de Blanchinis Ferrariensis e il suo carteggio scientifico col Regiomontano (1463–64)," in *Atti e memorie della deputazione ferrarese di storia patria*, **22**, fasc. 3, no. 2, (1915–1917), lvii.
9. Mohler, *op. cit.*, 312.
10. The total eclipse of the moon on 21 April 1464 was observed by Regiomontanus in Padua; see *Scripta clarissimi mathematici M. Ioannis Regiomontani* (Nuremberg, 1544), fol. 41v–42r; or Willebrord Snell, *Coeli et siderum . . . observationes Hassiacae* (Leiden, 1618), Ioannis de Monteregio . . . observationes, fol. 20v.
11. Maximilian Curtze, "Der Briefwechsel Regiomontan's mit Giovanni Bianchini, Jacob von Speier und Christian Roder," in *Abhandlungen zur Geschichte der Mathematik*, **12** (1902), 214.
12. Olaf Pedersen, "The Theorica Planetarum Literature of the Middle Ages," in *Ithaca, Proceedings of the Tenth International Congress of History of Science* (Paris, 1964), 617.
13. *Scripta . . . Regiomontani*, fol. 42r; Snell, fol. 21v.
14. The manuscript of Regiomontanus' *Tables of Directions* that Bylica presented to Cracow University is still preserved there; see Władysław Wisłocki, *Katalog rękopisów biblioteki uniwersytetu jagiellońskiego* (Cracow, 1877–1881), 188; and Jerzy Zathey *et al.*, *Historia biblioteki jagiellońskiej* (Cracow, 1966), 154, n. 64.
15. Curtze, *op. cit.*, 327. The lunar eclipse on 2 June 1471 was observed by Regiomontanus in Nuremberg—*Scripta*, fol. 42v; Snell, fol. 22r.
16. Samuel Eliot Morison, *Admiral of the Ocean Sea*, II (Boston, 1942), 400–403.
17. João de Barros, *Asia*, I, decade I, bk. 4, ch. 2 (Lisbon, 1945), 135. If Behaim's claim was correct, he was at most 16 years old when Regiomontanus left Nuremberg; see Richard Hennig, *Terrae incognitae*, 2nd ed., IV, (Leiden, 1944–1956), 434.
18. *Scripta*, fol. 27v; Snell, fol. 1v; Donald Beaver, "Bernard Walther: Innovator in Astronomical Observation," in *Journal for the History of Astronomy*, **1** (1970), 39–43.
19. Gassendi, *Tychonis Brahei . . . vita . . . accessit . . . Regiomontani . . . vita* (Paris, 1654), app., 92; and *Opera omnia*, V (Stuttgart-Bad Cannstatt, 1964; repr. of Lyons, 1658 ed.), 532.

BIBLIOGRAPHY

Regiomontanus' works were reprinted in *Joannis Regiomontari opera collectanea* (Osnabrück, 1972) with a biography by the ed. Felix Schmeidler. An older work is Ernst Zinner, *Leben und Wirken des Johannes Müller von Königsberg genannt Regiomontanus*, 2nd ed., rev. and enl. (Osnabrück, 1968). A recent trans. of Regiomontanus' *De triangulis omnimodis* is Barnabas Hughes, *Regiomontanus on Triangles* (Madison, Wis., 1967).

On Regiomontanus and his work, see the anonymous "Regiomontanus's Astrolabe at the National Maritime Museum," in *Nature*, **183** (1959), 508–509; and Edward Rosen, "Regiomontanus's Breviarium," in *Medievalia et Humanistica*, **15** (1963), 95–96.

EDWARD ROSEN

REGNAULT, HENRI VICTOR (*b.* Aix-la-Chapelle, France [now Aachen, Germany], 21 July 1810; *d.* Paris, France, 19 January 1878), *physics, chemistry*.

Regnault was left an orphan at an early age; his father, an officer in Napoleon's army, died in 1812 in the Russian campaign and his mother died shortly afterward. A close friend of his father supervised his education and then found employment for him in a draper's shop in Paris, where, despite considerable financial hardship, he took lessons for the entrance examination of the École Polytechnique. Admitted there in 1830, he graduated with distinction two years later and studied for two more years at the École des Mines in Paris before traveling extensively in Europe to study mining techniques and metallurgical processes. After short periods of research under Liebig at Giessen and Boussingault at Lyons, Regnault returned to the École Polytechnique as assistant to

Gay-Lussac in 1836 and in 1840 succeeded him in the chair of chemistry.

Also in 1840 he was elected to the chemical section of the Académie des Sciences, but his interests were already turning to physics and it was at the Collège de France, where he became professor of physics in 1841, that he performed his most important experimental work for the next thirteen years. From 1854 he lived and worked at Sèvres as director of the famous porcelain factory; and he was still engaged on research there when, in 1870, all his instruments and books were destroyed by Prussian soldiers. This blow, and the death of his son Henri late in the Franco-Prussian War, left Regnault a broken man; and his last years were clouded by grief and personal disability. Among his many honors were membership of the Legion of Honor (1850) and the Rumford and Copley medals of the Royal Society (1848 and 1869, respectively).

In his chemical work, nearly all of which dates from 1835 to 1839, Regnault followed no unified program of research. It was, however, his contributions in organic chemistry, most notably his studies of the action of chlorine on ethers and the research that led him to the discovery of vinyl chloride, dichloroethylene, trichloroethylene, and carbon tetrachloride which quickly won him an international reputation and the high esteem of such chemists as Liebig and Dumas.

Regnault's early work in physics arose directly from his chemistry. Encouraged by Dumas, who had long advocated the measurement of specific heats as a means of investigating the atomic composition of substances, he began with a systematic experimental study of the specific heats of a wide range of solids and liquids. In the course of this work, which occupied him from 1839 to 1842, he conclusively demonstrated the approximate nature of Dulong and Petit's law and confirmed the validity, within the same limits of accuracy, of F. E. Neumann's extension of the law from elements to compounds. In 1842, with his reputation as an experimenter established, Regnault was appointed by the minister of public works to redetermine all the physical constants involved in the design and operation of steam engines; it was then that he began the research on the thermal properties of gases for which he is now best known.

In fact Regnault did far more than was asked of him; and in nearly thirty years of painstaking experiments, all generously financed from government funds, he not only provided the standard data that practicing engineers were to use for several decades but also subjected the laws governing the physical properties of gases to a thorough and long-overdue reexamination. His most important conclusions, all

reached by 1853, concerned Dulong and Petit's law, which was shown to be only approximately true even for gases (thus contradicting Dulong's own conclusion on this point); the expansion coefficient, which Regnault measured to an unprecedented degree of accuracy and which he showed to vary with the nature of the gas; and Boyle's law, the deviations from which he investigated in great detail.

Speculation and discussions of theory are noticeably absent from Regnault's published work, as they are from his surviving notebooks. But in 1853 he declared his acceptance of the principle of the conservation of energy; and later he measured the mechanical equivalent of heat, although with only moderate success. In declaring his support for the principle, Regnault stated that he had subscribed to the mechanical theory of heat "for a long time" and that he had been led to it independently through his own experiments (*Comptes rendus . . . de l'Académie des sciences*, **36** [1853], 680). A comment made in 1854 by a former pupil, Ferdinand Reech, which suggests that Regnault had been won over to the new ideas only with great difficulty, seems more reliable, however (*Machine à air d'un nouveau système* [Paris, 1854], i). For there is certainly no evidence, other than Regnault's own testimony, to indicate that he arrived at the principle independently, although in many ways he was well poised to make the discovery.

As Dumas later regretted, Regnault was too cautious. In the quality of his meticulous experimental work, which he always conducted on the grand scale, he was unrivaled in his own day; and his compilations of data provided an indispensable tool for the pioneers of thermodynamics. Yet he lacked the intellectual capacity to participate in a truly creative way in the momentous developments in physics in the mid-nineteenth century.

BIBLIOGRAPHY

I. ORIGINAL WORKS. Regnault's accounts of his experiments in physics were collected in three vols. of the *Mémoires de l'Académie des sciences*, **21** (1847), **26** (1862), and **37**, pts. 1 and 2 (1868–1870)—which were also published, simultaneously and with identical pagination, as *Relation des expériences entreprises . . . pour déterminer les principales lois et les données numériques qui entrent dans le calcul des machines à vapeur*. His most important work in chemistry is contained in papers published in the *Annales de chimie et de physique*, but his *Cours élémentaire de chimie*, 3 vols. (Paris, 1847–1849), enjoyed a great success as a textbook, being translated into English, German, Dutch, Italian, and Spanish. Nearly all the surviving MS material, chiefly notebooks concerning his travel and early experiments, is in the library of the Institut de France, Paris.

II. SECONDARY LITERATURE. The fullest account of Regnault's life, given in the *éloge historique* that J. B. Dumas read to the Académie des Sciences in 1881, appears in *Mémoires de l'Académie des sciences*, **42** (1883), xxxvii–lxxv. Also useful are the tributes in *Comptes rendus . . . de l'Académie des sciences*, **86** (1878), 131–143; and two obituaries: by T. H. N. (T. H. Norton), in *Nature*, **17** (1878), 263–264; and J. H. Gladstone, in *Journal of the Chemical Society*, **33** (1878), 235–239. A critical discussion of Regnault's physics appears in R. Fox, *The Caloric Theory of Gases From Lavoisier to Regnault* (Oxford, 1971), 295–302.

ROBERT FOX

REICHENBACH, GEORG FRIEDRICH VON (*b.* Durlach [now part of Karlsruhe], Germany, 24 August 1771; *d.* Munich, Germany, 21 May 1826), *instrument making, mechanical engineering.*

Reichenbach was probably the greatest of the generation of engineers who initiated the industrialization of Germany. An original and versatile inventor, he contributed to many areas of technology and permanently shaped the design of scientific instruments and the construction of large hydraulic machinery.

Reichenbach grew up in Mannheim. His father, Johann Christoph Reichenbach, was a distinguished cannon borer and military engineer in the service of the elector palatine, who was later king of Bavaria; his mother was Helene Pfetsch. His education began in public school and in his father's workshop. From 1786 to 1790 Reichenbach attended the new School of Army Engineers at Mannheim, where he received a sound introduction into science. His free time was spent at the elector's observatory, where he learned a good deal of practical astronomy and perfected his skill in instrument making. After his graduation the elector, on Count Rumford's recommendation, sent young Reichenbach to England to study modern engineering (1791–1793). He met Matthew Boulton, Watt, and Ramsden; made secret sketches of Watt's latest engine; and supervised the construction of some large blast-furnace machinery. Upon his return Reichenbach was commissioned an engineering officer in the Bavarian army. In this function he improved ordnance design, installed cannon-boring facilities and musket factories, invented a rifled cannon, and occasionally served as assistant to Rumford. These duties did not prevent him, however, from engaging in a variety of other activities as well.

Reichenbach had recognized an acute demand for scientific instruments of high quality. Even the best German instruments, those of G. F. Brander and C. C. Höschel, were decidedly inferior to English-built ones. Since the precision of astronomical and geodetic instruments depended upon the exactness with which their scales were graduated, Reichenbach set out to invent a dividing machine that would be superior to the best existing one, that of Jesse Ramsden. Having succeeded in making that crucial invention (1800), he and a partner established a small instrument shop in Munich, which in 1804 was expanded into the Mathematisch-Mechanisches Institut von Reichenbach, Utzschneider, und Liebherr; the third partner, Joseph von Utzschneider, was a prominent Bavarian statesman and entrepreneur. Here Reichenbach built astronomical and surveying instruments that rapidly became famous for their precision and mechanical perfection, and were coveted and praised by such astronomers as Bessel, Gauss, and Laplace. Aware of the limitations of their optical systems, Utzschneider and Reichenbach searched for a competent optician; in 1806 they found young Joseph Fraunhofer. In 1809 Utzschneider, Reichenbach, and Fraunhofer founded a separate "optical institute" which was to supplement the older mechanical institute. Reichenbach and Fraunhofer's collaboration resulted in the design of astronomical instruments of unprecedented power and precision which, in turn, led to spectacular advances in observational astronomy. The two institutes also assumed the role of a national training center for instrument makers; the excellence of German scientific instruments in the nineteenth century is thus directly related to Reichenbach's initiative.

Reichenbach always had more than one project in hand at a time. While he was still building his most remarkable telescopes, he embarked on a venture that promised to fulfill an old ambition: the construction of large power machinery. In 1806 the Bavarian state, in order to increase the revenue from its saltworks, had decided to build a 108-kilometer pipeline to conduct brine from its Alpine origins to regions where fuel was more abundant. For the first stage of the project (the line from Bad Reichenhall to Traunstein) Reichenbach's institute had to build four water-powered pumping stations. Reichenbach solved the problem with "water pressure engines" (reciprocating piston engines similar to steam engines) of new design which were running smoothly in 1808, a year after he had received the order. In 1810 he completed the continuation of the line to Rosenheim. His crowning achievement was the third and final stage of the project, the line from Berchtesgaden to Bad Reichenhall. One of the two machines employed here had to overcome, in a single stage, a rise of over 1,200 feet. This machine, at the time considered the largest in the world, remained in continuous service from 1817 to 1958. Reichenbach had designed his water-pressure engines on the basis of mathematical calculation. During their construction, which presented some unfamiliar problems, he improvised various new machining and iron-casting techniques.

In 1811 Reichenbach resigned from the Bavarian army with the rank of captain and was appointed *Salinenrat* (councillor of saltworks). In 1814 he dissolved his association with Utzschneider and Fraunhofer and formed another partnership. In 1820 Reichenbach left instrument making entirely, to become director of the Bavarian Central Bureau of Highways and Bridges. In this capacity he was engaged in the construction of high-pressure steam engines, cast-iron bridges, city water supplies, and gas lighting systems, and in the establishment of a polytechnical institute; most of these projects were interrupted by his death.

Reichenbach's contributions were fully recognized during his lifetime. He was knighted, received numerous medals, and was elected to the Academy of Sciences of Paris and also to that of Munich.

BIBLIOGRAPHY

I. ORIGINAL WORKS. Reichenbach's few published writings are cataloged by Dyck (see below). His papers, as well as most of his surviving machines and instruments, are in the Deutsches Museum, Munich.

II. SECONDARY LITERATURE. Walther von Dyck, *Georg von Reichenbach* (Munich, 1912), is a definitive scientific biography with exhaustive documentation. In some biographical details it is supplemented in Ernst Kohler, *Georg von Reichenbach: Das Leben eines deutschen Erfinders* (Munich, 1933).

OTTO MAYR

REICHENBACH, HANS (*b.* Hamburg, Germany, 26 September 1891; *d.* Los Angeles, California, 9 April 1953), *philosophy of science, logic.*

Reichenbach was one of five children of Bruno Reichenbach, a prosperous wholesale merchant, and the former Selma Menzel. Both parents were members of the Reformed Church; his paternal grandparents were Jewish. The family was cultured, with a lively interest in music, chess, books, and the theater.

From 1910 to 1911 Reichenbach studied engineering in Stuttgart but, dissatisfied, turned to mathematics, physics, and philosophy, attending the universities of Berlin, Munich, and Göttingen. Among his teachers were Planck, Sommerfeld, Hilbert, Born, and Cassirer. He took his doctorate in philosophy at the University

of Erlangen in 1915, and another degree by state examination in mathematics and physics at Göttingen in 1916. His doctoral dissertation was on the validity of the laws of probability for physical reality. He wrote it without academic guidance, for he could find no professor interested in the topic. The completed dissertation consisted of an epistemological treatise and a mathematical calculus. After traveling in vain to several universities in search of a sponsor willing and able to read both parts, Reichenbach found at Erlangen a philosopher and a mathematician, each willing to sponsor the part within his competence and together willing to accept the dissertation as a whole. Decades later he would chuckle when he cited their decision as a fallacy of composition, adding: "But I did not point that out to the good professors at that time!"

Reichenbach served for two and a half years in the Signal Corps of the German army, contracting a severe illness at the Russian front. Throughout his life he regarded war as catastrophe and considered it a duty of intellectuals to combat the attitudes from which wars arise. From 1917 to 1920 he worked in the radio industry, continuing his studies in the evening. He was one of the five to attend Einstein's first seminar in relativity theory at the University of Berlin. From 1920 to 1926 he taught surveying, radio techniques, the theory of relativity, philosophy of science, and history of philosophy at the Technische Hochschule in Stuttgart.

In 1926 Reichenbach obtained a professorial appointment at the University of Berlin. Opposition to his appointment, due in part to his social activism during his student days and in part to his outspoken disrespect for many traditional metaphysical systems, was overcome by Einstein's persistent and witty pleading. In 1930 Reichenbach and Rudolf Carnap founded and edited the journal *Erkenntnis*, which for many years was the major organ of the Vienna Circle of logical positivists, of the Berlin Group for Empirical Philosophy, and of the International Committee for the Unity of Science. He also broadcast over the German state radio the lectures published in 1930 as *Atom und Kosmos*.

Within a few days of Hitler's election to power in 1933, Reichenbach was dismissed from the University of Berlin and from the state radio. Anticipating this action, he was on his way to Turkey before the dismissal notices were delivered. From 1933 to 1938 he taught philosophy at the University of Istanbul, where he was charged with reorganizing instruction in that subject. He was delighted to find that among his students there were many excellent young teachers, who had been given paid leave of absence by Ataturk's government so that they might profit from the presence in Turkey of German refugee professors.

In 1938 Reichenbach received his immigration permit to enter the United States, and from then until his death in 1953 was professor of philosophy at the University of California at Los Angeles, frequently lecturing at other universities and at congresses in the United States and Europe. Shortly before his death a volume was planned for the Library of Living Philosophers (edited by P. A. Schilpp) to include both Carnap and Reichenbach, but his death prevented fulfillment of the project.

As a teacher Reichenbach was extraordinarily effective. Carl Hempel, who studied under him at Berlin, stated: "His impact on his students was that of a blast of fresh, invigorating air; he did all he could to bridge the wide gap of inaccessibility and superiority that typically separated the German professor from his students." His pedagogical technique consisted of deliberately oversimplifying each difficult topic, after warning students that the simple preliminary account would be inaccurate and would later be corrected. Students who pursued advanced work with him found him kindly, witty, morally courageous, and loyal. Those whose interests or convictions differed from his sometimes found him arrogant and intolerant.

Reichenbach never substantially altered his epistemological stance, which can be briefly characterized as anti-Kantian, antiphenomenalistic empiricism. In his first book, *Relativitätstheorie und Erkenntnis apriori* (1920), he began his attack upon the Kantian doctrine of synthetic a priori knowledge, although at that time he still regarded the "concept of an object" as a priori. He declared, however, that this concept is a priori only in the sense of being a conceptual construction contributed by the mind to sense data, and not also, as Kant had believed, a priori in the sense of necessarily true for all minds. By 1930 Reichenbach had replaced this view with the thesis that the concept of a physical object results from a projective inductive inference. From that time on, he maintained that there is no synthetic a priori knowledge, defending this thesis by showing that every knowledge claim held by Kant to be a synthetic a priori truth could be classified as analytic a priori, synthetic a posteriori, or decisional. In accord with Helmholtz, Frege, and Russell, Reichenbach regarded the axioms and theorems of arithmetic as analytic a priori. He classified the parallel postulate and the theorems of Euclidean geometry as synthetic a posteriori if, in combination with congruence conventions, they are taken as descriptive of physical spatial relations. The Kantian principle of universal causality was also classified as synthetic a posteriori. In his later work,

after reformulating the principle of causality in terms of inductive inference, Reichenbach denied that it applies to the subatomic realm of quantum theory. As for the Kantian moral synthetic a prioris, he regarded them as volitional decisions, neither true nor false.

By 1924 Reichenbach had developed his theory of "equivalent descriptions," a central tenet of his theory of knowledge. It is formulated in his *Axiomatik der relativistischen Raum-Zeit-Lehre* (1924), in *Philosophie der Raum-Zeit-Lehre* (1928), in *Atom und Kosmos* (1930), in *Experience and Prediction* (1938), and in the less technical *Rise of Scientific Philosophy* (1951); and it is developed with new applications in his works on quantum mechanics and time. This theory attributes an indispensable role in physical theory to conventions but rejects the extreme conventionalism of Poincaré and his school. Reichenbach insisted that a completely stated description or physical theory must include conventional elements, in particular such "coordinating definitions" as equal lengths and simultaneous times. These definitions are not bits of knowledge, for such questions as whether or not two rods distant from each other have the same length are not empirically answerable. Hence such coordinations must be regarded as conventions, as definitions, as neither true nor false.

Physical theory contains much more than these conventional elements, however. The truth of a theory, the complete statement of which must include a set of coordinating definitions, is not a matter of convention but of empirical confirmation. Furthermore, one theory using one set of congruence conventions may be empirically equivalent to another theory using another set of conventions. For example, Riemannian geometry combined with the usual coordinating definitions of equal times, equal lengths, and straight lines yields a description of physical space equivalent to Euclidean geometry combined with coordinating definitions which attribute systematic changes to lengths of rigid rods.

This equivalence Reichenbach explicated as follows: When all possible observations confirm to the same degree two descriptions, one of which uses one set of congruence conventions and the other another set, the two descriptions are equivalent, that is, have the same knowledge content or cognitive meaning.

Theories of meaning had long been a focal concern of logical positivists, to whose work Reichenbach acknowledged indebtedness, offering his own theory of meaning as a development and correction of the positivists' verifiability theory. He disagreed with the positivists on two crucial points. First, their theory made complete verifiability (as true or false) a condition of cognitive meaningfulness. This, Reichenbach pointed out, denies that statements confirmed with probability have cognitive meaning, and hence consigns to the category "meaningless" all generalizations and all predictions of science. The strictness of their limitation on cognitive meaning had forced the positivists into a phenomenalism which equated the conclusions of physical science with statements about sensory data. Reichenbach proposed to give the neglected but all-important concept of probability the central role in theory of meaning which it actually plays in scientific method. He regarded the relation between observational data and physical theory as a probability inference, not a logical equivalence. This permits a "realistic" (as opposed to the positivists' phenomenalistic) view of the objects of scientific knowledge.

Reichenbach's second disagreement with the positivists' theory of knowledge involved the logical status of any criterion of cognitive meaningfulness. The positivists assumed that their theory was itself an item of knowledge, a description of the class of meaningful statements. Reichenbach declared that any definition of "knowledge" or of "cognitive meaning" is a volitional decision without truth character (see *Experience and Prediction*, ch. 1, esp. pp. 41, 62). He continued to use the label "theory of meaning" because each decision concerning what is to be accepted as cognitively meaningful is connected with two cognitive questions: whether the decision accords with the actual practice of scientists, and what subordinate decisions are logically entailed by the definition of meaning.

Reichenbach formulated his own decision concerning cognitive meaning in two principles: a proposition has meaning if it is possible to determine a degree of probability for it; and two sentences have the same meaning if they obtain the same degree of probability through every possible observation (for complete formulation, see *Experience and Prediction*, p. 54). The second of these principles he regarded as a modern version of Ockham's razor, core of the antimetaphysical attitude of every consistent empiricism. As examples of its application he cited Mach's criticism of the concept of force and Einstein's principle of the equivalence of gravitation and acceleration.

Throughout his life Reichenbach maintained that the mathematical or frequency concept of probability suffices and needs no supplementation by a priori equal probabilities or by such concepts as "degree of credibility." For prediction of individual events the probability was the "best wager," determined by the frequency of the narrowest class for which there were reliable statistics. He applied the frequency concept of probability to general hypotheses by regarding them as members of classes of hypotheses having known success ratios. These views were opposed by Russell in *Human Knowledge, Its Scope and Limits*, and were supported by Wesley Salmon in *Foundations of Scientific Inference*.

Reichenbach's work on induction was closely connected with his theory of probability, for it introduced the distinction between appraised and unappraised (or "blind") posits. The former have frequency probabilities attached to them; the latter admit of no probability estimate. One blind posit is involved in every inductive inference: the posit that frequencies of series of events converge toward limits. (Causal or one-to-one regularities are simply one case of statistical regularities.) With this thesis Reichenbach reopened the old question of the justification of induction. He accepted Hume's argument that inductive inferences admit of neither deductive (demonstrative) justification nor inductive justification (at pain of circularity). Thus there can be no proof of any sort that inductive inferences will ever succeed in the future, let alone succeed more often than they fail.

Nevertheless, Reichenbach offered the following "pragmatic justification" of our use of inductive inferences. He showed that if the world becomes such that inductive inferences usually fail, as would happen if no past regularities were to continue into the future, then no principles of predictive inference could succeed. Hence inductive procedures offer us our only chance of making successful predictions, although we cannot know whether they will succeed or not. If there are series of events with frequencies which converge toward limits, inductive methods will lead to increasingly successful predictions as observed frequencies approach those limits; if this condition does not obtain, no method whatsoever of making predictions will succeed. Since we cannot know that this necessary condition of successful prediction will not obtain, it would be unreasonable to renounce the method which will yield success if it does obtain. The choice is between certain cognitive failure and our only chance of success. Hence, Reichenbach concluded, it is reasonable to make inductive inferences—that is, to adopt and act on the blind posit that frequencies of series of events will converge toward limits (see *Experience and Prediction*, secs. 38–40).

In "Philosophy: Speculation or Science?" (1947) and in *The Rise of Scientific Philosophy* (1951) Reichenbach drew the corollaries for ethics of his theory of knowledge. His definition of cognitive meaning precluded any extrascientific kinds of knowledge. Hence moral principles and ethical aims are volitional decisions, not items of knowledge. He condemned traditional philosophical systems for conflating cognition and volition in the mistaken hope of establishing knowledge of ultimate values, and he also rejected John Dewey's attempt to test moral judgments by scientific methods. "There is no such thing as 'the good' in the sense of an object of knowledge" ("Philosophy: Speculation or Science?" p. 21). In his brief writings on the nature of moral judgments, his style and tone are as dogmatic as the style and tone of other ethical noncognitivists of the era. He does not mention that his classification of moral judgments as volitional decisions is a classification dependent upon his own decisional definition of cognitive meaning.

The Direction of Time, nearly completed before Reichenbach's death and published posthumously, is the culmination of his epistemological investigations of relativity physics and quantum theory. In it he applied his analyses of conventions, equivalent descriptions, probability inferences, and three-valued logic to subjective (experienced) time, to the time concepts of macrophysics, and to the possibility of establishing time order and time direction among subatomic events. He found that among the equivalent descriptions of the "interphenomena" of quantum theory, every possible description contains causal anomalies, reversals of time direction, or both. He concluded that both time order and time direction are statistical macrocosmic properties which cannot be traced to microcosmic events. To the question "Why is the flow of psychological time identical with the direction of increasing entropy?" his answer was "Man is a part of nature, and his memory is a registering instrument subject to the laws of information theory" (*The Direction of Time*, p. 269).

BIBLIOGRAPHY

I. ORIGINAL WORKS. A complete bibliography is in *Modern Philosophy of Science* (below). Reichenbach's earlier writings include *Der Begriff der Wahrscheinlichkeit für die mathematische Darstellung der Wirklichkeit* (Leipzig, 1915), his inaugural dissertation, also in *Zeitschrift für Philosophie und philosophische Kritik*, **161** (1916), 210–239, and **162** (1917), 98–112, 223–253, and summarized in "Der Begriff der Wahrscheinlichkeit für die mathematische Darstellung der Wirklichkeit," in *Naturwissenschaften*, **7**, no. 27 (1919), 482–483; *Relativitätstheorie und Erkenntnis apriori* (Berlin, 1920), trans. with an intro. by Maria Reichenbach as *The Theory of Relativity and A Priori Knowledge* (Berkeley–Los Angeles, 1965); *Axiomatik der relativistischen Raum-Zeit-Lehre*, no. 72 in the series Die Wissenschaft (Brunswick, 1924; repr. Brunswick, 1965), trans. by Maria Reichenbach with an intro. by W. C. Salmon as *Axiomatization of the Theory of Relativity* (Berkeley–Los Angeles, 1969); *Von Kopernikus bis Einstein*

(Berlin, 1927), trans. by R. B. Winn as *From Copernicus to Einstein* (New York, 1942; paperback ed., 1957); *Philosophie der Raum-Zeit-Lehre* (Berlin–Leipzig, 1928), trans. by Maria Reichenbach and John Freund, with intro. by Rudolf Carnap, as *The Philosophy of Space and Time* (New York, 1958); *Atom und Kosmos. Das physikalische Weltbild der Gegenwart* (Berlin, 1930), trans. by E. S. Allen, rev. and updated by Reichenbach, as *Atom and Cosmos. The World of Modern Physics* (London, 1932; New York, 1933; repr. New York, 1957); and *Experience and Prediction. An Analysis of the Foundations and the Structure of Knowledge* (Chicago, 1938).

Later works include "On the Justification of Induction," in *Journal of Philosophy*, **37**, no. 4 (1940), 97–103, repr. in Herbert Feigl and Wilfried Sellars, eds., *Readings in Philosophical Analysis* (New York, 1949), 324–329; *Philosophic Foundations of Quantum Mechanics* (Berkeley–Los Angeles, 1944); "Bertrand Russell's Logic," in P. A. Schilpp, ed., *The Philosophy of Bertrand Russell*, vol. 5 in Library of Living Philosophers (Evanston, Ill., 1944), 23–54; *Elements of Symbolic Logic* (New York, 1947); "Philosophy: Speculation or Science?" in *The Nation*, **164**, no. 1 (4 Jan. 1947), 19–22, repr. as "The Nature of a Question," in I. J. Lee, ed., *The Language of Wisdom and Folly* (New York, 1949), 111–113; *The Theory of Probability. . . .*, trans. by Maria Reichenbach and E. H. Hutten, 2nd ed. (Berkeley–Los Angeles, 1949); "The Philosophical Significance of the Theory of Relativity," in P. A. Schilpp, ed., *Albert Einstein: Philosopher-Scientist*, vol. 7 in Library of Living Philosophers (Evanston, Ill., 1949), 287–311, repr. in Herbert Feigl and May Brodbeck, eds., *Readings in the Philosophy of Science* (New York, 1953), 195–211, and in P. P. Wiener, *Readings in Philosophy of Science* (New York, 1953), 59–76; "Philosophical Foundations of Probability," in *Proceedings of the Berkeley Symposium on Mathematical Statistics and Probability* (Berkeley–Los Angeles, 1949), 1–20; *The Rise of Scientific Philosophy* (Berkeley–Los Angeles, 1951; 1954; paperback ed., 1956); *Nomological Statements and Admissible Operations*, in Studies in Logic and the Foundations of Mathematics (Amsterdam, 1954); *The Direction of Time*, Maria Reichenbach, ed. (Berkeley–Los Angeles, 1956); and *Modern Philosophy of Science: Selected Essays*, Maria Reichenbach, ed. and trans. (London, 1959).

II. SECONDARY LITERATURE. See A. Grünbaum, *Philosophical Problems of Space and Time* (New York, 1963), ch. 3; A. Grünbaum, W. C. Salmon, *et al.*, "A Panel Discussion of Simultaneity by Slow Clock Transport in the Special and General Theories of Relativity," in *Philosophy of Science*, **36**, no. 1 (Mar. 1969), 1–81; Ernest Nagel, "Review of Philosophical Foundations of Quantum Mechanics," in *Journal of Philosophy*, **42** (1945), 437–444; and "Probability and the Theory of Knowledge," in his *Sovereign Reason* (Glencoe, Ill., 1954); *Probability in Philosophy and Phenomenological Research*, V–VI (1945), which contains papers by Reichenbach, D. C. Williams, Ernest Nagel, Rudolf Carnap, Henry Margenau, and others; Hilary Putnam, review of *The Direction of Time*, in *Journal of Philosophy*, **59** (1962), 213–216; W. V. Quine, review of *Elements of Symbolic Logic*, *ibid.*, **45** (1948), 161–166; Bertrand Russell, *Human Knowledge, Its Scope and Limits* (New York, 1948), *passim*; W. C. Salmon, "Should We Attempt to Justify Induction?" in *Philosophical Studies*, **8** (1957), 33–48; and *Foundations of Scientific Inference* (Pittsburgh, 1966), 1–15 and *passim*.

CYNTHIA A. SCHUSTER

REICHENBACH, KARL (or **CARL**) **LUDWIG** (*b.* Stuttgart, Germany, 12 February 1788; *d.* Leipzig, Germany, 19 January 1869), *chemistry, speculative science.*

Reichenbach, whose father was librarian and archivist to the town of Stuttgart, was educated at the Stuttgart Gymnasium and at the University of Tübingen. While a student, he was arrested by the Napoleonic authorities and imprisoned in the fortress of Hohenasperg. After his release he received his doctorate (1811) with a thesis on a hydrostatic bellows.

Reichenbach worked for a time as a minor civil servant, but marriage made him financially independent; and he spent the years 1816–1818 touring industrial facilities, especially ironworks, in Germany, Austria, and France. At Hausach, in the grand duchy of Baden, he built charcoal ovens of a new type that produced more solid charcoal in less time and allowed the volatile products to be collected. In 1821, in partnership with Count Hugo zu Salm of Vienna, Reichenbach set up a group of metallurgical factories at Blansko in Moravia. They prospered and enabled him to purchase several estates, as well as interests in steelworks and blast furnaces at Terniz, Lower Austria, and Gaya, Moravia.

Shortly before Salm's death in 1836, he and Reichenbach had started a beet-sugar factory at Blansko, which soon ran into difficulties. Salm's sons quarreled bitterly with Reichenbach over the management of this enterprise; and the latter, already depressed by the death of his wife, was forced to engage in a long and exhausting lawsuit while coping with the problems of the factory. He eventually won the lawsuit; but in 1839, after the king of Württemberg had raised him to the rank of *Freiherr*, he disposed of most of his industrial interests and retired to his castle at Reisenberg, near Vienna, with the intention of devoting his time entirely to science.

During the 1830's Reichenbach published several papers on compounds isolated from the beechwood tar produced by his charcoal ovens. He gave them fanciful names of his own invention; the only ones to survive are creosote (a mixture of phenols) and paraffin (a mixture of solid aliphatic hydrocarbons). A deeply colored substance "pittacal" (later shown to be a triphenylmethane derivative) was sold as a dye stuff, but had little success. His other chemical interest was the analysis of meteorites. This work led to differences with Wilhelm Haidinger, head of the mineralogical section of the Natural History Museum in Vienna, who published on the same subject. Refused access to the museum's large collection of meteorites, Reichenbach retaliated by using his own resources to amass an equally important collection, which he later gave to the University of Tübingen.

About 1844 Reichenbach believed that he had discovered that certain "sensitive" people (mostly women) could detect physical stimuli, such as light, under conditions in which they were not detectable to the normal sense. From this time chemistry occupied second place in his interests. He initiated a long series of experiments which confirmed him in his beliefs, and Liebig was with difficulty persuaded to publish an account of them as a supplement to the *Annalen der Chemie*. Reichenbach was quite unaware of the elaborate precautions (to guard against suggestion, collusion, or minute sensory cues) which are now known to be essential if valid results are to be obtained from such experiments; his results undoubtedly were worthless, as most of his contemporaries suspected.

Reichenbach's experiments, however, led him to speculate about a universal force permeating all nature, differing from such known forces as gravitation and electricity. He called this "Od" and attempted to use it to explain the most diverse effects. He claimed to be able to photograph objects in total darkness by means of the "odic light" that they supposedly emitted; his results are difficult to explain—even on the assumption of fraud—unless the photographed objects were feebly radioactive. In 1862 Reichenbach thought he had demonstrated this effect to the satisfaction of seven professors at the University of Berlin, but they issued a statement to the press denying that they had been convinced. Meanwhile he continued to publish a stream of books and pamphlets elaborating his theories and refuting his numerous detractors.

Much of the wealth that Reichenbach had amassed during his industrial days was lost in unwise investment; his mode of life at Reisenberg gradually became one of seclusion and reduced circumstances. Locally he was regarded as an eccentric, if not a madman. His last months were spent on a visit to Leipzig with his elderly housekeeper, the only "sensitive" he could now rely on, in a pathetic effort to convince the psychologist Gustav Fechner, whom he regarded as his most fair-minded opponent; he died there, at the age of almost eighty-one.

BIBLIOGRAPHY

I. ORIGINAL WORKS. Besides his numerous short papers (the majority in Poggendorff's *Annalen der Physik*) on tar constituents and on meteorites, Reichenbach published a geological survey of part of Moravia, *Geologische Mittheilungen aus Mähren* (Vienna, 1834). He also wrote a very uncritical account of his work on beech-tar, *Das Kreosot, in chemischer, physischer und medicinischer Beziehung* (Leizpig, 1833: 2nd enlarged ed., 1835).

His first studies of "sensitives," *Untersuchung über den Magnetismus und damit verwandte Gegenstände*, were published as a supp. to Liebig's *Annalen der Chemie*, **53** (1845), separately paginated 1–270, as the result of unspecified (probably financial) pressure on Liebig. This supp. is not always to be found in bound series of the journal. Liebig, however, refused to publish the whole of the material, which finally appeared as *Physikalisch-physiologische Untersuchungen über die Dynamide des Magnetismus . . .*, 2 vols. (Brunswick, 1850). Two rival English translations soon appeared: William Gregory, *Researches on Magnetism* . . . (London–Edinburgh, 1850); and John Ashburner, *Physico-physiological Researches . . .* (London, 1851). Much of the same material was reworked into *Der sensitive Mensch und sein Verhalten zum Ode*, 2 vols. (Stuttgart, 1854).

Four papers on the photographic experiments were accepted by Poggendorff for publication in the *Annalen der Physik*, although only one actually appeared: "Zur Intensität der Lichterscheinungen," **112** (1861), 459.

The following is a list, probably incomplete, of Reichenbach's minor books or pamphlets on the "odic force": *Odisch-magnetische Briefe* (Stuttgart, 1852); *Köhlerglaube und Afterweisheit* (Vienna, 1855); *Odische Erwiderungen* (Vienna, 1856); *Wer ist sensitive, wer nicht?* (Vienna, 1856); *Die Pflanzenwelt in ihren Beziehungen. . . .* (Vienna, 1858); *Odische Begebenheiten* (Berlin, 1862); *Aphorisme über Sensitivität und Od* (Vienna, 1866); *Die odische Lohe . . .* (Vienna, 1867).

Mention should also be made of the curious affair of the supposed love letters between the Duke of Clarence (later King William IV) and Caroline von Linsingen, which were found among Reichenbach's effects after his death and published as *Ungedruckte Briefe und Abhandlungen . . .* (Leipzig, 1880). Seldom seriously considered genuine, these letters must have been composed as fiction either by Reichenbach himself or by one of the persons named in the document which he left, purporting to describe how the letters had come into his possession.

II. SECONDARY LITERATURE. The fullest account of Reichenbach's life and work is an anonymous obituary notice in *Almanach der Akademie der Wissenschaften in Wien*, **19** (1869), 326–369. The author, who can be identified as Johann Hoffer, professor of physics at Vienna, drew both on his own knowledge of Reichenbach, with copious extracts from their correspondence, and on biographical material supplied by the family. A slightly different version of the "family" material is printed in *Württembergische naturwissenschaftliche Jahreshefte*, **26** (1870), 62–75, and forms the basis of Ladenburg's entry in *Allgemeine deutsche Biographie*, XXVII, 670–671.

A very uncritical biography is prefaced (ix–lxxi) to an English version, by F. D. O'Byrne, of *Odisch-magnetische Briefe, Reichenbach's Letters on Od and Magnetism* (London, 1926). There is a more recent short article by Moritz Kohn in *Journal of Chemical Education*, **32** (1955), 188–189, with portrait on 170.

The declaration by the seven Berlin professors (C. G. Ehrenberg, Heinrich Magnus, Rose, Mitscherlich, Poggendorff, P. T. Riess, K. H. Schellbach) was published in *Allgemeine Zeitung* (4 June 1862). Gustav Fechner wrote an account of Reichenbach's last months in Leipzig in *Erinnerungen der letzten Tage der Odlehre und ihres Urhebers* (Leipzig, 1876). Fechner was not unsympathetic; he was clearly puzzled by some of the demonstrations, and could detect no fraud.

W. V. FARRAR

REICHERT, KARL BOGISLAUS (*b.* Rastenburg, East Prussia [now Kętrzyn, Poland], 20 December 1811; *d.* Berlin, Germany, 21 December 1883), *comparative anatomy, histology, embryology.*

Reichert made very important contributions to the knowledge of transformations of early structures during the embryo's development and to the understanding, from a comparative point of view, of the development of an individual organism. He also played an important role as editor of the leading German anatomical periodical. In his last years, however, his prestige diminished because of his opposition to the new biological theories.

Reichert's father, mayor of Rastenburg, died on the day his son was born. The boy's stepfather, a high school principal, saw to it that he had a good education. Reichert began the study of medicine at Königsberg, where K. E. von Baer aroused his interest in embryology. He soon went to Berlin, however, to continue his education at the military medical school (Friedrich-Wilhelm-Institut) and at the University of Berlin, where he became the protégé of Friedrich Schlemm and Johannes Müller. He graduated in 1836 with a dissertation on the gill arches of the vertebrate embryos. Alexander von Humboldt secured him a leave for scientific work and eventually freedom from his military obligations. As prosector of anatomy in Müller's department, Reichert was very active both in research and in preparing an annual review of histology for the *Archiv für Anatomie*. In 1843 he was called to the University of Dorpat as professor of anatomy. Ten years later he became professor of physiology at Breslau, succeeding K. T. E. von Siebold. After the death of Müller in 1858, Reichert occupied the chair of anatomy and remained in Berlin for the rest of his life.

Reichert's greatest accomplishment was the introduction into embryology of the cell theory soon after its formulation by Schwann. Studying the evolution of frog spawn, he described with great accuracy the consecutive stages of its development, demonstrating that globules formed during the segmentation of the vitellus are cells and that all subsequent parts of the embryonic organs derive from the cleavage cells of the embryo. Like other biologists, however, Reichert erroneously believed that the granules in the yolk of the egg also were individual cells. On the other hand, he carefully distinguished the formative and the nutritive parts of the yolk, calling them *Bildungsdotter* and *Nahrungsdotter*. He also made valuable observations on the evolution of the tadpole, especially of the head, and presented many general reflections on the formation of different structures by means of invagination which foreshadowed, to some degree, Haeckel's gastraea theory.

Reichert's discovery of true homologies between the middle-ear ossicles and primitive structures of the splanchnocranium of reptiles and other lower vertebrates has been confirmed by recent paleozoological research, which has found all the intermediary stages that Reichert had envisaged. This was a brilliant linking of embryology with comparative anatomy and physiology, an example of transformation of the original structure connected to a change of function. Another of Reichert's great contributions was his demonstration that all types of connective tissue are closely related, have many common features, and derive from the same primary structures.

In his last years Reichert, holder of a very important chair and editor of a major periodical, and thus formally a leader in his field, became completely isolated because of his stubborn opposition to Haeckel's theory of the homology of germinal layers throughout the animal kingdom, the concept of protoplasm and new developments of the cell theory, and Darwin's theory of the origin of species.

BIBLIOGRAPHY

I. Original Works. Reichert's works include *De embryonum arcubus sic dictis branchialibus* (Berlin, 1836); *Ueber die Visceralbogen der Wirbelthiere im Allgemeinen und deren Metamorphose bei Säugethieren und Vögeln* (Berlin, 1837); *Vergleichende Entwicklungsgeschichte des Kopfes der nackten Amphibien* (Königsberg, 1838); "Kritische Jahresberichte über die Fortschritte der mikroskopischen Anatomie," in *Archiv für Anatomie, Physiologie und wissenschaftliche Medizin* (1839–1858); *Das Entwick-lungsleben im Wirbelthierreiche* (Berlin, 1840); *Beiträge zur Kenntniss des Zustandes der heutigen Entwicklungsgeschichte* (Berlin, 1843); *Über die Entwicklung des befruchteten Säugethiereies* (Berlin, 1843); *Bemerkungen zur vergleichenden Naturforschung im Allgemeinen und vergleichende Beobachtungen über das Bindgewebe und die verwandten Gebilde* (Dorpat, 1845); *Die monogene Fortpflanzung* (Dorpat, 1852); *Der Bau des menschlichen Gehirns*, 2 vols. (Leipzig, 1859–1861); *Beitrag zur feineren Anatomie der Gehörschnecke des Menschen und der Säugethiere* (Berlin, 1864); and "Beschreibung einer frühzeitigen menschlichen Frucht im bläschenförmigen Bildungszustande nebst vergleichenden Untersuchungen über die bläschenförmigen Früchte der Säugethiere und des Menschen," in *Abhandlungen der Preussischen Akademie der Wissenschaften* (1873), a description of a 12-day-old human embryo, the earliest known at that time.

II. Secondary Literature. [G.] Br[oesike], "Carl Bogislaus Reichert," in *Berliner klinische Wochenschrift* (1884), 45–46; J. Pagel, "Reichert, Karl Bogislaus," in *Allgemeine deutsche Biographie*, XXVII (1888), 679–681; and Waldeyer, in *Biographisches Lexikon der hervorragenden Ärzte aller Zeiten und Völker*, IV (Berlin–Vienna, 1932), 752–753.

Vladislav Kruta

REID, HARRY FIELDING (*b.* Baltimore, Maryland, 18 May 1859; *d.* Baltimore, 18 June 1944), *geophysics.*

Reid, who may well be said to have been the first geophysicist in the United States, was the son of Andrew Reid and of Fanny Brooks, a grandniece of George Washington. His family spent some time in Switzerland when he was a child, and this led to a lifelong love of mountains and to his great interest in the mechanics of glaciers. In his early manhood glaciology was his principal interest.

Reid entered Johns Hopkins in 1876 and received his Ph.D. there in 1885. He proceeded to Case School of Applied Science as professor of mathematics and later was professor of physics. In 1894, after a year of teaching at the University of Chicago, he returned to his alma mater, where he taught until his retirement in 1930.

Reid's greatest contribution to geophysics was undoubtedly his masterful exposition of the "elastic rebound theory" of the immediate source of earthquake waves—the cause of earthquakes. The theory states that elastic strain accumulates slowly in the earth's rocky crust as a result of forces, presumably acting from below the crust, of uncertain origin. When this strain becomes too great for the crustal rocks to bear, they break along faults. The frictional grinding of the two sides of the fault against each other produces the elastic wave motion which we call an earthquake. The association of some earthquakes with surface fault breaks was recognized early, but many felt that the fault breaks were the result of an unspecified sudden catastrophe at depth.

The theory was generally adopted in the United States, but it took sixty years for it to be widely accepted in Japan and in Europe. Reid was on the committee appointed by the governor of California to report on the earthquake of 1906. He recognized the significance of the changes in position of various bench marks of the U.S. Coast and Geodetic Survey both before and during the earthquake. His careful analysis of this somewhat sparse data, plus his knowledge of mechanics and perhaps intuition, led him to the statement of the "elastic rebound" theory.

BIBLIOGRAPHY

A complete bibliography of Reid's writings, prepared by M. F. Alvey, is part of Edward W. Berry, "Memorial to Harry Fielding Reid," in *Proceedings of the Geological Society of America* (1944), 295–298. Among his works are "Mechanics of Glaciers," in *Journal of Geology*, **4** (1896), 912–918; "Geometry of Faults," in *Bulletin of the Geological Society of America*, **20** (1909), 171–196; *The California Earthquake of April 18, 1906. The Mechanics of the Earthquake*, vol. II of *Report of the State* [California] *Earthquake Investigation Commission*, which is Carnegie Institution of Washington, publication no. 87, vol. II (Washington, 1910); "The Elastic-Rebound Theory of Earthquakes," in *Publications of the University of California, Bulletin of the Department of Geology*, **6** (1911), 413–444; "Isostasy and Mountain Ranges," in *Proceedings of the American Philosophical Society*, **50** (1911), 444–451, also in *Bulletin of the American Geographical Society of New York*, **44** (1912), 354–360; and "The Mechanics of Earthquakes; the Elastic Rebound Theory; Regional Strain," in *Bulletin of the National Research Council. Washington*, **90** (1933), 87–103.

For information on Reid see the Berry article mentioned above.

PERRY BYERLY

REIDEMEISTER, KURT WERNER FRIEDRICH (*b.* Brunswick, Germany, 13 October 1893; *d.* Göttingen, Germany, 8 July 1971), *mathematics.*

The son of Hans Reidemeister and the former Sophie Langerfeldt, Reidemeister attended school in Brunswick. His student years at the universities of Freiburg, Munich, and Göttingen were interrupted by four years of military service during World War I. He passed the *Staatsexamen* in mathematics (Edmund Landau was his examiner), philosophy (at Freiburg H. Rickert had been his teacher), physics, chemistry, and geology in 1920. After having accepted an assistantship with E. Hecke at the University of Hamburg, he earned his doctorate in 1921 with a dissertation on algebraic number theory, "Über die Relativklassenzahl gewisser relativquadratischer Zahlkörper" (published in *Abhandlungen aus dem Mathematischen Seminar, Universität Hamburg*, **1** [1921]). At the same time he studied affine geometry, published several papers, and assisted Wilhelm Blaschke in editing the second volume of his *Vorlesungen über Differentialgeometrie*, entitled *Affine Differentialgeometrie* (Berlin, 1923).

In 1923 Kurt Reidemeister accepted an associate professorship in Vienna, where he came in close contact with Hans Hahn, with research on the foundations of mathematics, and with the Vienna philosophical circle. Two years later he accepted a full professorship at Königsberg, where he worked with other young mathematicians, notably Ruth Moufang, Richard Brauer, and Werner Burau. His interest at this time was in the foundations of geometry and combinatorial topology. He wrote books and articles in both fields. His *Knotentheorie* (Berlin, 1932; repr. New York, 1948) remained the standard work on knot theory for several decades.

In April 1933 Reidemeister was expelled from his Königsberg professorship because he opposed the Nazis. In 1934 he became professor at the University of Marburg, in the chair of Kurt Hensel. He remained there—except for a two-year visit to the Institute for Advanced Study at Princeton in 1948–1950—until he moved to the University of Göttingen in 1955. While at Marburg he collaborated with F. Bachmann, laying the foundations of a development which culminated in Bachmann's *Aufbau der Geometrie aus dem Spiegelungsbegriff* (1959), and with Helene Braun.

The foundations of geometry and topology, established on a purely combinatorial and group-theoretical basis without introduction of a limit concept, always held a prominent place in Reidemeister's mathematical research. He was convinced that problems in mathematics that are original should arise from vivid perception, and even from the beauty of geometrical objects, and that abstraction should only be the result of intensive thought, which justifies the lack of immediate visualization. In accordance with this view he was critical of the modern trend of replacing traditional geometry by linear algebra. He had worked out a modern course along the lines of Felix Klein's "Erlanger Programm," classifying the various geometries by their related groups. His book *Raum und Zahl* (Berlin–Göttingen–Heidelberg, 1957) gave an idea of this concrete approach to mathematics in which mathematical thinking and reflections on thought are to illuminate each other.

Besides mathematics it was the historical origin of mathematical and rational thought that fascinated Reidemeister most—the Greeks in particular and philosophy in general. Three of his historical articles were republished in 1949 under the title *Das exakte Denken der Griechen*. In several publications he expounded his own philosophical position, one of critical rationalism. Reidemeister was strongly opposed to existentialism, which came into vogue in Germany after 1945. He reproached it for lack of objectivity and logical reasoning (*Die Unsachlichkeit der Existenzphilosophie* [Berlin, 1954; 2nd ed., 1970]).

Although an advocate of enlightenment and rationality, Reidemeister was highly sensitive and responsive to beauty and symmetry. Among his publications are two small volumes of essays and poems: *Figuren* (Frankfurt, 1946) and *Von dem Schönen* (Hamburg, 1947). His last book was a memorial to Hilbert: *Hilbert-Gedenkband* (Berlin–Heidelberg–New York, 1971).

Reidemeister was married to Elisabeth Wagner, a photographer and the daughter of a Protestant minister at Riga.

BIBLIOGRAPHY

Reidemeister's publications are listed in Poggendorff, VI, 2144; and VIIa, 714.

Obituaries include R. Artzy, "Kurt Reidemeister, 13. 10. 1893–8. 7. 1971," in *Jahresbericht der Deutschen Mathematiker-Vereinigung*, **74** (1972), 96–104, with bibliography; and F. Bachmann, H. Behnke, W. Franz, "In Memoriam Kurt Reidemeister," in *Mathematische Annalen*, **199** (1972), 1–11.

CHRISTOPH J. SCRIBA

REIL, JOHANN CHRISTIAN (*b*. Rhaude, Germany, 20 February 1759; *d*. Halle, Germany, 22 November 1813), *medicine, physiology*.

Reil was one of the early and major figures of the transition period in German medicine widely labeled as "romantic." Reil's career as a leading medical educator and clinician reflects the gradual adoption of philosophical principles in the face of an inadequate and confused knowledge of vital phenomena. By education and training a hard-nosed clinician willing to interpret biology in physicochemical terms, he eventually succumbed to Schelling's comprehensive and speculative view of nature, abandoning his sober epistemological approach based on Kantian philosophy. Reil felt that *Naturphilosophie* offered an appropriate framework within which medicine could be viewed in its dual role of an empirical and a rational activity.

Reil was born the son of a Lutheran pastor in the small East Friesland town of Rhaude. After education at nearby Norden, he went to Göttingen to study medicine. In 1780 Reil transferred to the University of Halle, graduating two years later with a dissertation on biliary diseases. Among his teachers were the anatomist and surgeon Phillip F. T. Meckel (1756–1803) and the clinician Johann F. G. Goldhagen (1742–1788). After graduation Reil returned to Norden to practice medicine and published a highly popular manual of dietary instructions.

In 1787 Reil went to Halle as a clinical instructor, and then assistant professor, under the auspices of his former teacher Goldhagen. Following Goldhagen's death a year later, Reil was appointed clinical professor and director of the clinical institute. He also became Halle's municipal physician (*Stadtphysikus*) in 1789, a post which he retained throughout the difficult years of the Napoleonic occupation.

Reil witnessed the economic collapse of Halle in 1806 and the closing of its university. He spent much time caring for the wounded soldiers who crowded into the city's lazaretto. By 1807 he was involved in reorganizing Halle's institution of higher learning, which reopened in 1808 with Reil as dean of the medical school. He also promoted Halle as a center for balneotherapy.

In 1810 Reil was invited by Wilhelm von Humboldt to participate in the organization of the medical school at the University of Berlin. With the support of the clinician Christoph W. Hufeland (1762–1836), some of Reil's proposals were adopted and he himself was placed in charge of the university's medical clinic. Soon, however, he experienced difficulties because of his growing personal conflict with both Hufeland and Karl F. von Graefe (1787–1840), his former student who headed the surgical division, as well as with the Prussian bureaucracy.

With the renewal of hostilities against Napoleon in 1813, Reil volunteered for military duty and attempted to organize a private hospital in Berlin for the wounded, which would be controlled and staffed by a number of prominent citizens, including Princess Wilhelmina of Prussia. He strongly criticized the conditions prevailing in the larger military hospitals, advocating instead the creation of smaller and more manageable units. His pleas were, however, largely ignored by the established bureaucracy and actively opposed by von Graefe.

In April 1813, Reil appealed directly to the Prussian monarch, Friedrich Wilhelm III, and was appointed chief inspector of all lazarettos west of the Elbe. By

September of the same year, sanitary conditions had deteriorated alarmingly because of a tremendous increase in the number of casualties in Blücher's army; an epidemic of typhus among the poorly treated soldiers compounded the difficulties. Reil's efforts to halt the disease were greatly hindered by the battle of Leipzig. Reil was untiring in his efforts to evacuate these victims, organizing makeshift hospitals for them in Leipzig, Halle, and the surrounding villages. In the process he contracted typhus and died in his sister's house in Halle, a victim of his final humanitarian efforts.

Reil's career and achievements can be presented from three different viewpoints: as a famous physician with certain medical ideas and clinical competence; as an energetic medical educator and organizer of medical services; and as an innovator in psychiatric care.

When Reil returned to Halle in 1787, he was primarily concerned with clinical subjects. The four parts of his *Memorabilium clinicorum*, published between 1790 and 1795, portray a shrewd observer of human sickness, a keen diagnostician as well as a resourceful medical and surgical healer. Reil rejected the idea of a perennially beneficial healing force in nature, insisting that the physician take charge of the situation.

In 1795 Reil founded the first journal dealing with physiology in Germany, *Archiv für die Physiologie*, which was to present works in physics, chemistry, histology, biology, and comparative anatomy. One of the initial articles was a short monograph by Reil concerning the vital force of the organism. This subject was attracting great attention in contemporary medical circles, since the elucidation of the *Lebenskraft* was expected to provide the foundations for medical theory and practice.

Reil believed that the appearance and actions of the living organism were based on material and structural changes alone. For him *Lebenskraft* was only a term designating the special and characteristic manifestations of living matter. Generation, growth, nutrition, and reproduction all occurred according to chemical laws. He concluded that the physicochemical approach would be the most successful in achieving further understanding of living beings.

Reil published the first volume of his most important work on fevers in 1799. In the prologue he declared his opposition to the prevailing systems of medicine, stressing instead observation and experiment. In the organism he distinguished between "mixture"— chemical composition—and "form"—the overall result of chemical affinities and combinations—a theme also treated by his student Johann F. Meckel. There-

fore disease ought to be viewed as a deviation from the normal bodily "mixture" and "form." Such concepts led Reil to propose a "pathological chemistry" and to formulate plans for a new pharmacology based on the same premises.

Originally a follower of Kantian philosophy and epistemology, Reil gradually approached Schelling's philosophy of nature, especially after 1804. He had been concerned about the mind-body dualism, the uniqueness of the physicochemical reactions in living organisms, and the inadequacy of strictly mechanical modes of physiological explanation. In his final speech at the University of Halle, delivered on 8 September 1810, Reil revealed the transformation of his thought. He declared that his previous fondness for various explanations had finally given way to a "living perception of intuition." Instead of dealing with mechanical principles, medicine was now to be guided by certain fundamental ideas, and observation had therefore reached a higher level from which all objects could be seen in their natural relationships. Reil concluded that natural events could be traced back to laws which coincided with those of the thinking mind.

At Berlin, Reil's philosophical speculations did not endear him to Hufeland, as shown in the latter's critical remarks following the posthumous publication of Reil's last works. Reil's interpretation of cerebral anatomy in terms of polarity proved equally confusing, yet his pioneer anatomical research placed him in the forefront of contemporary neuroanatomy.

Reil was a visionary medical educator. Not only was he a stimulating teacher, but his emphasis on bedside teaching and research had a lasting impact. During his tenure at Halle, the medical school became the most prominent teaching center in Germany, boasting such firsts as clinical laboratories, better correlation between clinical findings and pathological anatomy, and instruction in psychology.

Reil attempted to close the gap between physicians and surgeons in Germany by proposing better educational standards for the latter. He was keenly interested in training paramedical personnel who could fill the unmet medical needs of the rural population; they would need to know only certain health regulations and procedures without possessing a thorough understanding of the bodily functions. By contrast, he insisted that the physician be properly educated and acquainted with the prevailing anatomical and physiological knowledge. Reil viewed physicians as individuals who apply certain known theoretical principles rather than as mere empirical technicians who operate purely at random.

Finally, Reil should be remembered for his contributions to the understanding and care of the mentally ill. To him psychiatry was but a branch of medicine, and he sought to explain psychological disturbances on the basis of cerebral malfunctions or "oscillations" of the brain. In *Rhapsodieen über die Anwendung der psychischen Curmethode auf Geisteszerrüttungen* (1803), he proposed the use of a psychological method for the treatment of mental disorders. Such an approach, to be carried out in a special hospital, implied a relatively more humane treatment for the mentally ill. Occupational therapy was among the more advanced ideas which he proposed. In this respect Reil's influence should be equated with that of Pinel in France.

BIBLIOGRAPHY

I. ORIGINAL WORKS. A complete bibliography of Reil's publications and book reviews is in Eulner (see below), 32–39. Among his best-known works is "Von der Lebenskraft," in *Archiv für die Physiologie*, **1** (1796), 8–162, repr. as vol. 2 of Sudhoffs Klassiker der Medizin (Leipzig, 1910).

Reil's most famous book was *Ueber die Erkenntniss und Cur der Fieber*, 5 vols. (Halle, 1799–1815), the last vol. published posthumously by Christian F. Nasse. Reil wrote on the principles of psychotherapy in *Rhapsodieen über die Anwendung der psychischen Curmethode auf Geisteszerrüttungen* (Halle, 1803). Following his death, Peter Krukenberg edited two of Reil's works: *Entwurf einer allgemeinen Therapie* (Halle, 1816) and *Entwurf einer allgemeinen Pathologie*, 3 vols. (Halle, 1815–1816), the latter with the assistance of C. F. Nasse.

Reil published many articles in his *Archiv* between 1796 and 1812, covering such diverse subjects as medical semeiology, the polarity of natural forces in the pregnant uterus, and the anatomy of the brain. Some of the articles were collected and published as *Kleine Schriften wissenschaftlichen und gemeinnützigen Inhalts* (Halle, 1817). Reil's clinical observations appeared as *Memorabilium clinicorum medico-practicorum*, 4 fascs. (Halle, 1790–1795).

II. SECONDARY LITERATURE. One of the more recent sources for Reil's life and work is a group of four articles edited by Rudolph Zaunick in *Nova acta Leopoldina*, **144** (1960), 5–159, commemorating Reil's 200th birthday. The first, by H. H. Eulner, is "Johann Christian Reil, Leben und Werk," 7–50, including an appendix with Reil's bibliography, a chronology of his life, and a complete list of secondary literature dealing with his accomplishments.

Another collection of writings about Reil is *Gedenkschrift zum 200. jährigen Geburtstag Dr. Johann C. Reil* (West Rhauderfehn, 1959). Somewhat more dated is Max Neuburger, *Johann Christian Reil* (Stuttgart, 1913), a monograph commemorating the centenary of Reil's death. In the same year another memorial speech appeared: Rudolf Beneke, *Johann C. Reil* (Halle, 1913). A contemporary account was written by Heinrich Steffens, *Johann Christian Reil, eine Denkschrift* (Halle, 1815).

A doctoral dissertation on Reil's romantic leanings is Liselotte Müller, *Johann Christian Reil und die Romantik* (Würzburg, 1935). Another useful source is I. Petzold, "Johann C. Reil, Begründer der modernen Psychotherapie?" in *Sudhoffs Archiv für Geschichte der Medizin und der Naturwissenschaften*, **41** (1957), 159–179. Reil's efforts to organize the military hospitals in 1813 is reflected in a series of his letters to the king and other authorities in Prussia: K. Sudhoff, "Johann Christian Reil im Befreiungsjahre 1813," in *Münchener medizinische Wochenschrift*, **60** (1913), 2578–2582.

Among the articles in English are W. A. White, "Critical Historical Review of Reil's *Rhapsodieen*," in *Journal of Nervous and Mental Diseases*, **43** (1916), 1–22; and Aubrey Lewis, "J. C. Reil: Innovator and Battler," in *Journal of the History of the Behavioral Sciences*, **1** (1965), 178–190.

GUENTER B. RISSE

REINHOLD, ERASMUS (*b.* Saalfeld, Germany, 22 October 1511; *d.* Saalfeld, 19 February 1553), *astronomy.*

Reinhold was, after Copernicus, the leading mathematical astronomer of the sixteenth century; and in computational ability he surpassed Copernicus himself. Nothing is known of his childhood: his father, Johann, was for a long time secretary to the last abbot of Saalfeld. He enrolled at the University of Wittenberg, and his name is inscribed in the dean's book for the winter term of 1530–1531. In May 1536 Philip Melanchthon appointed him professor of *mathematum superiorum* (astronomy) at the same time that Rheticus was named professor of lower mathematics. Reinhold was twice elected dean at Wittenberg: in the college of arts in the winter semester of 1540–1541 and in the college of philosophy in the summer semester of 1549. In the winter of 1549–1550 he became rector.

On 22 January 1537 he married Margareta Boner, daughter of a highly placed burgher in Saalfeld; she died in childbirth on 7 October 1548. In 1550 he remarried but he again lost his wife in childbirth, in 1552. He fled from Wittenberg in 1552 in an attempt to escape the plague, but he died the following year in Saalfeld. His brother Johann, who had become professor of mathematics of Greifswald, died there in 1553. Erasmus was survived by two daughters, Margareta and Katharina, and by a son, Erasmus, who became a physician and issued a series of annual prognostications in the 1570's.

Cheap printed university textbooks first became popular in the 1540's, particularly at Wittenberg; and Reinhold published a widely reprinted commentary on Peuerbach's *Theoricae novae planetarum* (1542) and one on the first book of Ptolemy's *Almagest* (1549). When Rheticus returned to Wittenberg in September

1541 from his visit to Copernicus, Reinhold was one of the first to examine the new astronomy; and in the preface to his commentary on Peuerbach he wrote: "I know of a modern scientist who is exceptionally skillful. He has raised a lively expectancy in everybody. One hopes that he will restore astronomy," and later, "I hope that this astronomer, whose genius all posterity will rightly admire, will at long last come to us from Prussia. . . ."

Reinhold's copy of Copernicus' *De revolutionibus* is painstakingly annotated; it is virtually impossible to detect an error in the printed text not already marked by Reinhold. The pattern of annotations suggests that he was primarily interested in the model-building aspects of the work, especially in the way Copernicus had used combinations of circles to eliminate the Ptolemaic equant, and that he considered the heliocentric arrangement simply as a mathematical hypothesis of secondary interest.

Although Copernicus' book includes tables as well as demonstrations, these were clumsy to use for calculations; and Reinhold therefore set out on "this huge and disagreeable task" (as Kepler called it) to cast them in a handier form. Already in January 1544 Reinhold wrote to his patron, Duke Albrecht of Prussia, about his intentions, but the actual work continued over many years and was interrupted by war in 1546–1547, when the university was closed. The resulting *Prutenic Tables*, named after both Copernicus and his patron, were finally printed in Tübingen in 1551; they rapidly became the most widely adopted astronomical tables. Reinhold systematically made small changes in the planetary parameters in order to have them conform more accurately with the observations recorded by Copernicus; he was apparently oblivious to the fact that this was an exercise in futility because of serious errors in the Copernican planetary positions. As for the arrangements of circles and epicyclets, Reinhold slavishly followed *De revolutionibus*, but the introduction to the tables, while praising Copernicus, was silent about the heliocentric cosmology.

The working manuscript in which Reinhold explored the effects of changing parameters on both the Ptolemaic and Copernican models still exists in Berlin. In 1957 A. Birkenmajer pointed out two short phrases which suggest that Reinhold had considered a proto-Tychonian arrangement of the planets, but this model was certainly not developed, and it is absent from other similar points in the manuscript. From hints in his printed works as well as in this manuscript we can at best conclude that Reinhold did not ascribe physical reality to any particular planetary system.

Although Tycho Brahe never met Reinhold, the latter's approach to Copernicus had a direct influence on the great Danish astronomer. Tycho came to Wittenberg on several occasions, and in 1575 he visited Reinhold's son in Saalfeld; there he copied many of the annotations from Reinhold's copy of *De revolutionibus* into his own. Reinhold's notes emphasized Copernicus' occasional uses of alternative arrangements of planetary circles, and it was in this framework that Tycho explored the various schemes that led to his own geocentric system.

The success of Reinhold's *Prutenic Tables* enhanced Copernicus' reputation, but his personal silence on the heliocentric world view fostered a pattern in astronomical lecturing at German universities that persisted for at least a generation after his own untimely death. Kaspar Peucer, his successor at Wittenberg, wrote:

> Of Erasmus Reinhold, my teacher, to whom I owe my eternal gratitude—a man well-versed not only in mathematics but in universal philosophy, and very careful besides—brilliant testimonies to this care exist and therefore his studies were correct and deserving the highest praise. He conceived of the greatest things, which he surely would have attacked and completed if a longer life had been granted him. Among others, he often promised us new hypotheses of motions, having grown weary of the Peuerbachians'. Unfortunately the other works that he was contemplating were impeded by the elaboration of the *Prutenic Tables*, which do exist, by their confirmation, which was somewhat weak, and by his premature death, which tore from us the fruits of his work that would have been handed down to posterity from his careful and unflagging study.

BIBLIOGRAPHY

I. ORIGINAL WORKS. Reinhold's astronomical writings are a short treatise on spherical astronomy, *Themata, quae continent methodicam tractationem de horizonte rationali ac sensili deque mutatione horizontium et meridianorum* (Wittenberg, 1541, 1544, 1545, 1550, 1553, 1558, 1561, 1578; the latter six eds. being appended to Sacrobosco's *Libellus de sphaera*); the commentary on Peuerbach, *Theoricae novae planetarum ab Erasmo Reinholdo Salueldensi pluribus figuris auctae et illustratae scholiis* (Wittenberg, 1542; revised ed. 1553; Paris, 1553, 1555; Wittenberg, 1556; Paris, 1558; Venice, 1562; Wittenberg, 1580, 1601, 1604, 1653); *Calendarium novum continens motum solis verum ex novis tabulis supputatum proprie ad annum XLII*, in Martin Luther, *Enchiridion piarum precationem* (Wittenberg, 1543); *Oratio de Joanne Regiomontano* (Wittenberg, 1549); and a commentary on Ptolemy, *Mathematicae constructionis liber primus, gr. et lat. editus* (Wittenberg, 1549, 1556; Paris, 1556[?], 1558, 1560, 1564, 1569)—the last ed. appeared under the variant title *Regulae artis mathematicae*.

See also *Ephemerides duorum annorum 50 et 51 supputatae ex novis tabulis astronomicis* (Tübingen, 1550); *Prutenicae tabulae coelestium motuum* (Tübingen, 1551, 1562), M.

Maestlin, ed. (Tübingen, 1571); C. Strubius, ed. (Wittenberg, 1585); J. Sturm, ed. (Rostock, 1598), pt. 1; and *Primus liber tabularum directionum* (Tübingen, 1554; Wittenberg, 1584, 1606), with the Wittenberg eds. reprinted in Regiomontanus' *Tabulae directionum*.

There exist also the *Oratio in promotione magistrorum recitata a M. Erasmo Reinhold, Salueldensi Decano* (Wittenberg, 1541), Uppsala Univ. Library, Obr 49:505; *Oratio de sophistica habita a magistro Erasmo Reinhold Salveldensi* (Wittenberg, 1541); *Oratio de Caspero Crucigero* (Wittenberg, 1549); and various official pronouncements in *Scriptorum publice propositorum a professoribus in Academia Witebergensi ab anno 1540 usque ad annum 1553* (Wittenberg, 1560), but in general these do not touch on astronomy.

An anonymous work sometimes attributed to Reinhold is *Hypotyposes orbium coelestium, quas appellant theoricas planetarium congruentes cum tabulis Alphonsinis et Copernici, seu etiam tabulis Prutenicis* (Strassburg, 1568), reissued with a variant title as *Absolutissimae orbium coelestium hypotyposes, quas planetarium theoricas vocant* (Cologne, 1573); much of this material must have come originally from Reinhold, but it is actually the work of his student Kaspar Peucer, who published it as *Hypotyposes astronomicae, seu theoriae planetarum* (Wittenberg, 1571).

Extracts from twenty-five informative letters to and from Reinhold are found in Johannes Voigt, *Briefwechsel der berühmtesten Gelehrnten des Zeitalters der Reformation mit Herzog Albrecht von Preuszen* (Königsberg, 1841), 514–546. The letters are preserved in the Herzogliche Briefarchiv (HBA, A4) in the Staatliches Archivlager Göttingen: Staatsarchiv Königsberg (Archivbestande Preuszischer Kulturbesitz). Two additional letters, to Johannes Crato, are in M246, nos. 399–400, Wrocław University Library.

Extant manuscripts include the "Commentarius in Opus Revolution Copernici" (cited in the text), Latin 2°391 in the Deutsche Staatsbibliothek, Berlin, ff. 1–63 and 187–259; a 204-leaf commentary on Euclid, Latin 4°32 in the same library; and Reinhold's annotated *De revolutionibus* in the Crawford Collection of the Royal Observatory in Edinburgh.

II. SECONDARY LITERATURE. The principal evaluation of Reinhold's influence is Owen Gingerich, "The Role of Erasmus Reinhold and the Prutenic Tables in the Dissemination of Copernican Theory," in J. Dobrzycki, ed., *Colloquia Copernicana* II (Wrocław, 1973), 43–62, 123–125. Aspects of his manuscript commentary on Copernicus are treated in Aleksander Birkenmajer, "Le commentaire inédit d'Erasmus Reinhold sur le *De Revolutionibus* de Nicolas Copernic," in *La science au seizième siècle* (Paris, 1960), 171–177; reprinted in Birkenmajer's *Études d'histoire des sciences en Pologne* (Wrocław, 1972), 761–766; and Janice Henderson, "On the Distances Between Sun, Moon, and Earth According to Ptolemy, Copernicus and Reinhold" (Ph.D. diss., Yale Univ., 1973). For the impact of Reinhold's philosophy at Wittenberg, see Robert Westman, "The Melanchthon Circle, Rheticus and the Wittenberg Interpretation of the Copernican Theory," in *Isis*, in press 1975, and Owen Gingerich,

"From Copernicus to Kepler: Heliocentrism as Model and as Reality," in *Proceedings of the American Philosophical Society*, **117** (1973), 513–522. See also Pierre Duhem, *To Save the Phenomena* (Chicago, 1969), 70–74, trans. by E. Doland and C. Maschler. Biographical details are found in Ernst Koch, "Magister Erasmus Reinhold aus Saalfeld," pp. 3–16 in *Saalfelder Weihnachtsbuchlein* (Saalfeld, 1908), a rare pamphlet found in the Zinner Collection, Malcolm Love Library, California State University, San Diego; and in *Fortsetzung und Ergänzungen zu C. G. Jöchers allgemeinen Gelehrten-Lexikon* (Bremen, 1819), **6**, cols. 1722–1723.

OWEN GINGERICH

RE'IS. See **Pirī Rais.**

REMAK, ROBERT (*b.* Posen, Germany [now Poznan, Poland], 30 July 1815; *d.* Kissingen, Germany, 29 August 1865), *histology, embryology, neurology.*

Remak's life has only recently been investigated in detail by Bruno Kisch. Remak was the oldest of the five children of Salomon Meyer Remak, who ran a tobacco shop and lottery office, and Friederike Caro. The family is generally thought to have been prosperous, although Alexander von Humboldt referred to them as being poor; both descriptions may well have been accurate, since changing political and economic conditions might have altered their circumstances, especially after the return of Poznan from Polish to Prussian sovereignty by the Congress of Vienna. The family were Orthodox Jews and maintained a close identification with Polish culture; Remak himself maintained both these allegiances, even after he moved to Prussia.

Remak received his earliest education at home, then attended a private school before entering the lower secondary school in Poznan. Illness forced him to interrupt his education for a year, but he returned to complete his secondary studies at the Poznan Polish Gymnasium. In 1833 he enrolled at the University of Berlin to study medicine. It was a propitious time, since Johannes Müller had just assumed the professorship of anatomy and physiology there, and Remak was able to profit from his instruction. Remak also studied under C. G. Ehrenberg, and Müller and Ehrenberg both allowed him to use their microscopes and otherwise assisted him in the independent research that he began while he was still an undergraduate. This work was itself given direction by Ehrenberg's observations on invertebrate ganglion cells and nerve fibers and by a remark by

Müller suggesting the still unproved existence of extremely fine primitive fibrils within the nerve fiber. Remak published his first studies on the fine structure of nerve tissue in 1836; this communication was reprinted with further reports in his dissertation of 1838, *Observationes anatomicae et microscopicae de systematis nervosi structura.*

The *Observationes* contained Remak's demonstration that the medullary nerve fibers are not hollow, as had been supposed, but rather surround a translucent substance—Remak thought that this core was flat, and therefore called it the "primitive band." (His work on this subject was parallel to that of Purkyně, and Purkyně's term for this central core —which he called the "axis cylinder"—became the one that was widely accepted.) Remak also reported his discovery of the marrowless nerve fibers in the sympathetic nervous system (which he called "organic" to distinguish them from the "animal" medullary fibers of the cerebrospinal nerves) and confirmed that these fibers originate in the ganglion cells. His findings were at first criticized, especially by G. G. Valentin and Jacob Henle, but were later proved to be correct. Remak translated his work into Polish, and in so doing created a new Polish nomenclature.

The *Observationes* gained Remak the M.D. from the University of Berlin in January 1838; he presumably took the state examination that enabled him to practice medicine shortly thereafter. Although he had no official position, Remak chose to stay in Müller's laboratory after he finished his studies; he supported himself by his medical practice and by giving private lessons in microscopy. (One of his first students was Albert Koelliker.) Although Remak wished to make a career in teaching, the way was barred to him, since in Prussia at that time Jews were not admitted to that profession. Remak considered emigrating to Paris, but was dissuaded by Humboldt, who urged him to continue his research. He therefore continued his investigations of nerve tissue, and in 1839 discovered ganglion cells in the human heart. This finding seemed to him to explain the relatively autonomous action of the heartbeat, which he knew to be independent of the central nervous system. He further demonstrated the small ganglia that occur in the gray nerve fibers of the lung, the larynx, the throat, and the tongue; he later found such ganglia in the wall of the urinary bladder.

In 1840 Remak turned his attention to the function of the "organic"—that is, the sympathetic—nervous system. He published articles on physiology and histology of the nervous system in general in the *Encyclopädisches Wörterbuch der medicinischen Wissenschaften* the following year. Other results of his most important microscopic examinations also appeared in 1841, under the title "Ueber die Entstehung der Blutkörperchen." The terminology that Remak employed in this work is often difficult to interpret, but the main body of his discussion is concerned with the *Zerschnürung* (literally, "splitting by constriction") of the nucleus with a subsequent division of the cell body—probably amitosis. Remak was to return to this subject in his later work on embryology, but it is interesting to note that as early as 1842 he was opposed to the notion that cells can be generated from a more or less homogeneous base substance.

In 1843 Remak again attempted to teach. He was falsely encouraged by the idea that the recent change of monarchs (Friedrich Wilhelm IV had become king in 1840) would be advantageous to the Prussian Jews, and against Humboldt's advice inquired of the ministry of education whether he might be made a *Dozent.* His request was refused, but in March of that year—with the consent of Müller and the somewhat hesitant assistance of Humboldt—he made a direct petition to the king, and was once again rejected. Remak then turned again to research, chiefly to clinical investigations carried out in Schönlein's laboratory, which he entered as an assistant in November.

Although pathology was now his chief concern, Remak continued to do important work on the nervous system and in embryology. In 1843 and 1844, he established the presence of extremely thin fibrils in the axis cylinder, while in the earlier year he conducted research on chicken embryos to demonstrate that the innermost portion of the germinal layer (later called the endoderm) is the site from which develops the epithelium of not only the gastrointestinal tract, but also that of the respiratory passages, as well as the parenchymas of the liver, pancreas, and thyroid. He also, in 1845, demonstrated the division of those cells in the embryo which develop into primitive muscle bundles. His chief work of pathological anatomy, *Diagnostische und pathogenetische Untersuchungen*, was published the same year.

Remak had by this time acquired some eminence; the introductory material to his book mentions his membership in the Leopoldina and the Senckenberg Scientific Society of Frankfurt, as well as his corresponding membership in the Warsaw Medical Society. Nevertheless, when he applied for the post of prosector of the Charité in Berlin in 1846, he was not granted it, and the position went to Rudolf Virchow, his junior by six years. At the end of 1847, however, Humboldt and Schönlein, who was physician in ordinary to the king, succeeded in obtaining a lectureship for Remak—who was disappointed because he had hoped for a full professorship. All the daily newspapers carried the account of Remak's first lecture, since it was the first time a Jew had taught at the University of Berlin. It may be assumed that his practice profited from the publicity; at any rate, in 1847 he married Feodore Meyer, the daughter of a Berlin banker and the next year gave up his position in Schönlein's laboratory.

In 1848 and 1849 Remak returned to his studies of the germinal layer. He demonstrated that the medullary canal arises from the central portion of the ectoderm, while its epithelium and associated glands develop from the periphery. He then stated that the body wall and the wall of the intestine develop from the mesoderm by cleavage. In 1850 Remak published the first of the three parts of his *Untersuchungen über die Entwicklung der Wirbelthiere*. In it he discussed the probability that the cells in fertilized chicken eggs divide continuously; he further remarked that the structural elements of the ectoderm and the endoderm become increasingly smaller as their numbers increase. He mistakenly asserted that the spinal ganglion and nerve stem in birds are formed from the mesoderm, and he was likewise mistaken in assuming the genetic deflection of the *chorda dorsalis*. He correctly observed the transformation of the primitive vertebra into the permanent vertebra in the chick, although he was able to offer only a partial interpretation of the role of the medullary plate as the site from which the mesenchyme is formed—for him, its importance lay in its supplying the material of which the oviduct is formed.

It was only in 1851 that Remak recognized that the sense organs are formed from the ectoderm. He reached this conclusion as a result of the fixation of blastodiscs with acetic acid, sublimate, or chromic acid. He did not achieve true staining, however, except in a series of preparations with tincture of iodine. He reported these findings, with others, in a second part of the *Untersuchungen*.

By 1852 Remak was able to announce certain conclusions concerning cell division. He set aside some of his cautious earlier formulations, and asserted that the cleavage of the frog egg is due to a continuous process of division that always begins with the nucleus. He posited similar divisions, likewise starting with the cell nucleus, as occurring in somewhat later developmental stages of embryonic cells, which are again produced by cleavage. These divisions are common to almost all types of tissue, including that of the muscles, nervous sytem, and cartilage, as well as the epidermis, connective tissue, intestinal epithelium,

and blood cells. Of particular importance was Remak's suggestion that, contrary to Schwann's view, the intercellular substances of the cartilage and connective tissue cannot be a cytoblastema. (It must be mentioned, however, that Remak made no specific note of mitotic cell division until 1858, when he presented an account of the formation of the blood in a five-day-old chick embryo.)

Publication of Remak's *Untersuchungen* was completed in 1855. A good deal of the third part was devoted to such supplementary studies as that of the ectodermal origin of the crystalline lens, which he demonstrated in the case of fish, batrachians, birds, and mammals. He also stated that feathers, hair, and nails are ectodermal formations. He examined the mesoderm with particular care, and noted that the portion of it that borders on the medullary canal represents a rudimentary form of the vertebral canal and its associated muscles. Remak was not at this time aware of the important role of the mesoderm as the rudimentary tissue from which the blood vessels arise.

In examining the endoderm, Remak confirmed the important differences that exist between holoblastic and meroblastic eggs. He nevertheless considered that their mode of development was fundamentally the same, and in particular he stressed the epithelial origin of the glands that empty into the intestine. "The whole substance of the liver," he wrote, "is therefore identical with the epithelium of the intestinal tube." Remak was unable to present a convincing demonstration of the derivation of the kidney tubules from the intestinal epithelium, and his emphasis upon the supposed origin of the nerves from the mesoderm is also unsatisfactory. On the whole, however, Remak's own assessment of the value of the *Untersuchungen* is an accurate one; he wrote that it established the position of histology among the sciences and "served as a foundation and stimulus for investigations of the interaction of the homologous formal constituents and heterologous tissues of the animal body."

Remak's first work on neurology, *Über methodische Electrisierung gelähmter Muskeln*, was also published in 1855. After 1856, when he lost the appointment to the chair of pathology at the University of Berlin to Virchow, Remak devoted himself entirely to his medical practice. He invented the technique of electrotherapy, which he applied successfully to his patients and described in his *Galvanotherapie der Nerven- und Muskelkrankheiten*, which he dedicated to Humboldt and published in 1858. In 1859 he was appointed an assistant professor at the university, but this belated recognition had no effect upon his subsequent career.

Remak was often ill; there are indications that even during his first years as an assistant in Schönlein's laboratory he had suffered from a chest ailment and had not expected to live much longer. To his sickness was added his frustration at being unable to win suitable academic appointment, so that he was often irritable and petulant in both his personal and his professional relationships. He died suddenly, while taking a rest cure.

BIBLIOGRAPHY

I. ORIGINAL WORKS. Remak's major scientific publications are "Vorläufige Mitteilung microscopischer Beobachtungen über den inneren Bau der Cerebrospinalnerven und über die Entwicklung ihrer Formelemente," in *Müllers Archiv für Anatomie, Physiologie und wissenschaftliche Medizin* (1836), 145–161; *Observationes anatomicae et microscopicae de systematis nervosi structura* (Berlin, 1838), his doctoral dissertation; "Ueber die Ganglien der Herznerven des Menschen und deren physiologische Bedeutung," in *Wochenschrift für die gesamte Heilkunde* (1839), 149–154; "Ueber die physiologische Bedeutung des organischen Nervensystems, besonders nach anatomischen Tatsachen," in *Monatsschrift für Medizin, Augenheilkunde und Chirurgie,* **3** (1840), 225–265; "Ueber die Entstehung der Blutkörperchen," in *Medizinische Zeitung,* **10** (1841), 127; "Ueber den Inhalt der Nervenprimitivröhren," in *Müllers Archiv für Anatomie, Physiologie und wissenschaftliche Medizin* (1843), 197–201; "Neurologische Erläuterungen," *ibid.,* (1844), 463–472; *Ueber ein selbständiges Darmnervensystem* (Berlin, 1847); and "Ueber die Funktion und Entwickelung des oberen Keimblattes im Ei der Wirbelthiere," in *Monatsberichte der Preussischen Akademie der Wissenschaften zu Berlin* (Oct. 1848), 362–365.

Also see *Untersuchungen über die Entwicklung der Wirbelthiere,* 3 pts. (Berlin, 1850–1855); "Ueber extracellulare Entstehung thierischer Zellen und über Vermehrung derselben durch Theilung," in *Müllers Archiv für Anatomie, Physiologie und wissenschaftliche Medizin* (1852), 47–57; *Galvanotherapie der Nerven- und Muskelkrankheiten* (Berlin, 1858), also in French trans. (Paris, 1860); and "Ueber die Theilung der Blutzellen beim Embryo," in *Müllers Archiv für Anatomie, Physiologie und wissenschaftliche Medizin* (1858), 178–188.

II. SECONDARY LITERATURE. See Arthur Hughes, *A History of Cytology* (London–New York, 1959), 58; Bruno Kisch, "Robert Remak," in "Forgotten Leaders in Modern Medicine," in *Transactions of the American Philosophical Society,* n.s. **44** (1954), 227–296; Leslie T. Morton, ed., *Garrison and Morton's Medical Bibliography,* 2nd ed. (London, 1961); and Julius Pagel, "Robert Remak," in *Allgemeine deutsche Biographie,* **28** (Leipzig, 1889), 191–192.

ERICH HINTZSCHE

REMSEN, IRA (*b.* New York, N.Y., 10 February 1846; *d.* Carmel, California, 4 March 1927), *chemistry, education.*

Remsen was educated in the New York public schools and, at the age of fourteen, entered the Free Academy (later the College of the City of New York). He did not complete the four-year course there but, at the urging of his father, James Vanderbilt Remsen, became an apprentice to a doctor who taught in a homeopathic medical school. Remsen was dismayed at the inadequacy of the instruction that he was offered and prevailed upon his father to permit him to enroll in the College of Physicians and Surgeons of Columbia University, from which he received the M.D. in 1867. Having completed his medical studies and reached his majority, Remsen decided to study chemistry—to which he had been attracted by the lectures of R. O. Doremus at the Cooper Union—and went to Munich to pursue that subject under Liebig. By that time, however, Liebig was no longer giving laboratory instruction, although Remsen was able to attend some of his lectures. Remsen therefore studied for a year with Jacob Volhard then, with Volhard's help, transferred in the autumn of 1867 to the University of Göttingen, where he worked with Rudolph Fittig. He was granted the Ph.D. in 1870 for his research on the structure of piperic and piperonylic acids.

When Fittig went to Tübingen as professor of chemistry later in 1870, he took Remsen with him as his laboratory and lecture assistant. Remsen held this post for two years, during which he worked independently on the oxidation of ortho- and parasulfotoluene. His investigations led him to Remsen's law that groups attached to the benzene ring in the ortho position protect paramethyl, paraethyl, and parapropyl groups from oxidation by nitric or chromic acid.

In 1872 Remsen returned to New York to seek an academic appointment. After some months, during which he translated Fittig's edition of Wöhler's *Organic Chemistry,* he was named professor of chemistry and physics at Williams College. He remained there for four years, performing his own laboratory research, although he was unable to institute a course of laboratory work for his students. In 1876, while still at Williams, he published his own *Principles of Theoretical Chemistry,* an influential text that emphasized Cannizzaro's determination of molecular weights through Avogadro's hypothesis.

Remsen's growing reputation attracted the attention of Daniel Coit Gilman, first president of the new Johns Hopkins University, and in 1876 Remsen accepted Gilman's offer of a professorship of chemistry. At Johns Hopkins, Remsen was able to introduce many of the teaching methods—especially the integration of laboratory research—with which he had become acquainted in Germany, and these practices had a profound influence on the teaching of chemistry in the United States, particularly at the graduate level, until World War II. Remsen attracted a large number of students from both the United States and Europe during his teaching career.

In 1879 Remsen invited Constantine Fahlberg, who had taken the Ph.D. at Leipzig, to continue the study of the oxidation of substituted benzene rings. Remsen had shown that ortho groups could be oxidized by potassium permanganate, and Fahlberg, working in Remsen's laboratory and at Remsen's suggestion, oxidized orthotoluene sulfamide by potassium permanganate to produce orthobenzoyl sulfimide. Fahlberg found the compound, later named saccharin, to be intensely sweet, and with the help of A. List patented the process for commercial manufacture. Although Remsen apparently felt some initial grievance about Fahlberg's behavior, he mastered his ill will and in 1907 acted impartially as head of the board appointed by Theodore Roosevelt to determine whether sodium benzoate (used as a food preservative) and saccharin were injurious to health.

Remsen founded, in 1879, the *American Chemical Journal*, the first continuing periodical devoted to American chemical research. He served as its chief editor until 1911, when it was incorporated into the *Journal of the American Chemical Society*. In 1887 he became secretary to the academic council of Johns Hopkins, and in 1901 he succeeded Gilman as president of the university. Despite the demands of his administrative duties, he remained as head of the chemistry laboratory until 1908. He retired as professor emeritus and president emeritus in 1913; he had previously refused to work for private firms (although he had taken active part in work for municipalities, notably Boston and Baltimore) but was then retained by Standard Oil as a laboratory consultant until his death in 1927. He was survived by his wife, Elizabeth Mallory Remsen, whom he had married in 1875, and by two sons; his ashes were placed in Remsen Hall at Johns Hopkins University.

BIBLIOGRAPHY

I. ORIGINAL WORKS. The Milton S. Eisenhower Library of Johns Hopkins University has more than 1,000 MS items by or related to Remsen. It also has several scrapbooks, including one entitled "Sewage Problems in Baltimore (1905-1912)" and another covering the resignation of Gilman from the university presidency and the

subsequent installation of Remsen. The John Work Garrett Library at Johns Hopkins has a collection of Remsen's medals and family memorabilia. MS material, chiefly letters, are scattered in other library collections, including the Edgar Fahs Smith collection at the University of Pennsylvania and the Lyman Churchill Newell collection in the chemistry department at Boston University. Remsen descendants hold others. E. Emmett Reid, formerly professor of organic chemistry at Johns Hopkins, assembled a scrapbook of personal recollections of Remsen recorded by the latter's students.

II. SECONDARY LITERATURE. The National Academy of Sciences devoted *Biographical Memoirs. National Academy of Sciences*, **14** (1932), 207–257, to Remsen. In that compilation W. A. Noyes and J. F. Norris assembled a biography from their papers in *Science*, **66** (1927), 243–246; *Journal of the Chemical Society* (1927), 3182–3189; and *Journal of the American Chemical Society. Proceedings* (1928), 67–79. A bibliography of papers published by Remsen and by his students, of Remsen's books, and of his addresses is repro. on 230–240 from *Journal of the American Chemical Society*, **50** (1928), 80.

Remsen's biography in *Dictionary of American Biography*, XV, 500–502, is by W. A. Noyes. Frederick H. Getman, *Life of Ira Remsen* (Easton, Pa., 1940), synthesizes all of the above material and adds Getman's own recollections of Remsen. The appendix lists some of Remsen's publications and is an abridgment of that in *Journal of the American Chemical Society* (1928) listed above. A brief biography of Remsen, by Aaron Ihde, is in E. Farber, *Great Chemists* (New York, 1961), 819–822. A summary of documents concerning Remsen in government archives may be found in *Current Literature*, **52** (1912), 304–305.

J. Z. FULLMER

RENARD, ALPHONSE FRANÇOIS (*b*. Renaix [now Ronse], Belgium, 26 September 1842; *d*. Brussels, Belgium, 9 July 1903), *geology, mineralogy.*

Renard's education was initially religious; and it was only in 1870, upon being sent to the Jesuit training college of Maria Laach in the Eifel, that he came in contact with the sciences and became interested in geology, mostly through the spectacular volcanic phenomena in that area.

In 1874, at the age of thirty, Renard was appointed professor of chemistry and geology at the Jesuit *collège* at Louvain. His teaching did not interrupt his theological education, and he was ordained a priest in 1877. Renard's growing scientific reputation led to his appointment as one of the curators of the Royal Natural History Museum at Brussels. In 1882 he abandoned teaching and devoted himself entirely to curatorial duties. In 1888, however, he accepted the chair of geology at the University of Ghent, a position he occupied until his death.

Renard's first contribution to geology was a monograph, written with Charles de La Vallée-Poussin in 1874, on the mineralogical and the stratigraphical characters of the "plutonic" rocks of Belgium and the French Ardennes. It is actually a study of the chemical, mineralogical, and structural aspects of the regional metamorphism of that area. This investigation was followed by several other papers on the same processes, as revealed by the petrography of particular rock types: phyllites, garnet schists, amphibolites, and dolomites. Renard at first considered these metamorphic rocks as the products of the mineralogical reorganization of the original sediments through intense tectonic deformation of the area. Several years later, having acquired a broader comprehension of metamorphism, he was inclined to doubt his first conclusions and to favor instead the action of contact metamorphism due to the intrusion of deep-seated igneous rocks not yet exposed by erosion. Recent studies have shown his original interpretation to be more correct.

This first group of publications established Renard's reputation throughout Europe as the unusual combination of a chemist, a petrographer, and a field geologist. It was therefore natural that upon completion of the *Challenger* expedition, the petrographic investigation of the samples should be entrusted to him. With Sir John Murray he published a series of preliminary papers on the collected materials; and eventually their work took its final form as the monumental *Deep-Sea Deposits*, a masterpiece in marine sedimentology which opened up an entirely new scientific field of oceanography.

The volume begins with a discussion of the methods of obtaining, examining, and describing deep-sea deposits. This introduction is followed by an account of the composition and geographical and bathymetrical distribution of deep-sea deposits in which abyssal red clay, radiolarian ooze, diatom ooze, Globigerina ooze, and pteropod ooze are exhaustively described. Then the organic constituents are examined, as are the mineral substances of terrestrial and extraterrestrial origin (cosmic dust). The volume ends with a description of the diagenetic products formed *in situ* on the ocean floor or within the muds, such as manganese nodules, phosphatic concretions, glauconite, and zeolites.

The development of Renard's scientific knowledge interfered with his religious beliefs, and he resigned from the Society of Jesus in 1884. In 1901 his separation from the Catholic Church became complete when he married. This emancipation, which he said was unavoidable, resulted in bitter and undeserved polemics that saddened the last years of his life.

BIBLIOGRAPHY

I. ORIGINAL WORKS. Renard's writings include "Mémoire sur les caractères minéralogiques et stratigraphiques des roches dites plutoniennes de la Belgique et de l'Ardenne française," in *Mémoires de l'Académie royale des sciences, des lettres et des beaux-arts de Belgique*, **40** (1874), also published as a book (Brussels, 1876), written with Charles de La Vallée-Poussin; "Report on the Petrology of the Rocks of St. Paul," in *Narrative*, II (London, 1882), app. B, and *Deep-Sea Deposits* (London, 1891), both vols. of *Report on the Scientific Results of the Exploring Voyage of H.M.S. Challenger During the Years 1873–76*; "Recherches sur la composition et la structure des phyllades ardennais," in *Bulletin du Musée royal d'histoire naturelle de Belgique*, **1** (1882), 1–54, and **2** (1883), 127–149; "On the Nomenclature, Origin, and Distribution of Deep-Sea Deposits," in *Proceedings of the Royal Society of Edinburgh*, **12** (1884), 495–529, also in French as "Notice sur la classification, le mode de formation et la distribution géographique des sédiments de mer profonde," in *Bulletin du Musée royal d'histoire naturelle de Belgique*, **3** (1884), 25–62, written with J. Murray; "On the Microscopic Characters of Volcanic Ashes and Cosmic Dust, and Their Distribution in Deep-Sea Deposits," in *Proceedings of the Royal Society of Edinburgh*, **12** (1884), 474–495, also in French as "Les caractères microscopiques des cendres volcaniques et des poussières cosmiques et leur rôle dans les sédiments de mer profonde," in *Bulletin du Musée royal d'histoire naturelle de Belgique*, **3** (1884), 1–23; and "Notice préliminaire sur les sédiments marins recueillis par l'expédition de la 'Belgica,'" in *Mémoires de l'Académie royale des sciences, des lettres et des beaux-arts de Belgique*, **61** (1901–1902), written with Henryk Arctowski.

II. SECONDARY LITERATURE. See two publications by A. Geikie: "Obituary of A. F. Renard," in *Geological Magazine*, 4th ser., **10** (1903), 525–527; and "Obituary of A. F. Renard," in *Quarterly Journal of the Geological Society of London*, **60** (1904), lix–lxiv.

ALBERT V. CAROZZI

RENAULT, BERNARD (*b.* Autun, France, 4 March 1836; *d.* Paris, France, 16 October 1904), *paleobotany*.

Renault came from a provincial middle-class family. His father, Lazare Renault, was a bailiff; his mother was Jeanne-Marie Goby. The family was of modest means and made considerable sacrifices to educate its gifted oldest son. He had to supplement the family income, and as early as 1855 he taught in a private school.

Renault's earliest scientific interest was in physics; his first doctoral dissertation (1867) was in physical chemistry and he published a number of short papers in that field, one of which was translated into German. In 1867 he was appointed teacher of chemistry and physics at the *lycée* at Cluny.

Through naturalist friends Renault early became interested in Carboniferous and Permian fossil plants, which were abundant around Autun. Some of them were silicified, so that even the microscopic details could be studied. By a happy coincidence Adolphe Brongniart, the founder of scientific paleobotany, was chief inspector of mines and had the official duty to inspect the *lycée* at Cluny. He met Renault, recognized his capacity, and encouraged him. Renault worked enthusiastically in his spare time, collected many new fossils, invented new and improved old methods for preparing the very tough material, and eagerly reported new discoveries to Brongniart. His first paleobotanical paper was published in 1869. During the Franco-Prussian War, Renault was given a high administrative position on the Committee of National Defense. In 1872 Brongniart summoned him to Paris, first as preparator and from 1876 as assistant naturalist at the Muséum d'Histoire Naturelle, a position he held until his death.

Renault obtained his second doctorate in 1879 with the dissertation "Structure comparée de quelques tiges de la flore carbonifère." A large work on silicified seeds was started with Brongniart and published under his name in 1881, even though most of the work was done by Renault. His lectures in paleobotany resulted in the four-volume *Cours de botanique fossile* (1881–1885), which he printed at his own expense.

Renault published more than 200 scientific papers, most of them on Carboniferous and Permian plants. His main contribution in this field was that he amassed an enormous amount of exact information on the microscopic and macroscopic anatomy of the fossil plants. He worked on material that was hard to prepare but which in detail could be observed extraordinarily well. This technique enabled him to combine the separate parts (roots, leaves, stems, fructifications, and spores) into real plants more precisely than ever before. In this respect Renault was certainly a worthy successor to his teacher Brongniart. He did not quite make the great discovery that many of the fernlike plants of the Carboniferous flora were really seed ferns, but the studies of that group are largely based on Renault's keen observations. He also worked with fossil microorganisms and thereby made a great contribution to the understanding of the formation of coal.

Renault was not recognized by his French colleagues, and official and academic honors came late or not at all. He had to do most of the preparation of his difficult material by himself; and the enormous amount of material he prepared, drew, and described is a monument not only to his scientific capacity but also to his diligence and perseverance.

BIBLIOGRAPHY

I. ORIGINAL WORKS. Renault's works include "Recherches sur les végétaux silicifiés d'Autun et de St. Étienne, *Bulletin de la Société Eduense* (1878); "Structure comparée de quelques tiges de la flore carbonifère," in *Nouvelles archives du Muséum d'histoire naturelle*, 2nd ser., **2** (1879); *Cours de botanique fossile*, 4 vols. (Paris, 1881–1885); *Les plantes fossiles* (Paris, 1888); and "Sur quelques microorganismes des combustibles fossiles," *Bulletin de la Société de l'industrie minérale de St-Étienne* (1899). A complete list of Renault's scientific papers is in Roche (see below).

II. SECONDARY LITERATURE. The best biography of Renault is by his friend A. Roche, "Biographie de Bernard Renault," in *Bulletin de la Société d'histoire naturelle d'Autun*, **18** (1905), 1–159; it includes a complete, annotated bibliography, and 10 plates of illustrations from Renault's scientific work. D. H. Scott, "Life and Work of Bernard Renault," in *Journal of the Royal Microscopical Society* (1906), 129–145, draws heavily on Roche but give a more balanced view of Renault's scientific importance.

NILS SPJELDNAES

RENAUT, JOSEPH-LOUIS (*b.* La Haye-Descartes, Indre-et-Loire, France, 7 December 1844; *d.* Lyons, France, 26 December 1917), *medicine, histology.*

Renaut began his medical studies at Tours in 1864, continued them in Paris in 1866 and in 1869 won the coveted position of *interne des hôpitaux*. He held this post until 1875, while studying histology under Cornil and Ranvier at the Collège de France.

Research never overshadowed Renaut's devotion to medicine. He completed his medical thesis on erysipelas (*Contribution à l'étude de l'érysipèle . . .*, 1874), and passed the *agrégation* with his paper *De l'intoxication saturnine . . .* (1875), earning two silver medals. In 1876 he accepted the directorship of the pathological anatomy laboratories at the Charité Hospital in Paris.

Many scholars in the French educational establishment felt that defeat in the Franco-Prussian War was due to the superiority of German education and technology. The ensuing French educational reforms decisively affected Renaut. He became a *répétiteur* in Claude Bernard's new histology laboratory at the Collège de France, dedicated by Ranvier, its director, to effective "competition with similar German establishments." In 1877 Renaut accepted the chair of general anatomy and histology at the new Medical Faculty of Lyons, which was created to emulate the German example of educational decentralization. He taught at Lyons for forty years and served as chief physician at the Croix-

Rousse, Perron, and Hôtel-Dieu hospitals, retiring from clinical service in 1900.

Renaut's research and writing focused on histology. In France his *Traité d'histologie pratique* (1889–1899) was considered, together with Ranvier's *Traité technique d'histologie*, as "the most important and original work on this science in the 19th century" (Mollard, "Le Professeur Renaut," p. 53). It emphasized comparative anatomy and revealed Renaut's interest in embryology and developmental physiology.

His contributions to histology and pathology include the study of the fibrohyaline membrane (known as "Renaut's layer") between the corium and epidermis; the secretory function of connective tissue and the intestinal epithelium; the ciliary epithelium in the lung; the diapedesis of blood cells across the intestinal epithelium; and the continuity of the lymphatic capillary system. He also investigated the syncytial nature of cardiac muscle fibers; the epithelial origin of neuroglia; the aggregation of lymphocytes to form lymphoid follicles; and the pathology of progressive muscular atrophy, of nephritis and myocarditis, and of the fibrous forms of tuberculosis in the lung.

He was a member of the Paris Academy of Medicine, the Société Anatomique, the Société de Biologie, and the Société de Dermatologie et de Syphiligraphie, which he helped found. He also founded the *Revue générale d'histologie* in 1904.

Renaut participated in many official functions at the University of Lyons and wrote poetry under the pseudonym Sylvain de Saulnay; one of his volumes, *Ombres colorées*, won a prize from the French Academy. He was one of France's most distinguished nineteenth-century histologists.

BIBLIOGRAPHY

I. ORIGINAL WORKS. Renaut's major works are *Contribution à l'étude anatomique et clinique de l'érysipèle et des oedèmes de la peau* (Paris, 1874); *De l'intoxication saturnine chronique* (Paris, 1875); *Note sur les lésions des faisceaux primitifs des muscles volontaires dans l'atrophie musculaire progressive et dans la paralysie saturnine* (Versailles, 1876), written with M. Debove; *Cours d'anatomie générale; Leçon d'ouverture, Faculté de médecine de Lyon* (Paris, 1877); *Note sur la tuberculose en général et sur ses formes fibreuses pneumoniques en particulier* (Lyons, 1879); *Travaux du laboratoire d'anatomie générale et d'histologie, 1880–1881* (Paris, 1882); *Instruction médicale sur le choléra, par la Société nationale de médecine de Lyon . . ., MM. Renaut, J. Teissier et Ferrand rapporteurs* (Lyons, 1884); *Titres et travaux scientifiques de J. Renaut* (Lyons, 1887); and *Traité d'histologie pratique*, 2 pts. in 4 vols. (Paris, 1889–1899)—*Le milieu intérieur et le tissu conjonctif lâche et modelé*, vol. I; *Tissus du squelette, tissu*

musculaire, système vasculaire, sanguin et lymphatique, vol. II; *Les épithéliums. L'ectoderme tégumentaire*, vol. III; *L'ectoderme neural. L'entoderme. Les reins. Les glandes génitales. La rate*, vol. IV.

Subsequent writings are *Note sur une nouvelle maladie organique du coeur: la myocardite segmentaire essentielle chronique* (Paris, 1890); *Note sur la structure des glandes à mucus du duodénum (glandes de Brunner)* (Versailles, 1879); and *Conseils à l'accouchée et à la jeune mère. Carnet Renaut, publication mensuelle* (Lyons, 1903). Renaut also edited *Travaux du laboratoire d'histologie annexé à la Chaire de médecine du Collège de France*, 3 vols. (Paris, 1875–1877); and *Revue générale d'histologie comprenant l'exposé successif des principales questions d'anatomie générale, de structure, de cytologie, d'histogénèse, d'histophysiologie et de technique histologique* (Paris–Lyons, 1904–1908). His vol. of poetry is *Ombres colorées* (Paris, 1906). He also wrote several articles in A. Dechambre, ed., *Dictionnaire encyclopédique des sciences médicales*— "Dermatoses (anatomie pathologique)," XXVIII, 141–267; "Epithélial, tissu," XXXV, 259–349; "Hémorragies," 4th ser., XIII, 335–475; Nerfs (anatomie)," 2nd ser., XII, 124–181; "Nerveux, système (anatomie)," 2nd ser., XII, 391–495; and "Sang (pathologie)," 3rd. ser., XIII, 501–591.

II. SECONDARY LITERATURE. The most complete biographical sketch is J. Mollard, "Le professeur Renaut," in *Lyon médical*, **127** (1918), 49–59. Shorter unsigned accounts are given in *France médicale. Revue d'études d'histoire de la médecine*, **53** (1906), 334–336, 347–348; *Médecine moderne*, 6th year, no. 53 (3 July 1895), 237; *Revue générale de clinique et de thérapeutique*, **32** (1918), 16; and *Biographisches Lexikon der hervorragenden Ärzte der letzten fünfzig Jahre*, II, 1284–1285. See also P. L. E. M. de Fleury, *Nos grands médecins d'aujourd'hui* (Paris, 1891), 371–376; G. Hayem, "Éloge de J.-L. Renaut, associé national," in *Bulletin de l'Académie de médecine*, **78** (1918), 795–798; and G. Linossier, "Le professeur J. Renaut," in *Paris médical*, 8th year, no. 2 (12 January 1918), iv.

DORA B. WEINER

RENEVIER, EUGÈNE (*b.* Lausanne, Switzerland, 26 March 1831; *d.* Lausanne, 4 May 1906), *geology, paleontology.*

Throughout his life Renevier was a man of enthusiasm and innovative spirit. He possessed the abilities, energy, motivation, and foresight to tackle the great geological challenges of his time. His accomplishments include contributions to the development of international geology as well as to the regional geology of his native Switzerland.

Renevier's mother died while he was young and, following the remarriage of his father, Charles Renevier, Eugène was sent to Stuttgart for his higher education. He studied paleontology under F. J. Pictet de la Rive at Geneva from 1851 to 1853 and continued his studies under Edmond Hébert at Paris in 1854. There he became interested in the nummulitic faunas of the Alps, in particular the partly overturned sequence at the Dent de Morcles–Diablerets massif in the canton of Vaud. In 1856 Renevier taught a course in zoology; and in 1859 he was elected professor of geology at the Académie de Lausanne, which became the University of Lausanne in 1890. As the authority on historical geology, he was relieved then from teaching courses in mineralogy and petrology. Renevier tended to speak over the heads of his audience, especially when treating matters of nomenclature.

Much of Renevier's scholarly work consisted of meticulous studies in stratigraphy, paleontology, and mineralogy. His structure sections across the High Calcareous Alps (Dent de Morcles) served mainly to illustrate the exposed stratigraphy, irrespective of "abnormal contacts"; tectonic conclusions were left for others to draw. Nevertheless, Renevier's exploration and detailed mapping in the High Calcareous Alps and Prealps were of great and permanent value and formed the basis for the later works of Hans Schardt, Maurice Lugeon, and Pierre Termier.

In addition to teaching, Renevier participated in and presided over a number of learned societies and associated working and planning commissions. He was a member of the federal geological commission of the Société Helvétique des Sciences Naturelles and the Commission Géologique du Simplon (1877, 1882), the agency responsible for planning the Simplon 19.7-kilometer rail tunnel, still the longest in the world.

Renevier was a prominent figure in the Swiss academic world, particularly in the French-speaking cantons. He was a founder of the Société Géologique Suisse in 1882 and served as its first president. In 1888 he began publishing the society's journal, *Eclogae geologicae Helvetiae*. Renevier developed the Musée Géologique Cantonale in Lausanne and served as its director for over four decades. He also founded the Missions Romandes, a missionary society in Africa. His Christian faith, however, was emphatically liberal and not in accord with the established local protestant church.

Although Renevier's contributions to the regional geology of Switzerland were outstanding, in retrospect it appears that his greatest achievement lay in his recognition of the need for international standards for geological nomenclature, classification, and graphics, and his subsequent efforts to have such standards established. At the first International Geological Congress (Paris, 1878), he was elected secretary-general of the newly formed Commission Géologique Internationale, which included subcommissions on the unification of classifications and nomenclature and on the unification of graphical procedures. Following Renevier's early proposal of using the array of the solar spectrum, the second International Geological Congress (Bologna, 1881) adopted the *gamme internationale de couleur*. The geologic column and corresponding color code as it appeared then is shown below:

Tertiary	yellow
Cretaceous	green
Jurassic	blue
Liassic	dark blue
Triassic	purple
Carboniferous	gray
Devonian	brown
Silurian and Cambrian	green-blue
Archean	pink

Within each period, darker and lighter tones were used to indicate older and younger stages, respectively. Sedimentary formations were designated by letters from the Latin alphabet in ascending order. As an alternative to the scheme of Latin letters, numerical ordering was accepted, with the lowest number representing the oldest formation. Characters of the Greek alphabet were assigned to different types of igneous rocks. At the second Congress, the assembly also adopted the modern dual classification of stratigraphic units (time units, time-rock units). A special commission was charged with the implementation of the new code on the planned Carte Géologique Internationale d'Europe (on a scale of 1:1,500,000). In 1894 Renevier presided over the sixth International Geological Congress at Zurich. At the following Congress in St. Petersburg (1897), he was elected president of the Commission Internationale de Classification Stratigraphique, which assumed the responsibilities of the previously established subcommissions of the Commission Géologique Internationale.

Renevier was active as rector of the University of Lausanne until his death. With the approach of the fiftieth anniversary of his teaching career, he ruled out personal gifts with the exception of two vertebrate fossils for his beloved museum. Days before the anniversary he died in an accident, apparently caused by his failing eyesight.

BIBLIOGRAPHY

I. ORIGINAL WORKS. Renevier's works include "Description des fossiles du terrain nummilitique supérieur," in *Bulletin de la Société géologique de France*, 2nd ser., **12** (1854), 589–604, written with Edmond Hébert; "Tableaux géologique" (10 stratigraphic tables printed on colored paper), in *Bulletin de la Société vaudoise des* *sciences naturelles*, **12**, no. 70–71 (1873–1874), portfolio; "Tableau des terrains sédimentaires qui représentent les époques de la phase organique," *ibid.*, **13**, no. 72 (1874), 218–252; "Structure géologique du Massif du Simplon," *ibid.*, **15**, no. 79 (1878), 281–304; "Sur l'emploi des couleurs et des termes désignant les subdivisions des terrains," in *Congrès géologique international, Paris, 1878* (Paris, 1880), 67–70; "Commission géologique internationale pour l'unification des procédées graphiques, IIme compte-rendu," in *Bulletin de la Société vaudoise des sciences naturelles*, **17**, no. 85 (1881), 165–188; "Étude géologique sur le nouveau project du tunnel coudé au travers du Simplon," *ibid.*, **19**, no. 89 (1883), 1–27; "Monographie des Hautes-Alpes Vaudoise," in *Matériaux pour la carte géologique de la Suisse*, **16** (1890), including *Carte spéciale* (1:50,000), no. 7 (1875), 563 pp.; "Chronographe géologique (2me édition du tableau des terrains sédimentaires aux couleurs conventionelles admise par les congrès géologiques internationaux)," in *Congrès géologique international, Zurich, 1894* (Lausanne, 1897), with 12 tables in portfolio; and "Commission internationale de classification stratigraphique," in *Eclogae geologicae Helvetiae*, **6** (1899), 35–46.

II. Secondary Literature. A short biographical notice on Renevier is in *Dictionnaire historique et biographique de la Suisse*, V (Neuchâtel, 1930), 443. A fairly complete biographical notice is the *éloge* by Maurice Lugeon, in *Actes de la Société helvétique des sciences naturelles*, **89** (1906), "Nécrologies," lxxxvii–cv, including a bibliography. Additional references are listed in L. Rollier, "Bibliographie géologique de la Suisse pour les années 1770 à 1900," in *Beiträge zur geologischen Karte der Schweiz*, **29** (1907–1908); E. Gogarten and W. Hauswirth, "Geologische Bibliographie der Schweiz 1900–1910," *ibid.*, n.s. **40** (1913); and "1817–1930 Bibliographie der Schweizerischen Naturforschenden Gesellschaft," in *Société helvétique des sciences naturelles* (Bern, 1934), 36, 207–208.

E. de Margerie, in *Critique et géologie*, I (Paris, 1943), 176–178, reviews Renevier's scheme of international geological colors; and E. B. Bailey, in his *Tectonic Essays, Mainly Alpine* (Oxford, 1935), 58–97, sets some of Renevier's works in historical perspective.

John Haller

RENNELL, JAMES (*b.* Upcott, near Chudleigh, Devon, England, 3 December 1742; *d.* London, England, 29 March 1830), *geography*.

Rennell, the son of an artillery officer, entered the Royal Navy in 1756 and served in the East Indies from 1760 to 1763. During this period he prepared charts of several harbors, having learned surveying on the voyage out from England. He then left the navy and in 1764 was appointed surveyor by the East India Company, first making a survey of the Ganges and then a general survey of Bengal. In 1767 the company made him surveyor general. In 1777,

suffering from ill health after wounds received in an affray with a band of fakirs, he returned to England with instructions to prepare a map of India from material at India House, London. He devoted the rest of his life to geographical research, maps, and memoirs.

Rennell was elected fellow of the Royal Society in 1781 and in 1791 was awarded its Copley Medal. He studied the works of classical geographers and wrote a commentary on Herodotus. He also acted as geographical adviser to the African Association, which was founded in 1788, and was rightly regarded as the most eminent British geographer of his period.

In addition to his regional work on Asia and North Africa, Rennell made several important contributions to physical geography. His detailed account of the Ganges, read to the Royal Society in 1781, was drawn on by James Hutton, John Playfair, and Lyell in their geological works. His work on ocean currents, which consisted of various papers from 1793 on and culminated in a posthumous book, *An Investigation of the Currents of the Atlantic Ocean* . . . (1832), was used by many subsequent writers. Humboldt visited Rennell in 1827 and consulted him on the subject of currents, and Rennell's map of the currents of the Atlantic appeared in several well-known atlases.

BIBLIOGRAPHY

I. Original Works. Rennell's principal works are *A Bengal Atlas* (London, 1779); *Memoir of a Map of Hindoostan* (London, 1783; 2nd ed., 1792; 3rd ed., 1793), with a map of all India dated 1782 and published separately; *The Geographical System of Herodotus Examined and Explained* . . . (London, 1800); and *An Investigation of the Currents of the Atlantic Ocean* . . . (London, 1832).

II. Secondary Literature. A detailed account of Rennell and his work in India is R. H. Phillimore, *Historical Records of the Survey of India*, I, (Dehra Dun, India, 1945), 369–378. A well-documented assessment of his geographical work, with bibliographical details and corrections of the errors of earlier authors, is J. N. Baker's "Major James Rennell, 1742–1830, and His Place in the History of Geography," in *The History of Geography* (Oxford, 1963), 130–157.

Joan M. Eyles

REPSOLD, ADOLF (*b.* Hamburg, Germany, 31 August 1806; *d.* Hamburg, 13 March 1871); **REPSOLD, JOHANN ADOLF** (*b.* Hamburg, 3 February 1838; *d.* Hamburg, 1 September 1919); **REPSOLD, JOHANN GEORG** (*b.* Wremen, near Bremerhaven, Germany, 19 September 1770; *d.* Hamburg, 14 January 1830), *instrument making*.

The family name derives from that of a manor house, Hrepesholt, in East Friesland, which is first recorded in connection with the establishment of a monastery in 983. In Johann Georg, Adolf, and Johann Adolf Repsold, the family produced three generations of outstanding designers and builders of astronomical instruments.

The first of these, Johann Georg Repsold, was the third child and eldest son of a minister, and was himself intended for a career in theology. He was more interested in technology, however, and, after attending the Latin school at Stade, in 1788 went to Cuxhaven to study mathematics and technical drawing with Reinhard Woltmann, a pilot on the Elbe River who later became director of the Hamburg waterworks. Repsold succeeded Woltmann as river pilot in 1795; in 1799 he married Eleonore Scharff, the daughter of a captain *(Spritzenmeister)* of the Hamburg fire department, which post he himself assumed in the same year.

At about the same time, Johann Georg Repsold met the Swiss physicist and astronomer Johann Kaspar Horner, who was engaged in measuring the estuaries of the Weser, Elbe, and Eider rivers, and discovered a mutual interest in the design of astronomical instruments. By 1800 Repsold had set up his own machine shop, and in 1803 he constructed for Horner a portable transit instrument, which the latter took with him on a world cruise with Adam Krusenstern. During this period he also made a meridian circle that Gauss purchased in 1815 for the Göttingen observatory; Repsold installed the instrument there in 1818, after making several improvements in it. These alterations, together with some later ones and with some new techniques for smelting optical glass, resulted in part from the technical correspondence that Repsold maintained with Gauss between 1807 and 1821. Gauss also suggested that European surveying methods might be improved through the use of trigonometric signals obtained from reflected sunlight, and for this purpose he built a heliotrope; in 1821 Repsold constructed one of these instruments, according to his own design, and it was subsequently used with considerable success by Heinrich Christian Schumacher.

Concomitant with his scientific work, Johann Georg Repsold continued to serve in the Hamburg fire department, having been appointed *Oberspritzenmeister* in 1808. On 14 January 1830, while directing his men in putting out a major fire, he was struck by a falling beam and fatally injured. He had, during his life, received a number of German and foreign honors, and following his death the Patriotic Society of Hamburg placed a bronze bust of him near the new Hamburg state astronomical observatory, which had itself been established in large part through Repsold's efforts.

Johann Georg Repsold was succeeded as fire captain by his third son, Adolf Repsold, who had served an apprenticeship in his father's workshop and, with his brother Georg, assumed direction of the family instrument business, renamed A. and G. Repsold. Adolf Repsold's first commissions were for a small transit instrument for Bessel and for a similar nine-foot (2.74-meter) instrument for the Edinburgh observatory, which was installed in 1831, the same year in which Repsold also designed a lamp system for the lighthouse on Wangerooge Island. In 1833 and 1834 he received orders for meridian circles for the Hamburg and Pulkovo observatories that obliged him to complete the large circular dividing machine that his father had begun. In 1838 he completed a transit instrument for use in the first vertical circle at Pulkovo; it was made to an innovative design whereby a system of levers was employed to compensate for axial deflection.

In 1836 Karl August von Steinheil came to Hamburg from Munich to assist Adolf Repsold in the manufacture of several standard measuring devices, including a rock-crystal kilogram weight and a glass meterstick, that had been ordered by the Bavarian state government. Steinheil contributed directly to one of the two great innovations that Repsold made during the 1840's—the cylindrical manipulation of the split objective and the apparatus by which scales could be read directly from the objective. It was Steinheil's suggestion that the scales, in the latter instance, be illuminated by means of electrically activated glowing platinum wires, a proposal that resulted in the practical application of the platinum-coil vacuum incandescent lamp that W. R. Grove had invented in 1840. Repsold may have also made use of a technique, patented by Frederick de Moleyns in 1841, to increase the brightness of the platinum coil by the injection of powdered charcoal.

Adolf Repsold also received constant help and encouragement from Bessel, who visited him in 1839 and ordered a meridian circle for the Königsberg observatory, which was installed in 1841, the same year in which Repsold completed an equatorial instrument with a clockwork mechanism for the Christiania (now Oslo) observatory. The latter instrument was novel not only in that it compensated for deviations in the axis of declination by a system of weights, but also in that it incorporated a fine adjustment, which was entirely independent of the clockwork, for the right ascension and a microscopic circular reading dial.

In 1842 Adolf Repsold was engaged in designing a heliometer for Oxford, a project in which he was particularly interested, when his work was interrupted by the great Hamburg fire of that year. He was stimulated to improve the design of the pump used in firefighting operations, and developed a vibrationless vane model that reduced the danger of breaking through the ice of the harbor during the winter months. Although Repsold's vane pump had only a single cog for each wheel, it was more powerful than the rotating reciprocating engine employed until that time. In 1849 he completed the Oxford heliometer.

Adolf Repsold built a new workshop in 1855, in which he manufactured an eight-foot (2.44-meter) refracting telescope for the observatory at Lisbon and an equatorial instrument for the Gotha observatory. From 1859 he was assisted by his eldest son, Johann Adolf Repsold, who had served an apprenticeship in the family laboratory, then worked for a year with C. A. F. Peters at the Altona observatory. In 1862 Johann Adolf Repsold became a partner in the family business; in 1867 another of Adolf Repsold's sons, Oskar, also joined the firm, and its name was changed again, to A. Repsold and Sons. Adolf Repsold simultaneously continued to work in the fire department, and in 1856 was promoted to *Oberspritzenmeister*, the post formerly held by his father. In 1858 he was made head of the central office of the Hamburg fire department, and worked toward increasing its effectiveness until his death from a heart attack in 1871.

Johann Adolf Repsold's own designs contributed significantly to the progress of contemporary astronomy. In 1879 he invented a spring pendulum for regulating timing mechanisms in parallactic mountings, and in 1890 he devised a micrometer that eliminated subjective bias. He also wrote important biographical and technical works on the history of astronomy. He was appointed a member of the board of trustees of the Physikalisch-Technischen Reichsanstalt in Berlin-Charlottenburg in 1887; he was elected to both the Leopoldina and the St. Petersburg Academy of Sciences. He received an honorary doctorate from Göttingen in 1911, and in 1918, shortly before his death, the senate of the city of Hamburg granted him the title of professor, and he was made an honorary member of the Hamburg Geographical Society.

BIBLIOGRAPHY

J. G. Repsold's writings are listed in Poggendorff, II, 607. Secondary literature includes R. Beneke, in *Allgemeine deutsche Biographie*, XXVIII (1889), 233–235; Friedrich Glitza, *Erinnerungen an J. G. Repsold's Leben* (Hamburg,

1870); J. A. Repsold, *Vermehrte Nachrichten über die Familie Repsold, nsbesondere über Joh. Georg Repsold* (Hamburg, 1896; 2nd ed., amended, 1915); and P. Riebesell, "Briefwechsel zwischen C. F. Gauss und J. G. Repsold," in *Mitteilungen der Mathematischen Gesellschaft in Hamburg*, 6, no. 8 (1928), 398–431.

A. Repsold's writings are listed in Poggendorff, II, 608, and III, 1108. Secondary literature is J. A. Repsold, *Nachrichten über Adolf Repsold, für die Familie zusammengestellt* (Hamburg, 1900); and F. Reuleaux, *Theoretische Kinematik. Grundzüge einer Theorie des Maschinenwesens* (Brunswick, 1875), 401–403, and table 8, fig. 9.

J. A. Repsold's writings are listed in Poggendorff, IV, 1232, and V, 1040. They include *Erinnerungen an Hermann Kaufmann's Jugendjahre. Zu seinem 100. Geburtstag* (Munich, 1908); *Zur Geschichte der astronomischen Messwerkzeuge von Purbach bis Reichenbach*, 2 vols. (Leipzig, 1908–1914); *Nachträge zu Geschichte der astronomischen Messwerkzeuge*, 2 vols. (Leipzig, 1908–1914); *Ludwig Friedrichsen. Ein Bild seines Lebens* (Hamburg, 1916); "Heron's Dioptra," in *Astronomische Nachrichten*, 206, no. 4931 (1918), 93–98; "Alte arabische Instrumente," *ibid.*, no. 4935, 125–135; "Über Schattenquadrate," *ibid.*, 135–138; "H. C. Schumacher," *ibid.*, 208, no. 4970–4971 (1918), 17–34; "Landgraf Wilhelm IV von Hessen und seine astronomischen Mitarbeiter," *ibid.*, 209, no. 5005–5006 (1919), 193–210; "Über vorgriechische Messwerkzeuge," *ibid.*, no. 5012, 305–307; "Friedrich Wilhelm Bessel," *ibid.*, 210, no. 5027–5028 (1919), 161–214; and "Über Instrumente aus der Repsold'schen Werkstatt," *ibid.*, 211, no. 5062 (1920), 405–414. An obituary is P. Harzer, "Nachruf auf Johann A. Repsold," in *Astronomische Nachrichten*, 209, no. 5007 (1919), 223–228.

More general information on the family is in L. Ambronn, *Handbuch der astronomischen Instrumentkunde*, 2 vols. (Berlin, 1899): J. G. Repsold, I, 286, 422; II, 513, 968; Adolf Repsold, I, 90; II, 562, 820; J. A. Repsold, II, 958; the family firm, I, 42, 74, 75, 303, 383, 398; II, 512, 553, 556, 560, 573, 582, 631, 653, 671, 677, 767, 865, 912, 930, 990; O. Heckmann, "Die Hamburger Sternwarte," in *Sterne und Weltraum*, 2, no. 1 (1963), 4–6; and L. Loewenherz, "Die Repsold'sche Werkstatt in Hamburg," in *Zeitschrift für Instrumentenkunde*, 7 (1887), 208–215.

Documents, family correspondence, shares of stock in the family firm, and diagrams and design sketches are in the Hamburg Staatsarchiv.

PAUL A. KIRCHVOGEL

RESPIGHI, LORENZO (*b.* Cortemaggiore, near Piacenza, Italy, 7 October 1824; *d.* Rome, Italy, 10 December 1889), *astronomy.*

Respighi studied at Parma and at the University of Bologna, where he took his degree in mathematics in 1847. Two years later he was appointed professor of mechanics and hydraulics at the University of

Bologna. His first works were purely mathematical, such as the well-known memoir on the principles of differential calculus, which Cauchy presented at the Academy of Sciences in Paris. His interest soon turned to astronomy, and in 1855 he succeeded Calandrelli as director of the astronomical observatory at the University of Bologna. Respighi moved easily from mathematics to observation, and he made an excellent determination of the latitude of the Bologna observatory. At the same time he worked on the reductions and discussion of meteorological and magnetic data accumulated by Calandrelli. In 1860 he published an exhaustive study of the comets observed during the years 1814–1843.

At Calandrelli's death in 1866, Respighi was appointed professor of astronomy at the University of Rome and was made director of the Campidoglio observatory. Here he devoted himself mainly to studying solar phenomena. During a three-year period he mapped more than 8,000 prominences, and his systematic solar studies continued for more than fifteen years. Especially important are the studies on the spectra of sunspots. He observed the splitting of the absorption lines, which was later explained by Hale as the result of a Zeeman effect of the magnetic field in the sunspots.

Respighi was the first to use the objective prism properly for the observation of stellar spectra. On 15 February 1869 he was able to show the French physicist Cornu excellent stellar spectra by placing a 12° prism in front of the equatorial telescope of the Campidoglio observatory.

Respighi contributed three catalogs to meridian astronomy: one of 285 stars (1877), one of 1,463 stars (1880), and one of 1,004 stars (1884). In the course of this work he also discussed the aberration of light, performing experiments with a water-filled telescope.

Respighi was a member of several academies, including the Accademia Nazionale dei Lincei, and a foreign member of the Royal Astronomical Society.

BIBLIOGRAPHY

A complete bibliography of Respighi's paper may be found in the obituary by P. Tacchini, in *Memorie della Società degli spettrocopisti italiani*, **18** (1889), 200–203. Accounts of his work are in C. André, G. Rayet, and A. Angot, *L'astronomie pratique et les observatoires*, V, *Observatoires d'Italie* (Paris, 1878), 82, 150; and M. Cimino, *Contributi scientifici dell'Osservatorio astronomico di Roma*, ser. 3, no. 25 (1964).

GUGLIELMO RIGHINI

RETZIUS, ANDERS ADOLF (*b.* Stockholm, Sweden, 13 October 1796; *d.* Stockholm, 18 April 1860), *anatomy, histology, anthropology.*

Retzius holds a distinguished place among nineteenth-century biologists for his contributions to comparative anatomy, histology, and anthropology, and as a pioneer of these disciplines in Scandinavia. He played a prominent role in the development of biological sciences there between 1820 and 1860, especially in the establishment of the Karolinska Institutet in Stockholm, and was one of those who guided Swedish biology and medicine into scientific channels against the strong influence of the speculative *Naturphilosophie* prevalent in many of the German universities.

Retzius' father, Anders Jahan Retzius (1742–1821), was professor of natural history at the University of Lund. One of his major accomplishments was to assemble an important collection of minerals and rocks for the university. Retzius was introduced to natural history, especially zoology, by his father; and when he entered the University of Lund, he came under the influence of Arvid Henrik Florman, professor of anatomy, who initiated him into the methods of dissection and careful observation. In 1816 he spent a year at Copenhagen with the anatomist Ludwig Levin Jacobson, the physicist Hans Christian Oersted, and zoologist J. H. Reinhard. The latter, a disciple of Cuvier, impressed young Retzius with his scholarly lectures on comparative anatomy, illustrated with his own preparations. After his return Retzius finished his medical studies with a dissertation on the anatomy of cartilaginous fishes, especially the dogfish and the ray, *Observationes in anatomiam chondropterygium praecipue squali et rajae generum* (Lund, 1819).

In 1823 Retzius was appointed professor of veterinary science at the Stockholm Veterinary Institution, and in 1824, sponsored by Berzelius, he also became professor of anatomy at the Karolinska Medico-Kirurgiska Institutet. In 1830 he was awarded the additional post of inspector of the Karolinska Institutet and was, besides Berzelius, the strongest personality in its early development. In 1840 he resigned his post at the Veterinary Institution and until his death devoted all his time to the two posts he held at the Karolinska Institutet.

Retzius' scientific work began with comparative anatomy, which he pioneered in Sweden and made a subject of medical training. During the studies for his inaugural dissertation he discovered the interrenal organ of elasmobranch fishes which is—as was shown later—homologous with the adrenal cortex of higher animals. A few years later (1822) he studied in detail a still more primitive vertebrate, the slime eel (*Myxine glutinosa)*, a curious animal placed by previous naturalists in such different groups as mollusks, fishes, and even amphibians. Retzius described in two brief but significant studies (1822 and 1824) its complicated cartilaginous cranium, digestive system, pronephric ducts (the pronephros itself was discovered later by J. Müller), and a gland shown later (also by Müller) to be homologous with the adrenal organ of higher forms. His descriptions of the vascular system and of the small but complex brain were almost complete. He also described the cranial, vagus, trigeminal, facialis, and statoacusticus nerves and the simply constructed internal ear, which has only one semicircular canal. Retzius' work on *Myxine* was a basis for further research by Johannes Müller, who described this creature—one of the few surviving links between vertebrates and the lowest chordates—more completely in a series of treatises (1834–1845) and usually receives credit for all the work on *Myxine*.

Retzius' work on the slime eel was related to his research on *Amphioxus*, the only link between vertebrates and invertebrates. *Amphioxus* had been described by Pallas as *Limax lanceolatus* (a mollusk) in 1774; in the 1830's it was found by O. G. Costa at Naples and by Lovén and Sunderwall on the Swedish coast. Costa recognized the kinship of the animal, which he named *Branchiostoma lubrum*, to the lowest fish; and W. Yarrell in 1836 identified the dorsal cord, named the creature *Amphioxus lanceolatus*, and classed it with the Cyclostomata. Retzius noticed several new features of this animal. He informed Müller of his findings and, because his vision was deteriorating, invited Müller to accompany him to Bohuslän, where they could investigate the living animals in detail. In just twelve days in September 1841 the investigation of *Amphioxus*' morphology was completed, and Müller presented the results to the Berlin Academy of Sciences on 6 December. Investigations of the *Amphioxus*, pursued mainly by A. Kovalevski, had great importance in the development of comparative anatomy and embryology.

In his other early anatomical work, Retzius described, with J. S. Billing, the ciliary and sphenopalatine ganglia in the horse and found the connections between the sympathetic trunk and the cerebrospinal nerves. Through his injection methods he discovered the peripheral canal of the cornea (later called Schlemm's canal) and demonstrated several previously unknown anastomoses in the vascular system. A study of the skeletons of birds in 1824 aroused his interest in the avian respiratory system and in the particular connections of the air sacs with the cavities of the long bones. He also compared avian and reptilian lungs.

In the period 1824–1835 Retzius made several journeys to other parts of Europe and to England, where he met many scientists and participated in several scientific meetings At a memorable meeting of German naturalists and physicians in Berlin in 1828 Retzius invited K. E. von Baer to use a dog to demonstrate the mammalian ovum, his famous discovery which had been made public at the beginning of that year but had not been mentioned to Baer by any other participant at the meeting. In 1833, after a journey to England, France, Germany, and Austria, Retzius attended one of the yearly congresses at Breslau. While there he worked with Purkyně, who introduced him to the use of the microscope and the techniques of preparing tissues for microscopic observations, in particularly hard tissues, bone, and teeth (grinding, decalcification).

This new knowledge and practical experience marked a turning point in Retzius' research, for after his return to Stockholm he began a series of microscopic studies—the most important of which were those of the structure of the teeth of several animal species. His work and that of Purkyně on this topic had a great effect and stimulated others, particularly in England (John Tomes, Owen, Huxley), to study the structure and development of teeth; A. Nasmyth quoted extensively from Retzius' and Purkyně's works. The term "Retzius' striae" for brown parallel lines crossing the enamel prisms bears witness to his precise observations and lasting contributions in this field.

During these studies, about 1840, it became evident that Retzius' eyesight had greatly deteriorated, and he had to abandon microscopic studies. For the last two decades of his life, he turned to gross anatomy, chiefly of the skeletal, circulatory, and nervous systems; to topographical anatomy; and to physical anthropology. The distinction of the pyloric antrum and the pyloric canal and the description, in the fundus of human and rodent stomachs, of the gastric canal, a gutterlike groove allowing a direct passage from cardia to the pylorus, are important original findings. The latter was forgotten, and was not confirmed and accepted until the beginning of the twentieth century. In topographical anatomy Retzius is known for Retzius' cavity, the prevesical space between the symphysis, the bladder, and the anterior abdominal wall that contains loose connective tissue and fat and affords the surgeon access to the bladder without opening the peritoneal cavity. "Retzius' ligament" commemorates his description of this structure, also called the fundiform ligament, on the ventral side of the ankle joint.

Retzius' most important work seems to have been in anthropology, where, following J. F. Blumenbach, he attempted to work out a way of classifying human ethnic groups according to their physical characters. After the discovery of numerous human remains in prehistoric graves in Scandinavia, Retzius noticed during their investigations considerable variation in the shape of the cranium. He extended his research to other European ethnic groups and found that human skulls could be divided, according to the proportion of length to breadth, into long (dolichocephalic) and short (brachycephalic), each race having a constant ratio between the breadth and length (cephalic index). Another division could be made according to the shape of the facial bones: orthognathous and prognathous.

The value of this work is not so much in Retzius' conclusions—in the division of populations according to their craniometric characters—but in his demonstrating the possibility of quantitative expression of different patterns of bodily forms and their mathematical treatment. Craniometric and anthropometric methods were soon widely adopted and developed, and new indexes were introduced. Thus there emerged a new branch of science—physical anthropology.

BIBLIOGRAPHY

I. ORIGINAL WORKS. Most of Retzius' writings are short articles or communications to scientific meetings written in Swedish and presented as factual reports of his findings, without theoretical reasoning or discussion of earlier work. A list of his published works is in S. Lovén, "Anders Adolf Retzius," in *Lefnadsteckningar öfver Svenska Vetenskaps akademiens ledamöter*, **2** (1878–1885), 20–36. Most of his important works were also published in German in *Archiv für Anatomie und Physiologie* (1826–1849) or in *Notizen aus dem Gebiete der Natur- und Heilkunde*. They include "Bemerkungen über den innern Bau der Zähne, mit besonderer Rücksicht auf den in Zahnknochen vorkommenden Röhrenbau, . . . mitgeteilt in Briefen an den Dr. Creplin in Greifswald . . .," in *Archiv für Anatomie, Physiologie, und wissenschaftliche Medizin* (1837), 486–566—extensive quotations in English are in A. Nasmyth, *Researches on the Development, Structure and Diseases of the Teeth*, I, *Historical Introduction* (London, 1839); "Ueber die Schädelformen der Nordbewohner," in *Archiv für Anatomie, Physiologie und wissenschaftliche Medizin* (1845), 84–129; "Bemerkungen über Schädel von Guarani-Indianern aus Brasilien," *ibid.* (1849), 543–553; "Kraniologisches," *ibid.*, 554–582; "Ueber das Ligamentum pelvioprostaticum oder den Apparat, durch welchen die Harnblase, die Prostata und die Harnröhre an der untern Beckenöffnung befestigt sind," *ibid.*, 182–190; and "Ueber die richtige Deutung der Seitenfortsätze an den Rücken- und Lendenwirbeln beim Menschen und bei den Säugethieren," *ibid.*, 593–685. After Retzius' death a collection of anthropological writings was published in Swedish, *Samlade skrifter at etnologiskt innehall* (Stockholm, 1864) and in German *Ethnologische Schriften von Anders Retzius, nach dem Tode des Verfassers gesammelt* (Stockholm, 1864). Later his son Gustaf published another collection, *Skrifter i skilda ämnen, jämte några bref, af Anders Retzius* (Stockholm, 1902), containing a history of the development of anatomy in northern Scandinavia, biographical sketches (mainly of Swedish scientists), and papers of more general interest. This volume includes also Retzius' letters to the Finnish anatomist E. J. Bonsdorff. See also C. M. Fürst, "Arvid Flormans bref till Anders Retzius," in *Lund Univ. Årsskrift*, **6**, no. 5 (1910). W. Haberling included many letters by Retzius in his *Johannes Müller, das Leben des rheinischen Naturforschers auf Grund neuer Quellen und seiner Briefe* (Leipzig, 1924). Other letters were published by V. Kruta in "Anders Retzius und Johannes Ev. Purkyně," in *Lychnos* (1956), 96–131, and (1959), 222–227; and B. Ottow, in *Ein Briefwechsel zwischen Anders Adolf Retzius und Karl Ernst von Baer* (Stockholm, 1963).

II. SECONDARY LITERATURE. Several biographies were published in Swedish, including one by Erik Müller, who had access to Retzius' papers and writings: "Anatomiska institutionen i Stockholm 1756–1910," in *Karolinska Mediko-kirurgiska Institutets historia*, III (Stockholm, 1910), 94–122. A short but comprehensive biography in English is O. Larsell, "Anders Adolf Retzius (1796–1860)," in *Annals of Medical History*, **6** (1924), 16–24.

VLADISLAV KRUTA

RETZIUS, MAGNUS GUSTAF (*b.* Stockholm, Sweden, 17 October 1842; *d.* Stockholm, 21 July 1919), *anatomy, histology, anthropology*.

Retzius came from a famous family of Swedish scientists. His grandfather, Anders Johan Retzius (1742–1821), was professor of natural history at the University of Lund and did distinguished work in chemistry, botany, zoology, mineralogy, and paleontology. His father, Anders Adolf Retzius, professor of anatomy at the Karolinska Institutet, achieved fame as an anatomist and anthropologist. His father's older brother, Magnus Christian Retzius (1795–1871), was a hygienist and professor of obstetrics at the Karolinska Institutet. Anders Retzius remarried in 1835; his second wife, the mother of Gustav Retzius, was Emilia Sofia Wahlberg, sister of the botanist and entomologist Peter Frederik Wahlberg.

After attending the Gymnasium in Stockholm, Retzius began to study medicine at age eighteen, first in Uppsala and later in Stockholm. He received his doctorate from the University of Lund in 1871 and in the same year became a *Dozent* in anatomy at the Karolinska Institutet. In 1862, 1869, and 1872–1873, he traveled in England, Germany, Switzerland, Italy, Belgium, France, Finland, and Russia to increase his knowledge. In 1877 a personal extraordinary professorship of histology was created for Retzius at the Karolinska Institutet. He was promoted to full professor of anatomy in 1889, but he resigned in 1890 to devote himself full time to pure research.

In 1876 Retzius married Anna Wilhelmina Hierta (1841–1924), daughter of the founder of the Stockholm *Aftonbladet*. She was an exceptionally active woman who, among other things, promoted medical education for women. The marriage not only brought Retzius the financial independence he sought for his scientific endeavors; it also provided him with an opportunity to demonstrate his wide-ranging interests in literature as temporary editor of the *Aftonbladet*. His literary efforts included travel descriptions, prize-winning collections of poems, and translations of the poems of Robert Burns into Swedish (1872).

The number and scope of Retzius' scientific publications were unique in his time. He presented the results of his research in more than 300 papers devoted to descriptive macroscopic and microscospic anatomy, comparative anatomy, embryology, anthropology, zoology, botany, and pathological anatomy. Retzius was very concerned with the presentation of his illustrations. Although the format he selected for his publications—large folio volumes—was very costly, it allowed him to furnish a synoptic view of his carefully executed drawings by means of unfolded plates. Most of his papers were in the new series of Biologische Untersuchungen (1890–1914). "These investigations constituted a kind of personal journal in which the editor was alone responsible for the costs, was the sole contributor, and usually the draftsman as well. Within its field this publication was unique. . . . As soon as Retzius had brought out one volume he began work on a new one" (C. M. Fürst, *Biologische Untersuchungen* [Jena, 1921], 7). He wrote in German "because at present the science of anatomy is studied most intensively in Germany, and as a result the terminology is most developed in this language," and "because in order to be of use, such specialized scientific studies must seek to reach a broader audience than works written in Swedish would find" (Retzius, in C. M. Fürst, *Biologische Untersuchungen* [Jena, 1921], 11).

Besides works that were accessible to only a few (predominantly foreign) specialists, Retzius coauthored a series of popular scientific works under the general title *Ur vår tids forskning* (from 1872). His collaborator was his friend Axel Key (1832–1901), a pathologist. The series filled the public's need for reliable information about science.

His research brought Retzius numerous honors, both in Sweden and abroad. The two-volume *Studien in der Anatomie des Nervensystems und des Bindegewebes* (1875–1876) that he wrote with Key won the Prix Montyon of the French Académie des Sciences. Through his election to the Swedish Academy and the Royal Swedish Academy of Sciences, Retzius became a member of the committee that awarded the Nobel prizes in literature and physiology or medicine. He was also an honorary member of the National Academy of Sciences in Washington, the Academy of Natural Sciences of Philadelphia, and the Royal Anthropological Institute. Retzius received honorary doctorates from Bologna (1888), Uppsala (1893), Harvard, Würzburg, and Budapest (1896).

Retzius' did work on the nervous system (central nervous system and its membranes, nerve cells and nerve fibers, sense organs—receptors of the external skin, odor, and taste receptors, the eye and ear); cells and cell division; bones, cartilage, connective tissue, and muscle tissue; the liver and the spleen; the ovum and its coverings; spermatozoa; embryology; anthropology and ethnography; methods in anatomical research; and on such miscellaneous topics as history of science, Swedenborg as anatomist and physiologist of the brain (1903), Linnaeus (1907), and an edition of the letters of Johannes Müller to Anders Retzius.

The bulk of Retzius' writings were devoted to neuroanatomy and neurophysiology. Their direct influence on contemporary work can be seen from the wealth of citations to them in, for example, the publications of Louis Ranvier (*Traité technique d'histologie* [1882]), S. Ramón y Cajal (1899), and K. Gegenbaur (in his comparative vertebrate anatomy [1898]). The development of experimental neurophysiology owed much to Retzius' study of microscopic structure, particularly of the conducting elements of the nerves and their sheaths (1876) but also of the sensory nerve endings. Given the limitations of the light microscope, Retzius advanced this study as far as was possible at the time. Even as late as 1950, when R. Lorente De Nó provoked a debate on the ineffectiveness of the nerve sheaths as diffusion barriers, most researchers found it necessary to refer to the still-authoritative investigations of Key and Retzius. Retzius' application of Golgi's silver nitrate method and, especially, of Paul Ehrlich's methylene blue dyeing method for nerve structures led to a further differentiation of the cellular elements and of the sensitive nerve endings in various classes of animals. On the basis of his own research Retzius became one of the early proponents of the neuron theory.

Further contributions that Retzius made to knowledge of the central nervous system include the description of the *eminentia saccularis* on the *tuber cinereum*, of the *corpus amygdaloideum*, amygdaloid nucleus, and of previously unnoticed convolutions in the rhinencephalon (in honor of his father he named these Retzius' *gyri*). Especially notable, according to Waldeyer, was Retzius' discovery of the significance of the Pacchionian bodies: through them, by means of a valve arrangement, the lymphatic fluids of the brain pass into the sinus of the dura mater.

Among Retzius' papers in anatomy and embryology, the most important were devoted to the process of bone formation (including the demonstration of the mitotic division of the cartilage cells) and to comparative studies. In the latter he dealt with the ape brain, the brains of many other types of animals, the comparative anatomy of the ear labyrinth, the development of the form of the human body during the fetal period, and the structure of spermatozoa (the many forms of which he documented in masterful illustrations).

Retzius' work in anthropology included observations of the Lapps of northern Finland; descriptions of ancient Swedish, Finnish, and Indian skulls; and anthropometric measurements of Swedish conscripts. He attempted to establish a correlation between brain structure and talent. As part of his effort to do this he described the brains of five people endowed with exceptional intellectual gifts: an unnamed statesman, the astronomer Hugo Gylden (1841–1896), the physicist Siljeström (1815–1892), the physiologist Otto Lovén (1835–1904), and the mathematician Sonya Kovalevsky (1850–1891). Finally, he investigated 100 brains to determine if those of men and women differ in macroscopic structure. Among the women's brains he found fewer deviations from the principal type; on the other hand, no arrangement of the furrows and convolutions was found to be more characteristic of one sex than of the other. At the request of his wife, Retzius' own brain was examined at the Karolinska Institutet.

BIBLIOGRAPHY

I. ORIGINAL WORKS. A list of "all the scientific works" of Retzius, grouped by subject, is in n.s. **19** of *Biologische Untersuchungen*, C. M. Fürst, ed. (Jena, 1921), 92–100. The volume also contains an excellent portrait of Retzius and a chronologically ordered synopsis of the papers he published in *Biologische Untersuchungen*, n.s. **1–18** (Jena, 1890–1914), 84–90.

Retzius' books include *Anatomische Untersuchungen. Das Gehörorgan der Knochenfische* (Stockholm, 1872); *Studien in der Anatomie des Nervensystems und des Bindegewebes*, 2 vols. (Stockholm, 1875–1876), written with Axel Key; *Finska kranier jämte några natur- och literaturstudier inom andra områden af finsk anthropologi* (Stockholm, 1878); *Finland i nordiska museet* (Helsinki, 1881); *Das Gehörorgan der Wirbelthiere*, 2 vols. (Stockholm, 1881–1884); *Das menschenhirn. Studien in der makroskopischen Morphologie*, 2 vols. (Stockholm, 1896); *Crania suecica antiqua* (Stockholm, 1899), in Swedish— also in German, with same Latin title (Stockholm, 1900); *Anthropologia suecica* (Stockholm, 1902), written with Carl M. Fürst; and *Cerebra simiarum illustrata. Das Affenhirn in bildlicher Darstellung* (Stockholm, 1906).

Retzius edited *Ethnologische Schriften von Anders Retzius* (Stockholm, 1864), for which he provided a foreword and notes; *Briefe von Johannes Müller an Anders Retzius* (Stockholm, 1900), which covers 1830–1857; and *Skrifter i skilda ämnen jämte några bref, af Anders Retzius* (Stockholm, 1902), for which he provided an introduction.

II. SECONDARY LITERATURE. See C. M. Fürst, "Gustaf Retzius," in *Mannus*, **11–12** (1919–1920), 433–435; and introduction to *Biologische Untersuchungen*, n.s. **19** (Jena, 1921); H. Hofberg, ed., *Svensk biografisk handlexikon*, II (Stockholm, 1906), 329–332; O. Larsell, "Gustaf Retzius 1842–1919," in *Scientific Monthly*, **10** (1920), 559–569; U. Nilsonne and S. Lindmann, "Magnus Gustaf Retzius," in *Svenska män och kvinnor*, VI (Stockholm, 1949), 296–297; E. Nordenskiöld, "Retzius," in V. Söderberg, ed., *Nordisk familjebok*, **16** (1932), 686–687; H. Östberg, "Om Gustaf Retzius och hans verk," in *Nordisk medicin-historisk årsbok* (1971), 200–208; L. Ribbing, "Scandinavian Anthropology in the 20th Century," in H. Lundborg and F. I. Linders, eds., *The Racial Characters of the Swedish Nation* (Uppsala, 1926), 5; H. Spatz, "Die vergleichende Morphologie des Gehirns vor und nach Ludwig Edinger," in *Edinger-Gedenkschrift* (Wiesbaden, 1959); J. H. Talbott, "Magnus Gustaf Retzius (1842–1919)," in *A Biographical History of Medicine* (New York–London, 1970), 934–936; and W. von Waldeyer-Hartz, "Gustaf Retzius," in *Deutsche medizinische Wochenschrift*, **45** (1919), 942–943.

GERHARD RUDOLPH

REULEAUX, FRANZ (*b.* Eschweiler, Germany, 30 September 1829; *d.* Berlin, Germany, 20 August 1905), *mechanical engineering.*

Reuleaux was significant in two respects. In engineering he is regarded as the founder of modern kinematics. In two highly original books on that subject he proposed a system of analyzing and classifying machinery that was philosophical in scope and that has proved remarkably durable. More generally, by virtue of a forceful and outgoing personality and a talent for publicity, he was recognized in Germany as the spokesman for engineers and for modern technology in general during the first two decades of the Second Empire, a period of rapid industrial growth.

Reuleaux was the fourth son of Johann Josef Reuleaux, one of the first steam engine manufacturers in the Rhineland and himself the son of a master engineer (*Kunstmeister*) from Liège. The elder Reuleaux died when Franz was a child, and in 1839 his widow moved the family to Koblenz, where Franz attended school through the intermediate grades. In 1844 his stepfather, named Scholl, chief engineer of an ironworks, apprenticed the boy to a Koblenz iron foundry and machine shop, while himself instructing him in theoretical subjects. In 1846 Reuleaux joined his father's factory at Eschweiler, now run by an uncle, first as a draftsman and later (1848) as a field engineer. In 1850 he enrolled at the Karlsruhe Polytechnische Schule, where he studied for two years under Redtenbacher, one of the foremost engineering teachers of the day. In 1852 and 1853 Reuleaux studied at the universities of Berlin and Bonn, where, besides attending lectures on science, mathematics, and philosophy, he wrote, in collaboration with a fellow student, C. L. Moll, a brief treatise on the strength of materials and a long handbook of machine design. Since the latter contained materials from Karlsruhe lectures, he was accused of plagiarism by Redtenbacher, his former teacher; most leading engineers, however, judged the work useful and forgave him. Reuleaux spent the years 1854 and 1855 in Cologne, first as the manager of a factory and then as an independent consulting engineer.

In 1856 Reuleaux was named professor of machine design at the newly founded Swiss Federal Polytechnical Institute. After eight years in Zurich, he became professor of mechanical engineering at the Gewerbe Institut in Berlin in 1864; he served as director of that school from 1867 until, after merging with the Berlin Bauakademie, it was reorganized in 1879 into the Technische Hochschule at Charlottenburg. He was elected rector of the institution for several terms. He retired from teaching in 1896 and remained in Berlin until his death.

Reuleaux's basic field was machine design. Besides a great number of minor publications on all aspects of this field, he wrote a handbook of machine design, *Der Constructeur*, highly popular at the time but later denounced as "a technological recipe book" (R. von Mises, 1929). Within the general field of machine design he discovered his own specialty in kinematics, with which his name is now permanently linked. Reuleaux had planned an exhaustive treatment of kinematics. The highly successful first volume on theoretical kinematics appeared in 1875; but a second

volume, on the more technical and practical aspects, was published when Reuleaux's influence was waning (1900) and received less notice. A projected third volume was never written.

The first volume, which had the greatest impact, is not an engineering book in the modern sense. Reuleaux saw its strongest points in logic and philosophy. Its subtitle, *Grundzüge einer Theorie des Maschinenwesens*, suggests the breadth of its ambitions. The volume consists of three parts. The first, in the tradition of Hachette, Borgnis, and Babbage, provides the logical and conceptual tools for analyzing and classifying machinery; prominent among them are the concepts of the kinematic pair and the kinematic chain, and a symbolic notation which Reuleaux hoped to employ algebraically in synthesizing mechanisms. The second part was devoted to the application of this conceptual apparatus to the task of "kinematic analysis," which consisted of breaking down given machines into chains of abstract components in order to identify mechanisms that were kinematically equivalent. The third and shortest section is hardly more than a veiled admission of failure in attaining its declared objective of "kinematic synthesis" by means of systematic-deductive methods.

Although falling short of its objectives, Reuleaux's kinematics was studied eagerly. It not only led to the cultivation of kinematics as an independent discipline, but also became particularly popular among non-technical readers. When its fatal weakness, a total disregard for dynamic phenomena, was recognized, the reaction was strong enough to cause kinematics to be struck from the Berlin curriculum after Reuleaux's retirement. Reuleaux's system of classifying mechanisms, however, proved definitive; and modern kinematics acknowledges a large debt to him.

Reuleaux affected German engineering and industry by means other than his books. He corresponded with leading industrialists, such as Eugen Langen, who valued his advice; he had an important role in the founding of the Mannesmann steelworks; he worked hard for the passage of comprehensive German patent legislation. Of special interest to him were world exhibitions. He was a member of the juries of the expositions at Paris (1867), Vienna (1873), and Philadelphia (1876); and he also served as the German commissioner at Philadelphia, Sydney (1879), and Melbourne (1881).

Reuleaux was a prolific writer, and covered a broad spectrum. He wrote not only on all aspects of engineering but also on technology and civilization (translation in the 1890 *Report of the Board of Regents of the Smithsonian Institution*), on the purity of the German language, and on technology and art; he also published a number of travel journals and a translation into German, in the original meter, of Longfellow's *Hiawatha*.

After enjoying great prestige in the 1870's and 1880's, Reuleaux saw his influence decline. A younger and scientifically more refined generation of engineers recognized the weaknesses of his technical teachings, weaknesses amplified by the all-too-broad scope of his interests and his love of bold formulation. The dominance of kinematics had to give way to a fuller consideration of dynamic problems. A controversy arose among academic engineers, the opposing parties being identified by their espousal of theory and of practice. The proponents of "practice" eventually gained control of most German institutes of technology, while Reuleaux, a prominent "theorist," spent his later years, especially after retirement, in increasing isolation.

BIBLIOGRAPHY

I. ORIGINAL WORKS. Bibliographies of Reuleaux's works are in Poggendorff, III, 1111; IV, 1234–1235; V, 1040; and VI, 2157; and in Carl Weihe, *Franz Reuleaux und seine Kinematik* (Berlin, 1925). His books include *Festigkeit der Materialien* (Brunswick, 1853), written with C. L. Moll; *Constructions-Lehre für den Maschinenbau* (Brunswick, 1854), written with C. L. Moll; *Construction und Berechnung der für den Maschinenbau wichtigsten Federarten* (Winterthur, 1857); *Der Constructeur* (Brunswick, 1861, 1865, 1872, 1889), English trans. by Henry H. Suplee as *The Constructor: A Handbook of Machine Design* (New York, 1893), also in French, Swedish, and Russian; *Theoretische Kinematik: Grundzüge einer Theorie des Maschinenwesens* (Brunswick, 1875), English trans. by Alexander B. W. Kennedy as *The Kinematics of Machinery: Outlines of a Theory of Machines* (London, 1876; repr. New York, 1963), also in French and Italian; *Briefe aus Philadelphia* (Brunswick, 1877); and *Theoretische Kinematik*, II, *Die praktischen Beziehungen der Kinematik zu Geometrie und Mechanik* (Brunswick, 1900).

II. SECONDARY LITERATURE. See Eugene S. Ferguson, "Kinematics of Mechanisms From the Time of Watt," in *Contributions from the Museum of History and Technology. U. S. National Museum Bulletin*, **228**, no. 27 (1962), 185–230; Wilhelm Hartmann, "F. Reuleaux," in *Zeitschrift des Vereins deutscher Ingenieure*, **49** (1905), 1481–1482; and "Gedenkrede bei der Enthüllung des Denkmals für Franz Reuleaux," *ibid.*, **57** (1913), 162–169; Heinrich Koch, "Franz Reuleaux und die Gründung der Mannesmannröhren-Werke," in *Tradition*, **5** (1960), 259–270; Alexander Lang, "Franz Reuleaux und die Maschinenwissenschaft," in *Zeitschrift für Socialwissenschaft*, **8** (1905), 804–809; Otto Mayr, "Symbolsprachen für mechanische Systeme im 19. Jahrhundert," in *Technikgeschichte*, **35** (1968), 223–240; R. von Mises, "Franz Reuleaux," in *Zeitschrift für angewandte Mathematik und Mechanik*, **9** (1929), 519; Wilhelm Oechsli, *Geschichte der Gründung des Eidg. Polytechnikums mit einer Übersicht seiner Entwicklung 1855–1905* (Frauenfeld, 1905), 179–182; Theodor Pöschl, "Franz Reuleaux," *Karlsruher akademische Reden*, no. 4 (1929); and Carl Weihe, *Franz Reuleaux und seine Kinematik* (Berlin, 1925); and "Franz Reuleaux und die Grundlagen seiner Kinematik," in *Abhandlungen und Berichte des Deutschen Museums*, **14** (1942), 83–104.

OTTO MAYR

REUSS, AUGUST EMANUEL (*b.* Bílina [German Bilin], Bohemia [now Czechoslovakia], 8 July 1811; *d.* Vienna, Austria, 26 November 1873), *micropaleontology, paleontology, stratigraphy*.

Reuss's father, Franz Ambrosius Reuss, was a physician at Bilin, a spa in northern Bohemia. He was also a serious student of geology and mineralogy who contributed considerably to the geological exploration of Bohemia. He tutored his son in these subjects and also in the standard curriculum.

Reuss completed his secondary schooling at the University of Prague, where he studied philosophy, science, and medicine from 1825 to 1833, receiving his M.D. degree on 10 November 1833. Following his father's death, Reuss took over the medical practice in Bilin. In his free time, however, he studied mineralogy, geology, and paleontology; and as early as 1837 he was able to report his first mineralogical and geological findings in northern Bohemia to the Vereinigung der Naturforscher in Prague. These findings inspired him to continue his research, and in 1849 he accepted the chair of mineralogy of the University of Prague. Thus he could give up his medical practice and devote himself entirely to science.

In 1863 Reuss was appointed professor of mineralogy at the University of Vienna, where he remained until his death. He accepted this post (as he had done at Prague) even though his principal areas of research were paleontology and stratigraphy—until the middle of the nineteenth century there were no chairs in paleontology, and geology and mineralogy still constituted a single field.

On 16 February 1841 Reuss married Anna Schubert, who also came from Bilin; they had two sons and three daughters. During the last years of his life, Reuss was forced because of ill health to curtail his academic activities. He died of a lung hemorrhage.

Reuss's scientific achievements brought him many honors, including honorary doctorates from the universities of Breslau and Vienna. He was rector of the University of Prague in 1859–1860, and he was knighted by the Emperor Franz Joseph in 1870.

Thereafter he published under the name of A. E. Ritter von Reuss.

In his first publications Reuss treated topics related to the geology and mineralogy of the region around Bilin; in fact, some sixty titles, just over half of his total production, were devoted to this, his native region. His interests soon turned to paleontology, however, the subject to which he made his greatest contribution. In the monograph *Die Versteinerungen der böhmischen Kreide-Formation* (1845–1846) he described and illustrated 776 species in fifty-one tables, arranged according to their stratigraphic occurrence. Even in this early work, which is still useful, he devoted particular attention to the micro-fossils (particularly Foraminifera and Ostracoda).

Reuss's studies covered a wide variety of topics: the stratigraphy of the Silurian near Prague; coprolites, Foraminifera, and Ostracoda of the Permian in Bohemia and Germany; corals and crabs of the Alpine Triassic; Bryozoa, corals, sponges, crabs, and snails of the Polish, Moravian, and Alpine Jurassic; and fishes and many invertebrate groups of the Cretaceous, not only of Bohemia but also from deposits in Poland, the Austrian Alps, and northern Germany. His studies of the Tertiary invertebrates were similarly wide-ranging, covering Foraminifera, corals, Bryozoa, mollusks, Ostracoda, and decapods, and including stratigraphic observations of the Tertiary in Bohemia, the Vienna and Mainz basins, the Austrian Alps, northern and central Germany, Galicia, Bessarabia, Hungary, and the Antwerp region. These investigations of marine invertebrates and the associated stratigraphic problems were supplemented by research on the Tertiary freshwater deposits of Bohemia and the fauna contained in them, especially mollusks.

Reuss was most interested in the Foraminifera of the Cretaceous and, particularly, of the Tertiary. In his monographs he enlarged a number of genera and groups (Peneroplidae, Lagenidae, and *Ataxophragmium*) and provided numerous descriptions of Foraminifera from various levels of the Cretaceous and Tertiary in northern and central Germany, Austria, and Belgium. His work contributed to an understanding of their taxonomy and biostratigraphy and exhibited many facts relating to their paleoecology and biogeography.

The classification of Foraminifera that Reuss proposed in 1861–1862 greatly influenced research on this order for decades, although his scheme is now considered to be an artificial one. He divided the fossil Foraminifera into twenty-one families encompassing 109 genera, arranged in accordance with the following criteria: the presence or absence of pores in the shell walls (Perforata, Imperforata); the composi-tion and structure of the casing shell; the shell form; and disposition of the chambers.

Reuss was among the first to practice "applied micropaleontology"—an approach that has played an important role since the 1920's in petroleum geology and in other branches of applied biostratigraphy. He was also the first to use Foraminifera to determine the age of Tertiary deposits in northern and central Germany, regions where macrofossils are rare.

Reuss's publications, especially those on the Foraminifera, are noted for their clarity and exactness of description and are still indispensable in the study of micropaleontology and other branches of paleontology and of biostratigraphy.

BIBLIOGRAPHY

I. ORIGINAL WORKS. A partial bibliography of Reuss's works is given in Poggendorff, III, 1112. His major works include *Geognostische Skizzen aus Böhmen. I. Die Umgebung von Teplitz und Bilin* (Prague–Litoměřice–Teplitz, 1840); *Die Versteinerungen der böhmischen Kreide-Formation* (Stuttgart, 1845–1846); "Neue Foraminiferen aus den Schichten des österreichischen Tertiärbeckens," in *Denkschriften der Akademie der Wissenschaften*, **1** (1849), 365–390, with 6 plates; "Beiträge zur Charakteristik der Kreideschichten in den Ostalpen," *ibid.*, **7** (1854), 1–156, with 31 tables; "Zur Kenntniss fossiler Krabben," *ibid.*, **17** (1859), 90 pp., with 24 plates; "Die marinen Tertiärschichten Böhmens und ihre Versteinerungen," in *Sitzungsberichte der Akademie der Wissenschaften in Wien*, **39** (1860), 207–288, with 8 plates; "Die fossilen Mollusken der tertiären Süsswasserkalke in Böhmen," *ibid.*, **42** (1860), 55–85, with 3 plates; "Entwurf einer systematischen Zusammenstellung der Foraminiferen," *ibid.*, **43** (1861); "Die Foraminiferen des norddeutschen Hils und Gaults," *ibid.*, **46** (1863), 5–100, with 13 plates; and "Die Foraminiferenfamilie der Lageniden," *ibid.*, **44** (1862), 308–342, with plates 1–7.

Subsequent writings are "Ueber die Paragenese der auf den Erzgängen von Přibram einbrechenden Mineralien," *ibid.*, **47** (1863), 13–76; "Beiträge zur Kenntniss der tertiären Foraminiferenfauna," *ibid.*, **48** (1863), 36–71, with 8 plates; *Kurze Übersicht der geognostischen Verhältnisse Böhmens* (Prague, 1854), with maps; "Die Bryozoen, Anthozoen und Spongiarien des braunen Juras von Balin bei Krakau," in *Denkschriften der Akademie der Wissenschaften*, **26** (1866), with 4 plates; "Beiträge zur Charakteristik der Tertiärschichten des nördlichen und mittleren Deutschlands," in *Sitzungsberichte der Akademie der Wissenschaften in Wien*, **18** (1855), 197–272, with plates 1–12; "Die fossile Fauna der Steinsalzablagerungen von Wieliczka in Galizien," *ibid.*, **55** (1866), with 8 plates; "Die fossilen Mollusken des Tertiärbeckens von Wien. Nr. 9, 10. Bivalven," in *Abhandlungen der Königlichen Kaiserlichen Geologischen Reichsanstalt*, **4** (1870), 431–479, plates 68–85; "Die fossilen Korallen des österreichisch-ungarischen Miozäns," in *Denkschriften der Akademie der Wissenschaften*, **31** (1871), with 21 plates; "Paläontologische Studien über die älteren Tertiärschichten der Alpen, III. Theil," *ibid.*, **33** (1872); "Die Foraminiferen, Bryozoen und Ostracoden des mittleren und oberen Pläner" in Geinitz, "Das Elbthal-Gebirge in Sachsen," in *Paläontographica. Beiträge zur Naturgeschichte der Vorzeit*, **20** (1874), 73–157, plates 20–28; and "Die Bryozoen des österreichisch-ungarischen Miozäns," in *Denkschriften der Akademie der Wissenschaften*, **33** (1874).

II. SECONDARY LITERATURE. On Reuss and his work, see H. Bartenstein, in *Paläontologische Zeitschrift*, **35** (1961), 248–250, with portrait; H. B. Geinitz, in *Leopoldina*, **9** (1874), 67–72, with bibliography; G. C. Laube, in *Mitteilungen des Vereins zur Geschichte der Deutschen in Böhmen*, **12**, no. 5 (1874), 193–205; A. Schrötter Ritter von Cristelli, in *Almanach der kaiserlichen Akademie der Wissenschaften*, **24** (1874), 129–151; and a bibliography in *Quarterly Journal of the Geological Society of London*, **30** (1874), xlvii–xlviii.

HEINZ TOBIEN
W. S. DALLAS

REUSS, FRANZ AMBROSIUS (*b*. Prague, Czechoslovakia, 3 October 1761; *d*. Bilina, Czechoslovakia, 9 September 1830), *mineralogy, geology, balneology.*

Reuss was the son of a tailor who had moved from southern Baden to Prague. He received his early education in that city, then entered its university, where he studied philosophy, natural science, and, afterwards, medicine. He took the M.D. on 4 October 1783 and established a practice in Prague; during the same period he also conducted mineralogical and geological investigations and went to Freiberg to hear Werner lecture at the Bergakademie. In 1784 Reuss became physician to Prince Lobkowitz and moved to the northwestern Bohemian city of Bilin, where, except for many field trips throughout Bohemia, he spent the rest of his life. On 3 September 1797 he married Katharina Scheithauer, the daughter of the manager of an estate in the vicinity; they had four daughters and four sons, of whom one, August Emanuel, became a well-known micropaleontologist and stratigrapher.

Bilin was already famous as a spa, and Reuss, in addition to his medical duties, supervised the exploitation of the mineral springs and mines belonging to his patron. In this connection he continued his mineralogical and geological researches and concurrently began an investigation of the spas of northern Bohemia. He was uniquely qualified for such an undertaking, since he was able to bring geological, petrological, chemical, and physical techniques to his descriptions of mineral springs, and to employ his knowledge of medicine in elucidating their therapeutic properties. His first balneological work, published in 1788, was devoted to the mineral springs of Bilin and contains the first scientific description of any spa, including the geological conditions that give rise to the springs, together with the physical and chemical characteristics and therapeutic effects of their waters. As a physician, Reuss suggested that patients ought actually to come to the springs for water cures (mineral waters were already extensively bottled and sold) and that spas should arrange facilities for bathing.

Reuss became a recognized expert in balneology, and visited a number of other Bohemian mineral springs. In 1818 he published a study of Marienbad, which aroused considerable interest. At about the same time he became acquainted with the work of Berzelius, who had also studied mineral springs, and began to advocate the use of Berzelius' analytical methods, which he wished to apply to the Bohemian springs.

Reuss's geological and mineralogical publications comprised works on the areas around Bilin and Carlsbad, on the Mittelgebirge of northwest Bohemia, and on the regions surrounding Leitmeritz, Bunzlau, and Kaurzim. His *Charakteristik der basaltischen Hornblende* was published in 1798 and his *Mineralogische und bergmännische Bemerkungen über Böhmen* appeared three years later. His most famous work, however, was the *Lehrbuch der Mineralogie*, of which eight volumes (six on mineralogy and two on geology) were published between 1801 and 1806. This text was of particular importance because it contained the most complete exposition of Werner's ideas, since Werner himself published little concerning his own neptunist theories. In it Reuss proved himself Werner's disciple, stating, for example, that basalt is of aquatic, rather than volcanic, origin. In general, Reuss's geological and mineralogical researches may be considered to have been the first comprehensive and reliable precursors for later work on the Tertiary sediments and volcanites of northern Bohemia, together with their pre-Tertiary basement.

Reuss carried out a number of official functions in Bilin in addition to his scientific and medical activities. He was supervisor of schools, and administered programs for vaccination and for public care of the sick and the elderly. His publications made him widely known in scientific circles; he corresponded with Goethe and Humboldt, and was elected to the Göttingen Academy of Sciences in 1800. In 1806 he was appointed *Bergrat*. He died of exhaustion attendant upon an abdominal injury.

BIBLIOGRAPHY

I. ORIGINAL WORKS. Reuss's medical writings include *Versuch einer Einleitung in die allgemeine Pathologie der Nerven* (Prague, 1788). His geological and mineralogical works include *Orographie des nordwestlichen Mittelgebirges in Böhmen, ein Beitrag zur Beantwortung der Frage: Ist der Basalt vulkanisch oder nicht?* (Dresden, 1790); *Mineralogische Geographie von Böhmen, I. Mineralogische Beschreibung des Leitmeritzer Kreises* (Dresden, 1793); *II. Mineralogische Beschreibung des Bunzlauer Kreises* (Dresden, 1797); *Mineralogische und bergmännische Bemerkungen über Böhmen. Mit einer Ansicht des Schlosses Rothenhaus* (Berlin, 1801); and *Lehrbuch der Mineralogie nach Karsten's mineralogischer Tabelle*, 8 vols. (Leipzig, 1801–1806).

His balneological works include *Naturgeschichte des Biliner Sauerbrunnens in Böhmen* (Prague, 1788); *Das Saidschützer Bitterwasser* (Prague, 1791); *Chemisch-medicinische Beschreibung des Franzbades oder Egerbrunnens nebst einer Literaturgeschichte der Quelle* (Dresden, 1794); *Die Gartenquelle zu Teplitz in chemisch und medicinischer Hinsicht untersucht* (Dresden, 1797); *Die Mineralquellen zu Bilin in Böhmen* (Vienna, 1808); *Das Marienbad bei Auschowitz auf der Herrschaft Tepl* (Prague, 1818); *Taschenbuch für die Badegäste zu Teplitz. Eine vollständige Beschreibung dieses Heilortes und seiner Umgebungen in topographischer, pittoresker, geschichtlicher, geognostischer und medicinischer Hinsicht* (Teplitz, 1823); and *Der Natron-Säuerling bei Bilin* (Prague, 1828).

II. SECONDARY LITERATURE. See A. C. P. Callisen, in *Medicinisches Schriftstellerlexicon*, XVI (Copenhagen, 1833), 3–7; vol. 31 (Copenhagen, 1843), 421; Poggendorff, II, 614; *Biographisches Lexikon des Kaiserthums Oesterreich*, XXV (Vienna, 1873), 354–356; *Allgemeine deutsche Biographie*, XXVIII, 307–308; F. Pemsel, *Dr. Franz Ambros Reuss. Zu seinem 100 Todestage am 9. September 1930* (Bilin, 1930); and K. A. von Zittel, *Geschichte der Geologie und Paläontologie bis Ende des 19. Jahrhunderts* (Munich-Leipzig, 1899), *passim*.

HEINZ TOBIEN

REY, ABEL (*b.* Châlon-sur-Saône, France, 29 December 1873; *d.* Paris, France, 13 January 1940), *philosophy*.

Rey studied classics at the *lycée* in Marseilles and in Paris, at the Lycée Louis-le-Grand and the Sorbonne. He won his *licence* in law, the *agrégation* in philosophy, and took the courses leading to a *licence ès sciences*. He studied philosophy and its history under Victor Brochard and Émile Boutroux and mathematics and the history of science under Émile Picard and Paul Tannery. He also shared the esthetic ideals of Gabriel Séailles and attended the lectures of Henri Poincaré. In addition he worked in the laboratories of Edmond Bouty and Lippmann

at a time when the alliance of philosophy and science was still a daring novelty in French academic life.

Rey taught philosophy at the *lycée* of Bourg-en-Bresse and then at Beauvais. The topics of his two dissertations for the *doctorat ès lettres* reflected his training in philosophy and science: *La théorie de la physique chez les physiciens contemporains* and *L'énergétique et le mécanisme au point de vue de la théorie de la connaissance* (Paris, 1907). In 1908 he was appointed to the Faculty of Letters of Dijon, where he established an experimental psychology laboratory equipped with the most advanced apparatus. The laboratory was the outcome of long reflection on psychology, which Rey had published in 1903 in *Leçons élémentaires de psychologie et de philosophie*. This work was accompanied by *Éléments de philosophie scientifique et morale* (1903) and *Leçons de morale fondées sur l'histoire des moeurs et des institutions* (1905), both of which were designed for use in the classics curriculum. He also wrote the chapter on invention for the first edition of Georges Dumas's *Traité de psychologie* (1923). In 1909 Rey published *La philosophie moderne*, which enjoyed a large audience, "curious about a philosophic panorama modernized by epistemology"—in the words of his student and biographer Pierre Ducassé.

In 1919, following the interruption of his research by World War I, Rey was named professor of the history and philosophy of science at the Sorbonne. In 1922 he published *Le retour éternel et la philosophie de la physique*, which contrasted scientific theories with "an intuition that has always recurred [in philosophy] from the Greeks until Nietzsche." In 1932 Rey was named first director of the new Institut d'Histoire des Sciences et des Techniques at the University of Paris. This post, which he held until his death, made him responsible for the publication of the first four volumes of *Thalès*, containing reports on the Institute's activities. He also supervised for the series Actualités Scientifiques et Industrielles, the collection of texts and translations "Pour Servir à l'Histoire de la Pensée Moderne" and the "Exposés d'Histoire et de Philosophie des Sciences."

Rey was associated with the Centre International de Synthèse from its beginning, and he organized its Section des Sciences de la Nature before becoming, in 1931, director of the Section de Synthèse Générale. In this latter capacity he collaborated with Henri Berr, Lucien Febvre, and Paul Langevin on the *Revue de synthèse* and the journal *Science*. A friend of Aldo Mieli, Hélène Metzger, and Pierre Brunet, he was elected one of the first members of the Académie Internationale d'Histoire des Sciences. Febvre invited Rey to collaborate on the *Encyclopédie française*, and with A. Meillet and P. Montel he edited the first volume, *L'outillage mental*, to which he contributed the opening pages: "De la pensée primitive à la pensée actuelle." Henri Berr, impressed by his skill in synthesizing, entrusted him with the four volumes of the series "Évolution de l'Humanité" that were devoted to the role of Greece in the origins of scientific thought: *La science dans l'antiquité*, I, *La science orientale avant les Grecs* (1930); II, *La jeunesse de la science grecque* (1933); III, *La maturité de la pensée scientifique en Grèce* (1939); IV, *L'apogée de la science technique grecque:* 1, *Les sciences de la nature et de l'homme, les mathématiques, d'Hippocrate à Platon* (1946); and 2, *L'essor de la mathématique* (1948).

A philosopher, Rey was fascinated by science. Free of dogmatism, he admired the rigor of Kant, the sincerity of Renouvier, and the intuitions of Comte. A remarkable teacher, he was sensitive to the strivings of individuals and the fate of society. In his thought a deep-rooted positivism was tempered by reason and sensitivity. As an epistemologist he was close to Gaston Milhaud, Émile Meyerson, and Léon Brunschvicg; but he maintained a personal orientation the governing idea of which is summarized in a remark he often made: "Real scientific truth lies in its historical curve. It never lies on one point of the curve."

BIBLIOGRAPHY

I. ORIGINAL WORKS. Rey's most important publications are cited in the text. He also contributed prefaces and articles to many periodicals, principally *Scientia, Revue de métaphysique et de morale, Revue philosophique de la France et de l'étranger*, and *Revue de synthèse*.

II. SECONDARY LITERATURE. See Léon Brunschvicg, "Abel Rey," in *Thalès*, **4**, for 1937–1939 (1940), 7–8; and Pierre Ducassé, "La vie et l'oeuvre d'Abel Rey (1873–1940)," in *Annales de l'Université de Paris*, **15**, no. 2 (Apr.–June 1940), 157–164.

SUZANNE DELORME

REY, JEAN (*b.* Le Bugue, France, *ca.* 1582/1583; *d.* 1645 or after), *chemistry*.

After graduating M.A. at Montaubin, Rey studied medicine at Montpellier, obtaining an M.D. in 1609. Few details are known of his life as a physician, except that he seems to have been highly regarded. His fame rests solely on his *Essays de Jean Rey docteur en médecine. Sur la recherche de la cause pour laquelle l'estain & le plomb augmentent de poids quand on les calcine* (Bazas, 1630), a reply to apothecary Pierre Brun's request for an explanation of why tin

and lead increased in weight when heated. Much of the *Essays* consists of argument based on "reason" and on analogy. For instance, one support for Rey's principal tenet—that the four "elements" of matter (earth, air, fire, and water) possessed the property of heaviness—was the ready separation, on standing, of a mixture of black enamel, solution of potassium acid tartrate, oil of turpentine, and aqua vitae, which, Rey said, represented the four elements.

The "heaviness" of air explained the increased weight on calcination; it resulted from the attachment of air that had been rendered denser, heavier, and in some measure adhesive by the vehement and long-continued heat of the furnace. Rey's only experiment appears in the sections of the *Essays* dismissing other possible explanations for the weight increase. He excluded furnace carbon partly on the ground that he obtained a weight increase when calcining tin on molten iron at his brother's ironworks.

The fascination of the *Essays* lies in its succinctness; and the matrix of ideas presented—at a time of far-reaching changes in science—appeals to reason, observation, and experiment, as well as to the skepticism about earlier writings. The *Essays* created some contemporary reaction, but its intrinsic interest lies in its anticipation of Lavoisier's recognition in 1772 that calcination involves combination with air. In 1775 Pierre Bayen drew attention to the *Essays*, and Lavoisier initially believed the work to be a forgery. Later, however, he spoke of it with admiration.

BIBLIOGRAPHY

There have been a number of reprints of the *Essays* since the first (Paris, 1777), with additional notes by Nicolas Gobet. The most recent, *The Essays of Jean Rey* (London, 1951), has an extensive introduction and full notes by Douglas McKie. McKie includes not only all the details available on Rey's life (from the studies of Gabriel Lafon and Henri Teulié) and on the impact of the *Essays*, but also details of its printing history. See also D. McKie, in *Ambix*, **6** (1957–1958), 136–139.

A full account of the work of Rey, with extensive bibliography, is in J. R. Partington, *A History of Chemistry*, II (London, 1961), 631–636.

J. K. CRELLIN

REYE, THEODOR (*b.* Ritzebüttel, near Cuxhaven, Germany, 20 June 1838; *d.* Würzburg, Germany, 2 July 1919), *mathematics*.

After schooling in Hamburg, Reye studied mechanical engineering and then mathematical physics at Hannover, Zurich, and Göttingen. He received his doctorate at Göttingen in 1861 with a dissertation on gas dynamics. After qualifying as lecturer at Zurich in 1863, he remained there until 1870 as a *Privatdozent* in mathematical physics. Following a short stay in Aachen came his most productive years 1872–1909, when he was professor of geometry at the University of Strasbourg. He remained in Strasbourg until after World War I, when he moved to Würzburg.

In his younger years Reye published works on physics and meteorology—for example, a book on cyclones (1872). The two-volume first edition of his *Geometrie der Lage* appeared in 1866 and 1868. He remained faithful throughout his life to the synthetic geometry presented in this work. His interest in geometry had been stimulated by analytical mechanics, and Culmann, the founder of graphic statics, had drawn his attention to Staudt's works on geometry. Staudt's books were considered very difficult to read; Reye's *Geometrie der Lage*, the fifth edition of which appeared in 1923, was easily comprehended.

Reye treated in detail the theory of conics and quadrics and of their linear systems, that of third-degree surfaces and some of the fourth degree, as well as many quadratic congruences and aggregates taken from line geometry. He was one of the leading geometers of his time, and he published a great deal on synthetic geometry. His name is linked to the axial complex of a second-degree surface, and he generalized the polarity theory of algebraic curves and surfaces, introducing the concept of apolarity.

Reye was the founder of that portion of projective geometry that E. A. Weiss later called point-series geometry. In a series of writings, Reye treated linear manifolds of projective plane pencils and of collinear bundles or spaces. Later these investigations were easily interpreted multidimensionally by means of the geometry of Segre manifolds. Reye refused to speak of true geometry when dealing with spaces of more than three dimensions. He was satisfied to interpret multidimensional relations in P_2 and P_3, that is, he treated the geometries of lines and spheres in P_3 as four-dimensional geometries. In 1878 Reye published a short work on spherical geometry, the only one of his mathematical writings, besides the *Geometrie der Lage*, to appear as a separate publication. An important configuration of twelve points, twelve planes, and sixteen lines in P_3 is named for Reye.

BIBLIOGRAPHY

I. ORIGINAL WORKS. Reye's writings include *Die Geometrie der Lage*, 2 vols. (Leipzig, 1866–1868), 5th ed., 3 vols. (Leipzig, 1923); *Synthetische Geometrie der Kugeln*

(Leipzig, 1879); "Über algebraische Flächen, die zueinander apolar sind," in *Journal für die reine und angewandte Mathematik*, **79** (1874), 159–175; and "Über lineare Mannigfaltigkeiten projektiver Ebenenbüschel und kollinearer Bündel oder Räume," *ibid.*, **104** (1889), 211–240; **106** (1890), 30–47, 315–329; **107** (1891), 162–178; **108** (1891), 89–124.

II. SECONDARY LITERATURE. See C. F. Geiser, "Zur Erinnerung an Theodor Reye," in *Vierteljahrsschrift der Naturforschenden Gesellschaft in Zürich*, **66** (1921), 158–160; C. Segre, "Cenno commemorativo di Reye," in *Atti dell' Accademia nazionale dei Lincei. Rendiconti*, 5th ser., **31** (1922), 269–272; and H. E. Timerding, "Theodor Reye," in *Jahresbericht der Deutschen Mathematiker-Vereinigung*, **31** (1922), 185–203.

WERNER BURAU

REYNA, FRANCISCO DE LA (*b.* Spain [?], *ca.* 1520), *physiology.*

Little is known of Reyna's life except that he was a farrier, veterinary surgeon, and "swine-doctor" of Zamora who cared for the horses of the Spanish nobility. The license for printing Reyna's popular *Libro de Albeyteria* was issued at Madrid in 1546, and the earliest dated edition known is that of Astorga, 1547. The manual, which went through at least thirteen editions, contains a controversial passage on the anatomy and physiology of the horse which has occasionally engaged the attention of historians since the eighteenth century, when Benito Feijoo y Montenegro claimed for Reyna the discovery of the circulation of the blood.

Certainly a number of William Harvey's predecessors made observations and articulated concepts which qualify them in some way as his precursors, and Reyna was among the lesser of these. The critical passage of the *Libro* (translated from the Spanish) is as follows (ed. 1547, f. 50b):

If you should be asked why, when a horse is bled from the front or hind legs, the blood comes from the lower part [of the opened vessel] and not the upper part, answer: In order to understand this problem, you must know that the principal veins proceed from the liver, and the arteries from the heart, and these principal veins are apportioned among the members of the body in this manner: in branches and webs of many small veins[1] through the external parts of the front and hind legs, going by means of the vessels. And from there these webs turn to pour through the principal veins which ascend from the hoofs through the front legs to the interior; so that the veins of the exterior parts have the function of carrying the blood downward, and the internal veins the function of carrying the blood upward. And so the blood goes around: and in a wheel,[2] through all the limbs and veins. It [the blood] has the function of carrying nourishment through the interior parts to the emperor of the body, the heart, which all the parts obey. This is the answer to this question.

Reyna's account was offered not as a discovery but as well-known information to be learned by the student. According to the Galenic system, when a vessel in a horse's leg is cut, it should bleed from the end nearest the body. Reyna explained why the reverse can be true by suggesting that blood flows into the leg through surface veins and out through deeper ones. Yet his narrative is so vague that the precise course is not clear, and he did not suggest a capillary structure.

Reyna's anatomy was essentially Galenic. The blood originates in the liver and is distributed through the veins. In no part of the *Libro* do we find the concept of the heart acting as a mechanical force-pump to motivate a circulatory system; Reyna merely says that some veins have the function of carrying nourishment to the heart. Because of this, and his statement about the liver, we must interpret the passage concerning the heart as the emperor of the body in an Aristotelian sense and not as an anticipation of Harvey, who although an Aristotelian himself, discovered the true function of the heart in the circulatory system. Reyna mentioned the arteries only once and not as part of his "wheel"; presumably their function is the Galenic one since none other was mentioned. Reyna did not suggest a passage of blood from the arteries to the veins. Galen had described an alternating movement of blood and pneuma occurring between the arteries and veins through the synanastomoses, but Reyna did not mention this.

Despite his traditionalism, the question posed at the beginning of the passage shows that Reyna could not square observations made during venesection with traditional physiology. His theory might be considered an attempt to reconcile observation (upward or centripetal flow of blood in equine leg veins) with Galenism (downward or centrifugal flow) by postulating both types of flow in differing veins. This interpretation would be more plausible had he not reported the exterior veins as carrying blood downward. Exterior veins were surely used for venesection, so that the theory is surprisingly discordant with the observation which inspired the passage.

Because of his lack of clarity, it is difficult to assess the precise nature of Reyna's contribution. His statement that "the blood goes around: and in a wheel" may be interpreted as implying some sort of venous "circulation," although one may argue that he meant merely that the various tissues are nourished in turn, or that he was only one of several pre-Har-

veians who used circular symbolism in writing of the blood. Despite these difficulties, it cannot be doubted that Reyna was among those before Harvey who saw that somehow there was more to the movements of the blood than Galen had taught. Yet it remained for Harvey to grasp the true nature of the circulatory system, and the *Libro de Albeyteria* in no way approaches the achievement of the *De motu cordis*.

NOTES

1. "Webs of many small veins," lit. *miseraycas*. The word usually refers to the portal vein and its branches.
2. Lit. *la sangre anda entorno* [*en torno* in later eds.]: *y en rueda*.

BIBLIOGRAPHY

I. ORIGINAL WORKS. The *Libro de Albeyteria* is the only work attributed to Reyna. Most of the known editions are listed in Antonio Palau y Dulcet, *Manual del Librero Hispanoamericano*, 2nd ed., XVI (Barcelona, 1964), 408–409. At least 17 eds. (to 1647) have been mentioned, but only 13 have been verified with certainty. A Burgos ed. known only from an imperfect copy in the Hispanic Society of America, with its presumably dated colophon missing, was called the 1st ed. by Keevil and Payne (see below). Palau suggested that it was the 2nd ed., but it is certainly later still; superficially it resembles the Burgos ed. of 1564, and may follow it.

The discussion of the movements of the blood differs slightly in later eds. from the text of 1647 given above, evidently through Reyna's emendation; for example, the passage ". . . through all the limbs and veins. It has the function of carrying nourishment..." becomes "...through all the limbs, and some veins have the function of carrying nourishment. . . ." Occasional minor changes in the text were also due to printers' variations. Reyna's seventeenth-century editor Fernando Calvo eventually changed the text and removed the passage about the "wheel" entirely, for reasons unknown.

II. SECONDARY LITERATURE. Benito Feijoo y Montenegro called attention to Reyna's passage in several parts of the *Cartas Eruditas, y Curiosas* (Madrid, 1742–1760), as did Samuel Taylor Coleridge in *Omniana* (London, 1812); for both see Dorothea Waley Singer, "S. T. Coleridge Suggests Two Anticipations of the Discovery of the Circulation of the Blood," in *Archeion*, **25** (1943), 31–39. Other discussions of Reyna are Cesareo Sanz Egaña, *Historia de la Veterinaria Espanola* (Madrid, 1941), 112–119; J. J. Keevil and L. M. Payne, "Francisco de la Reyna and the Circulation of the Blood," in *Lancet*, **260** (1951), 851–853; and Ronald S. Wilkinson, "The First Edition of Francisco de la Reyna's *Libro de Albeyteria*, 1547," in *Journal of the History of Medicine*, **23** (1968), 197–199. Harvey's precursors are examined by Walter Pagel, *William Harvey's Biological Ideas* (New York, 1967).

RONALD S. WILKINSON

REYNEAU, CHARLES RENÉ (*b.* Brissac, Maine-et-Loire, France, 11 June 1656; *d.* Paris, France, 24 February 1728), *mathematics.*

Reyneau is important historically as the author of a textbook, written at the request of Malebranche, that was designed to provide instruction in the new mathematics developed at the beginning of the eighteenth century. The son of a surgeon, he studied at the Oratorian *collège* in Angers. Attracted by the order, on 17 October 1676 he entered the Maison d'Institution in Paris, where, besides Malebranche, he met Jean Prestet, who had just published his *Élémens des mathématiques.* In 1679 Reyneau was sent to the Collège de Toulon, and in March 1681 he was ordained a priest there. In October 1682 he went to the University of Angers to replace Prestet as professor of mathematics, a post he held for twenty-three years. Suffering from deafness, he had former students substitute for him for several years but was finally obliged to give up teaching in 1705. Reyneau spent the rest of his life in Paris, at the Oratorian house on rue Saint-Honoré, and published his textbooks there. He was named an *associé libre* of the Académie Royale des Sciences on 12 February 1716.

Many surviving manuscripts reveal Reyneau's pedagogical ability and are a valuable source for the study of mathematics in France at the end of the seventeenth century. Reyneau was only slightly aware of the projects of Malebranche and L'Hospital in 1690–1691 and of the revolution resulting from Johann Bernoulli's stay in Paris in 1692. As late as 1694 all that Malebranche had for Reyneau to do was edit Prestet's posthumous *Géométrie.* But, after abandoning the last shred of Cartesian mathematics, Malebranche chose Reyneau to write the entirely new textbook required by this turnabout (1698).

Reyneau worked with two other Oratorians, Louis Byzance and Claude Jaquemet, who were better mathematicians than he. Reyneau had some difficulty in assimilating the differential and integral calculus and was very interested in the debates, beginning in 1700, provoked by Rolle on this subject. Reyneau's editorial efforts were frustrated in various ways, and the textbook was not published until 1708.

In 1705 Reyneau came into possession of Byzance's papers, which included a copy of the "Leçons" that Bernoulli had prepared for L'Hospital. Unfortunately, Reyneau lent some of the documents to Montmort, who lost them. On the whole, however, he preserved as well as possible the manuscripts of the group around Malebranche; and from them he drew the inspiration for a second didactic work, published in 1714. This work, which attempted to preserve the central conceptions of the Oratorian mathematics of the end of the preceding century, was less successful than the first.

Reyneau's most notable contribution to mathematical education was *Analyse démontrée* (1708). It was from the second edition of this work that d'Alembert learned the fundamentals of the subject.

BIBLIOGRAPHY

I. ORIGINAL WORKS. Reyneau's writings include *Analyse démontrée ou la méthode de résoudre les problèmes des mathématiques et d'apprendre facilement ces sciences, expliquée et démontrée . . . et appliquée . . . à découvrir les propriétés des figures de la géométrie simple et composée, à résoudre les problèmes . . . en employant le calcul ordinaire de l'algèbre, le calcul différentiel et le calcul intégral . . .,* 2 vols. (Paris, 1708; 2nd ed., enl., 1736–1738); *La science du calcul des grandeurs en général . . .,* 2 vols. (Paris, 1714–1735); *La logique ou l'art de raisonner juste à l'usage des dames* (Paris, 1744); and "Traité de la marine ou l'art de naviguer," MS no. 3729, Bibliothèque Mazarine, Paris.

II. SECONDARY LITERATURE. See an unsigned review of *L'analyse démontrée* in *Mémoires pour l'histoire des sciences et des beaux-arts,* **3** (1708), 1438–1452; Bernard de Fontenelle, "Éloge du Père Reyneau," in *Histoire de l'Académie royale des sciences pour l'année 1728 . . . avec les mémoires . . .* 112–116; and Pierre Costabel, "Deux inédits de la correspondance indirecte Leibniz–Reyneau," in *Revue d'histoire des sciences et de leurs applications,* **2** (1949), 311–332; "Rectification et compléments . . .," *ibid.,* **19** (1966), 167–169; and *Oeuvres de Malebranche,* XVII, pt. 2, *Malebranche et la réforme mathématique en France de 1689 à 1706* (Paris, 1968).

PIERRE COSTABEL

REYNOLDS, OSBORNE (*b.* Belfast, Ireland, 23 August 1842; *d.* Somerset, England, 21 February 1912), *engineering, physics.*

Reynolds was born into an Anglican clerical family. His father—like his grandfather and great-grandfather before him—was rector of Debach-with-Boulge in addition to having been a Cambridge wrangler in 1837, a fellow of Queens' College, principal of the Belfast Collegiate School, and headmaster of Dedham Grammar School, Essex.

Reynolds was educated first at Dedham and then privately before entering the service of Edward Hayes, a mechanical engineer, in 1861. He did an apprenticeship with Hayes in order to learn the mechanical arts before going like his father to Cambridge and eventually into a career in civil engineering. "In my boyhood," Reynolds later wrote, "I had the advantage of the constant guidance of my father, also a lover of mechanics, and a man of no mean attainments in mathematics and their application to physics."[1] At Cambridge, Reynolds was a successful mathematics student, passing the mathematical tripos in 1867 as seventh wrangler and receiving a fellowship, again like his father, at Queens' College. He thereupon entered the civil engineering firm of John Lawson.

In the following year a newly created professorship of engineering was advertised at Owens College, Manchester, at £500 per annum. Reynolds applied for the position and, despite his youth and inexperience, was awarded the post. Subsequently during the thirty-seven years of his tenure as professor, Reynolds investigated and contributed significantly to a wide variety of engineering and physics subjects.

From 1868 to 1873 Reynolds' attention focused largely on problems in electricity, magnetism, and the electromagnetic properties of solar and cometary phenomena. For the next two decades after 1873, his interests appear to have turned sharply toward mechanics, and especially toward the mechanics of fluids.

In an important paper of 1883, "An Experimental Investigation of the Circumstances which Determine whether the Motion of Water in Parallel Channels shall be Direct or Sinuous and of the Law of Resistance in Parallel Channels,"[2] Reynolds experimentally investigated the character of liquid flow through pipes and channels, and he demonstrated streamline and turbulent flow in pipes. He showed that there is a critical velocity, depending upon the kinematic viscosity, the diameter of the pipe, and a physical constant (the Reynolds number) for the fluid at which a transition between the two types of flow will occur. The "Reynolds stresses," resulting from his analysis, continue to play an important role in turbulence theory.

In 1886 Reynolds published "On the Theory of Lubrication,"[3] which became a classic paper on film lubrication, and which resulted in bearings that were capable of carrying high loads at speeds hitherto considered impossible.

Reynolds' analogy, which assumes that the rate of heat transfer between a fluid and its boundary is proportional to the internal diffusion of the fluid at and near the surface, was enunciated in a paper of 1874.[4]

Reynolds' most extensive experimental work concerned the mechanical equivalent of heat; specifically, he found the mean specific heat of water (in terms of work) between the freezing and boiling points. The results rank among the classic determinations of physical constants.

Reynolds also worked on the action of waves and currents in determining the character of estuaries, using models in the investigation; the development of turbines and pumps; studies on group-velocity, in which, according to Lamb, Reynolds was the first to show that group-velocity also provides the rate of transmission of energy; the theory of thermal transpiration; investigations of the radiometer; studies on the refraction of sound; and cavitation.

In 1885 Reynolds gave the name "dilatancy" to a peculiar property of a closely packed granular mass: it can increase the volume of its interstices when its shape is altered. Reynolds believed he saw in this phenomenon a possible aether-model explaining cohesion, light, and gravity. These speculations formed the basis for his 1902 Rede lecture, "On an Inversion of Ideas as to the Structure of the Universe," and afterward appeared in mathematical form as *The Sub-Mechanics of the Universe* (Cambridge, 1903; also as Volume III of Reynolds' *Papers on Mechanical and Physical Subjects*). In these papers Reynolds argued that, contrary to the vision of the kineticists, the universe is almost completely filled with absolutely rigid granules—as he insisted at the Rede lecture in 1902, "I have in my hand the first experimental model universe, a soft india rubber bag . . . filled with small shot."[5] He had, he said, "the fullest confidence that . . . ideas, such as I have endeavoured to sketch, will ultimately prevail."[6] George H. Bryan, reviewing *The Sub-Mechanics* in *Nature*, was impressed, and wrote, "It may be confidently anticipated that Prof. Osborne Reynolds' granular medium will play an important part in the physics of the future."[7]

In a sense, Reynolds' mechanical model—although moribund at its birth—was a suitable end to a distinguished career dedicated to the proposition that to "mechanical progress there is apparently no end: for as in the past so in the future, each step in any direction will remove limits and bring in past barriers which have till then blocked the way in other directions; and so what for the time being may appear to be a visible and or practical limit will turn out but a bend in the road."[8]

Reynolds was an active and dedicated member of the Manchester Literary and Philosophical Society, which he served as secretary for many years and as president for the term 1888–1889. Upon the death of Joule, he wrote an excellent biography, which was published in the Society's *Memoirs* for 1892. In 1877, Reynolds was elected a fellow of the Royal Society. In 1888 he also received a Royal Medal and in 1884 honorary LL.D from the University of Glasgow.

Because of ill health, Reynolds retired from active work in 1905. He spent his last years with greatly impaired mental and physical powers in Somerset. He left three sons and a daughter by his second marriage.

NOTES

1. Quoted in H. Lamb, "Osborne Reynolds," in *Proceedings of the Royal Society*, **88A** (1912–1913), xv.
2. *Philosophical Transactions of the Royal Society*, **174** (1883), 935–982.
3. *Ibid.*, **177** (1886), 157–234.
4. "On the Extent and Action of the Heating of Steam Boilers," in *Proceedings of the Manchester Literary and Philosophical Society*, **14** (1874–1875), 8.
5. *On an Inversion of Ideas as to the Structure of the Universe* (Cambridge, 1903), 28.
6. *Ibid.*, p. 44.
7. *Nature*, **68** (1903), 602.
8. O. Reynolds, *Report of the 57th Meeting of the British Association for the Advancement of Science* (1887), 861.

BIBLIOGRAPHY

I. ORIGINAL WORKS. Reynolds' most important papers are collected in Osborne Reynolds, *Papers on Mechanical and Physical Subjects*, 3 vols. (Cambridge, 1900–1903). Useful also is *On an Inversion of Ideas as to the Structure of the Universe* (Cambridge, 1903). His address to Section G of the British Association can be found in the Association's *Reports* (1887), 855–861. Good samples of Reynolds' popular style can be found in "The Two Manners of Motion of Water" and "Experiments Showing Dilatancy," in *Proceedings of the Royal Institution of Great Britain*, **11** (1884–1886), 44–52, 354–363. J. J. Thomson's views on Reynolds as a teacher are in his *Recollections and Reflections* (London, 1936). Reynolds' "Memoir of James Prescott Joule," *Memoirs and Proceedings of the Manchester Literary and Philosophical Society*, 4th ser., **6** (1892), is still the major book-length study of Joule.

II. SECONDARY LITERATURE. The Osborne Reynolds Centenary Symposium at the University of Manchester (September, 1968) has produced a valuable vol. about Reynolds and his contribution to engineering: D. M. McDowell and J. D. Jackson, eds., *Osborne Reynolds and Engineering Science Today* (Manchester, 1970), with a lively essay, Jack Allen, "The Life and Work of Osborne Reynolds" (pp. 1–82). To be consulted also are Horace Lamb, "Osborne Reynolds 1842–1912," in *Proceedings of the Royal Society*, **88A** (1912–1913), xv–xxi; A. H. Gibson, *Osborne Reynolds and His Work in Hydraulics and Hydrodynamics* (London, 1946); and "Prof. Osborne Reynolds, F.R.S." in *Nature*, **88** (1912), 590–591. See also R. W. Bailey, "The Contribution of Manchester Researches to Mechanical Science," in *Proceedings of the Institution of Mechanical Engineers*, **2** (1929), 613–683, with a discussion of Reynolds' mechanical work. On hydraulics, see H. Rouse and S. Ince, *History of Hydraulics* (New York, 1963), 206–212. On Reynolds at Owens College, see J. Thompson, *The Owens College: Its Foundation and Growth* (Manchester, 1886). For a brief description of Reynolds' construction of the universe and his own interesting reaction to it, see G. H. Bryan, "A New Mechanical Theory of the Aether," in *Nature*, **68** (1903), 600. A fuller bibliography on Reynolds can be found at the close of Allen's essay. I have had the privilege of consulting an unpublished biographical sketch by Hunter Rouse and a private communication by Prof. Stanley Corrsin of the Johns Hopkins University.

ROBERT H. KARGON

REY PASTOR, JULIO (*b*. Logroño, Spain, 16 August 1888; *d*. Buenos Aires, Argentina, 21 February 1962), *mathematics*.

Rey Pastor, a poet in his youth, studied science at the University of Zaragoza. In 1905 he published his first monograph, *Sobre los números consecutivos cuya suma es a la vez cuadrado y cubo perfecto*. Appointed professor of mathematical analysis at the University of Oviedo in 1911, Rey Pastor wrote the inaugural address for the academic year 1913–1914, *Los matemáticos españoles del siglo XVI* (enlarged and reprinted in 1925 and 1934). In this work he described the deplorable state of science in Spain under the Hapsburgs and, as a consequence, was accused of being unpatriotic. The following year he was professor at Madrid. A series of trips to Germany resulted in the monograph *Estudio geométrico de la polaridad* (Madrid, 1912) and his *Fundamentos de la geometría proyectiva superior* (Madrid, 1916). In the latter work Rey Pastor expounded the synthetic geometry of space in *n* dimensions, introducing concepts of great generality (for example, the definition of the curve) and developing them in all their consequences.

In 1915 Rey Pastor gave a series of lectures at the Institut d'Estudis Catalans in Barcelona on conformal mapping, in which he expounded and developed the work of H. A. Schwarz. Notes from those lectures by Esteban Terrades were published in Catalan. In 1917 Rey Pastor gave an extension course at the University of Buenos Aires and accepted a contract "to direct the advanced study of the exact sciences" in Argentina, spending half of the school year there and half in Spain.

Rey Pastor founded a mathematics laboratory, the Seminario Matemático de Madrid (1916), and the *Revista matemática hispanoamericana* (1919), and published the now-classic *Elementos de análisis algebraico* (Madrid, 1917), in which he introduced his own discoveries and innovations. Besides his mathematical work he studied the history of Spanish cartography, which led to publication of *La cartografía mallorquina*, written with E. García Camarero (Madrid, 1960).

BIBLIOGRAPHY

See Juan José González Covarrubia, *Julio Rey Pastor* (Buenos Aires, 1964); and Esteban Terrades, "Julio Rey Pastor como hombre e investigador," in *Homenaje a Rey Pastor*, I (Santa Fé, Argentina, 1945).

J. VERNET

RHAZES. See **Al-Rāzī.**

RHEITA, ANTON MARIA SCHYRLAEUS DE (**Antonín Maria Šírek z Vrajtu**) (*b*. Bohemia, 1597; *d*. Ravenna, Italy, 1660), *astronomy.*

Little is known of Rheita's life. He was a priest and a Capuchin, at first a member of the community in Vrajt (Rheita) in Bohemia. He apparently left that monastery during the Thirty Years' War, and by the 1640's was professor of theology at Trier. It is not certain when he went to Ravenna.

Rheita's work in observational astronomy and optics was carried out in Belgium in the 1640's. In 1643 he published at Louvain a tract of rather dubious scientific value entitled *Novem stellae circa Jovem visae, circa Saturnum sex, circum Martem nonnullae.* Two years later, at Antwerp, he brought out the work on which his scientific reputation rests, the *Oculus Enoch et Eliae, opus theologiae, philosophiae, et verbi dei praeconibus utile et iucundum.* This treatise contains, among a somewhat curious variety of topics, Rheita's description of an eyepiece for a Keplerian telescope that left the image reverted, his own invention; in it Rheita also made use of the terms "ocular" and "objective," which he himself had coined. Most interesting to historians of science, however, is Rheita's map of the moon, drawn according to his own observations. The map is eighteen centimeters in diameter, and although it is rather scanty in detail (and decidedly inferior to that published in Antwerp in the same year by M. F. van Langren), it is the first representation of the moon that places its southernmost part at the top, reproducing the image seen through an inverting astronomical telescope.

BIBLIOGRAPHY

Both of Rheita's surviving works have been cited in the text. There is no secondary literature.

ZDENĚK KOPAL

RHETICUS, GEORGE JOACHIM (*b*. Feldkirch, Austria, 16 February 1514; *d*. Kassa, Hungary [now Košice, Czechoslovakia], 4 December 1574), *mathematics, astronomy.*

Rheticus was the son of George Iserin, the town physician of Feldkirch, and Thomasina de Porris, an Italian lady. After Rheticus' father was beheaded for sorcery in 1528, his surname could no longer be used. Hence his widow reverted to her maiden name, de Porris, for herself and her two children. Our George Joachim de Porris tacked on "Rheticus" to indicate that he came from a place in what had been the ancient Roman province Rhaetia. Since he had not been born in Italy, he converted "de Porris" (meaning "of the leeks") into the German equivalent "von Lauchen." Then as a mature man he dropped both references to leeks, thereby transforming "Rheticus" from a geographical designation into an adopted surname. This fifth stage remained the name by which he is commonly known.

Rheticus' first teacher was his father. After the execution of his father Rheticus studied at Zurich, where Gesner was a schoolmate. He also met Paracelsus "and in the year 1532 had a conversation with him, a great man who published famous works."[1] In 1532 Rheticus matriculated at the University of Wittenberg, where he obtained his M.A. on 27 April 1536; ten days earlier he had publicly defended the thesis that Roman law did not absolutely prohibit all forms of astrological predictions, since predictions based on physical causes were permitted, like medical predictions.

In the same year Rheticus was appointed to teach elementary arithmetic and geometry at the University of Wittenberg. On 18 October 1538 he took a leave of absence for the purpose of visiting such leading astronomers as Johannes Schöner of Nuremberg, Peter Apian of Ingolstadt, and Philip Imser of Tübingen. At Feldkirch on 27 November 1538 he presented an edition of Sacrobosco (published earlier that year at Wittenberg) to Achilles Pirmin Gasser (1505–1577), who was his father's successor, twice removed, as town physician.[2] In the summer of 1539 Rheticus arrived in Frombork (Frauenburg) in order to learn from Copernicus himself about the rumored new and revolutionary cosmology.

The momentous meeting between Rheticus and Copernicus precipitated the beginning of modern astronomy. The reviver of the geokinetic system had long resisted friendly entreaties to release his masterpiece for publication, but permitted Rheticus to write a *Narratio prima* (*First Report*) about *De revolutionibus.* On 23 September 1539 Rheticus completed the *First Report*, which was published at Gdańsk in early 1540. The work was the earliest printed announcement to the educated public of a rival to the Ptolemaic system, which had dominated men's minds for fourteen hundred years. Rheticus immediately sent a copy of the *First Report* to Gasser, who promptly wrote a foreword for the second edition, which was published at Basel in 1541.[3] The first two editions of Rheticus' *First Report* did not detonate any such hostile explosion as Copernicus had feared would be the instant reaction to his geokineticism. Hence he finally made up his mind (perhaps by 9 June 1541) to let *De revolutionibus* be printed and began putting the final touches to his manuscript.

To the *First Report* Rheticus appended an *Encomium Borussiae*, a praise of Prussia based on his travels throughout that region. Presumably utilizing also Copernicus' earlier and incomplete geographical studies, Rheticus drew up a "Tabula chorographica auff Preussen und etliche umbliegende lender," which he presented to Duke Albert of Prussia on 28 August 1541. While Rheticus' "Topographical Survey of Prussia and Several Neighboring Lands" has not survived, it may have provided the foundation for the map of Prussia that was printed at Nuremberg in 1542 as the work of Rheticus' editorial assistant, Heinrich Zell. Rheticus' theoretical discussion of map-making, *Chorographia tewsch*, the first work he wrote in German, using his native Vorarlberg dialect,[4] was likewise dedicated to Duke Albert as a companion piece to the "Tabula chorographica." Since the duke had tried in vain to learn from other mathematicians how to anticipate the time of daily sunrise, Rheticus constructed a "small instrument for ascertaining the length of the day throughout the year." In transmitting his "Instrumentlin" to the duke on 29 August 1541, Rheticus asked Albert to recommend to both the Elector of Saxony and the University of Wittenberg that he be permitted to publish Copernicus' *De Revolutionibus*. Three days later Duke Albert complied, further requesting that Rheticus be retained in his professorship.

When Rheticus returned to Wittenberg for the opening of the winter semester, he was elected dean of the liberal arts faculty on 18 October 1541. In early 1542 he separately published—under the title *De lateribus et angulis triangulorum*[5]—the section on plane and spherical trigonometry in Copernicus' *De revolutionibus*. To this brief discussion of the *Sides and Angles of Triangles* Rheticus added a table of half-chords subtended in a circle. Such a half-chord is actually a sine, although both Copernicus and Rheticus studiously avoided the use of that term. The table of sines in the *Sides and Angles of Triangles* differs from the corresponding table in *De revolutionibus* by increasing the length of the radius from one hundred thousand to ten million and by diminishing the interval of the central angle from $10'$ to $1'$. Furthermore, by indicating the complementary angle

at the foot of the columns and at the right-hand side of the page, the 1542 table became the first to give the cosine directly, although that term is not mentioned. Rheticus did not ascribe the authorship of this table to Copernicus nor, presumably out of modesty, to himself. Nevertheless, the table was undoubtedly his doing. His independent place in the history of mathematics is due precisely to his computation of innovative and monumental trigonometrical tables.

Although such a purely technical work as Copernicus' *Sides and Angles of Triangles* could be published without opposition in Wittenberg, that citadel of Lutheran orthodoxy was no place to print Copernicus' *De revolutionibus*, with its far-reaching cosmological implications. Hence, shortly after the end of the winter semester on 30 April 1542, Rheticus left for Nuremberg, where on 1 August 1540 a printer had dedicated to him an astrological tract. Rheticus could not remain in Nuremberg long enough to supervise the entire printing of *De revolutionibus*, since he had been appointed professor of mathematics at the University of Leipzig, where he had to be present in mid-October 1542.

After teaching three years at Leipzig, Rheticus obtained a leave of absence. He went back to Feldkirch and then on to Milan, where he spent some time with Cardano. In Lindau, during the first five months of 1547, he suffered a severe mental disorder, which gave rise to rumors that he had gone mad and died. But he recovered well enough to teach mathematics at Constance for more than three months in the latter half of 1547. Then he moved to Zurich, where he studied medicine with his old classmate Gesner, who was now a widely recognized authority. On 13 February 1548 Rheticus reported to the University of Leipzig that on the advice of his doctors he would leave at Easter to undergo hydrotherapeutic treatment and thereafter return to his post.

At the beginning of the winter semester of 1548 Rheticus was back in harness, having been elected dean. In 1549 he became involved in a legal dispute with a goldsmith and then in April 1551 in a drunken homosexual encounter with a student, on account of which he had to run away from Leipzig.

Seeking to build a new career, Rheticus resumed the study of medicine at the University of Prague in 1551–1552. Although he was invited to teach mathematics at the University of Vienna in 1554, in the spring of that year he settled down at Cracow, where he practiced medicine for two full decades. On 12 April 1564 he wrote to a friend that he had not accepted an unofficial invitation by Peter Ramus to teach at the University of Paris. In Cracow, Rheticus' lifelong interest in astrology attained its greatest success. He had followed up his master's thesis of 1536 by inserting in 1539 an astrological section in his *First Report*, although Copernicus' astronomy was entirely free of that pathetic delusion. As late as 1 March 1562 Rheticus was still contemplating—on the basis of his astrological version of Copernicus[6]—the construction of a chronology of the world from creation to dissolution. But by correctly predicting in 1571 that the successor of King Sigismund Augustus of Poland "will reign only a very short time," Rheticus acquired immense renown as a seer.[7]

L. Valentine Otho, a student of mathematics at the University of Wittenberg, was deeply impressed by Rheticus' *Canon of the Doctrine of Triangles* (Leipzig, 1551), the first table to give all six trigonometric functions, including the first extensive table of tangents and the first printed table of secants (although such modern designations were eschewed by Rheticus as "Saracenic barbarisms"). Without any recourse to arcs, Rheticus' *Canon* defined the trigonometrical functions as ratios of the sides of a right triangle and related these ratios directly to the angles. By equating the functions of angles greater than 45° with the corresponding cofunctions of the complementary angles smaller than 45°, Rheticus reduced the length of his table by half.

When Otho went to visit Rheticus in 1574, he found him in Košice, where he had gone on the invitation of a local magnate. In the arrival of the youthful student to help him publish his life's work, Rheticus recognized a replay of the scenario he himself had enacted with Copernicus a generation earlier. But unfortunately the outcome was different, for Rheticus died on 4 December 1574, leaving his books and manuscripts to Otho, who faithfully promised to see his master's massive tables through the press.

These tables were a "labor of twelve years, while I always had to support a certain number of arithmeticians for these computations," Rheticus had informed Ramus in 1568.[8] Nevertheless Otho had to cope with enormous difficulties before he succeeded in fulfilling his promise to Rheticus. Through his deceased teacher's local patron, he obtained financial support from the Holy Roman Emperor, but within two years this ruler died. On 7 September 1576 Otho appealed from Košice to the Elector of Saxony, who consented to have him appointed as professor of mathematics at the University of Wittenberg. But in January 1581 Otho refused to sign a religious formula required of all the Wittenberg professors, and therefore he had to turn elsewhere.

He found his last patron in the count palatine, Frederick IV, with whom he signed a contract on 24 August 1587. Designated the count's official mathematician, permitted to eat at the table of the professors of the University of Heidelberg, and granted the aid of four students as computers, Otho was finally able to complete and publish in 1596 Rheticus' immense *Opus Palatinum de triangulis*, as Otho entitled it in gratitude to his backer.

The foundation of the Rheticus-Otho *Opus Palatinum* is the table of sines for the first quadrant 0° to 90°, the interval being 45″ and the radius 10^{15}. For purposes of interpolation, a process of successive halving was relentlessly pursued in order to find the small angle the sine of which is 1 in the fifteenth decimal place as the first significant figure. Then, with a radius of 10^{10}, the sines and cosines were computed for intervals of 10″. The functions of each degree occupy six full pages, so enormous was the labor expended in these computations.

After Otho's death, among his papers were found additional Rheticus manuscripts, which were published by Pitiscus in his *Thesaurus mathematicus* (Frankfurt am Main, 1613). These manuscripts included a table of sines for a radius of 10^{15} and intervals of 10″, but an interval of only 1″ for the two special cases of 1° and 89°. Although Rheticus' trigonometrical tables were understandably far from perfect, modern recomputations have found them accurate to a relatively high degree.

NOTES

1. Rheticus to Joachim Camerarius, 29 May 1569 (Burmeister, *Rhetikus*, III, 191).
2. *Bibliotheca apostolica vaticana, inventario dei libri stampati palatino-vaticani*, Enrico Stevenson, ed., vol. I, pt. 1 (Rome, 1886), libri latini, no. 2195.
3. Stevenson, no. 1532.
4. Part of it was translated into modern German, and the rest summarized by Heinz Balmer, *Beiträge zur Geschichte der Erkenntniss des Erdmagnetismus* (Aarau, 1956), pp. 279–286.
5. For the copy presented by Rheticus to Gasser on 20 June 1542 in Feldkirch, see Stevenson, no. 1528.
6. Burmeister, *Rhetikus*, III, 162.
7. *Ibid.*, III, 198.
8. *Ibid.*, III, 187.

BIBLIOGRAPHY

I. ORIGINAL WORKS. Rheticus' publications are listed in Karl Heinz Burmeister, *Georg Joachim Rhetikus 1514–1574, eine Bio-Bibliographie*, II (Wiesbaden, 1967–1968), 55–83; the extant MSS: II, 18–31; and correspondence: II, 32–39; III, 15–200.

II. SECONDARY LITERATURE. On Rheticus and his work, see the following references by K. H. Burmeister: *G. J. Rhetikus*, II, 84–92; III, 201; "G. J. Rhetikus und A. P. Gasser," in *Schriften des Vereins für Geschichte des Bodensees*, **86** (1968), 217–225; "G. G. Porro Retico," in *Archivio storico lombardo*, **7** (1968), 3–11; and "G. J.

Rheticus as a Geographer," in *Imago mundi*, **23** (1969), 73–76. See also Edward Rosen, "Rheticus's Earliest Extant Letter to Paul Eber," in *Isis*, **61** (1970), 384–386, with commentary by K. H. Burmeister; and "Rheticus as Editor of Sacrobosco" (in press).

EDWARD ROSEN

RIBAUCOUR, ALBERT (*b.* Lille, France, 28 November 1845; *d.* Philippeville [now Skikda], Algeria, 13 September 1893), *mathematics, engineering.*

Ribaucour was the son of Placide François Charles Ribaucour, a teacher of mathematics, and Angélique Françoise Devemy. In 1865 he entered the École Polytechnique in Paris and in 1867 began studying at the École des Ponts et Chaussées, which he left in 1870 to become an engineer at the Rochefort naval base. At Rochefort he showed an exceptional aptitude for engineering, which also distinguished him after transfer, in 1873, to Draguignan (Var), where from 1874 to 1876 he was in charge of road construction in Var. The bridges that he designed were remarkable because of their combination of maximum strength with minimum material. From 1878 to 1885 he stayed at Aix-en-Provence, where his skills displayed in the construction works on the canal of the Durance earned him a Légion d'Honneur and a gold medal at the Paris Exposition of 1889. Ribaucour's suspension bridge of Mallemort-sur-Corrège and his construction of the reservoir of Saint-Christophe (near Rognes, Bouches-du-Rhône) were especially praised.

After a short stay at Vesoul (Haute-Saône) in order to receive the title of chief engineer, Ribaucour was sent to Algeria, where from 1886 until his death he stayed at Philippeville and worked on the construction of railroads and harbor works.

Ribaucour's mathematical work—to which he dedicated himself especially under the influence of Mannheim—belonged to his spare time, except for a short period during 1873 and 1874, when he was *répétiteur* in geometry at the École Polytechnique. His main field was differential geometry, and his work was distinguished enough to earn him the Prix Dalmont in 1877 and a posthumous Prix Petit d'Ormoy in 1895, awarded by the Paris Academy. His most elaborate work was a study of minimal surfaces, *Étude des élassoïdes ou surfaces à courbure moyenne nulle*, presented to the Belgian Academy of Sciences in 1880 (in *Mémoires couronnés et mémoires des savants étrangers. Académie royale des sciences, des lettres et des beaux-arts de Belgique*, **44** [1881]). In the work he explained his method called *périmorphie*, which utilized a moving trihedron on a surface. The approach to minimal surfaces was to consider them as the envelope of the middle planes of isotropic congruences; this approach led Ribaucour to a wealth of results.

Many of Ribaucour's papers deal with congruences of circles and spheres. Special attention was devoted to those systems of circles that are orthogonal to a family of surfaces. Such systems form *systèmes cycliques*, and it is sufficient for the circles to be orthogonal to more than two surfaces for them to be orthogonal to a family. Ribaucour's research thus led him to envelopes of spheres, to triply orthogonal systems, cyclides, and surfaces of constant curvature.

BIBLIOGRAPHY

I. ORIGINAL WORKS. Ribaucour reported most of his results in *Comptes rendus hebdomadaires des séances de l'Académie des sciences*, **67** (1868), to **113** (1891); also in the *Nouvelles annales de mathématiques*, 2nd ser., **4** (1865), to **10** (1871); and the *Bulletin de la Société philomathique in Paris* (1867–1871). See also "Sur deux phénomenes d'hydrodynamique observés au bassin de Saint Christophe," in *Compte rendu de la 14ᵉ session de l'Association française pour l'avancement des sciences*, pt. 2 (1885), 252–255; and M. Salva, *Notice sur le port de Philippeville* (Paris, 1892), esp. chs. 5 and 6.

II. SECONDARY LITERATURE. A good approach to Ribaucour's work is through Gaston Darboux, *Leçons sur la théorie générale des surfaces et les applications géométriques du calcul infinitésimal*, 4 vols. (Paris, 1887–1896); see also L. Bianchi, *Lezioni di geometria differenziale*, 3rd ed., II (Pisa, 1923), esp. chs. 17, 19, 20, 21. Other works include P. Mansion, "Ribaucour," in *Mathésis*, 2nd ser., **3** (1893), 270–272; and P. M. d'Ocagne, "Un ingénieur et géomètre polytechnicien: Albert Ribaucour," *Bulletin de la Société des amis de l'École Polytechnique* (July, 1913). (A. Brunot in Paris and E. de Zelicourt in Aix have also provided data for this article.)

D. J. STRUIK

RIBEIRO SANTOS, CARLOS (*b.* Lisbon, Portugal, 21 December 1813; *d.* Lisbon, 23 December 1882), *engineering, geology.*

Ribeiro was the eldest of the five children of José Joaquin Ribeiro, who was employed in the silver foundry of the Lisbon mint, and Francisca Santos. The family was poor, and Ribeiro received only the rudiments of a primary education before he went to work in a haberdashery at the age of ten. In 1833, during the War of the Two Brothers, Ribeiro enlisted, on 4 August, in the artillery and on 5 September of that year volunteered for service in the constitutionalist forces, thus severing himself from the absolutist views of his parents. When the bloody civil conflict came to an end the following year, Ribeiro entered the Lisbon military college, from which he graduated as a commissioned officer in 1837. He remained in the army for the rest of his life, being promoted general a few weeks before he died.

Ribeiro began his geological researches in 1840. He was the first major Portuguese geologist, and gave expert advice on a number of subjects. In 1848 he served as director of a number of coal mines and in 1852, under a commission from the ministry of public works, he organized the section of mines, quarries, and geological works, founded the national mining service, and established the national geological survey. In 1854 he drew up a plan for supplying the city of Lisbon with water, and two years later he was concerned in linking the railroad between Portugal and Spain at the Elvas-Badajoz frontier. In August 1857 Ribeiro founded the commission for geological works and in November of the same year he organized the first topographical survey of Portugal, on a scale of 1:500,000. He himself had been the first to recognize the stratigraphic succession of the country, and the survey aided him in his own later geological studies. Ribeiro also served as secretary of the council on mines when it was created in 1859 and in 1865 was a councillor. In 1874 he became superintendent of the copper mines of Aljustrel.

In 1874 Ribeiro was elected a deputy to parliament, and took an active part in political affairs. In addition to his interest in national and geological matters, Ribeiro was concerned with early man; from 1866 to 1868 he undertook a study to ascertain the antiquity of man in Portugal, and he was one of the organizers of the Ninth International Congress of Anthropology and Prehistoric Archaeology that was held in Lisbon in 1880. He received many academic and other honors and at the time of his death belonged to a number of organizations, including the Lisbon Royal Academy of Sciences, the Coimbra Institute, and the Imperial and Royal Geology Institute of Vienna. He was survived by his wife, Ursula Damascio, whom he had married in 1846.

BIBLIOGRAPHY

I. ORIGINAL WORKS. For a partial list of Ribeiro's memoirs, see Royal Society Catalogue of Scientific Papers, V, 183; VIII, 742; XI, 165; XVIII, 166. His books include Indicazioni relative alla Commissione di geología nel Portogallo (Milan, 1865); Estudos geologicos (Lisbon, 1866); Memoria sobre o abastecimento de Lisboa com aguas de nascente e aguas de rio (Lisbon, 1867); Relatorio acerca da arborizacao geral do paiz (Lisbon, 1868); Relatorio sobre as minas de pyrite de ferro cuprica das cercanias da villa de Aljustrel e das minas do Sobral (Lisbon, 1873); and Estudos prehistoricos em Portugal, 2 vols. (Lisbon, 1878–1880).

II. SECONDARY LITERATURE. There are obituaries of Ribeiro by A. F. Loureiro, in Instituto, 2nd ser., 30 (1882–1883), 193–205; by J. F. Nery Delgado, in Neues Jahrbuch für Mineralogie, Geologie und Paläontologie, 2 (1883), 1–4, supp.; P. Choffat, in Bulletin de la Société géologique de France, 11 (1883), 321–329; the latter two also contain a bibliography of Ribeiro's works. See also C. Castello Branco, O general Carlos Ribeiro (recordaçoes da mocidade) (Porto, 1884); R. Severo, "Carlos Ribeiro," in Revista de sciencias naturaes e sociaes, 5 (1898), 153–177; S. A., "Rerum naturalium in Lusitania cultores— Carolus Ribeiro," in Broteria, 2 (1903), 93–97; J. S. Tavares, "Os naturalistas portugueses: Carlos Ribeiro," ibid., 98–103; J. F. Nery Delgado, "Elogio historico do general Carlos Ribeiro," in Revista de obras publicas e minas, 36 (1905), 1–59; and P. Choffat, "Biographies de geologes portugais. Carlos Ribeiro," in Comunicações Serviços geológicos de Portugal, 11 (1917), 275–281.

J. M. LÓPEZ DE AZCONA

RICCATI, JACOPO FRANCESCO (b. Venice, Italy, 28 May 1676; d. Treviso, Italy, 15 April 1754), mathematics.

Riccati was the son of a noble family who held land near Venice. He received his early education at the school for the nobility in Brescia then entered the University of Padua where, to please his father, he began to study law. He took the degree on 7 June 1696. At Padua he became a friend of Stefano degli Angeli, who encouraged him in the pursuit of mathematics; Riccati's detailed study of recent methods of mathematical analysis enabled him to solve, in 1710, a difficult problem posed in the Giornale de' letterati d'Italia. He soon embarked on the extensive series of mathematical publications that brought him contemporary fame. His renown was such that Peter the Great invited him to come to Russia as president of the St. Petersburg Academy of Sciences; he was also asked to Vienna as an imperial councillor and offered a professorship at the University of Padua. He declined all these offers, since he preferred to stay in Italy and devote himself to his studies privately, in his own family circle. He was often consulted by the senate of Venice, particularly on the construction of dikes along rivers and canals, and his expertise was deferred to on this and other topics. In addition to his works in mathematics and hydraulics, he published a number of studies on central forces that are marked by his enthusiastic advocacy of Newton's ideas.

Riccati's mathematical work dealt chiefly with analysis and, in particular, with differential equations.

He achieved notable results in lowering the order of equations and in the separation of variables. In 1722–1723 he was engaged to teach infinitesimal calculus to two young noblemen, Lodovico Riva and Giuseppe Suzzi, and his lectures to them, subsequently published, demonstrate the technique he employed. In expounding the known methods of integration of first-order differential equations, Riccati studied those equations that may be integrated with appropriate algebraic transformation before considering those that require a change of variables. He then discussed certain devices suggested by Johann I Bernoulli and expounded the method used by Gabriele Manfredi to integrate homogeneous equations. He further pointed out that in order to determine a curve endowed with an assigned property, it may at times be useful to relate it to some system of coordinates other than the usual one.

Riccati then discussed, with many examples, the integration methods that he himself had devised. Of these, one involves the reduction of the equation to a homogeneous one, while another, more interesting method is that of "halved separation," as Riccati called it. The technique of halved separation comprises three operations. In the first, the entire equation is multiplied or divided by an appropriate function of the unknown so that it becomes integrable; second, after this integration has been carried out, the result is considered to be equal to a new unknown, and one of the original variables is thus eliminated; and finally, the first two procedures are applied to the result until a new and desired result is attained. Riccati communicated this method to Bernardino Zendrini, a mathematician who was also superintendent of waterworks for the Venetian state; Zendrini passed it on to Leibniz, who considered it highly ingenious and wrote Riccati an encouraging letter. Riccati first published it as "Contrarisposta alle annotazioni del Sig. Niccolò Bernoulli" in *Giornale de' letterati d'Italia*, in 1715.

At a later point in his lectures, Riccati also dealt with higher-order equations, indicating how some of the techniques implicit in them may be further applied. He also took up the methods used by other mathematicians, of which he is occasionally critical.

In an earlier work, published in the *Giornale de' letterati d'Italia* in 1712, Riccati had already given the solutions to a number of problems related to plane curves determined by curvature properties, for which the integration of second-order differential equations is required. His results were widely known and used by other mathematicians (indeed, they were sometimes republished without mention of his name), and he himself repeated them in his lectures. The

earlier memoir also contains Riccati's important statement that the method he had used will lead to the integration of all differential equations of the type

$$f = \left(y, \frac{dy}{dx}, \frac{d^2y}{dx^2} \right) = 0,$$

which he was the first to consider in their generality.

Riccati also provided, in a memoir communicated to Zendrini in 1715 (and intended for publication in the *Giornale*, although it was not actually published until 1747, when it appeared in the *Comentarii* of the Bologna Academy of Sciences), the integration of an equation of the type $r = f(s)$, in which r is the radius of curvature and s the length of the arc. His result is significant because it bears upon the search for the Cartesian equation of a curve determined by its intrinsic equation. Riccati had already solved a general problem of the same type, in determining a curve of which the expression for the radius of curvature is known at any point whatever as a function of the radius vector ("Soluzione generale del problema inverso intorno ai raggi osculatori," published in the *Giornale* in 1712).

In his "Animadversiones in aequationes differentiales secundi gradus," published in *Acta eruditorum* in 1724, Riccati suggested the study of cases of integrability of the equation

$$X^m dx = dy + \frac{y^2\, dx}{X^n},$$

which is now known by his name. In response to this suggestion Nikolaus II Bernoulli wrote an important treatise on the equation and Daniel Bernoulli presented, in his *Exercitationes quaedam mathematicae*, the conditions under which it may be integrated by the method of separation of the variables. Euler also integrated it.

Riccati also drew upon the integration of differential equations in the context of his work in Newtonian mechanics. Thus, he studied the motion of cycloidal pendulums under the hypothesis that the force of resistance varies as the square of the velocity, and, in the results that he communicated to his former student Suzzi (on 5 March 1732), included the integration, by means of halved separation, of an equation in two variables, namely the arc s and the velocity u. In a memoir on the laws of resistance governing the retardation of the motion of bodies by a fluid medium he integrated, using a procedure different from that used by Manfredi, the homogeneous equation $ydy + 2\,budy = udu$.

In differential geometry, Riccati demonstrated that the segment lying between an arc and a tangent at point P of the ordinate at the point of the curve following P is an infinitesimal of the second order. In addition, he also studied a problem that had interested Descartes and Fermat—how to determine the algebraic curves of the minimum degree that must be employed to solve a geometric problem of a given order.

Riccati carried on an extensive correspondence with mathematicians all over Europe. His works were collected and published, four years after his death, by his sons, of whom two, Vincenzo and Giordano, were themselves eminent mathematicians.

BIBLIOGRAPHY

I. ORIGINAL WORKS. Riccati's complete works were published as *Opere del conte Jacopo Riccati*, 4 vols. (Lucca, 1761–1765). Vol. I contains an essay on the system of the universe and a treatise on the indeterminates in differential equations of the first order, and the reduction of those of the second and of higher orders. Vol. II deals with the principles and methods of physics; vol. III, with physiomathematical subjects; and vol. IV, with philosophical, ecclesiastical, rhetorical, practical, and scholarly topics.

II. SECONDARY LITERATURE. On Riccati and his work, see A. Agostini, "Riccati," in *Enciclopedia italiana*, XXIX (1936), 241; L. Berzolari, G. Vivanti, and D. Gigli, eds., *Enciclopedia delle matematiche elementari*, I, pt. 2 (Milan, 1932), 527; A. Fabroni, *Vitae italorum doctrina excellentium*, XVI (Pisa, 1795), 376 ff.; and G. Loria, *Curve piane speciali algebriche e trascendenti*, II (Milan, 1930), 168, 170; and *Storia delle matematiche*, 2nd ed. (Milan, 1950), 630, 631, 659 ff., 667, 701.

A. NATUCCI

RICCATI, VINCENZO

(*b.* Castelfranco, near Treviso, Italy, 11 January 1707; *d.* Treviso, 17 January 1775), *mathematics*.

Riccati was the second son of the mathematician Jacopo Francesco Riccati. He received his early education at home and under the auspices of the Jesuits, whose order he entered on 20 December 1726. In 1728 he went to teach literature in the Jesuit college in Piacenza; the following year he was sent to the college in Padua, where he remained until he was transferred to Parma in 1734. At some subsequent date he went to Rome to study theology, then, in 1739, returned to Bologna, where for the next thirty years he taught mathematics in the College of San Francesco Saverio. Like his father, Riccati was also skilled in hydraulic engineering and, under government commissions, carried out flood control projects along the Reno, Po, Adige, and Brenta rivers. He was much honored for this work, which saved the Venetian and Bolognan regions from disastrous flooding, and was made one of the first members of the Società dei Quarante. When Pope Clement XIV suppressed the Society of Jesus in 1773, Riccati retired to his family home in Treviso, where he died two years later.

Riccati further followed his father's example in studying the integration of differential equations, including some derived from geometrical problems. He, too, was well informed concerning pre-Eulerian mathematical analysis, and took his topics from other eminent mathematicians. Thus, a memoir by Johann I Bernoulli led him to consider the relationship between the lengths of two curves and a treatise by Jakob Hermann prompted him to suggest some methods whereby the conic equations of Cartesian coordinates might be discussed. He was also concerned with the rectification of conic sections and studied elliptic integrals as an introduction to the theory of elliptic functions.

Riccati's principal works in mathematics and physics were published in his two-volume *Opusculorum ad res physicas et mathematicas pertinentium* (1757–1762). He here introduced the use of hyperbolic functions to obtain the roots of certain types of algebraic equations, particularly cubic equations. He discussed this method further in the three-volume *Institutiones analyticae* (1765–1767), which he wrote in collaboration with Girolamo Saladini. In the latter work Riccati for the first time used the term "trigonometric lines" to indicate circular functions. He demonstrated that just as in a circle of radius 1, the coordinates of the extremity of an arc of φ length may be considered to be functions of twice the area of the sector determined by the arc $x = \cos \varphi$, $y = \sin \varphi$, so in an equilateral hyperbola, the coordinates of a point may be expressed as a function of twice the area of a hyperbolic sector w. He thus was able to make use of hyperbolic functions possessing properties similar to those of circular functions, obtaining (in modern notation)

$$\cosh w = \frac{e^w + e^{-w}}{2}, \qquad \sinh w = \frac{e^w - e^{-w}}{2},$$

from which the relations $\cosh^2 w - \sinh^2 w = 1$; $\sinh w + \cosh w = e^w$; and $\cosh 0 = 1$; $\sinh 0 = 0$.

Riccati and Saladini (who published a commentary on the process in Italian) then went on, in the second volume, to establish the formulas for the addition and subtraction of hyperbolic functions as well as the general formulas $2 \sinh nw = (\cosh w + \sinh w)^n - (\cosh w - \sinh w)^n$; $2 \cosh nw = (\cosh w + \sinh w)^n + (\cosh w - \sinh w)^n$. They were then able to calculate the derivatives of sinh w and cosh w, which they deduced from the geometric properties of the hyperbola. Riccati and Saladini thus anticipated Lambert in his study of hyperbolic functions, although Lambert, who published his findings in 1770, is often cited as having been the first to mention them.

Riccati and Saladini also considered the principle of the substitution of infinitesimals in the *Institutiones analyticae*, together with the application of the series of integral calculus and the rules of integration for certain classes of circular and hyperbolic functions. Their work may thus be considered to be the first extensive treatise on integral calculus, predating that of Euler. Although both Newton and Leibniz had recognized that integration and derivation are inverse operations, they had defined the integral of a function as a second function from which the former function is derived; Riccati and Saladini, on the other hand, considered differentiation to be the division of a quantity into its elements and integration to be the addition of these elements and offered examples of direct integrations.

Riccati's geometrical work includes a study, published in 1755 as "De natura et proprietatibus quarundam curvarum quae simul cum tractrice generantur, quaeque proinde syntractoriae nominabuntur," in which he examined the location of the points that divide the tangents of a tractrix in a certain relationship. Leibniz and Huygens also studied this curve, which may be defined as the locus of points so taken that the segment of the tangent between the point of contact and the intersection with a fixed straight line will be of constant length. Thus, given a cone of revolution with vertex V and a generator g, let t be a tangent perpendicular to g; each plane π passing through t will cut the cone in a conic, Γ, of which F_1 and F_2 are the foci. Rotating the plane π around the tangent t produces a curve called the strophoid, which had been discovered in France, possibly by Roberval. This curve was further studied by De Moivre, in 1715, and later by Gregorio Casali; Riccati and Saladini discussed it in the first volume of *Institutiones analyticae*.

Riccati and Saladini also considered the figure of the four-leaf rose, introduced by Guido Grandi, and further discussed the problem posed by Ibn al-Haytham in which given two points, A and B, it is required to find on a circular mirror a point C so located that a ray of light starting from A and reflected by the mirror at C passes through B. Ibn al-Haytham himself offered only a tortuous and confused solution to this problem, but in 1676 a simple geometrical solution was stated by Huygens, whose result Riccati and Saladini refined and further simplified. In the second volume of their work, they generalized a problem that was proposed to Descartes by Debeaune, then solved by Johann I Bernoulli and by L'Hospital.

BIBLIOGRAPHY

I. ORIGINAL WORKS. Riccati's works include *Opusculorum ad res physicas et mathematicas pertinentium*, 2 vols. (Bologna, 1757–1762); and *Institutiones analyticae*, 2 vols. (Bologna, 1765–1767), written with G. Saladini.

II. SECONDARY LITERATURE. On Riccati and his work, see Amedeo Agostini, "Riccati," in *Enciclopedia italiana*, XXIX (1936), 241; L. Berzolari, G. Vivanti, and D. Gigli, eds., *Enciclopedia delle matematiche elementari*, I, pt. 2 (Milan, 1932), 389, 478, 491; III, pt. 2 (1950), 826; and Gino Loria, *Curve piane speciali algebriche e trascendenti* (Milan, 1930), I, 72, 231, 427 and II, 153; and *Storia delle matematiche*, 2nd ed. (Milan, 1950), 663, 681, 706, 725.

A. NATUCCI

RICCI, MATTEO (*b*. Macerata, Italy, 6 October 1552; *d*. Peking, China, 11 May 1610), *mathematics, astronomy, geography, sinology.*

Ricci was the son of Giovanni Battista Ricci, a pharmacist, and Giovanna Angiolelli. In 1568 he went to Rome to study law, but in 1571 he joined the Jesuits and in 1572 was enrolled at the Collegio Romano, where he studied until 1577. One of his professors was the renowned Clavius. Ricci left Rome in 1577 when he was ordered to the missions in the Orient. He sailed from Lisbon for Goa, and from there moved on to Macao in 1582. In 1583 he entered the Chinese Empire, settling at Ch'ao-ching (Shiuhing), in Kwantung province. This expedition was the beginning of modern Catholic missions in China. After establishing missions in different parts of the empire, in 1601 Ricci finally settled in Peking, where, under the protection of the Emperor Wan-li, he remained until his death.

The success of Ricci's missionary activity was due not only to his personal high qualities and to his complete adaptation to China, both in customs and in language, but also to his authoritative knowledge of the sciences, especially mathematics, astronomy, and geography. He disseminated Western science by lecturing, publishing books and maps, and making instruments.

Besides his books in Chinese on religious and moral topics (including *Basic Treatise on God*; *Christian Doctrine*; *Treatise on Friendship*; and *Ten Paradoxes*),

Ricci is remembered for his Chinese works in the sciences, generally translations or shortened versions of works of Clavius. His Chinese pupils helped him with the Chinese literary style. These works comprised the *Astrolabe, Sphere, Arithmetic, Measures,* and *Isoperimeters.* But especially important was his Chinese version of the first six books of Euclid's *Elements,* also from the Latin text of Clavius. Entitled *A First Textbook of Geometry,* this work assures Ricci an important place in the history of mathematics. Written in collaboration with his pupil Hsu Kuang-ch'i, it was published at Peking in 1607. In about 1672 it was translated into Tatar at the suggestion of the Emperor K'ang Hsi. The work was completed in 1865, with the translation of the remaining books of Euclid, by the English Protestant missionary Alexander Wylie and the Chinese mathematician Li Shan-lan.

Ricci's map of the world is important in the history of geography. It was published at Ch'ao-ching in 1584 and at Nanking in 1600; later editions, one issued at the special request of the emperor, appeared at Peking. For the first time the Chinese had a complete idea of the distribution of the oceans and landmasses. Very few authentic copies of the map are known today. The copy at the Vatican Library (Peking, 1602) is entitled "Complete Geographical Map of All Kingdoms." It is an oval planisphere, on a folding screen of six panels, each seventy-and-a-half inches (1.79 meters) high and twenty-seven inches (0.69 meters) wide, with numerous illustrations and legends.

Ricci's other important contributions to geography were his calculation of the breadth of China in latitude (three-quarters the breadth assumed by Western geographers) and his identification of China and Peking with the Cathay and Cambaluc of Marco Polo. He shares the latter recognition with another Jesuit, Benedetto de Góis, who made a journey from India to China (1602–1605).

Ricci's life and activities are also documented in his letters, written in Italian and Portuguese, and in an extensive report, *Della entrata della compagnia di Giesù e Christianità nella Cina.* He was proposed for beatification in 1963 at the Second Ecumenical Vatican Council.

BIBLIOGRAPHY

I. Original Works. For a bibliography of Ricci's works, see Louis Pfister, *Notices biographiques et bibliographiques sur les Jésuites de l'ancienne mission de Chine 1552–1773,* 2 vols. (Shanghai, 1932–1934), I, 22–42; II, 9*–10*; Henri Bernard, "Les adaptations chinoises d'ouvrages européens: bibliographie chronologique depuis la venue des Portugais à Canton jusqu'à la mission française de Pekin 1514–1688," in *Monumenta serica,* **10** (1945), 1–57, 309–388; and Pasquale M. D'Elia, *Fonti Ricciane* (cited below), esp. III, 239–243.

Modern eds. of Ricci's works are Pietro Tacchi Venturi, *Opere storiche del P. Matteo Ricci S.I.,* 2 vols.: I. *I commentarj della Cina,* II. *Le lettere dalla Cina* (Macerata, 1911–1913), with intros., notes, tables, and a bibliography of Ricci's Chinese works compiled by Giovanni Vacca, II, 544–548; Pasquale M. D'Elia, *Il mappamondo cinese del P. Matteo Ricci S.I.* (Vatican City, 1938), a facs. ed. based on the 3rd ed. of the map (Peking, 1602), with trans., intro., and commentary; and Pasquale M. D'Elia, *Fonti Ricciane* (Rome, 1942–1949), the first three vols. of the planned national ed. of Ricci's works, which contain *Storia dell'introduzione del cristianesimo in Cina.*

Tacchi Venturi's ed. of the *Commentarj* and the *Storia* in D'Elia's ed. reproduce the autograph text of *Dell'entrata . . .,* cited in the text of the article. This MS discovered by Tacchi Venturi in 1910 (Archivio Romano della Compagnia di Gesù, *Jap.-Sin.,* n. 106a), was known in the Latin trans. of Nicolas Trigault, *De Christiana expeditione apud Sinas a Societate Iesu suscepta* (Augsburg, 1615). There are several eds. and trans. of this work, including L. J. Gallagher, *The China That Was: China As Discovered by the Jesuits at the Close of the Sixteenth Century* (Milwaukee, Wis., 1942).

D'Elia has edited other works by Ricci: "Il trattato sull'amicizia. Primo libro scritto in cinese da Matteo Ricci S.I. (1595)," in *Studia missionalia,* **7** (1952), 449–515, contains Ricci's Chinese text, an Italian trans., and commentary; "Musica e canti italiani a Pechino," in *Rivista degli studi orientali,* **30** (1955), 131–145, includes the Chinese text of eight songs by Ricci with Italian trans. and commentary; and "Presentazione della prima traduzione cinese di Euclide," in *Monumenta serica,* **15** (1956), 161–202, which gives an Italian trans., with commentary, of Chinese texts of Ricci and Hsu Kuang-ch'i.

II. Secondary Literature. One of the most prolific writers on Ricci and his work was Pasquale D'Elia; see the bibliography of his publications (1913–1959), in *Studia missionalia,* **10** (1960), 90–112. In his *Fonti Ricciane* D'Elia collected a rich bibliography on Ricci. See also Giovanni Vacca, "L'opera di Matteo Ricci," in *Nuova antologia,* 5th ser., **149** (1910), 265–275; and "Sull'opera geografica del P. Matteo Ricci," in *Rivista geografica italiana,* **48** (1941), 66–74.

Other sources are Arnaldo Masotti, "Sull'opera scientifica di Matteo Ricci," in *Rendiconti dell' Istituto lombardo di scienze e lettere,* **85** (1952), 415–445; Joseph Needham, *Science and Civilization in China,* 4 vols. (Cambridge, 1954–1971); and *Clerks and Craftsmen in China and the West* (Cambridge, 1970), 21, 205, 397. Two recent biographies are Vincent Cronin, *The Wise Man from the West* (Glasgow, 1961); and Fernando Bortone, *P. Matteo Ricci S.I., il "Saggio d'Occidente"* (Rome, 1965).

Arnaldo Masotti

RICCI, MICHELANGELO (*b.* Rome, Italy, 30 January 1619; *d.* Rome, 12 May 1682), *mathematics.*

Although he was never ordained, Ricci served the papal court in various capacities and on 1 September 1681 was made a cardinal by Pope Innocent XI. He was a member of the school of Galileo, although not a direct disciple; his teacher was Benedetto Castelli, whose students also included Torricelli. Torricelli himself was later a close friend of Ricci, and exerted a marked influence on Ricci's geometrical researches.

Ricci's only extant mathematical work is a nineteen-page printed booklet entitled *Geometrica exercitatio* (but more usually called by a later subtitle, *De maximis et minimis*), published in Rome in 1666. It enjoyed a wide circulation and was reprinted as an appendix to Nicolaus Mercator's *Logarithmo-technia*, issued in London in 1668. The work deals primarily with two problems: finding the maximum of the product $x^m(a - x)^n$, m and n being positive integers; and applying this result to the determination of the lines tangential to the parabolas $y^m = kx^n$. It thus represents a generalization of the property by which a tangent of the ordinary second-order parabola $y^2 = kx$ ($m = 2$, $n = 1$) meets the x-axis at a point of which the distance from the vertex (changing its sign) equals the abscissa of the point of contact. It has been suggested that Ricci's method anticipates the so-called method of induction from n to $n + 1$, since he begins with the values of $m = n = 1$, which he subsequently increases. The first explicit use of the method is, perhaps, the one set out in Pascal's posthumous *Traité du triangle arithmétique* of 1665, although some possibility exists that Ricci may not have been familiar with it.

Ricci's other mathematical contributions are contained in his numerous letters. These include his study of spirals (1644), his investigation of a family of curves more general than ordinary cycloids (1674), and the methods by which he recognized fairly explicitly that the treatment of tangents is an operation inverse to that of the calculation of areas (1668). His demonstrated competence in algebra was somewhat exceptional among the followers of Galileo, most of whom were more deeply concerned with geometrical speculation.

Ricci's extensive correspondence with both Italian and foreign scholars (including physicists and astronomers, as well as mathematicians) brought him considerable contemporary fame. Through such correspondence Ricci participated in the activities of the Florentine Accademia del Cimento, particularly in the final editing of its *Saggi*, published in 1667. He also served as an editor of the *Giornale dei letterati*, which was founded in Rome in the following year.

As a cardinal, he discussed with Vincenzo Viviani the life of Galileo that the latter was preparing, advising him on matters that the church felt to be of some delicacy.

A curious aspect of Ricci's career was his refusal to edit the manuscript remains of his friend and master Torricelli, who had in his will requested that Cavalieri and Ricci do so. Cavalieri died soon after Torricelli, so that the entire task devolved upon Ricci. Stating that he had too many other occupations—and that he had been away from mathematics too long—Ricci declined the undertaking. His action has been subjected to various interpretations; as a result of it, Torricelli's complete works were published only in the twentieth century.

BIBLIOGRAPHY

I. ORIGINAL WORKS. *Geometrica exercitatio* [*De maximis et minimis*] (Rome, 1666) was summarized in *Philosophical Transactions of the Royal Society*, **3** (1668), 738–740; and was reprinted in Nicolaus Mercator, *Logarithmotechnia* (London, 1668); and in Carlo Renaldini, *Geometra promotus* (Padua, 1670).

Ricci's correspondence was published in the following: *Bullettino di bibliografia e storia delle scienze matematiche e fisiche*, **18** (1885), see index; Raffaello Caverni, *Storia del metodo sperimentale in Italia*, V (Florence, 1898); C. R. Dati, *Lettera a Filaleti di Timauro Antiate della vera storia della cicloide . . .* (Florence, 1663), repr. in *Opere di Evangelista Torricelli*, G. Loria and G. Vassura, eds., I, pt. 2 (Faenza, 1919), 441–482; Angelo Fabroni, *Lettere inedite di uomini illustri*, 2 vols. (Florence, 1773–1775); Christiaan Huygens, *Oeuvres complètes*, 22 vols. (The Hague, 1888–1950); Ferdinando Jacoli, *Una lettera inedita del Cardinale Michelangelo Ricci a Gio. Domenico Cassini*, 1895, cited in P. Riccardi, *Biblioteca matematica italiana*, II, sec. 7, col. 82; Giovanni Lami, ed., *Novelle letterarie publicate in Firenze*, XIII (Florence, 1740–1769), col. 35; Giambattista Clemente de' Nelli, *Saggio di storia letteraria fiorentina del secolo XVII* (Lucca, 1759), 190; Carlo Renaldini, *Commercium epistolicum ab eodem cum viris eruditione* (Padua, 1682); Giovanni Targioni-Tozzetti, *Atti e memorie inedite dell'Accademia del cimento*, 3 vols. (Florence, 1780); Luigi Tenca, "Relazione fra Vincenzio Viviani e Michel Angelo Ricci," in *Rendiconti dell'Istituto lombardo di scienze e lettere*, Cl. di scienze, **87** (1954), 212–228; "M. A. Ricci," in *Atti e memorie dell'Accademia patavina di scienze, lettere ed arti*, **68** (1956), 1–8; and "Michel Angelo Ricci," in *Torricelliana*, **11** (1960), 5–13; Girolamo Tiraboschi, *Storia della letteratura italiana*, VIII (Venice, 1825), 554; and V. P. Zubov, "Iz perepiski mezhdu Evandzhelista Torrichelli i Mikelandzhelo Richi" ("From the Correspondence Between Evangelista Torricelli and Michelangelo Ricci"), in *Voprosy istorii estestvoznaniya i tekhniki*, **8** (1959), 95–101, which includes three letters, in Russian.

Fragments of Ricci's correspondence were also published in association with the following eds. of *Saggi di naturali esperienze fatte nell'Accademia del cimento:* Vincenzo Antinori, ed., 3rd ed. (Florence, 1841); and Giorgio Abetti and Pietro Pagnini, eds., *Le opere dei discepoli di Galileo Galilei, Edizione nazionale,* I, *L'Accademia del Cimento,* pt. 1 (Florence, 1942).

There does not appear to be any systematic study of Ricci's MS remains. There are quite possibly some fragments at the Bibliothèque Municipale, Toulouse; see the article by Costabel cited below. Other MSS are at the Biblioteca Apostolica Vaticana, Vatican City; Biblioteca Comunale and Museo Torricelliano, Faenza; and in the Galileiana MSS at the Biblioteca Nazionale Centrale, Florence. A substantial portion of the published correspondence derives from the Faenza and Florence collections; see esp.: Angiolo Procissi, "I Mss. Torricelliani conservati a Firenze," in *Evangelista Torricelli nel terzo centenario della morte* (Florence, 1951), 77–112. Indirect citations of Ricci and clues to the locations of other MSS can be found in various published collections of seventeenth-century correspondence, esp. B. Boncompagni, "Intorno ad alcune lettere di Evangelista Torricelli, del P. Marino Mersenne e di Francesco di Verdus," in *Bullettino di bibliografia e storia delle scienze matematiche e fisiche,* 8 (1875), 353–456; *Correspondance du P. Marin Mersenne* (Paris, 1932–); and M. C. Le Paige, "Correspondance de René-François de Sluse publiée pour la première fois et précédée d'une introduction," in *Bullettino di bibliografia e storia delle scienze matematiche e fisiche,* 17 (1884), 427–554, 603–726.

II. SECONDARY LITERATURE. On Ricci and his work, see the following: Amedeo Agostini, "Massimi e minimi nella corrispondenza di E. Torricelli con M. Ricci," in *Atti del IV Congresso dell'Unione matematica italiana,* II (Rome, 1953), 629–632; Davide Besso, "Sopra un opusculo di Michelangelo Ricci," in *Periodico di matematica per l'insegnamento secondario,* 8 (1892), 1–16; Pierre Costabel, "Un registre de manuscrits témoin de l'activité de Mersenne en Italie en 1645," in *Revue d'histoire des sciences et de leurs applications,* 22, no. 2 (1969), 155–162; Angelo Fabroni, *Vitae italorum doctrina excellentium,* II (Pisa, 1778), 200–221; Mario Gliozzi, "Origini e sviluppi dell'esperienza torricelliana," in *Opere di Evangelista Torricelli,* IV (Faenza, 1919), 231–294; Josef E. Hofmann, "Über die 'Exercitatio geometrica' des M. A. Ricci," in *Centaurus,* 9 (1964), 139–193; and Ferdinando Jacoli, "Evangelista Torricelli ed il metodo delle tangenti detto 'metodo del Roberval,' " in *Bullettino di bibliografia e storia delle scienze matematiche e fisiche,* 8 (1875), 265–304.

Two collections published in Faenza that deal primarily with Torricelli also include material on Ricci: *Torricelliana, pubblicate dalla commissione per le onoranze a Evangelista Torricelli, III centenario della scoperta del barometro,* 2 vols. (1945–1946); and the annual *Torricelliana, Bollettino della Società Torricelliana di scienze e lettere* (1949–). Every issue through 15 (1964) has articles mentioning Ricci.

Other sources on Ricci include Étienne Charavay, *Lettres autographes composant la collection de M. Bovet*

Alfred (Paris, 1885); Mario Guarnacci, *Vita et res gestae pontificum Romanorum,* I (Rome, 1751), cols. 189–194; Prospero Mandosio, *Biblioteca romana,* I (Rome, 1682), 344; Gabriel Maugain, *Étude sur l'évolution intellectuelle de l'Italie de 1657 à 1750 environ* (Paris, 1909); and Gaetano Moroni, *Dizionario di erudizione storico-ecclesiastica,* LVII (Venice, 1852), 177.

LUIGI CAMPEDELLI

RICCI, OSTILIO (*b.* Fermo, Italy, 1540; *d.* Florence [?], Italy, 15 January 1603), *mathematics, military engineering.*

Ricci was the son of Orazio Ricci and Elisabetta Gualteroni, patricians of Fermo. It is not known with any certainty where he studied mathematics, but some of his intellectual influences perhaps may be conjectured: Leonardo Olschki has likened his teaching to that of Niccolò Tartaglia, and Thomas Settle has pointed out a remarkable connection of Ricci with Leon Battista Alberti (see below). Ricci began, probably in 1580, to teach mathematics and likely military engineering to the pages of Francesco de' Medici, grand duke of Tuscany. In 1583, he also gave instruction to Galileo, the son of his friend Vincenzo Galilei, who was then nineteen years old and a medical student at the University of Pisa. Under Ricci's tutelage, Galileo studied Euclid and, later, Archimedes (a set of whose works Ricci gave to Galileo). Galileo also attended, with Ludovico Cardi da Cigoli and Giovanni de' Medici, the lessons on perspective that Ricci gave at the house of Bernardo Buontalenti in Florence, presumably around 1585. When Galileo applied for a chair at the University of Bologna in 1587, Ricci recommended him; he was also helpful in attaining for Galileo the chair of mathematics at the University of Pisa two years later.

Ricci was also active as an engineer. Around 1590 he was in Ferrara to study the courses of streams in that region and in the area of Bologna, a subject on which he wrote a report. He returned to Florence (from 1593 he taught at the Academy of Design) and in 1597, during a conflict between Tuscany and France, he directed the construction of fortifications on the islands off Marseilles. Later, according to his biographer Carlo Promis, he worked as a military engineer in Ferrara during the controversy between Pope Clement VIII and Cesare d'Este in 1597–1598.

Although he wrote on both mathematics and engineering, Ricci did not publish his works. Of the manuscripts that remain, two are of particular importance in establishing Ricci's influence on Galileo. One of these, a mathematical manuscript attributed to

Ricci and probably used by him in his teaching, has been identified by Thomas Settle as a copy of Alberti's *Ludi matematici.* Settle emphasizes Ricci's role as the spiritual intermediary between Alberti and Galileo; Alberti's influence, transmitted by Ricci, is apparent in Galileo's thought and experimental methods, and in some of his specific works, particularly *La bilancetta.* The other, a tract entitled "Libro primo delle fortificationi di M. Hostilio Ricci da Fermo," was discovered in Pesaro by Promis, who, in his biography of Ricci, noted the similarity between it and the treatise on fortifications by Galileo, and suggested that Galileo had probably been instructed in this subject by Ricci.

Of the few other manuscripts by Ricci, there may be mentioned a brief treatise taken from one of the Florentine manuscripts; it was published in 1929 by Federico Vinci under the title "L'uso dell'archimetro ovvero del modo di misurare con la vista." The manuscript is dated 1590, and deals with the use of the "archimeter," a simple instrument for the visual measurement of inaccessible distances, heights, and depths through the properties of similar triangles. Another manuscript, of which there is mention under the title "Intorno ad una leva ad argano," is now apparently lost. A manuscript of Giorgio Vasari (Rome, Biblioteca Angelica, n. 2220) mentions Ricci as solver of a peculiar geometrical question.

BIBLIOGRAPHY

I. ORIGINAL WORKS. "L'uso dell'archimetro" exists in two MSS at the National Library in Florence: Codici Magliabechiani II, 57, and VII, 380; the former was published in Federico Vinci, *Ostilio Ricci da Fermo, maestro di Galileo Galilei* (Fermo, 1929), 23–29, with nine facsimile figures. This is presumably the same work mentioned by Targioni-Tozzetti (p. 298) and Promis (p. 349) under the title *L'uso dell'aritmetica;* it is likely also the work described as concerning "il modo di misurare colla vista" by Nelli (p. 35), Fracassetti (pp. 30, 103), and Promis (p. 349), for which see below.

MS Gal. 10 (div. 1, anteriori, vol. X) in the Galilean Collection of the National Library, Florence, is entitled "Ricci Ostilio. Problemi geometrici." It is described in Angiolo Procissi, *La collezione Galileiana nella Biblioteca nazionale di Firenze,* I (Rome, 1959), 10. Fols. 1a–16a are derived from Alberti (Settle, pp. 121, 124). Settle also cites (p. 125, notes 6, 8, 9) various biographical documents on Ricci in the state archives of Florence, one of which gives the date of Ricci's death.

For Ricci's works on military fortifications, see Promis (pp. 341, 347–348) and the catalogs of the Campori collection, which is now at the Estense Library, Modena: Luigi Lodi, *Catalogo dei codici e degli autografi posseduti dal marchese Giuseppe Campori* (Modena, 1875), 273, art. 622; and Raimondo Vandini, *Appendice prima al catalogo dei codici e manoscritti posseduti dal marchese Giuseppe Campori* (Modena, 1886), 250, art. 753.

On Ricci's report on the waters of the Ferrara–Bologna region, see Frizzi, V, 28; and Promis, p. 343.

II. SECONDARY LITERATURE. Publications concerning Ricci's relations with Galileo are of particular interest. Mentions of Ricci in Galileo's application of 1587 and in the biographies of Galileo by Vincenzo Viviani and Niccolò Gherardini are in *Le opere di Galileo Galilei,* Antonio Favaro, ed., XIX (Florence, 1907), 36, 604–605, 636–638; Gherardini mistakenly called Ricci a priest, and Libri later called him an abbé. Ricci as a teacher is mentioned in a biography of Cigoli written in 1628 by his nephew Giovan Battista Cardi and published by Guido Battelli, *Vita di Lodovico Cardi Cigoli* (Florence, 1913), 14.

An early biography of Ricci is Giuseppe Santini, *Picenorum mathematicorum elogia* (Macerata, 1779), 51–52. In 1830 Giuseppe Fracassetti delivered to the Accademia Tiberina of Rome his *Elogio di Messer Ostilio Ricci da Fermo* (Fermo, 1830); Fracassetti later contributed a biography of Ricci to Antonio Hercolani, ed., *Biografie e ritratti di uomini illustri Piceni,* I (Forli, 1837), 97–106. See also Carlo Promis' biography of Ricci in "Gli ingegneri militari della Marca d'Ancona . . .," in *Miscellanea di storia italiana,* 6 (1865), 339–349. Vinci (see above) includes a biography of Ricci (pp. 7–21), his coat of arms, and notes on his family; there are also citations of biographical works on Ricci by Mistichelli (1844) and Giannini (1874), and of unpublished MSS by Eufemio Vinci on the nobility and leading men of Fermo, in the historical archives of the Vinci family in Fermo.

See also Giovan Battista Clemente de' Nelli, *Vita e commercio letterario di Galileo Galilei* (Florence, 1793), 35–36, 46, 797; Antonio Favaro, *Galileo Galilei e lo Studio di Padova* (Florence, 1883), I, 16–19, 23, 31—new ed. (Padua, 1966), I, 13–15, 18, 24; Antonio Frizzi, *Memorie per la storia di Ferrara,* V (Ferrara, 1809), 28; Riguccio Galluzzi, *Istoria del Granducato di Toscana sotto il governo della Casa Medici* (Florence, 1781), III, 291; also in Capolago, 1841 ed., V, 67; Guillaume Libri, *Histoire des sciences mathématiques en Italie,* IV (Paris, 1841; 2nd ed., Halle, 1865; repr. Bologna, 1967), 173–174; Ernan McMullin, ed., *Galileo, Man of Science* (New York, 1967), 53, 122, 234–235; Leonardo Olschki, *Geschichte der neusprachlichen wissenschaftlichen Literatur,* III, *Galilei und seine Zeit* (Halle, 1927; repr. Vaduz, 1965), 141–153; Thomas B. Settle, "Ostilio Ricci, A Bridge Between Alberti and Galileo," in *Acts of the Twelfth International Congress on the History of Sciences,* IIIB (Paris, 1971), 121–126; and Giovanni Targioni-Tozzetti, *Notizie sulla storia delle scienze fisiche in Toscana* (Florence, 1832), 298, 300.

ARNALDO MASOTTI

RICCI-CURBASTRO, GREGORIO (*b.* Lugo, Italy, 12 January 1853; *d.* Bologna, Italy, 6 August 1925), *mathematics, mathematical physics.*

Ricci-Curbastro[1] was the son of a noble family situated in the province of Ravenna. His father, Antonio Ricci-Curbastro, was a well-known engineer; his mother was Livia Vecchi. With his brother Domenico, Ricci received his elementary and secondary education from private teachers; he then, in 1869, entered the University of Rome to study philosophy and mathematics. After a year of study he returned home, and it was only in 1872 that he enrolled at the University of Bologna. The following year transferred to the Scuola Normale Superiore of Pisa, where he attended the courses of Betti, Dini, and Ernesto Padova. In 1875 Ricci defended a thesis entitled "On Fuchs's Research Concerning Linear Differential Equations," for which he received the degree of doctor of physical and mathematical sciences. The following year—in conformity with the then existing requirements for teaching—he presented a paper "On a Generalization of Riemann's Problem Concerning Hypergeometric Functions."[2] Betti then asked Ricci to write a series of articles on electrodynamics, particularly Maxwell's theory, for *Nuovo Cimento*. Under the influence of Dini, Ricci took up Lagrange's problem of a linear differential equation, on which he contributed a nineteen-page article to the *Giornale di matematiche di Battaglini*. Shortly afterward, having won a competition for a scholarship to study abroad, he spent a year (1877–1878) in Munich, where he attended the lectures of Felix Klein and A. Brill. Ricci greatly admired Klein, and his esteem was soon reciprocated; nevertheless, Ricci does not seem to have been decisively influenced by Klein's teaching. It was, rather, Riemann, Christoffel, and Lipschitz who inspired his future research. Indeed, their influence on him was even greater than that of his Italian teachers.

In 1879 Ricci worked as Dini's assistant in mathematics at Pisa. Then, on 1 December 1880, he was named professor of mathematical physics at the University of Padua, a position that he held without interruption for forty-five years. In 1891 he also began to teach higher algebra.

Ricci is best known for the invention of absolute differential calculus, which he elaborated over ten years of research (1884–1894). With this new calculus he was able to modify the usual procedures of the differential calculus in such a way that the formulas and results retain the same form whatever the system of variables used. This procedure requires the employment of systems of functions that behave, when a change of variables is made, like coefficients of expressions that are themselves independent (whether by nature or by convention) of the choice of variables. A further requirement is the introduction of an invariant element (called an absolute, from which the calculus takes its name), that is to say, an element that can also be used in dealing with other systems. The absolute that best lends itself to this operation is the quadratic differential form, which expresses, geometrically, the elementary distance between two points.

Ricci's attention was first drawn to the theory of the invariants of algebraic forms, which had been developed principally after Riemann wrote his thesis,[3] and to the works of Christoffel and of Lipschitz on the quadratic forms.[4] But it was essentially Christoffel's idea of covariant derivation that allowed Ricci to make the greatest progress. This operation, which possesses the characteristics of ordinary derivation, has the additional property of preserving, with respect to any change of variables, the invariance of the systems to which it is applied. Ricci realized that the methods introduced and utilized by these three authors required fuller development and were capable of being generalized. Their methods furnished the basis of Ricci's works on the quadratic differential forms (1884 and 1888) and on the parameters and the differential invariants of the quadratics (1886), which Ricci reduced to a problem of algebra. The method he used to demonstrate their invariance led him to the technique of absolute differential calculus, which he discussed in its entirety in four publications written between 1888 and 1892.

In 1893 Ricci revealed the first applications of his algorithm, to which he gave its specific name for the first time. Two years later, Klein urged him to make his methods more widely available in a complete exposition, but Ricci did not do so until five more years had passed. Meanwhile, he prepared a long paper on intrinsic geometry (published in the *Memorie dell'Accademia dei Lincei* in 1896), in which he examined the congruences of lines on an arbitrary Riemannian variety. He applied the absolute calculus to these problems by means of a special form given to the differential equations of the congruences, which appear with their covariant and contravariant systems and in this way arrived at the notion of a canonical orthogonal system of a given congruence. (In this case the coefficients of rotation replace the Christoffel symbols of absolute calculus.) Ricci next utilized the Riemann symbols to find the contract tensor (today called Ricci's tensor) that plays a fundamental role in the general theory of relativity. He also discovered invariants that occur in the theory of the curvature of varieties.[5]

This intrinsic geometry completed one stage in the development of absolute calculus, and Ricci was now in a position to fulfill Klein's earlier request. In collaboration with Levi-Civita, he published a seventy-seven-page memoir entitled "Méthodes de calcul différentiel absolu et leurs applications." The following brief discussion of the paper is, of necessity, limited to the simplest expressions used by Ricci and Levi-Civita.[6]

Given a change of variables $x_1, ..., x_n$ into

$$y_1, \cdots, y_n:$$
$$a_1 \, dx_1 + \cdots + a_n \, dx_n = b_1 \, dy_1 + \cdots + b_n \, dy_n,$$

one also has

$$b_i = \sum_j a_j \frac{\partial x_j}{\partial y_i}.$$

The system a_j is then said to be covariant of order 1. This will be the case if the a_j are the derivatives of a function $\varphi(x_1, ..., x_n)$. A system of arbitrary order m can then be generalized, from which a system

$$Y_{r_1 r_2 \cdots r_m}$$

may be obtained. The elements $dx_1, ..., dx_n$ form a contravariant system of order 1, which is written

$$dy_i = \sum_k \frac{\partial y_i}{\partial x_k} \cdot dx_k.$$

From this expression a system of order m,

$$Y^{(r_1 r_2 \cdots r_m)}$$

may be derived.

Next, a quadratic form is selected. Called the fundamental form,

$$\varphi = \sum a_{rs} \cdot dx_r \, dx_s,$$

this is an n-dimensional linear element of a variety V_n, with the a_{rs} forming a covariant system of order 2. The a^{rs} is established and generalized to

$$X^{(r_1 r_2 \cdots r_m)},$$

which is called reciprocal to the covariant system $X_{s_1 s_2 \cdots s_m}$ with respect to φ. With the equalities established by Christoffel,[7] it is possible to find formulas for deriving, from any covariant system of order m, a covariant system of order $m + 1$. This is what Ricci called covariant derivation based on φ. The contravariant derivation of a contravariant system is then defined by passing to the reciprocal system, which is derived, and returning again to the reciprocal system.

A chapter on intrinsic geometry as an instrument of computation deals with normal congruences, geodesic lines, isothermal families of surfaces, the canonical system with respect to a given congruence, and the canonical forms of the systems associated with the fundamental form. With regard to the last problem, Ricci started with a system X_r to which he associated a congruence defined by the equations

$$\frac{dx_1}{X^{(1)}} = \frac{dx_2}{X^{(2)}} = \cdots \frac{dx_n}{X^{(n)}},$$

whose covariant coordinated system will result from the elements $\lambda_{n/r} = x_r : \rho$ with $\rho^2 = \sum_1^n r X^{(r)} X_r$. The formulas $X_r = \rho \cdot \lambda_{n/r}$ furnish the canonical expressions of the X_r.

The authors then show how to proceed in order to arrive at general rules. The succeeding chapters are devoted to analytical, geometric, mechanical, and physical applications.

Analytical applications include classification of the quadratic forms of differentials; absolute invariants and fundamental invariants of the form φ; and differential parameters.

Geometric applications cover a study of two-dimensional varieties; remarks on surfaces of ordinary space; an extension of the theory of surfaces to linear spaces of n dimensions; groups of motions in an arbitrary variety; a complete study of the groups of motions of a three-dimensional variety; and comments on the relationship of this research with that done by Lie and Bianchi.

Mechanical applications include first integrals of the equations of dynamics. Here Ricci solved the Lagrange equations with respect to the second derivatives of the coordinates and found that

$$x_i'' = X^{(i)} - \sum_1^n rs \begin{Bmatrix} rs \\ i \end{Bmatrix} x_r' x_s'.$$

This is the form best suited to the question under examination. If, in seeking a function f of the x's and of the x'''s, it is desired that $f = $ constant be a first integral of the equations, then certain conditions must be satisfied. The latter, applied to the case in which there are no forces, yield the homogeneous integrals of the geodesics of the variety V_n, whose length ds^2 is expressed by $2T \, dt^2$.

Linear integrals, the quadratics, and the conditions of existence are then considered. Finally, Ricci and Levi-Civita took up surfaces whose geodesics possess a quadratic integral and the transformation of the equations of dynamics.

In their treatment of physical applications Ricci and Levi-Civita first examined the problem of the reducibility to two variables of the equation $\Delta u = 0$ (binary potentials), then went on to consider vector fields, and finally, equations in general coordinates of electrodynamics, of the theory of heat, and of elasticity.

The authors set forth a general statement of their work in their preface:

> The algorithm of absolute differential calculus, the *instrument matériel* of the methods,... can be found complete in a remark due to Christoffel. But the methods themselves and the advantages they offer have their *raison d'être* and their source in the intimate relationships that join them to the notion of an *n*-dimensional variety, which we owe to the brilliant minds of Gauss and Riemann.... Being thus associated in an essential way with V_n, it is the natural instrument of all those studies that have as their subject such a variety, or in which one encounters as a characteristic element a positive quadratic form of the differentials of *n* variables or of their derivatives.

In mechanics this is the case for kinetic energy, and it later proved to be the case, in general relativity, for the elementary interval between two events in space-time. Meanwhile, however, Ricci's methods—which Beltrami judged important, while adopting a prudent and reserved attitude toward them—were not known beyond the restricted circle of his students, and the memoir in the *Annalen* did not evoke a particularly enthusiastic response.

In 1911 Ricci and Levi-Civita sent to the *Bulletin des sciences mathématiques* a detailed exposition of the absolute calculus. The editors of the journal published it in abridged form with the comment that "essentially, it is only a calculus of differential covariants for a quadratic form," while adding that it was "very interesting."

Ricci was now almost sixty, and more than twenty-seven years had passed since he had begun his initial research. He probably was not aware that at the Zurich Polytechnikum, Marcel Grossmann, a colleague of Albert Einstein, had an intuition that only Ricci's methods could permit the expression of the quadrimensional metric of ds^2. And, indeed, it was by means of absolute differential calculus that Einstein was able to write his gravitational equations,[8] and on more than one occasion he paid tribute to the efficacy of this tool and to Ricci.[9]

In 1917 Levi-Civita, Ricci's brilliant student, introduced, with his new concept of parallel transport,[10] the geometric foundation of the algorithms of invariance, and Ricci's calculus gave rise to a series of developments and generalizations that confirmed its validity.

Ricci's other publications include a book on higher algebra (containing material from his course at Padua), a book on infinitesimal analysis, and papers on the theory of real numbers, an area in which he

extended the research begun by Dedekind. Between 1900 and 1924 he published twenty-two items, most of which dealt with absolute differential calculus. His last work was a paper on the theory of Riemannian varieties, presented to the International Congress of Mathematics held at Toronto in August 1924.

Ricci was a member of the Istituto Veneto (admitted in 1892, president 1916–1918), the Reale Accademia of Turin (1918), the Società dei Quaranta (1921), the Reale Accademia of Bologna (1922), and the Accademia Pontificia (1925). He became a corresponding member of the Paduan Academy in 1905 and a full member in 1915. The Reale Accademia dei Lincei, which elected him a corresponding member in 1899 and a national associate member in 1916, published many of Ricci's works.

In addition to his activities in research and teaching, Ricci held a number of civic posts. He served as provincial councillor and assisted in public works projects, including water supply and swamp drainage, at Lugo. He was elected communal councillor of Padua, where he was concerned with public education and finance, although he declined the post of mayor. In 1884 he married Bianca Bianchi Azzarani, who died in 1914; they had two sons and one daughter.

NOTES

1. This is his complete name. It is also the way in which he signed all his works, except for the one he published with his former student Levi-Civita in 1900 in the *Mathematische Annalen*, where he kept only the first part of his name. This memoir, written in French, made its senior author famous under the simple name of Ricci, and we shall keep to this usage.
2. These first two works by Ricci have never been published.
3. B. Riemann, "Ueber die Hypothesen, welche der Geometrie zu Grunde liegen," in *Gesammelte Werke*, 2nd ed. (Leipzig, 1892), 272–287.
4. See E. B. Christoffel, "Ueber die Transformation der homogenen Differentialausdrücke zweiten Grades," in *Journal für die reine und angewandte Mathematik*, **70** (1869), 46–70, 241–245; and R. Lipschitz, "Untersuchungen in Betreff der ganzen homogenen Funktionen von *n* Differentialen," *ibid.*, 71–102.
5. Concerning the curvature of surfaces in hyperspaces, Ricci mentions, in his "Méthodes de calcul différentiel absolu" of 1900 (p. 156), a paper by Lipschitz that he considers fundamental: "Entwickelungen einiger Eigenschaften der quadratischen Formen von *n* Differentialen," in *Journal für die reine und angewandte Mathematik*, **71** (1870), 274–295. Compare also, for these questions of intrinsic geometry, F. Schur, "Ueber den Zusammenhang der Räume constanten Riemann'schen Krümmungsmaasses mit den projectiven Räumen," in *Mathematische Annalen*, **27** (1886), 537–567.
6. It should be noted that Ricci puts the upper indices (of the contravariants) in parentheses and that he always uses the sign Σ for summations.
7. $$\frac{\partial a_{ik}}{\partial x_l} = \begin{bmatrix} i\,l \\ k \end{bmatrix} + \begin{bmatrix} k\,l \\ i \end{bmatrix} \quad \text{and} \quad \begin{bmatrix} i\,k \\ l \end{bmatrix} = \sum_p a_{lp} \begin{Bmatrix} i\,k \\ p \end{Bmatrix}$$

8. These are the well-known equations:

$$G_{\mu\nu} - \tfrac{1}{2}g_{\mu\nu}G = -\kappa T_{\mu\nu}.$$

Ricci's theorem shows that the covariant derivation cancels the effects of the variation of the metric tensor (Ricci does not use the term "tensor") and operates intrinsically on geometric entities. With the aid of this theorem and of the rules of tensor contraction one can write:

$$\nabla_\lambda \cdot S_\sigma^\lambda = 0,$$

where ∇_λ is the covariant derivation with respect to x^λ and S_σ^λ is the Einstein tensor. This relationship, which is fundamental in general relativity, serves to express the principle of the conservation of energy.

9. Compare, for example, "Entwurf einer verallgemeinerten Relativitätstheorie und einer Theorie der Gravitation. I. Physikalischer Teil von Albert Einstein. II. Mathematischer Teil von Marcel Grossmann," in *Zeitschrift für Mathematik und Physik*, **62** (1913), 225–261.

10. T. Levi-Civita, "Nozione di parallelismo in una varietà qualunque," in *Rendiconti del Circolo matematico di Palermo*, **42** (1917), 173.

BIBLIOGRAPHY

I. ORIGINAL WORKS. The obituary of Ricci by Levi-Civita contains a complete list of his scientific publications, running to sixty-one titles.

Ricci's early works include "Sopra un sistema di due equazioni differenziali lineari, di cui l'una è quella dei fattori integranti dell'altra," in *Giornale di matematiche di Battaglini* (Naples), **15** (1877), 135–153; "Sopra la deduzione di una nuova legge fondamentale di elettrodinamica," in *Nuovo cimento*, 3rd ser., **1** (1877), 58–72; "Sopra il modo di agire delle forze pondero- ed elettromotrici fra due conduttori filiformi secondo R. Clausius," *ibid.*, **2** (1877), 5–27; "Sulla teoria elettrodinamica di Maxwell," *ibid.*, 93–116; "Sulla funzione potenziale di conduttori di correnti galvaniche costanti," in *Atti del Istituto veneto di scienze, lettere ed arti*, 5th ser., **8** (1882), 1025–1048; "Sulla integrazione della equazione $\Delta U = f$," *ibid.*, 6th ser., **2** (1885), 1439–1444; "Sulla classificazione delle forme differenziali quadratiche," in *Atti dell' Accademia nazionale dei Lincei Rendiconti*, 4th ser., **4** (1888), 203–207.

Ricci laid the basis of absolute differential calculus in the following four articles: "Delle derivazioni covarianti e del loro uso nella analisi applicata," in *Studi editi dalla Università Padovana a commemorare l'ottavo centenario dalla origine della Università di Bologna*, III (Padua, 1888); "Sopra certi sistemi di funzioni," in *Atti dell' Accademia nazionale dei Lincei Rendiconti*, 4th ser., **5** (1889), 112–118; "Di un punto della teoria delle forme differenziali quadratiche," *ibid.*, 643–651; and "Résumé de quelques travaux sur les systèmes variables de fonctions," in *Bulletin des sciences mathématiques*, **16** (1892), 167–189.

On the applications of absolute calculus, intrinsic geometry, varieties, and groups see "Di alcune applicazioni del calcolo differenziale assoluto alla teoria delle forme differenziali quadratiche e dei sistemi a due variabili," in *Atti del Istituto veneto di scienze, lettere ed arti*, 7th ser., **4** (1893), 1336–1364; "Dei sistemi di coordinate atti a ridurre l'elemento lineare di una superficie alla forma $ds^2 = (U + V)(du^2 + dv^2)$," in *Atti dell'Accademia nazionale dei Lincei. Rendiconti*, 5th ser., **2** (1893), 73–81; "Sulla teoria delle linee geodetiche e dei sistemi isotermi di Liouville," in *Atti del Istituto veneto di scienze, lettere ed arti*, 7th ser., **5** (1894), 643–681; "Dei sistemi di congruenze ortogonali in una varietà qualunque," in *Atti dell' Accademia nazionale dei Lincei. Memorie*, 5th ser., **2** (1896), 275–322; and "Sur les groupes continus de mouvements d'une variété quelconque à trois dimensions," in *Comptes rendus . . . de l'Académie des sciences de Paris*, **127** (1898), 344–346, 360–361.

Ricci's articles on number theory are "Saggio di una teoria dei numeri reali secondo il concetto di Dedekind," in *Atti del Istituto veneto di scienze, lettere ed arti*, 7th ser., **4** (1893), 233–281; and "Della teoria dei numeri reali secondo il concetto di Dedekind," in *Giornale di matematiche di Battaglini*, **35** (1897), 22–74.

Ricci's memoir written with Levi-Civita, "Méthodes du calcul différentiel absolu et leurs applications," in *Mathematische Annalen*, **54** (1900), 125–201, appeared in a Polish trans., in *Praec matematyczno-fizyczne*, **12** (1901), 11–94, and was reprinted in Collection de Monographies Scientifiques Étrangères (Paris, 1923).

Ricci's books are *Lezioni di algebra complementare* (Padua–Verona, 1900); and *Lezioni di analisi infinitesimale. Funzioni di una variabile* (Padua, 1926).

His last publications include "Sulla determinazione di varietà dotate di proprietà intrinseche date a priori," in *Atti dell'Accademia nazionale dei Lincei. Rendiconti*, 5th ser., **19** (1910), 181–187 (first semester), and 85–90 (second semester); "Di un metodo per la determinazione di un sistema completo di invarianti per un dato sistema di forme," in *Rendiconti del Circolo matematico di Palermo*, **33** (1912), 194–200; "Sulle varietà a tre dimensioni dotate di terne principali di congruenze geodetiche," in *Atti dell'Accademia nazionale dei Lincei. Rendiconti*, 5th ser., **27** (1918), 21–28, 75–87; "Riducibilità delle quadriche differenziali e ds^2 della statica einsteiniana," *ibid.*, **31** (1922), 65–71; and "Di una proprietà delle congruenze di linee tracciate sulla sfera di raggio eguale ad 1," *ibid.*, **32** (1923), 265–267.

See also L'Unione Matematica Italiana, ed., *Opere de Ricci*, 2 vols. (1956–1957).

II. SECONDARY LITERATURE. The first account of Ricci's life and works is the excellent one by Levi-Civita, "Commemorazione del socio nazionale prof. Gregorio Ricci-Curbastro, letta dal socio T. L.-C. nella seduta del 3 gennaio 1925," in *Atti dell'Accademia nazionale dei Lincei. Memorie*, 6th ser., **1** (1926), 555–567. Angelo Tonolo, another disciple of Ricci, "Commemorazione di Gregorio Ricci-Curbastro nel primo centenario della nascita," in *Rendiconti del Seminario matematico della Università di Padova*, **23** (1954), 1–24, contains a beautiful portrait of Ricci and a partial bibliography. See also A. Natucci in *Giornale di matematiche di Battaglini*, 5th ser., **2** (1954), 437–442; and two articles in *Enciclopedia* *italiana,* on absolute differential calculus, XII, 796–798; and on Ricci's life and work, XXIX, 250.

Reports on almost all of Ricci's publications were published in *Bulletin des sciences mathématiques;* although sometimes very detailed, they often appeared only after a considerable delay. For the report on "Méthodes de calcul différentiel absolu," see **35** (1911), 107–111.

The most important works on absolute calculus and related questions are H. Weyl, *Raum, Zeit und Materie* (Berlin, 1918); G. Juvet, *Introduction au calcul tensoriel et au calcul différentiel absolu* (Paris, 1922); J. A. Schouten, *Der Ricci-Kalkül* (Berlin, 1924); A. S. Eddington, *The Mathematical Theory of Relativity,* 2nd ed. (Cambridge, 1924); L. P. Eisenhart, *Riemannian Geometry* (Princeton, N.J., 1926); and *Non-Riemannian Geometry* (New York, 1927); and É. Cartan, *Leçons sur la géométrie des espaces de Riemann* (Paris, 1928).

PIERRE SPEZIALI

RICCIOLI, GIAMBATTISTA (*b.* Ferrara, Italy, 1598; *d.* Bologna, Italy, 25 January 1671), *astronomy, geography.*

Riccioli entered the Society of Jesus when he was sixteen years old, and there received the comprehensive education that enabled him to teach Italian literature, philosophy, and theology, first at Parma and then at Bologna, while privately pursuing studies in astronomy and geography. He published extensively on the latter topics and these writings made him famous among his contemporaries, even though he rejected Galileo's example in using the vernacular and wrote most of his works in Latin. His commitment to church doctrine brought him into conflict with the ideas expressed by Galileo and his students and by the Florentine Accademia del Cimento. This attitude, together with the civil and religious pressures inherent in the Counter-Reformation, explains many of the apparent contradictions in Riccioli's scientific career. Following the Inquisition's condemnation of Galileo's astronomical theories, for example, Riccioli became one of the most ardent opponents of the Copernican system, which he tried to refute in every way. He nonetheless recognized the simplicity and the imaginative force of the Copernican theory, and acknowledged it as the best "mathematical hypothesis"— while striving to divorce it from any effective notion of truth.

In particular Riccioli designed a series of experiments by which he hoped to disprove Galileo's conclusions, but instead ratified them. This is especially true of his accurate and ingenious investigations of falling bodies. Although he was somewhat hampered by his reluctance to read Galileo's own works, his own skill as an experimenter served him well. With his fellow Jesuit, Francesco Maria Grimaldi, Riccioli succeeded in perfecting the pendulum as an instrument to measure time, thereby surpassing Galileo and his school and laying the groundwork for a number of important later applications.

Riccioli also made a number of significant astronomical measurements in an effort to expand and refine existing data. To this end he made measurements to determine the radius of the earth and to establish the ratio of water to land. His recourse to a mathematical treatment of these problems is noteworthy. He observed the topography of the moon and, in concert with Grimaldi, introduced some of the nomenclature that is still used to describe lunar features. Riccioli described sunspots, compiled star catalogues, and recorded his observation of a double star; he also noted the colored bands parallel to the equator of Jupiter and made observations of Saturn that, if he had had better instruments, might have led him to recognize its rings.

As a geographer, Riccioli set out to compose a single great treatise that would embrace all the geographical knowledge of his time. Although he did not complete this task, he published tables of latitude and longitude for a great number of separate localities, in which he corrected previous data and prepared the way for further developments in cartography. Despite the conservatism of the age in which he worked, Riccioli made honest and important contributions to science.

BIBLIOGRAPHY

I. ORIGINAL WORKS. Riccioli's writings are *Geographicae crucis fabrica et usus* (Bologna, 1643); *Almagestum novum astronomiam veterem novamque complectens* (Bologna, 1651, 1653); *Theses astronomicae de novissimo comete anni 1652* (Bologna, 1653), an anonymous work attributed to Riccioli by Lalande; *Geographiae et hydrographiae reformatae* (Bologna, 1661; Venice, 1672); *Astronomiae reformatae,* 2 vols. (Bologna, 1665); *Vindiciae kalendarii Gregoriani adversus Franciscum Leveram* (Bologna, 1666), published under the name of Michele Manfredi; *Argomento fisico-mattematico . . . contro il moto diurno della terra* (Bologna, 1668); *Apologia pro argumento physico-mathematico contra systema Copernicanum* (Venice, 1669); and *Chronologiae reformatae et ad certas conclusiones redactae,* 3 vols. (Bologna, 1669).

II. SECONDARY LITERATURE. An annotated bibliographic survey of Riccioli's writings is given in P. Riccardi, *Biblioteca matematica italiana,* I (Modena, 1893), cols. 370–374; Riccardi does not consider *De semidiametro terrae* (Bologna, 1655), sometimes attributed to Riccioli, to be by him. Sources on Riccioli cited by Riccardi include

P. Alegambe, *Bibliotheca scriptorum Soc. Jesu post exclusum* (Antwerp, 1643), 416; A. de Backer, *Bibliothèque des écrivains de la Compagnie de Jésus* (Liège, 1853–1861); J. S. Bailly, *Histoire de l'astronomie moderne*, II (Paris, 1779), 216; G. A. Barotti, *Memorie istoriche de' letterati ferraresi*, II (Ferrara, 1793), 270; Jacques Cassini, *De la grandeur et de la figure de la terre*, II (Paris, 1720 [1722]); G. B. Corniani, *I secoli della letteratura italiana commentario*, 9 vols. (Brescia, 1818–1819); A. Fabroni, *Vitae italorum doctrina excellentium*, 20 vols. (Pisa, 1778–1805); A. Libes, *Histoire philosophique des progrès de la physique*, 4 vols. (Paris, 1810–1813); J. E. Montucla, *Histoire des mathématiques*, 2nd ed., 4 vols. (Paris, 1799–1802); G. Tiraboschi, *Storia della letteratura italiana*, 27 vols. (Venice, 1822–1825); and Luigi Ughi, *Dizionario storico degli uomini illustri ferraresi*, II (Ferrara, 1804), 93.

LUIGI CAMPEDELLI

RICCÒ, ANNIBALE (*b.* Modena, Italy, 15 September 1844; *d.* Rome, Italy, 23 September 1919), *geophysics, astrophysics.*

After graduating in natural sciences from the University of Modena and in engineering from the Milan Polytechnic, Riccò became an assistant at the Modena observatory and then at the Palermo observatory. In 1885, under the sponsorship of the Accademia Gioenia, he founded the astrophysics observatory at Catania, and in 1890 he became professor of astrophysics at the university, the first chair of that subject in Italy. Riccò created an astronomic and meteorological station on Mt. Etna, and with A. Secchi and P. Tacchini he founded the Society of Italian Spectroscopists in 1872; he was editor of its memoirs until his death.

At the Palermo observatory he began a regular series of direct and spectroscopic observations of the sun, using a 25-centimeter Merz refractor and a direct-vision spectroscope. He continued this research in Catania for forty years, obtaining important results on the frequency, position, and development of sunspots and solar prominences and on their influence on terrestrial phenomena. Riccò collaborated with G. E. Hale in an unsuccessful attempt to photograph the corona in full sunlight, using a spectro-heliograph invented by Hale and taking advantage of the altitude of Mt. Etna.

In 1913, at the fifth conference of the International Union for Cooperation in Solar Research, Riccò presented the results of his observations made from 1880 to 1912 on solar prominences and their structure. He showed that the cycle of prominences lasts about as long as that of the sunspots, although there is a certain delay in the appearance of the former, in addition to other differences. Riccò noted two kinds of prominences: those associated with very active sunspots are extremely variable and are composed of hydrogen, helium, calcium, and other metals; quiescent prominences are composed almost exclusively of hydrogen and migrate toward the poles during the course of the eleven-year sunspot cycle. Riccò was one of the first to explain that the so-called filaments are merely prominences seen projected against the solar disk.

Riccò led three expeditions to study total solar eclipses, to Algeria (28 May 1900), Spain (30 August 1905), and the Crimea (August 1914). All of these expeditions achieved important results on the flash spectrum, on prominences and their relationship to the corona, and on the emission spectrum of the corona. In 1882, at Palermo, Riccò had pointed out the delayed occurrence of terrestrial magnetic storms with respect to the presence of extensive groups of sunspots. Ten years later he reported to the Académie des Sciences his discovery that magnetic storms begin on the earth forty to forty-five hours after the passage of the spots, or groups of spots, across the central meridian of the sun, and that consequently the presumed agent of the storms must travel to the earth at a speed of about 1,000 kilometers per second. This was the first observation of what was later called the solar wind.

For the international Carte du Ciel, Riccò organized and directed the study of those parts of the sky between 46° and 55° N. latitude that had been assigned to the Catania observatory. He observed the Daniel, Morehouse, and Halley comets (1908–1910) and discussed their constitution with Horn d'Arturo according to hypotheses proposed by Righi. In the fields of geodesy and geophysics he determined the gravitational anomalies and the terrestrial magnetic constants for Sicily, especially in relation to seismic activity, and carried out observations and studies on the crater of Mt. Etna during the 1910 eruption.

BIBLIOGRAPHY

A prolific writer, Riccò published mainly in the *Memorie della Società degli spettroscopisti italiani*. For a list of his papers (to 1900), see Royal Society *Catalogue of Scientific Papers*, VIII, 742; XI, 166–168; XII, 615; XVIII, 168–172. Subsequent writings are listed in Poggendorff, IV, 1241–1242; V, 1043; VI, 2165.

For a biography of Riccò, see G. Abetti, "Annibale Riccò, l'Accademia Gioenia e l'osservatorio astrofisico di Catania," in *Bollettino delle sedute dell' Accademia Gioenia di scienze naturali in Catania*, 4th ser., **3** (1955).

GIORGIO ABETTI

RICHARD SWINESHEAD. See **Swineshead, Richard.**

RICHARD, JULES ANTOINE (*b.* Blet, Cher, France, 12 August 1862; *d.* Châteauroux, Indre, France, 14 October 1956), *mathematics.*

Richard taught in several provincial *lycées,* including those at Tours, Dijon, and Châteauroux. He defended a doctoral thesis, on the surface of Fresnel waves, at the Faculté des Sciences of Paris on 22 November 1901. Of an eminently philosophical cast of mind, Richard published a work on the philosophy of mathematics at Paris in 1903. He collaborated on several scientific journals, most notably *Enseignement mathématique* (1905–1909), in which he was able to give free reign to his critical mind.

In an article published in *Enseignement mathématique,* "Sur une manière d'exposer la géométrie projective" (1905), Richard cited Staudt, David Hilbert, and Charles Méray. He based his exposition on the theorem of homological triangles, that is, on an implicit axiomatics very close to that of Staudt.

In a philosophical and mathematical article, "Sur la nature des axiomes de la géométrie" (1908), Richard distinguished four attitudes displayed by theoreticians and submitted them successively to critical analysis: (1) Geometry is founded upon arbitrarily chosen axioms or hypotheses; there are an infinite number of equally true geometries; (2) Experience provides the axioms; the basis of science is experimental, and its development is deductive; (3) Axioms are definitions —this third point of view is totally different from the first; (4) Axioms are neither experimental nor arbitrary; they force themselves upon us because without them experience would be impossible (this is the Kantian position). Richard found something unacceptable in each of these attitudes. He observed that the notions of the identity of two objects or of an invariable object are vague and that it is essential to make them precise; it is the role of axioms to do this. "Axioms are propositions the task of which is to make precise the notion of identity of two objects preexisting in our mind." Further on he asserted, "To explain the material universe is the goal of science."

Utilizing the group of anallagmatic spatial transformations and taking a subgroup that leaves a sphere invariant, Richard later remarks in the article that for a real sphere the subgroup is Lobachevskian, for a point sphere it is Euclidean, and for an imaginary sphere it is Riemannian. "One sees from this that, having admitted the notion of angle, one is free to choose the notion of the straight line in such a way that one or another of the three geometries is true." Hence, for Richard, difficulties persist, since "to study these groups we are obliged to assume that ordinary geometry has in fact been established." This article gave rise to several polemics, and Richard, having received a letter from Giuseppe Peano, returned to the question the following year.

In an article on mechanics, Richard took a mild swipe at Poincaré: "The consistent relativist will say not only that it is convenient to suppose that the earth revolves; he will say that it is convenient to suppose that the earth is round, that it has an invariable shape, and that it is greater than a billiard ball not contained in its interior."

"Richard's paradox or antinomy" was first stated in 1905 in a letter to Louis Olivier, director of the *Revue générale des sciences pures et appliquées.* Richard wrote, in substance:

The Revue has pointed out certain contradictions encountered in the general theory of sets.

It is not necessary to go as far as the theory of ordinal numbers to find such contradictions. Let E be the set of real numbers that can be defined by a finite number of words. This set is denumerable. One can form a number not belonging to this set.

"Let p be the nth decimal of the nth number of the set E; we form a number N having zero for the integral part and $p + 1$ for the nth decimal, if p is not equal to either 8 nor 9, and unity in the contrary case." This number does not belong to the set E. If it were the nth number of this set, its nth cipher would be the nth decimal numeral of this number, which it is not. I call G the group of letters in quotation marks [above]. The number N is defined by the words of the group G, that is to say by a finite number of words. It should therefore belong to the set E. That is the contradiction.

Richard then attempted to remove the contradiction by noting that N is not defined until after the construction of the set E. After having received some comments from Peano, he returned to the problem for the last time in 1907. Richard never presented his antinomy in any other form, although certain variants and simplifications falsely bearing his name are found in the literature.

BIBLIOGRAPHY

On Richard's paradox, see "Les principes des mathématiques et le problème des ensembles," in *Revue générale des sciences pures et appliquées,* **16**, no. 12 (30 June 1905), 541–543, which includes Richard's letter and Olivier's comments. The letter alone is reproduced in *Acta mathematica,* **30** (1906), 295–296. Richard returned to the question in "Sur un paradoxe de la théorie des ensembles et sur l'axiome de Zermelo," in *Enseignement mathématique,* **9** (1907), 94–98.

Sur la philosophie des mathématiques (Paris, 1903) was reviewed by P. Mansion, in *Mathésis*, 3rd ser., **3** (1903), 272; and *Notions de mécanique* (Paris, 1905) was reviewed by G. Combeliac, in *Enseignement mathématique*, **8** (1906), 90.

Articles published in *Enseignement mathématique* include "Sur une manière d'exposer la géométrie projective," **7** (1905), 366–374; "Sur les principes de la mécanique," **8** (1906), 137–143; "Considérations sur l'astronomie, sa place insuffisante dans les divers degrés de l'enseignement," **8** (1906), 208–216; "Sur la logique et la notion de nombre entier," **9** (1907), 39–44; "Sur la nature des axioms de la géométrie," **10** (1908), 60–65; and "Sur les translations," **11** (1909), 98–101.

JEAN ITARD

RICHARD, LOUIS PAUL ÉMILE (*b*. Rennes, France, 31 March 1795; *d*. Paris, France, 11 March 1849), *mathematics*.

Richard, the son of a lieutenant colonel in the artillery, was the eldest of four children. A physical impediment resulting from an accident prevented him from pursuing a military career, and he began teaching in 1814 as *maître d'étude* at the *lycée* in Douai. There he became friendly with the student A. J. H. Vincent, who became a historian of Greek mathematics and member of the Académie des Inscriptions et Belles-Lettres. The two friends later met again in Paris, where they held similar posts.

In 1815 Richard was appointed professor of the *sixième* at the Collège de Pontivy. He became professor of special mathematics the following year. In 1820 he was called to Paris to teach elementary mathematics at the Collège Saint-Louis. From there he went to the Collège Louis-le-Grand, and in 1822 he was given a chair of special mathematics, which he held until his death.

During this period virtually the only concern of secondary-school mathematics teachers in France was to prepare students for the entrance examination for the École Polytechnique. For this purpose, three classes were sufficient: preparatory, elementary, and, finally, special classes. Richard taught the latter class with extraordinary distinction. No program was imposed. Richard, rising above the routine, gave instruction in the principal modern theories, including the new geometry introduced by Poncelet. He was one of Poncelet's most fervent supporters, and when, in 1846, a chair of higher geometry was created at the Sorbonne for Michel Chasles, Richard was one of his most diligent auditors.

Richard stayed abreast of advances in mathematics, with which he constantly enriched his courses. The exercises he propounded were zealously investigated by his students. Of the many distinguished scientists whom he trained, the most famous, Evariste Galois, attended his class in 1828–1829. His students also included Le Verrier, J. A. Serret, and especially Hermite, to whom Richard entrusted the manuscripts of Galois's student exercises.

Richard never married.

BIBLIOGRAPHY

Despite the entreaties of his friends, Richard published nothing. On his life and work, see the notice by Olry Terquem, in *Nouvelles annales de mathématiques*, **8** (1849), 448–451.

JEAN ITARD

RICHARD OF WALLINGFORD (*b*. Wallingford, Berkshire, England, *ca*. 1292; *d*. St. Albans, Hertfordshire, England, 23 May 1336), *mathematics, astronomy, horology*.

Since Richard of Wallingford was for nine years abbot of St. Albans, the best chronicled monastery in England, much more is known about his personal life than about most medieval writers. From the *Gesta abbatum Monasterii Sancti Albani* (H. T. Riley, ed., Rolls Series, II [London, 1867], 181–183), we learn that he was the son of William, a blacksmith, and his wife, Isabella; the family was moderately prosperous. When his father died, Richard was adopted by the prior of Wallingford, William of Kirkeby, who in due course sent him to Oxford (*ca*. 1308).

Having determined in arts before he was twenty-three years of age, Richard left for St. Albans, where he assumed the monastic habit. He was ordained deacon (18 December 1316) and then priest (28 May 1317). His abbot sent him back to Oxford—probably to Gloucester College, a Benedictine establishment—where he studied philosophy and theology for nine years. Having determined as B.Th. in 1327, he returned once again to St. Albans to ask for the festive expenses of his graduation. While he was there the abbot died, and Richard was elected in his place. He visited Avignon for the papal confirmation, which was at length obtained, despite some legal difficulties.

Once back at St. Albans Richard found himself oppressed by three great burdens: the abbey was deeply in debt; the townsmen of St. Albans were in revolt, objecting to the abbot's feudal privileges; and he himself had contracted leprosy. Before he died of this disease, in 1336, he had cleared most of the debts and put down the revolt in the town. He had also kept control of a difficult internal situation, several of his monks having objected to his holding office, and one having gone so far as to instigate a papal inquisition. He skillfully negotiated these difficulties; and when he died in office, he left a reputation not only for moral firmness, but for intellectual and practical genius. He was especially remembered at St. Albans for the vast and intricate astronomical clock that he designed, and which in its final form was completed after his death.

Richard's first essay in mathematical or astronomical writing seems to have been a product of the early years in his second period at Oxford and was a set of instructions (canons) for the use of the tables that had been drawn up by John Maudith, the Merton College astronomer, in the approximate period 1310–1316. Richard followed this essay with *Quadripartitum*, a work on such of the fundamentals of trigonometry as were required for the solution of problems of spherical astronomy. The first part of this work has the appearance of a theory of trigonometrical identities, but at the time it was written it was regarded as a basis for the calculation of sines and cosines, and chords and versed chords. The next two parts of the *Quadripartitum* deal with a systematic and rigorous exposition of Menelaus' theorem, in the so-called "eighteen modes" of Thābit ibn Qurra. Finally, the work ends with an application of the foregoing principles to astronomy. The main sources of the work were Ptolemy's *Almagest*, the canons to the Toledan tables, and a short treatise that was possibly by Campanus of Novara. The *Quadripartitum* may reasonably be claimed as the first comprehensive medieval treatise on trigonometry to have been written in Europe, at least outside Spain and Islam. When Richard was abbot of St. Albans, he revised the work, taking into account the *Flores* of Jābir ibn Aflaḥ, but only one copy of the later recension is known.

Before finally leaving Oxford in 1327, Richard wrote three other works. The *Exafrenon pronosticacionum temporis* was a treatise on astrological meteorology. There is no good reason to doubt that it was his. Richard's most important finished treatise, *Tractatus albionis*, dealt with the theory, construction, and use of his instrument, the "Albion" ("all by one"), which was a highly original equatorium to assist in calculating planetary positions, together with ancillary instruments concerned with eclipse calculation, ordinary astrolabe practice, and a *saphea Arzachelis*. The Albion earned considerable renown in England, where Simon Tunsted produced a new version differing slightly from the original. John of Gmunden's recension included some new parameters, drawn from the Alphonsine tables, in place of the Toledan parameters. His version was much copied in southern Europe, and the instrument continued in vogue until the sixteenth century, influencing Schöner and Apian. The original treatise provides very few clues as to its sources, and many of its best parts are undoubtedly original. Regiomontanus drafted a much debased version.

While writing *Albion* (1326–1327) Richard composed a treatise on another new instrument he had designed, the "rectangulus." This instrument was meant as a substitute for the armillary sphere and was intended for observation and calculation. The chief advantage it was supposed to have was simplicity of construction: it was made from seven straight pivoted rods. Nevertheless, there were certain inherent disadvantages in its design. In connection with the treatise on the rectangulus, we note a table of an inverse trigonometrical function.

At St. Albans, where he was able to direct relatively large sums of money, Richard embarked on the task of building his astronomical clock. It seems that a sound mechanical escapement had been known for more than forty years when he began his work, but his clock is the first entirely mechanical clock of which we have detailed knowledge. It had a mechanism for hour-striking and an escapement older than (and in some ways superior to) the better-known verge and foliot as used with a contrate wheel. But the true originality of the design relates to its astronomical trains, with, for example, an oval wheel to give a variable velocity for the solar motion, correcting trains for the moon (leading to a theoretical error of only 7 parts in 10^6), and a lunar phase and eclipse mechanism. No mechanism of comparable complexity is known from any earlier time. The clock was lost to history after the dissolution of the monasteries in the sixteenth century. John Leland reported that it included planetary trains and a tidal dial. It had, of course, an astrolabe face, which, like the works, was probably ten feet across. The solitary treatise from which the details of the mechanism are known is now bound out of sequence, added to which it seems to have been copied from a pile of disordered drafts, with duplication of subject matter. It originally belonged to the subsacristan of the abbey.

BIBLIOGRAPHY

I. Original Works. Transcripts of parts of *Albion*, *Rectangulus*, and *Quadripartitum* have appeared in print. See R. T. Gunther, *Early Science in Oxford*, II (Oxford, 1926), 337–370, for parts of the first two works by H. H. Salter. J. D. Bond printed the first book only of *Quadripartitum*, from an inferior MS, in *Isis*, 5 (1923), 99–115, with English trans. (*ibid.*, 339–363). The bulk of Richard's writings are available only in MSS, but J. D. North has in the press a complete ed. of all the known writings, including translations and commentaries, and a discussion of conjectured and spurious works not mentioned above.

II. SECONDARY LITERATURE. For the best bibliographical guide before the work by J. D. North, see Thomas Tanner, *Bibliotheca Britannico-Hibernica* (London, 1748), 628–629. Bishop Tanner drew heavily on the sixteenth-century antiquary John Leland. The most fundamental biographical source is *Gesta abbatum*, referred to in the text above.

JOHN D. NORTH

BIBLIOGRAPHY

On Richards' life and work, see Helen K. Porter, "Francis John Richards 1901–1965," in *Biographical Memoirs of Fellows of the Royal Society*, **12** (1966), 423–436, with complete bibliography; and W. W. Schwabe, "Dr. F. J. Richards, F.R.S.," in *Nature*, **205** (1965), 853–854, and "Francis John Richards 1901–1965," in *Plant and Soil*, **22** (1965), 319–322.

A. D. KRIKORIAN

RICHARDS, FRANCIS JOHN (*b.* Burton-upon-Trent, England, 1 October 1901; *d.* Wye, Kent, England, 2 January 1965), *plant physiology.*

Richards was educated at the University of Birmingham, where he studied botany and biochemistry. In 1926 he joined F. G. Gregory at the Research Institute of Plant Physiology at Imperial College, London. In 1958, when Gregory retired, the Institute was dissolved, and Richards was made director of the new Agricultural Research Council Unit of Plant Morphogenesis and Nutrition at Rothamsted. This unit was later moved to Wye College, Kent. In 1954 he was elected a member of the Royal Society.

Richards is best known for his research on the mineral nutrition of cereal crops, especially the role of potassium and phosphorus. His international reputation has been established by his detailed studies and analyses of growth, respiration, photosynthesis, water content, carbohydrate and nitrogen metabolism in cereals under varying levels of nutrient supply, and his investigations of the substitution or partial replacement of the essential role of potassium by rubidium. This work emphasized that certain nutritional requirements are not absolute but may be a function of the environmental conditions under which plants grow—a view that has led to and still continues to yield important concepts in plant growth and development.

Richards' studies also disclosed many features of the metabolic consequences associated with mineral nutrient deficiency, such as the accumulation of the amide putrescine under potassium deficiency conditions. Richards appreciated that many factors of the environment interact with nutritional variables. He was one of the first to apply the then newly developed statistical methods to physiological and ecological data. His mathematical skill found expression in the devising of new methods of describing growth rates and leaf pattern production and arrangement in growing points (phyllotaxis).

RICHARDS, THEODORE WILLIAM (*b.* Germantown, Pennsylvania, 31 January 1868; *d.* Cambridge, Massachusetts, 2 April 1928), *chemistry.*

Richards was the son of gifted parents: William Trost Richards, a noted painter of seascapes, and Anna Matlack Richards, a Quaker author and poet. Because his mother felt that public education was geared to the slowest student in the class, Richards received his elementary and secondary schooling at home. At the age of six he became friendly with Josiah Parsons Cooke, Jr., professor of chemistry at Harvard University, who by showing the child Saturn through a telescope, awakened Richards' interest in science. By the time he joined the sophomore class at Haverford College at the age of fourteen, his only formal education had been attendance at some chemistry lectures at the University of Pennsylvania. In June 1885 he graduated at the head of his class with a degree in chemistry; and, eager to study under Cooke, he entered the senior class at Harvard the following fall. The youngest member of the class, Richards graduated in June 1886 with highest honors in chemistry.

As a graduate student at Harvard, Richards undertook—under Cooke's direction—the difficult laboratory problem of accurately determining the composition of water in order to obtain the relative weights of hydrogen and oxygen. He found that the ratio 1:15.96 disproved the validity of the statement (Prout's hypothesis) that atomic weights of the elements are integer multiples of that of hydrogen. In this investigation he learned the techniques that he later used so skillfully in his own laboratory: care, precision, and especially patience. In 1888 he received the doctorate; and because of the merit of his dissertation, he was awarded the Parker fellowship, which enabled him to spend a year abroad making many important professional friendships.

Upon his return in the autumn of 1889, Richards joined the Harvard faculty as an assistant in quantitative analysis and never severed his connection with the university (although he spent one year at the

University of Berlin as an exchange professor in 1907). Promotions came rapidly for Richards: instructor in 1891; assistant professor in 1894; and full professor in 1901 (this last promotion coming as a result of his being offered the chair of physical chemistry at the University of Göttingen). He was chairman of the chemistry department from 1903 until 1911 and director of the Wolcott Gibbs Memorial Laboratory from its opening in 1912 until his death in 1928. Holding the prestigious Erving professorship of chemistry from 1912 until 1928, he remained active in teaching and research until less than a month before his death. Richards married Miriam Stuart Thayer, the daughter of a professor at the Harvard Divinity School, in 1856; they had three children.

Richards' best-known studies were his determinations of the atomic weights of twenty-five elements, including those used to determine virtually all other atomic weights. For this work he was awarded the 1914 Nobel Prize in chemistry, the first chemist in the United States to be so honored. He also received the Davy Medal of the Royal Society (1910), the Faraday Medal of the Chemical Society (1911), and the Willard Gibbs Medal of the American Chemical Society (1912), among numerous awards, honorary degrees, and memberships in foreign societies. About one-half of his nearly three hundred published papers deal with atomic weights.

When Richards began publishing his work in 1887, the accepted values of atomic weights were based upon those determined by Stas in the 1860's. Stas's research was characterized by lengthy and careful procedures utilizing large quantities of materials and achieved an accuracy far exceeding that of earlier workers. These values were so well received that until 1905 no investigator seriously questioned them nor attempted to check his work.

After accurately determining the oxygen-hydrogen ratio, Richards turned to the atomic weights of several metallic elements. These early studies showed that copper from widely separated sources has exactly the same atomic weight. The studies also verified that the atomic weight of cobalt is greater than that of nickel, although cobalt precedes nickel in the periodic table. In his studies of the alkaline-earth strontium, he developed two important experimental devices: a bottling apparatus for transferring weighed samples without contact with moist air, and a nephelometer for accurately determining the amount of silver halide precipitate causing turbidity in samples. Most of Richard's analyses involved the precipitation of silver halide from solutions of halide salts of the desired element.

By 1905 Richards had become aware of serious errors in Stas's classical studies. Theory of precipitation had progressed far enough for him to see that his predecessor had neglected the slight but important solubility of silver chloride, that he had added solid silver nitrate to his solutions of metal halide salts (which caused silver nitrate to be included as an impurity in the precipitate), and that he had used such large quantities that impure samples dramatically increased the errors. Consequently, the Harvard group redetermined the atomic weights of several major elements previously studied by Stas: silver, nitrogen, chlorine, sodium, and potassium. In all cases Richards' work produced significant changes in the accepted values.

Richards' other major contribution to the field of atomic weights was a comprehensive study of the atomic weight of radioactive lead from uranium minerals. Although radioactivity was actively studied in many laboratories and the isotope concept had been suggested in 1907, the only experimental verification that chemically identical substances could have different atomic weights came from the analytical laboratory. The study of Richards and Lembert, published in 1914, was one of the first confirmations that the lead from radioactive minerals does have a different atomic weight from normal lead. Until Aston developed the mass spectrograph in 1919, tedious chemical analysis continued to provide the only experimental confirmation of various isotopes.

Richards was indirectly responsible for determining the atomic weights of thirty additional elements, since they were investigated by two of his former students, Gregory Baxter at Harvard and Otto Hönigschmid at Munich. Baxter became the first Richards professor of chemistry, a chair endowed in Richards' honor in 1925.

Although best known for his atomic weight studies, Richards directed a vigorous research program in thermochemistry and electrochemistry. He became interested in these subjects in 1895 when, upon the death of Cooke, Harvard sent him to visit the laboratories of Wilhelm Ostwald at Leipzig and Walther Nernst at Göttingen in order to improve his qualifications to teach physical chemistry. Many of his later investigations were a direct result of his theory of the compressible atom, an attempt to explain physically the variation of the constant "b" in van der Waals's equation of state. He proposed that an atom had a changeable volume, the magnitude of which depends upon its chemical state. Thus the volume of a potassium atom in its chloride salt is much less than that of a free potassium atom. Although the hypothesis was never adopted by other investigators, and although Richards spent much of his last ten years unsuccessfully trying to place the theory upon a firm mathematical foundation, his efforts led to the accurate determination of the physical constants of many elements and compounds. In nearly thirty publications on the subject, he attempted to correlate compressibilities of substances with their densities, surface tensions, heats of reaction, and other properties. As a part of this study in 1902, while investigating the behavior of galvanic cells at low temperatures, he approached the discovery of the principles enunciated by Nernst in 1906 as the third law of thermodynamics.

While measuring thermodynamic values in his compressibility studies, Richards became aware of certain shortcomings in the calorimetric methods then in use, especially the need to apply a complex cooling correction to the calculation of his results, on account of heat transfer from the reaction vessel to the calorimeter jacket. Seeking to eliminate the problem, Richards, Lawrence J. Henderson, and George Shannon Forbes in 1905 devised an adiabatic calorimeter, in which the jacket temperature could be adjusted to that of the reaction vessel. Although a similar calorimeter had actually been invented in 1849 by the Frenchman Charles C. Person, Richards and his colleagues were the first to use such a calorimeter extensively.

Using continually improved versions of this calorimeter, Richards published sixty papers on thermochemistry—many of them containing data that are still standard among the accepted values in handbooks of physical constants. Notable in this work is an extensive study by Richards, Allan W. Rowe, and Frank T. Gucker on the heats of dilution of metals in acids; the specific heats of acids, bases, and salts; and the heats of neutralization of strong and weak acids and bases. Richards' series of electrochemical studies includes the observation that "Faraday's law is not a mere approximation, but is rather . . . among the most precise . . . laws of nature." The Harvard group also carried out an extensive investigation of the electrical and thermodynamic properties of amalgams.

During his years as a member of the faculty, Richards created at Harvard a mecca for physical and analytical chemical research. Over sixty young men studied with him and became renowned chemists in their own right. Gilbert N. Lewis, Farrington Daniels, Arthur B. Lamb, Gregory P. Baxter, James B. Conant, Hobart H. Willard, and Otto Hönigschmid among others were products of the "Richards school." Richards was also noted for the excellence of his courses in physical chemistry. Although he distrusted mathematics and taught a much less rigorous approach than is presently offered to undergraduates, he liked to include historical material. The first introduction to the history of science for Conant, Henderson, and Frederick Barry came from Richards' lectures.

BIBLIOGRAPHY

I. ORIGINAL WORKS. No comprehensive bibliography of Richards' works has ever been published; however, a list of Richards' 292 published papers is in Sheldon J. Kopperl, *The Scientific Work of Theodore William Richards*, Ph.D. diss., University of Wisconsin, (Madison, 1970), 333–359.

On the composition of water, see "The Relative Values of the Atomic Weights of Hydrogen and Oxygen," in *Proceedings of the American Academy of Arts and Sciences*, **23** (1887), 149, written with Josiah Parsons Cooke, Jr. On the nephelometer, see "The Nephelometer, an Instrument for Detecting and Estimating Opalescent Precipitates," in *American Chemical Journal*, **31** (1904), 235, written with R. C. Wells. On the first redetermination of Stas's values, see "A Revision of the Atomic Weights of Sodium and Chlorine," in *Journal of the American Chemical Society*, **27** (1905), 459, written with R. C. Wells. Richards' first paper on lead from uranium ores is "The Atomic Weight of Lead of Radioactive Origin," *ibid.*, **36** (1914), 1329, written with Max E. Lembert. The low-temperature work with galvanic cells is "The Significance of Changing Atomic Volume. III," in *Proceedings of the American Academy of Arts and Sciences*, **38** (1902), 291.

The adiabatic calorimeter is first discussed in "The Elimination of Thermometric Lag and Accidental Loss of Heat in Calorimetry," *ibid.*, **41** (1905), 1, written with Lawrence J. Henderson and George Shannon Forbes. The remarks on Faraday's law come from "The Universally Exact Application of Faraday's Law," *ibid.*, **38** (1902), 407, written with W. N. Stull. Richards' only book-length publication is a collection of his early papers on atomic weights, translated into German and published as *Experimentelle Untersuchungen über Atomgewichte, 1887–1908* (Hamburg, 1909).

II. SECONDARY LITERATURE. The only book-length treatment of Richards is Kopperl's Ph.D. diss. cited above. An extensive sketch is Harold Hartley, "Theodore William Richards Memorial Lecture," in *Journal of the Chemical Society* (1930), 1930–1968, which summarizes Richards' work in detail. A recent short study on Richards' atomic weight investigations is Aaron J. Ihde, "Theodore William Richards and the Atomic Weight Problem," in *Science*, **164** (1969), 647–651.

SHELDON J. KOPPERL

RICHARDSON, BENJAMIN WARD (*b.* Somerby, Leicestershire, England, 31 October 1828; *d.* London, England, 21 November 1896), *medicine, pharmacology.*

Richardson was an eminent physician who was active in many of the reform movements of his time, including temperance, public hygiene, and sanitation. He strongly advocated more humane treatment of laboratory animals; was an early enthusiast of bicycling; and wrote poems, plays, songs, biographies, and a novel. Richardson's intellectual credentials were impeccable: he was apprenticed to a surgeon in Somerby and in 1847 entered Anderson's University (now Anderson's College). In 1850 he became licentiate of the Faculty of Physicians and Surgeons of Glasgow and in 1854, M.A. and M.D. of St. Andrews. He became a fellow of the Royal College of Physicians of London in 1865, of the Royal Society in 1867, and of the Faculty of Physicians and Surgeons of Glasgow in 1878. He was knighted in 1893.

Richardson also contributed to the development of a vigorous, scientific pharmacology. In a remarkable series of experiments supported financially by the British Association for the Advancement of Science, and published in the British Association *Reports* between 1863 and 1871, Richardson studied the physiological effects of a number of families of organic compounds. He deliberately chose substances the chemical compositions of which he could determine. By altering their constitutions through the controlled addition or substitution of various radicals, Richardson sought to determine the "physiological significance" of these radicals. He hoped to be able to predict a compound's physiological effect from this knowledge of its constituent elements, and from the known effects of chemically related substances.

Richardson applied his method to the study of compounds in the amyl, methyl, and ethyl series; and of a number of alcohols, hydrides, iodides, and chlorides. In the course of his investigations he found several therapeutically useful drugs, such as amyl nitrite and methylene bichloride. He was unable to formulate any general laws relating chemical constitution to physiological action. Nevertheless, his assumption that only portions of a molecule enter into the actual "physiologic reaction," and his method of investigating clusters of similar compounds, were adopted by later pharmacologists with fertile results.

BIBLIOGRAPHY

The major sources of information about Richardson's life are his autobiography, *Vita medica* (London, 1897), and a life by his daughter, prefixed to Richardson's book, *Disciples of Aesculapius*, 2 vols. (London, 1900). A short modern work is Sir Arthur MacNalty, *A Biography of Sir Benjamin Ward Richardson* (London, 1950), with a moderately complete bibliography, which may be supplemented by the Royal Society *Catalogue of Scientific Papers*, V, 187–188; VIII, 743–744; XI, 168–169; XVIII, 178.

Two recent papers stress Richardson's pharmacological work: J. Parascandola, "Structure-Activity Relationships—the Early Mirage," in *Pharmacy in History*, **13** (1971), 3–10; and W. F. Bynum, "Chemical Structure and Pharmacological Action: A Chapter in the History of 19th Century Molecular Pharmacology," in *Bulletin of the History of Medicine*, **44** (1970), 518–538.

WILLIAM F. BYNUM

RICHARDSON, OWEN WILLANS (*b.* Dewsbury, Yorkshire, England, 26 April 1879; *d.* Alton, Hampshire, England, 15 February 1959), *physics, electronics, thermionics.*

Richardson was one of the outstanding pure scientists behind the application of physics to the development of radio, telephony, television, and X-ray technology. An experimentalist of first rank and a foremost exponent of electrons and quantum theory throughout the first half of this century in the English-speaking world, Richardson was awarded the Nobel prize in 1928 "for his work on the thermionic phenomenon and especially for the discovery of the law named after him." In 1939 he was knighted for his services. Between 1901 and 1953 he published over 133 papers and three books. The quality of his long and productive career as experimentalist, teacher, and administrator makes him one of the prominent physical scientists of this century.

The only son of Joshua Henry and Charlotte Maria Richardson, Owen grew up near Leeds and later in the small mining town of Askern, near Doncaster. His father was a salesman of industrial tools. By the age of twelve the boy's precocity was so evident from his performance on parish-school examinations and his avid interest in plant life that he was admitted on a full scholarship to Batley Grammar School. From 1891 to 1897 he was a model pupil at Batley, winning many contests and exhibitions, including a scholarship to Cambridge, where he entered Trinity College in 1897. Studying at the Cavendish Laboratory under J. J. Thomson, with the group of scholars that included Ernest Rutherford, C. T. R. Wilson, Paul Langevin, and Harold A. Wilson, Richardson achieved highest distinction in classes in the natural and physical sciences. He placed first in the tripos for physics, chemistry, and botany and received his bachelor's degree in 1900.

Invited to remain at the Cavendish Laboratory after graduation, Richardson quickly became interested in the implications of Thomson's work on "cathode rays" and subatomic electrical "corpuscles." Neither the physical atom nor the concept of the electron was as yet widely accepted within the physics profession; but since 1897 Thomson among others had steadily built an edifice of evidence for the particulate nature of electricity, demonstrated through the use of high vacuum techniques applied to glass tubes impregnated by electrodes. While Rutherford and others were studying the ionic forms and processes related to Röntgen rays, radioactive elements, and radiation more generally, Richardson turned to the problem of describing how the metals of heated filaments emitted their streams of charged radiant energy. In 1901 he read his first two scientific papers before the Cambridge Philosophical Society, officially announcing in the second of them (25 November 1901) an empirical law regarding the emission behavior of electrical "corpuscles" per unit time from heated platinum surfaces in a vacuum. Its immediate consequence was to win young Richardson a promising reputation and election as a fellow of Trinity College in 1902, followed by the Clerk Maxwell scholarship and a D.Sc. from University College, London, in 1904. Also during these years Richardson collaborated with H. A. Wilson and H. O. Jones on other studies in physical and organic chemistry.

In 1906, the year that Thomson became a Nobel laureate for his work on the transmission of electricity through rarefied gases, Richardson accepted an appointment as professor of physics at Princeton University. Before leaving for New Jersey, he married Lillian Maud Wilson, the sister of his friend Harold. During their seven-year stay in the United States, their two sons and one daughter were born.

Having grown ever more concerned with both Boltzmann's statistical thermodynamics and Planck's quantum theory for electrodynamics, Richardson while at Princeton worked sometimes alone and sometimes in collaboration with F. C. Brown, Soddy, H. L. Cooke, and R. C. Ditto to perfect apparatus, to experiment, and to publish papers on photoelectricity, spectroscopy, X rays, and thermodynamics. In 1909 he coined the term "thermionics" as a title for an article (*Philosophical Magazine*, **17**, 6th ser., 813–833). He also then began the manuscript of his first book, *The Electron Theory of Matter* (1914), developed from lectures given to graduate students at Princeton, among whom were Robert H. Goddard and the brothers Arthur H. and Karl T. Compton. This book became a classic text for a generation of students interested in radio and electronics. In 1911 Richardson was elected a member of the American Philosophical Society and was thinking of becoming an American citizen. Then in 1913 the offer of the Wheatstone

professorship of physics at King's College, plus election as a fellow of the Royal Society, lured him back to the University of London.

Returning to England, Richardson left in America a legacy of rigorous teaching and research in what was beginning to be called electronic physics. Two of his sisters had married Princeton colleagues; and his brother-in-law, H. A. Wilson, had migrated from Canada to Texas. Thus family ties remained strong with the United States. Toward the end of his American sojourn, Richardson had published experimental proof that the electric current in tungsten is carried by electrons, and he had completed many other experiments on thermionic emission from various materials, on the photoelectric effect, and on positive ionization. He returned to Britain an acknowledged authority on metallic conduction, electrons, and heat.

The move back to London in 1914 was disruptive at first, yet Richardson was able to finish his first book and to start and complete his second, *The Emission of Electricity From Hot Bodies* (1916). Far more disruptive were the outbreak of World War I and the demands for secret military research into telecommunications. By now Richardson's expertise was a British asset. His reputation for having made possible the rapid development of J. A. Fleming's "thermionic valve" connected him with the growing industries for wireless telegraphy and telephony. During the war he managed, with C. B. Bazzoni and others, to publish a few works on spectroscopy, on tests of Bohr's theory of the atom, and on Einstein's analysis of the photoelectric effect. Also about this time his concern with the gyromagnetic effect led to his anticipating the electron-spin momentum later attributed to G. E. Uhlenbeck and S. A. Goudsmit. In the midst of the war Wilhelm Wien even published a Richardson chapter, "Glühelektroden," in his anthology *Kanalstrahlen und Ionizations* . . . (Leipzig, 1917).

After the war Richardson continued teaching in classroom and laboratory. In 1920 he received the Hughes Medal of the Royal Society, and in 1921–1922 he served as president of Section A (physics) of the British Association for the Advancement of Science. He relinquished all teaching duties in 1924, when he was given a dual appointment as Yarrow research professor of the Royal Society and as director for research in physics at King's College. In 1926–1928 he served also as president of the Physical Society. On 12 December 1929 he was awarded the 1928 Nobel prize for physics. The presentation speech stressed Richardson's originality both in advancing a theory for the electrical conductivity of metals and in pursuing strenuously for twelve years the experimental researches necessary to verify the nature of electronic flow in and emission from glowing filaments in vacuum tubes.

"Richardson's law," an abstruse empirical formula that allowed him to elaborate in detail the effect of heat on the interaction between electricity and matter, was by this time widely recognized as the scientific explanation for Fleming's rectifier and amplifier devices and for Lee de Forest's triode and Audion, as well as for modern X-ray and cathode-ray tubes of all sorts. It is significant that Richardson had entitled his first two papers in 1901 "On an Attempt to Detect Radiation From the Surface of Wires Carrying Alternating Currents of High Frequency" and "On the Negative Radiation From Hot Platinum" (see *Proceedings of the Cambridge Philosophical Society. Mathematical and Physical Sciences*, **11** [1901], 168–178, and 286–295, respectively). At the beginning of the century neither the concept of the electron nor of the physical atom was fully acceptable, and Richardson's work had centered on his intuition that both negatively and positively charged radiation emanate directly from the heated solid filaments themselves, rather than from interactions of neighboring gas molecules with hot bodies or with the electromagnetic ether.

In his Nobel acceptance lecture, "Thermionic Phenomena and the Laws Which Govern Them," Richardson recalled his predecessors briefly and his contemporaries at length. Remembering the difficulty of his early efforts to get rid of residual gas effects (by hand pumps; through weeks of baking; without the benefit of ductile tungsten, which only became available in 1913), Richardson recounted how he had struggled from 1901 to 1911 to test and perfect his thermionic theory of electron emission. Theories of metallic conduction advanced by Thomson, Drude, and Lorentz were maturing at the same time; but the main rival to the thermionic theory was the view that these emissions were not primary physical phenomena but rather secondary chemical reactions between hot filaments and residual gases. Richardson was proud to point to 1913 as the zenith of his confidence in this research:

> The advent of ductile tungsten enabled me, in 1913, to get very big currents under better vacuum conditions than had hitherto been possible and to show that the mass of the electrons emitted exceeded the mass of the chemicals which could possibly be consumed. This experiment, I think, ended that controversy so far as it could be regarded seriously [*Nobel Lectures . . .*, p. 227].

Meanwhile the progress of the rest of physics was so rapid that the canonical formulation of Richardson's law passed through several changes in notation. It is commonly known as the Richardson-Dushman equation, after Saul Dushman (1883–1954), a Russian-American electronics expert who, beginning in 1923, reinterpreted its implications in view of Langmuir's researches and in terms of Fermi-Dirac instead of Maxwell-Boltzmann statistics. Although Laue, Tolman, Sommerfeld, and Fowler, among others, also contributed to this ferment, Richardson himself remained acutely aware of the unsolved mysteries in electrodynamics, and he continued to lead his profession in thermionics for two decades after the quantum mechanical revolution of the 1920's. Since his own Nobel lecture was delayed a year and thus delivered on the same day that Louis de Broglie received his 1929 prize for the wave nature of electrons, Richardson alluded to the new matter waves of de Broglie and to the new quantum mechanics as promising new ways to understand old mysteries. Yet Richardson was pleased to note that the basic equation of thermionic emission "is still valid with the magnitude of the universal constant A unaltered in any essential way" (*Nobel Lectures . . .*, p. 233). Thus, we may give Richardson's law here as he modified it in 1911, reinterpreted it in 1915 with the universal constant A, and reexpressed it in 1929:

$$i = AT^2 e^{-w/kT}$$

where i is the thermionic current density; A is a universal constant based on the mass and charge of the electron as well as Boltzmann and Planck's constants; T is the absolute temperature in degrees Kelvin; e is the natural logarithmic base; w is a specific constant or the electronic work function of the metal used; and k is Boltzmann's constant.

Since first suggested in 1901 and extended in 1903, this equation had been based on the simple hypothesis that freely moving electrons in the interior of a hot conductor escape when they reach the surface, if the kinetic energy of their velocity is great enough to overcome the binding forces of the material. Thus, the central idea behind this equation ("that of an electron gas evaporating from the hot source") proved immensely fertile, suggesting all sorts of tests for electrons flowing against electromotive forces, electronic cooling of hot bodies by evaporation, and electronic heating of cool bodies independently of molecular chemical reactions. In short, Richardson said in 1929, "We could at that time [1900–1913] find out a great deal more about what the electrons in an electron gas were doing than we could about the molecules of an ordinary gas."

Richardson always found that his facility in mathematical calculations served him in good stead. He was dexterous and soft-spoken. His patience and precautions in the laboratory taught many students of electronics not to rush to rash conclusions about the nature of electricity and yet to espouse the experimentalists' faith that hunches for tests based on carefully controlled conditions will force nature to yield its secrets. Like Thomson, Richardson influenced a number of students and peers who became Nobel laureates, including A. H. Compton (1927), C. J. Davisson (1937), and Irving Langmuir (1932). Working with such collaborators as T. Tanaka, F. S. Robertson, P. M. Davidson, S. R. Rao, and A. K. Denisoff, Richardson's influence spread widely around the world.

During the 1920's and 1930's, Richardson published about three scientific papers each year, all meticulous and treating his chief interest—trying to understand the nexus between physics and chemistry. In 1930 he received the Royal Medal of the Royal Society, and in 1932 he was invited to deliver the Silliman memorial lectures at Yale University. When the lectures were published as his third and last book, *Molecular Hydrogen and Its Spectrum* (1934), it quickly became a classic exposition of experimental knowledge interacting with the new theories of quantum mechanics. More than a decade of intensive studies of the diatomic hydrogen molecule as revealed by thermionic and spectroscopic techniques were codified in this work.

During his years as foreign secretary for his section of the Royal Society, Richardson often played host to visiting scholars and scientists, including Lorentz, Bohr, Planck, Debye, and Sommerfeld. Richardson talked with them about hydrogen, the simplest of all atoms and molecules; why its structure had seemed inexplicable in the old quantum theory; and how its behavior might be better described or explained either by Heisenberg's matrix mechanics or by Schrödinger's wave mechanics. Dirac's theory of electrons, "holes," and positrons seems to have been of less interest to him than the excitation of soft X rays, the emission of electrons under the influence of chemical action, and the determination of Rydberg constants. Richardson's analyses of the multiplex lines of the H_2 spectrum and of the old question of why it happens that such an apparently simple structure should exhibit such complexity led him in 1934 to the judgment that "wave mechanics emerges triumphantly."

In 1939 Richardson was knighted. When World War II broke out, he reduced his worldwide correspondence on behalf of physics and concentrated on radar, sonar, electronic test instruments, and associated magnetrons and klystrons.

Richardson retired from the University of London in 1944 and moved to his country house, Chandos Lodge, near Alton, Hampshire. There he and his wife gardened, ran a dairy farm, and walked in the country until her death the next year. Their elder son, Harold, had joined the physics department at Bedford College, University of London, and their younger son, John, had entered the practice of medicine and psychiatry. In 1948 Richardson married Henrietta M. G. Rupp, a distinguished physicist in her own right who had been among the first to observe electron diffraction and who was an established authority on luminescence in solids.

Richardson was awarded honorary doctorates from the universities of Leeds, St. Andrews, and London. He maintained his membership in the American Philosophical Society and was elected a foreign member of academies of science in Norway, Sweden, Germany, and India. He was chosen an honorary fellow of the Institute of Physics in 1952. After retirement to northeast Hampshire, he worked with the new large reflection echelon spectrometer, designed by W. E. Williams, and published two papers with E. W. Foster on the fine structure of hydrogen lines. Answers about the ultimate nature of hydrogen were elusive, but Richardson seems to have enjoyed the quest more than the goal. His bookplate carried the simple sign for infinity, ∞, and his collection of 2,700 books on the atom remains intact as a monument to that quest.

BIBLIOGRAPHY

I. ORIGINAL WORKS. Richardson's books are *The Electron Theory of Matter* (Cambridge, 1914; 2nd ed., 1916); *The Emission of Electricity From Hot Bodies* (London, 1916; 2nd ed., 1921); and *Molecular Hydrogen and Its Spectrum* (New Haven, 1934).

Translations and 133 major articles are listed by William Wilson: "Owen Willans Richardson, 1879–1959," in *Biographical Memoirs of Fellows of the Royal Society*, **5** (1959), 207–215.

MS and memorabilia material is in the Sir Owen Richardson collection, consisting of about 25,000 items, in the Miriam Lutcher Stark Library at the University of Texas at Austin. A typescript "Catalog of the Sir Owen Richardson Collection," compiled by James H. Leech and Dessa Ewing, was completed in 3 vols. in Nov. 1967: I, letters sent and received; II, works, including unpublished papers, notes, editions; III, miscellaneous, family letters, papers, grades, referee reports.

II. SECONDARY LITERATURE. See E. W. Foster, "Sir Owen Richardson, F.R.S.," in *Nature*, **183** (4 Apr. 1959), 928–929; N. H. de V. Heathcote, *Nobel Prize Winners in Physics, 1901–1950* (New York, 1953), 278–286; *Nobel*

Lectures—Physics, Including Presentation Speeches and Laureates' Biographies, II, *1922–1941* (Amsterdam, 1965), 220–238; and an obituary in the *Times* (London) (16 Feb. 1959), 10, col. 3.

LOYD S. SWENSON, JR.

RICHE DE PRONY, GASPARD-FRANÇOIS-CLAIR-MARIE. See **Prony, Gaspard-François-Clair-Marie Riche de.**

RICHER, JEAN (*b.* 1630; *d.* Paris, France, 1696), *astronomy*, *physics*.

Richer's birthplace, like the day and month of his birth and death, is unknown. His education is likewise shrouded in mystery. Yet when the Académie Royale des Sciences was organized, he was admitted in 1666 as an *aide* or *élève astronome*. When the Academy conceived the project of sending a skillful observer to a place far away from Paris in order to compare corresponding or simultaneous observations made at two widely separated stations, the scientist chosen for this task was Richer.

According to the official financial accounts kept during the reign of Louis XIV, a maker of mathematical instruments was paid on 5 April 1670 for supplying Richer with the technical devices he needed to carry out his instructions. Meanwhile he had arrived at La Rochelle, on the Atlantic coast, where he measured the height of the tides in the harbor at the time of the vernal equinox. Bad weather kept his ship from sailing until 1 May. Even then it did not get very far before a severe storm stopped both his marine pendulum clocks, which had been constructed at royal expense, and the reliability of which he had been expected to test. Crossing the Atlantic Ocean, in July and August he observed the tides at two points on the coast of what was then French Canada (now part of New England). Presumably because he had a quadrant of the most recent design, equipped with cross hairs and with telescopic instead of open sights, he was able to determine the latitude of the French fort on Penobscot Bay in degrees, minutes, and seconds. Since previous determinations of latitude had been confined to the degree and the minute or, at best, the half-minute, Richer's was the most precise astronomical observation made up to that time anywhere in the western hemisphere.

On 17 September 1670 Richer was back in La Rochelle, where he measured the tides at the autumnal equinox. He reported the results of his North American expedition to the Academy of Sciences early in January 1671. Although the Academy's manuscript minutes for this period have not survived, a portion of Richer's report can be reconstituted from a letter written by Huygens. No similar means are available for the reconstruction of the log of the ship on which Richer sailed, and most of his correspondence has likewise perished.

For instance, in a message to Huygens, which has been lost, Richer explained to the celebrated inventor of the pendulum clock why his instruments had failed. For his part Huygens, in a letter dated 4 February 1671, blamed Richer for not taking proper care of the clocks; not applying a bit of oil where needed; and not restarting the clock that had stopped, instead of waiting for the second one to stop too.[1] No competent horologer had been on shipboard with Richer. In *Horologium oscillatorium* (Paris, 1673) Huygens complained that "where success was lacking, this ought to be imputed to the carelessness of those to whom the clocks had been entrusted rather than to the devices themselves."[2]

Huygens' dissatisfaction with Richer's performance was evidently not shared by the French. For when the Academy of Sciences decided to send an expedition to Cayenne Island—off the coast of French Guiana and a little less than 5° north of the equator—once more Richer was selected. As early as 30 January 1670 he had been listed as a *mathématicien* "designated to go to Cayenne to make astronomical observations useful for navigation." To prepare for the paired observations to be made at Paris and Cayenne, in September and October 1671 Richer observed jointly with his counterpart who was to remain behind in Paris. Richer's expense money was provided on 27 September, and two days later his passport was issued; he "left Paris by order of the king [Louis XIV] in the year 1671 in the month of October."[3] His ship set sail from La Rochelle on 8 February 1672[4] and on 22 April arrived at Cayenne. Richer remained there until 25 May 1673, when illness forced him to return to France earlier than he had planned.[5] After his departure his assistant died in Cayenne.

For unknown reasons, Richer was transferred from active service with the Academy of Sciences to fortifications and military construction with the title of *ingénieur du roi*. His written report on his South American expedition, "Observations astronomiques et physiques faites en l'isle de Caïenne," was published in the *Mémoires* of the Academy, of which he was designated a full-fledged member *(académicien)* in 1679. Thereafter, no information about him exists except that he died in Paris in 1696.

It was the publication of Richer's "Observations" that saved him from utter oblivion, for his South American report provided the basis for momentous conclusions concerning the earth and the other members of the solar system. He put the longitude of his primitive Cayenne observatory, built in the native style, "about 3 hours, 38 minutes, west of the Paris Observatory,"[6] only some three minutes in excess of the modern figure. From Richer's observations in 1672 of the eclipse of the moon on 7 November and of a satellite of Jupiter on 1 April, the difference in longitude between Paris and Cayenne was placed within fairly close range of the correct value.[7]

The near precision permitted the reliable reduction to the Paris meridian of Richer's observations of Mars. "During the months of August, September, October, and November in 1672 . . . the path of this planet was quite extraordinary,"[8] Mars being then in its perigee. A comparison of Richer's observations of perigean Mars with corresponding observations made elsewhere, yielded a reasonable value for the parallax of this planet and in turn for the parallax of the sun. From this calculation was derived a fairly close approximation of the fundamental astronomical unit, the distance from the sun to the earth. The dimensions of the solar system and of its constituent bodies were thus disclosed for the first time with substantial accuracy.

In addition to this dramatic enlargement of the size of the cosmos as traditionally conceived, Richer's physical observations led to an improved understanding of the shape of the earth. Among the numerous purposes for which the Academy had resolved to dispatch an observer to a location near the equator was the desire to acquire more definite knowledge about "the length of the seconds pendulum in this same place."[9] A pendulum requiring exactly one second for each swing in either direction turned out to be an immensely valuable research tool. In the report Richer said:

> One of the most important observations I have made is that of the length of the seconds pendulum, which has been found shorter in Cayenne than at Paris. For the same measurement marked on an iron rod in the former place in accordance with the length found necessary to make a seconds pendulum was transported to France and compared with the Paris measurement. The difference between them was found to be 1 1/4 lines, by which the Cayenne measurement falls short of the Paris measurement, which is 3 feet, 8 3/5 lines [1 line = 1/144 foot = 2 1/4 millimeters]. This observation was repeated during ten whole months, when no work passed without its being carefully performed several times. The vibrations of the simple pendulum which was used were very short and remained quite perceptible up to 52 minutes, and were compared with those of an extremely good clock whose vibrations indicated seconds.[10]

> For the measurement of time I had two pendulum clocks, one of which indicated seconds and the other half-seconds. They had been made by the . . . king's regular watchmaker who, by his precision and the refinement of his products, has up to the present surpassed all those who are busy making watches and pendulum clocks.[11]

Isaac Newton wrote in the *Principia*:

> Now several astronomers, sent into remote countries to make astronomical observations, have found that pendulum clocks do accordingly move slower near the equator than in our climates. And, first of all, in the year 1672, M. *Richer* took notice of it in the island of *Cayenne*; for when, in the month of *August*, he was observing the transits of the fixed stars over the meridian, he found his clock to go slower than it ought in respect of the mean motion of the sun at the rate of $2^m.28^s$. a day.[12]

Richer's empirical discovery that a pendulum keeping perfect time in Paris had to be shortened near the equator if the clock was to beat seconds there required a rational explanation. "The conclusion was . . . that the same pendulum moved more slowly at Cayenne than at Paris; consequently the effect of gravity was less at the equator than in our region . . . Therefore the earth must be higher at the equator than at the poles. Hence the earth is a spheroid flattened at the poles."[13] In this impressive chain of reasoning the first link was supplied by Richer as he patiently watched his pendulum clock in Cayenne Island.

NOTES

1. *Oeuvres complètes de Christiaan Huygens*, VII (The Hague, 1888–1950), 54–55.
2. *Ibid.*, XVIII, 116–117.
3. Richer, *Observations*, ch. 1 (repr., Paris, 1729), 235.
4. Not 6 February 1672, as in Alfred Lacroix, *Figures de savants*, III (Paris, 1938), 12.
5. J.-D. Cassini, *Élemens* (repr., Paris, 1730), 72.
6. Richer, *Observations*, ch. 2 (1729), 237–238.
7. Cassini, *Élemens* (1730), 69.
8. Richer, *Observations*, ch. 9 (1729), 278.
9. *Ibid.*, ch. 1, p. 235.
10. *Ibid.*, ch. 10, art. 1, p. 320.
11. *Ibid.*, ch. 2, p. 236.
12. Newton, *Principia*, bk. III, prop. 20.
13. D'Alembert, "Figure de la terre," in *Encyclopédie*.

BIBLIOGRAPHY

See "Observations astronomiques et physiques faites en l'isle de Caïenne," in *Mémoires de l'Académie royale des sciences, depuis 1666 jusqu'à 1699*, VII, pt. 1 (Paris, 1679), 231–326.

On Richer and his work, see J.-D. Cassini, "Les élemens de l'astronomie verifiez par Monsieur [Jean-Dominic] Cassini par le rapport de ses tables aux observations de M. Richer faites en l'isle de Cayenne," in *Mémoires de l'Académie royale des sciences, depuis 1666 jusqu'à 1699*, **8** (repr., Paris, 1730), 53–117; Alfred Lacroix, *Figures de savants*, III (Paris, 1938), 11–14; John W. Olmsted, "The Scientific Expedition of John Richer to Cayenne (1672–1673)," in *Isis*, **34** (1942–1943), 117–128; and "The Voyage of Jean Richer to Acadia in 1670," in *Proceedings of the American Philosophical Society*, **104** (1960), 612–634. See also *Index biographique des membres et correspondants de l'Académie des sciences du 22 décembre 1666 au 15 décembre 1967* (Paris, 1968), 468.

EDWARD ROSEN

RICHER DE BELLEVAL, PIERRE. See **Belleval, Pierre Richer de.**

RICHET, CHARLES ROBERT (*b.* Paris, France, 26 August 1850; *d.* Paris, 3 December 1935), *physiology, psychology*.

Richet was the son of Alfred Richet, a distinguished surgeon. While attending the Lycée Bonaparte, he was undecided whether to devote himself to literature or to science. He entered medical school but was bored with anatomy and surgery, and wrote poetry and drama to divert himself. Serving as an intern in 1873, he was placed in charge of a female ward, where he witnessed a hypnotic experiment. Over the next two years he produced numerous hypnotic trances in his patients, publishing a summary of his experiences in 1875. The characteristic behavior of hypnotic subjects, he argued, could not be explained away as simulation. The basic phenomena of a hypnotic trance followed as regular a course as a disease. The more often a person is placed in a hypnotic state, Richet observed, the more distinct the hypnotic phenomena become.

Richet's experience with hypnotism stimulated a lifelong interest in the associations between psychic and physiological phenomena, and helped persuade him to abandon a surgical career and turn to physiology. At the time he made that choice, physiology had passed through three decades of rapid, formative growth. Although debate over fundamental questions continued, the bulk of the research consisted of extensions and refinements of earlier investigations. Most of Richet's early physiological work followed that pattern. While discovering little that was startling, he made significant contributions to knowledge of gastric digestion, the nature of muscle contractions, the toxic effects of inorganic salts, and the production and regulation of animal heat.

Richet entered physiology at a time when the center of experimental activity had shifted to Germany and when French scientists were complaining of the lack of adequate support. Nevertheless, he gained access to an extraordinarily rich scientific milieu. Between 1876 and 1882 he worked extensively in the laboratories of Jules Marey and of Marcelin Berthelot at the Collège de France, and of Alfred Vulpian at the Faculty of Medicine. He also made histological examinations in the laboratory of Charles Robin, and studied digestion in fish at a marine biological station directed by Paul Bert. Richet never worked under the direct supervision of Claude Bernard, yet Bernard probably exerted a decisive influence, especially in terms of general orientation. He was influenced especially by Bernard's view of the relation between physiology and pathology, by his toxicological studies, and by his generalization that a poison introduced into the blood must act upon a specific tissue. Despite his switch from medicine to physiology, Richet continued to be involved in medical problems. In 1876 he served as assistant to the surgeon Aristide Verneuil. The work of Jean-Martin Charcot and others on hysteria and hypnotic therapy especially interested him. Richet seems to have been unusually suggestible; he absorbed ideas readily from those around him, and he eagerly followed up whatever was pointed out to him as a promising problem or approach. His familiarity with the literature of his field enabled him frequently to locate important gaps in the knowledge of problems that had been extensively investigated, and to find ways to fill them.

Richet entered Marey's laboratory primarily to learn the latter's methods of graphically recording physiological phenomena. Although the use of revolving drums to record muscle movements and pressure changes had been developed in the laboratory of Carl Ludwig, Marey had devised the most precise and versatile recording devices. Richet quickly found ways to apply Marey's recording methods in the context of his medical service. Recording the respiratory motions of tetanus patients, Richet showed that there were two forms of respiratory spasm: one occurred during inspiration, the other during expiration.

Verneuil recommended that for his doctoral dissertation Richet investigate the phenomena of traumatic pain. Richet began, in January 1876, by comparing the pain caused by electric stimulation of the healthy and affected sides of hemianesthetic hysteria patients; soon, however, he focused on the general function of sensory nerves. Utilizing frogs poisoned with strychnine, he demonstrated that sensory nerves deprived of their blood supply die gradually from the periphery toward the center. He measured the abnormal delay in the perception of an electric shock by ataxic patients, and showed that the lag was not proportional to the distance of the stimulus from the central nervous system. Subjecting normal humans to electrical stimuli, he found that stimuli too weak to be perceived if widely separated were felt if they were closer together. For equally spaced stimuli, the delay between their beginning and the first moment of perception was inversely proportional to the intensity. These effects were analogous to the known summation phenomena for muscle contractions.

In July 1876, Verneuil performed a gastric fistula operation on a young man named Marcelin R., whose esophagus had become blocked after he swallowed caustic potash. He urged Richet to use the opportunity to repeat William Beaumont's famous experiments on digestion. Although there had been dramatic progress in the experimental study of gastric digestion over the intervening period, some of the main features of the process remained obscure and controversial. Uncertainty still remained over whether the gastric acid was hydrochloric, lactic, or some other, for analysts could not be sure which of the acids recoverable from the complex gastric juice was in a free state.

Richet undertook this research in the laboratory of Berthelot, who gave him assistance, including a new method to ascertain the nature of the gastric acid. In 1872 Berthelot and E. C. Jungfleisch had shown that the relative quantities of a solute distributed between equal volumes of two immiscible solvents are constant, which they called its "partition coefficient." Berthelot advised Richet to mix an aqueous solution of gastric juice with ether. The high ratio of acid dissolved in water to that dissolved in ether indicated that the principal acid was a mineral acid and therefore undoubtedly hydrochloric. By an extension of the procedure Richet showed that an organic acid with a partition coefficient approaching that of an isomer of lactic acid forms during digestion, and suggested that the lactic acid derived from the fermentations of certain digesting aliments. This consideration led him to undertake experiments on the conditions affecting lactic fermentation, an investigation he pursued for many years.

At Bernard's suggestion Richet next studied the gastric juice of fish, which he found to provide very favorable conditions for investigation. It was highly active, strongly acidic, and, unlike the juice of mammals, was secreted as a coherent, mucilaginous mass. With it he confirmed that there is more chloride than the quantity contained in the salts of the juice, thereby supporting the opinion that the principal acid is hydrochloric. When he mixed sodium acetate with the juice, however, he discovered by means of the partition coefficient method that less than half as much acetic acid formed as the same quantity of hydrochloric acid ought to displace from the salt. He inferred that hydrochloric acid does not exist free in gastric juice, but in combination with an organic substance that he guessed might be a derivative of protein. In 1878 he published a comprehensive memoir on the properties of gastric juice in which he reviewed all aspects of the digestive process and its variations in mammals, fish, and invertebrates. He gave reasons for the disagreements among previous authors over the identity of the acid, and argued that the partition coefficient method was best suited to resolve the question because it did not alter the gastric juice. He noted that in order to obtain pure gastric juice from Marcelin R., he had him chew highly flavored foods, which produced an abundant flow. Since the food could not have entered Marcelin's stomach, Richet ascribed the secretion to a nervous reflex.

During 1878 Richet passed his *agrégation* examination and was named *professeur agrégé* at the Faculty of Medicine, where he began to work in the physiological laboratory directed by Vulpian. His first major investigation there was inspired mainly by Marey's research on muscle contraction. The earliest graphic recordings of muscle contractions were made in 1850 by Hermann von Helmholtz, who analyzed the resulting curves into a latent period, a short period of rapid ascent, and a more gradual descent, measuring the time consumed in each phase. The relatively large mass of his recording mechanism, however, limited its accuracy. Many physiologists labored to refine his method and to determine the influence of varied conditions on the amplitude and time relations of the contractions. During the 1860's Marey developed myographs that minimized the inertia of the system transmitting the muscle motions to the recording drum. He attained remarkably precise curves, distinguishing subtle differences in the characteristics of the movements according to the duration and intensity of the stimulus, and the condition of the muscle. Among the phenomena Marey helped to establish was that the extended contractions produced by direct electrical stimulations were composed of a series of brief "shocks," which fused to form a tetanic contraction if the stimuli were repeated at short enough intervals. He and others carried out most of these investigations with the gastrocnemius muscle of the frog because it was readily available and simple to prepare, although they were aware that the responses of different muscles varied widely.

Richet decided that a detailed study of an invertebrate muscle might help to distinguish general properties of muscle contraction from peculiarities of individual types. He therefore applied the current analytical techniques to the claw and tail muscles of the crayfish. They reacted quite differently. The tail muscle produced brief, strong contractions, did not go into tetanus unless the stimuli were very frequent, and quickly became fatigued. The contractions of the claw were much longer, and repeated stimuli easily produced a persistent tetanus. The claw muscle exhibited with remarkable clarity the phenomenon of latent summation.

Conflicting measurements of the latent period by other physiologists led Richet to examine more closely the length of the period in the crayfish claw. The time turned out to be variable, the length increasing especially with fatigue. If, however, while the muscle was still in a contracted state from one stimulus, he gave it a second identical stimulus, the second latent period was shorter than the first. As in his earlier study of hypnosis, Richet was particularly interested in the way in which a physiological response to a given influence may be conditioned by prior exposure to that influence. He also noticed that the relaxation of a detached claw muscle, after a contraction produced by a strong stimulus, was divided into two distinct periods—a short, quick descent followed by a plateau period, which he called contracture. In 1880, using a fresh, lightly weighted muscle, Richet was able to evoke a more distinctly demarcated contractive phase, consisting of a second contraction without an additional stimulus. He called the phenomenon the "secondary wave."

By 1883 Richet was well-established as an experimental physiologist and was directing students of his own. He also began to give the lectures in physiology at the Faculty of Medicine. For two years he concentrated his course on animal heat while carrying out experiments on the same topic. During the preceding decades attention had spread from the question of the source of animal heat to the problem of how warm-blooded animals can maintain a nearly constant temperature. Isidore Rosenthal, Eduard Pflüger and his students, Léon Fredericq, and others were seeking to answer experimentally such questions as what role changes in heat production and heat loss played in compensating for external changes; whether the stimulus to which such mechanisms respond is a peripheral reflex or a change in the internal temperature of the animal; and where the nervous center that coordinates this regulation is located. These problems had already been extensively investigated when Richet took them up; but the systems involved were so complex, and the conditions requiring investigation so extensive, that none of the questions had been conclusively answered.

The first question Richet treated was the role of the central nervous system in controlling temperature. Numerous observations by others had established that the temperature of a mammal rose when the cervical region of the spinal cord or the medulla oblongata was excited or injured. Richet and his two assistants, Eugene Gley and Pierre Rondeau, found in March 1884 that they could obtain the same effect by exciting the anterior lobe of the cerebral cortex of rabbits. This result, however, left Richet in the same dilemma that others had faced in analogous situations: Was the effect due to "a greater production of heat or a lessened loss at the periphery?" To resolve the question decisively, he realized, he would have to measure precisely the quantities of heat that the animals produced. After the pioneering studies of Lavoisier, and of Dulong and C. M. Despretz in the 1820's, few direct calorimetric measurements had been made; physiologists usually calculated the output indirectly from the products of respiratory combustion. In May 1884 Richet thought of constructing a very sensitive calorimeter in which the heat formed within a chamber would expand the air in a closed surrounding chamber, forcing a corresponding volume of water to overflow from the open end of a siphon connected with the air chamber. After calibrating the apparatus, he could measure the heat produced by the quantity of water collected. At almost the same time Arsène d'Arsonval also invented a calorimeter that utilized the expansion of air, but with a different measuring method. From d'Arsonval, Richet received several suggestions for his own instrument.

Meanwhile Richet showed that the characteristic rapid breathing of dogs was a cooling device, increasing the heat loss by evaporation of moisture from the tongue. In 1887 he demonstrated that the process was not a means to increase the respiratory gaseous exchanges; it could, in fact, take place only under conditions in which the dogs did not require immediate normal respiration—that is, when the blood was low in carbonic acid and saturated with oxygen. For this reason he named the effect thermal polypnea. Although thermal polypnea is a specialized adaptation in animals that do not perspire through their skins, it provided a very convenient means for elucidating the general character of temperature-regulating mechanisms, because its occurrence was directly observable.

By November 1884, Richet had perfected his siphon calorimeter. With it he demonstrated that the lesions of the cerebral cortex of rabbits that caused their temperatures to rise always produced a substantial increase in the heat evolved. These experiments led him to believe that changes in heat production were more important than vasomotor control of heat losses in regulating the general temperature of animals. His work also provided strong evidence for the involvement of the higher centers of the brain in the control of heat production.

Richet took advantage of his new calorimeter to examine the effects of other conditions on the heat output of animals. The most important factor, he found, was size. For rabbits, dogs, and guinea pigs he established that the larger the animal, the less heat it produced per unit weight. This was not a surprising result—Carl Bergmann had predicted it in 1848 on the basis of the mathematical relation between surface and volume—but it was a satisfying verification of the conformity of animals to physical principles. Calculating the surface area of the animals tested from their weights, by means of the simplifying assumption that they were approximately spherical, Richet demonstrated that their heat production was nearly proportional to their surface areas. In Germany, Max Rubner established the same relationship independently, using different methods.

Between 1887 and 1891 Richet extended his investigations of animal heat. He supported his view that thermal polypnea takes place only when the carbonic acid concentration of the blood is low, by showing that the breathing of a mixture of oxygen and carbon dioxide precludes its appearance. He confirmed his law of the proportionality of heat and surface area for several mammals, using the indirect method of measuring heat production by the quantity of carbon dioxide exhaled. In 1889 Richet showed that dogs anesthetized by chloral no longer produced heat in proportion to their surface area, from which he inferred that the different rates of heat production normally maintained by animals of different sizes must depend on the control of the central nervous system. Nevertheless, he reasoned, the great difference between the heat output of large and small animals must derive from differences in the rates of activity of muscle tissue—that explained why large dogs tend to be lethargic, whereas small ones are continually active and shiver frequently. The last observation led him to investigate shivering as a means of thermal regulation; he regarded it as the mechanism that compensates for coldness in the same way that polypnea compensates for heat. Like polypnea, he showed, shivering can be brought on either by a peripheral reflex or by a direct action on the central nervous system. To demonstrate the latter he anesthetized dogs with chloral. The internal temperatures of

the animals gradually decreased, until at a certain point the animals characteristically began to shiver. As soon as that happened, their temperatures started to rise. The heat production in the chloralized dogs was much less than normal, but returned almost to normal when they shivered.

The research by Richet described so far dealt mostly with what had already become classical problems in physiology. Closely attuned to current trends, he reflected in his diversified investigations a representative cross section of the dominant concerns of that maturing science. Increasingly after 1880, however, his work was influenced by the spreading evidence of the pervasive biological effects of microorganisms. These phenomena apparently first impinged on his activities in 1877, when a critic remarked that the organic acids found during his artificial digestion experiments were probably produced by bacteria in the gastric juice. Richet accepted this criticism but replied that bacteria must also be present in the stomach, so that the phenomenon probably occurred physiologically. Soon afterward he was able to find a few bacteria in the stomachs of fish. In 1882 and 1883 he discovered *Bacillus* and other types of motile bacteria in the peritoneal fluid, blood, and lymph of fish, and in the fluids of marine invertebrates. Using a medium of sterile beef bouillon, he and Louis Olivier succeeded in developing cultures from some of them. In 1881 Richet examined the transformation of urea to ammonium carbonate that he observed in the stomachs of dogs rendered uremic, and ascribed the reaction to the organized ferment discovered by Louis Pasteur. These studies involved the methods Pasteur had developed for growing pure cultures of specific microorganisms in suitable nutrient media. It was, therefore, probably in order to learn those methods that Richet carried out some of these investigations in the laboratory of Philippe van Tieghem, a former student of Pasteur's who had helped to perfect the techniques.

The event that was decisive in drawing Richet into the new field was Pasteur's announcement, in February 1880, that he had discovered a way to immunize chickens against the fatal fowl cholera by inoculating them with attenuated microbes cultured in a broth made from chicken muscle. Pasteur brought some of his preparations of the fowl cholera microbes to Vulpian's laboratory. Richet, who watched him demonstrate there, was deeply impressed. He was particularly attracted to the idea that the microbes might secrete a chemical toxin that caused the disease, and that there might also be a chemical substance opposing the toxin in immune animals. During his physiology course in 1881 he mentioned an observation by Auguste Chauveau that Algerian sheep are resistant to anthrax, whereas French sheep are susceptible to it. Richet proposed the hypothesis that the difference might arise from some extractive substance in the blood of the Algerian variety. That idea suggested to him that he might be able to protect animals vulnerable to a disease by giving them transfusions of blood from resistant animals. He was not able to follow up this plan, however, for several years.

In 1888 Richet and Jules Héricourt discovered a new type of staphylococcus bacterium in an epithelial tumor of a dog. Adopting the strategy that had become standard since Pasteur's investigation of fowl cholera, they grew a pure culture of the bacteria. Rabbits inoculated from the culture died, but in dogs the injection produced only a large abscess. This difference paralleled what Pasteur had found for the fowl cholera microbe, which caused the death of chickens but left guinea pigs healthy except for an abscess that contained the microorganisms. Still following Pasteur's methods, Richet and Héricourt attenuated the new microbe by cultivating it for several days or at a temperature above the optimum for its growth. Rabbits vaccinated with the attenuated microorganisms became immune to inoculations with the virulent form. Recalling Richet's earlier speculation that the blood of resistant animals might contain some special substance, they now sought to attain the same result in another way, by transfusing blood from dogs into rabbits. When injected directly into a vein, the dog's blood was toxic to the rabbits; but they obviated this difficulty by transfusing instead into the peritoneal cavity, from which the blood was gradually absorbed. Transfusions made from dogs that had previously been inoculated with the staphylococcus conferred immunity on the rabbits.

Following this success, Richet and Héricourt attempted to apply their new principle of "hemotherapy" to tuberculosis. They were able to retard the progress of the disease in rabbits by transfusing dog's blood, but were unable to prevent the death of the rabbits. Nevertheless, in December 1890 they transfused dog blood serum into a human tuberculosis patient, again attaining only a delay in the course of the disease. By inoculating dogs with bacilli of avian tuberculosis, a form of disease against which the dogs were resistant, they were able to make them immune to human tuberculosis, to which they normally were sensitive. Next they transfused blood from dogs vaccinated in this way into dogs infected with human tuberculosis and were able to slow, sometimes to arrest, the disease. Despite these hopeful effects, their goal of developing an effective vaccine for human use

eluded them. They were also unsuccessful in their attempts to apply serum therapy to human cancer.

After his hopes for serum therapy in tuberculosis had been disappointed, Richet turned to dietary factors, long known to have some effect. In 1899 he found that a diet of raw meat brought about a remarkable improvement in the physical condition of tubercular dogs. In 1900 he determined that the effect was due principally to "muscle plasma," that is the fluids extracted from meat. He ascribed the result not to the alimentary value of the plasma but to some kind of immunizing action, and called the method of treatment "zomotherapy." For many years he urged its application to human tuberculosis.

By 1900 Richet had completed a quarter-century of fruitful, prolific experimental activity. He had attained an eminent professional position, having been named to the chair of physiology at the Faculty of Medicine in 1887, and had attracted many able students. His resourceful research had contributed substantially to several important areas of physiology and to a new field of medicine; but he had made few discoveries that strayed beyond the bounds of the normal development of these fields. Most of his investigations were typical of a large, well-organized science; that is, they clustered around the fundamental work of other men. In the first decade of the twentieth century, when he had passed his fiftieth year, Richet finally entered the charmed circle of scientists who have discovered major unexpected phenomena.

The investigations from which Richet's principles of hemotherapy and zomotherapy emerged involved him in subsidiary toxicological problems. In 1889 he was confronted with the fact that the blood of one animal species is toxic to another. In 1900 he found that muscle plasma was toxic if injected directly into a vein. During the following year he tried to establish the toxic dose of muscle plasma for dogs, defined as the quantity per kilogram of the animal that would cause it eventually to die. His approach resembled efforts he had made twenty years earlier to define toxic doses of various metallic salts. Later in the year, as a member of a scientific expedition on the yacht of Prince Albert of Monaco, he had an unforeseen opportunity to utilize his toxicological experience. The prince and another scientist on the voyage advised him to study the toxic properties of the tentacles of the Portuguese man-of-war. While still on board, Richet and Paul Portier established that an aqueous or glycerin extract of the tentacles was extremely toxic. After returning to France, Richet decided to compare the properties of the poison with those of the poison obtainable from members of another class of coelenterates, the sea anemones. The effects of a glycerin extract of this toxin were similar to that from the Portuguese man-of-war. In his customary manner he set out with Portier to determine the toxic dose. They began by giving intravenous injections to a large number of dogs, in doses ranging from 0.05 to over 0.30 cc. of the extract per kilogram of the recipient. Those that received more than .30 cc. died after four or five days. Richet and Portier kept the surviving dogs in order to use them again for a similar experiment after they had recovered. When they reinjected these animals two to three weeks later, all those receiving doses of 0.08 to 0.25 cc. quickly died. This extra-ordinary outcome led them to reflect that the poison must have properties that are the opposite of the immunizing properties of serums, attenuated bacterial cultures, and other toxins. Instead of reinforcing the resistance of an animal to later injections, a sublethal dose diminished it. For this reason they called the property anaphylactic ("contrary to protection"). On 15 February 1902 they demonstrated the effect at a meeting of the Société de Biologie by injecting two dogs, one of which had been injected previously.

Richet and Portier next pursued the question of how long the anaphylactic effect of the poison would last, and found that the condition persisted for at least two and a half months. In 1904 and 1905 Richet observed effects lasting, with gradually diminishing intensity, for nearly eight months. The complexity of the symptoms of the anaphylactic animals led him and Portier, however, to suspect that the toxin contained several active substances. With the collaboration of Auguste Perret they were able, by February 1903, to isolate from the poison a whitish powder that reproduced most of the effects of the original glycerin extract. Because of the strong action on the vasomotor system, they called the substance congestin. During the next two years Richet refined the experiments, using a purer, more potent form of congestin and repeating with rabbits the results he had obtained with dogs. Suspecting from the beginning that many other poisons would produce analogous actions, he now began to study other cases. In 1905 he obtained an increased sensitivity by repeated injections of morphine, and in 1907 he extracted from mussels a poison similar in its effects to congestin. Meanwhile other investigators were finding that nontoxic albuminous substances, such as blood serum, could also cause anaphylactic phenomena. Inoculations of bacteria produced highly specific anaphylactic reactions to further inoculations with the same microbe. The anaphylactic response to blood serum turned out also to be specific for the species of the donor animal. Thus anaphylaxis was soon revealed to be a phenom-

enon of broad scope with potentially crucial consequences for medical practice.

In 1907 Richet began to construct a general theory of anaphylaxis based on his accumulating observations and those of his colleagues. The opposite, yet obviously analogous, effects of anaphylactic and immunizing toxins induced him to envision mechanisms patterned after the concepts of antigen-antibody interactions that Paul Ehrlich, Jules Bordet, and others had proposed to explain immunization. Reasoning along lines similar to theirs and also to the reasoning that he had followed a decade earlier when he transferred immunity by transferring the blood of a resistant animal, Richet inferred that some special substance must form in the blood of anaphylactic animals and must react with the toxin later injected. He confirmed this surmise when he was able to make dogs anaphylactic by injecting them with serum from an anaphylactic animal. The heightened sensitivity of an anaphylactic animal, he concluded, must derive from the presence within it of a substance that is not toxic but that reacts with a toxin such as congestin to produce a third, highly toxic substance. He designated the first and last of these substances "toxogenin" and "apotoxine," respectively. An anaphylactic poison, he asserted, produces a true disease. The temperature changes and denutrition that follow an injection of congestin are entirely equivalent to the evolution of a microbial infection. A better way to put it, he thought, was that a disease is really a slow intoxication in which the toxin is produced within the animal by the microbes and then provokes the formation of an immediately toxic substance.

The fact that anaphylaxis appeared to be a reaction harmful to the animal itself must have posed an awkward dilemma for Richet, who had often stressed that all physiological actions help to conserve the life of the organism. In some of his 1907 experiments he found that the anaphylactic period was followed by one of relative immunity, so that he was able to view the phenomena as an "admirable adaptation," in which anaphylaxis hastens the defensive reaction of an animal to feeble doses of microbial poisons. In subsequent experiments this effect did not appear consistently, however, so he was forced to try to reconcile anaphylaxis with biological finalism in a broader sense. He admitted that, on the whole, the defense of the organism must be weakened when its sensitivity to poisoning is augmented; but he claimed that the violent reaction that is harmful to the individual is essential to protect the stability of the species. If the albuminoid substances of other species that produce anaphylactic responses were permitted to penetrate repeatedly into the blood and tissues and remained there, the chemical identity of the species would be endangered; the somatic constitution acquired by natural selection would be continually at the mercy of accidental circumstances. That these explanations were teleological did not embarrass Richet; despite his faith in the physicochemical foundation of physiological phenomena, he also reiterated frequently that the teleological view "must always serve as a guideline in every biological doctrine."

In 1911 Richet reviewed the development of research on anaphylaxis in a monograph in which he assimilated the work of other investigators into the framework set by his own. He could already claim with justification that "the domain of anaphylaxis is very vast," and he could realistically portray that domain as emerging principally from his discovery. By 1913 its scope had become well enough recognized so that Richet was awarded a Nobel prize for his part in it.

Richet had many interests beyond physiological research and writing. Attracted to aviation through Marey's experiments on bird flight, he participated in the design and construction of one of the first airplanes to leave the ground under its own power. Through friends he became interested in spiritualistic phenomena, coming to believe that true premonitions sometimes occurred, and exerted a formative influence on the early development of what he called metapsychics. Richet was a dedicated pacifist and wrote a number of general history books in order to demonstrate the malevolent effects of war. He also composed poetry, novels, and drama.

Richet continued his research on anaphylaxis, lactic acid fermentation, zomotherapy, and other subjects after 1913. He spent part of World War I at the front investigating problems in the transfusion of blood plasma. In 1926 he received the Cross of the Legion of Honor.

BIBLIOGRAPHY

I. ORIGINAL WORKS. Books by Richet include *Recherches expérimentales et cliniques sur la sensibilité* (Paris, 1877); *Des circonvolutions cérébrales* (Paris, 1878), translated by E. Fowler as *Physiology and History of the Cerebral Convolutions* (New York, 1879); *Du suc gastrique chez l'homme et les animaux* (Paris, 1878); *Physiologie des muscles et des nerfs* (Paris, 1882); *L'homme et l'intelligence: Fragments de physiologie et de psychologie* (Paris, 1884); *La chaleur animale* (Paris, 1889); *Essai de psychologie générale*, 2nd ed. (Paris, 1891); *Dictionnaire de physiologie*, 10 vols. (Paris, 1895–1928), with the collab-

oration of P. Langlois, L. Lapicque, *et al.*; *L'anaphylaxie* (Paris, 1911); *Traité de métapsychique* (Paris, 1922); *La nouvelle zômothérapie* (Paris, 1924); *L'intelligence et l'homme* (Paris, 1927); and *L'avenir et la prémonition* (Paris, 1931).

Among Richet's most significant research articles are "Du somnambulisme provoqué," in *Journal de l'anatomie et de la physiologie normales et pathologiques de l'homme et des animaux*, **11** (1875), 348–378; "Des propriétés chimiques et physiologiques du suc gastrique," *ibid.*, **14** (1878), 170–333; "Contribution à la physiologie des centres nerveux et des muscles de l'écrevisse," in *Archives de physiologie normale et pathologique*, 2nd ser., **6** (1879), 262–284, 522–576; "Étude sur l'action physiologique comparée des chlorures alcalins," *ibid.*, 2nd ser., **10** (1882), 115–174, 366–387; "Recherches de calorimétrie," *ibid.*, 3rd. ser., **6** (1885), 237–291, 450–497; "Des conditions de la polypnée thermique," in *Comptes rendus . . . de l'Académie des sciences*, **105** (1887), 313–316; "Sur un microbe pyogène et septique et sur la vaccination contre ses effets," *ibid.*, **107** (1888), 690–692, written with J. Héricourt; "De la transfusion péritonéale, et de l'immunité qu'elle confère," *ibid.*, 748–750, written with Héricourt; "Le frisson comme appareil de régulation thermique," in *Archives de physiologie*, 5th ser., **5** (1893), 312–326; "De la formation d'urée dans le foie après la mort," in *Comptes rendus . . . de l'Académie des sciences*, **118** (1894), 1125–1128; "Du traitement de l'infection tuberculeuse par le plasma musculaire, ou zômothérapie," *ibid.*, **130** (1900), 605–609, written with Héricourt; "De l'action anaphylactique de certains venins," in *Comptes rendus de la Société de Biologie*, **54** (1902), 170–172, written with P. Portier; "De l'anaphylaxie en général et de l'anaphylaxie par le mytilo-congestine en particulier," in *Annales de l'Institut Pasteur*, **21** (1907), 497–524; "De l'anaphylaxie et des toxogénines," *ibid.*, **22** (1908), 465–495. The chronological evolution of his research can most easily be followed through the many short notes he published in *Comptes rendus de la Société de Biologie*.

Biographical material for this article is based mostly on an autobiographical essay in L. R. Grote, ed., *Die Medizin der Gegenwart in Selbstdarstellungen*, VII (Leipzig, 1928), 185–220, and Richet's *L'oeuvre de Pasteur* (Paris, 1923).

II. SECONDARY LITERATURE. See Jean-Louis Fauré, "Discours: Funérailles de Charles Richet," in *Notices et discours. Académie des sciences*, **1** (1937), 626–633; Marilisa Juri, *Charles Richet physiologiste (1850–1935)*, which is no. 34 of Zürcher Medizingeschichtliche Abhandlungen, E. H. Ackerknecht, ed. (Zurich, 1965); André Mayer, "Notice nécrologique sur M. Charles Richet (1850–1935)," in *Bulletin de l'Académie de médecine*, **115** (1936), 51–64; Paul Painlevé, "Charles Richet et l'aviation," in *A Charles Richet: Ses amis, ses collègues, ses élèves* (Paris, 1926), 61–63; A. Pi-Suner, "Charles Richet et la physiologie interfonctionnelle," *ibid.*, 68–72; and F. Saint Girons, "Charles Richet (1850–1935)," in *Revue générale des sciences pures et appliquées* (1935), 677–679, with portrait. There are also obituaries in *Nature*, **136** (1935),

1017–1018; *Proceedings of the Royal Society of Edinburgh*, **56** (1935), 276–278; and *Revue métapsychique*, **17** (1936), 1–42.

FREDERIC L. HOLMES

RICHMANN, GEORG WILHELM (*b.* Pernau, Estonia [now Pärnu, U.S.S.R.], 11 July 1711; *d.* St. Petersburg, Russia [now Leningrad, U.S.S.R.], 26 July 1753), *physics*.

Richmann was the posthumous son of Wilhelm Richmann, a German in the Swedish administration at Dorpat (Tartu), who had fled to Pernau to avoid the advancing armies of Peter the Great. Pernau itself fell to the Russians just before Richmann's birth; he consequently spent his youth, as he did his entire professional career, in the Russian empire. He received his early education in Reval (Tallinn), doubtless in one of the schools operated by the large German colony there. He subsequently attended the universities of Halle and, especially, Jena. There he studied mathematics and physics under G. E. Hamberger (1697–1755), who conceived of physics in the manner of Descartes and encouraged the search for what he called "natural laws," phenomenological relations between measurable physical quantities. Richmann firmly adopted this program, to which he had to recall Hamberger, who once negligently admitted special innate forces as the cause of cohesion.

About 1735 Richmann went to St. Petersburg as tutor to the sons of the powerful foreign minister A. I. Osterman. Richmann continued his own studies at the University staffed by members of the St. Petersburg Academy of Sciences. He worked primarily under G. W. Krafft (1701–1754), a facile mathematician much abler than Hamberger and no less interested in applying his art to quantifiable physical problems. In 1739—having enjoyed the unusual distinction of attending the Academy's meetings while yet a student ("because of his remarkable erudition and other good qualities")—Richmann entered it as an adjunct; two years later he was named extraordinary professor of physics; and in 1745 he succeeded Krafft, who had returned to Germany, as ordinary professor and director of the physical laboratory.

Richmann's first significant work concerned the determination of the temperature T produced by mixing a quantity of water m_1 at temperature T_1 with a quantity m_2 at temperature T_2. The problem, an old chestnut in medical literature, had earlier engaged Krafft, who concluded by experiment that

$$T = (11m_1T_1 + 8m_2T_2)/(11m_1 + 8m_2), T_1 > T_2.$$

This result disagreed with some measurements of Richmann's and, more importantly, with his excellent physical intuition, which assured him that Fahrenheit's thermometer was approximately linear and that (to use an anachronism) the specific heat of water was scarcely altered with temperature.[1] The problem then became, as many medical writers had glibly assumed, a simple case of averaging; and one should expect

$$T = (m_1T_1 + m_2T_2)/(m_1 + m_2).$$

Very careful measurements,[2] taking into account what we would call the water equivalent of the thermometer and container, confirmed Richmann's equation and revealed the cause of Krafft's error. "The whole business," he said, "shows clearly that physics should avoid mathematical abstractions with all diligence and whenever possible, and attend to every circumstance in individual cases."[3] This by no means meant defection from Hamberger's goals. Richmann explicitly made it his "business" to establish the quantitative phenomenological laws that form the basis of sound physics;[4] "only if we have accurately determined the properties of bodies can we legitimately infer other truths with certainty."[5]

Heat phenomena lent themselves preeminently to this program. Richmann painstakingly investigated how a body's rate of heating, of cooling, and (in the case of water) of evaporation, depended upon its nature and upon the difference between its temperature and that of the surrounding air. He succeeded in confirming Newton's law of cooling[6] and in demonstrating—independently of Nollet and in contradiction to most physicists—that the rate of gain or loss of heat bears no evident relation to density, elasticity, or hardness.[7] The former result seemed to him so secure that he later proposed using it and his own law of evaporation to define (and, suitably implemented, to measure) the time average of the ambient temperature;[8] as for the latter, it was just another "proof that in physics nothing can be established securely without experiments."[9]

Throughout these experiments Richmann assumed that heat was "a certain motion of certain corporeal particles," and could no more be known "absolutely" than could motion itself. Since, on this theory, the "cohesion of substances"—including the mercury in Fahrenheit's instruments—could be expected to decrease with increasing temperature, it eventually caused Richmann to question the linearity of the thermometer and to undermine the assumptions on which most of his work had been based.[10]

Although Richmann flirted with many subjects besides heat (for example, artificial magnets and several types of balance barometers),[11] the only other study that claimed his continuous attention was electricity, which he took up early in 1745. He commenced with the experiments of S. Gray, Dufay, C. A. Hausen, and G. M. Bose, which he found in Doppelmayr's admirable *Neu-entdeckte Phaenomena* (1744). Naturally Richmann quickly felt the need for a measuring instrument. He first took as a measure of electrical force the weight required to bring into equilibrium a balance one pan of which hung above the electrified object. But the device proved unsatisfactory, and he changed to the angle formed between a vertical rule attached to the electrified body and a thread suspended from the top of the rule. With this "index," as he called it, he found that the "capacity" of an insulated conductor (*corpus derivativae electricitatis finitae connexionis*) depended not only on its total mass but also upon its shape.[12] These experiments, which date from 1745, are perhaps the earliest of their kind. Richmann then examined the relative conductivity of silk rubbers of various colors and measured, as one might expect, the rate of leak of electricity from insulated conductors as a function of humidity. Such problems occupied him until 1752, when he learned of the lightning experiments at Marly and acquired a copy of Franklin's *Experiments and Observations on Electricity.*

Richmann repeated the experiments, pointing out exaggerations and errors which the Philadelphians would not have made, he said, had they bothered to construct a decent electrometer, for example, by utilizing the principle of the otherwise frivolous "electrical jack."[13] For his part he preferred his index, which he took to measure the "agitation" of "a certain electrical matter which surrounds electrified bodies to a certain distance."[14] He could not bring himself to accept negative electricity ("unless it is agreed that, as in mechanics, motion made in a contrary direction is negative")[15] or the impenetrability of glass, against which he designed an interesting experiment,[16] which Aepinus later showed to be in perfect, even in quantitative, agreement with Franklin's principles. But Richmann enthusiastically embraced the new doctrine about lightning and installed in his home an insulated rod with which to probe the agitated electrical matter of thunder clouds. He was more aware than Franklin of the risks involved. Fear of lightning, he wrote, is quite natural and will only be overcome when one understands how and why its stroke can be averted. That, of course, will require many observations and experiments. "Evidently in these times even the physicist has an opportunity to display his fortitude."[17] Richmann and his friend and colleague Lomonosov let no storm go by without carefully following its effects on the indi-

cators attached to their insulated poles. On 26 July 1753 it thundered in St. Petersburg. Richmann ran home from the Academy and bent to read his index. At that instant lightning hit the pole and struck him dead, a martyr to his mania for measurement.

NOTES

1. Richmann, "De quantitate caloris . . .," in *Novi commentarii*, **1** (1747–1748), 152–154.
2. "Formulae pro gradu excessus . . .," *ibid.*, 168–173.
3. "De quantitate caloris . . .," *ibid.*, 166.
4. "De indice electricitatis . . .," *ibid.*, **4** (1752–1753), 301.
5. "Inquisitio in decrementa et incrementa . . .," *ibid.*, 241–242.
6. "Inquisitio in legem . . .," *ibid.*, **1** (1747–1748), 174–197.
7. "De argento vivo . . .," *ibid.*, **3** (1750–1751), 309–339; and "Inquisitio in decrementa et incrementa . . .," *ibid.*, **4** (1752–1753), 241–270.
8. "Usus legis decrementi . . .," *ibid.*, **2** (1749), 172–178.
9. "De argento vivo . . .," *ibid.*, **3** (1750–1751), 309.
10. "Tentamen rationem calorum . . .," *ibid.*, **4** (1752–1753), 277–300.
11. "De barometro . . .," *ibid.*, **2** (1749), 181–209.
12. "De electricitate . . .," in *Commentarii*, **14** (1744–1746), 299–324.
13. "De indice electricitatis . . .," in *Novi commentarii*, **4** (1752–1753), 301–302, 323.
14. *Ibid.*, 305.
15. *Ibid.*, 323–324.
16. *Ibid.*, 324–325.
17. *Ibid.*, 335.

BIBLIOGRAPHY

I. ORIGINAL WORKS. Richmann's published work consists of 22 Latin memoirs, of which Poggendorff gives a full list, in the *Commentarii* and *Novi commentarii academiae scientiarum imperialis petropolitanae.* The most important memoirs are "De electricitate in corporibus producenda nova tentamina," in *Commentarii*, **14** (1744–1746), 299–324; "De quantitate caloris, quae post miscelam fluidorum certo gradu calidorum, oriri debet, cogitationes," in *Novi commentarii*, **1** (1747–1748), 152–167; "Formulae pro gradu excessus caloris supra gradum caloris mixti ex nive et sale ammoniaco, post miscelam duarum massarum aquarum, diverso gradu calidarum, confirmatio per experimenta," *ibid.*, 168–173; and "De indice electricitatis et ejus usu in definiendis artificialis et naturalis electricitatis phaenomenis, dissertatio," *ibid.*, **4** (1752–1753), 301–340.

Also of interest are "Inquisitio in legem, secundum quam calor fluidi in vase contenti, certo temporis intervallo, in temperie aëris constanter eadem decrescit vel crescit . : .," *ibid.*, **1** (1747–1748), 174–197; "Tentamen, legem evaporationis aquae calidae in aëre frigidiori constantis temperiei definiendi," *ibid.*, 198–205; "Usus legis decrementi caloris ad definiendam mediam certo temporis intervallo temperiem aëris ostentus . . .," *ibid.*, **2** (1749), 172–178; "De barometro cuius scala variationis insigniter augeri potest . . .," *ibid.*, 181–209; "De argento vivo calorem celerius recipiente et celerius perdente, quam multa fluida leviora experimenta et cogitationes," *ibid.*

3 (1750–1751), 309–339; "Inquisitio in decrementa et incrementa caloris solidorum in aëre," *ibid.*, **4** (1752–1753), 241–270; and "Tentamen rationem calorum respectivorum lentibus et thermometris definiendi," *ibid.*, 277–300.

Many of Richmann's scientific MSS and some correspondence, preserved in the Archives of the Russian Academy of Sciences, have been published in G. W. Richmann, *Trudy po fizike*, A. A. Eliseev, V. P. Zubov, A. M. Murzin, eds. (Moscow, 1956).

II. SECONDARY LITERATURE. Most Western sources, for example, Jöcher; Poggendorff; J. G. Meusel, in *Lexikon*, **11** (1811), 261–263; J. F. von Riecke and K. E. Napiersky, in *Allgemeines Schriftsstellerslexikon*, **3** (1831), 531–534; and L. Stieda, in *Allgemeine deutsche Biographie*, **28** (1899), 442–444, all copy one another and, ultimately, *Novi commentarii*, **4** (1753), 36. A valuable exception is F. C. Gadebusch, *Livländische Bibliothek*, **3** (1777), 22–29. Russian sources are much fuller: *Russkii biograficheskii slovar'*, **16** (1913), 233–240; and P. Pekarskii, *Istoriia imperatorskoi akademii nauk v Peterburge*, **1** (St. Petersburg, 1870), 697–717. See also *Protokoly zasdanii konferentsii imperatorskoi akademii nauk s 1725 po 1803 goda = Procès-verbaux des séances de l'académie impériale des sciences depuis sa fondation jusqu'à 1803*, I–II (St. Petersburg, 1897); *Materialy dlya istorii imperatorskoi akademii nauk (1716–1750)*, II–X (St. Petersburg, 1885–1900); and L. Euler, *Perepiska. Annotirovannyi ukazatel*, A. P. Youschkevitch, and V. I. Smirnov, eds. (Leningrad, 1967). A good portrait appears in *Materialy*, **4**, opposite p. 370.

A general account of Richmann's work is V. P. Zubov, "Die Begegnung der deutschen und der russischen Naturwissenschaft im 18. Jahrhundert und Euler," in *Die deutsch-russische Begegnung und Leonhard Euler* (Berlin, 1958), 19–48. On calorimetry, see D. McKie and N. H. de V. Heathcote, *The Discovery of Specific and Latent Heats* (London, 1935), 59–76; and V. P. Zubov, "La formule calorimétrique et ses origines," in *Mélanges Alexandre Koyré*, **1** (Paris, 1964), 654–661, and "Kalorimetricheskaya formula Rikhmana i ee predistorya," in *Trudy instituta istorii estestvoznaniia i tekhniki*, **5** (1955), 69–93. On the barometer, see W. E. Knowles Middleton, *The History of the Barometer* (Baltimore, 1964), 107, 376. On electricity, see B. S. Sotin, "Raboty G. V. Rikhmana po elektrichestvu," in *Trudy Instituta istorii estestvoznaniia i tekhniki*, **44** (1962), 3–42; A. G. Dorfman and M. I. Radovskii, "B. Franklin i russkie elektriki XVIII v.," *ibid.*, **19** (1957), 290–312; and B. G. Kuznetzov, "Razvitie ucheniia ob elektrichestve v russkoi nauke XVIII v.," *ibid.*, 313–385.

On Richmann's death, see "An Account of the Death of Mr. George Richmann," in *Philosophical Transactions of the Royal Society*, **44** (1755), 61–69; B. N. Menshutkin, *Russia's Lomonosov* (Princeton, 1952), 86–89; and D. Müller-Hillebrand, "Torbern Bergman as a Lightning Scientist," in *Daedalus. Tekniska museets årsbok* (1963), 35–76.

JOHN L. HEILBRON

RICHTER, JEREMIAS BENJAMIN (*b.* Hirschberg, Germany [now Jelenia Gora, Poland], 10 March 1762; *d.* Berlin, Germany, 4 April 1807), *chemistry*.

Richter graduated from the Hirschberg Gymnasium, and in 1778 joined the engineering corps of the Prussian army. He devoted his spare time to studying chemistry and, after seven years, left the army to enter the University of Königsberg, where he studied mathematics and philosophy and probably attended Kant's lectures. He was awarded the doctorate in 1789 with a dissertation, *De usu matheseos in chemia*, in which he set out the determinations of the specific gravities of a number of substances, both compounds and solutions, and attempted to determine the weight of phlogiston. He then went to Gross-Ober-Tschirnau, near Glogau, in Lower Silesia, where he established a laboratory and supported himself by chemical research and making aerometers. In 1795 he became secretary and assayer to the *Oberbergamt* at Breslau, and in 1798 went to Berlin to become "second Arcanist," or chemist, at the Royal Porcelain Works. He never held an academic position, never married, and died of tuberculosis at the age of forty-five.

Despite the brevity of his life and chronic financial hardship, Richter nevertheless managed to maintain a program of experimental investigations that yielded significant results. He reported them in numerous memoirs, as well as in the three-volume *Anfangsgründe der Stöchyometrie oder Messkunst chymischer Elemente* (1792–1794) and eleven small volumes entitled *Ueber die neuern Gegenstände der Chemie*, published between 1791 and 1802. (The serial publication of the *Neuern Gegenstände* grew out of Richter's need to have his work read, and was prompted by the disappointing reception accorded his larger work.)

Throughout his career, Richter's chemistry was shaped by his firm conviction that all chemical processes are based upon mathematical laws. An apostle of Kant's axiom that all true science is applied mathematics, he was unable to accept Kant's corollary that chemistry is a "systematic art" rather than a true science for the supposed reason that its principles are only empirical and its laws experimental. Richter's contrary view led him to the new concept of stoichiometry, the germ of which is contained in the introduction to his dissertation, where he stated that mathematics can be extended to all areas of science and art in which something can be measured. Chemistry should, therefore, be especially accessible to mathematics because its basic problem is to determine the exact proportions of the components of every compound.

This idea also appears in the preface to the *Anfangsgründe der Stöchyometrie*, in which Richter wrote that "All sciences concerned with magnitudes belong to mathematics. The reason that so little progress is made in this branch is that chemists only rarely occupy themselves with mathematics and mathematicians feel no call to make conquests for the art of measurement in the field of chemistry." He then went on (Volume I, part 1, page 121) to define his new specialty, stoichiometry, as "the science of measuring the quantitative proportions or mass ratios in which chemical elements stand one to another." He gave his theory a theological basis in a quotation from the Wisdom of Solomon, which he set out on the title page of the fourth part of his *Neuern Gegenstände*: "But Thou [God] made all things, in measure, and number, and weight." Faithful to that word, Richter devoted his whole life to searching for the laws according to which the chemical numbers are combined by "measure, number, and weight." This work took priority over all his other research (which was largely concerned with the chemistry of metals) and was the subject of most of his writing.

The experiments that led Richter to the law of neutrality grew out of his interest in determining the combining proportions of compounds. He had observed that calcium acetate and potassium tartrate solutions remain neutral on being mixed, while calcium tartrate is precipitated and potassium acetate remains in solution. In 1791 he described this phenomenon, and stated that neutralization should occur in all chemical decompositions by double affinity, to the extent that compounds used in the decomposition are themselves neutral. The following year he committed himself further and wrote that, when two neutral solutions are mixed and decomposition follows, the resulting products are neutral almost without exception. He drew two conclusions from this—first, that the compounds (for which he used the word "elements," by which he meant acids and bases) must "have among themselves a certain fixed ratio of mass," so that the compositions of the resulting products can be calculated mathematically from those of the interacting substances; and, second, that

> If the weights of the masses of two neutral compounds that decompose each other to give a neutral product are A and B, and if the mass of the one element in A is a, and that of the same one in B is b, then the masses of the elements in B are $B - b$ and b. The ratios of the masses of the elements in the neutral compounds before the reaction are $(A - a) : a$ and $(B - b) : b$; after decomposition, however, the masses of the new products formed are $a + (B - b)$ and $b + (A - a)$, and the ratios of the masses of the elements are $a : (B - b)$ and

$b : (A - a)$. If, therefore, the ratio of the masses in the compounds A and B is known, that in the new products formed is also known [*Anfangsgründe der Stöchyometrie*, I, 1, 124].

Richter presented this law (which had been anticipated by Guyton de Morveau in 1787) more generally in 1795. It was, he stated,

> . . . a true touchstone of the experiments instituted with regard to the ratios of neutrality; for if the proportions empirically found are not of the kind that is required by the law of decomposition by double affinity, where the decomposition actually taking place is accompanied by unchanged neutrality, they are to be rejected without further examination as incorrect, since an error has then occurred in the experiments instituted [*Neuern Gegenstände*, IV, 69].

In searching empirical evidence for this theory, Richter made quantitative researches to determine the proportion in which a number of oxides (including those of aluminum, magnesium, calcium, strontium, and barium) and a number of bases (ammonia, potassium, and sodium hydroxides) mixed with hydrochloric, sulfuric, nitric, or hydrofluoric acids. (He later introduced a variety of organic acids into these experiments.) He also determined the equivalent weights of a number of metals and metallic oxides and of chromic, molybdic, and tungstic acids; in 1797 he established that metal salts are also subject to the law of neutrality on mutual decomposition.

From his experiments on metals that dissolve in fixed weights of acid to neutral salts, Richter was led to conclude that the amount of oxygen in any base is the same as that needed to saturate a constant given amount of an acid. He applied the principle of the maintenance of neutrality further to establish that, when one metal precipitates another from a neutral salt, the quantities of both metals that will dissolve in the same amount of acid will also unite with identical quantities of oxygen to form oxides. He made a clear distinction between absolute neutrality (as demonstrated by potassium sulfate) and relative neutrality (as demonstrated by silver nitrate), in which a compound can absorb an excess of one of its components, as, for example, metallic salts do when they react in acid solution.

Richter had published his early stoichiometric researches as general descriptions. From his basic assumption that chemistry is a branch of applied mathematics, he offered his results as unassailable mathematical truths. He did not include accounts of his experimental work until he brought out (in 1793) the second volume of the *Anfangsgründe*, in which he also offered his first speculations on the series of masses. He proceeded from the notion that the combining proportions in a compound form arithmetical or geometrical series, and when his data did not confirm such numerical relationships, he emended his results in an arbitrary manner. He stated his belief succinctly in the preface to this volume, taking as a given that "the double affinities proceed in arithmetical progression, and after exact observations it is hardly possible to resist the notion that the whole chemical system consists of such progressions."

Richter made a number of determinations toward constructing such progressions. He established the ratio in which the alkaline earth oxides are neutralized by hydrochloric or sulfuric acid—finding, for example, that 1,000 parts by weight of hydrochloric acid are neutralized by 734 parts of alumina, 858 parts of magnesia, 1,107 parts of lime, or 3,099 parts of baryta—and, through an elaborate procedure, he developed his idea of the series of their masses. In a simplified form, his logic held that, given the examples chosen above, the numerical proportions of which are respectively 734, 858, 1,107, and 3,099, the mutual differences between these numbers are 124, 249, and 1,992. He then divided these differences to obtain $249/124 = 2 + 1/124$ and $1992/124 = 16 + 8/124$. Since the second result is eight times greater than the first, Richter altered the first difference to produce the series $124\frac{1}{2}$, 249, 1,992, or $249/2$, $249/2 \times 2$, $249/2 \times 16$. He then took $734 = a$ and $249/2 = b$ to find the arithmetical series

$$a = 734 \qquad a + 3b = 1,107\tfrac{1}{2}$$
$$a + b = 858\tfrac{1}{2} \qquad a + 19b = 3,099\tfrac{1}{2}.$$

From this he concluded, as he wrote in the second volume of the *Anfangsgründe*, that "The mass ratios in which the hitherto known alkaline earths assert neutrality with muriatic [hydrochloric] acid are therefore terms of a true arithmetical progression, which arises when to the first term is added a product of a certain magnitude with an odd number, except that many intermediate odd numbers are left out" (p. 31). The missing terms, Richter believed, must correspond to unknown bases.

He likewise tried to establish the numerical series for alkaline earths capable of neutralizing 1,000 parts by weight of sulfuric acid, but discovered that he had to modify his procedure. Having found that "The quantities of real and possible elements which belong to 1,000 parts of muriatic [hydrochloric] acid belong also to 1,394 parts of vitriolic [sulfuric] acid," Richter was able to derive a numerical series, which he compared to the earlier one to conclude that there must be a complete series consisting of a constant number a/b, to which the succeeding odd numbers

should be added. From this he derived a general series of the form

$$a,$$
$$a + b,$$
$$a + 3b,$$
$$a + 5b - 3,$$
$$a + 6b - 13,$$
$$a + 7b - (3 + 1),$$
$$a + 9b - (3 + 3), \text{ and so on.}$$

Therefore, he stated, "it is absolutely certain" that the alkaline earths reach neutrality with both hydrochloric and sulfuric acids according to quantities that are terms of an infinite series that increases by the product of a determinate quantity plus consecutive odd numbers. The terms missing from the series must represent undiscovered alkaline earths, or such anomalous alkaline earths as alumina. In 1797 Richter added strontium to his system, with a value of $a + 9b$, and in 1802 beryllium, with a value of $a + 6b$.

Richter's further researches on stoichiometry were published in the *Neuern Gegenstände*, beginning with Volume VI (1796). He developed series of masses for hydrofluoric, hydrochloric, sulfuric, and nitric acids, in accordance with their ability to neutralize magnesia, lime, and baryta, and established a geometric progression from his results. He also published a geometric series encompassing carbonic, sebacic, oxalic, formic, succinic, acetic, citric, and tartaric acids, of which he wrote that

> Without exposing myself to the charge of juggling figures, I found, guided by the analogy of experience, that carbonic acid, as well as the seven acids containing carbon that have been examined from the point of view of stoichiometry, are terms of a geometric progression, which differs from the former progression in that the exponents of the powers increase in the usual order of numbers; while on the contrary the arithmetical progression that the alkalies produce with those acids retain their form unchanged [*Neuern Gegenstände*, VI, v-vi].

There are again terms missing in the series of masses of these acids, and Richter assumed that they correspond to unknown acids containing carbon.

Richter generalized the differences that he found in the series of masses of certain groups of bases and acids to apply to all known bases and acids. He concluded that the equivalent weights of bases follow an arithmetical series, while the equivalent weights of acids progress geometrically, although he never discussed the reason for this relationship. He also found a mathematical relation for the quantity of oxygen with which a number of nonmetals combine in their highest state of oxidation, to which he gave the general form $1,381 + 119a$, in which a represents the series of triangular numbers 1, 3, 6, 10, 15. . . .

For all his systematization, Richter never gave a fully generalized statement of his law of equivalent proportion. It remained for Ernst Gottfried Fischer, professor of physics and mathematics at the Gymnasium zum Grauen Kloster of Berlin, to give a clear summary of Richter's work. In his 1802 translation of Berthollet's *Recherches sur les lois de l'affinité* (which was entitled *Claude Louis Berthollet über die Gesetze der Verwandtschaft*), Fischer collected and collated Richter's numerical values and combined them into a single table of equivalent weights. This table contained thirteen acids and eight bases, all referred to the single standard of 1,000 parts of sulfuric acid, the same standard that Richter himself had consistently used. At the same time, however, Fischer criticized Richter's series of masses as unacceptable hypotheses, a criticism that echoed that of the Kantian philosopher J. F. Fries.

Fischer's table was, nevertheless, responsible for the dissemination of Richter's work. Berthollet used it in his *Essai de statique chimique* of 1803, and Thomas Thomson incorporated it into the 1807 edition of his *System of Chemistry*. Richter himself recognized the value of Fischer's compilation, and in 1803 published a more complete table, containing the equivalent weights of eighteen acids and thirty bases. Although Berthollet's and Thomson's books received wide circulation, Richter's own did not, and this lack of recognition caused him considerable bitterness. Nor was Fischer his only critic; Gehlen, L. W. Gilbert, Schweigger, Bischoff, and Berzelius all found fault with the speculative aspects of his work, while praising his experimental methods. Others of Richter's contemporaries were critical of his verbosity, his obscurity, and his careless arithmetic.

By the time of Richter's death in 1807, Berzelius had realized the significance of his stoichiometry, although he incorrectly ascribed the discovery of the law of neutrality to Karl Friedrich Wenzel. G. H. Hess, in 1840, pointed out Berzelius' error, but Hermann Kopp perpetuated it in his *Geschichte der Chemie* of 1844 (although he, in turn, acknowledged his own mistake in 1873). Berzelius' textbook, nevertheless, brought Richter's work to the attention of a wide audience although he stated that Richter's experiments were not always correct. Specifically, Berzelius cited Richter's use of aluminum carbonate, which, he said, "does not exist"; he was here mistaken, since Richter had indeed worked with aluminum carbonate, which he obtained by precipitating an aluminum salt with an alkali carbonate, and then

drying the precipitate, which he carefully tested for its alumina content.

Richter performed other laboratory work besides the experiments directed toward validating his theory. In Volume XI of the *Neuern Gegenstände*, published in 1802, he reported his isolation of a new earth from beryl, which he called *Agust-Erde*. (This substance had, in fact, already been extracted two years previously by Trommsdorff, and a number of chemists, including Vauquelin, Klaproth, Bucholz, and, in 1803, Trommsdorff himself showed independently that it was merely impure calcium phosphate.) The same volume contains Richter's important researches on colloidal gold, in which he recognized that the purple substance precipitated out of a solution of gold and tin is "but an intimate mixture of extremely finely divided gold with tin calx." Richter went on to prepare colloidal gold in a number of ways, for example, from very dilute solutions of iron sulfate and gold, and from a water solution of the product formed by heating fulminating gold with borax. He described the colors of different gold preparations in detail and, by means of experiments on fulminating gold and ammonium nitrate, calculated the composition of ammonia to be 80.89 percent nitrogen and 19.11 percent hydrogen.

In addition to his own publications, Richter contributed, between 1803 and 1805, important articles on oxygen, light, neutrality (in which he gave his table of equivalent weights based on Fischer's), stoichiometry, and oxidation to David Bourguet's *Chemisches Handwörterbuch nach den neuesten Entdeckungen entworfen*, which he edited from Volume III on. Among his other later works, Richter also, in 1805, announced his discovery of a new metal in nickel ores, which he called "niccolanum"; it turned out, however, that this substance was not new, but was a rather impure nickel, containing cobalt, iron, and arsenic. Trommsdorff, Gehlen, Hisinger, and John Murray all demonstrated that such was the fact.

Since Richter was an eighteenth-century chemist, it is interesting to examine his attitude toward the doctrine of phlogiston. He was in this respect a product of his time. Even in his doctoral dissertation, he had attempted to determine the weight of phlogiston, and declared himself an opponent of Lavoisier's oxidation theory, which he thought to be insufficiently based upon experimental data. He believed metals to be composed of calx and phlogiston. Either phlogiston was freed from a solution of the metal and an acid like inflammable air (hydrogen) or else it combined with the acid. For example, he thought sulfur to be a compound of sulfuric acid and phlogiston; phosphorus, a compound of phosphoric acid and phlogiston; light, a compound of matter of fire and phlogiston; and so on.

After reading Christoph Girtanner's *Anfangsgründe der antiphlogistischen Chemie* (1792), however, Richter adopted most of the tenets of the antiphlogiston theory, although with some modifications. He continued to regard phlogiston as the principle of combustibility, and, in his new view, considered a metal to consist of phlogiston and a substratum. In combustion phlogiston combined with the matter of heat from oxygen gas (which Richter, like Lavoisier, believed to be a compound of "oxygen base" and the matter of heat), light and heat were emitted, and the remainder of the oxidized metal was dissolved in the matter of oxygen. Richter elaborated these concepts in a paper of 1793, "Entwurf eines Systemes der Phlogologie oder kurzgefassete Theorie der Phlogurgie," which, like much of the rest of his work, is characterized by complex nomenclature and diffuse style.

Richter was a member of the Gross-britannische Societät of Göttingen and of the Munich and St. Petersburg academies. Despite his imaginative theorizing and skillful laboratory work, his chemical achievements passed almost unnoticed among his contemporaries. The significance of his stoichiometry was not recognized for a number of years, and his results had virtually no influence on the development of chemistry until after the acceptance of Dalton's atomic theory.

BIBLIOGRAPHY

I. ORIGINAL WORKS. Richter's dissertation was *De usu matheseos in chemia. Dissertatio quam consentiente amplissima facultate philosophica pro receptione in eandem* (Königsberg, 1789).

Ueber die neuern Gegenstände der Chemie (Breslau–Hirschberg–Lissa, 1791–1802), was published in 11 parts: I, *Das ohnlängst entdeckte Halbmetall Uranium* (1791), contains 15 articles, mainly on the separation and purification of metals; II, *Ueber das Wasserbley und den daraus entstehenden blauen Carmin* (1792), deals with Carmine blue (a mixture of tin molybdate and blue molybdic oxide); III, *Den Versuch einer Critik des antiphlogistischen Systemes nebst einem Anhange* (1793), presents Richter's intermediate oxidation theory; IV, *Ueber Flussspathsäure und die neuentdeckte Ordnung chymischer Elemente* (1795); V, *Ueber Antiphlogistik, bequeme Scheidungswege und einige physische Parthien* (1795), has a number of tables of specific gravities, material on the construction of aerometers, and considerations on oxidations and thermometry; VI, *Ueber die Neutralitäts-Ordnung verbrennlicher Säuren nebst chymischen, insbesondere pharmaceutischen und metallurgischen Handgriffen* (1796), gives the methods of preparing organic

acids; VII, *Beyträge zur Antiphlogistik in Bezug auf die Göttlingischen Versuche* (1796), deals with the oxidation theory; VIII, *Ueber die Verhältnisse der Stronthian-Erde und quantitative Ordnung der Metalle* (1797), concerns the preparation of pure alcohol; IX, *Ueber die besonder Ordnung der Metalle und ihrer Verhältnisse* (1798); X, *Ueber das Chromium, Titan, Tellur, Wolfram und andere Metalle, nebst fernerer Entwickelung der quantitativen Ordnung* (1800); and XI, *Ueber die Glucine, Agust-Erde und einige besondere Eigenschaften des Goldes* (1802).

Anfangsgründe der Stöchyometrie oder Messkunst chymischer Elemente appeared in 3 vols.: I, pt. 1, . . . *welcher die reine Stöchyometrie enthält* (Breslau–Hirschberg, 1792); I, pt. 2, . . . *enthaltend die reine Thermimetrie und Phlogometrie* (Breslau–Hirschberg–Lissa, 1794); II, . . . *welcher die angewandte Stöchyometrie enthalt; für Mathematiker, Chymisten, Mineralogen und Pharmaceuten* (Breslau–Hirschberg, 1793); III, . . . *welcher der angewandten Stöchyometrie dritten Abschnitte und einen Anhang zu dem ersten und zweiten Theil enthält* (Breslau–Hirschberg, 1793; fascs. ed., Hildesheim, 1968).

Richter edited D. L. Bourguet's *Chemisches Handwörterbuch nach den neuesten Entdeckungen*, III–VI (Berlin, 1803–1805).

His essay on the phlogiston theory is "Entwurf eines Systemes der Phlogologie oder kurzgefassete Theorie der Phlogurgie," in *Neuern Gegenstände*, III (1793), 179–197. Another important essay is "Beyträge zur metallurgischen Chemie. Niccolanum, ein neu entdecktes, dem Nickel in manchem Betracht sehr ähnliches, Metall," in *Neues allgemeines Journal der Chemie*, 4 (1805), 392–401.

II. SECONDARY LITERATURE. See the following, listed chronologically: G. H. Hess, "On the Scientific Labours of Jeremias Benjamin Richter," in *Philosophical Magazine*, 21 (1842), 81–96, trans. from *Journal für praktische Chemie*, 24 (1841), 420–438; R. A. Smith, *Memoir of John Dalton and History of the Atomic Theory up to His Time* (London, 1856), 186–215; C. Löwig, *Jeremias Benjamin Richter. Der Entdecker der chemischen Proportionen. Eine Denkschrift* (Breslau, 1874); P. Schwarzkopf, "Jeremias Benjamin Richter. Anlässlich der hundertsten Wiederkehr seines Todestages," in *Chemikerzeitung*, 31 (1907), 471–475; G. Lockemann, "Jeremias Benjamin Richter in seiner Bedeutung für Naturwissenschaft und Technik," in *Technikgeschichte*, 30 (1941), 107–115; J. R. Partington and D. McKie, "Richter's Theory of Combustion," in "Historical Studies on the Phlogiston Theory," in *Annals of Science*, 4 (1939), 130–135; and J. R. Partington, "Jeremias Benjamin Richter and the Law of Reciprocal Proportions," *ibid.*, 7 (1951), 173–198, and 9 (1953), 289–314.

H. A. M. SNELDERS

RICHTHOFEN, FERDINAND VON (*b.* Karlsruhe, Silesia [now Poland], 5 May 1833; *d.* Berlin, Germany, 6 October 1905), *geology, geomorphology, geography.*

Richthofen was the scion of a landed Silesian family. After completing his secondary education at the Catholic Gymnasium in Breslau (now Wrocław), in 1850 he took up the study of geology at the university there and two years later went to the University of Berlin, where he graduated in 1856. The following year he joined a party of notable geologists, including C. W. Gümbel, on a geological tour of North Tirol and the Vorarlberg Alps. He was assigned the task of compiling the combined report and continuing the survey. Richthofen produced an admirable exposition of the Triassic succession in those areas. More important for Alpine geology was his independent publication. *Geognostische Beschreibung der Umgegend von Predazzo . . .* (1860), which, according to Zittel (1901, p. 476) "was greeted on its appearance with the highest recognition from all sides, and the author, who was little over twenty at the time, was looked upon as one of the first Alpine geologists."

In this work Richthofen successfully elaborated upon the Triassic succession in the South Tirol and the conditions under which it was formed. In opposition to the more catastrophic concepts then prevalent, he attributed most of the changes in the form of the ground and the tectonic disturbances to slow crustal movements. He also attributed the dolomitic masses and some other parts of the Triassic limestones in the southern Alps to reef-building corals upon a slowly subsiding sea floor.

Richthofen, with the aid of the Austrian Imperial Geological Institute, extended his research to the trachytic mountain ranges of the Carpathians, particularly in Transylvania. In 1860 he served as geologist with a Prussian government mission to Southeast Asia and the Far East, where his travels included an overland journey from Bangkok to Moulmein. But little came of this mission—most of his notes and collected materials were lost. In June 1862 he left for California and stayed there for the next six years, working as a journalist for German newspapers to which he reported on the mineral wealth and gold strikes (including the Comstock Lode). In the Rocky Mountains and the Sierra Nevada he recognized a definite sequence of igneous rocks—from propylite to trachyte and basalt—that was later confirmed by American geologists. In September 1868, four years after the Taiping Rebellion, he was able to realize his chief ambition and visit China. His trip was financed first by the Bank of California and later by the Chamber of Commerce of Shanghai in return for reports in English on the economic resources of the areas he visited. By May 1872 he had made seven long journeys and had traversed every province of the Chinese Empire, except Kansu and Yunnan. His reports,

published as *Letters on China* (Shanghai, 1870–1872), gave the first indications of the importance of the Shantung coalfield and emphasized the commercial potential of Tsingtao, a port later occupied by the Germans.

In 1872 Richthofen returned to Germany and spent the next thirty-three years mainly in writing and lecturing on China and promoting the study of geography in German universities. His written accounts of China, under the liberal patronage of the government, were planned on a monumental scale; but they progressed slowly because of his numerous academic commitments. He was president of the Gesellschaft für Erdkunde in Berlin from 1873 until his death. In 1875 he was elected professor of geology at the University of Bonn but delayed assuming the duties for four years while he compiled the first and part of a second volume on China.

In 1883 Richthofen received the chair of geography at Leipzig; his inaugural address was "Aufgaben und Methoden der heutigen Geographie." In 1886 he was persuaded to return to Berlin as professor of geography, and he held this post until he died suddenly of a stroke. In his later years he almost completed the establishment of the Museum für Meereskunde at the University of Berlin (it was finished by his successor, Albrecht Penck) and acted as rector of the university in 1903, when he delivered a notable address entitled "Triebkräfte und Richtungen der Erdkunde im neunzehnten Jahrhundert."

Richthofen's chief contributions were to Alpine stratigraphy; the geology and geography of China; geomorphology; and geographical methodology. The first volume of his monumental *China: Ergebnisse eigener Reisen und darauf gegründeter Studien* appeared in 1877. It dealt largely with the morphology and geology of Inner Asia and China and their influence on the movements of peoples. The next parts, published between 1882 and 1885, discussed North China and were based mainly on Richthofen's field observations and collections. They included special analyses by August Schenk of the fossil floras and an atlas of twenty-seven hypsographical and twenty-seven geological maps compiled largely from fieldwork and instrumental (aneroid) measurements. Richthofen concluded that the planes of unconformity in the rock series in China were due to marine abrasion on a subsiding landmass. He also described the masses of loess, which he attributed to subaerial deposition by wind, except in some localities where a "lake loess" indicated an association with water. When Richthofen died, the volume on southern China and the second part of the atlas were unfinished; but the text was completed from his copious notes by Ernst Tiessen and the maps were completed by Max Croll. These works were published in 1912 at the expense of the Prussian Kulturministerium. Tiessen also edited a full summary of the original notebooks, with a selection of the many admirable field sketches of people and landscapes, in *Ferdinand von Richthofen's Tagebücher aus China.*

Richthofen's geomorphological studies formed part of his geology and the fundamental basis of his geography. In 1875 he contributed a paper on geology to G. von Neumayer's *Anleitung zu Wissenschaftlichen Beobachtungen auf Reisen.* In 1886 he published a greatly enlarged version of this paper under the somewhat too comprehensive title *Führer für Forschungsreisende.* The first part of this guide deals with the techniques of field location and observation. The second part discusses at length the interrelationships of geology and surface forms, with considerable detail given to the physical processes involved. The final part contains accounts of soils, rocks, and mountain structures and classifies the main kinds of landforms *(Bodenplastik)* according to the process or dominant process in their formation. Thus the genetic aspect predominates, and external forms are used as sub-classificatory indices only where unavoidable. The text is illustrated by more than 100 small line blocks and in its classification of material and approach was the first truly successful compilation of genetic geomorphology. It immediately became the standard work in Germany for the systematic treatment of landforms and strongly influenced Albrecht Penck's *Morphologie der Erdoberfläche.*

Besides his large additions to the factual knowledge of China and elsewhere, Richthofen made outstanding contributions to geographical methodology and to the advancement of geography as an autonomous science. From about 1875 he devoted most of his time to geographical matters and took an interest in all branches of the discipline, although his geological training led him to emphasize the influence of the nature of the land surface upon its inhabitants. Richthofen believed that geography was concerned with the causal interrelationships of all formations and phenomena related to the surface of the earth (that is, *Erdoberflächenkunde,* rather than the more comprehensive *Erdkunde,* or earth science). It was a science based on field observations and measurements and was always concerned with the assembly of spatial distributions upon a physical background. The method of geographical investigation, however, varied with the scale of the project and aim of the prospector. There were two main fields of geography: special and general. Special geography was descriptive and synthetic and itself fell into two categories:

chorography and chorology. Chorography comprised the encyclopaedic registering, within the confines of any area *(Erdraum)*, of the systematic assembly of the phenomena and features belonging to the six realms of nature: land, water, atmosphere, plants, animals, and man. Analysis was not required except to divide the whole or the bigger areas into smaller components or unit areas. Chorology, although descriptive, went beyond chorography because it tried to explain the areal distribution of phenomena by studying their causal and dynamic (spatial) relationships.

The chorological method of special geography led to the second main field of geography—general geography, which dealt primarily with the general study of earthbound phenomena in an abstract or analytical way. It proceeded from the particular to the general and examined phenomena from four points of view or principles: morphology; material nature; dynamic or spatial interconnections; and development (forces and causes of change). Each of these principles would provide a distinctive aspect of general geography, while the last, or genetic, principle would serve to interpret the other three. But Richthofen preferred to apply all four aspects to the study of the six realms of nature. Thereby he brought the analytical approach into closer relationship with chorological studies and unified the numerous branches of geography within a broad physical framework. His scheme, however, was obviously two-sided; and while many of his disciples analyzed spatial arrangements of phenomena on a wide scale, others carried out research in depth on small areas.

Richthofen is generally regarded by geologists today as a stratigrapher who became the "Prince of *Forschungsreisende.*" Geographers and geomorphologists rightly acclaim him as one of the greatest forces in the modern development of their disciplines. Many of his students and followers held important chairs in geography in central Europe until the mid-1930's.

BIBLIOGRAPHY

I. ORIGINAL WORKS. Richthofen's major works are "Die Kalkalpen von Vorarlberg und Nordtirol," in *Jahrbuch der Geologischen Reichsanstalt,* **10** (1859–1861); *Geognostische Beschreibung der Umgegend von Predazzo . . .* (Gotha, 1860); "Die Metall-Produktion Californiens und der Angrenzenden Länder," in *Petermanns Mitteilungen,* Supp. **3**, no. 14 (1864); *Natural System of Volcanic Rocks* (San Francisco, 1867); *Letters on China* (Shanghai, 1870–1872, repr. 1900); and *China: Ergebnisse eigener Reisen und darauf gegründeter Studien,* 3 vols.—I (Berlin, 1877); II (1882–1885), which is pt. 3 (1882), on North China; pt. 4 (1883), on paleontology; and atlas (1885); III (1912),

E. Tiessen, ed. The work has been reprinted (Graz, 1968–1969).

Subsequent works include *Aufgaben und Methoden der heutigen Geographie* (Leipzig, 1883), his inaugural address; *Führer für Forschungsreisende* (Berlin, 1886); "Geomorphologische Studien aus Ostasien," in *Sitzungsberichte der Preussischen Akademie der Wissenschaften zu Berlin. Phys.-math. Klasse*, **11** (1900), 888–925; **36** (1901), 782–808; **38** (1902), 944–975; **40** (1903), 867–918; *Triebkräfte und Richtungen der Erdkunde im neunzehnter Jahrhundert* (Berlin, 1903), his rectorial address; *Ferdinand von Richthofen's Tagebücher aus China*, E. Tiessen, ed., 2 vols. (Berlin, 1907); and *Mitteilungen des Ferdinand von Richthofen*, E. Tiessen, ed. (Berlin, 1912).

II. SECONDARY LITERATURE. The *Festschrift* (Berlin, 1893) presented to Richthofen on his sixtieth birthday lacks a bibliography. On Richthofen and his work, see E. von Drygalski, "Von Richthofen und die deutsche Geographie," in *Zeitschrift der Gesellschaft für Erdkunde zu Berlin* (1933), 88–97; A. Hettner, "Ferdinand von Richthofens Bedeutung für die Geographie," in *Geographische Zeitschrift*, **12** (1906), 1–11; A. Penck, "Richthofens Bedeutung für die Geographie," in *Berliner geographische Arbeiten*, no. 5 (1933), 1–17, and *Deutsche Rundschau*, **225** (Nov. 1930), 154–157; E. G. Ravenstein, "Ferdinand Freiherr von Richthofen," in *Geographical Journal*, **26** (1905), 679–682; G. Wegener and H. von Wissmann, "Ferdinand von Richthofen," in *Die Grossen Deutschen*, V (Berlin, 1935–1936), 390–398; and Karl von Zittel, *History of Geology and Palaeontology*, M. M. Ogilvie-Gordon, trans. (London, 1901), *passim*.

ROBERT P. BECKINSALE

RICHTMYER, FLOYD KARKER (*b*. Cobleskill, New York, 12 October 1881; *d*. Ithaca, New York, 7 November 1939), *physics*.

Richtmyer's chief contributions to science were his enthusiastic teaching of physics, his zeal for administration in the community of physicists, and his experimental researches in X-ray spectra. He was brought up in the country, where he attended local public schools. After receiving the Ph.D. at Cornell University in 1910, he remained on the physics staff there until his death, becoming a full professor in 1918 and dean of the graduate school in 1931.

Richtmyer's first researches were in photometry, specifically, the application of photoelectricity to this field. After 1918, however, he turned his attention to X-ray spectroscopy and worked in this domain for the rest of his life. He became particularly interested in low-intensity satellite lines in X-ray spectra and developed high-precision methods for studying them. This research led to a broader program of investigation of widths and shapes of X-ray lines and absorption limits. This program was carried on at Cornell for many years, largely by Richtmyer's graduate students under his guidance.

Richtmyer's interest in teaching is reflected not only in the courses he gave and in his supervision of graduate students but also in his celebrated book, *Introduction to Modern Physics* (1928), which had considerable influence on the teaching of atomic physics and, with revisions and additions by colleagues after his death, remained a valuable work.

Richtmyer devoted much attention to professional societies and served as president of the American Physical Society, the Optical Society of America, and the American Association of Physics Teachers. In 1932 he was elected a member of the National Academy of Sciences. For several years he was also a member of the executive committee of the American Institute of Physics and was a long-time editor of the *Review of Scientific Instruments*. All who met Richtmyer were impressed with his active and unselfish devotion to the advancement of the profession of physics.

BIBLIOGRAPHY

I. ORIGINAL WORKS. Richtmyer's complete bibliography includes one book and seventy-three articles. A complete list is given by Ives, below. His book is *Introduction to Modern Physics* (New York, 1928). His major articles include "Dependence of Photoelectric Current on Light Intensity," in *Physical Review*, **29** (1909), 71–78; "Photoelectric Effect With Alkali Metals. II," *ibid.*, 404–408; "Photoelectric Cells in Photometry," in *Transactions of the Illuminating Engineering Society*, **8** (New York, 1913), 459–469; "Comparison of Flicker and Equality-of-Brightness Photometer," in *Bulletin of the Bureau of Standards*, **14** (1918), 87–113, written with E. C. Crittenden; "The Mass-Absorption Coefficient of Water, Aluminum, Copper and Molybdenum for X-rays of Short Wavelength," in *Physical Review*, **15** (1920), 547–549, written with Kerr Grant; "Absorption of X-rays," *ibid.*, **18** (1921), 13–30; and "The Size of the Electron as Determined by the Absorption and Scattering of X-rays," *ibid.*, **20** (1922), 87–88.

Subsequent writings include "The Structure of the K Lines of Molybdenum," *ibid.*, **23** (1924), 550–551, written with R. C. Spencer; "The Apparent Shape of X-ray Lines and Absorption Limits," *ibid.*, **26** (1925), 724–735; "Absorption of X-rays in Various Elements," in *Nature*, **120** (1927), 915–916; "Multiple Ionisation and Absorption of X-rays," in *Philosophical Magazine*, **6** (1928), 64–88; "Satellites of X-ray Lines $L\alpha$, $L\beta_1$, and $L\beta_2$," in *Physical Review*, **34** (1929), 574–581, written with R. D. Richtmyer; "Hyperfine Structure of X-ray Lines," *ibid.*, **36** (1930), 1017, written with S. W. Barnes and K. V. Manning; "Intensity of X-ray Satellites," *ibid.*, **36** (1930), 1044–1049, written with L. S. Taylor; "X-rays and Their Uses," in

Scientific Monthly, **32** (1931), 454–463; and "The Romance of the Next Decimal Place," in *Science*, **75** (1932), 1–5.

Later works are "Relative Intensities of Certain L-series X-ray Satellites in Cathode-ray and in Fluorescence Excitation," in *Physical Review* **44** (1933), 955–960, written with F. R. Hirsh, Jr.; "The Widths of the L-series Lines and of the Energy Levels of Au(79)," *ibid.*, **46** (1934), 843–860, written with S. W. Barnes and E. Ramberg; "Determination of the Shape, Wavelength and Widths of an X-ray Absorption Limit," *ibid.*, **43** (1933), 754, written with S. W. Barnes; "L-satellites in the Atomic Number Range $73 < Z < 79$," *ibid.*, **51** (1936), written with C. H. Shaw and R. E. Shrader; and "Determination of Widths of Energy States: Argon K Absorption Limit," *ibid.*, **52** (1937), 678–679, written with L. G. Parratt.

II. SECONDARY LITERATURE. On Richtmyer and his work, see Herbert E. Ives, "Floyd Karker Richtmyer 1881–1939," in *Biographical Memoirs. National Academy of Sciences*, **22** (1941), 71–81, with portrait and bibliography; and F. R. Hirsh's notice in *Dictionary of American Biography*, XXII, supp. 2 (New York, 1958), 556–557.

R. B. LINDSAY

RICKETTS, HOWARD TAYLOR (*b.* Findley, Ohio, 9 February 1871; *d.* Mexico City, Mexico, 3 May 1910), *pathology.*

In his short and brilliant career Ricketts pioneered research on a group of diseases that were later named after him. The rickettsial diseases are those caused by obligate, intracellular, parasitic, and pleomorphic microorganisms. Ricketts brought them to the attention of microbiology. He provided the basis for a rich understanding of *Rickettsia* by working intensively with Rocky Mountain spotted fever and Mexican typhus fever. The genus now includes all the very small, gram-negative, bacterium-like organisms found in arthropods and capable of transmission to some vertebrates. Difficult to stain, with Machiavello's technique, *Rickettsia* appears red against a bluish background. Henrique da Rocha-Lima named them in honor of Ricketts in 1916, and fundamental explorations of the genus were made by S. B. Wolbach shortly thereafter.

Ricketts, the son of Andrew Duncan and Nancy Jane Ricketts, was born on a farm. A husky and vigorous youth, he was motivated by deep Methodist conviction. In 1890, after completing preparatory school, he entered Northwestern University. Two years later he transferred to the University of Nebraska. The panic of 1893 destroyed his family's modest fortune, and from that time on Ricketts worked his way through school. In 1894 he entered Northwestern Medical School. His patron, W. H.

Allport, helped him secure employment there in the medical museum.

During his third year of medical school, Ricketts suffered a nervous breakdown. After his recovery he interned at Cook County Hospital, and in 1898 he received a pathology fellowship at Rush Medical College. In 1900 he married Myra Tubbs.

At the suggestion of L. Hektoen, head of the pathology department at the University of Chicago, Ricketts studied in Berlin, where his son, Henry, was born. He later went to the Pasteur Institute. These experiences in Europe perfected Ricketts' laboratory techniques and gave him a broader appreciation for theoretical microbiology.

In 1902 Ricketts became associate professor of pathology at the University of Chicago. His research there on blastomycosis was published in *Infection, Immunity, and Serum Therapy* (1906). When Ricketts traveled to Missoula, Montana, in the late spring of 1906, his vacation in the Rockies led to unexpected and exciting work. Accounts of a deadly spotted fever intrigued him, and he immediately began examining patients. He learned that the disease was geographically quite restricted and that 80 to 90 percent mortality rates were not uncommon.

The infective agent of spotted fever was thought to be *Piroplasma*, and several modes of transmission were suspected. Over the next three years Ricketts showed that the tick *Dermacentor andersoni* was responsible for transmission, but he failed to find *Piroplasma* and thus began to formulate a new theory. His work upset several real estate agents who were concerned lest land prices drop. One entomologist from the U.S. Department of Agriculture denied emphatically that the tick was responsible. By 1909 Ricketts thought he had seen the causative bacillus, and he outlined a control program that he hoped would destroy the reservoir of the disease. He based his ideas on the successful model of Texas tick fever that had been worked out by Theobald Smith and others. Although there were too many natural hosts for control to be effective, Ricketts' techniques and ideas were bringing him closer to the truth.

In 1910 Montana suffered an outbreak of typhoid fever and smallpox, and money ordinarily spent on preventing spotted fever was diverted to fight these other diseases. Ricketts thus accepted an invitation to examine typhus in the Valley of Mexico. Similarities between typhus and spotted fever would perhaps provide him with the key to the etiology of the diseases. He discovered that lice transmitted the fever, and he worked closely and rapidly with severely infected patients. He and two associates contracted the disease and Ricketts died. He had shown, however, that typhus (tabardillo) was distinct from spotted fever. He was buried in Kirkwood, Illinois.

Ricketts also pioneered the use of laboratory animals for inoculation experiments and disease identification. His work on immunity and serums became the basis for further advances in vaccine development. Wolbach of the Harvard Medical School wrote an apt eulogy: "Ricketts brought facts to light with brilliance and accuracy and indicated by the methods he used, most of the major lines of development subsequently employed in the study of rickettsial diseases."

BIBLIOGRAPHY

The Howard Taylor Ricketts Memorial Collection in ten boxes is in the Harper Memorial Library, University of Chicago. Important photographs, letters, drawings, and laboratory notes and protocols are reproduced in a family scrapbook on loan to the archives of the Rocky Mountain Laboratory (National Institute of Allergy and Infectious Diseases) in Hamilton, Montana. This scrapbook was assembled by the family in 1941 and materials relating to typhus were added in 1947. A permanent collection of Ricketts' work is: *Contributions to Medical Science by Howard Taylor Ricketts 1870–1910* (Chicago, 1911).

Secondary citations are in Frank L. Horsfall, Jr., and Igor Tamm, *Viral and Rickettsial Infections of Man*, 4th ed. (Philadelphia, 1965), 1059–1129. Also see *Selected Papers on Pathogenic Rickettsia*, Nicholas Hahon, ed. (Cambridge, Mass., 1968), xv–xxiii, 27–36, 41–46.

PIERCE C. MULLEN

RIDGWAY, ROBERT (*b.* Mt. Carmel, Illinois, 2 July 1850; *d.* Olney, Illinois, 25 March 1929), *ornithology*.

Ridgway received his love for the outdoors and most of his education from his father, David Ridgway, a small-town pharmacist. Having come to the attention of Baird of the Smithsonian Institution because of his bird drawings, young Ridgway was appointed, in 1867, zoologist of the U.S. Geological Survey directed by Clarence King. On his return he was employed by the Smithsonian, serving as curator of birds at the U.S. National Museum until he retired.

After the death of Baird and Coues, Ridgway was considered America's leading professional ornithologist. From 1887, when he published *Manual of North American Birds*, until 1919, when the eighth part of "Birds of North and Middle America" appeared, his publications were the standard reference works on North American birds. Ridgway is best described as a superb technician. He represented the acme of descriptive taxonomy. During his lifetime he described far more new genera, species, and subspecies of American birds than any other ornithologist. His bibliography of some 550 titles clearly delineates the scope of his interests. It includes reports on collections, the description of new taxa, records of new distributional facts, notes on geographic variation and nomenclature, generic revisions, as well as occasional popular papers. There is an almost total absence of generalized or philosophical papers, nor was his considerable knowledge of the living bird reflected in his publications.

In his lifetime Ridgway was universally admired, having set standards of accuracy and reliability that others could safely use as a basis for their researches. A warm advocate of trinomials, he was considered a "progressive" in the 1880's. The dominance of his ideas came to an end when polytypic species were much more broadly conceived by others than by Ridgway himself and when the genus was no longer defined purely morphologically. Consequently, most of the genera described by him are now considered synonyms.

Ridgway was a modest man and was so shy that he was virtually unknown except to his close friends. His capacity for work was amazing. His published works amount to approximately 13,000 printed pages, all written without the aid of typewriter or stenographer. Second to ornithology, Ridgway was best known for his color key. This work, consisting of fifty-three plates showing 1,115 colors in small rectangles, was published in an edition of 5,000 copies. Each color was hand-mixed according to a careful formula, and to ensure absolute uniformity each color was produced at a single time in sufficient quantity for the entire edition. The work was widely used not only among scientists but also among florists; manufacturers of paints, chemicals, and wallpapers; and in many government offices. When a careful description of colors was necessary, no better method for objective description was known, until fairly recently, except by comparison with Ridgway's Color Standard.

BIBLIOGRAPHY

I. ORIGINAL WORKS. Ridgway's works include *A History of North American Birds: Land Birds*, 3 vols. (Boston, 1874), written with S. F. Baird and T. M. Brewer; *A Manual of North American Birds* (Philadelphia, 1887); "The Birds of North and Middle America," in *Bulletin. United States National Museum*, no. 50, pt. 1 (1901); pt. 2 (1902); pt. 3 (1904); pt. 4 (1907); pt. 5 (1911); pt. 6 (1914); pt. 7 (1916); pt. 8 (1919); and *Color Standards and Color Nomenclature* (Baltimore, 1912). A bibliography accompanies both of the biographies listed below.

II. SECONDARY LITERATURE. On Ridgway's life and work, see A. Wetmore, in *Biographical Memoirs. National Academy of Sciences*, **15** (1932), 57–101; and H. Harris, "Robert Ridgway," in *Condor*, **30** (1928), 5–118.

ERNST MAYR

IBN RIḌWĀN, ABŪ'L-ḤASAN ʿALĪ IBN ʿALĪ IBN JAʿAFAR AL-MIṢRĪ (*b.* El Gīzah, Egypt, A.D. 998; *d.* Cairo, Egypt, A.D. 1061 or 1069), *medicine*.

Most of what is known about Ibn Riḍwān's life is based upon the information about him given by Ibn al-Qifṭī and Ibn Abī Uṣaybiʿa, although his own autobiography offers some additional clues. His father was employed in a bakery, and Ibn Riḍwān himself referred to his own early poverty. Ibn Riḍwān early displayed a taste for study, and went to Cairo at the age of ten; when he was about fifteen he began to teach himself medicine while earning his living by casting horoscopes for passersby. He married when he was about thirty and adopted one child, an orphan girl who, around 1044, disappeared with his money and valuables. The shock is said to have affected Ibn Riḍwān's sanity. He was happier in his medical career, and it is reported that his advice was sought by the king of Makran, Abū'l-Muʿaskar, who had suffered a hemiplegia. His success was assured when a Fāṭamid caliph of Cairo named him chief of all the physicians in Egypt; J. Schacht identifies this ruler as al-Mustanṣir, who reigned from 1036 to 1094, rather than al-Ḥākim (whose name is given by Ibn Abī Uṣaybiʿa), who died in 1021, when Ibn Riḍwān was only twenty-three.

Ibn Riḍwān's short autobiography is preserved in the work of Ibn Abī Uṣaybiʿa. In it Ibn Riḍwān asserts that each man should practice the profession that most suits him. In his own case, it is clear that the configuration of the heavens on the day of his birth predestined him for medicine, a science that, he emphasizes, borders closely on philosophy and is equally pleasing to God. His piety is everywhere apparent; although he stipulated that the physician should be paid in cash for his consultations, and should not lend money except in cases of real necessity, he also stated that if borrowed money is not returned, the loan should be considered an act of charity, done for God, who also sustained him in more severe defeats.

Ibn Riḍwān subscribed to a code of scrupulous medical ethics, Hippocratic in inspiration. He pledged to be cleanly dressed, to perfume his clothes, to be pure of heart and chaste of eye, and not to desire women or seek riches. He was to keep professional secrets and to avoid bullying the patient, in whom he should rather inspire confidence and whom he must treat gently and without deception. He swore not to use dangerous remedies, to respect life, and not to perform abortions. Like al-Ghazālī, Ibn Riḍwān thought medicine as useful to the soul as to the body, and even wrote a treatise on how to attain happiness through medicine. He was interested in treating the soul, and made notes on the works of Posidonius and Philagrius, which might loosely be termed psychiatric and neurological. He saw a relationship between medicine and morality, and was further concerned with the survival of the soul after death, in which connection he drew upon the ideas of Plato and Aristotle.

Ibn Riḍwān's lofty conception of the art of medicine was complemented by his conscientious practice of it. He combined his considerable theoretical knowledge with careful clinical observation. He asserted that the physician should first consider the exterior aspect of the parts of the body and note the color, temperature, and texture of the skin; he should then examine the function of the internal and external organs, testing, for example, the acuteness of hearing and vision and the articulation of speech by the tongue. Ibn Riḍwān also advocated evaluating the muscles of the patient by having him lift weights, and ascertaining the force of his prehension by having him grasp objects. He recommended observing the walk of the patient, both forward and backward, palpating the abdomen to determine the state of the intestines, feeling the pulse to discover the "temperament" of the heart, and examining the urine and humors to determine the condition of the liver. He asked the patient questions to determine his state of mind, and carefully noted his responses, as well as aspects of his behavior and personal preferences. He attempted to discover whether the patient's affliction was of recent origin or of long duration, and he based his treatment on all of these factors.

At the end of his autobiography Ibn Riḍwān lists some of the books that he had read, including works of literature and religious law, Hippocrates and Galen, Dioscorides' treatise on simples, Rufus of Ephesus, Oribasius, Paul of Aegina, al-Rāzī's *Al-Ḥāwī*, writings about agriculture and pharmacopoeias, Ptolemy's *Almagest* and *Quadripartitum*, and Plato, Aristotle, Alexander of Aphrodisias, Themistius, and al-Fārābī. This list reflects the scope of Ibn Riḍwān's self-education, although he was frequently attacked as an autodidact who had only book learning. He defended the superiority of his training, however, and denounced his critics for the slightest deviation from ancient texts; he thus became involved in a controversy with Ibn Buṭlān, who offered both logical and psychological arguments to prove that verbal teaching facilitates learning. Ibn Riḍwān wrote several polemics against Ibn Buṭlān, on this and other subjects, and also attacked Ḥunayn ibn Isḥāq for his translation of Galen and al-Rāzī, for his doubts concerning Galen, and for impiety and the denial of prophecy. He wrote against Ibn al-Jazzār and Ibn al-Ṭayyib, whom he accused of sophistry.

Ibn Riḍwān's medical writings embraced a wide variety of topics. He was the author of about fourteen commentaries on, and summaries of, Hippocrates and Galen; his commentary on the latter's *Ars parva* was translated into Hebrew, as was his compendium of the principles of medicine. He also wrote a series of short treatises on the treatment of elephantiasis, purgatives, syrups and electuaries, classifications of fevers, tumors, recurrent fevers, and asthma, as well as a book on medical education. A longer book dealt with diseases prevalent in Egypt; Ibn Riḍwān there discussed preventive measures, sanitation, the rules of hygiene, and the causes of plague, a disease of which he had had firsthand experience during an epidemic in Cairo in 1044. He compiled notes on pharmacology and a dictionary of simples, arranged alphabetically.

Ibn Riḍwān's nonmedical works comprise treatises on Aristotelian physics (on the superiority of Aristotle's science, on heat, and on the physical existence of points and lines), metaphysical tracts (on prime matter and the eternal existence of the world), and books on the climatology of Egypt and the utility of Porphyry's *Isagoge*. His commentary on Ptolemy's *Quadripartitum* was translated into Latin and is extant in an edition printed in Venice. Although he championed the written word, Ibn Riḍwān also lectured; his students included the Fāṭimid prince al-Mubashshir ibn al-Fātik, himself a writer and philosopher, and the Jewish physician Afrāīm ibn al-Zaffar.

It is apparent from his writings and his activities that Ibn Riḍwān possessed a logical and calculating mind, coupled with a great love of his profession and of knowledge itself. His character must have been formed in part by the poverty of his early years, since he was economical, while eschewing both waste and parsimony, and constantly concerned with financial security and providing for his old age. He was nonetheless generous to others, an attitude fostered by his religious faith and perhaps by his natural goodness. He was cautious and regulated his life methodically. Although he was jealous of his fame, and sometimes intemperate in debate, he was modest about his own accomplishments, calling himself merely industrious

and criticizing his own lack of scientific rigor. It must, in fact, be admitted that Ibn Riḍwān did not advance the course of medicine. In his controversies, for example, he displayed a rather scholastic mentality, and his strict adherence to the Greek masters may be at least in part the result of his self-tuition. But he remains a good witness to the science of his time, and his was a complex personality, not without charm.

BIBLIOGRAPHY

On Ibn Riḍwān and his work, see Ibn al-Qifṭī, *Ta'rīkh al-ḥukamā'* ("History of the Philosophers"), J. Lippert, ed. (Leipzig, 1903), 443–444; Ibn Abī Uṣaybi'a, *'Uyūn al anbā' fī ṭabaqāt al-aṭibbā'*, II ("Sources of Information on the Categories of Doctors"; Cairo, 1884), 99–105; C. Brockelmann, ed., *Geschichte der arabischen Literatur*, I (Leiden, 1943), 637, supp. I, p. 886; M. C. Lyons, " 'On the Nature of Man' in 'Alī ibn Riḍwān's *Epitome*," in *Al-Andalus*, **30** (1965), 181–188; C. Gabrieli, in *Isis*, **6** (1924), 500–506; and M. Meyerhof and J. Schacht, *The Medico-philosophical Controversy Between Ibn Butlan of Baghdad and Ibn Riḍwan, a Contribution to the History of Greek Learning Among the Arabs* (Cairo, 1937).

A translation of the second part of *Risalā fī daf' maḍārr al-abdān bî-arḍ Miṣr* ("Treatise on Avoiding What is Harmful for the Body in Egypt"), is by M. Meyerhof, in *Sitzungsberichte der Physikalisch-medizinischen Sozietät in Erlangen*, **54** (1923), 197–214; and in *Comptes rendus du Congrès international de médecine tropicale et d'hygiène*, II (Cairo, 1929), 211–235.

ROGER ARNALDEZ

RIECKE, EDUARD (*b.* Stuttgart, Germany, 1 December 1845; *d.* Göttingen, Germany, 11 June 1915), *physics.*

Riecke received his early education in Stuttgart and in 1866 entered the University of Tübingen to study mathematics under Carl Neumann, son of Franz Neumann, the great theoretician. He also studied experimental physics with E. Reusch, who whetted his interest in theoretical studies in crystal physics. This combination of experimental and theoretical work was to be a hallmark of Riecke's later work. He completed his undergraduate studies in 1869 and took a teaching post in mathematics at a school in Stuttgart.

At the beginning of 1870 Riecke was offered a state scholarship to continue his studies at the University of Göttingen. He had hardly begun this work when he was conscripted to serve in the Franco-Prussian War. The following year he returned to Göttingen to study mathematics with Clebsch and experimental physics with W. Weber and Kohlrausch. In May 1871, under

Weber's guidance, he completed his dissertation on the magnetization numbers of iron in weak magnetic fields.

In 1873 Riecke was called to Göttingen as extraordinary professor of physics; and in 1881 he became ordinary professor, taking over Weber's laboratory and institute. When W. Voigt was named ordinary professor of theoretical physics (1883) at Göttingen, he and Riecke were promised a new physics institute. Although this institute did not actually begin functioning until 1905, Riecke carried out, under difficult conditions, both experimental and theoretical studies over a wide range of areas in physics and physical chemistry during his forty-five-year tenure at Göttingen. His competence in physics is reflected in his textbook, *Lehrbuch der Physik* (1896), which went through five editions. In 1899 Riecke became the first editor of the journal *Physikalische Zeitschrift*; he continued his close association with the journal until his death. Even at the age of seventy he remained an active worker in physics. His last work, published posthumously in June 1915, was a rigorous mathematical résumé of particle physics and line spectra in which he concentrated on Bohr's theory of series spectra in hydrogen and helium.

Early in his career Riecke studied ferromagnetism and showed (1871) that in the presence of weak magnetic fields the magnetization numbers for iron are not independent of the strength of the external field, as had been assumed by Franz Neumann. About 1885 Riecke conducted theoretical and experimental researches in hydrodynamics and later, about 1890, undertook theoretical studies in thermodynamics, concentrating on the concept of thermodynamic potentials, which he applied to problems in physical chemistry. Noteworthy among these studies was one, in 1893, in which he analyzed muscle contraction in living organisms by using thermodynamic potentials.

The chief thrust of Riecke's research was related to the revival and establishment of granular theories of electricity and crystal structure. The particulate nature of electricity had been proposed by Weber but had been under severe attack during the 1860's and 1870's. It was largely through the efforts of Helmholtz that the granular theory of electricity was revived in the 1880's. Helmholtz had stressed the fact that if one accepts the hypothesis that matter is atomistic, then one must also accept an atomistic conception of electricity. Riecke sought to confirm this atomic hypothesis of electricity: with the aid of (1) Geissler tubes, (2) studies of atmospheric electricity and the motion of electrified particles in the electromagnetic field, (3) mathematical and experimental studies on the behavior of crystals, and (4) researches in the metallic state and electrical conduction in metals. His researches in these areas were not done serially but were intertwined, stretching over the period 1880–1915.

The most noteworthy of his results in Geissler tubes was the identification of negatively charged particles projected from the cathode; and he demonstrated that this charge is the same regardless of the metal from which the cathode is constructed (1899). Riecke's analysis of the motion of electrified particles in electromagnetic fields (1902) had direct application to the theory of the aurora. Also, he made important contributions to the molecular theory of pyroelectric and piezoelectric phenomena in tourmaline and quartz (1891, 1914). But his most important and influential researches were undoubtedly on the theory of conduction in metals and a granular theory of the properties of metals.

Riecke's major paper on this subject was published in 1898. He envisaged the metal as being composed of neutral atoms bound together in a lattice. Provision was made for ionization of some of these atoms to explain positive metal ions and negative electrons. The properties of the metal were accounted for by hypothesizing relationships between the two types of charged particles and their environment. The theory attempted to analyze electrical conduction; heat conduction; the Wiedemann-Franz ratio; various contact effects (including the Peltier effect, contact potentials, and the Thompson effect); various phenomena associated with the presence of an external magnetic field (including the Hall effect, the Nernst effect, and the Leduc effect); and electrical and thermal conductivity in alloys.

Riecke's basic assumption was that the space between molecules contained not only negative but also positive particles. Each metal molecule was capable of discharging both positive and negative charges by collision. The discharged ions then drifted in straight lines until they came under the influence of metal molecules still bound in the lattice, under the influence of which the ions were bent into circular paths of various sizes. Thus the charges would move randomly through the metal. The model was that of a gas; and Riecke assumed that the average speed of the particles was proportional to the square root of the absolute temperature, thus making it possible to apply the mathematical techniques of the kinetic theory of gases to the properties of metals.

The chief success of the theory, the accurate prediction of the Wiedemann-Franz ratio, was short-lived since soon after Riecke produced his theory it was shown that the ratio of the thermal conductivity to the electrical conductivity for most metals is not strictly proportional to the temperature. Nevertheless, Riecke had broken new ground. Drude and Lorentz followed with more successful variants of the Riecke theory, which themselves were eventually supplanted in 1927 and 1928 by the Sommerfeld theory utilizing nonclassical, Fermi-Dirac statistics.

BIBLIOGRAPHY

I. ORIGINAL WORKS. Many of Riecke's papers were published in both the *Göttingen Nachrichten* and the *Annalen der Physik*. Where such duplication exists the more easily accessible *Annalen der Physik* has been cited. His papers include "Ueber die elektrischen Elementargesetze," in *Annalen der Physik*, **11** (1880), 278–315; "Ueber die sogenannte unipolare Induction," *ibid.*, 413–432; "Ueber die Bewegung eines elektrischen Teilchens in einem homogenen magnetischen Felde und das negative elektrische Glimmlicht," *ibid.*, **13** (1881), 191–194; "Messung der vom Erdmagnetismus auf einen drehbaren linearen Stromleiter ausgeübten Kraft," *ibid.*, 194–204; "Beiträge zur Lehre vom inducirten Magnetismus," *ibid.*, 465–507; "Ueber die elektromagnetische Rotation einer Flüssigkeit," *ibid.*, **25** (1885), 496–511; and "Ueber die Pyroelektricität des Turmalins," *ibid.*, **28** (1886), 43–80; and **40** (1890), 264–306.

Subsequent writings are "Ueber elektrische Ladung durch gleitende Reibung," *ibid.*, **42** (1891), 465–482; "Das thermische Potential für verdünnten Lösungen," *ibid.*, 483–501; "Thermodynamik des Turmalins und mechanische Theorie der Muskelcontraction," **49** (1893), 430–458; "Moleculartheorie der piëzoelectrischen und pyroelectrischen Erscheinungen," *ibid.*, 459–486; "Der Satz vom thermodynamischen Potential beim Gleichgewichte eines heterogenen Systems mit Anwendung auf die Theorie von van der Waals und das Gesetz des Siedepunktes," **53** (1894), 379–391; "Zur Lehre von der Quellung," *ibid.*, 564–592; "Ueber die Vertheilung der freien Elektricität in Innern einer Geissler'schen Röhre," *ibid.*, **63** (1897), 220–233; "Zur Theorie des Galvanismus und der Wärme," **66** (1898), 353–389, 545–581; "Ueber den Reactionsdruck der Kathodenstrahlen," *ibid.*, 954–979; and "Ueber die Vertheilung von freier Elektricität an der Oberfläche einer Crooks'schen Röhre," *ibid.*, **69** (1899), 788–800.

Later writings include "Zur Kinetik der Serienschwingungen eines Linienspektruns," *ibid.*, **1** (1900), 399–413; "Ueber das Verhältnis der Leitfähigkeiten der Metalle für Wärme und für Elektricität," *ibid.*, **2** (1900), 835–842; "Ueber charakteristische Curven bei der elektrischen Entladung durch verdünnte Gase," *ibid.*, **4** (1901), 592–616; "Ist die metallische Leitung verbunden mit einem Transport von Metallionen?" in *Physikalische Zeitschrift*, **2** (1901), 639; "Ueber die Zerstreuung der Electricitat in abgeschlossenen Räumen," in *Göttingen Nachrichten* (1903), 1–16, 32–38; "Ueber näherungweise gestättige Ströme zwischen plan-parallelen Platten," *ibid.*, 336–343; "Ueber einige Eigenschaften des Radiumatoms," *ibid.* (1907), 163–170; "Ueber die Bewegung der α-Ionen," in *Annalen der Physik*, **27** (1908), 797–818; "Zur Theorie des Interferenzversuches von Michelson," in *Göttingen Nachrichten* (1911), 271–277; "Zur molekularen Theorie der Piezoelektrizität des Turmalins," in *Physikalische Zeitschrift*, **13** (1912), 409–415; and "Bohrs Theorie der Serienspektren von Wasserstoff und Helium," *ibid.*, **16** (1915), 222–227.

II. SECONDARY LITERATURE. E. T. Whittaker, *A History of the Theories of Aether and Electricity*, 2 vols. (New York, 1960), esp. I, chap. 11. An obituary notice by W. Voigt appears in *Physikalische Zeitschrift*, **16** (1915), 219–221.

STANLEY GOLDBERG

RIEMANN, GEORG FRIEDRICH BERNHARD (*b*. Breselenz, near Dannenberg, Germany, 17 September 1826; *d*. Selasca, Italy, 20 July 1866), *mathematics, mathematical physics.*

Bernhard Riemann, as he was called, was the second of six children of a Protestant minister, Friedrich Bernhard Riemann, and the former Charlotte Ebell. The children received their elementary education from their father, who was later assisted by a local teacher. Riemann showed remarkable skill in arithmetic at an early age. From Easter 1840 he attended the Lyceum in Hannover, where he lived with his grandmother. When she died two years later, he entered the Johanneum in Lüneburg. He was a good student and keenly interested in mathematics beyond the level offered at the school.

In the spring term of 1846 Riemann enrolled at Göttingen University to study theology and philology, but he also attended mathematical lectures and finally received his father's permission to devote himself wholly to mathematics. At that time, however, Göttingen offered a rather poor mathematical education; even Gauss taught only elementary courses. In the spring term of 1847 Riemann went to Berlin University, where a host of students flocked around Jacobi, Dirichlet, and Steiner. He became acquainted with Jacobi and Dirichlet, the latter exerting the greatest influence upon him. When Riemann returned to Göttingen in the spring term of 1849, the situation had changed as a result of the physicist W. E. Weber's return. For three terms Riemann attended courses and seminars in physics, philosophy, and education. In November 1851 he submitted his thesis on complex function theory and Riemann surfaces (*Gesammelte mathematische Werke. Nachträge*, pp. 3–43), which he defended on 16 December to earn the Ph.D.

Riemann then prepared for his *Habilitation* as a *Privatdozent*, which took him two and a half years.

At the end of 1853 he submitted his *Habilitationsschrift* on Fourier series (*ibid.*, pp. 227–271) and a list of three possible subjects for his *Habilitationsvortrag*. Against Riemann's expectation Gauss chose the third: "Über die Hypothesen, welche der Geometrie zu Grunde liegen" (*ibid.*, pp. 272–287). It was thus through Gauss's acumen that the splendid idea of this paper was saved for posterity. Both papers were posthumously published in 1867, and in the twentieth century the second became a great classic of mathematics. Its reading on 10 June 1854 was one of the highlights in the history of mathematics: young, timid Riemann lecturing to the aged, legendary Gauss, who would not live past the next spring, on consequences of ideas the old man must have recognized as his own and which he had long secretly cultivated. W. Weber recounts how perplexed Gauss was, and how with unusual emotion he praised Riemann's profundity on their way home.

At that time Riemann also worked as an assistant, probably unpaid, to H. Weber. His first course as a *Privatdozent* was on partial differential equations with applications to physics. His courses in 1855–1856, in which he expounded his now famous theory of Abelian functions, were attended by C. A. Bjerknes, Dedekind, and Ernst Schering; the theory itself, one of the most notable masterworks of mathematics, was published in 1857 (*ibid.*, pp. 88–144). Meanwhile he had published a paper on hypergeometric series (*ibid.*, pp. 64–87).

When Gauss died early in 1855, his chair went to Dirichlet. Attempts to make Riemann an extraordinary professor failed; instead he received a salary of 200 taler a year. In 1857 he was appointed extraordinary professor at a salary of 300 taler. After Dirichlet's death in 1859 Riemann finally became a full professor.

On 3 June 1862 Riemann married Elise Koch, of Körchow, Mecklenburg-Schwerin; they had a daughter. In July 1862 he suffered an attack of pleuritis; in spite of periodic recoveries he was a dying man for the remaining four years of his life. His premature death by "consumption" is usually imputed to that illness of 1862, but numerous early complaints about bad health and the early deaths of his mother, his brother, and three sisters make it probable that he had suffered from tuberculosis long before. To cure his illness in a better climate, as was then customary, Riemann took a leave of absence and found financial support for a stay in Italy. The winter of 1862–1863 was spent on Sicily; in the spring he traveled through Italy as a tourist and a lover of fine art. He visited Italian mathematicians, in particular Betti, whom he had known at Göttingen. In

June 1863 he was back in Göttingen, but his health deteriorated so rapidly that he returned to Italy. He stayed in northern Italy from August 1864 to October 1865. He spent the winter of 1865–1866 in Göttingen, then left for Italy in June 1866. On 16 June he arrived at Selasca on Lake Maggiore. The day before his death he was lying under a fig tree with a view of the landscape and working on the great paper on natural philosophy that he left unfinished. He died fully conscious, while his wife said the Lord's Prayer. He was buried in the cemetery of Biganzole.

Riemann's evolution was slow and his life short. What his work lacks in quantity is more than compensated for by its superb quality. One of the most profound and imaginative mathematicians of all time, he had a strong inclination to philosophy, indeed, was a great philosopher. Had he lived and worked longer, philosophers would acknowledge him as one of them. His style was conceptual rather than algorithmic—and to a higher degree than that of any mathematician before him. He never tried to conceal his thought in a thicket of formulas. After more than a century his papers are still so modern that any mathematician can read them without historical comment, and with intense pleasure.

Riemann's papers were edited by H. Weber and R. Dedekind in 1876 with a biography by Dedekind. A somewhat revised second edition appeared in 1892, and a supplement containing a list of Riemann's courses was edited by M. Noether and W. Wirtinger in 1902. A reprint of the second edition and the supplement appeared in 1953. It bears an extra English title page and an introduction in English by Hans Lewy. The latter consists of a biographical sketch and a short analysis of part of Riemann's work. There is a French translation of the first edition of Dedekind and Weber. Riemann's style, influenced by philosophical reading, exhibits the worst aspects of German syntax; it must be a mystery to anyone who has not mastered German. No complete appreciation of Riemann's work has ever been written. There exist only a few superficial, more or less dithyrambic, sermons. Among the rare historical accounts of the theory of algebraic functions in which Riemann's contributions are duly reported are Brill and Noether's "Die Entwicklung der Theorie der algebraischen Functionen . . ." (1894) and the articles by Wirtinger (1901) and Krazer and Wirtinger (1920) in *Encyclopädie der mathematischen Wissenschaften*. The greater part of *Gesammelte mathematische Werke* consists of posthumous publications and unpublished works. Some of Riemann's courses have been published. *Partielle Differentialgleichungen* . . . and *Schwere, Electricität und Magnetismus* are fairly authentic but not quite congenial editions; H. Weber's *Die partiellen Differentialgleichungen* is not authentic; and it is doubtful to what degree *Elliptische Funktionen* is authentic.

People who know only the happy ending of the story can hardly imagine the state of affairs in complex analysis around 1850. The field of elliptic functions had grown rapidly for a quarter of a century, although their most fundamental property, double periodicity, had not been properly understood; it had been discovered by Abel and Jacobi as an algebraic curiosity rather than a topological necessity. The more the field expanded, the more was algorithmic skill required to compensate for the lack of fundamental understanding. Hyperelliptic integrals gave much trouble, but no one knew why. Nevertheless, progress was made. Despite Abel's theorem, integrals of general algebraic functions were still a mystery. Cauchy had struggled with general function theory for thirty-five years. In a slow progression he had discovered fundamentals that were badly needed but still inadequately appreciated. In 1851, the year in which Riemann defended his thesis, he had reached the height of his own understanding of complex functions. Cauchy had early hit upon the sound definition of the subject functions, by differentiability in the complex domain rather than by analytic expressions. He had characterized them by what are now called the Cauchy-Riemann differential equations. Riemann was the first to accept this view wholeheartedly. Cauchy had also discovered complex integration, the integral theorem, residues, the integral formula, and the power series development; he had even done work on multivalent functions, had dared freely to follow functions and integrals by continuation through the plane, and consequently had come to understand the periods of elliptic and hyperelliptic integrals, although not the reason for their existence. There was one thing he lacked: Riemann surfaces.

The local branching behavior of algebraic functions had been clearly understood by V. Puiseux. In his 1851 thesis (*Gesammelte mathematische Werke. Nachträge*, pp. 3–43) Riemann defined surfaces branched over a complex domain, which, as becomes clear in his 1857 paper on Abelian functions (*ibid.*, pp. 88–144), may contain points at infinity. Rather than suppose such a surface to be generated by a multivalued function, he proved this generation in the case of a closed surface. It is quite credible that Riemann also knew the abstract Riemann surface to be a variety with a complex differentiable structure, although Friedrich Prym's testimony to this, as reported by F. Klein, was later disclaimed by the former (F. Klein, *Über Riemann's Theorie der*

algebraischen Funktionen und ihrer Integrale, p. 502). Riemann clearly understood a complex function on a Riemann surface as a conformal mapping of this surface. To understand the global multivalency of such mappings, he analyzed Riemann surfaces topologically: a surface T is called "simply connected" if it falls apart at every crosscut; it is $(m + 1)$ times connected if it is turned into a simply connected surface T' by m crosscuts. According to Riemann's definition, crosscuts join one boundary point to the other; he forgot about closed cuts, perhaps because originally he did not include infinity in the surface. By Green's theorem, which he used instead of Cauchy's, Riemann proved the integral of a complex continuously differentiable function on a simply connected surface to be univalent.

A fragment from Riemann's papers reveals sound ideas even on higher-dimensional homology that subsequently were worked out by Betti and Poincaré. There are no indications that Riemann knew about homotopy and about the simply connected cover of a Riemann surface. These ideas were originated by Poincaré.

The analytic tool of Riemann's thesis is what he called Dirichlet's principle in his 1857 paper. He had learned it in Dirichlet's courses and traced it back to Gauss. In fact it is due to W. Thomson (Lord Kelvin) ("Sur une équation aux dérivées partielles . . ."). It says that among the continuous functions u defined in a domain T with the same given boundary values, the one that minimizes the surface integral

$$\iint |\operatorname{grad} u|^2 \, dT$$

satisfies Laplace's equation

$$\Delta u = 0$$

(is a potential function); it is used to assure the existence of a solution of Laplace's equation which assumes reasonable given boundary values—or, rather, a complex differentiable function if its real part is prescribed on the boundary of T and its imaginary part in one point. (Since Riemann solved this problem by Dirichlet's principle, it is often called Dirichlet's problem, which usage is sheer nonsense.) Of course, if T is not simply connected, the imaginary part can be multivalued; or if it is restricted to a simply connected T', it may show constant jumps (periods) at the crosscuts by which T' was obtained.

In his thesis Riemann was satisfied with one application of Dirichlet's principle: his celebrated mapping theorem, which states that every simply connected domain T (with boundary) can be mapped one-to-one onto the interior of a circle by a complex differentiable function (conformal mapping). Riemann's proof can hardly match modern standards of rigor even if Dirichlet's principle is granted.

Riemann's most exciting applications of Dirichlet's principle are found in his 1857 paper. Here he considers a closed Riemann surface T. Let n be the number of its sheets and $2p + 1$ the multiplicity of its connection (that is, in the now usual terminology, formulated by Clebsch, of genus p). Dirichlet's principle, applied to simply connected T', yields differentiable functions with prescribed singularities, which of course show obligatory imaginary periods at the crosscuts. Riemann asserted that he could prescribe periods with arbitrary real parts along the crosscuts. This is true, but his argument, as it stands, is wrong. The assertion cannot be proved by assigning arbitrary boundary values to the real part of the competing functions at one side of the crosscut, since this would not guarantee a constant jump of the imaginary part. Rather one has to prescribe the constant jump of the real part combined with the continuity of the normal derivative across the crosscut, which would require another sort of Dirichlet's principle. No doubt Riemann meant it this way, but apparently his readers did not understand it. It is the one point on which all who have tried to justify Riemann's method have deviated from his argument to circumvent the gap although the necessary version of Dirichlet's principle would not have been harder to establish than the usual one.

If Riemann's procedure is granted, the finite functions on T (integrals of the first kind) form a linear space of real dimension $2p + 2$. By admitting enough polar singularities Riemann removed more or fewer periods. The univalent functions with simple poles in m given general points form an $(m - p + 1)$-dimensional linear variety. Actually, for special m-tuples the dimension may be larger—this should be recognized as Gustav Roch's contribution to Riemann's result.

The foregoing results stress the importance of the genus p, which Abel had come across much earlier in a purely algebraic context. By analytic means Riemann obtained the well-known formula that connects the genus to the number of branchings, although he also mentioned its purely topological character.

It is easily seen that the univalent functions w on T with m poles fulfill an algebraic equation $F(w, z) = 0$ of degrees n and m in w and z. It is a striking feature that these functions were secured by a transcendental procedure, which was then complemented by an algebraic one. In a sense this was the birth of algebraic

geometry, which even in the cradle showed the congenital defects with which it would be plagued for many years—the policy of stating and proving that something holds "in general" without explaining what "in general" means and whether the "general" case ever occurs. Riemann stated that the discriminant of $F(w, z)$ is of degree $2m(n - 1)$, which is true only "in general." The discriminant accounts for the branching points and for what in algebraic geometry were to be called the multiple points of the algebraic curve defined by $F(w, z) = 0$. The general univalent function on T with m poles, presented as a rational quotient $\varphi(w, z)/\psi(w, z)$, must be able to separate the partners of a multiplicity, which means that both φ and ψ must vanish in the multiple points—or, in algebraic geometry terms, that they must be adjoint. An enumeration shows that such functions depend on $m - p + 1$ complex parameters, as they should. In this way the integrands of the integrals of the first kind are presented by $\varphi/(\partial F/\partial w)$, where the numerator is an adjoint function.

The image of a univalent function on T was considered as a new Riemann surface T^*. Thus Riemann was led to study rational mappings of Riemann surfaces and to form classes of birationally equivalent surfaces. Up to birational equivalence Riemann counted $3p - 3$ parameters for $p > 1$, the "modules." The notion, the character, and the dimension of the manifold of modules were to remain controversial for more than half a century.

To prepare theta-functions the crosscuts of T are chosen in pairs a_j, b_j $(j = 1, \cdots, p)$, where b_j crosses a_j in the positive sense and no crosscut crosses one with a different subscript. Furthermore, the integrals of the first kind u_j $(j = 1, \cdots, p)$ are chosen with a period πi at the crosscut a_j and 0 at the other, a_k. The period of u_j at b_k is then called a_{jk}. By the marvelous trick of integration of $u_j \, dw_k$ over the boundary, the symmetry of the system a_{jk} is obtained; and integration of $w \, d\overline{w}$ with $w = \sum m_j u_j$ yields the result that the real part of $\sum a_{kl} m_k m_l$ is positive definite.

As if to render homage to his other master, Riemann now turned from the Dirichlet integral to the Jacobi inversion problem, showing himself to be as skillful in algorithmic as he was profound in conceptual thinking.

When elliptic integrals had been mastered by inversion, the same problem arose for integrals of arbitrary algebraic functions. It was more difficult because of the paradoxical phenomenon of more than two periods. Jacobi saw how to avoid this stumbling block: instead of inverting one integral of the first kind, he took p independent ones $u_1, \cdots, u_p$

to formulate a p-dimensional inversion problem—namely, solving the system ($i = 1, \cdots, p$)

$$u_i(\eta_1) + \cdots + u_i(\eta_p) = e_i \text{ mod periods.}$$

This problem had been tackled in special cases by Göpel (1847) and Rosenhain (1851), and more profoundly by Weierstrass (1856). With tremendous ingenuity it was now considered by Riemann.

The tool was, of course, a generalization of Jacobi's theta-function, which had proved so useful when elliptic integrals must be inverted. Riemann's insight into the periods of functions on the Riemann surface showed him the way to find the right theta-functions. They were defined by

$$\vartheta(v_1, \cdots, v_p) = \sum_m \exp(\sum_{jk} a_{jk} m_j m_k + 2\sum_j v_j m_j),$$

where the a_{jk} are the periods mentioned earlier and m runs through all systems of integer $m_1, \cdots, m_p$. Thanks to the negative definiteness of the real part of $\sum_{jk} a_{jk} m_j m_k$ this series converges. It is also characterized by the equations

$$\vartheta(v) = \vartheta(v_1, \cdots, v_j + \pi i, \cdots, v_p),$$

$$\vartheta(v) = \exp(2v_h + a_{hk}) \cdot \vartheta(v_j + a_{jk}).$$

The integrals of the first kind $u_j - e_j$ are now substituted for v_j. $\vartheta(u_1 - e_1, \cdots, u_p - e_p)$ is a function of $x \in T'$, which passes continuously through the crosscuts a_j and multiplies by $\exp(-2[u_j - e_j])$ at b_j. The clever idea of integration of $d \log \vartheta$ along the boundary of T' shows ϑ, if not vanishing identically, to have exactly p roots $\eta_1, \cdots, \eta_p$ in T'. Integrating $\log \vartheta \, du_j$ again yields

$$e_j = \sum_k u_j(\eta_k)$$

up to periods and constants that can be removed by a suitable norming of the u_j. This solves Jacobi's problem for those systems $e_1, \cdots, e_p$ for which $\vartheta(u_j - e_j)$ does not vanish identically. Exceptions can exist and are investigated. In *Gesammelte mathematische Werke. Nachträge* (pp. 212–224), Riemann proves that $\vartheta(r) = 0$ if and only if

$$r_j = \sum_{k=1}^{p-1} u_j(\eta_k) \text{ mod periods}$$

for suitable systems $\eta_1, \cdots, \eta_{p-1}$ and finds how many such systems there are. Riemann's proofs, particularly for the uniqueness of the solution of $e_j = \sum_k u_j(\eta_k)$, show serious gaps which are not easy to fill (see C. Neumann, *Vorlesungen über Riemann's Theorie. . .*, 2nd ed., pp. 334–336).

The reception of Riemann's work sketched above would be an interesting subject of historical study. But it would not be enough to read papers and books related to this work. One can easily verify that its impact was tremendous and its direct influence both immediate and long-lasting—say thirty to forty years. To know how this influence worked, one should consult other sources, such as personal reminiscences and correspondence. Yet no major sources of this sort have been published. We lack even the lists of his students, which should still exist in Göttingen. One important factor in the dissemination of Riemann's results, if not his ideas, must have been C. Neumann's *Vorlesungen über Riemann's Theorie . . .*, which, according to people around 1900, "made things so easy it was affronting"—indeed, it is a marvelous book, written by a great teacher. Riemann needed an interpreter like Neumann because his notions were so new. How could one work with concepts that were not accessible to algorithmization, such as Riemann surfaces, crosscuts, degree of connection, and integration around rather abstract domains?

Even Neumann did not fully succeed. Late in the 1850's or early 1860's the rumor spread that Weierstrass had disproved Riemann's method. Indeed, Weierstrass had shown—and much later published—that Dirichlet's principle, lavishly applied by Riemann, was not as evident as it appeared to be. The lower bound of the Dirichlet integral did not guarantee the existence of a minimizing function. Weierstrass' criticism initiated a new chapter in the history of mathematical rigor. It might have come as a shock, but one may doubt whether it did. It is more likely that people felt relieved of the duty to learn and accept Riemann's method—since, after all, Weierstrass said it was wrong. Thus investigators set out to reestablish Riemann's results with quite different methods: nongeometric function-theory methods in the Weierstrass style; algebraic-geometry methods as propagated by the brilliant young Clebsch and later by Brill and M. Noether and the Italian school; invariant theory methods developed by H. Weber, Noether, and finally Klein; and arithmetic methods by Dedekind and H. Weber. All used Riemann's material but his method was entirely neglected. Theta-functions became a fashionable subject but were not studied in Riemann's spirit. During the rest of the century Riemann's results exerted a tremendous influence; his way of thinking, but little. Even the Cauchy-Riemann definition of analytic function was discredited, and Weierstrass' definition by power series prevailed.

In 1869–1870 H. A. Schwarz undertook to prove Riemann's mapping theorem by different methods that, he claimed, would guarantee the validity of all of Riemann's existence theorems as well. One method was to solve the problem first for polygons and then by approximation for arbitrary domains; the other, an alternating procedure which allowed one to solve the boundary problem of the Laplace equation for the union of two domains if it had previously been solved for the two domains separately. From 1870 C. Neumann had tackled the boundary value problem by double layers on the boundary and by integral equations; in the second edition (1884) of his *Vorlesungen über Riemann's Theorie* . . . he used alternating methods to reestablish all existence theorems needed in his version of Riemann's theory of algebraic functions. Establishing the mapping theorem and the boundary value theorem for open or irregularly bounded surfaces was still a long way off, however. Poincaré's *méthode de balayage* (1890) represented great progress. The speediest approach to Riemann's mapping theorem in its most general form was found by C. Carathéodory and P. Koebe. Meanwhile, a great thing had happened: Hilbert had saved Dirichlet's principle (1901), the most direct approach to Riemann's results. (See A. Dinghas, *Vorlesungen über Funktionentheorie*, esp. pp. 298–303.)

The first to try reviving Riemann's geometric methods in complex function theory was Klein, a student of Clebsch's who in the late 1870's had discovered Riemann. In 1892 he wrote a booklet to propagate his own version of Riemann's theory, which was much in Riemann's spirit. It is a beautiful book, and it would be interesting to know how it was received. Probably many took offense at its lack of rigor; Klein was too much in Riemann's image to be convincing to people who would not believe the latter.

In the same period Riemann's function theory first broke through the bounds to which Riemann's broad view was restricted; function theory, in a sense, took a turn that contradicted Riemann's most profound work. (See H. Freudenthal, "Poincaré et les fonctions automorphes.") Poincaré, a young man with little experience, encountered problems that had once led to Jacobi's inversion problem, although in a different context. It was again the existence of (multivalent) functions on a Riemann surface that assume every value once at most—the problem of uniformization, as it would soon be called. Since the integrals of the first kind did not do the job, Jacobi had considered the system of p of such functions, which should assume every general p-tuple of values once. Riemann had solved this Jacobi problem, but Poincaré did not know about Jacobi's artifice. He knew so little about what had happened in the past

that instead of trying functions that behave additively or multiplicatively at the crosscuts, as had always been done, he chose the correct ones, which at the crosscuts undergo fractional linear changes but had never been thought of; when inverted, they led to the automorphic functions, which at the same time were studied by Klein.

This simple, and afterward obvious, idea rendered Jacobi's problem and its solution by Riemann obsolete. At this point Riemann, who everywhere opened new perspectives, had been too much a slave to tradition; nevertheless, uniformization and automorphic functions were the seeds of the final victory of Riemann's function theory in the twentieth century. It seems ironic, since this chapter of function theory went beyond and against Riemann's ideas, although in a more profound sense it was also much in Riemann's spirit. A beautiful monograph in that spirit was written by H. Weyl in 1913 (see also J. L. V. Ahlfors and L. Sorio, *Riemann Surfaces*).

The remark that nobody before Poincaré had thought of other than additive or multiplicative behavior at the crosscuts needs some comment. First, there were modular functions, but they did not pose a problem because from the outset they had been known in the correctly inverted form; they were linked to uniformization by Klein. Second, Riemann was nearer to what Poincaré would do than one would think at first sight. In another paper of 1857 (*Gesammelte mathematische Werke. Nachträge*, pp. 67–83) he considered hypergeometric functions, which had been dealt with previously by Gauss and Kummer, defining them in an axiomatic fashion which gave him all known facts on hypergeometric functions with almost no reasoning. A hypergeometric function $P(x; a, b, c; \alpha, \beta, \gamma; \alpha', \beta', \gamma')$ should have singularities at a, b, c, where it behaves as $(x - a)^\alpha Q(x) + (x - a)^{\alpha'} R(x)$, and so on, with regular Q and R; and between three arbitrary branches of P there should be a linear relation with constant coefficients.

Riemann's manuscripts yield clear evidence that he had viewed such behavior at singularities in a much broader context (*ibid.*, pp. 379–390). He had anticipated some of L. Fuchs's ideas on differential equations, and he had worked on what at the end of the century became famous as Riemann's problem. It was included by Hilbert in his choice of twenty-three problems: One asks for a k-dimensional linear space of regular functions, with branchings at most in the points $a_1, \cdots, a_l$, which undergoes given linear transformations under circulations around the $a_1, \cdots, a_l$. Hilbert and Josef Plemelj tackled this problem, but the circumstances are so confusing that it is not easy to decide whether it has been solved more than partially. (See L. Bieberbach, *Theorie der gewönlichen Differentialgleichungen*, esp. pp. 245–252.)

If there is one paper of Riemann's that can compete with that on Abelian functions as a contributor to his fame, it is that of 1859 on the ζ function.

The function ζ defined by

$$\zeta(s) = \sum_{n=0}^{\infty} n^{-s}$$

is known as Riemann's ζ function although it goes back as far as Euler, who had noted that

$$\sum n^{-s} = \prod (1 - p^{-s})^{-1}$$

where the product runs over all primes p. This relation explains why the ζ function is so important in number theory. The sum defining ζ converges for $Re\ s > 1$ only, and even the product diverges for $Re\ s < 1$. By introducing the Γ function Riemann found an everywhere convergent integral representation. That in turn led him to consider

$$\eta(s) = \zeta(s)\ \Gamma(\tfrac{1}{2}s)\ \pi^{-\frac{1}{2}s},$$

which is invariant under the substitution of $1 - s$, for s. This is the famous functional equation for the ζ function. Another proof via theta-functions gives the same result.

It is easily seen that all nontrivial roots of ζ must have their real part between 0 and 1 (in 1896 Hadamard and de la Vallée-Poussin succeeded in excluding the real parts 0 and 1). Without proof Riemann stated that the number of roots with an imaginary part between 0 and T is

$$\frac{1}{2\pi}\ T \log T - \frac{1 + \log 2\pi}{2\pi}\ T + O(\log T)$$

(proved by Hans von Mangoldt in 1905) and then, with no fuss he said that it seemed quite probable that all nontrivial roots of ζ have the real part 1/2, although after a few superficial attempts he had shelved this problem. This is the famous Riemann hypothesis; in spite of the tremendous work devoted to it by numerous mathematicians, it is still open to proof or disproof. It is even unknown which arguments led Riemann to this hypothesis; his report may suggest that they were numerical ones. Indeed, modern numerical investigations show the truth of the Riemann hypothesis for the 25,000 roots with imaginary part between 0 and 170,571.35 (R. S. Lehman, "Separation of Zeros of the Riemann Zeta-Functions"); Good and Churchhouse ("The Riemann Hypothesis and the Pseudorandom Features of the Möbius Sequence") seem to have proceeded

to the 2,000,000th root. In 1914 G. H. Hardy showed that if not all, then at least infinitely many, roots have their real part 1/2.

Riemann stated in his paper that ζ had an infinite number of nontrivial roots and allowed a product presentation by means of them (which was actually proved by Hadamard in 1893).

The goal of Riemann's paper was to find an analytic expression for the number $F(x)$ of prime numbers below x. Numerical surveys up to $x = 3,000,000$ had shown the function $F(x)$ to be a bit smaller than the integral logarithm $Li(x)$. Instead of $F(x)$ Riemann considered

$$f(x) = \sum \frac{1}{n}\ F(x^{1/n})$$

and proved a formula which, duly corrected, reads

$$f(x) = \log 1/2 + Li(x) - \sum_{\alpha} Li(x^{\alpha})$$
$$+ \int_{x}^{\infty} \frac{dx}{x(x^2 + 1) \log x},$$

where α runs symmetrically over the nontrivial roots of ζ. For $F(x)$ this means

$$F(x) = \sum \mu(n) n^{-1}\ Li(x^{1/n}),$$

where μ is the Möbius function.

For an idea of the subsequent development and the enormous literature related to Riemann's paper, one is advised to consult E. Landau, *Handbuch der Lehre von der Verteilung der Primzahlen* (esp. I, 29–36) and E. C. Titchmarsh, *The Theory of the Zeta-Function*.

Riemann taught courses in mathematical physics. A few have been published: *Partielle Differentialgleichungen und deren Anwendung auf physikalische Fragen* and *Schwere, Electricität und Magnetismus*. The former in particular was so admired by physicists that its original version was reprinted as late as 1938. Riemann also made original contributions to physics, even one to the physics of hearing, wherein no mathematics is involved. A great part of his work is on applications of potential theory. He tried to understand electric and magnetic interaction as propagated with a finite velocity rather than as an *actio in distans* (*Gesammelte mathematische Werke. Nachträge*, pp. 49–54, 288–293; *Schwere, Electricität und Magnetismus*, pp. 326–330). Some historians consider this pre-Maxwellian work as important (see G. Lampariello, in *Der Begriff des Raumes in der Geometrie*, pp. 222–234). Continuing work of Dirichlet, in 1861 Riemann studied the motion of a liquid mass under its own gravity, within a varying ellipsoidal surface (*Gesammelte mathematische Werke. Nachträge*, pp.

182–211), a problem that has been the subject of many works. One of Riemann's classic results deals with the stability of an ellipsoid rotating around a principal axis under equatorial disturbances. A question in the theory of heat proposed by the Académie des Sciences in 1858 was answered by Riemann in 1861 (*ibid.*, pp. 391–423). His solution did not win the prize because he had not sufficiently revealed his arguments. That treatise is important for the interpretation of Riemann's inaugural address.

Riemann's most important contribution to mathematical physics was his 1860 paper on sound waves (*ibid.*, pp. 157–175). Sound waves of infinitesimal amplitude were well-known; Riemann studied those of finite amplitude in the one-dimensional case and under the assumption that the pressure p depended on the density ρ in a definite way. Riemann's presentation discloses so strong an intuitive motivation that the reader feels inclined to illustrate every step of the mathematical argumentation by a drawing. Riemann shows that if u is the gas velocity and

$$\omega = \int_{\rho_0}^{\rho} (dp/d\rho)^{1/2} \rho^{-1} \, d\rho,$$

then any given value of $\omega + u$ moves forward with the velocity $(dp/d\rho)^{1/2} + u$ and any $\omega - u$ moves backward with the velocity $-(dp/d\rho)^{1/2} + u$. An original disturbance splits into two opposite waves. Since phases with large ρ travel faster, they should overtake their predecessors. Actually the rarefaction waves grow thicker, and the condensation waves thinner—finally becoming shock waves. Modern aerodynamics took up the theory of shock waves, although under physical conditions other than those admitted by Riemann.

Riemann's paper on sound waves is also very important mathematically, giving rise to the general theory of hyperbolic differential equations. Riemann introduced the adjoint equation and translated Green's function from the elliptic to the hyperbolic case, where it is usually called Riemann's function. The problem to solve

$$L(u) = \frac{\partial^2 w}{\partial x \, \partial y} + a \frac{\partial w}{\partial x} + b \frac{\partial w}{\partial y} + cw = 0,$$

if w and $\partial w/\partial n$ are given on a curve that meets no characteristic twice, is reduced to that of solving the adjoint equation by a Green function that fulfills

$$\partial G/\partial y = aG, \qquad \partial G/\partial x = bG,$$

along the characteristics $x = \xi$ and $y = \eta$ and assumes the value 1 at $\ulcorner \xi, \eta \urcorner$.

Riemann's method was generalized by J. Hadamard (see *Lectures on Cauchy's Problem in Linear Partial Differential Equations*) to higher dimensions, where Riemann's function had to be replaced by a more sophisticated tool.

A few other contributions, all posthumous, by Riemann to real calculus should be mentioned: his first manuscript, of 1847 (*Gesammelte mathematische Werke. Nachträge*, pp. 353–366), in which he defined derivatives of nonintegral order by extending a Cauchy formula for multiple integration; his famous *Habilitationsschrift* on Fourier series of 1851 (*ibid.*, pp. 227–271), which contains not only a criterion for a function to be represented by its Fourier series but also the definition of the Riemann integral, the first integral definition that applied to very general discontinuous functions; and a paper on minimal surfaces—that is, of minimal area if compared with others in the same frame (*ibid.*, pp. 445–454). Riemann noticed that the spherical mapping of such a surface by parallel unit normals was conformal; the study of minimal surfaces was revived in the 1920's and 1930's, particularly in J. Douglas' sensational investigations.

Riemann left many philosophical fragments—which, however, do not constitute a philosophy. Yet his more mathematical than philosophical *Habilitationsvortrag*, "Über die Hypothesen, welche der Geometrie zu Grunde liegen" (*ibid.*, pp. 272–287), made a strong impact upon philosophy of space. Riemann, philosophically influenced by J. F. Herbart rather than by Kant, held that the a priori of space, if there was any, was topological rather than metric. The topological substratum of space is the n-dimensional manifold—Riemann probably was the first to define it. The metric structure must be ascertained by experience. Although there are other possibilities, Riemann decided in favor of the simplest: to describe the metric such that the square of the arc element is a positive definite quadratic form in the local differentials,

$$ds^2 = \sum g_{ij} \, dx^i dx^j.$$

The structure thus obtained is now called a Riemann space. It possesses shortest lines, now called geodesics, which resemble ordinary straight lines. In fact, at first approximation in a geodesic coordinate system such a metric is flat Euclidean, in the same way that a curved surface up to higher-order terms looks like its tangent plane. Beings living on the surface may discover the curvature of their world and compute it at any point as a consequence of observed deviations from Pythagoras' theorem. Likewise, one can define curvatures of n-dimensional Riemann spaces by noting the higher-order deviations that the ds^2 shows from a Euclidean space. This definition of the curvature tensor is actually the main point in Riemann's inaugural address. Gauss had introduced curvature in his investigations on surfaces; and earlier than Riemann he had noticed that this curvature could be defined as an internal feature of the surface not depending on the surrounding space, although in Gauss's paper this fundamental insight is lost in the host of formulas.

A vanishing curvature tensor characterizes (locally) Euclidean spaces, which are a special case of spaces with the same curvature at every point and every planar direction. That constant can be positive, as is the case with spheres, or negative, as is the case with the non-Euclidean geometries of Bolyai and Lobachevsky—names not mentioned by Riemann. Freely moving rigid bodies are feasible only in spaces of constant curvature.

Riemann's lecture contains nearly no formulas. A few technical details are found in an earlier mentioned paper (*ibid.*, pp. 391–423). The reception of Riemann's ideas was slow. Riemann spaces became an important source of tensor calculus. Covariant and contravariant differentiation were added in G. Ricci's absolute differential calculus (from 1877). T. Levi-Civita and J. A. Schouten (1917) based it on infinitesimal parallelism. H. Weyl and E. Cartan reviewed and generalized the entire theory.

In the nineteenth century Riemann spaces were at best accepted as an abstract mathematical theory. As a philosophy of space they had no effect. In revolutionary ideas of space Riemann was eclipsed by Helmholtz, whose "Über die Thatsachen, die der Geometrie zum Grunde liegen" pronounced his criticism of Riemann: facts versus hypotheses. Helmholtz' version of Kant's philosophy of space was that no geometry could exist except by a notion of congruence—in other words, geometry presupposed freely movable rigid bodies. Therefore, Riemann spaces with nonconstant curvature were to be considered as philosophically wrong. Helmholtz formulated a beautiful space problem, postulating the free mobility of solid bodies; its solutions were the spaces with constant curvature. Thus Helmholtz could boast that he was able to derive from facts what Riemann must assume as a hypothesis.

Helmholtz' arguments against Riemann were often repeated (see B. Erdmann, *Die Axiome der Geometrie*), even by Poincaré, who later admitted that they were entirely wrong. Indeed, the gist of Riemann's address had been that what would be needed for metric geometry is the congruence not of solids but of (one-dimensional) rods. This was overlooked by almost everyone who evaluated Riemann's address

philosophically. Others did not understand the topological substrate, arguing that it presupposed numbers and, hence, Euclidean space. The average level in the nineteenth-century discussions was even lower. Curvature of a space not contained in another was against common sense. Adversaries as well as champions of curved spaces overlooked the main point: Riemann's mathematical procedure to define curvature as an internal rather than an external feature. (See H. Freudenthal, "The Main Trends in the Foundations of Geometry in the 19th Century.")

Yet there was more profound wisdom in Riemann's thought than people would admit. The general relativity theory splendidly justified his work. In the mathematical apparatus developed from Riemann's address, Einstein found the frame to fit his physical ideas, his cosmology, and cosmogony; and the spirit of Riemann's address was just what physics needed: the metric structure determined by physical data.

General relativity provoked an accelerated production in general differential geometry, although its quality did not always match its quantity. But the gist of Riemann's address and its philosophy have been incorporated into the foundations of mathematics.

According to Riemann, it was said, the metric of space was an experience that complemented its a priori topological structure. Yet this does not exactly reproduce Riemann's idea, which was infinitely more sophisticated:

> The problem of the validity of the presuppositions of geometry in the infinitely small is related to that of the internal reason of the metric. In this question one should notice that in a discrete manifold the principle of the metric is contained in the very concept of the manifold, whereas in a continuous manifold it must come from elsewhere. Consequently either the entity on which space rests is a discrete manifold or the reason of the metric should be found outside, in the forces acting on it [Neumann, *Vorlesungen über Riemanns Theorie*].

Maybe these words conceal more profound wisdom than we yet can fathom.

BIBLIOGRAPHY

I. ORIGINAL WORKS. Riemann's writings were collected in *Gesammelte mathematische Werke und wissenschaftlicher Nachlass*, R. Dedekind and H. Weber, eds. (Leipzig, 1876; 2nd ed., 1892). It was translated into French by L. Laugel as *Oeuvres mathématiques* (Paris, 1898), with a preface by Hermite and an essay by Klein. A supplement is *Gesammelte mathematische Werke. Nachträge*, M. Noether and W. Wirtinger, eds. (Leipzig, 1902). An English version is *The Collected Works*, H. Weber, ed., assisted by R. Dedekind (New York, 1953), with supp. by M. Noether and Wirtinger and a new intro. by Hans Lewy; this is based on the 1892 ed. of *Gesammelte . . . Nachlass* and the 1902 . . . *Nachträge*.

Individual works include *Partielle Differentialgleichungen und deren Anwendung auf physikalische Fragen. Vorlesungen*, K. Hattendorff, ed. (Brunswick, 1896; 3rd ed., 1881; repr., 1938); *Schwere, Electricität und Magnetismus, nach Vorlesungen*, K. Hattendorff, ed. (Hannover, 1876); and *Elliptische Funktionen. Vorlesungen mit Zusätzen*, H. Stahl, ed. (Leipzig, 1899).

See also H. Weber, *Die partiellen Differentialgleichungen der mathematischen Physik. Nach Riemann's Vorlesungen bearbeitet* (4th ed., Brunswick, 1901; 5th ed., 1912); and P. Frank and R. von Mises, eds., *Die Differential- und Integralgleichungen der Mechanik und Physik*, 2 vols. (Brunswick, 1925), the 7th ed. of Weber's work (see above)—the 2nd, enl. ed. (Brunswick, 1930) is the 8th ed. of Weber's work.

II. Secondary Literature. Reference sources include J. L. V. Ahlfors and L. Sario, *Riemann Surfaces* (Princeton, 1960); L. Bieberbach, *Theorie der gewöhnlichen Differentialgleichungen* (Berlin, 1953), esp. pp. 245–252; A. Brill and M. Noether, "Die Entwicklung der Theorie der algebraischen Functionen in älterer und neuerer Zeit," in *Jahresbericht der Deutschen Mathematiker Vereinigung*, **3** (1894), 107–566; E. Cartan, *La géométrie des espaces de Riemann*, Mémorial des sciences mathématiques, no. 9 (Paris, 1925); and *Leçons sur la géométrie des espaces de Riemann* (Paris, 1928); R. Courant, "Bernhard Riemann und die Mathematik der letzten hundert Jahre," in *Naturwissenschaften*, **14** (1926), 813–818, 1265–1277; A. Dinghas, *Vorlesungen über Funktionentheorie* (Berlin, 1961), esp. pp. 298–303; J. Douglas, "Solution of the Problem of Plateau," in *Transactions of the American Mathematical Society*, **33** (1931), 263–321; B. Erdmann, *Die Axiome der Geometrie* (Leipzig, 1877); H. Freudenthal, "Poincaré et les fonctions automorphes," in *Livre du centenaire de la naissance de Henri Poincaré, 1854–1954* (Paris, 1955), pp. 212–219; and "The Main Trends in the Foundations of Geometry in the 19th Century," in *Logic, Methodology and Philosophy of Science* (Stanford, Calif., 1962), pp. 613–621; I. J. Good and R. F. Churchhouse, "The Riemann Hypothesis and Pseudorandom Features of the Möbius Sequence," in *Mathematics of Computation*, **22** (1968), 857–862; J. Hadamard, *Lectures on Cauchy's Problem in Linear Partial Differential Equations* (New Haven, 1923; repr. New York, 1952); H. von Helmholtz, "Über die Thatsachen, die der Geometrie zum Grunde liegen," in *Nachrichten von der Gesellschaft der Wissenschaften zu Göttingen* (1868), 193–221, also in his *Wissenschaftliche Abhandlungen* II (Leipzig, 1883), 618–639; F. Klein, *Über Riemann's Theorie der algebraischen Funktionen und ihrer Integrale* (Leipzig, 1882), also in his *Gesammelte mathematische Abhandlungen*, III (Leipzig, 1923), 501–573; and "Riemann und seine Bedeutung für die Entwicklung der modernen Mathematik," in *Jahresbericht der Deutschen Mathe-*

matiker-vereinigung, **4** (1897), 71–87, also in his *Gesammelte mathematische Abhandlungen*, III, 482–497; A. Krazer and W. Wirtinger, "Abelsche Funktionen und allgemeine Funktionenkörper," in *Encyklopädie der mathematischen Wissenschaften*, IIB, **7** (Leipzig, 1920), 604–873; E. Landau, *Handbuch der Lehre von der Verteilung der Primzahlen*, 2 vols. (Leipzig, 1909), esp. I, 29–36; R. S. Lehman, "Separation of Zeros of the Riemann Zeta-Function," in *Mathematics of Computation*, **20** (1966), 523–541; J. Naas and K. Schröder, eds., *Der Begriff des Raumes in der Geometrie—Bericht von der Riemann-Tagung des Forschungsinstituts für Mathematik*, Schriftenreihe des Forschungsinstituts für Mathematik, no. 1 (Berlin, 1957), esp. K. Schröder, pp. 14–26; H. Freudenthal, pp. 92–97; G. Lampariello, pp. 222–234; and O. Haupt, pp. 303–317; C. Neumann, *Vorlesungen über Riemann's Theorie der Abelschen Integrale* (Leipzig, 1865; 2nd ed., 1884); and "Zur Theorie des logarithmischen und des Newton'schen Potentials," in *Mathematische Annalen*, **11** (1877), 558–566; M. Noether, "Zu F. Klein's Schrift 'Über Riemann's Theorie der algebraischen Funktionen,'" in *Zeitschrift für Mathematik und Physik*, Hist.-lit. Abt., **27** (1882), 201–206; and "Übermittlung von Nachschriften Riemannscher Vorlesungen," in *Nachrichten von der Gesellschaft der Wissenschaften zu Göttingen*, Geschäftliche Mitteilungen (1909), 23–25; E. Schering, "Bernhard Riemann zum Gedächtnis," in *Nachrichten von der Gesellschaft der Wissenschaften zu Göttingen* (1867), 305–314; and *Gesammelte mathematische Werke*, R. Haussner and K. Schering, eds., 2 vols. (Berlin, 1902–1909); H. A. Schwarz, *Gesammelte mathematische Abhandlungen*, II (Berlin, 1890), 108–210; W. Thomson, "Sur une équation aux dérivées partielles qui se présente dans plusieurs questions de mathématique physique," in *Journal de mathématiques pures et appliquées*, **12** (1847), 493–496; E. C. Titchmarsh, *The Theory of the Zeta-Function* (Oxford, 1951); H. Weyl, *Die Idee der Riemannschen Fläche* (Leipzig, 1913; 2nd ed., 1923); and *Raum, Zeit und Materie. Vorlesungen über allgemeine Relativitätstheorie* (Berlin, 1918); and W. Wirtinger, "Algebraische Funktionen und ihre Integrale," in *Encyklopädie der mathematischen Wissenschaften*, IIB, **2** (Leipzig, 1901), 115–175.

Hans Freudenthal

RIES (or **RISZ, RIESZ, RIS**, or **RIESE**), **ADAM** (*b*. Staffelstein, upper Franconia, Germany, 1492; *d*. Annaberg-Buchholz, Germany, 30 March 1559), *mathematics, mining*.

The son of Contz and Eva Riese, Adam (who always signed himself simply "Risz" or "Ries"[1]) came from a wealthy family. Little is known about his youth and nothing about his education. In 1509 he was at Zwickau, where his younger brother Conradus was attending the famous Latin school, and in 1515 he was living in Annaberg, a mining town. Ries finally settled at Erfurt in 1518, working there until 1522 or 1523 as a *Rechenmeister*. He benefited greatly from his contact with the university humanists, who gathered at the house of Georg Sturtz, a rich physician from Annaberg.

Ries wrote his first two books while at Erfurt: *Rechnung auff der linihen* (1518), of which no copy of the first edition is known, and *Rechenung auff der linihen vnd federn* (1522), which had gone through more than 108 editions by 1656. Sturtz encouraged Ries to study algebra—or *Coss*, as it was then called—which had slowly become known in Germany during the second half of the fifteenth century. Sturtz also· recommended certain authors: Johann Widman, who had given the first lecture on algebra in Germany (at Leipzig in 1486), and Heinrich Schreiber (or Grammateus, who taught Rudolff at Vienna), whose arithmetic book of 1518 contained sections on algebra incorporating an improved symbolism. Among the books that Sturtz made available to his friend was an old one—and, as Ries stated, "cast-off" (in the sense of "disordered" or "uncared-for")—containing a group of essays on algebraic topics.[2] Ries was therefore able to compose a *Coss* while still at Erfurt. He completed the book in Annaberg, where he had resettled about 1523, and dedicated the manuscript to Sturtz.[3]

In 1525 Ries married Anna Leuber, by whom he had eight children. He then purchased his own home and became a citizen of Annaberg. He held important positions in the ducal mining administration: *Rezess-schreiber* (recorder of mine yields, from 1525), *Gegenschreiber* (recorder of ownership of mining shares from 1532), and *Zehnter auf dem Geyer* (calculator of ducal tithes, 1533–1539). While fulfilling his official responsibilities he still found time to continue teaching arithmetic. He ran a highly regarded school, and improved and revised his books. During this period he wrote a comprehensive work, *Rechenung nach der lenge, auff den Linihen vnd Feder*, which far surpassed his books written at Erfurt, especially in the number of examples. Most of the work had been completed by 1525; but it was not published until 1550, after Elector Maurice of Saxony had advanced the printing costs. Because the expense was so great, the book was reprinted only once, in 1616.

The year 1539 was decisive for Ries. Duke Georg, an intransigent defender of Catholicism, was succeeded by his brother Heinrich, who favored the Lutherans. The change in rulers ended the religious troubles with which Annaberg, like so many other German cities, had been afflicted.[4] In the same year Ries received the title "Churfürstlich Sächsischer Hofarithmeticus." During this period Ries prepared

a revised edition of his *Coss*, in which he referred to the achievements of his contemporaries Rudolff, Stifel, and Cardano.[5] This was his last work.

In all his arithmetic books (but with greatest detail in the one of 1550) Ries described how the computations were done, both on the abacus and with the new Indian methods. He employed the rule of three to solve many problems encountered in everyday life. While asserting that he had found "proper instruction in only a few places" in the arithmetic of his predecessors, Ries failed to set forth the logical foundations of the subject. Instead, he simply presented formulas with the command "Do it this way."

Ries did, however, furnish the student with a great number of exercises. The steps to be followed were presented in detail, and the reader could check the correctness of answers by following the procedure used to obtain them. Ries surpassed his predecessors in the presentation of his material: it was clear and orderly, and proceeded methodically from the simpler to the more difficult.

Besides the section on gauging, the *Rechenung nach der lenge* contains an extensive section entitled "Practica," in which Ries solves problems according to the "Welsh practice" through the use of proportional parts.[6] In addition he treats problems taken from recreational mathematics, solving them according to the *regula falsi*. Particularly noteworthy is the fact that in his table of square roots the fractions are repeated in a manner that prepared the way for the use of decimal fractions.[7]

In the *Coss*, too, Ries proves to be a good mathematician. He recognized that the Cossists employed a superfluous number of distinct types of equations. He knew about negative quantities and did compute with them.

Ries composed a work that was commissioned by the city of Annaberg, *Ein Gerechnet Büchlein auff den Schöffel, Eimer vnd Pfundtgewicht . . .* (1533), which contains tables of measures and prices from which one could immediately determine the cost of more than one of an item for which a unit price was given, and a *Brotordnung*. From the latter one could directly read off the correct weight for loaves of bread when grain prices varied and the price of an individual loaf was held constant.

It is not known how Ries learned Latin. While in the Erzgebirge he gained a thorough knowledge of mining and of mining problems that lend themselves to computation. At Erfurt he obtained the mathematics books of Widman, Köbel, and Grammateus, and he also saw the book from which Widman had taken his examples, which ultimately stem from the *Algorismus Ratisbonensis*.[8]

Ries furnished precise information on the sources of his *Coss*. As early as 1515 he had solved algebraic problems in Annaberg with a coin assayer named Conrad.[9] Later he studied a revision of al-Khwārizmī's *Algebra* prepared by Andreas Alexander.[10] His principal source, however, was the old book he had received from Sturtz.[11] The contents of this work included al-Khwārizmī's *Algebra* in the translation by Robert of Chester, the *De numeris datis* of Jordanus de Nemore, the *Liber augmenti et diminutionis* of "Abraham," and a Latin and a German algebra (1481).

Because Ries's *Coss* was never printed it had little influence on the development of mathematics.[12] His arithmetic books enjoyed a different fate. Between 1518 and 1656 they went through more than 100 editions in cities from Stettin to Augsburg and from Breslau to Zurich.[13] Ries did more than any previous author to spread knowledge of arithmetic, the branch of mathematics most useful in arts and trade. He was a pioneer in the use of Indian numerals. Ries soon became synonymous with "arithmetic"; to this day, "nach Adam Ries" signifies the accuracy of a calculation.

NOTES

1. See F. Deubner, *Nach Adam Ries*, p. 109; and Roch, *Adam Ries*, p. 79.
2. The book, now Codex Dresdensis C 80, once belonged to Widman.
3. It was partially published by Berlet in 1860 and again in 1892.
4. Ries's name appeared on the list of Lutheran citizens of Annaberg that Duke Georg requested city officials to prepare. He was not persecuted for his religion, however; obviously his services were needed. See Roch, *op. cit.*, p. 21.
5. No study has yet been made of the extent to which Ries utilized the new knowledge in preparing this revision, which exists, in Ries's own handwriting, in an incomplete MS volume in the Erzgebirgsmuseum, Annaberg-Buchholz. This volume also contains an introduction to arithmetic, the old *Coss*, and a German translation of the *Data*. See Berlet, p. 27.
6. In *Regula proportionum*, Ries alludes to their relationship to the rule of three. See 121r.
7. For example, $\sqrt{19}$ = "4 gantze und 358 tausend-teil." See the arithmetic book of 1550, fols. Aa IIIv and Bb IIr ff, and the arithmetic book of 1574 (available in facsimile), fols. 84v ff.
8. See Berlet, p. 34. Widman also took examples from the Bamberg arithmetic book of 1483, which had borrowed problems from the *Algorismus Ratisbonensis*. See Kaunzner, pp. 26, 102 ff.
9. See Berlet, p. 53.
10. On Andreas Alexander, see *Neue deutsche Biographie*, I (1953), 195 f. The original MS of this revision is lost, but there are four copies done by Ries and his sons Jakob and Abraham. See Curtze's ed. in *Abhandlungen zur Geschichte der Mathematik*, 13 (1902), 435–651.
11. For the marginal notations made by Ries, see Kaunzner, p. 35.
12. Three extracts written later can be found in the Sächsische

Landesbibliothek, Dresden. On this matter see Roch, *Adam Ries*, p. 62. Stifel admired Ries's examples and incorporated many of them in his *Deutsche Arithmetica*.
13. For the printing history, see Smith, p. 139; and F. Deubner, "Adam Ries."

BIBLIOGRAPHY

A complete bibliography of Ries's works, including MSS, is in F. Deubner, "Adam Ries" (see below); most of them also are listed in D. E. Smith, *Rara arithmetica* (see below).

Secondary literature includes Bruno Berlet, *Adam Riese, sein Leben, seine Rechenbücher und seine Art zu rechnen . . .* (Leipzig–Frankfurt, 1892); Moritz Cantor, *Vorlesungen über Geschichte der Mathematik*, II, pt. 2 (1900), 420–429; Dorothy I. Carpenter, "Adam Riese," in *Mathematics Teacher*, 58 (Oct. 1965), 538–543; Fritz Deubner, *Nach Adam Riese. Leben und Wirken des grossen Rechenmeisters* (Leipzig–Jena, 1959); and "Adam Ries, der Rechenmeister des deutschen Volkes," in *Zeitschrift für Geschichte der Naturwissenschaften, der Technik und der Medizin*, 1, no. 3 (1964), 11–44; Hildegard Deubner, "Adam Ries—Rechenmeister des deutschen Volkes. Teil II, 1," in *Schriftenreihe für Geschichte der Naturwissenschaften, Technik und Medizin*, 7, no. 1 (1970), 1–22; Wolfgang Kaunzner, "Über Johannes Widmann von Eger . . .," in *Veröffentlichungen des Forschungsinstituts des Deutschen Museums für die Geschichte der Naturwissenschaften und der Technik*, ser. C, no. 4 (1968); Willy Roch, *Adam Ries. Ein Lebensbild des grossen Rechenmeisters* (Frankfurt, 1959); and *Die Kinder des Rechenmeisters Adam Ries*, vol. I of Veröffentlichungen des Adam-Ries-Bundes (Staffelstein, 1960); D. E. Smith, *Rara arithmetica*, 4th ed. (New York, 1970), 138–140, 250 ff.; F. Unger, *Die Methodik der praktischen Arithmetik in historischer Entwicklung* (Leipzig, 1888), 48–53; Kurt Vogel, "Adam Riese, der deutsche Rechenmeister," in *Deutsches Museum. Abhandlungen und Berichte*, 27, no. 3 (1959), 1–37; and "Nachlese zum 400. Todestag von Adam Ries(e)," in *Praxis der Mathematik*, 1 (1959), 85–88; and H. E. Wappler, "Zur Geschichte der deutschen Algebra im 15. Jahrhundert," in *Programm Gymnasium Zwickau* (1887).

Additional articles on Ries are cited in all works on the history of mathematics; for literature from the period 1544–1900, see especially the article by H. Deubner cited above.

KURT VOGEL

RIESZ, FRIGYES (FRÉDÉRIC) (*b.* Györ, Hungary, 22 January 1880; *d.* Budapest, Hungary, 28 February 1956), *mathematics.*

Riesz's father, Ignacz, was a physician; and his younger brother Marcel was also a distinguished mathematician. He studied at the Polytechnic in Zurich and then at Budapest and Göttingen before taking his doctorate at Budapest. After further study at Paris and Göttingen and teaching school in

Hungary, he was appointed to the University of Kolozsvár in 1911. In 1920 the university was moved to Szeged, where, in collaboration with A. Haar, Riesz created the János Bolyai Mathematical Institute and its journal, *Acta scientiarum mathematicarum*. In 1946 he went to the University of Budapest, where he died ten years later after a long illness.

Riesz's output is most easily judged from the 1,600-page edition of his writings (cited in the text as *Works*). He concentrated on abstract and general theories connected with mathematical analysis, especially functional analysis. One of the theorems for which he is best remembered is the Riesz-Fischer theorem (1907), so called because it was discovered at the same time by Emil Fischer. Riesz formulated it as follows (*Works*, 378–381; cf. 389–395). Let $\{\phi_i(x)\}$ be a set of orthogonal functions over $[a, b]$ of which each member is summable and square-summable. Associate with each ϕ_i a real number a_i. Then $\sum_{i=1}^{\infty} a_i^2$ is convergent if and only if there exists a function f such that

$$a_i = \int_a^b f(x)\, \phi_i(x)\, dx, \qquad i = 1, 2, \cdots. \qquad (1)$$

In this form the theorem implies that the $\{a_i\}$ are the coefficients of the expansion of f in terms of the $\{\phi_i\}$ and that f itself is square-summable. This result, the converse of Parseval's theorem, immediately attracted great interest and soon was being re-proved.

Riesz had been motivated to discover his theorem by Hilbert's work on integral equations. Under the influence of Maurice Fréchet's abstract approach to function spaces, such studies became associated with the new subject of functional analysis. Riesz made significant contributions to this field, concentrating on the space of L^p functions (functions f for which $|f^p|$, $p > 1$, is Lebesgue integrable). He provided much of the groundwork for Banach spaces (*Works*, esp. 441–489) and later applied functional analysis to ergodic theory.

Riesz's best-known result in functional analysis has become known as the Riesz representation theorem. He formulated it in 1909, as follows (*Works*, 400–402). Let A be a linear (distributive, continuous) functional, mapping real-valued continuous functions f over [0,1] onto the real numbers. Then A is bounded, and can be represented by the Stieltjes integral

$$A(f) = \int_0^1 f(x)\, d\alpha(x), \qquad (2)$$

where α is a function of bounded variation. The theorem was a landmark in the subject and has proved susceptible to extensive generalizations and applications.

Another implication of Hilbert's work on integral equations that Riesz studied was its close connection with infinite matrices. In *Les systèmes d'équations linéaires à une infinité d'inconnues* (1913; *Works*, 829–1016), Riesz tried not only to systematize the results then known into a general theory but also to apply them to bilinear and quadratic forms, trigonometric series, and certain kinds of differential and integral equations.

Functional analysis and its ramifications were Riesz's most consistent interests; and in 1952 he published his other book, a collaboration with his student B. Szökefnalvy-Nagy, *Leçons d'analyse fonctionnelle*. A classic survey of the subject, it appeared in later French editions and in German and English translations.

In much of his work Riesz relied on the Lebesgue integral, and during the 1920's he reformulated the theory itself in a "constructive" manner independent of the theory of measure (*Works*, 200–214). He required only the idea of a set of measure zero and built up the integral from "simple functions" (effectively, step functions) to more general kinds. He also re-proved some of the basic theorems of the Lebesgue theory.

In the topics so far discussed, Riesz was a significant contributor in fields that had already been developed. But a topic he created was subharmonic functions. A function f of two or more variables is subharmonic if it is bounded above in an open domain D; is continuous almost everywhere in D; and, on the boundary of any subdomain D' of D, is not greater than any function F that is continuous there and harmonic within. The definition is valuable for domains in which the Dirichlet problem is solvable and F is unique, for then $f \leqslant F$ within D and $f = F$ on its boundary. By means of a criterion for subharmonicity given by

$$f(x_0, y_0) \qquad\qquad\qquad\qquad\qquad (3)$$
$$\leqslant \frac{1}{2\pi} \int_0^{2\pi} F(x_0 + r\sin p, y_0 + r\cos p)\, dp,$$

where r is the radius and (x_0, y_0) the center of a small circle within D, Riesz was able to construct a systematized theory (see esp. *Works*, 685–739) incorporating applications to the theory of functions and to potential theory.

Among Riesz's other mathematical interests, some early work dealt with projective geometry. Soon afterward he took up matters in point set topology, such as the definition of continuity and the classification of order-types. He also worked in complex variables and approximation theory.

BIBLIOGRAPHY

I. ORIGINAL WORKS. Riesz's writings were collected in *Összegyűjtött munkái—Oeuvres complètes—Gesammelte Arbeiten*, Á. Császár, ed., 2 vols. (Budapest, 1960), with illustrations and a complete bibliography but little discussion of his work. *Leçons d'analyse fonctionnelle* (Budapest, 1952; 5th ed., 1968), written with B. Szökefnalvy-Nagy, was translated into English by L. F. Boron as *Functional Analysis* (New York, 1955).

II. SECONDARY LITERATURE. On Riesz's work in functional analysis and on the Riesz-Fischer theorem, see M. Bernkopf, "The Development of Functional Spaces With Particular Reference to Their Origins in Integral Equation Theory," in *Archive for History of Exact Sciences*, **3** (1966–1967), 1–96, esp. 48–62. See also E. Fischer, "Sur la convergence en moyenne," in *Comptes rendus . . . de l'Académie des sciences*, **144** (1907), 1022–1024; and J. Batt, "Die Verallgemeinerungen des Darstellungssatzes von F. Riesz und ihre Anwendungen," in *Jahresbericht der Deutschen Mathematiker-vereinigung*, **74** (1973), 147–181.

I. GRATTAN-GUINNESS

RIGHI, AUGUSTO (*b*. Bologna, Italy, 27 August 1850; *d*. Bologna, 8 June 1920), *physics.*

Righi studied in Bologna at the Technical School (1861–1867), then took the four-year mathematics at the University, and after another year graduated from the School of Engineering in 1872, with a dissertation in physics; the previous year he had been appointed assistant to the chair of physics. From 1873 to 1880 he was the physics teacher at the Technical School, and in November 1880 he won the competition for the newly established chair of experimental physics at the University of Palermo. He was professor of physics at the University of Padua from November 1885 to 1889, when he returned to Bologna as professor at the Institute of Physics of the University. He taught there until his death.

Righi is remembered for his studies on electric oscillations, which contributed to wireless telegraphy; his early studies, of greater importance to basic and applied research, are almost forgotten. His graduation thesis (1872) concerned the invention of an induction electrometer that permitted him to investigate weak electrostatic phenomena, including the Volta effect. The device could not only amplify and measure an initially minute electric charge, but also served as an induction electrostatic generator and thus constituted a precise, small-scale model of the van de Graaff accelerator (1933). Righi's interest in the development of experimental devices as well as the mathematical approach to the interpretation of data led to an important analytical paper on the composition of

vibrational motion (1873), described by Lissajous some months before. In it he presented original ideas on the composition of two harmonic orthogonal motions (not necessarily of the same period) in a plane, and the resulting curves. Righi also considered the same problem in three dimensions and defined the trajectories that result from three harmonic motions orthogonal to each other. In a subsequent work (1875), taking as his point of departure Helmholtz' studies in physiological optics, he described the *polystereoscope*, of his own invention, and proposed a new theorem of projective geometry in order to offer a physical and mathematical solution of the problems of binocular vision and the stereoscopic effect.

At the Technical School of Bologna, Righi turned increasingly to applied research. In 1880 he discovered and described magnetic hysteresis, a few months before Warburg, who is credited with the discovery. Although he patented a microphone using conductive powder and a loudspeaker, his inventions elicited little interest.

Righi's work at Palermo centered on the Hall and Kerr effects. He discovered that the Hall effect is several thousand times greater in bismuth than in gold and that the magnetic field in bismuth also causes a variation of electric and—as S.-A. Leduc also discovered—thermal resistance. Continuing his research in Padua, Righi also began studying the photoelectric effect, inspired by Hertz's casual observation (1887) that light that is rich in high-frequency radiation is conducive to a discharge between two electrodes. In a preliminary note of March 1888, Righi demonstrated that when two electrodes are exposed to radiation, rich in ultraviolet rays, they act like a voltaic couple, and he called this phenomenon the photoelectric effect. He also described the connection on series of multiple couples forming a photoelectric battery, pointing out that the maximum effect is obtained with selenium. Wilhelm Hallwachs had published a memoir on this subject less than two months earlier and is credited with priority of discovery, although he had not clarified the phenomenon so completely.

In his lectures at Bologna from 1892 on, Righi demonstrated Hertz's recent experiments on electromagnetic waves, and in 1893 he divulged his own preliminary findings on their nature. Unlike Marconi, who was attempting to apply Hertzian waves in wireless telegraphy, Righi wished to use them to prove the laws of classical optics. In order not to resort to mirrors, prisms, and lenses of prohibitive dimensions, he reduced the wavelength used in his experiments to only 26 mm (May 1894), thereby opening the new field of microwaves to subsequent research and technology. In this way, Righi had demonstrated that Hertzian waves not only interfere with each other and are refracted and reflected, but that they are also subject to diffraction, absorption, and double refraction, like the waves of the visible spectrum. The results of his experiments were published in the widely read *L'ottica delle oscillazioni elettriche* (1897), which is still considered a classic of experimental electromagnetism.

Righi reached the zenith of his activity toward the turn of the twentieth century, with important contributions to the study of X rays and of the Zeeman effect. In 1901 he demonstrated that the solution of Maxwell's equations can be reduced to the solution of one equation with only one vector (in two modes). The following year he described the production of nonlinear relaxation oscillations and discussed their theory. With B. Dessau he wrote the first work on wireless telegraphy, *La telegrafia senza fila* (1903). In the meantime, he had begun experiments on the conduction of gases under various conditions of pressure and ionization, inside tubes with several electrodes, and under the influence of magnetic fields. He continued these experiments until his death.

From 1918 Righi concentrated on the Michelson-Morley experiment, criticizing it and suggesting modifications. He was fascinated by the theory of relativity, even though he thought it had still not been demonstrated definitely in an experimental way.

BIBLIOGRAPHY

I. ORIGINAL WORKS. Righi was a prolific writer; more than 130 papers written before 1900 are listed in the Royal Society *Catalogue of Scientific Papers*, VIII, 751; XI, 181–182; XII, 619; and XVIII, 206–208. See also Poggendorff, III, 1123–1124; IV, 1251–1253; V, 1052–1053; and VI, 2179–2180. His books include *L'ottica delle oscillazioni elettriche* (Bologna, 1897), trans. into German by B. Dessau as *Die Optik der elektrischen Schwingungen* (Leipzig, 1898); *La telegrafia senza fila* (Bologna, 1903), written with B. Dessau, also in German trans. as *Die Telegraphie ohne Draht* (Brunswick, 1903, 1907); *Modern Theory of Physical Phenomena, Radioactivity, Ions, Electrons*, A. Trowbridge, trans. (New York–London, 1904); *Sur quelques expériences connues considerées au point de vue de la théorie des électrons* (Paris, 1906); and *I fenomeni elettro-atomici sotto l'azione del magnetismo* (Bologna, 1918), with bibliography of Righi's writings.

II. SECONDARY LITERATURE. On Righi and his work, see (listed chronologically) *Le feste giubilari di Augusto Righi, per l'inaugurazione del nuovo Istituto di fisica* (Bologna, 1907); B. Dessau, *L'opera scientifica di Augusto Righi* (Rome, 1907), repr. in *Giornale di fisica*, **11** (1970), 53–73,

with intro. by G. Tabarroni; *Arduo* (July 1920)—an entire issue devoted to Righi; L. Amaduzzi, "Augusto Righi, necrologia," in *Archiginnasio*, **15** (1920), 222–226; O. M. Corbino, *Commemorazione di Augusto Righi* (Rome, 1920); L. Amaduzzi, "Commemorazione di A. Righi," in *Elettrotecnica*, **8** (1921), 62–68; P. Cardani, "In memoria di Augusto Righi," in *Nuovo cimento*, 6th ser., **21** (1921), 53–186, with bibliography; L. Donati, *Commemorazione di Augusto Righi* (Bologna, 1923); G. Diaz de Santillani, "Righi," in *Enciclopedia italiana*, XXIX, 328–329; L. Imperatori, *Augusto Righi* (Milan, 1940); P. Veronesi, "Augusto Righi, scienziato," in *Emilia*, **1** (1950), 310–311; G. C. Dalla Noce and G. Valle, *Scelta di scritti di Augusto Righi* (Bologna, 1950), with complete bibliography; G. Valle, "Discorso commemorativo di Augusto Righi," in *Elettrotecnica*, **37** (1950), 483–487; G. Tabarroni, *Bologna e la storia della radiazione* (Bologna, 1965); and "La formazione di Augusto Righi nella Bologna di un secolo fa," in *Strenna storica bolognese*, **19** (1969), 271–292; A. M. Angelini, "Rievocazione di Augusto Righi," in Elettrotecnica, **58** (1971), 57–75; and D. Graffi, "Nel 50° anniversario della morte di A. Righi," in *Atti dell'Accademia delle scienze dell' Instituto di Bologna. Rendiconti*, 12th ser., **8** (1971), 34–42.

GIORGIO TABARRONI

RIMA, TOMMASO (*b.* Mosogno, Ticino, Switzerland, 11 December 1775; *d.* Venice, Italy, 26 February 1843), *medicine*.

After studying grammar and rhetoric at Locarno and Lugano, Rima went to Rome in 1793 for further studies, which he completed in 1798 with degrees in medicine and surgery. In 1799 he enlisted in the army as surgeon major, continuing his long military career until 1820.

From 1806 to 1810 he was head surgeon of the Sant' Ambrogio Military Hospital in Milan, where in 1808 he became also acting professor of military surgery. On 1 January 1811 he was transferred to Mantua as head surgeon and professor of surgery; it was in the hospital there that he conducted his first experiments in the radical treatment of varices of the lower limbs. He was obliged to interrupt this research in 1812 when he was transferred to Ancona. This move marked the beginning of a long period during which his activities of health official took priority over those of surgeon. Discharged from military service in 1820, he became head surgeon at Ravenna and, in 1822, at Venice, where he remained until his death.

Rima's fundamental contributions are the discovery of the blood reflux in varicose veins of the leg and the surgical procedure for the treatment of such veins, which he devised himself and used successfully. His innovations, rediscovered in 1891 by Friedrich Trendelenburg, still underlie the pathology and therapy of varicose veins. He also studied contagion of infectious diseases, in particular contagious ophthalmia.

BIBLIOGRAPHY

I. ORIGINAL WORKS. "Sulla cura radicale del varicocele," in *Giornale per servire ai progressi della patologia e della materia medica*, **4** (1836), 398–416; and "Sulla causa prossima delle varici alle estremità inferiori, e sulla loro cura radicale," *ibid.*, **5** (1836), 265–301; were separately reprinted as *Cura radicale delle varici dedotta dalla causa prossima scoperta e dimostrata in due memorie lette nell'-Ateneo dal socio ordinario Tommaso Dr. Rima* (Venice, 1838).

II. SECONDARY LITERATURE. On Rima's life and work, see Luigi Belloni, "Una ricerca del contagio vivo agli albori dell'Ottocento," in *Gesnerus*, **8** (1951), 15–31; "L'autobiografia del chirurgo Tommaso Rima (Necrologia del Dr. T. Rima)," *ibid.*, **10** (1953), 151–186; and "Valvole venose e flusso centrifugo del sangue. Cenni storici," in *Simposi clinici*, **5** (1968), xlix–lvi; and Davide Giordano, *Nel centenario della dottrina di Tommaso Rima su le varici* (Venice, 1925).

LUIGI BELLONI

RINGER, SYDNEY (*b.* Norwich, England, 1835; *d.* Lastingham, Yorkshire, England, 14 October 1910), *medicine, physiology.*

Ringer's entire professional career was associated with the University College Hospital, London. In 1854 he entered the medical school of University College, where he graduated M.B. in 1860 and M.D. in 1863. He was appointed assistant physician to the University College Hospital in 1863, physician in 1866, and consulting physician in 1900. Ringer also served successively as professor of materia medica, pharmacology and therapeutics, and the principles and practice of medicine at the University College faculty of medicine. In 1887 he was named Holme professor of clinical medicine, a chair he held until his retirement in 1900. In 1870 he became a fellow of the Royal College of Physicians and in 1885 a fellow of the Royal Society.

Ringer was an outstanding bedside teacher who continued the high standard of clinical instruction that had been established at the University College Hospital by Sir William Jenner, Sir John Russell Reynolds, and Walter Hayle Walshe. Ringer's clinical orientation may be clearly seen from his popular *Handbook of Therapeutics*, which passed through thirteen editions between 1869 and 1897. This book was originally commissioned as a revision of Jonathan Pereira's massive *Elements of Materia Medica*, but Ringer was little concerned with the minutiae of traditional medical botany and materia medica. He offered instead a thoroughly practical treatise in which the actions and indications of drugs were concisely summarized. Pharmacology remained one of Ringer's lifelong interests, and he incorporated into the successive editions of the *Handbook* the essence of a burgeoning literature on the specific actions of medicines. Ringer's own contribution to pharmacology included papers on the actions of various substances, including digitalis, atropine, muscarine, and pilocarpine. He was also the first person to investigate the direct effects of anesthetics on cardiac tissue.

Among Ringer's other medical publications were a short work *On the Temperature of the Body as a Means of Diagnosis in Phthisis and Tuberculosis* (1865) and the sections on parotitis, measles, and miliaria in Reynolds' *System of Medicine* (1866–1879).

Patient care, clinical teaching, and writing thus occupied most of Ringer's career, but for many years he also maintained a small laboratory in the department of physiology; at that time there was no pharmacology laboratory at University College. With the aid of a series of collaborators, including E. G. A. Morshead, William Murrell, Harrington Sainsbury, and Dudley Buxton, Ringer published between 1875 and 1895 more than thirty papers devoted to the actions of inorganic salts on living tissues.

Carl Ludwig in his 1865 inaugural address at Leipzig had stressed the importance of studying isolated organs, and with various pupils he had developed some perfusion techniques. From the beginning the heart had served as the principal organ for these "extravital" investigations, and most of Ringer's physiological work relied on Ludwig's experimental model. In a classic series of experiments performed between 1882 and 1885, Ringer began with the isolated heart of the frog suspended in a 0.75 percent solution of sodium chloride. He then introduced additional substances (for example, blood and albumin) to the solution and observed the effects on the beating heart. He demonstrated that the abnormally prolonged ventricular dilatation induced by pure sodium chloride solution is reversed by both blood and albumin. Ringer was able to identify the active substance as potassium. He also showed that small amounts of calcium in the perfusing solution are necessary for the maintenance of a normal heartbeat, a discovery he made after realizing that the distilled water he was using actually contained traces of calcium. Ringer thus gradually perfected Ludwig's perfusion technique by proving that if small amounts of potassium and calcium are added to the normal solution of sodium chloride, isolated organs can be kept functional for long periods of time. "Ringer's Solution" became an immediate necessity for the physiological laboratory.

Ringer subsequently broadened his investigations on the effects of inorganic salts on living tissues. He studied the influences of solutions on striated muscles, on the growth of the tadpole, and on the contractile tissue of fishes. He was also the first to show conclusively that the presence of calcium is necessary for the normal operation of the blood clotting mechanism.

Ringer's researches were carried out independently of Arrhenius' contemporary formulation of the theory of electrolyte dissociation. Consequently, Ringer's papers record a series of well-planned and executed experiments devoid of a coherent theoretical framework. This deficiency diminished Ringer's immediate influence, but by the time of his death the wider significance of his physiological work was more generally recognized.

BIBLIOGRAPHY

I. ORIGINAL WORKS. Most of Ringer's physiological papers were published in the early volumes of the *Journal of Physiology*. Some of his more important papers in that journal include "Concerning the Influence Exerted by Each of the Constituents of the Blood on the Contraction of the Ventricle," in *Journal of Physiology*, **3** (1880–1882), 380–393; and **4** (1883–1884), 29–42, 222–225; "The Influence of Saline Media on Fishes," *ibid.*, **4** (1883–1884), vi–viii; and **5** (1884–1885), 98–115; "On the Mutual Antagonism Between Lime and Potash Salts in Toxic Doses," *ibid.*, **5** (1884–1885), 247–254; and "Further Observations Regarding the Antagonism Between Calcium Salts and Sodium, Potassium and Ammonium Salts," *ibid.*, **18** (1895), 425–429. Ringer's solution may be said to date from his paper "Regarding the Action of Hydrate of Soda, Hydrate of Ammonia, and Hydrate of Potash on the Ventricle of the Frog's Heart," *ibid.*, **3** (1880–1882), 195–202.

With Buxton, Ringer wrote two important papers: "Concerning the Action of Calcium, Potassium, and Sodium Salts Upon the Eel's Heart and Upon the Skeletal Muscles of the Frog," *ibid.*, **8** (1887), 15–19; and "Upon the Similarity and Dissimilarity of the Behaviour of Cardiac and Skeletal Muscle When Brought Into Relation With Solutions Containing Sodium, Calcium and Potassium Salts," *ibid.*, **8** (1887), 288–295.

Ringer's major paper on calcium and blood clotting was written in collaboration with Sainsbury: "On the Influence of Certain Salts Upon the Act of Clotting," *ibid.*, **11** (1890), i–ii, 369–383.

A fuller bibliography of Ringer's scientific papers may be found in the Royal Society *Catalogue of Scientific Papers*, V, 209; VIII, 752; XI, 185; XVIII, 214.

II. SECONDARY LITERATURE. There is no full assessment of Ringer's work. Contemporary obituaries include those in *Lancet* (1910), **2**, 1386–1387; *British Medical Journal* (1910), **2**, 1384–1386; *Proceedings of the Royal Society*, **84B** (1912), i–iii; and the *Biochemical Journal*, **5** (1911), i–xix.

WILLIAM F. BYNUM

RINMAN, SVEN (*b.* Uppsala, Sweden, 12 June 1720; *d.* Ekskilstuna, Sweden, 20 December 1792), *metallurgy.*

For generations, the Rinman family produced scientists who worked in mining and metallurgy. Sven Rinman was the son of Gustaf Rinman, county treasurer in Uppsala, and Magdalena Leijonmarck.

After studies at the University of Uppsala, he devoted his working life to the metal industries. As director of mining and metallurgy and as adviser to proprietors and managers of mines and iron works all over Sweden—duties he performed on behalf of the Swedish Ironmasters' Association and the Royal Mines Authority—Rinman made substantial theoretical and practical contributions to the improvement of iron and steel production methods, especially of charcoal blast furnaces. His works, both published and unpublished, are highly esteemed sources for knowledge of iron- and steelmaking during the eighteenth century, both in Sweden and in the countries he visited.

Rinman was the first to identify carbon—plumbago, at least—as the causative agent in manufacturing steel from iron. He is also regarded as one of the foremost pioneers in using the blowpipe for analyses. A first-rate inventor and a skilled scientist, he is known as the "father of the Swedish iron industry."

BIBLIOGRAPHY

I. ORIGINAL WORKS. Rinman's books include *Anledningar til kunskap om den gröfre jern- och stålförädlingen och des förbättrande* (Stockholm, 1772); 2nd ed.., rev. by Carl Rinman, with biography by Gustaf Broling (Falun, 1829), also in German (Vienna, 1790); *försök til järnets historia, med tillämpning för slögder och handtverk*, 2 vols. (Stockholm, 1782), also in German (Berlin, 1785); *Bergwerks lexicon*, 2 vols. (Stockholm, 1788–1789), a descriptive handbook, alphabetically arranged of technical terms in Swedish, German, and French; and *Afhandling rörande mechaniquen med tillämpning i synnerhet til bruk och bergwerk*, II (Stockholm, 1794).

A fellow of the Royal Swedish Academy of Sciences, Rinman published 26 essays in its *Handlingar* (1745–1781).

II. SECONDARY LITERATURE. See C. Forsstrand, *Sven Rinman. Minnesteckning till 200-årsdagen av hans födelse* (Stockholm, 1920); and G. Malmborg, ed., *Sven Rinmans tjänsteberättelser rörande den grövre järnförädlingen 1761–1770*, Jernkontorets Bergshistoriska Skriftserie, no. 4 (Stockholm, 1935).

TORSTEN ALTHIN

RÍO, ANDRÉS MANUEL DEL (*b.* Madrid, Spain, 10 November 1764; *d.* Mexico City, Mexico, 23 May 1849), *mineralogy, geology.*

Del Río was the son of José del Río and María Antonia Fernández. He studied at the Colegio de San Isidro in Madrid, distinguishing himself in the classics, then entered the Universidad de Alcalá de Henares, where he studied experimental physics under José Solano. He graduated in 1781 and then continued his work at the Real Academia de Minas de Almadén under a subsidy from Charles III, who was attempting to bring Spain into the mainstream of western science. Del Río attracted the notice of Diego Gardoquí, the minister of mines, and through his interest received a further subsidy to study in the great intellectual centers of Europe. Del Río thus spent four years in Paris, where he studied first medicine and then chemistry (with d'Arcet). He next attended the Bergakademie at Freiberg, where he heard Werner's lectures on mineralogy and J. F. Lempe's on mining science and became a friend of Humboldt. He continued his studies at the Bergakademie at Schemnitz (now Baňskà Štiavnica, Czechoslovakia) and at mines in Saxony and England. Del Río next returned to Paris and studied chemistry with Lavoisier. When Lavoisier was arrested and imprisoned during the Terror, Del Río fled to England in disguise.

In 1794 Del Río went to Mexico to take up a post as professor of mineralogy at the newly founded Colegio de Minería in Mexico City. The Colegio was the first institution of technical education in the New World; its graduates, who had completed a comprehensive four-year curriculum and a two-year apprenticeship in one or more mining districts, were to serve as inspectors of mines throughout Mexico, Central America, and the Philippine Islands. Del Río assumed his professorship in 1795; his *Elementos de Orictognosia*, the first textbook of mineralogy to be published in the Americas, was published in the same year and represented the first critical exposition of Werner's system of mineralogy to be written in Spanish. Del Río's course in mineralogy was the first formal instruction in the subject in the New World and he was largely responsible for introducing modern science and modern engineering methods into the mining industry of Mexico; Humboldt, visiting the Colegio in 1803, was favorably impressed by its students, its laboratory facilities, and by Del Río's textbook.

When Humboldt left Mexico after this visit, he took with him a sample of the mineral vanadinite from Zimapán, Mexico, in which Del Río had discovered a new metallic element in 1801. He also took Del Río's account of his chemical observations of it, for possible publication; this chemical description was, however, lost in a shipwreck. The new element was vanadium (which Del Río called panchromium or, later, erythronium), and Humboldt gave the sample to Collet-Descotils, who mistakenly concluded that it was the element chromium, which had been discovered in 1797. Although Del Río's chemical investigation had indicated the distinct character of the new substance, he nevertheless accepted Collet-Descotils's evaluation, and claimed to be the first to discover chromium in the Americas. Nils Sefström subsequently found vanadium in magnetite from Falun, Sweden, and Wöhler demonstrated that it was identical to the material found in Zimapán. Del Río was accordingly resentful of having been denied priority in its discovery, and his friendship with Humboldt cooled. He had no better luck with other new minerals he described, for in every case they had been described previously by some other worker, or were later found to be a combination of already known minerals, or were never confirmed.

In the meantime, Del Río had become deeply involved in Mexican affairs, and took an active part in the scientific and cultural life of Mexico City. Following the War of Independence, he was named a deputy to the revolutionary Cortes in Spain, then returned to Mexico to take a small part in the masquerade imperial court of Agustín I (Itúrbide). When the republic was again in power, all Spaniards were expelled from Mexico; although Del Río was specifically exempted from the final ban of 1829, he nonetheless chose to go into exile, and went to Philadelphia. He remained there for five years, during which he participated in meetings of the Philadelphia Academy of Sciences and the American Philosophical Society and was elected president of the new Geological Society of Pennsylvania. Returning to Mexico, he resumed his teaching in the face of the almost insuperable difficulties imposed by the turbulent political situation; the Colegio had suffered from the prevalent instability and disorganization, and all of its facilities had fallen into a state of decay. Despite the ruin that surrounded him, Del Río continued to teach a few students, make new chemical analyses of minerals, publish papers, and revise his textbook. He did work on the origin of mineral veins, the paragenesis of sulfide minerals, and the effects of trace elements on physical properties and polymorphism. He also published critical comments on works by Karsten, Klaproth, Haüy, Breithaupt, and Berzelius. Del Río's contributions to science were substantial in spite of the situation under which he worked, but conditions in Mexico allowed only a few of his students to continue in the field. Del Río was honored as a member of a number of learned societies, both in Mexico and abroad, but he died a pauper.

BIBLIOGRAPHY

I. ORIGINAL WORKS. Most of Del Río's publications are listed in Rafael Aquilar y Santillán, "Bibliografia geologica y minera de la Republica Mexicana," in *Boletín del Instituto Geologica de México*, **10** (1898), 101–102, and **17** (1908), 202–205. Several papers that he published in the United States are cited in J. M. Nickles, "Bibliography of Geology of North America," *Bulletin of the United States Geological Survey*, no. **746** (1923), 878.

The most important of Río's works is *Elementos de Orictognosia*, pt. 1 (Mexico City, 1795) and pt. 2 (Mexico City, 1805); and 2nd ed. (Mexico City, 1832–1846), with *Suplemento . . . de mi mineralogia impresa en Filadelphia en 1832 . . .* (Mexico City, 1848). Río also translated, with important additions, L. G. Karsten, *Tablas mineralogicas* (Mexico City, 1804), and J. Berzelius, *Nueva sistema mineral . . .* (Mexico City, 1827).

See also Río's discussions of contemporary developments in mineralogy in "Carta dirigida al señor abate Haüy . . .," in *Seminario politico y literario de México*, **2** (1821), 173–182, and "Ein Paar Anmerkungen zu dem Handbuche der Mineralogie von Hoffman, fortgesetzt von Breithaupt," in *Annalen der Physik*, **71** (1822), 7–12.

II. SECONDARY LITERATURE. The best general biography of Río is Santiago Ramírez, "Biografía del Sr. D. Andrés Manuel del Río," in *Boletín de la Sociedad de Geografia y Estadistica Mexicana*, **2** (1890), 205–251, and repr. separately (Mexico City, 1891). See also Arturo Arnaíz y Freg, *Andrés Manuel del Río* (Mexico City, 1936), repr. without illustrations in *Revista de historia de America*, **25** (1948), 27–68, which includes a list of documents relating to Río in archives in Mexico and Spain.

The discovery of vanadium is discussed by Mary Elvira Weeks and Henry M. Leicester, *Discovery of the Elements*, 7th ed. (Easton, Pa., 1968), 351–364. Other helpful works include J. L. Amorós, "Notas sobre la historia de la cristalografía y mineralogía. V. La mineralogía española en 1800; La 'Orictognosia' de Andrés del Río," in *Boletín del Real Sociedad Española Historia Natural*, **62** (1964), 199–220; and "The Introduction of Werner's Mineralogical Ideas in Spain and in the Spanish Colonies of America," in *Freiberger Forschungshefte*, **223**C (1967), 231–236; Modesto Bargalló, "Homenaje a . . . Del Rio . .," in *Ciencia* (*Mexico*), **10** (1950), 270–278; and Walter Howe, *The Mining Guild of New Spain* . . . (Cambridge, Mass., 1949).

WILLIAM T. HOLSER

RÍO-HORTEGA, PÍO DEL (*b.* Portillo, Valladolid province, Spain, 5 May 1882; *d.* Buenos Aires, Argentina, 1 June 1945), *neurohistology.*

Río-Hortega received his medical degree in 1905 from the University of Valladolid, where he became assistant professor of histology. In 1915, after two years of study in Paris and London, he returned to Madrid and joined the laboratory staff of Nicolás Achúcarro, the histologist known for his studies of neuroglia and whose team worked closely with that of Santiago Ramón y Cajal.

Between 1914 and 1916 Río-Hortega worked on various histological topics, including the structure of the ovary and the fine texture of cancer cells. He did not begin his major work on the interstitial cells of the nervous system until 1916. Both Ramón y Cajal and Achúcarro had studied the cytology of neurons and astrocytes, but the nature of the cells that the former called "corpuscles without processes" or the "third element" remained unclear.

By 1918 Río-Hortega had developed a silver carbonate stain enabling him to explicate the fine structure of the "third element," which he found to consist of two different cytological types: microglia, small cells of mesodermal origin with spiny processes, dispersed throughout the central nervous system; and interfascicular glia or oligodendroglia, cells of ectodermal origin that follow and surround the nerve fibers. Río-Hortega's demonstration that these cells do not lack processes—as Ramón y Cajal had thought—led to heated controversy that strained relations between the two researchers.

The next phase of Río-Hortega's career unfolded at the National Institute of Cancer in Madrid, where he headed the research division from 1928 to 1936. There he produced the basic cytological descriptions necessary for the classification of gliomatous and other tumors of the central nervous system. Soon after the outbreak of the Spanish Civil War he left Spain to work in neuropathology laboratories in Paris and Oxford. In 1940 he removed to Argentina. There he organized a laboratory in Buenos Aires and founded the *Archivos de histología normal y patológica*, in which he published his last studies on the neuroglial character of the satellite cells surrounding the neurons of sensory ganglia.

BIBLIOGRAPHY

An account of Río-Hortega's writings in Spanish is in J. M. Ortiz Picón, "La obra histoneurológica del doctor Pío del Río-Hortega (1882–1945)," in *Archivos de neurobiología*, **34** (1971), 39–70.

For the work on microglia, see Río-Hortega, "Microglia," in Wilder Penfield, ed., *Cytology and Cellular Pathology of the Nervous System*, 3 vols. (New York, 1932), III, 483–534; Jorge Bullo, "Contribuciones de la escuela de Cajal sobre histopatología de la neuroglia y microglia," in *Archivos de histología normal y patológica*, **2** (1945), 425–445; and the following articles by Penfield, an American disciple of Río-Hortega: "Oligodendroglia and Its Relation to Classical Neuroglia," in *Brain*, **47** (1924), 430–452; "Microglia and the Process of Phagocytosis in Gliomas," in *American Journal of Pathology*, **1** (1925), 77–89; and "Neuroglia: Normal and Pathological," in *Cytology and Cellular Pathology of the Nervous System*, II, 423–479.

For biographical details, see the article by Ortiz Picón cited above and Penfield's obituary in *Archives of Neurology and Psychiatry*, **54** (1945), 413–416. On Río-Hortega's place in the school of Spanish histologists, see José María López Piñero, "La contribución de la escuela histológica española a la ciencia universal," in *Tercer programa* (Madrid), no. 5 (Apr.–June 1967), 39–59. His relationship with Achúcarro is discussed by several contributors to the memorial volume *Nicolás Achúcarro. Su vida y su obra* (Madrid, 1968). Elements of an unpublished autobiographical MS describing Río-Hortega's relationship with Ramón y Cajal appear in Dorothy F. Cannon, *Explorer of the Human Brain. The Life of Santiago Ramón y Cajal* (New York, 1949).

THOMAS F. GLICK

RIOLAN, JEAN, JR. (*b*. Paris, France, 15 February 1580; *d*. Paris, 19 February 1657), *anatomy, medicine*.

Jean Riolan, Sr. (1539–1606), came from Amiens and was a leading member of the Paris Medical Faculty, serving as dean in 1585–1586. He published a series of commentaries on the works of Fernel as well as many other medical and philosophical works. His wife was of the Piètre family, which was very prominent in Parisian medicine during the sixteenth and seventeenth centuries; its members included several deans of the faculty.

After studying chiefly under his uncle Simon Piètre, Jean, Jr., took his M.D. in 1604. He was soon named to a new chair of anatomy and botany at the University of Paris and also to a chair of medicine at the Collège Royal. From 1640 until his death he was dean of the college; and although he never held the formal position of dean of the faculty, in his later years he was called its doyen in an honorific sense.

He enjoyed the close friendship of his younger colleague Guy Patin, although the latter's correspondence depicts the elderly Riolan as a harsh, embittered, and unforgiving personality. His unhappy marriage produced three sons and two daughters.

Riolan was born into a very conservative medical establishment, among the chief concerns of which were the suppression of new currents in medicine that threatened to undermine the classical tradition and (closely related to this) the protection of its privileges against rival practitioners. Riolan participated vigorously in these activities of the faculty. From 1603 to 1606 he took part in a pamphlet war that his father had begun against Duchesne and other Paracelsians, and he later published attacks on Parisian surgeons and on graduates of the rival faculty of Montpellier who were practicing in Paris.

But almost alone among his colleagues in the faculty, Riolan stood for more than just a smug defense of the past. From his youth he was a dedicated student of anatomy, and through his efforts he restored to Paris its eminence, lost after the middle of the sixteenth century, as a center of teaching and research in that field. He had an abiding respect for Galen as the surest guide to anatomy; but in determining matters of structural detail, he placed ultimate reliance on his own thorough and critical experience in dissection, which commanded the sincere respect even of less conservative contemporaries. He established his reputation through a series of textbooks, the most important being the second edition of *Anthropographia* (1626); these works reveal a mastery not only of original anatomical observation and of the classical and modern anatomical literature but also of classical learning in general. In his later *Encheiridium* (1648) he included a systematic presentation of both morbid and normal anatomy. He regarded himself, and to some extent was regarded by others, as an international arbiter in anatomical matters; and many of his other publications were concerned with assessing and criticizing the work of contemporaries.

Riolan's anatomical studies were interrupted when, in 1633, he became the principal physician of the queen mother, Marie de Médicis. He accompanied her on the foreign travels that were enforced upon her because of her political machinations. Thus he was with her in England from 1638 to 1641, and he attended her final illness at Cologne in 1642. Upon his permanent return to Paris, he resumed his anatomical pursuits to the extent that his advancing age and declining health would permit, producing a spate of new publications beginning in 1648. The harshly polemical character of many of these later works, however, reflects the profound change that had occurred in anatomy and physiology in the interval. Earlier, Riolan had assumed that new discoveries in anatomy could only enhance and confirm the ancient medical doctrines; but now he was confronted with a number of discoveries that (as he saw perhaps more clearly than the proponents of such discoveries) threatened the very foundations of traditional medical practice. Riolan felt that the value of these practices, and hence of the theories on which they were based, had been established beyond question by the experience of countless physicians, and that this evidence must be taken into account along with anatomical and vivisectional evidence in judging new physiological theories. He therefore tried to accept the new discoveries without undermining established practices and to show that, to the extent that the innovations were incompatible with tradition, they were not based on sound observation.

Harvey's discovery of the circulation of the blood was the most important threat that Riolan sought to disarm. He had met Harvey during his stay in London; and although his initial reaction to the circulation was favorable, he eventually concluded that the idea would subvert much of humoral medicine. He therefore developed his own concept of circulation, according to which the blood circulates very slowly (perhaps once or twice a day), and only through the larger arteries and veins of the body, which he claimed were connected directly by anastomoses. In the smaller vessels the blood moves gradually outward as it is assimilated by the parts of the body, so that local variations in the humors can develop. Above all, the blood does not circulate through the mesenteric vessels, where the incoming nutriment must remain until it has been properly concocted and purified by the liver, the central role of which in determining the overall humoral balance of the body is thus preserved.

Riolan admitted that some vivisectional evidence seems to support the general, rapid Harveian circulation; but he maintained that under normal physiological conditions the movements of the blood are much more tranquil than those observed in vivisection. Riolan's publication of these ideas in his *Encheiridium* (1648) occasioned Harvey's first and only formal reply to his critics (1649). Riolan developed his own concept of circulation and his criticisms of Harvey in a further series of tracts, in one of which (1652) he added a second "Hippocratic circulation" involving only the arms and legs.

The discovery of the thoracic duct by Pecquet (1651) and of the lymphatic vessels by Thomas Bartholin (1653) provoked another series of responses by Riolan (1652, 1653). He accepted the purely anatomical aspects of this work but bitterly opposed the interpretations of Pecquet and Bartholin, who reduced the liver to insignificance in the overall economy of the body. While his position was chiefly motivated by his concern with protecting Galenic medicine, he defended it with concrete evidence regarding the importance of the liver that the innovators had chosen to ignore.

BIBLIOGRAPHY

I. ORIGINAL WORKS. Riolan's first anatomical work was *Schola anatomica novis et raris observationibus illustrata. Cui adiuncta est accurata foetus humani historia* (Paris, 1608). A revised and enlarged version was published as *Anatome* and appended to his edition of his father's *Opera omnia, tam hactenus edita quam, postuma . . .* (Paris, 1610). *Anthropographia ex propriis et novis observationibus collecta* (Paris, 1618) was a further expansion of this text, while *Anthropographia et osteologia* (Paris, 1626) also incorporated other earlier works: *Osteologia, ex veterum et recentium praeceptis descripta* (Paris, 1614), and *Anatomica humani foetus historia. Adjectae sunt viventis animalis observationes anatomicae* (Paris, 1618). A final revised ed. of *Anthropographia* was included in *Opera anatomica vetera, recognita, & auctiora, quam-plurima nova* (Paris, 1649). *Encheiridium anatomicum, et pathologicum* (Paris, 1648) was a shorter compendium.

Riolan's numerous tracts on the circulation and lymphatics were published in four main collections: *Opuscula anatomica nova* (London, 1649), also included in the *Opera anatomica* (1649); *Opuscula anatomica, varia, & nova* (Paris, 1652); *Opuscula nova anatomica* (Paris, 1653); and *Responsiones duae, prima ad experimenta nova Pecqueti, altera ad Pecquetianos* (Paris, 1655). Most of his many other controversial tracts are listed in the *Catalogue général des livres imprimés de la Bibliothèque national. Auteurs*, CLII (Paris, 1938), cols. 348–352.

II. SECONDARY LITERATURE. No full biographical study exists. R. Tabuteau, *Deux anatomistes français: Les Riolan* (Paris, 1929), is short but useful. Additional biographical details are provided by T. Vetter, "Jean Riolan, second du nom, qui ne fut pas doyen des écoles de Paris," in *Presse médicale*, **73** (1965), 3269–3274; and G. Whitteridge, *William Harvey and the Circulation of the Blood* (London–New York, 1971), 175–200. For general background, including some discussion of Riolan, see J. Lévy-Valensi, *La médecine et les médecins français au XVIIe siècle* (Paris, 1933); and J. Roger, *Les sciences de la vie dans la pensée française du XVIIIe siècle*, 2nd ed. (Paris, 1971), 7–48.

The best general assessment of Riolan's work is N. Mani, "Jean Riolan II (1580–1657) and Medical Research," in *Bulletin of the History of Medicine*, **42** (1968), 121–144. For his views on the heart and circulation, see Whitteridge, *op. cit.*; W. Pagel, *William Harvey's Biological Ideas* (Basel–New York, 1967), 74–76, 216–218; and K. Rothschuh, "Jean Riolan jun. (1580–1657) im

Streit mit Paul Marquart Schlegel (1605–1653) um die Blutbewegungslehre Harveys," in *Gesnerus*, **21** (1964), 72–82.

JEROME J. BYLEBYL

RISNER, FRIEDRICH (*b.* Herzfeld, Hesse, Germany; *d.* Herzfeld, *ca.* 1580), *mathematics, optics*.

Risner spent most of his scholarly life as the protégé and colleague of Peter Ramus, the renowned anti-Aristotelian humanist and educational reformer, who was accustomed to collaborating with younger scholars on mathematical subjects. Risner's mathematical ability is evident from Ramus' reference to his "assistant in mathematical studies, . . . Friedrich Risner, so well versed in mathematics" (Hooykaas, *Humanisme*, p. 45). In his will Ramus established a chair in mathematics at the Collège Royal de France and specified that Risner should be its first occupant. When legal disputes over the chair were settled in 1576, Risner accepted the salary but never lectured; he resigned a few months later and returned to his native Hesse, where he died after a long illness.

The most noteworthy result of the collaboration between Ramus and Risner was the immensely influential edition (1572) of the optical works of Ibn al-Haytham and Witelo. Risner produced the *editio princeps* of Ibn al-Haytham's *Optics (De aspectibus)* from two manuscripts discovered by Ramus, adding citations and subdividing the book into propositions. Witelo's *Perspectiva* had already been published twice; but Risner improved the text by comparing several manuscripts, redrafted the figures, and added citations to corresponding propositions in Ibn-al Haytham's *Optics*.

The collaboration of Ramus and Risner also resulted in the posthumous publication of Risner's *Opticae libri quatuor*, which later influenced Snell. It seems that the book was begun during the early years of Ramus and Risner's association; and it is probable that the basic outline was Ramus', while Risner was given the task of providing appropriate demonstrations and discussion. Only the first of the four books is complete, however, and the final two consist of little more than the enunciations of the propositions. The work depends primarily on Witelo, although other ancient and medieval authors are also cited.

BIBLIOGRAPHY

I. ORIGINAL WORKS. Risner edited *Opticae thesaurus Alhazeni Arabis libri septem, nunc primum editi. Eiusdem liber de crepusculis et nubium ascensionibus. Item Vitellonis*

thuringopoloni libri X. Omnes instaurati, figuris illustrati et aucti, adiectis etiam in Alhazenum commentariis, a Federico Risnero (Basel, 1572). Risner's own *Optica* has been published as *Opticae libri quatuor ex voto Petri Rami novissimo per Fridericum Risnerum ejusdem in mathematicis adjutorem, olim conscripti* (Kassel, 1606); and *Risneri optica cum annotationibus Willebrordi Snellii*, J. A. Vollgraff, ed. (Ghent, 1918).

II. SECONDARY LITERATURE. On Risner's life and works, see Bernardino Baldi, "Vite di matematici Arabi," in *Bullettino di bibliografia e di storia delle scienze matematiche e fisiche*, V (1872), 461–462; R. Hooykaas, *Humanisme, science et réforme: Pierre de la Ramée* (Leiden, 1958); J. J. Verdonk, *Petrus Ramus en de wiskunde* (Assen, 1966), 66–73; and Charles Waddington, *Ramus (Pierre de la Ramée), sa vie, ses écrits et ses opinions* (Paris, 1855).

DAVID C. LINDBERG

RISTORO (or **RESTORO**) **D'AREZZO** (Arezzo, Italy, possibly *ca.* 1210–1220; *d.* after 1282), *natural history*.

The scanty information available on Ristoro's life derives entirely from his *Composizione del mondo*, which he completed in Arezzo in 1282. The work is in the vernacular and is thus the oldest surviving scientific text in Italian. Ristoro mentions that he wrote his treatise in a monastery, which suggests that he was a monk, but of which order is not known. His interest in the science of the stars was paramount, but he was also an artist, skilled in drawing, painting, and illuminating. An indication of his refined artistic taste and of his keen spirit of observation is his description of the decorative patterns of a number of Etruscan and Roman vases (*arretina vasa*), in which he admired, besides other features, the fine representation of plants, animals, mountains, valleys, rivers, and forests. His critical judgment is all the more striking in a time when archaeological remains were poorly understood and ignored.

Convinced that "it is a dreadful thing for the inhabitant of a house not to know how it is made" (Book I, ch. 1), Ristoro collected in the *Composizione del mondo* all available knowledge of cosmology. The work begins with a detailed treatment of astronomy and astrology; he made no distinction between them. He described in particular the eight heavens of the fixed stars and the known planets and the four spheres corresponding to the four elements. For Ristoro the importance of the celestial bodies lay in their influence upon the earth, and he likened the action of the stars on terrestrial phenomena to that of a seal that gives form and nobility to wax, itself a shapeless and vulgar material. Every known celestial and terrestrial phenomenon was described by Ristoro, from the distribution of earth and seas, to the creation and disintegration of mountains, from meteorology to the circulation of waters, from plants and animals, to the consideration of man, "the most noble of all the animals" (Book I, ch. 1).

Ristoro's work was essentially a compilation from ancient authors, available to him either in the original, like Aristotle's *De coelo*, or through Arabic translations and adaptations that were subsequently rendered into Latin, such as the *Liber de aggregationibus stellarum* of al-Farghānī. The compilation, however, contains some startling critical comments: for instance, he denied that the motion of the spheres caused heavenly music; and he attributed the twinkling of the stars to the eye, rather than to the stars themselves. There are also noteworthy personal observations, including the description of the total solar eclipse that he observed in Arezzo in 1239. "The sky was clear and without clouds, when the air began to turn yellow, and I saw the body of the sun being covered little by little until it became obscured and it was night; and I saw Mercury near the sun; and all the stars could be seen. . . . All the animals and the birds became frightened, and the wild beasts could be captured with ease. . . . And I saw the sun remain covered the length of time a man can walk 250 paces; and the air and the earth began to grow cold" (Book I, ch. 15).

Of considerable interest are observations that Ristoro made on the natural history of the regions surrounding Arezzo and Siena. While traveling through the marine Pliocene terrains of those regions, he recognized the organic nature of the fossil remains and the great variety of their shapes, and attributed their presence in the hills to the universal deluge. He made other observations on the calcareous encrusting power of certain hydrothermal springs and on emanations of natural gas. In a passage that is fairly representative of his thought, Ristoro points out that both natural and human environments must change with time, even though he was mistaken in considering the two kinds of phenomena on the same level and in taking as his example an unrealistically brief period: "If a man returned to his region after less than a thousand years, he would not recognize the places he had known, because he would find the mountains worn down and changed, and in like manner the valleys, streams, rivers, springs, cities, hamlets and villages, and even the language. Where he had left a city, he would find woods, and vice versa; where a mountain had stood, a valley would now be and vice versa; and he would find his region everywhere deeply changed, so that it would never seem to him that he had been there before" (Book VII, ch. 4).

Opinions on Ristoro's position in the history of the sciences are divided; some consider him to be a mere medieval compiler, his ideas dominated by Aristotelian and Scholastic conceptions of the world; others, emphasizing his keen powers of observation, see him as a direct precursor of the Renaissance. In fact Ristoro was in every way a child of his times, even if he was a scholar of great intelligence and wide knowledge. Although he insisted that his observations of nature were made personally and in specific places, he names none of the places in question; much of the work of his commentators has thus become a task of identification. Observations of nature rapidly acquired in his mind an abstract character and were not regarded as meaningless when detached from reality. The general tenet of recent historical criticism can well be applied to Ristoro: medieval artists and writers had a keen sense of observation, but their contact with the reality of the senses was limited to the specific and the episodic. Hence their depiction of particular details was realistic, but their conception of the whole was not, a generalization of which the *Composizione del mondo* provides a perfect example.

BIBLIOGRAPHY

I. ORIGINAL WORKS. There is unfortunately no critical ed. of the *Composizione del mondo*. The most reliable text is Biblioteca Riccardiana, Florence, Codex 2164, which, according to some scholars is the autograph of Ristoro himself or at least a very close copy. Only the first book of this text has been transcribed: *Il primo libro della Composizione del mondo di Ristoro d'Arezzo, a cura di G. Amalfi* (Naples, 1888). E. Narducci's ed. of the complete work, *Ristoro d'Arezzo, Della composizione del mondo, testo italiano del 1282* (Rome, 1859; Milan, 1864), is very poor and is based on a corrupt mid-fifteenth-century codex.

II. SECONDARY LITERATURE. On Ristoro and his work, see H. D. Austin, "Accredited Citations in Ristoro d'Arezzo's Composizione del mondo," in *Studi medievali*, **4** (1912–1913), 335–382; U. Losacco, "Pensiero scientifico ed osservazioni naturali di Ristoro d'Arezzo," in *Rivista geografica italiana*, **50** (1943), 31–61; A. Michel, *Die Sprache der Composizione del mondo* (Halle, 1905); F. Rodolico, "Commento ad alcuni passi di Ristoro d'Arezzo," in *Rivista di storia delle scienze mediche e naturali*, 8th ser., **38** (1947), 17–23; and A. Zancanella, *Scienza e magia ai tempi di Ristoro d'Arezzo e di Dante* (Perugia, 1935). For useful notes on Ristoro's work, see M. Baratta, *Leonardo da Vinci e i problemi della terra* (Turin, 1903).

FRANCESCO RODOLICO

RITCHEY, GEORGE WILLIS (*b.* Tupper's Plains, Ohio, 31 December 1864; *d.* Azusa, California, 4 November 1945), *astronomy.*

The son of an amateur astronomer and instrument maker, Ritchey acquired a practical knowledge of astronomy as a youth. After completing his formal education at the University of Cincinnati in 1887, he became an instructor at the Chicago Manual Training School, remaining there until 1896. Because his ability had impressed George Ellery Hale, he was asked to become chief optician at the newly founded Yerkes Observatory, in which capacity he adapted the 40-inch refractor for photography. From 1901 to 1906 he was on the astronomy faculty at the University of Chicago, and in 1904 he was elected an associate of the Royal Astronomical Society. In 1906, again at the request of Hale, Ritchey became the head of instrument construction at another new observatory, Mt. Wilson. There he almost single-handedly produced the 60-inch telescope and did most of the work on the 100-inch disk for the Hooper telescope.

During World War I, Ritchey trained over 100 people to make optical parts for gunsights for the U.S. Ordnance Department. At the end of the war, following a bitter controversy at Mt. Wilson, Ritchey was dismissed from the staff because of a health problem and for allegedly exceeding his authority. He spent the next five years at his private laboratory in California, continuing the work he had started at Mt. Wilson, particularly on the development of cellular mirrors.

In 1923 Ritchey was invited by the National Observatory in Paris to discuss some of his recent innovations; later he became director of the observatory's astrophotographic laboratory. While in Paris, Ritchey, with Henri Chrétien, perfected a design for an aplanatic reflector. Upon completion of the first such telescope in 1930, Ritchey was made a knight of the Legion of Honor. He returned to the United States in 1931 to become director of photographic and telescopic research at the U.S. Naval Observatory, where he began construction of a 40-inch Ritchey-Chrétien reflector. When it was completed in 1936, he retired to Azusa, California.

Ritchey not only fashioned some of the largest instruments of his era but also used them to make important astronomical contributions; moreover, his meticulous care in observing was almost unequaled. He demonstrated the similarity between the dark lanes in the Milky Way and the Andromeda nebula (M31), and he resolved the outer portions of M31 into individual stars. In 1917 Ritchey photographed novae (and supernovae) in M31, thereby facilitating the first reasonably accurate measurement of the distance to a spiral nebula.

BIBLIOGRAPHY

I. ORIGINAL WORKS. Ritchey's chief articles include "Celestial Photography With 40-Inch Visual Telescope of the Yerkes Observatory," in *Astrophysical Journal*, **12** (1900), 352–360; "On Some Methods and Results in Direct Photography With the 60-Inch Reflecting Telescope of the Mount Wilson Solar Observatory," *ibid.*, **32** (1910), 26–35; and "The Thomas Young Oration: The Modern Reflecting Telescope and the New Astronomical Photography," in *Transactions of the Optical Society*, **29** (1927–1928), 197–224. Ritchey's longest work, which was published late in his professional life and summarized much of his career, laid plans for an ambitious eight-meter telescope: *The Development of Astro-Photography and the Great Telescope of the Future* (Paris, 1929), published by the Société Astronomique de France.

II. SECONDARY LITERATURE. For biographical information on Ritchey, see J. McKeen Cattell and Jacques Cattell, *American Men of Science* (New York, 1938), 1189; and G. Edward Pendray, *Men, Mirrors and Stars* (New York, 1935), 127–135, 222–228, 235–237, 259, 267–268, also in rev. ed. (New York, 1946), 223–238, 266, 269, 305–306. There are brief obituaries by Dorrit Hoffleit, in *Sky and Telescope*, **5** (1946), 11; and by F. J. Hargreaves, in *Monthly Notices of the Royal Astronomical Society*, **107** (1907), 36–38. Private, polemical correspondence both with and about Ritchey is in "The George Ellery Hale Papers, 1882–1937," microfilm ed. prepared at California Institute of Technology (Pasadena, 1968).

RICHARD BERENDZEN

RITT, JOSEPH FELS (*b.* New York, N.Y., 23 August 1893; *d.* New York, 5 January 1951), *mathematics.*

After two years of study at the College of the City of New York, Ritt obtained the B.A. from George Washington University in 1913. He received the Ph.D. from Columbia University in 1917 for a work on linear homogeneous differential operators with constant coefficients. He was colloquium lecturer of the American Mathematical Society (1932), a member of the National Academy of Sciences, and vice-president of the American Mathematical Society from 1938 to 1940.

Ritt's early work was highly classical. Papers entitled "On Algebraic Functions Which Can Be Expressed in Terms of Radicals," "Permutable Rational Functions," and "Periodic Functions With a Multiplication Theorem" were the result of a thorough study of classic masters. His work on elementary functions took its inspiration directly from Liouville.

Ritt also investigated the algebraic aspects of the theory of differential equations, considering differential polynomials or forms in the unknown functions $y_1, \ldots, y_n$ and their derivatives with coefficients that are functions meromorphic in some domain. Given a system Σ of such forms, he shows that there exists a finite subsystem of Σ having the same set of solutions as Σ. Furthermore, if a form G vanishes for every solution of the system of forms $H_1, \ldots, H_r$, then some power of G is a linear combination of the H_i and their derivatives. These arguments lead to the statement that every infinite system of forms has a finite basis.

In considering reducibility Ritt concluded that the perfect differential ideal generated by a system of forms equals the intersection of the prime ideals associated with its irreducible components. The purpose of this work was to advance knowledge of "general" and "singular" solutions, which in the preceding literature (Laplace, Lagrange, and Poisson, for example) had been very unsatisfactory.

Contributions to algebraic differential equations and algebraic difference equations were made by many of Ritt's students after 1932, in particular, by E. R. Kolchin, W. Strodt, H. W. Raudenbush, and H. Levi. Differential algebra, a new branch of modern algebra, also owes much to these early researches.

BIBLIOGRAPHY

The most important aspects of Ritt's work are summed up in two books published in the Colloquium Publications series of the American Mathematical Society: *Differential Equations From the Algebraic Standpoint* (New York, 1932) and *Differential Algebra* (New York, 1950).

Complete bibliographies of Ritt's writings are included in the notices by E. R. Lorch, in *Bulletin of the American Mathematical Society*, **57** (1951), 307–318; and by Paul A. Smith, in *Biographical Memoirs. National Academy of Sciences*, **29** (1956), 253–264.

EDGAR R. LORCH

RITTENHOUSE, DAVID (*b.* Paper Mill Run, near Germantown, Pennsylvania, 8 April 1732; *d.* Philadelphia, Pennsylvania, 26 June 1796), *technology, astronomy, natural philosophy.*

Rittenhouse, the son of Matthias and Elizabeth Williams Rittenhouse, was raised on his father's farm in Norriton, about twenty miles north of Philadelphia. His paternal ancestry was German Mennonite and maternal, Welsh Quaker, but no strong denominational loyalty was encouraged in his home. In his mature years, Rittenhouse main-

tained a Presbyterian Church membership. His education was largely informal, and he was regarded as self-taught. On 20 February 1766 he married Eleanor Coulston, by whom he had two daughters; following her death, he married Hannah Jacobs toward the end of 1772. Rittenhouse's health was seldom good but seldom seriously impaired, the primary difficulty probably being a duodenal ulcer. Except for the wartime occupation, surveying expeditions, and summers in Norriton, he spent most of his life after 1770 in Philadelphia.

As one of the leading Philadelphia mechanics, Rittenhouse played an important role in the American Revolution. He participated in formulating the radical Pennsylvania Constitution of 1776 and served on the Board of War and as vice-president of the Council of Safety—with occasional responsibility for executive leadership of the state. From 1777 through 1789 he was treasurer of Pennsylvania.

Basically, Rittenhouse was a maker of clocks and mathematical instruments. His long-case clocks were novel not in their mechanism but in their fine workmanship. Three included small orreries, one had only a single hand, and his astronomical clock used a compensation pendulum of his own design. His masterpieces in clockwork were two large orreries in which he achieved beauty and a high degree of precision. Rittenhouse's instruments, many of which have been preserved, were superior to those previously produced in America. He made surveyors' compasses, levels, transits, telescopes, and zenith sectors as well as thermometers, barometers, at least one hygrometer, and occasional eyeglasses. He made early use of spider webs for cross hairs in telescopes, and he erected a collimating telescope in his observatory. Rittenhouse's fine surveying instruments were used for laying out national boundaries years after his death. Because he constructed vernier compasses, they became known in America as Rittenhouse compasses; and his improvement on Franklin's Pennsylvania fireplace was called the Rittenhouse stove.

Most of Rittenhouse's science and much of his nonpolitical service were closely related to his making of clocks and instruments. Starting as a manufacturer of surveying instruments, he became the most celebrated American surveyor, serving on commissions that marked portions of Pennsylvania's boundaries with Maryland, New Jersey, New York, and what became the Northwest Territory, and portions of New York's boundaries with New Jersey and Massachusetts. Beginning in 1773, Rittenhouse supplied the astronomical calculations for almanacs in Pennsylvania, Maryland, and Virginia. During the

Revolution he helped to design the Delaware River defenses and worked on the production of saltpeter and guns—including experimentation with telescopic sights for rifles and rifled cannon. He provided informed advice when Thomas Jefferson was working out his report on weights and measures; and as first director of the U.S. Mint (1792–1795), he produced and put into operation machinery that was new to him but for which his clockmaking career had prepared him.

Astronomy was Rittenhouse's primary scientific study, a pursuit to which he moved easily from his orreries, telescopes, and surveying. He began to study mathematics and science at an early age, first attaining recognition in the observation of the transit of Venus of 1769, which was important because of worldwide efforts to establish the sun's parallax. On this occasion Rittenhouse emerged as the leading figure in the American Philosophical Society's observations and in its initial volume of *Transactions*. He made many of the instruments and assembled all of those used by the Norriton observation group; he carried through key related observations and projections and contributed to the best American calculation of the parallax. Rittenhouse established a Philadelphia observatory, where he kept daily records and conducted regular observations, publishing data and calculations on meteors, comets, Jupiter's satellites, Mercury, Uranus (following its discovery), and various eclipses. In calculating planetary orbits and positions, he worked out some solutions of his own. Rittenhouse's mathematical work was largely related to the study of astronomy, his best paper being an original solution for finding the place of a planet in its orbit. He also devised an arithmetical method for calculating logarithms and published a paper on the sums of the several powers of the sines—an offshoot of a study of the period of a pendulum.

Rittenhouse's other work in science was experimental, except for descriptive accounts of the effects of lightning and a few papers on meteorology, geology, and aspects of natural history that he published in magazines. He experimented with various compensation pendulums, with the expansion of steel, and with the expansion of wood—for which he obtained good values. After familiar experiments with magnetism, Rittenhouse produced a very clear statement of the concept of magnetic dipoles. In an investigation of diffraction, he constructed plane transmission gratings, using fine wire across a frame. With one of these gratings he observed six orders of spectra, obtained good values for their angular displacement and clearly stated the law governing their displacement. Rittenhouse also studied and correctly reported the primary causes of the illusion of reversible

relief. This work was picked up and passed into subsequent literature after his death. Even this paper, however, was characteristic of most of his work, in that it represented an isolated study, not part of a continuing dialogue or investigation.

Rittenhouse succeeded Franklin as president of the American Philosophical Society, which he strengthened in its role as a platform of science; nearly all of his scientific papers were published in its *Transactions*. His craftsmanship supported science by providing precise scientific instruments. More directly, he contributed data and new information as well as several fruitful ideas in fields now widely separated. Although he wrote occasional speculative and mathematical pieces and others on the history of science, his scientific contributions were almost wholly observational and experimental.

BIBLIOGRAPHY

I. ORIGINAL WORKS. All of Rittenhouse's papers, unless otherwise noted, are in *Transactions of the American Philosophical Society*. His papers on the transit of Venus are "Projection of the Ensuing Transit of Venus," in *Transactions*, **1** (1771), 4; "Observations at Norriton," *ibid.*, 13–22; "Account of the Contacts," *ibid.*, 26–28; "Delineation of the Transit of Venus," *ibid.*, 36–38; and [Method of Finding Parallax], *ibid.*, appendix, 59–60.

Other astronomical papers are "Observations on the Comet," in *Transactions*, **1**, 37–45; "An Easy Method of Deducing the True Time of the Sun's Passing the Meridian," *ibid.*, 47–49; "Account of a Meteor," *ibid.*, **2** (1786), 173–176; "New Method of Placing a Meridian Mark," *ibid.*, 181–183; "Observations of a Comet," *ibid.*, 195; "Astronomical Observations," *ibid.*, 260–263; "Astronomical Observations," *ibid.*, **3** (1793), 153–155; and "An Account of a Comet," *ibid.*, 261.

His mathematical papers are "A Method of Finding the Sum of the Several Powers of the Sines," in *Transactions*, **3**, 155–156; "To Determine the True Place of a Planet," *ibid.*, **4** (1799), 21–26; and "Method of Raising the Common Logarithm," *ibid.*, 69–71.

His experimental papers are "Explanation of an Optical Deception," in *Transactions*, **2**, 37–42; "An Account of Some Experiments in Magnetism," *ibid.*, 178–181; "An Optical Problem," *ibid.*, 202–206; "On the Improvement of Timekeepers," *ibid.*, **4**, 26–28; and "On the Expansion of Wood by Heat," *ibid.*, 28–31. The orrery paper is "A Description of a New Orrery," in *Transactions*, **1** (1771), 1–3.

On lightning and electricity, he wrote "Account of Several Houses in Philadelphia Struck With Lightning," in *Transactions*, **3**, 119–122, with John Jones; "An Account of the Effects of a Stroke of Lightning," *ibid.*, 122–125, with Francis Hopkinson; and "Experiments on the *Gymnotus electricus*," in *Philadelphia Medical and Physical Journal*, **1**, pt. 2 (1805), 96–100, 159–161.

His papers on natural history are in *Columbian Magazine*, **1** (1786–1787), 49–53, 284, 301–303. His essay on the history of astronomy is *An Oration* (Philadelphia, 1775). Locations of MSS and of clocks and instruments are in Hindle (below).

II. SECONDARY LITERATURE. The most recent biography is Brooke Hindle, *David Rittenhouse* (Princeton, 1964). William Barton, *Memoirs of the Life of David Rittenhouse* (Philadelphia, 1813), is of continuing value for its insights and for excerpts of letters now lost. The most important evaluations of aspects of Rittenhouse's science are Thomas D. Cope, "The Rittenhouse Diffraction Grating," in *Journal of the Franklin Institute*, **214** (1932), 99–104; and W. Carl Rufus, "David Rittenhouse as a Mathematical Disciple of Newton," in *Scripta mathematica*, **8** (1941), 228–231.

BROOKE HINDLE

RITTER, JOHANN WILHELM (*b.* Samitz, near Haynow [now Chojnów, Poland], Silesia, 16 December 1776; *d.* Munich, Germany, 23 January 1810), *chemistry, physics, physiology.*

Ritter was the son of Johann Wilhelm Ritter, a Protestant pastor, and Juliana Friderica, née Decovius. After attending Latin school until the age of fourteen, he was sent to Liegnitz (now Legnica) by his father to serve as an apothecary's apprentice. During the next five years, while learning and practicing the trade, he found ample opportunity to pursue his own studies from chemical texts and independent experiments, often to the neglect of his professional assignments. His yearning for more intellectual stimulation and independence was satisfied in 1795 by a modest inheritance. Thus, at the age of nineteen, Ritter entered the University of Jena in April 1796. Encouraged by Alexander von Humboldt, he soon began independent galvanism studies.

During the period 1797–1804, Ritter engaged in empirical research and writing, mainly on electrochemistry and electrophysiology. In these years he attained recognition and respect throughout Europe. In 1801 the duke of Gotha and Altenburg, Ernst II, appointed him to his court in Gotha, where Ritter lived in early 1802; and in that year he carried out electrical experiments on a grand scale. During the winter semester of 1803–1804, he lectured on galvanism at the University of Jena. A disagreement, however, with university officials about the nature of his future appointment disrupted further association with that institution. Subsequently Ritter sought a new position, and in November 1804 he was called to serve as a full member of the Bavarian Academy of Sciences in Munich. He moved to the new location the following spring with his young wife and daughter.

In June 1804 he had married Johanna Dorothea Munchgesang, with whom he had previously lived for some time; they had three more children in Munich.

Severe financial stress, heavy domestic responsibilities, and family illnesses, as well as disagreements with professional contemporaries, characterized the period in Bavaria from 1805 to 1810. Nevertheless Ritter continued to conduct a full program of research, although an increasing amount of his writing had philosophical overtones. The influence and reliability of his later empirical works were marred by a growing suspicion within the larger scientific community that Ritter's conclusions and even facts were affected adversely by the *Naturphilosophie* then in vogue in several German intellectual centers, including Munich. Embittered, deeply in debt, and drawing solace only from his family and a few friends, Ritter died shortly after his thirty-third birthday from a pulmonary disease incurred in his impetuous, unregulated life style.

During his thirteen-year career of active research, Ritter wrote five scientific works (some multivolume) and two philosophical treatments of his views on science. These represent a total of thirteen separate volumes. In addition, he published some twenty journal articles not collected in the above volumes. Despite this prodigious outpouring, little knowledge of Ritter was to be found, particularly outside of Germany, even in the next generation. Many of his unique discoveries in electrophysiology and electrochemistry were of little influence; more often than not, they were independently rediscovered later by other men. To understand the lack of Ritter's lasting influence, one must recognize something of his particular philosophical view of nature and its origin. Such a background also provides a basis for appreciating the wide and apparently disparate range of Ritter's research interests.

The intellectual climate at Jena provided young Ritter with a receptive sounding board for his discoveries and speculations. In turn the philosophical views of Romanticism and *Naturphilosophie* prevalent in Jena reinforced Ritter in many of his own ideas about nature. One of the chief notions, widely held at the time, notably by Schelling, Goethe, Herber, and other Romantic writers whom he knew, was that of a basic unity in nature. The universe as a whole was believed to possess a world-soul and all natural processes within it were thought to be interrelated and interdependent. A second general principle was that of polarity. Examples of natural contrasting polarities abounded—frictional electricity ($+$ and $-$), magnets (N and S), decomposition of water (H_2 and O_2), and the composition of air (O_2 and N_2).

Contrary to the deductive-speculative system of *Naturphilosophie* propounded by Schelling, with whom Ritter broke off association in 1799 because of a priority quarrel, Ritter was a firm advocate of the empirical approach in studying nature. Nonetheless, when he speculated about the interpretation of observed facts, as he often did in his papers, Ritter customarily referred to a conceptual framework of unity and polarity in nature.

In his first paper, read to a Jena natural history society in 1797 (although not published until 1806), Ritter argued for unity between organic and inorganic realms. His basis was galvanic studies of muscle contraction when a cell of two dissimilar metals is used to excite a muscle. In further researches he pursued both the chemical processes and the physiological responses aroused by the galvanic cell, but he always held to a clear recognition that a single principle was involved. Following the chemical approach, he used the cell in 1799 to produce electrolysis of water and thus demonstrated the identity between frictional electricity and the galvanic fluid. The production of electricity simultaneously with chemical phenomena, in itself showing interrelatedness, led Ritter to experiment with polarization in the cell. He built the first dry cell in 1802 and a secondary charging battery, or accumulator, in 1803. In various publications from 1800 to 1806 Ritter announced detailed investigations on the electrical potential series corresponding to affinity tables, the electrical conductibilities of metals, the generation of thermal currents, and the dependency of resistance on dimensions.

Besides these inorganic investigations, Ritter continued to seek results from the electrical excitation of muscle and sensory organs. In his attempt to reconcile the opposing voltaic and galvanic views of galvanism, Ritter used his concept of the "galvanic chain," or cells within the organic body, to explain such discoveries as "creeping in" at low currents to desensitize muscles, the diminished sensitivity of nerve responses when various chemicals are used, and the electrophysiology of plants (electronastism). Much of the success of Ritter in his studies of the excitability of sensory perception by electrical impulses was due to his use of his own body, even at very high voltages.

Ritter was motivated to extend his investigations to other areas of nature largely because of his philosophical views. These efforts met with mixed success. In one case the electrical polarity of the voltaic cell suggested to him, in analogy with a bar magnet, that the cell might be magnetized. He announced in 1801 (without subsequent confirmation by his peers) that a two-part needle of zinc and silver always oriented itself like a compass needle in space. A second case arose from Herschel's discovery in 1800 that invisible thermal rays are detectable beyond the red end of a prismatic spectrum. Ritter immediately hypothesized a possible polarity in the spectrum and looked for invisible radiation beyond the violet end. Using paper soaked with silver chloride in the spectral region, as Scheele had done earlier, he succeeded in 1801 in finding the greatest amount of blackening (reduction to silver) just beyond the visible violet. He called this effect a polarity between "deoxydizing rays" near the violet and "oxydizing rays" near the red, thus drawing an analogy to electrolysis produced by the voltaic cell.

Under the influence of the Munich mystic F. von Baader, and also along the lines of his own predispositions, Ritter spent much time from 1806 to 1809 investigating occult practices like water divining, metal witching, and sword swinging. Imagining that there exists a subterranean electricity analogous to geomagnetism, Ritter believed that he had successfully uncovered a general principle governing the interdependencies of inorganic nature and human phenomena. He gave the general name of siderism to this subject and before his death edited the first and only issue of a newly founded periodical by that title. Except for a small circle of colleagues, scientists in general were skeptical of the authenticity of the research. A posthumous note in the *Annales de chemie* in 1810 asserted that Ritter himself repudiated his belief in siderism just before his death.

Ritter's philosophical views were communicated to only a few scientists who were intimate friends, including Oersted and G. H. Schubert. The reception of his scientific work was damaged by his abtruse style, his tendency to mix philosophical implications with scientific observations, and a frequent delay of several years between a cryptic first announcement in some journal and the detailed account in a privately published collection.

BIBLIOGRAPHY

I. ORIGINAL WORKS. Ritter's scientific style and philosophical views first appeared in *Beweis, dass ein beständiger Galvanismus den Lebensprocess in dem thierreich begleite* (Weimar, 1798). He also edited a series of collected articles, mostly written by himself, on the subject of galvanism: *Beiträge zur nähern Kenntniss des Galvanismus und der Resultate seiner Untersuchung*, 2 vols. (Jena, 1800–1805); Further electrical experiments primarily on inorganic bodies are described in great detail in *Das Elektrische System der Korper, ein Versuch* (Leipzig, 1805).

Ritter's notes, articles, and letters, along with various unpublished papers and letters, were published in *Physisch-chemische Abhandlungen in chronologischer Ordung*, 3 vols. (Leipzig, 1806). Of particular historical interest are Ritter's hitherto unpublished first paper on galvanism addressed to the Naturforschende Gesellschaft zu Jena in Oct. 1797 (vol. I); his paper on his thought process in the discovery of untraviolet radiation delivered to the same society in 1801 (vol. II); and a dictation of the first part of his 1803–1804 lectures on galvanism given at Jena (vol. III). In all, 36 separate papers and letters are printed. His address at the founding ceremony of the Royal Bavarian Academy of Sciences was published as *Die Physik als Kunst, ein Versuch die Tendenz der Physik aus ihrer Geschichte zu deuten* (Munich, 1806; repr. Berlin, 1940).

Ritter founded the journal *Der Siderismus, Neue Beiträge zur nähern Kenntnis des Galvanismus*; only one volume appeared (1808). He also collected a series of aphorisms concerning his views on life and science in *Fragmente aus dem Nachlass eines jungen Physikers; Ein Taschenbuch für Freunde der Natur*, 2 vols. (Heidelberg, 1810).

The Royal Society *Catalogue of Scientific Papers*, V, 217–219, lists some twenty other journal articles.

II. SECONDARY LITERATURE. Although Ritter has not received a full-length biographical treatment, various aspects of his scientific influence and work have been repeatedly analyzed since the mid-nineteenth century. In an influential work R. Haym, *Die Romantische Schule* (Berlin, 1870), 612–619, discredited the scientific merit of Ritter's research, even though E. Du Bois-Reymond, *Untersuchungen über die tierische Elektrizität*, II (Berlin, 1849), 320 ff., had assessed clearly the empirical value of Ritter's work in electrophysiology. More recently, D. Huffmeier, "Johann Wilhelm Ritter. Naturforscher oder Naturphilosoph?" in *Sudhoffs Archiv für Geschichte der Medizin und der Naturwissenschaften*, **45** (1961), 225–234 has reinforced the latter view.

The scientific authenticity of Ritter's research on electrochemistry has been established by W. Ostwald, *Abhandlungen und Vorträge*, (Leipzig, 1904), 359–383; and H. Schimank, "Johann Wilhelm Ritter; Der Begründer der wissenschaftlichen Elecktrochemie," in *Abhandlungen und Berichte des Deutschen Museums*, **5** (1933), 175–203. C. von Klinckowstroem explored in detail the nature of Ritter's association with the Jena circle: "Goethe und Ritter; mit Ritters Briefe an Goethe," in *Jahrbuch der Goethe Gesellschaft*, **8** (1921), 135–151; "Drei Briefe von Johann W. Ritter," in *Der grundgescheute Antiquarius*, **1** (1921), 120–130; "Johann Wilhelm Ritter und der Electromagnetismus," in *Archiv für die Geschichte der Naturwissenschaften und der Technik*, **9** (1922), 68–85.

Three recent collections of writings by Ritter, with extensive historical commentary, provide the best historical material available: F. Klemm and A. Hermann, eds., *Briefe Eines Romantischen Physikers; Johann Wilhelm Ritter an Gotthilf Heinrich Schubert und an Karl von Hardenberg* (Munich, 1966), contains not only an informative set of 8 hitherto unpublished letters with commentary but also an excellent bibliography of secondary works about Ritter; K. Poppe, ed., *Johann Wilhelm Ritter:*

Fragmente aus dem Nachlass eines jungen Physikers (Stuttgart, 1968); and A. Hermann, ed., "Begründung der Elektrochemie und Entdeckung der ultravioletten Strahlen von Johann Wilhelm Ritter; eine Auswahl aus den Schriften des romantischen Physikers," in *Ostwalds Klassiker der exakten Wissenschaften*, n.s. **2** (1968), provides an excellent introduction, commentary, and bibliography of primary source material.

The archives of the Bavarian Academy of Sciences, Munich, contain much of Ritter's personal notes and library. The Deutsches Museum, Munich, has a collection of experimental apparatus used by Ritter.

ROBERT J. MCRAE

RITZ, WALTER (*b.* Sion, Switzerland, 22 February 1878; *d.* Göttingen, Germany, 7 July 1909), *theoretical physics.*

Ritz, the second of five children of Raphaël Ritz, a well-known landscape painter, was inclined from his youth toward science and mathematics. Following graduation from the *collège cantonal* in Sion (1895), he attended the *cours technique* at the *lycée cantonal* there for two years, studying calculus in his spare time. In the fall of 1897 he passed the entrance examinations to the Eidgenössische Technische Hochschule and the family moved to Zurich. Just prior to this, in September 1897, Ritz suffered a trauma that he regarded as the origin of the ill health from which, soon after, he began to suffer.[1] Climbing Mont Pleureur with friends, he looked back to see a group of them slip on fresh snow and plunge over a cliff; the emotional stress was compounded by physical overexertion and overexposure in the rescue efforts.

At Zurich, Ritz soon abandoned his original intention of becoming an engineer, ostensibly because of poor health, and joined the score of students— among them Albert Einstein—in the heavily mathematical "pure" section (Abteilung VI).[2] Here he studied especially Riemann's works and the two thick volumes of Woldemar Voigt's recently published *Compendium der theoretischen Physik* (1895–1896). At Easter 1901 following a severe case of pleurisy that the humid Zurich climate was presumed to aggravate, Ritz transferred to Göttingen. There he studied principally with Voigt (also with Max Abraham, T. Des Coudres, David Hilbert, Walter Kaufmann, Felix Klein, Georg E. Müller, and E. Riecke).

It was consistent with the interests and methods of the professors of physics, Voigt and Riecke, that Ritz chose as his dissertation topic a theory of spectral series. Moreover, he approached the problem not in terms of the avant garde conception of electrons but, rather, by conceiving atoms to be elastic continua.

Ritz postulated that line spectra—for which the doubly infinite generalized Balmer formula for the frequencies of the spectral lines emitted by hydrogen, $\nu = N(1/n^2 - 1/m^2)$, m, $n = 1, 2, 3,...$, was taken as the paradigm—originate in the proper vibrations of two-dimensional structures (indeed, in the transverse normal modes of a plane square plate).[3]

Since it was well known from the researches of Poincaré, Rayleigh, and others that the wave equation resulting from the action-by-contact of the theory of elasticity cannot give a Balmer-like distribution of frequencies, Ritz's object was to invent an interaction that would. "Leaning as fully as possible on mechanics and electrodynamics, one puts forward physically visualizable mathematical operations of which the interpretation as the vibrations of an appropriate 'model' leads to the laws of series spectra."[4] Ritz obtained the desired result, $\nu = N(1/n^2 - 1/m^2)$, from a tenth-order partial differential equation.[5] In constructing that equation he was obliged to put forward, for the energy of deformation of his elastic plate, mathematical expressions of which the physical interpretation as action-at-a-distance forces increasing with the separation between the points of the plate certainly exceeded the limits of physical plausibility, and perhaps the limits of visualizability as well.

Among spectroscopists there was evidently general agreement that Ritz's "theoretical foundation is not worth much."[6] If, nevertheless, they studied his dissertation "with great interest" in the spring of 1903, that was because of the novel series laws that Ritz was able to extract from his mathematical operations, largely by means of his thorough familiarity with the empirical data and his talent for generalizing empirical regularities. These operations themselves were of interest to mathematicians and stimulated research by Ivar Fredholm and Jacques Hadamard into the theory of integral equations with an infinite number of eigenvalues (frequencies) within a finite interval.[7]

A solution of his tenth-order equation with a Balmer distribution of frequencies had of course required appropriate boundary conditions; altering these gave the Balmer formula but with m and n replaced by infinite series $m + a + b/m^2 + c/m^4...$, $n + a' + b'/n^2 + c'/n^4 + \cdots$. The first approximation, $\nu = N/(n + a')^2 - N/(m + a)^2$ was Rydberg's formula, which was known to represent at least roughly the series in other elements. Ritz maintained that in the atoms of all elements the spectral vibrator is a face of one and the same cube, with appropriate differences in the constraints at its edges; that N, being expressed in his theory in terms of the size, mass, and "elastic" constants of the face itself, was

a rigorously universal constant; and that the Rydberg formula was thus to be improved not by adjusting N but by adding the next term in the series, b/m^2. Ritz showed that such a formula represented the spectral series, particularly the lines in the red and infrared corresponding to low values of m and n, with unprecedented accuracy. Moreover, the structure of his formula allowed him to adopt all of Rydberg's proposed interrelations between the several series of a given element. [Anticipating the notation Ritz himself introduced five years later:[8]

principal series: $\nu = (1.5, s, \sigma) - (m, p, \pi)$
sharp series: $\nu = (2, p, \pi) - (m, s, \sigma)$
diffuse series: $\nu = (2, p, \pi) - (m, d, \delta)$,

where $(m, a, a) \equiv N/(m + a + Na/m^2 + \cdots)^2$.] He achieved this impressive agreement with experiment and "put into the shade" all previous formulas,[9] even though he used substantially fewer adjustable parameters. And in fact, as emerged twelve years later from Sommerfeld's work, Balmer's, Rydberg's, and Ritz's formulas are successive approximations to the energy of a Rutherford-Bohr atom.[10]

Ritz's dissertation was accepted by Voigt late in 1902, and the oral doctoral examination was passed *summa cum laude* on 19 December. By the beginning of March 1903 the dissertation was ready for submission to the *Annalen der Physik*; and at the end of March, Ritz set off for Leiden (via Hannover, in order to discuss some spectroscopic questions with Carl Runge).[11] At Leiden, Ritz, with Paul Ehrenfest, attended lectures and seminars of H. A. Lorentz, whose electron theory was then coming to occupy the central position in theoretical physics. The six-week visit was made jointly by these two young theorists, who had formed a close friendship during the preceding three semesters at Göttingen. Ehrenfest, Ritz's junior by almost two years, enormously admired the cool, restrained personality; the quick, sarcastic wit; the fertile imagination; the rigorous mathematical talent; and, above all, the penetrating criticism of the physical theories of others. Although their scientific predilections soon diverged widely, Ritz exerted a great influence upon Ehrenfest's intellectual development.[12]

Ritz left Leiden with, if anything, even less of an inclination toward Lorentz's views. By late May or early June 1903 he was at work in Bonn, where Heinrich Kayser's institute offered the best spectroscopic facilities in Germany. Within a few weeks Ritz had found the missing $m = 4$ diffuse series line of potassium precisely where, in his dissertation, he had predicted it.[13] But after this initial success the work progressed slowly—August Hagenbach, then

Kayser's assistant, thought that "for experimental work he [Ritz] had neither the physical strength nor the necessary patience."[14] During the summer holidays Ritz was in Zurich, where Ehrenfest visited him for a week in September.[15] Shortly after returning to Bonn in November 1903, unable to obtain a piece of apparatus he wanted, Ritz abandoned his work and went to Paris.

With a recommendation from Pierre Weiss, Ritz was received into Aimé Cotton's and Henri Abraham's laboratory at the École Normale Supérieure, where he labored through that winter and into the following summer on a process for preparing infrared-sensitive photographic plates. Apparently it was primarily with an eye to an academic career that Ritz undertook experimental work at all—it was necessary in order to qualify for other than the exceedingly few positions available to pure theorists. His choice of this particular problem resulted from the crucial importance for his series formulas (and, four years later, for his "combination principle") of spectral lines in the red and infrared, a region that, because of inadequate means of detection, had been neglected by spectroscopists. Ritz had begun to obtain very promising results when, in July 1904, his health failed and he was compelled to retire to Zurich.

For the next three years Ritz resided outside the scientific centers, seeking the restoration of his health in various reputedly salubrious climes: St. Blasien in the Black Forest, Waldkirch in the canton of St. Gall, Sion, and Nice. During this period he published nothing, and during the first two years accepted the medical opinion that he should work as little as possible. In 1906 Ritz began to work intensively again, and that winter, despairing of recovering his health, he resolved to deprive himself of scientific companionship no longer.[16]

Late in September of 1907 Ritz settled in Tübingen, his mother's family's home but also a center of spectroscopic research.[17] Hoping that a few years remained to him, and anxious about his financial situation, in the spring of 1908 he moved to Göttingen, resolved to qualify as a lecturer. The *Habilitation* took place in February 1909, though Ritz was no longer strong enough to exercise the privilege of delivering a course of lectures or to join in any other way in the social-scientific life of the town. In mid-May he entered the Göttingen medical clinic, where, seven weeks later, he died.[18]

In Ritz's last year and a half (early 1908 through mid-1909), the fruits of six years of thinking and three years of working came forth in a spate: eighteen publications running to some 400 pages. These papers fall under three general topics: theoretical spectros-

copy, the foundations of electrodynamics, and a method for the numerical solution of boundary-value problems—with each group of investigations largely conceptually independent of the others.

In December 1904 the Paris Academy of Sciences had announced as the subject of the Prix Vaillant (4,000 francs), to be awarded three years hence, "to perfect in some important respect the problem in analysis relating to the equilibrium of an elastic plate in a rigid frame, that is, the problem of integrating the equation $\partial^4 u/\partial x^4 + 2\partial^4 u/\partial x^2 \partial y^2 + \partial^4 u/\partial y^4 = f(x, y)$ with the condition that the function u and its derivative normal to the boundary of the plate be zero. To examine more particularly the case of a rectangular boundary."[19] Ritz's dissertation had given him extensive experience closely related to this question. Moreover, in his financial dependence and professional isolation, this type of competition and reward seemed especially attractive. Returning to work in 1906, he threw himself into the problem of developing a rigorous yet practicable procedure for directly constructing the solution by successive approximations.

Ritz recast the problem from the integration of this partial differential equation, $\Delta\Delta u = f$, to the variation of an integral $J \equiv \int [\frac{1}{2}(\Delta w)^2 - f(x) \cdot w] \, dx$. (In order to avoid inessential complexities the one-dimensional case will be discussed here.) The object then was to find among all possible deformations $w(x)$ of the system that particular one, $u(x)$, for which this integral, expressing the potential energy as a function of the deformation and the distribution of applied forces $f(x)$, assumes its minimum value. Introducing a complete orthonormal set of functions $\psi_i(x)$—the choice of which was guided by the geometry of the case—the solution u was expressed formally as $\sum_{i=1}^{\infty} a_i\psi_i$, and the problem became: to construct a sequence of functions

$$w_1 \equiv a_1^{(1)}\psi_1, \ w_2 \equiv a_1^{(2)}\psi_1 + a_2^{(2)}\psi_2, \ldots$$
$$w_n \equiv a_1^{(n)}\psi_1 + a_2^{(n)}\psi_2 + \cdots + a_n^{(n)}\psi_n$$

that would converge to u. Ritz simply substituted these linear combinations of basis functions ψ_i with undetermined coefficients $a_i^{(n)}$ into J, the integrand of which thus became an explicit function of x. The integration could then be carried through, yielding $J^{(n)}$ as a function of the n parameters $a_i^{(n)}$. Their values were now ingeniously fixed as the roots of the n equations $\partial^{(n)}J/\partial a_i^{(n)} = 0$—that is, the condition that w_n be that linear combination of the first n basis functions for which J is a minimum. Drawing heavily upon Hilbert's recent work in the calculus of variations, Ritz proved rigorously that for the equation of elasticity, $\Delta\Delta u = f$ (and also for Dirichlet's problem, $\Delta u = f$), the w_n constructed in this way converge to the solution u.[20]

Ritz then proceeded to show that although this proof could be carried through only for the case of static equilibrium, the corresponding equation for standing waves, $\Delta u = k^2 u$, could be formally subjected to the same procedure, and that the sequence of approximate solutions thus obtained converged with extraordinary rapidity.[21] Two years later, as a striking demonstration of the power of his method, Ritz calculated the sequence of Chladni figures for a square plate up to the thirtieth harmonic. "That has fatigued me, but there is no relation at all between the fatigue and the celebrity of this problem which remained insoluble despite all efforts."[22]

Ritz's memoir, one of a dozen submitted for the prize, did not win even an honorable mention from the distinguished jury, for which H. Poincaré, E. Picard, and P. Painlevé were the *rapporteurs*.[23] In April 1909, however, Ritz had the satisfaction of being sought out at Göttingen by Poincaré, who apologized in the name of the Academy for the injustice and promised Ritz a prize that year as "reparation."[24] For indeed, following Ritz's publications in 1908 the significance of his work was immediately recognized— and his method was immediately adopted—especially by theoretical engineers, notably Hans Lorenz, Aurel Stodola, Theodore von Kármán, Arpad Nadai, and S. Timoshenko.[25]

The prize memoir was completed before the 31 December 1906 deadline; after repaying the overdraft upon his reserves of strength with some weeks of fever, early in 1907, while still at Nice, Ritz returned to the question of the mechanism responsible for the production of line spectra. His recent experience with elastic plates persuaded him that the approach he had taken since 1902—construing the spectral frequencies as those of a quasi-elastic body of which the potential energy of deformation has an anomalous form—"must, after trying for years in every direction, be discarded as a misguided idea."[26]

But Ritz did not abandon his methodologic predilections and thus "came *a priori* finally to the result that vortical processes must be involved here, and with it the idea of the magnetic field immediately arose," namely that the frequency of circulation of a charged particle under the influence of magnetic forces, although proportional to the square root of the force in accord with Rayleigh's rule, is directly proportional to the magnetic field giving rise to these forces.[27] Ritz set himself to inventing mechanisms for producing magnetic fields that followed the Balmer

formula. The first of these (March 1907) utilized a rather Victorian apparatus of particles suspended by inextensible wires. The second, definitive mechanism (June 1907) was of a remarkable simplicity. Observing the similarity between the Balmer formula and the field of a thin bar magnet at a point on its axis, $H \propto (1/r_1^2 - 1/r_2^2)$, Ritz proposed to construe $\nu = N(1/2^2 - 1/m^2)$ as the frequency of circular vibration (Larmor frequency) of a charged particle situated at a distance $2a$ beyond the end of a row of $m - 2$ identical linear magnets, of length a, arranged end to end. Shifting the position of the charged particle immediately gave the Rydberg formula, but unfortunately Ritz's own series formula did not follow so naturally.[28] Late in 1907, at Tübingen and in "resonance" with Paschen, Ritz applied his model to the Zeeman effect; although the predictions differed from those of the electron theory and from Runge's rule, Ritz maintained that his results gave the better agreement with the data.[29]

Ritz was ambivalent about the ontologic status of his model, now arguing that it was physically more plausible than the Lorentz electron theory, now seeing it as possessing "the advantage from the epistemological standpoint that it need make no assumptions about the form of the elements out of which the atoms are thought to be built up, but operates only with rigid lengths or separations."[30] The theory seems to have been viewed most favorably, and the model taken quite seriously, among spectroscopists.[31] As a mechanism for emission of radiation, it differed from the vibrating continuous-media models in one very important respect: the various lines of a series, rather than all being emitted simultaneously *qua* harmonics, were each emitted by a different system or a different state of the system. As Ritz duly emphasized,[32] recent experiments on resonance radiation seemed to support such a view— which proved to be at least half the truth.

A far more important contribution to theoretical spectroscopy was Ritz's "combination principle," which presented itself to him at Göttingen in April or May 1908. Undoubtedly a central topic in the discussions between Ritz and Paschen the preceding winter had been the relation of the series that Arno Bergmann had recently found (in the infrared spectra of the alkalies) to the three types of series previously recognized: principal, first subordinate, and second subordinate. Ritz concluded that Bergmann's series were analogous to the hypothetical

$$\nu = N(1/3^2 - 1/m^2)$$

series of hydrogen—which in a very limited sense they are—therefore analogous to the Balmer series, and therefore a type of first subordinate or diffuse series.[33] (Before Bohr, the Balmer series was generally regarded as the diffuse series of hydrogen.)

Carl Runge, on the other hand, had concluded that the limits of Bergmann's series stand in the same relation to the limits of the diffuse series as the limits of the second subordinate (sharp) series stand to the limits of the principal series—which indeed they do—and therefore surmised erroneously that the parallel was complete, that Bergmann's series are a type of principal series.[34] After vigorously attacking Runge's interpretation, Ritz came to see that the question of whether the Bergmann series were diffuse or principal had no clear meaning, that Runge had pointed out a valid relation—Ritz gave only niggardly acknowledgment and no apology—and that the important question was how the interconnections between the series formulas, which Rydberg had propounded, ought to be extended and generalized.[35]

Ritz's answer, the combination principle, although amply exemplified, was none too clearly stated. Essentially Ritz emphasized that in Rydberg's series formulas, and in his own improvement upon them, the frequency of a spectral line was given as the difference between two terms, a running term and a constant term or series limit, and that the latter was a particular value of the running term in some other series formula. Ritz generalized this relation along the lines of the generalized Balmer formula: that apart from certain restrictions on the minimum values of the integer in the running term of a given type (p terms, 2; d terms, 3; b terms, 4)—which restriction was an important generalization of a circumstance that had puzzled Ritz since 1902, and could now be explained as a necessary consequence of his "atomic magnets" theory—the subtractive (or additive) "combination" of any two terms from any two series, or even from one and the same series, gives the frequency of an actually existing spectral line. Subsequently, in a semipopular article, Ritz gave two brief statements of his law: "that by the addition or subtraction of two observed lines or series one obtains the frequency of a new line or series of lines" and "that each of the two terms of the formula has, in some manner, an independent existence, and that one obtains the lines of a spectrum by combining such terms with each other in various ways."[36]

Although Ritz had thus already gone too far, he went still further, reifying not merely the terms themselves but also the symbols in his formulas for the terms. In this erroneous way Ritz "derived," for example, the Bergmann terms from the doublet p terms, $(m, b, \beta) = (m, p_1 - p_2, \pi_1 - \pi_2)$, and boasted "that almost the whole of the new lines and series

recently discovered in the alkalies by Lenard, Konen and Hagenbach, Saunders, Moll, Bergmann, and others can be precisely derived from the known series formulas for these elements—without introducing any new constant whatsoever."[37] Moreover, on the basis of his principle Ritz predicted a great many lines in the infrared spectra of other elements. In his first publication he was able to announce the confirmation of a considerable number of these lines by Paschen, whose experiments in the spring of 1908—including the search for the "Paschen series" of hydrogen, $\nu = N(1/3^2 - 1/m^2)$—Ritz had been prompting by mail.[38]

Ritz's work on theoretical spectroscopy and boundary-value problems was fully appreciated by his contemporaries and was incorporated in the cognitive edifices they were constructing. It was otherwise with his work on the foundations of electrodynamics, to which he gave his most continued effort, about which he expressed himself with the greatest vigor, and for which he vainly wished a few more years of life. While Ritz was coming of age scientifically, electrodynamics as reformed by Lorentz was "being made the pivot of a new conception of nature, replacing the older mechanical conception." But instead of enthusiastically espousing this new conception as did so many others at Göttingen, Ritz felt a certain antipathy toward it and directed his critical attention to "the logical fundaments of this vast intellectual edifice."[39] To Ritz's critical eye the structure seemed ill-founded and poorly suited to experience. Maxwell's equations admitted far too large a class of solutions; and even after this class had been arbitrarily restricted by additional equations (such as the "Lorentz condition"), there still remained unphysical advanced potentials and convergent spherical waves, as well as the not directly observable electric and magnetic fields.

Ritz, like Einstein, was particularly concerned with the conflict between the classical principle of the relativity of motion and the Maxwell-Lorentz theory, with its immobile ether as a privileged reference system (which experiment seemed unable to put into evidence). But Ritz concluded that the incompatibility ought to be resolved in the opposite way from that which Lorentz, Poincaré, and Einstein were then choosing and which the community of physicists would soon adopt.[40] Instead of altering kinematics and dynamics through the Lorentz transformations in order to accord with Maxwell's field equations, Ritz insisted—invoking his critique of Maxwell's theory—that it was electrodynamics and optics that required modification in order to satisfy the classical principle of relativity: "The only conclusion which . . . appears possible to me is that the ether does not exist, or more exactly, that it is necessary to renounce the use of that image; that the motion of light is a relative motion like all the others; that only relative velocities play a role in the laws of nature; finally, that it is necessary to renounce partial differential equations and the notion of a field. . . ."[41]

In Ritz's positivist program not merely the ether but also the electromagnetic potentials, the fields derived from them, and the equations governing them were to be replaced by "elementary actions" between spatially separated charged particles. The expression for these elementary actions was to involve not the prior times $t = r/c$, as in the Lorentz-Einstein theory, but the times $t = r/(c + v_r)$, where r is the radial distance between the charges and v_r is their relative radial velocity at the prior time. This, Ritz stressed, was thus an "emission" theory; "luminous energy is to be regarded as *projected*, and not as *propagated*." "Fictitious" particles of light are emitted with velocity c from radiating charges; "the waves of the ether will be replaced by a distribution of the emanation, periodic in time and in space."[42]

In principle there were an infinite number of different forms for the elementary action satisfying this condition and compatible with existing experiments; Ritz the positivist regarded this as the most important advantage of his approach: simply by adding further terms to the interaction, further phenomena—particularly gravity—might be accounted for.[43] In practice, Ritz was never able to find a fully satisfactory form even for the electrodynamic interactions alone.

Returning to Göttingen in the spring of 1908, Ritz found insuperable resistance to his electrodynamic ideas; they were simply "monstrous."[44] Although there was no experimental fact that told squarely against Ritz until 1924 (when the Michelson experiment was performed with astronomical light sources),[45] the point of view that he brought forward never received the critical attention or sympathetic extension it deserved.

NOTES

1. It is clear that Ritz died of tuberculosis (specifically, lung hemorrhage)—Mrs. L. Ritz to P. Ehrenfest, 20 July 1909—but it is unclear just when the disease was diagnosed.
2. C. Seelig, *Albert Einstein. Eine dokumentarische Biographie* (Zurich, 1954), 29, 159; *Eidgenössische technische Hochschule 1855–1955* (Zurich, 1955). The vita appended to Ritz's Göttingen dissertation, *Zur Theorie der Serienspektren* (Leipzig, 1903), lists the following as his teachers at Zurich: Wilhelm Fiedler, Franel, K. F. Geiser, Hirsch, Lacombe, Minkowski, Pernet, H. F. Weber.

3. Ritz, *Zur Theorie der Serienspektren*; *Oeuvres*, 17–18, 78–80.
4. *Ibid.*; *Oeuvres*, 3.
5. Riecke had previously constructed a tenth-order differential equation that gave the Runge formula,
$$\nu = a + b/m^2 + c/m^4 + \cdots.$$
Physikalische Zeitschrift, **1** (1899), 10–11.
6. I. Runge, *Carl Runge*, 110. Runge to H. Kayser, 11 May 1903 (Staatsbibliothek Preussischer Kulturbesitz, Berlin, Darmst. H (11) 1885).
7. J. Hadamard, in *Société française de physique, Bulletin des séances*, **35** (1907), 73*.
8. Ritz, "Über ein neues Gesetz der Serienspektren," in *Physikalische Zeitschrift*, **9** (1908), 521–529; *Oeuvres*, 142.
9. Arthur Schuster, "Spectroscopy," in *Encyclopaedia Britannica*, 11th ed. (1911), XXV, 625a.
10. J. L. Heilbron, *A History of the Problem of Atomic Structure . . .* Ph.D. diss., Univ. of California, Berkeley, 1964, available on University Microfilms (Ann Arbor, Mich., 1965), 359.
11. Ritz to Runge, 25 Mar. 1903 (Deutsches Museum).
12. M. J. Klein, *Ehrenfest*, 41, 45, 165–166, 182, 188, 204.
13. Ritz, "Über das Spektrum von Kalium," in *Annalen der Physik*, 4th ser., **12**, 444–446, dated 4 July 1903; *Oeuvres*, 85–87.
14. Hagenbach, "J. J. Balmer und W. Ritz," 454.
15. Klein, *op. cit.*, 47.
16. Ritz to R. Fueter, Winter 1904–05, quoted in *Verhandlungen der Schweizerischen naturforschenden Gesellschaft*, **92** (1909), 100; Ritz to Fueter, Winter 1906–1907, quoted *ibid.*, p. 102.
17. Ritz to Ehrenfest, 19 Oct. [1907].
18. Ritz to Ehrenfest, 22 Dec. [1908], 19 Feb. 1909; Ritz to Runge, 29 May 1909.
19. *Comptes rendus . . . de l'Académie des sciences*, **139** (1904), 1135.
20. Ritz, "Über eine neue Methode zur Lösung gewisser Variationsprobleme der mathematischen Physik," in *Journal für die reine und angewandte Mathematik*, **135** (1908), 1–61; *Oeuvres*, 192–250. Summarized in "Über eine neue Methode zur Lösung gewisser Randwertaufgaben," in *Nachrichten von der Gesellschaft der Wissenschaften zu Göttingen*, Math.-phys. Klasse (1908), 236–248, presented 16 May 1908; *Oeuvres*, 251–264.
21. *Oeuvres*, 246–250, 263–264.
22. Ritz to P. Weiss, 15 Dec. 1908, quoted in *Oeuvres* xvi–xviii. Ritz, "Theorie der Transversalschwingungen einer quadratischen Platte mit freien Rändern," in *Annalen der Physik*, **28** (1909), 737–786, dated Jan. 1909; *Oeuvres*, 265–316.
23. *Comptes rendus . . . de l'Académie des sciences*, **145** (2 Dec. 1907), 983–984. Among the five memoirs that received some mark of distinction from the jury, not one appears to have treated the question as anything more than one in pure "analysis."
24. Ritz to Ehrenfest, 17 Dec. 1908; Ritz to R. Fueter, quoted in *Verhandlungen der Schweizerischen naturforschenden Gesellschaft*, **92** (1909), 101. Ritz was posthumously awarded a prize of 2,000 francs from the Fondation Leconte "for his works in mathematical physics and mechanics." *Comptes rendus . . . de l'Académie des sciences*, **149** (20 Dec. 1909), 1291.
25. C. Runge and F. A. Willers, in *Encyklopädie der mathematischen Wissenschaften*, II, pt. 3 (1915), ch. c2, par. 21. Ritz's method became known as the "Rayleigh-Ritz method" because it had been employed in a particular case thirty years earlier by Rayleigh in his *Theory of Sound*, a work that Ritz knew well.
26. Ritz to Ehrenfest, 19 Oct. [1907].
27. Loc. cit.
28. Ritz, "Sur l'origine des spectres en séries," in *Comptes rendus . . . de l'Académie des sciences*, **144** (18 Mar. 1907),

634–636; *Oeuvres*, 91–94. Ritz, "Sur l'origine des spectres en séries," *Comptes rendus . . . de l'Académie des sciences*, **145** (16 July 1907), 178–180; *Oeuvres*, 95–97.

29. Ritz to Ehrenfest, 19 Oct. [1907]; Ritz, "Magnetische Atomfelder und Serienspektren," in *Annalen der Physik*, **25** (1908), 660–696, dated Tübingen, Jan. 1908; *Oeuvres*, 98–136.

30. *Ibid.*; *Oeuvres*, pp. 110–111.

31. Ritz to Ehrenfest, 19 Oct. [1907], reporting Paschen's reaction; Runge to Ritz [n.d., draft, *ca.* October 1907] (Deutsches Museum); H. Deslandres, *Société française de physique, Bulletin des séances*, **37** (Jan. 1909), 3*; Pierre Weiss, *Journal de physique théorique et appliquée*, **9** (1910), 986–988; H. Poincaré, *ibid.*, **2** (1912), 352; A. Cotton, *Revue générale des sciences pures et appliquées*, **22** (1911), 597; Anon., *Archives des sciences physiques et naturelles*, **26** (1908), 425–427.

32. *Oeuvres*, 112–113.

33. Ritz, "Über die Spektren der Alkalien. Bemerkungen zu der Arbeit des Herrn C. Runge," in *Physikalische Zeitschrift*, **9** (1908), 244–245, dated Zurich, Mar. 1908; *Oeuvres*, 137–140.

34. C. Runge, "Über die Spektren der Alkalien," in *Physikalische Zeitschrift*, **9** (1908), 1–2.

35. Ritz, "Über ein neues Gesetz der Serienspektren," *ibid.*, 521–529, dated Göttingen, June 1908; *Oeuvres*, 141–162. I. Runge, *Carl Runge*, 134–135.

36. Ritz, "Les spectres de lignes et la constitution des atomes," in *Revue générale des sciences*, **20** (1909), 171–175; *Oeuvres*, 170–180, on 173.

37. Ritz, "Über ein neues Gesetz . . .," *Oeuvres*, 141. In an undated MS abstract of this work Ritz stated explicitly the principle involved: "By additive or subtractive combination, be it of the series formulas themselves, or be it of the constants that enter into them, new formulas are formed, which allow the new lines in the alkalies discovered in recent years by Lenard and others to be calculated completely from those previously known, and which admit of extensive applications to other elements also, in particular He." *Oeuvres*, 162.

38. Letters from Ritz to Paschen, Apr.–July 1908, quoted in *Oeuvres*, 521–525. See also P. Forman, "Paschen," in *Dictionary of Scientific Biography*.

39. Ritz, "Recherches critiques sur l'électrodynamique générale," in *Annales de chimie et de physique*, **13** (1908), 145–275; *Oeuvres*, 317–426. Summarized in "Recherches critiques sur les théories électrodynamiques de Cl. Maxwell et de H.-A. Lorentz," in *Archives des sciences physiques et naturelles*, **26** (Aug. 1908), 209–236; *Oeuvres*, 427–446. Both quotations on 429.

40. Ritz to R. Fueter, late 1905, quoted in *Verhandlungen der Schweizerischen naturforschenden Gesellschaft*, **92** (1909), 101.

41. *Oeuvres*, 369; also 429–430, 436–437.

42. *Oeuvres*, 459–460. Ritz never made, and would have repudiated, any connection between his fictitious light particles and Einstein's heuristic light quanta. Consistent with the fundamentally reactionary character of Ritz's scientific and methodologic inclinations, he was most unsympathetic to the quantum theory and regarded the failure of the Maxwell-Lorentz theory to yield a nonabsurd blackbody radiation formula as but one more example of the demonstrated capacity of the theory to yield unphysical and indeterminate solutions. Ritz, "Über die Grundlagen der Elektrodynamik und die Theorie der schwarzen Strahlung," *Physikalische Zeitschrift*, **9** (1908), 903–907; *Oeuvres*, 493–502.

43. *Oeuvres*, 517.

44. Ritz to P. Weiss, Whit Monday 1908, quoted *Oeuvres*, xx.

45. Loyd S. Swenson, Jr., *The Ethereal Aether; A History of the Michelson-Morley-Miller Aether-Drift Experiments, 1880–1930* (Austin, Tex., 1972), 205.

BIBLIOGRAPHY

I. Original Works. Walter Ritz, *Gesammelte Werke. Oeuvres* (Paris, 1911), edited, with a biographical introduction, by Pierre Weiss, includes Ritz's published scientific papers, a few brief MSS, and some extracts from his correspondence. Not included or mentioned by Weiss is Ritz's anonymous feuilleton, "Die N-Strahlen," in *Neue Züricher Zeitung*, no. 259 (18 Sept. 1906), 1–2.

The manuscript of Ritz's entry in the competition for the Prix Vaillant of 1907, 38 pages in folio, together with a rapporteur's summary, is in the archives of the Académie des Sciences, Paris.

Eleven cards and letters from Ritz to Paul Ehrenfest are in the Ehrenfest Scientific Correspondence, Rijksmuseum voor Geschiedenis der Natuurwetenschappen, Leiden, and the Archive for History of Quantum Physics: 27.10.03, 27.7.07, 17.9.07, 19.10 [07], 10. 12. 07, 17.12.07, 17.12.08, 22.12.[08] and draft of E.'s reply, 12.2.09, 19.2.09, 18.6.09. Further a dozen letters from Ritz's mother, 1909–1912, are in the Ehrenfest Personal Correspondence in the same archives.

Five cards and letters from Ritz to Carl Runge are in the Handschriftensammlung of the Deutsches Museum, Munich (signature 1948/49): 23 May [1902], 25 Mar. 1903, 28 Mar. [1903], Tübingen [n.d., *ca.* Oct. 1907; with draft of Runge's reply], 29 May 1909. The disposition of the MSS and correspondence extracted in the *Oeuvres* is unknown.

II. Secondary Literature. The principal biographical sources are Weiss's preface to the *Oeuvres* and Rudolf Fueter, "Walter Ritz," in *Neue Züricher Zeitung* (1 Sept. 1909), repr. in *Verhandlungen der Schweizerischen naturforschende Gesellschaft*, **92** (1909), Nekrologe, 96–104. All biographical data in the present article for which no other source is given derive from these or from the vitae published in Ritz's doctoral dissertation and in the *Chronik der Georg-August-Universität zu Göttingen* (1908–1909), 11–12, upon his *Habilitation*. (Weiss's insertion of an "h" in Ritz's given name is at variance with all contemporary sources. Both Weiss and Fueter give spring 1900 as the date of Ritz's transference to Göttingen, but I have followed the vitae. The dissertation, and it alone, gives Ritz's birthday as 28 February.) August Hagenbach, "J. J. Balmer und W. Ritz," in *Naturwissenschaften*, **9** (1921), 451–455, is largely derivative.

Valuable biographical information is in Iris Runge, *Carl Runge und sein wissenschaftliches Werk* (Göttingen, 1949), including an extract from Runge's funeral oration, July 1909 (p. 135); and Martin J. Klein, *Paul Ehrenfest*, I, *The Making of a Theoretical Physicist* (Amsterdam, 1970).

PAUL FORMAN

RIVA-ROCCI, SCIPIONE (*b.* Almese, Piedmont, Italy, 7 August 1863; *d.* Rapallo, Liguria, Italy, 15 March 1937), *medicine*.

After graduating in medicine and surgery in 1888 from the University of Turin, Riva-Rocci became assistant lecturer at the Propaedeutic Medical Clinic in Turin directed by Carlo Forlanini. In 1894 he became lecturer in special medical pathology, and in 1898 he followed Forlanini to the University of Pavia. From 1900 to 1928 he was director and head physician of the Ospedale di Varese, and from 1908 to 1921 he also lectured at the Pediatric Clinic of Pavia University, where he introduced the subject. A typical representative of the mechanically based clinical approach successfully advocated by Forlanini, inventor of the technique of artificial pneumothorax for the treatment of pulmonary tuberculosis, contributed to the development of this method through original physiopathological research. He demonstrated the importance of the eccentric pressure of the pulmonary alveolus as a phthisiogenic factor, and he showed that the respiratory function is not substantially endangered in individuals suffering from a reduction of respiratory lung area, particularly in patients with tuberculosis of the lung during pneumothoracic treatment.

His fundamental contribution (1896) was the mercury sphygmomanometer, which is easy to use and gives sufficiently reliable results. This device, the standard instrument for measuring blood pressure, led to many new developments in the therapy of hypertension disease. A fundamental role in spreading the use of the instrument was played by Harvey Cushing, who on a visit to Pavia in 1901, found Riva-Rocci's sphygmomanometer a valuable means of reducing mortality from anesthesia, especially during intracranial surgery. The main feature of the instrument is the pneumatic armband which, by distributing compression uniformly around the arm, eliminates the pressural hyperelevation obtained with previous apparatus, which exerted compression only on the artery. In addition, pressure is measured on the humeral artery, which, compared with more distal arteries, better reflects pressure conditions at the level of the aorta and is less subject to other disturbing factors.

BIBLIOGRAPHY

I. ORIGINAL WORKS. Riva-Rocci's main publication was "Un nuovo sfigmomanometro," in *Gazzetta medica di Torino*, **47** (1896), 981–996, 1001–1017; a partial trans. is in Arthur Ruskin, *Classics in Arterial Hypertension* (Springfield, Ill., 1956), 103–125.

II. SECONDARY LITERATURE. See Luigi Belloni, "Gli inizi della sfigmomanometria clinica," in *Symposium Ciba*, **3** (1955), 56–57 (only in the Italian and Spanish ed.); and "Scipione Riva-Rocci e il suo sfigmomanometro (1896)," in *Simposi clinici*, **10** (1973), i–viii; Enrico Benassi, "Scipione Riva-Rocci," in *Minerva medica*, **54** (1963), 3766–3771; John F. Fulton, *Harvey Cushing, A Biography* (Springfield, Ill., 1946), 212; and Luigi Ponticaccia, "La vita e l'opera di Scipione Riva-Rocci," in *Bollettino della Società medico-chirurgica della provincia di Varese*, **14** (1959), 7–21, with a bibliographical note on Riva-Rocci's works.

LUIGI BELLONI

RIVETT, ALBERT CHERBURY DAVID (*b.* Port Esperance, Tasmania, Australia, 4 December 1885; *d.* Sydney, Australia, 1 April 1961), *chemistry, scientific administration.*

As chief executive officer and deputy chairman (1927-1945) and later chairman (1945-1949) of the Council for Scientific and Industrial Research, Rivett was a principal leader of Australia's first government organization of science. He shaped the direction of a national policy of scientific research.

Rivett was the son of a Congregational clergyman, Albert Rivett, who immigrated to Australia from Norwich, England, in 1879, and Elizabeth Cherbury. He was educated at Wesley College, Melbourne, and at the University of Melbourne, where he majored in chemistry in 1906. He was elected Rhodes Scholar for Victoria the following year. At Oxford, where he shared a laboratory bench with Henry Tizard, he studied physical chemistry under Sidgwick and obtained the B.A. in 1909 and the B.Sc. in 1910. During 1910 he worked under Arrhenius at the Nobel Institute of Physical Chemistry in Stockholm, where he broadened his research interest in equilibria within heterogeneous systems.

In 1911 Rivett returned to Australia as lecturer in the chemistry department of the University of Melbourne. In the same year he married Stella Deakin, daughter of Alfred Deakin, a former prime minister of Australia. During World War I, Rivett worked on the production of ammonium nitrate at the government munition works in Swindon, England. He later became associate professor of chemistry at Melbourne (1921) and succeeded to the chair of chemistry (1924). He published *Phase Rule and the Study of Heterogeneous Equilibria* (1923). In 1927 he left the university to take up a full-time post at the newly formed Council for Scientific and Industrial Research.

Throughout the nineteenth century the Australian colonial legislatures gave scant support to science. But federation at the turn of the century and the influence of the war encouraged the concept of national scientific development, and the Institute of Science and Industry was founded by act of Parliament in 1920. The plan proved ineffectual; and it was not until 1926 that a new, liberally endowed and politically independent organization, the Council for Scientific and Industrial Research, was established by the federal government to direct and encourage scientific research in the agricultural and pastoral industries, to train scientific workers, and to serve as a liaison with governments and scientific institutes abroad.

As chief executive officer and deputy to the chairman, Sir George Julius, a practical engineer, Rivett exercised responsibility for initiating programs of research in animal health and nutrition; parasites and pests; economic botany; soil and pasture improvement; food preservation; forests and fisheries; and, with the decision of the government in 1937 to extend scientific and technical assistance to secondary industry, develop a national standards laboratory and establish divisions of industrial chemistry, aeronautics, and radiophysics. Rivett's own scientific background influenced his approach. He placed a marked emphasis on original and fundamental research; gave broad responsibility to his division chiefs; and sought to ease the deep-rooted tensions between the modestly funded universities and the growing colossus of government science. His conviction that science could flourish only in an atmosphere of free and open inquiry led to the ultimate separation of the aeronautical division from the council in the late 1940's. Conceding little, Rivett was thus able to give his organization a remarkable measure of independence from the normal machinery of government.

Rivett received a number of honorary degrees and titles. He was created K.C.M.G. in 1935 and elected fellow of the Royal Society in 1941. He also helped to establish the Australian National University. In 1954 he was a founding fellow of the Australian Academy of Science.

BIBLIOGRAPHY

I. ORIGINAL WORKS. Rivett's publications are listed in *Biographical Memoirs of Fellows of the Royal Society*, **12** (1966), 454–455. See also *Application of Science to Industry* (Brisbane, 1944), J. M. Macrossan lecture.

II. SECONDARY LITERATURE. On Rivett and his work, see George Currie and John Graham, *The Origins of the C.S.I.R.O. Science and the Commonwealth Government, 1901–1926* (Melbourne, 1966); Hedley Marston, "Albert Cherbury David Rivett," in *Biographical Memoirs of Fellows of the Royal Society*, **12** (1966), 437–455; D. P. Mellor, *The Role of Science and Industry*, Australia in the War of 1939–1945, ser. 4, civil vol. V (Canberra, 1958); and Rohan Rivett, *David Rivett: Fighter for Australian Science* (North Blackburn, Victoria, 1972).

ANN MOZLEY

RIVIÈRE DE PRÉCOURT, ÉMILE-VALÈRE (*b.* Paris, France, 22 April 1835; *d.* Paris, 25 January 1922), *speleology, anthropology.*

Rivière, son of a general practitioner, contemplated a career in medicine after his graduation from the Lycée Bonaparte and for a while was an intern at the Asile de Vincennes at Le Vésinet.

Because of ill health Rivière journeyed to Cannes in 1868 and to Menton in 1869. In 1870, having settled at Menton, he began to explore the nine caves of the Baoussé-Roussé, a promontory of Jurassic limestone just across the Italian border in the commune of Grimaldi. He sought and received permission from the Italian government to excavate the first four caves and to control access to the fifth, sixth, and seventh.

On 26 March 1872 Rivière uncovered, at a depth of 6.55 meters and about seven meters from the entrance of the fourth cave (Grotte de Cavillion), a nearly perfect adult male skeleton. It lay on its left side with its legs slightly flexed and its arms folded upward so that the left hand cradled its cheek. The skull bore a headdress made of more than two hundred perforated *Nassa* shells interspersed with twenty-two stag canine teeth. A garter of forty-one *Nassa neritea* adorned the left knee. On the forehead there rested a needle or dagger-like weapon made from a deer radius and under the back of the skull lay two broken triangular silex knives. The skull and implements buried with the skeleton were covered with powdered hematite, a substance also found in a small trench cut near the mouth.

During February 1873 Rivière excavated a skeleton from the sixth cave, and in June of the same year he unearthed two more. The first two, remains of adults, were sprinkled with the ferruginous powder and were accompanied by shell, flint, and bone grave goods. The last, a youth of approximately fifteen, was interred face-down without ornaments and without the red coloration.

Explorations in the first cave in July 1875 yielded skeletons of two children, ages four to six. They were lying together in extended position, arms at sides, feet outward. They were covered by a mantle of shells, a single flint interred between them. The children's bones lacked the red powder.

The human remains were found in deposits also containing the bones of cave bear, hyena, rhinoceros, and stag. Rivière contended that they were contemporary with the animals and that they therefore dated from the Pleistocene. He argued, further, that the peroxide of iron sprinkled on the bones of the adults and the weapons buried with them were evidence that the Cro-Magnon peoples, of which he was convinced his discoveries were examples, practiced funereal rites. A description of his finds and the conclusions drawn from them are in his *Paléontologie: De l'antiquité de l'homme dans les Alpes-Maritimes* (1887).

Rivière's claims were vigorously contested at the time by most prehistorians. Gabriel de Mortillet, William Boyd Dawkins, and others believed that the skeletons had been buried in Pleistocene deposits by Neolithic peoples. The controversy over the geological age of the Baoussé-Roussé skeletons continued until an archaeological team commissioned by Albert I of Monaco re-explored the Grimaldi caverns and produced indisputable evidence that Rivière had been correct.

Late in 1887 Rivière turned his attention to the caves of the Dordogne. In June 1887 he discovered a previously unknown chamber at the grotto of La Mouthe. Carved into the walls were representations of bison, ibex, reindeer, horse, and mammoth. The authenticity of the La Mouthe cave art was challenged, as the polychromes at Altamira, Spain, had been in 1879. Rivière presented evidence that the carvings were partially covered by debris from both the Paleolithic and Neolithic and that therefore they had to be dated at least from the Paleolithic; but his discoveries remained controversial until after publication by Breuil and L. Capitan of their discoveries at Les Combarelles and Font-de-Gaume in 1901.

BIBLIOGRAPHY

I. ORIGINAL WORKS. Rivière's main publication is *Paléontologie: De l'antiquité de l'homme dans les Alpes-Maritimes* (Paris, 1887). Brief notices and articles on the finds near Menton appear in the *Comptes rendus* of the Paris Academy and the International Congress of Anthropology and Prehistoric Archaeology, as well as of the Association Française pour l'Avancement des Sciences, between 1872 and 1900. Reports of his investigations of the Dordogne caves are in the *Comptes rendus* of the Paris Academy: **119** (1894), 358–361; **122** (1896), 1563–1565; **123** (1896), 714–715; **124** (1897), 731–734. The engravings of the La Mouthe grotto are reproduced and described in *Report of the Board of Regents of the Smithsonian Institution* for 1901 (1902), 439–449, a trans. from the *Bulletin et mémoires de la Société d'anthropologie de Paris*, 5th ser., **2** (1901), 509–518.

II. SECONDARY LITERATURE. See J. Bossavy's "Nécrologie," in *Bulletin de la Société préhistorique française*, **19** (1922), 257–264; and M. Boule and H. Valois, *Fossil Men* (London–New York, 1957), translated by Michael Bullock from *Les hommes fossiles*, 5th ed.

MARTHA B. KENDALL

RIVINUS, AUGUSTUS QUIRINUS. See **Bachmann, Augustus Quirinus.**

ROBERT OF LINCOLN. See **Grosseteste, Robert.**

ROBERTS, ISAAC (*b.* Groes, near Denbigh, North Wales, 27 January 1829; *d.* Crowborough, Sussex, England, 17 July 1904), *astronomy.*

The son of a farmer, Roberts was brought up in Liverpool and after an elementary education was apprenticed in 1844 to a local firm of builders and lime burners. He became manager of the firm but in 1859 set up in Liverpool, independently, the firm of Roberts and Robinson, which secured many important contracts in the city.

Gradually scientific interests occupied more and more of Roberts' attention. In 1870 he became a fellow of the Geological Society; and in 1878 he purchased a seven-inch refractor, made by Cooke, for visual astronomy. In 1883, however, he began to experiment with stellar photography and eventually obtained a twenty-inch reflector, with a 100-inch focal length, from Howard Grubb of Dublin. (The photographs were taken in the focus of the mirror.) In January 1886 Roberts announced to the Royal Astronomical Society that during the previous year he had taken 200 photographs of stars that might be measured for position and also long-exposure photographs of the Orion and Andromeda nebulae and of the Pleiades cluster. Later in 1886 he exhibited a three-hour exposure of the Pleiades that revealed the astonishing and unsuspected nebulosity that surrounds these stars. In 1887 he attended the Conference of Astronomers in Paris; this conference planned an international photographic chart of the sky, and thereafter Roberts concentrated his own efforts on objects of special interest.

In 1888 Roberts retired from business to devote himself to his scientific work; and in 1890 he settled in Crowborough, in the south of England, where the skies offered better viewing than near Liverpool. At Crowborough he supervised the work of his assistant, W. S. Franks, in the photography of stars, star clusters, and nebulae. The dramatic photographs that resulted were for many years exhibited regularly at the Royal Astronomical Society. In 1893 Roberts published a volume containing photographs and descriptions of his observatory, his telescope, and fifty-one celestial objects. Although his photographs of nebulae lacked the definition that Keeler attained at the Lick Observatory at the close of the century, they offered an objective record to replace the subjective sketches on which earlier generations had been forced to rely in their efforts to establish the structure of the nebulae. Indeed, "the detection of changes in the structure of nebulae," which would establish such nebulae as small-scale objects and not as island universes of stars, was the first use Roberts foresaw for his book. But the photographs were often misinterpreted. William Huggins (*Scientific Papers* [London, 1909], 173) considered the Andromeda nebula "a planetary system at a somewhat advanced stage of evolution." Roberts, in a second volume published six years later, expressed his belief that this nebula had rotated. The photographs, in short, were more significant for the possibilities of nebular photography that they demonstrated than for the inferences drawn from them.

Roberts was twice married but had no children. The honors that his astronomical photography earned included the 1895 Gold Medal of the Royal Astronomical Society.

BIBLIOGRAPHY

I. ORIGINAL WORKS. Roberts published his photographs in *Photographs of Stars, Starclusters and Nebulae*, 2 vols. (London, 1893–1899); and Mrs. Isaac Roberts (née Dorothea Klumpke), his widow, published a further portfolio, *Isaac Roberts' Atlas of 52 Regions, A Guide to Herschel's Fields* (Paris, 1929). His articles appeared principally in *Monthly Notices of the Royal Astronomical Society, Astrophysical Journal, Astronomische Nachrichten,* and *Proceedings of the Liverpool Geological Society*; lists of these are given in *Poggendorff*, IV, 1259; and V, 1057; and in a posthumous article edited by Mrs. Isaac Roberts, "The Nebula H. V. 20 Ceti," in *Monthly Notices of the Royal Astronomical Society*, **65** (1915), 191–200.

II. SECONDARY LITERATURE. Useful, short biographies are in *Monthly Notices of the Royal Astronomical Society*, **65** (1905), 345–347; *Observatory*, **27** (1904), 300–303; and *Dictionary of National Biography, 1901–1911*, 209–211. See also Gérard de Vaucouleurs, *Astronomical Photography* (London, 1961).

MICHAEL A. HOSKIN

ROBERTS-AUSTEN, WILLIAM CHANDLER (*b.* Kennington, Surrey, England, 3 March 1843; *d.* London, England, 22 November 1902), *metallurgy.*

Roberts-Austen was the eldest son of George Roberts and Maria Louisa Chandler. (In 1885, as heir of his uncle Nathaniel Lawrence Austen and at his request, he adopted by royal license the additional surname Austen. His early work is published under the name Roberts.) After private education he entered the Royal School of Mines, South Kensington,

London, at the age of eighteen, becoming an associate in 1865. Although he intended to be a mining engineer, his career was determined by his becoming private assistant to Thomas Graham, then master of the mint. Roberts is said to have done some of Graham's experimental work on colloids; and although he was not officially connected with the mint it was in its work that he became most adept. After Graham's death in 1869, the Treasury gave him an official post as part of a long-planned reorganization, placing him in charge of the assay of coin (bullion was assayed by another officer). In 1882 the two posts were combined and Roberts-Austen became "chemist and assayer of the mint," a position he occupied until his death.

In 1870, with the deputy-master of the mint, Sir Charles Fremantle, Roberts-Austen toured thirteen European mints to study their methods, a journey that marked the beginning of his international interests and influence. He conferred with Jean Stas, who was head of the Brussels mint, and as a result introduced into the British mint the volumetric, argentometric assay devised by Gay-Lussac for the Paris mint in 1832. Official confidence in him was shown by his being entrusted with the preparation in 1873 of trial plates (the legal standards of reference at the annual "trial of the pyx," for checking gold and silver coinage).

Roberts-Austen's study of the means of preparing plates of fine gold and silver led him to a fundamental study of the constitution of alloys, into which he introduced new physicochemical theories and new experimental techniques. He applied calorimetric methods to the measurement of the solidification of copper-silver alloys, expressing his results for the first time as "freezing-point" curves.

In 1876, in conjunction with Joseph Lockyer, Roberts-Austen studied the possibility of using the spectroscope for quantitative analysis. Although the method was not adopted at the mint, his interest stimulated its progress. This is characteristic of the way in which research aimed at supplementing the traditional methods of assay yielded scientific results of wider significance. He was elected a fellow of the Royal Society in 1875 and was an original member of the Physical Society at its formation in 1874.

In 1880 John Percy resigned his post of professor of metallurgy at the Royal School of Mines, following his disagreement with a proposal to move the whole school to South Kensington. Roberts-Austen was appointed to succeed him while still retaining his post at the mint. The report of the deputy master of the mint stated: "This Department could not fail to derive advantage from the appointment of the Chemist of the Mint to a position in which, as Professor, every advance in metallurgical science must of necessity come under his notice." A worthy successor to Percy, Roberts-Austen was an excellent teacher and enlarged the scope of research at the school.

In 1880 Roberts-Austen began a long study of the hardening of steel, important to the mint because of the use of steel for dies. The mint's needs were soon met; but the fundamental study continued, particularly under the encouragement given by the Institution of Mechanical Engineers. It led to the important series of Alloys Research Reports.

For the rest of his career Roberts-Austen succeeded in maintaining the difficult balance between the administrative demands of a post and the effective pursuit of the scientific inquiries it generates. In 1883 he toured the principal mints and assay offices in the United States, and in 1885 he began a study of the effect of small amounts of impurity on the tensile strength of standard gold. This work was developed in the next few years and was considered in a paper of 1888 that was an early example of the interpretation of the properties of a range of elements in terms of the periodic classification of the elements.

Roberts-Austen's work had been so successful up to this time that he might well have been content to continue to use the same methods. In 1890, however, he not only adopted methods and attitudes that were very advanced for the time but also initiated new ones. Experimentally he recognized the inadequacy of the calorimetric methods he had used for the study of fused alloys, and adopted Le Chatelier's thermo-couple pyrometer. He added greatly to the utility of the mirror galvanometer that detected the thermo-couple response by recording the movements of the reflected beam photographically. The ability to record rates of cooling in masses of high melting point extended Roberts-Austen's knowledge of eutectics and gave him the basis of a theory of alloy composition. In 1891 he described an alloy of gold and aluminum that he recognized as the first true intermetallic compound. His automatic recording pyrometer later proved of great industrial use.

Roberts-Austen also developed the use of photo-micrography in the study of metal crystal structures, his apparatus providing the pattern for similar installations in industry, the Royal Arsenal, and the engineering laboratory at Cambridge. He served on many committees, the most important of which was that leading to the establishment of the National Physical Laboratory. He was honored as a K.C.B. in 1899.

Roberts-Austen studied historical metallurgy, investigating with the sculptor Alfred Gilbert the revival of fifteenth-century methods of casting. He lectured widely on the colors of metals for artwork and even referred to the subject in official mint reports. His principal publication, the textbook *Introduction to the Study of Metallurgy* (1891), ran to six editions and was very influential.

BIBLIOGRAPHY

I. ORIGINAL WORKS. Roberts-Austen's largest single work was *Introduction to the Study of Metallurgy* (London, 1891, 1893, 1894, 1897, 1902, 1908). Over 80 papers dealing with metallurgical subjects are listed in Smith (see below), together with 12 written in collaboration on related subjects, such as spectroscopic analysis. Eight papers deal with the metallurgy of art objects.

II. SECONDARY LITERATURE. Roberts-Austen's most important and characteristic papers were assembled by his colleague S. W. Smith in *Roberts-Austen: A Record of His Work* (London, 1914). They are presented in groups and interspersed with biographical matter, his several parallel careers being given separate chronological treatment. More orthodox summary accounts of his life are given in London *Times* (24 Nov. 1902), 9; *Engineering*, **74** (1902), 717–718, with portrait; *Journal of the Iron and Steel Institute*, **62** (1902), 361–364; and *Proceedings of the Royal Society*, **75** (1905), 192–198.

FRANK GREENAWAY

ROBERVAL, GILLES PERSONNE (or **PERSONIER**) **DE** (*b.* near Senlis, France, 10 August 1602;[1] *d.* Paris, France, 27 October 1675), *mathematics, mechanics, physics.*

Very little is known about Roberval's childhood and adolescence. His parents seem to have been simple farmers. He stated that he was born and educated among the people (*inter multos*). J.-B. du Hamel reports that he devoted himself to mathematical studies beginning at the age of fourteen. Having left his family at an unknown date, Roberval traveled through various regions of the country, earning a living from private lessons and continuing to educate himself. In 1628 he arrived in Paris and put himself in touch with the scientists of the Mersenne circle: Claude Mydorge, Claude Hardy, and Étienne and Blaise Pascal. Mersenne, especially, always held Roberval in the highest esteem. In 1632 Roberval became professor of philosophy at the Collège de Maître Gervais. On 24 June 1634, he was proclaimed the winner in the triennial competition for the Ramus chair (a position that he kept for the rest of his life) at the Collège Royal in Paris, where at the end of 1655 he also succeeded to Gassendi's chair of mathematics. In 1666 Roberval was one of the charter members of the Académie des Sciences in Paris.

Roberval's tendency to keep his own discoveries secret has been attributed to his desire to profit from them in order to retain the Ramus chair. But this habit also resulted in his tardy and rather frequent claims to priority. He himself published only two works: *Traité de méchanique* (1636) and *Aristarchi Samii de mundi systemate* (1644). A rather full collection of his treatises and letters was published in the *Divers ouvrages de mathématique et de physique par messieurs de l'Académie royale des sciences* (1693), but since few of his other writings were published in the following period, Roberval was for long eclipsed by Fermat, Pascal, and, above all, by Descartes, his irreconcilable adversary. Serious research on Roberval dates from approximately the end of the nineteenth century, and many of his writings still remain unpublished.

In the field of elementary geometry, a collection of Roberval's manuscripts includes some remarkable constructions of isoperimetric figures,[2] of which at least two are earlier than October 1636.[3] In addition, in 1644 Mersenne reported, without indicating the procedure employed, several problems solved by Roberval under the condition of the *extrema*.[4]

Roberval was one of the leading proponents of the geometry of infinitesimals, which he claimed to have taken directly from Archimedes, without having known the work of Cavalieri. Moreover, in supposing that the constituent elements of a figure possess the same dimensions as the figure itself, Roberval came closer to the integral calculus than did Cavalieri, although Roberval's reasoning in this matter was not free from imprecision. The numerous results that he obtained in this area are collected in the *Divers ouvrages*, under the title of *Traité des indivisibles*. One of the first important findings was, in modern terms, the definite integration of the rational power, which he most probably completed around 1636, although by what manner we are not certain. The other important result was the integration of the sine, and he formulated by virtue of it the problem of which he was so proud: Trace on a right cylinder, with a single motion of the compass, a surface equal to that of a given square, or, except for the bases, of an oblique cylinder. Furthermore, in 1644 Mersenne recounts —but again without saying how—that Roberval was the first to square the surface of the oblique cone. Yet the most famous of his works in this domain concerns the cycloid. Roberval introduced the "compagne" ("partner") ($x = r\theta$, $y = r - r \cos \theta$) of the original cycloidal curve and appears to have succeeded, before the end of 1636, in the quadrature of the latter and in the cubature of the solid that it generates in turning around its base. But the cubature of the solid of revolution around its axis (presented in the treatise *Ad trochoidem, ejusque solida*),[5] must have been achieved between May 1644 and October 1645. Roberval, moreover, knew how to extend all these results to the general case:

$$x = a\theta - b \sin \theta, \quad y = a - b \cos \theta.$$

On account of his method of the "composition of movements" Roberval may be called the founder of kinematic geometry. This procedure had three applications—the fundamental and most famous being the construction of tangents. "By means of the specific properties of the curved line," he stated, "examine the various movements made by the point which describes it at the location where you wish to draw the tangent: from all these movements compose a single one; draw the line of direction of the composed movement, and you will have the tangent of the curved line."[6]

Roberval conceived this remarkably intuitive method during his earliest research on the cycloid (before 1636). At first, he kept the invention secret, but he finally taught it between 1639 and 1644; his disciple François du Verdus recorded his lessons in *Observations sur la composition des mouvemens, et sur le moyen de trouver les touchantes des lignes courbes*.[7] Jean-Marie-Constant Duhamel's criticism of this method (1838) applied only to the abuse some others had made of the parallelogram or parallelepiped of velocities; Roberval himself employed, in his own fashion, the rule advanced by Duhamel. In the last analysis, the latter failed to recognize that Roberval's method had to do with the moving system of coordinates.

In the second place, he also applied this procedure to comparison of the lengths of curves, a subject almost untouched since antiquity. In the winter of 1642–1643, Roberval equated not only the spiral and the parabola in their ordinary forms, but also the curves $r = k\theta^n$ and $y^{n+1} = k'x^n$ (n being any whole number).[8] He accomplished this equation by purely kinematic considerations, probably without making any computations at all (according to an unpublished work, preserved in the archives of the Academy of Sciences). In addition, he declared that he had carried out the rectification of the simple cycloid before 1640, by reducing it kinematically to the integration of the sine—a serious claim that would deprive Torricelli of the glory of having first rectified a curve (1644 or 1645), but for which there is not yet any objective proof. It is possible, however, that Roberval discovered before August 1648 the equality in length of the generalized cycloid and the ellipse, which was established by Pascal in 1659.[9]

The third application consisted in determining extrema, and four problems of this type are solved in an unpublished manuscript.[10] The solution is certainly ingenious, but we know of no other writings by Roberval that treat more difficult problems of the same type in this fashion. It was quadrature that, in accord with the general trend of mathematics in his century, Roberval pursued as the principal goal of his kinematics; but his efforts in this direction were not fruitful. Lacking the aid of analysis, his kinematics was still far from the Newtonian method of fluxions.

Roberval composed a treatise on algebra, *De recognitione aequationum*, and another on analytic geometry, *De geometrica planarum et cubicarum aequationum resolutione*.[11] Before 1632, he had studied the "logistica speciosa" of Viète; but the first treatise, which probably preceded Descartes's *Géométrie*, contains only the rudiments of the theory of equations. On the other hand, in 1636 he had already resorted to algebra in search of a tangent. By revealing the details of such works, he would have assured himself a more prominent place in the history of analytic geometry, and even in that of differential calculus. But the second treatise cited above was written later than the *Géométrie*, and therefore contributed nothing of particular interest, except for a thorough discussion of the "ovale optique," the style of which, however, is that of elementary geometry.

Turning now to mechanics, his *Traité* of 1636 led to the law of the composition of forces through a study—like the one in the *Beghinselen der weegconst* of Stevin—of the equilibrium of a body supported at first on an inclined plane, then suspended by two cords. But when Roberval reduced the equilibrium on the inclined plane to that of the balance, he was very close to Galileo, whose *Le meccaniche* he no doubt knew through the efforts of Mersenne. This treatise of Roberval contains, however, a clear notion of the pressure that the body exerts on the plane, and of the equivalent resistance that the latter opposes to the former. Moreover, Roberval stated that the treatise "is only a sample of a greater work on Mechanics which cannot so soon make its appearance." Perhaps he meant the French version of a booklet in Latin, now lost, which he had written before 1634, and from which he had just taken this treatise on statics.

Roberval's ambition did not cease to grow. In 1647 he wrote to Torricelli: "We have constructed a mechanics which is new from its foundations to its roof, having rejected, save for a small number, the ancient stones with which it had been built."[12] Roberval indicated the materials for this work: book I, on the center of action of forces *(de centro virtutis potentiarum)* in general; book II, on the balance; book III, on the center of action of particular forces;

book IV, on the cord; book V, on instruments and machines; book VI, on the forces that act within certain media; book VII, on compound movements; and book VIII, on the center of percussion of moving forces. This great treatise has not come down to us; the most we can do is to find some traces of its content among Roberval's surviving papers. The "Tractatus mechanicus, anno 1645,"[13] treating of the composition of parallel forces, might have been the beginning of book I. The "demonstratio mechanica,"[14] establishing the law of the balance in the manner of Stevin and Galileo,[15] was undoubtedly destined for book II. Since book III was to deal with the center of gravity, the "Theorema lemmaticum"[16] would have served as its basis; this unpublished manuscript demonstrates the general equation of moments in space of three dimensions. As for book IV, the booklet mentioned above would already have furnished the material for it. The "Proposition fondamentale pour les corps flottants sur l'eau"[17] would have been included in book VI. With regard to book VII, we may refer to the preceding paragraphs on kinematic geometry. And yet around 1669, Roberval wrote *Projet d'un livre de mechanique traitant des mouvemens composez*.[18] Book VII would therefore not have been completed; Roberval dreamed, certainly with too great temerity, of a vast physical theory based uniquely on the composition of motions. Concerning book VIII, a part of it is in the unpublished manuscript "De centro percussionis,"[19] as well as three texts of 1646 in the *Oeuvres de Descartes*.[20] The problem of the center of oscillation of the compound pendulum provoked in 1646 a new polemic between Descartes and Roberval. Although Descartes had a better idea of the center of oscillation, nevertheless he was wrong in neglecting the directions of the forces. Roberval well knew how to rectify this error. But he did not consider it necessary to locate the center of oscillation on the right line linking the point of suspension and the center of gravity of the body under consideration; he doubted that, even located on this right line, the point that he determined was precisely the center of oscillation.

On 21 August 1669 Roberval presented to the Royal Academy of Sciences the plans for a particular type of balance, which today bears his name. Although the notion of virtual work is clearly contained and expressed therein under the name "momentum," he probably considered only the finite path of the weight.

In mechanics, no less than in mathematics, Roberval displayed a great concern for rigor. For example, he began the "Tractatus mechanicus, anno 1645" by postulating the possibility of the movement of a point along an arbitrary curve. It is not the case, however, that he failed to appreciate the importance of experiment in mechanics; and positivism is even more evident in his work in physics. In this regard, it is also useful to consult his philosophical reflections, such as "Les principes du debvoir et des cognoissances humaines,"[21] *L'évidence—le fait avéré—la chymère*,[22] and the unpublished manuscript "Quelle créance l'homme doit avoir à ses sens et à son entendement" (in Archives of the Academy of Sciences).

Roberval's positivism appears in a particularly nuanced form in the book *De mundi systemate* of 1644, where he claimed to have translated an Arabic manuscript of Aristarchus, to which he had added his own notes, all of them favorable to the author. Yet he did not adhere to the system of Aristarchus to the exclusion of those of Ptolemy and Tycho Brahe. In the dedication of the work, Roberval wrote: "Perhaps all three of these systems are false and the true one unknown. Still, that of Aristarchus seemed to me to be the simplest and the best adapted to the laws of nature." It is with this reservation that Roberval expressed his opinion on the great system of the world (the solar system), the minor systems (planetary), the motions of the sun and the planets, the declination of the moon, the apogees and perigees, the agitation of the oceans, the precession of the equinoxes, and the comets. Despite this reservation, Roberval appeared convinced of the existence of universal attraction, which—under the inspiration of Kepler—he put forth as the foundation of his entire astronomy: "In all this worldly matter [the fluid of which the world is composed, according to our author], and in each of its parts, resides a certain property, or accident, by the force of which this matter contracts into a single continuous body."[23]

On the problem of the vacuum, which had been agitating French scientific circles since 1645, Roberval composed two *Narrationes*.[24] In the first of them (dated 20 September 1647), he reported Pascal's experiments at Rouen and the experiments he himself subsequently had undertaken. Roberval agreed with his friend, concluding that if the space at the top of the barometric tube was not absolutely empty, it was free of all the elements alleged by the philosophers. But the second *Narratio*—probably composed in May–June and October 1648—is much more important. In this work Roberval proved himself to be a very skillful and scrupulous experimenter. He explained the suspension of the mercury in the tube by the pressure of the air on the exterior mercury. Moreover, a very ingenious apparatus that he had invented to support this thesis later served as the prototype for the one in Pascal's experiment of "the vacuum in the vacuum," described in the *Traité de l'équilibre des liqueurs et de la pesanteur de la masse de l'air*. Roberval

thus remained in agreement with his friend, save that he attributed the equilibrium of the liquids to the universal attraction mentioned in the discussion of his astronomical work. But, on the other hand, he did deliberately assert the existence of rarefied air in the top of the tube. He showed in particular that an exhausted carp bladder placed in the empty space of the tube became inflated by virtue of the spontaneous dilation of the air. And that, in principle, is all he wished to do. He refrained from tackling the ancient question of whether a vacuum existed in nature. While ironically returning the problem to the schools, he did not at all tolerate the Cartesian confusion of space and matter.

The same positivism is evident in book II of *L'optique et la catoptrique de Mersenne* (1651), in which proposition 4 is particularly remarkable for its disdainful rejection of all speculation on the nature of light. Roberval promised himself to "join experiment to reasoning" in the study of the phenomenon of reflection. In the same spirit, he rewrote the "Livre troisiesme de la dioptrique, ou des lunettes,"[25] where we find again his cherished geometric dissertation on the oval.

NOTES

1. According to Pierre Desnoyers, secretary to the queen of Poland and correspondent of Roberval.
2. Bibliothèque Nationale, fonds latin, nouvelle acquisition 2340, 1_r–6_v, 196_r–214_r.
3. Given two cones of unequal bases or heights of two cones of equal volume and surface area (the bases included or excluded), find the cones. See *Oeuvres de Fermat*, II, 82–83.
4. M. Mersenne, "Phaenomena hydraulica," in *Cogitata physico-mathematica*, 55–77.
5. *Divers ouvrages*, 257–274.
6. *Ibid.*, 80.
7. *Ibid.*, 69–111.
8. One determines k, k' in such a manner that the two curves will have equal subtangents for any given pair of equal values of r and y.
9. The arc corresponding to the interval $[\theta_1, \theta_2]$ of the generalized cycloid (see end of fourth paragraph) is double the arc corresponding to the interval $[\theta_1/2, \theta_2/2]$ of the ellipse $x = (a + b) \cos \theta$, $y = (a - b) \sin \theta$.
10. Bibliothèque Nationale, fds. fr. 9119, 464_v–470_r.
11. *Divers ouvrages*, 114–189.
12. The original is in Latin, in *Divers ouvrages*, 301.
13. Bibliothèque Nationale, fds. lat. 7226, 2_r–27_r.
14. *Ibid.*, 31_r–33_v.
15. *The Principal Works of Simon Stevin*, I (Amsterdam, 1955), 116–125; *Le opere di Galileo Galilei*, VIII, 152–153.
16. Bibliothèque Nationale, fds. lat. 7226, 59_r–82_r.
17. *Ibid.*, 207_v–210_r.
18. *Divers ouvrages*, 112–113.
19. Bibliothèque Nationale, fds. lat. nouv. acq. 2341, 41_r–45_r.
20. *Oeuvres de Descartes*, IV, 420–428, 502–508.
21. In Victor Cousin, *Fragments de philosophie cartésienne*, 242–261.
22. *Oeuvres de Blaise Pascal*, II, 49–51.

23. *Op. cit.*, 2–3.
24. *Oeuvres de Blaise Pascal*, II, 21–35, 310–340.
25. Bibliothèque Nationale, fds. fr. 12279, 1r–108r.

BIBLIOGRAPHY

I. ORIGINAL WORKS. Works published during Roberval's lifetime include *Traité de mechanique. Des poids soustenus par des puissances sur les plans inclinez à l'horizon. Des puissances qui soustiennent un poids suspendu à deux chordes* (Paris, 1636), repr. in Mersenne, *Harmonie universelle* (see below); and *Aristarchi Samii de mundi systemate, partibus et motibus ejusdem libellus. Adjectae sunt AE. de Roberval . . . notae in eundem libellum* (Paris, 1644), repr. with some modifications in Mersenne, *Novarum observationum . . .* (see below). The extract of a letter to Mersenne and of another to Torricelli is in Dati (see below), pp. 8, 12–14. Two anti-Cartesian letters and an annexed fragment were published by C. Clerselier in *Lettres de M. Des-Cartes*, III (Paris, 1667), 313–321, 498–505. See also *L'optique et la catoptrique du R. P. Mersenne, mise en lumière après la mort de l'autheur*, (Paris, 1651), of which propositions 4–16 (pp. 88–131) of bk. II are actually the work of Roberval.

Posthumous publications before the end of the nineteenth century include 4 letters to Fermat, in *Varia opera mathematica D. Petri de Fermat Senatoris Tolosani* (Toulouse, 1679), 124–130, 138–141, 152–153, 165–166; 6 treatises (one of which is completely fragmentary) and 2 letters in *Divers ouvrages de mathématique et de physique par messieurs de l'Académie royale des sciences* (Paris, 1693), 69–302, repr. in *Mémoires de l'Académie royale des sciences, depuis 1666 jusqu'a 1699*, VI (Paris, 1730), 1–478; "Avant-propos sur les mathématiques," in Cousin (see below), 236–239; and "Les principes du debvoir et des cognoissances humaines," *ibid.*, 242–261.

Letters and other minor writtings of Roberval have been published or reprinted in the following works, in which one may also find various information about him: Charles Henry, *Huygens et Roberval* (Leiden, 1880), 35–41; *Oeuvres de Blaise Pascal*, I–II (see below); *Oeuvres complètes de Christiaan Huygens, publiées par la Société Hollandaise des Sciences*, I (The Hague, 1888); *Oeuvres de Fermat*, Charles Henry and Paul Tannery, eds., II, IV (Paris, 1894, 1912), with supp. by Cornelis de Waard (Paris, 1922); *Oeuvres de Descartes*, C. Adam and P. Tannery, eds., II, IV (Paris, 1898, 1901); *Opere di Evangelista Torricelli*, G. Loria and G. Vassura, eds., III (Faenza, 1919); Mersenne's *Correspondance*, C. de Waard, R. Pintard, and B. Rochet, eds., III–XII (Paris, 1946–1973) and L. Auger, *Un savant méconnu* (see below), 179–202, which presents two new documents on gravity (*pesanteur*) and on the so-called Roberval balance.

The Bibliothèque Nationale in Paris possesses numerous unpublished MSS of Roberval, cataloged as follows: fonds latins 7226, 11195, 11197; fds lat. nouvelles acquisitions 2338, 2340, 2341; fds. français 9119, 9120, 12279; and fds. fr. nouv. acq. 1086, 2340, 5161, 5175, 5856. The Archives of the Académie des Sciences also preserve a good number of Roberval's papers, almost all of which are unpublished and poorly classified. Besides the writings dealing with mathematics and mechanics, one may find items such as "Cours d'astronomie," "L'aere chrestienne," "Tractatus de architectura," "Geographie physique sur les golfes," "L'arpentage," and "Du nivelage." The Bibliothèque Sainte Geneviève possesses, according to Auger, "Trois tables de la grandeur des parties d'une fortification royale, dédiées à M. le Duc de Buckingham par M. Roberval, Paris, 1645."

Many of Roberval's papers preserved in the archives of the Académie des Sciences have been classified by Alan Gabbey and are described in a catalog (deposited in the archives of the Academy), which he prepared in 1966, and which covers MSS and documents by or relating to the founding members of the Academy, other than Huygens. The present author was not able to consult the catalog at the time of the composing of this article.

II. SECONDARY LITERATURE. On Roberval and his work, see Léon Auger, "Les idées de Roberval sur le système du monde," in *Revue d'histoire des sciences et de leurs applications*, 10, no. 3 (1957), 226–234, and *Un savant méconnu: G. P. de Roberval, son activité intellectuelle dans les domaines mathématique, physique, mécanique et philosophique* (Paris, 1962); Adrien Baillet, *La vie de Monsieur Des-Cartes*, 2 vols. (Paris, 1691); Le Marquis de Condorcet, "Éloge de Roberval" (1773), republished in *Oeuvres de Condorcet*, II (Paris, 1847), 5–12; Pierre Costabel, "La controverse Descartes–Roberval au sujet du centre d'oscillation," in *Revue des sciences humaines*, n.s. 61, fasc. 61 (Lille–Paris, 1951), 74–86; Victor Cousin, "Roberval philosophe," in *Fragments de philosophie cartésienne* (Paris, 1845), 229–261; Carlo Dati, *Lettera a Filaleti di Timauro Antiate della vera storia della cicloide, e della famosissima esperienza dell'argento vivo* (Florence, 1663); Jean-Marie-Constant Duhamel, "Note sur la méthode des tangentes de Roberval," in *Mémoires présentés par divers savants à l'Académie des sciences de l'Institut de France*, 5 (1838), 257–266; Pierre Duhem, *Les origines de la statique*, 2 vols. (Paris, 1905–1906), and *Études sur Léonard de Vinci*, I (Paris, 1955); Marie-Antoinette Fleury and Georges Bailhache, "Le testament, l'inventaire après décès . . . de Gassendi," in *Tricentenaire de Gassendi*, Actes du Congrès de Digne, 1955 (Paris–Digne, 1957), 21–68; Kokiti Hara, "Étude sur la théorie des mouvements composés de Roberval" (*thèse de troisième cycle*, defended in 1965 at the Faculté des lettres et sciences humaines de l'Université de Paris), and "Remarque sur la quadrature de la surface du cône oblique," in *Revue d'histoire des sciences et de leurs applications*, 20, no. 4 (1967), 317–332; J. E. Hofmann and P. Costabel, "A propos d'un problème de Roberval," *ibid.*, 5, no. 4 (1952), 312–333; Jean Itard, "Autre remarque sur la quadrature de la surface du cône oblique," *ibid.*, 20, no. 4 (1967), 333–335; Robert Lenoble, "Roberval éditeur de Mersenne et du P. Niceron," *ibid.*, 10, no. 3 (1957), 235–254; and M. Mersenne, *Harmonie universelle, . . .* 2 vols. (Paris, 1636), reprinted in facsimile (Paris, 1963),

Cogitata physico-mathematica (Paris, 1644), and *Novarum observationum physico-mathematicarum tomus III* (Paris, 1647); Blaise Pascal, *Histoire de la roulette . . .* (10 Oct. 1658), also in Latin, *Historia trochoidis* (same date), republished in *Oeuvres de Blaise Pascal*, L. Brunschvicg, P. Boutroux, and F. Gazier, eds., VIII (1914), 195–223; Bernard Rochot, "Roberval, Mariotte et la logique," in *Archives internationales d'histoire des sciences*, **22** (1953), 38–43; Paul Tannery, *La correspondance de Descartes dans les inédits du fonds Libri* (Paris, 1893), republished in *Mémoires scientifiques*, **6** (1926), 153–267; Cornelis de Waard, "Une lettre inédite de Roberval du 6 janvier 1637 contenant le premier énoncé de la cycloïde," in *Bulletin des sciences mathématiques*, 2nd ser., **45** (1921), 206–216, 220–230; and Evelyn Walker, *A Study of the Traité des indivisibles de G. P. de Roberval* (New York, 1932).

KOKITI HARA

ROBIN, CHARLES-PHILLIPE (*b.* Jasseron, Ain, France, 4 June 1821; *d.* Jasseron, 6 October 1885), *biology, histology.*

Robin's father's family were rich bourgeoisie with a high regard for scholarship. He was greatly influenced by his mother, Adelaïde Tardy, whose family included several physicians. Robin attended the boarding-school at Menestruel, near Poncin (Ain), where as a young boy he lost an eye while playing with his fellow students. He had to wear a glass prosthesis, and monocular vision may have played some role in his later predilection for working with the microscope.

After studying the classics at the Collège Royal of Lyons, Robin enrolled in 1838 at the Faculty of Medicine of Paris. From the start of his medical studies, anatomy and biological research attracted him much more than clinical medicine. In 1845, while still a student, Robin traveled with Hermann Lebert to the coast of Normandy and to the Channel Islands to collect specimens for the museum of comparative anatomy that M. J. B. Orfila wished to establish in Paris. Robin subsequently published (partly in collaboration with Lebert) a series of notes on such topics as the lymphatic and venous systems of marine animals, the reproductive mechanism of the squid, the comparative anatomy of the genitalia, the structural elements of the fibroplastic tissue, and the morphology of various animal and vegetable parasites (1845–1846). His initial research displayed two recurrent characteristics, skillful use of the microscope and a comparative approach to his subject matter.

Robin received his medical degree on 31 August 1846 with a thesis on the topographical anatomy of the region of the groin. In 1847 he defended two theses for the *doctorat ès sciences naturelles*: an original and

important zoological investigation of the electric organs of Rajidae, and a biological examination of parasitic vegetable growths of man and animals. He was the first to describe *Oidium albicans (Candida albicans)* and to explain thrush as a parasitosis caused by this microscopic fungus.

Having won the *agrégation* in natural history (1847), Robin began a private course in pathological anatomy, set up a laboratory of comparative anatomy, and in 1849 replaced Achille Richard in the chair of natural history at the Faculty of Medicine of Paris. The primary subjects of his publications were now the histology of the vertebrate nervous system and the microscopic structure of tumors. Through Pierre-François Rayer, Robin met Émile Littré and was introduced to Auguste Comte, whose lectures on positivist philosophy captivated him. In positivism Robin found the possibility of giving to his specialized studies a doctrinal unity. Lamarck introduced the term *biologie* into France and Comte and Littré had popularized it. But Robin was responsible for its final acceptance and for the later elaboration of the concept of general biology in French scientific circles.

Robin was the leading proponent of the Société de Biologie; he urged its creation, wrote up its statutes, and, with Rayer, Claude Bernard, and Brown-Séquard, directed its initial activities (1848). At the first meeting of this society, Robin presented a memoir entitled "Sur la direction que se sont proposé, en se réunissant, les membres fondateurs de la Société de Biologie pour répondre au titre qu'ils ont choisi." This credo of positivist biology exercised a significant influence on the orientation of physiological, medical, and zoological research in France.

Robin set forth in detail his own ideas on biology in two books, published in 1849 and 1851, *Du microscope* and *Tableau d'anatomie*. In this view of general anatomy Robin went beyond Bichat, asserting that the anatomical element itself, independent of the tissue of which it is a part, ought to be the subject of both morphological and physiological research. At the same time, Robin asserted that life did not depend on a rigid structure but on a "state of organization" —in fact, on "a particular molecular state." The notion of the blastema, central to Schwann's cell theory, fully corresponded to Robin's ideas, but he was never able to adopt the cell theory in its newest phase, as formulated by Virchow. Robin never accepted the view that the cell could be the single fundamental component of organized beings. For him the real seat of life was constituted by the humoral parts of the organism. Beyond the fixed anatomical elements, there must be, he thought, a molecular organization that explained the morphology. In his opinion, therefore, microscopic investigation was only a stage of biological research and must be followed by chemical analysis. In collaboration with a chemist, F. Verdeil, Robin studied the chemical compounds of which the organism is composed. Despite its display of useful information, the resulting *Traité de chimie anatomique et physiologique, normale et pathologique* (1852–1853), showed that research oriented in this direction led at that time to a dead end and that, given the contemporary state of chemical knowledge, the superiority of a morphological approach was undeniable.

In carrying out his ambitious program of histological and biochemical research Robin made serious errors but also a number of important discoveries, including the description of the osteoclasts in regard to bone formation (1851), study of the change in the uterine mucus during pregnancy and some new facts on the microscopic structure of ganglions and of neuroglia.

At Littré's request he helped him to revise Nysten's *Dictionnaire de médecine*. Far from being a simple guide to semantics and orthography, Littré and Robin's *Dictionnaire* became "the medical code of the positivist doctrine" (E. Chauffard). A chair of histology was created for Robin in 1862 at the Paris Faculty of Medicine. In 1864 he founded, with Brown-Séquard, the *Journal de l'anatomie et de la physiologie normales et pathologiques de l'homme et des animaux.*

With Robin's election to the Académie des Sciences on 15 January 1866, politics, teaching, and administrative tasks began to take precedence over scientific research. During the Franco-Prussian War he held the post of director of the army medical corps, and in 1873 he was named director of the marine zoology laboratory at Concarneau. In 1875 he was elected to the Senate as deputy from the Ain. In politics Robin was a partisan of the left, a "free-thinker," heavily engaged in anticlerical activity.

Although Robin's early micrographical research was a valuable contribution to science, and although his conception of general anatomy was historically useful as a transition between Bichat's histology and cellular biology, a similarly favorable judgment cannot be rendered on the work and ideas of the last period of his life. He opposed Virchow's cellular pathology, refused to accept such modern histological methods as slices and stainings, and opposed Pasteur's microbiological discoveries. When Robin's lifework, *Anatomie et physiologie cellulaires*, was published in 1873, even his students were abandoning him. His teaching no longer reflected contemporary scientific thought.

Robin never married. He led a frugal life, wholly dedicated to his work. He was a brilliant but peremptory teacher. Serving as medical counselor to many French

writers (Mérimée, Sainte-Beuve, Taine, Flaubert, A. Dumas *fils*, Michelet, About, the Goncourt brothers), Robin deserves to be labeled the "*éminence grise* of naturalism" (P. Voivenel).

BIBLIOGRAPHY

I. Original Works. Robin's principal books are *Recherches sur un appareil qui se trouve sur les poissons du genre des Raies* (Paris, 1847); *Du microscope et des injections dans leurs applications à l'anatomie et à la pathologie* (Paris, 1849; rev. ed., Paris, 1877); *Tableaux d'anatomie* (Paris, 1851); *Traité de chimie anatomique et physiologique, normale et pathologique*, 3 vols. (Paris, 1852–1853), written with F. Verdeil; *Histoire naturelle des végétaux parasites qui croissent sur l'homme et sur les animaux vivants* (rev. ed., Paris, 1853); *Dictionnaire de médecine, de chirurgie, de pharmacie, des sciences accessoires et de l'art vétérinaire*, 10th ed. of P. H. Nysten, *Dictionnaire de médecine*, 2 vols. (Paris, 1855; 11th ed., 1858; 12th ed., 1865), written with E. Littré; *Leçons sur les humeurs normales et morbides du corps de l'homme* (Paris, 1867; 2nd ed., 1874); *Anatomie et physiologie cellulaires* (Paris, 1873); and *Nouveau dictionnaire abrégé de médecine* (Paris, 1886).

Robin was a prolific writer; for a bibliography of about 300 of his articles (1844–1885), see *Journal de l'anatomie et de la physiologie normales et pathologiques de l'homme et des animaux*, **22** (1886), clxvi–clxxxiv.

II. Secondary Literature. The best biographical study of Robin is undoubtedly V. Genty, *Un grand biologiste: Charles Robin (1821–1885), sa vie, ses amitiés philosophiques et littéraires* (Lyons, 1931), but the treatment of his purely scientific work is superficial and excessively indulgent. The basic work remains the *éloge* of his disciple G. Pouchet, "Charles Robin (1821–1885), sa vie, son oeuvre," in *Journal de l'anatomie et de la physiologie . . .*, **22** (1886), i–clxv. Also useful is his autobiographical *Notice sur les travaux d'anatomie et de zoologie de M. Charles Robin*, 2 pts. (Paris, 1860–1865); and a collective publication, "L'oeuvre scientifique de Charles Robin," in *Annales de la Société d'émulation de l'agriculture, sciences, lettres et arts de l'Ain*, **21** (1888), 59–305. For insights into Robin's personality, see G. Variot, "Quelques souvenirs anecdotiques sur Charles Robin," in *Bulletin de la Société française d'histoire de la médecine*, **19** (1925), 8–15.

M. D. Grmek

ROBINET, JEAN-BAPTISTE RENÉ (*b.* Rennes, France, 23 June 1735; *d.* Rennes, 24 March 1820), *literature, philosophy, natural history.*

Robinet, who began his career as a writer, worked for booksellers in Holland and then in Bouillon. Besides translating works by Hume (*Essais de morale* [1760]) and by several English writers—including

Frances Sheridan, John Langhorne, and Sir Charles Morell—he collaborated on a number of periodicals, dictionaries, and other joint works. He also wrote French and English grammars, an *Analyse raisonnée de Bayle* (London, 1755–1770), and a few works in economics and politics. Returning to Paris in 1778, he was appointed a royal censor. At the outbreak of the French Revolution he withdrew to Rennes, where he spent the rest of his life.

Robinet's most important work from the point of view of the history of science is his four-volume treatise *De la nature* (1761–1766). It was followed in 1768 by *Considérations philosophiques de la gradation naturelle des formes de l'être, ou les essais de la nature qui apprend à faire l'homme* and in 1769 by *Parallèle de la condition et des facultés de l'homme avec la condition et les facultés des autres animaux.*

Robinet's principal intention in these works was to demonstrate that there is an equal quantity of good and evil in all the conditions and creatures of the universe. A staunch proponent of the idea of "the chain of beings," he believed that it was continuous, its elements separated only by imperceptible gradations. At the base of the chain, creatures are characterized by very little good or evil. At the top stands man, in whom good and evil, always present to an equal degree, are engaged in a convulsive struggle.

This vision of the moral world is based on a vision of the physical world, according to which all things—even stones—are sentient, living, and organized. Robinet argued that the phenomenon of nutrition—in which matter circulates from the earth to plants, then to herbivores, and finally to carnivores and man—proves that matter is organized and sentient. If this were not the case, how could inorganic matter constitute and nourish organs? The difference between beings, from crystals to man, is therefore only a difference in the degree of organization. The sponges and polyps constitute, respectively, the transitions between stones and plants, and between plants and animals. In this scheme Robinet carried to an extreme the ideas of Buffon and Charles Bonnet.

Robinet's metaphysical ideas, strongly influenced by Leibniz, are of a certain interest; they even attracted the attention of Hegel. According to Robinet, nature, although created, is, like God, eternal; but it exists in succession and in time. The created universe contains the germ of every being that will develop—in the literal sense of the term—in the course of its history. Robinet followed Leibniz in adopting the theory of the preexistence of germs, but he did not think that these germs, which were created at the beginning of the world, were encased ("*emboîtés*") inside each other. He held that the progressive development of these germs could give rise to completely unknown creatures, unexpected variations of the initial prototype. Accordingly, in certain of his statements Robinet appears to be heralding the theory of transformism and the idea that all living beings are constructed according to a single basic plan. In fact, he was only elaborating ideas he had borrowed from Leibniz and perhaps also, directly or indirectly, from Paracelsus.

The proof that Robinet did not propose a genuine theory of transformism is furnished by *Considérations philosophiques.* Here, claiming to exhibit "Nature, who teaches us how to construct man," he discusses a collection of stones, roots, and animals possessing more or less imaginary resemblances to parts of the human body. The total absence of a critical sense, or even of simple common sense, shows that Robinet was much more a metaphysician than a naturalist.

Nevertheless, Robinet's work illustrates several important elements in the scientific thinking of the second half of the eighteenth century: the unity of nature, the chain of beings, universal dynamism and sensibility, and—at this early date—vitalism. It also illustrates the role of Leibniz in the development of Enlightenment ideas on living nature. Robinet's writings influenced philosophers rather than scientists, but they cast an interesting light on the genesis of the theory of transformism.

BIBLIOGRAPHY

Robinet's major works have been cited in the text. On his life and work, see Jacques Roger, *Les sciences de la vie dans la pensée française du XVIIIe siècle* (Paris, 1963), 642–651; and Corrado Rosso, "Il paradosso di Robinet," in *Filosofia,* **5**, no. 1 (1954), 37–62.

JACQUES ROGER

ROBINS, BENJAMIN (*b.* Bath, England, 1707; *d.* Fort St. David, India, 29 July 1751), *mathematics, military engineering.*

Robins probably is best known as the inventor of the ballistic pendulum. Today the device is used to demonstrate conservation of momentum as well as for the purpose to which Robins put it: to determine the muzzle velocity of bullets. Robins needed experimental confirmation of his theoretical computations.

Born of Quaker parents, Robins never showed an inclination for pacifism. Trained as a teacher, he soon left that profession for mathematics and for ballistics and fortifications. In 1727 he published an article in the *Philosophical Transactions of the Royal Society,* a demonstration of the eleventh proposition of Newton's *Treatise on Quadratures*. Although the *Dictionary of National Biography* asserts that he accomplished the work without help, it is doubtful that Newton missed passing on it. Robins became one of Newton's most adamant defenders, often to the point of indelicacy. Much of his writing was devoted to attacks on Newton's enemies—Leibniz, the Bernoullis, Berkeley, and James Jurin. Robins took part in the celebrated *vis viva* controversy, the subject of most of his polemics.

Robins' best-known work, *New Principles of Gunnery*, appeared in 1742. Euler translated it into German in 1745, adding his own commentary. It was also translated into French in 1751. It was there that Robins described the ballistic pendulum. His other work on ballistics was far from trivial, including studies of the resistance of fluid media to high-speed objects, pressures on projectiles inside a gun barrel, the rifling of barrel pieces, and the shape of actual, as opposed to ideal, trajectories. For his service he was awarded the Copley Medal in 1747. His last work consisted of investigations of rockets for the purpose of military signaling.

Robins never married. He died in India, where he had gone to assist the British East India Company in renovating fortifications.

BIBLIOGRAPHY

I. Original Works. Robins published only two articles in the *Philosophical Transactions of the Royal Society*: "Demonstration of the Eleventh Proposition of Sir I. Newton's Treatise of Quadratures," **34** (1727), 230–236; and "On the Height to Which Rockets Will Ascend," **46** (1749), 131–133; and participated in the research for John Ellicott, "An Account of Some Experiments . . . to Discover the Height to Which Rockets May Be Made to Ascend," **46** (1750), 578–584. His major work remains *New Principles of Gunnery* (London, 1742). The most valuable collection of his writings was published by a friend ten years after his death: *Mathematical Tracts of the Late Benjamin Robins*, James Wilson, ed., 2 vols. (London, 1761). This collection contains Robins' book on gunnery, the polemics on the *vis viva* controversy and other articles read to the Royal Society but until then unpublished, reprints of the published articles, and Wilson's personal comments on the life and character of his old friend.

II. Secondary Literature. There is almost nothing except occasional mention of Robins in general histories. There are accounts of his life in British biographical series, most notably the *Dictionary of National Biography*. No full biography has been published.

J. Morton Briggs, Jr.

ROBIQUET, PIERRE-JEAN (*b.* Rennes, France, 13 January 1780; *d.* Paris, France, 29 April 1840), *chemistry, pharmacy.*

As a youth Robiquet was caught up in the turbulence of Revolutionary France. After serving an apprenticeship and working in pharmacies in Lorient, Rennes, and Paris, he became Fourcroy's *préparateur* in the chemical laboratory shared with Vauquelin on the rue des Bourdonnais in Paris. Here Robiquet had the opportunity to assist the two chemists with their analyses of urinary calculi and to become friends with Vauquelin's assistant, Thenard. Conscripted into the army in 1799 as a military pharmacist, he experienced many hardships in the Italian campaign but was able to attend the lectures of Volta on physics and of Scarpa on anatomy while he was stationed in Pavia.

After the French victory at Marengo, Robiquet was assigned to the military teaching hospital in Rennes in 1801 and several years later to the Val-de-Grâce in Paris, from which he resigned in 1807 to work in Vauquelin's private laboratory. Needing more money, he left Vauquelin's employ and established his own pharmacy, to which he added facilities for the manufacture of chemicals. Robiquet's teaching career began in 1811 with his appointment as *répétiteur* in chemistry at the École Polytechnique and assistant professor of the natural history of drugs at the École de Pharmacie, where he became full professor in 1814. Poor health forced him to resign his professorship in 1824 and to accept the post of treasurer of the pharmacy school, which he held for the rest of his life. He was elected to the Academy of Medicine in 1820 and to the Academy of Sciences in 1833.

Robiquet took a leading role in the expanding search during the first decades of the nineteenth century for new constituents in natural products, termed *principes immédiats*, *principes prochains* (Fourcroy), or "proximate principles." The Scottish physician and toxicologist Robert Christison, who came to Paris in 1820 to study analytical chemistry in Robiquet's laboratory on the rue de la Monnaie, later recalled: "My own foremost desire was to practice Proximate Organic Analysis. This branch of chemistry had been cultivated for a few years, . . . and nowhere with such energy as in Paris." Robiquet's earliest research dealt with the analysis of asparagus juice (1805) and was followed by the joint publication with Vauquelin of their isolation of asparagine (1806). In succeeding years Robiquet discovered glycyrrhizin in licorice (1809); analyzed cantharides (1810) and kermes (1812); discovered caffeine (1821) independently of Pierre Pelletier, Caventou, and F. Runge; discovered narcotine (1817) and codeine (1832) in opium; with J. J. Colin isolated alizarin and purpurin from madder (1826–1827); and discovered orcinol in lichens (1829).

One of Robiquet's most significant investigations, carried on with Antoine Boutron-Charlard in 1830 on bitter almonds, led to their discovery of amygdalin. They were not able, however, to account for the production of oil of bitter almonds (benzaldehyde) in some of their experiments nor to grasp the theoretical implications of their discovery of amygdalin, which was elucidated by Wöhler and Liebig in their remarkable work on the benzoyl radical in 1832. Robiquet also produced rufigallic acid (1836) and discovered citraconic acid (1837). Also noteworthy were his early investigations in inorganic chemistry, notably his study of carbon disulfide (1807) and his purification of baryta (1807) and nickel (1809).

Recognized by his contemporaries as an outstanding teacher and one of the most eminent analytical and experimental chemists, Robiquet was also a person of considerable charm. In the words of Christison: "He was a man of middle stature, very like a handsome, shapely, English gentleman, with a sharp, lively, amiable expression, and a fair share of French quickness of temper, but under admirable control."

BIBLIOGRAPHY

I. ORIGINAL WORKS. Among Robiquet's most important writings are "Essai analytique des asperges," in *Annales de chimie*, **55** (1805), 152–171; "Découverte d'un nouveau principe végétal dans le suc des asperges," *ibid.*, **57** (1806), 88–93, written with Vauquelin; "Sur le soufre liquide de Lampadius avec une note de Vauquelin," *ibid.*, **61** (1807), 145–152; "Sur la préparation de la baryte pure," *ibid.*, **62** (1807), 61–64; "Sur la purification de nickel par l'hydrogène sulfuré," *ibid.*, **69** (1809), 285–292; "Analyse de la réglisse," *ibid.*, **72** (1809), 143–159; "Expériences sur les cantharides," *ibid.*, **76** (1810), 302–322; "Observations sur la nature du kermès," *ibid.*, **81** (1812), 317–331; "Observations sur le mémoire de M. Sertuerner relatif à l'analyse de l'opium," *ibid.*, 2nd ser., **5** (1817), 275–278; and "Nouvelles expériences sur l'huile volatile d'amandes amères," *ibid.*, **21** (1822), 250–255.

See also "Sur un nouveau principe immédiat des végétaux (l'alizarin) obtenu de la garance," in *Journal de pharmacie et des sciences accessoires*, 2nd ser., **12** (1826), 407–412, written with Colin; "Nouvelles recherches sur la matière colorante de la garance," in *Annales de chimie et de physique*, 2nd ser., **34** (1827), 225–253, written with Colin; "Essai analytique des lichens de l'orseille," *ibid.*, **42** (1829), 236–257; "Nouvelles expériences sur les amandes amères, et sur l'huile volatile qu'elles fournissent," *ibid.*, **44** (1830), 352–382, written with Boutron-Charlard; "Nouvelles observations sur les principaux produits de l'opium," *ibid.*, **51** (1832), 225–267; "Nouvelles observa-

tions sur l'orcine," *ibid.*, **58** (1835), 320–335; "Notice sur l'acide gallique," in *L'Institut* (Paris), **4** (1836), 179–180; and "De l'action de la chaleur sur l'acide citrique," in *Annales de chimie et de physique*, 2nd ser., **65** (1837), 68–86. For a more complete listing of Robiquet's papers, see Royal Society *Catalogue of Scientific Papers*, V, 240–243.

II. SECONDARY LITERATURE. See A. Balland, *Les pharmaciens militaires français* (Paris, 1913), 181–182; A. Bussy, "Éloge de Pierre Robiquet," in *Journal de pharmacie et des sciences accessoires*, **27** (1841), 220–242; *Centenaire de l'École supérieure de pharmacie* (Paris, 1904), 282–283; P. Crété, "Pierre Robiquet," in *Figures pharmaceutiques françaises* (Paris, 1953), 47–52; E. Pariset, *Histoire des membres de l'Académie royale de médecine* (Paris, 1850), 584–587; and J. R. Partington, *A History of Chemistry*, IV (London–New York, 1964), 241–242, 327–328, *passim*. An interesting personal account of Robiquet and his laboratory is in R. Christison, *The Life of Sir Robert Christison, Bart.*, I (Edinburgh–London, 1885), 267–274. For a discussion of the theoretical significance of the discovery of amygdalin by Robiquet and Boutron-Charlard, see E. V. McCollum, *A History of Nutrition* (Boston, 1957), 49–50.

ALEX BERMAN

ROBISON, JOHN (*b.* Boghall, Stirlingshire, Scotland, 1739; *d.* Edinburgh, Scotland, 30 January 1805), *physics, applied mechanics.*

Robison was professor of natural philosophy at the University of Edinburgh at a time when it had become a flourishing center of science and learning, and he was an influential and prolific author of scientific works in which he often blended practical and theoretical knowledge.

The son of a prosperous merchant, Robison attended a Glasgow elementary school and, at the age of eleven, entered the University of Glasgow, where he received the M.A. in 1756. Although he was encouraged by his father to pursue a clerical career, his strongest interests were in mathematics and mechanics. Accordingly, in 1759 he accepted an offer to tutor a son of Admiral Charles Knowles in mathematics and navigation, thus becoming involved in the naval affairs that occupied him intermittently over the next fourteen years. Robison went with the younger Knowles to Canada, where he served as midshipman, performed surveys, and took part in James Wolfe's assault on Quebec. Later he sailed to Portugal. During these periods of service at sea he gained considerable knowledge of seamanship and naval technology. In 1761 Robison was appointed by the Board of Longitude to represent it in the testing of the timekeeper John Harrison had constructed for determining longitude at sea; he observed the device on a trip to Jamaica. By 1762, however, the prospects of a naval career had dimmed and, after briefly reconsidering the Church, he returned to Glasgow to resume his studies.

Before leaving Glasgow in 1758, Robison had become acquainted with Joseph Black and James Watt. Indeed, it was Robison, who had already published a note on an improvement of the Newcomen engine, who first turned Watt's attention to the steam engine. He revived these friendships and became Black's student. During the next four years Robison was closely associated with both men; and in 1766, when Black transferred to the University of Edinburgh, his recommendation led to Robison's appointment as lecturer in chemistry at the University of Glasgow. The appointment was renewed annually, and Robison seemed to be established in an academic career. In 1770, however, he left Glasgow again, this time to accompany Admiral Knowles to St. Petersburg, where Knowles served as president of the Russian Board of Admiralty and Robison acted as his secretary. He worked on plans to improve the construction and navigation of Russian warships until 1772, when he accepted the post of inspector general of the corps of marine cadets at Kronstadt, an appointment that carried the rank of lieutenant colonel. When Robison was offered—again on Black's recommendation— the position of professor of natural philosophy at the University of Edinburgh in 1773, he abandoned a naval career for the second time and returned to Scotland. The following year he took up his duties at the university, where he remained until his death. One of the founders of the Royal Society of Edinburgh, he was elected its first general secretary in 1783.

Although he published three articles in the early volumes of the *Transactions of the Royal Society of Edinburgh*, Robison's scientific career remained undistinguished until he became the principal contributor to the third edition of the *Encyclopaedia Britannica*. From 1793 to 1801 he composed a remarkably wide-ranging series of articles that, according to Thomas Young (who revised several of them for the fourth edition), "taken together, undeniably exhibit a more complete view of the modern improvements of physical science than had ever before been in the possession of the British public . . . ." Their strongest influence was in applied structural mechanics, where the articles "Strength of Materials," "Roof," "Arch," "Carpentry," and "Centre" (for bridges) essentially constituted a unique course of instruction deliberately presented in a didactic manner for the benefit of the artisans and craftsmen who filled the ranks of British engineering.

In preparing these articles Robison consulted the engineer John Rennie (the elder), who had attended his courses at the University of Edinburgh and whose practical knowledge of building he valued greatly. His approach to these technical topics was, however, essentially theoretical. In his discussion of the theory of flexure, Robison emphasized the soundness of the neglected analyses that Parent and Coulomb had formulated and that now appeared for the first time in English. And in his presentation of column theory he called attention to Euler's error in assuming that when a column bends, the cross sections sustain only tension. An indication of the influence of these articles is that through them the terms "strength of materials" and "neutral point" (to designate the position on the cross section of a beam where the stresses are zero) became established. After his death many of Robison's *Britannica* articles were edited by David Brewster, who had been one of his pupils, and were published as *A System of Mechanical Philosophy* (1822).

Public recognition of the need for applied mechanics came only late in Robison's life. In 1800, when Thomas Telford proposed a cast-iron bridge over the Thames in the form of a single arch, there were no theoretically informed engineers who could analyze and evaluate the design; instead Parliament sought advice from a committee divided into "theoreticians" and "practitioners," Robison being included among the former. Despite a few helpful suggestions by Robison and several of the "practitioners," the general inability of the committee to analyze Telford's design directed attention to the importance of furthering the application of mathematics and mechanics to the problems of structural engineering. Half a century later, applied mechanics had become a distinct field of study and professorships of engineering had been established. In 1855 W. J. Rankine, in his inaugural address at the University of Glasgow, paid tribute to Robison for his role in uniting theoretical science and the practical arts.

Robison made only the slightest original contributions to scientific research, the most notable being his determination, on strictly experimental grounds, that electrical attraction and repulsion follow an inverse-square law. Using an electrometer of his own design, he found the repulsion to be inversely proportional to the 2.06th power of the distance and the attraction slightly less than the second power. Although he read a paper on his results in 1769 (two years after Priestley had arrived at the inverse-square law through an elegant analogy between electricity and gravitation), he failed to publish them until 1801.

Robison was deeply religious and politically conservative. He became favorably impressed with Bošković's theory of point atoms both because he considered it to be an elaboration of Newtonian principles (the action between the points is accounted for by attractive and repulsive forces) and because he saw it as a rejoinder to the materialism that he believed was corrupting natural philosophy. He contributed a long article on Bošković's system to the *Britannica* and lectured on point atomism to his students at Edinburgh.

Robison's most widely read work, however, was a fiercely anti-Jacobin tract in which he attributed the French Revolution largely to Continental Freemasonry (into which he had been initiated during his journey to Russia), materialism, and the influence of the "German Union" and the short-lived "Order of Illuminati." The book, an intemperate expression of the Tory politics that dominated Edinburgh during the 1790's, was later soundly criticized as credulous and tendentious by John Playfair, a steadfast and outspoken Whig, who succeeded Robison in the chair of natural philosophy at Edinburgh. Published in 1797, it went through several editions in two years and, despite the implausibility of the argument, was well received in anti-Jacobin and some sectarian circles.

In 1798 Robison was awarded an honorary LL.D. by the College of New Jersey (now Princeton University), more for his political and religious views than for his contributions to science.

BIBLIOGRAPHY

I. ORIGINAL WORKS. Robison contributed the following articles to the 3rd ed. of the *Encyclopaedia Britannica*, 18 vols. (Edinburgh, 1797): "Optics," "Philosophy" (jointly with the editor, George Gleig), "Physics," "Pneumatics," "Precession," "Projectiles," "Pumps," "Resistance," "Rivers," "Roof," "Rope-making," "Rotation," "Seamanship," "Signal," "Sound," "Specific Gravity," "Statics," "Steam," "Steam Engine," "Steelyard," "Strength" (of materials), "Telescope," "Tide," "Trumpet," "Variation," and "Waterworks"; and to the *Supplement*, 2 vols. (Edinburgh, 1801): "Arch," "Astronomy," "Boscovich," "Carpentry," "Centre" (for bridges), "Dynamics," "Electricity," "Impulsion," "Involution," "Machinery," "Magnetism," "Mechanics," "Percussion," "Piano-forte," "Position," "Temperament" (in music), "Thunder," "Trumpet," "Tschirnhaus," and "Watchwork." Many of these (slightly edited), along with several of Robison's other works, are most readily accessible in the vols. published posthumously under his name as *A System of Mechanical Philosophy*, David Brewster, ed., 4 vols. (Edinburgh, 1822). "Steam" and "Steam Engine" were published separately as *The Articles Steam and Steam Engines*, with notes by James Watt (Edinburgh, 1818).

Robison edited the notes for Joseph Black's chemistry lectures as *Lectures on the Elements of Chemistry*, 2 vols. (Edinburgh, 1803); the comments he supplied reflect his reserved attitude toward Lavoisier and the French school of chemistry. He intended to produce a complete treatise on what he generally termed "mechanical philosophy" (approximately equivalent to what was then known as "natural philosophy" and what is now designated as physics), but at his death he had completed only *Outlines of a Course of Lectures on Mechanical Philosophy* (Edinburgh, 1797) and the first vol. of *Elements of Mechanical Philosophy* (Edinburgh, 1804).

The *Transactions of the Royal Society of Edinburgh* contain three papers by Robison: "The Orbit and Motion of the *Georgium Sidus*, Determined Directly From Observations," **1**, pt. 2 (1788), 305–332; "Observations on the Places of the Georgian Planet, Made at Edinburgh With an Equatoreal Instrument," **2**, pt. 2 (1790), 37–38; and "On the Motion of Light, as Affected by Refracting and Reflecting Substances, Which Are Also in Motion," *ibid.*, 83–111. (Robison has sometimes been confused with his son, Sir John Robison, who also was general secretary of the Royal Society of Edinburgh. Even the Society's index to its *Transactions* attributes one of the son's articles to the father.)

Robison's early note on the Newcomen engine appeared in *Universal Magazine of Knowledge and Pleasure,* **21** (1757), 229–231 (signed J—n R—n). Thomas Young credited Robison as the anonymous author of a critical review of one of George Atwood's books on arch theory: "Atwood on the Construction of Arches," in *British Critic*, **23** (1804), 6–14. The report Robison prepared for Parliament on Telford's bridge design is in *Report From the Select Committee Upon the Improvement of the Port of London*, British Sessional Papers (House of Commons, 1801), III. One of his MSS, "Professor Robison's Narrative of Mr. Watt's Invention of the Improved Engine Versus Hornblower and Maberley 1796," is reproduced in Eric Robinson and A. E. Musson, *James Watt and the Steam Revolution* (New York, 1969), 23–38. Many of Robison's letters, mainly to Watt and Black, are in Eric Robinson and Douglas McKie, *Partners in Science* (Cambridge, Mass., 1970).

His book on the French Revolution is *Proofs of a Conspiracy Against All the Religions and Governments of Europe* (Edinburgh, 1797).

II. SECONDARY LITERATURE. The fullest biographical work on Robison is John Playfair, "Biographical Account of the Late John Robison," in *Transactions of the Royal Society of Edinburgh*, **7** (1815), 495–539. Three years before Robison's death George Gleig, the editor of the *Encyclopaedia Britannica*, who knew him well and shared his political views, published a biographical article: "Dr. Robison," in *Anti-Jacobin Review and Magazine*, **11** (1802), 91–97, repr. in *Philosophical Magazine*, **13** (1802), 386–394. Thomas Young, "Life of Robison," in *Miscellaneous Works of the Late Thomas Young*, George Peacock, ed., 3 vols. (London, 1855), II, 505–517, contains additional information.

Two recent articles throw new light on Robison's place in the intellectual and political life of late eighteenth-century Scotland: Richard Olson, "The Reception of Boscovich's Ideas in Scotland," in *Isis*, **60** (1969), 91–103; and J. B. Morrell, "Professors Robison and Playfair, and the *Theophobia Gallica:* Natural Philosophy, Religion and Politics in Edinburgh, 1789–1815," in *Notes and Records. Royal Society of London*, **26** (1971), 43–63.

HAROLD DORN

ROCHE, ÉDOUARD ALBERT (*b*. Montpellier, France, 17 October 1820; *d*. Montpellier, 18 April 1883), *celestial mechanics, geophysics, meteorology.*

Roche spent nearly his whole life in his native city. Several members of his family had been professors at the University of Montpellier, where he earned his *docteur ès sciences* in 1844. While in Paris for three years in order to increase his knowledge of analysis and celestial mechanics, he engaged in scientific discussions with Cauchy and Le Verrier. Arago, who had taken notice of Roche's observations of the solar eclipse of 1842, welcomed him at the Paris observatory as an independent student. In 1849 Roche was appointed *chargé de cours* at the Faculté des Sciences of Montpellier, and in 1852 he was named professor of pure mathematics. He was elected a corresponding member of the Académie des Sciences in December 1873. Roche had suffered from delicate health since his youth, and, exhausted by work, he was obliged to take a leave of absence from Montpellier in 1881. Eighteen months later he died of an inflammation of the lungs.

Roche's investigations concerned primarily the internal structure and form of the free surface of celestial bodies, a subject he had treated in his doctoral dissertation. The law of the differential variation of terrestrial density, which he proposed in 1848, is still used. Roche studied the equilibrium figures of a rotating fluid mass subjected, in addition to internal forces, to an external attractive force or to a central attractive force; Roche's limit, the maximum value that the distance of a satellite imposes on its diameter (stated in 1849), is an essential criterion in cosmogony. He also considered the form of cometary envelopes and analyzed the effect of a repulsive force originating in the sun. The shape of comets was thus correctly explained in 1859, before the physical discovery of radiation pressure.

The elements permitting the study of two fundamental problems were now conjoined. In 1873 Roche undertook a critical examination of Laplace's cosmogonic hypothesis, which had never been the subject of thorough mathematical study. Roche provided important additions to it in order to render it coherent. In 1881 he analyzed hypotheses concerning the structure of the earth and was led to propose and study the first "earth model" with a solid nucleus.

Roche investigated various areas of pure mathematics and meteorology, and his generalization of Taylor's formula has become classical. He also definitively solved the historic problem of solar obfuscations (temporary diminutions in the solar radiation) by showing that each of the cases cited was the result of an eclipse or of a local atmospheric phenomenon.

BIBLIOGRAPHY

I. ORIGINAL WORKS. Almost all of Roche's work is contained in thirty articles in *Mémoires de l'Académie des sciences et lettres de Montpellier. Section des sciences*—cited henceforth as *Mémoires*—from 1848 to 1882.

On fluid masses in rotation, see *Mémoires*, **1** (1849), 243–262; **1** (1850), 333–348; and **2** (1851), 21–32. On the atmospheres of the planets and the shape of comets, see *Mémoires*, **2** (1854), 399–439; *Annales de l'Observatoire de Paris*, **5** (1859), 353–393; *Mémoires*, **4** (1860), 427–478; and **5** (1862), 263–302. On the interior of the earth and gravity, see *Mémoires*, **1** (1848), 117–128; **2** (1853), 251–264; **3** (1855), 107–124; and especially **10** (1881), 221–266. Laplace's cosmogonic hypothesis is analyzed in "Essai sur la constitution et l'origine du système solaire," in *Mémoires*, **8** (1873), 235–327. On Taylor's formula, see *Mémoires*, **4** (1858), 125–130; and **5** (1864), 419–430; and *Journal de mathématiques pures et appliquées*, 2nd ser., **9** (1864), 129–134. On solar obfuscations, see *Mémoires*, **6** (1868), 385–469. Two other articles on meteorology, published in vols. **10** (1882) and 2nd ser., **2** (1898), of the *Mémoires*, established the invariableness of Montpellier's climate since the eighteenth century.

II. SECONDARY LITERATURE. There are reports on Roche's work in *Comptes rendus . . . de l'Académie des sciences:* on comets (by J. Babinet) in **51** (1860), 417–419; on the origin of the solar system (by H. Faye) in **77** (1873), 957–962; and on his work as a whole (by F. Tisserand) in **96** (1883), 1171–1179. H. Poincaré analyzed the Laplace-Roche cosmogonic hypothesis in *Leçons sur les hypothèses cosmogoniques* (Paris, 1911), 15–68. In his *Traité de mécanique céleste*, 4 vols. (Paris, 1889–1896), F. Tisserand set forth Roche's results on the shapes of fluid masses and on the interior of the earth (II, 110–116, 237–244) as well as on comets (IV, 245–257).

The speeches delivered at Roche's funeral appeared in *Mémoires*, **10** (1883). See also J. Boussinesq, "Notice sur la vie et les travaux de M. Roche," in *Mémoires de la Société des sciences, de l'agriculture et des arts de Lille*, 4th ser., **14** (1883), 17–35; and "Professor A. Roche," in *Nature*, **28** (1883), 11–12; and *Comptes rendus . . . de l'Académie des sciences*, **96** (1883), 1171–1179.

JACQUES R. LÉVY

RODRIGUES, JOÃO. See **Lusitanus, Amatus.**

ROEBUCK, JOHN (*b.* Sheffield, England, 1718; *d.* Borrowstounness, West Lothian, Scotland, 17 July 1794), *chemistry, technology.*

The third son of John Roebuck, a prosperous Sheffield cutler, manufacturer, and merchant, and Sarah Roe, Roebuck was educated at Sheffield Grammar School and Dr. Doddridge's Academy, Northampton. He did not join the family business but elected to study medicine at Edinburgh, where he met William Cullen, David Hume, Joseph Black, and other Scottish intellectuals. From Edinburgh he went to Leiden, where he graduated M.D. in 1742. His dissertation, dated 1743, is entitled "An Investigation Into the Effects of Rarefied Atmosphere on the Human Body."

Following graduation Roebuck settled as physician in Birmingham. He disliked medical practice, however, and turned progressively to chemistry as manufacturer and consultant, finally becoming a refiner (recoverer) of gold and silver for the Birmingham jewelry trade. By 1746, in conjunction with the Birmingham merchant Samuel Garbett, Roebuck was established as a metal refiner and consultant chemist. Later he began to produce sulfuric acid, made in lead instead of glass vessels, which greatly reduced the cost.

In 1749 Roebuck and Garbett opened a second vitriol works at Prestonpans, near Edinburgh; the market for sulfuric acid there was probably as a "sour" in bleaching. An export trade soon built up with the Low Countries. Roebuck and Garbett did not register a patent until 1771, but their application in Scotland was rejected by the Court of Session on the ground that the process had been used in England since 1756. Despite setbacks, the Prestonpans vitriol works prospered. Roebuck added ceramic production and, most important, after carefully weighing the prospects, founded the Carron ironworks near Falkirk, Stirlingshire, in 1759. His partners were his brothers Benjamin, Thomas, and Ebenezer Roebuck; Samuel Garbett; and the Cadells of Cockenzie, shipowners and timber traders who had already attempted iron production. Carron was the real foundation of the Scottish iron industry. The first furnace, in which coke replaced charcoal, was blown on 1 January 1760; and the same year John Smeaton installed a blowing engine. Malleable iron was produced from 1762 (B.P. 1762 no. 780). In 1773 George III granted the Carron works a royal charter, and from 1779 it made the ordnance known as "carronades."

The success of the Carron works led Roebuck to lease coal mines and saltworks at Borrowstounness, West Lothian. The pumping engines there were inadequate; Roebuck brought in James Watt, and they experimented with Watt's improved engine. Watt owed Joseph Black £1,200 and Roebuck took over the debt in exchange for a two-thirds share in Watt's patent (B.P. 1769 no. 913); but this transaction, coupled with other activities, overtaxed his financial resources. He was forced to withdraw capital from his Birmingham, Prestonpans, and Carron firms. Matthew Boulton of Birmingham canceled a loan in exchange for the two-thirds share in Watt's patent and so led to the foundation of the firm of Boulton and Watt. Roebuck remained at Borrowstounness, where he managed the coal mines and saltworks and engaged in various scientific experiments, including an attempt to produce synthetic alkali.

Roebuck was a fellow of the Royal Society of Edinburgh and of London and a freeman of Edinburgh.

BIBLIOGRAPHY

I. ORIGINAL WORKS. Two of Roebuck's articles are "A Comparison of the Heat of London and Edinburgh," in *Philosophical Transactions of the Royal Society,* **65** (1775), 459–462; and "Experiments on Ignited Bodies," *ibid.,* **66** (1776), 509–510. Copies of his M.D. dissertation, "An Investigation Into the Effects of Rarefied Atmosphere on the Human Body," are in the University of Leiden's archives and at St. Bartholomew's Medical College, London.

II. SECONDARY LITERATURE. General Register House, Edinburgh, contains the Customs House returns from Prestonpans and specifications of patents and drawings for the period 1767–1787. Birmingham Reference Library has Soho MSS, the correspondence of Boulton and Watt; documents relative to a suit by the Carron Ironworks Company against Samuel Garbett (1777–1779); and a thesis by P. S. Bebbington, "Samuel Garbett, 1717–1803." The Patent Office Library, London, has preserved abridgments of specifications relating to various inventions.

See also A. Clow and N. L. Clow, "John Roebuck (1718–1794)," in *Chemistry and Industry,* **61** (1942), 497–498; and "Vitriol in the Industrial Revolution," in *Economic History Review,* **15** (1945), 44–55; Henry Hamilton, "The Founding of Carron Ironworks," in *Scottish Historical Review,* **25** (1928), 185–193; R. Jardine, "An Account of Dr. John Roebuck, M.D., F.R.S.," in *Transactions of the Royal Society of Edinburgh,* **4** (1798), 65–87; and *Journal of the House of Lords,* **34** (1774–1776), 76, 217.

For additional information, see Henry M. Cadell, *The Story of the Forth* (Glasgow, 1913), 143–194; A. Clow and N. L. Clow, *The Chemical Revolution* (London, 1952), 93–95, 133–143, 181, 333–341; Henry Hamilton,

An Economic History of Scotland (Oxford, 1963), 140–141, 180, 193–214; Samuel Parkes, *Chemical Essays*, 4 vols. (London, 1815), II, 377–378, 399; IV, 17; Richard B. Prosser, *Birmingham Inventors and Inventions* (Birmingham, 1881), 16; Arthur W. Roebuck, *The Roebuck Story* (Don Mills, Ontario, 1963), 6–18; and Samuel Smiles, *Industrial Biography* (London, 1863), 135.

ARCHIBALD CLOW

ROEMER. See also **Römer.**

ROEMER, FERDINAND (*b.* Hildesheim, Germany, 5 January 1818; *d.* Breslau, Germany [now Wrocław, Poland], 14 December 1891), *geology, paleontology.*

Roemer's father was councillor of the High Court of Justice in Hildesheim; his mother was the daughter of the mayor of that city. Like his brothers Friedrich Adolph and Hermann, both of whom also became geologists, Roemer attended secondary school in Hildesheim, then studied law at Göttingen and Heidelberg. He also attended lectures on the natural sciences, particularly geology and mineralogy, and developed a strong interest in them. Political difficulties prevented him from taking the final law examination, and Roemer thereupon decided to devote himself completely to a scientific career.

In the spring of 1840 Roemer moved to Berlin, where he completed his studies. His dissertation, written in Latin and concerning the fossil species of the genus *Astarte*, was published in 1843. Roemer then went to Bonn, where he spent two years investigating the geology of the Rhenish Mountains. He incorporated his results into his first major work, *Das Rheinische Übergangsgebirge*, published in 1844. In the spring of 1845 Roemer went to the United States on behalf of the Society for the Protection of German Emigrants. He spent most of his time in Texas, where he remained until April 1847. His vividly written accounts of his travels comprised a pioneering study on the physiography of Texas, together with a report on the society and culture of its inhabitants, remarks on political and economic conditions, and a short, clear description of the geology of the region.

Roemer returned to Germany in the fall of 1847 to become *Privatdozent* in geology and mineralogy at the University of Bonn. In 1852 he brought out a major monograph on the fauna of the marine Cretaceous period of Texas, with an appendix on Paleozoic and Tertiary fossils; in this, as in his other works on the Cretaceous, Roemer was the first to point out the differences between the northern (boreal) and southern (mediterranean) faunas, which he saw as an expression of climatic variation. In the same year Roemer also published a monograph on the extinct Blastoidea, one of a series of works on Paleozoic fossils, especially echinoderms. In 1854 he returned to the Cretaceous, devoting a fundamental paper to the Cretaceous of Westphalia.

In 1855 Roemer was appointed full professor of geology and paleontology at the University of Breslau, where he stayed for the rest of his life. Among his first works there was his new, enlarged edition of the volume dealing with the Paleozoic in H. G. Bronn's *Lethaea geognostica*, published in 1856. Roemer contributed a comprehensive survey of Paleozoic floras and faunas, with the divisions and correlations of the era throughout the world. He projected a second part, which was to comprise a systematic description of stratigraphically important fossils, but did not carry it out (it was finished by Fritz Frech). Roemer's own work is noteworthy because it constitutes the first full statement of his researches in this field and because he applied Murchison's views on the English Paleozoic to the conditions existing in Germany. In a further, separate study (published in 1860) Roemer described the Silurian fauna of western Tennessee and compared it with European faunas of the same period.

Throughout his career in Breslau, Roemer made a series of thorough investigations of the geology and paleontology of Silesia. His research on the geology of Upper Silesia, begun in 1862, resulted in a number of geological maps and short communications. Eight years of fieldwork by Roemer and his colleagues culminated in his *Geologie von Oberschlesien* of 1870, a book that was for many years indispensable to the study of this area, rich in coal, iron, zinc, and lead. During this period Roemer also studied Pleistocene mammals, especially those of Silesia and Poland, and, starting in 1872, published a number of papers on the woolly rhinoceros, elasmothere, musk-ox, cave bear, and mammoth, as well as the Bovidae.

Roemer also systematically examined the boulders that the northern glaciers had transported, in moraines, to Silesia during the Pleistocene. He determined their fossil content and established that they had originated in the Baltic area, charting the course that they had followed. He published his results individually, then summed them up in his long article "Lethaea erratica," published in 1885.

Roemer's scientific accomplishments brought him a number of honors, among them membership in the academies of Berlin, St. Petersburg, and Munich and the Murchison Medal of the Geological Society of London. He traveled widely, mastered many modern languages, and was highly esteemed by his colleagues both in and out of Germany. He maintained a lifelong interest in literature and art, and was an engaging companion and witty raconteur. He died suddenly, of a heart attack, and was survived by his wife, Katharina Schäfer, whom he had married in 1869.

BIBLIOGRAPHY

I. ORIGINAL WORKS. Roemer's writings include *Das Rheinische Übergangsgebirge. Eine paläontologisch-geognostische Darstellung* (Hannover, 1844); *Texas. Mit besonderer Rücksicht auf deutsche Auswanderung und die physischen Verhältnisse des Landes nach eigener Beobachtung geschildert* . . . . (Bonn, 1849); *Die Kreidebildungen von Texas und ihre organische Einschlüsse, mit einem die Beschreibung von Versteinerungen aus paläozoischen und tertiären Schichten enthaltenden Anhange* (Bonn, 1852); "Monographie der fossilen Crinoideenfamilie der Blastoideen und der Gattung Pentatrematites," in *Archiv für Naturgeschichte*, **17** (1852), 323–397; "Die Kreidebildungen Westfalens. Mit einer geognostischen Übersichtskarte," in *Verhandlungen des Naturhistorischen Vereins der preussischen Rheinlande, Westfalens und des Regierungsbezirks Osnabrück*, **11** (1854), 29–180; *H. G. Bronn's Lethaea geognostica oder Abbildung und Beschreibung der für die Gebirgsformationen bezeichnendsten Versteinerungen*; I, pt. 2, *Palaeo-Lethaea*, 3rd. ed. (Stuttgart, 1856); *Die silurische Fauna des westlichen Tennessee* (Breslau, 1860); *Geologie von Oberschlesien* (Breslau, 1870); *Lethaea palaeozoica oder Beschreibung und Abbildung der für die einzelnen Abtheilungen der palaeozoischen Formation bezeichnendsten Versteinerungen*, 3 vols. (Stuttgart, 1876, 1880, 1883); "Die Knochenhöhlen von Ojcow in Polen," in *Palaeontographica*, **29** (1883), 193–233; and "Lethaea erratica oder Aufzählung und Beschreibung der in der norddeutschen Ebene vorkommenden Diluvial-Geschiebe nordischer Sedimentärgesteine," in *Paläontologische Abhandlungen*, **2** (1885), 250–420.

II. SECONDARY LITERATURE. See the following, listed chronologically: W. Dames, "Ferdinand Roemer†," in *Neues Jahrbuch für Mineralogie, Geologie und Paläontologie*, **1** (1892), 1–32, with complete bibliography of 344 titles; C. Struckmann, "Ferdinand Roemer," in *Leopoldina*, **28** (1892), 31–32, 43–46, 63–67, with partial bibliography; F. W. Simonds, "Dr. Ferdinand Roemer, the Father of the Geology of Texas; His Life and Work," in *American Geologist*, **29** (1902), 131–140, with 30 titles relating to North America, also, without bibliography, in *Geological Magazine*, 4th ser., **9** (1902), 412–417; and H. Bartenstein, "125 Jahre deutsche Unterkreide-Stratigraphie—ein historischer Rückblick auf das geologisch-paläontologische Wirken der drei Brüder Roemer aus Hildesheim," in *Neues Jahrbuch für Geologie und Paläontologie. Monatshefte*, **10** (1966), 595–602, with portrait.

HEINZ TOBIEN

ROEMER, FRIEDRICH ADOLPH (*b*. Hildesheim, Germany, 14 April 1809; *d*. Clausthal, Germany, 25 November 1869), *stratigraphy, paleontology*.

Like his younger brother Ferdinand, Roemer attended the Gymnasium in Hildesheim, where his father was councillor of the High Court of Justice. From 1828 to 1831 he studied law at Göttingen and Berlin, and then was a judicial official in Hildesheim and in Bovenden, near Göttingen. In 1843 he was transferred to the Mining Office at Clausthal in the Harz Mountains. A few years later Roemer was placed in charge of teaching geology and mineralogy at the Mining School in that city, and in 1862 he was appointed its director. He resigned from the government service in 1867 because of poor health. In the last years of his life Roemer gave large sums of money and donated his extensive collections of minerals and fossils to the city museum of Hildesheim, which had been founded by his younger brother Hermann; the museum still bears the family name.

Roemer's independent study of geology and paleontology dates from his appointment as judicial officer in Hildesheim, and his interest was inspired by geological conditions in the area. Lacking both formal training and research experience, he began his investigations in the northwestern German Jurassic. He obtained the necessary paleontological literature from libraries in Göttingen and Hannover and copied it, including the illustrations, in his own hand.

Roemer's first extensive work, *Die Versteinerungen des norddeutschen Oolithen-Gebirges*, appeared in 1836. It provided the first insight into the fossil riches (over 250 new species) and stratigraphic subdivisions of the northwestern German Jurassic, as well as a comparison with the southern German and English Jurassic. The results of his subsequent investigations, which were extended to the Cretaceous, appeared in 1841 as *Die Versteinerungen des norddeutschen Kreidegebirges*; nine species of plants, seven hundred fifty of Metazoa, thirty-three of Foraminifera, and seven of Ostracoda were described. Until then knowledge of the Cretaceous fossils in northwestern Germany was incomplete, and the stratigraphic division contained many errors. Beyrich, a severe critic, recognized the importance of these studies in 1849 and praised them.

In the meantime, the publications of Murchison and Sedgwick on the Silurian and Devonian in England had attracted Roemer's attention. They led to his concern with the Harz Mountains, on the northwest border of which Hildesheim lies and in the middle of which Clausthal is located. In six long papers published between 1843 and 1866, Roemer described primarily the Devonian and Lower Carboniferous of the northwestern Harz with regard to its fossil contents and detailed stratigraphic division. These works represented major progress in knowledge of the Devonian and Lower Carboniferous. When Roemer began his investigations in the Harz, only a few fossils from three or four localities—most of them incorrectly interpreted—were known. When he completed them, he had described more than 500 species, from many localities, of mostly varied stratigraphic position within the Devonian and Lower Carboniferous. Moreover, he published monographs on the Tertiary Bryozoa and Anthozoa, as well as on the Cretaceous sponges of northwestern Germany. His serious interest in botany, which dated from his school days, is attested to by a large work on the algae of Germany (1845).

In addition to specialized studies Roemer produced comprehensive summary presentations of his field of study. Thus in 1853 he composed—at the request of a friend, the biologist Johannes Leunis—the third part (geology and mineralogy) of the latter's *Synopsis der drei Naturreiche*, a popular and widely disseminated work. The *Synopsis* also presented contemporary knowledge in zoology and botany. Roemer's talent for synopsis was due in no small measure to his teaching, over a period of twenty-four years, at the Mining School (now the Technical University) of Clausthal. The extensive mineral collection there—one of the largest in Germany—is the result of his work.

Roemer's monographs on the Jurassic and Cretaceous in northwestern Germany and on the Paleozoic of the northwestern Harz provided the foundation for knowledge of the faunas and for the present stratigraphy of these geological periods in northern Germany. In addition, many of the fossil forms that he described have become supraregional guide fossils: for example, approximately twenty of his Foraminifera and Ostracoda species from the northern German Cretaceous today have European, and in part worldwide, stratigraphic significance.

BIBLIOGRAPHY

I. ORIGINAL WORKS. Roemer's writings include *Die Versteinerungen des norddeutschen Oolithen-Gebirges* (Hannover, 1836); *Die Versteinerungen des norddeutschen Kreidegebirges* (Hannover, 1841); *Die Versteinerungen des Harzgebirges* (Hannover, 1843); *Die Algen Deutschlands* (Hannover, 1845); "Beiträge zur geologischen Kenntnis des nord-westlichen Harzgebirges," in *Palaeontographica*, **3** (1850), 1–67; **3** (1852), 69–112; **5** (1855), 109–156; **9** (1860), 153–202; **13** (1866), 201–236; "Beschreibung der norddeutschen tertiären Polyparien," *ibid.*, **9** (1862), 199–246;

and "Die Spongitarien des norddeutschen Kreidegebirges," *ibid.*, **13** (1864), 1–64.

II. SECONDARY LITERATURE. See the following, listed chronologically: Ferdinand Roemer, "Nekrolog von Friedrich Adolph Roemer," in *Zeitschrift der Deutschen geologischen Gesellschaft*, **22** (1870), 96–102, with partial bibliography; E. Böckh, H. J. Martini, and A. Pilger; "Friedrich Adolph Roemer (1809–1869)," in *Geologisches Jahrbuch*, **76** (1959), xxi–xxviii, with complete bibliography; and H. Bartenstein, "125 Jahre deutsche *Unterkreide-Stratigraphie*—ein historischer Rückblick auf das geologisch-paläontologische Wirken der drei Brüder Roemer aus Hildesheim," in *Neues Jahrbuch für Geologie und Paläontologie, Monatshefte*, **10** (1966), 595–602, with portrait.

HEINZ TOBIEN

ROENTGEN, WILHELM. See **Röntgen, Wilhelm Conrad.**

ROESEL VON ROSENHOF, AUGUST JOHANN (*b.* Arnstadt, Germany, 30 March 1705; *d.* Nuremberg, Germany, 27 March 1759), *painting, engraving, natural science.*

Roesel von Rosenhof came from an Austrian noble family. His grandfather Franz Roesel and his brother Wolf were granted a patent of nobility by Emperor Ferdinand II. In 1753 August Johann had the patent officially validated and from that date added "von Rosenhof" to his name. Following the death of his father, his godmother, the reigning princess of Arnstadt-Schwarzburg, Auguste Dorothea, assumed responsibility for his education.

In 1720 he was apprenticed to his uncle, the painter Wilhelm Roesel. Four years later he was recalled to the court at Arnstadt by his godmother. A planned trip to Italy to further his artistic development failed to materialize; and he obtained permission to travel to Nuremberg with his brother and sister to continue his training at the city's academy under the supervision of Johann Daniel Preisler.

Through an aunt who lived in Copenhagen and was lady-in-waiting to the crown prince (later King Christian VI), Roesel went to the Danish court in mid-1726. His portraits and miniatures were so well received that he was asked to settle in Denmark. Unwilling to take this step, he left after only two years, taking with him presents and a letter of introduction from the crown prince. A high fever forced him to interrupt his return journey with a four-week stay in Hamburg. While he was there, an acquaintance brought him a copy of an illustrated book entitled

Metamorphosis insectorum Surinamensium (1705) by Maria Sybilla Merian. After leafing through this splendid volume Roesel decided to devote special attention to insects and their metamorphoses, and to publish illustrated books.

After his recovery Roesel returned to Nuremberg, became a citizen, and married Elisabeth Maria Rosa, the daughter of a surgeon, on 3 June 1737. He began to study insects and, despite frequent admonitions that he should not waste "precious time on the depiction of such harmful and revolting creatures," he tirelessly gathered living caterpillars and butterflies in the course of excursions in the Nuremberg region.

After years of preparation the first installment of *Der monatlich-herausgegebenen Insecten-Belustigung* finally appeared in 1740. The work was praised by Johann Philipp Breyne and Réaumur, in large part for its outstanding illustrations. When the first volume was completed in 1746, Roesel wrote a preface in which he attempted to define the insects as a systematic unity and to divide this unity into classes, orders, and families. In the entire work he devoted the most space to the butterflies; beetles, grasshoppers, crickets, gnats, flies, and dragonflies were not considered until the second volume. The third volume contains (along with descriptions of butterflies) essays on crayfish and the natural history of the polyps, which Roesel studied for more than a year, until he succeeded in finding them for the first time. Roesel's son-in-law and closest collaborator, Christian Friedrich Carl Kleemann, took charge of the publication of the last installments of the third volume and of all of the fourth volume.

During his research on insects, Roesel studied amphibians and reptiles. This work culminated in the publication of *Historia naturalis ranarum* in several installments between 1753 and 1758. When all of them had appeared, Albrecht von Haller offered to contribute a preface to the work. The text of the *Historia*, printed in both Latin and German, contains descriptions of all the German frogs and toads. The twenty-four plates in large folio format are presented in pairs: those on the left-hand page show simple outlines of their subjects; those on the right-hand page are produced in unusually sumptuous and graphic colors. Along with illustrations of animal habitats and of copulating animals, the book contains drawings of anatomical preparations, individual organs, skeletons, and various stages of larval development. There are also extant six preliminary watercolors from a further work in which Roesel planned to treat tailed amphibians and lizards in a similar fashion.

In the biography of his father-in-law, Kleemann wrote: "The consequence of this ardent research and

exploration was a painful joint ailment and finally, indeed, a serious apoplectic fit." Further strokes led to the paralysis of Roesel's left side. Several weeks before his death Roesel was admitted to the Altdorfer Deutsche Gesellschaft (16 February 1759).

BIBLIOGRAPHY

I. ORIGINAL WORKS. Roesel's main work, *Monatlich-herausgegebenen Insecten-Belustigung*, appeared in 4 vols.: I, . . . *in welchem die in 6 Classen eingetheilte Papilionen mit ihrem Ursprung, Verwandlung und allen wunderbaren Eingeschaften* . . . (Nuremberg, 1746); II, . . . *welcher 8 Classen verschiedener, sowohl inländischer, als auch ausländischer Insecten* . . . (Nuremberg, 1749); III, . . . *worinnen ausser verschiedenen, zu denen in den beeden ersten Theilen enthaltenen Classen, gehörigen Insecten, auch mancherley Arten von 8 neuen Classen* . . ., with new observations and comments by C. F. C. Kleemann (Nuremberg, 1755); IV, . . . *in welchem ausser verschiedenen in- und ausländischen Insecten, auch die hiesige grosse Kreutz-Spinne nach ihrem Ursprung* . . ., C. F. C. Kleemann, ed. (Nuremberg, 1761). His other noteworthy book is *Historia naturalis ranarum* . . .— *Die natürliche Historie der Frösche hiesigen Landes* . . . (Nuremberg, 1758), with pref. by Albrecht von Haller.

II. SECONDARY LITERATURE. See L. Baege, "Unbekannte Vogelbilder Rösels von Rosenhof aus der Frühzeit ornithologischer Prachtillustrationen," in *Journal für Ornithologie*, **105** (1964), 464–467; W. Hess, "August Johann Rösel von Rosenhof," in *Allgemeine deutsche Biographie*, XXIX (1889), 188–189; C. F. C. Kleemann, "Ausführliche und zuverlässige Nachricht von dem Leben, Schriften und Werken des verstorbenen Miniaturmahlers und scharfsichtigen Naturforschers August Johann Roesels von Rosenhof," in *Monatlich-herausgegebenen Insecten-Belustigung*, IV (Nuremberg, 1761), with a copper-plate portrait; F. Leydig, "Herpetologische Zeichnungen aus dem Nachlass Rösels von Rosenhof," in *Verhandlungen des Naturhistorischen Vereins des Preussischen Rheinlandes, Westfalens und des Regierungsbezirks Osnabrück*, **35** (1878), 1–41; and A. J. Ziegeler, "Rösel von Rosenhof," in *Natur und Haus*, **12** (1904), 220–234, 245–251.

A. GEUS

ROGER OF HEREFORD (*fl.* England, second half of the twelfth century), *astronomy, astrology.*

Very little is known with certainty about Roger of Hereford's career. There has been considerable speculation seeking to identify him with several other contemporaneous Englishmen named Roger. At the beginning of his *Compotus*, which is dated 9 September 1176, Roger refers to himself as "young" but adds that he has devoted many years to learning. The period of his activity probably lies in the decade from 1170 to 1180.

Roger of Hereford wrote several astronomical works. The *Compotus*, which consists of five books of twenty-six chapters, is critical of other Latin computists. The work is dedicated to Gilbert Foliot, bishop of Hereford until 1163, and then bishop of London. Roger also composed a set of astronomical tables for the latitude of Hereford, dated 1178, based on the Toledan and Marseilles tables. His other astronomical treatises include *De ortu et occasione signorum* and *Theorica planetarum*. Taking into consideration the period during which the latter work was probably written and the dates of the availability of Ptolemy's *Almagest* in the Latin West (1160, 1175), Roger's *Theorica* is likely one of the earliest works in that genre in the post-Latin Ptolemy period. The Digby manuscript of the *Theorica* (Digby MS 168, fols. 69 f.) is entitled "Incipit theorica Rogeri Herefordensis"; a later hand has added "floruit A.D. 1170 sub Henrico 2⁰." In the *Theorica*, Roger describes the "Hindu" procedure for the determination of planetary latitudes, a technique that entered the West in the Toledan Tables as well as from other Arabic sources. He also provides the Ptolemaic method for latitudes, which he calls "more likely."

In addition to these astronomical contributions, Roger of Hereford wrote several works on astrology. One of these, *Liber de divisione astronomiae*, in the Bibliothèque Nationale manuscript, begins, "In the name of God the pious and merciful . . .," which has a decided Arabic flavor and would indicate that the work might be a translation. Other astrological treatises by Roger include *De quatuor partibus iudicorum astronomie*, *De iudiciis astronomie*, *Iudicia Herefordensis*, and *De tribus generalibus iudiciis astronomie*. Several of these treatises apparently are extracts from the four-part work on astrology. Roger also wrote *De re metallicis*.

Roger of Hereford can be placed in that group of twelfth-century Englishmen who were instrumental in bringing Arabic scientific materials to the Latin West, either through direct translation or in the form of adaptations of Arabic sources. The group includes Robert of Chester, Daniel of Morley, Alfred of Sarashel (Alfred Anglicus), and Adelard of Bath. Whether Roger knew Arabic or traveled to Spain is unknown. Alfred of Sarashel, who translated the Arabic version of the *De vegetabilis* or *De plantis*, a work attributed to Aristotle but written by Nicholas of Damascus, dedicated his translation to Roger.

Whether Roger's activity continued into the 1180's is unknown. He may have been the Roger, clerk of Hereford, who served as itinerant justice with Walter Map in 1185. It has also been suggested that he died as a monk at the abbey of Bury St. Edmunds.

BIBLIOGRAPHY

See Pierre Duhem, *Le système du monde*, III (Paris, 1958), 222–223, 520–523; C. W. Haskins, *Studies in the History of Mediaeval Science* (Cambridge, Mass., 1927), 87, 123–128; Josiah C. Russell, "Hereford and Arabic Science in England," in *Isis*, **18** (1932), 14–25; and Lynn Thorndike, *A History of Magic and Experimental Science*, II (New York, 1923), 181–187, 260.

CLAUDIA KREN

ROGERS, HENRY DARWIN (*b*. Philadelphia, Pennsylvania, 1 August 1808; *d*. Shawlands, near Glasgow, Scotland, 29 May 1866), and **ROGERS, WILLIAM BARTON** (*b*. Philadelphia, 7 December 1804; *d*. Boston, Massachusetts, 30 May 1882), *geology*.

Henry and William Rogers, along with their brothers James (1802-1852) and Robert (1813-1884), were important in the mid-nineteenth-century American scientific community. Their father had fled Ireland in 1798 because he had publicly expressed strong sympathy for the leaders of the rebellion of that year. He settled in Philadelphia, where he obtained an M.D. from the University of Pennsylvania. Later he became professor of natural philosophy and chemistry at the College of William and Mary. Each of the sons studied medicine or chemistry and became professor at William and Mary, the University of Virginia, or the University of Pennsylvania. William and Henry also secured appointments as state geologists of Virginia and Pennsylvania, respectively; the other brothers, who were more exclusively chemists, together wrote one of the earliest textbooks of chemistry in the United States. The Rogers brothers, especially William and Henry, were deeply involved in the organization of the Association of American Geologists in 1840 and in its transformation into the American Association for the Advancement of Science in 1848. Both served as chairman of the former association, Henry in 1843 and William in 1845 and 1847; William served as president of the latter association in 1876.

During much of this period, William and Henry were developing plans for organizing a more rigorous and practical kind of scientific education involving sustained laboratory work. After moving to Boston (Henry in 1845, William in 1853), they pursued the idea vigorously until the state was induced to charter the Massachusetts Institute of Technology. William became its first president (1862–1870, 1878–1881) and is regarded as one of the principal architects of modern scientific and engineering curricula. In 1855 Henry moved to Scotland, and in 1857 he was appointed regius professor of natural history at the University of Glasgow—probably the first native American to be appointed to a chair anywhere in Europe. William and Robert were charter members of the U.S. National Academy of Sciences in 1863, and William became its president in 1879.

In 1832 Henry, fascinated by the social theories of Robert Owen, went to England for a year with Owen's son to give scientific lectures to workmen. There he was introduced into scientific and especially geological circles, then in great ferment. When he returned to the United States he quickly infected William with his new enthusiasm for geology. It was an auspicious moment; many of the states were launching large-scale geological surveys, and for the most part collaboration among the state geologists (and Logan in Canada) was both cordial and fruitful. The Rogers brothers and their assistants mapped New Jersey, Pennsylvania, and Virginia (which then comprised West Virginia), including the largest sector of the Appalachian Mountains then being studied. They showed extraordinary insight in unraveling concomitantly the stratigraphy and the structure of the mountain chain. They made little overt use of paleontology in stratigraphy, although they were in close communication with the geologists contemporaneously mapping New York State, where James Hall and Lardner Vanuxem were demonstrating the great usefulness of William Smith's principle of faunal succession in strata continuous with those of Pennsylvania. Whether because of this communication or not, the Rogers brothers made no serious blunders in the stratigraphy like those made by previous workers unacquainted with the use of paleontology.

Concerning structure, the brothers made evident the beautiful simplicity and elegance of the Appalachian fold system and provided the first adequate understanding of the geologic structure of any large mountain belt.

The period of tight money that followed the panic of 1837 halted the brothers' survey work and prevented the prompt publication of their final reports; Henry's was published about two decades later, but William's remained unpublished. Nevertheless, in a joint paper presented in 1842, they described with great clarity and elegance and illustrated in excellent cross sections, with no vertical exaggeration, the great regular, curvilinear, asymmetrical folds that characterize the Valley and Ridge province of the Appalachians, especially in Pennsylvania and northern Virginia, and some of the major thrust faults of southwest Virginia. By demonstrating that any explanation of such folding and faulting must provide for a remarkable uniformity of forces over a large region, they greatly weakened current theories of vertical uplift and strengthened ideas of tangential forces operating perpendicular to the trend of linear mountain belts. Their paper attracted wide and favorable attention both in America and in Europe, especially in Britain, and was recognized as the first major American contribution to geological theory. Henry also showed that slaty cleavage is spatially and geometrically bound to folding and hence must be produced by the same regional forces.

The Rogers' dynamic explanation of these geological facts leaned toward catastrophism and met with little acceptance, but their work paved the way for James Dwight Dana's more uniformitarian theory of tangential contraction, which dominated thinking on orogeny well into the twentieth century. Although this theory was seriously challenged in the first half of the present century by a renewal of theories based on vertical uplift (coupled with lateral gravitational sliding to explain the evident tangential shortening) and by theories based on assumed convection inside the Earth, it was not superseded by a new general theory of orogeny until the 1960's. At that time geophysical evidence was added to the geological facts to establish the existence of slow but very large-scale motions within the Earth's mantle, of which the tangential shortening observed in mountain belts is simply a side effect (the "new global tectonics" or plate tectonics).

BIBLIOGRAPHY

I. ORIGINAL WORKS. The Rogers brothers' writings include William and Henry Rogers, "On the Physical Structure of the Appalachian Chain, as Exemplifying the Laws Which Have Regulated the Elevation of Great Mountain Chains Generally," in *Reports of the Meetings of the Association of American Geologists and Naturalists* (1843), 474–531, abstracts in *American Journal of Science*, **43** (1842), 177–178; in *Reports of the British Association for the Advancement of Science* for 1842, pt. 2, 40–42 (also see the 1884 work by William, cited below); Henry Rogers, "On the Direction of the Slaty Cleavage in Strata of the Southeastern Belts of the Appalachian Chain, and the Parallelism of the Cleavage Dip With the Planes of Maximum Temperature," in *Proceedings of the American Association of Geologists and Naturalists*, **6** (1845), 49–50; and *The Geology of Pennsylvania*, 2 vols. (Edinburgh–Philadelphia, 1858); and William Rogers, *A Reprint of Annual Reports and Other Papers on the Geology of the Virginias* (New York, 1884).

II. SECONDARY LITERATURE. See J. W. Gregory, *Henry Darwin Rogers* (Glasgow, 1916); Mrs. W. B.

Rogers, *Life and Letters of William Barton Rogers*, 2 vols. (Boston, 1896); and W. S. W. Ruschenberger, "A Sketch of the Life of Robert E. Rogers, M.D., LL.D., With Biographical Notices of His Father and Brothers," in *Proceedings of the American Philosophical Society*, **23** (1886), 104–146, which is not entirely accurate.

JOHN RODGERS

ROHAULT, JACQUES (*b.* Amiens, France, 1620; *d.* Paris, France, 1675), *natural philosophy, scientific methodology.*

Rohault was the leading advocate and teacher of Descartes's natural philosophy among the first generation of French Cartesians. Little is known of the details of his life, especially the early years. Most accounts of his career derive from the largely retrospective reports of his father-in-law, Claude Clerselier.[1] Rohault was the son of Ambroise Rohault, a wealthy wine merchant, and Antoinette de Ponthieu. He received his early education in Amiens, most likely a scholastic training at the Jesuit *collège* there.[2] He completed his studies in Paris, where, apart from his formal academic routine, he reportedly haunted the shops of artisans, exchanging information on mechanical contrivances and procedures. Rohault also began to teach himself mathematics, eventually gaining enough mastery to establish himself as a private tutor.

The circumstances under which Rohault embraced Cartesianism are far from clear. Alexandre Saverien, an eighteenth-century biographer, claimed that Rohault's philosophical studies ultimately attracted him to Cartesianism and that he subsequently found his way into the Cartesian circle led by Clerselier, an *avocat* of the Paris Parlement and later editor of Descartes's correspondence.[3] Clerselier reported that he had studied mathematics under Rohault (although he does not date the origin of the relationship),[4] and it is therefore possible that Clerselier recruited his young tutor for the Cartesian school. As if to consolidate the association, Rohault married Clerselier's daughter Geneviève, probably in 1648.[5]

Rohault's contemporary fame rested on the very popular weekly lectures he held at his house in Paris, beginning sometime in the mid-1650's. He rapidly became the leading Cartesian practitioner of the scientific lecture with experimental illustration. His fluent style of lecturing and lucid restatements of Descartes's physical theories made Cartesian science intelligible to large segments of the educated Parisian public for the first time.[6] Since his predominantly lay audience shared the nondogmatic, hypothetical, and experimental ideology of the new science at mid-

century, Rohault played down the more dogmatic tendencies of Descartes's work and sought to join Cartesian explanatory principles to experimental practice and to give a probabilistic interpretation of the truth value of specific explanations.

The lectures covered, one by one, the major problems of natural philosophy. Rohault began each session with a discourse on the general nature of the subjects under discussion, inviting interruptions by questioners and opening the floor to debate at the end of his exposition. Then, starting from the basic mechanical principles of Descartes's philosophy, he moved to explanations of the particular phenomena under examination, confirming the explanations by experiments. The most famous of the experiments were those on the weight of the air, including a simplified apparatus for performing Roberval's and Auzout's demonstrations of "a void within a void"; the production of artificial rainbows; and a meticulously ordered series of experiments illustrating the Cartesian explanation of magnetism. These demonstrations, scheduled in advance for a given subject, attracted large crowds of courtiers, bourgeois, administrative officers, and foreign virtuosos.

During the 1660's Rohault emerged as the arbiter of Cartesian scientific affairs in Paris. While his lecturing and tutoring continued, he also became an active participant in the Montmor Academy and other circles of leading natural philosophers. In 1665 Rohault recruited Pierre-Sylvain Régis to the Cartesian movement. After several months of instruction in Cartesian science and the arts of the *conférencier*, Régis was sent by Rohault to spread the doctrine in Toulouse. Rohault also organized the ceremonies marking the return of Descartes's remains to Paris from Stockholm in 1667.

Rohault's masterwork, the *Traité de physique* (1671), became the era's leading textbook on natural philosophy. Intended as an elementary synthesis of Cartesian science, it was largely based on the material and pedagogical approach that Rohault had developed in his *conférences*. As Paul Mouy observed, Rohault did not wish to appear as a mere Cartesian partisan in the *Traité*, but as a sympathetic arbiter between the systems of Aristotle and Descartes.[7] Hence, adopting the standard scholastic division of the subject matter of natural philosophy (no doubt to ease acceptance of the work in the schools), he strove to separate the supposed views of Aristotle from the bastardizations of the medieval commentators.[8] The *Traité* usually presents Aristotle as having been generally correct in his approaches and broad conclusions,[9] and often introduces Cartesian views as more complete elaborations of Aristotelian foundations. The revolutionary implications of Cartesian metaphysics and epistemology are somewhat played down.

In the *Traité* Rohault accepted Descartes's principles of natural philosophy: that the essence of matter is extension, that the universe is a plenum, that the quantity of motion in the universe is conserved, that three kinds of matter or elements exist, and that the mechanical contact of bodies is the only cause of change of motion. His discussion of the essential topics in Cartesian science—dioptrics, theory of colors, cosmology and vortex celestial mechanics, meteorological phenomena, and mechanistic physiology—follows the Cartesian line as set down in the *Principia* and in the posthumously published *Traité de l'homme*. Although Rohault also recounted the Cartesian laws of motion,[10] he had surprisingly little to say about the "rules of collision," which had become a central problem in natural philosophy with the appearance of Descartes's *Principia* in 1644. Rohault presented only two rules for cases of inelastic collisions.[11] Usually well informed about the latest developments in natural philosophy, he made no mention of the recent work of Wren, Wallis, Huygens, and Borelli on such laws of collision. In addition, despite his calls in the preface to the *Traité* for a quantitative approach to natural philosophy, Rohault (like Descartes before him) made little use of mathematical argument to establish his positions.

The strength of the *Traité* and its contemporary appeal lay in Rohault's ability to weave new experimental findings, as well as his knowledge of craft and chemical processes, within a verbal web of Cartesian mechanistic discourse. Even hostile critics, such as the anti-Cartesian Lagrange, noted that Rohault's presentations were fuller, more systematic, and better integrated with experiments than comparable sections in the *Principia*.[12] Perhaps the best examples of Rohault's procedures were his discussions of the experiments concerning the void and his analysis of the nature of liquids.

Recasting most of Pascal's experiments on the void in Cartesian terms, Rohault maintained, against Pascal and Torricelli, that the space at the top of the Torricellian tube is only apparently void and in fact is filled with Cartesian subtle matter. The space cannot be void, because it manifests the physical properties of transmitting light and responding to changes in temperature by expanding or contracting.[13] Nevertheless, Rohault accepted the explanation of the experiments on the void on the basis of a concept of the weight of the superincumbent air.[14] In several instances, however, he conflated this concept with the properly Cartesian conception of vortex flows of subtle matter, such as he conceived to be caused by pulling out the plunger of a syringe used to draw air or water.[15] This conflation allowed Rohault to claim that when one pulls the plunger of a syringe of which the opposite end is closed, the motion of the plunger squeezes the subtle matter out of the surrounding air, through the pores in the tube, and into the supposedly void space thus created.[16]

In his chapter on the nature of liquidity, Rohault made use of the putative mechanical properties of the air and of the existence in bodies of pores of various sizes, in order to formulate Cartesian explanations of the behavior of the famed Batavian glass drops, the shape of the meniscus in wetted and unwetted glasses, and some examples of capillary action in narrow tubes.[17] He seems to have been the first to observe these capillarity phenomena systematically.[18] As in his analysis of the void, Rohault was cognizant of the latest experimental findings and attempted an explanation employing Cartesian terms accompanied by a compelling array of experimental manipulations that seemed to confirm his views. One notes the same mode of presentation even in the rather cursory fourth book of the *Traité*, where his sketch of Cartesian physiology is enriched by reference to the work of Aselli on the lacteals, Pecquet on the thoracic duct, and Steno on the mechanism of muscle contraction.[19]

The *Traité* reflects Rohault's explicit view that natural philosophical explanations are probable at best and liable to falsification by one experimental counterinstance. For Rohault, an explanation is more probable to the degree that it has been formulated by consideration of fewer properties of the *explicandum* and that it can be extended to cover new experimental phenomena, which may or may not have been suggested by manipulation of the explanatory schema itself.[20] This probabilistic interpretation of scientific explanation is similar in its main lines to the contentions of such important contemporaries as Pascal, Huygens, and Mariotte. Hence, partisan devotion to Descartes did not isolate Rohault from the most sophisticated lines of contemporary thought on scientific method. Indeed, one may view him as having systematized the hints toward a probabilistic interpretation present in the latter portions of Descartes's *Principia*, and thus as having remained within the Cartesian school by consistently reorienting its methodological assumptions along lines only vaguely suggested by Descartes.[21]

The success of the *Traité de physique* was immediate. A favorable review appeared in the *Philosophical Transactions of the Royal Society* for 17 April 1671.[22] The even more laudatory anonymous reviewer in the *Journal des sçavans* praised Rohault's avoidance of metaphysical disputation, his recourse to experimental apparatus, and his familiarity with a variety of useful arts.[23] Given Rohault's sympathetic treatment of the subject matter, neither reviewer had reason to view his Cartesianism as a constricting or dogmatic position.

New editions quickly followed at a pace unprecedented for a textbook of natural philosophy. Within five years three new or revised editions were published in Paris, and two in Amsterdam; by 1730 there had been ten separate publications of subsequent Paris editions or revisions thereof.[24] A Latin translation (Geneva, 1674) by Théophile Bonet made possible the use of the *Traité* as a university text. In 1697 Samuel Clarke, then a young Cambridge B.A. and a confirmed Newtonian, published a new Latin translation, adding notes based on Newton's views in order to counter Rohault's Cartesian text. He thus opened the second phase of the history of the *Traité*.

At first Clarke seems to have been motivated as much by a desire to improve Bonet's clumsy translation as to undermine the *Traité* from within by insertion of Newtonian views. In subsequent versions (1702, 1708, 1710) Clarke expanded the notes, widened their subject matter, and sharpened their pro-Newtonian edge to nearly its final form. With the 1710 edition Clarke was able to utilize large portions of Newton's *Opticks*, which he had translated into Latin in 1706.[25] This version was the first in which Clarke's notes systematically refuted the text on most important issues in order to advance the Newtonian world view.[26] The last edition of Clarke's Latin translation appeared at London in 1739.

One should not suppose that Clarke's Newtonian notes were the sole reason for continued use of the text until the mid-eighteenth century. As a comprehensive natural philosophy Rohault's Cartesianism was still being taught in colleges and read in lay society. In addition, the availability in one text of the two leading interpretations of natural philosophy no doubt contributed to the longevity of the work.

Rohault's last years were troubled by a rising political and theological reaction to Cartesianism in France. The growing popular influence of Cartesianism, the scientific wing of which was led by Rohault, elicited both official government repression and private literary attacks.[27] Cartesianism was especially suspected of endangering public morals and undermining the tenets of the Catholic faith. Rohault's last work, *Entretiens sur la philosophie* (1671), sought to reverse this argument by establishing that only the Cartesian interpretation of the Eucharist, as opposed to the scholastic view, provides an unimpeachable basis for the admission of the real presence as an article of faith. Carrying the argument to the scholastic

camp, Rohault went on to claim that Cartesian animal automatism is more conducive to a proper understanding of the human soul than the school teachings that attribute immaterial forms to animate and inanimate entities alike. Nevertheless, the *Entretiens* did little to dampen anti-Cartesian sentiment, and Rohault was still suspected of heresy by some at the time of his death.

In 1670, at the height of his career, Rohault had obtained the *privilège du roi* for the publication of a collection of treatises on practical subjects, including elementary arithmetic, mechanics, perspective, and military architecture, in addition to a French translation of the first six books of Euclid's *Elements*. The writing of the *Traité* and the *Entretiens*, as well as the political difficulties of Cartesianism, delayed completion of the project until Clerselier published Rohault's *Oeuvres posthumes* in 1682. Clerselier added an important preface to the work in which he attempted to justify Rohault's Cartesianism. Like the *Traité*, the treatises grew out of Rohault's teaching and testify to the wide range of subjects he covered and to the diversity of his students. His "Traité de méchanique" in the *Oeuvres* is notable for its Archimedean approach and the avoidance of any systematic attempt to link the science of mechanics with the principles of Cartesian physics.[28]

NOTES

1. See Clerselier's preface to *Oeuvres posthumes de M. Rohault* (1682), unpaginated; and his preface to vol. II of the correspondence of Descartes (1659), repr. in the Adam and Tannery ed. of the *Oeuvres de Descartes*, V (1903), 630.
2. Paul Mouy, *Le développement de la physique Cartésienne*, p. 108.
3. Alexandre Saverien, *Histoire de philosophes modernes*, VI, *Histoire des physiciens* (Paris, 1758), 7.
4. Clerselier, preface to *Oeuvres posthumes*.
5. See Mouy, *op. cit.*, 110.
6. See L. L. Laudan's intro. to a reprint (New York, 1969) of John Clarke's 1723 English trans. of the *Traité*, *A System of Natural Philosophy*, I, xiii. (Hereafter cited as *System* according to bk., ch., and para. for those employing other eds. of the *Traité*.)
7. Mouy, *op. cit.*, 116.
8. See *System*, pt. I, ch. 27, para. 10, on the scholastics' invention of intentional species; or pt. I, ch. 7, paras. 10–13, where Aristotle's conception of matter is differentiated from that of the "Aristotelians" and sympathetically compared with the Cartesian view.
9. *Ibid.*, Rohault's preface (unpaginated), on Aristotle's correct contention that there is no void, and also on Aristotle's supposed use of mechanical considerations in explanation.
10. *Ibid.*, pt. I, ch. 5, para. 8, on conservation of state; ch. 10, para. 13, on conservation of quantity of motion; ch. 11, para. 1, on the law of inertia.
11. *Ibid.*, ch. 11, paras. 5–6.
12. J.-B. Lagrange, *Les principes de la philosophie, contre les nouveaux philosophes*, I (Paris, 1684), 30–31; cited in Mouy, *op. cit.*, 113–114.

13. *System*, pt. I, ch. 12, paras. 25–26.

14. *Ibid.*, paras. 17–23.

15. *Ibid.*, para 14. See also ch. 22, para. 69, where Rohault attributes the concave shape of the meniscus in a wetted glass to the difficulty encountered by the air in circulating into and out of the top of the glass in order to depress the water level; and paras. 81–82, where capillary phenomena in thin tubes are explained by the inability of the air to flow freely into the top of the tube and, hence, to depress the water level as much as usual.

16. *Ibid.*, ch. 12, paras. 8–9.

17. *Ibid.*, ch. 22, paras. 47–54, on the Batavian drops; paras. 68–74 on the meniscus; paras. 81–82 on capillarity phenomena.

18. See *Oeuvres posthumes*, 594; and Florin Périer's "Avertissement" to his preface to Pascal's *Traitez de l'équilibre des liqueurs et de la pesanteur de la masse de l'air* (1663) in 'Pascal's *Oeuvres complètes* ..., L. Brunschvicg, P. Boutroux, eds., III (Paris, 1908), 280. As the eds. observe (p. 280, n.1), Rohault's observations date from at least 1659.

19. *System*, pt. IV, ch. 6, para. 2, on Aselli; para. 4 on Pecquet; and ch. 3, para. 6, on Steno.

20. *Ibid.*, pt. I, ch. 3, para. 4.

21. See Descartes, *Principes*, pt. IV, prin. 204, in *Oeuvres de Descartes*, Adam and Tannery, eds., IX, pt. 2, 322–323; or Latin *Principia, ibid.*, VIII, 327.

22. *Philosophical Transactions of the Royal Society*, no. 70 (17 Apr. 1671), 2138–2141.

23. *Journal des sçavans* (22 June 1671), 624–625.

24. For a full summary of the various eds. see M. Hoskin " 'Mining All Within,' Clarke's Notes to Rohault's *Traité de physique*."

25. *Ibid.*, 360.

26. *Ibid.*, 361.

27. See Francisque Bouillier, *Histoire de la philosophie cartésienne*, 3rd ed., I (Paris, 1868), chs. 21, 22.

28. But see prop. XXVI, cor. 1, of the "Traité de méchanique" in *Oeuvres posthumes*, 568–569, where Rohault attempts to justify the resolution of components of velocity in a collision on the basis of prop. XXVI, which deals with the resolution of static forces involved when a heavy sphere is at rest on two mutually intersecting inclined planes.

BIBLIOGRAPHY

I. ORIGINAL WORKS. Rohault's chief work is the *Traité de physique* (Paris, 1671). Hoskin's article, cited below, gives a full list of the numerous subsequent eds. The *Traité de physique* is most accessible in the recent reprint of John Clarke's original English trans. (London, 1723), *A System of Natural Philosophy*, 2 vols. (New York, 1969), with an intro. by L. L. Laudan. Other works of Rohault are *Entretiens sur la philosophie* (Paris, 1671) and *Oeuvres posthumes de M. Rohault* (Paris, 1682). Mention should also be made of an anonymous "Discours des fièvres" appended to the edition of Descartes's *Traité de la lumière* published in 1664 in Paris. Mouy first noted the striking similarity between this treatise and the last chapter of the *Traité de physique*, in which Rohault presented a corpuscular-mechanical theory of fevers along Cartesian lines (Mouy, *op. cit.*, pp. 65, 126). The order and content of the main arguments of the two works are identical. In addition, the "Avis du libraire au lecteur" states that the "Discours" was first presented by a "philosopher and mathematician" at one of the weekly meetings of *savants* at the home of M. Montmor. There can be little doubt therefore

that this brief text constitutes the first published work of Rohault.

II. SECONDARY LITERATURE. There is relatively little secondary literature on Rohault. Aside from the reports of Clerselier mentioned in the notes, the following are of most value: A. G. A. Balz, *Cartesian Studies* (New York, 1951), 28–41; M. Hoskin, " 'Mining All Within,' Clarke's Notes to Rohault's *Traité de physique*," in *Thomist*, **24** (1961), 357–363; Paul Mouy, *Le développement de la physique cartésienne* (Paris, 1934), esp. 108–138, containing the most comprehensive available summary and analysis of Rohault's life and work. On Rohault's views on methodology and their relation to contemporary currents in French methodological thought, see L. L. Laudan's unpublished doctoral dissertation, "The Idea of a Physical Theory From Galileo to Newton: Studies in Seventeenth Century Methodology" (Princeton, 1966), ch. 8.

JOHN A. SCHUSTER

ROHN, KARL (*b.* Schwanheim, near Bensheim, Hesse, Germany, 28 January 1855; *d.* Leipzig, Germany, 4 August 1920), *mathematics.*

Rohn entered the Polytechnikum at Darmstadt in 1872, studying engineering and then mathematics. He continued his work in the latter at the universities of Leipzig and Munich, receiving his doctorate at Munich in 1878 and qualifying as lecturer at Leipzig a year later. In 1884 he became an assistant professor at Leipzig, and in 1887 full professor of descriptive geometry at the Technische Hochschule in Dresden. From 1904 until his death he was full professor at the University of Leipzig.

In his dissertation and in his *Habilitationsschrift* Rohn, stimulated by F. Klein, examined the relationship of Kummer's surface to hyperelliptic functions. In these early writings he demonstrated his ability to work out the connections between geometric and algebraic-analytic relations. In the following years Rohn further developed these capacities and became an acknowledged master in all questions concerning the algebraic geometry of the real P_2 and P_3, where it is possible to overlook the different figures. This concerns forms of algebraic curves and surfaces up to degree 4, linear and quadratic congruences, and complexes of lines in P_3. Gifted with a strong spatial intuition, Rohn possessed outstanding ability to select geometric facts from algebraic equations.

In several instances no decisive advance has been made on the results that Rohn obtained. This is especially true of his investigations on fourth-degree surfaces having one triple point or having finitely many isolated singular points. Most later studies concerning fourth-degree surfaces with only isolated singularities have been devoted to Kummer surfaces, which possess the greatest number of singular points (sixteen). Early in his career Rohn also constructed spatial models of the surfaces and space-curves he was studying. Rohn was the first to solve the difficult problem concerning the possible positions of the eleven ovals that the real branch of a sixth-degree plane curve can maximally possess. These problems were of great interest to Hilbert, but he did not succeed in resolving them.

BIBLIOGRAPHY

I. ORIGINAL WORKS. Rohn's writings include "Über Flächen 4. Ordnung mit dreifachem Punkte," in *Mathematische Annalen*, **24** (1884), 55–151; "Über Flächen 4. Ordnung mit 8–16 Knotenpunkten," in *Berichte über die Verhandlungen der Sächsischen Akademie der Wissenschaften zu Leipzig*, **36** (1884), 52–60; *Lehrbuch der darstellenden Geometrie*, 2 vols. (Leipzig, 1893–1896), written with E. Papperitz; and "Die ebene Kurve 6. Ordnung mit 11 Ovalen," in *Berichte über die Verhandlungen der Sächsischen Akademie der Wissenschaften zu Leipzig*, **63** (1911), 540–555.

II. SECONDARY LITERATURE. See O. Hölder, "Nekrolog für K. Rohn," in *Leipziger Berichte*, **72** (1920), 107–127; and F. Schur, "Karl Rohn," in *Jahresbericht der Deutschen Mathematiker-vereinigung*, **32** (1923), 201–211.

WERNER BURAU

ROLANDO, LUIGI (*b.* Turin, Italy, 16 June 1773; *d.* Turin, 20 April 1831), *medicine, anatomy, physiology, zoology.*

After the death of his father Rolando was entrusted to the priest Antonio Maffei, a paternal uncle, who attended to his education. He enrolled at the Faculty of Medicine in Turin, where he showed particular interest in the anatomy courses of G. F. Cigna, who considered Rolando his most promising pupil. This bent, however, did not prevent him from also devoting time and studies to comparative anatomy and zoology. His thesis, an anatomical and physiological study of the lungs in various classes of animals, also dealt with pleuropulmonary ailments of tubercular origin.

In 1802 Rolando began to practice medicine, and his increasing fame led to his appointment, on 15 November 1804, as professor of practical medicine at Sassari University. On his way to Sardinia, Rolando remained in Florence for three years, becoming friends with Paolo Mascagni and Felice Fontana, continuing his studies of anatomy, and practicing anatomical drawing. In 1807 he assumed the chair assigned to him in Sassari where he was also given the post of chief physician. Rolando's anatomico-physiological studies from this period dealt mainly with the structure and function of the nervous system in man and animals, using the comparative method with which he had become familiar. In the meantime he also continued zoological research.

In 1814 Rolando accompanied the royal family on its return from Sardinia to Turin and became professor of anatomy at Turin University, in addition to his various posts in scientific and health organizations. Despite this combination of engagements, which considerably weakened his health, he published an exceptional number of articles and works on entomology, zoology, general physiology, and pathology. He died of cancer of the pylorus.

Rolando's most important studies were devoted to the anatomical, physiological, and embryological examination of the brain. In particular, he examined the gray matter, which he considered to be different from that of the striate bodies; and he discovered the cerebral branches and fibrous processes, which he studied by means of serial sections. Contradicting existing opinions, Rolando asserted that in the first stages of development of the central nervous system of the embryo two vesicles appear, representing the medulla oblongata, from which the cerebral hemispheres are then developed; to the latter he attributed the intellectual faculties.

Experimental research on the cerebellum led him to consider, before Flourens—with whom he had an argument on the priority of the discovery—that this organ governs muscular movements. Rolando had observed that in animals in which lesions of this organ are induced, such movements progressively decrease until they disappear entirely, parallel with the extent and seriousness of the damage. Flourens then correctly restricted the functions of the cerebellum to the coordination of movement, but he refused to acknowledge that Rolando had preceded him along this course.

Rolando's name is linked to the Rolandic fissure, the sulcus separating the frontal lobe from the parietal.

BIBLIOGRAPHY

I. ORIGINAL WORKS. Rolando's principal writings include *Anatomico-physiologica-comparativa disquisitio in respirationis organa* (Turin, 1801); *Phtiseos pulmonalis specimen theoretico-practicum* (Turin, 1801); *Observations anatomiques sur la structure du sphinx nerii et autres insectes* (Sassari, 1805); *Sulle cause da cui dipende la vita negli esseri organizzati* (Florence, 1807); *Saggio sopra la vera struttura del cervello dell'uomo e degli animali e sopra le funzioni del sistema nervoso* (Sassari, 1809; 2nd ed., Turin, 1828, is the more important); *Humani corporis fabricae ac functionum analysis adumbrata* (Turin, 1817);

Osservazioni sulla pleura e sul peritoneo (Turin, 1818); *Anatomes physiologica* (Turin, 1819); *Cenni fisico-patologici sulle differenti specie d'eccitabilità e d'eccitamento, sull'irritazione e sulle potenze eccitanti, debilitanti e irritanti* (Turin, 1821); *Riflessioni e sperimenti tendenti allo scioglimento di alcune questioni riguardanti la respirazione e la calorificazione* (Turin, 1821); *Description d'un animal nouveau qui appartient à la classe des échinodermes* (Turin, 1822); "Recherches anatomiques sur la moelle allongée," in *Atti dell' Accademia delle scienze di Torino*, **29** (1825), 1–78; and "Osservazioni sul cervelletto, *ibid.*, 163–188.

II. SECONDARY LITERATURE. On Rolando's life, see P. Capparoni, *Profili bio-bibliografici di medici e naturalisti celebri italiani*, II (Rome, 1928), 97–101, with portrait.

CARLO CASTELLANI

ROLFINCK, GUERNER (*b.* Hamburg, Germany, 15 November 1599; *d.* Jena, Germany, 6 May 1673), *medicine, chemistry, botany.*

Rolfinck was named for his father, the rector of the Johanneum in Hamburg. He studied medicine at Wittenberg under Daniel Sennert and then at Leiden. He continued his education at Oxford and Paris, and received the M.D. degree at Padua on 7 April 1625. It is said that he was offered a professorship at Padua but declined because he wished to return to Germany. In 1629 Rolfinck went to Jena as professor of anatomy, surgery, and botany. He remained there until his death forty-four years later.

In 1629 Rolfinck built the first anatomical theater at Jena and gave lectures involving dissection, an innovation that aroused controversy. Two recently executed criminals were the subjects of dissection, which for a time became known as "Rolfincking" and—by implication—served as an additional deterrent to the "criminal classes." Rolfinck was the first German to accept Harvey's theory of the circulation of the blood, which he taught (in 1632) with such enthusiasm that he compared Harvey to Columbus. He collected surgical instruments, made a personal contribution to ophthalmology (he was the first to demonstrate the location of cataracts in the lens of the eye), and is credited with revolutionizing medicine at Jena.

Although botanical excursions and chemical lectures had been held earlier at Jena, Rolfinck was the first professor in both fields. Botany was part of his original professorship, and he founded the botanical garden in 1631. In 1638 he established the chemical laboratory, and in the following year was appointed *director exercitii chymia*, which became a professorship in 1641.

The expression "transitional figure" fits Rolfinck

well. Despite his renovation of the Jena medical school, he was grounded in an earlier time and produced both a commentary on Hippocrates and an epitome of the medicine of al-Rāzī (he was competent in both Greek and Arabic). He also wrote on botany, but from 1655 he was increasingly occupied with chemistry. In the latter field Rolfinck showed himself the enemy of alchemy and "superstition." His austere approach is revealed in the preface to his *Chemia in artis formam redacta*, where he says that the book is not universal but particular, not for transmutation but for medicine, and solely "occupied with the resolution of mixts for the benefit of human health." He even wrote a book on "chemical nonentities," chemical works and operations that, since they are unnatural, cannot be accomplished, notwithstanding the "contrary clamoring" of the ordinary chemists. The "nonentities" include such old favorites as oils from precious stones, "sweet" oils of acids and bitter salts, drinkable gold, and "fixed" (solidified) mercury.

Rolfinck's research contributions to science were minor but those to the organization of science for a new age clearly were major. In addition to the activities already mentioned, he maintained his personal medical practice, trained 104 doctoral candidates, and was six times rector of the University of Jena.

BIBLIOGRAPHY

I. ORIGINAL WORKS. Rolfinck's major writings are *Epitome methodi cognoscendi et curandi particulares corporis effectis, secundum ordinem Abubatri Rhazae ad Regun Mansorum* (Jena, 1655); *Dissertationes anatomicae methoda synthetica* (Nuremberg, 1656); *Chemia in artis formam redacta* (Jena, 1661); *Non ens chimicum* (Jena, 1670); and *De vegetabilibus, plantis, suffructibus, fructibus, arboribus in genere* (Jena, 1670).

II. SECONDARY LITERATURE. See the following, listed chronologically: G. W. Wedel, *Oratorio in funere Rolfinkii* (Jena, 1673); J. Günther, *Lebenskizzen der Professoren der Universität Jena, seit 1558 bis 1858* (Jena, 1858); F. Chemnitius, *Die Chemie in Jena von Rolfinck bis Knorr* (Jena, 1929); Ernst Giese and Benno von Hagen, *Geschichte der medizinischen Fakultät der Friedrichs-Schiller-Universität Jena* (Jena, 1958); Max Steinmetz, ed., *Geschichte der Universität Jena*, 2 vols. (Jena, 1958); and J. R. Partington, *A History of Chemistry*, II (London, 1961), 312–314.

R. P. MULTHAUF

ROLLE, MICHEL (*b.* Ambert, Basse-Auvergne, France, 21 April 1652; *d.* Paris, France, 8 November 1719), *mathematics.*

The son of a shopkeeper, Rolle received only a very elementary education. He worked first as a transcriber for a notary and then for various attorneys in his native region. At the age of twenty-three he moved to Paris. Married early and burdened with a family, he had difficulty earning sufficient money as master scribe and reckoner. But by independent study he learned algebra and Diophantine analysis. In the *Journal des sçavans* of 31 August 1682 Rolle gave an elegant solution to a difficult problem publicly posed by Ozanam: to find four numbers the difference of any two of which is a perfect square as well as the sum of the first three. Ozanam had stated that the smallest of the four numbers would have at least fifty figures. Rolle provided a solution in which the four numbers were expressed by homogeneous polynomials in two variables and of degree twenty. The smallest numbers found in this fashion each had only seven figures.

This brilliant exploit brought Rolle public recognition. Colbert took an interest in him and obtained for him a reward and, it was said, a pension. Rolle later enjoyed the patronage of the minister Louvois. He gave lessons in elementary mathematics to the latter's fourth son, Camille Le Tellier, abbé de Louvois (1675–1718). Rolle even received an administrative post in the ministry of war, from which he soon resigned.

Rolle entered the Académie des Sciences in 1685 with the title—rather disconcerting for us—of *élève astronome*. When the Académie was reorganized in 1699 he became *pensionnaire géomètre*, a post that assured him a regular salary. In 1708 he suffered an attack of apoplexy. He recovered, but a second attack in 1719 proved fatal.

Although it was his skill in Diophantine analysis that made Rolle's reputation, his favorite area was the algebra of equations, in which he published *Traité d'algèbre* (1690), his most famous work. In this book he designated, following Albert Girard (1629), the *n*th root of a number a, $\sqrt[n]{a}$, not as $\sqrt{@a}$, as was usually done before him. His notation soon became generally accepted. He retained the Cartesian equality sign ∞ until 1691, when he adopted the equal sign ($=$), which originated with Robert Recorde (1557).

In 1691, Rolle adopted, in advance of many of his contemporaries and in opposition to Descartes, the present order relation for the set of the real numbers: "I take $-2a$ for a greater quantity than $-5a$."

Rolle's *Algèbre* contains interesting considerations on systems of affine equations. Following the techniques established by Bachet de Méziriac (1621), Rolle utilized the Euclidean algorithm for resolving Diophantine linear equations. He employed the same algorithm to find the greatest common divisor of two polynomials, and in 1691 he was able to eliminate one variable between two equations.

The *Traité d'algèbre*, the language of which is so special, has remained famous, thanks notably to the method of "cascades." Rolle used this method to separate the roots of an algebraic equation. He justified it by showing (1691) that if $P(x) = 0$ is the given equation, and if it admits two reals roots a and b, then $P'(b) = (b - a) Q(b)$, where Q is a polynomial. $P'(x)$, a polynomial derived from $P(x)$, is what Rolle called the "first cascade" of the polynomial $P(x)$. The second cascade is our second derivative, and so on.

Arranging the real roots in ascending order, Rolle showed that between two consecutive roots of $P(x)$ there exists a root of $P'(x)$. His methods of demonstration are elaborations of the method utilized by Jan Hudde in his search for extrema (1658).

In 1846 Giusto Bellavitis gave Rolle's name to the present theorem: if the function $f(x)$ is defined and continuous on the segment $[ab]$, if $f(a) = f(b)$, and if $f'(x)$ exists in the interior of the segment, then $f'(x)$ is equal to zero at least once in the segment.

In 1699 the three pensionary geometers of the Academy were the Abbé Jean Gallois, a partisan of Greek mathematics; Rolle, an autodidact but very well versed in Cartesian techniques; and Pierre Varignon, who favored the ideas of Leibniz. L'Hospital was an honorary academician; in 1696 he had published *Analyse des infiniment petits*. The Academy was very divided over infinitesimal analysis. Rolle—incited, it was said, by influential persons—vigorously attacked infinitesimal analysis and strove to demonstrate that it was not based on solid reasoning and led to errors. Among the examples he chose were the curves

$$y - b = (x^2 - 2ax + a^2 - b^2)^{2/3} \text{ and}$$

$$y = 2 + \sqrt{(4 + 2x)} + \sqrt{(4x)}.$$

Varignon defended the new methods and pointed out the paralogisms that Rolle displayed in discussing these examples. The latter, too plebeian to control himself, created an uproar. A commission established to resolve the matter was unable to come to a decision. The dispute lasted from 1700 to 1701, and then continued in the *Journal des sçavans* in the form of exchanges between Rolle and a newcomer, Joseph Saurin. The Academy again intervened, and in the fall of 1706 Rolle acknowledged to Varignon, Fontenelle, and Malebranche that he had given up and fully recognized the value of the new techniques.

Rolle also displayed a certain vigor in the field of Cartesian geometry. In 1693, in the *Journal des sçavans*, he offered a prize of sixty pistoles for the solution, without the use of his methods, of the following problem: construct the roots of an equation by utilizing a given arc of an algebraic curve. Before leaving Paris, Johann I Bernoulli had given a solution to this problem in Latin to L'Hospital and had requested him to submit a French translation to the *Journal*. The solution did not meet with Rolle's approval, and the resulting polemic lasted for five numbers of the *Journal*; the sixty pistoles remained in the donor's coffers.

Another of Rolle's achievements is an observation that, though initially paradoxical, was recognized as correct by Saurin: two arcs of algebraic curves the convexity of which is in the same direction can have a large number of common points.

Rolle was a skillful algebraist who broke with Cartesian techniques; and his opposition to infinitesimal methods, in the final analysis, was beneficial.

BIBLIOGRAPHY

I. ORIGINAL WORKS. Rolle's books are *Traité d'algèbre, ou principes généraux pour résoudre les questions de mathématique* (Paris, 1690); *Démonstration d'une méthode pour resoudre les egalitez de tous les degrez; suivie de deux autres méthodes dont la première donne les moyens de résoudre ces mêmes égalitez par la géométrie, et la seconde pour résoudre plusieurs questions de Diophante qui n'ont pas encore esté resoluës* (Paris, 1691); and *Méthode pour resoudre les équations indéterminées de l'algèbre* (Paris, 1699). See also two papers in *Mémoires de l'Académie royale des Sciences:* "Règles pour l'approximation des racines des cubes irrationnels" (read 31 Jan. 1692), **10**; and "Méthode pour résoudre les égalités de tous degrés qui sont exprimés en termes généraux" (read 15 Mar. 1692), *ibid.*, 26–33.

Rolle's later papers in the Mémoires of the Académie des Sciences include "Remarques sur les lignes géométriques" (1702), 171–175, (1703), 132–139; "Du nouveau système de l'infini" (1703), 312–336; "De l'inverse des tangentes" (1705), 222–225; "Méthode pour trouver les foyers des lignes géométriques de tous les genres" (1706), 284–295; "Recherches sur les courbes géométriques et mécaniques, où l'on propose quelques règles pour trouver les rayons de leurs développées" (1707), 370–381; "Éclaircissements sur la construction des égalitez" (1708), 339–374 (1709), 320–350; "De l'évanouissement des quantités inconnues" (1709), 419–451; "Règles et remarques pour la construction des égalités" (1711), 86–100; "Remarque sur un paradoxe des effections géométriques" (1713), 243–261, read 12 July 1713; and "Suite des remarques sur un paradoxe des effections géométrique" (1714), 5–22, read 10 Jan. 1714.

Rolle also published many articles in the *Journal des sçavans* beginning in 1682.

II. SECONDARY LITERATURE. See the *éloge* by Fontenelle, read at the Académie des Sciences on 10 Apr. 1720, Niels Nielsen, *Géomètres français du dix-huitième siècle* (Copenhagen, 1935), 382–390; Gino Loria, *Storia delle matematiche*, 2nd ed. (Milan, 1950), 670–673; J. E.

Montucla, *Histoire des mathématiques*, II (Paris, 1758), 361–368, 2nd. ed., III (Paris, 1802; repr. 1960), 110–116, (which contains a detailed account of the polemic over the infinitesimal calculus—Montucla had access to Varignon's MSS); Cramer, ed., *Virorum Celebrium G. G. Leibnitii et Johan Bernoullii commercium philosophicum et mathematicum*, 2 vols. (Lausanne, 1745)—see index and II, 148 for Cramer's note containing a list of the articles in the *Journal des sçavans* concerning the dispute between Rolle and Saurin; O. Spiess, ed., *Der Briefwechsel von Johann I Bernoulli* (Basel, 1955)—see index and p. 393 for a note on the articles in the *Journal des sçavans* concerning the prize offered by Rolle and claimed by Johann I Bernoulli; Malebranche, *Oeuvres complètes*, XVII, pt. 2 (Mathematica) (Paris, 1968); Pierre Costabel, *Pierre Varignon (1654–1722) et la diffusion en France du calcul différentiel et intégral*, in the series Conférences du Palais de la Découverte (Paris, 1965); Petre Sergescu, *Un episod din batalia pentru triumful calculului diferential: polemica Rolle-Saurin 1702–1705* (Bucharest, 1942), repr. in *Essais scientifiques* (Timisoara, 1944); D. E. Smith, *A Source Book in Mathematics* (1929; repr. New York, 1959), 253–260, which contains Rolle's theorem and partial English translations of the works of 1690 and 1691; L. E. Dickson, *History of the Theory of Numbers*, II (New York, 1934), 45 (on linear Diophantine equations) and p. 447 (on the problem posed by Ozanam that Rolle solved); Dickson nowhere cites the passage of the 1691 work concerning Diophantine analysis; Jean Prestet, *Nouveaux élémens des mathématiques*, II (Paris, 1689), 238, a solution, by a different procedure, of Ozanam's problem and a criticism of Rolle's method—this work is not cited in Dickson; and Jakob Hermann, "Observationes in schediasma quod Dn. Rolle cum hac inscriptione: Éclaircissements sur la construction des égalitez," in *Commentariis Academiae Regiae Scientiarum* (1708, published in 1727), *Miscellanea Berolinensia*, III (Berlin, 1927), 131–146.

JEAN ITARD

ROLLESTON, GEORGE (*b.* Maltby, near Rotherham, Yorkshire, England, 30 July 1829; *d.* Oxford, England, 16 June 1881), *comparative anatomy, zoology, archaeology, anthropology.*

Linacre professor of anatomy and physiology at Oxford from 1860 until his death, Rolleston pioneered in the teaching of elementary zoology by means of the still-dominant "type" system, in which a few representative organisms are selected for the student to dissect. He was the second son of the Rev. George Rolleston, vicar and squire of Maltby Hall, from whom he received his early education at home. In 1839 he entered the grammar school at Gainsborough and two years later the collegiate school at Sheffield. In 1846 he won an open scholarship at Pembroke College, Oxford, which he entered early in 1847. He graduated B.A. in 1850 with a first class in classics. Following his election on 27 June 1851 to the Sheppard fellowship (established at Pembroke in 1846 to promote the study of law and medicine), Rolleston shifted his attention from classics to medicine; and in October 1851 he enrolled at the medical school attached to St. Bartholomew's Hospital in London. He graduated M.A. from Oxford in 1853, M.B. in 1854, and M.D. in 1857. In 1856 he became a member, and in 1859 a fellow, of the Royal College of Physicians, where he delivered the Harveian oration in 1873. He was elected fellow of the Royal Society in 1862 and of Merton College, Oxford, in 1872.

In 1855, near the end of the Crimean War, Rolleston was appointed one of the physicians to the British Civil Hospital in the Turkish seaport of Smyrna (now İzmir), where he remained after the war to write a report for the war secretary on the sanitary and other conditions of the city (1856). He returned to England in 1857, serving briefly as assistant physician at the Hospital for Sick Children in London. Later in the same year he moved back to Oxford, where he had been appointed physician to the Radcliffe Infirmary and Lee's reader in anatomy at Christ Church College. On 21 September 1861 he married Grace Davy, daughter of Dr. John Davy and niece of Sir Humphry Davy. His wife survived him, as did seven children, including Humphry Davy Rolleston, regius professor of physics at Cambridge from 1925 to 1932.

Of Rolleston's research papers, the most interesting bear on the celebrated Victorian controversy over man's place in nature. More specifically, Rolleston supported T. H. Huxley against Richard Owen on the most crucial issues in their debate over the similarities and differences between human and simian brains. Like Huxley, he denied Owen's claim that the human brain contained anatomical structures (notably the hippocampus) not found in simian brains. At the same time, however, Rolleston emphasized the large differences of degree (if not of kind) between human and simian brains, noting in particular the greater absolute weight and height of the human brain and the greater complexity of its cerebral convolutions. He also insisted that his support of Huxley's anatomical position should not be extended to the materialist implications sometimes drawn from Huxley's results. In fact, Rolleston's support of Huxley was expressed in a manner so cautious and contorted, and occasionally so obscure, that some inattentive readers placed him among Owen's supporters.[1]

More generally, Rolleston's verbose and convoluted prose style made it extremely difficult to follow his train of thought on any topic. It is doubtful that he made any important original contributions to the scientific literature, despite learned and wide-ranging papers on mammalian placental structures; the development of the enamel in mammalian teeth; the Stone, Bronze, and Iron ages; and prehistoric pigs and cats. As the latter two topics perhaps suggest, Rolleston shared the contemporary enthusiasm for archaeology and anthropology, an enthusiasm that carried him into the fields of craniology and anthropometry. Several of his papers concern the classification of human skulls and skeletons excavated from prehistoric burial mounds in various parts of England. If his work in these areas now seems marred by ethnocentricity and male complacency, it was at least representative of the age and laid the basis for the superb collection of human skulls in the Oxford Museum.

Rolleston introduced the type system of zoological instruction at Oxford in the early 1860's:

> His "types" were the Rat, the Common Pigeon, the Frog, the Perch, the Crayfish, Blackbeetle, Anodon, Snail, Earthworm, Leech, Tapeworm. He had a series of dissections of these mounted, also loose dissections and elaborate MS. descriptions. The student went through the series, dissecting fresh specimens for himself. After some ten years' experience Rolleston printed his MS. directions and notes as a book, called *Forms of Animal Life*.[2]

From the beginning the type system has had its critics, who lament its restricted focus on dissection and on a small number of animals. Also from the beginning its defenders have insisted on its value as the only possible means of giving the beginning student real exposure to dissection and research techniques. In any case, Rolleston must share credit (or blame) for the type system with Huxley, who had been chiefly responsible for his appointment as first incumbent of the Linacre chair and whose advice he reportedly followed in developing the system.[3] Certainly Huxley played a crucial role in the elaboration and wide diffusion of the type system.[4]

Rolleston took an active part in meliorist causes, notably the temperance movement, and in university and municipal politics. As a member of the Oxford local municipal board, he pressed for the isolation of smallpox cases during an epidemic in 1871 and did much to improve the municipal drainage system, water supply, and sanitary regulations. In testimony before the Royal Commission on Vivisection in 1875, he steered a middle course between the community of research physiologists and the more rabid antivivisectionists. The provisions of the Vivisection Act of 1876 reportedly satisfied him, but it is an exaggeration to claim that the act "was in great measure framed on his recommendations."[5] In general, Rolleston was a transitional figure, not quite ready to embrace the new experimental movement emanating mostly from Germany but attuned to the importance of most of the new currents in biological thought, including Darwinian evolutionary theory.

NOTES

1. See Leonard Huxley, *Life and Letters of Thomas Henry Huxley*, 2 vols. (London, 1900), I, 191. For a general discussion of the debate over man's place in nature, see William F. Bynum, "The Problem of Man in British Natural History, 1800–1863" (Ph.D. diss., Cambridge, 1974).
2. Huxley, *op. cit.*, 377–378, quoting a letter from E. Ray Lankester.
3. See Cyril Bibby, "Thomas Henry Huxley and University Development," in *Victorian Studies*, **2** (1958), 97–116, on 100. See also Huxley, *loc. cit.*
4. See G. L. Geison, "Michael Foster and the Rise of the Cambridge School of Physiology, 1870–1900" (Ph.D. diss., Yale, 1970), 275–300.
5. E. B. Tylor, "Life of Dr. Rolleston," lx. For an antidote to this simplistic assessment, see Richard D. French, *Antivivisection and Medical Science in Victorian Society* (Princeton, 1975).

BIBLIOGRAPHY

I. ORIGINAL WORKS. Rolleston's most influential work, and his only book, was *Forms of Animal Life: A Manual of Comparative Anatomy With Descriptions of Selected Types* (Oxford, 1870); 2nd ed., rev. and enl. by W. Hatchett Jackson (1888). Forty-eight of his publications are gathered in George Rolleston, *Scientific Papers and Addresses*, arranged and edited by William Turner, with a biographical sketch by Edward B. Tylor, 2 vols. (Oxford, 1884). This work also contains a bibliography of 79 items, including several book reviews and brief notes (pp. lxvii–lxxvi), and a digest of three previously unpublished archaeological notes (II, 937–944). The Royal Society *Catalogue of Scientific Papers*, V, 260–261; VIII, 771; XI, 210–211; XII, 626; lists 30 papers by Rolleston and two of which he was coauthor. Twenty-eight letters exchanged between Rolleston and T. H. Huxley are preserved in the Huxley papers, Imperial College of Science and Technology, London; there is a microfilm copy at the library of the American Philosophical Society, Philadelphia. See Warren R. Dawson, *The Huxley Papers: A Descriptive Catalogue of the Correspondence, Manuscripts and Miscellaneous Papers of Thomas Henry Huxley* (London, 1946), 135–136.

II. SECONDARY LITERATURE. For obituary sketches of Rolleston, see Edward B. Tylor, "Life of Dr. Rolleston," in *Scientific Papers and Addresses*, I, ix–lxv; D'Arcy Power, in *Dictionary of National Biography*, XVII, 167–169; and W. H. Flower, in *Nature*, **24** (1881), 192–193, repr. in *Proceedings of the Royal Society*, **33** (1882), xxiv–xxvii.

GERALD L. GEISON

ROLLET, JOSEPH-PIERRE-MARTIN (*b*. Lagnieu, Ain, France, 12 November 1824; *d*. Lyons, France, 2 August 1894), *venereology*.

After graduation from the *lycée* at Lyons, Rollet studied medicine in Paris, became *interne des hôpitaux* in 1845, and worked on the surgical services of Beaujon Hospital under Stanislas Laugier, at St.-Antoine under Auguste Bérard, and at the Pitié under Jacques Lisfranc. In 1848 he published his doctoral thesis on traumatic hemorrhages in the skull and received a bronze medal for tending the victims of the June Days. Having failed in the competitive examination at Lyons for chief-of-service in surgery at the Hôtel-Dieu in 1849, Rollet won the competition at the Antiquaille Hospital, where patients with venereal disease predominated. The position was not available until 1855; he spent the waiting period in study and private practice. His research, teaching, and writing on syphilis during his nine years at the Antiquaille made him famous.

In French syphilography confusion reigned. The fashionable Paris professor Philippe Ricord, who dominated the field, taught that secondary syphilis is not contagious. Rollet soon showed that two diseases were being confused: "Rollet's chancre" is a mixed infection consisting of what are now called Schaudinn's bacillus (*Treponema pallidum*) and Ducrey's bacillus (*Haemophilus ducreyi*). Rollet succeeded in differentiating between the two bacilli, in those days before the germ theory of disease was generally accepted, by painstaking clinical observation, establishing that syphilis has a mean incubation period of twenty-five days. He published his work in *Recherches . . . sur la syphilis* (1861) and *Traité des maladies vénériennes* (1865). He also contributed articles to Amédée Dechambre's *Dictionnaire encyclopédique des sciences médicales*.

Important practical and legal consequences derived from the fact that hereditary and secondary syphilis were recognized as contagious. Secondary syphilitic infection was widespread among glassblowers: three workers usually shared one blowing iron. Rollet identified this tool as the carrier of the "virus"; and the glass industry soon mechanized the operation, substituting compressed air for human breath.

Wet nurses were blamed for infecting infants with syphilis. Rollet showed that the reverse often was the case: babies with congenital syphilis transmitted the infection through their mouths. He also cautioned that Jennerian vaccination often propagated syphilis and urged that animal serum be substituted for arm-to-arm vaccination. He summarized his work at the termination of his stewardship at the Antiquaille in *Coup d'oeil rétrospectif . . .* (1864).

Appointed to membership, and soon the presidency, of the Conseil d'Hygiène et de Salubrité du Département du Rhône, Rollet became the delegate from Lyons to the International Congress of Medicine at Paris (1867), where he presented a report on general prophylactic measures against venereal disease. The Congress empowered him to approach the French foreign ministry to explore possibilities for the international control of venereal disease, but the Franco-Prussian War permanently interrupted this work.

In 1877 Rollet was appointed professor of hygiene at the new Medical Faculty of Lyons. He was by then a member of the Société Anatomique, the Société de Médecine Publique et d'Hygiène Professionnelle, an associate member of the Paris Academy of Medicine, and a corresponding member of the Academy of Sciences. He died at Lyons while presiding over a congress of the Society of Dermatology and Syphilography, which he had helped to found.

BIBLIOGRAPHY

I. ORIGINAL WORKS. Rollet's major writings are *Des agents contagieux des maladies de la peau* (Lyons, 1855); *Annuaire de la syphilis et des maladies de la peau* (Paris, 1858), written with P. Diday; *Mémoire sur le sarcocèle fongueux syphilitique* (Lyons, 1858); *Nouvelles recherches sur le rhumatisme blennorhagique* (Lyons, 1858); *Inoculation, contagion et confusion en matière de syphilis* (Lyons, 1860); *De la pluralité des maladies vénériennes* (Paris, 1860); *Recherches cliniques et expérimentales sur la syphilis, le chancre simple et la blennorrhagie* (Lyons, 1861); *Recherches sur plusieurs maladies de la peau, réputées rares ou exotiques, qu'il convient de rattacher à la syphilis* (Paris, 1861); *De la transmission de la syphilis entre nourrissons et nourrices au point de vue de la médecine légale* (Paris, 1861); *De la contagion de la syphilis secondaire* (Lyons, 1862); *Coup d'oeil rétrospectif sur la syphilis et les maladies de la peau; compte-rendu d'un exercice de neuf années, de 1855 à 1864* (Paris, 1864); *Traité des maladies vénériennes* (Paris, 1865); *Nouvelles conjectures sur la maladie de Job* (Paris, 1867); *De la prophylaxie générale des maladies vénériennes* (Lyons, 1867); *Prophylaxie internationale des maladies vénériennes* (Paris, 1869); *De la nécessité de l'adjonction des médecins au conseil d'administration des hospices* (Lyons, 1871); *Des caractères particuliers et du traitement de la blessure d'Alexandre le Grand, reçue dans le combat contre les Malliens* (Lyons, 1877); *Des applications du feu à l'hygiène dans les temps préhistoriques* (Lyons, 1879); *Influence des filtres naturels sur les eaux potables* (Lyons, 1882); *Rapport sur les mesures sanitaires applicables à Lyon en prévision du choléra* (Lyons, 1883); and *Épidémie de fièvre typhoïde à l'École normale et au collège de Cluny* (Lyons, 1887).

Rollet also wrote the following articles for A. Dechambre, ed., *Dictionnaire encyclopédique des sciences médicales:* "Acné syphilitique," I, 571–574; "Balanite," VIII, 263–276; "Blennorrhagie," IX, 638–696; "Bouche (maladies vénériennes)," X, 249–267; "Bubon," XI, 256–275; "Chancre," XV, 224–286; "Lichen syphilitique," 2nd ser., II, 529–531; "Mal de la baie de St. Paul," 2nd ser., IV, 205–207; "Mal de Brunn," *ibid.*, 211–212; "Mal de Chavanne-Lure," *ibid.*, 212–214; "Mal de Fiume ou de Scherlievo," *ibid.*, 214–217; "Mal de Ste. Euphémie," *ibid.*, 227; "Mamelles (maladies syphilitiques)," *ibid.*, 436–455; "Mercurielles (maladies)," 2nd ser., VII, 96–107; "Rupia syphilitique," 3rd ser., V, 624–627; "Syphilides," 3rd ser., XIV, 212–247, written with E. Chambard; "Syphilis," *ibid.*, 255–501; and "Syphilisation," *ibid.*, 678–691.

II. SECONDARY LITERATURE. See L. Bonnet, "Les fêtes du centenaire de J. Rollet," in *Lyon médical*, **134** (1924), 740–743; J. Gaté and J. Rousset, "La dermato-vénérologie lyonnaise," in *Lyon et la médecine*, special issue of *Revue lyonnaise de médecine* (1958), 346–347; M.A. Horand, "Notice biographique sur J. Rollet," in *Lyon médical*, **83** (1896), 577–586, and **84** (1897), 11–19, 44–50, 84–92; É. Jeanselme, "Le centenaire de Rollet à Lyon," in *Bulletin de la société française d'histoire de la médecine*, **18** (1924), 351–356; L. Jullien, "J. Rollet, 1824–1894," in *Annales de dermatologie et de syphiligraphie*, 3rd ser., **5** (1894), 1209–1219; J. Nicolas, "Joseph Rollet," in *Bulletin médical*, **38** (1924), 1411–1414; and G. Thibierge, "Rollet et son oeuvre," in *Gazette des hôpitaux civils et militaires*, **97** (1924), 1601–1606.

There are also short notices in *Archives provinciales de chirurgie*, **3** (1894), 601–604; A. Hirsch, ed., *Biographisches Lexikon der hervorragenden Ärzte aller Zeiten und Völker*, 2nd ed. (Berlin–Vienna, 1932), IV, 864–865; *British Medical Journal* (1894), **2**, 452; *Gazette hebdomadaire de médecine et de chirurgie*, **31** (1894), 392; and *Revue générale de clinique et de thérapeutique*, **39** (1925), supp. 395–409.

DORA B. WEINER

ROMANES, GEORGE JOHN (*b.* Kingston, Ontario, 2 May 1848; *d.* Oxford, England, 23 May 1894), *physiology, comparative psychology, evolution.*

Romanes was the third son of Rev. George Romanes, a classical scholar and theologian, and Isabella Gair Smith. In the year of Romanes' birth his father inherited a considerable fortune, resigned his post as professor of Greek at Queen's University in Kingston, and moved his family to Britain, finally settling in Regent's Park, London. Romanes grew up in comfortable circumstances, accompanying his parents on several trips to the Continent. Except for a brief period in a preparatory school (cut short by illness), he was educated at home. This early education was casual and unsystematic, and Romanes apparently showed no early signs of intellectual promise. At the age of seventeen he was sent to a tutor to read in preparation for the university, and in October 1867 he entered Gonville and Caius College, Cambridge.

The six years (1867–1873) that Romanes spent at Cambridge were decisive both in drawing out his talents and in determining the direction of his future work. After reading mathematics for a time, his attention was drawn to science by fellow students; and in 1868 he competed successfully for a natural science scholarship. By the time he received a second class on the natural science tripos of 1870, Romanes had abandoned his intention of entering the Church; and shortly thereafter he began to study physiology under Michael Foster. Foster, at that time just beginning his great reform of British physiology, was a pioneer in the evolutionary approach to the subject; and it was while working with him that Romanes read Darwin's works for the first time. In the exciting atmosphere of Foster's laboratory, Romanes' interest grew rapidly; and a personal note from Darwin praising a piece of his in *Nature* reinforced his decision to devote himself to scientific research.

In 1874 Romanes took his M.A. and moved to London, where he continued his physiological studies under William Sharpey and John Burdon-Sanderson at University College, and extended the circle of his scientific acquaintances. In the same year he visited Darwin, beginning a personal friendship that remained close until the latter's death.

Romanes received his training and began his scientific career at a time when evolutionary thinking dominated British scientific circles. Working under Foster and Burdon-Sanderson, and befriended by Darwin himself, he was thoroughly imbued with evolutionary modes of thought. This influence is clearly reflected in each of the three areas of study to which he devoted himself: physiology, comparative psychology, and the theory of evolution.

In the twenty years following his move from Cambridge and before his death in 1894, Romanes proved to be one of the most brilliant of the second generation of British Darwinists. Freed from professional obligations by family wealth, uninterested in politics, and endowed with tremendous energy and enthusiasm, he was able to put all his resources at the disposal of science. A clear and forceful writer, he was a prolific contributor not only to scientific journals but also to the monthly reviews that then helped to shape educated opinion. His dialectical skill and unshakable good humor made him a formidable opponent in scientific debate and a popular lecturer. Most of his books were translated into French and German, and his work was widely read in the United States. In addition to his writing and lecturing, Romanes also served as zoological secretary

of the Linnean Society and secretary of the Physiological Society, and was a member of the Council of University College, London. In 1891 he founded at Oxford the series of lectures that bears his name.

Romanes' most important early scientific work was in invertebrate physiology. Working at a small marine laboratory set up at his summer home on the Scottish coast, he began by trying to determine whether the medusa (jellyfish) possessed nerve tissue. This was still an open question and one with definite evolutionary interest, since nerve tissue had already been found in all higher organisms.

Excision experiments showed that the marginal tissue of the rim of the medusa's swimming bell was able to continue its rhythmic contractions when separated from the bell, while the bell itself was paralyzed by the same operation. Romanes concluded that the centers of spontaneous contraction of the medusa were localized in the marginal tissue. Further excision experiments indicated that such centers of spontaneity were concentrated in the small marginal bodies, or lithocysts. At the same time the excised bell retained the capacity to contract in response to mechanical or electrical stimulation. Severe spiral sectioning of the bell with interdigitating cuts failed to destroy its contractile properties. If carried far enough, however, sectioning would eventually produce a "block" of the waves of contraction at a single point. To explain these physiological results, Romanes at first postulated the existence of a fine plexus or dense crisscrossing grid of "lines of nervous discharge," without, however, being able to describe this plexus histologically. These experiments and their results formed the subject of Romanes' first paper for the Royal Society, published in 1876.

At Romanes' request the medusa was examined histologically in 1877 by his friend Edward A. Schäfer (later Sharpey-Schafer), who, using a gold chloride stain, was able to demonstrate the existence of an "interlacement" of nerve fibers on the undersurface of the bell. Schäfer concluded that the individual nerve fibers, though everywhere closely spaced, were anatomically discontinuous. This was a startling result at the time, since physiological continuity was then thought to demand the anastomosis of nervous elements, and Romanes at first hesitated to accept it. In his later papers he did acknowledge it, however, and by 1885 he was arguing that structural discontinuity of nervous elements was entirely compatible with their physiological continuity. This result of the work of Romanes and Schäfer, together with Romanes' general comparative approach and his skillful coordination of behavioral, physiological, and histological data, had a significant impact on Charles

Sherrington's development of the synapse concept.

As a student of Michael Foster and inspired by his evolutionary approach to physiology, Romanes had been alert from the beginning to the possible comparative significance of his work on the medusa, especially in relation to Foster's work on the vertebrate heart. At the time that his first paper on the medusa was published, he thought that both the vertebrate heart and the bell of the medusa might be cases of an intermediate evolutionary stage between primitive, irritable protoplasm and fully differentiated nerve tissue. When Schäfer's histological results forced him to abandon this idea, he turned, in his second and third papers, to the elucidation of functional parallels. Probably under Foster's influence, Romanes now gave special attention to rhythmic motion. He began to speak of the lithocysts of the medusa as ganglia, and in his third paper he gave a theory of ganglionic action almost identical to Foster's theory of the vertebrate heart. According to this theory, the ganglia normally released a continuous impulse to which the muscle tissue could respond rhythmically. At the same time that Romanes was working toward these conclusions, Foster was directing the attention of another student, Walter Gaskell, to the problems of rhythmic motion in the heart. Romanes' clear and decisive conclusions and the analogy he drew between the cases of the medusa and the heart were important elements in the background of Gaskell's later demonstration of the myogenic origin of the heartbeat.

While he was carrying out this important work in physiology, Romanes was gathering observations and corresponding with Darwin on the subject of animal intelligence. To both Darwin and Romanes it appeared that the theory of evolution required a fundamental continuity in the spectrum of mental life, extending from the lowest organisms up to and including man. Moreover, the ascending stages of mental development should be susceptible of explanation in terms of natural causes. Romanes set himself the task of demonstrating the fact of this continuous development and of giving an account of psychological processes in the light of their probable historical origins. The means to this end was to be a new science of animal intelligence, which he named, in a hopeful allusion to its successful anatomical counterpart, comparative psychology.

Romanes published three books on the subject. The first and most successful, *Animal Intelligence* (London, 1882), was entirely devoted to a more or less systematic presentation of a large number of observations of animal activities presumed to be indicative of intelligence. Anecdotal material gathered from diverse sources was included, together with the results of Romanes' own observations. This was followed by two explicitly theoretical works, *Mental Evolution in Animals* (London, 1883) and *Mental Evolution in Man* (London, 1888). Romanes took as part of the goal of comparative psychology the accurate description of the mental states of animals based on inference from their observed behavior. His own account of these mental states was imbedded in the categories of the British associationist tradition. Like Spencer before him, however, Romanes adapted the associationist scheme to allow for the phylogenetic development of successively higher mental levels, outlining a continuous historical evolution culminating in human intelligence.

Romanes' theoretical books on comparative psychology never exerted much influence. It was rather the observational and experimental side of his work that led to further developments in the field. Romanes' systematic and relatively critical approach marked a definite advance in the study of animal behavior, hitherto a subject for casual and often sentimental anecdotes. Nevertheless, newer workers in the field, led by his younger friend and colleague C. Lloyd Morgan, criticized the observational methods of *Animal Intelligence* as lacking in rigor and Romanes' inferences as too often anthropomorphic. The tendency of later workers in comparative psychology, such as Morgan and Jacques Loeb, was toward greater objectivity in the study of animal behavior, and to reduce any inferences regarding consciousness to the minimum entailed by experiment. The same tendency played a prominent role in the thinking of the early behaviorists in the United States. Yet although his work was soon superseded, Romanes occupies a crucial place in the origins of the modern study of animal behavior.

Romanes was a major participant in the controversies that arose among the Darwinists in the two decades following Darwin's death. The most general question at issue was whether natural selection was the only factor at work in evolution or whether there were other, subordinate factors, without which many phenomena remained unexplained. The question of the inheritability of acquired characters—especially the effects of use and disuse—became a central issue after August Weismann published his theory of the continuity of the germ plasm in 1883. Weismann denied to the inheritability of acquired characters even the restricted role that Darwin had allotted it. This denial drew a spirited protest from Herbert Spencer, who argued for its retention. Romanes generally concurred with Spencer in rejecting a single-factor approach to evolution and called for further experimental investigation to decide the particular question of the inheritance of acquired characters. Romanes' own experimental investigations undertaken to decide the question proved inconclusive, as had his earlier attempts to provide experimental evidence for Darwin's hypothesis of pangenesis.

A second controversy arose around a proposal made by Romanes himself. In May 1886 he read a lengthy paper to the Linnean Society, "Physiological Selection; an Additional Suggestion on the Origin of Species." In it he argued that an account of evolution relying solely on natural selection failed to explain three classes of facts: the seeming nonutility of many specific characters, interspecific sterility, and the need for varieties to escape the swamping effects of intercrossing if they were to become established as permanent species. To meet these difficulties Romanes suggested that mutual infertility between two or more portions of a species population might sometimes arise prior to morphological or other distinctions. The varieties thus reproductively isolated would then be free to develop different (and possibly nonuseful) characters independently of one another, even though mixing freely in the same geographical area.

Romanes' paper provoked a lively but generally negative response from British biologists who, although sometimes acknowledging the difficulties in question, could not accept Romanes' proposed solution. Interest in the issue waned and was revived only in 1888 with the publication of a paper by the American naturalist John Thomas Gulick. Gulick argued that the main task of evolution theory was to give an adequate account of the causes and conditions of evolutionary divergence. Such an account could not be given in terms of natural selection alone, but required recognition of a wide range of mechanisms by which varieties or incipient species became reproductively isolated from one another. In Gulick's scheme Romanes' "physiological selection" appeared as one possible mode of varietal isolation.

Romanes enthusiastically adopted Gulick's more general approach, and in volume III of *Darwin, and After Darwin* (1897) he argued vigorously for recognition of the role of isolation in evolution. Although few evolutionists accepted all of Romanes' arguments and conclusions, the possible roles of isolation and physiological selection had become issues to be debated; and evidence of concern with these issues is reflected in much of the literature on evolution of the first decade of the twentieth century.

BIBLIOGRAPHY

I. ORIGINAL WORKS. A nearly complete bibliography of Romanes' publications may be assembled by consulting four sources: the *British Museum Catalogue of Printed Books*, CCV, 787–789; the Royal Society *Catalogue of Scientific Papers*, VIII, 772; XI, 211–212; and XVIII, 281; the *Cumulative Author Index for Poole's Index to Periodical Literature*, 373; and *Wellesley Index to Victorian Periodicals 1824–1900*, 1069.

His most important papers in invertebrate physiology are "Preliminary Observations on the Locomotor System of Medusae," in *Philosophical Transactions of the Royal Society*, **166** (1876), 269–313; "Further Observations on the Locomotor System of Medusae," ibid., **167** (1877), 659–752; and "Concluding Observations on the Locomotor System of Medusae," ibid., **171** (1880), 161–202. An additional study (not mentioned in the text) done with James Cossar Ewart was published as "Observations on the Locomotor System of Echinodermata," ibid., **172** (1881), 829–885. The substance of these papers was brought together and published for a wider audience as *Jelly-Fish, Star-Fish, and Sea Urchins* (London, 1885).

The three books cited in the text contain the major part of Romanes' work in comparative psychology. The numerous letters to *Nature* and articles that he contributed on this subject were largely derived from material in these volumes.

Romanes' skill as an expositor of the theory of evolution for the educated layman is best represented by "The Scientific Evidence of Organic Evolution," in *Fortnightly Review*, **36** (1881), 739–758; and *Darwin, and After Darwin*, I, *The Darwinian Theory* (London, 1892). Romanes' principal contributions to the Spencer-Weismann controversy were a review of Spencer's *The Factors of Organic Evolution*, in *Nature*, **35** (1886–1887), 362–364; "The Factors of Organic Evolution," ibid., **36** (1887), 401–407; "Weismann's Theory of Heredity," in *Contemporary Review*, **57** (1890), 686–699; "Are the Effects of Use and Disuse Inherited?" in *Nature*, **43** (1890–1891), 217–220; "Mr. Herbert Spencer on 'Natural Selection,'" in *Contemporary Review*, **63** (1893), 499–517; "The Spencer-Weismann Controversy," ibid., **64** (1893), 50–53; "A Note on Panmixia," ibid., 611–612; *An Examination of Weismannism* (London, 1893); and *Darwin, and After Darwin*, II, *Post-Darwinian Questions, Heredity and Utility* (London, 1895). Chapter 1 of this book consists of an important introductory essay, "The Darwinism of Darwin and of the Post-Darwinian Schools," in which Romanes surveys the issues from his own point of view. Romanes' original paper on physiological selection appeared in *Journal of the Linnean Society*, Zoology, **19** (1886), 337–411. His further contributions to the debate on physiological selection and isolation as factors in evolution include "Physiological Selection," in *Nineteenth Century*, **21** (1887), 59–80; "Before and After Darwin," in *Nature*, **41** (1889–1890), 524–525; "Mr. A. R. Wallace on Physiological Selection," in *Monist*, **1** (1890–1891), 1–20; and *Darwin, and After Darwin*, III, *Post-Darwinian Questions, Isolation and Physiological Selection* (London, 1897).

Although there is no single repository of Romanes' MS material, most of it is concentrated in London: archives of the Linnean Society and the Zoological Society; University College (the Burdon-Sanderson papers and the Galton papers); the British Museum (the Wallace papers, the Gladstone papers, and the Croll papers); the Royal Botanic Gardens, Kew (the Hooker papers and the Thiselton-Dyer papers); The Wellcome Institute of the History of Medicine (the Sharpey-Schafer papers); Imperial College of Science and Technology (the T. H. Huxley papers and the Armstrong papers); the Royal Society (paper relating to the early physiological work); and the Passmore Edwards Museum (the Meldola papers). There are several important items in the C. Lloyd Morgan papers at the University of Bristol (relating to Morgan's editing of *Darwin, and After Darwin*, III). The American Philosophical Society possesses a large number of letters from Charles Darwin to Romanes (some unpublished) as well as 12 letters from Romanes to James Paget among the latter's papers and approximately 100 letters from Romanes to John Thomas Gulick.

II. SECONDARY LITERATURE. The standard biographical source for Romanes is Ethel Romanes, *The Life and Letters of George John Romanes* (London, 1896). Contemporary obituary notices were published by John Burdon-Sanderson, in *Proceedings of the Royal Society*, **57** (1894–1895), vii–xiv; E. Ray Lankester, in *Nature*, **50** (1894), 108–109; C. Lloyd Morgan, in *Dictionary of National Biography*, XLIX, 177–180; and E. B. Poulton, letter to the *Times* (London), 19 June 1894.

Romanes' work in invertebrate physiology and its evolutionary context have been examined in two articles by Richard D. French: "Darwin and the Physiologists, or the Medusa and Modern Cardiology," in *Journal of the History of Biology*, **3** (1970), 253–274; and "Some Concepts of Nerve Structure and Function in Britain, 1875–1885: Background to Sir Charles Sherrington and the Synapse Concept," in *Medical History*, **14** (1970), 154–165; and in Gerald L. Geison, "Michael Foster and the Rise of the Cambridge School of Physiology, 1870–1900" (Ph. D. diss., Yale, 1970), esp. 459–475.

Literature on Romanes' work in comparative psychology is of uneven quality and must be used with discretion. Useful standard histories of psychology are Edwin G. Boring, *A History of Experimental Psychology*, 2nd ed. (New York, 1950), 240–244, 468–476, 620–631; J. C. Flügel, *A Hundred Years of Psychology*, 2nd ed. (London, 1951), 111–125; L. S. Hearnshaw, *A Short History of British Psychology 1840–1940* (London, 1964), 34–46, 86–100; and R. S. Peters, ed., *Brett's History of Psychology*, rev. ed. (London, 1962), esp. 694–699, 737–738. Two relevant articles by Philip H. Gray are "The Morgan-Romanes Controversy: A Contradiction in the History of Comparative Psychology," in *Proceedings of the Montana Academy of Sciences*, **23** (1963), 225–230; and "Prerequisite to an Analysis of Behaviorism: The Conscious Automaton Theory From Spalding to William James," in *Journal of the History of the Behavioral Sciences*, **4** (1968), 365–376.

Romanes' theory of physiological selection and the reactions that it provoked are discussed in John E. Lesch, "The Role of Isolation in Evolution: George J. Romanes

and John T. Gulick," *Isis* (in press). An account of the life and scientific work of John T. Gulick, including much of the Romanes-Gulick correspondence, may be found in Addison Gulick, *Evolutionist and Missionary: John Thomas Gulick* (Chicago, 1932). Some of the problems involved in the recognition of the role of isolation in evolution are treated from the point of view of contemporary biology in Ernst Mayr, "Isolation as an Evolutionary Factor," in *Proceedings of the American Philosophical Society*, *103* (1959), 221–230.

Romanes' lifelong preoccupation with the relationship between religion and science, a side of his thought not considered in this article, is discussed in Frank Miller Turner, *Between Science and Religion: The Reaction to Scientific Naturalism in Late Victorian England* (New Haven–London, 1974), 134–163.

JOHN E. LESCH

ROMÉ DE L'ISLE (or **DELISLE**), **JEAN-BAPTISTE LOUIS** (*b.* Gray, France, 29 August 1736; *d.* Paris, France, 7 March 1790), *crystallography, mineralogy.*

Romé, the son of a lieutenant in the cavalry, studied humanities at the College Ste. Barbe in Paris. In 1756 he entered the Royal Corps of Artillery and Engineering, which he accompanied, as a secretary, to the French Indies in the following year. From 1758 until 1761 he was in the enclave of Pondicherry, French India. When it fell to the English in 1761 Romé was taken prisoner and transported to China, where he stayed until 1764, when he returned to France.

In Paris, Romé met the chemist and mineralogist B. G. Sage and attended his course in chemistry. In 1766 he published his first work, dealing with the then fashionable topic of freshwater polyps. Without making any observations, Romé set forth a hypothesis on the polyp. The work was a false start; Romé afterward confined himself to the study of mineralogy and chemistry.

In 1767 Romé was employed, on Sage's recommendation, to draw up a catalogue of the curiosities that had been collected by Pedro Francisco Davila, who wished to sell his cabinet of natural history before returning to Peru. The work ran to three volumes, in the second of which Romé, in agreement with Linnaeus, stressed the importance of crystalline form in mineralogical description. While engaged in this project, Romé met Michelet d'Ennery, an avid collector of coins and medals. Until Michelet d'Ennery's death in 1786, Romé lived in his house and dedicated himself to the study of minerals. He earned money by cataloguing at least fourteen mineral collections, according to his own, probably incomplete, list in the bibliography to his *Cristallographie* of 1783. Three of the catalogues remained in manuscript.

In 1772 Romé published the *Essai de cristallographie*, in which he identified 110 crystal forms (drawing upon Linnaeus, who had listed forty) and described in minute detail the minerals that exhibited them. He subdivided the various substances into salts, stones, pyrites, and metallic minerals, stating that he agreed with Linnaeus that geometrical form is the chief characteristic by which minerals may be classified. Also like Linnaeus, he held that saline principles imprinted their own geometrical form upon the earthy constituent of each mineral. In his description of the "primitive" form of each substance, and of the more complex forms derived from them, Romé did not depend on the exact measurement of crystalline angles; he gave values only for plane angles, and those were not consistent. Indeed, he has the plane angles of quartz varying between 70° and 75°, and it seems that in this early work he was not particularly concerned with the idea that such measurements exhibit strict constancy. In 1773 he brought out a description of the metallic ores of his own mineral cabinet, in which he discussed the origin, metamorphosis, and paragenesis of each.

In 1779 Romé became involved in a controversy concerning the theory of a central terrestrial fire and the eventual cooling of the earth. His opinion that all terrestrial heat derived from the sun brought him into opposition with Buffon (and with Bailly, who in a "Lettre à M. Voltaire" asked that author to support Buffon's view, as he had been the apologist of Newton's theory). In rejecting the idea of the central fire, Romé refrained from criticizing the geological conclusions that Buffon drew from it, although it is nonetheless clear from other of his writings (and especially from his warm praise of Werner and Saussure) that he favored the neptunist view.

Romé also opposed Buffon in the matter of methodology. In this controversy he took the side of the *nomenclateurs* (as they were disparagingly called), who followed Linnaeus in considering classification to be one of the chief ends of the natural sciences. He defended this aim against the *systémateurs*, who, like Buffon, were more concerned with building general systems on the basis of hypotheses that were not necessarily confirmed by empirical research. In addition, Romé disagreed with Buffon's opinion that crystal forms can be explained by "organic molecules"; he believed that Buffon undervalued the role of crystallography in refusing to recognize the geometrical form of crystals as a specific characteristic of minerals.

Romé's major work, the *Cristallographie* (1783), was first advertised as a second edition of his *Essai*, but instead it was expanded and comprised three volumes and an atlas describing more than 450 crystal forms. In this book, rather than using any physical basis, Romé followed both Linnaeus and Domenico Guglielmini in classifying crystals by arbitrary primitive forms—the regular tetrahedron, the cube, rectangular octahedrons, parallelepipeds, rhomboidal octahedrons (that is, rhombic dipyramids), and dodecahedrons with triangular planes (hexagonal dipyramids). Each crystal described was measured precisely. In the course of making terracotta models, Romé's assistant, Arnould Carangeot, had discovered the constancy of interfacial angles; and, using a contact goniometer invented for the purpose, he had made measurements of the interfacial angles (exact to about half a degree) of each mineral that Romé listed. Both these aspects—the tabulation of primitive forms and the measurement of interfacial angles—were of central importance to Romé's crystallography.

The fundamental law of the constancy of interfacial angles—that the faces of a crystal may vary in their relative dimensions, but the respective inclination of these same faces is constant and invariable in each species—implied that each species must have a characteristic primitive form with characteristic constant angles (a thesis that neither Buffon nor Bergman was willing to accept). Romé believed the chief task of crystallography to be the derivation of secondary crystal forms from primary ones, by means of truncation of the solid angles or edges. The term "truncation" had already been used by Cappeller, Guglielmini, Bergman, and, especially, Romé's disciple Jean Démeste, who had pointed out that it should be considered a purely geometrical device. Romé considered the way in which secondary forms arise in nature "a most impenetrable mystery."

Although Romé considered truncation to be a defect of the crystal, he maintained that the constancy of interfacial angles is equally valid for secondary and for primitive crystal forms. In his classification of mineral species, he recognized that although the most regular geometrical forms (cube, regular octahedron, regular tetrahedron) may be characteristic of more than one species, the properties of density and hardness would still make identification possible. On the basis of analogy, Romé believed that the form of the integrant molecules of a crystal must be identical with the primitive form and must therefore be constant and characteristic for each species. Since these integrant molecules had to be of the same magnitude and to consist of the same elementary particles (or constituent molecules), Romé concluded that they must exist in fixed proportions; the characteristic molecular form of a substance necessarily, therefore, depends on the (unknown) forms of all its constituent molecules, and on their arrangement and proportion within the integrant molecule.

While at work on his *Cristallographie*, Romé had become convinced, on the basis of chemical evidence, that Linnaeus' doctrine—according to which a small quantity of a salt was necessary to evoke the primitive crystal form out of the passive, earthy principle of a stony or metallic mineral—could not possibly be correct, since all solid mineral substances must have a specific and particular crystal form. He also recognized that, despite advances in chemistry, constant geometrical crystal form was a better criterion for classification than constant chemical composition. Romé's belief that each mineral species possesses both a characteristic primitive crystal form and a characteristic chemical composition obviated the concept of similarity of form with different substances (isomorphism). In rejecting isomorphism—as, for example, of zinc spar (smithsonite) and iron spar (siderite) with calcspar (calcite)—Romé was led to explain the formation of those minerals by analogy with the formation of fossils, wherein organic material is replaced by pyrite. He consequently concluded that these substances show a form that is "alien and accidental," since it does not correspond to that of their own constituent molecules.

Because Romé had only vague notions of symmetry relations, he sometimes separated forms that belong together and brought together forms that are essentially different. He considered the cube and the regular octahedron to be different primitive forms, even when they occurred in the same substance, a discrepancy that he explained by assuming a slightly different proportion of acid and base, for example, in alum crystals exhibiting these two forms. In the case of galena he was forced to attribute the change of its primitive form from a cube to an octahedron to "inversion"—an opposite arrangement—of the same molecules. On the other hand, the forms that Romé derived from the rhomboidal parallelepiped proved to have different degrees of symmetry, as may be seen in his evoking, by elongation, a monoclinic parallelepiped of iron sulfate from a trigonal calcspar rhombohedron. Somewhat vaguely, he attributed the truncation of the primitive form to a difference in molecular form, which itself might result from either a difference in the proportions of its chemical constituents or from a disturbance of the normal arrangement of the molecules. In any case, he held to his belief that the act of truncation does not occur in nature, although his opponents charged that he did in fact believe that it did (an injustice that Romé returned in falsely imputing to Haüy the belief that the cleavage form is a true kernel within the crystal).

Romé's relationship with Haüy was at best strained. He was sharply critical of the speculative nature of Haüy's theory, which was then in its early stages. Haüy, in turn, was staunchly defended by Buffon's collaborator Daubenton, and chose to ignore Romé's work insofar as it was possible for him to do so. Romé professed unwavering empiricism, an unwillingness to substitute "the dreams of our imagination for the majestic silence of Nature." Like Lavoisier he refused to speculate about the nature of molecules and atoms, and he further criticized not only Haüy but also Bergman for trying to demonstrate mathematically the internal structure of crystals before complete observational data had been gathered. The morphology of crystals was, he believed, so far from being completely understood that the study of the anatomy of crystals (or "cristallotomie," as he called it) should properly be deferred.

All the same, Romé himself could not resist the temptation occasionally to pronounce a hypothesis concerning structure in some particular case. Some of these hypotheses were, like his ideas about symmetry, vague and contradictory, particularly because he believed that the molecular arrangement of secondary crystal forms can differ from that of primitive ones. For example, he held that molecules of calcspar were always in the form of the cleavage rhombohedron but were differently arranged in crystals of different form.

In 1784 Romé brought out a book on the external characteristics of minerals, a supplement to the *Cristallographie*. In this work, *Des caractères extérieurs des minéraux, ou réponse à cette question*: *Existe-t-il dans les substances du règne minéral des caractères qu'on puisse regarder comme spécifiques*; *et au cas qu'il en existe, quels sont ces caractères?*, he stated his firm belief that form, density, and hardness were sufficient criteria to permit the identification of any mineral species.

The following year Romé was granted a pension from the public treasury and in 1789 Louis XVI added a further stipend from the royal treasury. The latter was especially welcome, since Romé had been in straitened financial circumstances since the death of Michelet d'Ennery in 1786. Romé was executor of D'Ennery's estate and thus helped to catalogue the extensive collection of coins and medals. He made a comparison between the weight of the Roman pound and the weight of the Paris pound which provided further material for his last great work, the *Métrologie* (1789), in which he compared a number of the weights and measures of antiquity with modern counterparts. He advised the States-General to introduce the Roman system of weights and measures, or at least to unify the system and standards of France.

Romé's chief scientific goal was the establishment of mineralogy on a firm basis of crystallography. His major contribution toward this end was the formulation of the law of the constancy of interfacial angles, which became the cornerstone of crystallography. Although earlier investigators—including Hooke, Erasmus Bartholin, Steno, Huygens, Philippe de la Hire, and Guglielmini—had made incidental statements about such a constancy in one or two substances, Carangeot and Romé were the first to enunciate it as a general law of nature. Romé was nevertheless a poor theoretician. His solution of the geometrical relation of the crystal forms within the same species remained much inferior to Haüy's; although Haüy's crystallochemical conception—geometrical form and chemical composition as the definition of a species, and pseudomorphoses as the explanation of isomorphous crystals—was much the same as Romé's.

In his general chemical views, Romé largely followed Sage, remaining faithful to the phlogiston theory. He also adhered to J. F. Meyer's theory of "acidum pingue" and energetically rejected Lavoisier's "absurd" theory of combustion.

It was Sage who made Romé's work known in France, through his lectures at the Mint (beginning in 1778) and at the École des Mines (beginning in 1783). In addition, Romé had a number of correspondents throughout Europe who provided him with mineral specimens. Among his pupils, his favorite was Jacques-Louis de Bournon, who wrote a number of geological and crystallographical studies, of which the most important was the *Traité complet de la chaux carbonatée*, published in London in 1808. A favorite correspondent was Jean Démeste, a physician of Liège, whose 1779 *Lettres au Dr. Bernard* were wholly based on Romé's crystallography and Sage's chemistry. Romé also encouraged Fabien Gautier d'Agoty to publish a splendid set of colored engravings of crystals and mineral groups in the first volume of his *Histoire naturelle* (1781). Romé wrote the explanatory captions.

Romé's friendship with Sage also brought him membership in a number of learned societies, including the academies of Mainz, Stockholm, Berlin, and St. Petersburg. But it did him no service with the Paris Académie des Sciences, which rejected him on the ostensible grounds that he was a mere "catalogue maker." It is likely that Romé's controversies with Buffon also played a part in his rebuff by the Academy.

BIBLIOGRAPHY

I. ORIGINAL WORKS. Romé's first publication was *Lettre de M. Deromé Delisle à M. Bertrand sur les polypes d'eau douce* (Paris, 1766). The foreword by P. F. Davila to the *Catalogue systématique et raisonné des curiosités de la nature et de l'art qui composent le cabinet de M. Davila*, 3 vols. (Paris, 1767), acknowledges Romé's share in the composition of the part on natural history and the description of some of the artificial curiosities. Romé himself claimed only to have written the section on natural history. He lists the other catalogues he wrote in the bibliography of his *Cristallographie*, which contains 14 catalogues, including those of the collections of Jacob Forster and Claude-Marc-Antoine Varennes de Béost.

The work on the central fire of the earth was first published under the initials M. D. R. D.: *L'action du feu central bannie de la surface du globe, et le soleil rétabli dans ses droits; contre les assertions de MM. le Comte de Buffon, Bailly, de Mairan . . .* (Stockholm–Paris, 1779); the 2nd ed., which includes answers to objections made to the first, is entitled *L'action du feu central démontrée nulle à la surface du globe, contre les assertions de MM. le Comte de Buffon, Bailly, de Mairan . . .* (Stockholm–Paris, 1781). His name was now given in full.

Romé's crystallographical writings, however, are by far the most important part of his work: *Essai de cristallographie, ou description des figures géométriques, propres à différens corps du règne minéral, connus vulgairement sous le nom de cristaux* (Paris, 1772), German trans. by C. E. Weigel (Greifswald, 1777); the much enlarged 2nd ed., *Cristallographie, ou description des formes propres à tous les corps du règne minéral, dans l'état de combinaison saline, pierreuse ou métallique*, 4 vols. (Paris, 1783); and the supp. to the latter, *Des caractères extérieurs des minéraux* (Paris, 1784).

Additional writings on mineralogy are *Description méthodique d'une collection de minéraux, du cabinet de M.D.R.D.L. Ouvrage où l'on donne de nouvelles idées sur la formation et la décomposition des mines . . .* (Paris, 1773); "Mémoire ou observations sur les altérations qui surviennent naturelles à différentes mines métalliques et particulièrement aux pyrites martiales," in *Observations sur la physique*, **16** (1780), 245–256; and *Observations sur les rapports qui paroissent exister entre la mine dite cristaux d'étain et les cristaux de fer octaèdres* (Erfurt, 1786).

Romé identified the alabaster of the ancients in "De antiquorum alabastrite et variis quibusdam lapidibus quos recentiores alabastri nomine appellaverunt disquisitiones historico-physico-criticae," in *Nova acta physico-medica exhibentia ephemerides*, **6** (1778), 186–199. His antiquarian interests also inspired his last work: *Métrologie, ou tables pour servir à l'intelligence des poids et mesures des anciens, et principalement à déterminer la valeur des monnoies grecques et romaines, d'après leur rapport avec les poids, les mesures et le numéraire actuel de la France* (Paris, 1789), German trans. by Gottfried Grosse (Brunswick, 1792). The dedication of this work, "To my Fatherland, which is undergoing a rebirth under Louis XVI," pays homage to the minister Necker and to the National Assembly, and thus unambiguously demonstrates his political feelings.

Romé also wrote the explanatory notes to Fabien Gautier d'Agoty's *Histoire naturelle ou exposition générale de toutes ses parties, gravées et imprimées en couleurs naturelles, avec des notes historiques,* pt. 1 (Paris, 1781); and he revised and enlarged the MS of the "delilio-sagiano-linnean letters" sent by his pupil Jean Démeste from Liège to Paris to be printed under the supervision of the Marquis d'Aoust, *Lettres du docteur Démeste au docteur Bernard, sur la chymie, la docimasie, la cristallographie, la lithologie, la minéralogie et la physique en général,* 2 vols. (Paris, 1779).

II. SECONDARY LITERATURE. Biographical articles on Romé are J. C. Delamétherie, "Notice sur la vie et les ouvrages de M. de Romé de l'Isle," in *Observations sur la physique, sur l'histoire naturelle et sur les arts,* **36** (1780), 315–323; and C. S. Weiss, in *Biographie universelle, ancienne et moderne* (Michaud), XXXVIII (1824), 521–523. Further details are in A. Birembaut, "Les préoccupations des minéralogistes français au 18e siècle," in *Actes de la 72e session de l'Association française pour l'avancement des sciences* (1953), 534–538, in which Carangeot's claims to the discovery of the law of constant angles also are maintained. On his relations with Démeste, see M. Florkin, "Vie de Jean Démeste, médecin et minéralogiste," in *Revue médicale de Liège,* **10** (1955), 543–555.

Romé's crystallographical work is dealt with in C. M. Marx, *Geschichte der Crystallkunde* (Karlsruhe–Baden, 1825), 120–131; and H. Metzger, *La genèse de la science des cristaux* (Paris, 1918), 65–75, 189–192. A more detailed analysis of his crystallography and his crystal chemistry, also in comparison with modern conceptions, has been given by R. Hooykaas in the following publications: "De kristallografie van J. B. de Romé de l'Isle," in *Chemisch weekblad,* **47** (1951), 848–855; "The Species Concept in 18th Century Mineralogy," in *Archives internationales d'histoire des sciences,* no. 31 (1952), 45–55; and *La naissance de la cristallographie en France au XVIIIe siècle* (Paris, 1953), 8–12, 23–24. His sparse and wavering ideas on crystal structure are analyzed in R. Hooykaas, "Romé de l'Isle en de structuur theorie," in *Chemisch weekblad,* **47** (1951), 909–914. On the priority of the discovery of the law of constancy of angles, see Birembaut, *op. cit.*; R. Hooykaas, "De oudste kristallografie," in *Chemisch weekblad,* **46** (1950), 438–440, also in *Revue d'histoire des sciences,* **12** (1959), 182–185; and J. G. Burke, *Origins of the Science of Crystals* (Berkeley, 1966), 69–71.

R. HOOYKAAS

ROMER, EUGENIUSZ MIKOŁAJ (*b.* Lvov, Poland [now U.S.S.R.], 3 February 1871; *d.* Cracow, Poland, 28 January 1954), *geography, cartography.*

Romer's father, Edmund Romer, was an official in the Galician administration; his mother was Irena Körtvelyessy de Asguth. Both were members of the impoverished nobility. Romer attended Jagiellonian University in Cracow from 1889 until 1891, specializing in history and geography. In 1891 he went

to Halle, where he studied with the geographer A. Kirchhoff, then to Lvov, where he spent two years working with the geobotanist A. Rehman. His doctoral thesis was concerned with climatology, specifically the distribution of heat over the face of the earth. In 1895 he studied geomorphology with A. Penck in Vienna, then the following year went to Berlin to continue his work on that subject with F. von Richthofen and to study meteorology with J. F. von Bezold, R. Assman, and A. Berson.

Returning to Poland, Romer taught geography in Lvov. In 1908 he published his first text, *Szkolny atlas geograficzny* ("School Geographical Atlas"), a hypsometric synthesis of the face of the entire earth, which went through a number of editions. He became professor of geography at the University of Lvov in 1911 and in 1916 brought out the encyclopedic *Geograficzno-statystyczny atlas Polski* ("Geographical-Statistical Atlas of Poland"). In 1919 Romer served as an expert on territorial affairs at the Paris Peace Conference, and at about the same time he organized the Lvov Institute of Geography, where he taught a large number of students. In 1921 he founded the Lvov Institute of Cartography and in 1924 a cartographical publishing house which, together with the periodical *Polski przegląd kartograficzny* ("Polish Cartographic Revue"; founded by him in 1923); expedited the publication of the large number of atlases, maps, and globes that he brought out over the years. He made a major contribution to cartography, improving maps for both scientific and didactic purposes, and strongly advocated the superiority of contour maps to hachured ones.

Romer also did significant work in geomorphology. He viewed geomorphological processes as climatic phenomena, while using both the concepts of W. M. Davis and his deductive model. He rejected any interpretation of geomorphological processes that relied on tectonics; in his chief work on the subject, *Tatrzańska epoka lodowa* ("The Tatra Glacial Epoch"; 1929), he demonstrated the specific nature of the glaciation that had occurred in Poland and posited cycles consisting of four glacial and three interglacial periods.

Climatology itself was one of Romer's earliest interests, as is evident in his choice of subject for his doctoral thesis. His later work included studies of specific climatological problems and syntheses of the climate of Poland in which he incorporated the geographical principles set out in his other writings. His most notable publications of this sort were *Pogląd na klimat Polski* ("A View of the Climate of Poland"; 1938) and *O klimacie Polski* ("On the Climate of Poland"; 1939), which he based upon an analysis of isothermal and isohyetal maps, devoting particular attention to the trend of closely spaced lines and the mutual intersections of isotherms, as well as the migration of isarithms and the duration of the seasons.

World War II interrupted Romer's work on his *Wielki powszechny atlas geograficzny* ("Great World Geographical Atlas"), which he had begun in 1936. During the conflict and the following occupation he spent most of his time in retreat in the monastery of the Resurrectionists in Lvov. In 1945 he became professor of geography at Jagiellonian University in Cracow.

Romer lived modestly with his wife, Jadwiga Rossknecht, and their two sons, Witold and Edmund. He was active in community affairs and research, and from 1908 on had participated in international scientific meetings and congresses. He was a member of a number of scientific societies, both Polish and foreign. In 1913 the government of the United States named a glacier at Glacier Bay, Alaska, in his honor, in recognition of the studies he had made of Alaskan fjords in that year.

BIBLIOGRAPHY

I. ORIGINAL WORKS. A number of Romer's writings were collected in *Wybór prac* ("Selected Writings"), 4 vols. (Warsaw, 1960–1964).

II. SECONDARY LITERATURE. Vol. I of *Wybór prac* includes a comprehensive study by Julian Czyżewski, "Życie i działalność Eugeniusza Romera" ("The Life and Work of Eugeniusz Romer"), and a bibliography of more than 500 titles. See also Łucja Mazurkiewicz-Herzowa, *Eugeniusz Romer* (Warsaw, 1966), which also contains an extensive bibliography.

JOZEF BABICZ

RÖMER. See also **Roemer.**

RÖMER, OLE CHRISTENSEN (or **ROEMER, OLAUS**) (*b.* Aarhus, Denmark, 25 September 1644; *d.* Copenhagen, Denmark, 19 September 1710), *astronomy.*

Römer came from a family of small merchants. In 1662 he was sent to the University of Copenhagen, where he studied with both Thomas Bartholin, professor of medicine, and his brother Erasmus Bartholin, a physician who was better known for his discovery of double refraction of light in Iceland spar.

He lived in Erasmus Bartholin's house and studied astronomy and mathematics under his direction; Bartholin was so impressed with Römer's abilities that he entrusted him with the editing of the unpublished manuscripts of Tycho Brahe.

In 1671 Erasmus Bartholin was visited by Jean Picard, who had been sent by the Académie des Sciences to measure precisely the position of Tycho Brahe's observatory, the Uraniborg, on the island of Hven. In September of that year Bartholin and Römer accompanied Picard to Hven, where, in order to redetermine the longitude of the observatory, they made observations of a series of eclipses of the first satellite of Jupiter, while G. D. Cassini carried out the same work in Paris. When Picard returned to Paris he took with him a notebook containing eight months' observations, the original manuscripts of Tycho Brahe's observational works, and Römer, whom he had persuaded to work there under the auspices of the Academy.

Upon his arrival in Paris, Römer was assigned lodgings in the new Royal Observatory building designed by Charles Perrault; he lived there for nine years. He was appointed by Louis XIV to be tutor to the dauphin and also made a number of astronomical observations, both in Paris and in other parts of France to which he was sent on behalf of the Academy. He constructed clocks and other devices, in which he displayed great mechanical skill and ingenuity, and invented a micrometer for differential measurement of position that was so superior to previous instruments that it was speedily adopted into general use.

Römer's greatest work, however, grew out of the problem that he had initially considered with Picard, the times of the occultations of the satellites of Jupiter. These measurements were of considerable practical use, since it was recognized soon after the discovery of the Jovian satellites that their frequent occultations—particularly those of the first satellite, Io—by the planet itself represent well-defined moments of celestial time, which may be compared with time at the place of observation to establish geographical longitude. This knowledge was of particular use to mariners, and astronomers began to concern themselves with drawing up ephemerides predicting the times of eclipses at a fixed meridian, for example at Paris or Greenwich. Galileo had attempted to construct such an ephemeris, without notable success, and the task was assigned to the astronomers of the new Paris observatory by Colbert. G. D. Cassini and his nephew Maraldi discovered the first large inequality in the periodic times of the minima, that caused by the eccentricity of the orbit of Jupiter around the sun; their second discovery, announced by

Cassini in August 1675, was more interesting, since the inequality seemed to depend on the position of the earth relative to Jupiter.

Cassini considered, but discarded, the idea that the fluctuation of periodic times might be caused by the finite speed of light; it remained to Römer to demonstrate that such was indeed the case. With rare exceptions, previous astronomers, both ancient and more recent—including Aristotle, Kepler, and Descartes—had held that light propagated itself instantaneously. Galileo, on the other hand, was not only convinced of its finite velocity, but also designed an experiment (although not an adequate one) by which the speed of light might be measured. These divergent views were discussed among the Paris academicians, and were well known to Römer.

In his observational work Römer noticed that the eclipses of Io occurred at longer intervals as the earth receded from Jupiter, but happened in closer sequence as the earth and that planet came closer together. Beginning from the point at which the earth and Jupiter were closest to each other, Römer tried to predict the time of occurrence of an eclipse of Io at a later date, when the earth and Jupiter had drawn further apart. In September 1676 he announced to the members of the Academy that the eclipse predicted for 9 November of that year would be ten minutes later than the calculations made from previous eclipses would indicate. Observations confirmed his hypothesis, and Römer correctly interpreted this phenomenon as being the result of the finite velocity of light. He was thereupon able to report to the Academy that the speed of light was such as to take twenty-two minutes for light to cross the full diameter of the annual orbit of the earth; in other terms, that the light from the sun would reach earth in eleven minutes (a time interval now measured to be about eight minutes and twenty seconds). The speed of light was thus established scientifically for the first time, with a value of about 140,000 miles per second—a reasonable first approximation to the currently accepted value of 186,282 miles per second.

Römer's results were not immediately accepted by everyone. The Cassinis remained unconvinced for some time, and Descartes's view concerning the instantaneous propagation of light retained some currency. It was only after James Bradley, in 1729, discovered the periodic annual displacements in the positions of all stars in respect to the ecliptic—their aberration—that Römer's interpretation prevailed. The value for the time of the passage of light from the sun to the earth, deduced from Bradley's aberration constant, was eight minutes and twelve seconds, a more accurate approximation than that originally obtained by Römer, whose result had incorporated additional perturbations of motion that he had not recognized.

In 1679 Römer undertook a scientific mission to England, where he met Newton, Flamsteed, and Halley. In 1681 he returned to Denmark to become professor of mathematics at the University of Copenhagen. He left France at a propitious time, since four years after his departure Louis XIV revoked the Edict of Nantes, and as a Protestant Römer would surely have been forced to leave the country, as was Christiaan Huygens. Christian V of Denmark appointed Römer his astronomer royal and director of the observatory and gave him a number of technical and advisory duties; at one time or another Römer served as master of the mint, harbor surveyor, inspector of naval architecture, ballistics expert, and head of a commission to inspect highways. He performed all these tasks with distinction, and in 1688 was made a member of the privy council. In 1693 he became first judiciary magistrate of Copenhagen and in 1694 he was made chief tax assessor, in which office he devised an efficient and equitable system of taxation.

Despite the press of other official duties, Römer did not neglect his work as astronomer royal and director of the Copenhagen observatory. He performed a large number of astronomical observations (as many, in fact, as had Tycho Brahe) and designed and constructed astronomical instruments, particularly transit circles. The Copenhagen observatory was one of the oldest in Europe; at the time that Römer became its director, it was housed in the "Round Tower" built by Christian IV for Longomontanus in 1637. Römer began his observations there but soon found the site unsatisfactory. He therefore converted his own house into an auxiliary observatory, and in 1704 created another observatory, the Tusculaneum, located between Copenhagen and Roskilde and equipped with excellent and innovative instruments. (Indeed, Römer would seem to have been the first astronomer to have attached a telescope to a transit circle.)

Concomitant to his work in astronomical instrumentation, Römer invented a new thermometer. He was the first to recognize that the scales of thermometers designed to give concordant results must be based upon two fixed points. For this purpose he chose the boiling point of water and the melting point of snow—the same points later used by Celsius. Fahrenheit met Römer in 1708 and (according to his letter to Boerhaave in 1729) adopted so many of Römer's ideas and techniques that the Fahrenheit thermometer should really have been named the Römer thermometer.

In 1705 Römer became mayor of Copenhagen; he was soon thereafter named prefect of police as well. The same year Frederick IV, who had succeeded Christian V, made Römer a senator and in 1707 he named him head of the state council of the realm. In each of these capacities Römer served faithfully and well. He died at the age of sixty-five, survived by his second wife, whom he had married in 1698. His first wife, his mentor Erasmus Bartholin's daughter Anna Maria, whom he married in 1681, died in 1694. He had no children.

BIBLIOGRAPHY

I. ORIGINAL WORKS. Most of Römer's astronomical MSS were lost in the great fire that destroyed Copenhagen in 1728. The only surviving observations are the "Triduum," published in J. G. Galle, *O. Roemeri triduum observationum astronomicarum a. 1706 . . . institutarum* (Berlin, 1845), which includes Römer's notes on observations made 21–23 Oct. 1706; and a subsequently discovered work, the "Adversaria," published as *Ole Römers Adversaria*, T. Eibe and K. Meyer, eds. (Copenhagen, 1910). Memoirs on astronomical observations, written with J. D. Cassini and Jean Picard, appeared in *Mémoires de l'Académie contenant les ouvrages adoptés . . . avant . . . 1699*, I (The Hague, 1731); see also "A More Particular Account of the Last Eclipse of the Moon," in *Philosophical Transactions of the Royal Society*, **10** (1675), 257–260; "Observationes lunares," *ibid.*, 388–389; and "Démonstration touchant le mouvement de la lumière," in *Journal des savants* (7 Dec. 1676), 233–236, translated into English in *Philosophical Transactions of the Royal Society*, **12** (1677), 893–894. Römer's astronomical instruments were described in *Machines et inventions approuvées par l'Académie des sciences*, I (Paris, 1735), 81–89; and in *Miscellanea Berolinensia*, III (Berlin, 1727), 276–278.

II. SECONDARY LITERATURE. The chief source of information concerning Römer's methods and ideas was written by his disciple, P. Horrebow, *Basis astronomiae sive astronomiae pars mechanica* (Copenhagen, 1735). I. B. Cohen, "Roemer and the First Determination of the Velocity of Light," in *Isis*, **31** (1940), 327–379, is a valuable study and includes a bibliography of works on Römer. See also G. van Biesbroeck and A. Tiberghien, "Études sur les notes astronomiques contenues dans les Adversaria d'Ole Römer," in *Oversigt over det K. Danske Videnskabernes Selskabs Forhandlinger* (1913), 213–324; M. C. Harding, *Ole Römer som ingeniør* (Copenhagen, 1918); M. Pihl, *Ole Römers videnskabelige liv* (Copenhagen, 1944); E. Strömgren, *Ole Römer som astronom* (Copenhagen, 1944); and the notice by K. Meyer in *Dansk biografisk leksikon*, XX (Copenhagen, 1951), 329–400, with bibliography of secondary literature. See also William Derham, *The Artificial Clock-Maker; to Which Is Added a Supplement Containing . . . Monsieur Römer's Satellite-Instrument*, 2nd ed., enl. (London, 1700).

ZDENĚK KOPAL

RONDELET, GUILLAUME (*b.* Montpellier, France, 27 September 1507; *d.* Réalmont, Tarn, France, 30 July 1566), *ichthyology, medicine, anatomy.*

Although he was active in several branches of biology, Rondelet's reputation effectively depends on his massive compendium on aquatic life, which covered far more species than any earlier work in that field. Despite its theoretical limitation, it laid the foundations for later ichthyological research and was the standard reference work for over a century.

Rondelet was the son of a drug and spice merchant who died while Guillaume was a child, leaving him to be brought up by an elder brother. In 1525 Rondelet went to study humanities at Paris but in June 1529 transferred to the Medical Faculty at Montpellier, where he was procurator in 1530 and became friendly with Rabelais, whose character Rondibilis may be based on him. After gaining practical experience as physician and schoolteacher in Pertuis (Vaucluse), he returned to Paris in the mid-1530's to study anatomy under Johannes Guinter. Rondelet then practiced for a time at Maringues, Puy de Dôme, after which he returned to Montpellier, where he graduated M.D. in 1537 and married Jeanne Sandre in 1538. He was effectively supported by his wife's elder sister until June 1545, when he was appointed regius professor of medicine at Montpellier.

In addition to this post, during the 1540's Rondelet was personal physician to François Cardinal Tournon, whom he accompanied on visits to Antwerp and to Bordeaux, Bayonne, and other towns on the southwest coast of France, where he learned something of the local whaling industry. In 1549 he went by sea to Rome with the cardinal, for the election of Pope Julius III. On the way home he visited Venice and the university towns of northern Italy. In 1551 Rondelet left the cardinal's service and returned to Montpellier, where he was elected chancellor of the university in 1556. It was probably at his initiative that the university set up its first anatomy theater in the same year.

In 1554–1555 Rondelet published *Libri de piscibus marinis in quibus verae piscium effigies expressae sunt*; the second part is entitled *Universae aquatilium historiae pars altera*. A French translation, *L'histoire entière des poissons*, possibly by his pupil Laurent Joubert, appeared in 1558. In his own day Rondelet was almost as well known as an anatomist as a zoologist. A popular lecturer, Rondelet attracted scholars from all over Europe: Coiter and Bauhin; L'Écluse; L'Obel, who inherited his botanical manuscripts; and Daleschamps. Gesner and Aldrovandi also studied briefly under him.

The main title of Rondelet's great work is a misnomer, and that of the second part should be applied to the whole; in fact the book covers the whole of freshwater as well as marine zoology, and it is not restricted to fish. All aquatic animals are included: marine mammals, arthropods and mollusks, riverine amphibians, and even beavers. The first four books are devoted to general considerations: how fish can be distinguished by their ways of life, parts, actions, "manners and complections"; they constitute, in effect, a treatise on comparative anatomy and physiology. No definite system of classification is adopted, however. Rondelet begins with what he calls "instrumental parts": the head, then such internal organs as the gills, heart, liver, and kidneys, organs of locomotion and reproduction, and processes such as Garbels. Next he deals with "mixture of elements," as revealed by taste, smell, color, and "idiosyncrasies," such as that of the torpedo's power of stunning all it touches. In book IV Rondelet's observations on differing systems of digestion, reproduction, and respiration are related to differences in physiological "activities." He expresses a vague concept of homology: "all parts correspond in proportion to those which have the same use, the same situation, notwithstanding they may be diverse in substance and form." Aristotelian ideas on the correlation of parts are exploited, as well as teleological attitudes that led him to try to relate form to function and environment. Thus, he notes that scaly fish, without lungs, have only three chambers in their hearts, in contrast with marine mammals, which have four, and tries to explain the association. But Rondelet goes beyond Aristotle—for instance, when he argues that fish need air for their "animal spirits," taking in water and air (dissolved in water) and expelling the water through their gills—and points to experimental evidence: if fish are kept in a vessel full of water whose lid is closed, they will suffocate. If a small amount of air remains between surface and lid, they appear to struggle to get close to the surface. He attacks those who will not accept the superior authority of experience and tries to base himself on anatomical investigations. He is at his best describing unusual structures or reporting examinations of stomach contents—for instance of a large starfish found on the beach at Maguelonne.

The rest of the work is an encyclopedia of over 300 aquatic animals, almost all of which are illustrated. Each section opens with the subject's names in several languages, including local variants, and then outlines its way of life, feeding habits, and characteristic anatomical features, both external and internal (gastronomic notes are sometimes added). For those fish he could inspect on the coast of Languedoc, Rondelet is thorough and usually accurate; the work long remained the basic guide to the region. For the rest, not surprisingly, it is less valuable. Although the concept of correlation enabled him to dispose of a "marine lion," Rondelet is often credulous about sea monsters. Marine mammals are quite well treated; he had dissected the respiratory organs of some smaller Cetacea and gives apparently the first zoological accounts of the sperm whale and the manatee.

After his first wife died in 1560, Rondelet married Tryphène de La Croix. In later years he seems to have been attracted by ideas of religious reform, and by 1563 he was reckoned a member of the Protestant community. A detailed account of Rondelet's terminal illness in the summer of 1566 reports not only on his symptoms but also on the company and the prayers offered, which have a Huguenot tinge to them.

But Rondelet's character, as portrayed by Joubert, was hardly Calvinistic: he sounds more Rabelaisian—a great lover of good food, especially sweets and cakes, grapes, and cherries. A keen musician, he constructed a fiddle for himself while a student. When he grew too stout for dancing, he loved to watch dances and to give balls. Like so many of his day, he had an enthusiastic interest in agriculture and building, and spent much on alteration of his houses. Above all, he was generous and fond of good company, "very merry and a lover of jokes and fun."

BIBLIOGRAPHY

I. ORIGINAL WORKS. Rondelet's major work is *Libri de piscibus marinis in quibus verae piscium effigies expressae sunt: Universae aquatilium historiae pars altera* (Lyons, 1554–1555). A minor medical work, *De ponderibus*, was published during his lifetime (Lyons, 1560), as were a few short treatises in medical compendiums. The principal ed. of his medical writings is that by L. Joubert, *Methodus curandorum omnium morborum corporis humani . . .* (Paris, 1573).

II. SECONDARY LITERATURE. The fundamental source is L. Joubert, "Vita Gulielmi Rondeletii," written in 1568 and published in his *Opera latina*, II (Lyon, 1582), 186–93. See also P. Delaunay, *La zoologie au seizième siècle* (Paris, 1962); and J. M. Oppenheimer, "Guillaume Rondelet," in *Bulletin of the Institute of the History of Medicine*, **4** (1936), 817–834; and C. Dulieu, "Guillaume Rondelet," in *Clio medica*, 1 (1965), 89–111. *Monspeliensis Hippocrates*, **12** (summer 1961), has a portrait (probably not from life) on the cover, with a brief curriculum vitae and genealogical table, and two articles: H. Harant and D. Jarry, "Oeuvre zoologique de Guillaume Rondelet," pp. 5–10; and A. Pagès, "L'observation clinique d'un médecin malade: Guillaume Rondelet," pp. 13–19.

A. G. KELLER

RÖNTGEN (ROENTGEN), WILHELM CONRAD

(*b*. Lennep im Bergischen [now part of Remscheid], Rhine Province, Germany, 27 March 1845; *d*. Munich, Germany, 10 February 1923), *physics*.

Röntgen was the only son of Friedrich Conrad Röntgen, a cloth manufacturer and merchant of Lennep, who belonged to an old Lutheran Rhineland family. His wife, Charlotte Constanze Frowein, was born in Holland, although her family, too, came originally from Lennep. When their son was three, they moved to Appeldoorn in Holland, and here Röntgen attended a private boarding school, the Institute of Martinus Herman van Doorn.

He seems not to have been a particularly studious boy, preferring to be out-of-doors and to use his hands. There is some doubt concerning the exact course of Röntgen's education until he entered the Utrecht Technical School in December 1862, at the age of sixteen. Apparently he was expelled from a school in Utrecht for refusing to identify a classmate who had caricatured one of the masters. The continuity of his formal progress toward the university was thus broken, and he was never accepted as a regular student by the University of Utrecht—to his own and his parents' distress. After two-and-a-half years at the Technical School, and nine months' attendance at the philosophy classes of the university, he passed an examination to enter the Polytechnic at Zurich, as a student of mechanical engineering.

Röntgen was extremely happy in Switzerland, both in his work and in his social life. He received his diploma as a mechanical engineer in 1868 and his doctor of philosophy degree a year later. With these qualifications, he became assistant to the professor of physics, August Kundt, whose friendship and support greatly furthered Röntgen's career. While working in Zurich, Röntgen met his future wife, Anna Bertha Ludwig, the daughter of a German exile. In 1871 Röntgen accompanied Kundt to the University of Würzburg, and the following year he married Bertha Ludwig. The couple had no children, but they adopted Bertha Röntgen's niece in 1887.

Würzburg saw the real beginning of Röntgen's academic career, although at first he was disappointed because the university refused to give him any academic position, since he lacked the formal educational requirements. Shortly after his marriage, he moved to Strasbourg with Kundt, where he became tutor in the very fine Physical Institute. He spent the year 1875 as professor at the Agricultural Academy of Hohenheim, but he missed the excellent equipment at Strasbourg and soon returned there to teach theoretical physics. The series of papers he produced during the next two years resulted in his being offered the chair of physics at the University of Giessen, in Hesse. From 1879 to 1888 he worked at Giessen, building such a reputation that he was offered professorships at both Jena and Utrecht. He was not tempted to move, however, until the Royal University of Würzburg offered him the joint posts of professor of physics and director of the Physical Institute. In 1894 he became rector of the University of Würzburg. The following year Röntgen made his momentous discovery of X rays, which brought him international fame. He was made an honorary doctor of medicine of Würzburg in 1896, an honorary citizen of his birthplace, Lennep, and a corresponding member of the Berlin and Munich academies. On 30 November 1896 the Royal Society of London awarded jointly to Röntgen and Lenard the Rumford Medal. In 1900 Columbia University awarded Röntgen the Barnard Medal. The final accolades for this unassuming scientist were the erection of his statue on the Potsdam Bridge in Berlin, and the award, in 1901, of the first Nobel Prize for physics. He gave his prize money to further scientific studies at the University of Würzburg. In 1900, at the request of the Bavarian government, Röntgen moved from Würzburg to the chair of physics and the directorship of the Physical Institute at Munich.

Röntgen's last years were shadowed by the distresses and privations of World War I. His wife died after a long illness in 1919, and in 1920 he retired from his chair at Munich. He spent a great deal of his time at his country house at Weilheim, near Munich, where he had an extensive library. He continued to work and to enjoy long country walks until the year before his death, which followed a short illness.

Röntgen's early training as an engineer and his years as Kundt's assistant at Würzburg, where there was no laboratory mechanic, formed his lifelong habit of making his own apparatus. He was, indeed, the meticulously conscientious experimenter. Röntgen invariably worked alone in the laboratory, and with nothing to disturb his concentration, he was able to develop acute powers of observation. He was able to detect and measure extremely small effects, for example, the compressibility of liquids and solids and the rotation of the plane of polarization of light in gases. His reticence caused him to shun public engagements, and he never acquired the requisite lecturer's skills. He even declined to give the expected lecture when he won the Nobel Prize. Röntgen was well known for his assiduous reading of the scientific literature, yet he never allowed his retiring and studious nature to interfere with his university administrative duties. His attitude to his profession is clearly defined in the address that he gave in 1894, when he became rector of Würzburg University:

The University is a nursery of scientific research and mental education, a place for the cultivation of ideals for students as well as for teachers. Her significance as such is much greater than her practical usefulness, and for this reason one should endeavour, in filling vacant places, to choose men who have distinguished themselves as investigators and promotors of Science, and not only as teachers; for every genuine scientist, whatever his line, who takes his task seriously, fundamentally follows purely ideal goals and is an idealist in the best sense of the word. Teachers and students of the University should consider it a great honour to be members of this organization. Pride in one's profession is demanded, but not professional conceit, snobbery or academic arrogance, all of which grow from false egotism [Glasser, 1933, p. 100].

In all, Röntgen wrote fifty-eight papers, some with collaborators. Most of them were published in *Annalen der Physik und Chemie*. The fifteen Strasbourg publications covered such topics as the ratio of the specific heats of gases, the conductivity of heat in crystals, and the rotation of the plane of polarization of light in gases. Four papers on this last subject were the result of his joint work with Kundt. It was only because of the very high level of their experimental skill that the phenomenon was able to be observed and measured, something that Faraday had not been able to achieve. During Röntgen's professorship at Giessen, he published eighteen papers. Work on the relation between light and electricity was being done by Röntgen at much the same time as by Kerr, who discovered the effect that bears his name. As part of his lifelong interest in crystals, he studied pyro-electrical and piezoelectrical phenomena. Having constructed a very sensitive air thermometer, he was able to measure the absorption of heat in water vapor, and his flair for experiment was also shown by his work on the compressibility of liquids and solids.

Röntgen's fame rests on two pieces of work, both of which were far outside his normal field of research. The discovery of X rays is the more famous, but the earlier one concerned the magnetic effects produced in a dielectric, such as a glass plate, when it is moved between two electrically charged condenser plates. He set himself to test the electromagnetic theory of Maxwell, which implies that there will be a magnetic field in a dielectric whenever the electric field changes. In 1878 Rowland claimed to have detected the magnetic effect caused by the motion of electrostatic charges, but others could not repeat the experiments. For Röntgen here was a challenge. In a paper published in 1888 he demonstrated beyond doubt both the reality of the effect and the ability of Maxwell's theory to explain it quantitatively. H. A. Lorentz named the effect the "roentgen current," and Röntgen

himself considered it as having as much importance as his discovery of X rays because it led to the theories of Lorentz and is the basis for modern theories of electricity.

During the eleven years which he spent at Würzburg, Röntgen published eighteen papers, the final three embodying the discovery of X rays. The earlier papers dealt with the effects of pressure on the physical properties of solids and liquids. While professor at Munich, administrative work took up so much of his time that only seven papers were produced between 1900 and 1921. These were concerned with the physical properties of crystals, their electrical conductivity, and the influence of radiation on them. The investigations published in 1914 on pyroelectricity and piezo-electricity proved of particular significance in clarifying the real nature of these effects.

It is, of course, for the discovery of X rays, as he called them, that Röntgen is known to the general public. In Germany, the name given to the rays is more usually *Röntgenstrahlen*. It is now known that X rays are part of the electromagnetic spectrum, as is light. The wavelengths of X rays are very short, occupying the region 0.01 to 50 angstroms.

On Friday, 8 November 1895, Röntgen first suspected the existence of a new phenomenon when he observed that crystals of barium platinocyanide fluoresced at some distance from a Crookes tube with which he was experimenting. Again, this investigation into gas discharges was outside his normal field of interest. Hertz and Lenard had published on the penetrating powers of cathode rays (electrons), and Röntgen thought that there were unsolved problems worth investigation. He found time to begin his repetition of their experiments in October 1895. As a preliminary to viewing the cathode rays on a fluorescent screen, Röntgen completely covered his discharge tube with a black card, and then chanced to notice that such a screen lying on a bench some distance away was glowing brightly. Although others had operated Crookes tubes in laboratories for over thirty years, it was Röntgen who found that X rays are emitted by the part of the glass wall of the tube that is opposite the cathode and that receives the beam of cathode rays. He spent six weeks in absolute concentration, repeating and extending his observations on the properties of the new rays. He found that they travel in straight lines, cannot be refracted or reflected, are not deviated by a magnet, and can travel about two meters in air. He soon discovered the penetrating properties of the rays, and was able to produce photographs of balance-weights in a closed box, the chamber of a shotgun, and a piece of nonhomogeneous metal; he also noticed the outlines of the bones in his fingers on these photographs. The apparent magical nature of the new rays was something of a shock even to Röntgen, and he, naturally, wished to be absolutely sure of the repeatability of the effects before publishing. On 22 December he brought his wife into the laboratory and made an X-ray photograph of her hand. It was no doubt the possibility of seeing living skeletons, thus pandering to man's morbid curiosity, that contributed to the peculiarly rapid worldwide dissemination of the discovery. The first communication on the rays, on 28 December, was to the editors of the Physical and Medical Society of Würzburg, and by 1 January 1896 Röntgen was able to send reprints and, in some cases, photographs to his friends and colleagues. Emil Warburg displayed some of the photographs at a meeting of the Berlin Physical Society on 4 January. The *Wiener Presse* carried the story of the discovery on 5 January, and on the following day the news broke around the world. The world's response was remarkably swift, both the general public and the scientific community reacting in their characteristic ways. For the former, the apparent magic caught the imagination, and for the latter, Crookes tubes and generators were promptly sold in great numbers.

After a royal summons, Röntgen demonstrated the effects of X rays to the Kaiser and the court on 13 January. He was immediately awarded the Prussian Order of the Crown, Second Class.

In March 1896, a second paper on X rays was published, and there followed a third in 1897, after which Röntgen returned to the study of the physics of solids. He had shown clearly the uses of the new rays for medicine and metallurgy, and so founded radiology, but the discovery of the nature of the rays and other applications he left to others. The hypothesis that X rays were transverse electromagnetic rays was proved by the experiments of Friedrich and Knipping, based on Laue's idea of using a crystal as a diffraction grating. The possibility of an X-ray spectrometer was developed brilliantly by Moseley, whose papers of 1913 and 1914 showed the physical significance of atomic numbers and predicted the existence of three undiscovered elements.

X rays must have been produced by others long before Röntgen, probably with some of the electrical apparatus used during the eighteenth century. Crookes himself, in 1879, complained of fogged photographic plates that happened to be stored near his cathode-ray tubes. A. W. Goodspeed and W. N. Jennings in Philadelphia in 1890 noticed a peculiar blackening of photographic plates after having demonstrated a Crookes tube, but they failed to follow up their observation. Lenard and some other German physicists had noticed the fluorescence near Crookes tubes, but since they were concentrating on studying the properties of cathode rays, the strange side effects were not examined.

BIBLIOGRAPHY

Röntgen's publications are listed in Otto Glasser, *Wilhelm Conrad Röntgen und die Geschichte der Röntgenstrahlen* (Berlin, 1931; 2nd ed. 1959); published in English as *Wilhelm Conrad Röntgen and the History of the Roentgen Rays* (London, 1933). This work lists some three dozen books and articles concerning Röntgen, and over 1,000 books and pamphlets on X rays published in the year 1896.

The Deutsches Röntgen Museum at Remscheid-Lennep contains Röntgen's personal possessions, library, photographic slides, private correspondence, and reminiscences of him recorded on magnetic tape.

W. Robert Nitske, *The Life of Wilhelm Conrad Röntgen. Discoverer of the X Ray* (Tucson, 1971), although it was intended as a popular biography, contains a six-page bibliography that includes papers concerned with the history of the discovery.

G. L'E. TURNER

ROOD, OGDEN NICHOLAS (*b.* Danbury, Connecticut, 3 February 1831; *d.* New York, N.Y., 12 November 1902), *physics.*

The son of Anson Rood, a Congregational minister, and Alida Gouverneur Ogden, Rood graduated from the College of New Jersey (Princeton) in 1852 and did postgraduate work from 1852 to 1858 at Yale, Munich, and Berlin. Married in 1858 to Matilde Prunner of Munich, he then joined the staff of the University of Troy, a newly founded but short-lived denominational institution in Troy, New York, as professor of chemistry. In 1863 Rood was appointed professor of physics at Columbia University, where he served until his death, and in 1865 he was elected to the National Academy of Sciences.

A talented and prolific experimentalist, Rood managed to measure the duration of a spark discharge in a Leyden jar and the high resistance of dielectrics. He also found a way to produce and measure the magnitude of vacuums in the range of 10^{-9} atmospheres. Perhaps his most significant contribution to science was his technique of flicker photometry for comparing the brightness of different colors. The photometry of brightness depends upon the direct visual comparison of two adjacent fields of illumination, and the eye cannot make the judgment if the two fields are unequal in wavelength. Rood pointed out that since one saw a flicker when two differently colored surfaces were alternatively illuminated by lights of unequal brightness, the intensities were the same when the flicker disappeared.

An accomplished painter, Rood had a specially keen interest in physiological optics and theories of color. In addition to his research in this area, in 1879 he published *Modern Chromatics*, a popular summary of the field addressed to both physicists and artists. The book was widely read by painters in both Europe and the United States and was known as "the impressionist's Bible." Ironically, Rood himself disliked the impressionists and once said, "If that is all I have done for art, I wish I had never written that book."

BIBLIOGRAPHY

Edward L. Nichols, "Ogden Nicholas Rood," in *Biographical Memoirs. National Academy of Sciences*, **6** (1909), 447–472, is a useful résumé and contains a complete bibliography of Rood's writings. About 1,000 items, including papers, correspondence, sketchbooks, drawings, etchings, photos, and memorabilia, are in the Columbia University Library.

DANIEL J. KEVLES

ROOKE, LAWRENCE (*b.* Deptford [now part of London], England, 23 March 1622; *d.* London, England, 7 July 1662), *astronomy, natural philosophy.*

Rooke was educated at Eton and at King's College, Cambridge, receiving the B.A. in 1643 and the M.A. in 1647. In 1650 he became a fellow-commoner of Wadham College, Oxford, in order to enjoy the company and instruction of John Wilkins, Seth Ward, and the circle of virtuosi gathered about them. While at Oxford he occasionally assisted Robert Boyle in his chemical experiments. In 1652 Rooke was named professor of astronomy at Gresham College in London, exchanging his position in 1657 for the professorship of geometry. After 1658 many of his Oxford associates joined him at regular meetings in his and Christopher Wren's rooms after their respective weekly Gresham lectures. It was in his room that the organization was founded that, shortly after his death, became the Royal Society.

Although universally esteemed for the breadth and solidity of his learning, Rooke wrote no systematic treatises and his work was primarily empirical and practical. He observed the comet of 1652 and performed a series of experiments with Wren on the collision of elastic bodies and, with Jonathan Goddard, on the effect of radiant heat on oil in a tube. His interest in practical maritime problems led him to draw up a list of systematic observations to be made by seamen that would be useful for the improvement of navigation. To help solve the problem of determining longitude at sea, he undertook a series of observations of the eclipses of Jupiter's satellites and proposed the systematic telescopic observation of lunar eclipses. Rooke suggested that several of Hevelius' lunar landmarks be accepted as a general standard against which to measure the position of the earth's shadow and that the altitudes of certain especially bright stars be taken as a standard measure of time for the observations.

BIBLIOGRAPHY

I. ORIGINAL WORKS. Rooke's writings include "Directions for Sea-Men, Bound for Far Voyages," in *Philosophical Transactions of the Royal Society*, **1** (1666), 140–143; "A Method for Observing the Eclipses of the Moon," *ibid.*, 388–390, repr. in Thomas Sprat, *History of the Royal Society of London* (London, 1667), 180–182; "Discourse Concerning the Observations of the Eclipses of the Satellites of Jupiter," in Sprat, *op. cit.*, 183–189; and "Observationes in cometam qui mense Decembri anno 1652 apparuit," in Seth Ward, *De Cometis . . .* (Oxford, 1653), 39.

II. SECONDARY LITERATURE. See John Ward, *Lives of the Professors of Gresham College* (London, 1740), 90–95; and Colin A. Ronan, "Laurence Rooke (1622–1662)," in *Notes and Records Royal Society of London*, **15** (1960), 113–118.

WILBUR APPLEBAUM

ROOMEN, ADRIAAN VAN (*b.* Louvain, Belgium [?], 29 September 1561; *d.* Mainz, Germany, 4 May 1615), *mathematics, medicine.*

Roomen's father, for whom he was named, was a merchant; his mother was Maria van den Daele. According to the dedication of his *Ideae mathematicae* (1593), he studied mathematics and philosophy at the Jesuit College in Cologne. In 1585 he spent some time in Rome, where he met Clavius. From about 1586 to 1592 van Roomen was professor of medicine and mathematics at Louvain. He then became professor of medicine at Würzburg, where on 17 May 1593 he gave his first lecture. From 1596 to 1603 he was also "mathematician" of the chapter in Würzburg; his duties included drawing up the calendar each year. In 1598 van Roomen was at Prague, where the Emperor Rudolf II very probably bestowed the titles of count palatine and imperial court physician upon him. In 1601 he was in France for three months to recover his health, and during his stay he visited Viète. Between 1603 and 1610 he lived in both Würzburg and Louvain;

he was ordained a priest in the latter city at the end of 1604 or the beginning of 1605.

In 1610 van Roomen was invited to teach mathematics in Zamosc, Poland; his pupil was most likely Thomas Zamojski, son of the founder of the college in that town. During his sojourn there (September 1610–July 1612) he became acquainted with the Polish mathematician Jan Brożek, whom he met several times and with whom he conducted a correspondence. In one of his letters to van Roomen, Brożek posed two questions. The first concerned the dispute between the astronomers Giovanni Antonio Magini and David Origanus, and the second concerned a theorem on isoperimetric figures from the *Geometria* of Petrus Ramus. Van Roomen's answer to the latter question forms the most interesting part of the correspondence and was published by Brożek in his *Epistolae* (Cracow, 1615) and in his *Apologia* (Danzig, 1652).

An important part of van Roomen's works dealt with mathematical subjects, especially trigonometry and the calculation of chords in a circle. His first known work, *Ouranographia* (Louvain or Antwerp, 1591), is a speculative consideration on nature, specifically the number and the motion of the heavenly spheres. His *Ideae mathematicae pars prima* (Antwerp, 1593), dedicated to Clavius, was intended to be the first part of a great work on the calculation of chords in a circle and on the quadrature of the circle. In it van Roomen hoped to publish his discoveries on regular polygons; but except for some fragments, the remainder of the work did not appear. In the introduction van Roomen states that for some years he had tried to find a general rule to calculate the sides of all regular polygons. He discovered three methods, one of which used algebraic equations. For all regular polygons from the triangle up to the eighty-sided polygon he derived the equations and sent them to Ludolph van Ceulen, to whom he left the calculation of the solutions. In his work van Roomen gives, without any proof, the calculation to thirty-two decimal places of the sides of regular three-, four-, five-, and fifteen-sided polygons and of the polygons arising from the preceding by a continuous doubling of the number of the sides. He continued his calculations up to the polygon with $15 \cdot 2^{60}$ sides, and with the help of the side of the regular 251,658,240-sided polygon he calculated π to sixteen decimal places.

At the beginning of his treatise van Roomen propounded to all the geometers the famous equation of the forty-fifth degree. An ambassador from the Netherlands told Henry IV that France did not possess a single geometer capable of solving the problem. Henry sent for Viète, who at once gave a solution and, the next day, twenty-two more. In his turn Viète proposed to van Roomen the Apollonian problem: to draw a circle touching three given circles. Van Roomen published his answer in *Problema Apolloniacum* (Würzburg, 1596). He solved the problem by the intersection of two hyperbolas, but he did not give a construction in the proper sense. Viète published his own geometrical solution in his *Apollonius Gallus* (Paris, 1600).

In 1594 Scaliger published his *Cyclometrica elementa duo*, in which he tried to prove that Archimedes' approximation of π was incorrect. At once he was attacked by several mathematicians, among them Viète and van Ceulen, as well as van Roomen in his *In Archimedis circuli dimensionem* (Geneva, 1597). The first part of this tract contained a reedition of the Greek text of Archimedes' *On the Measurement of the Circle*, with a Latin translation and an elaborate analysis. In the second part, "Apologia pro Archimede ad clarissimum virum Josephum Scaligerum," van Roomen refuted Scaliger's objections to Archimedes' tract. In the third part he refuted, in ten dialogues, the quadratures of the circle of Oronce Finé, Simon van der Eycke, Raymarus Ursus, and Scaliger.

Van Roomen also wrote a commentary on al-Khwārizmī's *Algebra*, "In Mahumedis Algebram prolegomena," which is now lost, the copy at the University of Louvain having been destroyed in 1914 and that at Douai in 1944. Van Roomen was partial to extensive calculations, as can be seen in his *Chordarum arcubus circuli* (Würzburg, 1602). In this work he gave, to 220 or 300 decimal places, the square roots needed for the calculation of the side of the regular thirty-sided polygon. He also wrote several works on plane and spherical trigonometry, including the *Speculum astronomicum* (Louvain, 1606) and the *Canon triangulorum sphaericorum* (Mainz, 1609). These works contain the first systematic use of a trigonometric notation.

In his terminology van Roomen imitated Viète, using the expressions "prosinus" and "transinuosa" for tangent and secant, respectively. The tables for sines, tangents, and secants, together with their cofunctions, in the *Canon triangulorum* were borrowed from Clavius. A last contribution to the project developed in his *Ideae mathematicae* is in van Roomen's *Mathematicae analyseos triumphus* (Louvain, 1609). In this work he calculated the sides of the nine-sided and eighteen-sided regular polygons to 108 decimal places.

Besides his printed works there were manuscripts, now lost, containing unpublished works by van Roomen: the "Tractatus de notatione numerorum" and the "Nova multiplicandi, dividendi, quadrata componendi, radices extrahendi ratio." The last, dealing with his methods for calculating with large numbers, was published in 1904 by H. Bosmans.

BIBLIOGRAPHY

The best survey of van Roomen's life and works is the article by P. Bockstaele in *Nationaal biografisch woordenboek* (Brussels, 1966), cols. 751–765, which also contains an extensive bibliography.

H. L. L. BUSARD

ROOZEBOOM, HENDRIK WILLEM BAKHUIS (*b.* Alkmaar, Netherlands, 24 October 1854; *d.* Amsterdam, Netherlands, 8 February 1907), *physical chemistry*.

After completing high school in Alkmaar (1872), Roozeboom worked in a laboratory that did research on the composition of foods and water. In 1878 he became assistant to J. M. van Bemmelen, professor of chemistry at the University of Leiden, where he continued his studies in chemistry (1878–1882). Roozeboom received a doctorate in 1884 with a dissertation on the hydrates of sulfur dioxide, chlorine, bromine, and hydrogen chloride; in it he studied the relationship among the three states of matter (solid, liquid, gas) at different temperatures and pressures. In 1886 van der Waals brought to Roozeboom's attention J. W. Gibbs's phase rule (1876), which defines the conditions of equilibrium as a relationship between the number of components of a system C and the number of coexisting phases P, according to the equation

$$F = C + 2 - P,$$

where F is the degrees of freedom or variability of the system. This rule gave Roozeboom the theoretical guide to his investigations on heterogeneous equilibriums, which he began to study in 1882. He became the founder of the scientific doctrine of heterogeneous equilibriums, important not only in theoretical chemistry but also for its practical applications in metallurgy and geology.

After receiving the doctorate, Roozeboom remained at Leiden as assistant to van Bemmelen. From 1881 to 1896 he also taught at a girls' high school in Leiden, and in 1890 he was appointed university lecturer in physical chemistry. In 1896 Roozeboom succeeded van't Hoff as professor of general chemistry at the University of Amsterdam, where he remained for the rest of his life. In 1880 he became a member of the Koninklijke Nederlandse Akademie van Wetenschappen. Besides his scientific work, Roozeboom was very interested in social problems; in 1895 he was one of the founders of the Christelijke Vereniging van Natuur- en Geneeskundigen in Nederland.

Roozeboom's importance in science stems from his application of the then little-known phase rule to the study of heterogeneous equilibriums. In his first important publication, "Sur les différentes formes de l'équilibre chimique hétérogène" (1887), he systematically arranged all the known dissociation equilibriums on the basis of the phase rule according to the number of components and the number and nature of the phases. In many publications Roozeboom used two- and three-dimensional figures to illustrate his theory. He always studied simple, generally applicable systems, such as calcium chloride–water and ferric chloride–water, which provide an almost complete illustration of the theory of salt hydrates (1889). With his pupil F. A. H. Schreinemakers, Roozeboom investigated the ternary system hydrogen chloride–ferric chloride–water (1894). Alone he studied astrakanite ($Na_2SO_4 \cdot MgSO_4 \cdot 4H_2O$) as an example of a double salt in a four-phase system (three solid and one gas).

In Amsterdam, Roozeboom made especially systematic studies of heterogeneous equilibriums. He gave much attention to mixed crystals, which provided insight into the homogeneous solid phase. Roozeboom represented all phase equilibriums with pressure-temperature-concentration diagrams. He and his pupils gave pseudosystems (such as tautomeric compounds) a phase-theoretical treatment—for example, the system acetaldehyde-paraldehyde and the system sulfur-chlorine.

Roozeboom's application of the concept of a homogeneous phase to mixed crystals, gas mixtures, and liquid mixtures, from which a general theoretical treatment of equilibriums became possible, was very important. From it he derived five types of fusion curves, which were experimentally confirmed by his pupils. Both fusion lines and transition phenomena were observed. Roozeboom's work profoundly stimulated the systematic and comprehensive investigation of alloys through practical application of the phase rule to the study of alloys of cadmium and tin and binary and ternary alloys of tin, bismuth, cadmium, and lead. Systems of two optical antipodes and liquid mixed crystals were also studied. All these investigations confirmed Roozeboom's theory, which was also applied to technical problems, such as the system iron-steel, and to metallurgical and experimental petrographical subjects.

In volume I of *Die heterogenen Gleichgewichte vom Standpunkte der Phasenlehre* (1901) one-component heterogeneous equilibrium systems were systematically treated. In volume II, part 1 (1904), Roozeboom treated two-component systems that contain only one liquid phase. After his death the work was completed by his pupils Schreinemakers, E. H. Büchner, and A. H. W. Aten (1911–1918). W. D. Bancroft described Roozeboom as having done "far more than any one else to show the importance and significance of Gibbs's Phase Rule" (*The Phase Rule*, p. iii).

BIBLIOGRAPHY

I. ORIGINAL WORKS. A complete bibliography of Roozeboom's publications is given by J. M. A. van Bemmelen, W. P. Jorissen, and W. E. Ringer in *Berichte der Deutschen chemischen Gesellschaft*, **40** (1907), 1570–1574.

Among his works are "Recherches sur quelques hydrates de gaz," in *Recueil des travaux chimique des Pays-Bas et de la Belgique*, **3** (1884), 29–104, his dissertation; "Sur les conditions d'équilibre de deux corps dans les trois états, solide, liquide et gazeux, d'après M. v. d. Waals," *ibid.*, **5** (1886), 335–350; "Sur les différentes formes de l'équilibre chimique hétérogène," *ibid.*, **6** (1887), 262–303; "Sur l'astrakanite et les sels doubles hydratés en général," *ibid.*, 333–355; "Étude expérimentale et théorique sur les conditions de l'équilibre entre les combinaisons solides et liquides de l'eau avec des sels, particulièrement avec le chlorure de calcium," *ibid.*, **8** (1889), 1–146; *Die Bedeutung der Phasenlehre* (Leipzig, 1900); "Eisen und Stahl vom Standpunkte der Phasenlehre," in *Zeitschrift für physikalische Chemie*, **34** (1900), 437–487, English trans. in *Journal of the Iron and Steel Institute* (1900), no. 2, 311–316; and *Die heterogenen Gleichgewichte vom Standpunkte der Phasenlehre* (Brunswick): I (1901); II, pt. 1 (1904); II, pt. 2, by E. H. Büchner (1918); II, pt. 3, by A. H. W. Aten (1918); III, pt. 1, by F. A. H. Schreinemakers (1911); III, pt. 2, by Schreinemakers (1913).

II. SECONDARY LITERATURE. See J. M. van Bemmelen, in *Chemisch weekblad*, **4** (1907), 249–285; J. M. van Bemmelen, W. P. Jorissen, and W. E. Ringer, in *Berichte der Deutschen chemischen Gesellschaft*, **40** (1907), 1541–1574; A. F. Holleman, in *Chemisch weekblad*, **4** (1907), 119–132; R. Hooykaas, in *Geloof en wetenschap*, **53** (1955), 68–77; W. P. Jorissen and W. E. Ringer, *H. W. Bakhuis Roozeboom*, Mannen en vrouwen van beteekenis in onze dagen, no. 37, sec. 4 (Haarlem, 1907), 155–218; and W. Stortenbeker, in *Recueil des travaux chimiques des Pays-Bas et de la Belgique*, **27** (1908), 1–51.

H. A. M. SNELDERS

ROSA, DANIELE (*b.* Susa, near Turin, Italy, 29 October 1857; *d.* Novi Ligure, Italy, 30 April 1944), *zoology.*

Rosa studied science and medicine at Turin, receiving the doctorate in 1880. He completed his training under Ernst Ehlers at the zoology institute of the University of Göttingen, then was an assistant at the zoology museum in Turin. Later he taught zoology and comparative anatomy at the universities of Sassari, Modena, Florence, Turin, and, again, Modena. He retired in 1932.

Rosa published works on the morphology and systematics of the oligochaetes (Annelida). In two monographs that appeared in 1899 and 1918 he set forth his own theory of the origin of species and of their transformations. He held that the extinction of species results primarily from a steady decrease in variability and postulated a "law of progressively diminished variation": the longer a species has been in existence, the less it varies. Since the lower species supposedly possessed a better-preserved capacity for variation, in the course of time they replaced the higher species. The law was concerned essentially with causes, which were unknown. Rosa therefore changed its basis to what he termed a "law of progressively diminished variability." The effect of this law was slowed, however, because not all parts of an animal become modified—and therefore reach an end point—simultaneously. Nevertheless, a consequence of the law was that the emergence of new forms eventually ceased. Rosa's law also asserted the existence of orthogenesis in nature. This orthogenesis, he claimed, was not affected by individual variations, since the latter, which were "Darwinian" as opposed to "phylogenetic," had, in his view, no influence on the transformation of species. The theory found no supporters.

BIBLIOGRAPHY

Rosa's main works are *La riduzione progressiva della variabilità ed i suoi rapporti coll'estinzione e l'origine della specie* (Turin, 1899), translated into German as *Die progressive Reduktion der Variabilität und ihre Beziehungen zum Aussterben und zur Entstehung der Arten* (Jena, 1903); and *Ologenesi. Nuova teoria dell'evoluzione e della distribuzione geografica dei viventi* (Florence, 1918), translated into French as *L'ologenèse. Nouvelle théorie de l'évolution et de la distribution géographique des êtres vivants* (Paris, 1931). An obituary on Rosa is Celso Guareschi, in *Dall'Annuario dell'Università di Modena Anni acc.* (1942–1944), 269–270.

HANS QUERNER

ROSANES, JAKOB (*b.* Brody, Austria-Hungary [now Ukrainian S.S.R.], 16 August 1842; *d.* Breslau, Germany [now Wrocław, Poland], 6 January 1922), *mathematics.*

Rosanes was the son of Leo Rosanes, a merchant. From 1860 until 1865 he studied at the universities of Berlin and Breslau; having taken the Ph.D. at the latter in 1865, he remained there for the rest of his career. In 1870 he became *Privatdozent*, in 1873 professor extraordinarius, and in 1876 ordinary professor; he also served the university as its rector during the academic year 1903–1904.

Rosanes' mathematical papers concerned the various questions of algebraic geometry and invariant theory that were current in the nineteenth century. One of his first papers, written with Moritz Pasch, discussed a problem on conics in closure-position. In 1870 he provided a demonstration that each plane Cremona transformation can be factored as a product of quadratic transformations, a theorem that M. Noether also proved independently at about the same time. Both demonstrations were, however, incomplete and were put into final form by G. Castelnuovo some thirty years later.

Rosanes' contributions to the theory of invariants were made in the 1870's and 1880's. He gave conditions for a form to be expressed as a power-sum of other forms, then, in a series of papers, treated linearly dependent point systems in a plane and in space. In later years his scientific productivity declined, but his rector's lecture of 1903, on the characteristic features of nineteenth-century mathematics, is noteworthy. Like a number of other mathematicians, Rosanes was also interested in chess and published a book on *Theorie und Praxis des Schachspiels*. He retired from the university in 1911 and spent the rest of his life in Breslau. In 1876 he married Emilie Rawitscher.

BIBLIOGRAPHY

Rosanes' works, cited in the text, are "Über das einem Kegelschnitt umbeschriebene und einem anderen einbeschriebene Polygon," in *Journal für die reine und angewandte Mathematik*, **64** (1865), 126–166, written with M. Pasch; "Über diejenigen rationalen Substitutionen, welche eine rationale Umkehrung zulassen," *ibid.*, **73** (1871), 97–111; "Über linear abhängige Punktsysteme," *ibid.*, **88** (1880), 241–273; and "Charakteristische Züge in der Entwicklung der Mathematik des 19. Jahrhunderts," in *Jahresbericht der Deutschen Mathematische-Vereinigung*, **13** (1904), 17–30.

WERNER BURAU

ROSCOE, HENRY ENFIELD (*b*. London, England, 7 January 1833; *d*. Leatherhead, Surrey, England, 18 December 1915), *chemistry*.

Roscoe was the son of Henry Roscoe, a Liverpool barrister and judge. He was educated at the high school of the Liverpool Institute, where his interest in chem-

istry and natural philosophy was awakened by W. H. Balmain, who provided a chemical laboratory for the boys. Roscoe fitted up a room in his house at Liverpool, in which he performed experiments and gave illustrated lectures to his cousins and friends.

In 1848 Roscoe, a Dissenter, entered University College, London, where he studied chemistry with Thomas Graham. He decided upon chemistry as a career and at the beginning of his second session at University College affiliated with the Birkbeck Chemical Laboratory under A. W. Williamson, who soon set him to work at original investigation and who chose him as his private assistant. Roscoe took his B.A. degree in 1853 and chose to continue his chemical studies at Heidelberg, with Bunsen, who trained him in quantitative techniques and gas analysis and encouraged him in original researches, some of which appeared in *Justus Liebigs Annalen der Chemie* (1854).

Roscoe received his Ph.D. by oral examination "six months after I first went to Heidelberg." He returned there in the fall of 1855 and began joint work with Bunsen on the chemical action of light. When Williamson was appointed Graham's successor as professor of chemistry in 1855, he appointed Roscoe to the post of lecture assistant. Roscoe returned to University College, London, for the winter session of 1855-1856, bringing with him W. Dittmar as his private assistant, to aid him in his researches.

In London, like many other chemists, Roscoe had to juggle several positions to scratch out a career. He taught chemistry at an army school; was a consultant to a government committee on ventilation that was determining the amount of carbon dioxide in various enclosed places; and attempted to develop a private consulting laboratory. When, however, the chair of chemistry at Owens College in Manchester, recently vacated by Edward Frankland, was advertised in the summer of 1857, Roscoe applied for the position, supported by testimonials from Bunsen, Liebig, Williamson, and Graham. There were fifteen applicants, including Robert Angus Smith and Frederick Crace-Calvert; but Roscoe was selected at an annual salary of £150 plus fees. He took up his new duties as soon as he was able to close his London laboratory and make arrangements in Manchester.

Owing to its local unpopularity, the college was almost at the point of extinction. The *Manchester Guardian* had pronounced the Owens experiment a "mortifying failure," and Roscoe was even refused lodgings in town when the landlord learned of his affiliation. There were only thirty-four students, of whom fifteen worked in the chemical laboratory. Roscoe set about altering the situation by demonstrating to the community the potential of Owens College to aid the economic life of the region. With the new principal, J. G. Greenwood, he prodded the college toward a greater emphasis upon science and what were considered practical subjects. Roscoe was instrumental in bringing Robert Clifton (later professor at Oxford) to teach natural philosophy and, by revising the chemical curriculum, was able to build a reputation for sound chemical teaching and for developing research students. Steadily, over a period of twenty-five years, he convinced manufacturers of the necessity for chemical training at Owens. In the short run the results were dramatic: by 1863 the number of day students at Owens had risen to 110, including 38 in the chemical laboratory.

Part of Roscoe's success must be laid to his active role in the city's scientific community. He joined and ultimately became an officer of the Manchester Literary and Philosophical Society; and during the cotton scarcity beginning in 1862, he was instrumental in organizing evening cultural events, including science lectures by himself. The success of these first attempts spurred Roscoe to expand his efforts along similar lines; and in 1866 he instituted Science Lectures for the People, which were given for eleven years by such luminaries as Huxley, William Carpenter, William Spottiswoode, Tyndall, and Huggins. The lectures were published and sold widely for a penny.

Roscoe's local reputation was further enhanced by his services as civic expert; he performed, for example, numerous analyses for local gas and water boards. He served the national government as well, both as consultant and as royal commissioner, on the Commission on Noxious Vapours (1876) and the Commission on Technical Instruction (1881). In the decade following his service, he lobbied extensively for technical education, his efforts bearing fruit in the Technical Instruction Act of 1889.

Without question, however, Roscoe's historical importance lay squarely in his great success as teacher and founder of what became known as the Manchester school of chemistry. While the debt to Bunsen is considerable and abundantly acknowledged, much of the character, style, and success of Roscoe's chemical school must be attributed to his own talent for organization and to the exigencies of regional requirements and support.

Like Bunsen, Roscoe began the working day with lecturing; then, after some time in his office attending to correspondence and permitting the students to commence their laboratory work, he made his rounds, counseling the young men, giving directions, and assessing their progress. The program for students began with the elementary laboratory, intended to inculcate the principles of method and accuracy in both practical work and theory. The advanced students continued with quantitative analysis, closely supervised by the professor and the demonstrators; Roscoe insisted that even students intended for chemical works in which the processes were admittedly cruder should be thoroughly trained in the "exacter processes." Finally, certain advanced students were encouraged, by precept and example, to embark upon original research; during Roscoe's thirty years at Owens, students and demonstrators published over 120 original papers, a record unequaled in Great Britain. The very best students were sent to Germany to complete their training.

Roscoe responded to the increasing specialization of chemistry in a creative way unmatched in Britain. He chose for his assistants and demonstrators men of unusual ability and worth, and permitted them to follow their special interests, ensuring that institutional flexibility would encourage them further. Carl Schorlemmer, his assistant, was named professor of organic chemistry in 1874, the first—and for many years the sole—professor of that subject in Great Britain. Assistant lecturers were employed for thermal chemistry, gas analysis, crystallography, and advanced organic chemistry; and lectureships in technological chemistry and metallurgy were introduced in 1885.

The results of these efforts were striking. Unquestionably, Owens College (later Victoria University) became, under Roscoe's tutelage, the leading chemistry school in Britain, providing staff for numerous academic and industrial enterprises. Students of Owens College went on to teach at Owens itself and at many of the new institutions for higher education in Britain and the colonies.

Roscoe's own researches, wide-ranging and of considerable number, centered largely upon inorganic chemistry. His first important work, on the laws of photochemical action, was undertaken jointly with Bunsen and lasted from 1855 to 1862. Investigating the gradual combination of chlorine and hydrogen under light, they demonstrated that the amount of photochemical action produced by a constant source varies inversely as the square of the distance and that the absorption varies directly as the intensity. They also measured the chemical action of the parts of the solar spectrum, describing the existence of several maximums of chemical intensity.

Roscoe's most important original researches concerned the metal vanadium. On a visit to Cheshire copper mines in 1865, he learned of the presence of the rare metal there and was able to prepare oxides of vanadium on which he and his assistant T. E. Thorpe began work. Roscoe demonstrated that Berzelius had been incorrect in viewing vanadium as analogous to chromium (Roscoe pointed to its relationship to the phosphorus-arsenic group) and had set the atomic weight too high. Berzelius had, he showed, conflated the oxychloride with the trichloride. Roscoe was the first to prepare the metal itself in a pure form by reducing the dichloride. Additional work on vanadium was completed under his direction by senior students at Owens.

In his own lifetime it was recognized that Roscoe's pedagogical publications carried even greater weight than his original researches. His *Lessons in Elementary Chemistry* (1866) and *Chemistry* in the Science Primer Series for Macmillan (1872), classics of their type, were widely adopted and were translated into nine languages. His substantial *Treatise on Chemistry*, the success of which largely depended upon his coauthor Schorlemmer, appeared in 1877–1884. Through his translation of Bunsen and Kirchhoff's classic work on spectrum analysis (*Chemische Analyse durch Spectralbeobachtungen*) and through his lectures, Roscoe was instrumental in calling attention to the new and revolutionary subject. His lectures before the Society of Apothecaries, published as *Spectrum Analysis* (1869), went through several editions, the last amply augmented by Arthur Schuster. Roscoe also contributed to the history of chemistry a popular life of Dalton and *A New View of the Origin of Dalton's Atomic Theory* (1896), a classic reevaluation of the atomic theory, written with Arthur Harden.

Roscoe was active in numerous scientific organizations and served several as an officer. He was elected a fellow of the Royal Society in 1863 and was subsequently elected to the Council and vice-presidency; he was president of the chemical section of the British Association for the Advancement of Science in 1870 and 1884, and president of the entire Association at its Manchester meeting in 1887; he was elected president of the Chemical Society in 1881 and was a founder-member and first president of the Society of Chemical Industry in 1881.

Roscoe was always closely concerned with the chemical industry, both in the capacity of consultant while at Owens and later as an industrialist himself, although these areas are curiously omitted from his autobiography. He was associated with the Aluminium Company of Oldbury and London as early as 1889 and was chosen a board member when the company was reconstituted as the Castner-Kellner Alkali Company, Ltd., in 1895. Even his researches on vanadium were spurred by the promise of industrial application. He wrote to Bunsen in 1876, for example, that "a friend of mine has made about *200 kilos* of pure Vanadic acid! I spent £6,000 on it! The best of the thing is that Vanadium will turn out to be a most valuable sub-

stance for Calico-Printers and Dyers—as by its means an aniline-black can be prepared which is far superior to that obtained with copper salts." After 1876 "vanadium blacks" became widely used but were found not to stand washing as well as blacks obtained with copper salts.

In 1884 Roscoe was knighted for his services on behalf of technical education. In the following year he was pressed to stand for election to Parliament for the upper-middle-class district of South Manchester and served as a Liberal member of Parliament from 1885 until his defeat in the election of 1895. In 1896 he became vice-chancellor of the University of London, which position he held until 1902. He was instrumental in the founding and direction of what is now the Lister Institute in Chelsea, modeled upon the Institut Pasteur in Paris, and served as a Carnegie trustee after 1901. In 1909 he was selected for the Privy Council.

BIBLIOGRAPHY

I. ORIGINAL WORKS. Roscoe's major books include *Lessons in Elementary Chemistry* (London, 1866; 7th ed., enl., 1906); *Spectrum Analysis* (London, 1869; 4th ed., with A. Schuster, 1885); *Chemistry* (London, 1872); *Record of Work Done in the Chemical Department of the Owens College* (Manchester, 1887); *Inorganic Chemistry for Beginners* (London, 1893; 2nd ed., 1912); *John Dalton and the Rise of Modern Chemistry* (London, 1895); and *The Life and Experiences of Sir Henry Enfield Roscoe, D.C.L., LL.D., F.R.S.* (London, 1906).

With Arthur Harden he wrote *A New View of the Origin of Dalton's Atomic Theory* (London, 1896); and *Inorganic Chemistry for Advanced Students* (London, 1899; 2nd ed., 1910). With Carl Schorlemmer he wrote *A Treatise on Chemistry*, 3 vols., (London 1877–1884).

Roscoe wrote a large number of scientific papers individually and jointly with colleagues and students; these are listed in the Royal Society *Catalogue of Scientific Papers*, V, 273–274; VIII, 778–779; XI, 216–217; XVIII, 291. There are MSS at the University of Manchester, correspondence at the Chemical Society, London, and scrapbooks at the John Rylands Library, Manchester. Letters to and from Roscoe can also be found in the Bunsen collection at Heidelberg.

II. SECONDARY LITERATURE. The standard biography is still T. E. Thorpe's *The Right Honourable Sir Henry Enfield Roscoe* (London, 1916). Thorpe also contributed a long *éloge* to *Proceedings of the Royal Society*, **93** (1916), i–xxi. See also H. B. Dixon's sketch in *Dictionary of National Biography* and the obituaries by Francis Jones and A. W. Waters in *Memoirs and Proceedings of the Manchester Literary and Philosophical Society*, **60** (1916), lii–lxiii. G. N. Burkhardt, "Schools of Chemistry in Great Britain and Ireland—XIII: The University of Manchester," in *Journal of the Royal Institute of Chemistry*, **78** (Sept.

1954), 448–460, discusses the Roscoe-Schorlemmer period. On technical education, see D. Thompson, "The Influence of Sir H. E. Roscoe on the Development of Scientific and Technical Education During the Second Half of the 19th Century" (M.Ed. thesis, Univ. of Leeds, 1957–1958). On Roscoe at Owens, see E. Fiddes, *Chapters in the History of Owens College and of Manchester University 1851–1914* (Manchester, 1937); H. B. Charlton, *Portrait of a University 1851–1951* (Manchester, 1951); P. J. Hartog, *The Owens College Manchester* (London, 1900); and J. Thompson, *The Owens College, Its Foundation and Growth* (Manchester, 1886).

ROBERT H. KARGON

ROSE, GUSTAV (*b.* Berlin, Germany, 18 March 1798; *d.* Berlin, 15 July 1873), *mineralogy, crystallography.*

Rose's family had a strong tradition in science. His grandfather, Valentin Rose the elder, invented the low-melting alloy still known as Rose's metal. Klaproth, a close friend of the family until his death in 1817, had earlier (1771–1780) been in charge of the family pharmacy before Gustav's father, Valentin Rose the younger, an original contributor to the methodology of inorganic chemical analysis, came of age. When he was only seventeen, Gustav and his brothers fought in the campaign against Napoleon in 1815. The following year his apprenticeship at a mine in Silesia was interrupted by illness. He returned to Berlin, where he studied mineralogy under C. S. Weiss. His dissertation, *De sphenis atque titanitae systematae crystallino*, was presented at the University of Kiel in December 1820. In this work, the first monograph on crystal morphology of a mineral species based on accurate measurements with a reflecting goniometer, Rose established the identity of sphene and titanite.

Following his elder brother Heinrich, later professor of chemistry at Berlin, and Eilhard Mitscherlich, whose discovery of isomorphism, announced in 1819, had been supported by Rose's accurate goniometric measurements, he then spent several years in Berzelius' laboratory at Stockholm. Rose returned to Berlin in 1823 to become a *Dozent* under Weiss and extraordinary professor in 1826. He became *ordentlicher* professor in 1839, succeeded Weiss as director of the Mineralogy Museum in 1856, and remained active in these posts until his death.

In 1829 Rose, with C. G. Ehrenberg, was chosen to accompany Humboldt on a scientific journey commissioned by the czar to the Urals, the Altai, and the Caspian Sea. This took him as far as the frontier of China. Rose's two-volume chronicle of the journey includes extensive observations on geology, mineralogy, and mineral resources of the regions traversed that were quoted widely and for a long time were the chief source of information on these matters.

Rose published about 125 papers, touching nearly all aspects of mineralogy known in his time. Much of his work was concerned with particular minerals or mineral groups. He discovered about fifteen new minerals, all still regarded as valid species, the most important being anorthite; and he also made significant contributions in many other fields. Through his meticulous goniometric measurements he contributed to the development of the concept of isomorphism, adding some important examples. With Riess (1843), in his only paper with a coauthor, he introduced the still-current terms "analogous pole" and "antilogous pole" in connection with the correlation of pyroelectric effects with morphology. He properly distinguished between positive and negative rhombohedrons in quartz and established its correct crystal class. Rose made important contributions to the study of meteorites, to the crystallography of the brittle metals and the noble metals, and to experimental petrology, repeating and extending James Hall's experiments on marble. One of his last papers (1871) dealt with the relations between thermoelectricity and morphology in pyrite.

Rose's *Elemente der Krystallographie*, in the first and second editions, represented the latest advances of the science at the time. Yet the first volume of the third edition, which was prepared by Alexander Sadebeck under Rose's direction and appeared just after Rose's death, shows practically no sign of the progress of the science in the thirty-five years following the appearance of the second edition. By contrast, Rose's *Mineralsystem* (1852) was strictly modern and most influential. It put an end to the "natural classifications" that had been a hindrance to the progress of mineralogy and became a model for later classifications.

With a dozen others, among them Humboldt and Mitscherlich, Rose founded the Deutsche Geologische Gesellschaft in July 1848, just at the time that publication of the 1 : 100,000 geologic map of Silesia, by Rose and Beyrich, was begun. Rose was very active in the society, serving as secretary and later repeatedly as president. In 1852 he presented fifty thin sections of rocks at a meeting of the society, seven years before the appearance of Sorby's classic paper, which usually is considered to mark the beginning of microscopic petrography.

Rose's biographers all emphasize that he was exceptionally modest and gentle and enjoyed the lasting esteem of his colleagues and students. Among the more distinguished of the latter were the explorer Ferdinand von Richthofen, G. vom Rath, Paul von Groth, and his successor, C. F. M. Websky.

BIBLIOGRAPHY

I. Original Works. A complete bibliography of Rose's publications is in Poggendorff, II, 692–694; and III, 1141–1142. His first major book was *Elemente der Krystallographie* (Berlin, 1833), also in French trans. (Paris, 1834); 2nd ed. (1838); a nominal 3rd ed., *Gustav Rose's Elemente der Krystallographie*, consists of 3 vols.: I (1873), by Alexander Sadebeck, Rose's assistant during his last years, covers the same ground as the earlier eds.; II (1876), by Sadebeck, is entitled *Angewandte Krystallographie*; and III (1887), by C. F. M. Websky, is *Berechnen der Krystalle*. Other books are *Mineralogisch-geognostische Reise nach dem Ural, dem Altai und dem Kaspischen Meere*, 2 vols. (Berlin, 1837–1842); and *Das krystallochemische Mineralsystem* (Leipzig, 1852).

Among his articles is "Ueber die Entdeckung der Isomorphie. Eine Ergänzung der Gedächtnissrede auf E. Mitscherlich," in *Zeitschrift der Deutschen geologischen Gesellschaft*, **20** (1868), 621–630. Adverse comments led Rose to write this supplement to his memorial address for Mitscherlich (1864). Here he details his contribution to the discovery of isomorphism and, incidentally, gives a partial account of his early scientific career, especially his relations with Mitscherlich and the circumstances of his first meeting with Berzelius.

II. Secondary Literature. See C. Rammelsberg, "Zur Erinnerung an Gustav Rose," in *Zeitschrift der Deutschen geologischen Gesellschaft*, **25** (1873), i–xix; and G. von Rath, "Gustav Rose, Nekrolog," in *Annalen der Physik*, **150** (1873), 647–652, by Rose's son-in-law.

A. Pabst

ROSE, HEINRICH (*b.* Berlin, Germany, 6 August 1795; *d.* Berlin, 27 January 1864), *chemistry.*

Rose was born into a family of scientists. His father and grandfather—both of whom were named Valentin Rose—were pharmacists who wrote on chemical and pharmaceutical subjects. His brother Gustav became a well-known mineralogist, and cousins and nephews later distinguished themselves in medicine and industrial chemistry. His was an established bourgeois professional family; and this fact helps to explain the even placid course of Rose's career, which after the Napoleonic period proceeded without drama and lacked development. He had no need, as had some of his more brilliant colleagues, such as Dumas and Liebig, for scientific entrepreneurship.

Rose's first training was in pharmacy, at Danzig. The war intervened, and with his brothers he joined the Prussian forces for the last campaign against Napoleon. He was in Paris in 1815 with the occupying armies, and while there met some of the foremost French scientists—Gay-Lussac, Biot, Vauquelin, and especially Berthollet, with whom he had a number of friendly conversations and for whose point of view on chemical dynamics he gained (and kept) a respect unusual for the time. On his return to Berlin he continued his studies, working for a time in the summer of 1816 with Martin Heinrich Klaproth, whom, in a sense, he succeeded as the purest and narrowest German chemical analyst. (Klaproth had been long and intimately associated with the Rose family: he had worked as assistant to the elder Valentin Rose and had become the guardian of his children after Rose's death in 1771.) Heinrich Rose next was apprenticed to a pharmacist in Mitau (near Riga), where he spent much of his spare time in discussion with Theodore von Grotthus, who had an estate nearby. Rose's earliest published writing appeared in a work by Grotthus. In 1819 Rose traveled via St. Petersburg and Finland to Stockholm to work with Berzelius. The great Swedish chemist had him continue some researches he had already begun on mica, and started him on the investigations of the properties of titanium, which became the subject of his dissertation.

Eilhard Mitscherlich came to Stockholm in 1819 and Gustav Rose followed in 1821. Rose, Mitscherlich, and Wöhler became Berzelius' main disciples in Berlin. Rose left Stockholm in the autumn of 1821 and proceeded to Kiel, where he submitted his dissertation on the oxygen and sulfur compounds of titanium. The doctorate was presently awarded. He then returned to Berlin. In 1822 he became *Privatdozent* in chemistry at the University of Berlin; in the following year he was made *extraordinarius*, and in 1835 *ordinarius*. Although he traveled some in later years, his life after 1822 centered on the university routine of teaching and research and on the round of activities of scientific Berlin. He became a member of the Prussian Akademie der Wissenschaften in 1832. He was twice married. To his great grief he survived his second wife and a daughter by that marriage.

Rose's contribution to chemistry was a piecemeal cumulative lifelong effort that can be divided into two aspects: (1) the training he gave to students directly at the University of Berlin and indirectly through his great textbook on analytical chemistry; and (2) the scores of analyses of mainly inorganic substances and minerals, the reports on which were published with unflagging regularity from 1820 until several years after his death. Most of them appeared in Poggendorff's *Annalen*, in the *Berichte* of the

German Chemical Society, and in the *Monatsberichte* of the Prussian Academy, and many were translated and published in other major European journals of chemistry.

A bare list of Rose's papers presents at first glance a confusing picture of analytical results without plan or direction. Indeed, the bulk of his papers consists of miscellaneous analyses of minerals that he collected (he went with Alexander von Humboldt and Ehrenberg on an expedition to the Urals in 1829), or that his brother Gustav submitted to him for analysis, or that sundry mineralogists, both amateur and professional, sent to him from all over the world. There were also analyses of a few compounds of practically the whole range of metals, earths, and alkaline metal earths. Rose also conducted several series of systematic investigations, some lasting several years. He was almost always working on several projects at the same time and thus the investigations overlap. He amplified his first researches on titanium with a number of papers on this element in the 1820's and one in 1844 on titanic acid. Starting in 1826 he examined the properties of phosphorus and its acids; the reports of this work (twenty-five papers or more) continued until 1849, and ran concurrently with research on ammonia compounds, since Rose thought that ammonia and phosphoretted hydrogen were "analogous" substances. In 1844 he began to investigate the properties of the mineral columbite. That led him to his discovery of niobium and to his classic papers, which continued until his death, on the properties of niobium and tantalum. Intermittently he presented the results of experiments on the compounds of chlorine and sulfur—especially the metallic and alkaline-metal-earth compounds. Information on these compounds was useful for analytical purposes, of course, but was also important because they were central to chemical theory.

In 1851 Rose began investigating the behavior of water in chemical compounds and its influence on chemical decomposition, particularly among the metal salts of weak acids. Berzelius had noted that the so-called law of neutrality did not always hold in reactions involving these compounds. Precipitates from solutions of earth and metal salts by alkaline carbonates, for example, sometimes resulted in basic hydrated salts rather than in carbonates corresponding to the original alkaline one. Rose was able to show the influence of temperature and concentration on reactions of this kind, and it reaffirmed his belief that Berthollet's insights into the influence of physical circumstances on chemical reactions were better founded than was generally supposed. Rose had found numerous instances of the law of mass, but he made no attempt to generalize or quantify it, as Guldberg and Waage did a short time later.

Rose also contributed about fifteen papers on organic chemistry. In 1839 he briefly joined the European debate on the theory of the subject with a paper on etherification but most of his work in this area was again analytical, although he seems to have been intrigued by the ways in which living things incorporate and use inorganic substances: hence his analyses of iron in blood, silica and iron in infusoria, and the series of papers (1848–1850) entitled "The Inorganic Components of Organic Bodies."

The *Handbook of Analytical Chemistry*, first published in 1829, was a modest work in one volume intended for beginners. Demand for it grew, however, and it went through several editions, becoming over the years more encyclopedic and comprehensive, until for a time it stood as the standard reference work on the subject. The seventh (and last) edition was prepared after Rose's death by one of his students, Rudolph Finkener. The theoretical framework and nomenclature in the *Analytical Chemistry* was that of Berzelius, from whose general dualistic atomic theory Rose, like Berzelius' other great students Mitscherlich and Wöhler, never strayed.

Rose the man is best conveyed in the following description by a French student, Adolphe Remelé, who said of the master's lectures:

> He looked upon the various substances that he was manipulating, as well as their reactions, under a thoroughly familial point of view: they were like so many children entrusted to his tutelage. Every time he explained simple, clear, well-defined phenomena, he assumed a jovial and smiling countenance; on the other hand, he almost got angry at certain *mischievous* [Remelé's italics] bodies, the properties of which did not obey ordinary laws and troubled general theoretical views; in his eyes, this was unruly behavior.

BIBLIOGRAPHY

I. ORIGINAL WORKS. Rose's works include *Dissertatio de titanio ejusque connubio cum oxygenio et sulphure* (Kiel, 1821), his doctoral thesis; *Handbuch der Analytischen Chemie* (Berlin, 1829); 5th ed. entitled *Ausführliches Handbuch der Analytischen Chemie*, 2 vols. (Brunswick, 1851); 6th ed. published in French as *Traité complet de chimie analytique*, 2 vols. (Paris, 1859–1861), with notes by Peligot; final ed., 2 vols. (Leipzig, 1867–1871); English trans. by J. Griffin, *A Manual of Analytical Chemistry* (London, 1831); and by A. Normandy from the 4th German ed., *A Practical Treatise of Chemical Analysis*, 2 vols. (London, 1848); *Gedächtnissrede auf Berzelius* (Berlin, 1852).

The list of Rose's papers in the Royal Society *Catalogue of Printed Papers* is accurate and, as far as is known, complete.

II. SECONDARY LITERATURE. See "Heinrich Rose," in *American Journal of Science*, 2nd ser., **38** (1864), 305–330, signed "D"; J. R. Partington, *A History of Chemistry*, IV, 185–190; Karl Rammelsberg, "Gedächtniss-rede auf Heinrich Rose," in *Abhandlungen der Königlichen Akademie der Wissenschaften zu Berlin* (1865), 1–31; and Adolphe Remelé, "Notice biographique sur le Professeur Henri Rose," in *Moniteur scientifique*, 2nd ser., **6** (1864), 385–389.

There is much about Rose in the published correspondence of Berzelius, Wöhler, Liebig, and Mitscherlich.

STUART PIERSON

ROSENBERG, HANS OSWALD (*b.* Berlin, Germany, 18 May 1879; *d.* Istanbul, Turkey, 26 July 1940), *astronomy*, *astrophysics*.

The son of a merchant, Rosenberg attended the Wilhelms-Gymnasium in Berlin. In 1899 and 1900 he studied astronomy and other sciences at the Universities of Munich, Berlin, and Strasbourg. He completed his doctoral dissertation at Strasbourg in 1904. The subject, which he chose at the suggestion of Gustav Müller of Potsdam, was an investigation of the variation of the long-period variable star χ Cygni.

At Strasbourg in 1904 Rosenberg observed the occultation of α Tauri. He also developed a method for determining the paths of meteors and applied it to the bright meteor of 21 March 1904. He then turned his attention exclusively to photometry, fitting a recording wedge-photometer of his own design to the Strasbourg observatory's six-inch refractor in order to measure the brightness of the comet 1903 IV. Using the same arrangement, Rosenberg determined the zero point of stepwise estimations of brightness measurements of the variable star Wx Cygni. This method was used by C. Wirtz in 1906 and 1907 to measure the brightness of Saturn. Rosenberg also developed a new design for the wedge photometer that lightened and simplified it.

To acquire more experience in the techniques of spectrophotometry, Rosenberg went in 1907 to Göttingen, where K. Schwarzschild placed the university observatory's ultraviolet prism camera at his disposal. First Rosenberg made spectrographs of the comet 1907 IV. From 1907 to 1909 he observed the sun and seventy bright stars up to the third magnitude. In the course of this program he recorded 378 spectra and measured them with the Hartmann microphotometer. A short report on this work appeared in 1913, and the complete results were published the following year. From these data Rosenberg derived the color

temperatures of many stars. In general his values were higher than those that J. Scheiner and J. G. Wilsing at Potsdam had obtained on the basis of visual observations. This discrepancy stimulated further study of the question.

At Göttingen, Rosenberg also investigated the relationship between brightness and spectral type among the Pleiades. With this study, which had been suggested by Schwarzschild, Rosenberg qualified as lecturer at the University of Tübingen in 1910. Rosenberg erected a private observatory on the Österberg, furnished with a 4.5-meter dome and a 130-mm. refractor fitted with a double camera. Among his first projects there was a spectrographic study of the nova Geminorum 2 of 1912.

From 1913 to 1921 (except during the war) Rosenberg concentrated on photoelectric photometry. In 1919, with P. Goetz, he made a photometric study of the lunar surface, using a potassium cell. In connection with this work he also developed a new type of polarization photometer. Other studies dealt with photographic star disks and with the brightness of the atmosphere in the neighborhood of brighter stars (Sirius, 1919); with the "influence of focusing on photographically effective wavelengths"; and with "the establishment of a normal sequence for the determination of effective wavelengths" (in collaboration with O. Bergstrand).

In 1921 Rosenberg began to study ways of increasing the strength of photocells by means of electron tubes. He also investigated previously unknown forms of photoelectric fatigue in alkali-metal cells. In 1925 Rosenberg designed an electromicrophotometer for making photometric measurements of the smallest focal star disks. With this instrument not only the photocell but also the electrometer could be used as a zero-point indicator. While working on this problem Rosenberg also determined the general reliability of the recording photometer (1925).

On 1 April 1926 Rosenberg accepted a position at the University of Kiel; as part of the agreement he was named director of its astronomical observatory. He had his instruments brought from Tübingen and supervised a complete refitting of the observatory that, in effect, adapted it for astrophysical research. In addition, he devoted considerable time to teaching. Once again Rosenberg concentrated primarily on photometric and spectrophotometric studies, and he organized two expeditions to measure solar eclipses, to Lapland in 1917 and to Thailand in 1929. Appointed dean of the Philosophy Faculty in 1930, he continued to produce scientific monographs.

Forced to leave Kiel for political reasons in 1934, Rosenberg accepted a visiting professorship at the

University of Chicago. While working at the university's Yerkes Observatory, he developed a double filter that made possible simultaneous photographs and cleanly separated focal images and extrafocal images. Rosenberg continued to make photographic-photometric observations of variable stars until 1937. In 1938 he went to Istanbul as university professor and director of the observatory. He was given responsibility for erecting a new observatory, but he succumbed to heat stroke in 1940 and was unable to complete the project. The year before his death he had been elected to the International Astronomical Union.

BIBLIOGRAPHY

I. ORIGINAL WORKS. Rosenberg's writings include *Der veränderliche Stern* χ *Cygni* (Halle, 1906); "Photographische Untersuchung der Intensitätsverteilung in Sternspektren," in *Nova acta Leopoldina*, **101**, no. 2 (1914); "Strahlungseigenschaften der Sonne," in *Handbuch der Physik*, XIX (Berlin, 1928); "Lichtelektrische Photometrie," in *Handbuch der Astrophysik*, II, pt. 1 (Berlin, 1929), 380–430, and VII (Berlin, 1936), 84–89; and *Die Entwicklung des räumlichen Weltbildes der Astronomie*, Kieler Universitätsreden, no. 11 (1930), 3–27.

II. SECONDARY LITERATURE. See W. Gleissberg, "Hans Rosenberg," in *İstanbul üniversitesi fen fakültesi mecmuasi*, **5**, fasc. 1–2 (1940), 36–39, with portrait and bibliography, also in *İstanbul üniversitesi Orman fakültesi yayinlari*, **13** (1940); C. Schmidt-Schönbeck, "Hans Rosenberg," in *300 Jahre Physik und Astronomie an der Kieler Universität* (Kiel, 1965), 196–200, 260; and C. Schönbeck, "Physik und Astronomie," in K. Jordan, ed., *Geschichte der Mathematik, der Naturwissenschaften und der Landwirtschaftswissenschaften* (Neumünster, 1968), 59–93; and *Handbuch der Experimentalphysik*, XXVI, *Astrophysik*, B. Strömgren, ed. (Leipzig, 1937), 40; and J. G. Wilsing, "Über effektive Sterntemperaturen," in *Astronomische Nachrichten*, **204** (1917), 153–159.

DIEDRICH WATTENBERG

ROSENBERGER, JOHANN KARL FERDINAND

(*b.* Lobeda, Germany, 29 August 1845; *d.* Oberstdorf, Germany, 11 September 1899), *history of physics.*

Rosenberger had already begun work as an elementary school teacher when he belatedly decided to pursue mathematics and physics at the University of Jena. After completing his Ph.D. in 1870, he taught mathematics and natural history at a series of private schools in Hamburg while preparing for the *Staatsexamen*, which he passed in 1876. In 1877 he moved to the Musterschule in Frankfurt am Main, where he taught until his death from a stroke. (From

1893 he was professor of mathematics and physics.) In 1892 he was elected a member of the Leopold-Carolinische Akademie der deutschen Naturforscher.

Except for his first work, an algebraic study of the basic laws of arithmetic (1876), Rosenberger's scholarly career was devoted entirely to the history of the physical sciences. In the course of that career he produced two major studies and had embarked on a third when he died. The first study, a history of physics from Greek antiquity to 1880, appeared in three parts over the period 1882–1890. Meant to "present the historical development of physics so that one can easily discern both the momentary state of the science at any point in time and the tendency of its path of development" (*Die Geschichte der Physik*, pt. 1, v), the work focused thematically on the emergence and articulation of the modern scientific method.

For Rosenberger, that method consisted of "hypothetico-deductive" reasoning grounded in a delicate balance and interplay of experiment, mathematics, and philosophy of nature. Only in his own century had the full ideal been achieved. The philosophical and mathematical genius of the Greeks could not compensate for their lack of experimental data and procedures. During the Middle Ages, speculative philosophy so dominated scientific thought that it became "corrective of experience" when not avoiding experience altogether. If seventeenth-century scientists introduced the experimental method, they were for a time (1650–1690) so overwhelmed by it as to equate science with crude empiricism. Under the impact of Newton's *Principia*, the pendulum swung toward mathematics, until the electrical research of the eighteenth century revived the prestige of experimental physics while retaining its mathematical articulation.

Rosenberger argued, however, that if mathematics and experiment had found their point of methodological equilibrium, a more lasting victim of Newton's success and influence was philosophy, to which the scientists' aversion was only strengthened by the futility of early nineteenth-century *Naturphilosophie*. One aim of Rosenberger's *Geschichte* was to remind scientists of the role of philosophy in scientific thought and of the dangers in ignoring it. He stated:

> An independent philosophy of nature without experimental and mathematical basis is impossible as real science, history teaches us that; but a pure empiricism without philosophical schooling, without a general philosophical science to set goals, yields at best a conglomerate of knowledge or otherwise, if it cannot do without hypothesis, slips as easily into cloudland as does pure philosophy of nature [*Die Geschichte der Physik*, pt. 2, p. 219].

Growing interest in the creative scientific process, which Rosenberger discussed in his pamphlet of 1885, and in the confrontation of opposing scientific theories ostensibly grounded in the same "correct" method reinforced Rosenberger's attitude toward philosophy and lent new importance in his mind to the history of science as a discipline. In his work on Newton (1895), and especially on the clash between Newtonianism and (Leibnizian) Cartesianism, he shifted his focus from scientific method to the function of authority within the scientific community and sought the personal and cultural factors that give scientists their authority and ability to attract followers. Because Rosenberger could compare the authority of Newton's "right" (but, by 1895, beleaguered) system only to that of Aristotle's "wrong" system, he was led to view scientific "truth" in more relative terms than before.

> All our fundamental scientific concepts have only a certain temporal truth, which in time changes to falsehood (the earlier truth of which can nevertheless not be ignored). . . . It is, for example, certain that, for his time, Aristotle created a correct, comprehensive system of natural explanation, the temporal truth of which one cannot properly deny [*Newton*, 528].

But the once healthy authority and model of Aristotle degenerated into rigid scholasticism, as Newtonianism threatened to do in the eighteenth and early nineteenth centuries. It was not authority *per se* but, rather, rigid authority that posed a danger to the progress of science. For Rosenberger, as he set forth in his last article (1899), the antidote to such rigidity lay in a critical history of science that, by demonstrating the temporal truth of contemporary science, undercut those claims to final insight that supported modern scientific scholasticism.

In the introduction to part three of his *Geschichte*, Rosenberger concluded that "we stand [in 1887] not at the end of a period of physics, but in the midst of one . . ., the duration of which cannot yet be seen" (*Die Geschichte der Physik*, pt. 3, p. 12). He was referring to electromagnetic theory; and his last research on the history of electricity in the eighteenth and nineteenth centuries, published as a series of lectures (1898), aimed at elucidating an ongoing scientific issue through historical perspective.

BIBLIOGRAPHY

I. ORIGINAL WORKS. Rosenberger's books are *Die Buchstabenrechnung. Eine Entwicklung der Gesetze der Grundrechnung-Arten* (Jena, 1876); *Die Geschichte der Physik in Grundzügen mit synchronistischen Tabellen der Mathematik, der Chemie und beschreibenden Naturwissen-schaften sowie der allgemeinen Geschichte*, 3 pts.—pt. 1, *Altertum und Mittelalter* (Brunswick, 1882); pt. 2, *1600–1780* (Brunswick, 1884); pt. 3, *1780–1880* (Brunswick, 1887–1890); *Die Genesis wissenschaftlicher Entdeckungen und Erfindungen* (Brunswick, 1885); *Isaac Newton und seine physikalischen Principien* (Leipzig, 1895); and *Die moderne Entwicklung der elektrischen Principien* (Leipzig, 1898).

His articles include "Übergang von metaphysischen Anfangsgründen der Naturwissenschaften zur Physik," in *Berichte des Freien Deutschen Hochstiftes*, **2** (1886); "Zum Gedächtnis Otto v. Guericke," *ibid.*, **3** (1887), 110–131. "Über Irrlichter," *ibid.*, **5** (1889), 2–12; "Geschichtliche Entwicklung der Theorie der Gewitter," *ibid.*, **7** (1891), 10–27. "Orientierung des Menschen im Raume," *ibid.*, **8** (1892), 89–109; "Fortschreitende Entwicklung des Menschengeschlechtes," in *Wissenschaftlich-Philosophische Viertel-Jahres-Schriften* (1892). "Über die erste Entwicklung der Elektrisiermaschinen," in *Verhandlungen der 68. Versammlung deutscher Naturforscher und Ärzte*, pt. 2, 1 (Frankfurt am Main, 1896), 66 ff; "Die erste Entwicklung der Elektrisiermaschinen," in *Abhandlungen zur Geschichte der Mathematik*, **8** (1898), 69–88; "Die ersten Beobachtungen über elektrische Entladungen," *ibid.*, 89–112; and "Die Geschichte der exakten Wissenschaften und der Nutzen ihres Studiums," *ibid.*, **9** (1899), 361–381.

II. SECONDARY LITERATURE. On Rosenberger and his work, see Sigmund Günther, in *Bibliotheca mathematica*, 3rd ser., 1 (1900), 217–224.

MICHAEL S. MAHONEY

ROSENBERGER, OTTO AUGUST (*b.* Tukums, Latvia, 10 August 1800; *d.* Halle, Germany, 23 January 1890), *astronomy*.

Rosenberger was the son of a physician who moved with his family to Königsberg in 1811. He attended the Königsberg Gymnasium, then entered the university, where he studied mathematics and astronomy and formed the tie with Bessel that determined the direction of his professional life. While still a student he worked with Bessel in making observations and computations, and from 1821 on was regularly mentioned in the *Astronomische Beobachtungen auf der K. Universitätssternwarte zu Königsberg*, which Bessel edited. On 18 June 1821 Bessel wrote to H. C. Schumacher that the observations of the chief stars that he had made with the Cary circle had been "very accurately computed" and reduced to the year 1815 by his students Rosenberger and H. F. Scherk. Rosenberger and Scherk had also computed the elements of the Pons comet of 1818, and Rosenberger determined the parabolic elements of it; his results were published in the *Berliner astronomisches Jahrbuch* for 1824.

Bessel further entrusted Rosenberger with computing the elements and ephemerides of the comet of 1821; these findings appeared in the first issue of the *Astronomische Nachrichten* in the following year. "Rosenberger," Bessel wrote of this work, "has been able to represent all the observations—the European ones made before the perihelion as well as the American ones made after it—in the form of a parabola; and the errors do not exceed the uncertainties of the observations." Indeed, Bessel thought so highly of Rosenberger's promise that when his assistant Argelander left the Königsberg observatory in 1823, Bessel named Rosenberger to replace him.

Bessel first assigned Rosenberger to compute a number of occultations of the Pleiades so as to obtain more accurate values for the distances between the observatories at which the observations had been made. Rosenberger also participated in Bessel's zone observations (an error in this work became known only in 1861, when Eduard Schönfeld demonstrated that in an observation of zone 285, performed on 23 April 1825, Rosenberger had confused the asteroid Pallas with a fixed star). In 1824 Rosenberger computed a new catalogue, which appeared in the *Astronomische Beobachtungen* as "Verzeichnis der geraden Aufsteigungen der 36 Fundamentalsterne für 1825." This catalogue was based upon the observations that he had made with the Königsberg meridian circle between 1821 and 1824; he also took part in the observation of the comet of the latter year and of the "moon stars."

On 23 May 1826 Rosenberger was named, on Bessel's recommendation, to succeed J. F. Pfaff at the University of Halle with the positions of extraordinary professor of mathematics and director of the observatory. Having completed the requirements for the doctorate, which was awarded him by the University of Königsberg on 16 July of the same year, Rosenberger went to Halle to take up his new duties the following October. He found that the observatory there, which had been erected in 1790 for exclusively pedagogical purposes, was sadly underequipped, and offered him no opportunity to conduct systematic observations. He therefore turned to theoretical studies, and in 1827 published an examination of the results reported by an expedition sent to Lapland by the Paris Académie des Sciences in 1736–1737. The purpose of this venture, which was supervised by Maupertuis, was to measure the length of a degree of longitude; the resulting measurements had provoked considerable criticism, but Rosenberger, after carefully weighing the merits of the undertaking, concluded that the work was valid.

In 1830 a small meridian circle, a Fraunhofer telescope with a focal length of seventy-two inches, and two astronomical clocks were installed in the Halle observatory at Rosenberger's request. He was thus able to carry out occasional astronomical observations as part of his teaching. On 5 July 1831 he was promoted full professor.

In 1834, shortly before the expected reappearance of Halley's comet, Rosenberger, working part of the time with Olbers, made new computations of the comet's orbital elements and ephemerides. His results were in close accord with the later actual observations. In a series of separate publications, he investigated the elements of the comet during its appearances in 1682 and 1759, and gave an account of its perturbations during the intervening period. These were his last publications; he devoted the rest of his career to his teaching and administrative duties, and continued to lecture enthusiastically even after his retirement in 1879.

BIBLIOGRAPHY

A bibliography of Rosenberger's writings is in *General-Register der Astronomischen Nachrichten, Bände 1 bis 40* (Kiel, 1936), 94. Thirty-seven letters from Rosenberger to Bessel (1821–1835) and seventeen letters from Rosenberg to H. C. Schumacher (1827–1835) are in the archives of the Akademie der Wissenschaften der DDR, Berlin.

Biographical material is in the obituary by A. Wangerin, "Otto August Rosenberger," in *Astronomische Nachrichten*, **123** (1890), 415–416.

DIEDRICH WATTENBERG

ROSENBLUETH, ARTURO (*b.* Ciudad Guerrero, Chihuahua, Mexico, 2 October 1900; *d.* Mexico City, Mexico, 20 September 1970), *neurophysiology*.

Rosenblueth studied at the Franco-English College in Mexico City and at the Medical School of the University of Mexico, then continued his medical studies at Berlin, and then at Paris, where he obtained his medical degree. In 1927 he returned to the Medical School of the University of Mexico and devoted himself to physiological research and teaching. In 1930 he was offered a research fellowship under Walter B. Cannon at Harvard, which led to a long and productive collaboration. Cannon had shown that the "constancy of the internal medium," discovered by Claude Bernard, reflected a general regulatory function or "homeostasis," which presumably was operated essentially by the autonomic nervous system. He and Rosenblueth demonstrated this assumption by extirpation of both sympathetic chains of a cat without impairing the animal's survival. Through many of these delicate operations they elucidated the details of the sympathetic regulatory action, which results from successive nerve impulses that conduct a quantum of chemical mediator into the terminal organ. These minimal doses summate and elicit various effects: vasomotor action, visceral muscle contraction, and hormonal secretion. These experiments decisively confirmed the theory of "chemical mediation," then frequently questioned.

At the same time Rosenblueth interested Norbert Wiener in the functional analysis of the nervous system. Their first paper (1943) was the starting point of the work that led Wiener to edify the new science of cybernetics. Rosenblueth was elected assistant professor at the Harvard Medical School in 1934. In 1944 he returned to Mexico City as director of research at the new Institute of Cardiology. The collaboration with Wiener had been maintained for nearly ten years. The neurophysiologist brought his knowledge of the coding information carried by the nerve impulse, and the mathematician demonstrated that the information theory adequately describes the coding in every detail. Rosenblueth and Wiener published papers on the mathematical expression of the conduction of impulses in a network of nerve cells, a statistical analysis of synaptic transmissions, psychology and cybernetics, and even the aesthetics of science.

In 1961 Rosenblueth founded, at the National Polytechnical Institute of Mexico, the center for advanced studies. There he successfully promoted interdisciplinary and international research, uniting the Anglo-Saxon and Latin civilizations of America. He was also a philosopher of science. In his last work, *Mind and Brain* (1970), he showed how all our knowledge of the material universe is based upon coded nerve impulses. All other features ascribed to the universe are essentially mental. Thus Rosenblueth gave a new and clear expression of the classical dualism between mind and brain.

BIBLIOGRAPHY

Rosenblueth wrote three monographs, *Autonomic Neuro-Effector Systems* (New York, 1937), written with W. B. Cannon; *The Supersensitivity of Denervated Structures. A Law of Denervation* (New York, 1949), written with W. B. Cannon; and *The Transmission of Nerve Impulses at Neuroeffector Junctions and Peripheral Synapses* (Cambridge, Mass., 1950).

Among his more important papers are "The Electric Responses of the Tail Pilomotors and Nictitating Membrane of the Cat," in *American Journal of Physiology*, **137** (1942), 263–279, written with D. D. Bond and W. B.

Cannon; "The Control of Clonic Responses of the Cerebral Cortex," *ibid.*, 681–694; "The Action of Electrical Stimuli on the Turtle's Ventricle," *ibid.*, **138** (1942), 50–64, written with W. Daughaday and D. D. Bond; "The Influence of Interelectrodal Distance in Electrical Stimulation of Nerve and Striated and Ventricular Muscle," *ibid.*, **138** (1943), 583–586, written with G. H. Acheson; "The Centrifugal Course of Wallerian Degeneration," *ibid.*, **139** (1943), 247–254; "Behavior, Purpose and Teleology," in *Philosophy of Science*, **10** (1943), 18–24, written with N. Wiener and J. Bigelow; "The Interaction of Myelated Fibers in Mammalian Nerve Trunks," in *American Journal of Physiology*, **140** (1944), 656–670; "Recruitment of Mammalian Nerve Fibers," *ibid.*, **141** (1944), 196–204; "The Role of Models in Science," in *Philosophy of Science*, **12** (1945), 316–321, written with N. Wiener; "The Mathematical Formulation of the Problem of Conduction of Impulses in a Network of Connected Excitable Elements, Specifically in Cardiac Muscle," in *Archivos del Instituto de cardiologia, Mexico*, **16** (1946), 205–265, written with N. Wiener; "An Account of the Spike Potential of Axons," in *Journal of Cellular and Comparative Physiology*, **32** (1948), 275–318, written with N. Wiener, W. Pitts, and J. Garcia Ramos: "The Functional Refractory Period of Axons," *ibid.*, **33** (1949), 405–440, written with J. Alanis and J. Mandoki; "A Statistical Analysis of Synaptic Excitation," *ibid.*, **34** (1949), 173–206, written with N. Wiener, W. Pitts, and J. Garcia Ramos; and "Purposeful and Non-Purposeful Behavior," in *Philosophy of Science*, **17** (1950), 318–326, written with N. Wiener; "The Local Responses of Axons," in *Ergebnisse der Physiologie*, **47** (1952), 23–69.

Later papers are "Functional Refractory Period of Cardiac Tissues," in *American Journal of Physiology*, **194** (1958), 171–183; "Two Processes for Auriculo-Ventricular and Ventriculo-Auricular Propagation of Impulses in the Heart," *ibid.*, **194** (1958), 495–498; "Ventricular Echoes," *ibid.*, **195** (1958), 53–60; "Some Properties of the Mammalian Ventricular Muscle," in *Archives internationales de physiologie et de biochimie*, **67** (1959), 276–293, written with J. Alanis and R. Rubio; "The Adaptation of the Ventricular Muscle to Different Circulatory Conditions," *ibid.*, **67** (1959), 358–373, written with J. Alanis, E. Lopez, and R. Rubio; "The Two Staircase Phenomena," *ibid.*, **67** (1959), 374–383; "Tetanic Summation in Isotonic and Isometric Responses," *ibid.*, **68** (1960), 165–180, written with R. Rubio; "The Relations Between Isometric and Isotonic Contraction," *ibid.*, **68** (1960), 181–189, written with R. Rubio; "The Accessory Motor Innervation of the Diaphragm," *ibid.*, **69** (1961), 19–25, written with J. Alanis and G. Pilar; "Relations Between Coronary Flow and Work of the Heart," in *American Journal of Physiology*, **200** (1962), 243–246, written with J. Alanis, R. Rubio, and G. Pilar; "Some Phenomena Usually Associated with Spreading Depression," in *Acta physiologica latino-americana*, **16** (1966), 141–179; "Slow Potential Changes in the Spinal Cord," *ibid.*, **16** (1966), 324–334, written with J. Garcia Ramos; "The Relations Between the Impedance and the emf Changes in the Cerebral Cortex," *ibid.*, **17** (1967), 76–87, written with J. Garcia Ramos and L. F. Nims; and "Slow Potential and Impedance Changes in the Medulla of the Cat," *ibid.*, **16** (1966), 212–219.

Rosenblueth also wrote a philosophical monograph, *Mind and Brain. A Philosophy of Science* (Cambridge, Mass., 1970).

A. M. MONNIER

ROSENBUSCH, HARRY (KARL HEINRICH FERDINAND) (*b*. Einbeck, Germany, 24 June 1836; *d*. Heidelberg, Germany, 20 January 1914), *geology*.

Rosenbusch was the son of a schoolteacher. His father was chronically ill, and died when Rosenbusch was young; the family was left in difficult circumstances, and it was only with considerable effort that Rosenbusch's mother managed to send the boy to secondary school. Rosenbusch there made a good record in languages, but a poor one in mathematics and physics; he therefore began to study classical languages when he entered the University of Göttingen. Financial difficulties caused him to leave the university soon thereafter, and he went to Brazil as private tutor to a rich family. Five years later he returned to Germany, and entered the University of Heidelberg, where a lecture by Robert Bunsen, together with observations he had made in Brazil, stimulated his interest in chemistry and geology.

In 1869, when he was nearly thirty-three years old, Rosenbusch graduated from the University of Freiburg with a thesis on the nephelinite of the Katzenbuckel, a mountain near Heidelberg which is still of great interest to petrographers. He was appointed *Privatdozent* in the following year. In 1873 Rosenbusch was called to the recently reorganized University of Strasbourg as professor extraordinarius of mineralogy and petrography. In 1878 he returned to Heidelberg as professor of mineralogy, a post that he held until his retirement in 1908.

Rosenbusch was greatly influential in establishing petrography as a true geological and historical science. He was particularly effective in advocating the use of the polarizing microscope in investigating rocks (although he neither invented this technique—thin sections had been introduced in England by H. C. Sorby in 1860 and the method had then been brought to Germany by F. Zirkel and H. Vogelsang—nor created essentially new procedures). His students came from all over the world, and a number of his geological ideas were almost universally accepted and propagated.

Rosenbusch published the first edition of his monumental textbook, of which the first volume was entitled *Mikroskopische Physiographie der petro-graphisch wichtigsten Mineralien* and the second *Mikroskopische Physiographie der massigen Gesteine*, while he was still in Strasbourg. The two books went through four enlarged editions, and became the standard works on igneous rocks. In later versions, the first volume was put into a systematic mathematical and physical form by Rosenbusch's student E. A. Wülfing, while the second volume was comprehensive and descriptive in nature. The subsequent editions of each work also incorporated Rosenbusch's own recent results in a number of areas; he published only a few of his discoveries separately, as for example sagvandite, monchiquite, and eukolite and a few discussions of discrete geological ideas.

Among Rosenbusch's individually published works, the papers that he wrote while in Strasbourg concerning the gradual contact metamorphism of the Steiger Schiefer, near Barr-Andlau, are especially significant. In these papers he carefully described the alteration of the mineral content of the slates of this region, and went on to prove that no chemical alteration except the loss of water took place in them. His data were so convincing that for some fifty years his ideas of contact-metamorphism without chemical change were generalized and widely accepted, although in fact the case he described was that of a special type of high-intensity intrusion into a particularly inert rock.

In 1878 Rosenbusch recognized the relationship between melanocratic and leucocratic dyke-rocks (Ganggestein) and the plutonic parent body. Although his conclusions and interpretations were sometimes misleading, as in the instance of the lamprophyres, his descriptions and observations were of high quality. The opposition to Rosenbusch's "Kerntheorie" (nucleus theory)—whereby $NaAlSi_2$ was the characteristic "nucleus" of rocks of the Atlantic series, as was $CaAl_2Si_2$ for rocks of the Pacific series—was likewise directed against the form of the interpretation, rather than on the assembled data.

Rosenbusch's most important contribution, however, lay in his establishment of a fundamentally genetic and mineralogical (as opposed to chemical) classification of rocks. In his textbook, he first stressed the importance of the mineralogical characteristics of rocks in classification then, by the second edition, developed a system in which igneous rocks are classified by geological position, texture, and finally their mineralogical and chemical composition.

BIBLIOGRAPHY

Rosenbusch's important publications include *Der Nephelinit von Katzenbuckel* (Freiburg, 1869); Über einige vulkanische Gesteine von Java," in *Berichte der*

Naturforschenden Gesellschaft zu Freiburg im Breisgau, **6** (1872), 36 p.; *Mikroskopische Physiographie der petrographisch wichtigen Mineralien* (Stuttgart, 1873); "Über die Kontaktzone von Barr-Andlau," in *Neues Jahrbuch für Mineralogie* (1875); "Ein neues Mikroskop für mineralogische und petrographische Untersuchungen," *ibid.* (1876); "Die Steiger Schiefer und ihre Kontaktzone an den Granititen von Barr-Andlau und Hohwald," in *Abhandlungen zur geologischen Spezialkarte von Elsass-Lothringen*, **1**, XII-XIX (1877), 79–393; "Die Gesteinsarten von Ekersund," in *Nyt magazin for naturvidenskaberne*, **27** (1883), 8p.; "Zur Auffassung der chemischen Natur des Grundgebirges," in *Mineralogische und petrographische Mitteilungen*, **12** (1891), 49–61; "Über Struktur und Klassifikation der Eruptivgesteine," *ibid.*, 351–396; and "Studien im Gneisgebirge des Schwarzwaldes: Einleitendes: I. Kohlenstoffführende Gneisgesteine des Schwarzwaldes," in *Mitteilungen aus der Grossherzoglichen Bad. geologischen Landesanstalt*, IV (1899), 9–48; and "II. Die Kalksilikatfelse im Rench- und Kinzigitgneis. 1. Die Paraaugitgneise, 2. die Paraamphibolgneise," *ibid.* (1901), 369.

PAUL RAMDOHR

ROSENHAIN, JOHANN GEORG (*b.* Königsberg, Prussia, 10 June 1816; *d.* Königsberg, 14 May 1887), *mathematics*.

Rosenhain studied at the University of Königsberg, where he earned the Ph.D. In 1844 he qualified as lecturer at the University of Breslau and remained there as a *Privatdozent* until 1848. His participation in the revolutionary activities of 1848 deprived him of any chance to further his career at Breslau. He therefore qualified as lecturer again in 1851, this time at the University of Vienna. In 1857 he returned to Königsberg, where he was an associate professor until a year before his death.

While studying at Königsberg, Rosenhain was especially close to Jacobi; and while still a student in the 1830's he edited some of Jacobi's lectures. His own scientific activity was mainly inspired by Jacobi, who had enriched the theory of elliptic functions with many new concepts and had formulated, on the basis of Abel's theorem, the inverse problem, named for him, for an Abelian integral on a curve of the arbitrary genus p. The next step was to solve this problem for $p = 2$.

In 1846 the Paris Academy had offered a prize for the solution of that problem, and Rosenhain won it in 1851 for his work entitled "Sur les fonctions de deux variables à quatre périodes, qui sont les inverses des intégrales ultra-elliptiques de la première classe." Göpel had solved this problem at almost the same time, but he did not enter the competition. Rosenhain's

work followed Jacobi even more closely than did Göpel's.

In his unpublished dissertation Rosenhain had already treated triple periodic functions in two variables. The solution of the inverse problem for $p = 2$ presented him with considerable difficulties, as can be seen in his communications to Jacobi published in Crelle's *Journal*. It was not until chapter 3 of his prize essay that he introduced, in the same manner as Göpel, the sixteen θ functions in two variables and examined in detail their periodic properties and the algebraic relations.

Most important, Rosenhain demonstrated (in modern terminology) that the squares of the quotients of these sixteen θ functions can be conceived of as functions of the product surface of a hyperelliptic curve of $p = 2$ with itself. Starting from this point and employing the previously derived addition theorem of the θ quotients, Rosenhain succeeded in demonstrating more simply than Göpel that these quotients solve the inverse problem for $p = 2$. Rosenhain never fulfilled the expectations held for him in his younger years, and published nothing after his prize essay.

BIBLIOGRAPHY

Extracts from most of Rosenhain's letters to Jacobi concerning hyperelliptic transcendentals are in *Journal für die reine und angewandte Mathematik*, **40** (1850), 319–360. Rosenhain's prizewinning work is "Sur les fonctions de deux variables à quatre périodes, qui sont les inverses des intégrales ultra-elliptiques de la première classe," in *Mémoires présentés par divers savants*, 2nd ser., **11** (1851), 361–468; also translated into German as *Abhandlung über die Functionen zweier Variabler mit vier Perioden*, Ostwalds Klassiker der Exakten Wissenschaften, no. 65 (Leipzig, 1895).

WERNER BURAU

ROSENHAIN, WALTER (*b.* Berlin, Germany, 24 August 1875; *d.* Kingston Hill, Surrey, England, 17 March 1934), *metallurgy*.

Rosenhain was the son of Moritz Rosenhain, a businessman. His mother was the daughter of a rabbi. The family immigrated to Australia in 1880 so that the son would not have to serve in the Prussian military. After attending Wesley College, Melbourne, and then Queen's College, Melbourne University (1892–1897), where he received the bachelor of civil engineering, Rosenhain went on to Cambridge University as 1851 Exhibition scholar (1897–1900). He received the D.Sc. from Melbourne in 1909. A fellow of the Royal Society (1913), Rosenhain was also a founder-member, president, and fellow of the Institute of Metals and Carnegie and Bessemer medalist of the Iron and Steel Institute. At Cambridge, Rosenhain studied under James Alfred Ewing, professor of mechanics. Initially he was assigned to a project concerning the dynamics of steam jets. This problem proved uncongenial, and in November 1898 Rosenhain abandoned it to undertake an investigation, suggested by Ewing, that was to shape his career.

Rosenhain applied the micrographic technique pioneered by Henry Sorby to a study of metal strips that had been polished and then deformed. In this way he discovered slip lines, which indicate that plastic deformation has taken place by the sliding of crystalline lamellae over each other. This discovery, which formed the subject of the Bakerian lecture in 1899, was of major importance for two reasons: it confirmed, thirteen years before the advent of X-ray diffraction, that metals consist of crystalline grains (an opinion still contested at that time); and it showed how plastic deformation, the most useful distinguishing mark of the metallic state, is possible without disruption of crystalline order. This research led Rosenhain to specialize in metallurgy and created his enduring reliance on microscopic techniques.

In 1900 Rosenhain left Cambridge. No opening in metallurgical industry was available, and he accepted a post with Chance Brothers Ltd., a Smethwick glass-manufacturing firm. He described himself as "a tame scientist kept on the premises."[1] His engineering knowledge also found extensive application, however, and he retained a permanent interest in the technology of glass. His experiences led him to the study of refractory crucible materials of high purity. While continuing research in a private metallurgical laboratory Rosenhain first adopted George Beilby's hypothesis that a thin metallic layer between slip lamellae is reduced to the amorphous state; this layer was taken to be very hard, like glass, thus explaining the mystery of work-hardening during plastic deformation. (Recovery of work-hardened metal was attributed to crystallization of the amorphous layer.)

Rosenhain soon extrapolated this notion to form the hypothesis that the boundaries between metal grains consist of thin (liquidlike) amorphous layers that, by analogy with work-hardened metal, he believed to be very hard at low temperatures but so soft at high temperatures as to favor intergranular rupture. Later he came to believe that the hardening of steel is due to the presence in quenched steel of amorphous layers that (by analogy with work-hardened metal) he took to be very hard at low temperatures. This complex

of ideas, the "amorphous hypothesis," became the scientific mainspring of Rosenhain's standpoint and he devoted his exceptional powers as a controversialist to its defense. Modern techniques of X-ray diffraction and electron microscopy, used to establish the nature of hardened steel and the role of dislocations in plastic deformation, have proved the amorphous hypothesis wrong in all three of its aspects. But Rosenhain's impassioned advocacy of the hypothesis did lead him to undertake valuable experimental work, particularly on plastic deformation.

A particularly informative account of the "β-iron controversy," concerning the basic mechanism of the hardening of steel and including an account of Rosenhain's early role in it, was published by Morris Cohen and James M. Harris.[2] (Before conceiving the amorphous interpretation, Rosenhain was a firm defender of the fallacious β-iron theory of hardening.)

In 1906 Rosenhain was offered the post of superintendent of the recently established department of metallurgy and metallurgical chemistry at the National Physical Laboratory, Teddington. Rosenhain accepted the post, considering it a stepping-stone[3] to better things, but he became absorbed by the work and remained there for twenty-five years. Under Rosenhain it became one of the world's largest and most renowned metallurgical laboratories.

When Rosenhain arrived at Teddington, metallurgy (as the department's quondam name illustrates) was virtually a branch of chemistry. Rosenhain steered metallurgy in the direction of physics, and through his influence the new science of physical metallurgy emerged. His 1914 book *Introduction to Physical Metallurgy* was widely influential. This reorientation of the aims and methods of metallurgy was essential to rapid progress in the understanding of the structure and behavior of metals and alloys. Rosenhain, trained as an engineer, maintained close connections with the metallurgical industry; for him, the later separation of advanced metallurgical science and technology would have been unthinkable. His industrial outlook and connections enabled him to leave Teddington in 1931, before compulsory retirement from the civil service, to become a free-lance metallurgical consultant in London.

In 1923 Rosenhain toured American industrial and academic metallurgical installations, and his series of eleven articles in *Engineer* provides an expert impression of American metallurgy at that time.[4]

At Teddington, Rosenhain participated especially in the development of instruments for physical metallurgical research, such as his gradient furnace and the plotting thermograph for registering thermal anomalies during the cooling of alloys.[5] He also improved the metallurgical microscope and invented a recording dilatometer. He directed a long series of researches on the constitution of steels and on the constitution and age-hardening of aluminum alloys, which included the important aluminum-nickel-magnesium alloy known as "Y alloy." Rosenhain established new standards of accuracy, paying particular attention to the purity of the constituent metals and—equally important—of the refractories used for making the melting crucibles. He also studied copper alloys and dental amalgams (the first instance of subzero metallography). He was one of the first to study the regularities governing the properties of series of solid solutions. Rosenhain influenced metallurgy both as an experimentalist and as a catalyst for the work of others. During a period when the conceptual basis of quantitative treatment of problems in physical metallurgy was not yet available, he contributed little of permanent importance as a theorist, except as a forceful controversialist who spurred others to fruitful attempts to prove him wrong.

NOTES

1. The source of this information is an unpublished biographical MS prepared by Mrs. Nancy Kirsner of Melbourne, Rosenhain's daughter, who kindly placed it at the writer's disposal, together with supplementary comments by the late Daniel Hanson, Rosenhain's senior collaborator in his second decade at the National Physical Laboratory.
2. *Sorby Centennial Symposium*, 209–233.
3. Kirsner, *op. cit.*
4. Rosenhain for some years edited and frequently contributed anonymously to a special supplement, entitled *Metallurgist*, of the journal *Engineer*.
5. Rosenhain, "Some Methods of Research in Physical Metallurgy," in *Journal of the Institute of Metals* (1929).

BIBLIOGRAPHY

I. ORIGINAL WORKS. A complete bibliography of books, papers, and occasional articles by Rosenhain is in John Haughton's obituary of him (see below), 28–32. His more important or characteristic publications include "The Crystalline Structure of Metals," in *Philosophical Transactions of the Royal Society*, **193**A (1900), 353–375, written with J. A. Ewing; *Glass Manufacture* (London, 1908); "The Crystalline Structure of Iron at High Temperatures," in *Proceedings of the Royal Society*, **83**A (1909), 200–209, written with J. C. Humfrey; "The Fatigue and Crystallization of Metals," in *Journal of the West of Scotland Iron and Steel Institute*, **16** (1909), 129–146; "Metallographic Investigations of Alloys," in *Journal of the Institute of Metals*, **1** (1909), 200–226; "Ninth Report to the Alloys Research Committee on the Properties of Some Alloys

of Copper, Aluminium and Manganese," in *Proceedings of the Institution of Mechanical Engineers* (1910), 119–292, written with F. C. Lantsberry; "The Constitution of the Alloys of Aluminium and Zinc," in *Philosophical Transactions of the Royal Society*, **211**A (1911), 315–343, written with S. L. Archbutt; "The Intercrystalline Cohesion of Metals," in *Journal of the Institute of Metals*, **10** (1913), 119–139, written with D. Ewen; "The Tenacity, Deformation, and Fracture of Soft Steel at High Temperatures," in *Journal of the Iron and Steel Institute*, **87** (1913), 219–271, written with J. C. Humfrey; *An Introduction to Physical Metallurgy* (London, 1914); and "Some Appliances for Metallographic Research," in *Journal of the Institute of Metals*, **13** (1915), 160–183.

Later works are "Aluminium and Its Alloys," in *Journal of the Royal Society of Arts*, **68** (1920), 791–798, 805–817, 819–827; *Eleventh Report to the Alloys Research Committee on Some Alloys of Aluminium* (London, 1921), summarized in *Proceedings of the Institution of Mechanical Engineers* (1921), 699–725, written with S. L. Archbutt and D. Hanson; "The Hardness of Solid Solutions," in *Proceedings of the Royal Society*, **99**A (1921), 196–202; "The Inner Structure of Alloys," in *Journal of the Institute of Metals*, **30** (1923), 3–26; "Science and Industry in America," in *Engineer*, **136** (1923), 270–271, 298–299, 312, 330–331, 358–359, 384–385, 412–413, 440–441, 468–469, 494–496, 522–524; "Solid Solutions," in *Transactions of the American Institute of Mining and Metallurgical Engineers*, **69** (1923), 1003–1034; "The Present Position of the Amorphous Theory," in *Metallurgist*, **1** (1925), 2–4; "The Metallography of Solid Mercury and Amalgams," in *Proceedings of the Royal Society*, **113**A (1926), 1–6, written with A. J. Murphy; "Presidential Address," in *Journal of the Institute of Metals*, **39** (1928), 27–51; "Some Methods of Research in Physical Metallurgy," *ibid.*, **42** (1929), 31–68; "The Development of Materials for Aircraft Purposes," in *Journal of the Royal Aeronautical Society*, **34** (1930), 631–642; and "Physik und Metallkunde," in *Zeitschrift für Metallkunde*, **22** (1930), 73–78.

II. SECONDARY LITERATURE. See Cecil Desch, in *Obituary Notices of Fellows of the Royal Society of London*, **1** (1932–1935), 353–359; Daniel Hanson, in *Journal of the Institute of Metals*, **54** (1934), 313–315; and John Haughton, "The Work of Walter Rosenhain," *ibid.*, **55** (1934), 17–32, with full bibliography. See also C. S. Smith, ed., *The Sorby Centennial Symposium on the History of Metallurgy* (New York, 1965), 221–222, 317–320.

R. W. CAHN

ROSENHEIM, ARTHUR (*b.* New York, N.Y., 17 August 1865; *d.* Berlin, Germany, 21 March 1942), *chemistry.*

Rosenheim was the son of Wilhelm Rosenheim and Maria Hallgarten. When he was eight years old the family returned to Germany and settled in Berlin.

Later he claimed German citizenship. He was educated at the Wilhelm-Gymnasium in Berlin, from which he graduated in the autumn of 1884. He then entered the philosophy faculty of the University of Berlin. In 1886 he went for one semester to the University of Heidelberg and then, from 1887 to 1888, wrote his Ph.D. dissertation, "Über Vanadinwolframsäure," at Berlin. His teachers at Heidelberg were Bunsen, Königsberger, and Rosenbusch; at Berlin they included Gabriel, Geiger, Helmholtz, Hofmann, Kundt, Liebermann, Rammelsberg, and Tiemann. Rosenheim received the Ph.D. *cum laude* on 24 November 1888. The following year he went to Munich but in 1891 returned to Berlin as assistant at the II. Chemical University Laboratory. In March 1896 Rosenheim qualified as *Dozent* with his thesis "Über die Einwirkung anorganischer Metallsäuren auf organische Säuren."

From 1891 Rosenheim worked at the Wissenschaftlich Chemisches Laboratorium, which he shared with Friedheim and later with Richard Meyer. In 1903 he was named professor of chemistry at Berlin, where he lectured until 1920 and then, for three years, supervised practical exercises in inorganic chemistry. Later he taught graduate courses for advanced students. A Jew, Rosenheim's teaching license was revoked by the Nazi minister in 1933. But he remained in Berlin until his death, possibly of apoplexy. He was buried in the Jewish cemetery in Berlin-Weissensee.

Except for a review article on heteropoly acids in Richard Abegg's *Handbuch* (1905), all of Rosenheim's chemical contributions appeared in journal articles published between 1870 and 1934. His findings were based on careful analyses and often rectified the errors of earlier chemists. Landolt and Emil Fischer remarked that his investigations chiefly repeated and extended the research of others and were scientifically reliable.

Rosenheim's papers are not easily grouped by distinct periods: he usually worked on several projects simultaneously and often returned to previous ones years later. Certain of his publications contain carefully specified directions for conducting analyses, and these constitute penetrating explanations of the chemical processes involved. Noteworthy are his papers on methods of determining vanadic acid (1890–1892); the separation of nickel and cobalt, of nickel and zinc, and of zinc and cobalt from complex thiocyanates; the analyses of hypophosphoric, phosphorous, hypophosphorous, and phosphoric acids (1909); on the gravimetric determination of tellurium and the alkalimetric determination of telluric acid (1911); and the determination of thorium by means of sodium phosphate (1912). Using his own preparations and analyses, Rosenheim also explained the reactions between ferric ions and thiocyanate and corrected the erroneous views of earlier researchers (1901).

Rosenheim's thoroughness is evident in a series of studies on the molecular weight of hypophosphoric acid. On the basis of conductivity measurements and ebullioscopic molar mass determinations (1906, 1908, 1910), he assigned to this acid the incorrect formula H_2PO_3. He then compared his results with those of other chemists, at first misinterpreting their findings. Later, in 1928, using kinetic measurements, he obtained the correct molecular weight, $H_4P_2O_6$. In the same paper, he also formulated the correct structure with P—P bonding.

Generally Rosenheim's arguments are supported by exhaustive experimental evidence and are so thoroughly conceived that they need only minor correction in view of current research. Although the application of physical methods has yielded considerable additional data about complex compounds, for example, heteropoly acids, Rosenheim's achievements are in no way diminished.

Rosenheim's publications dealt primarily with six areas of research:

1. The complex chemistry of nitrogenous, phosphorous, arsenical, and sulfurous ligands, including amines, cyanides, phosphites, hypophosphites, hypophosphates, phosphates, diphosphates, arsenates, diarsenates, thiocyanates, sulfates, and thiosulfates.

2. The complex compounds formed by halides and organic compounds, especially acids and ketones. In this connection he also investigated thio acids.

3. The salts of elements of adjacent groups and the coordination complexes to which they give rise. Notable are his studies on the halides of the elements of adjacent lower valence groups—his observations match the cluster structure that has since been established for these substances.

4. Internal complex borates and beryllates, as well as bismuth, antimony, and arsenic compounds.

5. Peroxy compounds of elements from adjacent groups.

6. Isopoly and heteropoly acids.

In all of these studies Rosenheim used Werner's coordination theory and arrived at conclusions that are still valid. Current refinements concern primarily structural questions.

With Miolati, Rosenheim laid the foundations for later research on isopoly and heteropoly acids. His last publication (1934) was a comprehensive survey of contemporary knowledge of heteropoly compounds, and it offered a critical evaluation of the most important experimental findings. Rosenheim also wrote many works on the nomenclature of inorganic compounds and on bibliographic matters. With J.

Koppel, he prepared the cumulative index of the first one hundred volumes of *Zeitschrift für anorganische Chemie*. He also wrote obituary notices on several scientists, including Carl Friedheim (1911), Leopold Spiegel (1927), and Fritz Raschig (1929). In 1925 a special volume of the *Zeitschrift für anorganische und allgemeine Chemie* (no. 147) was a *Festschrift* for Rosenheim on the occasion of his sixtieth birthday.

BIBLIOGRAPHY

Rosenheim's major works are "8 Seiten über Elektroreduktion der Wolframsäure nach experimentellen Untersuchungen von R. Bernhardi-Grisson," in *Proceedings of the International Congress of Applied Chemistry* (1909); and "88 Seiten über Heteropolysäuren," in R. Abegg and F. Auerbach, eds., *Handbuch der anorganischen Chemie*, IV (Leipzig, 1921). An extensive bibliography is given in Poggendorff, IV, 1271; V, 1066–1067; VI, 2220–2221; and VIIa, 813.

On Rosenheim and his work, see Kürschner's *Deutscher Gelehrtenkalender* (1937).

L. KOLDITZ

ROSENHOF, AUGUST JOHANN ROESEL VON.
See **Roesel von Rosenhof, August Johann.**

ROSS, FRANK ELMORE (*b*. San Francisco, California, 2 April 1874; *d*. Altadena, California, 21 September 1960), *astronomy*.

Ross was one of the most versatile astronomers of the twentieth century. He obtained a B.S. degree from the University of California at Berkeley in 1896 and for several years thereafter gained valuable experience through one-year appointments as teacher (mathematics and physics at Mt. Tamalpais Military Academy), on fellowships (a year each at Berkeley and Lick Observatory), and as an assistant professor of mathematics at the University of Nevada. Ross then returned to Berkeley, where he received the Ph.D. in 1901. For the next two years he computed perturbations of Watson asteroids, then spent a year as an assistant in the *Nautical Almanac Office*, where he attracted the attention of Simon Newcomb.

For two years Ross was research assistant at the Carnegie Institution, working on planetary and lunar problems under Newcomb. At Newcomb's suggestion, he computed a definitive orbit of Phoebe, the ninth satellite of Saturn. This was his first major independent

publication in astronomy. The differential coordinates of Phoebe given in the *American Ephemeris* beginning with 1909 are derived from the elements and tables published by Ross.

In 1905 Ross became director of the International Latitude Observatory at Gaithersburg, Maryland, where he took part in the observations and reductions for the precise determination of latitude required to study the motion of the earth's axis of rotation. In 1909 the U.S. Coast and Geodetic Survey received a grant from the International Geodetic Association for the construction and operation of a photographic instrument to be used in determining the latitude variation. The instrument, known as the photographic zenith tube, or PZT, was designed by Ross.

The optical system of the PZT is patterned after the visual instrument known as the zenith tube, which had been designed by Airy and installed at Greenwich in 1851. Airy's instrument was no longer in use, but Ross recognized the merits of its basic principle and was successful in developing an improved telescope of extremely high precision. He modified the optical system by locating the second nodal point of the lens in the focal plane, and he constructed it to operate as a photographic instrument. The PZT then had two major improvements over Airy's instrument: the first made the PZT insensitive to tilt of the lens; the second was effective in smoothing out short-period oscillations caused by refraction and also made the instrument impersonal.

The completed instrument was installed at Gaithersburg in June 1911, and by October 1914 Ross had photographed 6,944 stars on a total of 450 nights. Thereafter the program was discontinued by the Coast and Geodetic Survey and the instrument was sold to the U.S. Naval Observatory, where it was designated as PZT no. 1 and was in continuous operation from 1915 until 1955. In 1934 the instrument was modified to determine time as well as the variation in latitude. Its use for the determination of time proved to be superior to visual methods, and copies of the instrument are used by the world's major time services.

While at Gaithersburg, Ross completed the computations of the orbits of the sixth and seventh satellites of Jupiter that he had begun in Washington. His elements and tables were used to derive the differential coordinates of these satellites, which were published in the *American Ephemeris* from 1912 through 1947.

In 1912 Ross turned his attention to the cause of the discordance between the positions of Mars deduced from observations and those computed from Newcomb's theory of Mars in *Astronomical Papers*; this discordance amounted to six seconds of arc in right ascension in 1905. Ross's corrections to Newcomb's theory, published in *Astronomical Papers* (1917), have been used in the *American Ephemeris*, beginning with the volumes for 1922.

Ross is listed in the *American Ephemeris* as a part-time member of the staff of the *Nautical Almanac Office* from 1907 to 1919. In 1915 he was appointed physicist at the Eastman Kodak Company laboratories, where for nine years he studied and perfected the techniques of photography and lens design that he later successfully applied in astronomy. In 1924 he joined the staff of the Yerkes Observatory, where he continued astronomical observations until his retirement in 1939. While at Yerkes he was on the faculty of the University of Chicago as associate professor (1924–1928) and professor (1929–1939) of astronomy.

One of Ross's early projects at Yerkes was to rephotograph the stellar fields that had been photographed by Barnard. In comparing the new plates with the older ones, he found many variable stars and stars with large proper motions. The *Astronomical Journal* reported these discoveries, and by 1931 Ross was credited with the discovery of 379 variable stars and 869 proper-motion stars.

Ross photographed Mars in different colors with the Mount Wilson sixty-inch reflector and with the Lick thirty-six-inch refractor at the planet's opposition in 1926. The next year his photographic observations of Venus at Mount Wilson Observatory revealed unusual temporary markings, or shadings, in the planet's atmosphere.

Ross spent the remainder of his life in California after his retirement from the Yerkes Observatory in 1939. He was furnished an office at the Mount Wilson Observatory, although he was not an official member of the staff. Although retired, Ross contributed to optics by designing and computing lenses. His designs included wide-angle camera lenses and elements to correct coma in the largest reflectors of his time. He is credited with the design of the twenty-inch astrograph at the Lick Observatory. Many of his lens designs were used outside astronomy.

Ross was a member of the American Astronomical Society and the National Academy of Sciences, and was an associate of the Royal Astronomical Society.

BIBLIOGRAPHY

Ross's longer works are *Latitude Observations With Photographic Zenith Tube at Gaithersburg, Md.*, U.S. Coast and Geodetic Survey special publication no. 27 (Washington, D.C., 1915); *New Elements of Mars and Tables for Correcting the Heliocentric Positions Derived From Astronomical Papers, Vol. VI, Part IV*, which is *Astronomical Papers*, 9, pt. 2 (1917); and *The Physics of the Developed Photographic Image* (New York–Rochester, N.Y., 1924).

Articles in *Astronomische Nachrichten* include "New Elements of Jupiter's 7th Satellite," **174** (1907), 359–362; "The Instrumental Constants of a Zenith Telescope," **190** (1912), 19–22; "The Kimura Term in the Latitude Variation and the Constant of Aberration," **192** (1912), 133–142; and "Magnitudes and Colors of the Eros Comparison Stars," **239** (1930), 289–301, written with R. S. Zug.

In *Astronomical Journal*, Ross published "Observation of Asteroids," **19** (1899), 194–195; "The Moon's Mean Longitude and the Eclipse of August 21, 1914," **28** (1914), 153–156; "The Moon's Mean Longitude and the Eclipse of Feb. 3, 1916," **29** (1915), 65–68; "The Sun's Mean Longitude," **29**, (1916), 152–156; "Investigations on the Orbit of Mars," *ibid.*, 157–163; "New Proper-Motion Stars," listed in **36–41** (1925–1931) and **45–48** (1936–1939); and "New Variable Stars," listed in **36–41** (1925–1931).

Ross's contributions to *Astrophysical Journal* are "Photographic Photometry and the Purkinje Effect," **52** (1920), 86–97; "Image Contraction and Distortion on Photographic Plates," *ibid.*, 98–109; "Photographic Sharpness and Resolving Power," *ibid.*, 201–231; "The Mutual Action of Adjacent Photographic Images," **53** (1921), 349–374; "Astronomical Photographic Photometry and the Purkinje Effect," **56** (1922), 345–372; "Film Distortion and Accuracy of Photographic Registration of Position," **57** (1923), 33–48; "Mensurational Characteristics of Photographic Film," **59** (1924), 181–191; "Characteristics of Photographic Desensitizers and Distortions on Plates due to Local Desensitizing," **61** (1925), 337–352; "Photographs of Mars, 1926," **64** (1926), 243–249; "Photograph of the Orion Nebulosities," **65** (1927), 137–139; "Nebulosities in Monoceros, Taurus, and Perseus," **67** (1928), 281–295; "Photographs of Venus," **68** (1928), 57–92; "An Abnormal Phenomenon of Photographic Plates," **73** (1931), 54–55; "Photographs of the Milky Way in Cygnus and Cepheus," **74** (1931), 85–90; "Correcting Lenses for Refractors," **76** (1932), 184–201; "Astrometry With Mirrors and Lenses," **77** (1933), 243–269; "Lens Systems for Correcting Coma of Mirrors," **81** (1935), 156–172; "Photographic Photometry," **84** (1936), 241–269; "Limiting Magnitudes," **88** (1938), 548–579; "The 48-Inch Schmidt Telescope for the Astrophysical Observatory of the California Institute of Technology," **92** (1940), 400–407; and "Parabolizing Mirrors Without a Flat," **98** (1943), 341–346.

Popular Astronomy contains "Planetary Photography," **31** (1923), 21; "Accuracy of Photographic Position Registration," *ibid.*, 22; "Distortions on Spotted Photographic Plates," **32** (1924), 619–620; and "Photographs of Venus," **35** (1927), 492.

Among Ross's contributions to *Publications of the Astronomical Society of the Pacific* are "Lenses and Their Focal Adjustment in Relation to Photometry," **38** (1926),

312–314; "The Optics of Reflecting Telescopes," **46** (1934), 339–345; and "Photographic Measures of a Close Double Star," **48** (1936), 221–222.

Other articles are "Differential Equations Belonging to a Ternary Linearoid Group," in *American Journal of Mathematics,* **25** (1903), 179–205; "Semi-Definitive Elements of Jupiter's 6th Satellite," *University of California Publications. Astronomy. Lick Observatory Bulletin,* **4** (1906), 110–112; "Empirical Short Period Terms in the Moon's Mean Longitude," in *Monthly Notices of the Royal Astronomical Society,* **72** (1911), 27; "A Wide-Angle Astronomical Doublet," in *Journal of the Optical Society of America,* **5,** no. 2 (Mar. 1921), 123–130; and "Limiting Magnitudes With Red Sensitive Plates," in *Publications of the American Astronomical Society,* **9** (1939), 270–271.

Works of which Ross was a coauthor include *Investigation of Inequalities in the Motion of the Moon Produced by the Action of the Planets,* Carnegie Institution publication no. 72 (Washington, D.C., 1907), written with Simon Newcomb; *Tables of Minor Planets Discovered by James C. Watson* (Washington, D.C., 1910), to which A. O. Leuschner was the major contributor; *Atlas of the Northern Milky Way* (Chicago, 1934), written with Mary R. Calvert; and *Magnitudes and Colors of Stars North of +80°,* Carnegie Institution publication no. 532 (Washington, D.C., 1941), also *Papers of the Mount Wilson Observatory,* **6** (1941), written with F. H. Seares and M. C. Joyner.

RALPH F. HAUPT

ROSS, JAMES CLARK (*b.* London, England, 15 April 1800; *d.* Aylesbury, England, 3 April 1862), *polar navigation, geomagnetism.*

Ross, the third son of George Ross, entered the Royal Navy as a midshipman in April 1812 and served under his uncle, John Ross, in the Baltic and North seas and the Downs. He subsequently accompanied him on surveys of the North Sea and the coast of the White Sea, in the course of which the longitude of Archangel was determined by observations of Jupiter's satellites.

After service from 1815 to 1817 in Scottish waters, in 1818 Ross accompanied his uncle on the latter's attempt, with William Edward Parry, to discover the Northwest Passage via Davis Strait.

In 1819–1820 Ross sailed with Parry when a further attempt was made to force the passage via Baffin Bay, Lancaster Sound, and Bering Strait. The partial success of this voyage in reaching Melville Island brought the expedition a reward of £5,000 from the Board of Longitude and encouraged the next expedition (1821–1823).

Promoted to lieutenant on 26 December 1822, Ross sailed with Parry on the latter's third voyage on

H.M.S. *Fury,* which was wrecked in Prince Regent Inlet on 30 July 1825. The *Fury* was abandoned, and all her crew were taken aboard H.M.S. *Hecla.* In 1827 Ross again sailed with Parry, this time to Spitsbergen in an effort to reach the North Pole over the ice. On the return of this venture, Ross was promoted to commander on 8 November 1827.

From 1829 to 1833 Ross accompanied his uncle, Sir John Ross, on the private expedition promoted by Felix Booth (1775–1850), a wealthy distiller who contributed £17,000 toward the cost. Sailing in a small vessel, James Ross carried out the sledging operations on the coasts of the Boothia Peninsula (named after the sponsor) and King William Island. On 1 June 1831 he discovered the north magnetic pole at latitude 70°05'17" north and longitude 96°45'48" west. Promoted to post captain on 28 October 1834, Ross commanded H.M.S. *Cove* on a voyage to Baffin Bay in 1836 for the relief of iced-in whalers.

Ross's accumulated expertise in magnetic observations led to his employment by the Admiralty in 1838 on a magnetic survey (declination and dip) of the United Kingdom. The success of these operations was followed by his appointment in 1839 to command an expedition, sponsored by the British Association for the Advancement of Science and the Royal Society, for magnetic and geographical discovery in the Antarctic.

In September 1839 Ross sailed on what was to be a four-year voyage. On 1 January 1841 he crossed the Antarctic Circle, discovering Victoria Land, the 12,000-foot volcano at latitude 77°35' south that he named Mount Erebus, and "the marvellous range of ice cliffs barring the approach to the Pole."

Ross returned to England in 1843 with a large accumulation of observations on magnetism and other branches of natural sciences, including geology and marine life at great depths. He had carried out in his survey the greatest work of its kind yet performed— and, remarkably, with the loss of only one man through illness. This was due in no small measure to the great attention given to the selection of supplies for a mixed diet.

In 1847 Ross published a comprehensive two-volume account of his voyage. The following year he was placed in command of H.M.S. *Enterprise* on the first search expedition to relieve the expedition headed by Sir John Franklin. The main purpose was not achieved, but no opportunity was lost in accumulating observations.

Although Ross saw no further naval service after 1849, he was thereafter regarded as the leading authority on all aspects of Arctic navigation. His magnetic observations were reduced and published over a period of more than twenty years by the geomagnetician Sir Edward Sabine, in the *Philosophical Transactions of the Royal Society.*

Ross was elected to the Linnean Society in 1824 and to the Royal Society on 11 December 1828, and was awarded the gold medals of the geographical societies of London and Paris in 1842 for his Antarctic exploration. In 1844 he was knighted, received the degree of Doctor of Civil Laws, and was named to the French Legion of Honor. He was a corresponding member of the Paris Academy of Sciences and other foreign scientific societies for many years.

BIBLIOGRAPHY

I. Original Works. Ross published the account of his voyage to the southern hemisphere as *A Voyage of Discovery and Research in the Southern and Antarctic Regions During the Years 1839 to 1843*, 2 vols. (London, 1847), which includes observations on geology, marine life, and social conditions. His papers include "Observations to Determine the Amount of Atmospheric Refraction at Port Bowen, in the Years 1824–1825," in *Philosophical Transactions of the Royal Society*, **116** (1826), pt. 4, 206–230; "On the Position of the North Magnetic Pole," *ibid.*, **124** (1834), 47–52, written with H. Foster and W. E. Parry; "Observations on the Direction and Intensity of the Terrestrial Magnetic Force in Ireland," in *Report of the British Association for the Advancement of Science* (1835), 116–162, written with E. Sabine and H. Lloyd; and "On the Effect of the Pressure of the Atmosphere on the Mean Level of the Ocean," in *Philosophical Transactions of the Royal Society*, **144** (1854), 285–296.

II. Secondary Literature. Most of Ross's magnetic observations were discussed and published by E. Sabine as "Contributions to Terrestrial Magnetism," in *Philosophical Transactions of the Royal Society*, **132** (1842), 9–41; **133** (1843), 145–232; **134** (1844), 87–224; **136** (1846), 429–432; and **156** (1866), 453–543. See also *Dictionary of National Biography*; an obituary notice in *Proceedings of the Royal Society*, **12** (1862–1863), lxi–lxiii; and Royal Greenwich Observatory, Board of Longitude papers.

P. S. Laurie

ROSS, RONALD (*b.* Almora, Nepal, 13 May 1857; *d.* Putney, London, England, 16 September 1932), *medicine.*

Ross, the eldest of ten children of a British army officer serving in India, received an English dame and boarding school education. Subsequently he studied medicine at St. Bartholomew's Hospital, London, but with little enthusiasm. Although he had bowed to his father's wish that he not become an artist, his passionate interest in the arts took up much of his time. Ross published plays, short dramas, romances, fables, and poetry, much of which received the approbation of John Masefield. He was married in 1889 to Rosa Bloxam and had two sons and two daughters. He was elected to the Royal Society in 1901 and knighted in 1911.

Apart from the arts, Ross had an abiding interest in mathematics, much of his self-education in the subject being undertaken while he was serving in the Indian Medical Service (1881–1888). Ross, whose complex personality can be surmised though hardly delineated from his manifold activities, apparently became more and more conscious of medical problems the longer he remained in India. Later he wrote: "I was neglecting my duty in the medical profession. I was doing my current work, it was true; but what had I attempted towards bettering mankind by trying to discover the causes of those diseases which are perhaps mankind's chief enemies?"

Unlike most of his colleagues in the Indian Medical Service, Ross was research-minded. In 1888, during his first furlough in England, he earned the newly established Diploma of Public Health and took a course in bacteriology under E. Emanuel Klein. On returning to India, he studied malaria, initially believing that it was caused by intestinal auto-intoxication.

During Ross's next furlough in England (1894) his successful, far-reaching malarial studies were initiated. This was due in large measure to the influence of Patrick Manson, for three reasons a key figure in Ross's contributions to the unraveling of the life history of the malarial parasite. First, Manson demonstrated convincingly to a skeptical Ross the correctness of Alphonse Laveran's pioneering observations of 1880: that the blood of malarial patients contained pigmented bodies of parasites. Second, Manson propounded a theory that mosquitoes transmitted malaria ("On the Nature and Significance of the Crescentic and Flagellate Bodies in Malarial Blood," in *British Medical Journal* [1894], **2**, 1306–1308). Third, through an extensive exchange of correspondence with Ross, he helped to sustain the latter's researches in India during more than three years of difficulties that arose not only from problems of technique and of obtaining volunteer patients, but also from regimental duties and unsympathetic superiors.

In essence the problem Ross set himself—to prove Manson's hypothesis that mosquitoes transmitted malaria—was enormous, for he had to contend with two variables: a variety of mosquitoes and a variety of parasites. The questions were which mosquito was the vector and which parasites were the malarial parasites (now known to be species of *Plasmodium*). The main points in Ross's contributions to the problem during 1895–1898 can be summarized as follows.

First, Ross demonstrated that volunteers who drank water contaminated with infected mosquitoes (including larvae) failed to contract the disease. This, along with his earlier doubts that aerial and water contamination provided a ready explanation for the epidemiology of malaria, did much to direct his attention to the possibility that transmission might be via mosquito bites, a point of view expressed by A. F. A. King in 1883. Ross apparently was ignorant of King's work until 1899; and in fact he met continual problems because of a shortage of scientific literature in India, above all in connection with identifying and classifying mosquitoes.

Second, Ross's studies on the parasites in mosquitoes involved learning how to identify mosquitoes and to dissect their internal organs. From the beginning Ross was especially concerned with the "motile" parasitic filaments found in mosquito stomachs. The question was what happened to them. While he failed to recognize that the filaments were gametes (a point first appreciated by W. G. MacCallum in 1897), Ross's supposition that they developed into another stage stood him in good stead, for it ensured that he spaced out his examination of individual mosquitoes from groups that had fed on malarial patients. Even so, it was not until 20 August 1897 (later to be called "Mosquito day" by Ross) that he observed in the stomach wall of a type of mosquito he had not hitherto encountered (a malarial vector *Anopheles*, rather than the *Culex* and *Stegobium* he had been investigating for over two years) a cyst containing granules of black pigment similar to the pigmented bodies initially observed by Laveran.

Third, owing to the administration of the Indian Medical Service, Ross was unable to continue his studies on this stimulating find, which he had immediately recognized should lead him to unravel the complete life cycle. Some months later, however, he was able to study malaria in caged birds and to demonstrate the parasite life cycle, including stages in mosquito salivary glands. He also was able to demonstrate that mosquitoes could transmit malaria directly from infected to healthy birds.

The demonstration that the life cycle of the human malarial parasite was identical was accomplished by Italian workers, but a bitter priority dispute arose. The award of the Nobel Prize to Ross in 1902 laid to rest much of the recrimination.

Ross's experimental career ended in 1899, when he retired from the Indian Medical Service. His autobiography, published in 1922, entitles the remaining part of his life "The Fight for Life," reflecting bitterness at the lack of help and recognition he received in India, and also perhaps an egotistical facet of his personality. When he returned to England, he soon became lecturer at the new school for tropical medicine at Liverpool, where his influence was important in its success in pioneering tropical medicine education in Britain. Later Ross held the chair in tropical medicine at Liverpool, and in 1912 he moved to London to take up consulting practice.

Much of the rest of Ross's life was concerned with public health programs against malaria. They were of outstanding importance, for, at least initially, he had to fight a widespread belief that malaria was caused by effluvia or miasmas, especially those emanating from marshes. His inaugural lecture at Liverpool, "The Possibility of Extirpating Malaria From Certain Localities by a New Method" (destruction of mosquitoes), was an initial broadside that was followed by a range of publications, most of them written for laymen. In 1899 the first monograph of the Liverpool School of Tropical Diseases appeared, a short pamphlet of fourteen pages entitled *Instructions for the Prevention of Malarial Fever for the Use of Residents in Malarious Places*. It succinctly set out information on such topics as how to avoid being bitten and how to destroy mosquitoes. Other publications on the same theme were *Mosquito Brigades and How to Organise Them* (London, 1902) and the much more extensive *The Prevention of Malaria* (London, 1910), which included surveys of conditions in many countries by various contributors. Ross himself traveled fairly extensively to undertake malarial prevention campaigns. During World War I he became consultant in malaria to the War Office. His missionary zeal stood him in good stead in this work, though a pugnacious streak in his personality often aroused hostility and created difficulties which others might have avoided.

Ross's name and work were perpetuated during his lifetime through the opening in 1926 of the Ross Institute of Tropical Hygiene, the aim of which was to promote research and to stimulate malarial control measures. Ross was its first director, and continued to write on malaria and to encourage others. One of the more interesting notes he wrote while at the Institute was the foreword to M. E. Macgregor's *Mosquito Surveys. A Handbook for Anti-malarial and Anti-mosquito Field Workers* (London, 1927). The Ross Institute still exists, adding further interest to the famous emotional verses Ross penned on 21 August 1897, almost immediately after his discovery of the pigmented cyst in the mosquito wall. These verses appeared in his long poem "Exile."

Before Thy feet I fall,
Lord, who made high my fate;
For in the mighty small
Thou showed'st the mighty great.
Henceforth I will resound
But praises unto Thee;
Tho' I was beat and bound
Thou gavest me victory.

BIBLIOGRAPHY

I. ORIGINAL WORKS. The main printed source of information is Ross's *Memoirs With a Full Account of the Great Malaria Problem and Its Solution* (London, 1923); the MS material from which the book was partly compiled is at the Ross Institute and the London School of Hygiene and Tropical Medicine. The *Memoirs* include a full list of Ross's writings up to 1922; a complete bibliography has been compiled by M. J. Rees in an unpublished dissertation for the Diploma in Librarianship and Archives (University of London, 1966).

Apart from Ross's publications noted in the text, three others deserve mention: "On Some Peculiar Pigmented Cells Found in Two Mosquitoes Fed on Malarial Blood," in *British Medical Journal* (1897), **2**, 1786–1788 (observations were added by "Dr. Thin, Mr. Bland Sutton and Dr. Patrick Manson"); *Report on the Cultivation of Proteosoma, Labbe, in Grey Mosquitoes* (Calcutta, 1898); for a more accessible report, see P. Manson, "Surgeon-Major Ronald Ross's Recent Investigations on the Mosquito-Malaria Theory," in *British Medical Journal* (1898), **1**, 1575–1577; and "Researches on Malaria," his Nobel Prize lecture, in *Journal of the Royal Army Medical Corps*, **4** (1905), 450–474, 541–579, 705–729, also privately reprinted separately. See also *Nobel Lectures Including Presentation Speeches and Laureates' Biographies. Physiology or Medicine 1901–1921* (London, 1967), 21–119. Some of the points in the Nobel lecture are corrected by Ross in his *Memoirs*.

II. SECONDARY LITERATURE. Many articles and notices have appeared on Ross, but a scholarly account written against a background of the times is still lacking. Obituary notices of particular value are in *Obituary Notices of Fellows of the Royal Society of London*, **1** (1932–1935), 108–114; *British Medical Journal* (1932), **2**, 609–611; and *Lancet* (1932), **2**, 695–697.

An interesting account of Ross's malarial work is in P. H. Manson-Bahr and A. Alcock, *The Life and Work of Sir Patrick Manson* (London, 1927). It treats some of the questions that were raised as to whether Manson deserved as much credit as Ross, insofar as he suggested the mosquito hypothesis and sustained Ross with ideas. Manson, however, always gave Ross entire credit for overcoming the tremendous practical problems and for refining the hypothesis.

J. K. CRELLIN

ROSSBY, CARL-GUSTAF ARVID (*b.* Stockholm, Sweden, 28 December 1898; *d.* Stockholm, 19 August 1957), *meteorology, oceanography.*

Rossby was the son of Arvid Rossby, a construction engineer, and Alma Charlotta Marelius. He entered the University of Stockholm in 1917, and studied mathematics, mechanics, and astronomy. His career in meteorology began in 1919, when he joined the scientific staff at the Geophysical Institute at Bergen, Norway. Here the polar-front theory of cyclones was being developed under the leadership of V. F. Bjerknes. In 1921 Rossby studied hydrodynamics at the Geophysical Institute of the University of Leipzig and worked at the Prussian Aerological Observatory at Lindenberg. In the same year he returned to Stockholm and entered the Swedish Meteorological and Hydrological Service. He took part in several meteorological and oceanographic expeditions. He also continued his studies of mathematical physics at the University and received his licentiate in 1925. In 1926 he won a one-year fellowship to the U.S. Weather Bureau in Washington. He remained in the United States for more than twenty years, marrying Harriot Alexander in 1929 and obtaining American citizenship in 1939.

Rossby was instrumental in bringing American meteorology to a position of world leadership. At first, however, his attempt to introduce the polar-front theory and other innovations was met with hostility at the U.S. Weather Bureau, which was largely bureaucratic rather than a center of scientific research. In 1927 Rossby accepted the chairmanship of the Committee on Aeronautical Meteorology of the Daniel Guggenheim Fund for Promotion of Aeronautics, in which capacity he established a model weather service for civil aviation in California. He became associate professor of meteorology at the Massachusetts Institute of Technology in 1928 and rapidly established a strong department.

Rossby was appointed assistant chief of the Weather Bureau, in charge of research and education, in 1939. During the next two years he worked at reorganizing the Bureau and strengthening its scientific mission. In 1941 Rossby became chairman of the newly founded department of meteorology at the University of Chicago, where he brought together an outstanding group of scientists from many different countries. During the early 1940's, when the needs of warfare made meteorology a key science, Rossby organized a very effective military educational program in meteorology and worked for the establishment of an adequate global observing and forecasting service. Recognizing the importance of tropical meteorology, he was instrumental in founding the Institute of

Tropical Meteorology at the University of Puerto Rico. He also reorganized the American Meteorological Society and established the *Journal of Meteorology* and later, in Sweden, the geophysical journal *Tellus*.

Rossby became increasingly active in promoting international collaboration in the late 1940's. At the request of the Swedish government, he returned to Sweden in 1950 and organized the International Meteorological Institute. He continued to make extended visits to the United States during the 1950's, mainly in connection with his work at the Woods Hole Oceanographic Institute.

Despite his strenuous work as organizer, director, and promoter, Rossby still found time and energy for high-quality research. His original contributions to meteorology reveal a deep insight into the fundamental processes taking place in the atmosphere and oceans. He produced many new ideas that he submitted for discussior to the group of scientists working around him, often approaching a complex problem by introducing bold simplifications to be accounted for later.

During the 1920's Rossby worked on atmospheric turbulence and the theory of atmospheric pressure variations. While at MIT he continued his research on atmospheric and oceanic turbulence and introduced the concepts of mixing length, the roughness parameter, and the logarithmic wind profile. During his early years at MIT, Rossby applied thermodynamics to air mass analysis, a subject first systematically studied by T. Bergeron in 1928. Rossby designed a graphical method, the Rossby diagram, for the identification of air masses and the processes that give rise to their formation and modification (1932).

Pursuing a method advocated by W. N. Shaw during the 1920's, Rossby and his collaborators developed the technique of isentropic analysis. Using this technique, which allowed the tracing of large-scale air currents, Rossby started his studies of the dynamics of the general circulation of the atmosphere. In connection with a project on long-range forecasting started in 1935, he began to investigate the circumpolar system of long waves, now called Rossby waves, in the westerly winds of the middle and upper troposphere. These waves exert a controlling influence on weather conditions in the lower troposphere. Rossby developed a dynamic theory of these long waves on the basis of the theorem of conservation of absolute vorticity (propounded by Helmholtz in 1858) and derived a simple formula, now called the Rossby equation, for their propagation speed (1939, 1940). This formula became perhaps the most celebrated analytic solution of a dynamic equation in meteorological literature.

Rossby's theoretical analysis profoundly influenced both applied and theoretical meteorology and oceanography during subsequent decades. The applicability of his theoretical results convinced Rossby that the principal changes in the atmospheric circulation could be predicted by considering readjustments of the horizontal velocity field without taking into account changes in the vertical structure of the atmosphere. In 1940 he and his collaborators carried out the first numerical predictions for a "one-layer" barotropic atmosphere in which vorticity was conserved. These calculations and the introduction of high-speed electronic computers during the 1940's set the stage for the simultaneous development of forecasting techniques and theory through comparisons of calculated and observed atmospheric states, as Bjerknes had envisioned in 1904.

Under Rossby's leadership the research group at Chicago became engaged in synoptic, theoretical and experimental studies of the general circulation and developed most of the basic concepts of the jet stream, the core of high-speed winds embedded in the upper long waves; it was first revealed in the late nineteenth century by observation of the drift of cirrus clouds, but was systematically encountered and investigated only as a result of the establishment of a worldwide network of upper-air sounding stations during World War II. In analogy to his work on ocean currents (1936), Rossby found a partial explanation for the existence and maintenance of the observed latitudinal wind distribution that was based on the concepts of large-scale lateral mixing and conservation of absolute vorticity (1947).

Back in Stockholm, Rossby continued to work on problems of atmospheric and oceanic circulation and their interactions. In addition he began to study geochemistry in general and atmospheric chemistry in particular, seeing it as an opportunity to broaden the scope of meteorology. He organized an international network for the investigation of the distribution of trace elements in the atmosphere.

BIBLIOGRAPHY

I. Original Works. Bibliographies of Rossby's works are in B. Bolin, ed., *The Atmosphere and Sea in Motion*, the Rossby memorial volume (New York, 1959), 60–64; and in the biography by H. R. Byers (see below). His writings include "Thermodynamics Applied to Air Mass Analysis," *Massachusetts Institute of Technology, Papers in Physical Oceanography and Meteorology*, **1** (1932), no. 3; "A Generalization of the Theory of the Mixing Length With Application to Atmospheric and Oceanic Turbulence," *ibid.*, no. 4; "The Layer of Frictional

Influence in Wind and Ocean Currents," in *Papers in Physical Oceanography and Meteorology*, **3** (1935), no. 3, written with R. B. Montgomery; "Dynamics of Steady Ocean Currents in the Light of Experimental Fluid Mechanics," *ibid.*, **5** (1936), no. 1; and "Relation Between Variations in the Intensity of the Zonal Circulation of the Atmosphere and the Displacements of the Semi-Permanent Centers of Action," in *Journal of Marine Research*, **2** (1939), 38–55, written with collaborators.

Later works are "Planetary Flow Patterns in the Atmosphere," in *Quarterly Journal of the Royal Meteorological Society* (Toronto proceedings supp.), **66** (1940), 68–87; "The Scientific Basis of Modern Meteorology," in U. S. Department of Agriculture, *Yearbook of Agriculture* (1941), 599–655; "Kinematic and Hydrostatic Properties of Certain Long Waves in the Westerlies," in *University of Chicago, Department of Meteorology, Miscellaneous Reports*, no. 5 (1942); "On the Propagation of Frequencies and Energy in Certain Types of Oceanic and Atmospheric Waves," in *Journal of Meteorology*, **2** (1945), 187–204; "On the Distribution of Angular Velocity in Gaseous Envelopes Under the Influence of Large-Scale Horizontal Mixing Processes," in *Bulletin. American Meteorological Society*, **28** (1947), 53–68; "On the Dispersion of Planetary Waves in a Barotropic Atmosphere," in *Tellus*, **1** (1949), 54–58; and "On the Vertical and Horizontal Concentration of Momentum in Air and Ocean Currents," *ibid.*, **3** (1951), 15–27.

II. SECONDARY LITERATURE. See T. Bergeron, "The Young Carl-Gustaf Rossby," in B. Bolin, ed., *The Atmosphere and Sea in Motion* (New York, 1959), 51–55; B. Bolin, in *Tellus*, **9** (1957), 257–258; H. R. Byers, "Carl-Gustaf Rossby, the Organizer," in B. Bolin, ed., *op. cit.*, 56–59; and "Carl-Gustaf Arvid Rossby," in *Biographical Memoirs. National Academy of Sciences*, **34**, no. 11 (1960), 249–270; and G. W. Platzman, "The Rossby Wave," in *Quarterly Journal of the Royal Meteorological Society*, **94** (1968), 225–248.

GISELA KUTZBACH

ROSSE, WILLIAM. See **Parsons, William, Third Earl of Rosse.**

ROSSETTI, FRANCESCO (*b.* Trento, Italy, 14 September 1833; *d.* Padua, Italy, 20 April 1885), *physics.*

Rossetti studied physics at the University of Vienna, where he obtained his doctorate. At the age of twenty-four he went to Venice, to teach physics and mathematics at the Liceo Santa Caterina. Some years later he went to Paris, where he worked in Regnault's laboratory. When he returned to Italy, he was appointed to the chair of physics at the University of Padua, where he remained until his death.

Rossetti was interested in astrophysics, but his main work concerned electrical piles and electrical generators. His studies on the temperatures of an electric arc led him to shift his interest to the temperature of the sun. In this respect he was a pioneer in the use of a thermocouple to measure the solar constant. Rossetti concluded that the solar temperature had to be about 10,238° K.

BIBLIOGRAPHY

Rossetti published 43 scientific papers in the *Atti dell'Istituto veneto . . ., Atti dell'Accademia dei Lincei, Nuovo cimento*, and *Journal de physique*. His work on the temperature of the sun, which was awarded a special prize by the Accademia dei Lincei, was also published in *Philosophical Magazine*, 5th ser., **8** (1879).

GUGLIELMO RIGHINI

ROSTAN, LÉON LOUIS (*b.* St.-Maximin, Var, France, 16 March 1790; *d.* Paris, France, 4 October 1866), *medicine.*

The son of a wealthy bourgeois family, Rostan received a good education in Marseilles and at boarding schools in Paris. In 1809 he became an *interne des hôpitaux* at La Salpêtrière in Paris, serving under the surgeon A. M. Lallement and then under Philippe Pinel. He received his medical degree on 13 May 1812 with the dissertation *Essai sur le charlatanisme*, dealing with the sociology and ethics of medicine. His thesis defense was presided over by Pinel, whose influence remained paramount in Rostan's thought.

Named an inspector in the Service de Santé at La Salpêtrière, Rostan performed outstanding service during the epidemic of exanthematic typhus in 1814. (He himself suffered a severe attack of the disease.) Appointed *médecin-adjoint* at La Salpêtrière in 1818, he organized courses in clinical medicine. These lectures, held at the bedside, were presented in a straightforward manner but with great erudition and a highly developed clinical sense. They were immensely popular and influenced auditors as celebrated as J. M. Charcot and K. R. Wunderlich.

During this period Rostan conducted original research on encephalomalacia, myocardial hypertrophy, and cardiac asthma. His scientific publications were notable for the precision of their clinical observations and for the clarity of their anatomico-pathological explanations. Following his appointment in 1883 as professor of clinical medicine at the Hôtel-Dieu in Paris, Rostan's teaching became more dogmatic. An opponent of vitalism and of the medical system of Broussais, he developed and defended a doctrine that was termed "organicism" by his adversary, Frédéric-Joseph Bérard of Montpellier. Rostan married late in life and subsequently divided his time between work at the hospital and increasingly longer stays at his country house in Vauxcelles, Provence. He died after a long illness, rendered hemiplegic by cerebral apoplexy.

Rostan was a typical representative of the anatomicoclinical school of Paris. Confident of its sensualist and positivist approach and able to draw on the rich material for observation furnished by the great urban hospitals, this school constituted a bridge between Hippocratic and modern experimental medicine. In all of his scientific work, Rostan pursued the same goal: to explain clinical symptoms by specific organic lesions. His studies of the vascular disorders of the central nervous system and his definition of *ramollissement cérébral* (encephalomalacia) (1820–1823) were fundamental advances in the knowledge and interpretation of apoplexy. He also demonstrated that *asthme des vieillards* is not a functional disorder of the bronchia or lungs but is the clinical manifestation of an organic lesion, most often of cardiac origin. Rostan's "organicism" was a materialist-oriented medical theory, according to which life was defined as an ensemble of functions arising entirely from a specific structure. The so-called vital qualities are determined by the arrangement of the organs, and diseases are malfunctions resulting from material organic modifications.

BIBLIOGRAPHY

I. ORIGINAL WORKS. Rostan's principal books are *Recherches sur le ramollissement du cerveau* (Paris, 1820; 2nd ed., rev., Paris, 1823); *Cours élémentaire d'hygiène*, 2 vols. (Paris, 1822; 2nd ed., 1828); *Traité élémentaire de diagnostic, de pronostic, d'indications thérapeutiques, ou cours de médecine clinique*, 3 vols. (Paris, 1826; 2nd ed., 1830); and *Exposition des principes de l'organicisme* (Paris, 1831; 2nd ed., 1846; 3rd ed., 1864). His most important article is "Mémoire sur cette question: l'asthme des vieillards est-il une affection nerveuse?" in *Nouveau journal de médecine, chirurgie, pharmacie*, **3** (1818), 3–30; there is an English trans. by S. Jarcho in *American Journal of Cardiology*, **23** (1969), 584–587.

II. SECONDARY LITERATURE. On Rostan's life, see J. Béclard, *Notices et portraits* (Paris, 1878), 137–166; and M. Genty, "Rostan," in *Les biographies médicales*, **9** (1935), fasc. 3, 81–96. On his work in clinical neurology, see A. Rousseau, "L'analyse diagnostique de l'apoplexie par les médecins de l'école de Paris au début du XIXe siècle," in *Castalia*, **21** (1965), 11–24. His work in cardiology is analyzed in M. D. Grmek and A. Rousseau, "L'oeuvre cardiologique de Léon Rostan," in *Revue d'histoire des sciences*, **19** (1966), 29–52. For a brief account of "organicism," see T. S. Hall, *Ideas of Life and Matter* (Chicago, 1969), II, 251–254.

M. D. GRMEK

ROTH, JUSTUS LUDWIG ADOLPH (*b.* Hamburg, Germany, 15 September 1818; *d.* Berlin, Germany, 1 April 1892), *geology, petrology.*

Roth was the son of a pharmacist. He studied pharmacology and geology at the universities of Berlin, Tübingen, and Jena. His professors included Heinrich and Gustav Rose and Quenstedt. He obtained his Ph.D. in 1844 at Jena. From 1845 to 1848 he directed the Roth'sche Apotheke in Hamburg, which he had inherited from his father, but then leased it and moved to Berlin to dedicate himself entirely to the earth sciences.

During an assistantship at a pharmacy in Dresden, Roth published his first treatise, *Die Kugelformen im Mineralreich . . .* (1844), a publication to which the painter Ludwig Richter contributed some figures. In 1848 he helped to organize the Deutsche Geologische Gesellschaft and was its secretary from 1849 to 1866. Following a stay in Naples (1855–1856), he published his monograph *Der Vesuv und die Umgebung von Neapel* (1857).

Roth combined his knowledge of chemistry and petrography to produce a collection of all available rock analyses—*Die Gesteins-Analysen in tabellarischer Übersicht* (1861)—and a critical evaluation of analytical methods and results. In this classification of rocks he tried to combine their chemical, mineralogical, and geological properties. In 1861 he obtained his *Habilitation* at the University of Berlin, and he illustrated his lectures there with the carefully collected specimens from his private collection.

Excursions to the volcanic provinces of Germany gave Roth an opportunity to apply his knowledge of recent volcanism to the fossil volcanoes of the Eifel. Following the death of Mitscherlich, his mentor at Berlin, Roth edited his manuscripts on volcanology. In 1867 Roth was appointed extraordinary professor at the University of Berlin. In the same year (during which his first wife died) he was made a full member of the Royal Prussian Academy of Sciences.

Roth's lifelong goal was to write an advanced textbook on general and chemical geology. His *Beiträge zur Petrographie der plutonischen Gesteine* (Berlin 1869, 1873, 1879, 1884) were milestones in this project; and with various research reports, especially his comprehensive study "Über die Lehre vom Metamorphismus . . .," he contributed original work on metamorphic rocks and the processes of metamorphism. In 1879 he published the first volume of his masterpiece, *Allgemeine und chemische Geologie*.

In 1887, when the last part of the second volume of his monumental handbook of petrography, *Allgemeine und chemische Geologie*, appeared, he was appointed full professor of petrography and general geology at Berlin. Only the first part of Roth's third volume was published (1890) during his life. Following his death, his daughter Elisabeth edited the second part, which Roth had prepared, and wrote supplements to previous volumes. Roth was representative of his time in that his publications were of three types: original papers on many topics, but primarily on petrography, his preferred subject; papers of a monographic or textbook nature; and popular books or articles.

Roth's main contribution was in the field of systematic petrography. He was for two or three decades the leading specialist in this field, and thus a predecessor of Rosenbusch.

BIBLIOGRAPHY

I. ORIGINAL WORKS. Roth's major works are *Die Kugelformen im Mineralreiche und deren Einfluss auf die Absonderungsgestalten der Gesteine. Ein Beitrag zur geognostischen Formenlehre mit Rücksicht auf Landschaftsmalerei* (Dresden–Leipzig, 1844); *Der Vesuv und die Umgebung von Neapel. Eine Monographie* (Berlin, 1857); *Die Gesteins-Analysen in tabellarischer Übersicht und mit kritischen Erläuterungen* (Berlin, 1861); "Über die Lehre vom Metamorphismus und die Entstehung der krystallinischen Schiefer," in *Abhandlungen, Akademie der Wissenschaften, Berlin*, Phys. Cl. (1871), 151–232; *Allgemeine und chemische Geologie*, 3 vols. (Berlin, 1879–1893), I, *Bildung und Umbildung der Mineralien. Quell-, Fluss- und Meerwasser. Die Absätze* (Berlin, 1879); II, pt. 1, *Allgemeines und ältere Eruptivgesteine* (Berlin, 1883); II, pt. 2, *Jüngere Eruptivgesteine* (Berlin, 1885); II, pt. 3, *Krystallinische Schiefer und Sedimentgesteine* (Berlin, 1887); III, pt. 1, *Die Erstarrungskruste und die Lehre vom Metamorphismus* (Berlin, 1890); III, pt. 2, *Verwitterung, Zersetzung und Zerstörung der Gesteine* (Berlin, 1893); and "Die Eintheilung und die chemische Beschaffenheit der Eruptivgesteine," in *Zeitschrift der Deutschen geologischen Gesellschaft*, **43** (1891), 1–42.

II. SECONDARY LITERATURE. On Roth and his work, see Ahlenstiel, "Justus Roth," in *Jahresheft des Naturwissenschaftlichen Vereins für das Fürstenthum Lüneburg*, **12** (1890–1892), 42–44; T. Liebisch, "Justus Roth," in *Neues Jahrbuch für Mineralogie, Geologie und Paläontologie*, **2** (1893), 1–20; and A. Rothpletz's article in *Allgemeine deutsche Biographie*, LIII, 533–534.

G. C. AMSTUTZ

ROTHMANN, CHRISTOPH (*b.* Bernburg, Anhalt, Germany; *d.* Bernburg, 1599–1608), *astronomy*.

The exact dates of Rothmann's birth and death are not known. He studied theology in Wittenberg and also attended lectures in mathematics and astronomy (maybe under Johann Praetorius). In 1577 William IV of Hesse persuaded Rothmann to come to Kassel as his *mathematicus* and assist him in his observatory and with the publication of his star catalog. The clocks and other instruments were in the care of Bürgi, an outstanding instrument maker and a competent mathematician. Rothmann and Bürgi worked on "prosthaphairesis," a method for converting multiplication into addition by means of trigonometric functions. Rothmann, who seems to have been a conceited and difficult person, tried to claim priority for the discovery, although the very modest Bürgi probably contributed more.

In 1590 Rothmann visited Tycho Brahe at Hven. He stayed one month and then, instead of returning to Kassel, went back to his native town, where he became occupied with theological controversies. He remained there until his death, which must have occurred between 1599 (when, according to a letter from Tycho to Severin, he was still living) and 1608 (when a notice *Bericht von der Tauffe* was published in Goslar as posthumous).

Rothmann's importance derives mainly from his correspondence with Tycho Brahe. Tycho had visited William IV in Kassel in 1575; but they did not communicate again until the comet of 1585, which led to an exchange of letters between Tycho in Hven and William IV and Rothmann in Kassel that lasted for six years. A last letter was written by Rothmann from Bernburg in 1594. This correspondence covered all aspects of contemporary astronomy: instruments and methods of observing, the Copernican system (which Rothmann supported against Tycho's system), comets, and auroras. These letters were eventually published by Tycho in *Tychonis Brahe Dani Epistolarum astronomicarum . . .* (Uraniborg, 1596).

Rothmann did not publish anything himself. According to Rudolf Wolf, his manuscripts are preserved in Kassel and consist of several star catalogs, a description of the Kassel observatory and its instruments, a general work on astronomy, and trigonometric tables.

BIBLIOGRAPHY

Willebrord Snell published part of Rothmann's MSS in his *Coeli et siderum in eo errantium observationes hassiacae* (Leiden, 1618).

The contents of the MSS are extensively described by R. Wolf, *Astronomische Mittheilungen*, **5**, no. 45 (1876–1880). See also his *Geschichte der Astronomie* (Munich, 1877); and J. L. E. Dreyer, *Tycho Brahe, a Picture of Scientific Life and Work in the Sixteenth Century* (Edinburgh, 1890; repr. New York, 1963).

LETTIE S. MULTHAUF

ROUELLE, GUILLAUME-FRANÇOIS (*b.* Mathieu, Calvados, France, 15 September 1703; *d.* Passy, Seine, France, 3 August 1770), *chemistry, geology.*

Born in a village now on the outskirts of Caen, Rouelle attended the University of Caen, where he reportedly earned a *maitre-ès-arts*. About 1730 he went to Paris and became apprenticed to an obscure German pharmacist named J. G. Spitzley. By 1740 he was giving lectures on chemistry and pharmacy in the Place Maubert, not far from the Jardin du Roi; and he soon attracted the attention of Buffon, who, in 1742, appointed him to the post of demonstrator in chemistry at the Jardin. In 1746 Rouelle moved his laboratory to the rue Jacob (faubourg St.-Germain), where he taught private courses for the rest of his active career. Once admitted to the Company of Apothecaries of Paris in 1750—even though he had not completed the ten years of training required of licensed pharmacists—he added a pharmacy shop to his laboratory. This shop, still in existence, was later operated by Hilaire-Marin Rouelle and eventually passed into the hands of chemist and pharmacist Bertrand Pelletier.

Rouelle's duties at the Jardin required him to perform experiments to illustrate the theories propounded by the professor of chemistry, at that time Louis-Claude Bourdelin, a chemist of no particular distinction. Since, in an arrangement without parallel, Rouelle's title was actually demonstrator "sous le titre de Professeur," he also delivered lectures on theory; and his courses were thus complete and independent of Bourdelin's. Like all courses at the Jardin, Rouelle's were open to the public; those he taught in the rue Jacob were more detailed and were attended by a fee-paying clientele.

The content of Rouelle's lectures was not wholly original; many of his experiments resembled those in Lemery's popular *Cours de chymie* (1675 and subsequent editions). But he did improve upon traditional techniques in organic analysis by moderating the temperatures employed so that reagents and distillates were not destroyed. He also lectured on the classification of salts, the subject of his most important publications. Examining both crystal form and chemical composition, he distinguished neutral salts from those with "an excess of acid" and those with "very little acid." Not only did his definitions of acid, alkali, and salt bring precision to an area then in a state of confusion, but his published memoirs also revealed an experimental skill and clarity of thought that were generally not the qualities remarked upon by contemporaries who attended his lectures.

Among Rouelle's innovations was his adoption, with modifications, of the phlogiston theory of Stahl and the conclusion put forth by the English physiologist Hales (denied by Stahl) that air can be a chemical constituent of matter. These two ideas became part of Rouelle's own synthesis, the fundamental tenet of which was that earth, air, fire (phlogiston), and water all serve as both chemical elements and physical "instruments" that assist in the process of chemical change. Most contemporaries did not fully recognize the originality of this theory, assuming instead that Rouelle was transmitting the ideas of his predecessors or that he was simply reviving the four elements of Aristotle. It was often assumed, too, that phlogiston in particular had entered into French chemistry early in the century. The evidence, however, suggests that the phlogiston theory was accepted in France only in mid-century and then only as part of a broader chemical theory; the isolation of phlogiston from this context seems to date from the 1770's.

How Rouelle became familiar with the work of Stahl and other German chemists remains mysterious. Contemporaries sometimes described him as untutored and even semiliterate, but such reports are undoubtedly exaggerated and may mean only that his manners were boorish and provincial. One pupil, Antoine Monnet, believed that Rouelle gleaned much information from his many foreign students. Certainly Rouelle was sufficiently aware of the value of German (and Latin) works to advocate their translation into French, and one student attributed to his influence the translation by d'Holbach of several treatises by Stahl.[1] Other pupils, including P.-F. Dreux, J.-F. DeMachy, and A.-A. Cadet de Vaux, reveal that Rouelle inspired or even specifically requested the translations that they themselves produced.

Early in his career, Rouelle introduced into his lectures some discussion of the structure of the crust of the earth. He classified all geological formations into two groups: the unfossiliferous, largely granitic masses forming a primitive core; and the more recent, fossiliferous, sedimentary strata superimposed upon the primitive. During the 1760's, he modified this scheme to include an intermediate series of strata corresponding to the Coal Measures. Like other aspects of his teaching, this classification was uncommon when he first taught it—reputedly in about

1740—and almost a cliché when he retired, since it was to be suggested, independently, by many of his French and foreign contemporaries. Rouelle's importance as a geologist lies in the fact that he offered elementary instruction, inspiration, and some fruitful ideas to two talented pupils, Lavoisier and Desmarest.

No reading or summary of the content of Rouelle's lectures can quite account for the impact he had upon his contemporaries. He was universally considered to be an extraordinary teacher, even in an era of such teachers as Joseph Black, Boerhaave, and Bernard de Jussieu. But he differed from these men in that his style was unusually stirring and flamboyant. The following description by Vicq d'Azyr is not atypical:

> His eloquence was not a matter of words; he presented his ideas the way nature does her productions, in a disorder which was always pleasing and with an abundance which was never wearisome. . . . When he exclaimed: Listen to me, for I am the only one who can prove these truths for you, one knew that this was not evidence of vanity, but rather the transport of a soul fired by boundless zeal.[2]

Less typical in its emphasis upon Rouelle's competence as a chemist is a passage by Cadet de Vaux: "I am proud of having learned from him the principles of my profession, [which I did] on the advice of my brother [L.-C. Cadet] who, although he could have taught me very well indeed, thought he could do no better than to turn me over to Rouelle."[3]

Rouelle's more distinguished pupils included Lavoisier, Desmarest, Macquer, Venel, D'Arcet, and Bayen. As impressive is the list of nonscientists known to have attended his lectures: Diderot, d'Holbach, Jean-Jacques Rousseau, Malesherbes, and Turgot.

Among the distinctions awarded Rouelle were memberships in the Académie Royale des Sciences (1744) and in the academies of Stockholm and Erfurt. He also served on royal commissions charged with the investigation of such subjects as a cure for distemper in cattle, improvement of the refining of saltpeter, and the examination of alloys used in coinage. In 1753 he succeeded C. F. Geoffroy as inspector general of pharmacy at the Hôtel-Dieu. As Rouelle's health deteriorated, his lectures at the Jardin du Roi were sometimes delivered by his younger brother, who was appointed demonstrator in chemistry in 1768.

Rouelle's personal life and professional career in many ways remain obscure because of the limited nature of the available sources; this problem is so pervasive and so crucial that it deserves some analysis. Rouelle himself published little and seems to have left no manuscripts; and the information about him that is recorded by his contemporaries is of varying relia-

bility, largely anecdotal, and far from complete. The content of his chemistry lectures is known solely in the manuscript versions left by his students. Many manuscripts are, in fact, traceable to the notes of a single pupil, Diderot; and they therefore record Rouelle's teachings only for the period 1754–1758. While texts of other dates exist, they are not numerous (and those later than 1758 are often derivatives of the Diderot tradition) and do little to reveal any development in Rouelle's ideas during his long career. Furthermore, contemporaries sometimes attributed to Rouelle ideas not to be found in the manuscripts, and the accuracy of such reports cannot always be tested. Rouelle intended to publish a textbook based on his lectures—and a group of disciples, including his younger brother, had similar plans—but no such work ever appeared.

The sources for his activities in areas other than chemistry are even more slender and indirect. He is said to have frequented the Café Procope and Holbach's salon, but little is known about his relations with the *philosophes* he met there. A few extant manuscripts supply virtually all that is known about the courses in pharmacy that he taught for about thirty years. Weekly gatherings of colleagues and selected students are said to have taken place at his laboratory, but there is no record of conversations and almost none of participants.

NOTES

1. Bibliothèque de l'Institut de France, Paris, MS 6036, fol. 4r.
2. Félix Vicq d'Azyr, *Oeuvres*, J.-L. Moreau de la Sarthe, ed., I (Paris, 1805), 280.
3. J. R. Spielmann, *Instituts de chymie*, trans. by A.-A. Cadet de Vaux, I (Paris, 1770), xx–xxi.

BIBLIOGRAPHY

Standard bibliographies, including the published catalogues of the Bibliothèque Nationale and the British Museum, normally confuse Rouelle and his brother; a good working rule in these cases is to assume that items published after 1770 are by Hilaire-Marin rather than Guillaume-François. A bibliography of original and secondary works and some discussion of MSS are in R. Rappaport, "G.-F. Rouelle: An Eighteenth-Century Chemist and Teacher," in *Chymia*, 6 (1960), 68–101, and the sequel, "Rouelle and Stahl—The Phlogistic Revolution in France," *ibid.*, 7 (1961), 73–102; and Jean Mayer, "Portrait d'un chimiste: Guillaume-François Rouelle (1703–1770)," in *Revue d'histoire des sciences et de leurs applications*, 23 (1970), 305–332. The following are addenda to these articles.

There is no published checklist of MSS, but most are in France and are readily found listed in the series

Catalogue général des manuscrits des bibliothèques publiques de France. Excellent copies in England are at the Wellcome Historical Medical Library, London, and the Science Library, Clifton College, Bristol. The MS once owned by Denis I. Duveen is now at Cornell University, Ithaca, New York. Many MSS do not contain the lectures on geology. MSS in the Diderot tradition stem from Rouelle's private courses, as is shown by the copy at the Bibliothèque Municipale of Aire-sur-la-Lys (Pas-de-Calais). An unusually complete version of the public lectures is at the Bibliothèque Centrale, Muséum National d'Histoire Naturelle, Paris, MS 2542. There are few MSS of the pharmacy lectures; these include copies at the Bibliothèques Municipales of Lille and Arras, the Wellcome Historical Medical Library, and the Faculté de Pharmacie, Paris.

Two periodicals, the *Avantcoureur* and the *Journal de médecine, chirurgie, pharmacie, &c.*, are valuable for notices and sometimes descriptions of the content of courses offered by Rouelle, his brother, and their rivals. The papers of Antoine Monnet (École des Mines, Paris, MSS 4678, 4685) supply impressions and details by a disciple of questionable reliability. Anonymous notes for a eulogy of Rouelle are in the Bibliothèque de l'Institut de France, Paris, Fonds Cuvier, tome 182, pièce 9.

RHODA RAPPAPORT

ROUELLE, HILAIRE-MARIN (*b.* Mathieu, Calvados, France, February 1718; *d.* Paris, France, 7 April 1779), *chemistry.*

Known as Rouelle *le jeune* or *le cadet*, Hilaire-Marin was less famous, original, and influential than his older brother Guillaume-François; but their careers were parallel to such an extent that the two men are often confused. Hilaire-Marin learned chemistry from his brother and then assisted him in most of his activities, which ranged from the study of saltpeter to the delivery of lectures at the Jardin du Roi, where in 1768 the younger Rouelle became demonstrator in chemistry. As a result of this close collaboration the possible contributions of Hilaire-Marin to work credited to his brother remain unknown.

After the death of Guillaume-François in 1770, Hilaire-Marin began to teach private courses in chemistry and pharmacy at their shop on the rue Jacob. The chemistry lectures seem to have been patterned consciously on those of his brother, whose pedagogical methods he admired. As a teacher, Hilaire-Marin is said to have had more dexterity than his brother—he was certainly less flamboyant and less absentminded—and this may help to explain his ability to proceed still further with organic chemistry, adding to the work of his brother several fine analyses of animal material. Some indication of the content of his courses can be found in his *Tableau de l'analyse chimique* (1774). His other publications consist of articles, the most significant of which discuss fixed air and the combustion of the diamond (the latter in collaboration with D'Arcet).

During the period 1762–1775, Hilaire-Marin held a post of apothecary to Louis Philippe, Duc d'Orléans, eventually resigning in favor of his nephew. Never elected to the Académie Royale des Sciences, although he was at least twice a candidate, he was a member of the academies of Madrid and Erfurt and the Royal Society of Arts (London).

BIBLIOGRAPHY

I. ORIGINAL WORKS. Rouelle's major work is *Tableau de l'analyse chimique* (Paris, 1774). Articles are in the *Journal de médecine, chirurgie, pharmacie, &c.* and in *Observations sur la physique, sur l'histoire naturelle et sur les arts.* His article "Observations . . . sur l'air fixe et sur ses effets dans certaines eaux minerales" was reprinted in Lavoisier, *Opuscules physiques et chymiques* (Paris, 1774), 154–170. Interfoliated notes in one copy of G.-F. Rouelle's lectures often report the work of Hilaire-Marin, Bibliothèque Municipale de Bordeaux, MSS 564–565, pp. 558a–558d and *passim.*

II. SECONDARY LITERATURE. On Rouelle and his work, see J.-P. Contant, *L'Enseignement de la chimie au Jardin royal des plantes de Paris* (Cahors, 1952); and [J.-A. Mongez?], "Eloge de M. Rouelle le jeune," in *Observations sur la physique, sur l'histoire naturelle et sur les arts,* **16** (September 1780), 165–174.

A short eulogy by P.-J. Macquer, Hilaire-Marin's colleague at the Jardin du Roi, is in the Bibliothèque Nationale, Paris, MS fr. 9132, fols. 175–176. Most works on Guillaume-François also deal briefly with Hilaire-Marin.

RHODA RAPPAPORT

ROUELLE, JEAN (*b.* probably Douzy [Ardennes], France, 1751 or 1753; *d.* unknown), *chemistry, natural history.*

Nephew of the two Rouelle brothers, Jean was probably the same nephew described by contemporaries as laboratory assistant to Guillaume-François at the Jardin du Roi. He was a physician and in 1775 succeeded Hilaire-Marin as apothecary to the Duc d'Orléans. He sailed in 1788 to the United States, having been named to a ten-year post as mineralogist in chief and professor and demonstrator of natural history, chemistry, and botany at the new Académie des Sciences et Beaux-Arts des États-Unis (Richmond, Virginia). The Academy, established by

A.-M. Quesnay de Beaurepaire, soon foundered, but Rouelle fulfilled part of his contract by forming a collection of animals, plants, and minerals which he brought back to France in 1798. He then taught chemistry at the École Centrale in Charleville (Ardennes). He was a member of the American Philosophical Society (20 January 1792).

BIBLIOGRAPHY

John [*sic*] Rouelle, *A Complete Treatise on the Mineral Waters of Virginia* (Philadelphia, 1792); MS "Observations sur les cultures coloniales et en particulier sur celle de la canne a sucre . . .," Bibliothèque Centrale, Muséum National d'Histoire Naturelle, Paris, MS 1981, I, no. 2505, Archives Nationales, Paris, AJ[15]551; H. E. Sigerist, "Rise and Fall of the American Spa," in *Ciba Symposia*, **8** (1946), 313–326; H. M., "L'Enseignement à l'École centrale de Charleville," in *Études ardennaises*, no. 14 (July, 1958), pp. 21–23.

RHODA RAPPAPORT

ROUGET, CHARLES MARIE BENJAMIN (*b.* Gisors, France, 19 August 1824; *d.* Paris, France, 1904), *physiology, histology*.

The son of a surgeon in the armies of Napoleon, Rouget attended the Collège Sainte-Barbe and received medical training at the teaching hospitals of Paris. In 1860 he was appointed to the chair of physiology at the University of Montpellier, where, it is said, he was an efficient and inspiring teacher. In 1879 he was one of two candidates whose names were submitted by the Académie des Sciences for the vacant chair of physiology in the Muséum d'Histoire Naturelle. Although unsuccessful on this occasion, he was appointed to the professorship when it fell vacant again shortly afterward, and held it until his retirement in 1893. His colleagues specially remembered him as a brilliant histologist.

In summarizing Rouget's achievement one is faced with the same difficulty that beset the Academy when they were preparing his citation for the La Caze Prize of 1887. On account of the extent and variety of his published writings it was almost impossible to decide what work was most representative of his contribution to science. There are few instances in which his research is not characterized by some degree of originality. His interest in animal starch, on which he first wrote in 1859, was obviously informed by Claude Bernard's discovery that glycogen functions in the nutritive process by making sugar available; but the view that diabetes mellitus was caused by some internal nutritive disorder was largely Rouget's own speculation.

Animal reproductive organs and functions were among Rouget's recurring interests. Although not wholly original, one of his finest early papers, published in 1856, described the function of the gubernaculum testis in guiding the descent of the testes through the inguinal canal. He gave clear and tolerably accurate accounts of vascular engorgement of erectile tissue and of muscle contraction of the seminal glands in sexual phenomena. He also gave plausible explanations of ovulation and menstruation; of how the ovules, fertile or otherwise, find their way into the funnel of the oviduct; and of why menstrual hemorrhage occurs only in humans and other primates. His interest in sexual phenomena was partly an outgrowth of his speculation on the mechanism of erectility, and in 1868 he published "Erectile Movements," in *Archives de physiologie*, in which he explained that erectility is not, like contraction, a "simple act of the tissue," but the capacity of the tissue to hold unusual quantities of blood forced into its reservoirs by the constriction of blood vessels. As a result of his microscopic research on the female gonads, especially during 1879, he confirmed the presence of primordial egg cells in the ovaries of newly born mammals.

Rouget was especially interested in three topics: the eye, contractile tissue, and nerve endings. He wrote extensively on the eye, finally submitting a "new theory of vision." One of his more original contributions concerned visual accommodation, which is the automatic adjustment of the curvature of the lens so that a distinct image will be formed on the retina. The nearer the object, the greater the convexity of the lens. The lens is suspended from a ring of thickened tissue (the ciliary process) whose circular contraction reduces the tension on the suspensory ligament and enables the lens, with its natural elasticity, to assume a more convex shape. Within the ciliary process are two sets of muscle fibers; one set runs meridionally and the other circularly. Rouget was not the first to discover the circular muscles, and he lost no time in refuting the suggestion that he had plagiarized some of the work of Johannes Müller. Rouget was more original in explaining how the circular fibers are responsible for reducing the diameter of the ciliary margin for visual accommodation, and on the whole his account is correct. Rather less plausibly, he also assigned a role to the blood vessels. As the circular muscles contract, he argued, the constriction of the veins, returning from the iris, induces an engorgement in the ciliary tissue, and the turgidity assists in the reduction of the diameter of the annular rim.[1] (Elsewhere Rouget used the same explanation of engorgement as part of his explanation of glandular contraction.)

Rouget's interest in the phenomenon of muscle contraction led him to put forward a strange theory. In 1861 he challenged Brücke's finding that because the alternate bands of striped muscle fiber exhibit different optical properties, the cause of muscle contraction is probably chemical. Rouget suggested that these effects were due to the physical arrangement of the inner parts of the fiber. In 1863 Rouget wrote, "If smooth fibers can become striped in consequence of the foldings, which are the basis of their contraction, striped fibers can become completely smooth by forcible extension, and the mechanical conditions of their contraction are fulfilled by foldings similar to [those] seen in contracted smooth fibers." Briefly, his developed theory of muscle contraction, first published in 1866, envisaged the striped fibrils—that is, the longitudinal myofibrils within voluntary muscle fibers—as resilient helices, the contraction of which was due to simple elasticity but the extension of which was a function of the energy derived ultimately from nutrition.[2] As a model for this physiological theory, he chose the mechanism of the coiled, ciliated process of the Vorticella.[3] Obviously he was drawn into untenable views about muscle action in relation to nervous excitation.

Capillary contractility is phenomenon closely linked with Rouget. Although people were aware of the periodic changes in the caliber of blood capillaries, there was no evidence of any contractile apparatus, such as are found in the middle coat of arteries. Examining the capillaries of the hyaloid membrane of the eye of a frog, Rouget found on the outside, endothelial surface of the capillary, isolated nuclei surrounded by protoplasm, the ramifications of which extend like rings around the vessel.[4] Having observed what he took to be constriction of these ramified extensions on the capillaries of newt larvae, he concluded that they are modified smooth muscle cells. Since then, there has been a continued controversy on the function of the "cells of Rouget," and the topic was still a living issue in the 1960's, the available information indicating that although they were prototypes of smooth muscle-fiber, it was very unlikely that they were actively contractile.

Rouget's research on nerve endings has also been of permanent value. In 1866 and 1868 he published findings on the sensory receptors of the skin, particularly the receptors associated with Filippo Pacini and Wilhelm Krause, in which he adumbrated, though in terms that would be rejected today, that the capsule of the latter is not independent tissue, as Koelliker

had maintained, but a development of the myelin sheath of the supplying sensory nerve.[5] One of Rouget's last papers was on the termination of sensory nerve fibers in skeletal muscle.[6] Some of his finest microscopic research was more particularly concerned with end plates.[7] His descriptions of motor plates are mainly elaborations of the histological research of Doyère, Quatrefages, and Kühne, although he added considerable detail to their accounts. Thus he described how the "nerve tube" (axon) with its contour (Schwann's sheath) reaches the surface of the muscle, and then opens out, the nerve substance putting itself into contact with the contractile fibers of the muscle, and the contour of the "nerve-tube" mingling with the sarcolemma. He identified the end of the nerve plate as a disk of granular matter (cytoplasm), with several nuclei, which was extended on the surface of the muscle fibrils. He failed, however, to appreciate the significance of the synaptic cleft.

One outgrowth of Rouget's interest in nerve endings is seen in his papers on the electric organs of certain fishes. Thus he recorded the resemblance between the pattern of nerve threads that supply the electroplates, and those seen in motor nerve endplates. While not going so far as to suggest that the electroplates of the Torpedo are modified muscle tissue, he maintained that the electrical discharge produced by the nerves was merely an outflow of the energy that normally would lead to muscle movement.

NOTES

1. "Recherches anatomiques et physiologiques sur les appareils érectiles, . . ." in *Comptes rendus hebdomadaires des séances de l' Académie des sciences*, **42** (1856), 937–941.
2. "Note sur des photographies microscopiques relatives à la structure des muscles et aux phénomènes de la contraction musculaire," *ibid.*, **62** (1866), 1314–1317.
3. "Note sur les phénomènes de contraction musculaire chez les vorticelles," *ibid.*, **64** (1867), 1128–1132.
4. "Note sur le développement de la tunique contractile des vaisseaux," *ibid.*, **79** (1874), 559–562.
5. "Sur la structure intime des corpuscules nerveux de la conjonctive et des corpuscules du tact chez l'homme," *ibid.*, **66** (1868), 825–829.
6. "Terminaison des nerfs musculaire sur les faisceaux striés," *ibid.*, **123** (1896), 127–128.
7. "Note sur la terminaison des nerfs moteur dans les muscles," *ibid.*, **62** (1866), 1377–1381.

BIBLIOGRAPHY

Material on Rouget is remarkably scant. See the obituary, M. Nestor Gréhant, "Charles Rouget, notice necrologique," in *Nouvelles archives du Muséum d'histoire naturelle*, **6** (1904), iii. Information on Rouget's earlier career is in the autobiographical *Notice à l'appui de la candidature de M. le Dr. Charles Rouget à la chaire de* *physiologie vacante à la Faculté de Médecine de Montpellier* (Paris, 1860). For an appraisal of Rouget's contribution to medicine, see the citation "Prix L. La Caze," in *Comptes rendus hebdomadaires des séances de l'Académie des sciences*, **105** (1887), 1372. A catalog of Rouget's more important publications is "Liste des ouvrages et mémoires publiés par M. Ch. Rouget," in *Nouvelles archives du Muséum d'histoire naturelle*, **6** (1904), viii–xiii.

A. E. BEST

ROUILLIER, KARL FRANTSOVICH (*b.* Nizhny-Novgorod [now Gorky], Russia, 20 April 1814; *d.* Moscow, Russia, 21 April 1858), *biology, paleontology, geology.*

Rouillier's father, a shoemaker, was a Frenchman who had settled in Russia; his mother was a midwife. He was educated in Nizhny-Novgorod and, from 1829 to 1833, at the Moscow Medical and Surgical Academy. He was a military physician until 1840; from then until his death he held the chair of zoology at Moscow University. He was secretary of the Moscow Society of Natural Scientists (1840–1851), and from 1854 until his death he was editor of the journal *Vestnik estestvennykh nauk* ("Herald of the Natural Sciences").

Rouillier's early works (1840–1848) were devoted to classical studies of the Jurassic, Carboniferous, and Quaternary deposits of the Moscow basin. His proposed schema for subdividing these deposits in the Srednerussky (Mid-Russian) basin is still significant. He explained the uniqueness of the two upper levels of the Upper Jurassic surrounding Moscow in terms of the existence of various climatic zones and of isolated faunal areas during the Jurassic period; he was thus the first to raise the question of the zoogeography and paleoclimatology of the Jurassic seas.

Rouillier's comparative-historical approach was fundamental to his method of investigation. To understand the interrelationships and evolution of a species, it is insufficient, he believed, merely to establish morphological affinity between various groups of fossilized and modern forms. Rather, it is necessary to demonstrate that certain forms were transformed into others by "a series of gradual changes," that is, to establish their actual lineage. It is also necessary to analyze the constantly changing relationships of an organism to its environment that occur during its entire period of existence.

Rouillier's zoological work centered mainly on the influence of environmental conditions on animal life and included the study of the migration of birds and fishes, seasonal phenomena, the modifiability of animals under domestication, and the nature and origin of instincts. As early as 1841 he had rejected the concept of the unchangeability of species, and his profound conviction in the historical development of the organic world remained unshakable. He stressed the occurrence of slow and incessant changes in function, as well as in organization of animal life.

Rouillier was guided by the ideas of Lamarck and Geoffroy Saint-Hilaire in his attempt to explain the causes and laws of heredity, and his well-formulated theory of biological phenomena was the most advanced of its time. Each characteristic of an organism, according to Rouillier, is merely the historically established form of the relationship between the organism and its environment; it is innate "only to the extent that it was innately possible and necessary for it to appear at a certain time." Despite "the principle of final causes" there is no absolute expediency or predetermined adaptation of an organism to its environment. In opposition to Lamarck, Rouillier demonstrated the possibility of proving evolution without invoking the "internal efforts" of an organism to its "will to improvement," and he rejected Lamarck's mechanistic notion of "orgasm." Rouillier ascribed great significance to the complete extinction of separate systematic groups and often returned to this question.

Under the general concept of external conditions Rouillier included not only the abiotic environment but also the mutual influence of organisms. He distinguished between intraspecies and interspecies relations, noted the supplanting of one species by another, and considered the "competition" between species. The discrepancy between the number of offspring produced and the number reaching maturity also interested Rouillier, and such ideas as "war in nature" and nature as "a natural theater of war" recur in his work.

Although the concept of natural selection was missing in Rouillier's thought, his results represented a major step in the study of the development of organic forms from the simple to the complex, determined by the interaction of an organism with its environment. He was a founder of the first school of evolutionists. His historical explanation of the origin and development of animal instincts crowned his teaching, emphasized its breadth, and testified to his deep understanding of the universality and fruitfulness of the principle of evolution.

Rouillier suffered persecution by the authorities practically from the beginning of his career, and he labored under intolerable conditions forced upon him by the minister of education. He died of a brain hemorrhage at the age of forty-four.

BIBLIOGRAPHY

I. ORIGINAL WORKS. Rouillier's selected biological works were published as *Izbrannye biologicheskie proizvedenia*, L. S. Davitashvili and S. R. Mikulinsky, eds., in the series Klassiki Nauki (Moscow, 1954). His papers include "Les principales variations de Terabratula acuta dans l'oolite de Moscou," in *Bulletin de la Société impériale de naturalistes de Moscou*, n.s. **17** (1844), 889–894; "Über die Fauna des Moskauer Gouvernements und ihre Veränderungen in den einzelnen Epochen der Erdbildung," in *Archiv für wissenschaftliche Kunde von Russland*, **5** (1847), 443–482; "Études progressives sur la paléontologie des environs de Moscou," in *Bulletin de la Société impériale de naturalistes de Moscou*, **20** (1847), 371–447, written with A. Vosinsky; and "Études progressives sur la géologie de Moscou," *ibid.*, **21** (1848), 263–268; and **22** (1849), 3–17, 337–399, written with A. Vosinsky.

II. SECONDARY LITERATURE. On Rouillier and his work, see A. P. Bogdanov, *K. F. Rulie i ego predshestvenniki po kafedre zoologii Moskovskogo universiteta* ("K. F. Rouillier and His Predecessors in the Department of Zoology at Moscow University"; Moscow, 1885); S. R. Mikulinsky, *K. F. Rulie i ego uchenie o razvitii organicheskogo mira* ("K. F. Rouillier and His Work in the Development of the Organic World"; Moscow, 1957), with a detailed bibliography of Rouillier's works and of secondary literature; S. V. Petrov, *Vydayushchysya russky biolog K. F. Rulie* ("The Eminent Russian Biologist K. F. Rouillier"; Moscow, 1949); and B. E. Raykov, "Russky biolog-evolyutsionist K. F. Rulie" ("The Russian Biologist-Evolutionist K. F. Rouillier"), in *Russkie biologi-evolyutsionisty do Darvina*, III ("Russian Biologist-Evolutionists Before Darwin"; Moscow–Leningrad, 1955).

S. R. MIKULINSKY

ROUTH, EDWARD JOHN (*b.* Quebec, Canada, 20 January 1831; *d.* Cambridge, England, 7 June 1907), *mechanics.*

Born in Canada, the son of a high-ranking army officer, Routh was sent to England in 1842 to attend University College School, London. He subsequently studied at University College, London, where he fell under the influence of Augustus de Morgan, and after graduating B.A. there, he matriculated as a pensioner at Peterhouse, Cambridge. James Clerk Maxwell entered Peterhouse at the same time but transferred after two terms to Trinity College (it is said because he saw in Routh a strong competitor for the academic honors in the gift of the fellows of Peterhouse). If Maxwell moved for that reason, then his judgment was sound, for in the mathematical tripos of 1854, Routh was senior wrangler and Maxwell was second. In the examination for the Smith's prizes of the same year the names of Routh and Maxwell were bracketed as first equal; it was the first time on record that the examiners had divided the honors equally between two candidates.

Elected to a fellowship of Peterhouse in 1855, Routh thenceforth dedicated himself to the task of preparing undergraduates for the public examinations of the University of Cambridge. He was an inspiring teacher and became the most famous of the great Cambridge "coaches." His success in coaching students for the mathematical tripos may be gauged by the fact that twenty-seven of his students were senior wranglers and more than forty of them were Smith's prizemen. His gift for lucid exposition is also evidenced by his authorship of a set of advanced treatises which were destined to become the standard texts of classical applied mathematics. The first, and probably the most famous, of these was the two-volume *A Treatise on Dynamics . . . of Rigid Bodies* (1860); by the time of the author's death this had gone to a seventh edition and, in German translation, had the distinction of a foreword by Felix Klein. This was followed by *A Treatise on Analytical Statistics . . .* (1891) and *A Treatise on the Dynamics of a Particle . . .* (1898). Routh was coauthor (with H. W. Watson) of *Solutions of Senate House Problems* (1860).

Although he regarded himself primarily as a teacher, Routh through his original papers made a distinctive contribution to classical mechanics. His theorem of the "modified Lagrangian function" was one of the most significant contributions to the mechanics of his time. It was contained in the *Treatise on the Stability of a Given State of Motion, Particularly Steady Motion* (1877) for which he was awarded the Adams Prize for that year. For this and for his work on dynamical stability, interest in which has recently revived, he was elected a fellow of the Royal Society of London. In 1883 he shared with his friend W. H. Besant the distinction of being the first to graduate Sc.D. in the University of Cambridge.

BIBLIOGRAPHY

I. ORIGINAL WORKS. Routh's books were *Analytical View of Sir Isaac Newton's Principia* (London, 1855), written with H. Brougham; *A Treatise on the Dynamics of the System of Rigid Bodies* (London, 1860); *A Treatise on the Stability of a Given State of Motion Particularly Steady Motion* (London, 1877); *A Treatise on Analytical Statics with Numerous Examples*, 2 vols. (Cambridge, 1891–1892); and *A Treatise on the Dynamics of a Particle With Numerous Examples* (Cambridge, 1898). He also wrote numerous papers, the results of which are reported fully in the treatises cited above.

II. SECONDARY LITERATURE. Short notices of Routh's life are contained in obituaries by W. W. Rouse Ball, in *Cambridge Review*, **28** (1907), 480–481; by S. Larmor, in *Nature* (27 June 1907), 202; A. J. Forsyth, in *Proceedings of the London Mathematical Society*, **5** (1907), xiv–xx; and H. H. T., in *Monthly Notices of the Royal Astronomical Society*, **68** (1908), 239–241. See also the obituary notice, J. Larmor, in *Proceedings of the Royal Society*, **84A** (1910–1911), xii–xvi; and J. D. Hamilton Dickson, in *Dictionary of National Biography*, 2nd. supp., III, 233–235.

I. N. SNEDDON

ROUX, PIERRE PAUL ÉMILE (*b.* Confolens, Charente, France, 17 December 1853; *d.* Paris, France, 3 November 1933), *bacteriology*.

Roux was one of the principal founders of medical bacteriology, both through his collaboration with Pasteur and his own achievements. After attending secondary school in Aurillac and in Puy, he began his medical studies at Clermont-Ferrand. There he met Émile Duclaux, who imparted to Roux his enthusiasm for Pasteur. To complete his studies, Roux went to Paris, where in November 1878 he was accepted as an assistant at Pasteur's laboratory. With Chamberland and Thuillier he became associated in Pasteur's research on the etiology of anthrax and then on the attenuation of viruses, the basis for the preparation of Pasteur's vaccines.

After the perfection of vaccines against chicken cholera and anthrax, research on the prevention of rabies was begun. The difficulties were particularly great, since the pathogenic agent, or rabies virus, remained unknown. Roux succeeded in producing experimental cases of rabies (especially in rabbits), which displayed a regular sequence of development (by selecting a *virus fixe*). The virulence of a rabic medulla could then be diminished under certain conditions.

When the Institut Pasteur was created in 1888, Roux was placed in charge of instruction in microbiology. At the same time, he became director of the Service de Microbie Technique and began his most important original work. First, he confirmed the pathological role of the diphtheria bacillus, discovered shortly before in Germany by Klebs and Loeffler. With the bacillus he was able to reproduce paralysis experimentally in guinea pigs. Finally, and perhaps most important, he demonstrated with A. E. J. Yersin that the pathogenic power of this bacillus depends not merely on its presence but, rather, on a poison, or toxin, that it produces. This toxin spreads throughout an organism, and, in Roux's words, "snake venoms themselves are not as deadly." Soon afterward an analogous toxin, produced by the tetanus bacillus, was discovered by Knud Faber in Denmark.

In the course of their research to transform bacterial toxins into vaccines, a fundamental discovery was made in Berlin by Behring and Kitasato: a counterpoison (or antitoxin) forms in the serum of animals which have received quantities of toxin too weak to kill them and this serum can be used to protect other animals against an injection of a quantity which would otherwise certainly be fatal. It seemed curious, however, that among human beings the protection derived in this manner proved to be comparatively weak.

For that reason, Roux began to study the subject once again. Putting aside for the moment antitetanus serotherapy, he devoted all his efforts to investigating diphtheria. Working with horses, he determined the best conditions (1892–1893) for obtaining an antidiphtheritic serum; and in 1894, with the aid of Louis Martin and Auguste Chaillou, he treated diphtheritic children with this serum. The results, presented later that year to the Tenth International Congress of Hygiene in Budapest, evoked a response famous for its enthusiasm. The disease was completely conquered after the discovery of Gaston Ramon's vaccine, the *anatoxin*.

Roux's personal work was almost completely interrupted when, upon the death of Duclaux (1904), he became director of the Institut Pasteur. He held this post until he died, subordinating his own work in order to be of more help to those under him.

BIBLIOGRAPHY

I. ORIGINAL WORKS. Roux wrote very little. His memoirs written with Pasteur before 1890 were published in the *Comptes rendus hebdomadaires des séances de l'Académie des Sciences et de l'Academie de médecine*. Several later papers (1904–1933) can also be found there. His most important investigations were presented in the *Annales* and the *Bulletin de l'Institut Pasteur* (until 1903). The most important are "Contribution à l'étude de la diphtérie," in *Annales de l'Institut Pasteur*, and "Trois cent cas de diphtérie traités par le serum antidiphtérique," *ibid.*

II. SECONDARY LITERATURE. Numerous studies have been devoted to Roux. A list of these is given in A. Delaunay, *L'Institut Pasteur, des origines à aujourd'hui* (Paris, 1962). Two biographies should also be mentioned: M. Cressac, *Le docteur Roux, mon oncle* (Paris, 1950); and E. Lagrange, *Monsieur Roux* (Brussels, 1954).

A. DELAUNAY

ROUX, WILHELM (*b*. Jena, Germany, 9 June 1850; *d*. Halle, Germany, 15 September 1924), *embryology, developmental mechanics, anatomy.*

Roux single-mindedly devoted his life to science. Even in his autobiography he gave only the scantiest details about his family and extrascientific activities. The obituaries and tributes by admirers fail to provide us with more.

Roux was the fourth child and only son of F. A. Wilhelm Ludwig Roux, a well-known university fencing master, and Clotilde Baumbach. The paternal side of the family stemmed from a Huguenot line that fled France after the revocation of the Edict of Nantes. Roux himself described his youth as being *freudearm* and explained that by inclination he remained aloof from most of his school comrades. Instead, possessing an early interest in science, he immersed himself "secretly" in Johann Müller's *Pouillet's Lehrbuch der Physik und Meteorologie.* At the age of fourteen he attended the Oberrealschule in Meiningen, where its director encouraged his scientific bent. The Franco-Prussian War interrupted his first year at the University of Jena. Upon his return from military service, and after additional preparatory studies, Roux matriculated in 1873 in the medical faculty. He attended the lectures of Preyer, Haeckel, and Gustav Schwalbe, and Rudolf Eucken's seminar on Kant, an experience Roux always cherished. Two semesters in Berlin working under Virchow, a semester listening to Friedrich von Recklinghausen in Strasbourg, and a dissertation completed in 1878 under the supervision of Schwalbe completed Roux's formal education. During the winter of 1877 he passed his state medical examinations.

Roux's first employment was in Leipzig as an assistant at Franz Hofmann's hygienic institute, a position which he later confessed taught him the value of exacting laboratory techniques. Relief came from the drudgery of analyses when in 1879 Carl Hasse invited him to his anatomical institute in Breslau (Wrocław). Roux remained there for ten years (1879–1889), first as *Dozent*, then as associate professor, and finally, after April 1889, as director of his own Institut für Entwickelungsgeschichte. In August of 1889 Roux became professor of anatomy in Innsbruck but returned to Prussia in 1895 as director of the anatomical institute at the University of Halle, a position he held until April 1921. By the time of his retirement he had seen himself honored on many occasions as the founder of *Entwicklungsmechanik.* At Halle a prize had been established in his name for contributions in experimental embryology, and the University of Leipzig made him an honorary doctor at its tercentenary. Roux was a member of thirty-seven professional societies and an honorary member of the Deutsche Anatomische Gesellschaft and the American Society of Naturalists. At his death Roux was survived by his wife, Thusnelda Haertel, two sons, Erwin and Wilhelm, and a daughter, Irmgard.

Any attempt to evaluate Roux's long and prolific scientific career is made difficult by the task of sifting the real achievements from the reams of proclamations and self-assertions. That Roux had a substantial impact upon his peers and upon contemporary embryology there can be little doubt, but it is equally clear that he was a ferocious propagandist for his own accomplishments.

Jena, where Roux studied and did his first research, was an exciting and intense place for an aspiring embryologist. Haeckel, at the height of his career, had just refashioned the zoological institute. While Roux was a student, Haeckel lectured on general zoology, vertebrate natural history, embryology, anthropology, and human histology. Roux attended some of these courses and confessed that he was influenced by Haeckel's style and monistic philosophy; he insisted, however, that he never worked directly under Haeckel's tutelage. The physiologist Preyer was closer to his own age. Roux knew him personally and must have been familiar with his many physiological experiments on developing chicks. Later Preyer's *Spezielle Physiologie des Embryo* (1885) revealed the extent to which his physiological experiments had become oriented to the problems of development, if not differentiation. Roux's own supervisor, Schwalbe, was an anatomist who had a lively concern for the mechanics of growth. It was Schwalbe who first directed Roux to the problem of relating form to function in the embryo.

Roux's dissertation grew out of his studies on the form of branching blood vessels in muscle mesoderm. He developed a technique of injecting wax into the vessels, and upon dissolving the surrounding tissues, he was left with only a naked casting of the branches. His intent was to generalize the shapes and angles of vessel branching into rules of development. Thus, for example, Roux concluded that the axis of a tributary vessel lay on the plane which contained the longitudinal axis of the main vessel and the center of the ellipse formed by the juncture of both vessels. It is interesting to note that even at this early stage in his career Roux did not confine his generalizations to a descriptive equation. He drew a parallel between the vessel branches and the shape and direction of flowing water and thereupon arrived at a tentative conclusion that blood pressure had a bearing on the patterns of branching. By making an analogy between hydrodynamics and hemodynamics, Roux implied a search

The results were much more dramatic than his earlier *Anstichversuche* had led him to anticipate. Upon preserving and staining his specimens at gastrulation, Roux found that he had produced half-embryos: semiblastulae, and semigastrulae with clearly identifiable single neural folds, and anterior and posterior semigastrulae. It seemed to him that the undisturbed blastomeres had continued to develop as though nothing had happened. Roux inferred that development was "a mosaic of at least four vertical pieces each developing substantially independently."[3] Roux also noted that the damaged half occasionally became reorganized at a later stage as nuclei from the developing half migrated into the cytoplasm of the disturbed blastomere—a "post-generation," as he called it. Both phenomena, the normal development of the undamaged blastomeres and the regeneration of the pierced blastomeres, demonstrated the interplay between Roux's basic factors of independent and dependent differentiation. His experimental technique had isolated a formative action in the nucleus and had seemingly shown that the nucleus of each blastomere contained both the necessary and sufficient developmental components to direct a specific independent line of differentiation.

Although this mosaic pattern of development in the frog was shown in 1910 to be an inaccurate description of true amphibian development, it accorded well with a highly speculative suggestion made by Roux himself as early as 1883. At that time Roux became interested in Strasburger's and Flemming's work on karyokinetic figures. Obviously, Roux rationalized, such a complex and indirect division of the nucleus must serve some functional purpose; here perhaps was a mechanism for dividing the nucleus into two qualitatively unequal halves and for segregating these differences to the daughter blastomeres. During this speculative foray, Roux also indicated that he had become wedded to a belief that the nucleus itself was composed of a complex of macromolecules or hereditary particles; it was these units, he suggested, which were "purposefully" parceled out during the mitotic divisions of development. The 1888 "half-embryo" work, with its dramatic demonstration on the somatic level of independent differentiation of the first four blastomeres, corroborated perfectly with the notion of unequal nuclear division. Roux forgot the fanciful overtones of his 1883 suggestion and began to speak of mosaic development as a germinal phenomenon as well as a description of somatic events.

In 1893, in a monograph on mosaic development, Roux underscored the connection he had inferred between the events of early cleavage and the patterns of nuclear division. There is no doubt that in this general statement Roux saw inheritance and development as opposite sides of the same coin. The material that passed from generation to generation had to be so constituted as to initiate and carry on nearly identical processes of embryogenesis. At the same time the somatic cells, those resultants of differentiation, must contain very different genetic arrangements. Roux felt the paradox could only be resolved by accepting an isolated germinal track and his earlier suggestion of qualitatively unequal nuclear division during development. Moreover, in accepting the proposition that there existed hereditary particles, Roux insisted that he was not resurrecting the discredited theory of preformation; after all, he still recognized a correlative interaction between cells at most stages of development and during the process of regeneration. Only in 1902, after Theodor Boveri had marshaled strong evidence against qualitative nuclear division, did Roux relinquish his belief that a mosaic development was directly caused by the mechanics of mitotic division. The concept of a mosaic development was restored to a description of somatic events. After Boveri's demonstration Roux reasoned that the cytoplasm and the nucleus must continually interact with one another during development. His primary distinction between independent and dependent differentiation, however, remained a basic dichotomy and organizing theme for subsequent embryological research.

It was claimed that Roux was a propagandist as well as an investigator. The banner under which he marched was *Entwicklungsmechanik*, and the domains he defended were experimentalism and mechanism. Roux maintained a parochial view of the advent of experimentalism in embryology. His concern to establish priority over Pflüger when they simultaneously published experimental work on the determination of the longitudinal axis in amphibians was a typical reaction. On a number of occasions Roux drew up a four-tier hierarchy of scientific method in an effort to distinguish his own accomplishments from those of his contemporaries and predecessors. He explained that the (1) descriptive and (2) comparative methods were the most primitive and least informative about the causes of development. The (3) descriptive experiment, Roux claimed, was characteristic of the endeavors of His, Rauber, and Balfour, all of whom had searched for causes and even on occasion performed laboratory manipulations. Their experiments, however, were not systematic and were so constructed as to reveal only analogies between mechanical and embryonic phenomena. Roux maintained that only (4) a causal analysis, that combination of mental and laboratory isolation and manipulation of biological factors, gave the scientist a certainty about the causes of development. He contended that he alone had first moved beyond casual manipulations to a causal analysis of development; this was the core of his *Entwicklungsmechanik*.

His insistence was, of course, absurd, for it overlooked the work of his contemporaries Hertwig, Pflüger, and Born, to say nothing of the older tradition of experimental teratology in France. Nevertheless, Roux's assertions had some basis. He had dramatized experimentation on the embryo in a way that his peers had not, and in a unique manner he had made explicit that a true experiment was not just a manipulation of the egg but included a mental dissection of development into a number of hypothesized factors. In this respect his analysis of development into dependent and independent factors promoted a causal understanding. Roux's model for experimental embryology was physiology, and it was befitting that the physiologist Rudolf Heidenhain suggested to Roux the term *Entwicklungsmechanik* for his program of research.

Entwicklungsmechanik denoted more than a method of research; Roux made it clear that he was ascribing a mechanistic interpretation to the embryological events he discovered. It was not a simple and crude mechanism which read into every vital phenomenon the direct action of matter in motion. Instead, it entailed a commitment to the forces and matter legitimized by physicists and chemists and accepted these as the appropriate and sole ontological entities to be employed in a causal understanding of life. Despite such a metaphysical commitment, Roux recognized that biologists might never succeed in the task of explaining vital activities in physical and chemical terms. This nevertheless was the challenge and the only legitimate pursuit for the biologist.

Finally, *Entwicklungsmechanik* was a declared program of research. Roux recognized that its methods were akin to physiology yet the object was the explanation of form; it was therefore a new approach to a traditional morphological problem. In a revealing passage, Roux explained that he employed the term because no matter how experimentally inclined his students might be, they would never be eligible for academic chairs in physiology. The appellation helped them maintain a claim on positions in anatomy.

To further his program, Roux promoted a number of publications. His *Archiv für Entwicklungsmechanik der Organismen* (after his death retitled *Roux Archiv für Entwicklungsmechanik*) was founded in 1894 and continues to be published. For many years it was the most prestigious journal in experimental embryology and served as the model for its American counterpart, *Journal of Experimental Zoology*. Roux founded two

monograph series, in 1905 and 1909. He initiated the first of these with his *Entwicklungsmechanik, ein neuer Zweig der biologischen Wissenschaft*, which came closer than any of his other works to serving as a textbook for his program. A dictionary, *Terminologie der Entwicklungsmechanik der Tiere und Pflanzen*, compiled in collaboration with two botanists, Correns and Küster, and an anatomist, Fischel, presented a valuable compendium of definitions and historical notes for the general area of experimental morphology. Roux himself is cited as the originator or discoverer of so many of the entries, however, that the work today is little more than a mirror of a past age, reflecting the tight and egocentric hold Roux kept on his program. A *Festschrift*, published on his seventieth birthday, contained both accolades and thoughtful papers devoted to examining Roux's impact on other fields. Its collective message left the impression that many leading biologists identified with a tradition headed by Roux. Roux produced in his autobiography a long list of biologists (forty-four Germans, seventeen Austrians, and twenty-three Americans, among others) who he claimed were all representatives of *Entwicklungsmechanik*. Despite the implications, they can hardly be considered to have constituted a school of Roux students and devotees, for the list contains nearly all the important embryologists and cytologists of the day.

Experimental embryology was hardly the invention of Roux. He excelled in it at a time when it made dramatic strides; he made substantial contributions to theories of development and inheritance; he proselytized for and philosophized about it; and he produced the journals and terminology for its dissemination. To the degree that he performed these tasks more zealously than his contemporaries, he was the titular sire of modern experimental embryology.

NOTES

1. Eduard Pflüger, "Ueber den Einfluss der Schwerkraft auf die Theilung der Zellen," in *Pflüger's Archiv für die gesammte physiologie des Menschen und der Tiere*, **32** (1883), 64.
2. Wilhelm Roux, "Beiträge zur Entwickelungsmechanik des Embryo. Nr. 1). Zur Orientierung über einige Probleme der embryonalen Entwickelung," repr. in *Gesammelte Abhandlungen*, II, 154–155.
3. Wilhelm Roux, "On the Artificial Production of Half-Embryos by the Destruction of One of the First Two Blastomeres . . .," Benjamin H. Willier and Jane Oppenheimer, eds., *Foundations of Experimental Embryology*, p. 36.

BIBLIOGRAPHY

I. ORIGINAL WORKS. *Gesammelte Abhandlungen über Entwickelungsmechanik der Organismen*, 2 vols. (Leipzig, 1895), contains reprints of all the significant articles by

Roux published prior to 1895. Later major works include "Für unser Programm und seine Verwirklichung," in *Archiv für Entwickelungsmechanik der Organismen*, **5** (1897), 1–80, 219–342; "Ueber die Ursachen der Bestimmung der Hauptrichtung des Embryo im Froschei," in *Anatomischer Anzeiger*, **23** (1903), 65–91, 113–150, 161–183; *Die Entwickelungsmechanik, ein neuer Zweig der biologischen Wissenschaft (Vorträge und Aufsätze über Entwickelungsmechanik der Organismen)* (Leipzig, 1905); Roux, *et al.*, eds., *Terminologie der Entwicklungsmechanik der Tiere und Pflanzen* (Leipzig, 1912); "Die Selbstregulation, ein charakteristisches und nicht notwendig vitalistisches Vermögen aller Lebewesen," in *Nova acta. Abhandlungen der Kaiserlichen Leopoldinisch-Carolinischen Deutschen Akademie der Naturforscher*, **100**, no. 2 (1914), 1–91. Roux's autobiography, L. R. Grote, ed., "Wilhelm Roux in Halle a.S.," in *Die Medizin der Gegenwart in Selbstdarstellungen*, 2 vols. (Leipzig, 1923), I, 141–206, contains a full bibliography of all his writings. There are very few English translations of Roux's works. Two easily accessible ones are "The Problems, Methods, and Scope of Developmental Mechanics," William Morton Wheeler, trans., in *Biological Lectures Delivered at the Marine Biological Laboratory of Woods Hole in the Summer Session of 1894* (Boston, 1895), 149–190; and "Contributions to the Developmental Mechanics of the Embryo. On the Artificial Production of Half-Embryos by Destruction of One of the First Two Blastomeres, and the Later Development (Postgeneration) of the Missing Half of the Body," Hans Laufer, trans., in Benjamin H. Willier and Jane M. Oppenheimer, eds., *Foundations of Experimental Embryology* (Englewood Cliffs, N.J., 1964), 2–37.

II. SECONDARY LITERATURE. Kurt Altmann, "Zur kausalen Histogenese des Knorpels. W. Roux's Theorie und die experimentelle Wirklichkeit," in *Ergebnisse der Anatomie und Entwicklungsgeschichte*, **37** (1964), 5–31; Dietrich Barfurth, "Wilhelm Roux, ein Nachruf," in *Archiv für mikroskopische Anatomie und Entwicklungsmechanik*, **104** (1925), 1–22; Frederick B. Churchill, "Chabry, Roux, and the Experimental Method in Nineteenth-Century Embryology," in Ronald N. Giere and Richard S. Westfall, *Foundations of Scientific Method: The Nineteenth Century* (Bloomington, Ind., 1973), 161–205; Thomas Hunt Morgan, "Developmental Mechanics," in *Science*, n.s. **7** (1898), 156–158; Jane M. Oppenheimer, "Questions Posed by Classical Descriptive and Experimental Embryology" and "Analysis of Development: Methods and Techniques," in *Essays in the History of Embryology and Biology* (Cambridge, Mass., 1967), 62–91, 173–205; Eduard Pflüger, "Ueber den Einfluss der Schwerkraft auf die Theilung der Zellen," in *Pflüger's Archiv für die gesammte physiologie des Menschen und der Tiere* (1883) **31**: 311–318 and **32**: 1–79; E. S. Russell, *Form and Function, a Contribution to the History of Animal Morphology* (London, 1916), 314–334; E. S. Russell, *The Interpretation of Development and Heredity* (Oxford, 1930), 95–111; J. S. Wilkie, "Early Studies of Biological Regulation: An Historical Survey," in H. Kalmus, ed., *Regulation and Control in Living Systems* (London, 1966), 259–289.

"Wilhelm Roux zur Feier seines siebzigsten Geburtstages," in *Naturwissenschaften*, **8** (1920), 431–459, contains separate articles by Dietrich Barfurth, Hermann Braus, Hans Driesch, Ernst Küster, Georg Magnus, and Hans Spemann. For an excellent contemporary introduction to the problems of experimental embryology of the day see Thomas Hunt Morgan, *The Development of the Frog's Egg* (New York, 1897). There exists a privately printed genealogy of the Roux family, Oskar Roux, *Louis Roux aus Grenoble in Südfrankreich und seine Nachkommen in Deutschland und Amerika* (Jena, 1912).

FREDERICK B. CHURCHILL

ROVERETO, GAETANO (*b.* Mele, near Voltri, Genoa, Italy, 15 November 1870; *d.* Genoa, 23 November 1952), *geology, geomorphology*.

Rovereto was born into an aristocratic Genoese family, the son of Giuseppe Rovereto and Teresa Picardo. His formal education ended in the first years of secondary school. His early attraction to the natural sciences led him to attend, as an amateur, scientific meetings and geographical excursions. He thus met Arturo Issel, whom he eventually succeeded at the University of Genoa, and Senofonte Squinabol. They soon recognized his ability and encouraged him to study the geological sciences, in which he rapidly attracted notice. Rovereto had the advantages and defects of the self-taught: independence of judgment and originality in interpreting observed facts, but also a certain critical deficiency and an excessively personal and careless literary style that often rendered his arguments difficult to understand. He lived in Genoa, apart from a period in Argentina from 1910 to 1913, during which he conducted geological explorations from the Chaco to Patagonia. At the University of Genoa he was an assistant in the geological museum and then, from 1922, professor of geology, physical geography, and applied geology.

His first study (1891), of the serpentine regions of Liguria, was published after several years of research and was rich in original observations. Rovereto's continuing interest in the complex and disputed geology of Liguria culminated, after a half century of research, in the monumental "Liguria geologica" (1939), a model of analytic and synthetic treatment. He was fascinated by new ideas, and in these studies, as in others on the Apuan Alps, Monte Circeo, Capri, and Corsica, he applied the nappe theory for the first time in Italy, arriving at results which, although often vigorously contested, nevertheless contributed to the progress of modern tectonics. The first such observations (1904) were on the Apuan Alps; his last work, on the origin and mobility of the ophiolites of the Alps and Apennines, appeared in 1951, shortly before his death.

A skilled geologist and a brilliant geomorphologist, Rovereto published three extensive memoirs between 1902 and 1906: on coastlines, with special reference to Liguria; on the valleys of Liguria; and on the Gran Paradiso, one of the largest massifs of the Italian Alps. A volume of geomorphological studies on various regions of Italy (1908) was followed by a series of five memoirs (1911–1921) under the comprehensive title "Studi di geomorfologia argentina." Rovereto began his research at a time when the theories of W. M. Davis were becoming known; they differed from those of Rovereto, who had met him during a scientific excursion in northern Italy. In place of the concept of repetitive cycles—fundamental in Davis' theory—Rovereto proposed a theory of continuous development, as much for fluviatile as for karstic and marine erosion. In 1923–1924 he published a geomorphological treatise entitled *Forme della terra*.

Rovereto's interest was not confined to pure science. He dedicated a considerable part of his activity to applied geology, in Italy (railway planning) and in Argentina (hydraulic works for the irrigation of arid regions). He stimulated interest in the applications of geology by founding, in 1903 with Paolo Vinassa de Regny, the *Giornale di geologia pratica*. He also considered the practical aspects of geomorphology, especially in its influence on the character of towns.

BIBLIOGRAPHY

I. ORIGINAL WORKS. Rovereto's writings include "La serie degli scisti e delle serpentine antiche in Liguria," in *Atti della Società ligustica di scienze naturali e geografiche*, **2** (1891), 325–346; **4** (1893), 97–141, his first study; "Geomorfologia delle coste, ossia appunti per spiegare la genesi delle forme costiere," *ibid.*, **13–14** (1902–1903), 1–189; "Geomorfologia delle valli liguri," in *Atti della Università di Genova*, **18** (1904), 1–226; "Geomorfologia del gruppo del Gran Paradiso," in *Bollettino del Club alpino italiano*, **38** (1906), 1–75; *Studi di geomorfologia* (Genoa, 1908); "Studi di geomorfologia argentina," in *Bollettino della Società geologica italiana:* 1. "La Sierra di Cordova," **30** (1911), 1–19; 2. "Il Rio della Plata," *ibid.*, 313–350; 3. "La valle del Rio Negro," **31** (1912), 181–237; **32** (1913), 101–142; 4. "La Pampa," **33** (1914),75–128;**39** (1920), 1–48; 5. "La Penisola Valdéz," **40** (1921), 1–47; *Forme della terra*, 2 vols. (Milan, 1923–1924); and "Liguria geologica," in *Memorie della Società geologica italiana*, **20** (1939), 1–744.

II. SECONDARY LITERATURE. On Rovereto and his work, see S. Conti, "Gaetano Rovereto," in *Annali, ricerche e studi di geografia*, **8** (1952), 173–182, with complete bibliography; M. Gortani, "Commemorazione di Gaetano Rovereto," in *Rendiconti dell'Accademia dei Lincei. Classe di scienze fisiche*, 8th ser., **14** (1953), 336–337; and A. Sestini, "Gaetano Rovereto," in *Rivista geografica italiana*, **60** (1953), 54–55; and "Gli studi di geomorfologia argentina del prof. Gaetano Rovereto," in *Bollettino della Società geografica italiana*, 5th ser., **11** (1922), 123–125.

FRANCESCO RODOLICO

ROWE, ALLAN WINTER (*b*. Gloucester, Massachusetts, 31 July 1879; *d*. Boston, Massachusetts, 6 December 1934), *physiological chemistry*.

Rowe was the son of Arthur Howard Rowe and Lucy Haskell Rowe. In 1901 he received a B.S. in chemistry from the Massachusetts Institute of Technology; a M.S. from Wesleyan University in 1904; and in 1906 a Ph.D. from the University of Göttingen, where he studied under Nernst. He returned to the United States and undertook further graduate study with Theodore William Richards at Harvard University from 1907 until 1914 (receiving no degree). Late in life he was awarded the M.A. from Harvard as of 1908 in recognition of the value of his thermochemical studies. He described himself as an unmarried Republican, who although not a church member preferred Unitarian services.

Rowe's teaching career began in 1902 at Wesleyan, where he was assistant in chemistry for two years while working for his master's degree. Shortly after receiving his Ph.D. in 1906, he joined the faculty of the Boston University School of Medicine as a lecturer. Two years later he was promoted to professor of chemistry, a position he held until his untimely death. For most of his adult life Rowe was affiliated with the Robert Dawson Evans Memorial for Clinical Research and Preventive Medicine. Since he was one of the first research associates chosen by the founding director of the memorial in 1910, his advice was frequently sought in many matters; and in 1921 he was appointed director of research, a position he held until his death. He was responsible for drastically enlarging the staff of the memorial and also the scope of its research.

Physiological chemistry, specifically the disorders of the ductless glands, was his major area of research during his last twenty years. An orderly, logical survey of normal individuals and of those known to have abnormal endocrine functions led him to design a system of objective tests to recognize these disorders. Forty-seven papers, some written with collaborators, appeared between 1929 and 1934. Included among these publications were three series: "Studies of the Endocrine Glands," "The Metabolism of Galactose," and "Vital Function Studies." He also published papers on changes caused by pregnancy, the chemistry of urine, and behavior problems of the young. In 1932 his monograph *The Differential Diagnosis of Endocrine Disorders* appeared and was favorably reviewed by his professional colleagues. The book contains a survey of work done after 1912, but it is primarily a summary of his publications of 1929 and 1930.

Although not a physician, Rowe was honored by the medical profession. He was president of the Society for the Study of Internal Secretions (1933), vice-president and trustee of the Memorial Foundation for Neuro-Endocrine Research, and on the editorial staff of *Endocrinology*. He was a member of the American Academy of Medicine, honorary member of the American Medical Association and the Massachusetts Medical Society, and an honorary secretary of the International Anesthesia Society.

Rowe and Richard's studies of the thermochemistry of electrolytes resulted in the publication of standard data, which still appear in handbooks. Using the adiabatic calorimeter in their measurements (thus eliminating the need for complex cooling corrections in their calculations and yielding more accurate results), Rowe and Richards studied heats of dilution and specific heats of numerous simple acids, bases, and salts. The results were of interest in connection with the then new theories of electrolytic dissociation, although Richards cautioned against the premature attempt to explain the data solely on those bases.

The most significant studies of the series were determinations of the heat of neutralization of strong acids and bases. Checking earlier observations that this value is nearly independent of the acid and base used, Rowe observed that upon dilution of the solutions the heat decreases regularly and a nearly constant value is indeed observed for different samples. Extrapolation to zero concentration yielded a value for the heat of dissociation of water of about 13.65 calories per mole. These preliminary results later enabled Richards and his students to determine the value with greater accuracy and to investigate other systems such as acetic acid, which shows an increase with dilution in the heat of neutralization. Although Rowe had completed the experimental work by 1914, publication was delayed until after the end of World War I.

Shortly before the war, Rowe sustained a badly fractured ankle and was rejected for active military service. He performed, however, a great service to his country by organizing and helping to maintain Base Hospital 44, which was regarded as a model field hospital. He took pride in his nickname "Mother of the Unit."

Rowe's knowledge of different areas of chemistry was essential to his teaching at Boston University, where at different times he gave courses in inorganic, organic, biological, physiological, and pathological chemistry; qualitative and quantitative analysis; toxicology; and dietetics. He was an accomplished lecturer. During the winter of 1932–1933, he gave twenty-five speeches to learned associations. A number of graduate students earned doctorates under his direction.

BIBLIOGRAPHY

I. ORIGINAL WORKS. Rowe's monograph *Differential Diagnosis of Endocrine Disorders* (Baltimore, 1932) lists in its bibliography many of his significant articles on that subject. His thermochemical studies are cited in his last paper on that subject, "The Heats of Neutralization of Potassium, Sodium, and Lithium Hydroxides With Hydrochloric, Hydrobromic, Hyriodic, and Nitric Acids, at Various Dilutions," in *Journal of the American Chemical Society*, **44** (1922), 684, written with Theodore W. Richards.

II. SECONDARY LITERATURE. A short, anonymous obituary notice with a portrait is in "Allan Winter Rowe," in *Bostonia* (Jan. 1935), 2–8. A more readily accessible abstract of this notice is W. Goodwin, "Allan Winter Rowe, 1879–1934," in *Journal of the Chemical Society* (1935), 863–864. The only detailed treatment of his thermochemical studies is Sheldon J. Kopperl, *The Scientific Work of Theodore William Richards*, Ph.D. dissertation, University of Wisconsin (Madison, 1970), 218–222.

SHELDON J. KOPPERL

ROWLAND, HENRY AUGUSTUS (*b.* Honesdale, Pennsylvania, 27 November 1848; *d.* Baltimore, Maryland, 16 April 1901), *physics.*

Rowland was the son of Harriette Heyer, the daughter of a New York merchant, and Henry Augustus Rowland, Sr., who, like his father and grandfather before him, had gone to Yale and entered the Protestant clergy. In the spring of 1865 Rowland's mother (now a widow) enrolled him in the Phillips Academy at Andover, evidently as a first step toward Yale and the ministry; but Rowland had been an avid chemical and electrical experimenter as a boy, and he wanted to study engineering. In the fall of 1865 he went to the Rensselaer Polytechnic Institute, where, after developing a distaste for a career in the business world, he resolved to devote himself to science. In 1870 he graduated with a degree in civil engineering and then spent a year as a railroad surveyor and another year as a teacher at the College of Wooster in Ohio. In 1872 he returned to Rensselaer as an instructor of physics.

Rowland's first major research was an investigation of the magnetic permeability of iron, steel, and nickel. In order to determine this quantity, he set up toroidal transformers made of each of the three metals in question, broke or reversed the direct current in the primary windings, and measured the charge that flowed in the secondary circuit. Plotting the permeability against what he thought was the induced magnetic field, B, for each of the metals, Rowland found that a general mathematical function could be fitted to all the curves; but by using the toroidal arrangement, he had actually measured ΔB rather than B. His data were distorted by effects of hysteresis unknown to him, and it is clear from modern theories of ferromagnetism that his mathematical function had no physical significance. Rowland had proved that—contrary to the assumption then common in the literature on ferromagnetism—magnetic permeability varied with the "magnetizing force," H and, hence, with B. His work on the subject won the praise of Maxwell and established his reputation as one of the most promising young experimental physicists in the United States.

In 1875 Rowland accepted the chair in physics at the new Johns Hopkins University and went to Europe for a year to inspect various laboratories and purchase apparatus. He traveled widely, discussed contemporary physics with many leading practitioners of his discipline, including Maxwell, with whom he became good friends, and Helmholtz, in whose laboratory he spent four months. Rowland spent about $6,000 on apparatus for the Hopkins physics laboratory, emphasizing equipment suited for research rather than for teaching demonstrations. By the late 1870's the Hopkins facility was far better equipped than any other American or even many European laboratories. In part because of his European trip, Rowland kept in touch all his life with developments in physics abroad, which was unusual for an American physicist of his day. He regarded himself as midway between an experimental and mathematical physicist and often tried to focus his experimental efforts on problems of theoretical import.

In 1868, stimulated by his study of Faraday's *Electrical Researches*, Rowland conceived an experiment to test whether the magnetic effect produced by electric current was the direct result of charge moving through space or of some interaction between the current and the conducting body. Performing the experiment while in Helmholtz' laboratory, Rowland used a charged vulcanite disc with an astatic needle suspended above it to register magnetic effects. He found no magnetic effects in the arrangement for the interactive case, in which the charge was held stationary while the disc was made to rotate. But he did detect magnetic effects when the charge was allowed to rotate with the disc, and the motion of the astatic needle correlated with the rotational sense of the current. Though not decisively in favor of one or another of the prevailing electrical theories, Rowland's experiment was the first, as Helmholtz reported to the Berlin Academy of Science, to demonstrate that the motion of charged bodies produced magnetic effects.

Like his studies of permeability, Rowland's subsequent work was characterized by meticulous attention to experimental detail and remarkable mechanical ingenuity. In the late 1870's he established an authoritative figure for the absolute value of the ohm. At the opening of the 1880's, he painstakingly redetermined the mechanical equivalent of heat and conclusively showed that the specific heat of water varied with temperature. Then Rowland turned to the work for which he is best known, the invention and ruling of the concave spectral grating.

The range, resolving power, and accuracy of a grating were determined respectively by the number, density, and regularity of its rulings. Lewis Rutherfurd, an amateur astronomer in New York City, had managed to rule up to two square inches of metal with thirty thousand lines per linear inch, but his gratings were inaccurate. Rowland recognized that to make a grating of highly uniform line-spacings, one needed an exceedingly regular drive screw in the ruling engine. He found that he could manufacture a nearly perfect drive screw from a roughly cut screw simply by grinding it in an eleven-inch-long nut which, split parallel to its axis, was clamped over the screw. With the problem of the screw overcome, Rowland could rule up to forty-three thousand lines per inch on more than twenty-five square inches of metal and, hence, construct gratings of unprecedented accuracy and resolving power.

Rowland saw numerous advantages in ruling his gratings on a spherically concave grating rather than on a flat surface. Since such gratings were self-focusing, they eliminated the need for lenses, in which the glass absorbed infrared and ultraviolet radiation. More important, the optical properties of a concave grating permitted a vast simplification in the observation of spectra. Consider a circle drawn tangent to the inner face of the grating with a radius equal to half its curvature. Wherever on the circle the source was placed, its spectrum would come to a normal focus at an eyepiece placed at the opposite end of a diameter from the grating. If the eyepiece was fixed at that point, the focus of the apparatus had to be set only once. By moving the source, one could quickly read off the wavelengths of numerous spectral lines from a scale on a chord of the circle, easily photograph the spectrum of one element superimposed on that of another element, or reliably determine line intensities even in the infrared region.

The concave grating reduced the work of days to a few hours, and Rowland sold over 100 of them at cost to physicists throughout the world. In the 1880's Rowland remapped the solar spectrum; his wavelength tables, which were ten times more accurate than their best predecessors, became the standard for over a generation. At the Paris Exposition of 1890 his gratings and map of the solar spectrum received a gold medal and a grand prize. Rowland's numerous other professional honors included appointment as a delegate of the United States government to various international congresses on the determination of electrical units. He became a foreign member of the Royal Society of London and the French Academy of Sciences and was elected to the National Academy of Sciences, which awarded him its Rumford and Draper medals. He was a founder and also the first president of the American Physical Society.

In 1883, as vice-president of the American Association for the Advancement of Science, Rowland delivered a celebrated address, "A Plea for Pure Science," in which he disparaged technological invention and called upon his fellow countrymen to do more to foster basic research. But near the end of his life Rowland became an inventor himself. In 1890 he married Henrietta Troup Harrison. Not long afterward it was discovered that he had diabetes, and, eager to assure the future of his family, he worked on the development of a multiplex telegraph. Although technically successful, the system had not proved to be feasible commercially by the time of his death. In accordance with his express wish, Rowland's ashes were interred in the wall of the basement laboratory, where the engine with which he ruled his gratings was housed.

BIBLIOGRAPHY

A sizable collection of Rowland's personal papers and scientific notebooks (1868–1901) is at the Johns Hopkins University, and he left about 100 letters (1865–1884) with his daughter, Harriette H. Rowland. Numerous letters from Rowland exist in the papers of his good friend, Edward C. Pickering, in the archives of Harvard University.

Rowland's published writings and addresses are in *The Physical Papers of Henry Augustus Rowland* (Baltimore, 1902), which includes a biographical introduction by Thomas Corwin Mendenhall. John David Miller, "Rowland and the Nature of Electrical Currents," in *Isis*, **63** (1972), is indispensable. Other useful treatments of Rowland include J. S. Ames, "Henry Augustus Rowland," in *Dictionary of American Biography*, **8** (1935), 198–199; and Samuel Rezneck, "The Education of an American Scientist: H. A. Rowland, 1848–1901," in *American Journal of Physics*, **28** (1960), 155–162, and "An American Physicist's Year in Europe: Henry Rowland, 1875–1876," *ibid.*, **30** (1962), 877–886.

DANIEL J. KEVLES

ROWNING, JOHN (*b.* Ashby, Lincolnshire, England, 1701[?]; *d.* London, England, November 1771), *mathematics, natural philosophy.*

Educated in local schools, Rowning may then have worked with his father, also John Rowning (probably a watchmaker, as another son entered that trade, and John was credited with mechanical abilities). He was admitted as sizar to Magdalene College, Cambridge, in 1721. He gained the B.A. in 1724, a fellowship in 1725, and the M.A. in 1728. A college tutor for some years, he joined William Deane, an instrument-maker in London, about 1733 in giving courses of lectures in experimental philosophy. In 1733 he wrote a paper describing a barometer with a changeable scale of variation.

In 1734 Rowning became rector of Westley Waterless, Cambridgeshire, and by 1738, rector of Anderby, Lincolnshire, one of six livings in gift of Magdalene College. He became a member of the Gentleman's Society of Spalding, which was founded in 1710 and was the oldest provincial learned society in England. Under his urging, the Society temporarily forgot its antiquarian pursuits in the study of experimental philosophy. In 1756 he published the preliminaries to a projected text (never printed) in which he outlines his method of teaching fluxions and denies the "Analyst's" (George Berkeley) objections to the subject. His second, and last, mathematical work was a paper describing an analogue machine for the graphical solution of equations.

His most significant work was the *Compendious System of Natural Philosophy*, one of the most popular texts throughout the eighteenth century. The work was used at Cambridge and Oxford, at the College of William and Mary in Virginia, at many dissenting academies, and by John Wesley as a text for his itinerant preachers, it was also mentioned in the correspondence of people as various as John Adams, William Beckford, and Joseph Priestley.

Chiefly distinguished for its clarity, the work should also be noted for its explicit rejection of Newtonian ether, its explanation of forces as the continuing action of God upon matter, and its proposal of alternating spheres of attraction and repulsion some twenty years before Bošković's *Philosophiae*.

Rowning died at his London lodgings late in November 1771; he left a daughter, Mrs. Thomas Brown of Spalding, his heiress and executrix.

BIBLIOGRAPHY

I. ORIGINAL WORKS. Rowning's works are "A Description of a Barometer, Wherein the Scale of Variation May Be Increased at Pleasure," in *Philosophical Transactions of the Royal Society*, **38** (1733–1734), 39–42; the barometer is also illustrated and described in "Barometer," in *Encyclopaedia Britannica*, 3rd ed., III (1797), 25, with plate on facing page; *A Compendious System of Natural Philosophy: With Notes Containing the Mathematical Demonstrations, and Some Occasional Remarks*, pt. 1 (Cambridge, 1735); pt. 2 (London, 1736); pt. 3 (London, 1737); pt. 4 (London, 1742–1743). Each pt. was also revised and republished as successive pts. were issued, and any extant set may consist of varying eds. of pts. and secs. within pts. The 6th, 7th, and 8th eds. appear to be all of the same years: 1767, 1772, and 1779 respectively; *A Preliminary Discourse to an Intended Treatise on the Fluxionary Method* (London, 1756), reviewed by William Bewley, in *Monthly Review*, **14** (1756), 286–289; "Directions for Making a Machine for Finding the Roots of Equations Universally, With the Manner of Using It," in *Philosophical Transactions of the Royal Society*, **60** (1770), 240–256; and a copy of the syllabus of Rowning's course, *A Compleat Course of Experimental Philosophy and Astronomy*, is in the Science Museum, Oxford, MS Radcliffe 29.

II. SECONDARY LITERATURE. On Rowning and his work, see John Nichols, *Literary Anecdotes of the Eighteenth Century*, VI, pt. 1 (London, 1812), 109, 124; Robert E. Schofield, *Mechanism and Materialism* (Princeton, 1970), 34–39; and John Venn and J. A. Venn, *Alumni Cantabrigienses*, III, pt. 1 (Cambridge, 1927).

ROBERT E. SCHOFIELD

ROZHDESTVENSKY, DMITRY SERGEEVICH (*b.* St. Petersburg, Russia, 7 April 1876; *d.* Leningrad, U.S.S.R., 25 June 1940), *physics*.

Rozhdestvensky's father was a teacher in a Gymnasium. His parents encouraged him in his studies, and from childhood he studied foreign languages. He graduated in 1894 from the Gymnasium, with a silver medal, and entered St. Petersburg University. Rozhdestvensky graduated from the mathematical section of the university in 1900 and remained at the university to prepare for a teaching career. From 1901 to 1903 he worked with the physicist Paul Drude at the Institute of Physics in Giessen. On his return to Russia he became laboratory assistant at St. Petersburg University. He chose as the subject of his research the study of the course of anomalous dispersion near lines of absorption in sodium vapors. The lack of major specialists in optics at the university obliged Rozhdestvensky to work virtually independently, and he devised an original experimental method of research, which later was widely used. This project served as the theme for his master's dissertation, which he defended in 1912 and for which he was awarded the Mendeleev Medal of the Academy of Sciences. Extending his method to potassium, rubidium, and cesium vapors, Rozhdestvensky in 1915 defended his doctoral dissertation on simple relationships in the spectra of alkali metals. In 1916 he was elected professor and head of the Physics Institute of Petrograd University and began to work with a group of scientists and engineers on the preparation of optical glass, in which the Russian army was seriously lacking during World War I.

Soon after the October Revolution Rozhdestvensky presented a detailed project for the organization of a State Optical Institute, which was created in 1918 and of which he was director until 1932. At the Institute, Rozhdestvensky created a leading school, from which came important Soviet scientists including A. N. Terenin, V. A. Fok, Y. F. Gross, S. E. Frish, and V. K. Prokofiev. In 1925 he was elected corresponding member of the Academy of Sciences of the U.S.S.R. and, in 1929, academician. In the Academy Rozhdestvensky headed the Spectroscopic Commission until the end of his life.

Rozhdestvensky's research centered basically on three problems: anomalous dispersion near the lines of absorption in atomic spectra, the theory of atomic spectra, and the theory of the microscope. His quantitative research on anomalous dispersion was based on the experimental method that he developed of combining two spectral instruments, a diffraction grating and an interferometer, to obtain alternating light and dark bands. The visual demonstration of so-called hooks was an original way of showing refraction near the lines of absorption. This research made it possible for the first time to measure the numerical values of the ratios of intensity in the double lines of the alkali metals and the absolute values of the probability of quantum transitions, which play a very important role in the theory. The method was also successfully used for the study of discharge in gases.

Rozhdestvensky's work in the theory of atomic spectra (1920–1924) was important for interpreting spectra in the period before quantum mechanics. This research was especially useful in the development of quantum mechanical methods of calculating spectral terms worked out by V. A. Fok. In the theory of the microscope (1939–1940) Rozhdestvensky considered practical conditions of illumination of an object, especially the most effective use of the microscope for transparent biological objects.

Rozhdestvensky's organizational and scientific work in the State Optical Institute played a very important role in the development of the optical industry in the U.S.S.R.

BIBLIOGRAPHY

Rozhdestvensky's *Izbrannye trudy* ("Selected Works"; Moscow–Leningrad, 1964) includes a list of his writings. On his life and work, see K. K. Baumgart, "Dmitry Sergeevich Rozhdestvensky," in *Pyatdesyat let Gosudarstvennogo opticheskogo instituta* (Leningrad, 1968); S. E. Frish, *Dmitry Sergeevich Rozhdestvensky. Zhizn i deyatelnost* (". . . Life and Work"; Leningrad, 1954); and the notice in *Lyudi russkoy nauki. Matematika.* ("People of Russian Science. Mathematics"; Moscow, 1961), 303–313.

J. G. DORFMAN

RUBENS, HEINRICH (Henri Leopold) (*b.* Wiesbaden, Germany, 30 March 1865; *d.* Berlin, Germany, 17 July 1922), *physics.*

Rubens was the son of Barend Eliazer Rubens, a jeweler who had left Amsterdam to settle in Frankfurt am Main in 1859, and Bertha Kohn, who came from Speyer. On 17 March 1884 he matriculated from the Frankfurt Realgymnasium "Wöhlerschule," then entered the Darmstadt Technische Hochschule to study the new subject of "electro-technics" for one semester. He next attended the Polytechnic Institute in Berlin-Charlottenburg for a year. Having realized that his interest and talent "were preferentially directed to the theoretical study of mathematics and natural sciences,"[1] Rubens entered the University of Berlin to study physics in the winter of 1885. He next attended the University of Strasbourg, where he studied chiefly with the physicist August Kundt and the electrical engineer Franz Stenger. After four terms Rubens followed Kundt to the University of Berlin, from which he received the Ph.D. on 4 March 1889.

In 1890 Rubens became an assistant at the Königliches Physikalisches Institut of the University of Berlin; while he was doing research there he received the *venia legendi* on 17 February 1892. On 1 October 1896 he was given the title "professor" at the

Charlottenburg Technische Hochschule and appointed director of its physical laboratory. He was made full professor in 1900, and in 1904 he was named director of the Physikalische Sammlung. During this period he continued to carry out experimental work, chiefly at the Physikalisch-Technische Reichsanstalt. In 1906 Rubens succeeded Drude as professor of experimental physics at the University of Berlin and at the same time became director of its Königliches Physikalisches Institut. In 1908 he was elected to the Königlich Preussische Akademie der Wissenschaften zu Berlin.

Rubens' work was considerably influenced by that of Kundt, whose investigation of the optical properties of metals stimulated him to do research on the reflection of light from metals as a function of rays of great wavelength. From this early direction—and from the technology then at hand (including the principle of Svanberg's resistance thermometer)—Rubens' lifework, the exploration of the far infrared region, was determined. He had, moreover, favorable circumstances in which to work, since the University of Berlin was at that time becoming a center for European physical research. At the same time, he continued his theoretical work, as evidenced in his lectures on the "Theorie der Elasticität fester Körper mit besonderer Berücksichtigung einiger Probleme der Akustik und Optik" (winter 1892/1893 and 1893/1894) and "Theorie der Bewegung von Flüssigkeiten" (summer 1894 and 1895). "Die experimentellen Grundlagen der Maxwell'schen Theorie," a course that he announced for the winter terms of 1894 and 1895, grew out of his study of Maxwell's theory of electromagnetism and out of his correspondence with Hertz about it. In the summer of 1893 Rubens also gave a lecture entitled "Ausgewählte Capitel aus der Geschichte der Physik," one of several on the history of physics given at the University of Berlin.[2]

Rubens' interest in electricity and electrically generated waves is apparent in his first researches. In his doctoral thesis he set forth the result of an experiment in which he employed Carl Baur's tinfoil resistance thermometer to demonstrate that metallic reflection increases in the range from visible to infrared light. In collaboration with his former teacher Carl A. Paalzow, Rubens then repeated the experiment with a bolometer dynamometer that they had constructed to measure the intensities of electrical currents through the heat effect. Having confirmed the elastic theory of light, Rubens next turned to proving the existence of electromagnetic waves and, with R. Ritter, extended his earlier experiments by measuring bolometrically the polarization and reflection of electric waves. In 1890, in partial collaboration with

M. Arons, Rubens measured bolometrically the velocity of the propagation of electrically produced waves in isolating liquids and solid bodies and thereby tested the validity of Maxwell's law

$$n = \sqrt{\epsilon}.$$

Following a suggestion of Hertz, Rubens also used the bolometer to investigate standing electric waves in wires.

In 1892 Rubens extended the work of his former teacher Henri Du Bois, using both the bolometer and fine metal gratings to show the reversal of polarization of optical rays of sufficiently great wavelength—a first step toward establishing the electromagnetic theory in the far infrared region. By confirming and using such dispersion formulas as those of Eduard Ketteler (1887) and H. von Helmholtz (1893), Rubens was able by 1894 to measure wavelengths up to 8.95 microns, a value that Paschen had also just attained.

Rubens used a variety of substances in his refraction experiments, including rock salt and sylvite. With the American physicist E. F. Nichols,[3] who collaborated with him at Berlin from 1894 until 1896, Rubens began a series of experiments based on his observation that the refractive index of quartz cannot be measured for wavelengths greater than 4.20 microns because of the absorption properties of this substance. Their investigations of reflection, transmission, and dispersion indicated that in this region quartz became opaque, that it "[passed] over completely from a non-metallic to a metallic body." Drawing upon this data—and upon the phenomenon of selective absorption observed by G. Magnus (1870), together with the absorption bands assumed by the Ketteler-Helmholtz dispersion equation—Rubens and Nichols pointed out that by reflecting waves several times in the bands of the "same substance as the source advantageously emits . . . unassisted by either prism or grating, homogeneous rays of great wavelength may be obtained."

Using a sensitive bolometer of platinum, coated with an electrolytically deposited layer of platinum black, and the "Panzergalvanometer" (a galvanometer shielded against electromagnetic disturbances by an armor of cast steel, recently invented by Du Bois and Rubens), Rubens and Nichols then succeeded in detecting what Rubens in 1897 named *Reststrahlen* (or residual rays). The first substances that they tested were quartz, for which the value of the residual ray is about 0.0088 millimeter, and fluorite (about 0.0244 millimeter). In 1898 Rubens and Emil Aschkinass obtained the residual rays of rock salt (0.0512 millimeter), using Rubens' extremely sensi-

tive new thermopile of constantan and iron. In accordance with his close attachment to the idea of Hertzian waves, Rubens in 1896 produced a resonance of the fluorite rays on a grating of five-micron silver strips, finding it to be "in complete analogy" to the resonance of electric waves that Antonio Garbasso had demonstrated in 1893.

As early as 1898 Hermann Beckmann had, under Rubens' supervision, measured the variations with temperature of the residual rays reflected by fluorite to determine the exponential constant in Wien's law of energy distribution. He discovered that this constant was nearly twice as large as was consistent with the theory, a result that Rubens endeavored to save by a calculus of error.[4] On 7 October 1900, however, Rubens mentioned to Planck that his results concerning residual rays of great wavelength were actually consistent with a law that Lord Rayleigh had suggested in June of that year; inspired by this, Planck discovered his own radiation law on the same day. In their later work, neither Rubens nor his collaborator Kurlbaum ceased to stress the approximate character of Rayleigh's early law as well as the limitations of other formulas (including Planck's) on the distribution of radiation energy.

Rubens subsequently returned to the problem of relating properties of infrared waves to the electromagnetic theory of light. With Ernst Hagen, of the Physikalisch-Technische Reichsanstalt, he used different metals to determine experimentally the absorption, reflection, and emission of waves up to 25.5 microns. In 1903 they discovered that the coefficient of penetration of such waves, $100-R$, is inversely proportional to the root of the electrical conductance κ, a relationship for which Rubens' friend Planck immediately provided a theoretical basis. In 1910 Rubens and Hagen determined the region in which, according to Maxwell's theory, the heat radiation of metals perceptibly changes to a function of temperature to be between two and five microns. This result constituted the first confirmation of the validity of the electromagnetic theory as applied to the infrared.

From 1910 on, Rubens and his collaborators were concerned with bridging the gap between "optical" and "electric" waves. They found residual rays of greater length by using potassium bromide (88.3 microns) or, in 1914, thallium iodide (151.8 microns). Working from another direction, Rubens and Otto von Baeyer attempted in 1910 to close the gap by using small "electrical" waves of about two millimeters; their results were not unequivocal because of the inconstancy of the waves produced by the very small vibrators that they employed. The earlier method of isolating waves by virtue of the higher refractive index of quartz (first used in 1898) proved to be more successful; in 1910 Rubens and R. W. Wood obtained from quartz waves of 108 microns, measured interferometrically by means of the radio-micrometer. After having replaced the older zirconium-oxide burner by the Welsbach mantle (1905), which radiates selectively, Rubens and Baeyer in 1911 discovered the waves of 210 microns and 324 microns emitted by the quartz mercury lamp, a device that is still the only source of optical waves of great length.

The far-infrared rays that Rubens had obtained allowed him and Du Bois to prove conclusively the inversion of polarization of optical rays of great wavelength. Moreover, in 1914 Rubens and Schwarzschild demonstrated that solar rays of sufficiently great wavelength are absorbed by the atmosphere of the earth. Rubens' discoveries found further application in the atomic theories of Erwin Madelung and Frederick A. Lindemann in 1910, and, most importantly, in Niels Bjerrum's 1912 theory of a rotational-vibrational spectrum, a theory that was in turn applied in later work by Rubens and Georg Hettner. (Bohr's theory of equating the differences between atomic energies with the light quanta—rather than using the energies themselves, as Bjerrum did—was not then utilized by workers concerned with the far infrared.) In 1916 Planck saved the classical electrodynamics of Rubens and Hettner's "Bjerrum-quanta," while Einstein and Nernst had previously used Rubens' values to explain the temperature variation of specific heats. But in 1921 Planck's quantum hypothesis was confirmed by Rubens and by Gerhardt Michel, in response to questions raised by Nernst and by T. Wulf, who had in 1919 recognized certain differences between experimental data and the theoretical values embodied in the hypothesis. In these and other experiments Rubens proved his skill in the investigation of the far infrared and its applications.[5]

Rubens was also engaged in other areas of physics, including wireless telegraphy, radioactivity, acoustic waves, photoelectric effect, and the concept of potential. That he was open-minded toward the development of physical theories may be seen in the historical introductions to some of his papers.

NOTES

1. According to his own handwritten curriculum vitae, dated 19 Feb. 1892.
2. See H. Kangro, *Vorgeschichte des Planckschen Strahlungsgesetzes* (Wiesbaden, 1970), 126–127.
3. Rubens collaborated with Americans several times, for example, with Benjamin W. Snow of Henry, Illinois, and Augustus Trowbridge of New York.
4. See H. Kangro (1970), 176.
5. Rubens' monograph on this subject, although under contract with a publisher, was never printed.

BIBLIOGRAPHY

I. ORIGINAL WORKS. A nearly complete list of the papers that Rubens published in journals, compiled by G. Hettner, is in *Naturwissenschaften*, 10 (1922), 1038–1040; this may be supplemented by Poggendorff, IV and V. A bibliography of Rubens' papers on infrared radiation, compiled by E. D. Palik, is in K. D. Möller and W. G. Rothschild, *Far-Infrared Spectroscopy* (New York–London–Toronto, 1971), 680–689; the bibliography is nearly complete for this aspect of Rubens' work and lists reprints and translations of the original articles in addition to those in the above bibliographies, but it contains a number of errors (note that paper no. 9 of Du Bois and Rubens is correctly entitled "Einige neuere Galvanometerformen," in *Elektrotechnische Zeitschrift*, 15 [1894], 321–323; that no. 10 does not exist; that no. 50, "Recherches sur le spectre infrarouge . . .," in *Revue générale des sciences pures et appliquées*, 11 [1900], 7–13, is signed only by Rubens; and that no. 76, "Spectre d'émission des machons Auer," in *Radium*, 2 [1905], 397, is only an announcement of four lines by Léon Bloch).

Among the works of Rubens upon which this article is based, the most important are a diss. (2 March 1889), *Die selective Reflexion der Metalle* (Berlin, n.d. [1889]), which includes a "Vita" in Latin; "Über die Fortpflanzungsgeschwindigkeit electrischer Wellen in isolirenden Flüssigkeiten," in *Annalen der Physik*, 278 (1891), 581–592, written with L. Arons, which contains a new approach to Maxwell's relation; "Einige neuere Galvanometerformen," in *Elektrotechnische Zeitschrift*, 15 (1894), written with H. Du Bois, and containing an account of important new experimental devices; and "Über eine neue Thermosäule," in *Zeitschrift für Instrumentenkunde*, 18 (1898), 65–69.

Rubens' first paper on residual rays is "Über Wärmestrahlen von grosser Wellenlänge," in *Naturwissenschaftliche Rundschau*, 11 (1896), 545–549, written with E. F. Nichols. On the irregularities caused by selective molecular absorption, see "Beobachtungen über Absorption und Emission von Wasserdampf und Kohlensäure im ultraroten Spectrum," in *Annalen der Physik*, 300 (1898), 602–605, written with E. Aschkinass. "Isolirung langwelliger Wärmestrahlen durch Quarzprismen," *ibid.*, 303 (1899), 459–466, also written with Aschkinass, is concerned with chromatic aberration. "Über die Emission langwelliger Wärmestrahlen durch den schwarzen Körper bei verschiedenen Temperaturen," in *Sitzungsberichte der Königlich Preussischen Akademie der Wissenschaften* (1900), 929–941, written with F. Kurlbaum, sets out the decisive measurements that led to Planck's new radiation formula. "Über Beziehungen zwischen dem Reflexionsvermögen der Metalle und ihrem elektrischen Leitvermögen," *ibid.* (1903), 269–277, written with E. Hagen, gives proofs of the relationships of the properties of metals to the electromagnetic theory of light; see also "Änderung des Emissionsvermögens der Metalle mit der Temperatur," in *Verhandlungen der Deutschen Physikalischen Gesellschaft*, **10** (1908), 710–712, written with Hagen; and "Über die Änderung des Emissionsvermögens der Metalle mit der Temperatur im kurzwelligen Teil des Ultrarots," *ibid.*, **12** (1910), 929–941, written with Hagen.

Rubens' plan for a scientific life was presented in his inaugural lecture before the Berlin Academy and is in *Sitzungsberichte der Preussischen Akademie der Wissenschaften zu Berlin* (1908), 714–717. His remarks on the history of the Royal Physical Institute, founded in Berlin in 1862 by G. Magnus, are in *Geschichte der Königlichen Friedrich-Wilhelm-Universität zu Berlin*, III (Halle, 1910), 278–296.

Final proof of the polarization of rays in the far infrared, according to Maxwell's theory, is in "Polarisation langwelliger Wärmestrahlung durch Hertzsche Drahtgitter," in *Verhandlungen der Deutschen Physikalischen Gesellschaft*, **13** (1911), 431–444, written with H. Du Bois. Rubens' first application of Bjerrum's quanta is in "Das langwellige Wasserdampfspektrum und seine Deutung durch die Quantentheorie," in *Sitzungsberichte der Preussischen Akademie der Wissenschaften zu Berlin* (1916), 167–183, written with G. Hettner; a document on contemporary knowledge of atomism, published shortly before Bohr's work, is *Die Entwicklung der Atomistik* (Berlin, 1913). An excellent survey of heat radiation is "Wärmestrahlung," in P. Hinneberg, ed., *Die Kultur der Gegenwart, ihre Entwicklung und ihre Ziele*, **1**, pt. 3, sec. 3 (Leipzig–Berlin, 1915), 187–208. Rubens' only historical survey of his own work in infrared spectroscopy is "Das ultrarote Spektrum und seine Bedeutung für die Bestätigung der elektromagnetischen Lichttheorie," in *Sitzungsberichte der Preussischen Akademie der Wissenschaften zu Berlin* (1917), 47–63. For a final proof of Planck's radiation theory against justified criticism of systematical experimental deviations, see "Beitrag zur Prüfung der Planckschen Strahlungsformel," *ibid.* (1917), 590–610, written with G. Michel.

MSS (including letters) are listed in the T. S. Kuhn, J. L. Heilbron, P. Forman, L. Allen, eds., *Sources for History of Quantum Physics* (Philadelphia, 1967), 80, which contains some errors, and to which the following items may be added: in the Staatsbibliothek Preussischer Kulturbesitz, Berlin, Germany (not in Marburg) are two curricula vitae, dated 20 June 1890 and 19 Feb. 1892; a visiting card, dated 22 Nov. 1903; three letters to an unknown person, dated 2 June 1903, 17 June 1903, and 24 June 1903; a letter to Max Iklé, dated 31 Aug. 1908; a postal card to Hans Geitel, dated 8 Nov. 1907; a letter to Wilhelm Waldeyer, dated 17 July 1916; a letter to Eilhard Wiedemann, dated 25 Nov. 1916; a visiting card to Karl Flügge, dated 9 Dec. 1917; and a card to L. Darmstaedter, dated 27 Oct. 1909. In the Nachlass of Felix von Luschan are two postcards, dated 6 Apr. 1918 and 1 Mar. 1919, and one letter, dated 19 Dec. 1917, all directed to Felix von Luschan. In the Nachlass of Johannes

Stark are 10 letters to J. Stark, dated 15 Apr. (without year), 17 Nov. 1908, 4 Feb. 1909, 15 Oct. 1909, 3 Nov. 1909, 18 Jan. 1910 (xerographic copy), 16 May 1911, 13 Nov. 1912, 17 Feb. 1917 (xerographic copy), and 29 Mar. 1917; two letter-cards to J. Stark, dated 27 Nov. 1909 and 3 Nov. 1913; five postcards to J. Stark, dated 15 Feb. 1913, 18 Apr. 1913, 29 May 1913, 15 July 1913, and 14 Nov. 1913; two visiting cards to J. Stark, dated 15 May 1907 and 8 Dec. 1911; the copies of two letters of J. Stark, probably to Rubens, one dated 14 Nov. 1909, the other n.d. (not legible); one protocol of a discussion between Stark, Rubens, and Einstein (n.p., n.d.); and one letter of Marie Rubens to Stark, dated 12 Nov. 1909. The library of Erlangen mentioned in the *Sources for History of Quantum Physics* (1967) is in Germany (BRD); the 4 letters to H. Hertz in the library of the Deutsches Museum, Munich, are dated 3 June 1890, 13 Nov. 1890, 29 Oct. 1910, and 9 Feb. (without year); the letter to Leo Graetz is dated 21 July 1904.

A letter of Rubens to Arnold Sommerfeld, dated 18 June 1917, is also in the library of the American Philosophical Society, Philadelphia; and at the publishers F. Vieweg and Son, Braunschweig, Germany, are two letters of Rubens to the publishers, dated 28 June 1904 and 15 May 1906.

II. SECONDARY LITERATURE. A memorial collection, "Dem Andenken an Heinrich Rubens," is in *Naturwissenschaften*, **10** (1922), 1015–1040. It includes articles by W. Westphal (on Rubens' personality), E. Regener, G. Hertz, and O. von Baeyer (on two-millimeter electrical waves and the radiation of the mercury lamp), J. Franck and R. Pohl, and G. Hettner (on the history of Rubens' role in the immediate origin of Planck's radiation law), in addition to a bibliography of Rubens' papers in journals, compiled by Hettner.

A historical sketch of Rubens' scientific achievements is J. Franck, in *Verhandlungen der Deutschen physikalischen Gesellschaft*, 3rd ser., **3** (1922), 76–91. See also obituaries by J. Franck and R. Pohl, in *Physikalische Zeitschrift*, **23** (1922), 377–382; and "R. W. L.," and Joseph Larmor, in *Nature*, **110** (1922), 740–742. M. Planck, "Gedächtnisrede," in *Sitzungsberichte der Preussischen Akademie der Wissenschaften zu Berlin*, Phil.-hist. Klasse (1923), cviii–cxiii, emphasizes Rubens' work for the Academy. M. von Laue, "Heinrich Rubens," in *Deutsches Biographisches Jahrbuch*, IV (1929), 228–230, is somewhat superficial and contains some errors.

On Rubens' early research and the relationship between his residual-ray experiments and Planck's quantum physics, see H. Kangro, *Vorgeschichte des Planckschen Strahlungsgesetzes, Messungen und Theorien der spektralen Energieverteilung bis zur Begründung der Quantenhypothese* (Wiesbaden, 1970; English ed. in press), 49–60, 126–128, 160–164, 173–175, 200–208; see also H. Kangro, "Ultrarotstrahlung bis zur Grenze elektrisch erzeugter Wellen, das Lebenswerk von Heinrich Rubens, I. Experimenteller Beweis der elektromagnetischen Lichttheorie für das Ultrarot," in *Annals of Science*, **26** (1970), 235–259, and "II. Experimente zur Überbrückung der Spektrumslücke zwischen optischen und elektrischen Wellen, Verknüpfung mit der Quantentheorie," *ibid.*, **27** (1971), 165–170, which together give an account of Rubens' life and of his work on infrared radiation from about 1900.

Two portraits of Rubens are that in *Naturwissenschaften*, **10** (1922), facing page 1015, and that in *Physikalische Zeitschrift*, **23** (1922), 377.

Hans Kangro

RUBNER, MAX (*b*. Munich, Germany, 2 June 1854; *d*. Berlin, Germany, 27 April 1932), *physiology, hygiene.*

Rubner was the son of Johann Nepomuk Rubner, a locksmith, and of Barbara Duscher. His scientific inclinations were already dominant in his Gymnasium years in Munich, and he methodically pursued his interests outside of school. At the age of fifteen he had his own microscope and chemical apparatus. Rubner began studying medicine at Munich in 1873 and completed the course in 1877. When he received his doctorate in 1878, he already had several publications to his credit. While a medical student, he also worked in the chemical institute under Liebig's successor Adolf Baeyer and with Carl Voit, the director of the physiological institute.

After he had completed his studies, Rubner joined Voit's institute as an unpaid assistant. In his work on metabolism he developed a novel approach to bioenergetics. Dissatisfied with the purely chemical-material approach to nutritive substances and their effects, he investigated their energy values in the body. Voit did not agree with the direction his work was taking, and for a long time paid no attention to Rubner's *Habilitationsschrift*. In the meantime Rubner spent the academic year 1880–1881 in Carl Ludwig's physiological institute at Leipzig. There, where he hoped his physicalist approach would be more appreciated, he studied the physiology of the circulation, the muscles, and digestion and undertook an investigation of the metabolism of the mammalian muscle artificially supplied with blood.

In 1883 Rubner qualified as a lecturer in physiology at Munich, where in the next two years he conceived most of his novel views on energy maintenance, the validity of the law of the conservation of energy in the animal economy, the isodynamic relationship of nutrients, and the loss of energy through radiation and evaporation according to the law of surfaces. He later gradually deepened and demonstrated these views with extremely subtle methods and quantitative exactness. In 1880 Rubner published his medical dissertation, *Über die Ausnutzung einiger Nahrungsmittel im Darmkanal des Menschen.* Three years later he published "Die Vertretungswerthe der hauptsächlichsten organischen Nahrungsstoffe im Thierkörper." Setting forth his "law of isodynamics," which provided a method of calculating the quantity of each constituent (fats, proteins, starch) required to produce an equal amount of energy when consumed in the body. In 1885 he published the exact caloric values of nutritive substances.

In September 1885 Rubner was offered, on the same day, an assistant professorship in hygiene at Marburg and a chair in pharmacology at Munich. He chose Marburg, convinced that hygiene was simply applied physiology. In the same year an improvised institute of hygiene was created for him; in 1888 it was equipped with a lecture room and laboratories. Rubner remained at Marburg until the winter semester of 1891. While at Marburg he composed the works on heat regulation, body surface, and the exchange of substances that were published as "Biologische Gesetze" (1887).

In 1891 Rubner succeeded Robert Koch in the chair of hygiene in Berlin, with the explicit understanding that hygiene would be emphasized over bacteriology. A large new institute was built for him in 1905. Rubner moved to the chair of physiology on 1 April 1909, succeeding Engelmann. On 1 April 1913 he assumed additional duties as director of the Kaiser-Wilhelm-Institut für Arbeitsphysiologie, which he had been instrumental in creating. This organization undertook specialized research on industrial hygiene and industrial medicine.

Rubner published most of his works while at Berlin. They encompassed all aspects of the physiology of nutrition and metabolism, including the hygienic effects of clothing, climate, air and temperature, and the nutrition of entire populations. Among the most original results of his research in these years was his clarification of the specific dynamic effect of foodstuffs; begun between 1883 and 1885, this work was developed fully in *Gesetze des Energieverbrauchs* (1902). In 1894 Rubner definitively established the validity of the principle of the conservation of energy in living organisms—a goal toward which he and other physiologists had long been working. Between 1896 and 1903 he clarified the influence of cold on metabolism and the roles of radiation, conduction, and evaporation in heat loss. In 1892 Rubner had taken over the editing of the *Archiv für Hygiene und Bakteriologie.* He published many original articles in it, including those on the *Eiweissminimum*, the minimum daily protein intake that preserves a balance between nitrogen intake and nitrogen elimination, and on the attrition quota *Abnutzungsquote*, the amount of daily nitrogen loss through elimination without protein intake (1908). In a very original way Rubner sought to consider the lifespans of mammals in relation to their period of growth and the energy consumption of the protoplasm. When growth stops, the protoplasm can carry out only a constant magnitude of energy exchanges and ultimately loses even this capacity (1908). Lifespan is therefore a function of energy consumption.

During World War I Rubner became very active in the field of national nutrition. He investigated the shift in food consumption resulting from urbanization and other changes in German life (1913–1914). In 1918 Rubner studied in detail the disastrous effects on Germany's civilian population of the blockade maintained by the Allies beyond the end of the war. In his last years Rubner related the results of his research in nutrition and metabolism to very general human problems. He spoke before the Prussian Academy of Sciences in Berlin on world nutrition— past, present, and future (1928); on man's struggle for life (1928); and on the "enemies of mankind": malnutrition, disease, hunger, poor living conditions, and poor health care (1932).

Rubner possessed a rough manner, a notorious taciturnity, and a sarcastic sense of humor. As an investigator he was scrupulously exact in his measurements and inventive in the construction of proper apparatus (a nutrient calorimeter and an animal calorimeter, among many others). His achievements led to his election to the scientific academies of Munich, Vienna, Oslo, Washington, D.C., Stockholm, and Uppsala. In 1910–1911 he was rector of the University of Berlin.

BIBLIOGRAPHY

I. Original Works. Rubner's writings include "Die Vertretungswerthe der hauptsächlichsten organischen Nahrungsstoffe im Thierkörper," in *Zeitschrift für Biologie*, **19** (1883), 313–396; "Biologische Gesetze," in *Jahresberichte der Universität Marburg* (1887); "Die Quelle der thierischen Wärme," in *Zeitschrift für Biologie*, **30** (1894), 73–142; *Gesetze des Energieverbrauchs bei der Ernährung* (Leipzig–Vienna, 1902); *Das Problem der Lebensdauer und seine Beziehungen zu Wachstum und zu Ernährung* (Munich–Berlin, 1908); *Volksernährungsfragen* (Leipzig, 1908); *Kraft und Stoff im Haushalt der Natur* (Leipzig, 1909); *Die Ernährungsphysiologie der Hefezelle bei alkoholischer Gärung* (Leipzig, 1913); *Wandlungen in der Volksernährung* (Leipzig, 1913); *Über moderne Ernährungsformen* (Munich–Berlin, 1914); "Der Kampf des Menschen um das Leben," in *Sitzungsberichte der Preussischen Akademie der Wissenschaften zu Berlin* (1928), lxviii–cvi; "Die Welternährung in Vergangenheit, Gegenwart und Zukunft," *ibid.*, Phys.-math. Kl. (1928), 159–183; "Ernstes und Heiteres aus meinem Leben," in *Deutsche medizinische Wochenschrift*, **56** (1930), 1099–1101, 1138–1140; and "Die Feinde der Menschheit," in *Sitzungsberichte der Preussischen Akademie der Wissenschaften*, Phys.-math. Kl. (1932), 329–335.

II. Secondary Literature. See E. Atzler, "Worte des Gedenkens für Max Rubner," in *Arbeitsphysiologie*, **5** (1932), 497–499; R. Fick, "Gedächtnisrede auf Max Rubner," in *Sitzungsberichte der Preussischen Akademie der Wissenschaften zu Berlin* (1932), meeting of 30 June; E. Grafe, "Zum Tode von Max Rubner," in *Medizinische Klinik*, **28** (1932), 740–741; Karl Kisskalt, "Zu Rubners 100. Geburtstag," in *Archiv für Hygiene und Bakteriologie*, **138** (1954), 235; K. E. Rothschuh, *Entwicklungsgeschichte physiologischer Probleme in Tabellenform* (Munich, 1952), 33–41; *Geschichte der Physiologie* (Berlin–Heidelberg, 1953), 182–183; and *Physiologie. Der Wandel ihrer Konzepte, Probleme und Methoden* (Freiburg–Munich, 1968), 290–293; Oscar Spitta, "Max Rubner zum Gedächtnis," in *Zentralblatt für die gesamte Hygiene und ihre Grenzgebiete*, **27**, no. 3 (1932), 160a–160f; H. Steudel, "Max Rubner †," in *Forschungen und Fortschritte*, **8** (1932), 203–204; and "Zum Tode von Geheimrat Rubner," in *Die Medizinische Welt*, **6** (1932), 724; and K. Thomas, "Max Rubner," in *Deutsche medizinische Wochenschrift*, **50** (1924), 727; "Max Rubner," in *Mitteilungen der Max-Planck-Gesellschaft* (1954), 63–68; and "Erinnerungen an Max Rubner," *ibid.*, (1960), 91–111.

K. E. Rothschuh

RUDBECK, OLOF (*b*. Västerås, Sweden, 1630; *d*. Uppsala, Sweden, 17 September 1702), *medicine, anatomy, botany.*

Rudbeck was the son of Johannes Rudbeckius, bishop of Västerås, and the most important Swedish ecclesiastic of his time. He received his early schooling in Västerås, then, in 1648, entered the University of Uppsala to study medicine. Although the Uppsala Faculty of Medicine was not a distinguished one, Rudbeck learned at least the fundamentals of anatomy and botany from one of his professors there, Johannes Franckenius. Rudbeck then began to work on his own, particularly in animal anatomy.

Since Harvey had published his discovery of the circulation of the blood twenty years earlier, the attention of anatomists had turned to the systems of vessels in animals. Rudbeck was accordingly drawn to this subject and in the fall of 1650, when he was not yet twenty years old, reported on previously unknown vessels (lymphatic vessels) that carried a colorless fluid from the liver. At the same time, and independently of Pecquet, he discovered the thoracic duct, through which the lacteal vessels discharge chyle into the veins. Rudbeck performed a number of systematic dissections and vivisections of calves, sheep, cats, and dogs, and by the following year was able to elucidate the structure of this system and to demonstrate its connections with the lymph glands.

In 1652 Rudbeck demonstrated his anatomical discoveries, using a dog as his subject, before Queen Christina and her court at Uppsala. It might thus be considered that he had communicated them at this time, although he postponed publication. Instead, he brought out a preliminary dissertation on the circulation of the blood, *De circulatione sanguinis* (1652), in which he discussed Harvey's still controversial doctrine in terms similar to those employed by Johannes de Wale and Thomas Bartholin, but in addition presented arguments based on his own experiments. Only then, in the summer of 1653, did he publish his short work on the lymphatic system, *Nova exercitatio anatomica, exhibens ductus hepaticos aquosos et vasa glandularum serosa*, a clear and convincing description of the newly discovered vessels (the *vasa serosa*) and of their course and valves, the lymph glands, and the nature of the lymphatic fluid.

Rudbeck's delay in publishing his findings led to a bitter priority dispute. Thomas Bartholin and his assistant Michael Lyser had been conducting research in Copenhagen on the lymphatic system at about the same time that Rudbeck was doing his work, and in spring 1653, a few weeks before Rudbeck published his paper, Bartholin brought out one of his own, *Vasa lymphatica*. In 1654 Siboldus Hemsterhuis published both papers in his *Messis aurea triennalis*, and Rudbeck's originality came into question. While Bartholin remained aloof from the controversy, his student Martin Bogdan attacked Rudbeck's work. Rudbeck, who was in Holland at the time, began the dispute with some remarks directed against Bartholin, published in Hemsterhuis' work, and Bogdan immediately issued a separate pamphlet accusing Rudbeck of plagiarism. Rudbeck defended himself with *Insidiae structae* (1654), and Bogdan promptly responded with *Apologia pro vasis lymphaticis*, published in the same year. Rudbeck offered his final statement of the matter in 1657 in *Ad Thomam Bartholinum danum epistola*, in which he reiterated his account of his discoveries and passionately repudiated Bartholin's claims to priority.

Amid these disputes, Rudbeck moved from Holland, where he completed his medical education at the University of Leiden (under Johannes van Horne, among others), to Sweden, where he was in 1655 appointed assistant professor in the Medical Faculty of Uppsala University. In 1660 he was appointed full professor. An enthusiast in all that he did, Rudbeck set out to raise the Medical Faculty from the state of decay into which it had fallen. To this end, he used

his own money to establish a botanical garden, which soon became one of the best in Europe. In 1662–1663 he also built (partly with his own hands) a spacious anatomical theater, which he had designed after the Anatomicum in Padua. This theater (which is still extant) was erected on the roof of the university building called the Gustavianum.

Rudbeck extended his activities to encompass reform of the entire university. He designed a new building to house it, created an institute for the physical education of the sons of the nobility, and established a workshop with models of machines for technological instruction. He was also active as an architect, and built houses, bridges, and water conduits. He printed his own books, and organized the musical life of the university (he himself composed music and, upon request, sang, in a great bass voice that is said to have drowned out the trumpets and the drums). He was also a warm defender of scientific liberty and an active supporter of Cartesian philosophy, a subject of heated controversy at Uppsala in the 1660's and 1680's.

Rudbeck's interest in anatomy and the teaching of medicine gradually waned as he pursued these other projects. From the 1670's he devoted most of his efforts to patriotic historical works; inspired by early Swedish historians, Rudbeck took up the notion that Sweden was the cradle of civilization, Plato's lost Atlantis from which the Greeks, Phoenicians, and other early peoples had received their knowledge and gods. He developed this idea in his bizarre and overwhelming four-volume work, *Atland eller Manhem* (commonly called *Atlantica*), published in Swedish and Latin (1679–1702), which achieved considerable European notoriety.

At the same time, Rudbeck undertook his most important botanical work. About 1670 he began to plan, with typical lack of moderation, an illustrated book that would describe all known plants. He was inspired in this enterprise by the herbarium of Joachim Burser, which had been donated to Uppsala in 1666, and he made Gaspard Bauhin's *Pinax* his model in nomenclature and species definition. He employed a large number of draftsmen and wood engravers, including his daughters and his son Olof, and set them to work; by 1690 some 2,000 blocks had been cut, and another 1,200 were added in the next ten years. Rudbeck and an even larger staff concurrently worked on an equally comprehensive series of hand-drawn and hand-colored illustrations, all of them, like the woodcuts, life-size wherever possible. The first two parts of the woodcuts appeared in 1701 and 1702; entitled *Campus Elysius*, they contained rough but clear and beautiful woodcuts of grasses, lilies, and orchids. It is estimated the entire work should have comprised about 7,000 species, but the great fire that swept Uppsala in 1702 destroyed almost all of the finished blocks, together with Rudbeck's collections, books, and manuscripts. He died shortly thereafter.

Rudbeck's important work as an anatomist was, perhaps, marred by his dispute with Bartholin. It is now clear that both men worked independently and that both should properly be considered to have discovered the lymphatic system. Although it is certain that Bartholin published first, and that he gave the lymphatic glands their present name, it is equally certain that Rudbeck was the first to demonstrate the glands and to recognize their importance. Rudbeck's published description of the lymphatic system is, moreover, the more complete and clear one.

As a botanist, Rudbeck founded the tradition of natural history at Uppsala from which the young Linnaeus later profited. Rudbeck's son, Olof (1660–1740), was a more direct link, succeeding his father in 1691 as professor of medicine at Uppsala. The younger Rudbeck was a competent botanist and zoologist, whose *De fundamentali plantarum notitia rite acquirenda*, published at Utrecht in 1690, may be considered a precursor of Linnaeus' work in reforming the botanical system. (He also wrote a work, with colored illustrations, on the birds of Sweden, which is still extant.) About 1730 he became Linnaeus' patron and teacher and was instrumental in inspiring the fruitful journey to Swedish Lapland that Linnaeus undertook in 1732.

BIBLIOGRAPHY

I. ORIGINAL WORKS. Rudbeck's most important anatomical and botanical writings are cited in the text. His *Nova exercitatio* has been printed in facsimile (Uppsala, 1930) and translated into English, with an introduction by Göran Liljestrand, in *Bulletin of the History of Medicine*, **11** (1942), 304–339. He also published three catalogues of the plants in the botanical garden in Uppsala, the last and most extensive being *Hortus botanicus* (Uppsala, 1685). Ninety woodcuts of the *Campus Elysius* were published by J. E. Smith as *Reliquiae Rudbeckianae* (London, 1789). A complete bibliography of the works of both Rudbecks, father and son, is in Johannes Rudbeck, *Bibliotheca Rudbeckiana* (Stockholm, 1918).

II. SECONDARY LITERATURE. There is no modern biography of Rudbeck. His anatomical work, with a general background, is treated by Erik Nordenskiöld in *The History of Biology* (New York, 1928), 145–147. On his work on the circulation of the blood, see Sten Lindroth, "Harvey, Descartes, and Young Olaus Rudbeck," in *Journal of the History of Medicine*, **12** (1957), 209–219. The best recent discussions of Rudbeck's discovery of the

lymphatic vessels and his polemic with Bartholin are Nils von Hofsten, "Upptäckten av bröstgången och lymfkärlssystemet," in *Lychnos* (1939), 262–288; and Axel Garboe, *Thomas Bartholin*, I (Copenhagen, 1949), 120–173. For Rudbeck's botanical works, see Gunnar Eriksson, *Botanikens historia i Sverige intill år 1800* (Uppsala, 1969), 71–76, 135–138, and the bibliography.

STEN LINDROTH

RÜDENBERG, REINHOLD (*b*. Hannover, Germany, 4 February 1883; *d*. Boston, Massachusetts, 25 December 1961), *electrical engineering, electrophysics*.

Rüdenberg, the son of a manufacturer, studied electrical and mechanical engineering at the Technische Hochschule in Hannover. After passing his final degree examination (1906) and his doctoral examination with distinction, he was assistant at Göttingen to Ludwig Prandtl, who was engaged in aerodynamic research. In 1908 Rüdenberg entered the Siemens-Schuckert works in Berlin as a testing engineer for electrical machines. Later he was placed in charge of the development division of the Berlin plant and served as the firm's chief electrician. In 1916 he created the world's first 60-MVA turbine generator for the Goldenberg power station in the Rhineland.

In 1913 Rüdenberg became *Privatdozent* at the Berlin Technische Hochschule. His first lectures dealt with three-phase commutator motors. He was granted the title "professor" in 1919 and named an honorary professor in 1927.

In over 100 publications, including several books, Rüdenberg treated heavy-current engineering and, occasionally, light-current engineering as well. His textbook on electrical switching processes was a great success; the first edition appeared in 1923 and the fourth, in English, in 1950. His more than 300 patents record the many contributions he made to all areas of electrical engineering.

In 1936 Rüdenberg decided to leave Germany; he went to England, where he worked until 1938 as a consulting engineer for the General Electric Company, Ltd., in London. In 1939 he accepted the Gordon McKay professorship at Harvard. There he lectured on electric machines, on energy transfer, and on switching and compensation processes. In 1952, following his retirement, he was a visiting lecturer at the University of California at Berkeley and Los Angeles, at Rio de Janeiro, and at Montevideo.

Rüdenberg's most important honors were an honorary doctorate from the Technische Hochschule in Karlsruhe (1921), a medal from the Stevens Institute of Technology in Hoboken, New Jersey (for his work in 1931 on the development of the electron microscope), and an honorary degree from the Technical University of West Berlin.

BIBLIOGRAPHY

A complete bibliography of Rüdenberg's publications is in the Jacottet and Strigel article below.

Secondary literature includes Ekkehard Hieronimus, "Reinhold Rüdenberg," in *Leben und Schicksal*, a publication honoring the consecration of the synagogue in Hannover (Hannover, 1963), 143–149, with portrait; P. Jacottet and R. Strigel, "Reinhold Rüdenberg zum 75. Geburtstag," in *Elektrotechnische Zeitschrift*, **79**, no. 4 (1958), 97–100, including publications for his jubilee and a portrait; and A. Timascheff, "Reinhold Rüdenberg gestorben," in *Elektrotechnische Zeitschrift*, **83**, no. 8 (9 Apr. 1962), 283–284, with portrait.

SIGFRID VON WEIHER

RUDIO, FERDINAND (*b.* Wiesbaden, Germany, 2 August 1856; *d.* Zurich, Switzerland, 21 June 1929), *mathematics, history of science.*

Rudio completed his secondary schooling in Wiesbaden, then, in 1874, entered the Zurich Polytechnic to study physics and mathematics. From 1877 until 1880 he studied at the University of Berlin, from which he received the doctorate *magna cum laude* with a dissertation on Kummer's problem of determining all surfaces of which the centers of curvature form second-order confocal surfaces. Rudio's solution utilized reduction to a differential equation. He also worked in group theory, algebra, and geometry. In 1881 he returned to Zurich as lecturer at the Polytechnic; he was appointed professor of mathematics there in 1889 and served in that post until 1928. He also administered the Polytechnic's library from 1893 to 1919.

Rudio wrote on a number of topics in the history of mathematics, including the quadrature of the circle, Simplicius' work on quadratures, and Hippocrates' lunes. He also composed biographies of contemporary mathematicians, and wrote a history of the Zurich Naturforschende Gesellschaft for the years 1746 to 1896.

Of particular importance was Rudio's project for editing the collected works of Euler. He first proposed this edition in 1883, on the occasion of the centenary of Euler's death, then brought it up again before the meeting of the first International Congress of Mathematicians at Zurich in 1897, and finally suggested it as an appropriate memorial for the bicentennial of Euler's birth in 1907. His efforts bore fruit in 1909,

when the Naturforschende Gesellschaft decided to undertake the work, and named Rudio general editor. He himself edited two volumes (the *Commentationes arithmeticae*) and brought out an additional three in collaboration, including the *Introductio in analysin infinitorum*. In all, he supervised the production of some thirty volumes.

BIBLIOGRAPHY

I. ORIGINAL WORKS. Rudio's publications include "Zur Theorie der Flächen, deren Krümmungsmittelpunktsflächen confocale Flächen zweiten Grades sind," an abstract of his inaugural dissertation (Berlin, 1880), published in *Journal für die reine und angewandte Mathematik*, **95** (1883), 240–246; *Archimedes, Huygens, Lambert, Legendre. Vier Abhandlungen* (Leipzig, 1892); "Die Möndchen des Hippokrates," in *Vierteljahrsschrift der Naturforschenden Gesellschaft in Zürich*, **50** (1905), 177–200; and "Der Bericht des Simplicius über die Quadraturen des Antiphon und des Hippokrates," in *Bibliotheca mathematica*, 3rd ser., **3** (1902), 7–62.

II. SECONDARY LITERATURE. See G. Pólya, obituary, in *Vierteljahrsschrift der Naturforschenden Gesellschaft in Zürich*, **74** (1929), 329–330; Alice Rudio, "F. Rudio," in *Biographisches Lexikon verstorbener Schweizer*, II (Zurich, 1948), 230; and C. Schröter and R. Fueter, "Ferdinand Rudio zum 70. Geburtstag," in *Vierteljahrsschrift der Naturforschenden Gesellschaft in Zürich*, **71** (1926), 115–135, with portrait and bibliography of his works.

J. J. BURCKHARDT

RUDOLFF (or RUDOLF), CHRISTOFF (*b.* Jauer, Silesia [now Jawor, Poland], end of the fifteenth century; *d.* Vienna, Austria, first half of the sixteenth century), *mathematics.*

Virtually nothing is known about Rudolff's life.[1] It was formerly thought that he was born in 1499 and died in 1545, but these dates are not confirmed by any documentary evidence.[2] The earliest reliable information attests his presence in Vienna in 1525, the year in which he dedicated his *Coss* to the bishop of Brixen (now Bressanone, Italy). This book was the first comprehensive work in German on algebra, or *Coss*, as it was then called. Rudolff learned the subject from Grammateus, who taught at the University of Vienna from 1517 to 1521. In 1521 Grammateus went to Nuremberg and then Erfurt, not returning to Vienna until the summer semester of 1525.[3] Consequently, Rudolff must have been in Vienna before 1521. He supported himself by giving private lessons; and although he was not affiliated with the university, he was able to use its library. Some critics accused him of stealing the examples for his *Coss* from the Vienna Library, an accusation against which he was defended by Michael Stifel in the preface to the new edition of the *Coss*. In 1526 Rudolff published an arithmetic book entitled *Künstliche Rechnung mit der Ziffer und mit den Zahlpfennigen*, and in 1530 an *Exempelbüchlin*; which was reprinted as "exempelbüchle" in later editions of the *Künstliche Rechnung*. He stated his intention to publish an improved, Latin version of his *Coss* containing new examples, but he never did so.[4]

The *Coss* is divided into two parts. In the first, Rudolff devotes twelve chapters to the topics that the reader must master before taking up the study of algebra (the solution of equations). In chapters 1–4 he presents the basic operations and the rule of three, giving examples with whole numbers and fractions, and then treats the extraction of square and cube roots. In the section "Progredieren" he states in the style of recipes the summation formulas of arithmetic and geometric series. By relating the geometric series, the "Progredieren in Proportz" (*proportio dupla, tripla,* and so forth), to the series of natural numbers, he obtains the configuration

0	1	2	3	4	5	$\cdots$
1	2	4	8	16	32	$\cdots$

This procedure enables him to determine an arbitrarily high member of the geometric series.

In chapters 5 and 6 Rudolff carries out the four operations and the rule of three on algebraic polynomials,[5] after first setting forth the names and symbols of the powers of the unknowns. The schema

0	1	2	3	4	5	6	7	8	9
r	z	c	zz	β	zc	$b\beta$	zzz	cc	

shows that he considers the proper designation of a member without x to be $x^0 = 1$. (Grammateus still used N (for *numerus*) in such cases instead of 0.) Chapters 7–11 are devoted to roots, binomials, and residues. Rudolff distinguishes three types of roots: rational, irrational, and communicant. Two roots are communicant if they have a common rational factor: for example, $\sqrt[3]{16}$ and $\sqrt[3]{54}$. If a factor is brought under the radicals, then the "denominierte" number is formed; for example, $5\sqrt{18}$ gives rise to $\sqrt{450}$. Rudolff computes $\sqrt{a + \sqrt{b}} = \sqrt{c} + \sqrt{d}$, using letters instead of numbers.[6] The first part of the book concludes with a short explanation of the five types of "proportioned" numbers (*multiple, super-particular,* and so forth).

In part two of the *Coss* (which is divided into three sections) Rudolff discusses first- and second-degree equations and their variations of higher degree. He assumes the existence of only eight distinct *equationen*

or "rules of the coss," not the twenty-four distinguished by earlier cossists. In his presentation of the sixth rule ($ax^2 + c = bx$) he deliberately admits only the one solution that fits the conditions of the problem under study. Later he recognized his error.[7] The second section offers rules (*cautelae*) for solving equations, and the third is a collection of problems containing over 400 examples. Some of the problems involve abstract numbers; others, taken from daily life, are presented in fantastic forms similar to those of the *Enigmata* of recreational mathematics. In some of the problems Rudolff introduces a second unknown, q (for *quantitas*). If there are more unknowns than equations, the problem is considered indeterminate. For several such problems concerned with "splitting the bill" (*Zechenaufgaben*) Rudolff supplied all the possible solutions.[8]

The *Coss* ends with three cubic problems. Rudolff does not work out their solutions because, as he stated, he wanted to stimulate further algebraic research.[9] In an "addendum" he presents still another "verbal computation" and a drawing of a cube with edge $3 + \sqrt{2}$. According to Stifel, with this illustration Rudolff sought to hint at the solution of the cubic equation, which was then unknown.[10]

Rudolff's other major work, *Künstliche Rechnung mit der Ziffer und mit den Zahlpfennigen*, consists of three parts: a "Grundbüchlein," in which the beginner is introduced to computing with whole numbers and fractions; a "Regelbüchlein," which treats the rule of three and the "Welsh practice"; and an "Exempelbüchlein," some 300 problems that vividly evoke the commerce and manufacturing of the period. Rudolff also includes several *Enigmata*, termed "amusing calculations." Examples of this type are the rule of Ta-Yen, hound and hare, or the "horse sale," in which the price of thirty-two horseshoe nails is expressed in a geometric sequence. As in the *Coss*, the two number sequences are related to each other, with the first sequence again beginning with 0.[11]

The *Exempelbüchlin* that Rudolff published in 1530 contains 293 problems, as well as tables of measurements for many regions, a list of symbols used in gauging, and numerous hints for solving problems. Decimal fractions appear in the computation of compound interest.[12]

Rudolff's importance in the history of mathematics lies in his having written the first comprehensive book on algebra in German. In this work he went far beyond his teacher Grammateus, especially concerning calculation with rational and irrational polynomials. Rudolff was aware of the double root of the equation $ax^2 + b = cx$ and gave all the solutions to indeterminate first-degree equations. His writings are remarkable both for the occasional appearance of decimal fractions and for improvements in symbolism. Adding a diagonal stroke to the points used by earlier cossists, Rudolff introduced the signs $\sqrt{}$, $\sqrt[3]{}$, $\sqrt[4]{}$ for the second, third, and fourth roots.[13] His work also gives a hint of the beginnings of exponential arithmetic and the fundamental idea of logarithms—that is, setting x^0 equal to 1. His methodical hints on using the *Coss* are worth noting as well. In brief, Rudolff's role in the development of mathematical studies in Germany was analogous to that of Fibonacci in Italy.[14]

Rudolff learned arithmetic from early printed books on the subject, some of which he cited.[15] It is obvious that he thoroughly studied Johannes Widman's arithmetic book (1489).[16] He also obtained information from other writings on algebra. He mentioned earlier books in which the solutions to problems are introduced by the words "ponatur una res,"[17] and in which the second power of the unknown is designated by the word "substantia" instead of "zensus." These remarks indicate Robert of Chester's translation of the *Algebra* of al-Khwārizmī, which exists in a fourteenth-century manuscript.[18] In particular, Rudolff used an algebraic treatise included in a volume compiled at Vienna by Johann Vögelin.[19] Further, he was acquainted, directly or indirectly, with the Regensburg algebra (1461), from which he took a problem involving computation of compound interest; this problem, the solution to which was $\sqrt{600} - 20$, was earlier used by Widman in his arithmetic book of 1489.[20]

Rudolff's *Künstliche Rechnung* was widely read and was reprinted as late as 1588, at Augsburg. The importance of the *Coss* was recognized by Gemma Frisius and Stifel, but it soon went out of print. In 1553 Stifel brought out a new edition of the *Coss* containing supplementary material.

NOTES

1. J. E. Scheibel, who, like Rudolff, came from Silesia, sought information about his compatriot. See *Einleitung zur mathematischen Bücherkenntnis*, I (Breslau, 1769–1775), 313. As late as 1850 nothing was known in Vienna concerning the circumstances of Rudolff's life. See *Zeitschrift für Slawistik*, **1** (1956), 132.
2. R. Wolf, *Geschichte der Astronomie* (Munich, 1877), 340.
3. On Grammateus see W. Kaunzner, "Über die Algebra bei Heinrich Schreyber," in *Verhandlungen des Historischen Vereins für Oberpfalz und Regensburg*, **110** (1970), 227–239; and *Neue deutsche Biographie*, VI (1964), 738–739.
4. See *Coss*, $A_{II}r$ and *Künstliche Rechnung*, fol. s_4v.
5. The division by a polynomial was only an attempt at this operation; see *Coss*, fol. E_1r.
6. See *Coss*, fols. $G_{III}r$, P_7v, P_8r.

7. On the ambiguous wording of the sixth rule see *Künstliche Rechnung*, fol. s_4v f.
8. See *Coss*, $R_{VI}r$ f.
9. The third of these problems—cited by Cantor (*Vorlesungen über Geschichte der Mathematik*, II, pt. 2, 426) and Smith (*History of Mathematics*, II, 458)—and the suggested method for solving it both derive from Stifel.
10. The drawing is in P. Treutlein, p. 89.
11. See *Künstliche Rechnung*, fol. s_2v.
12. The decimal fractions are separated by a stroke; for example: 393|75. See *Exempelbüchlin*, fol. x_1v.
13. See H. E. Wappler, "Zur Geschichte der deutschen Mathematik im 15. Jahrhundert," p. 13, n. 1.
14. On this point see M. Terquem, "Christophe Rudolff," p. 326.
15. See *Coss*, $C_V r$.
16. The problem of "splitting the bill" (*Zechenaufgabe*) that Rudolff found in "an arithmetic book"—see *Coss*, R_{VIV}—appeared earlier in the work by Widman. A comparison between Rudolff's problems and Widman's is given by Treutlein, pp. 120 f.
17. The solutions to problems in both the Regensburg algebra of 1461 and the Latin algebra of Codex Dresdensis C 80 begin with words such as "Pono quod lucrum sit una res" or "Pono quod A 1 r[em] habeat." See, respectively, M. Curtze, "Ein Beitrag zur Geschichte der Algebra in Deutschland im fünfzehnten Jahrhundert," pp. 58 ff.; and Wappler, *op. cit.*, p. 19.
18. Vienna, Cod. Vind. 4770; see L. C. Karpinski, *Robert of Chester's Latin Translation of the Algebra of al-Khowarizmi* (Ann Arbor, Mich., 1915), repr. as pt. I of Karpinski and J. G. Winter, *Contributions to the History of Science* (Ann Arbor, 1930).
19. This volume is Cod. Vind. 5277. See Wappler, *op. cit.*, p. 3, n. 2. Cod. Lat. Mon. 19691 is a copy of Cod. Vind. 5277. The marginal notes in the Vienna MS were incorporated into the text of the Munich MS. They both contain the 24 old rules of the cossists, as did Widman's Latin algebra. Widman's work had many problems in common with Cod. Lat. Mon. 26639. The relationships among all these texts have not yet been determined; on this question see W. Kaunzner, *Über Christoff Rudolff und seine Coss*, p. 2 and n. 28.
20. Problem 16 of the "Fünfften regl" in Rudolff's *Coss* (fol. $X_{VI}r$) appears in the following works: Cod. Dresd. C 80, fol. 356v (see Wappler, *op. cit.*, p. 21); the Regensburg algebra in Cod. Lat. Mon. 14908, fol. 149v (see Curtze, *op. cit.*, p. 61); and Widman's arithmetic of 1489, 127v.

BIBLIOGRAPHY

I. Original Works. Rudolff's writings are *Behend und hübsch Rechnung durch die künstreichen Regeln Algebre so gemeincklich die Coss genent werden* (Strasbourg, 1525), new ed. by M. Stifel (Königsberg, 1553—the colophon is dated 1554); *Künstliche rechnung mit der ziffer vnd mit den Zal pfenningen sampt der Wellischen Practica vnnd allerley fortheil auff die Regel de Tri* (Vienna, 1526)—Smith, *Rara arithmetica*, p. 152, cites 11 eds. (there are some other eds., but here the 1550 ed. was used); and *Exempelbüchlin* (Augsburg, 1530, 1538, 1540), also included in later eds. of *Künstliche Rechnung*.

II. Secondary Literature. See *Allgemeine deutsche Biographie*, XXIX, 571–572; M. Cantor, *Vorlesungen über Geschichte der Mathematik*, II, 2nd ed. (Leipzig, 1913), 397–399, 425–429; M. Curtze, "Ein Beitrag zur Geschichte der Algebra in Deutschland im fünfzehnten Jahrhundert," in *Abhandlungen zur Geschichte der Mathematik*, **7** (1895), 31–74; A. Drechsler, *Scholien zu Christoph Rudolphs Coss* (Dresden, 1851); C. J. Gerhardt, *Geschichte der Mathematik in Deutschland* (Munich, 1877), 54–59; W. Kaunzner, *Über Christoff Rudolff und seine Coss*, no. 67 in Veröffentlichungen des Forschungsinstituts des Deutschen Museums für die Geschichte der Naturwissenschaften und der Technik, ser. A (Munich, 1970); and *Deutsche Mathematiker des 15. und 16. Jahrhunderts und ihre Symbolik*, no. 90 in the same series (Munich, 1971); C. F. Müller, "Henricus Grammateus und sein *Algorismus de integris*," in *Programm Gymnasium Zwickau* (1896); D. E. Smith, *Rara arithmetica* (Boston–London, 1908), 151 ff.; and *History of Mathematics*, 2 vols. (New York, 1923–1925), I, 328 f., and II, 721; M. Terquem, "Christophe Rudolf," in *Annali di scienze matematiche e fisiche*, **8** (1857), 325–338; P. Treutlein, "Die deutsche Coss," in *Abhandlungen zur Geschichte der Mathematik*, **2** (1879), 15 ff., 44 ff.; F. Unger, *Die Methodik der praktischen Arithmetik* (Leipzig, 1888), 238; and H. E. Wappler, "Zur Geschichte der deutschen Mathematik im 15. Jahrhundert," in *Programm Gymnasium Zwickau* (1887); and "Zur Geschichte der deutschen Algebra," in *Abhandlungen zur Geschichte der Mathematik*, **9** (1899), 537–554.

KURT VOGEL

RUDOLPHI, KARL ASMUND (*b.* Stockholm, Sweden, 14 July 1771; *d.* Berlin, Germany, 29 November 1832), *anatomy, physiology, helminthology*.

Rudolphi was the son of the vice-rector of the German school in Stockholm, where he spent his early years. He studied philosophy, natural sciences, and medicine at Greifswald, from which he graduated in philosophy in 1793 and in medicine, with a dissertation on intestinal worms, in 1794. In 1801, having finished a course at the Berlin Veterinary School, he returned to Greifswald, first as professor in the veterinary institute, then, in 1808, as professor of medicine in the medical faculty. In 1810 he was appointed to the chair of anatomy and physiology at the newly established University of Berlin; he became one of the most influential members of the university, and remained there for the rest of his life.

Some of Rudolphi's early work was in botany. In 1805 he shared an award from the Göttingen science society with his friend H. F. Link, with whom he established a new direction in the study of plant morphology. Rudolphi carried out a careful examination of all parts of a number of species of plants at all stages of their growth and concluded that without exception each plant consists wholly of cellular tissue in the early stages of its development and in large part of cellular tissue at the later stages of its growth. This constituted an important first step toward the recognition of the cell as the basic unit of the structure of plants. Rudolphi was, on the other hand, criticized for his erroneous assertion that mushrooms differ so greatly in their structure from other plants that they do not belong to the plant kingdom.

Rudolphi also did pioneering work in helminthology, which he developed as a special branch of zoology. In the course of making several thousand observations he identified and described a large number of new species of worms parasitic in both animals and man. (In 1803 Zeder had listed 391 species; Rudolphi in 1819 catalogued 993.) In his masterful and detailed description of the structure and life cycle of parasitic worms, Rudolphi established the basis for all subsequent systematic research on these animals. He nonetheless believed that these parasites are generated by disease in the body of the host, even though his predecessor P. S. Pallas had tried to demonstrate that their eggs enter the body of the host from outside. And although his work on helminthology was widely influential, Rudolphi's attempt at more general zoological classification was both unsuccessful and soon forgotten.

Among Rudolphi's works in comparative anatomy, his studies of intestinal villi in vertebrates were especially important because they constituted both a contribution toward Bichat's tissue theory and a demonstration of the utility of the microscope in the investigation of animal morphology. These examinations were among the first in the new field of comparative histology.

Rudolphi's textbook on physiology, *Grundriss der Physiologie*, was also widely influential, although he had completed only two volumes at the time of his death. It was based upon Rudolphi's own experience, particularly in comparative anatomy, and served as a useful counterforce to the romantic and speculative physiology then prevailing in Germany. (Müller even mentioned the brevity with which Rudolphi treated any subject to which he could not add critical comment or new information at first hand.) In the general section of his textbook Rudolphi took an anthropological view of physiology, stressing the differences between man and the great apes. His simple and concise grouping of tissues was valuable. He gave a clear and realistic account of the functions of various parts of the body, which in general agreed with modern conceptions.

Rudolphi repudiated the fantastic conceptions of life put forward by Schelling and Oken, and likewise dismissed speculations about symmetry and Stahl's notion of the soul as being the cause of bodily phenomena. The idea of the soul or mind, he pointed out, does not contribute to the understanding of physiology; in a study of the cerebral cavities he cited the entire brain as being the organ of intelligence—in opposition to Sömmering, who located intellect in the cerebrospinal fluid—and indicated that such complex processes as thought and reasoning must arise from a complex structure. Although he rejected many of the more mystical notions of science—including Gall's phrenology, Blumenbach's *nisus formativus*, and Cuvier's theory of catastrophes—Rudolphi supported the "animal chemistry" of Berzelius, with whom he was in personal touch, and included his discoveries and views in his own physiology.

Rudolphi was to some degree hampered in his physiological investigations by his reluctance to perform experiments upon living animals. He advocated more humane experimentation and based most of his own work on his studies in comparative anatomy, emphasizing the relationships between structure and function. He stressed the importance of chemistry as well and insisted upon the primary role of the exact sciences in medical training and practice, thereby giving impetus to the inductive and empirical physiology and medicine that had begun to develop at the University of Berlin. He was an inspiring teacher and was often of aid to younger scientists, both in Germany and abroad. He traveled in a number of European countries and corresponded with foreign scientists, particularly Scandinavians; he influenced and helped Baer, Purkyně, Müller, Siebold, Nordmann, and Lovén, among others. He was an important figure in the transition from romantic science to the modern approach in biology and medicine.

Rudolphi was also a collector of medallic portraits of physicians and scientists. His index of his collection, especially in C. L. Duisburg's final enlarged 1862 edition, became a standard work in this special branch of numismatics.

BIBLIOGRAPHY

I. Original Works. Johannes Müller, "Gedächtnisrede auf Carl Asmund Rudolphi," in *Abhandlungen der Königlichen Akademie der Wissenschaften zu Berlin*, Physikalische Klasse (1837), xvii–xxxviii, contains a bibliography of sixty-five titles of Rudolphi's work, including *Entozoorum, sive vermium intestinalium, historia naturalis*, 2 vols. (Amsterdam, 1808–1810); *Beiträge zur Anthropologie und allgemeine Naturgeschichte* (Berlin, 1812); *Entozoorum synopsis* (Berlin, 1819); *Grundriss der Physiologie*, 2 vols. (Berlin, 1821–1828); and *Index numismatum in virorum de rebus medicis vel physicis meritorum memoriam percussorum* (Berlin, 1823).

II. Secondary Literature. In addition to Müller, cited above, on Rudolphi and his work see his colleague H. F. Link, "Nachricht von dem Leben des Königl. geh. med. Raths und Professors K. A. Rudolphi," in *Medizinische Zeitung*, **2** (1833), 17–20, the first appreciation of his career. More recent works are M. Dittrich, "Die Bedeutung von Karl Asmund Rudolphi (1771–1832) für die Entwicklung der Medizin und Naturwissenschaften im 19. Jahrhundert," in *Wissenschaftliche Zeitschrift der Universität Greifswald*, Math.-naturwiss. Reihe, **16** (1967), 249–277; E. Nordenskiöld, *The History of Biology* (New York, 1928), 352–355; A. Waldeyer, "Carl Asmund Rudolphi und Johannes Müller," in *Forschen und Wirken. Festschrift zur 150 Jahr-Feier der Humboldt-Universität zu Berlin*, I (Berlin, 1960), 97–115; and W. Waldeyer, in *Biographisches Lexikon der hervorragenden Ärzte*, 2nd ed., IV (Berlin–Vienna, 1932), 911–913.

Vladislav Kruta

RUEDEMANN, RUDOLF (*b.* Georgenthal, Germany, 16 October 1864; *d.* Albany, New York, 18 June 1956), *paleontology, geology.*

Ruedemann, America's foremost graptolite specialist, was one of the three children of Albert and Franziska Seebach Ruedemann. The family was extremely poor (which may explain Ruedemann's later frugality), but Ruedemann nevertheless was able to enroll in the University of Jena, where he studied mathematics with Gustav Steinmann and biology with Johannes Walther. He soon became interested in geology. His doctoral dissertation, *Contact Metamorphose an der Reuth*, was written in response to a problem in petrography assigned by Steinmann's successor, Ernst Kalkowsky. Ruedemann received the degree *magna cum laude* in 1887; Kalkowsky had recommended that he be given a *summa cum laude*, but Ruedemann, in an early exhibition of the tactlessness that marked his career, unwisely opposed a favorite theory of his examining professor in chemistry. Haeckel, a friend of both, intervened to calm the antagonists.

From 1887 until 1892 Ruedemann had tenure at the University of Strasbourg. In the latter year he immigrated to the United States, where he first taught science in the high schools of Lowville and Dolgeville, New York. When John Mason Clarke succeeded James Hall as state paleontologist, however, he hired Ruedemann as his assistant, and Ruedemann began working at the New York State Museum in Albany in March 1899. Ruedemann himself succeeded Clarke as state paleontologist in 1925, and held this position until his retirement in 1937, although he continued to do research until 1942.

Ruedemann investigated the geology and paleontology of the principal valleys of New York, especially the Mohawk, where the Ordovician shales and limestones offered abundant and varied invertebrate fossils, many of which were previously undescribed. In particular, his interest was attracted by the enigmatic graptolites, which he began to study while he was teaching in Dolgeville. The material that he collected included complete growth series of *Diplograptus*—the first known for a graptolite (1897). Ruedemann was able to establish a graptolite zonation for the New York Ordovician black shales. These shales, older eastward, were replaced westward by shelf limestones. Graptolites increasingly became Ruedemann's chief paleontological concern, and his monumental "Graptolites of North America," published in 1947, marked the culmination of his career. He also attained an international reputation as a paleontologist through his researches on eurypterids (extinct arthropods unique to New York State), and did notable work on radiolarian cherts, nautiloid cephalopods, Paleozoic plankton, and a number of problematic fossils.

In structural geology, Ruedemann's study of the exotic Rysedorph conglomerate fauna (1901) led him to suggest, in 1909, that the present position of the deformed Taconic rocks had been the result of a far-ranging westward thrust—a view now universally accepted. He thus introduced into American geology the concepts of nappe and thrust that had been so successful in Alpine geology, as demonstrated in the work of M.-A. Bertrand, Hans Schardt, M. Lugeon, and A. Heim. Ruedemann's studies, together with those of Arthur Keith, therefore led to the ultimate resolution of the complex controversy surrounding the Taconic rocks which had long vexed American geologists. He also published stratigraphic and areal studies of the Thousand Islands (1910), Saratoga Springs (1914), the Capital District (1930), and the Catskills (1942). As a technical innovation he introduced copper electroplating of fragile gutta-percha castings of fossils, from which molds might be obtained.

Ruedemann was vice-president (1916) of the Geological Society of America and vice-president (1911) and president (1916) of the Paleontological Society. He was elected to the National Academy of Sciences in 1928.

He was married to Elizabeth Heitzmann and they had one daughter and six sons.

BIBLIOGRAPHY

I. ORIGINAL WORKS. Among Ruedemann's 163 scientific articles are "Development and Mode of Growth of *Diplograptus* M'Coy," in *New York State Geologist. Annual Report*, no. 14 (1897), 217–149; "Trenton Conglomerate of Rysedorph Hill, Rensselaer Co., N. Y., and Its Fauna," in *Bulletin of the New York State Museum*,

no. 49 (1901), 3–114; "Guelph Fauna in the State of New York," in *Memoir of the New York State Museum*, no. 5 (1903), 195, written with J. M. Clarke; "Graptolites of New York," *ibid.*, no. 7 (1904) and no. 11 (1908); "Cephalopods of the Beekmantown and Chazy Formations of the Champlain Basin," *Bulletin of the New York State Museum*, no. 90 (1906); "Types of Inliers Observed in New York," *ibid.*, no. 133 (1909); "Geology of the Thousand Islands Region," *ibid.*, no. 145 (1910), written with H. P. Cushing, H. L. Fairchild, and C. H. Smyth, Jr.; "The Eurypterida of New York," *Memoir of the New York State Museum*, no. 14 (1912), written with J. M. Clarke; "Geology of Saratoga Springs and Vicinity," *Bulletin of the New York State Museum*, no. 169 (1914); "The Utica and Lorraine Formations of New York. I. Stratigraphy," *ibid.*, no. 258 (1925); and "II. Systematic Paleontology," *ibid.*, no. 262 (1925) and no. 272 (1926); "Geology of the Capital District," *ibid.*, no. 285 (1930); "Paleozoic Plankton of North America," *Memoir of the Geological Society of America*, no. 2 (1934); and "Eastern New York Ordovician Cherts," in *Bulletin of the Geological Society of America*, **47** (1936), 1535–1586, written with T. Y. Wilson.

Later publications include "Geology of the Catskill and Kaaterskill Quadrangles. Part 1. Cambrian and Ordovician Geology," *Bulletin of the New York State Museum*, no. 331 (1942); and especially, "Graptolites of North America," *Memoir of the Geological Society of America*, no. 19 (1947).

II. SECONDARY LITERATURE. On Ruedemann and his work see Winifred Goldring, "Memorial to Rudolf Ruedemann (1864–1956)," in *Proceedings of the Geological Society of America. Annual Report*, no. 1957 (1958), 153–162.

DONALD W. FISHER

RUEL, JEAN, also known as **RUELLIUS** (*b.* Soissons, France, 1474; *d.* Paris, France, 24 September 1537), *medicine, botany.*

According to some biographers, Ruel (sometimes spelled du Ruel or de la Ruelle, on account of confusion with the fifteenth-century engraver Jean des Ruelles) was born in 1479 (Georges Gibault thinks that 1479 is the wrong date, probably the result of a typographical error).

Ruel was self-taught and fluent in Greek and Latin. Although very few details of his life are available, it is known that he studied medicine and that he received the M.D. in 1508. The following year he became physician to Francis I. Thoroughly absorbed in his own works, Ruel refused to follow the court on its frequent travels. He married and was a devoted father, but when his wife died he entered holy orders so that he could devote himself entirely to study. Étienne Porcher, bishop of Paris, obtained for Ruel a canonry at the cathedral of Notre-Dame: Ruel was thus freed of material concerns. He died of a stroke, which was probably brought on by overexertion after too sedentary a life. He was buried in Notre-Dame.

Although Ruel's works are compilations of the works of earlier authors, they are notable examples of the first attempts to popularize botany. Ruel's *De natura stirpium* (Paris, 1536) is elegantly written and furnishes many vernacular plant names. The work begins with general elements of botany borrowed from Theophrastus, whom Ruel considered the father of botany. Instead of a botanical classification, Ruel used alphabetical order, which rendered the book of great practical value. In each chapter, only one subject is discussed; for example, stalk, leaves, bark, flowers, germination, grafting, vegetables, cereals, and medicinal properties. He also provided information on the odors and tastes of plants. Unfortunately, the information on etymology is not very reliable, since it was copied from the ancients. *De natura stirpium* was dedicated to Ruel's patron, Francis I, who paid the cost of printing. Only a few copies of the work can still be found; the rarity of the book is attested by its absence from the libraries of Jussieu and Joseph Decaisne (according to Georges Gibault), both of whom were informed bibliophiles. Undertaking extensive research to find the best manuscripts, Ruel—whom Guillaume Budé called the "eagle of interpreters"—produced a number of excellent translations. In 1530 Ruel published in Paris *De medicina veterinaria*, a Latin compilation of everything on veterinary medicine that had been written in Greek. In 1516 he published his translation of Dioscorides' *De materia medica*. Several other editions were published, some posthumously, among which the 1549 edition is perhaps the best. Ruellia, an ornamental plant of the family Acanthaceae, was named by Plumier in honor of Ruel.

BIBLIOGRAPHY

I. ORIGINAL WORKS. Ruel's works include *De natura stirpium libri tres* (Paris, 1536; Basel, 1537, 1543); *Prima (-secunda) pars de natura stirpium libri tres*, 3 vols. (Venice, 1538); *In Ruellium de stirpibus epitome, cui accesserunt volatilium, gressibilium, piscium & placentarum magis frequentium apud Gallias nomina, per Leodegarium a Quercu* (Paris, 1539, 1542, 1543, 1544; Rouen, 1539); *Veterinariae medicinae libri II, Johanne Ruellio, interprete* (Paris, 1530); *Veterinariae medicinae libri duo, a Johanne Ruellio . . . olim quidem latinitate donati, nunc vero iidem sua, hoc est graeca, lingua primum in lucem aediti* (Basel, 1537); and *In P. Virg. Maronis Moretum scholia ex prestantissimis quibusque scriptoribus, maxime ex Jo. Ruelli lucubrationibus huc transposita, per H. Sussannaeum* (Paris, 1542).

Ruel translated the following works: Dioscorides, *De medicinali materia libri quinque* (Paris, 1516); *Libri octo graeca & latine*, Jacobus Goupy, ed. (Paris, 1549); Johannes Actuarius, *De medicamentorum compositione* (Paris, 1539); *Operum tomus II. De medicamentorum compositione* (Lyons, 1556); and Pollidore Vergile, *Le hystoriographe (dont le bibliographe Brunet se demande s'il faut vraiment l'attribuer à Ruel)* (Paris, 1544).

Ruel supervised the reprint of *Scribonius largus. De compositionibus medicamentorum liber unus* (Paris, 1528); *Scribonius largus: De compositione medicamentorum liber* (Basel, 1529); and *Celse (Aurelius Cornelius). De re medica libri octo* (Paris, 1529).

II. SECONDARY LITERATURE. On Ruel and his work, see Henri Ernest Baillon, *Dictionnaire de botanique*; Nicolas François Joseph Eloy, *Dictionnaire historique de la médecine contenant son origine, ses progrès, ses révolutions, ses sectes, et son état chez différens peuples, ce que l'on a dit des Dieux on héros anciens de cette science, l'histoire des plus célèbres médecins . . . et le catalogue de leurs principaux ouvrages*, IV (Liège–Frankfurt am Main, 1755), 132–133; Georges Gibault, "Notice biographique sur Jean Ruel médecin et botaniste au XVIe siècle," in *Bulletin de la Société archéologique, historique et scientifique de Soissons*, 15 (1908), 1001–1001, with portrait; Michaud, *Biographie universelle*; and Scévole 1 er de Sainte-Marthe, *Virorum doctrina illustrium qui hoc seculo in Gallia floruerunt, elogia, authore Scaevola Sammarthano. Augustoriti Pictonum* (Poitiers, 1598).

PAUL JOVET
J. C. MALLET

RUFFER, MARC ARMAND (*b.* Lyons, France, 1859; *d.* eastern Mediterranean, *ca.* 2 May 1917), *paleopathology.*

Ruffer was the son of Baron Jacques de Ruffer, a banker. He was first educated in Paris, where he was later a pupil of Pasteur and Metchnikoff. Then, after additional study in Germany, he went to England, where he graduated B.A. from the University of Oxford in 1882 and M.B. from University College, London, in 1887. He later served as director of the British Institute of Preventive Medicine. After contracting a laboratory infection there, he convalesced in Egypt, where he became professor of bacteriology at the Cairo Medical School. He was knighted in 1916 for his contributions to bacteriology and hygiene and for his service to the Red Cross. Ruffer was drowned while returning to Egypt from Salonika, when his ship was torpedoed during World War I.

Ruffer's reputation rests securely on his pioneering work in paleopathology. Although anticipated in paleohistological studies by Czermak, D. M. Fouquet, H. M. Wilder, and Samuel Shattock and in the osseous pathology of ancient remains by Elliot Smith and Wood Jones, Ruffer's influence in paleopathology is unparalleled. Few paleopathological works are published without extensive reference to him.

Because of the ample material supplied by his colleagues in the field of Egyptology, Ruffer was able to revolutionize techniques for the microscopic examination of ancient human tissues. He published important papers on the normal and pathological histology of Egyptian mummies of all periods. Many normal tissues, even from Predynastic bodies, were often wonderfully preserved. Ruffer's most important pathological observations revealed the presence of eggs of *Schistosoma bilharzia* in mummy kidneys, of degenerative arterial disease in bodies ranging from the New Kingdom to Coptic Christian times, and of probable smallpox lesions. Contemporary paleopathology utilizes more sophisticated tinctorial, histochemical, and electron microscope techniques, but these are all derived from Ruffer's original methods.

In the field of morbid anatomy, Ruffer's major observations concerned osseous and dental pathology. These observations complemented and consolidated important earlier work by Elliot Smith and Wood Jones on human remains uncovered by the Archaeological Survey of Nubia (1907–1908). His studies proved conclusively that osteoarthritis was a common degenerative disease over a period of three millennia in ancient Egypt. Ruffer showed also that dental disease was very common. Among the other pathological lesions that he described, the most important was a unique example of Pott's disease of the spine in a New Kingdom mummy. Because the soft tissues had survived, the diagnosis of tuberculosis could be made with great confidence. This report is of considerable significance since it supports tentative diagnoses of tuberculosis in other purely skeletal remains from ancient Egypt.

Ruffer showed conclusively that certain common and important maladies have affected man for almost 5,000 years, a valuable observation for both the physician and the historian. In 1921 Ruffer's papers were published posthumously in a collected volume. The book has continued to stimulate interest in the field of paleopathology.

BIBLIOGRAPHY

Ruffer's papers were published in *Studies in the Palaeopathology of Egypt,* R. L. Moodie, ed. (Chicago, 1921).

Useful secondary sources are D. Brothwell and A. T. Sandison, eds., *Diseases in Antiquity* (Springfield, Ill., 1967); S. Jarcho, ed., *Human Palaeopathology* (New Haven, 1966); and A. T. Sandison, "Sir Marc Armand Ruffer (1859–1917)–Pioneer of Palaeopathology," in *Medical History,* **11** (1967), 150–156.

A. T. SANDISON

RUFFINI, ANGELO (*b.* Pretare, near Arquata del Tronto, Italy, 17 July 1864; *d.* Baragazza, near Castiglione dei Pepoli, Italy, 7 September 1929), *histology, embryology.*

Ruffini, son of Giacomo Ruffini, a magistrate, and of Vincenza Saladini, received his secondary education in Ascoli Piceno. In 1884 he began to study medicine at Bologna, graduating in 1890. While a student Ruffini worked on histology in the laboratory of comparative anatomy, and in 1888 he set up a small histology laboratory in the workhouse at Bologna. In 1890 the clinician Augusto Murri appointed him manager of the microscopy laboratory of the Bologna Medical Clinic. Four years later Ruffini had qualified himself to teach histology. Ironically, just when his aim of teaching seemed to be attained, financial needs forced him to accept a job as a country physician: in January 1895 he was director of the little hospital of Lucignano, near Arezzo. Ruffini remained there for six years; and in this hospital he equipped, at his own expense, a laboratory for histological research. Thus he was able to carry out and publish some of his most important work. While at Lucignano, Ruffini gave a free course in histology at the Institute of Anatomy of the University of Siena; and in 1901 he was appointed professor of embryology at Siena. In 1912, having won a national competition, he was appointed professor of histology and general physiology at the University of Bologna, a post he held until his death.

From 1890 to 1906 Ruffini worked on the structures of proprioceptive sensibility. In 1888, while still a student, he had begun research on nerve receptors, and he was thus the first to see the ultraterminal plates, formed at the extremity of a nerve stem that already has a motor nerve ending. In 1891, shortly after receiving his degree, he discovered in man a new form of terminal nerve expansion now known as Ruffini's corpuscle; he published his discovery in 1894 as "Di un nuovo organo nervoso terminale e sulla presenza dei corpuscoli Golgi-Mazzoni nel connettivo sottocutaneo dei polpastrelli delle dita dell'uomo." Ruffini described two new forms of nerve expansion in the dermal papillae that also are named for him — the papillary brushes and the interlaced ansae —

thereby demonstrating that all the papillae, even those lacking Meissner's corpuscles, were innervated. In addition, he discovered minute amyelinic networks in the subpapillary stratum of the skin.

Ruffini also was the first to observe Golgi-Mazzoni corpuscles in the subcutaneous cellular structure, and he illustrated (1900) the intermediate forms that represent links between the classical Pacinian corpuscles and the more typical forms of the Golgi-Mazzoni corpuscles. Of great importance was his contribution (1905) to the study of the thin sheath interposed in the sensory nerve fiber between the myelinic sheath and Henle's sheath, today known as the subsidiary sheath of Ruffini. Ruffini also made a valuable contribution to knowledge of the minute structure of the organs of Golgi. In addition, he worked diligently to present the first complete, exhaustive description of the expansions in the neuromuscular spindles. His discovery in 1900 and 1902 of the ultraterminal fibrils and of fibrils that establish anastomosis between nerve corpuscles gave him the fundamental basis with which to oppose the too rigid outlines, then accepted, of the neuron doctrine.

Ruffini's fame is still essentially linked with his illustration of the nerve expansions of proprioceptive sensibility. Nevertheless, his observations on the earliest stages of the development of the fertilized amphibian egg were far more important and original. This research, begun in 1906 and pursued throughout his life, increased knowledge of embryogeny. Ruffini's embryological work, collected in his book *Fisiogenia*, may be placed between the work of Wilhelm Roux and that of Hans Spemann, who received the 1935 Nobel Prize for medicine or physiology. Ruffini published his *Fisiogenia* in 1925, one year after the death of Roux, the celebrated theorist of *Entwicklungsmechanik*; and in the year of Ruffini's death (1929) Waldemar Schleip published a book on the same subject, the determinism of the embryo's early development.

Ruffini was convinced that classical embryology, such as that of Wilhelm His, had become inadequate to understanding of the formation of · the germ layers. On the other hand, he was convinced that Roux's views were excessively one-sided. Roux's imposing work was based on the action of mechanical forces exclusively (see Castaldi 1925); but Ruffini demonstrated, by means of a very precise technique, that the phenomena of gastrulation are dominated by physiological processes: cellular secretion and a peculiar form of cellular movement that he called "stichotropismus" ("a file movement of the cells"). He also first demonstrated that the cellular fields from which the germ layers develop are already well distinguished in the blastula of Amphibia. Yet Ruffini, strongly convinced of potential cellular specificity, neglected the importance of morphogenetic correlations, which were emphasized by Spemann. Although Ruffini's *Fisiogenia* is thus preformist, it is nevertheless a wide-ranging and exhaustive treatise on the problems of general embryology.

Ruffini, who was awarded a gold medal in 1910 by the Accademia Nazionale dei Quaranta and (on the suggestion of Sherrington) a prize by the Royal Society, was a great teacher and observer; the Ruffini Institute of Histology and of General Embryology at Bologna is named in his honor.

BIBLIOGRAPHY

I. Original Works. Ruffini's writings include "Su due casi di anastomosi diretta fra i prolungamenti protoplasmatici delle cellule gangliari del cervello," in *Bolletino delle scienze mediche*, 6th ser., **24** (1889); "Di una particolare reticella nervosa e di alcuni corpuscoli del Pacini che si trovano in connessione cogli organi muscolo-tendinei del gatto. Nota preventiva," in *Atti dell' Accademia nazionale dei Lincei. Rendiconti*. Classe di scienze fisiche, matematiche e naturali, 5th ser., 7, no. 1 (1892), 442–446; "Sulla terminazione nervosa nei fusi muscolari e sul loro significato fisiologico. Nota preventiva," *ibid.*, no. 2 (1892), 31–38; "Considerazioni critiche sui recenti studi dell'apparato nervoso nei fusi neuromuscolari," in *Anatomischer Anzeiger*, **9** (1894), 80–88; "Di un nuovo organo nervoso terminale e sulla presenza dei corpuscoli Golgi-Mazzoni nel connettivo sottocutaneo dei polpastrelli delle dita dell'uomo," in *Atti dell'Accademia nazionale dei Lincei. Memorie*. Classe di scienze fisiche e naturali, 4th ser., **7** (1894), session of 12 Nov. 1893, also in *Monitore zoologico italiano*, **6** (1895), 196–203; "Sulla fine anatomia dei fusi neuromuscolari del gatto e sul loro significato fisiologico," in *Monitore zoologico italiano*, **7** (1896), 49–52, also in *Journal of Physiology*, **23**, no. 3 (1898), 190–208; and *Sulla presenza di nuove forme di terminazioni nervose nello strato papillare e subpapillare della cute dell'uomo* (Siena, 1898).

See also "Sulle fibrille nervose ultraterminali nelle piastre motrici dell'uomo," in *Rivista di patologia nervosa e mentale*, **5** (1900), 433–444, with notes by the Hungarian histologist István Apáthy; "Contributo allo studio della cute umana," in *Monitore zoologico italiano*, **11** (1900), 117–118, 282–289; "Le fibrille nervose ultraterminali nelle terminazioni nervose di senso e la teoria del neurone," in *Rivista di patologia nervosa e mentale*, **6** (1901), 70–82; "Di una nuova guaina (*Guaina sussidiaria*) nel tratto terminale delle fibre nervose di senso dell'uomo," in *Zeitschrift für wissenschaftliche Zoologie*, **79** (1905), 150–170; "Le espansioni nervose periferiche alla luce dell'analisi moderna," in *Monitore zoologico italiano*, **17** (1906), 16–33, 68–87; "Contributo alla conoscenza della ontogenesi degli anfibi anuri e urodeli," in *Atti dell'Accademia dei fisiocritici in Siena*, 4th ser., **18** (Nov. 1906), and **19** (Nov. 1907), also in *Archivio italiano di anatomia e di embriologia*, **6** (1907), 129–156; and *Fisiogenia. La biodinamica dello sviluppo e i fondamentali problemi morfologici dell' embriologia generale* (Milan, 1925).

II. SECONDARY LITERATURE. See L. Castaldi, "La vita e l'opera di Wilhelm Roux: 1850–1924," in *Rivista di biologia*, **7** (1925), 97–104, with a complete list of Roux's works; M. Clara, "Neue Untersuchungen zur Frage der Teilung bei den Talgdrüsen. Zugleich ein Beitrag zur Frage des 'Stichotropismus' in der Formbildung," in *Zeitschrift für mikroskopisch-anatomische Forschung*, **18** (1929), 487–519, which summarizes Ruffini's views on stichotropismus and reproduces his drawings concerning gastrulation; G. Cotronei, "Angelo Ruffini," in *Rivista di biologia*, **12** (1930), 198–202; E. Giacomini, "La vita e l'opera di Angelo Ruffini," in *Monitore zoologico italiano*, **40** (1929), 277–292, with a complete list of Ruffini's 85 works; G. Lambertini, "L'opera neurologica di Angelo Ruffini," in *Essays on the History of Italian Neurology. Proceedings of the International Symposium on the History of Neurology: Varenna 30 VIII/1 IX, 1961* (Milan, 1963), 195–202; G. Levi, "La Fisiogenia di Angelo Ruffini," *in Monitore zoologico italiano*, **36** (1925), 176–180; and "Determinazione e specificità dei tessuti," in *Archivio italiano di anatomia e istologia patologica*, **1** (1930), 55–84; and W. Schleip, *Die Determination der Primitiventwicklung* (Leipzig, 1929).

PIETRO FRANCESCHINI

RUFFINI, PAOLO (*b.* Valentano, Italy, 22 September 1765; *d.* Modena, Italy, 10 May 1822), *mathematics, medicine, philosophy.*

Ruffini was the son of Basilio Ruffini, a physician, and Maria Francesca Ippoliti. While he was in his teens, his family moved to Modena, where he spent the rest of his life. At the University of Modena he studied medicine, philosophy, literature, and mathematics, including geometry under Luigi Fantini and infinitesimal calculus under Paolo Cassiani. When Cassiani was appointed councillor of the Este domains, Ruffini, while still a student, was entrusted with his course on the foundations of analysis for the academic year 1787–1788. Ruffini obtained his degree in philosophy and medicine on 9 June 1788 and, soon afterward, that in mathematics. On 15 October 1788 he was appointed professor of the foundations of analysis, and in 1791 he replaced Fantini, who had been obliged by blindness to give up teaching, as professor of the elements of mathematics. Also in 1791 Ruffini was licensed by the Collegiate Medical Court of Modena to practice medicine. His exceptional versatility was reflected in his simultaneous activity as physician and researcher and teacher in mathematics—especially at a time when scientific specialization predominated.

Following the occupation of Modena by Napoleon's troops in 1796, Ruffini was appointed, against his wishes, representative from the department of Panaro to the Junior Council of the Cisalpine Republic. Relieved of these duties, he resumed his scientific activity at the beginning of 1798. His subsequent refusal, on religious grounds, to swear an oath of allegiance to the republic resulted in his exclusion from teaching and from holding any public office. Ruffini accepted the experience calmly, continuing to practice medicine and to pursue mathematical research. It was during this period that he published the mathematical theorem known as the Abel-Ruffini theorem: a general algebraic equation of higher than the fourth degree cannot be solved by means of radical-rational operations.

A preliminary demonstration of this result appeared in *Teoria generale delle equazioni* (1799). Discussions with mathematicians such as Malfatti, Gregorio Fontana, and Pietro Paoli led to publication of the theorem in refined form in *Riflessioni intorno alla soluzione delle equazioni algebriche generali* (1813). Ruffini's results were received with extreme reserve and suspicion by almost every leading mathematician. Only Cauchy accorded them full credence, writing to Ruffini in 1821: "Your memoir on the general resolution of equations is a work that has always seemed to me worthy of the attention of mathematicians and one that, in my opinion, demonstrates completely the impossibility of solving algebraically equations of higher than the fourth degree." Following its independent demonstration by Abel in 1824, the theorem eventually took its place in the general theory of the solubility of algebraic equations that Galois constructed on the basis of the theory of permutation groups.

Ruffini's methods began with the relations that Lagrange had discovered between solutions of third- and fourth-degree equations and permutations of three and four elements; and Ruffini's development of this starting point contributed effectively to the transition from classical to abstract algebra and to the theory of permutation groups. This theory is distinguished from classical algebra by its greater generality: it operates not with numbers or figures, as in traditional mathematics, but with indefinite entities, on which logical operations are performed.

Ruffini also developed the basic rule, named for him, for determining the quotient and remainder that result from the division of a polynomial in the variable x by a binomial of the form $x - a$. He treated the problem of determining the roots of any algebraic equation with a preestablished approximation by means of infinite algorisms (continuous fractions, development in series).

Ruffini was a staunch advocate of rigor in infinitesimal processes, a requirement that had assumed special importance toward the turn of the nineteenth century. Despite the success obtained following the algorismic systematization of calculus by Newton and Leibniz, there was an increasing awareness of the uncertainty of the foundations of infinitesimal analysis and of the lack of rigor of demonstrations in this field. A critical detail of the issue concerned the use of divergent and indeterminate series. As president of the Società Italiana dei Quaranta, Ruffini refused to approve two papers by Giuliano Frullani, presented by Paoli, because they used series of which the convergence had not been demonstrated. Although Frullani cited Euler and Laplace as having remained unconcerned about convergence in treating similar problems, Ruffini remained firm in his own demand for rigor. His position was supported by Cauchy in his *Analyse algébrique* (1821) and by Abel in a letter to Holmboe in 1826.

The application of Ruffini's mathematical outlook to philosophical questions is reflected in *Della immaterialità dell'anima* (1806), in which he enunciated the "theorem" that a being endowed with the faculty of knowledge is necessarily immaterial. His extremely detailed argument is developed by showing irresolvable differences between the properties of material beings and of beings endowed with the faculty of knowledge—such as the human soul. In another philosophical work, *Riflessioni critiche sopra il saggio filosofico intorno alla probabilità del signor Conte Laplace* (1821), Ruffini attempted to refute certain theses in Laplace's *Essai philosophique sur les probabilités* (1812) that he considered contrary to religion and morality. He began by rejecting the conception of Laplace's intelligence, which was inspired by the hypothesis of a rigid universal determinism. Ruffini argued from the basis of man's direct psychological experience of the exercise of his free will, which effects a change not only in states of consciousness but also in the physical world. Citing Jakob Bernoulli's theorem on probability and frequency, Ruffini developed a criticism of the applicability of the urn model to problems concerning the probability of natural events and attempted to determine to what extent the analogy between the two types of considerations holds true. In contrast with Laplace, who attempted to apply his calculus indiscriminately to moral actions, Ruffini observed that since the faculties of the soul are not magnitudes, they cannot be measured quantitatively.

The mathematician and physician converged in Ruffini to consider the probability that a living organism is formed by chance. He examined probability in relation to the truthfulness of evidence, showing that Laplace's solution applied to a different problem than that under consideration and that it represented a faulty application of Bayes's theorem. Ruffini thus anticipated the thinking of certain modern writers on the calculus of probability (see G. Castelnuovo, *Calcolo della probabilità*, I [Bologna, 1947], 150).

With the fall of Napoleon and the return of the Este family to Modena, Ruffini was appointed rector of the restored university in 1814. The contemporary political climate rendered his rectorship especially difficult, despite his enthusiasm, discretion, and honesty. He also held the chairs of applied mathematics and practical medicine until his death, but poor health forced him to relinquish the chair of clinical medicine in 1819.

Ruffini's patients included the destitute, as well as the duchess, of Modena. While tending to the victims of the typhus epidemic of 1817–1818 he contracted a serious form of the disease. In "Memoria del tifo contagioso" (1820), written after his recovery, he dealt with the symptoms and treatment of typhus on the basis of his own experience. Despite advice that he moderate his activities, he resumed his scientific and medical work. His strength gradually ebbed; and in April 1822, after a visit to one of his patients, he was struck by a raging fever, which obliged him to give up his activities. This last illness (chronic pericarditis) led to his death.

He was almost completely forgotten after his death, because of political and ideological reasons as well as the difficulty of interpreting his writings. His research bore precious fruit, however, largely through the work of Cauchy.

BIBLIOGRAPHY

I. ORIGINAL WORKS. Ruffini's writings include *Teoria generale delle equazioni in cui si dimostra impossibile la* soluzione algebrica delle equazioni generali di grado superiore al quarto, 2 vols. (Bologna, 1799); "Della soluzione delle equazioni algebriche determinate particolari di grado superiore al quarto," in *Memorie di matematica e di fisica della Società italiana delle scienze*, **9** (1802), 444–526; "Riflessioni intorno alla rettificazione, ed alla quadratura del circolo," *ibid.*, 527–557; "Della insolubilità delle equazioni algebriche generali di grado superiore al quarto," *ibid.*, **10**, pt. 2 (1803), 410–470; *Sopra la determinazione delle radici delle equazioni numeriche di qualunque grado* (Modena, 1804); "Risposta . . . ai dubbi propostigli dal socio Gianfrancesco Malfatti sopra la insolubilità delle equazioni di grado superiore al quarto," in *Memorie di matematica e di fisica della Società italiana delle scienze*, **12**, pt. 1 (1805), 213–267; "Riflessioni . . . intorno al metodo proposto dal consocio Gianfrancesco Malfatti per la soluzione delle equazioni di quinto grado," *ibid.*, 321–336; *Della immaterialità dell'anima* (Modena, 1806); and "Della insolubilità delle equazioni algebriche generali di grado superiore al quarto qualunque metodo si adoperi algebrico esso siasi o trascendente," in *Memorie dell'Istituto nazionale italiano*, Classe di fisica e di matematica, **1**, pt. 2 (1806), 433–450.

Subsequent works are "Alcune proprietà generali delle funzioni," in *Memorie di matematica e di fisica della Società italiana delle scienze*, **13**, pt. 1 (1807), 292–335; *Algebra e sua appendice*, 2 vols. (Modena, 1807–1808); "Di un nuovo metodo generale di estrarre le radici numeriche," in *Memorie di matematica e di fisica della Società italiana delle scienze*, **16**, pt. 1 (1813), 373–429; *Riflessioni intorno alla soluzione delle equazioni algebriche generali* (Modena, 1813); "Memoria del tifo contagioso," in *Memorie della Società italiana delle scienze*, Phys. sec., **18**, pt. 1 (1820), 350–381; "Intorno al metodo generale proposto dal Signor Hoêne Wronscki onde risolvere le equazioni di tutti i gradi," *ibid.*, Math. sec., **18**, fasc. 1 (1820), 56–68; and "Opuscolo I e II della classificazione delle curve algebriche a semplice curvatura," *ibid.*, 69–142, 269–396.

See also *Riflessioni critiche sopra il saggio filosofico intorno alla probabilità del signor Conte Laplace* (Modena, 1821); "Elogio di Berengario da Carpi," in *Fasti letterari della città di Modena e Reggio*. III (Modena, 1824); "Alcune proprietà delle radici dell'unità," in *Memorie dell'I.R. Istituto del regno lombardo-veneto*, **3** (1824), 67–84; "Riflessioni intorno alla eccitabilità, all'eccitamento, agli stimoli, ai controstimoli, alle potenze irritative, alle diatesi sì ipersteniche che iposteniche," in *Memorie della R. Accademia di scienze, lettere ed arti in Modena*, **1** (1833), 1–55; "Osservazioni intorno al moto dei razzi alle Congreve," *ibid.*, 56–78; "Intorno alla definizione della vita assegnata da Brown," *ibid.*, 319–333; and *Opere matematiche*, E. Bortolotti, ed., 3 vols. (Rome, 1953–1954).

MSS, letters, and documents relating to Ruffini are in the Academy of Science, Letters, and Arts of Modena.

II. SECONDARY LITERATURE. See the following, listed chronologically: A. Lombardi, *Notizie sulla vita e sugli scritti del prof. Paolo Ruffini* (Modena, 1824); H. Burkhardt, "Die Anfänge der Gruppentheorie und Paolo Ruffini," in *Zeitschrift für Mathematik und Physik*, **37** (1892), supp., 119–159; and "Paolo Ruffini e i primordi della teoria dei gruppi," in *Annali di matematica*, 2nd ser., **22** (1894), 175–212; E. Bortolotti, "Influenza dell'opera matematica di Paolo Ruffini sullo svolgimento delle teorie algebriche," in *Annuario della R. Università di Modena* (1902–1903), 21–77; "Un teorema di Paolo Ruffini sulla teoria delle sostituzioni," in *Rendiconti dell' Accademia dei Lincei*, ser. 5a, **22** (1913), 1st sem., pp. 679–683; and "I primordi della teoria generale dei gruppi di operazioni e la dimostrazione data da Paolo Ruffini della impossibilità di risolvere con funzioni trascendenti esatte le equazioni generali di grado superiore al quarto," in *Memorie della R. Accademia di scienze, lettere ed arti in Modena*, ser. 3a, **12** (1913), 179–195; G. Barbensi, *Paolo Ruffini nel suo tempo* (Modena, 1955), with complete bibliography of Ruffini's writings; G. Varoli, "Su un'opera pressochè sconosciuta di Paolo Ruffini," in *Statistica* (Bologna) (July–Sept. 1957), 421–442; and E. Carruccio, "Paolo Ruffini matematico e pensatore," in *Memorie della R. Accademia di scienze, lettere ed arti in Modena*, 6th ser., **8** (1966), liii–lxix.

ETTORE CARRUCCIO

RUFFO, GIORDANO (*b.* Gerace [?], Calabria, Italy; *fl.* Italy, middle of the thirteenth century), *veterinary medicine.*

Ruffo was either the son or the nephew of Pietro Ruffo, count of Catanzaro and viceroy of Sicily. He was himself the lord of Valle di Crati, and in 1239 was named governor of Cassino. He spent a part of his life with other members of his family at the court of the emperor Frederick II, who learned that Ruffo was expert in caring for horses and made him his farrier. Ruffo's treatise *De medicina equorum* is dedicated to the emperor's memory and must therefore have been finished after Frederick's sudden death in 1250. It is almost equally certain that the work was put into final form before the death of Frederick's successor, Conrad, in 1254, for Ruffo's involvement in the subsequent political turmoil would otherwise have precluded its completion. In the internecine struggle between Conradin, Frederick's grandson, and Manfred, Frederick's illegitimate son, Ruffo opposed Manfred, who took him prisoner at the beginning of the war. Following the Assembly of Barletta, in February 1256, Ruffo's eyes were put out. It is probable that he died in captivity.

The *De medicina equorum* consists of six parts: "De generatione et nativitate equi," "De captione et domatione equi," "De custodia et doctrina equi," "De cognitione pulcritudinis corporis equi," "De egritudinibus naturalibus venientibus," and "De accidentalibus infirmitatibus et lesionibus equorum." As their titles suggest, the first four sections deal with horses in general, while the last two are more specifically concerned with veterinary medicine.

It is likely that Ruffo knew the Byzantine *Hippiatrica*, perhaps in the compilation by Hierocles. His pathology derives from the humoral theory that Galen elaborated from the work of Hippocrates, and his work, moreover, refers to the Galenic theory of various tumors. The *De medicina equorum* is, however, essentially the product of Ruffo's personal experience and acute observation (certain passages, for example, suggest that he performed autopsies). Although he was of course ignorant of the circulation of the blood, he distinguished between veins and arteries, and he offered a method of differential diagnosis for cases of lameness. It is noteworthy that there is no astrology in his book.

The *De medicina equorum* was widely disseminated in both manuscript and printed form. It was probably written in Latin and was translated into Sicilian, Italian, French, Provençal, and Catalan. Its influence in the development of veterinary medicine was considerable.

BIBLIOGRAPHY

The only modern ed. of the *De medicina equorum*—H. Molin, *Jordani Ruffi Calabriensis hippiatrica* (Padua, 1818)—is so rare that it is virtually inaccessible. G. Beaujouan, Y. Poulle-Drieux, and J.-M. Dureau-Lapeyssonie, *Médecine humaine et vétérinaire à la fin du moyen âge* (Geneva–Paris, 1966) (Centre de recherches d'histoire et de philologie de la IVe section de l'Ecole pratique des hautes études, V, *Hautes études médiévales et modernes, 2*), pp. 17–21, contains references to the older eds., a list of the MSS of the various versions, and a bibliography on Ruffo and on his treatise; it contains also (pp. 51–114) a study on the veterinary portion of the *De medicina equorum* from a methodological point of view.

YVONNE POULLE-DRIEUX

RUFINUS (*fl.* Italy, second half of thirteenth century), *botany, medicine.*

All that is known of Rufinus is what can be gleaned from his only known work, *Liber de virtu-* *tibus herbarum.* He was an Italian monk and priest of the second half of the thirteenth century who, having studied astronomy among the liberal arts at Naples and Bologna, turned to examine "the lower realm" of herbs; the *De virtutibus herbarum*, finished after 1287, apparently was composed late in his career. This work lists nearly a thousand medicinal materials, mostly vegetable simples, presenting for each a brief summary of its description by earlier authorities (most commonly Macer Floridus, Dioscorides, and the early medieval herbal called *Circa instans*). Fully one-fifth of the text, however, consists of Rufinus' own contributions, which are outstanding for including a great many careful botanical descriptions of a detail quite unknown to earlier medical writers. He regularly describes the stem, flower, and leaves of a plant under consideration and contrasts it with similar plants. Much of his knowledge comes from direct acquaintance with these plants or from the lore of the practicing herbalists of Naples, Bologna, and Genoa; Rufinus often supplies vernacular names as well as Latin synonyms for the plants he discusses.

For all its interest in descriptive botany, Rufinus' book was intended as an aid to medical practice and is much closer in spirit to the Salernitan sources on which he drew than to the contemporary Scholastic botany of Albertus Magnus—to which, indeed, it makes no reference whatsoever. Perhaps because it lay outside the direction taken by much of medicine in the later Middle Ages, it seems to have been very little used; only Benedetto Rinio, in the early fifteenth century, has been shown to have had a knowledge of it.

BIBLIOGRAPHY

The *Liber de virtutibus herbarum* is known in only one MS, Florence, Biblioteca Laurenziana, Fondo Ashburnham 116 (189–121), of 118 fols. It has been edited, with an introduction, by Lynn Thorndike, as *The Herbal of Rufinus* (Chicago, 1946). Thorndike's introduction is in part an expansion and correction of his earlier article "Rufinus: A Forgotten Botanist of the Thirteenth Century," in *Isis*, **18** (1932), 63–76, but it does not retain the article's translations into English of sample passages from Rufinus' text.

MICHAEL MCVAUGH

RUFUS OF EPHESUS (*fl.* late first century B.C. to mid-first century A.D. [?]), *medicine.*

The dates of Rufus' birth and death are not known. According to the *Suda Lexicon* he lived "under Trajan"; yet he was also mentioned by the physician Damocrates, who lived during the reigns of Nero and Vespasian (see *Claudii Galeni Opera Omnia,* Kühn, ed., XIV, 119). Thus the report "under Trajan" appears to give too late a date for Rufus. Wellmann[1] claims that the *Suda Lexicon* is correct and that Damocrates was referring to another Rufus, namely, "Menius Rufus" (cited by Galen, XIII, 1010). But the existence of this latter Rufus is highly problematical. The supposition was made solely to account for the chronological discrepancy and is not convincing. (The very form of the name "Menius" is curious and suggests a corruption in the text.) Since no *terminus ante quem* for Rufus is known, it is even possible that he lived part of his life—or at least was born—in the first century B.C. The Latin personal name "Rufus" (that is, "red-blond") is documented for the republican period.

It is certain that the physician Rufus was Greek; and all his writings are in Greek. His name, however, strongly indicates that he was in some way connected with Rome, although it is not known whether he was ever in Rome or even in Italy.[2] It has even been suggested that because of Rufus' character and basic scientific outlook he deliberately stayed away from Rome, that "constantly sensation-seeking world capital."[3] However that may be, we know that Rufus studied and practiced medicine in his native city of Ephesus.[4] Beyond this, it can be concluded from many statements that he must have lived for a long time in Egypt, mainly in Alexandria.[5]

The available evidence does not indicate for certain whether Rufus belonged to a medical school. Ilberg stated that he was a member of the school of the "dogmatists."[6] But this designation—which was already employed in antiquity—implied nothing more than those who did not belong explicitly either to skeptical philosophy or to skeptical-empirical medicine (that is, to the medical school of the empiricists), and consequently it in no way implied allegiance to a "school" in the strict sense.[7] To describe Rufus' medical views in general terms, one must emphasize both his Hippocratism—which was far from uncritical[8]—and his eclecticism.[9]

Rufus' works are notable for the exceptional richness of their clinical observations. (One branch of his clinical studies, the investigations on melancholy, has recently been explored in detail.[10]) His works are further characterized by the care with which he evaluated his observations, for example, in the anamnestic treatise *Questions of the Physi-cian (to the Patients).* In certain areas Rufus' knowledge undoubtedly exceeded even Galen's, although, to be sure, Galen was the greater systematist. Yet, however thoroughly Rufus knew a subject, he remained cautious in his pronouncements, although he did not fall into the skepticism that was fashionable in his age. He never engaged in polemics, and his criticisms were extremely restrained and objective. In short, even in the brilliant intellectual world of the Hellenistic-Roman period, Rufus was undoubtedly a striking and independent medical figure.

The breadth of Rufus' knowledge and interests is reflected in the titles of his writings, of which ninety-six genuine works (or independent sections of works) are known.[11] He wrote some of his works in verse (hexameters)[12] in accordance with the tradition of didactic medical poems. This tradition was still honored during Nero's reign by physicians like Andromachus and Damocrates, who were following such Hellenistic predecessors as Nicandros.

A number of Rufus' genuine works have been preserved. His treatise *On the Naming of the Parts of the Human Body* appears, from the form in which it has come down to us, to be a compilation, which perhaps was only partially written by (or taken from) Rufus.[13] It is important to note that in this work anatomy is viewed primarily from the perspective of medical education and that Rufus deplored the fact that dissection of human corpses was no longer permitted.[14] His first monograph on medical anamnesis, *Questions of the Physician (to the Patients),* is in H. Gärtner's critical, annotated edition (see Bibliography). His *On Kidney and Bladder Ailments* will shortly appear in an edition prepared by A. Sideras; the textual criticisms have already been separately published. He also wrote *Satyriasis and Spermatorrhoea* and *On Joint-Diseases.* The genuineness of *Synopsis Concerning the Pulse* is questionable, and it is fairly certain that *Anatomy of the Parts of the Human Body* and *On the Bones* are not genuine works.

A series of fragments from Rufus' other writings can be found in the great medical compilations of Oribasius and Aëtius of Amida of late antiquity. Ilberg showed that more information can be derived from these sources than even Daremberg and Ruelle were able to obtain for their great edition of Rufus' works.[15] Valuable material can also be culled from the late Latin and Arabic traditions. Here, too, it is possible to go beyond Daremberg and Ruelle, as Flashar has recently shown for the particular case of Rufus' treatise on melancholy.[16]

The Arabic sources give the most detailed list of Rufus' works.[17] An examination of Ilberg's list of titles (presently the most comprehensive list, but one he himself admitted is subject to expansion and correction) makes it clear that Rufus never addressed himself in his writings to nonmedical subjects. Thus he differed from Galen and Soranus, who wrote on philosophy, philology, and medicine. Yet, within the field of medicine the scope of Rufus' interests was enormous. To cite only a few examples, it encompassed such topics as *Living at Sea, The Purchase of Slaves,* and problems of coition and potency. (He is cited as the author of a work entitled *Ointment for a Powerful Erection.*)

Several of Rufus' works were translated into Late Latin, including *On Joint-Diseases (De podagra).* Other works were cited by Western medieval physicians, for example, Constantine the African, who mentions the treatise on melancholy. In Byzantium, Rufus was counted (along with Hippocrates, Galen, and Cheiron, the mythical centaur and teacher of physicians) among "the quieting foursome of diseases" (ἡ τετράριθμος τῶν παθῶν γαληνότης), as it is expressed in the almost untranslatable baroque Greek of Byzantium.[18] The Arabs esteemed Rufus and frequently cited his work.

Although Rufus has been somewhat neglected by modern historians of medicine, at least some of his writings have appeared in new critical editions (see Bibliography). But a new complete edition (including all fragments and references) is still lacking. For this, a definite decision would be necessary with regard to the genuineness of certain treatises. A comprehensive modern book on Rufus is likewise lacking.

NOTES

1. Quoted by Ilberg, *Rufus von Ephesos,* 36.
2. See *ibid.,* 3.
3. *Ibid.,* 51.
4. *Ibid.,* 2.
5. *Ibid.,* 2 f.
6. *Ibid.,* 3 f.
7. See F. Kudlien, "Dogmatische Ärzte," in Pauly-Wissowa, *Real-Encyclopädie der classischen Altertumswissenschaft,* supp. X (1965), col. 197 f.
8. See Gärtner, *Rufus von Ephesos,* 102 ff.
9. Cf. Ilberg, *Rufus von Ephesos,* 4.
10. See H. Flashar, *Melancholie und Melancholiker in den medizinischen Theorien der Antike* (Berlin, 1966), 84–104.
11. See Ilberg, *Rufus von Ephesos,* 47–50.
12. *Ibid.,* sec. 4, title no. 2, p. 49.
13. *Ibid.,* 7–12.
14. Cf. L. Edelstein in O. and L. Temkin, eds., *Ancient Medicine. Selected Papers of Ludwig Edelstein* (Baltimore, 1967), 250 and 270 f.
15. Ilberg, *Rufus von Ephesos,* 25 ff.
16. Cf. Flashar, op. cit., 88 ff.
17. Cf. Ilberg, *Rufus von Ephesos,* 43 ff.
18. See H. Gossen, "Rufus no. 18," in Pauly-Wissowa, *Real-Encyclopädie der classischen Altertumswissenschaft,* IA1 (1914), col. 1212.

BIBLIOGRAPHY

Rufus' major writings are collected in C. Daremberg and E. Ruelle, eds., *Oeuvres de Rufus d'Ephèse* (repr., Amsterdam, 1963). See also H. Gärtner, "Rufus von Ephesos: Die Fragen des Arztes an den Kranken," in *Corpus medicorum Graecorum,* supp. 4 (Berlin, 1962); and A. Sideras, "Textkritische Beiträge zur Schrift des Rufus von Ephesos 'De renum et vesicae morbis,'" in *Abhandlungen der geistes- und sozialwissenschaftlichen Klasse der Akademie der Wissenschaften und der Literatur,* no. 3 (1971).

On Rufus and his work, see J. Ilberg, "Rufus von Ephesos, Ein griechischer Arzt in trajanischer Zeit," in *Abhandlungen der philologisch-historischen Klasse der Sächsischen Akademie der Wissenschaften,* **41,** no. 1 (1930).

FRIDOLF KUDLIEN

RÜHMKORFF, HEINRICH DANIEL (*b.* Hannover, Germany, 15 January 1803; *d.* Paris, France, 20 December 1877), *technology.*

As one of ten children of a postal ostler, Rühmkorff was encouraged to support himself at an early age. He was apprenticed to a mechanic named Wellhausen in Hannover until he was eighteen, after which he wandered about Germany for two years. He spent a year and a half in Paris assisting a physics lecturer, and in 1824 he went to England where he found a position in the workshop of Joseph Brahmah, known as the inventor of the hydraulic press. He returned home in 1827 and then went to Paris, where, after working in several shops, he found employment with Charles Chevalier, renowned for his optical instruments. In 1855 he founded his own shop, a "dingy little bureau in the rue Champollion" that became well-known to physicists throughout the world, especially for the quality of the electrical apparatus it produced.

Rühmkorff is most widely known as the inventor of an induction coil capable of producing sparks a foot or more in length. In 1864 he was presented with a 50,000-franc prize for his work. Established by the emperor in 1852, it was to be awarded for the most important discovery in the application of electricity (although the prize was to be given after five years, a postponement was deemed necessary by the committee.)

Rühmkorff's coil was widely used as a convenient power source for the operation of Geissler and Crookes tubes as well as for various other high-voltage laboratory needs, and it had practical application in detonating explosives. But if its importance is not open to question, the extent of Rühmkorff's role in its design is. Basic knowledge of induction originated in the work of Faraday and Joseph Henry, and an induction coil capable of producing small sparks was constructed by Charles G. Page in 1838. The interrupter mechanism was devised by Auguste de La Rive. On the basis of these accomplishments, Rühmkorff in 1851 constructed a coil that was capable of producing sparks of moderate length. Fizeau then introduced the use of a condenser that moderated the opposing effect of the secondary current on the primary current. More important was the development of a winding technique that would prevent the destruction of the coil by internal arcing. One method was suggested by Poggendorff in 1854; in another, employed by E. S. Ritchie in 1857, the wire was wound in parallel spiral disks.

Rühmkorff's electrical apparatus was widely used in physical laboratories, and the many surviving examples testify to his workmanship. It is reported that he died nearly impoverished, having given his money to science or to charitable causes.

BIBLIOGRAPHY

The best account of Rühmkorff's life and work appears in a 36-page volume by Emil Kosack: *Heinrich Daniel Rühmkorff, ein deutscher Erfinder* (Leipzig–Hannover, 1903). There are brief but relatively informative unsigned sketches in *Nature,* **17** (1877), 16, repr. in *Journal of the Franklin Institute,* **105** (1878), 133–134; and in *Scientific American,* **38** (1878), 81. Discussions of the history of the induction coil include E. S. Ritchie, "On Electrical Machines," in *Journal of the Franklin Institute,* **73** (1862), 58–60; an unsigned article in *Scientific American,* **12** (1865), 6; one signed S. H. W. in the same issue, 69; Charles G. Page, *A History of Induction* (Washington, D.C., 1867), esp. 115–121; T. du Moncel, *Exposé des applications d'électricité,* 3rd ed. (Paris, 1873), II, 238–262, and *Notice sur l'appareil d'induction électrique de Rühmkorff,* 5th ed. (Paris, 1867); Florian Cajori, *A History of Physics* (New York, 1899), 245–246; and George Shiers, "The Induction Coil," in *Scientific American,* **224** (1971), 80–87.

Bernard S. Finn

RUINI, CARLO (*b.* Bologna, Italy, *ca.* 1530; *d.* Bologna, 1598), *anatomy, veterinary surgery.*

A Bolognese aristocrat, senator, and high-ranking lawyer, Ruini is—somewhat surprisingly—remembered chiefly for the two-volume *Anatomia del cavallo, infermità et suoi rimedii.* The first edition appeared after Ruini's death (Bologna, 1598) and was followed throughout the seventeenth century by other editions and translations; and in 1706–1707 the first edition was reissued.

In the introduction to book 1 of the first volume Ruini stresses the importance of "artful instruction" concerning the body of the horse, which leads to knowledge of its constitution and of the means of prolonging its life. It was from riding the horse, Ruini points out, that man derived the title *cavaliere* (knight) to denote his own valor and nobility. After recounting the horse's role in both work and recreation, Ruini concludes the introduction by explaining that his aims are to describe each part of the horse's body, the nature of the ailments that afflict them, and the means of curing this worthy and noble animal.

In the first volume, which deals mainly with anatomy, Ruini includes notes on physiology that reflect his teleological Galenic approach. In the first book the morphology of the head is described in detail. The second book deals with the neck and its organs, the lungs, the heart, and the thoracic muscles, blood vessels, and nerves. The third book covers the liver, spleen, kidneys, stomach, intestines, peritoneum, and bladder. The structures of these organs and their positions are described, as are the lumbosacral region and its muscles, blood vessels, and nerves. The fourth book describes the genital system, and the fifth deals with the extremities.

Volume II deals specifically with equine diseases and their cures. Explaining that he has followed the methods used by Aristotle, Hippocrates, and Galen to describe the human body, Ruini considers equine pathology, beginning with conditions of a general nature, such as fever, before progressing to descriptions of specific diseases. He considered it necessary to place pathology on a constitutional foundation because he believed that from knowledge of the horse's physical disposition one could more easily understand the nature of disease; also, from knowing the age of the horse, one could determine the appropriate treatment at any phase of an illness. At the beginning of the first book, Ruini discusses at length the four Galenic humors (choleric, sanguine, phlegmatic, and melancholic) and ways of telling a horse's age. He then offers a detailed analysis of fever, distinguishing three types, giving general causes and a general cure, and discussing fevers of various origins.

The second book considers various types of horse in regard to humoral pathology, using criteria based essentially on the concept of the four qualities (hot, cold, moist, dry). Ruini then examines a series of "affections" of the brain: frenzy, rage or fury, and insanity, leading to convulsions and paralysis. The book concludes with the diseases of the neck. In the third book Ruini describes the diseases of the heart and the lungs; in the fourth, the afflictions of the digestive tract, from diarrhea to jaundice; and in the fifth, hernia, diseases of the testicles and penis, and problems of obstetrics. The sixth book deals with the diseases of the legs.

On the whole, Ruini's treatise was still closely bound to the Scholastic tradition. It does, however, show the effort made by its author, who must certainly have known the work of Vesalius, to produce a work that would manifest the new direction being taken by sixteenth-century anatomy. Because it was so traditional, his treatment of pathology, although minutely detailed, is less valuable than his study of anatomy. A pioneer in the latter field, Ruini deserves to be ranked among the founders of comparative anatomy, along with Vesalius, Belon, Rondelet, and Coiter.

Some scholars believe that Ruini was active in the discovery of the lesser and greater circulatory systems, and would therefore place him in the group that included Colombo and Cesalpino. A full discussion of this hypothesis would be out of place here, but it is probable that Ruini was one of many who had an inkling of the circulation of the blood.

BIBLIOGRAPHY

I. ORIGINAL WORKS. Ruini's work was *Anatomia del cavallo, infermità et suoi rimedii. Opera nuova degna di qualsivoglia prencipe, et cavaliere, et molto necessaria a filosofi, medici, cavallerizzi et marescalchi,* 2 vols. (Bologna, 1598; Venice, 1599), reprinted in Venice (1706–1707).

II. SECONDARY LITERATURE. On Ruini and his work, see *Dizionario enciclopedico italiano,* X (Rome, 1959), 622; G. B. Ercolani, *Carlo Ruini—Curiosità storiche e bibliografiche intorno alle scoperte della circolazione del sangue* (Bologna, 1873); Pagel, "Ruini," in *Biographisches Lexicon der hervorragenden Ärzte aller Zeiten und Völker,* 2nd ed., IV (Berlin–Vienna, 1932), 921; and C. Singer, *A Short History of Anatomy From the Greeks to Harvey* (New York, 1957), 153.

LORIS PREMUDA

RUIZ, HIPÓLITO (*b.* Belorado, Burgos province, Spain, 8 August 1754; *d.* Madrid, Spain, 1816), *botany.*

As a youth Ruiz was apprenticed to the pharmacy of an uncle in Madrid, where he learned chemistry and other sciences. In 1772 he began work in the botanical garden at Migas Calientes, where he attracted the attention of the director, Casimiro Gómez Ortega. As a result he was chosen first botanist of a joint French-Spanish expedition to Peru.

The expedition, led by Ruiz, José Antonio Pavón y Jiménez, and Joseph Dombey, reached Lima on 8 April 1778. The objectives of the project were more utilitarian than theoretical, and much of Ruiz' energy was directed toward the investigation of cinchona. His first publication after returning to Spain in February 1785 was *Quinología* (1792), which described seven species of cinchona and praised the medicinal qualities of a quina extract that he had developed. (In 1801 Ruiz and Pavón published a supplement to the *Quinología* to respond to charges made by Francisco Antonio Zea, who questioned the probity of their descriptions and asserted the priority of discoveries by José Celestino Mutis.)

Ruiz next turned to the major portion of the expedition's work, the *Flora Peruviana.* A slight introductory volume, the *Prodromus,* appeared in 1794. It generated an intense dispute with Antonio Cavanilles, who claimed to have already described genera that Ruiz and Pavón announced as newly discovered.

The first three volumes of the richly illustrated *Flora Peruviana,* published between 1798 and 1802, described 751 plant species through the first seven Linnaean classes. Ruiz spent the rest of his life in futile attempts to raise enough money to finish the project.

Tireless collectors and classifiers of plants, Ruiz and Pavón were extremely conservative theoretical botanists, defending the Linnaean system against all innovations, particularly the reduction of the number of classes.

BIBLIOGRAPHY

I. ORIGINAL WORKS. Works written by Ruiz alone include *Quinología, o tratado del árbol de la quina o cascarilla* (Madrid, 1792), with translations into Italian (Rome, 1792), German (Göttingen, 1794), and English (London, 1800); *Relación del viaje hecho a los reynos del Perú y Chile,* Agustín J. Barreiro, ed. (Madrid, 1931), translated into English by B. E. Dahlgren as *Travels of*

Ruiz, Pavón and Dombey in Peru and Chile (1777–1778) (Chicago, 1940). Another ed. of the *Relación* was prepared by Jaime Jaramillo-Arango, 2 vols. (Madrid, 1952). On the controversy with Cavanilles, see *Respuesta para desengaño del público a la impugnación que ha divulgado prematuramente el presbítero don Josef Antonio Cavanilles, contra el pródromo de la Flora del Perú* (Madrid, 1796).

The major botanical descriptions were published in collaboration with Pavón: *Florae Peruvianae et Chilensis prodromus* (Madrid, 1794; 2nd ed., Rome, 1797); *Flora Peruviana et Chilensis,* 3 vols. (Madrid, 1798–1802); and *Suplemento a la Quinología* (Madrid, 1801).

II. SECONDARY LITERATURE. A definitive study of the Ruiz-Pavón expedition is Arthur R. Steele, *Flowers for the King* (Durham, N.C., 1964). See also Enrique Alvarez López, "Algunos aspectos de la obra de Ruiz y Pavón," in *Anales del Instituto Botánico A. J. Cavanilles,* **12** (1953), 5–110.

THOMAS F. GLICK

RULAND, MARTIN (*b.* Lauingen, Germany, 11 November 1569; *d.* Prague, Bohemia, 3 April 1611), *medicine, iatrochemistry.*

Ruland's father was Martin Ruland the elder (1532–1602), who, in the last years of his life, was physician to Emperor Rudolf II. Ruland received the M.D. from the University of Basel in 1587. His interest in alchemy and iatrochemistry may have developed while he was at Basel or may have been entirely due to the influence of his father, who favored Paracelsian reforms and the use of chemically prepared medicines.

Nothing is known of Ruland again until 1594, when he was a physician at Regensburg. His alchemical interests first emerged in two works of 1595 and 1597, which argued that the gold tooth reportedly cut by a Silesian boy was genuine and could have been naturally generated, a conclusion of some alchemical significance. In 1600 Ruland published an extensive discussion of the nature, causes, symptoms, and treatment of the *morbus hungaricus,* which was in all probability typhus. He also proposed a number of remedies, many of which involved chemical preparations.

The major aspects of Ruland's thought were his alchemical philosophy of nature and his advocacy of chemical medicines. In 1606, while at Regensburg, he was charged with dispensing poisonous medicines. (Even as late as the turn of the seventeenth century, Paracelsian medical reforms and iatrochemical remedies had little official approval in Germany.) Ruland's attitude was moderate; he did not entirely reject the traditional Galenic posi-

tion, but his concept of medicine was generally based upon Paracelsian theories. Because nature and man are primarily chemical in composition and function, he believed, the physician should study chemistry in order to understand nature and should use chemicals to aid nature in curing diseases. He therefore concluded that chemically prepared remedies were safe and legitimate. Ruland's works illustrate the tension between traditional Galenist medical theory and the reformist iatrochemists, as well as between individuals of similar outlook, such as himself and his major opponent, Johann Oberndorfer, thereby indicating the complexity of medical controversies at that time. In 1607 Ruland was appointed physician to Emperor Rudolf II and settled in Prague.

Ruland's cosmology was derived from Renaissance Neoplatonic Hermeticism, according to which the cosmos is a unity modeled on divine archetypes. All aspects of the universe are interconnected by spiritual forces, and nature is strictly chemical in its operation. Salt, sulfur, and mercury, the three principles of Paracelsus, are the basis of all things; and the principal instrument for the study of nature is fire and the alchemical processes involving it. On the basis of this cosmology, Ruland argued that the transmutation of metals into gold was possible with the aid of the philosophers' stone, which was also the universal medicine capable of curing all diseases. *Lexicon alchemiae,* including much Paracelsian terminology, was issued posthumously in 1612.

Ruland's work is significant as an illustration of the process of the assimilation of Paracelsian reforms in medicine and chemistry, which had an important impact on the development of those fields in the late sixteenth and early seventeenth centuries. Further, before the establishment of a mechanical and mathematical conception of nature in the seventeenth century, many natural philosophers, like Ruland, found in the Hermetic and alchemical approach to nature a stimulating alternative to traditional Aristotelian natural philosophy.

BIBLIOGRAPHY

I. ORIGINAL WORKS. There has been some confusion between the works of Ruland and those attributed to his father. The following are undoubtedly by Martin Ruland the younger: *Nova et in omni memoria omnino in audita historia de aureo dente qui nuper in Silesia puero cuidam septenni succrevisse magna omnium admiratione animadversus est et eiusdem de eodem judicium* (Frankfurt, 1595); *Demonstratio judicii de dente aureo pueri Silesii adversus responsionem M. Ioh. Ingolstetteri* (Frankfurt, 1597); *De perniciosae luis ungaricae* (Frankfurt, 1600); *Progymnasmata alchemiae . . .* (Frankfurt, 1607); *Propugnaculum chymiatriae: Das ist, Beantwortung und Beschützung der alchymistischen Artzneyen . . .* (Leipzig, 1608); *Alexicacus chymiatricus . . .* (Frankfurt, 1611); and *Lexicon alchemiae sive dictionarium alchemisticum . . .* (Frankfurt, 1612).

II. SECONDARY LITERATURE. There has been no serious study of Ruland, and his life will not be covered well until archival material, as well as his works, can be studied in depth. The works below give what information there is and some relevant aspects of the background. For earlier secondary works, consult the bibliography in John Ferguson, *Bibliotheca chemica,* 2 vols. (London, 1954), II, 304; F. H. Garrison, *An Introduction to the History of Medicine* (Philadelphia, 1929), 243; J. R. Partington, *A History of Chemistry,* II (London, 1961), 161–162; Wolfgang Schneider, "Die deutschen Pharmakopöen des 16. Jahrhunderts und Paracelsus," in *Pharmazeutische Zeitung,* **106** (1961), 1141–1145; and "Der Wandel des Arzneischatzes im 17. Jahrhundert und Paracelsus," in *Sudhoffs Archiv für Geschichte der Medizin und der Naturwissenschaften,* **45** (1961), 201–215; and Lynn Thorndike, *A History of Magic and Experimental Science,* 8 vols. (New York, 1922–1948), VII, 159–160, and VIII, 371–372.

N. H. CLULEE

RÜLEIN, ULRICH (usually called **ULRICH RÜLEIN VON CALW**) (*b.* Calw [?], Germany, 1465/1469; *d.* Germany, 1523), *geology, mining, medicine.*

Although Rülein was one of the leading figures of the German Renaissance, very little is known about him. Mentioned in the 1556 edition of Agricola's *De re metallica* and in the dedicatory letter to the 1550 edition of that work, he was scarcely heard of again for nearly 350 years, until his place in the history of science was rediscovered toward the end of the nineteenth century. As a result of subsequent research, his name is permanently linked with the birth of modern geology and mining science and especially with the creation of the modern science of mineral deposits.

The year of Rülein's birth, let alone the exact date, has not been established. Even the place of his birth cannot be determined from the surviving records, although his family's place of origin can be asserted with confidence. The year of his death has been established with virtual certainty, but the sources yield no exact date or place.

The earliest confirmed mention of Rülein, his matriculation at the University of Leipzig in 1485, is the sole basis for making even an approximate calculation of the year of his birth. From the desig-

nation "molitoris" in the matriculation register, and from other information on the Rülein family, it is known that Rülein's father and grandfather were millers. Accordingly, the family must have been relatively prosperous.

The designation "de Calb" formed part of Rülein's Latinized name, and in German this element was written "Kalb." It was therefore long assumed that he was born either in the city of Calbe (on the Saale River) or in Kalbe (on the Milde River), both of which are near Magdeburg. Recent research, however, has revealed that his family never lived in either city but, rather, in Calw, in Württemberg. Since the church records of Calw were burned in 1693, the date of Rülein's birth probably will never be known.

Rülein may have attended Calw's Latin school or he may have received his earliest instruction at the school in nearby Kloster Hirsau, which greatly emphasized mathematics. He studied the liberal arts at the University of Leipzig and earned his master's and doctor's degrees in 1490. His principal interest must have been in natural science and mathematics. He then turned to the study of medicine, obtaining his M.D. degree in 1496 or 1497. It is not known whether he completed his medical studies at Leipzig or at another university. It is very probable that Rülein was identical with an "Ulrich Kalb, of the University of Leipzig, master of the liberal arts and of philosophy and professor of mathematics." A student of the latter's, Balthasar Licht, dedicated an arithmetic book, *Algorithmus linealis* (1500), to him. It may, therefore, be supposed that Rülein was a *Dozent* in mathematics at Leipzig between 1490 and 1496 or 1497—while he was studying medicine—a situation not at all unusual at the time.

By then Rülein must already have enjoyed a reputation as a scholar and a man of practical experience. This is the only explanation for his being asked by the duke of Saxony in 1496 to plan the building of the city of Annaberg. Silver ore had been discovered there in 1492, and within a few years its mines had become the most productive in Europe. The immense influx of people made it necessary to create a comprehensive city plan as quickly as possible. Rülein's success in this difficult assignment can be inferred from his having been commissioned in 1521 to plan another Saxon mining city, Marienberg, where silver had been found in 1520.

Rülein's closest ties, however, were to the most famous German mining city, Freiberg. He was appointed its municipal physician in 1497, and in the same year he purchased a house there. (Hence, Agricola called him "Kalbius Fribergius.") Rülein married in 1500. His wife, whose name and origins are unknown, died probably in 1524 or 1525; they had four sons and one daughter.

For more than twenty years Rülein was active in Freiberg as a physician, mining expert, surveyor, and politician. He often served as city councillor and twice as mayor (1514–1515 and 1517–1519). During his first term as mayor he loosened the dominant hold of the Church on education through the founding in 1515 of one of the earliest municipal Latin schools in Germany. (Later called a Gymnasium, the school is today a renowned high school, the Geschwister Scholl.) Rülein brought such famous scholars as Johann Rhagius Aesticampianus and Petrus Mosellanus to the school as teachers. In 1521 Rülein wrote two works on precautions to be taken against the plague. Highly respected as a physician, he was equally well regarded for the encouragement he gave to education and culture while holding public office. In short, he was the guiding spirit of the intellectual life of Freiberg in the period immediately preceding the Reformation.

Rülein's most important achievement was a brief work entitled *Ein nützlich Bergbüchlein*. It appeared anonymously about 1500 (the date of publication of the first edition cannot be precisely determined), but no less an authority than Agricola attributed it to Rülein. Comprising ten chapters, the *Bergbüchlein* deals with the formation of ores, the nature of ore deposits in general, surveying techniques useful in mining, and particular characteristics of deposits of gold and mercury as well as of ores of silver, zinc, copper, iron, and lead. The book is structured as a dialogue between an experienced miner (Daniel) and a young man (Knappius) eager to learn about the industry.

In the introductory chapter on the formation of ores, Rülein repeats the traditional alchemical views but adopts a critical stance toward them. What is really new and important in the book is his description of ore deposits and his instructions for surveying a mine. Rülein was the first to examine fully the personal experiences of miners concerning ores and ore deposits, which he then generalized and introduced into the literature. Thus the publication of his book was a crucial first step in the development of the science of mining. The *Bergbüchlein* constituted a bridge between a centuries-old practice and the theory required by the opening of major new mines and the vastly increased demands for their yields.

The *Bergbüchlein* is also notable for having been written in German—although Rülein was fluent in Latin. In this respect it inaugurated an era of transition from the universal use of Latin in scholarly discourse to the publication of scientific works in the vernacular.

One of Freiberg's wealthy citizens, Rülein owned several houses in the city and shares in a number of silver mines. His efforts to raise the level of education in Freiberg led to heated conflicts in the city council, from which he resigned in 1519. Rülein spent some of his remaining years in Leipzig, but it is not known whether he died there or in Freiberg.

BIBLIOGRAPHY

I. ORIGINAL WORKS. *Ein nützlich Bergbüchlein*, Rülein's principal work, was first published (without indication of place or year) about 1500. The 2nd and 3rd eds. appeared under the title *Ein wohlgeordnetes und nützliches Büchlein, wie man Bergwerk suchen und finden soll . . .* (Augsburg, 1505; Worms, 1518). The 4th (Erfurt, 1527) bore the original title. Further eds. were sometimes accompanied by a *Probierbüchlein*—not written by Rülein (Frankfurt, 1533; Augsburg, 1534; Frankfurt, 1535). The *Bergbüchlein* also appeared as part of a collection by Johann Haselberger entitled *Der Ursprung gemeiner Bergrecht* (n.p., ca. 1538; Augsburg, 1539). It was subsequently reprinted in *Ursprung und Ordnungen der Bergwerke* (Leipzig, 1616); in *Magazin für die Bergbaukunde* (Dresden), no. 9 (1792), 219–274; and in *Zeitschrift für Bergrecht* (Bonn), **26** (1885), 508.

The *Bergbüchlein* was partially translated into French by Adolf Gurlt in Gabriel Auguste Daubrée, "La génération des minéraux métalliques, dans la pratique des mineurs du moyen âge, d'après le *Bergbüchlein*," in *Extrait du Journal des savants* (Paris, 1890), 379–392, 441–452. It was translated into English by Anneliese G. Sisco and Cyril S. Smith in *Bergwerk- und Probierbüchlein. A Translation From the German* (New York, 1949). See also Judica I. M. Mendels, "Das Bergbüchlein. A Text Edition" (Ph.D. diss., Johns Hopkins, 1953). The most recent ed. of the work, containing a facs. of the original ed. and a new German trans., is in Wilhelm Pieper, *Ulrich Rülein von Calw und sein Bergbüchlein* (Berlin, 1955), no. D7 in the series Freiberger Forschungshefte—the only ed. based on the text of the 1st ed. It also contains the most comprehensive biography of Rülein and a critical discussion of much of the information presented in previous accounts.

In a *Pestschrift* published in 1521 Rülein described the plague and gave instructions on how to protect against it. This work has not survived, and we possess only a seven-page extract entitled *Eine Unterweisung wie man sich zu Zeiten der Pestilenz verhalten soll* (1521), which appeared anonymously, with no indication of place or year of publication.

II. SECONDARY LITERATURE. The earliest biographical account of Rülein is in Hieronymus Weller, *Analecta Welleriana. In lateinischer Sprache zussamengelesen und verdeutscht durch Michael Hempel*, II (Freiberg, 1596), chs. 30 and 31, 67b–68b. After 300 years Rülein was rediscovered by Georg Heinrich Jacobi, who examined his relationship with Agricola in *Der Mineralog Georgius Agricola und sein Verhältnis zur Wissenschaft seiner Zeit* (Werdau, Saxony, 1889).

The first modern biographical study is Constantin Täschner, "Der Arzt, Bürgermeister und Bergbauschriftsteller Ulrich Rülein von Kalbe," in *Mitteilungen des Freiberger Altertumsvereins*, no. 50 (1915), 21–27. Further details are in Otto Clemen, "Der Freiberger Stadtphysikus Ulrich Rülein von Kalbe," in *Neues Archiv für sächsische Geschichte und Altertumskunde*, **41** (1920), 135–139; Ernst Darmstaedter, *Berg-, Probier- und Kunstbüchlein*, which is Münchener Beiträge zur Geschichte und Literatur der Naturwissenschaft und Medizin, nos. 2–3 (Munich, 1926); Otto E. Schmidt, "Ulrich Rülein von Kalbe," in *Mitteilungen des Landesvereins sächsischer Heimatschutz*, **26** (1932), 111–114; and Karl Lüdemann, "Ulrich Rülein von Kalbe, der Verfasser des ersten deutschen Buches über den Bergbau," in *Mitteilungen des Freiberger Altertumsvereins*, no. 64 (1934), 67–75.

HANS BAUMGÄRTEL

RUMFORD. See Thompson, Benjamin.

AL-RŪMĪ. See Qāḍī Zāda al-Rūmī.

RUMOVSKY, STEPAN YAKOVLEVICH (*b.* Stary Pogost, near Vladimir, Russia, 9 November 1734; *d.* St. Petersburg, Russia, 18 July 1812), *astronomy, mathematics, geodesy.*

Rumovsky was the son of a priest. In 1739, after the family moved to St. Petersburg, he entered the Aleksander Nevsky Seminary, where he studied for nine years. In 1748 Rumovsky was one of four students chosen to study at the university of the St. Petersburg Academy of Sciences, where he heard Lomonosov lecture in chemistry, Richmann in mathematics and physics, and Nikita Popov in astronomy. In 1750 Rumovsky decided to specialize in mathematics and two years later presented a work that dealt with the use of tangents to find a straight line equal to the arc of an ellipse. In 1753 Rumovsky became adjunct of the Academy of Sciences after having submitted to it a work in which he offered a solution to the problem, posed by Kepler, of finding a semiordinate for a given sector. In his review of this paper, Euler noted the author's gift for mathematics. Rumovsky continued his mathematical education under Euler for

two years, living in his house in Berlin and working under his direction.

After returning to St. Petersburg, Rumovsky held various posts at the Academy of Sciences. In 1760 he was sent to A. N. Grischow. He lived with Grischow in St. Petersburg and in his well-furnished observatory "became skilled in astronomical practice." In 1761 he took part in an expedition to Selenginsk in Transbaikalia, to observe the transit of Venus.

After Lomonosov's death, Rumovsky succeeded him as director of the geographical department of the St. Petersburg Academy, holding the post from 1766 to 1805; he also headed the astronomical observatory of the Academy from 1763. An active participant in the expedition to observe the 1769 transit of Venus, Rumovsky supervised the preparations of the observers, the choice of sites, and the building of temporary observatories. In connection with the latter problem he corresponded with James Short and with Euler. Rumovsky himself observed the transit on the Kola Peninsula. Having analyzed all the observations from both transits, he calculated the value of the solar parallax as 8.67"—he came nearer than any of his contemporaries to the presently accepted value (8.79").

Participating in a number of expeditions, Rumovsky determined the longitude and latitude of various sites, then compiled the first summary catalog of the astronomically determined coordinates of sixty-two sites in Russia. The *Berliner astronomisches Jahrbuch* for 1790 published fifty-seven determinations from Rumovsky's catalog—thirty-nine for European Russia and eighteen for Siberia. Prior to Rumovsky's work, only seventeen complete determinations had been made. According to W. Struve, Rumovsky's determinations were "distinguished by a precision remarkable for that time, the probable error in longitude not exceeding 32 seconds or 8' of arc."

Rumovsky's more than fifty basic scientific works cover astronomy, geodesy, mathematics, and physics. Like Lomonosov, his interests included a profound study of the Russian language and literature. In 1783 he became a member of the Russian Academy. Especially valuable was his activity in the translation and compilation of the first etymological dictionary of the Russian Academy (6 vols., 1789–1794), for which he received its gold medal.

Rumovsky began teaching in 1757 at the university of the St. Petersburg Academy, where he lectured in mathematics and in theoretical and practical astronomy. In 1763 he was appointed extraordinary professor of astronomy and, in 1767, professor of astronomy and an honorary member of the Academy of Sciences. From 1800 to 1803 he was a vice-president of the Academy of Sciences. In 1803 he became a member of the Main School Administration Board, which was charged with introducing educational reforms. In the same year Rumovsky also became superintendent of the Kazan educational district. In his effort to create a university at Kazan and, more specifically, to establish a physics and mathematics faculty, he recruited J. J. von Littrow from Austria for the department of astronomy. The university was opened in 1804.

BIBLIOGRAPHY

I. ORIGINAL WORKS. Rumovsky's basic writings are "Rassuzhdenie o kometakh (v svyazi s poyavleniem komety Galleya)" ("Reflections on Comets [in Connection With the Appearance of Halley's Comet]"), in *Ezhemesyachnye sochinenia k polze i uveseleniyu sluzhashchie*, 6 (1757), 40–53; *Sokrashchenia matematiki . . . (rukovodstvo dlya gimnazistov)* ("Abridged Mathematics [Text for Gymnasium Students]"; St. Petersburg, 1760); *Rech o nachale i prirashchenii optiki do nyneshnikh vremen . . .* ("Speech on the Beginning and Growth of Optics to Our Time . . ."; St. Petersburg, 1763); *Investigatio parallaxeos solis ex observatione transitus Veneris per discum solis Selenginski habita . . .* (St. Petersburg, 1764); the translation of Euler's *Pisma o raznykh fizicheskikh i filosofskikh materiakh . . .* ("Letters on Various Physical and Philosophical Matters . . ."), 3 vols. (St. Petersburg, 1768–1774); and "Nablyudenia nad prokhozhdeniem Venery cherez disk solntsa 23 maya v Kole . . ." ("Observations on the Transit of Venus Across the Sun's Disk 23 May in Kola . . ."), in *Novye Kommentarii Peterburgskoy Akademii nauk* (1769), 111–153.

See also *Yavlenia Venery v solntse v Rossyskoy imperii v 1769 godu uchinennye s istoricheskim preduvedomleniem* ("Observations on the Phenomenon of Venus and the Sun, Made in the Russian Empire in 1769, With Historical Notice"; St. Petersburg, 1771); *Tablitsy s pokazaniem shiroty i dolgoty mest Rossyskoy imperii cherez nablyudenia [astronomicheskie] opredelennye . . .* ("Tables With Indications of the Latitude and Longitude of Places in the Russian Empire Determined Through [Astronomical] Observations . . ."; St. Petersburg, 1780); translation of F. T. Schubert, *Rukovodstvo k astronomicheskim nablyudeniam, sluzhashchim k opredeleniyu dolgoty i shiroty mest* ("Handbook for Astronomical Observations Used for the Determination of Longitude and Latitude of Places"; St. Petersburg, 1803); and *Letopis Kornelia Tatsita* ("Chronicles of Cornelius Tacitus"), 4 pts. (St. Petersburg, 1806–1809), containing a Latin text and Russian trans. of Tacitus'

Annals and two articles by Rumovsky: "Izvestia o zhizni Tatsita" ("Notes on the Life of Tacitus") and "Kratkoe izyasnenie nekotorykh slov, vstrechayushchikhsya v *Letopisi* Kornelia Tatsita" ("Brief Explanation of Certain Words Encountered in the *Annals* of Cornelius Tacitus").

II. SECONDARY LITERATURE. See V. V. Bobylin, "S. Y. Rumovsky," in *Russky biografichesky slovar* ("Russian Biographical Dictionary"), XVII (1918), 441–450; V. L. Chenakal, "James Short i russkaya astronomia XVIII v." ("James Short and Russian Astronomy in the Eighteenth Century"), in *Istoriko-astronomicheskie issledovania* (1959), no. 5, 76–82; S. F. Ogorodnikov, "Tri astronomicheskie observatorii v Laplandii" ("Three Astronomical Observatories in Lapland"), in *Russkaya starina*, 33 (Jan. 1882), 177–187; V. E. Prudnikov, *Russkie pedagogi-matematiki XVIII– XIX vekov* ("Russian Teacher-Mathematicians of the Eighteenth and Nineteenth Centuries"; Moscow, 1956), 84–101; W. Struve, "Obzor geograficheskikh rabot v Rossii" ("Survey of Geographical Work in Russia"), in *Zapiski Russkogo geograficheskogo obshchestva* (1849), nos. 1–2, 23–35; and A. P. Youschkevitch, "Euler i russkaya matematika v XVIII veke" ("Euler and Russian Mathematics in the Eighteenth Century"), in *Trudy Instituta istorii estestvoznania*, 3 (1949), 104–108.

P. G. KULIKOVSKY

RUNGE, CARL DAVID TOLMÉ (*b*. Bremen, Germany, 30 August 1856; *d*. Göttingen, Germany, 3 January 1927), *mathematics, physics*.

Runge was the third son of Julius Runge and his wife Fanny. His father, of a Bremen merchant family, had accumulated a comfortable capital during some twenty years in Havana, then retired to Bremen a few years before his early and unexpected death in 1864. Fanny Runge was herself the daughter of a foreign merchant in Havana, an Englishman of Huguenot descent, Charles David Tolmé. English was the language of choice between Runge's parents, and three of his four elder siblings eventually settled in England. There was thus a strong British element in his upbringing, particularly an emphasis upon sport, self-reliance, and fair play that, in combination with the civic traditions of the Hanseatic town, influenced his political and social views. All three of his brothers pursued commercial careers; but Runge, the most closely attached to his mother and an excellent student, pointed from his youth toward a more intellectual career.

At nineteen, after completing the Gymnasium, Runge spent six months on a pilgrimage with his mother to the cultural shrines of Italy. On his return at Easter of 1876, he enrolled at the University of Munich, registering for four courses on literature and philosophy and only one in science (taught by Jolly). But six weeks after the start of the semester, Runge had made up his mind to concentrate upon mathematics and physics. In his three semesters at Munich he attended several courses with Max Planck; they became warm friends and remained in close personal contact throughout Runge's life.[1] In the fall of 1877 Planck and Runge went to Berlin together; but Runge, not much attracted by the lectures of Kirchhoff and Helmholtz that he heard, turned to pure mathematics, becoming one of Weierstrass' disciples. In the winter semester 1878–1879 Runge also attended Friedrich Paulsen's seminar course on Hume.[2] A close and lasting personal friendship developed with Paulsen, whom Runge, upon the completion of his doctorate, placed alongside Weierstrass as one of the two men "to whom I owe the best of my knowledge and ability." The dissertation, submitted in the spring of 1880, dealt with differential geometry, a topic unrelated to Weierstrass' interest (or his own subsequent work). It stemmed from an independent study of Gauss's *Disquisitiones generales circa superficies curvas* and was stimulated by the discussion of these questions in the Mathematischer Verein, the student mathematical society, in which Runge played an active role.

Although he had resolved upon an academic career, Runge, as was customary, spent the year following his doctorate preparing for the *Lehramtsexamen* for secondary school teachers. In the fall of 1881 he returned to Berlin to continue his education, largely transferring his allegiance to Leopold Kronecker. In his *Habilitationsschrift* (February 1883), influenced by Kronecker, Runge obtained a general procedure for the numerical solution of algebraic equations in which the roots were expressed as infinite series of rational functions of the coefficients, and the three traditional procedures for numerical solution of Newton, Bernoulli, and Gräffe were derived as special cases from a single function-theoretic theorem. This problem, which he treated as one in pure mathematics, was indeed to become Runge's characteristic *Fragestellung*— but only after his defection (1887) to "applied mathematics," and then from the diametrically opposite perspective, namely as a problem in numerical computation.

Meanwhile, accepted into Kronecker's circle in Berlin, Runge continued to work on a variety of problems in algebra and function theory. The feeling of being at the very center of the mathematical world dampened the urge to publish, and it was only after the promising young pure mathematician visited Mittag-Leffler in Stockholm for two weeks in September 1884 that his results were released in a spate in Mittag-Leffler's *Acta mathematica* (1885).

Runge—tall, lean, with a large and finely sculptured head—had developed exceptional skill as an ice skater in his youth; and in Berlin in the early 1880's, when that activity was becoming extremely fashionable, he cut a striking figure. He drew the attention of the children of Emil du Bois-Reymond; and, after three years of close friendship with that sporty clan, in 1885 Runge was betrothed to one of the daughters, Aimée. The precondition of the marriage—which took place in August 1887 and produced two sons and four liberated daughters—was a professorship. The first call, to the Technische Hochschule at Hannover, came in March 1886. Runge took up his duties immediately and remained there for eighteen years.

Within a year of his arrival at Hannover, Runge had undergone a thorough reorientation in his research interests and his attitude toward mathematics, a reorientation viewed by his former teachers and fellow students almost as treason. The initial step was Runge's immersion in the problem of constructing formulas, analogous to that which J. J. Balmer had recently found for hydrogen, giving the wavelengths of the spectral lines of other elements. Curiously, the stimulus for this investigation originated only very indirectly from the spectroscopist Heinrich Kayser, who had come to Hannover as professor of physics in the fall of 1885 and who was then lunching daily with Runge. Rather, Runge's attention was first drawn to these questions late in 1886 by his future father-in-law, who had been stimulated by a lecture and subsequent conversation with Kayser in Berlin in June 1885.[3] All three men became interested in the problem primarily because of its fundamental physical importance: "affording a much deeper insight into the composition and nature of the molecules [atoms] than any other physical process."[4] Runge set himself the goal of finding for each element a single formula giving all its spectral lines; "then the constants of this formula would be just as characteristic of the element as, let's say, the atomic weight."[5]

Runge began his investigations by using published data, especially those of G. D. Liveing and J. Dewar on the spectra of lithium, sodium, potassium, calcium, and zinc; he found many series of lines that could be represented by adding to Balmer's formula—$1/\lambda = A - B/m^2$—a third term, either C/m or C/m^4.[6] The inaccuracy of the available measurements made it uncertain what significance to attach to these formulas. Kayser, who had abandoned his spectroscopic researches in favor of the expansion of gases when he left Berlin, now responded to Runge's passion. He proposed "to make no use whatsoever of the available data, and . . . to determine anew the spectra of the elements from one end to the other"[7] with at least an order of magnitude greater accuracy.

This proposal was feasible with the photographic techniques and Rowland gratings that had become available in the preceding five years. Kayser's first spectrograms were made in May 1887; and for the next seven years, until his call to Bonn, he and Runge worked together at this task—Runge doing all the calculations of series and gradually taking a large role in the experimental work. They were aided by a number of grants for equipment from the Berlin Academy of Sciences (through the influence first of Helmholtz and subsequently of Planck).[8] Their results, published in seven *Abhandlungen der Preussischen Akademie der Wissenschaften*, ran to more than 350 pages.[9]

As the work progressed, the ultimate goal of unraveling atomic structure receded into the background, and was replaced by an overriding concern with precision of the data, of the methods of data reduction, and of the determination of the constants in the series formulas, without regard to their physical interpretation. Although Runge's approach to the problem of spectral series was thus far more "scientific" than that of Rydberg, whose treatise appeared simultaneously with Kayser and Runge's, by the turn of the century it was clear that Runge's formulas were physically rather barren while Rydberg's proved ever more fruitful.

After Kayser's departure Runge struggled on alone for six months. His second seven-year collaboration began early in 1895, when, following Ramsay's discovery of terrestrial helium, Runge induced Friedrich Paschen to join him in an investigation of the spectrum of that substance. Paschen, an experimentalist of extraordinary virtuosity, had come to Hannover in 1891 as Kayser's teaching assistant but had thus far not participated in spectroscopic work. With great speed and accuracy Runge and Paschen now identified all the chief lines due to helium and, surprisingly, were able to arrange them all into two systems of spectral series.[10] This was the first instance of either achievement. The latter was taken as evidence that helium was a mixture of two elements until 1897, when Runge and Paschen, continuing the Kayser-Runge program of measurements, showed that oxygen too had more than one system of series.[11]

The final substantial collaboration with Paschen, and Runge's most important contribution to theoretical spectroscopy, occurred in 1900–1902, after Thomas Preston had presented evidence for a close connection between the type of splitting of spectral lines in a magnetic field and the type of series to which they belong. In part through contact with Paschen and in part through recognition of Rydberg's unreasonable success in extracting the "right" formulas from "inadequate" data, but perhaps also in part reflecting the changing methodological ideals in the exact sciences circa 1900, Runge had gradually come to allow a freer rein to fantasy and speculation in his own work. Now, analyzing their magnetic splitting data, Runge found not only that the splitting was characteristic of the series, and quantitatively as well as qualitatively identical for analogous series in the spectra of different elements, but also that all the splittings were rational fractions of the "normal" splitting given by the Lorentz theory of the Zeeman effect.[12] This last result, known as Runge's rule, brought him great applause and for twenty years remained an incitement to both theoretical and experimental spectroscopists. Eventually, however, it proved to be largely misleading.

The exceedingly solid work with Kayser and the brilliant work with Paschen drew the attention and approval of the numerous British and American spectroscopists and astrophysicists (but not of the German physicists, who on the whole showed remarkably little interest in spectroscopy). On visits to England (1895) and America (1897), Runge became acquainted with, and was found particularly congenial by, many physicists—including the two contemporaries whom Runge most admired, Lord Rayleigh and A. A. Michelson. Following Runge's visit to Yerkes Observatory, George E. Hale was moved to offer him a research professorship there.

After Paschen's departure for Tübingen in 1901, Runge continued his experimental spectroscopic work in collaboration with Julius Precht. When Runge transferred to Göttingen in October 1904, it was intended that his work on the Zeeman effect be continued there; and for this purpose extensive facilities were provided in Woldemar Voigt's new physical institute.[13] In fact, however, Runge never used them until after his retirement in 1925; his only experimental work in those two decades was performed with Paschen at Tübingen in October 1913.

First Kayser, then Paschen had been called to chairs at first-class universities, but Runge remained stranded at the Technische Hochschule.

Planck tried more than once to arrange a call to Berlin, but could never persuade his colleagues to propose a man whom the mathematicians refused to recognize as a mathematician, nor the physicists as a physicist. In 1904, however, Klein, doubtless seconded by Woldemar Voigt, managed to persuade his Göttingen colleagues to include Runge among three nominees, albeit in last place. He then used his great influence with Friedrich Althoff, the head of the university section of the Prussian Ministry of Education, in order to have the position offered to Runge on most generous terms: 11,000 marks per year income and an independent institute comprising some fifteen rooms.[14]

Runge went to Göttingen as the first (and last) occupant of the first full professorship for "angewandte Mathematik" in Germany. It was as the leading practitioner—indeed, in a sense as the inventor and sole practitioner—of this discipline that he was best known among his contemporaries. Although the bulk of his publications had been in spectroscopy, Runge had never ceased to regard himself as a mathematician. The laborious reductions of spectroscopic observations and computations with spectral formulas, as well as his preparation of courses for engineering students, had led him to conceive an "executive" branch of mathematics to be joined to the "legislative," or pure, branch of the discipline.

"Applied mathematics" as understood and practiced by Runge was not at all concerned with the rigorous mathematical treatment of models derived from the physical world, and very little concerned with the mathematical methods useful in physics and technology. Primarily it treated the theory and practice of numerical and graphical computation— with a great deal of emphasis on the teaching of the practice.[15] Some of the methods that Runge developed, notably the Runge-Kutta procedure for the numerical integration of differential equations,[16] have remained current or have gained in currency because they are suited for execution upon modern digital computers. On the whole, however, Runge's work belongs rather more to mathematical *Zeitgeschichte* than to mathematical history: it formed one wing of a broad movement in pre-World War I Germany toward applied mathematics, of which Felix Klein was the chief ideologist and strategist, but which did not survive Germany's defeat.

Runge, whose talent and pleasure in grasping and discussing the other fellow's problems had been largely frustrated since leaving Berlin,[17] threw himself fully into the lively scientific (and sporting) life of Göttingen. The number and importance of his publications declined. His interest and energies were absorbed in the development of an instructional program in "applied mathematics"; in regularly attending the mathematical, physical, and astronomical colloquiums, as well as the Academy of Sciences; in mediating between the younger mathematicians and physicists and king Klein; and in service as Klein's lieutenant in the movement for reform of mathematical curricula in Germany. Although Runge made no contributions to the quantum theory of spectra built upon Bohr's model, he followed this work fairly closely and sympathetically.

Despite his liberal political views—open opposition to the annexationists during World War I and membership in the Democratic Party afterward— Runge retained the confidence of his colleagues and his influence within the university. When Peter Debye vacated his chair in the spring of 1920 without a successor having been appointed, he urged that Runge, "because of his authority and his great knowledge of physics, is in my opinion the only person in Göttingen" capable of managing the affairs of the physical institute.[18] .And late in that year, when the Göttingen Academy was charged with forming the physics review committee of the Notgemeinschaft der Deutschen Wissenschaft, it elected Runge, its presiding secretary, as chairman—which immediately brought protests from physicists outside Göttingen, to whom he remained a "mathematician."[19]

Runge reached the obligatory retirement age of sixty-eight in 1923, but continued to examine and to administer his institute until his successor, Gustav Herglotz, arrived in 1925. The chair then ceased to be one of applied mathematics in any sense, least of all in Runge's sense; Runge had never even had any really talented students who had wished to be applied mathematicians in his sense. His scientific activity and his self-conception were too idiosyncratic, too heedless of conventional disciplinary boundaries and ideologies, too fully the free expression of his own broad mind and pleasure in scientific exchange. In excellent health at his seventieth birthday—doing handstands to amuse his grandchildren—he had several ambitious projects under way when he died suddenly of a heart attack six months later.

NOTES

1. Max Planck, "Persönliche Erinnerungen aus alten Zeiten" (1946), in his *Vorträge und Erinnerungen* (Stuttgart, 1949), 4.
2. F. Paulsen, *An Autobiography*, Theodor Lorenz trans. and ed. (New York, 1938), 278.

3. H. Kayser, "Erinnerungen aus meinem Leben," typescript (dated 1936, presented by Kayser to W. F. Meggers and by Meggers to the Library of the American Philosophical Society, Philadelphia), 144–147, 164, represents Runge as having been "inoculated" by du Bois-Reymond before his arrival in Hannover, but otherwise gives an account agreeing essentially with Iris Runge, *Carl Runge* (1949). I have drawn upon Kayser's "Erinnerungen" for information about the origins and style of the Kayser-Runge collaboration.

4. Kayser and Runge, "Über die Spectren der Elemente" [1st part], in *Abhandlungen der Preussischen Akademie der Wissenschaften* (1888), 4–5.

5. *Ibid.*, 7.

6. Runge, "On the Harmonic Series of Lines in the Spectra of the Elements," in *Report of the British Association for the Advancement of Science* (1888), 576–577.

7. Kayser and Runge, *op. cit.*, 9.

8. *Abhandlungen der Preussischen Akademie der Wissenschaften* (1891), xxi; (1893), xxii; Runge to Kayser, 22 Dec. 1894. Subsequently the Academy supported Runge's researches with Paschen (*Sitzungsberichte der Preussischen Akademie der Wissenschaften zu Berlin* [1900], 928) and Precht (*ibid.* [1903], 648).

9. The principal discussion of Runge's series formulas is in the "Dritter Abschnitt," dealing with the first column of the periodic table: *Abhandlungen der Preussischen Akademie der Wissenschaften* (1890); Runge to E. du Bois-Reymond, 27 May 1890.

10. Runge and Paschen, "Über das Spectrum des Heliums," in *Sitzungsberichte der Preussischen Akademie der Wissenschaften zu Berlin* (1895), 593, 639–643, presented 20 June 1895; "Über die Bestandtheile des Cleveit-Gases," *ibid.*, 749, 759–763, presented 11 July 1895; Runge to Kayser, 15 May 1895 and 13 July 1895.

11. Runge and Paschen, "Über die Serienspectra der Elemente. Sauerstoff, Schwefel und Selen," in *Annalen der Physik*, 61 (1897), 641–686.

12. Runge to Kayser, 5 June 1900, 1 May 1901, 17 July 1901; Runge and Paschen, "Über die Strahlung des Quecksilbers im magnetischen Felde," in *Abhandlungen der Preussischen Akademie der Wissenschaften* (1902), presented 6 Feb. 1902; "Über die Zerlegung einander entsprechender Serienlinien im magnetischen Felde," in *Sitzungsberichte der Preussischen Akademie der Wissenschaften* (1902), 349, 380–386, presented 10 Apr. 1902; and "Über die Zerlegung . . . Zweite Mitteilung," *ibid.*, 705, 720–730, presented 26 June 1902. All three papers were translated immediately in *Astrophysical Journal*, 15 (1902), 335–351, and 16 (1902), 123–134. In Apr. 1907 Runge extended his rule to the particularly multifarious Zeeman effects of neon: "Über die Zerlegung von Spektrallinien im magnetischen Felde," in *Physikalische Zeitschrift*, 8 (1907), 232–237.

13. Runge to Voigt, 10 July 1904; Göttinger Vereinigung . . ., *Die physikalischen Institute der Universität Göttingen . . .* (Leipzig, 1906), 43–47, 197–198.

14. Personalakten Runge, Universitätsarchiv. Göttingen; *Die physikalischen Institute der Universität Göttingen*, 95–111.

15. The most important of Runge's textbooks of "applied mathematics" are *Graphical Methods* (New York, 1912), his lectures as exchange professor at Columbia in 1909–1910; and *Vorlesungen über numerisches Rechnen* (Berlin, 1924), written with Hermann König. See Runge, "Was ist 'angewandte Mathematik'?" in *Zeitschrift für den mathematischen und naturwissenschaftlichen Unterricht*, 45 (1914), 269–271, and Wilhelm Lorey, *Das Studium der Mathematik an den deutschen Universitäten seit Anfang des 19. Jahrhunderts* (Leipzig–Berlin, 1916), 253–254, and *passim*, for the applied movement in general.

16. Runge, "Über die numerische Auflösung von Differentialgleichungen," in *Mathematische Annalen*, 46 (1895), 167–178.

17. Important exceptions are Runge's suggestions to Paschen and Planck for, respectively, the experimental discovery and the theoretical deduction of the blackbody radiation law. See Hans Kangro, *Vorgeschichte des Planckschen Strahlungsgesetzes* (Wiesbaden, 1970), esp. the letter from Planck to Runge, 14 Oct. 1898, located by Kangro in the Stadtbibliothek, Dortmund.

18. Debye to Kurator, 29 Mar. 1920 (Personalakten Runge).

19. Steffen Richter, "Forschungsförderung in Deutschland, 1920–1936" (Ph.D. diss., University of Stuttgart, 1971), 37.

BIBLIOGRAPHY

I. ORIGINAL WORKS. The most complete bibliography of Runge's publications is that in the biography by I. Runge (see below), 201–205. The following further items include the additional citations in Poggendorff, IV, 1286–1287; V, 1078–1079; VI, 2244; and in the bibliography published by Runge's son-in-law, Richard Courant, in *Zeitschrift für angewandte Mathematik und Mechanik*, 7 (1927), 416–419; "On the Line Spectra of the Elements," in *Nature*, 45 (1892), 607–608, and 46 (1892), 100, 200, 247; "On a Certain Law in the Spectra of Some of the Elements," in *Astronomy and Astrophysics*, 13 (1894), 128–130; "Spektralanalytische Untersuchungen," in *Unterrichtsblätter für Mathematik und Naturwissenschaften*, 5 (1899), 69–72; "Über das Zeemansche Phänomen" (abstract only), in *Sitzungsberichte der Preussischen Akademie der Wissenschaften* (1900), 635, written with F. Paschen; "Schwingungen des Lichtes im magnetischen Felde," in H. Kayser, ed., *Handbuch der Spectroscopie*, II (Leipzig, 1902), 612–672; "Über die spektroskopische Bestimmung des Atomgewichtes," in *Verhandlungen der Deutschen physikalischen Gesellschaft*, 5 (1903), 313–315; *Rechnungsformular zur Zerlegung einer empirisch gegebenen Funktion in Sinuswellen* (Brunswick, 1913), written with F. Emde; and "Method for Checking Measurements of Spectral Lines," in *Astrophysical Journal*, 64 (1926), 315–320.

When preparing the biography of her father, Iris Runge had available to her an exceedingly full *Nachlass*, including correspondence, diaries, programs of meetings, and newspaper clippings. In 1948 she deposited at the Deutsches Museum, Munich, a collection of letters, almost all to Runge, which cannot have been more than a small fraction of even the scientific correspondence in that *Nachlass*. The other materials presumably no longer exist (letter to the author from Wilhelm T. Runge, 8 May 1972), and may not have existed in 1948. The letters by Runge at the Deutsches Museum, and some six letters by Runge in other collections, are listed in T. S. Kuhn *et al.*, *Sources for History of Quantum Physics* (Philadelphia, 1967), 80.

The largest group of letters by Runge is in the Darmstädter collection, H 1885, at the Staatsbibliothek Preussischer Kulturbesitz, Berlin-Dahlem: to Emil du Bois-Reymond, 12 Jan. 1885, 24 June 1885, 27

May 1890, 30 Apr. 1892; to René du Bois-Reymond, 29 Dec. 1903, 15 Jan. 1906; to Hans Geitel, 1 July 1899, 29 July 1899; to Heinrich Kayser, 22 Dec. 1894, 20 Jan. 1895, 17 Mar. 1895, 15 May 1895, 13 July 1895, 22 July 1895, 27 July 1895, 24 Nov. 1895, 28 Nov. 1895, 8 May 1897, 30 Oct. 1899, 5 June 1900, 9 Oct. 1900, 28 Dec. 1900, 1 May 1901, 17 July 1901, 9 Jan. 1902, 17 Jan. 1902, 22 Jan. 1902, 14 June 1902, 2 May 1903, 11 May 1903, 15 Nov. 1903, 15 Sept. 1905, 12 Apr. 1913, 29 July 1913, 31 July 1913; to Johannes Knoblauch, 15 Sept. 1886. The Nachlass Stark in the same depository includes letters from Runge to Stark of 10 Aug. 1906, 31 Aug. 1906, 25 Apr. 1907, 27 June 1907, 6 July 1910, 29 Oct. 1911, 7 Dec. 1911; and copies of Stark's letters to Runge of 25 July 1906, 2 Aug. 1906, 27 Aug. 1906, 20 Nov. 1906, 25 Apr. 1907.

Runge's MS "Ausarbeitungen" of Weierstrass' lectures on the calculus of variations and on elliptic functions are in Stockholm: I. Grattan-Guinness, "Materials for the History of Mathematics in the Institut Mittag-Leffler," in *Isis*, **62** (1971), 363–374. Three letters regarding a plaque for Runge on his seventieth birthday are in the Hale Papers, California Institute of Technology Archives, Pasadena; a letter to William F. Meggers, 8 Dec. 1921, is in the Meggers Papers, American Institute of Physics, New York.

II. SECONDARY LITERATURE. Iris Runge, *Carl Runge und sein wissenschaftliches Werk* (Göttingen, 1949), which is *Abhandlungen der Akademie der Wissenschaften zu Göttingen*, Math.-phys. Kl., 3rd ser., no. 23, is by far the fullest and most authoritative source. Biographical data of which the source is not otherwise indicated are derived from this work. Runge's early spectroscopic work is discussed in William McGucken, *Nineteenth-Century Spectroscopy* (Baltimore, 1969); and his later magneto-optic work in James Brookes Spencer, *An Historical Investigation of the Zeeman Effect (1896–1913)* (Ann Arbor, Mich., 1964), issued by University Microfilms; and P. Forman, "Alfred Landé and the Anomalous Zeeman Effect, 1919–1921," in *Historical Studies in the Physical Sciences*, **2** (1970), 153–262.

Useful evaluations of Runge's work and personality are Ludwig Prandtl, "Carl Runge," in *Jahrbuch der Akademie der Wissenschaften zu Göttingen* (1926–1927), 58–62; and "Carl Runge," in *Naturwissenschaften*, **15** (1927), 227–229; Richard Courant, "Carl Runge als Mathematiker," *ibid.*, 229–231, and 473–474 for the ensuing exchange with Richard von Mises over the status of "angewandte Mathematik" in Germany; Friedrich Paschen, "Carl Runge als Spectroskopiker," *ibid.*, 231–233; and "Carl Runge," in *Astrophysical Journal*, **69** (1929), 317–321; Walther Lietzmann, "Carl Runge," in *Zeitschrift für den mathematischen und naturwissenschaftlichen Unterricht*, **58** (1927), 482–483; Hans Kienle, "Carl Runge," in *Vierteljahrsschrift der Astronomischen Gesellschaft* (Leipzig), **62** (1927), 173–177; Erich Trefftz, "Carl Runge," in *Zeitschrift für angewandte Mathematik und Mechanik*, **6** (1926), 423–424; H. L., "Prof. Carl Runge," in *Nature*, **119** (1927),

533–534; and Oliver Lodge, "Prof. Carl Runge," *ibid.*, 565.

PAUL FORMAN

RUNGE, FRIEDLIEB FERDINAND (*b.* Billwärder, near Hamburg, Germany, 8 February 1794; *d.* Oranienburg, Germany, 25 March 1867), *chemistry*.

At the age of fifteen Runge was apprenticed to an apothecary in Lübeck. He studied medicine at Berlin and then at Göttingen, and took his medical degree at the University of Jena in 1819. His dissertation dealt with the physiological action of the belladonna alkaloids. Runge then returned to the University of Berlin, where he received his doctorate in chemistry in 1822. After three years of travel across Europe, visiting chemical factories and laboratories, Runge became *Privatdozent* at the University of Breslau. In 1828 he was appointed extraordinary professor of technical chemistry.

In 1831 Runge moved to Berlin and was offered a position as a chemist in a chemical factory at Oranienburg owned by the Royal Maritime Society. In this industrial laboratory he carried out his important study of synthetic dyes. Through the distillation of coal tar and subsequent extraction of the fractions, Runge isolated and named carbolic acid (phenol), leucol (a mixture of quinoline, isoquinoline, and quinaldine), pyrrol, and cyanol (aniline). He also produced aniline black from cyanol, noted its value as a dye, and obtained a patent in 1834.

A pioneer in the use of paper chromatography, Runge published two books in 1850 explaining this technique of analysis: volume III of *Farbenchemie* and *Zur Farbenchemie: Musterbilder für Freunde des Schönen*. . . . The latter contained a collection of chromatograms showing concentric zones of different substances present in a solution that had radiated from the spot of application. After retiring from the chemical firm in 1852, Runge worked as a consultant until his death.

BIBLIOGRAPHY

I. ORIGINAL WORKS. Lists of Runge's works are in G. Kränzlein, "Zum 100-jährigen Gedächtnis der Arbeiten von F. F. Runge" (see below); and Royal Society *Catalogue of Scientific Papers*, V, 336–337; VIII, 799; XII, 640. Among his papers is "Ueber einige Produckte der Steinkohlendestillation," in *Annalen der Physik*, **31** (1834), 65–77, 513–524; and **32** (1834), 308–332. His major books include *Farbenchemie*, 3 vols. (Berlin, 1834–1850); *Einleitung in die technische Chemie für*

Jedermann (Berlin, 1836); *Zur Farbenchemie: Muster-bilder für Freunde des Schönen und zum Gebrauch für Zeichner, Maler, Verziehrer und Zeugdrucker* (Munich, 1850); and *Der Bildungstrieb der Stoffe* (Oranienburg, 1855).

II. SECONDARY LITERATURE. For more information on Runge, see Berthold Anft, *F. F. Runge, sein Leben und sein Werk* (Berlin, 1937); and "Friedlieb Ferdinand Runge: A Forgotten Chemist of the Nineteenth Century," in *Journal of Chemical Education,* **32** (1955), 566–574; and G. Kränzlein, "Zum 100-jährigen Gedächtnis der Arbeiten von F. F. Runge," in *Angewandte Chemie,* **48** (1935), 1–3.

DANIEL P. JONES

RUSH, BENJAMIN (*b.* Byberry, Pennsylvania, 4 January 1746; *d.* Philadelphia, Pennsylvania, 19 April 1813), *chemistry, medicine, psychiatry.*

Rush was born the fourth child of seven. He lost his father, John, when he was five but was fortunate in having a sturdy, emotionally and religiously steadfast mother (Susanna Hall Harvey), who opened a grocery store to support her children. At the age of eight Rush was sent to the school conducted by his uncle, Rev. Samuel Finley, and there came under the sway of the "Great Awakening" sweeping the colonies. His religious views were extended and polished under President Samuel Davies at the Presbyterian College of New Jersey (later Princeton), where he received the bachelor's degree in 1760. He remained devout throughout his life, viewing the world as a great unity, so constructed by a benevolent God that everything was comprehensible, meaningful, and existing for a purpose.

Under Davies' influence Rush contemplated the law as a career but decided instead in favor of medicine. He apprenticed himself to Dr. John Redman in Philadelphia for the next five years and during that time also took courses at the newly founded College of Philadelphia. He was exposed to some chemistry in the lectures on materia medica by John Morgan, who encouraged him to continue his medical education at Edinburgh, with the prospect of an appointment to the chair of chemistry on his return.

Enrolling in the medical program at the University of Edinburgh late in 1766, Rush furthered his chemical career by attending the lectures of Joseph Black for two consecutive years. In preparing his doctoral dissertation Rush applied his chemical bent to a study of the digestive processes in the human stomach. After vigorous self-experimenta-tion that included induced vomiting of special meals, he decided that the acidity of the stomach contents was caused by fermentation. Rush erred in his conclusion and realized it only in 1804, when faced with the new experimental evidence produced by his student John R. Young.

After his graduation Rush toured factories in England, investigating their use of chemical reactions, and visited leading French chemists: Baumé, Macquer, and Augustin Roux. On his return to America, he was appointed professor of chemistry on 1 August 1769 at the College of Philadelphia (today the University of Pennsylvania Medical College). The following year he issued the now-scarce *Syllabus of a Course of Lectures on Chemistry.* With the lectures presented in a medical context, it is not surprising that he devoted a quarter of this slim volume to pharmaceutical chemistry.

The appointment of Rush ushered in the formal beginnings of chemistry in America. Rush was pleased to accept this responsibility, not only for the increase in professional stature but also for the satisfaction he derived from following in the footsteps of two leaders of eighteenth-century medicine who also had been professors of chemistry: Herman Boerhaave of Leiden and William Cullen. He was determined that chemistry should be useful to the larger community, and therefore offered a course to the educated public in 1775 and to the students of the Young Ladies Academy of Philadelphia in 1787. In his teaching Rush closely followed the outlines of Joseph Black's course but, unfortunately, did not follow his teacher's penchant for experimental demonstrations.

Experimentation in general was not Rush's forte; he employed his chemical training only to expose the true nature of a quack cancer cure and to study the chemical composition and therapeutic effectiveness of various local mineral waters. He did use his knowledge to good avail during the Revolution, when he served on a governmental committee promoting the local manufacture of gunpowder; at that time his instructions for the manufacture of saltpeter were widely reprinted.

Rush's teaching of chemistry ended in October 1789, when his early mentor, John Morgan, died and Rush took over his position as professor of the theory and practice of medicine. Rush never lost interest in the selection of his successors, all of whom were his students: Caspar Wistar (1789–1791), James Hutchinson (1791–1793), James Woodhouse (1795–1809), and John Redman Coxe (1809–1818). Another student whom Rush encouraged was John Penington, who organized the first chemical society in the United States in 1789.

Rush began his practice of medicine in 1769. At first largely among the poor, it gradually grew to include a wide spectrum of society. Rush had been trained by Redman to honor the clinical observations and insights of Sydenham and to accept Boerhaave's theoretical system; but at Edinburgh he enthusiastically shifted his allegiance to Cullen's theory. With his professorship of 1789 he again began to modify his theoretical focus, and a collegiate reorganization that made him professor of the institutes of medicine (physiology) and clinical practice in late 1791 forced further reconsideration of his views of basic physiological processes. Developed in his teaching during these academic years and in his medical experiences with the yellow fever epidemic of 1793, his ideas were fixed by 1795. Whereas Cullen had made the nervous system (its overenergetic or underenergetic reactions) the center of his theory, Rush narrowed his focus to the responsiveness of the arterial system. Using fever as his paradigm, he said that a state of motion (or what he called the convulsive or irregular action) in the arteries was the sole cause of disease. Since the majority of illnesses appeared to him to arise from increased tension, he logically but overenthusiastically applied bleeding and other depleting remedies to his patients. History has roundly but often excessively condemned him for the vigor of this treatment.

As a physician Rush also must be seen as a successful and popular teacher of some 3,000 students during the forty-four years of his career. Many did not accept his theories, and the doctoral dissertations written under Rush in his later years leave a clear impression that his pupils had outstripped him in their ability to appreciate the growing experimentation in the medical sciences. More important, however, he inspired them; remained their medical consultant for life; and taught them to be observant, dedicated to their patients, and aware of the nuances of the doctor-patient relationship.

Rush's restless mind explored many spheres: theory and practice, medical jurisprudence, the physiology of balloon ascents, transcultural and especially Indian medicine, geriatrics, dentistry, veterinary medicine. Although active in many areas, he was concerned primarily with medicine and became widely recognized as the leading physician in the United States.

In 1787 Rush was placed in charge of the insane at the Pennsylvania Hospital. Psychiatric reform was accelerating throughout the Western world; and Rush was in step with such leaders as Vincen-

zo Chiarugi of Italy, Philippe Pinel of France and the Tuke family of Great Britain. Recognizing the need to see man as a whole, with body and mind "intimately united," Rush was deliberately unorthodox in devoting a large part of his physiological lectures to a discussion of the operations and functions of the mind. As he passed from his physical theories to psychology, he developed a complex body of theory based on a mixture of associationism and faculty psychology. Rush's practice and teaching of psychiatry culminated in the publication of *Medical Inquiries and Observations Upon the Diseases of the Mind* (1812), the first book on psychiatry by a native American. In this work he discussed, among many other topics, "moral derangements," a concept that had concerned him as early as 1786, when he published *An Enquiry Into the Influences of Physical Causes Upon the Moral Faculty*. He realized that not only intellect but also behavior and the emotions can be disturbed, and his attempts at understanding these phenomena represent his most creative contribution to psychiatric thought.

A man of the Enlightenment, Rush evinced the best qualities of the age—humanism, optimism, and a fervent belief in the progress of knowledge. These traits were visible in his political reform activities: he signed the Declaration of Independence and fought for the federal constitution. He helped found Dickinson College, supported greater education for women, and called for a network of colleges culminating in a national university. He opposed slavery and capital punishment, supported temperance and penal reform. As a proselytizer and inspirational teacher Rush had a great impact on the American scientific scene. But for all his confidence in the clarity of his observations, Rush's usual mode of validating his hypotheses was through analogy; and he never came to appreciate the experimental method for its true worth. As a medical theorist he belonged much more to the eighteenth-century system builders. For the field of science his importance lies, as Lyman Butterfield has so aptly said, in his role as "an evangelist of science."

BIBLIOGRAPHY

I. ORIGINAL WORKS. Rush's chemistry is in the reprint of his *Syllabus of a Course of Lectures on Chemistry* (Philadelphia, 1954), which contains a thoughtful introduction by L. H. Butterfield, and in the student notes of his lectures. Rush collected what he considered to be his most important writings in *Medical Inquiries and*

Observations, 2nd ed., 4 vols. (Philadelphia, 1805). To this should be added his *Medical Inquiries and Observations Upon the Diseases of the Mind* (Philadelphia, 1812; repr. New York, 1962) and his earlier *An Inquiry Into the Influence of Physical Causes Upon the Moral Faculty* (Philadelphia, 1786). The *Autobiography of Benjamin Rush,* G. W. Corner, ed. (Princeton, 1948); and *Letters of Benjamin Rush,* L. H. Butterfield, ed., 2 vols. (Princeton, 1951), provide a rich background to his life and associations and contain bibliographies listing all published works and MSS.

II. SECONDARY LITERATURE. Two major biographical studies exist: N. G. Goodman, *Benjamin Rush, Physician and Citizen, 1746–1813* (Philadelphia, 1934), is more of an intellectual history; C. Binger, *Revolutionary Doctor: Benjamin Rush, 1746–1813* (New York, 1966), is a warm, humanistic account of the man himself. D. J. D'Elia, "The Republican Theology of Benjamin Rush," in *Pennsylvania History,* **33,** no. 2 (1966), 187–203, provides important insights into Rush's philosophical stance. Essential to a study of his early career is W. Miles, "Benjamin Rush, Chemist," in *Chymia,* **4,** (1953), 33–77; and H. S. Klickstein offers additional information in "A Short History of the Professorships of Chemistry of the University of Pennsylvania School of Medicine, 1765–1847," in *Bulletin of the History of Medicine,* **27,** no. 1 (1953), 43–65. Rush's use of chemistry in his doctoral dissertation is discussed in D. F. Musto, "Benjamin Rush's Medical Thesis, 'On the Digestion of Food in the Stomach,' Edinburgh, 1768," in *Transactions and Studies of the College of Physicians of Philadelphia,* 4th ser., **33,** no. 2 (1965), 121–138.

A good evaluation of Rush the physician is given by R. H. Shryock in "Benjamin Rush From the Perspective of the Twentieth Century," in *Transactions and Studies of the College of Physicians of Philadelphia,* 4th ser., **14,** no. 2 (1946), 113–120. J. H. Powell, in *Bring Out Your Dead* (Philadelphia, 1949), shows Rush's dedicated but vehement behavior in the Philadelphia yellow fever epidemic. Rush's psychiatry is summarized in Shryock's "The Psychiatry of Benjamin Rush," in *American Journal of Psychiatry,* **101** (1945), 429–432; and in E. T. Carlson and M. M. Simpson, "The Definition of Mental Illness: Benjamin Rush (1746–1813)," *ibid.,* **121,** no. 3 (1964), 209–214. Carlson and Simpson have also shown the importance of his writings on the moral sense in their "Benjamin Rush's Medical Use of the Moral Faculty," in *Bulletin of the History of Medicine,* **39,** no. 1 (1965), 22–33.

ERIC T. CARLSON

DICTIONARY
OF
SCIENTIFIC BIOGRAPHY

PUBLISHED UNDER THE AUSPICES OF
THE AMERICAN COUNCIL OF LEARNED SOCIETIES

The American Council of Learned Societies, organized in 1919 for the purpose of advancing the study of the humanities and of the humanistic aspects of the social sciences, is a nonprofit federation comprising forty-five national scholarly groups. The Council represents the humanities in the United States in the International Union of Academies, provides fellowships and grants-in-aid, supports research-and-planning conferences and symposia, and sponsors special projects and scholarly publications.

MEMBER ORGANIZATIONS

AMERICAN PHILOSOPHICAL SOCIETY, 1743
AMERICAN ACADEMY OF ARTS AND SCIENCES, 1780
AMERICAN ANTIQUARIAN SOCIETY, 1812
AMERICAN ORIENTAL SOCIETY, 1842
AMERICAN NUMISMATIC SOCIETY, 1858
AMERICAN PHILOLOGICAL ASSOCIATION, 1869
ARCHAEOLOGICAL INSTITUTE OF AMERICA, 1879
SOCIETY OF BIBLICAL LITERATURE, 1880
MODERN LANGUAGE ASSOCIATION OF AMERICA, 1883
AMERICAN HISTORICAL ASSOCIATION, 1884
AMERICAN ECONOMIC ASSOCIATION, 1885
AMERICAN FOLKLORE SOCIETY, 1888
AMERICAN DIALECT SOCIETY, 1889
AMERICAN PSYCHOLOGICAL ASSOCIATION, 1892
ASSOCIATION OF AMERICAN LAW SCHOOLS, 1900
AMERICAN PHILOSOPHICAL ASSOCIATION, 1901
AMERICAN ANTHROPOLOGICAL ASSOCIATION, 1902
AMERICAN POLITICAL SCIENCE ASSOCIATION, 1903
BIBLIOGRAPHICAL SOCIETY OF AMERICA, 1904
ASSOCIATION OF AMERICAN GEOGRAPHERS, 1904
HISPANIC SOCIETY OF AMERICA, 1904
AMERICAN SOCIOLOGICAL ASSOCIATION, 1905
AMERICAN SOCIETY OF INTERNATIONAL LAW, 1906
ORGANIZATION OF AMERICAN HISTORIANS, 1907
AMERICAN ACADEMY OF RELIGION, 1909
COLLEGE ART ASSOCIATION OF AMERICA, 1912
HISTORY OF SCIENCE SOCIETY, 1924
LINGUISTIC SOCIETY OF AMERICA, 1924
MEDIAEVAL ACADEMY OF AMERICA, 1925
AMERICAN MUSICOLOGICAL SOCIETY, 1934
SOCIETY OF ARCHITECTURAL HISTORIANS, 1940
ECONOMIC HISTORY ASSOCIATION, 1940
ASSOCIATION FOR ASIAN STUDIES, 1941
AMERICAN SOCIETY FOR AESTHETICS, 1942
AMERICAN ASSOCIATION FOR THE ADVANCEMENT OF SLAVIC STUDIES, 1948
METAPHYSICAL SOCIETY OF AMERICA, 1950
AMERICAN STUDIES ASSOCIATION, 1950
RENAISSANCE SOCIETY OF AMERICA, 1954
SOCIETY FOR ETHNOMUSICOLOGY, 1955
AMERICAN SOCIETY FOR LEGAL HISTORY, 1956
AMERICAN SOCIETY FOR THEATRE RESEARCH, 1956
SOCIETY FOR THE HISTORY OF TECHNOLOGY, 1958
AMERICAN COMPARATIVE LITERATURE ASSOCIATION, 1960
AMERICAN SOCIETY FOR EIGHTEENTH-CENTURY STUDIES, 1969
ASSOCIATION FOR JEWISH STUDIES, 1969

DICTIONARY

OF

SCIENTIFIC BIOGRAPHY

CHARLES COULSTON GILLISPIE

Princeton University

EDITOR IN CHIEF

Volume 12

IBN RUSHD — JEAN-SERVAIS STAS

CHARLES SCRIBNER'S SONS . NEW YORK

Panel of Consultants

Contributors to Volume 12

The following are the contributors to Volume 12. Each author's name is followed by the institutional affiliation at the time of publication and the names of the articles written for this volume. The symbol† means that an author is deceased.

HANS AARSLEFF
Princeton University
SPRAT

GIORGIO ABETTI
Osservatorio Astrofisico di Arcetri
SCHIAPARELLI; SECCHI

FREDERIC J. AGATE, JR.
Columbia University
P. E. SMITH

GARLAND ALLEN
Washington University
SHULL

PIETRO AMBROSIONI
University of Bologna
SANARELLI

ADEL ANBOUBA
Institut Moderne de Liban
AL-SAMAW'AL

TOBY A. APPEL
Johns Hopkins University
SOULEYET

WILBUR APPLEBAUM
Illinois Institute of Technology
SCHICKARD

ROGER ARNALDEZ
IBN RUSHD

RICHARD P. AULIE
Loyola University
E. J. RUSSELL

OLIVER L. AUSTIN, JR.
University of Florida
SCLATER

ROBERT AYCOCK
North Carolina State University
E. F. SMITH

VASSILY BABKOFF
Academy of Sciences of the U.S.S.R.
SAKHAROV

LAWRENCE BADASH
University of California, Santa Barbara
E. RUTHERFORD

LUIGI BELLONI
University of Milan
SACCO

ENRIQUE BELTRÁN
Mexican Institute for the Conservation of Natural Resources
SIEDLECKI; SIGÜENZA Y GÓNGORA

RICHARD BERENDZEN
The American University
V. M. SLIPHER

WALTER L. BERG
SHALER

MICHAEL BERNKOPF
Pace University
E. SCHMIDT

P. W. BISHOP
Smithsonian Institution
STANTON

A. BLAAUW
European Southern Observatory, Hamburg
DE SITTER

L. J. BLACHER
Academy of Sciences of the U.S.S.R.
SCHMALGAUSEN

HERMANN BOERNER
University of Göttingen
SCHUR; SCHWARZ

UNO BOKLUND
Royal Pharmaceutical Institute, Stockholm
SCHEELE

MARTIN BOPP
University of Heidelberg
SACHS

FRANCK BOURDIER
École Pratique des Hautes Études
SERRES DE MESPLÈS

GERT H. BRIEGER
Duke University
SABIN

T. A. A. BROADBENT †
B. A. W. RUSSELL; SHANKS

HARCOURT BROWN
SALLO

K. E. BULLEN
University of Auckland
SEZAWA

IVOR BULMER-THOMAS
SERENUS

WERNER BURAU
University of Hamburg
SCHEFFERS; SCHROETER; SCHUBERT;
SCHWEIKART

JOHANN JAKOB BURCKHARDT
University of Zurich
SCHLÄFLI

JOHN G. BURKE
University of California, Los Angeles
SOHNCKE

WILLIAM F. BYNUM
University College London
H. W. SMITH

W. A. CAMPBELL
University of Newcastle Upon Tyne
SOLVAY

KENNETH L. CANEVA
University of Utah
SCHWEIGGER

MILIČ ČAPEK
Boston University
STALLO

ALBERT V. CAROZZI
University of Illinois at Urbana-Champaign
H. B. DE SAUSSURE; SCHLUMBERGER

CARLO CASTELLANI
University of Parma
SALVIANI

JAMES F. CHALLEY
Vassar College
M. SEGUIN

ALLAN CHAPMAN
University of Oxford
SHAKERLEY

CARLETON B. CHAPMAN
The Commonwealth Fund, New York City
E. SMITH; STARLING

ROBERT A. CHIPMAN
University of Toledo
C. W. SIEMENS

F. A. L. CLOWES
University of Oxford
SHARROCK

BRUCE C. COGAN
Johns Hopkins University
H. N. RUSSELL

EDWIN H. COLBERT
Museum of Northern Arizona
W. B. SCOTT

ALBERT B. COSTA
Duquesne University
H. J. SCHIFF

J. K. CRELLIN
Wellcome Institute of the History of Medicine
SCHULZE

JOHN F. DALY, S. J.
Saint Louis University
SACROBOSCO

GLYN DANIEL
University of Cambridge
SCHLIEMANN

UMBERTO MARIA D'ANTINI
SARPI

CONTRIBUTORS TO VOLUME 12

GAVIN DE BEER †
SLOANE

ALLEN G. DEBUS
University of Chicago
SEVERINUS

ALBERT DELAUNAY
Institut Pasteur, Paris
SERGENT

R. G. C. DESMOND
India Office Library of Records, London
SPRUCE

SALLY H. DIEKE
Johns Hopkins University
SCHWARZSCHILD; SEARES

HÂMIT DILGAN
Istanbul University
AL-SAMARQANDĪ

YVONNE DOLD-SAMPLONIUS
AL-SIJZĪ; SINĀN IBN THĀBIT IBN QURRA

CLAUDE E. DOLMAN
University of British Columbia
T. SMITH; SPALLANZANI

HAROLD DORN
Stevens Institute of Technology
SMEATON

SIGALIA DOSTROVSKY
Barnard College
SAGNAC; J. SAUVEUR; SAVART

J. M. EDMONDS
University of Oxford Museum
SOLLAS

VASILIY A. ESAKOV
Academy of Sciences of the U.S.S.R.
SARYCHEV

D. G. EVANS
National Institute for Biological Standards and Control, London
W. SMITH

JOSEPH EWAN
Tulane University
SETCHELL

JOAN M. EYLES
W. SMITH; SOWERBY

W. V. FARRAR
University of Manchester
SCHUNK

SISTER MAUREEN FARRELL, F.C.J.
University of Manchester
SCHUMACHER; SCHWABE; SOLDNER

VERA N. FEDCHINA
Academy of Sciences of the U.S.S.R.
SEDOV; SEMYONOV-TYAN-SHANSKY

JEAN FELDMANN
Pierre and Marie Curie University
SAUVAGEAU

MARTIN FICHMAN
Glendon College, York University
SIGORGNE

WALTHER FISCHER
SÉNARMONT

MARCEL FLORKIN
University of Liège
SCHWANN; SLUSE; STAS

PAUL FORMAN
Smithsonian Institution
SMEKAL; SOMMERFELD

PIETRO FRANCESCHINI
University of Florence
SCARPA

EUGENE FRANKEL
Trinity College, Hartford
SEEBECK

F. FRAUNBERGER
SCHUMANN

ARTHUR H. FRAZIER
SAXTON

H.-CHRIST. FREIESLEBEN
B. V. SCHMIDT; J. F. J. SCHMIDT; SEIDEL; SPOERER

HANS FREUDENTHAL
State University of Utrecht
SCHOENFLIES; SCHOTTKY

B. VON FREYBERG
University of Erlangen-Nuremberg
SCHLOTHEIM

DAVID J. FURLEY
Princeton University
SEXTUS EMPIRICUS

ELIZABETH B. GASKING †
A. E. R. A. SERRES

GERALD L. GEISON
Princeton University
SCHULTZE

A. GEUS
University of Marburg
SCHMIDEL; SIEBOLD

OWEN GINGERICH
Smithsonian Astrophysical Observatory
SHAPLEY

THOMAS F. GLICK
Boston University
SANCHEZ

J. B. GOUGH
Washington State University
SIGAUD DE LAFOND

I. GRATTAN-GUINNESS
Middlesex Polytechnic of Enfield
STÄCKEL

JOHN C. GREENE
University of Connecticut
B. SILLIMAN

A. T. GRIGORIAN
Academy of Sciences of the U.S.S.R.
SEGNER; SHATUNOVSKY; SOMOV

N. A. GRIGORIAN
Academy of Sciences of the U.S.S.R.
SAMOYLOV

M. D. GRMEK
Archives Internationales d'Histoire des Sciences
SANTORIO

J. GRUNOW †
SPRUNG

HENRY GUERLAC
Cornell University
SAGE

FRANCISCO GUERRA
SAINT-HILAIRE

JOHN S. HALL
Lowell Observatory, Flagstaff
E. C. SLIPHER

MARIE BOAS HALL
Imperial College
P. SHAW

THOMAS M. HARRIS
University of Reading
SEWARD

RICHARD HART
National Academy of Sciences
V. M. SLIPHER

MELVILLE H. HATCH
University of Washington
SCUDDER

THOMAS HAWKINS
Boston University
SAKS

KARL HEINIG
University of Berlin
SCHORLEMMER

ARMIN HERMANN
University of Stuttgart
SCHRÖDINGER; SOMMERFELD; STARK

H. M. HINE
University of Oxford
SENECA

ERICH HINTZSCHE
University of Bern
SOEMMERRING

E. DORRIT HOFFLEIT
Yale University
SCHLESINGER

J. E. HOFMANN †
SAINT VINCENT; SCHOOTEN

ZDENĚK HORNOF
ŠKODA

WŁODZIMIERZ HUBICKI
Marie Curie-Skłodowska University
SENDIVOGIUS

CONTRIBUTORS TO VOLUME 12

THOMAS PARKE HUGHES
University of Pennsylvania
E. W. VON SIEMENS; SPERRY

G. L. HUXLEY
Queen's University of Belfast
SOSIGENES

ALBERT Z. ISKANDAR
Wellcome Institute for the History of Medicine
IBN RUSHD

JEAN ITARD
Lycée Henri IV
SAINT-VENANT

DANIEL P. JONES
Oregon State University
SMITHSON

GISELA KANGRO
SALA

HANS KANGRO
University of Hamburg
SENNERT

ROBERT H. KARGON
Johns Hopkins University
SCHUSTER; R. A. SMITH

ALEX G. KELLER
University of Leicester
G. SCHOTT; O. DE SERRES

DANIEL J. KEVLES
California Institute of Technology
W. C. W. SABINE

DAVID A. KING
Smithsonian Institution Project in Medieval Islamic Astronomy
IBN AL-SHĀṬIR

LAWRENCE J. KING
State University of New York at Geneseo
C. K. SPRENGEL

LESTER S. KING
American Medical Association
STAHL

MARC KLEIN †
SCHLEIDEN

DAVID KNIGHT
University of Durham
SCHONLAND

H. KOBAYASHI
Shimane University
SATŌ FAMILY

AKIRO KOBORI
SEKI

ZDENĚK KOPAL
University of Manchester
SPENCER JONES

SHELDON J. KOPPERL
Grand Valley State Colleges
SCHRÖTTER; SERULLAS

HANS-GÜNTHER KÖRBER
Zentralbibliothek des Meteorologischen Dienstes der DDR, Potsdam
O. F. SCHOTT; SIEDENTOPF

EDNA E. KRAMER
Polytechnic Institute of New York
SOMMERVILLE

P. G. KULIKOVSKY
Moscow University
SHARONOV; SHAYN; SHIRAKATSÍ

KAZIMIERZ KURATOWSKI
Polish Academy of Sciences
SIERPIŃSKI

G. D. KUROCHKIN
Academy of Sciences of the U.S.S.R.
SEVERGIN

GISELA KUTZBACH
University of Wisconsin
W. N. SHAW

BENGT-OLOF LANDIN
University of Lund
SCHÖNHERR

M. G. LAROSHEVSKY
Academy of Sciences of the U.S.S.R.
SECHENOV

CHAUNCEY D. LEAKE
University of California, San Francisco
SANTORINI

HENRY M. LEICESTER
University of the Pacific
SØRENSEN

MARTIN LEVEY †
AL-SAMARQANDĪ

G. A. LINDEBOOM
Free University, Amsterdam
RUYSCH; SPIEGEL

MADELEINE LY-TIO-FANE
Sugar Industry Research Institute, Mauritius
SONNERAT

A. J. McCONNELL
University of Dublin
SALMON

WILLIAM McGUCKEN
University of Akron
J. SCHEINER

DUNCAN McKIE
University of Cambridge
L. J. SPENCER

ROGERS McVAUGH
University of Michigan Herbarium
SESSÉ Y LACASTA

KARL MÄGDEFRAU
University of Munich
SANIO

MICHAEL S. MAHONEY
Princeton University
SAURIN; STAMPIOEN

CLIFFORD L. MAIER
Wayne State University
RYDBERG

CARLTON MALEY
Wayne State University
P. E. SABINE

ERNST M. MANASSE
North Carolina Central University
SPEUSIPPUS

BRIAN G. MARSDEN
Smithsonian Astrophysical Observatory
ST. JOHN

KIRTLEY F. MATHER
Harvard University
SALISBURY

ERNST MAYR
Harvard University
SARS; SEMPER

KURT MENDELSSOHN
University of Oxford
SIMON

ROBERT K. MERTON
Columbia University
SARTON

MARKWART MICHLER
University of Giessen
SORANUS OF EPHESUS

WYNDHAM D. MILES
National Institutes of Health
E. F. SMITH

E. MIRZOYAN
Academy of Sciences of the U.S.S.R.
SEVERTSOF

A. M. MONNIER
University of Paris
A. SABATIER

EDGAR W. MORSE
California State College, Sonoma
R. SMITH

LETTIE S. MULTHAUF
SAMPSON; SCHRÖTER

SHIGERU NAKAYAMA
University of Tokyo
SHIBUKAWA; SHIZUKI

CLIFFORD M. NELSON
University of California, Berkeley
RÜTIMEYER

AXEL V. NIELSEN †
SCHJELLERUP

ALBERT NIJENHUIS
University of Pennsylvania
SCHOUTEN

CALVERT E. NORLAND
San Diego State College
SANDERSON

CONTRIBUTORS TO VOLUME 12

J. D. NORTH
University of Oxford
H. C. RUSSELL; H. J. S. SMITH

MARY JO NYE
University of Oklahoma
P. SABATIER

HERBERT OETTEL
SKOLEM

LEROY E. PAGE
Kansas State University
SCROPE

ELIZABETH C. PATTERSON
Albertus Magnus College
SOMERVILLE

J. D. Y. PEEL
University of Liverpool
H. SPENCER

STUART PIERSON
Memorial University of Newfoundland
A. SÉGUIN

VICENTE R. PILAPIL
California State University, Los Angeles
SERVETUS

P. E. PILET
University of Lausanne
N.-T. DE SAUSSURE; SCHEUCHZER;
SCHOPFER; SENEBIER

DAVID PINGREE
Brown University
ŚATĀNANDA; SPHUJIDHVAJA; ŚRĪDHARA;
ŚRĪPATI

A. F. PLAKHOTNIK
Academy of Sciences of the U.S.S.R.
SHOKALSKY; SHTOKMAN

EMMANUEL POULLE
École Nationale des Chartes
SIMON DE PHARES

VARADARAJA V. RAMAN
Rochester Institute of Technology
SAHA

P. RAMDOHR
University of Heidelberg
SCHNEIDERHÖHN

RHODA RAPPAPORT
Vassar College
SOULAVIE

ROY A. RAUSCHENBERG
Ohio University
SOLANDER

NATHAN REINGOLD
Smithsonian Institution
RUTHERFURD; E. SABINE; C. A. SCHOTT

P. W. RICHARDS
University College of North Wales
W. P. SCHIMPER

GUENTER B. RISSE
University of Wisconsin-Madison
SCHAUDINN; M. SCHIFF; SEMMELWEIS;
K. P. J. SPRENGEL

GLORIA ROBINSON
Yale University
SCHNEIDER; SCHÖNLEIN

FRANCESCO RODOLICO
University of Florence
SCILLA; SOLDANI

PAUL LAWRENCE ROSE
University of Cambridge
SCALIGER

EDWARD ROSEN
City University of New York
SCHÖNER

SIDNEY ROSS
Rensselaer Polytechnic Institute
SPRING

K. E. ROTHSCHUH
University of Münster/Westphalia
STANNIUS

M. J. S. RUDWICK
Free University, Amsterdam
SEDGWICK

DAVID RYNIN
University of California, Berkeley
SCHLICK

MORRIS H. SAFFRON
*New Jersey College of Medicine and
Dentistry*
SALERNITAN ANATOMISTS

A. P. M. SANDERS
State University, Utrecht
A. F. W. SCHIMPER; SCHOUW;
SCHWENDENER; SPIX

EBERHARD SCHMAUDERER
SERTÜRNER

F. SCHMEIDLER
University of Munich Observatory
SCHÖNFELD; SEELIGER

CHARLES B. SCHMITT
Warburg Institute
SCHEGK; SEVERINO

RONALD SCHORN
SEE

E. L. SCOTT
Stamford High School, Lincolnshire
D. RUTHERFORD

BENIAMINO SEGRE
Academia Nazionale dei Lincei
SEVERI

EDITH SELOW
SCHELLING

WILLIAM R. SHEA
McGill University
C. SCHEINER

N. P. SHIKHOBALOVA
Academy of Sciences of the U.S.S.R.
SKRYABIN

ELIZABETH NOBLE SHOR
Scripps Institution of Oceanography
SAY; SCHOOLCRAFT

DIANA M. SIMPKINS
Polytechnic of North London
J. E. SMITH; S. I. SMITH

NATHAN SIVIN
Massachusetts Institute of Technology
SHEN KUA

W. A. SMEATON
University College London
SENAC

CYRIL STANLEY SMITH
Massachusetts Institute of Technology
A. SAUVEUR; SORBY

IAN N. SNEDDON
University of Glasgow
SIMSON

H. A. M. SNELDERS
State University, Utrecht
SCHÖNBEIN; SCHROEDER VAN DER
KOLK; SMITS

E. SNORRASON
Rigshospitalet, Copenhagen
SALOMONSEN

Y. I. SOLOVIEV
Academy of Sciences of the U.S.S.R.
SHILOV

PIERRE SPEZIALI
University of Geneva
SEGRE

NILS SPJELDNAES
University of Aarhus
E. J. SCHMIDT; SCHREIBERS

F. STOCKMANS
University of Brussels
SAPORTA

D. J. STRUIK
Massachusetts Institute of Technology
SACCHERI; SCHOUTE; SERRET; SNEL

ROGER H. STUEWER
University of Minnesota
G. C. N. SCHMIDT

L. E. SUTTON
University of Oxford
SIDGWICK

JUDITH P. SWAZEY
Boston University
SHERRINGTON

CONTRIBUTORS TO VOLUME 12

MANFRED E. SZABO
Concordia University
SPORUS OF NICAEA

RENÉ TATON
École Pratique des Hautes Études
SERVOIS

DOUGLASS W. TAYLOR
University of Otago Medical School
SHARPEY; SHARPEY-SCHÄFER

M. TEICH
University of Cambridge
SKRAUP

ANDRZEJ A. TESKE †
SMOLUCHOWSKI·

ARNOLD THACKRAY
University of Pennsylvania
SARTON

K. BRYN THOMAS
Royal Berkshire Hospital
SNOW

ELIZABETH H. THOMSON
Yale University
B. SILLIMAN, JR.

VICTOR E. THOREN
Indiana University
SEVERIN

V. V. TIKHOMIROV
Academy of Sciences of the U.S.S.R.
SOKOLOV

HEINZ TOBIEN
University of Mainz
K. F. SCHIMPER

CAROLYN TOROSIAN
University of California, Berkeley
SHERARD

THADDEUS J. TRENN
University of Regensburg
SMITHELLS; SODDY

G. L'E. TURNER
University of Oxford
SHORT; SOLEIL

G. UBAGHS
University of Liège
SCHMERLING

GEORG USCHMANN
Deutsche Akademie der Naturforscher Leopoldina
SEMON

F. VAN STEENBERGHEN
University of Louvain
SIGER OF BRABANT

G. VERBEKE
University of Louvain
SIMPLICIUS

C. H. WADDINGTON
University of Edinburgh
SPEMANN

WILLIAM A. WALLACE, O.P.
Catholic University of America
SOTO

P. J. WALLIS
University of Newcastle Upon Tyne
SIMPSON

ANTHONY A. WALSH
Dickinson College
SPURZHEIM

C. W. WARDLAW
D. H. SCOTT

DEBORAH JEAN WARNER
Smithsonian Institution
SCHAEBERLE; SMYTH; SOUTH

DIEDRICH WATTENBERG
Archenhold Observatory
C. A. VON SCHMIDT

EUGENE WEGMANN
SCHARDT; SEDERHOLM

RONALD S. WILKINSON
Library of Congress
STARKEY

C. GORDON WINDER
University of Western Ontario
SELWYN

MARY P. WINSOR
University of Toronto
SAVIGNY

HANS WUSSING
Karl Marx University
SCHRÖDER

ELLIS L. YOCHELSON
SCHUCHERT

L. D. G. YOUNG
SEE

A. A. YOUSCHKEVITCH
Academy of Sciences of the U.S.S.R.
SLUTSKY

A. P. YOUSCHKEVITCH
Academy of Sciences of the U.S.S.R.
SEGNER; SHATUNOVSKY; SHNIRELMAN; SOKHOTSKY; SONIN

BRUNO ZANOBIO
University of Pavia
SERTOLI

DICTIONARY
OF
SCIENTIFIC BIOGRAPHY

DICTIONARY OF SCIENTIFIC BIOGRAPHY

IBN RUSHD — STAS

IBN RUSHD, ABŪ'L-WALĪD MUḤAMMAD IBN AHMAD IBN MUḤAMMAD, also known as **Averroës** (*b.* Cordoba, Spain, 1126; *d.* Marrakech, Morocco, 10 December 1198), *astronomy, philosophy, medicine.*

Ibn Rushd, who was called the Commentator in the Latin Middle Ages, came from an important family of jurists. His grandfather (who bore the same name as he, for which reason the philosopher is called the Grandson [*al-Ḥafīd*]) had been cadi (religious judge) and imam of the great mosque of Cordoba; he was also the author of a famous treatise on Malikite law, the *Kitāb al-Muqad dimāt al-mumahhidāt*, in which he set forth its principles with a view to facilitating its study. His father was also cadi. In this milieu the young Ibn Rushd received a very good Muslim education. His training was especially thorough in law, in which field his teacher was al-Ḥāfiẓ Abū Muḥammad ibn Rizq. He learned by heart the *Muwaṭṭa'* of the Imam Mālik. He was also initiated into the science of the traditions, but he was less interested in it than in the principles of law. In theology he worked through the Ashʿarite kalam, which, in Sunnite thought, represents a system of the *juste milieu* and of equilibrium between the extreme doctrines; it could not easily be defended except with dialectical arguments, which were inspired by controversies and often led to intellectually unsatisfactory compromises. Ibn Rushd later turned against this theology, attacking the most famous proponent of Ashʿarism, al-Ghazālī. Ibn Rushd was certainly well acquainted with the Muʿtazilite kalam, which sought to be more rational, and if he included it in his condemnation of the speculative methods of all the mutakallimun, he was not indifferent to the problems that occupied this school. But it is evident from his own works that he favored primarily the type of reasoning used by the jurists, which seemed to him much more solid than theological reasoning and, in the areas in which it finds application, much more in harmony with the requirements of pure logic.

Ibn Rushd studied medicine under Abū Jaʿfar Hārūn al-Tajālī (originally from Trujillo), a noted figure in Seville who was versed in the works of Aristotle and the ancient physicians. Thoroughly familiar with the principles (*uṣūl*) and various branches (*furūʿ*) of medical science, he was an excellent practitioner, and his cures were frequently successful. He was in the service of Abū Yaʿqūb Yūsuf (1163–1184), the father of al-Manṣūr Yaʿqūb ibn Yūsuf (1184–1199). This prince, during his stay in Seville, surrounded himself with philosophers, physicians, and poets. He patronized meetings of scientists, which were attended by men like Ibn Ṭufayl, Ibn Zuhr (Avenzoar), and Ibn Rushd himself. It is likely therefore that Abū Jaʿfar played an important role in the life of his student, teaching him not only medicine but also Aristotelian philosophy. It is worth noting, for a better understanding of Ibn Rushd's intellectual development, that he studied the Stagirite during his medical training. This explains why, later, while viewing Aristotle as the master of logic (*Ṣāḥib al-Manṭiq*) and the first *falāsifa*, he was particularly interested in the natural sciences and physics, which occupy such a prominent place in the thought of the Greek philosopher.

Ibn al-Abbār, a historian born in Valencia in the year following the Commentator's death, gives in his *Takmila* the name of another physician—a man of the first rank in the practice of his art—who was one of Ibn Rushd's teachers: Abū Marwān ibn Jurrayūl.

The biographers make no mention of Ibn Rushd's philosophical studies. Ibn Abī Uṣaibiʿa confirms that it was under Abū Jaʿfar that Ibn Rushd became interested in the philosophical sciences, and Ibn al-Abbār notes simply that he "inclined towards the sciences of the Ancients." These meager data are sufficient to substantiate the idea

1

that he approached philosophical problems with a scientific outlook, though without forgetting his early instruction in legal reasoning. To an important degree, therefore, it was his scientific and legal training that gave Ibn Rushd's thinking its particular cast.

It was still science that occupied Ibn Rushd's attention when he was in Marrakech, where, according to Renan, he supported the views of the Almohad ruler 'Abd al-Mu'min "in the erection of colleges that he was founding at this moment" (1153). We know, in fact, from his commentary on *De caelo* that Ibn Rushd conducted astronomical observations at Marrakech. He was undoubtedly referring to this period when he recalls, in the commentary on a book of the *Metaphysics*, his penchant for the study of astronomy in his younger years. It is possible that as early as this period he met Ibn Ṭufayl, who was to play a major role in his philosophical career by introducing him to Abū Ya'qūb Yūsuf. Now, Abū Bakr ibn Ṭufayl (Abubacer) was a philosopher, but also an astronomer. F. J. Carmody, in the introduction to his edition of al-Biṭrūjī's *De motibus celorum*, reports an interesting remark by the author of this work:

> You know, brother, that Abū Bakr ibn Ṭufayl, may God bless him, told us that he had been inspired with an astronomical system and with principles of motion other than those postulated by Ptolemy; these avoid use of eccentrics or epicycles. And he explained by this system all movements; and nothing impossible arose from this. He also promised to write on this matter; and his place in science is not unknown.

This declaration can explain the numerous similarities between the ideas of al-Biṭrūjī and those of Ibn Rushd, if it is assumed that they derived from a common source in the thought of Ibn Ṭufayl. More directly, perhaps, than medicine, astronomy posed metaphysical problems. This fact is brought out by the account of a meeting that Marrākushī, who reports it in his *Mu'jib*, had from the lips of one of Ibn Rushd's disciples. It concerns an encounter between Abū Ya'qūb, Ibn Ṭufayl, and Ibn Rushd. The prince asked if heaven is a substance that has always existed and will continue to exist throughout eternity, or if it has a beginning. Ibn Rushd, who was at first troubled, became more confident and took part brilliantly in the discussion. Henceforth he enjoyed the favor of the prince. This episode reveals the close relationship that existed at this time between the problems of astronomy and those of metaphysics.

Abū Ya'qūb, complaining of the obscurity of Aristotle's texts, asked Ibn Ṭufayl to make commentaries on them. The latter, thinking himself too old and too busy, may in turn have asked Ibn Rushd to undertake the project. This is perhaps what prevented him from pursuing the research and astronomical observations to which he would have preferred to devote his time.

Ibn Rushd remained in high favor throughout the reign of Abū Ya'qūb Yūsuf (1163–1184). In 1169 the philosopher became cadi of Seville, but he continued to work on his commentaries and paraphrases. In the latter year he completed his paraphrase of the *Parts of Animals*, and in the fourth book he stated that his task was made much more difficult by his official duties and by the absence of his books, which were still in Cordoba. He returned to Cordoba in 1171, still holding the office of cadi. Despite his many responsibilities, he managed to find even more time to devote to his commentaries. Between 1169 and 1179 he traveled through the Almohad empire, in particular to Seville, where he dated several of his works. In 1182 he went to Marrakech to replace Ibn Ṭufayl as chief physician to Abū Ya'qūb Yūsuf. He was then honored with an appointment as grand cadi of Cordoba.

During the reign of Ya'qūb al-Manṣūr Ibn Rushd enjoyed the prince's favor for ten years. It was only in 1195 that he fell into disgrace. It is possible, and even probable, that the Malikite *fuqahā'* — doctors of the law who in Spain were always the intransigent guardians of a legalistic form of Islam — had regained influence as a result of the struggles against the Christians. They may have then inspired a hardening of the attitude of the government toward all positions that could be suspected of weakening, at first doctrinally and then politically, the bastion of religion. Ibn Rushd was banished to Lucena, near Cordoba, and subsequently appeared before a high court of Cordoban notables who anathematized his doctrines. Edicts were issued ordering the burning of philosophy books and forbidding the study of philosophy.

When al-Manṣūr returned to Marrakech, to a Berber milieu, he canceled all these edicts and recalled the philosopher. But Ibn Rushd did not have long to enjoy his return to favor; he died at the end of 1198. He was buried in Marrakech near the Taghzut gate. Later his body was brought back to Cordoba. The mystic Ibn 'Arabī, who was still young, attended his funeral. He is supposed to have said, upon seeing the Commentator's remains placed on one side of the base of a monument and the books he had written placed on the other, that all these philosophical works were equal to no more than a corpse. Although another anecdote recounts a meeting between Ibn Rushd and Ibn 'Arabī, in which the old philosopher supposedly recognized the young man's genius, it is certain that Averroism in no way leads to mysticism, unlike Avicennism. Seen in this light, Ibn 'Arabī's judgment assumes its full significance. Moreover, in the anecdote in question, when Ibn 'Arabī finds himself in the presence of Ibn Rushd, he first says "yes": yes, no doubt, to the philosopher's intentions; then he says "no": no to the method, no to a system in which the immobile prime mover closes the universe in upon itself, leaving no prospect at all for a mystical life.

Astronomy. In his commentary on Aristotle's *Metaphysics*, Ibn Rushd wrote:

In my youth I hoped it would be possible for me to bring this research [in astronomy] to a successful conclusion. Now, in my old age, I have lost hope, for several obstacles have stood in my way. But what I say about it will perhaps attract the attention of future researchers. The astronomical science of our days surely offers nothing from which one can derive an existing reality. The model that has been developed in the times in which we live accords with the computations, not with existence.

These lines express the essence of Ibn Rushd's thinking on astronomy. He was interested in the subject and acquainted with the history of its theories. Capable of explaining what Aristotle said about the systems of Eudoxus and Callippus, he was just as well informed about the work of Ptolemy, and, through the latter, he had some knowledge of the ideas of the ancients who preceded Hipparchus. He also knew the writings of the Arab astronomers. In this connection, it should be recalled that whereas scientists like al-Battānī (Albategnius) and Ibn Yūnus remained faithful to Ptolemy, others, such as Farghānī, Zarqālī, and Biṭrūjī (who lived slightly later than Ibn Rushd, but whose conceptions are, in several respects, similar) altered more or less thoroughly the Ptolemaic explanations.

Certain authors returned to the vision of a world composed of homocentric spheres, while others took up again a theory that goes back to Thābit ibn Qurra, that of trepidations, or approach and recession (*al-iqbāl wa'l-idbār*), which Ibn Rushd briefly explains in his commentary. In this situation Ibn Rushd aligned himself with those astronomers who advocated a return to Aristotle, but in order to sort out his own ideas, he took into account the whole of the history of the subject that separates him from the Greek philosopher. In fact, the abundance and the weakness of the contending theories left him very perplexed. Although he treated the scientific aspects of these problems as an expert, he hesitated to offer definitive solutions.

Ibn Rushd was certainly influenced by the "moderns," but he did not follow them blindly. He remarked that if one considers the plurality of the planetary motions, one can distinguish three kinds: (1) those accessible to the naked eye; (2) those that can be detected only with the use of observational instruments—which sometimes take place over periods exceeding the lifetime of an individual and sometimes over shorter periods; and finally, (3) those whose existence is established only by reasoning. The first movements pose no problems, but their description is far from sufficient for astronomical investigation. The second kind require the continuous collaboration of several generations, during which time it is evident that the instruments used can undergo improvement. As for the movements postulated on the basis of reasoning, one cannot always be sure that they correspond to physical realities, although it is possible to criticize a hypothesis by appealing either to new observational data or to the requirements of physical principles. These considerations are responsible for the caution Ibn Rushd displays when judging theories based on a given state in the development of the science (for example, those concerning the number of planetary movements, or the theory of the spiral movements [*harakāt lawlabiyya*]). On the other hand, they account for his rigor when principles are at stake.

Widely varying figures had been proposed for the number of the planetary motions. Aristotle himself had counted fifty-five of them, which he reduced to forty-seven. Ibn Rushd relates that in his time the astronomers fixed this number at fifty, including the motions of the starry sphere. He himself admits forty-five: thirty-eight for the sphere of the fixed stars and the planets and seven for the diurnal motions of the planetary spheres. All the same, he wrote, "As to a profound examination of what is necessarily and really involved in this question, we leave it to those who devote themselves more completely to this art, those who dedicate themselves entirely to it and who concern themselves with nothing else." He expressed the same reserve regarding the spiral motions. They result from contrary motions, but they must be executed around different poles; for contrary motions around a single pole cancel themselves. It may be objected that if the sphere that carries the celestial body is situated between two spheres moved in opposite directions, its resultant motion will be a violent one (*haraka qasriyya*), an impossibility for such a body. The best recourse is to suppose that these spiral movements arise from contrary movements about different poles. On this supposition, the body is able to move sometimes directly and sometimes with a retrograde motion, sometimes more rapidly and sometimes less rapidly; and there can be differences of latitude with respect to the zodiacal sphere. This explanatory principle leads to no absurdity: the spiral motion is that which occurs in the heavens by the combination of the diurnal movement of the sphere of the planet with the movement of the planet in its oblique sphere (*al-falak al-mā'il*). Understood in this sense, spiral motions can be admitted.

Ibn Rushd found the system of eccentrics and epicycles, adopted and developed by Ptolemy, completely unacceptable. From the time of Plato the task of astronomy had been to save the phenomena by providing a rational account of the irregular apparent motions of the planets. The burden of Ibn Rushd's criticism of this type of explanation is that it is mathematical and not physical. Physics explains, and metaphysics confirms, that the motion of the celestial bodies should be uniform, contrary to what it appears to be to the sight. It is, therefore, necessary to construct a model (*hay'a*) of a planetary configuration in such a fashion that it yields the visible phenomena, without at the same time entailing physical impossibilities. Posed in this manner, the problem has only two conceivable solutions. But only one of them fulfills all the conditions: the one that furnishes a model corresponding to a physical reality and that considers the apparent motion as composed of several motions. In this conception the planet is moved by the motion of the sphere and thereby participates in the universal motion; but it also has its own peculiar motions within its sphere. Ibn Rushd gives the example of the government of a just city (*medīna fāḍila*, an expression borrowed from al-Fārābī); it is unique, and its unity is preserved to the degree that the various chiefs imitate the monarch in serving him. Each one has his own function, just like the monarch who has his function and whose activity is the noblest. Another example is the subordination of the arts and sciences, which aid each other in the execution of a single work. This is the case of the auxiliary arts of the science of equitation, such as the art of bridling a horse.

The second solution is to posit the existence of spheres the centers of which are exterior to the center of the world, the eccentrics (*al-aflāk al-khārijat al-marākiz*), as well as of epicycles having their centers on the deferent (called in Arabic *aflāk al-tadāwīr*). This option involves various kinds of constructions, a circumstance that explains the disagreement of the astronomers over the number of the movements of the heavenly bodies. Regarding the zodiacal anomaly (that is to say, the fact that the planets traverse equal arcs in unequal times), Ibn Rushd shows how the astronomer is led to multiply the movements: "When one calculates the movements of the heavenly bodies, the calculation requires that they be in definite places on the sphere of the zodiac. Now, observing them with instruments, one discovers them in other places, which requires the introduction of a new movement for the body in question." It was in this way that Ptolemy introduced new movements for the moon and the other planets. But it was impossible for him to base these upon a *hay'a*, a term designating a configuration that, according to Ibn Rushd, should not be a simple theoretical model, but a physical reality. Ibn Rushd raised particularly strong objections to the hypothesis of the equant (*circulus aequans*), to which he alludes in these terms: "The same is true regarding what he believed, that is to say, that the uniform movements of the planets on their eccentric spheres take place [in a uniform manner] with respect to centers other than those of the eccentric spheres."

It is clear that the center of the equant and the equant itself, on which absolutely nothing actually turns, are pure mathematical fictions without the least physical reality. The Ptolemaic theory does not accord with the nature of things. The existence of the epicycle is fundamentally impossible, for "the body that moves in a circle moves about the center of the universe and not exterior to it, since it is the movable body moved along a circle that determines the center." Thus, in contradistinction to geometry, in which a circle is defined with respect to its center, the physical method starts from the reality of the circular movement, which is what entails the position of a center, the earth. If, therefore, there were an eccentric, there would be another earth exterior to our own. Now, that is physically impossible; if there existed numerous centers other than the earth, heavy bodies would fall toward these centers out of their natural places. Moreover, these hypotheses imply the existence in the heavens of superfluous bodies (*faḍl*) with no utility "except that of producing a filling (ḥashw), as occurs, it is thought, in the body of animals." Elsewhere, concerning the theory of Eudoxus, Ibn Rushd writes: "There is no need to assume two movements of two celestial bodies [the second and third sphere for the moon and the sun]; for what their natures (ṭabāʾiʿ) can accomplish with a single instrument, they do not do with two."

At the end of these criticisms Ibn Rushd comes over to the opinions of the ancients: "They are exact," he asserts,

> . . . by virtue [of their conformity to] the principles of nature; they are established, according to me, on [the basis of] the movement of a sphere unique in itself, about a center unique in itself, with different poles, two or more, as a function of what is required by an application corresponding to the apparent motions; for it may happen accidentally that such motions will be more or less rapid, direct, or retrograde.

Ibn Rushd remained loyal to Aristotle because he considered the master's thought a coherent system that must be taken as a whole. Undoubtedly, metaphysics was not for Ibn Rushd a sovereign science that imposes its conceptions on the other sciences. On the contrary, it is to a certain degree tributary to the others; it draws its knowledge of mobile substances, whether corruptible or eternal, from physics; and from astronomy it derives everything that it knows concerning the motions of the heavens. Nevertheless, celestial phenomena can be understood only within the framework of a general theory of substance. The celestial bodies have only a single motion, eternal circular translation in space. Since motion has a contrary—rest— these bodies preserve in themselves the possibility of rest (imkān fī an taskana). Aristotle showed that this possibility remains in them a pure possibility that can never be realized. But the result is, that while this motion is eternal, it is not established as such in the celestial bodies themselves. Thus arises the necessity of a first mover, immaterial and immobile, which moves bodies "as the loved one moves the lover." There exists, among the heavenly bodies, a hierarchical order (tartīb): as the motor of all the rest, the first heaven is obviously anterior "by nature, by the place it occupies, and by its size," as well as by the great number of its stars and the rapidity of its movement. The order of the planets follows the order of their spheres with regard to position (makān), but for the velocities this order is inverted; those closest to the earth have the most rapid motion, whether because of the "nobility" of their motors or the smallness of their bodies (ajrām).

Ibn Rushd rejected the hypothesis of the "moderns" postulating a ninth sphere anterior to the first heaven. According to him, the reasoning that led to this doctrine was inspired by the Neoplatonic axiom that out of primary substance, which is one and simple, there can only proceed a being that is itself one. Now, what depends immediately on the prime mover is both the first heaven itself and the motor of the sphere that follows it. This, accordingly, is not a simple effect, and there must exist an anterior cause of this complexity. For Ibn Rushd this reasoning was pure fantasy; there is, at the level of the prime mover, neither procession, nor necessary dependence, nor action, since it moves while remaining immobile. Just as, for example, one and the same intelligible entity can be grasped by several knowing subjects from different points of view, the unique prime mover can be the end toward which several different mobile entities tend. The motive force that the first heaven receives from the prime mover is strictly analogous to the motive force that the sphere of Saturn receives from this same prime mover. "That is to say, the perfection of each sphere is given in the representation of the cause that is proper to it together with the first cause." It is in this manner that one must understand the motions of each of the heavenly bodies; they tend to a unique motion, which is that of the body itself, and which, in other terms, is their resultant. Similarly, the motions of the spheres tend toward the motion of the starry sphere, in the sense that they derive their perfection from the diurnal motion of the first heaven under the effect of the motion of the prime mover. Thus, the rejection of the Neoplatonic axiom allows the astronomer to unify the motions of the universe while justifying their diversity and plurality, above all at the level of the planets.

Philosophy. The philosophical writings of Ibn Rushd are divided into two groups, the commentaries on the works of Aristotle, and the personal writings, which are entitled *Faṣl al-Maqāl, Kitāb al-Kashf,* and *Tahāfut al-Tahāfut.*

As a commentator on Aristotle, Ibn Rushd attempted to restore the Stagirite's own thought, and to supplant the Neoplatonic interpretations of al-Fārābī and Ibn Sīnā. Ibn Rushd regarded Aristotelianism as the truth, inasmuch as truth is accessible to the human mind. Referring to a passage of the *Metaphysics*—which in the Arabic version reads, "The difficulty of metaphysics is shown by the fact that it has not been possible to grasp either the truth as a whole or one of its important parts" — Ibn Rushd wrote:

> Aristotle means that this [grasping of truth] has been impossible from the earliest times to his own age; it is as if he were hinting that he himself has grasped the truth, or at least most of the truth, and that what his predecessors grasped was very little in comparison, whether it be the whole or the most important part. The best thing is to assume that he comprehended the entire truth, and by the whole of the truth I mean that quantity which human nature—insofar as it is human—is capable of grasping [*Tafsīr mā baʿd at-Tabīʿat*, Bouyges, ed., I, p. 7, 6g].

The implication of this declaration should be carefully noted: Aristotle not only greatly advanced human knowledge, he brought it to the highest possible state of perfection. He enunciated all the truth that is accessible to man, that is to say, all that can be established by demonstrative proof (*burhān*).

Although Ibn Rushd had a more complete knowledge of the *corpus Aristotelicum* and analyzed it more carefully and more accurately than did his predecessors al-Fārābī and Ibn Sīnā, he continued to view Aristotle essentially as the master of logic (*Ṣāḥib al-Manṭiq*), and it was the logical rigor of the demonstrations in the Stagirite's philosophical and scientific writings that produced the greatest impression on him.

Ibn Rushd made an important qualification in his evaluation of Aristotle, however; he cautioned that while the Greek philosopher possessed the totality of the truth available to man, he did not possess the Truth itself. In other words, man is confronted with questions that cannot be answered by the strict application of logical reasoning. While following Aristotle in all his demonstrations, Ibn Rushd nevertheless allowed for faith in revealed truths. When the Koran touches on the same subject as philosophy, it is philosophy that must be heeded, and the sacred text must be interpreted so that it will agree with the requirements of demonstrative reason. But in those cases where philosophy is silent, then instruction must come from the word of God.

The obscurity of many Aristotelian texts permitted wide latitude in their interpretation. The Commentator (as Ibn Rushd was called in the Latin West) naturally did not always give the correct explanation, especially since he often had to work with defective and even incomprehensible translations. In any case, it is clear that he always interpreted the texts in such a way as to accomplish two things: emphasis on the opposition between Aristotle and Plato, and criticism and correction of the positions advanced by Ibn Sīnā. Ibn Rushd rejected the view of metaphysics as the universal science that gives to all the other sciences their goals and principles, as well as the corollary to this view, that all human knowledge can, in principle at least, be deduced from metaphysics. At the same time, he opposed a cosmology that claimed to deduce, by the process of emanation, the celestial world of the Intelligences and of the spheres from the existence of the First Principle (*al-Awwal*) or Necessary Being. Nor did Ibn Rushd accept the idea that the last of the Intelligences, that of the sphere of the moon (also called the Active Intellect), is the *dator formarum* (*wāhib al-ṣuwar*), which gives form to the material beings of the sublunary world. In short, he rejected the Avicennian world view that explained the universe as having started from above and as having then proceeded downward, moving from the superior to the inferior. In Ibn Rushd's eyes this was Platonism. Faithful in this regard to Aristotle, he considered that beings become what they are as a consequence of a movement of desire toward the First Unmoved Mover, which causes them to pass from a potential state to an active state. This movement is, therefore, from below to above. Similarly, metaphysics is not a primal science that abides in a region beyond physics, whence it projects its light on both thought and matter. Metaphysics is instead the keystone that supports the edifice of physics; but it is a keystone set into place only after the latter has been constructed. For Ibn Rushd, it is not metaphysics that gives physical science its subject: changeable substance. Without the study of physics, the human mind would lack even the idea of change or movement. Further, contrary to Ibn Sīnā's erroneous interpretation of a passage of the *Posterior Analytics* (I, 2, 72a), it is metaphysics that, far from supplying the answer to everything, presses physics with its own questions (*yuṣā-diru*). In discussing the main subject of metaphysics, being as being, Ibn Rushd stated that "being" is, first of all, a word that the metaphysician studies according to its different applications in order to show that it refers, first of all, to substance: "The nine categories relate to existence by the fact that their existence is in a real existent (*al-mawjūd al-ḥaqīqī*), which is substance." The metaphysician also investigates the word "being" from different points of view. "Aristotle noted these different points of view concerning such words [the analogues] in order to show that what is true of substance and the nine categories is also true of the word 'being' (*mawjūd*)." Thus, like Aristotle, Ibn Rushd accorded primacy to substance in his theory of being. This led him, in commenting on the *Metaphysics*, to hold that metaphysics in its entirety is a study of substance, corruptible or incorruptible, changeable or unchangeable. Such a study, of course, was not based on the fact that substance is qualified in this way, but rather on the fact that with these qualifications it is being. That is to say, metaphysics always considers substance under the aspect of being. Physics, on the other hand, studies substance as changeable, but it is, in any case, led to conceive of an unchanging substance. Thus the subject of this particular science is the same as that of metaphysics, but it is not examined from the point of view of being. Conversely, metaphysics is not unconcerned with the substances of our world or of the celestial world. It deals with the same substances discussed in Aristotle's *Physics* and *De caelo*, although it treats them from the point of view of being as being. Consequently, being as being is not a separate subject, distinct from all the others, and reserved to metaphysics in the way that, for example, changeable substance is the subject of physics and of no other science. Metaphysics studies being as being not in itself (that is impossible since it has no concept), but in all beings, and particularly in substance in all its forms.

Ibn Rushd propounded a theory of the intellect that is important both in itself and for its influence on the Latin Middle Ages. In order to understand it properly, one must constantly bear in mind that Ibn Rushd's main goal was to explain intellection without appealing to such separate intelligible entities as the Platonic Ideas. He observed, first, that man thinks by abstracting forms (called material forms) from the objects of perception. Apprehended by a process of abstraction, they are not intelligible entities perpetually *in actu*; rather, they are at first intelligible potentially, and only later actually. Thus, they are capable of being generated and of being destroyed; abstraction is not sufficient to prove that they exist separately. In fact, they are separable only in thought. Moreover, they consist of two elements, one of which plays the role of matter and the other that of form. This can be seen in the case of the concept "snub-nosed," in which concave is, as it were, the form, and the nose is the matter. Similarly, with the concept of man: it contains a quiddity that corresponds to its definition — this is the formal aspect; but there can be no man without flesh and bones — this is the material aspect

(cf. Aristotle, *De anima*, III, 4, 429b, 10 ff.). Consequently, in the intelligibles that we abstract there is a part that is liable to disappear (*fan*[in]) and a part that subsists (*baq*[in]). The latter resembles a purely immaterial "speculative intelligible" (*maʿqūl nazarī*). But such an intelligible is absolutely identical with the intellect that apprehends it. This being so, the material intelligibles that arise in us require an activating agent and a subject. Since it is evident that they are abstracted from perceived objects and, more immediately, from images of the imagination, it is permissible to suppose that in this faculty there is a disposition (*istiʿdād*) to produce and receive these intelligibles when they become actual. For Ibn Rushd, the forms of the imagination are, above all, the activating elements in the process. They are not subjects, except by virtue of the fact that the intelligibles are in them potentially, but not to the extent that they are *in actu*. Otherwise, there would be a mixture of the forms of the imagination and of the intelligible forms. Now, in order for it to be able to think all things, the intellect must be without mixture (cf. *De anima*, III, 4, 429a, 19). Ibn Rushd gives the name "first material intellect" to this "disposition to receive the intelligibles that are in the forms of the imagination." But this intellect cannot be the true subject; it can be generated and it can be destroyed, since the imagination is inseparable from the corporeal structure.

It is therefore necessary to introduce a second subject, the receptacle of the intelligible forms *in actu*. This is the material intellect (but not the first material intellect), also called the intellect *in potentia* (*al-ʿaql bi'l-quwwa*). It stands in the same relation to the intelligibles as the prime matter does to the perceptible forms. It is eternal (*azalī*), and it is called material because it plays a role analogous to that of prime matter. Like the latter, it cannot be generated and cannot be destroyed. As pure potentiality, it must receive from an intellect *in actu* the intelligibles *in actu*, without which it is nothing. But the individual human being, who, by his faculty of reason, not only conceives and apprehends but also exercises judgment (*hukm, tasdiq*), participates personally in these operations. The intelligibles that existed potentially in the forms of his imagination and are received in the material intellect common to all men constitute a kind of stockpile for each individual, to which he has free access whenever he wishes (*matā shā'a*). Thus arises the habitual intellect or intellect *en habitus* (*al-ʿaql bi'l-malaka*). An example of the knowledge it contains is that possessed by a professor at a time when he is not teaching, but which he can make actual at will when he begins to teach. Since this habitual intellect depends on the individual's decision, it appears that it is particular to each man and represents his personal store of intelligibles among all those that are received or can be received *in actu* in the common material intellect.

If the imagination plays an activating role in this process, then the agent upon which everything depends is an intellect perpetually in an active state, and is called, for this reason, the active intellect (*al-ʿaql al-faʿʿāl*). It bestows being upon the material intellect in the way that what is actual bestows being on what is potential. In order to accomplish this, it actualizes the disposition in the imagination by acting on the imaginative forms — which are potentially intelligible entities — causing them to pass into a state of actuality. But it is not only agent: in itself it is form, intellect perpetually *in actu*; and it is absolutely identical with the intelligible entity that it apprehends. It is this intelligible that is called speculative. It was seen above that by their formal aspect the material intelligibles we grasp are like speculative intelligibles. On this level, man can be said to think with a speculative intellect. "One may therefore suppose," wrote Ibn Rushd, "that it is possible for us to apprehend the active intellect." In this case we will have reached an intelligible that is itself eternal and that, in contradistinction to the material intelligibles, is not dependent for its existence on the act by which we conceive it. Man thus arrives at a state that is called conjunction (*ittiṣāl*) or union (*ittiḥād*). This is the path that the sufis sought to travel. But, according to Ibn Rushd, they did not really succeed. Turning to Ibn Bājja's theory on this subject, Ibn Rushd subjected it to thorough criticism, asking if such a state of union will be natural or divine. If it is divine, how could it be an ultimate perfection of nature? If it is natural, how could nature manage to produce a state in which she negates herself? If this conjunction does not occur as a result of natural perfection, then it must itself be a perfection in the sense that the separate forms are a perfection for the celestial bodies endowed with circular motion, which in itself is a perfect motion. "In short, it is a separate perfection for a natural relation of perfection that is in matter." The divine perfection exists only in the relationship (*iḍāfa*); that is to say, it is not there through a substantial presence. In conclusion, Ibn Rushd stated: "It is on account of this relationship [*nisba*] that the active intellect is called the acquired intellect [*al-ʿaql al-mustafād*].

It is evident, therefore, that while Ibn Rushd

employed the notion of an intellect common to all men, he did not infer from it that human immortality — which it alone can assure — is impersonal. He studiously modified Ibn Bājja's ideas so that they did not in fact lead to such a doctrine. If Ibn Rushd's conclusion is far from being clear and definite, it is owing to the difficulty of the problem. Moreover, the obscurity of even Aristotle on this very point is proof that the problem admits of no perfectly cogent demonstration. Under these conditions, one may let faith settle the issue.

In his personal writings Ibn Rushd sets forth his positions on the religious problems of his time, notably on the agreement between reason and faith and on the interpretation and speculative use that can be made of the verses of the Koran. A philosopher but also a believer, Ibn Rushd accepted the reality of Revelation. He maintained that the Koran and *hadith* encourage the study of nature. But the divine message takes into account the diversity of human capacities. All men are not equally responsive to rational demonstrations, and thus it is necessary to resort to dialectical or even rhetorical arguments. Whatever the means employed, it is our duty to understand both nature and the meaning of the language of Revelation. The role of the philosopher is either to furnish a demonstrative argument where it is appropriate but has not yet been formulated, or else to give the unsuitable literal meaning a metaphorical meaning (*majāz*) through a commentary containing figurative language (*ta'wīt*). The theologians (*al-mutakallimūn*) err in seeking to defend surface meanings that have no value as they stand with arguments that can only be dialectical if not sophistical. On the other hand, Ibn Rushd often appeals to the juridical reasoning of the *fuqahā'*. These methodological questions form the subject of Ibn Rushd's *Decisive Treatise and Exposition of the Convergence of the Religious Law and Philosophy*.

Ibn Rushd also wrote *Kitāb al-Kashf*, the full title of which may be rendered as *Exposition of the Methods of Demonstration Relative to the Religious Dogmas and to the Definition of the Equivocal Meanings and Innovations Encountered in the Process of Interpretation and which Alter the Truth and Lead to Error*. In this work Ibn Rushd examined the theories of the major theological sects, particularly the demonstrations of the existence of God, of His unicity, and of His attributes, and conceptions about the origin of the universe and the infinite chain of causation, as well as about predestination and human freedom. This treatise, too, is primarily methodological, but in it Ibn

Rushd stated his position on a number of issues, correcting errors based on false arguments and offering demonstrative proof wherever he can. Thus, on the subject of causes, he shows that God exercises an actual causality through his commandment (*amr*), and that it is not necessary to traverse in thought an infinite time to discover his creative act at the beginning of time. The question of human freedom, however, remains difficult to settle philosophically. One must trust in the Koran and accept its teaching of the existence of both divine omnipotence and human initiative, thus holding—as Bossuet was later to say—the two ends of the chain without knowing how they are joined. Finally, Ibn Rushd upholds the reality of a future life, stating that the dogma is not contrary to reason, even though reason cannot specify modalities of such an existence.

After the *Kashf*, which prepared the way for it, the *Tahāfut al-Tahāfut* may be considered the most complete exposition of Ibn Rushd's personal thought. It takes the form of a critique of the *Tahāfut al-Falāsifa*, in which the theologian al-Ghazālī refutes Ibn Sīnā in the name of religious dogma, using arguments that Ibn Rushd attacks because they are not demonstrative. While he considers al-Ghazālī's refutation worthless, he nevertheless thinks that Ibn Sīnā's ideas should be combated, and marshals a number of demonstrative proofs against the major themes of Avicennian thought. In the process, Ibn Rushd presents virtually an entire philosophical treatise. On the whole, he sought to replace Arab Neoplatonism with what he thought were Aristotle's real views, at the same time taking into account the demands of religious faith. Thus, while upholding the doctrine of the eternity of the creation, he explained that the First Mover moves the world not by a sort of attraction, but by his commandment (*amr*), like a king seated on his throne who has no need himself of moving in order to act. Offering an interpretation of what the Koran calls divine will, Ibn Rushd stated that it is the mode of action *ad extra* of a being perfectly transcendent to his own action and who thus can create a multiplicity of beings (contrary to the Neoplatonic doctrine, adopted by Ibn Sīnā, that the one can come only out of the one). In this perspective, Ibn Rushd demonstrated that God knows particular things in themselves and not in the universal—in this sense, that God's knowledge, which is creative, is closer to the knowledge we have of particulars than to our knowledge of the universal. With regard to the soul's destiny, Ibn Rushd, referring to Aristotle's *Nicomachean Ethics*, observed that the soul acquires not only the comtemplative virtues linked to the apprehension of the intelligible entities common to all men who think, but also personal moral virtues that it may preserve. A personal immortality is therefore possible.

ROGER ARNALDEZ

Medicine. The philosophical, religious, and legal works of Ibn Rushd have been studied more thoroughly than his medical books, since he was primarily a theologian-philosopher and scholar of the Koranic sciences. Among his teachers in medicine were ʿAlī Abū Jaʿfar ibn Hārūn al-Tarrajānī (from Tarragona) and Abū Marwān ibn Jurrayūl (or Ḥazbūl, according to al-Ṣafadī). Ibn Rushd's major work in medicine, *al-Kulliyyāt* ("Generalities"), was written between 1153 and 1169. Its subject matter leans heavily on Galen, and occasionally Hippocrates' name is mentioned. It is subdivided into seven books: *Tashrīḥ al-aʿḍāʾ* ("Anatomy of Organs"), *al-Ṣiḥḥa* ("Health"), *al-Maraḍ* ("Sickness"), *al-ʿAlāmāt* ("Symptoms"), *al-Adwiya wa 'l-aghdhiya* ("Drugs and Foods"), *Ḥifẓ al-ṣiḥḥa* ("Hygiene"), and *Shifāʿ al-amrāḍ* ("Therapy"). Ibn Rushd requested his close friend Ibn Zuhr to write a book on *al-Umūr al-juzʾiyya* (particularities, i.e., the treatment of head-to-toe diseases), which he did, and called his book *al-Taisīr fi 'l-mudāwāt wa 'l-tadbīr* ("An Aid to Therapy and Regimen"). Ibn Rushd's *al-Kulliyyāt* and Ibn Zuhr's *al-Taisīr* were meant to constitute a comprehensive medical textbook (hence certain printed Latin editions present these two books together), possibly to serve instead of Ibn Sīnā's *al-Qānūn*, which was not well received in Andalusia by Abu 'l-ʿAlāʾ Zuhr ibn ʿAbd al-Malik ibn Marwān ibn Zuhr (Ibn Zuhr's grandfather). Two Hebrew versions of *al-Kulliyyāt* are known, one by an unidentified translator, another by Solomon ben Abraham ben David. The Latin translation, *Colliget*, was made in Padua in 1255 by a Jew, Bonacosa, and the first edition was printed in Venice in 1482, followed by many other editions. Ibn Rushd wrote a *talkhīṣ* (abstract) of Galen's works, parts of which are preserved in Arabic manuscripts. He showed interest in Ibn Sīnā's *Urjūza fi 'l-ṭibb* ("Poem on Medicine," *Canticum de medicina . . .*), on which he wrote a commentary, *Sharḥ Urjūzat Ibn Sīnā*. It was translated into Hebrew prose by Moses ben Tibbon in 1260; a translation into Hebrew verse was completed at Béziers (France) in 1261 by Solomon ben Ayyub ben Joseph of Granada. Further, a Latin translation of the same work was made by Armengaud, son of Blaise, in 1280 or 1284, and a printed edition was published at Venice in 1484. Another revised Latin translation was made by Andrea Alpago, who translated Ibn Rushd's *Maqāla fī 'l-Tiryāq* ("Treatise on Theriac," *Tractatus de theriaca*).

So far, no evidence has been provided to support, or refute, the claim that Ibn Rushd is quoted as saying, "He who is occupied with the science of anatomy will have more faith in God." In 1182, he succeeded Ibn Ṭufayl, who retired on account of his advanced age from the post of court physician to the caliph Abū Yaʿqūb Yūsuf, and continued to be favored by his son and successor al-Manṣūr Yaʿqūb ibn Yūsuf until the year when Ibn Rushd fell out of favor and his philosophical works (but not his medical and other strictly scientific books) were banned or burned. In the East, the writings of al-Ghazālī against the principles of Greek philosophy probably led to changes in the medical curriculum, whereby Greek philosophy was gradually supplanted by Islamic theology, which included some aspects of philosophy, and particularly logic. Furthermore, the massacre of Herāt in 1222 and the Mongol invasion that led to the eradication of the eastern caliphate in 1258, and (in the West) the period that followed Ibn Rushd's unsuccessful attempts to defend philosophers against theologians paved the way for a decline in Arabic medicine. The great image of the *ḥakīm* (physician-philosopher), which culminated in the persons of al-Rāzī and Ibn Sīnā, has been superseded by that of *faqīh mushārik fī 'l-ʿulūm* (a jurist who participates in sciences), among whom were physician-jurists and theologian-physicians.

ALBERT Z. ISKANDAR

BIBLIOGRAPHY

I. ORIGINAL WORKS. *Incipit translatio Canticor. Avic. cum commento Averrhoys facta ab Arabico in Latinum a mag. Armegando blassi de Montepesulano* (Venice, 1484); *Abhomeron Abynzohar Colliget Averroys* (containing the *Taisīr* and *Antidotarium* of Avenzoar and the *al-Kuliyyāt* of Ibn Rushd, edited by Hieronymus Surianus) (Venice, 1496); *Collectaneorum de Re Medica Averrhoi philosophi . . . Sectiones tres. I. De sanitatis functionibus. ex Aristot. et Galeno. II. De sanitate tuenda, ex Galeno. III. De curandis morbis, a J. Bruyerino Campegio . . . nunc primum Latinitate donatae* (Lyons, 1537); *Quitab el culiat. Libro de las generalidades* (Publ. del Inst. Gen. Franco para la investigación hispano-árabe) (Larache, 1939), in Arabic, published by manuscript photo-reproduction; *Talkhīṣ k. al-ḥummayāt* (Abstract of the book: "Fevers," Escurial, MS 884, i.); *Talkhīṣ k. al-ʿilal wa al-aʿrāḍ* (Abstract of the book: "Diseases and Symptoms," Escurial, MS 884, iii.); *Talkhīṣ k. al-mizāj* (Abstract of the book: "Tempera-

ment," Escurial, MS 881, ii.); *Talkhīṣ k. al-quwā al-ṭabīʿiyya* (Abstract of the book: "Natural Faculties," Escurial, MSS 881, iii.; 884, ii.); *Talkhīṣ k. al-usṭuqussāt* (Abstract of the book: "Elements," Escurial, MS 881, i.); *Sharḥ urjūzat Ibn Sīnā fī al-ṭibb* (Commentary on Ibn Sīnā's "Poem on Medicine," Dār al-Kutub al-Miṣriyya Ṭibb 1239; Yale University Library, Landberg Collection, MS 157, n. 1513).

The Mediaeval Academy of America has begun publication of a critical ed. of the collected commentaries (medieval Latin and Hebrew trans.), H. A. Wolfson, *et al.*, eds., in the series Corpus Commentariorum Averrois in Aristotelem (Cambridge, Mass., 1949–); see H. A. Wolfson, "Revised Plan for the Publication of a *Corpus Commentariorum* . . .," in *Speculum*, **38** (1963), 88–104, which includes detailed bibliographical information. E. I. J. Rosenthal, *Averroes' Commentary on Plato's Republic*, 2nd ed., rev. (Cambridge, 1966), is based on a Hebrew trans. of the lost Arabic original.

For bibliographical details of Ibn Rushd's works, see also C. Brockelmann, *Geschichte der arabischen Literatur*, I (Leiden, 1943), 604–606, and supp. I (Leiden, 1937), 833–836; G. Sarton, *Introduction to the History of Science*, II, pt. 2 (Baltimore, 1931), 355–360; and M. Steinschneider, *Die hebräischen Übersetzungen des Mittelalters und die Juden als Dolmetscher* (Berlin, 1893; repr. Graz, 1956). M. J. Müller, ed., *Philosophie und Theologie des Averroes*, 2 vols. (Munich, 1859–1875), is a basic work.

II. SECONDARY LITERATURE. L. Leclerc, *Histoire de la médecine arabe*, II (Paris, 1876), 97–109; Ibn Abī ʾUṣaybiʿa's *ʿUyūn al-anbāʾ fī ṭabaqāt al-aṭibbāʾ*, A. Müller ed., II (Cairo-Königsberg, 1882–1884), 75–78; *The Encyclopaedia of Islam*, II (Leiden-London, 1913–1938), 410–413; new ed., III (Leiden-London, 1960–), 909–920; al-Yāfiʿī's *Mirʾāt al-janān*, III (Hyderabad, 1918–1920), 479; P. M. Bouyges, "Inventaire des textes arabes d'Averroés," in *Mélanges de l'Université Saint-Joseph*, **8**, 1 (1922), 3–54; *ibid.*, **9**, 2 (1924), 43–48 (additions and corrections to note V. 1); D. Campbell, *Arabian Medicine and Its Influence on the Middle Ages*, I (London, 1926), 92–96; G. Sarton, *Introduction to the History of Science*, II, pt. 1 (Baltimore, 1927–1948), 355–361; Y. A. Sarkīs, *Muʿjam al-maṭbūʿāt al-ʿarabiyya wa al-muʿarraba . . . .* I (Cairo. 1928–1931), 108–109; H. Ritter and S. Dedering, eds., *Das biographische Lexikon des Ṣalāḥaddīn Halīl Ibn Aibak aṣ-Ṣafadī*, II (Istanbul-Damascus, 1931–1970), 114–115; Ibn al-ʿImād's *Shadharāt al-dhahab fī akhbār man dhahab . . .*, IV (Cairo, 1931–1932), 320; C. Brockelmann, *Geschichte der arabischen Litteratur* (Leiden, 1943–1949), supplement I (Leiden, 1937–1942), 604, 833; H. P. J. Renaud, *Les manuscrits arabes de l'Escurial, décrits d'après les notes de H. Derenbourg*, II (Paris, 1941); *Publications de l'École Nationale des Langues Orientales Vivantes*, 5th ser., 91–92; 94–95; ʿA. M. al-ʿAqqād, *Nawābigh al-fikr al-ʿarabī, Ibn Rushd: Ibn Rushd al-ṭabīb*, Dār al-Maʿārif, ed. (Cairo, 1953), 96–112; Kh. al-Ziriklī, *al-Aʿlām . . .*, 2nd ed., VI

(Cairo, 1954–1959), 212–213; L. Nemoy, "Arabic Manuscripts in the Yale University Library," in *Transactions of the Connecticut Academy of Arts and Sciences*, XL (New Haven, 1956), 160 (n. 1513); F. X. Rodríguez Molero, "Averroes, médico y filósofo," in *Archivo Ibéro Americano de Historia de la Medicina*, 8 (1956), 187–190; S. Muntner, "Averrhoes (Abu-el-Walid ibn Ahmed ibn Rushd). Le médecin dans la littérature hébraïque," in *Imprensa Médica* 21, 4 (1957), 203–208: ʿU. R. Kaḥḥāla, *Muʿjam al-muʾallifīn* . . . , VIII (Damascus, 1957–1961), 313; Ṣ. el-Munajjed, "Maṣādir jadīda ʿan tārīkh al-ṭibb ʿind al-ʿarab," in *Majallat Maʿhad al-Makhṭūṭāt al-ʿArabiyya*, 5, 2 (1959), 257 (ns. 66–68); R. Walzer, *Greek into Arabic. Essays on Islamic Philosophy*, Oriental Studies, I (Oxford, 1962), 26–28; S. Hamarneh, "Bibliography on Medicine and Pharmacy in Medieval Islam. Mit einer Einführung Arabismus in der Geschichte der Pharmazie von Rudolf Schmitz," in *Veröffentlichungen der Internationalen Gesellschaft für Geschichte der Pharmazie*, e.V., n. s., 25 (1964), 92; S. Hamarneh, *Index of Manuscripts on Medicine, Pharmacy, and Allied Sciences in the Ẓāhiriyya Library* (Damascus, 1969), 175–178; A. Dietrich, "Medicinalia Arabica. Studien über arabische medizinische Handschriften in türkischen und syrischen Bibliotheken," in *Abhandlungen der Akademie der Wissenschaften zu Göttingen*, 3rd series, no. 66 (1966), 99–100 (n. 39); J. C. Bürgel, "Averroes, 'Contra Galenum.' Das Kapitel von der Atmung im Colliget des Averroes als ein Zeugnis mittelalterlich-islamischer Kritik an Galen, eingeleitet, arabisch herausgegeben und übersetzt," in *Nachrichten der Akademie der Wissenschaften in Göttingen*, I (1967), 9, 263–340; A. Z. Iskandar, *A Catalogue of Arabic Manuscripts on Medicine and Science in the Wellcome Historical Medical Library* (London, 1967), 37; B. S. Eastwood, "Averroes' View of the Retina—a Reappraisal," in *Journal of the History of Medicine*, 24 (1969), 77–82; A. Amerio, "Spunti di rinascimento scientifico negli averroisti latini del XIII secola," in *Med. Secoli*, 7 (1970), 13–18, refs.; M. Ullmann, *Die Medizin im Islam*, Handbuch der Orientalistik, supp. 6 (Leiden-Cologne, 1970), 166–167; R. Y. Ebied, *Bibliography of Mediaeval Arabic and Jewish Medicine and Allied Sciences*, Wellcome Institute of the History of Medicine (London, 1971), 107–108.

On Ibn Rushd's life and philosophical thought, see the following: L. Gauthier, *La théorie d'Ibn Rochd (Averroès) sur les rapports de la religion et de la philosophie* (Paris, 1909), with bibliography; and *Ibn Rochd (Averroès)* (Paris, 1948); M. Grabmann, *Der lateinische Averroismus des 13. Jahrhunderts und seine Stellung zur Christlichen Weltanschauung* (Munich, 1931); R. de Mendizábal Allende, *Averroes, un andaluz para Europa* (Madrid, 1971); F. W. Muller, *Der Rosenroman und der lateinische Averroismus des 13. Jahrhunderts* (Frankfurt, 1947); S. Münk, *Mélanges de philosophie juive et arabe* (Paris, 1859), 418–458; G. Quadri, *La filosofia degli arabi nel suo fiore*, II (Florence, 1939); and *La philosophie arabe dans l'Europe médiévale* (Paris, 1947), 198–340; E. Renan, *Averroès et l'averroïsme* (Paris, 1852; repr. 1949); and K. Werner, *Der Averroismus in der Christlich-peripatetischen Psychologie des späteren Mittelalters*, new ed., (Amsterdam, 1964).

RUSSELL, BERTRAND ARTHUR WILLIAM (*b.* Trelleck, Monmouthshire, England, 18 May 1872; *d.* Plas Penrhyn, near Penrhyndeudraeth, Wales, 2 February 1970), *mathematical logic.*

The Russell family has played a prominent part in the social, intellectual, and political life of Great Britain since the time of the Tudors; Russells were usually to be found on the Whig side of politics, with a firm belief in civil and religious liberty, as that phrase was interpreted by the Whigs. Lord John Russell (later first earl Russell), the third son of the sixth duke of Bedford, was an important figure in nineteenth-century politics: He was a leader in the struggle to establish the great Reform Act of 1832, held several high offices of state, and was twice prime minister in Whig and Whig-Liberal administrations. His eldest son, known by the courtesy title of Viscount Amberley, married Katherine Stanley, of another famous English family, the Stanleys of Alderley. The young couple were highly intelligent and were in strong sympathy with most of the reforming and progressive movements of their time, a stance that made them far from popular with the conservative section of the aristocracy. Unhappily, neither enjoyed good health; the wife died in 1874 and the husband in 1876. There were two children, Frank and Bertrand, the latter the younger by about seven years.

The Russell family did not approve of the arrangements made by Viscount Amberley for the upbringing of the two children in the event of his death. When this occurred, the boys were made wards in chancery and placed in the care of Earl Russell and his wife, who were then living at Pembroke Lodge in Richmond Park, a house in the gift of the Crown. Bertrand's grandfather died in 1878, but his grandmother lived until 1898 and had a strong influence on his early life.

Like many Victorian children of the upper class, the boy was educated at home by a succession of tutors, so that when he entered Trinity College, Cambridge, as a scholar in 1890, he had had no experience of communal life in an educational establishment save for a few months in a "cramming" school in London. At Trinity he was welcomed into a society that for intellectual brilliance could hardly have been bettered anywhere at that time. He obtained a first class in the mathematical tripos and in the moral sciences tripos, although the formality of examinations seems not to have appealed to him. He remarks in his autobiography that the university teachers "contributed little to my enjoyment of Cambridge," and that "I derived no benefits from lectures."

A great stimulus to Russell's development was his election in 1892 to the Apostles. This was a small, informal society, founded about 1820, that regarded itself—not without some justification—as composed of the intellectual cream of the university; its main object was the completely unfettered discussion of any subject whatsoever. One member was A. N. Whitehead, then a mathematical lecturer at Cambridge, who had read Russell's papers in the scholarship examination and had in consequence formed a high opinion of his ability. Through the society Russell acquired a circle of gifted friends: the philosophers G. E. Moore and Ellis McTaggart, the historian G. M. Trevelyan and his poet brother R. C. Trevelyan, the brilliant brothers Crompton and Theodore Llewelyn Davies, and later the economist J. M. Keynes and the essayist Lytton Strachey.

In the latter part of the nineteenth century, progressive opinion at Cambridge had begun to maintain that university dons should regard research as a primary activity, rather than as a secondary pursuit for leisure hours after teaching duties had been performed. This opinion was particularly strong in Trinity, where A. R. Forsyth, W. W. Rouse Ball, and Whitehead encouraged the researches of younger men such as E. W. Barnes, G. H. Hardy, J. H. Jeans, E. T. Whittaker, and Russell; these themselves exercised a great influence on the next generation, the remarkable set of Trinity mathematicians of the period 1900–1914.

One mode of encouragement was the establishment of prize fellowships, awarded for original dissertations; such a fellowship lasted for six years and involved no special duties, the object being to give a young man an unhindered opportunity for intellectual development. Russell was elected in 1895, on the strength of a dissertation on the foundations of geometry, published in 1897. During the later part of his tenure and after it lapsed, he was not in residence; but in 1910 the college appointed him to a special lectureship in logic and the philosophy of mathematics.

During World War I pacifism excited emotions much more bitter than was the case in World War II. Russell's strongly held views made him unpop-

ular in high places; and when in 1916 he published a leaflet protesting against the harsh treatment of a conscientious objector, he was prosecuted on a charge of making statements likely to prejudice recruiting for and discipline in the armed services, and fined £100. The Council, the governing body of Trinity, then dismissed him from his lectureship, and Russell broke all connection with the college by removing his name from the books. In 1918 another article of his was judged seditious, and he was sentenced to imprisonment for six months. The sentence was carried out with sufficient leniency to enable him to write his very useful *Introduction to Mathematical Philosophy* in Brixton Prison.

Many members of Trinity felt that the Council's action in dismissing Russell in 1916 was excessively harsh. After the war the breach was healed: in 1925 the college invited Russell to give the Tarner lectures, later published under the title *The Analysis of Matter*; and from 1944 until his death he was again a fellow of the college.

In the prologue to his *Autobiography* Russell tells us that three strong passions have governed his life: "the longing for love, the search for happiness, and unbearable pity for the suffering of mankind." His writings and his actions testify to the perseverance with which he pursued his aims from youth to extreme old age, undeterred by opposition and regardless of obloquy. Russell's perseverance did not necessarily imply obstinacy, for his mind was never closed; and if his acute sense of logic revealed to him a fallacy in his argument, he would not cling to a logically indefensible view but would rethink his position, on the basis of his three strong principles. But it was also his devotion to logic that led him frequently to reject the compromises so often forced on the practical politician.

Russell—not surprisingly, in view of his ancestry—was always ready to campaign for "progressive" causes. About 1907 he fought hard for women's suffrage, a cause that provoked more opposition and rowdyism in the United Kingdom than any other political question during this century—even more than the pacifism for which Russell was prosecuted in World War I. After the war Russell continued his search for a genuine democracy, in which freedom for the individual should be compatible with the common good; his experiments in education were designed to contribute to this end. Never insular, he would expose what he saw as the faults of his own country or of the English-speaking nations as caustically as he would those of the totalitarian regimes; but that the growth of the latter could be met only by war was a conclusion to which he came very reluctantly. At the close of World War II, his vision of humanity inevitably destroying itself through the potency of nuclear weapons caused him to lead a long campaign for nuclear disarmament.

A long list of books bears witness to Russell's endeavor to encourage human beings to think clearly, to understand the new scientific discoveries and to realize some of their implications, and to abhor injustice, violence, and war. *The Impact of Science on Society, History of Western Philosophy, Common Sense and Nuclear Warfare, Marriage and Morals, Freedom and Organisation,* and *Prospects of Industrial Civilisation,* to name only a few, show how earnestly he sought to promote his ideals. All were written in an English that was always clear and precise, and often beautiful. Critics might disagree with his opinions but seldom could misunderstand them. An occasional didactic arrogance might offend, but it could be forgiven in view of the author's manifest sincerity.

A few of Russell's many honors were fellowship of the Royal Society in 1908, the Order of Merit in 1949, and the Nobel Prize for literature in 1950.

Russell has told us in his autobiography how he began the study of geometry, with his elder brother as tutor, at the age of eleven. Like almost every other English boy of his time, he began on Euclid; unlike almost every other English boy, he was entranced, for he had not known that the world contained anything so delicious. His brother told him that the fifth proposition of book I, the notorious *pons asinorum,* was generally considered difficult; but Russell found it no trouble. Having been told, however, that Euclid proved things, he was disappointed at having to begin by assimilating an array of axioms and would not accept this necessity until his brother told him that unless he did so, his study of geometry could not proceed, thus extorting a reluctant acceptance. The anecdote is not irrelevant; Russell's mathematical work, which occupied him until he was over forty, was almost entirely concerned with probing and testing the foundations of mathematics, in order that the superstructure might be firmly established.

Russell's fellowship thesis was revised for publication in 1897 as *An Essay on the Foundations of Geometry.* Its basic theme was an examination of the status assigned to geometry by Kant in his doctrine of synthetic a priori judgments. Analytic propositions are propositions of pure logic; but synthetic propositions, such as "New York is a large city," cannot be obtained by purely logical processes. Thus all propositions that are known through experience are synthetic; but Kant would not accept the converse, that only such propositions are synthetic. An empirical proposition is derived from experience; but an a priori proposition, however derived, is eventually recognized to have a basis other than experience.

Kant's problem was to determine how synthetic a priori judgments or propositions are possible. He held that Euclidean geometry falls into this category, for geometry is concerned with what we perceive and thus is conditioned by our perceptions. This argument becomes dubious when the full implication of the existence of non-Euclidean geometries is appreciated; but although these geometries were discovered about 1830, the philosophical implications had hardly been fully grasped by the end of the nineteenth century. One considerable step was taken by Hilbert when he constructed his formal and abstract system based on his epigram that in geometry it must be possible to replace the words "points," "lines," and "planes" by the words "tables," "chairs," and "beer mugs"; but his *Grundlagen der Geometrie* was not published until 1899.

To Russell, non-Euclidean spaces were possible, in the philosophical sense that they are not condemned by any a priori argument as to the necessity of space for experience. His examination of fundamentals led him to conclude that for metrical geometry three axioms are a priori: (1) the axiom of free mobility, or congruence: shapes do not in any way depend on absolute position in space; (2) the axiom of dimensions: space must have a finite integral number of dimensions; (3) the axiom of distance: every point must have to every other point one and only one relation independent of the rest of space, this relation being the distance between the two points.

For projective geometry the a priori axioms are (1) as in metrical geometry; (2) space is continuous and infinitely divisible, the zero of extension being a point; (3) two points determine a unique figure, the straight line. For metrical geometry an empirical element enters into the concept of distance, but the two sets are otherwise equivalent. In the light of modern views on the nature of a geometry, these investigations must be regarded as meaningless or at least as devoted to the wrong kind of question. What remains of interest in the *Foundations of Geometry* is the surgical skill with which Russell can dissect a corpus of thought, and his command of an easy yet precise English style.

Following the publication of the *Foundations of Geometry,* Russell settled down to the composition of a comprehensive treatise on the principles of mathematics, to expound his belief that pure mathematics deals entirely with concepts that can be discussed on a basis of a small number of fundamental logical concepts, deducing all its propositions by means of a small number of fundamental logical principles. He was not satisfied with his first drafts, but in July 1900 he went to Paris with Whitehead to attend an International Congress of Philosophy. Here his meeting with Peano brought about, in his own phrase, "a turning point in my intellectual life." Until then he had had only a vague acquaintance with Peano's work, but the extraordinary skill and precision of Peano's contributions to discussions convinced Russell that such mastery must be to a large extent due to Peano's knowledge of mathematical logic and its symbolic language.

The work of Boole, Peirce, and Schröder had constructed a symbolic calculus of logic; and their success contributed to Peano's systematic attempt to place the whole of mathematics on a purely formal and abstract basis, for which purpose he utilized a symbolism of his own creation. This enabled him and his disciples to clarify distinctions hitherto obscured by the ambiguities of ordinary language, and to analyze the logic, basis, and structure of such mathematical concepts as the positive integers. The apparently trivial symbolism that replaces "The entity x is a member of the class A" with "$x \epsilon A$" leads not only to brevity but, more importantly, to a precision free from the ambiguities lurking in the statement "x is A." One result of Peano's work was to dispose of Kant's synthetic a priori judgments.

Russell rapidly mastered Peano's symbolism and ideas, and then resumed the writing of his book on principles; the whole of the first volume was completed within a few months of his meeting with Peano. Some sections were subjected to a thorough rewriting, however; and volume I of *Principles of Mathematics* was not published until 1903. The second edition (1937) is perhaps more valuable for the study of the development of Russell's ideas, for it both reprints the first edition and contains a new introduction in which Russell gives his own opinion on those points on which his views had changed since 1903; but in spite of Hilbert and Brouwer, he is still firm in his belief that mathematics and logic are identical.

The second volume of the *Principles,* to be written in cooperation with Whitehead, never appeared because it was replaced by the later *Principia Mathematica.* It was to have been a completely symbolic account of the assimilation of mathematics to logic, of which a descriptive version appears in volume I. The main sections of volume I treat indefinables of mathematics (including a description of Peano's symbolic logic), number, quantity, order, infinity and continuity, space, and matter and motion. On all these topics Russell's clarity of thought contributed to the establishment of precision; thus his analysis of the words "some," "any," "every," for instance, is very searching. In two places, at least, he was able to throw light on familiar but vexed topics of mathematical definition and technique.

That the positive integer 2 represents some property possessed by all couples may be intuitively acceptable but does not supply a precise definition, since neither existence nor uniqueness is guaranteed in this way. Russell's definition of a number uses a technique of equivalence that has had many further applications. The definition had already been given by Frege, but his work was not then known to Russell;[1] indeed, it was known to hardly any mathematician of the time, since Frege's style and symbolism are somewhat obscure. Two classes, A and $B,$ are said to be similar ($A \sim B$) if each element of either class can be uniquely mated to one element of the other class; clearly A is similar to itself. Similarity is a transitive and symmetrical relation; that is, if $A \sim B$ and $B \sim C,$ then $A \sim C,$ and if $A \sim B,$ then $B \sim A.$ The (cardinal) number of a class A is, then, the class consisting of all classes similar to $A.$ Thus every class has a cardinal number, and similar classes have the same cardinal number. The null class φ is such that $x \epsilon \varphi$ is universally false, and its cardinal is denoted by 0. A unit class contains some term x and is such that if y is a member of this class, then $x = y$; its cardinal is denoted by 1. The operations of addition and multiplication are then readily constructed. We thus have a workable definition with no difficulties about existence or uniqueness; but—and this is a considerable concession—the concept of "class" must be acceptable.

A similar clarification of the notion of a real number was also given by Russell. Various methods of definition were known, one of the most popular being that of Dedekind. Suppose that p and q are two mutually exclusive properties, such that every rational number possesses one or the other. Further, suppose that every rational possessing property p is less than any rational possessing property $q.$ This process defines a section of the rationals, giving a lower class L and an upper class R. If L has a greatest member, or if R has a least member, the section corresponds to this rational number. But if L has no greatest and R no least member, the section does not determine a rational. (The case in which L has a greatest and R has a least member cannot arise, since the rationals are dense.) Of the two possible cases, to say that the section in one case corresponds to a rational number and in the other it corresponds to or represents an irrational number is — if instinctive — not to define the irrational number, for the language is imprecise. Russell surmounts this difficulty by simply defining a real number to be a lower section L of the rationals. If L has a greatest member or R a least member, then this real number is rational; in the other case, this real number is irrational.

These are two of the outstanding points in the *Principles*. But the concept of "class" is evidently deeply involved, and the "contradiction of the greatest cardinal" caused Russell to probe the consequences of the acceptance of the class concept more profoundly, particularly in view of Cantor's work on infinite numbers. Cantor proved that the number of subclasses that can be formed out of a given class is greater than the number of members of the class, and thus it follows that there is no greatest cardinal number.

Yet if the class of all objects that can be counted is formed, this class must have a cardinal number that is the greatest possible. From this contradiction Russell was led to formulate a notorious antinomy: A class may or may not be a member of itself. Thus, the class of men, mankind, is not a man and is not a member of itself; on the other hand, the class consisting of the number 5 is the number 5, that is, it is a member of itself. Now let W denote the class of all classes not members of themselves. If W is not a member of itself, then by definition it belongs to W — that is, it is a member of itself. If W is a member of W, then by definition it is not a member of itself — that is, it is not a member of W. In the main text this contradiction is discussed but not resolved. In an appendix, however, there is a brief anticipation of an attempt to eliminate it by means of what Russell called the "theory of types," dealt with more fully in *Principia Mathematica*.

A commonsense reaction to this contradiction might well be a feeling that a class of objects is in a category different from that of the objects themselves, and so cannot reasonably be regarded as a member of itself. If x is a member of a class A, such that the definition of x depends on A, the definition is said to be impredicative; it has the appearance of circularity, since what is defined is part of its own definition. Poincaré suggested that the various antinomies were generated by accepting impredicative definitions,[2] and Russell enunciated the "vicious circle principle" that no class can contain entities definable only in terms of that class. This recourse, however, while ostracizing the antinomies, would also cast doubt on the validity of certain important processes in mathematical analysis; thus the definition of the exact upper bound is impredicative.

To meet the difficulty, Russell devised his theory of types. Very crudely outlined, it starts with primary individuals; these are of one type, say type 0. Properties of primary individuals are of type 1, properties of properties of individuals are of type 2, and so on. All admitted properties must belong to some type. Within a type, other than type 0, there are orders. In type 1, properties defined without using any totality belong to order 0; properties defined by means of a totality of properties of a given order belong to the next higher order. Then, finally, to exclude troubles arising from impredicative definitions, Russell introduced his axiom of reducibility: for any property of order other than 0, there is a property over precisely the same range that is of order 0; that is, in a given type any impredicative definition is logically equivalent to some predicative definition.

Whitehead and Russell themselves declared that they were not entirely happy about this new axiom. Even if an axiom may well be arbitrary, it should, so one feels, at least be plausible. Among the other axioms this appears as anomalous as, for instance, the notorious axiom of parallels seems to be in Euclidean geometry.

F. P. Ramsey showed that the antinomies could be separated into two kinds: those which are "logical," such as Russell's, and those which are "semantical," such as that involved in the assertion "I am lying."[3] The first class can then be eliminated by the simpler theory of types, in which the further classification into orders is not required. But even so, Ramsey did not regard this reconstruction of the Whitehead-Russell position as altogether satisfactory. Weyl pointed out that one might just as well accept the simpler axiomatic set theory of Zermelo and Fraenkel as a foundation, and remarked that a return to the standpoint of Whitehead and Russell was unthinkable.[4]

Before leaving the *Principles*, some other matters are worthy of note. First, the important calculus of relations, hinted at by De Morgan and explored by C. S. Peirce, is examined in detail.[5] Second, Russell was not concerned merely with foundations; he took the whole mathematical world as his parish. There is thus a long examination of the nature of space and of the characteristics of projective, descriptive, and metrical geometries. Third, there is an analysis of philosophical views on the nature of matter and motion. Finally, he draws, possibly for the first time, a clear distinction between a proposition, which must be true or false, and a propositional function, which becomes a proposition, with a truth value, only when the argument is given a determinate value; the propositional function "x is a prime number" becomes a proposition, which may be true or false, when x is specified.

In the three volumes of *Principia Mathematica* (1910 – 1913) Whitehead and Russell took up the task, attempted in Russell's uncompleted *Principles,* of constructing the whole body of mathematical doctrine by logical deduction from the basis of a small number of primitive ideas and a small number of primitive principles of logical inference, using a symbolism derived from that of Peano but considerably extended and systematized.

Associated with elementary propositions p, q, the primitive concepts are (1) negation, the contradictory of p, not-p, denoted by $\sim p$; (2) disjunction, or logical sum, asserting that at least one of p and q is true, denoted by $p \vee q$; (3) conjunction, or logical product, asserting that both p and q are true, denoted by $p \cdot q$; (4) implication, p implies q, denoted by $p \supset q$; (5) equivalence, p implies q and q implies p, denoted by $p \equiv q$. If a proposition is merely to be considered, it may be denoted simply by p; but if it is to be asserted, this is denoted by $\vdash p$, so that $\vdash$ may be read as "It is true that. . . ." The assertion of a propositional function for some undetermined value of the argument is denoted by $\exists$, which may be read as "There exists a . . . such that. . . ."

Dots are used systematically in place of brackets; the rule of operation is that the more dots, the stronger their effect. An example will show the way in which the dots are used. The proposition "if either p or q is true, and either p or 'q implies r' is true, then either p or r is true" may be written as

$$\vdash : \cdot p \vee q : p \cdot \vee \cdot q \supset r : \supset \cdot p \vee r.$$

The five concepts listed above are not independent. In the *Principia* negation and disjunction are taken as fundamental, and the other three are then defined in terms of these two. Thus conjunction, $p \cdot q$, is defined as $\sim (\sim p \vee \sim q)$; implication,

$p \supset q$, as $\sim p \vee q$; and equivalence, $p \equiv q$, as $(p \supset q) \cdot (q \supset p)$, which can of course now be expressed entirely in terms of the symbols $\sim$ and $\vee$. Another elementary function of two propositions is incompatibility; p is incompatible with q if either or both of p and q are false, that is, if they are not both true; the symbolic notation is p/q. Negation and disjunction are then definable in terms of incompatibility:

$$\sim p = p/p, p \vee q = p / \sim q.$$

Thus the five concepts (1) − (5) are all definable in terms of the single concept of incompatibility; for instance, $p \supset q = p/(q/q)$. This reduction was given by H. M. Sheffer in 1913,[6] although Willard Quine points out that Peirce recognized the possibility about 1880.

The primitive propositions first require two general principles of deduction — anything implied by a true proposition is true, and an analogous statement for propositional functions: when $\phi(x)$ and "$\phi(x)$ implies $\psi(x)$" can be asserted, then $\psi(x)$ can be asserted. There are then five primitive propositions of symbolic logic:

(1) Tautology. If either p is true or p is true, then p is true:

$$\vdash : p \vee p \cdot \supset \cdot p.$$

(2) Addition. If q is true, then "p or q" is true:

$$\vdash : q \cdot \supset \cdot p \vee q.$$

(3) Permutation. "p or q" implies "q or p":

$$\vdash : p \vee q \cdot \supset \cdot q \vee p.$$

(4) Association. If either p is true or "q or r" is true, then either q is true or "p or r" is true:

$$\vdash : p \vee (q \vee r) \cdot \supset \cdot q \vee (p \vee r).$$

(5) Summation. If q implies r, then "p or q" implies "p or r":

$$\vdash : \cdot q \supset r \cdot \supset : p \vee q \cdot \supset \cdot p \vee r.$$

Following up Sheffer's use of incompatibility as the single primitive concept, Nicod showed that the primitive propositions could be replaced by a single primitive proposition of the form

$$p \cdot \supset \cdot q \cdot r : \supset \cdot t \supset t \cdot s/q \supset p/s,$$

which, since $p \supset q = p/(q/q)$, can be expressed entirely in terms of the stroke symbol for incompatibility.[7]

The second edition of the *Principia* (1925 – 1927) was mainly a reprint of the first, with small errors corrected; but its worth to the student is considerably increased by the addition of a new introduction, of some thirty-four pages, in which the authors give an account of modifications and improvements rendered possible by work on the logical bases of mathematics following the appearance of the first edition—for instance, the researches of Sheffer and Nicod just mentioned. The authors are mildly apologetic about the notorious axiom of reducibility; they are not content with it, but are prepared to accept it until something better turns up. In particular, they refer to the work of Chwistek and of Wittgenstein, without, however, being able to give wholehearted approval.[8] Much of the introduction is devoted to Wittgenstein's theory and its consequences. Here they show that the results of volume I of the *Principia* stand, although proofs have to be revised; but they cannot, on Wittgenstein's theory, reestablish the important Dedekindian doctrine of the real number, nor Cantor's theorem that $2^n > n$, save for the case of n finite. The introduction was also much influenced by the views of Ramsey, whose death in 1930, at the age of twenty-seven, deprived Cambridge of a brilliant philosopher.

The publication of the *Principia* gave a marked impulse to the study of mathematical logic. The deft handling of a complicated but precise symbolism encouraged workers to use this powerful technique and thus avoid the ambiguities lurking in the earlier employment of ordinary language. The awkwardness and inadequacy of the theory of types and the axiom of reducibility led not only to further investigations of the Whitehead-Russell doctrine but also to an increased interest in rival theories, particularly Hilbert's formalism and Brouwer's intuitionism. Perhaps because none of these three competitors can be regarded as finally satisfactory, research on the foundations of mathematics has produced new results and opened up new problems the very existence of which could hardly have been foreseen in the early years of this century. Whitehead and Russell may have failed in their valiant attempt to place mathematics once and for all on an unassailable logical basis, but their failure may have contributed more to the development of mathematical logic than complete success would have done.

The *Introduction to Mathematical Philosophy* (1919), written while Russell was serving a sentence in Brixton Prison, is a genuine introduction but certainly is not "philosophy without tears"; it may perhaps best be described as *une oeuvre de haute vulgarisation*. The aim is to expound work

done in this field, particularly by Whitehead and Russell, without using the complex symbolism of *Principia Mathematica*. Russell's mastery of clear and precise English stood him in good stead for such a task, and many young students in the decade 1920–1930 were first drawn to mathematical logic by a study of this efficient and readable volume.

To explain the arrangement of the book, Russell remarks, "The most obvious and easy things in mathematics are not those that come logically at the beginning: they are things that, from the point of view of logical deduction, come somewhere near the middle." Taking such things as a starting point, a close analysis should lead back to general ideas and principles, from which the starting point can then be deduced or defined. This starting point is here taken to be the familiar set of positive integers; the theory of these, as shown by Peano, depends on the three primitive ideas of zero, number, and successor, and on five primitive propositions, one of which is the principle of mathematical induction. The integers themselves are then defined by the Frege-Russell method, using the class of all similar classes, and the relation between finiteness and mathematical induction is established. Order and relations are studied next, to enable rational, real, and complex numbers to be defined, after which the deeper topics of infinite cardinals and infinite ordinals can be broached. It is then possible to look at certain topics in analysis, such as limit processes and continuity.

The definition of multiplication when the number of factors may be infinite presents a subtle difficulty. If a class A has m members and a class B has n members, the product $m \times n$ can be defined as the number of ordered couples that can be formed by choosing the first term of the couple from A and the second from B. Here m or n or both may be infinite, and the definition may readily be generalized further to the situation in which there is a finite number of classes, $A, B, \cdots, K$. But if the number of classes is infinite, then in defect of a rule of selection, we are confronted with the impossible task of making an infinite number of arbitrary acts of choice.

To turn this obstacle, recourse must be had to the multiplicative axiom, or axiom of selection: given a class of mutually exclusive classes (none being null), there is at least one class that has exactly one term in common with each of the given classes. This is equivalent to Zermelo's axiom that every class can be well-ordered, that is, can be arranged in a series in which each subclass (not being null) has a first term.[9] This matter is dealt with in *Principia Mathematica,* but here Russell offers a pleasant illustration provided by the arithmetical perplexity of the millionaire who buys a pair of socks whenever he buys a pair of boots, ultimately purchasing an infinity of each. In dealing with the number of boots, the axiom is not required, since we can choose, say, the right boot (or the left) from every pair. But no such distinction is available in counting the socks, however, and here the axiom is needed.

The last six chapters of the *Introduction* are concerned with the theory of deduction and the general logical bases of mathematics, including an analysis of the use and nature of classes and the need, in Russell's theory, for a doctrine of types.

Among the essays collected in *Mysticism and Logic* (1921) are some that deal, in popular style, with Russell's views on mathematics and its logical foundations. One of these, "Mathematics and the Metaphysicians," written in 1901, had appeared in *International Monthly.* The editor had asked Russell to make the article "as romantic as possible," and hence it contains a number of quips, some now famous, in which the air of paradox masks a substantial degree of truth. To say that pure mathematics was discovered by George Boole in 1854 is merely Russell's way of stating that Boole was one of the first to recognize the identity of formal logic with mathematics, a point of view firmly held by Russell. In emphasizing that pure mathematics is made up of logical steps of the form "If p, then q" —that is, if such and such a proposition is true of anything, then such and such another proposition is true of that thing—Russell remarks that it is essential not to discuss whether the first proposition is really true, and not to mention what the anything is, of which it is supposed to be true. He is thus led to his oft-quoted description of mathematics as "the subject in which we never know what we are talking about, nor whether what we are saying is true." He comments that many people may find comfort in agreeing that the description is accurate.

Russell's gifts as a popularizer of knowledge are shown in a number of his other books, such as *The Analysis of Matter* and *The ABC of Relativity,* in which problems arising from contemporary physics are discussed. He never wholly divorced mathematics from its applications; and even his first book, on the foundations of geometry, had its origin in his wish to establish the concept of motion and the laws of dynamics on a secure logical basis.

In these later books his critical skill is exercised on the mathematical foundations of physics and occasionally is used to provide alternatives to suggested theories.

For instance, the advent of relativity, bringing in the notion of an event as a point in the space-time continuum, had encouraged Whitehead to deal with the definition of points and events by the application of his principle of extensive abstraction, discussed in detail in his *An Enquiry Concerning the Principles of Natural Knowledge* (1919). This principle has a certain affinity with the Frege-Russell definition of the number of a class as the class of all similar classes. To state the application very crudely, a point is defined as the set of all volumes that enclose that point; Whitehead is of course careful to frame the principle in such a way as to avoid the circularity suggested in this crude statement. This idea is then used to define an event. In *The Analysis of Matter,* Russell argues that while logically flawless, this definition, in the case of an event, does not seem genuinely to correspond to the nature of events as they occur in the physical world, and that it makes the large assumption that there is no minimum and no maximum to the time extent of an event. He develops an alternative theory involving an ingenious application of Hausdorff's axioms for a topological space.[10]

Whatever the final verdict on Russell's work in symbolic logic may be, his place among the outstanding leaders in this field in the present century must be secure.

NOTES

1. G. Frege, *Die Grundlagen der Arithmetik* (Breslau, 1884), also in English (Oxford, 1953); *Grundgesetze der Arithmetik, begriffsschriftlich abgeleitet,* 2 vols. (Jena, 1893–1903), vol. I also in English (Berkeley–Los Angeles, 1964).
2. H. Poincaré, "La logique de l'infini," in *Scientia,* **12** (1912).
3. F. P. Ramsey, "The Foundations of Mathematics," in *Proceedings of the London Mathematical Society,* 2nd ser., **25** (1926).
4. H. Weyl, "Mathematics and Logic," in *American Mathematical Monthly,* **53** (1946); and "David Hilbert and His Mathematical Work," in *Bulletin of the American Mathematical Society,* **50** (1944).
5. A. De Morgan, "On the Logic of Relations," in *Transactions of the Cambridge Philosophical Society,* **10** (1864); C. S. Peirce, "On the Algebra of Logic," in *American Journal of Mathematics,* **3** (1880).
6. H. M. Sheffer, "A Set of Five Independent Postulates of Boolean Algebra," in *Transactions of the American Mathematical Society,* **14** (1913).
7. Jean Nicod, "A Reduction in the Number of the Primitive Propositions of Logic," in *Proceedings of the Cambridge Philosophical Society. Mathematical and Physical Sciences,* **19** (1919).
8. Leon Chwistek, "The Theory of Constructive Types," in

Annales de la Société mathématique de Pologne (1924); L. Wittgenstein, *Tractatus Logico-Philosophicus* (London, 1922).

9. E. Zermelo, "Beweis, dass jede Menge wohlgeordnet werden kann," in *Mathematische Annalen*, **59** (1904); see also *ibid.*, **65** (1908).

10. F. Hausdorff, *Grundzüge der Mengenlehre* (Leipzig, 1914).

BIBLIOGRAPHY

I. ORIGINAL WORKS. Russell's contributions to mathematical logic are best studied through *An Essay on the Foundations of Geometry* (London, 1897); *The Principles of Mathematics*, I (London, 1903; 2nd ed. with new intro., 1937); *Principia Mathematica*, 3 vols. (London, 1910–1913; 2nd ed., with new intro., 1925–1927), written with A. N. Whitehead; and *Introduction to Mathematical Philosophy* (London, 1919).

The Autobiography of Bertrand Russell, 3 vols. (London, 1967–1969), contains a nontechnical account of his mathematical work in vol. I and early chs. of vol. II. *Mysticism and Logic* (London, 1917) contains some essays of a popular nature.

II. SECONDARY LITERATURE. The primary authority for Russell's life in his *Autobiography*, cited above. G. H. Hardy, *Bertrand Russell and Trinity* (London, 1970), clears away some misconceptions concerning Russell's relations with the college. An able and witty criticism of Russell's logical ideas is presented by P. E. B. Jourdain, given in *The Philosophy of Mr. B*rtr*nd R*ss*ll* (London, 1918).

For useful surveys of the doctrine of *Principia Mathematica*, see S. K. Langer, *An Introduction to Symbolic Logic* (New York, 1937); S. C. Kleene, *Introduction to Metamathematics* (New York, 1952), which has a valuable selected bibliography; and, by F. P. Ramsey, several items in *The Foundations of Mathematics and Other Logical Essays* (London, 1931). *The Philosophy of Bertrand Russell*, P. A. Schilpp, ed. (Chicago, 1944), contains a section by Gödel on Russell's mathematical logic.

To trace the many publications related directly or indirectly to Russell's work, see A. Church, "A Bibliography of Symbolic Logic," in *Journal of Symbolic Logic*, **1** (1936) and **3** (1938); it is also available as a separate volume. For the literature since 1935, the reader is advised to consult the volumes and index parts of the *Journal of Symbolic Logic*.

T. A. A. BROADBENT

RUSSELL, EDWARD JOHN (*b.* Frampton-on-Severn, Gloucestershire, England, 31 October 1872; *d.* Goring-on-Thames, Oxfordshire, England, 12 July 1965), *agricultural chemistry, agronomy.*

For a complete study of his life and work, see Supplement.

RUSSELL, HENRY CHAMBERLAINE (*b.* West Maitland, New South Wales, Australia, 17 March 1836; *d.* Sydney, Australia, 22 February 1907), *astronomy, meteorology.*

For a complete study of his life and work, see Supplement.

RUSSELL, HENRY NORRIS (*b.* Oyster Bay, New York, 25 October 1877; *d.* Princeton, New Jersey, 18 February 1957), *astrophysics, spectroscopy.*

Russell was educated at home until the age of twelve before attending a preparatory school in Princeton, the home of his maternal grandparents. His father, Alexander Russell, a Scottish-Canadian immigrant to the United States, was a Presbyterian minister in Oyster Bay. Both Russell's mother and maternal grandmother had had some advanced formal education and an outstanding gift for mathematics—a trait especially strong in Russell himself. He graduated *insigne cum laude* from Princeton University in 1897 and remained there to obtain his doctorate (1900). His dissertation was entitled "The General Perturbations of the Major Axis of *Eros*, by the Action of *Mars*."

After completing graduate work, Russell suffered a serious breakdown of his health and spent much of the following year at Oyster Bay. In the fall of 1901 he returned to Princeton and the following autumn began a three-year stay at Cambridge University. During the first year he was a student at King's College and also worked at the Cavendish Laboratory. The last two years were spent at the Cambridge observatory as a research assistant supported by the Carnegie Institution of Washington. There he worked in association with Arthur Hinks on a program of determining stellar parallaxes by photographic means. In September 1904 Russell was again taken seriously ill, and in his absence the remaining observations were made by Hinks.

In 1905 Russell accepted a post as instructor in astronomy at Princeton. In 1911 he was appointed professor of astronomy and, in 1912, director of the observatory, positions he held until his retirement in 1947. From June 1918 to early 1919 he was a consulting and experimental engineer in the Bureau of Aircraft Production of the Army Aviation Service. His chief responsibility was a study of problems in aircraft navigation, which included making observations in open aircraft at altitudes of up to 16,000 feet.

In 1921 Russell began his association with the Mt. Wilson Observatory, where he was a research associate until his retirement. In this capacity he usually spent two months of each year at the California observatory. Following his retirement he held research appointments at Lick and Harvard observatories.

Russell was a member of the American Astronomical Society (president, 1934–1937), the American Philosophical Society (president, 1931–1932), the National Academy of Sciences, the American Association for the Advancement of Science (president, 1933); an associate of the Royal Astronomical Society; a foreign member of the Royal Society; and a correspondent of the academies in Paris, Brussels, and Rome. He was president of the commissions of the International Astronomical Union on stellar spectroscopy and on the constitution of stars. Russell was awarded the Draper, Bruce, Rumford, Franklin, Janssen, and Royal Astronomical Society gold medals, and the Lalande Prize. In 1946 the American Astronomical Society established the annual Henry Norris Russell lectureship in his honor.

In the course of nearly sixty years of research, Russell concerned himself with most of the major problems of astrophysics. His principal contributions, however, can be summarized into four general categories. First, Russell presented (in 1912) the earliest systematic analysis of the variation of the light received from eclipsing binary stars; he later pointed out the importance of the motion of the periastron of the orbit in providing information about the internal structure of the component stars. Second, on the basis of his parallax studies, Russell developed a theory of stellar evolution that at the time was in good agreement with the known data. This work stimulated other astrophysicists, especially Arthur Eddington, and was the original context in which he introduced the Hertzsprung-Russell diagram. Third, in the 1920's Russell began a series of quantitative investigations of the absorption-line spectrum of the sun that resulted in a reliable determination of the abundance of various chemical elements in the solar atmosphere. This work provided clear evidence of the predominance of hydrogen in the sun and, by inference, in most stars. Fourth, Russell carried out, with various co-workers, extensive analyses of the spectra of a number of elements, those of calcium, tita-

nium, and iron being the more important. In this work he developed empirical rules for the relative strengths of lines of a given multiplet, and with F. A. Saunders he devised the theory of L-S coupling to explain spectra produced by atoms with more than one valence electron.

Russell's first research papers were published while he was a student at Princeton. Several of them, and most notably his dissertation, dealt with problems in celestial mechanics and orbit determinations. In light of his later work, however, the most interesting of these early studies was a short paper (1899) that showed how an upper limit to the densities of Algol-variable stars could be obtained. At this time the idea that these were binary stars, the components of which eclipsed each other, was not universally accepted. Only in a few particular cases had orbits been derived from the light variation. Recognizing that an upper limit to the sum of the diameters of the two stars (relative to the size of the orbit) could be determined from the duration of the eclipse, Russell derived limits for the mean densities of seventeen systems. After considering the possible systematic errors in his method, he concluded that Algol-variables were, as a class, much less dense than the sun—a determination that became important in his later work on stellar evolution.

At Cambridge, Russell learned about astrometric methods from Hinks, and in 1903 they embarked upon a program of photographically determining stellar parallaxes. Their two main objectives were to find the most suitable compromise between the amount of work done and the accuracy achieved, and to eliminate all known sources of systematic error. Because photographic techniques in astrometry were still plagued with difficulties and were generally considered inferior to visual observations, Hinks and Russell found it desirable to reconsider the entire observing procedure. The technique they developed was similar to that devised at about the same time by Frank Schlesinger in the United States and was one of the first modern parallax programs.

Russell completed the work of measuring all of the photographic plates and reducing the data in 1910. To him the most interesting result was the correlation between the spectral types and absolute magnitudes of different stars. For those of known parallax, the absolute magnitude decreased systematically from type B to type M—that is, from stars of high surface temperature to those of low surface temperature. This conclusion seemed contrary to the general opinion that many cool, red stars were at great distances and thus were of high luminosity. Russell pointed out that such stars were systematically excluded from parallax studies (since the parallax was undetectably small) and that all the known data could be accounted for under the assumption of two distinct groups of red stars.

To explain the existence of these two types of red stars, Russell adapted a theory of stellar evolution proposed by August Ritter and modified by Sir Norman Lockyer. According to this theory the stars first appeared as highly luminous cool objects that contracted and grew hotter until the high density of the gas caused a significant reduction in the compressibility. Thereafter the star decreased in brightness and in surface temperature. Thus the two kinds of red stars were representatives of the first and last stages of stellar evolution. This finding was in striking contrast with the prevalent view that stars evolved continuously from class B to class M.

To support his ideas, Russell returned to the study of binary stars. In several short papers presented at meetings of the American Astronomical Society between 1910 and 1912, he provided observational evidence that the basic distinguishing feature of the two groups of red stars was their density—as his theory demanded. He also saw other evidence for the correctness of his theory in the orbits of binary stars. Developing the ideas of G. H. Darwin (whose lectures he had attended at Cambridge) on the formation of binary stars from the fission of a single, rapidly rotating star (1910), Russell argued that the youngest stars were single and that binary stars with short orbital periods did not form until the density of the contracting star had become fairly high. The empirical evidence was that bright red stars were rarely members of binary systems and that when hot type-B stars were members of such systems the orbital periods were quite short. Thus Darwin's picture of the development of binary systems agreed quite satisfactorily with Russell's theory of the evolution of the individual stars.

The complete account of his theory of stellar evolution that Russell gave in December 1913 served to make his work more widely known. In this lecture he presented graphs plotting absolute magnitudes of stars against their spectral types (now known as Hertzsprung-Russell diagrams), which he used to illustrate the empirical evidence for his theory. It was also at this time that the terms "giant" and "dwarf" came into use, largely through his papers, to describe the two groups of

stars, although it is not clear who actually coined them.

While developing his ideas on evolution, Russell began a systematic study of the interpretation of the variations in intensity of the light from eclipsing binary systems. In the decade since he had written his paper on the densities of Algol-variables, two developments had made such a study desirable: the availability of data of much greater precision and completeness—primarily from the observations of Raymond S. Dugan at Princeton and from the photoelectric observations of Joel Stebbins at the University of Illinois—and Russell's need for the densities of individual stars as further supporting evidence for his theory of stellar evolution. These densities could be obtained only by detailed analysis of eclipsing binaries.

In the first of four papers (1912) on the subject, Russell stated his objective of determining both the orbital elements of the system and the dimensions and brightnesses of the component stars from the observed light curve. The problem was first reduced to one of the simplest cases—two spherical stars, seen as uniformly illuminated disks, moving in a circular orbit. Russell then showed under what circumstances a complete solution could be obtained and gave tabular values for the special functions that were required. His solution emphasized the importance of accurate observations of the binary system at all phases, not only at the primary eclipse.

The remaining papers showed how the basic solution could be extended to more realistic representations of the binary system. In the second paper Russell introduced the refinements of elliptical orbits of small eccentricity and of stars distorted into ellipsoids by their mutual gravitational attraction. Russell handled the latter effect by what he called "rectification"—a transformation of the observed light curve to remove the effects of ellipticity and to reduce the problem to the previously studied case of spherical stars.

At the end of this paper Russell briefly outlined a means of handling the problem of limb darkening—the decrease in brightness of the stellar disk near its apparent edge. The extensive calculations for this part of the theory were assigned to Harlow Shapley, who had recently arrived at Princeton as a graduate student. The results of their collective efforts were presented in the third and fourth papers of the series, in which the systematic treatment of eclipsing binaries was extended to those with limb-darkened components. Under Russell's direction Shapley later applied the new techniques to ninety eclipsing binaries, thereby providing a large number of new density determinations.

For nearly thirty years the standard techniques of dealing with eclipsing binary systems were essentially those introduced by Russell, and much of his nomenclature and notation became a permanent part of the subject. The wide acceptance of this work can in part be attributed to the very practical manner in which the analysis was presented. There were references to how much time certain calculations required and comments that certain refinements were not worth the work. Indeed, this was a characteristic feature of many of Russell's papers, in which a balance was struck between ease of computation and precision of results.

From 1914 to 1921 Russell worked on various subjects, some of which were a continuation of his study of stellar evolution. He published several papers on the orbits of visual binaries and the determination of the masses of the component stars. He continued a project on the photographic determination of the position of the moon, carried out jointly by the Harvard, Yale, and Princeton observatories. A review of the determination of the albedoes and magnitudes of planets and satellites was also conducted. In 1921 Russell showed that the age of the earth's crust was about 4×10^9 years, basing his statement on the radioactive decay of uranium and the abundance of its end products, lead and thorium.

Two developments in 1921 marked a major shift in Russell's career. The first was the publication of M. N. Saha's theory of the ionization of atoms in stellar atmospheres; the second was Russell's appointment as a research associate at the Mt. Wilson observatory, which brought him into close association with Walter S. Adams and other astronomers and spectroscopists in California.

During his first visit to Mt. Wilson, in the summer of 1921, Russell investigated the application of Saha's formula to the sun. To do so it was necessary to generalize Saha's original theory, which described a gas composed of a single atomic species. Russell pointed out that in a mixed gas, such as the sun's atmosphere, the ionization relationships were more complex because one of the products of the ionization process, the free electron, was common to the ionization reactions of all atoms. Thus the equilibrium state of ions and electrons could be determined only for all elements simultaneously. The result of this analysis was that the degree of ionization of an atom depended not only upon the pressure and temperature of the gas and the ionization potential of the atom, but also upon the relative abundances of other atomic species and their ionization potentials.

Russell undertook a critical test of the expanded theory by comparing the spectrum of the normal solar photosphere with that of sunspots. Since the temperatures of both the photosphere and the sunspots were well known and the pressures, although not well determined, were assumed to be equal, the relative strengths of absorption lines in the two spectra provided the desired comparison of theory and observation.

The most exact comparisons were for the alkali metals, since their ionization potentials were known. In particular Saha had predicted that the lines of neutral potassium would be stronger in the spot spectrum and that lines of neutral rubidium would be faintly visible in the sunspots, although they had not been detected in the normal solar spectrum. Russell's examination of spectra taken at Mt. Wilson confirmed these predictions and, in general, supported the Saha theory.

This success made Russell keenly aware of the tremendous possibilities that the new theory offered. The spectra of stars could now be used to give quantitative information about the state of the atmosphere where the lines were formed. In concluding his paper he wrote:

> The possibilities of the new method appear to be very great. To utilize it fully, years of work will be required to study the behavior of [the alkali earths, scandium, titanium, vanadium, manganese, and iron] and of others, in the stars, in laboratory spectra, and by direct measurement of ionization, but the prospect of our knowledge, both of atoms and of stars, as a result of such researches, makes it urgently desirable that they should be carried out ["Theory of Ionization," pp. 143–144].

Although he continued to investigate problems relating directly to the atmospheres of stars, such as the theoretical determination of the pressure at the solar photosphere (carried out jointly with John Q. Stewart), Russell soon turned to the determination of atomic structure through the study of spectra. His first major effort in this direction was a study, in collaboration with the Harvard spectroscopist F. A. Saunders, of the spectra of the alkali earths—calcium, scandium, and barium (1925). In 1923, when this work was being carried out, atomic theory was unable to explain "complex" spectra—the spectra of elements other than hydrogen, helium, and the alkali metals. Of the remaining elements, the spectra of the alkali earths were partially understood in terms of the Bohr atom; but it was clear that energy levels existed that did not fit into the regular series of terms.

Russell and Saunders had found several groups of lines in the ultraviolet spectrum of calcium that led to the identification of three new "anomalous" triplet terms for this element. With the additional data they were able to find some systematic relationships among the anomalous terms. The most important discovery was that the energies of some of these terms were greater than the ionization potential of the atom, a fact also recognized by Gregor Wentzel at about the same time. They interpreted this result as evidence that the anomalous terms were produced by an excitation of both optical electrons. This idea explained not only why an atom can absorb energy greater than its ionization potential, but also why the alkali metals—which have only one valence electron—do not have any anomalous terms.

From this basic concept of excitation of more than one electron (which had also occurred to Bohr), Russell and Saunders proceeded to extend Alfred Landé's vector model to account for the quantum numbers and energies of the anomalous terms. Landé's model identified the azimuthal quantum number with the orbital angular momentum of the electron, the multiplicity of the spectroscopic term with the angular momentum of the rest of the atom (the *Rumpf*), and the inner quantum number with the vectoral sum of the two. Russell and Saunders assumed that the quantized angular momentum of the individual excited electrons could be combined first, and the resultant combined with that of the *Rumpf*. This technique of handling complex spectra, which later became known as L-S coupling, proved quite successful in predicting both the energy levels for the anomalous terms and the observed transitions to those levels resulting from the excitation of only a single electron.

The final section of Russell's and Saunders' paper is of some interest in the history of spectroscopy for its attempt to introduce uniformity into the chaotic state of spectroscopic notation. The proposed system became the basis for the modern notation, although it was later refined by Russell, Allen G. Shenstone, and L. A. Turner (1929).

Russell subsequently turned to the problem of finding formulas that could represent the relative intensities of the spectral lines of a particular atom (1926). He accomplished this for the lines of a given multiplet, that is, all the lines arising from transitions between the various levels of two spectroscopic terms. (It is an indication of the vigorous

activity in this field in 1926 that two other spectroscopists, R. Kronig and Sommerfeld, derived similar formulas at the same time.) Determined without any detailed theory of atomic structure, the intensity formula was based upon Bohr's correspondence principle and a rule for the sum of the intensities of the lines having a given initial or final level. Thus, this work was carried out entirely within the framework of the "old" quantum mechanics, as was essentially all of Russell's spectroscopic work.

The motivation for these spectroscopic studies was, as Russell had indicated in 1921, not only atomic but also astrophysical. By 1928 he was able, in collaboration with Walter Adams and Charlotte Moore, to bring this new knowledge to bear on stellar spectra. The first problem to be solved was one of calibration: how could one deduce the number of atoms in the solar atmosphere that were responsible for producing a particular absorption line? Two methods appeared possible—a measurement of contours of the lines, followed by a theoretical interpretation in terms of atomic physics, or a direct calibration of the empirical Rowland intensity scale of spectral lines in terms of numbers of atoms. The latter method was chosen. Russell, Adams, and Moore assumed that the intensity of the lines, as derived from Russell's multiplet formulas, was proportional to the number of atoms acting to produce the line. The problem then reduced to that of calibrating the Rowland scale in terms of the theoretical one. Since the Mt. Wilson observatory was then revising Rowland's table of absorption lines in the solar spectrum, abundant information was available. A comparison of these data with the multiplet intensity formulas gave a relative scale of intensities—relative in the sense that although the shape of the curve was determined, the zero point was not; the zero point, in fact, proved to be different for each multiplet.

Yet even this relative calibration was of considerable interest. Adams and Russell used the results in conjunction with the Saha-Boltzmann relationship to compare the atmospheres of different stars (1928). Using the sun as a standard, they analyzed seven stars on the basis of high-dispersion spectra taken with the 100-inch telescope. In calculating the relative populations of excited states of atoms in different stars, they found—as expected—that in hotter stars the population of higher states was greater. In the cooler stars, however, the dependence on excitation potential was not what was expected (the Adams-Russell effect), leading them to believe that the atmospheres of cool stars were not in thermodynamic equilibrium. Their temperature determinations of these stars were in substantial agreement with the results of other methods, and the values they found for the partial pressure of free electrons emphasized the extremely low densities in red giant stars.

Russell continued this analysis in 1929 with the help of Albrecht Unsöld's measurements of line profiles. Unsöld's work provided an absolute calibration of the number of atoms involved in producing an absorption line. Because of the amount of work involved in his procedure, only a relatively small number of lines had been so analyzed. Using this work to provide the zero point for his own scale, Russell developed an absolute calibration scale for the Rowland intensities. He then showed that the total abundances of elements in the sun could be calculated by taking into account the atoms in various states of excitation and ionization. In this manner Russell determined the abundances of fifty-five elements and several molecules in the solar atmosphere. In many instances the abundance ratios between elements was similar to that in the earth's crust—with one notable exception. Hydrogen proved to be by far the most abundant element. This discovery was not completely unexpected, for Cecilia Payne had earlier found high abundances of hydrogen in giant stars but had dismissed the numerical values as "spurious." Russell's analysis had proceeded on more solid footing, however, and he was also able to show that the high abundance of hydrogen actually removed several other apparent difficulties in the analysis of the sun. Thus, there was clear evidence of the dominant abundance of hydrogen in the sun and, therefore, in most stars. It is difficult to overestimate the importance of this result in the development of astrophysics, since much of the subsequent progress has depended upon recognizing the predominant role of hydrogen in astrophysical processes.

Although this discussion of Russell's work in the 1920's might suggest that he proceeded single-mindedly toward the goal of a quantitative analysis of stellar spectra, Russell also investigated related matters. In atomic spectroscopy he carried out detailed analyses of several elements (most notably titanium, iron, and scandium); and on the basis of Friedrich Hund's theory of complex spectra, he found systematic similarities in the spectra of the elements of the iron group (those in the periodic table from potassium through zinc).

Eddington's discovery (1924) that the ideal gas law was applicable to the interiors of stars on the main sequence led Russell back to his old theory of stellar evolution, which had now been rendered untenable. Recognizing the chief problem to be the source of energy, he postulated highly temperature-sensitive processes of transforming matter to energy (1925). Thus, as a star contracted and grew hotter, the energy source would become activated and contraction would cease. The main sequence and giant branch were thus interpreted as stages in which different processes were active. As one of these processes proceeded, mass was converted into energy and the total mass of the star decreased. With the decrease, the temperature and luminosity changed as, consequently, did the position of the star in the Hertzsprung-Russell diagram. The main sequence and giant branches were again seen to be the evolutionary paths taken by stars. This theory was challenged by James Jeans, who claimed that Russell's stars would be unstable and who preferred an energy source the rate of which was independent of temperature or pressure. No substantial progress in unraveling the complexities of stellar evolution occurred, however, until the identification, fifteen years later, of the particular nuclear reactions occurring inside stars.

Russell also contributed to the theory of stellar structure. He suggested that instead of calculating models with a specific opacity formula or equation of state, an attempt should be made to postulate only very general principles and to search for distinctive relationships among stars (1931). The most important result of this approach, now known as the Vogt-Russell theorem, was that on very general grounds the properties of a star can be expected to be completely determined by its mass and chemical composition. Heinrich Vogt had derived a similar result, but apparently his work was not well known in England and the United States, where most of the research in stellar structure was being conducted.

Eddington's work had made the question of mass distribution inside a star an important one, and Russell realized that there was an empirical method of estimating the ratio of the mean density to the central density of a star (1928). In close binary systems the interaction of the distorted stars would result in an advance of the periastron of the orbit, and this advance could be detected from the light curve of an eclipsing binary system. Although Russell's initial results were not very satisfactory, his method was sound; and later investigators were able to derive better results.

During the 1930's and 1940's, Russell continued to work on most of the subjects that had occupied him in the previous twenty years. He made detailed analyses of the spectra of several more elements. His study of the orbits of visual binary stars and the masses derived therefrom led to the publication, with Charlotte Moore, of a monograph on the subject (1940). Russell enlarged his work on the chemical composition of the sun to include a study of molecular abundances (1933). He also returned to the analysis of eclipsing binary systems, extending his methods to more complicated systems and considering further the effects of the internal structure of the stars upon the advance of the periastron (1939, 1942).

While Russell continued to make significant contributions to astrophysical research during the latter part of his career, his role as an adviser and consultant to other astronomers became increasingly important. His yearly trips from Princeton to Mt. Wilson afforded many opportunities to visit other American observatories; and because his own interests covered such a wide range, he was often quite familiar with the problems on which others were working.

Russell's role as a critic and reviewer of contemporary research began at least as early as 1919, when, at the request of the National Academy of Sciences, he wrote a comprehensive review of current research in sidereal astronomy that pointed out to his colleagues some of the most important problems to be solved. Another indication of his interest in analyzing the work of others is his review of the dispute between Jeans and Eddington over the mass-luminosity relation for stars (1925). By placing their arguments within the framework of a more general theory, Russell resolved the apparent discrepancy between their results.

Russell's interest in teaching began fairly early in his career. By 1911 he had started a revision of the general astronomy textbook written by his predecessor at Princeton, Charles A. Young, but other work delayed its completion. With the collaboration of his Princeton colleagues Raymond Dugan and John Q. Stewart, it was finally published in 1926. The first volume was the originally projected revision of Young's book; the second contained largely new material, most of it written by Russell. This textbook was widely used for thirty years, and many American astronomers trained in this period were introduced to the subject through studying it.

After 1930 especially, Russell gave a number of lectures in which he reviewed the progress in various areas of astrophysics. Although new results were rarely presented in these talks, they did serve the important function of summarizing and organizing recent work.

These lectures, and even his textbook, were aimed principally at the scientist or potential scientist. To reach a larger, more general audience, Russell wrote a monthly article in *Scientific American*. Beginning in 1900 as a column describing the appearance of the evening sky for the coming month, the articles soon included information on recent research in astronomy. By 1911 Russell was regularly including a short essay on some astronomical subject, and this section of the article soon came to be the dominant feature. By 1943, when the last one appeared, he had written 500 short articles discussing all phases of astronomy.

Although in the vast majority of his writings Russell kept strictly to scientific matters, he did on several occasions discuss his ideas concerning science and religion. The fullest exposition was in a series of lectures given at Yale University in 1925 and published two years later as *Fate and Freedom*. The title reflects one of his central concerns: the conflict between the concept of a deterministic universe and the belief in free will. Although Russell concluded that the universe was completely mechanistic, he felt that the observed behavior of men should be considered a kind of statistical phenomenon and that consequently free will was as real as (to use his analogy) statistical phenomena in physics, such as the pressure of a gas. These ideas were formulated prior to the introduction into quantum mechanics of the uncertainty principle, which Russell in his later writings does not seem to have considered of central philosophical importance.

Over a period of fifty years Russell's work showed a continuous effort to provide a clear understanding of the physics of stars. Early in his career he focused on stellar evolution and the related problems of determining masses, radii, temperatures, luminosities, and densities of stars. A result of this effort was his series of investigations of eclipsing binary systems. His interest later turned to stellar atmospheres, the problems of determining pressures and temperatures, and the quantitative measurement of chemical abundances. An outgrowth of this work was his extensive work in the theory of atomic spectra.

BIBLIOGRAPHY

I. Original Works. Bibliographies of Russell's published writings are in Poggendorff, V, 1081–1082, and VI, 2249–2250, and at the end of the biographical essays by Shapley and Seaton (see below). The most extensive is that following Shapley's article, although it does not include abstracts of certain papers presented at meetings of the American Astronomical Society or Russell's articles in *Scientific American*.

Russell's principal publications include "The Densities of the Variable Stars of the Algol Type," in *Astrophysical Journal*, **10** (1899), 315–318; "The General Perturbations of the Major Axis of *Eros*, by the Action of *Mars*," in *Astronomical Journal*, **21** (1900), 25–28; "On the Origin of Binary Stars," in *Astrophysical Journal*, **31** (1910), 185–207; *Determinations of Stellar Parallax* (Washington, D.C., 1911); "On the Determination of the Orbital Elements of Eclipsing Variable Stars," in *Astrophysical Journal*, **35** (1912), 315–340, and **36** (1912), 54–74; "On Darkening at the Limb in Eclipsing Variables," *ibid.*, **36** (1912), 239–254, 385–408, written with H. Shapley; "Relations Between the Spectra and Other Characteristics of the Stars," in *Nature*, **93** (1914), 227–230, 252–258, 281–286; and "Some Problems of Sidereal Astronomy," in *Proceedings of the National Academy of Sciences of the United States of America*, **5** (1919), 391–416.

See also "A Superior Limit to the Age of the Earth's Crust," in *Proceedings of the Royal Society*, **99A** (1921), 84–86; "The Theory of Ionization and the Sun-Spot Spectrum," in *Astrophysical Journal*, **55** (1922), 119–144; "New Regularities in the Spectra of the Alkaline Earths," *ibid.*, **61** (1925), 38–69, written with F. A. Saunders; "The Intensities of Lines in Multiplets," in *Proceedings of the National Academy of Sciences*, **11** (1925), 314–328; "Note on the Relations Between the Mass, Temperature, and Luminosity of a Gaseous Star," in *Monthly Notices of the Royal Astronomical Society*, **85** (1925), 935–939; *Astronomy, a Revision of Young's Manual of Astronomy*, 2 vols. (Boston, 1926–1927), written with R. S. Dugan and J. Q. Stewart; *Fate and Freedom* (New Haven, 1927); "On the Advance of Periastron in Eclipsing Binaries," in *Monthly Notices of the Royal Astronomical Society*, **88** (1928), 641–643; "A Calibration of Rowland's Scale of Intensities for Solar Lines," in *Astrophysical Journal*, **68** (1928), 1–8, written with W. S. Adams and C. E. Moore; "Preliminary Results of a New Method for the Analysis of Stellar Spectra," *ibid.*, 9–36, written with W. S. Adams; "On the Composition of the Sun's Atmosphere," in *Astrophysical Journal*, **70** (1929), 11–82; "Notes on the Constitution of the Stars," in *Monthly Notices of the Royal Astronomical Society*, **91** (1931), 951–966, and **92** (1931), 146; *The Solar System and Its Origin* (New York, 1935); and *The Masses of the Stars, With a General Catalog of Dynamical Parallaxes* (Chicago, 1940), written with C. E. Moore.

II. SECONDARY LITERATURE. The most extensive biographical essays are those by F. J. M. Stratton in *Biographical Memoirs of Fellows of the Royal Society*, **3** (1957), 173–191; and by Harlow Shapley in *Biographical Memoirs. National Academy of Sciences*, **32** (1958), 354–378. Obituary notices include those by

Donald H. Menzel in *Yearbook. American Philosophical Society* (1958), 139–143; and Otto Struve, in *Publications of the Astronomical Society of the Pacific*, **69** (1957), 223–226. See also Axel V. Nielsen, "Contributions to the History of the Hertzsprung-Russell Diagram," in *Centaurus*, **9** (1964), 219–253; and the following articles in *Vistas in Astronomy*, **12** (1970): Katherine G. Kron, "Henry Norris Russell (1877–1957): Some Recollections," 3–6; Bancroft W. Sitterly, "Changing Interpretations of the Hertzsprung-Russell Diagram, 1910–1940: A Historical Note," 357–366; and R. Szafraniec, "Henry Norris Russell's Contribution to the Study of Eclipsing Variables," 7–20.

BRUCE C. COGAN

RUTHERFORD, DANIEL (*b.* Edinburgh, Scotland, 3 November 1749; *d.* Edinburgh, 15 December 1819), *chemistry.*

An uncle of Sir Walter Scott's, Rutherford is mentioned frequently in works devoted to the life and letters of the famous novelist; his place in the history of chemistry depends solely on his discovery of nitrogen. He was a son of John Rutherford, professor of medicine at the University of Edinburgh from 1726 to 1765, and his second wife, Anne Mackay; Scott's mother was Daniel Rutherford's stepsister.

A pupil of William Cullen and Joseph Black at Edinburgh, Rutherford traveled in Europe for about three years after obtaining his M.D. and before beginning practice at Edinburgh in 1775. Shortly afterward he became a member of the Philosophical Society of Edinburgh (later the Royal Society of Edinburgh), and in 1786 succeeded John Hope, father of Thomas Charles Hope, as regius professor of botany at the university. In the same year he married Harriet Mitchelson; they had two sons and three daughters.

Rutherford's M.D. dissertation, dated 12 September 1772, was devoted mainly to the discoveries regarding the gas that Black had called "fixed air" (carbon dioxide), but which Rutherford preferred to call "mephitic air." About two-thirds of the way through the work he wrote that, having dealt with the air from calcareous bodies, he would say something of the air rendered malignant by animal respiration. He noted the contraction of air in which animals had been confined, and said that by their respiration good air became in part "mephitic" but also suffered a further change: the separation of the mephitic air by means of a caustic alkaline solution did not render the remaining air wholesome. Although it gave no precipitate with limewater, it nonetheless extinguished flame and life.

This is the earliest published account of the awareness of a gas (nitrogen) that, although unable to support life and combustion, was clearly not Black's "fixed air." There can be little doubt about Rutherford's priority of publication, but priority of discovery seems attributable to Cavendish, Priestley, or Scheele. Cavendish, in a manuscript published thirty years after his death by W. V. Harcourt (*Report of the British Association for the Advancement of Science, 1839* [1840], 64–65), wrote: "Air which has passed thro' a charcoal fire contains a great deal of fixed air . . . but . . . consists principally of common air, which has suffered a change in its nature from the fire." He removed fixed air from air in which charcoal had been burned and found the remaining air unfit for combustion and of a density slightly less than that of common air.

The paper was undated but was marked "communicated to Dr. Priestley." The latter gave an inaccurate account of it in his classical paper "Observations on Different Kinds of Air," read (over four meetings) to the Royal Society in March 1772 (*Philosophical Transactions of the Royal Society,* **62** [1772], 147–252, see 225; the volume was not published until 1773). A few pages earlier in the same paper Priestley described how, in one of his own experiments, air in which moist iron filings and sulfur had been confined had decreased in volume by approximately a quarter; the remaining air gave no precipitate with limewater and was less dense than common air. Scheele's observation that air consisted of two gases, his names for which are usually translated as "vitiated air" (nitrogen) and "fire air" (oxygen), was not published until 1777 but may have been made as early as 1771.

BIBLIOGRAPHY

I. ORIGINAL WORKS. Rutherford's *Dissertatio inauguralis de aere fixo dicto, aut mephitico* (Edinburgh, 1772) is his only work of importance. An English translation by A. Crum Brown (communicated by L. Dobbin) is "Daniel Rutherford's Inaugural Dissertation," in *Journal of Chemical Education,* **12** (1935), 370–375. Rutherford also published a short account of a maximum and minimum thermometer, the design of which has been attributed to him but which he clearly attributes to his father, in "A Description of an Improved Thermometer," in *Transactions of the Royal Society of Edinburgh,* **3** (1794), 247–249. The authorship of two works ascribed to Rutherford by M. E. Weeks is uncertain.

II. SECONDARY LITERATURE. M. E. Weeks, "Daniel Rutherford and the Discovery of Nitrogen," in *Journal of Chemical Education,* **11** (1935), 101–107, repr. with additions in M. E. Weeks and H. M. Leicester, *Discovery of the Elements,* 7th ed. (Easton, Pa., 1968), 191–205, gives numerous references to Rutherford in the literature about Sir Walter Scott. D. McKie, "Daniel Rutherford and the Discovery of Nitrogen," in *Science Progress,* **29** (1935), 650–660, gives a translation of the relevant sections of the dissertation.

E. L. SCOTT

RUTHERFORD, ERNEST (*b.* between the settlements of Brightwater and Spring Grove, near Nelson, New Zealand, 30 August 1871; *d.* Cambridge, England, 19 October 1937), *physics, chemistry.*

Both of Rutherford's parents were taken as youngsters to New Zealand in the mid-nineteenth century. His father, James, from Perth, Scotland, acquired the skills of his wheelwright father and brought this technological inclination to his work: flax farming and processing, railroad-tie cutting, bridge construction, and small-scale farming. Although he was moderately successful in this range of endeavors, his family of a dozen children necessarily learned hard work and thrift. Rutherford's mother, Martha Thompson, accompanied her widowed mother to New Zealand from Hornchurch, Essex, England, and a few years later took over her mother's teaching post when she remarried. One need look no further than his parents for the source of Rutherford's characteristic traits of simplicity, directness, economy, energy, enthusiasm, and reverence for education.

Success in the local schools brought Rutherford a scholarship to Nelson College, a nearby secondary school. Until this time he had tinkered with clocks, made models of the waterwheels his father used in his mills, and at the age of ten had a copy of Balfour Stewart's science textbook; but he had not exhibited intellectual precocity or a predilection for a scientific career. At Nelson he excelled in nearly every subject, particularly mathematics, in which he was given a solid grounding by W. S. Littlejohn. Another scholarship took Rutherford in 1889 to Canterbury College, Christchurch, where he came under the influence of A. W. Bickerton, a man of contagious scientific enthusiasm whose cosmological theories were never taken seriously, and C. H. H. Cook, a rigorous and orthodox mathematician. At the conclusion of the three-year course Rutherford received his B.A. and a mathematical scholarship that enabled him to remain for another year. For his postgraduate work he obtained the M.A. in 1893, with double first-class honors in mathematics and mathematical physics and in physical science.

By this time Rutherford's special talents must have been apparent, for he was encouraged to stay at Canterbury for yet another year, during which he began research on the magnetization of iron by high-frequency discharges, work that earned him the B.Sc. in 1894. His activities until mid-1895 are not known for certain; but he seems to have continued this line of research under Bickerton, taught briefly at a boys' school, and fallen in love with his future wife, Mary Newton, the daughter of the woman in whose house he lodged.

In this first research Rutherford examined the magnetization of iron by a rapidly alternating electric current, such as the oscillatory discharge of a Leyden jar, and showed it to occur even with frequencies of over 10^8 cycles per second. Heinrich Hertz, less than a decade before, had caused a sensation by detecting the radio waves predicted by Maxwell's electromagnetic theory; and Rutherford, always interested in the latest scientific advances, probably was drawn to his own investigation involving high frequencies by Hertz's work. More important than Rutherford's initial observation—Joseph Henry had discovered the effect half a century earlier—was his finding that the alternating field diminished the magnetization of a needle that was already magnetized. This discovery enabled him to devise a detector of wireless signals before Marconi began his experiments, and during the next year or two Rutherford endeavored to increase the range and sensitivity of his device.

In 1895 Rutherford was awarded a scholarship established with the profits from the famous 1851 Exhibition in London. The terms of this award required attendance at another institution, and Rutherford chose Cambridge University's Cavendish Laboratory, of which the director, J. J. Thomson, was the leading authority on electromagnetic phenomena. The university had just altered its rules to admit graduates of other schools, thereby enabling Rutherford to become the laboratory's first research student. He brought with him to England his wireless wave detector and soon was able to receive signals from sources up to half a mile away. This work so impressed a number of Cambridge dons, J. J. Thomson included, that Rutherford quickly made a name for himself. Upon the discovery of X rays, Thomson asked Rutherford in early 1896 to join him in studying the effect of this radiation upon the discharge of electricity in gases.

Although he might have hesitated, for Rutherford was anxious to earn enough to marry his fiancée in New Zealand and saw a limited use for his detector in lighthouse or lighthouse-to-shore communication, he could not refuse the honor of Thomson's offer or the opportunity to investigate the most recently discovered physical phenomenon.

Out of this collaboration came a joint paper famous for its statement of a theory of ionization. The idea—that the X rays created an equal number of positive and negative carriers of electricity, or "ions," in the gas molecules—presumably was Thomson's, while much of the experimentation that placed this formerly descriptive subject on a quantitative basis was Rutherford's. The latter continued this work through 1897, measuring ionic velocities, rates of recombination, absorption of energy by gases, and the electrification of different gases while Thomson independently determined the existence of the particle later called the electron. Rutherford logically next examined the discharge of electricity by ultraviolet light, then conducted a similar study of the effects of uranium radiation. Again his inclination to pursue the most recent—and significant—problems led to a more detailed study of radioactivity. This was his field of endeavor for the next forty years; his work and that of his students was to make this the most significant area of physical science as radioactivity evolved into atomic physics and then into nuclear physics.

Radioactivity had been chanced upon in 1896 by Henri Becquerel, had enjoyed a brief period of moderate attention, and had then been abandoned even by its discoverer because it seemed relatively uninteresting among the numerous radiations being studied at the end of the nineteenth century. In early 1898 interest was somewhat revived when G. C. Schmidt and Marie Curie independently showed that not only uranium, but also thorium, exhibited this property. When Pierre and Marie Curie, with Gustave Bémont, announced later in 1898 the discovery of two new radioactive elements, polonium and radium, world scientific attention finally crystallized. Rutherford did not jump on this bandwagon, for his investigations had begun earlier, even before the discovery of thorium's activity. It is likely, in fact, that his own work alone would have served the same purpose as radium in creating the science of radioactivity, if somewhat more slowly; for within a short time Rutherford, not Becquerel or the Curies, was the dominant figure in the field.

He began by examining the Becquerel rays from

uranium. Indeed, until about 1904 the emissions received far more attention than the emitters. Passage of the radiation through foils revealed one type that was easily absorbed and another with greater penetrating ability; these Rutherford named alpha and beta, "for simplicity." While this work was in progress, Rutherford was seriously considering his future prospects. A lectureship or, even better, a fellowship at Trinity College would allow him to marry. But either a Trinity regulation about length of residence or, as he felt, the prevailing Cambridge snobbery toward those who had been undergraduates elsewhere, especially in the colonies, prevented the offer of a fellowship. With little hope of success, for older men with far greater teaching experience had also applied, Rutherford entered the competition for the professorship of physics at McGill University. The Montreal authorities, however, were looking for someone to direct work in their well-equipped laboratory and were convinced by Thomson's testimonial: "I have never had a student with more enthusiasm or ability for original research than Mr. Rutherford."

Arriving at McGill in September 1898, Rutherford found a warm welcome; perhaps the best laboratory in the western hemisphere (it was financed by a tobacco millionaire who considered smoking a disgusting habit); widespread skepticism that he would measure up in research ability to his predecessor, H. L. Callendar; and a department chairman, John Cox, who soon voluntarily assumed some of Rutherford's teaching duties when he recognized his colleague's genius. While in Cambridge, Rutherford's work in radioactivity had been solely with uranium minerals; in Montreal his first inclination was to examine thorium substances, since the activity of this element had been noticed only half a year earlier. When a colleague obtained erratic ionization measurements, Rutherford succeeded in tracing the irregularity to a gaseous radioactive product escaping from the thorium; and because he was uncertain of the nature of this product, in 1900 he gave it the deliberately vague name "emanation." Within a short time the emanations from radium and actinium also were found, by Ernst Dorn and F. Giesel, respectively.

The number of known radioelements was increasing. Rutherford added several more to the list, the next being thorium active deposit, which in time was resolved into thorium A, B, C, and so on. The active deposit, or excited activity, which was laid down on surfaces touched by the decaying emanation, was found by Rutherford because of the apparent breakdown of good insulators and

was described in *Philosophical Magazine* just one month after his announcement of the emanation. A curious feature he immediately noticed was that, unlike uranium, thorium, and radium, such materials as thorium emanation, radium emanation, their active deposits, and polonium lost their activities over periods of time. Moreover, the rate of this decrease was unique for each radioelement and thus an ideal identifying label. This meant that an exponential curve could be plotted for the half-life of each radioelement with a discernible decay period, and theory could thereby be compared with experiment.

Sir William Crookes, among others, doubted that uranium and thorium were intrinsically active; he believed, rather, that the active materials were only entrained with the atoms of these long-known elements. In 1900 he succeeded, through repeated dissolution and recrystallization of uranium nitrate, in preparing uranium that left no image on a photographic plate and in isolating the active constituent, called uranium X. But the confidence thereby generated in the stability of uranium was shaken little more than a year later, when Becquerel reexamined his materials, prepared by Crookes's method, and found that his uranium X was inactive, while his uranium had regained its activity. By this time Rutherford had recognized the need for skilled chemical assistance in his radioactivity investigations and had secured the services of a young chemistry demonstrator at McGill, Frederick Soddy. Together they removed most of the activity from a thorium compound, calling the active matter thorium X; but they too found that the X product lost its activity and that the thorium recovered its original level in a few weeks. Had Becquerel's similar finding for uranium not been immediately at hand, they might have searched for errors in their work. In early 1902, however, they began to plot the activities as a function of time, seeing evidence of a fundamental relationship in the equality of the time for thorium X to decay to half value and thorium to double in activity.

This work led directly to Rutherford's greatest achievement at McGill, for with Soddy he advanced the still-accepted explanation of radioactivity. Becquerel for several years had considered the phenomenon a form of long-lived phosphorescence, although by the first years of the twentieth century he spoke vaguely of a "molecular transformation." Crookes, in the British tradition of visualized mechanical models, had suggested a modified Maxwell demon sitting on each uranium atom and extracting the excess energy from faster-moving air molecules, this energy then appearing as uranium radiation. The Curies had considered several possibilities but inclined strongly toward the concept of an unknown ethereal radiation the existence of which is manifested only through its action on the heaviest elements, which then emit alpha, beta, and gamma rays as secondary radiations. Perhaps the most prescient idea was offered by Elster and Geitel—that the energy exhibited by radioactive substances comes not from external sources but from within the atoms themselves—but it was left to Rutherford and Soddy to add quantitative evidence to such speculation.

Their iconoclastic theory, variously called transformation, transmutation, and disintegration, first appeared in 1902 and was refined in the following year. Although alchemy had long been exorcised from scientific chemistry, they declared that "radioactivity is at once an atomic phenomenon and the accompaniment of a chemical change in which new kinds of matter are produced." The radioactive atoms decay, they argued, each decay signifying the transmutation of a parent into a daughter element, and each type of atom undergoing its transformation in a characteristic period. This insight set the course for their next several years of research, for the task was then to order all the known radioelements into decay series and to search for additional members of these families.

The theory also explained the experimental decay and rise curves as a measure of a radioelement's quantity and half-life. At equilibrium the same number of atoms of a parent transform as the number of atoms of its daughter and its granddaughter, and so on until a stable end product is obtained. But when a chemical process separates members of a series, the parent must regain its former activity as it produces additional daughters while its own numbers are maintained constant, unless it is the very first member of the family— whose numbers can only decrease. The daughter side of a chemical separation, however, is destined only to decay, for there is no means of replenishing its stock of transformed atoms.

Rutherford and Soddy saw that the apparently constant activities of uranium, thorium, and radium were due to half-lives that are long compared with human lives. This understanding overcame the puzzle at the core of all previous theories; for if the total radioactivity in the universe was growing smaller and tending to disappear, the law of conservation of energy would not be violated. They considered radioactivity a fundamental property of nature, fit to join the select group of electricity,

magnetism, light, and gravity. Not the least remarkable thing about this theory which proclaimed that the atom was not indestructible was the uncontroversial way in which it was accepted. Aside from the elderly and unalterable Lord Kelvin and the constantly contentious Henry Armstrong, the transformation theory encountered little opposition. Chemists, especially, although it violated views about the unchangeability of atoms that they "absorbed with their mothers' milk," could not refute the evidence and at most could adopt a wait-and-see attitude.

To a large degree Rutherford spent the next years mining this rich vein of interpretation. Working with Soddy and using the new liquid-air machine given to McGill by its wealthy benefactor, he condensed emanation at low temperatures, proving that it is a gas. Other tests convinced them that emanation belonged to the family of inert gases found not long before by Sir William Ramsay. Soddy then left Montreal in 1903 for London, where he and Ramsay proved spectroscopically that helium is produced during transformations from radium emanation. Such work was highly important, for there were numerous radioelements of which the chemical identity and place in the decay series were uncertain.

Helium, while not a radioelement, was of particular interest because of Rutherford's certainty that, as a positive ion, it was identical with the alpha particle. And the alpha particle, being of ponderable mass, he saw as the key in the change from an element of one atomic weight to an element of another. It fascinated Rutherford also because he could appreciate the enormous speed and energy with which it is ejected from a decaying atom. In 1903 he was able to deflect it in electric and magnetic fields, thereby showing its positive charge, but his charge-to-mass ratio measurement lacked the precision required to distinguish between a helium atom with two charges and a hydrogen atom with one charge. The proof of the particle's identity awaited Rutherford's transfer to Manchester, although he determined many useful facts about the alpha particle, such as the number emitted per second from one gram of radium, a constant that is the basis for several other important quantities, including the half-life of radium, and in 1906 made another assault upon the e/m ratio.

Halfway between Soddy's departure in 1903 and Otto Hahn's arrival at McGill in 1905, Rutherford found another chemist of comparable skill upon whom he could rely. This was Bertram Boltwood, who had proved circumstantially that uranium and radium are related, thus linking two previously separate decay series, and who in 1907 discovered ionium, the immediate parent of radium, which went far in proving the uranium-radium connection directly. Since Boltwood remained in New Haven, Connecticut, his collaboration with Rutherford was conducted through the mails. This work extended from determination of the quantity of radium present per gram of uranium in minerals to Rutherford's suggestion that, if quantity and rate of formation of a series' end product were known, it would be possible to calculate the age of the mineral. R. J. Strutt in England followed up this idea, using the helium found in radioactive substances; but the variable amount of this gas that escaped permitted only minimum age determinations. Boltwood showed the universal occurrence of lead with uranium minerals; considered this the series' final product; and, using Rutherford's value for radium's half-life and their figure for the amount of radium in a gram of uranium, was able to calculate the rate of formation for lead. The ages of some of his rock samples were over a billion years, furnishing for the first time quantitative proof of the antiquity of the earth.

Many other problems in radioactivity were pursued by Rutherford, sometimes alone, sometimes with one of the research students in the strong school he established. Among the projects in his laboratory were measurements of radiation energy, studies of beta- and gamma-ray properties, attempts to change rates of decay by extreme conditions of temperature, efforts to place actinium in a decay series, and investigations of the radioactivity of the earth and atmosphere. Few advances in this science throughout the world failed to be reflected in the work at McGill. Nor, single-minded though he was, did Rutherford entirely abandon other areas of science; radio, the conduction of electricity in gases, and N rays received some attention.

Rutherford's nine years at McGill, filled with the great work that brought him the Nobel Prize for chemistry in 1908, were no less replete with other professional activities. He was in great demand as a speaker and traveled frequently to distant parts of the United States and to England, to give a lecture, a series of talks, or a summer-session course. While he could not be expected to refuse the honor of speaking at the Royal Institution, the Bakerian lecture to the Royal Society (1904), or the Silliman lectures at Yale University (1905), some well-wishers urged him to limit his outside engagements. His time was also consumed in writing *Radio-Activity*, the first textbook on the subject

and recognized as a classic at its publication in 1904. So fast did the science progress, however, that Rutherford prepared a second edition the following year that was 50 percent larger. No sooner was this done than he faced the task of fashioning the Silliman lectures into a book. Small wonder that he confined his writing to journals for the next several years.

A veritable fallout of honors began to descend upon him, continuing for the rest of his life. Rutherford thoroughly enjoyed this recognition, for, while not vain, he was fully aware of his own worth. The Royal Society offered him fellowship in 1903 and the Rumford Medal the next year, while various universities presented both honorary degrees and job offers. While he was happy at McGill, Rutherford desired to return to England, where he would be closer to the world's leading scientific centers. Thus when Arthur Schuster offered to resign from his chair at the University of Manchester on the condition that Rutherford succeed him, the post and the laboratory were sufficiently attractive for Rutherford to make the move in 1907.

If the Cavendish, under Thomson, was the premier physics laboratory in England, Manchester, under Rutherford, was easily the second. Schuster had built a fine structure less than a decade earlier and bequeathed to his successor a strong research department, his assistant, Hans Geiger, and a personally endowed readership in mathematical physics, filled in turn by Harry Bateman, C. G. Darwin, and Niels Bohr. Rutherford's great and growing fame attracted to Manchester (and later to Cambridge) an extraordinarily talented group of research students who made profound contributions to physics and chemistry.

On his return to England, Rutherford had only a few milligrams of radioactive materials, a quantity insufficient for even his own research. In a generous gesture the Austrian Academy of Sciences sent, from the Joachimsthal uranium mines under its control, about 350 milligrams of radium chloride, as a joint loan to Rutherford and Ramsay. Unfortunately, Ramsay wished to retain possession indefinitely, while both saw the wisdom of leaving the supply undivided; so until the Vienna authorities sent another comparable radium supply for Rutherford's exclusive use, he was limited to work with the "draw" of emanation that Ramsay sent periodically from London. To a degree this determined most of Rutherford's initial investigations at Manchester, an extensive study of radium emanation; but he always found emanation and its active deposit decay products more convenient sources than radium itself.

The emanation could easily be purified in liquid air, and Rutherford soon determined the volume of this gas in equilibrium with one gram of radium. This corrected earlier results by Ramsay and A. T. Cameron, and, by confirming his calculated amount, removed some doubt cast on the accuracy of radioactive data and theory. With the spectroscopist Thomas Royds, Rutherford next photographed the spectrum of emanation, not examined since Ramsay and Collie's visual observations in 1904. Such work involved him in scientific controversy, which he usually sought to avoid; but after Soddy left Ramsay's laboratory, the latter's contributions to radioactivity were noted for their almost uniform incorrectness. Although an expert at handling minute quantities of rare gases, Ramsay never took the trouble to learn well the techniques of radioactivity. His imprecise work, coupled with a strong desire to gain priority, led him to publish quickly numerous results that Rutherford and others in this field felt compelled to correct. Further contributions to emanation studies included Rutherford's examination in 1909 of its vapor pressure at different low temperatures, and, with Harold Robinson in 1913, measurement of its heating effect.

Never one to limit the scope of his investigations—he preferred to advance across radioactivity in a wide path—Rutherford pursued "his" alpha particles in 1908. These were his favorites; the beta particles were too small and, being electrons, too common. The alphas, however, were massive, of atomic dimension; and he could clearly visualize them hurtling out of their parent atoms with enormous speed and energy. Certainly these would be the key to the physicist's classic goal: an understanding of the nature of matter. Until that time nothing had changed Rutherford's early conviction that the alpha particle was a doubly charged helium atom, but he had not succeeded in proving that belief. In 1908 he and Geiger were able to fire alpha particles into an evacuated tube containing a central, charged wire and to record single events. Ionization by collision, a process studied by Rutherford's former colleague at Cambridge, J. S. E. Townsend, caused a magnification of the single particle's charge sufficient to give the electrometer a measurable "kick." By this means they were able to count, for the first time accurately and directly, the number of alpha particles emitted per second from a gram of radium.

This experiment enabled Rutherford and Geiger to confirm that every alpha particle causes a faint but discrete flash when it strikes a luminescent zinc sulfide screen, and thus led directly to the widespread method of scintillation counting. It was also the origin of the electrical and electronic methods of particle counting in which Geiger later pioneered. But at this time the scintillation technique, now proved reliable, was more convenient. This counting work also led Rutherford and Geiger to the most accurate value of the fundamental electric charge e before Millikan performed his oil-drop experiment. They measured the total charge from a radium source and divided it by the number of alphas counted to obtain the charge per particle. Since this figure was about twice the previous values of e, they concluded that the alpha was indeed helium with a double charge. But Rutherford still desired decisive, direct proof; and here his skilled glassblower came to his aid. Otto Baumbach in 1908 was able to construct glass tubes thin enough to be transparent to the rapidly moving alpha particles yet capable of containing a gas. Such a tube was filled with emanation and was placed within a larger tube made of thicker glass. In time, alpha particles from the decaying emanation penetrated into and were trapped in the space between inner and outer tubes; and when Royds sparked the material in this space, they saw the spectrum of helium.

As in Montreal, Rutherford found chemical help in Manchester of the highest quality. Boltwood spent a year with him, during which time they redetermined more accurately the rate of production of helium by radium. By combining these results with those from the counting experiments mentioned above, they obtained Avogadro's number more directly than ever before. There were new researchers too—Alexander Russell, Kasimir Fajans, and Georg von Hevesy—fitting radioelements into the periodic table, generating information and ideas on which displacement laws and concept of isotopy would be based, and working on branching of the decay series, periods of the short-lived elements, and other radiochemical problems.

Rutherford's greatest discovery at Manchester—in fact, of his career—was of the nuclear structure of the atom. In retrospect, its origins can be seen in the slight evidences of alpha particle scattering in thin metal foils or sheets of mica, which he noticed while at McGill, and in similar scattering by air molecules in his later electrical counting experiments with Geiger. With a view to learning more about this scattering, both because it introduced experimental difficulties leading to less precise results and because it bore upon the perplexing question of the nature of alpha and beta absorption in matter, Geiger made a quantitative study of the phenomenon. Counting the scintillations produced by scattered alphas, he found that they increased with the atomic weight of the target foil and, until the particles could no longer penetrate the foil, with its thickness. Only very small angular deflections from the beam were measured and, as expected, fewer particles were bent through the larger angles.

In 1909 Rutherford and Geiger decided that Ernest Marsden, who had not yet taken his bachelor's degree, was ready for a real research problem. Much has been said of Rutherford's great insight in suggesting that Marsden look for large-angle alpha particle scattering; but, inspired though it was, it came logically from knowledge of the "diffuse" scattering still interfering with Geiger's measurement of small-angle scattering. On the other hand, Rutherford was aware that the alpha particle, being very fast and massive, was not likely to be scattered backward by the accumulated effect of a number of small deflections. His urging the experiment upon Marsden may well have been an example of his characteristic willingness to try "any damn fool experiment" on the chance that it might work. This one worked magnificently. Geiger then joined Marsden, and the two measured the exceedingly small number of particles that were deflected not only through ninety degrees, but more. Rutherford's reaction on learning of this—rather embellished over the years—has become a classic: "It was almost as incredible as if you fired a fifteen-inch shell at a piece of tissue paper and it came back and hit you."

Rutherford pondered long over the implications of this experiment, for it was early 1911 before he announced that he knew what the atom looked like. The small deflections investigated by Geiger could be reasonably explained by the theory of multiple scattering then current. This was based on the "plum pudding" model of the atom—a sphere of positive electrification in which electrons (plums) were regularly positioned—proposed by Lord Kelvin and highly refined by Thomson. The alpha particle was believed to suffer numerous collisions with the atoms of the target foil, each collision resulting in a small deflection; and a probability distribution for each angle could be calculated to compare with experiment. But for large angles the comparison failed; multiple scattering theory predicted virtually no deflections, while Geiger and Marsden found a measurable few.

Thomson's multiple-scattering theory, moreover, was challenged regarding beta particle encounters, its area of special competence: John Madsen, in Australia, obtained data on beta deflections that suggested that this type of scattering was done in a single collision. Other experiments, conducted at Manchester by William Wilson, showed that beta particles suffered inelastic collisions in their passage through matter; like the alpha particles, therefore, they gradually lost their energy, and it was possible to think that both particles experienced the same type of encounters.

By the end of 1910, Rutherford began tying these several factors into a new atomic model and theory of scattering. The alpha projectile, he said, changed course in a single encounter with a target atom. But for this to occur, the forces of electrical repulsion (or attraction—it made no difference for the mathematics) had to be concentrated in a region of 10^{-13} centimeters, whereas the atom was known to measure 10^{-8} centimeters. This meant that the atom consisted largely of empty space, with a very tiny and very dense charged nucleus at the center and opposite charges somehow placed in the surrounding void. Rutherford next calculated the probability of such single scattering at a given angle and found his predictions confirmed experimentally by Geiger and Marsden. The scientific community, however, was not impressed; this novel theory of the atom was not opposed, but largely ignored.

There were some, though, whose scientific orientation made them more likely than Rutherford to see the implications of a nuclear atom. One was Niels Bohr, who first met Rutherford in 1911 and later spent extended periods at Manchester. To Bohr it was apparent that radioactivity must be a phenomenon of the nucleus, while an element's chemical and physical properties were influenced by the electrons about this core. He brilliantly fitted chemical, radioactive, and spectroscopic data into the nuclear atom. His success in 1913 in applying quantum considerations to the orbital electron of hydrogen, thereby explaining its optical spectrum, eventually drew deserved attention to Rutherford's model of the atom. Bohr also treated heavier elements, and his attention to their electron arrangements brought the nuclear atom into chemistry.

H. G. J. Moseley was another of Rutherford's students whose work showed the fertility of the nuclear concept. In 1913, immediately after Max von Laue proved the wave nature of X rays and Rutherford's good friend at Leeds, W. H. Bragg, and his son, W. L. Bragg, who succeeded Rutherford at both Manchester and Cambridge, showed how to measure X-ray wavelengths by reflecting them from crystals, Moseley determined the wavelength of a particular line in the X-ray spectra of a large number of elements. When he organized his data according to each element's place in the periodic table, the wavelength (or frequency) of each line varied in regular steps. Only one thing, Moseley said, could change by such a constant amount: the positive charge on the atom's nucleus. Previously the organization of elements by atomic weights into the periodic table was seen by some as nothing more than fortuitous and by others as signifying a profound law of nature. It was Moseley's contribution to show that the profundity lay in the ordering of elements not by their weights but by their atomic numbers or nuclear charges, for it was precisely these charges that determined the number of orbital electrons and, hence, the chemical nature of the atom.

More was yet to come from Rutherford's school concerning the nucleus. Fajans and Soddy, both "alumni," and Russell, still at Manchester, each proposed a scheme to place the numerous radioelements into the periodic table. Russell's suggestion, a few months before Moseley's work mentioned above, and the more accurate versions that followed from the other two, stated simply that the daughter of an alpha-emitting element was two places to the left of the parent in the table, while the daughter of a beta-emitter was one place to the right. Moseley's work, showing that each place in the periodic table corresponds to a change of one nuclear charge, allowed further insight to these displacement laws, for the alpha particle bore a charge of $+2$ and the beta particle a charge of -1. That an alpha decay followed by two beta decays would lead back to the same place in the periodic table but, with a loss of about four atomic weight units, soon indicated the concept of isotopy.

Other important work was accomplished at Manchester, such as the Geiger-Nuttall rule connecting the range of an alpha particle with the average lifetime of the parent atom, beta- and gamma-ray spectroscopy, and the measurement by Rutherford and E. N. da C. Andrade in 1914 of gamma-ray wavelengths by the crystal technique. In all of these investigations Rutherford either played a direct part or kept closely abreast of developments during his daily rounds of the laboratory. At Manchester these rounds were possible, for Rutherford was largely spared time-consuming administrative duties and other chores. But he was

increasingly busy, and service on the Council of the Royal Society, the presidency of Section A of the British Association for the Advancement of Science in 1909, attendance at several overseas conferences, and his numerous lectures took him more and more away from the Midlands.

The outbreak of World War I caught Rutherford in Australia at a meeting of the British Association for the Advancement of Science. On his return to England, he found his laboratory virtually empty, for before governments found scientists useful in wartime and before conscription was introduced, many young scientists felt it was their duty to enlist for action. Rutherford himself was called upon to serve as a civilian member of the Admiralty's Board of Invention and Research committee dealing with submarine problems.

As the war progressed and hydrophone research became centralized at a naval base, Rutherford found time to return to his more customary line of investigation. A few years before, Marsden had noticed scintillations on a screen placed far beyond the range of alpha particles when these particles were allowed to bombard hydrogen. Rutherford repeated the experiment and showed that the scintillations were caused by hydrogen nuclei or protons. This was easily understood, but when he substituted nitrogen for the hydrogen, he saw the same proton flashes. The explanation he gave in 1919 stands beside the transformation theory of radioactivity and the nuclear atom as one of Rutherford's most important discoveries. This, he said, was a case of artificial disintegration of an element. Unstable, or radioactive, atoms disintegrated spontaneously; but here a stable nucleus was disrupted by the alpha particle, and a proton was one of the pieces broken off.

This line of work was to be the major theme for the remainder of Rutherford's career, which he spent at Cambridge from 1919. In that year Thomson was appointed master of Trinity College and decided to resign as director of the Cavendish Laboratory. The postwar period saw great activity in the game of professorial "musical chairs," but to no one's surprise Rutherford was elected as Thomson's successor. With him came James Chadwick, a former research student at Manchester who had spent the war years interned in Germany; he was to become Rutherford's closest collaborator. During the 1920's they determined that a number of light elements could be disintegrated by bombardment with swift alpha particles; as a corollary, they measured the distance of closest approach between projectile and target to ascertain both the size of the nucleus and that the inverse-square force law applied at this small distance.

There was no doubt that the alpha particle caused such elements as nitrogen, boron, fluorine, sodium, aluminum, and phosphorus to disintegrate. But did the alpha merely bounce off the target nucleus, which then emitted a proton, or did it combine with this nucleus? While these two reactions would form different elements, the number of atoms was too small for chemical tests. C. T. R. Wilson, who still worked independently in the Cavendish Laboratory, had perfected a cloud chamber before the war that was an ideal instrument to resolve this problem. If the alpha bounced off its target, there would be three tracks diverging from the collision point: the alpha, the proton, and the recoil nucleus. If, on the other hand, the alpha and the target formed a compound nucleus, there would be only two trails: the proton and the compound nucleus. From the photographs of some 400,000 alpha-particle tracks, P. M. S. Blackett in 1925 showed that it was the latter process which occurred, for he found eight doubly branched tracks from nuclear collisions. Later work at the Cavendish by Blackett and G. P. S. Occhialini, with cosmic rays triggering coincidence counters and thus photographing themselves in a cloud chamber, confirmed the discovery of the positron, made shortly before by Carl D. Anderson in California.

With its charge of +2, the alpha particle was too strongly repelled by the large numbers of positive charges on the nuclei of the heavier elements to cause disintegrations. To overcome this potential barrier, Rutherford recognized that projectiles might be accelerated but disregarded the alpha and proton because electrical engineers in the 1920's could not furnish the voltage required even to match the energy of alphas from natural radioactive sources. Instead, he inclined toward the idea of electron acceleration, with the thought that this projectile, once past the orbital electrons of similar charge, would be attracted to the nucleus. In the last few years of the 1920's, T. E. Allibone, one of a growing number at the Cavendish with engineering training—a new trend in physics—attempted disintegrations with accelerated electrons, but without success. George Gamow, on one of his several visits to Cambridge, then pointed out that the new wave mechanics predicted that a small number of particles of relatively low energy could tunnel through the potential barrier around a nucleus instead of climbing over it. This put the matter of effecting disintegrations by accelerated particles

of positive charge (and with more mass than the electron) back into the range of laboratory possibilities.

John Cockcroft and E. T. S. Walton built an apparatus capable of accelerating protons through several hundred thousand volts, with which, in 1932, they succeeded in bombarding lithium and producing alpha particles. By measuring the energy of these products, they further offered experimental proof of Einstein's famous relationship $E = mc^2$. The mass values were furnished with great accuracy by another long-time Cavendish member, F. W. Aston, who, like C. T. R. Wilson, worked to a large degree independently. In 1919 Rutherford had produced artificial disintegrations by natural means, that is, by alphas from naturally decaying radioactive materials. The Cockcroft-Walton work of 1932, artificial disintegrations by artificial means—that is, by accelerated particles—made Rutherford a believer in the quantum mechanics on which it was based. Although he was not generally interested in highly mathematical physical theories, especially ones difficult to visualize, these new ideas of the late 1920's worked—and that was Rutherford's criterion.

An instance of his willingness to use, if not fully understand, quantum mechanics was Rutherford's construction with Marcus Oliphant of a special discharge tube that generated a far more copious supply of protons than the Cockcroft-Walton apparatus and at lower voltages. His faith in the ability of these protons to tunnel through the potential barrier at these lower energies was rewarded with a number of disintegrations. But the heavier elements still resisted such bombardments; and it was clear that, for them, projectiles of greater energy were required. Ernest Lawrence, in California, had built the cyclotron a few years before and generously shared his plans with Rutherford. A new high-voltage laboratory was also planned for the Cavendish, to house a two-million-volt, commercially built apparatus; but neither of these heralds of the new age of "big science and big money" was in significant use by the time of Rutherford's death in 1937.

There were other important activities at the Cavendish, some with direct connections to Rutherford's main interest in disintegrations and others with no connection at all. Among the latter were the work of G. I. Taylor on problems in classical physics, E. V. Appleton on radio waves, and Peter Kapitza on phenomena in intense magnetic fields and at low temperatures. More in the mainstream of the laboratory's orientation was the long series of investigations by C. D. Ellis on beta- and gamma-ray spectra. Since these rays, as well as the alpha, come from the decaying nucleus, they offered a view of the energy levels in this nucleus and, hence, an insight to nuclear structure.

Not long after the discovery of heavy water in the United States, Rutherford obtained a small quantity of this precious fluid and in 1934, with Oliphant and Paul Harteck, bombarded deuterium with deuterons. This reaction was notable for the first achievement of what is now called fusion, as well as for the production of tritium. Another major advance at the Cavendish, the significance of which is not sufficiently appreciated, was the application by C. E. Wynn-Williams of Heinrich Greinacher's ideas for electronic amplification of ionization. With the Geiger-Müller tube, which was based on a different principle, and especially with Wynn-Williams' tubes and associated electronics, research workers in the laboratory were able to count particles at much higher rates than with scintillations and with other benefits.

Rutherford recognized the value to his experimentalists of contact with theoretical physicists and encouraged their presence in the laboratory. Some, such as Gamow, came as visitors for a period. The Cambridge theoreticians, however, were by administrative fiat in the mathematics department and were somewhat isolated from the Cavendish. A notable exception was Ralph Fowler, Rutherford's son-in-law, whose advice was eagerly sought during the nearly two decades that Rutherford directed the laboratory. Along with the Cockcroft-Walton experiment, the most important discovery during this period by one of Rutherford's colleagues was Chadwick's proof, in 1932, of the existence of the neutron. Rutherford had long considered the neutron a possibility and in his 1920 Bakerian lecture to the Royal Society had predicted its likely properties. Chadwick made several attempts to detect the neutral particle, but none was successful until he learned of experiments by the Joliot-Curies in Paris, in which, they said, extremely penetrating gamma rays were emitted. As he suspected, Chadwick found the rays were not gammas but neutrons; and not long afterward Norman Feather, also at the Cavendish, showed that neutrons were capable of causing nuclear disintegrations.

Even by the beginning of Rutherford's second period at Cambridge, he was a public figure. Increasingly beset with outside calls upon his time, he had less and less opportunity for his own research and for keeping abreast of his students'

work. Yet, with the tradition of enthusiasm for research that he had established earlier, his still frequent rounds to "ginger up" his "boys," and Chadwick's invaluable assistance, the laboratory's output remained far more than respectable.

From 1921, when he succeeded Thomson, until his death, Rutherford was professor of natural philosophy at the Royal Institution in London, a post that entailed several lectures each year. There were numerous other public lectures to which great honor was attached, such as his presidential address to the British Association for the Advancement of Science in 1923. Between 1925 and 1930 he was president of the Royal Society, and following this he became chairman of the advisory council to the British government's Department of Scientific and Industrial Research. Both posts involved many public appearances, such as opening conferences and new laboratories, in addition to administrative and policy-making chores. Although liberal-minded, Rutherford customarily side-stepped political issues. Yet he felt he could not remain idle when Nazi Germany expelled hundreds of Jewish scholars; and from 1933 he was president of the Academic Assistance Council, which sought to obtain positions and financial aid for these refugees.

In work that may be characterized as radioactivity at McGill, atomic physics at Manchester, and nuclear physics at Cambridge, Rutherford, more than any other, formed the views now held concerning the nature of matter. It is to be expected that numerous honors would come to such a man, called the greatest experimental physicist of his day and often compared with Faraday. In 1922 he received the Copley Medal, the highest award given by the Royal Society. Dozens of universities and scientific societies awarded him honorary degrees and memberships. For the fame Rutherford brought to the British Empire—not for his relatively minor services to his government—he was made a knight in 1914 and a peer (Baron Rutherford of Nelson) in 1931. King George V personally honored him in 1925 by conferring on him the Order of Merit, which is limited to a handful of the most distinguished living Englishmen.

BIBLIOGRAPHY

I. ORIGINAL WORKS. Approximately two dozen boxes of Rutherford's correspondence and miscellaneous papers are preserved in the Cambridge University Library. Most of his published papers have been reprinted, under the scientific direction of Sir James Chadwick, in *Collected Papers of Lord Rutherford of Nelson*, 3 vols. (London, 1962–1965). Rutherford's books are *Radio-Activity* (Cambridge, 1904; 2nd ed., 1905); *Radioactive Transformations* (London, 1906); *Radioactive Substances and Their Radiations* (Cambridge, 1913); *Radiations From Radioactive Substances* (Cambridge, 1930), written with J. Chadwick and C. D. Ellis; and *The Newer Alchemy* (Cambridge, 1937). A portion of his correspondence is reproduced in Lawrence Badash, ed., *Rutherford and Boltwood, Letters on Radioactivity* (New Haven, 1969). Badash has also compiled the *Rutherford Correspondence Catalog* (New York, 1974).

II. SECONDARY LITERATURE. There are three biographies written by Rutherford's former students: A. S. Eve, *Rutherford* (Cambridge, 1939); Norman Feather, *Lord Rutherford* (London, 1940); and E. N. da C. Andrade, *Rutherford and the Nature of the Atom* (London, 1964). Other biographies and partial biographies include Ivor Evans, *Man of Power, the Life Story of Baron Rutherford of Nelson, O.M., F.R.S.* (London, 1939); John Rowland, *Ernest Rutherford, Atom Pioneer* (London, 1955); Robin McKown, *Giant of the Atom, Ernest Rutherford* (New York, 1962); John Rowland, *Ernest Rutherford, Master of the Atom* (London, 1964); D. Danin, *Rutherford* (Moscow, 1966), in Russian; O. A. Staroselskaya-Nikitina, *Ernest Rutherford, 1871–1937* (Moscow, 1967), in Russian; E. S. Shire, *Rutherford and the Nuclear Atom* (London, 1972); and, especially valuable for personal information, Mark Oliphant, *Rutherford, Recollections of the Cambridge Days* (Amsterdam, 1972).

Collections of articles about Rutherford include *Rutherford by Those Who Knew Him*, which is the first five Rutherford lectures of the Physical Society, by H. R. Robinson, J. D. Cockcroft, M. L. Oliphant, E. Marsden, and A. S. Russell, reprinted from *Proceedings of the Physical Society*, 1943–1951; a series of Rutherford memorial lectures, by J. D. Cockcroft, J. Chadwick, E. Marsden, C. Darwin, E. N. da C. Andrade, P. M. S. Blackett, T. E. Allibone, and G. P. Thomson, in *Proceedings of the Royal Society*, from 1953 on; J. B. Birks, ed., *Rutherford at Manchester* (London, 1962); Albert Parry, ed., *Peter Kapitsa on Life and Science* (New York, 1968); *Notes and Records of the Royal Society*, 27 (Aug. 1972), an issue devoted to Rutherford, with articles by M. L. Oliphant, H. Massey, N. Feather, P. M. S. Blackett, W. B. Lewis, N. Mott, P. P. O'Shea, and J. B. Adams; P. L. Kapitza, ed., *Rutherford—Scholar and Teacher, On the Hundredth Anniversary of his Birth* (Moscow, 1973), which is in Russian.

Among the wide range of articles written about Rutherford during his lifetime, obituary notices, recollections, and historical studies are the following: A. S. Eve, "Some Scientific Centres. VIII. The Macdonald Physics Building, McGill University, Montreal," in *Nature*, 74 (1906), 272–275; J. A. Harker, "Some Scientific Centres. XI. The Physical Laboratories of Manchester University," in *Nature*, 76 (1907), 640–642; "The Ex-

tension of the Physical and Electrotechnical Laboratories of the University of Manchester," in *Nature*, 89 (1912), 46; N. Bohr, "Sir Ernest Rutherford, O.M., P.R.S.," in *Nature Supplement*, 118 (1926), 51–52; O. Hahn and L. Meitner, "Lord Rutherford zum Sechzigsten Geburtstag," in *Die Naturwissenschaften*, 19 (1931), 729; M. de Broglie, "Scientific Worthies: The Right Hon. Lord Rutherford of Nelson, O.M., F.R.S.," in *Nature*, 129 (1932), 665–669; and J. G. Crowther, "Lord Rutherford, O.M., F.R.S.," in *Great Contemporaries* (London, 1935), pp. 359–370.

Obituary notices are in the *Times* (London) 20, 21, 22, and 26 Oct. 1937; *New York Times*, 20 Oct. 1937 and 21 Jan. 1938; *Nature*, 140 (1937), 717, 746–755, 1047–1054; and 141 (1938), 841–842.

See also A. S. Russell, "More About Lord Rutherford," in *The Listener*, 18 (1937), 966; A. N. Shaw, "Rutherford at McGill," in the *McGill News* (Winter 1937), no pagination; C. M. Focken, "Lord Rutherford of Nelson, a Tribute to New Zealand's Greatest Scientist," a 19-page brochure (privately printed in New Zealand, n.d., but *ca.* 1938); the obituary notice by R. A. Millikan, in *Yearbook of the American Philosophical Society for 1938*, 386–388; F. R. Terroux, "The Rutherford Collection of Apparatus at McGill University," in *Transactions of the Royal Society of Canada*, 32 (1938), 9–16; the obituary notice by E. F. Burton, in *University of Toronto Quarterly*, 7 (1938), 329–338; A. George, "Lord Rutherford ou l'Alchimiste," in *La Revue de France*, (1938), 525–533; the obituary notice by G. Guében, in *Revue des Questions Scientifiques*, 113 (1938), 5–19; the obituary notice by A. S. Eve and J. Chadwick, in *Obituary Notices of the Royal Society of London*, 2 (1938), 395–423; H. Geiger, "Memories of Rutherford in Manchester," in *Nature*, 141 (1938), 244; H. Geiger, "Das Lebenswerk von Lord Rutherford of Nelson," in *Die Naturwissenschaften*, 26 (1938), 161–164; and obituary notices in *Proceedings of the Physical Society*, 50 (1938), 441–466.

Other articles are an obituary notice by E. Marsden, in *Transactions and Proceedings of the Royal Society of New Zealand*, 68 (1938), 4–16, to which is appended a partial bibliography compiled by C. M. Focken, pp. 17–25; "51 Years as Laboratory Steward," an interview with W. Kay, in the *Manchester Guardian*, 27 Dec. 1945; H. Tizard, "The Rutherford Memorial Lecture," in *Journal of the Chemical Society* (1946), 980–986; "Rutherford Commemoration, Paris, 7 and 8 November 1947," in *Notes and Records of the Royal Society*, 6 (1948), 67–68; H. Dale, "Some Personal Memories of Lord Rutherford of Nelson," in *Cawthron Lecture Series*, no. 25 (1950); P. M. S. Blackett, "Rutherford and After," in the *Manchester Guardian Weekly*, 63 (14 Dec. 1950), 13; E. N. da C. Andrade, "The Birth of the Nuclear Atom," in *Scientific American*, 195 (1956), 93–104; C. P. Snow, "The Age of Rutherford," in *Atlantic Monthly*, 202 (Nov. 1958), 76–81; C. D. Ellis, "Rutherford; One Aspect of a Complex Character," in *Trinity Review* (Lent 1960), 13–15; J. E. Geake, "Rutherford in

Manchester," in *Contemporary Physics*, **3** (1961), 155–158; "The Jubilee of the Nuclear Atom," in *Endeavour*, **21** (1962), 3–4; N. Feather, an essay-review of volume one of Rutherford's collected papers, in *Contemporary Physics*, **4** (1962), 73–76; W. A. Kay, "Recollections of Rutherford. Being the Personal Reminiscences of Lord Rutherford's Laboratory Assistant, Here Published for the First Time. Recorded and Annotated by Samuel Devons," in *The Natural Philosopher*, **1** (1963), 127–155; W. E. Burcham, "Rutherford at Manchester, 1907–1919," in *Contemporary Physics*, **5** (1964), 304–308; T. H. Osgood and H. S. Hirst, "Rutherford and His Alpha Particles," in *American Journal of Physics*, **32** (1964), 681–686; P. L. Kapitza, "Recollections of Lord Rutherford," in *Proceedings of the Royal Society*, **A294** (1966), 123–137; M. L. Oliphant, "The Two Ernests" [Rutherford and Lawrence], in *Physics Today*, **19** (Sept. 1966), 35–49, (Oct. 1966), 41–51; L. Badash, "How the 'Newer Alchemy' Was Received," in *Scientific American*, **215** (1966), 88–95; L. Badash, "Rutherford, Boltwood, and the Age of the Earth: The Origin of Radioactive Dating Techniques," in *Proceedings of the American Philosophical Society*, **112** (1968), 157–169; J. L. Heilbron, "The Scattering of α and β Particles and Rutherford's Atom," in *Archive for History of Exact Sciences*, **4** (1968), 247–307; L. Badash, "The Importance of Being Ernest Rutherford," in *Science*, **173** (1971), 873; T. Trenn, "Rutherford and Soddy: From a Search for Radioactive Constituents to the Disintegration Theory of Radioactivity," in *RETE Strukturgeschichte der Naturwissenschaften*, **1** (1971), 51–70; and T. Trenn, "The Geiger-Marsden Scattering Results and Rutherford's Atom, July 1912 to July 1913: The Shifting Significance of Scientific Evidence," in *Isis*, **65** (1974), 74–82.

For information about some of Rutherford's colleagues, see the various articles in the *DSB* and Albert Parry, ed., *Peter Kapitsa on Life and Science* (New York, 1968); *Sir Ernest Marsden. 80th Birthday Book* (Wellington, New Zealand, 1969); and Robert Reid, *Marie Curie* (New York, 1974).

Various aspects of the history of radioactivity may be found in the following selections: T. W. Chalmers, *A Short History of Radio-Activity* (London, 1951); Alfred Romer, *The Restless Atom* (Garden City, New York, 1960); A. Romer, ed., *The Discovery of Radioactivity and Transmutation* (New York, 1964); L. Badash, *The Early Developments in Radioactivity, With Emphasis on Contributions From the United States* (Ph.D. diss., Yale University, 1964); L. Badash, "Radioactivity Before the Curies," in *American Journal of Physics*, **33** (1965), 128–135; L. Badash, "Chance Favors the Prepared Mind: Henri Becquerel and the Discovery of Radioactivity," in *Archives Internationales d'Histoire des Sciences*, **18** (1965), 55–66; L. Badash, "The Discovery of Thorium's Radioactivity," in *Journal of Chemical Education*, **43** (1966), 219–220; L. Badash, "Becquerel's 'Unexposed' Photographic Plates," in *Isis*, **57** (1966), 267–269; L. Badash, "An Elster and Geitel Failure:

Magnetic Deflection of Beta Rays," in *Centaurus*, **11** (1966), 236–240; A. Romer, ed., *Radiochemistry and the Discovery of Isotopes* (New York, 1970); Marjorie Malley, "The Discovery of the Beta Particle," in *American Journal of Physics*, **39** (1971), 1454–1460; and Thaddeus J. Trenn, *The Rise and Early Development of the Disintegration Theory of Radioactivity* (Ph.D. diss., University of Wisconsin, 1971).

LAWRENCE BADASH

RUTHERFURD, LEWIS MORRIS (*b.* New York, N.Y., 25 November 1816; *d.* Tranquility, New Jersey, 30 May 1892), *astrophysics*.

As early as his student days at Williams College, Rutherfurd displayed an interest in science, assisting in the chemistry course. Initially he was destined for the law and studied with William H. Seward. He came from a prominent family and his independent means, subsequently augmented by marriage into the wealthy Stuyvesant family, must have made him seem like a few other members of what was an emerging American patriciate, a dilettantish amateur. Samuel Ward and J. P. Morgan the elder, for example, both gave up youthful interests in mathematics for more lucrative careers. The Rutherfurd who was on the yacht *America* during its challenge of the British must have appeared to superficial observers an unpromising candidate for honors in astrophysics. But he had already published his first scientific paper.

Freed from the need to practice law by his marriage, Rutherfurd traveled abroad for seven years, partly because of his wife's health. In Florence he associated with Amici, who was known for his work in both microscopy and optical astronomy. When Rutherfurd returned to the United States, he had an observatory constructed in 1856 on the Stuyvesant family estate, in what is now New York City's Lower East Side. Here he worked on astronomical photography and spectroscopy. Rutherfurd sometimes made his own instruments, work at which he was very skilled; more frequently the instruments were constructed by others according to his specifications.

In 1858 Rutherfurd started working on astronomical photography, using an 11.5-inch achromatic refracting telescope made by Henry Fitz. Although placing the plates at the actinic focus produced fine lunar photographs as well as images of Jupiter, Saturn, the sun, and stars of the fifth magnitude, Rutherfurd tried various expedients to obtain better photographs before he completely omitted the visual element from the telescope. Starting in December 1864 he employed a new 11.5-inch objective lens useful solely for photography. The resulting pictures of the moon were widely admired. In 1865 Rutherfurd began to photograph star clusters in order to map the heavens. His friend Benjamin Apthorp Gould collaborated in reducing the data for the Pleiades and the Praesepe. Rutherfurd also devised a micrometer for measuring the stellar photographs. When doubts were expressed concerning the stability of the photographic plates, particularly in connection with the proposed observations of the transits of Venus, Rutherfurd published results of tests of albuminized glass plates with wet collodion film (1872). After 1868 Rutherfurd used a thirteen-inch refractor with an exterior photographic corrective lens.

In 1861 O. W. Gibbs called Rutherfurd's attention to the spectroscopic work of Bunsen and Kirchhoff. By the next year Rutherfurd had made spectroscopic studies of the sun, moon, Jupiter, Mars, and sixteen fixed stars. From the last he independently gave a stellar spectra classification quite similar to Secchi's. Rutherfurd also used the spectroscope for color correction of telescope lenses.

His first spectroscopic observations used a cylindrical lens between a prism and the objective of the 11.25-inch Fitz telescope. In the winter of 1862–1863, Rutherfurd developed a spectroscope using glass prisms filled with carbon disulfide and ingenious devices for maintaining equal density of the disulfide and for adjusting the prisms. At the January 1864 meeting of the National Academy of Sciences, Rutherfurd displayed a never-published photograph of the solar spectrum 15 centimeters wide and 78.7 centimeters between lines H and F (according to Gould), with three times the lines given by Bunsen and Kirchhoff.

Encouraged by Gibbs and Ogden Rood of Columbia University, who were also working in this area, Rutherfurd turned in 1863 to diffraction gratings. An early ruling engine proved inadequate; in 1867 Rutherfurd devised another, in which a screw, rather than levers, powered by a turbine run by tap water, moved the plate. The gratings he produced were superior to the best then made (by Nobert), and by 1877 they were available with up to 17,296 lines to the inch. The earliest gratings were on glass; later ones had rulings on speculum metal. Rutherfurd freely and widely distributed his gratings, which were unsurpassed until the work of Rowland.

When his health began to fail in 1877, Rutherfurd started to dismantle his observatory, making his last observations in 1878. The growth of the city, in any event, made precision work at that location extremely difficult. A trustee of Columbia University, in 1881 Rutherfurd helped the university found its department of geodesy and practical astronomy; and in 1883 he donated the equipment of his observatory. In 1890 he transferred twenty volumes of plate measures and a large collection of his photographic plates that provided intellectual employment for the university observatory, named in his honor, for many years.

BIBLIOGRAPHY

I. ORIGINAL WORKS. Rutherfurd's writings include "Observations During the Lunar Eclipse, 12 September 1848," in *American Journal of Science*, n.s. **6** (1848), 435–437; "Astronomical Observations With the Spectroscope," *ibid.*, n.s. **35** (1863), 71–77; "Letter on a Companion to Sirius, Stellar Spectra, and the Spectroscope," *ibid.*, 407–409; "Observations on Stellar Spectra," *ibid.*, n.s. **36** (1863), 154–157; "On the Construction of the Spectroscope," *ibid.*, n.s. **39** (1865), 129–132; "Astronomical Photography," *ibid.*, 304–309; "On the Stability of the Collodion Film," *ibid.*, 3rd ser., **6** (1872), 430–433; "A Glass Circle for the Measurement of Angles," *ibid.*, 3rd ser., **12** (1876), 112–113. See also the correspondence of O. W. Gibbs at the Franklin Institute, Philadelphia, and the Ogden Rood correspondence at the Columbia University Library, New York City.

II. SECONDARY LITERATURE. The recent and enlightening article by Deborah Jean Warner, "Lewis M. Rutherfurd: Pioneer Astronomical Photographer and Spectroscopist," in *Technology and Culture*, **12** (1971), 180–216, is the best introduction to his work. Still useful are Benjamin Apthorp Gould, "Memoir of Lewis Morris Rutherfurd," in *Biographical Memoirs. National Academy of Sciences*, **3** (1895), 417–441; and John K. Rees, "Lewis Morris Rutherfurd," in *Astronomy and Astro-Physics*, **11** (1892), 689–697. The results of the work of the Columbia University's Rutherfurd Observatory on the materials presented by Rutherfurd are in John K. Rees, Harold Jacoby, Herman S. Davis, and Frank Schlesinger, *Lewis Morris Rutherfurd, a Brief Account of His Life and Work . . . ,* 2 vols. (New York, 1898–1919).

NATHAN REINGOLD

RÜTIMEYER, KARL LUDWIG (*b.* Biglen, Bern Canton, Switzerland, 26 February 1825; *d.* Basel, Switzerland, 25 November 1895), *vertebrate paleozoology, geography.*

The scion of an ancient cantonal family, Rüti-meyer was intended for the ministry. His parents, Albrecht Rütimeyer, a pastor in Biglen, and Marie Margaretha Küpfer, subsequently moved to Bern, where Rütimeyer studied theology and then medicine at the University of Bern. His early interests in natural history and geology were stimulated under the fascinating influence of Studer, who taught geology at the university. Rütimeyer's field studies in the 1840's of the Bernese Oberland and of the Solothurn Jura, as well as his correspondence with Murchison, led to his dissertation on the Swiss nummulitic terrain. In 1850 he received the doctorate in medicine.

Following three lively years of geological, zoological, and clinical study and experience in Paris, London, Leiden, and Italy, Rütimeyer accepted (1853) an extraordinary professorship in comparative anatomy at the university in Bern. Apparently his reception at Bern was somewhat mixed—perhaps because of his stance during the Sonderbund. He subsequently resigned from this post and spent part of 1854 with Murchison in London, where a Himalayan expedition reached the planning stages. Late in 1855 he was named ordinary professor of zoology and comparative anatomy at the University of Basel, thus beginning a distinguished teaching and research career that spanned almost four decades. He was later named rector (1865) and then professor in the medical and philosophical faculties, having received in 1874 a doctorate in philosophy *honoris causa*.

Rütimeyer made significant contributions to the natural history and evolutionary paleontology of ungulate mammals, especially the artiodactyls. His comparative odontography of ungulates (1863) was perhaps the first serious attempt after Darwin's *Origin* to interpret fossil mammals as parts of evolutionary lineages by showing the gradual change in dentitions. Rütimeyer discovered that ungulate milk teeth are conservative and thus closer in character to those of their nearest known ancestors than to permanent dentitions that are dissimilar in series. The significance that Rütimeyer attached to dental characters for phylogenetic interpretations preceded the more explicit series of horses and other ungulates proposed in the 1870's by Huxley, Kovalevsky, and Marsh.

A cautious developmentalist, Rütimeyer did not accept the Darwinian explanation of natural selection. It was likely too mechanistic a concept for a theologically trained biologist who held the widest view of natural history and whose writings reflect early influences of Humboldt's *Kosmos,* vertebrate body plans, and something of Karl von Baer's *Naturphilosophie.*

Rütimeyer's work advanced the study of mammalian evolution and biogeography. His investigation of the fauna of the Swiss lake dwellings (1862) appears to have been as significant for his subsequent researches as Darwin's interest in variation under domestication was for his ideas of transmutation of species. Rütimeyer's researches into the natural history, comparative osteology, evolutionary patterns, and paleozoogeography of perissodactyls, suids, and ruminants, as well as his studies of the diverse Eocene fauna of Egerkingen and of fossil turtles, earned him world renown. Many of his findings were used by Karl von Zittel in the latter's contemporary, general paleontological treatises.

In 1869 Rütimeyer published a significant analysis of Swiss valley and lake origin. To the earlier fluvialist explanations of Hutton and Lyell, he added an actualistic concept of valley development by headward stream erosion. Rütimeyer emphasized that varying rates of erosion, acting over long time intervals on differing geographies and rock types along stream courses, will produce diverse landforms. These forms might then be classified by relative age, a concept foreshadowing the Davisian tradition in geomorphology.

Rütimeyer traveled extensively in Europe during the 1870's and 1880's, adding to his experience and collections. In Basel he actively and influentially promoted and served both national and civic academic, natural history, paleontological, anthropological, and mountaineering societies and museums. With the brothers Sarasin, he aided Swiss conservation efforts. Rütimeyer retired from the university faculty late in 1893, but he retained his membership in the public libraries commission of Basel and also his directorship of its natural history museums. He continued his researches until his death. Rütimeyer was survived by his wife, Charlotte Laura Fankhauser, whom he had married in 1855, and their only child, Ludwig Leopold, who later became extraordinary professor of anthropology at Basel.

BIBLIOGRAPHY

I. Original Sources. Rütimeyer's autobiographical sketch is in *Gesammelte Kleine Schriften allgemeinen Inhalts aus dem Gebiete der Naturwissenschaft . . . ,* Hans Georg Stehlin, ed., I (Basel, 1898), in which several of his zoological papers are reprinted. Vol. II contains four of his earlier physiographical works, a quartet of memorials, and a bibliography. The Staatsarchiv in Basel contains complete correspondence between Rütimeyer and the cantonal government concerning his position as university professor and head of the Vergleichend-anatomische Sammlungen in the Museum of Natural History.

The Archiv also holds correspondence between Rütimeyer and Felix, Paul, and Fritz Sarasin, and Rudolf Staehelin-Stockmeyer. Most of his important letters, which are held by his granddaughter Dr. Elizabeth Rütimeyer and others of the family, are in *Brief von Ludwig Rütimeyer (1825–1895) als Manuscript gedruckt* (Basel, 1902) and "Ludwig Rütimeyer: Brief und Tagebuchblätter," both edited by L. L. Rütimeyer; the latter was published in L. E. Iselin and P. Sarasin, *Einleitung: Lebens- und Charakterbild Rütimeyers* (Frauenfeld, 1906), The University of Basel library holds a few letters and student notes from several of Rütimeyer's courses.

Rütimeyer's paleontological and zoological writings are cited in Alfred Sherwood Romer, *et al., Bibliography of Fossil Vertebrates Exclusive of North America, 1509–1927.* II, *L-Z*, which is *Memoirs. Geological Society of America,* **87** (1962), 782, 1191–1193. Major works include: "Die fauna der Pfahlbauten der Schweiz," in *Neue Denkschriften der Allgemeinen Schweizerischen Gesellschaft der gesammten Naturwissenschaften,* **19,** no. 1 (1862); "Eocaene Saugethier aus dem Gebiet des schweizerischen Jura," *ibid.,* **19,** no. 3 (1862); "Beiträge zur Kenntniss der fossilen Pferd und zur vergleichenden Odontographie der Hufthiere überhaupt," in *Verhandlungen der Naturforschenden Gesellschaft in Basel,* **3** (1863 [1862]), 558–696, pls. I–IV; *Die Grenzen der Thierwelt. Eine Betrachtung zu Darwin's Lehre* (Basel, 1868), dedicated to Karl von Baer; *Ueber die Art des Fortschrittes in den organischen Geschöpfen* (Basel, 1876); "Ueber einige Beziehungen zwischen den Säugetierstämmen alter und neuer Welt. Erster Nachtrag . . . Egerkingen," in *Abhandlungen der Schweizerischen Paläontologischen Gesellschaft,* **15** (1888), 1–63, pl. 1; "Uebersicht der eocaenen Fauna von Egerkingen. Nebst einer Erwiderung von Prof. E. D. Cope. Zweiter Nachtrag . . . Egerkingen (1862)," in *Verhandlungen der Naturforschenden Gesellschaft in Basel,* **9** (1890), 331–362; and, "Die eocaenen Säugethierwelt von Egerkingen. Gesammtdarstellung und dritter Nachtrag . . . 'Eocaenen Säugethieren . . . (1862),'" *op. cit.,* **18** (1891), 153 p., 8 pls.

Rütimeyer's writings in other fields are *Vom Meer bis nach den Alpen. Schilderungen von Bau, Form und Farbe unseres Continentes auf einem Durchschnitt von England bis Sicilien* (Basel, 1854), also published as *Van de Zee tot de Alpen* (Doesburgh, 1857); *Ueber Thal- und Seebildung. Beiträge zum Verständniss der Oberfläche der Schweiz* (Basel, 1869); "Ein Blick auf die Geschichte der Gletscherstudien in der Schweiz," in *Jahrbuch der Schweizer Alpenclub,* **16** (1881), 377–418; and *Entstehung und Verlauf der Vermessung des Rhonegletschers* (Basel, 1896).

Rütimeyer's verse is collected in *Gedichte von Ludwig Rütimeyer,* L. L. Rütimeyer, ed. (Basel, 1901).

II. SECONDARY LITERATURE. An evaluation of Rütimeyer's papers is Antoine Wahl, *L'Oeuvre géographique de L. Rütimeyer. Une analyse critique* (Fribourg, 1927). Of the ten memorials and obituaries listed in *Gesammelte Kleine Schriften,* II, the most extensive are L. E. Iselin, "Carl Ludwig Rütimeyer," in *Basler Jahrbuch 1897* (1897), 1–47; and Carl Schmidt, "Ludwig Rütimeyer," in *Basler Nachrichten,* 3–7 December 1895, repr. with modifications in *Verhandlungen der Schweizerischen Naturforschenden Gesellschaft, 78 Jahresversammlung . . . 1895* (1896), 213–256.

Brief sketches and evaluations are Eduard His, "Karl Ludwig Rütimeyer 1825–1895," in His, ed., *Basler Gelehrte des 19. Jahrhunderts* (Basel, 1941), 202–212; Adolf Portmann, "Ludwig Rütimeyer 1825–1895 . . . ," in Andrew Staehlin, ed., *Professoren der Universität Basel aus fünf Jahrhunderten. Bildnisse und Würdigungen . . .* (Basel, 1960), 160–161; H. G. Stehlin, "Karl Ludwig Rütimeyer aus Bern, 1825–1895," in Eduard Fueter, ed., *Grosse Schweizer Forscher . . .* (Zurich, 1941), 270–271; and Ewald Wust, "Ludwig Rütimeyer (1825–1895) als Begründer der historischen Paläontologie," in *Palaeontologische Zeitschrift,* **8,** no. 1/2 (1927), 34–39.

CLIFFORD M. NELSON

RUYSCH, FREDERIK (*b.* The Hague, Netherlands, 23 March 1638; *d.* Amsterdam, Netherlands, 22 February 1731), *botany, obstetrics, anatomy, medicine.*

Ruysch was descended from an old and notable family whose members held posts in various city governments, including those of Utrecht and Amsterdam. His great-great-grandfather was councillor to the bishop of Liège; his great-grandfather was councillor to the duke of Arensberg and, later, pensionary of Amsterdam; and his grandfather was secretary of the audit office. The Dutch war with Spain seems to have brought a change in 1576 in the fortunes of the family.

Ruysch was the son of Hendrik Ruysch, a secretary in the service of the state, and Anna van Berchem. Probably he attended the grammar school in The Hague. The early death of his father may explain Frederik's apprenticeship as a boy in an apothecary's shop. In 1661, even though not yet admitted to the apothecaries' guild, he prepared drugs and opened a shop in The Hague. The board of the guild ordered him to close the shop and forbade him to sell remedies to anyone until he had successfully passed the necessary examination, as he did on 16 and 17 June 1661. He was then admitted *confrater* in the guild and quickly reopened his shop. In the same year he married Maria Post, daughter of Pieter Post, the well-known architect of Frederik Henry, prince of Orange. Of this marriage many (perhaps twelve) children were born; but only two are known: Hendrik and Rachel, who married the painter Jurriaan Pool and was herself a well-known painter. Her works were bought by Johann Wilhelm, the elector palatinate. Rachel helped her father make anatomical preparations in his old age.

In his youth Ruysch had a passion for anatomy; and he himself told how he would ask grave diggers to open graves so that he could make anatomical investigations. Soon after his marriage, he began his medical studies at Leiden. To attend the lectures there, he had to travel from The Hague, where he lived and managed a chemist's shop. His teachers included Johannes van Horne, who was professor of anatomy and surgery, and Franciscus Sylvius, who taught practical medicine. On 28 July 1664 he received the M.D. at Leiden (not, as is sometimes reported, at Franeker) for his thesis *De pleuritide*, which was written under van Horne's guidance. He then established a medical practice in The Hague and almost immediately was overwhelmed with plague victims. During this time, he also conducted serious anatomical studies and continued them in his spare time. But he remained somewhat aloof from the experiments on live dogs that were then being carried out by de Graaf and Swammerdam (and attended by Steno) in the Leiden laboratories.

In 1665 Ruysch settled a dispute between van Horne and Louis de Bils, a self-taught, unqualified anatomist who claimed to be able to preserve corpses for years, when the latter cited van Horne's name in his fantastic theory on the course of the lymph in the lymph vessels. Although Bils firmly denied the existence of valves in these vessels, Ruysch succeeded in demonstrating their presence. His research, published as *Dilucidatio valvularum in vasis lymphaticis et lacteis* (1665), ended the controversy.

On 29 December 1666 Ruysch was named praelector of anatomy for the surgeon's guild in Amsterdam. He attended the session of the guild on 12 January 1667 and soon moved to Amsterdam. Ruysch held this post until his death in 1731. He was annoyed when, in 1727, at the age of eighty-nine, the burgomasters appointed Willem Roëll as his assistant without consulting him. As *praelector anatomiae*, Ruysch taught anatomy to the surgeons and performed the public dissections in the winter months. In this role he was painted twice with the masters of the guild: in 1670 by Adriaan Backer and in 1683 by Johan van Neck. The latter painted Ruysch dissecting a child while his son Hendrik (then about age twenty) is pictured as a boy holding the skeleton of a child. Also, Pool twice made a portrait of Ruysch. In 1672, after the death of Hendrik von Roonhuyse, Ruysch was also appointed city obstetrician. Thus he contributed to the education of midwives, giving one lesson a month and four demonstrations a year on female corpses. Ruysch held this post for forty years; in 1712 he retired in favor of his son Hendrik.

In 1679 Ruysch was appointed doctor of the court of justice. He reported on persons wounded or killed in robberies or quarrels—rather frequent occurrences in the great port of Amsterdam. Ruysch thus gathered extensive experience in forensic medicine. On 24 March 1685 he was appointed professor of botany at the Athenaeum Illustre and thus became supervisor of the botanical garden. He delivered three botanical lectures a week: to the surgeons, to the apothecaries' apprentices, and to the apothecaries. From 1692 he was assisted at the garden by Petrus Hotton and, later, by Caspar Commelin, who lectured on exotic plants. With F. Kiggelaar, Ruysch wrote a description of the rare plants in the garden.

Although Ruysch ably fulfilled these varied responsibilities for many years, he considered himself primarily an anatomist. He gave private courses in anatomy to foreign students and devoted himself throughout his life to making anatomical preparations. His skill in this art remains unsurpassed. The technique of injecting had already been used by Swammerdam and de Graaf during his student years at Leiden. De Graaf had invented a special syringe for this purpose, but Ruysch developed his own method and was thus able to prepare various organs (for example, the liver and the kidneys) and to preserve entire corpses for years. In the summer of 1696 he announced the dissection of bodies "which appear still to be alive but which have been dead for about two years."

Ruysch himself never disclosed the composition of the fluids he used, but in 1743 J. C. Rieger revealed that he used a mixture of talc, white wax, and cinnabar for injecting vessels, whereas his embalming fluid (liquor balsamicus) consisted of alcohol (prepared from wine or corn) to which some black pepper was added. Ruysch drew on his art not only for strict medical science but also for flights of fancy. He often made up preparations in a rather romantic, dramatic way. He prepared the corpse of a child as if it were alive so that Peter the Great was inclined to kiss it. A hydrocephalic child was prepared, seated on a cushion and with a placenta in its hands.

Ruysch displayed these preparations in several small rented houses in Amsterdam, and this "cabinet" became a major attraction for foreign visitors. He frequently added appropriate inscriptions referring to the brevity of life. Ruysch wrote a description (in both Dutch and Latin) of his collection in a series of ten books: *Thesaurus anatomicus primus* through *Thesaurus anatomicus decimus*. In 1715 he announced the sale of his collection. But no buyers presented themselves before 1717, when Peter the Great bought it for 30,000 guilders. It was carefully packed and transported by boat to Russia. The tale that the collection was destroyed by the sailors drinking the embalming fluid seems not to be true, or at least only partly so. Several pieces of the collection (for example, skeletons of children) are held by the Museum of the Academy of Sciences in Leningrad; the collection originally contained 935 items. Immediately after the sale, the energetic Ruysch, age seventy-nine, began to set up a new collection, which, after his death, was sold publicly. The greater part of it went to the king of Poland, John Sobieski, who entrusted it to the University of Wittenberg.

Ruysch had many friends and admirers, but also several critics with whom he became involved in scientific polemics—namely, G. Bidloo, J. J. Rau, and R. Vieussens. Boerhaave, his junior by thirty years, was a close friend. Ruysch visited Boerhaave at Leiden, and the latter seems to have passed several summer holidays with Ruysch. The friends' opinions diverged on some points of anatomy, including the structure of the glands. Ruysch rejected the view of Malpighi that the liver contains glandular tissue (parenchyma). Boerhaave supposed that Ruysch was misled by injecting the embalming fluid under too great a pressure. In 1721 the two friends published together two letters on the subject—*Opusculum anatomicum de fabrica glandularum.*

In addition to the valves in the lymph vessels, Ruysch described, independently of others, the arteria bronchialis. He also studied the eye and described a thin layer behind the retina (formerly called tunica Ruyschiana), as well as a circular muscle in the fundus uteri.

Foreign honors came relatively late to Ruysch. In 1705 he became a member of the Academia Leopoldo-Carolina. He was also elected to the Royal Society of London (1720) and to the Acadé-

mie des Sciences (1727) in Newton's place. At the end of his life Ruysch suffered a fracture of the collum femoris. The site of his grave is not known.

BIBLIOGRAPHY

I. ORIGINAL WORKS. Ruysch's major works are *Disputatio medica inauguralis de pleuritide* (Leiden, 1664); *Dilucidatio valvularum in vasis lymphaticis et lacteis* (The Hague, 1665; Leiden, 1667; Amsterdam, 1720, 2nd ed. 1742); a facs. of the 1665 ed., with intro. by A. M. Luyendijk-Elshout, appeared in Dutch Classics on the History of Science, no. 11 (Nieuwkoop, 1964); *Epistolae anatomicae problematicae,* 14 vols. (Amsterdam, 1696–1701); *Museum anatomicum Ruyschianum, sive catalogus rariorum quae in Authoris aedibus asservantur* (Amsterdam, 1691; 2nd ed., 1721; 3rd ed. 1737); *Thesaurus anatomicus . . . ,* 10 vols. (Amsterdam, 1701–1716), all with Dutch trans.; *Curae posteriores seu thesaurus anatomicus omnium precedentium maximus* (Amsterdam, 1724); and *Curae renovatae seu thesaurus anatomicus post curas posteriores novus* (Amsterdam, 1733).

Other writings are *Observationum anatomico-chirurgicarum centuria* (Amsterdam, 1691; 2nd ed., 1721; 3rd ed., 1737); *Thesaurus animalium primus* (Amsterdam, 1728); *Responsio ad Godefridi Bidloi libellum cui nomen vindicias inscripsit* (Amsterdam, 1697; 2nd ed., 1738); *Adversariorum anatomico-medico-chirurgicorum decas prima* (Amsterdam, 1717; 2nd ed., 1729), *decas secunda* (1720), *decas tertia* (1728); *Opusculum anatomicum de fabrica glandularum in corpore humano* (Leiden, 1722; Amsterdam, 1733), written with Boerhaave; *Tractatio anatomica de musculo in fundo uteri* (Amsterdam, 1723), with Dutch trans. by A. Lambrechts as *Over de baarmoeder-, of de ronde spier van de lijfmoeder* (Amsterdam, 1726; 2nd ed., 1731); *Opera omnia,* 4 vols. (Amsterdam, 1721); *Opera omnia anatomico-medicochirurgica huc usque edita,* 5 vols. (Amsterdam, 1737); *Alle de ontleed- genees- en heelkundige werken van Fr. Ruysch* (Amsterdam, 1744), Dutch trans. by Y. G. Arlebout; and *Horti medici Amstelodamensis rariorum descriptio . . .* (Amsterdam, 1697), written with F. Kiggelaar.

II. SECONDARY LITERATURE. On Ruysch and his work, see Bernard Fontenelle, "Éloge de M. Ruysch," in *Histoire de l'Académie royale des sciences pour l'année 1731 avec les mémoires de mathématique et physique pour la même année, tirés des registres de cette Académie;* N. T. Hazen, "Johnson's Life of Frederic Ruysch," in *Bulletin of the History of Medicine,* 7 (1939), 324; J. G. de Lint, "Frederik Ruysch," in *Nieuw Nederlandsch Biographisch Woordenboek,* III, 1108–1109; A. M. Luyendijk-Elshout's intro. to the facs. ed. of the *Dilucidatio valvularum* (see above); P. Scheltema, "Het leven van Frederik Ruysch" (M.D. diss., Univ. of

Leiden, 1886); and V. F. Schreiber's intro. to the *Opera omnia* (see above), also in the Dutch translation.

G. A. LINDEBOOM

RYDBERG, JOHANNES (JANNE) ROBERT (*b.* Halmstad, Sweden, 8 November 1854; *d.* Lund, Sweden, 28 December 1919), *mathematics, physics.*

Rydberg was the son of Sven R. and Maria Anderson Rydberg. After completing the Gymnasium at Halmstad in 1873, he entered the University of Lund, from which he received a bachelor's degree in philosophy in 1875. He continued his studies at Lund and was granted a doctorate in mathematics in 1879 after defending a dissertation on the construction of conic sections. In 1880 Rydberg was appointed a lecturer in mathematics. After some work on frictional electricity, he was named lecturer in physics in 1882 and was promoted to assistant at the Physics Institute in 1892. Rydberg married Lydia E. M. Carlsson in 1886; they had two daughters and a son. After provisionally occupying the professorship in physics at Lund from 1897, he was granted the appointment permanently in March 1901 and held it until November 1919. He was elected a foreign member of the Royal Society in 1919.

Rydberg's most significant scientific contributions were to spectroscopy; but his involvement with spectra had its origin in his interest in the periodic system of the elements, an interest that endured throughout his professional life. His earliest published papers in physics dealt with the periodic table. In the introduction to his major work on spectra (1890), he stated that he considered it only a part of a broader investigation, the goal of which was to achieve a more exact knowledge of the nature and constitution of the chemical and physical properties of the elements. He held that the effective force between atoms must be a periodic function of their atomic weights and that the periodic motions of the atoms, which presumably gave rise to the spectral lines and were dependent on the effective force, thus might be a fruitful study leading to a better knowledge of the mechanics, nature, and structure of atoms and molecules and to a deeper understanding of the periodic system of the other physical and chemical properties of the elements. In line with contemporary conceptions, Rydberg's view was that each individual line spectrum was the product of a single fundamental system of vibrations.

His major spectral work, "Recherches sur la constitution des spectres d'émission des éléments chimiques," published in 1890, mapped out Rydberg's total approach with remarkable clarity. He conceived of the spectrum of an element as composed of the superposition of three different types of series—one in which the lines were comparatively sharp, one in which the lines were more diffuse, and a third that he called principal series even though they consisted mostly of lines in the ultraviolet. The first lines were located in the visible spectrum and were usually the most intense. The members of each series might be single, double, triple, or of higher multiplicity. Any particular elementary spectrum might contain any number (even zero) of series of each of the basic types.

While Rydberg observed and measured some spectral lines on his own, he was not particularly noted as an experimental physicist and did not publish any of his experimental investigations or spectroscopic measurements. Most of the data he needed were already available in the voluminous literature. While T. R. Thalén and Bernhard Hasselberg, Rydberg's major Swedish contemporaries in spectral studies, concentrated upon accurate measurements of the spectra of the elements, Rydberg's major spectral contributions were to theory and mathematical form, and those to form were the ones of enduring value.

Unlike most others, Rydberg used wave numbers (the number of waves per unit length) instead of a correlated reciprocal, the directly measured wavelengths. This enabled him to manipulate his final formula into a particularly useful form.

Rydberg concluded that each series could be expressed approximately by an equation of the form

$$n = n_0 - \frac{N_0}{(m+\mu)^2},$$

where n was the wave number of a line; $N_0 = 109,721.6$, a constant common to all series and to all elements; n_0 and μ constants peculiar to the series; and m any positive integer (the number of the term). The lines of a series were generated by allowing m to take on integer values sequentially; n_0 defined the limit of the series that the wave number n approached when m became very large.

Just when he became occupied with confirming this relationship, Rydberg learned about Balmer's formula, which represented the observed lines of the hydrogen spectrum with extraordinary accuracy. He arranged Balmer's formula into its wave-number form and noted that, with appropriately selected constants, it was then a special form of his own more general formula. He felt that the success of Balmer's formula strengthened the justification of his own form. Thus encouraged, Rydberg proceeded to use the latter with sufficient success to propose it as the general formula for all series in all elementary line spectra, and to conclude that N_0 was indeed a universal constant, which has since become known as Rydberg's constant.

Spectroscopy had been a major developed field of physical study for several decades, but its most pressing need near the end of the nineteenth century was for the organization of its vast amount of data into some mathematically ordered form that theoreticians might find useful in their attempts to understand the underlying significance of spectra. Rydberg's general formula was the most important presentation of this type. Many others groped in the same general direction, mostly with ephemeral results. Rydberg's most significant competitors in this regard were Heinrich Kayser and Carl Runge, but their general formulas were of significantly different form.

The scope and structure of Rydberg's formula allowed him to note some important relationships. For example, he found not only that certain series with different values of μ exhibited the same value of n_0 but also that the value of the constant term n_0 in any series coincided with a member of the sequence of variable terms in some other series of the element. In particular, he discovered that the difference between the common limit of the diffuse and sharp series and the limit of the corresponding principal series gave the wave number of the common first-member term of the sharp and principal series, a relationship independently noted by Arthur Schuster and commonly known as the Rydberg-Schuster law.

Along this same line, Rydberg speculatively suggested as a comprehensive formula for every line of an element the relationship

$$\pm n = \frac{N_0}{(m_1 + \mu_1)^2} - \frac{N_0}{(m_2 + \mu_2)^2},$$

with which he hoped to represent a series according to whether he assumed either m_1 or m_2 to be variable. Thus, he viewed every spectral series as a set of differences between two terms of the type $N_0/(m + \mu)^2$ — that is, every spectral line would be expressed as $n = T_1 - T_2$, where T_1 and T_2 are two members of a set of terms characteristic of the element. This aspect, little appreciated at the time, was stated independently in 1908 by Walther Ritz and is commonly known as the Ritz combination principle.

The combination principle revealed several significant features about spectra. First, the wave number of each line could be conveniently represented as the difference between two numbers, called terms. Second, the terms could be naturally grouped into ordered sequences—the terms of each sequence converging toward zero. Third, the terms could be combined in various ways to give the wave numbers of the spectral lines. Fourth, a series of lines all having similar character resulted from the combination of all terms of one sequence taken in succession with a fixed term of another sequence. Thus, fifth, a large number of spectral lines could be expressed as the differences of a much smaller number of terms that in some way were characteristic of the atom and therefore, from a theoretical perspective, were more important than the lines themselves when speculating on atomic structure. Now it was these terms, rather than the lines, for which a direct physical interpretation should be found. This last point was widely overlooked by most contemporary physicists, including Rydberg.

As deeply as the notion of the existence of some fundamental mechanism might be stimulated by them, all the regularities noted by Rydberg were in themselves only empirical generalizations. His own theoretical concepts on atomic structure were still based on an analogy to acoustics. Therefore, Rydberg did not reach the final goal he had set for his work: an adequate insight into the nature and structure of the atom. His work did, however, provide a basis for the later development of successful ideas on atomic structure.

Some radically new ideas concerning the structure of the atom resulted from the development of other lines of evidence. In 1913 Niels Bohr proposed his theory of atomic structure based on Ernest Rutherford's nuclear atomic model and on Max Planck's quantum theory of radiation. These conceptions led to the first reasonably successful theoretical account of spectral data.

Bohr's view provided an immediate interpretation of the combination principle by identifying each Rydberg spectral term multiplied by hc (Planck's constant times the speed of light) with the energy of an allowable stationary state of the atom. The difference between two such states equaled the energy in the light quantum emitted in the transition from a higher allowable atomic-energy state to a lower one.

On this basis, spectral series were used to determine the excitation energies and ionization potentials of atoms. The further elaboration of these views led to a classification of the states of electron binding in a shell structure of the atoms that accounted for the periodic relationships of the properties of the elements, thereby fully justifying Rydberg's earlier faith that spectral studies could assist in attaining this goal. Rydberg played no role in this elaboration, however.

But earlier, along similar lines, Rydberg's study of the periodic properties of the elements led him in 1897 to suggest that certain characteristics of the elements could be more simply organized by using an atomic number instead of the atomic weights. This atomic number was to be identified with the ordinal index of the element in the periodic table. In 1906 Rydberg stated for the first time that 2, 8, and 18 (that is, $2n^2$, where $n = 1,2,3$) represented the number of elements in the early periods of the system. In 1913 he went further, correcting an earlier error about the number of rare earths from 36 to 32, thus allowing the $n = 4$ group to be included in the pattern.

Rydberg presented a spiral graph arrangement of the periodic table in which earlier holes in his system were corrected so that atomic numbers from helium on were two greater than at present. He maintained that there were two elements, nebulium and coronium, between hydrogen and helium in the system, supporting their existence by evidence from both spectra and graphical symmetry.

In 1913, H. G. J. Moseley published his paper based on researches on the characteristic X-ray spectra of the elements that strongly supported the fundamental importance of atomic numbers and Rydberg's basic expectations about the lengths of the periods of the periodic table. The physical reality that underlay Rydberg's atomic-number proposal was later interpreted as the positive charge on the atomic nucleus expressed in elementary units of charge.

Rydberg received a copy of Moseley's paper in manuscript form before publication. In a note written in 1914, he expressed satisfaction at the confirmation of his ideas on atomic numbers and the details of the periodic system, but he still maintained his conviction of the existence of the two elements between hydrogen and helium and the resulting difference of two in most atomic numbers. Later the nebulium spectrum was attributed to ionized oxygen and nitrogen, and the coronium lines to highly ionized iron.

Rydberg's health did not permit him to follow subsequent developments. In 1914 he became seriously ill. He went on an extended leave of absence that lasted until his formal retirement in 1919, a month before his death.

BIBLIOGRAPHY

I. ORIGINAL WORKS. Rydberg's most important spectral publication was "Recherches sur la constitution des spectres d'émission des éléments chimiques," in *Kungliga Svenska vetenskapsakademiens handlingar*, n.s. **23**, no. 11 (1890). Some of his other spectral works of significance are "On the Structure of the Line-Spectra of the Chemical Elements," in *Philosophical Magazine*, 5th ser., **29** (1890), 331–337; "Contributions à la connaissance des spectres linéaires," in *Ofversigt af K. Vetenskapsakademiens förhandlingar*, **50** (1893), 505–520, 677–691; "The New Elements of Cleveite Gas," in *Astrophysical Journal*, **4** (1896), 91–96; "The New Series in the Spectrum of Hydrogen," *ibid.*, **6** (1897), 233–238; "On the Constitution of the Red Spectrum of Argon," *ibid.*, 338–348; and "La distribution des raies spectrales," in *Rapports présentés au Congrès international de physique, Paris*, II (1900), 200–224.

Concerning his other work related to the periodic table, significant articles are "Die Gesetze der Atomgewichtszahlen," in *Bihang till K. Svenska vetenskapsakademiens handlingar*, **11**, no. 13 (1886); "Studien über die Atomgewichtszahlen," in *Zeitschrift für anorganische Chemie*, **14** (1897), 66–102; *Elektron der erste Grundstoff* (Berlin, 1906); "Untersuchungen über das System der Grundstoffe," in *Acta Universitatis lundensis*, Avd. 2, n.s. **9**, no. 18 (1913); and "The Ordinals of the Elements and the High-Frequency Spectra," in *Philosophical Magazine*, 6th ser., **28** (1914), 144–148.

II. SECONDARY LITERATURE. A short biography of value is Manne Siegbahn, in *Swedish Men of Science 1650–1950*, Sten Lindroth, ed., Burnett Anderson, trans. (Stockholm, 1952), 214–218. Siegbahn was a student at the University of Lund from 1906 to 1911 and an assistant at the Physics Institute from 1911 to 1914 while Rydberg was there. In the autumn of 1915 Siegbahn was appointed to fulfill Rydberg's duties while the latter went on an extended leave. In early 1920 Siegbahn permanently succeeded Rydberg in the chair of physics at Lund.

On the centenary of Rydberg's birth, an important collection of papers was presented at Lund: "Proceedings of the Rydberg Centennial Conference on Atomic Spectroscopy," in *Acta Universitatis lundensis*, Avd. 2, n.s. **50**, no. 21 (1954). Biographically, the two most significant articles are Niels Bohr, "Rydberg's Discovery of the Spectral Laws," 15–21; and Wolfgang Pauli, "Rydberg and the Periodic System of the Elements," 22–26.

Another biographical essay of merit is Sister St. John Nepomucene, "Rydberg: The Man and the Constant," in *Chymia*, **6** (1960), 127–145. Two brief biographical obituaries are in *Physikalische Zeitschrift,* **21** (1920), 113; and *Nature,* **105** (1920), 525.

C. L. Maier

SABATIER, ARMAND (*b.* Ganges, Hérault, France, 14 January 1834; *d.* Montpellier, Hérault, France, 22 December 1910), *comparative anatomy, philosophy.*

Sabatier's parents were dedicated Protestants. They closely supervised Armand's early education in the schools of his native town. Later he was admitted to the study of medicine at the nearby University of Montpellier, where the Faculty of Medicine was one of the oldest in Europe. His early interest in anatomy continued throughout his life.

In 1855 Sabatier was appointed assistant in anatomy at Montpellier. In 1858 he obtained an internship in Lyons. He subsequently returned to Montpellier to present his doctoral thesis. He worked in the department of anatomy and in 1869 was made associate professor. During the Franco-Prussian war, he served with distinction in command of a field ambulance. In 1873 he was associate professor and in 1876 professor of zoology in the Faculty of Sciences at Montpellier.

Sabatier's book on the heart and circulation in vertebrates (1873) immediately established his reputation as an anatomist. He made a detailed investigation of the cardiac morphology and physiology of amphibians and reptiles and from this research established the general laws that govern the functional evolution of the heart from fishes to mammals. He showed that these laws apply not only to the zoological series but also to the developing embryo. But he insisted that such parallelism between phylogenesis and ontogenesis should not be viewed too strictly. He did not intend the eventual deviation from this parallelism as evidence against the evolutionary doctrine.

Sabatier also compared the thoracic and pelvic girdles in the vertebrate series. He based this comparison on muscle insertions, which appear to be similar along the vertebrate series and to exhibit a certain similarity among bony structures. This work resolved several lengthy debates, including that on the significance in man of the coracoid process. He demonstrated that this structure is analogous to the pubis of the pelvic girdle.

Throughout his life Sabatier was interested in comparative osteology. He was also an excellent zoologist and cytologist. In such a commonplace mollusk as the mussel, he elucidated many unknown features of the circulatory system. He also investigated egg and sperm formations in various invertebrate groups and in the lower vertebrates. Although some of his conclusions have been challenged, many of his observations remain valid, and in some cases might (with modern techniques) be the starting point of fertile investigations.

In 1879 Sabatier founded one of the earliest marine laboratories—the Station Zoologique de Sète, which he installed for some years in the modest surroundings of a fisherman's cabin. Public support of marine stations was difficult to obtain in those days, but Sabatier's persistent efforts to win funds succeeded seventeen years later, when the station was finally given a well-equipped laboratory. But even when its facilities were modest, the laboratory was an active institution, and Sabatier trained many young marine biologists there.

Sabatier's mind inclined toward philosophy. He was a Christian of firm beliefs and a biologist of equally firm adherence to evolutionary doctrines. In two important books he brilliantly defended the compatibility of these two positions. In his *Philosophie de l'effort* . . . he maintained that man's striving toward a saintly or simply moral life furthers the survival of the species.

BIBLIOGRAPHY

I. Original Works. Sabatier's works include *Études sur le coeur et la circulation centrale dans la série des vertébrés. Anatomie, physiologie comparée, philosophie naturelle* (Paris–Montpellier, 1873); "Sur quelques points de l'anatomie de la Moule commune," in *Comptes rendus hebdomadaires des séances de l'Académie des sciences,* **79** (1874), 581–584; "Sur les cils musculoïdes de la Moule commune," *ibid.,* **81** (1875), 1060–1063; "Études sur la Moule commune (Mytilus edulis)," in *Mémoires de l'Académie de Montpellier, section des sciences,* **8** (1879), 413–506; "Comparaison des ceintures et des membres antérieurs et postérieurs dans la série des vertébrés," *ibid.,* **9** (1878), 277–336, **9** (1879), 337–709; "Appareil respiratoire des ampullaires," in *Comptes rendus hebdomadaires des séances de l'Académie des sciences,* **88** (1879), 1325–1328; "Formation du blastoderme chez les aranéides," *ibid.,* **92** (1881), 200–204; "La spermatogenèse chez les Annélides et les vertébrés," *ibid.,* **94** (1882), 172–175; "De la spermatogenèse chez les plagiostomes et les amphibiens," *ibid.,* **94** (1882), 1097–1100; "De l'ovogenèse chez les ascidiens," *ibid.,* **96** (1883), 799–802; "Sur les cellules du follicule de l'oeuf et sur la nature de la sexualité," *ibid.,* **96** (1883), 1804–1807; "Sur le noyau vitellin des aranéides," *ibid.,* **97** (1883), 1570–1573.

Later writings are "Sur la spermatogenèse des crustacés décapodes," in *Comptes rendus hebdomadaires des séances de l'Académie des sciences*, **100** (1885), "Sur la morphologie de l'ovaire chez les insectes," *ibid.*, **102** (1886), 61–64; "Recherches sur l'oeuf des ascidiens," in *Mémoires de l'Académie de Montpellier, section des sciences*, **10** (1885), 429–480; "Recueil de mémoires sur la morphologie des éléments sexuels et sur la nature de la sexualité," in *Travaux du laboratoire de zoologie de la Faculté des sciences de Montpellier et de la station zoologique de Sète*, **5** (1886), 1–271; "Sur les formes de spermatozoïdes de l'elédone musquée," in *Comptes rendus hebdomadaires des séances de l'Académie des sciences*, **106** (1888), 954–957; "Sur la station zoologique de Sète," *ibid.*, **109** (1889), 388–391; "La Spermatogenèse chez les locustides," *ibid.*, **111** (1890), 797–800; and "Sur quelques points de la spermatogenèse chez les sélaciens," *ibid.*, **120** (1895), 47–50.

See also "De la spermatogenèse chez les poissons sélaciens," in *Mémoires de l'Académie de Montpellier, section des sciences*, **2** (1896), 53–237; "Morphologie des membres des poissons osseux," in *Comptes rendus hebdomadaires des séances de l'Académie des sciences*, **122** (1896), 121–124; "Morphologie du sternum et des clavicules," *ibid.*, **124** (1897), 805–808; *ibid.*, 932–935; "Sur la signification morphologique des os en chevrons des vertébres caudales," *ibid.*, 932–935, written with Ducamp and Petit; "Etude des huîtres de Sète au point de vue des microbes pathogènes," *ibid.*, **125** (1897), 685–688, written with Ducamp and Petit; "Sur la genèse des épithéliums," *ibid.*, **127** (1898), 704–707, written with M. E. de Rouville; "Morphologie des ceintures et des membres pairs et impairs des sélaciens," *ibid.*, 928–932; "Morphologie de la ceinture pelvienne des amphibiens," *ibid.*, **130** (1900), 633–637; "Sur les mains scapulaires et pelviennes des poissons," *ibid.*, **137** (1903), 893–894; "Sur les mains scapulaires et pelviennes chez les poissons chondroptérygiens," *ibid.*, 1216–1220; and "Sur les mains scapulaires et pelviennes des poissons holocéphales et dipneustes," *ibid.*, **138** (1904), 249–253.

Sabatier's philosophical works are *Essai sur l'immortalité au point du vue de naturalisme évolutionniste*, (Paris, 1895); and *Philosophie de l'effort. Essais philosophiques d'un naturaliste* (Paris, 1903).

A. M. MONNIER

SABATIER, PAUL (*b.* Carcassonne, France, 5 November 1854; *d.* Toulouse, France, 14 August 1941), *chemistry*.

Sabatier achieved scientific distinction for his pioneering work in catalysis. From a family of modest means, he had his secondary education at Carcassonne and then at Toulouse. Admitted to the École Normale Supérieure in 1874, he received the *agrégé* in the physical sciences in 1877 and was first in his class. He taught briefly at the *lycée* of Nîmes, then, encouraged by Berthelot, entered the latter's laboratory at the Collège de France in 1878 and received his doctorate in the physical sciences in 1880. After a year at Bordeaux, Sabatier taught at Toulouse, where he was named to the chair of chemistry in 1884, when he was thirty, the minimum age for the post.

In 1907 Sabatier was offered Moissan's chair at the Sorbonne and that of Berthelot at the Collège de France. Although he realized that all candidates for the Académie des Sciences were required to be residents of Paris, he nevertheless chose to remain at Toulouse. In 1912 he shared the Nobel prize in chemistry with Victor Grignard, and in 1913 he became the first scientist elected to one of six chairs newly created by the Academy for provincial members. At this time Sabatier also was dean of the Faculty of Sciences at Toulouse, a post he held officially from 1905 to 1929. He was instrumental in founding three schools of applied science at Toulouse—in chemistry, electrical engineering, and agriculture. Both by personal example and by administrative action, Sabatier was throughout his life an important influence in steps toward the decentralization of scientific institutions in France.

Sabatier's initial researches were inorganic studies within the thermochemical tradition of Berthelot's laboratory. They included analyses of metallic and alkaline-earth sulfides and of chlorides, the preparation of hydrogen disulfide by vacuum distillation, the isolation of selenides of boron and silicon, the definition of basic cupric salts containing four copper atoms, and preparations of the deep blue nitrosodisulfonic acid and the basic mixed argentocupric salts. He studied the partition of a base between two acids, using the spectrophotometric change of coloration of chromates and dichromates as an indicator of acidity, and analyzed the velocity of transformation of metaphosphoric acid. In 1895 Sabatier had begun the preparation of metals by reduction of their oxides with hydrogen, when he noted with interest British chemists' preparation of nickel carbonyl by the direct action of carbon monoxide on finely divided nickel. Wondering if other unsaturated gaseous molecules might behave analogously to carbon monoxide, he succeeded in 1896 in fixing nitrogen peroxide on copper, cobalt, nickel, and iron.

Sabatier then learned that Moissan and Charles Moureu had failed to achieve a similar result with acetylene. Assured that they did not intend to pursue the experiment, he repeated it with the less violent hydrocarbon ethylene, heating an oxide of nickel to 300°C. in a current of hydrogen gas and then directing a current of ethylene upon the slivers of reduced nickel. He found that the resulting gaseous product was not hydrogen, as Moissan had assumed, but mostly ethane resulting from the hydrogenation of ethylene. Sabatier then succeeded in oxidizing acetylene to ethylene and ethane, and in 1901 attempted the transformation of benzene into cyclohexane. Berthelot had failed to do this with a hydriodic-acid hydrogenation agent, but Sabatier succeeded with benzene vapors and hydrogen over reduced nickel at 200°C.

In the next years Sabatier continued this work on hydrogenating organic compounds in the presence of finely disintegrated metals, for which he was awarded the 1912 Nobel Prize. Assisted by his student J. B. Senderens, Sabatier demonstrated the general applicability of his method to the hydrogenation of nonsaturated and aromatic carbides, ketones, aldehydes, phenols, nitriles, and nitrate derivatives. He synthesized methane from carbon monoxide, and demonstrated that at higher temperatures his hydrogenation procedures would lead to dehydrogenation, applying this principle to the production of aldehydes and ketones from their corresponding primary and secondary alcohols. Sabatier established that certain metallic oxides, particularly manganous oxide, behave analogously to metals in hydrogenation and dehydrogenation, although at slower rates; and that powdered oxides such as thoria, alumina, and silica possess hydration and dehydration properties. For example, reduced copper acts as a catalyst for splitting alcohol vapors into hydrogen and aldehyde, whereas replacing copper with alumina results in a division of alcohol into water and ethylene.

Sabatier's *La catalyse en chimie organique* first appeared in 1913, its utility enhanced by a principally empirical and analogical approach. His theory of catalytic mechanism, later termed "chemisorption," strongly opposed that of most nineteenth-century chemists. Berzelius, Ostwald, and others had assumed that known catalyzed reactions—such as the effect of platinum on the combustion of hydrogen and oxygen—resulted from an absorption of gases in the cavities of the porous metal, where compression and local temperature elevation led to chemical combination.

In contrast, Sabatier believed that in both homogeneous and heterogeneous systems, a temporary, unstable intermediary between the catalyst and one of the reactants forms on the surface of the catalyst. The intermediary's combination with the second reactant regenerates the catalyst. Like his predecessors' theory, Sabatier's view implied that the activity of a catalyst increases with its granular surface area; he thus also accounted for poisoning of a catalyst by impurities and for fatigue by surface modifications. But unlike his predecessors, Sabatier indicated that the course of a reaction would depend upon the chemical as well as the physical nature of the catalyst, a contention supported by his ability to manipulate the products of a reaction by substituting one catalyst for another (an oxide for a metal, for example). His view also predicted the empirically verified facts that a catalyst of hydrogenation will be equally one of dehydrogenation, and that promoters of catalysis are often the same types of material as inhibitors or poisons.

Although his work laid the foundation for many of the giant industries of the twentieth century, Sabatier paid little or no attention to the practical applications of his discoveries. He had no interest in liquid-phase hydrogenation and avoided high-pressure hydrogenation techniques. He obtained a few French patents, including one of 1909, which envisioned means of cracking heavy fractions of petroleum on a metal catalyst and then hydrogenating the volatile products.

BIBLIOGRAPHY

I. Original Works. Sabatier's most important publication was *La catalyse en chimie organique* (Paris, 1913; 2nd ed., 1920). E. Emmet Reid's translation, *Catalysis in Organic Chemistry* (New York, 1923), has been revised and reprinted in *Catalysis Then and Now* (Palisades Park, N.J., 1965), which contains numerous references to Sabatier's papers on catalysis. His two other major publications were his thesis, *Recherches thermiques sur les sulfures* (Paris, 1880) and *Leçons élémentaires de chimie agricole* (Paris, 1890). His 1926 address before the American Chemical Society in Cincinnati, Ohio, records his recollections about his work: "How I Have Been Led to the Direct Hydrogenation Method by Metallic Catalysts," in *Industrial and Engineering Chemistry*, **18** (Oct. 1926), 1005–1008.

II. Secondary Literature. There is no biography of Sabatier other than Lucien Babonneau's "Paul Sabatier," in *Génies occitans de la science* (Toulouse, 1947), 167–189. Other discussions of his life and work are in Gabriel Bertrand, Charles Camichel, *et al., Cérémonies du centenaire de la naissance de Paul Sabatier à Toulouse* (Hendaye, 1954); Charles Camichel, Gaston Dupouy *et al., Centenaire Paul Sabatier. Prix Nobel. Membre de l'Institut. 1854–1954* (Toulouse, 1956); and J. R. Partington, "Paul Sabatier," in *Nature*, **174** (1954), 859–860.

Mary Jo Nye

SABIN, FLORENCE RENA (*b.* Central City, Colorado, 9 November 1871; *d.* Denver, Colorado, 3 October 1953), *anatomy, immunology.*

The second daughter of Serena M. and George K. Sabin, who had given up the study of medicine to work in the Colorado mines, Florence Sabin attended schools in Denver and Vermont. She followed her sister Mary to Smith College and concentrated on mathematics and science, receiving the B.S. degree in 1893. While at Smith she became interested in women's rights and also decided to study medicine. The year she received her Smith degree, the Johns Hopkins Medical School opened with the financial help of a group of Baltimore women who stipulated that women be admitted on the same basis as men. Sabin taught mathematics for two years in Denver and zoology for a year at Smith to earn sufficient money to continue her education. She then matriculated at Johns Hopkins Medical School in the fall of 1896.

Sabin began her career in medical research under the stimulus and guidance of Franklin P. Mall, professor of anatomy at the Johns Hopkins Medical School. While still a student she undertook to construct a three-dimensional model of the mid- and lower brain. Her work was published as *An Atlas of the Medulla and Midbrain* in 1901 and quickly became a popular text. Also in 1901, a year after receiving the M.D. degree, she was appointed to a fellowship in anatomy after completion of an internship. She thus began a twenty-five-year academic association with Johns Hopkins. When she became an assistant in anatomy in 1902, Sabin had the distinction of becoming the first woman faculty member at the school. She was the first woman to achieve professorial rank at the Hopkins Medical School when she was promoted to professor of histology in 1917. Three similar distinctions occurred in 1924–1925, when she became the first woman elected president of the American Association of Anatomists, the first woman elected to the National Academy of Sciences, and the first woman to become a full member of the Rockefeller Institute.

In her early Johns Hopkins years, Sabin worked on the origins of blood cells and the lymphatics. Understanding of the anatomy of the lymph channels was vague at the time. Particularly under debate were the relationship of the smallest lymphatics to the tissue spaces and the embryonic origins of the lymph vessels. One theory held that the lymphatics arose from the tissue spaces and grew toward the veins. The opposing view was that they arose from the veins directly, by a series of small endothelial buds. Using small pig embryos rather than the larger ones used by earlier investigators, Sabin was able to show that the latter view was the correct one.

On numerous summer trips to German laboratories, Sabin learned and brought back techniques. One of the most important of these was supravital staining, the staining of living cells. She used this method in studies on the cellular reaction in tuberculosis and work on the site of antibody production. In 1925 Sabin left her teaching position in Baltimore to establish a laboratory at the Rockefeller Institute in New York that was devoted to the cellular aspects of the immune response. She worked especially with the large mononuclear white blood cells (monocytes), showing their role in the antigen-antibody reaction, a subject not yet fully understood fifty years later.

After thirteen years as a member of the Rockefeller Institute, Sabin became emeritus in 1938 and retired to Denver. She remained active on several national boards; but not until 1944, when she was asked to head a subcommittee on health for the governor's Post War Planning Committee, did she again work with her usual vigor and efficiency. Colorado had long prided itself as a health resort, and thus it came as a shock when the Sabin Committee began to publicize its findings. The state health department was an ineffective, politically controlled body, and its lack of efficiency and power was reflected in some of the worst health statistics of any state in the nation. Sabin worked tirelessly for the passage of new health laws. In 1947, the year of their passage, she was appointed chairman of the Interim Board of Health and Hospitals of Denver, a post she held until 1951, her eightieth year.

BIBLIOGRAPHY

I. ORIGINAL WORKS. Sabin's numerous scientific papers appeared for the most part in anatomical journals, *Science,* and publications of the institutions where she worked. Besides *An Atlas of the Medulla and Midbrain* (Baltimore, 1901), she contributed a widely cited chapter on the lymphatics to Franz Keibel and Franklin P. Mall's *Manual of Human Embryology,* II (Philadelphia, 1912), 709–745, that also appeared as *The Origin and Development of the Lymphatic System* (Baltimore, 1913). *Franklin Paine Mall, the Story of a Mind* (Baltimore, 1934) is the standard biography of Mall and reveals much about Sabin's attitudes as well.

II. SECONDARY LITERATURE. The most complete book about Sabin is Elinor Bluemel's *Florence Sabin, Colorado Woman of the Century* (Boulder, Colo.,

1959). Mary K. Phelan, *Probing the Unknown* (New York, 1969), is for younger readers. The best treatment of Sabin's scientific contributions is Philip D. McMaster and Michael Heidelberger, "Florence Rena Sabin," in *Biographical Memoirs. National Academy of Sciences,* **34** (1960), 271–305. See also Vincent T. Andriole, "Florence Rena Sabin—Teacher, Scientist, Citizen," in *Journal of the History of Medicine and Allied Sciences,* **14** (1959), 320–350; George W. Corner, *A History of the Rockefeller Institute 1901–1953* (New York, 1964), 238–239; Lawrence S. Kubie, "Florence Rena Sabin, 1871–1953," in *Perspectives in Biology and Medicine,* **4** (1961), 306–315; John H. Talbott, *A Biographical History of Medicine* (New York, 1970), 1181–1183; and Edna Yost, *American Women of Science* (Philadelphia, 1943), 62–79.

GERT H. BRIEGER

SABINE, EDWARD (*b.* Dublin, Ireland, 14 October 1788; *d.* Richmond, Surrey, England, 26 June 1883), *geophysics.*

An artillery officer, Sabine was a graduate of the Royal Military Academy, Woolwich. While retaining his commission—Sabine eventually reached the rank of general—he started scientific work at the close of the Napoleonic Wars. On the recommendation of the Royal Society, he accompanied John Ross on an expedition to seek the Northwest Passage in 1818 and was with William Edward Parry on his 1819–1820 Arctic expedition. From the latter voyage, he said, came the idea of a great shipborne expedition of "physical discovery" to the southern hemisphere.

The Royal Society next sent Sabine on a pendulum expedition in 1821–1822 around the Atlantic to determine the true figure of the earth, a project that brought him the Copley Medal. A pattern was developing in his work, clearly of a Humboldtian nature—the gathering and analysis of geophysical data on a large, even global, scale. While the range of Sabine's interests was wide, terrestrial magnetism attracted most of his attention. In 1826 he and Babbage worked jointly on the subject in the British Isles, an ironic collaboration in view of subsequent events. In the 1830's Sabine, Humphrey Lloyd, James Clark Ross, and others completed the magnetic survey of the British Isles; Sabine repeated the survey in 1858–1861.

Sabine was distinguished from his many contemporaries who collected similar data by his successful promotion and administration of a world-wide effort to gather terrestrial magnetism observations, designated the "magnetic fever" or the "magnetic crusade" by observers. Basic to an understanding of his accomplishments are the scientific viewpoints he embodied and the strategic position he came to occupy in the politics of British science for nearly four decades.

As a follower of Christopher Hansteen, in contradiction to Gauss's later theories on terrestrial magnetism, Sabine believed in the existence of two magnetic poles in each hemisphere and that terrestrial magnetism was essentially the same as, or closely related to, atmospheric phenomena; the latter view was widely held in the first third of the nineteenth century. Humboldt and Arago, for example, assumed a connection between the earth's central heat, volcanic eruptions, and atmospheric electricity. Seebeck's work on thermoelectricity reinforced the belief in the relationship to meteorological phenomena. Sir John Herschel and Charles Babbage assumed that the atmospheric electricity arose from a thermoelectric interaction of sky and earth that, in turn, produced terrestrial magnetism by a kind of induction. Gauss (1839) flatly limited the origins of terrestrial magnetism to the surface or interior of the planet, much to Sabine's dismay. Yet when Faraday published a theory similar to Herschel's (1851), Sabine informed him that the data gathered in the "crusade" disclosed none of the predicted correlations. Sabine ruefully admitted to Faraday that he had consulted William Thomson, who verified Gauss's mathematics. Yet Gauss's views did not dampen Sabine's interest in atmospheric phenomena or stop his search for extraterrestrial effects.

Sabine was infuriated by the confusion of his scientific aims with what he considered the lesser goal of geographic discovery. His was the widely shared tradition of viewing the earth as a heavenly body, the physical processes of which required study with the spirit and precision devoted to other astronomical phenomena. In tacit opposition were other views challenging both Sabine's methods and his order of priorities. In the early years of the British Association, for example, the study of "magnetism" was included in the stated scopes of two separate committees, one on the chemical sciences and the other dominated by astronomy and meteorology.

The distinction was, thus, between experimental and observational sciences. When the Royal Society launched the "magnetic crusade," it had two committees (physics and meteorology) combined under the direction of Sir John Herschel. Faraday accidentally came to one subcommittee meeting on instrumentation for the magnetic observations but thereafter was conspicuously absent from its deliberations. Herschel dealt with the distinction between observational and experimental fields in the *Preliminary Discourse on the Study of Natural Philosophy,* giving preference to the former because they lacked the opportunity of simply recreating an artificial situation. Herschel further elevated geophysical problems above astronomical problems because the former were not simply cyclical but undergoing complex secular changes. A missed observation might be literally unrepeatable. Sabine was in complete agreement.

In 1839 the British dispatched an expedition to the southern hemisphere and established a network of magnetic and meteorological observatories. Sabine had a key role in both the origins and the consequences of these events. The "magnetic crusade" did not originate, as is sometimes stated, in an 1836 suggestion of Humboldt's or in a desire to test Gauss's *Allgemeine Theorie des Erdmagnetismus.* As early as 1805, on his return from the Americas, Humboldt disclosed the important fact that the intensity of terrestrial magnetism varied at different points on the earth, thereby stimulating further interest in what many regarded as the great remaining physical mystery since Newton's work on gravitation. In 1828 Humboldt suggested a worldwide system of observations, and by 1830 a rather rickety one existed, stretching from Germany to Peking. Although Gauss's interest in the subject went back many years, the 1833 announcement of the first method of obtaining an absolute measurement of magnetic intensity brought him into prominence among British magneticians. By 1835, at least, Gauss and Weber were in contact with G. B. Airy, Humphrey Lloyd, and Sabine. The last two were aware as early as 1837 that Gauss was working on a general theory; it appeared early in 1839, in volume III of the *Resultate aus den Beobachtungen des Magnetischen Vereins,* in time to add testing of the theory to the goals of the worldwide effort. Of greater impact were the new method and the fact that by 1836–1838 Gauss and Weber's Magnetische Verein had a sixteen-station net of observatories stretching from Dublin to St. Petersburg from east to west and from Uppsala to Catania in the north-south axis.

Both intellectual curiosity and nationalistic zeal motivated Sabine and his associates. They looked back to Halley's work for precedent and spoke frankly about a great scientific prize slipping into foreign hands. In 1834 Arago, clearly Humboldtian in his views, wrote to the British Association, suggesting a British global effort, apparently unaware of Gauss's work. Having learned soon afterward of Gauss's method and the work of the Magnetische Verein, Sabine and others hesitated about siding with Gauss. His precise, large equipment was unsuitable for magnetic mapping of the oceans or the uses of scientific travelers. Sabine was also critical of the lack of observations of dip and inclination in the Gauss system. Unlike Herschel and Gauss, he was as interested in these aspects as in the intensity. Unlike Gauss and Herschel, Sabine believed that the routine periodic variations were as important as the readings for magnetic storms.

Originally an outsider to the British Association, Sabine was now very active there; and at the Dublin meeting of 1835 a resolution was passed calling on the government to send an expedition to the southern hemisphere and to open magnetic and meteorological observatories in the colonies. When this proved to no avail, Sabine convinced Humboldt in 1836 to write a letter to the duke of Sussex, the president of the Royal Society, calling for British action. For two years little happened outwardly, the Royal Society apparently being as unsuccessful as the British Association. From the surviving correspondence of participants, this was a period of intense politicking. Sabine was particularly anxious to avoid all pressures limiting the venture either to a voyage of discovery or to a series of fixed observatories. James Clark Ross was to head the former, and Humphrey Lloyd was to have charge of the theoretical work. Fellow artillery officers would staff the fixed observatories. From a scientific standpoint Sabine's most notable move was the publication of known intensity observations in the world (1837), which enabled Gauss to do the requisite calculations for the *Allgemeine Theorie.*

Yet the venture remained dormant in the Royal Society until John Herschel's return from South Africa in 1838. Although he once admitted to never having taken a magnetic reading, Herschel was generally interested in terrestrial magnetism. More important, he and Sabine were in complete agreement on the desirability of seizing this occasion to advance meteorology. Temperature and pressure readings were necessary, because they were sometimes responsible for greater effects on the compass needle than the earth's magnetism. Herschel also had great popular esteem and much influence with members of the government. At the British Association meeting at Newcastle-on-Tyne in 1838, he deftly swayed the crowd. Lloyd and Sabine went that year to Göttingen and Berlin to coordinate their coming venture with Gauss and Humboldt. On cue from Sabine and Herschel, Humboldt wrote a final letter to British officials, assuring the "crusade's" launching.

Originally the magnetic crusade was for three years, but Sabine very adroitly manipulated British and foreign opinion to get two successive three-year renewals. But disenchantment soon spread among former supporters. As early as 1839 Sabine aroused Lloyd's ire by taking over processing of the data, causing the latter to withdraw. Although the Royal Society at first refused to recognize Sabine's role, from 1841 to 1861 he maintained a staff at Woolwich for data reduction. Sabine's ambitions next clashed with Herschel's beliefs. The King's Observatory at Kew had been unused since Rigaud's death in 1839. Sabine wanted the facility to be the basic geophysical observatory for the empire, providing standard data and equipment for the colonial observatories. Neither Airy's Greenwich nor Lloyd's Dublin observatory would do. For three years the issue remained before the Royal Society while Sabine unsuccessfully tempted Herschel, who had long favored the founding of a facility combining geophysical observations, determination of standards, and physical experimentation. By 1842 the Royal Society had declined the offer of Kew because Herschel saw it as a limited observatory too narrowly tied to a particular venture. Sabine took the proposal to the British Association, which acquired the site in 1842 and managed the observatory until its 1871 transfer to the Royal Society. Sabine was very active in the management of Kew, which became a leading center for work in geophysics. From 1849 he was on the Kew Observatory Committee of the British Association, being particularly close to John Gassiot in its work.

Herschel early took exception to Sabine's seemingly endless compilation of data. The production of charts showing "lines of iso-x" aroused Herschel's ire; chartism, he called it. Not all facts were equally important, he insisted; and the data were not the ends, but merely the preliminaries to theory. Sabine, however, relished facts and was dubious of theoreticians' speculations.[1] In his view, the magnetic work was following the precedent of astronomy, a Baconian science in which accumulations of facts yielded sound theory. When the data from around the globe did not wholly validate Gauss, Sabine undoubtedly was pleased. Even more impressive was his discovery in 1851 of the relation between Schwabe's sunspot cycle and the periodicity of magnetic storms, even though it was marred by a priority squabble with Johann von Lamont. It was a vindication of a long-held belief. In the same year Sabine announced his important finding that the daily magnetic variation consists of two superimposed variations, one deriving from within and the other from outside the earth.

Much of science in Victorian Britain, not merely geomagnetism and related topics, is explicable in terms of Sabine's career. He was the artful dodger of the British scientific establishment. Bright, energetic, shrewd, he could have been the very model of Gilbert and Sullivan's modern major general. Although his publications are properly specialized, Sabine's range of interests was quite broad, as is evident from his association with Gassiot and the work at Kew of Francis Ronalds, John Welsh, and Balfour Stewart. He even published pieces on Arctic ornithology and Eskimos. As a proper Humboldtian or even a Herschelian, Sabine considered geomagnetism an aspect of an interconnected nature. Such a man commanded respect, loyalty, and even affection. Yet his enthusiasm, verging on the fanatic, and his intellectual limitations became increasingly tyrannical as the climate of ideas changed and aging took its toll.

Quite early, Sabine disclosed a talent for influencing the influentials and slipping into strategic positions. This talent produced the greatest embarrassment of his career. A member of the Royal Society Council in 1828, he was named an adviser to the Admiralty, an appointment that aroused Babbage's ire. Two years later Babbage, in *Reflections on the Decline of Science in England*, accused Sabine of falsifying data. If we are to believe his anonymous necrologist, Sabine was at least guilty of great naïveté in handling numbers, hardly an auspicious start for a career involving vast quantities of worldwide data. Babbage's attack also placed Sabine in the awkward position of being seemingly outside the wave of the future in the politics of British science—in the camp of the old guard.

Ever resilient, Sabine became active in Babbage's creation, the British Association, when furthering terrestrial magnetism called for that move; and soon he was an officer. The pattern was fixed and quite simple. Sabine became entrenched in both the Royal Society and the British Association, shifting programs adroitly from one to another to gain objectives, as in the cases of the magnetic crusade and the Kew observatory. Obstacles were evaded or removed. One suspects that more than coincidence was involved in Sabine's election as one of the general secretaries of the British Association at Newcastle in 1838 and the simultaneous resignation of Babbage as a trustee. Sabine remained a general secretary until 1859 with the exception of 1852, when he was president of the British Association. At various times in that period (1841–1861), in addition to Council membership, Sabine was foreign secretary, vice-president, and treasurer of the Royal Society. From 1861 until 1871 he was its president.

His unpublished correspondence discloses that Sabine, distressed by the disputes over reforming the Royal Society, viewed the magnetic crusade as a happy opportunity for the scientific community to present a united front. Abhorring divisions in the ranks of science, he and Grove tried to have the scientific societies unite when Burlington House became available. In the late 1840's Sabine, again with Grove, played a leading role in carrying out the reforms of the Royal Society that largely answered the earlier complaints of Davy and Babbage about the election of fellows.

Relying upon the support of the like-minded, Sabine quietly ensured that his intellectual interests received the lion's share of British Association funds. This is evident from its annual reports through the time of his resignation from the presidency of the Royal Society. Kew Observatory was the largest single recipient, but the number of grants to geophysical and related areas is notable. A suspicion arises that Sabine backed the move for the £1,000 parliamentary grant partly to quiet criticism from relatively neglected disciplines. Even allowing for his sincerity and for the general quality of the research, this allocation of resources was dubious, especially in the face of developments in Germany.

In time, Sabine's became a dead hand at the tiller, frozen on an old course. When Faraday reported his experimental demonstration of the relationship of light and magnetism in 1845, Sabine argued against giving him the Rumford Medal. James Clark Ross in 1834, Sabine asserted, had already described the effect as naturally observed on an Arctic voyage. Sabine was overruled.[2] When Grove was proposed for the Copley Medal in 1871, Sabine trotted out procedural quibbles to deny the honor.[3] It was Sabine who in 1863 answered the clamor of the younger naturalists for awarding the Copley Medal to Darwin by seeing that Adam Sedgwick was chosen.[4] (Note how well Sabine's later actions accorded with Babbage's earlier complaints about the Royal Society's distri-

bution of awards.) Accused by Tyndall of neglecting natural history, the octogenarian oligarch resigned the Royal Society presidency in 1871 when he realized the days of artful dodging were over.[5]

NOTES

1. Sabine to Tyndall, 24 Apr. [1855], vol. IV, 1307 of Tyndall Papers, Royal Institution: "I notice that he [Secchi] gives me credit for abstaining from all such attempts at combining facts and hypothetical connections or views. I have adhered to this quite as a duty—but have made my writings far less *interesting* than they might have been otherwise, thereby. For to many men speculations are far more attractive than facts."
2. Royal Society, *Minutes of Council*, printed ser., I, 512–513, 530–531. In Royal Society Library.
3. Gassiot, recommendation of Copley Medal for Grove, 18 Oct. 1871 (copy), with undated note by Grove on the events in the Royal Society Council: "My chance never recurred." Grove Papers, Royal Institution.
4. Sabine to John Phillips, 12 Nov. 1863, Miscellaneous MSS Collection, Library of the American Philosophical Society.
5. J. P. Gassiot, *Remarks on the Resignation of Sir Edward Sabine, K.C.B., of the Presidency of the Royal Society* (London, 1871).

BIBLIOGRAPHY

I. PRIMARY SOURCES. This account is largely derived from unpublished sources. The best collection of Sabine documents is in the archives of the Meteorological Office, Bracknell, Berkshire, as part of the records of the Kew Observatory. A smaller but valuable body of Sabine letters is in the library of the Royal Society, which also contains the papers of John Herschel, an essential source. The papers of Humphrey Lloyd, Sabine's colleague, are divided between the Royal Society and the archives of the Royal Greenwich Observatory. Unfortunately, few records of James Clark Ross survive: some are included in the two bodies of Sabine papers, and there is a smaller batch on sunspots at the Royal Greenwich Observatory. The Airy Papers at Greenwich are useful in presenting the views of an opponent of Sabine. Particularly valuable are the correspondence and minutes of the Royal Society's Committee on Physics (Including Meteorology) and its predecessors, in the archives of the Society. The Grove and Tyndall collections at the Royal Institution are very pertinent. Of the non-British collections, the Hansteen MSS at the University of Oslo are a rich, still largely unexplored source. The few items in the Gauss *Nachlass* at Göttingen are useful; the Quetelet Papers in the Académie Royale de Belgique are an extensive, rich source. Sabine materials at the American Philosophical Society and in the correspondence of Sabine's American friends A. D. Bache and Joseph Henry are also valuable.

Sabine's numerous articles (more than 100) are well covered in the Royal Society *Catalogue of Scientific*

Papers, V, 351–354; VIII, 805–806; XI, 251. In addition, see the following: *Remarks on the Account of the Late Voyage of Discovery to Baffin's Bay, Published by Captain J[ohn] Ross* (London, 1819) and the rejoinder by Ross, *An Explanation of Captain Sabine's Remarks on the Late Voyage of Discovery to Baffin Bay* (London, 1819); *North Georgia Gazette and Winter Chronicle* (1821); *An Account of Experiments to Determine the Figure of the Earth . . .* (London, 1825); the article on magnetism in the three eds. of the Admiralty's *A Manual of Scientific Enquiry . . .*; Sabine edited the 3rd ed. (London, 1859); *Observations on the Days of Unusual Magnetic Disturbances Made at the British Colonial Magnetic Observatories*, 2 vols. (London, 1843–1851); and 10 vols. of observations at the observatories. Sabine helped prepare the translations of Gauss and Weber's "Results of the Observations Made by the Magnetic Association in the Year 1836," in Taylor's *Scientific Memoirs*, **2** (1841), 20–25. Under his "superintendence," his wife, Elizabeth Julian Sabine, translated works by Humboldt and Arago.

II. SECONDARY SOURCES. See the following, arranged chronologically: John Ross, *A Voyage of Discovery in H. M. Ships Isabella and Alexander* (London, 1819); William Edward Parry, *Journal of a Voyage for the Discovery of a North-West Passage* (London, 1821); C. Babbage, *Reflections on the Decline of Science in England* (London, 1830); "Memoir of General Sir Edward Sabine, F.R.S., K.C.B.," in *Proceedings of the Royal Artillery*, **12** (1883), 381–396, unsigned, but obviously written by a military associate in the magnetic work; S. Chapman and J. Bartels, *Geomagnetism*, 2 vols. (Oxford, 1940); Johannes Georgi, "Edward Sabine, ein grosser Geophysiker des 19 Jahrhunderts," in *Deutsche hydrographische Zeitschrift*, **11** (1959), 225–239; and Nathan Reingold, "Babbage and Moll on the State of Science in Great Britain . . .," in *British Journal for the History of Science*, **4** (1968), 58–64.

NATHAN REINGOLD

SABINE, PAUL EARLS (*b.* Albion, Illinois, 22 January 1879; *d.* Colorado Springs, Colorado, 28 December 1958), *acoustics*.

Sabine was the son of a Methodist minister. After graduating from McKendree College in 1899 he went to Harvard University, where his cousin, Wallace Clement Sabine, was on the physics faculty. At Harvard he earned the baccalaureate (1903), the master's (1911), and the doctorate in physics (1915). Sabine taught at Worcester Academy (1903–1910), served as assistant in physics at Harvard (1915–1916), and was an assistant professor at Case School of Applied Science in Cleveland (1916–1918). In 1919 he became director of acoustical research at the Wallace Clement Sabine Laboratory of Acoustics, better known as the Riverbank Laboratory, in Geneva, Illinois. He remained there until his retirement in 1947, except for a period of war work at the Harvard Underwater Sound Laboratory (1942–1945). Sabine was a charter member of the Acoustical Society of America (1929), served as its fourth president (1935–1937), and was elected to honorary membership in 1954. After his retirement, while continuing to be active as a consultant, he turned his thoughts to the reconciliation of Christianity with the results of modern physical science and psychology. Sabine published his conclusions in *Atoms, Men and God*, his second book.

The bulk of Sabine's scientific work followed directly from Wallace Sabine's recognition of the reverberation time as the most significant variable affecting the acoustical quality of listening rooms and his subsequent discovery of the empirical relationship between reverberation time and total absorption. Total absorption was a summation of the sound-absorbing power of the various constituents of the room, a quantity that Wallace Sabine defined and showed how to measure. The Riverbank Laboratory was built for Wallace Sabine's use in determining the sound-absorptive properties of architectural materials as well as the sound absorption and transmission characteristics of architectural elements and types of construction. Wallace Sabine died just as the Riverbank Laboratory was completed, and Paul Sabine was appointed to carry out the research program that would make it possible to design acoustical environments on a scientific basis. The results of this research were incorporated in his textbook on acoustical design, *Acoustics and Architecture* (1932).

Sabine was also active as an acoustical consultant to architects. He was consulted in the design of the Radio City Music Hall in New York and of the Fels Planetarium in Philadelphia. He was most proud, however, of his work in the planning of the remodeling of the House and Senate Chambers of the United States Capitol Building after World War II. The acoustical design of the remodeled chambers was a notable success.

BIBLIOGRAPHY

Sabine's books are *Acoustics and Architecture* (New York–London, 1932) and *Atoms, Men and God* (New York, 1953). His scientific papers may be located through the cumulative indexes of the *Journal of the*

Acoustical Society of America and through the citations in *Acoustics and Architecture*. Several of his papers, as well as extensive sections of *Acoustics and Architecture,* offer retrospectives of the development of architectural acoustics as it grew from the work of Wallace Sabine.

CARLTON MALEY

SABINE, WALLACE CLEMENT WARE (*b*. Richwood, Ohio, 13 June 1868; *d*. Cambridge, Massachusetts, 10 January 1919), *physics*.

Sabine's parents, Hylas Sabine and Anna Ware, were both college-educated and had a strong interest in literature and science. His father, at one time a member of the Ohio State Senate and state commissioner of railways and telegraphs, was a farmer and landowner who lost most of his holdings in the panic of 1873. His mother, eager to see her two children do better, raised them both under a stern moral and educational regimen. The elder Sabines were practicing Protestants, but as an adult Wallace belonged to no church and professed no religious faith.

After earning an A.B. at Ohio State University in 1886, Sabine went to Harvard, where in 1888 he was awarded an M.A. in physics and in 1890 appointed to an instructorship. Neglecting to take a Ph.D., he devoted himself to teaching, and his courses were among the most popular in the department. A full professor in 1905, he was instrumental in the creation of the Harvard graduate school of applied science, which he administered as dean from 1906, the year of its founding, until 1915.

Following the United States's declaration of war in 1917, Sabine held various administrative posts in what became the Army Air Service. In June 1918 he became director of the Department of Technical Information of the Bureau of Aircraft Production, and that September he was appointed to the National Advisory Committee for Aeronautics by President Woodrow Wilson. At his death Sabine was a member of the American Physical Society and the National Academy of Sciences and a fellow of the American Academy of Arts and Sciences and the American Association for the Advancement of Science. He was married in 1900 to Jane Downes Kelly, a physician of Cambridge, Massachusetts.

As a research physicist Sabine is known for having turned architectural acoustics from a qualitative, rule-of-thumb practice into a quantitative engineering science. He started work in this field in 1895, when Charles William Eliot, the president of Harvard, asked him to do something about the very poor acoustics of the lecture hall in the university's new Fogg Art Museum. Measuring the time during which a given sound reverberated within the hall, Sabine found that a single syllable of speech persisted long enough to overlap confusingly with those that followed it. By hanging sonically absorptive materials on the walls, he reduced the reverberation time and, hence, improved the acoustical quality of the room.

In 1898, at Eliot's urging, the architectural firm of McKim, Mead, and White turned to Sabine for advice on the design of Symphony Hall in Boston. Using the raw data from his Fogg Museum experiments, Sabine managed, with the ingenious use of graphs, to derive an acoustical law of general applicability. He showed that the product of the reverberation time and the summed absorptive power of the walls, furnishings, and materials of appointment equaled a constant; and that this constant was directly proportional to the volume of the room. The formula enabled Sabine to predict the acoustical properties of an auditorium in advance of construction. The practical value of his law was confirmed by the acoustical success of Symphony Hall, and its essential scientific validity was demonstrated by a later analysis of reverberation that employed statistical methods from the kinetic theory of gases.

In 1900, in a comprehensive paper on reverberation, Sabine set down what have since been accepted as the three basic criteria for good acoustical quality in any auditorium: sufficient loudness, minimal distortion, and maximum distinctness. In subsequent years Sabine, who made his expertise available free to numerous architects, investigated how interference and resonance affect acoustics and the best way of sonically insulating a room. In honor of Sabine's seminal significance in architectural acoustics, the unit of sound-absorbing power is called the sabin.

BIBLIOGRAPHY

Sabine's *Collected Papers on Acoustics* (Cambridge, Mass., 1922) contains almost all of his important articles. A useful introduction to his life and work is Edwin H. Hall, "Wallace Clement Sabine," in *Biographical Memoirs. National Academy of Sciences,* **11,** no. 13 (1926), 1–19. William Dana Orcutt, *Wallace Clement Sabine: A Study in Achievement* (Norwood, Mass., 1933), apparently was commissioned by his widow and emphasizes Sabine's personal life and character.

DANIEL J. KEVLES

SACCHERI, (GIOVANNI) GIROLAMO (*b.* San Remo, Italy, 5 September 1667; *d.* Milan, Italy, 25 October 1733), *mathematics.*

Saccheri is sometimes confused with his Dominican namesake (1821–1894), a librarian at the Bibliotheca Casanatense of Rome. In 1685 Saccheri entered the Jesuit novitiate in Genoa and after two years taught at the Jesuit college in that city until 1690. Sent to Milan, he studied philosophy and theology at the Jesuit College of the Brera, and in March 1694 he was ordained a priest at Como. In the same year he was sent to teach philosophy first at Turin and, in 1697, at the Jesuit College of Pavia. In 1699 he began teaching philosophy at the university, where until his death he occupied the chair of mathematics.

One of Saccheri's teachers at the Brera was Tommaso Ceva, best known as a poet but also well versed in mathematics and mechanics. Through him Saccheri met his brother Giovanni, a mathematician living at the Gonzaga court in Mantua. This Ceva is known for his theorem in the geometry of triangles (1678). Under Ceva's influence Saccheri published his first book, *Quaesita geometrica* (1693), in which he solved a number of problems in elementary and coordinate geometry. Ceva sent this book to Vincenzo Viviani, one of the last surviving pupils of Galileo, who in 1692 (*Acta eruditorum*, 274–275) had challenged the learned world with the problem in analysis known as the window of Viviani. Although it had been solved by Leibniz and others, Viviani published his own solution and sent it to Saccheri in exchange for the *Quaesita.* Two letters from Saccheri to Viviani (1694) are preserved, one containing Saccheri's own solution (without proof).

While in Turin, Saccheri wrote *Logica demonstrativa* (1697), important because it treats questions relating to the compatibility of definitions. During his years at Pavia he wrote the *Neo-statica* (1708), inspired by and partly a polemic against T. Ceva's *De natura gravium* (Milan, 1669). This book seems of little importance now, being well within the bounds of Peripatetic statics. *Euclides ab omni naevo vindicatus* (1733), also written at Pavia, contains the classic text that made Saccheri a precursor of the discoverers of non-Euclidean geometry.

Saccheri's two most important books, the *Logica* and the *Euclides,* were virtually forgotten until they were rescued from oblivion—the *Euclides* by E. Beltrami in 1889 and the *Logica* by G. Vailati in 1903. They show that Euclid's fifth postulate (equivalent to the parallel axiom) intrigued Saccheri throughout his life. In the *Logica* it led him to investigate the nature of definitions and in the *Euclides* to an attempt to apply his logic to prove the correctness of the fifth postulate. Although the fallacy in this attempt is now apparent, much of Saccheri's logical and mathematical reasoning has become part of mathematical logic and non-Euclidean geometry.

The *Logica demonstrativa* is divided into four parts corresponding to Aristotle's *Analytica priora, Analytica posteriora, Topica,* and *De sophisticis Elenchis.* It is an attempt, probably the first in print, to explain the principles of logic *more geometrico.* Stress is placed on the distinction between *definitiones quid nominis* (nominal definitions), which simply define a concept, and *definitiones quid res* (real definitions), which are nominal definitions to which a postulate of existence is attached. But when we are concerned with existence, the question arises whether one part of the definition is compatible with another part. This may be the case in what Saccheri called complex definitions. In these discussions he was deeply influenced by Euclid's *Elements,* notably by the definition of parallelism of two lines. He warned against the definition, given by G. A. Borelli (*Euclides restitutus* [Pisa, 1658]), of parallels as equidistant straight lines. Thus Saccheri was one of the first to draw explicit attention to the question of consistency and compatibility of axioms.

To test whether a valid proposition is included in a definition, Saccheri proposed reasoning seemingly analogous to the classical *reductio ad absurdum,* using for his example *Elements* IX, 12: if $1, a_1, a_2, \cdots, a_n$ form a geometric progression and a_n has a prime factor p, then a_1 also contains this factor. There was a difference in Saccheri's proposal, however: his demonstration resulted from the fact that, reasoning from the negation, we obtain exactly the proposition to be proved, so that this proposition appears as the consequence of its own negation (an example of his reasoning is seen below). As Vailati observed, Saccheri's reasoning had much in common with that of Leibniz (see L. Couturat, *Opuscules et fragments inédits de Leibniz* [Paris, 1903]); but whereas Leibniz's primary inspiration came from algebra and the calculus, Saccheri's came from geometry.

In the *Euclides* Saccheri applied his logical principle to three "blemishes" in the *Elements.* By far the most important was his application of his type of *reductio ad absurdum* to Euclid's parallel axiom. He took as true Euclid's first twenty-six propositions and then assumed that the fifth postulate was false. Among the consequences of this hypothesis he sought a proposition to test the postulate itself. He found it in what is now called the quadri-

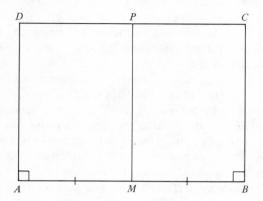

lateral of Saccheri, an isosceles birectangular quadrilateral consisting of a side *AB* and two sides of equal length, *AD* and *BC* at right angles to *AB*. Then without the fifth postulate it cannot be proved that the angles at *C* and *D* are right. One can prove that they are equal, since if a line *MP* is drawn through the midpoint *M* of *AB* perpendicular to *AB*, it intersects *DC* at its midpoint *P*. Thus there are three possibilities, giving rise to three hypotheses:

1. that of the right angle: $\angle C = \angle D = 1$ right angle;

2. that of the obtuse angle: $\angle C = \angle D > 1$ right angle;

3. that of the acute angle: $\angle C = \angle D < 1$ right angle.

Saccheri proceeded to prove that when each of these hypotheses is true in only one case, it is true in every other case. Thus in the first case the sum of the angles of a triangle is equal to, in the second it is greater than, and in the third case it is less than, two right angles.

For the proofs Saccheri needed the axiom of Archimedes and the principle of continuity. Then came the crucial point: he proved that for both the hypothesis of the right angle and that of the obtuse angle the fifth postulate holds. But the fifth postulate implies the hypothesis of the right angle; hence the hypothesis of the obtuse angle is false. (This argument is not now cogent because in the case of the obtuse angle the existence of the finite length of lines is accepted.) He could not dispose of the hypothesis of the acute angle in this way, but he was able to show that it leads to the existence of asymptotic straight lines, which, he concluded, was repugnant to the nature of the straight line. Saccheri thus thought that he had established the truth

of the hypothesis of the right angle and, hence, of the fifth postulate and of Euclidean geometry as a whole.

Several other theorems resulted from Saccheri's three hypotheses, some of which are now established as part of non-Euclidean geometry. The three types of quadrangles had already been studied by al-Khayyāmī and Nasīr-al-Dīn al-Tūsī; the latter was cited by John Wallis (1693) in a book known to Saccheri.

Saccheri's *Euclides,* although it had little direct influence on the subsequent discovery of non-Euclidean geometry, was not so forgotten as is sometimes believed. (See Segre, below.)

BIBLIOGRAPHY

I. ORIGINAL WORKS. Saccheri's writings include *Quaesita geometrica a comite Rugerio De Vigintimilliis* . . . (Milan, 1693), included in *Sphinx geometra, seu quesita geometrica proposita et solida* . . . (Parma, 1694); *Logica demonstrativa quam una cum thesibus ex tota philosophiae decerptis defendendam proposuit J. F. Caselette Graveriarum Comes* (Turin, 1697), 2nd ed. entitled *Logica demonstrativa auctore Hieronym. Saccherio Societatis Jesu* . . . (Pavia, 1701), 3rd ed. entitled *Logica demonstrativa, theologicis, philosophicis et mathematicis disciplinis accomodata* . . . (Cologne, 1735); *Neo-statica* . . . (Milan, 1708); and *Euclides ab omni naevo vindicatus: Sive conatus geometricus quo stabiliuntur prima ipsa universae geometriae principia* . . . (Milan, 1733). Theological works are listed in P. C. Sommervogel, *Bibliothèque des écrivains de la Compagnie de Jésus, VII* (Brussels–Paris, 1897), 360.

Letters by Saccheri to Viviani, Ceva, and Grandi are in A. Favaro, "Due lettere inedite del P. Girolamo Saccheri d. C. d. G. a Vincenzo Viviani," in *Rivista di fisica, matematica e scienze naturali,* 4 (1903), 424–434; A. Pascal, "Sopra una lettera inedita di G. Saccheri," in *Atti del R. Istituto Veneto di scienze, lettere ed arti,* 74 (1914–1915), 813–820; and A. Agostini, "Due lettere inedite di G. Saccheri," in *Memorie della R. Academia d'Italia,* Cl. di Scienze Matematiche e Naturale, 2, no. 7 (1931), 31–48.

II. SECONDARY LITERATURE. The full text of the *Logica demonstrativa,* with English trans., is in A. F. Emch, "The *Logica demonstrativa* of Girolamo Saccheri" (Ph.D. diss., Harvard, 1933), with a life of Saccheri by F. Gambarana from a MS at the Biblioteca Estense in Modena. The *Logica* is discussed by Emch in articles of the same title as his dissertation in *Scripta mathematica,* 3 (1935–1936), 51–60, 143–152, 221–233. On the *Logica,* see also G. Vailati, "Di un' opera dimenticata del P. Gerolamo Saccheri," in *Rivista filosofica,* 4 (1903), 528–540; with other papers on the book, it is also in *Scritti di G. Vailati (1863–1909)* (Leipzig–Florence, 1911), 477–484, see also 449–453.

Also of value is F. Enriques. *Per la storia della logica* (Bologna, 1922), 94–99, also available in French, German, and English.

The *Euclides ab omni naevo vindicatus* has been partially translated (only bk. I with the discussion of the parallel axiom) by P. Stäckel and F. Engel in *Die Theorie der Parallellinien von Euclid bis auf Gauss* (Leipzig, 1895), 31–136, and into English by G. B. Halsted in *Girolamo Saccheri's Euclides vindicatus* (Chicago–London, 1920). See also ten articles by Halsted on non-Euclidean geometry in *American Mathematical Monthly,* **1** (1894), see index, p. 447. There is an unsatisfactory Italian trans. by G. Boccardini, *L'Euclide emendato . . .* (Milan, 1904). Further literature on the *Euclides* (listed chronologically) includes E. Beltrami, "Un precursore italiano di Legendre e di Lobatschewsky," in *Atti della Reale Accademia Lincei, Rendiconti,* 4th ser., **5**, no. 1 (1889), 441–448, also in Beltrami's *Opere,* IV (Milan, 1920), 348–355; P. Mansion, "Analyse des recherches du P. Saccheri S.J. sur le postulatum d'Euclide," in *Annales de la Société scientifique de Bruxelles,* **14** (1889–1890), pt. 2, 46–59, also in *Mathesis,* 2nd ser., **1** (1891), supp. 15–29; C. Segre, "Congettare informo all'influenza di Girolamo Saccheri sulla formazione della geometrica non-euclidea," in *Atti dell'Accademia delle scienze* (Turin), **38** (1902–1903), 535–547; A. Pascal, "Girolamo Saccheri nella vita e nelle opere," in *Giornale di matematica di Battaglini,* **52** (1914), 229–251; and H. Bosmans, "Le géomètre Jérome Saccheri S. J.," in *Revue des questions scientifiques,* 4th ser., **7** (1925), 401–430.

Saccheri's contribution to non-Euclidean geometry is discussed in R. Bonola, *La geometria non-euclidea* (Bologna, 1906), also in English (Chicago, 1911; repr. 1955) and in German (Berlin, 1908). See also article on Saccheri by E. Carruccio, *Enciclopedia italiana di scienze, lettere ed arte,* **30** (Rome, 1936), 389–390.

D. J. STRUIK

SACCO, LUIGI (*b.* Varese, Lombardy, Italy, 9 March 1769; *d.* Milan, Italy, 26 December 1836), *medicine.*

Sacco obtained his degree in medicine and surgery in 1792 at the University of Pavia, where he was a pupil of Johann Peter Frank, the founder of social medicine. Sacco subsequently established a medical practice in Milan, where he became a friend of the physician Pietro Moscati. In 1778 Moscati had given the first public demonstrations in Milan of smallpox inoculation. He later became one of the foremost political personalities in Jacobinic and Napoleonic Milan.

Following Jenner's publication (1798) of his work on cowpox inoculation, Sacco had the good fortune, in September 1800, to find a spontaneous cowpox stock in the neighborhood of Varese. He used this stock to inoculate first himself and then a group of children on the farm where he was staying. From these and numerous other inoculations, he recognized the advantages of inoculating cowpox rather than human smallpox. He decided to publicize this new prophylactic practice, and he realized the importance of giving his work a social and political flavor in accordance with the new times.

Sacco persuaded the government of the Cisalpine Republic to set up a general vaccination department, which was entrusted to him. This department allowed Sacco to extend his work to many other regions of Italy besides Lombardy. By 1809 he had succeeded in reaching "a million and a half vaccinated people, five hundred thousand of whom I have had the satisfaction of vaccinating myself." In the same year he left his post as general director of vaccination and gave to the press his *Trattato di vaccinazione.* Shortly afterward this treatise was translated into French and German, thus presenting the important conclusions that he had been able to draw from his ample statistics. Sacco became a major advocate of cowpox vaccination, and his stock from Lombardy was sent to Jean de Carro in Vienna. The latter sent it in 1802 to Baghdad, and it was with this stock that the first vaccinations were given in the East Indies.

Sacco subscribed to the theory of *contagium vivum,* but he believed that the leukocytes in pus were the *animalcula* that caused disease and transmitted it from one individual to another by contagion. From 1803 he was chief physician at the Ospedale Maggiore in Milan and for several years served as director.

BIBLIOGRAPHY

I. ORIGINAL WORKS. Sacco's major works are *Osservazioni pratiche sull'uso del vajuolo vaccino, come preservativo del vajuolo umano* (Milan, 1801); *Omelia sopra il Vangelo della XIII. Domenica dopo la Pentecoste, in cui si parla dell'utile scoperta dell'innesto del vajuolo vaccino, recitata dal Vescovo di Goldstat, dalla Tedesca nell'Italiana lingua transportata* (Brescia, 1802; Parma, 1805; Pistoia, 1805); *Memoria sul vaccino unico mezzo per estirpare radicalmente il vajuolo umano diretta ai Governi che amano la prosperità delle loro nazioni* (Milan, 1803); and *Trattato di vaccinazione con osservazioni sul giavardo e sul vajuolo pecorino* (Milan, 1809), with French and German trans. as *Traité de vaccination . . . ,* Joseph Daquin, trans. (Chambéry, 1811), and *Neue Entdeckungen über die Kuhpocken . . . ,* Wilhelm Sprengel, trans. (Leipzig, 1812), respectively.

II. SECONDARY LITERATURE. On Sacco and his work, see Luigi Belloni, "L'innesto del vaccino," in *Storia di Milano*, XVI (1962), 960–971, with full bibliography; "Una ricerca del contagio vivo agli albori dell'Ottocento," in *Gesnerus*, **8** (1951), 15–31; and "Per la storia dell'innesto del vaiuolo a Milano," in *Physis*, **2** (Florence, 1960), 213–222.

LUIGI BELLONI

SACHS, JULIUS VON (*b.* Breslau, Silesia [now Wrocław, Poland], Germany, 2 October 1832; *d.* Würzburg, Germany, 29 May 1897), *botany, plant physiology.*

Sachs was the eighth of nine children of Christian Gottlieb Sachs, an engraver, and the former Maria-Theresia Hofbauer, who were quite poor. From 1840 to 1845 the gifted boy attended the poorly run seminary school in Breslau and, from 1845 to 1850, the Gymnasium, where he frequently was first in his class. After the deaths of his father (1848) and mother (1849), he had no means of support and was forced to leave school. Sachs had met the physiologist Purkyně, who took him to Prague as his personal assistant, where his principal duties were those of draftsman. In Prague, Sachs took his final secondary school examination (1851), after which he studied at the University of Prague until 1856. The most lasting intellectual influence upon him in this period was that of the philosopher Zimmermann. The botany and zoology lectures failed to hold his interest, but research that he conducted on his own led to eighteen publications on botanical and zoological topics in this period. These papers, general treatments designed for a popular audience, were translated into Czech and published in Purkyně's journal *Živa*.

Sachs received the Ph.D. in 1856; and in the same year he attended a scientific congress at Vienna, where he met many prominent botanists. The next year he qualified at Prague—although the faculty did not wish the new subject to be taught—as a lecturer in plant physiology. Thus, Sachs, who later became the leader in the field, was the first to teach a whole course of the subject at a German university. At this time plant physiology encompassed the whole of botany except systematics.

In 1859 Sachs became assistant in plant physiology at the Agricultural and Forestry College in Tharandt, near Dresden. With Wilhelm Hofmeister, he began to edit the *Handbuch der physiologischen Botanik* in 1860; and the following year he became a botany teacher at the Agricultural College in Poppelsdorf, near Bonn. During these extremely productive years at Poppelsdorf he laid the foundations of all his later scientific work. Sachs succeeded Anton de Bary as professor at Freiburg im Breisgau in 1867 but left the following year to become full professor of botany at Würzburg, a post he held for the rest of his life. Except for a trip to Norway and several visits to Italy, Sachs never left Würzburg. He refused offers that followed in rapid succession from Jena (1869), Heidelberg (1872), Vienna (1873), and Berlin (1877), as well as later ones from the Agricultural College of Berlin, from Bonn, and finally one from Munich (1891) to succeed Naegeli.

A brilliant lecturer and a highly imaginative experimenter, Sachs won fellow scientists to his views through the persuasive logic of his arguments. These talents, joined with his position as leader of the rapidly developing science of plant physiology, earned him an international reputation.

Sachs received many honors. The University of Würzburg elected him rector in 1871, and in 1877 he was named privy councillor and awarded the Order of Maximilian and the Order of the Bavarian Crown. This was accompanied by a grant of personal nobility, entitling Sachs to place "von" before his name. He also received honorary doctorates from the universities of Bonn, Bologna, and London. He was, in addition, a member or honorary member of many scientific societies and academies, including those of Frankfurt, Munich, Turin, and Amsterdam, as well as the Linnean Society of London and the Royal Society. Many important scientists were his students or worked for a time in his institute: Francis Darwin, Goebel, Klebs, F. Noll, Pfeffer, Stahl, De Vries, S. H. Vines, and Appel.

Throughout his life Sachs displayed an enormous appetite for work. Unfortunately, the intense inner restlessness that constantly drove him to new efforts and achievements also severely damaged his health, to such an extent that the letters from the last fifteen years of his life constitute one protracted health report (Goebel). He suffered from nervous disorders and excruciating pains—probably neuralgic—accompanied by insomnia and aggravated by extensive damage to his liver and kidneys.

Like many other outstanding scientists, Sachs was often overbearing and unfair. In scientific controversies and in many letters he occasionally adopted a harsh and implacable tone—which, to be sure, was not unusual in scholarly disputes of the nineteenth century. He was extremely reserved toward those around him, including most of his students, and had close ties with only a few people.

When Sachs began his scientific research, plant physiology was a totally neglected field; it became developed only through his work, which extended to nearly all branches of the subject. Even Sachs's earliest independent investigations aroused general admiration and are still of value. In the course of this research on the metabolism of stored nutrients during the germination of seeds (1858–1859), he discovered the transformation of oil into starch in *Ricinus* seeds. His work was characterized by a combination of microscopic and microchemical methods, by means of which he provided a clear picture of the catabolism and transport of stored nutrients. Another early investigation dealt with the culture of plants in pure nutrient solution (1860).

Pursuing research begun by Liebig, Sachs solved both practical and theoretically important problems regarding the mineral requirements of plants. In this connection he discovered the corrosive action of roots on marble slabs, indicating their ability to sequester minerals (1860) and the toxicity of solutions containing a single salt. He studied the influence of temperature on life processes (1860), especially the effects of freezing. He discovered the law of "cardinal points," according to which each vital process has a minimum, an optimum, and a maximum temperature that are mutually related.

Particularly important was Sachs's demonstration, beginning in 1861, that the starch in the chloroplasts is the first visible product of assimilation and that carbon dioxide assimilation (photosynthesis) actually occurs in chloroplasts. These discoveries, like many others that he made, are cornerstones of modern plant physiology.

Further experiments dealt with the effect of light and, above all, with the origins of etiolation (1862) and the formation of flowers and roots (from 1865). His highly significant studies of growth and its mechanisms in roots (from 1872) and shoots led to the discovery of the "great period of growth." Sachs also demonstrated that the formation of plants depends more on processes of cell enlargement than on those of cell division in the meristem. From about 1873 Sachs devoted increasing attention to the physiology of stimuli: geotropism, "heliotropism" (phototropism), and hydrotropism.

In his later investigations and theoretical papers, which lacked much of the experimental ardor of the earlier ones, Sachs sought evidence for his theory of "specific organ-forming substances." This theory took as its starting point the fact that although plants can grow in the dark, they cannot form flowers there (1865). From this, he contended, it follows that specific substances necessary for the formation of flowers are produced in the leaves and that these substances are essential for this development. Sachs claimed that similar substances cause differentiation in the shoots and roots. In his last publications he gave a detailed account of this theory of "matter and form," which was simultaneously a challenge to idealistic morphology as advocated by Alexander Braun.

Although modern plant physiology would be inconceivable without Sachs's contributions in these areas, he was, however, less successful in dealing with certain other questions, especially those concerning the transport of nutrients and water. He stubbornly held to the theory—not original with him—that water is conducted in the cell walls of wood (imbibition theory). Insisting that the sieve tubes play no role in transporting carbohydrates, he maintained instead that the latter are transported in the form of "wandering starch." Moreover, his great authority long delayed discovery of the real answers to these questions. Equally untenable were the attacks he made in the last years of his life against the mechanisms proposed by Darwin in his theory of evolution and against other writings of Darwin.

Sachs's skill in experimentation was astonishing, especially in view of the rudimentary methods available when he began his research. He was constantly concerned to point out the independence of physiology from physics and chemistry: "More and more I find that physiology achieves its most important results when it goes its own way entirely, without concerning itself very much with physics and chemistry" (letter of 15 May 1879). This is not to imply that Sachs did not use chemical and physical knowledge in his research. That he did so can be clearly seen, for example, in the number of microchemical demonstration methods he used, many of which he himself devised. Further examples are his proof of the existence of starch in the whole leaf by the iodine test, still used as a lecture experiment, and the gas bubble method for demonstrating the formation of oxygen in photosynthesis, also still used in laboratory classes.

Sachs invented or at least substantially improved many of the devices that were long prominent in botany laboratories: the hanging sieve, for demonstrating hydrotropism; the root box, for making visible the growth and branching of roots; the auxanometer, for automatically recording the processes of growth; a hand spectroscope, for measuring the light absorbed through the leaves; the clinostat, for compensating for gravity; centrifuges, for experiments involving centrifugal force; and thermostats and boxes for unilateral illumination. Although much of this apparatus now appears simple, even primitive, it was revolutionary in Sachs's day, when scarcely any apparatus other than the microscope was used. Sachs also was responsible for innovations in experimental technique, having introduced, for example, the use of seedlings in order to obtain a large number of uniform plants.

Many of the experimental papers written by Sachs and his students were published in *Arbeiten des Botanischen Instituts in Würzburg.* The most important were collected as *Gesammelte Abhandlungen über Pflanzenphysiologie.*

Sachs also exerted a major influence through his books. In *Geschichte der Botanik vom 16. Jahrhundert bis 1860* he went far beyond a dry historical description, presenting with great skill the basis of his own scientific work. This masterful presentation records all the fundamental elements upon which Sachs built his many theories. His immense contribution to the subject is most clearly apparent in *Lehrbuch der Botanik* (which went through four editions) and in *Vorlesungen über Pflanzenphysiologie* (which appeared in two editions). The *Vorlesungen* in particular, by virtue of the freer style permitted by the lecture form, illustrates Sachs's achievements in an especially vivid manner. The material it contains, much of which Sachs himself had elaborated, meshes harmoniously with contemporary knowledge in related branches of science and with it forms a unified whole.

Sachs's books were long the definitive works in plant physiology. Through them the results both of his own research and of that of many contemporaries, such as Hofmeister and Naegeli, became widely known. In fact, so complete was this process that it is frequently difficult to isolate Sachs's personal contributions. The numerous illustrations in these books, the majority of which derived from Sachs's drawings, often were incorporated, without his knowledge, in the textbooks of other authors, and this practice was continued until quite recent times.

In reviewing Sachs's work it becomes clear that he was little interested in making narrow observations or in answering highly specific questions. He always sought major laws of universal applicability. Even when he failed to solve a problem he attempted to outline a comprehensive framework which might lead to the relevant physiological aspects. To the extent permitted by contemporary knowledge, he was highly successful in this endeavor.

BIBLIOGRAPHY

I. ORIGINAL WORKS. Sachs's writings include *Handbuch der Experimentalphysiologie der Pflanzen,* which is vol. IV of *Handbuch der physiologischen Botanik,* W. Hofmeister, A. de Bary, T. Irmisch, N. Pringsheim, and J. Sachs, eds. (Leipzig, 1865); *Lehrbuch der Botanik* (Leipzig, 1868, 1870, 1872, 1874); *Geschichte der Botanik bis 1860* (Munich, 1875); *Vorlesungen über Pflanzenphysiologie* (Leipzig, 1882, 1887); and *Gesammelte Abhandlungen über Pflanzenphysiologie,* 2 vols. (Leipzig, 1892–1893). Sachs also edited *Arbeiten des Botanischen Instituts in Würzburg,* **1–3** (1871–1888).

II. SECONDARY LITERATURE. See K. Goebel, "Julius Sachs," in *Flora,* **84** (1897), 101–130; R. B. Harvey, "Julius von Sachs," in *Plant Physiology,* **4** (1929), 155–157; P. Hauptfleisch, "Julius von Sachs," in *Münchener medizinische Wochenschrift,* **26** (1897); F. Noll, "Julius v. Sachs," in *Naturwissenschaftliche Rundschau,* **12** (1897), 495–496; E. G. Pringsheim, *Julius Sachs der Begründer der neuen Pflanzenphysiologie 1832–1897* (Jena, 1932); W. Ruhland, "Julius Sachs," in *Handwörterbuch der Naturwissenschaften,* VIII (Jena, 1913), 529; and S. H. Vines, obituary in *Proceedings of the Royal Society,* **62** (1897–1898), xxiv–xxix.

MARTIN BOPP

SACROBOSCO, JOHANNES DE (or **JOHN OF HOLYWOOD**) (*b.* Holywood, Yorkshire, England, end of twelfth century; *d.* Paris, France, 1256 [1244?]), *astronomy.*

Sacrobosco (also called John or Johannes Halifax, Holyfax, Holywalde, Sacroboscus, Sacrobuschus, de Sacro Bosco, or de Sacro Busto) has been called a Scot, an Irishman, a Frenchman, a Brabançon, a Catalan, and a Jewish convert—all unfounded attributions. Some put his birthplace at Holywood near Dublin, or even at Nithsdale, Scotland. Very little is known of his life. English biographies maintain, and it is commonly held, that he was educated at Oxford.

After his studies Sacrobosco entered into orders and became a canon regular of the Order of St. Augustine at the famous monastery of Holywood in Nithsdale. About 1220 he went to Paris, where he spent most of his life and where he was admitted as a member of the university, on 5 June 1221, under the syndics of the Scottish nation. Elected professor of mathematics soon afterward, he won wide and enduring renown and was among the first exponents in the thirteenth century of the Arab arithmetic and algebra. By 1231 he was the outstanding mathematician and astronomer. He died in either 1244 or 1256 and was buried in the cloisters of the Fathers of Mercy, convent of St. Mathurin, in Paris.

The ambiguity of the year of Sacrobosco's death comes from the epitaph engraved on his tombstone in the convent cloisters: "M. Christi bis C. quarto deno quater anno De Sacro Bosco discrevit tempora ramus Gratia cui dederat nomen divina Joannes." If *quater* modifies only *deno*, then four times ten equals forty, plus four gives forty-four. If *quater* modifies *quarto deno*, then *quarto deno* is fourteen, and four times fourteen is fifty-six. The second interpretation is preferred by Johannes Fabricius, Christopher Saxius, Montucla, Bossut, and G. J. Vossius. An astrolabe decorates the stone, identifying the science to which he was most dedicated.

Sacrobosco's chief extant works are elementary textbooks on mathematics and astronomy: *De algorismo, De computo*, and *De sphaera*. All three are frequently found in the same manuscript, or at least bound together, and may be his only extant books. Prosdocimo de Beldemandis, however, wrote in his fifteenth-century commentary on the *De sphaera* that there were "many other works that it would take too long to enumerate here." A second arithmetic on fractions, *Algorismus de minutiis*, is attributed to Sacrobosco by Prosdocimo, but the opening words are those commonly occurring in a treatise usually ascribed to a Richard of England. Other works dubiously ascribed to him are a brief tract on physical or philosophical (that is, sexagesimal or astronomical) rather than common fractions, two tracts on the quadrant (one of which is more often ascribed to Campanus of Novara), an *Arithmetica communis*, and perhaps some commentaries on Aristotle.

Sacrobosco's fame rests firmly on his *De sphaera*, a small work based on Ptolemy and his Arabic commentators, published about 1220 and antedating the *De sphaera* of Grosseteste. It was quite generally adopted as the fundamental astronomy text, for often it was so clear that it needed little or no explanation. It was first used at the University of Paris.

There are only four chapters to the work. Chapter one defines a sphere, explains its divisions, including the four elements, and also comments on the heavens and their movements. The revolutions of the heavens are from east to west and their shape is spherical. The earth is a sphere, acting as the middle (or center) of the firmament; it is a mere point in relation to the total firmament and is immobile. Its measurements are also included.

Chapter two treats the various circles and their names—the celestial circle; the equinoctial; the movement of the *primum mobile* with its two parts, the north and south poles; the zodiac; the ecliptic; the colures; the meridian and the horizon; and the arctic and antarctic circles. It closes with an explanation of the five zones.

Chapter three explains the cosmic, chronic, and heliacal risings and settings of the signs and also their right and oblique ascensions. Explanations are furnished for the variations in the length of days in different global zones, namely, the equator, and in zones extending from the equator to the two poles. A discussion of the seven climes ends the chapter.

The movement of the sun and other planets and the causes of lunar and solar eclipses form the brief fourth chapter.

During the Middle Ages the *De sphaera* enjoyed great renown, and from the middle of the thirteenth century it was taught in all the schools of Europe. In the sixteenth century it gained the attention of mathematicians, including Clavius. As late as the seventeenth century it was used as a basic astronomy text, but after 1700 it was completely forgotten.

After Manilius' *Astronomica, The Sphere* was the first printed book on astronomy (Ferrara, 1472). Twenty-four more editions appeared in the following twenty-eight years, and more than forty editions from 1500 to 1547, the last being issued at Leiden. For eighty years after Barocius in 1570 had pointed out some eighty-four errors, *The Sphere* was still studied, and in the seventeenth century it served as a manual of astronomy in some German and Low Countries schools. Often it appeared with commentaries by the most distinguished scholars of the time. There were three Italian editions: by Maurus (Florence, 1550); by Dante de'Rinaldi (Florence, 1571), and by Francesco Pifferi at Sienna in 1537, 1550, 1572, 1579, 1604. Two French editions, by Martin Perer and Guillaume Desbordes, were printed at Paris in 1546 and 1576.

Some claim that *The Sphere* merely paraphrases the elementary ideas of Ptolemy and the Arab astronomers al-Battānī and al-Farghānī, but this is a great exaggeration. Sacrobosco's *Sphere* is far superior in structure and order to that of al-Farghānī. Resemblances should not be surprising, however, since both writers summarized the *Almagest*. Actually Sacrobosco omitted much of what is found in al-Farghānī, condensed what little he did use of it, and restated and rearranged the matter into a more effective plan. Indications are equally strong for his using Macrobius' *Commentary on the Dream of Scipio* as for al-Farghānī's *Elements*, although Macrobius is rarely mentioned.

Commentaries on *The Sphere* were written by Michael Scot between 1230 and 1235; by John Pecham, a Franciscan, sometime before 1279; by the Dominican Bernard of Le Treille between 1263 and 1266; and by Campanus of Novara between 1265 and 1292. Robertus Anglicus gave a course of lectures on Sacrobosco's *Sphere* at either the University of Paris or the University of Montpellier in 1271, which helps to date his own commentary. He stated that for the most part *The Sphere* was so clear it needed no further explanation.

Cecco d'Ascoli was the earliest of the fourteenth-century commentators on *The Sphere* (probably before 1324). The Dominican Ugo de Castello (Ugo di Città di Castello) wrote a commentary on *The Sphere* that was begun at Paris and finished at Florence in 1337. On 19 October 1346, at Ghent, Henry of Sinrenberg, an Augustinian friar, completed a commentary on *The Sphere*, in which he addressed friends and fellow Augustinians in the *studium* of the monastery of Milan. This indicates the wide geographical spread of the use of *The Sphere*, beyond universities to convent schools. Blasius of Parma commented on *The Sphere* in the late fourteenth century.

Fifteenth-century commentators included Prosdocimo de Beldemandis in 1418 and Franciscus Capuanus de Manfredonia about 1475. Their works are quite detailed and concern any subject that might be even remotely connected with *The Sphere*. Other commentators of this period were Jacques Lefèvre d'Étaples, Wenceslaus Faber of Budweis, and Pedro Ciruelo of Daroca (his commentary was published in February 1498). Incunabula editions of Sacrobosco's *Sphere* were brought out by Gasparino Borro (*d.* 1498), a Servite; and by George of Montferrat. Fausto Andrelini, the Italian humanist, lectured on *The Sphere* while at the University of Paris in 1496.

Sixteenth- and seventeenth-century commentators included Joannes Baptista Capuanus, Bartolomaeus Vespuccius, Erasmus Oswald (Schreckenfuchs), Maurus Florentinus, and Christoph Clavius, the Jesuit astronomer whose huge commentary was held in high regard.

Although the university curriculum of the time was based heavily on logic, Sacrobosco's *Sphere* was required reading, and properly so, first at Paris. The Faculty of Arts at Vienna in 1389 made *The Sphere* and the *De algorismo* required for the A.B. degree. At Erfurt in 1420 the matter of the quadrivium consisted solely of *The Sphere*. At Bourges in the fifteenth century, Euclid, various works of Aristotle, and *The Sphere* were required for the licentiate. In 1409 at Oxford it was required for the A.B. degree. At Bologna the Faculty of Arts required the *De algorismo* in the first year of their premedical course and *The Sphere* in the second year.

The *De computo ecclesiastico* or *De anni ratione*, written about 1232 and probably after *The Sphere*, points out the increasing error in the Julian calendar and suggests a remedy markedly similar to the one actually used under Gregory XIII, some 350 years later, in the revision of the calendar. It was printed at Paris in 1538(?), 1550, and 1572, and at Antwerp in 1547 and 1566.

Sacrobosco's *De computo ecclesiastico* deals with the division of time—marked out by the movements of the sun and moon and their interrelationship—into days and years. A discussion of the sun and its influence on the lengths of days, months, and years forms the first half of the work. In the second half, similar considerations are applied to the moon. The solar and lunar years, the solar and lunar cycles, intercalary days, and the movable feasts are examined.

De algorismo discusses the art of calculating with nonnegative integers. The work contains eleven chapters and examines numeration, addition, subtraction, mediation, duplication, multiplication, division, progression, and extraction of both the square and cube roots of numbers. Mediation is the halving of a given number, and duplication is the doubling of a number. Six rules of multiplication are included, depending on what combination of digits and articles are to be multiplied. Natural progression begins with the number one and the difference between adjacent numbers is one. Broken progression begins with the number two and the difference between adjacent terms of the progression is two. Two rules are given for finding the sum of terms in a natural progression, depending on whether the last term is an odd or even number. In the section on extracting roots, lineal, superficial, and solid numbers are first defined and their properties listed. Lastly, methods of extracting the square and cubic roots of a number are discussed.

Throughout the work examples of various arithmetic processes are included, and diagrams are supplied to facilitate understanding the method.

The *De algorismo* or *De arte numerandi* was first printed without date or place (1490[?]). It was later printed at Vienna by Hieronymus Vietor in 1517, at Cracow in 1521 or 1522, at Venice in 1523, and on occasion printed with *The Sphere*. It was the most widely used manual of arithmetic in the Middle Ages. Copies abound in manuscript form: some forty-three are in the Vatican holdings alone. Peter Philomenus of Dacia wrote a commentary on *De algorismo* toward the close of the thirteenth century, a copy of which is included in Holliwell's *Rara mathematica* of 1841.

BIBLIOGRAPHY

I. ORIGINAL WORKS. Maximilian Curtze's ed. of Sacrobosco's *Algorismus vulgaris*, with the commentary of Peter Philomenus of Dacia, is *Petri Philomeni de Dacia in Algorismus vulgarem Johannis de Sacrobosco commentarius una cum Algorismo ipso* (Copenhagen, 1897). Robert R. Steele translated Sacrobosco's *De arte numerandi* as "The Art of Nombryng," in R. Steele, ed., *The Earliest Arithmetics in English*, Early English Text Society Extra Ser. no. 118 (London, 1922), 33–51. MSS are McGill University MS 134, fols. 1r–20v; Vaticana Latina 3114, fols. 28–33; and Vaticana Rossi 732, fols. 76–80.

II. SECONDARY LITERATURE. See Pierre Duhem, *Le système du monde,* 5 vols. (Paris, 1913–1917); J. O. Halliwell-Phillips, *Rara arithmetica* (London, 1839; 2nd ed., 1841); *Histoire littéraire de la France,* XIX (1838), 1–4; and Lynn Thorndike, *History of Magic and Experimental Science,* IV (1941), 560; and *The Sphere of Sacrobosco and Its Commentators* (Chicago, 1949).

JOHN F. DALY, S.J.

AL-ṢADAFĪ. See Ibn Yūnus, Abu'l-Ḥasan ʿAlī ibn ʿAbd al-Raḥmān ibn Aḥmad ibn Yūnus al-Ṣadafī.

SAGE, BALTHAZAR-GEORGES (*b.* Paris, France, 7 May 1740; *d.* Paris, 9 September 1824), *metallurgy, assaying, chemistry.*

Except as founder of the Paris École des Mines,

there is little reason to rate Sage as an important scientific figure. He published extensively, often hurriedly, and—toward the dismal end of his life—usually for self-serving purposes. The chief modern study of Sage describes his scientific work as totally without value.[1] Another historian has called him a "faux savant."[2] Partington, on the whole, is somewhat more charitable.[3]

The younger son of a Paris pharmacist of limited means (who was the son of a notary of St.-Jean-de-Maurienne in Savoy named Sapienti), Sage was sent as a day student to the famous Collège des Quatre Nations.[4] His interest aroused in science, he became an accredited pharmacist and attended the public lectures on experimental physics of the Abbé Nollet and the chemical lectures of G. F. Rouelle, whose experiments he repeated in a small laboratory he was able to equip. In 1760 Sage opened at the family pharmacy a public course of lectures in chemistry and the techniques of assaying. Here, too, he housed the beginnings of an extensive mineralogical collection. In 1769 he published his first work, *Examen chymique de différentes substances minérales.*[5]

Skilled at finding and bemusing powerful patrons, Sage made connections at court and obtained the direct support of Louis XV, who put pressure on the Academy of Sciences (even sending his orders by special courier) to admit Sage, even if it meant installing him above the lowest rank.[6] The Academy evidently compromised, for it elected Sage in 1770 but ranked and promoted him according to its established procedures. At the death of Rouelle in that year, Sage replaced Cadet de Gassicourt as *adjoint,* when Cadet was promoted to *associé.* In 1777 Cadet became *pensionnaire,* and Sage succeeded him as *associé.*[7]

Sage's dubious qualifications as a chemist were soon evident. In his *Examen chymique* and his more successful *Élémens de minéralogie docimastique* (1772), Sage had claimed that a white lead ore found at the famous mine of Poullaouen in Brittany was a compound of lead and marine acid (hydrochloric acid), even asserting in the latter book that the ore contained twenty percent by weight of marine acid. Late in 1772 a master apothecary of Paris, one Laborie, reported to the Academy of Sciences an analysis of this ore which controverted Sage's results; accordingly, a committee composed of the other chemists in the Academy carried out experiments in Baumé's laboratory to decide between the contestants. They concluded unanimously that the ore contained no trace of marine acid.[8] It is perhaps significant that while both Laborie and Sage were invited to observe the experiments, only Laborie accepted the invitation.

Yet in these early years Sage took an active part in the Academy's proceedings, often joining other chemists, among them Lavoisier, on special investigating committees. In 1774, for example, Sage, Brisson, and Lavoisier demonstrated that the rival preparations of mercuric oxide (mercure précipité per se) by Baumé and Cadet were identical and — this was to play a part in the earliest preparation of oxygen gas in France — were reducible without the addition of substances supposed to contain phlogiston.[9] In 1778 Sage joined Macquer and Lavoisier in analyzing a sample of water from the Dead Sea.[10] He also was appointed one of the commissioners to award a prize in 1778 for the best method of increasing the production of saltpeter in the kingdom; and, with Macquer, Lavoisier, Baumé, and d'Arcy, he helped assemble the Recueil de mémoires et d'observations sur la formation & sur la fabrication du salpêtre (1776).[11] Later he contributed rare mineral specimens and precious stones from his collection for Lavoisier's experiments on the effect of intense heat produced by combustion with oxygen.[12]

The verdict on Sage as a chemist is almost universally negative. Partington, to be sure, ascribes to him a small number of minor chemical discoveries: the preparation, independently of Guyton de Morveau, of solid potassium ferrocyanide; the discovery of hypophosphoric acid; the demonstration (1776) that silver chloride can be reduced by iron and that phosphorus can reduce metallic salts to metals.[13]

If Lavoisier often cited Sage, it was not infrequently to correct his manifest errors. Perhaps the most notorious of Sage's stubborn mistakes was his report that he had obtained substantial amounts of gold from ashes. Others were unable to duplicate his results. The chemists of the Academy carried out a careful analysis and found that the amount of gold was not only infinitely small, but was derived from the minium used in treating the ashes.[14] Polemics on this matter shook the Academy, taking on the dimensions of a scandal. Indeed, according to one observer, censure of Sage by his colleagues was avoided only by personal antagonisms within the Academy.[15]

Most contemporaries viewed Sage as a successful teacher, esteemed by his students, but as a very poor chemist. This judgment was common well before he expressed his opposition to Lavoisier's new chemistry. In 1772 Turgot wrote to Condor-

cet that it was foolishness (sottise) for the Academy to have passed over the younger Rouelle and Jean Darcet and to elect Sage.[16] On 28 March 1775, Macquer — the dean of French chemists — wrote to Bergman that Sage

> . . . totally lacks a gift for chemistry [n'a point du tout l'espirit de la chimie] and does not understand that science at all. This is what makes him so bold in making hasty inferences from the slightest evidence [les moindres apparences]. In the end, he will seriously damage himself with the real chemists. But since he has as much confidence as he has little knowledge, many people who know no more than he, think of him as a remarkable discoverer, and these are highly-placed persons [grands personnages] and very numerous.[17]

A recent history of chemical mineralogy has little to say on Sage's behalf, describing his analytical technique as "most unsatisfactory" except when he was using traditional methods in assaying metallic ores. Because of his peculiar theories "he became hopelessly entangled whenever he attempted to use 'wet' methods of analysis."[18] In the light of his dubious competence and Macquer's opinion of him, it is ironic that Sage was promoted to pensionnaire in the Academy upon the latter's death in 1784.

Two years after his promotion Sage's Analyse chimique et concordance des trois règnes was published (1786). Fulsomely dedicated to Louis XVI, to whom he owed "tout ce que je suis," Sage described it as the result of twenty-five years of work. Here, as well as in his Mémoires historiques et physiques (1817), he set forth the doctrine that formed the basis of his teaching. Sage could not deny the experimental demonstration by his contemporaries that air is a mixture of gases. Yet he continued to believe not only in the reality of phlogiston but also in what he called l'acide igné, a substance imagined to be "the essence of the different aeriform fluids."[19] From the union of this acid with phlogiston arose the different sorts of air, notably oxygen, as well as electricity, oils, and the "principle" of metallicity. Acids were derived from this acide igné élémentaire, which was combined in compound bodies (corps organisés) with different amounts of phlogiston, earth, and water. In his Analyse chimique Sage noted that two celebrated scientists (Lavoisier and Laplace) had advanced the view that water is composed of inflammable air (hydrogen) and dephlogisticated air (oxygen). A series of ingenious experiments seemed to support their theory, but to Sage such experiments merely proved that water is one of the constituents of the different sorts of aeriform fluids.[20]

In 1787 there appeared, with the sanction of the Academy of Sciences, the famous *Nomenclature chimique*, the work of Guyton de Morveau, Lavoisier, Berthollet, and Fourcroy. To appraise this work, the Academy appointed a committee consisting of Baumé, Cadet de Gassicourt, Darcet, and Sage. The *rapporteurs* recognized without hesitation that a new theory of chemistry was implicitly being arrayed against the old.[21] They gave high praise to the authors of this radical work, pointed out some of its advantages, and raised certain doubts (about, among other matters, the conclusions drawn from the synthesis of water). Not only was it difficult, they wrote, to abandon all the principles of one's chemical education, but it was even more difficult suddenly to admit that a host of substances, which analogy suggested must be compound, should henceforth be regarded as simple substances. And they concluded:

> We therefore believe that one must submit this new theory, as well as its nomenclature, to the test of time, to the impact of experiments, to the consequent weighing of opinions; finally to the judgment of the public as the only tribunal by which it ought or can be judged.[22]

Although Sage's opposition was muted by the views of his more discreet coauthors, he never abandoned his idiosyncratic position. As late as 1810 he expressed his disapproval of the new chemistry and the new nomenclature by publishing his *Exposé des effets de la contagion nomenclative et réfutation des paradoxes qui dénaturent la physique*. When Sage published his *Mémoires historiques et physiques* in 1817 he still praised the old theory of Johann Friedrich Meyer and rejected Joseph Black's demonstration of some sixty years earlier that chalk is composed of quicklime and "fixed air" (carbon dioxide). Instead, Sage wrote, *terre calcaire* (calcium carbonate) is a salt composed of his cherished *acide igné* (obviously a descendant of Meyer's *acidum pingue*) with a specific earth, some oleaginous matter, and water. Yet, he continued, Black's erroneous theory (he described Black as a *chimiste anglais*!) had been adopted by the "secte lavoisienne" as if it were a demonstrated truth.[23]

The creation of the Paris École des Mines is Sage's major claim to distinction. The idea of such a school had first occurred to Daniel Trudaine, the *intendant des finances* who had founded in 1747 the École des Ponts et Chaussées. Yet it did not seem feasible at that time, in view of the primitive state of the French mining industry, to establish a parallel school of mines. Instead, on the advice of Jean Hellot, chief assayer at the mint, Trudaine reserved a few places at the Ponts et Chaussées for young persons recommended by the directors and owners of mines. Their professional instruction was to be completed by a course in chemistry and practical experience at one of the better-managed mines. In this fashion were trained the earliest inspectors and advisors to the mining industry, notably Gabriel Jars and J. P. F. Guillot-Duhamel (1730–1816).[24]

The first steps toward the creation of the École des Mines were taken under the auspices of Henri-Léonard-Jean-Baptiste Bertin. A protégé of Mme de Pompadour, he had served as *contrôleur-général des finances* from 1759 to 1763. On leaving that office, he retained the title of minister and secretary of state and put together a catch-all department, responsible for several diverse activities. Among these, and of particular interest to him, were the mines of France.[25]

Soon after taking over his new post, Bertin received memoirs proposing the creation of two schools of mines in the provinces. Consulted by Bertin, Antoine Monnet, recently appointed to the post of inspector general of mines, strongly opposed this suggestion and urged instead that courses in mineralogy and metallurgy be established at the Jardin des Plantes and that students, after being examined on these subjects (as well as on chemistry and mathematics), should complete their training at one of the better-run mines. Soon after, Monnet's colleague Jourdan, together with Sage, who then held the post of supervisor of assaying at the mint, urged Bertin to create a school of mines at Paris, to be supported by a tax levied on the mine owners. Monnet opposed this project with enthusiasm enhanced by his dislike of Sage.[26]

Yet Sage gained at least a partial victory. By letters patent of 11 June 1778 a chair of mineralogy and docimastic metallurgy was established at the mint. Sage received the royal appointment with a stipend of 2,000 livres per annum and with Guillot-Duhamel as his assistant. He installed his already substantial collection of minerals in the great hall of the building recently taken over for the mint and completely redesigned by the architect Jacques-Denis Antoine (1733–1801). Here Sage taught mineralogy, in what he later called his first École des Mines, to students among whom, we are told, were Chaptal (the future industrial chemist), the crystallographer Romé de l'Isle, and the surgeon

Jean Demeste (1743–1783), who described Sage as the most distinguished chemist in France.[27]

When Bertin left the government in 1780, the special department was abolished and the Service des Mines was put under four *intendants de commerce*. On 21 May 1781 the supervisory administration for the mines was further modified by the creation of a special *intendant* for the mines, Douet de La Boullaye. Under Douet, despite the predictable opposition of Monnet, three decrees (*arrêts*) dated 19 March 1783 were issued, one of which established a full-fledged school of mines, modeled after the highly successful École des Ponts et Chaussées. After stressing the backwardness of France's exploitation of its mineral resources and the difficulty of finding qualified personnel, this *arrêt* provided for the establishment of two professorships, one for mineralogy and assaying (*la chimie docimastique*) and the other for the practical aspects of mining (*l'exploitation des mines*). The school was to be located at the Hôtel de la Monnaie on the Quai de Conti; and the chair of mineralogy and assaying was filled by Sage, who was also to serve as director. The other professorship went to Guillot-Duhamel. Secondary posts were established for the teaching of mathematics (Prud'homme), drawing (Miché), and foreign languages (the Abbé Pierre-Romain Clouet). At least for a brief time Jean-Henri Hassenfratz, who in 1785 became a deputy inspector of mines, taught physics at the school, as did J. A. Charles.[28] Lesser staff included a curator of the mineralogical cabinet, the core of which was Sage's collection, ceded to the king in exchange for a lifetime annuity (*rente viagère*) of 5,000 livres.[29]

To be admitted to the École des Mines, a student had to be sixteen years old; the course of studies, in principle at least, lasted three years; and lectures were delivered from November to June. The number of students was never large: eight were admitted in 1783, and the entering class of 1786 numbered twenty-one. Of these, twelve were called *élèves titulaires* or *stipendiés,* for they received a government scholarship of 200 livres.[30] No new students were admitted after 1787.

Sage served as director and professor for at least seven years. The audience for his lectures may have been substantial, since besides those registered in the École des Mines, students of the École des Ponts et Chaussées were expected to attend his lectures on chemistry and mineralogy. His teaching, if the content can be judged from his *Analyse chimique des trois règnes,* was probably not very scientific or profound. He taught mineralogy, remarked Aguillon, as a simple display of minerals and rocks distinguished by inadequate and superficial characters, with no trace of the classification made possible by recent advances in chemistry and by the new crystallography of Haüy, both of which he totally ignored.[31]

The school, although small, was elegantly housed in the mint. Sage records that he received a substantial gratuity from the king for having salvaged gold worth some 440,000 francs from gilded decorations (*vieilles dorures*), and had received permission to apply his *gratification* of 40,000 francs to decorating the school's quarters according to the plans of the architect Antoine.[32] With the onset of the Revolution, and the increasingly desperate state of French finances, this extravagance was held against Sage and the institution he had founded.

Reporting in 1790 for the finance committee of the Constituent Assembly, Charles-François Lebrun, the future duke of Piacenza, concluded that the budget of the École des Mines was better suited to a country like Sweden, where mining was a major source of the national income, than to an agricultural country like France. The small size of the school, moreover, did not justify the sums expended. In June of the same year, Lebrun set forth a proposal of the finance committee that the Corps des Mines and its school should be amalgamated with the Corps des Ponts et Chaussées and its famous school.[33] This proposal fitted well with a plan for the reorganization of the Jardin des Plantes, which included the recommendation that the mineralogical collection of the mint should be transferred there and that its professor of chemistry should give a course in metallurgy. Neither proposal was implemented.[34] Indeed, the Constituent Assembly disbanded without having legislated any reorganization of the Corps des Mines and its school.

As a teaching institution the École des Mines continued a tenuous existence; its collections, although sought by various rival bodies, remained undisturbed. A law of 1792 continued the salaries of the mine inspectors, the teachers, and the students. Sage had played a major part in defending his school, but his much-vaunted contacts with the court came more and more to endanger his position. In 1793–1794 he was arrested and imprisoned, regaining his freedom only upon payment of a substantial fine.[35]

In 1794 the Committee of Public Safety, now

reformed but weakened in authority, created the Agence des Mines, to be composed of three members selected by the Committee.[36] Of the three appointees, two had been members of the first class at the École des Mines; all three continued to dominate mining affairs until the basic reorganization in 1810.[37]

Sage was simply ignored during these changes. The Agence des Mines established a new school, in the Hôtel de Mouchy in the rue de l'Université, and courses began there in November 1794. The school had a library, an assaying laboratory, and its own mineralogical collection (the core of which was Guettard's cabinet). The scientific level of the teaching staff was markedly raised: Vauquelin taught what had been Sage's course in assaying, mineralogy and physical geography were taught by Hassenfratz, and crystallography by Haüy. Two of Sage's original associates were retained: practical mining was again taught by Guillot-Duhamel, and the Abbé Clouet gave his course in German.[38]

In all this Sage played no part; but in 1797 the Directory put him once again in charge of his collections at the mint with a stipend of 6,000 francs. In 1801 he was elected a full member of the Institut de France.[39] A few years later, however, he became totally blind and soon after was deprived of his pension. His last years were spent in penury, in attempts to win back public favor, and in writing pamphlets protesting his ill treatment and enumerating his real and imaginary achievements. He died in his eighty-fifth year, an unregenerate royalist, a scientific fossil, and a pathetic hangover of the *ancien régime*.

NOTES

1. Paul Dorveaux, "Apothicaires membres de l'Académie royale des sciences, XI, Balthazar-Georges Sage," in *Revue d'histoire de la pharmacie*, 23 (1935), 152–166, 216–232. For this evaluation, see p. 232.
2. Arthur Birembaut, "L'enseignement de la minéralogie et des techniques minières," in René Taton, ed., *Enseignement et diffusion des sciences en France au XVIIIe siècle* (Paris, 1964), 387 and n. 2. Édouard Grimaux describes Sage as "expérimentateur maladroit, imagination fantaisiste, qui a beaucoup publié, beaucoup écrit, entassé erreurs sur erreurs, et n'a pas laissé dans la science un seul fait bien observé." See his *Lavoisier*, 2nd ed. (Paris, 1896), 122.
3. J. R. Partington, *A History of Chemistry*, III (1962), 97–98.
4. "Sage," in Didot-Hoefer, *Biographie générale*. The biographical sketch in John Ferguson, *Bibliotheca chemica*, II (Glasgow, 1906), 312–313, appears to draw uncritically upon the *Biographie générale*.
5. A copy of this work is cited in Denis I. Duveen, *Bibliotheca alchemica et chemica* (London, 1949), 523.
6. Dorveaux, *op. cit.*, 155–159.
7. The article in the *Biographie générale* errs in stating that in 1768, at the age of twenty-eight, Sage replaced Rouelle in the Academy of Sciences. This error reappears in Ferguson, *op. cit.*, p. 312. A reliable source is the *Index biographique des membres et correspondants de l'Académie des sciences* (Paris, 1939, and later eds.).
8. An account of this episode is given by Leslie J. M. Coleby in *Chemical Studies of P. J. Macquer* (London, 1938), 78–79. See *Oeuvres de Lavoisier*, IV (1868), 159–175. For his later admission of this error, see Sage, *Analyse chimique et concordance des trois règnes*, I, "Préface," v. Here, too, he admits his mistake in stating that manganese ore (pyrolusite) contains zinc; but he adds that he corrected himself by repeating the experiments of Gahn, Bergman, and Scheele. *Ibid.*, p. vi.
9. "Rapport sur une contestation entre MM. Cadet et Baumé sur le précipité per se," in *Oeuvres de Lavoisier*, IV (1868), 188–190.
10. "Analyse de l'eau du lac asphaltite," *ibid.*, II (1862), 234–237.
11. *Oeuvres de Lavoisier*, V (1892), 464, 481.
12. *Ibid.*, II (1862), 445–446, 473.
13. Partington, *op. cit.*, p. 97. This last discovery was cited with approval by Lavoisier in his important "Mémoire sur l'affinité du principe oxygine," in *Oeuvres de Lavoisier*, II, 554.
14. Coleby, *op. cit.*, pp. 114–115.
15. Arthur Birembaut, "L'Académie royale des sciences en 1780 vue par l'astronome suédois Lexell (1740–1784)," in *Revue d'histoire des sciences et de leurs applications*, 10 (1957), 156, 163–164.
16. Charles Henry, *Correspondance inédite de Condorcet et de Turgot* (Paris, 1883), 123.
17. *Torbern Bergman's Foreign Correspondence*, Göte Carlid and Johan Nordström, eds., I (Stockholm, 1965), 246. For similar skeptical allusions by Guyton de Morveau and by the Irish chemist Kirwan, see *ibid.*, pp. 112, 188. During a visit to Paris in 1781–1782, Alessandro Volta followed a course in chemistry given by Sage, probably the course at the mint. See *Epistolario di Alessandro Volta*, II (Bologna, 1951), 79.
18. David Roger Oldroyd, "From Paracelsus to Haüy: The Development of Mineralogy in Its Relation to Chemistry" (Ph.D. diss., Univ. of New South Wales, 1974). I owe this reference to Professor Seymour Mauskopf of Duke University.
19. Sage, *Analyse chimique*, I, "Préface," i–ii.
20. *Ibid.*, p. 29.
21. *Méthode de nomenclature chimique, proposé par MM. de Morveau, Lavoisier, Bertholet [sic], & de Fourcroy* (Paris, 1787), 244–249.
22. *Ibid.*, p. 251. Sage's open attack on the new nomenclature appeared in a series of letters published in Rozier's *Journal de physique* (1788, 1789). See Dorveaux, *op. cit.*, p. 223, n. 30.
23. Sage, *Mémoires historiques et physiques*, 50–51.
24. Aguillon, "L'École des mines . . .," 441–443, and Rouff, *Les mines de charbon*, 481. On Guillot-Duhamel, see H. Guerlac, "Some French Antecedents of the Chemical Revolution," in *Chymia*, 5 (1959), 93; and Alfred Lacroix, *Figures des savants*, I (Paris, 1932), 19–23.
25. Aguillon, *op. cit.*, p. 446. In 1766 Bertin officially commissioned Guettard to continue, under government sponsorship, his work on a mineralogical atlas of France, a project with which Lavoisier, and later Monnet, were closely associated.
26. On Jourdan, also called Jourdan de Montplaisir, who was an inspector of mines based at Lyons, see Aguillon, *op. cit.*, pp. 447–448. On Monnet see Rhoda Rappaport, "Guettard, Lavoisier, and Monnet: Geologists in the Service of

the French Monarchy" (Ph.D. diss., Cornell Univ., 1964), and her short article in this *Dictionary* with an excellent bibliography. According to Sage, Monnet was not the only influential opponent of his schemes. He charges Buffon with delaying his nomination to the chair for a year, hoping to have it awarded to his disciple Daubenton at the Jardin des Plantes. See Sage, *Mémoires historiques et physiques*, p. 73.

27. Aguillon, *op. cit.*, pp. 449–451. Partington, *op. cit.*, III, 65, describes Demeste as "a Liège surgeon who made chemistry a hobby." Sage deserves credit for his support of Romé de l'Isle, a distinguished crystallographer whose *Essai de cristallographie* first appeared in 1772. In turn Romé de l'Isle, in his *L'action du feu central bannie de la surface du globe* (2nd ed., 1781) quotes from Sage's experiments.

28. On Hassenfratz and Charles as teachers of physics in the early École des Mines, see Aguillon, *op. cit.*, p. 463. The *Almanach Royal* for 1789 lists only Sage, "Professeur de Minéralogie Docimastique, Directeur Général des Etudes, & Commissaire du Conseil pour les Essais des Mines"; Guillot-Duhamel, "Professeur de Géométrie Souterraine, et Démonstrateur de Toutes les Machines Servant à l'Exploitation des Mines"; the Abbé Clouet, "Professeur de Langues Étrangères"; and "M. de Voselles, Secrétaire Général & Garde du Cabinet du Roi." The same names appear in the *Almanach* through 1792. According to the *Rapport du Comité des finances – Département des Mines. Par M. Le Brun*, p. 4, the chair of foreign languages and the chair of physics were established in 1785. This and other related pamphlets can be consulted in the French Revolution collection of Cornell University.

29. In 1784 Sage published his *Description méthodique du cabinet de l'École royale des mines*.

30. In his *Fondation de l'École royale des mines*, 6, Sage writes that the twelve *élèves stipendiés* were paid 500 francs as their *traitements* and 200 francs for their travel, presumably to visit mines.

31. Aguillon, *op. cit.*, pp. 461–462, who adds that "en chimie et docimasie [Sage gave only] l'indication des recettes alors connues dans les laboratoires sans aucune véritable théorie scientifique pour les expliquer et les relier entre elles."

32. Sage, *Fondation de l'École royale des mines*, 7.

33. Aguillon, *op. cit.*, pp. 465–466.

34. Sage fought both proposals. In reply to Lebrun he published *Remarques de M. Sage, directeur de l'École royale des mines, sur l'extrait raisonné des rapports du Comité des finances de l'Assemblée nationale* (Paris, n.d. [probably 1790]). See also his *Observations sur un écrit qui a pour titre, vues sur le Jardin royal des plantes et le cabinet d'histoire naturelle; à Paris, chez Baudouin, imprimeur de l'Assemblée nationale, 1789* (Paris, 1790). Both pamphlets can be consulted in the Lavoisier Collection at Cornell University. In his *Mémoires historiques et physiques* Sage accused Lebrun of abolishing the 2,000 francs that was his stipend as professor of *minéralogie docimastique* at the mint.

35. Sage rather absurdly blamed his arrest and imprisonment on Guyton de Morveau and Fourcroy, who, while members of the Committee of Public Safety, "se vengèrent du ridicule que j'avais répandu sur leur vocabulisme insignifiant et sans euphonie, et de ce que j'avais prouvé que la nouvelle doctrine lavoisienne était une métachimie . . . ." *Mémoires historiques et physiques*, 74.

36. Aguillon, *op. cit.*, pp. 471–473.

37. The two who had studied at the École des Mines were Lelièvre and Lefebvre d'Hellancourt. Both were among those commissioned (*brevetés*) after a single year of study.

38. Aguillon, *op. cit.*, pp. 482–484.

39. According to the *Index biographique des membres et correspondants de l'Académie des sciences*, Sage had been elected *associé non-résident* of the section of natural history and mineralogy of the Première Classe of the Institut on 5 March 1796.

BIBLIOGRAPHY

I. Original Works. The following list, with no pretense at completeness, may give some idea of the range of Sage's interests. *Examen chymique de différentes substances minérales* (Paris, 1769) would seem to be his earliest publication. It was followed by *Élémens de minéralogie docimastique* (Paris, 1772; 2nd ed., 2 vols., 1777). *Mémoires de chimie* (Paris, 1773) contains, besides other papers, memoirs read to the Academy of Sciences from 1766 to 1772, many of them describing mineral analyses. These works were followed by *Analyse des blés, et expériences propres à faire connaître la qualité du froment* (Paris, 1776); and *Observations nouvelles sur les propriétés de l'alkali fluor ammoniacal* (Paris, 1778). His *L'art d'essayer l'or et l'argent* (Paris, 1780) contains his first description of the production of potassium ferrocyanide in solid form. This was followed by his ambitious *Analyse chimique et concordance des trois règnes de la nature*, 3 vols. (Paris, 1786). After the Revolution and in his later years he published various speculative works on electricity and natural philosophy, as well as his attack on Lavoisier's chemistry and the new chemical nomenclature: *Exposé des effets de la contagion nomenclative et la réfutation des paradoxes qui dénaturent la physique* (Paris, 1810). Of autobiographical interest are his *Mémoires historiques et physiques* (Paris, 1817); *Fondation de l'Ecole royale des mines à la monnaie* (Paris, 1817); and *Notice autobiographique* (Paris, 1818). Other of Sage's publications are cited in the notes and still others are listed in the catalog of printed books of the Bibliothèque Nationale.

II. Secondary Literature. Although a member of the Paris Academy of Sciences and (at the time of his death) of the Première Classe of the Institut de France, Sage was never accorded the customary honor of an *éloge* composed by the perpetual secretary. There is a passable, if not always accurate, sketch by E. M. [Ernest Mézières?] in the Didot-Hoefer *Nouvelle biographie générale*, based largely on Sage's autobiographical writings and an obituary notice in the *Journal de la librairie* of 1824. Sage is judged by E. M. as "ayant créé la docimasie en France," which would seem to be an exaggeration. The best modern evaluation is the two-part study by Paul Dorveaux (see Note 1): Sage's work in chemical metallurgy is briefly mentioned by Hermann Kopp in his *Geschichte der Chemie*, 2 vols. in 4 pts. (Brunswick, 1843), pt. II, 87–88, but is not discussed in F. Hoefer's *Histoire de la chimie*, 2 vols. (Paris, 1866). The most favorable account of Sage's chemical work (perhaps too favorable) is by J. R. Partington in his *History of Chemistry*, III (1962), 97–98. On the Paris École des Mines and Sage's role as its founder and first director, see Louis Aguillon, "L'École des mines de Paris—notice historique," in *Annales de mines*, 8th ser., **15** (1889); 433–686; and Marcel Rouff, *Les mines de charbon en France au XVIIIe siècle, 1744–1791* (Paris, 1922), 480–488.

HENRY GUERLAC

SAGNAC, GEORGES M. M. (*b*. Périgueux, France, 14 October 1869; *d*. Meudon, France, 26 February 1928), *physics.*

Sagnac studied mainly the radiation produced by X rays and the optics of interference. He is remembered today primarily for his design of a rotating interferometer and for the experimental results it provided. He was very interested in the theoretical analysis of optics of moving systems; his approach was classical, and he interpreted the results of his optical experiments as contradicting Einstein's theory of relativity.

Sagnac came from an old bourgeois family. His father was a lawyer who directed the Assurances Générales, and his mother was the daughter of a notary. From 1889 to 1894 Sagnac studied at the École Normale Supérieure and at the University of Paris. In 1894 he received his *agrégé* and became an assistant in physics at the university. In 1900 he was awarded the *docteur ès sciences* degree and became professor of physics at the University of Lille. From 1911 he was professor of physics at the University of Paris. He was twice a candidate for the physics section of the Academy, and he received the Pierson-Perrin, the Wilde, and the Lacaze prizes.

Sagnac's earliest research (and the subject of his dissertation) was motivated by the newly observed X rays.[1] From 1896, the year following Röntgen's discovery, to 1900 Sagnac studied the radiations produced by X rays. He observed that secondary radiation is produced when X rays fall on heavier metals. He showed that this radiation contained secondary X rays (of lower frequency than the original X rays) and negatively charged rays. He suggested that it was through the action of the negative rays that X rays produce ionization in air.

Sagnac next worked on the propagation of light in moving systems (he was particularly interested in the Fizeau effect) and with the optics of interference. From about 1910 he was especially concerned with designing and using a rotating interferometer. In this instrument all the components—light source, mirrors, and photographic plate—are on a disk that can be rotated at various velocities.

Light travels around the disk along a polygonal circuit determined by successive reflections from four mirrors placed around it. Light from the source is split into two beams that travel around the disk in opposite directions and, on recombination, produce interference fringes on the photographic plate. In 1913 Sagnac measured the shift in the position of the fringes when the direction of rotation was reversed.

Sagnac interpreted the shift in the position of the interference fringes in terms of an "ether wind" that gave the light beams traveling in opposite directions different velocities. He was convinced that the phenomenon observed with the rotating interferometer demonstrated the existence of an immobile ether. Sagnac later tried to develop a theory of electrodynamics that would retain classical ideas of space, time, and ether by analyzing the propagation of energy statistically and separately from the propagation of motion. He retained a lifelong dislike for relativity, and in 1923 he interpreted some observations of stellar color shift as being due to an ether wind, rather than as supporting the general theory of relativity.

The results of Sagnac's interferometer experiments were used by some scientists in France as an argument against the theory of relativity. As late as 1937 Dufour and Prunier repeated the experiment in a modified form for that purpose.[2] The experiment does not contradict relativity theory in any way, however, and in 1921 and 1937 Langevin responded with an explanation of how it should be interpreted.[3]

As Langevin explained it, the experiment can be understood equally well in terms of both classical and relativistic theories (the phenomenon involves only a first-order effect in v/c). The fringe shift is produced by the difference in optical path for light beams traveling in opposite directions around the rotating disk. In fact, the experiment provides a neat way to demonstrate the rotation of a system without reference to anything external to it. As Pauli observed, "It is essentially the optical analogue of the Foucault pendulum."[4] In terms of general relativity the experiment can be understood by thinking of the light in the coordinate system of the disk as being subjected to a gravitational field directed away from the center.

An analysis of the rotating interferometer experiment existed in Germany even before Sagnac performed it. Max von Laue discussed the theory of the experiment in 1911 in response to Michelson's suggestion that it be used to demonstrate the earth's motion with respect to the ether.[5] He showed that both relativity theory and ether theory predicted the same result for the experiment. Laue discussed Sagnac's experiment explicitly in 1919 and the similar experiment of F. Harress in 1920.[6]

NOTES

1. *De l'optique des rayons de Röntgen et des rayons secondaires qui en dérivent* (Paris, 1900).
2. A. Dufour and F. Prunier, "Sur l'observation du phénomène de Sagnac par un observateur non entraîné," in *Comptes rendus hebdomadaires des séances de l'Académie des sciences,* **204** (1937), 1925–1927.
3. Paul Langevin, "Sur la théorie de relativité et l'expérience de M. Sagnac" and "Sur l'expérience de M. Sagnac," in his *Oeuvres scientifiques* (Paris, 1950), 467–472.
4. W. Pauli, *Theory of Relativity* (New York, 1958), 19.
5. Max von Laue, "Über einem Versuch zur Optik der bewegten Körper," in his *Gesammelte Schriften und Vorträge,* I (Brunswick, 1961), 154–161.
6. Max von Laue, *Das Relativitätsprinzip,* 3rd ed. (Leipzig, 1919); and "Zum Versuch von F. Harress," in his *Gesammelte Schriften,* I. 526–541.

BIBLIOGRAPHY

I. ORIGINAL WORKS. Sagnac's papers include "Sur la transformation des rayons X par la matière," in *Éclairage électrique,* **18** (1899), 41–48; "Théorie nouvelle des phénomènes optiques d'entraînement de l'éther par la matière," in *Comptes rendus hebdomadaires des séances de l'Académie des sciences,* **129** (1899), 818–821; "Electrisation negative des rayons secondaires issus de la transformation des rayons X," in *Journal de physique théorique et appliquée,* 4th ser., **1** (1902), 13–21, with P. Curie; "Sur la propagation de la lumière dans un système en translation et sur l'aberration des étoiles," in *Comptes rendus hebdomadaires des séances de l'Académie des sciences,* **141** (1905), 1220–1223; "Sur les interférences de deux faisceaux superposés en sens inverses le long d'un circuit optique de grandes dimensions," *ibid.,* **150** (1910), 1302–1305; "Les systèmes optiques en mouvement et la translation de la terre," *ibid.,* **152** (1911), 310–313; "L'éther lumineux demontré par l'effet du vent relatif d'éther dans un interféromètre en rotation uniforme," *ibid.,* **157** (1913), 708–710; "Effet tourbillonnaire optique. La circulation de l'éther lumineux dans un interférographe tournant," in *Journal de physique théorique et appliquée,* 5th ser., **4** (1914), 177–195; "Les deux mécaniques simultanées et leurs liaisons réelles," in *Comptes rendus hebdomadaires des séances de l'Académie des sciences,* **171** (1920), 99–102; and "Sur le spectre variable périodique des étoiles doubles: Incompatibilité des phénomènes observés avec la théorie de la relativité générale," *ibid.,* **176** (1923), 161–173.

II. SECONDARY LITERATURE. Sagnac's scientific work was summarized in connection with the Academy's award of the Pierson-Perrin Prize in *Comptes rendus hebdomadaires des séances de l'Académie des sciences,* **169** (1919), 1227–1232. See also Sagnac's *Notice sur les titres et travaux scientifiques de M. Georges Sagnac* (Paris, 1920). For discussions of Sagnac's rotating interferometer experiment, see the works by Langevin, Pauli, and Laue listed in the notes and André Metz, "Les problèmes relatifs à la rotation dans la théorie de la relativité," in *Journal de physique et le radium,* 8th ser., **13** (1952), 224–238.

SIGALIA DOSTROVSKY

SAHA, MEGHNAD (*b.* Scoratali, Dacca district, India [now Bangladesh], 6 October 1894; *d.* New Delhi, India, 16 February 1956), *theoretical physics.*

Saha was the fifth child of Jagannath Saha, a small shopkeeper, and Bhubaneswari Debi. The family lived in modest circumstances, and Saha was able to go to primary school through the kindness of a local patron. He entered the Government Collegiate School on a scholarship when he was eleven, but was soon expelled for participating in a political demonstration; he then attended a private school and, in 1911, enrolled in the Calcutta Presidency College. He received the M.A. in applied mathematics in 1915 and, in the following year, was appointed lecturer in mathematics on the all-Indian faculty of the University College of Science, which had just been founded by Sir Astosh Mukherjee.

A personality conflict with the chairman of the mathematics department soon led Saha to transfer to the department of physics, where he taught and did research. He received the D.Sc. in 1918. In the meantime, he had become interested in astrophysics, and had begun a systematic study of twenty-five years of the *Monthly Notices of the Royal Astronomical Society.* Since he was also teaching thermodynamics on the graduate level (although he had not previously studied the subject in depth), he often considered the relationship between thermodynamics and astrophysics. Thus, when he read J. Eggert's paper "Über den Dissoziationzustand der Fixsterngase," published in *Physikalische Zeitschrift* in 1919, Saha was prepared to begin the work on thermal ionization that won him a permanent place in the history of science.

By boldly applying thermodynamics and quantum theory to stellar matter and by drawing an analogy between chemical dissociation and atomic ionization, Saha derived a formula by which the degree of ionization in a very hot gas could be expressed in terms of its temperature and electron pressure. He set out his results in a paper entitled "On Ionization in the Solar Chromosphere," published in *Philosophical Magazine* in 1920. There is some dispute concerning where this paper was written, since Saha left India in September 1919 on a scholarship that permitted him to visit both A. Fowler's laboratory at Imperial College, London, and later Nernst's laboratory in Berlin. At any rate, in London Saha worked not only with Fowler but with E. A. Milne, who developed important extensions of Saha's theory; while in Berlin he met Planck, Einstein, Laue, and Sommerfeld and set up experiments to confirm his theory, of which he gave an account in an article, "Versuch einer Theorie der physikalischen Erscheinungen bei höhen Temperaturen . . .," published in *Zeitschrift für Physik* in 1921.

Saha's theory may be considered to be the starting point of modern astrophysics. The long-noted absence of rubidium and cesium in the solar spectrum could be understood for the first time, since it follows from Saha's formula that the degree of ionization increases with temperature and decreases with pressure, and is therefore smaller for elements with lower ionization potentials. It is thus possible to predict whether or not the spectrum of an element may be expected in a region with a given temperature and pressure if its ionization potential is known. It was also soon confirmed that other lines existed in the cooler regions of the solar atmosphere, as the theory predicted, and H. N. Russell was thereby able to identify rubidium in the spectra of sunspots. (It must be noted, however, that in the years immediately following the publication of Saha's theory, enthusiasm led some astrophysicists to apply the equation rather rashly, and some significant errors were made.) But in general, important knowledge of stellar atmospheres could be obtained by an analysis of their spectra through Saha's theory.

Saha returned to India in 1921 to accept a professorship at the University of Calcutta, but lack of funds and encouragement prompted him to look for a position elsewhere. In 1923 he went to the University of Allahabad, where he taught for the next fifteen years. As confirmations and extensions of his theory brought him fame he received a number of honors. He was elected president of the physics section of the Indian Science Congress Association in 1925, and two years later he was named a member of the Royal Society. In 1938 he left the University of Allahabad to become professor of physics at the University of Calcutta, where he worked to create an institute of nuclear physics

(which was officially opened by Irène Joliot-Curie, and is now named in Saha's honor). He also initiated researches on cosmic rays, the ionosphere, and geophysics, and traveled widely in connection with his increasing organizational responsibilities. He was directly or indirectly responsible for the establishment of a number of scientific bodies and institutions, including the United Provinces Academy of Sciences (later the National Academy of Sciences), the Indian Science News Association, and the National Institute of Sciences.

Saha was further active in planning scientific and industrial projects for India. In particular he was involved in river valley projects (the result of having in his youth witnessed devastating floods), calendar reform, and geophysical explorations. In 1935 he founded the influential socioscientific journal *Science and Culture;* but although he was interested in the social implications of science, and although he was sympathetic to the Indian struggle for independence, he remained aloof from politics in general. He was opposed to the *khadi* movement of Ghandi and his followers because he thought that it would impede Indian industrialization. Following India's political independence, Saha devoted more time to social, political, and economic problems, and in 1951 he was elected, as an independent, to Parliament. He died during a trip to New Delhi to discuss some matters pertaining to the administration of his nuclear physics institute.

BIBLIOGRAPHY

A complete bibliography of Saha's works may be found in S. N. Sen, ed. (see below).

For further details on the life and works of Saha, consult S. N. Sen, ed., *Professor Meghnad Saha, His Life, Work and Philosophy* (1954); *Science and Culture,* **19** (1954), 442; *Science and Culture,* **22** (1956), a complete issue devoted to Saha; D. S. Kothari, "Meghnad Saha," in *Biographical Memoirs of Fellows of the Royal Society,* **4**, 217–236; and an obituary by F. J. M. Stratton in *Nature,* **177** (1956), 917.

For discussions of Saha's theory of thermal ionization, any work on astrophysics may be consulted. In particular, a nonmathematical treatment of the ideas as they stood in the early years of the theory may be found in H. Dingle, *Modern Astrophysics* (New York, 1924), 197–217. See also Giorgio Abetti, *The Sun,* J. B. Sidgwick, trans. (London, 1955), ch. 5. For a more technical discussion, see, A. Unsöld, *Physik der Sternatmosphären,* 2nd ed. (Berlin, 1955), 40–48; and S. Rosseland, *Theoretical Astrophysics* (Oxford, 1936), 170–183.

V. V. RAMAN

IBN AL-ṢAʾIGH. See Ibn Bājja, Abū Bakr Muḥammad ibn Yaḥyā ibn al-Ṣāʾigh.

SAINTE-MESME, MARQUIS DE. See L'Hospital (L'Hôpital), Guillaume-François-Antoine de.

SAINT-HILAIRE, AUGUSTIN FRANÇOIS CÉSAR PROUVENÇAL (usually known as **AUGUSTE DE**) (*b.* Orléans, France, 4 October 1779; *d.* La Turpinière, near Sennely, Loiret, France, 30 September 1853), *natural history.*

Saint-Hilaire was interested in entomology and botany from an early age; and despite his appointment as accountant in the civil service, in 1816 left for Rio de Janeiro, accompanying M. de Luxembourg, the French ambassador. He intensively surveyed the flora and fauna of Brazil from Jequitinhonha to the Río de la Plata for six years. In August 1822 Saint-Hilaire returned to Paris with 24,000 plants, 2,000 birds, 16,000 insects, 135 quadrupeds, and many reptiles, fishes, and minerals that he intended to classify. He fell seriously ill, however, lost his voice, and almost lost his sight while preparing the publication of the Brazilian flora. Under the care of Michel Dunal and Claude Lallemand at Montpellier, he recovered.

In 1830 Saint-Hilaire was elected a member of the Academy of Sciences and became professor at the Faculty of Sciences in Paris. On all his travels he made interesting anthropological, botanical, and pharmacognostic observations. His work on Brazilian botany, however, was superseded by that of Martius.

BIBLIOGRAPHY

Saint-Hilaire's first works appeared in the *Bulletin des sciences physiques, médicales, et d'agriculture d'Orléans.* Shortly after his return from Brazil he published "Aperçu d'un voyage dans l'intérieur du Brésil, la province cisplatine et les missions dites du Paraguay," in *Mémoires du Muséum d'histoire naturelle,* **9** (1822), 307–380. *Histoire des plantes les plus remarquables du Brésil et du Paraguay* (Paris, 1824) and *Plantes usuelles des brasiliens* (Paris, 1824) were followed by his most important work, *Flora Brasiliae meridionalis,* 3 vols. (Paris, 1825–1833), written with A. de Jussieu and J. Cambessèdes. Saint-Hilaire published the diaries of his travels in Brazil as *Voyage dans les provinces de Rio de Janeiro et de Minas Geraes* (Paris, 1830); *Voyage dans le district des diamants et sur le littoral du Brésil* (Paris, 1833); *Voyage aux sources du Rio de S. Francisco et dans la province de Goyaz,* 2 vols. (Paris, 1847–1848); and *Voyage dans les provinces de Saint Paul et de Sainte Catherine* (Paris, 1851). Saint-Hilaire's *Voyage à Rio Grande do Sul (Brésil)* (Orléans, 1887) was published posthumously, edited by R. de Dreuzy.

Anna Eliza Jenkins wrote an introductory essay to Saint-Hilaire's "Esquisse de mes voyages au Brésil et Paraguay," in *Chronica botanica,* **10** (1946), 1–62, with an interesting biographical study.

Francisco Guerra

ST. JOHN, CHARLES EDWARD (*b.* Allen, Michigan, 15 March 1857; *d.* Pasadena, California, 26 April 1935), *astronomy.*

The youngest in the family of Hiram Abiff St. John, a millwright, and Lois Amanda Bacon, St. John graduated from the Michigan Normal School in 1876. Overwork had weakened his health, however, and he could not resume work until 1885, when he became an instructor of physics at the normal school. He received the B.S. degree from Michigan Agricultural College in 1887; and after graduate study at the University of Michigan and at Harvard, including also a stay in Berlin and Heidelberg, he was awarded the Ph.D. by Harvard in 1896. After a year as an instructor at the University of Michigan, St. John was appointed associate professor of physics and astronomy at Oberlin College, becoming professor in 1899 and dean of the College of Arts and Sciences in 1906. Beginning in 1898, he spent several summer vacations working at the Yerkes Observatory. His association there with George Ellery Hale led him, in 1908, to resign his post at Oberlin and to join the staff of the Mount Wilson Observatory; after his retirement in 1930 he was appointed a research associate there.

St. John's research was mainly in solar physics. One of his earliest investigations (1910–1911) was of the H and K lines of ionized calcium in the solar spectrum. After measuring the absolute wavelengths of these lines in terrestrial sources, he compared them with those of the corresponding emission and absorption lines observed in the solar spectrum, making a particular study of the spectral lines over sunspots and flocculi. In 1913 St. John confirmed John Evershed's discovery concerning the displacement of Fraunhofer lines in the spectra of the penumbras of sunspots, and he made a detailed examination of the flow of gases in the spots. He concluded that ionized calcium exists at the highest level in the solar atmosphere, followed by hydrogen, with the metals and rare elements confined to the lower regions; his analysis of the flash spectrum obtained by S. A. Mitchell during the

1905 eclipse reinforced this conclusion. At the same time St. John refuted W. H. Julius' idea that the displacements in the spectral lines were due to anomalous dispersion in the solar atmosphere.

St. John's principal contribution was his revision of Rowland's table of the wavelengths in the solar spectrum. In collaboration with Charlotte Moore, Louise Ware, Edward Adams, and Harold Babcock, he produced this monumental work in 1928. Two independent series of measurements were made, one with the Mount Wilson tower spectrographs and the other with an interferometer; the differences rarely exceeded 0.002 Å. From the violet limit of 2,975 Å, the lines listed ranged beyond Rowland's extreme of 7,330 Å to about 10,200 Å. The revision listed the intensities of the spectral lines in spots as well as in the disk, and the lines were classified according to their furnace spectra and behavior under pressure; excitation potentials were also provided. In connection with this work St. John served for a time as president of the International Astronomical Union commission devoted to wavelength standards, and he was also president of the solar physics commission.

St. John paid particular attention to the problem of measuring the relativistic deflection of the lines in the solar spectrum. An initial study (1917), based on some forty lines, gave a negative result. By 1923, however, he was satisfied that the effect, which at the center of the sun amounts to about 0.01 Å, was detectable; and his final publication on the subject (1928), using 1,537 lines, illustrated it quite convincingly. Although St. John regarded this as his most important work, the motion of the perihelion of Mercury and the systematic displacement observed beyond the limb of the sun are generally regarded as more definitive tests of relativity.

In collaboration with Walter Adams and Seth Nicholson, St. John made spectroscopic observations of Mars and Venus, mainly in the hope of detecting the presence of oxygen and water vapor in their atmospheres.

BIBLIOGRAPHY

I. ORIGINAL WORKS. St. John's *Revision of Rowland's Preliminary Table of Solar Spectrum Wavelengths, With an Extension to the Present Limit of the Infra-Red*, written with C. E. Moore, L. M. Ware, E. F. Adams, and H. D. Babcock, is Carnegie Institution of Washington Publication no. 396 (Washington, D.C., 1928) and *Papers of the Mount Wilson Solar Observatory*, 3 (1928). Most of his other publications appeared in *Astrophysical Journal* and were reprinted as *Contribu-*

tions from the Mount Wilson Solar Observatory. They include "The Absolute Wave-Lengths of the H and K Lines of Calcium in Some Terrestrial Sources," in *Astrophysical Journal*, 31 (1910), 143–156; "The General Circulation of the Mean and High-Level Calcium Vapor in the Solar Atmosphere," *ibid.*, 32 (1910), 36–82; "Motion and Condition of Calcium Vapor Over Sun-Spots and Other Special Regions," *ibid.*, 34 (1911), 57–78, 131–153; "Radial Motions in Sun-Spots," *ibid.*, 37 (1913), 322–353, and 38 (1913), 341–391; "On the Distribution of the Elements in the Solar Atmosphere as Given by Flash Spectra," *ibid.*, 40 (1914), 356–376; "The Accuracy Obtainable in the Measured Separation of Close Solar Lines: Systematic Errors in the Rowland Table for Such Lines," *ibid.*, 44 (1916), 15–38, written with L. W. Ware; "The Elimination of Pole-Effect From the Source for Secondary Standards of Wave-Length," *ibid.*, 46 (1917), 138–166, written with H. D. Babcock; "On Systematic Displacements of Lines in Spectra of Venus," *ibid.*, 53 (1921), 380–391, written with S. B. Nicholson; "Evidence for the Gravitational Displacement of Lines in the Solar Spectrum Predicted by Einstein's Theory," *ibid.*, 67 (1928), 195–239; "Elements Unidentified or Doubtful in the Sun: Suggested Observations," *ibid.*, 70 (1929), 160–174; and "Excitation Potential in Solar Phenomena," *ibid.*, 319–330.

II. SECONDARY LITERATURE. Obituary notices appeared in *Astrophysical Journal*, 82 (1935), 273–283; *Publications of the Astronomical Society of the Pacific*, 47 (1935), 115–120; and *Popular Astronomy*, 43 (1935), 611–617. See also Walter S. Adams' notice in *Biographical Memoirs. National Academy of Sciences*, 18, no. 12 (1938), 285–304, with bibliography of St. John's works.

BRIAN G. MARSDEN

SAINT-VENANT, ADHÉMAR JEAN CLAUDE BARRÉ DE (*b.* Villiers-en-Bière, Seine et-Marne, France, 23 August 1797; *d.* St.-Ouen, Loir-et-Cher, France, 6 January 1886), *mechanics, geometry*.

Saint-Venant entered the École Polytechnique in 1813. Upon graduating he joined the Service des Poudres et Salpêtres and in 1823 transferred to the Service des Ponts et Chaussées, where he served for twenty years. He devoted the remainder of his life to teaching and, especially, to scientific research. In 1868 he was elected to the mechanics section of the Académie des Sciences, succeeding Poncelet.

Saint-Venant's investigations deal chiefly with the mechanics of solid bodies, elasticity, hydrostatics, and hydrodynamics. Closely related to engineering, they frequently had immediate applications to road- and bridge-building, to the control of streams, and to agriculture. On the basis of his work on the torsion of prisms or cylinders of any base and on the equilibrium of elastic beams, Saint-Venant presented a memoir to the Académie des Sciences in 1844 dealing with gauche curves. In it he introduced the term "binormal," which is still used: "This line is, in effect, normal to two consecutive elements at the same time."

In "Mémoire sur les sommes et les différences géométriques et sur leur usage pour simplifier la mécanique" (1845), Saint-Venant set forth a vector calculus displaying certain analogies with the conceptions of H. G. Grassmann. In a subsequent priority dispute Saint-Venant asserted in a letter to Grassmann, written in 1847, that his ideas dated from 1832.

Saint-Venant used this vector calculus in his lectures at the Institut Agronomique, which were published in 1851 as *Principes de mécanique fondés sur la cinématique*. In this book Saint-Venant, a convinced atomist, presented forces as divorced from the metaphysical concept of cause and from the physiological concept of muscular effort, both of which, in his opinion, obscured force as a kinematic concept accessible to the calculus. Although his atomistic conceptions did not prevail, his use of vector calculus was adopted in the French school system.

BIBLIOGRAPHY

I. ORIGINAL WORKS. The Royal Society *Catalogue of Scientific Papers*, I, 189–191; VIII, 812–814; and XI, 262, lists 111 works by Saint-Venant and four of which he was coauthor. Among his some 170 published writings are *Leçons de mécanique appliquée faites à l'École des ponts et chaussées* (Paris, 1838); "Mémoire et expériences sur l'écoulement de l'air," in *Journal de l'École polytechnique*, 16 (1839), 85–122, written with Laurent Wantzel: "Mémoire sur les courbes non planes," *ibid.*, 18 (1845), 1–76; "Mémoire sur les sommes et les différences géométriques et sur leur usage pour simplifier la mécanique," in *Comptes rendus . . . de l'Académie des sciences*, 21 (1845), 620–625; *Principes de mécanique fondés sur la cinématique* (Paris, 1851); "De l'interprétation géométrique des clefs algébriques et des déterminants," in *Comptes rendus . . . de l'Académie des sciences*, 36 (1853), 582–585; *Mécanique appliquée de Navier, annotée par Saint-Venant* (Paris, 1858); "Deux leçons sur la théorie générale de l'élasticité," in Chanoine Moigno, *Stratique* (Paris, 1868), lessons 21 and 22; R. Clebsch, *Theorie de l'élasticité des corps solides*, translated by Saint-Venant and Alfred Flamant, with notes by Saint-Venant (Paris, 1883), and "Résistance des fluides: Considérations historiques phy-

siques et pratiques relatives au problème de l'action dynamique mutuelle d'un fluide et d'un solide, spécialemente dans l'état de permanence supposé acquis dans leurs mouvements," in *Mémoires de l'Académie des sciences*, 44 (1888), 1–192, 271–273.

II. SECONDARY LITERATURE. See J. Boussinesq and A. Flamant, "Notice sur la vie et les travaux de M. de Saint-Venant," in *Annales des ponts et chaussées*, 6th ser., 12 (1886), 557–595, which includes a very comprehensive bibliography; Michel Chasles, *Rapport sur les progrès de la géométrie* (Paris, 1870), 197–199; Michael J. Crowe, *A History of Vector Analysis* (Notre Dame, Ind., 1967), 81–85; René Dugas, *Histoire de la mécanique* (Paris, 1950), 421–422; and E. Phillips, "Notice sur M. de Saint-Venant," in *Comptes rendus . . . de l'Académie des sciences*, 102 (1886), 141–147.

JEAN ITARD

ST. VICTOR, HUGH OF. See **Hugh of St. Victor.**

SAINT VINCENT, GREGORIUS (*b.* Bruges, Belgium, 8 September 1584; *d.* Ghent, Belgium, 27 January 1667), *mathematics, astronomy.*

Nothing is known of Gregorius' origins. He entered the Jesuit *collège* of Bruges in 1595 and from 1601 studied philosophy and mathematics at Douai. In 1605 he became a Jesuit novice at Rome and in 1607 was received into the order. His teacher Christoph Clavius recognized Gregorius' talents and arranged for him to remain in Rome to study philosophy, mathematics, and theology. When Galileo compared his telescope with those of the Jesuits in 1611, Gregorius hinted that he had doubts about the geocentric system, thereby displeasing the scholastically oriented philosophers.

After Clavius died in 1612, Gregorius went to Louvain to complete his theological studies, and in 1613 he was ordained priest. After being assigned to teach Greek for several years, first in Brussels, then in Bois-le-Duc ('s Hertogenbosch, Netherlands [1614]), and in Courtrai (1615), he served for a year as chaplain with the Spanish troops stationed in Belgium. He then became lecturer in mathematics at the Jesuit college in Antwerp, succeeding François d'Aguilon (*d.* 1617). Gregorius' *Theses de cometis* (Louvain, 1619) and *Theses mechanicae* (Antwerp, 1620) were defended by his student Jean Charles de la Faille, who later made them the basis of his highly regarded *Theoremata de centro gravitatis* (1632).

Established as a mathematician at Louvain in 1621, Gregorius elaborated the theory of conic sections on the basis of Commandino's editions of Archimedes (1558), Apollonius (1566), and Pappus (1588). He also developed a fruitful method of infinitesimals. His students Gualterus van Aelst and Johann Ciermans defended his *Theoremata mathematica scientiae staticae* (Louvain, 1624); and two other students, Guillaume Boelmans and Ignaz Derkennis, aided him in preparing the *Problema Austriacum,* a quadrature of the circle, which Gregorius regarded as his most important result. He requested permission from Rome to print his manuscript, but the general of the order, Mutio Vitelleschi, hesitated to grant it. Vitelleschi's doubts were strengthened by the opinion that Christoph Grienberger (Clavius' successor) rendered on the basis of preliminary material sent from Louvain.

Gregorius was called to Rome in 1625 to modify his manuscript but returned to Belgium in 1627 with the matter still unsettled. In 1628 he went to Prague, where he suffered a stroke. Following his recovery, his superiors granted his request that a former student, Theodor Moret, be made his assistant. His poor health forced Gregorius to decline an offer from the Madrid Academy in 1630. The following year he fled to Vienna just ahead of the advancing Swedes, but he was obliged to leave behind his scientific papers, including an extensive work on statics. A colleague, Rodrigo de Arriaga, rescued the studies on the conic sections and on methods of quadrature. Gregorius, who meanwhile had become a mathematician in Ghent (1632), did not receive his papers until 1641. He published them at Antwerp in 1647 as *Opus geometricum quadraturae circuli et sectionum coni.* His *Opus geometricum posthumum ad mesolabium* (Ghent, 1668) is an unimportant work, the first part of which had been printed at the time of his death.

Gregorius' major work is the *Opus geometricum* of 1647, misleadingly entitled *Problema Austriacum;* it is over 1,250 folio pages long and badly organized. It treats of four main subjects. Book I contains various introductory theorems on the circle and on triangles as well as geometrically clothed algebraic transformations. Book II includes the sums of geometric series obtained by means of transformation to the differences of the terms. Among the applications presented in this book is the step-by-step approximation of the trisection of an angle through continuous bisection, corresponding to the relationship $1/2 - 1/4 + 1/8 \mp \cdots = 1/3$. Another is the skillful treatment of Zeno of Elea's paradox of Achilles and the tortoise. In book VIII it is shown that if the horn an-

gle is conceived as a quantity, the axiom of the whole and the parts no longer holds.

Books III–VI are devoted to the circle, ellipse, parabola, and hyperbola, and to the correspondence between the parabola and the Archimedean spiral (today expressed as $x = r$, $y = r\phi$). These books contain various propositions concerning the metric and projective properties of conic sections. Their scope far exceeds that found in older treatments, but their presentation is unsystematic. Common properties are based on the figure of the conic section pencil $y^2 = 2\,px - (1 - \epsilon^2)x^2$, where ϵ is the parameter. (This figure had appeared in 1604 in a work by Kepler.) By inscribing and circumscribing rectangles in a geometric series in and about a hyperbola, Gregorius developed a quadrature of a segment bound by two asymptotes, a line parallel to one of them, and the portion of the curve contained between the two parallels. The relation between this procedure and logarithms was first noted by Alfonso Antonio de Sarasa (1649).

Book VII contains Gregorius' remarkable quadrature method. It is a summation procedure— the so-called *ductus plani in planum*—related to the method of indivisibles developed by Bonaventura Cavalieri, although the two are mutually independent. Gregorius' method, however, is somewhat better founded. In modern terms it amounts to the geometric interpretation of cubatures of the form $\int y(x) \cdot z(x) \cdot dx$. It touches on considerations related to the then-unknown *Method* of Archimedes, considerations that, in book IX, are applied to bodies of simple generation. A section of book VII deals with "virtual" parabolas, expressible in modern notation as $y = \sqrt{ax + b} + \sqrt{cx + d}$.

Book X is devoted to the quadrature of the circle, which here is based on cubatures of the following type:

$$X_1 = 2a\int_{x-c}^{x+c} \sqrt{a^2 - t^2}\,dt, \quad X_2 = \int_{x-c}^{x+c} (a^2 - t^2)\,dt,$$

$$X_3 = \int_{x-c}^{x+c} (a^2 - t^2)^2 dt/4a^2$$

$$Y_1 = 2a\int_{y-c}^{y+c} \sqrt{a^2 - t^2}\,dt, \quad Y_2 = \int_{y-c}^{y+c} (a^2 - t^2)\,dt,$$

$$Y_3 = \int_{y-c}^{y+c} (a^2 - t^2)^2 dt/4a^2,$$

given that $x \neq y$ and $0 \leq x - c < x + c \leq a$ and $0 \leq y - c < y + c \leq a$. The crucial element of the argument is the false assertion that from $X_2/Y_2 = (X_1/Y_1)^n$ it follows that $X_3/Y_3 = (X_2/Y_2)^n$. The result is the appearance in the calculations of an error of integration, first detected by Huygens (1651). The error arose from the geometric presentation of the argument, which made it extraordinarily difficult to get an overall grasp of the problem. This error considerably damaged Gregorius' reputation among mathematicians of the following generation. But their reaction was unfair, for his other results show that he was a creative mathematician with a broad command of the knowledge of his age. Although Gregorius basically despised algebraic terminology, he was, as his students recognized, one of the great pioneers in infinitesimal analysis.

BIBLIOGRAPHY

I. Original Works. Gregorius' unprinted posthumous papers are in the Bibliothèque Royale Albert Ier, nos. 5770–5793. Nos. 5770–5772, which are partially illustrated with remarkable figures, are discussed in E. Sauvenier-Goffin, "Les manuscrits de Saint-Vincent," in *Bulletin de la Société r. des sciences de Liège*, **20** (1951), 413–436, 563–590, 711–737. See also P. Bockstaele, "Four Letters from Gregorius a S. Vincentio to Christophe Grienberger," in *Janus*, **56** (1969), 191–202. A portrait of Gregorius is in *Opus geometricum posthumum* (1668).

II. Secondary Literature. For a biography of Gregorius, see H. Bosmans, "Saint Vincent (Grégoire de)," in *Biographie nationale belge*, XXI, cols. 141–171, which contains an extensive bibliography.

For a discussion of Gregorius' work on conic sections, see K. Bopp, "Die Kegelschnitte des Gregorius a S. Vincentio," in *Abhandlungen zur Geschichte der Mathematik*, no. 20 (1907), 83–314.

Gregorius' work on infinitesimal analysis and its influence on other mathematicians is discussed in J. E. Hofmann, "Das *Opus geometricum* des Gregorius a S. Vincentio und seine Einwirkung auf Leibniz," in *Abhandlungen der Preussischen Akademie der Wissenschaften*, Math. Naturwiss. Kl. (1941), no. 13.

On the debate over the quadrature of the circle, see A. A. de Sarasa, *Solutio problematis a M. Mersenno propositi* (Antwerp, 1649); C. Huygens, *Exetasis cyclometriae Gregorii a S. Vincentio*, which was appended to *Theoremata de quadratura hyperboles, ellipsis et circuli ex dato portionum gravitatis centro* (Leiden, 1651), repr. in Huygens' *Oeuvres complètes*, XI (The Hague, 1908), 315–337; and *Ad. Fr. X. Ainscom epistola* (The Hague, 1656), repr. in his *Oeuvres complètes*, XII (1910), 263–277; Alexius Sylvius, *Lunae circulares periodi . . ., adjunctum quoque est examen quarundam propositionum quadraturae circuli Gregorii a S. Vincentio* (Lesna, 1651), 374–418; Gottfried Alois Kinner von Löwenthurn, *Elucidatio geometrica problematis Austriaci* (Prague, 1653); Vincent Léotaud, *Examen circuli quadraturae . . .* (Lyons, 1654); and *Cyclomathia . . .* (Lyons, 1663), I, *Quadraturae examen confirmatur ac promovetur*; Marcus Meibom, *De proportionibus dialo-*gus (Copenhagen, 1655); and Franz X. Aynscom, *Expositio ac deductio geometrica quadraturarum circuli Gregorii a S. Vincentio* (Antwerp, 1656).

J. E. Hofmann

SAKHAROV, VLADIMIR VLADIMIROVICH (*b.* Simbirsk [now Ulyanovsk], Russia, 28 February 1902; *d.* Moscow, U.S.S.R., 9 January 1969), *genetics.*

Sakharov's research was basically in experimental chemical mutagenesis, polyploids, radiation genetics, and human genetics. His father, Vladimir Matveevich Sakharov, was an agronomist; his mother, Maria Antonovna Ponyatovskaya, was the daughter of a Moscow physician and taught French in the Simbirsk Gymnasium.

In 1919 the family moved to Moscow, and in 1920 Sakharov entered the Pedagogical Faculty of the Second Moscow State University, from which he graduated in 1926, defending as dissertations two works done at the Institute of Experimental Biology (IEB) from 1922 to 1924: "Novaya mutatsia u drozofily" ("New Mutation in Drosophila") and "Razbor muzykalnykh genealogy" ("Analysis of Musical Genealogies"). From 1925 through 1929 he taught soil science in a Moscow school.

In 1929 Sakharov entered the IEB, where he worked until 1948 under the immediate direction of its organizer and director, N. K. Koltsov. During the 1930's and early 1940's Sakharov did research on experimental mutagenesis. In 1932 he showed for the first time the mutagenic action of chemical agents (iodine, methylcholanthrene). Continuing this line of work, Sakharov formulated the idea of the "specific action of mutagenic factors" and showed the difference in the nature of spontaneous and induced chemical and physical factors in mutations. This research was completed with the discovery of the role of factors inherent in the mutational process (the aging of sperm, hibernation, inbreeding, and hybridization). At the beginning of the 1930's Sakharov also studied medical anthropogenesis, for instance, the distribution and character of inherited endemic goiter and of blood types in Uzbekistan (1929 and 1930).

In 1941 Sakharov began research on polyploids. With S. L. Frolova and V. V. Mansurova, he obtained by the colchicine method a highly fertile variety of tetraploid buckwheat, which by 1948 successfully competed with the best diploid variety. Sakharov continued to combine research with teaching in the department of general biology (headed by V. F. Natali of the Moscow Pedagogical Institute).

In 1950–1956 Sakharov worked at the Moscow Pharmaceutical Institute in the department of botany, which was headed by A. P. Zhebrak. Besides lecturing on plant genetics, he continued his research on polyploids (now on medicinal plants) and began experiments on chemical mutagenesis in plants. With B. M. Griner he created a botanical garden of medicinal plants. In 1956 he organized a section of genetics at the Moscow Society of Experimenters with Nature, which became the basis of the Vavilov All-Union Society of Geneticists and Selectionists, created in 1965.

In 1956 Sakharov moved to the Laboratory of Radiation Genetics of the Institute of Biophysics of the Soviet Academy of Sciences, where he headed a group for the study of polyploids; in 1966–1967 he headed the Laboratory of Polyploids of the Institute of General Genetics of the Soviet Academy of Sciences. In 1967 he moved his laboratory to the Institute of Biology of Development of the Soviet Academy. From 1956, combining his research on mutagenesis and polyploids, Sakharov began a comparative study of sensitivity of diploid and autotetraploid forms to the action of radiation and chemical mutagenesis (for example, on buckwheat and meadow brown butterflies), discovering the physiological protection of polyploids against the influence of mutagenesis. This work led Sakharov to pose the question and prove the possibility of selection for radiation-resistance on a theoretical level, and later in direct experiments on diploid and tetraploid buckwheat. At the end of the 1960's he became professor at the Timiryazev Institute of Plant Physiology, where he lectured on genetics.

Sakharov was a scientist with a broad general biological and philosophical viewpoint, as well as an author who both posed problems and indicated the means for solving them in an original way. His scientific credo was stated in a small book, *Organizm i sreda* ("Organism and Environment," 1968). Besides his personal scientific contribution, he played a large role in the development of genetics in the Soviet Union through his scientific administration and teaching and through propaganda for genetic knowledge, especially in the periods of conflict over genetics and its restoration.

BIBLIOGRAPHY

I. ORIGINAL WORKS. Sakharov's writings include "Novye mutatsii *Drosophila melanogaster*" ("New Mutations of *Drosophila melanogaster*"), in *Zhurnal eksperimentalnoi biologii i meditsiny*, ser. A, **1**, nos. 1–2 (1925), 75–91, written with A. S. Serebrovsky; "Yod kak khimichesky faktor deystvuyushchii na mutatsionny protsess u *Drosophila melanogaster*," in *Biologicheskii zhurnal*, **1**, nos. 3–4 (1932), 1–8, with German summary, "Erregung des Mutationprozesses bei *Drosophila melanogaster* durch Jodbehandlung," 7–8; "Jod als chemischer Faktor, der auf dem Mutationprozess von *Drosophila melanogaster wirkt*," in *Genetica* (The Hague), **18**, nos. 3–4 (1936), 193–216; "Spetsifichnost deystvia mutatsionnykh faktorov," in *Biologicheskii zhurnal*, **7**, no. 3 (1938), 595–618, with English summary, "The Specificity of the Action of the Factors of Mutation," 617–618; and "Vlianie inbridinga i gibridizatsii na temp mutatsionnogo protsessa," in *Zhurnal obshchei biologii*, **3**, nos. 1–2 (1942), 99–123, written with K. V. Magrzhikovskaya, with an English summary, "Effect of Inbreeding and Hybridization on the Mutation Rate," 120–123.

See also "Tetraploidy in Cultivated Buckwheat (*Fagopyrum esculentum*)," in *Doklady Akademii nauk SSSR*, **43**, no. 5 (1944), 213–216, written with S. L. Frolova and V. V. Mansurova; "Cytological Basis of High Fertility in Autotetraploid Buckwheat," in *Nature*, **158**, no. 4015 (1945), 520, written with S. L. Frolova and V. V. Mansurova; "Chuvstvitelnost diploidnykh i autotetraploidnykh rastenii k gamma-izlucheniyu," in *Botanicheskii zhurnal SSSR*, **43**, no. 7 (1958), 989–997, written with V. V. Mansurova and V. V. Khvostova, with English summary, "The Sensitivity of Diploid and Autotetraploid Plants to Gamma Radiation," 997; "Otbor na radioustoichivost diploidnykh i autotetraploidnykh form grechiki posevnoi (*Fagopyrum esculentum* Moënh)" ("Selection for Radio-Resistance in Diploid and Autotetraploid Forms of Buckwheat"), in *Radiobiologia*, **2**, no. 4 (1962), 595–600, written with R. N. Platonova; and *Organizm i sreda* ("Organism and Environment"; Moscow, 1968).

II. SECONDARY LITERATURE. See B. L. Astaurov, A. A. Malinovsky, and V. S. Andreev, "Vladimir Vladimirovich Sakharov," in *Genetika* (Moscow), **5**, no. 2 (1969), 177–182, with a bibliography of 84 titles; and Y. I. Polansky, "Vladimir Vladimirovich Sakharov (1902–1969)," in *Citologia*, **2**, no. 3 (1969), 398–400.

VASSILY BABKOFF

SAKS, STANISŁAW (*b.* Warsaw, Poland, 30 December 1897; *d.* Warsaw, November 1942), *mathematics*.

Saks was a member of the Polish school of mathematics that flourished between the two world wars. The son of Philip and Ann Łabedz Saks, he received his secondary education in Warsaw. In the autumn of 1915 Saks entered the newly established Polish University of Warsaw, from which he received his doctorate in 1921 with a dissertation in topology. From 1921 to 1939 he was an assistant at the Warsaw Technical University and, from 1926 to 1939, he also lectured at the University of Warsaw. In 1942 he was arrested by the Nazi authorities and killed, allegedly while attempting to escape from prison.

Most of Saks's research involved the theory of real functions, such as problems on the differentiability of functions and the properties of Denjoy-Perron integrals. His work also touched upon questions in such related fields as topology and functional analysis. The two mathematicians at the University of Warsaw who exerted the greatest influence upon Saks were the topologist Stefan Mazurkiewicz, from whom Saks acquired a sensitivity to topological problems and methods, and Wacław Sierpiński. Saks, in turn, considerably influenced the development of real analysis within the Polish school.

Saks's contributions to mathematics included two important books. The first, *Théorie de l'intégrale* (1933), grew out of his lectures at the University of Warsaw and appeared as the second volume in the series Monografie Matematyczne. A thoroughly revised English edition was published in 1937 as the seventh volume of the series. In this highly original work Saks systematically developed the theory of integration and differentiation from the standpoint of countably additive set functions. Widely read outside Poland, it is now considered a still useful classic. In 1938 Saks collaborated with Antoni Zygmund to produce the ninth volume of Monografie Matematyczne, *Funkcje analityczne,* which received the prize of the Polish Academy of Sciences that year. An English edition, published by Zygmund in 1952, helped make its contents known to a larger audience, for whom it has become a standard reference work on complex analysis.

BIBLIOGRAPHY

I. ORIGINAL WORKS. Saks's papers have not been published in collected form, but many appeared in *Fundamenta mathematicae* and, to a lesser extent, in *Studia mathematica*. His books are *Zarys teorii całki* (Warsaw, 1930); *Théorie de l'intégrale* (Warsaw, 1933), rev. as *Theory of the Integral*, L. C. Young, trans. (Warsaw–Lvov–New York, 1937; repr. New York, 1964); and *Funkcje analityczne* (Warsaw, 1938), written with A. Zygmund, rev. as *Analytic Functions*, E. J. Scott, trans. (Warsaw, 1952; 2nd English ed., 1965).

II. SECONDARY LITERATURE. Apparently nothing has been written on Saks. (The author is indebted to Professor A. Zygmund of the University of Chicago for much helpful information.) For further references concerning the Polish school of mathematics, see M. G. Kuzawa, *Modern Mathematics: The Genesis of a School in Poland* (New Haven, 1968).

THOMAS HAWKINS

SALA, ANGELO (Angelus) (*b.* Vicenza, Italy [?], 1576; *d.* Bützow, Germany, 2 October 1637), *pharmaceutics, chemistry, medicine.*

Sala was the son of Bernhardino Sala, a spinner; nothing is known about his mother. After Sala's grandfather, Angelo Sala, immigrated to Geneva, Sala followed him (at an unknown date), as did his brother Domenico. There the family converted to Calvinism. Sala probably had already begun his "chymical" studies in northern Italy (1593), perhaps during his years in Vicenza, which was close to Padua and Venice, then the centers of the manufacture and sale of medicines.

The years from 1602 to about 1612 were Sala's *Wanderjahre.* He is reported working as a physician in a number of cities, including Dresden (1602), Sondrio and Ponte (1604), Nuremberg (1606), Frauenfeld (1607), and Geneva (1609). In 1610 he served as physician to the Protestant troops under the command of Prince Johann of Nassau in the Upper Palatinate (for example, in Amberg and Neumark). Finally, in 1612, he settled in The Hague, the Netherlands, where for five years he practiced medicine and gave instruction in chemistry to medical students from various countries. He had already published two early works, *De variis tum chymicorum, tum galenistarum erroribus* . . . (n.p., 1608) and the famous *Anatomia vitrioli* (Geneva, 1609), but in The Hague he engaged in much more scientific research, which is reflected in a number of new works he published during this period.

From 1617 to 1620 he worked as physician in ordinary to Count Anton Günther of Oldenburg, living partly in the city of Oldenburg and partly in Jever. In addition, he supervised all the pharmacies in the count's territory. In Oldenburg, Sala met Anton Günther Billich, who became his student and an enthusiastic proponent of his chemical theories. In 1625 Billich married Sala's only daughter, Maria. (It is not known when Sala married his wife, Katherine von Brockdorf, who survived him.)

Sala continued to publish books while at Oldenburg and also at his next place of residence, Hamburg (1620–1625). Beginning in this period he wrote mainly in German (although he continued to give most of his works Latin titles). He made this decision after joining, in 1617, the Fruchtbringende Gesellschaft in Weimar, which promoted the use of the German language. In Hamburg, where he served as *Chymiater*, Sala published for the first time a list of the medications that he had prepared (Wandsbek [now part of Hamburg], 1624).

At the recommendation of the Landgrave Moritz of Hesse (Kassel), Sala transferred to the service of the rulers of Mecklenburg. Very active henceforth as both a physician (principally in connection with the use of drugs) and a researcher, he lived in Güstrow until 1636 and then, for the rest of his life, in Bützow. To be sure, during the troubles caused by the Thirty Years' War, he was obliged to move around several times, spending the years from 1628 to 1631 in Bernburg and Harzgerode in Anhalt. On his return trip he spent several days in Lübeck as a guest of the physician Hermann Westhoff. As early as 1622 Westhoff had reported on Sala's medicinal preparations to his friend Joachim Jungius (also from Lübeck) and had himself brought samples of them to Rostock. In his own theoretical research Jungius therefore drew indirectly on Sala's chemical studies and perhaps also met him.

Sala recorded his original chemical observations chiefly in his later works, notably those written after about 1620. Of his nineteen books, only a few have attracted the attention of historians. There exists no direct evidence concerning his medical practice. Early in his career, from about 1608, he followed the teachings of Paracelsus and agreed with the principle that "similia similibus curantur." From about 1625, he prescribed medicines in conformity with the basic principle "that they have a force . . . that resists the disease"; and he adopted a skeptical attitude toward the followers of Paracelsus, as previously he had done toward the Galenic physicians. Frequently attacked by representatives of both these medical traditions, Sala said that none of them was willing "to deviate a hair's breadth from his opinions, no matter whether he was right or wrong."[1]

Sala's chemical ideas proved to be historically influential. He performed the earliest known experiment in which a synthesis was confirmed by analysis ("anatomia"). Sala wished to establish "what blue *vitriolum* is, and out of what kind of pieces or materials it is composed or put together by nature."[2] As proof that this *vitriolum* consists of "copper ore, sulfur fumes, and water," Sala observed

that out of it one can distill, successively, water and "sulfur-bearing vapor" ("schweffelischen Dampff," that is to say, sulfuric acid), with the result that there remains "a reddish-brown substance similar to burned copper" (cupric oxide). Going beyond this qualitative analysis, Sala made quantitative observations, notably that these three substances can "exist in combination only in their appropriate proportion." He further remarked that after distillation of the water, the remainder "cannot weigh more than two-thirds with respect to the previous weight of the raw vitriol." This weight is now determined to be 64 percent; thus Sala was not yet able to arrive at exact results.[3]

Closely connected with Sala's conception of the constitution of vitriol out of constituent parts is his notion that the possibility of *reductio* (that is to say, of a return to the earlier state) is proof that no genuine *transmutatio* has occurred. Although Sala thought that the theory of transmutation was valid for salts in general, he did not think it applied to vitriol, since he did not consider it to be a salt. Another example of *reductio*, he thought, was the precipitation of copper by means of iron. Iron draws (*attrahit*) copper out of copper or blue vitriol solution (copper sulfate) but not out of iron vitriol solution or out of sulfuric acid. The *reductio* consists, according to Sala, of a joining together of the copper particles dispersed in the blue vitriol solution. Previously, in 1603, Nicolas Guibert had also interpreted *praecipitatio* as a process of *attractio*, but not from a corpuscular point of view. Both authors went beyond Libau, who still accepted the concept of *transmutatio*. Still, they did not perceive, as did van Helmont in 1624, that the iron goes into solution. Nor did they realize, as did Jungius in 1630, that an exact exchange of the two metals takes place, both at the metallic precipitate and in the substance in solution.[4]

Sala did not succeed in carrying out the *Anatomia antimonii* (that is, of antimony [V] sulfide).[5] It was too difficult at that time to distinguish between synthesis and analysis.[6] In the case of sal ammoniac, the synthesis was successful but not the analysis.[7] Sala considered sulfur to be a *compositum*, because he perceived corrosive "smoke" being liberated from it in the process of burning. It should be remembered that for the pharmacists of the period, preparation, and therefore synthesis, was of primary interest. Accordingly, Sala's attempts to conduct analyses are all the more praiseworthy. It is worth noting that in 1614 Sala reported the blackening of silver nitrate by sunlight.[8]

Considering his work as a whole, it is evident

that Sala was above all a practitioner. In his view, demonstrations could be carried out only through manual operations (*inventionibus manualibus*), that is to say, only with the aid of experimental examples, which he clearly distinguished from argumentation. For him, chemistry was still a handicraft (*ars*).

NOTES

1. R. Capobus, *Angelus Sala* (Berlin, 1933), 46.
2. A. Sala, *De natura, proprietatibus et usu spiritus vitrioli fundamentalis dissertatio. Oder Gründliche Beschreibung, was Spiritus Vitrioli eigentlich sey* (Hamburg, 1625), 4–5.
3. R. Hooykaas, *Het begrip element in zijn historisch-wijsgeerige ontwikkeling* (Utrecht. 1933), 150.
4. On the history of metallic precipitates, see H. Kangro, *Joachim Jungius' Experimente und Gedanken zur Begründung der Chemie als Wissenschaft* (Wiesbaden, 1968), 159–173.
5. Hooykaas (1933), 158.
6. Hooykaas, "The Experimental Origin of Chemical Atomic and Molecular Theory Before Boyle," in *Chymia*, 2 (1949), 72–73; H. Kangro, "Ein allgemeines Prinzip, mit dessen Hilfe im 17. Jahrhundert chemische Reaktionen ohne quantitative Analyse gedeutet worden sind," in *Beiträge zum XIII. Internationalen Kongress für Geschichte der Wissenschaft* (Moscow, 1974), 7th section, 225–231.
7. R. P. Multhauf, *The Origins of Chemistry* (London, 1966), 333.
8. J. R. Partington, *A History of Chemistry*, II (London, 1961), 278; Multhauf, *op. cit.*, 330.

BIBLIOGRAPHY

I. ORIGINAL WORKS. A complete bibliography is in Robert Capobus, *Angelus Sala* (Berlin, 1933), 53–55, although the titles are abbreviated and not always exact. Part of Sala's work is listed in the *Catalogue of the British Museum* and in the *Catalogue général des livres imprimés de la Bibliothèque Nationale* (Paris). Sala's Latin *Opera medico-chymica* appeared at Frankfurt am Main (1647), then at Rouen (1650), and again at Frankfurt am Main (1682).

Single works besides the two early works mentioned in the article and the work in n. 2 are *Septem planetarum terrestrium spagirica recensio* (Amsterdam, 1614); *Emetologia, ou enarration du naturel et usage des vomitoires* (Delft, 1613); *Opiologia* (The Hague, 1614); *Ternarius Bezoarticorum, ou trois souverains medicaments Bezoardiques* (Leiden, 1616); *Anatomia antimonii* (Leiden, 1617); *Anatomia vitrioli*, enlarged by the famous *Brevis demonstratio, quid sit vitriolum . . .* (Leiden, 1617); *Traicté de la peste* (Leiden, 1617); *Aphorismorum chymiatricorum synopsis* (Bremen, 1620); *Descriptio brevis antidoti pretiosi* (Marburg, 1620); *Chrysologia, seu examen auri chymicum* (Hamburg, 1622); *Von etlichen kräfftigen vnd hochbewerthen Medicamenten* (Wandesbeck, 1624); *De natura . . .* (Hamburg, 1625; see n. 2); and *Processus de auro potabili* (Strasbourg, 1630).

See also *Essentiarum vegetabilium anatome, darinnen von den fürtrefflichsten Nutzbarkeiten der Vegetabilischen Essentzen in der Artzney . . . gelehret vnnd gehandelt wird* (Rostock, 1630); *Tartarologia. Das ist: Von der Natur vnd Eigenschafft des Weinsteins* (Rostock, 1632); *Hydrelaeologia. Darinnen, wie man allerley Wasser, Oliteten vnd brennende Spiritus der Vegetabilischen Dingen, durch gewisse Chymische Regeln, vnd manualia destilliren vnd rectificiren soll . . . gehandelt wird* (Rostock, 1633); *Spagyrische Schatzkammer, Darinnen von unterschiedlichen, erbrechenmachenden . . . spagyrischen Medicamenten . . . treulich erwiesen vnd gelehret wird* (Rostock, 1634); *Saccharalogia. Darinnen erstlich von der Natur, qualiteten, nützlichem Gebrauch, vnd schädlichem Missbrauch des Zuckers: Darnach, wie von demselben ein Weinmässiger starker Getrank, Brandwein vnd Essig, als auch medicamenten . . . können bereitet werden . . . angezeiget wird* (Rostock, 1637). The last ed. of a work of Sala is the *Tractatus II de variis tum chymicorum, tum Galenistarum erroribus* (Frankfurt am Main, 1702).

II. SECONDARY LITERATURE. Apart from the literature mentioned in notes 3, 6, 7, and 8 there can only be recommended G. F. A. Blanck, *Angelus Sala, sein Leben und seine Werke* (Schwerin, 1883); and a short biographical sketch in *Die Mecklenburgischen Ärzte von den ältesten Zeiten bis zur Gegenwart* (Schwerin, 1929), 83.

GISELA KANGRO

SALERNITAN ANATOMISTS (*fl.* eleventh to thirteenth centuries, Salerno, Italy), *anatomy.*

In order to assess the contribution made by the school of anatomists at Salerno to the revival of their science in the West, it is essential to review briefly the arid millennium in the history of anatomy that began after the death of Galen (A.D. 199/200). Even during the latter's lifetime, dissection of the human body was no longer permitted at Alexandria; and the early Christian aversion to such studies is clearly shown by the opprobrium heaped on the memory of Herophilus by Tertullian, who castigated the great anatomist as "more of a butcher than a physician." Although Galen based many of his studies on the Barbary ape and rhesus monkey, he frequently made use of the pig (among other domestic animals), not only because of its ready availability but also because the internal organs were thought to show a remarkable similarity to those of man. Indeed, it was while vivisecting a pig that Galen discovered the function of the recurrent laryngeal nerve in voice production.

Although the written records do not say so, there is reason to believe that during the Dark Ages an occasional intrepid soul investigated the interior of the human body. A recent find in a fourth-century Roman catacomb shows a rather apathetic-looking group of master and pupils observing the dissection of a male corpse. Cassiodorus (sixth century) stated that human dissection was strictly forbidden by law and added a warning—presumably for the more intrepid—about cemetery guards and the harsh penalties for grave desecration. The various barbarian legal codes also contained sections on the violation of sepulchers. Some three centuries later George Teofano (d. 818) related that some Greek soldiers stationed in Bulgaria turned over a condemned prisoner to the physicians before consigning his body to the flames.

It is generally held that none of the anatomical works of Hippocrates, Aristotle, or Galen were in use in Europe before the end of the eleventh century. The meager anatomical literature available at Salerno before the arrival of the Constantinian translations included the *Schema anatomica* of Vindicianus (late fourth century), which survives in a solitary fragment written in the Benedictine script used at Salerno; the book on anatomical terminology in the *Differentiae* and the *Origines* of Isidore of Seville (*ca.* 560–636); and the anatomical chapters—derived ultimately from Celsus and Galen—in the *Epitome* of Paul of Aegina (seventh century). This situation changed radically with the arrival at Salerno of Constantine the African (d. 1087), translator of numerous works from the Arabic. Most important from an anatomical viewpoint was the *Kitāb al-Mālikī* of Haly Abbas (ᶜAlī ibn al-ᶜAbbās, d. 994), titled *Pantechne* by Constantine, who rendered parts of it into Latin and claimed it as his own original work. The *Kitāb al-Mālikī* was later (1127) translated in its entirety by Stephen of Antioch as the *Liber regalis*. The two chapters on anatomy in this book, although derived entirely from Galen and showing no evidence of direct observation, exerted a strong influence on the later anatomical writings produced at Salerno.

By the mid-eleventh century Salerno was rapidly approaching its zenith as the undisputed center of medical teaching in the Western world. The earliest writings of the school, such as the *Passionarius* of Gariopontus and the *Practica petrocelli*, were compiled for students and practitioners from late Greek and Byzantine works and contain little of anatomical interest. Yet there is every reason to believe that before the end of the century an annual public demonstration of porcine anatomy had become a traditional occurrence in the *civitas Hippocratica,* as Salerno came to be known.

The text that is considered to give the earliest account of such a dissection has for many centuries been attributed—although without any definite proof—to a Master Copho (*fl. ca.* 1080–1115), the author of *De modo medemdi* and other writings. The most primitive version of the *Anatomia porci,* as it is now called, treats only the neck, chest, and abdomen; the brief sections on the uterus and brain seem to have been added later, possibly by Stephen of Antioch, who called himself a pupil of Copho. In the cervical region the author identified the larynx, trachea, esophagus, epiglottis, and thyroid, and demonstrated to his pupils the function of the recurrent laryngeal nerve. He then proceeded to the contents of the thorax, identifying the pleura, pericardium, heart, and diaphragm as well as demonstrating the hollowness of the lung. He also described the course of the *vena concava* from the liver (as he thought) through the diaphragm into the inferior (right) auricle, where it becomes the artery "from which all other arteries arise." In the abdomen he enumerated eight subdivisions of the gastrointestinal tract, described the five-lobed liver, spleen, chylous vein, kidneys, ureters, omentum, and peritoneum. In the description of the uterus, allusion is made to seven cells or chambers that accommodate the fetuses, and to the placenta and membranes (secundines).

Throughout this brief tract (it contains less than a thousand words) the demonstrator speaks with assurance and clarity; his terminology has been described as transitional but is, in fact, almost entirely Greco-Latin or late Latin; and the three words of Arabic derivation may well have been interpolated during a revision. One recent student, noting that the order of dissection differs from the *de capite ad calcem* of Vindicianus and Isidore, suggests the influence of Celsus and Aristotle, although the means of transmission of the latter remains unclear. Certainly there is an agreement between the description of the large vessels of the heart in the *Historia animalium* and the *Anatomia porci.* The author seems to have been unfamiliar with the Galenic concept of blood passing from right to left ventricle through pores in the interventricular septum and depended on the earlier tradition (common to all twelfth-century Salernitan anatomies) that had the superior vena cava entering directly into the aorta.

Summing up the evidence for dating the earliest version of the *Anatomia porci,* I would suggest a year between 1080 and 1090, the decade just before the Constantinian translations had begun to circulate freely in Salerno.

Regardless of its antecedents and meagerness of content, the value of the *Anatomia porci* to the historian of science cannot be denied. Any eleventh-century investigation at which students from all over Europe could observe their master verify or criticize statements of other authorities must have had more than ordinary significance as marking the beginning of a new era. The text itself has survived in at least five manuscripts; it also has a remarkable printing history, having served as a working tool in book form from 1502 to 1655. Few works in the history of medicine can claim use over more than five centuries; long after the works of Mondino de' Luzzi and Vesalius were freely available, students continued to memorize the ancient text "as a preparatory exercise to the noble art of anatomy." Indeed, Mondino himself must have known and relied on this earlier work, since he continued to describe the human liver as five-lobed and the human uterus as bicornuate and multichambered.

The exact number of anatomies attributed to Salerno has varied with authorities, but Sudhoff and Corner agree on four. The second pig dissection, the *Demonstratio anatomica,* is related to the first as far as basic anatomical facts are concerned but differs greatly in style, methodology, and terminology. This work is approximately five times as long as the *Anatomia porci* and can best be described as an elaborate, polemic, early Scholastic commentary on the short primitive text. The author frequently wandered from descriptive anatomy to elaborate discussions of humoral physiology and made repeated references to the teleological concepts of Galen. Apparently this master had fallen completely under Arabic influence, and he made no attempt to conceal his indebtedness to Constantine. Indeed, he mentioned the latter by name on eight occasions, but only in reference to the terminology of various subdivisions of the gastrointestinal tract. Corner has clearly shown that vocabulary, phrases, and even entire paragraphs have been lifted almost verbatim from the *Pantechne.*

This second demonstration lacks the spontaneity of the first, and because of its prolixity never gained a student following. Since the author referred, somewhat pompously, to his own commentaries on several texts of the still-growing *Articella,* I have tentatively identified him with Master Bartholomaeus (*fl.* first half of the twelfth century), who is now known to have written precisely such commentaries as well as an *Anatomia* (all still unpublished). In his occasionally critical side remarks, the author confirmed the tradition of an annual dissection at Salerno; emphasized to his students the importance of committing his remarks to memory; and described the professional rivalry and competition for students in the still loosely organized school. He refused flatly to accept the *lateralia* "described in a recent booklet"—probably a revision of the *Anatomia porci*— because he "had never discovered these in animals," nor would he admit the usage of the term *faringes* proposed by a predecessor, because he had "not found it written in any book or heard it from any teacher." On the other hand, he did not hesitate to describe the human liver (which he obviously had never studied) as being five-lobed, "since certainly the same number [as in the pig] occurs in man."

The authorship of the third dissection, or *Anatomia mauri,* which has survived in four manuscripts (two of which explicitly name the author) has been attributed to Master Maurus, the *optimus physicus* of the school, who flourished in the latter half of the twelfth century and is known to have died in 1214. Although this brief tract shows familiarity with Arabic terminology, the descriptions are not so obviously dependent on the *Pantechne* as those of the second demonstration; and the author seems deliberately to have followed the *Anatomia porci* in format and method. In addition to the two hypochondria below the diaphragm mentioned by earlier authors, Maurus postulated, for teaching purposes, the existence of two superior hypochondria, one containing the lung and the other the heart. In his description of the circulation Maurus was somewhat more explicit than his predecessors: "Then you will see the cover of the heart and the heart itself and that *vena concava* which rises from the convexity of the liver through the diaphragm to the right auricle of the heart and then emerges through the left, where from the substance of the heart it acquires another coat and is thus transformed into an artery which is called the aorta, the name given to the Chief of all arteries." Again there is no mention of the interventricular pores. For the Latin term *peritoneum* Maurus also used the Arabic *siphac* and the Greek *epigasunta hymenon,* thus recalling the multilingual character of the Salernitan medical milieu. Nevertheless, the text in general is rather unoriginal and, aside from some changes in terminology, offers little improvement on its predecessors.

The fourth and last surviving porcine anatomy that can definitely be related to Salerno was discovered by Sudhoff and attributed by him to Urso of Calabria, a contemporary of Maurus, who fell entirely under the spell of the "new" Aristotelian logic. Basically a theoretician and philosopher rather than a physician, Urso was concerned only vaguely with morphological data; and the Scholastic method and terminology that dominated his work exclude any practical approach to the subject. Urso provided one historical note: he recalled that in his youth he had witnessed several dissections performed by his master, Matthaeus Platearius (d. ca. 1150).

In 1224 Frederick II issued a decree that surgeons must study anatomy for a year and be examined in that subject before they could practice; whether this involved actual dissection of the human body or simple observation is not known, nor has it ever been definitely proved whether a further edict of 1231, ordering that a human cadaver be publicly dissected at least once every five years in the presence of all practicing physicians and surgeons, was ever carried out. The political turmoil that threatened the latter years of Frederick's reign makes this unlikely.

For the sake of completeness we mention briefly an anatomy attributed to a Ricardus Salernitanus (d. 1252). He had undoubtedly studied at Salerno, for he mentioned many of its masters and reverentially compared Bartholomaeus with Hippocrates. He seems, however, to have taught at Montpellier; and his anatomy, not of the pig but of the human body, is thoroughly pseudo-Galenic and unoriginal. The work of Ricardus was edited and enlarged by a pupil, and the *Anatomia nicolai* is simply an elaboration of his master's text. Neither of these works can be considered as directly related to Salerno.

The earliest Salernitan anatomical demonstration had some influence on disciplines other than medicine. This is particularly evident in the writings of the philosopher-mystic Hugh of St. Victor (d. 1141), the great rival of Abelard at Paris. Three of Hugh's writings—"De bestiis et aliis rebus," "De hominis membris ac partibus," and "De natura hominis"—bear witness to the pervasive influence of the small but provocative *Anatomia porci* on the intellectual life of Europe.

BIBLIOGRAPHY

See George W. Corner, *Anatomical Texts of the Earlier Middle Ages* (Washington, D.C., 1927); Salvatore De Renzi, *Storia documentata della scuola di medicina di Salerno*, 2nd ed. (Naples, 1857), 254–255, 334–335; Thomas Haviland, " 'Anatomia porci,' a Twelfth Century Anatomy of the Pig Used in Teaching Human Anatomy," in *Wiener tierärztliche Monatsschrift*, Festschrift (1960), 246–265; Ynez Violé O'Neill, "William of Conches and the Cerebral Membranes," in *Clio Medica*, **2** (1967), 13–21; and "Another Look at the 'Anatomia porci,' " in *Viator*, **1** (1970), 115–124; Werner L. H. Ploss, ed., *Anatomia mauri* (Leipzig, 1921); Morris H. Saffron, *Maurus of Salerno* (Philadelphia, 1972), 13, 62; and Karl Sudhoff, "Die erste Tieranatomie von Salerno und ein neuer Salernitanischer Anatomietext," in *Archiv für Geschichte der Mathematik, der Naturwissenschaften und der Technik*, **10** (1927), 137–154; "Die vierte Salernitaner Anatomie," in *Archiv für Geschichte der Medizin*, **20** (1928), 33–50; and "Codex Fritz Paneth," in *Archiv für Geschichte der Mathematik, der Naturwissenschaften und der Technik*, **12** (1929), 2.

MORRIS H. SAFFRON

SALISBURY, ROLLIN DANIEL (*b.* Spring Prairie, Wisconsin, 17 August 1858; *d.* Chicago, Illinois, 15 August 1922), *geology, physical geography.*

Salisbury is best known for his collaboration with T. C. Chamberlin in the writing of geological textbooks that profoundly influenced the growth of the earth sciences during the first third of the twentieth century. His field studies, especially in Wisconsin, New Jersey, and Greenland, contributed notably to the then new science of glaciology.

Salisbury had come to Chamberlin's attention as a student at Beloit College, where Chamberlin was a part-time professor of geology. When he withdrew from that assignment to give full time to other duties, Salisbury was selected to succeed him, becoming professor of geology in 1884, only three years after receiving his bachelor's degree. His work at Beloit was interrupted for a year (1887–1888) by a trip to Europe, where he studied under Rosenbusch at Heidelberg and traced a previously unidentified belt of glacial moraines from Denmark to Russia. Chamberlin had employed him earlier in the glacial division of the United States Geological Survey, and his first important scientific contribution was their joint paper published in 1885. In 1891, four years after Chamberlin had become president of the University of Wisconsin, Salisbury resigned his post at Beloit to become professor of geology at the larger institution. Only a year later, however, Chamberlin accepted an invitation to organize the department of geology at the new University of Chicago and took Salisbury with him as professor of geographic geology. Both men remained at Chicago for the rest of their lives.

From 1899 until his death in 1922, Salisbury was dean of the Ogden Graduate School of Science at the University of Chicago, devoting much of his time to the manifold duties of that office. In addition, when the burgeoning program in geography was separated from that in geology to form a new department in 1903, he became its head. In 1918, when Chamberlin retired, Salisbury succeeded him as head of the department of geology, leaving the administration of the geography department to others. During his thirty years at the University of Chicago, Salisbury also assumed much of the editorial responsibility for the *Journal of Geology,* founded at Chicago during his first year there.

Salisbury never married. All his energy and devotion were concentrated on his students, his administrative responsibilities, and his science. He had a reputation for being brusque in speech and gruff in manner, but beneath the mask was a warm and kindly heart.

BIBLIOGRAPHY

I. ORIGINAL WORKS. Salisbury's writings include "Preliminary Paper on the Driftless Area of the Upper Mississippi Valley," in *Report of the United States Geological Survey,* **6** (1885), 205–322, written with T. C. Chamberlin; "On the Relationship of the Pleistocene to the Pre-Pleistocene Formations of Crowley's Ridge and Adjacent Areas South of the Limit of Glaciation," in *Report of the Arkansas Geological Survey* (for 1889), **2** (1891), 224–248; "The Drift of the North German Lowland," in *American Geologist,* **9** (1892), 294–318; "Distinct Glacial Epochs and the Criteria for Their Recognition," in *Journal of Geology,* **1** (1893), 61–84; "Salient Points Concerning the Glacial Geology of North Greenland," ibid., **4** (1896), 769–810; *The Physical Geography of New Jersey,* vol. IV of *Final Report of the [New Jersey] State Geologist* (Trenton, 1898); "The Geography of the Region About Devil's Lake and the Dalles of the Wisconsin," in *Bulletin of the Wisconsin Geological and Natural History Survey,* no. 5 (1900), 1–151, written with W. W. Atwood; *Geology,* 3 vols. (New York, 1904–1906), written with T. C. Chamberlin; *Physiography* (New York, 1907; 3rd ed., 1919); *Elements of Geography* (New York, 1912), written with H. H. Barrows and W. S. Tower; and *Introductory Geology* (New York, 1914), written with T. C. Chamberlin.

II. SECONDARY LITERATURE. See the memorial by R. T. Chamberlin, in *Bulletin of the Geological Society of America,* **42** (1931), 126–138, with bibliography of 95 titles; and T. C. Chamberlin's obituary in *Journal of Geology,* **30** (1922), 480–481.

KIRTLEY F. MATHER

SALLO, DENYS DE (*b.* Paris, France, 1626; *d.* Paris, 14 May 1669), *scientific journalism.*

Denys de Sallo, Seigneur de la Coudraye, was the son of Jacques de Sallo, conseiller of the grandchambre of the Parlement of Paris. He was educated at the Collège des Grassins, winning awards in Greek and Latin, before studying law. Admitted to the Paris bar in 1652, he won respect for his solid judgment, intelligence, and wit. In 1655 he married Elisabeth Mesnardeau, daughter of a colleague; they had one son and four daughters, the latter entering religious orders. Details of Sallo's life are extremely scarce; in 1657 he was in Frankfurt for the preliminaries to the election of a successor to Ferdinand III. E. Bigot records at this time that Sallo carried books from J. Boecler of Strasbourg to Paris for the astronomer Bouillaud.[1] During his last years Sallo was unable to walk, a condition attributed to his constant reading; although competent analysis of recorded symptoms now suggests a diabetic condition, then undiagnosed. In 1664 he began preparing the first scholarly periodical, *Journal des sçavans.* His 4,000 books and 200 manuscripts were sold after his death for 6,000 francs.

A *privilège* for the printing of the *Journal des sçavans* for twenty years was granted to Sallo on 8 August 1664; it was ceded on 30 December to Jean Cusson, printer and publisher; and on 5 January 1665 the first number was put on sale, priced at five sous. Thirteen weekly issues were published under Sallo's editorship; after the last, that of 30 March, there was a nine-month gap before resumption under Jean Gallois on 4 January 1666. A member of Sallo's household, Gallois was competent in languages and mathematics, and probably was one of three or four unidentified persons actively engaged in writing for the *Journal des sçavans* from the beginning.

Even before publication started, the *Journal des sçavans* had been actively promoted: by J. Chapelain, poet, critic, and correspondent of many French and foreign scholars; by Henri Justel, acquainted with innumerable travelers and men of letters; and by Émeric Bigot, well-known in foreign centers of learning. These and others had assured the *Journal* of a favorable reception abroad. Henry Oldenburg, secretary of the Royal Society, and Christiaan Huygens promised their assistance, which is apparent in the early issues. Sallo's active collaborators, apart from Gallois, for whom circumstantial evidence is persuasive, cannot be identified with assurance. Amable Bourzeis, the novelist Martin Le Roy de Gomberville, and Chapelain have been named on the basis of a rumor repeated by Gui Patin, who was not a friend of the circle in which the *Journal des sçavans* had originated. Of these only Chapelain probably was involved.

The *Journal des sçavans* responded to several aspects of contemporary life. Scientists, historians, philosophers, and others were finding that new facts, theories, and techniques posed issues that changed the basis of their thought. A rising skepticism about traditional views was transforming all disciplines, and the usual methods of exchanging information by letter were inadequate to the challenge. The *Gazette,* founded at Paris in 1631 by Théophraste Renaudot, reported competently on politics and military affairs; and it was logical that a similar periodical should chronicle events of intellectual interest. François de Mézeray, author of a comprehensive history of France and aware of the need for a record of cultural events as a basis for future histories of the sciences and arts, drafted a program (in the form of a *privilège*), outlining the aims of such a journal.[2] Mézeray lived in the same house in Paris as Sallo, a coincidence that cannot be passed over.

Mézeray's project emphasized the need for a record of discoveries and inventions, with only secondary attention to publications; as finally realized, the *Journal des sçavans* was a record of new books, a readable and critical account of current writings, and a marketable production in which the hand of the publisher is evident. The prospectus in the first pages of the first number resembles the program that Mézeray had outlined; but the accent was now on utility to the reader, whom the *Journal des sçavans* would inform of the content and value of the books reviewed.

In its first three months some eighty publications were discussed, sometimes at length. Almost all were in French or Latin, four in Italian or Spanish, and one in English—the first issue of the *Philosophical Transactions,* listed because of the repute of the newly founded Royal Society. A capable translator had been found who promised reviews of books in English. The *Journal des sçavans* was international from the outset: about half the books reviewed were published in Paris, while the rest came from London, Amsterdam, Rome, and other French and German cities. A quarter of the space was devoted to scientific material, some identified as trivial. Several important works were reviewed with insight: Thomas Willis' *Cerebri anatome,* at length; Steno's *De musculis et glandulis,* also in detail; G. Campani's work on his new lenses and telescopes; Clerselier's edition of Descartes's *De l'homme;* and G. Huret on the geometry of columns, treated with sharp criticism.

In addition there were reports of current scientific and technological developments: William Pet-

ty's double-hulled vessel and Robert Holmes's use of Huygens' clocks on Atlantic voyages. The most important scientific article offered an account[3] of a learned conference on comets held at the college of the Jesuits, followed by a detailed report on Adrien Auzout's calculation of the path of the comet then visible, with predictions for following weeks. While the views set forth before the distinguished audience at the conference were treated with formal respect, a discreet emphasis showed that traditional lore on the causes and nature of comets was not significant in comparison with accurate observation and sound mathematics, which demonstrated that the course followed by comets was much more regular than the lecturers, including Roberval, had asserted.

Various reasons have been given for the interruption of the *Journal des sçavans.* Sallo had offended certain persons of importance who were unaccustomed to seeing their work criticized in public print; the editor had commented critically on papal policy and the decrees of the Congregation of the Index concerning French publications; and the ironical view toward traditional attitudes in science and philosophy had indicated a leaning to excessive freedom of thought. The fact that the weekly numbers of the *Journal des sçavans* were not submitted in page proof for official approval was given as the cause of suppression; and Sallo, who could not tolerate constraints on speech and thought, was glad to turn his work over to others, although he seems to have acted as an intermediary for Gallois in dealing with Oldenburg in London. There is no record of his association with the *Journal des sçavans* after early 1668.

NOTES

1. Leonard E. Doucette, *Emery Bigot: Seventeenth-Century French Humanist* (Toronto, 1970), 11.
2. The papers of François Eudes de Mézeray in the Bibliothèque Nationale contain a leaf on which is drafted a *privilège* for a "Journal littéraire général" (Fonds français 20792, between ff. 112 and 113). This undated document was first published by the critic C. A. Sainte Beuve in 1853 (reprinted in *Causeries de lundi,* VIII, 183–184) and again from the original text by W. H. Evans in *L'Historien Mézeray et la conception de l'histoire en France au dix-septième siècle* (Paris, 1930), pp. 63–64.
3. *Journal des sçavans,* "du Lundy, 26 janvier, M.DC.LXV."

BIBLIOGRAPHY

A Paris *thèse de l'université* by Betty Trebell Morgan, *Histoire du Journal des Savants depuis 1665 jusqu'en 1701* (Paris, 1928), offers a rapid summary of the available information on De Sallo and his periodical; the book contains a full bibliography of earlier material. See also R. Birn, "Le *Journal des Savants* sous l'ancien régime," in *Journal des Savants* (1965), 15–35; Harcourt Brown, *Scientific Organizations in Seventeenth-Century France* (Baltimore, 1934; New York, 1967), 185–207; and "History and the Learned Journal," in *Journal of the History of Ideas*, **33**, no. 2 (1972), 365–378. For De Sallo, see Louis Moreri, *Le grand dictionaire* [sic] *historique ou le Mélange curieux de l'histoire sacrée et profane*, IV (Paris, 1699), 392. The fullest article on De Sallo is Dugast-Matifeux, "Débuts du journalisme littéraire en France: Denis de Sallo, fondateur du Journal des Savants," in *Annales de la Société d'Emulation de la Vendée (1883)*.

HARCOURT BROWN

SALMON, GEORGE (*b*. Cork, Ireland, 25 September 1819; *d*. Dublin, Ireland, 22 January 1904), *mathematics*.

Salmon's father, Michael Salmon, was a linen merchant; his mother, Helen, was the daughter of the Reverend Edward Weekes. After early schooling in Cork he entered Trinity College, Dublin, in 1833 to read classics and mathematics. He graduated in 1838 as first mathematical moderator. He was elected a fellow of Trinity College in 1841 and, as required by college statutes, took Holy Orders in the Church of Ireland. In 1844 he married Frances Anne, the daughter of the Reverend J. L. Salvador; they had four sons and two daughters.

The main burden of teaching in Trinity College was then borne by the fellows, and Salmon spent twenty-five years as a lecturer and tutor—mainly in the mathematical school, but also to a lesser extent in the divinity school. During this period he published some forty papers in various mathematical journals and wrote four important textbooks.

Over the years Salmon became frustrated by the heavy load of tutoring and lecturing, much of it of an elementary kind, and was disillusioned because he was not made a professor, a promotion that would have relieved him of most of this load and given him more time for his research. It must have been this which influenced him, in about 1860, to turn away from mathematics toward the theological studies in which he had always been interested—and which appeared to offer better prospects of promotion. In fact, in 1866 he was appointed regius professor of divinity and head of the divinity school, a post that he held for twenty-two years. During this period he published four more

books; they earned him a reputation as a theologian that was as great as the one he already had as a mathematician.

In 1888 Salmon was appointed provost of Trinity College. He remained an administrator for the rest of his life. He was a good and much-loved head of his college, although he had become a strong conservative in his old age, so that his provostship was a period of consolidation in the college rather than one of reform.

When Salmon joined the staff of Trinity College in 1841, its mathematical school was already internationally known and his colleagues included the well-known scholars Rowan Hamilton, James MacCullagh, Charles Graves, and Humphrey Lloyd. There was a strong bias toward synthetic geometry in the school, and it was in this field that Salmon began his research work, although he shortly became interested in the algebraic theories that were then being developed by Cayley and Sylvester in England and by Hermite and later Clebsch on the Continent. Salmon soon joined their number, and played an important part in the applications of the theory of invariants and covariants of algebraic forms to the geometry of curves and surfaces. He became a close friend of both Cayley and Sylvester and exchanged a voluminous mathematical correspondence with them for many years. His chief fame as a mathematician, however, rests on the series of textbooks that appeared between 1848 and 1862. These four treatises on conic sections, higher plane curves, modern higher algebra, and the geometry of three dimensions not only gave a comprehensive treatment of their respective fields but also were written with a clarity of expression and an elegance of style that made them models of what a textbook should be. They were translated into every western European language and ran into many editions (each incorporating the latest developments); they remained for many years the standard advanced textbooks in their respective subjects.

Salmon's own most important contributions to mathematics included his discovery (with Cayley) of the twenty-seven straight lines on the cubic surface, his classification of algebraic curves in space, his investigations of the singularities of the ruled surface generated by a line meeting three given directing curves, his solution of the problem of the degree of a surface reciprocal to a given surface, his researches in connection with families of surfaces subjected to restricted conditions, his conditions for repeated roots of an algebraic equation, and his theorem of the equianharmonic ratio of the four tangents to a plane cubic curve from a variable point on it.

When his investigations called for it, Salmon was an indefatigable calculator. The most famous example of this was his calculation of the invariant E of the binary sextic, which he published in the second edition (1866) of his treatise on modern higher algebra and which occupied thirteen pages of text.

BIBLIOGRAPHY

I. ORIGINAL WORKS. The Royal Society *Catalogue of Scientific Papers*, V, 381–382; VII, 819, lists forty-one memoirs by Salmon published between 1844 and 1872. His mathematical textbooks are *A Treatise on Conic Sections* (Dublin, 1848); *A Treatise on the Higher Plane Curves: Intended as a Sequel to a Treatise on Conic Sections* (Dublin, 1852); *Lessons Introductory to the Modern Higher Algebra* (Dublin, 1859); and *A Treatise on the Analytic Geometry of Three Dimensions* (Dublin, 1862).

His most important theological writings are *A Historical Introduction to the Study of the Books of the New Testament* (London, 1885); *The Infallibility of the Church* (London, 1888); *Some Thoughts on the Textual Criticism of the New Testament* (London, 1897); and *The Human Element in the Gospels*, N. J. D. White, ed. (London, 1907), posthumously published.

II. SECONDARY LITERATURE. On Salmon and his work, see the obituary by J. H. Bernard in *Proceedings of the British Academy*, 1 (1903–1904), 311–315; by R. S. Ball in *Proceedings of the London Mathematical Society*, 2nd ser., 1 (1903–1904), xxii–xxviii; the unsigned obituary in *Nature*, 69 (1903–1904), 324–326; and the obituary by C. J. Joly in *Proceedings of the Royal Society*, 75 (1905), 347–355. See also *The Times* (London) (23 Jan. 1904), 13; and *Dictionary of National Biography*.

A. J. McCONNELL

SALOMONSEN, CARL JULIUS (*b.* Copenhagen, Denmark, 6 December 1847; *d.* Copenhagen, 14 November 1924), *medicine, bacteriology.*

The only child of a wealthy and cultured family, Salomonsen obtained an excellent education in both the sciences and the classics. He was the son of Martin Salomonsen, a physician who had published scientific works on epidemiologic problems, and Eva Henriques, whose family included many eminent physicians in Denmark.

After undergraduate work at the Metropolitan school in Copenhagen, Salomonsen received the M.D. (1871) at the University of Copenhagen. He also took a great interest there in zoological studies, which later inspired his works in experimental pathology and parasitology. Through his father's friendships with the archaeologist Worsaae and the physiologist Peter Panum, Salomonsen met Virchow and became his private secretary, assisting him in his Copenhagen measurements of ancient skulls (1869).

During his first years as a physician, Salomonsen shared a large practice with his father. He also worked as an assistant in the pathology department of the Almindelig Hospital and, later, at the Kommunehospital, where Fritz Valdemar Rasmussen and Panum inspired his interest in the relationship between bacteria, which had recently been discovered, and pyemic processes. Pasteur and Koch had recently investigated this problem, and Panum had published his investigations concerning the putrefaction of blood and the role of microbes in the fermentative effects of putrefaction.

In 1877 Salomonsen defended his thesis "Studier over Blodets Forraadnelse"—some opponents finding it a book more of botanical than of medical interest. Nevertheless, it became the fundamental starting point for the study of bacteriology in Denmark. He showed that the colorshift in putrid blood at 5–10°C is produced by microbes that have begun to putrefy. With the aid of long capillary tubes he isolated live microbes and developed a method for growing them in pure cultures. (This method of cultivation was soon surpassed by Koch's use of transparent solidifiable media.) In his thesis Salomonsen suggested dyeing bacteria in diluted fuchsine solutions to preserve the form of the microbes better than was possible with the more complicated dyeing in hardened preparations. His thesis also gave effective support to the school of Ferdinand Cohn, who maintained that the bacteria were distinct species, against the school of Billroth, who considered them different forms of a single species of coccobacteria septica.

After defending his thesis, Salomonsen traveled to Germany and Paris for half a year. In Breslau (now Wrocław) he worked with Julius Cohnheim; and together they demonstrated the specificity of tuberculosis. That was in 1878, four years before Koch's discovery of the tubercle bacillus. They used an aseptic method of inoculating fresh tuberculous material into the camera anterior bulbi of rabbits. In Breslau Salomonsen also met many scientists, including Weigert, Welch, Neisser, Koch, and Ehrlich. In Paris he met Pasteur—an encounter that later proved of great significance for young Danish bacteriologists. On his return to Copenhagen, Salomonsen was appointed prosector at the

Copenhagen Municipal Hospital. He also lectured privately, and thus began his lifelong interest in education.

In 1878 Salomonsen published "Notits om Forekomsten af Bakterier i metastatiske Pusansamlinger hos Levende," an investigation of bacteria from pus accumulations in live human beings. He identified streptococci from various suppurations and, by inoculating pyogenic material in rabbits, developed a fatal peritonitis with streptococci in pure cultivation.

In 1883 Salomonsen was named lecturer in medical bacteriology at Copenhagen. His was the first such chair in Europe. In the cellar of the Botanical Museum of Copenhagen he assembled a circle of physicians and veterinarians who diligently followed his demonstrations and lectures. Among this group were the physicians Vilhelm Jensen; Thorvald Madsen, later director of the State Serum Institute; J. Christmas; and the veterinary surgeons Bernhard Bang and C. O. Jensen. In 1885 Salomonsen published his important textbook *Ledetraad i Bakteriologisk Teknik,* which was translated into English, French, and Spanish.

In 1893 Salomonsen was named professor of pathology at the University of Copenhagen and for nearly two decades used the new chemistry and physiology laboratory facilities that had been installed in a building in Ny Vestergade. In 1910 he moved to the Rigshospital, where a special building had been constructed for him.

During these years, Salomonsen published his investigations on immunity (1880), diphtheria (1891), and anthrax (1899). In 1884 he published, with Christmas, his studies on pseudoinfection and the ophthalmia produced by the jequirity. In this work he demonstrated that the jequirity microbe is nonvirulent: the venom promoted a morbid predisposition in test animals that enabled the microbes to develop in the blood, thus producing a pseudoinfection. When serum therapy was developed, Salomonsen quickly acknowledged both its practical and its scientific value; and in Ny Vestergade he built up a small department for the production of antidiphtheric serum. Because of the widespread use of this serum, both in hospitals and in research on immunity, the department soon sought better facilities, and Salomonsen proposed (1896, 1899) the construction of an independent serum institute. Despite political difficulties and economic restrictions, he succeeded with his plans and the institute was inaugurated in 1902. He had hoped to have it affiliated with the university, but it was taken over by the Ministry of Health. Denmark being a small country, the institute has been able to sponsor many centralized investigations on immunity serum reactions in infectious diseases. These surveys have made its research world famous.

Salomonsen was director of the State Serum Institute until 1909, when Madsen, his collaborator, succeeded him. They investigated antitoxin formation in horses that were diphtheria-immunized (1896) and demonstrated the fluctuating progress that led to studies of the basis of antitoxin formation. They showed that the cells of an organism that is actively immunized have the unique ability to produce antitoxins. They also observed a constant formation and destruction of such substances and identified the stability of the antitoxin quantity by bleeding the animals and replacing the evacuated blood with transfusions of fresh blood (1897). Because of his interest in hygienic problems, Salomonsen made certain that vaccination procedures in Denmark were reformed; only animal vaccine from healthy calves was used—human vaccine was prohibited.

With Georges Dreyer, Salomonsen demonstrated the physiological effects of radium on amoebas and trypanosomes. He also showed (1904) that radium produces severe hemolysis. Besides his inspiring work as a bacteriologist, Salomonsen was interested in the improvement of education. In 1917 he proposed the construction of a tuition-free college for needy students—an idea that was realized shortly before his death. His many addresses on science, biography, university education, and the history of medicine were published in 1917.

Salomonsen was a cofounder (1907) of the Danish Museum for the History of Medicine and from 1917 president of the Danish Society for the History of Medicine. As dean of the University of Copenhagen, he published (1910) a historical essay on epidemiologic theories during the first half of the nineteenth century; and he wrote (1914) on the sanctuary of Asclepius on Cos. In 1921 he published a book of silhouettes of many outstanding scientists of his time, taken from his personal collection. He also published two polemics on dysmorphism in art, believing that art forms produced just after World War I reflected a deranged state of mind.

In 1891 Salomonsen was elected to the Royal Danish Academy of Sciences. He was made honorary doctor of medicine in 1905 by Victoria University of Manchester and in 1911 by the University of Christiania (Oslo). In 1880 he married Ellen Henriques. His last years were complicated by crippling rheumatoid arthritis and the death of his wife and only daughter.

BIBLIOGRAPHY

I. ORIGINAL WORKS. Salomonsen's major works are *Studier over Blodets Forraadnelse* (Copenhagen, 1877), his M.D. diss.; "Notits om Forekomsten af Bakterier i metastatiske Pusansamlinger hos Levende," in *Nordiskt medicinskt archiv*, **10** (1878), 1–10; "Versuche über künstliche Tuberculose," in *Jahresberichte der Schlesischen Gesellschaft für vaterländische Kultur*, **56** (1878), 222–223, written with J. Cohnheim; "Om Indpodning af Tuberculose i Kaniners Iris," in *Nordiskt medicinskt archiv*, **11**, no. 12 (1879), 1–29, and no. 19, 1–38; "Eine einfache Methode zur Reinkultur verschiedener Fäulniss bacterien," in *Botanische Zeitung*, **38** (1880), 481–489; "Nyeste experimentelle Undersøgelser over Immunitet," in *Hospitalstidende*, **71** (1880), 861–872, 881–894; "Uber die Aetiologie des Jequerity-ophtalmie," in *Fortschritte der Medizin*, **2** (1884), 78–87, written with J. Christmas; "Ueber Pseudo-Infektion bei Fröschen," *ibid.*, 617–631; *Ledetraad for Medicinere i Bakteriologisk Teknik* (Copenhagen, 1885); and *Bakteriologisk Teknik for Medicinere* (Copenhagen, 1889; 4th ed., 1906); with French trans. by R. Durand-Fardel, *Technique élémentaire de bactériologique à l'usage des médecins* (Paris, 1891), and English trans., *Bacteriological Technology for Physicians* (New York, 1891).

Later works include "Difterilaerens nuvaerende Standpunkt," in *Ugeskrift for Laeger*, **24** (1891), 63–79; "Redegørelse for Virksomheden ved den serumtherapeutiske Anstalt," in *Hospitalstidende*, **4**, no. 1 (1896), 225–226; "Studier over Antitoxindannelse," in *Nordiskt medicinskt archiv*, **30** (1897), 1–21, written with T. Madsen; "Forslag om Oprettelse af et Seruminstitut," *ibid.*, **7**, no. 4 (1899), 142; "The Rise and Growth of the State Serum Institute," in the *Festskrift ved Indvielsen af Statens Seruminstitut* (Copenhagen, 1902), 1–20; "Recherches sur les effets physiologiques du radium," in *Comptes rendus hebdomadaires des séances de l'Académie des sciences*, **138** (1904), 1543–1545, written with G. Dreyer; *Erindringsord og Notesbog for Deltagere i de medicinskbakteriologiske Øvelser ved Københavns Universitet* (Copenhagen, 1906; 2nd ed., 1911); and *Epidemiologiske Teorier i den første Halvdel af det 19. Aarhundrede* (Copenhagen, 1910).

See also *Erindringsord og Notesbog for Deltagere i de experimental-pathologiske Øvelser ved Københavns Universitet* (Copenhagen, 1912; 2nd ed., 1919); *Menneskets Snyltere* (Copenhagen, 1913), with atlas (1913); *Asklepios' Helligdom på Kos* (Copenhagen, 1914); *Maa-Arbejder* (Copenhagen, 1917); *Smitsomme Sindslidelser for og nu med Henblik paa de nyeste Kunstretninger* (Copenhagen, 1919); *Om Dysmorfismens syge-* *lige Natur* (Copenhagen, 1920); and *Medicinske Silhouetter* (Copenhagen, 1921).

II. SECONDARY LITERATURE. On Salomonsen and his work, see E. Gotfredsen, *Medicinens Historie*, 3rd ed. (Copenhagen, 1973), *passim*; T. Madsen, "Carl Julius Salomonsen," in *Oversigt over glet Kd Danske Videnskabernes Selskabs Forhandlinger* (1925), 59–75; and Oluf Thomsen, "Carl Julius Salomonsen," in *Copenhagen University Program* (Nov. 1925), 75–81.

E. SNORRASON

SALVIANI, IPPOLITO (*b.* Citta di Castello [?], Italy, 1514; *d.* 1572), *medicine, natural history.*

Little is known of Salviani's life: according to some, he was born in Rome, to others, in Citta di Castello. The latter is more likely since Pope Julius III, by a *motu proprio*, made him a Roman citizen—a pointless action if he had been born in Rome.

It is certain that Salviani studied medicine and was closely connected with the Vatican. He was personal physician to Julius III, Paul IV, and Cardinal Cervini, who later became Pope Marcellus II. It is also known that he was a professor of practical medicine, from 1551 until at least 1568, at the Sapienza, the Roman university of the Renaissance. The Vatican gave him many honors: in 1564 the cardinal in charge sent Salviani in his place to supervise the degree sessions in medicine; he was made principal physician of the medical college of Rome; and in 1565 he was nominated *conservatore* ("registrar") of Rome, which was more an administrative than a medical post. As *conservatore* he arranged for the transportation to the Campidoglio of the two famous statues of Caesar Augustus and Julius Caesar, which until then had been in the keeping of Alessandro Ruffini, bishop of Melfi.

Salviani enjoyed considerable renown, probably because of his privileged position as papal physician; he also had many rich clients and, in time, became very wealthy.

Salviani published only one medical work, *De crisibus ad Galeni censuram* (1556). Its success is evident from a second edition that was published only two years later. He is better known, however, for his monumental work on natural science, *De piscibus tomi duo*, the publication date of which is uncertain; it has been conjectured by various biographers to be 1554, 1555, or 1558.

The work had been encouraged and supported financially by Cardinal Cervini. It describes, in two folio volumes, the fishes of the Mediterranean and is accompanied by beautiful copper engravings by various contemporary artists. Despite Aldrovandi's unconditional praise for the illustrations, more recent critics have pointed out that their merits are artistic rather than scientific. Many species of fishes are so approximately represented that they cannot be identified with precision.

The work was dedicated not to Cardinal Cervini, as Salviani had intended, but to the new pope, Paul IV; Cervini died before publication was complete. A slight bibliographical controversy has ensued from this alteration. Polidorus Vergilius, who is never very reliable, went so far as to reproduce the dedication that Salviani would have written for Cervini. But no such dedication has been found in any copy of the work, the one to Pope Paul IV being the only known dedication.

Salviani's *De piscibus tomi duo* was reprinted in Venice in 1600–1602, but with the title changed to *De aquatilium animalium . . . formis,* which led some biographers to believe that it was a different book. Salviani also wrote poetry and a play, *La ruffiana.* The latter was popular during his life and went through several editions, but it is completely ignored by modern literary critics.

BIBLIOGRAPHY

Salviani's two scientific works are *De crisibus ad Galeni censuram liber* (Rome, 1558), and *De piscibus tomi duo, cum eorumdem figuris aere incisis* [Rome, 1554], reprinted as *De aquatilium animalium curandorum formis* (Venice, 1600–1602). His play was entitled *La ruffiana* (Rome, 1554).

On Salviani and his work, see Polidorus Vergilius, *De inventoribus rerum* (Basel, 1563); G. Marini, *Degli archiatri pontifici,* 2 vols. (Rome, 1784); I, 402–405; II, 306–307, 314–317; and G. Tiraboschi, *Biblioteca Modenese,* VII, pt. 2 (Modena, 1781–1786), 119.

CARLO CASTELLANI

AL-SAMARQANDĪ, NAJĪB AL-DĪN ABŪ ḤĀMID MUḤAMMAD IBN ᶜALĪ IBN ᶜUMAR (*d.* Herat, Afghanistan, 1222), *medicine, materia medica.*

Al-Samarqandī, who flourished at the time of the philosopher Fakhr al-Dīn al-Rāzī (*d.* 1210), died during the pillage of Herat by the Mongols. Ibn Abī Uṣaybiᶜa states that al-Samarqandī was a famous physician and gives his name as Najīb al-Dīn Abū Ḥāmid Muḥammad ibn ᶜAlī ibn ᶜUmar al-Samarqandī. Nothing more is known of his life.

The most important of his medical works is *al-*

Asbāb wa'l-ᶜalāmāt ("Etiology and Symptoms [of Diseases]"). It is described in the work of Nafīs ibn ᶜIwaḍ al-Kirmānī (*d.* 1449), who wrote *Sharḥ al-asbāb wa'l-ᶜalāmāt* ("Commentary on Etiology and Symptoms"). According to Ibn Abī Uṣaybiᶜa, al-Samarqandī also wrote a book on the treatment of diseases by diet and two medical formularies. The treatise *Uṣūl tarkīb [al-adwiya]* ("On the Principles of Compounding Drugs") also is ascribed to him.

Works still extant are *al-Adwiya al-mufrada* ("Simple Drugs"), *Aghdhiyat al-marḍā* ("Diet for the Ill"), *al-Aghdhiya wa'l-ashriba wa-mā yattaṣil bihā* ("Food and Drink and What Relates to Them"), *Fī mudāwāt wajaᶜ al-mafāṣil* ("On the Cure of Pain in the Joints"), *Fi 'l-ṭibb* ("On Medicine"), *Fī kayfiyyat tarkīb ṭabaqāt al-ᶜayn* ("On the Mode of Composition of the Layers of the Eyes"), *Tractatus de medicamentis repertu facilibus, Aqrābādhīn* ("Medical Formulary"), *Fī ᶜilāj man suqiya 'l-sumūm aw nahashahu 'l-hawāmm waghayruhā* ("On the Treatment for One Who Has Been Poisoned or Has Had a Poisonous Bite, and Similar Cases"), *Ghāyat al-gharaḍ fī mu ᶜālajat al-amrād* ("The Last Word on Treating the Ill"), *Fī ittikhādh māʾ al-jubn wamanāfi ᶜihi wakayfiyyat istiᶜmālihi* ("On Administration of Water of Cheese, Its Benefits, and Its Various Uses"), and *Fi 'l-adwiya al-mustaᶜmala ᶜinda 'l-sayādila* ("On Drugs Prepared by Pharmacists").

It is significant that al-Samarqandī did not rely entirely on the old humoral pathology. In fact, he displayed originality in not considering the theory of humors of decisive importance in therapeutics. He dealt with the many "accidents" of drugs and conditions of the body in the much broader framework of medicine and the prescribing of drugs.

BIBLIOGRAPHY

Nafīs' *Commentary* on al-Samarqandī's *al-Asbāb wa'l-ᶜalāmāt* was published by Mawlawi ᶜAbd al-Majīd (Calcutta, 1836). The *Ṭibb-i Akbarī* of Muḥammad Arzānī, completed in 1700–1701, includes a Persian translation of this *Commentary.* Printed and lithograph eds. of the *Ṭibb-i Akbarī* are listed in G. Sarton, *Introduction to the History of Science,* II (Baltimore, 1962), 661.

See also G. M. Anawati, *Drogues et médicaments* (Cairo, 1959), 117–118, text in Arabic; Carl Brockelmann, *Geschichte der arabischen Literatur,* I² (Leiden, 1943), 491, and supp. I (Leiden, 1937), 895–896; S. Hamarneh and G. Sonnedecker, *A Pharmaceutical View of Abulcasis* (Leiden, 1963); L. Leclerc, *Histoire de la médecine arabe,* II (Paris, 1876), 128–129; Martin Levey, *Chemistry and Chemical Technology in Ancient Mesopotamia* (Amsterdam, 1959), introduction; *The Medical Formulary or Aqrābadhīn of Al-Kindī* (Madison, Wis., 1966); *Medieval Arabic Toxicology* (Philadelphia, 1966); and *The Medical Formulary of al-Samarqandī and the Relation of Early Arabic Simples to Those Found in the Indigenous Medicine of the Near East and India* (Philadelphia, 1967); P. Sbath and C. D. Averinos, *Deux traités médicaux* (Cairo, 1953); and Ibn Abī Uṣaybiᶜa, *Kitāb ᶜuyūn al-anbāʾ fī ṭabaqāt al-aṭibbāʾ*, A. Müller, ed., II (Cairo, 1882), 31; and (Beirut, 1950), pt. 2, 47–48.

MARTIN LEVEY

AL-SAMARQANDĪ, SHAMS AL-DĪN MUHAMMAD IBN ASHRAF AL-ḤUSAYNĪ (*b.* Samarkand, Uzbekistan, Russia, *fl.* 1276), *mathematics, logic, astronomy.*

Al-Samarqandī was a contemporary of Naṣīr al-Dīn al-Ṭūsī (1201–1274) and Quṭb al-Dīn al-Shīrāzī (1236–1311). Al-Samarqandī was not among the scientists associated with al-Ṭūsī at the observatory at Marāgha. A noted logician, al-Samarqandī was best known to mathematicians for his famous tract *Kitāb Ashkāl al-taʾsīs* ("Book on the Fundamental Theorems"). This work of twenty pages, probably composed around 1276, summarizes with their abridged demonstrations thirty-five fundamental propositions of Euclid's geometry. To write this short work, Samarqandī consulted the writings of Ibn al-Haytham, ᶜUmar al-Khayyāmī, al-Jawharī, Naṣīr al-Dīn al-Ṭūsī, and Athīr al-Dīn al-Abharī. Several mathematicians, notably Qāḍī Zāda, commented on this work by al-Samarqandī.

It was chiefly with his book on dialectics that al-Samarqandī became famous. This valuable work, entitled *Risāla fī ādāb al-baḥth* ("Tract on the Methods of Enquiry"), was the subject of several commentaries. Two other works on logic by al-Samarqandī are known: *Mīzān al-Qusṭās* and *Kitāb ᶜAyn al-naẓar fīᶜilm al-jadal.* Al-Samarqandī was also interested in astronomy. He wrote *Al-Tadhkira fī 'l-hay'a* ("Synopsis of Astronomy") and a star calendar for 1276–1277. His *Ṣaḥāʾif al-ilāhiyya* and his *ᶜAqāʾid* are two works on dogmatic theology.

BIBLIOGRAPHY

MSS of the works of al-Samarqandī are listed in C. Brockelmann, *Geschichte der arabischen Literatur,* I (Weimar, 1898), 486; and *ibid.,* supp. 1 (Leiden, 1937), 860. See also H. Suter, *Die Mathematiker und Astronomen der Araber* (Leipzig, 1900), 157; and "Nachträge und Berichtigungen zu 'Die Mathematiker . . .,'" in *Abhandlungen zur Geschichte der Mathematik,* **14** (1902), 176.

Also helpful are Ḥājjī Khalīfa's *Kashf al-ẓunūn,* G. Flügel, ed. (Leipzig, 1835–1855), I, 322; Carra de Vaux's article "Baḥth," in *Encyclopaedia of Islam,* 1st ed., I (1911), 587; and G. Sarton, *Introduction to the History of Science,* II (Baltimore, 1962), 1020–1021.

For a demonstration of Euclid's fifth postulate attributed to al-Samarqandī, see H. Dilgan, "Démonstration du Vᵉ postulat d'Euclide par Shams-ed-Dīn Samarkandī," in *Revue d'histoire des sciences et de leurs applications,* **13** (1960), 191–196. For the attribution of this demonstration to Athīr al-Dīn al-Abharī, see A. I. Sabra, "Thābit ibn Qurra on Euclid's Parallels Postulate," in *Journal of the Warburg and Courtauld Institutes,* **31** (1968), 14, note 9.

HÂMIT DILGAN

AL-SAMAWʾAL, IBN YAHYĀ AL-MAGHRIBĪ (*b.* Baghdad, Iraq; *d.* Marāgha, Iran [?], *ca.* 1180), *mathematics, medicine.*

Al-Samawʾal was the son of Yehuda ben Abun (or Abu'l-ᶜAbbās Yaḥyā al-Maghribī), a Jew learned in religion and Hebrew literature who had emigrated from Fez (Morocco) and settled in Baghdad. His mother was Anna Isaac Levi, an educated woman who was originally from Baṣra (Iraq). Al-Samawʾal thus grew up in a cultivated milieu; a maternal uncle was a physician, and after studying Hebrew and the Torah the boy was encouraged, when he was thirteen, to take up the study of medicine and the exact sciences. He then began to study medicine with Abu'l-Barakāt, while taking the opportunity to observe his uncle's practice. At the same time he started to learn mathematics, beginning with Hindu computational methods, *zījes* (astronomical tables), arithmetic, and *misāḥa* (practical techniques for determining measure, for use in surveying), then progressing to algebra and geometry.

Since scientific study had declined in Baghdad, al-Samawʾal was unable to find a teacher to instruct him beyond the first books of Euclid's *Elements* and was therefore obliged to study independently. He finished Euclid, then went on to the *Algebra* of Abū Kāmil, the *al-Badīᶜ* of al-Karajī, and the *Arithmetic* of al-Wasīṭī (most probably Maymūn ibn Najīb al-Wasīṭī, who collaborated in making astronomical observations with ᶜUmar al-Khayyāmī between 1072 and 1092). By the time he was eighteen, al-Samawʾal had read for himself all of the works fundamental to the study of mathematics and had developed his own mode of mathematical thinking.

In science, this independence of thought led al-Samaw'al to point out deficiencies in the work of al-Karajī (whom he admired) and to challenge the arrangement of the *Elements*; in religion he was similarly inclined to test the validity of the claims of the various creeds and came to accept those of Islam, although he postponed his conversion for a number of years to avoid distressing his father. His autobiography states that he reached his decision at Marāgha on 8 November 1163 as a result of a dream; four years later he wrote to his father, setting out the reasons for which he had changed his religion, and his father immediately set out to Aleppo to see him, dying en route. Al-Samaw'al himself spent the rest of his life as an itinerant physician in and around Marāgha. His earlier travels had taken him throughout Iraq, Syria, Kūhistān, and Ādharbayjān.

His biographers record that al-Samaw'al was a successful physician, and had emirs among his patients. In his autobiography al-Samaw'al recorded that he had compounded several new medicines, including an almost miraculous theriac, but no other account of them has survived. His only extant medical work is his *Nuzhat al-aṣḥāb* (usually translated as "The Companions' Promenade in the Garden of Love"), which is essentially a treatise on sexology and a collection of erotic stories. The medical content of the first and longer section of the book lies chiefly in descriptions of diseases and sexual deficiencies; the second, more strictly medical, part includes a discussion of states of virile debility and an account of diseases of the uterus and their treatment. This part is marked by al-Samaw'al's acute observation and his interest in the psychological aspects of disease; he provides a detailed description of the condition of being in love without recognizing it, and gives a general prescription for the anguished and melancholic that comprises well-lighted houses, the sight of running water and verdure, warm baths, and music.

It is, however, chiefly as a mathematician that al-Samaw'al merits a place in the history of science. His extant book on algebra, *Al-bāhir* ("The Dazzling"), written when he was nineteen years old, represents a remarkable development of the work of his predecessors. In it al-Samaw'al brought together the algebraic rules formulated by, in particular, al-Karajī and, to a lesser extent, Ibn Aslam and other authors, including al-Sijzī, Ibn al-Haytham, Qusṭā ibn Lūqā, and al-Ḥarīrī. The work consists of four parts, of which the first provides an account of operations on polynomials in one un-

known with rational coefficients; the second deals essentially with second-degree equations, indeterminate analysis, and summations; the third concerns irrational quantities; and the fourth, and last, section presents the application of algebraic principles to a number of problems.

It is apparent in the first section of this work that al-Samaw'al was the first Arab algebraist to undertake the study of relative numbers. He chose to treat them as if they possessed an identity proper to themselves, although he did not recognize the significance of this choice. He was thus able, in a truly bold stroke, to subtract from zero, writing that

> If we subtract an additive [positive] number from an empty power ($0 \cdot x^n - a \cdot x^n$), the same subtractive [negative] number remains; if we subtract the subtractive number from an empty power, the same additive number remains ($0 \cdot x^n - [-ax^n] = ax^n$). . . . If we subtract an additive number from a subtractive number, the remainder is their subtractive sum: $(-ax^n) - (bx^n) = -(a + b) x^n$; if we subtract a subtractive number from a greater subtractive number, the result is their subtractive difference; if the number from which one subtracts is smaller than the number subtracted, the result is their additive difference.

These rules appear in the later European work of Chuquet (1484), Pacioli (1494), Stifel (1544), and Cardano (1545); it is likely that al-Samaw'al reached them by considering the extraction of the square root of a polynomial.

Al-Karajī had conceived the algorithm of extraction, but did not succeed in applying it to the case in which the coefficients are subtractive. His failure may have stimulated al-Samaw'al's abilities, since the problem in which these rules are stated in the *Al-bāhir* is that of the extraction of the square root of

$$25x^6 + 9x^4 + 84x^2 + 64 + 100(1/x^2) + 64(1/x^4)$$
$$- 30x^5 - 40x^3 - 116x - 48(1/x) - 96(1/x^3).$$

Since al-Karajī's algebra lacked symbols, so that the numbers had to be spelled out in letters, this operation would have presented an insurmountable obstacle. Al-Samaw'al was able to overcome this trial of memory and imagination by using a visualization in which he assigned to each power of x a place in a table in which a polynomial was represented by the sequence of its coefficients, written in Hindu numerals. This technique, a major step in the development of symbolism, was requisite to the progress of algebra because of the increasing complexity of mathematical computations.

Al-Samaw'al's rules of subtraction were also

important in the division of polynomials; in the interest of obtaining better approximations he pursued division up to negative powers of x, and thereby approached the technique of development in series (although he overlooked the opportunity of unifying the various cases of the second-degree equation and of computation by double error). He computed the quotient of $20x^2 + 30x$ by $6x^2 + 12$, for example, to obtain the result $3 \ 1/3 + 5(1/x)$ $-6 \quad 2/3(1/x^2) - 10(1/x^3) + 13 \quad 1/3(1/x^4) + 20(1/x^5)$ $-26 \ 2/3(1/x^6) - 40(1/x^7)$. He then recognized that he could apply the law of the formation of coefficients $a_{n+2} = -2a_n$, which allowed him to write out the terms of the quotient directly up to $54,613 \cdot 1/3$ $(1/x^{28})$.

Al-Samaw'al further applied the rules of subtraction to the multiplication and division of the powers of x, which he placed in a single line on both sides of the number 1, to which he assigned the rank zero. The other powers and other constants are displayed on each side of zero, in ascending and descending order:

...	4	3	2	1	0	1	2	3	4	...
						$\frac{1}{x}$	$\frac{1}{x^2}$	$\frac{1}{x^3}$	$\frac{1}{x^4}$	
	x^4	x^3	x^2	x	1					

The rules of multiplication and division that al-Samaw'al enunciated are, except for their notation, those still in use.

The second part of the Al-bāhir contains the six classical equations ($ax = b$, $ax^2 + bx = c$, and so forth) that were set out by al-Khwārizmī. Interestingly, however, al-Samaw'al gave only geometrical demonstrations of the equations, although their algebraic solutions were known to his predecessor al-Karajī, who dedicated a monograph, 'Ilal ḥisāb al-jabr wa 'l-muqābala, to them. Al-Samaw'al then presented a remarkable calculation of the coefficients of $(a + b)^n$, which al-Karajī discovered after 1007 (for the dating of this and other of al-Karajī's works, see A. Anbouba, ed., L'algèbre al-Badī' d'al-Karagi [Beirut, 1964], p. 12). Since al-Karajī's original computation has been lost, al-Samaw'al's work is of particular interest in having preserved it; these coefficients are arranged in the triangular table that much later became known in the west as Tartaglia's or Pascal's triangle.

A further, and equally important, part of the second section of the book deals with number theory. This chapter contains about forty propositions, including that among n consecutive integers there is one divisible by n; that

$$1 \cdot 2 + 3 \cdot 4 + \cdots + (2n - 1) \, 2n = 1 + 3 + \cdots$$
$$+ (2n - 1) + 1^2 + 3^2 + \cdots + (2n - 1)^2;$$

and that $1^2 + 2^2 + \cdots + n^2 = n(n + 1)(2n + 1)/6$. Al-Samaw'al was especially proud of having established the last, since neither Ibn Aslam nor al-Karajī had been successful in doing so.

The chief importance of the second part of the Al-bāhir lies, however, in al-Samaw'al's use of recursive reasoning, which appears in such equations as

$$(n - 1) \sum_1^n i = (n + 1) \sum_1^{n-1} i \text{ and}$$
$$(n - 1) \sum_1^n i = (n - 1) \sum_1^{n-3} i + 3 \, (n - 1)^2.$$

The third part of Al-bāhir is chiefly concerned with the classification of irrationals found in Book X of the Elements. Al-Samaw'al's account is complete and clear, but contains nothing new or noteworthy except his rationalization of $\sqrt{30} : \sqrt{2} + \sqrt{5} + \sqrt{6}$, which had eluded al-Karajī.

The final section of the work contains a classification of problems by the number of their known solutions, a device used by earlier writers. Al-Samaw'al was led by this procedure to solve a varied group of these problems and to master a prodigious system of 210 equations in ten unknowns, a result of his having undertaken the determination of ten numbers of which are given their sums, taken six at a time. He further elucidated the 504 conditions necessary to the compatibility of the system. He overcame the lack of symbolic representation by the expedient of designating the unknown quantities 1, 2, 3, $\cdots$, 10, and was then able to draw up a table that started from

$$123456 \cdots 65$$
$$123457 \cdots 70$$
$$123458 \cdots 75$$
$$123459 \cdots 80.$$

Al-Samaw'al's intention in writing the Al-bāhir was to compensate for the deficiencies that he found in al-Karajī's work and to provide for algebra the same sort of systematization that the Elements gave to geometry. He wrote the book when he was young, then allowed it to remain unpublished for several years; it seems quite possible that he reworked it a number of times. It is difficult to ascertain the importance of the book to the development of algebra in the Arab world, but an indirect and restricted influence may be seen in the Miftaḥ-al-ḥisab ("Key of Arithmetic") of al-Kāshī, published in 1427. The book was apparently altogether unknown in the west.

The mathematical counterpart of the Al-bāhir,

the *Al-zāhir* ("The Flourishing"), has been lost, and of al-Samawꞌal's mathematical writings only two almost identical elementary treatises remain. These are the *Al-tabṣira* ("Brief Survey") and *Al-mūjiz* ("Summary"). The influence of the Arabic language is seen in their classification of fractions as deaf, Arab, or genitive, and both contain sections on ratios that are clearly derived from the work of al-Karajī. The ratio 80:3·7·9·10, for example, is expressed as a sum of fractions with numerator 1 and the respective denominators 3·10, 3·7·9, and 3·9·10; this use of the sexagesimal system reflects the continuing importance of the concerns of the commercial and administrative community of Baghdad, since this system was still favored by merchants and public servants. In his account of division, al-Samawꞌal noted the periodicity of the quotient 5:11, also calculated in the sexagesimal system. The last section of *Al-mūjiz* exists only in mutilated form, but what remains would indicate that it contained interesting material on abacuses.

In an additional work related to his mathematics al-Samawꞌal again demonstrated the independence of his thinking. In this, the *Kashf ꞌuwār al-munajjimīn* ("The Exposure of the Errors of the Astrologers"), he refuted the pronouncements of scientific astrology by pointing out the multiple contradictions in its interpretation of sidereal data, as well as the errors of measurement that he found in astrological observations. He then, for the sake of argument, assumed astrology to be valid, but showed that the astrologer could scarcely hope to make a valid prediction since, by al-Samawꞌal's count, he would have to take into simultaneous consideration 6,817 celestial indicators, a computation that would surely exceed his abilities.

BIBLIOGRAPHY

I. ORIGINAL WORKS. Al-Samawꞌal wrote at least eighty-five works, of which most have been lost. (It must be remembered, however, that the word *kitāb* may be used to designate either a brief note or a full volume.) A brief autobiographical writing may be found in *Ifḥām ṭā ꞌifat al-yahūd*, Cairo MS (Cat. F. Sayyid, I, p. 65), fols. 25–26, and in *Al-ajwiba al fākhira raddan ꞌala'l milla 'l-kāfira*, Paris MS 1456, fols. 64–65.

His mathematical works that have been preserved include *Al-bāhir* ("The Dazzling"), Istanbul MS Aya Sofya 2718, 115 ff., and Esat Ef. 3155; extracts are in A. Anbouba, "Mukhtārāt min Kitāb al-Bāhir," in *al-Mashriq* (Jan.-Feb. 1961), 61–108, while a complete modern ed., with notes, a detailed analysis, and French introduction is S. Ahmad and R. Rashed, eds., *Al-bāhir en algèbre*

d'as-Samawꞌal* (Damascus, 1972). *Al-ṭabṣira fi'l ḥisāb* ("Brief Survey") is Berlin MS 5962, 29 ff., Bodleian (Oxford) I, 194, and Ambrosiana (Milan) C. 211 ii; *Al-mūjiz al-mūḍawī* [?] *fi'l ḥisāb* ("Summary") is Istanbul MS Fatih 3439, 15 (consisting of thirty-three pages of thirty-five lines each); and *Kashf ꞌuwār al-munajjimīn wa ghalaṭihim fī akthar al-aꞌmāl wa'l-ahkām* ("The Exposure of the Errors of the Astrologers") is Leiden MS Cod. Or 98, 100 ff., and Bodleian I, 964, and II, 603, while its preface has been translated into English by F. Rosenthal, "Al-Asṭurlābī and al-Samawꞌal on Scientific Progress," in *Osiris*, 9 (1950), 555-564.

Al-Samawꞌal's only surviving medical work, *Nuzhat al-aṣhāb fī muꞌāsharat al-aḥbāb* ("The Companions' Promenade in the Garden of Love"), is preserved in Berlin MS 6381, Paris MS 3054, Gotha MS 2045, Istanbul MS Aya Sofya 2121, and Escorial MS 1830. His other extant works are apologetics, and include *Ifḥām ṭā'ifat al-yahūd* ("Confutation of the Jews"), Cairo MS Cat. F. Sayyid, I, p. 65, Cairo MS VI, ii, and Teheran MS I, 184, and II, 593, of which the text and an English trans. with an interesting introduction by M. Perlmann are *Proceedings of the American Academy for Jewish Research*. 32 (1964); and *Ghāyat al-maqṣūd fī 'l-radd ꞌala 'l-naṣārā wa 'l-yahūd* ("Decisive Refutation of the Christians and the Jews"), Istanbul MS As'ad 3153 and Asir 545. A last MS, *Badhl al-majhūd fī iqnāꞌ al-yahūd* ("The Effort to Persuade the Jews"), formerly in Berlin, has been lost since World War II. See also Perlmann (above), pp. 25–28, 127.

II. SECONDARY LITERATURE. Recent editions of Arabic sources on al-Samawꞌal and his work include Ibn Abī Uṣaybiꞌa, *ꞌUyūn al anbāꞌ*, II (Cairo, 1882), 30–31; Ḥājjī Khalīfa, *Kashf al-zunūn*, I (Istanbul, 1941), col. 664, and II (1943), col. 1377; Ibn al ꞌIbrī [Barhebraeus], *Taꞌrīkh mukhtaṣar al-duwal* (Beirut, 1958), 217; Ibn al-Qifṭī, *Ikhbār al-ꞌulama* (Cairo, 1908), 142; Ibn Ṣāꞌid al-Akfānī, *Irshād al-Qāṣid* (Beirut, 1904), 123, 125, 127; and Tash Kupri Zadeh, *Miftāḥ al-Saꞌāda*, I (Hyderabad, 1910), 327, 329.

Works by later authors include A. Anbouba, "Al-Karajī," in *Al-Dirāsāt al-Adabiyya* (Beirut, 1959); and "Mukhtārāt min Kitāb al-Bāhir," in *al-Mashriq* (Jan.-Feb. 1961), 61–108; C. Brockelmann, *Geschichte der arabischen Literatur*, I² (Leiden, 1943), 643, and Supp. I (Leiden, 1937), 892; H. Hirschfeld, "Al-Samawꞌal," in *Jewish Encyclopedia*, I (New York–London, 1901), 37–38; Lucien Leclerc, *Histoire de la médecine arabe*, II (Paris, 1876), 12–17; S. Perez, *Biografias de los mathematicos arabos que florecieron en España* (Madrid, 1921), 137; F. Rosenthal, "Al-Asṭurlābī and al-Samawꞌal on Scientific Progress," in *Osiris*, 9 (1950), 555–564; G. Sarton, *Introduction to the History of Science*, II (Baltimore, 1953), 401, and III (1953), 418, 596; M. Steinschneider, *Mathematik bei den Juden* (Hildesheim, 1964), 96; and H. Suter, *Die Mathematiker und Astronomen der Araber und ihre Werke* (Leipzig, 1900), biography, 24.

See also introductions and notes in S. Ahmad and

R. Rashed, eds., *Al-bāhir en algèbre d'as-Samaw'al* (Damascus, 1971); and in A. Anbouba, ed., *L'algèbre al-Badīᶜ d'al-Karagi* (Beirut, 1964).

ADEL ANBOUBA

SAMOYLOV, ALEKSANDR FILIPPOVICH (*b.* Odessa, Ukraine, Russia, 7 April 1867; *d.* Kazan, U.S.S.R., 22 July 1930), *physiology, electrophysiology, electrocardiography.*

After early education at the Gymnasium in Odessa, Samoylov studied for two years in the natural history section of the faculty of physics and mathematics of Novorossysk University. In 1891 he graduated from the medical faculty of the University of Dorpat, where he conducted his first experimental research under the guidance of the well-known pharmacologist E. R. Kobert. His dissertation for the M.D. degree was entitled "O sudbe zheleza v zhivotnom organizme" ("On the Fate of Iron in the Living Organism"; 1891).

From 1892 to 1894 Samoylov worked with Pavlov in the physiology section of the Institute of Experimental Medicine; and from 1894 to 1903 he was I. M. Sechenov's assistant and docent in the department of physiology of the Moscow University. Samoylov worked in the laboratories of Ludimar Hermann in Königsberg, Wilhelm Nagel in Berlin, and Johannes von Kries in Freiburg im Breisgau. On 3 October 1903 he became professor of physiology of Kazan University, where he remained until his death. In 1924 he became chairman of the department of physiology of Moscow University.

Samoylov conducted more than 120 original experimental and theoretical investigations in electrophysiology, electrocardiography, the physiology of sense organs, clinical physiology, and the history of science. His methodological research was important in the development and training of specialists in electrophysiology. His improvements in methods of string galvanometry include a special compensator and string indicator, and a special form of immersed electrodes. He was generally recognized as a leading electrophysiologist, and among his students at Kazan were I. S. Beritov, M. A. Kiselev, V. V. Parin, and M. N. Livanov. With Einthoven, Samoylov laid the foundations for clinical and theoretical electrocardiography.

In 1924 he published his hypothesis of the chemical nature of the transfer of nerve excitation in "O perekhode vozbuzhdenia s dvigatelnogo nerva na myshtsu" ("On the Transfer of Excitation From the Motor Nerve to the Muscle"). He also carried out electrophysiological research on the central nervous system and on the mechanisms of coordination of complex reflex actions. He added to the theory of reflexes new concepts of the existence of circular forms of excitation, and closed reflex arcs. This research was published in "Koltsevoy ritm vozbuzhdenia" ("The Circular Rhythm of Excitation"; 1930).

A gifted scientific administrator, Samoylov received the Lenin Prize in 1930 and for several years was a deputy of the Tatar Republic.

BIBLIOGRAPHY

Samoylov's major works are collected in *Izbrannye stati i rechi* ("Selected Articles and Speeches"; Moscow–Leningrad, 1946), with a bibliography of Samoylov's works, pp. 308–313; and *Izbrannye trudy* ("Selected Works"; Moscow, 1967).

On Samoylov and his work, see N. A. Grigorian, *Aleksandr Filippovich Samoylov* (Moscow, 1963).

N. A. GRIGORIAN

SAMPSON, RALPH ALLEN (*b.* Skull, County Cork, Ireland, 25 June 1866; *d.* Bath, England, 7 November 1939), *astronomy.*

Sampson studied at St. John's College, Cambridge, and in 1888 graduated third wrangler in the mathematical tripos. In 1891 he received the first Isaac Newton studentship in astronomy and physical optics at Cambridge, where he studied astronomical spectroscopy with Newall. In 1893 Sampson published "On the Rotation and Mechanical State of the Sun," in which he discussed the distribution of temperature by radiation and absorption. In the same year he was elected professor of mathematics at the University of Durham and also started his great work on the four large satellites of Jupiter.

The advent of the Harvard photometric eclipse observations of the satellites of Jupiter stimulated Sampson to reexamine previous observations, and as a result he published in 1909 "A Discussion of the Eclipses of Jupiter's Satellites 1878–1903." He developed a new theory for the motions of the four satellites and in 1910 published *Tables of the Four Great Satellites of Jupiter.* (Sampson's tables have since formed the basis for computing the phenomena for the national ephemerides.) The new theory itself was not published until 1921 in "Theory of the Four Great Satellites of Jupiter."

In 1910 Sampson was appointed astronomer

royal for Scotland and professor of astronomy at the University of Edinburgh, where he interested himself in problems concerning the determination of time. He encouraged the development of the Shortt free-pendulum clock, which became standard equipment in many observatories. Sampson also carried out photometric research and introduced the concept of spectrophotometric . gradients. He retired from the observatory in 1937.

BIBLIOGRAPHY

I. ORIGINAL WORKS. Sampson's major works are "On the Rotation and Mechanical State of the Sun," in *Memoirs of the Royal Astronomical Society,* **51** (1893), 123–183; "Description of Adam's Manuscripts on the Perturbation of Uranus," *ibid.,* **54** (1901), 143–170; "A Discussion of the Eclipses of Jupiter's Satellites 1878–1903," in *Annals of Harvard College Observatory,* **52** (1909), 149–343; "The Old Observations of the Eclipses of Jupiter's Satellites," in *Memoirs of the Royal Astronomical Society,* **59** (1909), 199–256; *Tables of the Four Great Satellites of Jupiter* (London, 1910); *The Sun* (Cambridge–New York, 1914); "A Census of the Sky," in *Observatory,* **38** (1915), 415–426; and "Theory of the Four Great Satellites of Jupiter," *Memoirs of the Royal Astronomical Society,* **63** (1921).

Sampson also edited the unpublished MSS of John C. Adams in *Scientific Papers of John Couch Adams,* II, pt. 1 (Cambridge, 1896–1900).

II. SECONDARY LITERATURE. On Sampson and his work, see W. M. H. Greaves, "Ralph Allen Sampson," in *Monthly Notices of the Royal Astronomical Society,* **100** (1940), 258–263.

LETTIE S. MULTHAUF

SANARELLI, GIUSEPPE (*b.* Monte San Savino, Italy, 24 April 1864; *d.* Rome, Italy, 6 April 1940), *hygiene.*

Sanarelli obtained his degree in medicine and surgery at the University of Siena in 1889. He then studied under Pasteur at the Institut Pasteur in Paris. In 1893 he received the chair of hygiene at the University of Siena but two years later went to Uruguay, at the invitation of the government, to set up an institute of experimental hygiene at the University of Montevideo. In 1898 he returned to Italy and became professor of hygiene at the University of Bologna and then at the University of Rome, where he remained until 1935, serving as vice-chancellor in 1922–1923.

Sanarelli was made doctor *honoris causae* of the universities of Paris and Toulouse; and he was a member of the Académie des Sciences, the Medical Academy of Paris, and the Royal Medical Academy of Belgium.

In 1892–1894 Sanarelli published his research on the pathogenesis of typhoid fever: he was the first to propose the theory of general infection with a secondary, localized infection in the intestine. (In 1925 he extended this theory to paratyphoid infections and to bacillary dysentery.) In a series of papers that appeared between 1916 and 1924, and as a book in 1931, Sanarelli discussed his investigations on the pathogenesis of cholera, demonstrating that the bacillus of cholera has a marked gastric enterotropism. He then studied (1924–1925) the so-called hematic carbuncle in the intestine and showed that cholera is hematogenetic rather than enteric.

From his studies on cholera, and especially those on the pathogenesis of choleraic algidity, Sanarelli discovered (1916) the hemorrhagic allergy, or "Sanarelli's phenomenon." He was also the first to propose the concept of hereditary immunity to tuberculosis, thus opposing the traditional concept of a hereditary predisposition to the disease. In Montevideo, Sanarelli had the opportunity to investigate the cause of yellow fever. From his research on diseased patients and corpses, he was able to isolate *Bacterium icteroides,* the earliest discovered (1897) human exit-paratyphoid.

In 1898 Sanarelli discovered the myxomatosis virus of the rabbit, the first known tumor disease in animals caused by a filterable virus. He was also the first to use the method of ultrafiltration through colloidal membranes. This method enabled him to study (1930) the germination and development of the tubercular ultravirus. In other important studies he showed that spirillum and fusiform associations of bacteria are not real microbic associations, but are, instead, a single germ, which can appear as either a rod or a spiral according to the environmental conditions. He called this microorganism *Heliconema vincenti.* Sanarelli was also a pioneer in the field of nasal vaccination (1924).

BIBLIOGRAPHY

I. ORIGINAL WORKS. Sanarelli's major works are "Die Ursachen der natürlichen Immunität gegen den Milzbrand," in *Zentralblatt für Bakteriologie und Parasitenkunde,* **9** (1891), 467; "Études sur la fièvre typhoïde expérimentale," in *Annales de l'Institut Pasteur,* **6** (1892), 721; **8** (1894), 193, 353; *Etiologia e patogenesi della febbre gialla* (Turin, 1897); *Etiología y patogenia de la fiebre amarilla* (Montevideo, 1897); "Das myxomatogene virus," in *Zentralblatt für Bakteriologie und Parasitenkunde*, **23** (1898), 865; *Tubercolosi ed evoluzione sociale* (Milan, 1913); *Manuel d'igiene generale e coloniale* (Florence, 1914); "La patogenesi del colera. Nota preventiva," in *Annali d'igiene sperimentale*, **26** (1916), 685; and "Pathogénie du choléra," in *Comptes rendus hebdomadaires des séances de l'Académie des sciences*, **163** (1916), 538. Later writings are "Le gastro-entérotropisme des vibrions," *ibid.*, **168** (1919), 578, repr. in *Annali d'igiene sperimentale*, **29** (1919), 129; **31** (1921), 1; "De la pathogénie du choléra . . . Voies de pénétration et de sortie des vibrions cholériques dans l'organisme animal," in *Annales de l'Institut Pasteur*, **37** (1923), 364, repr. in *Annali d'igiene sperimentale*, **33** (1923), 457; "L'algidité cholérique," in *Annales de l'Institut Pasteur*, **37** (1923), 806, repr. *Annali d'igiene sperimentale*, **34** (1924), 1; "Sulle vaccinazioni per via nasale," in *Annali d'igiene sperimentale*, **34** (1924); 861; "Sur le charbon dit 'intestinal,'" in *Comptes rendus hebdomadaires des séances de l'Académie des sciences*, **179** (1924), 937; *Nuove vedute sulle infezioni dell'apparato digerente* (Rome, 1925); and "Sulla patogenesi del carbonchio detto 'interno' o 'spontaneo,'" in *Annali d'igiene sperimentale*, **35** (1925), 273, repr. *Annales de l'Institut Pasteur*, **39** (1925), 209.

See also "Identité entre spirochètes et bacilles fusiformes. Les Héliconèmes 'vincenti,'" in *Annales de l'Institut Pasteur*, **41** (1927), 679; "Dimostrazione in vivo e in vitro delle forme filtranti del virus tubercolare," in *Annali d'igiene sperimentale*, **40** (1930), 589, written with A. Alessandrini; "Il fattore ereditario nella tubercolosi," in *Romana medica* (Rome, 1930); and *Il colera* (Milan, 1931).

Pietro Ambrosioni

SANCHEZ (or **SANCHES**), **FRANCISCO** (*b.* diocese of Braga, Portugal, 1550 or 1551; *d.* Toulouse, France, November 1623), *philosophy, medicine.*

An exponent of an extreme nominalist skepticism, Sanchez was probably born near Valença, Portugal, and was probably of Jewish descent. In 1562 his family immigrated to Bordeaux, a frequent refuge for fleeing *conversos*. In Bordeaux, Sanchez studied at the reformist Collège de Guyenne; but in 1569 he departed for Rome to study anatomy. In 1574 he completed his doctorate in medicine at the University of Montpellier. The following year, in order to avoid religious strife in heavily Protestant Montpellier, he settled in Toulouse.

In Toulouse, Sanchez held a variety of posts: professor of philosophy from 1585 and of medicine from 1612; rector of the university; and, for thirty years, director of the Hospital of Saint Jacques,

where he is said to have sequestered himself at night in order to study cadavers. His first writings were nevertheless on mathematics. In 1575 he wrote a letter to Clavius, who the previous year had published an edition of Euclid's *Elements*, questioning the certainty of geometrical proofs. In Sanchez' view, mathematics deals with ideal, rather than real, objects and therefore can say nothing certain about actual experience.

In 1576, at the age of twenty-six, Sanchez composed his major philosophical work, *Quod nihil scitur.* Beginning with a rigorously nominalist critique of scholastic epistemology, Sanchez attacked, first, the Aristotelian concept of science as concentrating too much on abstract categories rather than on real objects, and, second, the syllogistic method of doing science as self-fulfilling: anything can be proved by starting with the correct premises. He concluded that Aristotelian science was not science at all, since one cannot reach certitude through definitions nor in the endless search for causes. True science, according to Sanchez, must be the study of particulars, but is in fact, even with this limitation, beyond man's reach because of the imperfection of his senses. In order to replace scholastic methodology, Sanchez proposed a commonsensical, empirical approach, which would not seek vainly to gain knowledge, but would entertain the more circumscribed aim of dealing realistically with experience.

The very next year (1577) Sanchez extended his skeptical critique into the realm of astrology, in a long, versified comment on the hysterical reactions to the comet of that year. In the poem he ridiculed the beliefs of astrologers such as Francesco Giuntini, who asserted that all comets must be evaluated as the source of prognostication. Sanchez wrote that such beliefs were untenable philosophically because they seemed to limit man's free will and practically because such predictions were based on inconsistent correlations between the appearance of comets and human events. (In a later treatise he condemned, in a similar vein, the practice of divination through dreams.)

The practice of medicine was Sanchez' lifelong occupation and primary source of income. He was a shrewd clinical observer in the Hippocratic tradition, although he was careful to point out errors in the works of past medical writers. In his anatomical works, the influence of Colombo, Vesalius, and Falloppio is explicit. Moreover, his diagnostic experience, convincing him of the limitations of the human senses, played an important role in the formulation of his skeptical philosophy.

BIBLIOGRAPHY

I. ORIGINAL WORKS. See *Quod nihil scitur* (Lyons, 1581; 2nd ed., Frankfurt, 1618); Spanish trans., *Que nada se sabe,* with an intro. by Marcelino Menéndez y Pelayo (Buenos Aires, 1944). This major work, along with three shorter treatises *(De longitudine et brevitate vitae, In lib. Aristotelis physiognomicon commentarius,* and *De divinatione per somnum ad Aristotelem),* is also included in Sanchez' collected works: *Opera medica* (see below) and *Tractatus philosophici* (Rotterdam, 1649). Modern eds. are *Opera philosophica,* Joaquim de Carvalho, ed. (Coimbra, 1955); and *Tratados filosóficos,* A. Moreira de Sá, ed., 2 vols. (Lisbon, 1955).

The *Carmen de cometa* was originally published in Lyons (1578); a recent ed. is *O cometa do ano de 1577,* A. Moreira de Sá, ed. (Lisbon, 1950). The letter to Clavius, which originally circulated in MS, was first published by J. Iriarte, "Francisco Sánchez el Escéptico disfrazado de Carneades en discusión epistolar con Christóbal Clavio," in *Gregorianum,* **21** (1940), 413–451; for a Portuguese trans., see *Revista portuguesa de filósofia,* **1** (1945), 295–305.

His medical works are in *Opera medica* (Toulouse, 1636), a posthumous compilation of his classroom notes. The most representative treatises are *De morbis internis, Observationes in praxi,* and a *Summa anatomica.*

II. SECONDARY LITERATURE. For Sanchez' contributions to the philosophy of science, see João Cruz Costa, *Ensaio sôbre a vida e a obra do filósofo Francisco Sanchez* (São Paulo, 1942); J. Iriarte, "Francisco Sanchez . . . a la luz de muy recientes estudios," in *Razón y Fe,* **110** (1936), 23–42, 157–181; Evaristo de Moraes Filho, *Francisco Sanches na renasença portuguesa* (Rio de Janeiro, 1953); Artur Moreira de Sá, *Francisco Sanchez, filósofo e matemático,* 2 vols. (Lisbon, 1947); and Emilien Senchet, *Essai sur la méthode de Francisco Sanchez* (Paris, 1904).

General works setting Sanchez in historical perspective are John Owen, *The Skeptics of the French Renaissance* (London, 1893), and Richard H. Popkin, *The History of Scepticism From Erasmus to Descartes,* 2nd ed. (Assen, Netherlands–New York, 1963).

On the medical works of Sanchez, see Luis de Pina, "Francisco Sanches, médico," in *Revista portuguesa de filósofia,* **7** (1951), 156–191.

THOMAS F. GLICK

SANCTORIUS. See **Santorio, Santorio.**

SANDERSON, EZRA DWIGHT (*b.* Clio, Michigan, 25 September 1878; *d.* Ithaca, New York, 27 September 1944), *entomology, rural sociology.*

Sanderson was the son of John P. and Alice Sanderson. He attended Michigan Agriculture College, receiving the B.S. degree in 1897 and, a year later, the same degree in agriculture from Cornell University. His first position was that of assistant entomologist at Maryland Agricultural College, a post he held until the summer of 1899, when he accepted an assistantship with the federal Division of Entomology. During this year he married Anna Cecilia Blanford, by whom he had a daughter.

In the fall of 1899 Sanderson became entomologist for the Delaware Agricultural Experiment Station at Newark and associate professor of zoology at Delaware College, where he remained until 1902. He then accepted the post of state entomologist of Texas and professor of entomology at Texas Agricultural and Mechanical College, where he remained for two years. He was subsequently appointed professor of zoology at New Hampshire College and, three years later, director of the agricultural experiment station of that institution. In the autumn of 1910 he became dean of the College of Agriculture at West Virginia University and from 1912 to 1915 was director of its agricultural experiment station. He resigned from this position in order to pursue graduate studies in sociology at the University of Chicago, from which he received the Ph.D. in 1921.

In 1918 Sanderson became head of the Rural Social Organization àt Cornell, an institution that furthered the cause of rural sociology. He attracted many graduate students to this new discipline, and their joint efforts resulted in the provision of guidelines for ameliorating the plight of many rural communities. Sanderson's entire attention from 1918 until his retirement in 1943 was directed to these sociological activities.

Sanderson's active career falls into three fairly definite categories: economic entomology, academic administration, and rural sociology. Although he attained real distinction in each area, his enduring scientific reputation probably will rest on his earlier entomological contributions. The results of these fundamental and pioneering researches appeared in the early issues of the *Journal of Economic Entomology,* which has become the most important periodical dealing with applied entomology. Sanderson was influential in establishing it, as well as its parent sponsoring agent, the American Association of Economic Entomologists. His research articles are considered landmarks in the development of the recent discipline of insect ecology. Sanderson early realized the value of applying statistical analysis to the problems of applied entomology, an area to which he also contributed through his participation in establishing the standardization of insecticides, work that culminated in the Federal Insecticide Act of 1910. In many respects Sanderson was an "ideas" man who stimulated original thinking in entomology, and his early departure from the field was a distinct loss.

The central concept in Sanderson's approach to rural sociology appears to have been the building of the natural rural community, in which the centralized school became a major factor in community unification.

BIBLIOGRAPHY

Sanderson's entomological writings include *Insects Injurious to Staple Crops* (New York, 1902); *A Statistical Study of the Decrease in the Texas Cotton Crop Due to the Mexican Cotton Boll Weevil . . .* (Austin, Tex., 1905); "The Relation of Temperature to the Hibernation of Insects," in *Journal of Economic Entomology*, **1** (1908), 56–65; "The Influence of Minimum Temperatures in Limiting the Northern Distribution of Insects," *ibid.*, 245–262; "The Relation of Temperature to the Growth of Insects," *ibid.*, **3** (1910), 113–140; and *Insect Pests of the Farm, Garden, and Orchard* (New York, 1912; 3rd ed., rev. and enl. by L. M. Peairs, 1931).

His sociological works include *The Farmer and His Community* (New York, 1922); *The Rural Community; the Natural History of a Sociological Group* (New York, 1932), which contains ecological considerations far ahead of its time; *Rural Community Organization* (New York–London, 1939); and *Rural Sociology and Rural Social Organization* (New York–London, 1942).

An obituary is E. F. Phillips, "Dwight Sanderson, 1878–1944," in *Journal of Economic Entomology*, **37** (1944), 858–859.

C. E. NORLAND

SANDERSON, JOHN S. B. See **Burdon-Sanderson, John Scott.**

SANIO, KARL GUSTAV (*b.* Lyck, East Prussia [now Elk, Poland], 5 December 1832; *d.* Lyck, 28 January 1891), *botany.*

The son of Johann Sanio, a landowner, Sanio began studying the flora and fauna around Lyck while attending the Gymnasium. After passing the final secondary school examination in 1852, he studied science and medicine at the University of Königsberg and, from 1855 to 1857, at the University of Berlin, where his teachers were Alexander Braun and Nathanael Pringsheim. In the fall of 1857 Sanio returned to Königsberg, where he earned the

Ph.D. for the dissertation "Florula Lyccensis" in 1858. In the same year Sanio qualified as lecturer in botany with an essay entitled "Untersuchungen über die Epidermis und die Spaltöffnungen der Equisetaceen." The university administration objected to his manner of living, however, and he was ultimately compelled to give up teaching. He returned to his family in Lyck, where he conducted botanical research for the rest of his life.

Although Sanio's floristic examination of the region around Lyck essentially ended with his dissertation, he occasionally returned to floristics and systematics. He was mainly concerned, however, with other areas of botany, including the ferns, the Characae, and, above all, the moss genus *Drepanocladus* (*Harpidium*).

Sanio had begun microscopic studies of plant anatomy as a student, and his first publication was on the development of the spores of *Equisetum*. He recognized that the spiral bands (haptera) arise in the outer layer of the spore wall and exhibit an oblique structure on the exterior. Sanio was the first to point out that the stomata of *Equisetum* are composed of two pairs of cells: an external, upper pair and an inner, lower one with thickening stripes. The upper pair lies either at the same level as the epidermis or deeper. On the basis of this finding, Julius Milde later (1865) divided the genus *Equisetum* into two groups of species, *Equiseta phaneropora* and *Equiseta cryptopora*.

Sanio made his most important contributions in the anatomy of wood, a subject that had attracted his attention when he was a student. In his first studies on this topic, he dealt with the wood parenchyma, which is found in almost all woody plants. He found that it consists of rows of cells that are joined into a spindle-shaped unit and that arise from a fiber cell through oblique division. In autumn the wood parenchyma contains starch, which is dissolved in the spring.

Sanio explained the structure and development of cork so completely that later writers could add nothing essential to his account. He showed that every radial row of cork cells is derived from the action of a single cambium cell, which divides only centripetally or carries out one or two divisions in a centrifugal direction and then divides uninterruptedly only in a centripetal sequence. In the first case only cork is produced; in the second, a thin phelloderm is generated on the inside at the start of the process. The formation of cork occurs either in the epidermis or in the uppermost cortical layers, rarely in deeper tissue.

Sanio thoroughly examined 166 European and

exotic trees and shrubs. Through this research he resolved many uncertainties concerning the terminology and origin of various types of cells. He also explained the formation of annual rings and established the first table for identifying woods on the basis of anatomical features (1863).

Sanio's last works were devoted to the anatomy of the wood of the Scotch pine (*Pinus silvestris*). He showed that the cells of the wood (tracheids) in trunks and branches grow in length and breadth from the inside toward the outside for a number of annual rings, until a final size is reached. Sanio refuted Hartig's view that every radial row of wood and phloem cells derives from two mother cells, one of which transmits only wood cells toward the inside, while the other yields only phloem cells that move toward the outside. He demonstrated that each radial wood and phloem cell row in the cambium derives from only a single mother cell, which alternately sends wood cells toward the inside and phloem cells toward the outside.

Sanio owed the success of his scientific works to outstanding skill in preparing anatomical thin sections (with a razor he could produce series of sections of .03 mm) and to his great care in observation and drawing. His work also benefited from his critical attitude toward the literature and from examination of the greatest possible number of genera and families, which prevented hasty conclusions based on a few samples.

Sanio's findings and his precise terminology became widely known through their inclusion in H. A. de Bary's *Vergleichende Anatomie der Vegetationsorgane der Phanerogamen und Farne* (1877). Almost all the terms that he coined are still in use.

BIBLIOGRAPHY

I. Original Works. Sanio's writings include "Über die Entwickelung der Sporen von *Equisetum*," in *Botanische Zeitung*, **14** (1856), 175–185, 193–200; "Untersuchungen über diejenigen Zellen des Holzkörpers, welche im Winter assimilierte Stoffe führen," in *Linnaea*, **29** (1857), 111–168; "Florula Lyccensis," *ibid.*, 169–264; "Untersuchungen über die Epidermis und die Spaltöffnungen der Equisetaceen," *ibid.*, 385–416; "Vergleichende Untersuchungen über den Bau und die Entwickelung des Korkes," in *Jahrbücher für wissenschaftliche Botanik*, **2** (1860), 39–108; "Vergleichende Untersuchungen über die Elementarorgane des Holzkörpers," in *Botanische Zeitung*, **21** (1863), 85–91, 93–98, 101–111, 113–118, 121–128; "Vergleichende Untersuchungen über die Zusammensetzung des Holzkörpers," *ibid.*, 357–363, 369–375, 377–385, 389–399, 401–412; "Über die Grösse der Holzzellen

bei der gemeinen Kiefer (*Pinus silvestris*)," in *Jahrbücher für wissenschaftliche Botanik*, **8** (1872), 401–420; and "Anatomie der gemeinen Kiefer (*Pinus silvestris*)," *ibid.*, **9** (1873), 50–126.

II. Secondary Literature. A detailed obituary is by P. Ascherson, in *Verhandlungen des Botanischen Vereins der Provinz Brandenburg*, **34** (1893), xli–xlix. Shorter biographies are J. T. Ratzeburg, *Forstwissenschaftliches Schriftsteller-Lexikon* (Berlin, 1872), 449–450; and *Allgemeine deutsche Biographie*, LIII (1907), 709–711. Sanio's work is discussed in H. A. de Bary, *Vergleichende Anatomie der Vegetationsorgane der Phanerogamen und Farne* (Leipzig, 1877); K. Mägdefrau, *Geschichte der Botanik* (Stuttgart, 1973); Julius Sachs, *Geschichte der Botanik* (Munich, 1875); and T. Schmucker and G. Linnemann, "Geschichte der Anatomie des Holzes," in H. Freund, ed., *Handbuch der Mikroskopie in der Technik*, V, pt. 1 (Frankfurt, 1951), 1–78.

Karl Mägdefrau

SANTORINI, GIOVANNI DOMENICO (*b.* Venice, Italy, 6 June 1681; *d.* Venice, 7 May 1737), *medicine, anatomy.*

The son of an apothecary, Santorini studied medicine at Bologna, Padua, and Pisa, receiving the doctorate in 1701. He began anatomical dissection in 1703 and was demonstrator in anatomy at Venice from 1706 to 1728, when he became physician to the Spedaletto in that city.

Santorini was generally acknowledged as the outstanding anatomist of his time, carefully dissecting and delineating many difficult and complex gross features of the human body, such as facial muscles involved in emotional expression, accessory pancreatic ducts, and duodenal papillae. His name has been given to some of these structures, such as the arytenoid cartilages (1724), the risorius muscle, and the *plexus pudendalis venosus*.

Santorini's contributions began with *Opuscula medica de structura* (1705). His most important work was *Observationes anatomicae* (1724), a valuable exposition of details of human anatomy that contains "De musculis facies," "De aure exteriore," "De cerebro," "De naso," "De larynge," "De iis," "De abdominae," "De virorum naturalibus," and "De mulierum partis procreationes datis." Santorini was a popular teacher and a pioneer in teaching obstetrics.

BIBLIOGRAPHY

See *Opuscula medica de structura* (Venice, 1705; Rotterdam, 1718), later repr. in Georgius Baglivi, *Opera*

(Leiden, 1710); *Observationes anatomicae* (Venice, 1724; Leiden, 1739), his most important work; *Istoria d'un feto* (Venice, 1727); *Istruzioni intorno alla febbre* (Venice, 1734, 1751), and his *Opera* (Parma, 1773). His *Opuscula quatuor: De structura et motu fibrae, De nutritione animali, de haemorrhoidibus*, and *De catameniis* appeared in Baglivi's *Opera omnia medico-practica et anatomica* (Leiden, 1745).

CHAUNCEY D. LEAKE

SANTORIO, SANTORIO, also known as **Sanctorius** (*b.* Justinopolis, Venetian Republic [now Koper, Yugoslavia], 29 March 1561; *d.* Venice, Italy, 6 March 1636), *medicine, physiology, invention of measuring instruments.*

Santorio was the son of Antonio Santorio, a nobleman from Friuli who had come to Justinopolis (also named Capraria and Capodistria) as a high official of the Venetian Republic. While serving there, he married Elisabetta Cordonia, the heiress of a local noble family.

Santorio, the eldest of four children, received at his baptism his family name as his given name—a practice fashionable at that time in Istria. He was educated at Justinopolis and Venice, where his father's friendship with an illustrious nobleman, Morosini, enabled Santorio to share the same tutors as Morosini's sons, Paolo and Andrea. (The latter became a famous historian and a reformer of the University of Padua.) Thus Santorio acquired a thorough knowledge of classical languages and literature.

In 1575 Santorio enrolled at the Archilyceum of Padua, where he followed the traditional sequence of studying philosophy and then medicine. He obtained his doctor's degree in 1582. At the very beginning of his medical practice, Santorio carried out a systematic study of the changes in weight that occurred in his own body following the ingestion of food and the elimination of excrement.

Most biographers assert that, following an invitation from King Maximilian, Santorio lived for a long time in Poland. The only document supporting this statement is late testimony quoting a copy of a letter written by Nicolò Galerio, vicar of the University of Padua. In 1587 an important person requested a qualified physician from the medical faculty at Padua; the faculty replied through Galerio that Santorio—the most suitable choice because of his learning, his loyalty, and his zeal—was ready to undertake the journey. It is claimed that Galerio's response was addressed to the king of Poland, but this is an unproved conjecture. The original

document has disappeared. Capello was Santorio's only biographer to have seen the draft (also lost) of Galerio's letter, which, according to Capello's testimony, was written "for a certain Polish prince." Therefore, it can be assumed that the draft did not bear the name of the intended recipient. The Polish archives contain no material indicating that Santorio ever stayed in Poland, but there are several pieces of evidence attesting his presence in Croatia, particularly in Karlovac, during the last decade of the sixteenth century. The invitation of 1587 would thus seem to have come from a leading Croatian nobleman, probably Count Zrinski. From 1587 to 1599 Santorio spent most of his time, not in Poland, but among the South Slavs. It was on the Adriatic coast, at or near Senj, that Santorio made the first trials with both his wind gauge and a new apparatus designed to measure the force of water currents.

Santorio often returned to his native city and maintained cordial relations with the scientists of Venice and Padua. In 1599 he finally left Croatia and established a medical practice in Venice. One reason for his departure was perhaps a desire to escape the plague that in 1598 began to ravage the valley of the Sava River. Also, he was attracted by the exchange of new ideas and the intellectual fervor that Venice was then enjoying. The home of the Morosinis had become a true meeting place for the proponents of the new science: Santorio met Galileo and became friendly with Paolo Sarpi, Girolamo Fabrici, Giambattista Della Porta, and Francesco Sagredo, among others. In 1607 Santorio treated Sarpi after the latter had been wounded during an attempted political assassination.

Santorio's first book appeared in 1602, *Methodi vitandorum errorum omnium qui in arte medica contingunt*. It is a comprehensive study on the method to be followed in order to avoid making errors in the art of healing. Without breaking with the Galenic tradition, Santorio ventured certain criticisms of classical physiology and expressed ideas that prefigured the mechanistic explanations of the iatrophysical school. According to Santorio, the properties of the living body do not depend only on the four elementary qualities, nor even on the secondary and tertiary qualities, as they were defined by Galen, but also on "number, position, form, and other accidental factors." Santorio employed the analogy (later used by Descartes) between an organism and a clock, the movements of which depend on the number, the form, and the disposition of its parts. Nevertheless, he remained faithful to traditional humoral pathology and tried to explain all internal diseases as particular cases of humoral dyscrasia.

Although Santorio accepted the ancient scheme that attributed diseases to a bad mixture of the four humors, he modified it profoundly by some quantitative attributions. In the pathology of Hippocrates and Galen there is no discontinuity between eucrasia and the innumerable possible pathological deviations. Santorio, starting from certain remarks of Galen on the "degrees" of dyscrasia, defined the discontinuity of morbid states and proposed to make known, by mathematical deduction, all the possible diseases. (According to his calculations, their total number was about 80,000.) In *Methodi vitandorum errorum,* Santorio mentioned a few measuring instruments (the scale and the "pulsilogium") but did not seem to attach special importance to them. It should be emphasized, however, that theoretical considerations form only the background to this medical textbook, which is of an eminently practical orientation. The book contains several good descriptions of diseases and offers model cases of differential diagnostics, for example, the clinical distinction between an abscess of the mesentery and intestinal ulcerations.

On 6 October 1611 Santorio was appointed professor of theoretical medicine at the University of Padua. He owed his appointment to the success of his book, to his growing reputation as a practitioner, and to the support of his friends in the upper Venetian nobility. Originally he was supposed to have held the position for six years; but at the end of that period, in 1617, the Venetian Senate extended his contract for six more years and granted him an exceptionally high salary. Santorio's talents as an orator, the originality of his ideas, and his demonstrations of new methods of clinical examination made his courses very popular. As professor of theoretical medicine, Santorio was required to present and comment upon the *Ars parva* of Galen, the *Aphorisms* of Hippocrates, and the first book of the *Canon* of Ibn Sīnā. This obligation led him to publish most of his own views in the restricted form of scholarly commentaries on these three works. Cautious and introverted, Santorio preferred to express himself through allusions and to envelop his bold and original ideas in a thick layer of conventional erudition. Still, as early as 1602, he clearly set forth his personal creed: "One must believe first in one's own senses and in experience, then in reasoning, and only in the third place in the authority of Hippocrates, of Galen, of Aristotle, and of other excellent philosophers" (*Methodi vitandorum errorum,* 215).

In 1612 Santorio published *Commentaria in artem medicinalem Galeni,* a large volume of commentaries on Galen that contains the first printed mention of the air thermometer. In 1614 he published *De statica medicina,* a short work on the variation in weight experienced by the human body as a result of ingestion and excretion. The latter work made him famous. Filled with incisive and elliptic aphorisms, *De statica medicina* dazzled his contemporaries, although its style is rather irritating to modern readers. The book briefly describes the results of a long series of experiments that Santorio conducted with a scale and other measuring instruments. Its success stemmed chiefly from the simplicity and apparent precision of the methods by which he promised to preserve health and to direct all therapeutic measures.

On 9 February 1615 Santorio sent a copy of *De statica medicina* to Galileo. In an accompanying letter he explained that his work was based on two principles: first, Hippocrates' view that medicine is essentially the addition of what is lacking and the removal of what is superfluous; and second, experimentation. The origin of "static medicine" was, in fact, the Hippocratic conception that health consists in the harmony of the humors. One expression of this harmony is the equilibrium between the substances consumed by the organism and those rejected by it. According to this view, pathological conditions should be accompanied by a quantitative disequilibrium of the exchanges between the living body and its surroundings. To verify this supposition, Santorio turned to quantitative experimentation. With the aid of a chair scale, he systematically observed the daily variations in the weight of his body and showed that a large part of excretion takes place invisibly through the skin and lungs *(perspiratio insensibilis).* Moreover, he sought to determine the magnitude of this invisible excretion; its relationship to visible excretion; and its dependence on various factors, including the state of the atmosphere, diet, sleep, exercise, sexual activity, and age. Thus he invented instruments to measure ambient humidity and temperature. From this research he concluded (1) that *perspiratio insensibilis,* which had been known since Erasistratus but which was considered imponderable, could be determined by systematic weighing; (2) that it is, in itself, greater than all forms of sensible bodily excretions combined; and (3) that it is not constant but varies considerably as a function of several internal and external factors; for example, cold and sleep lessen it and fever increases it.

The invention of the thermometer gave rise to a

priority dispute between Santorio, Galileo, Sagredo, and several other scientists. It seems probable that Galileo invented the first open-air thermoscope. Santorio built a similar device, and whether or not he knew of Galileo's, he was the first to add a scale, thereby transforming the thermoscope into the thermometer. The exact date of this invention is unknown, but it falls between 1602 and 1612. The fixed reference points that Santorio employed to create a thermometric scale—namely, the temperature of snow and the temperature of a candle flame—are mentioned in his commentaries on Galen, although not in the first edition of that work. Galileo made no use at all of the thermoscope. In contrast, Santorio attempted to measure body temperature in health and illness and employed his thermometer in physiological experiments and in medical practice. It is not possible to estimate the accuracy of the figures that Santorio obtained with his apparatus. In any case, the instrument did have the great drawback of being subject to variations in barometric pressure, a factor that was still unknown.

Santorio also invented other measuring instruments and medical devices, including a hygrometer, a pendulum for measuring the pulse rate, a trocar, a special syringe for extracting bladder stones, and a bathing bed. He spoke of these inventions in his lectures and demonstrated their uses. For example, he publicly used the trocar for abdominal and thoracic paracentesis and even for a tracheotomy. Although Santorio promised to reveal the methods of construction of his apparatus in a book entitled *De instrumentis medicis,* the work never appeared.

Santorio was an advocate of the Copernican system and a fierce opponent of astrology. These stands involved him in difficulties with certain of his colleagues, who accused him, among other things, of neglecting his professional duties. An adept of the occult sciences, Ippolito Obizzi, professor at Ferrara, violently attacked *De statica medicina*; and Santorio felt himself obliged to respond in a new edition of his book. Many of the criticisms were justified (Santorio had attributed to static medicine a practical significance that proved to be grossly exaggerated), but Obizzi was wrong in attacking Santorio's experimental method.

In 1624, at the end of his second term as professor at Padua, Santorio wished to retire from his post. His request was granted by the Venetian Senate, which, as a sign of its regard for his services, also awarded him an annual pension and the permanent title of professor. Subsequently Santorio declined offers from the universities of Bologna, Pavia, and Messina and returned to Venice.

In 1625 he published there his commentaries on Ibn Sīnā's *Canon (Commentaria in primam fen primi libri Canonis Avicennae).* The chief value of this work lies in its revelation (still quite hesitant, although accompanied by diagrams for the first time) of the principles of construction of various instruments. In 1630 Santorio was given the task of organizing measures against an epidemic of plague in Venice. In the same year he was elected president of the Venetian College of Physicians.

A misogynist, Santorio never married. Although frugal and little concerned with personal comfort, he was eager for gain and in fact assembled a considerable fortune. He was restrained and prudent, but his style occasionally ran to incisive irony. Above all, Santorio was endowed with a highly critical intelligence. He quickly accepted the ideas of Galileo on mechanics and on the nature of the celestial bodies as well as those of Kepler on optics. Curiously, despite his penchant for mechanistic explanations, he did not grasp the significance of Harvey's discovery of the circulation of the blood.

Santorio died from a disease of the urinary tract. He was buried in the Church of the Servi in Venice, but his skeleton was removed when the church was demolished in 1812. His skull is now at the museum of the University of Padua.

Throughout the seventeenth century and the first half of the eighteenth, physicians sympathetic with the doctrines of iatrophysics praised Santorio as one of the greatest innovators in physiology and practical medicine. Many scientists agreed with Baglivi that the new medicine was based on two pillars: Santorio's statics and Harvey's discovery of the circulation of the blood. Boerhaave wrote of *De statica medicina* that "no medical book has attained this perfection." The exaggerated praise accorded to Santorio's little book detracted from its author's reputation in the nineteenth century, by which time scientists had rejected as illusory the medical content of his teaching, namely, the claim that knowledge of the quantity of *perspiratio insensibilis* provides essential information for hygiene, diagnostics, and therapeutics. Although static medicine is no longer a viable medical doctrine, the method through which it emerged is no less fruitful on that account. Santorio's great achievement was the introduction of quantitative experimentation into biological science. Undoubtedly inspired by the ideas of Galileo, Santorio opened the way to a mathematical and experimental analysis of physiological and pathological phenomena.

BIBLIOGRAPHY

I. ORIGINAL WORKS. Santorio's most famous work is *Ars de statica medicina sectionibus aphorismorum septem comprehensa* (Venice, 1614); the second, greatly revised ed. is *De medicina statica libri octo* (Venice, 1615). This second ed. was reprinted about forty times, either by itself or with commentaries by M. Lister, Baglivi, J. Gorter, D. L. Rüdiger, or A. C. Lorry. It was also translated into English (1676), Italian (1704), French (1722), and German (1736).

Santorio's other writings are *Methodi vitandorum errorum omnium qui in arte medica contingunt* (Venice, 1602); *Commentaria in artem medicinalem Galeni* (Venice, 1612); *Commentaria in primam fen primi libri Canonis Avicennae* (Venice, 1625); and *Liber de remediorum inventione* (Venice, 1629). These works are contained in his *Opera omnia quatuor tomis distincta* (Venice, 1660).

II. SECONDARY LITERATURE. The most helpful old biographies are A. Capello, *De vita cl. viri Sanctorii Sanctorii* (Venice, 1750); and J. Grandi, *De laudibus Sanctorii* (Venice, 1671). Further details can be found in M. Del Gaizo, *Ricerche storiche intorno a Santorio Santorio ed alla medicina statica* (Naples, 1889) and *Alcune conoscenze di Santorio intorno ai fenomeni della visione ed il testamento di lui* (Naples, 1891); and in P. Stancovich, *Biografie degli uomini illustri dell'Istria*, II (Trieste, 1829).

More recent studies of Santorio's life and work include A. Castiglioni, *La vita e l'opera di Santorio Santorio Capodistriano* (Bologna–Trieste, 1920), with English trans. by E. Recht in *Medical Life*, **38** (1931), 729–785; M. D. Grmek, *Istarski liječnik Santorio Santorio i njegovi aparati i instrumenti* (Zagreb, 1952); and R. H. Major, "Santorio Santorio," in *Annals of Medical History*, n.s. **10** (1938), 369–381.

For an appraisal of Santorio's role in the history of science, see M. D. Grmek, *L'Introduction de l'expérience quantitative dans les sciences biologiques* (Paris, 1962) and H. Miessen, *Die Verdienste Sanctorii Sanctorii um die Einführung physikalischer Methoden in die Heilkunde* (Düsseldorf, 1940). On static medicine see E. T. Renbourn, "The Natural History of Insensible Perspiration," in *Medical History*, **4** (1960), 135–152. The best work on the invention of the thermometer is W. E. K. Middleton, *A History of the Thermometer and Its Use in Meteorology* (Baltimore, 1966).

M. D. GRMEK

SAPORTA, LOUIS CHARLES JOSEPH GASTON DE (*b.* St. Zacharie, France, 28 July 1823; *d.* St. Zacharie, 26 January 1896), *paleobotany.*

Saporta's ancestors came from Zaragoza, Spain, and included physicians, botanists, and entomologists. He himself was at first more inclined toward literature, but following the death of his wife in 1850, when he was twenty-seven, he sought diversion in botany. His interest in paleobotany developed after he had by accident found some plant fossils from a nearby gypsum mine in an antique shop in Aix. He related this discovery to Adolphe Brongniart, a friend of his grandfather's, and Brongniart encouraged him to make a systematic study of the deposits around Aix.

Saporta devoted his initial research to the Tertiary flora of France, on which he published a number of monographs, illustrated with detailed drawings that he executed directly from nature. These included a study of a sequence of local floras separated by short intervals that he discovered in lacustrian formations from the Upper Eocene to the Lower Miocene. The remarkable state of preservation of the fossils that he studied allowed him to examine the outlines and rib networks of leaves and to make more exact determinations of species than are usually possible.

Saporta next turned his attention to the Mesozoic. Among his publications dealing with that period was an extensive examination of Jurassic flora that comprised four volumes each of text and illustrations. This work brought his services into demand, and he was invited to Belgium (to collaborate with A. F. Marion in a work on the Gelinden flora), to Portugal (to study Lower Cretaceous and Cretaceous floras), to Greece (to study the Miocene flora of Koumi), and to the United States (to assist Lesquereux in his work on the Cretaceous flora of the Dakotas). He published other major works, including *Le monde des plantes avant l'apparition de l'homme* (Paris, 1879), which represented revised versions of articles that Saporta had already brought out in nontechnical journals, *L'évolution du règne végétal* (Paris, 1881–1885), written with Marion, and *Origine paléontologique des arbres cultivés ou utilisés par l'homme* (Paris, 1888). The first of these gave a synoptic view of the stages through which vegetation on earth had passed, while the last two were more specialized treatments of the same subject.

In his work in general Saporta was a patient and meticulous researcher who attempted both to give a precise description of a species and to relate it to historical developments so as to clarify its origin. He was particularly concerned with elucidating the climatic conditions of an era and wrote a number of works on historic climate. He emphasized the difficulty of determining species, and advised caution; while he has been accused of erroneously increasing the number of species, such accusations are unjustified, and recent studies have frequently served to confirm his general views, although with some qualifications.

Saporta's scientific reputation was widespread. He was elected to the Académie d'Aix in 1866, and served as its president on several occasions; from 1886 he was permanent secretary of its science section. In 1872 he was admitted to the Académie de Marseille, and in 1876 he became a corresponding member of the Académie des Sciences. He was also an associate foreign member of the Royal Academy of Belgium and of the Madrid Academy of Science. In addition to his paleontological research, Saporta was interested in history. He wrote *La famille de Madame de Sévigné en Provence,* of which the first chapter served as his last presidential address to the Académie d'Aix.

BIBLIOGRAPHY

I. ORIGINAL WORKS. In addition to the works cited, see Saporta's own *Notice sur les travaux scientifiques* (Paris, 1875); Poggendorff lists a selection of his articles in periodicals, and R. Zeiller, cited below, gives a further bibliography. His book on the history of Provence is *La famille de Madame de Sévigné en Provence* (Paris, 1889).

II. SECONDARY LITERATURE. On Saporta and his work, see J. A. Henriques, "Luiz Carlos José Gaston, Marquez de Saporta," in *Boletim da Sociedade Broteriana,* **13** (1896), 1–10; A. Pons, "Contribution palynologique à l'étude de la flore et de la végétation pliocènes de la région Rhodanienne," in *Annales des sciences naturelles,* Botanique, 12th ser., **5** (1964), 499–722; R. Zeiller, "Le marquis G. de Saporta. Sa vie et ses travaux," in *Revue générale de botanique,* 7 (1895), 353–388, with bibliography; and Y. Conry, *Correspondance entre Charles Darwin et Gaston de Saporta* (Paris, 1972).

F. STOCKMANS

SARPI, PAOLO (*b.* Venice, Italy, 14 August 1552; *d.* Venice, 16 January 1623), *natural philosophy, theology.*

Sarpi was the son of Francesco Sarpi and Isabella Morelli. The financial straits of the family, after the death of his father, shifted the responsibility for Sarpi's education to his uncle, Ambrogio Morelli, the head titular priest of St. Hermangora. Impressed with the boy's precocity, Don Ambrogio placed him under the tutelage of the Servite friar Giammaria Capella, who instructed him in philosophy, theology, and logic, and was influential in his entry into the Servite Order at the age of fourteen. At the end of his novitiate, in 1570, Sarpi displayed such argumentative skill in publicly defending 318 philosophical and theological theses that the duke of Mantua, Guglielmo Gonzaga, appointed him court theologian and professor of positive theology.

In 1574, as a bachelor in theology, Sarpi worked with Cardinal Carlo Borromeo in Milan and, later that year, returned to Venice to teach philosophy in the Servite monastery. In 1578 he was awarded the degree of doctor of theology by the University of Padua and the following year was elected provincial of his order. As procurator general from 1585 to 1588, Sarpi resided in Rome, where he enjoyed the friendship of Pope Sixtus V, the Jesuit theologian Robert Bellarmine, and Giambattista Della Porta. He was appointed state theologian by the Venetian Senate in 1606 and counseled defiance of the bull of interdict and excommunication launched against Venice by Paul V. Having failed to appear before the Roman Inquisition to answer charges of heresy, he was excommunicated in January 1607. The reconciliation between the papacy and the Republic of Venice did not lessen the hostility harbored in Rome against Sarpi, and on 5 October 1607 he was the object of an attempted assassination that he accused the Roman Curia of engineering.

As adviser to the Venetian Senate, an office he continued to hold until his death sixteen years later, Sarpi arbitrated the dispute between Galileo and Baldassar Capra, who had claimed the invention of the proportional compass as his own. In July 1609, when offered the opportunity to purchase one of the earliest telescopes, the Senate referred the matter to Sarpi for his opinion. Sarpi, who as early as November 1608 had been the first in Italy to learn of the invention of the Flemish spectacle-maker Hans Lippershey, recommended that the offer be refused, confident that his friend Galileo could construct an instrument of comparable if not superior quality. This Galileo did, and presented it to the government as a gift in August 1609, in return for which he received a lifetime appointment to the University of Padua.

Chiefly remembered now for his highly biased *Istoria del Concilio Tridentino* (1619), Sarpi was well versed in the works of all the Scholastic philosophers, especially Ockham, whom he held in great esteem. His *Arte di ben pensare,* in which he distinguishes between sensation and reflection, and examines the relationship of the senses to cognition, has been credited with anticipating Locke's *Essay Concerning Human Understanding*.

Although Sarpi was highly praised for his mathematical and speculative abilities by such contemporaries as Galileo, Della Porta, and Acquapendente, all that remains by which one can judge the originality of his scientific thought are some letters and the notebooks containing his philosophical, physical, and mathematical thoughts. Extant are reliable copies of the originals that perished in a fire in 1769. They consist of more than 600 numbered paragraphs, which were written over a period of three decades but date chiefly from 1578 to 1597. Rather than forming a consistent philosophical system, the notebooks are a collage of disparate thoughts about the nature of the physical world—a chronicle of the intellectual evolution of a man in the mainstream of the scientific developments of his age. In Venice, Sarpi religiously frequented the *ridotto* at the home of the historian Andrea Morosini; while in Padua he regularly attended colloquiums sponsored by Giovanni Vincenzio Pinelli and, in Rome, at the Accademia dei Lincei. There he met and exchanged views with some of the most celebrated scientists of the day and, it appears, recorded in his notebooks their ideas in addition to his own.

The entries, chronologically annotated, touch upon every aspect of contemporary science, from a discussion of the corpuscular nature of light and a refutation of the Peripatetic denial of its passage through a vacuum, to an enumeration of the properties of conic sections. Of particular interest are Sarpi's ideas on optical relativity, his negation of the concept of absolute rest, and his refutation of the Aristotelian doctrine of an essential difference between natural and violent motion together with its corollary—that the two types of motion cannot be simultaneously operative in the same body. In anatomy, to which he devoted himself from 1582 to 1585, Sarpi has been credited with correctly interpreting the function of the venous valves and the discovery of the circulation of the blood, for which Harvey later provided experimental proof. The subject of magnetism is one in which the notebooks reflect a constant interest. Several entries suggest (as, a decade later, did William Gilbert, whom Sarpi knew) that all bodies fall to earth not because it is their nature to do so but because they are drawn to it "as iron is to a magnet." Other entries deal with the reflection of light by curved mirrors, centers of gravity, the relative weights of

bodies immersed in water, speeds of descent of freely falling bodies, the motion of projectiles, cause and effect, and a short discussion of the inconclusiveness of the arguments marshaled by opponents of the Copernican system.

BIBLIOGRAPHY

I. ORIGINAL WORKS. See F. Griselini, ed., *Opere di F. Paolo Sarpi Servita*, 8 vols. (Verona, 1761–1768). R. Amerio, *Pensieri naturali, metafisici e matematici del Padre Maestro Paolo Sarpi Servita* (Bari, 1951), consists of excerpts from the notebooks, including the *Arte di ben pensare*. Copies of the notebooks made by Configlio Capra in 1740 are in the Marciana Library in Venice, MSS Ital. Cl.II, no. 129 (Provenienza Giacomo Morelli), Collocazione 1914: "Opuscoli e frammenti del Padre M.ro Paola Servita, in varie materie filosofiche, volume primo, an. 1740, Pensieri naturali, metafisici e matematici, opusc. I," 1–142; *Arte di ben pensare*, 275–292. G. Gambarin edited *Istoria del Concilio Tridentino,* 3 vols. (Bari, 1935; repr. Florence, 1966).

II. SECONDARY LITERATURE. See G. Abetti, *Amici e nemici di Galileo* (Milan, 1945); D. Bertolini, ed., *Enciclopedia italiana*, XXX (1949), 877–879; A. G. Campbell, *The Life of Fra Paolo Sarpi* (London, 1869); P. Cassani, *Paolo Sarpi e le scienze matematiche e naturali* (Venice, 1882); A. Favaro, "Fra Paolo Sarpi, fisico e matematico, secondo i nuovi studi del prof. P. Cassani," in *Atti del R. Istituto veneto di scienze, lettere ed arti,* 6th ser., **1** (1882–1883), 893–911; G. Getto, *Paolo Sarpi* (Florence, 1967); Fulgenzio Micanzio, *Vita del Padre Paolo del Ordine dei Servi* (Leiden, 1646), also in English (London, 1651); A. Robertson, *Fra Paolo Sarpi the Greatest of the Venetians* (London, 1894); and T. A. Trollope, *Paul the Pope and Paul the Friar* (London, 1861).

UMBERTO MARIA D'ANTINI

SARS, MICHAEL (*b.* Bergen, Norway, 30 August 1805; *d.* Christiania [Oslo], Norway, 22 October 1869), *marine biology.*

Sars's father, for whom he was named, was born in Bremen but later settled in Bergen, as sea captain and merchant. The ancestors of his mother, Divert Henriche Heilmann, came from Narva, Estonia.

Sars's early enthusiasm for natural history was much encouraged by his teachers. He studied theology and received the candidate's degree in 1828. From 1828 to 1854 he served as teacher, vicar, and later rector of seashore communities in western Norway, often traveling by boat to visit parishioners and devoting much of his time to zoological studies. Some of his most important work was done in this period. Following his appointment on 7 August 1854, Sars was professor extraordinarius of zoology at the University of Christiania (Oslo) until his death; but his excessive concern with details of morphology made him neither a successful nor a popular teacher.

His marriage in 1831 to Maren Welhaven produced twelve children, one of whom, Georg Ossian, became a noted zoologist.

Sars traveled widely in his younger years, beginning with a journey to the northern seas of Norway. A large grant supported a six-month natural history trip in 1837 to Holland, France, Germany, Prague, Denmark, and Sweden. Sars often visited the Mediterranean, the Adriatic in 1851, and Naples and Messina in the winter of 1852–1853. These study trips not only greatly increased his knowledge but also brought him in contact with the leading zoologists of Europe.

One of the fathers of marine zoology, Sars from 1830 to 1860 made what is perhaps the greatest single contribution to the elucidation of the life cycles of marine invertebrates. His findings on the alternation of generations in coelenterates were among the most important evidence for Steenstrup's classical work on *Generationswechsel* (1842). Because the larval stages of most marine organisms are so different from the adults, their connection cannot be discovered until a series of intermediate stages is established; it was Sars who discovered many of them. Simultaneously with the Swedish zoologist Sven Lovén he found and described the first trochophore larvae (annelid). Likewise, Sars was the first to describe the veliger larvae of the mollusks (1837, 1840) and the bipinnaria larva (1835), which he later (1844, 1846) identified as a stage in the development of starfish.

From 1830 on, Sars used the dredge invented by O. F. Müller in his collecting. His numerous discoveries of deep-sea organisms by dredging were a sensation in his day because—until he proved otherwise—it had been universally assumed that the depths of the ocean where light did not penetrate were without life. In 1864 his son Georg, who participated in the collecting, dredged up from a depth of 300 fathoms (near Lofoten) the first living stalked crinoids (sea lilies), a group of organisms then known only from fossils and thought to have been extinct since the Mesozoic. The description of *Rhizocrinus lofotensis* (1864) was followed by intensified work at greater ocean depths, and in 1868 Sars published a memoir on 427 species of invertebrates collected off Norway at depths of from more than 200 to 450 fathoms. These exciting results induced him to promote deep-sea expeditions and thus led to the conception and organization of the *Challenger* expedition.

Although a competent taxonomist and morphologist, Sars was interested mainly in life cycles, larval stages, parental care through brood pouches, seasons of reproduction, cyclic phenomena, and migrations of marine organisms. His own research yielded major contributions to knowledge of annelids, ascidians, coelenterates, crustaceans, echinoderms, and mollusks. Actively interested in fossils, he also published a number of contributions to paleontology.

Sars published ninety-five papers (six posthumous). He was an honorary or corresponding member of more than twenty foreign academies and societies and was awarded honorary doctorates from the universities of Zurich (1846) and Berlin (1860). Most of his original publications appeared in Norwegian; but some were subsequently republished, at least in abstract, in French, English, or German.

BIBLIOGRAPHY

I. ORIGINAL WORKS. A complete bibliography of Sars's works is found in Økland (below); they include *Beskrivelser og iagttagelser over nogle . . . ved den Bergenske kyst levende dyr* (Bergen, 1835), French abstract in *Annales d'anatomie et de physiologie,* **2** (1838), 81–90; "Beitrag zur Entwicklungsgeschichte der Mollusken und Zoophyten," in *Archiv für Naturgeschichte,* **3** (1837), 402–407; **6** (1840), 196–219; "Mémoire pour servir à la connaissance des crinoïdes vivants," in University Program, first semester 1867 (Christiania, 1868); and "On Some Remarkable Forms of Animal Life From the Great Deeps of the Norwegian Coast I," *ibid.,* first semester 1869 (Christiania, 1872), written with G. O. Sars.

II. SECONDARY LITERATURE. Information on Sars's life and work is in Fridthjof Økland, *Michael Sars et Minneskrift* (Oslo, 1955), which has a full bibliography; the obituary by Gwynn Jeffreys in Nature, **1** (1870), 265–266; and J. J. S. Steenstrup, *Ueber den Generationswechsel* (Copenhagen, 1842).

ERNST MAYR

SARTON, GEORGE ALFRED LÉON (*b.* Ghent, Belgium, 31 August 1884; *d.* Cambridge, Massachusetts, 22 March 1956), *history of science.*

Sarton was the only child of Alfred Sarton, a chief engineer and director of the Belgian State

Railways. His mother, Léonie Van Halmé, died when he was a few months old. An isolated child, surrounded by servants, George had a prosperous but lonely upbringing.

Secure in a setting as bourgeois as a Balzac novel, Sarton followed a course of schooling that normally led to the study of philosophy at the University of Ghent. But he soon abandoned philosophy in disgust for the natural sciences. He studied chemistry (for which he won a gold medal), crystallography, and then mathematics. His 1911 D.Sc. dissertation on "Les principes de la mécanique de Newton" provided an early indication of the direction his interests were taking under the philosophical influence of Comte, Duhem, and Tannery. A visit to London at this time led to the systematic exploration of the works of Wells, Shaw, and the Fabians, whose ideas Sarton experienced as a refreshing contrast to the doctrinaire Marxism that he and his friends youthfully espoused. Socialism rather than communism thus came to seem the necessary and inevitable prelude to the final achievement of benevolent anarchism.

Sarton graduated from the university in 1911. In May of the same year he married Eleanor Mabel Elwes of London, who had experienced a similar lonely childhood, after being boarded out when her parents (her father was a civil and mining engineer) traveled abroad. Among other similarities in their backgrounds, her father was a Fabian and agnostic, while his was liberal, anticlerical, and a Mason. The small private income Sarton enjoyed was not enough to sustain a family, while all the assets of his wife's father had been lost in speculative mining stocks. Sarton thus found it necessary to seek employment. A 1910 note in his diary had already indicated his intention—before trying to get a post at the university—to "become the pupil, if I prove worthy, of Henri Poincaré: the most intelligent man of our time." He continued, in a passage both prophetic and revealing:

> It is almost certain that I shall devote a great part of my life to the study of "natural philosophy." There is great work to be accomplished in that direction. And—from that point of view—*living* history, the passionate history of the physical and mathematical sciences is still to be written. Isn't that really what history is, the evolution of human *greatness,* as well as its weakness?

This liberal and characteristically enlightened faith was to guide most of the remaining forty-six years of his life. The transition from a dawning conviction of the importance of a passionate history of the physical sciences to the systematic work of equipping a new discipline with tools and standards, and more especially the transition to paid employment in an as-yet-nonexistent profession, was to prove slow and complex.

Using the proceeds from the sale of his deceased father's wine cellar (the sale itself was a typically outrageous act of the confident and iconoclastic young man), Sarton bought a pleasant country house in Wondelgem, near Ghent. Here his only surviving child, May Sarton, was born in May 1912. At about this time Sarton made the bold decision to found *Isis,* his "*Revue consacrée á l'histoire de la science.*" Displaying the single-minded and disinterested opportunism that marks the actions of a man wholly convinced of his mission, he recruited a distinguished editorial board. By September 1912 he had secured the patronage of his idol Poincaré and Arrhenius, Durkheim, Heath, Jacques Loeb, Ostwald, Ramsay, and David Eugene Smith. Sarton's methodical placing of these names in categories shows he was already convinced that the history of science subsumed under its wider heading the histories of mathematics, technology, chemistry, medicine, biology, physics, and astronomy and required besides the expert advice of scientists, historians, sociologists, and historians of philosophy. The methodical division of his field also displays Sarton's passion for tidiness and classificatory order. This passion was to inform all his efforts and may in part explain his attraction to the work of Comte, just as it lay behind his particular admiration for Linnaeus among men of science.

The decision to found a journal was crucial. In retrospect we can see how *Isis* provided Sarton with the first of the institutional tools he needed, if a long-continued but still incoherent area of inquiry was to be transformed under his leadership into an articulated discipline, with agreed critical standards and a definitive cognitive identity. Sarton himself conceived of *Isis* as having far wider aims. His overarching vision and evangelical belief were announced to the world in a series of explanatory passages in the early numbers of the journal. As he pointed out, it was not the chosen domain of activity which made *Isis* unique, but the fact that no other journal would systematically and holistically connect methodological, sociological, and philosophical perspectives with the purely historical and thus allow historical inquiry to "attain its full significance."

Sarton always insisted that the history of science was by nature an encyclopedic discipline, that is, a discipline devoted to summation, comparison, and synthesis. Indeed his own interest in the history of science was "dominated . . . by a philosophical conception." As he wrote to a correspondent in 1927, "I am anxious to prove inductively the unity of knowledge and the unity of mankind." The immensity of the task did not daunt him. Rather it provided the rationale for a lifetime pursuit of difficult linguistic skills and wide-ranging historical and scientific knowledge. He eventually mastered fourteen languages.

Sarton was above all a man of the nineteenth century. He was culturally oriented toward universal history and the progressivist philosophies that found their basis in positive science and their end in the imminent and universal brotherhood of man. Yet in his thinking he was indebted to Condorcet as well as to Comte. The lines of English thought that led from Spencer to the Webbs and Shaw were also important in defining his developed view of the goals to be served by that new synthesis of knowledge to which the history of science was the essential key. High theory and a rigorous consistency were less urgent to him than sustained, appropriate action. Thus throughout his life, Sarton enjoyed the role of propagandist. His evangelizing on behalf of his chosen subject inevitably calls to mind the way Francis Bacon served as apostle for the field of science itself. And, like Bacon, Sarton had his most enduring impact in this vital, although little-acknowledged capacity. Other roles were more nearly central to his mission. With a discipline to be created, a world to be won, the provision of tools, techniques, methodologies, and intellectual orientations lay uppermost in his mind and at the forefront of his actions. A cognitive identity for his new discipline was the primary goal, and his own pattern of work was the self-exemplifying model of appropriate scholarship. Sarton was also well aware of the real, if less immediate, need for professional as well as cognitive identity, without which his field of learning could never be secure let alone accepted as crucial to man's intellectual quest. Appropriate exhortations poured from his pen. The need for career positions and institutes for the history of science were matters to which he often returned. Once again, Sarton provided the self-exemplifying models. He was to "invent" for himself both a research institute and a full-time career, when the discipline barely possessed a cognitive, let alone a professional identity.

While he would on occasion write in an avowedly pragmatic and relativist vein, it was the heritage from positivism, progressivism, and utopian so-cialism that more often controlled Sarton's argument and guided his actions. He would repeatedly present a "theorem on the history of science," which ran as follows:

> *Definition.* Science is systematized positive knowledge, or what has been taken as such at different ages and in different places. *Theorem.* The acquisition and systematization of positive knowledge are the only human activities which are truly cumulative and progressive. *Corollary.* The history of science is the only history which can illustrate the progress of mankind.

The scholar who until his last years largely devoted himself to critical bibliography could also say that "The quest for truth and beauty is indeed man's glory. This is certainly the highest moral certainty which history allows. . . . History itself is of no concern to us. . . . To build up [the] future, to make it beautiful [is rather the aim]." Whatever the contradictions and gaps between words and deeds, the fundamental belief was the one expressed in some words from the first volume of his monumental *Introduction* to the discipline. "The history of science is the history of mankind's unity, of its sublime purpose, of its gradual redemption."

In these statements we may perceive some reasons behind a paradox in Sarton's career. From one perspective his major achievement was that of the discipline builder: providing a key journal; establishing an identity for a field; encouraging the formation of a discipline-based learned society with its potential for sanction and reward (the History of Science Society, 1924); locating and mobilizing scarce resources of men and money in pursuit of crucial scholarly objectives; and seeking to furnish reference works, general surveys, advanced monographs, and teaching manuals. To create the necessary infrastructure for a coherent discipline was a task that demanded a lifetime of devotion.

To Sarton himself such work was only preliminary and minor compared with achieving the "new humanism," the holistic and all-embracing synthesis which would be based on a just appreciation of science in history. The yearning for this synthesis made his contributions to his new subject less than complete. Partly because his vision was so catholic, he could not communicate to others that sense of either the problematics or the conceptual and analytic *schema* necessary if his chosen field were to become a coherent, fully articulated discipline. The paradox is acute. Ambitious for the total vision, it is rather for bibliography, documentation, and the establishment of historical standards and facts that Sarton is most readily remembered.

In the early days of *Isis* all these matters, of course, lay in an unknown and surely unforeseen future. But the events that would crack, erode, and finally destroy the progressive, bourgeois confidence of which Sarton was such a supreme exemplar were already under way. The German invasion of Belgium in August 1914 immediately and dramatically rendered Sarton's private world precarious. Abandoning Wondelgem, his library, and his notes (which he buried in the garden), he fled to London with his family. For the next four years he lived a life of great uncertainty and occasional despair as he sought to find a context in which to pursue his ideals, and his as yet uninvented discipline. Prospects of an English position proved deceptive. Early in 1915 Sarton sailed to the United States, temporarily leaving his family behind. His hopes lay with the greater range and diversity of American institutions, with the progressive spirit, and with Robert S. Woodward.

Woodward was the second president and successful organizer of the Carnegie Institution of Washington. He was also a man with a personal interest in the history of science, and Sarton was in touch with him even before the forced retreat from Belgium. Woodward was initially unsympathetic but slowly softened. The universities in the United States provided sustenance both more immediate and more limited. When Sarton landed, the history of science was actually well established as an activity, although far from being an intellectual discipline and almost unthought of as a profession. A 1915 review article in *Science* makes this plain. It details no fewer than 162 courses in the history of particular sciences as well as fourteen general courses in the history of science, spread among 113 institutions.

Through the help of friends and acquaintances, Sarton managed to arrange a frenetic but sustaining round of guest lectures, seminars, and temporary appointments in American universities. One of those most involved in the promotion of the history of science as a new area of pedagogy was L. J. Henderson, a polymathic biological chemist and junior but influential member of the Harvard faculty. Since 1911 Henderson had himself been teaching at Harvard a regular course on the history of science. He was no doubt early aware of Sarton's program, of which he was to become such a willing supporter. As a member of the inner circle, he was in a position to advance the institutionalization of

the field at Harvard and in the process to have a profound effect on Sarton's life. By 3 May 1916 Henderson was able to write his new friend with the glad news that "from several different sources we have been able to put together $2,000 for your first year [at Harvard]. The second year is not fully arranged, but I have not much doubt that we shall be successful." On the strength of this encouraging letter, Sarton understandably wrote Woodward the optimistic interpretation that "I have been appointed lecturer on the history of science at Harvard University for two years. A new chair has been endowed for the purpose." As it turned out, Sarton was first featured as "lecturer in philosophy," with the bulk of his teaching being listed under the auspices of the department of philosophy. Partly because of the financial problems Harvard faced as a result of the entry of the United States into World War I (but perhaps also partly because his courses did not draw many students), his two-year appointment was not extended.

At this juncture, Woodward's informed aid was to prove decisive. In response to Sarton's renewed appeals and with the help of Andrew Dickson White, historian of the warfare between science and theology and a trustee of the Carnegie Institution, an appointment was created as research associate in the history of science, initially for two years from 1 July 1918.

Thus began an association with the Carnegie Institution that endured throughout Sarton's professional life. Although employed on a full-time basis from Washington, he remained in Cambridge, pleading the uncertainty of war and the great value of a study in the then-new Widener Library. When the end of his Harvard appointment raised the prospect of eviction from Widener, Woodward's solicited intervention proved crucial in allowing him to remain unmolested. The unexpected end of the war allowed Sarton to plan an expedition to Wondelgem to recover his library and notes in the summer of 1919. Following this journey he was supposed to settle in Washington. Instead he quickly returned to Cambridge. There, through the good offices of Henderson, he once more secured a room in Widener Library on the basis of an annual unpaid appointment at Harvard as lecturer in the history of science.

Secure in one of the world's great libraries, with his salary guaranteed by Carnegie, and with no specific duties other than those he fashioned for himself, Sarton was at last free to develop his own mission and his own life-style. It was against this background that the idea of an *Introduction to the History of Science* gradually matured. Aside from *Isis*, Sarton's immediate plans were often vague and shifting during the precarious years between 1912 and 1920. In 1915 he expressed the intention to sail for China and Japan. By 1918, when he had been at Harvard twenty months, his interests had somewhat changed. In reply to Woodward's request for specific information on the work he was undertaking, Sarton mentioned a study of Leonardo da Vinci's scientific manuscripts, which would take "about six months" to finish, and a book on *The Teaching of the History of Science*, which would take a further six months. A little later he was confiding to Woodward that throughout his life he planned "to carry on simultaneously research on ancient science and on XIXth century science." This same letter referred specifically to "my book on XIXth century physics."

In further exchanges Sarton set out one of those ambitious overarching programs that continually recurred in his thinking. The plan was "to lay the foundation of an empirical philosophy of science, to evidence the unity of science." The means included *Isis*; an annual series of studies in the history and method of science to be jointly edited with Charles Singer of Oxford; a "General History of Science and Civilization, to be written on an extensive scale by a large group of scholars"; a history of physics in the nineteenth century; a complete account of the [ancient] beginnings of science; facsimile copies of scientific books and manuscripts; a Chinese encyclopedia; an exhibition of the progress of science; and a catalogue of scientific instruments down to 1900. In all this, the cooperation of Charles Singer was described as essential; and if he could be brought to Washington (where Sarton still intended to go), the city would become an international center for the history of science.

The work on Leonardo gradually made Sarton aware of his own historical naiveté and lack of training. In common with all founders of disciplines, he was perforce an autodidact. His very anxiety to escape from dilettantism and to establish critical standards also protracted his work. Another reason for slow progress was Sarton's incurable tendency to project and begin several studies at once. His first annual report to the Carnegie Institution refers to the Leonardo studies, the accumulation of materials for a history of physics in the nineteenth century, "activity in behalf of the new humanism," and plans for a retrospective survey on the occasion of the twenty-fifth anniversary of the Carnegie Institution. Characteristically, this last would "consider the Institution not as an isolated unit, but as part of the scientific organization of the world."

As if these activities were not enough, early the next year Sarton was planning a general history of science. The work was to be edited jointly with Singer, written by the scholars best qualified, and published in ten or twelve volumes over the next decade by the Oxford University Press. As late as 1934 Sarton was still projecting what had by then become *The Harvard History of Science* in eight illustrated volumes "comparable to the Cambridge Medieval History and other Cambridge and Oxford standard collections." All this was in addition to his published plans for a research institute in the history of science, an incessant round of travel and public lecturing, the detailed personal editing and directing of *Isis,* and, throughout, work on the *Introduction* itself. This vast labyrinth of labor left no time for a private social life (although an hour of classical music remained a daily anodyne). And only by heroic effort could Sarton keep track of his many commitments and produce that controlled order he loved and which his critical bibliographies exemplify.

Sarton's tendency to project programs in the history of science had ramified intellectual roots. Never having been trained for historical work, he long underestimated its difficulty and slow-moving character, until hard-won experience taught him otherwise. Then again he was burdened with a sense of how many different things wanted doing, and all quickly, if the history of science were to become an academically reputable subject. Journals, teaching manuals, standard histories, source books, and, above all, critical bibliographies of what already existed and was being produced — each was desperately needed and would, if necessary, be produced by Sarton himself. In thinking so, he was of course displaying his emerging orientation as a visionary, almost solipsistic scholar rather than as administrative entrepreneur. But beyond such reasons, Sarton embraced global projects because he passionately believed in the unity of knowledge, in the integrity of experience, and in the need for a holistic philosophy that embraced art and science. "The moral failure which the [First World] War implied" made this philosophy all the more urgent. Only what he came to call "the new humanism" could afford the necessary "mixing of the historical with the scientific spirit, of life with knowledge, or beauty with truth." Specialization everywhere threatened such broader insight and reinforced Sarton's belief in "the necessity of synthetic or encyclopedic studies, to keep alive the pure spirit of science." In holding these views he manifested one French-language response to Cassirer and other neo-Kantian cultural generalists, just as Sarton's own idols had earlier reacted to the then-ruling German Kantians and Hegelians.

It was out of this complex background that Sarton's most ambitious and significant work gradually took shape. The essential impulse came in 1919, when he returned to Belgium and retrieved his private notes. With these notes at last safely ensconced in Widener Library, his second annual report to the Carnegie Institution was able to reintroduce what he described as "an old design." It was "the writing of an introduction to the history and philosophy of science, a sort of compendium of the sources of information to which the student . . . may have to refer." By January 1921 the project had grown greatly in scope. The work was now to be in two parts, dealing with the history and philosophy of science as a whole and the history and philosophy of special sciences and their branches. The fundamental aim remained that of establishing the history of science as an independent discipline, with its own tools and methods. The project continued to grow and develop in Sarton's mind. By 1927, with the appearance of the first volume of what was by then to be a multivolume *Introduction to the History of Science*, Sarton envisaged the project as containing "A purely chronological survey . . . which will require seven or eight more volumes," "Surveys of different types of civilization, e.g. Jewish, Muslim, Chinese . . . [in] seven or eight volumes," and a "Survey of the evolution of special sciences . . . [in] some eight or nine volumes."

Sarton was forty-three years of age, with ample time, he felt, to finish the first series down through the eighteenth century, write parts of the Semitic and Far Eastern volumes for the second, and write the whole of the physical sciences volume for the third series. In actuality, work on the *Introduction* progressed far more slowly than his optimism allowed. It was to be 1948 before the third (and final) volume of this magisterial work appeared. Even then its contents reached only as far as A.D. 1400.

The public events of the 1930's and 1940's, together with the growing realization of the immensity of the task he had undertaken, served to erode the beliefs that lay behind Sarton's *Introduction*. As he wrote many times, "the day of Munich was

the nadir of my life." Another blow from quite a different direction was the decision of the Carnegie Institution in 1941 "not to continue the work on the history of science after I am gone." The failure of Harvard to support his plans for an institute was equally hard to accept. J. B. Conant did, however, show his sympathy for the man, if not his ambitious plans, in 1940. In his capacity as Harvard president he arranged Sarton's transfer from lecturer on annual appointment to professor with tenure. Even after this new arrangement, Sarton still drew the major part of his salary from the Carnegie Institution. Yet he was moved to write Conant that "I hope that the day may come when I would serve only Harvard, which appreciates the humanities, including the humanities of science, and not the Carnegie Institution, which considers them 'irrelevant.' " But Conant preferred to have Carnegie continue to pay.

After all, Sarton was, in the special calculus of Harvard, a marginal although illustrious man. In 1940 he had still to produce his first successful Ph.D. candidate, his undergraduate courses remained small, and he almost completely avoided all committee service and routine academic administration. The difference in attitude of the research scholar and the budget-conscious university president was highlighted in Conant's politely negative response to one of Sarton's articles on an institute for the history of science: "I can sum up my point of view by saying that I feel Macaulay was necessary in the development of scholarly and reputable political history, although I understand that now he is not considered as being at all scholarly and hardly reputable. I feel the history of science badly needs a Macaulay, indeed several of them."

Sarton was manifestly no Macaulay, and his enduring monument lies not in narrative accounts that have shaped the thinking of a generation, nor in lectures and students. It lies rather in the creation of tools, standards, and critical self-awareness in a discipline. The *Introduction* was foremost among these tools. In form somewhat like an "inspired dictionary" (as Singer was to label it), the *Introduction* deals with the emergence and growth of positive knowledge by means of contemporaneous surveys across all disciplines, races, and cultures, and by systematic division into half-century time units. Critical bibliography was another essential basis of the work. Hence the deliberate cross-linking of information in the *Introduction* with that contained in *Isis* (and its later occasional fellow-journal *Osiris*). As Sarton himself expressed it in retrospective summary:

The materials contained in the *Introduction, Isis* and *Osiris* are integrated by means of thousands of cross references. Thus we may say that volume 1 [of the *Introduction*] was built on a foundation of 8 volumes; volume 2, on a foundation of 15; volume 3 on a foundation of 42. . . . the *Introduction* is the most elaborate work of its kind, and by far, in world literature. This statement can be made without falling under the suspicion of boasting, for it is objective, controllable, and obviously correct.

Besides such heroic achievement, Sarton's many other publications appear almost lighthearted. In his later years, especially, he occupied named lectureships at a number of American universities. From them came such studies as *The History of Science and the New Humanism* (New York, 1931) and *Appreciation of Ancient and Medieval Science During the Renaissance* (Philadelphia, 1955). The same desire to make his message known and his discipline accessible lay behind a bibliographic venture like *Horus: A Guide to the History of Science* (Waltham, Massachusetts, 1952). His last years were largely devoted to work on the projected eight-volume *History of Science,* from antiquity to the present, which was to emerge from his lectures at Harvard from 1916 to 1951. He lived to complete two volumes: *Ancient Science Through the Golden Age of Greece* (Cambridge, Massachusetts, 1952) and the posthumously published *Hellenistic Science and Culture in the Last Three Centuries B.C.* (Cambridge, Massachusetts, 1959).

In what amounted to his last testament, Sarton continued to express his faith that "the only road to intellectual progress was scientific research." Yet paradoxically, as various historians of science in antiquity observed in their reviews, these readable volumes, rich in factual detail and lacking in the synthesis of ideas, were primarily devoted to cultural history rather than narrowly scientific development.

George Sarton came to epitomize the history of science to both European and American audiences. In the years immediately after World War II, when the first small cluster of teaching positions in the discipline were appearing in the United States, he automatically served as the central reference point—the continuing propagandist for, and the ideal type of, the historian of science as researcher, scholar, and teacher. The international and the intensely personal aspects of his achievement also won increasing recognition. He was honored with a rich variety of awards, including the Charles Homer Haskins Medal of the Mediae-

val Academy of America (1949) and the George Sarton Medal of the History of Science Society (1955). In 1934 he was elected to the American Philosophical Society.

Sarton's immediate influence in the postwar years was that of the catalyst rather than that of the reactant. He had only a limited intellectual impact on the discipline he did so much to create. His holistic concerns, his ambition for full comprehension, and his emphasis on the moral virtues of historical inquiry all ran counter to the preference for depth, particularity of pertinent detail, and detached analysis that have increasingly characterized historical scholarship (not just the history of science) over the last half-century. The positivistic cast of his thought ("I have tried to name the people who were first to do this or that; to take the first step in the right direction. . . .") and his belief in the uniqueness of science were antithetical to the idealistic, intuitionistic, and relativistic currents so powerful in recent Western thought. Finally, his emphasis on historical approaches through biography and bibliography, necessary and useful as they are, could not capture the imagination of scholars or provide a powerful technique of analysis around which a research school could form.

Instead Sarton opened the way for Koyré to have a major impact on the first generation of American "professional" historians of science, when shifts in the larger society created a demand for university teachers of this new discipline in the later 1950's and early 1960's. Equipped with a growing range of reference aids and a profound sense of the importance of their task, they could not find in Sarton's work coherent general ideas or theoretically derived, finite problems and techniques of investigation.

Sarton's influence on the discipline he labored so faithfully to create has thus been further obscured. His monument is also of its nature becoming progressively less visible to newcomers to the field. In founding a journal, in emphasizing critical bibliography, in essaying broad surveys, and above all in writing the *Introduction,* he was creating elements required by the discipline, not methods to be emulated or finished products for display. Yet as all those who turn to his work are aware, he also possessed, preached, and put into practice a range of insights that reveal a mind of remarkable range, catholicity, and tenacity. And of course his presence at Harvard was crucial to the later creation and legitimation of a department that is now one of the world's major centers of the history of science.

The limits of Sarton's influence on the history of

science reveal by default how the cognitive identity of a discipline is a matter of theoretical orientation and worldview as well as tools and techniques. His inability to engineer the careers and train the disciples who would create a professional identity for his subject also demonstrates how much this latter aspect of discipline-building depends on factors beyond the control of any individual. The history of science is now a firmly institutionalized field of learning. At first glance it shows little trace of Sarton's influence. Yet he not only created and assembled the necessary building materials through heroic feats of labor, but he also saw himself as—and he was—the first deliberate architect of the history of science as an independent and organized discipline.

BIBLIOGRAPHY

Sarton wrote 15 books and over 300 articles and notes, besides editing *Isis* for almost 40 years and personally producing 79 critical bibliographies of the history of science (containing perhaps 100,000 of his brief analyses). For a complete bibliography of his publications, with a number of illuminating essays by colleagues, pupils, and friends, see the memorial issue of *Isis*, **49** (1957). There are many other obituary notices in scholarly journals.

More immediately helpful as a source for the study of Sarton's achievements is his daughter's delightful "Sketches for an Autobiography," in May Sarton, *I Knew a Phoenix* (New York, 1969). She has also written "An Informal Portrait of George Sarton," in *Texas Quarterly,* **5** (1962), 101–112. For a sampling of Sarton's varied writings, see Dorothy Stimson, ed., *Sarton on the History of Science* (Cambridge, Mass., 1962). The most recent study of his work is Arnold Thackray and Robert K. Merton, "On Discipline-Building: the Paradoxes of George Sarton," in *Isis,* **63** (1972), 673–695. Sarton's university education is treated by J. Gillis, "Paul Mansion en George Sarton," in *Mededelingen van der Koninklijke Academie . . . van België, Klasse der Wetenschappen,* **35** (1973), 3–21.

Some of Sarton's correspondence now preserved in Belgium is in Paul Van Oye, *George Sarton, de Mens en zijn Werk uit Brieven aan Vrienden en Kennissen* (Brussels, 1965). Most of the letters are in French or English and date from his youth. The commentary is in Flemish. Further letters of this period (in French) are in Suzanne Delorme, "La naissance d'*Isis*," in *Actes du XIᵉ Congrès international d'histoire des sciences,* **2** (Warsaw, 1967), 223–232. Lynn Thorndike's contribution to the memorial issue of *Isis* also includes some of his and Sarton's correspondence over the years.

Much Sarton correspondence undoubtedly exists in other private archives. Several collections of his letters are publicly available. Somewhere between 20,000 and 30,000 letters from 2,108 and to 788 correspondents are preserved and indexed in the Houghton Library of Harvard University. Small but important collections are at California Institute of Technology (George Ellery Hale papers) and Columbia University (David Eugene Smith papers). An unknown number of letters (perhaps 500 to 1,000) to and from the president of the Carnegie Institution of Washington are also available in an uncatalogued state in the Institution library. Sarton's work for and in relation to the Carnegie Institution over the years may be followed in his reports, published annually in the Institution *Yearbook* from 1919 to 1949.

Arnold Thackray
Robert K. Merton

SARYCHEV, GAVRIIL ANDREEVICH (*b.* St. Petersburg, Russia, 1763; *d.* St. Petersburg, 11 August 1831), *hydrography.*

In 1775 Sarychev entered the Naval Cadet Corps in Kronstadt, and was commissioned in 1781. From 1785 to 1793 he and Joseph Billings participated in an extensive surveying expedition to northeastern Siberia, the Aleutian Islands, and the shores of North America, the most important research on the northern Pacific and its shores by Russian scientists since the Bering-Chirikov expedition. The results of the expedition were described in *Puteshestvie flota kapitana Sarycheva . . .* ("The Journey of the Fleet of Captain Sarychev . . .," 1802) and in Billings' account of his trip along the Chukchi Peninsula (1811).

Beginning in 1802 Sarychev spent many years on surveying expeditions to the Baltic Sea. An atlas (1812) and *Lotsia . . .* ("Sailing Directions," 1817) for this sea were published under his direction. His handbook on surveying shores and on compiling marine charts (1804) played an important role in the development of methods of marine surveying from sail- and oar-powered vessels. Sarychev's methods of describing ocean shores, compiling charts, and writing up the results of expeditions served as models for many researchers. In 1808, as hydrographer-general, he directed the hydrographical service in Russia; and in 1829 he became an admiral.

Sarychev initiated and led many hydrographic expeditions during the first third of the nineteenth century and was teacher and guide of a large group of Russian seafaring scientists (F. P. Litke, M. F. Reyneke, A. E. Kolodkin, E. P. Manganeri, P. F. Anjou, F. P. Wrangel). No expedition was sent without his knowledge and supervision, either to inland seas or to the Far East and America. His great achievement was *Atlas severnoy chasti Vostochnogo okeana* ("Atlas of the Northern Part of the Eastern Ocean," 1826), compiled under his direction. Sarychev was elected an honorary member of the Academy of Sciences and of many university and scientific societies.

BIBLIOGRAPHY

I. ORIGINAL WORKS: Sarychev's writings include *Puteshestvie flota kapitana Sarycheva po severo-vostochnoy chasti Sibiri, Ledovitomu moryu i Vostochnomu okeanu. . .* ("The Journey of the Fleet of Captain Sarychev Through the Northeastern part of Siberia, the Arctic Sea and the Eastern Ocean . . ."), 2 vols. (St. Petersburg, 1802; 2nd ed., Moscow, 1952), with atlas and drawings; *Pravila, prinadlezhashchie k Morskoy Geodezii . . .* ("Rules Appropriate for Ocean Geodesy . . ."; St. Petersburg, 1804; 2nd ed., 1825); *Dnevnye zapiski plavania . . . Gavrily Sarycheva po Baltyskomu moryu i Finskomu zalivu v 1802, 1803, 1804 godakh. . . .* ("Daily Notes on the Voyage of . . . Sarychev on the Baltic Sea and the Gulf of Finland in 1802, 1803, and 1804. . . ."; St. Petersburg, 1808); *Puteshestvie kapitana Billingsa cherez Chukotskuyu zemlyu ot Beringova proliva do Nizhnekolymskogo ostroga . . .* ("Voyage of Captain Billings Across the Chukchi Peninsula From the Bering Strait to the Lower Kolyma Fortress . . ."; St. Petersburg, 1811); *Morskoy atlas vsego Baltyskogo morya s Finskim zalivom i Kattegotom . . .* ("Marine Atlas of the Entire Baltic Sea With the Gulf of Finland and Kattegat . . ."; St. Petersburg, 1812); *Lotsia ili Puteukazanie k bezopasnomu korablevozhdeniyu po Finskomu zalivu, Baltyskomu moryu i Kattegat* ("Sailing Directions or Voyage Instructions for Safely Piloting Ships in the Gulf of Finland, the Baltic Sea and Kattegat"; St. Petersburg, 1817); and *Atlas severnoy chasti Vostochnogo okeana . . .* ("Atlas of the Northern Part of the Eastern Ocean . . ."; St. Petersburg, 1826).

II. SECONDARY LITERATURE. See the following, listed chronologically: A. P. Sokolov, *Russkaya morskaya biblioteka. 1701–1851* ("Russian Marine Library. 1701–1851"), 2nd ed. (1883), which contains a biography of Sarychev and annotations to his work; N. N. Zubov, *Otechestvennye moreplavateli-issledovateli morey i okeanov* ("Native Sailors and Investigators of the Seas and Oceans"; Moscow, 1954), 123–131; M. I. Belov, *Arkticheskoe moreplavanie s drevneyshikh vremen do serediny XIX veka* ("Arctic Sea Voyages From Ancient Times to the Middle of the Nineteenth Century"; Moscow, 1956); E. E. Shvede; "G. A. Sarychev," in *Otechestvennye fiziko-geografy i puteshestvenniki* ("Native Physical Geographers and Travelers"; Moscow, 1959), 116–125; V. A. Esakov, A. F. Plakhotnik, and A. I. Alekseev, *Russkie okeanicheskie i morskie issledovania v XIX – nachale XX v* ("Russian Oceanic and Marine Research in the Nineteenth and Beginning of the Twentieth Centuries"; Moscow, 1964); and A. I. Alekseev, *Gavriil Andreevich Sarychev* (Moscow, 1966).

VASILIY A. ESAKOV

ŚATĀNANDA (*fl.* India, 1099), *astronomy.*

The only certain biographical data concerning Śatānanda are that he was the son of Śaṅkara and Sarasvatī and that he wrote the *Bhāsvatī* in 1099 on the basis of Varāhamihira's (*fl. ca.* 505) summary in the *Pañcasiddhāntikā* (I, 14; IX; X; XI [?]; and XVI) of Lāṭadeva's (*fl.* 505) recension of the *Sūryasiddhānta* according to the Ārdharātrikapakṣa (see essay in Supplement) of Āryabhaṭa I (*b.* 473). The last verse of the *Bhāsvatī*, it is true, refers to the divine utterance of Puruṣottama, and some have mistakenly concluded therefrom that Śatānanda lived at Puruṣottama or Puri, in Orissa; but the only Indian locality referred to by Śatānanda is Ujjayinī. In any case, the *Bhāsvatī*, containing only eighty-one verses, was instrumental in spreading this version of the *Sūryasiddhānta* throughout northern and, especially, eastern India, as can be seen from the existence of numerous manuscripts (nearly a hundred), commentaries, and editions. The commentators include the following:

1. Aniruddha (*b.* 1463), who wrote a *ṭīkā* in Benares in 1495 (see D. Pingree, *Census of the Exact Sciences in Sanskrit*, ser. A, I [Philadelphia, 1970], 43b).

2. Acyuta (*fl.* 1505–1534), who wrote a *Bhāsvatīratnadīpikā* in Bengal (see D. Pingree, *op. cit.,* I, 36a–36b).

3. Gaṇapati Bhaṭṭa (*fl.* 1512), who wrote a *vivṛti* in Bengal (see D. Pingree, *op. cit.,* vol. II [Philadelphia, 1971], 89a).

4. Mādhava Miśra (*fl.* 1525), who wrote a *vivaraṇa* at Kanauj.

5. Balabhadra (*b.* 1495), who wrote a *Bālabodhinī* in 1543 in Bengal.

6. Kuvera Miśra (*fl.* 1685), who wrote a *ṭīkā*, probably in Bengal (see D. Pingree, *op. cit.,* II, 47b).

7. Gaṅgādhara (*fl.* 1685), who wrote a *ṭīkā*, probably in Rājasthān (see D. Pingree, *op. cit.,* II, 85a).

8. Rāmakṛṣṇa (*fl.* 1738), who wrote a *Tattvaprakāśikā*.

9. Kamalanayana (*fl.* 1740?), who wrote an *udāharaṇa* in Mithilā (see D. Pingree, *op. cit.,* II, 20a).

10. Yogīndra (*fl.* 1742), who wrote an *udāharaṇa*, probably in Mithilā.

Those commentators who cannot be dated include Cakravartin *(Bhāsvatīpaddhati);* Cakravi-pradāsa *(ṭīkā);* Dharmāditya *(Bhāsvatītilaka);* Gopāla *(vivaraṇa;* see D. Pingree, *op. cit.,* II, 130b); Gopīnātha Sudhī *(Bhāsvatīprakāśikā;* see D. Pingree, *op. cit.,* II, 132b); Govinda Miśra *(ṭīkā;* see D. Pingree, *op. cit.,* II, 137a); Keśava (?) *(udāharaṇa;* see D. Pingree, *op. cit.,* II, 64a); Madhusūdana *(Subodhinī);* Rāmeśvara *(ṭīkā);* Vanamālin (Hindī *ṭīkā);* Viśveśvaranātha *(vyāk-hyā);* and Vṛndāvana *(udāharaṇa).*

BIBLIOGRAPHY

The *Bhāsvatī* has been published several times: at the Akhavāra Press (Benares, 1854); with *vivaraṇa* of Mād-hava Miśra in *Aruṇodaya,* I (1890–1891); with his own Sanskrit *ṭīkā, Chātrabodhinī,* and his own Hindī explanation by Mātṛprasāda Pāṇḍeya (Benares, 1917); and with his own *ṭīkā, Manoramā,* and several appendices by Ṭīkārāma Dhanañjaya (Vārāṇasī, N.D.).

There are short biographical notices on Śatānanda in S. Dvivedin, *Gaṇakataraṅgiṇī* (Benares, 1933), 33–34, repr. from *The Pandit,* n.s. **14** (1892); and in S. B. Dīkṣita, *Bhāratīya Jyotiḥśāstra* (Poona, 1896, repr. Poona, 1931), 243–245.

DAVID PINGREE

SATŌ NOBUHIRO (*b.* Ugo [now Akita] prefecture, Japan, 1769; *d.* Edo [now Tokyo], Japan, 6 January 1850); **SATŌ NOBUKAGE** (*b.* Nishimo-nai, 1674; *d.* Nishimonai, 1732); **SATŌ NOBUSUE** (*b.* 1724; *d.* 1784), *mining, agriculture, economics.*

The Satō family served the feudal lords of Ugo (now Akita) prefecture as physicians. Of the five generations so employed, there is little information about the first two, Satō I, Nobukuni (whose pen name was Kan'an) and Satō II, Nobutaka (Gen'an), save that they were father and son. Of Satō No-butaka's son Satō Nobukage (Fumaiken) it is known that while originally a physician, he studied agricultural administration, natural history, and natural science in order to find a means of helping farmers who had experienced crop failures. He spent the years 1688 to 1703 on the island of Yes-so (now Hokkaido) and drew upon his experiences there to compose the twelve-volume *New Theory of Developing the Country.* He also wrote a five-volume work entitled *Features of the Soil,* and a shorter, two-volume book called *Secrets of the Mountain Phase* and managed the Matsuoka mine in Ugo with notable success.

Satō IV, Nobusue (Genmeika), was, like his fa-ther Satō Nobukage, a physician. He also studied agricultural management and economics and wrote a work entitled *How to Preserve Fishermen's Villages* and another, in four volumes, called *Illustration of the Secret of the Mountain Phase.* His son, Satō V, was Satō Nobuhiro (Yūsai), who in 1782 accompanied him to Yesso, where he remained for a year. Satō Nobuhiro was present when his father died at the Ashio mine; he obeyed his last wish, and went to Edo to study science under Genzui Udagawa and Gentaku Ōtsuki. He there learned astronomy, mensuration, geography, and surveying. In 1787 he traveled to Kyushu and through western Honshu. In 1808 he studied literature with Atsu-tane Hirata and about 1839 he became closely associated with the leading members of Bansha, the Association of Foreign Learning. These includ-ed Watanabe Kazan, a painter and minister of the Tawara clan, and the physician Takano Chōei; Satō Nobuhiro was imprisoned with them in the same year, when Bansha was suppressed through the efforts of conservative scholars, but he was soon released. During the Tempo reformation (1841–1843) he acted as adviser to prime minister Mizuno Tadakuni.

The accomplishments of the elder Satōs can be understood only in the context of the secrecy in which science and technology were held in feudal Japan. Knowledge was jealously guarded and, for the most part, passed on orally from father to son. Even when a formal school existed (and it may be assumed that the works of the earlier Satōs were composed for such a school), its members were sworn to hold what they had learned in confidence. Even Satō Nobuhiro notes this tradition in his book *Laws of the Mine,* reminding his patron that "Al-though the laws of the mine are essentially the most precious secret of our family, I will transmit them to you, not revealing that fact even to the staff of my school, since I was impressed by your enthusiasm and patriotism. In consequence, you must not show this work even to your parents or your brother."

The knowledge accumulated by the Satō family and published by Satō Nobuhiro was concerned with politics, agricultural administration, eco-nomics, natural history, natural science, mining techniques, metallurgy, the exploration of ore de-posits, mine management, geography, education, law, and military science. Satō Nobukage and Satō Nobusue both wished to alleviate the lot of the farmers, which they had observed in their travels; Satō Nobuhiro, in addition, had in his youth expe-rienced the severe famine of the Temmei era (in

1782) and later observed the great famine of the Tempo period (in 1836). Since all the Satōs served a feudal lord, their concern with agricultural subjects was doubly legitimate.

The deep interest of the Satō family in mining was in part the result of the changing economic conditions of Japan in the seventeenth and eighteenth centuries. As trade developed rapidly, a merchant class rose to power and the warrior aristocracy, the samurai class, correspondingly declined. As Japan gradually began to emerge from feudalism, it became necessary for the feudal lords to find an economic basis other than agriculture, and some of them began to develop a mining industry; by the beginning of the eighteenth century the Ugo district, the seat of the Satōs, produced more copper than any other part of Japan. The interest of the Satōs in mining was therefore predictable.

Indeed, Satō Nobuhiro's best-known work concerns mining. Part of his *Secret of the Mountain Phase* is devoted to describing a method for predicting the presence of ores. The method given was, however, an unscientific one, based on the shape of the mountain that might contain the ores, together with the characteristics of the "spirit," or moisture, that evaporates from the ore body. A large part of the rest of the book describes methods for refining gold, silver, copper, lead, tin, iron, mercury, and sulfur ores, and discusses the management of mines, with particular emphasis on the well-being of miners. Satō Nobuhiro set out a system for the division of the operations of a mine into departments, and considered the role of each department, as well as its physical location at the site of a mine. His discussion includes the daily supply of food and other necessary goods, as well as the need for a recreation area for the miners and the part to be played by religion in their lives.

Satō Nobuhiro also presented a system for the management of the whole civil state in his *Elements of Economics* and *Government and Reactionism*. He designed an authoritarian, ideal state in which the class distinctions between the samurai, farmers, manufacturers, merchants, and peasants would be abolished and all Japan would be united under a single ruler. All land and all the means for production would be owned by the state, and the state would administer all commerce and foreign trade. Satō Nobuhiro's plan also included a system of free education, up to and including the university level.

In drawing up his scheme of government, Satō Nobuhiro was influenced by western political science and science, taking from them in particular the notion of the equality of men. He was a popularizer of western thought (as a young man he had attempted to learn Dutch, since the Dutch were the only westerners permitted in Japan at that time) and wrote the first Japanese works on western science and history. He spent much of his later life composing such studies and died in retirement.

H. KOBAYASHI

SAURIN, JOSEPH (*b.* Courthézon, Vaucluse, France, 1 September 1659; *d.* Paris, France, 29 December 1737), *mathematics, mechanics, cosmology.*

The youngest son of Pierre Saurin, a Calvinist minister of Grenoble, Saurin was educated at home and in 1684 entered the ministry as curate of Eure. Outspoken in the pulpit, he soon had to take refuge in Switzerland, where he became pastor of Bercher, Yverdon. No less combative in exile, he refused at first to sign the *Consensus* of Geneva (1685). The pressure brought to bear on him as a result apparently weakened his Calvinist persuasion; after discussions with elders in Holland, he had an audience with Bishop Bossuet in France and shortly thereafter, on 21 September 1690, embraced Roman Catholicism. After an adventurous[1] return to Switzerland to fetch his wife, the daughter of a wealthy family named de Crouzas, Saurin settled in Paris for the rest of his life.

Forced to find a new career, Saurin turned to mathematics, which he studied and then taught. By 1702, as mathematics editor for the *Journal des sçavans*, he was again involved in dispute, most notably with Rolle, over the infinitesimal calculus. Failing to get a satisfactory response from Rolle, Saurin appealed to the Academy of Sciences, of which Rolle was a member. The Academy avoided a direct decision in favor of an outsider by naming Saurin an *élève géomètre* on 10 March 1707 and a full *pensionnaire géomètre* on 13 May 1707.

Even this rise to prominence could not keep Saurin out of trouble. Accused by the poet Jean-Baptiste Rousseau of having written libelous poems against him, Saurin spent six months in jail before an *arrêt* of Parlement (7 April 1712) exonerated him and sent Rousseau into exile. Thereafter, Saurin appears to have retired to his scientific research, working all night and sleeping all day.[2] His active career ended with his being named a *vétéran* of the Academy in 1731. He died of lethargic fever in 1737, leaving at least one son, Bernard-Joseph (1706–1781), who earned some fame as a dramatic poet.

Saurin made no original contributions to mathematics. Rather, firmly committed to the new infinitesimal calculus, he explored the limits and possibilities of its methods and defended it against criticism based on lack of understanding. Rolle, for example, assumed that the new method of tangents could not handle singularities of multivalued curves where dy/dx took the form 0/0. In reply (1702, 1703, 1716), Saurin explicated the nature and treatment of such indeterminate expressions on the basis of L'Hospital's theorem (*Analyse des infiniment petits* [1969], section 9, article 163), by which, for a $f(x) = g(x)/h(x)$ of the form 0/0 at $x = a$, one determines $f(a)$ by differentiating $g(x)$ and $h(x)$ simultaneously until one of them is nonzero at $x = a$. His further study of multivalued curves (1723, 1725) became the basis for correcting Guisnée's and Crousaz's misunderstanding of the nature of extreme values and of their expression in the new calculus.

Saurin's two papers (1709) on curves of quickest descent represent a solution of a problem first posed by Jakob I Bernoulli—to find which of the infinitely many cycloids linking a given point as origin to a given line is the curve of quickest descent from the point to the line—and then an extension of the problem to any family of similar curves. Saurin followed the differential methods of Johann I Bernoulli, although he studiously avoided taking a position in the brothers' famous quarrel.[3]

Combining his command of infinitesimal methods with a firm understanding of the new dynamics, Saurin offered (1722) a sensitive and subtle explanation of why the infinitesimal path of a simple pendulum must be approximated by the arc of a cycloid rather than by the chord subtending the arc of the circle. Thus he defended Huygens' theory of the pendulum against the attacks of Antoine Parent and the Chevalier de Liouville. Saurin had already provided (1703) rather neat algebraic demonstrations of Huygens' theorems on centrifugal force and the cycloidal path and had done an experimental and theoretical study (1720) of the damping and driving effects of the escapement and weight in a pendulum clock.

Huygens himself became the target of Saurin's rebuttals on the issue of Descartes's vortex theory of gravity. Saurin's first effort (1703)—to explain how a terrestrial vortex with lines of force parallel to the equatorial plane could cause bodies to fall toward the center of the earth—was patently clumsy. In 1709, however, to counter Huygens' objection that the necessarily greater speed of the vortex would sweep objects off the earth, Saurin proposed an attenuated ether that, on the basis of Mariotte's experimental findings on the force of moving fluids, made the ether all but nonresisting while still accounting for gravity by its greater speed of rotation. In Johann I Bernoulli's opinion, it was the best theory of gravity devised up to that time. Although, as Aiton points out,[4] it offered the chance for a reconciliation of Cartesian and Newtonian cosmology, Saurin himself felt that Newtonianism threatened a return to "the ancient shadows of Peripateticism" (*Mémoires de mathématique et physique* [1709], 148).

NOTES

1. The biographical note in Didot speaks of an outstanding charge of theft, while other notices recount the dangers of religious persecution only.
2. Fontenelle, "Éloge de M. Saurin," 120, makes this point and then adds that Saurin had few friends.
3. Saurin did, however, sarcastically reject Johann's claims of priority over L'Hospital in the matter of indeterminate expressions; cf. his 1716 paper and Joseph E. Hofmann, *Geschichte der Mathematik*, III (Berlin, 1957), 11.
4. *Vortex Theory of Planetary Motion*, 176. On Bernoulli's judgment, *ibid.*, 188.

BIBLIOGRAPHY

I. ORIGINAL WORKS. Saurin's works include "Démonstration des théorèmes que M. Hu(y)gens a proposés dans son Traité de la Pendule sur la force centrifuge des corps mûs circulairement," in *Mémoires pour servir à l'histoire des sciences et des beaux-arts* (*Mémoires de Trevoux*), 1702 (Addition pour . . . Novembre et Decembre), 27–60; "Réponse à l'écrit de M. Rolle de l'Académie Royale des Sciences inséré dans le Journal du 13. Avril 1702, sous le titre de Règles et Remarques pour le Problème général des Tangentes par M. Saurin," in *Journal des sçavans* (Amsterdam ed.), **30** (3 Aug. 1702), 831–861; "Solution de la principale difficulté proposée par M. Hu(y)gens contre le système de M. Descartes, sur la cause de la pesanteur," *ibid.*, **31** (8 Jan. 1703), 36–47; "Remarques sur les courbes des deux premiers exemples proposés par M. Rolle dans le Journal du jeudi 13. Avril 1702," *ibid.* (15 Jan. 1703), 65–73, (22 Jan. 1703), 78–84; and "Manière aisée de démontrer l'égalité des temps dans les chutes d'un corps tombant par une cycloide de plus ou de moins haut, et de trouver le rapport du temps de la chute par la cycloide au temps de la chute perpendiculaire par son axe," *ibid.* (4 June 1703), 563–570.

In the *Mémoires de mathématique et physique* of the Paris Academy of Sciences, see "Solutions et analyses de quelques problèmes appartenants aux nouvelles méthodes" (1709), 26–33; "Examen d'une difficulté considérable proposée par M. Hu(y)ghens contre le système cartésien sur la cause de la pesanteur" (1709), 131–148; "Solution générale du problème, où parmi une infinité de courbes semblables décrites sur un plan verticale, et ayant un même axe et un même point d'origine, il s'agit de déterminer celle dont l'arc compris entre le point d'origine et une ligne donnée de position, est parcouru dans le plus court temps possible" (1709), 257–266, with addendum in 1710, pp. 208–214; "Remarques sur un cas singulier du problème général des tangentes" (1716), 59–79, 275–289; and "Problème" (1718), 89–92.

In the same journal, see "Démonstration d'une proposition avancée dans un des mémoires de 1709. Avec l'examen de quelques endroits de la *Recherche de la vérité*, qui se trouvent dans la dernière edition, et qui ont rapport à ce mémoire" (1718), 191–199; "Démonstration de l'impossibilité de la quadrature indéfinie du cercle. Avec une manière simple de trouver une suite de droites qui approchent de plus en plus d'un arc de cercle proposé, tant en dessus qu'en dessous" (1720), 15–19; "Remarques sur les horloges à pendule" (1720), 208–230; "Éclaircissement sur une difficulté proposé aux mathématiciens par M. le Chevalier de Liouville" (1722), 70–95; "Sur les figures inscrites et circonscrites au cercle" (1723), 10–11; "Dernières remarques sur un cas singulier du problème des tangentes" (1723), 222–250; "Observations sur la question des plus grandes et des plus petites quantités" (1725), 238–260; and "Recherches sur la rectification des baromètres" (1727), 282–296.

II. SECONDARY LITERATURE. Bernard Fontenelle's "Éloge de M. Saurin," in *Histoire de l'Académie royale des sciences . . .* (1737), 110–120, is the basis for the account in Joseph Bertrand's *L'Académie des sciences et les académiciens de 1666 à 1793* (Paris, 1869), 242–247. The entry in the Didot *Nouvelle biographie générale* provides some additional details from Swiss sources. Saurin earns only passing mention in histories of mathematics, but his vortex theory of gravity receives considerable attention from Eric J. Aiton in *The Vortex Theory of Planetary Motions* (London–New York, 1972), 172–176 and *passim*.

MICHAEL S. MAHONEY

SAUSSURE, HORACE BÉNÉDICT DE (*b.* near Geneva [Conches], 17 February 1740; *d.* Geneva, 22 January 1799), *geology, meteorology, botany, education.*

Saussure's ancestors emigrated to Geneva from France and Italy to escape the religious persecutions of the sixteenth century. (The name "Saulxures" is still to be found in several villages in Lorraine.) The family contained a number of writers and scientists. Saussure may have inherited his early interest in botany from his father, the agricultural author Nicolas de Saussure; from his mother, an invalid, he inherited an ability to endure hardship, a philosophical turn of mind, and a delicate constitution. He was sent to the *Collège* of Geneva when he was six and entered the university there when he was fourteen. During his early years he also took long walking trips in the vicinity of Geneva and in the Salève, the Voirons, and the Jura, and was strongly influenced by two naturalists, his uncle Charles Bonnet and the physician and botanist Albrecht von Haller.

Saussure completed a degree in philosophy in 1759 with a dissertation on the transmission of heat from the rays of the sun. In 1760 he made the first of a number of trips to Chamonix, on this occasion for the specific purpose of collecting plant specimens for Haller. He wrote a lyrical description of the mountains and glaciers of the area and climbed the Brévent with Pierre Simon, who served as his guide on a number of subsequent occasions. In addition, Saussure placed a notice in each of the surrounding parishes, offering a handsome reward to the first person to climb Mont Blanc.

In 1761 Saussure was a candidate for the chair of mathematics at the Academy of Geneva, but he was not elected; he therefore turned for consolation to the classical authors and continued his botanical investigations. His treatise *Observations sur l'écorce des feuilles et des pétales,* published in 1762, was dedicated to Haller. He presented this, together with a philosophical thesis entitled "Principal Causes of Errors Arising from the Qualities of the Mind" and a work "On Rainbows, Haloes, and Parhelia," when the chair of philosophy at the Academy became vacant in the same year. Haller strongly supported his candidacy, and he was elected. He lectured in French on physics and natural history and in Latin on metaphysics in alternate years.

Saussure made another trip to Chamonix in March 1764, and at about this time he began to concentrate on geology, although he did not abandon botany completely. In the following year he married Albertine Amélie Boissier, who brought him a considerable fortune and a beautiful house in Geneva. His passion for mountains (he had decided both to climb Mont Blanc and to take an annual alpine tour) nevertheless remained unabated and was to become a source of anxiety to his family. In July 1767 he completed a tour of Mont Blanc, where he carried out experiments on heat and cold, on the weight of the atmosphere, and on electricity and magnetism. In these he employed what was probably the first electrometer, an instrument that he himself had developed. He returned to Geneva for the birth of his son, Nicolas-Théodore, in October.

Saussure held liberal views that set him apart from most of his fellow patricians and professors, and in 1768 presented as a private citizen a plan for reform that advocated both a more democratic constitution and popular education. It was not accepted, and he thereupon decided to escape recurrent political crises between the aristocracy and the people of Geneva by immediately going on a "grand tour" with his wife. He spent some time in Paris, where he met Buffon (their dislike was mutual) and held discussions with a number of French geologists, including a long dialogue on the basalts of the Auvergne with Desmarest and Guettard. He also took every opportunity to attend plays. Saussure was in England by mid-summer of the same year, and visited coal, tin, and lead mines, as well as quarries, from the Midlands to Cornwall. In London he met with members of the Royal Society and discussed electricity with Benjamin Franklin. He and his wife returned to Geneva in February 1769, and Saussure refused any participation in local politics, although strongly sympathetic to the popular parties.

In the summer of 1771 Saussure made an expedition to study the lakes and flora of northern Italy; on his return trip through the Great St. Bernard he felt the first symptoms of a serious gastric illness, and by the fall of the following year his health had so declined that he was advised to spend the winter in a warmer climate. He therefore returned to Italy, where he investigated the iron mines of Elba, then proceeded to Rome and to Naples, where Sir William Hamilton acted as his guide on a visit to Vesuvius and the Campi Phlegraei. Saussure then went on to Sicily and on 5 June 1773 climbed Mt. Etna. He found his health improved and was enthusiastic about the attractions that Italy offered to the naturalist, although he published only a few papers on it instead of the larger work that he considered.

Returning to Geneva, Saussure was appointed rector of the university, in which post he served for two years, from 1774 until 1776. He took advantage of this position to present a series of proposals for the reform of the Geneva *Collège;* sensitive to the popular unrest that was enveloping all Europe, Saussure felt that its effects might be dissipated by more adequate public education. His proposals led to a violent controversy, with political overtones, particularly since he had drawn on some of the ideas set out in Rousseau's *Émile,* a work recently condemned at Geneva. Since his proposed reforms were not to be put into effect, Saussure tried another approach, founding the Société des Arts, in which different social classes and professions were to meet to apply the arts and sciences to industry. In 1776 Saussure served as the society's first president.

At the same time, Saussure also began to make the extensive alpine investigations that he described in the four volumes of his *Voyages dans les Alpes, précédés d'un essai sur l'histoire naturelle des environs de Genève,* published between 1779 and 1796. Although he resigned his professorship in January 1786, citing ill health, the years from 1774 through 1789 marked the period of his most strenuous alpine activity. This period also coincides with an increasing involvement of Saussure in local politics. Until 1781, his proposals for political and educational reforms were personal endeavors; after that date his sense of civic duty led him to accept numerous public offices. They ranged from his election to several councils of the patrician governing system before the revolution to membership in numerous committees involved in successive changes of the form of government in Geneva and in foreign affairs immediately before, during, and after the revolution, until his voluntary withdrawal from public life in 1794 after the famous Massacre of the Bastions.

In the summer of 1787 Saussure climbed Mont Blanc, a feat that had been accomplished for the first time only in the preceding August, by Jacques Balmat and Michel Gabriel Paccard. On his own ascent, Saussure was accompanied by eighteen guides and carried a number of scientific instruments, many of them of his own design and construction, with which he undertook appropriate experiments during the four and one-half hours that he spent at the summit. The great elevation also allowed him to observe directly a number of geological features about which he had previously only been able to speculate. The success of his expedition caught the popular imagination (at one point it was even suggested that the mountain be renamed in his honor) and won Saussure an international reputation. He was elected a fellow of the Royal Society of London in April 1788, under the sponsorship of Hamilton, and in 1791 he became one of the eight foreign members of the French Academy of Sciences.

Since Saussure had had only limited time on top of Mont Blanc, he sought further opportunities to perform high-altitude experiments and observations. In July 1788, with his son, Nicolas-Théodore, he camped for fifteen days at a base 11,000 feet high on the Col du Géant. Despite precarious conditions, the two men completed a full series of observations of the daily variations of winds and other meteorological phenomena. Saussure was also able to observe geological and topographical details during the hours of dawn and sunset, the conditions under which structure is most apparent.

In July 1789, again in the company of his son, Saussure set out to measure the height of Monte Rosa and to investigate its geological structure. He made a series of observations that confirmed his neptunist ideas about the formation of granite and was greatly impressed by the sight of the Matterhorn, which he decided to measure next. While he was on this expedition, Saussure received news of the fall of the Bastille. His letters show that he maintained his liberal attitudes—indeed, he had taken part in drafting legislation designed to alleviate in Geneva the unrest similar to that which produced events in France—but he was unfortunately too optimistic about the fate of the French securities in which most of his money was invested.

Saussure continued his scientific work as the French Revolution began to make itself felt in Geneva. In August 1792, during a short period of political calm, he returned with his son to the Matterhorn to measure its altitude. This was the last trip that he recorded in his *Voyages*. He returned to Geneva to learn that he had lost most of his fortune and that his wife's income had been drastically reduced. As the revolution spread to Geneva in the fall, many patrician families fled the country; Saussure remained, however, and participated in several committees that were attempting to draw up a new revolutionary constitution. In 1793 he took the voluntary oath to Liberty, Equality, and Fraternity.

Saussure once again took the opportunity to present a project for national public education, but his health had again begun to fail. He had a series of seizures that left him partially paralyzed in March 1794, yet he continued under the necessity of finding a position that would increase his income. In particular, Saussure hoped to find a professorship at a French university, perhaps combined with an inspectorship of mines, but no such position was offered him. He also considered the universities of Göttingen, Berlin, and St. Petersburg, and even an exile in the United States, where Thomas Jefferson was looking for a faculty for the new university he was organizing at Charlottesville, Virginia

(Jefferson had suggested to Washington that some of the professors at the University of Geneva, especially Saussure, might be glad to find refuge and employment in America). Nonetheless, no offer was actually made to him, and Saussure remained in Geneva until his death in 1799. By 1796, when the third and fourth volumes of his *Voyages* were published, largely under the direction of his son, he had been entirely incapacitated by a second stroke.

Although the *Voyages* are a record of Saussure's expeditions between 1774 and 1787, he did not set out the descriptions of his travels in chronological order. Rather, he abstracted and grouped the results of his journeys into three sections, of which one concerned his work in the area of Mont Blanc, another his trips through the Mont Cenis pass to the Italian and French Rivieras and his return through Provence, and the last his numerous expeditions to the area of the Gries, the St. Gotthard, and the Italian lakes. Subjects not subsumed under these general headings were given specific chapters, arranged chronologically. His book incorporated the observations that Saussure made on the spot, which he recorded instantly, then wrote up in a more finished draft within twenty-four hours. The work also reflects his method of exploration, whereby he returned to an area several times in order to complete and verify his observations.

The *Voyages* demonstrate Saussure's approach toward a theory of the earth, to which end he collected data tirelessly. He had arrived at a tentative theory as early as 1774, when he had investigated only a few alpine passes. In the same year he delivered a lecture on mountain structure; although the lecture itself is not preserved, an abstract of it appears in the *Voyages*.

As Saussure originally conceived of the structure of the Alps, the primitive and central chain of the mountains consist of vertical strata, while the marginal or secondary mountains adjacent to the primitive mountains consist of steeply inclined beds. These progressively approach the horizontal as they reach the margins of the chain, where they are partially surrounded by tertiary mountains that are composed of the debris of all earlier deposits. It was therefore apparent to Saussure that both the primitive and secondary mountains of the chain must consist of distinct strata that display a transition in both their composition and their structure.

Saussure had accepted the notion that the secondary mountains had been formed on the bottom of the sea; he therefore drew upon the similarities between the primitive and secondary mountains to argue that the primitive mountains were also deposited on the ocean floor. The earth must therefore have been covered by a universal sea, which through crystallizations and successive deposits generated first the primitive mountains, then, around them in a series of concentric shells, the secondary ranges. The fire—or other "elastic fluids"—enclosed within the earth later lifted and ruptured the entire crust of the earth; in this process, the primitive mountains were lifted up, while the secondary ones became tilted against them. The waters of the universal ocean then rushed into the fissures left along the margins of the Alps, transporting for great distances huge boulders that were eventually scattered across the plains. Following the retreat of the waters, the germs of plants and animals, made fecund by their exposure to the air, began to develop on the newly exposed ground and in the residual water that remained in depressions of the earth.

Although Saussure stated specifically that he had arrived at his ideas without giving any thought to any particular system, it is easy to recognize in his approach the neptunist ideas associated with the contemporary school of Werner, as well as a plutonist idea of vertical movements as the result of internal fire of Cartesian origin. The concept of the universal waters rushing away from the Alps prevented Saussure from understanding the systems of frontal moraines that he observed on both sides of the mountains, although he had actually seen the recent fluctuations of glaciers in the valley of Chamonix.

After some thirteen years of accurate observations and thinking, and particularly after having studied the St. Gotthard area, Saussure concluded that the dislocation, distortion, and even overturning of the alpine rocks had been caused by processes of horizontal compression, as well as by uplifting by internal explosions. He thus came close to an accurate understanding of the structure of the Alps.

Despite his early formulation of a theory of the earth, and despite his further work toward emending this theory, Saussure never presented his material in a final, synthetic form. Why he did not do so remains a matter of some controversy. Although Cuvier, in his *Éloge* of Saussure, suggested that Saussure deliberately refrained from developing any theory, an analysis of the *Voyages* indicates that, on the contrary, Saussure fully intended to write such a synthesis until the time that he became physically disabled. It must not be forgotten that Saussure became incapacitated when he was

only fifty-four; having completed his travels, he must have anticipated ten or twenty years in which he could put his data into order. Since his illness did not permit him to do so, he substituted, in the last volume of the *Voyages,* an "Agenda or General Compendium of the Results of the Observations and Investigations Which Are to Serve as a Basis for the Theory of the Earth," which he characterized as a list of the problems to be solved by his followers. He thereby placed in the hands of his successors the fruits of his thirty-six years of travel and work, and left to them the task of reaching a general theory.

Saussure envisioned a theory of the earth that was to be based on uniformitarian principles, in which the present state of the earth would be carefully studied both to elucidate its past history and to predict its future. He proposed that such a theory should begin with a thorough review of the results of all previous workers and a description of the character of the various rocks (including their fossils) that compose mountain ranges. A discussion of all the fundamental laws, both physical and chemical, that affect the atmospheric envelope of the earth and play a significant role in shaping its features should then follow; finally, the successive stages of the evolution of the crust of the earth through geological time should be demonstrated by relating observations to laws.

Whether or not Saussure himself could have constructed an adequate theory within this outline remains open to question. It seems as if he was perhaps too timid in drawing conclusions and in discarding the theories of others when they were in conflict with his own observations. It is also the case that the structure of the Alps, his chief concern, was simply too complicated to be understood in the light of the geological knowledge available to the eighteenth century. Saussure himself seems to have known this intuitively, since at the end of the *Voyages* he noted that the only constant feature of the Alps was their variety.

Nonetheless, Saussure's dedicated work was of great importance in the development of geology, since, among other things, it provided James Hutton with fundamental documentation. In addition, Saussure devised a number of useful instruments, among them a hair hygrometer that utilized a degreased human hair to measure humidity, and he also performed some experiments on the fusion of granites and porphyries that entitled him to be considered the first experimental petrologist. Finally, he popularized the very term "geology," which replaced "geognosy" in the 1770's and 1780's.

BIBLIOGRAPHY

I. ORIGINAL WORKS. Saussure's major work is *Voyages dans les Alpes, précédés d'un essai sur l'histoire naturelle des environs de Genève*, 4 vols. (Neuchâtel–Geneva, 1779–1796, and later eds.), repr. in facsimile (Bologna, 1969).

See also *Dissertatio physica de igne* (Geneva, 1758); *Observations sur l'écorce des feuilles et des pétales des plantes* (Geneva, 1762); "Description des effets du tonnerre, observés à Naples dans la maison de Mylord Tilney," in *Observations sur la physique*, **1** (1773), 442–450; *Projet de réforme pour le Collège de Genève* (Geneva, 1774); *Eclaircissemens sur le projet de réforme pour le collège de Genève* (Geneva, 1774); *Essais sur l'hygrométrie* (Neuchâtel, 1783); "Lettre à Son Excellence M. le Chevalier Hamilton . . . sur la géographie physique de l'Italie," in *Observations sur la physique*, **7** (1776), 19–38; "Lettre de M. de Saussure à M. Faujas de Saint-Fond," in Faujas de Saint-Fond, *Description des expériences de la machine aérostatique de M. M. de Montgolfier*, II (Paris, 1784), 112–127; "Lettre de M. de Saussure à M. l'Abbé J. A. Mongez le jeune, sur l'usage du chalumeau," in *Observations sur la physique*, **26** (1785), 409–413; *Relation abrégée d'un voyage à la cîme du Mont-Blanc* (Geneva, 1787); "Défense de l'hygromètre à cheveu," in *Observations sur la physique*, **32** (1788), 24–45, 98–107, and repr. separately (Geneva, 1788); "Description d'un cyanomètre ou d'un appareil destiné à mesurer la transparence de l'air," in *Memorie della R. Accademia delle scienze di Torino*, **4** (1788–1789), 409–424; and "Description d'un diaphanomètre ou d'un appareil destiné à mesurer l'intensité de la couleur bleue du ciel," *ibid.*, 425–453.

See further "De la constitution physique de l'Italie," in J. J. F. de Lalande, *Voyage en Italie . . .*, I (Geneva, 1790), 45–48; "Description de deux nouvelles espèces de trémelles douées d'un mouvement spontané," in *Observations sur la physique*, **37** (1790), 401–409; *Éloge historique de Charles Bonnet* (Geneva, 1793); *Rapport et projet de loi du Comité d'Instruction Publique. Lu à l'Assemblée Nationale le 9 août 1793 par les citoyens Dessaussure et Bourrit fils* (Geneva, 1793); "Sur les collines volcaniques du Brisgau," in *Journal de physique*, **1** (1794), 325–362; "Notice sur la mine de fer de Saint-George en Maurienne," in *Journal des mines*, **1**, no. 4 (1794), 56–61; "Agenda ou tableau général des observations et des recherches dont les résultats doivent servir de base à la théorie de la terre," *ibid.*, **4**, no. 20 (1796), 2–70, and repr. in last vol. of *Voyages* and separately as *Agenda du voyageur géologue tiré du 4ème volume des "Voyages dans les Alpes"* (Geneva, 1796); "Mémoire sur les variations de hauteur et de température de l'Arve," in *Journal de physique*, **4** (1798), 50–55; and *Voyages dans les Alpes, partie pittoresque des ouvrages de H. B. Saussure*, with intro. by A. Sayous (Geneva-Paris, 1834).

Saussure's letters to his wife have been published as *Lettres de H. B. Saussure à sa femme*, annotated by E. Galliard and H. F. Montagnier (Chambéry, 1937).

II. SECONDARY LITERATURE. On Saussure and his work, see Georges Cuvier, "Éloges historiques de Charles Bonnet et H. B. de Saussure," in *Recueil des éloges historiques lus dans les séances publiques de l'Institut Royal de France*, II (Strasbourg-Paris, 1819), 383–430; Douglas W. Freshfield and H. F. Montagnier, *The Life of Horace Bénédict de Saussure* (London, 1920); and Jean Senebier, *Mémoire historique sur la vie et les écrits de Horace Bénédict de Saussure, pour servir d'introduction à la lecture de ses ouvrages* (Geneva, 1801).

ALBERT V. CAROZZI

SAUSSURE, NICOLAS-THÉODORE DE (*b*. Geneva, Switzerland, 14 October 1767; *d*. Geneva, 18 April 1845), *chemistry, plant physiology.*

Saussure was the son of the scientist Horace-Bénédict de Saussure (1740–1799) and Albertine-Amélie Boissier. His father supervised his initial studies and Saussure aided his father in his research. During the famous ascent of Mont Blanc on 3 August 1787, Nicolas was assigned by his father to make all the meteorological and barometric observations. In 1788 Nicolas accompanied his father on the expedition to the Col du Géant. They remained for seventeen days and nights in the snow fields.

During other expeditions, Saussure made observations on the composition of the atmosphere, the density of the air, and the geodesic features of the area around Geneva. In July 1789 he climbed Monte Rosa, where he pursued his experiments on the weight of the air. Using new techniques, he corroborated, with great precision, the observations of Mariotte, which had given rise to the Boyle-Mariotte law.

At this time Saussure became passionately interested in chemistry and in plant physiology, and he accumulated original observations, particularly on the mineral nutrition of plants. He later published this work. When the Revolution broke out, Saussure left Geneva for England. He returned in 1802 to occupy a promised chair in plant physiology at the Geneva Academy. Instead, he was named honorary professor of mineralogy and geology, a title he held until 1835. Disappointed at not being able to teach plant chemistry, the subject that had absorbed his attention since his first publication in 1797, he requested a leave of absence several days after his nomination. He never gave a course at the Academy.

In 1797, in *Annales de Chimie*, Saussure published three remarkable articles on carbonic acid and its formation in plant tissues. These works, followed in 1800 by an important study of the role of soil in the development of plants, brought him the recognition of his fellow scientists.

Saussure published his *Recherches chimiques sur la végétation* in Paris in 1804. An immediate success, it was translated into German in 1805 by F. S. Voigt and went through many editions. This work, which took seven years to write, laid the foundations of a new science, phytochemistry. Saussure examined the chief active components of plants, their synthesis, and their decomposition. He specified the relationships between vegetation and the environment and here, too, did pioneering work in what became the fields of pedology and ecology.

Starting in 1808, Saussure published a series of important articles, most of them devoted to a rigorous analysis of biochemical reactions occurring in the plant cell. The first dealt with the phosphorus content of seeds (1808). Then came two works on the conversion of starch into sugars by the action of air and water (1814 and 1819). These were followed by a study of the oil stored in fruits (1820). Saussure also investigated the biochemical processes that take place during the maturation of fruits (1821) and flowers (1822). Later he turned to the chemistry of germination and was the first to note the influence of desiccation on several food grains (1826). He then examined the formation of sugar during the germination of wheat (1833) and compared germination to fermentation reactions (1833). Several of the publications in which he reported his analysis of fermentation were regarded highly by Pasteur.

Saussure next studied the action of fermentation on the oxygen and hydrogen of air (1839) and also alcoholic fermentation (1841). Toward the end of his life, he took up again his research on plant nutrition (1841). His last paper was on the germination of oilseeds (1842).

Saussure received many honors. He was named a corresponding member of the Institut de France (1805), member of the Conseil Représentatif de Genève (1814), and was a founding member of the Société Helvétique des Sciences Naturelles (1815). By 1825 he was an associate member of virtually all the great European academies, and in 1842 he was elected president of the Congrès Scientifique of Lyons. In 1837 A. P. de Candolle named a genus of composite flower *Saussurea* and a section of the genus *Theodora*.

BIBLIOGRAPHY

I. ORIGINAL WORKS. Saussure's works include "Essai sur cette question: La formation de l'acide carbonique est-elle essentielle à la végétation?" in *Annales de chimie*, **24** (1797), 135–149, 227–228, 336–337; *Recherches chimiques sur la végétation* (Paris, 1804); "Sur le phosphore que les graines fournissent par la distillation et sur la décomposition des phosphates alcalins par le carbone," in *Annales de chimie*, **65** (1808), 189–201; "Observations sur la décomposition de l'amidon à la température atmosphérique par l'action de l'air et de l'eau," *ibid.*, **11** (1819), 379–408; "Observations sur la combinaison de l'essence de citron avec l'acide muriatique et sur quelques substances huileuses," in *Archives des sciences physiques et naturelles*, **13** (1820), 20–42, 112–135; "De l'influence des fruits verts sur l'air avant leur maturité," in *Mémoires de la Société de physique et d'histoire naturelle de Genève*, **1** (1821), 245–287; "De l'action des fleurs sur l'air, et de leur chaleur propre," in *Annales de chimie et de physique*, **21** (1822), 279–304; "De l'influence du dessèchement sur la germination de plusieurs graines alimentaires," in *Mémoires de la Société de physique et d'histoire naturelle de Genève*, **3** (1826), 1–28; "De la formation du sucre dans la germination du froment," *ibid.*, **6** (1841), 239–256; "Faits relatifs à la fermentation vineuses," in *Archives des sciences physiques et naturelles*, **32** (1841), 180–256; and "Sur la nutrition des végétaux," *ibid.*, **36** (1841), 340–355.

II. SECONDARY LITERATURE. On Saussure and his work, see C. Borgeaud, *Histoire de l'Université de Genève*, II (Geneva, 1909), 83; J. Briquet, "Biographies des botanistes à Genève de 1500 à 1931," in *Bericht der Schweizerischen botanischen Gesellschaft*, **50** (1940), 425–428; M. Macaire, "Notice sur la vie et les écrits de Théodore de Saussure," in *Bibliothèque universelle de Genève, Nouvelle série*, **57** (1845).

P. E. PILET

SAUVAGEAU, CAMILLE-FRANÇOIS (*b.* Angers, Maine-et-Loire, France, 12 May 1861; *d.* Vitrac [near Sarlat], Dordogne, France, 5 August 1936), *botany*.

Sauvageau came from a family of landowners and lawyers that had been established for several centuries in Anjou. After completing his secondary education at a *collège libre* in Angers, he entered the University of Montpellier. He obtained his *licence ès sciences physiques* and *licence ès sciences naturelles* in 1884 and immediately became an assistant to Charles Flahault, who held the chair of botany at the university.

Having successfully passed the *agrégation*, in 1888 Sauvageau was named professor at the *lycée* in Bordeaux. He left this post the following year to prepare a dissertation at the Muséum National d'Histoire Naturelle in Paris. He worked in the laboratory of P. van Tieghem, who encouraged most of his students to investigate problems in plant anatomy. Thus it is not surprising that Sauvageau's thesis (1891) was entitled "Sur les feuilles de quelques Monocotylédones aquatiques." In this work Sauvageau emphasized marine species; and because his research had taken this direction, Flahault introduced him to Édouard Bornet. The latter started him on the study of marine algae, and Sauvageau soon gave up his anatomical research to concentrate on this group.

Shortly after earning his doctorate, Sauvageau was appointed *maître de conférences* at Lyons and then professor at the Faculty of Sciences of Dijon. In 1901 he became professor at the Faculty of Sciences of Bordeaux, a post that he held until his retirement in 1932. Sauvageau was married to Marie-Louise Michelot, by whom he had one daughter.

Sauvageau's scientific work brought him many honors. He was elected corresponding member of the Académie des Sciences (1918), and he was a member of numerous foreign scientific academies and societies.

In his initial research, Sauvageau studied the anatomy of aquatic monocotyledons, mycology, phytopathology, and, particularly, diseases of the grape. But the bulk of his work was devoted to marine algae and in fact was concentrated almost exclusively on the study of Phaeophyceae. The few papers that were not concerned with this group dealt with the marine flora of the Gulf of Gascony, the coloration of oysters by the blue diatom, gelose (agar), the commercial uses of marine algae, and the iodine-containing cells (*ioduques*) of Bonnemaisoniaceae.

Sauvageau's first research was concerned with Ectocarpaceae. He showed in particular the great variety of their reproductive organs (1892–1897). He observed the isogamous and anisogamous copulation of certain flagellate cells produced by these organs and also noted the frequency with which they develop parthenogenetically.

He then studied Myrionemaceae (1898) and Cutleriaceae (1899). In the course of this research he confirmed and extended the observations of Johannes Reinke and Paul Falkenberg on the heteromorphic life-cycle of Cutleriaceae and demonstrated that the alternation of generations of *Cutleria-Aglaozonia* is not absolutely constant.

In "Remarques sur les Sphacelariacées" (1900–

1914), a series of papers totaling more than six hundred pages, Sauvageau described the complex anatomical structure of the various genera of this family, discussed their reproduction and development, and established a classification for them. Also, he published works on the various forms of *Fucus* and on the *Cystoseira* of the Mediterranean and the Atlantic (1913); these works are model systematic and ecological studies.

It was known that some certain Phaeophyceae produce only unilocular organs and appear to be deprived of sexuality because their zoospores are incapable of copulation. This condition is the case of the Laminariaceae, whose mode of reproduction was still unknown as late as 1915. In that year Sauvageau announced that the zoospores of *Saccorhiza bulbosa* give rise to filamentous microscopic plantlets. He further stated that some of these plantlets produce antherozoids while others produce oospheres, and he concluded that the plantlets are actually forms of prothallia. He then extended his investigations to other Laminariaceae found on the coasts of France and showed that this family, type of the order of Laminariales, is characterized by a type of alternation of generations that was previously unknown among the Phaeophyceae but is comparable to that encountered among the ferns.

Sauvageau also showed that this type of alternation of generations is exhibited by other Phaeophyceae, namely, *Dictyosiphon* (in which the prothallium produces isogamous gametes) and *Carpomitra* (in which the female prothallium is regularly apogamic). These two genera became the model types of two new orders: Dictyosiphonales and Sporochnales.

Sauvageau showed that among other types of Phaeophyceae, the discoid or filamentous microscopic plantlets emerging from the developing zoospores are not sexed. Thus he concluded that this stage of development, which he called adelophyceae in contrast to that of macroscopic plants (delophyceae), is not that of a prothallium. Instead, this stage produces numerous zoospores that, if placed in culture, are capable of reproducing several generations of microscopic plantlets before reaching (under favorable conditions) the delophycean stage. Sauvageau gave the name *plethysmothalli* to these microscopic plantlets, which have no analogues among the other plants.

Sauvageau's discoveries, which were made with simple culture techniques, greatly elucidated the extremely complex cycles of the brown algae.

More concerned with facts than with theories, Sauvageau did not attempt to derive general conclusions from his many discoveries. Yet, these discoveries form the basis of all our present knowledge concerning Phaeophyceae—its classification and the evolution of its reproductive cycles.

BIBLIOGRAPHY

I. Original Works. Among Sauvageau's works are "Remarques sur la reproduction des Phéosporées et en particulier des *Ectocarpus*," in *Annales des sciences naturelles,* 8th ser., **2** (1896); "Sur quelques Myrionémacées (premier mémoire)," *ibid.,* **5** (1898); "Les Cutlériacées et leur alternance de générations," *ibid.,* **10** (1899); 265–362; *Remarques sur les Sphacélariacées,* published in parts in *Journal de botanique,* **14–18** (1900–1904) and separately (Bordeaux, 1904, 1914); "A propos de *Cystoseira* de Banyuls at de Guethary," *Bulletin de la Station biologique d'Arcachon,* **14** (1912); "Recherches sur les Laminaires des côtes de France," *Mémoires de l'Académie des sciences,* **56** (1918); and *Utilisation des Algues marines* (Paris, 1920). A large number of papers was published in *Bulletin de la Station biologique d'Arcachon* from 1905 to 1936.

II. Secondary Literature. See the obituary by P. Dangeard in *Bulletin. Société botanique de France,* **84** (1937), 13–18; and *Bulletin de la Station biologique d'Arcachon,* **34** (1937), 5–59; and J. Feldmann, "L'oeuvre de C. Sauvageau (1892–1936)," in *Histoire de la botanique en France* (Paris–Nice, 1954), 212–217.

Jean Feldmann

SAUVEUR, ALBERT (*b.* Louvain, Belgium, 21 June 1863; *d.* Cambridge, Massachusetts, 26 January 1939), *metallurgy, metallography.*

Sauveur was the son of Lambert and Hortense Franquin Sauveur. He was educated at the Athénée Royale, Brussels (where his father was *préfet*), and at the École de Mines in Liège (1881–1886). Sauveur came to the United States in 1887 and enrolled as an advanced student at the Massachusetts Institute of Technology, where he graduated in 1889 with a thesis on copper smelting. For a year and a half he worked for the Pennsylvania Steel Company at an undemanding job as an analyst, which left him time to study the metallurgy of steel on his own.

In 1891 Sauveur married Mary Prince Jones, by whom he had three children. In the same year he went to the Illinois Steel Company in order to set up a laboratory, and he became the first in the United States effectively to study the microscopy of steel. Sauveur showed how grain-size was affected by mechanical and thermal treatment, and how, in turn, the properties of steel depend on grain-size. In an 1896 paper Sauveur critically summarized all the diverse current theories of steel-hardening mechanisms. He attributed the hardness of quenched steel to a fine dispersion of carbide (Fe_3C) particles. This paper attracted international notice and led to a debate at a meeting of the American Institute of Mining Engineers that occupied a hundred printed pages (*Transactions,* **27** [1898], 846–944). Using quantitative methods for the first time in metallography, he related the grain-size and the volume-fraction of the principal microconstituents in steel (ferrite, cementite, austenite and its decomposition products, martensite and pearlite) to the carbon content and to the temperature of quenching. In the same year, the laboratory of the steel works was disbanded; Sauveur returned to Massachusetts, where he established a commercial testing laboratory and began a consulting practice that continued throughout his life.

In 1898 Sauveur founded a quarterly journal, *Metallographist,* which he edited until its failure in 1906; in 1904 the title was changed to the *Iron and Steel Magazine.* Published in the most active period of metallographic discovery, the journal is a prime (if not always a primary) source for metallurgical history and contains biographies by Sauveur of all the principal workers in metallography of the time. In association with H. M. Boylston, Sauveur started a correspondence course in metallography, which was followed by 1,500 students and did much to spread knowledge of the new techniques throughout American industry. Sauveur joined the Harvard faculty as instructor in metallurgy in 1899, became professor in 1905, and remained there until his death.

With his lucid book, *The Metallography and Heat Treatment of Iron and Steel* (1912), and as an urbane man of the world with wide international contacts in both scientific and industrial spheres, Sauveur had a great influence on metallurgical institutions and received many honors. Nevertheless, he made few important scientific discoveries after joining Harvard, and he had few eminent research students. His main contribution after 1896 was that of a gifted advocate and interpreter of the work of others.

BIBLIOGRAPHY

I. ORIGINAL WORKS. A list of Sauveur's 160 papers is in R. A. Daly, in *Biographical Memoirs. National Academy of Sciences*, **21** (1943), 26–33. Sauveur's most important papers are "The Microstructure of Steel," in *Transactions of the American Institute of Mining Engineers*, **22** (1893), 546–557; "The Microstructure of Steel and the Current Theories of Hardening," *ibid.*, **26** (1896), 863–906, with German trans. by Hanno von Juptner (Leipzig, 1898); and "Current Theories of Hardening Steel, Thirty Years Later," in *Transactions of the American Institute of Mining and Metallurgical Engineers*, **73** (1926), 859–908. Sauveur was editor of *Metallographist*, **1–6** (1898–1904); and its successor, *Iron and Steel Magazine*, **7–11** (1904–1906).

Sauveur's books include *Laboratory Experiments in Metallurgy* (Cambridge, Mass., 1908); *The Metallography of Iron and Steel* (Cambridge, Mass., 1912; 2nd ed., retitled *The Metallography and Heat Treatment of Iron and Steel*, Cambridge, Mass.–New York, 1916, 1918, 1920; 3rd ed., 1926; 4th ed. 1935). Sauveur also wrote two political pamphlets: *Germany and the European War* (Boston, 1914), and *Germany's Holy War* (Cambridge, Mass., 1915); and a brief autobiography, *Metallurgical Dialogue* (Cleveland, Ohio, 1935), reissued as *Metallurgical Reminiscences* (New York, 1937).

II. SECONDARY LITERATURE. On Sauveur and his work, see R. A. Daly, "Albert Sauveur," in *Biographical Memoirs. National Academy of Sciences*, **22** (1943), 121–133; and Cyril S. Smith, *A History of Metallography* (Chicago, 1960), ch. 16.

CYRIL STANLEY SMITH

SAUVEUR, JOSEPH (*b.* La Flèche, France, 24 March 1653; *d.* Paris, France, 9 July 1716), *physics.*

Sauveur worked on early problems in the physics of sound, especially beats, harmonics, and the determination of absolute frequency; and he was influential as a teacher of practical mathematics.

Sauveur was the son of Louis Sauveur, a notary, and Renée des Hayes. Born with a voice defect, he did not begin to speak until the age of seven and retained a lifelong difficulty with his speech. He first attended the famous Jesuit school of La Flèche, where arithmetic intrigued him. Hoping to learn science, Sauveur went to Paris in 1670, where he studied mathematics and medicine and attended the physics lectures of Jacques Rohault. Despite his speech problem, Sauveur became well known as a good teacher and was a tutor at the court of Louis XIV. Interested primarily in practical mathematics, he prepared tables for simplifying calculations and for converting weights and measures. He also worked on problems of engineering and in 1681 conducted hydraulic experiments with Mariotte at Chantilly. In 1691 he visited the town of Mons while it was under active siege, in connection with his plan to write a treatise on fortification. In 1703 he replaced Vauban as examiner for the Engineering Corps. When he obtained the chair of mathematics at the Collège Royal in 1686, Sauveur was sufficiently well known that he dared to read the required public speech (earlier, he had dropped a plan to apply for the post because the speech seemed too difficult for him). In 1696 he became a member of the Paris Academy of Sciences.

Like Mersenne and others in the seventeenth century, Sauveur used musical experience to obtain information on sound and vibration. According to Fontenelle, Sauveur was fascinated by music, even though he had no ear for it, and consulted frequently with musicians. Despite the musical foundation of his work, Sauveur proposed the development of a new subject, which he named *acoustique*,[1] dealing with sound in general rather than with the *son agréable* of music.

Sauveur began his work in acoustics by developing a method of classifying temperaments of the musical scale. He divided the octave into forty-three equal intervals, or *merides,* each of which was divided into seven *eptamerides.*[2] Sauveur's intention was to indicate the size of any musical interval, at least to a reasonable approximation with respect to the ability of the ear to discriminate pitch, in terms of an integral number of *eptamerides.* These divisions made it simple to describe and compare different tuning systems. For example, the fifths and thirds given by an integral number of *merides* approximate those of the one-fifth comma meantone tuning used in the sixteenth and seventeenth centuries.[3]

Sauveur's first work on the physics of vibration, originally presented to the Paris Academy in 1700, concerned the determination of absolute frequency. The problem of pitch standardization was a natural successor to the problem of temperament standardization. Sauveur wanted to use a *son fixe* of 100 cycles per second. Pitch had been identified with frequency early in the seventeenth century, and the ratio of the frequencies of a pair of tones was known, ultimately, from the inverse relative string lengths.[4] Sauveur was the first to use beats to determine the frequency difference and was therefore able to calculate the absolute frequencies. Since he correctly interpreted beats, it appears that Sauveur may have been the first to have an understanding of superposition.

To determine absolute frequency, Sauveur used a pair of organ pipes a small half-tone apart in just intonation (frequency ratio 25:24). This interval is sufficiently small that the beats can be counted, for low pitches. Furthermore, the interval can be obtained accurately by tuning through thirds and perfect fifths (for example, by tuning up two major thirds and then down a fifth). As a result of experiments done with Deslandes, an organ builder, Sauveur found that the frequency of an open organ pipe five Paris feet long was between 100 and 102 cps. Sauveur claimed to have obtained consistent results from experiments done with other pipes. Newton made a rough check of Sauveur's results: knowing that the velocity of sound is the product of frequency and wavelength, he knew the wavelength of a tone of 100 cps to be just over twice the length of Sauveur's pipe.[5] Since the exact dimensions of the pipe are not known, there is no way to determine how accurate Sauveur was in his determination of its frequency. However, its pitch was an A of the time, in agreement with a later determination of the frequency of an eighteenth-century Paris tuning fork.[6] If each of the three tunings made in the process of obtaining the small halftone is accurate to within half a cent (1/200 of an equal tempered semitone), it would be possible, in principle, to find the absolute frequency to within 2.5 percent.

Later, in work presented in 1713, Sauveur derived the frequency of a string theoretically. He treated the string, stretched horizontally and hanging in a curve because of the gravitational field, as a compound pendulum and he found the frequency of the swinging motion, assumed to have small amplitude. His result agrees with the modern one except for a factor of $\sqrt{10/\pi}$.

In 1701, at a lively meeting of the Paris Academy, Sauveur explained the basic properties of harmonics. Harmonics had seemed paradoxical ever since the early seventeenth century, when the identification of pitch with frequency had implied the seemingly impossible phenomenon of a single object vibrating simultaneously at several frequencies. Sauveur argued that a string can vibrate at additional, higher frequencies, which he called *sons harmoniques,* by dividing up into the appropriate number of equal shorter lengths separated by stationary points, which he called *noeuds.* Sauveur apparently did not know of the paper rider demonstration, reported by Wallis (1677)[7] and Robartes (1692),[8] that nodes are associated with higher modes. However, Wallis' paper was mentioned by a member of the audience, and Sauveur's argument for the existence of nodes culminated with Wallis' demonstration. In a discussion of the organ, presented in 1702, Sauveur stated explicitly (he was the first to do so) that harmonics are components of all musical sound and that they affect the timbre of a tone.

Among Sauveur's interests, the subject of harmonics proved the most important for later developments—in mathematics, physics, and music. In the eighteenth century, analysis of the vibrating string was inspired, in part, by knowledge of the higher vibrational modes. The composer and theorist Rameau used harmonics to provide a physical basis for his theory of harmony, and a century later Helmholtz emphasized the effect of harmonics on timbre. It was through Sauveur and the Paris Academy that ideas about harmonics became well known in the early eighteenth century.[9] Sauveur's terminology, including "harmonics" and "node," was adopted and is still current.

NOTES

1. The word had already been used occasionally in connection with sound.
2. Other systems of multiple division were already in use; the most important, developed by Huygens, was the division of the octave into 31 intervals.
3. J. Murray Barbour, *Tuning and Temperament, an Historical Survey* (East Lansing, Mich., 1953), 122.
4. Mersenne made the first known estimate of vibrational frequency by extrapolating from the countable vibrations of a long string to the frequency of a short section of it; see *Harmonie universelle* (Paris, 1636; facs. ed., 1963), I, Bk. 3, Prop. VI, 169–170.
5. In the 2nd and 3rd eds. of the *Principia,* Bk. 2, Prop. L; cf. Isaac Newton, *Mathematical Principles of Natural Philosophy,* F. Cajori, ed. (Berkeley, Calif., 1962), I, 383–384.
6. Alexander J. Ellis, "On the History of Musical Pitch," in *Journal of the Society of Arts,* **28** (1879–1880), 318, col. 1.
7. John Wallis, "On the Trembling of Consonant Strings, a New Musical Discovery," in *Philosophical Transactions of the Royal Society,* **12** (1677), 839–842.
8. Francis Robartes, "A Discourse Concerning the Musical Notes of the Trumpet, and the Trumpet Marine, and of Defects of the Same," *ibid.,* **17** (1692), 559–563.
9. In 1809, when Chladni's experimental demonstration of nodal lines inspired the Academy competition for theoretical analysis of vibrating surfaces, Prony remarked that Sauveur's work had led to important research on the vibrating string; see *Procès-verbaux de l'Académie des sciences,* IV (Hendaye, 1913), 175.

BIBLIOGRAPHY

Sauveur's papers include "Système général des intervalles des sons," in *Mémoires de l'Académie royale des sciences,* 1701 (Paris, 1704), 297–460; "Application des sons harmoniques à la composition des jeux d'orgues," *ibid.,* 1702 (Paris, 1704), 308–328; "Rapport des sons des cordes d'instruments de musique aux flèches des cordes; et nouvelle détermination des sons fixes," *ibid.,* 1713 (Paris, 1716), 324–348.

Fontenelle is the main source of information on Sauveur's life and the reception of his work. His *éloge* is in *Histoire de l'Académie royale des sciences,* 1716 (Paris, 1718), 79–87. His discussions of Sauveur's work include "Sur la détermination d'un son fixe," *ibid.,* 1700 (Paris, 1703), 131–140; "Sur un nouveau système de musique," *ibid.,* 1701 (Paris, 1704), 123–139; "Sur l'application des sons harmoniques aux jeux d'orgues," *ibid.,* 1702 (Paris, 1704), 90–92; and "Sur les cordes sonores, et sur une nouvelle détermination du son fixe," *ibid.,* 1713 (Paris, 1716), 68–75.

See also Léon Auger, "Les apports de J. Sauveur (1653–1716) à la création de l'acoustique," in *Revue d'histoire des sciences et de leurs applications,* 1 (1948), 323–336; and the article on Sauveur by F. Winckel in F. Blume, ed., *Die Musik in Geschichte und Gegenwart,* XI (Kassel, 1963), cols. 1437–1438.

Sigalia Dostrovsky

SAVART, FÉLIX (*b.* Mézières, France, 30 June 1791; *d.* Paris, France, 16 March 1841), *physics.*

Savart made experimental studies of many phenomena involving vibration. With Biot he showed that the magnetic field produced by the current in a long, straight wire is inversely proportional to the distance from the wire. In most of his vibrational studies Savart observed the nodal lines of vibrating surfaces and solids, and he thereby obtained information on vibrational modes and elastic properties.

Savart was the son of Gérard Savart, an engineer at the military school of Metz. His brother, Nicolas, who studied at the École Polytechnique and was an officer in the engineering corps, also did work on vibration. Savart studied medicine, first at the military hospital at Metz and then at the University of Strasbourg, where he received his medical degree in 1816. At this time Savart presumably was already interested in the physics of the violin, for he built an experimental instrument in 1817 and in 1819 presented a memoir on the subject to the Paris Academy of Sciences. Biot, one of Savart's first contacts in Paris, was interested in his work and helped him to find a position teaching physics there. In 1827, Savart replaced Fresnel as a member of the Paris Academy; and in 1828 he became a professor of experimental physics at the Collège de France, where he taught acoustics.

In 1820, a few months after Oersted's discovery of the magnetic field produced by a current, Biot and Savart determined the relative strength of the field by observing the rate of oscillation of a magnetic dipole suspended at various distances from a long, straight wire. In some measurements the earth's field was canceled by an appropriately placed magnet, while in others the apparatus was oriented so that the field produced at the dipole by the current was in the magnetic north-south direction.

In his earliest work Savart gave the first explanation of the function of certain parts of the violin. To learn how vibrations are transmitted from the strings to the rest of the instrument, he induced vibrations in a free wood plate by passing a vibrating string over a bridge at its center; he also used Chladni's sand-pattern technique to observe the resulting nodal lines. Savart showed that the bridge transmits the string's vibrations; that the plate can be made to vibrate at any frequency; and that the corresponding mode is a modification of an unforced mode. He demonstrated that the sound post also serves to transmit vibrations, and he explained that it therefore should not be placed under a nodal line. Thinking that symmetry and regularity would produce the best tone, Savart built a trapezoidal violin with rectangular sound holes. When the instrument was played before a committee that included Biot, the composer Cherubini, and other members of the Academy of Sciences and the Académie des Beaux-Arts, its tone was judged as extremely clear and even, but somewhat subdued.

Over a period of twenty years, Savart performed numerous experiments in acoustics and vibration. He generalized his work on the violin to analyze the vibrations of coupled systems. He also greatly extended Chladni's observations of the modes of plates: adding a dye to the sand, he made prints of the nodal patterns for brass plates in the shapes of circles, ellipses, and polygons. Savart was able to locate directly the nodes of a vibrating air column by lowering a light membrane covered with sand into a vertical pipe.

On the basis of his experience observing vibrational modes, Savart introduced a new way to learn about the structure of materials. The variation of nodal patterns for laminae cut along different planes of a nonisotropic material indicated the orientational dependence of the material's elasticity. Savart sought the axes of elasticity of various substances, including certain crystalline ones. His papers on this subject were translated and reprinted.

Savart also studied aspects of the voice and of hearing. In connection with determining the lower frequency limit of hearing, he devised and used the rotating toothed wheel for producing sound of any frequency. Savart was highly regarded as an experimenter, and his results were relevant for the contemporary analyses of vibration and elasticity made by Poisson, Cauchy, and Lamé.

BIBLIOGRAPHY

I. ORIGINAL WORKS. Savart's published works include *Mémoire sur la construction des instruments à chordes et à archet* (Paris, 1819); "Mémoire sur la communication des mouvements vibratoires entre les corps solides," in *Annales de chimie et de physique*, 2nd ser., **14** (1820), 113–172; "Note sur le magnétisme de la pile de Volta," *ibid.*, **15** (1820), 222–223, written with J. B. Biot; "Recherches sur les vibrations de l'air," *ibid.*, **24** (1823), 56–88; "Mémoire sur les vibrations des corps solides, considérées en général," *ibid.*, **25** (1824), 12–50, 138–178, 225–269; "Recherches sur l'élasticité des corps qui cristallisent regulièrement," *ibid.*, **40** (1829), 5–30, 113–137; "Recherches sur la structure des métaux," *ibid.*, **41** (1829), 61–75; and "Sur les modes de division des plaques vibrantes," *ibid.*, **73** (1840), 225–273. More of Savart's papers are listed in the Royal Society *Catalogue of Scientific Papers*, V (London, 1871), 419–420.

Savart began a book on acoustics but did not complete it. See *Comptes rendus . . . de l'Académie des sciences*, **12** (1841), 651–652.

II. SECONDARY LITERATURE. There are brief biographies of Savart in Michaud's *Biographie universelle*, XXXVIII, 104–105; and in *Nouvelle biographie générale* (Paris, 1969), 387–389. The report of the committee that studied Savart's *mémoire* on the violin is in *Annales de chimie et de physique*, 2nd ser., **12** (1819), 225–255. The determination of the Biot-Savart law is discussed in detail by J. B. Biot in *Précis élémentaire de physique*, 3rd ed., II (Paris, 1824), 707–723; this section is translated by O. M. Blunn in R. A. R. Tricker, *Early Electrodynamics* (Oxford, 1965), 119–139. Some of Savart's work on vibration and elasticity is discussed by I. Todhunter in *A History of the Theory of Elasticity*, K. Pearson, ed. (New York, 1960), 167–183.

SIGALIA DOSTROVSKY

SAVASORDA. See **Abraham bar Ḥiyya ha-Nasi**.

SAVIGNY, MARIE-JULES-CÉSAR LELORGNE DE (*b.* Provins, Brie, France, 5 April 1777; *d.* Versailles, France, 5 October 1851), *biology.*

Savigny used his full name only on legal documents. His parents, Jean-Jacques Lelorgne de Savigny and Françoise Josèphe de Barbaud, had one other child, Amable Eléanore Louise Josèphe. Savigny studied at the local Collège des Oratoriens; his education there was supplemented by visits to the Génovéfains de Saint-Jacques. Besides classical languages, he was introduced to botany, the use of the microscope, and tales of travel. He was studying chemistry and botany with a local apothecary when, in 1793, he was chosen by a government commission to study in Paris. He enrolled at the École de Santé, attended lectures at the Muséum d'Histoire Naturelle, and became known to Lamarck, Daubenton, Cuvier, and Étienne Geoffroy Saint-Hilaire. Savigny's mother, widowed and impoverished, joined him in Paris and died when he was twenty.

Savigny was about to begin a teaching career in botany when Cuvier urged him to join Napoleon's expedition to Egypt as a zoologist: he would study invertebrates while Geoffroy Saint-Hilaire studied vertebrates. With many other scholars, they remained in Egypt from June 1798 until January 1802. Savigny then returned to France and spent the rest of his life in Paris and Versailles, with the exception of an excursion to Italy between February and November of 1822. (Pallary first put this trip during the period 1817–1822 ["Vie," 45] but later found letters that show the visit to have been briefer ["Documents," 90, 101].) Beginning about 1817, Savigny suffered from a severe nervous affliction (not blindness) that made visual work, and even thought, impossible. His life as a scientist had virtually ended by the time he was elected to the Académie des Sciences in 1821. When the commission responsible for publishing the results of the Egyptian expedition finally despaired of receiving his promised descriptions, they corresponded with Savigny's devoted mistress, Agathe-Olympe Letellier de Sainteville. She seems to have helped effect a compromise: plates drawn under Savigny's supervision were eventually published in *Description de l'Égypte*, with notes written by Victor Audouin.

Savigny's *Mémoires sur les animaux sans vertèbres* (1816) deeply impressed his contemporaries and was cited as a model of fine zoology for the next fifty years. By example rather than by precept, he demonstrated the value of comparative morphology. Presumably it was his botanical training that led him to seek, in the insects and crustacea collected in Egypt, "perfectly Linnaean characters, that is, where the same organs are always disposed in the same order, and can be compared without interruption" (*Mémoires sur les animaux sans vertèbres,* iii).

Just as flowers were analyzed into calyx, corolla, stigma, and style, so Savigny analyzed insect mouthparts into labrum, mandibles, maxillae, and labium. Previous zoologists had distinguished Lepidoptera, for example, by the absence of regular mouthparts and the presence of a two-part coiled tubule. Savigny interpreted the tubule as a highly modified pair of maxillae and said that the other mouthparts, however minute, were present. Savigny extended his comparison to the appendages of crustacea and other arthropods, to the point of seeing the mouthparts of the horseshoe crab as homologous to the legs of insects. In the second part of the *Mémoires,* he compared the anatomy of solitary ascidians with that of various zoophytes, demonstrating that the latter should be regarded as colonial ascidians. He even suggested some homologies between ascidians and salps.

While Savigny was showing unsuspected homologies among these invertebrates, Geoffroy Saint-Hilaire was doing the same with the bones of reptiles and fish. Savigny's work encouraged Geoffroy Saint-Hilaire to search for homologies between vertebrates and arthropods. Savigny did not discuss principles or suggest a "unity of plan," but his work became a model for the morphological zoology that flourished in the nineteenth century.

BIBLIOGRAPHY

I. ORIGINAL WORKS. Savigny's principal works are *L'histoire naturelle et mythologique de l'ibis* (Paris, 1805); *Mémoires sur les animaux sans vertèbres* (Paris, 1816); *Description de l'Égypte, publiée par les ordres de sa majesté l'Empereur Napoléon-le-Grand. Histoire naturelle* (Paris, 1809–[1826]), which includes "Système des oiseaux de l'Égypte et de la Syrie" (1809), "Système des annélides, principalement de celles des côtes de l'Égypte et de la Syrie" (1820), and reprs. of his already published material on the ibis and on ascidians; and "Remarques sur certain phénomènes dont le principe est dans l'organe de la vue, ou fragments du journal d'un observateur atteint d'une maladie des yeux," in *Mémoires de l'Académie des sciences de l'Institut de France,* **18** (1842), 385–416.

II. SECONDARY LITERATURE. On his life and work, see Paul Pallary, "Marie Jules-César Savigny: sa Vie et son Oeuvre: la Vie de Savigny," in *Mémoires. Institut d'Égypte,* **17** (1931), 1–111; "L'oeuvre de Savigny," *ibid.,* **20** (1932), 1–112, with bibliography, 88–92; and "Documents," *ibid.,* **23** (1934), 1–203.

MARY P. WINSOR

SAXTON, JOSEPH (*b.* Huntingdon, Pennsylvania, 22 March 1799; *d.* Washington, D.C., 26 October 1873), *scientific instrumentation.*

Saxton was the son of James Saxton and Hannah Ashbaugh. After leaving school at the age of twelve and working at his father's nail factory in Huntingdon, he became apprenticed to a local watchmaker, David Newingham, from whom he apparently acquired his taste for precision craftsmanship. Newingham died soon afterward and in 1818 Saxton went to Philadelphia, in the hope of using his talents for precision work.

His earliest major achievement in Philadelphia was a clock with a unique temperature-compensating pendulum and an oil-less escapement. The Franklin Institute awarded him a silver medal for it in 1824. He also invented a device for shaping clock gear teeth into the ideal (epicycloidal) configuration and a "reflecting pyrometer and comparator" for checking the precision of pendulums; the city of Philadelphia awarded him the John Scott legacy medal for this device on 11 November 1841. He also made a cane gun in 1824; it was not muzzle-loaded but employed a metal-jacketed cartridge.

Saxton's ingenuity led Isaiah Lukens, a leading mechanic and tower clockmaker in Philadelphia, to make him an associate. Together they built the clock that occupied the steeple of Independence Hall from 1828 until 1876, when it was moved to the town hall in Germantown, where it is still in operation. All of Saxton's innovations had been installed in this particular clock.

In 1829 Saxton moved to London, where he built many of the permanent exhibits for the newly constructed Adelaide Gallery. An early visitor was Michael Faraday, whose electromagnetic discoveries inspired Saxton to build a highly regarded electric generator and electric motor. Other visitors included the physicist Charles Wheatstone and William Cubitt and Thomas Telford, two of London's best-known engineers. Saxton built a water-current meter for Cubitt and the apparatus that Wheatstone used for determining the speed of current electricity. He also built the apparatus that Telford and his assistant John Macneill used to study water resistance, at various speeds, of canal boats.

In January of 1835 Franklin Peale commissioned Saxton to build a precision assay balance for the Philadelphia Mint. When it was completed he returned to Philadelphia to take charge of the construction and maintenance of the balance scales at the mint. He occupied this post from 1837 until 1844. One of his balances is still on exhibit there.

In Philadelphia, Saxton pioneered in daguerreotype photography (1839) with a view of the state arsenal and the old Central High School. He also engraved a diffraction grating that enabled John William Draper to make the first photograph (a daguerreotype) of the diffraction spectrum (1844).

From 28 February 1844 until his death, Saxton was director of the U.S. Coast Survey Office of Weights and Measures, the forerunner of the National Bureau of Standards. Every state that had not previously been furnished with sets of federally approved weights, measures, and balance scales was supplied with sets built, checked, and in many instances personally installed at the various state capitals by Saxton. His skill in making precision balances reached its zenith during these years. A set of them, exhibited at the 1851 Great Exhibition in London, brought him a gold medal—the highest award.

Saxton designed and built for the Coast Survey a gauge for recording tide levels, a current meter like the one he had made in London, a unique hydrometer, and a maximum-minimum thermometer for deep-sea observations. Each of these instruments was the first of its kind manufactured in America.

For engraving degrees on a circle, he and William Wurdemann devised an automatic dividing engine that eliminated the errors caused by the body temperature of the operator.

Saxton married Mary H. Abercrombie in 1850. He was a member of the American Philosophical Society, the Franklin Institute, and a charter member (1863) of the National Academy of Sciences.

BIBLIOGRAPHY

I. ORIGINAL WORKS. Saxton's patents are Great Britain, 6351 (1832) and 6682 (1834); U.S., 3806, 22982, 23046, and 44460. See also National Archives Patent No. 5446X (1829). For descriptions of his exhibits, see the quarterly catalogs of the Society for the Illustration and Encouragement of Practical Science, Adelaide Gallery (London, 1832–1835); and the Society's *Magazine of Popular Science and Journal of the Useful Arts,* 1–4 (1836–1840), which contains material on Saxton's work, some probably written by him.

See also "Notice of Electro-Magnetic Experiments," in *Journal of the Franklin Institute,* 14 (1832), 66–72; "Description of a Revolving Keeper Magnet for Producing Electrical Currents," *ibid.,* 17 (1834), 155–156; and "On the Application of the Rotating Mirror to the Aneroid Barometer," in *Proceedings of the American Association for the Advancement of Science,* 12 (1858), 40–42.

A MS scrapbook and diary of his life in London are in the Smithsonian Institution archives; letters are in the U.S. National Archives, Record Gp. 104. A MS letter (2 Oct. 1825) is in Proctor papers, University of Virginia library.

Saxton's daguerreotype of the state arsenal and the old Central High School is at the Historical Society of Pennsylvania, Philadelphia. The Historical Society of York County, York, Pennsylvania, has a ruling machine of his construction.

II. SECONDARY LITERATURE. On Saxton and his work, see Joseph Henry's notice in *Biographical Memoirs. National Academy of Sciences,* 1 (1877), 287–316; J. Saxton Pendleton, *Joseph Saxton, 1799–1873* (Reading, Pa., 1935); C. W. Mitman, in *Dictionary of American Biography,* XVI (New York, 1943), 400; and Thomas Coulson, in *Journal of the Franklin Institute,* 259 (1955), 277–291.

ARTHUR H. FRAZIER

SAY, THOMAS

SAY, THOMAS (*b.* Philadelphia, Pennsylvania, 27 July 1787; *d.* New Harmony, Indiana, 10 October 1834), *entomology, conchology.*

One of the generation of self-taught naturalists, Say was an indifferent scholar in the Quaker boarding school he attended until the age of fifteen. His father, Benjamin Say, and his grandfather, Thomas Say, had both been physician-apothecaries, public-spirited philanthropists, and founders of hospitals; they were also noted as "fighting Quakers" for the colonial cause. Say's mother, Anna Bonsall Say, died when he was six years old. She was a granddaughter of John Bartram, and through this relationship Say became acquainted with the beetle and butterfly collections of William Bartram, for whom he collected specimens.

Say's father tried to discourage his son's early interest in natural history but inadvertently opened the door wider by putting him into partnership in the apothecary business with John Speakman. The partners' lack of business acumen soon brought the shop to failure; but the meetings of a congenial group of friends, occasionally in the back of the shop, led to the founding (1812) of the Academy of Natural Sciences of Philadelphia, with Say among the charter members. His father having died, Say lived frugally in the Academy building, tended the small museum, and studied his own collections. Here his friendship developed with William Maclure, president of the Academy from 1817.

In 1818, with Maclure, George Ord, and T. R. Peale, Say went on an expedition to the Sea Islands of Georgia and Spanish Florida, where they were thwarted by hostile Indians. The next year Say became chief zoologist on Major Stephen H. Long's exploration of the tributaries of the Missouri River, and in 1823 he was zoologist on Long's trip to the headwaters of the Mississippi River. Say declined the offer to be the expedition historian, but in addition to the many hitherto unknown animals he collected, he also gathered a great many plants. The explorers concluded that the treeless expanse between the Mississippi and Missouri rivers offered no possibilities for future settlement and would indefinitely remain the home of numberless bison and a few Indians.

Maclure persuaded Say to accompany him to New Harmony, Indiana, in 1825. This idealistic community, the dream of Scottish industrialist Robert Owen, had been established as an escape from the harshness of clamoring cities and as a proof that beauty, culture, and science could flourish where all worked willingly together. It failed. Say was among its victims, for, although hopeless at financial matters, he stayed as Maclure's agent after the latter's departure; and the malarial climate on the Wabash River contributed to Say's early death.

Say effectively did scientific work at New Harmony, and there he continued the study of mollusks that he had begun in Philadelphia. In 1816 he had published the first paper on American shells by an American. At New Harmony he completed and printed *American Conchology.* He also produced the third volume of *American Entomology,* which had been well under way. These works were illustrated by Charles-Alexandre Lesueur and by the talented Lucy Way Sistaire, whom Say had married at New Harmony. He also published descriptions of more than a thousand new species of insects.

New Harmony attracted the attention of scientists throughout the world, many of whom visited the community. Say's reputation led to his becoming a foreign member of the Linnean Society of London and the Zoological Society of London, and a correspondent of the Société Philomathique of Paris.

Say's studies of insects, on which he spent the greater time, had few predecessors in the United States. The first book on American insects appeared in 1797, based on notes of John Abbot, and studies of individual insects or groups went back as far as Paul Dudley's account of bees in 1723 and Moses Bartram's work on the seventeen-year cicada in 1767. The emphasis was on economic insects, as evidenced by Thomas Jefferson's participation on a committee to study the Hessian fly in 1792. Amateurs in the United States customarily sent specimens to Europe for identification, especially after the impetus given to the subject by Réaumur in the six-volume *History of Insects* (1734–1742). Say's entry into entomology changed that, for his published descriptions were accurate and readily usable by others. Although entirely too trustful of others and excessively modest, he had a delightful personality, and he readily identified specimens for a growing list of American collectors. He was familiar with American and European literature on insects and was a natural taxonomist, showing excellent judgment in selecting the significant features of each species so that his descriptions did not leave taxonomic confusion. Say described many important economic insects, which bear his name. Although he urged others to study also the habits of insects, he did little of that himself.

Unfortunately, after his death, Say's collection of insects was long neglected before it was finally established at the Academy of Natural Sciences and many of the type specimens were hopelessly damaged by dermestids.

BIBLIOGRAPHY

I. ORIGINAL WORKS. Say's two major works were *American Conchology,* 6 vols. (New Harmony, 1830–1834), and *American Entomology; or Descriptions of the Insects of North America,* 3 vols. (Philadelphia, 1817–1828); both are classics in their fields. His first publication of shells (the first by an American) was "Conchology," in *Nicholson's British Encyclopedia,* American ed. (1816–1817). Say's papers on insects were gathered into *The Complete Writings of Thomas Say on the Entomology of North America,* John L. LeConte, ed., 2 vols. (New York, 1859), with biography by George Ord. Some of Say's descriptions of specimens and some of his narratives are included in the reports of Long's two expeditions.

II. SECONDARY LITERATURE. For accounts of Say's prominence in early American science, see William J.

Youmans, *Pioneers of Science in America* (New York, 1896), 215–222; and Bernard Jaffe, *Men of Science in America* (New York, 1944), 130–153. See also William H. Dall, "Some American Conchologists," in *Proceedings of the Biological Society of Washington*, **4** (1888), 98–102; and E. O. Essig, *A History of Entomology* (New York, 1931), 750–756.

Elizabeth Noble Shor

SCALIGER (BORDONIUS), JULIUS CAESAR (*b.* Padua, Italy, 23/24 April 1484; *d.* Agen, France, 21 October 1558), *natural philosophy, medicine, botany.*

An evident desire to claim noble descent led Julius Caesar Bordonius (later called Scaliger) and his son, the great classical scholar Joseph Justus Scaliger, to trace their origins to the della Scala family, sometime rulers of Verona. Conveniently for the Scaliger claim, Julius Caesar's family died out around 1512, which thus made it difficult for contemporaries to verify the alleged genealogy. Although the Scaliger account has been widely accepted, recent research has exposed the elaborate camouflage surrounding Julius Caesar's birth and early life.

According to the Scaligers' version, Julius Caesar was born at Riva on Lake Garda on 23/24 April 1484. The parents were held to be Benedetto and Berenice Scaliger (della Scala). Julius was named Caesar at the insistence of Paul of Middelburg, the noted astronomer and astrologer, who had cast the infant's horoscope. Styling himself Count of Burden, Julius Caesar fought in the Imperial army against the Venetians, hereditary enemies of the della Scala. Afterward he toyed with monastic life at Bologna in the hope of becoming pope and of regaining the family property but, disillusioned, quit the monastery for the University of Bologna, where he studied Aristotelian philosophy and physics, before reentering the army. During his second tour, Julius Caesar studied medicine and collected medicinal herbs in northern Italy. Following his second military retirement, Scaliger in 1524 accompanied, as personal physician, Bishop Antonio della Rovere to Agen in southern France.

After the departure for Agen in 1524, the details of Scaliger's life are not in dispute, but the account of the period before 1524 is open to doubt. Although Scaliger was probably born in 1484, it was not at Riva but at Padua. His father was Benedetto Bordon, an expert illuminator of manuscripts and a graphic artist, and also an astronomer and geographer who had perhaps known Paul of Middelburg, when the latter was teaching at Padua in 1479. The Bordon family was of Paduan, not Veronese, stock, although it is possible that Scaliger's father held dual Paduan and Veronese citizenship. Certainly there is no possibility that Scaliger was a true descendant of the della Scala.

Scaliger seems to have grown up in his father's household at nearby Venice and may have entered a Franciscan convent there. It is conceivable that he knew the architect and mathematician Fra Giovanni Giocondo, also a Franciscan, who was at Venice after 1506. If Scaliger's tales of military service are largely true, then he was fighting from 1509 to 1515 and may, as he says, have fought at the great battle of Ravenna in 1512. After leaving the army, Scaliger did not go to the University of Bologna but to the University of Padua, where, as "Giulio, son of Benedetto Bordon," he received the doctorate in arts in 1519 and the following year seems to have been appointed lecturer in logic, a post that he did not accept. The refusal of the appointment may have been due to his pursuit of a doctorate in medicine, although there is no proof of Scaliger's medical degree. There is evidence, however, that Scaliger was working at Venice (1521–1524) on a translation of Plutarch that was published there in 1525. Interestingly, Joseph Justus Scaliger implied that his father had studied at Padua (despite his banishment from the city as a della Scala), when he said that his father had been taught mathematics by Pomponazzi as well as by Luca Gaurico. Scaliger himself claimed that his instructors in philosophy had included Marc'Antonio Zimara, Nifo, and Pomponazzi, all of whom were teachers at Padua.

After his arrival at Agen in 1524, Scaliger entered a new life as "Julius Caesar de l'Escale de Bordons." He became a well-known and respected physician and in March 1528 was naturalized as a French citizen. In April 1529 he married Andiette de la Roque Lobejac, who bore him fifteen children and was reckoned a restraining influence on Scaliger's combativeness. Active in civic life, Scaliger served as consul of Agen (1532–1533). In 1538, as a Huguenot sympathizer, he was summoned before the Inquisition but was acquitted. (Scaliger's *Poetics* and his commentaries on Theophrastus were later placed on the *Index of Prohibited Books*.) The Huguenot historian Théodore de Bèze regarded Scaliger's career as a paradigm of the relation between enlightened intellectual views and Protestantism. After prolonged attacks of gout, Scaliger died at Agen in 1558, possessed of a European reputation.

Scaliger first established his fame by a savage literary attack (Paris, 1531) on the satire of Erasmus against the Ciceronian stylists (1528). Later Scaliger wrote the more lasting *Poetics*, which represents a reworking of Aristotelian aesthetics in order to form an important early statement of neoclassicism. Among Scaliger's literary friends were Pierre de Ronsard, François Rabelais (for a time), and George Buchanan. (Scaliger's *De causis linguae Latinae* [1540] has been termed the first modern scientific attempt at a Latin grammar.)

The creative approach to classical thought is manifest in Scaliger's writings on botany. Scaliger sought to advance botany and simples by his admirable editions of three ancient treatises: the *De plantis* of pseudo-Aristotle (Nicolaus of Damascus) and the two works of Theophrastus on plants. All three works benefit from the editor's knowledge of actual specimens. The dedication of the pseudo-Aristotelian treatise (all were published posthumously) remarks that seasonal and regional variations make it difficult to identify European plants with classical descriptions, many of the regional variations also being vague or erroneous. Scaliger tried to effect a new and more consistent classification of plants but feared that ignorant physicians would continue to adhere to the older descriptions. Elsewhere he remarked: "It is necessary to submit everything to examination [and] not to embrace anything with servile adulation. The ancients must not put a brake on us."

In medicine Scaliger considered himself an empirical Averroist, who relied upon observation and experience rather than system. "I should like men of learning to become un-complex again and no longer consider themselves members of systematic schools." Scaliger's medical skills secured his appointment as physician to the king of Navarre (1548–1549). On account of Scaliger's reputation, many students eager for instruction, including Nostradamus and Rabelais, were attracted to Agen. The relationship between Scaliger and Rabelais was exacerbated by their conceit and by Rabelais's preference for systematic "ancient" medicine. Although Rabelais left Agen in 1530 to study under Scaliger's rival at Montpellier, the Rabelais-Scaliger hostility continued for decades. Rabelais's lampoon of "entelechy" and "endelechy" in *Gargantua and Pantagruel* (book V, chapter 19) seems directed at his former mentor's proneness to philosophize.

Scaliger was proud of his disputatious nature. In the *Exotericarum exercitationum* (1557) he wrote: "Vives maintains that silent meditation is more profitable than dispute. This is not true. Truth is brought forth by a collision of minds, as fire by a collision of stones. Unless I discover an antagonist, I can do nothing successfully." As Scaliger made his reputation by an attack on Erasmus, so he confirmed it with a spirited critique of Cardano's *De subtilitate libri XXI*. The *Exotericarum exercitationum* runs to well over 1,200 pages. When Cardano failed to reply immediately, Scaliger, believing a false rumor that Cardano had died, was stricken with remorse and wrote a funeral oration in which he repented for the onslaught on his late opponent. Ironically, Cardano published his reply two years after Scaliger's death.

Scaliger based his critique on a reprint (Lyons, 1554) of the first edition of *De subtilitate*, rather than the revised second edition (Basel, 1554) (perhaps because of difficulty of access to the latter). The full title of the *Exotericarum exercitationum* implies that the critique is merely the fifteenth book of Scaliger's philosophical exercises (the first fourteen remained unpublished). Following its target, the work ranges over the whole of natural philosophy. In astronomy Scaliger ridiculed Cardano's stress on the astrological significance of comets; and he denied that the world's decay is proven because the apse of the sun was thirty-one semidiameters nearer the earth than in Ptolemy's time. Scaliger also rejected several of Cardano's beliefs in natural history: that the swan sings at its death; that gems have occult virtues ("a flea has more virtue than all the gems"); that there exist corporeal spirits that eat; that the bear forms its cub by licking; and that the peacock is ashamed of its ugly legs. Like Cardano, Scaliger was aware that lead and tin gain in weight during calcination, although he preferred to explain the increase as a result of the addition of particles of fire to the metal.

In order to refute Cardano's theory of the origin of mountain springs, Scaliger used the strange argument that the sea is not in its natural place, since earth should be nearer than water to the center of the earth. Consequently, seawater presses upward, emerging sometimes through superior earth as a mountain spring. This view, of course, failed to account for the difference in salinity between sea and mountain water.

Scaliger casts aside Cardano's Aristotelian view that the medium is a motive force. This view is refuted experimentally when a thin wooden disk, cut from a plank, is set to spin within the plank. According to Scaliger, the air between the disk and the surrounding plank is insufficient to act as a motive force, as postulated by the Aristotelians. Instead, as an admirer of Parisian dynamics, Scaliger preferred to use the impetus theory (which he called *motio*). Following Albert of Saxony and Jean Buridan, Scaliger stated that accelerated motion is a result of a persisting gravity within the moving body. This gravity generates from instant to instant a new impetus, which, intensifying, produces acceleration. The impetus, although evanescent, is an efficient cause, and as such need not be coterminous with the effect.

The *Exotericarum exercitationum* won a celebrity that survived its author's death. Lipsius, Bacon, and Leibniz were among its later admirers; and Kepler, who read it as a young man, accepted its Averroist doctrine of attributing the movement of each star to a particular intelligence.

BIBLIOGRAPHY

I. ORIGINAL WORKS. Scaliger's works and commentaries include *Hippocratis liber de somniis cum J. C. Scaligeri commentariis* (Lyons, 1539, 1561, 1610, 1659); *In libros duos, qui inscribuntur de plantis, Aristotele autore, libri duo* (Paris, 1556; Geneva, 1566; Marburg, 1598); *Exotericarum exercitationum liber XV. De subtilitate ad Hieronymum Cardanum* (Paris, 1557); *Commentarii et animadversiones in sex libros de causis plantarum Theophrasti* (Geneva, 1566); *M. Manilii astronomicon cum commentariis J. C. Scaligeri* (Paris, 1579, 1590, 1599, 1600, 1655); *Animadversiones in Theophrasti historias plantarum* (Lyons, 1584, 1644); and *Aristotelis de animalibus historia J. C. Scaligeri interprete cum eiusdem commentariis* (Toulouse, 1619).

The funeral oration for Cardano is in *Epistolae aliquot nunc primum vulgatae,* Joseph Justus Scaliger, ed. (Toulouse, 1620–1621), 63–66. Other writings of Scaliger are in *Epistolae et orationes,* F. Donsa, ed. (Leiden, 1600). Autograph codices by Scaliger in the Bibliothèque Universitaire, Leiden, include MSS Scaligerani 18 (Galen in Greek); 27 (Latin miscellany); 34 (Aristotle, *De animalibus,* dated 17 Dec. 1538); and MS Graecus 44 (autographed letters). For the Scaligerani from the collection of Joseph Justus, see *Bibliotheca Universitatis Leidensis: codices manuscripti. II, codices Scaligerani* (Leiden, 1910). Cardano's reply to Scaliger, "Actio prima in calumniatorem librorum de subtilitate," was first published in *De subtilitate,* 3rd ed. (Basel, 1560), 1265–1426; cf. Girolamo Cardano, *De vita propria,* (Paris, 1643), ch. 48.

II. SECONDARY LITERATURE. On Scaliger and his work, see Joseph Justus Scaliger, *Epistolae* (Leiden, 1627), which includes the misleading life of Julius Caesar; and V. Hall, Jr., "The Life of Julius Caesar Scaliger (1484–1558)," in *Transactions of the American Philosophical Society,* n.s. **40** (1950), 85–170. The Scaliger version is refuted by P. O. Kristeller, in his review of Hall in *American Historical Review,* **107** (1952), 394–396; and by J. F. C. Richards, "The Elysium of Julius Caesar Bordonius (Scaliger)," in *Studies in the Renaissance,* **9** (1962), 195–217. Much new material appears in Myriam Billanovich, "Benedetto Bordon e Giulio Cesare Scaligero," in *Italia medioevale e umanistica,* **11** (1968), 187–256. For the botanical commentaries, see Charles B. Schmitt, "Theophrastus," in *Catalogus Translationum: Mediaeval and Renaissance Latin Commentaries and Translations,* P. O. Kristeller and F. Edward Cranz, eds., II (Washington, D.C., 1971), 239–322, 269–271, 274–275. Remarks on Scaliger's natural philosophy appear in Pierre Duhem, *Études sur Léonard de Vinci,* 3 vols. (Paris, 1906–1913), esp. I, 240–244; III, 198–204.

PAUL LAWRENCE ROSE

SCARPA, ANTONIO (*b.* Motta di Livenza, near Treviso, Italy, 19 May 1752; *d.* Pavia, Italy, 31 October 1832), *anatomy, neurology.*

Scarpa was an eminent anatomist, a skilled surgeon, and one of the powerful teachers at Pavia University during its period of greatest renown. The child of Giuseppe Scarpa, a boatman, and Francesca Corder, he was taught by his paternal uncle, Canon Paolo Scarpa. It was only with great financial difficulty that in 1766 he entered the University of Padua, where he became a favorite of Morgagni. After graduating on 31 May 1770, he assisted Morgagni until his death in December 1771. Scarpa was next helped by Girolamo Vandelli, physician to the duke of Modena; and by October 1772 he had been appointed professor of anatomy at Modena University and chief surgeon of the military hospital in that city. He worked at Modena for eleven years, the happiest of his academic life; it was there that he laid the foundations for his major work, which was carried out at Pavia. Highly esteemed for his brilliance, Scarpa had a true anatomical amphitheater built at Modena (it is still in existence); and in 1781 he obtained permission and funds for a study trip to Paris and London. In Paris he met the anatomist Félix Vicq d'Azyr and J. A. von Brambilla (1728–1800), surgeon-superintendent of the Imperial Austrian army; in London, the brothers William and John Hunter.

In 1783, through the assistance of Brambilla, Scarpa was named to the chair of anatomy at Pavia, where he gave his inaugural lecture on 25 November 1783, "Oratio de promovendis anatomicarum administrationum rationibus." At Pavia, Scarpa was responsible for the construction of a new, enlarged anatomical amphitheater. To show his gratitude to Brambilla, Scarpa, accompanied by Volta, went in the summer of 1784 to Vienna, where he performed some anatomical demonstrations and repeated his blood-transfusion experiments on sheep. From 1787 to 1812 he was in charge of teaching clinical surgery; in 1803, after thirty years, he gave up the teaching of anatomy. His fame was so great that in 1805 he was personally complimented by Napoleon; and in 1815 the restored Austrian government appointed him director of the medical school at Pavia, a post he held until 1818.

Scarpa also wrote and edited many works; from 1772 to 1825 he spent much of the money earned from his profession on the printing of his works. He retained an admirable clearness of mind even at a very advanced age, as is shown in his two letters *De gangliis* (1831) to Ernst Heinrich Weber. Possessor of the highest honors (he was a member of the Paris Academy, the Leopoldina, and the Royal Society), he died, unmarried, in his own home and was buried in the basilica of San Michele in Pavia.

All of Scarpa's work bears the unmistakable mark of his exacting personality. In his description of surgical procedures (amputation, the removal of cataracts, perineal cutting for the urinary calculi), the technique is always related to precise and very detailed anatomical description. In his monumental atlas on hernia (1809), he masterfully described the exact structure of the inguinal canal ("Memoria prima," sections 2–10) and of the crural ring ("Memoria terza," sections 2–4), as well as the disposition of the parts today known as the "triangle of Scarpa." His essay (1784) on freemartins was a pioneer study of hermaphroditism. His pathological works on diseases of the eye and on aneurysm were remarkable. Scarpa's greatest works, however—those that established him as a scientist— were in descriptive anatomy. His great skill in the use of the microscope is shown in his microscopical observations on nerve ganglia (1799) and on bones (1799). But above all he was a fine dissector and made his own anatomical drawings, which were engraved in copper by Faustino Anderloni.

Scarpa began his scientific activity with comparative investigation of the ear, suggested to him by Morgagni, *De structura fenestrae rotundae auris, et de tympano secundario* . . . (1772); for man and for the hen and pig he gave a more accurate and complete description of the osseous labyrinth and demonstrated the true function of the round window. In 1789 Scarpa made his historic observations on the membrane labyrinth, which he discov-

ered together with its endolymph (the perilymph had been discovered in 1761 by Domenico Cotugno). He also discovered the vestibule, which is admirably depicted by Anderloni in the 1794 edition of *Anatomicae disquisitiones de auditu* . . . . Scarpa precisely described the membrane semicircular canals with their ampullae and the utricle, and discovered the vestibular nerve and its ganglion (named for him [Herrick, 1928]). He was even able to observe the microscopical structure of the ampullae and identify the origin of the fibers of the vestibular nerve. Probably Scarpa had also observed the neurosensorial structure of the otolithic membrane. He accurately illustrated the course of the human acoustic nerve from the cochlea to the rhombencephalon. Enthusiasm for his own discoveries, however, led Scarpa to the mistaken affirmation that the semicircular canals are the organ of hearing. As early as 1672 Thomas Willis had correctly stated that the cochlea is the essential organ of hearing.

At about the same time Scarpa conducted research on the olfactory apparatus. His *De organo olfactus praecipuo* . . . (1785) presented the first illustration of the human olfactory nerves, olfactory bulbs, and olfactory tracts, as well as of the sphenopalatine ganglion (described by Johann Meckel the elder in 1748) and of the interior nasal nerves. He also documented his discovery of the human nasopalatine nerve. Scarpa was the first to provide a clear comparative anatomical illustration of the olfactory apparatus in the dogfish, reptiles, birds, and mammals.

These classic works on the auditory and olfactory apparatus were part of a broad plan of research on the nervous system, the premise of which had been set forth in Scarpa's *De nervorum gangliis et plexubus* (1779), the first accurate analysis of these nerve structures. Scarpa also was the first to distinguish the spinal (*ganglia simplicia*) from the sympathetic ganglia (*ganglia composita*) and to demonstrate that the spinal ganglia are formed only on the dorsal roots of the spinal nerves. Further, he stated that the ganglia of thoracic-lumbar sympathetic nerves are connected to the ventral roots of the spinal nerves only.

By 1779 Scarpa had established the foundation upon which he developed his great neurological work. In 1787 he studied the connection between the vagus nerve and its accessory, as acknowledged by Claude Bernard. According to Willis, the accessory nerve originated from the cervical spinal cord only, but Scarpa was the first to demonstrate that the accessory is also formed by fibers arising from the medulla oblongata. Scarpa's masterpiece was his *Tabulae . . . neurologicae* (1794), with seven life-size plates engraved in copper by Anderloni that illustrate the human glossopharyngeal, vagus, and hypoglossal nerves. Of particular interest are the two plates on the cardiac nerves, which were also a reply to arbitrary statements made by the physician Johann Bernard Behrends, who supported the views of Albrecht von Haller on the "irritability" of the heart. The latter was considered to be an autonomous contractile property and hence independent of sensibility; contractility and irritability were, in Haller's opinion, identical. This view formed one of the points of disagreement for Felice Fontana, who published his *De irritabilitatis legibus* in 1767. But Behrends also denied the existence of cardiac nerves and in 1792 published a book to show that the heart lacks nerves. In his plates Scarpa very accurately demonstrated the number, origin, and course of cardiac nerves. He also stated that the heart is richly supplied with its own nerve structures, that all of its nerves have ganglia, and that the terminal ramifications of the cardiac nerves are directly connected to cardiac muscular fibers. Scarpa thus decisively first demonstrated cardiac innervation.

BIBLIOGRAPHY

I. Original Works. Earlier writings by Scarpa are *De structura fenestrae rotundae auris, et de tympano secundario anatomicae observationes* (Modena, 1772); *Anatomicarum annotationum liber primus. De nervorum gangliis et plexubus* (Modena, 1779); "Osservazione anatomica sopra un vitello-vacca detto dagli Inglesi Freemartin," in *Memorie di matematica e di fisica della Società italiana delle scienze,* **2**, pt. 2 (1784), 846–852; *Anatomicarum annotationum liber secundus. De Organo olfactus praecipuo deque nervis nasalibus interioribus e pari quinto nervorum cerebri* (Pavia, 1785); "Abhandlung über den zum achten Paar der Gehirnnerven hinlaufenden Beynerven der Rückgräte," in *Abhandlungen der K. K. Medizinisch-chirurgischen Josephs-Akademie,* **1** (1787), 15–45, translated into Italian by Albrecht von Schoenberg as *Trattato sopra il nervo accessorio decorrente all'ottavo paio de' nervi cerebrali* (Naples, 1817); *Anatomicae disquisitiones de auditu et olfactu* (Pavia, 1789; Milan, 1794); *Tabulae ad illustrandam historiam anatomicam cardiacorum nervorum, noni nervorum cerebri, glossopharyngaei et pharyngaei ex octavo cerebri* (Pavia, 1794); and *De penitiori ossium structura commentarius* (Leipzig, 1799), repr. as *De anatome et pathologia ossium commentarii* (Pavia, 1827).

Later works are *Saggio di osservazioni e di esperienze sulle principali malattie degli occhi* (Pavia, 1801), 5th ed., enl., repr. as *Trattato delle principali malattie degli occhi,* 2 vols. (Pavia, 1816); *Memoria chirurgica sui piedi torti congeniti dei fanciulli e sulla maniera di correggere questa deformità* (Pavia, 1803; 2nd ed., enl., Pavia, 1806); *Osservazioni anatomico-chirurgiche sull'aneurisma* (Pavia, 1804), translated into English by John Henry Wishart as *A Treatise on the Anatomy, Pathology and Surgical Treatment of Aneurism* (Edinburgh, 1808); *Memorie anatomico-chirurgiche sulle ernie* (Milan, 1809; Pavia, 1819), translated into French, German, and English; *Elogio storico di Leon Battista Carcano 1536–1606 professore di notomia nella Università di Pavia* (Milan, 1813); *Memoria sulla legatura delle principali arterie degli arti* (Pavia, 1817); *Sull'ernia del perineo* (Pavia, 1821); *Memoria sull'idrocele del cordone spermatico* (Pavia, 1823); *Memoria sulla gravidanza susseguita da ascite* (Pavia, 1825); *Osservazioni sul taglio retto-vescicale per l'estrazione della pietra della vescica orinaria* (Milan, 1826); *De gangliis nervorum deque origine et essentia nervi intercostalis ad illustrem virum Henricum Weber anatomicum Lipsiensem epistola* (Pavia, 1831); *De gangliis deque utriusque ordinis nervorum per universum corpus distributione ad illustrem virum Henricum Weber anatomicum Lipsiensem epistola altera* (Pavia, 1831); and *Opuscoli di chirurgia,* 3 vols. (Pavia, 1825–1832), reprint of all his minor surgical notes.

Posthumously published works are *Opere del Cavaliere Antonio Scarpa,* Pietro Vannoni, ed., 5 pts. in 2 vols. (Florence, 1836–1838); *Atlante di tutte le opere del Professore Antonio Scarpa,* (Florence, 1839); and *Epistolario: 1772–1832* (Pavia, 1938), which contains Scarpa's autobiography and a collection of his 659 letters, reprinted in their original text.

Portraits of Antonio Scarpa are in his *Opere* (1836) and *Epistolario*; and in Favaro, *Antonio Scarpa*; Franceschini; Ovio; Politzer; and Putti.

II. Secondary Literature. See J. B. Behrends, *Dissertatio anatomico-physiologica qua demonstratur cor nervis carere* (Mainz, 1792); Claude Bernard, *Leçons sur la physiologie et la pathologie du système nerveux,* II (Paris, 1858), 271; M. Brazier, "Felice Fontana," in *Essays on the History of Italian Neurology. International Symposium on the History of Neurology . . .* (Milan, 1963), 107–116; P. Capparoni, *Spallanzani* (Turin, 1941), 101–114; G. Chiarugi, "Triangolo femorale dello Scarpa," in his *Istituzioni di anatomia dell'uomo,* II (Milan, 1924), 226–228; G. Favaro, "Antonio Scarpa e l'Università di Padova," in *Atti del Istituto veneto di scienze, lettere ed arti,* **91** (1931), 1–22; "Antonio Scarpa e i Caldani," *ibid.* 23–37; "Il 'Publicum doctoratus privilegium' di Antonio Scarpa," in *Rivista di storia delle scienze mediche e naturali,* **23** (1932), 193–204; "Antonio Scarpa e Michele Girardi," in *Valsalva,* **8** (1932), 742–748; *Antonio Scarpa e l'Università di Modena* (Modena, 1932); "I primi periodi della vita e della carriera di Antonio Scarpa descritti da un suo curriculum autografo," in *Bollettino dell'Istituto storico italiano dell'arte sanitaria,* **13** (1933), 29–32;

and "Antonio Scarpa nella storia dell'anatomia," in *Monitore zoologico italiano*, **43** (1933), supp., 29–43; P. Franceschini, *L'opera nevrologica di Antonio Scarpa* (Florence, 1962); C. J. Herrick, *An Introduction to Neurology* (Philadelphia, 1928), 399; G. Levi, *I gangli cerebrospinali* (Florence, 1908); G. Ovio, *L'oculistica di Antonio Scarpa e due secoli di storia*, 2 vols. (Naples, 1936); A. Politzer, "Antonio Scarpa," in his *Geschichte der Ohrenheilkunde*, I (Stuttgart, 1907), 260–271; and V. Putti, "Opere dello Scarpa riguardanti argomenti di anatomia e chirurgia dell'apparato motore," in *Biografie di chirurghi del XVI e XIX secolo* (Bologna, 1941), 24–28.

PIETRO FRANCESCHINI

SCHAEBERLE, JOHN MARTIN (*b.* Württemberg, Germany, 10 January 1853; *d.* Ann Arbor, Michigan, 17 September 1924), *practical astronomy*.

Schaeberle's most important work was with astronomical instruments. He devised new apparatus, particularly for astronomical photography; he figured mirrors and constructed telescopes; he investigated instrumental errors and atmospheric conditions; and he used instruments to good advantage. For this he was well trained, first as a machinist's apprentice and later as a civil engineer at the University of Michigan. Following graduation in 1876, Schaeberle spent twelve years at the university observatory before moving to Mt. Hamilton, California, as one of the original staff members of the Lick Observatory. For a brief period in 1897 Schaeberle served as acting director of Lick; then, after traveling around the world, he retired to Ann Arbor.

Schaeberle's early studies of comets included computations of their orbits and, with the aid of reflecting telescopes of his own construction, the discovery of two comets in 1880 and 1881. At Lick, Schaeberle turned to photography of stars, planets, nebulae, and solar eclipses. His visual observations led to his discovery in 1896 of the thirteenth-magnitude companion of Procyon.

During four eclipses—in California in 1889; at Cayenne, French Guiana, in 1889; at Mina Bronces, Chile, in 1893; and in Japan in 1896—Schaeberle took excellent large-scale photographs. From these he developed a mechanical theory of the solar corona (as opposed to the magnetic theories then widely discussed), according to which ejection of matter from the sun and its subsequent movement in a conic section accounted for the apparent structure of the corona.

Other theories expounded by Schaeberle, with perhaps more vigor than reason, concerned the history of the earth. He argued, for instance, that the inherent heat of the earth, not of the sun, controls the temperature of the earth, and that the sun is the parent body of the sidereal as well as of the solar system.

BIBLIOGRAPHY

The most extensive list of Schaeberle's published articles is in the Royal Society *Catalogue of Scientific Papers*, XI, 297; XVIII, 477–479.

Secondary literature includes W. J. Hussey, "John Martin Schaeberle," in *Publications, Astronomical Society of the Pacific*, **36** (1924), 309–312; a biography in *Dictionary of American Biography*, XVI, 412; and John A. Eddy, "The Schaeberle 40-ft. Eclipse Camera of Lick Observatory," in *Journal for the History of Astronomy*, **2** (1971), 1–22.

DEBORAH JEAN WARNER

SCHAFER. See **Sharpey-Schäfer, Edward Albert.**

SCHARDT, HANS (*b.* Basel, Switzerland, 18 June 1858; *d.* Zurich, Switzerland, 3 February 1931), *geology*.

Schardt was interested in natural history from the time he was a schoolboy in Basel. He entered the University of Lausanne to prepare himself for a career in pharmacy, but during his practical training at Yverdon (Vaud) he became acquainted with Édouard Desor and Auguste Jaccard, professors of geology at the University of Neuchâtel, who diverted his attention to the earth sciences. The structure and stratigraphy of the nearby Jura mountains provided Schardt with an excellent field for his researches, so that by the time he returned to Lausanne, geology had begun to become an increasingly important part of his studies. He received the diploma in both pharmacy and science in 1883, then, in the following year, presented at the University of Geneva a thesis entitled "Étude géologique sur le Pays-d'Enhaut Vaudois," for which he was awarded the D.Sc. He then took a job as science master at the *collège* at Montreux, which was admirably situated to allow him to continue his study of the Prealps. Schardt won four academic prizes between 1879 and 1891, the year in which he became a lecturer in geology at the University of Lausanne. In order to increase his knowledge of the subject, he undertook further studies, in 1892–1893, at the University of Heidelberg, where his teachers included Harry Rosenbusch.

In 1897 Schardt was called to the University of Neuchâtel to succeed Léon du Pasquier as professor of geology and paleontology. He was given only modest means with which to create a department, and devoted a considerable amount of his time to this task, spending Sundays and holidays on field trips. He soon began to take an active part in Neuchâtel scientific circles, and published a number of his observations in the series Mélanges Géologiques, edited by the Société Neuchâteloise des Sciences Naturelles. In 1911 Schardt left Neuchâtel for Zurich, where he had accepted an appointment as professor at both the Swiss Federal Polytechnical Institute and at the university. He also served as director of the university geological collections and, for several years, was concerned in organizing a department of geology and supervising the construction of new buildings to house it. He remained at Zurich for seventeen years; upon his retirement in 1928 he was succeeded by Rudolf Staub. He continued to do field research and to climb mountains until the year before he died of a stroke at the age of seventy-three.

Schardt's research encompassed tectonics, hydrology, stratigraphy, and engineering geology. His most important work, however, lay in his discovery of the older, rootless exotic complexes of the Alps, which, floating on younger series, led him to the hypothesis concerning the great alpine mass displacements that became known as the nappe theory.

Schardt's early research had made him familiar with the puzzling problem of the Prealps in which, from the Stockhorn in the east to the Chablais in the west, the Mesozoic series exhibits quite different facies from those of the surrounding mountains, so that it is obvious that they cannot have been laid down in the same basin. He made the complex flysch series of this region, which contains exotic blocks, the subject of his 1891 prize paper "Versuch einer Bahnbrechung zur Lösung der Flyschfrage und zur Entdeckung der Herkunft der exotischen Blöcke im Flysch."

In 1891–1893 Schardt, drawing upon his extensive knowledge of the geology of the area and influenced by the ideas of Marcel Bertrand, became convinced that the stratigraphical series of the Prealps had been deposited in basins far to the south, and had slid into their present position. This view was at a considerable variance with the geological thought of his contemporaries, and was greeted with, at best, condescension. The nappe theory found greater acceptance around 1903, when it had been developed and extended by Lugeon. It is interesting to note that it was not Schardt's original sliding theory (nappes du charriage) that then gained currency, but rather the thrusting theory (tectonique de poussée); Schardt's own hypothesis, as he himself had predicted, was fully accepted only after a number of years. In the meantime, Argand drew upon Schardt's studies of Jurassic folding, thrusting, and strike-slip faults for his own theory of cover folds (nappes de récouvrement).

Schardt did other important work in the exploration of the Simplon area, where he advised on the construction of the Simplon tunnel. He made observations of the recumbent folds of this metamorphosed area and described the springs that the tunnelers encountered, showing that these waters circulated in a highly complex manner, since both hot and cold springs can simultaneously flow from the same fissure. He applied his structural knowledge and analytical methods to a number of other hydrological problems, and he distinguished a number of modes of water circulation. He was particularly interested in karst hydrology and subterranean exsolution, and acted as consultant to a number of European countries on improving their water supplies. Schardt further demonstrated his expertise in both hydrology and engineering geology by advising on dam sites, landslides, and related phenomena; he was also concerned with discovering the correlations among geological structure, lithology, and the regime and chemical qualities of waters.

Schardt was vigorous and tenacious, a combination that allowed him to weather the disappointing years in which his nappe theory was scorned. His students were impressed by the force of his will and by his unfailing optimism; if he had decided to take them on a field trip, it was never postponed on account of bad weather. Schardt instead looked doggedly for a rift in the clouds, and, in deepening darkness, sought fossils by the light of matches. While reserved indoors, in the field Schardt became warm and even enthusiastic. He influenced his students by his toughness and capacity for hard work; he taught in both French and German, and supervised some sixty theses. He retained his early interest in plants and their properties, probably a relic of his pharmacological studies, and although not overly fond of alcoholic beverages nevertheless liked the bitter spirits prepared from the roots of the yellow gentian plant.

BIBLIOGRAPHY

I. ORIGINAL WORKS. Schardt published almost 200 papers and nine major geological maps. His most important writings include "Théorie des plis déjetés et couchés des Dents du Midi et des Tours Saillères," in *Bulletin de la Société neuchâteloise des sciences naturelles* (1890); "Sur la géologie du massif du Simplon," *ibid.*, 27 (1891); "L'origine des Alpes du Chablais et du Stockhorn, en Savoie et en Suisse," in *Comptes rendus hebdomadaires des séances de l'Académie des sciences,* 117 (1893); "Compte rendu de l'excursion au travers des Alpes de la Suisse occidentale," in *Comptes rendus du VIe Congrès géologique international à Zurich* (1896); "L'origine des régions exotiques et des Klippes du versant N. des Alpes suisses et leurs relations avec les blocs exotiques et les brèches du Flysch," in *Archives des sciences physiques et naturelles de Genève* (1897); "Note sur le profil géologique et la tectonique du massif du Simplon, comparés aux travaux antérieurs," in *Eclogae geologicae helvetiae,* 8 (1904); "Les causes du plissement et des chauvements dans le Jura," *ibid.,* 10, no. 4 (1908); "Neue Gesichtspunkte der Geologie. Antrittsrede als Professor der Geologie an der Universität Zurich," in *Mitteilungen der Naturforschenden Gesellschaft in Winterthur,* 9 (1911); "Unsere heutigen Kenntnisse vom Bau und von der Entstehung der Alpen. (Autorreferat)," in *Sitzungsberichte der Naturforschenden Gesellschaft in Zürich,* 71 (1926); and "Zur Kritik der Wegenerschen Theorie der Kontinentenverschiebung. (Autorreferat)," *ibid.* (1928).

II. SECONDARY LITERATURE. On Schardt and his works, see J. Leuba, "Le professeur Hans Schardt," in *Bulletin de la Société neuchâteloise des sciences naturelles*, n.s. 56 (1932), 103–119; and Hans Suter, "Professor Dr. Hans Schardt zu seinem 70. Geburtstag am 18 Juni, 1928," in *Vierteljahrsschrift der Naturforschenden Gesellschaft in Zürich,* 73 (1928), 375–388; and "Prof. Dr. Hans Schardt," in *Verhandlungen der Schweizerischen naturforschenden Gesellschaft,* 112 (1931), 411–422. All contain portraits and bibliographies, and the last contains a complete list of Schardt's works.

EUGENE WEGMANN

SCHAUDINN, FRITZ RICHARD (*b.* Röseningken, Germany, 19 September 1871; *d.* Hamburg, Germany, 22 June 1906), *zoology, medicine.*

Schaudinn's brief but highly successful career in the late nineteenth century and early twentieth century revealed the growing importance of protozoology for an understanding of various contemporary medical problems. His research on the etiologic agents of malaria and amebiasis was precise and accurate. The culmination of Schaudinn's investigations was the discovery of the microorganism responsible for venereal syphilis, first named *Spirochaeta pallida,* later *Treponema pallidum* or Schaudinn's bacillus. This important accomplishment greatly facilitated the subsequent development of an effective method of treating the disease.

The only son of an old East Prussian family of farmers, Schaudinn demonstrated in early childhood an interest in the natural world around him by systematically collecting plants, insects, and small animals. During his education in the cities of Insterburg (now Chernyakhovsk, U.S.S.R.) and Gumbinnen (now Gusev, U.S.S.R.), he showed considerable affection for both physics and chemistry. Schaudinn was a voracious reader with broad interests, especially in German philology, which led him to matriculate in philosophy at the University of Berlin in 1890. Soon he shifted his attention to the natural sciences, especially zoology; and under the direction of Schulze, he concentrated on the study of protozoa.

In 1894, after receiving a doctorate in the natural sciences from the university (he wrote a dissertation on a new genus of Foraminifera), Schaudinn participated in a highly successful expedition to Bergen, Norway, to study Arctic fauna. Upon his return he was appointed assistant at the Zoological Institute of the University of Berlin, where he intensified his protozoological studies. During the next four years, he published a number of papers dealing with the reproductive cycles of lower organisms such as the Coccidia and Haemosporidia.

In 1898 Schaudinn successfully defended his *Habilitationsschrift,* which was concerned with the importance of protozoological research on the cell theory, and he became a *Privatdozent* at the University of Berlin. In the same year he made another expedition to the Arctic. Unlike France, which had the Pasteur Institute, and Italy, with a Malaria Society, the zoologists, pathologists, and physicians of Germany worked in relative isolation from one another. Schaudinn vigorously promoted contacts between his discipline and medicine. He also stressed the importance of certain protozoa both as parasites of the human organism and as etiologic agents for certain diseases.

As a result of this interest, in 1901 Schaudinn was appointed director of a German-Austrian zoological station located in the town of Rovigno (now Rovinj, Yugoslavia) on the Dalmatian coast. The assignment was the most productive period in Schaudinn's life. He conducted several field studies on malaria in the nearby village of San Michele di Leme, observing the entrance of the sporozoite into the red blood cell. Moreover, Schaudinn clearly revealed the amoebic nature of tropical dysentery. His distinction between the harmless *Entamoeba coli* and the disease-producing *Entamoeba histolytica* was achieved by experimental self-infection with these organisms. Other studies carried out on the trypanosomes and Coccidia led in 1903 to Schaudinn's receiving the Tiedemann Award, although he had published only short abstracts on his findings. In the same year he was joined at Rovigno by Prowazek.

In 1904 Schaudinn was called to Berlin in order to assume the direction of the newly established Institute for Protozoology at the Imperial Ministry of Health. His first assignment was a study of the hookworm disease that affected German miners, and he was one of the first to confirm that the larvae of the parasite gained entrance to the body by actively penetrating the skin of the feet or legs.

During the spring of 1905, Schaudinn was given the task of verifying some experimental work on syphilis carried out by John Siegel under the direction of his former mentor Schulze. The conclusions of this rather dubious research pointed toward a common etiologic agent *(cytorrhyctes luis)* for scarlet fever, smallpox, hoof-and-mouth disease, and syphilis. The joint clinical and parasitological investigation of syphilis was conducted with the help of the dermatology clinic at the Charité Hospital in Berlin. The dermatologist Erich Hoffmann became Schaudinn's clinical consultant and the bacteriologist Fred Neufeld, his assistant.

Soon thereafter, Schaudinn detected a pale-looking spiral-shaped rod among a myriad of other microorganisms contained in a fresh microscopical preparation derived from the fluid of an eroded syphilitic papule. Schaudinn was excited about his finding, although he believed that the organism, named *Spirochaeta pallida*, was probably a saprophyte. Other types of spirochetes were found in subsequent preparations of nonsyphilitic lesions such as condylomas.

Thus, the first report of Schaudinn and Hoffmann dated 10 March 1905 merely pointed out the existence of *Spirochaeta pallida* in syphilitic lesions without ascribing to it any importance as a possible causal factor for the disease in question. Sometime later, Schaudinn was able to differentiate between the coarse and fine spirochetes and observe the *Spirochaeta pallida* in Giemsa-stained microscopical preparations obtained from syphilitic lymph nodes. These more significant findings, reported 17 May 1905 to the Berlin Medical Society, more clearly implicated the microorganism as the etiologic agent of syphilis, since it had been observed in both primary and secondary syphilitic lesions.

Schaudinn remained extremely cautious and, although he was confident of the causal relationship between the *Spirochaeta pallida* and syphilis, he was unconvincing in his presentation of his findings to the Berlin Medical Society on 17 and 25 May. He faced strong opposition from Schulze's followers, and the presiding officer's closing remarks at the meeting reflected the deep skepticism with which Schaudinn's discovery was greeted. The distinguished surgeon Ernst von Bergmann declared the session adjourned "until another agent responsible for syphilis engages our interest."

Soon various investigators in other countries were able to verify the repeated and exclusive presence of the *Spirochaeta pallida* in syphilitic lesions, but no pure cultures of the microorganism could be obtained immediately. Schaudinn received a series of offers to work and teach abroad, notably at the universities of London and Cambridge, but he finally took a leave of absence from his post with the Ministry of Health in order to work in the new protozoology laboratory of the Research Institute for Naval and Tropical Diseases at Hamburg.

The post was confirmed by the senate of Hamburg in early 1906, but Schaudinn's career at the new location was short-lived. From 1904 onward he had suffered from furunculosis, which caused an anal fistula. He died of a rectal abscess and general sepsis shortly after returning from the International Medical Congress in Portugal. Among the honors bestowed upon him were corresponding memberships in the Senckenbergische Naturforschende Gesellschaft (1903), the Imperial Academy of Sciences in St. Petersburg (1905), and the Berlin Society of Internal Medicine (1906).

BIBLIOGRAPHY

I. ORIGINAL WORKS. Schaudinn's most important articles, in a number of journals, have been collected in *Fritz Schaudinns Arbeiten* (Hamburg–Leipzig, 1911), published under the auspices of the Hamburgische Wissenschaftliche Stiftung; with biographical sketch by S. von Prowazek and with a list of all the material that Schaudinn published in his short career. Schaudinn's first paper on the discovery of *Spirochaeta pallida* was dated April 1905 and appeared as "Vorläufiger Bericht über das Vorkommen von Spirochaeten in syphilitischen Krankheitsprodukten und bei Papillomen," in *Arbeiten an das K. Gesundheitsamte,* **22,** no. 2 (1905), 527–534.

The article was shortly thereafter published with the same title in book form (Berlin, 1905). A second paper, also co-authored by Erich Hoffman, reported the findings of *Spirochaeta pallida* in syphilitic lymph nodes: "Ueber Spirochaetenbefunde im Lymphdrüsensaft Syphilitischer," in *Deutsche medizinische Wochenschrift*, **31**, no. 18 (1905), 711–714. See also Schaudinn's report to the Berlin Medical Society, "Ueber Spirochaeta pallida bei Syphilis und die Unterschiede dieser Form gegenüber anderer Arten dieser Gattung," in *Berliner klinische Wochenschrift*, **42**, no. 22 (29 May 1905), 673–675; no. 23 (5 June 1905), 726.

II. Secondary Literature. The best biographical account is Christel Kuhn, *Aus dem Leben Fritz Richard Schaudinns* (Stuttgart, 1949). Although brief, the work is based on extensive archival material and personal interviews with Schaudinn's widow and sister; it contains a small bibliography of secondary sources on Schaudinn and his work.

Shorter biographical sketches are in two obituaries: W. Loewenthal, in *Medizinische Klinik*, **2**, no. 26 (1906), 693–694; and S. von Prowazek, in *Wiener klinische Wochenschrift*, **19** (1906), 880–882. On the occasion of the 25th anniversary of Schaudinn's discovery, the personal recollections of one of his assistants appeared: F. Neufeld, "Zum 25 jährigen Gedenktage der Entdekkung des Syphiliserregers," in *Deutsche medizinische Wochenschrift*, **56** (1930), 710–712. Similar source material is in Erich Hoffmann, *Vorträge und Urkunden zur 25 jährigen Wiederkehr der Entdeckung des Syphiliserregers* (Berlin, 1930), and in A. Schuberg and H. Schlossberger, "Zum 25 Jahrestag der Entdeckung der Spirochaeta Pallida," in *Klinische Wochenschrift*, **9** (1930), 582–586.

See also O. T. Schultz, "Fritz Schaudinn, a Review of His Work," in *Johns Hopkins Hospital Bulletin*, **19** (1908), 169–173; J. H. Stokes, "Schaudinn, a Biographical Appreciation," in *Science*, **74** (1931), 502–506; and a short biographical sketch "Fritz Richard Schaudinn (1871–1906)," in John H. Talbott, ed., *A Biographical History of Medicine* (New York, 1970), 796–798.

Guenter B. Risse

SCHEELE, CARL WILHELM (*b.* Stralsund, Swedish Pomerania, 19 December 1742; *d.* Köping, Sweden, 21 May 1786), *pharmacy, chemistry.*

Scheele was the seventh of eleven children of Jochim Christian Scheel(e) and Margaretha Eleonora Warnekros. Like his oldest brother, Johan Martin, he became interested in pharmacy at an early age and chose it as his career. It is said that while still a boy he was taught how to read prescriptions and to write chemical symbols by two friends of the family in Stralsund, a physician named Schütte and a pharmacist named Cornelius.

After finishing school, which did not include a Gymnasium course, Scheele went to Göteborg in 1757 and began his training in the pharmacy of Martin Bauch, where his brother also had been an apprentice.

Bauch's great influence on Scheele's development has been confirmed by previously unused source material and is supported by the general reputation for competence that Bauch enjoyed among contemporaries as knowledgeable as Linnaeus. Although Bauch was sixty-three years old when Scheele came to him, he was by no means set on outdated alchemical theories—on the contrary, he was aware of new developments in his profession.

Scheele remained with Bauch beyond his six years of apprenticeship, until the pharmacy was sold in 1765. Then he left Göteborg and for the next ten years traveled as a journeyman. In Malmö he found work in a pharmacy run by Peter Magnus Kjellström, who fully understood Scheele's preference for experimental work and allowed him to work in the laboratory of the pharmacy. His stay in Malmö was especially important because of its proximity to the university city of Lund. It gave Scheele his first contact with the academic world through his friendship with Anders Retzius, lecturer in chemistry at the university and the same age as Scheele. The proximity of Copenhagen, a center of culture and trade, made it possible to buy recently published chemical literature.

Scheele remained in Malmö for three years, until tempted by the better facilities available in Stockholm: the Royal Swedish Academy of Sciences was located there, and Uppsala with its famous university was nearby. In the spring of 1768 he found a position in a Stockholm pharmacy run by Johan Scharenberg, but was allowed only to prepare prescriptions. Since this was hardly to his liking, in the summer of 1770 he moved to the establishment of Christian Ludwig Lokk in Uppsala. There he had a workbench in the laboratory, was soon recognized as an able chemist, and met Johan Gottlieb Gahn. Gahn was then an assistant to Torbern Bergman, and he soon brought together the unusually capable apprentice experimenter and the outstanding theorist. This contact eventually developed into a lifelong friendship.

Scheele made important discoveries during his five years in Uppsala. On 4 February 1775, still a *studiosus pharmaciae*, he was elected a member of the Swedish Academy of Sciences and began to publish in its *Handlingar* ("Transactions"). Also while in Uppsala he began his major work: to combine the many and varied experiments with fire and air made during the previous years into an integrated book.

Since after fulfilling his apprenticeship Scheele had for thirteen years contented himself with subordinate positions offering limited possibilities for experimental research, it is not surprising that he was eager to seize an opportunity that held the promise of independence. Sara Margaretha Sonneman, the daughter of a councilman in Köping, had married a pharmacist named Herman Pohl in 1772. He died in 1775, and now she was looking for someone to carry on the pharmacy privilege that she had inherited. In the summer of that year Scheele reached an agreement with the widow that he would manage the pharmacy independently for one year and that after nine months he could negotiate for its purchase. This agreement nearly fell through when another prospective buyer secretly rented the pharmacy. In less than a year, however, Scheele had become so popular and respected that the citizens of the province demanded that he continue to be the city's pharmacist. He remained in the small town until his death, disregarding all offers to leave it, including that to succeed Marggraf at Berlin.

Scheele left Köping only once, for a few days. After the dispute about the ownership of the pharmacy had been settled, he traveled to Stockholm, where he took his long-postponed pharmacy examination and swore the pharmacist's oath on 11 November 1777. Two weeks earlier he had finally taken his seat as a member of the Royal Swedish Academy of Sciences, to which he had been elected in 1775. On 23 November Scheele returned to Köping. On 18 May 1786, on his deathbed, he married the widow Pohl and made her his heiress.

The earlier periods of Scheele's scientific career have not yet been elucidated; but from the extant letters of his employers Bauch and Kjellström, as well as from statements of his co-workers in Göteborg, it is clear that his curiosity was as boundless as his persistence. A surviving inventory of 1755 shows that the two most famous chemical handbooks of the time, Kunckel's *Laboratorium chymicum* and Caspar Neumann's *Praelectiones chemicae*, were available in Bauch's pharmacy; and Scheele undertook to repeat the numerous experiments in these books. The critical mind that Scheele had developed in his teens soon led him to conclude that in many instances he knew better than the authorities of his time. This belief explains his boldness in declaring his opposition to an important detail of the phlogiston theory. According

to this theory all combustible material contained a special substance, phlogiston, that escaped as the material burned. Scheele, however, thought that the phlogiston went into the heat that was formed, so that it could be liberated from the heat or fire. Thus, while denying that fire was an element, he simultaneously changed the phlogiston theory by stating that some reductions can occur without the addition of the reducing medium, phlogiston, but solely by providing heat. As a demonstration he cited the reduction of silver nitrate to metallic silver by heating it to redness (in which case the necessary phlogiston was provided by the fire).

Scheele's refusal to accept phlogiston as an element of combustibility, regardless of what the leading authorities said, showed an unusual instinct for research in a teen-ager; and one could expect that such exceptional abilities would soon produce written results, perhaps as laboratory notes. Among the extant papers of this kind, however, nothing can be ascribed with certainty to the early years in Göteborg. For the last several years there has been a renewed attempt in Sweden to make an inventory and a more thorough examination of Scheele's manuscripts, and it is anticipated that this effort will lead to a more precise dating. The preserved laboratory notes are quite detailed but difficult to interpret, for they are jotted down in an ungrammatical Old German and in a script that is extremely hard to read. The words are often so abbreviated as to be incomprehensible and the text filled with variations of now obsolete chemical symbols, or with their Latin equivalents.

The first to try to decipher Scheele's manuscripts was the mineralogist and polar scientist Adolf Erik Nordenskiöld, who, in connection with the commemoration of the 150th anniversary of Scheele's birth (1892), published a selection of Scheele's correspondence with excerpts from his laboratory notes. Nordenskiöld himself described this as a pioneering work and pointed out that through the publication of Scheele's manuscripts it would be possible to learn more of his development, evolution of ideas, and work habits. The ambitious plan, however, yielded to an aim that seemed more in harmony with the patriotic feelings of the time: to secure Scheele's priority for the discovery of oxygen. When Nordenskiöld had accomplished this task, he abruptly halted deciphering of the manuscripts.

In 1942, in connection with the festivities held on the bicentennial of Scheele's birth, a new edition of the laboratory notes from which Nordenskiöld had presented excerpts was published. The work, edited by the physicist C. W. Oseen, still did not satisfy the demand for clear, scientific interpretation; and the dating was rejected by experts. Neither Nordenskiöld nor Oseen had exhausted the existing manuscript material, however. Therefore the chemist and historian of science Uno Boklund has undertaken the publication of all available laboratory notes and Scheele's correspondence, complete with commentary.

Among the items of Scheele's correspondence published by Nordenskiöld in 1892 are several letters to contemporary lecturers in chemistry at the University of Lund. Anders Retzius was a correspondent between 1 December 1767 and 26 April 1768. These letters are the oldest dated documents written by Scheele that have so far been found. The contents are concerned entirely with chemistry and thus are of great value in following the course of thought that led to Scheele's opposition to the leading theories and to his discovery of oxygen. It would be almost impossible to obtain a thorough understanding of Scheele and his work without an intimate knowledge of these temperamental and stimulating letters. They reveal a number of experiments on nitrous acid, which Scheele named "volatile acid of niter," which for that time were outstanding scientific achievements.

As is so often the case, Scheele's great experimental abilities were overlooked by scholars because of insufficient understanding of eighteenth-century chemistry. They did not realize that the young apprentice was facing a serious conflict, having reached a point in his research where the evidence of his experiments no longer agreed with the classic phlogiston theory. Relying on his experiments, he felt compelled to speak out, a course of action that led not only to the formulation of his theories of combustion and calcination but also to the discovery of oxygen. By comparing his experimental results with reports to Retzius of the same period, it can be seen that in connection with experiments with the volatile nitrous acid, Scheele had already observed that saltpeter becoming red-hot in a crucible seemed to "boil" and that above the crucible sparks flamed up and burned with a vivid brilliance. He did not know that this was caused by a gas emanating from the saltpeter that was identical with a component of the air, but his comments concerning the experiments are noteworthy:

> I realized the necessity to learn about fire. . . . But I soon realized that it was not possible to form an opinion on the phenomena of fire as long as one did not understand air. And after I had made a certain number of experiments, I . . . saw that the air penetrated into the burning material and constituted a component of flames and sparks.

This marked a milestone in the development of Scheele's thought, for if air penetrated into the burning material, fire could not be an element; it must be a compound.

It is not known how and when Scheele completed his research on the phenomena of combustion. According to Retzius, in 1768–1770, while in Stockholm, Scheele undertook a series of experiments that laid the basis for his book on air and fire. No attempt has ever been made to find out what these experiments signified, but a manuscript dating from the time before his stay in Stockholm shows that Scheele was already occupied with highly advanced gas experiments. It is of importance, but has gone unnoticed, that his experiments often had a physiological background. He examined how plants react to gases and gas mixtures and had already perfected a technique to isolate and collect in ox bladders the air he wanted to examine. It is especially interesting that he began to examine exhaled air and understood that lime removes *aer fixus* (carbon dioxide) from it and, thus, that air is separated into its various components through physiological processes. On the whole the above-mentioned manuscripts support Retzius' statements and, in combination with other known facts, give validity to the assumption that it was in Stockholm that Scheele definitively rejected the theory of air as an element.

Among the many problems that occupied Scheele in the 1760's, his examinations of volatile nitrous acid are undoubtedly the most interesting for tracing the evolution of his thought. But in letters to Retzius he mentions the excellent results of his work with Prussian blue, which many years later led to the discovery of prussic acid, and also of a long series of experiments with boric acid and the first experiments with hydrogen sulfide. The isolation of tartaric acid also was part of this early work; but Scheele's account of it, sent to the Academy of Sciences, went unnoticed. Retzius, who in the fall of 1768 had come to Stockholm, enlarged it with some additional experiments and published it in the *Handlingar* of the Academy in 1770, with due acknowledgment of Scheele's priority. While in Stockholm, Scheele also undertook research on the chemical reactions of light on silver salts and discovered that different parts of the solar spectrum reduce with different strength. According to

Retzius it was also in Stockholm that Scheele found that *aer fixus* was an acid.

The fields of interest mentioned so far reveal the youthful Scheele's imposing versatility, but they are greatly overshadowed by the veritable catalog of important chemical observations and discoveries that appeared under the heading "P. M. hördt af Herr Scheele. År 1770 om Våhren" ("Memorandum Heard from Herr Scheele. Spring 1770"). This document, preserved in the Scheele archives of the library of the Academy of Sciences in Stockholm, was written in Swedish by Johan Gottlieb Gahn.

By way of introduction this memorandum presents Scheele's observations concerning the chemical composition of *terra animalis* (bone ash), which later, after Gahn had proved the presence of phosphorus, led to Scheele's method of producing phosphorus from animal bones instead of from urine (as had been customary until then). Other observations concern tartaric acid, pyrotartaric and pyruvic acids, oxalic acid, gallic and pyrogallic acids, and citric acid. A number of notes about gases deal with the production of ammonia gas and its combustibility, with hydrochloric acid gas, and with *aer fixus*. One experiment concerns the change of water into earth, and there are many observations about different chemical compounds. Of the greatest interest are some notes on the solution of iron in acid, for they make it clear that Scheele already knew about the different oxidation grades of iron.

This detailed memorandum covering Scheele's earliest years at Uppsala sheds no light on the research in minerals that then attracted his interest. His work with fluorspar led to the discovery of a new mineral acid, hydrofluoric acid (mixed with hydrofluosilicic acid). The discovery of "the Swedish acid" attracted great attention and was the subject of Scheele's first independent contribution to the *Handlingar* of the Academy (1771). In a letter of 10 April 1772 Macquer wrote about it to Torbern Bergman and requested more information.

When the experiments on fluorspar were finished, Bergman suggested that Scheele study the chemistry of pyrolusite. Although chemists had been interested in the mineral for many years, they had discovered little more than properties necessary to differentiate it from other minerals. By contrast, Scheele began his examination by determining the solubility of pyrolusite, under various conditions, in all the known acids. Thus he was able to show that it has a very strong attraction for phlogiston (in modern terms, it is a strong oxidizing agent) and could place it in a phlogistic reducing system, which gave him the key to many of the mineral's properties that were not then understood.

The most important result of these experiments, however, was Scheele's discovery of chlorine, which occurred in connection with the attempts to dissolve pyrolusite in hydrochloric acid. In a communication to Gahn of 28 March 1773, Scheele said that by dissolving pyrolusite, hydrochloric acid loses its phlogiston and becomes a yellowish gas that smells like aqua regia and can dissolve gold but is barely soluble in water. In a subsequent letter he reported additional properties of the new gas, among them the observation that it reacts as a bleach with textile dyes, becoming hydrochloric acid in the process, and that the original colors cannot be restored.

From the same communication to Gahn it is clear that Scheele, after the successful dephlogistication of hydrochloric acid that resulted in the discovery of chlorine, conceived the idea of placing arsenic in a hypothetical phlogistic reducing system. In an article on arsenic, published in 1775, he says that shortly after he had discovered the phlogiston content of arsenic and that it could be dephlogisticated (oxidized), he decided to investigate arsenic without phlogiston and found that it contained an acid (arsenic acid). The detailed article describes more than 100 experiments and contains the first mention of the later famous "Scheele's green." Before the article was published, Scheele added a short note concerning a newly discovered material, "earth of heavy spar," which was later renamed baryta. A separate article about this earth was published in 1779.

Scheele's experimental activity in Uppsala was quite extensive. In a letter of 1 March 1773 he stated that he was the first to discover that gases are absorbed by recently ignited charcoal, and, in another letter, that he had analyzed *sal microcosmicum*, which could be isolated from urine, and found it to be sodium ammonium phosphate. In addition, he had worked out a method to produce *sal benzoes* by keeping it wet. This communication was read to the Academy of Sciences in 1774 and was later printed in the *Handlingar*. Another pharmaceutical preparation discussed in these letters was rhubarb with mercurous chloride.

Further studies of Scheele's correspondence from this period show that he began to make notes on still greater discoveries. Ironically, in foreign countries, especially in France, he and his work were much better known than in Sweden. Thus, when Lavoisier published his *Opuscules physiques et chymiques* at the beginning of January 1774 and sent a copy to the Academy of Sciences in Stockholm, he showed his great respect for Scheele's scientific work by stating that he would forward an additional copy to him.

In a letter to Lavoisier of 30 September 1774 Scheele freely disclosed his discovery of oxygen and, to demonstrate his gratitude for the book, gave him instructions on how to make pure oxygen. Until then no chemist who had produced or released it had been aware that it was a completely new gas or had been fully familiar with all its properties. Scheele demonstrated his superiority over his scientific contemporaries by giving Lavoisier information on both the chemical and the physiological properties of oxygen. This letter from Scheele is the earliest known written description of the detailed method of producing oxygen, together with complete information on its chemical nature and physiological properties. It was discovered among Lavoisier's papers by his biographer Édouard Grimaux, who published the letter in a way that clearly showed his awareness of its importance; as part of a communication entitled "Une lettre inédite de Scheele à Lavoisier," it adorns the first page of the initial issue, dated 15 January 1890, of the *Revue générale des sciences pures et appliquées*. The important content of the letter was not understood, however; it was ignored in the research on the chemical revolution and was not even reprinted by Grimaux. This letter has not been included in the most recent French edition of Lavoisier's correspondence.

In a letter of 2 August 1774 from Scheele to the secretary of the Stockholm Academy, Per Wargentin, in which he said that he had studied air and fire for many years, Scheele mentioned that until recently he had thought he was the only one to know of certain phenomena but that "some Englishman had gone very far in his researches." It is conceivable that this realization that he had rivals impelled him to summarize his own results, for the manuscript of *Chemische Abhandlung von der Luft und dem Feuer* was ready for the printer on 22 December 1775. In it, with the assistance of numerous elegant experiments, Scheele proved that air liberated from aerial acid (carbon dioxide) and water vapor consists of two gases: fire air (oxygen), which can support combustion, and vitiated, foul air (nitrogen), which cannot.

The printing of the book was delayed for two years, partly because Torbern Bergman did not deliver his promised preface until 1 July 1777; the work appeared the following month. By then oxy-gen was already known, and some of the observations in the book had been published by others. Priestley, during a visit to Paris in October 1774, informed Lavoisier, Le Roy, and many other scientists of his discovery of a new remarkable gas on 1 August 1774; and that date was long accepted as the "birthday" of oxygen. Now it is known from Scheele's laboratory notes that the discovery was made at least two years before Priestley's. Even the notion that Priestley should have published it earlier is negated when it is noted (Partington) that Bergman had published a summary of Scheele's discovery of oxygen and of his theory of heat in *Nova acta Regiae societatis scientiarum upsaliensis* many months before Priestley revealed his discovery.

In the last chapter of *Luft und Feuer*, Scheele wrote about his experiments with "fetid sulfurous air" (hydrogen sulfide), the first correct description of its properties; he was also the first to record a synthesis of the gas through the heating of sulfur in hydrogen. Scheele's view that this gas consisted of sulfur, heat, and phlogiston was correct, since he thought that hydrogen was a combination of heat and phlogiston. Further, Scheele provided the first description of hydrogen polysulfides, which he had already mentioned in a letter to Retzius in 1768.

The minerals plumbago (graphite) and molybdena (molybdenite, MoS_2) are so similar in appearance and physical properties that they are often confused. In an effort to differentiate them, Scheele thickened and boiled molybdenite with all known acids, but even arsenic and nitric acids had no effect. But the oxidizing nitric acid eventually yielded a white powder that was soluble in water and gave an acid reaction. Scheele named it *terra molybdaenae* (MoO_3) and observed that "the earth" gave a blue solution with concentrated sulfuric acid, a reaction that led him to consider it "not reluctant to attract phlogiston" — in other words, it could be reduced. Since Scheele did not possess a furnace that could reduce the earth to metal, he asked his friend Peter Jacob Hjelm, a mineralogist, to do this for him. In 1781, with improved ovens, Helm produced the metal and suggested the name molybdenum.

In 1770–1771 Scheele had examined plumbago (manuscript no. 2 in the *Brown Book*) but had obtained no definite results. He now took up the problem again and showed that the end product obtained by detonation with saltpeter or heating with arsenic acid was aerial acid (carbon dioxide). From this he concluded that graphite is carbon dioxide combined with a great amount of phlogiston—that carbon is a reduction product of carbon dioxide. In a letter to Bergman of 18 August 1780, Scheele discussed his thorough study of tungsten (*lapis ponderosus*, now called scheelite, $CaWO_4$). He had found that it consists of lime and a new acid, tungstic acid; but his efforts to reduce it to metal failed.

Scheele's contributions to inorganic chemistry should not overshadow his research in organic chemistry, which may be considered more imposing, since he had no precedent. In his preliminary attempts to isolate from plants or animals the delicate materials he planned to examine, he had to avoid the destructive methods (calcination, distilling until dry) common in inorganic chemistry and to proceed with greater caution, working with lower temperatures and learning to extract with water or some other solvent. Scheele was especially successful in his method of isolating organic acids by precipitating the acid as an insoluble calcium salt or potassium ferrocyanide and then separating it with diluted sulfuric acid.

Scheele had obtained excellent results in inorganic chemistry by oxidation, and the same methods in organic chemistry led to the discovery of many new acids. When this work is added to his researches in protein and fat, it is clear that Partington's judgment that Scheele's influence in organic chemistry was fundamental was fully justified.

Scheele had isolated tartaric acid during his stay in Malmö and had discussed its property of forming both acid and neutral salts. Later he observed the formation of pyrotartaric acid. Scheele documented his interest in plant acids in his first known letter (1 December 1767), in which he wrote to Retzius: "So far as the essential salts [acids] of plants are concerned, I have crystallized the juices of *Aconitum, Stramonium* and *Mentha crispa*." In manuscript no. 1 of the *Brown Book* he began to examine "rhubarb earth," which he thought consisted of citric acid bound with lime (also in manuscript no. 2). About 1770 he changed his mind and thought it was phosphoric acid, then finally found that "rhubarb earth" was the calcium salt of acid of sorrel. Until then no chemist had been able to prepare free acid of sorrel. At the beginning of 1776 Scheele had succeeded in isolating it through the use of baryta, but it was not absolutely pure. In 1784 he obtained the pure acid by precipitating it out with potassium ferrocyanide and then changing it with a calculated amount of sulfuric acid. At the same time he learned that the pure acid of sorrel was identical with acid of sugar that he had discov-

ered earlier and had produced through the oxidation of sugar with nitric acid. Both acids were later called oxalic acid.

Scheele's outstanding experimental ability also led to his solution of the difficult problem of obtaining pure crystallized citric acid from lemon juice. If he could do so, he could determine the properties of the acid and also compare it with acids of other fruits and berries. Scheele investigated twenty-one kinds of fruit and berries, as well as fifteen other materials of vegetable origin and some animal material (glue, egg white, egg yolk, blood), all of which were analyzed after oxidation with nitric acid. He finally found an organic acid (which he named malic acid) in the apple. The same process with milk sugar had already yielded yet another new acid, *acidum sacchari lactis*, later called mucic acid.

Valentine, Paracelsus, and Helmont had tried without success to analyze urinary calculi, and their successors had not fared any better. Scheele attempted to solve the problem in a different manner, and in 1776 he reported that he had separated from both kidney stones and urine the acid of calculus (uric acid). He added that it could be recognized by the red spots it produced on the skin after being dusted with nitric acid.

According to one of his letters to Retzius, Scheele's work with the coloring principle of Prussian blue dated from 1765. This intricate problem was one of the most difficult of Scheele's career, and it took eighteen years for him to reach his goal: the discovery of prussic acid. He reported that this *acidum berolinense* (hydrocyanic acid) had "a peculiar but not disagreeable smell, a taste somewhat approaching sweet, and warm in the mouth, at the same time exciting a cough." It is difficult, says Partington, to understand how he escaped with his life.

The capstones of Scheele's gigantic chemical edifice were his discovery of glycerol and of the art of preserving vinegar by heating the vessel containing it in a kettle with boiling water—pasteurization a century before Pasteur. He also found that vinegar is an oxidation product of alcohol and that "an ether of exquisite smell" (acetaldehyde) is an intermediate product during the oxidation.

The list of Scheele's discoveries does not tell the whole story of his work. For this, more thorough source studies are needed, especially editions of Scheele's collected manuscripts with expert commentary. When such material is available, it will be possible to consider his work as that of an indefatigable seeker after truth who was driven to test and retest the validity of contemporary answers to the great chemical controversies of the time, without regard for theoretical attitudes. Scheele's thousands of experiments seem to be random, but closer examination reveals that they are ordered into groups, each of which has a characteristic background that connects it to central theoretical chemical questions: the concept of phlogiston, the value concept, the concepts of elements, the concepts of alkali and acid, and the controversy over *acidum pingue*.

Also needed is a study of the dissemination of Scheele's ideas and experimental findings to other countries and their influence on the development of chemistry. It is therefore important to examine his foreign correspondence. Since Scheele informed Torbern Bergman in great detail of his discoveries, it is no less vital to determine the ways in which Bergman disseminated these chemical reports, which were given to him without any request for secrecy.

BIBLIOGRAPHY

I. ORIGINAL WORKS. MSS in the Library of the Royal Academy of Sciences in Stockholm are *Scheele's Brown Book*, a collection described as extremely difficult to decipher, that includes laboratory notes, drafts of papers, and drafts of letters (the draft of Scheele's letter to Lavoisier among them); a collection of laboratory notes, difficult to read, on sheets of paper of different shapes and sizes; and part of Scheele's extensive correspondence.

MSS in the library of the University of Uppsala are mainly Scheele's important letters to Torbern Bergman.

A selection of the above material has been published in the following editions: A. E. Nordenskiöld, ed., *Carl Wilhelm Scheele. Efterlemnade bref och anteckningar* (Stockholm, 1892), also in *Carl Wilhelm Scheele. Nachgelassene Briefe und Aufzeichnungen* (Berlin, 1893); C. W. Oseen, ed., *Carl Wilhelm Scheele, Manuskript 1756–1777* (Uppsala, 1942); and a selection of these MSS deciphered by C. W. Oseen, also published as *Carl Wilhelm Scheele, Manuskript 1756–1777* (Uppsala, 1942); Otto Zekert, ed., *Carl Wilhelm Scheele. Sein Leben und seine Werke*, 7 vols. (Mittenwald, 1931–1935).

None of these eds., however, offers exact deciphering and critical clarity. Nordenskiöld arbitrarily modernizes Scheele's seventeenth-century German and omits important letters, such as the correspondence with J. C. Wilcke, secretary of the Academy of Sciences. (These letters can be found in J. C. Oseen, *Johan Carl Wilcke. Experimentalfysiker* [Uppsala, 1939], 312–341). Oseen's ed. of . . . *Scheele, Manuskript 1756–1777* must be used with great care, since the work abounds with mis-

takes due to incorrect reading of the text and his dating of the MSS is extremely questionable. Johan Nordström's review, "En edition av Scheeles efterlämnade manuskript" ("An Edition of Scheele's Remaining Manuscripts"), in *Lychnos* (1942), 254–277, is recommended for its detailed and clear criticism. It is a thorough examination of Oseen's work and also severely criticizes Zekert's mistakes in his work on Scheele.

A new ed. of Scheele's work has now been started, supported by the Bank of Sweden Tercentenary Fund and other foundations. It is complete, with commentary on all his laboratory notes, letters to and from him, and English trans. of all of the material: *Carl Wilhelm Scheele. His Work and Life*, Uno Boklund, ed., 8 vols. Thus far 2 vols. have appeared: I–II, *The Brown Book* (Stockholm, 1969). In preparation are III–IV, *Laboratory Notes*, facs. ed. with intro., decipherment, English trans., and commentary; V, *Correspondence 1767–1777*, with intro. and parallel English trans.; VI–VII, *Correspondence 1778–1786*, with parallel English trans., commentary, and index; and VIII, which will cover Scheele's life and scientific achievements, his influence on contemporary European chemistry, and his role in the chemical revolution, plus a bibliography and general index.

A complete bibliography of Scheele's numerous printed works is in Nordenskiöld, *op. cit.*, xxxii–xxxviii. Partington includes a list of Scheele's most important works in his *A History of Chemistry*, III (London, 1962), 210–211.

II. SECONDARY LITERATURE. There is no standard biography of Scheele; but since the first obituary was published by Lorenz von Crell in *Chemische Annalen*, 1 (1787), 175–192, many works of varying reliability have appeared. Nordenskiöld, who published an interesting description of Scheele's life (*op. cit.*, vii–xxxi) gives a list of the most important biographies (pp. xli–cl). An indispensable guide for the study of the abundant literature about Scheele is Bengt Hildebrand, "Scheeleforskning och Scheelelitteratur," in *Lychnos* (1936), 76–102. An American contribution is Georg Urdang, *The Apothecary Chemist Carl Wilhelm Scheele, A Pictorial Biography* (Madison, Wis., 1942; 2nd ed., 1958). Important for questions of priority is J. R. Partington, "The Discovery of Oxygen," in *Journal of Chemical Education*, 39 (1962), 123–125.

Among the Swedish contributions are the following, listed chronologically: J. Nordström, "Några bortglömda brev och tidskriftsbidrag av Carl Wilhelm Scheele" ("Some Forgotten Letters and Journal Contributions by Carl Wilhelm Scheele"), in *Lychnos* (1942); and "Två notiser till Scheeles biografi" ("Two Notes to Scheele's Biography"), *ibid.*, 280–284; Uno Boklund, "A Lost Letter From Scheele to Lavoisier," *ibid.* (1957–1958), 39–62; *Carl Wilhelm Scheele. Bruna Boken. Utgiven i faksimil med deschiffrering och innehållsanalys jämte inledning och kommentar* ("Carl Wilhelm Scheele. The Brown Book. Published in Facsimile With Deciphering and an Analysis of the Contents With an Introduction and Commentary"; Stockholm, 1961); "Varför Scheele måste börja tala engelska" ("Why Scheele Had to Begin Speaking English"), in *Svensk farmaceutisk tidskrift*, **68** (1964), 967–979; and "Die Rolle Carl Wilhelm Scheeles in der chemischen Revolution des 18. Jahrhunderts," in *Ruperto-Carola*, **18** (1966), 306–317; and Hugo Olsson, *Kemiens historia i Sverige intill år 1800* ("The History of Sweden Until the Year 1800"; Uppsala, 1971), 136–151, 221–234, 282–297, and *passim*.

UNO BOKLUND

SCHEFFERS, GEORG (*b.* Altendorf, near Holzminden, Germany, 21 November 1866; *d.* Berlin, Germany, 12 August 1945), *mathematics*.

Scheffers studied mathematics and physics from 1884 to 1888 at the University of Leipzig, where his father was professor at the Academy of Art. He received the doctorate from Leipzig in 1890 and qualified as a lecturer there the following year. In 1896 he became extraordinary professor at the Technische Hochschule in Darmstadt, and in 1900 he was promoted to full professor. In 1907 he succeeded Guido Hauck as full professor at the Technische Hochschule in Charlottenburg, where he remained until his retirement in 1935.

As a student Scheffers was greatly influenced by Sophus Lie, who was professor at the University of Leipzig from 1886 to 1898. He followed Lie's suggestions in choosing topics for both his doctoral dissertation and his *Habilitationsschrift*, which dealt respectively with plane contact transformations and complex number systems. Scheffers' most important independent research inspired by Lie was his 1903 paper on Abel's theorem and translation surfaces.

In later years Scheffers' reputation was based largely on his own books. These writings, which grew out of his wide-ranging activities at technical colleges, were directed at a broader audience than the books he edited with Lie; and they all went through several editions. Scheffers' two-volume *Anwendung der Differential- und Integralrechnung auf Geometrie* (1901–1902) was a popular textbook of differential geometry. Also widely used was his revision of Serret's *Lehrbuch der Differential- und Integralrechnung*, the last edition of which appeared in 1924; subsequently it was superseded by books written in a more modern style. Scheffers also published *Lehrbuch der darstellenden Geometrie* and, in 1903, an article entitled "Besondere transzendente Kurven" in the *Encyklopädie der mathematischen Wissenschaften*.

Scheffers' favorite field of study was geometry and, more specifically, the differential geometry of intuitive space. In this area he was a master at discovering many properties of particular curves and surfaces and their representation; he also possessed a gift for giving an easily understandable account of them—although in a much wordier style than is now customary. His exceptional talent for vividly communicating material is also apparent in a later work on the grids used in topographic maps and stellar charts.

BIBLIOGRAPHY

Scheffers' original works are as follows: "Bestimmung einer Klasse von Berührungstransformationsgruppen," in *Acta mathematica*, **14** (1891), 117–178; *Zurückführung komplexer Zahlensysteme auf typische Formen* (Leipzig, 1891); *Anwendung der Differential- und Integralrechnung auf Geometrie*, 2 vols. (Leipzig, 1901–1902; 3rd ed., 1922–1923); "Das Abelsche Theorem und das Lie'sche Theorem über Translationsflächen," in *Acta mathematica*, **28** (1902), 65–91; "Besondere transzendente Kurven," in *Encyklopädie der mathematischen Wissenschaften*, III, pt. 3 (Leipzig, 1903), 185–268; *Lehrbuch der darstellenden Geometrie*, 2 vols. (Berlin, 1919–1920; 2nd ed., 1922–1927); and *Wie findet und zeichnet man Gradnetze von Land- und Sternkarten?* (Leipzig–Berlin, 1934).

Scheffers edited the following volumes by Lie: *Vorlesungen über Differentialgleichungen mit bekannten infinitesimalen Transformationen* (Leipzig, 1891); *Vorlesungen über kontinuierliche Gruppen mit geometrischen und anderen Anwendungen* (Leipzig, 1893); and collaborated with Lie on *Geometrie der Berührungstransformationen*, I (Leipzig, 1896). He also revised J. A. Serret's *Lehrbuch der Differential- und Integralrechnung*, A. Harnack, trans., 5th ed., 3 vols. (Leipzig, 1906–1914; I, 8th ed., 1924; II, 7th ed., 1921; III, 6th ed., 1924).

WERNER BURAU

SCHEGK, JAKOB (in Latin, **Jacobus Schegkius** or **Scheggius;** also sometimes called **Jakob Degen**) ((*b.* Schorndorf, Germany, 7 June 1511; *d.* Tübingen, Germany, 9 May 1587), *medicine, natural philosophy, methodology of science*.

Schegk was the son of a well-to-do burgher, Bernhard Degen; it is not known why he changed his name to Schegk. As a boy he was taught Latin, Greek, Hebrew, and rhetoric by Johann Thomas, a student of Johann Reuchlin's, before entering the University of Tübingen in 1527 to study philosophy. A year later he received the baccalaureate

150

and, in 1530, a master's degree. He also studied theology and medicine, taking a doctorate in the latter in 1539, and taught philosophy, logic, and medicine at Tübingen for forty-five years before retiring in 1577. During five decades Schegk published more than thirty works, including many very long ones, on philosophy, theology, and medicine. He was generally Aristotelian in orientation and wrote numerous commentaries and treatises on Aristotle's works, in addition to becoming involved in polemics with Theodore Beza, Petrus Ramus, and Simone Simoni. Besides many compendia of natural philosophy, he wrote *De demonstratione libri XV* (1564), in which he attempted to reassert the validity of Aristotle's scientific methodology, and in *De plastica seminis facultate libri tres* (1580) he argued in behalf of the Peripatetic doctrine of the formative power of the semen.

One of the most prominent sixteenth-century spokesmen for German Scholasticism, Schegk approached natural philosophy through a strong emphasis on a return to a study of the Greek text of Aristotle. Thus he followed in large measure the humanistic tradition of Renaissance Aristotelianism.

BIBLIOGRAPHY

I. ORIGINAL WORKS. Schegk's writings include *De demonstratione libri XV* (Basel, 1564); and *De plastica seminis facultate libri tres* (Strasbourg, 1580). The most complete list of his works is in the article by Sigwart (below). See also C. Sigwart, "Ein Collegium logicum im XVI. Jahrhundert," *Tübinger Universitätsschriften* for 1889–1890 (1890).

II. SECONDARY LITERATURE. The basic source for information on Schegk's life is Georg Liebler, *Oratio de vita . . . Jacobi Schegki . . .* (Tübingen, 1587). See also N. W. Gilbert, *Renaissance Concepts of Method* (New York, 1960), 158–162; W. Pagel, "William Harvey Revisited, Part II," in *History of Science*, 9 (1970), 1–41, esp. 26–30; P. Petersen, *Geschichte der Aristotelischen Philosophie im protestantischen Deutschland* (Leipzig, 1921; repr. Stuttgart–Bad Canstatt, 1964); C. Sigwart, "Jakob Schegk, Professor der Philosophie und Medizin," in his *Kleine Schriften*, I (Freiburg im Breisgau, 1889), 256–291, the best general survey; and C. Vasoli, *La dialettica e la retorica dell'umanesimo* (Milan, 1968), *passim*.

CHARLES B. SCHMITT

SCHEINER, CHRISTOPH (*b.* Wald, near Mindelheim, Swabia, Germany, 25 July 1573; *d.* Neisse, Silesia [now Nysa, Poland], 18 June 1650), *astronomy*.

Scheiner attended the Jesuit Latin school at Augsburg and the Jesuit College at Landsberg before he joined the Society of Jesus in 1595. In 1600 he was sent to Ingolstadt, where he studied philosophy and, especially, mathematics under Johann Lanz. From 1603 to 1605 he spent his "magisterium," or period of training as a teacher, at Dillingen, where he taught humanities in the Gymnasium and mathematics in the neighboring academy. During this period he invented the pantograph, an instrument for copying plans on any scale; and his results were published several years later in the *Pantographice, seu ars delineandi* (1631). He returned to Ingolstadt to study theology, and after completing his second novitiate or "third year" at Edersberg, he was appointed professor of Hebrew and mathematics at Ingolstadt in 1610.

The following year Scheiner constructed a telescope with which he began to make astronomical observations, and in March 1611 he detected the presence of spots on the sun. His religious superiors did not wish him to publish under his own name, lest he be mistaken and bring discredit on the Society of Jesus; but he communicated his discovery to his friend Marc Welser in Augsburg. In 1612 Welser had Scheiner's letters printed under the title *Tres epistolae de maculis solaribus*, and he sent copies abroad, notably to Galileo and Kepler. Scheiner believed that the spots were small planets circling the sun; and in a second series of letters, which Welser published in the same year as *De maculis solaribus . . . accuratior disquisitio*, Scheiner discussed the individual motion of the spots, their period of revolution, and the appearance of brighter patches or *faculae* on the surface of the sun. Having observed the lower conjunction of Venus with the sun, Scheiner concluded that Venus and Mercury revolve around the sun.

Welser had concealed Scheiner's identity under the pseudonym of *Apelles latens post tabulam*. Galileo, however, identified Scheiner as a Jesuit and took him to task in three letters addressed to Welser and published in Rome in 1613. Galileo claimed priority in the discovery of the sunspots and hinted darkly that Scheiner had been apprised of his achievement and was guilty of plagiarism. This criticism was unfair, for the sunspots were observed independently not only by Galileo in Florence and Scheiner in Ingolstadt, but also by Thomas Harriot in Oxford, Johann Fabricius in Wittenberg, and Domenico Passignani in Rome.

In Ingolstadt, Scheiner trained young mathematicians and organized public debates on current issues in astronomy. Two of these "disputations" were subsequently published. In the first, the *Disquisitiones mathematicae de controversis et novitatibus astronomicis,* Scheiner upheld the traditional view that the earth is at rest at the center of the universe but praised Galileo for his discovery of the phases of Venus and the satellites of Jupiter. In the second, *Exegeses fundamentorum gnomonicorum,* Scheiner discussed the theory behind sundials and explained their construction. In the *Sol ellipticus* (1615) and the *Refractiones caelestes* (1617), which he dedicated to Maximilian, the archduke of Tirol, Scheiner also called attention to the elliptical form of the sun near the horizon, and he explained the form as the effect of refraction.

In 1616 Scheiner accepted an invitation from Maximilian and took up residence at the court in Innsbruck. The following year he was ordained to the priesthood. He performed several experiments on the physiology of the eye. In the *Oculus, hoc est: fundamentum opticum* (1619) he showed that the retina is the seat of vision.

In 1620 the University of Freiburg im Breisgau was entrusted to the Jesuits; Scheiner was one of the first seven Jesuits to be assigned to the university, but the following year he was recalled to Innsbruck. In 1622 he accompanied the Archduke Charles, the bishop of Neisse, to that city; and in 1623 Scheiner was appointed superior of the Jesuit College to be erected there. In 1624 he left with the Archduke Charles on a journey to Spain; but they parted ways at Genoa, and Scheiner proceeded to Rome to settle matters concerning the foundation of the college. He was detained in Rome until March 1633. When not occupied with administrative problems, he busied himself with astronomical observations and the writing of his major work, the *Rosa ursina sive sol,* which was printed at Bracciano between 1626 and 1630. In the *Rosa ursina,* Scheiner confirmed his method and criticized Galileo for failing to mention the inclination of the axis of rotation of the sunspots to the plane of the ecliptic, which Scheiner determined as 7° 30' (modern value 7° 15').

Scheiner does not appear to have played an active role in the trial and condemnation of Galileo, and his refutation of the Copernican system, *Prodromus de sole mobili et stabili terra contra Galilaeum,* was published only posthumously in 1651.

From 1633 to 1639 Scheiner lived in Vienna and then in Neisse, where he was active in pastoral work until his death in 1650.

BIBLIOGRAPHY

I. Original Works. Scheiner's works include *Tres epistolae de maculis solaribus scriptae ad Marcum Velserum* . . . (Augsburg, 1612) and *De maculis solaribus et stellis circa Jovem errantibus, accuratior disquisitio* . . . (Augsburg, 1612), which were reedited as an appendix to Galileo Galilei, *Istoria e dimostrazioni intorno alle macchie solari* . . . (Rome, 1613).

See also *Disquisitiones mathematicae de controversis et novitatibus astronomicis* (Ingolstadt, 1614), written with his pupil Johann Georg Locher; *Sol ellipticus; hoc est novum et perpetuum solis contrahi soliti phaenomenon* (Augsburg, 1615); *Exegeses fundamentorum gnomonicorum* (Ingolstadt, 1615), written with his pupil Johann Georg Schoenberg; *Refractiones caelestes, sive solis elliptici phaenomenon illustratum* (Ingolstadt, 1617); *Oculus, hoc est: fundamentum opticum* . . . (Innsbruck, 1619; 2nd ed., Freiburg im Breisgau, 1621; 3rd ed., London, 1652); *Rosa ursina sive sol ex admirando facularum et macularum suarum phaenomeno varius necnon circa centrum suum et axem fixum, ab occasu in ortum annua, circaque alium axem mobilem ab ortu in occasum conversione quasi menstrua, super polos proprios, libris quatuor mobilis ostensus* (Bracciano, 1626–1630); and *Pantographice, seu ars delineandi res quaslibet* . . . (Rome, 1631; Italian eds: Padua, 1637; Bologna, 1653).

Two sets of lecture notes are preserved in the Bayerische Stadtsbibliothek in Munich: *Commentarius in Aristotelis libros de caelo et de meteorologicis. 1614–1615,* Pg. 128 (*Catal. MSS latin. Monach.,* n. 11878); and *Euclidis liber V dictatus an. 1615* (*ibid.,* n. 12425). There are also unpublished letters of Scheiner in the Bibliothèque Nationale in Paris.

II. Secondary Literature. On Scheiner and his work, see Augustin and Alois de Backer, *Bibliothèque de la Compagnie de Jésus,* VIII (Louvain, 1960), cols. 734–740; Anton von Braunmuehl, *Christoph Scheiner als Mathematiker, Physiker und Astronom* (Bamberg, 1891); Bellino Carrara, *"L'unicuique suum" a Galileo, Fabricius e Scheiner nella scoperta delle macchie solari* (Rome, 1900); Stillman Drake, *Galileo Studies* (Ann Arbor, 1970); Antonio Favaro, "Oppositori di Galileo. III. Cristoforo Scheiner," in *Atti del R. Istituto veneto di scienze, lettere ed arte,* 78 (1919), 1–107; Johann Schreiber, "P. Christoph Scheiner, S.J. und seine Sonnebeobachtungen," in *Natur und Offenbarung,* 48 (1902), 1–20, 78–98, 145–158, 209–221; and William R. Shea, "Galileo, Scheiner, and the Interpretation of Sunspots," in *Isis,* 61 (1970), 498–519.

William R. Shea

SCHEINER, JULIUS (*b.* Cologne, Germany, 25 November 1858; *d.* Potsdam, Germany, 20 December 1913), *astrophysics, astronomy.*

Scheiner was the son of Jacob Scheiner, a painter of landscapes and architectural subjects. In 1878 he entered the University of Bonn to read mathematics and natural science. While there he developed an interest in astronomy, which led to his becoming an assistant at the Bonn observatory. In 1882 he obtained the doctorate with a dissertation on the observations of Algol made by E. Schonfeld, then director of the observatory. Scheiner continued to work at the observatory until moving, early in 1887, to the Royal Astrophysical Observatory at Potsdam, where he remained for the rest of his life, rising from assistant to permanent assistant to senior observer (1900). In 1894 Scheiner was appointed extraordinary professor of astrophysics at the University of Berlin.

While at the Bonn observatory Scheiner was engaged primarily in assisting in zone observations. On moving to Potsdam he immediately set to work in astrophysics, the latest, and flourishing, branch of astronomy. He collaborated closely with Potsdam's director, Hermann Vogel, in applying the new instrument that Vogel had designed and named the spectrograph. Together they inaugurated the era of accurate measurement of stellar radial velocities. Their average probable error was only 2.6 kilometers per second, an improvement over earlier results by a factor of ten.

From the early 1890's Scheiner was also much occupied with celestial photography and with the preparation of the international astrographic chart. In connection with the latter he represented the Potsdam observatory at meetings in Paris in 1891, 1896, and 1900, and supervised the publication of six large volumes during the period 1899–1912. Work on the chart also benefited in several respects from Scheiner's considerable practical and experimental skills. For example, he tested the previously employed law of photographic photometry and showed it to be incorrect.

Around the turn of the century the close relations between Scheiner and Vogel became impaired, and afterward Scheiner worked in collaboration with J. Wilsing. Among other things they made a photometric determination of the relative intensities of the three principal lines in the nebular spectra and measured visually the radial velocities of nine of the brighter gaseous nebulae. Availing themselves of recent advances in the study of blackbody radiation, they also made determinations of the temperatures of more than 100 stars.

Scheiner's strengths lay in the experimental and practical areas of research. Drawing on his rich knowledge of both laboratory and workshop, he could quickly devise an experiment for settling a debated point. Scheiner was also an excellent teacher and enjoyed giving numerous popular accounts of astrophysical matters, both in lectures and in writing.

BIBLIOGRAPHY

I. ORIGINAL WORKS. Scheiner's more important books are *Die Spectralanalyse der Gestirne* (Leipzig, 1890), translated into English by E. B. Frost (Boston, 1890), a textbook on stellar physics; *Photographie der Gestirne* (Leipzig, 1897), at the time considered indispensable to those interested in any branch of the subject; *Strahlung und Temperatur der Sonne* (Leipzig, 1899), valuable for discussion of the temperature of the sun, in light of contemporary studies of blackbody radiation; *Populäre Astrophysik* (Leipzig–Berlin, 1908), Scheiner's 1906 lectures at Berlin; and *Spectralanalytische und photometrische Theorien* (Leipzig, 1909), meant for those with a general interest in astrophysics.

Scheiner's more important papers are discussed by J. Wilsing (see below).

II. SECONDARY LITERATURE. See E. B. Frost, "Julius Scheiner," in *Astrophysical Journal*, **41** (1915), 1–9; Hector Macpherson, Jr., *Astronomers of Today and Their Work* (London, 1905), 234–239; and J. Wilsing, "Julius Scheiner," in *Vierteljahrsschrift der Astronomischen Gesellschaft*, **49** (1914), 22–36.

WILLIAM McGUCKEN

SCHELLING, FREDERICK WILHELM JOSEPH (later **von Schelling**) (*b.* Leonberg, Württemberg, Germany, 27 January 1775; *d.* Bad Ragaz, Switzerland, 20 August 1854), *philosophy.*

Schelling was the son of Joseph Friedrich Schelling (1737–1812), a deacon, preacher, and theological writer, and the former Gottliebin Maria Cless. Both parents were the children of pastors. Schelling early displayed extraordinary gifts, and in 1790, at age fifteen, he entered the Protestant theological foundation at the University of Tübingen. There he studied theology and philosophy and was friends with Hegel and Hölderlin. Like most of his fellow students, he embraced the ideas of the French Revolution. Consequently, he forfeited the patronage of Duke Karl of Württemberg.

In 1791 Schelling began an intensive study of the works of Kant and of Fichte, whose importance for the further development of the Kantian philosophy he immediately recognized. In 1792 he obtained a master's degree in philosophy for a paper dealing with biblical subjects and, in 1795, a master's degree in theology. At this time (1794–1795) he wrote several essays on philosophy and theology.

After finishing his studies, Schelling took a post as a tutor, traveling with two aristocratic pupils to Leipzig in 1796. In his two years there he studied natural science and medicine and published several works on philosophy, which attracted the attention of Fichte and Goethe. At their suggestion, Schelling, in 1798, was offered a position as extraordinary professor at Jena, then a center of German cultural life. Around this time he spent several weeks in Dresden, where he was introduced into the circle of the Romantic school, notably to the brothers August Wilhelm and Friedrich Schlegel and their wives, and to Novalis and Tieck. At Jena, Schelling lectured primarily on Kant's transcendental idealism and Fichte's theory of science (*Wissenschaftstheorie*), but soon he began to develop his own ideas on the philosophy of nature (*Naturphilosophie*). In 1803 the Bavarian government invited him to Würzburg as a full professor. As at Jena, the first part of his stay was successful, but then he encountered increasing opposition, in this case from both Catholics and "dogmatic Kantians." He was accused of materialism, atheism, obscurantism, and mysticism.

Schelling left Würzburg in 1806 and moved to Munich. For the next thirty-five years he remained in the service of the Bavarian state, becoming a member of the Bavarian Academy of Sciences and general secretary of the Academy of Fine Arts. Granted a leave of absence from 1820 to 1827, Schelling went to Erlangen, where he gave private lectures on philosophy, mythology, and the history of philosophy. At Erlangen he enjoyed the respect and admiration of a circle of sympathetic friends.

In 1827 Schelling returned to Munich. The University had been moved from Landshut to the Bavarian capital, and King Ludwig I had made a great effort to attract outstanding faculty members. Schelling was appointed head curator of the scientific collections and president of the Academy of Sciences. He also made a major contribution to a project for reforming the Bavarian school system, but the project was abandoned after a reactionary turn in government policy.

Meanwhile, Hegel's philosophy had triumphed in Prussia, and it appeared that the schools of thought growing out of his teaching would soon be dominated by materialistic and antimonarchical ideas. Looking to Schelling to achieve the "destruction of the dragon's teeth" of materialism, King Friedrich Wilhelm IV of Prussia, encouraged

by the Humboldt brothers, offered the philosopher a post in Berlin. Despite his advanced age, Schelling accepted the offer, for he viewed it as an obligatory mission. He taught in Berlin for only a short time (1841–1846). After vehement disputes with one of his opponents, Heinrich Eberhard Gottlob Paulus, Schelling gave up his lectures, which initially had been very popular but then attracted fewer and fewer listeners. He withdrew increasingly from public life and devoted himself exclusively to perfecting his philosophy. He lived long enough to witness the victory of Hegelian philosophy and the rise of materialism.

Schelling's first marriage was to Dorothea Caroline Michaelis, daughter of the famous Orientalist Johann David Michaelis. Schelling met her in Dresden while she was still married to A. W. Schlegel. Extremely intelligent and witty, she had led a very eventful life. Her early death left Schelling deeply shaken. Although it may not have been the reason that he virtually ceased to publish (and even withdrew works already printed), the two events occurred at about the same time. In 1812 Schelling married Pauline Gotter, the daughter of a friend of his late wife. They enjoyed a harmonious marriage and had three sons and three daughters.

Unlike Kant or Hegel, Schelling left behind no finished, logically constructed system. While still young, he published a number of brilliant works, and whenever he embarked on a new subject, he aroused the highest expectations, which, however, he was not able to fulfill. Curious, enthusiastic, and receptive to a broad range of influences, he grasped at new ideas, reworked them, and integrated them into his own philosophy. He repeatedly started over from the beginning, used new terminology, invented new schemata, ignored or altered previous statements, and gave new content to earlier concepts. Most of his early writings were derived from or were composed at the same time as his lectures, often under pressure of time; consequently the writings contain material that is immature or insufficiently thought out. Journals that he founded in order to publish essays soon went out of existence. Much in his late philosophy is difficult to understand and is even difficult to enter into at all, laden as it is with mythological notions.

Earlier historians of philosophy distinguished definite stages in the development of Schelling's thinking. Recent research (especially that of Horst Fuhrmans and Manfred Schröter) has shown that, viewed from the standpoint of his late philosophy, Schelling's work as a whole developed organically and without breaks and that certain fundamental concepts and images are evident throughout. For the sake of convenience, Schelling's philosophy can be presented—with little violence to its contents—in a strictly chronological order: philosophy of nature and transcendental philosophy, philosophy of identity, and philosophy of religion.

Schelling attentively followed the discoveries of the great scientists of his time: Galvani, Volta, Lavoisier, John Brown, and Kielmeyer. Like Goethe, Schelling was temperamentally opposed to a purely mechanistic conception of nature. He viewed nature as a living organism, an active force. He was, however, just as little a vitalist as he was a mechanist, for he rejected the notion of an autonomous life-force. Rather he supposed that nature contains the organic within itself, inorganic matter being matter that has ceased to live.

In his treatment of Fichte's *Wissenschaftslehre*, Schelling expressed dissatisfaction with its assignment of a passive role to nature as the nonego (*Nicht-Ich*). In order to balance Fichte's primarily ethical concerns, he sought to create a "speculative physics." His starting point was Kant's concept of natural purpose as developed in the *Critique of Judgment* and Fichte's theory of the unconscious creative ego. In a work of 1797 Schelling developed Kant's idealism in accord with Fichte's position, in that he wished to make no distinction between the thing as represented and the actual thing. According to Schelling, nature institutes in the faculty of intuition a dynamic process based on attraction and repulsion, and it is this process that appears in the intuition as an object, recognizable as such and accessible to the understanding.

In order that the individual products of nature can come to exist as enduring objects at all, the perpetually active nature-force must be opposed by an obstructing or checking force. These mutually opposed forces, the formative drive (*Bildungstrieb*) and the obstructing force, arise through the spontaneous division of the original force (*Urkraft*) of nature. The entire realm of natural occurrences is permeated by polarity (a universal principle to which the more modern term "differentiation" may be given) and dualism (conflict of forces). Their effectiveness is determined by quantity (in the mechanical natural process) or quality (in the chemical natural process). In the dead body the opposing forces are neutralized; in the chemical process the equilibrium of the forces is destroyed; and in the life process these forces are engaged in a perpetual struggle.

Out of the original forces of attraction and repulsion there emerges the primary phenomenon (*Urphänomen*) of light as the duplexity of ether (repulsive) and oxygen (attractive). Just as in magnetism, which Schelling considered the primary phenomenon of polarity, a north and a south pole stand opposed to each other; in electricity, a negative and a positive pole; and in chemistry, acids and bases—so, too, he thought, the dynamic natural process consists, universally, in the unification of the opposites on a higher plane. Accordingly, the world Soul, the organizing principle of the entire universe, is a unity of mutually conflicting forces. Its existence explains the progress observable from the lowest to the highest forms.

In his *Naturphilosophie* Schelling dealt with all the important physical, chemical, and biological phenomena and processes that occupied the scientists of his day: ether, light, weight, heat, air, gravitation, the atom (which he conceived of as a center of force), matter, combustion, electricity, magnetism, and evolution. To Kant's mechanistic theory of the formation of the cosmos, Schelling opposed an organic theory according to which the universe came into being through the expansion and contraction of the primary matter (*Urmaterie*).

Schelling's philosophy of nature was well received not only by the poets and writers of the Romantic school, but also by L. Oken, H. Steffens, K. E. von Baer, and Karl Friedrich von Burdach. The physicians of the Brownian school, who viewed man as the unity of body and soul, also welcomed Schelling's ideas. In 1802 the faculty of medicine of the University of Landshut awarded Schelling an honorary doctorate of medicine.

In his transcendental philosophy, which is partly a reworking and partly a major revision of Fichte's *Wissenschaftslehre*, Schelling treated the problem of nature from the point of view of consciousness. With the philosophy of nature, it completed the theoretical part of his doctrines. In the practical part, Schelling took up questions concerning the freedom of the will, the moral law and natural law, and the philosophy of history. The latter field, he asserted, displays the realization of the unity of necessity and freedom. For Schelling, the summit of subjective activity is attained, not in morality, as Kant and Fichte held, but in the free creative act of the artist. He alone is capable of representing the infinite in the finite, for he brings the identity of the real and the ideal, toward which philosophy aspires, to concrete representation in the work of art.

With the philosophy of identity, Schelling went one step further: real and ideal, and nature and mind are seen to be identical when conceived with sufficient understanding. Mind, and life as the bearer of mind, can be understood only on the assumption that nature is not a conglobation of dead matter, but rather, in its essence, a living primary force, capable of infinite activity. The secret of the unity of nature and mind in the Absolute, however, can ultimately be grasped only in the completion of the creative act, which leads to the product of nature. This occurs in intellectual intuition, which, according to Schelling, affords an immediate, concrete, intuitive apprehension of the Absolute. (Kant, on the other hand, reserved this intuition to the divine intellect.) Such intuition cannot be taught, but it becomes immediately evident in the contemplation of art.

Schelling was a natural philosopher, not a scientist. His philosophy of nature must be assessed in terms of the historical situation in which it was created. Along with Fichte and Hegel, he propounded an epistemological idealism the starting-point of which was the philosophy of Kant. Schelling shared a number of traits with the literary figures, scientists, and philosophers who had embraced Romanticism in reaction against the rationalism of the Enlightenment: a sense of historical development and an emphasis on feeling, the irrational, fantasy, the creative individual, and intuition. All the same, Schelling—in contradistinction to many of his followers—retained a modicum of sobriety and openness to empirical findings. Even considering that the experimental basis of the science of his time was relatively narrow, Schelling applied his concepts to nature without sufficient regard for scientific rigor and thereby brought the speculative philosophy of nature into discredit. (To be sure, some of his disciples were even less disciplined in this respect.) In Schelling's own lifetime, readers found it difficult to follow his expositions because of his penchant for esoteric formulations. The problem was especially acute for those not intimately familiar with the history of philosophy, in which Schelling had immersed himself with characteristic enthusiasm. Not surprisingly, contemporary scientists and philosophers who favored mechanistic explanations were repelled by Schelling's notions, particularly since his notions were heavily burdened with analogies. Their skeptical attitude undoubtedly hardened in response to the new knowledge steadily accumulating in all fields of biology, chemistry, and physics as a result of intensive exact research and the application of more refined methods.

Although critics were correct in denouncing Schelling's all too facile use of concepts and images, they forgot the magnitude and profundity of Schelling's project and its fruitful stimuli to further thinking. A good example of the latter is the notion of a basic type (*Bauplan*), which appears in nature in limitless variations. Another is the concept of biological evolution, although Schelling's formulation of it is closer to the Romantic ideas of evolution as a product of the creative, active force in nature than it is to Darwin's origin of species.

The impact of Schelling's *Naturphilosophie* was very great. Aside from the direct influence that his theory of nature had on such scientists as L. Oken, H. Steffens, G. H. Schubert, and Franz von Baader, its fundamental ideas entered into the thinking of the age and became the common property of educated men. Traces of them can be found in Johannes Müller, K. E. von Baer, and C. G. Carus. Only this pervasiveness can account for the violent reaction of leading scientists like Virchow, Helmholtz, and du Bois-Reymond against speculative philosophy of nature at a time when idealism and Romanticism had long ceased to be vital intellectual movements and when scientists believed that all change in nature could be explained causally through reduction to mechanics.

Beginning in 1806 Schelling occupied himself increasingly with the philosophy of religion. His own thinking on this subject was stimulated by his encounter with the ideas of Jacob Boehme, which had been brought to his attention by Franz von Baader. Schelling's conceptions, which he termed "positive" philosophy, were frequently misunderstood and vigorously attacked as a form of gnostic theosophy. Schelling himself was partially responsible for this, since he refused to publish the themes he treated in his lectures (at Munich, Erlangen, Stuttgart, and Berlin). (Some works did reach the printer, but Schelling always recalled them at the last minute to make further revisions.) Many thousands of pages of handwritten manuscripts attest to this decades-long struggle with the ultimate problems of God, freedom, and the universe. Overcoming the impasse of Romanticism, and freeing himself, in this domain, from the thinking of his age, he arrived at a new realism—that of struggling, suffering man, the fundamental concept of existentialism. A direct line can be drawn from this formulation to the work of Kierkegaard, the founder of this new philosophy and for a time Schelling's student at Berlin. It has rightly been remarked (at a congress celebrating the hundredth anniversary of Schelling's death) that Schelling's philosophy of religion held more of the future within it than of the past.

The entire corpus of Schelling's manuscripts, which were preserved at the Bavarian State Library in Munich, was destroyed during World War II. Fortunately, in 1943 H. Fuhrmans encouraged M. Schröter to examine this material. Schröter either analyzed or in some cases copied portions of the "Weltalter," which he later published. The studies undertaken in connection with or in the wake of this project have given a new impetus to research on Schelling.

BIBLIOGRAPHY

I. ORIGINAL WORKS. Schelling's collected works were edited by K. F. A. von Schelling (his second eldest son) as *Sämtliche Werke* (Stuttgart–Augsburg, 1856–1861); they appeared in two sections: the first, in 10 vols., was published between 1856 and 1861; the second, which is made up of 4 vols. taken from the posthumous MSS, appeared between 1856 and 1858. Sec. 1, vol. I (writings from 1792 to 1797) contains the following works: Schelling's master's diss. on the origin of evil, based on Genesis 3. The full Latin title is "Antiquissimi de prima malorum humanorum origine philosophematis Genesis III explicandi tentamen criticum et philosophicum" (1792). Other writings are *Über Mythen, historische Sagen und Philosopheme der ältesten Welt* (1793); *Über die Möglichkeit einer Form der Philosophie überhaupt* (1794), Schelling's first purely philosophical work, which is a criticism of Kant in accord with Fichte's *Wissenschaftslehre*, and which sets forth the task of his future work; diss. for his theology degree, *Marcion* (1795), a study in the history of religion; a commentary on the *Wissenschaftslehre vom Ich als Prinzip der Philosophie, oder über das Unbedingte im menschlichen Wissen* (1795; repr., 1809), which shows the influence of Spinoza, whose absolute substance becomes, in Schelling, the active ego; *Philosophische Briefe über Dogmatismus und Kriticismus* (1795; repr., 1809), which first appeared anonymously, is directed against the dogmatic Kantians and exhibits the influence of Spinoza (Schelling develops Kantian doctrines in accord with Fichte's views); *Neue Deduktion des Naturrechts*, first published in *Philosophisches Journal*, 4 (1796), but completed in 1795 (it leads into the subject of practical philosophy); and *Abhandlungen zur Erläuterung des Idealismus der Wissenschaftslehre* (1796–1797; repr., 1809), an apology of Fichte's views, appeared in *Philosophisches Journal* as "Allgemeine Übersicht der neuesten philosophischen Literatur."

Vol. II, which has Schelling's first writings on the philosophy of nature, contains *Ideen zu einer Philosophie der natur* (1797), with subtitle in 2nd ed. of *Als Einleitung in das Studium dieser Wissenschaft* (here Schelling extends the *Wissenschaftslehre* into the speculative theory of nature); *Von der Weltseele, eine Hypothese der höheren Physik zur Erklärung des allgemeinen Organismus* (1798), preceded in the 2nd ed. (1806) by an essay "Über das Verhältniss des Realen und Idealen in der Natur oder Entwicklung der ersten Grundsätze der Naturphilosophie an den Principien der Schwere und des Lichts."

Schelling's first systematic works are in vol. III: *Erster Entwurf eines Systems der Naturphilosophie* (1799), with additions from Schelling's MSS; *Einleitung zu dem Entwurf eines Systems der Naturphilosophie oder über den Begriff der speculativen Physik und der inneren Organisation eines Systems der Wissenschaft* (1799); and *System des tranzendentalen Idealismus* (1800), an attempt to unite philosophy of nature, transcendental philosophy, and ethics.

Vol. IV contains *Allgemeine Deduction des dynamischen Processes oder der Kategorien der Physik* (1800), an essay on the boundary between philosophy of nature and the doctrine of identity, which originally appeared in *Zeitschrift für spekulative Physik*, 1 (1800); *Über den wahren Begriff der Naturphilosophie und die richtige Art ihre Probleme anzufassen* (1801); *Darstellung meines Systems der Philosophie* (1801) (Philosophy of nature is Plato's theory of ideas. This work is a fragment, only first part of the theory of nature, directed at Fichte); *Bruno oder über das göttliche und natürliche Prinzip der Dinge* (1802; 2nd, unaltered ed., 1842), a dialogue in which Bruno—named for Giordano Bruno—upholds Schelling's position and convinces Lucian (Fichte) of the validity of his position; and *Fernere Darstellungen aus dem System der Philosophie* (1802).

Vol. V is *Vorlesungen über die Methode des akademischen Studiums* (1803), a literary work, which includes Schelling's farewell offering to Würzburg and also a unified account of his worldview: philosophy is the doctrine of ideas (*Ideenlehre*); the 2nd ed. (1803) has additions of the *Ideen*.

Vol. VI contains *Immanuel Kant* (1804), an obituary, which, according to the editor, was unearthed from "an obscure journal" following up an allusion made by Schelling; *Philosophie und Religion* (1804), in which philosophy is coordinated with, and even subordinated to, religion; this work shows the influence of Neoplatonism and mysticism; *Propaedeutic to Philosophy* (ca. 1804), taken from the posthumous MSS; and *System der gesammten Philosophie und der Naturphilosophie insbesondere* (1804), a development of the first three potencies of the ideal side—taken from the posthumous MSS.

Vol. VII includes the following works: *Darlegung des wahren Verhältnisses der Naturphilosophie zu der verbesserten Fichteschen Lehre, eine Erläuterungsschrift der ersten* (1806), which marks Schelling's settling with and break from Fichte; various essays in the *Jahrbücher der Medicin als Wissenschaft* (1806–1807), including "Aphorismen zur Einleitung in die Naturphilosophie," "Aphorismen über die Naturphilosophie," "Kritische Fragmente," and "Vorläufige Bezeichnung des Standpunktes der Medicin nach Grundsätzen der Naturphilosophie"; *Über das Verhältnis der bildenden Künste zu der Natur* (1807), a lecture before the Academy of Sciences at Munich on the occasion of the birthday (12 Oct. 1807) of the Bavarian king; *Philosophische Untersuchungen über das Wesen der menschlichen Freiheit und die damit zusammenhängenden Gegenstände* (1809), taken from the posthumous MSS, still borders on the philosophy of nature but goes beyond both it and the theory of identity. In the winter of 1809–1810, following the death of Caroline, Schelling was in Stuttgart, where at the request of several friends he gave private lectures. In them he anticipated his later work in the field of the philosophy of religion. Other works in vol. VII are essays and book reviews (1807–1809) from *Jenaer, Erlanger Literaturzeitung,* and *Morgenblatt; Die Weltalter* (1814–1815), which, according to the editor, is from the posthumous MSS composed in 1814 or 1815, and which is the most complete of several existing versions of the work, which was to include three books, one each on the past, the present, and the future; *Über die Gottheiten von Samothrake* (1815), a lecture that is a supp. to the *Weltalter,* was delivered at a public session of the Bavarian Academy of Sciences on the king's name day in 1815; and short essays (1811) from the posthumous MSS (one of them was on "weather-shooting," that is, on the influence that cannon shots, for example, might have upon the weather [thunderstorms, hail]).

Vol. IX includes *Über den Zusammenhang der Natur mit der Geisterwelt* (ca. 1816–1817), a dialogue taken from the posthumous MSS; lectures delivered (1821–1825) at Erlangen, from the posthumous MSS; Schelling's first lecture in Munich (1827); and various lectures, addresses, and speeches given from 1828 to 1841 at the Academy of Sciences or at the University.

Works in vol. X are *Vorrede zu einer philosophischen Schrift des Herrn Cousin (Fragments philosophiques),* 2nd ed. (Paris, 1833), which contains an attack on Hegel; *Darstellung des philosophischen Empirismus* (1836), from the posthumous MSS of the introductory lectures on philosophy delivered at Munich for the last time in 1836; *Anthropologisches Schema* (no date), a brief outline, written at Munich and devoted to psychological and characterological questions; Schelling's first lecture at Berlin (1841); *Darstellung des Naturprozesses* (1843–1844), a fragment of the posthumous MSS of a lecture held at Berlin on the principles of philosophy; and *Vorwort zu Henrik Steffens nachgelassenen Schriften* (1845), from a public lecture—with a few additions in Steffen's memory—held on 24 April 1845.

In the preface of sec. 2, vol. I (1856), the editor K. F. A. von Schelling explains that he undertook to prepare this edition with the assistance of his brothers at the explicit request of his father, and that he was especially aided in the first volume by his younger brother Her-

mann Schelling. Part 1, vol. I, with a historical-critical introduction and lectures 1–10, contains a philosophical criticism of the possible ways of elucidating mythology. Revised by Schelling during his last years at Munich and again at Berlin from 1842 to 1845, pt. 1 was printed 30 years previously but was never published. Part 2 of vol I, with a philosophical introduction and lectures 11–24 (Schelling's last work), is an account of rational (negative) philosophy, which is presented, however, in the form he gave to it after he had fully developed his positive philosophy.

Philosophie der Mythologie, II (1857), which includes "Buch. Der Monotheismus" and "Buch. Die Mythologie" (these works were compiled by Schelling's son after his death); *Philosophie der Offenbarung,* III (1858), which consists of "Stellung der Aufgabe, nämlich der philosophischen Religion," "Lösung der Aufgabe," and a 3rd part, which includes "Einleitung in die Philosophie der Offenbarung, 1.–8. Vorlesung," "Der Philosophie der Offenbarung erster Teil," and "Der Philosophie der Offenbarung zweiter Teil."

Other editions of Schelling's works are *Werke,* M. Schröter, ed., 6 vols. (Munich, 1927–1928), with a supp. vol. containing the original versions of the *Weltalter* (1946); and *Werke,* Otto Weiss, ed., 3 vols. (Leipzig, 1907), with a foreword by A. Drews. See also *Clara oder Zusammenhang der Natur mit der Geisterwelt,* 2nd ed. (Stuttgart, 1862), and an expanded ed. prepared by M. Schröter (Munich, 1948). For Schelling's correspondence, see *Aus Schellings Leben in Briefen, 1775–1803,* I (Leipzig, 1869), II, III (Leipzig, 1870); "Briefwechsel," in *Maximillian II, König von Bayern und Schelling,* L. Trost and F. Leist, eds. (Stuttgart, 1891); and *Caroline, Briefe* (to her brothers and sisters, her daughter Auguste, the Gotter family, F. L. W. Meyer, A. W. and F. Schlegel, and J. Schelling), G. Waitz, ed. (Schelling's son-in-law).

II. SECONDARY LITERATURE. On Schelling and his work, see Kuno Fischer, "Geschichte der neueren Philosophie," in *Schellings Leben und Lehre,* 4th ed., VII (Heidelberg, 1923), with an appendix that contains H. Falkenheim's thorough list of literature on Schelling up to 1922; Horst Fuhrmans, *Schellings Philosophie der Weltalter. Schellings Philosophie in den Jahren 1806–1821. Zum Problem des Schellingschen Idealismus* (Düsseldorf, 1954); Eduard von Hartmann, *Schellings philosophisches System* (Leipzig, 1897); Hinrich Knittermeyer, *Schelling und die romantische Schule* (Munich, 1928); Wolfgang Pfeiffer-Belli, "Schelling und seine Weltalter," in *Philosophisches Jahrbuch,* **58,** pt. 1 (1948), 65–68; G. Schneeberger, *F. W. J. von Schelling* (Berne, 1954); Manfred Schröter, *Kritische Studien. Über Schelling und zur Kulturphilosophie* (Munich, 1971); and Wilhelm Szilasi, *Philosophie und Naturwissenschaft* (Berne, 1961).

See also Bernhard Taureck, *Mathematische und transzendentale Identität* (Vienna–Munich, 1973); Friedrich Ueberweg, *Grundriss der Geschichte der Philosophie* (Tübingen, 1951), 35–67; John Watson, *Schelling's*

Transcendental Idealism (Chicago, 1882); Wolfgang Wieland, "Schellings Lehre von der Zeit," in *Grundlagen und Voraussetzungen der Weltalterphilosophie* (Hamburg, 1956).

Schelling founded or cofounded the following periodicals: *Zeitschrift für speculative Physik*, 2 vols. (1800–1801), followed by *Neue Zeitschrift für speculative Physik*, pt. 1 (1802); *Kritisches Journal der Philosophie* (1802–1803), edited with Hegel; and *Jahrbücher der Medizin als Wissenschaft* (1806–1808), edited with A. F. Marcus.

EDITH SELOW

SCHEUCHZER, JOHANN JAKOB (*b.* Zurich, Switzerland, 2 August 1672; *d.* Zurich, 23 June 1733), *medicine, natural history, mathematics, geology, geophysics.*

A diligent pupil at the age of three, Scheuchzer later became a brilliant student at the Carolinum in Zurich. Devoted to the natural sciences, he decided to study medicine and, having won a scholarship in 1691, was able to enroll in both science and medicine courses at the Altdorf Academy, near Nuremberg. He remained there for two years, then went to Utrecht, where he was awarded the doctorate in 1694. The fossil collection that he began assembling in 1690 soon became famous and brought Scheuchzer to the attention of the scholarly world. In 1694 he returned to Zurich and began systematic exploration of the Alps. His first writings for the Collegium der Wohlgesinnten (1695) were a scientific study of the Helvetic Alps. Scheuchzer then went to Nuremberg, where he studied for a diploma in mathematics, intending to teach this subject. But he was recalled to Zurich to become assistant municipal physician and medical supervisor of the orphanage. A few years later he became head of the Bibliothèque des Bourgeois, a post that he occupied while serving as director of the Museum of Natural History (then called the Kunsthammer).

By the age of thirty Scheuchzer had become prominent in Zurich and was carrying on a voluminous correspondence with many European scholars that has become of great interest to historians of science. A grant from the Zurich government in 1702 enabled him to resume his Alpine excursions, which provided the subject for numerous communications on geology, geophysics, natural sciences, and medicine. The results of his annual excursions to the Alps are presented in *Helvetiae stoicheiographia* (1716–1718), his greatest work in natural history and geophysics. In 1716 he became profes-

sor of mathematics at the Carolinum, and a few months before his death he was named *premier médecin* of Zurich, professor of physics at the Academy, and *Chorherr.*

Scheuchzer left the municipal library of Zurich more than 260 folio volumes, which he wrote in less than forty years. The moving force in the establishment of paleontology in Switzerland, he is also considered the founder of paleobotany and his *Herbarium diluvianum* remained a standard through the nineteenth century. His work on a great variety of fossils and notably on *Homo diluvii testis* of Oensingen (1726) makes him generally considered the founder of European paleontology. Scheuchzer became famous for his medical studies on the effects of altitude, published a remarkable topographic map of Switzerland, and took an active part in the military life of his canton as an army doctor.

In addition to his scientific accomplishments, Scheuchzer compiled a twenty-nine-volume *Histoire suisse* and a critical collection of deeds and other documents, entitled *Diploma Helvetiae.*

BIBLIOGRAPHY

I. ORIGINAL WORKS. A complete bibliography of Scheuchzer is in the Steiger article (below) with a list of his correspondence. Among his works are his medical diss., *De surdo audiento* (Zurich, 1694); "De generatione conchitarum," in *Miscellanea curiosa Academiae naturae curiosorum*, IV (Zurich, 1697); *Helvetiae stoicheiographia, orographia et oreographia* (Zurich, 1716); *Homo diluvii testis* (1726); and *Physica sacra*, 3 vols. (Zurich, 1731–1733).

II. SECONDARY LITERATURE. The most complete account of Scheuchzer is R. Steiger, "Johann Jakob Scheuchzer (1672–1733)," in *Beiblatt zur Vierteljahrsschrift der Naturforschenden Gesellschaft in Zurich*, **21** (1933), 1–75, with a complete bibliography. See also C. Walkmeister, "J. J. Scheuchzer und seiner Zeit," in *Bericht der St. Gallischen naturwissenschaft Gesellschaft* (1896), 364–401; F. X. Hoeherl, "J. J. Scheuchzer, der Begrunder der physischen Geographie des Hochgebirges" (diss., University of Munich, 1901); and B. Peyer, "J. J. Scheuchzer im europaischen Geistleben seiner Zeit," in *Gesnerus*, **2** (1945), 23–33.

P. E. PILET

SCHIAPARELLI, GIOVANNI VIRGINIO (*b.* Savigliano, Cuneo province, Italy, 14 March 1835; *d.* Milan, Italy, 4 July 1910), *astronomy.*

After receiving his degree in civil engineering at Turin in 1854, Schiaparelli taught mathematics and studied modern languages and astronomy at the University of Turin. As a result of his increasing interest in astronomy, in 1857 he was sent by the Piedmont government to continue advanced studies at the observatories of Berlin and Pulkovo. On his return to Italy in 1860, Schiaparelli was appointed astronomer at the Brera Observatory in Milan; and in 1862, when his work had already brought him a certain renown, he succeeded Francesco Carlini as director. He retired voluntarily in 1900 and spent the rest of his life in Milan. In 1889 Schiaparelli became senator of the kingdom of Italy, he also was a member of the Lincei and of many other Italian and foreign academies. He received gold medals from the governments of Italy, England, and Germany, and twice won the Lalande Prize of the Institut de France.

At the beginning of his observations at Brera, Schiaparelli, using primitive instruments, discovered the asteroid Hesperia. In 1860 he became interested in comets and undertook a theoretical study on the initial direction of their tails in which he demonstrated the existence of a repulsive force that tends to pull parts of the tail away from the part opposite the sun. This force, combined with gravity, generates in the tail a parabola, similar to that described by terrestrial projectiles.

The appearance of the bright comet of 1862 stimulated Schiaparelli's interest in these celestial objects; and while observing it, he reflected on the forces that determine the features of comets in general. His accurate study of the shape and position of the tails of comets led him to new theories on the repulsive action exerted by the sun and to classify the tails according to this action.

Schiaparelli assiduously continued his observations of physical position and calculations of the orbits of comets while developing the idea that comets give rise to meteors. It was discovered that Biela's comet, when it appeared in 1845, had split in two; and on its next appearance, in August 1852, Secchi found its larger fragment. The smaller fragment was discovered about a month later at a far greater distance from the first than had previously been calculated. It became clear that the comet was disintegrating as it approached the sun. In April 1862, still using the modest instruments available at Brera, Schiaparelli observed the large comet of that year and saw that the nucleus—the luminous head of the comet followed by a long tail—was emitting a luminous jet that rapidly increased in size and assumed the shape of a clearly outlined pear. Its mass was much greater than the nucleus proper; the form was that of a small cloud

in which, over a clear background, more luminous points flared at intervals, like small stars visible in the field of a telescope.

The hypothesis had been advanced that meteor showers, observed over the centuries as originating from a well-defined point in the sky, could be related to the disintegration of comets. Schiaparelli's hypothesis was based chiefly on his observations of the meteor swarms of August 1866. He stated the hypothesis more definitely in five letters to Secchi "concerning the course and probable origin of the meteoric stars." Secchi published the letters in *Bullettino meteorologico dell'Osservatorio del Collegio romano* in 1866. To confirm his hypothesis Schiaparelli had to prove that if meteors are subject to the attraction of the sun, they must move in elliptical or parabolic orbits around it and that these orbits must be identical with or similar to those of the comets that cause meteor swarms. The latter become visible when the earth, in its course around the sun, meets either of the swarms that extend along its orbit. In the second letter Schiaparelli stated that in the planetary spaces the meteors must form a multitude of continuous currents that, on meeting the earth in its orbit, become visible in the form of luminous showers falling from a determined direction of the celestial sphere. Secchi accurately observed these directions (the "radiants") and proved that the orbit of the periodic stars of August is practically identical to that of the large comet of 1862.

Schiaparelli published a complete elaboration of his theory as *Entwurf einer astronomischen Theorie der Sternschnuppen* (1871). Following the extraordinary meteor shower of 27 November 1872, he explained the phenomena and the theory behind them in three letters. In the third letter he stated:

> The meteor showers are the product of the dissolution of the comets and consist of very minute particles that they . . . have abandoned along their orbit because of the disintegrating force that the sun and the planets exert on the very fine matter of which they are composed.

Schiaparelli's hypothesis on the origin of meteors, elaborated in depth and with elegance, has been fully confirmed in the several cases listed: these include the relation he discovered between the Perseids of 10 August and the great comet of 1862; C. H. F. Peters' observations concerning the November Leonids in relation to the comet of 1866; and Galle's and Weiss's studies of the first comet of 1861 and the meteor shower of 20 April. A fourth case deals with Biela's comet, the relations of which to certain previously observed meteors had been noted as early as 1867 by d'Arrest and Weiss and were confirmed by the meteor shower of 27 November 1872. The meteor showers of 1933 and 1946 are related to the Giacobini-Zinner comet; the swarm of the Taurids, to Encke's comet. Thus, Schiaparelli's theory was confirmed.

In 1877, using a Merz refractor far superior to the antiquated instruments he had previously used, Schiaparelli turned his attention to the study of Mars. This exceptional instrument, with an objective aperture of twenty-two centimeters, revealed "a large quantity of minute objects, which in the earlier oppositions had been overlooked by the gigantic telescopes in which foreign countries justly take pride." During the great opposition of Mars in 1877 Schiaparelli observed the planet thoroughly, detecting even the smallest surface features. He began to determine the orientation of Mars in the sky, publishing a first memoir in *Astronomische Nachrichten*: "Sur l'axe de rotation et sur la tâche polaire australe de Mars." It was followed by *Osservazioni astronomiche e fisiche sull'asse di rotazione e sulla topografia del pianeta Marte durante l'opposizione del 1877.*

In drawing a complete picture of the areographic positions of the fundamental points for the construction of an accurate map, Schiaparelli stated that the interpretation of the phenomena observed on Mars was still largely hypothetical, varying among observers even when they saw the same details. He was the first to classify the features as "seas" and "continents"; the term "canal" had been used by Secchi in his observations of 1859. Schiaparelli's was an original nomenclature, and he observed that "the names I adopted will in no way harm the cold and rigorous observations of facts." Although he understood that the features he observed on Mars were stable, like their terrestrial counterparts, he was cautious in drawing conclusions on the nature of the surface and atmosphere until he could establish that the seas, continents, and canals were identifiable with analogous terrestrial formations.

During the opposition of 1879–1880 Schiaparelli continued his observations, from which he prepared a catalog of the positions of all visible topographical features. In it he sought to provide a less geometric interpretation than that of his first map of 1877–1878. Schiaparelli noted that because Mars was moving away from the earth, its diame-

ter gradually decreased during subsequent oppositions. In 1879 he had observed that certain canals seemed to be splitting into two parts. In the 1881–1882 opposition he noted the increasing clarity of the geminations of canals, which he thought would greatly change current opinions on the physical constitution of the planet. His areographic map of this opposition apparently is a more geometric representation, perhaps in order to stress the gemination of the canals, which also appeared in the 1883–1884 oppositions. In the latter nearly all of them were split.

In his observations of the 1886 opposition, Schiaparelli used a new refractor with an aperture of fifty centimeters, among the largest at that time. The disk of the planet then measured only ten seconds in diameter and Schiaparelli, continuing to make increasingly geometric drawings, marked only one large gemination: the Nilus-Hydrae Fons. In the 1888 opposition, which occurred under good atmospheric conditions, he found it impossible to represent adequately all the detailed features and their colors. Observing the geminations that had been absent from the preceding opposition, Schiaparelli thought that their reappearance constituted a strictly periodic phenomenon related to the solar year of Mars, and that it was necessary to follow it closely in successive, and more favorable, oppositions. He noted that the split canals appeared and remained visible for a few days or weeks before again becoming simple canals or disappearing entirely.

Schiaparelli's observations of Mars ended with the 1890 opposition. This cycle included seven oppositions that present all conceivable varieties of inclinations of the axis, of the apparent diameter, and of geocentric declination, and it occurred at points along the zodiac almost equidistant from each other. Schiaparelli recalled that three astronomers of the Lick observatory, using the refractor with a ninety-two-centimeter aperture, insisted they saw the same details differently through the same telescope and, one might say, at the same instant. Other observers, using less powerful instruments, saw a thick web of lines (the canals proper) so clearly that they could be recognized with good telescopes having an aperture of only ten centimeters. The last areographic map, which Schiaparelli drew at the conclusion of his observations, is the most geometric of all and depicts most of the canals as split. E. Antoniadi, another well-known observer of Mars, using a telescope with an eighty-three-centimeter aperture at the Meudon (Paris) observatory, noted in 1930 that Schiaparelli, with instruments of equal power, had surpassed everyone with his numerous observations.

Schiaparelli also observed Saturn and, for eight years, the few dark spots visible on Mercury in the form of shadowy bands, difficult to recognize in full daylight. He concluded that Mercury revolves about the sun in the same manner that the moon does around the earth and Iapetus around Saturn — with the same side always turned to the sun. He also tried to solve the problem of the rotation of Venus on its axis. In outlining the history of the subject, which also concerned Jean-Dominique Cassini, he recalled the observations made at Rome in 1726 by Francesco Bianchini. On the basis of the diffused and indefinite shadows visible on the surface, Bianchini had concluded that it completed one rotation on its axis in about twenty-four days and eight hours. Schiaparelli observed Venus in 1877 and 1878, noting luminous oval spots that were quite visible but transitory, perhaps resulting from variations in the planet's atmosphere. He concluded that the rotation of Venus was "very slow . . . , much slower than Bianchini had supposed." From observations of well-defined spots, he obtained as a very probable result that the rotation occurs in 224.7 days, a period exactly equal to that of the sidereal revolution of the planet around an axis almost coincident with the perpendicular to the plane of the orbit.

Schiaparelli also made numerous observations of double stars. For many of the more interesting binaries, the measurements were continued for several years, in order to deduce the orbital elements of the systems.

Schiaparelli's work in the history of astronomy was noteworthy. The orientalist C. A. Nallino was sent by the Brera Observatory to the Escorial to copy and translate into Latin the only existing Arab text of al-Battānī's *Opus astronomicum*. Schiaparelli collaborated with Nallino to complete the translation, which was published by the Brera Observatory between 1899 and 1907. He also contributed many explanatory notes to several chapters. He noted that much more was known about Arab astronomy as a result of the translation of this vast work, which is not limited to the works of al-Battānī. A comparison between the *Opus astronomicum* and Ptolemy's *Almagest*, with which there are many points of similarity, is of special value to an understanding of the development of astronomy in the Arab world.

Schiaparelli had intended to compile a major work on the history of ancient astronomy, In preparation for it he read the original texts of the He-brews, Assyrians, Greeks, and Romans. During his lifetime he published several monographs on the subject. These and many similar works that Schiaparelli was not able to complete were published by his pupil Luigi Gabba in 1925–1927.

The first volume deals with the astronomy of Babylonia, of the Old Testament lands, and of Greece. In the second volume he treats later Greek astronomy: the homocentric spheres of Eudoxus of Cnidus, Callippus, and Aristotle. Schiaparelli next considers the origin of the Greek heliocentric planetary system with Aristarchus of Samos, who, as Schiaparelli demonstrates, must have developed the system through the hypothesis proposed centuries later by Tycho Brahe. According to Schiaparelli's research, the Tychonic system was known to the Greeks at the time of Heraclides Ponticus (fourth century B.C.). The second volume ends with minor writings on "parapegmata," the astrometeorological calendars of the ancients. The third volume contains writings designed to integrate the history of ancient astronomy: his studies on the calendar of the ancient Egyptians; on the observations and ephemerides of the Babylonians; on the phenomena of Venus, according to the discoveries made in the ruins of Nineveh; on the discovery of the precession of the equinoxes; and on the astronomy of Hipparchus. These three volumes, even though they may not contain all that Schiaparelli had wanted to include in the project, nevertheless provide data and information of inestimable value on the history of ancient astronomy.

BIBLIOGRAPHY

I. ORIGINAL WORKS. Schiaparelli's writings have been brought together in *Le opere di G. V. Schiaparelli*, 11 vols. (Milan, 1929–1943); and *Scritti sulla storia della astronomia antica*, Luigi Gabba, ed., 3 vols. (Bologna, 1925–1927).

II. SECONDARY LITERATURE. See G. Abetti, *Storia della astronomia* (Florence, 1963), 224–228; and the obituary by E. Millosevich, in *Memorie della Società degli spettroscopisti italiani*, **39** (1910), 138–140, with photograph.

GIORGIO ABETTI

SCHICKARD, WILHELM (*b*. Herrenberg, Germany, 22 April 1592; *d*. Tübingen, Germany, 23 October 1635), *astronomy, mathematics, natural philosophy.*

Schickard, a brilliant student, received the B.A.

in 1609 and the M.A. in 1611 from the University of Tübingen, where he continued with the study of theology and oriental languages until 1613. He then served as deacon or pastor in several nearby towns. In 1617 he befriended Kepler, who reawakened in him an interest in mathematics and astronomy and with whom he maintained an active correspondence for several years. In 1619 he was named professor of Hebrew at the University of Tübingen. Upon the death in 1631 of his former teacher, Michael Mästlin, Schickard succeeded to the chair of astronomy but continued to lecture on Hebrew.

Schickard was a polymath who knew several Near Eastern languages, some of which he taught himself. He was a skilled mechanic, cartographer, and engraver in wood and copperplate; and he wrote treatises on Semitic studies, mathematics, astronomy, optics, meteorology, and cartography. He invented and built a working model of the first modern mechanical calculator and proposed to Kepler the development of a mechanical means of calculating ephemerides. Schickard's works on astronomy include a lunar ephemeris, observations of the comets of 1618, and descriptions of unusual solar phenomena (meteors and the transit of Mercury in 1631). He also constructed and described a teaching device consisting of a hollow sphere in three segments with the heavens represented on the inside.

Schickard was an early supporter of Kepler's theories; his treatise on the 1631 transit of Mercury called attention to some of Kepler's ideas and works and to the superiority of the *Rudolphine Tables*. Schickard also mentioned Kepler's first two laws of planetary motion; the second law, however, was given only in the inverse-distance, rather than in the correct, equal-areas formulation.

BIBLIOGRAPHY

I. ORIGINAL WORKS. Schickard's unpublished MSS are in the Österreichische Nationalbibliothek of Vienna and in the Württembergische Landesbibliothek in Stuttgart. His chief works (all published in Tübingen) are *Astroscopium pro facillima stellarum cognitione noviter excogitatum* (1623); *Ignis versicolor e coelo sereno delapsus et Tubingae spectatus* (1623); *Weiterer Bericht von der Fliegenden Liecht-Kugel* (1624); *Anemographia, seu discursus philosophicus de ventis* (1631); *Contemplatio physica de origine animae rationalis* (1631); and *Pars responsi ad epistolas P. Gassendi . . . de mercurio sub sole viso et alijs novitatibus uranicis* (1632). Useful collections of his correspondence are in *Epistolae W. Schickarti et M. Berneggeri mutuae* (Strasbourg, 1673); *Johannes Kepler Gesammelte Werke*, **17–18**, Max Caspar, ed. (Munich, 1955, 1959); and the appendix to Schnurrer's biography (see below), pp. 249–274.

II. SECONDARY LITERATURE. The standard biographies remain Johann C. Speidel, in his ed. of Schickard's *Nova et plenior grammatica Hebraica* (Tübingen, 1731) and Christian F. Schnurrer, in *Biographische und litterarische Nachrichten von ehmaligen Lehrern der hebräischen Litteratur in Tübingen* (Ulm, 1792), 160–225. Recent accounts of Schickard's invention of the calculating machine are Franz Hammer, "Nicht Pascal, sondern der Tübinger Professor Wilhelm Schickard erfand die Rechenmaschine," in *Büromarkt-Bibliothek*, **13** (1958), 1023–1025; and René Taton, "Sur l'invention de la machine arithmétique," in *Revue d'histoire des sciences et de leurs applications*, **16** (1963), 139–160.

WILBUR APPLEBAUM

SCHIFF, HUGO JOSEF (*b.* Frankfurt, Germany, 26 April 1834; *d.* Florence, Italy, 8 September 1915), *chemistry.*

Schiff, the last surviving representative of the Karlsruhe Congress (1860), was the brother of the distinguished physiologist Moritz Schiff. He studied under Wöhler at Göttingen and was awarded the doctorate in 1857. Shortly thereafter he left Germany because of his strong liberal views and became *Privatdozent* at the University of Bern. In 1863 the physicist Carlo Matteucci invited Schiff to Florence. He taught chemistry at the Museo di Storia Naturale until 1876, when he became professor of general chemistry at the University of Turin. In 1879 he returned to Florence to assume the chair of chemistry at the Istituto di Studi Superiori, remaining there for the rest of his life.

In addition to his prolific research, Schiff devoted himself to disseminating chemical knowledge and to continuing the tradition represented by Berzelius and Wöhler, on whose teaching methods he modeled his own. At Florence he transformed the chemistry laboratory into one of the best in Europe. In 1871, with Cannizzaro and Francesco Selmi, he founded the journal *Gazzetta chimica italiana*. Schiff wrote the widely used *Introduzione allo studio della chimica* (1876) and also contributed many articles to Francesco Selmi's *Enciclopedia di chimica* (1868–1883). A man of many interests, he also published writings in history and literary criticism.

Schiff's chemical studies were predominantly in organic chemistry. His earliest noteworthy work was his isolation and investigation in 1857 of thionyl chloride from the action of sulfur dioxide on phosphorus pentachloride. In 1864 he discovered the condensation products of aldehydes and amines, later known as "Schiff bases." In 1866 Schiff introduced the fuchsine test for aldehydes, in which decolorized fuchsine regains its color in the presence of aldehydes, the color reaction being specific for aldehydes and serving to distinguish them from ketones. His studies on the color bases derived from furfural led to his discovery of the sensitive xylidine-acetic acid reagent for furfural.

Schiff published many papers on the constitution of natural glucosides, examining esculin, amygdalin, arbutin, helicin, and phlorizin. By fusing salicin with benzoic anhydride, he synthesized populin, a glucoside present in the bark and leaves of aspen trees. Other researches dealt with metal-ammonium compounds, the biuret reaction, esters of boric acid, and the constitution of tannins. A resourceful experimentalist, Schiff in 1866 devised the nitrometer that bears his name, an improved version of the Dumas method for the determination of nitrogen.

Before Emil Fischer, Schiff attempted to obtain high-molecular-weight polymers of amino acids having the properties of proteins. By condensing aspartic acid molecules, he prepared tetraaspartic and octoaspartic acids. His formulas did not, however, include the peptide linkages that Fischer later proposed to explain how amino acid molecules were joined in proteins.

BIBLIOGRAPHY

I. ORIGINAL WORKS. Schiff's books include *Untersuchungen über metallhaltige Anilinderivate und über die Bildung des Anilinroths* (Berlin, 1864); *Introduzione allo studio della chimica* (Turin, 1876); and *Empirismo e metodo nell'applicazione di chimica alle scienze naturale e biologiche* (Turin, 1877). The more important of his almost 300 published papers include "Über die Einwirkung des Phosphorsuperchlorids auf einige anorganische Säuren," in *Justus Liebigs Annalen der Chemie,* **102** (1857), 111–118; "Eine neue Reihe organischer Basen," *ibid.,* **131** (1864), 118–119; "Eine neue Reihe organischer Diamine," *ibid.,* **140** (1866), 92–137; "Zur Azotometrie," in *Zeitschrift für analytische Chemie,* **7** (1868), 430–432; "Untersuchungen über Salicinderivate," in *Justus Liebigs Annalen der Chemie,* **150** (1869), 193–200; and **154** (1870), 1–39; "Zur Stickstoffbestimmung," in *Berichte der Deutschen chemischen Gesellschaft,* **13** (1880), 885–887; "Über Polyaspartsäuren," *ibid.,* **30** (1897), 2449–2459; and "Intorno a composti poliaspartici," in *Gazzetta chimica italiana,* **28**, pt. 1 (1898), 49–64; **29**, pt. 1 (1899), 319–340; and **30**, pt. 1 (1900), 8–25.

II. SECONDARY LITERATURE. For accounts of Schiff's life and work, see M. Betti, in *Berichte der Deutschen chemischen Gesellschaft,* **48** (1915), 1566–1567; and *Journal of the Chemical Society,* **109** (1916), 424–428; I. Guareschi, in *Atti dell' Accademia delle scienze* (Turin), **52** (1917), 333–351; and William McPherson, in *Science,* **43** (1916), 921–922.

ALBERT B. COSTA

SCHIFF, MORITZ (*b.* Frankfurt, Germany, 28 January 1823; *d.* Geneva, Switzerland, 6 October 1896), *zoology, physiology.*

Schiff was one of the eminent biologists who pioneered the experimental method in the new science of physiology. Following in the steps of his teacher Magendie, he approached the subject matter from a biological point of view instead of carrying out reductionist physicochemical studies like those of du Bois-Reymond and Helmholtz. Schiff's often controversial vivisections uncovered details in spinal cord physiology, clarified the role of the autonomic nervous system, and revealed the importance of certain internal secretions.

Schiff was descended from a family of Jewish merchants. His interest in the natural sciences emerged in early childhood, during which he established in his attic a veritable museum of small animals. Following the family tradition, Schiff was sent to study the textile business, but his utter incompetence convinced his father that his talents lay elsewhere.

Thus, in the late 1830's Schiff began to study the natural sciences at the Senckenberg Institute, transferring to Heidelberg in 1840 in order to pursue a career in medicine. After some anatomical work with Friedrich Tiedemann, Schiff moved to Berlin, where he studied morphology with Johannes Müller. Finally, in 1843, he went to Göttingen as a student of Rudolf Wagner and received his medical degree there a year later.

After his graduation Schiff traveled to Paris in order to conduct zoological research at the museum of the Jardin des Plantes. He also visited local hospitals and studied experimental techniques with Magendie and served briefly as a research assistant to both François Longet and Carlo Matteucci, who were studying the physiology of the nervous system.

In 1845 Schiff returned to Frankfurt to practice medicine. Instead of seeing many patients, he spent most of his time in a small homemade laboratory, where he conducted physiological experiments. In 1846 he was appointed a member of the ornithology section of the Senckenberg Museum, where he cataloged South American birds.

Schiff's early research dealt with cardiac contraction and its possible relation to nerve-mediated action. By 1849 he had concluded that the diastole was a reflection of nervous exhaustion. Following studies on digestion and the influence of neural centers on bodily heat, Schiff received the Montyon Prize of the French Academy in 1854 for his work on bone physiology.

A year later Schiff decided to begin his academic career at Göttingen, but his petition to become *Privatdozent* in zoology was rejected by the university authorities without explanation. The true reasons for the decision were Schiff's Jewish ancestry and his past membership in the revolutionary medical corps during the 1848 uprising in Baden. Undaunted, Schiff accepted an appointment in 1856 as assistant professor of comparative anatomy and zoology at the University of Bern. In 1859, after rejecting an offer to teach physiology at Jena, he was appointed professor of physiology at the Istituto di Studii Superiori of the University of Florence in 1862. During the ensuing years, the most productive of his scientific career, he studied the pathways of pain and touch in the spinal cord.

In 1874 Schiff made a comparative study of the two anesthetic agents then in vogue, chloroform and ether. He succeeded in proving the toxicity of the former and branded its use dangerous. A systematic campaign aimed at his vivisection experiments forced Schiff to leave Florence in 1876, and he returned to Switzerland to assume the chair of physiology at the University of Geneva. Before his death from diabetes, Schiff studied primarily the functions of the thyroid, establishing the foundations for the surgical removal of goiters. He was a corresponding member of the Royal Academy of Rome and Paris Academy of Medicine, and a coeditor of the *Schweizerische Zeitschrift für Heilkunde.*

Schiff pioneered research on the vasomotor functions of the autonomic nervous system and especially the innervation of the heart. He also studied thyroid function, artificially inducing myxedema through the surgical removal of the gland and reversing that condition through thyroid transplants in dogs. Schiff wrote on the formation of glycogen in the liver in experimentally induced diabetes, using Claude Bernard's puncture of the fourth ventricle in the brain.

BIBLIOGRAPHY

I. ORIGINAL WORKS. Schiff's collected papers in German and French were published as *Moritz Schiff's gesammelte Beiträge zur Physiologie,* 4 vols. (Lausanne, 1894–1898). In vol. I Schiff himself rearranged some of his articles on the centers in the nervous system that are related to respiration. The papers are arranged by topic and then presented chronologically. Vol. IV was edited by Alexandre Herzen and Émile Levier.

Schiff's earlier books include *Untersuchungen zur Physiologie des Nervensystems mit Berücksichtigung der Pathologie* (Frankfurt, 1855); and *Untersuchungen über die Zuckerbildung in der Leber, und den Einfluss des Nervensystems auf die Erzeugung der Diabetes* (Würzburg, 1859). He began a textbook of physiology, completing the first volume on muscle physiology and neurophysiology: *Lehrbuch der Physiologie des Menschen* (Lahr, 1858–1859).

Several of his Italian lectures were translated into French by R. Guichard de Choisity as *Contribution à la physiologie: De l'inflammation et de la circulation* (Paris, 1873). Two works written in Italian are *Lezioni di fisiologia sperimentale sul sistema nervoso encefalico,* compiled by P. Marchi (Florence, 1865); and *La pupilla come estesiometro* (Florence, 1875).

II. SECONDARY LITERATURE. Among the short biographical sketches of Schiff is an obituary note by Arthur Biedl in *Wiener klinische Wochenschrift,* 9, no. 44 (1896), 1008–1010; the article by J. R. Ewald in *Allgemeine deutsche Biographie,* LIV, 8–11; and the entry in August Hirsch, ed., *Biographisches Lexikon der hervorragenden Ärzter,* 2nd ed., V, 72–73. More recent is P. Riedo, "Der Physiologe Moritz Schiff (1823–1896) und die Innervation des Herzens," *Zürcher medizingeschichtliche Abhandlungen,* n.s. no. 85 (1970).

Schiff's formal application for an academic position at the University of Göttingen has been published together with some biographical details by H. Friedenwald, "Notes on Moritz Schiff (1823–1896)," in *Bulletin of the Institute of the History of Medicine,* 5 (1937), 589–602; the document furnishes a chronological account of Schiff's physiological research before 1855. A recent brief editorial containing some paragraphs of Schiff's articles in English translation is "Moritz Schiff (1823–1896), Experimental Physiologist," in *Journal of the American Medical Association,* 203 (1968), 1133–1134.

GUENTER B. RISSE

SCHIMPER, ANDREAS FRANZ WILHELM (*b.* Strasbourg, France, 12 May 1856; *d.* Basel, Switzerland, 9 September 1901), *botany.*

Schimper was the son of Wilhelm Philipp Schimper, professor of natural history and geology at the University of Strasbourg and director of the city's museum of natural history. From 1864 to 1874 he attended the Strasbourg Gymnasium. His father allowed him to take part in the excursions he conducted for his students. His mother, Adèle Besson, who was greatly interested in her husband's botanical activities, stimulated the boy's interest in natural history.

In 1874 Schimper entered the University of Strasbourg, where he came under the influence of Anton de Bary. He received the doctorate in natural philosophy in November 1878. While at Strasbourg he studied the origin and development of starch grains. Following the death in 1880 of his father, whose assistant he had been, the trustees of the museum of natural history elected him director. De Bary opposed the appointment, although he acknowledged Schimper as one of his best students. As a result, Schimper accepted a post at the Lyons botanical garden but soon returned to Germany to work with Julius Sachs at Würzburg.

In the autumn of 1880 Schimper was appointed a fellow of the Johns Hopkins University in Baltimore. The results of his further observations on the growth of starch grains induced him to abandon Naegeli's intussusception theory. In the spring of 1881 he went to Florida and, the following winter, to the West Indies. These trips awakened Schimper's interest in plant geography. During the summer he visited the zoological summer laboratory at Annisquam, Massachusetts, where he studied insectivorous plants. Schimper returned to Germany in January 1882 and worked in the laboratory of Eduard Strasburger at Bonn until 1898. Strasburger, who ranked Schimper as one of his outstanding students, was instrumental in keeping him in Germany when an attractive post in the United States was offered to him in 1889.

From December 1882 to August 1883 Schimper traveled in Barbados, Trinidad, Venezuela, and Dominica, studying the morphology and biology of the epiphytes. On 16 November 1883 he was appointed lecturer in physiological botany at the University of Bonn. He lectured on plant geography, historical and geographical distribution of important cultivated plants, and, after 1885, on medicinal plants, pharmacognosy, and microscopic research on drugs and food products. He also made botanical excursions with his students. On 12 February 1886 he was named extraordinary professor.

During these years Schimper wrote very important medicopharmaceutical books. In August 1886 he traveled to Brazil, where he studied the mangrove vegetation. In both the West Indies and Brazil he made observations and physiological experiments to determine the influence of high salt con-

centrations on the marine littoral vegetation. To study the vegetation of tropical beaches Schimper visited Ceylon and Java in 1889–1890. During this voyage he visited the Buitenzorg (now Bogor) botanic garden, near Batavia (now Jakarta) and made excursions to several volcanoes with solfataras and halophytic vegetation.

In July 1898 Schimper joined the important German marine expedition on board the *Valdivia,* during which he studied the oceanic plankton flora and the vegetation of the Canary Islands, Kerguelen, the Seychelles, Cameroon, the Congo and eastern Africa, Sumatra, and the Cape of Good Hope. In October, near Cameroon, he suffered a severe attack of malaria, from the effects of which he never recovered.

In June 1898 Schimper had been appointed professor of botany at the University of Basel, where he took up his duties in April 1899. The following February he delivered his inaugural oration on marine plankton but his deteriorating health (he had suffered from diabetes since 1899) prevented him from carrying out his duties for long.

Schimper, who was able to penetrate quickly to the core of scientific problems, preferred to work independently. Not physically strong, he had enormous energy and enthusiasm, and a very deep love for nature. He was greatly interested in literature and the arts. Schimper did not care for large groups, preferring to spend the evenings with a few close friends. Although solid and accurate in his research, he was impulsive in thought and speech.

BIBLIOGRAPHY

The most important of Schimper's 27 books and articles are *Untersuchungen über die Proteinkrystalloide der Pflanzen* (Strasbourg, 1878), his dissertation; "Untersuchungen über die Entstehung der Stärkekörner," in *Botanische Zeitung,* **38** (1880), 881–902, in English in *Quarterly Journal of Microscopical Science,* **21** (1881), 291–306; "Untersuchungen über das Wachsthum der Stärkekörner," in *Botanische Zeitung,* **39** (1881), 185–194, 201–211, 217–228; "The Growth of Starch Grains," abstract, in *American Naturalist,* **15** (1881), 556–558; "Die Vegetationsorgane von *Prosopanche burmeisteri,*" in *Abhandlungen der Naturforschenden Gesellschaft zu Halle,* **15** (1882), 21–47; *Anleitung zur mikroskopischen Untersuchung der Nahrungs- und Genussmittel* (Jena, 1886); *Taschenbuch der medicinisch-pharmaceutischen Botanik und pflanzlichen Drogenkunde* (Strasbourg, 1886); *Schimper's botanische Mittheilungen aus den Tropen,* 3 vols. (Jena, 1888–1891): I, *Die Wechselbeziehungen zwischen Pflanzen und Ameisen im tropischen Amerika;* II, *Die epiphy-*

tische Vegetation Amerikas; III, *Die indomalayische Strandflora;* "Ueber Schutzmittel des Laubes gegen Transpiration, besonders in der Flora Java's," in *Sitzungsberichte der Preussischen Akademie der Wissenschaften zu Berlin,* Phys.-math. Kl., **40** (1890), 1045–1062; and *Pflanzengeographie auf physiologischer Grundlage* (Jena, 1898), trans. as *Plant Geography Upon a Physiological Basis* (Oxford, 1903).

A good biography, with complete bibliography, is H. Schenk, "A. F. W. Schimper," in *Berichte der Deutschen botanischen Gesellschaft, XIX, Generalversammlungsheft,* **1** (1901), 54–70.

A. P. M. SANDERS

SCHIMPER, KARL FRIEDRICH (*b.* Mannheim, Germany, 15 February 1803; *d.* Schwetzingen, near Heidelberg, Germany, 21 December 1867), *botany, geology, zoology, meteorology.*

Schimper's father was a mathematics teacher and later a government engineer; his mother, Meta, Baroness Furtenbach, came from a noble family of Nuremberg. Their uncongenial marriage ended in divorce. Schimper and his younger brother thus had an unhappy childhood, all the more so since their father had a very modest income. Schimper gained such a broad knowledge of plants during his school days in Mannheim that his teacher, F. W. L. Succow, asked him to collaborate on his *Flora Manhemiensis.* In 1822 Schimper began to study theology at the University of Heidelberg because it was the only subject for which he could obtain a scholarship. In 1826 he turned to the study of medicine and became friends with Alexander Braun and Louis Agassiz. The three continued their studies at Munich in 1827–1828, and in 1829 they received their doctorates. Braun and Agassiz returned home, but Schimper remained in Munich until 1841.

Schimper hoped to embark on an academic career in Munich but was unable to obtain a post. After a short stay in Zweibrücken (Rhenish Palatinate) he returned, disappointed and poor, to Mannheim. His situation did not improve until 1845, when the grand duke of the Rhenish Palatinate awarded him a small annuity. He also encountered difficulties in Mannheim, and in 1849 he moved to nearby Schwetzingen, where he spent the rest of his life. Lacking a permanent position and a regular income, he was constantly in financial difficulties. This period was interrupted only by a stay at Jena in 1854–1855. In 1856 Schimper's friends and admirers sought unsuccessfully to procure a professorship of botany for him. He died of dropsy in 1867, following months of confinement to bed. He was engaged twice, the second time to a sister of Alexander Braun. Neither engagement led to marriage.

In botany Schimper's principal concern was to formulate a theory of phyllotaxy. He pointed out that leaves that grow in spiral formations are arranged in regular, cyclic patterns and that each species has a characteristic pattern. He also dealt with the unequal, eccentric growth in thickness of branches of deciduous and coniferous trees (epinasty and hyponasty), with the morphology and physiology of plant roots, with the heterophylly of certain aquatic plants (variation in the forms of the leaves, depending on the depth of the water), and with water transport in mosses. Since his school days Schimper had had a good knowledge of floristics and systematics. Besides the Mannheim flora, he collaborated on the *Flora Friburgensis* with F. C. L. Spenner. A catalog of the mosses of Baden that Schimper planned to publish in the last years of his life never appeared, as was the case with many of his projects.

Since his days in Munich, Schimper had been interested in prehistoric animals. He did not, however, undertake specific research on particular groups of fossils, desiring instead to formulate an overall view. In "Eintheilung und Succession der Organismen" he proposed the existence of a succession of different faunas. They, and the floras that accompanied them, were related to each other; but phases of development *(Belebungen)* were, he held, separated by periods of desolation *(Verödungen).* Schimper expressed his ideas, which contradicted Cuvier's catastrophist theory, in a schema similar to a genealogical tree. Much of this account was similar to the theory of evolution, but Schimper completely rejected the theory of natural selection later formulated by Darwin.

In geology Schimper was especially interested in the traces of Pleistocene glaciers that he detected in the northern Alpine foothills, notably in Bavaria, Switzerland, and the Black Forest. This study led him to the concept of the "Ice Age," which he discussed in a paper presented in 1837 to the congress of Swiss scientists held at Neuchâtel. The paper gave rise to an unpleasant priority dispute with Louis Agassiz that was not decided in Schimper's favor until much later. Through his investigations in the Bavarian Calcareous Alps, Schimper recognized in 1840 that the Alps had not been raised by a force acting from below but had emerged as the result of horizontal pressure—in the same way that a range of folded mountains is created. The pressure was generated, he speculated, by the shrinking of the earth's core. These views anticipated the contraction theory proposed by Suess in 1875.

In 1843 Schimper presented a paper on prehistoric climatic conditions and postulated an alternation of warm and cold periods in the earth's history. Accordingly, his Ice Age was preceded by warmer weather. Further, he interpreted desiccation cracks and raindrop impressions in the Triassic of southern Germany as evidence of the existence, during a given period, of a dry or a moist climate. From the evidence of postglacial calcareous tuffs in Upper Bavaria and the annual rings in Triassic silicified trees, Schimper deduced the existence of change of seasons in the period he studied. Finally, he contended that the snails occurring in the Pleistocene loess indicated their emergence in a cool climate.

In hydrology and meteorology Schimper made a series of new and stimulating observations for which he proposed ingenious explanations. Among the topics he treated were the refraction and reflection of light in inland waters and their effect on plant life, the attenuation of the reflection of light from the surface of bodies of water because of the presence of pollen, and the formation of ice on rivers.

Schimper expressed many of his scientific ideas in poems. Several hundred of these are known, and they encompass the most varied verse forms. Their artistic merit has been quite variously judged. Schimper was an extraordinarily versatile scientist, but almost all of his publications have a preliminary or fragmentary character. He frequently planned to publish large-scale works but never carried out his intentions. Consequently, many of his results fell into neglect or were taken over and developed by others, with their origin often forgotten. This circumstance, together with his volcanic temperament and a certain lack of objectivity and of consideration for his colleagues, surely contributed to the difficulties that marked Schimper's personal life.

BIBLIOGRAPHY

I. ORIGINAL WORKS. Schimper's writings include "Vorträge über die Möglichkeit eines Verständnisses der Blattstellung," in *Flora*, **18**, no. 1 (1835), 145–192; "Vortrag über Blattstellungstheorie," in *Verhandlungen der Versammlung der Schweizerischen naturfor-schenden Gesellschaft*, **22** (1836), 114–117; "Auszug aus dem Schreiben des Herrn Dr. Schimper über die Eiszeit . . .," *ibid.*, **23** (1837), 38–51; *Gedichte* (Erlangen, 1840); "Über den Bau der bayerischen Kalkalpen," in *Amtlicher Bericht über die 18. Versammlung der Gesellschaft deutscher Naturforscher und Ärzte zu Erlangen* (1841), 93–100; *Über die Witterungsphasen der Vorwelt, Entwurf zu einem Vortrage bei Gelegenheit der zehnten Stiftungsfeier und Generalversammlung des Mannheimer Vereins für Naturkunde* (Mannheim, 1843); *Gedichte 1840–1846* (Mannheim, 1847); "Über hyponastische, epinastische und diplonastische Gewächse . . .," in *Amtlicher Bericht über die Versammlung deutscher Naturforscher und Ärzte zu Göttingen* (1854), 87–88; *Natursonette . . . eine Weihnachtsgabe für Gebildete* (Jena, 1854); "Nützliches Allerlei von der ganzen Pflanze; Auswahl förderlichster Thatsachen aus der Morphologie," in *Amtlicher Bericht über die Versammlung deutscher Naturforscher und Ärzte zu Bonn* (1857), 129–132, 137–138; "Wasser und Sonnenschein oder die Durchsichtigkeit und der Glanz der Gewässer betrachtet nach ihrem Einfluss auf die Entwickelungen organischer und geologischer Art am Äussern des Erdballs," in *Festschrift der Naturforschenden Gesellschaft zu Emden* (Emden, 1865), 37–66; and "Uber Eintheilung und Succession der Organismen," L. Eyrich, ed., in *Jahresbericht des Mannheimer Vereins für Naturkunde* for 1878–1882 (1882), 1–36.

II. SECONDARY LITERATURE. See L. Eyrich, "Nachrede zum Vortrage von Dr. K. F. Schimper . . .," in *Jahresbericht des Mannheimer Vereins für Naturkunde* for 1878–1882 (1882), 37–64, with biographical data, poems, and bibliography; R. Lauterborn, "Karl Friedrich Schimper, Leben und Schaffen eines deutschen Naturforschers," in "Der Rhein. Naturgeschichte eines deutschen Stromes," in *Berichte der Naturforschenden Gesellschaft in Freiburg i. Br.*, **33** (1934), 269–324, with bibliographical data and portrait; and G. H. O. Volger, *Leben und Leistungen des Naturforschers K. Schimper* (Frankfurt, 1889).

HEINZ TOBIEN

SCHIMPER, WILHELM PHILIPP (GUILLAUME PHILIPPE) (*b.* Dossenheim, Alsace, France, 12 January 1808; *d.* Strasbourg, France [then part of Germany], 20 March 1880), *botany.*

Schimper, the cousin of two notable botanists, Karl Friedrich (1803–1867) and Wilhelm Schimper (1804–1878), and father of the plant geographer A. F. W. Schimper, was the son of Franz Schimper, Lutheran pastor of Offweiler. From 1826 to 1833 he studied at Strasbourg University, first philosophy, philology, and mathematics, and later, to please his father, theology. But from an early age he had been attracted to the study of natural history, in which he was encouraged by his cousins, the elder of whom often visited Offweiler. Soon after graduating, Schimper decided to devote his life to science and made a long journey in the Alps, studying plants and especially mosses, in which his interest was stimulated by the apothecary Philipp Bruch (1781–1847) of Zweibrücken. In 1835 Schimper was appointed assistant in the geological section of the Strasbourg Natural History Museum. He remained with this institution in various capacities throughout his life and eventually became its director.

In 1845 Schimper received a degree in natural sciences and in 1848 obtained a doctorate for his *Recherches anatomiques sur les mousses*. In 1849 he married Adèle B. Besson, of Swiss origin; they had two daughters and one son. The even course of his busy life was temporarily disrupted by the Franco-Prussian War, as a result of which Alsace was ceded to Germany. Faced with the painful choice of leaving his native province for a post in Paris or staying in Strasbourg, he decided to remain at the museum and to accept the chair of geology and paleontology at the reorganized German university.

With Bruch, Schimper in 1836 began publication of the *Bryologia Europaea*, his most famous work, which set a new standard in the description and delineation of mosses. Publication of this work (which appeared in parts) was continued by Schimper after Bruch's death, for a time with the assistance of Theodor Gümbel. Among Schimper's other publications on mosses, his study of the structure and development of the sphagna is particularly valuable. Although best known as a botanist, he also worked and wrote on zoology and geology. He made important contributions to paleobotany, especially on the Triassic flora of the Vosges. His work covered a vast field, but he was a supremely competent observer and describer rather than an originator of new ideas.

Schimper traveled extensively throughout his life and visited many European countries; one of his most productive journeys was to the Sierra Nevada in Spain (1847), from which he brought back specimens of a previously undescribed species of ibex as well as many interesting plants.

Schimper's many-sided activity, maintained almost to the end of his life, was made possible by powers of physical endurance hardly suggested by his tall, emaciated figure and delicate appearance.

BIBLIOGRAPHY

I. ORIGINAL WORKS. Schimper's principal writings are *Bryologia Europaea seu genera muscorum Europaeorum monographice illustrata*, 6 vols. (Stuttgart, 1836–1855), also *Corollarium* (1856) and *Supplementum*, 2 vols. (1864–1866), written with P. Bruch and T. Gümbel, reprinted and rearranged in 3 vols. with intro. by P. A. Florschütz and W. D. Margadant (Amsterdam, 1971); *Monographie des plantes fossiles du Grès Bigarré de la chaine des Vosges* (Leipzig, 1844); *Recherches anatomiques sur les mousses* (Strasbourg, 1848), reissued in *Mémoires de la Société d'histoire naturelle de Strasbourg*, 4, no. 1 (1850), 1–69; *Mémoire pour servir à l'histoire naturelle des sphaignes (Sphagnum L.)* (Paris, 1857), reissued in *Mémoires présentés par divers savants*, Cl. sci. math. et phys., 15 (1858), 1–97, also published as *Versuch einer Entwicklungsgeschichte der Torfmoose (Sphagnum) und einer Monographie der in Europa vorkommenden Arten dieser Gattung* (Stuttgart, 1858); *Synopsis muscorum Europaeorum praemissa introductione de elementis bryologici tractante* (Strasbourg, 1860; 2nd ed., Stuttgart, 1876); *Traité de palaeontologie végétale ou la flore du monde primitif dans ses rapports avec les formations géologiques et la flore du monde actuel*, 3 vols. and atlas (Paris, 1869–1874); and *Palaeophytologie*, vol. II of K. A. von Zittel, ed., *Handbuch der Palaeontologie* (Munich–Leipzig, 1879), completed by A. Schenk.

II. SECONDARY LITERATURE. See "dBy" (Anton de Bary), "Wilhelm Philip Schimper," in *Botanische Zeitung*, 38, no. 26 (1880), 443–450; E. Desor, "Philipp Wilhelm Schimper," in *Neues Jahrbuch für Mineralogie, Geologie und Palaeontologie*, 2 (1880), 1–7; C. Grad, "Guillaume-Philippe Schimper, sa vie et ses travaux, 1808–1880," in *Bulletin de la Société d'histoire naturelle de Colmar*, 20–21 (1880), 351–392; T. Gümbel, "Dr. Philipp Wilhelm Schimper," in *Allgemeine deutsche Biographie*, XXXI (1890), 277–279; S. O. Lindberg, "W. Ph. Schimper," in *Meddelanden af Societas pro fauna et flora fennica*, 6 (1881), 268; and W. D. Margadant and P. Florschütz, introduction to repr. of *Bryologia Europaea* (above).

P. W. RICHARDS

SCHJELLERUP, HANS CARL FREDERIK CHRISTIAN (*b.* Odense, Denmark, 8 February 1827; *d.* Copenhagen, Denmark, 13 November 1887), *astronomy.*

Schjellerup was trained as a watchmaker, but through the interest and help of H. C. Oersted he was admitted to the Polytechnic School of Copenhagen, where he passed the final examination in applied mathematics and mechanics. In 1851 he was appointed senior astronomer at the Copenhagen observatory, a post he held until his death. In 1857 he received the doctorate at Jena with a dissertation deriving the orbit of the comet of 1580 by means of Tycho Brahe's observations.

In 1861–1863 Schjellerup observed 10,000 positions of faint stars in declinations between −15° and +15° using the new Pistor-Martin meridian circle. A catalog, inspired by Bessel's zone observing about forty years earlier, was published in 1864 by the Royal Danish Academy. Schjellerup's catalog, outstanding at the time for its completeness and accuracy, was used as recently as 1952 for determining the constant of precession.

Schjellerup compiled two catalogs of colored stars (1866, 1874); the second, an extension of the first, contains 400 stars. During the same period several astronomers started a survey of the spectra of such stars, and Schjellerup's compilations appeared in time to be used in this work.

Schjellerup also made a signal contribution to the history of astronomy. At an advanced age he studied oriental languages, especially Arabic and Chinese. He had access to a pair of Arabic manuscripts, one in the Royal Library at Copenhagen and the other in the Imperial Library at St. Petersburg, which contain a description of the sky elaborated about 950 by the Persian astronomer al-Ṣūfī. Al-Ṣūfī had made a careful comparison between Ptolemy's catalog from about A.D. 150, which contains the positions and magnitudes of about 1,000 stars, and the sky itself; his specifications of relative magnitudes are the only series of this kind of observation that remains from the years between antiquity and modern times. Schjellerup's French translation was published in 1874 by the Academy of St. Petersburg.

Another result of his rare combination of linguistic and scientific ability was Schjellerup's control of different moon tables of his time, taking into account three solar eclipses from 708, 600, and 548 B.C. that he found mentioned in ancient Chinese literature.

BIBLIOGRAPHY

I. ORIGINAL WORKS. Schjellerup's writings include "Tycho Brahes Original-Observationer, benyttede til Banebestemmelse af Cometen 1580," in *Kongelige Danske Videnskabernes Selskabs Skrifter*, Naturv.-math. Afd., 5th ser., **4** (1856), 1–39; *Stjernefortegnelse indeholdende 10000 Positioner af teleskopiske Fixstjerner imellem −15 og +15 Graders Deklination* (Copenhagen, 1864); "Catalog der rothen, isolirten Sterne, welche bis zum Jahre 1866 bekannt geworden sind," in *Astronomische Nachrichten*, **67** (1866), 97–112; "Eine Uranometrie aus dem zehnten Jahrhundert," ibid., **74** (1869), 97–104; "Bidrag til Bedømmelsen af de moderne Maaneelementers Paalidelighed," in *Oversigt over det K. Danske Videnskabernes Selskabs Forhandlinger* (1874), 64–95; *Description des étoiles fixes composée au milieu du dixième siècle de notre ère par l'astronome persan Abd-al-Rahman al-Süfi* (St. Petersburg, 1874); "Zweiter Catalog der rothen, isolirten Sterne, vervollständigt und fortgeführt bis zum Schluss des Jahres 1874," in *Vierteljahrsschrift der Astronomischen Gesellschaft*, **9** (1874), 252–287; and "Recherches sur l'astronomie des anciens," in *Copernicus* (Dublin), **1** (1881), 25–39, 41–47, 223–236.

II. SECONDARY LITERATURE. Obituaries are J. L. E. Dreyer, in *Monthly Notices of the Royal Astronomical Society*, **48** (1888), 171–174; V. Hjort, in *Tidsskrift for Mathematik*, 5th ser., **5** (1887), 148–153; and J. Holetschek, in *Deutsche Rundschau für Geographie*, **10** (1888), 381–382; and in *Sirius*, **21** (1888), 161–163. See also J. E. Gordon, "Derivation of the Constant of Precession From a Comparison of the Catalogues of Schjellerup (1865.0) and Morin-Kondratiev (1900.0)," in *Izvestiya Glavnoi astronomicheskoi observatorii v Pulkove*, **19**, pt. 1(1952), 72–121.

AXEL V. NIELSEN

SCHLÄFLI, LUDWIG (*b*. Grasswil, Bern, Switzerland, 15 January 1814; *d*. Bern, Switzerland, 20 March 1895), *mathematics.*

Schläfli, the son of Johann Ludwig Schläfli, a citizen of Burgdorf, and Magdalena Aebi, attended primary school in Burgdorf. With the aid of a scholarship he was able to study at the Gymnasium in Bern, where he displayed a gift for mathematics. He enrolled in the theological faculty at Bern but, not wishing to pursue an ecclesiastical career, decided to accept a post as teacher of mathematics and science at the *Burgerschule* in Thun. He taught there for ten years, using his few free hours to study higher mathematics. In the autumn of 1843 Jakob Steiner, who was traveling to Rome with Jacobi, Dirichlet, and Borchardt, proposed that they take Schläfli with them as interpreter. Schläfli thus had an opportunity to learn from the leading mathematicians of his time. Dirichlet instructed him daily in number theory, and Schläfli's later works on quadratic forms bear the mark of this early training. During this period Schläfli translated two works by Steiner and two by Jacobi into Italian.

In 1848 Schläfli became a *Privatdozent* at Bern, where, as he expressed it, he was "confined to a stipend of Fr. 400 and, in the literal sense of the word, had to do without (*darben musste*)." His nomination as extraordinary professor in 1853 did not much improve his situation, and it was not until he became a full professor in 1868 that he was freed from financial concerns. Schläfli's scientific achievements gained recognition only slowly. In 1863 he received an honorary doctorate from the University of Bern; in 1868 he became a corresponding member of the Istituto Lombardo di Scienze e Lettere in Milan and was later accorded the same honor by the Akademie der Wissenschaften in Göttingen (1871) and the Accademia dei Lincei (1883). He won the Jakob Steiner Prize for his geometric works in 1870.

While at Bern, Schläfli was concerned with two major problems, one in elimination theory and the other in *n*-dimensional geometry, and he brought his results together in two extensive works. The first problem is discussed in "Ueber die Resultante eines Systems mehrerer algebraischer Gleichungen. Ein Beitrag zur Theorie der Elimination" (published in *Denkschriften der Akademie der Wissenschaften*, **4**). Schläfli summarized the first part of this work in a letter to Steiner:

> For a given system of *n* equations of higher degree with *n* unknowns, I take a linear equation with literal (undetermined) coefficients *a, b, c,* ··· and show how one can thus obtain true resultants without burdening the calculation with extraneous factors. If everything else is given numerically, then the resultant must be decomposable into factors all of which are linear with respect to *a, b, c,* ···. In the case of each of these linear polynomials the coefficients of *a, b, c,* ··· are then values of the unknowns belonging to a *single* solution.

Drawing on the works of Hesse, Jacobi, and Cayley, Schläfli presented applications to special cases. He then developed the fundamental theorems on class and degree of an algebraic manifold, theorems that attracted the interest of the Italian school of geometers. The work concluded with an examination of the class equation of third-degree curves. Through this publication Schläfli became acquainted with Arthur Cayley, whose paper "Sur un théorème de M. Schläfli" begins: "In §13 of a very interesting memoir by M. Schläfli one finds a very beautiful theorem on resultants." The acquaintance led to an extensive correspondence and opened the way for Schläfli to publish in English journals. In his obituary of Schläfli, F. Brioschi wrote:

> While rereading this important work recently it

occurred to me that it displays the outstanding characterisics of Schläfli's work as a whole. These are, first, deep and firsthand knowledge of the writings of other authors; next, a desire and ability to generalize results; and, finally, great penetration in investigating problems from very different points of view.

The second of the two major works, "Theorie der vielfachen Kontinuität," was rejected by the academies of Vienna and Berlin because of its great length and was not published until 1901 (in *Neue Denkschriften der Schweizerischen naturforschenden Gesellschaft*). For many years only sections of it appeared in print—in the journals of Crelle and Liouville and in the *Quarterly Journal of Mathematics*. The core of this work consisted of the detailed theory of regular bodies in Euclidean space R_n of *n* dimensions and the associated problems of the regular subdivision of the higherdimensional spheres. Schläfli based his investigation of regular polytopes on his discovery that such objects can be characterized by certain symbols now known as Schläfli symbols. His definition was recursive: $\{k_1\}$ is the symbol of the plane regular k_1-gon. $\{k_1, \cdots, k_{n-1}\}$ is the Schläfli symbol of that regular polytope the boundary polytopes of which have the symbol $k_1, \cdots, k_{n-2}$ and the vertex polytopes of which have the symbol $\{k_2, \cdots, k_{n-1}\}$.

Schläfli discovered a way of finding all regular polytopes by calculating the numbers $k_1, \cdots, k_{n-1}$ in the following manner: In the plane, for every k_1 there exists a $\{k_1\}$. In considering *n*-space he started from Euler's theorem on polyhedrons, which he formulated and proved for R_n: assume a polytope with a_0 vertexes, a_1 edges, a_2 faces and so on in higher dimensions until a_{n-1} boundary polytopes of dimension $n-1$, and $a_n = 1$. Then it is true that

$$a_0 - a_1 + a_2 - \cdots (-1)^{n-1} a_{n-1} + (-1)^n a_n = 1.$$

For $n = 3$ the Euler theorem on polyhedrons becomes $a_0 - a_1 + a_2 = 2$. Since, further, for $\{k_1, k_2\}$ it is true that $k_2 \cdot a_0 = 2a_1 = k_1 \cdot a_2$, it follows that

$$a_0 : a_1 : a_2 : 1 = 4k_1 : 2k_1 k_2 : 4k_2 : [4 - (k_1 - 2)(k_2 - 2)].$$

The nature of the problem requires a positive value for $[4 - (k_1 - 2)(k_2 - 2)]$; therefore $(k_1 - 2)(k_2 - 2)$ can take only the values 1,2,3. This yields the following possibilities: $\{3,3\}$ tetrahedron, $\{3,4\}$ octahedron, $\{3,5\}$ icosahedron, $\{4,3\}$ cube, and $\{5,3\}$ dodecahedron. For $n = 4$ the Euler equation $a_0 - a_1 + a_2 - a_3 = 0$ becomes homogeneous and yields only the ratios of the a_i. Schläfli therefore determined the radius of the circumscribed sphere of a $\{k_1, k_2, k_3\}$ of edge length 1. If this radius is to be real, then it must be true that $\sin\frac{\pi}{k_1} \cdot \sin\frac{\pi}{k_3} >$ $\cos\frac{\pi}{k_2}$. This condition yields the six bodies {3,3,3}, {4,3,3}, {3,3,4}, {3,4,3}, {5,3,3}, and {3,3,5}. Schläfli proved further that in every R_n with $n > 4$ there are only three regular solids: {3, 3, $\cdots$, 3}, regular simplex; {4, 3, $\cdots$, 3}, n-dimensional cube; and {3, 3, $\cdots$, 3, 4}, regular n-dimensional octahedron.

Schläfli achieved another beautiful result by considering the unit sphere in R_n and n hyperplanes through the origin $(1) = 0, \cdots, (n) = 0$. Specifically, he found that the inequalities $(1) \geq 0, \cdots, (n) \geq 0$ determine a spherical simplex with surface S_n. Schläfli proved $dS_n = \frac{1}{n-2}\{\Sigma S_{n-2}\, d\lambda\}$, where S_{n-2} is the surface of a boundary simplex of two dimensions less and λ is a suitable angle between two such simplexes, and the summation extends over all such boundary simplexes. Let O_n be the surface of the sphere; then the Schläfli function f_n is defined by $S_n = \frac{1}{2^n} O_n f_n$. It can be proved that $f_{2m+1} = a_0\Sigma f_{2m} - a_1\Sigma f_{2m-2} + \cdots$, where a_k is proportional to the $k + 1$ Bernoulli number B_k. This equation states that the Schläfli function in a space of odd dimension can be reduced to Schläfli functions in spaces of even dimension. Concerning this discovery Schläfli wrote to Steiner: "I believe I am not overestimating the importance of this general theorem if I set it beside the most beautiful results that have been achieved in geometry."

Besides the theory of Schläfli functions the second section of the paper included a detailed treatment of the decomposition of an arbitrary spherical simplex into right-angled simplexes. The section concluded with a theorem on the sum of the squares of the projections of a ray on the vertex rays of a regular polytope, a question that has interested researchers in recent times.

The third section, headed "Verschiedene Anwendungen der Theorie der vielfachen Kontinuität, welche das Gebiet der linearen und sphärischen übersteigen," contains both applications of theorems of Binet, Monge, Chasles, and Dupin to quadratic continua in R_n and Schläfli's own discoveries. After first determining the midpoint, major axes, and conjugate diameters for a quadratic continuum, he demonstrates the law of inertia of the quadratic forms by means of continuity considerations. Among other results presented is a generalization of a theorem of Binet for a system of conjugate radii: the sum of the squares of all m-

fold parallelepipeds constructed out of the conjugate radii of a system is equal to the sum obtained when the system is formed from the major axes. Schläfli then divided the quadratic continua into two classes and generalized Monge's theorem on the director circle, or great circle, of a central conic section. He also examined confocal systems and showed that in R_3 their determination depends on a third-order linear differential equation.

After "Theorie der vielfachen Kontinuität" had appeared in its entirety, P. H. Schoute wrote in 1902:

> This treatise surpasses in scientific value a good portion of everything that has been published up to the present day in the field of multidimensional geometry. The author experienced the sad misfortune of those who are ahead of their time: the fruits of his most mature studies cannot bring him fame. And in this case the success of the division of the cubic surfaces was only a small compensation; for, in my opinion, this achievement, however valuable it might be, is far from conveying the genius expressed in the theory of manifold continuity.

Steiner communicated to Schläfli Cayley's discovery of the twenty-seven straight lines on the third-degree surface. Schläfli thereupon found the thirty-six "doubles sixes" on this surface and then the division of the cubic surface into twenty-two species according to the nature of the singularities. Schläfli also solved problems posed by the Italian school of geometers. He gave a condition under which a manifold has constant curvature: its geodesic lines must appear as straight lines in a suitable coordinate system. He also investigated the space of least dimension in which a manifold can be imbedded; his conjecture on this question was demonstrated by M. Janet and E. Cartan (1926–1927). Schläfli's work on the division of third-order surfaces led him to assert the one-sidedness of the projective plane in a letter to Felix Klein in 1874.

Schläfli wrote a work on the composition theory of quadratic forms in which he sought to provide the proof of the associative law that was lacking in Gauss's treatment of the subject. Schläfli's posthumous papers contain extensive tables for the class number of quadratic forms of both positive and negative determinants.

Schläfli's geometric and arithmetical studies were equaled in significance by his work in function theory. Stimulated by C. G. Neumann's investigations (1867) and following up the representation of the gamma function by a line integral, Schläfli gave the integral representation of the Bessel function $J_n(z)$ for arbitrary n, even where n is not integral. He also wrote an outstanding work on elliptic modular functions (1870) that gave rise to the designation "Schläfli modular equation." An examination of his posthumous manuscripts reveals that in 1867, ten years before Dedekind, Schläfli discovered the domain of discontinuity of the modular group and used it to make a careful analysis of the Hermite modular functions from the analytic, number theoretic, and geometric points of view. As early as 1868, moreover, Schläfli employed means that Weber discovered only twenty years later and termed f-functions or class invariants.

Besides his mathematical achievements, Schläfli was an expert on the flora of the canton of Bern and an accomplished student at languages. He possessed a profound knowledge of the *Veda*, and his posthumous manuscripts include ninety notebooks of Sanskrit and commentary on the *Rig-Veda*.

BIBLIOGRAPHY

I. ORIGINAL WORKS. Schläfli's writings were brought together as *Gesammelte mathematische Abhandlungen*, 3 vols. (Basel, 1950–1956). His correspondence with Steiner is in *Mitteilungen der Naturforschenden Gesellschaft in Bern* for 1896 (1897), 61–264. That with Cayley is in J. H. Graf, ed., *Briefwechsel von Ludwig Schläfli mit Arthur Cayley* (Bern, 1905); and that with Borchardt (1856–1877) is in *Mitteilungen der Naturforschenden Gesellschaft in Bern* for 1915 (1916), 50–69. Graf also edited the following: "Lettres de D. Chelini à L. Schläfli," in *Bullettino di bibliografia e di storia delle scienze matematiche e fisiche*, **17** (1915), 36–40; "Correspondance entre E. Beltrami et L. Schläfli," *ibid.*, 81–86, 113–122; and "Correspondance entre Luigi Cremona et Ludwig Schläfli," *ibid.*, **18** (1916), 21–35, 49–64, 81–83, 113–121, and **19** (1917), 9–14. Two letters from Schläfli to P. Tardy (1865) are in G. Loria, "Commemorazione del socio Prof. Placido Tardy," in *Atti dell'Accademia nazionale dei Lincei. Rendiconti*, Cl. fis., **24** (1915), 519–531.

II. SECONDARY LITERATURE. See J. J. Burckhardt, "Der mathematische Nachlass von Ludwig Schläfli, mit einem Anhang: Ueber Schläflis nachgelassene Manuskripte zur Theorie der quadratischen Formen," in *Mitteilungen der Naturforschenden Gesellschaft in Bern* for 1941 (1942), 1–22; and "Ludwig Schläfli," supp. no. 4 of *Elemente der Mathematik* (1948); J. H. Graf, "Ludwig Schläfli," in *Mitteilungen der Naturforschenden Gesellschaft in Bern* for 1895 (1896), 120–203; A. Häusermann, *Ueber die Berechnung singulärer Moduln bei Ludwig Schläfli* (inaugural diss., Zurich, 1943); W. Rytz, "Prof. Ludwig Schläfli als Botaniker," in *Mitteilungen der Naturforschenden Gesellschaft in Bern* for 1918 (1919), 213–220; and O. Schlaginhaufen, "Der Schädel des Mathematikers Ludwig Schläfli," *ibid.* for 1930 (1931), 35–66.

JOHANN JAKOB BURCKHARDT

SCHLEIDEN, JACOB MATHIAS (*b*. Hamburg, Germany, 5 April 1804; *d*. Frankfurt am Main, Germany, 23 June 1881), *botany*, *natural science*, *scientific popularization*.

Schleiden came from a well-to-do family; his father was municipal physician of Hamburg. After legal studies, culminating in a doctorate, at the University of Heidelberg (1824–1827), Schleiden established a legal practice in Hamburg. He was dissatisfied, however, and, after a period of deep depression, he abandoned this profession. In 1833 he began to study natural science at Göttingen and then transferred to Berlin. He devoted himself enthusiastically to the subject of botany, in which he was encouraged by his botanist uncle, Johann Horkel, to whom he remained forever grateful. During these years Alexander von Humboldt and Robert Brown were resident at Berlin. Schleiden worked in the laboratory of the celebrated physiologist Johannes Müller, and there he met Theodor Schwann. In this inspiring milieu, Schleiden worked intensively and produced noteworthy publications. He obtained his doctorate in 1839 at Jena and was then able to give free reign to his pedagogical fervor. He lectured and wrote both technical and popular scientific works on the widest range of topics.

Schleiden's lectures drew enthusiastic, overflow audiences; his numerous articles appeared in highly respected journals, or in collections that were often reprinted and translated. He declined an offer from the University of Giessen in 1846, but in 1850 he accepted nomination as titular professor of botany at Jena. He also received many honors from learned societies. In spite of his success, Schleiden decided to leave Jena. His combative personality probably contributed to this decision: he was often involved in polemics with leading figures of the day. Also, he had an insatiable desire to study problems going beyond the confines of botany and natural history. He soon became a highly regarded popular lecturer and writer; indeed, he was one of the most successful popularizers of the age—no small achievement at a time when scientists like Virchow, Helmholtz, Liebig, Moleschott, Alexander von Humboldt, and Ludwig Büchner, among others, were addressing the general public. Following his departure from Jena in 1862 and a stay at Dresden, Schleiden became professor of anthropology at Dorpat. Even though he soon left Dorpat, the Russian government granted him a pension. He became a *Privatgelehrter* and thereafter frequently moved from one city to another.

In 1838 Schleiden published "Beiträge zur Phytogenesis" in *Müller's Archiv*, one of the most respected journals of the time. This article, which was immediately translated into French and English, fixed his name in the history of biology. According to a well-known tradition, the cell theory was conceived in a conversation between Schleiden and Schwann on the subject of phytogenesis. In fact, however, historical investigation has shown that Schleiden's article, like Schwann's book, represents only one stage—although admittedly an important one—in the evolution of the search for the elementary unit common to the animal and plant kingdoms. (On this subject see the publications of Studnička, Klein, Baker, and Florkin; the biographies in this Dictionary of Mohl, Oken, Raspail, and Schwann, among others, should also be consulted.)

Schleiden reprinted the article in a collection of studies and dealt at length with its contents in his textbook on botany. In the article, which evoked wide interest and sparked violent debates, Schleiden starts from Robert Brown's discovery of the cell nucleus (1832), which Schleiden called the cytoblast, and then indicates its role in the formation of cells. According to Schleiden, as soon as the cytoblast reaches its final size, a fine, transparent vesicle forms around it: this is the new cell. The cell then crystallizes within a formative liquid. The best statement of this interpretation can be found in his botany textbook, *Grundzüge der wissenschaftlichen Botanik* (1842): "Since the elementary organic cells present a marked individualization and since they are the most general expression of the concept of the plant, it is necessary, first of all, to study this cell as the foundation of the vegetable world. We have therefore produced a study of the vegetable cell." This clearly announced the advent of plant cytology. This subject became the starting point of all subsequent botany textbooks. As Schleiden stated, the cells can form only in a liquid containing sugar, gum, and mucus (cytoblastema). This phenomenon occurs in the following manner: the mucous portion condenses into more or less round corpuscles (cytoblastus). On its surface, a part of the liquid is transformed into jelly, a relatively insoluble substance; thus there is created a closed gelatinous vesicle that is penetrated by the external liquid. . . . During the progressive expansion of the vesicle, the jelly of the wall is generally transformed into a membranous substance and the formation of the cell (cellula) is completed.

Schleiden's interpretation merits careful attention. The idea that cells are crystallized inside an amorphous primary substance is as old as the study of the cell itself. The idea can be traced back to the writings of Grew (1675), who compared the process with the fermentation of a paste or liquid. It appears again, independently, in Raspail, as a vesicular crystallization. Mohl observed cell division but was undecided in his views on the existence of "the free formation of cells," as this type of process was long called. Despite increasing evidence of nuclear activity during division, and despite Virchow's definitive aphorism, "omnis cellula a cellula," the notion of the formative blastema long survived, owing mainly to the support of Charles Robin. An eminent figure in the development of the subject of microscopic anatomy, Robin remained faithful to the notion, granting the cell a position of no special distinction among the anatomical elements of the higher organisms (see Klein, 1936; 1960).

From the start of his career, Schleiden showed a predilection for the microscope, and he contributed greatly to its introduction in biological research. He engaged in long and sometimes bitter disputes with Amici, one of the outstanding micrographers and opticians of the period. Schleiden is thought to have played an active role in the establishment of the Zeiss optical works in Jena.

Schleiden based his description of cytogenesis on an examination of the pollen tube. Ironically, his interpretation of this tube was fundamentally wrong, both morphologically and biologically: he considered it a female reproductive element in the plant. This error, like the one concerning the free genesis of cells, gave rise to much further research and to violent controversies; but one may truly call both these mistakes fruitful.

Schleiden's botany textbook merits an extensive methodological discussion, but we shall restrict our comments to a few essential points. A number of Schleiden's articles contain virulent criticism of the botanists of the first half of the nineteenth century, many of whom still upheld the ideas of nature philosophy, against which his textbook was a frontal attack. More important, however, it introduced new pedagogical standards that were to dominate the teaching of botany for years. Beginning with the second edition, the work bore the subtitle *Botanik als inductive Wissenschaft.*

Schleiden considered the inductive method the only valid one in biology, and the first part of his book constitutes an important document for the study of the methodology of natural history. He declared himself an enemy of all philosophical speculation, while at the same time adhering to the views of Kant and rejecting the label of materialist. He completed his attacks against the philosophers with a brief polemical monograph against the philosophy of Schelling and of Hegel. The entire structure of Schleiden's textbook was fundamentally new. The lengthy work begins with a study of the material elements of the plant. Next there is a large section on plant cytology, and then a treatment of morphology and organology. The book, which established the teaching of botany on a completely new basis, was often reprinted and appeared in various translations and adaptations. To appreciate the enthusiasm it aroused and its influence in turning young men to the study of botany, it is necessary to read the testimony of contemporaries, particularly of Julius Sachs, a famous botanist and author of a well-known history of botany.

Schleiden's pedagogical genius manifested itself in other publications as well. From the time of its founding in 1857, Schleiden was an assiduous contributor to Westermann's *Monatshefte,* a periodical that maintained high literary and scientific standards. His lectures, delivered to vast audiences, were occasionally published in book form and met with great success. Among the best known of these collections was *Die Pflanze und ihr Leben,* which was handsomely reproduced and reprinted many times. Another, somewhat more difficult series, *Wissenschaftliche Studien,* covered a wide range of topics in natural history. Later, Schleiden devoted entire monographs to subjects of apparently limited scope—for example, one to the rose and another to salt, in which he discussed its history, symbolism, and economic and social importance to human life. In other writings he dealt with the Isthmus of Suez and with anthropological questions. Among his last publications were scholarly studies on the fate of the Jews in the Middle Ages, on their martyrdom, and on their importance in transmitting knowledge to the Occident. These works, which were reprinted and translated, stimulated much interest; they also testify to the liberality of Schleiden's thinking in a period that witnessed the first anti-Semitic campaigns in the universities of Wilhelmine Germany.

This very liberality, however, joined with his combative nature, constantly involved Schleiden in debates and harsh polemics with the most eminent scientists and thinkers of the age, among whom were Amici, Fechner, Liebig, Mohl, Nees von Esenbeck, and Schelling. A few words may be said about his controversy with Fechner. The celebrated founder of modern psychophysiology, convinced of the existence of a soul in all living creatures, had published a work entitled *Nanna or the Soul of the Plants.* Schleiden violently attacked it, and Fechner responded with a book that is still delightful to read, *Professor Schleiden and the Moon.* But it should be noted that Schleiden often cut polemics short by the simple expedient of silence.

One of his early biographers, L. Errera, has a neat epitome of his career: "As a popularizer he was a model, as a scientist an initiator."

BIBLIOGRAPHY

I. ORIGINAL WORKS. A list of Schleiden's writings is given by Möbius (see below). The following works are cited in the text: "Beiträge zur Phytogenesis," in *Archiv für Anatomie, Physiologie und wissenschaftliche Medicin* (1838), 137–176, with French trans. in *Annales des sciences naturelles.* Botanique, **11** (1839), 242–252, 362–370, and English trans. in *Scientific Memoirs,* **2** (1841), 281–312; *Grundzüge der wissenschaftlichen Botanik,* 2 vols. (Leipzig, 1842–1843), 2nd ed., *Die Botanik als inductive Wissenschaft behandelt* (Leipzig, 1845–1846), 3rd ed., *Die Botanik als inductive Wissenschaft. Grundzüge der wissenschaftlichen Botanik nebst einer Einleitung als Anleitung zum Studium der Pflanzen* (Leipzig, 1849): vol. I, *Methodologische Grundlage. Vegetabilische Stofflehre. Die Lehre von der Pflanzenzellen,* vol. II, *Morphologie, Organologie,* with 153 figs. and 4 plates, 4th ed. (Leipzig, 1861); English trans. by E. Lankester as *Principles of Scientific Botany as an Inductive Science* (London, 1849; 2nd ed., 1868), facs. ed. by Lorch (see below).

See also *Beiträge zur Botanik. Gesammelte Aufsätze* (Leipzig, 1844); *Schelling's und Hegel's Verhältniss zur Naturwissenschaft* (Leipzig, 1844); *Die Pflanze und ihr Leben. Populäre Vorträge* (Leipzig, 1848; 5th ed., 1858), also translated into English (1848), French (1859), and Dutch (1873); *Studien. Populäre Vorträge* (Leipzig, 1855; 2nd ed., 1857); "Über den Materialismus unserer Zeit: Zerstreute Gedanken," in *Westermanns Monatshefte,* **1** (1857), 37–45; "Die Landenge von Suez und der Auszug der Israeliten aus Egypten," *ibid.,* **4** (1858), 262–273; "Ueber die Anthropologie als Grundlage für alle übrigen Wissenschaften, wie überhaupt der Menschenbildung," *ibid.,* **11** (1862), 49–58; *Das Salz. Seine Geschichte. Seine Symbolik und seine Bedeutung im Menschenleben. Eine monographische Skizze* (Leipzig,

1875); "Die Bedeutung der Juden für Erhaltung und Wiederbelebung der Wissenschaften im Mittelalter," in *Westermanns Monatshefte*, **41** (1877), 52–60, 156–169; and "Die Romantik des Martyriums bei den Juden im Mittelalter," *ibid.*, **44** (1878), 62–79, 166–178.

II. SECONDARY LITERATURE. On Schleiden and his work, see J. R. Baker, "The Cell Theory, a Restatement. History and Critique," in *Quarterly Journal of Microscopical Science*, **89** (1948), 103–125; **90** (1949), 87–108, 331; **93** (1952), 157–190; **94** (1953), 407–440; **96** (1955), 449–481; L. Errera, "J. M. Schleiden," in *Revue scientifique de la France et de l'étranger*, 3rd ser., **2** (1882), 289–298; G. T. Fechner, *Professor Schleiden und der Mond* (Leipzig, 1856); M. Florkin, *Naissance et déviation de la théorie cellulaire dans l'oeuvre de Théodore Schwann* (Paris, 1960), 57, 62; E. Hallier, "Mathias Jacob Schleiden. Seine Bedeutung für das wissenschaftliche Leben der Gegenwart geschildert," in *Westermanns Monatshefte*, **51** (1882), 348–358; M. Klein, *Histoire des origines de la théorie cellulaire* (Paris, 1936), 36–39; *A la recherche de l'unité élémentaire des organismes vivants. Histoire de la théorie cellulaire* (Paris, 1960), 18; J. Lorch, *Introduction to Principles of Botany as an Inductive Science by Mathias Jacob Schleiden*, Sources of Science, no. 40 (New York–London, 1969), i–xxxiv, a facs. ed. of the London 1849 ed.; M. Möbius, *Mathias Jacob Schleiden* (Leipzig, 1904); E. Nordenskjöld, *Die Geschichte der Biologie* (Jena, 1926), 396–401; C. Robin, *Anatomie et Physiologie cellulaires* (Paris, 1873), 565 ff.; J. Sachs, *Geschichte der Botanik* (Munich, 1875), 202–207, 349; F. L. Studnička, "Mathias Jacob Schleiden und die Zelltheorie von Theodor Schwann," in *Anatomischer Anzeiger*, **76** (1933), 80–95; and A. Wartenberg, "Mathias Jacob Schleiden," H. Freund and A. Berg, eds., in *Geschichte der Mikroskopie*, **1** (1963), 299–302.

MARC KLEIN

SCHLESINGER, FRANK (*b.* New York, N.Y., 11 May 1871; *d.* Lyme, Connecticut, 10 July 1943), *astronomy.*

Schlesinger was the youngest of seven children of William Joseph Schlesinger and Mary Wagner, both of whom were German immigrants. He was first educated in the public schools of New York City. He graduated in 1890 from the City College of New York and was awarded the Ph.D. from Columbia University in 1898. His dissertation, which was based on measurements of star positions on plates photographed many years before by Lewis Rutherfurd, was a forerunner in his distinguished career in astrometry.

During this era, determinations of stellar parallaxes (which are inversely proportional to the distances of the stars) were carried out largely by the visual methods of Bessel. Experimentation with photographic techniques was progressing, and Schlesinger was eager to test the performance of the Yerkes forty-inch refractor for photographic parallax work. The parallax of a star is defined as the angle at the star subtended by the radius of the earth's orbit. It is manifested by an apparent slight change in the direction of the star observed at intervals of six months, during which the earth moves in its orbit from one side to the other of the sun. In practice these parallactic displacements are measured relative to very distant faint stars, the motions of which are negligibly small. The uncertainties of the measurements are reduced photographically when there is a minimum disparity in the sizes of the images of the parallax star and the faint comparison stars. In order to achieve that, Schlesinger designed a rotating sector to occult the image of the bright star intermittently, while the faint stars were exposed continuously; this technique is still used extensively. By devising his time-saving "method of dependences," Schlesinger also improved and simplified the mathematical and numerical procedures for the reductions of the measurements.

After two years at Yerkes Observatory (1903–1905), Schlesinger was called to the Allegheny Observatory in Pittsburgh as its director. Here he expended considerable effort on spectroscopic studies of eclipsing and spectroscopic binary stars, and also on the improvement of instrumentation for parallax work. In addition, he began his first investigations for the preparation of "zone catalogues" to provide accurate positions and proper motions (that is, apparent changes in position) of many thousands of stars to the ninth and fainter magnitudes. These early beginnings he pursued vigorously at the Yale University Observatory, where he was director from 1920 until his retirement in 1941. With the enthusiastic collaboration of Ida Barney, Schlesinger published ten volumes of zone catalogues, including some 150,000 stars, between declinations −30° and +30° of declination.

At Yale he extended his work on parallaxes to the southern hemisphere, where the Yale-Columbia Southern Station began operation in Johannesburg, South Africa, in 1925. Before Schlesinger began his investigations of stellar parallaxes, only a few hundred were known; during his lifetime, and primarily owing to his direct influence, the number grew to four thousand.

In 1908 E. C. Pickering, director of the Harvard Observatory, had published his "Revised Harvard Photometry," giving the positions, magnitudes, and spectral classes of all stars of magnitude 6.5 and brighter. Using this as a basis, Schlesinger in 1924 published his first edition of the "Bright Star Catalogue," vastly extending its usefulness by adding proper motions, radial velocities, and other relevant data. The second edition, published with Louise Jenkins in 1940, became probably the most widely used of all astronomical catalogues.

BIBLIOGRAPHY

A list of 262 of Schlesinger's papers is in Dirk Brouwer, "Biographical Memoir of Frank Schlesinger," in *Biographical Memoirs. National Academy of Sciences,* **24** (1945), 105–144. These works by Schlesinger include "Photographic Determinations of Stellar Parallaxes," in *Probleme der Astronomie: Festschrift für Hugo v. Seeliger* (1924), 422–437; "Some Aspects of Astronomical Photography of Precision," in *Monthly Notices of the Royal Astronomical Society,* **87** (1927), 506–523, which is the first George Darwin Lecture; and *General Catalogue of Stellar Parallaxes,* 2nd ed. (New Haven, 1935), compiled with Louise Jenkins.

E. DORRIT HOFFLEIT

SCHLICK, (FRIEDRICH ALBERT) MORITZ (*b.* Berlin, Germany, 14 April 1882; *d.* Vienna, Austria, 22 June 1936), *theory of knowledge, philosophy of science.*

Schlick, the son of Albert Schlick and Agnes Arndt, studied at Heidelberg and Lausanne, and took his doctorate under Max Planck at Berlin in 1904 with a dissertation on the physics of light. An early and abiding interest in philosophy (he published his *Lebensweisheit* at the age of twenty-six) led him to abandon science in favor of a philosophical career; and in due course, following the necessary preparation, he entered academic life as a teacher of philosophy, first at Rostock and then at Kiel. After the publication of his *Allgemeine Erkenntnislehre* (1918), he was called to the chair of the philosophy of the inductive sciences at the University of Vienna (1922), formerly held by such eminent philosopher-scientists as Ernst Mach and Ludwig Boltzmann. Apart from two visits to the United States, where he held visiting professorships at Stanford University and the University of California at Berkeley, he retained this post until he was murdered by a deranged former student.

At Vienna, Schlick soon became the center of a group of thinkers with predominantly scientific and mathematical backgrounds, who were devoted to the cultivation and development of a scientific philosophy, as opposed to the then prevailing metaphysical orientation of Continental, and especially German, philosophy. The group came to be known as the Vienna Circle and later, in England and the United States, as the logical positivists. In addition to producing much solid philosophical work, published for the most part in the journal *Erkenntnis,* the Vienna Circle engaged in a fair amount of antimetaphysical crusading, earning enthusiastic support from younger philosophical "radicals" and at least as much hatred from the conservative representatives of "academic" philosophy.

Schlick, although the leading figure of the circle, was far from being a typical exponent of some of the views generally held to be representative of the group. While he shared their interest in logic as a tool of philosophical analysis and their repudiation of metaphysics as a viable discipline, as well as their rejection of synthetic a priori propositions, he was much more sympathetic than some members to the great figures of the Western philosophical tradition, despite strong criticism of what he considered to be their errors. Honoring science as man's highest intellectual achievement, he nevertheless deemed the problems of culture and of life to be of far greater significance. Like Wittgenstein, who greatly influenced him in his Vienna period, Schlick saw philosophical activity as meaning clarification, and as no less important for conduct and enlightening life's goals than in preparing the way for scientific ascertainment. In his ethical and related writings—*Lebensweisheit, Vom Sinn des Lebens* (a pamphlet of 1927), and *Fragen der Ethik* (1930)—his profound understanding of life often finds expression in the elevated language of the poet or the philosopher as man of wisdom. His main reputation will, however, probably rest on his work in theory of knowledge and philosophy of science.

Schlick's first work in this area was a brief expository, interpretive book (one of the earliest) on Einstein's theory of relativity, published in German in 1917 and in English translation as *Space and Time in Contemporary Physics* (1920). In it he stresses the profound philosophical implications of Einstein's work, to which, in his later writings, he frequently refers as a paradigm of philosophical activity conceived as meaning clarification—Einstein's main achievement having been his clarification of the hitherto vague concept of simultaneity at a distance.

Allgemeine Erkenntnislehre (1918), his major work, examines a very wide range of problems and concepts relating to scientific knowledge. Schlick begins with the all-important concept of knowledge itself, the analysis of which sets the tone and determines the special character of the work as a whole. Despite the half century since its appearance, the work remains perhaps the most comprehensive and valuable treatment of the general theory of knowledge. Schlick's central idea is that knowledge is discursive, not intuitive: it yields true descriptions of the object to be known as a special case of something already known; it is not simply awareness of something confronting us. In German the point can be made perspicuous by distinguishing between *Erkenntnis* (recognition, knowledge) and *Kenntnis* (cognition, in the narrow sense of acquaintance with something; experience of it). The whole of Schlick's epistemology is permeated and affected by this distinction. In its simplest terms the work may be seen as a polemic against Kant, who denied that we can have knowledge of what lies beyond the phenomenal, and against Hume and his modern followers, who tended to deny meaning to talk of the transcendent and identify reality with the immediately given. The latter view is sometimes expressed by the slogan "Only the given exists" and is closely connected with the view that what cannot be sensed cannot be known.

Since knowledge is not simple acquaintance with or awareness of some datum but, rather, true description of class membership, causal connections, or governing laws of objects to be known, knowledge is in no way restricted to what can be sensed. The relatively few electrical phenomena accessible to the senses do not give us knowledge of electricity but only furnish occasions for inquiring into their causes and consequences and, ultimately, for investigating the laws governing electrical events; for the true concern of knowledge is with laws, not things or appearances. Knowledge in its higher levels seeks ever more general laws from which can be derived those on lower levels. The ideal of knowledge would be maximum descriptive power using a minimum of concepts; then our picture of the universe would be of a single tightly knit system in which each thing (event) stood in some known relationship to all others. The business of science is not simply to report sense data, but to formulate and test hypotheses that have consequences ascertainable in the given.

Schlick brought this conception of knowledge to bear on the errors of traditional metaphysics, which he located in the vain effort to capture and express in language the immediate quality of life and experience—a task for which it is wholly unfit.

Only form or structure is expressible, communicable—can enter into knowledge. The immediate is ineffable, to be enjoyed or experienced (*erlebt*), a matter of feeling, while knowledge is a matter of intellect, of thought; the two are incommensurable.

Schlick's broad conception of science by no means excluded systematic knowledge relating to questions of life and values. He was, in fact, most deeply concerned with the meaning of life and the path to happiness. In his *Lebensweisheit* he explores the pleasure-happiness value of the senses, the instincts, and of personal and social relations and institutions, and he returned to this topic in the unfinished work *Natur und Kultur*. The richest source of happiness ultimately proves to be the social instincts and love. This theme plays a central role in Schlick's *Fragen der Ethik*, where it culminates in the view that in the long run virtue (moral conduct) and happiness go hand in hand, a fact which explains why man acts morally. Such explanation is the explicit topic of the book as a whole, which however also contains much rich psychological material bearing on human conduct and happiness, as well as passages of profound wisdom and eloquence. In *Lebensweisheit* Schlick develops Schiller's theme of man's finding his highest vocation in "play"—free, joyous activity pursued for its own sake; and in answer to the question "What is the meaning of life?" he answers "Youth!" "Preserve the spirit of youth," he urges, "for it is the meaning of life!" Youth alone makes sense of life, so that for the youthful spirit the question of the meaning of life does not arise.

BIBLIOGRAPHY

I. ORIGINAL WORKS. Schlick's early works are *Lebensweisheit* (Munich, 1908); "Das Grundproblem der Ästhetik in entwicklungsgeschichtlicher Beleuchtung," in *Archiv für die gesamte Psychologie,* **14** (1909), 102–132; "Die Grenze der naturwissenschaftlichen und philosophischen Begriffsbildung," in *Vierteljahrsschrift für wissenschaftliche Philosophie und Soziologie,* **34** (1910), 121–142; "Das Wesen der Wahrheit nach der modernen Logik," *ibid.,* 386–477; "Gibt es intuitive Erkenntnis?" *ibid.,* **37** (1913), 472–488; "Die philosophische Bedeutung des Relativitätsprinzips," in *Zeitschrift für Philosophie und philosophische Kritik,* **159** (1915), 129–175; "Idealität des Raumes, Introjektion und psychophysisches Problem," in *Vierteljahrsschrift für wissenschaftliche Philosophie und Soziologie,* **40** (1916), 230–254; *Raum und Zeit in der gegenwärtigen Physik* (Berlin, 1917; 4th ed., 1922), English trans. by H. L. Brose as *Space and Time in Contemporary Physics*

(Oxford, 1920); and *Allgemeine Erkenntnislehre* (Berlin, 1918; 2nd ed., 1925).

Subsequent works include "Naturphilosophische Betrachtungen über das Kausalprinzip," in *Naturwissenschaften*, **8** (1920), 461–474; "Naturphilosophie," in M. Dessoir, ed., *Lehrbuch der Philosophie*, II (Berlin, 1925), 397–492; "Erleben, Erkennen, Metaphysik," in *Kantstudien*, **31** (1926), 146–158; *Vom Sinn des Lebens* (Berlin, 1927); "Erkenntnistheorie und moderne Physik," in *Scientia*, **45** (May 1929), 307–316; *Fragen der Ethik* (Vienna, 1930), English trans. by D. Rynin as *Problems of Ethics* (New York, 1939; repr., New York, 1962); "Die Kausalität in der gegenwärtigen Physik," in *Naturwissenschaften*, **19** (1931), 145–162, English trans. by D. Rynin as "Causality in Contemporary Physics," in *British Journal for the Philosophy of Science*, **12** (1961–1962), 177–193, 281–298; and "Positivismus und Realismus," in *Erkenntnis*, **3** (1932), 1–31, English trans. by D. Rynin as "Positivism and Realism," in A. J. Ayer, ed., *Logical Positivism* (Glencoe, Ill., 1959), and in *Synthèse*, **7** (1948–1949), 478–505.

Later works include "Über das Fundament der Erkenntnis," in *Erkenntnis*, **4** (1934), 79–99, English trans. by D. Rynin as "The Foundation of Knowledge," in A. J. Ayer, ed., *Logical Positivism* (Glencoe, Ill., 1959); "Philosophie und Naturwissenschaft," in *Erkenntnis*, **4** (1934), 379–396; "Facts and Propositions," in *Analysis*, **2** (1935), 65–70; "Sind die Naturgesetze Konventionen?" in *Actes du congrès international de philosophie scientifique, Paris, 1935* (Paris, 1936), 8–17, English trans. by H. Feigl as "Are Natural Laws Conventions?" in H. Feigl and M. Brodbeck, eds., *Readings in the Philosophy of Science* (New York, 1953), 181–188; "Meaning and Verification," in *Philosophical Review*, **45** (1936), 339–369; "Quantentheorie und Erkennbarkeit der Natur," in *Erkenntnis*, **6** (1937), 317–326; "L'école de Vienne et la philosophie traditionnelle," in *Travaux du 9ᵉ congrès international de philosophie* (Paris, 1937); and "Über die Beziehung zwischen den psychologischen und den physikalischen Begriffen," in Schlick's *Gesammelte Aufsätze* (Vienna, 1938; repr. Hildesheim, 1969), English trans. by W. Sellars as "On the Relation Between Psychological and Physical Concepts," in H. Feigl and W. Sellars, ed., *Readings in Philosophical Analysis* (New York, 1949), and in French trans. by J. Haendler as "De la relation entre les notions psychologiques et les notions physiques," in *Revue de synthèse*, **10** (1935), 5–26—both the English and German versions have dropped lines.

Miscellaneous works are *Grundzüge der Naturphilosophie*, W. Hollitscher and J. Rauscher, eds. (Vienna, 1948), a version of his lectures at the University of Vienna on the philosophy of culture; *Natur und Kultur*, J. Rauscher, ed. (Vienna–Stuttgart, 1952), an unfinished work on the philosophy of nature; and *Aphorismen*, B. H. Schlick, ed. (Vienna, 1962), a privately printed selection of philosophic reflections and aphorisms taken from Schlick's writings.

II. SECONDARY LITERATURE. V. Kraft, *Der Wiener*

Kreis (Vienna, 1950), English trans. by A. Pap as *The Vienna Circle* (New York, 1953, 1969), has numerous references to, and comments on, Schlick's views during the last decade of his life. Biographical information together with impressions of Schlick as man, teacher, and philosopher may be found in the obituary articles by H. Reichenbach, in *Erkenntnis*, **6** (1936), 141–142; and H. Feigl, *ibid.*, **7** (1937), 393–419; and in F. Waisman's intro. to Schlick's *Gesammelte Aufsätze*.

See also Béla Juhos, "Moritz Schlick," in *Encyclopedia of Philosophy*, VII (New York, 1967), 319–324; and D. Rynin, "Moritz Schlick," in *International Encyclopedia of the Social Sciences*, XIV (New York, 1968), 52–56.

For critical evaluations, see comments in K. R. Popper, *The Logic of Scientific Discovery* (London–New York, 1959); and the critical discussion by D. Rynin prefixed to his trans. of "Positivism and Realism," in *Synthèse*, **7** (1948–1949), 466–477.

DAVID RYNIN

SCHLIEMANN, HEINRICH (*b.* Neu Buckow, Mecklenburg-Schwerin, Germany, 6 January 1822; *d.* Naples, Italy, 26 December 1890), *archaeology.*

Schliemann was the son of a poor Protestant minister, who encouraged his interest in classical antiquity. A picture of Troy in flames, in a copy of Jerrer's *Universal History* that his father had given him as a Christmas present, captured his imagination and fortified his belief in the reality of the events described by Homer; the picture remained in his memory throughout his youth and during his later career in business. Unable to continue his education past the age of fourteen, Schliemann became an apprentice to a grocer in 1836; in 1841 he decided to immigrate to America, and signed on as cabin boy on a ship that was wrecked shortly thereafter. He then settled in Amsterdam, and was employed by a Dutch business firm for five years, during which he learned almost all the European languages. In 1846 he was sent to St. Petersburg as the firm's agent there, but he soon started his own business, dealing chiefly in indigo, and became rich from it. In 1850 he was in California; his business continued to prosper, and he became an American citizen. He then returned to Russia, where he married, and, at the age of thirty-six, retired from business to devote his time and his great fortune to the study of prehistoric archaeology, and especially to finding the remains of Troy.

To this end, Schliemann studied ancient and modern Greek, traveled extensively in Europe, Egypt, Syria, and Greece, and studied archaeology in Paris. In 1864 he traveled around the world, then in 1868 visited archaeological sites in Greece and Asia Minor. In 1869 he published *Ithaka, der Peloponnes und Troja*, based upon his own excavations at Ithaca and Mycenae, in which he argued that the tombs of Agamemnon and Clytemnestra were to be sought in the citadel of Mycenae, rather than in the treasuries in the lower town. Troy, he went on to state, was not a myth, nor was it located, as some had claimed, at Burnarbashi; rather, it was to be found in the mound of Hissarlik, the site of historic Ilion, and there Schliemann decided to dig.

Some isolated discoveries concerning prehistoric Greek archaeology had been made before Schliemann began to dig at Troy. Chief among these was F. Fouqué's 1862 excavation of painted pottery and frescoed walls at Santorin; since these artifacts were covered by twenty-six feet of pumice deposited by the volcanic eruption of about 2000 B.C., there could be little doubt that they indicated a prehistoric Aegean culture. Schliemann's goal was specific—he wished to prove, through archaeology, the truth of Homer—but he in fact achieved much more; his work at Hissarlik led him to discover the archaeological record of centuries of pre-Homeric, prehistoric, and pre-Hellenic culture.

With his young Greek second wife, Sophia Engastromenos, whom he had married following his divorce, Schliemann began to dig at Hissarlik in 1871. Within the mound he found evidence of seven heavily fortified settlements, which he designated Troy and distinguished by Roman numerals, the deepest being Troy I. Troy II held the greatest interest for him; he found fortress walls, evidence of violent overthrow, and indications that the city had traded in gold, silver, ivory, amber, and jade. Because Troy II had been totally destroyed by fire, Schliemann called it the "burnt layer"; it was succeeded by the small villages of Troy III, Troy IV, and Troy V, and then by the grand Mycenaean city Troy VI. Since a considerable interval must have elapsed between Troy II and Troy VI, it was clear to Schliemann that Troy II must have existed well before the first Olympiad of 776 B.C., traditionally the earliest date in Greek history. He identified Troy II as Homeric Troy, "the citadel of Priam," and, the day before he finished the dig in 1873, found "Priam's treasure," a magnificent cache of gold objects that he hastily smuggled out of Turkey.

Schliemann's attempt to keep the treasure together, against the claims of the Ottoman government to a share of it, precluded his immediate

return to Hissarlik. He prepared his *Trojanische Altertümer* for publication in 1874 (his long business experience had made him assiduous in publishing his work immediately), then returned to Mycenae, where he dug for the tombs of Clytemnestra and Agamemnon at the site of his earlier prediction. Excavating within a circle of stones inside the Lion Gate of the citadel, he found the tombs he was looking for—the now-famous shaft graves. The contents of these graves (Schliemann discovered five, and a sixth was unearthed after he left the site) far surpassed "Priam's treasure" in richness, and included gold and silver vases, inlaid swords of gold, silver, copper, and bronze, gold ornaments for the clothing of the dead, and gold masks. In addition to his work within the citadel, Schliemann excavated two tholoi, the treasury of Atreus (or Agamemnon), and the treasury of Clytemnestra.

Schliemann's *Mycenae*, published in 1877, was written in eight weeks and represented a daily record of his excavations. Like the rest of his books, it was written in German and almost immediately translated into French and English. This book, together with the Mycenaean treasure itself (established in the National Museum in Athens), brought Schliemann considerable fame; the English translation carried a preface by W. E. Gladstone, himself a Homeric scholar as well as a statesman.

Following a short and not very productive visit to Ithaca, Schliemann, in 1879, returned to Hissarlik. He was assisted in this new series of digs by a classical archaeologist, Émile Burnouf, and by Rudolf Virchow, the founder of the Berlin Society for Anthropology, Ethnology, and Prehistory, and organizer of the Berlin Museum für Volkerkunde. The results of this expedition, including new evidence to identify Hissarlik with ancient Troy, are set out in *Ilios* (1881). In 1880 Schliemann went to Orchomenus, where he excavated the treasury of Minyas, a structure similar to the treasuries of Agamemnon and Clytemnestra at Mycenae. His book on the subject, *Orchomenos*, was also published in 1881. He returned to Hissarlik the following year, accompanied by Wilhelm Dörpfeld, who gave him expert assistance on an extensive dig that lasted until 1883. Dörpfeld, a practical architect who had worked with Ernst Curtius at Olympia, brought to the work at Troy the systematization and efficiency of the new German archaeology; he was able to expose the stratigraphy of Hissarlik with precision, and he revolutionized Schliemann's technique.

In 1884 Schliemann went to Tiryns, where he

uncovered the royal palace. In 1886 he traveled to Egypt with Virchow and visited the excavations being conducted by William Petrie (who characterized Schliemann as "dogmatic but always ready for facts"). During the next two years Schliemann also worked at Cythera and Pylos, then, in 1889, returned with Dörpfeld to Hissarlik. He was, throughout these last few years of his life, greatly afflicted by an ear ailment, and made a number of trips to Europe seeking a cure; it was on one such that he collapsed and died while in Naples.

Dörpfeld continued to work at Troy until 1894. Three years after Schliemann's death, he identified Troy VI, rather than Troy II, as the Homeric city, and established that the treasure that Schliemann had found at Troy II predated Priam's time. In his work at Mycenae, Orchomenus, and Tiryns, Schliemann also attributed to the Homeric Greeks artifacts of a much older civilization. The Greeks themselves had always regarded Mycenae and Tiryns as Homeric sites, but the scholarly world was deeply divided over the nature of Schliemann's discoveries. Some scholars willingly accepted his claims, while others argued that the artifacts were Byzantine in origin, or perhaps the work of Celts, Goths, Avars, Huns, or unspecified "orientals." Nonetheless, a number of Schliemann's contemporaries were certain that the Mycenaean civilization that Schliemann had found was not Homeric, but pre-Homeric, as is now known to be the case—what Schliemann had in fact discovered, in both Greece and western Anatolia, was the great pre-Hellenic civilization of the eastern Mediterranean, and this marks his chief contribution to prehistoric archaeology.

Schliemann's contribution to the development of archaeological technique and method has also been vigorously disputed. Stanley Casson, for example, called him (in *The Discovery of Man*, p. 221) the inventor of "a proper archaeological method which could be followed in any land," and added that Schliemann's techniques "constituted an innovation of the first order of importance in the study of the antiquity of man by archaeological methods." A. Michaelis, on the other hand, characterized Schliemann as "a complete stranger to every scientific method of treatment of his subject" and accused him of having "no idea that a method and a well-defined technique existed" (*A Century of Archaeological Discoveries*, p. 217).

It is, however, certain that Schliemann's excavation of Hissarlik was the first such operation conducted upon a tell and was, as Sir John Myres wrote, "the first large-scale dissection of a dryland settlement unguided by the remains of great monuments such as simplified the task in Babylon and Nineveh" (*The Cretan Labyrinth*, p. 273). Schliemann's discovery of seven occupation levels at Troy further gave a considerable impetus to the application of the principles of stratigraphy to archaeology, although it is necessary to note that he himself understood the stratigraphy of Hissarlik only slowly and with the assistance of Dörpfeld. (Indeed, he came to recognize the strata only slowly, thinking at one time that the whole mound covered Priam's city; and for a while the recurrence of stone tools puzzled him, so that he wrote that he could not understand "how it is that I am unearthing stone implements throughout the length of my excavations.")

Schliemann's archaeological work was of interest to the non-scientific world as well. He kept the public informed of his discoveries through his books and through his dispatches to the London *Times* and *Daily Telegraph*, as well as a number of other newspapers, so that, as A. T. White wrote, "every person of culture and education lived through the drama of discovering Troy" (*Lost Worlds*, p. 27). His readers were excited by the romance of his undertaking and rejoiced in Schliemann's incredible good luck in finding exactly what he had set out to find—the physical evidence of Homer's Troy, and a buried hoard of golden treasure. Schliemann also provided inspiration to a whole generation of professional archaeologists and ancient historians; although Emil Ludwig described him as "monomaniacal" and as perhaps of "a mythomaniacal nature which at times overstepped the limits of the normal," Sir John Myres wrote that upon the news of Schliemann's death it seemed to many that "the spring had gone out of the year" (*The Cretan Labyrinth*, p. 272).

BIBLIOGRAPHY

I. ORIGINAL WORKS. Schliemann's writings include *Ithaka, der Peloponnes und Troja* (1869); *Trojanische Altertümer* (1874); *Troja und seine Ruinen* (1875); *Mycenae* (1877); *Ilios* (1880); *Orchomenos* (1881); *Troja* (1884); and his autobiography, edited by his wife Sophia, *Selbstbiographie bis zu seinem Tode vervollständigt* (Leipzig, 1892; 9th ed., Wiesbaden, 1961).

II. SECONDARY LITERATURE. On Schliemann and his work, see Stanley Casson, *The Discovery of Man* (London, 1939); Emil Ludwig, *Schliemann: The Story of a Gold-Seeker* (London, 1931); Sir John Myres, *The Cretan Labyrinth* (London, 1933); Sir John Sandys, *A History of Classical Scholarship* (Cambridge, 1908);

and Carl Schuchardt, *Schliemann's Excavations and Archaeological and Historical Studies* (London, 1891).

GLYN DANIEL

SCHLOTHEIM, ERNST FRIEDRICH, BARON VON (*b.* Almenhausen, Thuringia, Germany, 2 April 1765; *d.* Gotha, Germany, 28 March 1832), *geology, paleontology.*

Schlotheim was the son of Ernst Ludwig von Schlotheim and Friederike Eberhardine von Stange. After receiving his basic education at home from a tutor, he attended the Gymnasium in Gotha from 1779 to 1781. He then studied public administration, and the natural sciences under Blumenbach, at Göttingen. Since he was especially interested in the geological sciences, he next went to Freiberg to study under Werner; he also became a friend of Alexander von Humboldt. In 1792 Schlotheim entered the Gotha civil service as an assessor, rising to minister and lord high marshal by 1828.

From 1822 Schlotheim also served as superintendent of the ducal art, natural history, coin, and book collections in Gotha. At an early age Schlotheim had started his own geological and especially paleontological collections; and he now began to publish his observations of the countryside. Reporting on the stratigraphy of the calcareous tufa at Gräfentonna, in which a complete fossil elephant skeleton had been found in 1695, Schlotheim recognized that in addition to indigenous stones, it contained other field stones, predominantly granite. He was the first to trace this combination to Scandinavia.

Schlotheim's later investigations were concerned primarily with paleontology. He studied the plants found in the bituminous schists of the Lower Permian in Thuringia and realized (1801, 1804) that they belonged to extinct species and could not be given — as had been customary — contemporary names. He also concluded that during the Lower Permian epoch Thuringia must have had a warmer climate. In his article of 1813 Schlotheim was the first to insist that the species of the fossils must be determined in order to distinguish the various formations. He called for the establishment of a nomenclature in paleontology analogous to that provided by Linnaeus for existing organisms. Paleontology would thereby become a tool for determining the age of strata — a principle advocated independently by William Smith (1816). Schlotheim's article also presented primarily a survey of fossils according to geological formations. His *Petrafaktenkunde* (1820) employed the binomial nomenclature systematically and thus put the science of paleontology onto a rigorous basis. It was the first major advance in that science since 1762, when J. G. Walch had comprehensively classified fossils within the zoological system; and it marked the beginning of an era of rapid and important growth.

Schlotheim also realized that the distribution of fossils, their association, their degree of preservation, the facies of the neighboring rock, and the mixture of marine with terrestrial forms made it possible to draw important inferences concerning the history of the earth. The study of these factors was not taken up systematically until a century later, as biostratinomy. Schlotheim was also critical of catastrophism, which was challenged for the first time in 1822 by his friend K. E. A. von Hoff, an early uniformitarian.

BIBLIOGRAPHY

I. ORIGINAL WORKS. Schlotheim's most important works are "Mineralogische Beschreibung der unteren Herrschaft Tonna," in J. C. W. Voigt, *Mineralogische und bergmännische Abhandlungen,* III (Leipzig, 1791), 182–200; "Beiträge zur nähern Kenntniss einzelner Fossilien," in *Magazin für die gesamte Mineralogie, Geognosie und mineralogische Erdschreibung,* 1 (1801), 143–172; "Über die Kräuterabdrücke im Schieferton und Sandstein der Steinkohlenformation," *ibid.; Beschreibung merkwürdiger Kräuterabdrücke und Pflanzenversteinerungen* (Gotha, 1804); "Beiträge zur Naturgeschichte der Versteinerungen in geognostischer Hinsicht," in *Taschenbuch für die gesammte Mineralogie,* 7 (1813), 3–134; "Die Versteinerungen im Höhlenkalkstein von Glücksbrunn," in *Magazin für die neuesten Entdeckungen in der gesammten Naturkunde,* 7 (1816), 156–158; "Beiträge zur Naturgeschichte der Versteinerungen in geognostischer Hinsicht," in *Denkschriften der K. Akademie der Wissenschaften zu München,* 6 (1816–1817), 13–36; "Der Kalktuff als Glied der aufgeschwemmten Gebirgsformation," in *Taschenbuch für die gesammte Mineralogie,* 12 (1818), 315–345; *Die Petrafaktenkunde auf ihrem jetzigen Standpunkte* (Gotha, 1820; supps., 1822, 1823); and *Der thüringische Flözmuschelkalkstein in besonderer Beziehung auf seine Versteinerungen* (Gotha, 1823).

II. SECONDARY LITERATURE. See C. Credner, "Ernst Friedrich von Schlotheim," in *Neuer Nekrolog der Deutschen,* 10, no. 1 (1832), 246–250; B. von Freyberg, "Ernst Friedrich Baron von Schlotheim," in "Aus der Heimat," *Naturwissenschaftliche Zeitschrift,* 45, no. 10 (1932); and *Die geologische Erforschung Thüringens in älterer Zeit* (Berlin, 1932); W. von Gümbel, "Ernst Friedrich Freiherr von Schlotheim," in *Allgemeine deutsche Biographie*, XXXI, 550–551; and K. A. von Zittel, *Geschichte der Geologie und Paläontologie* (Munich–Leipzig, 1899).

B. VON FREYBERG

SCHLUMBERGER, CHARLES (*b*. Mulhouse, France, 29 September 1825; *d*. Paris, France, 13 July 1905), *micropaleontology*.

After completing his education at the École Polytechnique, Schlumberger joined the navy corps of engineers at Toulon in 1849. His transfer in 1855, to Nancy, where he was in charge of purchasing timber, enabled him to make many field trips and to develop an interest in natural history, which dated from his youth. The collecting of fossils, combined with his acquaintance with Olry Terquem, led to his career as a micropaleontologist specializing in Foraminifera.

In 1879 Schlumberger went to Paris; and although he was promoted a few years later to the rank of chief engineer, he requested early retirement and became a guest scientist at the laboratory of paleontology of the Muséum d'Histoire Naturelle and later at the École des Mines. About 1882 he began his collaboration with E. Munier-Chalmas, who had just recognized dimorphism in *Nummulitids*. By means of a new thin-section technique developed by Schlumberger, both men were able to discover the same character among Miliolidae, as well as important features pertaining to their apertures.

By 1894 dimorphism in Foraminifera had been recognized by many as an extremely widespread character and had been explained by alternating phases of sexual and asexual reproduction. Unfortunately, Schlumberger was not active in this final development and interpretation of some of his original discoveries because Munier-Chalmas's reluctance to write papers in final form had gradually brought their collaboration to a standstill. Consequently Schlumberger continued his research on other groups of Foraminifera alone. He wrote in particular a series of important papers on the stratigraphic distribution of *Orbitoids*. After the death of Munier-Chalmas and during the last years of his life, he resumed his studies on Miliolidae.

Despite his lack of formal scientific training, Schlumberger contributed in a fundamental manner to the solution of one of the most puzzling problems of micropaleontology.

BIBLIOGRAPHY

Schlumberger's writings include "Note sur les Foraminifères," in *Feuille des jeunes naturalistes*, 12 (1882), 83–112; "Nouvelles observations sur le dimorphisme des Foraminifères," in *Comptes rendus . . . de l'Académie des sciences*, 96 (1883), 862–866, 1598–1601, written with E. Munier-Chalmas, also in English as "New Observations on the Dimorphism of the Foraminifera," in *Annals and Magazine of Natural History*, 5th ser., 11 (1883), 336–340; "Sur les Miliolidées trématophorées, lère partie," in *Bulletin de la Société géologique de France*, 3rd ser., 13 (1885), 273–323; "Révision des Biloculines des grands fonds, expéditions du *Travailleur* et du *Talisman*," in *Mémoires de la Société zoologique de France*, 4 (1891), 155–191; "Monographie des Miliolidées du golfe de Marseille," *ibid.*, 6 (1893), 199–222; "Première note sur les Orbitoïdes (*Orbitoides*, s. str.)," in *Bulletin de la Société géologique de France*, 4th ser., 1 (1901), 459–467; "Deuxième note sur les Orbitoïdes (*Orbitoides*, s. str.)," *ibid.*, 2 (1902), 255–261; "Troisième note sur les Orbitoïdes (*Orthophragmina* discoïdes)," *ibid.*, 3 (1903), 273–289; "Quatrième note sur les Orbitoïdes (*Orthophragmina* étoilés)," *ibid.*, 4 (1904), 119–135; and "Deuxième note sur les Miliolidées trématophorées," *ibid.*, 5 (1905), 115–134.

A complete list of Schlumberger's publications is in H. Douvillé, "Charles Schlumberger, notice nécrologique," in *Bulletin de la Société géologique de France*, 4th ser., 6 (1906), 340–350.

ALBERT V. CAROZZI

SCHMERLING, PHILIPPE-CHARLES (*b*. Delft, Netherlands, 24 February 1791; *d*. Liège, Belgium, 6 November 1836), *paleontology*.

Schmerling's paternal ancestors came from Austria. After completing his secondary education in Delft, he studied medicine at Leiden for two years and then went to The Hague. Appointed a health officer in 1812 and a military physician the following year, he left the army in 1816 in order to establish a civilian practice. In 1821 he married Sara Henriette Caroline Douglas, a descendant of a noble Scots family, one branch of which had immigrated to the Netherlands. A few months later Schmerling and his wife moved to Liège, where he continued his medical studies and, after receiving the doctorate in 1825, began to practice.

Schmerling became a paleontologist by chance. In 1829 he went to Chokier, a small village near Liège, to treat a poor quarry worker. He was amazed to see the man's children playing with very unusual bones that had been unearthed at a nearby quarry. Stimulated by this find, the first known excavation of fossil bones in Belgium, Schmerling traveled extensively in the region and within less than four years located more than forty similar sites. He collected the remains of some sixty animal species. Those that made him famous were human bones in an indisputably fossil state.

Following a carefully formulated plan, Schmerling studied first the caves, then the animal remains they contained, and finally the human bones. He observed that the stalagmitic floors covered deposits containing the remains of extinct species, such as the mammoth, and of apparently contemporaneous species still in existence, such as the wolf and the boar. He noted that the human remains were in the same state of preservation as the animal bones unearthed with them. They were of the same color, were dispersed in similar patterns, and sometimes were so abraded and scattered that the hypothesis of their deliberate burial in the caves had to be excluded. Therefore, Schmerling asserted, the human bones had undoubtedly been buried at the same time and by the same causes as the bones of the extinct species. He also discovered chipped stones and carved bones in the same conditions; their very presence, he asserted, demonstrated the existence of man during the "antediluvian period."

The most famous of the caves that Schmerling explored was that of Engis, located on the left bank of the Meuse about fifteen kilometers upstream from Liège. There, in 1830, he exhumed two skulls at different levels. One, of a child (today called Engis I), was lying at the base of the deposits, next to a mammoth's tooth; the other, of an adult (Engis II), was found at a somewhat higher level in a hole containing rhinoceros teeth and the bones of horses, reindeer, and several ruminants. (It was later shown that Engis I was of the Neanderthal type—the first such example ever found, whereas Engis II belonged to a variety of Cro-Magnon man.) Others before Schmerling (E. J. C. Esper in 1771 and Buckland in 1823) and at about the same time (P. Tournal) had discovered human bones associated with the remains of extinct animals, but Schmerling was the first to demonstrate the existence of fossil man by means of irrefutable stratigraphic arguments.

Schmerling's medical training led to his interest in the pathological lesions on the Quaternary mammalian bones found in the Belgian caves, and his 1836 memoir on paleopathology was one of the first of its kind. His recognition of the importance of this discipline is a further indication of both the originality and the scope of his thought.

Although Schmerling's five published works on paleontology (1832–1836) were characterized by exemplary scientific rigor, they generated little enthusiasm (except for the last one, on paleopathology, which stimulated much discussion in Germany). Indeed, his work fell into such neglect that many copies of his principal study (1833–1834) were destroyed after his death. Lyell visited Schmerling in 1834 and cited his discoveries in the third edition of *Principles of Geology* (1834, p. 161); but he did not grasp their importance until much later, as he himself admitted (1863). Schmerling's contemporaries were not, however, completely unaware of his scientific ability. He was elected a corresponding member of the Royal Academy of Brussels (1834) and a corresponding member of the Royal Institute of the Low Countries (1836).

BIBLIOGRAPHY

I. ORIGINAL WORKS. In addition to his doctoral dissertation (1825) and a note on dyeing with colchicum (1832), Schmerling published five works on paleontology. The two principal ones are *Recherches sur les ossements fossiles découverts dans les cavernes de la province de Liège*, 2 vols. and 2 vols. of plates (Liège, 1833–1834), translated or analyzed in Italy, France, Germany, Russia, England, and the United States, and containing the description of the Engis men; and "Notice sur quelques os de pachydermes découverts dans le terrain meuble près du village de Chokier," in *Bulletin de l'Académie royale des sciences et belles-lettres de Bruxelles*. Classe de sciences, **3** (1836), 82–87, his contribution to paleopathology.

II. SECONDARY LITERATURE. See the following, listed chronologically: C. Morren, "Notice sur la vie et les travaux de Philippe Charles Schmerling," in *Annuaire de l'Académie royale des sciences et belles-lettres de Bruxelles*, **4** (1838), 130–150; A. Le Roy, *L'Université de Liège depuis sa fondation* (Liège, 1869), 550; C. Lyell, *The Geological Evidences of the Antiquity of Man*, 1st ed. (London, 1863), 70–71; R. L. Moodie, *Paleopathology. An Introduction to the Study of Ancient Evidence of Disease* (Urbana, Ill., 1923), 64–65; C. Fraipont, "Les hommes fossiles d'Engis," *Archives de l'Institut de paléontologie humaine*, no. 16 (1936); and K. P. Oakley, "The Problem of Man's Antiquity. An Historical Survey," in *Bulletin of the British Museum (Natural History)*, Geology, **9**, no. 5 (1964), 91–93.

G. UBAGHS

SCHMIDEL (or **Schmiedel**), **CASIMIR CHRISTOPH** (*b.* Bayreuth, Germany, 21 November 1718; *d.* Ansbach, Germany, 18 December 1792), *medicine, natural history.*

Schmidel was the son of Georg Cornelius Schmidel, a Brandenburg financial councillor and physician-in-ordinary to the margrave in Bayreuth. Schmidel was best known for his studies of the morphology of the cryptogams and for his editing of Konrad Gesner's posthumous botanical publications. He also lectured and wrote many essays on general medicine and anatomy.

Following the early death of his parents, Schmidel left Bayreuth in 1728, going first to Arnstadt and then, in 1733, to Gera. He began medical studies at Jena in 1735 and continued them a year later at Halle, where he attended the lectures of Friedrich Hoffmann and Johann Heinrich Schulze. He returned to Jena in 1739 and on 17 February 1742 received the M.D. for his *Dissertatio inauguralis de exulceratione pericardii et cordis exemplo illustrata.* His teachers included Georg Erhard Hamberger, Simon Paul Hilscher, Karl Friedrich Kaltschmied, Hermann Friedrich Teichmeyer, Johann Adolph Wedel, and Johann Bernhard Wiedeburg. Schmidel learned natural history on his own and in the company of other amateurs.

After completing his studies, Schmidel became professor of pharmacology at the newly opened Friedrichs-Akademie and simultaneously established a medical practice in Bayreuth. When the university was moved the following year from Bayreuth to Erlangen, Schmidel assumed the second professorship of medicine at its new quarters. He was also appointed the first dean of the medical faculty. In 1745 he was named Brandenburg court councillor, and in 1750 he became a member of the Kaiserliche Akademie der Naturforscher in Halle. He took the cognomen Oribasius II.

From 1756 to 1758 Schmidel studied botany and geology in Saxony, Holland, and Switzerland. In 1760 Schmidel obtained the post of first full professor at Erlangen, a position left vacant by the death of Johann Friedrich Weismann. Schmidel's teaching responsibilities were chiefly in the fields of physiology, where he drew on Boerhaave's *Institutiones medicae,* and natural history, where he followed Linnaeus' *Systema naturae.* He also lectured on anatomy, surgery, dietetics, pathology, semiotics, therapeutics, and legal medicine. Because of scientific disagreements with his colleague Heinrich Friedrich von Delius, Schmidel left Erlangen and in 1763 went to Ansbach to serve as physi-

cian-in-ordinary to Margrave Carl Alexander. A few years later he had temporarily to give up this appointment because of a dispute with the sovereign. Thus he was left with sufficient leisure both to conduct extensive research in botany and to enjoy a career as a respected physician. In recognition of his services, Carl Alexander made him a privy councillor and president of the board of health.

In 1773 and 1774 Schmidel accompanied the ailing daughter of Margrave Friedrich of Bayreuth on a journey to Lausanne, where she consulted Simone-André Tissot, and then to Dieppe, in Normandy. (Schmidel's account of the trip was first published in 1794 in an edition prepared by Johann Christian Daniel Schreber.) Shortly thereafter, Friedrich requested that Schmidel serve as physician on a tour through France and Italy in 1775–1776. On 16 July 1783 Schmidel was awarded an honorary M.D. by the philosophy faculty of the University of Erlangen. During the last four years of his life, Schmidel suffered from mental disorders.

BIBLIOGRAPHY

I. ORIGINAL WORKS. Schmidel's works include *Dissertatio inauguralis de exulceratione pericardii et cordis exemplo illustrata* (Jena, 1742), his diss., "Anmerkungen über die bisherige Eintheilung der Schwämme, besonders nach ihren Arten," in *Erlangische gelehrte Anzeigen,* **19** (1746), 145–152; *Icones plantarum et analyses partium aeri incisae atque viuis coloribus insignitae, adjectis indicibus nominum necessariis, figurarum explicationibus et breuibus animaduersionibus* (Nuremberg, 1747); "Von der Grösse und Einrichtung der erschaffenen Erde," in *Fränkische Sammlungen von Anmerkungen aus der Naturlehre, Arzneygelahrtheit, Oekonomie,* **23** (1761), 195–208; *Demonstratio vteri praegnantis mulieris e foetu ad partum maturo in tabulis sex ad naturae magnitudinem post dissectionem depictis et ea methodo dispositis . . .* (Nuremberg, 1761); *Fossilium metalla et res metallicas concernentium, glebae suis coloribus expressae* (Nuremburg, 1762); "Beschreibung eines Seesterns mit rosenförmigen Verzierungen," in *Der Naturforscher,* **16** (1781), 1–7; *Vorstellung einiger merkwürdigen Versteinerungen mit kurzen Anmerkungen versehen* (Nuremberg, 1781); and *Descriptio itineris per Heluetiam, Galliam et Germaniae partem 1773 et 1774 instituti, mineralogici, botanici et historici argumenti* (Erlangen, 1794), Johann Christian Daniel Schreber, ed.

Schmidel also edited *Conradi Gessneri opera botanica, per duo secula desiderata, quorum pars prima prodromi loco figuras continet vltra 400 minoris formae, partim ligno excisas, partim aeri insculptas omnia. Ex bibliotheca D. Ch. Jac. Trew nunc primum in lucem edidit et praefatus est* (Nuremberg, 1753); *Conradi Gessneri historiae plantarum fasciculus quam ex bibliotheca D. Ch. Jac. Trew edidit et illustravit* (Nuremberg, 1759); and *Conradi Gessneri opera botanica,* 2 vols. (Nuremberg, 1764–1771).

II. Secondary Literature. On Schmidel and his work, see G. W. A. Fickenscher, *Gelehrtes Fürstenthum Baireut* (Nuremberg, 1804), 7, 112–127; W. Hess, "Kasimir Christoph Schmidel," in *Allgemeine Deutsche Biographie* (Berlin, 1890), 31, 700; H. Krauss, *Die Leibärzte der Ansbacher Markgrafen* (Neustadt an der Aisch, 1941); F. Leydig, "Kasimir Christoph Schmidel, Naturforscher und Arzt 1716–1792," in *Abhandlungen der Naturhistorischen Gesellschaft zu Nürnberg,* 15 (1905), 325–355; J. A. Vocke, *Geburts- und Todten-Almanach Ansbachischer Gelehrten, Schriftsteller und Künstler* (Augsburg, 1797), 2, 326–329; and T. Wohnhaas, "Miscellanea anatomica zu Kasimir Christoph Schmidel," in *Sitzungsberichte der Physikalisch-medizinischen Sozietät in Erlangen,* 82 (1963), 27–32.

A. Geus

SCHMIDT, BERNHARD VOLDEMAR (*b.* Naissaar, Estonia [now Estonian S.S.R.], 30 March 1879; *d.* Hamburg, Germany, 1 December 1935), *optics.*

Schmidt studied in Göteborg, Sweden, and then engineering in Mittweida, Saxony, where he established a small workshop for the manufacture of astronomical mirrors (up to about twenty centimeters in diameter). Their perfection was much appreciated. In 1905 Schmidt constructed for the Potsdam Astrophysical Observatory the first reflector with an aperture of forty centimeters and with a focal length of about one meter. Schmidt already had used for Cassegrain reflectors a spherical mirror that corrected spherical aberration by an adequate deformation of the second mirror.

In addition to the construction of astronomical instruments, Schmidt also improved and himself used some of the great objectives (for example, at the Hamburg observatory, Bergedorf). In 1909 he constructed for his own small observatory at Mittweida a new horizontal reflector, later on named "Uranostat," which consisted of a parabolic mirror of forty-centimeter aperture and of eleven-meter focal length. The reflector was mounted so that the axis was in the north-sourth direction and caught the light of the object to be observed from two plane mirrors rotatable around two perpendicular axes.

Later Schmidt constructed two similar arrangements of several mirrors for the Bergedorf and Breslau observatories. In 1926 Schmidt moved to Bergedorf, where with the first arrangement he himself made excellent photographs of Jupiter, Saturn, and the moon.

Schmidt always was an odd man, who neither married nor was willing to lead a normal life. He once said that he got his best ideas on awaking slowly after some days of complete intoxication. Drink was certainly the cause of his early death. At the Bergedorf observatory he had no regular duties and was free of economic worries, receiving generous aid from R. Schorr, the director of the observatory. Schmidt thus had the time and the resources to carry out his most famous work—the construction of a reflector without coma. He used a correction plate shaped as a very small curved circular torus, which compensates for spherical aberration and coma. Photographs could now be taken which yielded undistorted star images over a large field: formerly only objects near the optical axis could be delineated.

Schmidt had lost his right arm in an accident in his youth, but he nevertheless did all of his work alone and unaided. He polished his mirrors by hand, using glass instead of metal disks.

Schmidt spent the last year of his life in a mental hospital and died there.

BIBLIOGRAPHY

Schmidt's only work is "Ein lichtstarkes komafreies Spiegelsystem," in *Zentralzeitung für Optik und Mechanik, Elektrotechnik,* 52 (1931), 25–26. For works on Schmidt and his work, see B. Strömgren, "Das Schmidt'sche Spiegelteleskop," in *Vierteljahrsschrift der Astronomischen Gesellschaft,* 70 (1935), 65–86; and A. A. Wachmann, "From the Life of Bernhard Schmidt," in *Sky and Telescope,* 15 (1959), 4–9.

H.-Christ. Freiesleben

SCHMIDT, CARL AUGUST VON (*b.* Diefenbach, Württemberg, Germany, 1 January 1840; *d.* Stuttgart, Germany, 21 March 1929), *geophysics, astrophysics.*

Schmidt, the son of a schoolteacher, studied Protestant theology at the seminary in Tübingen from 1858 to 1862. In 1863 he obtained a doctorate in philosophy from the University of Tübingen, where during the same year he studied mathematics and natural science. From 1864 to 1866 he studied chemistry in Paris and then in Stuttgart at the polytechnic school until 1868. At this time he also took the examination to qualify for teaching mathematics and science in secondary school.

From 1868 to 1871 Schmidt was a student-teacher and from 1872 to 1904, professor at the Realgymnasium in Stuttgart, where he taught chemistry, physics, and mathematics. In addition, in 1896 he was appointed a full member of the Württemberg office of statistics and director of the central weather bureau in Stuttgart, which he headed until 1912. From 1902 to 1912 he served on the board of directors of the earthquake research center in Strasbourg and from 1906 to 1912, on the board of the weather station on Lake Constance. The latter institution, founded by H. Hergesell, employed kites in its meteorological research.

One of Schmidt's notable works (most of which dealt with geophysics and astrophysics) was his "Wellenbewegung und Erdbeben" (1888), in which he demonstrated that seismic waves do not spread rectilinearly from the focus of an earthquake but in curved paths. This phenomenon partly arises from Snell's law of refraction, and partly because the quotient of density to modulus of elasticity is not constant over the entire surface of the earth. Schmidt also established the law, named for him, concerning the turning point in the apparent propagation velocity of seismic waves; and he introduced the time-distance curve into seismology. In order to measure the vertical movement of the earth he devised the bifilar or trifilar gravimeter.

In 1886, in Württemberg, Schmidt made the first seismic measurements, and in 1892 he and K. Mack established an earthquake observatory in Hohenheim. As early as 1894 he pointed out the separation of seismic waves into longitudinal and transverse components, thus anticipating the subsequent findings of Wiechert. In 1896 Schmidt became director of the entire earthquake bureau in Württemberg, a position he held until 1912. During this period the geomagnetic survey of Württemberg, which Schmidt had promoted, was completed by K. G. F. Haussmann; Schmidt wrote the preface to the publication of the results of the survey. He also reported on the results in a separate work, as well as on the geomagnetic measurement of the Riesen Gebirge (1906)—a region in which current research indicates the presence of an impact crater of a large meteorite.

In other works Schmidt treated terrestrial magnetism and the shape of the earth (1895), the displacement of the terrestrial poles (1896), continental tides (1897), and plumb-line deflection (1898). He contributed to meteorology through works on the application of thermodynamics and the kinetic theory of gases to the study of the atmosphere (1889), on the fostering of aerological observations at the kite station on Lake Constance, and on the discussion of problems of climatology. Furthermore, he introduced the concept of barometric tendency into weather forecasting and investigated the mechanism of thunderstorms (1895).

Schmidt directed his astronomical research to questions of refraction in the solar atmosphere and chromosphere (beginning in 1891), and also to general problems of solar physics (rotation of the sun, sources of solar energy, the spectrum of the chromosphere, and solar flares). He also investigated phenomena on the planet Mars (1893) and commented on the stability of the rings of Saturn (1894).

Schmidt's importance lay in the breadth of his interests and in his knowledge of the interrelated aspects of geophysics and astrophysics.

BIBLIOGRAPHY

A list of Schmidt's works is in Poggendorff, III (1898), 1200; IV (1904), 1335–1336; and in *Beiträge zur Geophysik,* 22 (1929), 235–238.

On Schmidt and his work, see K. Kleinschmidt, *ibid.,* 22 (1929), 233–235; in *Meteorologische Zeitschrift,* 64 (1929), 265–267; and in *Unterrichtsblätter für Mathematik und Naturwissenschaften,* 35 (1929), 129.

DIEDRICH WATTENBERG

SCHMIDT, ERHARD (*b.* Dorpat, Germany [now Tartu, E.S.S.R.], 13 January 1876; *d.* Berlin, Germany, 6 December 1959), *mathematics.*

Schmidt's most significant contributions to mathematics were in integral equations and in the founding of Hilbert space theory. Specifically, he simplified and extended David Hilbert's results in the theory of integral equations; and he formalized Hilbert's distinct ideas on integral equations into the single concept of a Hilbert space, in the process introducing many geometrical terms. In addition, he made contributions in the fields of partial differential equations and geometry. The most important of these discoveries were the extensions of the isoperimetric inequality, first to *n*-dimensional Euclidean space and then to multidimensional hyperbolic and spherical spaces. Although his methods were classical rather than abstractionist, nevertheless he must be considered a founder of modern functional analysis.

The son of Alexander Schmidt, a medical biologist, Erhard studied at Dorpat, Berlin, and finally at Göttingen, where he was a doctoral candidate under Hilbert. His degree was awarded in 1905 after the presentation of his thesis, "Entwicklung willkürlicher Funktionen nach Systemen forgeschriebener." After short periods as a teacher in Bonn, Zurich, Erlangen, and Breslau (now Wrocław), in 1917 he went to the University of Berlin, where he was to remain the rest of his life. In 1946 he became the first director of the Research Institute for Mathematics of the German Academy of Sciences, a post he held until 1958. He was also one of the founders and first editors of *Mathematische Nachrichten* (1948).

The integral equation on which Schmidt's reputation is based has the form

$$f(s) = \phi(s) - \lambda \int_a^b K(s, t) \phi(t) dt. \qquad (1)$$

In (1), $K(s, t)$ — called the kernel — and $f(s)$ are known functions and ϕ is an unknown function that is to be found. This equation has a long history. Interest in it stemmed from its many applications; for example, if (1) can be solved, then the partial differential equation $\Delta u = \partial^2 u / \partial x^2 + \partial^2 u / \partial y^2 = 0$ with the prescribed condition $u(x, y) = b(s)$ of arc length s on the boundary of a given region of the plane can also be solved. This differential equation arises in many problems of physics.

From the early nineteenth century on, there were many attempts to solve equation (1), but only partial results were obtained until 1903. In that year Ivar Fredholm was able to present a complete solution to (1), although in his theory the parameter λ plays no significant role. Fredholm showed that for a fixed λ and K either (1) has a unique solution for every function f, or the associated homogeneous equation

$$\phi(s) - \lambda \int_0^1 K(s, t) \phi(t) dt = 0 \qquad (2)$$

has a finite number of linearly independent solutions; in this case (1) has a solution ϕ only for those f that satisfy certain orthogonality conditions.

In 1904 Hilbert continued the study. He first used a complicated limiting process involving infinite matrices to show that for the fixed but symmetric kernel K ($K(s, t) = K(t, s)$), there would always be values of λ — all real — for which (2) had nontrivial solutions. These λ's he called the *eigenvalues* associated with K, and the solutions he called *eigenfunctions*. He also proved that if f is such that there exists g continuous on [0, 1] with

$$f(s) = \int_a^b K(s, t) g(t) dt,$$

then $f(s)$ can be expanded in a series in eigenfunctions of K, that is,

$$f(s) = \sum_{p=1}^\infty a_p \phi_p$$

where $\{\phi_p\}$ is an orthonormal[1] set of eigenfunctions of K.

A year later Hilbert introduced the concept of infinite bilinear forms into both the theory of integral equations and the related topic of infinite matrices. He discovered the concept of complete continuity[2] for such forms and then showed that if $\{a_{ij} : i, j = 1, 2, \cdots\}$ are the coefficients of a completely continuous form, then the infinite system of linear equations

$$x_i + \sum_{j=1}^\infty a_{ij} x_j = a_i, \qquad i = 1, 2, \cdots \qquad (3)$$

either has a unique square summable[3] solution $\{x_i : i = 1, 2, \cdots\}$ for every square summable sequence $\{a_i\}$ or the associated homogeneous system $x_i + \Sigma a_{ij} x_j = 0$ has a finite number of linearly independent solutions. In the latter case, (3) will have solutions only for those sequences $\{a_i\}$ that satisfy certain orthogonality conditions. Hilbert then went on to prove again Fredholm's result converting equation (1) to equation (3) by using Fourier coefficients.

Schmidt's paper on integral equation (1) appeared in two parts in 1907. He began by reproving Hilbert's earlier results concerning symmetric kernels. He was able to simplify the proofs and also to show that Hilbert's theorems were valid under less restrictive conditions. Included in this part of the work is the well-known Gram-Schmidt process for the construction of a set of orthonormal functions from a given set of linearly independent functions.

Schmidt then went on to consider the case of (1) in which the kernel $K(s, t)$ is no longer symmetric. He showed that in this case, too, there always will be eigenvalues that are real. The eigenfunctions, however, now occur in adjoint pairs; that is, ϕ and ψ are adjoint eigenfunctions belonging to λ if ϕ satisfies

$$\phi(s) = \lambda \int_a^b K(t, s) \psi(t) dt,$$

which is called an eigenfunction of the first kind, and ψ satisfies

$$\psi(s) = \lambda \int_a^b K(s, t) \phi(t) dt,$$

an eigenfunction of the second kind. Moreover, if $\phi = \phi_1 + i\phi_2$, $\psi = \psi_1 + i\psi_2$, then ϕ_1 and ψ_1 are an adjoint pair of eigenfunctions, as are ϕ_2 and ψ_2. Thus, it is only necessary to consider real pairs of eigenfunctions.

Other extensions of the symmetric to the unsymmetric case were also developed by Schmidt. As a broadening of Hilbert's result, Schmidt proved (Hilbert-Schmidt theorem) that if f is such that there is a function g continuous on $[a, b]$ with

$$f(s) = \int_a^b K(s, t) g(t) dt,$$

then f can be represented by an orthonormal series of the eigenfunctions of the first kind of K; and if

$$f(s) = \int_a^b K(t, s) g(t) dt,$$

then f has a representation in a series of the second kind of eigenfunctions. He also proved a type of diagonalization theorem: If $x(s)$ and $y(s)$ are continuous on $[a, b]$, then

$$\int_a^b \int_a^b K(s, t) x(s) y(s) ds dt =$$

$$\sum_v \frac{1}{\lambda_v} \int_a^b x(s) \phi_v(s) ds \int y(t) \psi_v(t) dt$$

where $\{\phi_p\}$ and $\{\psi_p\}$ are orthonormal sets of eigenfunctions of the first or second kinds and λ_v are the associated eigenvalues.

The idea behind Schmidt's work is extremely simple. From the kernel $K(s, t)$ of equation (1) he constructed two new kernels:

$$\overline{K}(s, t) = \int_a^b K(s, r) K(t, r) dr$$

and

$$\underline{K}(s, t) = \int_a^b K(r, s) K(r, t) dr,$$

which are both symmetric. Then ϕ and ψ are an adjoint pair of eigenfunctions belonging to λ if and only if

$$\phi(s) = \lambda^2 \int_a^b \overline{K}(s, t) \, \phi \, (t) dt,$$

and

$$\psi(t) = \lambda^2 \int_a^b \underline{K}(s, t) \psi(t) dt;$$

that is, ϕ is an eigenfunction belonging to λ^2 of $\overline{K}$ and ψ is an eigenfunction belonging to λ^2 of $\underline{K}$. Thus Schmidt could then apply much of the earlier theory of symmetric kernels.

Schmidt's contributions to Hilbert space theory stem from Hilbert himself. Before Hilbert there had been some attempts to develop a general theory of infinite linear equations, but by the turn of the twentieth century only a few partial results had been obtained. Hilbert focused the attention of mathematicians on the connections among infinite linear systems, square summable sequences, and matrices of which the entries define completely continuous bilinear forms. These equations were of importance since their applications were useful not only in integral equations but also in differential equations and continued fractions.

In 1908 Schmidt published his study on the solution of infinitely many linear equations with infinitely many unknowns. Although his paper is in one sense a definitive work on the subject, its chief importance was the explicit development of the concept of a Hilbert space and also the geometry of such space—ideas that were only latent in Hilbert's own work.

A vector or point z of Schmidt's space H was a square summable sequence of complex numbers, $\{z_n\}$. The inner product of two vectors z and w—denoted by (z, w)—was given by the formula

$$(z, w) = \sum_{p=1}^{\infty} z_p w_p$$

and a norm—denoted by $\|z\|$—was defined by $\|z\| = \sqrt{(z, \bar{z})}$. The vectors z and w were defined to be perpendicular or orthogonal if $(z, w) = 0$, and Schmidt showed that any set of mutually orthogonal vectors must be linearly independent. The Gram-Schmidt orthogonalization process was then developed for linearly independent sets, and from this procedure necessary and sufficient conditions for a set to be linearly independent were derived.

Schmidt then considered convergence. If $\{z^n\}$ is a sequence of vectors of H, then $\{z^n\}$ is defined to converge strongly in H to z if $\lim_{h \to \infty} \|z^n - z\| = 0$, and $\{z^n\}$ is said to be a strong Cauchy sequence if $\lim \|z^p - z^n\| = 0$ independently in p and n. He then showed that every strong Cauchy sequence in H converges strongly to some element of H. Then the nontrivial concept of a closed subspace A of H was introduced. Schmidt showed how such subspaces could be constructed and then proved the projection theorem: If z is a vector H and A is a closed subspace of H, then z has a unique representation $z = a + w$ where a is in A and w is orthogonal to every vector in A. Furthermore, $\|w\| = \min \|z - y\|$ where y is any element of A, and this minimum is actually assumed only for $y = a$. Finally, these results were used to establish necessary and sufficient conditions under which the infinite system of equations

$$\sum_{p=1}^{\infty} a_{np} z_p = c_n$$

has a square summable solution $\{z_p\}$ where $\{c_n\}$ is a square summable sequence and, for each n, $\{a_{np}\}$ is also square summable. He then obtained specific representations for the solutions.

Schmidt's work on Hilbert space represents a long step toward modern mathematics. He was one of the earliest mathematicians to demonstrate that the ordinary experience of Euclidean concepts can be extended meaningfully beyond geometry into the idealized constructions of more complex abstract mathematics.

NOTES

1. The set $\{\phi_p\}$ is orthonormal if

$$\int_a^b (\phi_p(s))^2 \, ds = 1 \quad (p = 1, 2, \cdots)$$

and

$$\int_a^b \phi_p(s)\phi_q(s) \, ds = 0 \quad (p \neq q).$$

2. The form $K(x, x)$ is completely continuous at a if

$$\lim_{h\to\infty} \epsilon_i^{(h)} \to 0 \quad (i = 1, 2, \cdots)$$

implies that

$$\lim_{h\to\infty} K(a + \epsilon^{(h)}, a + \epsilon^{(h)}) = K(a, a)$$

where $a = (a_1, a_2, \cdots)$ and $\epsilon^{(h)} = (\epsilon_1^{(h)}, \epsilon_2^{(h)}, \cdots)$. In a Hilbert space this is stronger than ordinary continuity (in the norm topology).

3. The sequence of (complex) numbers $\{b_n\}$ is square summable if

$$\sum_{n=1}^{\infty} |b_n|^2 < \infty.$$

BIBLIOGRAPHY

I. ORIGINAL WORKS. A complete bibliography of Schmidt's works can be found in the obituary by Kurt Schröder in *Mathematische Nachrichten,* **25** (1963), 1–3. Particular attention is drawn to "Zur Theorie der linearen und nichtlinearen Integralgleichungen. I," in *Mathematische Annalen,* **63** (1907), 433–476; "Zur Theorie . . . II," *ibid.,* **64** (1907), 161–174; and "Über die Auflösung linearen Gleichungen mit unendlich vielen Unbekannten," in *Rendiconti del Circolo matematico di Palermo,* **25** (1908), 53–77.

II. SECONDARY LITERATURE. On Schmidt and his work, see Ernst Hellinger and Otto Toeplitz, "Integralgleichungen und Gleichungen mit unendlichvielen Unbekannten," in *Encyklopädie der Mathematischen Wissenschaften,* IIC, 13 (Leipzig, 1923–1927), 1335–1602. This article, also published under separate cover, is an excellent general treatise, and specifically shows the relationship between integral equation theory and the theory of infinite linear systems.

MICHAEL BERNKOPF

SCHMIDT, ERNST JOHANNES (*b.* Copenhagen, Denmark, 2 January 1877; *d.* Copenhagen, 21 February 1933), *marine biology.*

Schmidt was the son of Ernst Schmidt, an estate inspector, and Camilla Ellen Sophie Johanne Kjeldahl. His father died when he was seven. Schmidt eventually entered the University of Copenhagen to study botany and in 1898 received the M.Sc. and in 1903 the Ph.D. for a paper on the prop roots of the mangrove. From 1899 to 1909 he was attached to the Botanical Institute of the university, and from 1910 he served as director of the Carlsberg Physiological Laboratory in Copenhagen. Schmidt published numerous botanical papers, particularly on tropical faunas and marine plants; and with F. Weis he wrote a textbook on bacteria.

From 1899 Schmidt was a member of the Danish Commission for the Investigation of the Sea. He went on several marine biological expeditions in the North Atlantic, and his fame rests chiefly on his research there. Schmidt's interest soon turned to the larval development of fishes, and his first paper on this subject was published in 1904. In the same year he investigated eel larvae and was led into the study of the breeding of eels.

At that time the life cycle of eels was a complete mystery. The larvae until only recently had been described as a separate species; and the breeding grounds were unknown. From his research in the North Atlantic and in the Mediterranean (1908–1910), Schmidt hypothesized that the European eel has a common breeding ground in the Atlantic and that it belongs to a single population. To prove this theory, Schmidt led an expedition to the North Atlantic from 1920 to 1922. By tracing the area where the youngest larvae were found, he eventually located the breeding grounds in the Sargasso Sea. He was able to show that adult eels from all of western Europe and the Mediterranean migrate to this place and die after spawning. The larvae then migrate to their parents' adult habitats. (He showed also that American eels have a similar pattern and a nearby breeding ground.) This peculiar pattern had profound implications not only in biology but also in the discussion of continental drift.

To extend his theory to all eels, Schmidt received a grant from the Carlsberg Foundation to lead an expedition around the world (1928–1930). During this time, he collected material on other species of fishes as well, especially deepwater fishes.

Schmidt also studied the eel-like *Zoarces viviparus.* He showed that this species breeds locally and identified differences between the various populations. Later he extended these observations to other species and made a number of important studies of environmental effects on the size and shape of many populations. Schmidt's biometric findings contributed to the concept of the inter-breeding populations as a fundamental unit. Since most of this research was done on economically important species of fishes, his work was also of great value to fisheries.

Schmidt's research was highly appreciated during his lifetime, and he received many academic and public honors. He also took an active role in several international organizations concerned with the sea, especially the International Permanent Council for the Study of the Sea, which had its seat in Copenhagen.

In 1903 Schmidt married Ingeborg Kühle, daughter of the director of the Old Carlsberg Breweries. Their home was both a social and scientific center.

BIBLIOGRAPHY

I. ORIGINAL WORKS. A complete list of Schmidt's scientific publications is in Regan (see below). His major works are *Bakterierne. Naturhistorisk Grundlag for det baktereologiske Studium* (Copenhagen, 1899–1901), written with F. Weis, German ed., M. Porsild, trans. (Jena, 1902); "Danish Researches in the Atlantic and Mediterranean on the Life-History of the Fresh-Water Eel (*Anguilla vulgaris*)," in *Internationale Revue der gesamten Hydrobiologie u Hydrographie*, **5** (1912), 317–342; and "The Breeding Places of the Eel," in *Philosophical Transactions of the Royal Society*, **205** (1922), 179–208.

II. SECONDARY LITERATURE. The best biography is C. Tate Regan, "Johannes Schmidt," in *Journal du Conseil*, **8** (1933), 145–160, with complete bibliography.

NILS SPJELDNAES

SCHMIDT, GERHARD CARL NATHANIEL (*b.* London, England, 5 July 1865; *d.* Münster, Germany, 16 October 1949), *physical chemistry.*

Although born in England, Schmidt was of German ancestry. From 1886 he obtained his higher education at Tübingen, Berlin, Strasbourg, Greifswald, and Basel. In 1891, under Georg Kahlbaum's guidance, he received the Ph.D. at Basel. Schmidt's subsequent work on solutions, mixtures, and adsorption led to a lifelong interest in physical chemistry. In 1895 he received a *Dozentur* in Eilhard Wiedemann's small but exceptionally lively institute at the University of Erlangen. Schmidt often worked closely with Wiedemann himself; he studied luminescence, phosphorescence, and photoelectric phenomena. During a brief

excursion into another area in late 1897 or early 1898, he made the discovery for which he is most famous—the radioactivity of thorium. (Marie Curie soon made the same discovery independently.)

Schmidt made this discovery while examining "many elements and compounds" in an effort to determine whether any of the rays that were emitted bore a resemblance to those that Henri Becquerel had found emerging from uranium and uranium compounds. He located only one such element, thorium, and immediately conducted absorption, ionization, reflection, refraction, and polarization studies to determine the characteristics of its rays. Having combined a misinterpretation of Becquerel's with one of his own, Schmidt concluded that thorium rays most resembled Röntgen rays—a conclusion that soon required revision in view of the researches of Marie Curie and Ernest Rutherford.

In 1900 Schmidt became professor ordinarius of physics at the Forstakademie in Eberswalde but soon moved to Erlangen as professor extraordinarius (1901–1904). He then went to Königsberg as professor ordinarius and director of the physical institute (1904–1908) and finally to Münster, where he occupied the chair once held by Hittorf, whom Schmidt admired and commemorated in several addresses. Schmidt retired from this post in 1935.

During these years, Schmidt studied canal-ray and cathode-ray phenomena, the electrical conductivity of salt vapors, solid electrolytes, adsorption, passivity, and luminescence. He was unusually vigorous and healthy until the last year of his life, when he fractured his hipbone and was hospitalized. Shortly after he was released, he suffered a stroke and died.

BIBLIOGRAPHY

I. ORIGINAL WORKS. Schmidt's writings include "Ueber die von den Thorverbindungen und einigen anderen Substanzen ausgehende Strahlung," in *Annalen der Physik*, **65** (1898), 141–151; "Wilhelm Hittorf," *Schriften der Gesellschaft zur Förderung der Westfälischen Wilhelms-Universität zu Münster*, no. 4 (1924); and "Eilhard Wiedemann," in *Physikalische Zeitschrift*, **29** (1928), 185–190.

II. SECONDARY LITERATURE. On Schmidt and his work, see Lawrence Badash, "The Discovery of Thorium's Radioactivity," in *Journal of Chemical Education*, **43** (1966), 219–220. For obituary notices see A. Kratzer, *Physikalische Blätter*, **6** (1950), 30; and K. Kuhn,

Naturwissenschaftliche Rundschau, **4** (1951), 41. Schmidt's portrait is in J. A. Barth, *Deutsche Senioren der Physik* (Leipzig, 1936).

ROGER H. STUEWER

SCHMIDT, JOHANN FRIEDRICH JULIUS (*b.* Eutin, Germany, 26 October 1825; *d.* Athens, Greece, 7 February 1884), *astronomy, geophysics.*

Schmidt was the son of Carl Friedrich Schmidt, a glazier, and Maria Elisabeth Quirling. He received his early education in Eutin and Hamburg, where his interest in the sciences and his aptitude for drawing were encouraged. From 1842 he studied practical astronomy under Carl Rümker at the Hamburg observatory. Then, in 1845, he went to Benzenberg's private observatory in Bilk, near Düsseldorf. The following year, he assisted Argelander at the latter's observatory in Bonn, where he became an accomplished astronomer.

With Argelander's recommendation, Schmidt was named observator at the observatory of the canon and provost E. von Unkrechtsberg in Olmütz (now Olomouc), where he remained from 1853 to 1858. He was then named director (1858) of the new observatory founded by Baron Sica in Athens. Schmidt worked there until his death. He continued the observations that he had begun in Olmütz and also observed comets, variable stars, nebulae, sunspots, and the zodiacal light. He later studied volcanic and seismic phenomena but is known chiefly for his selenographic observations.

Although Schmidt's research was voluminous, he lacked the means to publish it. But through the intervention of the German ambassador to Greece, this work was placed in the Potsdam observatory.

Schmidt never married. He was well liked and respected at the Greek court; and the evening before his death, he attended the usual social gathering at the German Embassy in Athens. Schmidt received many honors and in 1868 was awarded the Ph.D. *honoris causa* by the University of Bonn.

BIBLIOGRAPHY

Schmidt's major works are *Das Zodiakallicht* (Brunswick, 1856); *Resultate aus 11-jährigen Beobachtungen der Sonnenflecken* (Vienna, 1857); *Physikalische Geographie von Griechenland* (Athens, 1869); *Vulkanstudien* (Leipzig, 1874); *Studien über Erdbeben* (Leipzig, 1875; 2nd ed., 1879); *Karte der Gebirge der Mondes nach eigenen Beobachtungen 1840–1874* (Berlin, 1878); and *Erläuterungsband* (Berlin, 1874).

The only useful secondary source is A. Krueger's obituary in *Astronomische Nachrichten,* **108** (1884), 129.

H.-CHRIST. FREIESLEBEN

SCHMIEDEL. See Schmidel, Casimir Christoph.

SCHNEIDER, FRIEDRICH ANTON (*b.* Zeitz, Germany, 13 July 1831; *d.* Breslau, Germany [now Wrocław, Poland], 30 May 1890), *zoology, comparative anatomy, cytology.*

Schneider's zoological interests were in morphology and systematization. After years of studying the roundworms and flatworms, he reported in an 1873 paper his various laboratory observations of the life history of the Platyhelminthes. The paper contains the first description of the process of cell division and the visible changes during its successive stages—a detailed account of Schneider's microscopic investigations, with drawings of the nucleus and the chromosomal strands as he had seen them in the flatworm *Mesostomum ehrenbergii.* Following Schneider's discovery, the phenomena of division were independently observed and reported on by Fol and Bütschli. Hertwig saw the fusion of the maternal and paternal nuclei in fertilization, and over the next years there developed a new understanding of the cell, the process and significance of fertilization, and the role of the chromosomes in inheritance.

Schneider was the son of Karl Friedrich Schneider, a merchant, and Friederike Wilhelmine Müller. He was frail and occasionally ill, and his schooling at the small Gymnasium at Zeitz was somewhat irregular. His mother died when he was young, and his father remarried. Schneider's father often took him on business trips and imbued his son with his own lively interest in literature and the arts. Country visits intensified the youth's love of nature, and at home there was a well-rounded library.

At the age of eighteen Schneider entered the University of Bonn, where he concentrated at first in mathematics and the natural sciences, but he increasingly leaned toward zoology, stimulated by the lectures he heard in that field. In 1851 he continued his studies at Berlin, where he came under the influence of a great teacher, Johannes Müller, whose laboratory provided a formative experience for so many students. In 1854 Schneider received the doctorate in philosophy at the University of

Berlin, but much of the next year was spent in Zeitz. His father had died and he had to assume family responsibilities.

Müller often took his students on field trips, and in 1855 Schneider accompanied him to Norway via Copenhagen. During the return voyage, their ship sank after a collision and explosion. The two were rescued, but another student of Müller's was drowned. Schneider visited Italy in 1856–1857; Naples and Messina provided unique opportunities for marine biological studies. In 1859 he habilitated at the University of Berlin and then taught as *Privatdozent*. He also served as a custodian in the zoological museum, where he worked especially on the nematode collections. He made many friends among his colleagues at the university and became well acquainted with Nathaniel Pringsheim, while both were working with the small marine forms in Helgoland, in the North Sea.

Schneider succeeded Leuckart as professor at the University of Giessen in 1869. He truly enjoyed lecturing and teaching, and he filled his laboratory with freshwater specimens from the nearby lakes and streams. Students gathered about him, and he also gave a number of lectures on various topics at meetings of the local scientific society, the Oberhessische Gesellschaft für Natur- und Heilkunde. In the society's *Bericht* appeared Schneider's "Untersuchungen über Plathelminthen," with the observations of cell division. Although he may not then have fully realized the implications of his studies of cell division, Schneider did recognize, even in 1873, that the phenomena he was describing were significant and that they opened up a new understanding of the cell.

The years he spent at Giessen were full. He even volunteered to care for the sick during the Franco-Prussian War and actually spent some time in France. He was rector of the university until 1881, when he was appointed professor of zoology and comparative anatomy at the University of Breslau. He became rector at Breslau in 1886, and it was also there that he married. Schneider founded the *Zoologische Beiträge,* directed the zoological institute, and taught actively until his death in 1890.

Schneider's major contribution grew out of his work on the flatworm *Mesostomum,* an ideal subject because of its transparency. He not only followed the life cycle of the living specimen, but using acetic acid to fix his microscopic sections he rendered the changes in the nucleus and chromosomes visible. Staining with carmine, for which the chromosomes showed an affinity, made them still more visible. The earliest mention of his using this stain is 1880, but how much earlier he had employed it is unknown.

Using acetic acid, Schneider saw that, contrary to accepted belief, the outline of the nucleus did not disappear. He saw the nucleolus disappear and a mass of filaments take form; the filaments arranged themselves on the equatorial plane of the cell, seemed to thicken and increase, and then—in an orderly and typical arrangement—one part of the filaments went to one pole; the other part, to the other pole. The cell was divided, and each of the two new cells again exhibited a fluid-filled nucleus and a nucleolus. Schneider then suggested that the nucleus might persist similarly in other forms in which it had been thought to disappear during division, and that the transitions of the chromosomes might be quite general in occurrence. He concluded that cell division might take place with or without these "metamorphoses."

Schneider's observations of mitosis remain outstanding, and he made some lasting contributions to morphology; but his intuitive conclusions were sometimes less sure. A friend later recalled that Schneider preferred the spoken to the written word; and he was known to like working out problems in his head. He eschewed writing whenever possible, and the various papers he did publish provide an incomplete view of the extent of his work and record errors that he later corrected.

In the years after his description of division, Schneider seemed to his colleagues to oppose the very advances in the understanding of the cell and its life processes that he had foretold. The priority of his 1873 observations was acknowledged, but he had received little notice then because he published his results in a paper on observations of the flatworm and systematics, in a journal that was not widely read. Then, as new researches clarified the process of fertilization and the division stages that Schneider had seen in the summer eggs of *Mesostomum,* he persisted in his own interpretations. Schneider thought that the spermatozoon breaks up or disappears after its entrance into the ovum; thus he did not agree with Hertwig that fertilization signifies a fusion of nuclei. The implications for the understanding of heredity of the view Schneider expressed are apparent, and his contemporaries in cytological investigation felt his concept to be regressive. Schneider in turn took on other investigations, and it is not known whether he ever changed his mind.

He left an important monograph on the nematodes, papers on a range of zoological subjects in the area of morphology, and numerous descriptions that were cited by other comparative anatomists who followed him.

BIBLIOGRAPHY

I. ORIGINAL WORKS. Schneider's most important writings are "Untersuchungen über Plathelminthen," in *Bericht der Oberhessischen Gesellschaft für Natur- und Heilkunde,* **14** (1873), 69–140; see esp. 113–116 and pl. V, fig. 5, a–f, for description of cleavage as Schneider saw it in the summer eggs of *Mesostomum.* See also "Über Befruchtung," in *Zoologischer Anzeiger,* **3** (1880), 252–257; "Über Befruchtung der thierischen Eier," *ibid.,* 426–427; *Das Ei und seine Befruchtung* (Breslau, 1883); and *Monographie der Nematoden* (Berlin, 1866; Farnborough, Hampshire, England, 1968).

II. SECONDARY LITERATURE. On Schneider and his work, see the following articles by Wulf Emmo Ankel: "Anton Schneider, 1831–1890," in Hugo Freund and Alexander Berg, eds., *Geschichte der Mikroskopie, Leben und Werk Grosser Forscher,* I (Frankfurt am Main, 1963), 303–311; "Anton Schneider, ein Bild und ein Nachruf," in *Bericht der Oberhessischen Gesellschaft für Natur- und Heilkunde zu Giessen, naturwissenschaftliche Abteilung,* n.s. **28** (1957), 163–185; and "Zur Geschichte der wissenschaftliche Biologie in Giessen," in *Ludwigs-Universität Justus Liebig Hochschule 1607–1957 . . . Giessen, Festschrift zur 350 Jahrfeier* (Giessen, 1957), 327–328.

Friedrich Keller, "Anton Schneider und die Geschichte der Karyokinese" (inaugural M.D. diss., Freiburg im Breisgau: Albert-Ludwigs University, 1926), presents biographical material and describes Schneider's studies and the conclusions other investigators were reaching meanwhile on cell division. For an evaluation of Schneider's work by a contemporary whose own researches were basic to cytology, see Oskar Hertwig, *Dokumente zur Geschichte der Zeugungslehre. Eine historische Studie als Abschluss eigener Forschung* (Bonn, 1918), 7–8, 58–59; see p. 9 for reproductions of Schneider's figures of division in *Mesostomum.* M. J. Sirks, "The Earliest Illustrations of Chromosomes," in *Genetica,* **26** (1952), 65–76, shows the chromosomes as they appeared to nineteenth-century investigators, and it discusses Schneider's observations in this context, including the above-cited drawings of egg cleavage and those of spermatocyte division. Schneider's contribution is assessed also in John R. Baker, "The Cell Theory: A Restatement, History and Critique, Part V, the Multiplication of Nuclei," in *Quarterly Journal of Microscopical Science,* 3rd ser., **96** (1955), 463; John A. Moore, *Heredity and Development* (New York, 1963), 22–23; (the last two sources have Schneider's landmark illustrations); and William Coleman, "Cell, Nucleus, and Inheritance: An Historical Study," in *Proceedings of the American Philosophical Society,* **109** (1965), 131.

GLORIA ROBINSON

SCHNEIDERHÖHN, HANS (*b.* Mainz, Germany, 2 June 1887; *d.* Sölden, near Freiburg im Breisgau, Germany, 5 August 1962), *geology.*

Schneiderhöhn possessed great knowledge of ore deposits and, with Joseph Murdoch and Rudolf Willem van der Veen, was one of the classical authors on ore microscopy. After graduating from the Gymnasium in Mainz, he studied geology and mineralogy at Freiburg, Munich, and especially Giessen, under the petrologist Erich Kaiser. After receiving his doctorate and spending a short time as an assistant to Kaiser, he was for several years the assistant to Theodor Liebisch, then professor of mineralogy at the University of Berlin. In 1913 Schneiderhöhn became the mineralogist for the Tsumeb mine in South-West Africa, which was then working the boundary between the spectacular oxidation zone and the enriched and primary sulfides. Forced by the war to remain in South-West Africa from 1914 to 1918, he began his major works on ore microscopy. He had only extremely primitive equipment, but the difficulties he experienced made him familiar with many technical and theoretical problems.

Schneiderhöhn made very careful studies of the primary mineral content of the Tsumeb deposit and its relations to the secondary enrichment. While doing this he discovered "Rosa Erz," now known as germanite, and made important observations of the unusual oxidation minerals of Tsumeb, as well as of the karst phenomena and the petrology of Otavi highlands. In addition, he explored other parts of South-West Africa, working with Ernst Reuning, and gained a profound knowledge of the extremely varied types of ore deposits.

In 1919 Schneiderhöhn was appointed to the chair of mineralogy at Giessen, and five years later he succeeded F. Klockmann at Aachen. In 1926 he accepted the chair of mineralogy at Freiburg im Breisgau, where he taught and worked until his retirement in 1955. He later moved to Sölden, in the Black Forest, where he continued his scientific work.

Schneiderhöhn examined many deposits in Europe, Africa, North America, and Turkey. His experience, memory, knowledge of the literature, and sharp and critical intellect gave him an intuitive sense of similarities, differences, and weak points in earlier opinions. He sometimes offered his interpretations and opinions—and published them—after a very short time, perhaps a visit of one or two days. Mistakes and oversimplifications were unavoidable; but generally, and especially in the most important cases (such as the Northern Rhodesian [now Zambian] copper deposits) he was right and his statements were later confirmed.

Friendships with Paul Niggli and (since his Berlin years) with Max Berek, who made major contributions to the optics of reflected light, were very fruitful for all concerned. Niggli's *Versuch einer natürlichen Klassifikation der im weiteren Sinne magmatischen Erzlagerstätten* was greatly influenced by discussions with Schneiderhöhn, and Berek's papers often dealt with topics suggested by Schneiderhöhn.

Careful study of the Manfeld copper shale convinced Schneiderhöhn that it was of sedimentary origin and had been formed in a manner similar to that of all black shales. The tiny globules of framboidal form were explained as "mineralized bacteria." This idea gave rise to strong discussion; but whether "bacteria" or other primitive organisms, the globules are surely organic. His discussion of these "Schwefelkreislauf" became a basic idea of geology.

Schneiderhöhn gradually became convinced that the purist magmatic ideas, such as those of Niggli and Louis Caryl Graton, were untenable in many cases where undoubtedly "hydrothermal" deposits could not be connected with magmatism. He suggested that such deposits could be derived from superficial waters, heated by some means, from much older deposits. This idea explains many enigmatic deposits—in northern Algeria, Tunisia, and Morocco, and lead-zinc-silver veins surrounding the much older Broken Hill—but probably fails elsewhere.

Schneiderhöhn presented his experience and tremendous knowledge of the literature in many books. *Die Lagerstätten der magmatischen Abfolge,* the first volume of his *Lehrbuch der Erzlagerstätten,* appeared in 1941 but had a very limited circulation; most of the stock, still in the publisher's office, was destroyed by the first bombing of Berlin. He then prepared *Erzlagerstätten Kurzvorlesungen,* a comprehensive introduction to the science of ore deposits. It appeared in several editions in German and has been translated into English and other languages. In the 1950's Schneiderhöhn began *Die Erzlagerstätten der Erde,* of which the first volume, *Die Erzlagerstätten der Frühkristallisation,* appeared in 1958. It contains excellent and critical descriptions and comparisons. The second volume, *Die Pegmatite,* appeared soon after his death. Much data had also been collected for the four remaining volumes that he had planned.

BIBLIOGRAPHY

Schneiderhöhn's works are *Lehrbuch der Erzlagerstättenkunde,* I (Jena, 1941); *Erzlagerstätten Kurzvorlesungen zur Einführung und zur Wiederholung* (Jena, 1944); *Die Erzlagerstätten der Fruhkristallisation* (Stuttgart, 1958); and *Die Erzlagerstätten der Erde,* II (Stuttgart, 1961).

On Schneiderhöhn and his work, see K. F. Chudoba, "Prof. Dr. phil. Hans Schneiderhöhn," in *Aufschluss,* **14** (1963), 106–107; D. Di Colbertaldo, "Nachruf für das Mitglied auf Lebenszeit Hans Schneiderhöhn," in *Rendiconti della Società mineralogica italiana.* **20,** 51–54; K. R. Mennert, "In Memoriam Hans Schneiderhöhn," in *Neues Jahrbuch für Mineralogie. Monatshefte* (1962), 245–246; and "Festband Hans Schneiderhöhn zum 70. Geburtstag," in *Neues Jahrbuch für Mineralogie. Abhandlungen,* **91** (1957).

P. RAMDOHR

SCHOENFLIES, ARTHUR MORITZ (*b.* Landsberg an der Warthe, Germany [now Gorzów, Poland], 17 April 1853; *d.* Frankfurt am Main, Germany, 27 May 1928), *mathematics, crystallography.*

Schoenflies studied with Kummer at the University of Berlin from 1870 to 1875 and received the Ph.D. in 1877. From 1878 he taught at a Gymnasium in Berlin and then, from 1880, in Colmar, Alsace. In 1884 he earned his *Habilitation* as *Privatdozent* at the University of Göttingen, where, in 1892, he was named professor extraordinarius and was given the chair of applied mathematics. (This chair had been created thanks to Felix Klein's initiative.) In 1899 Schoenflies was appointed professor ordinarius at the University of Königsberg and then, in 1911, at the Academy for Social and Commercial Sciences in Frankfurt am Main; this school became a university in 1914. He was later professor ordinarius (1914–1922) at the University of Frankfurt and in 1920–1921 served as rector of the university.

Schoenflies produced an extensive mathematical *oeuvre* consisting of about ninety papers and many reports and books. He started his scientific work with rather traditional geometry and kinematics. This research was published in 1886 (1) and was later translated into French (1*a*). In the same year, under Klein's influence, Schoenflies turned to Euclidean motion groups and regular space divisions. His investigations culminated in 1891 in his magnum opus (2). The result of this book, the 230 crystallographic groups, was at the same time obtained independently by E. S. Fedorov. During the last phase of this research, Schoenflies corresponded with Fedorov and was thus able to correct some minor errors that he had originally made in his classification. In 1923 Schoenflies reedited his 1891 publication under another title (2a). He also wrote a textbook on crystallography (9).

In the mid-1890's Schoenflies, by then in his forties, turned to topology and set theory. In 1898 he published an article (5) on this subject in the *Encyklopädie der mathematischen Wissenschaften*. He also published extensive reports in *Deutsche Mathematiker-Vereinigung*, which appeared in 1900 and 1908 (6) and were reedited in 1913 (6a). These reports were totally eclipsed by Hausdorff's *Grundzüge der Mengenlehre* (1914). The greater part of Schoenflies' original contributions to topology is contained in three papers (7) and is devoted to plane topology. He proved the topological invariance of the dimension of the square, and he invented the notions and theorems that are connected with the characterization of the simple closed curve in the plane by its dividing the plane into two domains of which it is the everywhere attainable boundary. There are numerous gaps and wrong statements in this part of Schoenflies' work, and these errors led L. Brouwer to some of his startling discoveries.

Schoenflies published four articles in the *Encyklopädie der mathematischen Wissenschaften* (on set theory, kinematics, crystallography, and projective geometry), in part with others (5). With W. Nernst, he wrote a textbook (1895) on calculus (3) that went through at least eleven editions and two Russian translations. He also wrote textbooks on descriptive geometry (8) and analytic geometry (10). In 1895 he edited the work of Julius Plücker (4). Schoenflies was elected a fellow of the Bayerische Akademie der Wissenschaften in 1918.

BIBLIOGRAPHY

I. ORIGINAL WORKS. Schoenflies' works are the following:

(1) *Geometrie der Bewegung in synthetischer Darstellung* (Leipzig, 1886), with French trans. by C. Speckel as (1a) *La géométrie du mouvement—exposé synthétique* (Paris, 1893);

(2) *Kristallsysteme und Kristallstruktur* (Leipzig, 1891); the 2nd ed. appeared as (2a) *Theorie der Kristallstruktur* (Berlin, 1923);

(3) *Einführung in die mathematische Behandlung der Naturwissenschaften—Kurzgefasstes Lehrbuch der Differential- und Integralrechnung* (Munich, 1895; 11th ed., 1931), written with W. Nernst;

(4) Julius Plücker, *Gesammelte Mathematische Abhandlungen*, Schoenflies, ed. (Leipzig, 1895);

(5) "Mengenlehre," in *Encyklopaedie der mathematischen Wissenschaften*, 184–207; "Kinematik," *ibid.*, IV, 190–278, written with M. Grübler; "Kristallographie," *ibid.*, V, pt. 7, 391–492, written with T. Liebisch and O. Mügge; "Projektive Geometrie," *ibid.*, III, pt. 5, 389–480;

(6) "Die Entwicklung der Lehre von den Punktmannigfaltigkeiten. I," in *Jahresbericht der Deutschen Mathematiker-Vereinigung*, 8 (1900). 1–250; "Die Entwicklung . . . II," supp. 2 (1908), 1–331;

(6a) *Entwicklung der Mengenlehre und ihrer Anwendungen* (Leipzig, 1913), written with H. Hahn;

(7) "Beiträge zur Theorie der Punktmengen," in *Mathematische Annalen*, 58 (1903), 195–234; 59 (1904), 152–160; 62 (1906), 286–326;

(8) *Einführung in die Hauptgesetze der zeichnerischen Darstellungsmethoden* (Leipzig, 1908);

(9) *Einführung in die Kristallstruktur—ein Lehrbuch* (Berlin, 1923);

(10) *Einführung in die analytische Geometrie der Ebene und des Raumes*, Grundlehren der Mathematischen Wissenschaften no. 21 (Leipzig, 1925), with 2nd ed. by M. Dehn (Leipzig, 1931).

II. SECONDARY LITERATURE. On Schoenflies and his work, see L. Bieberbach, "Arthur Schoenflies," in *Jahresbericht der Deutschen Mathematiker-Vereinigung*, 32 (1923), 1–6; J. J. Burckhardt, "Zur Entdeckung der 230 Raumgruppen," in *Archives for History of Exact Sciences*, 4 (1967), 235–246; "Der Briefwechsel von E. S. Fedorow mit A. Schoenflies, 1889–1908," *ibid.*, 7 (1971), 91–141; R. von Mises, "Schoenflies," in *Zeitschrift für angewandte Mathematik und Mechanik*, 3 (1923), 157–158; A. Sommerfeld, "A. Schoenflies," in *Jahrbuch der bayerischen Akademie der Wissenschaften* (1928–1929), 86–87; and K. Spangenberg, "A. Schönflies," in *Handwörterbuch der Naturwissenschaften*, 2nd ed., VIII (1933), 1108–1109.

HANS FREUDENTHAL

SCHÖNBEIN, CHRISTIAN FRIEDRICH (*b.* Metzingen, Swabia [now West Germany], 18 October 1799; *d.* Sauersberg, near Baden-Baden, Germany, 29 August 1868), *physical chemistry*.

His family's financial condition would not permit any advanced schooling, and at the age of fourteen Schönbein became an apprentice in a chemical and pharmaceutical factory in Böblingen. He acquired a profound knowledge of theoretical and applied chemistry, and he also privately studied Latin, French, English, philosophy, and mathematics.

In April 1820 Schönbein obtained a post at the chemical factory of J. G. Dingler in Augsburg, where he assisted Dingler in making German translations of French publications for the new *Dinglers polytechnisches Journal*. In his spare time Schönbein studied chemistry and later, in 1820, accepted a post in the chemical factory of J. N. Adam in Hemhofen, near Erlangen. He frequently visited the University of Erlangen, where he met the philosopher Schelling; J. W. A. Pfaff, professor of physics and mathematics; and G. H. Schubert, the *Naturphilosoph* and professor of zoology. He remained friends with Schelling until the latter's death in 1854. After a semester, Schönbein moved to Tübingen but in February 1823 returned to Erlangen.

Although largely a self-educated chemist, Schönbein taught (1823) chemistry, physics, and mineralogy at Friedrich Froebel's institute in Keilhau, a small town near Rudolfstadt, in Thuringia. In 1826 he was in England, where he taught mathematics and natural philosophy at a boys' school in Epsom (1826), and in 1827 he went to France, where he attended the lectures of Gay-Lussac, Ampère, César Despretz, and Thenard. In 1828 Schönbein moved to Basel. He received the Ph.D. *honoris causa* from the faculty of philosophy there and also lectured on physics and chemistry. In 1835 he was appointed professor of physics and chemistry at the University of Basel, where he remained until his death. In 1852, when the professorship was divided, Schönbein retained the chair of chemistry. From 1848 he was a member of the Basel parliament.

Schönbein was also interested in philosophy. The influence of *Naturphilosophie* (he was also a friend of Lorenz Oken) is evident in all of Schönbein's work, especially in his speculative views that lack a sufficient experimental basis. Schönbein published more than 350 works, mostly qualitative, covering a wide range of research—but especially ozone, autoxidation, induced reactions, guncotton, electrochemistry, and passive iron. Schönbein's speculative bent is evident even in his early studies (from 1835 on) on the passivity of iron. He started from the well-known fact that iron reacts with dilute nitric acid but not with concentrated nitric acid. He sought to explain this phenomenon as a type of polymerism (he himself spoke always of isomerism) and thus disagreed with Faraday's explanation of a layer of oxide on the iron. Schönbein assumed that a conversion of the metallic iron takes place. Thus he agreed with Friedrich Fischer, professor of philosophy at Basel, who accounted for the passivity of iron by means of polarization, through which the pure attracting chemical affinity is changed into attraction and repulsion.

Schönbein thought that the iron particles possessed two "poles," one that attracts and one that repels (1838). But under certain conditions the poles that attract oxygen are directed to the inside of the metallic iron while those that repel oxygen are directed to the outside, thus producing passive iron. To account for this passivity Schönbein assumed that iron in the passive state possesses on its surface the properties of a "noble" metal. Consequently, he questioned the status of iron as an element: if iron can be converted into a noble metal, would all other metals possess the same property? He claimed that too little is known about the nature of matter and the workable forces in it to give definitive answers to these questions. Schönbein's reasoning was based on the analogies he drew from his 1835 lecture on isomerism, in which he stated that all known examples of isomerism (for example, tartaric and racemic acids, fulminic and cyanic acids) are dimorphic, with the exception of sulfur. He concluded that all dimorphic substances must be composite, and that sulfur is a compound.

Schönbein is known primarily for his work on ozone. While conducting experiments on the decomposition of water (autumn 1839), he noticed that the oxygen obtained in the process had a peculiar odor similar to that produced when a large electrical machine is operating—a similarity first noted by the Dutch chemist van Marum (1785). Schönbein recognized that the substance is a gas, that it is produced at the anode, and that it resembles chlorine and bromine in its chemical and electric properties. He studied extensively the properties of this gas and found (1844) that it is produced when phosphorus glows in air. He also discovered that it bleaches litmus, frees iodine from potassium iodide, and changes potassium ferrocyanide into ferricyanide.

Schönbein's ideas concerning the nature of ozone were rather confused. At first, he thought that nitrogen is composed of ozone and hydrogen. But phosphorus cannot decompose nitrogen ("ozone-hydrogen") unless another substance is present that can combine with the hydrogen. That substance is oxygen. Thus nitrogen is decomposed by phosphorus only in the presence of oxygen. The hydrogen in the nitrogen reacts with the oxygen of the water and ozone partly liberates and partly is bound with the phosphorus to "ozone-phosphorus." Like phosphorus trichloride, "ozone-phosphorus" is decomposed by water, namely into phosphorous acid and nitrogen. Schönbein saw a strong analogy between ozone and the halogens chlorine and bromine. Because the electrical, chemical, and physiological reactions of ozone closely resemble those of chlorine and bromine, he concluded that ozone also forms a salt and that its chemical affinity must place it directly after chlorine.

In 1845 Marignac and Auguste Arthur de la Rive proved independently that ozone is formed by an electric spark in pure, dry oxygen. They regarded ozone as oxygen in a particular state of chemical affinity. Although Schönbein persisted in his belief that ozone is a compound, he held that its oxidation state is higher than that of hydrogen, or even more likely, that it is a particular compound of water and oxygen. He suggested that there are three forms of oxygen: ozone, antozone, and ordinary oxygen, the last being a neutralization product of the first two. Similarly, Schönbein concluded that chlorine is a compound. Only in 1851 did he declare that, in all probability, ozone is an allotropic form of oxygen.

In the field of physical chemistry Schönbein studied the phenomenon of autoxidation: the spontaneous oxidation of a substance by atmospheric oxygen, part of which combines with the substance while a second part is converted into ozone or "antozone" (hydrogen peroxide) or combines with another substance. From various reactions, he concluded that ordinary oxygen is converted into ozone and antozone. Oxidation of phosphorus yields ozone; oxidation of metals in the presence of water yields antozone. Schönbein also studied induced reactions. To investigate this phenomenon he first used compounds of sodium sulfite and sodium arsenite; only the former is oxidized when exposed to air. But both sulfite and arsenite are oxidized when mixed and exposed to air. Thus he stated (1858) that the oxidation of sulfite "induces" that of arsenite.

From 1836 on, Schönbein's publications on voltaic current attributed the origin of this form of electricity to chemical action. His 1838 tendency theory stated that the tendency of two substances to combine with each other is sufficient to disturb their chemical equilibrium and to produce an electric current.

At the 11 March 1846 meeting of the Naturforschende Gesellschaft in Basel, Schönbein announced his discovery of guncotton. He also discovered collodion, a solution of guncotton in ether, which early found applications in medicine and photography. Schönbein produced guncotton by dipping cotton-wool in a mixture of fuming nitric and sulfuric acids and then washing and drying the product. His method of preparation remained a secret until it was discovered independently in 1846 by Böttger and Friedrich Julius Otto. Schönbein was also interested in the rise of dissolved materials from filter paper. His technique of capillary chromatography was based on the specific low rate of individual substances in nonimpregnated slips of paper. The height of ascent under standardized conditions and the time required for the ascent were recognized as characteristics of each individual substance present in the mixture under investigation. Schönbein's pupil Friedrich Goppelsroeder greatly extended the knowledge of the technique of capillary analysis.

Schönbein's general ideas on chemistry, and particularly on the phenomenon of catalysis, are discussed in *Beiträge zur physikalischen Chemie* (1844). His dynamical ideas are also emphasized. Schönbein was opposed to the atomic theory: he rejected the explanation of chemical combination as the basis for the formation of chemical substances. He thought that the qualitative changes in the formation of chemical substances indicated that every particle of a chemical element is a system of continuously working molecular forces. Schönbein sent a copy of his *Beiträge* to Faraday; and in a covering letter he pointed out that for years he had doubted the validity of the atomic theory and that he considered the molecule of a compound to be the "centre of physical forces."

Although Schelling's influence is clearly evident in the work of Schönbein, he was not strictly a *Naturphilosoph*. More than once he expressed himself against the views of Schelling and Oken and he passionately denounced Hegel and his school. Nevertheless, Schönbein's work is filled with speculative remarks lacking an adequate experimental basis and with excessive recourse to analogy.

BIBLIOGRAPHY

I. ORIGINAL WORKS. Schönbein's works include "Ueber das Verhalten des Zinns und des Eisens gegen Salpetersäure," in *Annalen der Physik und Chemie*, 2nd ser., **37** (1836), 390–399; "Ueber das Verhalten des Eisens zum Sauerstoff," ibid., **38** (1836), 492–497; *Das Verhalten des Eisens zum Sauerstoff. Ein Beitrag zur Erweiterung electro-chemischer Kenntnisse* (Basel, 1837); "Further Experiments on the Current Electricity Excited by Chemical Tendencies, Independent of Ordinary Chemical Action," in *London and Edinburgh Philosophical Magazine and Journal of Science*, **12** (1838), 311–317; "Beobachtungen über das electromotorische Verhalten einiger Metallhyperoxyde des Platins und des Eisens," in *Annalen der Physik und Chemie*, 2nd ser., **43** (1838), 89–102; "Notiz über die Passivität des Eisens," *ibid.*, 103–104; and "Beobachtungen über den bei der Electrolyse des Wassers und dem Ausströmen der gewöhnlichen Electricität sich entwickelnden Geruch," *ibid.*, **50** (1840), 616–635.

Later writings are *Ueber die Häufigkeit der Berührungswirkungen auf dem Gebiete der Chemie* (Basel, 1843); *Beiträge zur physikalischen Chemie* (Basel, 1844); *Ueber die Erzeugung des Ozons auf chemischen Wege* (Basel, 1844); "On the Nature of Ozone," in *London, Edinburgh, and Dublin Philosophical Magazine and Journal of Science*, **27** (1845), 386–389; "Einige Bemerkungen über die Anwesenheit des Ozons in der atmosphärischen Luft und die Rolle, welche es bei langsamen Oxidationen spielen dürfte," in *Annalen der Physik und Chemie*, 3rd ser., **65** (1845), 161–172; *Chemische Beobachtungen über die langsame und rasche Verbrennung der Körper in atmosphärischer Luft* (Basel, 1845); *Denkschrift über das Ozon* (Basel, 1849); *Ueber den Einfluss des Sonnenlichtes auf die chemische Thätigkeit des Sauerstoffs und den Ursprüng der Wolken-electricität und des Gewitters* (Basel, 1850); and *Ueber den Zusammenhang der katalytischen Erscheinungen mit der Allotropie* (Basel, 1856).

Schönbein's letters are collected in *Letters of Faraday and Schönbein, 1836–1862*, G. W. A. Kahlbaum and F. V. Darbishire, eds. (Basel–London, 1899); *Letters of Jöns Jacob Berzelius and Christian Friedrich Schönbein, 1836–1847*, G. W. A. Kahlbaum, ed., trans. by F. V. Darbishire and N. V. Sidgwick (London, 1900); and *Justus von Liebig und Christian Friedrich Schönbein. Briefwechsel 1853–1868*, G. W. A. Kahlbaum and E. Thon, eds. (Leipzig, 1900).

II. Secondary Literature. The best study of the life and work of Schönbein is G. W. A. Kahlbaum and E. Schaer, *Christian Friedrich Schönbein, 1799–1868, Ein Blatt zur Geschichte des 19. Jahrhunderts*, 2 vols. (Leipzig, 1899–1901). See also E. Färber, "Christian Friedrich Schönbeins Werk," in *Prometheus*, **29** (1918), 413–416; E. Hagenbach, *Christian Friedrich Schönbein* (Basel, 1868); R. E. Oesper, "Christian Friedrich Schönbein," in *Journal of Chemical Education*, **6** (1929), 432–440, 677–685; and the chap. on Schönbein in W. Prandtl, *Deutsche Chemiker in der ersten Hälfte des neunzehnten Jahrhunderts* (Weinheim, 1956), 193–241.

H. A. M. Snelders

SCHÖNER, JOHANNES (*b.* Karlstadt, Germany, 16 January 1477; *d.* Nuremberg, Germany, 16 January 1547), *astronomy, geography.*

On or after 18 October 1494, Schöner paid the full fee when he enrolled in the University of Erfurt,[1] where he studied theology. He left the university before he took a degree, was ordained a Roman Catholic priest, and served in Bamberg, where an astrological tract was addressed to him in October 1506. In 1509 he bought manuscript ephemerides for the years 1464–1484, and on 20 March 1518 he was paid for binding a book for his bishop.

Schöner assembled a printing shop in his house in Bamberg. He himself set the type, carved the woodblocks for the illustrations, and bound the finished product. He also made his own globes. His earliest terrestrial globe named the recently discovered continental mass "America," the first printed globe to do so. This globe was issued with his *Luculentissima quaedam terrae totius descriptio* (Nuremberg, 1515), which Schöner dedicated to his bishop on 24 March 1515. He likewise dedicated to the bishop his *Solidi et sphaerici corporis sive globi astronomici canones usum et expeditam praxim ejusdem exprimentes* (Nuremberg, 1517). In 1521, again using his own press in Bamberg, Schöner published his *Aequatorium astronomicum*, with movable disks to represent the motions of the planets. Having neglected to celebrate mass, Schöner was relegated to officiate at early mass in Kirchehrenbach. In this small village near Forchheim he printed on his own press his *De nuper . . . repertis insulis ac regionibus*, with the gores (triangular segments) for his globe of 1523. On 24 April 1525 he finished the last of his Kirchehrenbach books, his correction of a faulty Latin translation of a work by al-Zarqālī. The threat of the rebellious peasants to kill all Roman Catholic clergymen ended Schöner's career as a priest in 1525.

Fortunately, in 1526 Nuremberg opened the Melanchthon Gymnasium, where Schöner taught mathematics for two decades. He turned Lutheran and married Anna Zelerin on 7 August 1527.[2] The appearance of a comet in August 1531 impelled him to publish Regiomontanus' *De cometae magnitudine . . . problemata XVI* (Nuremberg, 1531). Thus began Schöner's valuable editions of many previously unpublished works by the greatest astronomer of the fifteenth century. As a zealous defender of astrology against its critics, Schöner printed on his own press his *Horoscopium generale, omni regioni accomodum* (Nuremberg, 1535).

After Schöner's death his mathematical works, all of which had been placed on the *Index of Prohibited Books*, were published by his son Andreas, with a portrait of the father at the age of sixty-nine (*Opera mathematica* [Nuremberg, 1551; 2nd edition, revised and enlarged, 1561], signature B4v).

NOTES

1. H. J. C. Weissenborn, ed., *Acten der Erfurter Universität,* which is Geschichtsquellen der Provinz Sachsen und angrenzender Gebiete, vol. 8, II (Halle, 1881–1899), 185, left column, line 10.
2. Karl Schornbaum, *Ehebuch von St. Sebald in Nürnberg 1524–1543* (Nuremberg, 1949), 91, no. 3043. Our "Johann Schöner" was confused with a Hanns Schonner, who married on 15 May 1537 (Schornbaum, 80, no. 2617); see Johannes Kist, *Die Matrikel der Geistlichkeit des Bistums Bamberg 1400–1556* (Würzburg, 1955–1960), 367, no. 5585; Veröffentlichungen der Gesellschaft für fränkische Geschichte, IV. Reihe; Matrikeln fränkischer Schulen und Stände, 7.

BIBLIOGRAPHY

There is no full-length biographical study of Schöner. His astronomical publications are listed in Ernst Zinner, *Geschichte und Bibliographie der astronomischen Literatur in Deutschland zur Zeit der Renaissance,* 2nd ed. (Stuttgart, 1964). For the works that Schöner printed on his own press in Bamberg, Kirchehrenbach, and Nuremberg, see Karl Schottenloher, "Johann Schöner und seine Hausdruckerei," in *Zentralblatt für Bibliothekswesen,* **24** (1907), 145–155; and Henry Stevens, *Johann Schöner,* Charles H. Coote, ed. (London, 1888), 149–170, with the Latin text of Schöner's *De nuper . . . repertis insulis ac regionibus,* pp. 47–55, and an English trans., pp. 91–99. The Latin text of this letter was printed also by Franz Wieser; see "Der verschollene Globus des Johannes Schöner von 1523," in *Sitzungsberichte der philosophisch-historischen Classe der k. Akademie der Wissenschaften in Wien,* **117,** no. 5 (1888), 15–18. Wieser's *Magalhães-Strasse und Austral-Continent auf den Globen des Johannes Schöner* (Innsbruck, 1881) was recently reprinted (Amsterdam, 1967). Frederik Caspar Wieder, *Monumenta cartographica,* I (The Hague, 1925), 1–4, deals with Schöner; and the book was reviewed by George E. Nunn, "The Lost Globe Gores of Johann Schöner, 1523–1524," in *Geographical Review,* **17** (1927), 476–480.

EDWARD ROSEN

SCHÖNFELD, EDUARD (*b.* Hildburghausen, Germany, 22 December 1828; *d.* Bonn, Germany, 1 May 1891), *astronomy.*

Schönfeld was educated at the Gymnasium in Hildburghausen. He then attended the Technische Hochschule at Hannover, from which he was expelled in 1849 because of his involvement in the political unrest of 1848. Schönfeld continued his technical studies at Kassel but found them unsatisfactory and began to study chemistry and astronomy at Marburg University in the autumn of 1849. In 1852 he went to Bonn University, where he was introduced to Argelander and became his assistant in 1853; the following year Schönfeld received his doctorate. In 1859 he was appointed director of the observatory at Mannheim. When Argelander died in 1875, Schönfeld succeeded him as director of the Bonn observatory and as professor of astronomy at the university. An active member of the Astronomische Gesellschaft, he served as secretary from 1875 until his death.

Schönfeld's name is very closely connected with the *Bonner Durchmusterung,* a catalog of all stars down to the ninth magnitude begun by Argelander. Schönfeld was one of his most enthusiastic collaborators and made a great many of the necessary observations using a telescope having an aperture of not more than about three inches. Schönfeld also performed most of the preparatory work for the publication of the results. After Argelander's death, he started work to expand the *Durchmusterung* to include more southern declinations; within ten years he completed the *Schönfeld-Durchmusterung,* which extended Argelander's *Durchmusterung* from declination −2° down to −23°. All observations for this work were made by Schönfeld himself. The great importance of these *Durchmusterungen* is best shown by the fact that in 1967, almost a century later, they were completely reprinted.

Schönfeld also observed nebulae, variable stars, and comets. He published the results of his observations of nebulae at Mannheim in two catalogs that appeared in 1862 and 1875.

BIBLIOGRAPHY

Schönfeld collaborated with Argelander and A. Krüger on "Bonner Sternverzeichnis," *Astronomische Beobachtungen auf der Sternwarte zu Bonn,* **3** (1859); **4** (1861); and **5** (1862). He was sole author of "Bonner Sternverzeichnis," *ibid.,* **8** (1886).

See also E. von der Pahlen, *Lehrbuch der Stellarstatistik* (Leipzig, 1937), 182, 341.

F. SCHMEIDLER

SCHÖNHERR, CARL JOHAN (*b.* Stockholm, Sweden, 10 June 1772; *d.* Sparresäter, near Skara, Sweden, 28 March 1848), *entomology.*

Schönherr was completely self-taught. From his father he inherited a silk factory in Stockholm, but after a few years he gave it up to settle on his estate, Sparresäter. There he was in convenient proximity to his entomological teacher, Leonhard Gyllenhaal. and their close collaboration was important.

At the age of twelve Schönherr began an insect collection, which soon developed into an ardent hobby; and he taught himself Latin in order to understand entomological literature. He was an extremely systematic person, who wanted order and reason in everything, including entomological synonyms and nomenclature. He was obliged to undertake these topics in order to clear up the chaos that long held sway in the field, and between 1806 and 1817 he published *Synonymia insectorum*, concerning beetles. This series was continued later (1833–1845) under the same title with another work on the weevils, Curculionidae. The work ran to eight volumes.

Schönherr intended to work on the whole insect class, but like other similar projects, the result was the publication of an intensive and penetrating treatment of a single group: the beetles. Schönherr was especially interested in the large family of weevils, and his exemplary descriptions of the genera, and the analyses of synonyms and explanations of nomenclature have remained of lasting value. Besides Gyllenhaal, he also collaborated with C. H. Boheman and other experts.

Schönherr's private library was one of the greatest entomological book collections of his time and was also an unusually beautiful collection (now preserved in the library of the Royal Swedish Academy of Sciences). His insect collection was as rich; it is said to have included some 13,500 species in about 37,700 specimens. By his will, the collections were transferred to the Naturhistoriska Riksmuseet in Stockholm.

BIBLIOGRAPHY

Schönherr's works are *Synonymia insectorum oder Versuch einer Synonymie aller bisher bekannten Insecten nach Fabricius Syst. Eleutheratorum geordnet, mit Berichtigungen und Anmerkungen, wie auch Beschreibungen neuer Arten und illuminirten Kupfern*, 3 vols. and appendix (Stockholm–Skara, 1806–1817); "Entomologiska Anmärkningar och beskrifningar på några för Svenska Faunan nya Insekter," in *Kungliga Svenska vetenskapsakademiens handlingar*, **30** (1809), 48–58; "Pulex segnis, ny Svensk species," *ibid.*, **31** (1811), 98–102; *Curculionidum dispositio methodica cum generum characteribus, descriptionibus atque observationibus variis, seu prodromus ad Synonymiae insectorum Partem IV* (Leipzig, 1826); and *Synonymia insectorum oder Versuch einer Synonymie aller von mir bisher bekannten Insecten, mit Berichtigungen und Anmerkungen, wie auch mit Beschreibung neuer Arten*, 8 vols. (Paris, 1833–1845).

BENGT-OLOF LANDIN

SCHONLAND, BASIL FERDINAND JAMIESON (*b*. Grahamstown, South Africa, 5 February 1896; *d*. Winchester, England, 24 November 1972), *atmospheric electricity, scientific administration.*

Schonland's father received the Ph.D. from the University of Kiel and was the first professor of botany at Rhodes University, Grahamstown. His mother was the daughter of a botanist. Schonland was the eldest of three sons. After graduating at Grahamstown, he failed to win a Rhodes scholarship to Oxford, and thus, in 1915, went instead to Gonville and Caius College, Cambridge. He interrupted his studies there to serve in World War I, working on signals. In 1919 he returned to Cambridge and after graduating in 1920, registered for the Ph.D. at the Cavendish Laboratory. At Cambridge he met Isabel Marion Craib, a fellow South African, whom he married when he returned to South Africa in 1923. They had a son and two daughters. In 1927–1928 he returned to Cambridge.

From 1922 to 1936 Schonland taught physics at the University of Cape Town and from 1937 to 1954 at the University of Witwatersrand. He also served as deputy director of the Atomic Energy Research Establishment at Harwell (1954–1958) and from 1958 to 1960 was its director. He was elected a fellow of the Royal Society of London in 1938 and in 1960 was knighted.

At the Cavendish, Schonland studied cathode rays; but after returning to South Africa, he investigated thunderstorms. He designed apparatus to photograph lightning discharges and confirmed Charles Wilson's theory that positive ions are carried to the top rather than to the bottom of the thundercloud. Schonland described this work in *Atmospheric Electricity* (1932); *The Flight of Thunderbolts* (1950), a popularized account; and in numerous papers.

In 1938 Cockcroft invited Schonland to join a group of scientists investigating radar. During World War II, he continued this work, ultimately becoming scientific adviser to General Montgomery. In 1945 Schonland returned to South Africa to direct the Council for Scientific and Industrial Research; but after the 1948 elections he found South Africa less congenial. He resigned the post in 1950 and in 1954 returned to England.

BIBLIOGRAPHY

Schonland's writings include *Atmospheric Electricity* (London, 1932; 2nd ed., 1953); *The Flight of Thunderbolts* (Oxford, 1950; 2nd ed., 1964); and *The Atomists, 1805–1933* (Oxford, 1968).

On Schonland and his work, see T. E. Allibone, "Sir Basil Schonland," in *Biographical Memoirs of Fellows of the Royal Society*, **14** (1973), 629–653, with portrait and full bibliography.

DAVID KNIGHT

SCHÖNLEIN, JOHANN LUCAS (*b.* Bamberg, Germany, 30 November 1793; *d.* Bamberg, 23 January 1864), *medicine.*

Schönlein was his country's leading clinician, and his methods and teaching gave a new direction to German medicine. His brief paper on favus, which recognized for the first time a fungus parasite as the cause of a disease in man, contributed importantly to the understanding of contagious disease.

Schönlein was the only child of Thomas Schönlein, a successful ropemaker, and Margaretha Hümmer, who intended him at first to follow in his father's occupation. But his scholarly interests soon became apparent and, largely through his mother's intercession, he attended the local Gymnasium and proceeded to university studies.

A lifelong collector, Schönlein received early encouragement from a Gymnasium teacher as he amassed specimens ranging from stones to insects. During visits to the country, he gathered plants. He had a continuing interest in paleobotany and collected fossil plants while touring Switzerland on vacations. In his reading on botany he especially admired Linnaeus and Linnaeus' system of classification of plants. Zoology, too, drew Schönlein's curiosity, and independently he dissected frogs and lizards.

Schönlein first studied the natural sciences in 1811 at the University of Landshut but later undertook the study of medicine. Tiedemann was his teacher of comparative anatomy. In 1813 he transferred to Würzburg, where, as at Landshut, *Naturphilosophie* still exerted an influence. But Döllinger, professor of anatomy and physiology, who provided Schönlein with a wealth of material illustrating developmental anatomy, stressed observation. Schönlein's dissertation on the metamorphosis of the brain, "Von der Hirnmetamorphose" (1816), reflected both approaches. After visits to Göttingen and Jena and after an interval during which he practiced medicine, Schönlein became

Privatdozent (1817) at Würzburg, lecturing in pathological anatomy.

In 1819 Schönlein was named provisional head of the medical clinic at the Julius Hospital in Würzburg. By 1824 he had been appointed ordinary professor of special pathology and therapy and was director of the clinic. He soon attracted students from throughout Europe. At the clinic percussion and auscultation, which had been introduced in France, were first routinely used in Germany as diagnostic aids. Blood, urine, and various secretions were examined under the microscope and chemical reagents were utilized; autopsies provided still further information. Schönlein's bedside teaching emphasized direct observation and careful reasoning. Each student followed the patients' symptoms and the courses of their diseases.

Schönlein's respect for the systems of classification used in the natural sciences led him to develop his "natural historical school." He distinguished his approach from the "natural philosophical school," and set forth his own "system." He was convinced that, in much the same way that botanists and zoologists applied their systems, he could classify pathological conditions according to their characteristics and symptoms and thus establish their relationships: three classes—morphae, haematoses, and neuroses—were subdivided into families, then groups.

At Würzburg, Schönlein married Theresa Heffner in 1827. Although he had been made an honorary citizen of Würzburg, he was forced to leave because of his liberal beliefs and associations. His academic appointment was rescinded in 1832. Having refused a post at Passau, he established a medical practice in Frankfurt but again had to leave.

In 1833 Schönlein became professor of medicine at Zurich, where a new hospital was built, providing him with fine facilities, and his fame grew. At Zurich he wrote two papers, which, besides his thesis, were his only publications. (His lectures were published by his pupils.) The first paper, on crystals in the urine of typhoid fever patients, was of little moment; but the second, a letter to Johannes Müller, occupying only a page in Müller's *Archiv* for 1839, described the cause of *porrigo lupinosa,* or favus: a minute parasitic fungus.

Schönlein's paper related the sources of his interest in favus, a disease typified by crusts that form on the scalp. Following Bassi's discovery, confirmed by Audouin, that the silkworm disease *muscardine* is caused by a microscopic parasitic fungus (Bassi suggested that certain diseases in man might be similarly caused), Schönlein examined infected silkworms, again confirming Bassi's findings. Schönlein also noted the botanist Franz Unger's *Die Exantheme der Pflanzen* (1833), where Unger described plant diseases in which fungi are present and suggested a parallel between plant exanthems "and like diseases of the animal organism." In this connection Schönlein recalled his view "of the plant nature of many an impetigo," and his examination of fragments of the "pustules" of favus obtained from his own patients revealed "the fungous nature" of the disease. His communication implicated a living parasitic plant organism as the cause of a disease in man, an important step toward the understanding of contagious disease; but he never published on the further investigations he promised. Henle noted Schönlein's finding in 1840 in his paper "Von den Miasmen und Contagien und von den miasmatisch-contagiösen Krankheiten" but pointed out that it had not yet been shown whether the fungus actually caused the pustules or merely grew on them.

Meanwhile, Schönlein had left Zurich in 1839 for political and personal reasons—he had not been made a citizen there because he was a Catholic. In 1840 he became professor of medicine at Berlin and director of the clinic at the Charité. His assistant, Robert Remak, who in 1837 had seen the organism that caused favus but had not then recognized it as a fungus, carried on the experiments and later named it *Achorion schoenleinii*. Gruby also discovered it independently, not knowing at first of Schönlein's 1839 paper.

At Berlin, Schönlein's lectures and clinical work made him a guiding influence in medicine. His many distinguished students had included Schwann at Würzburg; Billroth and Virchow were among his students at Berlin.

Schönlein left Berlin in 1859 to retire with his daughters in Bamberg; his only son had died the previous year on a botanical expedition, and his wife had died in 1846. At Bamberg he devoted time to his books and collections. Although he was interested in ethnology and other subjects, his fine collection of writings on epidemiology is evidence of his lasting concern with contagious disease.

BIBLIOGRAPHY

I. ORIGINAL WORKS. Schönlein's works are "Ueber Crystalle im Darmcanal bei Typhus abdominalis," in *Archiv für Anatomie, Physiologie und wissenschaftliche Medicin* (1836), 258–261; "Zur Pathogenie der Impetigines," *ibid.* (1839), 82, on favus.

II. SECONDARY LITERATURE. Schönlein's lectures, published by his pupils, appear in various eds., some noted in the sources below (see also the catalogs of the Library of the Surgeon General); they include *Allgemeine und specielle Pathologie und Therapie*, 4th ed. (St. Gallen, 1839). Rudolf Virchow, *Gedächtnissrede auf Joh. Lucas Schönlein* (Berlin, 1865), is a comprehensive biography and appreciation; he added notes in "Aus Schönlein's Leben," in *Archiv für pathologische Anatomie und Physiologie und für klinische Medicin,* **33** (1865), 170–174. A. Göschen, "Johann Lucas Schönlein," in *Deutsche Klinik,* **17** (1865), 29–32; and W. Griesinger, "Zum Gedächtnisse an J. L. Schönlein," in *Aerztliches Intelligenz-Blatt,* **11** (1864), 445–451 and in *Berliner klinische Wochenschrift,* **1** (1864), 276–279, stress his contributions as a clinician. J. Pagel, "Johann Lucas Schoenlein," in *Biographisches Lexikon hervorragenden Ärzte des neunzehnten Jahrhunderts,* J. Pagel, ed. (Berlin–Vienna, 1901), 1522–1524, and his "Johann Lucas Schoenlein," in *Allgemeine deutsche Biographie.* XXXII (Leipzig, 1891), 315–319, are general accounts.

Erwin H. Ackerknecht, "Johan Lucas Schoenlein, 1793–1864," in *Journal of the History of Medicine and Allied Sciences,* **19** (1964), 131–138, has a facs. of the paper on favus with a trans.

See also Walther Koerting, "Zum hundertjährigen Todestag von Johann Lukas Schoenlein," in *Bayerisches Ärzteblatt,* **19** (1964), 58–60; and W. Löffler's detailed study, "Johann Lucas Schönlein (1793–1864, Zürich 1833–1839) und die Medizin seiner Zeit," in *Zürcher Spitalgeschichte,* Regierungsrat des Kantons Zürich, II (Zurich, 1951), 2–89; Friedrich von Müller, "Johann Lukas Schönlein, Professor der Medezin, 1793–1864," in *Lebensläufe aus Franken,* Gesellschaft für Frankische Geschichte (Erlangen, 1936), 332–339; and Henry E. Sigerist, *The Great Doctors,* trans. by Eden and Cedar Paul (New York, 1958), 295–299, 336. W. Schönfeld, "Aus der Frühzeit der Pilzerkrankungen des Menschen, Jean Victor Audouin (1797–1839), Agostino Bassi (1773–1856), Franz Unger (1800–1870)," in *Deutsche medizinische Wochenschrift,* **82,** pt. 2 (1957), 1235–1237, reviews the background of the investigation of favus. For Schönlein's description of peliosis rheumatica, or Schönlein's disease, see R. H. Major, *Classic Descriptions of Disease,* 3rd ed. (Springfield, Ill., 1945), 225–227, which is taken from the lectures published in 1841.

GLORIA ROBINSON

SCHOOLCRAFT, HENRY ROWE (*b.* Albany County, now Guilderland, New York, 28 March 1793; *d.* Washington, D.C., 10 December 1864), *ethnology.*

Schoolcraft's father, Lawrence Schoolcraft, a descendant of English settlers in Canada and New York, was a glass manufacturer, who served in the

Revolutionary War and the War of 1812. Schoolcraft's mother was Margaret Ann Barbara Rowe Schoolcraft, who was of English descent. Schoolcraft was a natural scholar, who quite young gathered a small library of good books. He collected rocks and minerals, painted, and wrote poems and essays that were locally published. With private instruction he prepared for Union College, and attended it for a time. He later studied informally under Frederick Hall at Middlebury College, Vermont.

Devoted to his father, Schoolcraft readily joined him in the manufacture of glass in 1809; both men superintended factories at several localities in New York and New England. Under the influx of cheaper foreign imports, all of these businesses collapsed soon after the War of 1812.

In 1818 Schoolcraft set out westward to tour the lead mines of Missouri, where he hoped to become superintendent by federal appointment. His six-thousand-mile tour of the Mississippi valley, the Ozark Mountains, and the lead district, published in 1819 as *A View of the Lead Mines of Missouri*, brought him to the attention of Secretary of War John C. Calhoun, who in 1820 assigned Schoolcraft as naturalist on the Lewis Cass expedition from Detroit to the source of the Mississippi River. The expedition turned back before reaching its goal, but twelve years later, at the request of Cass, who was by then secretary of war, Schoolcraft continued the search to Lake Itasca, considered the river's source by the accompanying Indian guides.

Schoolcraft's major scientific accomplishment resulted from his appointment as Indian agent in 1822 at Sault Ste. Marie, Michigan Territory. The first and only agent in that post, for nineteen years he gathered notes on the Indian tribes, especially the Chippewa and their customs, languages, myths, songs, and history. In this work Schoolcraft was greatly aided by his wife, Jane Johnston Schoolcraft, granddaughter of an Ojibwa chief; her father, a cultured fur trader, also provided valuable help. From 1836 Schoolcraft was also acting superintendent of Indian affairs for the northwest. He negotiated a number of treaties on land and mineral rights, and, devoutly religious himself, encouraged missionaries and their Indian schools.

In 1841 Schoolcraft left Michigan for New York in order to supervise his many publications. In 1847 he returned to the Office of Indian Affairs in Washington to compile information on all American Indian tribes. For over ten years he edited the six-volume *Historical and Statistical Information Respecting the History, Condition and Prospects of the Indian Tribes of the United States*, an unassorted, lavishly illustrated compendium of material gathered by agents and staff of the Office of Indian Affairs and by himself. His second wife, Mary Howard Schoolcraft, helped him considerably in continuing the work despite his increasing paralysis. These books and Schoolcraft's other publications on American Indian lore were Longfellow's major reference sources for *The Song of Hiawatha*.

Schoolcraft was elected to the Lyceum of Natural History of New York and to the New-York Historical Society in 1820, upon the publication of his report on the lead district. He was a founder of the American Ethnological Society in 1842 and was awarded the LL.D. by the University of Geneva in 1846.

Schoolcraft's reports on mineral occurrences in both the Missouri and the Lake Superior region were among the earliest descriptions of mineral resources in the United States. He presented assays of lead ore, descriptions of the Missouri mines, distribution of outcrops, local geography, and notes on other resources. From his first report in 1819, privately published, he learned the value of publication. Henceforth he prodigiously contributed notes, observations, poems, and articles to newspapers, magazines of his own founding, and established journals; he also published books and government reports. He often repeated material, and he made little effort to organize his observations. His name became widely known, and he developed correspondences with distinguished scientists, including Benjamin Silliman (1779–1864), and with literary figures. His extended reports on the two upper Mississippi expeditions considerably influenced new settlement into Michigan Territory.

Almost the only predecessor of Schoolcraft in making ethnological notes on American Indians was Governor Lewis Cass of Michigan Territory, who provided the agents in his area with questionnaires on Indian observations, and who undoubtedly sparked Schoolcraft's interest. The latter's own fondness for poetry led him to record especially the American Indian songs, myths, and legends. His contributions to ethnology resulted chiefly from his having been an early observer in the field and his presenting widespread, written results from which later workers could synthesize, after the westward expansion and conflict made observations less reliable.

BIBLIOGRAPHY

I. ORIGINAL WORKS. Schoolcraft's monument to early ethnology is *Historical and Statistical Information . . . of the Indian Tribes of the United States,* I–V (Washington, D.C., 1851–1855), VI (Philadelphia, 1857), to which Frances S. Nichols compiled an index, which is *Bulletin of the Bureau of American Ethnology,* **152** (1954). In addition to Schoolcraft's numerous shorter works, many of his observations of American Indians were also included in *Personal Memories of a Residence of Thirty Years With the Indian Tribes on the American Frontiers* (Philadelphia, 1851). *Algic Researches,* 2 vols. (New York, 1839), includes Indian oral legends, compiled over many years. For the state of New York, Schoolcraft compiled *Notes on the Iroquois* (New York, 1846).

The descriptive accounts of his Mississippi River and Missouri trips are in *A View of the Lead Mines of Missouri* (New York, 1819), *Narrative Journal of Travels Through the Northwestern Regions of the United States . . . to the Sources of the Mississippi River* (New York, 1821), and *Narrative of an Expedition Through the Upper Mississippi to Itasca Lake the Actual Source of the Mississippi* (New York, 1834).

II. SECONDARY LITERATURE. For an effusive biography of Schoolcraft, a long list of shorter biographical sources, and an extensive bibliography of his writings, see Chase S. Osborn and Stellanova Osborn, *Schoolcraft-Longfellow-Hiawatha* (Lancaster, Pa., 1942).

ELIZABETH NOBLE SHOR

SCHOOTEN, FRANS VAN (*b.* Leiden, Netherlands, *ca.* 1615; *d.* Leiden, 29 May 1660), *mathematics.*

Schooten's father, Frans van Schooten the Elder, succeeded Ludolph van Ceulen at the engineering school in Leiden. The younger Schooten enrolled at the University of Leiden in 1631 and was carefully trained in the tradition of the Dutch school of algebra. In early youth he studied Michael Stifel's edition of Christoph Rudolff's German *Coss.* He was also acquainted with the Dutch and French editions of the works of Simon Stevin, with van Ceulen's *Arithmetische en geometrische Fondamenten,* and with Albert Girard's *Invention nouvelle en l'algèbre.* Schooten studied Girard's edition of the mathematical works of Samuel Marolois and his edition of Stevin's *Arithmétique.* He was of course familiar with Commandino's editions of Archimedes, Apollonius, and Pappus, and with Cavalieri's geometry of indivisibles.

It was probably through his teacher, the Arabist and mathematician Jakob Gool, that Schooten met Descartes, who had just come to Leiden from Utrecht to supervise the printing of the *Discours de la méthode* (1637). Schooten saw the proofs of the *Géométrie* (the third supplement to the *Discours*) by the summer of 1637 at the latest. He recognized the utility of the new notation, but he had difficulty in mastering the contents of the work. He therefore undertook a more intensive study of literature on the subject and sought to discuss the work with colleagues.

Armed with letters of introduction from Descartes, Schooten went to Paris. Although he was a convinced Arminian, he received an extremely cordial welcome from the Minimite friar Marin Mersenne and his circle. In Paris, Schooten was able to read manuscripts of Viète and Fermat; and, on a commission from the Leiden printing firm of Elzevier, he gathered all the printed works of Viète that he could find. He went next to England, where he met the leading algebraists of the day, and finally to Ireland.

Schooten returned home in 1643 and served as his father's lecture assistant, introducing a number of young people—including Jan de Witt—to mathematics. He also prepared a collected edition of the mathematical writings of Viète (1646). Although Schooten generally followed the original texts closely, he did change the notation in several places to simplify the mathematical statements and to make the material more accessible, for Viète's idiosyncratic presentation and the large number of Greek technical terms rendered the originals quite difficult to read. Unfortunately, because he misunderstood a remark that Viète made concerning the unsuccessful edition of his *Canon mathematicus* (1579), Schooten omitted this work and the interesting explanatory remarks that accompanied it from his edition. Schooten had also brought back copies of Fermat's papers, but he was unable to convince Elzevier to publish them, especially since Descartes had expressed an unfavorable opinion of Fermat's work.

In 1645 Christiaan Huygens and his elder brother Constantijn began to study law at Leiden. They attended Schooten's general introductory course (published by Erasmus Bartholin in 1651), and in advanced private instruction became acquainted with many interesting questions in mathematics. A close friendship developed between Schooten and Christiaan Huygens, as their voluminous correspondence attests. The letters reveal how quickly Huygens outgrew the solicitous guidance of his teacher to become the leading mathematician and physicist of his time.

Schooten's first independent work was a study of the kinematic generation of conic sections (1646). In an appendix he treated the reduction of higher-order binomial irrationals $\sqrt[n]{a} + \sqrt{b}$ to the form $x + \sqrt{y}$ in cases where this is possible, using a development of a procedure of Stifel's. An interesting problem that Schooten considered was how to construct a cyclic quadrilateral of given sides, one of which is to be the diameter—a problem that Newton later treated in the lectures on *Arithmetica universalis* (*Mathematical Papers*, V, 162–181).

After the death of his father in 1645, Schooten took over his academic duties. He also worked on a Latin translation of Descartes's *Géométrie*. Although Descartes was not completely satisfied with Schooten's version (1649), it found a broad and receptive audience by virtue of its more carefully executed figures and its full commentary. It was from Schooten's edition of the *Géométrie* that contemporary mathematicians lacking proficiency in French first learned Cartesian mathematics. In this mathematics they encountered a systematic presentation of the material, not the customary, more classificatory approach that essentially listed single propositions, for the most part in unconnected parallel. Further, in the Cartesian scheme the central position was occupied by algebra, which Descartes considered to be the only "precise form of mathematics."

The great success of Schooten's edition led him to prepare a second, much enlarged one in two volumes (1659–1661), which became the standard mathematical work of the period. A third edition appeared in 1683, and an appendix to the fourth edition (1695) contained interesting remarks by Jakob Bernoulli. In the second edition Schooten not only greatly expanded his commentary, but also added new material including an example of Fermat's extreme value and tangent method (with a reference to Hérigone's *Cursus mathematicus* [*Supplementum*, 1642]) and a peculiar procedure for determining the center of gravity of parabolic segments. Since Fermat was not mentioned in the latter connection, it is likely that Schooten came upon the procedure independently, for he usually cited his sources very conscientiously.

In the first edition (1649) Schooten inserted Debeaune's rather insignificant *Notae breves* to the *Géométrie*. The commentary of the second edition contained valuable contributions by Huygens dealing with the intersections of a parabola with a circle and certain corollaries, as well as on an improved method of constructing tangents to the conchoid. Schooten also included longer contributions by his students: Jan Hudde's studies on equations and the rule of extreme values and Hendrik van Heuraet's rectification method.

Volume II of the second edition of the *Geometria* (1661) commences with a reprinting of Schooten's introductory lectures. This material is followed by Debeaune's work on the limits of roots of equations and then by de Witt's excellent tract on conic sections. The volume concludes with a paper by Schooten's younger half brother Pieter on the algebraic discussion of Descartes's data. This edition shows the great effort Schooten devoted to the training of his students and to the dissemination of their findings. This effort can be seen even more clearly in his wide-ranging correspondence, most of which is reprinted in Huygens' *Oeuvres complètes*. (Unfortunately, not all of Schooten's correspondence has been located.)

Schooten made an original contribution to mathematics with his *Exercitationes mathematicae* (1657). Book I contains elementary arithmetic and geometry problems similar to those found in van Ceulen's collection. Book II is devoted to constructions using straight lines only and Book III to the reconstruction of Apollonius' *Plane Loci* on the basis of hints given by Pappus. Book IV is a revised version of Schooten's treatment of the kinematic generation of conic sections, and Book V offers a collection of interesting individual problems. Worth noting, in particular, is the restatement of Hudde's method for the step-by-step building-up of equations for angular section and the determination of the girth of the folium of Descartes: $x^3 + y^3 = 3\,axy$. Also noteworthy is the determination of Heronian triangles of equal perimeter and equal area (Roberval's problem) according to Descartes's method (1633). As an appendix Schooten printed Huygens' *De ratiociniis in aleae ludo*, which was extremely important in the development of the theory of probability.

Schooten possessed an excellent knowledge of the mathematics of both his own time and earlier periods. Besides being an extraordinarily industrious and conscientious scholar, a skillful commentator, and an inspiring teacher, he was a man of rare unselfishness. He recognized his own limitations and did not seek to overstep them. Fascinated by the personality and ideas of Descartes, he worked hard to popularize the new mathematics; his highly successful efforts assured its triumph.

BIBLIOGRAPHY

I. ORIGINAL WORKS. Schooten's writings include his ed. of Viète's *Opera mathematica* (Leiden, 1646); *De organica conicarum sectionum in plano descriptione* (Leiden, 1646); *Geometria a Renato Descartes anno 1637 gallice edita, nunc autem . . . in linguam latinam versa* (Leiden, 1649; 2nd ed., 2 vols., Amsterdam, 1659–1661; 3rd ed., 1683; 4th ed., Frankfurt, 1695): *Principia matheseos universalis*, E. Bartholin, ed. (Leiden, 1651), also included in the 2nd ed. of *Geometria*; *Exercitationum mathematicarum libri quinque* (Leiden, 1657), also in Flemish as *Mathematische Oeffeningen* (Amsterdam, 1660); and "Tractatus de concinnandis demonstrationibus geometricis ex calculo algebraico," Pieter van Schooten, ed., in the 2nd ed. of *Geometria*.

II. SECONDARY LITERATURE. See J. E. Hofmann, "Frans van Schooten der Jüngere," in *Boethius*, II (Wiesbaden, 1962), with portrait; and C. de Waard, "Schooten, Frans van," in *Nieuw Nederlandsch biographisch woordenboeck*, VII (1927), 1110–1114.

J. E. HOFMANN

SCHOPFER, WILLIAM-HENRI (*b.* Yverdon, Switzerland, 8 May 1900; *d.* Geneva, Switzerland, 19 June 1962), *biology, microbiology, biochemistry.*

Schopfer received all of his higher education in Geneva and was awarded two *licences* in the natural sciences, one in 1923 and another in 1925. He won the Prix Davy for his first works, which were concerned essentially with parasitology and protozoology and dealt particularly with the molecular concentration of the juices of parasites (trematodes and cestodes). While working in the general botany laboratory, Schopfer prepared a dissertation under the direction of Robert Chodat. In 1928 he received his doctorate from the University of Geneva after presenting his remarkable work on the sexuality of mushrooms, in which he treated a completely new problem, the comparative biochemistry of sexual reproduction. Schopfer studied abroad on several occasions, notably at the Pasteur Institute in Paris, at the biological station at Roscoff, and in Berlin (1929–1930), where he worked with Hans Kniep, an outstanding specialist in the sexuality of microorganisms.

Schopfer advanced quickly in his academic career. In 1929 he was a *Privatdozent* in general physiology at the Faculty of Sciences of Geneva, and in 1933 he accepted the chair of botany and general biology at the University of Bern. In the same year he became director of the university botanical institute and garden. He remained there until his death, occupying the post of dean of the Faculty of Sciences in 1941–1942 and that of rector in 1948–1949. Schopfer was president of the Swiss Society of Microbiology (1942–1943) and of the Swiss Society for the History of Medicine and Natural Sciences (from 1946), and vice-president of the International Union of the History and Philosophy of Science. In addition he was actively involved with the scientific journals *Archiv für Mikrobiologie, Enzymologia, Excerpta medica,* and *Internationale Zeitschrift für Vitaminforschung.* Schopfer was awarded honorary doctorates by the Paris Faculty of Pharmacy (1949), the Nancy Faculty of Pharmacy (1950), the Lyons Faculty of Sciences (1950), and the Besançon Faculty of Sciences (1956). Between 1923 and 1962 he published 299 scientific works and directed twenty-two doctoral dissertations.

It was in 1927 that Schopfer, after about twenty works devoted primarily to protozoology, published his first memoir on the sexuality of the Mucoraceae. Henceforth the path of his research was clearly marked. All of Schopfer's works and the majority of those by his students were devoted to the study of the role of organic factors controlling the growth of microorganisms. Schopfer opened the immense area of research on microbial vitamins, in which plant and animal biochemists and physiologists met.

In 1930 he inaugurated a series of experiments, which became classics, on the mold *Phycomyces blakesleeanus.* Following an apparently fortuitous observation, in 1931 Schopfer made a series of studies that were decisive for the new science of vitaminology. This *Phycomyces* cannot grow on a synthetic medium containing purified maltose, but it will grow if unpurified maltose is used. Schopfer then demonstrated that the impurity, linked to maltose, was vitamin B_1. In this way it was demonstrated for the first time that an animal vitamin was necessary to the growth of a fungus. In many subsequent papers Schopfer demonstrated the roles of riboflavin, biotin, and mesoinositol—to cite only the well-known compounds—in the growth of microorganisms. To the classic distinction between autotrophic and heterotrophic individuals, Schopfer added that between auxoautotrophic species, capable of producing the vitamins necessary for their growth, and the auxoheterotrophic species incapable of synthesizing them. He devoted a series of studies to what he called artificial symbiosis. He showed, for example, that certain mushrooms that need only one of the constituents of the vitamin B complex, such as *Rhodotorula rubra*, the growth of which requires pyrimidine, and *Mucor ramanninanus*, which needs thiazole, are able to live in association without these substances.

But Schopfer employed mainly *Phycomyces* as a biological test of the quantitative dosage of vitamin B_1, and it permitted him to determine precisely a group of questions concerning the biosynthesis of vitamins, their role in the soil, and their importance for the higher organisms. Schopfer also concentrated on the mechanisms of action of these vitamins and demonstrated the effects of structurally similar substances that behave as antagonists, which he called antivitamins. All the investigation of vitamins conducted by Schopfer and his students were joined with those of other scientists in a remarkable work published in 1939, then reworked and completed in English as *Plants and Vitamins* (1943). Schopfer also conducted research on nitrogen metabolism, the in vitro culture of plant organs, graftings, and morphogenesis.

In addition to research, Schopfer was greatly interested in the history of science; and his studies in the history of biology are valuable for the abundance of their documentation and for his critical analysis of the facts discussed. Among his writings in this field are "L'histoire des théories relatives à la génération au 18me et au 19me siècle" (1944), *La recherche de l'unité en biologie* (1948), and a study of the work of Jules Raulin (1949). He also produced *Situation de la biologie dans le système des sciences* (1951), *L'évolution de la méthode en biologie du point de vue de l'histoire des sciences* (1952), and an edition of the letters of Leeuwenhoek (1955).

BIBLIOGRAPHY

I. ORIGINAL WORKS. A list of 299 scientific papers by Schopfer was published in *Mitteilungen der Naturforschenden Gesellschaft in Bern*, **20** (1964), 86–102. They include "Vitamine und Wachstumsfaktoren bei den Mikroorganismen, mit besonderer Berücksichtigung des Vitamins B_1," in *Ergebnisse der Biologie*, **16** (1939), 1–172; revised and enlarged as *Plants and Vitamins*, N. N. Noecker, trans. (Waltham, Mass., 1943; 2nd ed., 1949); "Les vitamines, facteurs de croissance pour les microorganismes," in *Schweizerische Zeitschrift für Pathologie und Bakteriologie*, **7** (1944), 303–345; "L'histoire des théories relatives à la génération au 18me et au 19me siècle," in *Actes de la Société helvétique des sciences naturelles* (1944), 192–193; "Les tests microbiologiques pour la détermination des vitamines," in *Experientia*, **1** (1945), 183–194, 219–229; *La recherche de l'unité en biologie* (Bern, 1948); *Les répercussions hors de France de l'oeuvre de Jules Raulin relative au zinc* (Lyons, 1948); "Remarque bibliographique sur l'histoire du terme Cambium," in *Archives internationales d'histoire des sciences*, **28** (1949), 457–458; "Situation de la biologie dans le systeme des sciences; les relations de la biologie avec les autres sciences," in *Actes de la Société helvétique des sciences naturelles* (1951), 68–80; "Le méso-inositol en biologie," in *Bulletin de la Société de chimie biologique*, **83** (1951), 1113–1146; "L'évolution de la méthode en biologie du point de vue de l'histoire des sciences," in *Actes scientifiques et industrielles Herrman*, **8** (1952), 117–125; "Les cultures d'organes et leurs applications en physiologie végétale," in *Actes de la Société helvétique des sciences naturelles* (1952), 61–73; "La botanique et le bien-être humain," in *Actes du 8me congrès international botanique, 1954* (1959), 53–57; and "Recherches sur le role du méso-inositol dans la biologie cellulaire de *Schizosaccharomyces pombe* Lindner," in *Archiv fur Mikrobiologie*, **44** (1962), 113–151, his last work.

II. SECONDARY LITERATURE. See K. H. Erismann, "William-Henri Schopfer (1900–1962)," in *Verhandlungen der Schweizerischen naturforschenden Gesellschaft*, **142**, pt. 1 (1962), 252–258; and "Die wissenschaftlichen Arbeiten von Prof. Dr. W.-H. Schopfer: Chronologisch geordner von 1923 bis 1962," in *Mitteilungen der Naturforschenden Gesellschaft in Bern*, n.s. **20** (1963), 86–102; and A. Tronchet, "Rapport sur l'attribution du Dr. H. C. de l'Université de Besançon à M. le professeur Schopfer," in *Annales scientifiques de l'Université de Besançon*, 2nd ser., **2** (1958), 1–4.

P. E. PILET

SCHORLEMMER, CARL (*b.* Darmstadt, Germany, 30 September 1834; *d.* Manchester, England, 27 June 1892), *organic chemistry, history of chemistry.*

Schorlemmer was the son of Johann Schorlemmer, a master carpenter. After attending the *Volksschule* and the *Realschule*, he enrolled at what is now the Technische Universität of Darmstadt. In 1853 he became an apprentice pharmacist in Umstadt (Hesse) and then an assistant pharmacist in Heidelberg. In the latter city Schorlemmer attended Bunsen's lectures, gave up pharmacy, and began to study chemistry. In 1858 he enrolled at the University of Giessen, where he studied chemistry for a semester under Heinrich Will and Hermann Kopp. On the advice of his friend William Dittmar he went to Manchester the following year. He began as a private assistant to H. E. Roscoe at Owens College, became an assistant instructor in 1861, lecturer in 1872, and in 1874, on Roscoe's recommendation, England's first professor of organic chemistry, at Owens College.

In the years before he became a professor, Schorlemmer experimented with simple hydrocarbons. During this period he made important contributions to the development of modern organic chemistry, including investigations of the compounds "methyl" (CH_3-CH_3) and "ethyl hydride" (C_2H_5H), which he recognized as equivalent (that is, ethane) and not isomeric, as had been suggested. Industrial research on coal tar distillate and American petroleum (naphtha) constituted the first petrochemical investigations.

Kopp's influence can be seen in Schorlemmer's experimental investigations, especially his studies on the relations between boiling point and chemical constitution, and it is unmistakable in Schorlemmer's works on the history of chemistry. A close friend of Friedrich Engels and Karl Marx, Schorlemmer became a member of the International Workingmen's Association and of the German Social Democratic party. A believer in scientific socialism, and in dialectical and historical materialism, Schorlemmer presented the history of chemistry from a sociological point of view and discovered important relations between chemistry, economics, and philosophy.

Schorlemmer was a member of many scientific societies and academies. The technical college at Merseburg, German Democratic Republic, is named for him.

BIBLIOGRAPHY

I. Original Works. Schorlemmer's writings include *Lehrbuch der Kohlenstoffverbindungen* (Brunswick, 1874), also in English, *A Manual of the Chemistry of the Carbon Compounds or Organic Chemistry* (London, 1874); *Treatise of Chemistry* (London, 1877), also in German, *Ausführliches Lehrbuch der Chemie* (Brunswick, 1877), written with H. E. Roscoe; and *The Rise and Development of Organic Chemistry* (Manchester, 1879; 2nd ed., London, 1895), also in German, *Der Ursprung und die Entwicklung der organischen Chemie* (Brunswick, 1889).

Schorlemmer also translated two works by H. E. Roscoe: *Kurzes Lehrbuch der Chemie* (Brunswick, 1868) and *Die Spektralanalyse* (Brunswick, 1870).

II. Secondary Literature. Appreciations and obituaries are in *Berichte der Deutschen chemischen Gesellschaft*, **25** (1892), 1106 ff.; *Festschrift der Technische Hochschule für Chemie "Carl Schorlemmer"* (Merseburg, 1964), 12 ff.; *Journal of the Chemical Society. Transactions*, **63** (1893), 756–763, with a complete bibliography of Schorlemmer's papers; *Nature*, **46**, no.

1191 (25 Aug. 1892), 394–395; and "Vorwärts 3.7. 1892," in Karl Marx and Friedrich Engels, *Werke*, XXII (Berlin, 1963), 313–315. See also Karl Heinig, "Carl Schorlemmer—der erste marxistische Chemiker. Darstellung seines wissenschaftlichen und gesellschaftlichen Wirkens" (Ph.D. diss., Humboldt University, Berlin, 1968); and *Carl Schorlemmer—Chemiker und Kommunist ersten Ranges* (Leipzig, 1974), with bibliography of Schorlemmer's works and secondary literature.

Karl Heinig

SCHOTT, CHARLES ANTHONY (*b.* Mannheim, Germany, 7 August 1826; *d.* Washington, D.C., 31 July 1901), *geophysics.*

Schott studied for six years at the Technische Hochschule in Karlsruhe, graduating as a civil engineer. The revolution of 1848, in which he participated briefly, and poor career prospects led him to emigrate to the United States in that year. On 8 December 1848 he received an appointment in the U.S. Coast Survey and remained in its service until his death. His initial post was in the agency's computing division; in 1855 he became its chief, relinquishing the title in 1899.

Although Schott had various field assignments, his career was principally in the Washington office. His division was responsible for processing the data gathered by Survey parties. Before the age of electronic computers these calculations were arduous, if not tedious, often requiring great ingenuity in devising shortcuts and methods of approximation. But Schott's importance in the Coast Survey far transcended his industry and cleverness in computing. On him, perhaps more than any other individual, depended the precision and the theoretical adequacy of the Survey's work. This involved not only the study of the instruments, observational techniques, and data but also appraisals of proposed innovations, including theoretical changes.

Evaluating Schott's role is awkward, since the Coast Survey was a team effort and its publications often did not identify particular contributors. Yet his bibliography and the esteem accorded by informed contemporaries are clues to his stature. Before John F. Hayford, Schott was the leading geodesist in the Survey. His work on the great triangulation across the continent was a high point in the older style of determining the figure of the earth, yielding results falling between those of Bessel and Alexander Ross Clarke. Like others in the Coast Survey, Schott was greatly interested in terrestrial magnetism and was responsible for several studies in this area. When the French Academy awarded him the Wilde Prize in 1898 for his contributions, it was recognition for nearly fifty years of collecting and reducing data, constructing new apparatus, studying the influence of the aurora, and investigating the relations of sunspots and magnetic storms.

Schott was also well-known for his climatological studies. The principal evidence of his scientific competence remains unpublished and unstudied in the records of the Coast and Geodetic Survey in the United States National Archives, which contain nearly 150 volumes of his reports on scientific topics, as well as similar documents dispersed in many of the agency's series of records.

BIBLIOGRAPHY

Cleveland Abbe's memoir of Schott in *Biographical Memoirs. National Academy of Sciences,* **8** (1915), 87–133, contains a very good bibliography of Schott's writings but is not very enlightening on his life. The greatest source is Record Group 23 of the U.S. National Archives, the Records of the Coast and Geodetic Survey. They are described by N. Reingold, in *Preliminary Inventory of the National Archives,* no. 105 (Washington, D.C., 1958). Of particular relevance are the Computing Division Reports (entry 45) and Geodetic Reports (entry 83).

Nathan Reingold

SCHOTT, GASPAR (*b.* Königshofen, near Würzburg, Germany, 5 February 1608; *d.* Würzburg, 22 May 1666), *mathematics, physics, technology.*

Apart from the place and date of his birth, nothing is known of Schott's origins; almost the only childhood recollection in his works is of a suction pump bursting at Paderborn in 1620, which suggests an early interest in machinery. In 1627 he entered the Society of Jesus and was sent to Würzburg University, where he studied philosophy under Athanasius Kircher. The Swedish invasion of the Palatinate in 1631 forced teacher and pupils to flee. Schott may have first accompanied Kircher to France, for he mentions his travels in that country; but he certainly completed his studies in theology, philosophy, and mathematics at Palermo. He remained in Sicily for twenty years, mostly teaching at Palermo, although he spent two years at Trapani. Nevertheless he was anxious to satisfy a strong thirst for knowledge and to resume his connection with Kircher, whom he always revered as his master. Schott was able to satisfy his desire in 1652, when he was sent to Rome, where for

three years he collaborated with Kircher on his researches. Schott decided that since Kircher did not have time to publish all that he knew and all the information communicated to him by Jesuits abroad, he himself would do so. While compiling this material, he returned to Germany in the summer of 1655, first to Mainz and then to Würzburg, where he taught mathematics and physics.

Schott first published what had originally been intended as a brief guide to the hydraulic and pneumatic instruments in Kircher's Roman museum, expanding it into the first version of his *Mechanica hydraulico-pneumatica.* But he added as an appendix a detailed account of Guericke's experiments on vacuums, the earliest published report of this work. This supplement contributed greatly to the success of Schott's compendium; and as a result he became the center of a network of correspondence as other Jesuits, as well as lay experimenters and mechanicians, wrote to inform him of their inventions and discoveries. Schott exchanged several letters with Guericke, seeking to draw him out by suggesting new problems, and published his later investigations. He also corresponded with Huygens and was the first to make Boyle's work on the air pump widely known in Germany. Schott repeated Guericke's experiments, and later those of Boyle, at Würzburg, as well as some medical experiments on the effects of intravenous injections. He does not, however, seem to have attempted any original investigations.

During the last years of his life, Schott was engaged in publishing this mass of material, besides what he had brought with him from Rome, adding his own commentaries and footnotes: he produced some eleven titles over eight years (1658–1666). But although his industry was impressive, these books consist largely of extracts from communications he had received or from books he had used. Schott was so determined to include all possible arguments on every side that it is often hard to discover what he himself thought. While he maintained that the experiments of Guericke, Torricelli, Boyle, and others had not produced a true vacuum, the space exhausted of air being filled with "aether," he accepted the assumption that the phenomena previously attributed to the effects of *horror vacui* were really due to atmospheric pressure or to the elasticity of the air. In a treatise on the then very popular theme of the origin of springs, his own opinion, when finally expressed, amounted to saying that everyone was right: some springs are due to precipitation, some to underground condensation, and some are connected directly to the sea.

Schott's chief works, the *Magia universalis* and the two companion volumes, *Physica curiosa* and *Technica curiosa,* are huge, uncritical collections, mines of quaint information in which significant nuggets must be extracted from a great deal of dross. Like many of his time, Schott believed that the principles of nature and art are best revealed in their exceptions. This makes him a useful source on the history of scientific instruments and mechanical technology; a treatise on "chronometric marvels" (which may be his own, since it is ascribed to "a friend" and often quotes his earlier writings) contains the first description of a universal joint to translate motion and a classification of gear teeth. Although the "natural curiosities" include some useful matter (such as on South American mammals), his syncretic attitude and taste for the abnormal made him far readier than most of his contemporaries to credit tales of ghosts, demons, and centaurs. All this writing about magic, both natural and supernatural, involved him in slight difficulties with the censors.

Schott apparently yearned for the intellectual delights of Rome, and after twenty-five years in Italy he suffered from German winters and had to have his own hypocaust installed. He visited Rome in 1661, and in 1664 he applied for a post to teach mathematics at the Jesuits' Roman college; this was rejected, and instead he was offered the headship of the college at Heiligenstadt, which he rejected, feeling himself unsuited to administration. Exhausted, it was said, by overwork on his books, he died in 1666.

Undoubtedly Schott was extraordinarily productive. But his contribution was essentially that of an editor who prepared the researches of others for the press without adding much of consequence. Still, he did much to popularize the achievements of contemporary physicists, especially—but not exclusively—in Catholic Germany.

BIBLIOGRAPHY

I. ORIGINAL WORKS. Schott's most important writings are *Mechanica hydraulico-pneumatica* (Würzburg, 1657); *Magia universalis,* 4 vols. (Würzburg, 1657–1659); *Physica curiosa,* 2 vols. (Würzburg, 1662); *Anatomia physico-hydrostatica fontium ac fluminum* (Würzburg, 1663); and *Technica curiosa* (Würzburg, 1664).

II. SECONDARY LITERATURE. All later articles are based on N. Southwell [N. Bacon], *Bibliotheca scriptorum Societatis Jesu* (Rome, 1682), 282; and A. de Backer, in *Bibliothèque des écrivains de la Compagnie de Jésus,* K. Sommervogel, ed., VII (Paris, 1896), 904–912. The only later biographer to add further information is G. Duhr, *Geschichte der Jesuiten in den Ländern deutscher Zunge,* III (Munich–Regensburg, 1923), 587–592.

A. G. KELLER

SCHOTT, OTTO FRIEDRICH (*b.* Witten, Germany, 17 December 1851; *d.* Jena, Germany, 27 August 1935), *glass chemistry, glass manufacture.*

A leading pioneer in glass chemistry, Schott created new types of glass of outstanding quality for use in optics, in the laboratory, and in industry. He came from a family of glassmakers: his father, a master in the making of window glass, became co-owner of a glassworks in Westphalia in 1853. Schott attended the *Realschule* in Witten and the trade school in Hagen. In 1869 he volunteered for service in chemical factories in Haspe and abroad. From 1870 to 1873 he studied chemistry and chemical technology at the technical college in Aachen and at the universities of Würzburg and Leipzig. He received his doctorate from the University of Jena in 1875 for a work on defects in window glass manufacturing. He then returned to Haspe as an industrial chemist and made study trips to England and France. In 1877–1878 he established an iodine and saltpeter factory in Oviedo, Spain, and in 1880 was responsible for the renovation of two Spanish glassworks.

In May 1879 Schott sent a sample of his newly smelted lithium glass to Ernst Abbe, requesting him to test its optical properties. The ensuing close collaboration between the two researchers led Schott to move to Jena in 1882. Glasses with high refractive power had always possessed high dispersive power and consequently could not satisfy the theoretical requirements for optical systems that Abbe had set forth. But the glass samples that Schott smelted finally made possible "a large variety in the gradation of optical characteristics," as Abbe wrote in 1881, and far surpassed the existing types of glass.

Putting to use his energy and talent for the technical application of knowledge, Schott, along with his associates Abbe, Carl Zeiss, and the latter's son Roderich, founded a glass technology laboratory in 1883 and, in 1884, the Glastechnische Versuchsanstalt at Jena. The latter, which subsequently became the Jena glassworks of Schott and Associates, soon achieved world fame for its Jena standard glass 16 III (thermometer glass, 1884), the laboratory glasses (beginning in 1892), Supraxglass

(1890's), uviol glass (1903), and apparatus glass 20 (1920). Schott directed the factory—which in 1919 became part of the Carl Zeiss Foundation—until 1927 and was the recipient of many honors.

BIBLIOGRAPHY

I. ORIGINAL WORKS. Schott's writings and patents include *Beiträge zur Theorie und Praxis der Glasfabrikation* (Jena, 1875), his dissertation; "Chemische Vorgänge beim Schmelzen des Glassatzes," in *Dinglers polytechnisches Journal*, **215** (1875), 529–538; "Ueber Abkühlung des Glases und vom sogenannten Hartglase," *ibid.*, **216** (1875), 75–78, 288; "Ueber die Constitution des Glases," *ibid.*, 346–353; "Ueber Krystallisations-Produkte im gewöhnlichen Glase," in *Annalen der Physik und Chemie*, **155** (1875), 422–442, also in *Dinglers polytechnisches Journal*, **218** (1875), 151–165; "Gewinnung des Schwefels aus Gyps und Glaubersalz bei der Glasfabrikation," *ibid.*, **221** (1876), 142–146 (Prussian patent of 3 Dec. 1875); "Studien über die Härtung des Glases," in *Verhandlungen des Vereins zur Beförderung des Gewerbefleisses*, **58** (1879), 273–305; "Lithiumglas," *ibid.*, **59** (1880), 130–135; *Beiträge zur Kenntniss der unorganischen Schmelzverbindungen* (Brunswick, 1881); "Über Glasschmelzerei für optische und andere wissenschaftliche Zwecke," in *Verhandlungsberichte des Vereins zur Beförderung des Gewerbefleisses*, **67** (1888), 162–180; and "Verfahren zur Herstellung von Verbund-Hartglas," *Patentschrift* no. 51, 578, Kl. 82, Glas, issued 11 Mar. 1892 (German patent issued 5 Apr. 1891). His scientific papers on the physical properties of new glasses written with A. Winkelmann were published in *Annalen der Physik und Chemie*, n.s. **51** (1894), 698–720, 730–746; **61** (1897), 105–141; and "Über elektrisches Kapillarlicht," *ibid.*, **59** (1896), 768–772.

Scientific correspondence includes *Der Briefwechsel zwischen Otto Schott und Ernst Abbe über das optische Glas 1879–1881*, H. Kühnert, ed. (Jena, 1946), with bibliography of Schott's scientific papers on p. xxxv; and *Briefe und Dokumente zur Geschichte des VEB Optik Jenaer Glaswerk Schott & Genossen*, I, *Die wissenschaftliche Grundlegung (Glastechnisches Laboratorium und Versuchsglashütte) 1882–1884*, H. Kühnert, ed. (Jena, 1953), with bibliography of Schott's scientific papers on pp. lxxix–lxxxi.

There are also bibliographies in Poggendorff, III, 1209; IV, 1346; and VI, 2364.

II. SECONDARY LITERATURE. Obituaries and other biographical notes are listed in Poggendorff, VIIa, 240. See especially E. Berger, in *Zeitschrift für technische Physik*, **17** (1936), 6–11; G. Keppler, in *Glastechnische Berichte*, **14** (1936), 49–54; H. Kühnert, in *Zeitschrift für technische Physik*, **17** (1936), 1–6; H. Schimank, in *Glastechnische Berichte*, **25** (1952), 18–24; A. Silverman, in *Bulletin of the American Ceramics Society*, **15** (1936), 169–175; and W. E. S. Turner, in *Journal of the Society of Glass Technology*, **20** (1936), 84–94.

Other literature is E. Berger, "50 Jahre Jenaer Glas," in *Zeiss Nachrichten*, no. 8 (Jan. 1935), 1–7; H. Kühnert, *Urkundenbuch zur Thüringischen Glashüttengeschichte* (Jena, 1934), 271–281; and *Otto Schott. Eine Studie über seine Wittener Zeit bis zur Gründung des Jenaer Glaswerkes* (Witten, 1940); M. von Rohr, "Zu Otto Schotts siebzigsten Geburtstag (17. Dez.)," in *Naturwissenschaften*, **9** (1921), 999–1010; and "Die Entwicklungsjahre der Kunst, optisches Glas zu schmelzen," *ibid.*, **12** (1924), 781–797; and E. Zschimmer, *Die Glasindustrie in Jena—ein Werk von Schott und Abbe*, 2nd ed. (Jena, 1923).

Also see E. Abbe, *Gesammelte Abhandlungen*, IV, pt. 1, *Arbeiten zum Glaswerk zwischen 1882 und 1885. Die Entstehung des Glaswerkes Schott & Gen.*, M. von Rohr. ed. (Jena, 1928); F. Auerbach, *Ernst Abbe* (Leipzig, 1918); and N. Günther, "Ernst Abbe," in *Dictionary of Scientific Biography*, I, 6–9.

HANS-GÜNTHER KÖRBER

SCHOTTKY, FRIEDRICH HERMANN (*b.* Breslau, Germany [now Wrocław, Poland], 24 July 1851; *d.* Berlin, Germany, 12 August 1935), *mathematics.*

After attending the Humanistisches Gymnasium St. Magdalenen in Breslau, Schottky studied mathematics and physics at Breslau University from 1870 to 1874 and continued his studies at Berlin with Weierstrass and Helmholtz. He received the Ph.D. in 1875, was admitted as a *Privatdozent* at Berlin in 1878, and in 1882 was appointed a professor at Zurich—at the university, according to one source, and at the Eidgenössische Technische Hochschule, according to another. In 1892 Schottky was appointed to a chair at Marburg University and in 1902 to one at Berlin, where he remained until 1922. In 1902 he was elected a fellow of the Preussische Akademie der Wissenschaften and, in 1911, a corresponding member of the Akademie der Wissenschaften in Göttingen.

Schottky's thesis [1,3] was an important contribution to the conformal mapping of multiply connected plane domains and was the origin of the famous mapping of a domain bounded by three disjoint circles, which, continued by mirror images, provides an example of an automorphic function with a Cantor set boundary. The dissertation also dealt with the conformal mapping of domains bounded by circular and conic arcs.

A contribution to the realm of Picard's theorem, known as Schottky's theorem [5], is an absolute estimation $C(f(0), |z|)$ for functions $f(z)$ defined in

$|z| < 1$ and omitting the values 0,1. Schottky also initiated the study of the oscillation, at the boundary, of regular functions defined in the unit circle [4].

The greater part of Schottky's work concerned elliptic, Abelian, and theta functions, a subject on which he wrote a book [2]. He published some fifty-five papers, most of them in *Journal für die reine und angewandte Mathematik, Mathematische Annalen,* and *Sitzungsberichte der Preussischen Akademie der Wissenschaften zu Berlin.* His work is difficult to read. Although he was a student of Weierstrass, his approach to function theory was Riemannian in spirit, combined with Weierstrassian rigor.

BIBLIOGRAPHY

I. ORIGINAL WORKS. Schottky's writings include [1] "Ueber die conforme Abbildung mehrfach zusammenhängender ebener Flächen," in *Journal für die reine und angewandte Mathematik,* **83** (1877), 300–351, his dissertation; [2] *Abriss einer Theorie der Abel'schen Functionen von drei Variablen* (Leipzig, 1880); [3] "Ueber eine specielle Function, welche bei einer bestimmten linearen Transformation ihres Arguments unverändert bleibt," in *Journal für die reine und angewandte Mathematik,* **101** (1887), 227–272; [4] "Ueber die Werteschwankungen der harmonischen Functionen," *ibid.,* **117** (1897), 225–253; [5] "Ueber den Picardschen Satz und die Borelschen Ungleichungen," in *Sitzungsberichte der Preussischen Akademie der Wissenschaften zu Berlin* (1904), 1244–1262; and "Bemerkungen zu meiner Mitteilung . . . ," *ibid.* (1906), 32–36.

II. SECONDARY LITERATURE. See [6] L. Bieberbach, "Friedrich Schottky zum 80. Geburtstage," in *Forschungen und Fortschritte,* **7** (1931), 300; and [7] "Gedächtnisrede auf Friedrich Schottky," in *Sitzungsberichte der Preussischen Akademie der Wissenschaften zu Berlin,* Math.-phys. Kl. (1936), cv–cvi; and the [8] obituary in *Nachrichten von der Gesellschaft der Wissenschaften zu Göttingen* (1935–1936), 6–7.

Portraits of Schottky are in *Acta mathematica 1882–1913, Table générale des tomes 1–35* (Uppsala, 1913), 168; and *Journal für die reine und angewandte Mathematik,* **165** (1931), frontispiece.

HANS FREUDENTHAL

SCHOUTE, PIETER HENDRIK (*b.* Wormerveer, Netherlands, 21 January 1846; *d.* Groningen, Netherlands, 18 April 1923), *mathematics.*

Schoute, whose family were industrialists on the Zaan near Amsterdam, studied at the Polytechnical School at Delft, from which he graduated in 1867 as a civil engineer. He continued his study of mathematics at Leiden, where he received his Ph.D. in 1870 with the dissertation "Homography Applied to the Theory of Quadric Surfaces." While teaching at high schools in Nijmegen (1871–1874) and The Hague (1874–1881), he published two textbooks on cosmography. From 1881 until his death he was professor of mathematics at the University of Groningen.

Schoute was a typical geometer. In his early work he investigated quadrics, algebraic curves, complexes, and congruences in the spirit of nineteenth-century projective, metrical, and enumerative geometry. From 1891 he turned to geometry in Euclidean spaces of more than three dimensions, then a field in which little work had been done. He did extensive research on regular polytopes (generalizations of regular polyhedrons). Some of his almost thirty papers in this field were written in collaboration with Alice Boole Stott (1860–1940), daughter of the logician George Boole.

Schoute was an editor of the *Revue semestrielle des publications mathématiques* from its founding in 1893, and in 1898 he became an editor of the *Nieuw archief voor wiskunde.* He held both positions until his death. In 1886 he became a member of the Royal Netherlands Academy of Sciences.

BIBLIOGRAPHY

I. ORIGINAL WORKS. Much of Schoute's work appeared in *Verhandelingen der Koninklyke nederlandsche akademie van wetenschappen,* Afdeeling Natuurkunde, 1st section; see esp. "Regelmässige Schnitte und Projektionen des Hundertzwanzigzelles und des Sechshundertzelles im vierdimensionalen Raume," **2,** no. 7 (1894); and **9,** no. 4 (1907). Writings on other polytopes are in **2,** no. 2 and 4 (1894), which deal with the 8-cell, the 16-cell, and the 24-cell. See also "Het vierdimensionale prismoïde," **5,** no. 2 (1896); and "Les hyperquadratiques dans l'espace à quatre dimensions," **7,** no. 4 (1900). Several articles appeared in *Archives néerlandaises des sciences exactes et naturelles,* 2nd ser. **5–9** (1896–1904). Many of Schoute's results were collected in his *Mehrdimensionale Geometrie,* 2 vols. (Leipzig, 1902–1905).

II. SECONDARY LITERATURE. See H. S. M. Coxeter, *Regular Polytopes* (New York–London, 1948; 2nd ed., 1963), *passim;* and D. J. Korteweg, "P. H. Schoute," in *Zittingsverslagen der Koninklyke nederlandsche akademie van wetenschappen,* **21** (1912–1913), 1396–1400. Also: H. Fehr, *Enseignement mathématique,* **35** (1913), 256–257.

D. J. STRUIK

SCHOUTEN, JAN ARNOLDUS (*b*. Nieuweramstel [now part of Amsterdam], Netherlands, 28 August 1883; *d*. Epe, Netherlands, 20 January 1971), *tensor analysis*.

A descendant of a prominent family of shipbuilders, Schouten grew up in comfortable surroundings. He became not only one of the founders of the "Ricci calculus" but also an efficient organizer (he was a founder of the Mathematical Center at Amsterdam in 1946) and an astute investor. A meticulous lecturer and painfully accurate author, he instilled the same standards in his pupils.

After studying electrical engineering at what is now the Technische Hogeschool at Delft, Schouten practiced this profession for a few years and then returned to study in Leiden when an inheritance gave him the necessary independence. Upon completion of his doctoral dissertation in 1914, his first contribution to the foundations of tensor analysis, he was appointed professor at Delft. In 1943 Schouten resigned the post, divorced his wife, and remarried. From then on, he lived in semiseclusion at Epe. Although he was a professor at the University of Amsterdam from 1948 to 1953, the Mathematical Center had replaced teaching as his first commitment. He served the Center until 1968 and was its director for about five years.

Schouten attained numerous distinctions during his lifetime, including membership in the Royal Netherlands Academy of Sciences, the rotating position of *rector magnificus* at Delft, the presidency of the 1954 International Congress of Mathematicians at Amsterdam, several terms as president of the Wiskundig Genootschap (the society of Netherlands mathematicians), and a royal decoration.

Schouten's scientific contributions comprise some 180 papers and six books, virtually all related to tensor analysis and its applications to differential geometry, Lie groups, relativity, unified field theory, and Pfaffian systems of differential equations. Having entered the field when it was in its infancy, he helped develop and perfect the basic techniques of local differential geometry and applied them in numerous ways. He discovered connections ("geodesic displacements") in Riemannian manifolds in 1919, independently of, although later than, Levi-Civita; and he also discovered basic properties of Kähler manifolds in 1931, two years before Kähler. Under the influence of Weyl and Eddington he was led to general linear connections and investigated affine, projective, and conformal manifolds.

Schouten's approach to differential geometry was strongly influenced by Felix Klein's "Erlanger Programm" (1872), which viewed each geometry as the theory of invariants of a particular group. This approach led him to a point of view that handled geometric problems more formally than most other prominent differential geometers of his time, notably Levi-Civita, E. Cartan, Veblen, Eisenhart, and Blaschke. This same point of view underlies his "kernel-index method," a notation of great precision, which he and his pupils used masterfully, but which gained favor elsewhere only in less extreme forms.

Schouten inspired numerous co-workers, including D. J. Struik, D. van Dantzig, J. Haantjes, E. R. van Kampen, V. Hlavaty, S. Gołab, Kentaro Yano, E. J. Post, and A. Nijenhuis. His influence extended as far as Russia and Japan.

BIBLIOGRAPHY

I. ORIGINAL WORKS. Most of Schouten's work on tensor analysis and differential geometry can be found or is referred to in *Der Ricci Kalkül* (Berlin, 1924); *Einführung in die neueren Methoden der Differential-geometrie*, 2 vols. (Groningen, 1934–1938), I, *Uebertragungslehre*, by Schouten, II, *Geometrie*, by D. J. Struik, also translated into Russian (Moscow, 1939, 1948); and *Pfaff's Problem and Its Generalisations* (Oxford, 1948), written with W. van der Kulk. *Ricci Calculus* (Berlin, 1954), the 2nd ed. of *Der Ricci Kalkül*, is completely rewritten and contains all that Schouten considered relevant in differential geometry at the end of his career. *Tensor Analysis for Physicists* (Oxford, 1951) is an attempt to spread to sophisticated physicists the subtleties of tensor analysis and its implications for field theory and elasticity.

A collection of Schouten's papers and correspondence has been deposited at the library of the Mathematical Center in Amsterdam.

II. SECONDARY LITERATURE. A short biographical article, concentrating on Schouten's scientific work, is D. J. Struik's *Levensbericht* on Schouten, in *Jaarboek der K. Nederlandsche akademie van wetenschappen* for 1971, pp. 94–100, with portrait. A. Nijenhuis, "J. A. Schouten: A Master at Tensors," in *Nieuw archief voor wiskunde*, 3rd ser., **20** (1972), 1–19, contains a complete list of publications.

ALBERT NIJENHUIS

SCHOUW, JOAKIM FREDERIK (*b*. Copenhagen, Denmark, 7 February 1789; *d*. Copenhagen, 28 April 1852), *plant geography, climatology*.

Schouw was the eldest of seven children born to Poul Schouw, a wine merchant, and Sara Georgia Liebenberg. Since he had to work at his father's business, he was educated by a tutor. His interest in botany began while he was attending the lectures on cryptogamic botany given in the winter of 1803–1804 by Martin Vahl, who assisted him in making a herbarium. Because working in his father's wine cellar had undermined his weak health, Schouw became a clerk in a lawyer's office in 1804; seven years later he passed the examination for the candidate's degree in law.

In 1812 Schouw took part in a botanical expedition to Norway headed by the Norwegian botanist Christian Smith and was strongly impressed by the conspicuous division of the vegetation into zones. After his return Schouw obtained a civil-service post in 1813. He pursued his interest in botany by studying all the available literature on plant geography, especially the works of Humboldt and Wahlenberg.

In 1816 Schouw received the Ph.D., along with a grant and a three-year leave to make a botanical trip to Italy, where he visited the Alps, the Apennines, and Sicily. On his return trip he visited P. de Candolle in Geneva and Adrien de Jussieu and Humboldt in Paris. Schouw returned to Copenhagen in 1820, where he was appointed extraordinary professor of botany, especially phytogeography, at Copenhagen University. During 1823–1824 Schouw had meteorological observations made in several Danish towns, the results of which he published in *Tidsskrift for Naturvidenskaberne*. In 1829, en route to Italy to complete his observations, he met Martius, Joseph Gerhard Zuccarini, and Mohl in Munich. Schouw was greatly interested in popularizing science and in improving the teaching of natural history. From 1831 he was editor of *Dansk Ugeskrift*, in which many of his popular-science lectures were printed. Through this work he became well known, and during the summer of 1839 he was invited to take part in the preparations for the meeting of Scandinavian naturalists at Göteborg.

In 1841 Schouw was appointed curator of the botanical gardens of Copenhagen, and four years later he became a full professor. During his last years his health deteriorated, and on 1 April 1853 he resigned.

While in Italy in 1817, Schouw met Susette Dalgas. They were married in 1827 and had a son and a daughter.

BIBLIOGRAPHY

I. ORIGINAL WORKS. Schouw's writings include *Dissertatio de sedibus plantarum originariis. Sectio prima.*

De pluribus cujusvis speciei individuis originariis statuendis (Copenhagen, 1816), his dissertation; "Einige Bemerkungen über zwei, die Pflanzengeographie betreffende Werke des Herrn von Humboldt," in *Jahrbücher der Gewächskunde*, 1 (1818), 6–56, unsigned; *Grundtraek til en almindelig Plantegeographie* (Copenhagen, 1822), translated by Schouw into German as *Grundzüge einer allgemeinen Pflanzengeographie* (Berlin, 1823), one of his principal works on plant geography; *Skildring af Vejrligets Tilstand i Danmark* (Copenhagen, 1826), his major work on climatology; *Specimen geographiae physicae comparativae . . .* (Copenhagen, 1828); and *Europa, En letfattelig Naturskildring* (Copenhagen, 1832).

Collections of his popular lectures are *Natur-Skildringer. En Raekke af almeenfattelige Forelaesniger* (Copenhagen, 1837, 1845) and *Naturskildringer. En Raekke populaere Foredrag. Ny foroget Udgave. Med Forfatterens Biographie* (Copenhagen, 1856).

Many of Schouw's letters are in the library of the botanical gardens, Copenhagen, and in the Royal Library, Copenhagen.

II. SECONDARY LITERATURE. Detailed biographies, in Danish, are Carl Christensen, "Joachim Frederik Schouw," in *Botanisk Tidsskrift*, 38, no. 1 (1923), 1–56; and *Den Danske Botanisk Historie*, I, pt. 1 (Copenhagen, 1924), 253–276; with a bibliography of Schouw's works, II (1926), 165–179.

A. P. M. SANDERS

SCHREIBERS, KARL (or CARL) FRANZ ANTON VON (*b.* Pressburg, Hungary [now Bratislava, Czechoslovakia], 15 August 1775; *d.* Vienna, Austria, 21 May 1852), *zoology*.

Schreibers came from a noble family that had supplied numerous civil servants and scientists to the Austrian Empire. His father, Joseph Ludwig von Schreibers, was in the military administration; and the family moved to Vienna when Schreibers was a boy. After graduating from the Gymnasium, he entered the University of Vienna and was soon attracted to natural science and medicine. In 1793 he published a two-volume work on mollusks. Five years later he received the M.D. degree and began to practice medicine with his uncle, Johann Ludwig von Schreibers.

Although Schreibers used and campaigned for inoculation against smallpox, he did not remain in the medical profession for long. He made an extensive study tour through most of western Europe; and in 1800 he was named professor at the University of Vienna, becoming director of both the zoological and the mineralogical museums there in 1806. Schreibers proved to be an effective museum organizer, improving the museums and enlarging the collections, partly through expeditions, notably to Brazil (1817–1822). In 1809 he was put in charge of removing the Vienna treasury and archives from the reach of the advancing French army, and in 1815 he led the Austrian commission to retrieve the art treasures confiscated by the French during the Napoleonic Wars.

The author of many scientific papers and a large general work on meteorites, Schreibers was active mainly in zoology. Especially in his younger days he readily adopted new ideas and put them to effective use. He was the first to teach Cuvier's system of comparative anatomy in German, and even after he had left the University of Vienna he remained an important teacher and adviser to younger scientists. Schreibers made notable contributions to ornithology and entomology, and was an expert on arachnids. His main field, however, was reptiles and amphibians, and he considerably extended the knowledge of the central European faunas. Schreibers' best-known work in this field was his description of the salamander *Proteus anguinus*, a blind white amphibian that lives only in dark caves. Schreibers understood the biological importance of the find, which was the introduction to a long series of studies on cave faunas and their development and distribution.

In 1848 the zoological and mineralogical museums caught fire during a thunderstorm; and Schreibers, who lived in the museum buildings, lost not only the collections and the library, but also his private library, manuscripts, and other belongings. He tried to rebuild the museums, but his age prevented it. He was obliged to retire and died shortly afterward. Schreibers was a highly respected scientist who received a number of titles and decorations, both for his work with the museums and for his scientific achievements. His most lasting works are the large monographs on the fauna of Austria and his investigations of *Proteus*.

BIBLIOGRAPHY

I. ORIGINAL WORKS. Schreibers' writings include *Versuch einer vollständigen Conchylienkenntniss nach Linne's System*, 2 vols. (Vienna, 1793); "A Historical and Anatomical Description of a Doubtful Amphibious Animal of Germany Called by Laurenti *Proteus anguinus*," in *Philosophical Transactions of the Royal Society* (1801), 241–264; *Nachrichten von den Kaiserlichen Naturforschern in Brasilien*, 2 vols. (Brünn [Brno], 1818–1820); and *Beiträge zur Geschichte und Kenntniss meteorische Stein- und Metalmassen* (Vienna, 1820).

II. SECONDARY LITERATURE. See A. F. G. Marshall, "Nekrolog des K. K. Hofrathes Carl, Ritter von Schreibers," in *Verhandlungen der Zoologisch-botanischen Gesellschaft in Wien*, **2** (1852), 46–51; and C. von Wurzbach, "Carl Franz Anton, Ritter von Schreibers," in *Biographisches Lexicon der Kaiserthum Osterreich*, XXXI (Vienna, 1876), 283–287.

NILS SPJELDNAES

SCHRÖDER, FRIEDRICH WILHELM KARL ERNST (*b.* Mannheim, Germany, 25 November 1841; *d.* Karlsruhe, Germany, 16 June 1902), *mathematics*.

Schröder was the son of Heinrich Schröder, who did much to foster the teaching of science in secondary and college-level schools and also strongly influenced his son to choose a scientific career. Schröder's mother, the former Karoline Walter, was the daughter of a minister. Her father tutored Ernst until he was fifteen, providing him with an excellent basic education, especially in Latin. In 1856 Schröder enrolled at the lyceum in Mannheim, from which he graduated in 1860.

Schröder then attended the University of Heidelberg, where he studied under Hesse, Kirchhoff, and Bunsen. He passed his doctoral examination in 1862 and spent the next two years studying mathematics and physics at the University of Königsberg under Franz Neumann and F. J. Richelot. Soon afterward, at Karlsruhe, he took the examination to qualify for teaching in secondary schools. He then went to the Eidgenössische Polytechnikum in Zurich, where he qualified as a lecturer in mathematics in 1865 and taught for a time. In 1874, after teaching at Karlsruhe, Pforzheim, and Baden-Baden, Schröder was offered, on the basis of his mathematical publications, a full professorship at the Technische Hochschule in Darmstadt. In 1876 he accepted a post at the Technische Hochschule in Karlsruhe, of which he became director in 1890. He most often lectured on arithmetic, trigonometry, and advanced analysis.

Schröder was described as kind and modest. A lifelong bachelor, he was an ardent mountain climber and cyclist, and learned to ski when he was sixty years old.

Schröder published more than forty mathematical works, including seven separately printed essays and books. They deal almost exclusively with the foundations of mathematics, notably with combinatorial analysis; the theory of functions of a real variable; and mathematical logic. Particularly noteworthy was his early support of Cantor's ideas on set theory, which he was one of the first to accept.

Through his writings on theoretical algebra and symbolic logic, especially *Algebra der Logik*, Schröder participated in the development of mathematical logic as an independent discipline in the second half of the nineteenth century. This is his real achievement, although his contribution was not recognized until the beginning of the twentieth century. Three factors accounted for the delay; the immature state of the field during his lifetime; a certain prolixity in his style; and, above all, the isolation imposed by his teaching in technical colleges. As a result he was an outsider, at a disadvantage in choosing terminology, in outlining his argumentation, and in judging what mathematical logic could accomplish.

Despite Schröder's relative isolation, his work was in the mainstream of the conceptual development of mathematical logic, the chief figures in which were Boole, de Morgan, and C. S. Peirce. Other new ideas that Schröder adopted and elaborated were Peano's formulation of the postulates of arithmetic (1889) and the abstract conception of mathematical operations vigorously set forth by Grassmann and Hankel. With respect to the philosophical problems raised in the formation of mathematical logic, Schröder was guided primarily by Lotze and Wundt, who closely followed Aristotle in questions of logic.

The terminology and contents of Schröder's "logical calculus" are now primarily of historical interest. His ideas, however, furnished the fundamental notion of mathematical logic: the partition of objects into classes. His work constituted a transitional stage that helped to prepare the way for the development of mathematical logic in the twentieth century.

BIBLIOGRAPHY

I. ORIGINAL WORKS. Schröder's writings are listed in Poggendorff, III, 1212–1213; IV, 1353–1354; V, 1131–1132. They include *Lehrbuch der Arithmetik und Algebra* (Leipzig, 1872); *Formale Elemente der absoluten Algebra* (Baden-Baden—Stuttgart, 1874); *Operationskreis des Logikkalküls* (Stuttgart, 1877); *Vorlesungen über die Algebra der Logik*, 3 vols. in 4 pts. (Leipzig, 1890–1905; 2nd ed., New York, 1966), II, pt. 2, edited by E. Müller; *Uber das Zeichen. Festrede bei dem Direktoratswechsel an der Technischen Hochschule zu Karlsruhe am 22. November 1890* (Karlsruhe, 1890); and *Abriss der Algebra der Logik*: pt. 1, *Elementarlehre* (Leipzig, 1909), and pt. 2, *Aussagentheorie, Funktionen,*

Gleichungen und Ungleichungen (Leipzig, 1910), both parts edited by E. Müller.

II. SECONDARY LITERATURE. See J. Lüroth, "Nekrolog auf Ernst Schrôder," in *Jahresbericht der Deutschen Mathematiker-Vereinigung,* **12** (1903), 249–265, with portrait and bibliography; and Lüroth's obituary and bibliography in Schröder's *Vorlesungen über die Algebra der Logik,* II, pt. 2 (1905), iii–xix.

H. WUSSING

SCHRÖDINGER, ERWIN (*b.* Vienna, Austria, 12 August 1887; *d.* Alpbach, Austria, 4 January 1961), *theoretical physics.*

Schrödinger's father, Rudolf Schrödinger, inherited an oilcloth factory, which, although run in an old-fashioned manner, was successful enough to free him of financial worries. After studying chemistry he turned to his real interests—painting and, later, botany—and published a series of scientific papers in the *Abhandlungen* and *Verhandlungen der Zoologisch-botanischen Gesellschaft in Wien.* He married the daughter of Alexander Bauer, professor of chemistry at the Technische Hochschule in Vienna; Erwin was their only child.

Schrödinger attended public elementary school only once, for a few weeks in Innsbruck, while his parents were on vacation. In Vienna an elementary school teacher came to his home twice a week to tutor him; but, in Schrödinger's opinion, his "friend, teacher, and tireless partner in conversation" was his father. In the fall of 1898 Schrödinger entered the highly regarded academic Gymnasium in Vienna. As was then customary, the curriculum emphasized Latin and Greek, the sciences being somewhat neglected. Schrödinger wrote: "I was a good student, regardless of the subject. I liked mathematics and physics, but also the rigorous logic of the ancient grammars. I hated only memorizing 'chance' historical and biographical dates and facts. I liked the German poets, especially the dramatists, but hated the scholastic dissection of their works."

As a student Schrödinger regularly attended the theater in Vienna and was a passionate admirer of Franz Grillparzer. He kept an album containing programs of the performances he had seen and made extensive annotations on them. He did not, however, neglect his studies. In 1907, during his third semester at the University of Vienna, he began to attend lectures in theoretical physics, which had just been resumed after a nearly two-year interruption following the death of Boltzmann. Friedrich Hasenöhrl's brilliant inaugural lecture on the work of his predecessor made a powerful impression on Schrödinger.

Schrödinger highly esteemed Hasenöhrl and attended his lectures on theoretical physics five days a week for eight successive semesters. He also was present at the mathematics lectures of Wilhelm Wirtinger and those on experimental physics of Franz Exner, whose laboratory assistant he later became.

In 1910 Schrödinger received the doctorate under Hasenöhrl, and the following year he became assistant to Exner at the university's Second Physics Institute, where he remained until the outbreak of war. During these years Egon von Schweidler was *Privatdozent* at the university; Schrödinger learned a great deal from him and called him his teacher, second only to Hasenöhrl and Exner. Schrödinger was obliged to supervise the large physics laboratory courses, a duty for which he was very thankful all his life because it taught him "through direct observation what measuring means."

Schrödinger served in World War I as an officer in the fortress artillery; and in the isolated areas where he was stationed, he often had time to study physics. In 1916, while at Prosecco, he learned the fundamentals of Einstein's general theory of relativity, which he at first found quite difficult to understand. Soon, however, he was able to follow Einstein's train of thought and the relevant calculations; he found much in the initial presentation of the theory that was "unnecessarily complicated."

As early as 1918 Schrödinger had a sure prospect of obtaining a position; he was to succeed Josef Geitler as extraordinary professor of theoretical physics at the University of Czernowitz (now Chernovtsy, Ukraine). "I intended to lecture there honorably on theoretical physics, at first on the model of the splendid lectures of my beloved teacher, fallen in the war, Fritz Hasenöhrl, and beyond this to study philosophy, deeply immersed as I then was in the writings of Spinoza, Schopenhauer, Mach, Richard Semon, and Richard Avenarius." The collapse of the Austro-Hungarian monarchy prevented this plan, and after the war he worked again at the Second Physics Institute in Vienna. As a result, Schrödinger's first scientific papers were in the experimental field. In 1913, at the summer home of Egon von Schweidler at Seeham, Schrödinger collaborated with K. W. F. Kohlrausch on a work that was awarded the Haitinger Prize of the Imperial Academy of Sciences and that was published as "Radium-A-Gehalt der Atmosphäre in Seeham 1913." At Seeham, Schrö-

dinger met Annemarie Bertel, whom he married on 6 April 1920.

Shortly after his marriage Schrödinger moved to Jena, where he was an assistant to Max Wien in the experimental physics laboratory. He left Jena after only four months, in order to accept an extraordinary professorship at the Technische Hochschule in Stuttgart. He remained there for only one semester; in the meantime he had received three offers of full professorships—from Kiel, Breslau, and Vienna. He would have preferred to succeed Hasenöhrl at Vienna, but the working conditions for university professors in Austria were then so poor that this alternative was unacceptable. Instead he went to Breslau, where a few weeks after his arrival he received and accepted an offer to assume the chair formerly held by Einstein and Max von Laue at Zurich.

While at Zurich, Schrödinger worked chiefly on problems related to the statistical theory of heat. He wrote papers on gas and reaction kinetics, oscillation problems, and the thermodynamics of lattice vibrations and their contribution to internal energy; in other works he elucidated aspects of mathematical statistics. In an article on the theory of specific heats and in a monograph on statistical thermodynamics he gave a comprehensive account of the latter subject.

Although Schrödinger published several contributions to the old quantum theory, he did not pursue that topic systematically. His first papers on relativity pointed to a second major field of interest. In addition to these works, and his early papers on relativity, Schrödinger made a detailed study, through both measurement and computation, of the metric of color space and the theory of color vision. The main results of his efforts were an article in J. H. J. Müller and C. S. M. Pouillet's *Lehrbuch der Physik* and the acceptance by physiologists of his interpretation of the relationship between the frequency of red-green color blindness and that of the blue-yellow type.

In the meantime, on 25 November 1924, Louis de Broglie defended his dissertation before the examining committee at the Sorbonne: "Recherche sur la théorie des quanta." The contents of the dissertation first became known through a direct communication from Paul Langevin to Einstein and then, more generally, through publication in the *Annales de physique*. At first no physicist—except Einstein—was willing to believe in the reality of the Broglie waves.

As in his first quantum papers, of 1905, Einstein at the end of 1924 again hypothesized "a far-reaching formal relationship between radiation and gas"; but by the latter year he was concerned primarily with the properties of the gas. Basing his analysis on what is today known as Einstein-Bose statistics, he obtained expression for the fluctuation in number of molecules that hinted at interference effects.

Schrödinger, who in 1925 was also investigating problems of quantum statistics, was "suddenly confronted with the importance of de Broglie's ideas" in reading Einstein's "Quantentheorie des einatomigen idealen Gases. 2. Abhandlung," which appeared on 9 February 1925 in *Sitzungsberichte der Preussischen Akademie der Wissenschaften zu Berlin*. He recognized that Einstein had introduced a fundamental new approach, but he sought "to recast it in a more pleasing form, to liberate it from Bose's statistics," which he deeply disliked.

Shortly before the middle of December, Schrödinger completed a paper on this topic, "Zur Einsteinschen Gastheorie," recorded as being received by *Physikalische Zeitschrift* on 15 December 1925. In an important and still unpublished letter to Einstein dated 28 April 1926, Schrödinger gave the following evaluation of his results: "I can . . . assert categorically that I have really achieved the liberation I mentioned above. . . . I stress the determination of the frequency spectrum in § 3. This whole conception falls entirely within the framework of 'wave mechanics'; it is simply the mechanics of waves applied to the gas instead of to the atom or the oscillator."

Schrödinger, who generally expressed his judgments in an intensely emotional way, termed the earlier Bohr-Sommerfeld quantum theory unsatisfactory, sometimes even disagreeable. Seeking to apply the new ideas to the problem of atomic structure, he "took seriously the de Broglie-Einstein wave theory of moving particles, according to which the particles are nothing more than a kind of 'wave crest' on a background of waves." As is evident in a letter of 16 November 1925, from Schrödinger to Alfred Landé, Schrödinger's conjectures on this topic date from the beginning of November 1925 and therefore from before the conclusion of his paper on Einstein's gas theory.

The intensity of Schrödinger's work on the problem increased as he saw that he was on the track of a "new atomic theory," and it reached a peak during his winter vacation in Arosa. On 27 December 1925 he wrote to Wilhelm Wien, editor of the *Annalen der Physik* in Munich that he was very optimistic: "I believe that I can give a vibrating system . . . that yields the hydrogen frequency *levels* as

its eigenfrequencies." The frequencies of the emitted light rays are then obtained, as Schrödinger observed, by establishing the differences of the two eigenfrequencies respectively.

> Consequently the way is opened toward a real understanding of Bohr's frequency calculation—it is really a vibration (or, as the case may be, interference) process, which occurs with the same frequency as the one we observe in the spectroscope.
>
> I hope that I will soon be able to report on this subject in a little more detail and in a more comprehensible fashion. In the meantime I must learn more mathematics, in order to fully master the vibration problem—a linear differential equation, similar to Bessel's, but less well known, and with remarkable boundary conditions that the equation 'carries within itself' and that are not externally predetermined.

The letter confirms what is already known from Schrödinger's publications and from other statements: that, as must have seemed logically consistent from the physics of the problem, he originally developed a relativistic theory. It must be emphasized, therefore, that Schrödinger worked out the relativistic version only at the end of 1925 and not, as historians of science had believed, in the middle of that year. The equation now known as the "Klein-Gordon equation" does yield the correct nonrelativistic Balmer term, but it gives an incorrect description of the fine structure. Schrödinger was deeply disappointed by this failure and must have thought at first that his whole method was basically wrong. Today it is known that the reason for the failure lay not in his bold initial approach but in the application of the theory of relativity, which, however, has itself been abundantly confirmed. The relativistic Schrödinger equation is obviously correct, but it describes particles without spin, whereas a description of electrons requires the Dirac equation. At the time, however, only the first steps had been taken toward an understanding of electron spin.

After a brief interruption Schrödinger took up his method again, but this time he treated the electron nonrelativistically. It soon became apparent that he had arrived at a theory that correctly represented the behavior of the electron to a very good approximation. The result was the emergence of wave mechanics in January 1926.

Schrödinger published the results of his research in a series of four papers in the *Annalen der Physik* bearing the overall title "Quantisierung als Eigenwertproblem." The first installment, sent on 26 January and received by Wien the next day, contains the first appearance in the literature of his famous wave equation, written out for the hydrogen atom. The solution of this equation follows, as Schrödinger put it, from the "well-known" method of the separation of variables. The radial dependency gives rise to the differential equation

$$\frac{d^2\chi}{dr^2} + \frac{2}{r}\frac{d\chi}{dr} + \left(\frac{2mE}{K^2} + \frac{2me^2}{K^2 r} - \frac{n(n+1)}{r^2}\right)\chi = 0.$$

In fulfilling the boundary conditions one obtains solutions only for certain values of the energy parameters, the stationary values. This seemed to Schrödinger to be the "salient point," but in Bohr's original theory—as its creator stressed from the beginning—it was one of the two fundamental postulates that had remained unexplained. Schrödinger emphasized that, in his theory,

> the ordinary quantization rule can be replaced by another condition in which the term "integral number" no longer appears. Rather, the integrality occurs in the same natural way as, say, the integrality in the modal numbers of a vibrating string. The new conception can be generalized and, I believe, penetrates very deeply into the true nature of the quantum rules.

In solving the differential equation for the radial function, Schrödinger received expert assistance from Hermann Weyl. A crucial element in their rapid success was the fact that the mathematical theory had already been completely worked out by Richard Courant and David Hilbert in their *Methoden der mathematischen Physik* (1924).

In his second paper (23 February 1926) Schrödinger gave a sort of "derivation" of his *undulatorischer Mechanik* in which he drew on the almost century-old work of William Rowan Hamilton. Hamilton was aware that geometrical optics was only a special case of wave optics valid for infinitely small wavelengths, and he showed how to make the transition from the characteristic (iconal) equation of geometrical optics to the differential equation of wave optics. Hamilton introduced the methods of geometrical optics into mechanics and obtained an equation similar to the iconal equation and now known as the Hamilton-Jacobi differential equation. In it the index of refraction is replaced, essentially, by the potential energy and the mass of the mechanical particle.

In Hamilton's work Schrödinger thus found an analogy between mechanics and geometrical optics. And, since geometrical optics "is only a gross approximation for light," he conjectured that the same cause was responsible for the failure of classical mechanics "in the case of very small orbital dimensions and very strong orbital curvature." Both would be only approximations for small wavelengths. Therefore, he said:

> Perhaps this failure is a complete analogy to the failure of geometrical optics, that is, the optics with infinitely small wavelengths; [a failure] that occurs, as is known, as soon as the "obstacles" or "openings" are no longer large relative to the real, finite wavelength. Perhaps our classical mechanics is the *complete* analogue of geometrical optics and, as such, false. . . . Therefore, we have to seek an "undulatory mechanics"—and the way to it that lies closest at hand is the wave-theoretical elaboration of Hamilton's model.

Consequently, Schrödinger introduced into his development of wave mechanics conceptions that differed completely from those underlying the quantum mechanics formulated by the Göttingen school. He himself stated: "It is hardly necessary to emphasize how much more agreeable it would be to represent a quantum transition as the passage of energy from one vibrational form into another, rather than to represent it as the jumping of electrons." In many passages Schrödinger (like Heisenberg) expressed his views in an almost polemical tone: "I . . . feel intimidated, not to say repelled, by what seem to me the very difficult methods [of matrix mechanics] and by the lack of clarity."

Despite his distaste for matrix mechanics, Schrödinger was "convinced of [its] inner connection" with wave mechanics. Hermann Weyl, to whom he had presented his purely mathematical problem, was unable to "provide the connecting link." Thereupon Schrödinger temporarily put aside his conjectures on the matter; but by the beginning of March 1926, much earlier than he had thought possible, he was able to show the formal, mathematical identity of the two theories.

The starting point for this analysis was the following observation:

> Given the extraordinary disparity, it is . . . odd that these two new quantum theories agree with each other even where they deviate from the old quantum theory. I note above all the peculiar "half-integrality" in the case of the oscillator and the rotator. This is truly remarkable, for the starting point, conception, method, and . . . entire mathematical apparatus appear to be fundamentally different for each theory.

Schrödinger remarked that Heisenberg's peculiar computational rules for functions of the $2n$

variables—q_1, q_2, $\cdots$, q_n, p_1, p_2, $\cdots$, p_n space and impulse coordinates—agree exactly with the computational rules that are valid in ordinary analysis for linear differential operators of n variables q_1, $\cdots$, q_n. The correspondence is of such a nature that each p_t in the function is replaced by the operator $\delta/\delta q_t$. As a result Schrödinger rewrote the equation $pq - qp = h/2\pi i$ (first formulated by Born) simply as $\dfrac{\delta}{\delta q} q - q \dfrac{\delta}{\delta q} = 1$, because the operator on the left side, applied to an arbitrary function of q, reproduces this function. On this basis Schrödinger proceeded to show the complete mathematical equivalence of the two theories. The matrices can be constructed from Schrödinger's eigenfunctions and vice versa.

With the demonstration of the mathematical identity of wave mechanics and matrix mechanics, physicists at last came into possession of the "new quantum theory" that had been sought for so long. In working with it they could use either of two mathematical tools: matrix computation or the method of setting up and solving a partial differential equation. Schrödinger's wave equation proved to be easier to handle. Moreover, physicists were more familiar with partial differential equations than with the new matrices. Therefore, Schrödinger's methods were more widely adopted for the mathematical treatment of the new theory. He contributed substantially to the elaboration of that treatment in his next two papers, especially through the development of his perturbation theory.

In his first publications Schrödinger had spoken of the wave function ψ as something that could be directly visualized—a vibration amplitude in three-dimensional space. He sought to interpret $\psi\bar{\psi}$ as electric charge density and hoped to establish physics on a thoroughgoing wave conception. Since, however, experiments clearly indicated the existence of strongly localized particles, he attempted to introduce the concept of the wave group: "One can try to construct a wave group of relatively small dimensions in all directions. Such a wave group presumably will obey the same laws of motion as an individual image point of the mechanical system."

Schrödinger attempted to develop this conception in "Der stetige Übergang von der Mikro- zur Makromechanik." It soon became apparent, however, that in almost all cases such a wave group disappears in infinitely short time and therefore cannot possibly represent a real particle. Schrödinger also observed that in the many-electron problem, the interpretation he originally had in mind is necessarily invalid in ordinary space: "$\psi\bar{\psi}$ is a sort of weight function in the configuration space of the system."

Shortly afterward Max Born interpreted $\psi\bar{\psi}$ as a probability, a view that Schrödinger considered a complete misinterpretation of his theory. From this time on, quantum theory developed in a way wholly different from the one Schrödinger had foreseen. In 1927 Heisenberg and Bohr succeeded in establishing, on a statistical foundation, an independent and consistent interpretation of quantum theory, the "Copenhagen interpretation." Schrödinger was "concerned and disappointed" that this "transcendental, almost psychical interpretation of the wave phenomena" had become "the almost universally accepted dogma." Schrödinger never changed his attitude on this subject, repeatedly defending the notion of "the electron as wave" and seeking to elaborate it without having recourse to the idea of "the electron as particle."

In 1927 Schrödinger accepted the prestigious offer, which had been declined by Arnold Sommerfeld, to succeed Max Planck in the chair of theoretical physics at the University of Berlin. At the same time he became a member of the Prussian Academy of Sciences. The University of Zurich vainly sought to persuade him to stay, offering him, among other inducements, a double professorship jointly with the Eidgenössische Technische Hochschule. Schrödinger was content in Zurich, despite occasional complaints; and his stay there had been very fruitful for the development of his scientific thought. Clearly, however, the city could not compete with Berlin, where, in the truest sense of the phrase, "physics was done." Berlin, with its two universities, the Kaiser Wilhelm Institute, the Physikalisch-Technische Reichsanstalt, and numerous industrial laboratories, offered the possibility of contact with a large number of first-rate physicists and chemists. Still, Schrödinger did not find it easy to make the decision. It was Max Planck who finally brought the vacillating Schrödinger to Berlin with the words: "It would make me happy"—as Schrödinger himself recorded in the Planck family guest book.

Although Schrödinger was extremely fond of nature, especially the Alps, and dreaded the prospect of living in a big city, he very much enjoyed his years in Berlin. He developed a close friendship with Planck, whose scientific and philosophical views were similar to his own. After the "wandering years from 1920 to 1927," this time of his life was "the very beautiful teaching and learning period."

In 1933 Schrödinger was deeply outraged at the new regime and its dismissal of outstandingly qualified scientists. Frederick A. Lindemann (later Viscount Cherwell) offered him the support of Imperial Chemical Industries; and after a summer vacation in Wolkenstein in the Grödnertal (Val Gardena), where he had a depressing meeting with Born and Weyl, Schrödinger moved to Oxford at the beginning of November. The fifth day after his arrival, he was accepted as a fellow of Magdalen College. At the same time the *Times* of London called his hotel to tell him that he had been awarded the Nobel Prize in physics for 1933, jointly with P. A. M. Dirac.

At Oxford, Schrödinger gradually became so homesick for Austria that he allowed himself to be persuaded to accept a post at Graz in the winter semester of 1936–1937. After the *Anschluss* he was subjected to strong pressure from the National Socialists, who had not forgotten his emigration from Germany in 1933. His friends at Oxford observed his difficulties with great concern.

As early as May 1938 Eamon de Valera, who had once been professor of mathematics at the University of Dublin, attempted to find a way of bringing Schrödinger to Ireland. By the time Schrödinger was dismissed, without notice, from his position at Graz on 1 September 1938, the first steps had already been taken. Fortunately, Schrödinger had been left his passport and was able to depart unhindered, although with only a small amount of baggage and no money. Passing through Rome and Geneva, he first returned to Oxford. De Valera had a law passed in the Irish Parliament establishing the Dublin Institute for Advanced Studies; but in order to keep busy until it opened, Schrödinger accepted a guest professorship at the Francqui Foundation in Ghent.

At the beginning of September 1939, Schrödinger, as a German émigré, suddenly found himself an enemy alien; but once more de Valera came to his assistance. Through the Irish high commissioner in Great Britain, he arranged for a letter of safe conduct to be issued for Schrödinger, who on 5 October 1939 passed through England on his way to Dublin with a transit visa valid for twenty-four hours. Schrödinger spent the next seventeen years in the Irish capital, where he was able to work in his new position undisturbed by external events. He later called these years of exile "a very, very beautiful time. Otherwise I would have never gotten to know and learned to love this beautiful island of Ireland. It is impossible to imagine what would have happened if, instead, I had been in Graz for these seventeen years."

The new Institute for Advanced Studies consisted of two sections, theoretical physics and Celtic languages, both located in a former townhouse on Merrion Square in Dublin. Young physicists from all over the world were given stipends enabling them to spend one or two years there. On the average there were ten to fifteen scholars in residence. Among them were Walter Thirring, Friedrich Mautner, Bruno Berdotti, and H. W. Peng. Like many of the others, Peng had previously worked with Max Born at Edinburgh. The yearly "summer school" in Dublin became famous as an informal gathering for the discussion of current problems of physics. Born and Dirac were frequent participants, and de Valera often came too.

In the years after his departure from Germany, Schrödinger published many works on the application and statistical interpretation of wave mechanics, on the mathematical character of the new statistics, and on its relationship to the statistical theory of heat. He also dealt with questions concerning general relativity, notably the relativistic treatment of wave fields, in contradistinction to the initial, nonrelativistic formulation of wave mechanics. In addition he wrote on a number of cosmological problems. Schrödinger, however, devoted an especially fervent effort, as did Einstein in his later years, to expanding the latter's theory of gravitation into a "unified field theory," the metric determination of which was to be established from a consideration of all the known forces between particles.

In his last creative period Schrödinger turned to a thorough study of the foundations of physics and their implications for philosophy and for the development of a world view. He wrote a number of studies on this subject in book form, most of them appearing first in English and then in German translation. It becomes particularly evident from the posthumously published *Meine Weltansicht* that Schrödinger was greatly concerned with the ancient Indian philosophy of life (Vedanta), which had led him to concepts that closely approximate Albert Schweitzer's "reverence for life." In "What Is Life?" Schrödinger points out why physics had amassed so little empirical evidence that might be applicable to the study of cell development: aperiodic crystals, in terms of which a gene should be considered, had not been investigated. But according to Delbrück's model, quantum physics made it possible to understand general persistence as well as the case of spontaneous mutation. Schrödinger was convinced that the biological process of growth could also be conceived on the basis of quantum theory according to the schema "order out of order." His analysis is outdated today; but during his lifetime it exerted enormous appeal among physicists (as Francis Crick corroborated) and induced many young people to study biology. Thus the great advances of molecular biology are indirectly linked to Schrödinger. He was a master of exposition, and Arnold Sommerfeld even spoke of a special "Schrödinger style." Schrödinger wrote and spoke four modern languages (as well as Greek and Latin), translated various items, and published a volume of poetry—while continuing to bestow great care on the preparation of his lectures, as is evident from their exceptional accuracy. To keep up this pace he required a marked alternation of intensely productive periods with creative pauses.

Soon after the end of the war, Austria tried to convince Schrödinger to return home. Even the president, Karl Renner, personally intervened in 1946; but Schrödinger was not willing to return while Vienna was under Soviet occupation. In the succeeding years he often visited the Tirol with his wife, but he did not return definitively until 1956, when he was given his own chair at the University of Vienna. A year later he turned seventy, the customary retirement age in Austria, but lectured for a further year (Ehrenjahr).

In his last years Austria honored Schrödinger with a lavish display of gratitude and recognition. Immediately after his return he received the prize of the city of Vienna. The national government endowed a prize bearing Schrödinger's name, to be awarded by the Austrian Academy of Sciences, and Schrödinger was its first recipient. In 1957 he was awarded the Austrian Medal for Arts and Science. He wrote that "Austria had treated me generously in every respect, and thus my academic career ended happily at the same Physics Institute where it had begun."

On 27 May 1957 Schrödinger was accepted into the German order Pour le mérite. He was also granted honorary doctorates from a number of universities and was a member of many scientific associations, including the Pontifical Academy of Sciences, the Royal Society of London, the Prussian (later German) Academy of Sciences in Berlin, and the Austrian Academy of Sciences. In 1957 Schrödinger survived an illness that threatened his life, and he never fully recovered his health. He died on 4 January 1961 and is buried in the small village of Alpbach, in his beloved Tirolean mountains.

BIBLIOGRAPHY

I. ORIGINAL WORKS. Schrödinger's important papers on wave mechanics are reprinted in *Abhandlungen zur Wellenmechanik* (Leipzig, 1927; 2nd ed., 1928); and *Die Wellenmechanik*, vol. 3 of Dokumente der Naturwissenschaft (Stuttgart, 1963), which contains an extensive bibliography compiled by E. E. Koch of Schrödinger's writings (pp. 193–199).

Some important correspondence is in Karl Przibram, ed., *Schrödinger. Einstein. Lorentz, Briefe zur Wellenmechanik* (Vienna, 1963), also translated into English by Martin J. Klein as *Letters on Wave Mechanics* (New York, 1967), which does not, however, contain Schrödinger's letter to Einstein (28 Apr. 1926). Unpublished letters to Arnold Sommerfeld are at the Sommerfeld estate in the library at the Deutsches Museum, Munich. Two letters to Hermann Weyl were published by Johannes Gerber in *Archive for History of Exact Sciences*, **5** (1969), 412–416. The sources of other letters to and from Schrödinger are in T. S. Kuhn *et al.*, *Sources for History of Quantum Physics. An Inventory and Report* (Philadelphia, 1967), 83–86.

II. SECONDARY LITERATURE. See Johannes Gerber, "Geschichte der Wellenmechanik," in *Archive for History of Exact Sciences*, **5** (1969), 349–416; Armin Hermann, "Erwin Schrödinger—eine Biographie," in *Die Wellenmechanik* (see above), 173–192; Max Jammer, *The Conceptual Development of Quantum Mechanics* (New York, 1966), 236–280; Martin J. Klein, "Einstein and the Wave-Particle Duality," in *Natural Philosopher*, **3** (1964), 1–49; V. V. Raman and Paul Forman, "Why Was It Schrödinger Who Developed de Broglie's Ideas?" in *Historical Studies in the Physical Sciences*, **1** (1969), 291–314; William T. Scott, *Erwin Schrödinger, an Introduction to His Writings* (Amherst, Mass., 1967); Robert Olby, "Schrödinger's Problem: What Is Life?" in *Journal of the History of Biology*, **4** (1971), 119–148.

ARMIN HERMANN

SCHROEDER VAN DER KOLK, JACOBUS LUDOVICUS CONRADUS (*b.* Leeuwarden, Netherlands, 14 March 1797; *d.* Utrecht, Netherlands, 1 May 1862), *medicine.*

Schroeder van der Kolk began the study of medicine in 1812 at the University of Groningen. As a student, he wrote two prize essays: one on the benefits accruing to the animal economy from the latent or combined caloric of air and water (1816) and the other on blood and its circulation (1819). In 1820 he received the M.D. for his dissertation on the coagulation of blood. He then established a medical practice in Hoorn but the following year was appointed resident physician at the Buitengasthuis in Amsterdam, where he treated about 400 patients daily. He also performed many postmortem examinations and did much on behalf of the 150 mental patients in the hospital. Thus he had abundant opportunity to acquire an extensive practical knowledge of various diseases. He prepared anatomical specimens and also formed a collection of pathological specimens. In 1826 he published his anatomical researches, *Observationes anatomico-pathologici et practici argumenti;* and in the same year he established a medical practice in Amsterdam. In 1827 he was appointed professor of anatomy and physiology at the University of Utrecht; he held this post until his death.

Schroeder van der Kolk's many articles on anatomy reveal both his skill in fine anatomical examination and his wide reading. At Utrecht he compared the anatomy of man with that of other vertebrates and often used the microscope to examine organs and tissues, a practice that was still uncommon. He also lectured on morbid anatomy and emphasized the value of anatomical investigations. In 1845 he discovered that tuberculous pulmonary tissue is easily recognized by the presence of elastic threads in the sputum. But he mistakenly concluded that these threads originate only from a tuberculous cavity.

To gain a better insight into brain disorders Schroeder van der Kolk closely examined the structure of the central nervous system. He conducted anatomical-physiological, pathological-anatomical, and clinical researches on the structure of the human brain and on that of higher animals. From 1855 he studied microscopically the spinal cord and the medulla oblongata. His most important discovery was the connection between the nervous fibers of the anterior roots of the medulla oblongata and the large branched cells of the anterior gray horns of the spinal cord.

Schroeder van der Kolk's studies are characterized by his accurate descriptions and fine illustrations. He published works on the anatomy of the tarsier *Stenops kukang* of the East Indies (1841); on the anatomy and physiology of the larva of the horse botfly *Gasterophilus intestinalis* (1845); on the structure of the lungs of birds (1858); on the liver of the elephant (1861); and, with Willem Vrolik, professor of anatomy at Amsterdam, on the comparative anatomy of the half-apes (1848). He

also wrote on the brain of the chimpanzee (1849) and the orangutan (1861), the vascular plexuses of the three-toed sloth *Bradypus tridactylus,* and on the limbs of birds (1848).

Schroeder van der Kolk was deeply influenced by vitalism, and he often proposed extreme teleological points of view, especially in his lectures on physiology (*ca.* 1840); he was resolute against the rising materialism of his time. Besides the primitive forces in nature (attraction and repulsion) and the imponderables (light, heat, electricity, galvanism, and magnetism), he saw in the vegetable and animal kingdoms various life-forces. In man, he believed that the nervous-force informs the sensibility with its impression of the outer world and that this force communicates the commands of the will to the muscles. Although he sought an empirical foundation for this philosophy, he was unable to escape from the concepts of *vis vitalis* and teleology. He was convinced that all events in the universe are focused on a good and just aim.

Besides his work in physiology and anatomy, Schroeder van der Kolk always strove for better care for the insane; his *Oratio de debita cura infaustam maniacorum sortem emendandi eosque sanandi in nostra patria neglecta* (1837) dealt with this subject. In 1827 he became a governor of the Utrechtsche Dolhuis and sought to improve both the treatment and housing of the insane. His reforms prompted legislation for general reform in the care of the mad; and after passage of the Lunacy Act in 1841, he was appointed inspector of lunatic asylums (1842–1862).

Schroeder van der Kolk's textbook on psychiatry was published posthumously by his pupil F. A. Hartsen. His clinical psychiatric concepts were also influenced by vitalism. He thought that body and soul interact in the life-force (or "brain-force"): in the insane the brain-force, rather than the soul (the "higher principle" in man), is ill. Thus the soul receives wrong data from the nervous-force and consequently reaches a wrong judgment.

BIBLIOGRAPHY

I. ORIGINAL WORKS. Schroeder van der Kolk's works include "Responsio ad quaestionem: quae sunt emolumenta praecipua, quae ex calorico latente, seu ligato, aëris et aquae ad oeconomiam animalem redundant," in *Annales Academiae Groninganae* (1815–1816); "Commentatio ad quaestionem, ab ordine medico anno 1818 propositam, de sanguinis vase effluentes coagulatione," *ibid.* (1818–1819); *Dissertatio physiologica-medica inauguralis, sistens sanguinis coagulantis historiam, cum experimentis ad eam illustrandam institutis* (Groningen, 1820); *Observationes anatomico-pathologici et practici argumenti* (Amsterdam, 1826); *Oratio de anatomiae pathologicae praecipue subtilioris studio utilissimo et ad morborum naturam investigandam maxime commendando* (Utrecht, 1827); and *Eene Voorlezing over het verschil tusschen doode natuurkrachten, levenskrachten en ziel* (Utrecht, 1835), with German trans. reprinted in *Opuscula selecta Neerlandicorum de arte medica*, **11** (1932), 283–359.

Later writings are *Oratio de debita cura infaustam maniacorum sortem emendandi eosque sanandi in nostra patria neglecta* (Utrecht, 1837), with English trans. reprinted in *Opuscula selecta Neerlandicorum de arte medica*, 7 (1927), 294–352; "Anatomisch-physiologisch onderzoek over het fijnere zamenstel en de werking van het ruggemerg," in *Verhandelingen der Koninklijke akademie van wetenschappen*, **2** (1855); and "Over het fijnere zamenstel en de werking van het verlengde ruggemerg en over de naaste oorzaak van epilepsie en hare rationele behandeling," *ibid.*, **6** (1858). The last two were translated by W. D. Moore as *On the Minute Structure and Functions of the Spinal Cord and Medulla Oblongata and On the Proximate Cause and Rational Treatment of Epilepsy* (London, 1859). See also *Handboek van de Pathologie en Therapie der Krankzinnigheid* (Utrecht, 1863), with trans. by J. T. Rudall as *The Pathology and Therapeutics of Mental Diseases* (London, 1870). Schroeder van der Kolk's lectures on physiology, "Physiologia corporis humani" (1840), are in G. ten Doesschate, *J. L. C. Schroeder van der Kolk als physioloog* (Utrecht, 1961).

II. SECONDARY LITERATURE. On Schroeder van der Kolk and his work, see C. A. Pekelharing, in *Nieuw Nederlandsch Biografisch Woordenboek*, II (Leiden, 1912), col. 700–705; W. Vrolik, in *Jaarboek van de Koninklijke Akademie van wetenschappen gevestigd te Amsterdam* (1862), 161–191; and P. van der Esch, *Jacobus Ludovicus Conradus Schroeder van der Kolk. 1797–1862. Leven en werken* (Amsterdam, 1954), with an extensive bibliography, pp. 95–119.

H. A. M. SNELDERS

SCHROETER, HEINRICH EDUARD (*b*. Königsberg, Germany [now Kaliningrad, R.S.F.S.R.], 8 January 1829; *d*. Breslau, Germany [now Wrocław, Poland], 3 January 1892), *mathematics*.

The son of a merchant, Schroeter attended the Altstädtische Gymnasium of his native city. In the summer of 1848 he began to study mathematics and physics at the University of Königsberg, and after his military service he continued his studies at Berlin for two years. He earned the doctorate at Königsberg in 1854 and qualified as lecturer in the fall of 1855 at the University of Breslau, where he became extraordinary professor in 1858 and full professor in 1861. He taught at Breslau until his death but was severely handicapped by paralysis during the final years of his life.

As a student at Königsberg, Schroeter attended the mathematics lectures of Friedrich Richelot, a follower of Jacobi. At Berlin his most important teachers were Dirichlet and Jakob Steiner. The influence of Steiner's ideas, on synthetic geometry in particular, was so strong that Schroeter later devoted almost all his research to this branch of mathematics. For his doctoral dissertation (under Richelot) and *Habilitationsschrift*, however, he chose topics from the theory of elliptic functions. Schroeter became more widely known through his association with Steiner—specifically, by editing the second part of Steiner's lectures on synthetic geometry.

The publication of Steiner's lectures ended with this second part, but Schroeter's extensive book of 1880 on the theory of second-order surfaces and third-order space curves can be considered a continuation of Steiner's work. Among the topics Schroeter treated were many metric properties of quadrics and cubic space curves; for unlike Staudt, for example, he did not confine himself to pure projective geometry. Schroeter pursued Steiner's fundamental aim of generating more complicated geometric elements from simpler ones (for instance, generating conic sections from the intersections of corresponding straight lines of projectively related pencils). Schroeter's name has been given to the generation of a third-degree plane curve c starting from six points of the plane, given that c should pass through six points of the plane and that further points are to be obtained using only a ruler; and to two generations of a third-degree surface when only one point and four straight lines in P_3 are given.

In 1888 Schroeter published a book in which he applied his approach to third-order plane curves. His last separately printed publication (1890) was devoted to fourth-order space curves of the first species, that is, to total intersections of two quadrics. Examining this topic from the viewpoint of synthetic geometry, Schroeter obtained many results on these curves, which are closely related to plane cubics. In his last years he studied various plane and spatial configurations, employing—as in all his writings—a purely elementary approach. In his view, all multidimensional considerations were not elementary, as were all those that were later designated by Felix Klein as belonging to higher geometry.

Schroeter's most important student in synthetic geometry was Rudolf Sturm.

BIBLIOGRAPHY

Schroeter's ed. of Steiner's work is . . . *Vorlesungen über synthetische Geometrie. Zweiter Teil: Die Theorie der Kegelschnitte, gestützt auf projectivische Eigenschaften* (Leipzig, 1867; 2nd ed., 1876). His own writings include *Die Theorie der Oberflächen 2. Ordnung und der Raumkurven 3. Ordnung als Erzeugnisse projectivischer Gebilde* (Leipzig, 1880); *Die Theorie der ebenen Kurven 3. Ordnung, auf synthetischem Wege abgeleitet* (Leipzig, 1888); and *Grundzüge einer rein-geometrischen Theorie der Raumkurven 4. Ordnung, I. Spezies* (Leipzig, 1890).

A biography is R. Sturm, "Heinrich Schroeter," in *Jahresberichte der Deutschen Mathematikervereinigung,* 2 (1893), 32–41.

WERNER BURAU

SCHRÖTER, JOHANN HIERONYMUS (*b.* Erfurt, Germany, 30 August 1745; *d.* Erfurt, 29 August 1816), *astronomy.*

Schröter studied law at Göttingen but also attended lectures in mathematics, physics, and astronomy, the last under Kästner. Upon completing his law studies he was appointed junior barrister in Hannover. Through his appreciation of music he met the Herschel family, who revived his interest in astronomy. In 1781 he became chief magistrate at Lilienthal, a post that left him free time to devote to astronomy. With the aid of the optician J. G. Schrader he built and equipped an observatory that subsequently became world-famous for the excellence of the instruments. Some were made in his own workshop; others he bought from Herschel, the latter including a reflector with a twenty-seven-foot focal length, the largest on the Continent. George III of England enabled Schröter to continue his astronomical work by buying all of his instruments, with the stipulation that they remain in Schröter's possession until his death, when they would become the property of the University of Göttingen. Schröter was also awarded a grant to hire an assistant. K. L. Harding and, later, F. W. Bessel were among those who held the post.

For thirty years the observatory at Lilienthal was a center of astronomical research and was visited by foreign astronomers. On 21 September 1800 it was the site of the congress organized to search the space between Mars and Jupiter for a planetary body.

Lilienthal was occupied during the Napoleonic Wars by the French, who looted and partly destroyed the observatory, although most of the instruments were saved. In the ensuing fire Schröter lost all copies of his own works, which he had published himself. He returned to Erfurt and built a new observatory, but his health failed and he did little observing. He died soon afterward.

Schröter was the first to observe the surface of the moon and the planets systematically over a long period. He made hundreds of drawings of lunar mountains and other features, and discovered and named the lunar rills. Unfortunately, his drawings were rough; and the standard of the images obtainable with the large reflectors was soon greatly improved by the refractors from the Munich workshops. *Selenotopographische Fragmente zur genauern Kenntniss der Mondfläche* was published at Lilienthal in 1791–1802. His observations of Venus appeared in *Aphroditographische Fragmente* . . . (Helmstedt, 1796), in which he estimated a rotation period of twenty-three hours and twenty-one minutes. He also thought that he observed mountains on the surface of Venus. In other works he noted lines on Mars (but did not call them canals), and he thought that the ring of Saturn was a solid body.

Schröter's reputation has been damaged by the many extravagant conclusions he drew from his observations. It may well be that his lasting influence on astronomy lies in the fact that he enabled Bessel and Harding to work in astronomy and that the selenographer J. F. J. Schmidt acquired his lifelong interest in the moon after he had read a copy of Schröter's work.

BIBLIOGRAPHY

A list of Schröter's publications can be found in Poggendorff, II, 846–847.

Secondary literature includes H.-B. Brenske, "Johann Hieronymus Schröter, der Amateurastronom von Lilienthal," in Walter Stein, ed., *Von Bremer Astronomen und Sternfreunden* (Bremen, 1958), 64–74; Gunther's article on Schröter in *Allgemeine deutsche Biographie,* XXXII, 570–572; and Dieter B. Herrmann, "Johann Hieronymus Schröter im Urtel seiner Zeit," in *Sterne,* **41** (1965), 136–143.

LETTIE S. MULTHAUF

SCHRÖTTER, ANTON VON (*b*. Olmütz, Austria [now Olomouc, Czechoslovakia], 26 November 1802; *d*. Vienna, Austria, 15 April 1875), *chemistry*.

Schrötter's father was an apothecary, and his mother was daughter of the mayor of Olmütz. In 1822, at his father's request, he entered the University of Vienna to study medicine but two years later turned to his true interest, mathematics and the natural sciences. On the advice of the mineralogist Friedrich Mohs, he devoted his attention to chemistry, learning applied analytical chemistry at the artillery school. Schrötter became assistant in physics and mathematics at the Technische Hochschule of Graz in 1827 and was promoted to professor of physics and chemistry three years later. Shortly thereafter he took a leave of absence and visited the important chemical laboratories in Germany and Paris. Using the experience gained on his trip, especially during his stay with Liebig at Giessen, he set up an impressive laboratory at Graz. In 1843 he was named professor of chemical technology at the Technische Hochschule in Vienna, and in 1845 he became professor of general chemistry. Appointed chief director of the mint in 1868, he held that position until his retirement in 1874.

Although he published over sixty papers on pure and applied chemistry, chiefly on the behavior of metals at very high and low temperatures, Schrötter is best known for his conclusive demonstration that red phosphorus (believed to be an oxide of white phosphorus) is truly an allotropic form of the element. In 1847 he demonstrated before the Vienna Academy that white phosphorus in a hermetically sealed bulb tube would turn red upon exposure to light although no oxygen or moisture was present. His suggestion of using amorphous phosphorus in the manufacture of matches led to the development of the safety match. This achievement brought him the Montyon Prize of the French Academy in 1856 and the Legion of Honor at the Paris Exhibition of 1855 for his great contribution to public safety. One of the active promoters of the Vienna Royal Academy of Sciences, Schrötter was a founding member and served as general secretary from 1850 until his death.

BIBLIOGRAPHY

I. ORIGINAL WORKS. Schrötter's important paper on red phosphorus appeared in several journals under several different titles. One is "Über einen neuen allotro-

pischen Zustand des Phosphors," in *Annalen der Physik und Chemie* (Poggendorff), **81** (1850), 276–298. An English abstract was published as "On the Allotropic Condition of Phosphorus," in *Report of the British Association for the Advancement of Science*, **19**, pt. 2 (1849), 42. His work on the safety match is described in his chapter on phosphorus and matches in August W. Hofmann, ed., *Bericht über die Entwickelung der chemischen Industrie während des letzten Jahrzehnts*, 2 vols. (Brunswick, 1875–1877). A bibliography of Schrötter's publications is in an unsigned obituary, "Anton Schrötter," in *Berichte der Deutschen chemischen Gesellschaft*, **9** (1876), 90–108.

II. SECONDARY LITERATURE. The obituary notice cited above is the most detailed of several short sketches of Schrötter's life. The most complete English obituary is an unsigned, untitled article in *Journal of the Chemical Society*, **29** (1876), 622–625. On his work with phosphorus, see Moritz Kohn, "The Discovery of Red Phosphorus (1847) by Anton von Schrötter (1802–1875)," in *Journal of Chemical Education*, **21** (1944), 522–554.

SHELDON J. KOPPERL

SCHUBERT, HERMANN CÄSAR HANNIBAL (*b*. Potsdam, Germany, 22 May 1848; *d*. Hamburg, Germany, 20 July 1911), *mathematics*.

Schubert, the son of an innkeeper, attended secondary schools in Potsdam and Spandau. He first studied mathematics and physics in 1867 at the University of Berlin and then went to Halle, where he received the doctorate in 1870. Soon afterward he became a secondary school teacher; his first post was at the Andreanum Gymnasium in Hildesheim (1872–1876). In 1876 he accepted the same post at the Johanneum in Hamburg. He remained there until 1908, having been promoted in 1887 to the rank of professor. Besides this school activity he was engaged by the Hamburg authorities to teach adult courses in which he dealt with various fields of mathematics for teachers already in the profession. In 1905 Schubert began to suffer from circulatory disorders that forced him to retire three years later. He died after a long illness that, toward the end, left him paralyzed. Schubert married Anna Hamel in 1873; they had four daughters.

Schubert published sixty-three works, including several books. His place in the history of mathematics is due chiefly to his work in enumerative geometry. He quickly established a reputation in that field on the basis of his doctoral dissertation, "Zur Theorie der Charakteristiken" (1870), and two earlier papers on the system of sixteen spheres that touch four given spheres. When he was only twenty-six, Schubert won the Gold Medal of the Royal Danish Academy of Sciences for the solution of a prize problem posed by H. G. Zeuthen on the extension of the theory of characteristics in cubic space curves (1874). A member of the Société Mathématique de France and honorary member of the Royal Netherlands Academy of Sciences, Schubert knew and corresponded with such famous geometers as Klein, Loria, and Hurwitz.

Schubert was content to remain in Hamburg, which had no university until 1919. Like Hermann Grassmann, he never became a university teacher and, in fact, declined offers that would have enabled him to do so. Mathematics in Hamburg centered in this period on the Mathematische Gesellschaft (founded in 1690 and still in existence), in the *Mitteilungen* of which Schubert published a number of papers.

In 1879 Schubert was able to present the methods and many individual results of his research in *Kalkül der abzählenden Geometrie*. Many further results were in papers he published until 1903.

Enumerative geometry is concerned with all those problems and theorems of algebraic geometry that involve a finite number of solutions. For example:

1. Bézout's theorem of the plane: two algebraic curves of orders a and b with no common elements have no more than ab points of intersection in common; this number can be reached.

2. Apollonius' theorem, according to which there are eight circles that simultaneously touch three given circles in the plane. Schubert's earliest works dealt with a spatial generalization of this theorem.

3. A somewhat more difficult result of enumerative geometry, Halphen's theorem: two algebraic linear congruences of P_3, one of order a and class b, and the other of order a' and class b', have in general $aa' + bb'$ straight lines in common.

Algebraically, the solution of the problems of enumerative geometry amounts to finding the number of solutions for certain systems of algebraic equations with finitely many solutions. Since the direct algebraic solution of the problems is possible only in the simplest cases, mathematicians sought to transform the system of equations, by continuous variation of the constants involved, into a system for which the number of solutions could be determined more easily. Poncelet devised this process, which he called the principle of continuity; in

his day, of course, the method could not be elucidated in exact terms. Schubert's achievement was to combine this procedure, which he called "the principle of the conservation of the number," with the Chasles correspondence principle, thus establishing the foundation of a calculus. With the aid of this calculus, which he modeled on Ernst Schröder's logical calculus, Schubert was able to solve many problems systematically.

In *Kalkül der abzählenden Geometrie* Schubert formulated his fundamental problem as follows: Let C_k be a given set of geometric objects that depend on k parameters. Then, on the model of Bézout's theorem, formulate theorems on the number of common objects of two subsets C_a and C'_{k-a} of C_k. Here C_a (and analogously C'_{k-a} are designated by certain characteristics, that is numbers $\rho_1, \cdots, \rho_s$ of objects that C_a has in common with certain previously designated elementary sets $E^1_{k-a}, \cdots, E^s_{k-a}$ of C_k of dimension $k - a$. The best known of Schubert's investigations are those for the case where C_k is the totality of all subspaces P_d of the projective P_n, where $k = (n - d)(d + 1)$. The appropiate elementary sets have since been known as Schubert sets, defined as follows: Let P_{a_i} ($i = 0, 1, \cdots, d$) be subspaces of P_n, each of them of dimension a_i with $0 \leqslant a_0 < a_1 < \cdots < a_d \leqslant n$ and $P'_{a_0} \subset P'_{a_1} \subset \cdots \subset P'_{a_d}$. Then Schubert designated as $[a_0, a_1, \cdots, a_d]$ the set of those P_d that intersect P'_{a_i} in at least i dimensions ($i = 0, 1, \cdots, d$). If the totality of all P_d in P_n is mapped into the points of the Grassmann-manifold $G_{n,d}$, there corresponds to $[a_0, a_1, \cdots, a_d]$ a subset of dimension $a_0 + a_1 + \cdots + a_d - \binom{d+1}{2}$ on $G_{n,d}$. Later investigations have shown that the Schubert sets are precisely the basic sets of $G_{n,d}$ in Severi's sense.

Another set that Schubert studied is the totality C_6 of all plane triangles. His results on this set were rederived and confirmed from the modern standpoint by J. G. Semple.

Schubert could not rigorously demonstrate the principle of the conservation of number with the means available in his time, and E. Study and G. Kohn showed through counterexamples that it could lead to false conclusions. Schubert avoided such errors through his sure instinct. In 1900, in his famous Paris lecture David Hilbert called for an exact proof of Schubert's principle (problem no. 15). In 1912 Severi published a rigorous proof, but it was little known outside Italy. B. L. van der Waerden independently established the principle in 1930 on the basis of the recently created concepts of modern algebra and topology.

Schubert was known to a broader public as the

editor of Sammlung Schubert, a series of textbooks in wide use before World War I. He wrote the first volume of the series, on arithmetic and algebra, and a subsequent volume on lower analysis. He also edited tables of logarithms and collections of problems for schools and published a simple method for computing logarithms.

Schubert was very interested in recreational mathematics and games of all kinds, including chess and skat, and in the mathematical questions that arise in connection with them. In 1897 he published the first edition of his book on recreational mathematics, *Mathematische Mussestunden;* the second edition, expanded to three volumes, appeared in 1900; and a thirteenth edition, revised by J. Erlebach, appeared in 1967. Schubert also was the author of the first article to appear in the *Encyklopädie der mathematischen Wissenschaften:* "Grundlagen der Arithmetik." His article, however, was subjected to severe criticism by the great pioneer in this area, Gottlob Frege.

BIBLIOGRAPHY

I. ORIGINAL WORKS. Schubert's writings include "Zur Theorie der Charakteristiken," in *Journal für die reine und angewandte Mathematik,* **71** (1870), 366–386; *Kalkül der abzählenden Geometrie* (Leipzig, 1879); "Abzählende Geometrie der Dreiecke," in *Mathematische Annalen,* **17** (1880), 153–212; *Mathematische Mussestunden* (Leipzig, 1897; 2nd ed., 3 vols., 1900; 13th ed., enl. by J. Erlebach, 1967); "Grundlagen der Arithmetik," in *Encyklopädie der mathematische Wissenschaften,* I, pt. 1 (1898), 1–29; *Arithmetik und Algebra* (Leipzig, 1898–1904); and *Niedere Analysis* (Leipzig, 1902).

II. SECONDARY LITERATURE. See W. Burau, "Der Hamburger Mathematiker Hermann Schubert," in *Mitteilungen der Mathematischen Gesellschaft in Hamburg,* 9th ser., **3** (1966), 10–20; G. Kohn, "Über das Prinzip von der Erhaltung der Anzahl," in *Archiv der Mathematik und Physik,* 3rd ser., **4** (1902), 312–316; J. G. Semple, "The Triangle as a Geometric Variable," in *Mathematica,* **1** (1954), 80–88; F. Severi, "Sul principio della conservazione del numero," in *Rendiconti del Circolo mathematico di Palermo,* **33** (1912), 313–327; "I fondamenti della geometria numerative," in *Annali di matematica pura ed applicata,* 4th ser., **19** (1940), 153–242; and *Grundlagen der abzählenden Geometrie* (Wolfenbüttel, 1948); and B. L. van der Waerden, "Topologische Begründung des Kalküls der abzählenden Geometrie," in *Mathematische Annalen,* **102** (1930), 337–362.

WERNER BURAU

(*b*. Cincinnati, Ohio, 3 July 1858; *d*. New Haven, Connecticut, 20 November 1942), *paleontology.*

Schuchert began his career as an untrained amateur collector of fossils and completed it as perhaps the most influential synthesizer of historical geology in North America. The oldest of six children of an impoverished immigrant Bavarian cabinetmaker, he received formal schooling only between the ages of six and twelve in a Catholic parochial school, which he was forced to leave to work in the family business. By the time he reached twenty, his father's health had failed and Schuchert was forced to support the entire household. In the meanwhile, fortunately, he had developed an intense interest in fossils and had amassed a significant collection.

Cincinnati, Ohio, lies in an area of Late Ordovician shales and limestone. These rocks are so crowded with fossils that before widespread urban paving specimens could even be collected from the street gutters after every rain. The city may well be built on the most fossiliferous spot on earth. For much of the nineteenth century, the "Cincinnati school" of enthusiastic amateurs was a vital part of the American study of paleontology.

In 1878 Schuchert joined forces with E. O. Ulrich, another local worker who rose to worldwide prominence as a paleontologist. Subsequently, Schuchert's business was destroyed by fire, but in the interim he had learned the art of lithographic illustration. Between 1884 and 1887 he was employed by Ulrich, and they drew more than 100 plates of illustrations of bryozoans—complex colonial organisms—for the geological surveys of Minnesota and Illinois. Concurrently he built up a magnificent collection of fossil brachiopods.

Both Schuchert's skill as an illustrator and the value of his collection induced James Hall, state geologist of New York, to employ him. Beginning in November 1888 he spent thirty months in Albany assisting in both the illustration and the writing of a classic text on brachiopods, but he received scant credit for his work. Schuchert worked for the Minnesota Geological Survey from 1891 to 1892, when available funds were exhausted. He then assisted C. E. Beecher at Yale in the preparation of fossils on large slabs that were exhibited at the Columbian Exposition in Chicago (1893).

C. D. Walcott, who had also worked under Hall's supervision, which bordered on the tyrannical, arranged in 1893 for Schuchert to join the U.S. Geological Survey in Washington. When Walcott became director of the Survey in 1894, most fossil collections were transferred to the U.S. National Museum, where Schuchert remained as curator for a decade. His summers were spent in fieldwork, including one season with R. E. Peary in western Greenland. During the winter he reorganized museum exhibits, curated collections, and studied fossils. He published about thirty papers during this decade, of which *Synopsis of American Fossil Brachiopoda* (1897) is the most enduring. He also prepared the section on Brachiopoda for the Zittel-Eastman *Text-Book of Paleontology* (1900).

Following Beecher's death in 1904, Schuchert was invited to join the Yale University faculty. At the age of forty-six he began a second career as a professor; the transition from museum to classroom apparently was painful but eventually highly successful. For the next twenty-one years he trained and influenced dozens of graduate students. His portion of *A Text-Book of Geology* (1915), written with L. V. Pirsson, went through several revisions and for at least three decades was the definitive American text of historical geology. For ten years Schuchert was chairman of the geology department of the Sheffield Scientific School and then served two additional years as head of a university-wide department.

A direct outgrowth of Schuchert's teaching efforts was the summarization of numerous stratigraphic details on maps to give a better picture of the changing distribution of land and seas during 600 million years. Although there is some doubt as to whether Schuchert or Ulrich was the first in North America to utilize this method of synthesizing data, there is no question as to the volume of work accomplished and the worldwide preeminence of Schuchert in this field. His 1910 "Paleogeography of North America" was a pioneer work and remains a classic. More than seventy-five other papers were written during this twenty-year interval. His work on older fossil starfish and a revision of the brachiopod section in the second edition (1913) of the Zittel-Eastman treatise are particularly noteworthy. He also directed substantial amounts of fieldwork with students in eastern Canada, a region in which he had special interest.

Schuchert completed his formal teaching career in 1923 but continued to advise and assist graduate students for almost two decades. His later research accomplishments were formidable, even for a bachelor "wedded to his science." With the aid of Clara Mae LeVene, he prepared another summary of published data on brachiopods, as well as a popular geology book and the definitive biography of the vertebrate paleontologist O. C. Marsh (1940).

Schuchert and Cooper (1932) is a standard reference for the study of two orders of brachiopods, but only one of many papers produced. In keeping his textbook on historical geology current, he became increasingly concerned with worldwide problems of correlation and methodology in the science, and this breadth of interest is reflected in the titles of his papers

The capstone of Schuchert's career was the publication of volume I of *Historical Geology of North America* (1935), on the Antillean-Caribbean region. A second volume, on the stratigraphy of the eastern and central United States, was published the year following his death; and a third was left partially completed. Without question Schuchert was the leader in synthesizing the geologic history of North America and was the last to grasp the entire literature encompassing details of 600 million years of change.

BIBLIOGRAPHY

Adolf Knopf, "Biographical Memoir of Charles Schuchert 1858–1942," in *Biographical Memoirs. National Academy of Sciences,* **27** (1952), 363–389, is the principal source that lists other memorials. In particular the memorial by C. O. Dunbar in *Proceedings of the Geological Society of America for 1942* (1943), 217–240, should be consulted. The esteem in which the Schuchert and Dunbar text on historical geology was held may be gathered from a review of the fourth edition (1941) by Cary Croneis in *Journal of Geology,* **49** (1941), 776–779.

ELLIS L. YOCHELSON

SCHULTZE, MAX JOHANN SIGISMUND (*b.* Freiburg im Breisgau, Germany, 25 March 1825; *d.* Bonn, Germany, 16 January 1874), *anatomy, microscopy.*

Schultze played a leading role in the movement to reform the cell theory as originally set forth by Schleiden and Schwann. Above all, Schultze and the other reformers disputed Schleiden and Schwann's emphasis on the cell wall and directed attention instead to the living substance (protoplasm) found within all cells, whether plant or animal.

Schultze was born to Frederike Bellermann and Karl August Sigismund Schultze, then professor of anatomy and physiology at the University of Freiburg. In 1830, when Schultze was five, his father became professor of anatomy at the University of Greifswald. After early education at home, where his interests in natural history, music, and drawing were nurtured and encouraged, Schultze attended the Gymnasium at Greifswald, and studied there from 1835 to 1845. In the summer of 1845 he entered the University of Greifswald as a medical student. He received all of his formal training there, except for the winter semester of 1846–1847, when he went to the University of Berlin. There he heard Johannes Müller lecture on anatomy and physiology and Ernst Brücke on the theory and use of the microscope. With a dissertation on the structure, function, and chemical composition of the arteries, Schultze graduated M.D. from Greifswald on 16 August 1849.

After another winter in Berlin, where he passed the state medical examination, Schultze returned to Greifswald as prosector in anatomy to his father. Later in 1850 he was also named *Privatdozent* in the faculty of medicine at Greifswald. He moved to Halle as assistant professor of anatomy in October 1854. In 1859 he became professor of anatomy and director of the anatomical institute at Bonn, where he remained despite offers from the universities of Strasbourg and Leipzig. At Bonn he planned and supervised the construction of a new anatomical institute, completed in 1872. Schultze was twice married; in 1854, to his cousin Christine Bellermann, who died of typhoid fever in 1865; and in 1868, to Sophie Sievers of Bonn. His death in 1874 was attributed to a perforated duodenal ulcer.

In 1851 Schultze found chlorophyll in the flatworm Turbellaria, thereby contributing to the recognition that animals as well as plants can contain that substance. His achievement won him an honorary Ph.D. from the University of Rostock in 1852 and the Blumenbach Traveling Scholarship from the University of Berlin in 1853. Schultze used this scholarship to go to Italy, where he studied marine zoology on the shores of the Adriatic Sea. Like Dujardin before him, he focused on the semifluid substance within the calcareous shells of the foraminifera. In a monograph of 1854 he described the results of this work and proposed the creation of a new class of shelled rhizopods, the Monothalamia, which lacked the internal partitions of Ehrenberg's Polythalamia. The monograph earned him Ehrenberg's lasting enmity.

In 1858 Schultze drew attention to the remarkable similarity between cyclosis in lower plants (notably the marine diatoms) and the streaming of granules in the pseudopodia of foraminifera and other lower animals. He also found that physical and chemical agents produced strikingly uniform effects on the contractile contents of plant and animal cells. By 1860 his studies of protozoa had led him to a generalization that implied a redefinition of the cell: "The less perfectly the surface of the protoplasm is hardened to a membrane, the nearer to the primitive membraneless condition does the cell find itself, a condition in which it exhibits only a small lump of protoplasm with nucleus."[1]

This definition of the cell as "ein nacktes Protoplasmaklümpchen mit Kern" became famous chiefly through Schultze's paper "Ueber Muskelkörperchen und das was man eine zelle zu nennen habe" (1861). At the time, controversy surrounded these "muscle corpuscles"—small, granular, spindlelike, nucleated masses of protoplasm found among the contractile fibers of striated muscle. Some histologists took these bodies to be complete cells while others supposed they were merely isolated nuclei. Schultze claimed that the argument stemmed mainly from disagreement over the definition of a cell. If histologists would only abandon the old "botanical" conception of the cell as a "bladderlike structure with membrane, contents, and nucleus," if they would recognize instead that a cell need not have a chemically distinct membrane, then they might agree with Schultze that the muscle corpuscles were wall-less cells that had fused to form a colonial muscle fiber. This particular conclusion eventually lost favor among histologists, who came to regard the muscle fiber as a single, multinucleated cell rather than as fused colony of many independent cells. But Schultze's redefinition of the cell and his emphasis on the cell substance won widespread support, despite the opposition of Remak and Reichert.

Even before Schultze entered the arena, Alexander Braun, Ferdinand Cohn, and Franz Leydig, among others, had sought to establish an identity between plant and animal cell substances and to insist that a cell need not have a distinct membrane. Schultze himself admitted that he intended only "to dress in words that which many have long perceived, though perhaps less definitely."[2] That his work nonetheless attracted so much attention can be ascribed to two factors: (1) unlike his predecessors, Schultze gave prominent attention to a tissue (muscle) characteristic of higher, differentiated animals; (2) he also campaigned for the adoption of a single word—protoplasm—to refer to the cell substance of both plants and animals. Following Dujardin, zoologists had generally used the name "sarcode" for the contractile contents of animal cells. Schultze urged them to adopt instead the name used by botanists for the plant cell substance and thereby to acknowledge "the complete correspondence that exists between plant and animal cells in all essential respects."[3]

This seemingly trivial suggestion helped to crystallize thinking about the substance of life, and the 1860's became "a heyday for speculation upon the nature of protoplasm and for the celebration of its amazing properties."[4] Gradually, as it became clear that protoplasm was not a unitary chemical substance but a dynamic emulsion of several substances, and as the quest for a substance of life focused increasingly on the nucleus, protoplasm lost much of its allure. Moreover, the detection of the plasma membrane—notably through Overton's plasmolytic studies of the 1890's—qualified Schultze's claim that a cell required no limiting boundary. Nonetheless, Schultze's critique of the original cell theory—and especially of the place of the cell wall in that theory—retained much of its cogency.

Apart from his role in the reform of the cell theory, Schultze did his most important work on the sense organs, particularly the retina, which was the subject of his inaugural lecture at Bonn in 1859. In a monograph of 1867 Schultze sought especially to elucidate the physiological role of the rods and cones. Emphasizing that the rods predominated in nocturnal animals (including the bat, the cat, and the owl), he suggested that these structures were better adapted than the cones for the simple perception of light. Partly because the sense of color in humans was proportional to the number of cones in a given region of the retina, he argued that the cones probably acted as the terminal nerve organs of the color sense, although they obviously served other visual functions as well.

Schultze also did valuable descriptive and taxonomic work, especially on rhizopods and sponges, although he shared the common skepticism of German naturalists toward Darwinian evolutionary theory. In 1864 he described prickles in the stratified squamous epithelium of mammalian tongue and skin, but without recognizing them as plasmodesmata. In 1865 he gave a clear description of the blood platelets. His studies of bioluminescence and of the electric organs of fishes also attracted considerable attention.

Schultze founded in 1865 and edited until his death the *Archiv für mikroskopische Anatomie und Entwicklungsmechanik*. This esteemed journal, in which he published most of his later papers, won the support of many of the leading German histologists and microscopists of the day. A consummate master of microscopic technique, Schultze introduced osmic acid as a fixative and iodized serum as a preservative "physiological fluid." He also designed a "hot stage" for the microscope, allowing the investigator to heat his preparations within reasonably precise temperatures. In the judgment of N. E. Nordenskiöld, Schultze "brought cytology to the farthest point possible" before the introduction of the microtome.[5]

NOTES

1. Max Schultze, "Die Gattung Cornuspira unter den Monothalamien und Bemerkungen über die Organisation und Fortpflanzung der Polythalamien," in *Archiv für Naturgeschichte*, **26** (1860), 299. Cf. *ibid.*, 305. See also Baker, "The Cell Theory," pt. 3, pp. 164–165.
2. Schultze, "Ueber Muskelkörperchen," 8.
3. *Ibid.*, 2, n. 2.
4. Coleman, "Cell, Nucleus, and Inheritance," 128.
5. Nils Erik Nordenskiöld, *The History of Biology: A Survey*, trans. from the Swedish by Leonard Bucknall Eyre (New York, 1949), 533–534.

BIBLIOGRAPHY

I. ORIGINAL WORKS. Schultze's published monographs include *Beiträge zur Naturgeschichte der Turbellarien. Erste Abtheilung* (Greifswald, 1851); *Ueber den Organismus der Polythalamien (Foraminiferen) nebst Bemerkungen über die Rhizopoden im Allgemeinen* (Leipzig, 1854); *Das Protoplasma der Rhizopoden und der Pflanzenzelle* (Leipzig, 1863); and *Zur Anatomie und Physiologie der Retina* (Bonn, 1867). For his famous paper on "muscle corpuscles," see "Ueber Muskelkörperchen und das was man eine Zelle zu nennen habe," in *Archiv für Anatomie, Physiologie und wissenschaftliche Medizin* (1861), 1–27.

In his obituary notice Schwalbe (see below) gives a chronological bibliography of eighty-two items by Schultze. The Royal Society *Catalogue of Scientific Papers*, V. 571–573; VIII, 894–895; XII, 66, lists eighty-eight papers by Schultze alone and one written with M. Rudneff. Nonetheless, Schwalbe's bibliography is probably complete, for the Royal Society *Catalogue* sometimes gives a separate listing to trans. or to multiple items that Schwalbe includes under one entry.

II. SECONDARY LITERATURE. G. Schwalbe's obituary notice, in *Archiv für mikroskopische Anatomie und Entwicklungsmechanik*, **10** (1874), i–xxiii, provides the basis for the article by Theodor H. Bast, in *Annals of Medical History*, n.s. **3** (1931), 166–178. Brief sketches, also derivative from Schwalbe, appear in *Allgemeine deutsche Biographie*, LIV (1908), 256–257; and in *Biographisches Lexicon der hervorragenden Ärzte aller Zeiten und Völker*, 2nd ed., V (1934), 162–163.

More generally, see William Coleman, "Cell, Nucleus, and Inheritance: An Historical Study," in *Proceedings* of the American Philosophical Society, **109** (1965), 124–158; G. L. Geison, "The Protoplasmic Theory of Life and the Vitalist-Mechanist Debate," in Isis, **60** (1969), 273–292; and "Towards a Substance of Life: Concepts of Protoplasm, 1835–1870" (M.A. thesis, Yale Univ., 1967), esp. ch. 4; and John R. Baker, "The Cell Theory: A Restatement, History, and Critique," in Quarterly Journal of Microscopical Science, in five parts: **89** (1948), 103–125; **90** (1949), 87–108; **93** (1952), 157–190; **94** (1953), 407–440; and **96** (1955), 449–481. References to Schultze will be found in pt. 2, pp. 95–96, and in pt. 3, pp. 164–165, 172, 176, 180, 189.

GERALD L. GEISON

SCHULZE, FRANZ FERDINAND (b. Naumburg, Germany, 17 January 1815; d. Rostock, Germany, 15 April 1873), *chemistry, microbiology.*

Schulze's career was centered principally around teaching agricultural chemistry at Eldena and chemistry and pharmacy at Rostock. Although his particular interests were wide-ranging, the core of much of his work was his expertise in analytical chemistry. He was not a great innovator but made a number of useful modifications to existing analytical techniques and equipment, such as in gas analysis and in the use of the blowpipe in the production of laboratory glassware (1).

Many of Schulze's activities were in the field of applied science, for example his long paper (1868) on the examination of well water for "those particles which are most relevant in hygiene"(2). In this study, prompted by an outbreak of cholera in Rostock in 1866, Schulze included a judicious summary of the difficulties of interpreting the nature and significance of airborne organic matter. He thus contributed to the current, far-reaching debate on whether microorganisms could be spontaneously generated from, for example, the floating organic matter that was widely believed also to cause fermentation and putrefaction. Schulze indicated that much organic matter was harmless, but the difficulty lay in identifying that which was undoubtedly poisonous. He felt, too, that the latter might be synonymous with the "mysterious domain" of microorganisms, as was being suggested by Pasteur. Schulze's interest in the subject of spontaneous generation of microorganisms extended back to 1836. At that time he demonstrated that after air was bubbled slowly through sulfuric acid, no growth of organisms occurred in a sterile culture medium through which the air was next passed (3). This carefully conducted experiment had consider-

able influence, for it was repeated frequently (sometimes with contradictory results) during the peak of controversy (from the late 1850's to the 1870's) over the question of spontaneous generation of microorganisms. It thus contributed significantly to the developing awareness of the experimental difficulties involved in handling microorganisms.

Although the title of Schulze's paper suggested that it was a preliminary communication, he published nothing more on spontaneous generation and immersed himself in agricultural and chemical topics. Much of his chemistry involved natural products, including the difficult areas of lignins and carbohydrates (4). Although current chemical techniques limited what he could accomplish, some of his results on the chemical similarities of lignins from various sources, and on the properties of starch, were sound. Apart from his laboratory achievements, Schulze also contributed to education, including his translation into German (5) of J. F. W. Johnston's *Elements of Agricultural Chemistry and Geology* (Edinburgh, 1841). Johnston was notably successful in stimulating interest in the application of science to agriculture, and Schulze wrote in his preface to the translation that "the more we [Germans] have reason to turn our attention to the practical sense and high level of development of [English] agriculture, the greater must be our trust in their judgment" (6). This comment—although apparently forgetting Liebig's agricultural studies—was just, and it also illustrates Schulze's own practical outlook.

BIBLIOGRAPHY

Schulze's writings include (1) "Die gasvolumetrische Analyse, als Hülfsmittel für wissenschaftliche agriculturchemische und technische Untersuchungen," in *Zeitschrift für analytische Chemie*, **2** (1863), 289–300; and "Beschreiben eines für chemische Laboratorien anwendbaren gebläse Apparates," in *Journal für praktische Chemie*, **43** (1848), 368–372; (2) "Ueber die Untersuchung der Brunnenwässer auf diejenigen Bestandtheile, welche für die Gesundheitspflege am meisten in betracht kommen," in *Dinglers polytechnisches Journal*, **188** (1868), 197–219; (3) "Vorläusige Mittheilung resultate einer experimentallen Beobachtung über generatio aequivoca," in *Annalen der Physik*, **39** (1836), 487–489; (4) "Beitrag zur Kenntniss des Lignins," in *Chemisches Zentralblatt*, n.s. 2 (1857), 321–325; and "Ueber die Metamorphose des Amylums," in *Annalen der Physik*, **39** (1836), 489–493; (5) *Anfangsgründe der praktischen Agricultur-Chemie und Geologie* (Neubrandenburg, 1845); and (6) *ibid.*, p. 4.

Schulze's many publications are in a variety of journals. No comprehensive list has been compiled, but most of his works are referred to in F. Ferchl, *Chemisch-pharmazeutisches Bio- und Bibliographikon* (Mittenwald, 1938), 490. Some background to his teaching career, particularly at Rostock, is in R. Schmitz, *Die Deutschen pharmaceutische chemischen Hochschulinstitute* (Stuttgart, 1969).

J. K. Crellin

SCHUMACHER, HEINRICH CHRISTIAN (*b*. Bad Bramstedt, Holstein, Germany, 3 September 1780; d. Altona, Germany, 28 December 1850), *astronomy, geodesy.*

Schumacher was the son of Andreas Schumacher, a magistrate, who died when the boy was nine. Schumacher was then placed under the care of a minister named Dörfer at the Lutheran church of Altona. He attended the Gymnasium at Altona, where he was introduced to mathematics, astronomy, and the use of astronomical instruments; studied jurisprudence at the universities of Kiel and Göttingen; and was awarded the LL.D. at Göttingen in 1806. In 1805–1806 he was a *Dozent* at the Faculty of Law of Dorpat and wrote two legal treatises. It was at Dorpat that Schumacher's interest in astronomy was revived by J. Pfaff.

In 1807 Schumacher obtained a salaried position at the University of Göttingen and studied astronomy under the direction of Gauss. His firm friendship with Gauss, begun at this time, was lifelong. From 1808 to 1810 Schumacher studied mathematics in Hamburg and translated Lazare Carnot's *La géométrie de position*. J. G. Repsold gave Schumacher access to the observatory at Hamburg, where he made a series of observations that served as the basis of a new star catalog.

In 1810 Schumacher was named extraordinary professor of astronomy at the University of Copenhagen; but he did not assume the duties connected with this post until after Thomas Bugge's death in 1815, serving in the meantime as director of the observatory at Mannheim (1813–1815). In 1817 the Danish government released Schumacher from his duties so that he could take part in the geodetic survey of Schleswig and Holstein.

During the years 1800–1825, the mapping of territory was in progress in many European states including Holland, Prussia, Hesse, and Bavaria. The work simultaneously in progress in many centers had to be coordinated. Schumacher was involved in the measurement of a degree between Skagen and Lauenburg, and also in the determination of the longitude along the arc between Copenhagen and the west coast of Jutland.

In 1821 Schumacher was appointed to direct the survey by the Royal Danish Academy of Copenhagen. In the same year the king, Frederick VI, arranged for the building and equipping of the observatory at Altona, where Schumacher worked for many years. His determination of the base line for the measurement of a degree between Skagen and Lauenburg was a masterpiece of accuracy and was in almost perfect agreement with the Hannoverian measurement by Gauss and the Hessian and Bavarian triangulations. Bessel used Schumacher's results in his calculations of the figure of the earth.

Schumacher rendered considerable service to astronomers of his day by the institution of various publications. Between 1820 and 1829 he published astronomical ephemerides and auxiliary tables in a form that did not necessitate reduction calculations. From 1829 the *Berliner astronomisches Jahrbuch*, edited by Bessel, continued to publish similar tables. In 1823 Schumacher edited the first volume of *Astronomische Nachrichten*, a journal to which astronomers of all nations could contribute. The founding of this journal, which is still published, is perhaps his greatest contribution to astronomy. He was at the center of a lively correspondence with the leading astronomers of his day, including Gauss, Olbers, and Bessel.

In conjunction with the English Board of Longitude, Schumacher in 1824 determined the difference in longitude between the observatories of Greenwich and Altona, using English and Danish chronometers. In 1830 he determined the length of a seconds pendulum at the castle of Güldenstein in Holstein, and between 1837 and 1839 he carried out experimental work for the Danish government on a comparison of the most important legal units of weight. Schumacher continued his topographical work until 1837, when the preparation of maps was taken over by the army.

Schumacher was elected to the Royal Society of London in 1821, and his portrait by H. Wolf was presented to the Society by the artist in 1847. Schumacher was honored by the Danish kings Frederick VI and Christian VIII.

BIBLIOGRAPHY

I. Original Works. Sixty-four titles of papers and scientific writings by Schumacher are listed in the Royal Society *Catalogue of Scientific Papers*, V, 576–577. See also Poggendorff, II, cols. 866–867.

II. Secondary Literature. A biographical account

of Schumacher is given in a lengthy obituary by C. F. R. Olufsen in *Astronomische Nachrichten,* **36**, supp. 864 (1853), cols. 393–404. Further biographical information appears in *English Cyclopaedia,* Biography Div., V (London, 1867), 343; *Allgemeine deutsche Biographie,* XXXIII, 32–33; and *Dansk biografisk Leksikon,* XXI, 429–432.

SISTER MAUREEN FARRELL, F.C.J.

SCHUMANN, VICTOR (*b.* Markranstädt, near Leipzig, Germany, 21 December 1841; *d.* Leipzig, 1 September 1913), *photography.*

Schumann is known among physicists as UV Schumann because of his development of photographic plates capable of registering ultraviolet light. The son of Karl Ferdinand Schumann, a physician, he attended the local elementary school and was then sent to a *Realschule* in Leipzig. Following a year of practical training in an engineering works, he studied mechanical engineering from 1861 to 1864 at the Royal Technical College at Chemnitz (now Karl-Marx-Stadt). In 1865 he was cofounder of an engineering works in Leipzig of which he served as technical director. Schumann was obliged to retire in 1893 because of overwork and damage to his eyes that appeared to be leading to blindness. His first wife, Auguste Baumgarten, daughter of a factory owner in Chemnitz, whom he married in 1871 after a youthful romance, died only seven years later. He remained a widower until 1909, when he married Elise Börner.

Schumann was an extremely successful engineer and designer, especially of machines for book production. It was, however, in photography that he won international recognition. Schumann took up photography as a hobby in 1872 and, after the death of his wife, spent all his limited leisure time on it. He became a member of the Berlin Photographic Society in 1882 (and later an honorary member), and in 1894 the University of Halle awarded him an honorary doctorate. During the last three years of his life he was also a full member of the Royal Saxon Academy of Sciences in Leipzig.

Schumann took up photography while it was still in its infancy. The necessity for long exposures was further complicated by the impossibility of representing various tints by different degrees of photographic density: a red or yellow object, for example, appeared as dark as a black one. Schumann found, however, that adding a small amount of silver iodide to the customary silver bromide emulsion not only permitted significantly shorter exposures, but also yielded markedly improved contrasts for red and yellow objects. This finding contradicted the publications of experts in England and on the Continent, who claimed that the addition of silver iodide produced still weaker contrasts. Accepting the challenge implicit in this disagreement, Schumann demonstrated that both sides were justified in their claims. The experts had prepared silver bromide and silver iodide emulsions separately and then mixed them, a technique that resulted in inferior plates. By reversing the procedure—dissolving the light-sensitive salts together in water and then adding the gelatin—Schumann produced plates that were far superior. Earlier, in 1873, he had increased sensitivity to red by allowing solutions of certain organic dyes to act on prepared plates. The announcement of this technique stimulated a wealth of further research by both amateurs and industry.

Schumann held that the most informative way of investigating the sensitivity of photographic plates to color was to expose them through a spectroscope. He was particularly interested in the ultraviolet region. As George Stokes had shown, this region could be made strikingly visible through a device in which the collector is formed by placing quartz lenses and calcite prisms on plates of fluorescent uranium glass. A problem remained, however: Why did shorter wavelengths of light darken the plates less than longer wavelengths? Schumann discovered two reasons for this. First, because the gelatin absorbed the light before it reached the silver bromide, he prepared plates in which these particles were not dispersed throughout the layer of gelatin but, rather, adhered to the top of it. Second, he established that the air in the spectrograph was highly disruptive, creating problems of both dispersion and, especially, absorption. Schumann discovered that a layer of air even 0.10 millimeter thick could completely absorb a spectral line.

Since no firm could furnish or build him a vacuum spectrograph, Schumann constructed one himself. The prisms and plates had to be adjusted precisely from outside the apparatus, and it took him a full year to devise an aperture that fulfilled his requirements. When finished, the device produced sharper spectra than any previously seen; with his plates it became possible to photograph spectra as low as 1,270 Å, whereas the limit had been 4,000 Å. (Theodore Lyman subsequently recorded spectra at 1,030 Å.) This further advance made it possible to demonstrate the Lyman series of hydrogen (predicted by J. J. Balmer), the first member of which lies at 1,215 Å. Since the majority of spectral lines of the elements lie in the ultraviolet region, the amateur photographer Schumann also became a pioneer of spectroscopy and of atomic physics.

BIBLIOGRAPHY

I. ORIGINAL WORKS. Schumann's writings include "Wirkung d. AgJ auf die Lichtempfindlichkeit d. AgBr-Gelatine," in *Photographisches Archiv* (1882); "Über die Photographie der Lichtstrahlen kleinster Wellenlänge," in *Sitzungsberichte der Akademie der Wissenschaften in Wien*, **102**, sec. 2a (1893), 415–475, 625–694; "Über ein neues Verfahren zur Herstellung ultraviolettempfindlicher Platten," *ibid.*, 994–1024; "New Method of Preparing Plates Sensitive to Ultraviolet Rays," in *Astrophysical Journal*, **3** (1896), 220–226, 387–395; and **4** (1896), 144–154; and "Verbessertes Verfahren zur Herstellung ultraviolettempfindlicher Platten," in *Annalen der Physik*, 4th ser., **5** (1901), 349–374. He also made many contributions to *Jahrbuch für Photographie und Reproductionstechnik* between 1887 and 1903.

II. SECONDARY LITERATURE. Obituaries are T. Lyman, in *Astrophysical Journal*, **39** (1914), 1–4, with portrait; and O. Wiener, "Victor Schumann," in *Berichte über die Verhandlungen der Sächsischen Gesellschaft der Wissenschaften*, **65** (1913), 409–413.

F. FRAUNBERGER

SCHUNCK, HENRY EDWARD (*b.* Manchester, England, 16 August 1820; *d.* Kersal, near Manchester, 13 January 1903), *organic chemistry.*

Schunck was of German descent, his father having settled in Manchester as the founder of the firm Schunck, Mylius and Co., textile merchants and calico-printers; his mother was a daughter of Johann Mylius. He was educated privately and received his first instruction in practical chemistry in the laboratory of William Henry, a family friend. He then studied under Heinrich Rose and H. G. Magnus at Berlin, and with Liebig at Giessen, before returning about 1842 to enter his father's printing works at Rochdale. After a few years, however, his increasing wealth enabled him to detach himself from the day-to-day management of the firm and to devote himself to research, mainly on the chemistry of the natural coloring matters.

Soon after returning to England, Schunck published the results of his work on the isolation and analysis of a pure crystalline compound (lecanorin) from the lichens that furnished the old dyestuffs archil and cudbear and the indicator litmus. Later (1847) he isolated the glycoside rubian, the precursor of alizarin in madder root, and studied other colored substances that accompany it. In 1853 he isolated the colorless precursor of indigo; this was indican, later shown to be a glycoside of indoxyl, which forms indigo as a result of oxidative hydrolysis in the dyebath.

Although a skillful and painstaking practical chemist and analyst, Schunck had little interest in theory. It was only while Hermann Roemer was serving as his assistant (1875–1879) that he began to relate his analyses to structural formulas. He and Roemer made important contributions to the study of the polyhydroxy anthraquinones, which were then becoming available synthetically; and they were among the first to use absorption spectroscopy as a tool for the identification of colored compounds. Later, assisted by Leon Marchlewski, Schunck made extensive studies of chlorophyll and its congeners, again using absorption spectroscopy. This problem was so difficult, however, that they achieved little, except to demonstrate a chemical relationship between chlorophyll and hemoglobin. Schunck's son Charles also participated in this work.

Although Schunck never held an academic post, he was the leading figure in the scientific life of Manchester for fifty years. He was four times president of the Manchester Literary and Philosophical Society, one of the earliest members of the Chemical Society, and a founder-member (later president) of the Society of Chemical Industry. He was elected fellow of the Royal Society in 1850. Shortly before his death he gave Manchester University the then huge sum of £20,000 for the endowment of research. His laboratory—reputedly the best private laboratory in the world—and his extensive library were bequeathed to the University, to which they were moved. He married in 1850 and had seven children, of whom only four survived him.

BIBLIOGRAPHY

I. ORIGINAL WORKS. Schunck and his assistants wrote nearly 100 papers. Of special interest is his address to the British Association, in *Report of the British Association for the Advancement of Science*, **57** (1887), 624–635, expressing the opinion that the future of organic chemistry lay in the elucidation of biological processes; and his presidential address to the Society of Chemical Industry, in *Journal of the Society of Chemical Industry*, **17** (1897), 586–594, in which he recalls the scientific personalities of Manchester in his younger days.

II. Secondary Literature. There are notices on Schunck in *Manchester Faces and Places*, IX (1898), 1–6, with portrait; and in *Dictionary of National Biography, 1901–1911*, 274–275. The main obituaries are in *Memoirs and Proceedings of the Manchester Literary and Philosophical Society*, 47 (1902–1903), xlix–liii; *Berichte der Deutschen chemischen Gesellschaft*, 36 (1903), 305; *Journal of the Society of Chemical Industry*, 22 (1903), 84; *Journal of the Society of Dyers and Colourists*, 19 (1903), 35–36, with portrait; *Nature*, 67 (1903), 274; and *Proceedings of the Royal Society*, 75 (1904–1905), 261–264. A short account of his scientific work is W. H. Perkin, Jr., "The Chemical Researches of Edward Schunck. . . .," in *Memoirs and Proceedings of the Manchester Literary and Philosophical Society*, 47, no. 6 (1902–1903), 1–8.

For an obituary, see *Berichte der Deutschen chemischen Gesellschaft*, 18 (1885), 285–289.

W. V. Farrar

SCHUR, ISSAI (*b.* Mogilev, Russia, 10 January 1875; *d.* Tel Aviv, Palestine [now Israel], 10 January 1941), *mathematics.*

Schur was one of the most brilliant Jewish mathematicians active in Germany during the first third of the twentieth century. He attended the Gymnasium in Libau (now Liepaja, Latvian S.S.R.) and then the University of Berlin, where he spent most of his scientific career. From 1911 until 1916, when he returned to Berlin, he was an assistant professor at Bonn. He became full professor at Berlin in 1919. Schur was forced to retire by the Nazi authorities in 1935 but was able to emigrate to Palestine in 1939. He died there of a heart ailment two years later. Schur had been a member of the Prussian Academy of Sciences before the Nazi purges. He married and had a son and daughter.

Schur's principal field was the representation theory of groups, founded a little before 1900 by his teacher Frobenius. Schur seems to have completed it shortly before World War I; but he returned to the subject after 1925, when it became important for physics. Further developed by his student Richard Brauer, it is in our time experiencing an extraordinary growth through the opening of new questions. Schur's dissertation (1901) became fundamental to the representation theory of the general linear group; in fact English mathematicians have named certain of the functions appearing in the work "S-functions" in Schur's honor. In 1905 Schur reestablished the theory of group characters—the keystone of representation theory. The most important tool involved is "Schur's lemma." Along with the representation of groups by integral linear substitutions, Schur was also the first to study representation by linear fractional substitutions, treating this more difficult problem almost completely in two works (1904, 1907). In 1906 Schur considered the fundamental problems that appear when an algebraic number field is taken as the domain; a number appearing in this connection is now called the Schur index. His works written after 1925 include a complete description of the rational and of the continuous representations of the general linear group; the foundations of this work were in his dissertation.

A lively interchange with many colleagues led Schur to contribute important memoirs to other areas of mathematics. Some of these were published as collaborations with other authors, although publications with dual authorship were almost unheard of at that time. Here we can only indicate the areas. First there was pure group theory, in which Schur adopted the surprising approach of proving without the aid of characters theorems that had previously been demonstrated only by that means. Second, he worked in the field of matrices. Third, he handled algebraic equations, sometimes proceeding to the evaluation of roots, and sometimes treating the so-called equations without affect, that is, with symmetric Galois groups. He was also the first to give examples of equations with alternating Galois groups. Fourth, he worked in number theory, notably in additive number theory; fifth in divergent series; sixth in integral equations; and lastly in function theory.

BIBLIOGRAPHY

Schur's writings are collected in *Gesammelte Abhandlungen*, A. Brauer and H. Rohrbach, ed., 3 vols. (Berlin, 1973). Moreover, two lectures have been published as *Die algebraischen Grundlagen der Darstellungstheorie der Gruppen: Zürcher Vorlesungen 1936*, E. Stiefel, ed.; and *Vorlesungen über Invariantentheorie*, H. Grunsky, ed. (Berlin, 1968).

On Schur and his work, see *Mathematische Zeitschrift*, 63 (1955–1956), a special issue published to commemorate Schur's eightieth birthday, with forty articles dedicated to his memory by leading mathematicians. See also Alfred Brauer, "Gedenkrede auf Issai Schur, gehalten 1960 bei der Schur-Gedenkfeier an der Humboldt-Universität Berlin," in *Gesammelte Abhandlungen*, I, which contains a detailed report of Schur's life and work.

H. Boerner

SCHUSTER, ARTHUR (*b.* Frankfurt, Germany, 12 September 1851; *d.* Yeldall, near Twyford,

Berkshire, England, 14 October 1934), *physics, applied mathematics.*

Schuster was the son of Francis Joseph Schuster, a well-to-do Jewish textile merchant with business connections in Great Britain. After the Seven Weeks' War the family firm moved to Manchester, England, when Frankfurt was annexed by Prussia. Schuster, baptized as a young boy, was educated privately and at the Frankfurt Gymnasium. He attended the Geneva Academy from 1868 until he joined his parents at Manchester in the summer of 1870.

By the age of sixteen Schuster had developed an interest in physical science, mainly through Henry Roscoe's elementary textbook on spectrum analysis. His parents saw at once that he lacked enthusiasm for business; and they consulted Roscoe, then professor of chemistry at Owens College, Manchester, who arranged for Schuster to enroll as a day student in October 1871. He studied physics under Balfour Stewart and was directed in research in spectrum analysis by Roscoe. Within a year he produced his first research paper, "On the Spectrum of Nitrogen." Again at Roscoe's suggestion, Schuster enrolled at Heidelberg under Kirchhoff and received his Ph.D. after a less-than-brilliant examination in 1873.

Schuster served at Owens in 1873 as unpaid demonstrator in the new physics laboratory and later, at the request of Lockyer, joined an eclipse expedition to the coast of Siam. Upon his return to England in 1875, Schuster remained at Owens for a semester and then joined Maxwell as a researcher at the Cavendish Laboratory, where he remained for five years, ultimately joining Lord Rayleigh in an absolute determination of the ohm.

In 1879 Schuster applied for a post at Mason Science College, Birmingham, but was rejected in favor of his friend J. H. Poynting. Two years later, when a professorship of applied mathematics was founded at Owens, Schuster was selected for the chair over his former student J. J. Thomson and Oliver Lodge. Subsequently he was rejected as Rayleigh's successor at the Cavendish in 1884; but after Balfour Stewart's death in 1887 he succeeded in the following year to the chair of physics at Manchester.

At the beginning of his Owens College career, Schuster resumed his interest in what was by then termed "spectroscopy." In an important paper, "On Harmonic Ratios in the Spectra of Gases" (*Proceedings of the Royal Society,* **31** [1881], 337–347), he refuted G. J. Stoney's explanation of spectral lines that used simple harmonic series

by demonstrating statistically that the spectra of five chosen elements conform more closely to a random distribution than to Stoney's "law." He concluded, however, that "Most probably some law hitherto undiscovered exists which in special cases resolves itself into the law of harmonic ratios." In 1897 Schuster independently discovered and published the relationship known as the Rydberg-Schuster law, which relates the convergence frequencies of different spectral series of the same substance.

Schuster's interests led him to investigate the spectra produced by the discharge of electricity through gases in otherwise evacuated tubes. Such electrical discharges were imperfectly understood, and he began a series of detailed investigations that led to his Bakerian lectures before the Royal Society in 1884 and 1890. Schuster's findings were of major importance: he showed that an electrical current was conducted through gases by ions and that once a gas was "dissociated" (ionized), a small potential would suffice to maintain a current.

Schuster was also the first to indicate the path toward determining the ratio e/m for cathode rays by using a magnetic field, a method that ultimately led to the discovery of the electron. In 1896, shortly after the appearance of Roentgen's researches, he offered the first suggestion that X rays were small-wavelength transverse vibrations of the ether.

Schuster's interests were too wide-ranging to give even a brief account here. His work on terrestrial magnetism, however, deserves special notice. In 1889 he showed that daily magnetic variations are of two kinds, internal and atmospheric. He attributed the latter to electric currents in the upper atmosphere, and the former to induction currents in the earth. In a later estimate of the ionization of the upper atmosphere he helped lay the groundwork for the studies of Heaviside and Kennelly.

In 1907 Schuster resigned his chair at Manchester and secured Ernest Rutherford as his successor, thus reinforcing Manchester's prominence in physical research.

Elected a fellow of the Royal Society in 1879, Schuster served twice on its Council and was secretary from 1912 to 1919. He was founder and first secretary of the International Research Council and served as president of the British Association in 1915. He was knighted in 1920.

A man of remarkable originality and ingenuity, Schuster often pointed the way toward novel areas but left the task of reaching research summits to others, a pattern perhaps inevitable in a period of exploding possibilities for one of such wide interests and perception.

BIBLIOGRAPHY

I. ORIGINAL WORKS. A record of Schuster's scientific papers from 1881 to 1906 is in *The Physical Laboratories of the University of Manchester* (Manchester, 1906), 45–60; papers published to 1901 are listed in the Royal Society *Catalogue of Scientific Papers*, VIII, 899; XI, 359–360; XVIII, 623–625. Schuster's major books include *Spectrum Analysis*, 4th ed. (London, 1885), written with H. E. Roscoe; *Introduction to the Theory of Optics* (London, 1904; 3rd ed., 1924); *The Progress of Physics During 33 Years (1875–1908)* (Cambridge, 1911); and *Biographical Fragments* (London, 1932). With Arthur Shipley he wrote *Britain's Heritage of Science* (London, 1917), a fascinating Victorian view of the history of science.

II. SECONDARY LITERATURE. On Schuster's life and work the following are of special value: G. C. Simpson, "Sir Arthur Schuster, 1851–1934," in *Obituary Notices of Fellows of the Royal Society of London*, 1 (1932–1935), 409–423; "Sir Arthur Schuster, FRS," in *Nature*, 134 (1934), 595–597; and his article in *Dictionary of National Biography*; G. E. Hale, "Sir Arthur Schuster," in *Astrophysical Journal*, 81 (1935), 97–106; R. S. Hutton, *Recollections of a Technologist* (London, 1964), pp. 103–106; and J. G. Crowther, *Scientific Types* (London, 1968), 333–358. See also *Manchester Faces and Places*, IV (1892–1893), 158–159; and *Commemoration of the 25th Anniversary of the Election of Arthur Schuster, F.R.S., to a Professorship in the Owens College* (Manchester, 1906).

On Schuster's work see Edmund Whittaker, *History of the Theories of Aether and Electricity*, I (New York, 1960), pp. 355–360; Norah Schuster, "Early Days of Roentgen Photography in Britain," in *British Medical Journal* (1962), 2, 1164–1166; D. L. Anderson, *The Discovery of the Electron* (Princeton, 1964), pp. 30, 42, 74; and William McGucken, *Nineteenth-Century Spectroscopy* (Baltimore, 1969), *passim*.

On Schuster at Owens, see P. J. Hartog, *The Owens College Manchester* (Manchester, 1900), pp. 54–59; and H. B. Charlton, *Portrait of a University* (Manchester, 1951), pp. 78–84.

ROBERT H. KARGON

SCHWABE, SAMUEL HEINRICH (*b.* Dessau, Germany, 25 October 1789; *d.* Dessau, 11 April 1875), *astronomy*.

Schwabe's father was a physician, and the family apothecary business was derived from his mother's family. After working from 1806 to 1809 as an assistant in the business, Schwabe continued his pharmaceutical studies at the University of Berlin in 1810–1812, under Klaproth and Hermbstädt. While at Berlin he became interested in astronomy and botany. Schwabe worked as an apothecary from 1812 until 1829, when he sold the business in order to give his time fully to his scientific interests.

On 17 December 1827 Schwabe rediscovered the eccentricity of Saturn's rings. In 1843 he made his first definite statement regarding the periodicity of sunspots, giving statistics for 1826–1843. He tabulated his results under four headings: the year, the number of groups of sunspots in the year, the number of days free from sunspots, and the number of days when observations were made. Schwabe realized that with the modest apparatus in his private observatory, numerical determination was difficult; for instance, on days when there was a large number of spots, he had probably underestimated them. His carefully compiled results demonstrated the existence of periodicity, although he wrongly estimated the period to be about ten years. His discovery remained unnoticed until Humboldt drew attention to it in 1851. Because Schwabe was an amateur astronomer, his discovery was all the more noteworthy, for the investigation of the occurrence of sunspots had been judged unprofitable by Lalande and Delambre.

After Schwabe's discovery, Rudolf Wolf collated all existing sunspot data and recalculated the period as just over eleven years. In 1857 the Royal Astronomical Society awarded Schwabe its gold medal. He was elected a member of the Royal Society of London in 1868.

BIBLIOGRAPHY

I. ORIGINAL WORKS. The Royal Society *Catalogue of Scientific Papers*, V, 582–585; and VIII, 901, lists 109 of Schwabe's printed papers and scientific works, including papers on the phenomena of frost patterns, haze, and mock suns. In 1865 he published a flora of Anhalt. Thirty-one volumes of Schwabe's astronomical observations were transferred after his death into the keeping of the Royal Astronomical Society. They cover his work from 1825 to 1867 and are held in the Society archives at Burlington House, London. See also Poggendorff, II, 871; III, 1223; VI, 2391.

II. SECONDARY LITERATURE. Biographical information is in *Allgemeine deutsche Biographie*, XXXIII, 159–161; T. Arendt, *Schwabe: Leben und Wirken* (Dessau, 1925), which I have not been able to see; and Gustav Partheil, "Samuel Heinrich Schwabe der dessauer Astronom und Botaniker," in *Jahrbuch, Heimat-*

liches, für Anhalt 1926. Accounts of his work appear in Alexander von Humboldt, *Kosmos: Entwurfeiner physischen weltbeschreibung,* III (Stuttgart–Augsburg, 1851), 379–405; and H. H. Turner, *Astronomical Discovery* (London, 1904), 155–176.

Sister Maureen Farrell, F.C.J.

SCHWANN, THEODOR AMBROSE HUBERT (*b.* Neuss, Germany, 7 December 1810; *d.* Cologne, Germany, 11 January 1882), *physiology.*

An account of Schwann's early life has some of the qualities of an edifying tale of saintly childhood. To his teachers and fellow pupils in primary school and at the progymnasium, Schwann was a cooperative child, diligent and modest. Little tempted by the delights of society, lacking self-confidence, and excessively shy, he withdrew into study, family life, and piety. Equally brilliant in all branches of learning, he showed a particular inclination for mathematics and physics. Given his lack of interest in the outside world, it was accepted that his vocation should be directed toward the Church when he left his native town in 1826 to enter the Jesuit College of the Three Crowns in Cologne.

Here Schwann came under the influence of an exceptional religious teacher, Wilhelm Smets. To Schwann, until then acquainted only with the strict aspects of piety, but also endowed with a brilliant intelligence and a lively sensibility, Smets's teaching of religion was the revelation of an entirely new aspect of God and especially of the singular fact of the liberty of man with regard to the whole of nature. It was from him that Schwann learned the lesson of the elevation of man by personal perfection.

Increasingly enamored with reason, Schwann renounced theology to take up medical studies. His philosophical position became that of a Christian rationalist whose personal philosophy was in the tradition of Descartes and Leibniz.

In October 1829 Schwann entered the University of Bonn, where he enrolled in the premedical curriculum and obtained his bachelor's degree in 1831. During this time, he attended Johannes Müller's lectures on physiology and began to assist him in the laboratory. In the autumn of 1831 he moved to Würzburg, where he studied for three semesters, attending clinical lectures. In April 1833 he left Würzburg for Berlin, where Müller had been appointed to teach anatomy and physiology. In Berlin, Schwann attended clinical demon-strations and, under Müller's guidance, prepared a dissertation on the necessity of air for the development of chicken eggs. He obtained the M.D. on 31 May 1834 and passed the state examination on 26 July. Schwann immediately became one of Müller's assistants and devoted all his time and efforts to research.

Although he remained a practicing Catholic, Schwann abandoned himself, especially after the death of his mother in 1835, to an extreme mechanistic tendency, which guided him in the impressive work he accomplished at Müller's laboratory in Berlin between 1834 and 1839. Schwann's conception of God at this time was the philosophical and impersonal God of Descartes.

During this period, Müller was working on the *Handbuch der Physiologie,* which introduced into Germany Magendie's experimental method in medical studies. Until his death Müller remained a convinced vitalist. Recourse to experimentation was for him (as it had been for Bichat) a means of studying the effects of the vital force peculiar to each organ. Restricted in his chemical and physical background, he progressively detached himself from physiology and devoted himself entirely to comparative morphology, in which field he acquired fame. On the other hand, from the beginning of his career as a researcher Schwann took a completely different position, which inaugurated the quantitative period of physiology.

Müller's *Handbuch* was not merely a compilation; he critically examined all the notions that he printed. Repeating the experiments of others, suggesting new ones, opening avenues not yet explored, this treatise is a work as unique in its conception as in its realization. In the section entrusted to him, Schwann enriched Müller's *Handbuch* with the results of extensive work and contributed numerous new notions. This book also contains an account of a study clearly showing Schwann's innovating tendency; his first experiments can be dated, on the basis of his laboratory notebooks, at 16 April 1835. In these Schwann envisaged experiments in which it would be possible to subject the physiological properties of an organ or of a tissue to physical measurement. One such method involved measuring the secretion of a gland. But it was the muscle that seemed to him likely to furnish the most rewarding results. He planned to measure the length of a muscle contracted by the action of the same stimulus for different loads or, further, to compare the intensity of the contraction with that of the stimulus. He accomplished this experiment by means of the "muscular balance" and in a sense established the first tension-length diagram.

The influence in physiological circles of this simple experiment is difficult for us to appreciate. "It was the first time," as du Bois-Reymond emphasized, "that someone examined an eminently vital force as a physical phenomenon and that the laws of its action were quantitatively expressed." In a milieu where the idealistic philosophy and the theories of Fichte and Hegel were still dominant, the *fundamental Versuch* came as a revelation and constituted the point of departure for a new physiology. Dissociating itself from the teaching of Müller and resolutely abandoning the notion of vital force for the study of molecular mechanisms, the school stemming from Schwann's experiment was distinguished particularly by the work of his successors at the Berlin laboratory, Emil du Bois-Reymond and Hermann von Helmholtz.

Parallel with his experiments on muscle, Schwann pursued the researches that led to his discovery of pepsin. About 1835, however, Gay-Lussac's observations, prompted by Nicolas Appert's experiments, made acceptable the notion that oxygen was the agent of both fermentation and putrefaction. This observation stimulated a recrudescence of the theories of spontaneous generation and a tendency to return to the ideas of Needham, for whom the effect of heat was to deprive the air of the oxygen necessary for the birth of "animalcules."

Having observed that neither infusorians nor the smell of putrefaction appeared in a maceration of meat that had been boiled, Schwann noted the appearance of both these phenomena when he used an unboiled maceration or unheated air. Convinced that it was the destruction of germs that prevented the development of infusorians and molds, and that prevented putrefaction, Schwann wished to make a counterproof by showing that the heating of air did not prevent the operation of a chemical process to which it contributed oxygen, and not germs. He demonstrated that a frog breathes normally in previously heated air; and he investigated alcoholic fermentation, which also depended, in the current opinion, on the presence of oxygen. To his great astonishment, Schwann observed that heating the air he bubbled through a boiled suspension of yeast in a sugary solution prevented fermentation in certain experiments. In January 1836 he noted in his laboratory notebook the conclusion that alcoholic fermentation is the work of a live organism.

The description of the multiplication of yeast cells appears in Schwann's laboratory notebook under the date 16 February 1836. The first public announcement of the relationship between alcoholic fermentation and the life cycle of yeast was by Cagniard de La Tour, who described the multiplication of yeast in the issue of *L'Institut* for 25 November 1836. Schwann's paper (1837), however, independently demonstrated the living nature of the agent of fermentation and presented arguments of a new sort.

Schwann was led to the idea that alcoholic fermentation was related to the metabolism of yeast by his conception that putrefaction was related to the metabolism of live organisms. The prevailing doctrine in Müller's laboratory was the vitalism derived from Paracelsus, and his principles were hostile to the Cartesian unity of natural forces. Schwann's mechanistic and unitarian antagonism toward this intellectual attitude had already been clearly manifested in his studies of muscle, of the mechanism of digestion, and of fermentation. His tendency to introduce a more exact mode of explanation than the current one in terms of the "vital force" culminated in the formulation of the cell theory.

The cell theory prolonged, on the biological terrain, the old debate over continuity and discontinuity in nature. The search for a common structural principle of live organisms, excluding such imaginary entities as Buffon's "molecules," has preoccupied many scientists. In his biography of Virchow, E. H. Ackerknecht distinguished several searches for a common principle. In the eighteenth century the principle was the "fiber." According to this view, which Ackerknecht designated as the first cell theory, the development of fibers began in little globules, like those recognized by Prochaska (1797). After these views were abandoned, a new theory appeared that John R. Baker termed "globulist," Ackerknecht's second cell theory. Its adherents included Lorenz Oken, Meckel, Mirbel, Dutrochet, Purkyně, Valentin, and Raspail. The notion of "globule" embraced a great variety of elementary units: particles and nuclei, as well as optical illusions. The globulists often included some form of cell among their "globules"; but none of them can be regarded as having conceived of the organism as composed *solely* of cells, of modified cells, or of products of cells. It was not until 1830 that the perfecting of the microscope permitted Robert Brown to recognize the presence of the nucleus as the essential characteristic of the plant cell.

In 1839, in his *Mikroskopische Untersuchungen*, Schwann formulated what Ackerknecht called the third cell theory, which insists on the common cellular origin of every living thing. By "cell" Schwann meant "a layer around a nucleus" that could differentiate itself: covered by a membrane, as the site of deposit of a more consistent substance; growing hollow, as a vacuole; or fusing itself with the "layer" of other cells. He also thought (incorrectly) that cells form around a nucleus within a "blastema," an amorphous substance that can be intracellular or extracellular. Ackerknecht's fourth cell theory, which remains current, is that of Remak and Virchow, the first part of which follows Schwann in acknowledging the cellular composition of organisms, with the cell as the vital element, the bearer of all the characteristics of life. The second part of this theory, expressed in the dictum "omnis cellula e cellula," contradicts Schwann's erroneous belief in the formation of cells within a "blastema."

The *Mikroskopische Untersuchungen* is composed of three parts. The first is devoted to a microscopic study of the *chorda dorsalis* in frog larvae. Studying that structure, Schwann found that it consists of polyhedral cells that have in or on the internal surface of their wall a structure corresponding to the nucleus of plant cells. New cells are formed within parent cells. He also found the structure of cartilage to be in accordance with the tissues of plants, and he believed that he had observed that the cartilage cells contain a nucleus and that they originate by formation of the nucleus, around which the cell develops. Schwann therefore was convinced that the cells of the *chorda dorsalis* and of cartilage were derived from structures of the same kind as the plant cells, with nucleus, membrane, and vacuole.

In the second part Schwann presented a demonstration of the same notion with regard to much more specialized elementary parts. He found that the varied forms of the "elementary parts" of tissues—be they epithelium, hoof, feather, crystalline lens, cartilage, bone, tooth, muscle tissue, fatty tissue, elastic tissue, nerve tissue—are products of cellular differentiation. The conclusion he drew from this observation was that "elementary parts," although quite distinct in a physiological sense, may be developed according to the same laws. The elementary parts of most tissues, when traced back from their state of complete development to their primary conditions, are only developments of cells.

In the third part, of a philosophical nature, Schwann, on the basis of his cell theory expounded in the first two parts, developed a theory of the cells that he presented as purely hypothetical. He stated that according to the cell theory, one may suppose that an organized body is not produced, as was accepted by theological theories, by a fundamental power guided in its operation by a definite idea, but is developed, according to blind laws of necessity, by powers that, like those of inorganic nature, are established by the very existence of matter. He believed that the source of life phenomena resided in another combination of the materials of the inorganic world, whether it be in a peculiar mode of union of the atoms to form molecules, or in the arrangement of these conglomerate molecules to form the parts of organisms.

Schwann stated that two groups of phenomena attend the formation of cells: those relating to the combination of molecules to form a cell (plastic phenomena) and those resulting from chemical changes in the component particles of the cells (metabolic phenomena). The cell attracts particles from its medium, which is not a mere solution of cell material but contains this material in other combinations, and it produces chemical changes in these particles. In addition, all the parts of the cell itself may be chemically modified during the process of its growth by a "metabolic power" that is an attribute of the cell itself; this is demonstrated by alcoholic fermentation, which provides a representation of the process that is repeated in all the cells of an organism. Metabolic changes occur not only in the cell contents but also in the solid parts, for example, the nucleus and the membrane. The metabolic processes, in which heat is evolved, are produced only at certain temperatures. All cells demonstrate respiration, a fundamental condition of metabolism; and each cell produces chemical changes in particular organic substances.

Schwann discerned a relation between the phenomena of cell formation, as he conceived it, and the phenomena of crystallization, a comparison developed at length, but only as a hypothesis.

Schwann defined his attitude toward the vital force, as it was accepted by Müller, who proposed the notion of the proper energy of tissues, thus:

A simple force different from matter, as it is supposed, the vital force would form the organism in the same way as an architect constructs a building according to a plan, but a plan of which he is not conscious. Furthermore, it would give to all our tissues that which is called their proper energy, that is, the properties that distinguish living tissues from dead tissues: muscles would owe it their contractility, nerves their irritability, glands their secretory function. Here, in a word, is the doctrine of the vitalist school. Never was I able to conceive the existence of a simple force that would itself change its mode of action in order to realize an idea, without, however, possessing the characteristic attributes of intelligent beings. I have always preferred to seek in the Creator rather than in the created the cause of the finality to which the whole of nature evidently bears witness; and I have also always rejected as illusory the explanation of vital phenomena as conceived by the Vitalist school. I laid down as a principle that these phenomena must be explained in the same way as those of inert nature [*Manifestation en l'honneur de M. le professeur Th. Schwann . . .*].

Schwann sought to replace theological explanation with physical explanation. For him the phenomena of life were not produced by a force acting according to an idea, but by forces acting blindly and with necessity, as in physics. Individual finality, as it was observed in each organism, was determined in the same manner as in inert nature: its explanation depended entirely upon the characteristics of matter and upon the blind forces with which it had been created by an infinitely intelligent being.

Schwann found the confirmation of this view in the notion of the uniformity of the texture and the growth of animals and plants, as he developed it in his cell theory. "The uniformity of this development demonstrated that it is the same force that everywhere unites molecules into cells, and that this force could be nothing but that of molecules or atoms: the fundamental phenomenon of life therefore had to have its *raison d'être* in the properties of atoms." The error suggested to Schwann by Schleiden—the formation of cells within a blastema, which Schwann tentatively compared to the phenomenon of crystallization—satisfied his chemical and physical preferences to such a high degree that one can understand a little more easily why he accepted it on the strength of arguments as weak as those he presented to demonstrate it: those concerning an alleged preexistence of the nucleus in the cartilages, for example.

The solution to the philosophical problem of finality proposed by Schwann transferred it from biology to the universe and its constituent particles, and from the vital force to the Creator. It continued to be influential philosophically, and Lotze was notably inspired by it in his celebrated article on the nature of life, "Leben, Lebenskraft" (1842). Schwann's cell theory can be regarded as marking the origin in biology of the school of mech-

anistic materialism that Brücke, du Bois-Reymond, Helmholtz, and Carl Ludwig made famous. According to Schwann, the theory that led from the molecule (the molecule of the chemist) to the organism by way of the universal stage of the cell, was inspired by an intellectual, mechanistic reaction to Müller's vitalism. Erroneous as it appears now in certain of its aspects, this theory led him to the inestimably significant discovery of the development of organisms through cellular differentiation.

Schwann's short and brilliant scientific career extended from 1834 to 1839, after which he abandoned rationalism and became a mystic. The scientist gave way to the professor, the inventor, and the theologian. The beginning of this transformation dates from the attacks directed at Schwann by the chemists. Having shown an exceptional insensitivity to epistemological obstacles during his years of fruitful work, he nevertheless succumbed to a particularly violent attack dictated by one of these obstacles. At the beginning of 1839, there appeared in the *Annalen der Pharmacie*, following a translation of a general paper by Turpin on the mechanism of alcoholic fermentation considered as a result of the activity of yeast, an article entitled "Das enträthselte Geheimnis der geistigen Gährung." The work of Wöhler, embellished by Liebig with some particularly ferocious touches, this satirical text presented a caricature of the views of Cagniard, Schwann, and Kützing on the role of yeast in alcoholic fermentation.

According to this facetious article, yeast suspended in water assumes the form of animal eggs that hatch with an unbelievable rapidity in a sugary solution. These animals, in the shape of an alembic, have neither teeth nor eyes but do have a stomach, an intestine, an anus (in the form of a pink dot), and urinary organs. Immediately upon leaving the eggs, they throw themselves on the sugar and devour it; it penetrates their stomachs, is digested, and produces excrements. In a word, they eat sugar, expelling alcohol through the anus and carbonic acid through the urinary organs. Moreover, their bladder has the shape of a champagne bottle.

Shortly afterward, a lengthy memoir by Liebig appeared in the same periodical. In it he formulated the theory of alcoholic fermentation as the result of instability produced in sugar by the instability of a substance formed through the access of air to the nitrogenous substances of plant juices. This theory enjoyed a long popularity among chemists, and it was not until Pasteur that justice was done to Cagniard, Schwann, and Kützing. The cruel treatment of Schwann by the scientific leaders of his time made it impossible for him to pursue a scientific career in Germany.

At the same time Schwann's ardent rationalism grew lukewarm; he became preoccupied with religious meditations, doubtless fostered by the influence of his brother Peter, a theologian. The brother was the author (under the pseudonym Dr. J. F. Müller) of an edition of *The Imitation of Christ*. His failure in his candidature for a chair at the University of Bonn, added to his other disappointments, sent Schwann into exile in 1839, when he became professor of anatomy at Louvain. But the mainspring of enthusiasm and discovery was broken. Like Pascal before him, he abandoned rationalism to return to the God of his childhood, the "God of the heart, not of reason." A conscientious professor at Louvain and at Liège (from 1848), Schwann spent the rest of his life in a solitary existence darkened by episodes of depression and anxiety.

Before he went to Liège in 1848, Schwann had been approached by his friend F. A. Spring, who presented the proposals of the Belgian government, including a substantial increase of salary and the promise, never fulfilled, of the construction of an institute of anatomy. When Schwann was appointed, he first received the chair of anatomy, which previously had been held by Spring, who also taught physiology. In 1858 Spring became professor of general pathology and clinical medicine, and Schwann of physiology as well as of general anatomy and embryology. In 1872 he abandoned general anatomy and in 1877 embryology, teaching only physiology until he retired in 1879.

During his stay at Louvain (1839–1848), Schwann developed a method of utilizing the biliary fistula for the study of the role of bile in digestion, and concluded that a lack of bile secretion in the digestive tract is incompatible with survival. He received the Sömmering Medal (1841); and in 1847 the Sydenham Society published an English version of his *Mikroskopische Untersuchungen*, translated by Henry Smith, who in his introduction presented the following judgment: "The treatise has now been seven years before the public, has been most acutely investigated by those best competent to test its value, and the first physiologists of our day have judged the discoveries which it unfolds as worthy to be ranked among the most important steps by which the science of physiology has ever been advanced."

The papers of 1844 and 1845 that record Schwann's work on the biliary fistula were his last physiological works. After that time, although he never ceased to work in the laboratory, his scientific inquiry lost its impact. Following his arrival in the prosperous industrial region of Liège, he became more of an inventor, developing a number of instruments used in mining technology, including pumps for the aspiration of water in coal mines and a respiratory apparatus for rescue operations. This instrument is the ancestor of the apparatus for measuring metabolism in man and of the devices used by divers.

Schwann's religious meditations occupied most of his time after his arrival at Liège. He composed what he intended to be a complement to the three parts of the *Mikroskopische Untersuchungen*, starting from the definition of the atom and extending the cell theory to a general system of organisms, including psychology and religion. His *Theoria* consisted of the three parts of the *Mikroskopische Untersuchungen* and of still unpublished chapters; a fourth part was two chapters on irritability and on brain function, and a fifth part concerned the theory of creation. The fourth and fifth parts are purely theological and philosophical. After his retirement Schwann remained in Liège, where he had formed many friendships. His life was troubled by very few incidents, the main one being the case of the "stigmatized" Louise Lateau. High Church authorities having wrongly interpreted what they considered to be a testimonial by him in favor of the miraculous nature of the phenomena, Schwann was forced to publish a statement of rectification.

During a Christmas visit to a brother and sister living in Cologne, he suffered a stroke and died on 11 January 1882, after two weeks of agony, during which he several times expressed the regret that he had not been able to publish the whole of his *Theoria*.

BIBLIOGRAPHY

I. Original Works. Schwann's earlier writings include *De necessitate aëris atmospherici ad evolutionem pulli in ovo incubito* (Berlin, 1834); "Uber die Nothwendigkeit der atmosphärischen Luft zur Entwicklung des Hühnchens in dem bebrüteten Ei," in *Notizen aus dem Gebiete der Natur- und Heilkunde*, **41** (1834), 241–245; Johannes Müller, *Handbuch der Physiologie des Menschen*, 2 vols. (Koblenz, 1834–1838), *passim*: reports by Schwann on his research on muscle structure, ends of motor nerves, laws of muscle contraction, walls of capillary vessels, contractility of arteries, division of primitive nerve fibers, regeneration of cut nerves, nerve conduction, movements of lymph in the mesentery, ciliary movements, erectile tissues; "Gefässe," "Hämatosis," "Harnsekretion," "Hautsekretion," in *Encyclopädisches Wörterbuch der medizinischen Wissenschaft*, XIV (Berlin, 1836); "Versuche über die künstliche Verdauung des geronnenen Eiweisses," in *Archiv für Anatomie und Physiologie* (1836), 66–90, written with Müller; "Über das Wesen des Verdauungsprocesses," *ibid.*, 90–119; "Über die feinere Nervenausbreitung im Schwanze von Froschlarven," in *Medizinische Zeitung*, 6 (1837), 169; "Verdauung," "Muskelkraft," "Generatio equivoca," in Oken's *Isis* (1837), no. 5, 509–510; no. 6, 523–524; and no. 7, 524, respectively; "Vorläufige Mittheilung betreffend Versuche über Weingährung und Fäulniss," in *Annalen der Physik und Chemie*, 41 (1837), 184–193; preliminary notes on the cell theory in *Neue Notizen aus dem Gebiete der Natur- und Heilkunde*, no. 91 (1838), 34–36; no. 103 (1838), 225–229; no. 112 (1838), 21–23; *Mikroskopische Untersuchungen über die Übereinstimmung in der Struktur und dem Wachstum der Thiere und Pflanzen* (Berlin, 1839), also in English (London, 1847) and Russian (Moscow, 1939), and repr. as no. 176 of Ostwald's Klassiker der Exakten Wissenschaften (Leipzig, 1910); and "Übersicht über die Entwicklung der Gewebe," in R. Wagner, *Lehrbuch der Physiologie*, I (Leipzig, 1839), 139–142.

During the 1840's and 1850's Schwann published "Instructions pour l'observation des phénomènes périodiques chez l'homme," in *Bulletin de l'Académie royale de Belgique*, 9 (1842), 120–137; "Recherches microscopiques sur la conformité de structure et d'accroissement des animaux et des plantes," in *Annales des sciences naturelles*, Zoologie, 2nd ser., 17 (1842), 5–19; "Mensuration d'organes," in *Nouveaux mémoires de l'Académie royale de Belgique*, 16 (1843), 51–52, and 18 (1845), 145; "Versuche um auszumitteln, ob die Galle im Organismus eine für das Leben wesentliche Rolle spielt," in *Archiv für Anatomie, Physiologie und wissenschaftliche Medicin* (1844), 127–159; "Expériences pour constater si la bile joue dans l'économie animale un rôle essentiel pour la vie," in *Nouveaux mémoires de l'Académie royale de Belgique*, 18 (1845), 3–29; a letter to Wagner concerning his last experiments on the biliary fistula, in R. Wagner, *Handwörterbuch der Physiologie*, III (Brunswick, 1846), 837; "Sur des graines tombées de l'air dans la Prusse rhénane," in *Bulletin de l'Académie royale de Belgique*, 19 (1852), 5–6; *Anatomie du corps humain*, 2 vols. (Brussels, 1852); *Rapport sur la situation exceptionnelle dans laquelle s'est trouvée la province de Liège à l'époque de l'épidémie cholérique de 1854 et 1855* (Liège, 1857); and *Erklärung der stöchiometrischen Tafel* (Cologne, 1858). Schwann's latest works include "Réponse à l'interpre-

tation de M. d'Omalius relative à la force vitale," in *Bulletin de l'Académie royale des sciences de Belgique*, 24 (1870), 683; "Notice sur Frédéric-Antoine Spring," in *Annuaire de l'Académie royale de Belgique* (1874); *Mein Gutachten über die Versuche die an der stigmatisirten Louise Lateau am 26. März 1869 angestellt wurden* (Cologne–Neuss, 1875); "Appareil permettant de pénétrer et de vivre dans un milieu irrespirable," in *Bulletin du musée de l'industrie de Belgique*, 21 (1877), 5–9; and *Description de deux appareils permettant de vivre dans un milieu irrespirable* (Liège, 1878), repr. in *Revue universelle des mines, de la métallurgie, des travaux publics, des sciences et des arts appliqués à l'industrie*, 7 (1880), 601–609. See also *Manifestation en l'honneur de M. le professeur Th. Schwann, Liège, 23 juin 1878. Liber memorialis publié par la Commission organisatrice* (Düsseldorf, 1879), which contains Schwann's reply to speeches of congratulation.

II. SECONDARY LITERATURE. See M. Florkin, *Théodore Schwann et les débuts de la médecine scientifique* (Paris, 1956), Conférences du Palais de la Découverte, ser. D, no. 43; *Naissance et déviation de la théorie cellulaire dans l'oeuvre de Théodore Schwann* (Paris, 1960); *Lettres de Théodore Schwann* (Liège, 1961); and "Théodore Schwann. 1810–1882," in *Florilège des sciences en Belgique* (Brussels, 1968); L. Fredericq, *Théodore Schwann, sa vie et ses travaux* (Liège, 1884); A. Le Roy, "Schwann, Théodore," in *L'Université de Liège depuis sa fondation* (Liège, 1869), 919–938; and R. Watermann, *Theodor Schwann. Leben und Werk* (Düsseldorf, 1960).

MARCEL FLORKIN

SCHWARZ, HERMANN AMANDUS (*b.* Hermsdorf, Silesia [now Sobiecin, Poland], 25 January 1843; *d.* Berlin, Germany, 30 November 1921), *mathematics.*

Schwarz, the son of an architect, was the leading mathematician in Berlin in the period following Kronecker, Kummer, and Weierstrass. He may be said to represent the link between these great mathematicians and the generation active in Germany in the first third of the twentieth century, a group that he greatly influenced. After attending the Gymnasium in Dortmund, he studied chemistry in Berlin at the Gewerbeinstitut (now the Technische Universität) but, under the influence of Kummer and Weierstrass, soon changed to mathematics. Schwarz received the doctorate in 1864 and then completed his training as a *Mittelschule* teacher. Immediately thereafter, in 1867, he was appointed assistant professor at Halle. In 1869 he became a full professor at the Eidgenössische

Technische Hochschule in Zurich and in 1875 assumed the same rank at the University of Göttingen. Schwarz succeeded Weierstrass at the University of Berlin in 1892 and lectured there until 1917. During this long period, teaching duties and concern for his many students took so much of his time that he published very little more. A contributing element may have been his propensity for handling both the important and the trivial with the same thoroughness, a trait also evident in his mathematical papers. Schwarz was a member of the Prussian and Bavarian academies of sciences. He was married to a daughter of Kummer.

Schwarz's greatest strength lay in his geometric intuition, which was brought to bear in his first publication, an elementary proof of the chief theorem of axonometry, which had been posed by Karl Pohlke, his teacher at the Gewerbeinstitut. The influence of Weierstrass, however, soon led Schwarz to place his geometric ability in the service of analysis; and this synthesis was the basis of his contribution to mathematics. Schwarz tended to work on narrowly defined, concrete, individual problems, but in solving them he developed methods the significance of which far transcended the problem under discussion.

Schwarz's most important contribution to the history of mathematics was the "rescue" of some of Riemann's achievements. The demonstrations had been justly challenged by Weierstrass. The question centered on the "main theorem" of conformal (similar in the least parts) mapping, which stated that every simply connected region of the plane can be conformally mapped onto a circular area. In order to prove it, Riemann had employed the relation of the problem to the first boundary-value problem of potential theory (Dirichlet's problem), which requires a solution of the partial differential equation $\Delta u = 0$ with prescribed values at the boundary of the region. Dirichlet believed he had disposed of this problem with the observation (Dirichlet's principle) that such a function yields an extreme value for a certain double integral; Weierstrass had objected that the existence of a function which can do that is not at all self-evident but must be demonstrated.

Schwarz first solved the mapping problem explicitly for various simple geometric figures—the square and the triangle—and then in general for polygons. He also treated the conformal mapping of polyhedral surfaces onto the spherical surface. These results enabled him to solve the two problems mentioned, that is, to present the first com-

pletely valid proofs for extended classes of regions by approximating the given region by means of polygons. These works contained the first statement of principles that are now familiar to all: the principle of reflection; the "alternating method," which provides a further method for the approximation of solution functions, and "Schwarz's lemma."

Schwarz also worked in the field of minimal surfaces (surfaces of least area), a characteristic problem of the calculus of variation. Such a surface must everywhere have zero mean curvature, and in general all surfaces with this property are termed minimal surfaces. The boundary-value problem requires in this case that a minimal surface be passed through a given closed space curve, a procedure that can be carried out experimentally by dipping a wire loop into a soap solution. Following his preference for concrete geometrical problems, Schwarz first solved the problem explicitly for special space curves, mostly consisting of straight sections, of which the curve composed of four out of six edges of a tetrahedron has become the best known.

In his most important work, a *Festschrift* for Weierstrass' seventieth birthday, Schwarz set himself the task of completely answering the question of whether a given minimal surface really yields a minimal area. Aside from the achievement itself, which contains the first complete treatment of the second variation in a multiple integral, this work introduced methods that immediately became extremely fruitful. For example, a function was constructed through successive approximations that Picard was able to employ in obtaining his existence proof for differential equations. Furthermore, Schwarz demonstrated the existence of a certain number, which could be viewed as the (least) eigenvalue for the eigenvalue problem of a certain differential equation (these concepts did not exist then). This was done through a method that Schwarz's student Erhard Schmidt later applied to the proof of the existence of an eigenvalue of an integral equation—a procedure that is one of the most important tools of modern analysis. In this connection Schwarz also employed the inequality for integrals that is today known as "Schwarz's inequality."

Algebra played the least role in Schwarz's work; his dissertation, however, was devoted to those surfaces developable into the plane that are given by algebraic equations of the first seven degrees. Much later he answered the question: In which cases does the Gaussian hypergeometric series represent an algebraic function? In approaching this matter, moreover, he developed trains of thought that led directly to the theory of automorphic functions, which was developed shortly afterward by Klein and Poincaré.

Of a series of minor works, executed with the same devotion and care as the major ones, two that involve criticism of predecessors and contemporaries remain to be mentioned. Schwarz presented the first rigorous proof that the sphere possesses a smaller surface area than any other body of the same volume. Earlier mathematicians, particularly Steiner, had implicitly supposed in their demonstrations the existence of a body with least surface area. Schwarz also pointed out that in the definition of the area of a curved surface appearing in many textbooks of his time, the method employed for determining the length of a curve was applied carelessly and that it therefore, for example, led to an infinitely great area resulting for so simple a surface as a cylindrical section.

BIBLIOGRAPHY

I. ORIGINAL WORKS. Schwarz's writings were collected in his *Gesammelte mathematische Abhandlungen*, 2 vols. (Berlin, 1890). He also compiled and edited *Nach Vorlesungen und Aufzeichnungen des Hrn. K. Weierstrass*, 12 pts. (Göttingen, 1881–1885), brought together in the 2nd ed. (Berlin, 1893), and also in French (Paris, 1894).

II. SECONDARY LITERATURE. See *Mathematische Abhandlungen Hermann Amandus Schwarz zu seinem fünfzigjährigen Doktorjubiläum gewidmet von Freunden und Schülern*, C. Carathéodory, G. Hessenberg, E. Landau, and L. Lichtenstein, eds. (Berlin, 1914), with portrait; L. Bieberbach, "H. A. Schwarz," in *Sitzungsberichte der Berliner mathematischen Gesellschaft*, **21** (1922), 47–51, with portrait and list of works not included in *Gesammelte Abhandlungen;* C. Carathéodory, "Hermann Amandus Schwarz," in *Deutsches biographisches Jahrbuch*, III (1921), 236–238; G. Hamel, "Zum Gedächtnis an Hermann Amandus Schwarz," in *Jahresberichte der Deutschen Mathematikervereinigung*, **32** (1923), 6–13, with portrait and complete bibliography; F. Lindemann, obituary in *Jahrbuch der bayerischen Akademie der Wissenschaften* (1922–1923), 75–77; R. von Mises, "H. A. Schwarz," in *Zeitschrift für angewandte Mathematik und Mechanik*, **1** (1921); and E. Schmidt, "Gedächtnisrede auf Hermann Amandus Schwarz," in *Sitzungsberichte der Preussischen Akademie der Wissenschaften zu Berlin* (1922), 85–87.

H. BOERNER

SCHWARZSCHILD, KARL (*b.* Frankfurt am Main, Germany, 9 October 1873; *d.* Potsdam, Germany, 11 May 1916), *astronomy.*

Schwarzschild was the eldest of five sons and one daughter born to Moses Martin Schwarzschild and his wife Henrietta Sabel. His father was a prosperous member of the business community in Frankfurt, with Jewish forebears in that city who can be traced back to Liebmann "of the Black Shield" (died 1594), and possibly even further, to one Elieser, also known as Liebmann, who came to Frankfurt from Cologne in 1450.

From his mother, a vivacious, warm person, Schwarzschild undoubtedly inherited his happy, outgoing personality; from his father, a capacity for sustained hard work. His childhood was spent in comfortable circumstances among a large circle of relatives, whose interests included art and music; he was the first to become a scientist.

After attending a Jewish primary school, Schwarzschild entered the municipal Gymnasium in Frankfurt at the age of eleven. His curiosity about the heavens was first manifested then; he saved his allowance and bought lenses to make a telescope. Indulging this interest, his father introduced him to a friend, J. Epstein, a mathematician who had a private observatory. With Epstein's son (later professor of mathematics in the University of Strasbourg), Schwarzschild learned to use a telescope and studied mathematics of a more advanced type than he was getting in school. His precocious mastery of celestial mechanics resulted in two papers on double star orbits, written when he was barely sixteen. They appeared in the *Astronomische Nachrichten* (1890).

In 1891 Schwarzschild began two years of study at the University of Strasbourg, where Ernst Becker, director of the observatory, guided the development of his skills in practical astronomy—skills that later were to form a solid underpinning for his masterful mathematical abilities.

At age twenty Schwarzschild went to the University of Munich. Three years later, in 1896, he obtained his Ph.D., *summa cum laude.* His dissertation, written under the direction of Hugo von Seeliger, was an application of Poincaré's theory of stable configurations in rotating bodies to several astronomical problems, including tidal deformation in satellites and the validity of Laplace's suggestion as to how the solar system had originated. Before graduating, Schwarzschild also found time to do some practical work: having read about Michelson's interferometer with two slits variably spaced, he devised a multislit instrument for himself. This instrument consisted of a coarse wire grating at a variable angle above the objective lens of a ten-inch telescope. He used it to measure the separation of close double stars; with a micrometer he found the distance between the tiny first-order spectra that resulted, and was able to detect separations as small as 0.88″ of arc. Later workers, notably Comstock and Hertzsprung, used this device to find the "effective wavelengths" of individual stars and so a clue to stellar surface temperatures.

After graduating, Schwarzschild became assistant at the Kuffner observatory in Ottakring (a western suburb of Vienna), where Leo de Ball was then director. Here Schwarzschild remained from October 1896 until June 1899, and here he began what became a lifetime avocation—giving lectures that conveyed to nonastronomers his own feelings about the excitement and significance astronomy holds for everyone.

During this period Schwarzschild published several papers on celestial mechanics, dealing with special cases of the three-body problem. But the main thrust of his work now became a coordinated attack on one of astronomy's most fundamental problems: stellar photometry. Measurement of the radiant energy reaching us from the stars was then still being done by eye, in principle just as Hipparchus had done two millennia earlier. Schwarzschild decided to try substituting a photographic plate for the human eye at the telescope. The many advantages of photography had been recognized (permanent record; coverage of a whole field of stars at once, and of invisible stars merely by increasing the time of exposure), but much work remained to be done before photography could even equal the eye of a trained observer. This work, both theoretical and practical, Schwarzschild now began. It culminated with the publication of his "Aktinometrie," so called because light producing a photochemical effect was then referred to as "actinic." Part A of "Aktinometrie," published in 1910, contains the earliest catalog of photographic magnitudes; it preceded Parkhurst's "Yerkes Actinometry" by several years.

Of the two likely ways of converting from black dots on a photographic plate to actual stellar magnitudes, Schwarzschild rejected measuring diameters of the images as inaccurate. He had investigated the theory of diffraction patterns as produced at various angular distances from the optic axis and found that the distribution of intensity in the concentric rings was not constant over an extended

field. He therefore decided to smear out the images by putting his plate inside the focal plane, and then to measure the density of the resulting blurs with a photometer. This technique had been suggested by Janssen in 1895, but Schwarzschild was actually the first to use it.

Next Schwarzschild investigated the response of the photographic emulsion in order to determine whether photographs taken at different exposure times could be directly compared. According to the photochemical law enunciated by Bunsen and Roscoe in 1862, the image of a given star should be identical with that produced for a star half as bright when exposed for twice the time. But other workers, such as Abney, noted deviations from such strict reciprocity. Schwarzschild was the first to quantify the particular "failure of reciprocity" that occurs under low levels of illumination such as from faint stars. He performed a series of laboratory experiments and concluded that the law must be modified by raising the exposure time to the power p, with p less than unity but a constant for any given combination of emulsion and development process. This relation is still known as Schwarzschild's law, and p as Schwarzschild's exponent, although subsequent work has shown that p is not as unvarying as Schwarzschild had thought.

Now Schwarzschild was ready to try out his techniques on the sky over Vienna. He photographed an aggregate of 367 stars, which included two that were known to vary in brightness. In following one of the variables, eta Aquilae, through several of its cycles, Schwarzschild found that its changes covered a considerably larger range of magnitude photographically than visually. He correctly attributed this difference to a rhythmic change in surface temperature and was therefore the first both to observe and to explain this phenomenon. It is one that occurs in all similar stars—the Cepheids.

These photometric results were presented to the University of Munich as Schwarzschild's monograph qualifying him to teach. He returned to Munich and served as *Privatdozent* for two years.

At a meeting of the German Astronomical Society held in Heidelberg in August 1900, Schwarzschild—once more under Seeliger's influence—discussed quite a different aspect of astronomy, namely the possibility that the geometry of space was non-Euclidean. He suggested two kinds of curvature: elliptic (positively curved and finite, like the surface of a sphere but with antipodal points considered identical) and hyperbolic (negatively curved and infinite). After considering the astronomical evidence then available, he concluded that, if space were curved, the radius of curvature of the universe must exceed the earth-sun distance by at least four million times for a hyperbolic universe, and by a hundred million times for an elliptic universe. The latter case carried the proviso that space absorption of about forty stellar magnitudes must occur, otherwise the returning light emitted from the far side of our sun would refocus into a visible countersun in the sky. These estimates of size, even as lower limits, are decidedly small; yet if the paper shows how small a universe Schwarzschild thought he lived in, it also demonstrates the unlimited boldness of his thought.

In another publication dating from this period, Schwarzschild considered the suggestion, published by Arrhenius a few months previously, that the tails of comets point away from the sun because the repulsive pressure of solar radiation outweighs gravity. Assuming that the tails were made up of completely reflecting spherical particles of a reasonable density, he found that the pressure of radiation could just barely exceed the gravitational attraction by the necessary twenty times, if the particles had diameters between 0.07 and 1.5 microns; below that size scattering would occur, and above it gravitation would predominate. He remarked that the occasional tail in which even greater repulsion seemed to be present could not be shaped by radiation pressure alone. Such is still believed to be the case.

In the fall of 1901 Schwarzschild was named associate professor at the University of Göttingen and also director of its observatory. The observatory, with the director's living quarters in one wing, had been built and equipped eighty years earlier by Gauss.

Less than a year later, at the age of twenty-eight, Schwarzschild was promoted to full professor. He remained eight years in Göttingen, the most productive and probably the happiest time in his life. He sharpened his ideas in discussions with a brilliant circle of colleagues; for example, during the summer semester of 1904 he was one of four men in charge of the mathematical-physical seminar, the other three being Klein, Hilbert, and Minkowski. He carried a heavy teaching load but each year managed to include a course entitled "Popular Astronomy." The observatory soon became a center for young intellectuals of all disciplines. But it did not at first offer Schwarzschild the instruments he needed to pursue his observational work.

Gauss's original meridian circle was obsolete, while the other main piece of equipment, a Repsold heliometer, was unsuitable. In 1904, however, A. Schobloch donated a seven-inch refractor. Schwarzschild and his collaborators used this telescope to photograph 3,522 stars, data on which appeared in the "Aktinometrie."

For this enterprise Schwarzschild had a new idea. Instead of using extrafocal images as before, he decided to use a special plateholder—his "Schraffierkassette"—held in the focal plane but moved mechanically during a three-minute exposure so that all the images came out as squares 0.25 mm. on a side. This plateholder, revised several times, was effective but cumbersome and was not used by subsequent investigators. The methods of reducing the plates were also refined. The final publication, *Aktinometrie, Teil B* (1912), contains for each star its fully corrected photographic magnitude and an indication of its surface temperature in the form of its color index (photographic minus visual magnitude). Schwarzschild used the Potsdam visual magnitudes and also those obtained at Harvard, for comparison.

To the tradition of geodetic measurements at Göttingen, initiated by Gauss, Schwarzschild contributed in 1903 a suspended zenith camera of his own design, to be used for photographic latitude determination. He also became interested in ballooning—in balloons filled with gas from the city plant, making it as hazardous a sport as Schwarzschild's other favorites, mountain climbing and skiing. To simplify the problem of navigating while in the swinging basket of a balloon, Schwarzschild developed a form of bubble sextant, with ancillary tables and nomograms.

While in Göttingen, Schwarzschild also became interested in the sun. He obtained a small grant and went by freighter from Hamburg to Algeria for the total solar eclipse of 30 August 1905. His companions were Carl Runge, an applied mathematician at the University of Göttingen, and Robert Emden, a physicist, then teaching in Munich, and Schwarzschild's brother-in-law. Their modest equipment, pictured in the published report (and including a rare photograph of Schwarzschild himself), was set up on the ruined stage of an old Roman amphitheater in Guelma, with a makeshift darkroom nearby. The ambient temperature was 108° F., but the planned program was carried out to the last detail. Of greatest interest were Schwarzschild's flash spectra: using a camera fitted with an objective prism (all the glass being transparent to near ultraviolet light) and roll film pulled through by hand, he got sixteen photographs of the solar spectrum in a period of thirty seconds, beginning ten seconds before second contact. Since the speed with which the moon passed across the solar disc was known, these spectrograms could be analyzed to give the chemical composition, both qualitative and quantitative, of regions at various heights on the sun. This ingenious method was later revived by Menzel in his jumping film spectrograph.

Returning to Germany, Schwarzschild enlisted the aid of his old friend Villiger (who had helped him to measure double star separations in 1896) to investigate how the intensity of ultraviolet light varied across the disc of the sun. The observations were made at Jena, in the Zeiss factory, using a thin film of silver on the objective of the telescope, to screen out all radiation longer than 3,200 Å. They found that the drop-off between center and limb was even more pronounced at these short wavelengths than in either the visible or the infrared.

In 1906 Schwarzschild published a theoretical work on the transfer of energy at and near the surface of the sun. In discussing the results of the 1905 eclipse he had assumed that equilibria in the different modes of energy transfer could occur simultaneously and therefore be treated separately. He had also remarked that the rapid decrease in density necessary to explain the observed sharpness of the solar disc favored the predominance of radiative over convective (adiabatic) transfer near the photosphere. He now developed a theory of radiative exchange and equilibrium that was quantitative, and therefore susceptible to experimental verification. He assumed that near its surface the sun was horizontally layered into regions, each of which received radiation from below and reemitted it outward. His first approximation to a solution of the integral equations involved was similar to one proposed by Schuster in 1905. It is therefore usually referred to as the Schuster-Schwarzschild model for a gray atmosphere, although they arrived at it independently. It represented a major step toward understanding stellar structure, but as it did not predict the correct flux it was superseded by Eddington's model, which did.

Other theoretical work that Schwarzschild did at Göttingen includes three papers on electrodynamics, written in 1903. In them he attempted to formulate the fundamental equations in terms of direct action at a distance, using Hamilton's principle of least action. This approach has been used again at least once, by Wheeler and Feynmann in 1945.

In 1905 Schwarzschild published three papers on geometrical optics, dealing exhaustively with the aberrations encountered in optical systems. Here he used Hamilton's "characteristic function" (called the "eikonal" by Bruns). In the first paper he showed how spherical aberrations originate, including those of higher order. In the second paper he demonstrated how, by combining two mirrors with aspherical surfaces, a telescope free of aberrations would result. In the third paper he provided formulas for computing a variety of compound optical systems. Max Born, writing in 1955, acknowledged that these papers formed the backbone of his own "Optik," first published in 1932.

Schwarzschild also made contributions to stellar statistics, at a time when the structure of our galaxy and the way it rotates were still mysterious. He considered, in two articles published in 1907 and 1908, the motions of nearby stars through space as related to estimates of their distances. The sparse observational material then available, including proper motions as tabulated in the Groombridge-Greenwich catalog, had already been analyzed by Kapteyn, who was surprised to find that peculiar motions (obtained from proper motions of the stars by correcting for the motion of the sun) were not random but seemed to favor two preferential directions. Kapteyn derived from this his "two stream hypothesis," which had stars moving past each other from opposite directions. Such a picture was intellectually unacceptable to Schwarzschild, who developed instead what he called a unitary picture, namely his hypothesis of an ellipsoidal velocity distribution, which he showed would fit the observed facts equally well. A third article, published in 1911 after he had left Göttingen, gave details of his methods and compared his results to those of Seeliger, Kapteyn, and Hertzsprung.

Schwarzschild's cosmological thoughts are available in a collection of four popular lectures, entitled "On the System of Fixed Stars." Perhaps of the greatest interest is the lecture read before the Scientific Society of Göttingen on 9 November 1907, dealing with Lambert's cosmological letters. Schwarzschild discussed the type of teleological arguments used (successfully) by Lambert to reach many of the same conclusions about the universe—including the plurality of inhabited worlds—that are adhered to even by those physical scientists to whom teleological arguments are anathema. Schwarzschild includes a wry comment that teleology is still fruitful in biological sciences, in the theory of evolution.

In 1909 Schwarzschild's life changed. He was appointed successor to Vogel as director of the Astrophysical Observatory in Potsdam. This was the most prestigious post available to an astronomer in Germany, but accepting it meant increased administrative burdens, and also giving up the academic surroundings so congenial to him. Potsdam had been an army town since the time of Frederick the Great, and the University of Berlin would be fully fifteen miles away. Nevertheless, he decided to go, and took up his duties on Telegraph Hill (where the observatory was located) late in 1909. With him he brought Else Rosenbach, the daughter of a professor of surgery at Göttingen, whom he had married on 22 October 1909. Their marriage was successful despite initial family misgivings (she was not Jewish) and was blessed with three children: Agathe, Martin (professor of astronomy in Princeton University), and Alfred.

Also making the move to Potsdam at this time was Ejnar Hertzsprung, who had come to Göttingen at Schwarzschild's invitation in April 1909.

In August 1910 Schwarzschild went to California to attend the Fourth Meeting of the International Union for Cooperative Solar Research, stopping off along the way to visit the major observatories in the United States. The published account of this trip sheds an interesting light on the differences between American and European astronomy just before World War I. Notwithstanding the better skies and the larger instruments, Schwarzschild's envy was directed mainly toward the installations both Harvard and Lick had in the Southern hemisphere. He came home ready to push for a German observatory south of the equator, and suggested Windhoek, then in German Southwest Africa, as a possible site.

The year 1910 also brought a return of Halley's comet. Schwarzschild, with E. Kron, measured photographs of this comet, taken by a Potsdam expedition to Tenerife. In their discussion of how the brightness diminished outward, there appeared for the first time the suggestion that fluorescent radiation occurs in comet tails (later amply verified).

At Potsdam, Schwarzschild's interests turned more toward spectroscopy. He designed a spectrographic objective, which was built by Zeiss. Appreciating the need for a quick and reliable way to determine the radial velocities of stars (to supplement proper motions for work in stellar statistics), he expanded, in 1913, upon the way E. C. Pickering had used an objective prism for this purpose. In 1914, anticipating his own later work in general

relativity, Schwarzschild attempted—unsuccessfully, as it turned out—to observe a gravitational red shift in the spectrum of the sun.

When World War I began in August 1914, Schwarzschild carried over into political life the unitary concepts that guided his scientific life. He volunteered for military service, feeling that loyalty to Germany should come ahead of professional ties and his personal background as a Jew. After an initial delay, because of his high government position, he was accepted and placed in charge of a weather station in Namur, Belgium. Subsequently he was commissioned as a lieutenant and attached to the headquarters staff of an artillery unit, serving first in France and later on the Eastern front. His assignment was to calculate trajectories for long-range missiles; a communication to the Berlin Academy in 1915 (not published until 1920 for security reasons) dealt with the effect of wind and air density on projectiles.

While serving in Russia, Schwarzschild wrote two papers on general relativity, presented to the Berlin Academy for him by Einstein, and published in 1916. The first paper, dealing with the gravitational field of a point mass in empty space, was the first exact solution of Einstein's field equations; Schwarzschild's comment is that this work of his "permits Mr. Einstein's result to shine with increased purity."

The well-known "Schwarzschild radius" appears in the second of these papers, which treated the gravitational field of a fluid sphere with constant density throughout. Such a simplification cannot, of course, represent any real star, but it does permit of an exact solution. This solution has a singularity at $R = 2MG/c^2$, where R is the (Schwarzschild) radius for an object of mass M, G the universal constant of gravitation, and c the velocity of light. Should a star, undergoing gravitational collapse, shrink down inside this radius, it becomes a "black hole" that emits no radiation and can be detected only by its gravitational effects. (As an example of how far a star has to shrink, R for our sun is less than two miles.) The black holes resulting from Schwarzschild's solution differ from those of Kerr's 1963 solution in that they have no angular momentum.

For some time—dating back to his studies of the solar atmosphere—Schwarzschild had been interested in the problem of relating spectral lines to the underlying structure of the atoms producing them. His last paper was an attempt to enlarge upon several recent papers on the quantum theory written by Planck and by Sommerfeld. This subject,

Schwarzschild said, awaits a Kepler—but he did not live to see who it would be.

While serving at the front in Russia, Schwarzschild developed symptoms of a rare, painful, and at the time incurable, malady called pemphigus. This is a metabolic disease of the skin, tentatively believed today to have an "auto-immune" mechanism. Schwarzschild was invalided home in March 1916 and spent the last two months of his life in a hospital. He was buried, according to his own wish, in the central cemetery in Göttingen.

The honors Schwarzschild received before he died at age forty-two were few, considering his accomplishments. He was elected to the Scientific Society of Göttingen in 1905, became a foreign associate of the Royal Astronomical Society of London in 1909, and a member of the German Academy of Science in Berlin in 1913. For his war work he was awarded an Iron Cross. In 1960 the Berlin Academy dedicated to him, as "the greatest German astronomer of the last hundred years," the Karl Schwarzschild telescope, a seventy-nine-inch reflector located at Tautenburg, a few miles from Jena, where the optical parts were made. In 1959 the German Astronomical Society established the Karl Schwarzschild lectureship in his honor, with invited lectures to be given by distinguished astronomers.

BIBLIOGRAPHY

I. ORIGINAL WORKS. Schwarzschild's schoolboy publications on celestial mechanics are "Zur Bahnbestimmung nach Bruns," in *Astronomische Nachrichten*, 124 (1890), cols. 211–216; and "Methode zur Bahnbestimmung der Doppelsterne," *ibid.*, cols. 215–218. His coarse wire objective grating is described in "Über Messung von Doppelsternen durch Interferenzen," *ibid.*, 139 (1896), cols. 353–360. His dissertation is "Die Poincaré'sche Theorie des Gleichgewichts einer homogenen rotierenden Flüssigkeitsmasse," in *Neue Annalen der K. Sternwarte in München*, 3 (1898), 231–299.

Schwarzschild's publications on the photographic determination of stellar magnitudes and color indices include "Die Beugungsfigur im Fernrohr weit asserhalb des Focus," in *Sitzungsberichte der Bayerischen Akademie der Wissenschaften zu München*, Math.-Phys. Kl., 28 (1898), 271–294, with figure following p. 362; "Über Abweichungen vom Reciprocitätgesetz für Bromsilbergelatine," in *Photographische Korrespondenz*, 36 (1899), 109–112, trans. as "On the Deviations from the Law of Reciprocity for Bromide of Silver Gelatine," in *Astrophysical Journal*, 11 (1900), 89–91; "Die Bestimmung von Sternhelligkeiten aus extrafokalen photographischen Aufnahmen," in *Publikationen der von Kuffnerschen Sternwarte*, 5 (1900), B1–B23; his Habilitationsschrift, "Beiträge zur photographischen Photometrie der Gestirne," *ibid.*, 5 (1900), C1–C135; "Über die photographische Vergleichung der Helligkeit verschiedenfarbiger Sterne," in *Sitzungsberichte der Akademie der Wissenschaften in Wien*, Math.-Natur. Kl., 109, sec. 2a (1900), 1127–1134; "Über eine Schraffierkassette zur Aktinometrie der Sterne," in *Astronomische Nachrichten*, 170 (1906), cols. 277–282, written with Br. Meyermann; "Über eine neue Schraffierkassette," *ibid.*, 174 (1907), cols. 297–300, written with Br. Meyermann; "Aktinometrie der Sterne der BD bis zur Grösse 7,5 in der Zone 0° bis +20° Deklination": Teil A in *Abhandlungen der K. Gesellschaft der Wissenschaften zu Göttingen*, Math.-Phys. Kl., n.s. 6, no. 6 (1910), written with Br. Meyermann, A. Kohlschütter, and O. Birck; Teil B, *ibid.*, n.s. 8, no. 3 (1912), written with Br. Meyermann, A. Kohlschütter, O. Birck, and W. Dziewulski; and "Über die Schlierkorrektion bei der Halbgittermethode zur Bestimmung photographischer Sterngrössen," in *Astronomische Nachrichten*, 193 (1912), cols. 81–84.

Publications by Schwarzschild dealing with comets include "Der Druck des Lichts auf kleine Kugeln und die Arrheniussche Theorie der Cometenschweife," in *Sitzungsberichte der Bayerischen Akademie der Wissenschaften zu München*, Math.-Phys. Kl., 31 (1901), 293–338; and "Über die Helligkeitsverteilung im Schweife des Halleyschen Kometen," in *Nachrichten von der Gesellschaft der Wissenschaften zu Göttingen*, Math.-Phys. Kl., 1911 (1911), 197–208, written with E. Kron, trans. as "On the Distribution of Brightness in the Tail of Halley's Comet," in *Astrophysical Journal*, 34 (1911), 342–352.

The suspended zenith camera devised by Schwarzschild is described in "Über Breitenbestimmung mit Hilfe eines hängenden Zenithkamera," in *Astronomische Nachrichten*, 164 (1903), cols. 177–182.

Schwarzschild discussed the sun and its atmosphere, including the formation of Fraunhofer lines, in papers among which are "Über die totale Sonnenfinsternis vom 30. August 1905," *Abhandlungen der Gesellschaft der Wissenschaften zu Göttingen*, Math.-Phys. Kl., n.s. 5, no. 2 (1907), with three halftone plates; "Über die Helligkeitsverteilung des ultravioletten Lichtes auf der Sonnenscheibe," in *Physikalische Zeitschrift*, 6 (1905), 737–744, written with W. Villiger, trans. as "On the Distribution of Brightness of the Ultra-Violet Light on the Sun's Disc," in *Astrophysical Journal*, 23 (1906), 284–305, which includes additional material; "Über das Gleichgewicht der Sonnenatmosphäre," in *Nachrichten von der Gesellschaft der Wissenschaften zu Göttingen*, Math.-Phys. Kl., 1906 (1906), 41–53; and "Über Diffusion und Absorption in der Sonnenatmosphäre," in *Sitzungsberichte der Preussischen Akademie der Wissenschaften zu Berlin*, 1914 (1914), 1183–1200.

Schwarzschild's three articles on electrodynamics, "Zur Elektrodynamik," appeared in *Nachrichten von der Gesellschaft der Wissenschaften zu Göttingen*,

Math.-Phys. Kl. (1903), I, 126–131; II, 132–141; III, 245–278. His three articles on optical aberrations appeared as "Untersuchung zur geometrischen Optik," in *Abhandlungen der K. Gesellschaft der Wissenschaften zu Göttingen*, Math.-Phys. Kl., n.s. 4 (1905): I as no. 1, II as no. 2, III as no. 3.

Schwarzschild's papers on stellar motions and stellar statistics include "Über die Eigenbewegung der Fixsterne," in *Nachrichten von der Gesellschaft der Wissenschaften zu Göttingen*, Math.-Phys. Kl., **1907** (1907), 614–631; "Über die Bestimmung von Vertex und Apex nach der Ellipsoidhypothese aus einer geringeren Anzahl beobachteter Eigenbewegung," *ibid.*, **1908** (1908), 191–200; "Zur Stellarstatistik," in *Astronomische Nachrichten*, **190** (1911), cols. 361–376; and "Beitrag zur Bestimmung von Radialgeschwindigkeiten mit dem Objektivprisma," in *Publicationen des Astrophysikalischen Observatoriums zu Potsdam*, **23**, pt. 1 (1913), no. 69.

Schwarzschild's publications dealing with space curvature and general relativity are "Über das zulässige Krummungsmass des Raumes," in *Vierteljahrsschrift der Astronomischen Gesellschaft*, **35** (1900), 337–347, with a summary of the discussion that followed his talk on 311–312; "Über die Verschiebung der Bande bei 3883 Å in Sonnenspektrum," in *Sitzungsberichte der Preussischen Akademie der Wissenschaften zu Berlin*, **1914** (1914), 1201–1213; "Über das Gravitationsfeld eines Massenpunktes nach der Einsteinschen Theorie," *ibid.*, **1916** (1916), 189–196; and "Über das Gravitationsfeld einer Kugel aus inkompressibler Flüssigkeit nach der Einsteinschen Theorie," *ibid.*, **1916** (1916), 424–434.

Schwarzschild's last paper was "Zur Quantenhypothese," *ibid.*, **1916** (1916), 548–568; his classified work on ballistics appeared as "Über den Einfluss von Wind und Luftdichte auf die Geschossbahn," *ibid.*, **1920** (1920), 37–63.

For Schwarzschild's general attitude toward astronomy, including its relation to other disciplines, see "Über Himmelsmechanik," in *Physikalische Zeitschrift*, **4** (1903), 765–773. See also the collection of his popular lectures, reprinted in pamphlet form in honor of Seeliger's sixtieth birthday, *Uber das System der Fixsterne* (Leipzig, 1909; 2nd ed., Leipzig, 1916) (the talk "Über Lamberts kosmologische Briefe," found in this pamphlet, first appeared in *Nachrichten von der Gesellschaft der Wissenschaften zu Göttingen, Geschäftliche Mitteilungen*, 1907, 88–102); "Die grossen Sternwarten der Vereinigen Staaten," in *Internationale Wochenschrift für Wissenschaft, Kunst und Technik*, **49** (1910), cols. 1531–1544; and his inaugural address to the Berlin Academy, in *Sitzungsberichte der Preussischen Akademie der Wissenschaften zu Berlin*, **1913** (1913), 596–600, followed by Planck's reply, 600–602.

A list of 116 publications by Schwarzschild is appended to the obituary written by O. Blumenthal. Apart from official director's reports from Göttingen and Potsdam, which appeared yearly in *Vierteljahrsschrift der Astrono-*

mischen Gesellschaft, and similar special reports in *Astronomische Nachrichten*, it seems to be complete, with the exception of a contribution, "Stationare Geschwindigkeitsverteilung in Sternsystem," printed posthumously (and marked "fragment") in *Probleme der Astronomie, Festschrift für Hugo v. Seeliger* (Berlin, 1924), 94–105.

II. SECONDARY LITERATURE. Obituary notices on Schwarzschild include O. Blumenthal, in *Jahresbericht der Deutschen Mathematikervereinigung*, **26** (1918), 56–75, with photograph, facs. signature, and list of publications starting on p. 70; A. S. Eddington, in *Monthly Notices of the Royal Astronomical Society*, **77** (1917), 314–319; A. Einstein, in *Sitzungsberichte der Preussischen Akademie der Wissenschaften zu Berlin*, **1916** (1916), 768–770; and E. Hertzsprung, in *Astrophysical Journal*, **45** (1917), 285–292, with portrait facing p. 285. S. Oppenheim wrote an essay commemorating the fiftieth anniversary of Schwarzschild's birth, which appeared in *Vierteljahrsschrift der Astronomischen Gesellschaft*, **58** (1923), 191–209. Max Born's reminiscences of Schwarzschild as a professor at Göttingen are included in his "Astronomical Recollections," in *Vistas in Astronomy*, **1** (1955), 41–44.

Author's Note: I am indebted to Professor Martin Schwarzschild for a number of facts about his parents that are not available in print.

SALLY H. DIEKE

SCHWEIGGER, JOHANN SALOMO CHRISTOPH (*b.* Erlangen, Bavaria, 8 April 1779; *d.* Halle, Prussia, 6 September 1857), *physics, chemistry*.

Schweigger was the son of Friedrich Schweigger, extraordinary professor of theology at the Protestant University of Erlangen and archdeacon of a parish in that city. In addition to attending the Gymnasium in Erlangen, he received a thorough education in classical and Semitic languages and in philosophy from his father and several of his father's learned friends. On 7 April 1800 he received the Ph.D. at Erlangen; his dissertation, "De Diomede Homeri," dealt with the unifying characteristics of Homer's heroes. Having also studied mathematics and physics, he lectured on mathematics and natural science as a *Privatdozent* at Erlangen until 1803, when he was called to Bayreuth as professor of mathematics and physics at the Gymnasium. He then was professor of chemistry and physics at the Physikotechnisches Institut (called also the Höhere Realschule) in Nuremberg from 1811 until its dissolution in 1816. A corresponding member of the Bavarian Academy of Sciences since 1813, Schweigger went to Munich in 1817 as an ordinary member but left that year probably to become ordinary professor of physics and chemistry at the University of Erlangen. His status at the Academy thereby reverted to corresponding member, although he later became a foreign member (probably in 1847). In 1819 he was called to the University of Halle as ordinary professor of physics and chemistry, and remained there until his death. In 1816 he had become a corresponding member of the Göttingen Gesellschaft der Wissenschaften and a member of the Kaiserliche Leopoldino-Carolinische Deutsche Akademie der Naturforscher, of which he was an *Adjunkt des Direktoriums* by 1817. Schweigger, who married late in life, had three sons and one daughter.

Schweigger is perhaps best known as founder of the *Journal für Chemie und Physik,* of which he edited fifty-four volumes between 1811 and 1828.[1] This important journal published both original articles and translations and, being somewhat less prestigious than the *Annalen der Physik,* served as the organ of publication of the first papers of a number of scientists, including Wilhelm Weber and Gustav Theodor Fechner. Schweigger's literary output consisted largely of review articles and commentaries on the papers of others.

At the start of his scientific career Schweigger published several papers (1806, 1808) that questioned the validity of Volta's contact theory of electricity by describing an active pile that, according to that theory, should not have been capable of producing any electricity. At first content to conclude merely that the liquid conductor must play a more important role in the pile than simply preventing metallic contact, he subsequently offered an explanation in terms of a polarization of the water by the metal plates. This work was extended in 1817 by the discovery that in certain piles chemical action caused a reversal in polarity of the normally positive zinc and negative copper. Schweigger did not, however, refine or extend his experiments and thus cannot be counted among the major protagonists of the chemical theory of the pile.

Schweigger shared with Poggendorff the honor of constructing the first simple galvanometer, his *Multiplicator,* or multiplier, which he demonstrated before the Naturforschende Gesellschaft of Halle on 16 September and 4 November 1820. Schweigger's device consisted of a figure-eight-shaped coil of wire, a construction he preferred to a simple loop because he wished to demonstrate the equal and opposite electromagnetic effects on a magnetic needle placed in either side. Indeed, his use of the device never went much beyond such simple demonstrations, nor did he show any inclination to refine it into a measuring instrument or to discover its laws of operation. Schweigger thought the operation of the multiplier provided vivid proof of the existence of a "double magnetic polarity" perpendicular to the current in a connecting wire (that is, a wire connecting the poles of a pile). He envisioned this polarity as a succession of magnetic axes oppositely directed along the top and bottom of the wire, so that a compass would point in opposite directions when held above or below it. That he regarded this hypothesis on the nature of electromagnetism as a direct expression of the simple facts was characteristic of Schweigger's concretely pictorial approach to physical theory.

NOTES

1. The journal was continued until 1833 (vol. **69**) by his adopted son, Franz Wilhelm Schweigger-Seidel, who had been coeditor since 1825 (vol. **45**). Johann Ludwig Georg Meinecke also had been a coeditor, from 1819 (vol. **26**) until 1823 (vol. **38**). From the beginning of 1821 (vol. **31**) to 1830 the journal also bore the title *Jahrbuch der Chemie und Physik,* with its own volume numbers. Finally, from 1825 (vol. **45**) to 1827 (vol. **51**) it was subtitled *als eine Zeitschrift des wissenschaftlichen Vereins zur Verbreitung von Naturkenntniss und höherer Wahrheit.*

BIBLIOGRAPHY

I. ORIGINAL WORKS. Extensive bibliographies are found in the Royal Society *Catalogue of Scientific Papers,* V, 589–592; and in Poggendorff, II, cols. 873–875. Many more of his commentaries on others' papers are listed in G. C. Wittstein, ed., *Autoren- und Sach-Register zu sämmtlichen neunundsechzig Bänden des Schweigger'schen Journals für Chemie und Physik. (Jahrgänge 1811–1833)* (Munich, 1848), 88–91. See also Snelders' article (below).

II. SECONDARY LITERATURE. On Schweigger's life and work, see Carl Friedrich Philipp von Martius, "Denkrede auf Joh. Salomo Christoph Schweigger," in *Gelehrte Anzeigen der k. Bayerischen Akademie der Wissenschaften,* **46,** Jan.–June 1858, cols. 81–99, repr. in Martius' *Akademische Denkreden* (Leipzig, 1866), 345–364. Also useful is the article by "K" in *Allgemeine deutsche Biographie,* XXXIII (1891), 335–339. For his contribution to the founding of the Gesellschaft Deutscher Naturforscher und Ärzte, see Heinz Degen, "Die Gründungsgeschichte der Gesellschaft deutscher Naturforscher und Ärzte," in *Naturwissenschaftliche Rundschau,* **8** (1955), 421–427, 472–480. An informative account of an important aspect of his work is H. A. M. Snelders, "J. S. C. Schweigger: His Romanticism and His Crystal Electrical Theory of Matter," in *Isis,* **62** (1971), 328–338, which should be read to supplement the information given here. Some of his work on the pile and the multiplier is discussed in Wilhelm Ostwald, *Elektrochemie, ihre Geschichte und Lehre* (Leipzig, 1896), 301–304, 371–373.

KENNETH L. CANEVA

SCHWEIKART, FERDINAND KARL (*b.* Erbach, Germany, 28 February 1780; *d.* Königsberg, Germany [now Kaliningrad, R.S.F.S.R.], 17 August 1859), *mathematics*.

Schweikart studied law at the University of Marburg from 1796 to 1798 and received his doctorate in law from Jena in the latter year. After practicing at Erbach from 1800 to 1803, he worked as a private tutor until 1809; his pupils included the prince of Hohenlohe-Ingelfingen. In 1809 Schweikart became extraordinary professor of law at the University of Giessen; from 1812 to 1816 he was full professor at the University of Kharkov; and from 1816 to 1820 he taught at Marburg. From 1821 until his death he was professor of law at Königsberg, where he also earned a doctorate in philosophy.

Schweikart published extensively in his principal field of endeavor, including a work on the relationship of natural and positive law (1801). Early in his life he also became interested in mathematics, and he holds an important place in the prehistory of non-Euclidean geometry. While a student at Marburg, the lectures of J. K. F. Hauff had stimulated him to consider the problem of parallel lines, which provided the subject of his only publication in mathematics (1807). His approach was still completely Euclidean; but later he arrived at the beginnings of a hyperbolic geometry, which he called astral geometry. He made this advance independently of Gauss, Bolyai, and Lobachevsky, as is proved by the correspondence cited by Engel and Stäckel in *Die Theorie der Parallellinien von Euklid bis Gauss*. The astronomer Christian Gerling, a student of Gauss, wrote in a letter to Wolfgang Bolyai, the father of János Bolyai, that in 1819 Schweikart had reported on the basic elements of his "astral geometry" to colleagues at Marburg. Schweikart also wrote on this topic to his nephew Taurinus in Cologne. Stimulated by his uncle's work, Taurinus had virtually discovered hyperbolic trigonometry; but unlike his uncle, he still believed in the sole validity of Euclidean geometry.

The three chief founders of hyperbolic geometry surpassed Schweikart only because of the thoroughness with which they examined specific topics of this subject. The demands of his legal career

undoubtedly prevented him from finding sufficient time to undertake similarly extensive research.

BIBLIOGRAPHY

Schweikart's only mathematical work is *Die Theorie der Parallellinien, nebst einem Vorschlag ihrer Verbannung aus der Geometrie* (Jena–Leipzig, 1807).

A secondary source is Friedrich Engel and Paul Stäckel, *Die Theorie der Parallellinien von Euklid bis Gauss* (Leipzig, 1895), 243–252.

WERNER BURAU

SCHWENDENER, SIMON (*b.* Buchs, St. Gallen, Switzerland, 10 February 1829; *d.* Berlin, Germany, 27 May 1919), *plant anatomy*.

Schwendener was the son of a farmer; he himself worked on his grandfather's farm during summer vacations from school. His father wished him to become a teacher, so in 1847 Schwendener began to prepare himself for this career, although he had little inclination for it. From 1849 to 1850 he studied science at the Academy of Geneva. His father could not afford to allow him to attend the University of Zurich for further study, as he had planned, and Schwendener then accepted a teaching position in Wädenswil. In 1853 his grandfather died and left him a small legacy, which he employed to support himself at the university. He graduated from the University of Zurich in 1854 and was awarded the Ph.D., *summa cum laude*, in 1856.

Schwendener's Ph.D. thesis, *Ueber die periodische Erscheinungen der Natur, insbesondere der Pflanzenwelt*, was inspired by Candolle and finished under the direction of Oswald Heer. Another early—and lasting—influence on Schwendener's work was that of Carl Naegeli, whose Zurich lectures Schwendener had attended. Both Naegeli and Schwendener were interested in plant morphology and ontogeny, and Schwendener attracted Naegeli's attention while attending his courses at the Zurich Eidgenössische Technische Hochschule. As a result, when Naegeli was later, in 1857, appointed professor at the University of Munich, he invited Schwendener to be his assistant, and thus initiated a close collaboration. Their two-volume work *Das Mikroskop* (published in 1865–1867) represented three years' mutual effort; in it Naegeli and Schwendener not only demonstrated a number of details of plant anatomy but also set out the principles that Abbe used in his optical work. Schwendener also did independent research on lichens, and published the first of his several works on that subject in 1860, the same year in which he became *Privatdozent*.

In 1867 Schwendener returned to Switzerland to take up an appointment as *professor ordinarius* and director of the botanic garden at the University of Basel. He here continued the work on lichens that culminated in his *Die Algentypen der Flechtengonidien. Programm für die Rektoratsfeier der Universität*, published in 1869. In this book Schwendener first stated that lichens are a composite of algae and fungi (later termed "symbiosis" by De Bary and confirmed by the work of Bornet and Stahl). In support of his thesis, Schwendener offered a considerable amount of histological evidence.

Schwendener then began to examine the mechanical properties of plants, seeking an analogue in plants to the animal skeleton or to the materials used in constructing a bridge. He found that the principles of mechanics govern the structure of the stems of plants, with maximum rigidity resulting from minimum materials; he presented his findings in *Das mechanische Prinzip im anatomischen Bau der Monocotylen,* in which, however, he gave no account of the causal development of the structures that he described. The book, published in 1874, was received unenthusiastically by his fellow botanists.

At Basel Schwendener worked in comparative isolation and taught only a few students; he was therefore happy to accept an appointment in 1877 to succeed Wilhelm Hofmeister at Tübingen. Here he studied phyllotaxy, again applying mechanical principles. Hofmeister had been the first to refute the spiral theory of leaf development, stating that each point on the growing region of a stem is a potential leaf and adding that new leaves occur within the largest gaps between existing ones. Schwendener went on to demonstrate, using mechanical models, that leaf arrangement was the result of displacement by contact between leaf primordia; his theory was at first fully accepted, then attacked. It remains an important one in the development of the theory of phyllotaxy, and was stated in *Mechanische Theorie der Blattstellungen,* published in 1878, the same year that Schwendener went to Berlin to succeed Alexander Braun.

Schwendener stayed in Berlin for thirty-one years, during which time he taught a great number of students and wrote articles in defense of his theory of phyllotaxy and on the movement of fluids in plants, the structure and mechanics of stomata, the theory of descent in botany, and torsion in plants. He was elected to the Berlin Academy of Sciences in 1880 and in 1882 was one of the founders of the Deutsche Botanische Gesellschaft. He served as rector of the university from 1887 and retired in 1909. He was, in addition, an honorary member of a number of foreign scientific societies.

Schwendener never married. He was interested in the arts and in literature, and composed several poems. Music, however, he found "a rather disagreeable noise."

BIBLIOGRAPHY

I. ORIGINAL WORKS. Schwendener's most important writings are his dissertation, *Ueber die periodische Erscheinungen der Natur, insbesondere der Pflanzenwelt* (Zurich, 1856); *Das Mikroskop,* 2 vols. (Leipzig, 1865–1867), written with Carl Naegeli; *Die Algentypen der Flechtengonidien. Programm für die Rektoratsfeier der Universität* (Basel, 1869); *Das mechanische Prinzip im anatomischen Bau der Monocotylen* (Leipzig, 1874); and *Mechanische Theorie der Blattstellungen* (Leipzig, 1878). A number of the results of his work at Berlin are collected in *Gesammelte botanische Abhandlungen* (Berlin, 1898).

II. SECONDARY LITERATURE. A short autobiography, written in 1900, and a bibliography are included in A. Zimmermann, "Simon Schwendener," in *Berichte der Deutschen botanischen Gesellschaft,* **40** (1922), 53–76. A detailed biography is G. Haberlandt, *ibid.,* **47** (1929), 1–20.

A. P. M. SANDERS

SCILLA, AGOSTINO (*b.* Messina, Sicily, 10 August 1629; *d.* Rome, Italy, 31 May 1700), *geology.*

Having shown an aptitude for painting since childhood, Scilla was sent to Rome to study art. After his return to Messina, he soon became well known and is still considered one of the best painters of the seventeenth-century Sicilian school. The failure of the Messina revolt against the Spanish (1674–1678), in which Scilla had taken part, forced him into exile, first at Turin and then at Rome. He is famous not only as a painter but also as a scholar, a man of culture well versed in science and the humanities; the latter field is illustrated by his knowledge of ancient Sicilian coins. In the domain of science he was a good mathematician, but he is particularly remembered as the author of *La vana speculazione disingannata dal* senso (1670), today considered one of the classics of geology.

In the seventeenth century most scientists still considered fossils to be *lusus naturae,* sports of nature, born within rocks through the influence of the stars or by other strange means. At the Accademia dei Lincei, in which the supporters of the new science gathered around Galileo, one of its first members, Francesco Stelluti, persisted in this view. It seems highly likely that Stelluti was the academician who moved Scilla to claim the right, in his book, of denouncing *vana speculazione* ("vain speculation"). Fabio Colonna, Scilla, and Niels Stensen put the study of geology in the seventeenth century on the right path, even if they still placed the universal deluge as the origin of those phenomena they were so carefully and objectively observing in Italy.

Convinced that "to doubt things is the best and only way to begin to know them, even approximately," Scilla described with admirable clarity and critical sense the observations he had made on the fossiliferous sedimentary terrains of both shores of the Strait of Messina, dealing with the succession of strata, the genesis of the rocks, and particularly the nature of the fossils. He anticipated the principle that the present is the key to the past when he explained the repetition of coarse, medium, and fine-grained materials in terms of what he could see actually happening in the same places under the action of rapid torrents. He considered fossils to be animal remains imprisoned in rocks that are now hard but were originally muddy or sandy soil.

Extending his researches to other parts of Sicily and to Malta, Scilla studied the zoological features of each fossil, comparing them with those of analogous living species. He did not limit himself to the study of mollusks, the origins of which seemed obvious to him, but tackled more difficult problems, recognizing the presence of fossil corals and echinoderms and showing that the much-discussed *glossopetrae* are the teeth of sharks. To deny the organic origin of fossils, he concluded, was to "commit the sin of disputing a known truth."

BIBLIOGRAPHY

La vana speculazione disingannata dal senso (Naples, 1670) includes 28 beautifully executed engravings of fossil and living marine animals that reveal the keen spirit of observation of the painter and the naturalist. The volume was later published in Latin as *De corporibus marinis quae defossa reperiuntur* (Rome, 1747).

The main secondary source is G. Seguenza, *Discorso intorno Agostino Scilla* (Messina, 1868). Other studies deal exclusively with Scilla as a painter.

                                    FRANCESCO RODOLICO

SCLATER, PHILIP LUTLEY (*b.* Tangier Park, Hampshire, England, 4 November 1829; *d.* Odiham, Hampshire, England, 27 June 1913), *ornithology.*

Sclater was the second son of William Lutley Sclater and Anne Maria Bowyer. His family were landed gentry, and his elder brother became a member of Parliament and later the first Lord Basing. Educated at Oxford, Sclater took the B.A. degree in 1849 and remained at Corpus Christi College for another two years, studying natural history. In 1851 he entered Lincoln's Inn and was later admitted to the bar. He practiced law for a number of years, but he constantly maintained his interest in natural history. Sclater often traveled abroad during this decade, both on the Continent, where he was a frequent visitor to the home of Charles Lucien Bonaparte, and to America, where he met Cassin, Leidy, and John LeConte.

Elected a fellow of the Zoological Society of London in 1850, Sclater became a member of its Council in 1857 and its secretary in 1860, a post he held for forty-three years. In 1858 he took a prominent part in founding *Ibis,* the journal of the British Ornithologists' Union. He became the first editor and continued to be for fifty-four years, except during the period 1865–1877.

This was the great age of zoological travel and exploration, of classification and anatomy. Among Sclater's first major contributions to zoology, and perhaps one of his greatest, was the 1858 paper "On the General Geographic Distribution of the Members of the Class Aves" (*Journal of the Linnean Society,* Zoology, **2,** 130–145), in which he classified the zoogeographical regions of the world on the basis of their bird life. The division into six distinct regions that he proposed was later adopted for all other animals, and is still used by students of zoogeography.

In 1862 Sclater married Jane Anne Eliza Hunter Blair, daughter of a baronet, by whom he had six children. His eldest son, William Lutley, also became an outstanding ornithologist and succeeded his father as editor of *Ibis* just before the latter's death in 1913; he continued as editor until 1930.

In 1861 Sclater was elected a fellow of the Royal Society and served two terms on its Council. He was a member of the British Association for the Advancement of Science (1847) and served as one of its two general secretaries for five years. He was also a life fellow of the Linnean and the Royal Geographical societies and the Geological Society of London, and a corresponding or honorary member of more than forty other scientific societies in Great Britain and abroad.

Sclater's most engrossing duties were with the Zoological Society of London. He spent a number of years reorganizing its affairs, increasing its membership and income, rebuilding the main buildings in the Zoological Gardens, repaying its debts, and keeping the society solvent. He also saw to it that the various publications of the society—the *Proceedings* (now titled *Journal of Zoology*), *Transactions, Zoological Record,* and sundry lists of animals and garden guides—were issued regularly and on schedule.

When his elder brother accepted (1874) a position as president of the Local Government Board in Disraeli's second administration, Sclater acted as his private secretary for two years and subsequently was offered a permanent position in the civil service. He declined this offer because it would take him from his work in natural history.

In connection with his work on zoogeography, Sclater wrote monographs on the tanager genus *Calliste* (1857) and on the jaçanas and puffbirds (1882). With Osbert Salvin, he published *Exotic Ornithology* (1869), which described many new and rare Central and South American birds, and *Nomenclator avium neotropicalum* (1873). A series of papers in the *Proceedings of the Zoological Society of London* in 1877 and 1878 reported on the birds collected by the *Challenger* expedition. In 1888 and 1889 Sclater issued his *Argentine Ornithology,* with notes by W. H. Hudson on the habits of the birds. He also wrote four volumes of the monumental *Catalogue of the Birds in the British Museum*—he prepared volume XI in 1886, XIV in 1888, XV in 1890, and half of XIX in 1891.

Sclater's close friends were the great zoologists of the time—Alfred Newton, who succeeded him as editor of *Ibis* (1865–1871); Salvin, who succeeded Newton until 1877; Canon Tristram; and Alfred Henry Garrod, the anatomist. One of his closest friends was T. H. Huxley, who was also one of his staunchest supporters on the Council of the Zoological Society. Charles Darwin often visited him in his office, bringing long lists of memoranda to discuss.

Sclater was a man of strong personality. Probably this quality is reflected best in his association with the British Ornithologists' Club, which he helped R. Bowdler Sharpe establish in 1892, and of which he was elected chairman. He chaired the monthly meetings and delivered an inaugural address at the beginning of each session.

After resigning as secretary of the Zoological Society in 1903, Sclater retired to his home in Odiham, where he was widely known as an active justice of the peace and a frequent rider with the Hampshire Hunt, of which he was by far the oldest member. He continued to visit the library of the Zoological Society and the collection of his birds at the British Museum (Natural History) until his death, at the age of eighty-three, following a carriage accident.

BIBLIOGRAPHY

I. ORIGINAL WORKS. Sclater was one of the most prolific writers of scientific papers, books, articles, and notes of his time. His publications, listed in the several bibliographies below, total almost 1,400 titles. Many of these were short notes of a few pages or less in periodical literature, some only a few lines announcing new acquisitions by the Zoological Society or exhibits at the British Ornithologists' Club, describing new taxa, or making corrections in the systematic literature. On the other hand, the several volumes he contributed to the *Catalogue of the Birds in the British Museum* are weighty tomes of substance and importance.

II. SECONDARY LITERATURE. A detailed biography is included in G. Brown Goode, "Bibliography of the Published Writings of Philip Lutley Sclater, F.R.S., Secretary of the Zoological Society of London," *Bulletin. United States National Museum,* no. 49 (1896). This lists 1,289 titles to that year, contains a portrait, and names the new families, genera, and species he described. This biography was abridged (with slight corrections) for *Ibis,* 9th ser., 11, jubilee supp. (1908), 129–137, commemorating the "original members" of the British Ornithologists' Union. His longest obituary, by A. H. Evans, in *Ibis,* 10th ser., 1 (1913), 672–686, lists 582 titles on birds alone.

OLIVER L. AUSTIN, JR.

SCOT. See **Michael Scot.**

SCOTT, DUKINFIELD HENRY (*b.* London, England, 28 November 1854; *d.* Basingstoke, England, 29 January 1934), *botany, paleobotany.*

Scott was the youngest son of George Gilbert Scott, an eminent architect, and Caroline Oldrid.

An unusually well-informed botanist, he was among those who laid the foundations of scientific botany and paleobotany.

He was educated privately at home, where he became an avid collector of the local flora. From an early age he used a standard work on systematic botany, and his interests and reading were surprisingly profound. At the age of fourteen he read Joseph Hooker's presidential address to the British Association for the Advancement of Science (1868) in support of Darwinism, and soon thereafter (using English translations), the books of the Continental botanists Alexander Braun, Hugo von Mohl, Carl Wilhelm von Naegeli, and Wilhelm Hofmeister. In addition, he studied Henfrey and Griffith's *Micrographic Dictionary*, with its many German citations. (German literature was ahead of the field in those days.)

Upon entering Christ Church, Oxford, Scott's botanical interests and scholarship, which had sustained his boyhood, slackened or failed, probably through lack of encouragement. In 1876 he received his B.A. and from 1876 to 1879 studied engineering. Upon the death of his father in 1878, Scott came into independent means. His old enthusiasm for botany reasserted itself, and in 1880 he went to Würzburg to study botany with Sachs.

In August 1882 Scott graduated as doctor of philosophy, *summa cum laude*, with a thesis on plant structure: "The Development of the Milk Vessels in Plants." Upon returning to England he succeeded Bower in 1883 as assistant to Daniel Oliver at University College and in 1885 as lecturer in the Royal College of Science under T. H. Huxley. During this period Bower and Scott, great friends, labored on their individual anatomical projects in the small Jodrell Laboratory in Kew Gardens, where Scott produced further evidence for his interpretation of the nature of milk vessels. Later he became honorary keeper of the laboratory and wrote (sometimes in collaboration) several excellent anatomical papers on the fleshy roots of *Sesbania;* on the stems of species *(Strychnos, Ipomoea, Yucca, Dracaena)* with peculiar or anomalous stem vascular tissues; and on the pitchers of *Dischidia*. Scott's associations with the many distinguished visiting botanists and his sojourn in Germany—his "spiritual home"—made him something of an internationalist.

In 1889–1890 the aging W. C. Williamson of the University of Manchester invited Scott and Bower to see his collection of fossil plants and sections thereof—in the hope that they might continue his novel investigations. As Scott himself said, he became an instant convert to the study of fossil plants. While he never ceased to emphasize the importance of comparative anatomical studies of fossil and living plants, he also kept in close touch with virtually the whole field of botany, including plant physiology and genetics. For example, although he did not work in physiology or genetics, both active fields of research, he was alert to their importance. Scott wrote two admirable textbooks, which together make up *An Introduction to Structural Botany:* pt. I, *Flowering Plants;* pt. II, *Flowerless Plants* (1896). Essentially based on the detailed study of selected "types," the textbooks contain many drawings by Scott's wife.

Williamson's pioneering work on Carboniferous fossils had been published in a series of memoirs in *Philosophical Transactions of the Royal Society* (1871–1893). Unfortunately many of his contemporaries had but little understanding of his remarkable achievements. In Scott, Williamson recruited as a co-worker a first-class scholar and plant anatomist, who had an eye for detail and possessed an enviable objectivity. In a joint memoir to the Royal Society on "Further Observations on the Organization of Fossil Plants of the Coal Measures," Williamson and Scott in 1894 gave an important general review of the morphological characters of *Calamites, Calamostachys,* and *Sphenophyllum,* including comparisons with recent plants. The complex strobili (cones) of some of these plants must have taxed their techniques and powers of interpretation. Scott pointed out, with convincing vigor, that studies of the diverse anatomy of extinct and of living species were essentially reciprocal in character and quite essential to any basic consideration of the evolution of plants. In 1895 (the year of Williamson's death) two further memoirs, which dealt with the roots of *Calamites, Lyginodendron,* and *Heterangium,* were published, again under joint authorship. In the memoirs the anatomy of fossil species, with relevant comparisons with recent ferns and cycads, was impeccably portrayed for the first time.

In a notable memoir, contributed to the Royal Society in 1897, Scott described *Cheirostrobus,* a new and exceptional type of cone from the Lower Carboniferous beds at Pettycur on the Firth of Forth. This paper affords a good example of Scott's skill in investigation and interpretation.

In 1900 Scott gave a course of lectures at University College, London, published in *Studies in Fossil Botany*. Because of the beautiful illustrations and engaging clarity of the book, it became, and has remained, a classic. In 1908 the original volume was so enlarged that it was necessary to issue the work in two volumes, and a third edition appeared in 1920–1923. These volumes and Scott's various public lectures did much to bring a proper awareness of the importance of the study of fossils to the botanical world.

During this period Scott also continued to work on selected living plants, for example, cycads and *Isoetes hystrix*. In 1901 he described the seedlike fructifications of *Lepidocarpon*, which were of special interest because they afforded an indication of how early seeds—"in a nascent stage of evolution"—might have originated. In 1902 he published an account of *Dadoxylon* and of its curious vascular structure. In the same year there appeared *The Old Wood and the New*, in which the common anatomical features of Paleozoic stems were closely and critically examined. Later he elucidated the structure and probable affinity of *Stauropteris*, *Botryopteris*, *Trigonocarpus*, *Mesoxylon*, and *Ankyropteris*.

Numerous investigations of ancient Pteridophyta, Cycadofilicales, and other groups, with discoveries by contemporaries, provided Scott with a wide range of information on ancient plants and enabled him to write *The Evolution of Plants* (1911). At the time, the book was of rather special interest, for Scott gave an unusually concise and clear account of the possible affinity of the Jurassic-Cretaceous Bennettitales, that is, of plants that vegetatively seem close to recent cycads but which, on the evidence of their reproductive structures, must be far apart. Scott raised the question: Could plants with these remarkable reproductive organs perhaps have prepared the way for the evolution of the flowering plants? In 1922 he included a wider range of materials in *Extinct Plants and Problems of Evolution*.

In 1887 Scott married Henderina Victoria Klaassen, who had written several botanical papers. In addition to illustrating some of Scott's books and papers, she also provided secretarial help. They had two sons who survived infancy: the younger died suddenly at school in 1914; the elder was killed in the British Ypres salient in 1917. Four daughters survived.

Scott received many academic honors. In 1894 he was elected a member of the Royal Society and in 1906 received the Royal Medal. From 1908 to 1912 he was president of the Linnean Society. In 1926 he was awarded the Darwin Medal and two years later, the Wollaston Medal of the Geological Society.

BIBLIOGRAPHY

Besides the works listed in the text, Scott wrote "The Present Position of Morphological Botany," in *Reports of the British Association for the Advancement of Science* (1896), 922–1010; and "The Present Position of the Theory of Descent, in Relation to the Early History of Plants," *ibid.* (1922), 170–186, his presidential addresses to the botany section of the Association. See also "Reminiscences of a Victorian Botanist," A. B. Rendle, ed., in *Journal of Botany, British and Foreign* (1934). A bibliography is given in *Annals of Botany,* **72** (1935), with portrait.

On Scott and his work, see F. O. Bower, *Sixty Years of Botany (1875–1935): Impressions of an Eyewitness* (London, 1938); and A. C. S., "Dukinfield Henry Scott," in *Obituary Notices of Fellows of the Royal Society of London,* **1** (1932), 205–227.

C. W. WARDLAW

SCOTT, WILLIAM BERRYMAN (*b.* Cincinnati, Ohio, 12 February 1858; *d.* Princeton, New Jersey, 29 March 1947), *vertebrate paleontology, geology.*

The son of William McKendree Scott and Mary Elizabeth Hodge, and a great-great-great-grandson of Benjamin Franklin, Scott lived virtually his entire life in Princeton, New Jersey, where his family moved when he was three years old. He attended the College of New Jersey (now Princeton University) and began his scientific career with field trips to Bridger Basin, Wyoming, in 1877 and 1878, accompanied by his lifelong friend, Henry Fairfield Osborn.

After completion of graduate studies in England and Germany (where he obtained the doctorate at the University of Heidelberg), Scott returned to Princeton in order to join the faculty, where he served for fifty years and then for seventeen years more, after his formal retirement. In 1883 he married Alice Adeline Post; they had seven children. In 1884, at the age of twenty-six, he became a full professor, and in 1904, first chairman of the department of geology, a post he held until his retirement in 1930.

Scott devoted a large part of his life to quiet teaching and research on fossil mammals. After a few expeditions at the beginning of his career, he abandoned the search for fossils in the field, although he was an inveterate traveler throughout most of his life. His research, almost entirely devoted to fossil mammals, produced some 177 published contributions. His viewpoint was pragmatic, and little of his attention was devoted to the more theoretical aspects of his subject. Perhaps his two greatest efforts were the editing and writing in part of the impressive *Reports of the Princeton University Expeditions to Patagonia* (1905–1912) and the writing of several large volumes on the Oligocene mammals of the beds of the White River, South Dakota (1936–1941). He also wrote a standard geology textbook and the widely used *History of Land Mammals in the Western Hemisphere* (1913).

BIBLIOGRAPHY

Scott wrote an autobiography, *Memoirs of a Palaeontologist* (Princeton, N.J., 1939). The full bibliography of W. B. Scott is G. G. Simpson, "Biographical Memoir of William Berryman Scott," in *Biographical Memoirs. National Academy of Sciences,* **25** (1948), 175–203.

EDWIN H. COLBERT

SCOTUS. See Duns Scotus, John.

SCROPE, GEORGE JULIUS POULETT (*b.* London, England, 10 March 1797; *d.* Fairlawn [near Cobham], Surrey, England, 19 January 1876), *geology.*

Scrope was the son of John Poulett Thomson, head of a firm engaged in trade with Russia, and Charlotte Jacob. Scrope's younger brother was Charles Edward Poulett Thomson, Lord Sydenham. Scrope was educated at Harrow; at Pembroke College, Oxford (1815–1816); and at St. John's College, Cambridge (1816–1821). Upon his marriage (22 March 1821) to Emma Phipps Scrope, heiress of William Scrope, of Castle Combe, Wiltshire, he assumed her name (which is pronounced Scroop).

Scrope was elected to the Geological Society in 1824 and served as secretary in 1825; he was later awarded the Society's Wollaston Medal (1867). He was elected also to the Royal Society (1826) and was founder and first president of the Wiltshire Archaeological and Natural History Society (1853–1855). From 1833 to 1868 he was a Liberal member of Parliament for Stroud, Gloucestershire. He was an advocate of free trade and various social and economic reforms, especially the poor law, but he took no part in parliamentary debate. He instead wrote extensively on a variety of political and economic subjects. He was said to have written nearly seventy anonymous pamphlets, earning him the nickname "Pamphlet Scrope." About 1867, after the death of his wife, who had been an invalid as the result of a riding accident soon after their marriage, he sold Castle Combe and moved to Fairlawn, near Cobham, Surrey. On 14 November 1867 he married Margaret Elizabeth Savage, who survived him. There were no children from either marriage.

Scrope's interest in geology was first aroused in Italy by the sight of Vesuvius in continual eruption during the winters of 1817–1818 and 1818–1819. In 1819–1820 he visited Sicily and the Lipari Islands, studying Mount Etna and Stromboli. On the advice of Edward Clarke and Adam Sedgwick at Cambridge, he studied the extinct volcanic region of Auvergne in central France in the summer and fall of 1821. He then went to northern Italy and eventually to Naples, where he arrived in time to observe the violent eruption of Vesuvius in October 1822. During his return to England in the summer of 1823, he visited the volcanic regions of the Eifel in Germany.

On the basis of his geological fieldwork, Scrope wrote two books. The first book, *Considerations on Volcanos* (1825), has been called "the earliest systematic treatise on vulcanology."[1] It was poorly received by geologists, since it put forth in a dogmatic fashion hypotheses about every phase of volcanic activity and concluded with a "new theory of the earth." At this time Scrope was an ardent advocate of the theory of a cooling earth, which implied that the frequency and intensity of earthquakes and volcanic eruptions had declined over geological time. He believed that the forces of heat within the earth were still capable of producing, on rare occasions, cataclysmic upheavals (perhaps of whole mountain ranges or continents), causing destructive diluvial waves. At the same time he argued that

> The laws or processes of nature we have every reason to believe invariable. Their *results* from time to time vary, according to the combinations of influential circumstances [p. 242]. . . . Until, after a long investigation, and with the most liberal allowances for all possible variations, and an unlimited series of ages, [present-day processes] have been found wholly inadequate to the purpose, it would be the height of absurdity to have recourse to any gratuitous and unexampled hypothesis [pp. v–vi].

These ideas were indebted in part to the writings

of James Hutton, John Playfair, and James Hall, but Scrope's restatement in combination with the theory of a cooling earth was to provide the basis for a catastrophist opposition to the uniformitarianism of Lyell.

Scrope's second book, *Memoir on the Geology of Central France* (1827, but written in 1822), was more uniformitarian in approach, being devoted to a small region during geologically recent times. Improving upon the work of French geologists in Auvergne, he showed that currents of lava, which had flowed into valleys at various times, appeared at different heights above the river beds, marking successive steps in the progress of erosion of the valleys by the rivers. Scrope refuted the arguments for a recent deluge in the region and showed the untenability of attempts to classify the volcanic cones into antediluvian and postdiluvian on the basis of the amount of erosion they had undergone. As additional evidence that rivers are capable of producing impressive valleys unaided by a deluge, he cited in 1830 the presence of entrenched meanders in the valleys of the Meuse and Moselle.[2] These arguments were effective in undermining the Cuvier-Buckland theory of a recent universal deluge, identical with the Biblical flood, that had deposited debris and had carved the major valleys. The theory was eventually abandoned by its remaining supporters in the Geological Society.

Scrope always believed that his two books had greatly influenced the development of Lyell's uniformitarian views; and Lyell's dependence is traceable in two articles that he wrote in 1826 and 1827 for the *Quarterly Review*. The latter was a very favorable review of Scrope's second book. Lyell soon after verified Scrope's observations in central France and Italy (1828–1829). Therefore, Scrope can, paradoxically, be regarded as a parent of both uniformitarianism and its catastrophist opposition in Great Britain. He never committed himself to either side and always occupied a middle ground.

As one of a number of younger geologists who wished, as he expressed it, to free geology "from the clutches of Moses," Scrope assisted Lyell in the completion of the first volume of his *Principles of Geology* (1830), which had as a principal objective the extermination of theological influence on geology. A favorable review was ensured, since the publisher also owned the *Quarterly Review*; and its editor, after consulting Lyell, chose Scrope as the reviewer. Scrope wrote the review with Lyell's advice, and he also reviewed Lyell's third volume for the same journal (1835). In these and later reviews, Scrope argued strongly against reli-

gious influence in geology, yet he asserted that there was no conflict between the Bible and geology and wrote enthusiastically about the great contributions of geology to natural theology. He put forth, in opposition to Lyell, the evidence for progressive change in the histories of the earth and of life; and he was unimpressed by Lyell's hypothesis of a chemical source of heat within the earth.

Scrope rejected Lyell's metamorphic theory, which asserted that the "stratified primary rocks," such as gneiss and schist, were normal sedimentary rocks altered by internal heat. He contended instead that the stratified primary rocks had originally been disintegrated granite deposited in hot agitated water when the primitive earth was hot and barren.[3] He later gave up this explanation but continued to believe that the foliation of these rocks was caused by differential movement. Scrope regarded the earth as essentially solid in its interior, kept in this state by intense pressure brought about by the expansive forces of heat and steam. The subterranean matter was thus in a state of tension so that any slight reduction of external pressure, caused perhaps by a change in atmospheric pressure or by the tidal action of the sun and moon, or an increase in internal temperature brought about by a local influx of heat would be sufficient to cause local melting accompanied by fracturing of the surrounding rock and the penetration of the fractures by the molten rock.

Scrope saw no essential difference between the causes of volcanoes and of earthquakes; a volcano was the result of fracturing at shallow depths, which allowed the fluid material to reach the surface. In 1856 he argued that this same cause could also produce a violent fracture and elevation of the overlying crust and the extrusion through a fissure of a ridge of crystalline matter. As the central axis of this protruded ridge or mountain range rose, the hot and partially fluid matter on the sides would be subject to friction, causing differential movement and the formation of parallel striations. In support of this theory he argued that mountain ranges typically show granite in the center, passing on the sides into gneiss (or "squeezed granite") and, further on, into schist.[4]

To the end of his life, Scrope continued to believe in the possibility of rare cataclysmic earth movements far in excess of anything that Lyell would allow, although Scrope's early catastrophism was soon modified to the extent that he believed that a mountain range had been formed by a succession of upheavals rather than by one grand upheaval, as advocated by Élie de Beaumont.

Scrope adopted a mean position between the two extremes following the analogy with volcanic eruptions, which vary greatly in violence. By 1872 he had also sufficiently changed his views so that he could regard the theory of a cooling earth as only a conjecture, belonging rather to astronomy than to geology. He therefore rejected T. H. Huxley's attempt to replace the uniformitarian interpretation of earth history by an evolutionary one, asserting that there was no evidence to show that the overall rate of earth movement had varied perceptibly during geological time. He did not, however, contest the evidence for biological evolution.[5]

Scrope's original scientific work virtually ceased for many years after he entered Parliament. He returned to geology in the mid-1850's, prompted by his desire to assist Lyell in combating a revival of the theory of "craters of elevation" of Humboldt, Buch, and Élie de Beaumont—a theory that he had first attacked in 1825. This theory regarded some volcanic cones as produced by a single explosive upheaval of strata rather than being built slowly by successive deposits of volcanic materials erupted from a vent, as Scrope and Lyell had always maintained. In two articles (1856, 1859) Scrope helped to refute the theory, pointing out its inconsistencies and the lack of agreement among its supporters as to the criteria for distinguishing these cones from cones formed by ordinary eruptions.[6]

After revisiting Auvergne in the summer of 1857, Scrope published a revised edition of his work on central France (1858); he dedicated the work to Lyell. This was followed by a greatly altered edition of his work on volcanoes (1862). During the last fifteen years of his life, he wrote many letters and short articles, which appeared mostly in the *Geological Magazine*. These writings served to correct the errors of others, to reaffirm or revive theories that Scrope had published in his first books, and to remind the world of his priority. He gave a general summary of his views in a new preface that accompanied the reissue of his work on volcanoes in 1872.

Among the "errors" attacked by Scrope in his last years were (1) the theory of a hot liquid earth with a thin crust, the contractions of which cause the earth to crumple, forming mountain ranges; (2) the tendency of German geologists to postulate a rigid law of succession of different types of lava from a volcanic vent; (3) the theory, held by Lyell, that the influx of seawater into the interior of the earth is the triggering cause of earthquakes and volcanic activity; and (4) any theory that ignored the primary importance of subterranean forces in the history of the earth. Thus, in a controversy with Jukes over the origin of valleys (1866), Scrope felt it necessary to remind the extreme fluvialists of the primacy of internal forces in creating the topography of the earth. Scrope believed that stream erosion was insignificant in comparison, yet he had long been among the leading advocates of "rain and rivers" as agencies of denudation in opposition to the tendency of Lyell and others to stress marine denudation.[7] Later (1872) he warned the fluvialists of the probability of vast denudational effects produced in the past by gigantic waves accompanying cataclysmic earth movements.

One of Scrope's favorite theories was that most lavas at the time of their appearance on the surface are not in a state of fusion but consist of solid crystals sliding over one another because of the expansive force of the steam mixed with them. In support of this theory he pointed to the rarity of glassy textures in most lavas. Despite the criticism his theory incurred from Lyell, after he first proposed it in 1825, Scrope continued to advance it, although without much success.

Like Lyell, Scrope was virtually blind during his last years. He generously encouraged young geologists to continue his investigations of volcanic phenomena, and his financial support enabled Archibald Geikie (1870) and John W. Judd (1876) to study Vesuvius and the Lipari Islands.

Scrope remained an amateur in geology in the sense that his knowledge never extended far beyond his principal fields of interest. He had little knowledge of paleontology, and his theories on tectonic mechanisms and on the origin of metamorphic rocks exhibit a deficient knowledge of chemistry and physics. His books of the 1820's showed considerable originality, and, principally by means of their influence on Lyell, helped steer geology into a more uniformitarian path. His later writings may have helped to keep geology on a middle course by combating extreme views from whatever side.

NOTES

1. See John W. Judd, *Volcanoes: What They Are and What They Teach,* 6th ed. (London, 1903), p. 5. Judd credits Scrope with a number of contributions to vulcanology.
2. *Philosophical Magazine,* **7** (1830), 210–211 (Geological Society Proceedings).
3. *Quarterly Review,* **43** (1830), 411–469, and **53** (1835), 406–448.
4. *Geological Society Quarterly Journal,* **12** (1856), 326–350; cf. *Volcanos* (1862), pp. 265ff.
5. *Volcanos* (London, 1872), Preface.

6. *Geological Society Quarterly Journal,* **12** (1856), 326–350, and **15** (1859), 505–549.
7. *Geological Magazine,* **3** (1866). Cf. *Geology and Extinct Volcanos of Central France* (1858), 208.

BIBLIOGRAPHY

I. ORIGINAL WORKS. Scrope's major geological works are *Considerations on Volcanos* (London, 1825), 2nd ed., *Volcanos* (London, 1862; reissued with new preface, 1872), and *Memoir on the Geology of Central France* (London, 1827), 2nd ed., *Geology and Extinct Volcanos of Central France* (London, 1858). The Royal Society *Catalogue of Scientific Papers* lists some 36 scientific articles by Scrope, omitting at least four articles in the *Geological Magazine,* two letters in *Nature* (1875), and four reviews of geological works in the *Quarterly Review* (listed in the *Wellesley Index to Victorian Periodicals,* vol. I). Besides many pamphlets, Scrope wrote a number of reviews on nonscientific subjects, a biography of his brother, a history of Castle Combe, and works on economics and other subjects. Some 32 of Scrope's letters to Lyell (1828–1874) are at the American Philosophical Society.

II. SECONDARY WORKS. There is no biography of Scrope. T. G. Bonney's article in the *Dictionary of National Biography,* XVII (1897), 1073–1074, is rather uninformative about Scrope's personal life, as are the various obituaries. Scrope's contributions to geology are listed in *Geological Magazine,* **13** (1876), 96. His geological work is discussed in Karl Alfred von Zittel, *History of Geology and Palaeontology* (London, 1901), 259–263, and in R. J. Chorley, A. J. Dunn, and R. P. Beckinsale, *History of the Study of Landforms,* I (London, 1964), 125–130, 146–147, 357, 390–391, 400.

L. E. PAGE

SCUDDER, SAMUEL HUBBARD (*b.* Boston, Massachusetts, 13 April 1837; *d.* Cambridge, Massachusetts, 17 May 1911), *systematic entomology.*

Scudder was the son of Charles Scudder and Sarah Lathrop Coit, both of whom were Congregationalists. He attended Williams College (1853–1857) and studied with Louis Agassiz at Harvard (1857–1864).

Scudder held various positions in Cambridge and Boston and was instrumental in founding the Cambridge Entomological Club and *Psyche* in 1874. Between 1858 and 1902 he published 791 papers. His noteworthy works include *Catalogue of Scientific Serials of All Countries . . . 1633–1876* (1879); *Nomenclator zoologicus,* 2 pts. (1882–1884), a list of the generic names proposed in zoology; and the three-volume *Butterflies of the Eastern United States and Canada* (1888–1889).

Scudder's main contributions were in the study of Orthoptera—of which he described 630 species—and fossil insects. He named about 1,144 species of fossil insects, mostly while employed as paleontologist in the United States Geological Survey between 1886 and 1892.

Scudder was one of the most learned and productive American systematic entomologists of his day. As recently as 1920 Willis Stanley Blatchley said that "to him more than to all his predecessors and contemporaries combined is due our present knowledge of the Orthoptera." Scudder drew whatever taxonomic and evolutionary conclusions his data indicated, although new methods, especially studies of genitalia, and more extensive series of specimens, have required the modification of some of his conclusions. It is also doubtful whether fossils like the Coleoptera, which he studied, can always be named with as much taxonomic precision and be integrated as precisely with living species as Scudder held that they could. The beetles, for instance, do not exhibit the characters on which contemporary classification depends. To date no one has attempted either to continue or to revise Scudder's work on fossil beetles, at least in anything like Scudder's detail.

Scudder's personal life was unfortunate. In 1867 he married Ethelinda Jane Blatchford, who died in 1872. Their son died at the age of twenty-seven in 1896, the same year that Scudder began to show signs of Parkinson's disease. Realistic about his prospects, Scudder ended his work in 1902. He transferred his personal collection to the Museum of Comparative Zoology at Harvard College and his library to the Boston Society of Natural History and to Williams College. He died in 1911 after years of seclusion and invalidism.

BIBLIOGRAPHY

On Scudder and his work, see J. S. Kingsley, William L. W. Field, T. D. A. Cockerell, and Albert P. Morse, "Appraisals of Scudder as a Naturalist," in *Psyche,* **18** (1911), 174–192, with portrait; and Alfred Goldsborough Mayor, "Samuel Hubbard Scudder 1837–1911," in *Biographical Memoirs. National Academy of Sciences,* **17** (1919), 79–104, with portrait and bibliography of 791 titles.

MELVILLE H. HATCH

SEARES, FREDERICK HANLEY (*b.* near Cassopolis, Michigan, 17 May 1873; *d.* Honolulu, Hawaii, 20 July 1964), *astronomy.*

Seares was associated with the Mount Wilson Observatory for thirty-six years, fifteen of them as assistant director. His principal contribution lay in the field of photographic photometry: he standardized the stellar magnitude system and extended it to include stars fainter than the eighteenth magnitude.

The son of Isaac Newton Seares and his wife, the former Ella Ardelia Swartwout, Seares was born on a farm in the southwest corner of Michigan. By the time he was fourteen, the family was living in California, where he attended the Pasadena high school. He received a B.S. degree (with honors) from the University of California in Berkeley in 1895. This was his highest earned degree, although he remained at Berkeley four more years, two with the title of fellow and two more as instructor in astronomy.

In 1896 Seares married Mabel Urmy. Their only child, Richard, was born in Paris in 1900, while Seares was studying at the Sorbonne; the previous year Seares attended the University of Berlin.

Returning to the United States in 1901, Seares served for eight years as professor of astronomy and director of the Laws Observatory at the University of Missouri in Columbia. Then, at the invitation of G. E. Hale, he returned to Pasadena as superintendent of the computing division of the Mount Wilson Solar Observatory, but also to do research with the then brand-new sixty-inch telescope.

In this latter connection Seares was soon photographing stars in "selected areas" of the sky that had been chosen by Kapteyn. This collaboration culminated, more than twenty years later, in a catalog listing 67,941 stars, located in 139 areas (none larger than 23′ of arc in diameter) systematically scattered over the sky north of declination −15°. Seares's problem was to intercompare, in a reliable way, light signals from widely separated stars that differed by as much as ten million times in brightness. His method was to photograph each area several times on the same plate, using absorbing wire gauze screens and reduced apertures to introduce known reductions in light, up to eleven magnitudes, between successive images of each star. Relative magnitudes could then be read off, and a common zero point applied to give absolute values.

The preliminary results of Seares's photometric program were so impressive that in 1922 he was elected president of the committee on stellar photometry of the International Astronomical Union. In 1932 the IAU adopted international magnitude standards based on his work, but Seares himself was not yet satisfied: he then began work on a catalog that appeared in 1941, providing data on 2,271 stars within 10° of the North celestial pole, to be used for both visual and photographic magnitude standards.

Seares also wrote a number of papers on galactic structure. His conclusion as to the relative brightness of our Milky Way compared to other "spiral nebulae" was influential in the stand taken by Shapley (his student at Missouri and his co-worker at Mount Wilson) in the Curtis-Shapley debate of 1920. Seares also investigated the way interstellar material obscures and reddens the light from distant stars.

Seares was elected to the American Philosophical Society in 1917, an associate of the Royal Astronomical Society in 1918, and to the National Academy of Sciences in 1919. He received an LL.D. degree from the University of California in 1930 and another from the University of Missouri in 1934. In 1940 he was awarded the Bruce Gold Medal of the Astronomical Society of the Pacific. That same year his wife died; two years later he married his longtime assistant Mary Cross Joyner, and with her eventually retired to Honolulu, where he died at age ninety-one.

BIBLIOGRAPHY

I. ORIGINAL WORKS. For a description of the Laws Observatory, see "Report of the Director of the Laws Observatory of the University of Missouri," in *Publications of the Astronomical Society of the Pacific*, **15** (1903), 167–168.

Seares's photometric techniques are described in "Photographic Photometry with the 60-inch Reflector of the Mount Wilson Solar Observatory," in *Astrophysical Journal*, **39** (1914), 307–340; his so-called exposure ratio method for direct determination of the color index (a number to be subtracted from the photographic magnitude of a star to get its visual magnitude) is outlined in "A Simple Method of Determining the Colors of the Stars," in *Proceedings of the National Academy of Sciences of the United States of America*, **2** (1916), 521–525. His paper entitled "The Surface Brightness of the Galactic System as Seen From a Distant External Point and a Comparison With Spiral Nebulae," appeared in *Astrophysical Journal*, **52** (1920), 162–182.

The *Mount Wilson Catalogue of Photographic Magnitudes in Selected Areas 1–139* is vol. IV of the Papers of the Mount Wilson Observatory and appeared in 1930 as Carnegie Institution of Washington Publication no. 402; the authors are given as F. H. Seares, J. C. Kapteyn, and P. J. van Rhijn (although Kapteyn had been dead almost eight years), assisted by Mary C. Joyner and Myrtle L. Richmond. *Magnitudes and Colors of Stars North of +80°*, written with Frank E. Ross and Mary C. Joyner, appeared as vol. VI of the same series and as Carnegie Publication no. 532 (1941); this work is Seares's final attempt to provide a set of absolute standard magnitudes in the North Polar region, replacing those printed in *Transactions of the International Astronomical Union,* **4** (1932), 140–152, that had been adopted as international standards.

Seares's ideas on interstellar obscuration are summarized in "The Dust of Space," based on his Bruce Gold Medal lecture, in *Publications of the Astronomical Society of the Pacific,* **52** (1940), 80–115.

A list of 161 publications by Seares follows Joy's 1967 biographical memoir (see below).

II. SECONDARY LITERATURE. For a contemporary discussion of Seares's work, see Alfred H. Joy, "The Award of the Bruce Gold Medal to F. H. Seares," in *Publications of the Astronomical Society of the Pacific,* **52** (1940), 69–79, with photograph facing p. 69. Joy also wrote the entry on Seares in *Biographical Memoirs, National Academy of Sciences,* **39** (1967), 417–431 with bibliography 432–444. An obituary by R. O. Redman appeared in *Quarterly Journal of the Royal Astronomical Society,* **7** (1966), 75–79.

SALLY H. DIEKE

SECCHI, (PIETRO) ANGELO (*b.* Reggio nell' Emilia, Italy, 18 June 1818; *d.* Rome, Italy, 26 February 1878), *astronomy, astrophysics.*

Secchi's father was a cabinetmaker, and the family was of modest circumstances. Secchi himself attended the Jesuit school in Reggio nell'Emilia, then, toward the end of 1833, when he was fifteen, entered the Jesuit novitiate in Rome. Two years later he entered the Collegio Romano, where he distinguished himself in a course that included physics and mathematics. In 1841 he was appointed instructor in these subjects at the Jesuit college in Loreto and soon became known for the originality of his lectures. Between 1844 and 1848 Secchi was obliged to return to his theological studies, but was able simultaneously to continue his astronomical work under Francesco de Vico, director of the observatory of the Collegio Romano and professor of astronomy at the Gregorian University in Rome.

When the Jesuits were expelled from Rome in 1848, Secchi went first to the flourishing Jesuit college at Stonyhurst, Lancashire, England. He then established himself at Georgetown University in Washington, D.C., where he acted as assistant to the director of the observatory, Father P. Curley, and made further studies in both theoretical and practical astronomy. He met the hydrographer M. F. Maury, and became acquainted with Maury's important meteorological works. In 1849 the ban against the Jesuits was lifted, and Secchi was able to return to Rome, where he took up an appointment as director of the observatory of the Collegio Romano.

Secchi recognized immediately that the observatory and its equipment were inadequate. He determined both to build a new observatory and to reshape the course of the research to be performed there, placing a new emphasis on astrophysics. His predecessor de Vico had already conducted important observations on comets, the rotation of Venus, and the satellites of Saturn, and Secchi, using the outmoded instruments of the old installation, began to examine the physical aspect of the ring of Saturn. In 1851 he began to study the sun, measuring the intensity of solar radiation with a thermoelectric pile, a technique previously employed by Joseph Henry to determine the temperature of a sunspot. Secchi first measured the radiation of the sun during the eclipse of 1851, then applied this method to the full solar disk in an attempt to establish the relationship between the temperature of the disk at its center and that at its edges. In the course of his work during the eclipse he made several daguerreotypes that must be reckoned among the earliest applications of photography in the study of the celestial bodies. His results indicated that the center of the solar disk emits almost twice as much radiation as does its borders.

The following year Secchi collated the two fragments into which Biela's comet had split during its appearance in 1846. He was able to observe a weak star through one of the segments, and was thereby able to demonstrate the thinness of the matter of which the comet was constituted. In 1853 Secchi discovered a comet with a multiple nucleus. He was, during this period in which astronomers were first beginning to understand the true nature of comets, more interested in determining their physical composition than in observing their position, and his interest extended to falling stars, which he introduced as a subject for study by his students at the Collegio Romano. His own observations led him to the conclusion that falling stars were of cosmic origin and thus paved the way for the work of Schiaparelli. At about the same time he made his first investigations of nebulae, by which he sought to confirm the results obtained by William Herschel, James Clark Ross, William Huggins, and Hermann Vogel.

Secchi was further engaged in preparing the new observatory that was being constructed over the church of St. Ignatius, part of the Collegio Romano. He installed a Merz equatorial telescope with an aperture of 24.5 centimeters and a focal distance of 4.3 meters, and with it resumed his observations of Saturn. With the new instrument, Secchi was able to determine the physical characteristics of the planet, including its polar flattening and the eccentricity of its ring. Having made further observations of Mars, during its opposition of 1859, Secchi announced that he had seen two permanent "canals" between the two reddish equatorial continents of that planet; he thus introduced a term which was widely taken up by his successors. In the same year he made studies of the moon, during which he measured and and made a detailed drawing of the crater of Copernicus for the purpose of noting seasonal variations, should such occur. He attempted to apply photography to the study of that body, but determined that his refractor was not suitable for photographic use, since its visual objective was corrected for the human eye, rather than for the special qualities of photographic emulsions.

About 1860 Secchi began to make observations and drawings of Jupiter. He noted that disturbances similar to terrestrial storms occurred in its atmosphere and, with the Merz objective, studied its satellites in some detail. In particular, he observed the spots on the Jovian moons, their periods of rotation about their axes, and the characteristics of their reflected light. He made spectroscopic studies of both Jupiter and Saturn and discovered the presence of special absorption bands in their spectra, which led him to conclude that the atmospheres of these planets contained elements different from terrestrial ones. He also made note of the almost permanent transparency of the atmosphere of Mars and proved that Uranus and Neptune do not have discontinuous spectra, but rather demonstrate bands that have the same qualities (although of much greater intensity) as those that he had found in the spectra of Jupiter and Saturn.

Secchi soon turned his attention to the sun. From the beginning he was convinced of the applicability of his research to a wide variety of celestial and terrestrial phenomena, a belief that he later summed up (in the introduction to his *Le soleil*), when he wrote that

Whatever our researches and [whatever] the knowledge that we acquire from them, it will not be in our power to regulate the influence of the sun. Nevertheless, the action of this star is too intimately related to the phenomena of life, heat, and light to render useless the studies that may enable us to investigate its nature. And, on the other hand, who knows whether or not an intimate relationship may exist between certain solar phenomena and some terrestrial ones that it would be important for us to predict with some degree of certainty?

Secchi made use of a good helioscopic eyepiece and projections to observe, on the photosphere, a great number of small luminous granules. These granules were of a variety of sizes and shapes, although the commonest were oval; they stood out upon a darker (although not entirely black) ground. This granulation was broken, most notably at the edges of the disk, by luminous tongues, which Secchi named "facole," and by the small black holes (which he called "pori") that are the points from which sunspots originate. Secchi noted that the luminous granules represented the extremities of columns of the warmer gases that arise from the cooler and less luminous solar surfaces. Having observed that the formation of sunspots, which generally appeared after a period of surface agitation, was usually accompanied by the appearance of less brilliant luminous tongues (now called "flares"), Secchi determined to investigate them. He concluded that these flares were, in fact, complex groups of gases with several nuclei (or dark central shadows), surrounded by half-shadows.

Secchi also observed the chromospheric eruptions that cross the nuclei of sunspots and split them into segments. It was clear to him that such wide, rapid, and complex surface movements could not occur in a solid substance, and he therefore suggested that the entire photosphere must be composed of an elastic fluid, similar to a gas, through which the sunspots move in a manner similar to terrestrial cyclones. He noticed that these vortices are more frequent during a period in which sunspots are being formed, when the surface movements that create the spots create currents that converge toward the nuclei. This represented further evidence of the gaseous nature of the sunspots and of the photosphere. Secchi lastly applied the law of diminution of angular velocity to the movement of sunspots from the equator to the poles of the solar surface to ascertain that the sun, or at least the photosphere, moves in accordance with the laws that govern a fluid mass.

Secchi wished to study the sun spectroscopically, and to that end commissioned Hofmann and

Merz to construct spectroscopes incorporating a series of prisms, while G. B. Amici constructed instruments for direct viewing. Taking up Kirchhoff's researches, Secchi demonstrated that the absorbing stratum of the sun, later identified as an inversion layer, must be very thin. With a large dispersion spectroscope, which he attached to the Merz equatorial, Secchi was able to observe the inversion of the hydrogen line on the chromosphere, which occurred an instant before the appearance of the dark D lines of Fraunhofer. Since the continuous spectra, with the exception of a few lines of sodium and magnesia, were also inverted, Secchi concluded that the stratum that in that instant partially inverts the dark lines of certain metals (such as the rose stratum of the chromosphere) also inverts the dark lines of hydrogen. This spectral analysis provided further confirmation of a solar atmosphere similar to the terrestrial one, although containing many lines of unknown origin, possibly those of elements that did not exist on earth. Secchi's spectroscopic examination of sunspots led him to recognize that the lines they exhibit are those of the solar atmosphere, although more or less widened, intensified, or weakened.

Secchi also realized the importance of observing the chromosphere and the corona during total eclipses of the sun. His first opportunity to do so occurred in 1860, when the totality of a solar eclipse was visible in Spain. Secchi traveled to Desierto de las Palmas, near Castellón de la Plana, where he made observations with a Cauchoix refractor; he then compared his results with those De la Rue had made with a Kew photoheliograph at Rivabellosa. Secchi thus concluded that the prominences seen during the eclipse were real, rather than a play of light as some had suggested; that they were solar in origin; and that the corona was also real and thicker at the equator than at the poles, and thickest at forty-five degrees. After the 1868 eclipse Secchi used the technique, simultaneously developed by Janssen and Lockyer, of enlarging the aperture of the spectroscope directed toward the solar border to observe the prominences and chromosphere in full sunlight. He applied this method further in 1869, when he began, with Respighi, the series of observations of the "spectroscopic images of the solar border" that he published in *Memorie della Società degli spettroscopisti italiani* in the following year.

Secchi also published a number of observations on solar prominences in his treatise *Le soleil*, issued in French in 1875–1877. This work was illustrated with magnificent color plates of the chromosphere and of the various types of prominences. Among those shown were the small flames (now called "spiculae") that he observed in the region of sunspots, converging toward the center of the eruption. These, he noted, became higher, more slender, and extremely luminous at the solar poles, where considerable activity manifested itself even during periods of relative calm on the solar surface. He classified the prominences as "quiescent" and "eruptive"—terminology that is still current.

Secchi made an especially careful study of the forces to which solar prominences are subjected, measuring the velocity of the masses expelled by the sun and their movement in the solar atmosphere. He used the spectroscope to observe the shifts, caused by the Doppler effect, of the lines present in the spectra of the prominences and observed that the expelled matter is not only launched upward but also frequently animated by vortical movements that give a spiral appearance to the luminous spouts. He speculated that the variations in velocity now attributed to the presence of magnetic fields must be caused by an as yet unrecognized periodic force. He observed that secular variations, of very short duration, also occurred at the diameter of the sun, and pointed out that irregularities in the shape of the sun were most apparent during periods of (and in the regions of) maximum photospheric agitation.

While conducting his solar studies, Secchi was further concerned with the study of the physical constitution of comets. During the appearance of the comet of 1861 he observed that the head of the comet emitted jets of discontinuous gases, which formed parabolic envelopes about it and, by a backward movement, produced the tail. In observing the comet in 1862, he was able to see that the jets of gases altered from night to night in correspondence to the comet's distance from the sun. He established the presence of carbon in the spectrum of the comet of 1866 and noted, in addition to the lines of emission, the continuous spectrum that indicated the existence of direct or reflected light. He found carbon associated with hydrogen or oxygen in the bands exhibited by comets of later years.

In 1862 Secchi, in furtherance of the early work on falling stars that had been performed at the Collegio Romano, carried out simultaneous observations at Rome and at Civitavecchia in order to determine the altitude of falling stars, which he calculated as being between seventy-five and 250 kilometers, with a falling velocity of ninety kilometers per second. From these and later observations he also established the similarity of falling stars, asteroids, and aerolites. Using the spectrosope he demonstrated that all these phenomena contained such metals as iron, magnesium, and sodium; he was particularly active during the rain of fire of 27 November 1872, "in which," he wrote, "the layers of distant light resemble the surge of snowflakes."

Secchi's spectroscopic research on luminous stars began in 1863, when he was visited in Rome by Janssen, who had a small spectroscope with him. Together they attached the spectroscope to a Merz equatorial and observed the stars, communicating their joint findings to the Paris Academy. Secchi and Janssen found Fraunhofer lines in the stellar spectra they examined, and identified some of these with terrestrial elements; Secchi then began to work with more sophisticated equipment to define the differences in solar spectra already noted by Donati and Huggins. He determined to investigate the spectra of a large number of stars, and set out his plan in a communication to the Pontificia Accademia Tiberina on 27 January 1868. He prefaced his report with the statement that: "In substance I wanted to see whether, just as the stars are countless, their composition is also proportionately varied. This was my query, and having been fortunate enough to perfect the observation instrument, the harvest was abundant, even more than I had hoped." The instrument to which he referred was a spectroscope equipped with a luminous aperture with which he could analyze even the weakest stars; with it Secchi was able to recognize a specific spectral type in a number of principal stars.

Secchi then determined to examine other stars by means of simple differential methods. Toward the end of 1869 he decided to adopt the apparatus that Fraunhofer had used, which consisted of a flint prism, fifteen centimeters in diameter, with a refracting angle of twelve degrees. He mounted this prism directly in front of the twenty-three-centimeter objective of the Merz equatorial telescope; he thus obtained greater luminosity and a larger dispersion than had been possible with the direct-vision prism, inserted between the objective and the ocular piece, that he had used previously. He next experimented with a circular prism, constructed by Merz, which he attached to the Cauchoix equatorial, the objective having an aperture equal to that of the prism. With this new combination he discovered that "whereas the stars are very numerous, their spectra are reduced to a few well-defined and distinct forms, which for the sake of brevity we may call types." He went on to note that he had examined at least 4,000 stars, and was able to divide them into five types, with the high-temperature white stars at one end of the scale and the low-temperature red stars at the other. This classification still bears Secchi's name; it was soon adopted almost universally.

Secchi also classified nebulae according to his spectroscopic examination, into planetary, elliptical, and irregular forms. He examined examples of each type and found spectral lines of emission produced both by hydrogen and by elements then unknown on earth. The presence of these unknown elements led him to deduce that some nebulae are masses of pure gas—that is, that not all of them are resolvable into stars—a confirmation of a theory put forth by Herschel and accepted by Secchi himself. Of the elliptical nebulae of the external galaxies he was particularly interested in Andromeda, of which he studied the continuous spectrum.

In his observations of Andromeda Secchi noticed two black canals transversing the nebula. It seemed to him that these must be zones of non-luminous matter, projected on the nebula proper and intercepting the light of the stars, a phenomenon that he discovered in other parts of the Milky Way. He noted that it was improbable that these canals should, in fact, represent apertures, particularly given the gaseous nature of nebulae; many years later observations confirmed the presence of dark masses dispersed in space, which are seen projected on the background of a sky made luminous by light-emitting cosmic matter.

Secchi had almost ceased to make observations by 1873, when the Jesuits were again expelled from the Collegio Romano. He was nonetheless allowed to remain at his observatory through the intervention of the government, and he spent his last years preparing his scholarly writings for their final editions. He also continued to publish notes in the *Memorie della Società degli spettroscopisti italiani*, of which he had been, with Tacchini, a founder in 1870.

BIBLIOGRAPHY

I. ORIGINAL WORKS. Secchi's works include *Le soleil* (Paris, 1875–1877); *Le stelle* (Milan, 1877); and *L'astronomia in Roma nel pontificato di Pio IX* (Rome, 1877). A complete list of his writings is given by Bricarelli, cited below.

II. SECONDARY LITERATURE. On Secchi and his works, see G. Abetti, *Padre Angelo Secchi* (Milan, 1928); and *Storia dell'astronomia* (Florence, 1963), 159, 187–203, 205–206, 228–229, 255, 259, 266, 270, 278,

299, 301, 304, 352, 364, 377, 383, 389–391, 395, 401; and Bricarelli, "Della vita e delle opere del P. Angelo Secchi," in *Memorie dell'Accademia pontificia dei Nuovi Lincei*, **4** (1888).

GIORGIO ABETTI

SECHENOV, IVAN MIKHAYLOVICH (*b.* Teply Stan [now Sechenovo], Simbirsk guberniya [now Arzamas oblast], Russia, 1 August 1829; *d.* Moscow, Russia, 2 November 1905), *physiology, physical chemistry, psychology.*

Sechenov's father was a landowner, and his mother was of peasant stock. He was educated at home and then attended the Military Engineering School in St. Petersburg (1843–1848). From 1848 to 1850 he was a military engineer at Kiev, and for the following six years he studied medicine at Moscow University. Upon graduation he went to Germany, where, until 1860, he studied and worked in the laboratories of Johannes Müller, E. du Bois-Reymond, Helmholtz, and Ludwig.

In 1860 Sechenov presented his dissertation *Materialy dlya budushchey fiziologii alkogolnogo opyanenia* ("Data for the Future Physiology of Alcoholic Intoxication") to the St. Petersburg Medico-Surgical Academy, at which he was appointed professor (1860–1870) and founded the first Russian school of physiology. After resigning to protest the rejection of Elie Metchnikoff, his candidate for the chair of zoology, Sechenov conducted chemical research in Mendeleev's laboratory in St. Petersburg. The following year he accepted the chair at the Novorossysk University at Odessa (1871–1876) and subsequently became professor at St. Petersburg (1876–1888) and Moscow (1891–1901). Sechenov was elected an honorary fellow of the Russian Academy of Sciences in 1904.

His first investigations were devoted to gaseous exchange (his dissertation) and electrophysiology *O zhivotnom elektrichestve* ("On Animal Electricity," 1862). In November 1862, while working in the Paris laboratory of Claude Bernard (who did not, however, collaborate on this work), Sechenov reported on "central inhibition"— the repressive effects of thalamic nerve centers on spinal reflexes. He thus inaugurated research on inhibition phenomena in the central nervous system.

His discovery led Sechenov to suggest the theory of cerebral behavior mechanisms, according to which all conscious and unconscious acts are reflexes in terms of their structure ("means of origin"). This theory provided the basis for the development of neurophysiology and objective psychology in Russia, including the investigations of Pavlov and Bekhterev. Although accepted by the intelligentsia as an aspect of a scientific view of the nature of man, the theory was interpreted in more conservative circles as a threat to moral principles and social order. In elaborating the theory of the functions of higher nerve centers, Sechenov established a principle of self-regulation that was set forth in works published in 1866, 1891, and 1898. The main role was attributed to the coordination between nerve centers. Sechenov emphasized not only "form but activity, not only topographical isolation of organs but the combination of central processes into natural groups"—an interpretation that contradicted conceptions based on "anatomical principle."

In 1881 Sechenov established the existence of periodic spontaneous fluctuations of bioelectric potentials in the brain. Proceeding from the conception of muscle as a "receptor" of sensory information, and having analyzed disorders of nerve and muscle activities (ataxia), he concluded that signals reflecting the muscle effects are involved in the regulation of motor activity in animals and man. This was a precursor of the conception of feedback as an essential factor in the organization of behavior. He introduced an essentially new interpretation of reflex by declaring that it consisted of sensation (a signal) and movement. Sensation, he believed, was determined not by information at the level of consciousness but on the basis of objective functions of vital activity and served to distinguish among conditions and to regulate actions.

Maintaining that the reality of sensation is rooted in the reality of the motor act, Sechenov developed a new approach to the functions of the sensory organs in *Fiziologia organov chuvstv* ("Physiology of the Sensory Organs," 1867). A receptor was responsible merely for a signal and thus constituted only half of the complete physiological mechanism, the other half being muscle activity. (This represented "the principle of coordinating movement and sensation, a unity between reception by and action of a muscle.") The signals of muscle sensation were the main source of information on the space-time characteristics of the environment.

On the basis of the new "reflective" scheme, Sechenov suggested a plan for reorganizing psychology into an objective natural science based on physiology (1873). Instead of being an adjunct of physiology, psychology should become a study of the psychic regulation of behavior using the methods and conceptions of natural sciences.

Sechenov also investigated the chemistry of respiration; the physiology of respiration, particularly at reduced atmospheric pressure; and the physics and chemistry of solutions. He designed a new type of absorptiometer and used it to establish the law of solution of gases in salt solutions with which they did not react. He also constructed a device for studying respiration when a man was moving or at rest. His *Ocherk rabochikh dvizheny u cheloveka* ("A Survey of the Working Movements of Man," 1901) laid the foundation for later investigations into the physiology of work in Russia.

Sechenov was active in the struggle for equal rights for women and for self-government of the universities. His students and followers included many outstanding Russian physiologists: Vvedensky, Pavlov, Ukhtomsky, I. R. Tarkhanov, Samoylov.

BIBLIOGRAPHY

I. ORIGINAL WORKS. Sechenov's main writings are *Materialy dlya budushchey fiziologii alkogolnogo opyanenia* ("Data for the Future Physiology of Alcoholic Intoxication"; St. Petersburg, 1860); *O zhivatnom elektrichestve* ("On Animal Electricity"; St. Petersburg, 1862); "Refleksy golovnogo mozga" ("Reflexes of the Brain"), in *Meditsinsky vestnik*, nos. 47–48 (1863); *Fiziologia nervnoy sistemy* ("Physiology of the Nervous System"; St. Petersburg, 1866); "Komu i kak razrabatyvat psikhologiyu" ("Who Must Investigate Psychology and How"), in *Vestnik Evropy* (1873), no. 4; "Elementi mysli" ("The Elements of Thought"), *ibid.*, (1878), nos. 3–4; *O pogloshchenii ugolnoy kisloty solyanymi rastvorami i krovyu* ("On Absorption of Carbon Acid by Salt Solutions and Blood"; St. Petersburg, 1879); *Fiziologia nervnykh tsentrov* ("Physiology of Nerve Centers"; St. Petersburg, 1891); and *Ocherk rabochikh dvizheny u cheloveka* ("A Survey of the Working Movements of Man"; Moscow, 1901).

Collections of his writings are available in French as *Études psychologiques* (Paris, 1889) and in English as *Selected Works*, A. A. Subkov, ed. (Moscow–Leningrad, 1935).

II. SECONDARY LITERATURE. See K. S. Koshtoyants, *I. M. Sechenov* (Moscow, 1950); A. F. Samoylov, "I. M. Sechenov i ego mysli o roli myshtsy v nashem poznanii prirody" ("Sechenov and His Ideas on the Role of Muscle in Our Knowledge of Nature"), in *Nauchnoe slovo* (1930), no. 5; N. E. Vvedensky, "I. M. Sechenov," in his *Polnoe sobranie sochineny* ("Complete Works"), VII (Leningrad, 1963); and M. G. Laroshevsky, *I. M. Sechenov* (Leningrad, 1968).

M. G. LAROSHEVSKY

SEDERHOLM, JOHANNES JAKOB (*b.* Helsinki, Finland, 20 July 1863; *d.* Helsinki, 26 June 1934), *geology, petrology.*

Sederholm was the third of eight children of Claes Theodor Sederholm and Maria Sofia Christina Blomquist. His father, owner of a printing plant, founded and published the major daily newspaper of Helsinki and also published books. In 1882 Sederholm entered the University of Helsinki to study philosophy. His health was weak, however, and his doctor, doubtful of his recovery, recommended that he abandon formal academic pursuits for a branch of the natural sciences that would get him outdoors. He chose geology. Sederholm had to fight against illness throughout his life.

In the department of natural history at Helsinki, Sederholm met Axel Palmén, later a renowned ornithologist, and Wilhelm Ramsay, who became a leading geologist of Fennoscandia; they became lifelong friends. After receiving his B.A. in 1885, Sederholm continued his studies from 1886 to 1888 at the University of Stockholm, where he met Brøgger, who was teaching mineralogy and geology there. His imposing, multifaceted personality inspired several generations of Fennoscandian geologists, and Sederholm always spoke of him as "my dear teacher and master"; they remained lifelong friends.

During the summers Sederholm served as assistant geologist (1883–1887), and in 1888 was appointed a member of the Geological Commission (Geological Survey). To become more familiar with modern trends in petrology, he studied with Rosenbusch at Heidelberg, where he met American geologists and thus began a lifelong link with America. Sederholm completed two important papers while at Heidelberg, on the eruptive rocks of southern Finland and on the rapakivi granites. In the first he showed that many of these rocks are metamorphosed and that two degrees of metamorphism can be distinguished, an idea that was developed (metamorphic depth zones) more than ten years later by Becke and Grubenmann. The paper on the rapakivi granites marked the starting point of research that remained a lifelong concern. The former was also presented as a dissertation on 30 May 1891 with F. J. Wiik, a declared antiactualist, as critic.

After having passed his examination in 1892, Sederholm received the doctorate at the age of twenty-nine, became director of the Geological Commission, and married Anna Ingeborg Mathilde von Christierson. They had two daughters and one son. As director of the Geological Commission for over forty years, he not only created a modern research institution for basic and economic geology but also established new standards for the study of the world's crystalline basements, the Precambrian continental shields. He retired during the summer of 1933 and died the following June.

Established in 1877 under the directorship of Karl Adolf Moberg, the Geological Commission was the first geological survey of Finland with geological mapping. The first sheet appeared in 1879 on a scale of 1:200,000; bedrock and superficial deposits were represented on the same sheet. Thirty-seven quadrangles of this atlas were subsequently published, but only a few principal petrological groups were distinguished within the basement. Under Sederholm's enthusiastic leadership, however, another type of map was adopted, with bedrock and superficial deposits published on separate sheets on a scale of 1:400,000. Whereas the earlier maps had been based on petrological distinctions, the new series employed principles proposed in Sederholm's paper of 1893. Age relationships were of primary importance, but the methods of classical stratigraphy were scarcely applicable; and new principles for determining the succession of geological events were needed. During the course of this long and exacting work, methods and classifications were constantly adjusted to conform with the growth of knowledge and were influenced by critics and exchanges with other geologists. The continuing feedback from field observations modified and improved the manner of mapping over the years. The dominant thread was always actualism.

One of the methods chiefly responsible for the final success of the survey was Sederholm's idea of mapping crucially important areas or outcrops on a finer scale, and the geology of Finland was thus represented at different enlargements. Some outcrops were mapped at 1:10, others at 1:50, 1:200, or 1:10,000. In the limit, photographs of ice-smoothed hillocks, of hand specimens, and of thin sections under the microscope completed the spectrum of scales.

With Sederholm's method of multiple levels of research, Finland became the classic example of a Precambrian basement; for in addition to various mapping scales, the Commission was also administered at various levels of organization. In this way the project, rather than becoming merely the sum of its parts, acquired new properties and the individual parts assumed new functions at different levels. Studies conducted at each level had their own methodology and governing principles. Because of qualitative differences, the same methods of study are best applied to closely related levels. This distinction between different levels was not understood by those who sought a unity of methods to study the entire spectrum: for instance, those who advocated mapping only petrographic differences without paying attention to formations having the same or a different geological history.

The large units of sedimentary and volcanic formations are separated by unconformities, breaks in geological history that are most evident if they are marked by conglomerates representing specimens of a former erosional surface. Sederholm's eagerness to find conglomerates was rewarded by his important discovery of such formations in the sedimentary series of Lake Näsijärvi, near Tampere, in 1899. Many other sedimentary structures are well preserved, such as the varves containing some of the very ancient fossils *Corycium enigmaticum*. This was a triumph for uniformitarian principles, and the area became a classic that was often visited by foreign geologists. Another important member of these series consists of basic volcanic rocks transformed into uralite-porphyrites.

In order to obtain a uniformitarian picture of the events that occurred in the Svecofennian area of the Precambrian basement, it was necessary to find the basement supporting the sedimentary series and the piles of volcanic effusions, as well as the source areas for the clastic rocks. But again, new obstacles appeared that could not be eliminated until the possibility of the reactivation of old basements was recognized. Observations of relevant exposures were described, and interpretive principles proposed, in 1897, when an official excursion of the Seventh International Geological Congress visited the shores of Näsijärvi; these views spread rapidly and soon appeared in textbooks.

Dikes of volcanic and granitic intrusions with their ramifications provided, by their intersections, still another means of establishing a chronological sequence; but continuing in this way, geologists soon became ensnared in riddles that at first sight appeared insoluble. Before further progress was possible, such problems had to be pursued on several levels. During the summer of 1906 Sederholm, then a member of the Diet, was obliged to remain near Helsinki and was unable to travel to the wilderness. He decided to study the *roches moutonnées* of southern Finland with his assistant Hans Hausen. The multitude of phenomena of these southern rocks, varying from one step to the next although the number of mineral components and their combinations were sharply limited, was initially extremely confusing. These phenomena were only partly intelligible through methods of classical petrology, for which reason they had not been observed earlier, even though they occupy vast areas in crystalline basements of any age. A new set of ideas and methods was needed to describe and interpret the observations and to enable others to see these phenomena. Thus a new level of research was intercalated between the petrographic level, based mainly on microscopic study, and large-scale mapping. After the variegated patterns visible on the flat rock surfaces of southern Finland were mapped on scales from 1:10 to 1:200, an unsuspected wealth of new phenomena was revealed and time sequences were established. The apparatus of descriptive and interpretive terms that was created gave petrologists of the old school cause to shudder. Nevertheless, a new domain, existing from unknown ages and hitherto invisible, was revealed; and a new model for mass circulation within the sialic crust was proposed.

The mixed rocks, previously considered impure, were named migmatites and became the center of petrologic interest. Different patterns of such mixed rocks were distinguished, from "agmatites" (with angular fragments) to "nebulites" (with cloudy appearance), and from ptygmatic veins to palingenetic and anatectic granites (older granites, gneisses, or schists grading into rocks that resembled granite). Sederholm considered these areas to be former "granite factories" or "granite works" (1907), arrested during their activity by cooling and rendered accessible by deep erosion, smoothed by glacial polishing, and washed bare by wave action. A new cyclic model was proposed, similar in some points to Hutton's hypothesis; but in this sequence every phase was represented by visible outcrops, so that the whole of the events could be observed in the field. It was a new victory for uniformitarianism. The granites in this model came not from "unknown depths," as a result of differentiation from a basic magma, but from reactivation *in situ* of older materials (anatexis and palingenesis). These mobilized granites could then rise in the crust.

The case of the "paradoxical inclusions" illustrates the sort of problem that confronted Sederholm and his method of solution. These inclusions are both older and younger than the enclosing rocks. The explanation was that basic dikes were intruded into former gneisses or schists, which were later reactivated and transformed into palingenetic granite enclosing the fragments of the former dikes. These phenomena made it possible to distinguish the chronological sequence of events in areas characterized as ultrametamorphic and to coordinate these sequences over more extended areas.

On another level, belts characterized by distinct historical evolution could be distinguished and mapped, some with definite limits (such as the Karelian belt, well discernible from the much older basement in eastern Karelia) or more indistinct (such as the Svecofennian belt in southern Finland, characterized by the Bottnian formation). Sederholm published a series of maps of the basement of Finland, perfected after the progressive mapping. Most of them were included in the *Atlas of Finland*, a work in which he was deeply interested.

The study of fracture patterns cutting the old basement was conducted on the scale of the old shield (1911, 1913, 1932). After the deep erosion that removed the upper zones and exposed the metamorphic and ultrametamorphic levels, the brittle crust was broken into a mosaiclike pattern, undergoing weathering, erosion, and glacial scouring, and thus producing some of the most characteristic features of the Fennoscandian landscape.

In "The Average Composition of the Earth's Crust in Finland" (1925), another study of old crystalline basements, Sederholm attempted to characterize the average geochemical composition of deeply eroded crustal areas. The investigation differed from the usual means of rock analyses (for example, Frank Wigglesworth Clarke's) because the analyses entering into the calculation were weighted according to the areas actually underlain by the corresponding rocks. By establishing the weighted average composition, the result offered a valuable example—the only one thus far—of an extended crystalline basement cut at this deep level and represented an important contribution to the geochemical knowledge of the sialic crust.

Sederholm was a keen observer for whom a granite was not an abstract concept but the embodiment of a distinct "personality." Appearance and behavior, related to composition and variability, formed the characteristic profile. Thus he was often able to predict other qualities invisible to the naked eye, such as a percentage of fluorine or other mineral components. These granite individuals were grouped into four families, which also were groups of different relative age that occupy special places in the evolution of the crustal sector.

To learn more about these individual characteristics, Sederholm studied special textures: orbicular, spotted, and nodular granites and the rapakivi

(1928). His paper on "synantectic" minerals (1916) was representative of work on the component material of the rocks. Here his main aim was to establish the sequence of events, and the phenomena of synantectic minerals are a good example. The interfaces of two adjacent minerals, formed in stable association, can begin to produce new minerals by interaction of the primary minerals; in this way two, and sometimes more, stages are marked in the evolution of the rock, characterized by different physicochemical conditions. More interested in transformations than in phase equilibriums, Sederholm noted the results of laboratory experiments but could hardly consider them final criteria. For him, they were useful only insofar as they could be integrated within the organic interdependence of different levels of apprehension.

In this way a spectrum of new methods of observation and interpretations, each set adapted to its own level, was created; it produced a hitherto unsuspected picture of crustal evolution in a historical perspective. Those accustomed to a taxonomic approach found it difficult to accept this manner of reasoning, which was unusual at the beginning of the twentieth century; and it became necessary to reply to the vigorous opposition.

Most of the old crystalline basements are more or less peneplained with low relief, and it was considered impossible to distinguish three-dimensional structures of higher orders of magnitude. Fortunately for Sederholm, a young Swiss geologist, C. E. Wegmann, appeared and showed how to obtain a three-dimensional interpretation, how to reconstruct ancient mountain chains with their different axial culminations and depressions, and how to discover the movements that they had undergone. The former belts became orogens with visible cross sections that were many kilometers in depth. Although in his sixties, Sederholm was still so open-minded and eager to complete his work that he became an ardent student of structural methods, even though their results contradicted some of his former hypotheses. On the other hand, they offered further evidence against the theory of the ubiquity of folding and furnished new conceptual tools for uniformitarian reconstruction.

Sederholm's outstanding scientific achievements brought him many honors and medals, and the respect accorded him was turned to the welfare of his native land. In 1906 he became a member of the Diet, representing the third estate. An active member of the Workers Institution, he lectured on scientific and other topics and was the author of semipopular and popular articles, papers, and books, on social questions as well as earth sciences. He was a founder of the Geographical Society of Finland and served five times as president, and was one of the most active members of the editorial staff of the *Atlas of Finland* (1899–1925). With the establishment of the Republic of Finland in July 1919, Sederholm was entrusted with several important missions to the League of Nations, including that concerning the issue of whether the Åland Islands should be Swedish or Finnish. The League also appointed him a member of the Commission for the Independence of Albania. He made two journeys to that country to inspect its frontiers. Unarmed and without military escort, he met the Serbian guerrillas and arranged for their peaceful withdrawal.

Sederholm directed much attention to economic questions, not only economic geology but also general problems. As a member, and twice chairman, of the Economic Society of Finland, he delivered several lectures, one of them on the reasons for North American industrial supremacy (1905). F. W. Taylor's system of scientific management interested him deeply, probably because of its multileveled reasoning, and in 1915 Sederholm published a noted monograph entitled *Arbetets vetenskap* ("The Science of Work").

Sederholm was greatly interested in America, first in the Precambrian shield and its chronology, but also in the politics, economics, industry, social conditions, hospitals, and many other facets of the nation's life. He traveled several times in Canada and the United States on lecture tours, accepting honors (including the Penrose Medal, in 1928), visiting important outcrops, collecting documents, and observing general conditions. He particularly enjoyed recalling his discussions with N. L. Bowen, the foremost leader of the diametrically opposed magmatist school of petrology. This great scientist admitted that many of the phenomena described by Sederholm were inexplicable by his own theory—in sharp contrast with other petrologists, who simply denied the possibility that they had occurred. One of Sederholm's favorite projects was a book on America for Europeans, to have been entitled *The Country of Brobdingnag*. But after retiring, so many matters awaited completion that the time left him was too limited to reap the entire harvest on the land he had tilled.

BIBLIOGRAPHY

I. ORIGINAL WORKS. There is a bibliography of all Sederholm's main geological publications and a selection of his more important nongeological work in Sederholm's *Selected Works* (New York, 1967), 589–594.

His earlier writings include "Über die finnländischen Rapakiwigesteine," in *Tschermaks mineralogische und petrographische Mitteilungen*, **12** (1891), 1–31; "Studien über archäische Eruptivgesteine aus dem südwestlichen Finnland," *ibid.*, 97–142; "Om bärggrunden i södra Finland. Deutsches Referat," in *Fennia*, **8**, no. 3 (1893), 1–137; "Les excursions en Finlande," in *Guide du VII Congrès géologique international*, XIII (St. Petersburg, 1897), 1–22, written with W. Ramsay; "Über eine archäische Sedimentformation im südwestlichen Finnland und ihre Bedeutung für die Erklärung der Entstehungsweise des Grundgebirges," in *Bulletin de la Commission géologique de la Finlande*, **6** (1899); "Über die Entstehung des Urgebirges," in *Förhandlingar vid Nordiska Naturforskare och läkaremötet* (Helsinki) (7 July 1902), 88–109; "Über den gegenwärtigen Stand unserer Kenntnis der kristallinischen Schiefer von Finnland," in *Comptes rendus du IX Congrès géologique international* (Vienna, 1903), 609–630; "Om granit och gneis, deras uppkomst, uppträdande och utbredning inom urberget i Fennoskandia," in *Bulletin de la Commission géologique de la Finlande*, **23** (1907); and "Einige Probleme der präkambrischen Geologie von Fennoskandia," in *Geologische Rundschau*, **1** (1910), 126–135.

Subsequent works include "Om palingenesen i den sydfinska skärgården samt den finska urbergsindelningen," in *Geologiska föreningens i Stockholm förhandlingar*, **34** (1912), 285–316; "Sur les vestiges de la vie dans les formations proterozoiques," in *Comptes rendus du XI Congrès géologique international* (1912), 515–523; "Die regionale Umschmelzung (Anatexis) erläutert an typischen Beispielen," *ibid.*, 573–586; "Subdivision of the Pre-Cambrian of Fennoscandia," *ibid.*, 683–698; "Hutton och Werner," in *Festskrift tillägnad Edvard Westermarck* (Helsinki, 1912), 279–291; "Über ptygmatischen Faltungen," in *Neues Jahrbuch für Mineralogie, Geologie und Paläontologie*, supp. **35** (1913), 491–512; "Über die Entstehung der migmatitischen Gesteine," in *Geologische Rundschau*, **4** (1913), 174–185; "On Regional Granitization or Anatexis," in *Comptes rendus du XII Congrès géologique international* (Toronto, 1913), 319–324; *Arbetets vetenskap* ("The Science of Work"; 1915); "Different Types of Pre-Cambrian Unconformities," *ibid.*, 313–318; "On Synantectic Minerals and Related Phenomena," in *Bulletin de la Commission géologique de la Finlande*, **48** (1916); "On Migmatites and Associated Pre-Cambrian Rocks of Southwestern Finland. I: The Pellinge Region," *ibid.*, **58** (1923); "The Average Composition of the Earth's Crust in Finland," *ibid.*, **70** (1925), also in *Fennia*, **45**, no. 18 (1925), 1–20; "On Migmatites and Associated Pre-Cambrian Rocks of Southwestern Finland. II: The Region Around the Barösundsfjärd West of Helsingfors and Neighbouring Areas," *ibid.*, **77** (1926); "On Orbicular Granites, Spotted and Nodular Granites and the Rapakivi Texture," *ibid.*, **83** (1928);

"On Migmatites and Associated Pre-Cambrian Rocks of Southwestern Finland. III: The Åland Islands," *ibid.*, **107** (1934); and "Ultrametamorphism and Anatexis," in *Pan-American Geologist*, **61** (1934), 241–250.

II. SECONDARY LITERATURE. See Pentti Eskola, "Outline of Sederholm's Life and Work," in Sederholm's *Selected Works* (New York, 1967), 577–594, with portrait and bibliography; Victor Hackman, "Jakob Johannes Sederholm," *Bulletin de la Commission géologique de la Finlande*, no. 117 (1935), with portrait and bibliography; Hans Hausen, "J. J. Sederholm," in *Svenska folkskolans vänners kalender* (1934), 151–159, with two portraits; and Hans Hausen, *The History of Geology and Mineralogy in Finland, 1828–1918* (Helsinki, 1968), with bibliography; Aarne Laitakari, "Geologische Bibliographie Finnlands 1555–1933," *Bulletin de la Commission géologique de la Finlande*, no. 108 (1934); and Väinö Tanner, *Jakob Johannes Sederholm* (Helsinki, 1937), read to the Swedish Academy of Technical Sciences in Finland on 22 Mar. 1935.

EUGENE WEGMANN

SEDGWICK, ADAM (*b*. Dent, Yorkshire, England, 22 March 1785; *d*. Cambridge, England, 27 January 1873), *geology.*

Sedgwick was the second son of Richard Sedgwick, vicar of the rural parish of Dent. From a local grammar school he went to Trinity College, Cambridge, where he graduated in 1808 with distinction in mathematics. He was elected a fellow of Trinity in 1810 and was ordained in 1817. He never married, and Trinity College remained his home for the rest of his life.

In 1818 Sedgwick was elected Woodwardian professor of geology at Cambridge and immediately took to his new work with boundless energy and enthusiasm (his later assertion of his complete ignorance of geology at this time probably was exaggerated). He made his first geological expedition that summer, and became a fellow of the Geological Society of London later in the year. In 1819 he played a leading part in the founding of the Cambridge Philosophical Society, which was designed to promote the study of the natural sciences at Cambridge; and the same year he gave the first of his annual courses of lectures on geology (a series that he continued without interruption until 1870). Although Sedgwick's lectures were, until the latter part of his life, optional and extracurricular, they were immensely popular; and their influence on successive generations of Cambridge students, and hence on the shaping of English educated opinion on geology, is hard to overestimate.

Sedgwick's lecturing style was clear and vivid,

and he had a direct and informal manner that made him accessible to and popular with students. But he found the formal composition of scientific papers irksome, and their completion repeatedly was delayed by recurrent ill health (dating from a serious breakdown in 1813) and by his political activities and administrative responsibilities within the college and the university. His published works, therefore, hardly reflect the full extent of his achievement.

Sedgwick inherited a geological collection somewhat enlarged from Woodward's original bequest; but he himself expanded it during his long tenure of the chair into one of the finest geological museums in the world, partly through his own and his students' collecting activities, and partly through purchasing fine specimens and collections with his own resources or with funds raised from public appeals. In 1841 a new museum building was opened to accommodate this rapidly growing collection; by the end of his life this in turn was inadequate, and the present Sedgwick Museum was erected as a memorial to him.

Sedgwick was president of the Geological Society of London from 1829 to 1831, and of the British Association at its first visit to Cambridge in 1833. In 1834 his appointment as a prebendary of Norwich gave him greater financial independence (the stipend of his chair was very small), but at the cost of requiring his residence at Norwich for part of every year for the rest of his life.

In politics Sedgwick was one of the most prominent Liberals at Cambridge, and was in the forefront of the movement for university reform; Prince Albert, on becoming chancellor of the university in 1847, chose him to act as his "secretary" at Cambridge, and Sedgwick was later (1850–1852) a member of the royal commission on the reform of the university.

Sedgwick began his geological work under the influence of William Conybeare, and his earliest major studies were on the stratigraphy of the problematical and poorly fossiliferous deposits of the New Red Sandstone. In a monograph published in 1829 he successfully used the distinctive Magnesian Limestone of northeastern England as a key to these strata, and was able to correlate them with the classic successions in Germany. This work showed Sedgwick to be a field geologist with an exceptional flair for grasping the regional significance of local details. The wider relevance of its conclusions lay in his interpretation of the strata as the products of long-continued processes, and in his emphasis on the strata as conformable "con-

necting links" from the Coal Measures below to the Lias (Jurassic) above. The same emphasis on stratigraphical continuity is evident in his joint work with Roderick Murchison on the eastern Alps at about the same period, in which he clearly was concerned to bridge the apparent faunal break between "Secondary" and Tertiary strata.

Sedgwick therefore naturally welcomed some aspects of Charles Lyell's work: in his presidential addresses to the Geological Society in 1830 and 1831 he agreed with Lyell that incalculably vast periods of time must be inferred for many geological events, and he retracted his earlier view (derived from William Buckland) that a single paroxysmal episode could account for all the "diluvial" deposits. But in reviewing the first volume of Lyell's *Principles of Geology* he sharply criticized Lyell's confusion of different meanings of "uniformity." He agreed that the "primary laws of matter" were "immutable," but he felt it was "a merely gratuitous hypothesis" to assume that the geological processes based on those laws must have been rigidly uniform in their intensity throughout earth-history: uniformity in the latter sense had to be tested by empirical observation, not assumed a priori.

In conformity with this empiricist program, Sedgwick approved Élie de Beaumont's theory of occasional paroxysmal elevation of mountain ranges, because it made explicable many common phenomena (local folding of strata and local unconformities) in terms of events that, although abrupt and uncommon in occurrence, were perfectly natural in their mechanism. Similarly, he rejected Lyell's assertion of steady-state uniformity in the organic realm, because the facts of the fossil record seemed to indicate unequivocally a gradual approach to the "present system of things." Above all, Sedgwick felt that the geologically recent appearance of man was "absolutely subversive" of Lyell's Huttonian conclusion.

Sedgwick's most important geological work, which led to the foundation of the Cambrian system, seems to have been motivated by a desire to penetrate the fossil record back to its farthest limits, and to demonstrate that life had indeed had a beginning in time. Some of his earliest fieldwork was an attempt to unravel the complex structure of the old rocks of Devon and Cornwall, and later he studied in detail those of the Lake District (where for many years he enjoyed the friendship of William Wordsworth). In the classic paper "Remarks on the Structure of Large Mineral Masses" (1835) he combined his mathematical training with his skill as a field geologist in the first clear analysis of the effects of diagenesis and low-grade metamorphism. In particular, his exposition of the distinction between stratification, jointing, and slaty cleavage provided the crucial technical key for the interpretation of the structure of regions with complex folding.

This had already enabled him, as early as 1832, to discover the essential structure and succession of the ancient rocks of North Wales (on his first expedition there, in 1831, he gave Charles Darwin his first training as a field geologist). In the same years Murchison was studying apparently younger "Transition" strata in the Welsh Borderland. During their first and only joint study (1834) of the relation between their two areas, Murchison assured Sedgwick that his strata (later termed "Silurian") lay wholly above Sedgwick's "Cambrian" (so named in 1835), although there was clear—and not unexpected—faunal similarity between Murchison's Lower Silurian strata and Sedgwick's Upper Cambrian Bala series.

Their collaboration continued fruitfully in their solution of the anomalous problem of the slaty rocks of Devonshire, where De La Beche's discovery of Carboniferous plants in rocks of ancient appearance had seemed for a time to threaten the validity of both Murchison's and Sedgwick's stratigraphy. They worked closely in discovering first that the anomalous fossils had come from a syncline of slaty rocks of Carboniferous age, and then that the older rocks in the region were not Silurian or Cambrian but the lateral equivalents of the Old Red Sandstone, which they termed "Devonian." This interpretation was confirmed by a joint expedition to the Continent in 1839.

Sedgwick then turned his attention back to the problem of the Cambrian strata in Wales, as part of a larger program for a synoptic work (never completed) on all the Paleozoic strata of Britain. Although he could not point to a Cambrian fauna as distinctive as Murchison's Silurian fauna, he rightly emphasized that the vast thickness of the Cambrian strata and the apparent beginning of the fossil record within them justified their status as a "system" of comparable importance. Indeed, he underlined their theoretical significance by including them with the Silurian in a broader category of "Protozoic." He was therefore disconcerted and eventually exasperated when Murchison claimed that the upper (and fossiliferous) part of his Cambrian was nothing other than Murchison's Lower Silurian. Murchison gradually extended this claim until he had annexed virtually the whole of the Cambrian and reduced it to a synonym of Lower Silurian.

Sedgwick later was angered by what he regarded as editorial tampering with a crucial paper on the subject submitted to the Geological Society. His stratigraphical nomenclature was altered, apparently in the interests of editorial uniformity, and possibly in innocence of the massive theoretical and personal implications. Reacting with characteristic vehemence, he later broke off all dealings with the Society.

Only gradually did Sedgwick detect the root cause of the controversy. Murchison earlier had misinterpreted the order of succession of his Lower Silurian strata in their type area, so that they were not in fact younger than Sedgwick's Upper Cambrian Bala series, but of the same age. Even more seriously, Sedgwick found in 1854 that Murchison had confused some Upper Silurian strata (May Hill sandstone) with these much earlier Lower Silurian strata (Caradoc series), thus giving the Silurian faunas a spurious uniformity down into Sedgwick's Upper Cambrian.

But Murchison naturally was reluctant to admit these errors, and his position as head of the Geological Survey (from 1855) allowed him to retain his interpretation in all the Survey's official publications. What had begun as a controversy with important implications for the understanding of the earliest part of the fossil record gradually degenerated into a priority dispute between two equally obstinate old men. The conflict was not settled until much later, when the discovery of earlier faunas in Wales rehabilitated the term Cambrian for what had been Sedgwick's Lower Cambrian; and an "Ordovician" system was proposed ironically for the disputed strata (that is, Sedgwick's "Upper Cambrian" and Murchison's "Lower Silurian") between the newly restricted Cambrian and Silurian systems.

The Geological Society acknowledged the outstanding value of Sedgwick's work in 1850 (before he was estranged from it) by awarding him their highest honor, the Wollaston Medal; and in 1863 the Royal Society, to which he had been elected in 1830, awarded him the Copley Medal.

Sedgwick was seventy-four when Darwin's *Origin of Species* was published, but his rejection of his pupil's evolutionary theory was not simply a consequence of old age. Thirty years earlier, while welcoming naturalistic explanations for geological events (including those responsible for "diluvial" deposits), he had felt that no purely natural mechanism could ever account for the origin of new organic species. This belief was grounded in his strong sense of the "designful" beauty of organisms: although he was not primarily a paleontologist, his favorite lecture subject was the adaptive significance of fossil mammals. But his concept of teleology was philosophically unsophisticated, and like Paley (whose work had influenced him deeply) he simply wished the contemplation of organic design to nourish the sense of wonder that would lead the mind toward God.

Any theory of the transmutation of species by natural means therefore seemed to Sedgwick to threaten this preparatory function of natural theology and to lead to a "train of monstrous consequences." Above all, since he saw clearly that whatever applied to other species ultimately must apply to man, the consequences of any evolutionary theory seemed to him to include the denial of the reality of the "moral" realm. Although a devout Christian, Sedgwick was no fundamentalist in religious matters: he was a prominent member of the Broad Church party within the Anglican Church, and was greatly concerned to see progressive science and enlightened theology working independently toward an ultimate synthesis of truth. But this meant that he was as much concerned to expose the pretensions of a naïve materialism that undermined man's moral responsibility as he was to attack simplistic "reconciliations" between geology and Scripture. Thus in his presidential addresses of 1830 and 1831 he criticized the popular "Scriptural geologists" even more vehemently than the proponents of transmutation. His influential *Discourse on the Studies of the University* (1833) linked a reassertion of the place of geology within the tradition of natural theology with a trenchant criticism of utilitarian moral philosophy; and the same concerns later were expressed vehemently in a major anonymous review (1845) of Robert Chambers' *Vestiges of the Natural History of Creation.*

Sedgwick was angered not only by the inaccuracies and pseudoscientific pretensions of Chambers' unsigned book: more fundamentally he feared that the specious plausibility of its all-embracing principle of naturalistic "development" would undermine the sense of personal responsibility that he believed was basic to the nature of man in society. He reiterated his views even more vehemently in a vastly enlarged fifth edition of the *Discourse* in 1850; by the time Darwin sent him a copy of the *Origin of Species,* Sedgwick's antipathy to the "materialistic" tendencies of evolutionary theories

had become so obsessive that his reaction was predictable.

By the end of his long life Sedgwick had in effect survived into a new period in the history of science; and although he was widely admired and even loved as a warmhearted and noble character, many of his scientific views seemed remote and antiquated. But in his prime he had been one of the most distinguished geologists within an exceptionally talented generation; and his concern for the broader implications of science had left an enduring mark on the place of science in university education.

BIBLIOGRAPHY

I. ORIGINAL WORKS. Sedgwick's more important earlier works include *A Syllabus of a Course of Lectures on Geology* (Cambridge, 1821); "On the Origin of Alluvial and Diluvial Formations," in *Annals of Philosophy,* n.s. **9** (1825), 241–247, and **10** (1825), 18–37; "On the Geological Relations and Internal Structure of the Magnesian Limestone, and on the Lower Portions of the New Red Sandstone Series . . .," in *Transactions of the Geological Society of London,* 2nd ser., **3**, pt. 1 (1829), 37–124; "Address[es] to the Geological Society, Delivered on the Evening of the Anniversary . . .," in *Proceedings of the Geological Society of London,* **1** (1830), 187–212, and (1831), 281–316; "A Sketch of the Structure of the Eastern Alps . . .," in *Transactions of the Geological Society of London,* 2nd ser., **3**, pt. 2 (1832), 301–420, written with Murchison; *A Discourse on the Studies of the University* (Cambridge, 1833; 5th ed., 1850; repr. Leicester, 1969); "Remarks on the Structure of Large Mineral Masses, and Especially on the Chemical Changes Produced in the Aggregation of Stratified Rocks During Different Periods After Their Deposition," in *Transactions of the Geological Society of London,* 2nd ser., **3**, pt. 3 (1835), 461–486; "Introduction to the General Structure of the Cumbrian Mountains . . .," *ibid.,* **4**, pt. 1 (1835), 47–68; and "A Synopsis of the English Series of Stratified Rocks Inferior to the Old Red Sandstone," in *Proceedings of the Geological Society of London,* **2** (1838), 675–685, and "Supplement [to the same]," *ibid.,* **3** (1841), 541–554.

In the 1840's he published "On the Physical Structure of Devonshire, and on the Subdivisions and Geological Relations of Its Older Stratified Deposits, &c.," in *Transactions of the Geological Society of London,* 2nd ser., **5**, pt. 3 (1840), 633–704, written with Murchison; "On the Distribution and Classification of the Older or Palaeozoic Deposits of the North of Germany and Belgium, and Their Comparison With Formations of the Same Age in the British Isles," *ibid.,* **6**, pt. 2 (1842), 221–301, written with Murchison; "Three Letters

Upon the Geology of the Lake District Addressed to W. Wordsworth, Esq.," in John Hudson, ed., *A Complete Guide to the Lakes . . .* (Kendal, 1842)—two further letters were published in later eds. (1846, 1853); "On the Older Palaeozoic *(Protozoic)* Rocks of North Wales," in *Quarterly Journal of the Geological Society of London,* **1** (1845), 5–22; and [Anonymous review of] "Vestiges of the Natural History of Creation," in *Edinburgh Review,* **82** (1845), 1–85.

His last works include *A Synopsis of the Classification of the British Palaeozoic Rocks With a Systematic Description of the British Palaeozoic Fossils in the Geological Museum of the University of Cambridge* (Cambridge, 1851–1855), the description of fossils by Frederick McCoy; "On the Classification and Nomenclature of the Lower Palaeozoic Rocks of England and Wales," in *Quarterly Journal of the Geological Society of London,* **8** (1852), 136–168; and "On the May Hill Sandstone, and the Palaeozoic System of England," in *Philosophical Magazine,* 4th ser., **8** (1854), 301–317, 359–370.

University Library, Cambridge, has a large collection of scientific letters to Sedgwick; his geological collections and field notebooks are in the Sedgwick Museum, Cambridge.

II. SECONDARY LITERATURE. The principal biographical source is John Willis Clark and Thomas McKenny Hughes, *The Life and Letters of Adam Sedgwick . . .,* 2 vols. (Cambridge, 1890), which contains balanced assessments of Sedgwick's geological work from a late nineteenth-century viewpoint but relatively little of his more strictly scientific correspondence.

For an important modern essay on the cultural milieu of Sedgwick's work, see Walter F. Cannon, "Scientists and Broad Churchmen: An Early Victorian Intellectual Network," in *Journal of British Studies,* **4** (1964), 65–88. The biography by Clark and Hughes contains a fairly full list of Sedgwick's work, excluding items published anonymously but including most of his political pamphlets and other writings as well as his scientific papers.

M. J. S. RUDWICK

SEDOV, GEORGY YAKOVLEVICH (*b.* Krivaya Kosa, on the Sea of Azov, Russia, 1877; *d.* during an expedition to the North Pole, 5 March 1914), *hydrography, polar exploration.*

The son of a fisherman, Sedov attended the local church school. In 1894 he served as a sailor on a trading ship, and the following year he entered the naval school at Rostov-on-Don. After qualifying as a long-distance navigator in 1898, he sailed on the Sea of Azov and the Black Sea as navigator and captain. In 1900 Sedov entered naval service; he passed the examinations for the naval corps the next year and, with the rank of ensign, was assigned to the Main Hydrographical Administration of the Naval Ministry in St. Petersburg. In 1902 he joined an expedition to Novaya Zemlya. His work included a survey of the island of Vaygach and soundings of Yugorsky Shar.

In 1909 Sedov headed an expedition to explore the mouth of the Kolyma River and to investigate the possibility of approaching it from the sea. The group made a plane-table survey and sounding of the river, conducted meteorological and hydrological observations, surveyed the seacoast, and, before beginning navigation, gave a complete description of the mouth of the Kolyma. At the same time Sedov gathered ethnographical and geological material and made astronomical observations. In 1910 he explored and mapped Krestovaya Gulf on the western shore of Novaya Zemlya.

In 1912 Sedov proposed a sled expedition to the North Pole. His project was not supported by the government, so he organized an expedition with voluntary contributions, collecting 108,000 rubles and chartering a ship. On 10 September 1912 the *St. Fok* sailed from Arkhangelsk with twenty-two crew members and scientists. At Novaya Zemlya she encountered a severe storm, began to leak, and shipped water. On the way to Franz Josef Land the route was blocked by heavy ice, and the expedition was forced to winter at Novaya Zemlya. During the winter Sedov organized expeditions by sleigh into the interior and along the shores; reached Cape Zhelania, on foot; and mapped the shore of the unexplored portion of Novaya Zemlya. The expedition reached Franz Josef Land on 13 September 1913 and began its second wintering at Bukhta Tikhaya on Hooker Island.

On 15 February 1914 Sedov, suffering from scurvy, headed for the North Pole with three dog teams and two sailors, G. Linnik and A. Pustoshny. He died en route and was buried at Cape Auk on Rudolf Island.

Sedov and the members of his expedition to the North Pole made meteorological and magnetic observations, investigated and mapped the northwestern and Karskiye Vorota shores of Novaya Zemlya, crossed the northern island of Novaya Zemlya, corrected maps of Novaya Zemlya, gave geological descriptions of Hooker Island and Franz Josef Land, made observations on the condition of the ice, and determined astronomical points. Energetic and, as his friends wrote, "bold to the point of madness," Sedov passionately strove to reach the North Pole.

BIBLIOGRAPHY

I. ORIGINAL WORKS. Sedov's writings include "Puteshestvie v Kolymu v 1909 g." ("Journey to the Kolyma in 1909"), in *Zapiski po gidrografii*, **41**, no. 2–3 (1917), 263–326; "Ekspeditsia po issledovaniyu guby Krestovoy na Novoy Zemle v 1910 g." ("Expedition to Investigate Krestovaya Bay on Novaya Zemlya in 1910"), *ibid.*, **43**, no. 1 (1919), 119–136; "Tsarskoe pravitelstvo i polyarnaya ekspeditsia G. Y. Sedova, 1912–1916 gg. Dokumenty" ("The Tsarist Government and the Polar Expedition of G. Y. Sedov. Documents"), in *Krasny arkhiv*, **88** (1938), 16–76, with intro. by S. Nagornow; and "Materialy rabot G. Y. Sedova za vremya pervoy zimovki polyarnoy ekspeditsii 1912–1913 gg." ("Materials From the Works of G. Y. Sedov on the First Wintering of the Polar Expedition of 1912–1913"), K. A. Bogdanov, ed., in *Izvestiya Akademii nauk SSSR*, geographical ser. (1957), no. 3, 85–90.

II. SECONDARY LITERATURE. See A. F. Laktionov, *Severny polyus* ("The North Pole"; Moscow, 1955); B. G. Ostrovsky, *Bezvremenno ushedshie (G. Y. Sedov, V. A. Rusanov, G. L. Brusilov i E. V. Tol)* ("Those Who Died Prematurely [G. Y. Sedov, V. A. Rusanov, G. L. Brusilov and E. V. Tol]"; Leningrad, 1934), 3–31; N. V. Pinegin, "Georgy Sedov," foreword to V. Y. Vize, *Glavny sevmorput* ("The Main Northern Sea Routes"), 2nd ed. (Moscow–Leningrad, 1953), 345; A. I. Soloviev and G. V. Karpov, "Georgy Yakovlevich Sedov," in *Otechestvennye fiziko-geografy* ("Native Physical Geographers"; Moscow, 1959), 434–439; and V. Y. Vize, "Posledny put Sedova" ("Sedov's Last Trip"), in *Sovetskaya arktika* (1939), no. 3, 86–93; and "Georgy Yakovlevich Sedov," in *Russkie moreplavateli* ("Russian Seafarers"; Moscow, 1953), 317–327.

VERA N. FEDCHINA

SEE, THOMAS JEFFERSON JACKSON (*b.* near Montgomery City, Missouri, 19 February 1866; *d.* Oakland, California, 4 July 1962), *astronomy.*

See graduated from the University of Missouri in 1889 and received his doctorate from Berlin in 1892. After three years at the University of Chicago, he spent two years directing the Lowell survey of southern double stars. In 1899 See was appointed U.S. Navy professor of mathematics. He spent three years at the Naval Observatory and one year at the Naval Academy before becoming director of the observatory at Mare Island, California, where he remained until his retirement in 1930.

See's early investigations of double stars led him to study stellar evolution and the evolution of the earth and of the solar system. He was the first to postulate that matter was expelled from stars by repulsive forces, the material condensing into multiple stars or, frequently, solar systems, and that the orbits of the planets become circular because of a resisting medium. Satellites, including the moon, were small planets captured by their primaries. He also believed that planets without dense atmospheres preserved evidence of their formation in their cratered surfaces. See was the first to perform an experimental study of craters formed by high-velocity projectiles. He noted the evidence for lunar erosion by impact and melting, blanketing by dust, and subsidence.

See did not accept the prevalent belief that mountains and ocean trenches resulted from cooling and shrinkage of the earth, postulating instead a dynamic relationship between these features that involved an exchange of material from beneath the continents.

He also formulated the wave theory of gravitation and suggested that the red shift of galaxies was due not to an expanding universe but to the interaction of light and gravity waves.

See's numerous publications were considered unorthodox and were dismissed by scientists of his time. Many of his ideas, however, are in striking agreement with current theories.

BIBLIOGRAPHY

I. ORIGINAL WORKS. See's voluminous publications may best be sampled in *Researches on the Evolution of Stellar Systems*, 2 vols. (Lynn, Mass., 1896–1910), esp. vol. II, *The Capture Theory of Cosmical Evolution*. A number of his earlier papers are reprinted by William Larkin Webb in his *Brief Biography and Popular Account of the Unparalleled Discoveries of T. J. J. See* (Lynn, Mass., 1913). See's "New Theory of the Aether" is presented in *Astronomische Nachrichten*, **211** (1920), 49–86, 137–190; **212** (1920), 233–302, 385–454; **214** (1921), 281–359; **215** (1922), 49–138; **217** (1922), 193–284; and **226** (1926), 401–497 (special ed.). See also *Wave Theory!*, 3 vols. (Lynn, Mass., 1941–1950).

II. SECONDARY LITERATURE. Obituary notices are by J. Ashbrook, in *Sky and Telescope*, **24** (1962), 193, 202; and New York *Times* (5 July 1962). The only biography of See is that by W. L. Webb (above), which, however, deals only with See's life before 1913. A classic demolition of See's "capture theory" is F. R. Moulton, "Capture Theory and Capture Practice," in *Popular Astronomy*, **20** (1912), 67–82. Moulton had previously shown that See's "discovery" of an invisible body in the system of 70 Ophiuchi was impossible in "The Limits of Temporary Stability of Satellite Motion, With an Application to the Question of the Existence of an Unseen Body in the Binary System F. 70 Ophiuchi," in *Astronomical Journal*, **20** (1899), 33–37. See's reply, and his dismissal from that journal, are *ibid.*, 56.

R. A. J. SCHORN
L. D. G. YOUNG

SEEBECK, THOMAS (*b.* Tallinn, Estonia, 9 April 1770; *d.* Berlin, Germany, 10 December 1831), *electricity, magnetism, optics.*

Thomas Seebeck, the discoverer of thermoelectricity and one of the most distinguished experimental physicists of the early nineteenth century, was born in Estonia, to a well-to-do merchant family. After graduating from a Gymnasium he studied medicine at Berlin and at the University of Göttingen, receiving an M.D. from the latter in 1802. As a student Seebeck developed a strong interest in the natural sciences and decided to devote himself to scientific research rather than to medical practice. He married in 1795 and shortly thereafter began the career of a rich scientific enthusiast.

As a natural philosopher Seebeck was attracted to Jena, where an important intellectual circle was developing in the early 1800's around the philosophers Schelling and Hegel, the scientists Ritter and Oken, and the poet-philosopher Goethe. Here, partly inspired by Goethe's anti-Newtonian theory of colors (the *Farbenlehre*), Seebeck undertook his first research in optics in 1806. Its goal was to investigate the heating and chemical effects of the different colors of the solar spectrum.

Several scientists before Seebeck—Marsiglio Landriani, A. M. de Rochon, William Herschel, and John Leslie—had examined the heating effects of rays of different colors, and Herschel had shown in 1800 that heating could be detected in the region beyond the red end of the visible spectrum. Seebeck confirmed Herschel's finding and added to it the discovery of a slight rise in temperature beyond the violet region. He also found that the position of the point of greatest heat in the spectrum varied with the nature of the prism producing the colors. Seebeck also repeated and expanded upon Ritter's experiments on the coloration effects of different rays on silver salts.

Malus's discovery of polarization in 1808 led Seebeck to his next research in optics. In 1812 and 1813 he observed the "entoptic" patterns in pieces of glass under stress or subject to uneven heating and viewed through a polarizer. He also discovered optical rotatory polarization in certain oils,

the power of tourmaline to produce a single polarized ray of light, and the system of colored rings produced in polarized light by Iceland spar cut orthogonal to its axis. All these discoveries, unfortunately, were partially anticipated by Brewster or Biot. Nonetheless, Seebeck shared the Paris Academy of Science's annual prize in 1816 for his work on polarization in stressed glass, and was elected to the Berlin Academy in 1814.

In the 1820's Seebeck, having moved to Berlin, became interested in the phenomena of magnetism. He repeated the discovery made independently by Arago and Davy, that an electric current can induce magnetism in iron and steel, and that a steel needle is strongly magnetized when drawn around a conductor. Next, he expanded upon Arago's discovery that the oscillations of a magnetized iron needle about the magnetic meridian can be damped by placing a slab of almost any material in the immediate vicinity. He performed a large number of experiments on the magnetizability of different metals and first noticed the anomalous behavior of magnetized red-hot iron, an early indication of the phenomenon known as hysteresis.

By far Seebeck's most significant discovery, however, was that of thermoelectricity—or thermomagnetism, as he called it—in 1822. While he was studying the influence of heat on galvanic arrangements, it occurred to him that heat might create magnetism in two different metals joined to form a closed circuit. He joined a semicircular piece of bismuth with a similar piece of copper and fastened the ends together to form a circle. When heat was applied to either of the bismuth-copper junctions, a magnetic needle placed nearby behaved as if the circle were a closed, current-carrying circuit. By repeating this experiment many times with different pairs of metals and other conductors, Seebeck was able to order the various conducting materials in a thermoelectric series with bismuth at the extreme negative end and tellurium at the extreme positive end. He did not, however, believe that an electric current was actually set up in the bimetallic rings and preferred to describe his effect as "thermomagnetism."

BIBLIOGRAPHY

I. ORIGINAL WORKS. Seebeck's studies on the heating effects of the solar spectrum are presented in "Über die ungleiche Erregung der Wärme in prismatischen Sonnenbilde," in *Abhandlungen der Preussischen Akademie der*

Wissenschaften (1818–1819), 305–350. His work on the chemical effects of different colored rays appears in *Journal für Chemie und Physik,* **2** (1811), 263–264. The two memoirs on polarized light are "Einige neue Versuche und Beobachtungen über Spiegelung und Brechung des Lichtes," *ibid.,* **7** (1812), 259–298, 382–384; and "Von den entoptischen Farbenfiguren und den Bedingungen ihrer Bildung in Gläsern," *ibid.,* **12** (1814), 1–16. His discovery of thermoelectricity is described in "Magnetische Polarisation der Metalle und Erze durch Temperatur-Differenz," in *Abhandlungen der Preussischen Akademie der Wissenschaften* (1822–1823), 265–373. Other works are listed in the Royal Society *Catalogue of Scientific Papers,* V, 620–621; and at the end of Poggendorff's "Gedachtnissrede auf T. J. Seebeck" (see below).

II. SECONDARY LITERATURE. The only substantial biography of Seebeck is J. C. Poggendorff, "Gedachtnissrede auf Thomas Johann Seebeck, gehalten in den öffentlichen Sitzung vom 7 Juli 1839," in *Abhandlungen der Preussischen Akademie der Wissenschaften* (1841), xix–xxvii. *Encyclopaedia Britannica,* 7th ed. (1842), contains brief discussions of Seebeck's principal discoveries in electromagnetism, VIII, 574, and XXI, 684; thermoelectricity, VIII, 574, and XXI, 695; the prismatic spectrum, XIII, 332; magnetism, XIII, 694, 709, 712; and optics, XIII, 370, 420, 464.

EUGENE FRANKEL

SEELIGER, HUGO VON (*b.* Biala, near Bielitz, Austrian Silesia [now Bielsko Biala, Poland], 23 September 1849; *d.* Munich, Germany, 2 December 1924), *astronomy.*

Seeliger's father was mayor of Biala. After graduating from the Gymnasium at Teschen, he studied astronomy at Heidelberg and Leipzig, where he took his doctorate under Bruhns in 1871. From 1871 to 1873 he was assistant at the Leipzig observatory and, from 1873 to 1878, observer at the Bonn observatory. He served as *Privatdozent* at Leipzig until 1881, when he became director of the Gotha observatory. The following year he succeeded Lamont as professor of astronomy and as director of the Munich observatory, retaining both posts until his death.

One of the most famous astronomers of his time, Seeliger was interested mainly in theoretical astronomy. His activities, however, also included observations, and in 1874 he led an expedition to the Pacific to observe the transit of Venus. His papers on stellar statistics were his most important contribution to theoretical astronomy. He was the first astronomer to develop the fundamental equations of the relations between various statistical functions of the stars. The immense number of stars in the Milky Way precludes a determination of its constitution by studying each star separately. Seeliger found that there are integral equations relating the function of stellar density in space with the distribution of stars of apparent magnitude; the first of these functions can be calculated after the distribution has been determined by observation. Although later improved and simplified in certain points, Seeliger's theory established the fundamental principles of stellar statistics.

Seeliger's important papers on the illumination of cosmic objects deal especially with Saturn and the zodiacal light. His speculations on the occurrence of novae were later abandoned by astronomers. His research on the law of gravitation provided a strong impetus to astronomical cosmology. He found that the Euclidean structure of space, nonvanishing mean density of matter, and overall validity of Newton's law of gravitation were incompatible, because a universe constructed on these principles would be unstable. In an attempt to resolve this difficulty, Seeliger assumed that Newton's law required small corrections. Later cosmological research, partly stimulated by his results, made it clear that no instability occurs if all matter in the universe moves—for instance, if the universe expands, as has been believed since about 1930.

In 1885 Seeliger married Sophie Stoeltzel, daughter of a professor at the Munich Technische Hochschule; they had two sons. Seeliger was president of the Astronomische Gesellschaft from 1896 to 1921 and of the Munich Academy of Sciences from 1918 to 1923. Many students took their doctorates under his supervision, the best known being Karl Schwarzschild. Seeliger's influence on the development of astronomy was thus much greater than many astronomers believe.

BIBLIOGRAPHY

I. ORIGINAL WORKS. Seeliger's papers on stellar statistics are in the *Abhandlungen* and *Sitzungsberichte der Bayerischen Akademie der Wissenschaften,* Math.-phys. Kl. (1884–1920). Some papers also appeared in *Astronomische Nachrichten;* three articles in the latter— **137** (1895), 129–136; and **138** (1895), 51–54, 255–258— deal with Newton's law of gravitation. See also "Zur Theorie der Beleuchtung der grossen Planeten, insbesondere des Saturn," in *Abhandlungen der Bayerischen Akademie der Wissenschaften,* Math.-phys. Kl., **16** (1887), 403–516; and "Theorie der Beleuchtung staubförmiger Massen, insbesondere des Saturnringes," *ibid.,* **18** (1893), 1–72.

A complete bibliography of Seeliger's publications and of all doctoral theses supervised by him is given by H. Kienle, in *Vierteljahrsschrift der Astronomischen Gesellschaft,* **60** (1925), 18–23.

II. Secondary Literature. See G. Deutschland, "Die Untersuchungen H. V. Seeliger's über das Fixsternsystem," in *Vierteljahrsschrift der Astronomischen Gesellschaft,* **54** (1919), 25–131; and E. von der Pahlen, *Lehrbuch der Stellarstatistik* (Leipzig, 1937), 370–409.

F. Schmeidler

SEGNER, JÁNOS-ANDRÁS (JOHANN ANDREAS VON) (*b.* Pressburg, Hungary [now Bratislava, Czechoslovakia], 9 October 1704; *d.* Halle, Germany, 5 October 1777), *mathematics, physics.*

Segner was the son of Miklós Segner, a merchant. He was educated in the Gymnasiums of Pressburg and Debrecen and then at the University of Jena (1725–1730), from which he received the M.D. Simultaneously he also studied physics and mathematics and in 1728 he published a work containing a demonstration of Descartes's rule of signs for the determination of the number of positive and negative roots of an algebraic equation when all roots are real; later he devoted another article to the derivation of this rule (1758). Segner practiced medicine in Pressburg and Debrecen for a short time. He went to Jena in 1732 as assistant professor and in 1733 became extraordinary professor of mathematics. In 1735 he was named ordinary professor of mathematics and physics at Göttingen; and from 1755 until his death he held the same chair at the University of Halle. He was a foreign member of the Berlin (1746) and St. Petersburg (1754) academies of science.

Segner's invention of one of the first reaction hydraulic turbines, named for him, was of outstanding importance. It consists of a wheel rotating under the action of water streaming from parallel and oppositely directed tubes. He wrote of this invention in a letter to Euler dated 11 January 1750, and in the same year he described in detail the construction and action of his machine, which he later improved; further improvements in construction were added by Euler. Segner's letters to Euler give detailed evidence of the progress of early work in the theory and construction of reaction hydraulic turbines (Euler's letters to Segner are lost). Segner's wheel is now used for horticultural irrigation and serves as a demonstration device in schools. Segner generally spent much time constructing and perfecting scientific devices, from a slide rule to clocks and telescopes.

While studying the theory of tubes, Segner introduced the three principal axes of rotation (axes of inertia) of a solid body and offered first considerations on this problem. Euler made considerable use of this discovery and, referring motion to principal axes of inertia, deduced his important equations of the motion of a solid body (1765).

Segner also wrote on various problems of physics and mathematics. He defended Newton's theory of the emanation of light (1740), developed an original graphic device for the construction of roots of algebraic equations (1761), and presented a recurrent solution of Euler's famous problem of the number of possible dissections of an *n*-gon into triangles by means of noncrossing diagonals (1761). In mathematical logic Segner developed Leibniz' ideas and was one of the first to make extensive use of an entire system of symbolic designations to formalize logical conclusions; he did not, however, confine himself to classical syllogistics.

Segner wrote a number of mathematical manuals, proceeding, to a certain extent, from Euler's works and using his advice, which were popular in their time.

BIBLIOGRAPHY

I. Original Works. Bibliographies of Segner's works are in Poggendorff, II, cols. 892–894; and C. von Wurzbach, ed., *Biographisches Lexikon des Kaiserthums Oesterreich,* XXXIII (1877), 318–320. His writings include *Dissertatio epistolica . . . qua regulam Harrioti, de modo ex aequationum signis numerum radicum eas componentium cognoscendi demonstrare conatur* (Jena, 1728); *Specimen logicae universaliter demonstratae* (Jena, 1740); *Programma in quo computatio formae atque virium machinae nuper descriptae* (Göttingen, 1750); *Programma in quo theoriam machinae cujusdam hydraulicae praemittit* (Göttingen, 1750); *Specimen theoriae turbinum* (Halle, 1755); *Cursus mathematicus,* 5 vols. (Halle, 1758–1767); and *Elementa analyseos infinitorum,* 2 vols. (Halle, 1761–1763).

II. Secondary Literature. See *Allgemeine deutsche Biographie,* XXXIII (1891), 609–610; M. Cantor, *Vorlesungen über Geschichte der Mathematik,* III–IV (Leipzig, 1900–1908), see index; *Leonhardi Euleri Opera omnia,* fourth series, *Commercium epistolicum,* I, A. Juškerič, V. Smirnov, and W. Habicht, eds. (Basel, 1975), 403–426; Jakusc István, "Segner András," in *Fizikai szemle,* **5** (1955), 56–65; N. M. Raskin, "Voprosy tekhniki u Eylera" ("Technical Problems of Euler"), in M. A. Lavrentiev, A. P. Youschkevitch, and A. T. Grigorian, eds., *Leonard Eyler. Sbornik statey . . .* (Moscow, 1958), 509–536; and F. Rosenberger, *Die*

Geschichte der Physik in Grundzügen, II (Brunswick, 1884), 345.

A. P. YOUSCHKEVITCH
A. T. GRIGORIAN

SEGRE, CORRADO (*b.* Saluzzo, Italy, 20 August 1863; *d.* Turin, Italy, 18 May 1924), *mathematics.*

Segre studied under Enrico D'Ovidio at the University of Turin, where he formed a long friendship with his fellow student Gino Loria. Segre submitted his doctoral dissertation in 1883, when he was only twenty; and in the same year he was named assistant to the professor of algebra and to the professor of analytic geometry. Two years later he became an assistant in descriptive geometry, and from 1885 to 1888 he replaced Giuseppe Bruno in the courses on projective geometry. In 1888 he succeeded D'Ovidio in the chair of higher geometry, a post he held without interruption until his death.

Segre was much influenced by D'Ovidio's course on the geometry of ruled spaces (1881–1882). D'Ovidio started from the ideas of Plücker, which had been taken up and developed by Felix Klein. According to these ideas, the geometry of ruled space is equivalent to the study of a quadratic variety of four dimensions imbedded in a linear space of five dimensions. In his lectures D'Ovidio examined the works of Veronese on the projective geometry of hyperspaces and those of Weierstrass on bilinear and quadratic forms. These topics inspired much of Segre's research, beginning with his thesis. The latter consists of two parts: a study of the quadrics in a linear space of arbitrary dimension and an examination of the geometry of the right line and of its quadratic series. Before completing his thesis, Segre collaborated with Loria on a twenty-two-page article in French that they sent to Klein, who published it in *Mathematische Annalen* (1883). A long and active correspondence between Segre and Klein then ensued.

Segre's mathematical work can be divided into four distinct areas, all of which are linked by a common concern with the problem of space. The first of these areas comprises Segre's articles on the geometric properties that are invariant under linear transformations of space. In this connection Segre showed the value of investigating hyperspaces in the study of three-dimensional space S_3. For example, a ruled surface of S_3, which is composed of right lines, can be represented by a curve in S_5; it thus becomes possible to reduce the classification of surfaces to that of curves. The insufficiencies of the earlier theories proposed by A. Möbius, Grassmann, Cayley, and Cremona were thus soon revealed.

According to Segre, a ruled surface in a space S_n can also be considered a variety of ∞^2 points distributed on ∞^1 right lines. Further, Segre generalized the theory of the loci formed by ∞^1 right lines of S_n to the theory of the loci formed by ∞^1 planes. He took as his point of departure certain problems on bundles of quadrics that Weierstrass and L. Kronecker had treated in a purely algebraic manner.

At this time it was known that the intersection of two quadrics of S_3 is a quartic the projection of which from a point exterior to it onto a plane is a quartic with two double points. John Casey and Gaston Darboux had shown that its study is useful for that of fourth-order surfaces, called cyclides. Segre reexamined and generalized the problem by placing the two quadrics in a space S_4. He also investigated the locus resulting from the intersection of two quadrics of S_5 and discovered that it is no longer a surface but rather a three-dimensional variety that can be interpreted as a complex quadratic of S_3. From this result he confirmed in an elegant manner the famous fourth-order surface with sixteen double points, which had been found by Kummer in 1864 and bear his name. Before Segre's findings, the study of this surface required the use of extremely complicated algebraic procedures.

Segre next began a series of works on the properties of algebraic curves and ruled surfaces subjected to birational transformations. Alfred Clebsch, Paul Gordan, Alexander Brill, and Max Noether had already studied these transformations with a view toward giving a geometric interpretation to the theory of Abelian functions. Segre showed the advantage gained by operating in a hyperspace. His article of 1896 on the birational transformations of a surface contains the invariant that Zeuthen had encountered under another form in 1871, now called the Zeuthen-Segre invariant.

Segre's interest in 1890 in the properties of the Riemann sphere led him to a third area of research: the role of imaginary elements in geometry. He laid the basis of a new theory of hyperalgebraic entities by representing complex points of S_n by means of the ∞^{2n} real points of one of the varieties V_{2n}. (This variety has since been named for Segre.) Certain of Segre's hyperalgebraic transformations possess invariant properties, and he was led to enlarge the concept of a point. To this end, he introduced points that he called bicomplex, which correspond to the ordinary complex points of the real image. Their coordinates are bicomplex numbers constructed with the aid of the two unities i and j, such that:

$$i \cdot j = j \cdot i$$

and

$$i^2 = j^2 = -1.$$

Later, in 1912, Segre returned to this subject, when, utilizing the works of Von Staudt, he studied another type of complex geometry.

Darboux's *Leçons sur la théorie générale des surfaces*, which Segre often used in his courses, inspired him to investigate (from 1907) infinitesimal geometry. Extending the work of Darboux, Segre studied a certain class of surfaces in S_n defined by second-order linear partial differential equations. These surfaces are described by a moving point of which the homogeneous coordinates — functions of two independent parameters u and v — are the solutions of a second-order partial differential equation. Among the surfaces of a hyperspace, Segre was particularly interested in those that lead to a Laplace equation. In an article of 1908 on the conjugate tangents of a surface, he established a relationship between the points of the tangent plane and those of the planes passing through the origin. To establish this relationship he employed infinitesimals of higher order in a problem concerning the neighborhood of a point. This procedure led him to introduce a new system of lines, analogous to those studied by Darboux, traced on the surface; they were named Segre lines, and their differential equation was established by Fubini. It may be noted that Segre's last publication dealt with differential geometry. Segre wrote a long article on hyperspaces for the *Encyklopädie der mathematischen Wissenschaften*, containing all that was then known about such spaces. A model article, it is notable for its clarity and elegance.

Segre became a member of the Academy of Turin in 1889. He long served on the editorial board of the *Annali di matematica pura ed applicata*, on which he was succeeded by his former student Severi of the University of Rome.

Through his teaching and his publications, Segre played an important role in reviving an interest in geometry in Italy. His reputation and the new ideas he presented in his courses attracted many Italian and foreign students to Turin. Segre's contribution to the knowledge of space assures him a place after Cremona in the ranks of the most illustrious members of the new Italian school of geometry.

BIBLIOGRAPHY

I. ORIGINAL WORKS. The complete works of Segre have been published as *Opere*, 4 vols. (1957–1963), but it does not contain the paper "Mehrdimensionale Räume" (see below). A complete list of Segre's publications, 128 titles, is given in G. Loria (see below). A. Terracini lists ninety-eight titles. See also Poggendorff, V, 1151–1152.

Segre's most important works include "Sur les différentes espèces de complexes du 2e degré des droites qui coupent harmoniquement deux surfaces du 2e ordre." in *Mathematische Annalen*, **23** (1883), 213–234, written with G. Loria; "Studio sulle quadriche in uno spazio lineare ad un numero qualunque di dimensioni," in *Memorie dell'Accademia delle scienze di Torino*, **36** (1883), 3–86; "Sulla geometria della retta e delle sue serie quadratiche," *ibid.*, 87–157; "Note sur les complexes quadratiques dont la surface singulière est une surface du 2e degré double," in *Mathematische Annalen*, **23** (1883), 235–243; "Sulle geometrie metriche dei complessi lineari e delle sfere e sulle loro mutue analogie," in *Atti dell'Accademia delle scienze*, **19** (1883), 159–186; "Sulla teoria e sulla classificazione delle omografie in uno spazio lineare ad un numero qualunque di dimensioni," in *Memorie della R. Accademia dei Lincei*, 3rd ser., **19** (1884), 127–148; and "Étude des différentes surfaces du quatrième ordre à conique double ou cuspidale (générale ou décomposée) considérées comme des projections de l'intersection de deux variétés quadratiques de l'espace à quatre dimensions," in *Mathematische Annalen*, **24** (1884), 313–444.

Later writings are "Le coppie di elementi imaginari nella geometria proiettiva sintetica," in *Memorie dell' Accademia delle scienze di Torino*, **38** (1886), 3–24; "Recherches générales sur les courbes et les surfaces réglées algébriques," in *Mathematische Annalen*, **30** (1887), 203–226, and **34** (1889), 1–25; "Un nuovo campo di ricerche geometriche," in *Atti dell'Academia delle scienze*, **25** (1889), 276–301, 430–457, 592–612, and **26** (1890), 35–71; "Su alcuni indirizzi nelle investigazioni geometriche," in *Rivista di matematica*, **1** (1891), 42–66, with English trans. by J. W. Young as "On Some Tendencies in Geometric Investigations," in *Bulletin of the American Mathematical Society*, 2nd ser., **10** (1904), 442–468; "Le rappresentazioni reali delle forme complesse e gli enti iperalgebrici," in *Mathematische Annalen*, **40** (1891), 413–467; "Intorno ad un carattere delle superficie e delle varietà superiori algebriche," in *Atti dell'Accademia delle scienze*, **31** (1896), 485–501; and "Su un problema relativo alle intersezioni di curve e superficie," *ibid.*, **33** (1898), 19–23.

See also "Intorno ai punti di Weierstrass di una curva algebrica," in *Atti dell'Accademia nazionale dei Lincei. Rendiconti*, 5th ser., **8** (1889), 89–91; "Gli ordini delle varietà che annullano i determinanti dei diversi gradi estratti da una data matrice," *ibid.*, **9** (1900), 253–260; "Su una classe di superficie degli iperspazi, legata con le equazioni lineari alle derivate parziali di 2° ordine," in *Atti dell'Accademia delle scienze*, **42** (1907), 1047–1079; "Complementi alla teoria delle tangenti coniugate di una superficie," in *Atti dell'Accademia nazionale dei Lincei. Rendiconti*, 5th ser., **17** (1908), 405–412; "Mehrdimensionale Räume," in *Encyklopädie der mathematischen Wissenschaften*, III, pt. 3, fasc. 7 (1918), 769–972; "Sulle corrispondenze quadrilineari tra forme di prima specie e su alcune loro rappresentazioni spaziali," in *Annali di matematica pura ed applicata*, 3rd ser., **29** (1920), 105–140; "Sui fochi di 2° ordine dei sistemi infiniti di piani e sulle curve iperspaziali con una doppia infinità di piani plurisecanti," in *Atti dell'Accademia nazionale dei Lincei. Rendiconti*, 5th ser., **30** (1921), 67–71; "Le superficie degli iperspazi con una doppia infinità di curve piane o spaziali," in *Atti dell'Accademia delle scienze*, **56** (1921), 143–157; "Sugli elementi lineari che hanno comuni la tangente e il piano osculatore," in *Atti dell'Accademia nazionale dei Lincei. Rendiconti*, 5th ser., **33** (1924), 325–329; "Le curve piane d'ordine n circoscritte ad un $(n + 1)$ = latero completo di tangenti ed una classe particolare di superficie con doppio sistema coniugato di coni circoscritti," in *Atti dell'Accademia delle scienze*, **59** (1924), 303–320.

II. Secondary Literature. Obituary notices are L. Berzolari, in *Rendiconti dell'Istituto lombardo di scienze e lettere*, **57** (1924), 528–532; G. Castelnuovo, in *Atti dell'Accademia nazionale dei Lincei. Rendiconti*, 5th ser., **33** (1924), 353–359; E. Pascal, in *Rendiconti dell'Accademia delle scienze fisiche e matematiche*, **30** (1924), 114–116; and V. Volterra, in *Atti dell'Accademia nazionale dei Lincei. Rendiconti*, 5th ser., **33** (1924), 459–461. See also Poggendorff, VI, 2407, for a list of notices.

On Segre and his work, see H. F. Baker's article in *Journal of the London Mathematical Society*, **1** (1926), 263–271, with trans. by G. Loria in *Bollettino dell'Unione matematica italiana*, **6** (1927), 276–284; J. L. Coolidge, in *Bulletin of the American Mathematical Society*, **33** (1927), 352–357; G. Loria, in *Annali di matematica pura ed applicata*, 4th ser., **2** (1924), 1–21; and A. Terracini, in *Jahresbericht der Deutschen Mathematikervereinigung*, **35** (1926), 209–250, with portrait. The articles by Loria, Terracini, and Baker are especially helpful.

Pierre Speziali

SÉGUIN, ARMAND (*b.* Paris, France, March 1767; *d.* Paris, 24 January 1835), *chemistry, chemical technology, physiology.*

Séguin's father was the treasurer (apparently corrupt) of the duke of Orléans. Although little is known of his education, it is evident that he had some training in the sciences, for by the mid-1780's he had become part of Lavoisier's circle. Séguin was Lavoisier's assistant from 1789 until the latter's death in 1794. During this time, Séguin produced those memoirs (some written with Lavoisier) on heat and respiration for which he is best known. In 1796 he assisted Mme Lavoisier in preparing her husband's memoirs for publication. But they soon ended this collaboration because Séguin was unwilling to denounce in print Lavoisier's executioners and because, in a preface prepared by Séguin, he took more credit for Lavoisier's later work than Mme Lavoisier rightly thought he deserved.

Meanwhile, Séguin had become director of a tanning works at Sèvres, an enterprise that was based on processes of his own devising and that won the contract to supply the revolutionary armies with boot leather. Although Séguin made no fundamental innovations in tanning, he found several ways to rationalize and accelerate the operation, being unhampered by craft mysteries. He shortened the time for tanning hides from more than a year to a few weeks. Understandably, he received support, encouragement, and subsidies from the Comité de Salut Public.

Séguin made money out of the wars, but Napoleon, who had no love for profiteers, later reduced his fortune by taxes and fines (even to the point of imprisoning him), but did not succeed in impoverishing him. Séguin survived the Empire and Restoration and lived thereafter the life of an eccentric, Balzacian *rentier*, devoting most of his intellectual energies after 1815 to the composition of pamphlets on government finances. He was also devoted to horseracing and wrote a number of works on that subject.

Séguin's scientific contributions were made chiefly in collaboration with others. In the work he published with Fourcroy and Vauquelin on the synthesis of water, and in the papers he wrote with, or under the tutelage of, Lavoisier, he wrote crisply and authoritatively. Thereafter he lost his grip, and, with the exception of his research on tannin, his independent investigations were trivial. Although he contributed a number of "first memoirs" on several subjects, mostly pharmaceutical, to the *Annales de chimie* in 1814, these papers had been written years before and were never followed up.

In May 1790 Séguin read to the Académie des

Sciences the report on the large-scale synthesis of water carried out in Fourcroy's laboratory by himself, Fourcroy, and Vauquelin. The purpose of this experiment was to establish finally that water is composed only of hydrogen and oxygen and that the weight of water is fully accounted for by the weights of the two gases. They also sought to determine accurately the combining ratio of the components of water, an especially important constant in oxygen chemistry. They found that the ratio hydrogen:oxygen is 2.052:1 by volume and 14.338:85.662 by weight. (The discrepancies from the true figures probably arose from the difficulties of weighing the gases.)

Séguin's papers on respiration, animal heat, and caloric are all derivative from Lavoisier's work and ideas, although the evidence suggests that Séguin was an enthusiastic collaborator and one who made the master's ideas his own. Séguin's series of papers on caloric competently summarized and systematized Lavoisier's later thinking on the subject and also, as Robert Fox has recently shown, thoroughly refuted the Irvinist theory of caloric. This work helped to consolidate the Lavoisier-Laplace version of caloric theory and thus paved the way for the developments of that theory in the following three decades.

BIBLIOGRAPHY

I. ORIGINAL WORKS. Séguin's works are listed in the Royal Society *Catalogue of Printed Papers*, V, 628–629; those which he contributed to Lavoisier's *Mémoires de chimie*, 2 vols. (Paris, 1805), are listed in Denis I. Duveen and Herbert S. Klickstein, *A Bibliography of the Works of Antoine Laurent Lavoisier 1743–1794* (London, 1954), 204–214. His papers written with Lavoisier are "Premier mémoire sur la respiration," in *Mémoires de l'Académie royale des sciences* (1789), 566–584; "Premier mémoire sur la transpiration des animaux," *ibid.* (1790), 601–612; "Second mémoire sur la transpiration," in *Annales de chimie*, **90** (1814), 5–28; "Second mémoire sur la respiration," *ibid.*, **91** (1814), 318–334; and "Sur la respiration des animaux," in Lavoisier, *Mémoires de chimie*, II, 52–64 (in the separately paginated last section). See also "Observations générales sur la respiration et sur la chaleur animale," in *Observations sur la physique, sur l'histoire naturelle et sur les arts*, **37** (1790), 467–472. Séguin's paper with Fourcroy and Vauquelin is "Mémoire sur la combustion du gaz hydrogène dans des vaisseaux clos," in *Annales de chimie*, **8** (1791), 230–307; reprinted in Lavoisier, *op. cit.*, II, 313–413.

The Bibliothèque Nationale *Catalogue des livres imprimés* lists Séguin's financial and miscellaneous writings.

II. SECONDARY LITERATURE. On Séguin and his work, see Michaud, ed., *Biographie universelle*; and J. R. Partington, *History of Chemistry*, III (London, 1962), 106–107. For his contribution to the theory of heat, see Robert Fox, *Caloric Theory of Gases* (Oxford, 1971), chap. 2, especially pp. 38–39; for his relations with Mme Lavoisier, see Édouard Grimaux, *Lavoisier* (Paris, 1888), 330–336, and Duveen and Klickstein, *op. cit.*, 199–204; and for his work on tanning, see C. H. Lelièvre and Pelletier, "Rapport au Comité de Salut Public, sur les nouveaux moyens de tanner les cuirs, proposés par le cit. Armand Séguin," in *Annales de chimie*, **20** (1797), 15–77.

STUART PIERSON

SEGUIN, MARC, also known as **SEGUIN AÎNÉ** (*b.* Annonay, France, 20 April 1786; *d.* Annonay, 24 February 1875), *engineering, physics.*

The eldest son of a small but prosperous manufacturer in the Ardèche, Seguin completed his formal education at an undistinguished Parisian boarding school. From his arrival in Paris in 1799 his interest in science and engineering was stimulated and decisively shaped through his informal contact with his granduncle Joseph Montgolfier, the famous balloonist. Closely related by birth and later by marriage to the Montgolfier family, Seguin always regarded himself as Montgolfier's leading disciple. From Montgolfier Seguin acquired several scientific perspectives seldom expressed by the contemporary scientific community, notably an emphasis on *vis viva* (mv^2) as the quantity that is conserved not only in heat engines and machines but also in the universe. Seguin also credited Montgolfier with rejecting the caloric theory of heat and suggesting, instead, that heat and motion, both manifestations of an unknown but common cause, were interconvertible. Montgolfier died in 1810, and shortly thereafter Seguin returned to Annonay.

Over the next two decades Seguin, aided by his younger brothers, executed a series of successful and innovative engineering projects. In 1825, at Tournon on the Rhone, he erected the first successful suspension bridge in France to use cables of iron wire. After a brief attempt to establish regular steamboat service on the Rhone, the Seguin brothers organized the company that built France's first modern railroad, a line between Saint-Étienne and Lyons completed in 1832. Finding the Stephenson locomotive unable to generate enough steam for high-speed operation, in 1827 Seguin invented the multitubular, or fire tube, boiler. After working briefly on the Left Bank railroad between

Paris and Versailles, Seguin retired from active engineering and moved to Fontenay in 1838. Marked both by numerous technical innovations and by shrewd economic planning, his engineering projects provide one of the earliest examples in France of large-scale civil engineering undertaken and financed by private companies. For these engineering achievements Seguin was elected a correspondent of the Académie des Sciences in 1845.

Except for some contact with J. B. Biot, who had invested heavily in the engineering enterprises, Seguin always remained on the periphery of professional French science. Through numerous visits to England, however, he became acquainted with several prominent British scientists, notably John Herschel, Michael Faraday, and Humphry Davy. Two letters from Seguin to Herschel and to Brewster were published in Edinburgh journals in 1824 and 1825. These contain his first statements, perhaps derived from Montgolfier, of an extreme form of Newtonianism in which all physical forces are to be explained as the result of the inverse-square law of attraction between molecules. Matter, Seguin argued, consists of small, dense molecules in constant motion in miniature solar systems. Magnetic, electrical, and thermal phenomena are the result of particular velocities and orbits. Faster molecules escape from orbit along tangential paths, producing light and radiant heat. Explicitly identifying heat as molecular velocity, Seguin added that when molecules transmit their velocity to external objects, a conversion of heat into mechanical effect occurs. The qualitative, synthetic style of Seguin's papers and his conceptions of heat and light were in sharp contrast with the mathematical caloric and wave theories then dominant in France. In spite of a close similarity to contemporary British science, the papers apparently had no influence.

In 1839 Seguin returned to the problem of heat in his most important publication, *De l'influence des chemins de fer,* a handbook for the design and construction of railroads. In the chapter on steam locomotive performance, he first rejected the caloric theory because its major premise, the existence of heat as a fluid conserved in all processes, would allow the reuse of heat in an engine and thus would imply perpetual motion. "It appears more natural to me to suppose that a certain quantity of heat disappears in the very act of the production of force or mechanical power, and conversely; and that the two phenomena are linked together by conditions that assign to them invariable relationships" (p. 382). This assumption was the basis for

his later claim to priority over Joule and Mayer for the statement of the convertibility and conservation of heat and work.

To determine the numerical relationship, Seguin imagined a unit weight of steam generated under pressure in a cylinder and allowed to expand adiabatically until its temperature fell twenty degrees Centigrade. Assuming that steam is a perfect gas that remains saturated during expansion, he used standard engineering formulas and steam tables to calculate the work performed against the piston. The results, given in a table for different twenty-degree intervals, showed a steady decrease in the amount of work produced as the intervals were located higher on the temperature scale, a result indicating either that the conversion of heat into work did not follow an invariable law or, as Seguin chose to conclude, that the heat loss of twenty degrees as measured on a thermometer was not a true measure of the heat lost by the steam. Unable to specify the relation between temperature loss and total loss of heat content, he made no attempt to define a unit of heat or to state its mechanical equivalent.

Joule's determination of the mechanical equivalent was published in France in 1847, and one month later Seguin responded with a note to the Academy supporting Joule's conclusions. Seguin called attention to his 1839 work as an earlier attempt to find the numerical equivalent and demonstrated that the average of his tabular results, converted to Joule's units, gave a value very close to the figure Joule had obtained by completely different methods. Conversely, Seguin regarded Joule's experiments as a confirmation of his own more general theory outlined in 1824, and his paper concluded with an announcement of his intention to explain the identity of heat and work through an extension of the law of universal attraction.

In the later priority controversy surrounding the principle of energy conservation, it became clear that Mayer, Joule, and Helmholtz had worked independently of Seguin and that, despite Seguin's repeated claims, only in retrospect could his 1839 suggestions be clearly interpreted as a mechanical equivalent of heat. Although Seguin's theory of forces implied a conservation of energy, he focused his attention on the motion of molecules and made no attempt to define a concept of energy or to state its conservation.

From 1848 until his death Seguin conducted a broad campaign to win acceptance for his program of reducing all physical forces to the single Newtonian law of molecular attraction. In 1853 he bought a bankrupt journal and began publishing a weekly scientific magazine, *Cosmos*. Under the editorship of the abbé François Moigno the journal served as an important vehicle for the popularization of science and as a forum for Seguin's theories. In 1856 Seguin commissioned a French translation of William Grove's *Correlation of Physical Forces* and added notes and commentary. Throughout Seguin's prolific publications the earlier themes of 1824 were repeated and elaborated. The only major addition was a strong emphasis, adopted from Biot, on crystal structure and polarized light. Seguin developed gravitational models for the structure of matter and presented particle theories for heat, light, electricity, and magnetism. These theories, together with his attacks on the ether hypothesis and on the reduction of physics to analytical mathematics, were completely contrary to contemporary scientific opinion and had no apparent subsequent influence.

BIBLIOGRAPHY

I. ORIGINAL WORKS. Seguin's thirty-six journal articles are listed in the Royal Society *Catalogue of Scientific Papers,* V and VIII. The most important are "Letter to Sir. J. Herschel: Observations on the Effects of Heat and of Motion," in *Edinburgh Philosophical Journal,* 10 (1824), 280–283; "Letter to Dr. Brewster on the Effects of Heat and Motion," in *Edinburgh Journal of Science,* 3 (1825), 276–281; and "Note à l'appui de l'opinion émise par M. Joule, sur l'identité du mouvement et du calorique," in *Comptes rendus . . . de l'Académie des sciences,* 25 (1847), 420–422. Seguin published nearly thirty books, many of them reprints of earlier journal articles. Of special interest are *Des ponts en fil de fer* (Paris, 1824; 2nd ed., 1826); *De l'influence des chemins de fer et de l'art de les tracer et de les construire* (Paris, 1839; repr. 1887); and *Mémoire sur les causes et sur les effets de la chaleur, de la lumière et de l'électricité* (Paris, 1865).

II. SECONDARY LITERATURE. The only biography, P. E. Marchal and Laurent Seguin, *Marc Seguin 1786– 1875: La naissance du premier chemin de fer français* (Lyons, 1957), is useful, but uncritical. Seguin is discussed in the context of the several men who approached energy conservation by various routes in T. S. Kuhn, "Energy Conservation as an Example of Simultaneous Discovery," in Marshall Clagett, ed., *Critical Problems in the History of Science* (Madison, Wis., 1959), 321–356.

JAMES F. CHALLEY

SEIDEL, PHILIPP LUDWIG VON (*b.* Zweibrücken, Germany, 24 October 1821; *d.* Munich, Germany, 13 August 1896), *astronomy, mathematics.*

Seidel was the son of Justus Christian Felix Seidel, a post office official, and Julie Reinhold. Because of his father's work he began school in Nördlingen, continued in Nuremberg, and finished in Hof. After passing the graduation examination in the fall of 1839 he took private lessons from L. C. Schnürlein, a teacher of mathematics at the Hof Gymnasium who had studied under Gauss. In the spring of 1840, Seidel entered Berlin University, where he attended the lectures of Dirichlet and Encke, for whom he subsequently carried out calculations at the astronomical observatory. In the fall of 1842 he moved to Königsberg, where he studied with Bessel, Jacobi, and F. E. Neumann. When Jacobi left Königsberg because of his health in the fall of 1843, Seidel, on Bessel's recommendation, moved to Munich and obtained the doctorate with the dissertation "Über die beste Form der Spiegel in Teleskopen" in January 1846. Six months later he qualified as a *Privatdozent* on the basis of his "Untersuchungen über die Konvergenz und Divergenz der Kettenbrüche."

These two works treat two fields investigated by Seidel throughout his life, dioptrics and mathematical analysis. He also produced works in probability theory and photometry, the latter stimulated by his collaboration with Steinheil. In his mathematical investigations he depended on Dirichlet but filled important gaps left by his teacher—for instance, introducing the concept of nonuniform convergence.

Seidel's photometric measurements of fixed stars and planets were the first true measurements of this kind. The precise evaluation of his observations by methods of probability theory, considering atmospheric extinction, are worthy of special mention. At Steinheil's suggestion Seidel derived trigonometric formulas for points lateral to the axis of an optical system; they soon became important for astronomical photography and led to the production of improved telescopes.

Besides the application of probability theory to astronomy, Seidel investigated the relation between the frequency of certain diseases and climatic conditions at Munich. His pioneer work in several fields was acknowledged by Bavaria. In 1847 he became assistant professor, in 1855 full professor, and later royal privy councillor; he also received a number of medals, one of them connected with nobility. Seidel was a member of the Bavarian Academy of Sciences (1851) and corresponding member of the Berlin and Göttingen academies, as well as a member of the Commission for the European Measurement of a Degree and of a group observing a transit of Venus.

Seidel suffered from eye problems and was obliged to retire early. A bachelor, he was cared for by his unmarried sister Lucie until 1889, and later by the widow of a clergyman named Langhans.

Seidel's lectures covered mathematics, including probability theory and the method of least squares, astronomy, and dioptrics. He never accepted Riemannian geometry.

BIBLIOGRAPHY

I. ORIGINAL WORKS. Seidel's earlier works include "Über die Bestimmung der Brechungs- und Zerstreuungs-Verhältnisses verschiedener Medien," in *Abhandlungen der Bayerischen Akademie der Wissenschaften*, Math.-phys. Kl., **5** (1848), 253–268, written with K. A. Steinheil; "Note über eine Eigenschaft der Reihen, welche discontinuierliche Functionen darstellen," *ibid.*, 381–393; "Untersuchungen über die gegenseitigen Helligkeiten der Fixsterne erster Grösse, und über die Extinction des Lichtes in der Atmosphäre," *ibid.*, **6**, no. 3 (1852), 539–660; "Zur Theorie der Fernrohr-Objective," in *Astronomische Nachrichten*, **35** (1852), 301–316; "Zur Dioptrik," *ibid.*, **37** (1853), 105–120; "Bemerkungen über den Zusammenhang zwischen dem Bildungsgesetze eines Kettenbruches und der Art des Fortgangs seiner Näherungsbrüche," in *Abhandlungen der Bayerischen Akademie der Wissenschaften*, Math.-phys. Kl., **7** (1855), 559–602; "Entwicklung der Glieder 3.ter Ordnung, welche den Weg eines ausserhalb der Ebene der Axe gelegenen Lichtstrahles durch ein System brechender Medien bestimmen," in *Astronomische Nachrichten*, **43** (1856), 289–332; "Über den Einfluss der Theorie der Fehler, mit welchen die durch optische Instrumente gesehenen Bilder behaftet sind, und über die mathematischen Bedingungen ihrer Aufhebung," in *Abhandlungen der naturwissenschaftlich-technischen Commission der Bayerischen Akademie der Wissenschaften*, **1** (1857), 227–267; and "Untersuchungen über die Lichtstärke der Planeten vergleichen mit den Sternen, und über die relative Weisse ihrer Oberfläche," in *Monumenta saecularum der Bayerischen Akademie der Wissenschaften* (1859), 1–102.

Among his later writings are "Resultate photometrischer Messungen an 208 der vorzüglichsten Fixsterne," in *Abhandlungen der Bayerischen Akademie der Wissenschaften*, Math.-phys. Kl., **9** (1863), 419–607; "Über eine Anwendung der Wahrscheinlichkeitsrechnung, bezüglich auf die Schwankungen in den Durchsichtigkeitsverhältnissen der Luft," in *Sitzungsberichte der Bayerischen Akademie der Wissenschaften zu München* (1863), pt. 2, 320–350; "Trigonometrische Formeln für den allgemeinsten Fall der Brechung des Lichtes an centrirten sphärischen Flächen," *ibid.* (1866), pt. 2, 263–284; "Über ein Verfahren, die Gleichungen, auf welche die Methode der kleinsten Quadrate führt, sowie lineäre Gleichungen überhaupt," in *Abhandlungen der Bayerischen Akademie der Wissenschaften*, Math.-phys. Kl., **11**, no. 3 (1874), 81–108; and "Über eine einfache Entstehungsweise der Bernoulli'schen Zahlen," in *Sitzungsberichte der Bayerischen Akademie der Wissenschaften*, Math.-phys. Kl., n.s. **7** (1877), 157–187.

II. SECONDARY LITERATURE. See F. Lindemann, *Gedächtnisrede auf L. Ph. von Seidel* (Munich, 1898).

H.-CHRIST. FREIESLEBEN

SEKI, TAKAKAZU (*b.* Huzioka[?], Japan, 1642[?]; *d.* Edo [now Tokyo], Japan, 24 October 1708), *mathematics*.

Knowledge of Seki's life is meager and indirect. His place of birth is variously given as Huzioka or Edo, and the year as 1642 or 1644. He was the second son of Nagaakira Utiyama, a samurai; his mother's name is unknown. He was adopted by the patriarch of the Seki family, an accountant, whom he accompanied to Edo as chief of the Bureau of Supply. In 1706, having grown too old to fulfill the duties of this office, he was transferred to a sinecure and died two years later.

Seki is reported to have begun the study of mathematics under Yositane Takahara, a brilliant disciple of Sigeyosi Mōri. Mōri is the author of *Warizansyo* ("A Book on Division," 1622), believed to be the first book on mathematics written by a Japanese, but little is known about Takahara. Of particular influence on Seki's mathematics was *Suan-hsüeh ch'i-mêng* ("Introduction to Mathematical Studies," 1299), compiled by Chu Shih-chieh, a collection of problems solved by a method known in Chinese as *t'ien-yuan shu* ("method of the celestial element"); this method makes it possible to solve a problem by transforming it into an algebraic equation with one variable. Kazuyuki Sawaguchi, allegedly the first Japanese mathematician to master this method, used it to solve 150 problems proposed by Masaoki Satō in his *Sanpô Kongenki* ("Fundamentals of Mathematics," 1667). He then organized these problems, together with their solutions, into the seven-volume collection *Kokon sanpōki* ("Ancient and Modern Mathematics," 1670). At the end of the last volume Sawaguchi presented fifteen problems that he believed to be unsolvable by means of *t'ien-yuan shu*.

Seki's solutions to these problems were published in 1674 as *Hatubi sanpō*. Because mathematicians of those days were unable to grasp how the solutions had been accomplished, Katahiro Takebe (1664–1739), a distinguished disciple of Seki, published *Hatubi sanpō endan genkai* ("An Easy Guide to the *Hatubi sanpō*," 1685), the work that made Seki's method known.

In Chinese mathematics operations were performed by means of instruments called *suan-ch'ou* ("calculating rods") and therefore could not treat any algebraic expressions except those with one variable and numerical coefficients. Seki introduced Chinese ideographs and wrote them to the right of a vertical—such as $|a$—where a is used, for typographical reasons, instead of a Chinese ideograph. Seki called this notation *bōsyohō* ("method of writing by the side") and used it as the basis of his *endan zyutu* ("*endan* method"). *Endan zyutu* enabled him to represent known and unknown quantities by Chinese ideographs and led him to form equations with literal coefficients of any degree and with several variables. The technique was later renamed *tenzan zyutu* by Yosisuke Matunaga (1693–1744), the third licensee of the secret mathematical methods of Seki's school.

The *Hatubi sanpō* did not include Seki's principal theorems, which were kept secret; but in order to initiate students he arranged the theorems systematically. Parts of these works and some more extended theorems were collected and published posthumously by his disciples as *Katuyō sanpō*—which, with *Sekiryū sanpō sitibusyo* ("Seven Books on the Mathematics of Seki's School," 1907), is sufficient to grasp his mathematics.

Seki first treated general theories of algebraic equations. Since his method of side writing was inconvenient for writing general equations of degree *n*, he worked with equations of the second through fifth degrees. But his treatment was so general that his method could be applied to equations of any degree. Seki attempted to find a means of solving second-degree algebraic equations with numerical coefficients but, having no concept of algebraic solution, directed his efforts at finding an approximate solution. He discovered a procedure, used long before in China, that was substantially the same as Horner's. The notion of a discriminant of an algebraic equation also was introduced by Seki. Although he had no notion of the derivative, he derived from an algebraic expression $f(x)$ another expression that was the equivalent of $f'(x)$ in modern notation. Eliminating *x* from the pair of equations $f(x) = 0$ and $f'(x) = 0$, he obtained what is

now called a discriminant. With the help of this expression Seki found double roots of the equation $f(x) = 0$. He also developed a method similar to Newton's by which an approximate value of the root of a numerical equation can be computed. Seki's *tenzan zyutu* was important in the treatment of problems that can be transformed into the solution of a system of simultaneous equations. In order to solve the problem of elimination, he introduced determinants and gave a rule for expressing them diagrammatically. For third-order equations his formula was basically similar to Sarrus'.

The method that Seki named *syōsahō* was intended to determine the coefficients of the expression $y = a_1 x + a_2 x^2 + \cdots + a_n x^n$, when *n* values of $x_1, x_2, \cdots, x_n$ of *x* and the corresponding *n* values of $y_1, y_2, \cdots, y_n$ of *y* are given. Since his notation was not suitable for the general case corresponding to an arbitrary *n*, Seki treated the case that corresponds to the special value of *n*. His solution was similar to the method of finite difference.

Another method, *daseki zyutu*, also is important in Seki's mathematics. Its purpose is to find values of $s_p = 1^p + 2^p + \cdots + n^p$ for $p = 1, 2, 3, \cdots$. Using the *syōsahō*, Seki calculated $s_1, s_2, \cdots, s_6$ for a particular value of *n*:

$$s_1 = \frac{n^2}{2} + \frac{n}{2}$$

$$s_2 = \frac{n^3}{3} + \frac{n^2}{2} + \frac{1}{6} \cdot \frac{2n}{2}$$

$$s_3 = \frac{n^4}{4} + \frac{n^3}{2} + \frac{1}{6} \cdot \frac{3n^2}{2}$$

$$s_4 = \frac{n^5}{5} + \frac{n^4}{2} + \frac{1}{6} \cdot \frac{4n^3}{2} - \frac{1}{30} \cdot \frac{4 \cdot 3 \cdot 2n}{2 \cdot 3 \cdot 4}$$

$$s_5 = \frac{n^6}{6} + \frac{n^5}{2} + \frac{1}{6} \cdot \frac{5n^4}{2} - \frac{1}{30} \cdot \frac{5 \cdot 4 \cdot 3n^2}{2 \cdot 3 \cdot 4}$$

$$s_6 = \frac{n^7}{7} + \frac{n^6}{2} + \frac{1}{6} \cdot \frac{6n^5}{2} - \frac{1}{30} \cdot \frac{6 \cdot 5 \cdot 4n^3}{2 \cdot 3 \cdot 4} +$$

$$\frac{1}{42} \cdot \frac{6 \cdot 5 \cdot 4 \cdot 3 \cdot 2n}{2 \cdot 3 \cdot 4 \cdot 5 \cdot 6}.$$

The numbers 1/6, 1/30, 1/42 are Bernoulli numbers, which were introduced in his *Ars conjectandi* (1713).

Enri ("principle of the circle"), one of Seki's important contributions, consists of rectification of the circumference of a circle, rectification of a circular arc, and cubature of a sphere. In the rectification of a circumference, Seki considered a circle of diameter 1 and an inscribed regular polygon of 2^n sides. He believed that the inscribed polygon gradually loses its angularities and finally becomes a circumference of the circle when the number of sides is increased without limit. He therefore calculated the perimeter c_i of the regular polygon of 2^i sides, where i represents any integer not greater than 17, and devised a method by which he was able to obtain a better result from c_{15}, c_{16}, c_{17}. The formula was

$$s = c_{16} + \frac{(c_{16} - c_{15})(c_{17} - c_{16})}{(c_{16} - c_{15}) - (c_{17} - c_{16})},$$

where

$$c_{15} = 3.14159264877698566708,$$
$$c_{16} = 3.14159265235659135571,$$

and

$$c_{17} = 3.14159265328890277755.$$

Therefore, $s = 3.14159265359$, where s is the circumference of the circle; this value is accurate except for the last figure. Also in connection with this problem Seki created a method of approximation called *reiyaku zyutu*, by which he theoretically obtained 355/113 as an approximate value of π, a value found much earlier in China.

In the rectification of a circular arc, Seki considered an arc of which the chord is 8 and the corresponding sagitta is 2. He considered an inscribed open polygon of 2^{15} sides and, using the above formula, calculated the approximate value of the arc as 9.272953. In the cubature of a sphere, Seki calculated the volume of a sphere of diameter 10 and obtained $523\frac{203}{339}$ as an approximate value. Using the approximate value of π as 355/113, Seki observed that if d is the diameter of this sphere,

$$\frac{\pi}{6} d^3 = \frac{355}{678} \times (10)^3 = 523\frac{203}{339}.$$

Certain other geometrical problems, which do not belong among the *enri*, concern ellipses. Seki believed that any ellipse can be obtained through cutting a suitable circular cylinder by a plane. In order to obtain the area of an ellipse, he cut off from a circular cylinder two segments with generatrices of equal length such that one has elliptical bases and the other has circular bases. These two cylinders have the same volume. By equating the volumes of these two pieces of the cylinder, Seki obtained the result that the area of an ellipse is $\pi/4 \times AA' \times BB'$, where AA' and BB' denote, respec-

tively, the major and minor axis of an ellipse. Among the problems of cubature, that of a solid generated by revolving a segment of circle about a straight line that lies on the same plane as the segment is noteworthy. His result in this area had generality, and the theorem that he established was substantially the same as that now called the theorem of Pappus and Guldin.

BIBLIOGRAPHY

I. ORIGINAL WORKS. Seki's published writings are *Hatubi sanpō* (Edo, 1674) and *Katuyō sanpō* (Tokyo–Kyoto, 1709), copies of which are owned by the Mathematical Institute, Faculty of Science, Kyoto University. *Sekiryū sanpō sitibusyo* ("Seven Books on the Mathematics of Seki's School"; Tokyo, 1907), is a collection of Seki's papers (1683–1685) that were transmitted from master to pupils, who were permitted to copy them. An important collection of MS papers, "Sanbusyō" ("Three Selected Papers"), is discussed by Fujiwara (see below). See also *Collected Papers of Takakazu Seki* (Osaka, 1974).

II. SECONDARY LITERATURE. See the following, listed chronologically: Katahiro Takebe, *Hatubi sanpō endan genkai* ("An Easy Guide to the Hatubi sanpō"; n.p., 1685), a copy of which is owned by the Mathematical Institute, Faculty of Science, Kyoto University, with a facs. of the text of *Hatubi sanpō* and an explanation of Seki's method of solution; Dairoku Kikuchi, "Seki's Method of Finding the Length of an Arc of a Circle," in *Proceedings of the Physico-Mathematical Society of Japan*, 8, no. 5 (1899), 179–198; and Matsusaburo Fujiwara, *Mathematics of Japan Before the Meiji Era*, II (Tokyo, 1956), 133–265, in Japanese. There are many other works that treat Seki's mathematics, such as Yoshio Mikami, *The Development of Mathematics in China and Japan* (Leipzig–New York, 1913; repr. New York, n.d.), but only those of scientific importance are cited here.

AKIRA KOBORI

SELWYN, ALFRED RICHARD CECIL (*b.* Kilmington, Somerset, England, 28 July 1824; *d.* Vancouver, British Columbia, 19 October 1902), *geology.*

Selwyn was born into an English upper-class family. His father was a canon in the church and his mother was the daughter of a nobleman. His education began at home under private tutors and was completed in Switzerland, although to what level is not known. He was interested in the natural sciences and especially in geology; at the age of twenty-one he was engaged by the Geological Survey of Great Britain. His mapping assignments were in the lower Paleozoic graptolitic sediments and black volcanics of western Shropshire and North Wales. Selwyn is credited with sixteen map sheets, which were described by his superior, and sometime field associate, Sir A. C. Ramsay, as "the perfection of beauty."

At the end of 1852 Selwyn accepted the post of director of the Geological Survey of Victoria, Australia, which he held for seventeen years. About the same time he married Matilda Charlotte Selwyn, daughter of the Rev. Edward Selwyn of Hemmingford Abbotts, and they had nine children. The area of Victoria is ten times that of Wales, but the rocks are generally similar in type and age. Although he had only a small staff of geologists, sixty-one geological maps (with accompanying sections at a scale of two inches to a mile) were produced. Since Victoria was the scene of intensive gold prospecting, particular attention was given to the distribution of auriferous veins and the extension of gold-bearing placer deposits beneath younger Tertiary lava flows. He administered the survey, actively participated in field mapping, and served as a commissioner to international exhibitions in London (1862), Dublin (1865), and Paris (1866). The survey was abruptly terminated in 1869, when the colonial legislature refused to vote a budget.

In December 1869 Selwyn became the second director of the Geological Survey of Canada, succeeding Sir William Logan, who had established a small but effective organization. The following year Selwyn investigated the gold fields of Nova Scotia. With the confederation of Canada in 1867 the jurisdiction of the survey over Ontario and Quebec had been expanded to include the Maritime Provinces. Moreover, negotiations for union were in progress with the western provinces, and such a unification would require the construction of a railroad to the Pacific coast. Thus, in 1871, at the request of the government, Selwyn went via San Francisco to British Columbia to traverse on horseback possible rail routes in the southern part of the province. The dense forest and raging streams made progress so difficult that a distance of only five miles a day was normal. At one point, a hungry horse seems to have eaten the records of two days' observations. Selwyn strengthened the assumption by feeding the horse blank pages from the same notebook.

In 1873 Selwyn made a geological reconnaissance of the western provinces, traveling by horse

cart from Winnipeg to the Rocky Mountains, and in 1875 he examined a proposed northern railroad route via the Peace River. On this latter trip he was accompanied by a botanist because the survey had been expanded to include all aspects of natural history. Thus within five years Selwyn had a broad knowledge of Canadian geology and the hardships of fieldwork in Canada. Later his responsibilities as director of the survey prevented him from conducting prolonged fieldwork; but he still visited in the field on a regular basis, usually to inspect mineral finds of economic importance.

During his directorship, the staff of the Geological Survey was increased to over thirty employees, including some of the most highly qualified geologists graduating from Canadian universities. Sections within the survey were set up for chemistry, mineralogy, paleontology, natural history, topographic mapping, and administration. The survey library was greatly expanded and was described as one of the great scientific libraries in North America. Selwyn initiated new sections for mines, borings, water supply, and statistics of mineral production. The survey was represented at international exhibits in Philadelphia (1876), Paris (1878), London (1886), and Chicago (1893), all of which drew attention to the mineral industry of Canada and credit to Selwyn's organization. The Survey Act of 1890 required that the geologists had to be graduates of recognized universities.

Selwyn's term as director, however, was marked by much internal dissension, possibly because of his authoritarian attitude, and agitation by malcontents both inside and outside the survey. A government inquiry in 1884 into the operations of the survey has been described as breaking open a hornet's nest. Some of the problems arose from the expansion of Canada, devoting too much time to one provincial area rather than another, low pay and different salaries for individuals with the same qualifications, slow promotions, the delay or suppression of publications, poor accounting, lack of public relations, pursuing a purely scientific (rather than practical) goal, and following in the footsteps of Logan. Although the final report was critical of the discord, the level of activity, and the delay and seeming suppression of publications, Selwyn remained as director and none of the employees resigned. Selwyn's superannuation at age seventy in 1894 proved embarrassing when the government, during his absence in England, issued an order for his immediate retirement and the appointment of G. M. Dawson as director, about which Selwyn learned upon his return.

BIBLIOGRAPHY

I. Original Works. A complete list of Selwyn's publications is given in the biographical sketch by H. M. Ami (see below). His most notable contributions are the sixty-one geological maps of Victoria, which, according to the present director of the Geological Survey of Victoria, are still held in the highest regard. His thirty-seven publications for Canada, listed by J. M. Nickles in "Geologic Literature of North America 1785–1918," in *Bulletin of the United States Geological Survey*, **746** (1923), 931–932, are mainly the annual summary reports of the Geological Survey of Canada. Selwyn admitted after retirement that he had "an antipathy to the mechanical labour of writing," but he was described as a master of clear English expression. His broad interests are evident in "Notes and Observations on the Gold Fields of Quebec and Nova Scotia," in *Canadian Geological Survey, Report of Progress 1870–71* (1872), 252–282; "The Stratigraphy of the Quebec Group," in *Canadian Naturalist and Geologist*, n.s. **9** (1879), 17–31; and "Tracks of Organic Origin in the Animikie Group," in *American Journal of Science*, 3rd ser. (1890), 145–147.

II. Secondary Literature. See H. M. Ami, "Memorial or Sketch of the Life of the Late Dr. A. R. C. Selwyn, C.M.G., LL.D., F.R.S., F.G.S., etc., etc., Director of the Geological Survey of Canada from 1869 to 1894," in *Transactions of the Royal Society of Canada*, Section IV, Geological Sciences and Mineralogy (1904), 173–205. An interpretative account of Selwyn as director of the Geological Survey can be found in M. Zaslow, *Reading the Rocks: The Story of the Geological Survey of Canada, 1842–1972* (Toronto, 1975), *passim*.

C. Gordon Winder

SEMMELWEIS, IGNAZ PHILIPP (*b.* Buda, Hungary, 1 July 1818; *d.* Vienna, Austria, 13 August 1865), *medicine*.

Semmelweis was one of the most prominent medical figures of his time. His discovery concerning the etiology and prevention of puerperal fever was a brilliant example of fact-finding, meaningful statistical analysis, and keen inductive reasoning. The highly successful prophylactic hand washings made him a pioneer in antisepsis during the pre-bacteriological era in spite of deliberate opposition and uninformed resistance.

Semmelweis was born in Tabán, an old commercial sector of Buda. The fifth child of a prosperous shopkeeper of German origin, he received his elementary education at the Catholic Gymnasium of Buda, then completed his schooling at the University of Pest between 1835 and 1837.

In the fall of 1837, Semmelweis traveled to Vienna, ostensibly to enroll in its law school. His father wanted him to become a military advocate in the service of the Austrian bureaucracy. Soon after his arrival, however, he was attracted to medicine; and seemingly without parental opposition he matriculated in the medical school.

After completing his first year of studies at Vienna, Semmelweis returned to Pest and continued at the local university during the academic years 1839–1841. The backward conditions in the school, however, caused his return to Vienna in 1841 for further studies at the Second Vienna Medical School, which became one of the leading world centers for almost a century with its amalgamation of laboratory and bedside medicine. During the last two years of study, Semmelweis came in close contact with three of the most promising figures of the new school: Karl von Rokitansky, Josef Skoda, and Ferdinand von Hebra.

After voluntarily attending seminars led by these teachers, Semmelweis completed his botanically oriented dissertation early in 1844. He remained in Vienna after graduation, repeating a two-month course in practical midwifery and receiving a master's degree in the subject. He also completed some surgical training and spent almost fifteen months (October 1844–February 1846) with Skoda learning diagnostic and statistical methods. Finally Semmelweis applied for the position of assistant in the First Obstetrical Clinic of the university's teaching institution, the Vienna General Hospital.

In July of 1846 Semmelweis became the titular house officer of the First Clinic, which was then under the direction of Johann Klein. Among his numerous duties were the instruction of medical students, assistance at surgical procedures, and the regular performance of all clinical examinations. One of the most pressing problems facing him was the high maternal and neonatal mortality due to puerperal fever, 13.10 percent. Curiously, however, the Second Obstetrical Clinic in the same hospital exhibited a much lower mortality rate, 2.03 percent. The only difference between them lay in their function. The First was the teaching service for medical students, while the Second had been selected in 1839 for the instruction of midwives. Although everyone was baffled by the contrasting mortality figures, no clear explanation for the differences was forthcoming. The disease was considered to be an inevitable aspect of contemporary hospital-based obstetrics, a product of unknown agents operating in conjunction with elusive atmospheric conditions.

After a temporary demotion to allow the reinstatement of his predecessor, who soon left Vienna for a professorship at Tübingen, Semmelweis resumed his post in March 1847. During his short vacation in Venice, the tragic death of his friend Jakob Kolletschka, professor of forensic medicine, occurred after his finger was accidentally punctured with a knife during a postmortem examination. Interestingly, Kolletschka's own autopsy revealed a pathological situation akin to that of the women who were dying from puerperal fever.

Prepared through his intensive pathological training with Rokitansky, who had placed all cadavers from the gynecology ward at his disposal for dissection, Semmelweis made a crucial association. He promptly connected the idea of cadaveric contamination with puerperal fever, and made a detailed study of the mortality statistics of both obstetrical clinics. He concluded that he and the students carried the infecting particles on their hands from the autopsy room to the patients they examined during labor. This startling hypothesis led Semmelweis to devise a novel system of prophylaxis in May 1847. Realizing that the cadaveric smell emanating from the hands of the dissectors reflected the presence of the incriminated poisonous matter, he instituted the use of a solution of chlorinated lime for washing hands between autopsy work and examination of patients. Despite early protests, especially from the medical students and hospital staff, Semmelweis was able to enforce the new procedure vigorously; and in barely one month the mortality from puerperal fever declined in his clinic from 12.24 percent to 2.38 percent. A subsequent temporary resurgence of the dreaded ailment was traced to contamination with putrid material from a patient suffering from uterine cancer and another with a knee infection.

In spite of the dramatic practical results of his washings, Semmelweis refused to communicate his method officially to the learned circles of Vienna, nor was he eager to explain it on paper. Hence, Hebra finally wrote two articles in his behalf, explaining the etiology of puerperal fever and strongly recommending use of chlorinated lime as a preventive. Although foreign physicians and the leading members of the Viennese school were impressed by Semmelweis' apparent discovery, the papers failed to generate widespread support.

During 1848 Semmelweis gradually widened his prophylaxis to include all instruments coming in contact with patients in labor. His statistically documented success in virtually eliminating puerperal fever from the hospital ward led to efforts by Skoda to create an official commission to investigate the results. The proposal was ultimately rejected by the Ministry of Education, however, a casualty of the political struggle between the defeated liberals of the 1848 movement and the newly empowered conservatives in both the university and the government bureaucracy.

Angered by favorable reports concerning the new methods that indirectly represented an indictment of his own beliefs and actions, Klein refused to reappoint Semmelweis in March 1849. Undaunted, he applied for an unpaid instructorship in midwifery. In the meantime he began to carry out animal experiments to prove his clinical conclusions with the aid of the physiologist Ernst Brücke and a grant from the Vienna Academy of Sciences.

Semmelweis was at last persuaded to present his findings personally to the local medical community. On 15 May 1850 he delivered a lecture to the Association of Physicians in Vienna, meeting under the presidency of Rokitansky. The following October he received the long-awaited appointment as a *Privatdozent* in midwifery, but the routine governmental decree stipulated that he could only teach obstetrics on a mannequin. Faced with financial difficulties in supporting his family, and perhaps discouraged, Semmelweis abruptly left the Austrian capital, returning to Pest without notifying even his closest friends. Such a hasty decision jeopardized forever his chances to overcome the Viennese skeptics gradually with the dedicated help of Rokitansky, Skoda, Hebra, and other colleagues.

In Hungary, Semmelweis found a backward and depressed political and scientific atmosphere following the crushing defeat of the liberals in the revolution of 1848. Despite the unfavorable circumstances, he managed to receive an honorary appointment and took charge of the maternity ward of Pest's St. Rochus Hospital in May 1851, remaining there until 1857. He soon was able to implement his new prophylaxis against puerperal fever, with great success, while building an extensive private practice.

Following the death of the incumbent, Semmelweis was appointed by the Austrian Ministry of Education to the chair of theoretical and practical midwifery at the University of Pest in July 1855, although he had been only the second choice of the local medical faculty. He subsequently devoted his efforts to improving the appalling conditions of the university's lying-in hospital, a difficult task in the face of severe economic restrictions. In 1855 Semmelweis instituted his chlorine hand washings in the clinic, and he gradually achieved good results despite initial carelessness by the hospital staff. His lectures, delivered in Hungarian by decree of the Austrian authorities, attracted large student audiences. Semmelweis also became active in university affairs, serving on committees dealing with medical education, clinical services, and library organization.

In 1861 Semmelweis finally published his momentous discovery in book form. The work was written in German and discussed, at length, the historical circumstances surrounding his discovery of the cause and prevention of puerperal fever. A number of unfavorable foreign reviews of the book prompted Semmelweis to lash out against his critics in a series of open letters written in 1861–1862, which did little to advance his ideas.

After 1863 Semmelweis' increasing bitterness and frustration at the lack of acceptance of his method finally broke his hitherto indomitable spirit. He became alternately apathetic and pathologically enraged about his mission as a savior of mothers. In July 1865 Semmelweis suffered what appeared to be a form of mental illness; and after a journey to Vienna imposed by friends and relatives, he was committed to an asylum, the Niederösterreichische Heil- und Pflegeanstalt. He died there only two weeks later, the victim of a generalized sepsis ironically similar to that of puerperal fever, which had ensued from a surgically infected finger.

Semmelweis' achievement must be considered against the medical milieu of his time. The ontological concept of disease insisted on specific disease entities that could be distinctly correlated both clinically and pathologically. Puerperal fever, however, exhibited multiple and varying anatomical localizations and a baffling symptomatology closely related to the evolution of generalized sepsis. The apparent connection between this fever and erysipelas further clouded the issue. Moreover, the idea of a specific contagion causing the disease was not borne out by the clinical experience.

In the face of such theoretical uncertainties and the profusion of causes attributable to the disease, Semmelweis displayed a brilliant methodology borrowed from his teachers at the Second Vienna Medical School. He partially solved the puzzle through extensive and meticulous dissections of those who had succumbed to the disease, eventu-

ally recognizing the crucial similarities of all septic states. The methodical exclusion of possible etiological factors—one variable at a time—followed Skoda's diagnostic procedure, while the employment of statistical data was transferred from therapeutic analysis to the elucidation of the decisive factor responsible for the disease. In finally arriving at his discovery, Semmelweis successfully seized upon his built-in control group of women at the Second Clinic, a fortunate situation unparalleled elsewhere.

The subsequent lack of recognition for Semmelweis' prophylaxis can be attributed to several factors. An initial lack of proper publicity among Viennese and foreign visiting physicians led to misunderstandings and an incomplete assessment of the intended procedure. Further, political feuds led to an identification of Semmelweis with the liberal and reform-oriented faction of the Viennese medical faculty, a group temporarily thwarted in their objectives by the crushing defeat of 1848. Finally, Semmelweis' abrupt departure from the arena robbed him of the possibility of eventually persuading his Viennese colleagues of the soundness of the chlorine washings. Operating from a politically suppressed and scientifically backward country with a second-rate university, Semmelweis was effectively hampered in the promulgation of his ideas. His later, rather violent and passionate polemics added little further credence to a somewhat cumbersome method that was difficult to implement among hospital staff members content with the status quo. Most important, however, was the lack of a good explanation for Semmelweis' empirically derived procedure, a development made possible only through the ensuing work of Pasteur.

BIBLIOGRAPHY

I. Original Works. Semmelweis' meager writings are listed in *Medical Classics*, **5** (1941), 340–341; and in *Bulletin of the History of Medicine*, **20** (1946), 653–707. His most important work was *Die Aetiologie, der Begriff und die Prophylaxis des Kindbettfiebers* (Pest–Vienna–Leipzig, 1861), reprinted with a new introduction by A. F. Guttmacher (New York–London, 1966). This work was translated into English by F. R. Murphy and appeared as "The Etiology, the Concept and Prophylaxis of Childbed Fever," in *Medical Classics*, **5** (1941), 350–773. However, Semmelweis first published his ideas in 1858 under the title "A gyermekagyi láz koroktana" ("The Etiology of Childbed Fevers") in *Orvosi hétilap* (1858): no. 1, 1–5; no. 2, 17–21; no. 5,

65–69; no. 6, 81–84; no. 21, 321–326; no. 22, 337–342; no. 23, 353–359.

In 1903 the Hungarian Society for the Publication of Medical Works collected and published Semmelweis' works in Hungarian while the Hungarian Academy simultaneously arranged for an edition of the collected writings in German: *Semmelweis gesammelte Werke*, edited and partially translated from the Hungarian by T. von Győry (Jena, 1905). This book contains articles dealing with gynecological subjects originally published in *Orvosi hétilap*. A short letter written by Semmelweis in English and dated at Pest, 21 April 1862, is "On the Origin and Prevention of Puerperal Fever," in *Medical Times and Gazette* (London), **1** (1862), 601–602.

II. SECONDARY LITERATURE. Bibliographies of works on Semmelweis are Emerson C. Kelly, in *Medical Classics*, **5** (1941), 341–347, and Frank P. Murphy, in *Bulletin of the History of Medicine*, **20** (1946), 653–707.

Among the numerous biographies, the following standard ones are mentioned in chronological order: Alfred Hegar, *Ignaz Philipp Semmelweis. Sein Leben und seine Lehre, zugleich ein Beitrag zur Lehre der fieberhaften Wundkrankheiten* (Freiburg–Tübingen, 1882); Fritz Schürer von Waldheim, *Ignaz Philipp Semmelweis, sein Leben und Wirken* (Vienna–Leipzig, 1905); William J. Sinclair, *Semmelweis, His Life and Doctrines* (Manchester, 1909); and, more recently, the English version of a Hungarian work initially published in 1965; György Gortvay and Imre Zoltán, *Semmelweis–His Life and Work,* translated by E. Róna (Budapest, 1968). Shorter biographical accounts are E. Podach, *Ignaz Philipp Semmelweis* (Berlin–Leipzig, 1947); and Henry E. Sigerist, in *The Great Doctors*, translated by Eden and Cedar Paul (Baltimore, 1933), ch. 41, 338–343.

A great number of articles and monographs concerning aspects of Semmelweis' life (especially the question of his terminal illness) and discovery have been published in Europe, especially in Hungary and Austria. Many of these papers were delivered during the celebrations commemorating the 150th anniversary of Semmelweis' birth in 1968 and appeared in two special numbers of the journal *Communicationes de historia artis medicinae Orvostörténeti közlemények*, nos. 46–47 (1968) and 55–56 (1970). Other articles appeared previously in *Orvosi hétilap*, **106** (1965); *Orszagos orvostörténeti* (Communicationes ex Bibliotheca historiae medicae hungarica), no. 42–43 (1967); and *Zentralblatt für Gynäkologie*, **87** (1965).

Among the main essays printed in both Hungarian and English in *Orvostörténeti közlemények*, no. 55–56 (1970), are I. Zoltán, "Semmelweis," 19–29; G. Regöy-Mérei, "The Pathological Reconstruction of Semmelweis' Disease on the Basis of the Catamnestic Analysis and Paleopathological Examination," 65–92; I. Benedek, "The Illness and Death of Semmelweis," 103–113; and L. Madai, "Semmelweis and Statistical Science," 157–174.

Other valuable contributions concerning Semmelweis' Viennese period are Erna Lesky, "Ignaz Philipp Sem-

melweis und die Wiener medizinische Schule," in *Sitzungsberichte der Österreichischen Akademie der Wissenschaften*, phil.-hist. Kl., **245** (1964), 3. Abh., 1–93, summarized in "Ignaz Philipp Semmelweis, Legende und Historie," in *Deutsche medizinische Wochenschrift*, **97** (1972), 627–632. The same author also published additional background material in "Wiener Aktenmaterial zur Berufung Semmelweis' im Jahre 1855," in *Orvostörténeti közlemények*, no. 46–47 (1968), 35–54; and *Die Wiener medizinische Schule im 19. Jahrhundert* (Graz–Vienna–Cologne, 1964), 209–220.

Selected recent articles on the subject are S. Fekete, "Die Geburtshilfe zur Zeit Semmelweis," in *Clio Medica*, **5** (1970), 35–44; H. Böttger, "Förderer der Semmelweisschen Lehre," in *Sudhoffs Archiv für Geschichte der Medizin und der Naturwissenschaften*, **39** (1955), 341–362; D. Tutzke, "Die Auswirkungen der Lehre von Semmelweis auf die öffentliche Gesundheitspflege," in *Medizinische Monatsschrift*, **20** (1966), 459–462; and S. D. Elek, "Semmelweis and the Oath of Hippocrates," in *Proceedings of the Royal Society of Medicine*, **59** (1969), 346–352.

GUENTER B. RISSE

SEMON, RICHARD WOLFGANG (*b.* Berlin, Germany, 22 August 1859; *d.* Munich, Germany, 27 December 1918), *zoology, anatomy*.

Semon's father, Simon Joseph Semon, a banker and stockbroker, and his mother, Henrietta Aschenheim, came from well-to-do Jewish families. His older brother Felix was a leading laryngologist in England, becoming physician in ordinary to King Edward VII and receiving a knighthood.

After reading the works of Charles Darwin and Ernst Haeckel, Semon became interested in biology while still at the Gymnasium. In 1879 he began to study zoology at Jena under Haeckel, whose views on natural philosophy had a lasting influence on Semon. He later stated that Haeckel's school was characterized by "the feeling for the connection of all branches of human knowledge [and] monism as a method of thinking and research." Beginning in 1881, Semon studied medicine at Heidelberg and at the same time prepared under Otto Bütschli's supervision a dissertation on "Das Nervensystem der Holothurien." He obtained the Ph.D. with this work at Jena in 1883, and a year later he passed the state medical examination at Heidelberg.

In 1885 Semon served as physician on an expedition to Africa led by Robert Flegel, but he had to withdraw from the expedition because of malaria. He then worked at the zoology station in Naples (1885–1886) before becoming an assistant at the

Jena anatomical institute in 1886. After receiving his medical degree that same year, Semon qualified as a university lecturer at Jena in 1887 with a work "Die indifferente Anlage der Keimdrüsen beim Hünchen und ihre Differenzierung zum Hoden" ("The Undifferentiated Rudiments of the Genital Glands of the Chick and Their Differentiation Into Testicles"). In 1891 Semon was named extraordinary professor at Jena.

From 1891 to 1893 Semon supervised an expedition to Australia. He published the results with the assistance of a number of co-workers.

In 1897 Semon gave up his teaching activities at Jena for personal reasons and began working as a private scholar in Munich. Two years later he married Maria Krehl (born Geibel), who became widely known for her translations of the works of Auguste-Henri Forel, L. Morgan, and the young Charles Darwin.

Semon based his zoological works on comparative anatomical and embryological studies of echinoderms, snails, fish, and birds. In Australia he was concerned mainly with the habitats, sexual reproduction, and development of the lungfish *Ceratodus forsteri*, as well as with the monotremata. His travel accounts contain remarks on animal geography, paleontology, and geology, as well as ethnographic and anthropological observations, and vivid descriptions of landscapes.

After 1900 Semon devoted himself primarily to an attempt to bring together into a unified concept "all those phenomena in the organic world that involve reproduction of any kind." Out of this attempt emerged his hypothesis of "the mneme as the enduring principle in changes occurring in organic life." Semon used the concept of the mneme (cell memory) to designate a particular property of organic substance, pursuing ideas first put forth in 1870 by Ewald Hering, who saw in memory "a universal function of organic matter." According to Semon, not only can a stimulus influence irritable organic substance temporarily, but it can also effect a persisting, latent change. In this case, the stimulus inscribes itself by means of an engram. These engrams can, he thought, be eliminated by certain influences, or ecphoria. The mneme consists of the sum of the engrams that an organism has individually acquired in the course of its life or that have been produced within it by heredity. In Semon's view, under favorable circumstances the individually acquired adaptations can be passed on to offspring (somatogenic inheritance).

Semon's Lamarckian concepts were rejected by, among others, August Weismann and Wilhelm Ludvig Johannsen, but were welcomed by Forel and his student Eugen Bleuler (the founder of Mnemismus). According to Jürgen Schatzmann, modern research on memory and on heredity provides partial justification for Semon's ideas, since "in both processes entirely similar chemical structures are involved." Nevertheless, the idea of an identity of the two processes is no more acceptable today than it was when Semon proposed it.

BIBLIOGRAPHY

I. ORIGINAL WORKS. A complete list of Semon's publications can be found in Otto Lubarsch, ed., *Bewustseinsvorgang im Gehirnprozess. Eine Studie von R. Semon* (Wiesbaden, 1920), xlv–xlviii.

Semon's principal works are *Zoologische Forschungsreisen in Australien und dem Malayischen Archipel-Denkschriften der medizinisch naturwissenschaftlichen Gesellschaft Jena*, **4–8** (Jena, 1893–1913)—this contains all Semon's works on the Ceratodus as well as on the monotremata and Marsupialia; *Im australischen Busch und an den Küsten des Korallenmeeres* (Leipzig, 1896); *Die Mneme als erhaltendes Prinzip im Wechsel des organischen Geschehens* (Leipzig, 1904; 3rd ed. 1911); *Das Problem der Vererbung 'erworbener Eigenschaften'* (Leipzig, 1912); and "Aus Haeckels Schule," in H. Schmidt, ed., *Was wir Ernst Haeckel verdanken* (Leipzig, 1914).

Semon's letters for Forel can be found in the archives of the Medizinhistorische Institut of the University of Zurich. Semon's letters to Haeckel are held at the Ernst-Haeckel-Haus of the Friedrich-Schiller-Universität in Jena; Haeckel's letters to Semon are held at the Bavarian State Library in Munich.

II. SECONDARY LITERATURE. See the article by Lubarsch in the book he edited under the title *Bewustseinsvorgang und Gehirnprozess von Richard Semon* (Wiesbaden, 1920). See also Auguste-Henri Forel, "Richard Semons Mneme als erhaltendes Prinzip im Wechsel des organischen Geschehens," in *Archiv für Rassen- und Gesellschaftsbiologie*, **2**, no. 2 (1905), and "Richard Semon," in *Journal für Psychologie und Neurologie*, **25** (1919); O. Lubarsch, "Richard Wolfgang Semon," in *Münchener medizinische Wochenschrift*, **66**, no. 11 (1919); August Weismann, "Semons 'Mneme' und die Vererbung erworbener Eigenschaften," in *Archiv für Rassen- und Gesellschaftsbiologie*, **3**, no. 1 (1906); Georg Uschmann, in *Geschichte der Zoologie und der zoologischen Anstalten in Jena* (Jena, 1959); and Jürgen Schatzmann, "Richard Semon (1859–1918) und seine Mnemetheorie," in *Zürcher medizingeschichtliche Abhandlungen*, n.s. no. 58 (1968).

GEORG USCHMANN

SEMPER, CARL GOTTFRIED

SEMPER, CARL GOTTFRIED (*b.* Altona, Germany, 6 July 1832; *d.* Würzburg, Germany, 29 May 1893), *zoology, ecology.*

Semper's father was a manufacturer; his uncle was the famous architect Gottfried Semper. Carl first studied engineering at Hannover (1851–1854) but eventually chose to become a naturalist and explorer. Thus he studied zoology, histology, and comparative anatomy (particularly of the invertebrates) at the University of Würzburg under Koelliker, Leydig, and Gegenbaur. His thesis (1856) dealt with the anatomy and physiology of the pulmonate snails. From 1857 to 1865 he traveled in Europe and then in the eastern tropics.

Except for a year (1862) on the Palau Islands of Babelthaup and Peleliu, Semper devoted the period from December 1858 to May 1865 to the exploration of the Philippines. During this time, he endured extreme hardship, frequent serious illness (malaria and dysentery), and great danger (hostile natives and travel in unseaworthy boats). Nevertheless, it was a period of outstanding achievement, chiefly because of Semper's dogged determination. He acquired magnificent zoological and ethnographic collections on the islands of Luzon, Bohol, Cebu, Leyte, and Mindanao, thus laying a permanent foundation for future research in the Philippines.

After his return to Germany, Semper was appointed *Privatdozent* (1866) at the University of Würzburg and later full professor of zoology and director of the zoological institute (1869). He was an active teacher with numerous students, but in 1887 a stroke left him a semi-invalid and forced him, in 1893, to retire.

Semper's bibliography of ninety titles reveals his broad interests. In addition to his journal articles, in which he reported his Philippine travels, he published two books on geography and ethnology—one dealing with the Philippines, the other with the Palau Islands. Mollusks were his favored group of animals and more than half of his publications are devoted to this group; but he also published on corals, holothurians, pycnogonids, ascidians, annelids, crustaceans, and sharks. His interests included taxonomy, anatomy, histology, and phylogeny. He was especially interested in evolution, and contributed his share to the construction of phylogenetic trees, basing his investigation in part on the (erroneous) belief that the urogenital system of the sharks could be homologized with that of the annelids. This assumption involved him in a heated controversy with Max Fürbringer and others.

The ten-quarto-volume *Reisen im Archipel der Philippinen*, which contains Semper's scientific reports on his collections, is a monument to his industry and determination. It remains, to this day, an important source book, particularly for the Philippine mollusks.

Probably Semper is best known for his two-volume *Die natürlichen Existenzbedingungen der Thiere*, the first textbook on animal ecology. The work was based on a series of twelve lectures given in 1877 at the Lowell Technological Institute in Boston. In this volume on physiological ecology he discussed the influence of the physical and biotic environment on the structure, distribution, and habits of organisms. Food, light and darkness, temperature, water, water currents, and various aspects of symbiosis, parasitism, and predator-prey relations are treated. Semper's emphasis on the living animal was of decisive importance in an age that placed an exaggerated value on the study of morphology.

BIBLIOGRAPHY

I. ORIGINAL WORKS. Semper's works include vols. I (Holothuria), III (Terrestrial Mollusks), V, and VI (Lepidoptera) of *Reisen im Archipel der Philippinen,* 10 vols. (Wiesbaden, 1868–1905) (for a complete collation of the 100 parts in which this work was issued, see R. I. Johnson, in *Journal of the Society for the Bibliography of Natural History,* 5 [1969], 144–147); and *Die natürlichen Existenzbedingungen der Thiere,* 2 vols. (Leipzig, 1880), with English trans., *Animal Life as Affected by the Natural Conditions of Existence* (London, 1880).

II. SECONDARY LITERATURE. On Semper's life and work, see A. Schuberg, "Carl Semper," in *Reisen . . . ,* X (1895), vi–xxi, with complete bibliography.

ERNST MAYR

SEMYONOV-TYAN-SHANSKY, PETR PETROVICH

SEMYONOV-TYAN-SHANSKY, PETR PETROVICH (*b.* near Urusov, Ryazan guberniya, Russia, 14 January 1827; *d.* St. Petersburg, Russia, 11 March 1914), *geography, statistics.*

Semyonov's father, Petr Nikolaevich Semyonov, was a landowner and well-known playwright; his mother, Aleksandra Petrovna Blank, came from a French family that had immigrated to Russia at the end of the seventeenth century. Semyonov was interested in botany and history as a child and was educated at home by private tutors. In 1842 he entered the school for guard cadets at St. Petersburg, from which he graduated with distinction. Three years later he enrolled in the natural sciences section of St. Petersburg University. After graduating in 1848 he was elected a member of the Russian Geographical Society, and by the end of his life he had held honorary memberships in seventy-three Russian and foreign scientific societies and institutions.

In addition to his scientific activity, Semyonov was an expert on Dutch painting and collected 700 pictures and 3,500 prints by Dutch and Flemish masters of the sixteenth and seventeenth centuries; the collection was presented to the Hermitage Museum in 1910. He also published works on seventeenth-century Dutch painting and in 1874 was elected an honorary member of the Academy of Arts in St. Petersburg. His abiding interest in entomology was reflected in his collection of 700,000 specimens, given to the Zoological Museum of the Academy of Sciences.

Semyonov began his geographical research in 1849, when, at the request of the Free Economic Society, he and Nikolai Danilevsky investigated the chernozem zone of European Russia. The botanical material that they collected provided the basis for his master's thesis, defended in 1851, on the flora of the Don basin in relation to the geographical distribution of plants in European Russia. In the same year, at the request of the Russian Geographical Society, he began work on a translation of Karl Ritter's *Die Erdkunde von Asien*, taking into consideration material obtained after 1830. This project stimulated his interest in the then almost unknown Tien Shan. From 1853 to 1855 he lived in Berlin and became acquainted with Ritter.

In preparation for an expedition to Tien Shan, Semyonov attended lectures at Berlin University, studied geography and geology, traveled on foot through the mountainous regions of western Europe and Switzerland, studied volcanic phenomena, and made seventeen ascents of Vesuvius. His acquaintance with Humboldt and study of his scientific work especially influenced his scientific outlook; Humboldt, in turn, enthusiastically encouraged Semyonov to explore Tien Shan.

In 1855 Semyonov returned to Russia, and the following year he published the first volume of his translation of Ritter's *Die Erdkunde von Asien*, devoted to Mongolia, Manchuria, and northern China. Semyonov's extensive additions constituted half the volume, and his edition subsequently acquired the significance of an independent work. In the introduction he emphasized the need for geography to deal with the particulars of nature as they related

to agriculture and discussed the importance of developing a Russian orthography—especially for Chinese place names—and geographical terminology; he himself introduced "upland" (*nagore*), "plateau" (*ploskogore*), "hollow" (*kotlovina*), and "foothills" (*predgore*). In his annotations he corrected Ritter's text and also disputed the author's assertion that the junctions of the Caspian with the Aral and Black seas could have occurred only in prehistoric times. This problem is still unresolved. He also corrected Humboldt and accounted for the Caspian depression "by the gradual drying up of the seas and not by a volcanic collapse, as in the Dead Sea."

In 1856 Semyonov reached Tien Shan. He had traveled through the cities of Ekaterinburg (now Sverdlovsk), Omsk, Barnaul, Semipalatinsk, and the fortress of Verny (now Alma-Ata); and en route he had met the geographers G. N. Potanin and C. C. Valikhanov, and the exiled F. Dostoevsky. From Verny he made two excursions to Lake Issyk Kul and one to Kuldja. In 1857 Semyonov wrote to the Russian Geographical Society: "My second long trip to the Chu River exceeded my expectations. I not only succeeded in crossing the Chu but even in reaching Issyk Kul by this route, that is, by its western extremity, on which no European had yet set foot and which no scientific research of any kind had touched."

In 1857 Semyonov crossed the northern chain of the Terskey Ala-Tau range; discovered the upper reaches of the Naryn River—the main source of the Syr Darya; and climbed along the canyon of the Dzhuuk, having observed the hill and valley topography of the elevated watersheds. He then crossed the Tien Shan to the basin of the Tarim, climbed the Khan Tengri group, and discovered broad glaciers in the upper reaches of the Sazydzhas River. On his return he studied the Trans-Ili Ala-Tau range, Dzungarian Ala-Tau, Lake Alakol, and the Tarbagatai range.

Semyonov's route enabled him to trace the overall configuration of the country and to discover the actual structure of the interior of Asia. He refuted Humboldt's assertion of the volcanic origin of the mountains of central Asia and of the presence of the north-south Bolor (Muztagh Ata) range. His observations provided a basis for refuting Ritter's and Humboldt's assertion that the Chu River arises from Lake Issyk Kul. The river, he discovered, only approaches the lake, which has no outlet and only provisional connections with the Chu along the channel of the Kutemalda. He pointed out the great altitude of the snow line in the mountains of central Asia (from 11,000 to 15,000 feet), convincing Humboldt that "the dryness of the climate elevates the snowline to an unusual extent." He established the existence of extensive glaciation in the Tien Shan, first suggested by Humboldt, and compiled the first orographical scheme of the area. He discussed the tectonics and geological structure, noting the line of east-west elevations and intermontane depressions, in which young sedimentary rock had developed. Semyonov also described the vertical division of the landscape in the mountains of the Trans-Ili Ala-Tau into five zones of vegetation and evaluated their agricultural potential. The vast collection of geological and botanical specimens that he amassed included insects, mollusks, and ethnographical material.

After his return to Russia in 1857 Semyonov became active as a scientific encyclopedist. His thirty-year study of Russian economics and statistics was reflected in publication of the five-volume *Geografichesko-statistichesky slovar* ("Geographical-Statistical Dictionary," 1863–1885), a basic reference work. As head of the Central Statistical Committee from 1864 Semyonov organized important studies and introduced the geographical method into the study of landed property, sowing area, and yields; the material was grouped by districts classified according to their natural and economic features. In 1870 he organized the First All-Russian Statistical Congress. He conducted a classic investigation of the peasant economy in 1880, and the first general census of Russia (1897) was carried out on his initiative. In 1882 Semyonov became a senator, and from 1897 he was a member of the State Council.

In 1860 Semyonov was elected president of the Section of Physical Geography of the Russian Geographical Society, and from 1873 he was vice-president of the Society. He organized many expeditions to central Asia, including those of N. M. Przhevalsky, M. V. Pevtsov, G. N. Potanin, P. K. Kozlov, and V. A. Obruchev, the results of which substantially altered existing ideas about Asia. He actively assisted the expedition of N. N. Miklukho-Maklai and G. Sedov; organized, with Shokalsky, the Kamchatka expedition of the Russian Geographical Society; and assisted scientists in political exile, including Potanin, A. L. Chekanovsky, and I. D. Chersky.

In 1888 Semyonov visited central Asia for the second time, passing through Ashkhabad and Bukhara to Samarkand and Tashkent. He traveled up the valley of the Zeravshan River and climbed the mountains of Turkestan and the Gissar range. His descriptions of the natural history and economy of the area are still of scientific value.

As a popularizer of geographical knowledge, Semyonov wrote and edited many works. He translated and supplemented Ritter's *Die Erdkunde von Asien* with sections on the Altai, Sayan, Baikal, and regions around Lake Baikal. He edited the multivolume *Zhivopisnaya Rossia* ("Scenic Russia") and wrote a three-volume history of the Russian Geographical Society. In 1906, on the fiftieth anniversary of his expedition to Tien Shan, the epithet "Tyan-Shansky" was officially added to the family name.

BIBLIOGRAPHY

I. ORIGINAL WORKS. Semyonov's earlier writings include "Neskolko zametok o granitsakh geologicheskikh formatsy v sredney i yuzhnoy Rossii" ("Some Notes on the Boundaries of Geological Formations in Central and Southern Russia"), in *Geograficheskie izvestiya* (1850), 513–518; *Pridonskaya flora v ee otnosheniakh s geograficheskim raspredeleniem rasteny v Evropeyskoy Rossii* ("Flora of the Don in Relation to the Geographical Distribution of Plants in European Russia"; St. Petersburg, 1851), his master's diss.; "Obozrenie Amura v fiziko-geograficheskom otnoshenii" ("Review of the Amur in Its Physical-Geographical Aspects"), in *Vestnik Russkogo geograficheskogo obshchestva*, no. 15 (1855), 227–255; *Zemlevedenie Azii K. Rittera* (Ritter's *Die Erdkunde von Asien*), translated and supplemented by Semyonov, 5 vols. (St. Petersburg, 1856–1879); "Pervaya poezdka na Tyan-Shan ili Nebesny khrebet do verkhoviev r. Yaksarta ili Syr-Dari v 1857 g." ("First Trip to the Tien Shan or Heavenly Range as Far as the Upper Reaches of the Jaxartes or Syr Darya River in 1857"), in *Vestnik Russkogo geograficheskogo obshchestva*, no. 23 (1858), 7–25; "Zapiska po voprosu ob obmelenii Azovskogo morya" ("Note on . . . the Shallowness of the Sea of Azov"), *ibid.*, no. 30 (1860); and *Geografichesko-statistichesky slovar Rossyskoy imperii* ("Geographical-Statistical Dictionary of the Russian Empire"), 5 vols. (St. Petersburg, 1863–1885).

Subsequent works are "Naselennost Evropeyskoy Rossii v zavisimosti ot prichin, obuslovlivayushchikh raspredelenie naselenia imperii" ("Population of European Russia in Relation to the Conditions That Determine the Distribution of the Population of the Empire"), in *Statistichesky vremennik Rossyskoy imperii*, 2, no. 1 (1871), 125–156; articles in *Statistika pozemelnoy sobstvennosti i naselennykh mest Evropeyskoy Rossii* ("Statistics on Landed Property and Settled Localities of European Russia"), pts. 1–2, 4–5 (St. Petersburg, 1880–1884); "O vozvrashchenii Amu-Dari v Kaspyskom more" ("On the Return of the Amu Darya Into the

Caspian Sea"), in *Moskovskie vedomosti*, nos. 45–46 (1881); "Oblast kraynego severa Evropeyskoy Rossii v eyo sovremennom ekonomicheskom sostoyanii" (". . . the Extreme North of European Russia in Its Present Economic Condition"), in Semyonov, ed., *Zhivopisnaya Rossia* ("Scenic Russia"), I (St. Petersburg–Moscow, 1881), 313–336; "Ozernaya oblast v eyo sovremennon ekonomicheskom sostoyanii" ("The Lake Region in Its Present Economic Condition"), *ibid.*, 817–834; "Ermitazh i kartinnye gallerei Peterburga" ("The Hermitage and Picture Galleries of St. Petersburg"), *ibid.*, 687–720; "Obshchy obzor ekonomicheskogo sostoyania Finlyandii" ("A General Survey of the Economic Conditions of Finland"), *ibid.*, II (St. Petersburg–Moscow, 1882), 119–128; "Belorusskaya oblast v eyo sovremennom ekonomicheskom sostoyanii" ("The Belorussian Region in Its Present Economic Condition"), *ibid.*, III (St. Petersburg–Moscow, 1882), 473–490; "Zapadnaya Sibir v eyo sovremennom ekonomicheskom sostoyanii" ("Western Siberia in Its Present Economic Condition"), *ibid.*, IX (St. Petersburg–Moscow, 1884), 349–370; and "Nebesny khrebet i Zailysky kray" ("The Heavenly Range and the Trans-Ili Region"), *ibid.*, X (St. Petersburg–Moscow, 1885), 333–376.

Among his later works are *Kratkoe rukovodstvo dlya sobirania zhukov ili zhestkokrylykh (Coleoptera) i babochek ili cheshuekrylykh (Lepidoptera)* ("A Short Guide to Collecting Beetles . . . and Butterflies . . ."; St. Petersburg, 1882; 2nd ed., 1893); "Turkestan i Zakaspysky kray v 1888 godu" ("Turkestan and the Transcaspian Region in 1888"); in *Izvestiya Russkogo geograficheskogo obshchestva*, **24**, no. 4 (1888), 289–347; *Istoria poluvekovoy deyatelnosti Russkogo geograficheskogo obshchestva 1845–1895* ("History of a Half-Century . . . of the Russian Geographical Society . . ."), 3 vols. (St. Petersburg, 1896); "Kharakternye vyvody iz pervoy vseobshchey perepisi" ("Characteristic Conclusions From the First General Census"), in *Izvestiya Russkogo geograficheskogo obshchestva*, **33** (1897), 249–270; "Sibir" ("Siberia"), in Semyonov, ed., *Okrainy Rossii. Sibir, Turkestan, Kavkaz, i polyarnaya chast Evropeyskoy Rossii* ("Outlying Districts of Russia. Siberia, Turkestan, the Caucasus, and the Polar Part of European Russia"; St. Petersburg, 1900); "Rastitelny i zhivotny mir" ("The Plant and Animal World"), II, ch. 3; and "Istoricheskie sudby srednerusskoy chernozemnoy oblasti i kulturnye eyo uspekhi" ("The Historical Fate of the Central Russian Chernozem Regions and Their Cultural Progress"), II, ch. 4, written with V. I. Lamansky, in *Rossia* (St. Petersburg, 1899–1914); and *Memuary* ("Memoirs"), 4 vols. (I, III, IV, Petrograd, 1916–1917; II, Leningrad, 1946–1947).

II. SECONDARY LITERATURE. See L. S. Berg, "Petr Petrovich Semyonov-Tyan-Shansky," in his *Ocherki po istorii russkikh geograficheskikh otkryty* ("Sketches in the History of Russian Geographical Discoveries"; Moscow–Leningrad, 1946), 232–272; and *Vsesoyuznoe geograficheskoe obshchestvo za sto let* ("The All-

Union Geographical Society for the Last Hundred Years"; Moscow–Leningrad, 1946), 57–77; V. I. Chernyavsky, *P. P. Semyonov-Tyan-Shansky i ego trudy po geografii* ("Semyonov . . . and His Work in Geography"; Moscow, 1955), 296; A. A. Dostoevsky, "P. P. Semyonov-Tyan-Shansky kak issledovatel, geograf i statistik" (". . . Semyonov . . . as Researcher, Geographer, and Statistician"), in *Pamyati P. P. Semyonov-Tyan-Shansky* ("Recollections of Semyonov . . ."; Petrograd, 1914), 9–22; G. Y. Grumm-Grzhimaylo, "P. P. Semyonov-Tyan-Shansky kak geograf" (". . . Semyonov . . . as Geographer"), in A. A. Dostoevsky, ed., *Petr Petrovich Semyonov-Tyan-Shansky, ego zhizn i deyatelnost* (". . . Semyonov . . ., His Life and Work"; Leningrad, 1928), 161–165; V. I. Lavrov, "Petr Petrovich Semyonov-Tyan-Shansky," in *Lyudi russkoy nauki* ("People of Russian Science"; Moscow, 1962), 460–468; S. I. Ognev, "P. P. Semyonov-Tyan-Shansky," in *Byulleten Moskovskogo obshchestva ispytateley prirody*, Biol. ser., **51**, no. 3 (1946), 122–137; and Y. K. Efremov, "Petr Petrovich Semyonov-Tyan-Shansky kak fiziko-geograf" (". . . Semyonov . . . as Physical Geographer"), in *Otechestvennye fiziko-geografy* ("Native Physical Geographers"; Moscow, 1959), 284–293.

VERA N. FEDCHINA

SENAC, JEAN-BAPTISTE (*b.* near Lombez, Gascony, France, *ca.* 1693; *d.* Paris, France, 20 December 1770), *anatomy, physiology, medicine, chemistry (?).*

Nothing definite is known either of Senac's family or of his early life; nor is it certain where he received the M.D.—although Montpellier and Reims are possibilities. He was elected to the Académie des Sciences in 1723 as an anatomist and in 1741 was made *associé vétéran*, a sign that he was no longer an active member. He served as a doctor with the army and became personal physician to the Maréchal de Saxe. In 1752 he succeeded François Chicoyneau as chief physician to Louis XV and was appointed a councillor of state.

In 1724 Senac published anonymously *L'anatomie d'Heister*, a detailed account of human anatomy and physiology in which he advocated mechanical rather than chemical explanations of bodily functions. Between 1724 and 1729 he wrote several anatomical memoirs, principally on the respiratory organs; and he established that, contrary to popular belief, little water enters the lungs and stomach of the drowned. After twenty years of research on the structure, action, and diseases of the heart, he published in 1749 his most important work, *Traité de la structure du coeur. . . .* He made important new observations on both healthy and diseased hearts, and the book remained authoritative for many years.

Two anonymous medical books have been attributed to Senac. *Traité . . . de la peste* (1744) includes observations made by Chicoyneau and others during the Marseilles plague of 1720; but a treatise on fevers, *De recondita febrium* (1759), is clearly based on the author's personal experience as a physician.

Senac has also been suspected of writing *Nouveau cours de chymie, suivant les principes de Newton et de Sthall* [sic] (1723), a book consisting partly of extracts from the writings of John Freind. It introduced early Newtonian chemical ideas to French readers—although with little immediate impact—and helped to spread knowledge of the phlogiston theory.

BIBLIOGRAPHY

I. ORIGINAL WORKS. There are four publications by Senac in *Histoire et Mémoires de l'Académie Royale des Sciences*; "Sur les organes de la respiration," in *Mémoires* for 1724 (1726), 159–175; "Sur les noyés," in *Histoire* for 1725 (1727), 12–15; "Sur les mouvements des levres," in *Histoire* for 1727 (1729), 13–15; and "Sur le diaphragme," in *Mémoires* for 1729 (1731), 118–134.

The only book with Senac's name on the title page is *Traité de la structure du coeur, de son action, et de ses maladies*, 2 vols. (Paris, 1749; 2nd ed., corrected and enlarged by Senac, with additions by Antoine Portal, 2 vols., Paris, 1774). Eds. dated 1777 and 1783 have also been recorded.

It is certain that Senac wrote *L'anatomie d'Heister, avec des essais de physique sur l'usage de parties du corps humain, & sur le méchanisme de leurs mouvemens* (Paris, 1724; 2nd ed., 1735; 3rd ed., 1753). When discussing drowning (2nd ed., p. 618), the author cites his own publication in the memoirs of the Académie for 1725, and this can refer only to Senac's "Sur les noyés" (see above). In the preface (2nd ed., pp. xii–xiii) Senac stated that the book was based on lessons that he had given to foreign students recommended to him by Freind; he had used Heister's *Compendium anatomicum* (Altdorf, 1717) as a text, but had added much new material, some of which contradicted Heister.

The *Nouveau cours de chymie suivant les principes de Newton et de Sthall* [sic] was attributed to Senac in the sale catalog of E. F. Geoffroy's library (*Catalogus librorum . . . Stephani-Francisci Geoffroy* [Paris, 1731], 94, item 1363). P. J. Macquer described it as "a work of Senac's youth which he has never acknowledged" (Macquer to T. O. Bergman, 22 February 1768, in G. Carlid and J. Nordström, eds., *Torbern Bergman's Foreign Correspondence*, I [Stockholm, 1965], p. 230). It

may be significant that *Nouveau cours* and *L'anatomie d'Heister* had the same publisher, Jacques Vincent, and appeared about the same time, and also that Senac was evidently in touch with Freind (see above), whose writings were utilized by the author of *Nouveau cours*. But in 1778 Eloy (see below) denied that Senac was the author, as did de la Porte and Renauldin (see below), and the attribution cannot be regarded as certain. There is an Italian trans., *Nuovo corso di chimica secondo i principe di Newton e di Sthall*, only the 2nd ed. of which (Venice, 1750) has been located (Wellcome Institute of the History of Medicine Library, London); the censorship license is dated 29 September 1737 and the date of registration is 9 April 1738 (p. 483), so the 1st ed. probably appeared in 1738.

According to Eloy, Senac wrote the anonymous *Traité des causes, des accidens, et de la cure de la peste . . .* (Paris, 1744). He also attributed to Senac *De recondita febrium intermittentium, tum remittentium natura, et de earum curatione . . .* (Amsterdam, 1759; enl. ed., Geneva, 1769); the Amsterdam ed. was translated into English, with a few notes, by Charles Caldwell: *A Treatise on the Hidden Nature, and the Treatment of Intermitting and Remitting Fevers . . . by Jean Senac* (Philadelphia, 1805).

II. SECONDARY LITERATURE. A short account of Senac is in N. F. J. Eloy, *Dictionnaire historique de la médecine, ancienne et moderne*, IV (Mons, 1778), 245–247; further information is given by H. de la Porte and L. J. Renauldin in Michaud's *Biographie universelle*, XLII (Paris, 1825), 1–2, and by G. Degris, *Étude sur Senac, premier médecin de Louis XV* (Paris, 1901).

Senac's work on the heart and its diseases is placed in its historical context by D. Guthrie, "The Evolution of Cardiology," in E. A. Underwood, ed., *Science, Medicine and History, Essays in Honour of Charles Singer*, II (London, 1953), 508–517, esp. 511; and by J. O. Leibowitz, *The History of Coronary Heart Disease* (London, 1970), 75–76. The contents of the *Nouveau cours de chymie . . .* are discussed by J. R. Partington, *History of Chemistry*, III (London, 1962), 58–59; and its importance in the development of Newtonian chemistry is assessed by A. Thackray, *Atoms and Powers* (Cambridge, Mass., 1970), 94–95.

W. A. SMEATON

SÉNARMONT, HENRI HUREAU DE (*b.* Broué, Eure-et-Loire, France, 6 September 1808; *d.* Paris, France, 30 June 1862), *crystallography, mineralogy.*

At Sénarmont's request, no obituaries appeared after his death; thus little is known of his private life. He was the son of Amédée Hureau Sénarmont, a landowner in Badonville in the commune of Broué, and Amélie Rey. After attending the Collège Rollin and the Collège Charlemagne in Paris, he entered the École Polytechnique in 1826.

Three years later he joined the state mining administration as *élève-ingénieur*. In this capacity he visited (1831) the arsenal in Toulouse and was assigned to the steel mills in Rive-de-Gier and Le Creusot. In 1833 he worked temporarily with the engineer Conte, who was carrying out special assignments in the Autun basin. In 1835 he became a mining engineer, second class.

Sénarmont's work brought him to Nantes and Angers in 1835–1836. In the summer of the latter year he was assigned to prepare geological maps of Aube and Seine-et-Oise. (Seine-et-Marne was added in 1837, and the map of Aube was canceled in 1839.) In 1843 he published *Sur la géologie des départements de Seine-et-Oise et Seine-et-Marne*, which appeared in two sections the following year.

From 1840 to 1847 Sénarmont was engaged in inspecting steam engines in the Seine department. He was promoted to mining engineer, first class, in 1841 and then, in 1848, to assistant chief engineer. In 1847 he was made an examiner at the École Polytechnique; but in the same year he was transferred to the École des Mines, where he received the chair of mineralogy. In 1849 Sénarmont joined the editorial staff of the *Annales des mines*. Later he was named dean of students, librarian, and secretary of the council at the École des Mines. In 1852, following the death of Beudant, Sénarmont became a member of the mineralogy section of the Académie des Sciences; he served as vice-president in 1858 and as president in 1859. In 1854 he became an editor of the *Annales de chimie et de physique* and held this post until his death. He also received the chair of physics at the École Polytechnique (1856–1862).

Sénarmont's enthusiasm for crystallography was first awakened by Fresnel. His importance is based on his demonstration of the directional dependency of the physical properties of crystals and on his experiments on the synthesis of minerals under conditions corresponding to those in nature. In his first paper, "Mémoire sur les modifications que la réflexion spéculaire à la surface des corps métalliques imprime à un rayon de lumière polarisée" (1840), Sénarmont introduced, for the purpose of measuring phase differences, a thin layer of mica known as the 1/4 λ plate. In "Mémoire sur la réflexion et la double réfraction de la lumière par les cristaux doués de l'opacité métallique" (1847), he discussed perpendicular incidence on calcite and stated (1) that a rotation of the plane of polarization can be observed and (2) that an analogous rotation of the plane of polarization occurs for re-

flected light on stibnite (antimony sulfide) and is evidence of double refraction. The difficulty of the problems that Sénarmont investigated is evident when one considers that it was only in 1887 that Drude attempted to determine quantitatively the optical properties of stibnite, and not until 1931 that problems of quantitative ore microscopy were clarified by Max Berek.

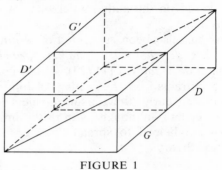

FIGURE 1

In 1850 Sénarmont described his polariscope (Figure 1), a plate composed of four quartz prisms with its upper and lower faces normal to the optic axis (D and D' are optically rotatory clockwise and G and G' counterclockwise):

> If this parallel-sided plate is set at right angles to a beam of polarized, parallel light—that is to say, in such a way that the light is traveling along the optic axis—the plate will be seen to be covered with straight fringes parallel to the refracting edges of the prisms. If the principal section of the analyzer is the same as the initial plane of polarization, the dark central fringe will be situated in the middle of the plate, where the thickness of each of the crossed prisms is the same; hence it will form a straight line along the front and back halves of the plate.[1]

Sénarmont made important contributions to isomorphism in "Recherches sur les propriétés optiques biréfringentes des corps isomorphes" (1851). Besides an exact description of his procedures, he furnished a considerable amount of crystallographic data and described many isomorphic compounds. To determine the character of the axes of optical elasticity, he utilized quartz plates and wedges: "The thickness is always that which is required to produce colors by compensation if the slightly prismatic laminae recommended by M. Biot are employed; and the axis of the plate parallel to the optical axis of the quartz will always be the one of greatest optical elasticity."[2]

In his "Recherches" Sénarmont also provided a method for measuring the binormal angle (optic

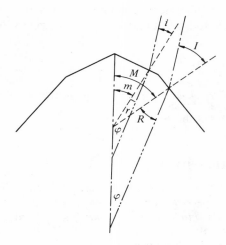

FIGURE 2

axial angle) in convergently polarized light (Figure 2). If $m = r + \varphi = i + \theta$ and $M = R + \varphi = I + \Theta$, it follows that

$$\sin(m - \theta) = l\sin(m - \varphi),\ \sin(M - \Theta) = l\sin(M - \varphi);$$

$$\tan\left(\frac{M + m}{2} - \varphi\right) = \tan\left(\frac{M - m}{2}\right) \frac{\left[\dfrac{M + m}{2} - (\theta + \Theta)\right]}{\left[\dfrac{M - m}{2} + (\theta - \Theta)\right]}.$$

Here, i and I are the corresponding angles of emergence, θ and Θ the apparent half-angles of the optic axes after this emergence, and l the index of refraction. Thus Sénarmont was able to determine the binormal angle of K_2SO_4 without knowing the index of refraction. In this work he also described the dispersion of the optic axes and the shift of the plane of these axes toward red in crystalline solid solutions of potassium Rochelle salt and ammonium Rochelle salt.

In "Recherches sur la double réfraction" (1856), Sénarmont provided a thorough account of all of the phenomena related to parallel and convergent polarized light beams. He also observed carefully the boundary cones of rays in total reflection. In "Sur la réflexion totale de la lumière extérieurement à la surface des cristaux biréfringents" (1856), he developed the formulas for conic sections in optically uniaxial crystals and in certain optically biaxial crystals. Virtually the only fluid refractive medium available to him was carbon disulfide. In "Note sur quelques formules propres à la détermination des trois indices principaux dans les cri-

staux biréfringents" (1857), he applied the method of minimal diffraction to prisms of doubly refractive crystals in order to determine the principal indexes of refraction.

In "Mémoires sur la conductibilité des substances cristallisées pour la chaleur" (1847), Sénarmont demonstrated that thermal conductivity is dependent on crystal symmetry. He bored through the middle of thin crystal plates, waxed them, and through the hole placed a silver tube, which he heated. The swelling of the wax indicated an isothermal corresponding to the melting temperature of the wax. On all the slices of isometric crystals and on all the basal slices of tetragonal and hexagonal crystals, the isotherms were circular; in all other cases they were elliptical. From these patterns of swelling in the wax he recognized the optically positive nature of quartz. In 1848 he showed that, when subjected to unilateral pressure, melting isotropic bodies (like glass) yield isotherms that resemble those that are characteristic of anisotropic materials.

Analogously, Sénarmont established the directional dependence of surface conductivity in crystals in "Mémoire sur la conductibilité superficielle des corps cristallisées pour l'électricité de tension" (1850). He discovered this dependence by wrapping a crystal in tinfoil (which had been grounded) and by placing a metal point on the crystal surface at a point where a circular piece of the tinfoil had been cut out; the point was then connected to the positive conductor of an electrostatic machine, and the whole apparatus was placed under the glass bell of an air pump. At reduced pressure he was able to observe, in the dark, circular or elliptical figures of light, depending on the nature of the crystal. He stated that: "The continuous and silent flux of the electricity of the rarefied air does not, it is true, leave permanent traces; but it does manifest itself in the darkness by a faint light that persists throughout the whole experiment and that makes all its details visible. . . ."[3]

In "Expériences sur la production artificielle de polychromisme dans les substances cristallisées" (1851), Sénarmont was the first to describe the production of artificial pleochroism in strontium nitrate pentahydrate, which had been prepared by saturating the substance with ammoniacal logwood extract and other organic dyestuffs. He also reported on less successful experiments involving rock candy (sugar), Rochelle salt, potassium nitrate, and sodium nitrate.

The syntheses that Sénarmont carried out in the

years 1849–1851 are recounted in "Expériences sur la formation des minéraux par la voie humide dans les gîtes métallifères concrétionnées" (1851) and are an essential contribution to the understanding of mineral formation. Since CO_2, H_2S, alkali salts, sulfides, and carbonates predominate in thermal springs, he assumed that the formation of ore veins from these components would necessarily occur at elevated temperatures and pressures. Thus he placed those components that he wished to have interact in sealed glass tubes, which were inserted into a sealed pipe filled with water. The apparatus was then embedded in coal dust and heated in the gas ovens of the steel mills at Ivry-sur-Seine. Through either double decomposition of a soluble salt with Na_2CO_3 or $CaCO_3$, or precipitation of a soluble salt using alkali carbonate in a supersaturated CO_2 solution, Sénarmont produced magnesite, siderite, rhodochrosite, cobalt carbonate, nickel carbonate, smithsonite, and malachite, as well as barite, fluorite, and quartz in crystalline form.

Sénarmont also synthesized pure silver, copper, arsenic, and hematite. From metallic salts and alkali sulfides he obtained mostly amorphous sulfides—including marcasite, pyrite, manganese sulfide, hauerite, NiS, CO_3S_4, sphalerite, galena, and chalcopyrite; and he obtained realgar, orpiment, stibnite, bismuthinite, arsenopyrite, proustite, and pyrargyrite in crystalline form. He was similarly successful in crystallizing PbS and ZnS in a supersaturated H_2S solution and in obtaining pyrite and chalcopyrite in the form of a granulated powder with metallic luster. If he wished to avoid an immediate reaction, he inserted into the glass tube a thin ampul containing a salt and a gas bubble. When heat was applied the gas bubble burst the ampul. Altogether Sénarmont succeeded in synthesizing twenty-nine vein minerals from the alkali sulfides and carbonates commonly found in thermal springs with metallic salts. In these syntheses the temperature rarely exceeded 350° C. He gave an exact crystallographic description of all the compounds he was able to crystallize. In addition, he described the effect of the solutions employed on the glass tubes, the flaking off of pieces of glass from the tubes, and the danger of explosion involved in the use of sealed tubes (bombs).

In 1851 Sénarmont described for the first time both rhombic antimony bloom (valentinite) from Sensa in Algeria and natural isometric antimony oxide from Minina (near Sensa), thus confirming the dimorphism of Sb_2O_3. Dana gave the name senarmontite to the isometric form of Sb_2O_3.

Sénarmont also demonstrated that common silicon carbide belongs to the isometric system (1856).

NOTES

1. *Annales de chimie et de physique*, 3rd ser., **28** (1850), 281.
2. *Ibid.*, **33** (1851), 401.
3. *Ibid.*, **28** (1850), 261.

BIBLIOGRAPHY

I. ORIGINAL WORKS. For many of Sénarmont's important papers written between 1840 and 1851, see *Annales de chimie et de physique*, 2nd and 3rd ser. The Royal Society *Catalogue of Scientific Papers*, V (1871), 641–643, lists thirty-eight titles by Sénarmont. Poggendorff, curiously, omits most of the papers that appeared in German journals, as well as "Extraits de minéralogie," which appeared in *Annales des mines*, 5th ser., **6–19** (1854–1861). Sénarmont's personal instructions from and reports to the French mining administration are recorded in *Annales des mines*, 3rd ser. (1833–1841) and 4th ser. (1842–1851).

II. SECONDARY LITERATURE. See J. L. F. Bertrand, "Éloge de Sénarmont, lu à la Société des amis des sciences, 16 avril 1863," in *Société de secours des amis des sciences* (1863), 27–56; Walter Fischer, *Gesteins- und Lagerssättenbildung im Wandel der wissenschaftlichen Anachauung* (Stuttgart, 1961); E. Hoppe, *Geschichte der Physik* (Brunswick, 1926); F. von Kobell, *Geschichte der Mineralogie 1650–1860* (Munich, 1864); T. Liebisch, *Physikalische Kristallographie* (Leipzig, 1891); Joseph Michaud, ed., *Biographie universelle ancienne et moderne*, XXXIX (1969), 56; F. Pockels, *Lehrbuch der Kristalloptik* (Leipzig–Berlin, 1906); and *Nouvelle biographie générale*, XLIII (Paris, 1864).

WALTER FISCHER

SENDIVOGIUS (SĘDZIMIR or SĘDZIWÓJ), MICHAEL (*b.* Skorsko or Łukawica, Poland, 2 February 1566; *d.* Cravar, Silesia, June [?] 1636), *alchemy.*

Sendivogius' parents, Jacob Sędzimir and Catherine Pielsz Rogowska, were both of noble families and had a small estate near Nowy Sącz, in the Cracow district. After studying in Italy, Sendivogius entered the University of Leipzig in 1590, moving a year later to the University of Vienna. In 1593 he entered the service of Emperor Rudolf II in Prague as a courier, and in 1594 he also became courier and later secretary to the Polish king Sigismund III; this dual service was made possible by the close friendship of the two rulers and their common enmity to Turkey. Also in 1594 Sendivogius married Veronica Stieber, a wealthy widow. He soon came to Rudolf's attention and participated in his alchemical experiments, becoming his favorite and trusted friend. In Cracow and Prague, and on his many official missions, Sendivogius met prominent political figures and scientists; his friends included the alchemists Joachim Tancke, Oswald Croll, J. Orthel, J. Kapr von Kaprstein, V. Lavinius, R. Egli, Martin Ruland, and Michael Maier.

Sendivogius' name appears in the rolls of the University of Altdorf for 1595—probably as an imperial official, rather than as a student. As mentioned by his biographer Carolides a Carlsperga (1598), he may also have visited the universities of Rostock, Ingolstadt, and Cambridge. In 1597 Sendivogius bought the Fumberg estate, near Prague, from the widow of the English alchemist Edward Kelley. Around this time, and under the influence of the Polish master of heraldry Bartłomiej Paprocki, he changed his family name to the nobler-sounding Sędziwój (latinized as Sendivogius) and began to sign his name Michael Sendivogius Liberbaro de Skorsko et Łukawica. In 1599 he was accused before the municipal court of Prague of being responsible for the death of a friend and fellow alchemist, a rich Bohemian merchant named Louis Koralek, and was sentenced to prison. He was released as a result of the diplomatic intervention of Sigismund III.

At this time Sendivogius' wife and two of their four children died. Offended by the noninterference of Rudolf in the Koralek affair and by his failure to defend him before the arrest, Sendivogius sold his estate at Fumberg and, with his surviving children, Veronica and Christopher, returned to Poland. In 1602 he was recalled to Prague and was appointed imperial privy councillor. In 1605, while on a diplomatic mission to France, Sendivogius was lured to the court of Duke Frederick of Württemberg at Stuttgart. Having claimed in *De lapide philosophorum* (1604) to be the "true possessor" of the "mystery of the philosophical stone," he was soon imprisoned. Sigismund III, Rudolf II, and several German princes intervened on his behalf; alarmed by this, Frederick arranged for Sendivogius to escape and laid the blame on his court alchemist, Heinrich Mühlenfells. Put to torture, Mühlenfells pleaded guilty and was condemned to the gallows.

In 1607 Sendivogius visited Cologne, where he published *Dialogus mercurii . . .,* a kind of satire on alchemy. On his return to Poland he became courtier to Queen Constantia, the second wife of Sigismund III. With crown marshall Mikołaj Wolski he established many smithies and iron and brass foundries in Krzepice, which later became a leading industrial center. His collaboration with Wolski was undoubtedly lucrative, for Sendivogius soon became the owner of several houses in Cracow. In 1615–1616 he visited Johannes Hartmann's laboratory in Marburg. Around 1619 Sendivogius transferred his allegiance to Emperor Ferdinand II, for whom he established lead foundries in Silesia. In 1626 he was appointed privy councillor and in 1631, as compensation for long-unpaid salaries, he received the estates of Cravar and Kounty in Crnow county, Moravia. In ruinous condition following the Thirty Years War, the estates proved to be the source of great expense.

Mysterious and intriguing, Sendivogius was undoubtedly a political double agent. His adventures with the Scottish alchemist Alexander Seton in Saxony seem to be a literary fiction, created years after his death; there is no mention of them in the materials of the Landesarchiv in Dresden. Sendivogius was considered by contemporary and succeeding generations of alchemists to be the true possessor of the "great mystery" and a member of the Rosicrucians. He was greatly admired and was frequently cited in seventeenth- and eighteenth-century alchemical treatises.

His alchemical writings had no influence on the development of chemistry, but his treatise *De lapide philosophorum* is of great value for the history of science. Besides recipes for the philosophers' stone, it contains interesting notes concerning the components of air. Sendivogius believed that the air contained a hidden life-giving and fire-supporting agent, the "invisible niter" or "philosophical saltpeter"—the food of life without which nothing could live or grow. This "invisible niter" was born in the rays of the sun and moon, from which it flowed down to the earth in rain or dew. During rainstorms the "niter" passed from the air into the earth, combining with its constituents to form "saltpeter." This process, Sendivogius maintained, occurred continuously and in plain view, although no one noticed or understood it. This argument contains the first idea of the existence of oxygen.

An exponent of the then fashionable Hermetic philosophy, Sendivogius was greatly influenced by Paracelsus and Alexander von Suchten. The *ignis naturae* or *balsamum vitae* of Paracelsus and the *spiritus vitae* of Suchten were redefined by Sendivogius as "philosophical niter"—with the important added explanation of its role in nature. His views on the fire-supporting and life-giving "saltpeter" apparently were derived from the current belief that fires in the salt mines at Wieliczka, near Cracow, were caused by the abundance of saltpeter in the air.

Several seventeenth-century alchemists, including Bathurst and Mayow, investigated Sendivogius' recipes for the philosophers' stone and as a first step searched for the "philosophical niter" in the air.

BIBLIOGRAPHY

I. ORIGINAL WORKS. Sendivogius' main writings are *De lapide philosophorum tractatus duodecim e naturae fonte et manuali experientia depromti* (n.p., 1604), since Jean Beguin's ed. (Paris, 1608) also known as *Cosmopolitani novum lumen chymicum; Dialogus mercurii, alchimistae et naturae* (Cologne, 1607); and *Tractatus de sulphure altero naturae principio* (Cologne, 1616). All three treatises are known in over eighty eds. and have been translated into German, English, French, Russian, and Polish. Sendivogius' *Processus super centrum universi seu sal centrale* was edited by Johann Becher in his *Chymischer Glückshafen* (Frankfurt, 1682), 231–240.

The fifty-five letters allegedly by Sendivogius, first published by J. Manget in *Bibliotheca chemica curiosa*, II (Geneva, 1702), 493 ff. and also published in German, were written in the second half of the seventeenth century, long after his death.

Bibliographies of Sendivogius' works are in the following (listed chronologically); John Ferguson, *Bibliotheca chemica*, II (Glasgow, 1906), 364–370; C. Zibrt, *Bibliographie česke historia*, III (Prague, 1906), 523–526; K. Estreicher, *Bibliografia polska*, XXVII (Cracow, 1929), 332–342; and R. Bugaj, *Michał Sędziwój* (Wrocław–Warsaw–Cracow, 1968), 280–304.

II. SECONDARY LITERATURE. Early biographies of Sendivogius include Georgius Carolides a Carlsperga, *Praecepta institutionis* (Prague, 1598); Bartłomiej Paprocki, *Jina czastka, nove kratochwile* (Prague, 1598); and J. Chorinus, *Illustris foeminae D. Dn. Veronicae Stiberiae* (Prague, 1604). Later biographies (listed chronologically) are P. N. Lenglet Dufresnoy, *Histoire de la philosophie hermétique*, I (The Hague, 1742), 332–333; A. Batowski, "List Poliarka Micigna," in *Rozmaitości— Gazeta Lwowska*, 19 (1858), 153–156; W. Szymanowski, "Michał Sędziwój," in *Tygodnik ilustrowany*, 5 (1862), 181–218; M. Wiszniewski, *Bakona metoda tłumaczenia natury* (Warsaw, 1876), 130–136; J. Brincken, "O życiu i pismach Michała Sędziwoja," in *Biblioteka Warszawska*, 2 (1846), 479–506; J. Svatek, *Culturhistorische Bilder aus Böhmen* (Vienna, 1879), 78–84; J. Read, *Humour and Humanism in Chemistry* (London, 1947), 52–65; and R. Bugaj, *W poszukiwaniu kamienia filozoficznego* (Warsaw, 1957). These works,

however, have little value in terms of history or the history of science.

Archival materials concerning the life of Sendivogius are presented in the following (cited chronologically): B. Peška, "Praski meštan a polsky alchymista," in *Svêtozor*, VI (1872), 471–495; C. R. Elvert, "Der Alchemist Sendivogius," in *Notizenblatt der k. und k. Mährischschlesischen Gesellschaft zur Beförderung des Ackerbaues*, Hist.-stat. Kl., 12 (1883), 20–22; J. Zukal, "Alchymista Michael Sendivoj," *Vestnik matice opavske*, 3 (1909), no. 17; O. Zachar, *Z dejin alchymie v zemich ceskych* (Kladno, 1910); W. Hubicki, "Michael Sendivogius's Theory, Its Origin and Significance in the History of Chemistry," in *Proceedings of the Tenth International Congress on the History of Science, Ithaca, 1962* (Paris, 1964), II, 829–833; "The True Life of Michael Sendivogius," in *Actes du XI Congrès international d'histoire des sciences*, IV (Warsaw, 1965), 31–35; and "Zapomniana teoria," in *Problemy*, 22 (1966), 98–103; and *Ossolineum PAN*, IV (Wrocław–Warsaw–Cracow, 1968), 41–45.

WŁODZIMIERZ HUBICKI

SENEBIER, JEAN (*b.* Geneva, Switzerland, 6 May 1742; *d.* Geneva, 22 July 1809), *physiology.*

Although Senebier, the son of Jean-Antoine Senebier, a merchant, and Marie Tessier, was interested in natural history, his family intended him to be a minister. After having presented a distinguished thesis on polygamy, he was ordained pastor of the Protestant church of Geneva in 1765. He then spent a year in Paris, where he became acquainted with more people in the scientific and theatrical worlds than in the church. In 1770 he published *Contes moraux* and became friends with Abraham Trembley, who influenced the young Protestant minister profoundly.

Charles Bonnet encouraged Senebier to work in the natural sciences and enabled him to perform his first experiments in plant physiology. Following Bonnet's advice, in 1768 Senebier answered a question on the art of observing posed by the Netherlands Society of Sciences at Haarlem. It received an honorable mention and was published in 1772. He became pastor of Chancy, near Geneva, in 1769 but four years later resigned his post to become librarian for the Republic of Geneva. In 1777 Senebier published the first volume of Spallanzani's *Opuscules de physique animale et végétale,* thereby introducing French readers to his work. He later translated most of the works of Spallanzani, with whom he maintained close contact. In 1779 Senebier began to publish his *Action de la lumière sur la végétation,* the study of photosynthesis that established his reputation as a physiologist. The first edition of this voluminous *Traité de physiologie végétale* appeared in 1800.

Senebier became the center of a group of young life scientists, and he taught them all: Pierre Huber (1777–1840); A.-P. de Candolle; Jean-Antoine Colladon (1758–1830), the pharmacist whom many consider to be one of Mendel's precursors; and N. T. de Saussure, who started photochemistry. He was particularly close to François Huber (1750–1831), with whom he conducted experiments on bees and published *Influence de l'air . . . dans la germination* (1801).

Two of Senebier's publications can still repay attention today: his research on photosynthesis and his works on the experimental method, which he defined with precision fifty years before Claude Bernard.

In several works, but especially in *Expériences sur l'action de la lumière solaire dans la végétation* (1788), Senebier paid particular attention to the gas exchanges of green plants exposed to light. He was the first to observe that in sunlight such plants absorb carbonic acid gas and emit oxygen while manufacturing a substance with a carbon base.

As an extension of his communication on the "art of observing" of 1769, Senebier published the two-volume *Art d'observer* in 1775. The three-volume *Essai sur l'art d'observer et de faire des expériences* (1802) sums up the fundamental theses of the experimental method. This work is impressive for the closeness of Senebier's thought to that of Claude Bernard and for the degree to which the ideas expressed in Bernard's *Introduction à l'étude de la médecine expérimentale* were formulated in the work of Senebier.

BIBLIOGRAPHY

I. ORIGINAL WORKS. Senebier's works include *Art d'observer* (Geneva, 1775); *Expériences sur l'action de la lumière solaire dans la végétation* (Geneva, 1788); *Physiologie végétale,* 5 vols. (Geneva, 1800); and *Essai sur l'art d'observer et de faire des expériences* (Geneva, 1802).

II. SECONDARY LITERATURE. See J. Briquet, "Bibliographie des botanistes à Genève," in *Bulletin de la Société botanique suisse,* **50** (1940), 433; J. P. Maunoir, *Éloge historique de M. Jean Senebier* (Geneva, 1810); P. E. Pilet, "Jean Senebier, un des précurseurs de Claude Bernard," in *Archives internationales d'histoire des sciences,* **15** (1962), 303–313; P. Revilliod, *Physiciens et naturalistes genevois* (Geneva, 1942), 48.

P. E. PILET

SENECA, LUCIUS ANNAEUS (*b.* Córdoba, Spain, *ca.* 4 B.C.–A.D. 1; *d.* near Rome, April A.D. 65), *physical science.*

Seneca came from a distinguished provincial family of Italian origin; his father, for whom he was named, wrote on history and rhetoric. The younger Seneca was educated at Rome and then for a time devoted himself to philosophy, particularly to the teaching of the eclectic Sextians and the Stoics. Ethics was his main concern; but his interests extended to physics, for in his youth he produced a book, now lost, on earthquakes. In accordance with his father's wishes he entered politics, beginning his senatorial career soon after A.D. 31 with the post of quaestor. During the next ten years he became established as one of Rome's leading orators and writers and won influential friends within the imperial family. In A.D. 41 he was implicated in a court intrigue and banished to Corsica, a grave setback to his career; but eight years later his fortunes were restored when Agrippina, wife of the Emperor Claudius, recalled him and appointed him tutor to her son Nero. In A.D. 54 Nero, then aged sixteen, became emperor; and for the next eight years he governed with the assistance of Seneca and Sextus Afranius Burrus, the commander of the Praetorian Guard. Toward the end of this period of generally sound government Nero turned to different, less scrupulous advisers, so that when Burrus died in A.D. 62, Seneca withdrew from the court, his influence with Nero at an end. Three years later he was accused of involvement in the abortive Pisonian conspiracy against Nero. The evidence against him was weak, but Nero ordered him to commit suicide.

Writing and philosophy occupied Seneca's leisure throughout his life. The tragedies and ethical works have always been his best-known writings, but also extant is one of his scientific books, *Naturales quaestiones,* written around A.D. 62. It is typical of Roman scientific writing, a popularizing work largely derived from Greek sources. Seneca shows the eclectic's independence in choosing between rival theories, but he has no original ideas to contribute. He writes of the need for further careful investigation of natural phenomena but did not conduct any fresh research, although casual observation did provide him with some valuable new information.

The extant part of the *Naturales quaestiones,* which has survived incomplete, deals with meteorological phenomena, rivers, earthquakes, meteors, and comets, topics that all belonged to "meteorology" in the ancient sense. Apart from Aristotle's

Meteorologica it is the longest extant ancient work on the subject; hence it is the main source for the history of Greek meteorology after Aristotle, since it draws heavily on Greek sources and mentions the theories of many individuals whose works are lost. Admittedly Seneca had little interest in the historical development of the subject; knowing few of his predecessors' works at first hand, he sometimes misunderstood or oversimplified their ideas and did not always sharply distinguish his own interpretations and comments. Furthermore, his characteristically terse and brilliant prose lacked the clarity and precision of expression needed for scientific writing. Yet despite these limitations the work greatly enlarges our knowledge of Greek meteorology after Aristotle.

The *Naturales quaestiones* owes more to the meteorological works of Posidonius than to any other single source, although the loss of these works prevents the extent of the debt from being known in detail. Posidonius had followed Aristotle closely, although he placed Aristotelian theories in the context of his own world system, a modification of the Stoic one. The main features of Seneca's world view were probably Posidonian. He thought that the stars and planets are nourished by vapors given off from the earth. An innate energy possessed by air, and the Aristotelian exhalations, account for most events in the atmosphere. (Aristotle had attributed most meteorological phenomena to the activity of moist and dry "exhalations" emitted from the earth's surface, roughly equivalent to water vapor and radiated heat.) To explain earthquakes and rivers, Seneca assumed that the earth is like a living creature, permeated by channels for water and air analogous to veins and arteries. But he disagreed with Posidonius and Aristotle about the nature of comets, effectively criticizing their theory that these are a variety of meteor and using his own observations of the comets that appeared in A.D. 54 and 60 to support the view that they are heavenly bodies like planets, with regular orbits. Based on good evidence and well argued, this part of the *Naturales quaestiones* is in sharp contrast with the rest, which, like most Greek meteorology, abounds in untested speculation and analogy.

Certain broader issues also interested Seneca. As a Stoic he rejected Epicurean physics, particularly the atomic theory of matter, and the denial that the world was created and ordered by a rational God. Like most Stoics he accepted the principles of astrology and divination, and attempted to answer some of the skeptical arguments against

them. The problem of relating science to moral life was of especial importance to him, for almost a third of the *Naturales quaestiones* is about ethical and theological subjects: Seneca thought that through the rational investigation of the universe, men may learn what their attitude toward the material world should be and may reach a true awareness of God's nature, free from all superstition.

After the immediate popularity enjoyed by all Seneca's writings, the scientific works were little read in the ancient world and never became established textbooks. The *Naturales quaestiones* survived the Middle Ages, contributing to the rediscovery of ancient science in Western Europe during the twelfth century, and was still read as a scientific work during the Renaissance. Today it gives an instructive picture of the state of Roman science in the first century A.D., and of the history of Greek meteorology, has considerable literary interest, and illuminates our knowledge of Seneca himself.

BIBLIOGRAPHY

I. ORIGINAL WORKS. Modern eds. of the *Naturales quaestiones* are by A. Gercke (Leipzig 1907; repr. Stuttgart, 1970); P. Oltramare, with French trans. and notes (Paris, 1929; repr. 1961); and T. H. Corcoran, with English trans., 2 vols. (London–Cambridge, Mass., 1971–1972). For recent eds. of other works, see Motto's bibliography (cited below).

II. SECONDARY LITERATURE. Bibliographies are by W. Schaub, of works since 1900 relating to the *Naturales quaestiones*, in the 1970 repr. of Gercke's ed., pp. xlvii–lxi; and by A. L. Motto, of works on all of Seneca's prose since 1940, in *Classical World*, **54** (1960–1961), 13–18, 37–48, 70–71, 111–112; **64** (1970–1971), 141–158, 177–186, 191. A few of the works are O. Gilbert, *Die meteorologischen Theorien des griechischen Altertums* (Leipzig, 1907; repr. Hildesheim, 1967); R. Waltz, *Vie de Sénèque* (Paris, 1909); and G. Stahl, "Die *Naturales Quaestiones* Senecas. Ein Beitrag zum Spiritualisierungsprozess der römischen Stoa," in *Hermes,* **92** (1964), 425–454. The following articles in Pauly-Wissowa, *Real-Encyclopädie der classischen Altertumswissenschaft* refer to Seneca: W. Gundel, "Kometen," XI, 1143–1193; W. Capelle, "Erdbebenforschung," supp. IV, 344–374; and "Meteorologie," supp. VI, 315–358; A. Rehm, "Nilschwelle," XVII, pt. 1, 571–590; and R. Böker and H. Gundel, "Winde," 2nd ser., VIIIa, pt. 2, 2211–2387.

H. M. HINE

SENNERT, DANIEL (*b.* Breslau, Germany [now Wrocław, Poland], 25 November 1572; *d.* Wittenberg, Germany, 21 July 1637), *medicine, chemistry*.

Sennert was the son of a shoemaker, Nicolaus Sennert, and Catharina Helmania, both of whom came from Silesia (from Lähn and Zopten, respectively). After attending the schools in his native Breslau, Sennert enrolled at the University of Wittenberg on 6 June 1593. He followed the basic course of study in the philosophy faculty and was awarded the master's degree on 5 April 1598. He originally intended to become a teacher, but studied medicine instead for three years at the universities of Leipzig, Jena, and Frankfurt an der Oder. In 1601 he entered medical practice in Berlin under the supervision of the physician Johann Georg Magnus. After a short stay at the University of Basel, he obtained the doctor of medicine degree from Wittenberg on 10 September 1601. Altering his plans immediately to open a medical practice, he successfully sought the professorship of medicine at Wittenberg, which he was named to on 15 September 1602 and held until his death.

In 1603 Sennert married Margarethe Schatt of Wittenberg, by whom he had seven children. One of them, Andreas, became a famous Orientalist. Margarethe died in 1622, and in 1624 Sennert married Helene Burenius of Dresden. Following her death, Sennert married, in 1633, Margarethe Kramer of the principality of Sachsen-Altenburg. The last two marriages were childless. Sennert died from what was diagnosed as "plague," after he himself had survived six epidemics of the disease while serving as physician in Wittenberg.

Sennert was a well-known teacher, physician, and scientific writer. His medical ideas have never received a thorough treatment, although his publishing activity began[1] in 1611 with the lengthy *Institutionum medicinae libri V* and continued with other large works on medicine, such as *De febribus libri IV* of 1619 and *Practicae medicinae*, which appeared successively in six books between 1628 and 1636. Closely related to these works were others in which he dealt in detail with *chymia*, a subject that had been placed in the service of medicine by Paracelsus. This group began with the revealingly titled *De chymicorum cum Aristotelicis et Galenicis consensu ac dissensu liber I* (Wittenberg, 1619). Sennert's views are generally difficult to judge, because he attempted to reconcile the theories of Aristotle, Galen, Paracelsus, and the supporters of the traditional atomic hypotheses. For this reason Lasswitz judged Sennert too one-sidedly from the point of view of atomism, while Thorndike and Partington gave equally dis-

torted assessments, but from the standpoints, respectively, of magic and modern chemistry. Even Ramsauer's attempt to describe Sennert's amalgamation of the various doctrines remained incomplete, in part because he lacked sufficient knowledge of the requisite languages.[2]

In medicine, Sennert defended Galen's humoral pathology. From the three Paracelsian principles, sulfur, salt, and mercury, he derived, like Jean Béguin, properties suitable for the chemical treatment of disease. Sennert thought that sulfur, which he considered to be the principle of burning, "phlogiston" (1619), was responsible for the heat of the heart. Salt supposedly served as the radical of the liver, and mercury as the spiritual principle of the brain. Sennert assumed that these three principles cannot be isolated; otherwise, their medicinal properties would be destroyed—an opinion still held in 1662 by Robert Boyle with respect to chemical action.[3] Sennert broke with Paracelsus, however, in rejecting the macroscopic influence of the celestial bodies. He also rejected Paracelsus' view that diseases are caused by the *ens deale*, *ens astrale*, *ens naturale*, *ens spirituale*, and *ens veneni*.[4] Further, unlike Paracelsus, Sennert thought that the *tria prima* are actually *prima mista* (first combinations), specifically, that they are composed of the four Peripatetic elements. He agreed with Paracelsus, however, that all natural bodies contain a *vis seminalis* that bestows life on them (and that, for example, causes the growth of metals). He joined this seminal force in a Neoplatonic fashion with the logos to produce a *principium plasticum* (formative principle).

From the above, it is obvious how strongly rooted the academic physician Sennert was in the ancient tradition. Even where he conceived from a corpuscular point of view the constituents joined together under the form, he retained the Peripatetic notion of form (which is independent of celestial influence).

In the theory proper of change in natural processes—which includes what we call "chemistry"—Sennert followed to some extent the views of the physicians Jean Fernel and Ibn Sīnā, according to whom the forms of the constituents persist under the new *forma superior* of the *mixtum* (for example, of the chemical compound) (1629). In 1619 Sennert, still following Ibn Rushd in this respect, held that the parts of the form persist and therefore that the form persists as such individually: *eadem numero*. Notwithstanding, in order to allow the new natural body (*mixtum*) to come into being, Sennert posited the existence of an *impetus* (*natura*, όρμή), which functions as a causal force. This is in addition to the *spiritus architectonicus*, which first creates the new forms, but then also the occult forms (or properties, as stated in the writings of Fernel). Sennert also retained until at least 1636 the notion he took from Albertus Magnus and Jacopo Zabarella of a multiplication of the forms responsible for the properties.

The fact that Sennert, like Ibn Sīnā, assumed the persistence of the forms of the constituents was made easy for him by notions dominant in the medical tradition since the time of Galen (for example, in the interpretation of the analysis of milk). To be sure, Sennert did not think it necessary to prove this theory through an actual separation of the constituents. This is evident from his theory of the elements. On the basis of J. C. Scaliger's principle, as *in essendo*, so *in causando*, Sennert inferred, in a direct, *a posteriori* manner, the existence of the four Peripatetic elements and the three Paracelsian principles from the perceptible properties of substances. For Sennert, as for Aristotle, this reasoning sufficed as a demonstration of the existence of the elementary constituents. The constituents must, of course, already be present in the body *potentia*, that is to say, with respect to their possible effect. Otherwise the formations of the new body (*mixtum*) would not be a qualitative change (*alteratio*) but a destruction or new creation. Combustibility, for example, demonstrates that the principle "sulfur" does have an effect. Sennert held that it was the concern of the wise, not of the common people, to discover these basic conceptions (the elements and the *tria prima*).

Sennert's fundamental principle of analysis was: *ex iis corpora naturalia constant, in quae resolvuntur* ("natural bodies consist of that into which they are decomposed"). He also correctly inverted this proposition, which, incidentally, he attributed to Hippocrates. He felt that he was faithful to the medical tradition not only in adopting this proposition, but also in accepting Galen's definition of the elements as very small particles (*minima*). In order to arrive at the elements, he added—true to the Aristotelian tradition—the necessity of the *ultima resolutio* (1635). In his view, the *prima mista* combined from the elements are the direct causes of the perceptible properties (*prima qualitates*: color, taste, smell). Thus, sulfur gives rise to the ability to burn and to smell, and salt to crystallization and to taste. Sennert was not sure what properties to assign to mercury.

Sennert held that the mechanism of reaction consisted of two stages: (1) the bodies split up into *minimae particulae*; (2) they then move about and reform as a new body. Although Sennert referred to a corresponding definition of Scaliger's, Galen had already put forth the idea of the comminution and reciprocal action of the fragments. The end product is no mere assemblage of the particles; rather, its properties are determined by the *natura quinta* associated with the form. In 1619 Sennert named this comminution and recombination *diacrisis* and *syncrisis*, respectively. (For Jungius' different terminology, see the article on him in this Dictionary.) A novel element in this account is that Sennert, an Aristotelian, went beyond the concepts of *actio* (influence) and *passio* (being influenced by) and raised the question of a *re-actio* and *re-passio*, perhaps on grounds of symmetry.

Sennert made use of an atomic theory properly so-called only in certain cases where he wished to show the persistence of the nature of a substance (1629). Specifically, he adopted this point of view in discussing distillation, sublimation, coagulation, the melting of the gold-silver alloy, and the solution of silver in nitric acid and of common salt in water. It was not until 1636, however, that he spoke of the unalterable persistence of the particles themselves.

Between 1611 and 1636 Sennert developed a compromise atomic hypothesis starting from the assumptions of the Averroistic school of Padua.

At first sight it appears surprising that in 1624 Sennert argued (falsely) for a *transmutatio* (that is, genuine transformation) of iron into copper during precipitation. He held that since many authors had confirmed it, it was a waste of time to continue the dispute.

For Sennert (as for A. G. Billich) chemistry is not an auxiliary art, but has its own inner goal (*finis internus*, 1629): to decompose natural substances and to prepare them for use. It thus possesses an independent character. Its external goals are healing and the metamorphosis of metals. An academic physician, Sennert incorporated the craft *chymia* into scientific medicine. First one conducts chemical experiments; the reasons that account for them will not fail to be discovered. In 1629 he stated that both medicine and chemistry must, because they are parts of a single knowledge of nature, seek to discover natural laws in such a way that the doctrine of nature (*Physica*) supplies *chymia* with the theory of natural principles, whereas *chymia* furnishes *Physica* with the experience acquired from chemical operations.

NOTES

1. Robert Multhauf, *The Origins of Chemistry* (London, 1966), 265, n. 27; and H. Kangro, *Joachim Jungius' Experiment und Gedanken zur Begründung der Chemie als Wissenschaft* (Wiesbaden, 1968). R. Hooykaas devotes a chapter to Sennert's chemical procedures in *Het begrip element in zijn historisch-wijsgeerige ontwikkeling* (Utrecht, 1933), 160–167.
2. Kangro, *op. cit.*, 144.
3. R. Boyle, *The Sceptical Chymist*, conclusion.
4. Multhauf, *op. cit.*, 265, notes that these *entia* are five in number, as are the eternal substances of al-Rāzī. It may be added that the notion of the *natura quinta* appeared in the third century B.C. and that Étienne de Clave (1624) employed five principles: *tria prima, phlegma, caput mortuum*.

BIBLIOGRAPHY

I. Original Works. There does not exist a complete bibliography of Sennert's works; some are listed in the catalogs of the Bibliothèque Nationale, Paris, and of the Library of the British Museum. See also J. R. Partington, *A History of Chemistry*, II (London–New York, 1961), 271–272; H. Kangro, *Joachim Jungius' Experimente und Gedanken zur Begründung der Chemie als Wissenschaft* (Wiesbaden, 1968), 407; the titles given in *Biographisches Lexikon der hervorragenden Ärzte aller Zeiten und Völker*, A. Hirsch, E. Gurlt, and A. Wernich eds., 2nd ed., V (Berlin–Vienna, 1934), 230, are not very exact.

The main works are *Epitome naturalis scientiae, comprehensa disputationibus viginti sex, in . . . Academia Witebergensi . . . propositis a M. Daniele Sennerto* (Wittenberg, 1600) (I have not seen this); *Institutionum medicinae libri V.* (Wittenberg, 1611); *Disputatio medicina, qua suam de occultis, seu totius substantiae quas vocant morbis sententiam defendit D. Sennertus* ([Wittenberg], 1616); *Epitome naturalis scientiae* (Wittenberg, 1618), including atomism, most probably for the first time and so different from the disputations in 1600; *De chymicorum cum Aristotelicis et Galenicis consensu ac dissensu liber I* (Wittenberg, 1619); *De febribus libri IV* (Wittenberg, 1619); *De scorbuto tractatus* (Wittenberg, 1624); *De dysenteria tractatus* (Wittenberg, 1626); *Disputatio physica de gustu et tactu* (Wittenberg, 1626); *Disputatio physica de auditu et olefactu* (Wittenberg, 1626); *Practicae medicinae, liber I, II, III, IV, V, VI* (appeared at different places from 1628 till 1636); *De chymicorum cum Aristotelicis et Galenicis consensu ac dissensu liber: cui accessit appendix de constitutione chymiae* (Wittenberg, 1629), very different from the work ed. in 1619; *De arthritide tractatus* (Wittenberg, 1631); *Epitome institutionum medicinae* (Wittenberg, 1631); *Epitome librorum de febribus* (Wittenberg, 1634); *Hypomnemata Physica* (Frankfurt am Main, 1636); *Paralipomena cum praemissâ methodo discendi medicinam, tractatus posthumus* (Wittenberg, 1642); Sennertus (Daniel), Culpeper (Nicholas), Cole (Abdiah), *Thirteen Books on Natural Philosophy* (London, 1660).

A work of unknown authorship (see Kangro [1968], pp. 126–127, note 178) is *Auctarium epitomes physicae . . . Danielis Sennerti* (edited simultaneously at Wittenberg and Hamburg, 1635).

Moreover Sennert has written many *Disputationes*, which still wait to be revealed among the occasional writings of various authors bound together in volumes of the seventeenth century.

II. Secondary Literature. No full presentation of Sennert's life and work is known. Biographical details are included in the works of August Buchner, *Dissertationum Academicarum volumen II* (Wittenberg, 1651); J. Graetzer, *Lebensbilder hervorragender schlesischer Ärzte* (Breslau, 1889); and *Allgemeine Deutsche Biographie*, R. V. Liliencron and F. X. von Wegele, eds., published by the Historische Commission bei der Königlichen Akademie der Wissenschaften (at Munich), XXXIV (Munich–Leipzig, 1892), 34–35.

On chemical views of Sennert, consult Reijer Hooykaas (1933: see note 1), 160–167; J. F. Partington, *A History of Chemistry*, XXIII (London, 1961), 271–276; Robert P. Multhauf (1966: see note 1), *passim*; Hans Kangro (1968: see note 1), *passim*; Allen G. Debus, "Guintherius, Libavius and Sennert: The Chemical Compromise in Early Modern Medicine," in *Science, Medicine and Society in the Renaissance*, A. G. Debus, ed. (New York, 1972), 157–165, and "The Paracelsians and the Chemists: the Chemical Dilemma in Renaissance Medicine," in *Clio medica*, 7 (1972), 195. Sennert's atomism is particularly treated by Kurd Lasswitz, "Die Erneuerung der Atomistik in Deutschland durch Daniel Sennert und sein Zusammenhang mit Asklepiades von Bithynien," in *Vierteljahrsschrift für wissenschaftliche Philosophie*, 3 (1879), 408–434; idem, in *Geschichte der Atomistik vom Mittelalter bis Newton*, 2 vols. (Hamburg–Leipzig, 1890), vol. I, 436–454 and *passim* in vols. I and II; and Rembert Ramsauer, *Die Atomistik des Daniel Sennert* (Kiel, 1935), the last not always reliable.

A few thoughts extracted from Sennert's large medical work are briefly dealt with by Walter Pagel, "Daniel Sennert's Critical Defence of Paracelsus," in *Paracelsus, An Introduction to Philosophical Medicine in the Era of Renaissance* (Basel–New York, 1958), 333–343; "William Harvey Revisited," in *History of Science*, 8 (1969), 10, and 9 (1970), 7, 11–20; and Peter H. Niebyl, "Sennert, van Helmont, and Medical Ontology," in *Bulletin of the History of Medicine*, 45 (1971), 115–137.

On older sources dealing with views of Sennert, see Ferguson.

Hans Kangro

SERENUS (*b.* Antinoupolis, Egypt, *fl.* fourth century A.D. [?]), *mathematics*.

Serenus was the author of two treatises on conic sections, *On the Section of a Cylinder* and *On the Section of a Cone,* which have survived, and a commentary on the *Conics* of Apollonius, which has not. From a subscription in a later hand to the Vatican archetype of the first-named work and from the title of the second as given in a Paris manuscript from Mount Athos, Serenus' birth can be placed at Antinoupolis, a city founded by Hadrian in A.D. 122. This birthplace gives an upper limit for his date. As Serenus reckoned Apollonius among the "ancient" writers on conics, and used two lemmas proved by Pappus to transform certain unequal proportions,[1] he is generally thought to have flourished in the fourth century. Certainly his surviving works belonged to an age when Greek geometry had passed its creative phase.[2]

On the Section of a Cylinder, dedicated to an otherwise unknown Cyrus, consists of an introduction, eight definitions, and thirty-three propositions. It counters what is said to have been a prevalent belief—that the curve formed by the oblique section of a cylinder differs from the curve formed by the oblique section of a cone known as the ellipse. In the final five propositions Serenus defended a friend Peithon, who, not satisfied with Euclid's treatment, had defined parallels to be such lines as are cast on a wall or a roof by a pillar with a light behind it. Even in the decline of Greek mathematics this description had been a source of amusement to Peithon's contemporaries.[3]

On the Section of a Cone, also dedicated to Cyrus, consists of an introduction and sixty-nine propositions. It deals mainly with the areas of triangular sections of right or scalene cones made by planes passing through the vertex. Serenus specified the conditions for which the area of a triangle in a certain class is a maximum, those for which two triangles in a particular class may be equal, and so on. In some instances he also evaluated areas.

Serenus himself bore witness to his lost commentary on Apollonius.[4] Certain manuscripts of Theon of Smyrna preserve a fragment that may have come from that work or from a separate collection of lemmas. It is introduced with the words, "From Serenus the philosopher out of the lemmas," and it lays down that if a number of rectilineal angles be subtended at a point on the diameter of a circle (not being its center) by equal arcs of the circle, an angle nearer the center is always less than an angle farther away; this is applied to angles subtended at the center of the ecliptic by equal arcs of the eccentric circle of the sun.

NOTES

1. *De sectione coni*, Prop. XIX, in J. L. Heiberg, *Sereni Antinoensis opuscula,* pp. 160.15–162.11; Pappus, *Collectio* VII. 45 and 47, in F. Hultsch, *Pappi Alexandrini Collectionis quae supersunt,* II (Berlin, 1877), pp. 684.20–686.4, 686.15–27.
2. Halley's beliefs (*Apollonii Pergaei Conicorum libri octo,* Praefatio *ad finem*) that Serenus was born at Antissa in Lesbos and that a lower date for his life is given by an apparent reference to him in the commentary of Marinus (*fl.* A.D. 425) on Euclid's *Data* (David Gregory, *Euclidis quae supersunt omnia* [Oxford, 1703], p. 457.3) have been shown philologically by Heiberg in his review of M. Cantor, *Vorlesungen über Geschichte der Mathematik* in *Revue critique d'histoire et de Littérature,* 11 (1881), 381, and by Menge (*Euclidis opera omnia,* J. L. Heiberg and H. Menge, eds., VI [Leipzig, 1896], p. 248.3–4, where there is no mention of Serenus) to be erroneous.
3. J. L. Heiberg, *Sereni Antinoensis opuscula,* p. 96.14–25.
4. *Ibid.,* pp. 26–27.

BIBLIOGRAPHY

I. ORIGINAL WORKS. It seems likely that from the seventh century the two surviving works of Serenus and the commentary of Eutocius were bound with the *Conics* of Apollonius—Theodorus Metochita certainly read them together early in the fourteenth century—and their survival is probably due to these circumstances. A Latin trans. of Serenus' *De sectione cylindri* and *De sectione coni* was published by F. Commandinus at the end of his *Apollonii conicorum libri quatri* (Bologna, 1566). The Greek text was first published by E. Halley in *Apollonii Pergaei Conicorum libri octo et Sereni Antissensis De sectione cylindri et coni libri duo* (Oxford, 1710). A definitive critical ed. with Latin trans. was published by J. L. Heiberg, *Sereni Antinoensis opuscula* (Leipzig, 1896). A German trans. was made by E. Nizze, *Serenus von Antissa: Ueber den Schnitt des Cylinders* (Stralsund, 1860) and *Ueber den Schnitt des Kegels* (Stralsund, 1861); and there is an excellent French trans. with intro. and notes by Paul Ver Eecke, *Serenus d'Antinoë: Le livre De la section du cylindre e le livre De la section du cône* (Paris–Bruges, 1929).

The fragment from the lemmas has been published by T. H. Martin, *Theonis Platonici Liber De astronomia* (Paris, 1849; repr. Groningen, 1971), 340–343, with a Latin trans. and by J. L. Heiberg, *Sereni Antinoensis opuscula,* XVIII–XIX.

II. SECONDARY LITERATURE. See Thomas Heath, *History of Greek Mathematics,* II (Oxford, 1921), 519–526; J. L. Heiberg, "Über der Geburtsort des Serenos," in *Bibliotheca mathematica,* n.s. 8 (1894), 97–98; Gino Loria, *Le scienze esatte nell' antica Grecia,* 2nd ed. (Milan, 1914), 727–735; T. H. Martin, *Theonis Platoni-*

ci *Liber De astronomia* (Paris, 1849; repr. Groningen, 1971), 79–81; and Paul Tannery, "Serenus d'Antissa," in *Bulletin des sciences mathématiques et astronomiques*, 2nd ser., **7** (1883), 237–244, repr. in *Mémoires scientifiques*, I (Paris–Toulouse, 1912), 290–299.

IVOR BULMER-THOMAS

SERGENT, EDMOND (*b.* Philippeville [now Skikda], Algeria, 23 March 1876; *d.* Andilly-en-Bassigny, France, 20 August 1969), *epidemiology, immunology.*

The son of a French career soldier stationed in North Africa, Sergent studied medicine at Algiers. In 1900 he began to study under Émile Roux, assistant director of the Institut Pasteur in Paris. Shortly before, Grassi had demonstrated the role of mosquitoes in the propagation of malaria, and at Roux's suggestion Sergent went to Algeria to investigate the implications of this discovery. From 1900 to 1910 Sergent spent the malaria season in Algeria and the rest of the year in Paris. In 1912 he became director of the Institut Pasteur at Algiers, a post he held until 12 April 1963, when he was suddenly removed from office for political reasons. He left his native country and spent the rest of his life in France.

Sergent produced an extensive body of scientific work. He often collaborated with other researchers, notably his brother Étienne. A large portion of this research dealt with malaria. For more than forty-five years, both in the laboratory and in the field, the Sergent brothers tirelessly studied the factors involved in the spread of the disease: the protozoan (pathogenic agent), the *Anopheles* (vector), and man (reservoir of the parasite). This work culminated in 1926 in the creation, near Algiers, of an experimental station known as Marais des Ouled Mendil. Henceforth men were able to live where previously the presence of malaria had precluded settlement.

Sergent demonstrated that malaria is not the only disease in which an insect acts as the vector of the pathogenic agent. He was able to show this by studying the mode of transmission of various diseases including relapsing fever (spread by lice), bouton d'Orient (sandflies of the genus *Phlebotomus*), and various types of babesioses (ticks). Sergent also devised the concept of "premunition," according to which an organism's immunity to certain infections (including malaria and tuberculosis) can be assured only if it permanently carries the pathogenic agent in an attenuated state.

Sergent had great administrative ability and was placed in charge of many missions. He was known for his warmth and kindness, qualities most memorably displayed in his filial relationship with Roux. Sergent belonged to many scientific societies and received a number of honors, including the coveted Manson Medal (1962).

BIBLIOGRAPHY

I. ORIGINAL WORKS. With his brother Étienne, Sergent wrote *Vingt-cinq années d'étude et de prophylaxie du paludisme en Algérie* (Algiers, 1928) and *Histoire d'un marais algérien* (Algiers, 1947). Alone he wrote *Les travaux scientifiques de l'Institut Pasteur en Algérie de 1900 à 1962* (Paris, 1964). Many of his works can be found in *Archives de l'Institut Pasteur d'Algérie.*

II. SECONDARY LITERATURE. See Albert Delaunay, *L'Institut Pasteur des origines à aujourd'hui* (Paris, 1962); and "Edmond Sergent (1876–1969)," in *Annales de l'Institut Pasteur*, **118** (May 1970), 593.

ALBERT DELAUNAY

SERRES, ANTOINE ÉTIENNE REYNAUD AUGUSTIN (*b.* Clairac, France, 12 September 1786; *d.* Paris, France, 22 January 1868), *comparative anatomy, embryology.*

Serres was trained in Paris and received his medical degree in 1810. From 1808 to 1822 he worked at the Hôtel Dieu. In 1820 he was awarded the prize for physiological research by the Académie des Sciences and the following year gained a special prize for his two-volume work on the comparative anatomy of the brains of vertebrate animals. In 1822 Serres was appointed chief medical officer at the Hôpital de la Pitié. He was elected to the Académie de Medecine in 1822 and to the Académie des Sciences in 1828. In 1839 he preceded Flourens as professor of comparative anatomy at the Jardin des Plantes and two years later became president of the Académie des Sciences. He was created an officer (1841) and a commander (1848) of the Légion d'Honneur.

Serres did research into the development of the bones and teeth in normal and abnormal fetuses and studied the comparative anatomy of a number of vertebrate organs. He noted that many organs start from a number of isolated centers, which eventually unite to form a single adult organ—an observation that he regarded as a complete confirmation of epigenesis. In his general approach to the nature of life and the harmony between the organs he was clearly influenced by Cuvier, who mentioned Serres's work with admiration.

Serres's theoretical position was more closely akin to that of Geoffroy Saint-Hilaire, who regarded Serres as his collaborator. Serres believed that there was only one underlying animal type and that in the course of their development, the organs of the higher animals repeated the form of the equivalent organs in lower organisms. These ideas were not new; the nature philosophers in Germany had suggested similar ideas, and Meckel claimed that the higher animals pass through developmental stages analogous to the adult forms of lower animals. The distinction between a repetition of the organs of lower forms and the repetition of the actual organisms is often blurred; the latter view is sometimes called the Serres-Meckel Law.

After 1828 belief in either version of the Serres-Meckel law was gradually abandoned as the result of Baer's criticism, but throughout the 1840's and 1850's Serres continued to write papers in which he maintained his original views. These publications brought him into conflict with his younger colleagues and especially with Milne-Edwards; nevertheless, in 1859 Serres produced a final memoir in which he still maintained his original views.

A careful and precise observer of anatomical detail, Serres's work was neglected by the next generation, who were converts to an evolutionary outlook and scorned Serres's type of speculation. Haeckel's biogenetic law (1866)—each animal in its development recapitulates its evolutionary history—has much in common with the Serres-Meckel law.

Historians tend to treat Serres as a mere disciple of Geoffroy Saint-Hilaire, which again is to do him less than justice.

BIBLIOGRAPHY

I. ORIGINAL WORKS. Serres's chief writings are *Anatomie comparée du cerveau dans les quatre classes des animaux vertébrés*, 2 vols. plus atlas (Paris, 1824–1826); and *Anatomie comparée, transcendante. Principes d'embryogénie de zoogénie et de teratogénie* (Paris, 1859). His early works appeared in *Annales des sciences naturelles*, **11**, **12** (1827); **16** (1829); **21** (1830). Other publications are listed in *Catalogue de la Bibliothèque Nationale*.

II. SECONDARY LITERATURE. On Serres and his work, see Ch. Coury, "L'identification de la fièvre 'entéro-mésentérique' (ou typhoide) par Petit et Serres (1813)," in *Semaine des hôpitaux de Paris*, **40** (1964), 3056–3064; Ch. Coury and R. Rullière, "Deux grands iniciateurs de la recherche clinique moderne—Petit et Serres—leur étude de la fièvre entéro-mésentérique (1811–1813)," in *Presse médicale*, **72** (1964), 2487–2490, and in *Aktuelle Probleme aus der Geschichte der Medizin* (Basel–New York, 1966), 596–603; P. Huard, "A propos du centenaire de la mort d'Étienne Serres (1786–1868)," in *Pagine di storia delle medicina*, **14**, no. 5 (1970), 11–15; E. J. Russell, *Form and Function* (London, 1916); and an inaugural diss. by M. M. Kraegel von der Heyden, *Étienne-Renaud-Augustin-Serres (1787–1868) Entdecker des Abdominaltyphus* (Zurich, 1972), which lists also most of the earlier literature.

ELIZABETH B. GASKING

SERRES, OLIVIER DE (or **DES**) (*b.* Villeneuve-de-Berg, Ardèche, France, 1539; *d.* Villeneuve-de-Berg, 2 July 1619), *agronomy.*

Serres's father, Jacques de Serres, and his mother, Louise de Leyris, came from families long established as small landowners and lawyers in Vivarais; it is uncertain at what point the estate of Pradel, which was to be Serres's home, passed into their hands. Since his father died while Olivier was still a boy, it is uncertain what his education was; he may have studied for a while at the University of Valence, but presumably he did not graduate. In 1559 he married Marguerite d'Arcons; they had seven children. Although he had to travel from time to time, possibly to Germany and Italy, Pradel and its improvement were his life's work. As a young man he was converted to Protestantism; as early as 1561 he seems to have been regarded as a leader of the local Huguenots, obtaining a preacher for them from Geneva.

Much of the little that is known of Serres's life relates to his position in the Reformed community in time of civil war. In 1562 the parish church vessels were entrusted to him for sale. He also commanded forces in local campaigns of the various wars of the 1560's and 1570's and played a leading role in the capture of Villeneuve by the Huguenots in 1573, although he should almost certainly be exonerated of all blame for the massacre that followed. In fact, he was driven from Pradel more than once during these years and participated at least three times in conferences to arrange a local peace. At the end of the century he spent some time at Paris, where he presented his plans for the expansion of sericulture and diffusion of the mulberry to Henry IV.

At the same time Serres saw to the publication of his book *Théâtre d'agriculture* (1600). Since he was by then over sixty, the work can be regarded as the fruit of his life's experience. His aim was to present a complete survey of all aspects of agriculture, starting with advice on the proper way to run a household and proceeding, by way of discussions of various types of soil, to describe all the domesticated animals and plants known to him and to give useful hints on their cultivation. He also was an enthusiastic advocate of the use of irrigation to improve meadows, of careful drainage, and of the conservation of water.

Unlike most books of that period on farming and estate management, which tended to limit themselves to the codification and dissemination of the best current practice, Serres's was among the first, at least north of the Alps, to argue for widespread innovation and experiment. By the time his book appeared, he had acquired a national reputation as an authority on the silkworm, so much so that the relevant portion of his book appeared as a kind of preprint, *La cueillette de la soye,* the year before the main body, and was translated into German and English. In it he gave one of the first detailed accounts of the life cycle of the silkworm, although he did not eliminate the possibility of spontaneous generation (but acknowledged that he had never observed it). He believed that the silkworm could be reared much farther north than was then accepted. Serres also devised, or promoted, a method of manufacturing coarse cloth from the bark of the mulberry trees, the leaves of which fed the worms. His work on this, *La seconde richesse du meurier-blanc,* also was printed independently and was translated. With this new motive added to the needs of the silk industry, Serres bore a large part of the responsibility for the mulberry craze of the next decades and, in particular, inspired the king to make extensive plantings at Paris.

Serres was also keen on introducing crops previously unknown to France. An advocate of the sowing of artificial grasses, he devoted a chapter of his *Théâtre* to sainfoin, the use of which was then spreading north from Italy and Spain. Serres introduced hops to France and was the first French agricultural writer to describe and encourage the cultivation of maize and potatoes. Among exotic barnyard fowl—which gave him some difficulty in distinguishing one species from another—he appears to have known the turkey.

But the proportion of novelty in the whole should not be exaggerated. Although it is a treatise of agricultural improvement, his *Théâtre* is also a depiction of the old ideal of the self-sufficient and patriarchal small estate, whose master, while trying to increase his patrimony, seeks above all to be content with his lot and holds fast to the simplicity of rural life.

BIBLIOGRAPHY

I. ORIGINAL WORKS. Serres's main work is *Théâtre d'agriculture et mesnage des champs* (Paris, 1600), of which at least nineteen eds. had appeared by 1675; there is also an annotated ed., 2 vols. (Paris, 1804–1805). Sections of it that were published separately are *La cueillette de la soye* (Paris, 1599) and *La seconde richesse du meurier-blanc* (Paris, 1603).

II. SECONDARY LITERATURE. See M. de Fels, *Olivier de Serres* (Paris, 1963); G. Lizerand, *Le régime rural de l'ancienne France* (Paris, 1942), 79–80; and H. Vaschalde, *Olivier de Serres, seigneur du Pradel, sa vie et ses travaux* (Paris, 1886).

A. G. KELLER

SERRES DE MESPLÈS, MARCEL PIERRE TOUSSAINT DE (*b.* Montpellier, France, 3 November 1780; *d.* Montpellier, 22 July 1862), *zoology, geology.*

Serres (or de Serres) came from a rich family of drapers that belonged to the nobility of the robe. His mother died when he was very young, and he became a rebellious and lazy student, indifferent to punishment. Following his adolescence, however, Serres became a tireless scientific author whose 300 writings are noteworthy both in number and in diversity and encompass the natural and physical sciences, technology, jurisprudence, social and economic statistics, and travel accounts.

Serres studied law and in 1805 became deputy public prosecutor at the court of Montpellier; he was nevertheless more attracted by scientific research. Although his family suffered sudden financial ruin, Serres was able to go to Paris to study in 1807 through the generosity of Count Pierre Daru, a close collaborator of Napoleon. In Paris he attended the lectures of Haüy, Cuvier, Lamarck, and Geoffroy Saint-Hilaire. He also had the support of Berthollet and became friendly with Alexandre Brongniart and Constant Prévost. At the request of Daru, Serres spent 1809–1811 in Austria and Bavaria, studying technical processes of possible value to French industry. His account of this mission filled sixteen volumes (1813–1823), three of which were devoted specifically to technology. In 1813 Serres translated into French a partially unpublished work by the physicist H. C. Oersted:

Recherches sur l'identité des forces chimiques et électriques.

In 1809 Daru arranged for Serres to receive the chair of mineralogy and geology at the reorganized University of Montpellier. Serres assumed the post in 1811 but continued to spend much time in Paris. After the fall of the Empire and Daru's disgrace (1814), he settled permanently in Montpellier and reentered the magistracy, remaining in office until 1852. This responsibility accounts for his authorship, in 1823, of a large *Manuel des cours d'assises.*

Among Serres's earliest works were some dealing with the anatomy of insects, particularly the Orthoptera. The most detailed and original of these studies concerned the organs of vision. In an extensive work published in 1842 Serres presented the first synthesis of knowledge of animal migrations. In geology he accepted the theory that present causes have always been sufficient, as proposed by Lamarck and Prévost; this view was under attack from Cuvier and had not yet been developed by Lyell into uniformitarianism. Serres applied his knowledge of chemistry to the study of rocks, particularly flints of the chalk formations (1850). Serres was apparently the first, in 1817, to consider dating fossil bones by their fluorine concentration.

Serres, who was friendly with William Buckland, stimulated interest on the Continent in cave excavations and discovered human bones that he believed to be contemporary with the semifossilized bones of extinct animal species. He was unaware, however, of the existence of chipped flint tools. In 1836 Serres published an excellent synthetic work on caves, and three years later he wrote a monumental monograph on the cave of Lunel-Viel. In 1829 Jules Desnoyers had proposed the word *Quaternaire* to designate the most recent geological period; Serres redefined it soon afterward in a more valid fashion and spread its use.

Influenced by Lamarck and Étienne Geoffroy Saint-Hilaire, Serres stated in his *Géognosie* (1829) that "extinct species [*générations*] appear to be linked by an uninterrupted chain to present species." An English summary of this book, published the following year, emphasized the increasing complexity of organisms during the course of geological time. Serres subsequently adopted a new approach to the problem, however, as did Brongniart, Deshayes, and E. R. A. Serres. They rejected the idea of the variability of species. At the same time they relaxed the notion of fixity of species by accepting the evidence for successive creations—an idea that seemed to be confirmed by stratigraphic geology.

Serres sought to reconcile science and the Bible in *Cosmogonie de Moïse comparée aux faits géologiques*, which went through three editions (1838–1859) and appeared in at least two translations (German and Spanish). In this work he asserted that only cultivated or domestic species are variable. In 1851, however, he again stressed the "gradual perfecting of organized beings," in effect adopting an unacknowledged evolutionary view.

BIBLIOGRAPHY

I. Original Works. The works listed in the bibliography given by Rouville (see below) and in Royal Society *Catalogue of Scientific Papers,* V, 651–659; and VIII, 937, total about 300. The major ones are *Mémoire sur les yeux composés et les yeux lisses des Orthoptères* . . . (Montpellier, 1813), also in German (Berlin, 1826); *Géognosie des terrains tertiaires* (Montpellier–Paris, 1829), summarized in *Edinburgh Journal of Natural and Geographical Science,* 2 (1830), 294–295; *Essai sur les cavernes à ossements et sur les causes qui les y ont accumulés* (Montpellier, 1836; 3rd ed., Paris, 1838); *Cosmogonie de Moïse comparée aux faits géologiques* (Paris, 1838; 3rd ed., 1859); *Des causes des migrations des animaux* (Haarlem, 1842; 2nd ed., enl., Paris, 1845); and "Du perfectionnement graduel des êtres organisés," in *Actes de la Société linnéenne de Bordeaux,* 17 (1851), 5–32, 85–117, 181–213, 389–421; 18 (1852), 5–37, 97–129, 193–257, 427–459; and 19 (1853), 1–37, 77–113.

II. Secondary Literature. See P. G. de Rouville, *Éloge historique de Marcel de Serres* (Montpellier, 1863); and Rumelin, "Marcel de Serres," in Michaud, *Biographie générale,* 2nd ed., XXXIX, 128–131.

Franck Bourdier

SERRET, JOSEPH ALFRED (*b.* Paris, France, 30 August 1819; *d.* Versailles, France, 2 March 1885), *mathematics.*

Serret is sometimes confused with Paul Joseph Serret (1827–1898), a mathematician at the Université Catholique in Paris. After graduating from the École Polytechnique in 1840, Serret decided on a life of science and in 1848 became an entrance examiner at the École. After several other academic appointments he was named professor of celestial mechanics (1861) at the Collège de France and then, in 1863, professor of differential and integral calculus at the Sorbonne. In 1873 he joined the Bureau des Longitudes.

With his contemporaries P.-O. Bonnet and J. Bertrand, Serret belonged to that group of mathematicians in Paris who greatly advanced differential calculus during the period 1840–1865, and the fundamental formulas in the theory of space curves bear his name and that of J. F. Frenet. Serret also worked in number theory, calculus, mechanics, and astronomy and wrote several popular textbooks, including *Cours d'algèbre supérieure* (1849) and *Cours de calcul différentiel et intégral* (1867–1868).

In 1860 Serret succeeded Poinsot in the Académie des Sciences. After 1871 Serret's health declined; and he retired to Versailles, where he lived quietly with his family until his death.

BIBLIOGRAPHY

I. ORIGINAL WORKS. The *Journal des mathématiques pures et appliquées* contains several of Serret's papers, including "Mémoire sur les surfaces orthogonales," **12** (1847), 241–254; "Sur quelques formules relatives à la théorie des courbes à double courbure," **16** (1851), 193–207; and "Mémoire sur les surfaces dont toutes les lignes de courbure sont planes ou sphériques," **18** (1853), 113–162. See also "Sur la moindre surface comprise entre des lignes droites données, non situées dans le même plan," in *Comptes rendus hebdomadaire des séances de l'Académie des sciences,* **40** (1855), 1078–1082.

Serret edited the *Oeuvres de Lagrange,* 14 vols. (Paris, 1867–1892), and the fifth ed. of Monge's *Application de l'analyse à la géométrie* (Paris, 1850), with annotations.

II. SECONDARY LITERATURE. An obituary notice is given in *Bulletin des sciences mathématiques,* 2nd ser., **9** (1885), 123–132.

D. J. STRUIK

SERTOLI, ENRICO (*b.* Sondrio, Italy, 6 June 1842; *d.* Sondrio, 28 January 1910), *physiology, histology.*

Sertoli studied medicine at the University of Pavia, where, with Giulio Bizzozero and Camillo Golgi, he was a pupil of the physiologist Eusebio Oehl, who systematically developed studies in microscopic anatomy and histology at Pavia. After graduating in 1865, Sertoli moved to Vienna to study physiology under Brücke; but he returned to Italy the following year to take part in the campaign against Austria. After the war, he went to Tübingen in 1867 to work in Hoppe-Seyler's laboratory there.

From 1870 to 1907 Sertoli was professor of anatomy and physiology, and after 1907 of physiology only, at the Advanced Royal School of Veterinary Medicine in Milan, which, at that time, enjoyed university status. There he founded the Laboratory of Experimental Physiology.

Sertoli was an outstanding exponent of microscopic anatomy. In his first scientific work (1865) he identified and described the branched cells in the seminiferous tubules of the human testicle, which are still known as Sertoli cells. This research, carried out under Oehl's direction at the Pavia Institute of Physiology, was the starting point for further investigations; Sertoli later studied the structure of the testicle and spermatogenesis.

With Hoppe-Seyler, Sertoli reported (1867) on the importance of the blood proteins, and especially of the globulins, as mediators of alkalies in the alternating process of fixation and clearance of carbon dioxide; they discovered that this process is carried out in the capillary beds of the greater and lesser circulation, respectively. Sertoli's findings in this field were subsequently confirmed by Zuntz and Otto Loewi.

In Milan, Sertoli studied (1882–1883) the persistent excitability and extreme sensitivity to thermal stimuli of the smooth muscles. He is also credited with the first leiomyogram, which he obtained by using the retractor muscle of the penis, which was particularly suitable for the purpose because of its length, uniformity, and parallel fibrocells.

BIBLIOGRAPHY

I. ORIGINAL WORKS. A list of Sertoli's publications is given in Pugliese (see below). His major works include "Dell'esistenza di particolari cellule ramificate nei canalicoli seminiferi del testicolo umano," in *Morgagni,* **7** (1865), 31–40, with one plate; "Ueber die Bindung der Kohlensäure im Blute und ihre Ausscheidung in der Lunge," in *Medicinisch-chemische Untersuchungen,* **2** (1867), 350–365; and in *Zentralblatt für die medizinischen Wissenschaften,* **6** (1868), 145–147; "Osservazioni sulla struttura dei canalicoli seminiferi del testicolo," in *Gazzetta medica lombarda,* 4 d.s. **6** (1871), 413–415; "Sulla struttura dei canalicoli seminiferi del testicolo studiata in rapporto allo sviluppo dei nemaspermi," *ibid.,* 2 d.s. **7** (1875), 401–403; "Di un pseudo-ermafrodismo in una capra," in *Archivio di medicina veterinaria,* **1** (1876), 22–33, written with G. Generali; "Sulla struttura dei canalicoli seminiferi dei testicoli studiata in rapporto allo sviluppo dei nemaspermi," in *Archivio per le scienze mediche,* **2** (1878), 107–146, 267–295, and plates 3–4; "Contribuzioni alla fisiologia generale dei muscoli lisci," in *Rendiconti dell'Istituto lombardo di scienze e lettere*, **15** (1882), 567–582; "Contribution à la physiologie général des muscles lisses," in *Archives italiennes de biologie*, **3** (1883), 78–94; "Della cariocinesi nella spermatogenesi," in *Rendiconti dell'Istituto lombardo di scienze e lettere*, **18** (1885), 833–839; and "Sur la caryokinèse dans la spermatogénèse," in *Archives italiennes de biologie*, **7** (1886), 369–375.

II. SECONDARY LITERATURE. On Sertoli and his work, see L. Belloni, "Enrico Sertoli in la medicina a Milano dal settecento al 1915," in *Storia di Milano*. Fondazione Treccani degli Alfieri, **16** (1962), 1028; A. Pugliese, "Henri Sertoli," in *Archives italiennes de biologie*, **53** (1910), 161–164; and F. Usuelli, "Enrico Sertoli (1842–1910)," in *Annuario veterinario italiano* (1934–1935), 455–461.

BRUNO ZANOBIO

SERTÜRNER, FRIEDRICH WILHELM ADAM FERDINAND (*b.* Neuhaus, near Paderborn, Germany, 19 June 1783; *d.* Hameln, Germany, 20 February 1841), *pharmacology.*

Sertürner's parents were Austrian. His father, Joseph Simon Serdinner (the spelling varies), married Marie Therese Brockmann and entered the service of Friedrich Wilhelm, prince-bishop of Paderborn, his son's godfather, as engineer and state building inspector. In 1798 Sertürner's father and his princely patron both died. The youth, then fifteen, was apprenticed to Cramer, the court apothecary, and in 1803 passed his assistant's test with excellent marks. In 1806 Sertürner became assistant to Hink, the town apothecary of Einbeck. In 1809 the French government of Westphalia licensed him to open his own pharmacy. Upon the return of the Hanoverian government, the license was revoked as an act of the French occupation forces. After a long, unsuccessful litigation, Sertürner took over the town pharmacy of Hameln in 1820. The following year he married Leonore von Rettberg. A capable assistant relieved him of the routine work in the prosperous pharmacy and thus, financially secure, he was able to devote himself to his scientific interests.

In later years Sertürner apparently suffered increasingly from mental disturbances, and his hypochondria became quite evident during his last years.

Aside from his chemical-pharmaceutical work, his passion was the construction of firearms. He designed a breechloader and tested new alloys for bullets. An arms manufacturer named Stürmer undertook the manufacture of these novelties and demonstrated them to the war ministries of Hannover and Prussia.

In his first scientific work Sertürner endeavored to isolate the "sleep-inducing factor" in opium and discovered a process whereby he could use ammonia to separate practically pure morphine from aqueous clarified opium extract. Until then Scheele's work had led to the assumption that all active substances in plants were acids. Sertürner called the newly discovered "sleep-inducing factor" a "vegetable alkali." He stated that this was most certainly the first representative of a new class of plant matter and called for the further search for other vegetable alkalies. Thus the foundation was laid for alkaloid chemistry. The introduction of morphine into pharmaceutics was later compared to the introduction of iron into metallurgy.

Sertürner's first publications on morphine in 1805 and 1806 failed to attract attention. In 1817, when Sertürner republished the results of his research in enlarged and more detailed form, the importance of his work was recognized.

Considering the obstacles confronting Sertürner's work on morphine, it is surprising that he succeeded at all. An autodidact with only sparse knowledge of the relevant literature, he conducted his research with the simplest equipment while performing the strenuous duties of an apothecary's assistant. Sometimes years elapsed between his research projects. Sertürner played an important part in that period of organic chemistry—between Scheele and F. F. Runge—when fundamental research was carried out by gifted investigators (mostly pharmacists) with only limited equipment.

Before Davy, Sertürner established in 1806 that caustic alkalies are not elements, but compounds of oxygen plus another combustible element similar to hydrogen. He failed, however, to interest a scientific journal in publishing his paper.

Sertürner's tendency toward speculation became so pronounced that even his most advanced colleagues often failed to comprehend him. His views on the life element "zoon," on the "cold nature of sunlight," on "atmospheric heat," and on "fire oxide" were unfounded and failed to stimulate productive thinking. Consequently, Sertürner acquired a dubious reputation, which partly explains why his two further important discoveries were ignored.

He developed a theory on the formation of ether from alcohol and sulfuric acid and established the formulas for three different "sulphovinic acids" (ethyl sulfuric acids). Although he came close to being correct, he met with universal rejection.

No attention was paid to his paper dealing with the cholera epidemics prevalent at the time. In this work he was the first to point out the real cause of the disease. On the basis of the known fact that objects exposed to severe cold or heat no longer transmitted the disease, he ascribed its cause to a toxic, self-reproducing living agent.

Only his work on morphine brought Sertürner numerous honors.

BIBLIOGRAPHY

I. Original Works. Sertürner's writings include *System der chemischen Physik*, 2 vols. (Göttingen, 1820–1822); *Annalen für das Universalsystem der Elemente*, 3 vols. (Göttingen, 1826–1829); *Dringende Aufforderung an das deutsche Vaterland, in Beziehung der orientalischen Brechruhr* (Göttingen, 1831); *Einige Belehrungen für das gebildete und gelehrte Publikum über den gegenwärtigen Zustand der Heilkunde und der Naturwissenschaften im allgemeinen, mit besonderer Rücksicht auf das gemeine Leben* . . . (Göttingen, 1838); and Franz Krömeke, ed., *Friedrich Wilhelm Sertürner, der Entdecker der Morphiums. Lebensbild und Neudruck der Original-Morphiumarbeiten* (Jena, 1925).

II. Secondary Literature. See Hermann Coenen, "Über das Jahr der Morphiumentdeckung Sertürners in Paderborn," in *Archiv der Pharmazie*, 287 (1954), 166–180; F. von Gizyki, "Die Aufnahme des Morphins in den Arzneischatz," in *Deutsche Apotheker-Zeitung*, 96 (1956), 583–584; P. J. Hanzlik, "125th Anniversary of the Discovery of Morphine by Sertürner," in *Journal of the American Pharmaceutical Association*, 18 (1929), 375–384; Georg Lockemann, "Friedrich Wilhelm Sertürner," in *Zeitschrift für angewandte Chemie*, 37, no. 30 (1924), 526–532; Hermann Trommsdorff, "Trommsdorff und Sertürner; Johann Bartholomä Trommsdorff und seine Zeitgenossen. Teil 2," in *Jahrbücher der Akademie gemeinnütziger Wissenschaften in Erfurt*, 55 (1941), 133–243; and J. Valentin, "Der erkenntnistheoretische Wandel Sertürners im Jahre 1804," in *Deutsche Apotheker-Zeitung*, 97 (1957), 573–574.

Eberhard Schmauderer

SERULLAS, GEORGES-SIMON (*b.* Poncin, Ain, France, 21 November 1774; *d.* Paris, France, 25 May 1832), *chemistry, pharmacy.*

Serullas was the son of a notary and seemed destined to follow his father's profession. After doing well in his early studies, however, he enrolled in a pharmacy course in 1793 and became a military pharmacist. During a campaign in the Alps he learned botany, physics, and chemistry. He spent several years in Italy, where, following the European blockade, he was put in charge of preparing a huge amount of grape syrup as a sugar substitute for consumption in the military hospitals. He was chief pharmacist under the command of Ney throughout the Italian, German, and Russian campaigns.

After the siege of Torgau, where he lived for a time, Serullas became chief pharmacist as well as the first professor of pharmacy at the military hospital in Metz. From then on, he devoted himself to intellectual pursuits; for example, at the age of forty-two he began to study Greek and mathematics. In 1825 Serullas became chief pharmacist and professor at the Val de Grâce and was named professor of chemistry at the Jardin des Plantes. In 1829 he was elected a member of the Paris Academy, succeeding Vauquelin. He died of cholera, which he contracted at the funeral of Georges Cuvier.

Serullas's earliest research, which involved sugar and sugar substitutes, was followed by investigations of alloys of sodium and potassium. He probably is best known, however, for his studies of iodine and bromine and their compounds. In 1823 he discovered iodoform (CHI_3), which he called *hydriodure de carbone*. Serullas's confusion, which reflects the state of organic chemistry of the period, is indicated by the fact that he called presumably the same compound *protoiodure de carbone* in 1823 and *periodure de carbone* in 1828. In 1824 he prepared cyanogen iodide (discovered by Humphry Davy in 1816) by a more efficient method.

His studies with bromine led to the preparation of ethyl bromide; cyanogen bromide; a selenium bromide; several compounds of bromine with arsenic, bismuth, and antimony; and an *éther hydrobromique*. Serullas found that the *hydrocarbure de brome* (bromoform) remains solid up to 7°C., a fact ignored during previous work. He also experimented with chlorine and in 1828 discovered cyanuric chloride. This work on halogen compounds led to the publication of the well-received book *Sur quelques composés d'iode, tels que le chlorure d'iode, sur l'action mutuelle de l'acide iodique et de la morphine ou de ses sels, sur l'acide iodique cristallisé* (1830), which became an important reference work in legal medicine. In 1827 he discovered cyanamide and cyanuric acid. His other investigations involved perchloric acid, phosphonium iodide (PH_4I), and ether.

BIBLIOGRAPHY

I. Original Works. J. R. Partington, *A History of Chemistry*, IV (London, 1964), 83–84, 89, 254, 325,

342, 349–350, 358, gives references to the papers describing most of Serullas's studies that were published in the *Annales de chimie* or the *Mémoires* of the Paris Academy. In addition, the following early books and memoirs are significant: *Mémoires pour le perfectionnement des moyens d'obtenir la matière sucrée des végétaux indigènes*, 2 vols. (Paris, 1810–1813); *Mémoire sur la conversion de la matière sucrée en alcool* (Paris, 1817); *Sur les fumigations chloriques* (Paris, 1817); *Observations physico-chimiques sur les alliages du potassium et du sodium avec d'autres métaux* (Metz, 1821); *Moyen d'enflammer la poudre sous l'eau* (Metz, 1822); and *Sur quelques composés d'iode . . .* (Paris, 1830).

II. Secondary Literature. A short, relatively accessible sketch of Serullas is in F. Hoefer, ed., *Nouvelle biographie générale*, XLIII (1867), 802–803. Longer obituary notices are Jean Antoine Lodibert, *Éloge historique de Serullas* (Paris, 1837); and Julien Joseph Virey, *Notice sur Serullas* (Paris, 1832). Partington (see above) makes several references to Serullas's investigations of organic halides in relation to the work of other French chemists.

Sheldon J. Kopperl

SERVETUS, MICHAEL (*b.* Villanueva de Sixena [?], Spain, 29 September 1511 [?]; *d.* Geneva, Switzerland, 27 October 1553), *biology, philosophy.*

There is controversy as to the date and place of birth of Michael Servetus. The conflicting data were supplied by Servetus himself during his trials at Lyons and Geneva, when he was anxious to mislead his inquisitors. The more likely date, 29 September 1511 as against the traditional 1509, is corroborated by two separate statements made by Servetus: that he was forty-two at the time of his trial in Lyons and that he was twenty when he published his first book. Villanueva de Sixena, in the province of Huesca, has been authenticated as the place where his family resided; hence Servetus' choice of Villanovanus as a pseudonym. He also had stated, however, that he was born at Tudela, Navarre, thus leading some historians to suppose that his family lived there at the time of his birth and later moved to Villanueva. Evidence is lacking to support this inference.

Servetus' parents were Antonio Serveto, alias Reves (a pseudonym Servetus also used), a notary, and Catalina Conesa. They were "Old Christian" nobles; and one of his brothers was a priest. Not much is known of his early education. Possibly, after completing church school, he attended the University of Zaragoza, which was not far from his home. There he learned Latin.

A combination of intellectual precocity and family connections led Servetus, at the age of fifteen, to enter the service of the learned Franciscan friar Juan de Quintana, who held a doctorate from the Sorbonne and was a member of the Cortes of Aragon. The influence of Erasmianism in Spain was well manifested in Quintana, who at the Diet of Augsburg told Melanchthon that he was unable to understand why Luther's doctrine of justification by faith should have aroused so much controversy.

Servetus temporarily left Quintana's employ to pursue legal studies at the University of Toulouse in 1528. This institution was considered preeminent in the field of law, and Servetus described it as "the mother of those skilled in law." (All quotations are from the O'Malley translation.) It was here that Servetus became interested in scriptural studies, and he may have studied Greek and Hebrew as linguistic aids. His stay in Toulouse was brief; for Quintana, having been named confessor to Emperor Charles V, recalled him to his service in 1529. In Quintana's train Servetus witnessed Charles V's coronation as Holy Roman Emperor at Bologna and later traveled to Germany, where the emperor hoped to settle the Protestant problem.

Servetus' earlier studies of the Bible had raised grave doubts in his mind. Nowhere in the Bible did he find the word "Trinity." He must have realized that he could no longer stay in Quintana's employ, for he left his patron in 1530. By July of that year he was in Basel. He hoped to meet Erasmus, but the latter had left Basel more than a year earlier. Johannes Oecolampadius was now the chief reformer of that city; and Servetus stayed in his house as a guest, probably for the ten months that he remained in Basel.

Oecolampadius, who was forty-eight, showed great forbearance toward the nineteen-year-old Servetus, who, aside from voicing anti-Trinitarian doctrines, was a contentious, vain, and stubborn young man. Oecolampadius made every effort to convert Servetus, but to no avail. At the conference of reformers in Zurich, he was driven to complain about this Spaniard who was spreading the Arian heresy in his city. On his return to Basel, Oecolampadius heeded Zwingli's admonition that every possible means be taken to prevent Servetus' heresy from spreading, and Servetus left Basel.

He chose to go to Strasbourg, for he had befriended Martin Bucer and the city was well-known for its tolerant attitude toward sectarian movements. At Hagenau, near Strasbourg, Servetus found a printer, Johannes Setzer, who had published some 150 titles, and convinced him to print his first work. In July 1531, *De trinitatis erroribus* went on sale.

In this work, published when he was only twenty, Servetus displayed a very wide range of reading. He cited many authors and pitted their views against the Bible in its original Greek and Hebrew texts. Thus he was able to show the discrepancy between later Scholastic theories and the original Biblical statements on the Trinity. Servetus denied the doctrine of three equal persons in Godhead and brought on himself the condemnation of both Catholics and Protestants.

Bucer refuted *De trinitatis* publicly, and the city magistrates banned its sale in Strasbourg. Servetus returned to Basel; but the reception of his work there was, if anything, harsher than in Strasbourg. Partly in order to allay criticism, he published *De trinitate* (1532). Although the tone of this second work was not as unrestrained, Servetus fundamentally stuck to his doctrines. It therefore was advisable for him to leave Switzerland.

Assuming the name of Michel de Villeneuve, Servetus moved to France. In 1533 he was studying in Paris at the Collège Calvi. At this time Paris was beginning to crack down on heretics, however; and a meeting between Servetus and Calvin did not take place, probably because of the former's fear of being apprehended. Servetus decided to go to Lyons, where he may have stopped briefly before going to Paris. Being a trade center, Lyons was a relatively more tolerant city; moreover, there were many great printing houses there where he could find work.

Servetus became a corrector and editor for the most famous publishers in Lyons, the brothers Trechsel. For them he prepared two editions of Ptolemy's *Geography* (1535, 1541) and three editions of the Bible (an octavo Bible and the Santis Pagnini Bible, both in 1542, and a seven-volume edition that appeared in 1542). His edition of the Santis Pagnini Bible is the best-known and is remarkable for its theory of prophecy.

Ptolemy's *Geography* had been the standard work on the subject since the second century and had often appeared in Greek and Latin editions. Servetus used Willibald Pirckheimer's edition of 1524 and compared it with the Greek text and other editions. With great relish he pointed out the many errors that had crept into the Pirckheimer edition. Subsequent geographers have acknowledged the validity of these corrections.

In the preparation of the text that accompanied the fifty maps, Servetus stated that he had consulted eighty works. In reference to the New World,

he wrote that "those err to high heaven . . . who contend that this continent should be called America, since Amerigo approached that land long after Columbus. . . ." Some historians have gone so far as to claim that he was the founder of comparative geography because of his comments on national characteristics and his interest in national psychology, which were new for the time. The success of this work was attested to by the fact that Servetus was commissioned to do another edition, with minor alterations, in 1541.

Servetus' interest in medicine was aroused in connection with his proofreading duties. Many medical works were published in Lyons, and the Trechsel firm published the writings of the distinguished medical humanist Symphorien Champier, with whom Servetus struck up a close friendship. Significantly, Servetus' first medical work, *In Leonardum Fuchsium apologia* (1536), was a defense of Champier, who had become involved in a controversy with Leonhard Fuchs. In the *Apologia* Servetus expounded his belief in the healing powers of certain herbs.

It was probably Champier who advised Servetus to return to Paris and study medicine. The preface of the *Apologia* was dated 12 November 1536, from Paris. There Servetus became part of a distinguished medical circle, and his teachers included Sylvius (Jacques Dubois), Fernel, and Johannes Guinter. The last singled him out, together with Vesalius, as his most able assistant in dissection. (Servetus and Vesalius may not have known each other personally; evidence suggests that the latter had returned to Louvain by the time Servetus reached Paris.)

In 1537 Servetus published what was essentially a continuation of the *Apologia,* the *Syruporum universa ratio,* which was so successful that it went through six editions and helped finance his stay in Paris. This fundamentally Galenic work centered on the use of syrups for curative purposes and contained a significant passage on the use of "correct" foods, especially citrus fruits, as an aid in the assimilative process of digestion. Servetus also maintained that sickness was the perversion of the natural functions of body organs and was not caused by the introduction of new elements into the body. The *Apologia* and the *Syruporum* were noteworthy contributions to modern pharmacology.

Contrary to university regulations, which at any rate were only laxly enforced, Servetus supplemented his dwindling funds by giving lectures, although he did not have a Master of Arts degree. His original subject was geography, in which his edition of Ptolemy had given him enough reputation. He then moved on to astronomy and became involved in judicial astrology or forecasting and its relation to medicine. He was charged before the Faculty of Medicine with lecturing on astrology. In his defense Servetus wrote the *Apologetica disceptatio pro astrologia* (1538), which he hastened to publish despite indications that the Parlement of Paris had been asked to issue an injunction against its publication. Too late to prevent its appearance, the Parlement confiscated all copies of the *Astrologia* and reprimanded its author. Because of the absence of any record that he received a degree in medicine, many authors have surmised that this incident prevented Servetus from completing his studies. Nevertheless, his 1541 contract to edit the Bible referred to him as a *docteur en médecine;* and, although this document was drafted in Lyons, there certainly were persons in that city who had known him well in Paris.

In 1538 Servetus returned to Lyons; then moved to Charlieu, where he practiced medicine for three years; then returned to Lyons. During most of the latter time Servetus lived at Vienne in the palace of Archbishop Pierre Palmier, the outstanding churchman of the region, who had attended his lectures in Paris and whom he now served as personal physician. He also practiced medicine at large for the next twelve years; and his colleagues elected him prior of the Confraternity of St. Luke, with responsibility to supervise the apothecaries and tend to the indigent hospital patients.

Although outwardly living as a Catholic, Servetus did not abandon his theological studies and his original doctrines. He had been at work on his *magnum opus,* the *Christianismi restitutio,* which was published on 3 January 1553. This theological treatise contained Servetus' imperishable contribution to science as the first man in the West to discover the lesser circulation of the blood. His primary concern, however, was theological: the problem of the introduction of the divine spirit into the blood and its dissemination throughout the body. He stated that the blood was not transmitted from the right ventricle of the heart to the left by way of the septum, for "that middle wall, since it is lacking in vessels and mechanisms, is not suitable for that communication and elaboration, although something may possibly sweat through." Rather, noting the size of the pulmonary artery, Servetus concluded that it was too large for simply transporting a small portion of the blood for the nutriment of the lungs, the function that Galen had ascribed to it. Servetus asserted that blood passed through the lungs for oxygenation. A further statement that the "vital spirit is then transfused from the left ventricle of the heart into the arteries of the whole body" showed that Servetus had arrived at the threshold of the complete circulation. Since his interest was primarily theological, however, he did not pursue this; and we have no way of knowing whether he could have done so.

Servetus' claim to the discovery of the lesser circulation has been questioned. Reference has been made to Realdo Colombo's *De re anatomica,* which, although published in 1559, was written earlier. Nonetheless, a manuscript of *Christianismi restitutio* in the Bibliothèque Nationale bears evidence of having been written before 1546. Ibn al-Nafīs described the lesser circulation in a work that dates from the mid-thirteenth century, but Servetus' finding was made independently.

The publication of *Christianismi restitutio* and his earlier letters to Calvin, including drafts of some chapters of the book, led to Servetus' undoing, for Calvin allowed these letters to be used to inform the authorities in Lyons as to the true identity of Michel de Villeneuve. On 4 April 1553, Servetus was arrested and imprisoned but managed to escape three days later. In the meantime most of the thousand copies of his work were confiscated and burned; it was not until 1694 that Servetus' discovery of the lesser circulation became known. When Harvey announced the discovery of the general circulation in 1628, he did not know of Servetus' contribution.

Servetus remained out of sight for four months in France and then decided to go to Italy. He chose the route that passed through Geneva, where on 13 August he was recognized and denounced by Calvin to the magistrates. He was sentenced to be burned at the stake; his last cry was a reaffirmation of his views on the Trinity.

BIBLIOGRAPHY

I. Original Works. A complete listing of Servetus' works is in John F. Fulton, *Michael Servetus: Humanist and Martyr* (New York, 1953), which also shows where copies may be found. An able translation of the scientific writings of Servetus is C. D. O'Malley, *Michael Servetus: A Translation of His Geographical, Medical and Astrological Writings* (Philadelphia, 1953).

II. Secondary Literature. The most readable biography is Roland H. Bainton, *Hunted Heretic: The Life and Death of Michael Servetus, 1511–1553* (Boston, 1953; repr., 1960), which also contains a good list of periodical literature, including the sixty-eight articles and six books of the assiduous nineteenth-century Servetus scholar Henri Tollin. The following emphasize various aspects of Servetus' life: B. Becker, ed., *Autour de Michel Servet et de Sebastien Castellion* (Haarlem, 1953); Eloy Bullón y Fernández, *Miguel Servet y la geografía del Renacimiento,* 3rd ed. (Madrid, 1945); Pierre Cavard, *Le procès de Michel Servet de Vienne* (Vienne, 1953); and Juan-Manuel Palacios Sánchez, *El ilustre aragonés Miguel Servet* (Huesca, 1956).

Two works that place the doctrines of Servetus in their wider theological context are Earl Morse Wilbur, *A History of Unitarianism,* I (Cambridge, Mass., 1947), 3–4 and *passim*; and George Huntston Williams, *The Radical Reformation* (Philadelphia, 1962).

VICENTE R. PILAPIL

SERVOIS, FRANÇOIS-JOSEPH (*b.* Mont-de-Laval, Doubs, France, 19 July 1767; *d.* Mont-de-Laval, 17 April 1847), *mathematics.*

Servois was the son of Jacques-Ignace Servois, a merchant, and Jeanne-Marie Jolliet. He was ordained a priest at Besançon at the beginning of the Revolution, but in 1793 he gave up his ecclesiastical duties in order to join the army. In 1794, after a brief stay at the artillery school of Châlons-sur-Marne, he was made a lieutenant. While serving in several campaigns as staff officer, he devoted his leisure time to the study of mathematics. With the support of Legendre, he was appointed professor of mathematics at the artillery school of Besançon in July 1801. A few months later he transferred to the school at Châlons-sur-Marne; in 1802, to the artillery school at Metz; and in 1808, to the school at La Fère. After a brief return to Metz as professor at the artillery and engineering school, he was appointed curator of the artillery museum at Paris in 1816. He held the post until 1827, when he retired to his native village.

Like a number of his colleagues who taught at military schools, Servois closely followed developments in mathematics and sought, at times successfully, to make an original contribution. His first publication was a short work on pure and applied geometry: *Solutions peu connues de différents problèmes de géométrie pratique . . .* (1805). Drawing upon Mascheroni's *Geometria del compasso* and upon Lazare Carnot's *Géométrie de position* (1803), Servois formulated some notions of modern geometry and applied them to practical problems. The book was well received, and Poncelet considered it a "truly original work, notable for presenting the first applications of the

theory of transversals to the geometry of the ruler or surveyor's staff, thus revealing the fruitfulness and utility of this theory" (*Traité des propriétés projectives* [Paris, 1822], xliv).

Servois presented three memoirs before the Académie des Sciences. The first was on the principles of differential calculus and the development of functions in series (1805; new version, 1810); the second, which was never published, was devoted to the elements of dynamics (1809; additions in 1811); the third, also never published, dealt with the "determination of cometary and planetary orbits." In 1810 Servois published a study on the principle of virtual velocities in the *Mémoires* of the Turin Academy, but most of his subsequent papers appeared in Gergonne's *Annales de mathématiques pures et appliquées*. In his first contribution to the latter, he solved two construction problems by projective methods and introduced the term *pôle*. His ability in geometry was recognized by Poncelet, who consulted him on several occasions while writing his *Traité des propriétés projectives*.

In a letter to Gergonne of November 1813, Servois criticized, in the name of the primacy of algebraic language, the geometric representation of imaginary numbers that had recently been proposed by J. R. Argand and J. F. Français: "I confess that I do not yet see in this notation anything but a geometric mask applied to analytic forms the direct use of which seems to me simple and more expeditious" (*Annales de mathématiques . . . ,* 4, no. 7 [January 1814], 230). This formalist conception of algebra made Servois one of the chief precursors of the English school of symbolic algebra. It can be seen still more clearly in his "Essai sur un nouveau mode d'exposition des principes du calcul différentiel," which contains the most important aspects of the memoir presented to the Academy in 1805 and 1810. Familiar with the work of Hindenburg's combinatorial school and with L. F. A. Arbogast's *Calcul des dérivations*, Servois sought in the "Essai" to provide differential calculus with a rigorous foundation. In the course of this effort he developed the first elements of what became the calculus of operations. Observing that this calculus is based on the conservation of certain properties of the operations to which it is applied, he introduced the fundamental notions of "commutative property" and "distributive property" (*Annales,* 5, no. 5 [November 1814], 98). He did not, however, always distinguish between "function" and "operation." Servois's memoir, which more or less directly inspired the work of

Robert Murphy and of George Boole, was followed by an interesting critique of the various presentations of the principles of differential calculus, particularly the theory of the infinitely small and the method of Wronski.

Although Servois did not produce a major body of work, he made a number of original contributions to various branches of mathematics and prepared the way for important later developments.

BIBLIOGRAPHY

I. ORIGINAL WORKS. Servois's book on geometry, *Solutions peu connues de différents problèmes de géométrie pratique pour servir de supplément aux traités de cette science* (Metz–Paris, 1805), was followed by *Lettre de S . . . à F . . . professeur de mathématiques sur le Traité analytique des courbes et surfaces du second ordre* (Paris, 1802). He also published "De principio velocitatum virtualium," in *Mémoires de l'Académie impériale des sciences de Turin,* **18** (1809–1810), pt. 2, 177–244.

The following articles appeared in Gergonne's *Annales de mathématiques pures et appliquées:* "Solutions de deux problèmes de construction," **1,** no. 11 (May 1811), 332–335, 337–341; "Démonstrations de quelques formules de trigonométrie sphérique," **2,** no. 3 (Sept. 1811), 84–88; "Remarques relatives à la formule logarithmique" (dated 2 Oct. 1811), **2,** no. 7 (Jan. 1812), 178–179; "Calendrier perpétuel," **4,** no. 3 (Sept. 1813), 84–90; "Sur la théorie des quantités imaginaires. Lettre de M. Servois" (dated 23 Nov. 1813), **4,** no. 7 (Jan. 1814), 228–235, also in J. R. Argand, *Essai sur une manière de représenter les quantités imaginaires . . .,* 2nd ed. (Paris, 1874), 101–109; "Essai sur un nouveau mode d'exposition des principes du calcul différentiel," **5,** no. 4 (Oct. 1814), 93–140; "Réflexions sur les divers systèmes d'exposition des principes du calcul différentiel, et, en particulier, sur la doctrine des infiniment petits" (La Fère, 10 Aug. 1814), **5,** no. 5 (Nov. 1814), 141–170. The last two articles were printed together in a pamphlet (Nîmes, 1814).

See also "Note de M. Servois (Sur la trigonométrie des Indiens)," in *Correspondance sur l'École polytechnique,* **3,** no. 3 (Jan. 1816), 265–266; "Mémoire sur les quadratures," in *Annales de mathématiques . . .,* **8,** no. 3 (Sept. 1817), 73–115; "Lambert (Henri-Jean)," in Michaud, ed., *Biographie universelle,* XXIII (1819), 46–51; "Trajectoire," in *Dictionnaire de l'artillerie,* G.-H. Cotty, ed. (Paris, 1822), 464–471; "Lettre sur la théorie des parallèles" (dated 15 Nov. 1825), in *Annales de mathématiques . . .,* **16,** no. 7 (Feb. 1826), 233–238. Royal Society *Catalogue of Scientific Papers,* V, 665, gives only a portion of this bibliography.

II. SECONDARY LITERATURE. The principal account of Servois's career and writings is J. Boyer, "Le mathématicien franc-comtois François-Joseph Servois, an-

cien conservateur du Musée d'artillerie d'après des documents inédits," in *Mémoires de la Société d'émulation du Doubs,* 6th. ser., **9** (1894), 5–37, also separately printed as a 26-page pamphlet (Besançon, 1895).

Comments on Servois's work are given by S. F. Lacroix, in *Procès-verbaux des séances de l'Académie des sciences,* V (Hendaye, 1914), 99–101; and in *Traité du calcul différentiel et du calcul intégral,* 2nd ed., III (Paris, 1819), see index; by J. V. Poncelet, in *Traité des propriétés projectives . . .* (Paris, 1822), v–vi, xliv; and in *Applications d'analyse et de géométrie,* II (Paris, 1864), 530–552; by M. Chasles, in *Aperçu historique . . .* (Paris, 1875), see index; by O. Terquem, in *Bulletin de bibliographie, d'histoire et biographie mathématique,* **1,** 84, 93, 110, 185, supp. to *Nouvelles annales de mathématiques,* **14** (1855); and by S. Pincherle, "Équations et opérations fonctionnelles," in *Encyclopédie des sciences mathématiques,* II, pt. 5, fasc. 1 (Paris–Leipzig, 1912), 4–5; and in *Intermédiaire des mathématiciens,* **2** (1895), 58, 220, and **23** (1916), 195. The most recent study, N. Nielsen, *Géomètres français sous la Révolution* (Copenhagen–Paris, 1929), 221–224, analyzes certain aspects of Servois's work in greater detail but contains a number of errors.

RENÉ TATON

SESSÉ Y LACASTA, MARTÍN DE (*b.* Baraguas, Aragón, Spain, 11 December 1751 [?]; *d.* Madrid, Spain, 4 October 1808), *botany.*

Sessé studied medicine, practiced in Madrid (1775–1776), then served as an army doctor in Spain and in Cuba. He moved to Mexico City in 1785. On 13 March 1787 he was named director of the Royal Botanical Expedition to New Spain and director of the Royal Botanical Garden in Mexico City. Early in 1789 Sessé gave up his medical work to devote full time to botany. He continued to hold his two directorships until he returned to Spain. In spite of the pressure of administrative detail involving the expedition and the garden, he took part in long field excursions to western Mexico (1789–1792), to the Atlantic slope of Mexico (1793), and to Cuba and Puerto Rico (1795–1798). Sessé left Mexico for the last time about April 1803 and reached Spain in November, having stopped in Cuba to arrange for the shipment of his West Indian collection, the bulk of which reached Madrid in June 1804. He seems to have accomplished little in his remaining years.

Sessé was a competent botanist, as is shown by his existing manuscript notes and botanical descriptions. He apparently enjoyed collecting and analyzing plants in the field and understood thoroughly the standard practices and concepts of his day. Sessé's contribution to botany is linked with that of José Mariano Mociño and can hardly be considered apart from it. The posthumous works *Plantae Novae Hispaniae* and *Flora Mexicana* were attributed to Sessé and Mociño as joint authors, but it is probable that a major part of the botanical study and writing was done by Mociño under the nominal direction of Sessé, whose principal contribution to science seems to have been related to his administrative and executive functions.

It was Sessé who originally conceived and proposed the Botanical Expedition to New Spain, and with Vicente Cervantes he helped plan and maintain the Botanical Garden in Mexico. He dealt with several viceroys in turn, keeping them informed of progress and trying to convince them of the continuing value of the botanical work. Under his direction the members of the expedition conducted or took part in major excursions to all parts of Mexico except the extreme north, to Central America as far as Costa Rica, to the Greater Antilles, and to the northern Pacific. For about fifteen years he kept a group of temperamental naturalists and artists occupied and relatively contented (with the conspicuous exception of Longinos Martínez), sometimes under very trying circumstances. Finally, he managed to return to Spain with all the expedition's collections, manuscripts, and paintings intact.

BIBLIOGRAPHY

I. ORIGINAL WORKS. Two volumes, attributed to Sessé and Mociño jointly, were published in Mexico between 1887 and 1897. These appeared first in parts, as supplements to the periodical *La Naturaleza*. *Plantae Novae Hispaniae* (1887–1891) was based on a MS written by Mociño, completed at Guadalajara, Jalisco, forwarded from there to the viceroy, the Conde de Revilla-Gigedo, in July 1791, and now in the archives of the Instituto Botánico "A. J. Cavanilles," Madrid. It is a complete flora, including the species of flowering plants studied by the Botanical Expedition up to about the beginning of 1791. A 2nd ed. was published in book form at Mexico City in 1893.

Flora Mexicana (1891–1897) was based on a very heterogeneous series of notes on individual plant-species, from many parts of Spanish America. These comprised a part, but by no means all, of the notes prepared by the members of the Botanical Expedition. Discovered, in no particular order, in the archives at Madrid, the notes were organized by the editor into the Linnaean classes and were published without careful study or collation. A 2nd ed. was published in book form at Mexico City in 1894, before the later parts of the first edition appeared in *Naturaleza*.

Original letters, memoranda, and other documents relative to the Botanical Expedition to New Spain are to be found in the Mexican National Archives, sec. "Historia," vols. 460–466, 527. A few documents apparently of similar origin are in the William L. Clements Library, University of Michigan, Ann Arbor. Sessé's official correspondence as director of the Expedition and of the Royal Botanical Garden is voluminous. The papers of Revilla-Gigedo, acquired in 1954 by a private collector in the United States, contain some information not accessible elsewhere but have been little studied in this connection.

The richest source of MS material in Spain is the archive of the Instituto Botánico "A. J. Cavanilles," Madrid. It contains most of the existing MSS dealing with strictly botanical matters of the expedition to New Spain: the MS of *Plantae Novae Hispaniae*, various botanical descriptions, fragments of unpublished floras (including a "Flora guatemalensis" by Mociño), inventories of paintings, and collections from the various excursions carried out in Mexico. Descriptions or copies of most of these inventories have been published by Arias Divito (see below) or in the papers cited by him. Arias Divito also lists (p. 307) the other major sources of MS material in Madrid and Seville.

II. SECONDARY LITERATURE. An extensively documented account of Sessé, Mociño, and their co-workers, based primarily upon materials in the Archivo General de la Nación, Mexico City, is H. W. Rickett, "The Royal Botanical Expedition to New Spain," in *Chronica botanica*, **11** (1947), 1–86. Juan Carlos Arias Divito, *Las expediciones científicas españolas durante el siglo XVIII* (Madrid, 1968), is based primarily on Spanish archival sources. It includes copies of many previously unpublished inventories of plants, animals, and paintings, and a considerable bibliography that supplements the references cited by Rickett. Additional information, especially relative to the members of the Malaspina Expedition who were in Mexico at the same time as the Royal Botanical Expedition, is in Iris Higbie Wilson, "Scientific Aspects of Spanish Exploration in New Spain During the Late Eighteenth Century" (Ph.D. diss., Univ. of Southern California, 1962).

The story of the disaffected naturalist Longinos Martínez, who left the Botanical Expedition after a long and bitter quarrel with Sessé, is told in Lesley Bird Simpson, *Journal of José Longinos Martínez* (San Francisco, 1961).

The botanical specimens collected in New Spain from about 1787 to 1799 number more than 10,000; perhaps 8,000 compose the "Sessé and Mociño" herbarium at the Instituto Botánico "A. J. Cavanilles," Madrid; and several thousand duplicates are scattered through the larger European herbaria, especially those in London, Paris, Geneva, Florence, and Oxford. After the death of Sessé and the flight of Mociño with the retreating French in 1812, the Spanish botanist José Antonio Pavón sold to collectors at least 15,000 duplicate specimens, including many of those collected by the expedition of Sessé and Mociño. As these found their way gradually into large public and private herbaria, they were much studied and cited by botanists, and thus ironically became of more scientific value than the original herbarium, which remained unstudied in Madrid. Much effort has been expended in recent years, as more has become known of the work of the Botanical Expedition, in documenting these specimens that constitute perhaps the most valuable part of the legacy of Sessé and Mociño. A part of the story of their sale and the dispersal of duplicate specimens from New Spain is told by Arthur Robert Steele in *Flowers for the King* (Durham, N.C., 1964), 291–315, which describes in detail Pavón's dealings with Aylmer Bourke Lambert and Philip Barker Webb. An uncataloged MS in the department of botany, British Museum (Natural History), lists the plants sold by Pavón to Lambert and later bought for the Museum.

ROGERS MCVAUGH

SETCHELL, WILLIAM ALBERT (*b.* Norwich, Connecticut, 15 April 1864; *d.* Berkeley, California, 5 April 1943), *botany, geography.*

Setchell was the acknowledged authority on marine algae of the northern Pacific, and on the role of crustaceous algae in coral reef formation. He also advanced knowledge of the role of temperature in delimiting plant distributions, and initiated studies on the genus *Nicotiana.*

Setchell was the son of George Case Setchell and Ann Davis Setchell. With George R. Case, Setchell published a catalog of local wild plants before he was twenty; and the botanist Daniel Cady Eaton encouraged his interest in botany while he was an undergraduate at Yale. His lifetime predilection for algae dated from his Harvard years, when, as a Morgan fellow, he came under the tutelage of William Gilson Farlow. For twenty-five years Setchell, with his associates F. S. Collins and Isaac Holden, issued *Phycotheca Boreali-Americana* (1895–1919), numbering 200,000 specimens of algae. In 1920 Setchell went to Samoa under the aegis of the Carnegie Institution of Washington, the first of many journeys to the South Pacific to study coral reef formation, ethnobotanical subjects, and patterns of plant distribution. From these and other trips he collected and donated hundreds of objects, including a notable pipe collection now at the Robert H. Lowie Museum of Anthropology. Meanwhile Setchell amassed a collection of living tobaccos and began an investigation of their morphology and hybridization. His colleague Thomas Harper Goodspeed took up this study and published a comprehensive monograph entitled *The Genus Nicotiana* (1954).

The history of botany was one of Setchell's pleasures, and his course in that subject attracted students and colleagues alike. He enjoined students to compare various topics among the herbals which he presented to the Biology Library of the University of California at Berkeley.

Setchell emphasized the critical role of temperature in the delimitation of algal and flowering plant ranges; this led to his "waves of anthesis" principle: species succeed one another in flowering with every rise of 5°F. in the vernal temperature. He also predicated the critical role of establishment in the distribution of organisms, as a biologic corollary of Liebig's law of the minimum.

Setchell's wife, Clara Ball Pearson Caldwell, whom he married in 1920, shared his enthusiasms for fourteen years. His foil in the laboratory preparation of algae was Nathaniel L. Gardner (1864–1937), with whom he wrote several revisions of Pacific marine algae. Setchell's close associates described him as "of commanding presence, magnetic personality, catholic taste, and congenial disposition" (Clausen, Bonar, and Evans, p. 39).

BIBLIOGRAPHY

I. ORIGINAL WORKS. Setchell's chief works are "Marine Algae of the Pacific Coast of North America," in *University of California Publications in Botany,* **8,** pts. 1–3 (1919–1925); "American Samoa," in *Publications. Carnegie Institution of Washington,* no. 341 (1924), 1–275; "Temperature and Anthesis," in *American Journal of Botany,* **12** (1925), 178–188; and "Geographic Elements of the Marine Flora of the North Pacific Ocean," in *American Naturalist,* **69** (1935), 560–577. Prophetic of his later phenological interests was *A Catalogue of Wild Plants Growing in Norwich and Vicinity, Arranged in Order of Flowering for the Year 1882* (Norwich, Conn., 1883), written with George R. Case. A bibliography of his publications, prepared by T. H. Goodspeed and Lee Bonar, was appended to D. H. Campbell's sketch (see below). Copies of 210 papers (1890–1935) are bound in 35 vols. in the Biology Library, University of California, Berkeley. His correspondence is preserved in the department of botany. His unpublished outline for his history of botany course was extensively utilized by Howard S. Reed in his *Short History of the Plant Sciences* (Waltham, Mass., 1942).

II. SECONDARY LITERATURE. The best obituary of Setchell, with interpretive background comment, was written by his Stanford colleague D. H. Campbell, in *Biographical Memoirs. National Academy of Sciences,* **23** (1945), 127–147, with portrait by Peter van Valkenburgh. *Essays in Geobotany in Honor of William Albert Setchell* (Berkeley, 1936) includes a biographical intro. by the editor, T. H. Goodspeed. Other sketches were published by Roy E. Clausen, Lee Bonar, and Herbert M. Evans, in *University of California In Memoriam* (Berkeley, 1943) 37–39; by A. D. Cotton, in *Proceedings of the Linnean Society of London*, **156** (1943–1944), 232–233; by Francis Drouet, in *American Midland Naturalist*, **30** (1943), 529–532; and by Herbert L. Mason, in *Madroño*, **7** (1943), 91–93, with portrait. A disparate estimate will be read in the private journals of his colleague Willis Linn Jepson when they are made available to the public.

JOSEPH EWAN

SEVERGIN, VASILY MIKHAYLOVICH (*b*. St. Petersburg, Russia, 19 September 1765; *d*. St. Petersburg, 29 November 1826), *mineralogy, chemistry, technology*.

The son of a court musician, Severgin was accepted, on his father's petition, in 1776 at the private Gymnasium of the St. Petersburg Academy of Sciences. In 1784 he enrolled at the Academy's university, choosing mineralogy as his specialty. The following year he was sent to the University of Göttingen, where he studied the outcrops of basalt near Göttingen and became involved in the controversy between the neptunists and the plutonists.

In 1789 Severgin returned to St. Petersburg and presented two scientific papers to the Academy, on the properties and formation of basalt and on alkaline salts. In the first paper, directed against the neptunists, he argued that basalt originated in a fiery liquid fusion. On 25 June 1789 Severgin was elected adjunct to the chair of mineralogy. His chief orientation was toward mineral chemistry, in which he applied the ideas of Lavoisier.

In his enlarged and supplemented Russian translation of Kirwan's *Elements of Mineralogy* (1791) Severgin classified and described minerals on the basis of their chemical composition. Two years later he was elected professor of mineralogy, a post that carried with it the title of academician. Much of his subsequent work was devoted to the study of Russian minerals, their regional distribution, and methods of extracting and processing them. Severgin disseminated chemical and mineralogical knowledge through his many textbooks and lectures at the Institute of Mines and at the Medical and Surgical Academy.

His *Pervye osnovania mineralogii* ("Foundations of Mineralogy," 1798) was the first textbook in Russian on the subject. Besides describing minerals and rocks, Severgin classified petrifactions as "simple" or "complex." Among the former

were marble, jasper, and flint (quartz), which lack foreign particles. He gave detailed descriptions of the physical and chemical properties of minerals, developing "wet" methods of analysis as well as methods using the blowpipe; his techniques of determining the external characteristics of minerals are still used, substantially unchanged. In 1804 Severgin founded *Tekhnologichesky zhurnal*. A frequent contributor to the journal, he was also its editor until 1824. His mineralogical dictionary (1807) contained detailed explanations and was an important contribution to the literature.

Concentrating on mineral chemistry rather than on crystallography, Severgin reported in 1798 on the significance of mineral associations—galena and sphalerite, for example—on the basis of which he developed a theory of the contiguity of minerals: "What I call the contiguity of minerals is the joint occurrence of two or more minerals in one place . . . for example, the association of quartz with mica, virgin gold, and others" (*Pervye osnovania mineralogii*, p. 85–86).

The mineral collections of the Institute of Mines, the Free Economic Society, and the many St. Petersburg amateur collectors, as well as his personal collection, served Severgin as the material for *Opyt mineralogicheskogo zemleopisania Rossyskogo gosudarstva* ("An Attempt at the Mineralogical Description of the Territory of the Russian State"). The first volume, a physical-geographical survey, describes structures and lithology as well as the hydrographic network; the second volume deals with the geographical distribution of minerals.

In 1819 Severgin published his translation of the book on minerals of Pliny's *Natural History*. He also contributed to the development of scientific terminology, introducing Russian terms still used for oxide, silicon dioxide, alkali, and splintery and conchoidal fracture of minerals. His chemical works had a practical orientation and dealt with the extraction of mineral salts, the testing of medicinal chemical substances, assaying, and the production of saltpeter.

BIBLIOGRAPHY

I. ORIGINAL WORKS. Severgin's writings include *Nachalnye osnovania estestvennoy istorii* . . . ("The Foundations of Natural History"), a trans. of Kirwan's *Elements of Mineralogy*, enl. and supp., 2 vols. (St. Petersburg, 1791); *Mineralogicheskie, geograficheskie i drugie smeshannye izvestia o Altayskikh gorakh, prinadlezhashchikh k Rossyskomu vladeniyu* ("Mineralogical,

Geographical, and Other . . . Information on the Altai Mountains, Which Belong to the Russian Domain"), trans. from the German of H. M. Renovanz (St. Petersburg, 1792); "Opisanie Dalgrenevoy payalnoy trubki, deystvuyushchey pomoshchyu mekha s pokazaniem upotreblenia onoy" ("Description of the Dahlgren Blowpipe, Which Works With the Aid of a Bellows, With Instructions for Its Use"), in *Trudy Volnogo ekonomicheskogo obshchestva* (1792), nos. 14 and 15; *Pervye osnovania mineralogii . . .* ("Foundations of Mineralogy"), 2 vols. (St. Petersburg, 1798); "O estestve i obrazovanii bazalta, ili stolbchatogo kamnya" ("On the Nature and Formation of Basalt or of Basaltiform Rock"), in *Akademicheskie sochinenia* (1801), no. 1, 332–359; *Probirnoe iskusstvo* ("Assaying"; St. Petersburg, 1801); *Zapiski puteshestvia po zapadnym provintsiam Rossyskogo gosudarstva . . . 1802 i 1803* ("Notes of a Journey Through the Western Provinces of the Russian State . . . in 1802 and 1803"; St. Petersburg, 1803).

They also include *Prodolzhenie zapisok puteshestvia po zapadnym provintsiam Rossyskogo gosudarstva* ("Continuation of the Notes of a Journey Through the Western Provinces of the Russian State"; St. Petersburg, 1804); *Obozrenie Rossyskoy Finlyandii* ("A Survey of Russian Finland"; St. Petersburg, 1804); *Podrobny slovar mineralogichesky . . .* ("A Detailed Mineralogical Dictionary . . ."), 2 vols. (St. Petersburg, 1807); *Opyt mineralogicheskogo zemleopisania Rossyskogo gosudarstva* ("An Attempt to Describe Mineralogically the Territory of the Russian State"), 2 vols. (St. Petersburg, 1809); *Slovar khimichesky* ("Chemical Dictionary"), trans. of the work of Charles-Louis Cadet de Gassicourt, 4 vols. (St. Petersburg, 1810–1813); "Obozrenie mineralnogo kabineta imperatorskoy Akademii nauk" ("A Survey of the Mineral Cabinet of the Imperial Academy of Sciences"), in *Tekhnologichesky zhurnal*, **11**, no. 1 (1814); *Novaya sistema mineralov, osnovannaya na naruzhnykh otlichitelnykh priznakakh* ("A New System of Minerals, Based on External Distinctive Characteristics"; St. Petersburg, 1816); and *Kaya Plinia sekunda—Estestvennaya istoria iskopaemykh tel*, trans. of the mineralogical section from Pliny the Elder's *Natural History* (St. Petersburg, 1819).

II. SECONDARY LITERATURE. See G. P. Barsanov, "V. M. Severgin i mineralogia ego vremeni v Rossii" ("Severgin and the Mineralogy of His Time in Russia"), in *Izvestiya Akademii nauk SSSR*, geol. ser. (1949), no. 5; *Bolshaya sovetskaya entsiklopedia* ("Great Soviet Encyclopedia"), 2nd ed., XXXVIII, 303; A. E. Fersman, "Mineralogia v Akademii nauk za 220 let" ("220 Years of Mineralogy in the Academy of Sciences"), in *Ocherki po istorii Akademii nauk* ("Essays on the History of the Academy of Sciences"; Moscow–Leningrad, 1945); D. P. Grigorev and I. I. Shafranovsky, "V. M. Severgin," in *Vydayushchiesya russkie mineralogi* ("Outstanding Russian Mineralogists"; Moscow–Leningrad, 1949); A. N. Ivanov, "Vasily Mikhaylovich Severgin," in *Lyudi russkoy nauki* ("People of Russian Science"; Moscow, 1962), 7–15; A. V. Nemilova, and

I. I. Shafranovsky, "Akademik Severgin V. M. i ego rol v istorii russkoy mineralogii (k 120 letiyu so dnya smerti, 1765–1826)" ("Academician Severgin and His Role in the History of Russian Mineralogy [on the 120th Anniversary of His Death]"), in *Priroda*, **36** (1947), no. 3, 72–75; D. P. Rezvoy, "Akademik Vasily Severgin— Russky mineralog i geognost (1765–1826)" ("Academician Severgin—Russian Mineralogist and Geognost"), in *Mineralogicheskii sbornik* (1953), no. 7; and M. I. Sukhomlinov, "Ocherk zhizni i deyatelnosti akademika Severgina" ("An Essay on the Life and Career of Academician Severgin"), in *Istoria Rossyskoy Akademii* ("History of the Russian Academy"), appendix to *Zapiski Imperatorskoi akademii nauk*, **32**, no. 1 (1878).

G. D. KUROCHKIN

SEVERI, FRANCESCO (*b*. Arezzo, Italy, 13 April 1879; *d*. Rome, Italy, 8 December 1961), *mathematics*.

From 1898 until his death, Severi published more than 400 books and papers on mathematics, history of science, education, and philosophy. His most outstanding contributions, however, were in the field of algebraic geometry. Severi acquired the taste for elegant synthetic arguments while studying with Segre at the University of Turin, from which he graduated in 1900 under Segre's guidance. At Turin he became interested in algebraic and enumerative problems and developed a broad geometric eclecticism and a formidable dexterity in the projective geometry of higher spaces.

In the latter field Segre published (1894) an interesting reworking of geometry on an algebraic curve. In Italy, Bertini and Castelnuovo also contributed to this field, while in Germany, Brill and Max Noether, in the footsteps of Riemann, made more studies using different methods. A more invariant view, derived from the transformations introduced by Cremona thirty years earlier, led Castelnuovo and Enriques to lay the foundations of a similar theory for algebraic surfaces. This theory anticipated the work of Picard in France on the same subject.

Having served as *assistente* to Enriques at Bologna in 1902 and to Bertini at Pisa in 1903, Severi was drawn to these new developments. He attempted, with great success, to explain important and still unsolved problems along with work in new areas. He perfected the theory of birational invariants of algebraic surfaces and created an analogous (but more complex) theory for algebraic varieties of arbitrary dimension. The completion of this work was to take him another fifty years. Severi's work on algebraic geometry can best be described by dividing it into five sections, rather than maintaining a chronological order.

1. Enumerative and Projective Geometry, Intersections, and Questions on the Foundations of Algebraic Geometry. The proof of the principle of the "conservation of number," established heuristically by Schubert in the nineteenth century, was listed by Hilbert at the Paris Congress of 1900 as one of the fundamental unsolved problems of mathematics. Severi subsequently found and proved the conditions under which this principle is true. Thus he refined Schubert's work and also advanced it through the theory of the base and the theory of characteristics.

Twenty years later this research by Severi inspired several mathematicians, including William V. D. Hodge, Wei L. Chow, and Bartel van der Waerden. Severi also introduced the important notion of the invariant order of an algebraic variety, which led to the theory of minimal models; and he studied improper double points of algebraic surfaces, characters of embedding of one variety in another, and generalizations of Bezout's theorem from intersections of plane curves to those of arbitrary varieties in higher projective spaces.

2. Series and Systems of Equivalence. This theory, created almost wholly by Severi, added to algebraic geometry many important entities, for example, the canonical varieties of arbitrary dimension—a theory (completed later by Beniamino Segre and J. A. Todd) that has had considerable connections and implications in both algebra and topology. He generalized the theory of linear equivalence to arbitrary subvarieties of a given variety and also made lengthy fundamental studies of rational equivalence, algebraic equivalence, and algebraic correspondences between varieties.

3. Geometry on Algebraic Surfaces. At the beginning of the twentieth century, the geometry of algebraic surfaces had reached a dead end. Although Castelnuovo and Enriques had defined the genera, irregularity, and plurigenera of surfaces and had characterized those surfaces that are birationally equivalent to a plane or to a ruled surface, there were still several unsolved problems; and Picard's introduction of three types of integrals on a surface suggested many additional questions, a large number of which were successfully explained by Severi. Also, Severi reduced Picard's three types of integrals to normal form and found conditions of integrability for certain linear differential equations on a surface.

Severi introduced the notion of semiexact differentials of the first type and, using Hodge's findings, surprisingly showed that they are always exact. An important property utilized in these investigations was the completeness of the characteristic series of a continuous complete system. Much effort was later required to establish this result in its correct generality.

4. Geometry on Algebraic Varieties. The extension from surfaces to varieties of three or more dimensions is no less difficult than that from curves to surfaces. In an early memoir (1909), Severi established the basis for the extended theory with his study of linear systems of hypersurfaces. He gave various definitions of the arithmetic genus of a variety and proved their equivalence, thus partially extending the Riemannn-Roch theorem and also Picard's theorem on the regularity of the adjoint.

Besides his work on the foundations of the general theory of algebraic varieties, Severi established the theory of irregularity and made important studies of continuous systems of curves in the plane and in higher projective spaces.

5. Abelian and Quasi-Abelian Varieties. The theory of Abelian varieties V_p, of dimension p, originated in 1889 with Picard, who investigated algebraic V_p possessing a continuous, transitive, Abelian group of ∞^p birational automorphisms. The infinitesimal transformations of the group led to p independent integrals. Picard maintained that these integrals were all of the first type, but Severi showed that this is not true for $p = 2$ if the group is not absolutely transitive.

The study of these V_p is connected with that of a particular type of functions of several complex variables—a generalization of elliptic functions. These functions are related to particular varieties, called Picard and Albanese varieties, to which Severi devoted several works. When the group of the V_p is transitive, but not necessarily absolutely transitive, V_p is called quasi-Abelian. Severi discussed these V_p in a lengthy paper written during the turbulent period October 1944 – May 1945.

Algebraic geometry has undergone several revolutionary changes in the twentieth century that have led to many schools and to several widely differing methods of approach. Severi's work remains not merely a monument to him but also a valuable source from which all algebraic geometers continue to draw ideas. He himself characterized his approach to mathematical research with the following admonition:

> Let us not pride ourselves too much on perfect rigour, which we today believe to be capable of reducing so large a part of mathematics to nothing, and let us not discard what does not appear quite as rigorous,

for tomorrow we will certainly find imperfections in our perfection and from some brilliant, intuitive thought which had not yet the blessing of rigour will be drawn unthinkable results [Severi, "Intuizionismo e astrattismo nella matematica contemporanea," in *Atti del congresso. Unione matematica italiana* (Sept. 1948), p. 30].

BIBLIOGRAPHY

I. Original Works. Severi's works include *Vorlesungen über algebraische Geometrie*, L. Löffler, ed. and trans. (Leipzig, 1921; repr., New York–London, 1968); "Geometria delle serie lineari," in *Trattato di geometria algebrica*, 1, pt. 1 (1926), 145–169; *Serie, sistemi d'equivalenza e corrispondenze algebriche sulle varietà algebriche*, F. Conforto and E. Martinelli, eds. (Rome, 1942); *Funzioni quasi abeliane* (Vatican City, 1947); *Memorie scelte*, B. Segre, ed. (Bologna, 1950); *Geometria dei sistemi algebrici sopra una superficie e sopra una varietà algebrica*, II (Rome, 1957), III (Rome, 1959); and *Il teorema di Rimann-Roch per curve, superficie e varietà* (Berlin, 1958). Severi's mathematical papers, collected in seven volumes, will soon be published by the Lincei Academy (vols. I and II have already appeared in 1971 and 1974).

II. Secondary Literature. On Severi and his work, see B. Segre, *L'opera scientifica di Francesco Severi* (Rome, 1962), with complete bibliography.

Beniamino Segre

SEVERIN, CHRISTIAN, also known as **Longomontanus** (*b.* Longberg, Jutland, Denmark, 4 October 1562; *d.* Copenhagen, Denmark, 8 October 1647), *astronomy.*

Severin was the son of Søren Poulsen and Maren Christensdatter, both of whom were humble peasants. Finding education an uncertain and intermittent luxury, especially after the early death of his father, Severin did not complete his basic education until 1588. At that time he entered the service of Tycho Brahe and stayed with him until 1597, when Tycho left Denmark. After his *Wanderjahre* in Germany, Severin received the M.A. at the University of Rostock and then returned home to begin his career. By 1607 he was professor of mathematics and astronomy at the University of Copenhagen, where he remained until his death.

When Tycho died in 1601, his program for the restoration of astronomy was unfinished. The observational aspects were complete, but two important tasks remained: the selection and integration of the data into accounts of the motions of the planets, and the presentation of the results of the entire program in the form of a systematic treatise. Severin, Tycho's sole disciple, assumed the re-sponsibility and fulfilled both tasks in his voluminous *Astronomia danica* (1622). Regarded as the testament of Tycho, the work was eagerly received and quickly won a place in seventeenth-century astronomical literature. Even after the appearance of Kepler's *Tabulae Rudolphinae* (1627), a rival work that bore the imprimatur of Tycho, Severin's *Astronomia danica* retained sufficient prestige (despite its staidness) to warrant reprinting in 1640 and 1663.

Unfortunately Severin found himself looking backward to Tycho, instead of forward into the seventeenth century. Although Severin worked and wrote in the era of Kepler and Galileo, he denounced ellipses, denied heliocentrism, denigrated the telescope, and ignored logarithms. Severin departed from Tycho in only one significant respect—he assumed diurnal rotation of the earth.

Because Severin's career was virtually determined by his unique status as the literary heir of Tycho, it is impossible to form an independent estimate of his contemporary reputation. He was highly esteemed by Tycho for his skill at manipulating observational data, and he may have played an important role in Tycho's remarkable research on the lunar theory. Regardless of his competence as a planetary theorist, Severin's reputation will always suffer in comparison with Kepler's achievements in the same task. In addition to his astronomical interests, Severin also displayed considerable enthusiasm for pure mathematics, but with notably less success. Concerned principally with the quadrature of the circle, he believed that he had solved the problem with a precise evaluation of π as equal to $78/43 \sqrt{3}$.

BIBLIOGRAPHY

A complete bibliography of Severin's works is in H. Ehreneron-Müller, *Forfatterlexicon*, V (Copenhagen, 1929), 181–185. His major works are *Cyclometria ex lunulis reciproce demonstrata* (Copenhagen, 1612); *Astronomia danica* (Amsterdam; 1622, 1640, 1663); *Inventio quadraturae circuli* (Copenhagen, 1634); and *Introductio in theatrum astronomicum* (Copenhagen, 1639).

On Severin and his work, see J.-B.-J. Delambre, *Histoire de l'astronomie moderne*, I (Paris, 1821), 262–287.

Victor E. Thoren

SEVERINO, MARCO AURELIO (*b.* Tarsia, Calabria, Italy, 2 November 1580; *d.* Naples, Italy, 12 July 1656), *biomedical sciences.*

Severino was the son of Beatrice Orangia and Jacopo Severino, a successful lawyer who died when his son was seven. Marco Aurelio's mother directed his early education, in Latin, Greek, rhetoric, poetry, and law, at various schools in Calabria. He then continued his studies at Naples, soon moving from law to medicine as his chosen field. At Naples he met Tommaso Campanella, who, although not officially one of Severino's teachers, was nevertheless an important influence in the formation of his thought. From Campanella, he learned the rudiments of Telesio's philosophical system, which formed the basis of the critical anti-Aristotelianism that marked Severino's later work. After taking a medical degree at Salerno in 1606 (although his studies had largely been at Naples), Severino returned to Tarsia to begin medical practice. Three years later he returned to Naples to study surgery with Giulio Iasolino. From 1610 Severino taught surgery and anatomy privately at Naples. When the university chair in these subjects fell vacant in 1615, Severino was named to fill it and shortly afterward was also named first surgeon at the Ospedale Degli Incurabili.

Severino's fame as a surgeon spread rapidly, and students came from all parts of Europe to study with him. Ultimately his published works were better known and more frequently published north of the Alps than in Italy. He corresponded with many of the important physicians and scientists of his time, including William Harvey and John Houghton in England; Thomas Bartholin and Ole Worm in Denmark; J. G. Volkamer and Joannes Vesling in Germany; and Campanella, Iasolino, and Tommaso Cornelio in Italy. Severino was called before the Inquisition for allegedly unorthodox religious and philosophical views but was eventually acquitted. He died of the plague in Naples and was buried without a marker in the church of S. Biagio de' Librai.

Severino's writings are marked by a general emphasis on observation and experience, which he traced back to a medical tradition stemming from Democritus. But he remained deeply influenced by metaphysical ideas and accepted teleology in nature, neo-Platonic hierarchical schemes, and the Paracelsian version of the microcosm-macrocosm relationship. His strongly anti-Aristotelian sentiments (which are especially evident in *Zootomia Democritea* and in *Antiperipatias*) derived in part from the native southern Italian intellectual heritage of Telesio, della Porta, and Campanella.

The bulk of Severino's printed works dealt with surgery and anatomy. He published both compre-

hensive treatises and specific detailed monographs on these subjects. His fame as a surgeon is illustrated by the broad distribution of works like *De efficaci medicina,* in which he championed surgery as a legitimate medical technique in opposition to the iatrochemical approach of some contemporaries.

Severino's permanent contributions, however, seem to lie in his anatomical works, especially *Zootomia Democritea.* This work might with some justification be called "the earliest comprehensive treatise on comparative anatomy" (Cole) and, indeed, it emphasized throughout an approach in which human anatomy is related to that of other animals. Severino viewed the study of anatomy as one way to uncover a clearer knowledge of divine creation. Since man, animals, and plants form a continuous hierarchical structure, the anatomy of all three must be studied in conjunction. Severino recognized a close similarity between the anatomy of man and of animals and considered important the detailed study of nonhuman anatomy. He himself dissected and studied a wide range of specimens, both vertebrate and invertebrate. He contended that even tiny animals and insects must be studied by the anatomist, if necessary with the aid of a microscope, although he does not seem to have made much use of that instrument himself.

The *Antiperipatias* illustrates Severino's critical attitude toward the Aristotelians. He argued against the Peripatetic view—that fish do not breathe air—by trying to demonstrate, following the atomistic philosophy of Democritus, that fish actually utilize the air that is dissolved in water.

Severino's full significance in the flourishing scientific culture of Naples during his time and also his importance as a figure of international renown in the seventeenth century have not been studied in detail. Particularly important are his relations to Harvey and his place in the discussions arising from the publication in 1628 of Harvey's *De motu cordis.* Most of the key documents touching on this aspect of his activities remain in manuscript and have never been properly evaluated, nor is there an adequate survey of his other unpublished writings. Thus his place in the development of seventeenth-century biomedicine is not firmly fixed.

BIBLIOGRAPHY

I. ORIGINAL WORKS. A more complete bibliography is in Schmitt and Webster (below). See also *Therapeuta Neapolitanus* (Naples, 1653), fols. §§2r – §§4r and Severino's trans. of Antonio Colmenero de Ledesma, *Chocolata Inda* (Nuremberg, 1644), 69–73.

Severino's major works include *De recondita abscessuum natura* (Naples, 1632); *Zootomia Democritea* (Nuremberg, 1645); *De efficaci medicina* (Frankfurt, 1646); *Vipera Pythia* (Padua, 1650); *Trimembris chirurgia* (Frankfurt, 1653); *Therapeuta Neapolitanus* (Naples, 1653); *Seilo-phlebotome castigata* (Hanau, 1654); *Quaestiones anatomicae quatuor* (Frankfurt, 1654); *Antiperipatias. Hoc est adversus Aristoteleos de respiratione piscium diatribe . . . De piscibus in sicco viventibus . . . Phoca illustratus*, 2 pts. (Naples, 1655, 1659); *Synopseos chirurgiae* (Amsterdam, 1644).

At his death Severino left numerous unpublished works, some of which appeared in print posthumously. Most of his MSS went first to Antonio Bulifon, then to Giammaria Lancisi; they remain in the Biblioteca Lancisiana in Rome. For some information see V. Ducceschi, "L'epistolario di Marco Aurelio Severino (1580–1656)," in *Rivista di storia delle scienze mediche e naturali*, **5** (1923), 213–223, and the article by Schmitt and Webster, which summarizes the extant MS sources thus far uncovered. For a list of the seventy-seven vols. of MSS in the Biblioteca Lancisiana, which contain Severino materials, see P. De Angelis, *Giovanni Maria Lancisi, La Biblioteca Lancisiana, L'Accademia Lancisiana* (Rome, 1965), 151–163.

II. SECONDARY LITERATURE. The anonymous *Vita*, prefaced by Severino's *Antiperipatias* (1659), fols. 3v–4v, remains the most important contemporary source for his life. Of the more recent works, see esp. L. Amabile, "Marco Aurelio Severino," in *Rivista critica di cultura calabrese*, **2** (1922); "Due artisti ed uno scienziato . . . Marco Aurelio Severino nel Santo Officio Napoletano," in *Atti della Reale Accademia di scienze morali e politiche* (Società reale di Napoli), **24** (1891), 433–503; N. Badaloni, *Introduzione a G. B. Vico* (Milan, 1961), 25–37; L. Belloni, "Severino als Vorläufer Malpighis," in *Nova acta Leopoldina*, n.s. **27** (1963), 213–224, and "La dottrina della circulazione del sangue e la Scuola Galileiana, 1636–61," in *Gesnerus*, **28** (1971), 7–33; and *Biographisches Lexikon der hervorragenden Ärzte*, V (Munich–Basel, 1962), 242–243.

See also P. Capparoni, *Profili biobibliografici di medici e naturalisti celebri italiani dal secolo XV al secolo XVIII*, II (Rome, 1925–1928), 75–78; F. J. Cole, *History of Comparative Anatomy* (London, 1949), 132–149; Pietro Magliari, *Elogio istorico di M. A. Severino* (Naples, 1815); A. Portal, *Histoire de l'anatomie et de la chirurgie*, II (Paris, 1770), 493–505; C. B. Schmitt and C. Webster, "Harvey and M. A. Severino: A Neglected Medical Relationship," in *Bulletin of the History of Medicine*, **45** (1971), 49–75; and "Marco Aurelio Severino and His Relationship to William Harvey: Some Preliminary Considerations," in A. G. Debus, ed., *Science, Medicine and Society in the Renaissance*, II (New York, 1972), 63–72; and J. C. Trent, "Five Letters of Marcus Aurelius Severinus . . . ," in *Bulletin of the History of Medicine*, **15** (1944), 306–323.

An immense amount of material concerning Severino was collected by Luigi Amabile, who was prevented by his death from publishing a major work on him. His notes, including transcriptions from manuscripts in various libraries, are preserved in Naples, Biblioteca nazionale, MSS XI.AA.35–37.

CHARLES B. SCHMITT

SEVERINUS, PETRUS (or **PEDER SØRENSON**) (*b.* Ribe, Jutland, Denmark, 1542 [or 1540]; *d.* Copenhagen, Denmark, July 1602), *chemistry, medicine.*

Severinus attended the University of Copenhagen, where he lectured on Latin poetry at the age of twenty. After studying medicine briefly in France, he returned to Copenhagen to take his Master of Arts degree. He was officially appointed Professor Paedagogicus, and, with the offer of financial support from the University of Copenhagen, he set out with Johannes Pratensis (also a noted sixteenth-century Paracelsist) to study abroad. From 1565 to 1571 Severinus traveled throughout Germany, France, and Italy, attending various universities. Although he first matriculated at the University of Padua, he took his M.D. degree in France. Later, in Florence, he completed his major work, *Idea medicinae philosophicae* (1571), which he dedicated to the Danish king, Frederick II, a monarch genuinely interested in the sciences. On his return home Severinus was appointed canon of Roskild and became a physician to the court, a post he held for the next three decades. In 1602 he was offered the chair of medicine at the University of Copenhagen, but he died of the plague before the appointment officially began.

Severinus was widely known in the iatrochemical circles of his time. In Denmark he was closely associated at court with Tycho Brahe, who was claimed by sixteenth- and seventeenth-century chemists as a leading authority in this field. The writings of Severinus attest to his close relationship with the Paracelsians Livinius Battus and Theodor Zwinger the elder. The English iatrochemist Thomas Moffett, who visited Denmark in 1582, dedicated his important *De jure et praestantia chemicorum medicamentorum* (1585) to Severinus.

Only two of Severinus' many papers were published: *Idea medicinae philosophicae* (1571) and *Epistola scripta Theophrasto Paracelso* (1572). The latter, a short panegyric, was written thirty years after the death of Paracelsus and it reached its greatest audience when it was included in the

Latin edition of Paracelsus' works (1658). Far more important is the *Idea medicinae philosophicae,* which purported to contain the "entire doctrine of Paracelsus, Hippocrates, and Galen." Although earlier syntheses of the Paracelsian corpus had been written by Leo Suavius (1568) and Albert Wimpenaeus (1569), the work of Severinus was immediately accepted as one of the most authoritative documents of the Paracelsian school.

The *Idea* was a defense of the Paracelsian doctrines in opposition to the traditional medicine. In his attack Severinus labeled Galen as little more than a compiler who had been forced to arrange the work of his predecessors into some sort of order; seeking a unifying principle for this task, Galen had chosen the methods of the geometricians. His attempt to make medicine a part of geometry, with its own principles, axioms, and mathematical explanations, had been finally disproved, Severinus argued, only in recent years, when a number of new diseases had ravaged the Continent. Because they could not be controlled by physicians trained only in the traditional methods, it was therefore proper to seek something new and more effective. The answer was to be found in the medicine that was the glory of Paracelsus, a scholar whose method was devoid of the "mathematical" approach of the Galenists. In contrast, the truths to be found in his work were based on the fresh observations of the chemists. In one of the most frequently quoted passages of the sixteenth-century scientific literature, Severinus tells his readers to discard their books and to seek a knowledge of nature through personal experience.

The *Epistola* indicates in a few pages those Paracelsian texts known to Severinus, while the *Idea* shows just how deeply steeped in those sources he was. He fully accepted Paracelsus' endorsement of the macrocosm-microcosm universe; and he wrote that man has within him rivers, seas, mountains, and valleys in a fashion analogous to the greater world. Severinus accepted the doctrine of signatures and firmly condemned the humoral pathology of the ancients. Again, like Paracelsus, he broke with traditional medicine in affirming that the harmony of nature requires that like must cure like. In contrast, the Galenists insisted that contraries cured.

Much of the *Idea medicinae philosophicae* is devoted to the elements and the principles. The influence of traditional alchemy may be seen in the acceptance of both material and insensible elements, while the inconsistencies of Paracelsus regarding the relationship of the Aristotelian elements and the Paracelsian principles also is mirrored. Severinus' view of the universe was vitalistic and he believed that naturalists should seek out the vital principle in all substances. He stated that the elements contained certain forces, or *astra,* that, in connection with the chemical principles, formed *semina.* These *semina* were to be found in all parts of a given body—in man, however, they were perfected in the generative organs. The seed was properly called "astral" in nature because it had the magisterial power of life, which could not be destroyed even through the processes of putrefaction and dissolution.

As Pagel has shown, Severinus was the most eloquent exponent of epigenesis in the period between Aristotle and Harvey. He believed that the *semen* could give rise to a complex organism, not by virtue of the matter present but through its internal endowment, an intrinsic "knowledge" within it. While his views were significant, they were not based on embryological observations. In his role as defender of Paracelsus, Severinus placed strong emphasis on the supremacy of the heart because of its relationship to the vital spirit. Nevertheless, to him the role of the heart was somewhat less important than that of the blood, since the essential life force reached all parts of the body through this vehicle. Thus, although Severinus adopted the hard line of the Paracelsians against the ancients, his views on the primacy of the heart and the blood—as well as his espousal of epigenesis—show an Aristotelian influence and also mark him as a significant precursor of Harvey.

As the first major synthesis of the Paracelsian corpus, the *Idea medicinae philosophicae* was highly influential. Printed three times between 1571 and 1660, it was widely quoted not only by adherents of the new Paracelsian medicine but also by its opponents. Thomas Erastus wrote against Severinus in his attack on Paracelsus (1572–1573), and Daniel Sennert similarly discussed the views of Severinus in his *De chymicorum cum Aristotelicis et Galenicis consensu ac dissensu liber* (1619). Francis Bacon thought highly of the ability of Severinus, and he regretted only that he had devoted his time and talent to supporting the "useless" opinions of Paracelsus. Perhaps the greatest impact of the *Idea medicinae philosophicae* is to be found in the work of William Davison, who was appointed the first lecturer in chemistry at the Jardin des Plantes in Paris. In 1660 Davison's *Commentariorum in . . . Petri Severini Dani Ideam medicinae philosophicae . . . prodromus* was published along with the much shorter original text of the *Idea*. A condensed commentary, also by Davison, appeared in 1663.

BIBLIOGRAPHY

I. ORIGINAL WORKS. *Epistola scripta Theophrasto Paracelso: In qua ratio ordinis et nominum, adeoque totius philosophiae adeptae methodus compendiose et erudite ostenditur a Petro Severino Dano, philosophiae et medicinae doctore* (Basel, 1572) is most conveniently found in the Latin *Opera omnia* of Paracelsus, Fridericus Bitiskius, ed., I (Geneva, 1658), 4v–2r. The *Idea medicinae philosophicae. Continens fundamenta totius doctrinae Paracelsicae Hippocraticae & Galenicae* (Basel, 1571; 2nd ed., Erfurt, 1616) was reprinted (The Hague, 1660) with a long commentary by William Davison, the *Commentariorum in . . . Petri Severini Dani Ideam medicinae philosophicae . . . prodromus.* Davison's brief commentary entitled *Commentaria in Ideam medicinae philosophicae Petri Severini Dani, medici incomparabilis & philosophi sublimis* (The Hague, 1663) is an entirely different work. A contemporary English trans., probably made by A. Bartlet *ca.* 1600, of the *Idea medicinae philosophicae* exists in MS at the British Museum (Sloane MS 11).

II. SECONDARY LITERATURE. The secondary literature on Severinus is scanty, and Kurt Sprengel, *Versuch einer pragmatischen Geschichte der Arzneikunde*, 2nd ed., 5 vols. (Halle, 1800–1803), III (1801), 408–413, remains of interest. Important material on the Scandinavian Paracelsians is in Sten Lindroth, *Paracelsismen i Sverige till 1600-talets mitt* (Uppsala, 1943), 21–25, and *passim*; more recent are Eyvind Bastholm, "Petrus Severinus (1542–1602). A Danish Paracelsist," in *Proceedings of the XXI International Congress of the History of Medicine* (Siena, 1968), 1080–1085; and "Petrus Severinus (1542–1602). En dansk paracelsist," in *Särtryck ur Sydsvenska medicinhistoriska sällskapets årsskrift* (1970), 53–72.

On the relationship of the work of Severinus to the Paracelsian corpus, see Walter Pagel, *Paracelsus. An Introduction to Philosophical Medicine in the Era of the Renaissance* (Basel–New York, 1958), *passim*. More specifically, Pagel has investigated Severinus as a precursor of William Harvey in *William Harvey's Biological Ideas. Selected Aspects and Historical Background* (Basel–New York, 1967), 239–347.

The relationship of Severinus' views on mathematics to those of other Paracelsians is discussed in Allen G. Debus, "Mathematics and Nature in the Chemical Texts of the Renaissance," in *Ambix,* 15 (1968), 1–28, 211; and a summary of his thought from the standpoint of the chemist is included in J. R. Partington, *A History of Chemistry,* II (London, 1961), 163–164.

The impact of the work of Severinus on the contemporary English medical literature is treated in Allen G. Debus, *The English Paracelsians* (London, 1965; New York, 1966), 20 and *passim*. The extent of his influence throughout European medical circles is reflected in the many references in John Ferguson, *Bibliotheca chemica* (Glasgow, 1906; repr. London, 1954), II, 378–379.

ALLEN G. DEBUS

SEVERTSOV, ALEKSEY NIKOLAEVICH (*b.* Moscow, Russia, 11 September 1866; *d.* Moscow, 16 December 1936), *comparative anatomy, evolutionary morphology.*

Severtsov spent his early childhood in the village of Petrov, Voronezh gubernia, with his father, N. A. Severtsov, a zoologist and explorer, and his mother, S. A. Severtsova. He received his secondary education in a private gymnasium. After his graduation in 1885, Severtsov entered the department of physics and mathematics at Moscow University. Severtsov's teachers were M. A. Menzbir, I. M. Sechenov, K. A. Timiryazev, and V. V. Markovnikov.

While a student Severtsov, with P. P. Sushkin, entered a department competition on the organization and taxonomy of the Apoda and was awarded a gold medal. In 1895 Severtsov defended his master's dissertation, "O razvitii zatylochnoy oblasti nizshikh pozvonochnykh v svyaz; s voprosom o metamerii golovy" ("On the Development of the Occipital Area of the Lower Vertebrates in Connection With Metameres of the Head"). For the next three years he worked in the marine biological stations at Banyuls-sur-Mer, Villefranche, and Naples, and in the zoological laboratories at Munich and Kiel. The research done abroad was included in his doctoral dissertation, "Metameria golov elektricheskogo skata" ("Metameres of the Head of the Torpedo Ray"), which he defended in 1898. Severtsov did scientific and administrative work at Dorpat (now Tartu, Estonian S.S.R.), where from 1898 to 1902 he occupied the chair of zoology, then at Kiev (1902–1911), and at Moscow (1911–1930).

In 1930 Severtsov founded at Moscow University a laboratory of evolutionary morphology that later became the Institute of Evolutionary Morphology (now the A. N. Severtsov Institute of Evolutionary Morphology and Animal Ecology). In recognition of his scientific contributions he was elected an academician of the Soviet Academy of Sciences and of the Academy of Sciences of the Ukrainian S.S.R.

Severtsov chose comparative anatomy as his specialty, for he early recognized the necessity of extending morphology and evolutionary theory. Rather than study morphological regularities of evolution however, for a long time he limited himself to problems of comparative anatomy and phylogeny. His classic research on metameres of the vertebrate head (1895) revealed the evolution of the head. Severtsov devoted a number of works to explaining the origin and evolution of the extremities of vertebrates; he believed that the pentadactylic limbs of terrestrial vertebrates arose from the many-rayed fins of fish. In 1926 and 1934 he supplemented this conclusion with the idea that the fins had originated from the lateral folds. Clarifying the course of vertebrate phylogenetic development also was facilitated by Severtsov's research on the origin of the osseous scales and the evolution of the osseous skulls of fish. He believed that the covering parts of the skull arose from the osseous rhombic scales of the cutaneous cover. Severtsov also investigated the origin of the maxillary apparatus, the branchial skeleton, and breathing organs in fish.

Using material obtained through his own work and that of his pupils, Severtsov attempted to restate the evolution of the lower vertebrates in three articles (1916, 1917, 1925). He considered the lower cartilaginous fish, the basic groups of osseous fish, and the ancestors of land vertebrates. Severtsov proposed uniting Ostracodermi with the Cyclostomata in one group of ancient agnathous vertebrates. He tried to re-create the structure of the ancestors of the vertebrates—the primary Acrania and Gnathostomata, ancestors of agnathous and gnathostomatous vertebrates—and to draw a family tree of the lower vertebrates. The first two parts of "Issledovania ob evolyutsii nizshikh pozvonochnykh" ("Research on the Evolution of the Lower Vertebrates") were awarded the K. E. Baer Prize in 1919.

In 1910 Severtsov presented a report at the Twelfth Congress of Russian Natural Scientists and Physicians that defined the purposes and methodology of the evolutionary morphology of animals. In it he pointed to the scarcely touched area of research opened after the discovery of the biogenetic law by Haeckel, the principle of change of function by A. Dohrn, the law of substitution by N. Kleinenberg, and the phyletic correlation and certain other regularities of evolution.

Supporting the materialistic view of the evolution of the organic world, Severtsov denied the concepts of vitalism and autogenesis. Without knowledge of the laws of evolution, he held, it is impossible to understand either the laws of individual development or the laws of life in general. Opposing a dogmatic interpretation of Darwin's theory, he favored a bold posing of new problems and the introduction of new ideas. Severtsov devoted twenty-five years to the realization of this program. The result was the strict theory of the morphological regularities of evolution, the nucleus of which was the theory of phyloembryogeny and the morphobiological theory of evolution. Having rehabilitated the Darwinist principle of the variability of all ontogenetic stages in the process of evolution, Severtsov investigated modes of phylogenetic transformation of ontogenesis and their influence on the character and tempo of evolution. The theory of phyloembryogeny helped to overcome the limitations of the two opposing points of view: that ontogeny is a function of phylogeny, and vice versa. Severtsov saw the evolution of form and the evolution of ontogeny as mutually interacting processes.

Severtsov first studied the relations of ontogeny and phylogeny. The lively polemic concerning the theory of the gastraea, the theory of embryonic layers, problems of homology, and other phylogenetic questions indicated that the biogenetic law and theory of recapitulation in the form given by Haeckel was unsatisfactory for explaining the correlation of the individual and historical development of organisms. Turning from isolated consideration of the biogenetic law, Severtsov tried to free it from the traditional Haeckelian treatment and to give a new basis to the phenomenon of recapitulation in terms of his own theory of phyloembryogeny.

According to the latter, evolutionary changes arise through changes in the first stages of ontogeny (archallaxis), changes in the intermediate stages (deviation), and the addition of new final stages (anabolism). Thus this theory rehabilitated and supplemented the Darwinian principle of variations in all stages of ontogeny during evolution with concrete morphological data.

According to the theory of phyloembryogeny, the biogenetic law is the consequence of evolution by means of anabolism. Archallaxis and deviation limit the completeness of recapitulation. Severtsov showed that Baer's law, which asserted that the characteristics of small systematic groups appear late in ontogeny while the signs of larger systematic groups appear early, was not a general rule. He related the development of characteristics of animals in a determinate order to anabolic evolution, and viewed Baer's statement that characteristics of large systematic groups appear in the early stages of ontogeny as the consequence of the anabolic divergent monophyletic evolution.

The theory of phyloembryogeny also can be viewed as a morphological theory of ontogenetic evolution. Severtsov proposed an addition to it in 1934: the hypothesis of ontogenetic evolution of many-celled animals was invoked as an aid in determining the origin and regularity of development of modes of phyloembryogeny. Severtsov considered anabolism the primary method of ontogenetic evolution and the reason for the origin of primary recapitulation. His hypothesis assumed that ontogenetic evolution was completed in the early stages, not only through "piecing" of stages but also by means of archallaxis, deviation, and heterochronism. These secondary modes changed the linear order of stages of ontogeny and caused a reduction in recapitulation. Severtsov also included cenogenesis in the regular processes of ontogenetic evolution, showing the irregularity of its opposite, polygenesis.

The study of the relationship of ontogeny and phylogeny on the level of the whole organism, of separate organs, and of tissues showed the universality of the theory of phyloembryogeny. First formulated on the basis of the study of vertebrates, this theory was later recognized in the morphology of vertebrates, the morphology of plants, histology, physiology, and anthropology.

Darwin's theory presented evolution as a gradual progressive increase in organization, a replacement of lower forms by higher. Darwin gave a basically correct but also highly general solution to the problem of the development of species. His followers—T. H. Huxley, M. Neumayr, Mechnikov, V. O. Kovalevsky—continued the analysis of this problem. Ideas on progress that were alien to Darwinism were widespread, however: the mechanical-Lamarckian theory of ontogeny and autogenetic views. There was no coherent theory showing the basic trends of progress in the organic world from a Darwinian point of view. Severtsov's morphological theory of evolution filled that lack.

According to Severtsov (1925, 1931), biological progress occurs by means of a general increase in the life activity of the organism (aromorphosis), individual adaptation (idioadaptation or adaptation in the narrow sense), embryonic adjustment, and morphophysiological regression (for example, the transition to parasitism). In all but the first case the level of organization does not rise (and in the case of morphophysiological regression it is lowered); but, because it is better adapted to the conditions of existence, the group of organisms that is retarded in development gains the opportunity to compete successfully with more highly organized forms. Thus the existence of both highly organized and primitive forms among contemporary fauna is explained.

Using the theory of phyloembryogeny and the studies of the main directions of the evolutionary process, Severtsov closely associated the methods of the phylogenetic transformation of organs ("methods of transition," according to Darwin) with his hypothesis of correlation. According to this theory, in evolution a few characteristics change at first through heredity; the remaining characteristics and the organism as a whole then change in correlation with these primary changes. His hypothesis served as the basis for the solution of the problem of mutual adaptation (coadaptation) of organs in phylogenesis.

With the appearance of evolutionary morphology, work proceeded on the problem of reduction. Severtsov provided a detailed concept of the courses of reduction, distinguishing the sequential shedding of the final stages of development during the decrease in original formation of the organ (rudimentation) and the reduction of a normally formed organ until it disappears completely (aphanisia).

Severtsov limited his investigations to morphological regularities of evolution and to the routes by which the evolution of form and structure was achieved. Nevertheless, he clearly saw and upheld the tendency to synthesize the data and generalizations achieved by descriptive and experimental biology. Thus he understood and stated that a complete theory of the relation of individual development and evolution cannot be constructed exclusively from morphological material. In his opinion such a theory could be created only by synthesizing the data of evolutionary morphology, genetics, mechanics of development, and ecology. His prognosis was justified: in the 1930's and 1940's Severtsov's student I. I. Shmalgauzen achieved such a synthesis, creating the theory of the organism as a whole in its individual and historical development and the theory of the course and regularities of the evolutionary process.

BIBLIOGRAPHY

I. ORIGINAL WORKS. Severtsov's works have been collected in *Sobranie Sochinenii*, I. I. Shmalgauzen and

E. N. Pavlovsky, eds., 5 vols. (Moscow–Leningrad, 1945–1951). His basic writings are "O razvitii zatylochnoy oblasti nizshikh pozvonochnykh v svyazi s voprosom o metamerii golovy" ("On the Development of the Occipital Area of the Lower Vertebrates in Connection With Metameres of the Head"), in *Uchenye zapiski Moskovskogo universiteta*, Nat.-hist. cl. (1895), no. 2, 1–95; *Ocherki po istorii razvitia golovy pozvonochnykh* ("Sketches in the History of the Development of the Vertebrate Head"; Moscow, 1898); "Evolyutsia i embriologia" ("Evolution and Embryology"), in *Dnevnik XII Sezda russkikh estestvoispytateley i vrachey* ("Daily Journal of the XII Congress of Russian Natural Scientists and Physicians"; Moscow, 1910), 262–275; *Etyudy po teorii evolyutsii* ("Studies in the Theory of Evolution"; Kiev, 1912); *Sovremennye zadachi evolyutsionnoy teorii* ("Contemporary Problems in Evolutionary Theory"; Moscow, 1914); "Issledovania ob evolyutsii nizshikh pozvonochnykh" ("Research on the Evolution of the Lower Vertebrates"), in *Russkii arkhiv anatomii, gistologii i embriologii*, **1**, no. 1 (1916), 1–114; no. 3 (1917), 503–656; and **3**, no. 2 (1924), 279–360 – the first two parts are available in French as "Études sur l'évolution des vertébrés inférieures. I. Morphologie du squelette et de la musculature de la tête des Cyclostomes" and "II. Organisation des ancêtres des vertébrés actuels," in *Archives russes d'anatomie, d'histologie et d'embryologie*, **1**, no. 1 (1916) and no. 3 (1917); *Glavnye napravlenia evolyutsionnogo protsessa* ("Main Trends in the Evolutionary Process"; Moscow, 1925); *Morphologische Gesetzmässigkeiten der Evolution* (Jena, 1931); and *Morfologicheskie zakonomernosti evolyutsii* ("Morphological Regularities in Evolution"; Moscow–Leningrad, 1939).

II. Secondary Literature. See B. S. Matveev and A. N. Druzhinin, "Zhizn i tvorchestvo A. N. Severtsova" ("Life and Work of Severtsov"), in *Pamyati akademika A. N. Severtsova* ("Memories of Academician Severtsov"), I (Moscow–Leningrad, 1939); L. B. Severtsova, *Aleksey Nikolaevich Severtsov* (Moscow–Leningrad, 1946; 1951); and I. I. Shmalgauzen, *Nauchnaya deyatelnost akademika A. N. Severtsova kak teoretika-evolyutsionista* ("Scientific Activity of Severtsov as Evolutionary Theorist"), in *Pamyati akademika A. N. Severtsova*, I.

E. Mirzoyan

SEWARD, ALBERT CHARLES (*b.* Lancaster, England, 9 October 1863; *d.* London, England, 11 April 1941), *paleobotany.*

Seward's interest in geology was first inspired by John Marr's lectures. At Cambridge he took his degree in geology and botany, and on the advice of Thomas McKenny Hughes, decided to work on fossil plants. Botany at Cambridge was then emerging from a period of neglect, and Seward stated that it was quite late in his career when he first heard of living cells and protoplasm – "a revelation." To further his training as a paleobotanist he spent a year at Manchester with W. C. Williamson and then traveled to European countries where work was being done in that field.

Seward worked on the whole range of fossil plants, but about half his papers dealt with the Mesozoic. He was appointed a lecturer at St. John's College, Cambridge, in 1890 and later was made a fellow of the college; he became professor of botany in 1906. His steady stream of papers began in 1888, but it was his great revision of the English Weald flora (1895) that brought him fame and election at the age of thirty-five to the Royal Society. Collections of fossil plants from all over the world poured into Seward's laboratory. He collected very little until late in his life, when he spent a summer on the western Greenland Cretaceous.

Seward's synthetic papers dealt with the history of floras, especially with past climates as deduced from fossil plants and with the ways in which changing climates altered the geographical distribution of vegetation. His anatomical descriptions were careful, but he preferred the gross specimen on a large slab to the minute detail. Seward was not interested in the subtleties of comparative and evolutionary morphology and would, if possible, lump species rather than divide them. Impatient with the intricacies of nomenclature that had arisen during his working life, he often proceeded on the basis of what he considered good sense.

Seward became professor of the large and growing department of botany in 1906 and master of Downing College in 1915, both posts that many treated as full-time. In addition he gave generous help to the British Museum and British Association. Although he undertook nearly all of the elementary teaching in botany in addition to giving advanced lectures, he was able to set aside time almost every day for research. Not until he took on yet another full-time post, that of vice-chancellor of the university, was his paleobotanical research suspended.

Seward's retirement years were as well organized as his working life. At seventy-three he retired voluntarily from his chair and the mastership, and settled down to work near the British Museum. He made it a rule to finish all work before Easter and before Christmas; and so on Good Friday eve of 1941 he left on his desk the finished manuscript of his final paper and answers to all his letters, and went to bed. He died in his sleep.

BIBLIOGRAPHY

Seward's major works include *The Wealden Flora*, 2 pts. (London, 1894–1895); *The Jurassic Flora*, 2 vols. (London, 1900–1904); *Fossil Plants*, 4 vols. (Cambridge, 1898–1919); and *Plant Life Through the Ages* (Cambridge, 1931). See also "A Petrified Williamsonia From Scotland," in *Philosophical Transactions of the Royal Society*, **B204** (1912), 201; and "The Cretaceous Plant Bearing Rocks of Western Greenland," *ibid.*, **218** (1926), 215.

A bibliography of Seward's works is in his obituary in *Obituary Notices of Fellows of the Royal Society of London* (1941).

TOM M. HARRIS

SEXTUS EMPIRICUS (*fl. ca.* A.D. 200), *medicine, skeptical philosophy.*

The rediscovery of Sextus' writings in the sixteenth century and the publication of his *Pyrrhonian Hypotyposes* (or *Outlines of Pyrrhonism*) in a Latin translation in 1562 led to an epistemological crisis at the time of the Reformation. About the man himself, almost nothing is known. The name Sextus is Latin, but he wrote in Greek; and in *Against the Grammarians* he used the first person plural: "We say . . . whereas the Athenians and Coans say. . . ."[1] The name Empiricus signifies that he was a member of the Empirical school of medicine, which claimed to understand and treat diseases without postulating theoretical causes like "humors" or "spirits." Presumably Sextus was a practicing physician of this school, although he observed that of the medical schools it was the Methodists rather than the Empiricists who were closest to skepticism.[2] He also mentioned a work of his called *Medical Notes* (Ιατρικὰ ὑπομνήματα),[3] but this does not survive. His historical importance is in the field of philosophy, not medicine.

Sextus' surviving works include *Pyrrhonian Hypotyposes*, in three books, of which the first outlines the skeptical position and the second and third criticize other philosophical schools, subject by subject. The criticisms contained in these latter books are expanded in the rest of his work, which is usually grouped under the single title Πρὸς μαθηματικούς (*Adversus mathematicos*, or *Against the Professors*). There are eleven books, referred to by individual titles in the English edition by R. G. Bury (and in this article), as follows: I, *Against the Grammarians*; II, *Against the Rhetoricians*; III, *Against the Geometers*; IV, *Against the Arithmeticians*; V, *Against the Astrologers*; VI, *Against the Musicians*; VII–VIII, *Against the Logicians, I–II*; IX–X, *Against the Physicists, I–II*; and XI, *Against the Ethicists.*

Sextus probably derived most of his knowledge of earlier philosophers from handbooks and compendia, rather than from original sources; but he is a valuable source of information nonetheless. Occasionally he transcribed verbatim; for example, the prologue of Parmenides' great philosophical poem,[4] some fragments of Empedocles,[5] and Critias' poem on the invention of the gods[6] are known to us only from the text of Sextus. More often he gave summaries, some of which are very important: his critique of Pythagoreanism[7] gives details of that theory that are otherwise unknown. (P. Wilpert, *Zwei aristotelische Frühschriften über die Ideenlehre* [Regensburg, 1949], has claimed that much of this was derived from Plato's lectures, but the claim has not found much agreement.) Sextus gives the only surviving account of Diodorus Cronus' "arguments against motion."[8]

The Stoic school, however, founded by Zeno of Citium in Athens about 300 B.C. and by Sextus' time the most firmly established of all the philosophical schools, was his chief target. The division of philosophy into logic, physics, and ethics was Stoic, and he organized his own books according to this scheme. Logic, for him and also for the Stoics, included epistemology, and consequently was the most important division. *Against the Logicians* and the summary in book II of *Pyrrhonian Hypotyposes* constitute the fullest source of information on Stoic logic and have been used extensively in the recent reconstruction of the logical theory of the Stoics.[9]

Sextus' own professed philosophy of skepticism is derived ultimately from Pyrrho of Elis, but more directly from Aenesidemus. In *Pyrrhonian Hypotyposes*[10] Sextus enumerated the ten "tropes" for bringing about suspension of judgment, which elsewhere[11] he attributed to Aenesidemus. These tropes constitute a systematic exposition of arguments that throw doubt on the ability of the senses to give knowledge of the external world. They are followed by a critique of the notions of cause and sign, also derived from Aenesidemus.

Sextus' own position, which has to be collected from statements scattered over his works, amounts to an extreme form of skepticism. He distinguished between the external object and the phenomenon, and asserted that our sense perceptions are of phenomena only, that the external object is unknowable, and that nothing is true.[12] To count as true a statement must be verifiable and be about a real object: the first condition eliminates statements about the external world, and the second those about sense impressions. It follows that not even the statements of the skeptical philosopher are true. Sextus used the famous image of the ladder: just as a man can climb up to a high place and then kick the ladder down, so the skeptical philosopher can use argument to reach his position and ultimately demolish his own argument.[13] The result, according to Sextus, should be *ataraxia*, or peace.[14]

NOTES

1. *Against the Grammarians*, 246.
2. *Pyrrhonian Hypotyposes*, bk. I, 237–241.
3. *Against the Logicians*, bk. I, 202.
4. *Ibid.*, 111.
5. *Ibid.*, 122–124; *Against the Physicists*, bk. I, 127–129.
6. *Against the Physicists*, bk. I, 54.
7. *Ibid.*, bk. II, 248–309.
8. *Against the Physicists*, bk. I, 85–120.
9. See esp. B. Mates, "Stoic Logic"; and William and Martha Kneale, *Development of Logic* (Oxford, 1962), chap. 3.
10. Bk. I, 36–163.
11. *Against the Logicians*, bk. I, 345.
12. *Pyrrhonian Hypotyposes*, bk. II, 88 ff.
13. *Against the Logicians*, bk. II, 481; cf. L. Wittgenstein, *Tractatus Logico-philosophicus* (London, 1922), prop. 6.54.
14. *Pyrrhonian Hypotyposes*, bk. I, 8.

BIBLIOGRAPHY

I. ORIGINAL WORKS. The main source is Sextus Empiricus, *Opera*, 4 vols. (Leipzig, 1912–1962), vols. I–II, H. Mutschmann, ed.; vol. III, J. Mau, ed.; vol. IV, K. Janáček, ed., is an index. See also *Sextus Empiricus*, R. G. Bury, ed. and trans., 4 vols. (London–Cambridge, 1933–1949); and *Scepticism, Man and God: Selections From the Major Writings of Sextus Empiricus*, P. Halle, ed., Sanford Etheridge, trans. (Middletown, Conn., 1964).

II. SECONDARY LITERATURE. The most recent and best source is Charlotte Stough, *Greek Scepticism: A Study in Epistemology* (Berkeley–Los Angeles, 1969). See also V. Brochard, *Les Sceptiques Grecs*, 2nd ed. (Paris, 1955); Roderick Chisholm, "Sextus Empiricus and Modern Empiricism," in *Philosophy of Science*, **8** (1941), 371–384; A. Goedeckemeyer, *Die Geschichte des griechischen Skeptizismus* (Leipzig, 1905); K. Janáček, "Prolegomena to Sextus Empiricus," in *Acta Universitatis Palackianae Olomucensis*, **4** (1948), "Sextus Empiricus en der Arbeit," in *Philologus*, **100** (1956), 100–107, and *Sextus Empiricus' Sceptical Methods*, Acta Universitatis Carolinae Philologica, Monographia XXXVIII (Prague, 1972); Benson Mates, "Stoic Logic and the Text of Sextus Empiricus," in *American Journal of Philology*, **70** (1949), 290–298, and *Stoic Logic*

(Berkeley–Los Angeles, 1961); A. Philip McMahon, "Sextus Empiricus and the Arts," in *Harvard Studies in Classical Philology*, **42** (1931), 79–137; and Richard Popkin, *History of Scepticism From Erasmus to Descartes* (Assen, 1960).

DAVID J. FURLEY

SEZAWA, KATSUTADA (*b.* Yamaguchi, Japan, 21 August 1895; *d.* Tokyo, Japan, 23 April 1944), *applied mathematics, theoretical seismology.*

The son of a judge, Sezawa entered the Imperial University of Tokyo in 1918 and graduated as a shipbuilding engineer in 1921. He was then appointed assistant professor of engineering at the university and became a full professor in 1928. He first worked on problems of shipbuilding engineering and of vibration, in which he acquired a considerable reputation. In 1925 he started a lifelong association with the Earthquake Research Institute of the university, becoming its director in 1943. In 1932 he visited Britain, Germany, and the United States to further his studies in theoretical seismology, the field in which he is now best known. Sezawa was associated with many Japanese research organizations, including the Seismological Society of Japan, the Aeronautical Research Institute, and the Aviation Council. During World War II his responsibilities were greatly extended: he was supervisor of research projects at the Admiralty and a member of the Air Force Weapons Research Committee and was closely connected with army research.

When Sezawa began his research career, the new science of seismology presented challenging problems requiring sophisticated mathematical analysis. Sezawa's pioneering work on many of these problems provided the basis for important later developments both in Japan and elsewhere, and his mathematical ability was a significant factor in raising the world standard of seismological research. Much of this work was carried out in collaboration with his brilliant former pupil Kiyoshi Kanai.

Sezawa is particularly noted for his contributions to the theory of seismic surface waves, one of the main wave types generated by earthquakes. These waves tend to become increasingly regular in form as they move away from an earthquake source. Mathematical analysis of this property throws light on the structure of the outer part of the earth. By this means Sezawa derived useful estimates of layering in the earth's crust and produced evidence indicating that the Pacific crust is thinner than the Eurasian. More profoundly, his work pointed the way to important developments of the existing mathematical theory of seismic surface waves.

Sezawa produced a body of theory important to such seismological problems as seiches in lakes and tsunami (seismic sea waves) generated by large earthquakes, and the mechanism of earthquake generation. He applied his earlier studies of vibration theory to the problems of vibrations excited in buildings and bridges by strong earthquakes, contributed to the theory of designing structures to withstand earthquakes, and participated in related experimental work.

Although he was often unwell, Sezawa drove himself hard and lived very austerely. His health deteriorated seriously during the war, and he died at the age of forty-eight. The Imperial Academy of Sciences of Japan awarded him its highest honor, the Imperial Order of Merit, in 1931 and elected him a member in 1943.

BIBLIOGRAPHY

Most of Sezawa's 140 research papers on seismology and earthquake engineering were published as *Bulletins* of the Earthquake Research Institute of Tokyo Imperial University. Forty-six of his more important papers are listed in W. M. Ewing, W. S. Jardetzky, and F. Press, *Elastic Waves in Layered Media* (New York, 1957). He also wrote 50 papers on aeronautics, published in *Reports of the Aeronautical Research Institute, Tokyo University*; and 20 papers on ship construction, published as Reports of the Society of Naval Architecture of Japan.

An obituary was published in Japanese in *Jishin,* **16,** no. 5 (May 1944), 1–2.

K. E. BULLEN

SHAKERLEY, JEREMY (*b.* Halifax, Yorkshire, England, November 1626; *d.* India, *ca.* 1655), *astronomy.*

Shakerley was the son of William Shakerley of North Owram, Halifax. After a childhood in Yorkshire and a visit to Ireland, he settled in Pendle Forest, Lancashire, the address on his first surviving astronomical letter, sent to the London astrologer William Lilly in January 1648. The correspondence with Lilly, spanning 1648–1650, comprises nine letters and is concerned largely with patronage and astrology. It also shows that Shakerley

first became interested in mathematics around 1646. Over the years Lilly supplied him with books, stationery, and other aids, although he never extended to him the complete patronage he craved.

In 1649 Shakerley was taken into the Towneley household at Carré Hall, Burnley, Lancashire, where he was encouraged in his scientific pursuits. The Towneleys were prominent patrons of learning in the north of England and assisted the astronomers Crabtree and Gascoigne as well as Shakerley. While in the household of the Towneleys, who were Royalist Catholics, Shakerley continued to correspond with Lilly, a Parliamentarian and a Protestant—seemingly with the full knowledge of Towneley. Nothing is known of Shakerley's own religious allegiance, although he appears to have been a latitudinarian Protestant.

Like other practitioners of astronomy in the region, Shakerley was self-educated, having acquired his knowledge from the works of Kepler and Boulliau, whose achievements he greatly admired, as well as other authors. He was remarkably well versed in the literature of the "new philosophy"; and in his printed works he took great delight in ridiculing more conservative astronomers.

Many of the early letters to Lilly deal with a mixture of astronomical and astrological matters. They show that although Shakerley had a completely physical explanation for celestial mechanics, he nonetheless accepted that the natural motions of the planets also contained a supernatural element that vindicated his astrological beliefs. A letter describing the appearance of two mock suns in 1648 clearly illustrates this harmony of physical and astrological beliefs, for after giving a purely optical explanation for the mock suns, he invited Lilly to pronounce "astrological judgment" upon them. But as the correspondence progressed, Shakerley became increasingly skeptical about the validity of astrology, despite his enthusiasm for Lilly's intention to reform the science.

Shakerley was the first mathematician to recognize the significance of the work of the Liverpool astronomer Jeremiah Horrocks, whose papers had been acquired by Christopher Towneley and had been taken to Carré Hall after Horrocks' death. In each of his three published works, Shakerley spoke of Horrocks' researches with the highest praise, especially his discoveries concerning the lunar theory. When attacking Vincent Wing's *Urania practica* (1649), he used material from Horrocks' surviving papers to frame his own critique of

Wing's faulty lunar theory. The debt to Horrocks was again recognized by Shakerley in his *Tabulae Britannicae* (1653).

As a supplement to his almanac for 1651, Shakerley predicted a transit of Mercury on 24 October 1651. He described how best to observe it, using the projection method, and enumerates the new astronomical data that could be gained by doing so. Horrocks, who first observed a transit of Venus in 1639, was again cited; and it is clear that Shakerley had access to his unpublished papers on the transit, then at Carré Hall. He also referred to Gassendi, who in 1631 had been the first astronomer to observe a transit of Mercury. As the 1651 transit would occur during the night in European latitudes, he stated that it would be observed most advantageously in eastern lands.

Probably as a result of his attack on Vincent Wing, Lilly withdrew his support from Shakerley in 1650. No doubt because of the bleakness of his prospects at home, Shakerley emigrated to India. It was from Surat that he observed the Mercury transit predicted in his almanac. The observation began at 6:40 A.M.; but the proximity of adjacent buildings along with other demands on his time prevented him from making more than one sighting, and even this was impeded by his lack of adequate instruments. Mercury appeared to be "brownish black" and was less than half a minute in diameter. This information, and a drawing of the solar disk with Mercury delineated, was sent to Henry Osborne in London. The letter gives no indication as to Shakerley's occupation in Surat; but it is certain that he did not make the voyage merely to see the transit, as Vincent Wing suggested in 1669. He was probably an employee of the East India Company, although no mention of his name occurs in the company's records for the early 1650's.

Apart from reporting the Mercury transit, the Osborne letter relates other astronomical observations, such as that of a comet in 1652. Scarcely any references to astrology occur in this letter, and Shakerley confined himself to reporting purely physical data. While in India he became deeply interested in Brahmin astronomy. By the time of his letter to Osborne in January 1653, he had already learned a considerable amount, especially about the Indian calendar, and promised to devote more time to its study.

Shakerley was the second man ever to witness a Mercury transit, was probably the first Englishman to undertake systematic astronomical observations in India, and was one of the earliest men of science to seriously interest himself in the astronomy of the Brahmins.

After his letter to Osborne in 1653, nothing more was heard of Shakerley, although his *Tabulae Britannicae* was published at London in the same year. John Booker wrote to him in India in 1655, but he does not record having received a reply. In 1675 Edward Sherburne laconically remarked that Shakerley had "dyed in the East Indies."

BIBLIOGRAPHY

I. ORIGINAL WORKS. Twelve of Shakerley's letters are extant. Eleven are in the Bodleian Library, Ashmole 242 and 423, including his correspondence with Lilly and Osborne. Shakerley's letter on astrology and the doubts that he entertained therein, sent to John Matteson on 5 Mar. 1649, is published in the *Historical Manuscripts Commission Report, Various Collections,* **8** (1913), p. 61. Shakerley's three published works are *Anatomy of "Urania Practica"* (London, 1649); *Synopsis compendiana* (London, 1651), his almanac for 1651, which includes a supplement predicting the impending Mercury transit; and *Tabulae Britannicae* (London, 1653).

II. SECONDARY LITERATURE. Shakerley obtained posthumous recognition in Vincent Wing's *Astronomia Britannica*, 2nd ed. (London, 1669); and in the appendix to Edward Sherburne's *The Sphere of Marcus Manilius* (London, 1675).

ALLAN CHAPMAN

SHALER, NATHANIEL SOUTHGATE (*b.* Newport, Kentucky, 20 February 1841; *d.* Cambridge, Massachusetts, 10 April 1906), *geology.*

Shaler was the son of Nathaniel Burger Shaler, a physician, and Anne Hinde Southgate, daughter of a prosperous lawyer. He entered Lawrence Scientific School of Harvard University; studied under Louis Agassiz, although he came to reject Agassiz's anti-Darwinism, and earned the S.B. *summa cum laude* in 1862. His Civil War service in Kentucky curtailed by illness, Shaler returned to Harvard in 1864 as a university lecturer and remained there until his death, at which time he had become professor of geology and dean of the Lawrence Scientific School. Shaler worked intermittently for the U.S. Coast Survey and was director of the Kentucky Geological Survey (1873–1880), undertaking the first systematic survey of that state. In

1880 he joined the U.S. Geological Survey on a part-time basis and was head of its Atlantic Coast division from 1884 to 1900.

Shaler made a notable contribution to reclamation geology. His reports on inundated lands of the eastern United States and their reclamation value appeared within a decade after John Wesley Powell's *Report on the Lands of the Arid Region of the United States* (1878). Shaler said that an interesting reclamation alternative to irrigation of arid western lands would be the drainage of swamps and marshes. His concern with inundated lands reflected his broader interest in relating geology to the effects of man's activity upon the earth's surface. Shaler's *Aspects of the Earth* (1889), similar in mood to George Perkins Marsh's *Nature and Man* (1864), was intended to show the necessity for human understanding of the environment. This theme was later applied in his *Nature and Man in America* (1891), an important work on the environmental interpretation of history.

Shaler attained recognition through his popularization of geology, the sheer volume of his writing, and his national prominence within the profession (he was elected president of the Geological Society of America in 1895). Optimistic, enthusiastic, and sometimes prone to exaggeration, he contributed to the dissemination of geological knowledge and linked geology to human progress.

BIBLIOGRAPHY

I. ORIGINAL WORKS. The following scientific writings provide some indication of the range of Shaler's geological interests: "On the Formation of Mountain Chains," in *Geological Magazine,* **5** (1868), 511–517; "Sea Coast Swamps of the Eastern United States," in United States Geological Survey, *Sixth Annual Report* (Washington, D.C., 1886), 359–398; "The Geology of the Island of Mount Desert, Maine," *ibid., Eighth Annual Report* (Washington, D.C., 1889), 987–1061; "General Account of the Fresh-Water Morasses of the United States, With a Description of the Dismal Swamp District of Virginia and North Carolina," *ibid., Tenth Annual Report* (Washington, D.C., 1890), 255–339; "The Origin and Nature of Soils," *ibid., Twelth Annual Report* (Washington, D.C., 1892), 213–345; "Preliminary Report on the Geology of the Common Roads of the United States," *ibid., Fifteenth Annual Report* (Washington, D.C., 1895), 255–306.

Representative of Shaler's effort to popularize geological knowledge are *Illustrations of the Earth's Surface: Glaciers* (Boston, 1881); *The First Book of Geology* (Boston, 1884); *Aspects of the Earth* (New York, 1889); *Sea and Land: Features of Coasts and Oceans* (New

York, 1892); and *Outlines of the Earth's History* (New York, 1898). The last work was castigated by the geologist Israel C. Russell, who termed it a "nature-novel" in his review in *Science,* n.s. **8** (1898), 712–715.

Shaler's application of geology and geography to history is revealed in *Kentucky: A Pioneer Commonwealth* (Boston, 1884) and in *Nature and Man in America* (New York, 1891).

Shaler's memoirs were published as *The Autobiography of Nathaniel Southgate Shaler With a Supplementary Memoir by His Wife* (Boston, 1909).

II. SECONDARY LITERATURE. Brief assessments of Shaler are "Professors N. S. Shaler and I. C. Russell," in *Nature,* **74** (1906), 226–227; and "Nathaniel Southgate Shaler," in *Science,* n.s. **23** (1906), 869–872. A lengthy account is Walter L. Berg, "Nathaniel Southgate Shaler: A Critical Study of an Earth Scientist," unpub. diss., University Microfilms (Ann Arbor, Mich., 1957). A full bibliography of Shaler's writings is in John E. Wolff, "Memoir of Nathaniel Southgate Shaler," in *Bulletin of the Geological Society of America,* **18** (1908), 592–608.

WALTER L. BERG

SHANKS, WILLIAM (*b.* Corsenside, Northumberland, England, 25 January 1812; *d.* Houghton-le-Spring, Durham, England, 1882), *mathematics.*

Shanks's contributions to mathematics lie entirely in the field of computation, in which he was influenced by William Rutherford of Edinburgh. From 1847 his life was spent in Houghton-le-Spring, a small town in the coal-mining area of County Durham. There he kept a boarding school, and carried out his laborious and generally reliable calculations, most of which concerned the constant π, the ratio of the circumference of a circle to its diameter.

Modern methods for the calculation of π rely mainly on the formula

$$\text{arc tan } x = x - \frac{x^3}{3} + \frac{x^5}{5} - \cdots ,$$

discovered independently by James Gregory (1670) and Leibniz (1673). With $x = 1$ it yields

$$\frac{1}{4}\pi = 1 - \frac{1}{3} + \frac{1}{5} - \cdots ;$$

but the series converges too slowly to be of use, and more rapid processes may be obtained by using the Gregory series in formulas derived from the addition theorem

$$\text{arc tan } x + \text{arc tan } y = \text{arc tan } \{(x+y)/(1-xy)\}.$$

By repeated application of this theorem, John Machin (1706) found the convenient formula

$$\frac{1}{4}\pi = 4 \text{ arc tan}\left(\frac{1}{5}\right) - \text{arc tan}\left(\frac{1}{239}\right),$$

and calculated π to 100 decimal places.

This and similar formulas encouraged more extended calculations: here it is enough to note that in 1853 Rutherford gave 440 decimal places; and in the same year Shanks, in conjunction with Rutherford, gave 530 places, which proved to be his most accurate value. Also in 1853 Shanks gave 607 places, and the value to 500 places was independently checked. Some errors were corrected in 1873; and by that year Shanks, using Machin's formula, carried his calculations to 707 decimal places. There the matter rested for a considerable period. Subsidiary calculations provided the natural logarithms of 2, 3, 5, 10 to 137 decimal places and the values of 2^n, with $n = 12m + 1$ for $m = 1, 2, \cdots, 60$. Shanks also computed the value of e and of Euler's constant γ to a great many decimal places, and prepared a table of the prime numbers less than 60,000.

In 1944 D. F. Ferguson of the Royal Naval College, Dartmouth, attracted by the formula

$$\frac{1}{4}\pi = 3 \text{ arc tan}\left(\frac{1}{4}\right) + \text{arc tan}\left(\frac{1}{20}\right) + \text{arc tan}\left(\frac{1}{1985}\right),$$

proceeded to calculate π and compare his value with that given by Shanks. At the 528th decimal place there was a disagreement that was not reduced when Ferguson rechecked his own work, which he eventually carried to 710 decimal places. This discrepancy was communicated to R. C. Archibald, editor of *Mathematical Tables and Aids to Computation*, who suggested to J. W. Wrench, Jr., and L. B. Smith that they might recalculate π by Machin's formula; their value, to 808 decimal places, confirmed Ferguson's result and his identification of two terms omitted by Shanks, which had caused the latter's errors. Modern computing machinery has carried the calculation of π to great lengths: in 1949 the first such determination, by ENIAC, went to 2,000 decimal places; by 1960 at least 100,000 places were known.

BIBLIOGRAPHY

I. ORIGINAL WORKS. Shanks's book was *Contributions to Mathematics, Comprising Chiefly the Rectification of the Circle . . .* (London–Cambridge–Durham, 1853). His papers are listed in Royal Society *Catalogue of Scientific Papers*, V, 672; VIII, 941; and XI, 401; and include nine memoirs published in *Proceedings of the Royal Society*, **6–22** (1854–1874).

II. SECONDARY LITERATURE. Poggendorff mentions Shanks briefly, in III, 1241; but his reference in IV, 1390, to an obituary by J. C. Hoffmann in *Zeitschrift für mathematischen und naturwissenschaftlichen Unterricht*, **26** (1895), is misleading; this item merely reproduces Shanks's 1873 figures, with a little comment. Local sources could supply only the information that Shanks kept a school. A. Fletcher, J. C. P. Miller, and L. Rosenhead, *An Index of Mathematical Tables*, 2nd ed. (London, 1962), gives full bibliographical details and critical notes. A concise history of the evaluation of π is given by E. W. Hobson, *Squaring the Circle* (London, 1913; repr. 1953).

Ferguson's two notes on his evaluation of π are in *Mathematical Gazette*, no. 289 (May 1946), 89–90; and no. 298 (Feb. 1948), 37. A note by R. C. Archibald, J. W. Wrench, Jr., L. B. Smith, and D. F. Ferguson, in *Mathematical Tables and Other Aids to Computation*, **2** (Apr. 1947), gives agreed figures (as in Ferguson's second note) and adds some details.

T. A. A. BROADBENT

SHAPLEY, HARLOW (*b.* Nashville, Missouri, 2 November 1885; *d.* Boulder, Colorado, 20 October 1972), *astronomy*.

Shapley was born in a farmhouse near Carthage, Missouri, the son of Willis Shapley, a farmer and schoolteacher, and Sarah Stowell. He received the equivalent of a fifth-grade education in a nearby rural school, and later took a short business course in Pittsburg, Kansas. At age sixteen he was a reporter on the Chanute, Kansas, *Daily Sun*, and a year later worked briefly as a police reporter in Joplin, Missouri. Determined to qualify for college, he and his younger brother John applied to the high school in Carthage, but they were turned down as unprepared. Instead, they attended the Presbyterian Carthage Collegiate Institute, from which Harlow graduated after two semesters. (Although Harlow's intellectual ambition was not shared by his twin brother, Horace, his younger brother John became an eminent art historian.)

In 1907 Shapley enrolled at the University of Missouri, intending to enter the projected School of Journalism, only to find that the school would not open for another year. Consequently, he took up astronomy almost by accident. Shapley's choice of astronomy was reinforced in his third year when Frederick H. Seares, director of the Laws Observatory, offered him a teaching assistantship. After three years at the university, Shapley received a B.A. with high honors in mathematics and physics in 1910 and an M.A. in 1911. When recommending him for the Thaw fellowship in astronomy at Princeton, Seares mentioned Shapley's "phenomenal industry," his "independence of thought and a certain originality," and his "diversity of interest."

Upon receiving the fellowship in 1911, Shapley began working on eclipsing binaries with Henry Norris Russell, who became one of his closest friends and confidants. Their joint work, based on the use of new computing methods, for the first time yielded extensive knowledge of the sizes of stars. Besides the new methods of computing, Shapley used the polarizing photometer with the 23-inch refractor at Princeton, obtaining nearly 10,000 measurements. Within two years he had completed his doctoral dissertation. In an expanded version of his thesis, eventually published as a 176-page quarto volume in the *Princeton University Observatory Contributions*, Shapley analyzed ninety eclipsing binaries; scarcely ten orbits had previously been computed. Otto Struve later called this "the most significant single contribution toward our understanding of the physical characteristics of very close double stars."

As an important by-product of his research, Shapley disproved the commonly accepted opinion that Cepheid variables were binary stars. He showed that if the Cepheids were indeed double stars, the two components of their prototype, Delta Cephei, would have to fall inside each other. He therefore concluded that the Cepheid variables are not double but single stars that pulsate, thus changing their brightness as they change in size. Arthur Eddington carried out the theoretical analysis that made the pulsation hypothesis creditable but, as the extant correspondence reveals, there was always a close interaction between Shapley, the observer, and Eddington, the theoretician.

In 1913, as Shapley was finishing his thesis, he inquired about job prospects with Seares, who had left Missouri for the Mount Wilson Observatory. Seares arranged for Shapley to have an interview with George Ellery Hale, and shortly thereafter Shapley obtained a post at the California observatory. He did not go west immediately, but first took a five-month European tour with his brother John, and then stayed several months longer in Princeton to complete his monograph on eclipsing binaries. En route to Pasadena, on 15 April 1914, he married Martha Betz, whom he had met in a mathematics class at Missouri. Later she collaborated with Shapley on several papers and eventually became an expert in her own right on eclipsing binaries. The Shapley family grew to include a daughter and four sons.

The nature and direction of Shapley's research at Mount Wilson was foreshadowed by a visit he made to the Harvard College Observatory shortly before completing his graduate work at Princeton. There he discussed his future plans with Solon I. Bailey, who suggested that Shapley use the Mount Wilson sixty-inch telescope to study variable stars in globular clusters. It was precisely this suggestion that led to Shapley's most remarkable discoveries.

The globular clusters that became the focal point of Shapley's work are extremely remote and highly concentrated stellar systems, arranged in a spherical form and consisting of tens of thousands of stars. Before Shapley began his research at Mount Wilson, Bailey had already detected a number of Cepheid variables in the globular clusters. In addition, Henrietta Leavitt of Harvard had identified many variable stars in the two Magellanic Clouds. Her investigations indicated that the longer the periodic cycle of light variation, the brighter the star.

Shapley enlarged on Bailey's and Leavitt's work first by discovering many new Cepheid variables in globular clusters and second by devising a method of measuring distances to these clusters based on the relationship between Cepheid brightness and period. Before he could exploit this so-called period-luminosity relationship, Shapley had to calibrate the absolute brightness or luminosity of at least one Cepheid. Because no Cepheids are close enough to be measured by direct trigonometric methods, he relied on an ingenious statistical procedure to establish the distance and hence the luminosity of a typical Cepheid variable.

With his newly calibrated standard candle for the measurement of stellar distances, Shapley established a radically altered conception of the size of the Milky Way system. It is difficult to convey a sense of the intensive amount of work required to set up the magnitude sequences and to obtain the multiple plates needed to determine the periods of the variable stars. The fact that Shapley produced a series of eleven papers on star clusters before reaching his remarkable conclusions on galactic structure is indicative of the extraordinary number of hours devoted exclusively to data gathering. On 6 February 1917, Shapley wrote to the Dutch astronomer Kapteyn, who had been a regular visitor at Mount Wilson, that "the work on clusters goes on monotonously—monotonous so far as labor is concerned, but the results are continual pleasure. Give me time enough and I shall get something out of the problem yet."

A year later, on 8 January 1918, Shapley wrote to Eddington about a new breakthrough:

I have had in mind from the first that results more important to the problem of the galactic system than to any other question might be contributed by the cluster studies. Now, with startling suddenness and definiteness, they seem to have elucidated the whole sidereal structure. . . .

The luminosity-period law of Cepheid variation—a fundamental feature in this work—is now very prettily defined. It is based upon 230 stars with periods ranging from about 100 days to five hours. The measurement of the magnitudes necessary for the determination of the distances and space distribution of the clusters took a painful amount of stupid labor, but I am forgetting that for now we have the parallaxes of every one of them. . . .

To be brief, the globular clusters outline the sidereal system, but they avoid the plane of the Milky Way. . . . All of our naked-eye stars, the irregular nebulae, eclipsing binaries—everything we know about, in fact, and call remote, [belong to this system] except those compactly formed globular clusters, a few outlying cluster-type variables, the Magellanic clouds, and perhaps, the spiral nebulae. The globular clusters apparently can form and exist only in the parts of the universe where the star material is less dense and the gravitational forces less powerful than along the galactic plane. This view of the general system, I am afraid, will necessitate alterations in our ideas of star distribution and density in the galactic system.

The widely accepted view of the Milky Way in 1918 had resulted largely from the statistical work of Kapteyn. According to those laborious studies, the sun lay near the center of a flat lens-shaped stellar aggregation with the great majority of stars encompassed within a disk about 10,000 light-years in diameter. In contrast, Shapley maintained that Kapteyn's system, containing most of the stars and clusters that we can see, constituted only a small part of a much larger galactic system that was centered within the remote congregation of globular clusters in the direction of Sagittarius. In writing to Eddington, Shapley indicated that the equatorial diameter of the system was about 300,000 light-years, with a center some 60,000 light-years distant.

Walter Baade later described Shapley's achievement in his own picturesque way:

I have always admired the way in which Shapley finished this whole problem in a very short time, ending up with a picture of the Galaxy that just about smashed up all the old school's ideas about galactic dimensions.

It was a very exciting time, for these distances seemed to be fantastically large, and the "old boys" did not take them sitting down. But Shapley's determination of the distances of the globular clusters simply demanded these larger dimensions [*Stars and Galaxies*, p. 9].

Among the other very exciting things then going on at Mount Wilson was the discovery of novae in the spiral nebulae. At that time it was uncertain whether the spiral nebulae were satellites of our own Galaxy or "island universes," that is, stellar systems comparable in form and structure to the Milky Way but located far beyond our galactic system. Shapley realized that if the luminosities of the novae were known, these stars could then provide a key to the distances of the spirals. In 1917 he suggested that the Andromeda nebula had a distance of some one million light-years, a measure close to the result now accepted. Yet, almost immediately, Shapley withdrew his statement. The reason for his action was twofold. First, Adriaan van Maanen at Mount Wilson had studied the proper motions of stars in spiral nebulae and had found that the spirals were rotating. His investigation of the spiral M 101 led to the conclusion that if this object were actually located at the distance indicated by its nova, it would be an enormous galaxy and hence its linear motions would be just incredibly large—an appreciable fraction of the speed of light. (Those who argued that the spirals lay outside the galactic system were obliged to consider van Maanen's measures spurious, and subsequent research proved their view correct.)

By 1918 Shapley had a second reason for questioning the validity of the island universe theory. The Milky Way, as he had begun to envision it, was an enormous and lumpy structure that seemed to bear little resemblance to the spirals. He was loath to believe that the spirals could be comparable to the immense galactic system.

Shapley's indefatigable researching, speculating, and publicizing of his own views eventually led to the now famous debate on the scale of the universe presented before the National Academy of Sciences in Washington in April 1920. Throughout the encounter, Shapley maintained a cordial relationship with his opponent, Heber D. Curtis of the Lick Observatory, even though Curtis had written that the two speakers should go after each other "hammer and tongs." Shapley outlined his findings on the large dimensions of the Galaxy, presenting his points mainly in a nontechnical fashion. Curtis, on the other hand, spoke rather technically, trying his best to demolish Shapley's hypotheses about the luminosity of stars in globular clusters and to deflate the concept of a large distance scale for the galactic system. At the same time, Curtis argued quite correctly about the great distances of the spiral nebulae, a topic that Shapley tried to ignore. Before the debate there had been little direct communication between Shapley and Curtis, but for the joint publication, they freely exchanged working drafts. Hence Shapley's published version differs greatly from his comparatively popular oral presentation.

Shapley's alert and inquisitive nature led him to still other investigations on Mount Wilson. For example, he discovered a quantitative linear relation between temperature and the running speed of ants on the mountain, and he was always particularly proud of his five technical papers on ants in *Proceedings of the National Academy of Sciences*, *Psyche*, and *Bulletin of the Ecological Society of America*. Altogether, during his seven years at Mount Wilson he published over 100 papers.

Shapley's enthusiastic researching and his penchant for speculation sometimes led him astray. In the spring of 1918 he became excited about an explanation for the phenomenon of star streaming that completely missed the correct reasons. Ironically enough, it was the analysis of star streaming in terms of the rotation of our Galaxy, set forth by Bertil Lindblad in the 1920's, that provided one of the convincing finishing touches to Shapley's picture of the Milky Way.

While at Mount Wilson, Shapley had recognized that his own interests in variable stars and clusters were closely akin to the main concerns of the research programs carried out at Harvard under Edward C. Pickering's directorship, and he sometimes contemplated the possibility of becoming Pickering's successor. In his reminiscences Shapley wrote:

> The day I heard that Pickering had died, on my way home for lunch, I stopped at the corner of two streets—I could name them now—and pondered on whether I should give up a research career. Should I, or should I not? Should I curb my ambition? Finally I said to myself, "All right. I'll take a shot at it" [*Through Rugged Ways to the Stars*, p. 82].

In fact, the directorship was first offered to Henry Norris Russell, who on 13 June 1920 wrote frankly to Hale at Mount Wilson Observatory:

> If they accept this plan, I will then propose Shapley for second in command . . . consider what Shapley and I could do at Harvard!

> Between us, we cover the field of sidereal astrophysics rather fully. We can both do some theory,— and I might keep Shapley from too riotous an imagination,—in print. Moreover, Shapley knows the field of modern photographic photometry and is familiar with big reflectors. He would have good ideas for the use of the 60-inch mirror which is at Harvard, but has never been utilized. . . .
> Shapley couldn't swing the thing alone, I am convinced of this after trying to measure myself with the job. . . . But he would make a bully second, and would be sure to grow—I mean in knowledge of the world and of affairs; if he grew intellectually he would be a prodigy!

In the end, Russell turned down the position, President Lowell of Harvard then offered Shapley a staff appointment, but not the directorship. When Shapley promptly declined the job, Hale informed Lowell that the Mount Wilson Observatory would grant Shapley a year's leave of absence if Harvard wished to make a trial arrangement. Lowell agreed, and in April 1921 Shapley took up residence in Cambridge; on 31 October he was awarded the appointment as full director.

At Harvard Observatory, Shapley immediately offered his encouragement for the completion and extension of the *Henry Draper Catalogue* of stellar spectral classifications, and with various collaborators, including Annie Jump Cannon and Lindblad, he began extensive researches into the distribution and distances of stars of various spectral types. Even while at Mount Wilson, Shapley had hoped to use the Harvard objective prism plates to determine spectrographically the distances of bright southern stars, but Walter S. Adams, the acting director, had made him return the plates on the grounds that it was inappropriate for a Mount Wilson staff member to use observational material from elsewhere.

At Harvard, Shapley seized the opportunity to study the Magellanic Clouds, the objects in the southern hemisphere in which the period-luminosity relation for Cepheids had first been established. Because Harvard Observatory had maintained a southern station for many years, photographic plates were already available, and Shapley in 1924 revised his earlier distance estimate for the Small Magellanic Cloud upward to 100,000 light-years, at that time the largest published distance for any object. Throughout his tenure as director, Shapley was always proud of the existence of a southern station, and with it he established a virtual monopoly on the study of the Magellanic Clouds. In 1927 the station was moved from Arequipa, Peru, to Bloemfontein, South Africa, and simultaneously Shapley persuaded the Rockefeller Foundation to provide a sixty-inch reflector for the new site. The giant emission nebula, 30 Doradus, in the Large Magellanic Cloud received special study, and in 1937 he and John S. Paraskevopoulos published photographs from the Rockefeller reflector that showed for the first time the obscured nuclear cluster of blue-white supergiants. They also identified the red (M-type) supergiants in the association.

In February 1924 Edwin Hubble wrote to Shapley about his discovery of two Cepheid variables in the Andromeda nebula, M 31. Shapley responded that the letter was "the most entertaining literature I have seen for a long time," and promised to send a revised period-luminosity curve. Shapley must have realized at once that the spirals were, after all, extremely distant objects. His research interests turned increasingly toward these nebulae, which he called galaxies. By the end of the decade a considerable rivalry developed with Hubble, who called the spirals "extragalactic nebulae," and these terms became shibboleths in an even broader competition between east coast and west coast American astronomy. The rivalry was exacerbated by the Rockefeller Foundation decision to sponsor a 200-inch telescope for the Mount Wilson Observatory; Shapley naturally had hoped for greater development of his southern station.

Shapley's principal work on galaxies took the form of vast surveys that recorded tens of thousands of these objects in both hemispheres of the sky. His work showed not only the enormous numbers of galaxies, but also their irregular distribution, a point he emphasized in contrast to Hubble, who tended to stress the homogeneity necessary for simple cosmological modeling. An early result of these surveys was the "Shapley-Ames Catalogue" of 1,249 galaxies, including 1,025 brighter than the thirteenth magnitude.

Shapley's major discovery of the 1930's, a consequence of the galaxy surveys, was the identification of the first two dwarf systems, in the southern constellations Sculptor and Fornax, both now firmly established as members of our local family of galaxies.

After leaving Mount Wilson, Shapley's greatest contribution was not so much any particular astronomical discovery, but rather the extraordinarily stimulating environment he created at the Harvard College Observatory. Cambridge, Massachusetts, in the 1920's became the crossroad through which nearly every major astronomer passed, a status that culminated in the congress of the International Astronomical Union there in 1932. Cecilia Payne and Donald Menzel came to Harvard to pursue pioneering astrophysical problems, and Payne's doctoral thesis on stellar atmospheres, published as the first of the *Harvard Observatory Monographs*, was pronounced by Henry Norris Russell as the best he had ever read with the possible exception of Shapley's. Previously there had been no graduate program in astronomy at Harvard. Shapley quickly set about building a distinguished department whose alumni in turn became the leaders in other graduate programs throughout the country. Among the staff members Shapley brought to Harvard in the late 1920's to assist in building a graduate program were H. H. Plaskett and Bart J. Bok. The first Radcliffe and Harvard astronomy Ph.D.'s after Miss Payne, were Frank Hogg, Emma Williams, and Helen Sawyer. Graduates in the 1930's included Peter Millman, Carl Seyfert, Frank Edmondson, Jesse Greenstein, and Leo Goldberg.

Under Shapley the Harvard Observatory became a mecca for young astronomers throughout the world. In his early days there he became a confirmed internationalist, and during the late 1930's he helped rescue European refugee scientists and bring them to the United States. Bok reports, "One of these who came to Harvard [Richard Prager of Berlin] told me quietly and seriously that every night at least a thousand Jewish scientists were saying a prayer of thanks for Harlow Shapley's humanitarian efforts to help save them and their families."

A brilliant and witty speaker, Shapley accepted numerous lecture assignments, including the Halley lecture in Oxford (1928), the Darwin lecture of the Royal Astronomical Society (London, 1934), and the Henry Norris Russell lecture of the American Astronomical Society (Haverford, 1950), as well as popular lectures in churches and small colleges. His original insights dramatized the vastness of the universe and the peripheralness of man. A confirmed agnostic, he nevertheless frequently participated in conferences on science and religion and edited the book *Science Ponders Religion* (New York, 1960).

Shapley was the recipient of many honors, beginning with his election to the National Academy of Sciences in 1924 and the Draper Medal awarded him by the Academy in 1926. His other numerous awards included the Rumford Medal of the American Academy of Arts and Sciences in 1933, the Gold Medal of the Royal Astronomical Society in 1934, and the Pope Pius XI Prize in 1941. He became an honorary national academician in a dozen foreign countries, and won even more honorary doctorates. Shapley served as president of the American Astronomical Society (1943–1946), of the Society of the Sigma Xi (1943–1947), and of the American Academy of Arts and Sciences (1939–1944), which he was particularly instrumental in revitalizing.

After World War II, Shapley gave increasing priority to national and international affairs. Consequently, his effectiveness as an astronomer began to decline, and Harvard began to lose the leading position it had reached in astronomy during the 1930's, according to Bok, in his "Biographical Memoir of Harlow Shapley," in *Biographical Memoirs. National Academy of Sciences.* Bok adds that in retrospect it seems a pity that Shapley did not resign his directorship to assume some important administrative post in science commensurate with his role as citizen of the world.

One of Shapley's proudest achievements during the late 1940's was his role in the formation of the United Nations Educational, Scientific, and Cultural Organization. Kirtley Mather has written, "Shapley almost singlehandedly prevented the deletion of the 'S' from UNESCO." In 1945 Shapley was one of the Americans sent to London by the State Department to write the UNESCO Charter, and he firmly believed the opening lines: "Since wars begin in the minds of men, it is in the minds of men that the defenses of peace must be constructed."

In 1945 Shapley was Harvard's representative at the celebration of the 220th anniversary of the Academy of Sciences in Moscow. One of the few Americans permitted to visit the Soviet Union in that era, he became an outspoken champion of cooperation with Soviet intellectuals when such a view was becoming increasingly unpopular. For several years Shapley served as chairman of the Independent Citizens Committee of the Arts, Sciences, and Professions, an organization that helped raise money to support liberal candidates for Congress. In November 1946 he was subpoenaed by the House Committee on Un-American Activities. Congressman John Rankin, who had been sitting behind closed doors as a one-man subcommittee, emerged to state, "I have never seen a witness treat a committee with more contempt." A month later Shapley was elected president of the American Association for the Advancement of Science, a move interpreted as a rebuke to the committee and a vote of confidence in the Harvard astronomer. In the late 1940's Shapley made headlines when he chaired several meetings of left-wing organizations to which Russian delegates were invited. In March 1950 he was named by Senator Joseph McCarthy as one of five alleged Communists connected with the State Department, but later in the year Shapley was completely exonerated by the Senate Foreign Relations Committee.

Shapley continued as director of the Harvard College Observatory until the fall of 1952. In his seventies, he was still very active, giving much time to the grants committees of the American Philosophical Society and the Society of the Sigma Xi, and thoroughly enjoying himself as he traveled far and wide on lecture tours. Following his eighty-fifth birthday, his strength began to fail rapidly. He moved to Boulder, Colorado, where his son Alan resided, and there he died in 1972.

A versatile and imaginative thinker with a vivid personality, Shapley made devoted allies and bitter enemies. His friends called him a Renaissance man and forgave his vanity, while even his detractors conceded that he was one of the most stimulating figures in twentieth-century science.

BIBLIOGRAPHY

I. ORIGINAL WORKS. A. *Books and Articles.* An extensive bibliography of Shapley's writings (about 600 items), prepared by Mildred Shapley Matthews and based on an earlier version by Thomasine Brooks, will appear in *Biographical Memoirs. National Academy of Sciences.* Only his most noteworthy publications are listed here.

Shapley's doctoral dissertation was published as "A Study of the Orbits of Eclipsing Binaries," in *Princeton University Observatory Contributions,* no. 3 (1915). A related and pioneering paper was "On the Nature and Cause of Cepheid Variation," in *Astrophysical Journal,* **40** (1914), 448–465. Two dozen additional articles on binaries and variable stars appeared in *Astrophysical Journal* and *Publications of the Astronomical Society of the Pacific* during the years 1913–1916.

Shapley's Mount Wilson studies of clusters and galactic structure appeared between 1916 and 1921 primarily in two long series, "Studies of the Magnitudes in Star Clusters" (13 parts in *Proceedings of the National Academy of Sciences*) and "Studies Based on the Colors and Magnitudes in Stellar Clusters" (19 parts mostly in the *Astrophysical Journal,* but the first three only in *Contributions from the Mount Wilson Solar Observatory*); the most important of these studies are "Sixth Paper: On the Determination of the Distances of Globular Clusters," in *Astrophysical Journal,* **48** (1918), 89–124; "Seventh Paper: The Distances, Distribution in Space, and Dimensions of 69 Globular Clusters," *ibid.,* **48** (1918), 154–181; and "Twelfth Paper: Remarks on the

Arrangement of the Sidereal Universe," *ibid.*, **49** (1919), 311–336. An excellent review of his early ideas on the arrangement of the Milky Way is "Star Clusters and the Structure of the Universe," in *Scientia*, **26** (1919), 269–276, 353–361, and **27** (1920), 93–101, 185–193.

His changing views on the nature of the spiral nebulae are revealed in "Note on the Magnitudes of Novae in Spiral Nebulae," in *Publications of the Astronomical Society of the Pacific*, **29** (1917), 213–217, and "On the Existence of External Galaxies," *ibid.*, **31** (1919), 261–268. The famous Shapley-Curtis debate, in a much revised form, appears as "The Scale of the Universe," in *National Research Council Bulletin*, **2**, no 11; Shapley's argument is found on pages 171–193 and H. D. Curtis' on pages 194–217. Shapley's work on clusters culminated in his monograph *Star Clusters* (New York, 1930), and in the extended summary "Stellar Clusters," in *Handbuch der Astrophysik*, **5**, part 2 (1933), 698–773. A later review of this work is "A Half Century of Globular Clusters," in *Popular Astronomy*, **57** (1949), 203–229.

While at Harvard, Shapley produced a series of eight articles on "The Magellanic Clouds," in *Harvard College Observatory Circulars* (1924–1925), and, with Adelaide Ames, another series of five parts on "The Coma-Virgo Galaxies," in *Harvard College Observatory Bulletin* (1929). He also prepared, with Miss Ames, "A Survey of the External Galaxies Brighter than the Thirteenth Magnitude," in *Annals of Harvard College Observatory*, **88**, no. 2 (1932), commonly called the "Shapley-Ames catalogue." For descriptions of Shapley's vast galaxy surveys recorded in *Annals of Harvard College Observatory*, **88**, no. 5 (1935), and **105**, no. 8 (1938), see "A Study of 7900 External Galaxies," in *Proceedings of the National Academy of Sciences*, **21** (1935), 587–592, and "A Survey of Thirty-six Thousand Southern Galaxies," *ibid.*, **23** (1937), 449–453. The discovery of the Sculptor and Fornax dwarf galaxies is announced in "Two Stellar Systems of a New Kind," in *Nature*, **142** (1938), 715–716. From 1939 to 1942 Shapley published the first sixteen pages of a series entitled "Galactic and Extragalactic Studies," in *Proceedings of the National Academy of Sciences*; he completed the series finally with paper 23 in 1955.

At least as early as 1935 Shapley had outlined a proposed book on the structure of the Milky Way and the Magellanic Clouds, and had by 1948 planned a monograph solely on the Magellanic Clouds. Although the work was never completed in this form, its development strongly influenced the scope of his *Galaxies* (Philadelphia, 1943; rev. ed., Cambridge, Mass., 1961), in the semipopular Harvard Books on Astronomy series, as well as his Henry Norris Russell lecture to the American Astronomical Society in 1950 (summarized in "A Survey of the Inner Metagalaxy," in *American Scientist*, **39** (1951), 609–628). The Russell lecture and a Sigma Xi lecture, "The Clouds of Magellan, a Gateway to the Sidereal Universe," *ibid.*, **44** (1956), 73–97, were later expanded into *The Inner Metagalaxy* (New Haven,

1957). Among the research works synthesized in this volume was a series of seventeen papers, mostly by Shapley, entitled "Magellanic Clouds," appearing in the *Proceedings of the National Academy of Sciences* between 1951 and 1955. See also "Comparison of the Magellanic Clouds with the Galactic System," in *Publications of the Observatory of the University of Michigan*, **10** (1951), 79–84.

Shapley wrote several distinguished popular books of essays on astronomy: *Starlight* (New York, 1926), *Flights from Chaos: A Survey of Material Systems from Atoms to Galaxies* (New York, 1930), *Of Stars and Men* (Boston, 1958), *Beyond the Observatory* (New York, 1967), and *The View from a Distant Star* (New York, 1963). The last of these volumes incorporates one of his most influential and widely reprinted essays, "A Design for Fighting," originally in *American Scholar*, **14** (1945), 19–32. Another essay that Shapley held as one of his most significant was "Cosmography: an Approach to Orientation," in *American Scientist*, **42** (1954), 471–486.

Shapley prepared several anthologies including *A Source Book in Astronomy*, with H. E. Howarth (New York, 1929), and *Source Book in Astronomy 1900–1950* (Cambridge, Mass., 1960).

B. *Manuscripts*. The Harvard University Archives contains 216 archival boxes of Shapley's correspondence, manuscripts, and memorabilia. Under shelf mark HUG 4773.10 are filed the so-called personal papers deposited by Shapley and his family, with access and literary rights under the discretion of the Archives. Boxes 1a to 5d contain manuscripts including scientific papers and associated research, book typescripts, and radio scripts; found here also is the original manuscript of the 1920 debate with Curtis, as well as numerous other lectures. Boxes 23b and 23c contain early correspondence from about 1910 to 1921. Later personal correspondence, mostly after Shapley's retirement from the Harvard Observatory directorship in 1952, is found in boxes 18a–23a. Other large blocks of correspondence and memoranda relating mostly to non-observatory committees and assignments (including political affairs) are found in boxes 10a–17d. Travel diaries and biographical materials are found in 25a–26c. The second group of boxes, ninety in all, shelf mark UA V 630.22, contain the observatory director's correspondence for 1921–1954, with access by permission of Harvard College Observatory. In this category can be found Shapley's voluminous correspondence with Henry Norris Russell, for example, plus letters to and from George Ellery Hale, Edwin Hubble, Adriaan van Maanen, and many others.

II. SECONDARY LITERATURE. Shapley's book of autobiographical reminiscences, *Through Rugged Ways to the Stars* (New York, 1969), was produced from an oral interview, although with reliance on memories of an earlier anecdotal book that he researched but never wrote. (The original tapes are in the Niels Bohr Library, American Institute of Physics, in New York.) "It is not the very best of autobiographies, but it does show the true

Harlow Shapley with all his wonderful ideals, his vanity, his compassion and his greatness," writes Bart J. Bok in "Biographical Memoir of Harlow Shapley," in *Biographical Memoirs. National Academy of Sciences*, **48** (1976). An unpublished book-length biography has been written by Shapley's daughter Mildred Shapley Matthews. See also Bart J. Bok, "Harlow Shapley, Cosmographer," in *American Scholar*, **40** (1971), 470–474, and "Harlow Shapley—Cosmographer and Humanitarian," in *Sky and Telescope*, **44** (1972). 354–357. An extensive obituary appeared in *The New York Times*, 21 October 1972; see also Hudson Hoagland, "Harlow Shapley—Some Recollections," in *Publications of the Astronomical Society of the Pacific*, **77** (1965), 422–430.

Earlier accounts include Frank Robbins, "The Royal Astronomical Society's Gold Medallist," in *Journal of the British Astronomical Association*, **44** (1934), 177–179, and the entry in Anna Rothe and Evelyn Lohr, eds., *Current Biography 1952* (New York, 1953), 533–535. For other aspects of Shapley's career, see Kirtley Mather, "Harlow Shapley, Man of the World," in *American Scholar*, **40** (1971), 475–481, and Don K. Price, "The Scientist as Politican," in *American Academy of Arts and Sciences Bulletin*, **26** (1973), 25–34.

Shapley's earlier work is reviewed by Owen Gingerich, "Harlow Shapley and Mount Wilson," *ibid.*, 10–24; Otto Struve, "A Historic Debate About the Universe," in *Sky and Telescope*, **19** (1960), 398–401; Bart J. Bok, "Harlow Shapley and the Discovery of the Center of Our Galaxy," in J. Neyman, ed., *The Heritage of Copernicus—Theories "Pleasing to the Mind"* (Cambridge, 1975); Helen Sawyer Hogg, "Harlow Shapley and Globular Clusters," in *Publications of the Astronomical Society of the Pacific*, **77** (1965), 336–346; and Bart J. Bok, "Shapley's Researches on the Magellanic Clouds," *ibid.*, 416–421. See also Richard Berendzen, Richard Hart, and Daniel Seeley, *Man Discovers the Galaxies: Case Studies on the Development of Modern Astronomy, 1900–1940* (in press), especially sections I and II and the Reader.

OWEN GINGERICH

Al-SHARĪF AL-IDRĪSĪ. See **Al-Idrīsī, Abū ʿAbd Allāh Muḥammad ibn Muḥammad ibn ʿAbd Allāh ibn Idrīs, al-Sharīf al-Idrīsī.**

SHARONOV, VSEVOLOD VASILIEVICH (*b*. St. Petersburg, Russia, 10 March 1901; *d*. Leningrad, U.S.S.R., 26 November 1964), *astronomy, geophysics.*

Sharonov was the son of an Imperial opera singer. After graduating from the Gymnasium in 1918, he entered the Faculty of Physics and Mathematics of Petrograd University, from which he graduated only in 1926 after having served in the Red Army (1919–1924). From his youth Sharonov systematically conducted various astronomical observations, particularly of sunspots. While still a student he headed the section of solar research of the Russian Society of Amateurs of Astronomy and published instructions for conducting observations throughout the country; the results were then sent to Sharonov. He later recounted his experiences in *Solntse i ego nablyudenia* ("The Sun and Observations of It"; 1948, 1953).

As a graduate student (1926–1929) at the Leningrad Astronomical Institute (now the Institute of Theoretical Astronomy of the Soviet Academy of Sciences) Sharonov conducted a substantial part of his experimental work at the State Optical Institute in Leningrad and at the Pulkovo, Simeiz, Tashkent, and Leningrad University observatories. He defended his dissertation, on the theory and application of the photometric wedge, in October 1929. After working at the Tashkent observatory, he returned to Leningrad in 1930 as senior scientific worker. In 1932 Sharonov organized a photometric laboratory at the University of Leningrad (later the Laboratory of Planetary Astronomy) and carried out important projects on absolute photometry, on the colorimetry of the moon and planets, and on atmospheric optics. He devised and tested a number of photometric instruments for solving the problem of "visibility of distant objects and sources of light." In December 1940 Sharonov defended his doctoral dissertation, on "indexes of visibility."

From 1938 to 1941 Sharonov was a docent at Leningrad University, and from 1941 to 1944 he directed the astrophysics laboratory of the part of Leningrad University evacuated to Yelabuga, Tatar A.S.S.R. From 1944 he was professor at Leningrad University and, from 1951, director of its astronomical observatory.

Having studied the photometric wedge as the most convenient instrument for astronomical and geophysical observations, Sharonov worked from 1930 to 1936 at the Institute of Air Surveys on problems of aerophotometry applied to the aerial photographic survey of landscapes under various conditions of illumination and visibility. The original instruments that he developed for this purpose included an aeroexponometer, a sensitoscope, a haze measurer, an epsilometer, a diaphanometer, a universal wedge photometer, and a visual colorimeter. He used them in his photometrical laboratory and on expeditions to measure the transparency of the atmosphere and the visibility of distant objects. One group of his works was devoted to methods of spectrophotometry applied to the measurement of the color of the clear sky and to the determination of the solar light constant, which he found to be 135,000 lux.

Sharonov applied the absolute methods of photometry and colorimetry to study the lunar surface and the solar corona, which he observed successfully during seven total solar eclipses (1936–1963). He published tables and graphs of the variation in several photometric properties of more than 100 objects on the lunar surface. Comparing them with analogous investigations of rock and mineral specimens, he found confirming evidence of the "meteor-slag theory" advanced by his wife, N. N. Sytinskaya. According to this theory, almost all lunar rocks are covered with porous, spongy layers formed by the caking of particles that result from the fall of meteoric bodies, the latter being accompanied by explosions and sharp local rises in temperature.

Sharonov observed the oppositions of Mars at Tashkent in 1939, 1956, and 1958. His photometric and colorimetric research on the planets was summarized in *Fotometricheskie issledovania prirody planet i sputnikov* ("Photometric Research on the Nature of the Planets and Satellites"; 1954), *Priroda planet* ("The Nature of the Planets"; 1958), and *Planeta Venera* ("The Planet Venus"; 1965). He also developed the hypothesis that the surface of Mars is covered with limonite dust.

Sharonov was in charge of the Soviet study of noctilucent clouds for the International Geophysical Year in 1957–1959.

BIBLIOGRAPHY

I. ORIGINAL WORKS. Sharonov's more than 200 basic writings include "Issledovanie fotometricheskogo klina" ("Research on the Photometric Wedge"), in *Trudy Tashkentskoi astronomicheskoi observatorii*, 3 (1930), 84–100; "O sisteme i nomenklature astronomicheskikh svetovykh velichin" ("On the System and Nomenclature of Astronomical Light Quantities"), in *Astronomicheskii zhurnal*, 9, nos. 1–2 (1932), 82–101; "O kharakteristikakh otrazhatelnoy sposobnosti nebesnykh tel" ("On the Properties of the Reflecting Capacity of Heavenly Bodies"), *ibid.*, 11, no. 5 (1934), 473–483; "On the Determination of the Absolute Reflectivity of the Moon and Planetary Surfaces," in *Trudy Astronomicheskoi observatorii Leningradskogo gosudarstvennogo universiteta*, 6 (1936), 26–33; "A Simple Method of Checking the Purkinje Effect in Variable Star Observations," in *Variable Stars*, 5, no. 3 (1936), 68–70; "Absolute Photographic Photometry of Saturn's Disk," in *Poulkovo Observatory Circular*, nos. 26–27 (1939), 37–51;

"Opyt izmerenia absolyutnykh znacheny koeffitsientov yarkosti razlichnykh uchastkov lunnoy poverkhnosti" ("Experiment in Measuring the Absolute Values of the Coefficients of Brightness of Various Parts of the Lunar Surface"), in *Trudy Astronomicheskoi observatorii Leningradskogo gosudarstvennogo universiteta*, 10 (1939), 28–60; and "Universalny klinovoy fotometr" ("Universal Wedge Photometer"), *ibid.*, 72–81, repr. in *Uchenye zapiski Leningradskogo universiteta*, no. 31, Math. Ser., no. 3.

Later works are "Absolute Photographic Photometry and Colorimetry of Jupiter's Disk at the Opposition of 1928," in *Poulkovo Observatory Circular*, no. 30 (1940), 48–64; "Problemy absolyutnoy fotometrii tel solnechnoy sistemy" ("Problems of Absolute Photometry of Bodies in the Solar System"), in *Uchenye zapiski Leningradskogo Universiteta*, no. 53 (1940), 5–36; "Osveshchennost v lunnye nochi" ("Illumination on Moonlit Nights"), in *Astronomicheskie zhurnal*, 20, no. 1 (1943), 21–33; *Vidimost dalekikh predmetov i ogney* ("The Visibility of Distant Objects and Sources of Light"; Moscow–Leningrad, 1944); *Izmerenie i raschet vidimosti dalekikh predmetov* ("Measurement and Calculation of Visibilities of Distant Objects"; Moscow–Leningrad, 1947); *Mars* (Moscow–Leningrad, 1947); "Diafanoskop, ego teoria, issledovanie i primenenie" ("The Diaphanometer, Its Theory, Research, and Use"), in *Trudy Glavnoi geofizicheskoi observatorii imeni A. I. Voeikova* (1948), no. 11, 73–110; and "Opyt absolyutnoy fotometrii serebristykh oblakov" ("Experiment on the Absolute Photometry of Noctilucent Clouds"), in *Nauchny byulleten Leningradskogo universiteta*, no. 22 (1948), 5–16.

See also "Issledovanie otrazhatelnoy sposobnosti lunnoy poverkhnosti" ("Research on the Reflecting Capacities of the Lunar Surface"), in *Trudy Astronomicheskoi observatorii Leningradskogo gosudarstvennogo universiteta*, 16 (1952), 114–154, written with N. N. Sytinskaya; "'Yavlenie Lomonosova' i ego znachenie dlya astronomii" ("'Lomonosov's Phenomenon' and Its Importance for Astronomy"), in *Astronomicheskii zhurnal*, 29, no. 6 (1952), 728–737; "Problema fotometricheskikh nablyudeny lunnykh zatmeny" ("The Problem of Photometric Observations During Lunar Eclipses"), in *Vestnik Leningradskogo Universiteta*, no. 2 (1953), 47–61; "Fotometricheskie issledovania prirody planet i sputnikov" ("Photometric Researchs on the Nature of Planets and Satellites"), in *Uspekhi astronomicheskikh nauk*, 6 (1954), 181–249, which has a bibliography of 256 titles; and "Fotometricheskie i kolorimetricheskie sravnenia poverkhnosti Marsa s obraztsami limonita i gornykh porod krasnotsvetnykh tolshch" ("Photometric and Colorimetric Comparisons of the Surface of Mars With Specimens of Limonite and of Rock From the Red Layers"), in *Izvestiya Komissii po fizike planet Astrosoveta Akademii nauk SSSR*, no. 2 (1960), 30.

II. SECONDARY LITERATURE. See the obituary in *Astronomicheskii tsirkulyar. Byuro astronomicheskikh soobshchenii, Akademiya nauk SSSR*, no. 311 (1964); V. A. Bronshten, "Leningradsky issledovatel planet"

("Leningrad Investigator of the Planets"), in *Zemlya i vselennaya*, no. 5 (1969), 70–71; *Izvestiya Komissii po fizike planet Astrosoveta Akademii nauk SSSR*, no. 5 (1965), 105–111; L. N. Radlova, in *Astronomicheskii Kalendar na 1966* (Moscow, 1965), 242–245, with portrait; and *Uchenye zapiski Leningradskogo universiteta*, no. 328 (1965), 175–177.

P. G. KULIKOVSKY

SHARPEY, WILLIAM (*b.* Arbroath, Scotland, 1 April 1802; *d.* London, England, 11 April 1880), *anatomy, physiology.*

Sharpey was the posthumous son and fifth child of Henry Sharpey, a shipowner of Folkestone who had settled in Arbroath, Scotland, and his wife Mary Balfour. He was educated at Arbroath and at the University of Edinburgh, where he graduated M.D. in 1823. He became a fellow of the Royal College of Surgeons of Edinburgh in 1830. From 1831 to 1836 he taught anatomy extramurally in Edinburgh and in the latter year was appointed to the chair of anatomy and physiology at University College, London, in which post he spent the remainder of his professional life, retiring in 1874. He never married.

Sharpey became a fellow of the Royal Society of Edinburgh in 1834 and five years later was elected a fellow of the Royal Society, of which he was secretary from 1853 to 1872. At various times he was a member of many other official and learned bodies. He appears, after his London appointment, largely to have given up original work in favor of teaching and administration.

Although many who knew Sharpey considered him the real founder of the British school of physiology, comparing him in this respect with Johannes Müller in Germany, he made no great impact as a young man, and his appointment to the London chair evoked much surprise. He was of course a product of the famous Edinburgh medical school, still deeply influenced by the tradition of Cullen, John Gregory, and the first two Monros. Little is known of his early life, other than that he traveled extensively in Europe, studying under Panizza, Rudolphi, and Tiedemann and became familiar with French and German science and medicine. He published little original work—a few papers on cilia and ciliary motion, a long note on decidual structure in the English translation of Müller's *Handbuch der Physiologie des Menschen*, and a description in Quain's *Elements of Anatomy* of what are still referred to as Sharpey's bone fibers.

Sharpey's considerable authority stemmed from his membership of what would now be called the scientific "establishment," and from the fact that he was a great and inspiring teacher, who from his chair and by virtue of his position in the Royal Society did all that he could to further the development of physiology. His pupils included Joseph Lister; Michael Foster, who founded the Cambridge school of physiology; and E. A. Schäfer, who many years later added Sharpey's name to his own. He also collaborated with Burdon-Sanderson, who succeeded him and later became first holder of the chair of physiology at Oxford. These men, writing in the atmosphere of the late nineteenth century, were inclined to attribute much of Sharpey's influence to his firm opposition to vitalism in biology; but he was not averse to using vitalistic concepts in certain contexts, and the philosophy underlying his physiology must be interpreted with some care.

In the last years of his life Sharpey was much disturbed by the increasing agitation in Britain against experiments on living animals, and he kept in close touch with the negotiations that culminated in the act of 1876 "to amend the Law relating to Cruelty to Animals." In 1876 he was elected, along with Darwin, to honorary membership of the newly founded Physiological Society, an expression of the widely held opinion that for many years he had been the mainstay of physiology in Britain.

BIBLIOGRAPHY

The outlines of Sharpey's career are given in the obituary notice in *Proceedings of the Royal Society,* **31** (1880), xi–xix; in the *Dictionary of National Biography;* and in E. A. Sharpey-Schäfer, *History of the Physiological Society* (London, 1927), esp. 17–19, 31. The only modern and the only full-length study, by D. W. Taylor in *Medical History,* **15** (1971), 126–153, 241–259, includes a bibliography of his published work and draws to a considerable extent on unpublished MSS: sets of lecture notes including those taken by Lister, miscellaneous papers, referees' reports to the Royal Society, and letters written toward the end of his life to E. A. Schäfer and others. Sharpey's presidential address to the Physiological Section of the British Medical Association, in *British Medical Journal* (1862), ii, 162–171, is of considerable interest because it sets out his views about the state of his science after a lifetime of teaching and from a position of eminence among his contemporaries.

DOUGLASS W. TAYLOR

SHARPEY-SCHÄFER, EDWARD ALBERT (*b.* London, England, 2 June 1850; *d.* North Berwick, Scotland, 29 March 1935), *histology, physiology.*

Schäfer was the third son of J. W. H. Schäfer, who was born in Hamburg but had become a naturalized Englishman and a merchant in the City of London, and Jessie Browne. He was educated at University College, London, where he came under the lasting influence of William Sharpey and where he qualified in medicine in 1874. When Sharpey retired from his chair in that year and Burdon-Sanderson was appointed his successor and first Jodrell professor of physiology, Schäfer became assistant professor, succeeding Burdon-Sanderson in 1883. He was appointed to the chair of physiology at the University of Edinburgh in 1899 and retired in 1933. In 1878 he married Maud Dixey, who died in 1896; they had two sons and two daughters. In 1900 he married Ethel Roberts. One daughter died young in tragic circumstances, and both sons were killed in World War I. In 1918 he took the name of Sharpey-Schäfer "partly on Jack's account [his elder son, John Sharpey Schäfer], partly because it was the name of my old teacher and master in Physiology—the best friend I ever had."

Schäfer received many distinctions. He was elected a fellow of the Royal Society in 1878, and was awarded a Royal Medal in 1902 and the Copley Medal in 1924. He was president of the British Association in 1912, of the Eleventh International Physiological Congress, which met at Edinburgh in 1923, and of the Royal Society of Edinburgh in 1933. He also received many honorary degrees and was knighted in 1913.

In the laboratory Schäfer's catholicity of interest was impressive. He edited and contributed six chapters on topics as diverse as the biochemistry of blood, the ductless glands, the neuron, and cerebral localization to the *Textbook of Physiology,* a work for advanced students, in which each chapter was written by a leading authority.

Schäfer's early work was histological and embryological. His *Essentials of Histology* (1885) was one of the most widely used books on the subject in English; the sixteenth edition was published in 1954. One paper from those years in particular deserves to rank as a classic although its importance was not appreciated at the time. Whether nerve cells were separate and individual units, or whether their processes anastomosed to form a nerve net, was an eagerly debated question in the latter part of the nineteenth century. The main protagonists were Golgi and Ramón y Cajal. As a result of the interest of his friend G. J. Romanes in the problem of locomotion in Medusae, Schäfer was led to study the structure of the subumbrellar nervous plexus in *Aurelia aurita.* He found that each nerve fiber was distinct from and nowhere structurally continuous with any other; he thought it reasonable to assume fiber-to-fiber transmission from "inductive action," possibly electric, the result being the same as if there were a real network.

With E. Klein and J. N. Langley, Schäfer was appointed in 1881 to adjudicate on the conflicting claims of Goltz, on the one hand, and of Ferrier and Yeo, on the other, about the effects of ablation of defined areas of the mammalian cerebral cortex. This aroused his interest in cerebral localization and led him to seek the collaboration of Victor Horsley in a series of experiments that appreciably added to the results of Ferrier. This interest was later extended to include the spinal cord, and until the end of his career he worked intermittently on neurophysiological problems.

Schäfer made his most notable contributions in the field of endocrinology, and his papers with George Oliver on the effects of suprarenal and of pituitary extracts are landmarks in the history of physiology. Oliver, a practicing physician, sought Schäfer's advice on the supposed effects of orally administered suprarenal and other extracts. The two men showed that while most were inactive, intravenous injections of suprarenal extract produced a dramatic rise in arterial blood pressure, arterial constriction, vagal stimulation (reflex, as we now know), and an increase in the rate and force of cardiac contraction in vagotomized animals. They also pointed out that these effects derived from the medulla, not from the cortex of the gland. They then demonstrated the pressor effect of pituitary extract given intravenously. After Howell's discovery that this depended only on the posterior lobe of the gland, Schäfer continued to investigate the problem, which was the subject of his Croonian lecture to the Royal Society in 1909. He found, however, that posterior lobe extract apparently caused an increase in renal volume and a diuresis, independent of its pressor effect. These observations on the kidney led to considerable confusion among workers in the field, which was dissipated only by E. B. Verney's unequivocal demonstration of the antidiuretic effect in 1926.

The importance of the pancreas in carbohydrate metabolism had been obvious since the experiments of Joseph von Mering and Minkowski in 1889. As early as 1894 Schäfer pointed out on morphological grounds that the islet tissue might act collectively as an organ of internal secretion by means of which the pancreas produced its effect on the blood sugar level, an illustration of his prescience and of his oft-stated belief in the value of histology to physiology. In 1913, in his Lane lectures, he suggested the name "insuline" for the still hypothetical substance and also introduced the terms "autacoid" and "chalone" into endocrinology. (He later pointed out [*The Endocrine Organs,* 2nd ed., II, 343] that he was not the first to use the word "insuline.")

Schäfer's practical turn of mind showed itself in the invention of new laboratory methods and useful modifications of existing procedures. His name became familiar to many who could have had only the haziest idea of his eminence as a physiologist following the publication in 1903 of his method of artificial respiration—now superseded by mouth-to-mouth inflation of the lungs—for which he was awarded the Distinguished Service Medal of the Royal Life Saving Society in 1909.

An uncompromising opponent of the antivivisection movement, Schäfer did not mince words in public about what he regarded as the hypocrisy of those opposed to experiments on living animals. His presidential address to the British Association in 1912 offended many lay people, who considered it a dogmatically materialistic explanation of the origins of life in terms of physics and chemistry. During World War I, when anti-German hysteria was at its height, his forthright but unavailing defense of his colleague W. Cramer did not increase his popularity, even in academic circles.

Schäfer was an original member of the Physiological Society; and in 1926 he was appointed to write its history for its jubilee, a task for which he was uniquely qualified—knowing personally almost all the members since its inception. In 1908 he founded the *Quarterly Journal of Experimental Physiology,* which he continued to edit until 1933; in that year a special number was dedicated to him, written entirely by his pupils, the number and worldwide distribution of whom testified to his influence as a teacher. He remained an active experimentalist almost until his retirement, and one of his last papers reported his observations on the results of nerve section and regeneration. He himself, at the age of seventy-seven, had been the experimental subject.

BIBLIOGRAPHY

I. Original Works. Schäfer's output of published work was very large. It includes numerous short contri-

butions to the *Proceedings* of the Physiological and Royal societies, about sixty full-length scientific papers, an impressive number of textbooks, and lectures and addresses.

His contributions to science, apart from his textbooks, are to be found mainly in the *Philosophical Transactions* and the *Proceedings of the Royal Society,* in *Brain,* in *Journal of Physiology,* and in *Quarterly Journal of Experimental Physiology.* Of those on the nervous system, the most important are "Observations on the Nervous System of *Aurelia aurita,*" in *Philosophical Transactions of the Royal Society,* **169** (1878), 563–575; "A Record of Experiments Upon the Functions of the Cerebral Cortex," *ibid.,* **B179** (1888), 1–45, written with Victor Horsley; "An Investigation Into the Functions of the Occipital and Temporal Lobes of the Monkey's Brain," *ibid.,* 303–328, written with S. M. Brown; and "The Nerve Cell Considered as the Basis of Neurology," in *Brain,* **16** (1893), 134–169.

His fundamental work with G. Oliver on the suprarenal gland was reported in preliminary form in *Journal of Physiology,* **16** (1894), i–iv, and **17** (1895), ix–xiv, and, in full, as "The Physiological Effects of Extracts of the Suprarenal Capsules," **18** (1895), 230–276. The results of his research with Oliver on the pituitary are to be found briefly in *Journal of Physiology,* **18** (1895), 277–279; more fully, with S. Vincent, in "The Physiological Effects of Extracts of the Pituitary Body," *ibid.,* **25** (1899), 87–97; and, with P. T. Herring, in "On the Action of Pituitary Extracts on the Kidney," in *Philosophical Transactions of the Royal Society,* **199** (1908), 1–29. His Oliver-Sharpey lectures, "The Present Condition of Our Knowledge Regarding the Suprarenal Capsules," in *British Medical Journal* (1908), **1,** 1277–1281, 1346–1351; and his Croonian lecture, "The Functions of the Pituitary Body," in *Proceedings of the Royal Society,* **B81** (1909), 442–468, also should be consulted.

Schäfer's early statements about the possible function of the islets of Langerhans are given in his address to the Physiological Section of the British Association, in *Report of the British Association for the Advancement of Science,* **64** (1894), 795–814; that to the British Medical Association, in *British Medical Journal* (1895), ii, 341–348; and in his *Textbook of Physiology,* I (Edinburgh–London, 1898), 930; this book is a well-documented and valuable historical guide to the physiological knowledge of the time.

Schäfer's Lane lectures, given at Stanford University in 1913, were published as *An Introduction to the Study of the Endocrine Glands and Internal Secretions* (Stanford, 1914) and include a lengthy discussion on the terminology of the subject. They were published in a revised form as *The Endocrine Organs: An Introduction to the Study of Internal Secretion* (London–New York, 1916; 2nd ed., enl., 2 vols., 1924–1926).

His work on artificial respiration was reported as "Description of a Simple and Efficient Method of Performing Artificial Respiration in the Human Subject Especially in Cases of Drowning. To Which Is Appended Instructions for the Treatment of the Apparently Drowned," in *Medico-Chirurgical Transactions*, **87** (1904), 609–623.

Schäfer's presidential address to the British Association, in *Report of the British Association for the Advancement of Science*, **82** (1912), 3–36, is an interesting exposition of his views on physiology in general; as is his Horsley memorial lecture, "The Relations of Surgery and Physiology," in *British Medical Journal* (1923), **2**, 739–745. In addition to his *History of the Physiological Society During Its First Fifty Years 1876–1926* (London, 1927), an invaluable source of material is the collection of Sharpey-Schäfer papers in the Library of the Wellcome Institute of the History of Medicine, London. These include his private diaries, from which the quoted explanation of his change of name was taken, letters from physiologists all over the world covering a period of sixty years, and many other documents and newspaper clippings.

II. SECONDARY LITERATURE. The only secondary sources appear to be the obituary notices in *Quarterly Journal of Experimental Physiology . . .*, **25** (1935), 99–104; *Nature*, **135** (1935), 608–610; *Lancet* (1935), **1**, 843–845; *British Medical Journal* (1935), i, 741–742; *Obituary Notices of Fellows of the Royal Society*, no. 4 (1935); *Dictionary of National Biography, 1931–1940*; and Sherrington's Sharpey-Schäfer lecture, in *Edinburgh Medical Journal*, n.s. **42** (1935), 393–406.

DOUGLASS W. TAYLOR

SHARROCK, ROBERT (*b.* Adstock, England, June [?] 1630; *d.* Bishop's Waltham, England, 11 July 1684), *botany.*

Sharrock, the son of a clergyman, was educated at Winchester College and became a perpetual fellow of New College, Oxford, on 5 March 1649. After taking the B.C.L. in 1654 and the D.C.L. in 1661, he was ordained and held several church benefices, becoming a canon in 1669 and archdeacon of Winchester in 1684.

Sharrock took a scientific interest in the cultivation of plants. He was well acquainted with the classical work on plants and with the prevalent myths and superstitions concerning agriculture and horticulture, and he tested these against his own observations. His *History of the Propagation and Improvement of Vegetables* (1660) was dedicated to Robert Boyle and shows Sharrock's experimental approach to botany as well as a profound knowledge of methods of propagating plants by seeds, vegetative reproduction, budding, and grafting, and of the improvement of soil by cultivation and by leguminous crops. He was skeptical about the transmutation of species at a time when some professors of botany believed in it, and he

showed Boyle how the belief arose from insufficient investigation. He demonstrated that shreds and ashes of plants could not grow into new plants, and that grafting a red rose onto a white did not produce the striped *Rosa mundi*. He also conducted experiments on phototropism and made observations on the morphology of seeds and on phyllotaxy.

Had he devoted his life to the study of plants, Sharrock's experimental approach might have made him one of the most important botanists. But he also wrote other books on law, religion, and political philosophy, in which he attacked Hobbes's views on ethics. He also contributed prefaces to three of Boyle's treatises on physics.

BIBLIOGRAPHY

Sharrock's book on botany is *The History of the Propagation and Improvement of Vegetables by the Concurrence of Art and Nature* (Oxford, 1660, 1666, 1672; London, 1694), the last under the title *An Improvement to the Art of Gardening.*

Secondary literature includes *Athenae Oxonienses*, II (London, 1692), 580–581; J. Britten and G. S. Boulger, *A Biographical Index of Deceased British and Irish Botanists* (London, 1931), 152; J. Foster, *Alumni Oxonienses* (Oxford, 1892), 1340; J. R. Green, *A History of Botany* (London, 1914), 56, 125; and B. Porter, "Robert Sharrock," in *Dictionary of National Biography*, XVII, 1368–1369.

F. A. L. CLOWES

IBN AL-SHĀṬIR, ꜥALĀꜣ AL-DĪN ABUꜣL-ḤASAN ꜥALĪ IBN IBRĀHĪM (*b.* Damascus, Syria, *ca.* 1305; *d.* Damascus, *ca.* 1375), *astronomy.*

Ibn al-Shāṭir (Suter, no. 416) was perhaps the most distinguished Muslim astronomer of the fourteenth century. Although he was head *muwaqqit* at the Umayyad mosque in Damascus, responsible for the regulation of the astronomically defined times of prayer, his works on astronomical timekeeping are considerably less significant than those of his colleague al-Khalīlī. On the other hand, Ibn al-Shāṭir shared the interest of his earlier contemporaries Ibn al-Sarrāj, Ibn al-Ghazūlī, and al-Mizzī (Suter, nos. 508, 412, and 406) in astrolabes and quadrants; and he constructed sundials. Nevertheless, Ibn al-Shāṭir's most significant contribution to astronomy was his planetary theory. In his planetary models he incorporated various ingenious modifications of those of Ptolemy. Also, with the reservation that they are geocentric, his models are the same as those of Copernicus. Ibn al-Shāṭir's planetary theory was investigated for the first time in the 1950's, and the discovery that his models were mathematically identical to those of Copernicus raised the very interesting question of a possible transmission of his planetary theory to Europe. This question has since been the subject of a number of investigations, but research on the astronomy of Ibn al-Shāṭir and his sources, let alone on the later influence of his planetary theory in the Islamic world or Europe, is still at a preliminary stage.

Only a few details of the life of Ibn al-Shāṭir are known. His father died when the boy was six years old; and he was brought up by his grandfather, who taught him the art of inlaying ivory. At the age of about ten he traveled to Cairo and Alexandria to study astronomy, and, presumably, his interest in spherical astronomy was fired by the extensive compendium on spherical astronomy and instruments compiled in Cairo about 1280 by Abū ʿAlī al-Marrākushī (Suter, no. 363). In his early work Ibn al-Shāṭir revealed something of his debt to al-Mizzī, who also had worked in Egypt. In his treatise on the "perfect" quadrant, he depended on a pair of distinctive parameters also used, but not derived, by al-Mizzī: 33;27° for the latitude of Damascus and 23;33° for the obliquity of the ecliptic. In A.H. 765 (1363/1364) he derived the new pair of values 33;30° and 23;31°, which he employed in his later works.

Planetary Astronomy. Ibn al-Shāṭir appears to have begun his work on planetary astronomy by preparing a *zīj*, an astronomical handbook with tables. Some two hundred *zījes* were compiled by the astronomers of medieval Islam, and several had been prepared in Damascus prior to the time of Ibn al-Shāṭir (for example, Kennedy, nos. 15/16, 89, 41, and 42). Ibn al-Shāṭir's first *zīj*, which has not survived, was inappropriately called *Nihāyat al-ghāyāt fī 'l-aʿmāl al-falakiyyāt* ("The Final Work on Astronomical Operations") and was based on strictly Ptolemaic planetary theory. In a later treatise entitled *Taʿlīq al-arṣād* ("Comments on Observations"), he described the observations and procedures with which he had constructed his new planetary models and derived new parameters. No copy of this treatise is known to exist in the manuscript sources. Later, in *Nihāyat al-sūl fī tashīh al-uṣūl* ("A Final Inquiry Concerning the Rectification of Planetary Theory"), Ibn al-Shāṭir presented the reasoning behind his new planetary models. This work has survived.[1] Finally, Ibn al-Shāṭir's *al-Zīj al-jadīd* ("The New Astronomical Handbook"), which survives in a number of manuscript copies, contains a new set of planetary tables based on his new theory and parameters.[2]

Ibn al-Shāṭir introduced this later *zīj* in the following way:

God granted me success in working on this science [astronomy] and made it easy for me after I had mastered arithmetic, surveying, geometry, instrument making, and had actually invented many kinds of astronomical instruments. I came across the books of certain of my predecessors among the noted scholars in this branch of science, and I found that the most distinguished of the later astronomers, such as al-Majrīṭī Abu'l-Walīd al-Maghribī [Ibn Rushd?], Ibn al-Haytham, Naṣīr al-Ṭūsī, Muʾayyad al-ʿUrḍī [assistant to al-Ṭūsī, fl. Damascus and Persia ca. 1250], Quṭb al-Shīrāzī, and Ibn Shukr al-Maghribī [Suter, no. 376], and others, had adduced doubts concerning the well-known astronomy of the spheres according to Ptolemy. These doubts were indisputable and [concerned matters] incompatible with the geometrical and physical models that had been established [by Ptolemy]. These scholars took pains to make models that would adequately represent the longitudinal and latitudinal motions of the planets, and not introduce inconsistencies. They were not granted success, however, and they admitted this in their writings.

I therefore asked Almighty God to give me inspiration and help me to invent models that would achieve what was required, and God—may He be praised and exalted, all praise and gratitude to Him—did enable me to devise universal models for the planetary motions in longitude and latitude and all other observable features of their motions, models that were free—thank God—from the doubts surrounding previous models. I described these new models and gave the necessary proof of their viability in my book called *Taʿlīq al-arṣād*, "Comments on Observations," and I gave a short description of the models themselves in my book called *Nihāyat al-sūl fī tashīh al-uṣūl*, "A Final Enquiry Concerning the Rectification of Planetary Theory." Then I asked God—may He be exalted—for guidance in compiling a book that would contain [rules for] the precise determination of planetary positions and motions and the secrets of the planetary attributes, according to the mean motions that I found by observation, the distances that I computed, and the tables that I compiled on the basis of the new corrected astronomy. This book should be a fundamental work for people to rely on, in which astronomical operations and problems are precisely formulated. . . .

Of the surviving works by the scholars mentioned by Ibn al-Shāṭir, only the *Tadhkira* of al-Ṭūsī and the *Nihāyat al-idrāk* and the *Tuhfa shāhīya* of Quṭb al-Dīn al-Shīrāzī describe non-Ptolemaic planetary models.[3] Quṭb al-Dīn remarked several times in his treatises that most contemporary astronomers preferred such-and-such a non-Ptolemaic model, suggesting that he was one of several scholars who tried to modify the Ptolemaic models.

The essence of Ibn al-Shāṭir's planetary theory is the apparent removal of the eccentric deferent and equant of the Ptolemaic models, with secondary epicycles used instead. The motivation for this was aesthetic rather than scientific; the ultimate object was to produce a planetary theory composed of uniform motions in circular orbits rather than to improve the bases of practical astronomy. In the case of the sun, no apparent advantage was gained by the additional epicycle. In the case of the moon, the new configuration to some extent corrected the major defect of the Ptolemaic lunar theory, since it considerably reduced the variation of the lunar distance. In the case of the planets, the relative sizes of the primary and secondary epicycles were chosen so that the models were mathematically equivalent to those of Ptolemy.[4]

Below is a brief outline of Ibn al-Shāṭir's new planetary theory. All numbers are expressed sexagesimally (see Kennedy, p. 139).

The Solar Theory. The mean sun, $\bar{S}$, is situated on the deferent circle, radius $r = 60$, which rotates from west to east about the center of the universe O. The apogee moves from west to east about O at a rate of one degree in sixty Persian years of 365 days each. (Ibn al-Shāṭir accepted the rate of precession as one degree in seventy Persian years.)

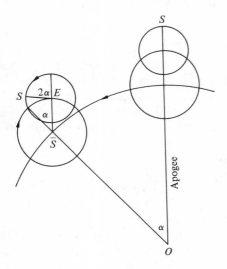

FIGURE 1. Ibn al-Shāṭir's solar model (α: apogee distance).

The primary epicycle has center $\bar{S}$ and radius $r_1 = 4;37$ and rotates with the motion of $\bar{S}$ relative to

the apogee and in the opposite sense. The radius $\bar{S}E$ thus remains parallel to the apsidal line. The true sun, S, is situated on the secondary epicycle, center E and radius $r_2 = 2;30$, which rotates with double the motion of $\bar{S}$ relative to the apogee and in the same sense.

The resultant maximum equation in this model is $2;2,6°$ and occurs when $\bar{S}$ is about 97° from the apogee, the position of which is given as Gemini 29;12° in December 1331. Ibn al-Shāṭir retained the Ptolemaic eccentricity 2;30 in his value of r_2; and his maximum equation corresponds to a resultant eccentricity of about 2;8, which is close to his value for $r_1 - r_2$. The solar distances at apogee and perigee are now 52;53 and 1,7;7, as against Ptolemy's 57;30 and 1,2;30. Ibn al-Shāṭir's new solar model appears to be the result of an attempt to make the variation of the solar distance correspond closely to the variation of the lunar distance in his new lunar model.

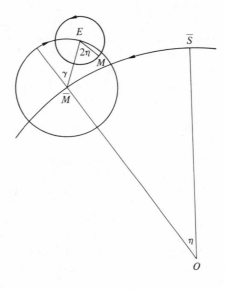

FIGURE 2. Ibn al-Shāṭir's lunar model (γ: mean anomaly; 2η: double elongation).

The Lunar Theory. The orbit of the moon is inclined at an angle of 5° to the plane of the ecliptic, and the nodes move from east to west with a constant motion. The mean moon, $\overline{M}$, is situated on the deferent, radius $r = 60$, which rotates about O from west to east in such a way that the resultant motion of $\overline{M}$ is the mean sidereal motion. The primary epicycle, center $\overline{M}$ and radius $r_1 = 6;35$, rotates with the mean anomaly in the opposite direction. The true moon, M, is situated on a secondary epicycle, centered at E on the first epicycle and having radius $r_2 = 1;25$, which rotates from west to east at twice the difference between the lunar and solar mean motions.

As a consequence of the resultant motion, the moon will always be at the perigee of the secondary epicycle at mean syzygies and at its apogee at quadrature. The apparent epicycle of radius $r_1 - r_2 = 5;10$ at syzygies accounts for the equation of center, and the gradual increase in its apparent radius to $r_1 + r_2 = 8;0$ as it approaches quadrature accounts for the evection. The maximum equation of the resultant epicycle is $7;40°$, which is Ptolemy's value. Also the lunar distance now varies between $r - (r_1 - r_2) = 54;50$ and $r + (r_1 - r_2) = 1,5;10$ at the syzygies and between $r - (r_1 + r_2) = 52;0$ and $r + (r_1 + r_2) = 1,8;0$ at the quadratures. Thus the major objection to the Ptolemaic model—in which the moon could come as close as $34;7$ to the earth at quadrature, so that its apparent diameter should be almost twice its mean value—was eliminated.

Planetary Theory. The mean planet, $\bar{P}$, here considered in the plane of the ecliptic, is situated on the deferent, radius $r = 60$, which rotates about the center of the universe from west to east with the mean longitudinal motion. The primary epicycle, radius r_1, rotates in the opposite direction at the same rate corrected for the motion of the apogees, again one degree in sixty Persian years. Thus the radius $\bar{P}E$ remains parallel to the apsidal line. The secondary epicycle, radius r_2, rotates about E from west to east at twice this rate. The true planet, P, is situated on the tertiary epicycle, radius r_3, which rotates with the mean anomaly about point F on the secondary epicycle. The anomaly is reckoned from the true epicyclic apogee, which is point G such that FG is parallel to $O\bar{P}$. In the case of the outer planets, FP remains parallel to the line joining O to $\bar{S}$. In the case of the inner planets, the direction $O\bar{P}$ defines $\bar{S}$.

In order to preserve the Ptolemaic distances in the apsidal line and at quadrature, the geometry of the models requires that

$$r_1 - r_2 = e \quad \text{and} \quad r_1 + r_2 = 2e,$$

where e is the Ptolemaic eccentricity, so that $r_1 = 3e/2$ and $r_2 = e/2$. At least in the case of the three outer planets, Ibn al-Shāṭir's values of r_1 and r_2 are precisely these. For Venus he takes

$$r_1 - r_2 = e \quad \text{and} \quad r_1 + r_2 = 2e',$$

where $e = 1;15$ is Ptolemy's eccentricity for Venus and $2e' = 2;7$ is the resultant double eccentricity of Ibn al-Shāṭir's own solar model.

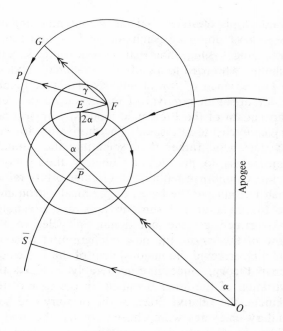

FIGURE 3. Ibn al-Shāṭir's model for the outer planets (α: apogee distance; γ: mean anomaly).

Because of the large eccentricity of the orbit of Mercury, Ibn al-Shāṭir's model was more elaborate than those for the other planets. Two additional epicycles placed at the end of r_3 have the effect of expanding and contracting its length in simple harmonic motion, with a period twice that of the mean longitudinal motion corrected for the motion of the apogee. Also, the sense of rotation of the epicycle with radius r_2 is the reverse of that for the other planets.

The solar, lunar, and planetary equation tables in *al-Zīj al-jadīd* were based on these new models. The accompanying mean motion tables, however, were based on parameters different from those stated in the *Nihāyat al-sūl*. Also, although in this treatise Ibn al-Shāṭir presented a new theory of planetary latitudes to accompany his new longitude theory, the latitude tables in *al-Zīj al-jadīd* were, with the exception of those for Venus, ultimately derived from Ptolemy's *Almagest*.[5]

Astronomical Timekeeping. Ibn al-Shāṭir compiled prayer tables, that is, a set of tables displaying the values of certain spherical astronomical functions relating to the times of prayer. The latitude used for these tables was $34°$, corresponding to an unspecified locality just north of Damascus. These tables, not discovered until 1974, display such functions as the duration of morning and evening twilight and the time of the afternoon prayer, as well as such standard spherical astronomical functions as the solar meridian altitude, the lengths of daytime and nighttime, and right and oblique ascensions. Values are given in degrees and minutes for each degree of solar longitude, corresponding roughly to each day of the year.

A more extensive set of tables for timekeeping at Damascus was compiled by al-Mizzī; but it was replaced by the corpus of tables compiled by al-Khalīlī, which were based on slightly different parameters.

Sundials. In A.H. 773 (1371/1372) Ibn al-Shāṭir designed and constructed a magnificent horizontal sundial that was erected on the northern minaret of the Umayyad mosque. The instrument now on the minaret is an exact copy made in the late nineteenth century by the astronomer al-Ṭanṭāwī, the last of a long line of Syrian *muwaqqits* working in the medieval astronomical tradition. Fragments of the original instrument are preserved in the garden of the National Museum, Damascus. Ibn al-Shāṭir's sundial, described for the first time in 1971 by L. Janin, consisted of a slab of marble measuring approximately one meter by two meters. A complex system of curves engraved on the marble enabled the *muwaqqit* to read the time of day in equinoctial hours since sunrise or before sunset and to reckon time with respect to daybreak and nightfall and with respect to the beginning of the interval during which the afternoon prayer should be performed, this being defined in terms of shadow lengths. The curves on this sundial probably were drawn according to a set of tables, compiled especially for the purpose, that displayed the coordinates of the points corresponding to the hours on the solstitial and equinoctial shadow traces. Tables of such coordinates for horizontal sundials to be used in Mecca, Medina, Cairo, Baghdad, and Damascus had been compiled early in the ninth century by al-Khwārizmī; and less than a century before Ibn al-Shāṭir's time, new sets of sundial tables for various latitudes had been compiled in Cairo by al-Marrākushī (Suter, no. 363) and al-Maqsī (Suter, no. 383). None of the several later sets of Islamic sundial tables that are still extant is attributed to Ibn al-Shāṭir. One such set based on his parameters survives, however, in MS Damascus Ẓāhirīya 9353, where it is attributed to al-Ṭanṭāwī.

A considerably less sophisticated sundial made by Ibn al-Shāṭir in A.H. 767 (1365/1366) is preserved in the Aḥmadiyya *madrasa* in Aleppo. It is contained in a box called *ṣandūq al-yawāqīt* ("jewel box"), measuring twelve centimeters by twelve centimeters by three centimeters. It could be used to find the times (*al-mawāqīt*) of the midday and afternoon prayers, as well as to establish the local meridian and, hence, the direction of Mecca.[6]

Astrolabes and Quadrants. Among the astronomers of Damascus and Cairo in the thirteenth, fourteenth, and fifteenth centuries, many varieties of quadrants rivaled the astrolabe as a handy analog computer. Certain instruments devised for solving the standard problems of spherical astronomy for any latitude were more of theoretical interest than of practical value. It should also be remembered that tables were available to Ibn al-Shāṭir for solving all such problems with greater accuracy than was possible using any of the several available varieties of quadrant.

Ibn al-Shāṭir wrote on the ordinary planispheric astrolabe and designed an astrolabe that he called *al-āla al-jāmiᶜa* ("the universal instrument").[7] Ibn al-Shāṭir also wrote on the two most commonly used quadrants, *al-rubᶜ al-muqanṭarāt* (the almucantar quadrant) and *al-rubᶜ al-mujayyab* (the sine quadrant). The first bore a stereographic projection of the celestial sphere for a particular latitude, and the second a trigonometric grid for solving the standard problems of spherical astronomy. A given instrument might have markings of each kind on either side.

Two special quadrants designed by Ibn al-Shāṭir were called *al-rubᶜ al-ᶜAlāʾī* (the ᶜAlāʾī quadrant, the appellation being derived from ᶜAlāʾ al-Dīn, part of his name) and *al-rubᶜ al-tāmm* (the "perfect" quadrant). Both quadrants were modifications of the simpler and ultimately more useful sine quadrant. No examples of either are known to survive. The ᶜAlāʾī quadrant bore a grid, like the sine quadrant, of orthogonal coordinate lines dividing each axis into sixty (or ninety) equal parts and also a family of parallel lines joining corresponding points on both axes. (In modern notation, if we denote the axes by $x = 0$ and $y = 0$, and the radius of the quadrant by $R = 60$, the grid consists of the lines $x = n$, $y = n$, $x + y = n$, for $n = 1, 2, \cdots, R$.) Ibn al-Shāṭir described how to use the instrument for finding products, quotients, and standard trigonometric functions, and for solving such problems as the determination of the first and second declination and right ascension for given ecliptic longitude, the length of daylight and twilight for a given terrestrial latitude and solar longitude, and the time of day for a given terrestrial latitude, solar longitude, and solar altitude. The "perfect" quadrant bore a grid of two sets of equispaced lines drawn parallel to the sides of an equilateral triangle inscribed in the quadrant with one axis as base. (In algebraic notation, the grid consisted of lines $y = \pm x \tan 60° + n$, for $n = 1, 2, \cdots, 60$.) The instrument could be used for solving the same problems as the ⁛Alā⁛ī quadrant. Ibn al-Shāṭir's treatise on the perfect quadrant concludes with a hundred questions and answers on topics relating to spherical astronomy.

Mechanical Devices. The Arab historian al-Ṣafadī reported that he visited Ibn al-Shāṭir in A.H. 743 (1343) and inspected an "astrolabe" that the latter had constructed. His account is difficult to understand, but it appears that the instrument was shaped like an arch, measured three-quarters of a cubit in length, and was fixed perpendicular to a wall. Part of the instrument rotated once in twenty-four hours and somehow displayed both the equinoctial and the seasonal hours. The driving mechanism was not visible and probably was built into the wall. Apart from this obscure reference we have no contemporary record of any continuation of the sophisticated tradition of mechanical devices that flourished in Syria some two hundred years before the time of Ibn al-Shāṭir.

Later Influence. There is no indication in the known sources that any Muslim astronomers after Ibn al-Shāṭir concerned themselves with non-Ptolemaic astronomy. The *zījes* of al-Kāshī and of Ulugh Beg (Kennedy, nos. 20 and 11), compiled in Samarkand in the first half of the fifteenth century, were the only astronomical works of major consequence prepared by Muslim astronomers after Ibn al-Shāṭir; and they are based on strictly Ptolemaic planetary theory following the tradition of the thirteenth century *Īlkhānī zīj* of al-Ṭūsī (Kennedy, no. 6). Nevertheless, later astronomers in Damascus and Cairo prepared commentaries on, and new versions of, Ibn al-Shāṭir's *Zīj al-jadīd.* His *zīj* was used in Damascus for several centuries, but it had to compete with adaptations of other works in which the planetary mean motion tables were modified for Damascus: a recension of al-Ṭūsī's *Īlkhānī zīj* prepared by al-Ḥalabī (*fl. ca.* 1425, Suter, no. 434); a recension of the *zīj* of Ulugh Beg prepared by al-Ṣāliḥī (*fl. ca.* 1500, Suter no. 454); and a recension of al-Kāshī's *Khāqānī zīj* prepared by Ibn al-Kayyāl (*fl. ca.* 1550, Suter, no. 474).

Another Damascus astronomer, Ibn Zurayq (*fl. ca.* 1400, Suter, no. 426), prepared an abridgment of Ibn al-Shāṭir's *zīj,* called *al-Rawḍ al-⁛āṭir,* that was very popular. Al-Ḥalabī, in one source (see Kennedy, no. 34) reported to have been a *muwaqqit* at the Hagia Sofia mosque in Istanbul but more probably to be identified with the Damascus astronomer mentioned above, compiled a *zīj* called *Nuzhat al-nāẓir,* based on that of Ibn al-Shāṭir. An

astronomer named al-Nabulusī (*fl. ca.* 1590), who may have worked in Damascus or Cairo, compiled a *zīj* called *al-Misk al-⁛āṭir* based on *al-Zīj al-jadīd.*

In Cairo, al-Kawm al-Rīshī (*fl. ca.* 1400, Suter, no. 428) adapted Ibn al-Shāṭir's planetary tables to the longitude of Cairo in his *zīj* entitled *al-Lum⁛a.* The contemporary Egyptian astronomer Ibn al-Majdī (Suter, no. 432; Kennedy, no. 36) compiled another set of planetary tables entitled *al-Durr al-yatīm,* from which planetary positions could be found with relative facility from a given date in the Muslim lunar calendar; he stated that the parameters underlying his tables were those of Ibn al-Shāṭir. Another Egyptian astronomer, Jamāl al-Dīn Yūsuf al-Khiṭāⁱī, prepared an extensive set of double-argument planetary equation tables based on those of Ibn al-Shāṭir.

Each of these works was used in Cairo for several centuries, alongside solar and lunar tables extracted from the *Ḥakimī zīj* of the tenth-century astronomer Ibn Yūnus (Kennedy, no. 14) and recensions of the *zīj* of Ulugh Beg prepared by Ibn Abi l-Fatḥ al-Ṣūfī (*fl. ca.* 1460; Suter, no. 447; Kennedy, no. 37) and Riḍwān ibn al-Razzāz (*fl. ca.* 1680; Kennedy, no. X209). The popularity of Ibn al-Shāṭir in Egypt is illustrated by the fact that a commentary on al-Kawm al-Rīshī's *zīj al-lum⁛a* was written in the mid-nineteenth century by Muḥammad al-Khuḍrī. There is evidence that Ibn al-Shāṭir's *zīj* was known in Tunis in the late fourteenth century but was replaced by a Tunisian version of Ulugh Beg's *zīj.* None of the numerous works purporting to be based on Ibn al-Shāṭir's *zīj* has been studied in modern times.

Ibn al-Shāṭir's principal treatises on instruments remained popular for several centuries in Syria, Egypt, and Turkey, the three centers of astronomical timekeeping in the Islamic world. Thus his influence in later Islamic astronomy was widespread but, as far as we can tell, unfruitful. On the other hand, the reappearance of his planetary models in the writings of Copernicus strongly suggests the possibility of the transmission of some details of these models beyond the frontiers of Islam.

NOTES

1. A critical edition of the Arabic text and an English translation have been prepared by V. Roberts but both are unpublished.
2. A brief summary of the contents of this *zīj* has been published by E. S. Kennedy.
3. These have been discussed in the secondary literature by E. S. Kennedy and W. Hartner, but more research is necessary before the extent of Ibn al-Shāṭir's debt to them and to other sources can be ascertained.

4. Ibn al-Shāṭir's planetary theory has been described in a series of four articles by E. S. Kennedy, V. Roberts, and F. Abbud. Before the theory may be more fully understood, however, the complete text of the *Nihāyat al-sūl* must be published with a translation and commentary, and also the relevant parts of the *Zīj al-jadīd*, including the planetary tables. It is also necessary to continue the search for his other works amidst the vast numbers of Arabic astronomical manuscripts which survive in libraries around the world untouched by modern scholarship. The manuscript (no. 66/5) in the Khālidiyya Library, Jerusalem, of a work attributed to Ibn al-Shāṭir that is entitled *Risāla fi 'l-hay²a al-jadīda* ("Treatise on the New Astronomy"), is unfortunately only a copy of his *Nihāyat al-su²l*.

5. For the other contents of *al-Zīj al-jadīd*, see the summary by E. S. Kennedy. The topics treated are the standard subject matter of *zījes*, although some of Ibn al-Shāṭir's tables for parallax and lunar visibility are of a kind not attested in earlier works.

6. A treatise on the use of Ibn al-Shāṭir's "jewel box" was written by the Egyptian astronomer Ibn Abi 'l-Fatḥ al-Ṣūfī (*fl. ca.* 1475; Suter, no. 447/7).

7. Two examples of this instrument, both made by Ibn al-Shāṭir in A.H. 738 (1337/1338) are preserved at the Museum of Islamic Art, Cairo, and the Bibliothèque Nationale, Paris, but have not been properly studied. Another astrolabe made by him in 1326 is preserved at the Observatoire National, Paris.

BIBLIOGRAPHY

I. ORIGINAL WORKS. For lists of Ibn al-Shāṭir's works and MSS thereof, consult H. Suter, no. 416; C. Brockelmann, *Geschichte der arabischen Literatur*, 2nd ed., II (Leiden, 1943–1949), 156, and supp., II (Leiden, 1937–1942), 157; and the much less reliable A. Azzawi, *History of Astronomy in Iraq* (Baghdad, 1958), 162–171, in Arabic.

The following titles are attributed to Ibn al-Shāṭir:

On planetary astronomy, *Nihāyat al-ghāyāt fi 'l-a*ᶜ*māl al-falakiyyāt*, an astronomical handbook with tables (not extant but mentioned in *Zīj Ibn al-Shāṭir*); *Nihāyat al-sūl f ʿī tashīh al-usūl*, on planetary theory (extant); *Taʿlīq al-arṣād*, on observations (not extant but mentioned in *Zīj Ibn al-Shāṭir*); and *Zīj Ibn al-Shāṭir* or *al-Zīj al-jadīd*, astronomical handbook with tables (extant).

His sole work on astronomical timekeeping is a set of prayer tables for latitude 34° (extant in MS Cairo Dār al-Kutub *mīqāt* 1170, fols. 11r–22v; intro. in MS Leiden Universitetsbibliothek Or. 1001, fols. 108r–113r).

Works on instruments are *al-Naf*ᶜ *al-*ᶜ*āmm fi 'l-*ᶜ*amal bi-l-rub*ᶜ *al-tāmm li-mawāqīt al-Islām*, on the "perfect" quadrant (extant); *Iḍāḥ al-mughayyab fi 'l-*ᶜ*amal bi-l-rub*ᶜ *al-mujayyab*, on the sine quadrant (extant); *Tuḥfat al-sāmi*ᶜ *fi 'l-*ᶜ*amal bi-l-rub*ᶜ *al-jāmi*ᶜ, on the "universal" quadrant (not extant, but see *Nuzhat al-sāmi*ᶜ *fi 'l-*ᶜ*amal bi-l-rub*ᶜ *al-jāmi*ᶜ); *Nuzhat al-sāmi*ᶜ *fī 'l-*ᶜ*amal bi-l-rub*ᶜ *al-jāmi*ᶜ, a shorter version of *Tuḥfat al-sāmi*ᶜ (extant); *al-Ashi*ᶜᶜ*a al-lāmi*ᶜ*a fī 'l-*ᶜ*amal bi-l-āla al-jāmi*ᶜ*a*, on the "universal instrument" (extant); *al-Rawḍāt al-muzhirāt fi 'l-*ᶜ*amal bi-rub*ᶜ *al-muqanṭarāt*, on the use of the almucantar quadrant (extant); *Risāla fi 'l-rub*ᶜ *al-*ᶜ*Alā²ī*, on the *ᶜAlā²ī* quadrant (extant); *Risāla fi 'l-asṭurlāb*, on the astrolabe (extant); *Risāla fi uṣūl* ᶜ*ilm al-asṭurlāb*, on the principles of the astrolabe (extant); and *Mukhtaṣar fi 'l-*ᶜ*amal bi-l-asṭurlāb wa-rub*ᶜ *al-muqanṭarāt wa-l-rub*ᶜ *al-mujayyab*, on the use of the astrolabe, almucantar quadrant, and sine quadrant (extant).

Miscellaneous writings are *Fi 'l-nisba al-sittīniya*, probably on sexagesimal arithmetic (extant); *Urjūza fi 'l-kawākib*, a poem on the stars (extant); *Risāla fi istikhrāj al-ta²rīkh*, on calendrical calculations (extant); and *Kitāb al-jabr wa-l-muqābala*, on algebra (Azzawi, p. 165, states that there is a work with this title by Ibn al-Shāṭir preserved in Cairo).

II. SECONDARY LITERATURE. References to Suter and Kennedy are to the basic bibliographical sources: H. Suter, *Die Mathematiker und Astronomen der Araber und ihre Werke* (Leipzig, 1900), and E. S. Kennedy, "A Survey of Islamic Astronomical Tables," in *Transactions of the American Philosophical Society*, n.s. **46**, no. 2 (1956), 121–177.

The only biographical study thus far is E. Wiedemann, "Ibn al Schâtir, ein arabischer Astronom aus dem 14. Jahrhundert," in *Sitzungsberichte der physikalisch-medizinischen Sozietät in Erlangen*, **60** (1928), 317–326, repr. in his *Aufsätze zur arabischen Wissenschaftsgeschichte*, II (Hildesheim, 1970), 729–738.

On the *Zīj* of Ibn al-Shāṭir, consult E. S. Kennedy, "A Survey of Islamic Astronomical Tables," in *Transactions of the American Philosophical Society*, n.s. **46**, no. 2 (1956), no. 11. See also A. Sayili, *The Observatory in Islam* (Ankara, 1960), 245.

On Ibn al-Shāṭir's planetary theory, see the following, listed chronologically: V. Roberts, "The Solar and Lunar Theory of Ibn al-Shāṭir: A Pre-Copernican Copernican Model," in *Isis*, **48** (1957), 428–432; E. S. Kennedy and V. Roberts, "The Planetary Theory of Ibn al-Shāṭir," *ibid.*, **50** (1959), 227–235; F. Abbud, "The Planetary Theory of Ibn al-Shāṭir: Reduction of the Geometric Models to Numerical Tables," *ibid.*, **53** (1962), 492–499; V. Roberts, "The Planetary Theory of Ibn al-Shāṭir: Latitudes of the Planets," *ibid.*, **57** (1966), 208–219; E. S. Kennedy, "Late Medieval Planetary Theory," *ibid.*, **57** (1966), 365–378; and W. Hartner, "Ptolemy, Azarquiel, Ibn al-Shāṭir, and Copernicus on Mercury: A Study of Parameters," in *Archives internationales d'histoire des sciences*, **24** (1974), 5–25.

The possible transmission of late Islamic planetary theory to Europe is discussed in W. Hartner, "Naṣīr al-Dīn's Lunar Theory," in *Physis: Rivista internazionale di storia della scienza*, **11** (1969), 287–304; E. S. Kennedy, "Planetary Theory in the Medieval Near East and Its Transmission to Europe," and W. Hartner, "Trepidation and Planetary Theories: Common Features in Late Islamic and Early Renaissance Astronomy," in *Accademia Nazionale dei Lincei*, **13°** *Convegno Volta*, (1971), 595–604 and 609–629, respectively; I. N. Veselovsky, "Copernicus and Naṣīr al-Dīn al-Ṭūsī," in *Journal for the History of Astronomy*, **4** (1973), 128–130; G. Rosinska, "Naṣīr al-Dīn al-Ṭūsī and Ibn al-

Shāṭir in Cracow?" in *Isis*, **65** (1974), 239–243; and W. Hartner, "The Astronomical Background of Nicolaus Copernicus," in *Studia Copernicana* (1975).

On the quadrants designed by Ibn al-Shāṭir, see P. Schmalzl, *Zur Geschichte des Quadranten bei den Arabern* (Munich, 1929). On his sundial, see L. Janin, "Le cadran solaire de la mosquée Umayyade à Damas," in *Centaurus*, **16** (1971), 285–298. Ibn al-Shāṭir's "jewel box" is described and illustrated in S. Reich and G. Wiet, "Un astrolabe syrien du XIVᵉ siècle," in *Bulletin de l'Institut français d'archéologie orientale du Caire*, **38** (1939), 195–202. See also L. A. Mayer, *Islamic Astrolabists and Their Works* (Geneva, 1956), 40–41.

DAVID A. KING

SHATUNOVSKY, SAMUIL OSIPOVICH (*b.* Znamenka, Melitopol district, Tavricheskaya guberniya, Russia, 25 March 1859; *d.* Odessa, U.S.S.R., 27 March 1929), *mathematics*.

Shatunovsky was the ninth child in the family of an impoverished artisan. In 1877 he graduated from a technological high school in Kherson and the following year completed a specialized supplementary course at Rostov. He then studied for a short time at the Technological College and the College of Transport in St. Petersburg. Shatunovsky, however, was interested in mathematics rather than technology; instead of following the curriculum at the college, he attended the lectures of Chebyshev and his disciples at St. Petersburg University. Unable to enroll at the university (he did not have the prerequisite diploma from a classical high school), Shatunovsky attempted to acquire a higher mathematical education in Switzerland. In 1887 lack of money forced him to return to Russia, where he was a private teacher in small towns in the south. One of his works that was sent to Odessa was well received by local mathematicians, who invited him to move there. He was elected a member (1897) and secretary (1898) of the mathematical department of the Novorossysky (Odessa) Society of Natural Scientists, and for some time taught school. In 1905 Shatunovsky passed the examinations for the master's degree and became assistant professor at Novorossysky (Odessa) University, where he worked until his death, becoming professor in 1920. In 1906–1920 he also taught at the Women's School for Higher Education.

Shatunovsky's principal works concern the foundations of mathematics. Independently of Hilbert he elaborated an axiomatic theory of the measurement of areas of rectilinear figures and reported on the subject to the Society of Natural Scientists in Odessa (1897) and at the Tenth Congress of the All-Russian Society of Natural Scientists and Physicians (1898). Publication of Hilbert's *Die Grundlagen der Geometrie* (1899) probably kept Shatunovsky from stating his theory, which was almost identical with Hilbert's, in print. From 1898 to 1902 Shatunovsky developed his theory for measuring the volumes of polyhedrons. In his theory of areas the principal concept is that of the invariant of one triangle (the product of base times corresponding height), and in the theory of volumes the principal notion is the invariant of the tetrahedron (the product of the area of some face times corresponding height). These studies led Shatunovsky to an axiomatic general theory of scalar quantities.

From 1906 Shatunovsky taught introduction to analysis; his lectures contain an original description of the theory of sets and functions, particularly of the definition of irrational and real numbers. The generalization of the concept of limit suggested in them is close to that introduced by E. H. Moore in 1915 ("Definition of Limit in General Integral Analysis," in *Proceedings of the National Academy of Sciences* [1915], no. 12). For a long time Shatunovsky's *Vvedenie v analiz* could be obtained only as a lithograph (Odessa, 1906–1907), however, and was not printed until 1923.

In a report to the Society of Natural Scientists in Odessa (1901), Shatunovsky critically approached the problem of applying the logical law of the excluded third to the elements of infinite sets. He discussed the subject in print in the introduction to his master's thesis, published in 1917. Pointing out the logical inadmissibility of the purely formal use of the logical law of the excluded third, the applicability of which needs special verification every time, Shatunovsky did not reach conclusions as radical as those presented by L. E. J. Brouwer in his works on intuitionism. Shatunovsky's thesis contains a new construction of Galois's theory that does not presuppose the existence of the roots of algebraic equations, which is demonstrated only in the final part of this work.

Shatunovsky also wrote articles and books on elementary mathematics. In them, for example, he stated a general principle for solving trigonometrical problems and a classification of problems connected with this principle.

BIBLIOGRAPHY

I. ORIGINAL WORKS. Shatunovsky's writings include "Ob izmerenii obemov mnogogrannikov" ("On the

Measurement of Volumes of Polyhedrons"), in *Vestnik opytnoi fiziki i elementarnoi matematiki* (1902), 82–87, 104–108, 127–132, 149–155; "Über den Rauminhalt der Polyeder," in *Mathematische Annalen*, **57** (1903), 496–508; "O postulatakh lezhashchikh v osnovanii ponyatia o velichine" ("On the Basic Postulates of the Concept of Quantity"), in *Zapiski Matematicheskago otdeleniya Novorossiiskago obshchestva estestvoispytatelei*, **26** (1904); *Algebra kak uchenie o sravnenyakh po funktsionalnym modulyam* ("Algebra as the Theory of Congruences With Respect to the Functional Modulus"; Odessa, 1917); *Vvedenie v analiz* ("Introduction to Analysis"; Odessa, 1923); and *Metody reshenia zadach pryamolineynoy trigonometrii* ("Methods of Solving Problems in Rectilinear Trigonometry"; Moscow, 1929).

II. SECONDARY LITERATURE. See .E. Y. Bakhmutskaya, "O rannikh rabotakh Shatunovskogo po osnovaniam matematiki" ("On Shatunovsky's First Research on the Foundations of Mathematics"), in *Istoriko–matematicheskie issledovaniya*, **16** (1965), 207–216; N. G. Chebotarev, "Samuil Osipovich Shatunovsky," in *Uspekhi matematicheskikh nauk*, **7** (1940), 316–321; V. F. Kagan, "S. O. Shatunovsky," in Shatunovsky's *Metody . . . trigonometrii* (above); and "Etudy po osnovaniam geometrii" ("Essays on the Foundations of Geometry"), in *Vestnik opytnoi fiziki i elementarnoi matematiki*, (1901), 286–292, also in Kagan's book *Ocherki po geometrii* ("Geometrical Essays"; Moscow, 1963), 147–154; *Matematika v SSSR za tridtsat let* ("Mathematics in the U.S.S.R. for Thirty Years"; Moscow–Leningrad, 1948), see index; F. A. Medvedev, "O formirovanii ponyatia obobshchennogo predela" ("On the Development of the Concept of the Generalized Limit"), in *Trudy Instituta istorii estestvoznaniya i tekhniki. Akademiya nauk SSSR*, **34** (1960), 299–322; *Nauka v SSSR za pyatnadtsat let. Matematika* ("Science in the U.S.S.R. During Fifteen Years. Mathematics"; Moscow–Leningrad, 1932), see index; J. Z. Shtokalo, ed., *Istoria otechestvennoy matematiki*, 4 vols. ("A History of [Russian] Mathematics"; Kiev, 1966–1970), see index; and A. P. Youschkevitch, *Istoria matematiki v Rossii do 1917 goda* ("History of Mathematics in Russia Until 1917"; Moscow, 1968), see index.

A. P. YOUSCHKEVITCH
A. T. GRIGORIAN

SHAW, PETER (*b.* Lichfield, Staffordshire, England, March or April 1694; *d.* London, England, 15 March 1764), *chemistry*.

Shaw's father was master of the Lichfield Grammar School, so it is probable that the boy received a good education, although nothing is recorded of his life between 1704, when his father died, and 1723, when his first publication appeared. Since he translated from Latin easily and well, he evidently had a good grounding in the classics. He also learned medicine and chemistry, and made his living by translating, writing, and editing books on these two subjects and by practicing medicine. His edition of Boyle's works, and his translations of Boerhaave and Stahl, were popular and influential. The most important of his own early writings was *A New Practice of Physic*, based on the teachings of Sydenham and Boerhaave.

Shaw's interest in chemistry evidently was deepened by his study of Boerhaave and Stahl. He welcomed Stahl's search for a "universal chemistry" but rejected his mysticism and his doctrine of phlogiston; indeed, he never translated any part of Stahl's writings on those subjects. His rational, experimental, and eclectic approach is revealed in the title he gave to his chemical lectures, the text of which he later published. From 1733 to 1737 he practiced at Scarborough and was active in promoting its spa. He became involved with the notorious Joanna Stephens' remedies, which, as she claimed, dissolved urinary calculi *in situ*. They were a complex mixture that included calcined snail shells and soap, and Shaw believed in their efficacy.

In 1740 Shaw was admitted licentiate of the College of Physicians and soon established an extensive and fashionable practice. He was made M.D. of Cambridge by mandamus in 1751, fellow of the Royal Society in 1752, candidate of the College of Physicians in 1753, and physician in ordinary to George III in 1760. Shaw was active in promoting the aims of the newly founded Society for the Encouragement of Arts, Manufactures and Commerce, and was highly influential in chemical and medical circles.

BIBLIOGRAPHY

I. ORIGINAL WORKS. Shaw's own writings are *A Treatise of Incurable Diseases* (London, 1723); *The Juice of the Grape, or Wine Preferable to Water* (London, 1724), published anonymously but traditionally ascribed to Shaw; *A New Practice of Physic* (London, 1726; 1728; 5th ed., 1753); *Three Essays in Artificial Philosophy, or Universal Chemistry* (London, 1731); *Chemical Lectures Publickly Read in London in . . . 1731 and 1732; and Since at Scarborough, in 1733, for the Improvement of Arts, Trades, and Natural Philosophy* (London, 1734; 1755; Paris, 1759); *An Enquiry Into the Contents, Virtues and Uses of the Scarborough Spaw-Waters* (London, 1734); *Examination of the Reasons for and Against the Subscription for a Medicament for the Stone* (London, 1738); *Inquiries on the Nature of Miss Stephens's Medicaments* (London, 1738); and

Essays for the Improvement of Arts, Manufactures and Commerce by Means of Chemistry (London, 1761).

With Francis Hauksbee the younger he wrote *An Essay for Introducing a Portable Laboratory, by Means Whereof All the Chemical Operations Are Commodiously Performed for the Purposes of Philosophy, Medicine, Metallurgy, and Family. With Sculptures* (London, 1731) and *Proposals for a Course of Chemical Experiments: With a View to Practical Philosophy, Arts, Trade and Business* (London, 1731).

His translations include *The Dispensatory of the Royal College of Physicians of Edinburgh* (London, 1727), 5th ed. (London, 1753); *A New Method of Chemistry,* 2 vols. (London, 1727), trans. of Hermann Boerhaave's *Institutiones chemiae* (Paris, 1724), incorporating student notes in collaboration with Ephraim Chambers; also new, rev. ed. (London, 1741, 1753); *Philosophical Principles of Universal Chemistry* (London, 1730), trans. of all but the last part of G. E. Stahl's *Collegium Jenense: The Philosophical Works of Francis Bacon,* 3 vols. (London, 1733; French ed. Paris, 1765); and B. Varenius, *A Compleat System of. . . Geography,* rev. and corrected by Shaw (London, 1733; 1734; 1736; 1765).

Shaw edited John Quincy, *Praelectiones pharmaceuticae; or a Course of Lectures in Pharmacy, Chymical and Galenical, Published From His Original Manuscript, With a Preface, . . .* (London, 1723); and *Philosophical Works of the Honourable Robert Boyle, Abridged, Methodised and Disposed Under the General Heads of Physics, Statics, Pneumatics, Natural History, Chemistry, and Medicine,* 3 vols. (London, 1725).

II. SECONDARY LITERATURE. The definitive account is F. W. Gibbs, "Peter Shaw and the Revival of Chemistry," in *Annals of Science,* 7 (1951), 211–237, which corrects the accounts in *Dictionary of National Biography* and in W. Munk, *The Roll of the Royal College of Physicians of London,* 2nd ed. (London, 1878), upon which all earlier accounts were based.

MARIE BOAS HALL

SHAW, WILLIAM NAPIER (*b.* Birmingham, England, 4 March 1854; *d.* London, England, 23 March 1945), *meteorology, physics.*

Shaw received his education at King Edward's School in Birmingham and Emmanuel College, Cambridge, where he studied mathematics and natural sciences. After his graduation in 1876 he was elected a fellow of his college. In 1879, after a semester of study under Helmholtz at Berlin, he was appointed demonstrator at the Cavendish Laboratory, jointly with his lifelong friend R. T. Glazebrook. He became lecturer in experimental physics in 1887 and assistant director of the laboratory in 1898. His publications during this time dealt with experimental physics. He also began work on problems of ventilating buildings.

Partly on the basis of his work on hygrometric methods and instruments, begun in 1879, Shaw was appointed a member of the Meteorological Council in 1897. With his appointment as secretary in 1900, he forsook the opportunity of a university career at Cambridge. He became director of the Meteorological Office in 1905 and held this post until his retirement in 1920. In 1907 he became reader in meteorology at the University of London, and he was first professor of meteorology at Imperial College from 1920 to 1924. He was president of the International Meteorological Committee from 1906 to 1923.

Shaw's contributions to meteorology were more far-reaching than his writings would indicate. Under his administration the Meteorological Office was transformed through the introduction of a trained scientific staff and the consequent emphasis on studies of the physics of the atmosphere. This activity complemented the customary statistical treatment of observations.

One of Shaw's most important publications, *The Life History of Surface Air-Currents,* pointed the way toward air-mass analysis and the concept of fronts (later developed by the Norwegian school of meteorologists) by showing that trajectories of air converging toward various parts of mid-latitude storms originated in widely different regions. Shaw did not pursue these results, however, and his work had little influence on meteorological practice and theory. In association with W. H. Dines, he subsequently turned to the study of the upper atmosphere by means of kites and balloons. Shaw introduced the principle of isentropic analysis, later developed by C.-G. Rossby and his collaborators, and devised a thermodynamic diagram (the tephigram) that is widely used in meteorology. His enthusiasm for the observational and diagrammatic approach was based on the conviction, dating from his student years under Maxwell, that atmospheric problems should be handled by determining the dynamics from the observations of motion.

Shaw took a particular interest in educating the public on meteorology and related subjects, and served on a number of advisory committees. After his retirement he completed his four-volume *Manual of Meteorology,* a unique account of the historical roots and the physical and mathematical basis of the subject. His writings in general reflect a deep interest and insight into the historical development of meteorology.

Shaw was knighted in 1915 and was a fellow of the Royal Society and honorary or foreign member of many academies and societies. He received the Symons Medal (1910), the Buys Ballot Medal (1923), and the Royal Medal (1923).

BIBLIOGRAPHY

I. ORIGINAL WORKS. A complete bibliography of Shaw's works is in *Selected Meteorological Papers of Sir Napier Shaw* (London, 1955). His publications included *Practical Physics* (London, 1885), written with R. T. Glazebrook; "Report on Hygrometric Methods," in *Philosophical Transactions of the Royal Society*, **A179** (1888), 73–149; "Ventilation and Warming," in T. Stevenson and S. Murphy, eds., *A Treatise on Hygiene and Public Health*, I (London, 1890), 31–148; *The Life History of Surface Air-Currents* (London, 1906), written with R. G. K. Lempfert; *Weather Forecasting* (London, 1911); *The Air and Its Ways* (Cambridge, 1923); and *Manual of Meteorology*, 4 vols. (Cambridge, 1926–1931). Shaw's MSS and correspondence are in the archives of the Meteorological Office, Bracknell, England.

II. SECONDARY LITERATURE. Obituaries are in *Obituary Notices of Fellows of the Royal Society of London*, **5** (1945), 202–230, with selected bibliography; and *Quarterly Journal of the Royal Meteorological Society*, **71** (1945), 187–194. See also D. Brunt, "A Hundred Years of Meteorology (1851–1951)," in *Advancement of Science*, **8** (1951), 114–124.

GISELA KUTZBACH

SHAYN, GRIGORY ABRAMOVICH (*b*. Odessa, Russia, 19 April 1892; *d*. Abramtsevo, near Moscow, U.S.S.R., 4 August 1956), *astrophysicist.*

The son of a joiner, Shayn completed only elementary school; but in 1911 he passed with distinction the examinations for the graduation certificate as an extramural student. He became interested in astronomy at the age of ten, and at fourteen or fifteen he seriously observed meteors with binoculars; in 1910 his first scientific work, "Vychislenie radianta Perseid" ("A Calculation of the Radiant of the Perseids"), was published in *Izvestiya Russkogo astronomicheskogo obshchestva* (**16**, no. 5, 194–197).

In 1912 Shayn entered the Faculty of Physics and Mathematics at Yurev (Dorpat) University. After serving in the army from 1914 to 1917, he completed his university education in 1919 at Perm, to which the university had been evacuated, and began his teaching career. The following year he passed the examinations for the master's degree

and became an assistant in the department of astronomy at Tomsk University. In 1921 he transferred to Pulkovo and devoted himself completely to scientific work.

In 1925 Shayn and his wife, Pelageya Fedorovna Sannikova, moved to the Simeiz section of the Pulkovo observatory, where Shayn supervised the installation of a 102-centimeter reflector, which had been ordered before the war from the British firm of Grubb. In January 1926 the first spectrogram was obtained with it. Shayn worked with this instrument until World War II. In 1935 he was awarded the doctorate in physical and mathematical sciences. Two years later he was elected foreign member of the Royal Astronomical Society, and in 1939 he became an academician of the Soviet Academy of Sciences.

During World War II part of the staff of the Simeiz observatory was evacuated to the Abastumani astrophysical observatory in Georgia, and there Shayn continued to study spectrograms evacuated from Simeiz. After the war Shayn participated in restoring the destroyed Simeiz observatory and in building a large modern astrophysical observatory in the mountains of the central Crimea. In 1945 he was named director of the Crimean Astrophysical Observatory of the Soviet Academy of Sciences. Seven years later, having asked, for reasons of health, to be relieved of his responsibilities as director, Shayn was named head of the section on the physics of nebulae and interstellar mediums. During the following four years he carried out important investigations of nebulae and galactic magnetic fields.

In the 1920's Shayn became interested in the evolution of stars and turned to double stars, correctly asserting that their components must be of the same age. He compared the most reliable data obtained from the literature on the components of double stars, in order to construct a "spectrum-luminosity" (Hertzsprung-Russell) diagram. Shayn studied the evolution of doubles, the changes in the proportions of the masses of the components, and the mass-luminosity and spectrum-luminosity relationships. He drew the important conclusion, later fully confirmed, that the evolution of the larger component must be more rapid than that of the smaller.

At the same time as O. Struve, and in partial collaboration with him by means of correspondence, Shayn discovered the rapid rotation of a number of stars of early spectral classes by analyzing the form of the spectral lines. Theoretically considering the forms of spectral lines of rotating and non-rotating stars, he provided a method of determining the velocity of rotation. With V. A. Albitsky, Shayn obtained precise determinations of the radial velocity of about 800 stars, discovered several dozen spectroscopic binaries, and computed the elements of the orbits of many of them.

Spectrophotometry was a natural continuation of Shayn's study of spectroscopic binaries. His aim was to investigate the behavior of the absorption lines and bands and their influence on the color of the stars, their apparent bolometric magnitudes and other properties, and the relation of the normal color to luminosity. Research on the spectra of the long-period variables was associated with the elucidation of all the peculiarities of the spectrum-luminosity and period-luminosity relationships. A number of Shayn's spectrophotometric investigations dealt with planetary nebulae, the integral spectrum of the Milky Way clouds, and the spectrum of the rings of Saturn. His observations of the total solar eclipse of 1936 provided material for the study of the physics of the solar corona.

At Abastumani, Shayn used spectrograms from Simeiz to offer an original interpretation of the coexistence of emission lines and lines of absorption (high- and low-temperature spectra) in the spectra of long-period variables such as Mira Ceti. A paradox was removed by the hypothesis that the physical obscuration of the source of high-temperature radiation was provided by the extended atmosphere of such a star. Shayn indicated the possible similarity of the turbulent phenomena of solar activity (chromospheric flares and similar processes) to phenomena that cause the outward motion of hot matter from a star and the appearance of emission lines in its spectrum.

Shayn conducted important research on isotopes in the atmospheres of stars. In 1940 he discovered that the isotope C^{13} content in several stars was very great: for the earth the proportion of C^{13} to C^{12} is approximately $1:70-1:90$, for the sun it is less than $1:10$, while on certain red so-called carbon stars it reaches $1:2$. Besides the two known bands of the heavy molecule of cyanogen $C^{13}N^{14}$, Shayn also demonstrated the presence of many bands in the red and violet range of the spectra of these stars. American physicists were thus led to make a new experimental determination of the cross section of the capture of protons by the nuclei of C^{12} and C^{13}. As a result it was unexpectedly shown that the amount of isotope C^{13} on earth and on the sun is abnormally small. For his work on isotopes in the atmospheres of carbon stars Shayn was awarded the State Prize in 1950.

By 1948 Shayn and his collaborator of many years, Vera F. Gaze, had developed a special method of photography based on the fact that gas nebulae radiate all their energy in certain bright emission lines. Using two meniscus cameras with objectives of 450 millimeters of high optical efficiency (1:1.4) and a field of about 4°, they discovered more than 150 new emission nebulae and published several catalogs of diffuse nebulae, as well as the photographic *Atlas diffuznykh gazovykh tumannostey* ("Atlas of Diffuse Gas Nebulae"; 1952). Using the same method, Shayn also found emission objects in other galaxies. All these nebulae appeared to be distributed along the branches of spirals. From 1950 Shayn used a more powerful 640-millimeter camera.

In the galactic emission nebulae he noted peculiarities in the distribution of the matter in the so-called peripheral nebulae, and he concluded that the nebulae expanded over the course of time. From this he formulated the important cosmogonic statement of the common formation of the association of hot stars and nebulae. Studying the continuous spectra of certain diffuse nebulae, he showed that in accord with the hypothesis of A. Y. Kipper (Tartu), the continuous spectrum of nebulae is related not to the scattering of light by interstellar dust but to the "two-quantum jump" in hydrogen atoms. Shayn concluded that the role of the dust had been exaggerated, for its density proved to be much less than that of the gaseous medium (for example, 1/100 in the Orion nebula).

Studying the filamentary nebulae, Shayn found that in most cases the filaments were oriented parallel to the galactic equator; to explain their elongation and orientation he posited the existence in the galaxy of powerful magnetic fields. He discovered that the direction of the elongated nebulae was similar to the direction of the plane of polarization of light. This made it possible to define the borders of the branches of our galaxy more precisely and to assert the presence of an "arm" in the direction of the constellation Sagittarius. The magnetic theory explained why the gas does not disperse beyond the branches and does not fill the space between the "arms" of the spirals.

BIBLIOGRAPHY

I. Original Works. A complete bibliography of 249 titles, compiled by N. B. Lavrova, is *Istoriko-astronomicheskie issledovaniya*, **3** (1957), 596–607. Among them are "On the Rotation of Stars," in *Monthly No-*

tices of the Royal Astronomical Society, **89** (1929), 222–239, written with O. Struve; "The Absorption Continuum in the Violet Region of the Spectra of Carbon Stars," in *Astrophysical Journal*, **106** (1947), 86–91, written with O. Struve; "Otnoshenie kontsentratsii isotopov C^{13} i C^{14} v atmosferakh zvezd" ("The Relative Concentration of the Isotopes C^{13} and C^{14} in the Atmosphere of Stars"), in *Uspekhi fizicheskikh nauk*, **43** (1951), 3–10, written with V. F. Gaze; "Certain Peculiar Structures in Interstellar Clouds," in *Gas Dynamics of Cosmic Clouds* (Amsterdam, 1955), 37–38; and "On the Groups of Diffuse Emission Nebulae," in *Vistas in Astronomy*, II (London, 1955), 1066–1069.

II. SECONDARY LITERATURE. See "Akademik Grigory Abramovich Shayn. 1892–1956," in *Vestnik Akademii nauk SSSR* (1956), no. 10, 84; "Akademik Grigory Abramovich Shayn," in *Voprosy kosmogonii*, **5** (1957), 3–5; P. P. Dobronravin, "Grigory Abramovich i Pelageya Fedorovna Shayn," in *Peremennye zvezdy*, **11**, no. 4 (1958), 321–324; "G. A. Shayn (1892–1956)," in *Astronomicheskii zhurnal*, **33**, no. 4 (1956), 465–468, with portrait; "Grigory Abramovich Shayn," in *Izvestiya Krymskoi astrofizicheskii observatorii*, **17** (1957), 3–10; "Grigory Abramovich Shayn (1892–1956)," in *Materialy k biobibliografii uchenykh SSSR*, Astron. ser. (1960), no. 2, intro. by P. P. Dobronravin and bibliography by O. V. Isakova; S. B. Pikel'ner, "G. A. Shayn," in *Astronomicheskii tsirkulyar. Byuro astronomicheskikh soobshchenii, Akademiya nauk SSSR*, no. 172 (1956), 1–2; and "G. A. Shayn (1892–1956)," in *Istoriko-astronomicheskie issledovaniya*, **3** (1957), 551–607, with complete bibliography of 249 titles compiled by N. B. Lavrova, 596–607; the article "Shayn" in *Bolshaya sovetskaya entsiklopedia* ("Great Soviet Encyclopedia"), 2nd ed., XLVII, 499; and O. Struve, "G. A. Shain and Russian Astronomy," in *Sky and Telescope*, **17**, no. 6 (1958), 272–274.

P. G. KULIKOVSKY

SHEN KUA[1] (*b.* 1031, registered at Ch'ien-t'ang[2] [now Hangchow, Chekiang province], China; *d.* Ching-k'ou, Jun prefecture[3] [now Chinkiang, Kiangsu province], China, 1095), *polymathy, astronomy.*

Shen was the son of Shen Chou[4] (*ca.* 978–1052) and his wife, whose maiden name was Hsu.[5] Shen Chou came of a gentry family with neither large landholdings nor an unbroken tradition of civil service. He spent his life in minor provincial posts, with several years in the capital judiciary. Shen Kua apparently received his early education from his mother. A native of Soochow (the region of which was known for its flourishing manufactures, commerce, and agriculture), she was forty-four or forty-five years old when he was born. Shen's background made possible his entry into the imperial bureaucracy, the only conventional road to advancement for educated people of his time. Unlike colleagues who came from the ancient great clans, he could count on few advantages save those earned by his striving and the full use of his intellectual talents. Shortly after he was assigned to the court, he became a confidant of the emperor and played a brilliant part in resolving the crises of the time. But within slightly over a decade his career in the capital was ended by impeachment. After a provincial appointment and five years of meritorious military accomplishment, he was doubly disgraced and politically burned out. The extremes of Shen's career and the shaping of his experience and achievement in science and technology become comprehensible only if the pivotal circumstances of his time are first considered.

Historical Background. Shen's time was in many senses the climax of a major transition in the Chinese polity, society, and economy.

Three centuries earlier the center of gravity in all these respects still lay in the north, the old center of civilization of the Han people. Wealth and power rested in the hands of the old aristocratic landowning families. Governmental institutions incorporated the tension between their private interests and the inevitable desire of their foremost peer, the emperor, to concentrate authority. The civil service examination system was beginning to give the central government a means to shape a uniform education for its future officials; but since birth or local recommendation determined who was tested, the mass of commoners remained uninvolved. The social ideals prevalent among the elite were static; the ideal past was cited to discourage innovation; and the moral example of those who ruled, rather than responsive institutions or prescriptive law, was held to be the key to the healthy state. The classicist's paradigm of a two-class society—self-sufficient agriculturalists ruled and civilized by humane generalists, with land as the only true wealth—did not encourage commerce, industry, or the exploitation of natural resources. The wants of the great families, whose civil servant members were becoming city dwellers by the middle of the eighth century, nonetheless gave momentum to all of these activities; but the majority of the population still took no part in the rudimentary money economy.

The T'ang order began a long, slow collapse about 750, until in the first half of the tenth century the empire of "All Under Heaven" was reduced to a succession of ephemeral and competing kingdoms. When the universal state was reconstructed in the Northern Sung (960–1126), its foundations were in many important respects different from those of the early T'ang. A new dynasty was not only, as classical monarchic theory had it, a fresh dispensation of the cosmos; it was also the occasion for institutionalizing a new distribution of power in society. The cumulative result of changes in taxation had been to make the old families accountable for their estates as they had not been earlier, and to encourage smaller landholdings—and, thus, a wider diffusion of wealth.

The center of vitality had moved southeast to the lower Yangtze valley, which had long before emerged as the major rice-yielding region. By this time its fertility, combined with its relative freedom from restrictive social arrangements, had bred a new subculture that was more productive in industry than elsewhere and hospitable to the growth of commerce and stable markets, the beginnings of a uniform money economy, and the great broadening of education that printing had just made possible. The new southern elite was, on the whole, small gentry, and lacked the military traditions of the ancient northern clans and of power holders in the period of disunion. Their families were often too involved in trade for them to despise it. Although conservative, as all Chinese elites have been, they were prepared to think of change as a useful tool. The novelties of attitude and value were often slighter or subtler than such a brief account can convey, but within the established limits of Chinese social ideals their consequences were very considerable.

In Shen Kua's time the old families still provided many of the very highest officials and thus wielded great influence, positive or obstructive, in discussions about the future of China. But they had become merely influential members of a new political constellation that brought a variety of convictions and interests to that perennial debate. An especially obvious new element was that many southern small gentry families like Shen's established traditions of civil service, either as a main means of support or to protect and further their other concerns. Once a family's social standing was achieved, one or more members could enter the bureaucracy freely because of experience as subordinates in local administration or because they were amply prepared by education for the examinations. Their sons could enter still more freely because special access to both direct appointment and examination was provided to offspring of officials.

Not sharing the old vision of a virtue-dominated social order fixed by precedent, men of the new elite were willing to sponsor institutional renovation in order to cope directly with contemporary problems. Dependent on their own talents and often needing their salaries, they were dedicated to building a rational, systematic, and in most respects more centrally oriented administration. They were willing to make law an instrument of policy, and insisted that local officials be rated not only on the moral example they set but also quantitatively—on how effectively they made land arable and collected taxes. In the name of efficiency they devoted themselves to removing customary curbs on imperial authority and (with only partial success in the Sung) to dismantling the structures of privilege that underlay regional autonomy. Only later would it become clear that they were completing the metamorphosis of the emperor from paramount aristocrat to autocrat. At the same time they were successfully demanding more policy-making authority as the emperor's surrogates, although at the cost to themselves of greater conformity than officials of the old type had willingly accepted.

This irreversible transition did not lead to a modern state, but only to a new and ultimately stagnant pattern. The most accelerated phase of change was the activity of what is called the New Policies[6] group (actually a shifting coalition) between 1069 and 1085. Its leader, Wang An-shih[7] (1021–1086), was brought to the capital in 1068 by the young emperor Shen-tsung, who had just taken the throne. Within two years Wang had become first privy councillor. He resigned for nine months in 1074, when pressure from his antagonists persuaded the emperor to be less permissive, and returned permanently to private life in 1076. The New Policies continued to be applied and extended, but with less and less attention to their founding principles, until Shen-tsung's death in 1085. Under the regency of the empress dowager, enemies of the reform attempted for eight years to extirpate Wang's influence and take revenge upon his adherents. When Emperor Che-tsung came of age in 1093, the New Policies were revived, but were so bent toward selfish ends and administered so disastrously that the word "reform" is hardly applicable.

Wang An-shih's opponents were many: the old aristocrats, career bureaucrats of the sort who would oppose any change as disruptive, officials whose individual or group interests ran in other directions —and men of high ideals who found his proposals ill-advised and his personal style too intolerant.*

No institution had evolved through Chinese history to work out and resolve conflicts of political viewpoint. This lack was filled by cliques, intrigues, and appeals to imperial intervention. Division and corruption among active supporters of the New Policies also had been a problem from the start. The scope of Wang's program was so large that he had to take competent support where he found it. The new access to power that he offered attracted ambitious men, many of whom had little real sympathy for his convictions and dedicated themselves primarily to manipulation and graft. Once Wang was gone, the leadership of his group tended to become a battleground for aspirations of this kind. The internal and external enemies of the New Policies left the program a shambles by the time the Chin Tartars drove the Sung south in 1127.

A primary aim of the reforms was financial security of the state, which prompted initiatives in water control and land reclamation, encouragement of extractive industries and agriculture, intervention in commerce, and rationalization of taxes. Another goal, particularly at the emperor's insistence, was military strength. There had been a long confrontation between the Chinese and the powerful Khitan empire, pastoral masters of mounted combat to the north (renamed Liao in 1066). Seventy years of fitful peace were punctuated by humiliating Chinese failures to recapture territory south of the Great Wall and maintained by large annual bribes. For three decades the Tangut people of the northwest had posed an almost equally unpalatable demand for appeasement. Victory or détente through strength, the emperor hoped, could be bought on both fronts with the wealth that the New Policies generated from man's exploitation of nature. Here too expertise was needed in cartography, strategic theory and tactical doctrine (both of which contained cosmological elements), design and manufacture of war matériel, fortification, troop organization and training, and development of a stable economy in border regions.

---

*In the successive reform movements of the Northern Sung there were considerable differences in the alignment of men with different beliefs and backgrounds. See the discussion in James T. C. Liu, "An Early Sung Reformer: Fan Chung-yen," in John K. Fairbank, ed., *Chinese Thought and Institutions* (Chicago, 1957), 105–131, esp. 107–109. The generalizations of the present article and of current scholarship as a whole are crude and tentative, pending the "comparative analysis of the interrelationships between ideology and family, class, status-group, and regional interests" that Robert M. Hartwell has called for in "Historical Analogism, Public Policy, and Social Science in Eleventh- and Twelfth-Century China," in *American Historical Review*, **76** (1971), 690–727.

Shen Kua contributed to nearly every field of New Policies activity, both civil and military. His social background and political commitments cannot be considered responsible for his scientific talent or curiosity; the antecedents and loyalties of other major contemporary scientific figures were very different from his. But a review of his career and of his work will show how regularly his involvement with particular technical themes and problems grew out of his activities in government.

Life. From about 1040 Shen traveled with his father to successive official posts from Szechwan in the west to the international port of Amoy. He was exposed not only to the geographical diversity of China but also to the broad range of technical and managerial problems—public works, finance, improvement of agriculture, maintenance of waterways—that were among the universal responsibilities of local administrators. Because his physical constitution was weak, he became interested in medicine at an early age.

Late in 1051, when Shen was twenty, his father died. As soon as the customary inactivity of the mourning period ended in 1054, Shen received the first of a series of minor local posts; his father's service exempted him from the prefectural examination. His planning ability became almost immediately apparent when he designed and superintended a drainage and embankment system that reclaimed some hundred thousand acres of swampland for agriculture. This was the first of a series of projects that established his reputation for skill in water control. In 1061, as a subprefect in Ning-kuo[8] (now Fu-hu,[9] Anhwei province), after a cartographic survey and a historical study of previous earthworks in the region, he applied the labor of fourteen thousand people to another massive land reclamation scheme that won the recognition of the emperor. In a series of floods four years later, Shen noted, it was the only such project not overwhelmed. He wrote characteristically that in the first year it returned the cost of the grain used, and that there was more than a tenfold profit on cash expended. In 1063 he passed the national examinations. Posted to Yangchow, he impressed the fiscal intendant (a post then equivalent to governor), Chang Ch'u[10] (1015–1080), who recommended him for a court appointment leading to a career in the professional financial administration.*

---

*The succession of fiscal posts that often led to a seat on the Council of State in the eleventh century has been documented by Robert M. Hartwell in "Financial Expertise, Examinations, and the Formulation of Economic Policy in Northern Sung China," in *Journal of Asian Studies*, **30**, 281–314.

Shen apparently used the time not occupied by his early metropolitan appointments, which were conventional and undemanding, to study astronomy. In reply to the informal questions of a superior he set down clear explanations, still extant, of the sphericity of the sun and moon as proved by lunar phases, of eclipse limits, and of the retrogradation of the lunar nodes. They demonstrate an exceptional ability to visualize motions in space, which were at best implicit in the numerical procedures of traditional astronomy and seldom were discussed in technical writing. In 1072 Shen was given an additional appointment as director of the Astronomical Bureau. With the collaboration of his remarkable commoner protégé Wei P'u[11] and the aid of other scholarly amateurs, using books gathered from all over the country, he undertook a major calendar reform. He planned an ambitious series of daily observations to extend over five years, using renovated and redesigned instruments. When he took office, the bureau was staffed with incompetents. He forced the dismissal of six whom he caught falsifying records of phenomena, but the obstruction of those who remained doomed his program of observations and kept his new system of ephemerides computation from being among the two or three most securely founded before modern times. Shen's personal involvement in later stages of the reform undoubtedly was limited by his gradual movement into the vortex of factional politics.

Shen was early known to Wang An-shih, who composed his father's epitaph while a young provincial official; Shen eventually came to be publicly identified by enemies of the New Policies as among the eighteen members of Wang's intimate clique. In late 1072, in support of Wang's program, Shen surveyed the silting of the Pien Canal near the capital by an original technique, dredged it, and demonstrated the value of the silt as fertilizer. Until mid-1075 he spent much time traveling as a troubleshooter of sorts, inspecting and reporting on water control projects, military preparations, and local administrations—and, it has been conjectured, providing encouragement to Wang's provincial supporters. Shen was put in charge of arsenal activities and, in 1075, was sponsored by Wang (then head of government) to revise defensive military tactics, a task the throne had proposed for Wang himself.

In 1074 the Khitan were pressing negotiations to move their borders further south. Incompetent and timorous Chinese negotiators were conceding unfounded Liao assertions about the language and substance of previous agreements. Shen built a solid Chinese case by going to the archives, as no one had bothered to do before. His embassy in mid-1075 to the camp of the Khitan monarch on Mt. Yung-an[12] (near modern P'ing-ch'üan,[13] Hopei) was triumphant. He described himself surrounded by a thousand hostile onlookers, calling on his staff, who had memorized the old documents of the Khitan themselves, to cite without pause or flurry the exact reference to refute one historical claim after another.

Shen returned to China—with biological specimens and maps of the territories he had passed through—to become a Han-lin academician, to be given charge of a large-scale water control survey in the Yangtze region, and then to become head of the Finance Commission. While in this very powerful position he untangled a variety of contradictory policies, producing in the process some of the most penetrating writings before modern times on the operation and regulation of supply and demand, on methods of forecasting prices in order to intervene effectively in the market, and on factors that affect the supply of currency (varying through hoarding, counterfeiting, and melting) as the value of the metal in it fluctuates about its controlled monetary value. In the autumn of 1077, just as his revision of critical fiscal measures was well launched, he was impeached by the corrupt and vindictive censor Ts'ai Ch'ueh[14] (1036–1093). The charge was that Shen had opposed a New Policies taxation measure in an underhanded, inconsistent, and improper way. It was credited by historians for centuries, but its truth has been refuted in every detail by recent Chinese research. His protector Wang An-shih had just left government; it is believed, given the mood of the time, that by threatening an established budget item in order to ease the burdens of the poor, Shen became an easy victim of factional maneuvering.

The emperor was not only the ritual synapse between the political and natural orders; he was a human being whose likes and dislikes were indulged within broad limits that could be further widened by force of his personal charisma and will. The closer to him an official penetrated, the more achievement and even survival became subject to imperial whim and the intrigue of colleagues. Although the record is fragmentary, it gives the impression that Shen Kua was maneuvered by Wang An-shih into the proximity of the throne because of his brilliance, judgment, and effectiveness at complicated tasks. Nothing indicates that he was adept at protecting himself. He attracted the most damaging animosity not from opponents of the

New Policies but from designing members of his coalition. Once the emperor qualified his support of the New Policies in 1074, the risk of debacle remained great and imminent. Many officials who had risen with Wang fought furiously for the power that would keep them afloat even though the program sank. They did not wish to be deterred by a colleague who judged issues on their own merits. They probably also felt, as others did, that a man of Shen's age and rank did not deserve the emperor's confidence.

Ts'ai Ch'ueh was rising into the vacuum that Wang's retirement had left. The emperor depended increasingly on Ts'ai's monetary counsel and could not easily disregard what he insisted upon. For three years it was impossible to overcome his objections and those of another censor, and to rehabilitate Shen. Finally Shen was sent to Yen-chou[15] (now Yenan, Shensi province), on the necessary route for military operations by or against the Tanguts, as commissioner for prefectural civil and military affairs.[16] The Tanguts were then divided and weakened, minor Chinese conquests around 1070 had set the stage for a war, and the treasury had ample funds. Shen played an important part in organizing and fortifying for the victorious offensive of the autumn of 1081. In extending Sung control he showed a practical as well as a theoretical mastery of the art of warfare. He was cited for merit and given several honorary appointments. It was probably at the same time that he was ennobled as state foundation viscount.[17] In his sixteen months at Yen-chou, Shen received 273 personal letters from the emperor. His standing at the court was in principle reestablished. Whether he had become shrewd enough to survive there was never tested.

Shen and a colleague followed up the victory by proposing fortifications to close another important region to the Tanguts. The emperor referred the matter to an ambitious and arrogant official who, ignoring the proposal, changed the plan to provide defenses for what Shen argued was an indefensible and strategically useless location. Shen was commanded to leave the vicinity of the new citadel so as not to share in the credit for the anticipated victory. When the Tangut attack came, the emissary's force was decimated while Shen, with imperial permission, was successfully defending a key town on the enemy invasion route to Yen-chou. The campaign thus provided the Tanguts with no opening for advance—but Ts'ai Ch'ueh was now a privy councillor. As titular military commander Shen was held responsible for the defeat and con-

siderable loss of life. At the age of fifty-one his career was over. The towns he saved were later abandoned by the anti–New Policies regime to no advantage, just as the lands he had saved from the Khitan through diplomacy had since been lost by another negotiator.

Shen spent six years in fixed probationary residence, forbidden to engage in official matters. He used at least two of these years to complete a great imperially commissioned atlas of all territory then under Chinese control. He had been working on this atlas intermittently since, as finance commissioner a decade earlier, he had had access to court documents. His reward included the privilege of living where he chose.

Ten years earlier Shen had bought, sight unseen, a garden estate on the outskirts of Ching-k'ou. In 1086, visiting it for the first time, he recognized it as a landscape of poignant beauty that he had seen repeatedly in dreams, and named it Dream Brook (*Meng ch'i*,[18] alternately read *Meng hsi*). He moved there in 1088. Despite a pardon and the award of sinecures to support him in his old age, he spent seven years of leisure, isolation, and illness until his death there.*

Shen's writings, of which only a few are extant even in part, include commentaries on Confucian classics, two atlases, reports on his diplomatic missions, a collection of literary works, and monographs on rituals, music, mathematical harmonics, administration, mathematical astronomy, astronomical instruments, defensive tactics and fortification, painting, tea, medicine, and poetry. Of three books compiled during his last years at Dream Brook, one, "Good Prescriptions" (*Liang fang*[19]), was devoted to medical therapy, theory, and philology; the other two belong to particularly Chinese genres. "Record of Longings Forgotten" (*Wang huai lu*[20]), a collection of notes on the life of the gentleman farmer in the mountains, contains useful information on implements and agricultural technique and, unlike more conventional agricultural treatises up to that time, on the culture of medicinal plants.

"Brush Talks From Dream Brook" (*Meng ch'i pi t'an*[21]) and its sequels, extant and well-edited in modern times, is by any reckoning one of the most remarkable documents of early science and technology. It is a collection of about six hundred recollections and observations, ranging from one or two sentences to about a page of modern print— "because I had only my writing brush and ink slab

*For a translation that conveys the flavor of Shen's autobiography, see Donald Holzman, "Shen Kua," 275–276.

to converse with, I call it Brush Talks." They are loosely grouped under topics (seventeen in all current versions), of which seven contain considerable matter of interest in the study of nature and man's use of it: "Regularities Underlying the Phenomena"[22] (mostly astronomy, astrology, cosmology, divination), "Technical Skills"[23] (mathematics and its applications, technology, medicine), "Philology"[24] (including etymology and meanings of technical terms), "Strange Occurrences"[25] (incorporating various natural observations), "Artifacts and Implements"[26] (techniques reflected in ancient objects), "Miscellaneous Notes"[27] (greatly overlapping other sections), and "Deliberations on Materia Medica"[28] (most of it on untangling historic and regional confusions in identities of medical substances).

Notices of the highest originality stand cheek by jowl with trivial didacticisms, court anecdotes, and ephemeral curiosities under all these rubrics; other sections were given to topics conventional in collections of jottings—memorable people, wisdom in emergencies, and so on. Shen's theoretical discussions of scientific topics employed the abstract concepts of his time—yin-yang, the Five Phases (wu-hsing[29]), ch'i,[30] and so on. A large fraction of the book's contents is devoted to fate, divination, and portents, his belief in which has been ignored by historians seeking to identify in him the prototype of the modern scientist. The author of "Brush Talks" has been compared with Leibniz; and in an era of happier relations with the Soviet Union, Hu Tao-ching, the foremost authority on Shen, referred to him as the Lomonosov of his day. But Shen was writing for gentlemen of universal curiosity and humanistic temperament; custom, wisdom, language, and oddity were as important themes as nature and artifice.

Because Shen's interests were multifarious, the record unsystematic, and its form too confining for anything but fragmentary insight, only accumulation can provide a fair impression of what constitutes his importance. What follows is the mere sample that space allows of his attempts to deepen the contemporary understanding of nature, his observations that directed the attention of his educated contemporaries to important phenomena or processes, and his own technical accomplishments. They are grouped to bring out contiguity of subject matter without interposing the radically different disciplinary divisions of modern science. These samples will become the basis of discussion—which, given the state of research, must be highly

tentative—of the epistemological underpinnings of Shen's work, and of the unity of his scientific thought with elements that today would be considered unscientific, primitive, or superstitious. Finally, it will be possible to evaluate Shen's life as a case study in the reconcilability of Confucianism and science, which the conventional wisdom among sinologists for over a generation has tended to place in opposition.

Quantity and Measure. Mathematics was not the queen of sciences in traditional China. It did not exist except as embodied in specific problems about the physical world. Abstract thought about numbers was always concerned with their qualities rather than their properties, and thus remained numerology. This art, although it blended into arithmetic, was only partly distinct from other symbolic means (in the anthropologist's sense: magical, ritual, religious, divinatory) for exploring the inherent patterns of nature and man's relation to it. Computation, on the other hand, was applied to a great variety of mensurational, accounting, and other everyday tasks of the administrator in a coherent tradition of textbooks. Occasionally curiosity and skill pushed beyond these pragmatic limits, but never very far. Some of the problems that Shen presented in "Brush Talks" had no application, but his enthusiasm for them was in no way qualified.

In addition to this accumulation of individual problems there were two exact sciences, in which mathematics served theory to advance knowledge of the patterns underlying the phenomena. One was mathematical harmonics (lü lü[31]), which explored the relations between musical intervals and the dimensions of instruments that produced them, in ways analogous to the Pythagorean art. Its appeal was much the same in both China and Greece: it demonstrated how deeply the power of number was grounded in nature. For this reason in China mathematical harmonics was often put into the same category as mathematical astronomy, which also had foundations in metaphysics. Astronomy, by far the more technically sophisticated of the two exact sciences, was normally employed on behalf of the monarch. Unpredictable phenomena and failures of prediction were either good or bad omens. Bad omens were interpreted as warnings that the emperor's mediating virtue, which maintained concord between the cosmic and political orders, was deficient. Successful prediction of celestial events was symbolic preservation or enhancement of the charisma of the ruling dynasty.

The annual calendar (or almanac) issued by authority of the throne was thus of great ceremonial importance. It encompassed all predictable phenomena, including planetary phenomena and eclipses. The utilitarian calendrical aspects—lunar months and solar years—had long since been refined past any practical demand for accuracy, but astronomical reinforcement of the Mandate of Heaven called forth endless attempts at greater precision of constants. As it became conventional to institute a complete new system for computing these ephemerides when a new emperor was enthroned, technical novelty was at a premium. When new ideas were unavailable, trivial recasting of old techniques was usually substituted. Repeated failures of prediction were another motive for reform of the astronomical system. In such cases too the system was in principle replaced as a unit rather than repaired. Most systems survived or fell on their ability to predict eclipses, particularly solar eclipses. These were the least amenable of all celestial phenomena to the algebraic, nongeometric style of mathematics. Prior to Shen's time little effort had gone into predicting the apparent motions of the planets, which lacked the immediacy of solar and lunar phenomena. This was, in fact, an omission that Shen seems to have been the first to confront.

General Mathematics. As wood-block printing became widespread, the government used it to propagate carefully edited collections of important ancient textbooks for use in education. This was being done in medicine at the time Shen entered the capital bureaucracy. In 1084 a collection of ten mathematical manuals, made four centuries earlier and reconstituted as well as extant texts allowed, was printed. The authority of these projects served both to fix textual traditions, preserving selected treatises from further attrition, and passively to encourage the fading into oblivion of books left out. Shen thus lived at a pivotal period in the development of mathematics, and his judgments on lost techniques and disused technical terms (such as 300, 306) have played an important part in later attempts to interpret them.*

*Numbers in parentheses are item numbers in the Hu Tao-ching edition of *Meng ch'i pi t'an* (the latter is referred to hereafter as "Brush Talks"). Roman volume numbers followed by page numbers refer to translations in Joseph Needham *et al.*, *Science and Civilisation in China*. Where my own understanding differs considerably from Needham's, an asterisk follows the page reference. All quotations below are from Shen, and all translations are my own. Full bibliographical data are given in the notes only for sources of too limited pertinence to be included in the bibliography. Chinese and Japanese family names precede personal names throughout this article.

"Brush Talks" is also an essential source for the study of pre-Sung metrology, currency, and other subjects related to computation.

Shen used mathematics in the formulation of policy arguments more consistently than most of his colleagues; examples are his critique of military tactics in terms of space required for formations (579) and his computation that a campaign of thirty-one days is the longest that can feasibly be provisioned by human carriers (205). But of the computational methods discussed in his "Technical Skills" chapter, those not related to astronomy are almost all abstractly oriented.

This original bent emerges most clearly in two problems. One departs from earlier formulas for computing the frustum of a solid rectangular pyramid. Shen worked out the volume of the same figure if composed of stacked articles (he mentioned *go* pieces, bricks, wine vats) that leave interstices (301). Since Shen intended this "volume with interstices" (*ch'i chi*[32]) method to be applicable regardless of the shape of the objects stacked, what he gave is a correct formula for the number of objects, which are thus to be considered of unit volume. His presentation has several interesting features. Needham has suggested that the concern with interstices (and, one would add, unit volumes) may have been a step in the direction of geometric exhaustion methods (III, 142–143)—although it was tentative and bore fruit only in seventeenth-century Japan. Second, instead of the worked-out problem with actual dimensions that is conventional in early textbooks, Shen simply gave a generalized formula: "double the lower length, add to the upper length, multiply by the lower width," and so on. Third, this was the earliest known case in China of a problem involving higher series. Built on earlier numerical approaches to arithmetical progressions, it provided a basis for more elaborate treatment by Yang Hui[33] (1261) and Chu Shih-chieh[34] (1303).

The second problem of interest was said "in a story" to have been solved by one of China's greatest astronomers, the Tantric Buddhist patriarch I-hsing[35] (682–727): the number of possible situations on a *go* board, with nineteen by nineteen intersections on which any number of black or white pieces may be placed. Whether I-hsing actually solved this problem we do not know; Shen's single paragraph was the first and last known discussion of permutations in traditional mathematics. It stated the order of magnitude of the answer—"approximately speaking one must write the character *wan*[36] (10,000) fifty-two times in succession"—adding exact answers for smaller arrays, three methods of solution, and a note on the limited traditional notation for very large numbers (304).*

Mathematical Harmonics. The Pythagoreans were fascinated by the relations of concordant intervals to the plucked strings that produced them, since the lengths between stops were proportionate to simple ratios of integers. The Chinese built up a similar science on a gamut of standard pipes. Beginning with a pipe eight inches long and 0.9 inch in diameter, they generated the lengths of subsequent pipes by multiplying the previous length alternately by 2/3 and 4/3, making twelve pipes within an approximate octave. The dozen were then related to such categories as the twelve divisions of the tropical year, in order to provide a cosmic basis for the system of modes that the pipes determined. A pentatonic scale, which could be used in any of the twelve modes, provided similar associations with the Five Phases. This basis was extended to metrology by defining the lengths and capacities of the pipes in terms of millet grains of standard dimensions. Shen provided a lucid and concise explanation of these fundamentals of mathematical harmonics, and corrected grotesque complications that had crept into a canonic source through miscopying of numbers (143, 549). He also experimentally studied stringed instruments. By straddling strings with paper figures, he showed that strings tuned to the same notes on different instruments resonate, as do those tuned an octave apart on the same instrument (537; cf. IV.1, 130). His two chapters on music and harmonics[37] are also a trove of information on composition and performance.

Astronomy. Shen's major contributions in astronomy were his attempts to visualize celestial motions spatially, his arc-sagitta methods that for the first time moved algebraic techniques toward trigonometry, and his insistence on daily observational records as a basis for his calendar reform. The first had no direct application in computation of the ephemerides, although it may well have inspired (and at the same time have been inspired by) the second, which grew out of traditional mensurational arithmetic. It has been suggested that the clarity of Shen's cosmological explanations led to his appointment to the Astronomical Bureau,

which provided opportunity for his contributions in the second and third areas. But circumstances that arose from the bureaucratic character of mathematical astronomy made these contributions futile in his lifetime.

Shen's discussions of solar, lunar, and eclipse phenomena (130–131; excerpts, III, 415–416) have been mentioned. By far the most remarkable of his cosmological hypotheses attempted to account for variations in the apparent planetary motions, including retrogradation. This concern is not to be taken for granted, since traditional astronomers preferred purely numerical approaches to prediction, unlike the spatial geometric models of Greek antiquity, and showed little interest in planetary problems. Noting that the greatest planetary anomaly occurred near the stationary points, Shen proposed a model in which the planet traced out a figure like a willow leaf attached at one side to the periphery of a circle (see Figure 1). The change in direction of the planet's motion with respect to the stars was explained by its travel along the pointed ends of the leaf (148).* The willow leaf, in other words, served one of the same functions that the epicycle served in Europe. It is characteristic that, having taken a tack that in the West was prompted entirely by geometric reasoning, Shen's first resort should have been a familiar physical object. Use of a pointed figure doubtless would not have survived a mathematical analysis of observational data, but this remained an offhand suggestion.

Another early outcome of Shen's service at the court was a series of proposals for the redesign of major astronomical instruments: the gnomon, which was still employed to measure the noon shadow and fix the solstices; the armillary sphere, with which angular measurements were made; and the clepsydra, used to determine the time of observations (and to regulate court activities). Shen's improved versions of the latter two apparently were not built until late 1073, after he had taken charge of the Astronomical Bureau. The armillary at least was discarded for a new one in 1082, a casualty of his personal disgrace.

Shen's clepsydra proposals represent a new design of the overflow-tank type (Needham's Type B; III, 315–319, 325), but the most significant outcome of his work on this instrument was a jotting on problems of calibration. Day and night were by custom separately divided into hours, the

---

*This was translated in part by Needham (III, 139). The extant text, even in Hu's edition, is very corrupt. It has been edited and considerably emended by Ch'ien Pao-tsung in *Sung Yuan shu-hsueh-shih lun-wen-chi*, 266–269.

*Translated by N. Sivin, *Cosmos and Computation in Early Chinese Mathematical Astronomy* (Leiden, 1969), also published in *T'oung Pao*, **55** (1969), 1–73 (see 71–73), from which the figure is reproduced with permission of E. J. Brill.

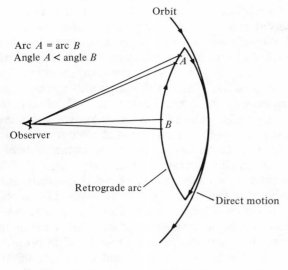

Arc *A* = arc *B*
Angle *A* < angle *B*

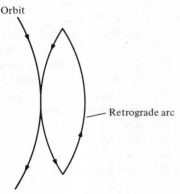

FIGURE 1. Shen Kua's explanation of planetary anomaly. He suggested that the "willow leaf" could be either inside or outside the orbit. No drawings appear with the text in "Brush Talks."

length of which varied with the season. The time was read off graduated float rods, day and night sets of which were changed twenty-four times a year. Shen pointed out that this crude and inadequate scheme amounted to linear interpolation, "treating the ecliptic as a polygon rather than a circle," and argued for the use of higher-order interpolation (128).

The best armillary sphere available in the central administration when Shen first worked there was based on a three-hundred-year-old design "and lacked ease of operation" (150). The most interesting of Shen's improvements was in the diameter of the naked-eye sighting tube. At least from the first millennium B.C. a succession of stars had been taken up and abandoned as the pole star. In the late fifth century of the current era Tsu Keng[38] discovered that the current polestar, 4339 Camelopardi, rotated about a point slightly more than a

degree away. This determination of the true pole was incorporated in subsequent instruments by making the radius of their sighting tubes 1.5 Chinese degrees (each 360/365.25°). The excursion of the pole star just inside the field of view thus provided a nightly check on orientation. Six hundred years later Shen found that the polestar could no longer be kept in view throughout the night. He gradually widened the tube, using plots of the pole-star's position made three times each night for three months to adjust aim, until his new calibration revealed that the distance of the star from "the unmoving place at the celestial pole" was now slightly over three degrees (127; III, 262). Shen's successors followed him in treating the distance as variable, although the relation of this secular change to the equinoctial precession was not explored. Aware of the periodic retrogradation of the lunar nodes, Shen also discarded the armillary ring representing the moon's path, which could not reflect this motion; it was never used again.

Calendar Reform. On the accession of Shen-tsung in 1068, a new computational system was expected. The inability of the incumbent specialists to produce one left Shen with a clear mandate when he took over the Astronomical Bureau in 1072. The situation became even more awkward when he was forced to bring in Wei P'u and others from outside the civil service, although few of the incompetents already in the bureau could be dislodged, in order to begin work on the calendar reform. It is not yet possible to tell what part of the work was done by Shen and what part by his assistants, although it is clear that Wei took responsibility for compiling the system as Shen became increasingly occupied elsewhere in government. Wei, a commoner whose connection with Shen was first reported in 1068, bore the brunt of fervent opposition within the bureau. He was even formally accused of malfeasance.

Shen knew that previous Sung astronomical systems had suffered greatly from reliance on old observations, and had a clear conception of what new data were needed for the first major advance in centuries. Unabating opposition within the bureau and his own demanding involvements outside it limited the number of innovations of lasting importance in his Oblatory Epoch (*Feng-yuan*[39]) system. It was the official basis of calendar computation from 1075, the year of its completion, to 1094, a period very close to the average for systems of the Northern Sung. That the system was not used longer has little to do with its merits, since except in cases of spectacular failure, Sung astronomical systems changed as rulers changed. Shen's was replaced when a new era was marked by the coming of age of Che-tsung. The immediate vicissitudes and long-term influence of three special features will give a general idea of the limits that historical actuality set upon Shen's astronomical ambitions.

The boldest aspect of Shen's program was the attempt to master the apparent motions of the planets—not merely their mean speeds and prominent phenomena—for the first time. This could not be done with a few observations of stationary points, occultations, and maximum elongations. Shen and Wei therefore planned a series of observations of a kind not proposed in Europe until the time of Tycho Brahe, five centuries later: exact coordinates read three times a night for five years. Similar records were to be kept for the moon's positions, since previous Sung systems had still used the lunar theory of I-hsing, which after 350 years had accumulated considerable error. These records were the most unfortunate casualty of the antagonism within the Bureau. Shen and Wei had no recourse but to produce a conventional planetary theory based mainly on old observations. They were able to correct the lunar error, but even this proposal provoked such an outcry that it could be vindicated only by a public demonstration using a gnomon (116).

A second issue was the central one of eclipse prediction. Previous attempts to add or subtract correction factors showed the futility of this approach. It was Wei P'u who "realized that, because the old eclipse technique used the mean sun, [the apparent sun] was ahead of it in the accelerated phase of its motion and behind it in the retarded phase." He therefore incorporated apparent solar motion into the eclipse theory (139). This had been done centuries earlier but abandoned.

A major obstacle in eclipse prediction, as well as in such workaday problems as the projection of observations in equatorial coordinates onto the ecliptic, was the absence of spherical geometry. Shen's evolution of arc-chord-sagitta relations out of some inferior approximations for segment areas given in the arithmetical classics was a first step toward trigonometry, making it possible in effect to apply sine relations and a fair approximation of cosine relations (301; III, 39, with diagram). The great remaining lack, as in planetary theory, was a mass of fresh observations on which to base new parameters. That this weakness could threaten the continuance of the system became clear the year after it was adopted (1076), when the failure of a predicted lunar eclipse to occur left Shen and his associates open to attack. Shen parried with a successful request that astronomical students at the Han-lin Academy observatory be ordered to carry out his observational program "for three or five years" and to communicate the results to the original compilers. Whether this attempt to bypass the stalemate at the Astronomical Bureau's observatory was well-conceived remains unknown, for in the next year Shen's impeachment aborted it.

In sum, the immediate outcome of the Oblatory Epoch calendar reform was undistinguished, and within half a century the official documents embodying it had been lost. It is impossible to be sure, for instance, to what extent arc-sagitta relations had been incorporated after Shen invented them. But enough information survived in proposals, reports, Shen's writings, and compendiums of various sorts for his astronomical system to play a considerable part in the highest achievement of traditional Chinese mathematical astronomy, the Season-Granting (*Shou shih*[40]) system of Kuo Shou-ching[41] (1280). Kuo carried out a sustained program of observation using instruments that incorporated Shen's improvements. He took up Shen's arc-sagitta formula, greatly improving the cosine approximation, and applied it to the equator-ecliptic transform. Aware of Shen's emphasis on the continuous variation of quantities in nature, and of his criticism of linear interpolation in clepsydra design, Kuo used higher-order interpolation to an unprecedented extent in his calendar reform.

Shen recorded another scheme for reform of the civil calendar that was most remarkable for his time and place. It almost certainly occurred to him in the last decade of his life. The traditional lunisolar calendar was a series of compromises in reconciling two incommensurable quantities. The modern value for the tropical year is 365.2422 days, and that for the synodic month 29.53059 days, so that there are roughly 12.37 lunar months per solar year. The practical problem was to design a civil calendar with an integral number of days each month, and an integral number of months each year, in such a way that the long-term averages approach the astronomical constants. Hardly two of the roughly one hundred computational systems recorded in early China solved this problem in exactly the same way, just as there was endless tactical variety in other traditional societies, but strategy was generally the same. Months of twenty-nine and thirty days alternated, with occasional pairs of long months to raise the average slightly. Intercalary thirteenth months were inserted roughly seven times every nineteen years, which comes to 0.37 additional months per year.

By a millennium before Shen's time the calendar was more than adequate in these respects for every civil need, although attempts to further refine the approximation led to endless retouching. The rhythms of administration, and to some extent of commerce, were of course paramount in the design of the lunisolar calendar, despite pieties about imperial concern for agriculture. It is most unlikely that Chinese peasants ever needed a printed almanac by which to regulate their activity; what they consulted, if anything, was its notations of lucky and unlucky days. Division of the year by lunar months is, in fact, useless for agriculture, since the seasons that pace the farmer's work vary with the sun alone. The Chinese calendar also incorporated twelve equal divisions of the tropical year (ch'i[30], like the Babylonian *tithis*), further subdivided into twenty-four periods with such names as Spring Begins, Grain Rains, and Insects Awaken. These provided a reliable notation for seasonal change in the part of northern China in which the series originated.

Shen's suggestion was a purely solar calendar, based on the twelve divisions of the tropical year (average 30.43697 days in his system) instead of on the lunation. The civil calendar would thus alternate months of thirty and thirty-one days, with pairs of short months as necessary to approach the average. This would provide truly seasonal months and at the same time do away with "that goitrous excrescence" the intercalary month. "As for the waxing and waning of the moon, although some phenomena such as pregnancy and the tides are tied to them, they have nothing to do with seasons or changes of climate; let them simply be noted in the almanac" (545). Shen was aware that because the lunisolar calendar went back to hoary antiquity "it is by no means appropriate to criticize it." He predicted that his discussion "will call forth offense and derision, but in another time there will be those who use my arguments." This proposal was in fact considered by later scholars the greatest blemish on Shen's astronomical talent. His posterity appeared in the mid-nineteenth century, with the even more radical solar calendar enacted for a few years by the T'ai-p'ing rebels.* His work was

*Kuo T'ing-i,[42] *T'ai-p'ing t'ien kuo li-fa k'ao-ting*[43] ("Review of the Calendrical Methods of the T'ai-p'ing Heavenly Kingdom," 1937; reprinted Taipei, 1963); Lo Erh-kang,[44] *T'ien li k'ao chi t'ien li yü yin yang li jih tui-chao-piao*[45] ("On the T'ai-p'ing Calendar, With a Concordance Table for the Lunar and Gregorian Calendars"; Peking, 1955).

cited to justify historically more respectable proposals between that time and the adoption of the Gregorian calendar in 1912.

Configuration and Change. Chinese natural philosophers, unlike the majority in the postclassical West, did not dismiss the possibility that terrestrial phenomena could conform to mathematical regularities. But given the strength of Chinese quantitative sciences in numerical rather than geometric approaches, the very late and partial development of mathematical generalization, and the complete absence of notions of rigor, it is only consistent that much of the effort to discover such regularities produced numerology. Thus the most obvious of Shen's contributions to understanding of the earth and its phenomena are qualitative.

Magnetism. For more than a millennium before Shen's time, south-pointing objects carved from magnetite had been used from time to time in ceremonial and magic, and in 1044 objects cut from sheet iron and magnetized by thermoremanence were recommended for pathfinding in a book on military arts. Shen took up the matter of needles rubbed against lodestone by contemporary magi, discussed floating and other mountings, recommended suspension, noted that some needles point north and some south, and asserted that "they are always displaced slightly east rather than pointing due south"—all in about a hundred characters (437; IV.1, 249–250). This recognition of magnetic declination depended not only on consideration of a suspended needle but also on the improved meridian determined by Shen's measurement of the distance between the polestar and true north; declination in his part of China at the time has been estimated as between five and ten degrees (Needham and Peter J. Smith, "Magnetic Declination in Mediaeval China," in *Nature* [17 June 1967], 1213–1214. See the historical table in *Science and Civilisation in China*, IV.1, 310).

Shen may have been anticipated by geomancers, who practiced a sophisticated protoscience of land configuration and siting, but the dates of texts on which such claims have been based are questionable. The use of compass needles in navigation is recorded shortly after Shen's death, and later descriptions provide enough detail to show that the twenty-four-point rose that Shen substituted for the old eight compass points (perhaps also under the stimulus of the better meridian, if not of geomantic practice) had become widely used. He apparently was unaware of the polarity of magnetite itself, since in another article he explained the difference between north-pointing and south-pointing needles as "perhaps because the character of the stone also varies" (588; IV.1, 250).

Cartography. It has been conjectured that Shen was the first to use a compass in mapmaking, although traditional methods would have sufficed. Neither his early maps of Khitan territory nor the atlas of China completed in 1087 have survived to answer this question. But in an enclosure to the latter he did separately record bearings between points using his twenty-four-point compass rose, as well as rectilinear distances rather than, as customary, distances along established routes (he calls the use of distances "as the bird flies" ancient, but we have no earlier record). "Thus although in later generations the maps may be lost, given my book the territorial divisions may be laid out according to the twenty-four directions, and the maps speedily reconstructed without the least discrepancy" (575; III, 576). His great atlas included twenty-three maps drawn to a uniform scale of 1:900,000; the general map was ten by twelve Chinese feet. There is no evidence that the handbook outlasted the maps.

Three-dimensional topographic maps go back at least to Hsieh Chuang[46] (421–466), who had a demountable wooden model carved, apparently on the basis of an ancient map. In 1075, while inspecting the Khitan border, Shen embodied information gathered from the commander and the results of his own travels in a series of relief maps modeled, for the sake of portability, in plastic media—wheat paste and sawdust until the weather turned freezing, then beeswax—on wooden bases. These were carried to the capital and duplicated in wood; similar models were thenceforth required from other frontier regions (472; III, 580).

Shen's regular use of both historical research and special on-the-ground surveys to solve such cartographic problems as tracing changes in watercourses also is noteworthy (431). Typical of his ingenious topographic survey methods were those used in 1072 to measure the slope of the Pien Canal near the capital. There he built a series of dikes in temporary, narrow parallel channels to measure incremental changes in water level (457; III, 577*).

Formation of the Earth. In 1074, in the T'ai-hang mountain range (Hopei), Shen noticed strata of "bivalve shells and ovoid rocks running horizontally through a cliff like a belt. This was once a seashore, although the sea is now hundreds of miles east. What we call our continent is an inundation of silt. . . . This mud year by year flows eastward, forming continental land." A similar stratum

had been observed long before by Yen Chen-ch'ing[47] (708–784), who vaguely suggested its origin in the sea; but Shen—whose duties had made him intimately familiar with the process of silting—opened a new line of investigation by proposing a mechanism (430; III, 604).

Probably on his southward drought survey earlier in the same year, Shen saw the Yen-tang range (Chekiang), a series of fantastic rock formations "invisible from beyond the ridgeline [opposite], but towering to the sky when seen from the valleys. If we trace the underlying pattern, it must be that great waters in the valleys have attacked and washed away all the sand and earth, leaving only the great rocks erect and looming up." His explanation proceeded to generalize the shaping role of erosion, and then to apply it to the hills that divide streams in the loess country of northwest China— "miniatures of the Yen-tang mountains, but in earth rather than stone" (433; III, 603–604).

Shen reported a variety of contemporary finds of petrified plants and animals (373–374; III, 614–618). He remarked particularly on a stony formation he identified as originally a grove of interconnected bamboo roots and shoots, found dozens of feet below ground level at Yenan (Shensi). He knew from his military service there that the climate was too dry to grow bamboo: "Can it be that in earliest times [literally, 'before antiquity'] the land was lower and the climate moister, suitable for bamboo?" (373). About a century later the great philosopher and polymath Chu Hsi[48] (1130–1200), who knew Shen's jottings well and often extended ideas from them in his teaching, suggested that the stone of certain mountains was itself petrified silt deposits. But Shen's notion of prehistoric climatic change, like that of the reshaping of land by erosion, was not pushed further soon after his lifetime.

Atmospheric Phenomena. Although Shen did not report important original discoveries of his own, he preserved a number of interesting observations not recorded elsewhere. Perhaps the most important is a vivid description of a tornado (385; translated in Holzman, "Shen Kua," 286), the veracity of which was questioned by modern meteorologists until, in the first decade of the twentieth century, the Sikawei Observatory in Shantung reported phenomena of the same kind, previously thought restricted to the western hemisphere. Shen was also responsible for transmitting an explanation of the rainbow by Sun Ssu-kung,[49] an elder contemporary in the court who was also considered one of the best mathematical astronomers of his era. "The rainbow is the image [literally, 'shadow'] of the sun in rain, and occurs when the sun shines upon it." This sentence does not, as often claimed, adduce refraction (pinhole or mirror images were regularly called "shadows"; see 44). Shen was prompted to determine by experiment that the rainbow is visible only opposite the sun (357). Later Chu Hsi, aware of Shen's account, added that by the time the rainbow appears "the rain ch'i[30] has already thinned out; this in turn is because sunlight has shone on and attenuated the rain ch'i."[30] Ch'i must mean vapor here; the notion of reflections off individual drops is, as in Sun's explanation, implicit at best. Shen also recorded the fall of a fist-sized meteorite in more detail and with less mystification than previous reports. The particulars of its fall came from a careful account by another of Wang An-shih's associates. The object was recovered and exhibited, but Shen did not claim that he himself had observed that "its color is like that of iron, which it also resembles in weight" (340; III, 433–434).

Products of the Earth. Responsibilities with respect to fiscal policy gave Shen a detailed knowledge of important commodities, their varieties, and the circumstances of their production, as may be seen from his descriptions of tea (208) and salt (221). Inflammable seepages from rock had been known a millennium before Shen's time, and for centuries had been used locally as lamp fuel and lubricant. While civil and military commissioner near Yen-chou, he noted the blackness of soot from petroleum and began an industry to manufacture the solid cakes of carbon ink used for writing and painting throughout China. Good ink was then made by burning pine resin, but Shen knew that North China was being rapidly deforested. He remarked that, in contrast with the growing scarcity of trees, "petroleum is produced inexhaustibly within the earth." The name Shen coined for petroleum[50a] is the one used today, and the source in Shensi province that he developed is still exploited. In the same article he quoted a poem of his that is among the earliest records of the economic importance of coal, then beginning to replace charcoal as a fuel (421; III, 609, partial).

Optical Phenomena. Shen's interest in image formation was not directly connected with his worldly concerns. His motivation is more plausibly traced to the play of his curiosity over old artifacts than to the improvement of naked-eye astronomical instruments.

In the canons of the Mohist school (*ca.* 300 B.C.) is a set of propositions explaining the formation of shadows and of optical images (considered a kind of shadow) in plane, convex, and concave mirrors. One proposition is widely believed to concern pinhole images, although textual corruption and ambiguity make this uncertain. These propositions are in many respects correct, although very schematic, and rays of light are not presupposed. Shen concerned himself with the single question of why a concave mirror forms an inverted image. He posited an "obstruction" (*ai*[50]), analogous to an oarlock, that constricts the "shadow" to a shape like that of a narrow-waisted drum—or, as we would put it, to form two cones apex to apex, the second constituting the inverted image. Like the Mohists, Shen clearly believed that inversion takes place before the image is reflected. He expressly likened the inverted image to that of a moving object formed on the wall of a room through a small opening in a paper window. Aware for the first time that there is a range of distances from a concave mirror within which no image is formed (that is, between the center of curvature and the focal point), he explained that this blank region, corresponding to the pinhole, is the locus of "obstruction" (44; translated in A. C. Graham and N. Sivin, "A Systematic Approach to the Mohist Optics," in S. Nakayama and N. Sivin, eds., *Chinese Science: Explorations of an Ancient Tradition* [Cambridge, Mass., 1973], 145–147). His pinhole observation was adventitious, but his approach to the burning-mirror was experimental in its details.

Two other observations of optical interest are found under the rubric "Artifacts and Implements." The first, in the "Sequel to Brush Talks," noted that when the ancients cast bronze mirrors, they made the faces just convex enough that, regardless of size, every mirror would reflect a whole face. By Shen's time this refinement had been abandoned and the reasoning behind the curvature forgotten, so that collectors were having the faces of old mirrors scraped flat (327; IV.1, 93).

The second jotting is the oldest record of a Far Eastern curiosity still being investigated: "magic mirrors," or, as Shen called them, "transparent mirrors." Shen described a bronze mirror with a smooth face and an integrally cast inscription in relief on its back (both conventional features). When the mirror was used to reflect the sun onto a wall, the inscription was duplicated within the image. Shen cited with approval an anonymous explanation: "When the mirror is cast, the thinner parts cool first; the raised design on the back, being thicker, cools later and the shrinkage of the bronze is greater. Although the inscription is on the obverse, there are imperceptible traces of it on the face, so that it becomes visible within the light." He then qualified this explanation as incomplete, because he had tried mirrors in his own and other collections that were physically indistinguishable from the "transparent" ones and found that they did not cast images (330; IV.1, 94*). His doubt was justified, although the approach taken by his informant was at least as good as those of some modern metallurgists. Although cooling rate plays no discernible part, the variation in thickness is indeed responsible for the image in this sort of mirror, the most common among several types extant. Filing considerable bronze off the face of the mirror after casting is the key. This releases tensions in the metal and gives rise to slight deformations that produce the image.

Productive Techniques and Materials. The technologies of Shen's time were not cumulative and linked to science, but independent artisanal traditions transmitted from master to pupil. Shen left so many unique and informative accounts of ancient and contemporary processes among his jottings that "Brush Talks" has become a major source for early technology. Shen's interests in contemporary techniques can in most cases be linked to broad concerns of his official career; but the exceptional richness of his record bespeaks a rare curiosity, and the trenchancy of his descriptions a seriousness about mechanical detail unusual among scholar-officials. His notes on techniques lost by his time reflect the application of this technical curiosity and seriousness to archaeology, which was just becoming a distinct branch of investigation in the eleventh century.

Most of Shen's cultured contemporaries had a keen appreciation for good workmanship but considered the artisans responsible for it beneath notice except for occasional condescension. Shen wrote about resourceful craftsmen and ingenious laborers with much the same admiration he gave to judicious statesmen. He did not lose sight of the social distance between himself and members of the lower orders, but in his writing there is no snobbishness about the concert of hand, eye, and mind.

Contemporary Techniques. The most famous example is Shen's account of the invention of movable-type printing by the artisan Pi Sheng[51] (*fl.* 1041–1048). Shen described the carving and firing of ceramic type and the method of imbedding and leveling them in a layer of resin, wax, and paper ash in an iron form, one form being set as a second is printed. As in xylography, water-base ink was used. Since the porous, thin paper took it up with little pressure, no press was needed. Shen also remarked, with his usual acumen, that the process could become faster than carving wood blocks only with very large editions[52] (the average then has been estimated at between fifty and a hundred copies). Unevenness of the surface and absorption of ink by the fired clay must have posed serious problems. Abandonment of the process after Pi died was probably due to the lack of economic incentive that Shen noted. The long series of royally subsidized Korean experiments in the fifteenth century that perfected cast-metal typesetting still began with Pi Sheng's imbedding technique as described by Shen. Whether he knew Pi is unclear, but Shen's cousins preserved Pi's original font (307; translated in full, but not entirely accurately, in T. F. Carter, *The Invention of Printing in China and Its Spread Westward*, L. C. Goodrich, ed., 2nd ed. [New York, 1955], 212).

Shen left a number of descriptions of metallurgical interest—for instance, an account of the recovery of copper from a mineral creek by replacement of iron, a process then being carried out on an industrial scale to provide metal for currency (455; II, 267); observations of two of the three steelmaking processes used in early China (56; translated in Needham, *The Development of Iron and Steel Technology in China* [London, 1958], 33–34; the book was reprinted at Cambridge, England, in 1964); and remarks on a little-known cold-working method used by smiths of the Ch'iang[53] people of western China to make extremely tough steel armor (333). Water control techniques of which he records details include pound-locks with double slipways (213; IV.3, 351–352), piles for strengthening embankments (210; IV.3, 322–323), and sectional gabions for closing gaps after embankment repairs (207; IV.3, 342–343).

Ancient Techniques. The concern for understanding ancient techniques began with the commentators on the Confucian and other classics more than a millennium earlier. Exegesis remained an important activity in China, and the productive methods of golden antiquity were investigated with the same assiduity as anything else mentioned in its literary remains. For various reasons—among them the recovery of ancient artifacts in large numbers for the first time, the growth of collecting, and the elaboration of a conscious aesthetics—archaeology began to emerge from the footnotes less than a century before Shen's time, especially in monographs on ancient implements and ritual institutions. He was familiar with this literature and responded to it critically. Much of his writing in the "Artifacts and Implements" chapter falls squarely in this tradition, drawing on the testimony of both objects and books.

Shen's vision of the past as a repertory of lost processes introduced an influential new theme. A constant concern in his writing was not only that the workmanship of the past be esteemed for its excellence, but also that the present be enriched through understanding what the practical arts had been capable of. Although the belief was still current that the inventions that first made civilization possible were all due to semidivine monarchs of archaic times, in a letter Shen saw the technological past as successful for just the opposite reason: "How could all of this have come from the Sages? Every sort of workman and administrator, the people of the towns and those of the countryside— none failed to take part" (*Ch'ang-hsing chi* [1718 ed.], 19:53b).

Shen's remarks on magic mirrors are typical of his effort to understand lost processes. Another example is his reconstruction (and personal trial) of ancient crossbow marksmanship, interpreting a gnomic aiming formula in an ancient footnote with the aid of a graduated sight and trigger assembly that he examined after it was unearthed (331; III, 574–575*). The most famous instance of Shen's use of literary sources for the study of techniques has to do with the remarkable modular system of architecture used in public buildings. The set of standard proportions is well-known from an official compilation printed about a decade after Shen's death. Shen, by describing the proportion system of the Timberwork Canon (*Mu ching*[54]), attributed to a great builder of about 1000 and already falling out of use, demonstrates the antiquity of this art (299; IV.3, 82–83).

Medicine. By Shen's time medicine, which from the start drew heavily upon natural philosophy for its conceptual underpinnings, had accumulated a classical tradition. Not only was each new treatise consciously built upon its predecessors, but a major goal of new work was restoring an understanding that medical scholars believed was deepest in the oldest writings. The revealed truth of the archaic canons was too concentrated for ordinary latter-day minds, who could hope to recapture it only as the culmination of a lifetime of study. Writers in the intervening centuries referred to the early classics as the ultimate source of significance even while aware that empirical and practical knowledge had considerably advanced since antiquity. The major contribution of the continuous tradition of medical writing was to fit new experiences into the old framework and, when necessary, to construct new frameworks in the spirit of the old. As woodblock printing became feasible, standard editions of the chief classics were compiled and disseminated by government committees. This increased the respectability of the curing arts as a field of study. Large numbers of men from the scholar-official class began to take up medicine, not in competition with those who made a living by it but as a means of self-cultivation allied to cosmology and occasionally useful. The initial motivations commonly were personal ill health and the desire to serve one's sick parents.

Shen, as noted earlier, began the study of medicine early, for the former reason. One of his two therapeutic compilations survives in somewhat altered form. Its preface is a long disquisition on the difficulty of adequate diagnosis and therapy, as well as on the proper selection, preparation, and administration of drugs. His criticisms of contemporary trends toward simplification remind us that the development of urban culture and education in Sung China had led to increased medical practice among ordinary people as well as study by the literati. As protoscientific medicine began to displace magico-ritual folk remedies (at least in the cities), there were more half-educated physicians to be criticized by learned amateurs such as Shen. Shen's most characteristic contribution was undoubtedly his emphasis on his own experience, unusual in a tradition whose literature in the Sung still tended to depend heavily on copying wholesale from earlier treatises. Shen not only omitted any prescription the efficacy of which he had not witnessed, but appended to most a description of the circumstances in which it had succeeded. He provided many precise descriptions of medicinal substances of animal, vegetable, and mineral origin. Although he had no more interest in general taxonomic schemes than other pharmacognostic scholars of his time, his concern for exact identification and for philological accuracy gave his critical remarks enduring value. Many were incorporated into later compilations on materia medica, and Shen's writing also served as a stimulus to the work a few decades later of the great pharmacognostic critic K'ou Tsung-shih[55] in his "Dilatations Upon the Pharmacopoeias" (*Pen-ts'ao yen i*,[56] 1116).

A recent discovery of considerable interest is that certain medical preparations from human urine collectively called "autumn mineral" (*ch'iu shih*[57]), which have a long history in China, contain high concentrations of steroid hormones and some protein hormones as well. In "Good Prescriptions" Shen gives one of the earliest accounts, in the form of detailed instructions for two such preparations that he performed in 1061 (other accounts by contemporaries are harder to date).*

Perhaps Shen's most famous writing on general medical matters is one in which he refutes the common belief that there are three passages in the throat—as shown, for instance, in the first book of drawings of the internal organs based directly on dissection (1045).† His supporting argument is not from independent dissection but from sufficient reason—"When liquid and solid are imbibed together, how can it be that in one's mouth they sort themselves into two throat channels?" He thus saw the larynx as the beginning of a network for distributing throughout the body the vital energy carried in atmospheric air, and the esophagus as carrying nutriment directly to the stomach cavity, where its assimilation begins. This was a significant increase in clarity as well as accuracy (480).

A passage that has been praised for its simple but beautiful language takes issue with the ancient principle that medicinal plants should be gathered in the second and eighth lunar months (when they were thought easiest to identify). In a few hundred words it epitomizes the variation of ripening time with the identity and variety of the plant; the part used in therapy; the physiological effect needed for the application; altitude; climate; and, for domesticated medicinal plants, variation with planting time, fertilization, and other details of horticulture. The sophistication of this passage reflects not only increasing domestication (exceptional in earlier eras) but also the integration of drugs from every corner of China into the expanding commercial network.

Conclusion. The expansiveness of Northern Sung society and its relative openness to talent, not to mention increasing government sponsorship

*See Lu Gwei-djen and Joseph Needham, "Medieval Preparations of Urinary Steroid Hormones," in *Medical History*, **8** (1964), 101–121; Miyashita Saburō,[58] *Kanyaku shūseki no yakushigakuteki kenkyū*[59] ("A Historical Pharmaceutical Study of the Chinese Drug 'Autumn Mineral' the *Ch'iu-shih*"; Osaka, 1969), esp. 9–12.

†Persons untrained in medicine performed the dissection upon executed bandits in 1045 and recorded what was found under the direction of an enthusiastic amateur. Another episode of the same kind, undertaken explicitly to correct the earlier drawings, took place at the beginning of the twelfth century. There is no reliable account of either in any European language, but see Watanabe Kōzō,[60] "Genson suru Chūgoku kinsei made no gozō rokufu zu no gaisetsu"[61] ("A Survey of Extant Chinese Anatomical Drawings Before Modern Times"), in *Nihon ishigaku zasshi*, **7** (1956), 88.

of learning, made this an important period in the history of every branch of science and technology. Shen was not the first polymath it produced. There was also Yen Su[62] (*fl.* 1016), who designed an odometer and south-pointing chariot (in which a differential gear assembly kept figures pointing in a constant direction as the chariot turned), improved the design of the water clock and other astronomical instruments, and wrote on mathematical harmonics and the tides. In Shen's lifetime there was Su Sung[63] (1020–1101), who was first privy councillor during the last part of the reaction against the New Policies (1092–1093). Through the 1060's he played a major part in a large imperially sponsored compilation of materia medica, and in the editing and printing of ancient medical classics. In 1088 a group that he headed completed a great water-driven astronomical clock incorporating an escapement device. Their detailed description of the mechanism included the oldest star map extant in printed form, based on a new stellar survey. (The book has been studied and translated in Wang Ling, Joseph Needham, Derek J. de Solla Price, *et al., Heavenly Clockwork* [Cambridge, England, 1960].) That Yen, Su, and Shen were all in the central administration is not surprising. The projects on which they were trained and those in which they worked out many of their ideas were of a scale that only the imperial treasury could (or at least would) support.

Breadth of interest alone does not account for Shen's importance for the study of the Chinese scientific intellect. Another aspect is his profound technical curiosity. A number of the phenomena he recorded were mentioned by others; but even when others' descriptions happen to be fuller, they usually are of considerably less interest because their subject matter is treated as a mere curiosity or as an occasion for anecdote rather than as a challenge to comprehension. Above all, one is aware in Shen, as in other great scientific figures, of a special directness. A member of a society in which the weight of the past always lay heavily on work of the mind, he nevertheless often cut past deeply ingrained structures and assumptions. This was as true in his program of astronomical observations and his audacious solar calendar as in his work on government policies. People in the Sung were aware that man's world had greatly expanded since antiquity, and questioning of precedent (in the name of a return to classical principles) was inherent in the New Policies. Shen's commitment to this political point of view can only have reinforced the sense of cumulative improvement of techniques and increasing accuracy over time that one finds in major Chinese astronomers. But given these predispositions and opportunities, Shen remains in many senses an atypical figure, even in his time and among his associates.

There certainly is much that a modern scientist or engineer finds familiar, not only in the way Shen went about making sense of the physical world but also in the temper of his discourse, despite the profoundly antique nature of the concepts he used. One comes away from his writings confident that he would see much of modern science as a culmination (not the only possible culmination) of his own investigations—more confident than after reading Plato, Aristotle, or St. Thomas Aquinas. But does Shen's special configuration of abilities and motivations suggest that a genetic accident produced, out of time, a scientific rationalist-empiricist of essentially modern type? To answer this question it is necessary to look at Shen's larger conception of reality, of which his scientific notions compose only a part but from which they are inseparable.

The Relation of Scientific Thought to Reality. The sense of cumulative enterprise in mathematical astronomy did not imply the positivistic conviction that eventually the whole pattern could be mastered. Instead, from the earliest discussions there was a prevalent attitude that scientific explanation—whether in terms of number or of abstract qualitative concepts, such as yin-yang—merely expresses, for human purposes, limited aspects of a pattern of constant relations too subtle to be understood directly. No one expressed this attitude more clearly than Shen. In instance after instance he emphasized the inability of secular knowledge to encompass phenomena: the reason for magnetic declination (437), why lightning striking a house can melt metal objects without burning the wooden structure (347), and so on.

Shen made this point most clearly in connection with astronomy. In one passage he discussed the fine variations that astronomers must, in the nature of their work, ignore. Every constant, every mean value obscures continuous variation of every parameter (123). In his official proposals on the armillary sphere,[64] he argued that measure is an artifact, that it allows particular phenomena to be "caught" (po[65]) in observational instruments, where they are no longer part of the continuum of nature. That Shen saw as the condition of their comprehensibility. This and similar evidence amount not merely to an appreciation of the role of abstraction in science, but also to the steady conviction that abstraction is a limited process incapable of producing universal and fundamental knowledge of the concrete phenomenal world. Nature is too rich, too multivariant, too subtle (wei[66]). This limitation did not detract from the interest or worth of theoretical inquiry, and did not lead intellectuals to question whether learning could contribute to the satisfaction of social needs; but the ambit of rationalism in traditional scientific thought was definitely circumscribed.

In this light Shen's explanatory metaphors become more comprehensible. In his remarkable suggestion that variations in planetary speed may be represented by a compounded figure, he chose to fasten to the periphery of his circle a willow leaf, whereas in Europe no figure but another circle was thinkable (148). When explaining optical image inversion in terms of converging and diverging rays, the images of the oarlock and waisted drum occurred to him (44). The variation in polarity of different magnetized needles was likened to the shedding of antlers by two species of deer in opposite seasons (588; IV.1, 250), and so on. Geometric figures, numbers, and quantities were useful for computation but had very limited value, not so great as cogent metaphors from the world of experience, in understanding the pattern inherent in physical reality.

Many Chinese thinkers, even in the Sung, did believe in number as a key to the pattern of physical reality; but their search was concentrated in numerology (especially as founded on the "Great Commentary" to the *Book of Changes*) rather than in mathematics. This is not to imply that numerology was a distraction from mathematics. The two were not considered alternate means to the same goal.

Other Kinds of Knowledge. Did Shen believe that other ways of knowing complemented and completed empirical and theoretical investigation? Aside from its scientific aspects, Shen's thought has been so little studied that only some tentative suggestions can be offered. Contemplation and disciplined self-examination were ancient themes in Confucianism, and by Shen's time illumination was widely considered among the learned as a source of knowledge complementary to that given by experience of the external world. The domestication and secularization of Buddhist and Taoist meditation were gradually leading to a more introspective and less ritualistic approach to self-realization. This tendency was later elaborated with great variety of emphasis and weight in the schools of neo-Confucianism.

To understand what part contemplation and meditation played in the thought of Shen Kua requires a clearer view than we now have of their currency and coloring in his time, of the considerable role of Wang An-shih's thought in his intellectual development, and of Shen's own attitudes as indirectly expressed in his literary remains. There is as yet no sound basis for evaluating his interest in Taoist arcana that seems to have peaked in his thirties, his public remarks that express sympathetic interest in illuminationist (Ch'an, Japanese "Zen") Buddhism, and his statement in an autobiographical fragment that Ch'an meditation was one of the things to which he turned his attention after retirement. In any case these involvements refract aspects of his epistemology that cannot be overlooked without badly distorting our recognition of the whole.

Teraji Jun has recently demonstrated this point in examining how strong a factor in Shen's motivation and individuality was his belief in destiny and prognostication. There are crucial passages, especially in his commentary on Mencius, where Shen spoke of the necessity for choosing what is true and holding to it, and called the rule of the heart and mind by sensory experience "the way of the small man." The basis of moral choice was an autonomous inner authority defined in an original way but largely in Mencian terms, a centeredness "filling the space between sky and earth," unquestionably linked with the self-reliance that marked his unhappy career.

It is not immediately obvious why someone who so valued individual responsibility should have been fascinated by fate and divination, which in fact are the themes of whole chapters of "Brush Talks." Shen does not seem to have viewed these enthusiasms as in conflict with his scientific knowledge. His delight in strange occurrences and his tendency to place matters of scientific interest under that rubric begin to make sense under the hypothesis that he accepted the odd, the exceptional, and the affront to common sense as a challenge for explanation at another time, or by someone else — without assuming that explanation was inevitable. In his hundreds of jottings on people, the person he chose to praise is most often the one who did not do the obvious thing, even when it seemed the sound thing to do.

At one point Shen provided a thoroughly rational explanation of the relations between fate and prognostication. The future can of course be foreknown by certain people, he said, but it is a mistake to conclude that all matters are preordained. The vision of the future is always experienced in present time; the years in the interim also become simultaneous. One can do nothing to avoid an undesirable future so glimpsed. Authentic foreknowledge would have witnessed the evasive measures; a vision that failed to see them could not be authentic foreknowledge (350).

In addition to the visionary ability of certain minds, Shen pondered universally accessible methods of divination, which (he seems to have believed) do not describe the future or the spatially distant so much as provide counsel about them or aid thought about them. In one of his chapters, "Regularities Underlying the Phenomena," he explained why the same divinatory technique gives different outcomes when used by different people, and thus has no inherent verifiability. He quoted the "Great Commentary" to the *Book of Changes* to the effect that understanding is a matter of the clarity and divinity (in a very abstract sense) within one's mind. But because the mind is never without burdens that hinder access to its divinity, Shen reasoned, one's communion with it may take place through a passive mediating object or procedure (144, 145). This divinity is, for Shen's sources, the moral center of the individual. Prognostication, however ritualized (as we would put it), thus draws indirectly upon the power of self-examination. Access to the future, whether by vision or by divination, is a perfectly natural phenomenon that is imperfectly distinct, on the one hand, from the moral faculties, the choices of which condition the future, and, on the other, from science, the rational comprehension of the natural order as reflected in all authentic experience.

Thus it appears that introspection supplemented by divinatory procedures was a legitimate means to knowledge in Shen Kua's eyes, just as painstaking observation and measurement of natural phenomena were another. He neither confused the two approaches nor attempted to draw a clear line between them. Nor was he inclined to assess the comparative importance of these ways of knowing.

The complementarity in Shen's attitudes toward knowledge is echoed by another in the external world of his work. Computational astronomy and divination of various kinds (including judicial astrology) were equally weighty functions carried out by the central government on the emperor's behalf, for both kinds of activity were established supports of his charisma. The need to combine science with ritual in this sphere is implied in an important memorial of Wang An-shih: because the monarch acts on behalf of the natural order, he can safeguard the empire and command the assent of the governed only through knowledge of nature. Ritually expressed awe of that order, without knowledge, is not enough (*Hsu tzu chih t'ung chien ch'ang pien*[67] ["Materials for the Sequel to the Comprehensive Mirror for Aid in Government"], presented to the throne 1168 [1881 ed.], 236:16b). Teraji has acutely pointed out that this is precisely the political justification for Shen's research, and the reason that traditional bureaucrat-scientists who were concerned mainly with maintaining ancient practices were not what Wang wanted.

Confucianism and Science. Recent attempts in both East and West to construct a historical sociology of Chinese science have in large part been built around a contrast between Confucianist and Taoist ideology. The values of the Confucian elite are often described as oriented toward stasis, hierarchy, bureaucracy, and bookishness. These characteristics are seen as perennially in tension with the appetite of socially marginal Taoists for novelty and change, their tendency to contemplate nature and the individual in it as a system, and their fascination with techniques, which kept them in touch with craftsmen and made them willing to engage in manual work themselves. It will no doubt be possible eventually to excavate a falsifiable, and thus historically testable, hypothesis from the mound of observations and speculations in this vein that have accumulated over the last half-century. For the moment, all one can do is point out how relentlessly unsociological this discussion has been.

Sociology is about groups of people. Doctrines are germane to sociology to the extent that their effect on what groups of people do, or on how they form, can be demonstrated. Generalizations about people who accept a certain doctrine have no sociological significance unless such people can be shown to act as a group, or at least to identify themselves as a group. The term "Confucian" is commonly used indifferently even by specialists to refer to a master of ceremonial, a professional teacher of Confucian doctrines, a philosopher who contributes to their elaboration, someone who attempts to live by Confucius' teachings, any member of the civil service, any member of the gentry regardless of ambition toward officialdom, or any conventional person (since it was conventional to quote Confucian doctrines in support of conventional behavior). A "Taoist" can be anyone from a hereditary priest ordained by the Heavenly Master to a retired bureaucrat of mildly unconventional tastes living on a city estate. Either group, by criteria in common use, includes people who would make opposite choices on practically any issue. This being so, the proposition "Taoists were more friendly toward science and technology than Confucians" reduces to "Educated individuals who hold unconventional sentiments are more inclined to value activities unconventional for the educated than are educated people who hold conventional sentiments." That is probably not quite a tautologous statement, but it is sociologically vacuous and historically uninteresting.

Unease of this sort is probably the most obvious outcome of reflection on Shen Kua's career. By sentimental criteria he can be assigned to Confucianism, Taoism, or Buddhism, to suit the historian's proclivities.* He was a member of the elite, a responsible official, a writer of commentaries on several of the Confucian classics, and a user of the concepts of Confucius' successor Mencius to explore the depths of his own identity. He spoke well and knowledgeably of Buddhism. He practiced arcane disciplines, such as breath control, that he called Taoist.

As for his allegiances, Shen was prominently associated with a powerful but shifting group of background very generally similar to his own. Social stasis and institutional fixity were impediments to their aims in reshaping government. At the same time, the new balance of power toward which they strove was more authoritarian than the old. Underlying their common effort was an enormous disparity of motivation, from the well-intentioned (Shen) to the simultaneously manipulative and corrupt (Ts'ai Ch'ueh).

Were these Confucians more or less Confucian than their Confucian opponents? Wang An-shih earned enduring stature for his commentaries on the classics and his thought on canonic themes. His followers seem to have found inspiration in the

---

*A new element was introduced in 1974 in a book issued as part of the "anti-Confucius anti-Lin Piao" campaign against current ideological trends. Two of its essays (pp. 118–140) portray Shen as a legalist and a relentless opponent of Confucianism. "Legalist" is a term applied to writers on government and administration concentrated in the last centuries before the Christian era, especially those who argued that polity must be built on law and regulation, in contrast with the traditionalist faith of Confucius in rites and moral example. Although the arguments in this book are too distorted and too selective in their use of sources to be of interest as history, they become intelligible when "legalism" and "Confucianism" are understood as code words for the political convictions of two contending power groups in China today, as portrayed by spokesmen for one of the two. The book is *Ju-Fa tou-cheng yü wo kuo ku-tai k'o-hsueh chi-shu ti fa-chan*[68] ("The Struggle Between Confucianism and Legalism and the Development of Science and Technology in Our Country in Ancient Times"; Peking, 1974). The first printing was 31,000 copies.

classics as often as their enemies and as those who avoided taking a political position. This is not to say that everyone understood the Confucian teachings in the same way. The latter were not, from the viewpoint of intellectual history, a set of tightly linked ideas that set fixed limits on change; rather, they were a diverse and fragmentary collection of texts reinterpreted in every age. They were understood differently by every individual and group who looked to them for guidance when coping with problems of the moment.

The major commentaries, which attempted to define the meanings of Confucian teachings philologically, carried enormous authority; and governments (that of Wang An-shih, for instance) repeatedly attempted to make one interpretation orthodox. But the urge to pin down meanings was always in conflict with precisely what made these books classic. Their unlimited depth of significance depended more on what could be read into them than on precisely what their authors had meant them to say. That depth made them applicable to an infinity of human predicaments and social issues, unprecedented as well as perennial. Late neo-Confucian philosophers striking out in new directions demonstrated again and again how little the bounded intellectual horizons and social prejudices of the classics' authors objectively limited what may be drawn from them.

In other words, the Confucian canon had the influence it did because it provided a conceptual language that over the centuries educated people used and redefined in thinking out decisions and justifying action and inaction. The classics were often cited as a pattern for static social harmony and willing subordination in arguments against the New Policies. Shen, on the other hand, used them to argue for flexibility in social relations and for greater receptivity toward new possibilities than was usual in his time. Either as a social institution or as an ideology, Confucianism is too protean and thus too elusive a base for generalizations about the social foundations of science and techniques in China.

Institutions also changed constantly, but at least they were tangible entities. It is essential to consider them when tracing the social connections of science. Very little is known about how scientists were educated in the Northern Sung period; the obvious next step is a collective study of a great many biographies. In Shen's case we can see a pattern that certainly was not unique. He was, so far as we know, self-educated in astronomy, but with many learned associates to draw upon. In medicine

and breath control he probably received teaching in the traditional master-disciple relationship. Defined in the ages before printing made possible access to large collections of books, this relationship involved the student's memorizing the classics (more often one than several) that the teacher had mastered. This verbatim transmission of a text was supplemented by the teacher's oral explanations. The relation was deepened by ceremonial formality; the master took on the obligation to monitor the disciple's moral as well as intellectual growth, and the disciple accepted the responsibility of becoming a link in an endless chain of transmission. Schools were largely communities of masters and disciples. The scale of government-sponsored elementary schools in the provinces was small in Shen's youth, and began to compete with private academies only in the New Policies period. The two sorts together did not serve more than a small minority of youth.

By the eighth century there were small schools in the central government to train technical specialists. The masters, usually several in number, were functionaries, representing the departments of the bureau that the disciples were being trained to staff. The schools for medicine and astronomy could not lead to the top of government, but guaranteed steady advancement between minor sinecures. Very few of the great physicians or astronomers of traditional China began in these schools.

In the absence of evidence to the contrary, there is no reason to believe that Shen Kua ever attended a school of any sort, nor does that make him untypical. His early education by his mother, his training in medicine by an obscure physician and others who remain unknown, and his catch-as-catch-can studies of most other matters do not set him apart from his contemporaries. With no knowledge of particulars one cannot even guess how his personal style in technical work was formed. But to say that we are ignorant is not to say everything. The intimate relations of master and teacher and the isolation of the autodidact were themselves important institutions in the Northern Sung, institutions of a sort that did not discourage the emergence of unforeseen abilities in the small number of people who had the opportunity to be educated. Shen did not have to cope with a standard curriculum, for better or worse. If we are searching for the decisive curriculum of science and technology, it is necessary to look outside the realm of education.

The Civil Service and Science. One institution above all others influenced the mature ideas and attitudes of the ruling stratum: the bureaucracy. What can be said about its influence on science and technology in the life of Shen Kua? First, like every bureaucracy, it depended upon science and technology. It supported both sorts of activity on a scale otherwise unattainable, and unheard of in Europe at the time. Shen's curiosity, experience, and skills were so largely shaped by the civil service that it is absurd to ask what he would have become had he lived as a country gentleman or a Taoist priest. On the other hand, as elsewhere, technicians were certainly less important to the priorities of the state than administrators. The responsibility of the former was to provide the emperor and his administrators with wealth and other tools for the realization of policy. Specialist positions in science and engineering did not often serve as the beginnings of great careers.

By the New Policies period a career stream for economic experts had been established. It could assimilate people who combined technological acumen with fiscal skills, and carry them to the central councils of the empire. Shen's early technical feats were performed in general administrative posts, but his talents came to be valued and he rose quickly through formal and informal structures. It is not irrelevant that his directorship of the Astronomical Bureau was never more than a concurrent position. His attempt to combine an effective voice in the shaping of change with scientific contributions ended in personal disaster. He was ruined by men of his own faction, apparently for his political seriousness and naïveté. His astronomical work was rendered futile by subordinates because of his professional demands upon them. The bureaucracy was not neutral; it was a two-edged sword.

The civil service provided a form for great projects in science and technology, and practically monopolized certain disciplines, such as mathematical astronomy and observational astrology. Printing gave it the wherewithal to determine much of the content of elementary technical education (as in medicine and mathematics). A man of Wei P'u's genius, who had not had the opportunity to enter the bureaucracy by a regular route, was looked down upon and deliberately frustrated. Had Shen himself chosen to be a mere technician, his standing in the civil service would have been sufficient to protect him from personal attack. He would have had more time but less power. It would be rash indeed to speculate that his calendar reform would not have failed. But there is a larger issue.

Shen's mind was shaped for the civil service, as were those of his ancestors and peers, by an early education centered in moral philosophy and letters. He was a generalist. The development of depth in thought and work was left to his own proclivities. Only a superficial knowledge of technical matters was expected of him as a youth—a situation not very different from that of the British civil service generalist of some decades ago. Shen's growing responsibilities in fiscal affairs were the one aspect of his career that we can be sure encouraged him to draw coherence out of his varied experiences and studies. For this reason and others of which we are still ignorant, the great breadth of his knowledge was accompanied by enough depth to let him write monographs of some importance and, even through his brief jottings, to reshape Chinese knowledge of certain phenomena. But distraction is a theme that runs through his writings: promising studies laid aside; endless skirmishes to defend administratively measures that spoke for themselves technically and strategically; proposals negated by political setbacks. Regardless of his capacity for scientific depth and his willingness to find his way to it, the sheer busyness of his career drastically limited him. The works of his final leisure, however valuable, were all superficial in form. Was this the result of habit, of distance necessitated by disillusion, or of an aesthetic choice of the style appropriate for conversing with one's brush and ink slab in a silent garden? That remains for deeper study to decide.

What, then, was responsible for Shen Kua's scientific personality? We do not know the answers to all sorts of prior questions. The greatest difficulty comes in learning what these questions should be—in isolating the important issues, in coming to terms with the paucity and partiality of the sources, and in doing justice to a rich mind that, despite its absorption in a quest that transcends people and eras, partook fully of its time and place. It is not a matter of mechanically juxtaposing the usual factors: intelligence, subjectivity, philosophical convictions, social background, career, and other experiences. We have already seen how problematic the last three are. The most conspicuous traits of Shen's consciousness were open curiosity, mental independence (without the intolerance for intellectual disagreement that was a major limitation of Wang An-shih), sympathy for the unconventional, ambition, loyalty, and lack of snobbishness. The first four are considered marks of promise among technical people today, although one often meets great scientists who lack one or more of them. Were these characteristics in Shen due to heredity, to early experiences and education, or to influences encountered in adult life? This is an example of the sort of question that bars understanding; surely Shen was the sum of all three. The secret of his uniqueness will not yield itself to historical method, however powerful, unless it is applied with imagination, artifice, and awareness of the springs of human complexity.

Attitudes Toward Nature. When examined closely, attitudes toward nature in the late eleventh century become as elusive as attitudes toward Confucian humanism. The richly articulated philosophic vision of man in harmony with his physical surroundings was proving quite incapable of preventing the deforestation of northern China, which was virtually complete a generation after Shen's death. One cannot even speak of the defeat of that vision in an encounter of ideas, for no intellectual confrontation is recorded. What happened? The most obvious part of the answer is that the people who were chopping down the trees for charcoal were not the people who were seeking union with the ineffable cosmic Tao. Since that social difference was of very long standing, however, it does not explain the crescendo of exploitation in the Northern Sung. The coincidence of that fateful shift with the rise of large-scale industry and market networks is again obvious enough.* What needs to be explained, in fact, is the survival of the naturalist ideal until modern times.

The dilemma emerges clearly in the attitudes of Shen Kua and Wang An-shih toward nature. The orientations that pervade "Brush Talks" are in most respects the same as those of literati thinking about nature a millennium earlier. Philosophical pigeonholes are largely beside the point. Some "Confucians" thought about nature a great deal, and some, convinced that human society is the sole proper object of reflection and action, as little as possible; but their perspectives were, on the whole, the ones common to all Chinese who could read and write. Nature was an organismic system, its rhythms cyclic and governed by the inherent and concordant pattern uniting all phenomena.

*It was made obvious in a brilliant series of papers by Robert M. Hartwell: "A Revolution in the Chinese Iron and Coal Industries During the Northern Sung, 960–1126 A.D.," in *Journal of Asian Studies*, **21** (1962), 153–162; "Markets, Technology, and the Structure of Enterprise in the Development of the Eleventh-Century Chinese Iron and Steel Industry," in *Journal of Economic History*, **26** (1966), 29–58; "A Cycle of Economic Change in Imperial China: Coal and Iron in Northeast China, 750–1350," in *Journal of the Economic and Social History of the Orient*, **10** (1967), 102–159.

It comes as a shock to see Shen's definition of salt in a memorial: "Salt is a means to wealth, profit without end emerging from the sea" (*Hsu tzu chih t'ung chien ch'ang pien*, 280:17b–21b). This was not a slip, nor is it difficult to find philosophical precedents. Shen saw the fiscal function of the state (for which he briefly had supreme responsibility) as the provision of wealth from nature. His recommendations encouraged extractive industries and manufactures, and mobilization of the popular strength for land reclamation, in order to increase national wealth. In that respect he was faithful to the priorities of Wang An-shih. This is a far cry from the senior civil servant in China in the 1960's designing a campaign to convince farmers that nature is an enemy to be conquered, tamed, and remolded to social ends. But neither is it the pastoral ideal.

Why this discrepancy between nature as the ideal pattern to which man adjusts and nature as a (still beneficent) means of enrichment? Why does Shen seem not to be conscious of it as contradictory? These are questions on which the research has yet to be done. But Shen Kua's career, considered in the round, suggests a working hypothesis. Such notions as yin-yang, the Five Phases, and certain related ideas associated with the *Book of Changes* are often considered to have been hindrances to an autochthonous scientific revolution in traditional China. This is, of course, an elementary fallacy, comparable to considering the railroad, because it filled a need satisfactorily for so long, an impediment to the invention of the airplane. The old Chinese world view had much in common with cosmological ideas practically universal in Europe until the consummation of the Scientific Revolution—the four elements and so on—but that gave way soon enough. Historically speaking, Chinese organismic naturalism was not a rigid framework of ideas that barred change; rather, it was the only conceptual language available for thinking about nature and communicating one's thoughts, new or old, to others. Like any language, it imposed form and was itself malleable. Its historical possibilities were less a matter of original etymology or definition than of the ambiguity and extensibility that let people in later ages read new and often drastically changed import into old words. There is no true paradox in appeals to the harmony of man and nature by Shen and others before and after him who favored the exploitation of nature in the interests of the state. Although such activist thinkers stretched the old pattern of understanding, its fabric remained seamless. Their definition of what they wanted could not transcend it. Only the more desperate urgencies of another time could finally stretch it until it tore.

BIBLIOGRAPHY

I. ORIGINAL WORKS. The best attempt at a complete list of Shen's writings is in an appendix to Hu Tao-ching's standard ed. of "Brush Talks," *Meng ch'i pi t'an chiao cheng*[69] ("Brush Talks From Dream Brook, a Variorum Edition"), rev. ed., 2 vols. (Peking, 1960 [1st ed., Shanghai, 1956]), 1151–1156. There are forty titles, including some only mentioned in early writings about Shen. A portion of the list belongs to parts or earlier versions of larger writings. It has been suggested that the high rate of attrition was due to the campaign of Ts'ai Ching[70] (1046–1126), virtual dictator during the revival of the New Policies in the first quarter of the twelfth century, to obliterate the literary remains of his predecessors as well as their enemies. (See Ch'en Teng-yuan,[71] *Ku-chin tien-chi chü-san k'ao*[72] ["A Study of the Collection and Dispersion of Classical Writings in Ancient and Modern Times"; Shanghai, 1936], 54.) Six works are extant, although only two appear to be substantially unaltered, and considerable fragments of four others exist. Those of scientific interest are described below:

1. *Meng ch'i pi t'an*[21] ("Brush Talks From Dream Brook"), written over the greater part of Shen's retirement and possibly printed during his lifetime. It was first quoted in a book dated 1095. Originally it consisted of thirty *chüan* (a chapterlike division); but all extant versions, descended from a xylograph of 1166, follow an unknown prior editor's rearrangement into twenty-six *chüan*. The editor of the 1166 reprint noted a number of errors already in the text that he could not correct for want of variants. There are 587 jottings.

The practically definitive ed. of this book and its sequels (items 2 and 3 below), and in many other respects the foundation of future studies, is the Hu Tao-ching recension mentioned two paragraphs above. It includes a carefully collated and corrected text with variorum notes and modern (but occasionally faulty) punctuation, based on all important printed versions and on five previous sets of notes on variants. It also provides exegetic and explanatory notes and generous quotations from documents concerning Shen, from his other books, from the reflections of other early writers on his subject matter, and from modern Chinese (and to some extent Japanese and Western) scholarship. Appendixes include thirty-six additional jottings or fragments that have survived only in the writings or compilations of others; all known prefaces and colophons; notes on eds. by early bibliographers and collators; a chronological biography; a list of Shen's writings; and an index to names and variant names of all persons mentioned in "Brush Talks" (a tool still very rare in Chinese publications). A 1-vol. version of the text with minimal apparatus was published by Hu as *Hsin chiao cheng Meng ch'i pi t'an*[73] ("Brush Talks From Dream Brook, Newly Edited"; Peking, 1957).

2. *Pu pi t'an*[74] ("Supplement to Brush Talks"), listed in most early bibliographies as two *chüan* but rearranged into three *chüan* with some alteration of order in the 1631 ed. Ninety-one jottings. Hu suggests that this and the next item were edited posthumously from Shen's notes. There is even stronger evidence for this hypothesis than he adduces, for some articles appear to be rejected drafts of jottings in "Brush Talks" (compare 588 with 437, 601 with 274).

3. *Hsu pi t'an*[75] ("Sequel to Brush Talks"), eleven jottings in one *chüan*, mostly on literature.

4. *Hsi-ning Feng-yuan li*[76] ("The Oblatory Epoch Astronomical System of the Splendid Peace Reign Period," 1075), lost, but listed in a Sung bibliography as seven *chüan*. This was the official report embodying Shen's calendar reform. It would have followed the usual arrangement, providing lists of constants and step-by-step instructions for computation, with tables as needed, so that the complete ephemerides could be calculated by someone with no knowledge of astronomy. Since a *Hsi-ning Feng-yuan li ching*[77] ("Canon of the Oblatory Epoch Astronomical System . . .") in three *chüan* is separately recorded, the remaining four *chüan* may have been, as in other instances, an official critique (*li i*[78]) outlining the observational basis of the system and reporting on tests of its accuracy. The Sung standard history also records a ready reckoner (*li ch'eng*[79]) in fourteen *chüan*, used to simplify calculations, and a detailed explanation of the mathematics with worked-out examples (*pei ts'ao*[80]) in six *chüan*. Surviving fragments of the basic document have been gathered by the great student of ancient astronomy Li Jui[81] (1765–1814) under the title *Pu hsiu Sung Feng-yuan shu*[82] ("Restoration of the Sung Oblatory Epoch Techniques"), printed in his *Li shih i shu*[83] ("Posthumous works of Mr. Li," 1823).

5. *Liang fang*[19] ("Good Prescriptions"), a work of ten or fifteen *chüan* compiled during Shen's retirement. In the Sung it was combined with a smaller medical miscellany by the greatest literary figure of Shen's time, Su Shih[84] (1036–1101), a moderate but influential opponent of the New Policies. The conflation is called *Su Shen nei-han liang fang*[85] ("Good Prescriptions by the Han-lin Academicians Su and Shen"), often referred to as *Su Shen liang fang*. The most broadly based text is that in the *Chih pu-tsu chai ts'ung-shu*[86] collection and modern reprints descended from it. One copy of an illustrated Ming ed. still exists. Shen's original compilation was lost sometime after 1500. There is some overlap between *chüan* 1 of *Su Shen liang fang* and jottings in *chüan* 26 of *Meng ch'i pi t'an*; see the comparison in Hu's *Chiao cheng*, pp. 880–882. A lost collection of prescriptions in twenty *chüan*, *Ling yuan fang*[87] ("Prescriptions From the Holy Garden"), is quoted in Sung treatises on mate-

ria medica. Hu has shown that it was written before *Liang fang* (*Meng ch'i pi t'an chiao cheng*, pp. 830–831).

6. *Wang huai lu*[20] ("Record of Longings Forgotten"), three *chüan*, compiled during Shen's retirement. It incorporates a lost book of observations on mountain living written (or at least begun) in Shen's youth and entitled *Huai shan lu*[88] ("Record of Longings for the Mountains"). His retirement to Dream Brook satisfied his early longings, hence the title of the later collection. It was lost soon after his death. The only well-known excerpts, in the *Shuo fu*[89] collection, are on implements useful to the well-born mountain dweller, but Hu Tao-ching in a recent study has shown that the book was correctly classified by early bibliographers as agricultural. See "Shen Kua ti nung-hsueh chu-tso *Meng ch'i Wang huai lu*"[90] ("Shen Kua's Agricultural Work . . ."), in *Wen shih*,[91] 3 (1963), 221–225. Hu's collection of all known fragments has not yet appeared.

7. *Ch'ang-hsing chi*[92] ("Collected Literary Works of [the Viscount of] Ch'ang-hsing"), originally forty-one *chüan*, almost certainly a posthumous compilation. Includes prose, poetry, and administrative documents prized for their language. By the time this work was reprinted in the Ming (*ca.* fifteenth century), only nineteen *chüan* of the Sung version remained. An additional three *chüan* were collected from other works and printed at the head of the recension in *Shen shih san hsien-sheng wen chi*[93] (1718). This is now the best ed. available. The collection includes important astronomical documents and a great deal of information on Shen's intellectual formation, in particular his commentary on Mencius (*Meng-tzu chieh*[94]) in *chüan* 23.

The only book in any Western language that translates more than a few examples of Shen's writings is Joseph Needham *et al.*, *Science and Civilisation in China*, 7 vols. projected (Cambridge, 1954–   ), particularly from vol. III on. The translations always occur in context, usually with fuller historical background than given in Chinese publications. Occasionally the English version is extremely free, as when "Meng ch'i" is translated "Dream Pool." Translations into modern Chinese are sprinkled through Chang Chia-chü,[95] *Shen Kua* (Shanghai, 1962). A complete Japanese trans. of "Brush Talks" and its sequels is an ongoing project of the History of Science Seminar, Research Institute for Humanistic Studies (Jimbun Kagaku Kenkyūsho[96]), Kyoto University. A representative selection of English translations will be included in a sourcebook of Chinese science being compiled by N. Sivin.

II. SECONDARY LITERATURE. There is no bibliography devoted to studies of Shen's life or work, but most primary and secondary sources in Chinese have been cited in Hu's ed. or in the footnotes to the biography of Shen by Chang Chia-chü (see above). The latter is the fullest and most accurate account of Shen's life, and pays attention to the whole range of his work. It is generally critical in method, but sometimes careless. Like

other recent Chinese accounts, it is extremely positivistic, patronizing toward "feudal" aspects of Shen's mentality, and inclined to exaggerate his sympathies toward the common people. A concise survey of Shen's life and positive contributions by a great historian of mathematics is Ch'ien Pao-tsung,[97] "Shen Kua," in Seminar in the History of the Natural Sciences, ed., *Chung-kuo ku-tai k'o-hsueh-chia*[98] ("Ancient Chinese Scientists"; Peking, 1959), 111–121. Another work of interest by Hu Tao-ching, overlapping to some extent the preface to his ed. of "Brush Talks," is "Shen Kua ti cheng-chih ch'ing-hsiang ho t'a tsai k'o-hsueh ch'eng-chiu-shang ti li-shih t'iao-chien"[99] ("Shen Kua's Political Tendencies and the Historical Conditions Bearing on His Scientific Accomplishments"), in Li Kuang-pi and Ch'ien Chün-yeh,[100] eds., *Chung-kuo li-shih jen-wu lun-chi*[101] ("Essays on Chinese Historical Figures"; Peking, 1957), 330–347. Its summary of scientific and technical accomplishments in the Northern Sung period from 960 to *ca.* 1100 is especially useful.

In addition to discursive biographical studies, Shen's life has been the subject of four chronologies (*nien-p'u*[102]), an old form in which individual events are simply listed year by year along with related data. The fullest in print (although obsolete in a number of respects) is Chang Yin-lin,[103] "Shen Kua pien nien shih chi"[104] ("A Chronicle of Shen Kua"), in *Ch'ing-hua hsueh-pao*,[105] 11 (1936), 323–358. That appended to the 2-vol. Hu Tao-ching ed. of "Brush Talks," 1141–1156, is especially handy because of its references to jottings and to sources cited in the book's notes. The most up-to-date and accurate chronology is the one at the end of Chang Chia-chü, *Shen Kua*, 235–259. Hu Tao-ching, in his colophon to the 1960 ed. of "Brush Talks," remarked that his own book-length chronology was in the press, but it has not yet appeared.

Yabuuchi Kiyoshi,[106] Japan's leading historian of science, has provided a characteristically reflective discussion of the historic circumstances of Shen's career in "Shin Katsu to sono gyōseki,"[107] ("Shen Kua and His Achievements"), in *Kagakushi kenkyū*,[108] 48 (1958), 1–6. The most stimulating contribution to the study of Shen in the past decade is Teraji Jun,[109] "Shin Katsu no shizen kenkyū to sono haikei"[110] ("The Natural Investigations of Shen Kua and Their Background"), in *Hiroshima daigaku bungakubu kiyō*,[111] 27, no. 1 (1967), 99–121. Rejecting the prevalent tendency to prove Shen's greatness by citing anticipations of European science and technology, the author has made a fruitful and original effort to grasp the inner coherence of his thought and work. This article provided a point of departure for the first two sections of the "Conclusion" of the present article.

The first, and so far the only, European introduction to Shen's life is Donald Holzman, "Shen Kua and his *Meng-ch'i pi-t'an*," in *T'oung Pao* (Leiden), 46 (1958), 260–292, occasioned by the first publication of Hu's ed. of "Brush Talks." In addition to providing a critical and well-proportioned biographical sketch, Holzman has paid more attention to Shen's humanistic scholarship than has any other author discussed in this section. He also considers some of the evidence for Shen's position in the history of science, but reaches no conclusion. He tends to ask whether Shen's ideas are correct from to-day's point of view rather than what they contributed to better understanding of nature in the Sung. The most re-liable and compendious introduction to the New Policies is James T. C. Liu, *Reform in Sung China. Wang An-shih (1021–1086) and His New Policies* (Cambridge, Mass., 1959). A full-length intellectual biography of Shen is under way by N. Sivin.

The first modern study of any aspect of Shen's inter-ests, largely responsible for the attention paid him by Chinese educated in modern science, is Chu K'o-chen,[112] "Pei Sung Shen Kua tui-yü ti-hsueh chih kung-hsien yü chi-shu"[113] ("Contributions to and Records Concerning the Earth Sciences by Shen Kua of the Northern Sung Period"), in *K'o-hsueh*,[114] **11** (1926), 792–807. Chu's erudite and broadly conceived article has influenced much of the later writing on the subject. A great number of observations on Shen's scientific and technical ideas are distributed through Needham *et al.*, *Science and Civilisation in China*, as well as through the topical studies by leading Japanese specialists in Yabu-uchi Kiyoshi, ed., *Sō Gen jidai no kagaku gijutsu shi*[115] ("History of Science and Technology in the Sung and Yuan Periods"; Kyoto, 1967).

There is no recent investigation in depth of Shen's as-tronomical activities, but a good technical description of what were traditionally considered his most important contributions is found in Juan Yuan,[116] *Ch'ou jen chuan*[117] ("Biographies of Mathematical Astronomers" [1799]; Shanghai, 1935), 20:238–243. Shen's most noteworthy mathematical problems have been studied in the various articles in Ch'ien Pao-tsung, ed., *Sung Yuan shu-hsueh-shih lun-wen-chi*[118] ("Essays in the History of Mathematics in the Sung and Yuan Periods"; Peking, 1966). The considerable portion of "Brush Talks" de-voted to music is evaluated and used in Rulan C. Pian, *Sonq [sic] Dynasty Musical Sources and Their Interpre-tation* (Cambridge, Mass., 1967), esp. 30–32. Shen's ideas concerning economic theory, the circulation of money, and similar topics have been related to traditions of thought on these subjects in an unpublished study by Robert M. Hartwell. A number of interesting ideas are found in Sakade Yoshinobu's[119] positivistic discussion of Shen's use of theory, "Shin Katsu no shizenkan ni tsuite"[120] ("On Shen Kua's Conception of Nature"), in *Tōhōgaku*,[121] **39** (1970), 74–87. Shen's remarks on ancient techniques are elucidated in Hsia Nai,[122] "Shen Kua ho k'ao-ku-hsueh"[123] ("Shen Kua and Archaeol-ogy"), in *K'ao-ku*,[124] no. 5 (1974), 277–289, also in *K'ao-ku hsueh-pao*,[125] no. 2 (1974), 1–14, with English summary, 15–17.

N. SIVIN

NOTES

1. 沈括	29. 五行	53. 羌	筆談	和他在科学成就	119. 坂出祥伸
2. 錢塘	30. 氣	54. 木經	74. 補筆談	上的历史条件	120. 沈括の自然觀に
3. 潤州京口	31. 律呂	55. 寇宗奭	75. 續筆談	100. 李光璧, 錢君曄	ついて
4. 沈周	32. 隙積	56. 本草衍義	76. 熙寧奉元曆	101. 中國歷史人物	121. 東方學
5. 許	33. 楊輝	57. 秋石	77. 曆經	論集	122. 夏鼐
6. 新法	34. 朱世傑	58. 宮下三郎	78. 曆議	102. 年譜	123. 沈括和考古学
7. 王安石	35. 一行	59. 漢薬秋石の薬	79. 立成	103. 張蔭麟	124. 考古
8. 寧國	36. 萬	史学的研究	80. 備草	104. 編年事輯	125. 学报
9. 蕪湖	37. 樂律	60. 渡辺幸三	81. 李銳	105. 清華學報	
10. 張藂	38. 祖暅	61. 現存する中国	82. 補修宋奉元術	106. 藪內清	
11. 衛朴	39. 奉元	近世までの五	83. 李氏遺書	107. 沈括とその業績	
12. 永安	40. 授時	蔵六府図の概	84. 蘇軾	108. 科学史研究	
13. 平泉	41. 郭守敬	説	85. 蘇沈內翰良方	109. 寺地遵	
14. 蔡確	42. 郭廷以	62. 燕蕭	86. 知不足齋叢書	110. 沈括の自然研	
15. 延州	43. 太平天国曆法	63. 蘇頌	87. 靈苑方	究とその背景	
16. 經略安撫使	考訂	64. 渾儀議	88. 懷山錄	111. 広島大学文学	
17. 開國子	44. 羅爾綱	65. 搏	89. 說郛	部紀要	
18. 夢溪	45. 天曆考及天曆	66. 微	90. 沈括的農學著	112. 竺可楨	
19. 良方	與陰陽曆日對	67. 續資治通鑑長編	作《夢溪忘懷錄》	113. 北宋沈括對於	
20. 忘懷錄	照表	68. 儒法斗争与我	91. 文史	地學之貢獻與	
21. 夢溪筆談	46. 謝莊	国古代科学技	92. 長興集	紀述	
22. 象數	47. 顏眞卿	术的发展	93. 沈氏三先生文集	114. 科學	
23. 技藝	48. 朱熹	69. 胡道靜, 校證	94. 孟子解	115. 宋元時代の科	
24. 辯證	49. 孫思恭	70. 蔡京	95. 張家駒	学技術史	
25. 異事	50. 礨	71. 陳登原	96. 人文科學研究所	116. 阮元	
26. 器用	50a. 石油	72. 古今典籍聚	97. 錢宝琮	117. 疇人傳	
27. 雜識	51. 畢昇	散考	98. 中国古代科学家	118. 宋元数学史论	
28. 藥議	52. 十百千	73. 新校證夢溪	99. 沈括的政治傾向	文集	

SHERARD, WILLIAM (*b*. Bushby, Leicestershire, England, 27 February 1659; *d*. London, England, 11 August 1728), *botany.*

William Sherard was the eldest son of George and Mary Sheerwood, or Sherwood. He received his secondary education at Merchant Taylors' School and in 1677 was elected to St. John's College, Oxford, where he developed a lasting interest in botany and established a close friendship with Jacob Bobart. In December 1683 Sherard took the bachelor's degree in common law and was elected law fellow of St. John's College. Granted leave for foreign travel, he studied three years with Tournefort in Paris and spent the summer of 1688 studying with Paul Hermann in Leiden. He collected plants in the Swiss Alps, Geneva, Rome, Naples, Cornwall, and Jersey, supplying lists of plants that were published by John Ray in his *Synopsis methodica stirpium Britannicarum* (1690) and *Stirpium Europaearum . . . sylloge* (1694).

Between 1690 and 1702, Sherard, as tutor to various young noblemen, made two more tours of the Continent and received the degree of doctor of common law from St. John's College on 23 June 1694. During this period, Sherard began a revision of Gaspard Bauhin's *Pinax*. This work, which occupied the remainder of his life, was never finished; the manuscript is with his library at Oxford.

After a brief appointment as "Commissioner for the Sick and Wounded, and for the Exchange of Prisoners" in 1702, Sherard received an appointment by the Levant Company as consul at Smyrna in 1703. While there he collected plants in Greece and Anatolia, copied antiquarian artifacts, and collected coins. He returned to England in 1717 with a considerable fortune. In 1718 he was elected a fellow of the Royal Society. His brother, James Sherard (1666–1737), a physician and apothecary, who had amassed a fortune in business, retired in 1720 and bought a country house at Eltham in Kent, where he established one of the finest botanical gardens in England. Sherard himself made three trips to the Continent between 1721 and 1727, bringing Dillenius from Giessen in August 1721 to assist with the *Pinax*. He was particularly impressed with Dillenius' knowledge of cryptogams and thought that bringing him to Oxford would enhance the department and the progress of the science.

On his death in 1728, Sherard left to Oxford his herbarium of 12,000 to 14,000 specimens, still preserved intact, and his library of more than 600 volumes. In addition, he bequeathed £3,000 to establish the Sherardian chair of botany, naming Dillenius the first Sherardian professor. Unfortunately, his brother James Sherard, who acted as executor, delayed settlement of his estate until 1734.

Sherard's contemporaries considered him an excellent and knowledgeable botanist. Nevertheless, he wrote and published little during his lifetime. He became the friend and correspondent of nearly every major botanist of his day, and his letters, which occupy four volumes, reveal his generosity in gifts of specimens, seeds, living plants, books, and subscriptions. In 1695 he edited Paul Hermann's manuscript of *Paradisus Batavus* for his widow's benefit. He assisted Pier Antonio Micheli and Paolo Boccone with subscriptions for publications and contributions of plants. In 1721 Sherard prevailed upon Boerhaave to edit the lifework of the ailing Sébastien Vaillant and assisted in cataloguing specimens for it. Sherard's assistance is acknowledged by Bobart in his *Historia oxoniensis* and by Ray in his *Historia plantarum*. Specimens with his notations are found in the herbaria of Tournefort, Ray, Dillenius, Vaillant, and Sloane. Concerning Sherard's problem in completing the *Pinax*, Clokie stated, "His difficulty seems to have been to concentrate on his work instead of helping his friends. His generosity to his friends seems to have known no limit."

BIBLIOGRAPHY

I. ORIGINAL WORKS. Sherard's most important work is *Schola botanica, sive catalogus plantarum quas ab aliquot annis in Horto Regio Parisiensi studiosis indigitavit vir clarissimus Joseph Pitton Tournefort, D.M., ut et Pauli Hermanni P.P. Paradisi Batavi prodromus, in quo plantae rariores omnes, in Batavorum Hortis hactenus cultae, et plurimam partem à nemine antea descriptae, recensentur* (Amsterdam, 1689). This publication is a list of the plants found in the Royal Garden in Paris, arranged according to the Tournefort system with the prodromus for Paul Hermann's *Paradisus Batavus* that Sherard subsequently edited. It was published under the initials S. W. A., which have been interpreted by various authors to stand for Simone Wartono Anglo or some variation thereof. Gorham (p. 12) gives the name as Sherardus Wilhelmus Anglus. Regardless of the name for which the initials stand, all authorities agree that it is the work of William Sherard. (See also Jackson, pp. 136–137 and Clokie, pl 18.)

Other writings are "The Way of Making Several China Varnishes Sent From the Jesuits in China to the Great Duke of Tuscany, Communicated by Dr. William Sherard," in *Philosophical Transactions of the Royal*

Society, **22** (1700), 525 (Sherard probably gained the information while in Rome with his pupil, the Duke of Beaufort): "An Account of the Strange Effects of the Indian Varnish, Wrote by Dr. Joseph del Papa, Physician to the Cardinal de Medices, at the Desire of the Great Duke of Tuscany. Communicated by Dr. William Sherard," *ibid.* (1701), 947; "An Account of a New Island Raised Near Sant-Erini in the Archipelago; Being Part of a Letter to Mr. James Petiver, F.R.S. From Dr. W. Sherard, Consul at Smyrna . . .," *ibid.*, **26** (1708), 67 (date of writing was 24 July 1707; the news came to Smyrna from the English consul at Milo); "An Account of the Poyson Wood Tree in New England. By the Honourable Paul Dudley, Esq., F.R.S. Communicated by John Chamberlain, Esq.," *ibid.*, **31** (1721), 145; "A Farther Account of the Same Tree. By William Sherard, LL.D., R.S.S.," *ibid.*, p. 147.

II. SECONDARY LITERATURE. A complete list of literature on Sherard and his work is in J. Britten and G. S. Boulger, *A Biographical Index of Deceased British and Irish Botanists*, 2nd ed. (London, 1931). See especially *Dictionary of National Biography*, XVIII, p. 67; G. Druce and S. Vines, *The Dillenian Herbaria* (Oxford, 1907); M. Epstein, *The Early History of the Levant Company* (New York, 1968); J. Green, *History of Botany* (New York, 1914); R. T. Günther, *Oxford Gardens* (Oxford, 1912); D. P. Micheli, *Targioni-Tozzetti* (Florence, 1858); R. Pulteney, *Pulteney's Sketches* (London, 1790); D. Richardson, ed., *Richardson Correspondence* (Yarmouth, 1835); and A. C. Wood, *A History of the Levant Company* (New York, 1964).

Sources to which specific reference has been made are H. N. Clokie, *An Account of the Herbaria of the Department of Botany in the University of Oxford* (Oxford, 1964); G. C. Gorham, *Memoirs of John and Thomas Martyn* (London, 1830); and B. D. Jackson, "A Sketch of the Life of William Sherard," in *Journal of Botany*, **12** (1874), 129–138.

CAROLYN D. TOROSIAN

SHERRINGTON, CHARLES SCOTT (*b.* London, England, 27 November 1857; *d.* Eastbourne, England, 4 March 1952), *neurophysiology.*

Sherrington was the son of Anne Brookes and James Norton Sherrington. After his father's death, in Sherrington's early childhood, his mother married Dr. Caleb Rose, Jr., of Ipswich. The Rose home, a gathering place for artists and scholars, helped to shape Sherrington's broad interests in science, philosophy, history, and poetry. After attending the Ipswich Grammar School from 1870 to 1875, Sherrington, encouraged by his stepfather, began medical training at St. Thomas's Hospital in London. In 1879 improved family finances enabled

him to enter Caius College, Cambridge, where he studied physiology in Sir Michael Foster's laboratory. He worked chiefly under John Newport Langley and Walter Gaskell, who imparted to him their dominant interest in how anatomical structure reflects, or is expressed in, physiological function. From 1884 to 1887 Sherrington completed his medical courses and did graduate study and research in Europe under Friedrich Goltz, Rudolph Virchow, and Robert Koch, gaining a superb grounding in physiology, morphology, histology, and pathology. In 1887 he was appointed a lecturer in systematic physiology at St. Thomas's, and from 1891 to 1895 he served as physician-superintendent of the Brown Institution, a London animal hospital. From 1895 to 1912 Sherrington held the Holt professorship of physiology at Liverpool and from 1913 to 1935 the Wayneflete chair of physiology at Oxford.

Sherrington married Ethel Wright of Suffolk, England, on 27 August 1891; their only child, Carr E. R. Sherrington, was born in 1897. He himself was a man of diverse interests. Outside of the laboratory his activities included sports (a feature event during his years in London was Sunday morning parachute jumping from the tower of St. Thomas's), work in many scientific organizations, academic affairs at Liverpool and Oxford, and studies for the government on such problems as industrial fatigue. After his retirement from Oxford, his pursuits included lecturing and writing, trusteeship of the British Museum, and service as governor of the Ipswich School and Ipswich town adviser on museums and health services. In a career that spanned sixty-nine years, Sherrington is remembered mainly for his scientific contributions but he was also the teacher who prepared *Mammalian Physiology. A Course of Practical Exercises* (1919), the poet who wrote *The Assaying of Brabantius* (1925), and the philosopher and historian whose writings included *Man on His Nature* (1941) and *The Endeavour of Jean Fernel* (1946). His numerous honors included the presidency of the Royal Society of London (1920–1925), Knight Grand Cross of the British Empire (1922), Order of Merit (1924), and the Nobel Prize for physiology or medicine (1932); at his death he was an honorary fellow, member, or associate of more than forty academies and had received honorary degrees from twenty-two universities.

Sherrington's classic investigations dealt primarily with reflex motor behavior in vertebrates, detailing the nature of muscle management at the spinal level. The data, terms, and concepts that he introduced have become such a fundamental part of the neurosciences that it is perhaps not surprising their authorship is often forgotten: such terms as proprioceptive, nociceptive, recruitment, fractionation, occlusion, myotatic, neuron pool, motoneuron, and synapse, and such concepts as the final common path, the motor unit, the neuron threshold, central excitatory and inhibitory states, proprioception, reciprocal innervation, and the integrative action of the nervous system.

Sherrington's scientific work may be broadly divided into two phases: from the 1880's to the publication of *The Integrative Action of the Nervous System* in 1906, and from 1906 to his receipt of the Nobel Prize in 1932. When he began his work, in the 1880's, the data and theories about the structure and function of the nervous system that had developed over the centuries were at best piecemeal. Controversy was rampant in almost every area and, apart from some textbook presentations, few attempts had been made to correlate structural and functional data within a given field of study, much less to interrelate the various separate channels of work on the nervous system.

The study of reflex actions, for example, went on almost independently of work on such problems as the structure and interconnection of nerve cells, the differentiation of the sensory and motor functions of the spinal cord, and the determination of brain structure and function. A fairly extensive fund of techniques, data, and theories about reflexes was available, but the field of reflex physiology was greatly in need of reorganization. Techniques were generally imprecise; sounder anatomical knowledge was needed; above all, experimentally based concepts—with which to interpret the known facts of reflex action and evaluate their role in the animal economy—were singularly lacking.

Sherrington decided to concentrate on neurophysiology rather than on pathology, his initial interest, when he returned to England in 1887 from Koch's laboratory in Berlin. He credited Gaskell with directing his attention from his first neurophysiological investigations into brain–spinal cord connections to the physiology of the spinal cord. Sherrington began by studying the little-understood phenomenon of the knee jerk, reporting the results of his first analyses (1891, 1892) of the muscles and nerves upon which the jerk depends. In these studies, however, he found that he could not deal satisfactorily with functional problems in the face of a major gap in neuroanatomical knowledge—the distribution of the sensory and motor fibers of the spinal cord. For a decade of what

seemed to him often "boring" and "pedestrian" research, Sherrington therefore surveyed the field of distribution of each spinal root, creating the anatomical foundation necessary for physiological work. His three major contributions to neuroanatomy were mapping motor pathways, chiefly those in the lumbosacral plexus (1892), establishing the existence of sensory nerves in muscles (1894), and tracing the cutaneous distribution of the posterior spinal roots (1894, 1898).

Concomitant with his anatomical work, and often deriving from it, was a profusion of ideas and observations on the reflex functions of the spinal cord. The two major, intertwined lines of these researches were the analyses of antagonistic muscle action and of larger pieces of reflex "machinery" such as the extension, flexion, and scratch reflexes of the hind limb. Out of these studies emerged Sherrington's conviction that the "main secret of nervous co-ordination . . . lies in the compounding of reflexes," a compounding built up by the play of reflex arcs about their "common paths." Behind this play lie the key processes of inhibitory and excitatory actions at the junctional regions between nerve cells—the synapses.

Like other investigators of the nervous system, Sherrington faced the task of devising techniques for reducing and controlling the complexity of the nervous system to the point where meaningful data could be obtained. His first steps were to concentrate upon the reflex functions of the cord rather than on the more complex field of the brain; to choose an appropriate experimental animal, the monkey; and to make parallel control and comparison experiments on lower forms to establish the necessary points of anatomical knowledge.

Sherrington's basic method was to study simple motor acts which could be made to occur in isolation, correlating his exacting analyses of input-output relations of reflex responses with anatomical and histological data. He used two types of experimental preparations: the classic spinal animal and the decerebrate animal. The effects of decerebration had been partially described by many earlier workers, such as Magendie, Bernard, and Flourens, but it was Sherrington who named decerebrate rigidity and, in a fundamental paper of 1898 and later publications, established it as a phenomenon in its own right and as a major tool for examining the reflex functions of the spinal cord, particularly the nature of inhibition.

The last decades of the nineteenth century saw the rise and fall of numerous theories about the nature of central inhibition, such as the controversial and influential center theory advanced by Johann Setchenov in 1863. It was against the background of these theories and the emergence of the neuron theory that Sherrington began to work out his ideas on the roles of inhibition and excitation in motor behavior and on the reflex nature of inhibition itself. The most important theme in Sherrington's functional researches up to 1900, for both his understanding of the operation of spinal reflexes per se and his comprehension of the mechanisms of nervous coordination, was his analysis of the reciprocal innervation of antagonistic muscles. It was the principle of reciprocal innervation, as Lord Adrian commented, "which opened the way to the further advance from the simple to the complex. It was the clue to the whole system of traffic control in the spinal cord and throughout the central pathways."[1]

The results of Sherrington's exhaustive study of reciprocal innervation, which stemmed from his observations on the knee jerk, are found chiefly in his fourteen classic "Notes" in the *Proceedings of the Royal Society* from 1893 to 1909. He first used the term "reciprocal innervation" in the title of the third "Note," read before the Royal Society on 21 January 1897; the term, he explained, denoted the "particular form of correlation" in which one muscle of an antagonistic couple is relaxed as its mechanical opponent actively contracts. Four months later, as the Royal Society's Croonian lecturer, he proposed his classic definition of reciprocal innervation as that form of coordination in which "inhibito-motor spinal reflexes occur quite habitually and concurrently with many of the excito-motor."

Another critical event of 1897 was Sherrington's introduction of the term and concept of synapse in Michael Foster's *Textbook of Physiology*. "So far as our present knowledge goes we are led to think that the tip of a twig of the [axon's] arborescence is not continuous with but merely in contact with the substance of the dendrite or cell body on which it impinges. Such a connection of one nerve-cell with another might be called a synapsis."

Sherrington's statement reflects the impact upon ideas of the structural and functional interrelations of the nervous system created by the neuron theory, introduced in 1889 by Ramón y Cajal. Prior to Ramón y Cajal's researches the dominant neurohistological view was the reticular theory, which held that nerve impulses are transmitted throughout the body over a continuous network or reticulum of anastomosing nerve processes. Ramón y Cajal's preparations showed that definitely limited conduction paths exist in the gray matter and that

nerve impulses are somehow transmitted by contact or contiguity, not by continuity. The significance of Sherrington's choice of the neuron theory and his coining of synapse has been well stated by Ragnar Granit: "When Sherrington decided in favor of nerve-cell contacts he refashioned thinking in this field along lines that determined its future course for all time and also tied it to the newly born science of electrophysiology. . . . Only a contact theory could bridge the gap between reflex transmission and electrophysiology; such is the power of a fundamental concept like the synapse."[2]

Between 1897 and 1900 Sherrington formulated a comprehensive picture of the motor functions of the spinal cord. His conception of these functions, of the rules that govern them, their mechanisms of control, and their role in the unitary functioning of the nervous system, were set forth in his Croonian and Marshall Hall lectures and in E. A. Schäfer's *Text Book of Physiology*. By 1900 Sherrington had assembled the major ingredients of the integrative action concept. From a study of the seemingly simple anatomy and physiology of the knee jerk he had become engaged in a series of broader problems, such as the nature and mechanisms of antagonistic muscle action, the production and maintenance of decerebrate rigidity, and the nature and significance of spinal shock. From these and other researches he developed a number of basic functional principles: reciprocal innervation; interaction between higher and lower level centers of motor control; and muscular sense, inhibition, and facilitation as the three key mechanisms of muscle management at the spinal level. Recognizing the import of the neuron theory, he had perceived that many of the characteristic properties of reflex pathways might be explicable by the events at the synapse.

The next phase of his work was to determine how reflex arcs combine to form successively larger and more complex reflex patterns. Sherrington's analysis of the scratch and other hind limb reflexes confirmed his earlier findings: the same functional principles obtain in both the simplest and most complex reflex actions. And, because of their very complexity, the hind limb reflexes further illuminated a wide range of phenomena underlying motor coordination, such as inhibition, facilitation, spinal induction, and the events at the synapse. In his definitive paper on the scratch reflex ("Observations on the Scratch Reflex in the Spinal Dog," *Journal of Physiology*, **34** [1906], 1–50), the properties of reflexes as found in the isolated spinal cord were described more minutely and fully than ever before in the annals of neurophysiology.

In 1904 Sherrington enunciated the essentials of the integrative action concept to the Physiological Section of the British Association for the Advancement of Science in a presidential address entitled "The Correlation of Reflexes and the Principle of the Common Path." It was his most important published conceptual statement before *The Integrative Action of the Nervous System* (1906). The main theme was the reflex chain of the synaptic system: the receptive neuron forms a private path into the brain or cord; within the "great central organ" many private paths converge at an internuncial neuron to form a public or common path which runs to the motor neuron; from the motor neuron, impulses travel over a final common path to converge upon the effector organ. "The singleness of action from moment to moment," assured by the principle of the common path, Sherrington declared, "is a keystone in the construction of the individual whose unity it is the specific office of the nervous system to perfect."

The immediate fruitfulness of the common path principle can be seen in Sherrington's papers of 1905 and 1906, in which he extended his analysis of the mechanisms controlling reflex actions. Working with the hind limb reflexes, he now focused on spinal induction, inhibition, and his fundamental concept of the proprioceptive system.

Sherrington journeyed to the United States to deliver ten lectures on "Integrative Action by the Nervous System" as the second Silliman Memorial Lecturer at Yale University in April 1904. His oral delivery, complex and difficult to follow even for those familiar with his work, left the majority of his steadily dwindling audience less than enthusiastic; but publication of the lectures in 1906 was recognized as an epochal event in the development of neurophysiology. The lasting value of *The Integrative Action of the Nervous System* for students of neurophysiology is reflected in the numerous reviews of its fifth reprinting, in 1947. F. M. R. Walshe, writing in the *British Medical Journal*, asserted:

> I have called it "an imperishable work," for it is one of those works, rare in science, the permanent value of which is unquestionable, and I believe that future generations of physiologists will so acclaim it. In physiology, it holds a position similar to that of Newton's *Principia* in physics. . . . For it is more than an orderly record of precise observations: it is a product of sustained thought upon what is essential-

ly—though only his genius revealed it as such—a single problem—namely, the mode of nervous action.[3]

The Integrative Action consists, in essence, of a synthesis of Sherrington's own researches and concepts and a chronicle of relevant work by other investigators. The structure and major concepts of the book may be divided into six parts: (1) Sherrington's definition of his topic in the first seven and one-half pages of lecture I; (2) lectures I–III, treating of coordination in the simple reflex; (3) lectures IV–VI, concerned with coordination between reflexes—their interaction and compounding by simultaneous and successive combination; (4) lecture VII, reflexes as adapted reactions; (5) lectures VIII–IX, the brain's role in integrative motor action; and (6) lecture X, sensual fusion. Sherrington's written analysis of the integrative action of the nervous system followed basically the same pattern as his research work. He began by delineating the characteristics of the simple reflex, the smallest functional unit of integrative action as seen in the spinal animal, and then built toward the complex patterns of reflex muscle management, guided by the brain, in the intact animal.

For a person unfamiliar with Sherrington's work, the first pages of lecture I provide a succinct statement of the meaning and scope of the concept of integrative action. In them he laid down three central propositions: (1) the nervous system is one, if not the only, major integrating agent in complex multicellular organisms; (2) the reflex is the unit reaction in nervous integration; (3) there are two grades of reflex coordination, that effected by the simple reflex and that effected by the simultaneous and successive combination of reflexes. Working from these premises he proceeded to demonstrate in meticulous detail the basic theme of the integrative action concept: "The nervous synthesis of an individual from what without it were a mere aggregation of commensal organs resolves itself into coordination by reflex action."

Sherrington continued his active life as researcher, teacher, writer, and prominent member of the international scientific community in the years after 1906, moving from Liverpool to Oxford in 1913. When his research was curtailed by World War I, Sherrington devoted himself to government war work with his customary drive and efficiency. Few episodes better illustrate his character than his activities during the summer of 1915, when he disappeared from home on a bicycle, presumably for a holiday, leaving no address. His whereabouts were finally revealed when he needed to replace a lost collar stud—having decided to study industrial fatigue *in situ,* he was working incognito as an unskilled workman at the Vickers-Maxim shell factory in Birmingham.

At the end of the war Sherrington resumed his extensive research program, and until he was almost seventy-five he performed at least one long experiment every week and spent many hours analyzing data. When he retired in 1935, the "Sherrington school" at Oxford had issued a series of influential papers on such topics as afterdischarge, summation, recruitment, postural contraction, and the motor unit, to illuminate the finer details of the reflex activity of the spinal cord. Two of Sherrington's last comprehensive reviews of muscle management at the spinal level, summarizing and synthesizing the work of the Oxford years, are his 1931 Hughlings Jackson lecture, "Quantitative Management of Contraction in Lowest Level Coordination," and chapter seven of *Reflex Activity of the Spinal Cord* (1932). His coauthors for the later work exemplify the continuing prominence of the Sherrington school in modern neurophysiology: R. S. Creed, D. Denny-Brown, J. C. Eccles, and E. G. T. Liddell.

Sherrington's tenure at Oxford may be characterized as a period of quantitation, testing and refining the concepts set forth in *The Integrative Action.* New techniques, in particular the development of isometric myography, made possible the accurate measurement of muscle tensions in various preparations. Sherrington could now measure and balance excitatory and inhibitory processes against each other, learning virtually everything about reflexes that was possible without the aid of the more sophisticated electronic methods that were developed as his career drew to a close.

Three fundamental publications of the Oxford period were Sherrington's papers on the stretch reflex (1924), central excitatory and inhibitory states (1925), and the motor unit (1930). These papers presented both the culmination of issues raised by his earlier work and the basis for many subsequent major advances in unraveling the operations of the nervous system. Sherrington and Liddell's analysis of the stretch reflex, the basic reflex used in standing, grew out of Sherrington's studies in the 1890's on the response of the "muscular sense organs" to stretching and contraction. By 1905 he had observed and described the stretch reflex, although its naming and definition awaited his and Liddell's later work. In their 1924 and 1925 papers, they reported the results of using individual isolated knee extensor muscles in a decerebrate preparation, with the free ends of the muscle attached to an isometric myograph. These researches, in turn, led to the definition of the nature of autogenetic excitation and inhibition and to our present understanding of muscle tonus, attitude, and posture.

Sherrington's work on central excitatory and inhibitory states and the motor unit were the culmination, for him, of the research he had begun forty years earlier on the functional anatomy of sensory and motor pathways, and led to the studies of the finer details of synaptic conduction, which now fill volumes.

After 1906 one of the chief problems occupying Sherrington's attention was inhibition. By 1925, in "Remarks on Some Aspects of Reflex Inhibition," he was ready to state his concept of central excitatory and inhibitory states and, as he had first suggested in 1908, of excitation and inhibition interacting algebraically at the synapse. In the 1925 paper, Sherrington marshaled the evidence developed in over twenty-five years of experiments, reasoning back from the phenomena of muscle contraction, as seen principally in the crossed extensor and ipsilateral flexor reflexes, to the events at the synapse. Inhibition, he demonstrated, is a distinct phenomenon although it is almost identical in its properties to excitation and obeys the same laws. The further testing and development of his ideas on central excitation and inhibition occupied much of Sherrington's time during his last decade of research. His concept of central excitatory and inhibitory states has been confirmed, expanded, and reformulated in terms of postsynaptic excitatory and inhibitory potentials by Sir John Eccles and others, using such techniques as intracellular recordings from motoneurons.

The concept of the motor unit, Sherrington's last major contribution to neurophysiology, can be seen as a more sophisticated, experimentally based development of the principle of the common path. The motor unit, in simple terms, is a spinal motoneuron (or motor cell in the ventral horn of the spinal cord) which, by the branching of its axon, controls and coordinates the actions of more than 100 muscle fibers. Years of exacting study by Sherrington and his colleagues went into the paper on the motor unit which he published with Eccles in 1930: the writing of the paper itself occupied Sherrington for over two years. A glimpse of Sherrington the scientist, at seventy-two, working on his last major paper, is offered by John Fulton's diary entry for 2 April 1930:[4]

> It is remarkable to see how at this time of his life he is beginning to correlate all the various aspects of his own work; structure of fibers, reflexes, series of central excitations, etc., etc., into a beautifully synthesized body of knowledge. For him the days simply weren't long enough, and there are all manner of things that need investigation. The rest of the world in his eyes is a little slow because it does not see all these things staring it in the face.

Sherrington received the Nobel Prize in 1932 specifically for his isolation and functional analysis of the motor unit. Somewhat ironically, the man who developed the single-unit concept never used the then newly available electrical recording techniques, but Sherrington can scarcely be held at fault for not making such studies when he was in his seventies. Sherrington shared the Nobel award with Edgar Douglas Adrian, Foulerton professor of physiology at Cambridge, for his analysis of the frequency discharge of single units. Both men had long sought to define the properties of the nerve cell as the functional unit of the central nervous system—Adrian through investigating the afferent input from sense organs, Sherrington the nature of motor input—and Adrian's work during the 1920's probably had the most immediate influence on the direction and development of Sherrington's ideas. To many, particularly those who had worked with him over the years, Sherrington's award was long overdue—a circumstance perhaps partly explained by the Nobel Committee's difficulties in citing a specific discovery on which to base the prize.

The work noted above is a small, but highly important, portion of the experiments and ideas generated by Sherrington and his pupils and colleagues during his twenty-three years at Oxford. It would be a difficult task indeed to select the most important single achievement from among the vast program of researches in which Sherrington was engaged from the 1880's to his retirement in 1935: brain-cord connections and spinal degenerations; the distribution of motor and sensory roots; the proprioceptive system; the characteristics of synaptic reflex arc conduction; reciprocal innervation and the nature of central inhibition; the reflex patterns of the spinal, decerebrate, and intact animal; and the nature of supraspinal control as seen in the functional organization of the motor cortex. Perhaps most significant are his inseparable analysis of reciprocal innervation and inhibition, his studies

of muscle tonus (posture), and his conceptual definition of the nature of synaptic action in effecting the unitary or integrated behavior of nerve cells.

In more general terms, by examining the antecedents to, and tracing the course and content of, Sherrington's researches, one sees how strikingly the new outweighed the old in his work: in instance after instance Sherrington himself "made the time ripe" for answering a given problem. He pioneered new techniques and apparatus, such as the method of successive degenerations, surgical procedures for mammalian decerebration, and the use of the myograph for reflex recordings, and established new methodological canons with his meticulously designed and executed experiments. Second, he marshaled extant facts and theories and added a host of new ones about each topic he studied, emphasizing particularly the correlation of structural and functional data.

The scope of his specific contribution clearly marks Sherrington as a major figure in the history of the neurosciences. The greater significance of his work, however, lies in his "synthetic attitude," his perception of the interrelatedness of his varied researches. One of his goals was to explain the functional unity of motor behavior, primarily by interpreting central nervous system function in histological terms. From the content of his 1897 Croonian lecture on the mammalian spinal cord as an organ of reflex action, it appears that he was moving toward the concept of integrative action by that date.

Sherrington's work is resolvable into a threefold study of reflex actions, using the nerve cell and its interconnections as his basic analytical unit: their gross and histological architecture, the spinal and higher level mechanisms controlling them, and their functions in vertebrate motor behavior. Although his most fundamental experiments were performed around the turn of the century, it was the work during the 1920's at Oxford which has shaped many facets of neurophysiological research up to the present.

Sherrington's name remains linked most closely, however, with his integrative action concept, although the idea of nervous integration did not originate with him. The fact of motor coordination and the participation of the nervous system in its operation had been recognized since antiquity, and prototypes of the integrative action concept may be found from the ancient idea of "sympathy" to Flourens's studies of how specific brain regions affect an animal's functional unity. The uniqueness and significance of Sherrington's work, epitomized by the integrative action concept, lies in the fact that it provided the first comprehensive, experimentally documented explanation of how the nervous system, through the unit mechanism of reflex action, produces an integrated or coordinated motor organism. It was this watershed achievement, synthesizing the work of one era and opening a new one, that led Sherrington's peers to designate him as "the main architect of the nervous system," "the supreme philosopher of the nervous system," "the author of the *Principia* of physiology," and the "man who almost singlehandedly crystallized the special field of neurophysiology."

Sherrington's work was motivated, in large measure, by a desire to explain organized, purposeful behavior, and he looked upon reflexes as very simple items of such behavior. Thus, in the Silliman lectures and in subsequent writings he voiced distinct reservations about the ability of his analysis of reflex action to account for the functional solidarity of vertebrates, especially among primate forms. These reservations stemmed, in part, from the realization that his work left unanswered many questions about the nature of reflex motor control that would be resolved by others using more sophisticated techniques and types of experiments.

His other reservations about his work stemmed from his belief that reflex action was only a small part of the integration of higher vertebrates. Sherrington himself might have written the words of the sixteenth-century physician Jean Fernel, quoted in *Man on His Nature*: " . . . our task, now that we have dealt with the excellent structure of the body, cannot stop there, because man is a body and a mind together." As the quotation suggests, this aspect of Sherrington's self-appraisal is intimately related to his position on the question of the relation of mind and body. In this realm Sherrington was a dualist. For Sherrington man is the product of natural forces, yet he encompasses a territory which neurophysiology cannot reach — the realm of mind and thought. One senses, however, that Sherrington was not a dualist by philosophical choice but, rather, felt constrained to adopt the position because the sciences of his day offered no evidence or means of bridging the gap between mind and brain.

In the foreword to the 1947 reissue of *The Integrative Action*, Sherrington summarized his years of study and thought about the roles and relations of body and mind in animal integration, distinguishing three systems or levels of integrative action. At the first level, physicochemical processes weld the body's organs into a "unified machine." This welding is exemplified by the integrative action of the nervous system, of which the unit mechanism, as Sherrington had shown, is reflex action.

Reflex action vanished completely as a mechanism in Sherrington's second system of integrative action, "the field of the psyche." At this level, he held, "the physical creates from psychical data a percipient, thinking, and endeavouring individual." For Sherrington "the physical is never anything but physical, or the psychical anything but psychical," yet the two systems "are largely complemental and life brings them co-operatively together at innumerable points." This mind-body liaison was, to Sherrington, the third and highest level of integrative action. "In all of those types of organisms in which the physical and psychical coexist, each of the two achieves its aim only by reason of a *contact utile* between them. And this liaison can rank as the final and supreme integration completing the individual."

To Sherrington, called the "supreme philosopher of the nervous system," the most baffling and challenging problem for both scientists and philosophers was how the mind-body liaison is effected. In 1947, he commented succinctly that the issue "remains where Aristotle left it 2,000 years ago." While Sherrington could not explain a dualistic interaction, neither could he find any valid basis for reducing mind to a manifestation of physical energy. Thus, as he explained in the closing sentence of the foreword, dualism seemed to be as reasonable an assumption as monism. "That our being should consist of *two* fundamental entities offers I suppose no greater inherent probability than that it should rest on one only."

Near the end of his life Sherrington made one of his final and most positive statements about the levels of integrative action that produce the totality of an animal such as man. During a conversation with Sir Russell Brain, he said "the reflex was a very useful idea, but it has served its purpose. What the reflex does is so banal. You don't think that what we are doing now is reflex, do you? No, no, no."[5]

Although Sherrington's dualistic philosophy was disturbing to many of his scientific colleagues, it did not diminish their estimate of his contributions to neurophysiology. Of the scores of tributes which marked his death on 4 March 1952, none more simply expressed the sentiments of those who knew him than the words of Henry Viets: "A great and good man has died. . . . We stand on mighty shoulders."[6]

NOTES

1. E. D. Adrian, "The Analysis of the Nervous System: Sherrington Memorial Lecture," in *Proceedings of the Royal Society of Medicine,* **50** (1957), 993.
2. Ragnar Granit, *Charles Scott Sherrington,* p. 43.
3. F. M. R. Walshe, "A Foundation of Neurology. *The Integrative Action of the Nervous System,*" in *British Medical Journal* (1947), **2**, 823.
4. John F. Fulton Papers, Historical Library, Yale University School of Medicine.
5. Quoted in G. E. W. Wolstenholme and C. M. O'Connor, eds., *Ciba Foundation Symposium on the Neurological Bases of Behaviour, in Commemoration of Sir Charles Scott Sherrington* (Boston, 1948), p. 24.
6. Henry Viets, "Charles Scott Sherrington, 1857–1952," in *New England Journal of Medicine,* **246** (1952), 981.

BIBLIOGRAPHY

I. Original Works. A complete Sherrington bibliography is in Fulton (1952); an extensive although not complete listing is in Swazey (1969). Both are cited below. For those wishing to explore the development of Sherrington's scientific ideas and his nonscientific works, the following are suggested: *The Central Nervous System,* vol. III of Michael Foster, *A Textbook of Physiology,* 7th ed. (London, 1897); "The Mammalian Spinal Cord as an Organ of Reflex Action. Croonian Lecture," in *Proceedings of the Royal Society,* **61** (1897) 220–221, an abstract that was printed in full as sec. 4 of Sherrington's "Experiments in Examination of the Peripheral Distribution of the Fibres of the Posterior Roots of Some Spinal Nerves (II)," in *Philosophical Transactions of the Royal Society,* **190B** (1898), 45–186; "Decerebrate Rigidity, and Reflex Co-ordination of Movements," in *Journal of Physiology,* **22** (1898), 319–332; "On the Spinal Animal (The Marshall Hall Lecture)," in *Medico-Chirurgical Transactions,* **82** (1899), 449–477; "The Spinal Cord," "The Parts of the Brain Below the Cerebral Cortex," "Cutaneous Sensations," and "The Muscular Sense," in E. A. Schäfer, ed., *Text Book of Physiology,* II (Edinburgh, 1900), 783–1025; "The Correlation of Reflexes and the Principle of the Common Path," in *Report of the British Association for the Advancement of Science,* **74** (1904), 1–14; *The Integrative Action of the Nervous System* (New Haven, 1906); "Reflex Inhibition as a Factor in the Co-ordination of Movements and Postures," in *Quarterly Journal of Experimental Physiology,* **6** (1913), 251–310; "Some Aspects of Animal Mechanism. Presidential Address, British Association for the Advancement of Science," in *Report of the British Association for the Advancement of Science* (1922), 1–15; "Reflexes in Response to Stretch (Myotatic Reflexes)," in *Proceedings of the*

Royal Society, **86B** (1924), 212–242, written with E. G. T. Liddell; *The Assaying of Brabantius and Other Verses* (Oxford, 1925); "Remarks on Some Aspects of Reflex Inhibition," in *Proceedings of the Royal Society,* **97B** (1925), 519–545; "Numbers and Contraction-values of Individual Motor-units Examined in Some Muscles of the Limb," *ibid.,* **106B** (1930), 326–357, written with J. C. Eccles; *Reflex Activity of the Spinal Cord* (Oxford, 1932), written with R. S. Creed *et al.; Inhibition as a Co-ordinative Factor* (Stockholm, 1932), the Nobel lecture delivered at Stockholm, 12 Dec. 1932; *The Brain and Its Mechanism* (Cambridge, 1933); *Man on His Nature* (Cambridge, 1941), the Gifford lectures, Edinburgh, 1937–1938; *The Endeavour of Jean Fernel* (Cambridge, 1946); and "Marginalia," in E. A. Underwood, ed., *Science, Medicine and History,* II (Oxford, 1954), 545–553.

II. Secondary Literature. See Edgar D. Adrian, "The Analysis of the Nervous System: Sherrington Memorial Lecture," in *Proceedings of the Royal Society of Medicine,* **50** (1957), 991–998; Mary Brazier, "The Historical Development of Neurophysiology," in John Field, H. W. Magoun, and V. E. Hall, eds., *Handbook of Physiology. Section 1: Neurophysiology* (Washington, D.C., 1959); Georges Canguilhem, *La formation du concept de réflexe au XVIIe et XVIIIe siècles* (Paris, 1955); Lord Cohen of Birkenhead, *Sherrington: Physiologist, Philosopher, Poet* (Liverpool, 1958), vol. IV of the University of Liverpool Sherrington Lectures; Derek Denny-Brown, "The Sherrington School of Physiology," in *Journal of Neurophysiology,* **20** (1957), 543–548; Franklin Fearing, *Reflex Action: A Study in the History of Physiological Psychology* (Baltimore, 1930; repr. New York, 1964); John F. Fulton, "Sir Charles Scott Sherrington, O. M.," in *Journal of Neurophysiology,* **15** (1952), 167–190; and "Historical Reflections on the Backgrounds of Neurophysiology: Inhibition, Excitation, and Integration of Activity," in Chandler M. Brooks and P. F. Cranefield, eds., *The Historical Development of Physiological Thought* (New York, 1959); Ragnar Granit, *Charles Scott Sherrington: An Appraisal* (London, 1966); E. G. T. Liddell, "Charles Scott Sherrington, 1857–1952," in *Obituary Notices of Fellows of the Royal Society of London,* **8** (1952), 241–259; and *The Discovery of Reflexes* (Oxford, 1960); Wilder Penfield, "Sir Charles Sherrington, Poet and Philosopher," in *Brain,* **80** (1957), 402–410; Carr E. R. Sherrington, *Memories* (privately printed, 1957), the Beaumont lecture, Yale University, 15 Nov. 1957; Judith P. Swazey, "Sherrington's Concept of Integrative Action," in *Journal of the History of Biology,* **1** (1968), 57–89; and *Reflexes and Motor Integration: Sherrington's Concept of Integrative Action* (Cambridge, Mass., 1969).

A major source of information about Sherrington and his life and work is contained in the papers of the late Dr. John F. Fulton, who had a long and close personal and professional association with him. Fulton first studied under Sherrington in 1921, when he went to Oxford as a Rhodes Scholar. The Fulton papers are housed at the Historical Library, Yale University School of Medicine, New Haven, Connecticut.

JUDITH P. SWAZEY

SHIBUKAWA, HARUMI (*b.* Kyoto, Japan, 27 December 1639; *d.* Edo [now Tokyo], Japan, 11 November 1715), *astronomy.*

Harumi was the son of Yasui Santetsu, a professional *go* player in the service of the Tokugawa shogunate. After his father's death he assumed his name and profession, becoming known as Yasui Santetsu II. Trained in *go,* Harumi studied Chinese and Japanese classics, Shintoism, and calendrical astronomy with various teachers. His distinguished service in calendar reform led to his appointment in 1685 as official astronomer. He later returned to his original name, Shibukawa.

The Chinese lunisolar Hsuan-ming calendar, adopted in Japan in 862, had not been reformed for more than eight hundred years. Over the centuries the discrepancy in the length of a solar year had increased so that by Harumi's time there was a two-day delay in the winter solstice. Moreover, it had become of little use for its traditional purpose, the precise prediction of solar and lunar eclipses. Although the error caused little inconvenience or confusion in daily life, reformation of the official calendar became an event of major political importance—to strengthen the prestige and authority of the imperial court vis-à-vis the shogunate by providing the people with an accurate calendar.

An able mathematician and skilled diplomat, Harumi urged calendar reform by pointing out faults in the Hsuan-ming calendar, seeking to prove the discrepancy by actual observation of winter and summer solstices and by demonstrating the possibility of establishing a calendar that would be more accurate—as confirmed by observation—in predicting eclipses.

In 1669 Harumi began conducting astronomical observations, probably the first systematic observations made in Japan. Following the procedures of traditional astronomy, he set up a gnomon and measured the lengths of shadows at various points before and after the winter solstice, in order to calculate the time of occurrence. He was especially interested in the Shou-shih calendar of the Yüan dynasty (1279–1368), a crowning achievement of calendrical astronomy adopted in China in 1282; and his observations were based upon its methods.

In 1673 Harumi proposed to the emperor the adoption of the Shou-shih calendar. In a report entitled "On the Eclipses" he compared the differing calculations, derived from both calendars, of solar and lunar eclipses, for the period 1673–1675. Of the six subsequently observed eclipses, however, the solar eclipse of 1675 was found in better agreement with the calculations based on the Hsuanming calendar.

Stunned and disheartened by this experience, Harumi improved his calculations and subsequently devised a calendar that agreed with the incorrectly predicted eclipse. In 1683 he again proposed a calendar revision to the emperor, the first calendar devised by a Japanese that was completely independent of the Chinese calendars.

However, the tendency of Japanese astronomers to follow Chinese practice was not easily overcome. In 1684 the fifteen-member Board of Astronomy decided to adopt a newer Chinese calendar, the Ta-t'ung. Appalled by this decision, Harumi engaged in some quiet behind-the-scenes maneuvering and finally succeeded in reversing the decision. Later that year his own calendar revision (the Jokyo calendar) was implemented.

Like the Shou-shih calendar, the Jokyo incorporated the secular variation term of the length of the tropical year. Such correctional terms were used to explain the records of winter solstices in the remote past. The Ta-t'ung, although following the substance of the Shou-shih, had abandoned secular variation terms in the belief that such minimal revisions could not be proved by observation; Harumi's own adoption of them was based on his conviction that it would render his calendar more profound.

The Jokyo and Shou-shih calendars can be considered the culmination of traditional Chinese calendrical astronomy, distinguished—like Babylonian astronomy—in their concentration on numerical and algebraic matters. Neither used the geometrical approach or the schematic model of Western astronomy, and thus, taken together, neither surpassed Ptolemy's *Almagest.*

During his career, the Shih-hsien calendar, compiled by Jesuits, had been introduced in 1644 in China, and Harumi frequently referred to it. There was, unfortunately, no way for him to learn about the system adopted in compiling it. The Japanese government's closed-door policy, begun in the 1630's, included a ban directed mainly against

Jesuit writings and prevented him from obtaining an important work on Sino-Jesuit astronomy, *Hsi-yang hsin-fa li-shu* ("Treatises on Calendrical Science According to New Western Methods," 1645).

In his references to Western theory, Harumi based his information exclusively on *T'ien-ching huo-wen* ("Queries on the Classics of Heaven," 1675), by Yu I. Rather than a scientific treatment of observational values or of the methods of calculation derived from them, the work was merely a popular explanation of various astronomical and cosmological theories. Harumi was especially impressed by its clear explanation, using a geometrical model, of eclipses, which he had never found in Chinese calendrical writings. Yet, as an astronomer focusing on useful parameters or numerical values that could be borrowed to improve the precision of a calendar, he found the work disappointing and came to regard Western astronomers as "barbarians who may have theories but cannot prove methods." It is regrettable that sufficient material for evaluating Western theories was not available to him.

During the eighteenth century Japanese astronomy altered its orientation from China to the West. Harumi belonged to the first generation of astronomers who, with only limited knowledge of Western astronomy, began evaluating the merits of both systems. He was a leader of those Japanese astronomers who, through their science, initiated the acknowledgment of Western superiority and the modernization of Japan.

BIBLIOGRAPHY

In English, see Shigeru Nakayama, *A History of Japanese Astronomy, Chinese Background and Western Impact* (Cambridge, Mass., 1969).

In Japanese, works on Harumi and his work are Endo Toshisada, "Shibukawa Harumi," MS preserved at the Japan Academy, begun in 1904; Shigeru Nakayama, "Shibukawa Harumi's Solstitial Observation and the Hsiao-chang Method," in *Tenmon Geppo*, no. 58 (1965); and *A History of Japanese Astronomy, Chinese Background and Western Impact* (Cambridge, Mass., 1969); Nishiuchi Masaru, *Study of Shibukawa Harumi* (Tokyo, 1940); Shibukawa Takaya, "Biography of Master Harumi," in *Nihon Kyoiku Shiryo*, **9** (1889); Tani Jinzan, "Shinro Menmei," in *Nihon Bunko*, **4** (1890); and *Jinzan Shu* (Tokyo, 1909); and Watanabe Toshio, *Shibukawa Harumi, the Pioneer of Japanese Calendrical Astronomy, and the History of Astronomy in the Edo Period* (Tokyo, 1965).

SHIGERU NAKAYAMA

SHILOV, NIKOLAY ALEKSANDROVICH (*b.* Moscow, Russia, 10 July 1872; *d.* Gagry, U.S.S.R., 17 August 1930), *chemistry.*

Shilov graduated from Moscow University in 1895. In 1896–1897 and 1901–1904 he worked in Ostwald's physics and chemistry laboratory at Leipzig, where he began research on chemical kinetics. From 1910 he was professor in the department of inorganic chemistry at the Moscow Technical College and from 1911, professor at the Moscow Commercial Institute. From 1915 to 1918 he headed the technical section on gas, attached to the headquarters on the Western Front. In 1919–1921 Shilov helped organize the Institute of Chemical Research in Moscow.

Shilov's master's thesis (1905) systematized a large amount of experimental material on conjugate oxidation reactions. He there offered several theoretical generalizations and developed terminology for all the processes and active components, namely chemical induction, induction factor, actor, inductor, and acceptor (terms used in the field of catalysis). Subsequently these terms became generally accepted in chemical literature. Shilov gave special attention to the study of self-induction and to transitional phenomena between induction and catalysis.

Shilov's research demonstrated the central role of intermediary products in the kinetics of conjugate oxidation reactions. Using the conjugate oxidation reactions as a model, he developed a theory of the action of photographic developers.

Shilov began his study of gas adsorption during World War I. Introducing Zelinsky's charcoal filter gas mask into the Russian army, Shilov established a laboratory at the front, where he conducted a broad investigation of the adsorption of gases in an air flow containing poisonous substances. He established the relationship between the length of the layer of adsorbent and the duration of its effectiveness.

In 1919 Shilov began to study the adsorption of substances from solutions and the distribution of substances between two liquid phases. He also constructed a formula for the distribution of a substance between two solvents.

Investigating hydrolytic adsorption, Shilov demonstrated that, according to the surface oxides of carbon, the latter element manifests the characteristics of either a positive or a negative adsorbent. Shilov's representation of active carbon, as an adsorbent having various surface functional groups that enter into exchange reactions with adsorbing substances, must be considered the first explanation of the principle of the action of ion exchangers (cation and anion exchange resins).

BIBLIOGRAPHY

I. ORIGINAL WORKS. Shilov's writings include "Zur Systematik und Theorie gekoppelter Oxydations-Reduktionsvorgänge," in *Zeitschrift für physikalische Chemie*, **46** (1903), 777–817, written with R. Lüther; *O sopryazhennykh reaktsiakh okislenia* ("On Conjugate Oxidation Reactions"; Moscow, 1905); "K teorii fotograficheskogo proyavitelya" ("Toward a Theory of Photographic Developers"), in *Sbornik posvyashchenny K. A. Timiryazevu ego uchenikami* ("A Collection Dedicated to Timiryazev by His Students"; Moscow, 1916), 111–129; "Adsorbtsia elektrolitov i molekulyarnye sily" ("Molecular Forces and the Adsorption of Electrolytes"), in *Vestnik Lomonosovskogo fiziko-khimicheskogo obshchestva v Moskve*, **1**, no. 1 (1919), 1–137, written with L. K. Lepin; "Raspredelenie veshchestva mezhdu dvumya rastvoritelyami i silovoe pole rastvora" ("The Distribution of a Substance Between Two Solvents and the Strength Field of a Solution"), *ibid.*, no. 2 (1920), 1–103, written with L. K. Lepin; "K voprosu ob adsorbtsii postoronnego gaza iz toka vozdukha" ("Toward the Question of the Adsorption of Foreign Gases From an Air Flow"), in *Zhurnal Russkogo fiziko-khimicheskogo obshchestva*, chem. sec., **61** (1929), 1107; written with L. K. Lepin and A. S. Voznesensky; and "Studien über Kohleoberflächenoxyde," in *Zeitschrift für physikalische Chemie*, **150** (1930), 31–36, written with E. G. Shatunovskaya and K. V. Chmutov.

II. SECONDARY LITERATURE. See "Krupny russky ucheny Nikolay Aleksandrovich Shilov" ("The Outstanding Russian Scientist Shilov"), in *Uspekhi khimii*, **15**, no. 2 (1946), 233–264; and N. N. Ushakova, *Nikolay Aleksandrovich Shilov* (Moscow, 1966).

Y. I. SOLOVIEV

SHIRAKATSÍ, ANANIA (also known as **Ananias of Shirak**) (*b.* Shirakavan [now Ani], Armenia, *ca.* 620; *d.* shortly after 685), *mathematics, geography, philosophy, astronomy.*

A representative of the progressive Armenian scholars of the seventh century and a follower of the best traditions of Hellenistic science and culture, Shirakatsí lived during the period when Armenia had lost her political independence; the western part being ruled by Byzantium and the eastern by Persia. He received his basic education at a local monastery school. After several journeys in search of a teacher of mathematics, which he considered the "mother of all sciences," Shirakatsí reached Trebizond and entered the school of the

Greek scientist Tychicus, who taught the children of many Byzantine nobles. During the next eight years he studied mathematics, cosmography, philosophy, and several other sciences, before returning to his native region of Shirak, where he opened a school. In addition to teaching, he conducted scientific research and wrote works on astronomy, mathematics, geography, history, and other sciences. He possessed truly encyclopedic knowledge and the ability to reach the essence of matters.

Shirakatsí produced his most important scientific work from the 650's through the 670's. In 667–669 he was concerned with the reform of the Armenian calendar, anticipating the modern desire for an "immovable" calendar.

Shirakatsí's scientific works are known through manuscripts of the eleventh through seventeenth centuries that are scattered in the Soviet Union, Italy, Great Britain, Austria, Israel, and perhaps other countries. His advanced philosophical and cosmological views brought him to the attention of official circles, and he was persecuted by both lay and ecclesiastical authorities.

Like the scientists of antiquity, Shirakatsí believed that the world consists of four elements: earth, water, air, and fire. The world, in which he included plants, animals, and man, is a "definite composition of intermixed elements."

All things in nature move and are subject to change. Old substances decompose in due course, and new forms arise in their place. Creation, wrote Shirakatsí, is the basis of destruction; and destruction in its turn is the basis of creation; "as a consequence of this harmless contradiction, the world acquires its eternal existence."

On the form of the earth, Shirakatsí wrote: "The earth seems to me to have an egg-shaped form: as the yolk in spherical form is in the middle, the white around it, and a shell surrounds it all, so the earth is in the center like the yolk, the air around it like the white, and the sky surrounds it all like the shell." In the early Middle Ages such ideas were very daring.

In connection with the spherical form of the earth, Shirakatsí spoke of the mountains and canyons "on the other side" of the earth. He wrote about the antipodes, of animals and people "like a fly, moving on an apple equally well on all sides. When it is night on one side of the earth, the sun lights the other half of the earth's sphere."

Shirakatsí believed that the earth is in equilibrium in space because the force of gravity, which pulls it down, is opposed by the force of the wind, which tries to raise it.

Criticizing numerous legends that explained the Milky Way, Shirakatsí gave an explanation that was correct and bold for the time: "The Milky Way is a mass of thickly clustered and weakly shining stars."

The moon, in Shirakatsí's opinion, does not emit its own light but reflects that of the sun. He associated the phases of the moon, which proceed from changes in the mutual positions of the sun and moon, with this reflection of the sun's light.

Shirakatsí gave a correct explanation for solar and lunar eclipses and composed a special table for calculating their occurrence, using the nineteen-year "lunar cycle."

In his *Geometry of Astronomy* Shirakatsí tried to determine the distance from the earth to the sun, the moon, and the planets, and to estimate the true dimensions of the sun. Such a problem was of course beyond the observational techniques of his time.

Shirakatsí's works on the calendar were of great importance. He studied and compared the calendar systems of fourteen nations, among them the Agvancians (ancient inhabitants of Azerbaidjan), who did not leave any written records, and the Cappadocians.

Of special interest is his *Tables of the Lunar Cycle*, the authorship of which was established in the mid-twentieth century. The Yerevan Matenadaran, a repository of ancient manuscripts and books, possesses ten records that contain this work, which was based on his own observations. He wrote in his foreword:

> I, Anania Shirakatsí, have faithfully studied the course and changes in the appearance of the moon through all the days of its passage and, noting them, have fixed this information in tables, wishing to lighten the work of those who are interested. And I have drawn first the newborn moon, and then the full moon, on what day it takes place, at what time—in the night or in the day, at what hour and what minute.

Shirakatsí considered not only the days of the various phases of the moon but also the hours, which had not been done in any previous calendar. A comparison of his lunar tables with modern data shows the former's great precision.

Shirakatsí's textbook of arithmetic is one of the oldest known Armenian textbooks. Its mathematical tables—of multiplication and of arithmetical and geometrical progression—also are the oldest.

BIBLIOGRAPHY

I. ORIGINAL WORKS. Shirakatsí's writings include *Ananiayi Širakunwoy mnatsordk' panic'* ("Collected Works of Anania Shirakatsí"), K. P. Patkanian, ed. (St. Petersburg, 1877), in classical Armenian; his autobiography in English, in F. C. Conybeare, "Ananias of Širak," in *Byzantinische Zeitschrift*, **11** (1897), 572–584, also translated into French by H. Berberian in *Revue des Études Arméniennes*, n.s. **1** (1964), 189; *T'uabanut'iwn* ("Arithmetic"), A. G. Abrahamean, ed. (Erevan, 1939); *Tiezeragitut'iwn ew Tomar* ("Cosmography and Chronology"), A. G. Abrahamean, ed. (Erevan, 1940), also in Armenian; *Tablitsy lunnogo kruga* ("Tables of the Motions of the Moon"), A. G. Abrahamean, ed. (Erevan, 1962), in Russian and Armenian; and a collection of Shirakatsí's other works, *Anania Širakac'u matenadrut'iwn* ("The Works of Anania Shirakatsí"), A. G. Abrahamean, ed. (Erevan, 1944), in Armenian.

II. SECONDARY LITERATURE. See A. G. Abrahamean and G. B. Petrosian, *Ananias Shirakatsy* (Erevan, 1970), in Russian; F. C. Conybeare, "Ananias of Shirak, 'On Christmas,'" in *Expositor* (London), 5th ser., **4** (1896), 321–337; R. H. Hewsen, "Science in VIIth Century Armenia: Ananias of Širak," in *Isis*, **59** (1968), 32–45; I. A. Orbely, *"Voprosy i reshenia" Ananii Shirakatsi* ("The Problems and Solutions of Anania Shirakatsi") (Petrograd, 1918); W. Petri, "Anania Shirakazi—ein armenischer Kosmograph des 7. Jahrhunderts," in *Zeitschrift der Deutschen morgenländischen Gesellschaft*, **114**, no. 2 (1964), 269–288, also in *Mitteilungen der Sternwarte München*, **1**, no. 14 (1964), 269–288; G. Ter-Mkrtchian, "Anania Shirakatsy," in *Ararat* (1896), 96–104, 143–152, 199–208, 292–296, 336–344; and B. E. Tumanian and R. A. Abramian, "Ob astronomicheskikh rabotakh Ananii Shirakatsi" ("On the Astronomical Works of Anania Shirakatsi"), in *Istoriko-astronomicheskie issledovaniya*, no. 2 (1956), 239–246.

P. G. KULIKOVSKY

AL-SHĪRĀZĪ. See **Quṭb al-Dīn al-Shīrāzī.**

SHIZUKI, TADAO (*b.* Nagasaki, Japan, 1760; *d.* Nagasaki, 22 August 1806), *natural philosophy.*

Tadao's surname by birth was Nakano, but he was adopted by the Shizuki family, whose head was a government interpreter from the Dutch. His nickname was Tadajiro, and professionally he was known as Ryuho. In 1776 he became an assistant interpreter, succeeding his adoptive father, but he retired from that post the following year because of ill health. After leaving public service, he spent the rest of his life in the private study of Dutch books and in contemplation. Although his health was delicate and he had a reputation for unsociability, he attracted brilliant followers. He wrote books about the Dutch language and partially translated the Dutch translation of Engelbert Kaempher's *Geschichte und Beschreibung von Japan*. But Tadao's major work was *Rekisho shinsho*, a compilation of his own theories on natural philosophy, inspired by his translation of the Dutch version of John Keill's *Introductio ad veram physicam* and the *Introductio ad veram astronomiam*.

Tadao's natural philosophy was unusual within the Japanese intellectual tradition. His approach to problems was close to that of modern Western natural philosophers: he attempted to find fundamental explanations for all natural phenomena rather than merely describe them as moral, aesthetic, or practical problems.

Western mechanistic philosophy must have seemed remote and uninteresting in an intellectual climate dominated by Confucius' moral supreme doctrine, but Tadao nonetheless absorbed himself for twenty years in translating Keill's book and constructing his own natural philosophy. Perhaps he was able to do so because he was removed from the intellectual tradition centered around the Japanese Confucians and because he was able to pursue his study in Nagasaki, which was then the only opening to the outside world. Japanese knowledge of western science was at that time very limited. While Chinese translations of Western books on science had just been released from governmental ban, these were the work of Jesuits stationed in China and covered only Tychonic astronomy. The only sources for Copernicanism and Newtonianism were books brought in by the Dutch, who were the only Westerners permitted to trade with Japan and who were permitted to land only at Nagasaki. Tadao was thus ideally situated.

Other interpreters and physicians had contributed to introducing modern Western science into Japan, but Tadao had a far better understanding of the Dutch language and a superior insight into philosophical concepts. It is therefore curious that he chose to spend so much of his life translating Keill's book. If he meant simply to introduce Western natural science, why did he not choose other popular, up-to-date books and well-arranged texts, such as, for example, Benjamin Martin's *Philosophical Grammar*, the Dutch translation of which must have been easy to obtain despite the limited availability of Dutch books?

Although he translated other works on request, Tadao probably concentrated on Keill's book because it was difficult, under the import restrictions, to find an equally sophisticated work and because

the other textbooks and popular books available to him did not give him the same intellectual satisfaction. In advocating Newtonianism, Keill's work had a polemical tone, and more importantly and quite unlike the readjusted interpretations of later authors, it dealt in abstractions and included a great many elements of natural philosophy. The book especially suited Tadao's inclination toward natural philosophy and Tadao thus became the first Newtonian in the East.

The earliest extant manuscript of Tadao's translation of Keill's work is "On Attractive Force" (1784). Considering the level of knowledge of Western science in the Orient at that time, it would seem that Tadao struggled excellently with a very difficult subject. Since Copernican theory can be grasped as problems of angular variations, it presented no conceptional difficulty to the Oriental mind; but Newtonianism was being introduced into the Orient for the first time, and there was no Japanese vocabulary to embrace the concepts of corpuscle, vacuum, gravity, or force. Tadao therefore had to invent such words, and some of his inventions became standard in countries in which Chinese characters are used. The concept of atomism was also unknown in the Orient, and Tadao attempted to adopt the neo-Confucian idea of *ch'i* as a corresponding notion; but since the concept of *ch'i* is based on the model of a continuously changing fluid, he found it difficult to combine the idea of discrete atoms with the monistic *ch'i* concept of condensation and rarefaction.

Atomism requires a clear distinction between atom and vacuum, but for Tadao vacuum was a rarefied state of *ch'i*; and atom and vacuum were continuously caught in the spectrum of rarefaction of *ch'i*. Keill, however, explained electrical phenomena by means of effluvia, and Tadao found the concept of effluvia much more congenial to interpretation by *ch'i* than that of the atom. Tadao also claimed that effluvia offered an explanation for not only electrical force but also gravity. Since he considered weight to be the accumulation of *ch'i*, both electromagnetic force and gravity could be explained thereby.

Tadao's commentary contains many original ideas. He applied Western principles of attractive force and the traditional idea of Yin-Yang to the explanation of the phenomenal world. His adaptation of Yin-Yang incorporated two different— positive and negative—forces, and he used the variation in the balance of these two forces to explain what Western scientists considered to be a variation in the magnitude of only one force. Had

Tadao developed this idea further he might have constructed his own system, which would probably have been significantly different from its Western counterpart. His method seems particularly suitable to the explanation of electromagnetic force.

Although Tadao and Keill were from completely different academic environments, Tadao's native aptitude seems superior to Keill's, and he himself even experimented with a barometer, at a time when a Japanese experimental science barely existed. The subjects that Tadao tried to explain by attractive force—other than the data he obtained from Dutch books—were, however, limited to spontaneously occurring phenomena—as, for example, atmospheric events and plant life. (He even tried to construct a primitive vegetable physiology based upon attractive force.).

In his work on attractive force Keill developed the ideas of Newton's *Opticks* and tried to explain chemical and other phenomena by means of homogeneous particles and their intermolecular force, using the inverse third or fourth powers of their distances. But he could neither measure nor test what he wanted to prove and thus failed to achieve his objective. Unlike Keill and other Newtonians who sought to prove quantitative change in chemical phenomena, Tadao lacked the social background that might have fostered an interest in chemistry; and no such academic discipline had yet developed in Japan. Thus his interest in attractive force remained purely that of a natural philosopher.

The theory of cosmic dual forces in Tadao's commentary "On Attractive Force" is compatible with modern science, and there is the possibility that one subsumes the other: it is an acceptable and permissible concept for modern scientists. But in his later studies Tadao's ideas conformed more closely to traditional natural philosophy, and they became more speculative. During this time he completed his *Rekisho shinsho*, in which he commented systematically on Newtonian dynamics—although he was concerned primarily with their metaphysical basis: "All things have the property of gravity. Although gravity originally emerged from the inexplicable process of creation, it can be comprehended by the intelligence and hence is not absolutely inexplicable. Yet the cause of gravity is quite inscrutable. Even with advanced Western instruments and mathematics, the fundamental cause is indeterminable."

Although the second volume of *Rekisho shinsho* was entitled *Immeasurable*, Tadao sought in Oriental thinking a solution for matters that in the West would have been explained as acts of God. He established unitary *ch'i* and its dual function (rarefaction and condensation) as the basis for his natural philosophy and thus wrote that

The space of the universe contains only one substance, *ch'i*, but it can also be either empty or full. Thus in one there is two, and in two, one. If there were only the one, there could be no difference between the rarefied and the condensed. Is there not then a difference between the rarefied and the condensed? By the existence of these two contrary principles the phenomena are caused in endless succession. Because of the oneness of the substance, the universe is monistic. The cause of these principles is beyond my comprehension, but the best way to comprehend the subtlety of these principles is to study the teaching of the *Book of Changes* [*I-Ching*].

If there were only the two, the *ch'i* of heaven and earth could not be transmitted from one to the other. The shining *ch'i* of the sun and heavenly bodies is reflected from one to the other, and goes to and from the extremities of heaven. Permeating space without a single gap, it rises and falls, undergoing countless transformations. We must, then, admit that there is only a single *ch'i* . . . , which differs in respect to condensation and rarefaction. Condensation and rarefaction are the same in causal origin as emptiness and plenitude. The extreme of rarefaction is emptiness. Perfect plenitude and perfect emptiness, combining with each other, make a single state. This is why Laotzu said, "Nonexistence occupies nonspace." Nevertheless, even in the extremities of heaven, there is not the slightest space of pure plenitude or nonemptiness.

We must acknowledge that the light of the stars can permeate the broad heavens, and that the *ch'i* of fire can penetrate rock and metal. For example although winter is cold, and summer hot, even in winter there is still some warm *ch'i* and even in summer there is some cold *ch'i*. Or take the case of soil, placed in water. As the soil is condensed and heavy, the water, being light and rarefied, is like heaven. The soil, mixing with the water, discolors it; and the water, permeating the soil, moistens it. Therefore in the water there is nowhere where the soil *ch'i* is absent, and in the soil, nowhere where the water *ch'i* is absent.

Accordingly, Tadao arranged the subjects of Newtonian dynamics, including gravity, into a series of chapters on divination, theory of monistic *ch'i*, and Newtonian dynamics. He was not as successful, however, in relating the natural philosophy that he derived from the theory of monistic *ch'i* to Newtonian dynamics.

Tadao was one of the first to introduce Copernican theory into the Orient, although his interest was not in heliocentric theory. The locational relationship between the sun and the universe was for him simply a problem of changing the coordinate system; and, from the natural-philosophical point of view, it was not a substantial problem. In traditional neo-Confucian cosmology the problem involved the dynamics of motion and inactivity. Tadao did not refer to Copernicanism as a heliocentric theory but, rather, as an earth-moving theory, in which heaven and earth are composed of the cosmic dual forces of *ch'i*. The fast and light *ch'i* ascend and become heaven while the heavy and slow *ch'i* gather together and become earth. This system represents Chu Hsi's dynamic cosmogony and, unlike Aristotelian celestial-terrestrial dichotomy, makes no sharp distinction between heaven and earth, motion and inactivity. Another neo-Confucian philosopher also thought that the earth did not have absolute fixation in the middle of the universe but was situated at one end of a continuous spectrum binding Yin-Yang polar concepts (for example, shade and light, inactivity and motion, slow and fast) and that it rotated in relation to its surroundings.

Tadao viewed Copernicanism as absolute relativism but could not decide whether earthly or heavenly motion was more correct. He also pointed out that he had found the word "earth moving" in an ancient Chinese book and went on to suggest that the Chinese first conceived the earth-moving theory. (The word "earth moving" as it appears in this ancient text can also be translated as "earthquake," however.) The term "earth moving," as Tadao used it, has been retained as a scientific, technical word, even though it does not distinguish rotation from revolution.

Tadao also raised the question of why the planets rotate and revolve in the same direction in planes not greatly inclined to the ecliptic. In a section entitled "Kenkon bunpan zusetsu" ("The Formulation of the Cosmos, Illustrated") he proposed a hypothesis concerning the formation of the planetary system. His hypothesis recalls immediately the celebrated hypotheses of Kant and Laplace. Because of the relative inaccessibility of Western treatises, however, it is unlikely that Tadao derived his idea from anyone else. In view of his background in neo-Confucian ideas, his hypothesis was not a titanic leap—many aspects of it were already present in neo-Confucian vortex cosmogony. Hence, an infusion of ideas concerning attraction and centrifugal force provided Tadao with a more elaborate mechanical hypothesis, which he formulated in accordance with the heliocentric system.

BIBLIOGRAPHY

Tadao's major works are *Rekisho Shinsho* ("New Treatise on Calendrical Phenomena"; Heibon-sha, 1956), 2 vols., Japanese Philosophical Thoughts no. 9; and *Kyurikiron* ("On Attractive Force"; Iwanami, 1972).

On Tadao and his work in Western languages, see Yoshio Mikami, "On Shizuki Tadao's Translation of Keill's Astronomical Treatise," in *Nieuw archief voor wiskunde*, **11** (1913), 1–11; Shigeru Nakayama, *History of Japanese Astronomy* (Cambridge, Mass., 1969); Ohmori Minoru, "A Study of the *Rekisho Shinsho*," in *Japanese Studies in the History of Science*, no. 2 (1963), 146–153, and no. 3 (1964), 81–88; S. Yajima, "Théorie nébulaire de Shizuki (1760–1806)," in *Archives internationales d'histoire des sciences*, **12** (1956), 169–173; and Tadashi Yoshida, "The Rangako of Shizuki Tadao: the Introduction of Western Science in Tokugawa Japan," Ph.D. thesis at Princeton University (1974).

In Japanese are Kanda Shigeru, "Translations by Shizuki Tadao," in *Rangaku shiryo Kenkyukai Hokoka*, no. 107 (1961); Hiroto Saigusa, "On Newton, Who Existed in Japan the Past Two Centuries," in *Yokohama Daigaku Ronso*, Social Science, no. 1 (1958); Watanabe Kurasuke, "Summary on Work of Dutch Translator, Shizuki," in *Nagasaki Gakkai Sosho*, 4th ed. (1957).

SHIGERU NAKAYAMA

SHMALHAUZEN, IVAN IVANOVICH (*b.* Kiev, Russia, 23 April 1884; *d.* Moscow, U.S.S.R., 7 October 1963), *biology.*

The son of a professor of botany at Kiev University, Shmalhauzen graduated from the natural sciences section of the Physics and Mathematics Faculty of Kiev University in 1909. He remained there as assistant to A. N. Severtsov until 1911, passed his master's degree examination in 1912, and in 1914 presented his thesis, "Neparnye plavniki ryb; ikh filogeniticheskoe razvitie" ("Unpaired Fins of Fish; Their Phylogenetic Development"), at Moscow University. In 1916 Shmalhauzen defended his doctoral dissertation, "Razvitie konechmostey amfiby i ikh znachenie v voprose proiskhozhdenia nazemnykh pozvonochnykh" ("The Development of Extremities in Amphibians and Their Significance in the Origin of Land Vertebrates"). He was professor at Yuriev (Tartu) University, from 1916 to 1920 and headed the department of embryology and dynamics of development at Kiev from 1920 to 1937. At Kiev he organized the Biological Institute (later the Institute of Zoology) at the Ukrainian Academy of Sciences and was its director from 1941. From 1936 to 1948 Shmalhauzen was director of the A. N. Severtsov

Institute of Evolutionary Morphology and of the Academy of Sciences of the U.S.S.R., and was head of the department of Darwinism at Moscow University from 1939 to 1948. During his last years he worked at the Zoological Institute of the Academy of Sciences. He was elected academician of the Ukrainian Academy of Sciences (1922) and of the Academy of Sciences of the U.S.S.R. (1935).

Beginning his scientific career as a comparative anatomist, Shmalhauzen later expanded his research to the study of individual development and directed research in experimental embryology. Studying the growth of animals, he formulated a theory of growth according to which that process has an exponential character exhibiting an inverse relationship between growth and differentiation: attenuation of growth is described by a parabolic curve, since the specific rate of growth decreases in inverse proportion to age. During various distinct periods of individual life the energy of growth can be expressed by the constant of growth C_v, for the computation of which Shmalhauzen proposed

the formula $C_v = \dfrac{\log v_1 - \log v_0}{0.4343\,(t_1 - t_0)} = \text{constant}$, where

v_0 and v_1 are the mass of the body at the moments t_0 and t_1. He later studied the correlations of growth in individual and historic development that are the basis of the integrity of the organism and arise through the effect of natural selection.

In specialized monographs (1938, 1939, 1946) Shmalhauzen examined the course and regularities of the evolutionary process. In particular he provided a basis for classifying the types of natural selection. He established, in addition to an "active" natural selection that stabilizes mutational changes favoring survival under changed conditions, the existence of "stabilizing" natural selection. The latter preserves "normal" characteristics and eliminates deviations as long as the environment remains relatively unchanged. In the process of stabilizing selection "under the cover of the normal phenotype," recessive or balancing genes accumulate; with environmental changes they can serve as material for "active" natural selection.

In 1938, while studying the relations that exist in ontogeny and in phylogeny, Shmalhauzen suggested the idea, later (1968) substantiated in detail, that methods of cybernetics might be applied to the study of ontogenetic and phylogenetic development. His many years of teaching resulted in textbooks on comparative anatomy and on Darwinism. During the last years of his life Shmalhauzen re-

sumed the study of the origin of land vertebrates and published a monograph on that subject (1964).

BIBLIOGRAPHY

I. ORIGINAL WORKS. Shmalhauzen wrote more than 200 works, including the following monographs: *Rost organizmov* ("The Growth of Organisms"; Kiev, 1932), in Ukrainian; *Organizm kak tseloe v individualnom i istoricheskom razvitii* ("The Organism as a Whole in Individual and Historical Development"; Moscow–Leningrad, 1938; 2nd ed., 1942); *Puti i zakonomernosti evolyutsionnogo protsessa* ("The Ways and Regularities of the Evolutionary Process"; Moscow–Leningrad, 1939); *Faktory evolyutsii. Teoria stabiliziruyushchego otbora* ("Factors of Evolution. Theory of Stabilizing Selection"; Moscow–Leningrad, 1946; 2nd ed., 1968); also in English trans. (Philadelphia, 1949); *Proiskhozhdenie nazemnykh pozvonochnykh* ("The Origin of Terrestrial Vertebrates"; Moscow, 1964), also in English trans. (New York, 1968); and *Kiberneticheskie voprosy biologii* ("Cybernetic Questions of Biology"; Novosibirsk, 1968).

II. SECONDARY LITERATURE. See A. A. Makhotin, "Ivan Ivanovich Shmalhauzen (1884–1963)," in *Zoologicheskii zhurnal*, **43**, pt. 2 (1964), 297–302; and L. P. Tatarinov and B. A. Trofimov, "Akademik Ivan Ivanovich Shmalhauzen (1884–1963)," in *Paleontologicheskii zhurnal* (1964), no. 2, 169–173.

L. J. BLACHER

SHNIRELMAN, LEV GENRIKHOVICH (*b.* Gomel, Russia, 2 January 1905; *d.* Moscow, U.S.S.R., 24 September 1938), *mathematics.*

The son of a teacher, Shnirelman displayed remarkable mathematical abilities even as a child. In his twelfth year he studied an entire course of elementary mathematics at home; and in 1921 he entered Moscow University, where he attended courses taught by N. N. Lusin, P. S. Uryson, and A. Y. Khinchin. While still a student he obtained several interesting results in algebra, geometry, and topology that he did not wish to publish, considering them of insufficient importance. After two and a half years Shnirelman graduated from the university, then remained for further study. Having completed his graduate work, he became professor and head of the department of mathematics at the Don Polytechnical Institute in Novocherkassk (1929). In the following year Shnirelman returned to Moscow and taught at the university. He was elected a corresponding member of the Soviet Academy of Sciences in 1933, and from 1934 he worked in the Mathematical Institute of the Academy.

In 1927–1929 Shnirelman, with his friend L. A. Lyusternik, made important contributions to the qualitative (topological) methods of the calculus of variations. Their starting point was Poincaré's problem of the three geodesics, which they first solved completely and generally by showing the existence of three closed geodesics on every simply connected surface (every surface homeomorphic to a sphere). For the proof of this theorem the authors used a method, which they broadly generalized, that had been devised by G. Birkhoff, who in 1919 showed the existence of one closed geodesic. Shnirelman and Lyusternik also applied their "principle of the stationary point" to other problems of geometry "im Grossen." They also presented a new topological invariant, the category of point sets.

In 1930 Shnirelman introduced an original and profound idea into number theory, using the concept of the compactness α of the sequence of natural numbers, $n_1, n_2, n_3, \ldots$ so that $\alpha = \inf. \dfrac{N(x)}{x}$ ($x \geq 1$) where $N(x)$ is a number of the members of the sequence not exceeding x, and proving that every natural number n is representable as the sum of a finite (and independent of n) number of members of the sequences with a positive compactness. This allowed Shnirelman to prove, in particular, that any natural number is the sum of a certain finite number k of prime numbers—the Goldbach hypothesis in a less rigid form. According to the Goldbach hypothesis $k = 3$; by Shnirelman's method it is now possible to show that k is not greater than 20. Shnirelman also stated several arithmetical propositions, among them a generalization of Waring's theorem.

BIBLIOGRAPHY

I. ORIGINAL WORKS. Shnirelman's writings include "Sur un principe topologique en analyse," in *Comptes rendus . . . de l'Académie des sciences*, **188** (1929), 295–297, written with L. A. Lyusternik; "Existence de trois géodésiques fermées sur toute surface de genre 0," *ibid.*, 534–536, written with Lyusternik; "Sur le problème de trois géodésiques fermées sur les surfaces de genre 0," *ibid.*, **189** (1929), 269–271, written with Lyusternik; "Ob additivnykh svoystavakh chisel" ("On the Additive Properties of Numbers"), in *Izvestiya Donskogo politekhnicheskogo instituta v Novocherkasske*, **14**, nos. 2–3 (1930), 3–28, also in *Uspekhi matematicheskikh nauk*, **6** (1939), 9–25; "Über eine neue kombinator-
ische Invariante," in *Monatshefte für Mathematik und Physik*, **37** (1930), 131–134; *Topologicheskie metody v variatsionnykh zadachakh* ("Topological Methods in Variational Problems"; Moscow, 1930), written with Lyusternik; "Über additive Eigenschaften von Zahlen," in *Mathematische Annalen*, **107** (1933), 649–690; "Ob additivnykh svoystvakh chisel" ("On the Additive Properties of Numbers"), in *Uspekhi matematicheskikh nauk*, **7** (1940), 7–46; and "O slozhenii posledovatelnostey" ("On Addition of Sequences"), *ibid.*, 62–63.

II. SECONDARY LITERATURE. See "L. G. Shnirelman (1905–1938)," in *Uspekhi matematicheskikh nauk*, **6** (1939), 3–8; *Matematika v SSSR za pyatnadtsat let* ("Mathematics in the U.S.S.R. for Fifteen Years"; Moscow–Leningrad, 1932); *Matematika v SSSR za tridtsat let* ("Mathematics in the U.S.S.R. for Thirty Years"; Moscow–Leningrad, 1948); and *Matematika v SSSR za sorok let* ("Mathematics in the U.S.S.R. for Forty Years"), 2 vols. (Moscow, 1959), esp. II, 781–782.

A. P. YOUSCHKEVITCH

SHOKALSKY, YULY MIKHAYLOVICH (*b.* St. Petersburg, Russia, 17 October 1856; *d.* Leningrad, U.S.S.R., 26 March 1940), *oceanography, geography.*

Fascinated by geography as a child, Shokalsky enrolled at the Naval College in St. Petersburg in 1873. Although it became apparent on his first student voyages that he was strongly inclined to seasickness, he did not abandon his chosen career. After graduating in 1877, he began service with the Baltic fleet; but realizing that his command responsibilities did not leave him time to pursue his scientific interests, he completed his education at the Naval Academy from 1878 to 1880.

In 1881 Shokalsky became head of the marine meteorology division at the Central Physics Observatory (now the A. I. Voeykov Central Geophysical Observatory in Leningrad). A year later his first scientific paper, on weather forecasting, was published. From 1883 to 1908 he taught geography and marine description at the Naval College; and, from 1908, physical geography, meteorology and, later, oceanography at the Naval Academy, where he was the first professor of this subject in Russia. Shokalsky was a member of the fleet's department of hydrography, directing its central naval library from 1890 and the hydrometeorological section from 1907.

From 1900 to 1915 Shokalsky developed a large-scale project for oceanographic research and participated in the preparations for the expedition of the icebreakers *Taymyr* and *Vaygach* to the

Arctic Ocean—the expedition that discovered Severnaya Zemlya. He was also one of the organizers of the Eleventh International Congress on Navigation, held in St. Petersburg in 1908. Shokalsky's plan, formulated in 1902, for the creation of a marine science center was realized in 1921 with the establishment in Monaco of the International Hydrographic Organization.

Elected president of the Russian Geographic Society in 1914, Shokalsky made a major contribution to cartography, geomorphology, terrestrial hydrology, glaciology, and geodesy. In *Okeanografia* (1917), his most important work, he postulated the mutual dependence of all marine phenomena; this concept was subsequently elaborated to include the idea of a "worldwide ocean," an unbroken totality of all interconnected bodies of salt water. In addition to discussing physical phenomena, Shokalsky gave a full description of the methodology and techniques of shipboard observation. The work, which became extremely popular both in the Soviet Union and elsewhere, was awarded prizes by the Russian Academy of Sciences (1919) and by the Paris Academy of Sciences (1923). It is still of value.

In 1909 the Hydrographic Administration had commissioned Shokalsky to plan and determine the methodology of a comprehensive oceanographic expedition to the Black Sea. The project was to include all branches of marine science and would extend to adjacent basins: the Kerchenskiy Proliv, Bosporus, and Dardanelles straits, and the seas of Azov and Marmara. Although Shokalsky completed plans for the expedition in 1914, it was delayed for nine years by the outbreak of World War I and the Soviet Revolution.

During the first year the study comprised only two hydrological sections taken from the narrowest part of the Black Sea—from Cape Sarych, in the southern Crimea, to Inebolu, on the Anatolian coast of Turkey. By 1925 Shokalsky had extended the survey to the entire Black Sea, and its work was completed in 1935 with a long winter cruise. During the twelve-year expedition fifty-three voyages were made, and over 1,600 hydrological samples and approximately 2,000 soil and biological samples were taken. Data obtained on the density of seawater in relation to depth discredited the concept of an upper layer, containing atmospheric oxygen in solution, distinct from a lower, hydrosulfidic layer. Instead, the information indicated a high mobility of both layers and a constant interexchange. Extremely long cores of earth were also taken from the seabed; one, 380 centimeters in length, was a record at the time.

Despite his advanced age, Shokalsky remained active during the 1930's. He served on the commission to develop the programs of the Second International Polar Year (1932–1933) and, with V. V. Shuleykin and N. N. Zubov, strove for the inclusion of oceanographic studies. He also was responsible for compiling a new physical map of the North Polar region. In 1933 he was a member of the organizational committee of the Fourth Hydrologic Conference of Baltic Nations, held at Leningrad; was a member of the presidium of the marine section; and the chairman of the Commission on the [Aquatic] Balance of the Baltic Sea. The following year, at the age of almost eighty, he was the Soviet delegate to the Fourteenth International Geographical Congress, in Warsaw.

From 1929 until his death Shokalsky taught oceanography at Leningrad University; his lectures in general oceanography were supplemented by a course in regional oceanography that was entirely of his own devising.

Shokalsky was a member of fifteen foreign geographic societies and received an honorary doctorate in geography from the University of Bordeaux. He was an honorary member of the Soviet Academy of Sciences, honorary president of the Russian Geographic Society, and received the title "honored scientist."

BIBLIOGRAPHY

I. ORIGINAL WORKS. The most important of Shokalsky's more than 1,300 publications are "O predskazanii veroyatnoy pogody i shtormov" ("On Forecasting Weather and Storms"), in *Morskoi sbornik*, **192**, no. 10 (1882), 87–125, his first scientific paper; "Ocherk razvitia fiziki okeanov" ("Essay on the Development of the Physics of Oceans"), in *Fiziko-matematichesky ezhegodnik*, **1** (1900), 158–207; "Vzglyad na sovremennoe sostoyanie okeanografii" ("View of the Contemporary State of Oceanography"), in *Izvestiya Russkogo geograficheskogo obshchestva*, **43** (1911), 503–520; *Okeanografia* (Petrograd, 1917; 2nd ed., 1959), his major work; *Geografichesky atlas v 32 tablitsakh* ("Geographic Atlas in 32 Tables"; Leningrad, 1930); *Fizicheskaya okeanografia* (Moscow, 1933); "Okeanografia Chernogo morya" ("Oceanography of the Black Sea"), in *Doklady sovetskoy delegatsii na Mezhdunarodnom geograficheskom kongresse v Varshave*, no. 14 (1934); "O rabotakh po okeanografii v Mirovom okeane" ("On Papers About Oceanography in the World Ocean"), in *Doklady prochitannye na Tretem* *plenume gruppy geografii i geofiziki Akademii nauk SSSR* ("Reports Read at the Third Plenum of the Geography and Geophysics Group of the Soviet Academy of Sciences"; Moscow, 1937), 5–18; "Okeanograficheskie issledovania Soyuza SSR za 20 let" ("Twenty Years of Oceanographic Investigation in the U.S.S.R."), in *Matematika i estestvoznanie v SSSR* ("Mathematics and Natural Science in the U.S.S.R."; Moscow, 1938), 901–917; and "Kartografirovanie morey Sovetskogo Soyuza," in *Dvadtsat let sovetskoy kartografii i geodezii, 1919–1939* (Moscow, 1939), pt. 2, 162–190.

II. Secondary Literature. See E. Andreeva, *Y. M. Shokalsky* (Leningrad, 1956); *Pamyati Yulia Mikhaylovicha Shokalskogo. Sbornik statey i materialov*, I (Moscow–Leningrad, 1946); and Z. Y. Shokalskaya, *Zhiznenny puty Y. M. Shokalskogo* ("The Life of . . . Shokalsky"; Moscow, 1960).

A. F. Plakhotnik

SHORT, JAMES (*b.* Edinburgh, Scotland, 10 June 1710; *d.* Stoke Newington, Essex, England, 15 June 1768), *optics.*

Short, whose father was an Edinburgh joiner and burgess, was orphaned at the age of ten. He was educated at Heriot's Hospital, a school for the sons of poor burgesses, and at the Edinburgh High School. In 1726 he entered the University of Edinburgh. His M.A., however, was awarded by the University of St. Andrews in 1753. Because of his excellent academic record, his grandmother hoped that Short would enter the ministry, but he became instead a protégé of Colin Maclaurin, professor of mathematics at Edinburgh. Encouraged by Maclaurin, Short started making mirrors for reflecting telescopes, first of glass, then of speculum metal. His skill in this work received immediate recognition and was the source of his subsequent success. Indeed, the perfecting of these metal mirrors was his lifework, for, unlike other instrument makers of his time, he was a specialist.

Short settled permanently in London in 1738, having been elected a fellow of the Royal Society the previous year. A bachelor, Short's devotion to his chosen profession, "Optician, solely for reflecting telescopes," brought him an international reputation and a fortune that was estimated at £20,000 at the time of his death. Short was one of forty-five founder members of the Philosophical Society of Edinburgh; was a founder member, in 1754, of the Society for the Encouragement of Arts, Manufactures and Commerce; and, in 1757, became a foreign member of the Royal Swedish Academy of Sciences.

From 1738 to the end of his life, Short's home and workshop were in Surrey Street, off the Strand. There he made astronomical observations that were frequently reported in the *Philosophical Transactions*. Short's portrait, painted by Benjamin Wilson, shows him in his observatory at Surrey Street, overlooking the Thames; behind him is the five-foot-focal-length reflecting telescope that he habitually used.

Short generously encouraged the work of his fellow telescope maker, and potential rival, John Dollond, by communicating to the Royal Society, in 1758, Dollond's experimental work on the correction of chromatic aberration in the refracting telescope. Because of Dollond's achievement, the refractor was, by the end of Short's life, beginning to supersede the reflector. Another friend to whom Short gave active support was John Harrison, the chronometer maker. Had Short not championed the notoriously difficult and unpopular Harrison in his disputes with the Board of Longitude, Short might well have become Astronomer Royal. According to Alexander Small, writing to Benjamin Franklin, who was a friend of Short, the help that Short gave to Harrison annoyed the Earl of Morton, who consequently opposed Short's appointment to the Royal Observatory.

The most reliable source of information about Short's method of work is the Plymouth physician John Mudge, who visited Short's workshop and who read two papers to the Royal Society on the founding and figuring of metal mirrors for reflecting telescopes. This art was partly in the composition and founding of the metal and partly in the polishing (or figuring) of the mirror. Apparently it was in the latter skill that Short particularly excelled; Mudge wrote of his mirrors, "they are all exquisitely figured." Recent studies show that Short made a total of 1,370 reflecting telescopes, of which 110 still exist.

The two most important astronomical events of the mid-eighteenth century were the transits of Venus in 1761 and 1769. The purpose of making worldwide observations of the transits was to calculate the solar parallax, and hence to discover the dimensions of the solar system. Astronomers throughout Europe were eager to acquire the best possible instruments for the task. They turned naturally to London, the reputation of whose instrument makers was unrivaled, and, for telescopes, to Short. A week before his death he was still concerned with the despatch of instruments to the Russian Imperial Academy. Short was also, as a mathematician, involved in the calculations of the parallax, and he published two long papers on this problem in the *Philosophical Transactions of the Royal Society* (1762, 1763). Short became a member of the council of the Royal Society in 1760 and was appointed to its special committee set up to study the 1769 transit. He died before he could take an active part in the plans he had helped to formulate, but his instruments were used to observe the second transit from stations throughout the world. Two of his instruments traveled with Cook on the *Endeavour* to Tahiti.

BIBLIOGRAPHY

I. ORIGINAL WORKS. Short's works include "Description and Uses of the Equatorial Telescope, or Portable Observatory," in *Philosophical Transactions of the Royal Society,* **46** (1749–1750), 241–246; "The Observations of the Internal Contact of Venus With the Sun's Limb, in the Late Transit, Made in Different Places of Europe, Compared With the Time of the Same Contact Observed at the Cape of Good Hope, and the Parallax of the Sun From Thence Determined," *ibid.,* **52** (1762), 611–628; "The Difference of Longitude Between the Royal Observatories of Greenwich and Paris, Determined by the Observations of the Transits of Mercury Over the Sun in the Years 1723, 1736, 1743, and 1753," *ibid.,* **53** (1763), 158–169; and "Second Paper Concerning the Parallax of the Sun Determined From the Observations of the Late Transit of Venus; in Which This Subject is Treated of More at Length, and the Quantity of the Parallax More Fully Ascertained," *ibid.,* 300–345. Short published twenty-eight additional papers in the *Philosophical Transactions of the Royal Society,* volumes **41** to **59,** most of these being brief accounts of astronomical observations. MS material is in the British Museum Add. MS 4434. Short's telescopes are housed in museums throughout the world; the Museum of the History of Science, Oxford, holds the longest (twelve-foot focal length) and also seven others.

II. SECONDARY LITERATURE. See D. J. Bryden, *James Short and His Telescopes. Royal Scottish Museum, Edinburgh, July 26th–September 7th, 1968* (Edinburgh, 1968); "Note on a Further Portrait of James Short, FRS," in *Notes and Records. Royal Society of London,* **24** (1969), 109–112; and "A Portrait Sketch of James Short by the 11th Earl of Buchan," in *Journal of the Royal Society of Arts,* **118** (1970), 792; V. L. Chenakal, "James Short and Russian Astronomy," in *Istoriko-astronomicheskie issledovaniya,* 5 (1959), 11–82; and G. L'E. Turner, "A Portrait of James Short, FRS, Attributed to Benjamin Wilson, FRS," in *Notes and Records. Royal Society of London,* **22** (1967), 105–112; "James Short, FRS, and His Contribution to the Construction of Reflecting Telescopes," *ibid.,* **24** (1969), 91–108; and "Mr James Short, Optician, Founding Member of the Society," in *Journal of the Royal Society of Arts,* **118** (1970), 788–792.

G. L'E. TURNER

SHTOKMAN, VLADIMIR BORISOVICH (*b.* Moscow, Russia, 10 March 1909; *d.* Moscow, 14 June 1968), *earth science, oceanography.*

In 1928 Shtokman entered the Faculty of Physics and Mathematics of Moscow University and chose to specialize in the geophysics of the earth's hydrosphere. The death of his parents in 1930 and his own illness in 1931 obliged him to interrupt his studies, and he began work as a laboratory assistant at the Institute of Oceanography in Moscow. By 1932 he had become a junior staff member as a result of his first independent scientific work, an investigation of the form of the curved underwater section of a cable carrying oceanographic instruments. He was promoted to senior staff member the following year, after having planned and led an oceanographic expedition in Motovskiy Gulf of the Barents Sea that yielded important results.

In 1934 Shtokman was sent to the Azerbaydzhan branch of the All-Union Scientific Research Institute of Ocean Fisheries and Oceanography (VNIRO) at Baku, where he organized and directed a laboratory of physical oceanography.

By applying probability theory and random functions to the study of ocean turbulence, Shtokman was a pioneer in introducing new statistical ideas and methods of studying this phenomenon into the U.S.S.R. He also developed direct techniques of measuring turbulent pulsation and current velocity at a series of points. These investigations were carried out in the central and southern Caspian Sea during a series of oceanographic expeditions, the majority of which Shtokman led personally. As a result he derived completely new results for turbulent pulsations in the horizontal velocity of marine currents. He also was the first to obtain the coefficients of horizontal turbulent exchange by direct methods.

In 1938 Shtokman was awarded the candidate's degree in physical-mathematical sciences, without defending a dissertation. In the late 1930's and early 1940's he worked in Moscow as director of the hydrology section of VNIRO and as senior staff member at the Institute of Theoretical Geophysics of the U.S.S.R. Academy of Sciences, and during the first years of World War II he was at Krasnoyarsk as senior staff member of the Arctic Institute, which had been evacuated there. During these years Shtokman conducted and published investigations on the theoretical bases of the computation of geostrophic currents from oceanographic measurements; on indirect methods for computing geostrophic currents in the Greenland Sea; on the analysis of water masses of the central part of the northern Arctic Ocean; and on the peculiarities of distribution of Atlantic waters in the Arctic Ocean. Shtokman employed temperature-salt curves in his critical examination of the widely used Jacobsen method of determining the intensity of intermixing of masses of water, and showed that the geometrical part of the method contained serious error. This work led him to research on the application of temperature-salt curves in the analysis of marine water masses, as a result of which a strict theory of these curves (geometry of temperature-salt curves) was developed.

In 1943 Shtokman returned to Moscow, defended his dissertation for the doctorate in physical-mathematical sciences, and began work at the newly created laboratory of oceanography of the Soviet Academy of Sciences, where he remained for the last twenty-five years of his life. During this time the laboratory of oceanography grew into the Institute of Oceanography of the Soviet Academy of Sciences; and Shtokman was director of the section of physical oceanography, the laboratory of ocean dynamics, and the theoretical section.

By the mid-1940's Shtokman had become widely known as an eminent Soviet specialist in the theory of ocean currents. He made an important step forward from the relations, proposed by Ekman, between currents and wind at a given point of the ocean and discovered the connections between currents and wind over the area of the ocean. Shtokman was the first to show clearly the important role in the dynamics of ocean currents of the transverse irregularities of tangential stress exerted by wind on water. In research published in 1945 he showed that the transverse irregularity of tangential stress of a following wind is an important reason for the horizontal circulation in the ocean. In 1941 Shtokman had shown that the presence of transverse irregularities of tangential stress of a wind over a closed sea inevitably results in countercurrents. The countercurrents exist not only in closed seas, however, but also in the open ocean, particularly in the equatorial zone.

In 1947 Shtokman published a special investigation, *Vozmozhny li protivotechenia v bezbrezhnom more, obuslovlennye lokalnoy neravnomernostyu vetra?* ("Are Countercurrents Caused by Local Irregularity of the Wind Possible in an Open Sea?"). The answer was affirmative. In a work published in 1948 Shtokman showed that the main peculiarities of equatorial countercurrents are explained by the dynamic effect of a stable zone irregularity (trade winds) in the lower latitudes.

In 1946 research on the influence of the wind field on currents led Shtokman to develop a theoretical method of computing mean velocities of wind currents as a function of depth—the method of full streams. The closed system of hydrodynamic equations used in this method proved easy to solve and stimulated significant progress in the development of theoretical models of ocean currents for many years. Soon after Shtokman's work appeared, a method of full streams was developed in the United States by Sverdrup (1947). Unlike Shtokman's work, which dealt with the basin of a limited area and thus did not consider extensive change in the Coriolis parameter (β effect), Sverdrup, by developing the same method for the open sea, did consider the β effect.

In subsequent research Shtokman sought an easier method of mapping the full streams of the seas of the U.S.S.R., and the theoretical model of density that he introduced made it possible to compute approximately the velocity of currents at various depths.

In the last years of his life, Shtokman did mathematical work on the circulation of water, especially around islands in a direction opposite to the circulation in the surrounding ocean. Such a situation occurs in the area around Taiwan in winter and around Iceland and the Kurile Islands throughout the year. In 1966 Shtokman showed that the reason for this phenomenon is the disturbance effect (violation of water exchange), introduced by the island into the horizontal circulation stimulated by the transverse irregularity of wind within the limits of a closed area.

BIBLIOGRAPHY

I. Original Works. Shtokman's major published writings are "Turbulentny obmen v rayone Agrakhanskogo poluostrova v Kaspyskom more" ("Turbulent Exchange in the Region of the Agrakhansky Peninsula in the Caspian Sea"), in *Zhurnal geofiziki*, 6, no. 4 (1936), 340–388; "O turbulentnom obmene v sredney i yuzhnoy chasti Kaspyskogo morya" ("On Turbulent Exchange in the Middle and Southern Parts of the Caspian Sea"), in *Izvestiya Akademii nauk SSSR*, geog.-geophys. sec. (1940), no. 4, 569–592; "O pulsatsiakh gorizontalnykh komponent skorosti morskikh techeny, obuslovlennykh turbulentnostyu bolshogo masshtaba" ("On the Pulsations of the Horizontal Components of Velocity of the Ocean Currents Caused by Large-Scale Turbulence"), *ibid.* (1941), nos. 4–5, 475–486; "Osnovy teorii temperaturno-solenostnykh krivykh, kak metoda izuchenia peremeshivania i transformatsii vodnykh mass morya" ("Bases of the Theory of Temperature-Salt Curves as a Method of Studying the Mixing and Transformation of Masses of Seawater"), in *Problemy arktiki* (1943), no. 1, 32–71; "Geometricheskie svoystva temperaturno-solenostnykh krivykh pri smeshenii trekh vodnykh mass v neogranichennom more" ("Geometrical Properties of Temperature-Salt Curves in the Mixing of Three Water Masses in the Open Ocean"), in *Doklady Akademii nauk SSSR*, 43, no. 8 (1944), 351–355; "Poperechnaya neravnomernost nagonnogo vetra kak odna iz vazhnykh prichin gorizontalnoy tsirkulyatsii v more" ("Transverse Irregularity of a Following Wind as One of the Important Reasons for Horizontal Circulation in the Ocean"), *ibid.*, 49, no. 2 (1945), 102–106; and "Teoria ekvatorialnykh protivotecheny v okeanakh" ("Theory of Equatorial Countercurrent in the Oceans"), *ibid.*, 52, no. 4 (1946), 311–314.

Subsequent works are "Uravnenia polya polnykh potokov, vozbuzhdaemykh vetrom v neodnorodnom okeane" ("Equations of the Fields of Full Streams Generated by Wind in the Nonuniform Ocean"), in *Doklady Akademii nauk SSSR,* 54, no. 5 (1946), 407–410; "Ispolzovanie analogii mezhdu polnym potokom v more i izgibom zakreplennoy plastiny . . ." ("Use of the Analogy Between a Full Stream in the Ocean and the Curve of a Strengthened Plate . . ."), *ibid.*, 54, no. 8 (1946), 689–692; "Novye dokazatelstva znachenia neravnomernosti vetra kak odnoy iz prichin tsirkulyatsii v more" ("New Proof of the Significance of the Irregularity of Wind as One of the Reasons for Circulation in the Ocean"), *ibid.*, 58, no. 1 (1947), 53–56; *Ekvatorialnye protivotechenia v okeanakh* ("Equatorial Countercurrents in the Oceans"; Leningrad, 1948); "Issledovanie vliania vetra i relefa dna na rezultiruyushchuyu tsirkulyatsiyu i raspredelenie mass v neodnorodnom okeane ili more" ("Research on the Influence of Wind and the Surface of the Bottom on the Resulting Circulation and Distribution of Masses in a Nonuniform Ocean or Sea"), in *Trudy Instituta okeanologii Akademii nauk SSSR,* 3 (1949), 3–65; "Vlianie vetra na techenia v Beringovom prolive . . ." ("Influence of the Wind on the Currents in the Bering Strait . . ."), *ibid.*, 25 (1957), 17–24; "Ob odnoy probleme dinamiki okeanicheskoy tsirkulyatsii" ("On One Problem of the Dynamics of Ocean Circulation"), in *Simpozium po matematicheskim i gidrodinamicheskim metodam izuchenia fizicheskikh protsessov v okeane, Moskva, 25–28 maya 1966, tezisy dokladov* ("Symposium on Mathematical and Hydrodynamical Methods of Studying the Physical Processes in the Ocean . . . Abstracts of Reports"; Moscow, 1966), 60–61; and "Razvitie teorii morskoy i okeanicheskoy tsirku-

lyatsii v SSSR za 50 let" ("Development of the Theory of Sea and Ocean Circulation in the U.S.S.R. for Fifty Years"), in *Okeanologia,* 7, no. 5 (1967), 761–773.

II. SECONDARY LITERATURE. See the obituaries by his editorial colleagues, "Vladimir Borisovich Shtokman," in *Izvestiya Akademii nauk SSSR*, Fiz. atmos. i okeana ser., 4, no. 9 (1968), 1012–1013; "Vladimir Borisovich Shtokman," in *Okeanologia,* 8, no. 4 (1968), 771, by a group of his colleagues.

A. F. PLAKHOTNIK

SHUJĀᶜ IBN ASLAM, AL-MIṢRI. See Abū Kāmil Shujāᶜ ibn Aslam ibn Muḥammad ibn Shujāᶜ

SHULL, AARON FRANKLIN (*b*. Miami County, Ohio, 1 August 1881; *d*. Ann Arbor, Michigan, 7 November 1961), *genetics, evolution.*

Shull was one of eight children born to Harrison and Catherine Ryman Shull. Although they had relatively little formal education, both parents encouraged intellectual activity among their children. The father was a farmer and lay minister; the mother had strong interests in serious reading, especially topics relating to natural history, and in her later years became an accomplished horticulturist with a firsthand knowledge of plants. Because the family moved from farm to farm in south-central Ohio, Shull's early schooling was largely informal. In 1904 he enrolled as an undergraduate at the University of Michigan, completing his A.B. degree in 1908. In the summer following his graduation he worked with the Michigan Biological Survey, and in the fall of that year he entered Columbia University for graduate work in zoology. Stimulated by the work of T. H. Morgan and E. B. Wilson, both at Columbia, Shull became interested in problems of heredity, particularly sex determination, on which he wrote his thesis. After obtaining his Ph.D. in 1911, he returned to the University of Michigan as an instructor in zoology (1911–1912), assistant professor (1912–1914), associate professor (1914–1921), and professor (1921–1951). His only break with this university occurred in the summer of 1938, when he was a visiting professor at the University of California, Berkeley. Although a researcher of considerable merit, Shull perhaps is best remembered in American biology for his teaching and writing. A stimulating lecturer, he was also a prolific writer of monographs and textbooks that introduced countless students to modern, experimental biology and to rigorous concepts in general biology, heredity, and evolution.

Shull studied in considerable depth the life cycle of and sex determination problem in rotifers, a subject that had been variously interpreted by workers in the last part of the nineteenth and first part of the twentieth centuries. A study of developmental physiology and sex determination in aphids, expanded the earlier work to the problem of sex determination in general. In the 1920's and 1930's he experimented with induction of crossing-over in *Drosophila* by physical factors, most notably heat. In all these studies, Shull's focus was on the relationship between heredity and environment in determining the phenotype of an organism. His interests centered on some of the highly controversial questions, which received so much prominence at the time, concerning the influence of environment on heredity. A Mendelian from his earliest days, Shull was interested not only in working out specific problems of heredity—was food a factor in determining sex, or does heat induce crossing-over in *Drosophila?*— but also in the relationships between these phenomena and the process of organic evolution. Recognizing earlier than many biologists that one side of the evolutionary coin was the problem of the origin of hereditary variation (the other side was the effect of selection on these variations), he concentrated much of his research on determining the relationship between genetic and environmental factors in producing the phenotypes on which selection acted.

In his studies and writings on evolution, Shull was incisive and penetrating in his analyses of such subsidiary problems as the nature and origin of mimicry. A fervent advocate of drawing conclusions only from the data at hand, he severely criticized the concept of mimicry advanced by G. D. Hale Carpenter. To Shull, phenotypic properties such as mimicry (in which one species comes to resemble another as a means of protection against predators) could not always be judged as being of survival value simply because, to man, the two species appeared to resemble each other. He pointed out that two forms resembling each other in human vision might appear quite distinct to some predator (such as a bird). Shull cited the work of W. L. McAtee, who, in 1912, had shown from analysis of the contents of birds' stomachs that mimics were eaten, like nonmimics, roughly in proportion to their availability. Thus, whatever resemblance the mimics may or may not have borne to the model appeared to be of little value in helping the prey escape its predators. Shull's opposition to the theory of warning coloration and mimicry was based largely on what he considered the tendency to misuse evidence, to interject anthropomorphism and subjective speculation into an area where the collection of solid facts would be more beneficial. A strong empiricist and experimentalist, Shull was one of a group of younger biologists who rejected the nonrigorous and speculative tradition in biology represented by the older morphologists and neo-Darwinians. Although a strong proponent of Darwin's theory of natural selection, Shull felt that among evolutionists there was a preponderance of purely descriptive, nonrigorous, and nonanalytical thinking that did not take into consideration the recent findings in such experimental areas as physiology, development, and genetics.

Shull carried his concern for new methods in biology into his teaching and educational writings. A strong advocate of the "principles" approach to biology, he organized his courses and teaching monographs in terms of generally applicable biological concepts; cell structure and function, transport, unity and control, embryonic development, genetics, systematics, and geographic distribution. To the modern reader this may appear to be a customary and routine type of organization; it was in fact quite novel in Shull's day. During the first three decades of the twentieth century, most biology courses were organized along phylogenetic and "type specimen" lines. Students were given a detailed, descriptive, and anatomically oriented survey of animal or plant types (poriferans, annelids, crustaceans, echinoderms, vertebrates, tracheophytes, angiosperms, and so on). To Shull, this method of organization not only was uninteresting for most students but also offered a limited view of biology. Why not, he asked, study the characteristics of cells by using examples of many cell types from many different species of animals and plants? Biological principles are what the student is going to remember most; the anatomical detail will be important if—and only if—the student decides to pursue some area of biology in depth. Besides, he argued, the old "type specimen" course did not even do what it claimed to do: give the student a picture of evolutionary development. Shull pointed out, for example, that most such courses involved detailed consideration of a modern echinoderm (starfish, sand dollar, sea urchin); yet echinoderms as a group represent an evolutionary offshoot that was by no means to be found in the mainstream of animal phylogeny. In *Principles* of Animal Biology (1934), Evolution (1936), and Heredity (1938), Shull gave this approach a concrete form that he hoped would aid others in developing a "principles" approach to biology at the undergraduate and graduate levels.

During his academic career Shull received numerous honors and was a member of many professional societies. He was a fellow of the American Association for the Advancement of Science and was appointed Russel lecturer for 1951. He was a member of the American Society of Naturalists, serving as secretary (1920–1926), vice-president (1929), and president (1934). He also belonged to the American Society of Zoologists, the Genetics Society of America, the Society for the Study of Evolution, the Eugenics Society (London), and the National Association of Biology Teachers; was a fellow of the Entomological Society of America; and served as president of the Michigan Academy of Sciences (1921–1922). In 1911 Shull married Margaret Jeffrey Buckley; they had four children.

BIBLIOGRAPHY

I. ORIGINAL WORKS. Shull wrote a large number and variety of scholarly papers on sex determination, heredity, evolution, and education. A few of the most notable are "Color Sport Among Locustidae," in Science, 26 (1907), 218–219; "Nutrition and Sex Determination in Rotifers," ibid., 38 (1913), 786–788; "Biological Principles in the Zoology Course," ibid., 48 (1918), 648–649; "Crossovers in Male Drosophila melanogaster Induced by Heat," ibid., 80 (1934), 103–104, written with Maurice Whittingill; "Weismann and Haeckel: One Hundred Years," ibid., 81 (1935), 443–452; and "Needs of the Mimicry Theory," ibid., 85 (1937), 496–498.

II. SECONDARY LITERATURE. There is no detailed biographical sketch of Shull's life or work. He is listed in American Men of Science, 9th ed. (Lancaster, Pa., 1955), 1030. Letters to and from Shull can be found at the American Philosophical Society, in the Jennings, Blakeslee, Demerec, Dunn, and Davenport papers.

GARLAND ALLEN

SIDGWICK, NEVIL VINCENT (b. Oxford, England, 8 May 1873; d. Oxford, 15 March 1952), chemistry.

Sidgwick was born into a family of unusual distinction. His father, William Carr Sidgwick, was a fellow of Merton College, Oxford. His uncles included Henry Sidgwick, professor of moral philosophy at Cambridge; Arthur Sidgwick, reader in

Greek at Oxford; and (by marriage) Edward White Benson, archbishop of Canterbury. From them he may well have inherited or absorbed his power of mental organization and his command of language. Until he was twelve years old Sidgwick was educated at home, mainly by his mother, Sarah Isabella Thompson. Her uncle was General Thomas Perronet Thompson, F.R.S., and it was to her that he owed his introduction to botany and natural history and his general love of science.

Sidgwick went to Rugby School when he was thirteen and, unusually for that time, studied both the classics and science. In 1892 he returned to Oxford as a scholar of Christ Church. His tutor was Vernon Harcourt, a pioneer in reaction kinetics. Sidgwick earned a first-class degree in chemistry in 1895. Then, reputedly because of a disparaging remark by a relative about science, he studied literae humaniores and gained a brilliant first in 1897, largely by his performance in philosophy. Next he went to Germany, where he studied physical chemistry under Georg Bredig in Ostwald's laboratory at Leipzig and then organic chemistry with Hans von Pechmann at Tübingen.

In 1900 Sidgwick was elected a fellow of Lincoln College, Oxford, and there spent the rest of his working life. Until 1920 he had published only eighteen papers. Most of these were concerned with the kinetics of organic reactions, the others mainly with the relation of solubility and chemical structure. These papers described good, careful work; but none was of great importance. He would have been unknown outside Oxford had he not written Organic Chemistry of Nitrogen (1910). This was his first essay in applying the ideas and quantitative methods of physical chemistry to the facts and systematics of descriptive chemistry, a task that gradually became his major interest. The book was a great success, not only because of his shrewd selection of topics and his clarity of thought but also because of the intellectual excitement conveyed by his style.

From 1920 Sidgwick's rate of publication increased rapidly, probably because the introduction of a year of research as a part of the chemistry course provided him with more research pupils. His major accomplishment was establishing that there can be a definite bond between a group containing a fairly acidic hydrogen atom (for example, hydroxyl group) and an oxygen-rich group (for example, nitro group). Such a bond via a hydrogen atom had already been postulated by various people; but the clear, systematic attack by Sidgwick and his pupils played a major part in gaining general acceptance of the idea. He was elected a fellow of the Royal Society in 1922.

In 1914, while traveling to Australia for a meeting of the British Association, Sidgwick met Ernest Rutherford and immediately came under his spell. This friendship was of crucial importance, for it inspired Sidgwick to try to explain chemical behavior in terms of atomic structure, as G. N. Lewis and Langmuir were also doing. The first major fruit of this new interest was the publication in 1927 of *Electronic Theory of Valency*, which was intended as an exposition of principles to be followed by a second volume applying them systematically. One of the most novel and important parts of this book was that concerned with the coordination compounds or complexes so extensively studied by A. Werner. Sidgwick showed that, by using the concept of the dative bond (wherein both bonding electrons are provided initially by one of the two atoms involved, instead of one by each), it is possible to rationalize these compounds more successfully than had been done previously. Lewis had already put forward the concept of this bond, but it was Sidgwick's systematic application of the idea that made chemists realize its value and wide importance. Sidgwick also emphasized that both ionic and covalent bonds exist, that generally they are sharply distinguished, and that a given bond might exist in either form, for example, when an acid ionizes. The book presented a brilliant discussion of a wide range of topics on a simple basis, and it had a profound effect.

Even as the book was written, however, the theoretical basis that Sidgwick had used, namely, the quantum mechanics of Bohr, Sommerfeld, and W. Wilson, was being discarded by physicists in favor of a much more general mechanics formulated in matrix form by W. Heisenberg and as a wave equation by Schrödinger. In 1927 W. Heitler and F. London produced an explanation of the covalent bond that was far more fundamental than anything previously advanced. A year later F. Hund produced an alternative treatment. Both theories were quickly taken up and developed. From the former, J. C. Slater and L. Pauling derived, in the early 1930's, a basis for stereochemistry; and they also introduced the concept of "resonance" in molecules; electrons holding a set of atoms together may not be localized as pairs between pairs of atoms but may be more generally distributed or "delocalized," so that each electron can be considered to play a part in holding several atoms together. From Hund's explanation R. S. Mulliken and E. Hückel developed an alternative treatment for many-atom molecules, the "molecular orbital" method.

Sidgwick went to the United States for the first time in 1931, as George Fisher Baker nonresident lecturer in chemistry at Cornell University, and in his travels met Pauling. They immediately became fast friends. Thereafter one of Sidgwick's main preoccupations was to expound the concept of resonance to British chemists. He did so in various articles and in his presidential addresses to the Chemical Society (1936, 1937). He had a flair for extracting the essence of a mathematical argument and expressing it verbally.

Sidgwick had become embarrassed by the success of his books. There had long been a demand for a new edition of his book on nitrogen compounds. Twenty-five years after publication secondhand copies were selling for four times the original price. But realizing that he could not prepare a new edition, he had his colleagues T. W. J. Taylor and W. Baker take over the task and prepare a reworked edition in 1937. Another edition was produced in 1966 by I. T. Millar and H. D. Springall. To meet the requirements of his Cornell appointment Sidgwick published *Some Physical Properties of the Covalent Link in Chemistry* (1933), which dealt mainly with the new experimental methods for investigating structure, such as heats of formation, lengths, and electric dipole moments of bonds. Because of its relatively limited scope, it did not have the importance of the book on valence theory.

Sidgwick then began writing volume II of his *Electronic Theory of Valency*. Curiously, the hardships created by World War II may have helped Sidgwick to complete this task, the size of which became clearer and more daunting as he proceeded. The limitations on travel meant fewer outside demands on his time; and still more important was the diminution in the world output of primary publications. He consulted 10,000 papers, mostly on his own but also with help from H. M. Powell and R. V. G. Ewens. Eventually, in 1950 when he was seventy-seven, the work appeared as *Chemical Elements and Their Compounds*, two large volumes of about 750,000 words. While it did not have the impact of volume I, it was still a landmark in the development of chemistry. Once again he had brought unification, this time to an even vaster body of fact. His clear critical mind and easy style illuminate every word. Although the book has become dated it still is a prime source of earlier references. The emphasis by chemists on the molecular orbital method began just as his writing finished. In the 1950's the ligand field treatment of complexes was developed at Oxford by L. E. Orgel, a pupil of a pupil. Accurate determinations of geometric structures of molecules began to pour forth after 1945. Completely unforeseen developments occurred, notably in the discovery of new organometallic compounds and of compounds of the once inert gases.

Today it is impossible for a large book about descriptive chemistry to remain in date for more than a few years after it is started; and no satisfactory technique for dealing with this situation has yet been devised. What is certain is that no one man can ever again attempt such a task with any hope of being effective; thus Sidgwick may be regarded as the last of a line. He showed shrewdness, or had luck, in applying his talents when they were particularly useful in the development of chemistry.

Sidgwick did not live long enough to enjoy his final success. His health had been slowly failing for more than ten years, and he had completed his writing only because of his indomitable will. But he was able to achieve again his dearest ambition, to revisit the United States, to meet his old friends, and to see the beauty of New England in the fall. On the return voyage he collapsed aboard ship. He was brought back to Oxford, where he died in his sleep.

Sidgwick's influence on those who knew him was as much due to his personality as to his writings. He never married, although he enjoyed the company of women—provided that they were intelligent. He lived in college—very simply save when he entertained. He was a generous and genial host, delighting in lively discussions and pungent repartee.

BIBLIOGRAPHY

I. ORIGINAL WORKS. A complete bibliography is given by Tizard (see below). Sidgwick's books are *Organic Chemistry of Nitrogen* (Oxford, 1910); *Electronic Theory of Valency* (Oxford, 1927); *Some Physical Properties of the Covalent Link in Chemistry* (Ithaca, N. Y., 1933); and *Chemical Elements and Their Compounds*, 2 vols. (Oxford, 1950).

II. SECONDARY LITERATURE. The most helpful source is Sir Henry Tizard's article in *Obituary Notices of Fellows of the Royal Society of London*, **9** (1954), 237–258, with bibliography. See also L. E. Sutton's obituary in *Proceedings of the Chemical Society* (1958), 310–319.

L. E. SUTTON

SIEBOLD, CARL THEODOR ERNST VON (*b.* Würzburg, Germany, 16 February 1804; *d.* Munich, Germany, 7 April 1885), *medicine, zoology.*

Siebold was the third child of Elias von Siebold, professor of medicine and midwifery at Würzburg, and Sophie von Schäffer. His happy childhood was punctuated by fondly remembered vacations in Regensburg at the home of his grandfather Jakob Christian Gottlieb von Schäffer, whose extensive natural history collection first stimulated Siebold's interest in the subject. Siebold accompanied his older brother Eduard (who later became professor of medicine and midwifery at Marburg and then at Göttingen) and their friend Ignaz Döllinger on entomological and botanical excursions in the woods near Würzburg. In a short autobiography Siebold wrote of his friendship with Döllinger:

> This relationship gave me the opportunity one day to enter the study of Döllinger's father, the famous founder of embryology. And it was there that I glimpsed a saucer filled with black wax placed on a desk near the window; a flea was affixed to the saucer with needles so that the arrangement of its intestines could easily be examined. This anatomical preparation made a deep and lasting impression on me.

Siebold began his schooling in Würzburg and continued it at the Gymnasium zum Grauen Kloster in 1816, when his father assumed a post at the University of Berlin. During these years Siebold collected and identified butterflies and dug up newts, snails, and mussels. He passed the final secondary school examination in the fall of 1823 and, acceding to his father's wishes, began to study medicine, although even at this early date he would have preferred to devote himself exclusively to zoology. He spent the first two semesters of his medical studies at Berlin, attending the lectures of Karl Martin Lichtenstein, Link, and Rudolphi, among others. In the fall of 1824 he went to Göttingen, where his interest in natural history was encouraged by Blumenbach and Johann Hausmann. Three years later he returned to Berlin, and on 28 April 1828 received the M.D. The sudden death of his father on 12 July 1828 obliged Siebold to find some means of support. Accordingly, he prepared to practice medicine, passing the two official qualifying examinations in 1829 and 1830. In the spring of 1831 he was named district physician in Heilsberg (now Lidzmark). On 10 April 1831, shortly before assuming his duties, he married the twenty-six-year-old Fanny Nöldechen.

In Königsberg, en route to Heilsberg, Siebold met Karl Ernst von Baer, who offered him guidance and assistance in his scientific studies during the next three years by sending him technical literature and information. Siebold's wish to be near a university again was fulfilled by his transfer in 1834 to the post of municipal physician in Königsberg. He still hoped for the opportunity to qualify as a university lecturer, and within a few months it seemed that he would have his chance. Both Baer, who had received a post in St. Petersburg, and Lichtenstein supported Siebold's candidacy at Albertus University, but their efforts were futile — a ruling dating to 1544 prohibited Catholics from teaching there.

Following this disappointment Siebold accepted a post in the same year (1834) in Danzig as municipal physician and director of a school of midwifery. His investigations during this period on the phenomena of generation in jellyfish, intestinal worms, and insects can be viewed as preliminary studies for Steenstrup's fundamental work on the alternation of generations (1842). Siebold also published extensively on invertebrates of Prussia.

Siebold's many publications soon attracted the attention of zoologists. No less a figure than Alexander von Humboldt—who was a guest at Siebold's house in Danzig on 12 and 13 September 1840—intervened successfully with King Ludwig I of Bavaria in favor of Siebold's appointment to the chair of zoology and comparative anatomy at the Friedrich Alexander University in Erlangen, left vacant by the departure of Rudolph Wagner. On 1 March 1841 Siebold returned with his family to his native Franconia. In addition to his regular lectures, Siebold also taught veterinary medicine, physiology, and histology, requiring his students to use the microscope. At the beginning of 1845 Siebold accepted an offer from the University of Freiburg, and on 28 October 1845 he was formally welcomed by the university senate.

Siebold developed a close friendship at Freiburg im Breisgau with the botanist Alexander Braun, who accompanied him in 1847 to the congress of Swiss scientists at Schaffhausen. On this occasion Siebold realized a long-cherished plan: he and Braun agreed to join with Naegeli and Koelliker in editing a new journal of botany and zoology. Various circumstances, including Braun's election to the office of vice-chancellor of the University of Berlin and Naegeli's departure for Freiburg, prevented the journal from appearing in the form originally envisioned. Instead, Siebold and Koelliker created the *Zeitschrift für wissenschaftliche Zoologie*, first published in 1848. In the same year Siebold completed his *Lehrbuch der vergleichenden*

Anatomie der wirbellosen Thiere, begun in 1845 and one of the most important systematic reforms since the work of Cuvier. In it Siebold divided the Radiata into groups and characterized the Protozoa as single-celled organisms.

The political disturbances of 1848 led Siebold, in the spring of 1850, to Breslau, where he succeeded Purkyně as professor of physiology. At Breslau, Siebold continued his research on the development of the Cestoda and discovered the parthenogenesis of the honeybee. Wagner considered the discovery of parthenogenesis to be "one of the most disconcerting obstacles impeding the formulation of so-called general laws of animal life processes." Siebold remained at Breslau for only two years. Disappointed with the university, he was delighted when the Bavarian ministry of education in June 1852 began negotiations concerning his assuming the professorship of physiology and comparative anatomy at Munich. The discussions lasted until the late fall, but on 26 November Siebold informed the dean of the Breslau medical faculty that he had accepted the post.

Siebold became a member of the medical faculty of the Ludwig Maximilian University in Munich on 13 April 1853; and on 18 January 1854 he was accepted as a member of the Bavarian Academy of Sciences, of which he had been a corresponding member since 1848. According to the terms of Siebold's contract, he was required to establish an institute of physiology in Munich. Thus he lectured on both zoology and physiology only until he was able to relinquish the chair of physiology, along with the post of curator of the anatomical institute, to Theodor Bischoff. Siebold's lectures encompassed zoology, comparative anatomy, the reproductive biology of man and animals, and parasitology. He also was curator of the Bavarian state collections of comparative anatomy, physiology, and zoology.

While in Munich, Siebold completed *Über die Band- und Blasenwürmer nebst einer Einleitung über die Enstehung der Eingeweidewürmer* and actively participated in the research of his student Bilharz. Siebold's most important topic of research, however, was parthenogenesis; and in 1856 he published *Wahre Parthenogenesis bei Schmetterlingen und Bienen*. In addition, he obtained a royal contract (dated 3 May 1854) to produce a monograph on the fishes of Central Europe, a task that required extended travel and an intensive study of the specialized literature. The result of nine years' work, *Die Süsswasserfische von Mitteleuropa*, illustrated with sixty-four woodcuts and two colorplates, finally appeared in 1863.

Siebold's wife died on 26 December 1854, one of the last victims of the cholera epidemic. A year later, in Göttingen, Siebold married her younger sister, Antoynie Nöldechen. Siebold was acquainted with artists and scholars, and she made their home a center of stimulating social gatherings. He also belonged to the Munich poets' circle, whose members were regularly guests of the king.

At a congress of German scientists and physicians in Königsberg in 1860, Siebold met the young Ernst Haeckel, with whom he felt closely united in discussions of the Darwinian theory of evolution. Their common enthusiasm provided the basis of a lifelong correspondence and friendship. In a letter to Haeckel on his fortieth birthday, the seventy-year-old Siebold wrote:

> Oh, how I wish I could see this reform carried through! For I must tell you that brilliant though it is, it is not easy for a zoologist trained in the old school. Instead of being able to relax in my later years, I have to learn just as much—no, even more—than I did during all my younger days. If you reflect that in old age it is much harder to learn than to forget, you will bear with me.

At the end of the winter semester of 1882–1883, during which he gave a two-hour lecture course, Siebold submitted his request for retirement. In a birthday letter to Haeckel dated 18 February 1883, he confessed: "I have been much upset recently by the fact that in my lectures, which I was accustomed to give without the aid of notes, I cannot always remember the scientific names of animals with their genus and species designations—this causes me the greatest embarrassment." Ludwig II granted him permission to retire on 11 March 1883, two years before his death.

BIBLIOGRAPHY

I. ORIGINAL WORKS. Siebold's works include *Observationes quaedam de Salamandris et Tritonibus* (Berlin, 1828), his M.D. thesis; "Über die Spermatozoen der Crustaceen, Insecten, Gastropoden und einiger anderer wirbelloser Thiere," in *Archiv für Anatomie, Physiologie und wissenschaftliche Medicin*, **3** (1836), 13–53; "Fernere Beobachtungen über die Spermatozoen der wirbellosen Thiere," *ibid.*, 232–255; "Fernere Beobachtungen über die Spermatozoen der wirbellosen Thiere. 3. Die Spermatozoen der Bivalven. 4. Die Spermatozoen in den befruchteten Insecten-Weibchen," *ibid.*, **4** (1837), 381–439; "Zur Entwicklungsgeschichte der Helminthen," in K. F. Burdach, ed., *Die Physiologie als Erfah-*

rungswissenschaft, II (Leipzig, 1837), 183–213; "Beiträge zur Naturgeschichte der wirbellosen Thiere. Über Medusa, Cyclops, Loligo, Gregarina und Xenos," in *Neueste Schriften der Naturforschenden Gesellschaft in Danzig*, **3** (1839), 1–94; *Observationes quaedam entomologicae de oxybelo uniglume atque miltogramma conica* (Erlangen, 1841); and *Viro summe reverendo collegae . . .* (Erlangen, 1844).

Later writings are "Parasiten," in R. Wagner, ed., *Handwörterbuch der Physiologie mit Rücksicht auf physiologische Pathologie*, II (Brunswick, 1844), 641–692; *Lehrbuch der vergleichenden Anatomie der wirbellosen Thiere* (Berlin, 1848); "Über den Generationswechsel der Cestoden nebst einer Revision der Gattung Tetrarhynchus," in *Zeitschrift für wissenschaftliche Zoologie*, **2** (1850), 198–253; *Über die Band- und Blasenwürmer nebst einer Einleitung über die Entstehung der Eingeweidewürmer* (Leipzig, 1854); *Wahre Parthenogenesis bei Schmetterlingen und Bienen. Ein Beitrag zur Fortpflanzungsgeschichte der Thiere* (Leipzig, 1856); *Über Parthenogenesis* (Munich, 1862); *Die Süsswasserfische von Mitteleuropa* (Leipzig, 1863); and *Beiträge zur Parthenogenesis der Arthropoden* (Leipzig, 1871).

A short autobiography, used by A. Koelliker for his biographical sketch, is still extant in MS in a private collection in Freiburg im Breisgau.

II. SECONDARY LITERATURE. On Siebold and his work, see E. Ehlers, "Carl Theodor Ernst von Siebold. Eine biographische Skizze," in *Zeitschrift für wissenschaftliche Zoologie*, **42** (1885), i–xxiii; R. Hertwig, *Gedächtnisrede auf Carl Theodor von Siebold gehalten in der öffentlichen Sitzung der k. Bayerischen Akademie der Wissenschaften . . .* (Munich, 1886); A. Koelliker, "Carl Theodor von Siebold, eine biographische Skizze," in *Zeitschrift für wissenschaftliche Zoologie*, supp. **30** (1878), v–xxix; H. Körner, "Die Würzburger Siebold. Eine Gelehrtenfamilie des 18. und 19. Jahrhunderts," in *Deutsches Familienarchiv*, nos. 34–35 (1967), 451–1080; G. Olpp, *Hervorragende Tropenärzte in Wort und Bild* (Munich, 1932), 378–379; and F. Winckel, "Carl Theodor Ernst von Siebold," in *Allgemeine deutsche Biographie*, XXXIV (1892), 186–188.

ARMIN GEUS

SIEDENTOPF, HENRY FRIEDRICH WILHELM (*b.* Bremen, Germany, 22 September 1872; *d.* Jena, Germany, 8 May 1940), *physics*.

Siedentopf was a distinguished practitioner of scientific microscopy and a pioneer in ultramicroscopy and microphotography. After graduating from the Gymnasium in Bremen, he studied at Leipzig and Göttingen and became assistant to the mineralogist and crystallographer Theodor Liebisch. In 1896 Siedentopf received the doctorate at Göttingen under Woldemar Voigt for the dissertation

"Über die Capillaritätsconstanten geschmolzener Metalle." He became assistant to Franz Richarz at Greifswald in 1898 but the following year accepted an offer from Ernst Abbe to join the Zeiss optical works in Jena. Siedentopf first worked in Abbe's laboratory, and from 1907 to 1938 he was director of the company's microscopy division. In 1918 he was named both titular and ordinary professor at the University of Jena, where he lectured on scientific microscopy. With Hermann Ambronn and August Köhler he gave courses at Jena every summer.

Siedentopf's most important achievement was the development of the "slit ultramicroscope," which he perfected in 1902–1903 in collaboration with Richard Zsigmondy. They constructed the instrument in order to make visible the gold particles in ruby glass, which Zsigmondy had unsuccessfully attempted to do alone. The device was based on the principle that under intense illumination with an electric arc lamp, the ultramicroscopic particles can be made to act as origins of small diffraction cones, which are visible in the objective. In the instrument the smallest particles were no longer illuminated from below, as in the ordinary microscope, but from the side; and the light that they refracted appeared in the ultramicroscope's field of view. Zsigmondy later developed this instrument into the immersion microscope, and Siedentopf created the cardioid ultramicroscope. From 1907 Siedentopf devoted his attention to microcinematography and in 1911 devised a stationary apparatus with a time-lapse camera and a slow-motion camera. Between 1919 and 1923 he constructed skin and capillary microscopes, and in 1922 he developed the photomicroscope with "Phoku" eyepiece and attached miniature camera. The latter device was a milestone in the development of microphotography.

BIBLIOGRAPHY

I. ORIGINAL WORKS. Siedentopf wrote about 50 scientific papers and obtained a number of patents for optical and microscopical devices. See Poggendorff, IV, 1394; V, 1162; VI, 2442; VIIa, 4, 405. His works include "Über die Sichtbarmachung und Grössenbestimmung ultramikroskopischer Teilchen mit besonderer Anwendung von Goldrubingläsern," in *Annalen der Physik*, 4th ser., **10** (1903), 1–39, written with R. Zsigmondy; "Bisphärische Spiegelkondensatoren für Ultramikroskopie," *ibid.*, **39** (1912), 1177–1186; *Übungen zur wissenschaftlichen Mikroskopie* (Leipzig, 1913); "Über den Nachweis der Form von Ultramikronen," in *Kolloid-*

Zeitschrift, supp. to **36** (1925), "Zsigmondy-Festschrift," 1–14; "Über die optische Abbildung von Nicht-Selbstleuchtern," in *Zeitschrift für Physik*, **50** (1928), 297–309; and "Mikroskopische Beobachtungen an Strichgittern mit periodischen Teilungsfehlern," *ibid.*, **107** (1937), 251–257; **108** (1938), 279–287; **109** (1938), 260–272; and (with similar title) **112** (1939), 704–726.

II. SECONDARY LITERATURE. Obituary notices are in *Deutsche optische Wochenschrift*, **61** (1940), 110; and *Kolloid-Zeitschrift*, **91** (1940), B218; a biography is being prepared by F. Stier for *Neue deutsche Biographie*. See also the following, listed chronologically: F. Hauser, "Die Entwicklung mikroskopischer Apparate bei der Firma Zeiss in dem ersten Jahrhundert ihres Bestehens," in *Jenaer Jahrbuch* (1952), 1–64, see 47–48, 55; R. Jobst, "120 Jahre wissenschaftlicher Gerätebau in Jena," in *Jenaer Rundschau* (1966), no. 4, supp., "Carl Zeiss 150. Geburtstag," 25; and H. Gause, "Das Spaltultramikroskop nach Siedentopf und Zsigmondy–eine historische und optische Betrachtung," *ibid.*, no. 6, 327–333.

HANS-GÜNTHER KÖRBER

SIEDLECKI, MICHAŁ (*b.* Cracow, Poland, 1873; *d.* Sachsenhausen, Germany, January 1940), *zoology, cytology.*

Siedlecki was born into a middle-class family. He studied at Jagiellonian Cracow University at Cracow and received the M.D. in 1895. The following year he went to Berlin to work with Schulze, whose assistant at that time was Schaudinn. From 1897 to 1899 Siedlecki worked at the Pasteur Institute under Metchnikoff and also at the Zoological Station in Naples under Anton Dohrn.

Siedlecki's first paper dealt with leukocytes of Urodela (1895). He later studied the phagocytes of Annelida (1903) and, with Caullery, Echinodermata (1903). With Kostanecki, he published an interesting paper on the cytology of *Ascaris*. In 1908–1909, Siedlecki went to Java, where he became interested in frogs of the genus *Rhacophorus* and in fishes and birds.

Siedlecki's outstanding scientific contributions, however, were in protozoology, which he first studied at Schulze's laboratory. When Siedlecki went there he intended to work under Schaudinn on Foraminifera. But Schaudinn was also interested in the life histories of Sporozoa and suggested to Siedlecki that they work together in this area, studying the Coccidia of *Lithobius forficatus*. In a joint, preliminary paper (1897), they described the life histories of *Coccidium schneideri* and *Adelea ovata*. This paper, a classic in protozoology, was the first to describe correctly the life cycle of Coccidia. Siedlecki's trip to Naples and Schaudinn's military obligations prevented them from publishing the full results of their researches.

Siedlecki was the first (1898) to describe the sexual cycle of *Klossia octospina* (or *Aggregata eberthi*) in *Sepia*, the only known host until 1908, when Léger and Duboscq found an asexual cycle in *Portunus*. The following year he published the first complete life cycle of a gregarine *(Adelea ovata)* living in *Lithobius forficatus*.

In 1900 Siedlecki was appointed lecturer of zoology at Cracow and in 1912 professor and also director of the Zoological Laboratory and Museum. From 1919 to 1921 he was rector at the University of Vilna, helping to plan its reconstruction after World War I. Later he returned to Cracow. He also served as the Polish representative to the Permanent Council for Marine Exploration and to the International Committee for Bird Preservation.

After the Nazi invasion of Poland, Siedlecki and many of his colleagues were arrested on charges of promoting Polish nationalism. He was jailed first at Cracow but was later transferred to a prison in Breslau, and finally to a concentration camp at Sachsenhausen-Oranienburg, where reportedly he died of heart failure and "ill treatment."

BIBLIOGRAPHY

I. ORIGINAL WORKS. Siedlecki's works include "Beiträge zur Kenntnis der Coccidien," in *Verhandlungen der Deutschen zoologischen Gesellschaft*, **7** (1897), 192–203, written with F. Schaudinn; "Étude cytologique et cycle évolutif de la coccidie de la seiche," in *Annales de l'Institut Pasteur*, **12** (1898), 799–836; and "Étude cytologique et cycle évolutif de *Adelea ovata* Schneider," *ibid.*, **13** (1898), 169–192.

II. SECONDARY LITERATURE. On Siedlecki and his work, see "Michael Siedlecki," in *Nature*, **145** (22 June 1940), 963; "Michał Siedlecki," in *Zoologica poloniae*, **4** (1948), 51; and Clifford Dobell, "Michał Siedlecki (1873–1940)," in *Parasitology*, **33** (1941), 1–7.

ENRIQUE BELTRÁN

SIEMENS, CHARLES WILLIAM (CARL WILHELM) (*b.* Lenthe, near Hannover, Germany, 4 April 1823; *d.* London, England, 19 November 1883), *engineering.*

The seventh son of Christian Ferdinand Siemens, a prosperous farmer, and Eleonore Deichmann, Siemens was naturalized as a British subject on 19 March 1859. He married Anne Gordon on

23 July of that year. Siemens was a member of the Institution of Civil Engineers, the Institution of Mechanical Engineers (president, 1872), the Iron and Steel Institute (president, 1877), the Society of Telegraph Engineers (first president, 1872), the British Association (president, 1882), and the Royal Society of Arts (chairman, 1882) and a fellow of the Royal Society.

Having received a sound German technical education, Siemens went at age twenty to England and profitably promoted an electroplating invention of his older brother Werner (who in 1847 founded the German company of Siemens and Halske). After several years of indifferent success with other inventions, including regenerative steam engines, an engine governor, and a printing technique, he became agent in Britain for his brother's telegraph equipment and later a partner in his subsidiary British company. During the same period (1850–1858) Siemens developed a highly successful meter for measuring water consumption. These activities, combined with his important invention (1861) of the regenerative gas furnace and its application to open-hearth steelmaking and other industrial processes, made him independently wealthy before 1870.

In 1874 Siemens designed the cable ship *Faraday* and assisted in the laying of the first of several transatlantic cables that it completed. During the last fifteen years of his life he actively supported the development of the engineering profession and its societies and stimulated public interest in the conservation of fuel, the reduction of air pollution, and the potential value of electric power in a wide variety of engineering applications.

BIBLIOGRAPHY

Siemens' writings are collected in *The Scientific Works of C. William Siemens, Kt.*, E. F. Bamber, ed., 3 vols. (London, 1889).

A biography is William Pole, *The Life of Sir William Siemens* (London, 1888). See also H. T. Wood in *Dictionary of National Biography*, XVIII (Oxford, 1922). Many incidental references are in Georg Siemens, *History of the House of Siemens*, 2 vols. (Freiburg, 1957), and in a collection of his brother's memoirs, *Werner von Siemens, Inventor and Entrepreneur* (Clifton, N.J., 1966).

ROBERT A. CHIPMAN

SIEMENS, ERNST WERNER VON (*b.* Lenthe, near Hannover, Germany, 13 December 1816;

d. Berlin-Charlottenburg, Germany, 6 December 1892), *electrical science, technology.*

Siemens' father, Christian Ferdinand, was a farmer and estate manager descended from a middle-class family long prominent in the affairs of Goslar. His mother, Eleonore Deichmann, bore fourteen children and, of the ten surviving, he was the oldest. In 1832 he entered a Gymnasium in Lübeck, where he gave early indication of an abiding interest in science. Although economic difficulties at home thwarted his plan to study at the *Bauakademie* in Berlin, Siemens won an appointment as an officer candidate at the Prussian artillery and engineering school in Berlin. From 1835 to 1838 he studied mathematics, physics, and chemistry under instructors who also lectured at the university.

Stationed as an officer at a provincial garrison, Siemens used his free time to apply science to practical inventions. After the death of his mother and of his father months later in 1840, he was spurred on by the financial need of his brothers and sisters. His first successful invention was an improved process for gold- and silverplating. Rights to the process were sold in England in 1843 by his brother Wilhelm (later Sir William) to Elkington of Birmingham. Transferred to the staff of the Berlin artillery works, he soon joined the circle of Gustav Magnus, professor of physics at the University of Berlin. The group, which included du Bois-Reymond, Clausius, and Helmholtz, heard Siemens lecture on his indicator telegraph in 1845.

After improving upon the indicator telegraph of Charles Wheatstone, Siemens developed an entire telegraph system, including a method of providing the wire with a seamless insulation of gutta-percha. In 1847, together with Johann Georg Halske, the university's scientific instrument maker, he founded the Telegraphenbauanstalt von Siemens & Halske to manufacture and construct telegraph systems.

The firm obtained government contracts to build a telegraph network in northern Germany, including the line that in 1849 carried the dramatic news from the revolutionary Frankfurt Parliament to the Prussian king, Frederick William IV, in Berlin, that he had been elected German emperor (a dubious honor he declined). Although disagreements cut off Prussian government contracts after 1850, Siemens, having left the army, visited Russia and planned an extensive telegraph network, including a line from St. Petersburg to the Crimea, used during the Crimean War. The Russian business was so extensive that Siemens' brother Carl was made resident Russian representative, and so profitable

that Siemens could conduct research that resulted not only in telegraph improvements but also in advances in underwater cable telegraphy.

Siemens became scientific consultant to the British government on underwater telegraphy; and Siemens Brothers in London, headed by William, manufactured and laid cable. For that company Siemens helped design the first special cable-laying ship, the *Faraday*, which, after 1875, laid five Atlantic cables in ten years. An even more dramatic achievement was Siemens' organization and construction of the Indo-European telegraph from London via Berlin, Odessa, and Teheran to Calcutta, completed in 1870.

Siemens' outstanding contribution to scientific technology was his discovery of the dynamo principle, announced to the Berlin Academy of Sciences in January 1867. Having already introduced the double-T armature, he found it possible to connect the armature, the electromagnetic field, and the external load of an electrical generator in a single circuit, thereby avoiding the costly permanent magnets previously used in the field. Other inventors and scientists—Sóren Hjorth, Anyos Jedlik, Alfred Varley, Charles Wheatstone, and Moses Farmer—discovered the dynamo principle at about the same time; but Siemens foresaw the consequences of his "dynamo" for heavy-current, or power, uses and developed practical applications. His company pioneered in using electricity for streetcars and mine locomotives, in electrolysis, and in central generating stations. In 1889 Siemens retired from active management of the family firm, which, including the daughter firms in London, St. Petersburg, and Vienna, employed about five thousand workers.

Unlike many major inventor-engineers of the nineteenth century, Siemens valued science highly, steadfastly advocating that technology not only should be based upon scientific theory but also should be analyzed to derive theory. His own efforts provided an excellent example, for he often published his analyses of telegraph and cable technology in *Dinglers polytechnisches Journal,* Poggendorff's *Annalen der Physik,* and in the reports of the Berlin Academy of Sciences. Siemens helped to establish scientific standards of measurement, designing among other things a universal galvanometer. In a period of sharp international competition, he advised the Prussian government that a nation would never gain and maintain international status if it did not excel in research and base its technology and science upon it. His determination and financial assistance resulted in the establishment of the Physikalische-Technische Reichsanstalt in Berlin (1887), a government-supported research institution first headed by Helmholtz.

Siemens received an honorary doctorate from the University of Berlin (1860), was a member of the Berlin Academy of Sciences (1873), and was ennobled in 1888. He died a few days after publication of the first edition of his *Lebenserinnerungen*, a memoir still in print.

BIBLIOGRAPHY

I. ORIGINAL WORKS. Siemens' autobiography was *Lebenserinnerungen* (Berlin, 1892; 17th ed. Munich, 1966), also available in English as . . . *Recollections* (London, 1893; 2nd ed., London–Munich, 1966). His papers were collected as *Wissenschaftliche und technische Arbeiten*, 2 vols. (Berlin, 1889–1891); a selection of his letters and a 190-page biography are in *Werner Siemens: Ein kurzgefasstes Lebensbild nebst einer Auswahl seiner Briefe*, Conrad Matschoss, ed., 2 vols. (Berlin, 1916). Six thousand letters of the Siemens brothers from 1842 to 1892 are in the Werner von Siemens Institut, Munich.

II. SECONDARY LITERATURE. A concise, well-informed biography by the head of the Siemens archives in Munich is Sigfrid von Weiher, *Werner von Siemens: Ein Leben für Wissenschaft, Technik und Wirtschaft* (Göttingen, 1970); the same author also contributed to the series of pamphlets published by the Deutsches Museum: *Werner von Siemens, ein Wegbereiter der deutschen Industrie* (Munich, 1966). Also useful are Karl Burhenne, *Werner Siemens als Sozialpolitiker* (Munich, 1932); Richard Ehrenberg, *Die Unternehmungen der Brüder Siemens* (Jena, 1906); Friedrich Heintzenberg, *Werner von Siemens in Briefen an seine Familie und an Freunde* (Stuttgart, 1953); and Conrad Wandrey, *Werner Siemens—Geschichte seines Lebens und Wirkens* (Munich, 1942).

THOMAS PARKE HUGHES

SIERPIŃSKI, WACŁAW (*b.* Warsaw, Poland, 14 March 1882; *d.* Warsaw, 21 October 1969), *mathematics.*

Sierpiński was the son of Constantine Sierpiński, a prominent physician, and Louise Łapińska. He entered the University of Warsaw in 1900 and studied under G. Voronoi, an outstanding expert on number theory who influenced his scientific career for the next decade or more. Sierpiński's important contributions to number theory (for instance, in the theory of equipartitions) were continued and developed in G. H. Hardy, Edmund Landau, and H. Weyl. In 1903 the university awarded

Sierpiński a gold medal for mathematics; his abilities in this area were evident from childhood. He received his degree the following year.

Sierpiński's most important work, however, was in set theory, and in 1908 he was the first to teach a systematic course on that subject. He investigated set theory and related domains (point-set topology, theory of functions of a real variable) for fifty years; he devoted the last fifteen to number theory. He also served as editor in chief of *Acta arithmetica*.

Sierpiński published some six hundred papers on set theory and a hundred on number theory. The most important of his books and monographs on set theory are *Hypothèse du continu* (1934) and *Cardinal and Ordinal Numbers* (1958). His chief work on number theory was *Elementary Theory of Numbers* (1964). His papers contained new and important theorems (some of which bear his name), geometrical constructions (Sierpiński curves), concepts, and original and improved proofs of earlier theorems. His findings stimulated further research by his students and by mathematicians throughout the world.

Sierpiński was a foreign member of twelve academies of science (among them the French, the Lincei, and Pontifical), and he received honorary doctorates from ten universities (including Paris, Moscow, and Amsterdam). He was also elected vice-president of the Polish Academy of Sciences and was awarded the scientific prize of the first degree (1949) and the Grand Cross of the Order of Polonia Restituta (1958).

Sierpiński's career spanned more than sixty years; he lectured at the University of Lvov until 1914 and then, after World War I, at the University of Warsaw. He was considered an excellent and stimulating teacher. About 1920 Sierpiński, Janiszewski, and Mazurkiewicz created a Polish school of mathematics centered on foundations, set theory, and applications, and also founded in 1919 a periodical to specialize in these areas, *Fundamenta mathematicae*. The first editor in chief was Janiszewski, and after his death in 1920 Sierpiński and Mazurkiewicz carried on the work for decades.

BIBLIOGRAPHY

I. ORIGINAL WORKS. Sierpiński's most important works are *Hypothèse du continu* (Warsaw, 1934); *Cardinal and Ordinal Numbers* (Warsaw, 1958); and *Elementary Theory of Numbers* (Warsaw, 1964).

II. SECONDARY LITERATURE. Works on Sierpiński

and his work are M. Fryde, "Wacław Sierpiński-Mathematician," in *Scripta mathematica*, **27** (1964), 105–111; S. Hartman, "Les travaux de W. Sierpiński sur l'analyse," in *Oeuvres choisies*, I (1974), 217–221; S. Hartman, K. Kuratowski, E. Marczewski, A. Mostowski, "Travaux de W. Sierpiński sur la théorie des ensembles et ses applications," *ibid.*, II (1975), 9–36; K. Kuratowski, "Wacław Sierpiński (1882–1969)," in *Acta arithmetica*, **21** (1972), 1–5; A. Schinzel, "Wacław Sierpiński's Papers on the Theory of Numbers," *ibid.*, 7–13.

KAZIMIERZ KURATOWSKI

SIGAUD DE LAFOND, JOSEPH-AIGNAN (*b.* Bourges, France, 5 January 1730; *d.* Bourges, 26 January 1810), *experimental physics, chemistry, medicine.*

Sigaud de Lafond was the son of a clockmaker who was an artisan and man of letters. He began his education by studying for the priesthood with the Jesuits in Bourges, but later he decided to become a physician instead. He went to Paris and enrolled as a medical student at the school of Saint-Côme.

While preparing for his medical degree, Sigaud attended the famous course of public lectures given by the Abbé Nollet, who aroused in him such a lively interest in experimental science that Sigaud became first a tutor in philosophy and mathematics and then a demonstrator in experimental science at the Collège Louis-le-Grand. In 1760 he succeeded the Abbé Nollet in his chair at Louis-le-Grand, where he taught courses in anatomy, physiology, and also those courses in experimental physics that had been taught by his famous predecessor.

In 1770 Sigaud became a professor of surgery at the school of Saint-Côme. In 1782 he returned to Bourges, where, after four years, he obtained a chair in physics at the local *collège*. The Revolution closed the *collèges*, making his position temporarily difficult; but with the reorganization of public instruction under the National Convention and Directory, he became in 1795 professor of physics and chemistry at Bourges at the École Centrale, which replaced the old *collège*. With the creation of the *lycées*, Fourcroy, a former student of Sigaud's and a member of the Council of State, appointed him *proviseur* (headmaster) of the school at Bourges. He resigned this position in 1808, two years before his death.

On 28 February 1796 Sigaud was elected a nonresident associate of the section of experimental physics of the National Institute. He belonged also to the academies of Montpellier, Florence, and St. Petersburg. In 1795 the Convention included him on a list of savants who were to receive a subsidy of 3,000 francs in gratitude for their services.

Sigaud was a prolific writer in the fields of experimental physics, chemistry, medicine, and (apparently as a consequence of his early Jesuit training) theology. Experimental science was a fashionable pursuit among the leisured classes in eighteenth-century France, and Sigaud was one of several illustrious popularizers who satisfied the intellectual appetites and curiosities of an ever-increasing number of amateurs of science. Popular interest tended toward the more spectacular examples of natural phenomena; and lectures accompanied by demonstrations, especially on electricity and on the newly discovered gases, always attracted large and enthusiastic crowds. As a follower of the Abbé Nollet, Sigaud was apparently quite successful in appealing to this group of virtuosi, and most of his publications were written for the enlightened layman rather than the professional researcher. As a result, his work was generally not profound, creative, or original. He avoided theoretical explanations and instead emphasized phenomenological aspects. There is something, too, in his writing of the vulgar catering to the "goût des merveilles"—the popular fascination with the strange, the unusual, the bizarre. He devoted an entire two-volume work to the "marvels of nature," which went through at least two French editions and was translated into German. His positive contributions to science were in the area of experimental technique. He is sometimes attributed with the invention of the glass insulator and the circular glass plate (to replace the glass globe) in electrical machines. (A. Wolf [see Bibliography] attributes the latter invention to Ingenhousz and Ramsden, and perhaps also to Planta.)

In the 1770's Sigaud collaborated with Macquer in investigating the aeriform fluids or "airs," newly discovered by Priestley. In 1776 they burned a quantity of the so-called "inflammable air" (hydrogen), and by holding a porcelain saucer over the flame they managed to collect a few drops of a colorless liquid that both researchers agreed was water. The experiment is often cited as an anticipation of some of the work later done by Cavendish, Lavoisier, and Monge on the synthesis of water, but neither Macquer nor Sigaud de Lafond fully recognized the significance of their observation.

In medicine, Sigaud achieved a certain notoriety for proposing, in a communication to the Royal Academy of Surgery, that section of the pubic symphysis could, in certain cases, be substituted for cesarean section. The Academy rejected the idea, but Sigaud was resolved to try it anyway. Specializing in midwifery, he established his medical practice in Paris, and finally in October 1777, he found an opportunity to put his new operation into practice. A pregnant woman, about forty years of age and deformed from rickets, came to him for help. She had already lost four babies, and the consensus of medical opinion was that she had no chance of bearing live children without a cesarean section. Sigaud, assisted by Alphonse Le Roy, performed instead a section of the pubic symphysis, and mother and child both survived the operation. Before the faculty of medicine in Paris, Sigaud read a memoir describing his procedure. The faculty ordered that the memoir be published in Latin and French and had a silver medal struck in honor of Sigaud and his assistant.

BIBLIOGRAPHY

I. Original Works. Sigaud's works include *Leçons de physique expérimentale*, 2 vols. (Paris, 1767); *Traité de l'électricité . . .* (Paris, 1771); *Description et usage d'un cabinet de physique expérimentale*, 2 vols. (Paris, 1775); *Élémens de physique théorique et expérimentale . . .*, 4 vols. (Paris, 1777); *Essai sur différentes espèces d'air, qu'on désigne sous le nom d'air fixe . . .* (Paris, 1779); *Dictionnaire des merveilles de la nature*, 2 vols. (Paris, 1781); *Dictionnaire de physique*, 5 vols. (1781–1782); *Précis historique et expérimental des phénomènes électriques . . .* (Paris, 1781); *Physique générale*, 5 vols. (Paris, 1788–1792); *Examen de quelques principes erronés en électricité* (Paris, 1796); and *De l'électricité médicale* (Paris, 1803).

II. Secondary Literature. On Sigaud and his work, see *Biographie universelle*, XLII (Paris, 1825), 316–318; H. Boyer, in *Nouvelle biographie générale*, XLIII (Paris, 1864), 966–967; Mechin-Desquins, *Notice historique sur Sigaud de Lafond* (Bourges, 1841); and A. Wolf, *History of Science, Technology, and Philosophy in the Eighteenth Century*, I (New York, 1961), 220. For details concerning Sigaud's course of public demonstrations, see Jean Torlais, "La physique expérimentale," in R. Taton, ed., *Enseignement et diffusion des sciences en France au XVIIIè siècle* (Paris, 1964), 619–645.

J. B. Gough

SIGER OF BRABANT (*b.* Brabant, *ca.* 1240; *d.* Orvieto, Italy, 1281/1284), *philosophy.*

Nothing is known of Siger's birthplace, his family, or his early education. He arrived in Paris probably between 1255 and 1260, was admitted to the Picard *nation* of the University of Paris, and became master of arts between 1260 and 1265. His name is first cited in a document dated 27 August 1266, in which he appears as a boisterous and pugnacious young teacher at the Faculty of Arts. He received a special rebuke in Thomas Aquinas' *De unitate intellectus* (1270); and on 10 December 1270 the bishop of Paris, Étienne Tempier, condemned thirteen heterodox propositions taken from the writings of Siger and his partisans. After 1270 Siger tempered his doctrinal positions, but remained the leader of the dissident minority party in the Faculty of Arts. Later he was summoned by the inquisitor of France but fled in late 1276 with two other teachers and took refuge at the papal court, the tribunal of which was reputedly more lenient than that of the inquisitors. (On 7 March 1277 Tempier, with Siger's teaching particularly in mind, condemned 219 propositions.) At the papal court, Siger was placed under house surveillance in the company of a cleric. Sometime during the pontificate of Martin IV (1281–1285), the cleric, in a fit of madness, stabbed Siger to death. John Peckham attests his death in a letter dated 10 November 1284.

Dante esteemed Siger and in his *Divine Comedy* consigned him to Paradise (canto X) beside Thomas Aquinas, in the crown of twelve sages, where he represented autonomous philosophy. A firm believer in the separation of the two powers in Christianity, Dante viewed Siger as the victim of attacks by conservative theologians. Although Siger's career ended prematurely, his historical role was nevertheless fundamental because of the reactions he provoked in university circles—and on such men as Bonaventure, Thomas Aquinas, Albert the Great, and John Peckham.

On the basis of his writings discovered to date, it is known that Siger was concerned primarily with metaphysics and psychology, and secondarily with logic and natural philosophy; several questions on ethics also have been discovered. His contributions to science can be found in his writings on psychology and natural philosophy, in which areas connections can be made between philosophy and science.

Siger's writings on psychology are devoted to problems of the intellective soul, and these are treated in an exclusively philosophical manner: the works contain only those few elements of descriptive psychology that are indispensable for formulating the philosophical problems.

Siger wrote several works on natural philosophy. Of the three *Quaestiones naturales* found in Paris, the first is purely philosophical and deals with the uniqueness of the substantial form; the second defends the Aristotelian principle "Everything that moves is moved by something else"; and the third discusses the problem of gravity, which is resolved in the spirit of Ibn Rushd.

Six other *Quaestiones naturales* have been found in Lisbon. The third and sixth are purely philosophical, and two others are patterned on Aristotle's pseudophilosophical hypotheses on the "natural place" of simple bodies (the first question) and on the influence of the heavens (*orbis*) in human generation (the fourth question), which are of no interest for experimental science. Only the second and fifth questions have some scientific taste. The second interprets an experiment in physics: "If a lighted candle is placed in a vessel put on water, why does the water then rise in the vessel?" Siger's answer is inspired by Aristotle. The candle warms the air, which then rises to the top of the vessel. Because a vacuum is impossible and water is fluid, the water rises (remaining in contact with the air), compresses the air, and thus increases its own ascending motion. But the vessel would shatter if placed mouth down on the earth, since the latter cannot rise because of its cohesion and weight. The fifth question is a brief, abstract discussion of the paradox of Achilles and the tortoise.

In *Impossibilia* the problem of gravity is again examined in chapter 4. After a long discussion, in which he rejects the opinions of Albertus Magnus and of Thomas Aquinas, Siger again adopts (but modifies) Ibn Rushd's thesis. The entire discussion is developed according to Aristotelian physics (with all its prejudices).

De aeternitate mundi treats the eternity of mankind in a purely philosophical context. *Compendium de generatione et corruptione*, a fragment of which has been discovered in a manuscript in Lilienfeld, Austria, is a brief, unoriginal analysis of Aristotle's treatise. Almost all of the twenty-four *Quaestiones super libros I et II Physicorum*, discovered at the Vatican, deal with purely philosophical problems. The only questions of possible interest from a scientific viewpoint are II, 1 and 2, on the natural movement of light and heavy bodies (which is explained in the same way as in the other writings), and II, 5, in which Siger explains that *musica*, *perspectiva*, and *astrologia* are intermediary between the purely natural and mathematical sciences.

Siger's contribution to experimental science seems insignificant. Even those problems that could have been treated scientifically were given a philosophical explanation and were solved without originality by relying on the principles of Aristotle and Ibn Rushd. If the commentaries of Munich MS 9559, which have been attributed to him by Martin Grabmann, were truly by Siger, this assessment would be different, for then Siger would be the author of an important series of commentaries on natural philosophy in which many scientific questions are discussed. Formerly the author accepted Grabmann's attribution; but serious difficulties have since been raised concerning the authenticity of several commentaries, and it seems preferable not to take them into account here.

BIBLIOGRAPHY

I. ORIGINAL WORKS. All the works of Siger quoted in the article are in B. Bazán, ed., *Siger de Brabant. Quaestiones in tertium de anima. De anima intellectiva. De aeternitate mundi* (Louvain, 1972), and B. Bazán, ed., *Siger de Brabant. Écrits de logique, de morale et de physique* (Louvain, 1974).

II. SECONDARY LITERATURE. See P. Mandonnet, *Siger de Brabant et l'averroïsme latin au XIIIe siècle*, 2nd ed., 2 vols. (Louvain, 1908–1911); F. Van Steenberghen, *Siger de Brabant d'après ses oeuvres inédites*, 2 vols. (Louvain, 1931–1942); *La philosophie au XIIIe siècle* (Louvain, 1966), 357–402; and *Introduction à l'étude de la philosophie médiévale* (Louvain, 1974), *passim* (see *Table onomastique*, p. 603).

F. VAN STEENBERGHEN

SIGORGNE, PIERRE (*b*. Rembercourt-aux-Pots, France, 24 October 1719; *d*. Mâcon, France, 10 November 1809), *physics, science popularization.*

The son of Pierre Sigorgne, a minor judicial official, and Marguerite du Moulin, Sigorgne received his theological degrees at the Sorbonne and assumed the chair of philosophy at the Collège Duplessis (Paris) in 1740. He quickly established himself as a gifted educator and popularizer of science, and was prominent in introducing Newtonian theories into the French university curriculum. His promising Paris career ended, however, when he was arrested in 1749 as the alleged author of satirical verses concerning Louis XV and Madame de Pompadour. He spent the remainder of his life in exile at Mâcon, where he continued his scientific work while proving himself a distinguished ecclesiastical administrator. Sigorgne maintained an active correspondence with many of the important scientists and *philosophes* of the period, winning a reputation as a genial and enlightened reconciler of science and theology. He was named *correspondant* of the Académie des Sciences in 1778 and *correspondant* of the Institut de France in 1803, and was one of the founders of the Mâcon Academy.

Sigorgne's *Institutions léibnitiennes* (1767), an accurate but critical account of Leibniz's cosmological theories, contributed to the more informed discussion of German philosophy in France; but his main importance for the history of science lies in his vigorous and effective popularization of Newtonian ideas. Although the introduction of Newton's theories into France was well advanced by 1740, Cartesian ideas still exerted a powerful influence. Sigorgne's courses of lectures at the Collège Duplessis provided a detailed and sophisticated treatment of recent physical theories, notably the Newtonian concept of universal gravitation; and his courses in philosophy included systematic instruction in mathematics and contributed to the spread of ideas on the calculus.

His cautious advocacy of Newtonian science was broadened with the publication of Sigorgne's *Examen et refutation des leçons de physique expliquées par M. de Molières* (1741). Primarily an attack on Privat de Molières's influential attempt to reconcile Cartesian and Newtonian theories, the *Examen* demolished the vortex theory as emended by Privat de Molières to obviate the major objections to Cartesian physics. Sigorgne forcefully demonstrated the Newtonian arguments for the physical instability of the hypothetical vortices and the mathematical incompatibility between vortex motion and Kepler's laws. His *Institutions newtoniennes* (1747), a clear introduction to Newtonian mathematical and physical principles, contributed to the acceptance of the attraction theory by the French scientific community. A Latin résumé of the *Institutions newtoniennes* (1748) was rapidly recognized as a standard Newtonian textbook in Western Europe.

The most successful of Sigorgne's efforts to apply the concept of universal gravitation is his explanation of capillary phenomena by the laws of attraction, which was awarded a prize by the Rouen Academy in 1748. His chemical theories, in contrast, were of little significance. Sigorgne shared the misguided vision of those eighteenth-century Newtonians who sought to explain observed chemical behavior on the basis of interparticulate forces, the operation of which would be subject to exact mathematical treatment. Thus, ironically, those ideas that had been skillfully exploited by Sigorgne in defense of the new physics at mid-century reappeared at the end of his long career in a series of ill-tempered attacks on modern chemistry.

A minor but respected *savant* of Enlightenment France, Sigorgne used his gifts of exposition to bring developing scientific ideas before a broad public.

BIBLIOGRAPHY

I. ORIGINAL WORKS. Sigorgne's first important writings, his polemics against Privat de Molières and his Cartesian supporters, include *Examen et refutation des leçons de physique expliquées par M. de Molières au Collège royal de France* (Paris, 1741); *Réplique à M. de Molières ou démonstration physico-mathématique de l'insuffisance et de l'impossibilité des tourbillons* (Paris, 1741); and "A Physico-Mathematical Demonstration of the Impossibility and Insufficiency of Vortices," in *Philosophical Transactions of the Royal Society*, **41,** no. 457 (1740), 409–435.

His major work is *Institutions newtoniennes, ou introduction à la philosophie de Newton* (Paris, 1747; 2nd ed., 1769). A Latin résumé of *Institutions newtoniennes, Astronomiae physicae juxta Newtoni principia breviarium, methodo scolastica ad usum studiosae juventutis* (Paris, 1748), was widely used in France and Germany as a standard Newtonian textbook and was translated into Italian by Giulio Carbonara as *Istituzioni neutoniane* (Lucca, 1757).

Sigorgne published a summary of his prize essay on the effects of attraction in capillary tube phenomena as an appendix to the 2nd ed. of *Institutions newtoniennes*; the essay had been submitted to the Rouen Academy in 1748 as "Dissertatio physico-mecanica de ascensu et suspensione liquorum intra tubos capillares." Sigorgne's critical exposition of Leibnizian cosmology is *Institutions léibnitiennes, ou précis de la monadologie* (Lyons, 1767). An example of Sigorgne's enduring fascination with Newtonian explanations is the merely curious *Examen nouveau de la chimie moderne, avec une dissertation sur la force* (Mâcon, 1807).

II. SECONDARY LITERATURE. The fullest recent account of Sigorgne's life and work is Martial Griveaud, "Un physicien oublié du XVIIIe siècle: L'Abbé Pierre Sigorgne de Rembercourt-aux-Pots," in *Annales de l'est*, 4th ser., 3 (1935), 77–107. J.-M. Guerrier, "Étude critique sur les oeuvres de l'Abbé Sigorgne," in *Annales de l'Académie de Mâcon*, 3rd ser., **14** (1909), 432–458, is concerned mainly with an analysis of Sigorgne's literary output but has some brief comments on his scientific writings. A short account of Sigorgne's life and work appears in F. Hoefer, ed., *Nouvelle biographie générale*, XLIII (1864), 988–989. Useful comments on the significance of Sigorgne's university courses appear in René Taton, *Enseignement et diffusion des sciences en France au XVIIIe siècle* (Paris, 1964), 142, 627.

MARTIN FICHMAN

SIGÜENZA Y GÓNGORA

SIGÜENZA Y GÓNGORA, CARLOS DE (*b.* Mexico City, 20 August 1645; *d.* Mexico City, 22 August 1700), *mathematics, astronomy, natural history.*

Sigüenza's father was tutor to Prince Baltazar before going to New Spain. After receiving his first education at home, Sigüenza entered the Jesuit Colegio de Tepozotlán and took his first vows in 1662. He continued his studies at the Colegio del Espíritu Santo at Puebla until 1667, when he was expelled for disciplinary reasons; he remained a secular priest. During the following years Sigüenza was a student at the University of Mexico and chaplain at Amor de Dios Hospital. In 1672 he was awarded the chair of astrology and mathematics at the university and occupied it for more than twenty years.

In 1680, to calm the fears aroused by a comet, Sigüenza wrote *Manifiesto filosófico contra los cometas* (1681), which drew a reply from Martín de la Torre the same year. To answer it Sigüenza wrote *El Belerofonte matemático* (now lost), which aroused the antagonism of Father Eusebio Kino, a Jesuit missionary who was a renowned mathematician and astronomer, leading him to publish a strong response to Sigüenza's arguments: *Exposición astronómica del cometa* (1681). Kino's book gave Sigüenza the opportunity to publish in 1690 *Libra astronómica y philosóphica*, a short book of great significance for its sound mathematical background, anti-Aristotelian outlook, and familiarity with modern authors: Copernicus, Galileo, Descartes, Kepler, and Tycho Brahe.

As royal cosmographer, Sigüenza made valuable observations and drew good charts. These included a general map of New Spain, probably the first by a Mexican, best known through a reproduction by Beaumont in 1873–1874; a map of the lakes of the Valley of Mexico, probably made in 1691, but not published until 1748, and reprinted in 1768, 1783, and 1786; and a map of the bay of Santa María de Galve (Pensacola), 1693. In 1692 the viceroy's palace was set on fire during a riot and Sigüenza risked his life to save valuable papers in the archives.

Sigüenza projected writing a history of ancient Mexico and collected much material, but little was published. His manuscripts, now lost, were considered by contemporaries of great value. He assembled a large library, said to be the best in the realm. In 1693 Sigüenza was sent with Admiral Andrés de Pez to reconnoiter Pensacola Bay; he kept an interesting diary and made valuable charts.

BIBLIOGRAPHY

I. ORIGINAL WORKS. Besides those works cited in text, Sigüenza wrote *Piedad heróica de don Fernando Cortes* (Mexico City, 1689); *Trofeo de la justicia española* (Mexico City, 1691); *Mercurio volante* (Mexico City, 1693), and several unpublished MSS.

II. SECONDARY LITERATURE. See F. Pérez Salazar, *Obras de Carlos de Sigüenza y Góngora con una biografía* (Mexico City, 1928); J. Rojas Garcidueñas, *Don Carlos de Sigüenza y Góngora. Erudito barroco* (Mexico City, 1945); I. A. Leonard, *Don Carlos de Sigüenza y Góngora. A Mexican Savant of the Seventeenth Century* (Berkeley, 1929).

ENRIQUE BELTRÁN

AL-SIJZĪ

AL-SIJZĪ, ABŪ SAʿĪD AḤMAD IBN MUḤAMMAD IBN ʿABD AL-JALĪL (*b.* Sijistān, Persia, *ca.* 945; *d. ca.* 1020), *geometry, astronomy, astrology.*

Al-Sijzī is also known as al-Sijazī, al-Sijizī, or al-Sijarī. The following evidence indicates that he was an older contemporary of al-Bīrūnī (973–*ca.* 1050): he is not mentioned in Ibn al-Nadīm's *Fihrist* (987), but al-Bīrūnī quoted him in his *Chronology*. Al-Bīrūnī wrote to al-Sijzī on the determination of the *qibla* (direction of Mecca, for prayer) and on a proof by his teacher Manṣūr ibn ʿIrāq for the theory of the transversal figure. Conversely, al-Sijzī quoted three propositions by al-Bīrūnī in his treatise on trisecting an angle, which he ended with five problems of al-Bīrūnī. Around 969 al-Sijzī had written and copied mathematical works at Shīrāz, a later version of which is in Paris (Bib. Nat. arabe 2457). Presumably around the same time (*ca.* 967) he composed his *Kitāb al-qirānāt* ("Book of the Conjunctions"), which contains references to an even earlier work of his, *Muntakhab Kitāb al-ulūf* ("Summary of the Thousands of Abū Maʿshar"). In 969–970 al-Sijzī assisted at the observations of the meridian transits in Shīrāz conducted by ʿAbd al-Raḥmān al-Ṣūfī.

Al-Sijzī may have spent some time in Khurāsān, since he answered questions by mathematicians of that region. He dedicated works to the Sayyid Amīr Abū Jaʿfar Aḥmad ibn Muḥammad, a prince of Balkh (*d.* 1019) (L. Massignon, *Opera omnia*, I [Beirut, 1963], 650–666), and to the Buwayhid Caliph ʿAḍud al-Dawla (Shīrāz–Baghdad, 949–983).

Al-Sijzī's main scientific activity was in astrology, and he had a vast knowledge of the older literature. He usually compiled and tabulated, adding his own critical commentary. Al-Sijzī summarized three works by Abū Maʿshar and wrote on the second of the five books ascribed to Zoroaster in his *Kitāb Zarādusht ṣuwar darajāt al-falak* ("The Book of Zoroaster on the Pictures of the Degrees of the Zodiac"). In his *Kitāb al-qirānāt*, which treats general astrology and its history, he used Sassanid material and sources from the time of Hārūn al-Rashīd and from the late Umayyad period. In *Zāʾirjāt*, a book on horoscopes, he gave tables based on Hermes, Ptolemy, Dorotheus, and "the moderns." Al-Sijzī's tables, together with those of Ptolemy, are quoted by Iḥtiyāzuᶜ l'Dīn Muḥammad in his *Judicial Astrology* (Trinity College, Cambridge). Al-Bīrūnī described in his *Kitāb fī istīᶜāb* three degenerate astrolabes constructed by al-Sijzī: one fish-shaped, one anemone-shaped, and one skiff-shaped.

Al-Sijzī's mathematical papers are less numerous but more significant than his astrological ones, and he is therefore better known as a geometer. He wrote original treatises on spheres and conic sections, the construction of a conic compass, and the trisection of an angle by intersecting a circle with an equilateral hyperbola. This method became widely accepted: Abū'l-Jūd, for example, describes it in the Leiden manuscript Or 168(13). Al-Sijzī mentioned several other methods for solving this problem, including one by "mobile geometry," which he ascribed to the ancients; but he omitted any reference to Pappus. His treatise on proportions in the transversal figure is especially useful for astronomy, and his emphasis on the position of the lines was new and important. Al-Sijzī constructed the regular heptagon according to the same principle as that used by al-Qūhī. He also wrote articles on subdividing segments and several letters on problems related to the work of Euclid and Archimedes.

BIBLIOGRAPHY

I. ORIGINAL WORKS. Al-Sijzī's available mathematical MSS are listed in F. Sezgin, *Geschichte des arabischen Schrifttums*, V (Leiden, 1974), 331–334. On the astrological MSS see M. Krause, "Stambuler Handschriften islamischer Mathematiker," in *Quellen und Studien zur Geschichte der Mathematik, Astronomie und Physik*, B.3 (1934), 468–472; and W. Thomson and G. Junge, *The Commentary of Pappus on Book X of Euclid's Elements* (Cambridge, 1930), 48–51. C. Brockelmann, *Geschichte der arabischen Literatur*, I (Leiden, 1943), 246–247; and supp. I (Leiden, 1937), 388–389, lists a few more MSS and additional copies. Neither mentions *Kitāb al-qirānāt wa tahāwīl sinī al-ᶜālam*, a MS that David Pingree dealt with in *The Thousands of Abū Maᶜshar* (London, 1968). In this work Pingree also

discusses the *Muntakhab Kitāb al-ulūf*, which was partly translated by E. S. Kennedy in "The World-Year of the Persians," in *Journal of the American Oriental Society*, **83**, no. 3 (1963), 315–327. Translations or discussions of mathematical treatises are found in: F. Woepcke, *L'Algèbre d'Omar Alkhayāmī* (Paris, 1851), 117–127; and "Trois traités arabes sur le compas parfait," in *Notices et extraits de la Bibliothèque nationale*, **22**, part 1 (1874), 112–115; C. Schoy, "Graecoarabische Studien," in *Isis*, **8** (1926), 21–40; H. Bürger and K. Kohl, "Thabits Werk über den Transversalensatz," in *Abhandlungen zur Geschichte der Naturwissenschaften und der Medizin*, **7** (1924), 49–53; and L. A. Sédillot, "Notice de plusieurs opuscules mathématiques," in *Notices et extraits de la Bibliothèque nationale*, **13** (1838), 136–145. Edited by the Osmania Oriental Publications Bureau is *Risāla fī 'l-shakl al-gaṭṭāᶜ* ("On the Transversal-Theorem"; Hyderabad, 1948).

II. SECONDARY LITERATURE. There are few biographical references to al-Sijzī. On the observations in Shīrāz see al-Bīrūnī, *Taḥdīd nihāyāt al-amākin li-taṣḥiḥ masāfāt al-masākin* (Cairo, 1962), 99; and E. S. Kennedy, *A Commentary Upon Bīrūnī's Kitāb Taḥdīd al-Amākin* (Beirut, 1973), 42. On al-Sijzī as an astrologer see Pingree (see above), 21–26, 55, 63–67, 70–127. On his mathematics consult Sezgin (see above), 46–47, 329–334; Thomson and Junge (see above), 43–51; and the notes of G. Bergsträsser in "Pappos Kommentar zum Xten Buch von Euklid's Elementen," in *Islam*, **21** (1938), 195–198. On his astrolabes see Josef Frank, "Zur Geschichte des Astrolabs," in *Sitzungsberichte der Physikalisch-medizinischen Sozietät in Erlangen*, **50–51** (1918–1919), 290–293; and al-Bīrūnī, *Al-Qānūn al Masᶜūdī*, I (Hyderabad, 1954), introduction, 17–18.

YVONNE DOLD-SAMPLONIUS

SILLIMAN, BENJAMIN (*b.* North Stratford [now Trumbull], Connecticut, 8 August 1779; *d.* New Haven, Connecticut, 24 November 1864), *chemistry, mineralogy, geology.*

Graduated from Yale College in 1796, Silliman was diverted from following his father and grandfather in the law when he was offered the newly established (1802) professorship of chemistry and natural history at Yale. Untrained in these subjects, Silliman went to Philadelphia to study, profiting greatly not only from formal course work in the medical school there but also from occasional visits with John Maclean at Princeton and from informal chemical experiments with his classmate and fellow boarder Robert Hare. In the spring of 1805 Silliman sailed for Britain to continue his scientific education and to purchase books and apparatus for Yale College. After visiting Liverpool, Manchester, London, Holland, and the mining districts of Cornwall, he settled in Edinburgh, where he spent the winter studying chemistry, geology, and medicine.

In the years following his return to the United States, Silliman established himself as a leading figure in American science less through his original research than through his teaching and educational statesmanship at Yale, his editorship of the *American Journal of Science*, his public lectures on chemistry and geology, his textbooks, and his role in founding and strengthening scientific organizations.

As an original investigator, Silliman made his chief contributions during the early part of his career, the best-known being his description and chemical analysis of the Weston meteor of 14 December 1807 and his experiments with the oxyhydrogen blowpipe and the deflagrator, both invented by his friend Robert Hare. Silliman's analysis of fragments of the Weston meteor was widely reprinted in Europe and won him election to the American Philosophical Society. His experiments on the fusion of refractory substances also attracted considerable attention abroad. Using his own improved version of Hare's blowpipe, Silliman added substantially to the list of substances proved capable of fusion by heat, including zircon, lime, magnesia, chalcedony, beryl, and corundum (1813). In his experiments with the deflagrator on the fusion of carbon (1822), Silliman noted the transfer of volatilized carbon from the positive electrode to the negative, an observation subsequently confirmed by César Despretz. Silliman's geological papers were mostly descriptive essays on New England localities, but George P. Merrill credits him with anticipating the aqueo-igneous theory of eruptive rocks in his views on rock crystallization.

By 1820 Silliman had made Yale College the leading center in the United States for training in chemistry, geology, and mineralogy. Through his friendship and collaboration with George Gibbs of Newport, Rhode Island, he secured for Yale the splendid collection of minerals that Gibbs had purchased in Europe during his travels there. Arranged according to Haüy's system, these specimens served as an invaluable teaching aid for Silliman's lectures. Meanwhile, Silliman published four American editions of William Henry's *Epitome of Chemistry* with notes and additions. In 1830–1831 he brought out his own *Elements of Chemistry*, a solid, up-to-date work that compared favorably with European textbooks of that day. Silliman also prepared three American editions of Robert Bakewell's *Introduction to Geology*, accompanied

by an appendix in which he outlined his own geological lectures and endeavored to demonstrate the harmony of geology and Genesis.

An excellent teacher, Silliman trained a generation of American chemists, geologists, and mineralogists, including Denison Olmsted, Amos Eaton, Edward Hitchcock, Chester Dewey, Oliver P. Hubbard, George T. Bowen, Charles U. Shepard, James Dwight Dana, and Benjamin Silliman, Jr. He also took the lead in establishing graduate and professional training in the sciences at Yale. He helped to found a medical school at Yale (1813) and served as professor of chemistry and pharmacy in that institution for nearly forty years. In 1846 Silliman joined with his son to establish a professorship of agricultural chemistry and plant and animal physiology at Yale, for the express purpose of providing graduate training in chemistry and its applications to agriculture. The eventual result was the Department of Philosophy and the Arts, from which grew the Graduate School of Yale University and the Sheffield Scientific School.

In 1818 Silliman launched the *American Journal of Science*, which quickly became the leading American scientific journal and gave Silliman an international reputation. By 1830 the *Journal* was self-sustaining and was drawing important contributions from all fields of American science, including applied science. Deeply interested in the applications of science, Silliman kept his readers posted on the progress of the industrial arts. As a consulting chemist and geologist, he inspected mining properties in New England, Pennsylvania, Maryland, and Virginia, and published excerpts of his reports in the *Journal*. In 1838 Silliman's son joined him in editing the journal, and nine years later his son-in-law James Dwight Dana began assisting in the work.

Meanwhile, Silliman had begun to carry the cause of science to the American public as a lecturer on chemistry and geology. Beginning in 1834, at Hartford, Connecticut, Silliman extended his lecturing activities to Boston and thence to New York, Baltimore, Washington, Pittsburgh, St. Louis, Mobile, and New Orleans. These lectures did much to generate interest in science throughout the country, as Charles Lyell noted during his American travels. They also served to allay religious opposition to science, since Silliman went out of his way to harmonize Genesis and geology.

As his reputation increased, Silliman became a member of a great many scientific societies both at home and abroad. Elected early to the Connecticut Academy of Arts and Sciences, the American Philosophical Society, and the American Academy of Arts and Sciences, he joined with George Gibbs in founding the short-lived American Geological Society in 1819. More successful was the Association of American Geologists, formed in 1840 by Edward Hitchcock and several geologists from the New York and Pennsylvania surveys. Silliman served as president of this organization in 1841–1842 and remained active in it as it evolved into the American Association for the Advancement of Science.

On his second trip abroad in 1851 Silliman, now well-known in the world of science, was warmly received by his European colleagues. He was a charter member of the National Academy of Sciences, established in 1863. His scientific work is commemorated in the name of the mineral sillimanite.

BIBLIOGRAPHY

I. ORIGINAL WORKS. Extensive Silliman MSS, including much correspondence, his student diary for 1795–1796, account books, daybook for 1840–1864, and a 9-vol. MS, "Origin and Progress in Chemistry, Mineralogy, and Geology in Yale College and in Other Places, With Personal Reminiscences," are at the Yale University Library. Numerous other letters may be found at the Historical Society of Pennsylvania, the Library of the American Philosophical Society, the Library Company of Philadelphia, and the New-York Historical Society. Silliman's scientific papers are listed in the Royal Society *Catalogue of Scientific Papers*, V, 694–697. Other writings are listed in F. B. Dexter, *Biographical Sketches of the Graduates of Yale College, With Annals of the College History*, V (New York, 1911), 220–227. Silliman's European travels are narrated in his *A Journal of Travels in England, Holland and Scotland . . . in the Years 1805 and 1806*, 2 vols. (New York, 1810) and *A Visit to Europe in 1851*, 2 vols. (New York, 1853).

II. SECONDARY LITERATURE. For a nineteenth-century view of Silliman's life and work, see George P. Fisher, *Life of Benjamin Silliman, M.D., LL.D.*, 2 vols. (New York, 1866), which contains extensive quotations from the MS sources. A more recent and very well balanced account of Silliman's career is John F. Fulton and Elizabeth H. Thomson, *Benjamin Silliman 1779–1864, Pathfinder in American Science* (New York, 1947), containing a useful section, "Bibliography and Sources." See also Alexis Caswell's article in *Biographical Memoirs. National Academy of Sciences*, 1 (1877), 99–112; R. H. Chittenden, *History of the Sheffield Scientific School of Yale University 1846–1922*, I (New Haven, 1928), chs. 1–2; E. S. Dana, Charles Schuchert, *et al.*, *A Century of Science in America, With Special Reference to the American Journal of Science, 1818–1918* (New Haven,

1918), ch. 1; *Memorial of the Centennial of the Yale Medical School* (New Haven, 1915), ch. 1; George P. Merrill, *The First 100 Years of American Geology* (New Haven, 1924), 157; and Margaret Rossiter, "Benjamin Silliman and the Lowell Institute: The Popularization of Science in Nineteenth-Century America," in *New England Quarterly*, **44** (1971), 602–626.

JOHN C. GREENE

SILLIMAN, BENJAMIN, JR. (*b.* New Haven, Connecticut, 4 December 1816; *d.* New Haven, 14 January 1885), *chemistry, geology.*

Silliman was the fourth child and second son of Benjamin Silliman, professor of chemistry at Yale College, and Harriet Trumbull. After graduation from Yale College in 1837, he studied and did research in his father's laboratory, earning an M.A. in 1840. He also worked briefly in the private laboratory of the well-known Boston chemist Charles T. Jackson.

In 1838 Silliman began to assist his father on the internationally known *American Journal of Science and Arts*. His name appeared on the masthead in 1841, and he continued in various editorial capacities until his death. This work brought him into contact with many foreign scientists, a number of whom visited his father's house and later his own, and gave him a broad knowledge of the progress of scientific research.

In the late 1830's Silliman also began to assist his father on lecture tours and mining surveys. These trips widened his acquaintance with American scientists, gave him experience in the effective presentation of science, of which his father was a master, and offered training in practical geology.

On 14 May 1840 Silliman married Susan Huldah Forbes. They had seven children, of whom a son (Benjamin) and four daughters lived to maturity. A man of great personal charm, tremendous energy, and enthusiasm, Silliman was warmhearted and trusting to a fault. He dispensed hospitality generously, sent his daughters to Europe to study, and in general lived on a scale beyond his professional salary.

Silliman was one of the fifty original members of the National Academy of Sciences, incorporated by act of Congress in 1863. He was an associate fellow of the American Academy of Arts and Sciences and a member of many other societies in the United States and abroad. He received an honorary M.D. from the Medical College of South Carolina in 1849 and an LL.D. from Jefferson Medical College in 1884.

Silliman's professional career may be divided into four areas: contributions as an editor of the *American Journal of Science*, as a teacher and author of textbooks, as an analytical chemist doing experimental work, and as a consultant in chemistry and geology.

Silliman's long and effective career as a teacher started in 1842, when he began providing laboratory experience for his father's students interested in advanced training. Many were only slightly his junior; and together they turned out creditable work in the small laboratory, several studies appearing in the *American Journal of Science and Arts*. One of these students, John Pitkin Norton, who later studied in Europe, succeeded on his return, with the help of the Sillimans, in persuading the Corporation of Yale College to establish two professorships in 1846: agricultural chemistry (to which Norton was appointed) and applied chemistry (to 'which Silliman was appointed). The need to provide a degree for graduates of the School of Applied Chemistry (later the Yale Scientific School and finally the Sheffield Scientific School) led to the establishment in 1847 of the Department of Philosophy and the Arts (subsequently the Graduate School) and of the degree of Doctor of Philosophy, the first to be awarded in the United States (1863).

Since he and Norton received no salary from Yale, Silliman had to leave Norton to carry on alone in 1849 and accepted the professorship of medical chemistry and toxicology in the medical department at the University of Louisville, Kentucky. He returned to Yale in 1854 as professor of general and applied chemistry when his father retired.

In 1847 Silliman published the first of two textbooks that were clearly written, well arranged, and justifiably popular for many years in American colleges. Some fifty thousand copies of the *First Principles of Chemistry*, to which T. Sterry Hunt contributed the section on organic chemistry, were sold in the first twenty-five years. His *First Principles of Physics or Natural Philosophy* (1859) was, according to J. D. Dana, long the best-known textbook in physics in the country. Both volumes showed broad knowledge and Silliman's remarkable ability to synthesize, explain, and extract from the work of others with great clarity and effectiveness, which made his books especially useful to students.

As a laboratory investigator Silliman never had a well-defined program. His choice of research projects reflected his own wide-ranging interests and the practical problems addressed to him by men who were concerned with economic development of the country's resources but had no scientific knowledge. Certain interests persisted, however—mineralogy, petroleum, coal, precious metals, and combustion of gases for illumination (he was a director of the New Haven Gas Works for many years).

Silliman's selected bibliography, which includes thirteen books and pamphlets and nearly a hundred papers, does not list many of the often extensive reports written for private clients. One of these was probably his most important publication, for it launched the world's petroleum industry. In a report dated 16 April 1855, he set forth his methods and results in a chemical analysis of rock oil from Venango County, Pennsylvania, and recommended uses of the several products discovered. Silliman used fractional distillation to break down the components—a method utilized in Europe but little employed in America for that purpose. He identified kerosene, an inexpensive and safe illuminant; paraffin, better than tallow for candles; lubricants, to replace animal grease; and, by passing crude petroleum through heated coke, an illuminating gas of high quality. For a low-boiling fraction (gasoline) he could propose no use. That had to await the development of the internal combustion engine, but for the next half century the petroleum industry utilized the other components and the methods (including steam distillation) that Silliman suggested for preparing and purifying them. This report showed his potential as an original, imaginative investigator—a potential not realized in subsequent work.

Another important publication, not supplanted to this day, was *American Contributions to Chemistry*, a biographical dictionary of American chemists including bibliographies of their work prepared for the "centennial of chemistry," a celebration of the hundredth anniversary of the discovery of oxygen, and presented in part on 1 August 1874 at Northumberland, Pennsylvania, Priestley's place of residence in America. It involved a tremendous amount of work and, on the whole, was remarkably complete and accurate.

Silliman had of necessity been augmenting his meager professorial salary for some time, as did his father[1] and most other members of the academic community, by outside commissions of various sorts; and in 1864 he made a year-long trip to California to seek new opportunities. It was a time of great excitement about the resources of the country, and Silliman shared the general curiosity about undeveloped lands in the West. On this and several subsequent trips he examined many properties for clients eager to capitalize on his reputation as a geologist—both potential oil-yielding sites and gold and silver mines. Enthusiasm excited by the promise of great oil and mineral wealth, coupled with his natural optimism, resulted in generally favorable reports, useful to promoters in the formation of companies with authorized capital in the millions. Never quoted were more guarded statements or conditions that Silliman specified must be met if results were to justify promise. Although they were based on sometimes brief and insufficient study, the majority of his predictions ultimately were realized, partly because of his sound knowledge of geology and an intuitive ability to sense unseen potentials.

Silliman's enthusiastic lectures on California's rich resources, with special mention of oil in the southern part of the state, brought him into conflict with the head of the California Geological Survey, Josiah D. Whitney, and his former assistant, Silliman's friend and student William H. Brewer, professor of agriculture in the Scientific School (1864–1903); both were on record as saying that there was no oil in southern California. Fearing that Silliman's opinion would jeopardize the Survey, Whitney mounted a vicious attack, one of the most acrimonious and bitter in the annals of science, that continued intermittently, with Brewer's help, until Silliman's death. He accused Silliman of deliberately swindling the public for large fees and thus of degrading all scientists. Whitney was aided by the failure of the oil company formed on the basis of Silliman's report and by the fact that the oil sample on which Silliman had based part of his judgment proved to be "salted."[2] Damaging, too, to Silliman was the subsequent failure of silver and gold mining companies the formation of which had also been assisted by his enthusiastic reports.

Silliman's enemies were unable to have him ousted from the National Academy of Sciences, the American Academy of Arts and Sciences, or Yale; but they did hurt his scientific reputation and they forced him to resign (1870) from Yale College (but not from the medical faculty) and to sever connections with the Sheffield Scientific School. They also turned friends against him and brought disgrace and anguish to him and his family.

Silliman amassed an enormous file of evidence to support his opinions in response to Whitney's charges before the National Academy of Sciences, printed the "salted" oil report in the *American Journal of Science and Arts,* and promised an investigation; but thereafter he maintained a dignified silence and outwardly cheerful mien. He did not gloat, publicly or privately, when improved methods and machinery yielded rich oil strikes in the late 1870's in southern California, or when the Bodie mine (California) produced gold beyond even his great expectations and a report of a Congressional investigation of the Emma mine (Utah) contained no criticism of his judgment.[3]

Silliman's excellent reputation as an editor, teacher, and author of useful books remains undiminished. As a laboratory investigator he was careful and methodical but showed originality only on rare occasions, as in the investigation of rock oil in 1855 and in the use of certain techniques—such as the production of daguerreotype pictures by the light of the carbon arc and the use of an improved goniometer, based on a modification of a European model, in his examination of American micas. Had his energies not been diverted into so many channels, Silliman might have made more notable contributions to chemistry. His geological work showed excellent training, extensive knowledge, and sound judgment. It was when he ventured from academic surroundings into the commercial world that his optimism and guilelessness helped to create circumstances that led to his undoing. The ultimate vindication of his judgment on the important issues could not erase his personal tragedy, but it did restore for the record his reputation as a scientist.

NOTES

1. See, for example, Margaret W. Rossiter, "Benjamin Silliman and the Lowell Institute: The Popularization of Science in Nineteenth-Century America," in *New England Quarterly,* **44** (1971), 602–626.
2. It was later thought that Silliman never explained what he discovered in his investigation of the salted sample because John B. Church, the husband of his eldest sister, might have been implicated.
3. Silliman's testimony before the Committee on Foreign Affairs of the House of Representatives (the investigation ran from February to May 1876) gave him the opportunity to state under oath that his fee for two trips to the mine in Utah was $25,000, less than half the amount that his adversaries had claimed.

BIBLIOGRAPHY

I. ORIGINAL WORKS. Silliman's most important books include *First Principles of Chemistry* (Philadelphia–Boston, 1847; rev. 1850, 1853); *First Principles of Physics, or Natural Philosophy* (Philadelphia, 1859; 2nd ed., 1861); *A Century of Medicine and Chemistry* (New Haven, 1871); and *American Contributions to Chemis-*

try (Philadelphia, 1874). Pamphlets are *Fuel for Locomotive Steam Use* (New York, 1855) and *Report on the Rock Oil, or Petroleum, From Venango Co., Pennsylvania* (New Haven, 1855).

His papers include "A Daguerreotype Experiment by Galvanic Light," in *American Journal of Science*, **43** (1842), 185–186, written with W. H. Goode; "On the Use of Carbon in Grove's Battery," *ibid.*, 393; "Report on the Intrusive Trap of the New Red Sandstone of Connecticut," *ibid.*, **47** (1844), 107–108; "On the Chemical Composition of the Calcareous Corals," *ibid.*, 2nd ser., **1** (1846), 189–199; "Optical Examination of Several American Micas," *ibid.*, **10** (1850), 372–383; "On the Existence of the Mastodon in the Deep-Lying Gold Placers of California," *ibid.*, **45** (1868), 378–381; "On Flame Temperatures in Their Relations to Composition and Luminosity," *ibid.*, **49** (1870), 339–347, written with Henry Wurtz (repr. in *Chemical News, Journal of the Franklin Institute, Philosophical Magazine*, and *Journal of Gas-lighting, Water-supply and Sanitary Improvement* [London]); "Researches on Water-Gas," in *Journal of Gas-lighting, Water-supply, and Sanitary Improvement*, **24** (1874), 544–545, 574–576, 608–610, 640–641, 675–677, written with Henry Wurtz—this paper appeared first as a book in 1869 and in the *American Gas-Light Journal and Chemical Repertory*, beginning with the issue of 16 Jan. 1874, p. 21.

There are important collections of letters, diaries, reports, and memorabilia pertaining to Silliman and his work at the Yale University Library, in the archives of the National Academy of Sciences, at the Bancroft Library (University of California), the Stanford University Library, the Huntington Library (San Marino, California), and the DeGolyer Foundation Library (Dallas, Texas). For details concerning these materials and further sources, see Gerald T. White (below).

II. SECONDARY LITERATURE. There is no full-length biography of Silliman. The best source for the period 1865–1885 is Gerald T. White, *Scientists in Conflict. The Beginnings of the Oil Industry in California* (San Marino, Calif., 1968), which contains the substance of a number of earlier papers by White and, through its bibliographical note and extensive footnotes, is an excellent guide to the sources by and about Silliman. Other sources are Russell H. Chittenden, *History of the Sheffield Scientific School of Yale University 1846–1922*, I (New Haven, 1928), esp. 38, 42, 45–51, 64, 66, 69, 110, 116, 122, 287; [James Dwight Dana], "Benjamin Silliman," in *American Journal of Science*, **29** (1885), 85–92; W. L. Kingsley, *Yale College: A Sketch of Its History*, II (New York, 1879), 81–83, 105–107; Louis I. Kuslan, "The Founding of the Yale School of Applied Chemistry," in *Journal of the History of Medicine and Allied Sciences*, **24** (1969), 430–451; and Arthur W. Wright, "Biographical Memoir of Benjamin Silliman 1816–1885," in *Biographical Memoirs. National Academy of Sciences*, **7** (1913), 115–141.

ELIZABETH H. THOMSON

SIMON, FRANZ EUGEN (FRANCIS) (*b.* Berlin, Germany, 2 July 1893; *d.* Oxford, England, 31 October 1956), *physics*.

The son of a wealthy estate dealer, Simon, and his two sisters, grew up in comparative affluence. Although he received a classical education, he developed a strong interest in science; and in 1912 he went to Munich to read physics. A year later he was called up for military service; and from 1914 to 1918 he served as lieutenant in the field artillery. He resumed his studies at the University of Berlin in 1919, and the following year he started work for his Ph.D. under the supervision of Nernst. His dissertation concerned the measurement of specific heats at low temperatures, a line of research that was closely connected with Nernst's heat theorem, now generally known as the third law of thermodynamics. The subject remained the basis of Simon's scientific interest throughout his life. After obtaining his doctorate in 1921, Simon remained at Berlin, where in 1924 he became *Privatdozent* and, three years later, associate professor. It was during this period that his school of low-temperature physics was founded and that he did his outstanding work combining low-temperature and high-pressure techniques.

In 1931 Simon was appointed to the chair of physical chemistry at the Technical University of Breslau, succeeding Eucken. With some of the former members of his Berlin school he began to assemble and to set up low-temperature equipment, but the economic depression and political uncertainty severely hampered these efforts. Simon spent part of 1932 as visiting professor at Berkeley. When, a few months after his return from America, Hitler assumed power, Simon realized that despite his war service, his days in Germany were numbered. He tendered his resignation in June 1933 and accepted the invitation of F. A. Lindemann (later Lord Cherwell) to work at the Clarendon Laboratory, Oxford, where a small helium liquefaction plant had been set up by one of Simon's former co-workers, K. Mendelssohn. Another member of his Berlin School, Nicholas Kurti, accompanied Simon to Oxford and worked with him until Simon's death.

At first Simon occupied no regular position in Oxford but received a grant from Imperial Chemical Industries, which, through Lindemann's efforts, was helping many scientific refugees to establish themselves. In 1935 Simon was appointed to the readership in thermodynamics, and in 1945 he became professor of thermodynamics. He held this post until 1956, when Lindemann retired as Lee professor of experimental philosophy and Simon became his successor. Simon died while making plans for the further development of the Clarendon Laboratory, only a few weeks after taking up his new appointment.

Except for the interruption by the war years, Simon devoted all his time at Oxford to building up a new research school in low-temperature physics. At his suggestion Kurti, while still at Berlin, had begun work on the low-temperature properties of paramagnetic salts; and after settling at Oxford, much of Simon's work was taken up with developing the method of cooling by adiabatic demagnetization and with investigating the properties of matter at temperatures below 1° K. During the last years of his life his interest shifted from the ordering of electron spins to that of nuclear spins, and in 1951 the first nuclear alignment was achieved by members of his research group. Although Simon was greatly interested in this combination of low temperatures and nuclear physics, it took second place to the aim that he and Kurti had pursued for many years, the cooling by adiabatic demagnetization of nuclear spins. After long preparations and often disappointing pilot experiments, the final goal was reached a few months before Simon's death. The nuclear spin system of copper had been cooled to a temperature of less than 20 microdegrees absolute.

In addition to his magnetic work, Simon continued his interest in the general properties of matter at liquid helium temperatures, such as specific heats and thermal conductivities. He also resumed the research at high pressures, but perhaps not quite with the vigor it deserved. During the war Simon took part in the work on the atomic bomb and was particularly concerned with isotope separation by gaseous diffusion. In his later years he was much concerned with the problems of utilizing scientific advances technologically; and in his writings on these subjects politicians, as well as industry, came in for a good deal of criticism.

In 1941 Simon was elected a fellow of the Royal Society, and in 1948 he received its Rumford Medal. Two years later he was the first recipient of the Kamerlingh Onnes Medal of the Dutch Institute of Refrigeration, and in 1952 he was awarded the Linde Medal; in the same year he was elected an honorary foreign member of the American Academy of Arts and Sciences. For his war work on atomic energy he was given the C.B.E. in 1946, and in 1955 he was knighted.

When assessing Simon's scientific achievement, it is tempting to rate the spectacular success of nuclear cooling highest, but this would not do justice to the large body of outstanding work directed toward the proof and elucidation of the third law of thermodynamics. When Simon came to Nernst in 1920, the latter had just written the famous monograph in which he had proved to his own satisfaction that the heat theorem was correct and that it should be regarded as a basic law of physics. There were many who did not share this conviction, however, and it was left to Simon to prove the validity of the law in which Nernst was now beginning to lose interest. The law requires that, as absolute zero is approached, any system must tend to a state of zero entropy (that is, to maximum statistical orderliness). It was pointed out that many systems, such as glasses, retained an obvious degree of disorder, even when cooled to the lowest possible temperatures. Simon noticed that these systems were not in thermodynamic equilibrium and that, if by magic or by waiting for immensely long times they could be guided into equilibrium, the process would be accompanied by the liberation of heat. In other words, none of these systems could be used to reach absolute zero and thereby infringe upon the third law.

Simon and his school also investigated systems, such as ortho-hydrogen, the chemical constant of which seemed to suggest the existence of a zero-point entropy. Their experiments proved that some ordering always occurs, sometimes at less than 1° K., which brings the system into agreement with the third law. Simon also provided the explanation for the very strange fact that under its own vapor pressure helium remains a liquid, even at absolute zero. He showed that this is due to the vibration of the atoms under the quantum-mechanical zero-point energy that again is a consequence of Nernst's theorem.

Simon's work on high pressure was an investigation of the melting curves of solidified gases, especially that of helium. It was known that up to the highest pressures, the equilibrium between the solid and the fluid phase of any pure substance follows a smooth curve with no indication of a critical point or of a change in the temperature function. He had the brilliant idea of extending investigations to the substance with the lowest critical data: helium. Thus, using it as a model system, Simon eventually was able to explore the melting curve up to ten times the liquid-gas critical temperature. Even under these extreme conditions, however, no solid-fluid critical point was found. In fact, parallel experiments on the thermodynamic properties showed that with rising pressure and temperature, the two phases become increasingly dissimilar.

Curiously, Simon was at first less known through his scientific results than through the methods of obtaining them. When he began his work, only three laboratories in the world commanded the expensive means of liquefying helium. It was Simon's great achievement to develop small-scale apparatus of novel and ingenious design that eventually permitted not only his but many other laboratories to experiment in this otherwise closed domain.

BIBLIOGRAPHY

Simon and his co-workers published more than 150 papers, only a few of which can be cited here: "Zur Frage der Nullpunktsenergie," in *Zeitschrift für Physik*, **16** (1923), 183–199, written with K. Bennewitz; "Die Bestimmung der freien Energie," in H. Geiger and K. Scheel, eds., *Handbuch der Physik*, X, pt. 7 (Berlin, 1926), 350; "Fünfundzwanzig Jahre Nernstscher Wärmesatz," in *Ergebnisse der exakten Naturwissenschaften*, **9** (1930), 222–274; "The Approach to the Absolute Zero of Temperature," in *Proceedings of the Royal Institution*, **28** (1935), 515–541; "On the Range of Stability of the Fluid State," in *Transactions of the Faraday Society*, **33** (1936), 65–73; "The Determination of Temperature Below 1° K.," in *Science Progress*, **34** (July 1939), 31–46; *The Neglect of Science* (Oxford, 1951); "Low Temperature Problems, A General Survey," in F. E. Simon *et al.*, *Low Temperature Physics* (London, 1952), 1–29; and "The Third Law of Thermodynamics—a Historical Survey (40th Guthrie Lecture)," in *Yearbook of the Physical Society* (1956), 1.

For Simon's life see N. Kurti's obituary in *Biographical Memoirs of Fellows of the Royal Society*; and Nancy Arms, *A Prophet in Two Countries: The Life of F. E. Simon* (London, 1966).

KURT MENDELSSOHN

SIMON BREDON. See **Bredon, Simon.**

SIMON DE PHARES (*b.* Meung-sur-Loire [?], France, *ca.* 1450; *d.* Paris, France, after 1499), *astrology.*

All that is known about Simon is what he wrote in his *Recueil des plus célèbres astrologues.*[1] Born perhaps in Meung-sur-Loire, he studied law in Orléans and then entered the Faculty of Arts in Paris, where he studied Sacrobosco's *De sphaera* and al-Qabīṣī's *Introductorium*. He then joined the service of Mathieu de Nanterre,[2] no doubt as astrologer, and later that of Duke John II of Bour-

bon. While serving the latter, Simon completed his training with the German astrologer Conrad Heingarter, whom he subsequently considered his master. After studying for two years at Oxford, he traveled in Scotland and Ireland; returned to France to take courses at the Faculty of Medicine in Montpellier; journeyed in Italy (Rome, Venice) and Egypt (Cairo, Alexandria); then traveled through the Alps of Savoy and Switzerland for four years, botanizing and examining rocks.

After the death in 1488 of John II, to whose service he had returned, Simon moved to Lyons, married, and raised a family. His house, near the cathedral, was furnished with a library of two hundred books and was his astrological office. He was so famous that King Charles VIII, while passing through Lyons on All Saints' Day of 1490, was anxious to consult him. Such glory and success were bound to arouse jealousy. Accused of sorcery before the episcopal court of Lyons, Simon was ordered to cease his activities and his library was confiscated. His appeal to the Parlement of Paris failed, and the books in his library were censured by the Faculty of Theology. He subsequently moved to Paris, where he was still living in 1499, apparently having failed to obtain royal intervention, although he had composed a work justifying his activities for Charles VIII.

This justification is known today by the title *Recueil des plus célèbres astrologues et quelques hommes doctes*, which was added at the end of the sixteenth century to the only existing manuscript, which appears to be the original. It is the first part of a work in which Simon had also planned to present the principles of astrology and of the divinatory arts, in order that the scientific contributions of the former could easily be separated from the charlatanism of the latter. That the planned parts probably were never written is unfortunate, for Simon's experience in a trade that he practiced rather like a liberal profession, his curiosity in the most varied areas of science, and his training in at least four universities and through extensive travels would have imparted considerable value to the work.

Composed in French, Simon's justification of astrology was conceived as a panegyric of astrologers who had honored their profession by the success of their predictions, and he should therefore be considered the first historiographer of astronomy. The *Recueil* is a series of accounts, arranged in theoretically chronological order, devoted to famous astrologers and their works. It is thus a priceless source of information, since Simon went so far as to give the incipits of some works of his colleagues.

The quality of the information obviously is highly variable, and the reports on Simon's contemporaries are of greater interest than the mythical biographies of the founders of astrology. Even for the fourteenth century the chronology is often defective: John of Murs and Firmin de Belleval are assigned dates that are too recent, as are Roger Bacon and "Barthelemy de Morbecha" (William of Moerbeke), who are placed in the middle of the fourteenth century. Because erroneous information is presented side by side with precise data, the *Recueil* must be read very critically.[3] Nevertheless, it is a tool of very great importance for historians of medieval astronomy and medicine.

NOTES

1. The information provided by the *Recueil* on the circumstances of Simon's trial before the Parlement has been completed by E. Wickersheimer with the aid of archival documents, and by the material published by Charles Du Plessis d'Argentré in his *Collectio judiciorum de novis erroribus*, I (Paris, 1728).
2. Mathieu de Nanterre was the premier president of the Parlement of Paris. The minutes of *étude* VIII in the *minutier central* of the Archives Nationales preserve several documents on Mathieu de Nanterre for 1480–1485.
3. Among the errors is Simon's dating of his work from the sixteenth year of the reign of Charles VIII, who actually died in April 1498, before having completed the fifteenth year of his reign.

BIBLIOGRAPHY

An excellent ed. of the only known work by Simon de Phares is *Recueil des plus célèbres astrologues et quelques hommes doctes faict par Symon de Phares du temps de Charles VIIIe, publié d'après le manuscrit unique de la Bibliothèque nationale par Ernest Wickersheimer* (Paris, 1929). The biography included in this ed. (pp. vi–xii) was reprinted in E. Wickersheimer, *Dictionnaire biographique des médecins en France au moyen âge* (Paris, 1936), 743–744. It also is the basis of the account by L. Thorndike in *A History of Magic and Experimental Science*, IV (New York, 1934), 545–557, which contains interesting speculations concerning the reasons for Simon's travels in the Alps and the relations of Simon and of Charles VIII to the Parlement of Paris.

EMMANUEL POULLE

SIMPLICIUS (*b.* Cilicia, *ca.* 500; *d.* after 533), *philosophy.*

Simplicius was one of the most famous representatives of Neoplatonism in the sixth century. An outstanding scholar, he was the author of extensive commentaries on Aristotle that contain much valuable information on previous Greek philosophy, including the pre-Socratics.

Very little is known of his life. According to Agathias (*History*, II,30,3), he was born in Cilicia. He received his first philosophical education in Alexandria at the school of Ammonius Hermiae,[1] the author of a large commentary on the *Peri Hermeneias* and on some other logical, physical, and metaphysical treatises of Aristotle. These works strongly influenced not only the commentaries of Simplicius but also those written by the philosophers of the Alexandrian School: Asclepius, Philoponus, and Olympiodorus.[2] Simplicius also studied philosophy at Athens in the school of Damascius,[3] the author of *Problems and Solutions About the First Principles*, known for his doctrine of the Ineffable First Principle. According to Damascius no name is capable of expressing adequately the nature of that Principle, not even the Plotinian name of "the One." Damascius was the last pagan Neoplatonist in the unbroken succession of the Athenian school, where he was teaching when Justinian closed it in 529. Simplicius, who at that time was a member of Damascius' circle, left Athens with him and five other philosophers and moved to Persia (531–532). Their exile was only temporary, for they returned to the empire after the treaty of peace between the Byzantines and the Persians (533). According to Agathias (*History*, II,31,4), the terms of the treaty would have guaranteed to the philosophers full security in their own environment: they were not to be compelled to accept anything against their personal conviction, and they were never to be prevented from living according to their own philosophical doctrine.[4]

There are grounds for supposing that Simplicius settled in Athens after returning from Persia.[5] Presumably he was not allowed to deliver public lectures and thus could devote all his time to research and writing. Hence his commentaries are not related to any teaching activity; rather, they show the character of written expositions that carefully analyze the Aristotelian text and interpret it in the light of the whole history of Greek philosophy. Simplicius always endeavored to harmonize and reconcile Plato and Aristotle by reducing the differences between them to a question of vocabulary, point of view, or even misunderstanding of some Platonic theories by the Stagirite.

Simplicius was not the first to take this approach. According to W. Jaeger, this trend can be traced to Posidonius and to Neoplatonic philosophy in general. The same method was certainly used by Ammonius, who always attempted to reduce the opposition between Plato and Aristotle to different viewpoints. For example, in dealing with Aristotle's criticism of the theory of ideas, Ammonius believed this criticism to concern not the authentic doctrine of Plato, but rather the opinion of some philosophers who attributed to the Ideas an independent subsistence, separate from the Intellect of the Demiurge (Asclepius, *In Metaphysicorum*, 69,24–27; 73,27).

Apparently Simplicius was persuaded that this approach was in agreement with the attitude of the φιλομαθεῖς and that it uncovered the true meaning of philosophical doctrines. At first glance, he said, some theories seem to be quite contradictory, but a more accurate inquiry shows them to be reconcilable (*In de Caelo*, 159,3–9). Moreover, in explaining a philosophical text, one should not be biased for or against its author. Hence Simplicius opposed the method of Alexander, who from the beginning is suspicious of Plato in the same way that others are inspired with prejudice against Aristotle (*In de Caelo*, 297,1–4). Since agreement on an opinion, even a prephilosophical one, has often been considered a criterion of truth, Aristotle and the Stoics frequently used the argument of universal agreement. Therefore, having to cope with the increasing influence of Christianity, late Neoplatonic philosophers wanted to argue against the presumed disaccord between the main representatives of Greek philosophy, Plato and Aristotle, in order to enhance their own doctrine. As a Christian, Philoponus did not have the same motives for harmonizing Plato and Aristotle; he firmly opposed attempts to reconcile them and called this interpretation a kind of mythology. Aristotle, he held, did not argue against those who misunderstood Plato but against the authentic Platonic doctrine.

As a commentator Simplicius did not overestimate his own contributions but was quite aware of his debt to other philosophers, especially to Alexander, Iamblichus, and Porphyry (*In Categorias*, 3, 10–13). He did not hesitate to call his own commentaries a mere introduction to the writings of these famous masters (*In Categorias*, 3,13–17), nor did he cling fanatically to his own interpretations; he was happy to exchange them for better explanations (*In Categorias*, 350,8–9). On the other hand, the work of a commentator is far from being a neutral undertaking or a question of mere erudition; it is chiefly an opportunity to become more familiar with the text under consideration and to elucidate some intricate passages (*In Enchiridion*, Praefatio, 2,24–29; *In de Caelo*, 102,15; 166,14–16; *In Categorias*, 3,4–6); hence Simplicius' constant concern to obtain reliable documents and to check the historical value of this information, as when he verified the information provided by Alexander about the squaring of the circle according to Hippocrates of Chios (*In Physicorum*, 60, 22–68, 32).

Simplicius adhered to the Aristotelian doctrine of the eternity of the world, as a theory that fits perfectly into the Neoplatonic ontology insofar as the eternal movement of the heavens is a necessary link between the pure eternity of the intelligible reality and the temporal character of material beings. With respect to this question, Simplicius strongly opposed Philoponus, who asserted the beginning of the world through divine creation. Philoponus, however, did not argue as a Christian, nor did he base his refutation of the Aristotelian doctrine on arguments drawn from his Christian faith. According to him, God is the principle of whatever exists: if time is infinite, nothing may ever come to be, because an infinite number of conditions of possibility are to be fulfilled before anything could begin to exist—which is clearly impossible. Simplicius' notion of "infinite" is different; it does not mean an infinity existing at once, but a possibility of transcending any boundary. Consequently the conception of time exposed by both authors is not the same. Simplicius professed a cyclical conception; Philoponus adhered to a linear view without regular return of the same events. Philoponus also substantiated divine creation in time, without preexisting matter; whereas Simplicius maintained that although heaven, the first and highest corporeal reality, is totally dependent upon God, it has never come to exist; it must be eternal, because it springs immediately from God.

In his *Corollarium de tempore* (*In Physicorum*, 773,8–800,25) Simplicius drew a general survey of the different theories about time, dealing with older as well as with more recent philosophers. According to his view, time is closely related to the life of the soul; but the activity of the soul does not merely coincide with time, because it occurs in time. Previous to all things existing in time, there is a time that makes them temporal and arranges the extension of their existence in an orderly fashion. The nature of that time, like the nature of the soul, is intermediary between being and becoming. Consequently the soul does not exist in time but is the principle of the temporal character of its own activity, as it is the origin of the time of the cosmos.

To a certain extent, soul and time may be identified, although the conceptual distinction must be maintained. Simplicius wondered whether this logical distinction may entail an ontological one; his reply was rather hesitant. In stressing the connection between time and soul, Simplicius approached Plotinus; on the other hand, he was also influenced by Iamblichus and Proclus insofar as he dealt not only with the numbered time, but also with the numbering time, that is, the regulating principle of movement.

The earliest preserved work of Simplicius seems to be his commentary on Epictetus' *Enchiridion*. K. Praechter was the first to believe that this work predates the Aristotelian commentaries or even that it was written during Simplicius' stay at Alexandria. Praechter argued mainly from the text that has been chosen for explication; from the absence of references to Iamblichus, Proclus, or Damascius; from the less intricate doctrine of the first principles; and finally from its kinship with the commentary of Hierocles on the *Carmen aureum*.[6] Praechter's thesis is certainly questionable. In two contributions I. Hadot has shown that the influence of Proclus and even of Damascius is undeniable.[7] In his discussion of the Manichaean cosmogony Simplicius also seemingly relied on information drawn from conversations with Manichaean sages; this is probably related to his stay in Persia with King Chosroes, who always showed a keen interest in philosophical problems.[8] The allusion to the "tyrannic circumstances" that afforded him an opportunity for dealing with Epictetus' *Enchiridion* (138,17–19) may suggest that Simplicius wrote his commentary after the Edict of Justinian (529). From these and other similar anti-Christian statements that occur in the commentary, and also from the way in which the duty of a philosopher in corrupt states is presented, A. Cameron argued that this work may have been written precisely during the years 529–531.[9]

Among the commentaries on Aristotle, the first to be mentioned according to chronological order seems to be the *In de Caelo*; some passages in the first book of this work, where the criticisms of Philoponus against Aristotle are refuted, are referred to in the commentary on the *Physics* (1118,3; 1146,27; 1169,7; 1175,32; 1178,36; 1330,2; 1335,1). In 529 Philoponus published his *De aeternitate mundi contra Proclum*. Between this work and his *De aeternitate mundi contra Aristotelem*, he completed two other works, one of

which is the commentary on the *Meteorologica*.[10] Hence the *In de Caelo* could hardly be dated before 535; presumably the work was written shortly after Simplicius' stay in Persia.

The commentary on the *Physics* is certainly later than the *In de Caelo* because of its references to this work already noted. On the other hand, it is prior to the commentary on the *Categoriae*, because it is referred to in this last work (*In Categorias*, 435,23–24). The commentary was written after the death of Damascius, but we do not know this date. According to A. Cameron,[11] Damascius was alive as late as 538; consequently the commentary could hardly have been written before 540.

Both the commentaries on the *Categoriae* and the *De anima* are to be dated after the *In Physicorum*. To date, the authenticity of the commentary on the *De Anima* has hardly been questioned; nevertheless, certain features may suggest that this work has been erroneously attributed to Simplicius. The solution of this problem is of some importance, because our information about some lost works of Simplicius depends on it. The author of the *In de Anima* six times refers to earlier writings: to an *Epitome Physicorum Theophrasti* (136,29), to a work on the *Metaphysics* (28,20; 217,26), and to a commentary on the *Physics* (35,14; 120,4; 198,5). If Simplicius is to be considered the author of the *In de Anima*, then perhaps only two not unimportant works, namely the *Epitome Physicorum Theophrasti* and a commentary on the *Metaphysics*, have been lost.

Only fragments have been preserved of the commentary on the *Premises* of the first book of the *Elementa Euclidis* (see *Anaritii in decem libros priores Elementorum Euclidis commentarii*). On the other hand, some scholia on Proclus' *In Platonis Timaeum commentaria* may have been composed by Simplicius.[12] As to the *Scholia in Hermogenis artem oratoriam* (see Fabricius-Harles, V,770, referring to Lambeck-Kollar, VII, 549–553), there is no reference to Simplicius in the description of the codex Vindob. Phil. gr. 15 by H. Hunger.[13] Fabricius-Harles (IX,567) also mentions a commentary on Iamblichus' *De Pythagorica secta libri tres* and a *Commentarius brevis de Syllogismis*; no trace of either work has ever been found.

Simplicius' work was very influential, especially his commentary on the *Categoriae*, translated into Latin by William of Moerbeke in 1266. Part of the commentary on the *De Caelo* was first translated by Robert Grosseteste;[14] and a complete Latin version was executed by William of Moerbeke in 1271.

NOTES

1. In his commentaries Simplicius frequently mentions Ammonius as his master: *In de Caelo*, 271,19; 462,20–21; *In Physicorum*, 59,23–24; 59,30–31; 183,18; 192,14; 198,17; 1363,8.
2. Cf. *Anonymous Prolegomena to Platonic Philosophy*, L. G. Westerink, ed. (Amsterdam, 1962), xi.
3. For references to Damascius as the master of Simplicius see *In Physicorum*, 462,17; 601,19; 630,35; 644,10; 774,28; 778,27; 795,14.
4. Some serious criticisms against the reliability of the information given by Agathias were recently raised by A. Cameron; see "The Last Days of the Academy at Athens," 18.
5. *Ibid.*, 22–26.
6. See K. Praechter, *Simplikios*, cols. 206–210.
7. See I. Hadot, *Le Système théologique de Simplicius dans son commentaire sur le manuel d'Epictète*, 270, 272–273, and 278–279.
8. See I. Hadot, *Die Widerlegung des Manichäismus im Epiktetkommentar des Simplikios*, 46, 56–57.
9. See A. Cameron, *op. cit.*, 13–17.
10. See É. Evrard, "Les convictions religieuses de Jean Philopon et la date de son commentaire aux 'Météorologiques,'" in *Bulletin de l'Académie royale de Belgique. Classe des lettres et des sciences morales et politiques*, 5th ser., **39** (1953), 345.
11. A. Cameron, *op. cit.*, 22.
12. Proclus, *Théologie Platonicienne*, H. D. Saffrey and L. G. Westerink, ed. and trans., Bk. 1 (Paris, 1968), clii–cliii.
13. H. Hunger, *Katalog der griechischen Handschriften der oesterreichischen Nationalbibliothek*, I (Vienna, 1961), 147–148.
14. See D. Allan, "Mediaeval Versions of Aristotle's 'De Caelo' and of the Commentary of Simplicius," in *Mediaeval and Renaissance Studies*, **2** (1950), 82–120.

BIBLIOGRAPHY

I. ORIGINAL WORKS. Simplicius' works are *Commentarius in Enchiridion Epicteti*, Jo. Schweighäuser, ed., in *Theophrasti Characteres*, . . . Fr. Dübner, ed. (Paris, 1840); *In Aristotelis physicorum libros quattuor priores commentaria*, H. Diels, ed., Commentaria in Aristotelem Graeca, vol. IX (Berlin, 1882); *In libros Aristotelis de Anima commentaria*, M. Hayduck, ed., *ibid.*, XI (Berlin, 1882); *In Aristotelis de Caelo commentaria*, J. L. Heiberg, ed., *ibid.*, VII (Berlin, 1894); *In Aristotelis physicorum libros quattuor posteriores commentaria*, H. Diels, ed., *ibid.*, X (Berlin, 1895); *In Aristotelis Categorias commentarium*, C. Kalbfleisch, ed., *ibid.*, VIII (Berlin, 1907); and *Commentaire sur les Catégories d'Aristote*, in *Corpus Latinum commentariorum in Aristotelem graecorum*, I (Louvain–Paris, 1971), II (Leiden, 1975), a critical ed. by A. Pattin of William of Moerbeke's translation.

II. SECONDARY LITERATURE. On Simplicius and his work, see *Anaritii in decem libros priores Elementorum Euclidis commentarii, ex interpretatione Gherardi Cremonensis in codice Cracoviensi 569 servata*, M. Curtze, ed. (Leipzig, 1899); A. Cameron, "The Last Days of the Academy at Athens," in *Proceedings of the Cambridge Philological Society*, n.s. **15** (1969), 7–29; E. Ducci, "In τὸ ἐὸν parmenideo nella interpretazione di Simplicio," in *Angelicum*, **40** (1963), 173–194, 313–327; I. Hadot, "Die Widerlegung des Manichäismus im Epiktetkommentar des Simplikios," in *Archiv für Geschichte der Philosophie*, **51** (1969), 31–57; and "Le système théologique de Simplicius dans son commentaire sur le manuel d'Épictète," in *Le néoplatonisme. Colloques internationaux du C.N.R.S.* (Royaumont, 1969), 265–279.

See also H. Meyer, *Das Corollarium de Tempore des Simplikios und die Aporien des Aristoteles zur Zeit* (Meisenheim am Glan, 1969); B. Nardi, "Il commento di Simplicio al De anima nelle controversie della fine del secolo XV e del secolo XVI," in *Saggi sull' Aristotelismo padovano dal secolo XIV al XVI* (Florence, 1958), 365–442; K. Praechter, "Simplikios," in Pauly-Wissowa, *Real-Encyclopädie der classischen Altertumswissenschaft*, 2nd ser., III, 204–213; A. I. Sabra, "Simplicius's Proof of Euclid's Parallels Postulate," in *Journal of the Warburg and Courtauld Institutes*, **32** (1969), 1–24; and W. Wieland, "Die Ewigkeit der Welt (Der Streit zwischen Joannes Philoponus und Simplicius)," in *Die Gegenwart der Griechen im neuerem Denken. Festschrift für H.-G. Gadamer zum 60. Geburstag* (Tübingen, 1960), 291–316.

G. VERBEKE

SIMPSON, THOMAS (*b.* Market Bosworth, Leicestershire, England, 20 August 1710; *d.* Market Bosworth, 14 May 1761), *mathematics.*

Simpson's father, a weaver, wanted him to take up the same trade. After limited education the son moved to Nuneaton, where he was influenced by the 1724 eclipse and by a visiting peddler, who lent him a copy of Cocker's *Arithmetic* and a work by Partridge on astrology. Young Simpson made such progress with his studies that he acquired a local reputation as a fortune-teller. He was able to leave his weaving and marry his landlady, a widow Swinfield, whose son was a little older than her new husband. About 1733 an unfortunate incident obliged him to move to Derby, where he taught at an evening school and resumed his trade as a weaver during the day.

By the beginning of 1736 Simpson had moved to London and settled in Spitalfields, where the Mathematical Society had flourished for two decades. In 1736 his first mathematical contributions were published in the well-known *Ladies' Diary*. One of these showed that he was already versed in the subject of fluxions, which had elicited a growing interest, as illustrated by the famous controversy sparked by Bishop George Berkeley in 1734.[1] In December 1735 Simpson had issued proposals[2] for publishing his first book, *A New Treatise of Fluxions*, which appeared in 1737. Although publication may have been delayed by the author's teaching duties, it indicated his success, which enabled him to bring his family from Derby, and his future career as a mathematics teacher, editor, and textbook writer.

Robert Heath's accusation of plagiarism probably brought Simpson useful publicity, which was supplemented by the publication in 1740 of *The Laws of Chance* and *Essays on Several Subjects*. They were rapidly followed by *Annuities and Reversions* (1742) and *Mathematical Dissertations* (1743), the latter being dedicated to Martin Folkes, then president of the Royal Society, with whom he had been in correspondence for some months. Apart from Francis Blake, Simpson's other correspondents were relatively humble philomaths. Largely through Folkes's support, Simpson was appointed second mathematical master at the Royal Military Academy, Woolwich, in August 1743 and was elected fellow of the Royal Society two years later.

Simpson seems to have been quite successful as a teacher, and his duties left him time for other activities. Three subsequent textbooks were bestsellers, partly because of his position and partly because of their scope: *Algebra*, with ten English editions in 1745–1826, besides American and German versions; *Geometry*, six London editions between 1747 and 1821, five at Paris, and one at Amsterdam; and *Trigonometry*, five London editions in 1748–1799, besides French and American versions. *Geometry*, which led to an argument with Robert Simson (whose editions of Euclid became very popular), represented a significant revision of the original Greek treatment along the lines of Clairaut and other Continental mathematicians.

Simpson's influence on English mathematics was extended by his editorship of the annual *Ladies' Diary* from 1754.[3] This post demanded an extensive correspondence with contributors throughout the country, in addition to the normal responsibility of seeing the work through the press; and Simpson seems to have worn himself out with his many activities and aged prematurely. In 1760 he became involved as a consultant on the best form for a new bridge across the Thames at Blackfriars.[4] The intense work on this project accelerated his death.

Simpson obtained a reputation as "the ablest Analyst (if we regard the useful purposes of Analytical Science) that this country [Britain] can boast of" and as author of one of the two best treatises "on the Fluxionary Calculus."[5] He was aware of the importance of Continental mathematicians, for the first book on the subject he read was a translation from the French of L'Hospital's *Analyse des infiniment petits*; and the final paragraph of the preface to his last work, *Miscellaneous Tracts* (1757), was by nature of a testament. Having mentioned in the latter that he had "chiefly adhered to the analytic method of Investigation," he warned that "by a diligent cultivation of the Modern Analysis, . . . Foreign Mathematicians have, of late, been able to push their Researches farther, in many particulars, than Sir Isaac Newton and his Followers here, have done. . . ."

Although Simpson clearly was more interested in the applications to problems in series and mechanics[6] than in the foundations of analysis, he avoided the difficulties of infinitesimals by his definition: "The Fluxions of variable Quantities are always measured by their Relation to each other; and are ever expressed by the finite spaces that would be uniformly described in equal Times, with the Velocities by which those Quantities are generated." F. M. Clarke has detailed the correspondence with Francis Blake, author of an anonymous but influential pamphlet, *Explanation of Fluxions* (1741), which clarified his and Simpson's ideas on the subject before the appearance, in an enlarged and revised form, of Simpson's *Doctrine and Applications of Fluxions* (1750); this work inspired another polemic.

Until his death in 1754, one of the leading mathematicians in England was Abraham De Moivre, whose well-known *Doctrine of Chances* (1718) included work on annuities. In *Laws of Chance* (1740) Simpson wrote approvingly of De Moivre but claimed to have investigated two problems in probability for which the latter had given only the results; two years later he issued *Annuities and Reversions*. The latter was criticized by De Moivre, and Simpson replied immediately with an appendix that seems to have effectively terminated the dispute. In 1752 Simpson issued a supplementary essay that included his much-quoted tables on the valuation of lives according to London bills of mortality. Paradoxically he is now best remembered for Simpson's rule, discovered long before him, for determining the area under a curve,

$$\frac{Aa + 4Bb + Cc}{3} \times AB,$$

obtained by replacing the curve by a parabola with vertical axis going through the points a, b, and c.[7] Fifty years after Simpson's death Robert Woodhouse and his disciples achieved Simpson's aim with the reform of mathematical analysis at Cambridge, which brought English mathematics once more into the front rank of European developments.

NOTES

1. Details are in F. Cajori, *A History of the Conceptions of Limits and Fluxions in Great Britain From Newton to Woodhouse* (Chicago, 1919), chs. 3, 4.
2. D. F. McKenzie and J. C. Ross, *A Ledger of Charles Ackers* (Oxford, 1968), no. 398, quotes 750 copies costing £1 each. The only known copy is that in the Simpson papers, IV.
3. Simpson also contributed to other periodicals. See R. C. Archibald, "Notes on Some Minor English Mathematical Serials," in *Mathematical Gazette*, 14, no. 200 (Apr. 1929), 379–400.
4. A brief account in F. M. Clarke, *Simpson and His Times*, can be supplemented by J. Nichols, *The History and the Antiquities of the County of Leicester*, IV (London, 1811), 510–514.
5. The assessments by R. Woodhouse and J. Playfair are quoted in Simpson's *Fluxions* (1823), iv.
6. See I. Todhunter, *A History of the Mathematical Theories of Attraction* (London, 1873), ch. 10, for an estimate of Simpson's contributions to this subject and (sec. 294) his estimate that Simpson was "at the head of the non-academical body of English mathematicians" and second only to Newton.
7. Given in *Mathematical Dissertations* (1743), p. 110, for the equidistant ordinates *Aa*, *Bb*, and *Cc*.

BIBLIOGRAPHY

I. ORIGINAL WORKS. Clarke (see below) gives the full titles, but not details of the eds., of Simpson's works except the last: *Miscellaneous Tracts on Some Curious and Very Interesting Subjects in Mechanics, Physical-Astronomy and Speculative Mathematics; Wherein the Precessions of the Equinox, the Nutation of the Earth's Axis, and the Motion of the Moon in Her Orbit, Are Determined* (London, 1757). Simpson's books reprinted many of his *Philosophical Transactions* articles, listed in Poggendorff, II, 937.

II. SECONDARY LITERATURE. The main source for this article is Frances M. Clarke, *Thomas Simpson and His Times* (New York, 1929), based on her 1929 Columbia University thesis but incompletely documented and unindexed. This often quotes from the 8 vols. of Simpson papers in Columbia University Library, which kindly sent a microfilm to the writer. Most other biographies depend on Charles Hutton, "Memoirs of the Life and Writings of the Author," prefixed to Simpson's *Select Exercises* (London, 1792), itself an extended version of an account in the *Annual Register* (1764), 29–38.

P. J. WALLIS

SIMSON, ROBERT (*b.* West Kilbride, Ayrshire, Scotland, 14 October 1687; *d.* Glasgow, Scotland, 1 October 1768), *geometry*.

Simson's father, Robert, was a prosperous merchant in Glasgow who had acquired the small estate of Kirktonhall in West Kilbride; his mother, Agnes, whose maiden name was also Simson, came from a family that had provided parish ministers for the Church of Scotland from the time of the Reformation. It was with the intention of training for the Church that Simson matriculated at the University of Glasgow in 1701. He followed the standard course in the faculty of arts (Latin, Greek, logic, natural philosophy) and then devoted himself to the study of theology and Semitic languages. During these years, one of his teachers was his maternal uncle, John Simson, professor of divinity. He also acquired a knowledge of natural history that was a source of pleasure to him throughout his life; it is interesting to note that until his death he was held by his contemporaries to be one of the best botanists of his time.

At this time no instruction in mathematics was given at the University of Glasgow. The chair of mathematics had been revived in 1691 and during the years 1691–1696 it was occupied by George Sinclair, a mathematician and engineer of some repute. On his death Sinclair was succeeded by his son, Robert, who flagrantly neglected the duties of his chair. Thus Simson had no formal tuition in mathematics. It would appear to have been through reading George Sinclair's *Tyrocinia Mathematica in Novem Tractatus* (Glasgow, 1661) that Simson's interest in the subject was first aroused and it was this work that encouraged him to read Euclid's *Elements* (in the edition of Commandinus). He soon became absorbed in the study of geometry and acquired such a reputation as an "amateur" mathematician that in 1710 the senate of the university, having relieved Sinclair of his office, offered Simson the chair of mathematics. Simson declined the invitation on the grounds that he had received no formal training in mathematics; when the senate reaffirmed its confidence in his ability to discharge the duties of the chair, Simson suggested that the appointment be left open for a year, during which he would devote himself entirely to the study of mathematics.

Simson chose to spend the academic year 1710–1711 in London. He had originally intended to study in Oxford, but his efforts to make contact with mathematicians there were unsuccessful; so he spent the year at Christ's Hospital (the Blue Coat school), where, under the aegis of Samuel

Pepys, a mathematical school had been founded for the purpose of training navigation officers for the Royal Navy. More important than the formal instruction that Simson received there were the personal relationships he established with several prominent mathematicians: John Caswell, James Jurin (secretary of the Royal Society), and Humphrey Ditton. He was most profoundly influenced by Halley, who had recently been appointed Savilian professor of geometry at Oxford while still a captain in the Royal Navy; not only was Halley regarded as second only to Newton in the field of scientific research, he was also a distinguished scholar (and editor) of the works of the Greek mathematicians.

While Simson was still in London the senate of the University of Glasgow elected him (on 11 March 1711) to the chair of mathematics on the condition that "he give satisfactory proof of his skill in mathematics previous to his admission." On his return to Glasgow he submitted to a simple test and was duly admitted professor of mathematics on 20 November 1711.

At Glasgow, Simson's first task was to design a proper course in mathematics. The course extended over two complete academic years, each of seven months' duration; to each class he lectured for five hours a week. Although his own interest was entirely in geometry, he lectured on Newton's theory of fluxions; on Cartesian geometry, algebra, and the theory of logarithms; and on mechanics and geometrical optics. Among his students were Maclaurin, Matthew Stewart, and William Trail, all of whom subsequently occupied chairs of mathematics in Scottish universities.

Simson lived the rest of his life in rooms within the College of Glasgow; outwardly his life gave every appearance of being uneventful—so much so that it was highlighted only by the conferment upon him in 1746 of the M.D. (*honoris causa*) by the University of St. Andrews. In 1761 John Williamson was appointed his assistant and successor.

Simson's lifework was devoted to the restoration of "lost" works of the Greek geometers and to the preparation of definitive editions of those works that had survived. Halley had encouraged this predilection for the works of the Greek geometers. (Simson's classical education and his knowledge of oriental languages were especially useful to him.) He first turned his attention to the restoration of Euclid's porisms, which are known only from the scant account in Pappus' *Mathematical Collections*. Although Fermat claimed to have restored Euclid's work, and Halley had edited the Greek

text of the preface to Pappus' seventh book, Simson is usually regarded as the first to have thrown real light on the matter. In a paper, "Two General Propositions of Pappus, in Which Many of Euclid's Porisms Are Included" (*Philosophical Transactions of the Royal Society*, **32** [1723], 330), Simson elucidated two general propositions of Pappus and showed that they contained several of the porisms as special cases. He continued to work on this topic throughout his life, but nothing further was published until *De porismatibus tractatus* appeared posthumously in 1776. Simson's only other genuine research paper, "An Explanation of an Obscure Passage in Albert Girard's Commentary on Simon Stevin's Works, p. 169, 170," appeared in 1753 (*Philosophical Transactions of the Royal Society*, **48**, 368).

Simson's book on conic sections (1735) used only geometrical methods. Although he was familiar with the methods of coordinate geometry—and lectured upon them—he developed the subject in the style of the classical Greek authors. His authoritative account of the *loci plani* of Apollonius appeared in 1749. But his most influential work was his definitive edition (1756) of Euclid's *Elements*. This edition was the basis of every subsequent edition of the *Elements* until the beginning of the twentieth century. Simson adopted the perhaps naive view that Euclid's treatise in its original form had been free from logical faults—any blemishes were regarded by him as being due to the bungling of editors such as Theon. Simson's restoration of Euclid's *Data* was added to his second edition of the *Elements* (1762).

A posthumous edition of Simson's unpublished mathematical works was published as *Opera Quaedam Reliqua R. Simson* (1776) at the expense of Philip Stanhope, second earl of Stanhope. It consists of four books: *De porismatibus tractatus*; *De sectione determinata*; *De logarithmis liber*; and *De limitibus quantitatum et rationum, fragmentum*. The last two books are based on his lectures to students; *De logarithmis* is a purely geometrical theory of logarithms. *De limitibus* is of great interest because it shows that Simson perceived that the fluxionary calculus of Newton rested on insecure foundations; accordingly, he attempted to place the theory of limits on a rigorous foundation. His failure lies probably in the fact that he tried to formulate the theory entirely in terms that would have been intelligible to a Greek geometer of the Alexandrian School.

Simson's manuscripts contain a great variety of miscellaneous geometrical propositions and many interesting reflections on various aspects of mathematical teaching and research, but none of it in a state for publication. He also prepared a draft of an edition of the complete works of Pappus that was based on material he had received from Halley many years earlier, and it is perhaps for this reason that a transcript was obtained by the Clarendon Press at Oxford.

On his death Simson bequeathed to the University of Glasgow his collection of mathematical books—at that time recognized as the most complete in the British Isles. They are preserved as the Simson Collection of the university library.

BIBLIOGRAPHY

I. ORIGINAL WORKS. Simson's works include *Sectionum Conicarum Libri V* (Edinburgh, 1735; 2nd ed., enlarged, 1750); *Apollonii Pergaei Locorum Planorum Libri II, restituti a R. Simson* (Glasgow, 1749); *Elements of Euclid* (Glasgow, 1756), of which the 2nd ed. (1762) contained Euclid's *Data*; and *Opera Quaedam Reliqua R. Simson*, James Clow, ed. (Glasgow, 1776).

II. SECONDARY LITERATURE. The best source is William Trail, *Life and Writings of Robert Simson* (Bath, 1812).

IAN N. SNEDDON

IBN SĪNĀ (or **Avicenna**), **ABU ALI AL-HUSSEIN IBN ABDALLAH** (*b.* Kharmaithen, near Bukhara, Persia [now Uzbekistan, U.S.S.R.], 980; *d.* Hamadān, Persia [now Iran], 1037), *philosophy, medicine, biology, astronomy.*

For a detailed study of his life and work, see Supplement.

SINĀN IBN THĀBIT IBN QURRA, ABŪ SAᶜĪD (*b. ca.* 880; *d.* Baghdad, 943), *medicine, astronomy, mathematics.*

The son of Thābit ibn Qurra al-Ḥarrānī (*ca.* 830–901), and the father of Ibrāhīm ibn Sinān ibn Thābit (908–946), Sinān belonged to the sect of the Sabians originating in Ḥarrān. One of the most famous physicians of his time, Sinān worked mainly in Baghdad. He was born probably around 880: al-Masᶜūdī mentions a description by Sinān of the life at the court of Caliph al-Muᶜtaḍid (892–902), his father's protector. Apparently, Sinān held no position before 908. He was then physician to the caliphs al-Muqtadir (908–932), al-Qāhir (932–934), and al-Rāḍī (934–940).

Under al-Muqtadir, Sinān brilliantly directed the hospitals and medical administration of Baghdad. He was not a Muslim, and he cared for the faithful and unfaithful without discrimination. In 931, after a fatal malpractice, every Baghdad doctor, except a few famous ones, had to pass a test before Sinān.

Under al-Qāhir, Sabians were persecuted, and Sinān had to become Muslim and later fled to Khurāsān, returning under al-Rāḍī. After the latter's death he served Amīr Abu 'L-Ḥusayn Baḥkam in Wāsiṭ, looking after his character and physical health.

None of Sinān's work is extant. As listed by Ibn al-Qifṭī, it can be divided into three categories: historical-political, mathematical, and astronomical; no medical texts are mentioned. A treatise of the first kind contained the already mentioned description of life at the court of al-Muᶜtaḍid, and, among other things, a sketch for a government according to Plato's *Republic*. Al-Maᶜsūdī criticizes it, adding that Sinān should rather have occupied himself with topics within his competence, such as the science of Euclid, the *Almagest*, astronomy, the theories of meteorological phenomena, logic, metaphysics, and the philosophical systems of Socrates, Plato, and Aristotle.

Four mathematical treatises are listed: one addressed to ᶜAḍud al-Dawla; a correction of a commentary on his entire work by Abū Sahl al-Qūhī, made on the latter's request; one connected with Archimedes' *On Triangles*; and a correction, with additions, of Aqāṭun's *On Elements of Geometry* (Is this the Aya Sofya MS 4830, 5, *Kitāb al-Mafrūḍāt* by Aqāṭun?). The first two treatises cannot be Sinān's, since the addressees were active in the second half of the tenth century.

As to the third category, only the content of the *Kitāb al-Anwāʾ* (dedicated to al-Muᶜtaḍid) is somewhat known through excerpts by al-Bīrūnī; it is probably identical with *Kitāb al-Istiwāʾ* listed in the *Fihrist* and Ibn al-Qifṭī. The *anwāʾ* are the meteorological qualities of the individual days. Scholars disagree on their cause. Some scholars deduce them from the rising and setting of the fixed stars, others by comparing the weather in the past. Sinān maintains the latter opinion and disapproves of Galen, who wants to decide between the two only after prolonged experimental examination. Sinān agrees on the difficulty of testing them in a short period. He advises to verify whether the Arabs and Persians agree on a *nawʾ* (singular form of *anwāʾ*); if they do, it is most probable. According to al-Bīrūnī, Sinān also relates an Egyptian theory and one by Hipparchus, on where to fix the beginnings of the seasons.

One of the other astronomical treatises, directed to the Sabian Abū Isḥāq Ibrāhīm ibn Hilāl (*ca.* 924–994), is on the assignment of the planets to the days of the week. The seven planets were important in Sabian religion; each one had its own temple. Ibn al-Qifṭī lists several works on Sabian rites and religion.

BIBLIOGRAPHY

I. ORIGINAL WORKS. Fuat Sezgin, *Geschichte des arabischen Schrifttums,* V (Leiden, 1974), 291; Ibn al-Qifṭī, *Ta'rīkh-al-ḥukamā',* J. Lippert, ed. (Leipzig, 1903), 195 and the Ibn Abī Uṣaybi'a, *Ṭabaqāt al-aṭibbā',* I. A. Müller, ed. (Cairo, 1882), 224, list Sinān's work, of which nothing is extant. Al-Bīrūnī gives excerpts from the *Kitāb al-anwā'* in his *Chronology of Ancient Nations,* C. E. Sachau, ed. (London, 1879), 232, 233, 262, 322; see on this subject also O. Neugebauer, "An Arabic Version of Ptolemy's Parapegma From the 'Phaseis,' " in *Journal of the American Oriental Society,* **91,** no. 4 (1971), 506. A translation of the Aya Sofya MS 4830, 5, of which Sinān might be the author, is in preparation by the writer of this article.

II. SECONDARY LITERATURE. Biographical references can be found in Ibn al-Qifṭī, *Ta'rīkh al-ḥukamā',* J. Lippert, ed. (Leipzig, 1903), 190–195; Ibn Abī Uṣaybi'a, *Ṭabaqāt al-aṭibbā',* I. A. Müller, ed. (Cairo, 1882), 220–224; and C. Brockelmann, *Geschichte der arabischen Literatur,* I (Leiden, 1943), 244–245 and supp. I (Leiden, 1937), 386. Ya'qub al-Nadīm, *Kitāb al-Fihrist,* G. Flügel, ed. (Leipzig, 1871–1872), 272, 302, mentions Sinān, without giving much information. D. Chwolson, *Die Ssabier und der Ssabismus,* I (St. Petersburg, 1856; repr., Amsterdam, 1965), 569–577, elucidates Sinān's biography and Sabian religion. L. Leclerc, *Histoire de la médecine arabe* (Paris, 1876), 365–368, emphasizes Sinān the physician. Al-Mas'ūdī's description and criticism is to be found in al-Mas'ūdī, *Murūj al-dhahab wa ma'ādin al-jawhar, Les prairies d'or,* I, Arabic text and French translation by C. Barbier de Meynard and Pavet de Courteille (Paris, 1861), 19–20.

YVONNE DOLD-SAMPLONIUS

SITTER, WILLEM DE (*b.* Sneek, Netherlands, 6 May 1872; *d.* Leiden, Netherlands, 20 November 1934), *astronomy.*

De Sitter was the son of L. U. De Sitter, a judge, who became president of the court at Arnhem, and T. W. S. Bertling. After preparatory education at Arnhem he entered the University of Groningen, where he studied mathematics and physics. Later he became interested in astronomy while participating (under Kapteyn's guidance) in the work of the astronomical laboratory. From 1897 to 1899 he worked under David Gill at the Royal Observatory in Cape Town, South Africa, and next served as an assistant to Kapteyn at Groningen until 1908, when he was appointed professor of astronomy at the University of Leiden. From 1919 until his death he also was director of the Leiden observatory.

De Sitter's main contributions to astronomy lie in the fields of celestial mechanics (particularly his research into the intricate problem of the dynamics of the satellites of Jupiter), the determination of the fundamental astronomical constants, and the theory of relativity applied to cosmology. He also contributed significantly, during his early years, to stellar photometry and to the measurement of stellar parallaxes in the context of Kapteyn's general program of researches on the structure of the Milky Way. Throughout his career De Sitter often acknowledged that his scientific approach was strongly influenced by Kapteyn and Gill.

Of the twelve satellites of Jupiter, four (Io, Europa, Ganymede, and Callisto) are of the fifth and sixth stellar magnitude, whereas the remaining ones are of thirteenth magnitude and fainter. These four, discovered by Galileo in 1610, were used by Römer in 1675 to determine the velocity of light. Their brightness enabled accurate determinations to be made of their projected positions on the sky with respect to Jupiter and to each other, first by heliometer observations and then photographically. De Sitter first participated at the Cape Observatory in the heliometer observations started by Gill and W. H. Finlay and then undertook their reduction and discussion, which led to his doctoral thesis at Groningen, "Discussion of Heliometer Observations of Jupiter's Satellites" (1901).

Satellites of a planet slightly disturb the motions of each other by their mutual attractions, and a study of these perturbations enables the mass of a satellite to be determined with its orbital elements. This problem was one of high mathematical complexity, requiring critical appreciation of both the value and limitations of the observations. De Sitter was particularly well prepared for the task. The satellites of Jupiter continued to interest him for the next thirty years. At his instigation a series of photographic observations were obtained at observatories in Cape Town, Greenwich, Johannesburg, Leiden, and Pulkovo. Their analysis, and a discussion of old observations of the eclipses of these satellites (by Jupiter) dating from 1668, led to an extensive series of publications by De Sitter in various journals and observatory publications, culminating in his "New Mathematical Theory of Jupiter's Satellites" (1925).

Shortly after Einstein's first publication on the restricted principle of relativity, De Sitter discussed its consequences for the small deviations in the motions of the moon and the planets; and after Einstein's paper on the generalized theory of relativity, De Sitter published (1916–1917) a series of three papers on "Einstein's Theory of Gravitation and Its Astronomical Consequences" in *Monthly Notices of the Royal Astronomical Society.* In the third of these papers he introduced what soon became known as the "De Sitter universe" as an alternative to the "Einstein universe." Because of his broad knowledge of dynamical astronomy, De Sitter was able to discuss fully the astronomical consequences of the theory of relativity, and he was among the first to appreciate its significance for astronomy. Apparently De Sitter's papers contributed uniquely to the introduction into the English-speaking countries of Einstein's theory during and shortly after World War I, and, for instance, led to Eddington's solar eclipse expeditions of 1919 to measure the gravitational deflection of light rays passing near the sun. De Sitter showed that in addition to the solution given by Einstein himself for the Einstein field equation (representing a static universe) a second model was possible with systematic motions—particularly the "expanding universe"—provided the density of matter could be considered negligible. Subsequent work by Georges Lemaître, Eddington, and De Sitter led to solutions satisfying more accurately both theory and observations, from which modern cosmology has emerged.

Closely related to De Sitter's work on the satellites of Jupiter were his investigations of the rotation of the earth and of the fundamental astronomical constants. Starting in 1915 with a discussion of the figure and composition of the earth, he tried to combine in a coherent system results from geodetic and gravity measurements with those from astronomical observations. In his paper "Secular Accelerations and Fluctuations of the Longitude of the Moon, the Sun, Mercury and Venus" (1927), De Sitter showed that these phenomena can be understood by assuming that varying tidal friction influences the rotation of the earth as well as the motion of the moon and by assuming internal changes in the moment of inertia of the earth. A comprehensive discussion of the fundamental, but observationally interrelated, astronomical constants—for example, the parallax of the sun, the constant of aberration, and the constant of nutation—appeared in "The Most Probable Values of Some Astronomical Constants; 1st Paper: Constants Connected With the Earth" (1927). An unfinished manuscript extending this work was edited and commented upon by D. Brouwer and was published posthumously in 1938 as "On the System of Astronomical Constants."

As director of the Leiden observatory, De Sitter successfully reorganized the institute, adding an astrophysical department and modern observing facilities. The latter included an arrangement with the Union Observatory in Johannesburg for the use of the telescopes there. This arrangement later led to the establishment of a Leiden station in Johannesburg that was equipped with a twin astrograph donated by the Rockefeller Foundation. These organizational efforts and nearly uninterrupted research were carried out by De Sitter despite repeated periods of illness. In 1933 he published a *Short History of the Observatory of the University at Leiden, 1633–1933* to commemorate its third centennial. In 1921 he created the *Bulletin of the Astronomical Institutes of the Netherlands.*

De Sitter was president of the International Astronomical Union from 1925 to 1928 and in that capacity did much to reestablish relations between scientists of formerly hostile countries. He received many honors, including the Gold Medal of the Royal Astronomical Society of London (1931), the Bruce Medal of the Astronomical Society of the Pacific, and honorary degrees from Cambridge, Cape Town, Oxford, and Wesleyan universities.

BIBLIOGRAPHY

I. ORIGINAL WORKS. A list of principal publications is given in C. H. Hins's obituary (see below). De Sitter's works include "Secular Accelerations and Fluctuations of the Longitude of the Moon, the Sun, Mercury and Venus," in *Bulletin of the Astronomical Institutes of the Netherlands,* **4,** no. 124 (1927), 21–38; "The Most Probable Values of Some Astronomical Constants; 1st Paper: Constants Connected With the Earth," *ibid.,* 57–61; "New Mathematical Theory of Jupiter's Satellites," in *Annalen van de Sterrewacht te Leiden,* **12,** pt. 3 (1925), 1–83, and in his George Darwin lecture "Jupiter's Galilean Satellites," which appeared in *Monthly Notices of the Royal Astronomical Society,* **91** (1931), 706–738; *Kosmos, a Course of Six Lectures on the Development of Our Insight Into the Structure of the Universe* (Cambridge, Mass., 1932); *The Astronomical Aspect of the Theory of Relativity* (Berkeley, Calif., 1933); and "On the System of Astronomical Constants," D. Brouwer, ed., in *Bulletin of the Astronomical Institutes of the Netherlands,* **8** (1938), 213–231.

II. SECONDARY LITERATURE. Extensive obituaries are given by C. H. Hins, in *Hemel en Dampkring*, **33** (1935), 3–18, with bibliography; H. Spencer Jones, in *Monthly Notices of the Royal Astronomical Society*, **95** (1935), 343–347; and J. H. Oort, in *Observatory*, **58** (1935), 22–27. De Sitter's wife, Eleonora De Sitter-Suermondt, wrote *Willem de Sitter, een Mensenleven* (Haarlem, 1948), a memoir.

A. BLAAUW

ŠKODA, JOSEF (*b.* Pilsen, Bohemia [now Czechoslovakia], 10 December 1805; *d.* Vienna, Austria, 13 June 1881), *internal medicine.*

Since Škoda was frequently ill during childhood, he entered high school in Pilsen only at the age of twelve. He graduated near the top of his class in 1825 and entered the Faculty of Medicine at Vienna. His dissertation, on the "De morborum divisione," may be considered the first evidence of his critical turn of mind.

After graduation on 18 July 1831, Škoda returned to Pilsen and established a medical practice. At this time the first pandemic of Asiatic cholera was approaching Czechoslovakia; and he became the district cholera specialist, first in the Chrudim region, then in Kouřim, and finally in Pilsen (1831–1832). He realized how little his formal training in medicine had prepared him for medical practice, and further that in the fight against cholera, more could be achieved through preventive hygienic measures than by means of many officially recommended medicines. Škoda therefore returned to Vienna for further study. Before obtaining an unsalaried post as a doctor in the internal department of the General Hospital (autumn of 1833), he worked in Karl Rokitansky's Pathology-Anatomy Institute and developed a close relationship with him. The collaboration did not cease when Škoda moved to clinical work; indeed, it was only when Škoda confirmed his clinical diagnoses on the dissection table and was able to perform experiments in the dissection room that his collaboration with Rokitansky achieved its real purpose.

Around 1836 Škoda began investigating the fundamentals of percussion and auscultation, two of the modern examination methods in clinical medicine. Both had been propagated in France (Corvisart, Laënnec, Gaspard Bayle) and in Great Britain (Charles Williams, Robert Graves, William Stokes) within clinics but had not penetrated into general medical practice because they were difficult to master. Škoda used his knowledge of physics, which he had learned at the high school in Pilsen and had studied further at the University of Vienna under Julius Baumgärtner.

Škoda based his research into percussion and auscultation on physical acoustics. He simplified and unified the terminology, defined concepts, and supplemented them with his own observations and experience. After his first publications in 1836 and 1837, he elaborated his doctrine and in 1839 published it formally as *Abhandlung über Perkussion und Auskultation*. Škoda improved each of the five new editions, responding to criticism and to new ideas. The book was also translated into English and French.

Škoda critically evaluated the doctrines of the French school of medicine, which distinguished percussion sounds according to the organ—the thigh, the liver, the intestine, the lung, or whatever—and substituted a physical classification of percussion sound in four categories: from full to empty, from clear to muffled, from tympanous to nontympanous, from high to deep. A part of modern diagnostics is Škoda's discovery of tympanous percussion in the presence of serous pleurisy.

In the theory of auscultation Škoda first distinguished reverberations (heart sounds) from cardiac murmurs. On the basis of comparative observations of healthy people and those known to have heart disease he learned to diagnose various heart illnesses from the presence of murmurs in individual valves. He also evaluated pulsations of the neck veins and accentuation of further reverberations in the pulmonary artery. Through his lucid account of functional changes and symptoms attendant upon various changes in valves of the heart or the pericardium, he established the principles of the clinical physiology of heart diseases.

By comparing manifestations of sickness in the body, and its physical and chemical signs, with pathological findings at autopsies, Škoda was able to make accurate diagnoses. His critical approach to therapy led him to replace obsolete, inefficient methods (venesection) with rational new methods (puncture of empyema of the pericardium and pleura); he also introduced effective medicines, such as chloral hydrate and salicylic acid. His principles formed the basis of diagnoses and a simpler and more humane therapy. He also evaluated the results of medical treatment by means of statistical methods. Because of this, he was unjustly regarded as a therapeutic nihilist.

Škoda was a born teacher. In high school he taught his classmates, and while studying medicine he earned money by giving private lessons. About 1836, while on a hospital staff, he began courses in percussion and auscultation for doctors, which were his sole source of income; he continued them until his appointment as professor of internal diseases at the Vienna Faculty of Medicine on 26 September 1846. These courses carried his doctrines to foreign universities.

Škoda almost entirely eliminated typhoid fever in Vienna by securing the construction of a water main from mountain springs at a time when the true cause of the disease was unknown. (He had already demonstrated preventive measures during the cholera epidemic in Czechoslovakia.) Thus he enlarged the sphere of his activities into public health and epidemiology.

BIBLIOGRAPHY

I. ORIGINAL WORKS. Škoda's most important works are "Ueber die Perkussion," in *Medizinischer Jahrbücher des K. K. osterreichischen Staates*, **20** (1836), 453–473, 514–566; "Ueber den Herzstoss und die durch Herzbewegungen verursachten Töne," *ibid.*, **22** (1837), 227–266; "Anwendung der Perkussion bei Untersuchung der Organe des Unterleibes," *ibid.*, **23** (1837), 236–262, 410–439; "Ueber Abdominaltyphus und desse Behandlung mit Alumen crudum," *ibid.*, **24** (1838), 5–46, written with A. Dobler; "Untersuchungsmethode zur Bestimmung des Zustandes des Herzens," *ibid.*, **27** (1839), 528–559; "Ueber Pericarditis in pathologischer und diagnostischer Beziehung," *ibid.*, **28** (1839), 55–74, 227–272, 397–433, written with J. Kolletschka; *Abhandlung über Perkussion und Auskultation* (Vienna, 1839; 6th ed., 1864); "Äuszug aus der Eintrittsrede," in *Zeitschrift der Gesellschaft der Aerzte in Wien*, **3** (1847), 258–265; "Fälle von Lungenbrand behandelt und geheilt durch Einathmen von Terpentinöldampfen," *ibid.*, **9** (1853), 445–447; and "Ueber die Funktion der Vorkammern des Herzens und über den Einfluss der Kontraktionskraft der Lunge und der Respirationsbewegungen auf die Blutzirkulation," *ibid.*, 193–213.

II. SECONDARY LITERATURE. See the following, listed chronologically: Constantin Wurzbach, *Biographisches Lexikon des Kaiserthums Oesterreich*, XXXV (Vienna, 1877), 66–72; Maximillian Sternberg, *Josef Skoda* (Vienna, 1924); Erna Lesky, *Die Wiener medizinische Schule im 19 Jahrhundert* (Graz–Cologne, 1965), 142–149; and Zdeněk Hornof, "Josef Škoda als Choleraarzt in Böhmen," in *Clio Medica*, **2** (1967), 55–62; and "The Study of Josef Škoda at the Medical Faculty in Vienna in the Period 1825–1831," in *Plzeňský lékařský sborník*, **31** (1968), 131–148, in Czech with English summary.

ZDENĚK HORNOF

SKOLEM, ALBERT THORALF (*b.* Sandsvaer, Norway, 23 May 1887; *d.* Oslo, Norway, 23 March 1963), *mathematics.*

Skolem was the son of Even Skolem, a teacher, and Helene Olette Vaal. He took his *examen artium* in Oslo in 1905 and then studied mathematics (his preferred subject), physics, chemistry, zoology, and botany. In 1913 he passed the state examination with distinction.

In 1909 Skolem became an assistant to Olaf Birkeland and in 1913–1914 traveled with him in the Sudan to observe the zodiacal light. Then, in 1915–1916, he studied in Göttingen. In the latter year he returned to Oslo, where he was made *Dozent* in 1918. He received his doctorate in 1926.

Skolem conducted independent research at the Christian Michelsens Institute in Bergen from 1930 to 1938, when he returned as full professor to the University of Oslo. He retired in 1950. On several occasions after 1938 he was a visiting professor in America. Skolem served as editor of various mathematical periodicals and was a member of several learned societies. In 1962 he received the Gunnerus Medal in Trondheim.

Skolem published more than 175 works. His main field of research was the foundations of mathematics; but he also worked on algebra, number theory, set theory, algebraic topology, group theory, lattice theory, and Dirichlet series. Half of his works are concerned with Diophantine equations, and in this connection he developed a p-adic method. In 1920 he stated the Skolem-Löwenheim theorem: If a finite or denumerably infinite sentential set is formulable in the ordinary predicate calculus, then it is satisfiable in a denumerable field of individuals.

Skolem freed set theory from Cantor's definitions. In 1923 he presented the Skolem-Noether theorem on the characterization of the automorphism of simple algebras. According to this theorem, it is impossible to establish within a predicate calculus a categorical axiom system for the natural numbers by means of a finite or denumerably infinite set of propositions (1929).

Most of Skolem's works appeared in Norway, although his monograph *Diophantische Gleichungen* (1938) was published in Berlin. With Viggo Brun, he brought out a new edition (1927) of Netto's textbook on combinatorial analysis, for which he wrote all the notes and an important addendum.

Skolem also investigated the formal feasibility of various theories and concerned himself with the discovery of simpler, more constructive demonstrations of known theorems. He was especially influenced by the mathematicians Sylow and Thue.

BIBLIOGRAPHY

For a bibliography of Skolem, see T. Nagell, "Thoralf Skolem in Memoriam," in *Acta mathematica*, **110** (1963), which lists 171 titles. On his life, see also Erik Fenstadt, "Thoralf Albert Skolem in Memoriam," in *Nordisk Mathematisk Tidsskrift*, **45** (1963), 145–153, with portrait; and Ingebrigt Johansson, "Minnetale over Professor Thoralf Skolem," in *Norske Videnskåps-Akademi i Oslo Arbok 1964* (1964), 37–41.

H. OETTEL

SKRAUP, ZDENKO HANS (*b.* Prague, Czechoslovakia, 3 March 1850; *d.* Vienna, Austria, 10 September 1910), *chemistry.*

Skraup came from a Czech family of musicians; his father composed church music and popular songs, and his uncle is remembered to this day for his composition of the Czech national anthem. Although musically gifted as well, Skraup turned to chemistry, which he studied at the German Technische Hochschule in Prague. He became a fervent German-Austrian patriot, abandoning the national allegiance of his Czech forebears. After completing his studies and briefly working in a porcelain factory near Karlovy Vary (Karlsbad) and in the mint in Vienna, Skraup became assistant under Friedrich Rochleder and then Adolf Lieben, who held professorships at the University of Vienna. In 1886 Skraup moved to Graz, where he was appointed professor of chemistry first at the Technische Hochschule and then a year later at the University of Graz. In 1906 he accepted the invitation to succeed Lieben at Vienna.

The development of science in Austria and Bohemia was closely linked during the years before 1918, as is clearly evident from an examination of the activities of the group of eminent chemists including Adolph Martin Pleischl, Jacob Redtenbacher, Rochleder, Heinrich Hlasiwetz, Lieben, and Skraup, in Prague and Vienna (and other Austrian university towns). It was because of the influence of Rochleder, one of the founders of modern phytochemistry, that Skraup became interested in quinine alkaloids, an area of study important to medical and structural chemistry. In turn it was Skraup who guided young co-workers, among them Fritz Pregl, in the field of physiological chemistry.

Skraup's most renowned scientific contribution was his synthesis of quinoline. The published account of his work resulted in the development of heterocyclid chemistry of the quinoline series. Although the relation of quinoline to various alkaloids was recognized at the time, there was no easy way to prepare the substance. From the investigations of Karl Graebe on alizarin blue and of Wilhelm Königs on quinoline, it became evident to Skraup that heating nitrobenzene and glycerol in the presence of sulfuric acid could produce the compound

$$C_6H_5NO_2 + C_3H_8O_3 = C_9H_7N + 3H_2O + O_2.$$

Skraup tried out the reaction and confirmed this conjecture but found that the yield of quinoline was rather low. He believed that this drawback was mainly the result of the evolution of oxygen, and in order to avoid it he proceeded to combine aniline with glycerol under the same conditions:

$$C_6H_7N + C_3H_8O_3 = C_9H_7N + 3H_2O + H_2,$$

only finding that the amount of quinoline was again small. After combining the two methods, thus effectively oxidizing oxygen to water, Skraup obtained a satisfactory yield of quinoline:

$$2C_6H_5NH_2 + C_6H_5NO_2 + C_3H_8O_3 = 3C_9H_7N + 11H_2O.$$

Skraup's synthesis became a general method for the preparation of quinolines, in which an aromatic primary amine is heated with glycerol and sulfuric acid in the presence of nitrobenzene or some other oxidizing agent.

BIBLIOGRAPHY

The fundamental paper "Eine Synthese des Chinolins" appeared in the *Sitzungsberichte der mathematischnaturwissenschaftlichen Classe der Kaiserlichen Akademie der Wissenschaften*, **81** (1880), pt. II,I–V. Skraup's works are listed in succeeding volumes of Poggendorff, III (1898), 1254; IV, pt. II (1904), 1402, and V, pt. II (1926), 1173. The obituary by H. Schrötter in *Berichte der Deutschen chemischen Gesellschaft*, **43** (1910), 3683–3702, is informative. See also the article by M. Kohn, "A Chapter of the History of Chemistry in Vienna," in *Journal of Chemical Education*, **20** (1943), 471–473.

M. TEICH

SKRYABIN, KONSTANTIN IVANOVICH (*b.* St. Petersburg, Russia [now Leningrad, U.S.S.R.], 7 December 1878; *d.* Moscow, U.S.S.R., 17 October 1972), *helminthology, public health.*

The son of a communications engineer, Skryabin attended the Dorpat (now Tartu) Veterinary Institute while auditing classes at the Faculty of Biology of Dorpat University. From 1905 he was a veterinarian in Kazakhstan. Sent abroad to continue his research on helminthology, he worked in the laboratories of Max Braun, Lue, Alcide Railliet, and Furman.

In 1917 Skryabin became professor in the first chair of parasitology in Russia, at Novocherkassk, and from 1920 he headed the chair of parasitology of the Moscow Veterinary Institute (now the K. I. Skryabin Moscow Veterinary Academy). Three important helminthological research institutes were organized in Moscow under his direction: the helminthological section of the State Institute of Experimental Veterinary Medicine, reorganized in 1931 as the All-Union Institute of Helminthology and named for Skryabin in 1939; the helminthological section of the Central Tropical Institute (now the I. E. Martsinovsky Institute of Medical Parasitology and Tropical Medicine), which Skryabin directed from 1921 to 1949; and a small parasitological laboratory of the Faculty of Physics and Mathematics of Moscow University that he headed for several years. In 1942 the Laboratory of Helminthology of the U.S.S.R. Academy of Sciences was established, and it was headed by Skryabin until his death.

Skryabin's more than 700 works are devoted to morphology, biology, phylogeny, taxonomy, the geography of helminths, epidemiology (epizootiology), helminthiasis, clinical pathogenesis, preventive and therapeutic measures, and the development of principles and radical methods for eliminating helminths in man and animals. The more than 340 expeditions organized under his general direction—and frequently with his participation—played a major role in disseminating knowledge of helminths of man and animals. Material gathered on these expeditions was amassed through the method of complete helminthological dissection, which Skryabin elaborated. In addition to his revisions of many taxonomic groups, Skryabin described more than 200 new species and 100 new genera of helminths.

Skryabin's methods of prophylaxis against helminthiases were widely used in the Soviet Union and in other countries, and his principles of the complete elimination of various species of pathogenic helminths led, in certain areas, to the total liquidation of a number of helminthiases, including ascariasis, ancylostomiasis, and taeniarhynchosis. Through his work such helminthiases of animals as echinococcosis, cysticercosis of cattle, dictyocaulosis of large horned animals, fascioliasis of cattle, and *Moniezia* infections have been virtually eliminated in many districts in the Soviet Union.

Skryabin's special interest in the trematode and the diseases it causes resulted in the important monograph *Trematody zhivotnykh i cheloveka* ("Trematodes of Animals and Man"), of which twenty-five of twenty-seven projected volumes were published. Each volume included a description of a given taxonomic group and presented a new system and tables for diagnostic determinations of trematodosis in animals. His concern to further the progress of helminthology led him to create and head a commission, established in 1922 at the zoological museum of the U.S.S.R. Academy of Sciences, to coordinate biological, medical, veterinary, and agronomical research. The commission was reorganized in 1940 as the All-Union Society of Helminthologists of the U.S.S.R. Academy of Sciences.

His emphasis on the maintenance of relations with foreign scientists led Skryabin to travel frequently after World War II and to present many reports to the International Epizootic Bureau in Paris (1930–1937) and to the Eleventh International Veterinary Congress in London in 1930. His proposal for the international coordination of research on measures against trichinosis, echinococcosis, and fascioliasis was accepted at a meeting of parasitologists in Hungary in 1959. The international journal *Helminthologia* was founded that year at his request, and he served as president of its editorial board.

Skryabin was an honorary or active member of many foreign academies and scientific societies. He was twice awarded the State Prize of the U.S.S.R. (1941, 1950), received the Lenin Prize in 1957, and in 1958 was awarded the honorary title Hero of Socialist Labor.

N. P. SHIKHOBALOVA

SLIPHER, EARL C. (*b.* Mulberry, Indiana, 25 March 1883; *d.* Flagstaff, Arizona, 7 August 1964), *planetary astronomy.*

After receiving his B.S. degree at Indiana University, Slipher joined the staff of the Lowell Observatory at Flagstaff, Arizona, in 1905; he worked there until the day before his death. Slipher was a pioneer in planetary photography, and the quality of his photographs has seldom been surpassed. He regularly observed the brighter planets during their favorable oppositions over a period of more than fifty-five years, and his photographic sequences that show long-term changes on Mars and Jupiter and the various aspects of Saturn are unique.

Slipher's special interest, like that of Percival Lowell, was the study of Mars. He obtained almost 200,000 images of the planet, nearly half of them through telescopes during expeditions to Chile in 1907 and to South Africa in 1939, 1954, and 1956. At these sites he obtained sharper photographs because Mars crossed the meridian close to the observer's zenith, where the turbulent effects of the atmosphere were minimal.

Slipher was one of the first to recognize that multiple-image printing could improve the quality of information extracted from a series of photographs taken within a sufficiently brief interval to avoid blurring because of planetary rotation. He would make a single print by accurately superimposing a number of unusually sharp individual images. He also was one of the first to standardize his plates for photometric measures; this practice, which he initiated in 1918, is now generally followed.

Slipher found that features on Mars's surface, which normally are invisible when photographed in the violet, sometimes stand out as clearly as they do in yellow light. He referred to this phenomenon as "blue clearing." His other discoveries about Mars include the "W" clouds, the appearance of a very large dust cloud in 1956, and many secular and transient changes.

The culmination of Slipher's work on Mars was *Mars, the Photographic Story* (1962), a compilation of 512 photographs that graphically illustrated many facts known about the planet. *The Brighter Planets* was published two years later. Only two weeks before his death he saw a preliminary copy of the deluxe edition, which contained photographic reproductions instead of halftones.

Photographs and scientific discoveries made from space vehicles have, since Slipher's death, drastically changed the thinking of many on the idea that canals, oases, and valleys of vegetation have been seen and sometimes photographed on Mars. Slipher shared this older point of view, which had been widely publicized many years before by Percival Lowell. These classical interpretations do not importantly detract, however, from the value of the unique and extensive series of photographs that is his legacy to planetary science. His photographic sequence of Venus, one image of which shows a complete halo around the planet,

has been frequently reprinted; and his photograph taken on 4 December 1911, showing Mars about to be occulted by the moon, is a striking illustration of the relative apparent size and surface brightness of these two objects.

Slipher's many honors included an honorary D.Sc. from the University of Arizona, Tucson, and an honorary LL.D. from Arizona State College (now Northern Arizona University in Flagstaff).

BIBLIOGRAPHY

Slipher published a number of short papers usually based on photographs. The best of these were reproduced in the two books mentioned above. An interim summary of his work is "The Planets From Observations at the Lowell Observatory," in *Proceedings of the American Philosophical Society*, **79** (1938), 441–470.

An obituary by A. P. Fitzgerald is in *Irish Astronomical Journal*, **6** (1964), 297.

JOHN S. HALL

SLIPHER, VESTO MELVIN (*b*. Mulberry, Indiana, 11 November 1875; *d*. Flagstaff, Arizona, 8 November 1969), *astronomy*.

Slipher, a son of David Clarke and Hannah App Slipher, perfected techniques in spectroscopy and achieved great advances in galactic astronomy. He earned his B.A. (1901), his M.A. (1903), and his Ph.D. (1909) degrees at Indiana University, and received honorary degrees from the University of Arizona (1923), Indiana University (1929), the University of Toronto (1935), and Northern Arizona University (1965).

Soon after receiving his B.A., Slipher was asked by Percival Lowell to join the staff of the Lowell Observatory at Flagstaff, Arizona. Lowell had selected the site because its high altitude was conducive to good visibility. He had obtained a twenty-four-inch Alvan Clark refractor and a John Brashear spectrograph. Slipher installed the spectrographic equipment in 1902 and began work under Lowell's enthusiastic, driving direction.

Slipher's main contributions were to spectroscopy, in which he both pioneered instrumental techniques and made major discoveries. His research can be divided into three areas: planetary atmospheres and rotations, diffuse nebulae and the interstellar medium, and rotations and radial velocities of spiral nebulae.

Shortly after arriving at Lowell Observatory, Slipher began studying the rotations of the planets.

Since Venus shows no surface markings, optical determinations of its rotation period proved difficult. Slipher oriented the slit of his spectrograph perpendicular to the terminator of Venus and measured the inclination of the spectral lines. In 1903, after taking twenty-six plates, he determined that the period was surprisingly long—certainly much greater than the twenty-four hours that it was commonly believed to be. And in the next issue of the *Lowell Observatory Bulletin*, Slipher announced his measurements for the rotation of Mars. These results, obtained in the same manner as for Venus, are close to presently accepted values. He continued spectrographic observations of planetary rotation periods, and by 1912 he had measured them for Jupiter, Saturn, and Uranus.

Slipher's spectrograms also clearly showed, for the first time, bands in the spectra of the Jovian planets. In 1934 Rupert Wildt identified some of the spectral features as being caused by ammonia and methane, and Slipher and Arthur Adel identified many of the remaining bands. For his work on planetary spectroscopy, Slipher was awarded the gold medal of the Royal Astronomical Society in 1933.

In the area of diffuse nebulae and interstellar material, in 1912 Slipher noticed that the diffuse nebulosity in the Pleiades shows a dark-line spectrum similar to that of the stars surrounding the Pleiades; he therefore concluded that the nebula shines by reflected light. This discovery, one of the first to give incontrovertible evidence of particulate matter in interstellar space, paved the way for the work of Hertzsprung and Hubble on emission and absorption nebulae.

In 1908, while studying the spectrum of a binary star, Slipher discovered a sharp calcium line that did not exhibit the oscillatory motion of its companions. Recalling that J. F. Hartmann had found a similar line in 1904, he studied more spectra and found several other such lines. To explain the phenomenon, he correctly reasoned that there must be gas between the stars and the earth.

Thus Slipher's research on interstellar space was extremely important, for he demonstrated the existence of both dust and gas. During the late 1920's and early 1930's, the studies of the interstellar medium by Eddington, Plaskett, Trumpler, and others were directly influenced by his work.

Probably the most significant aspect of Slipher's research, however, dealt with spiral nebulae. During the fall and winter of 1912, he obtained a series of spectrograms indicating that the Andromeda Nebula is approaching the sun at a mean velocity

of 300 kilometers per second, the greatest radial velocity that had been observed.

Slipher continued such observations; and by 1914, when he released his results, he had obtained Doppler shifts for fourteen spirals. Despite the initially enthusiastic response of the astronomical community, many questioned Slipher's findings. For over a decade—until others began to believe and understand the implications of his findings—he was virtually the only observer investigating the velocities of extragalactic nebulae.

By 1925 Slipher had measured thirty-nine of the forty-four known radial velocities of spirals, the majority of which showed large velocities of recession, as much as 1,125 kilometers per second. Although the nature of spirals was not definitely known until Hubble proved in 1924 that they are external galaxies similar to the Milky Way, Slipher's early results suggested to a few perceptive astronomers that spirals are exterior to our system. Since the radial velocities of the spirals are so extraordinarily great, they probably could not be contained within the Milky Way. On 14 March 1914, just weeks after Slipher's original announcement, Hertzsprung wrote to him:

> My harty [sic] congratulations to your beautiful discovery of the great radial velocity of some spiral nebulae.
>
> It seems to me, that with this discovery the great question, if the spirals belong to the system of the Milky Way or not, is answered with great certainty to the end, that they do not. . . .

Moreover, H. D. Curtis, the chief proponent of the revival of the "island universe" theory (before Hubble's discovery of Cepheids in spirals), appears to have been influenced by Slipher's findings.

Hubble's velocity-distance relationship, first presented in 1929, was made possible by Slipher's velocity measurements. To construct the relationship, Hubble used these velocities and the distance measurements available for eighteen isolated nebulae and four objects in the Virgo cluster. The relationship also was used to compute distances for the nebulae on Slipher's list in which no stars could be detected.

The possibility of a relationship between distances and velocities of galaxies had been considered for years; C. Wirtz, K. Lundmark, and others had attempted unsuccessfully to construct such a relationship. Reliable distances were needed, but they were unobtainable without large instruments (like Mount Wilson's Hooker telescope) and ingenious techniques (like Shapley's period-luminosity law). Credit deservedly belongs to Hubble for his work on measuring these distances and in recognizing their relationship to the velocities; nevertheless, Slipher's findings were crucial to the discovery. His work prepared the way for investigations of the motions of galaxies and for cosmological theories based on an expanding universe.

Slipher also measured rotations of spirals, using the technique he had developed in his studies of planetary rotation. He found rotational velocities on the order of a few hundred kilometers per second and the direction of motion to be such as to "wind up" the spirals. These results contradicted the controversial proper motion measurements of Adriaan van Maanen. This discrepancy was not entirely resolved until the 1930's, when it was demonstrated conclusively that van Maanen's measurements had been subject to systematic errors.

Other areas of Slipher's research included the determination of radial velocities of globular clusters, spectroscopic studies of comets and aurorae, and observations of bright lines and bands in night sky spectra.

Slipher was also an unusually competent administrator; indeed, in recognition of that ability, as well as for his research, the Astronomical Society of the Pacific awarded him the Bruce Medal in 1935. He received his first experience in administration in 1915, when Lowell made him assistant director of the observatory. He became acting director upon Lowell's death in 1916 and continued in that capacity until he was made director in 1926, a post he held until 1952. During his directorship he supervised the trans-Neptunian planet search, which culminated in 1930 in the discovery of Pluto by Clyde Tombaugh, a staff member at Lowell.

Slipher's other administrative experience included serving as president of the Commission on Nebulae (no. 28) of the International Astronomical Union (1925 and 1928), vice-president of the American Astronomical Society (1931), and vice-president of the American Association for the Advancement of Science (1933).

In his work with the I.A.U. commission, Slipher made another important contribution to astronomy. As its president he became the center for all information concerning nebulae, serving as coordinator and organizer during the mid- and late 1920's, when the nature of galaxies and their relationship to the universe as a whole were being discovered by Hubble, Lundmark, and others.

Slipher was a member of the American Academy of Arts and Sciences, the American Philosoph-ical Society, the Astronomical Society of France, Phi Beta Kappa, and Sigma Xi. He received the Lalande Prize of the Paris Academy of Sciences in 1919 and the Draper Gold Medal of the National Academy of Sciences in 1922.

BIBLIOGRAPHY

I. ORIGINAL WORKS. Slipher's most important publications include "A Spectrographic Investigation on the Rotational Velocity of Venus," in *Lowell Observatory Bulletin*, no. 3 (1903), 9–18; "On the Efficiency of the Spectrograph for Investigating Planetary Rotations and on the Accuracy of the Inclination Method of Measurement: Tests on the Rotation of the Planet Mars," *ibid.*, no. 4 (1903), 19–33; "The Lowell Spectrograph," in *Astrophysical Journal*, **28** (1908), 397–404; "Peculiar Star Spectra Suggestive of Selective Absorption of Light in Space," in *Lowell Observatory Bulletin*, no. 51 (1909), 1–2; "The Radial Velocity of the Andromeda Nebula," *ibid.*, no. 58 (1913), 56–57; "Spectrographic Observations of Nebulae," in *Popular Astronomy*, **23** (1915), 21–24; and "Spectroscopic Studies of the Planets," in *Monthly Notices of the Royal Astronomical Society*, **93** (1933), 657–668.

II. SECONDARY LITERATURE. Obituaries are in *Publications of the Astronomical Society of the Pacific*, **81** (1969), 922–923; and *New York Times* (10 Nov. 1969), 47. Two excellent biographies of Slipher have been prepared by John S. Hall: "V. M. Slipher's Trailblazing Career," in *Sky and Telescope*, **39** (1970), 84–86; and "Vesto Melvin Slipher," in *Yearbook. American Philosophical Society* (1970), 161–166. Comments on Slipher's research were published on the occasion of several of his awards: S. Einarsson, "The Award of the Bruce Gold Medal to Dr. Vesto Melvin Slipher," in *Publications of the Astronomical Society of the Pacific*, **47** (1935), 5–10; and "Gold Medal Award, President's Speech," in *Monthly Notices of the Royal Astronomical Society*, **93** (1933), 476–477. For additional, related information, see Otto Struve and Velta Zebergs, *Astronomy of the 20th Century* (New York, 1962); A. Pannekoek, *A History of Astronomy* (London, 1961); and J. D. Fernie, "The Historical Quest for the Nature of the Spiral Nebulae," in *Publications of the Astronomical Society of the Pacific*, **82** (1970), 1189–1230. Slipher's private papers and correspondence (including the Hertzsprung letter cited above) are cataloged and are on microfilm at Lowell Observatory, Flagstaff, Arizona.

RICHARD HART
RICHARD BERENDZEN

SLOANE, SIR HANS (*b.* Killyleagh, County Down, Northern Ireland, 16 April 1660; *d.* Chelsea, London, England, 11 January 1753), *medicine, natural history.*

Sloane was the youngest of seven sons born to Alexander Sloane and Sarah Hickes, daughter of the chaplain to Archbishop Laud. The Sloane family emigrated to Ireland from Scotland during the reign of James I (VI), and the name was originally written Slowman or Slowan. Hans Sloane's first name was a compliment to the Hamiltons, earls of Clanbrassill, a family in which it was common.

After the Restoration, Sloane's father, receiver-general of taxes from County Down for the earl of Clanbrassill, became one of the commissioners of array. In a census of the previous year he is shown as having twenty-two people on his land, so he must have been a man of standing and property. He died in 1666.

In his youth Sloane turned his interest toward natural history: "I had from my youth been very much pleas'd with the Study of Plants, and other Parts of Nature, and had seen most of those Kinds of Curiosities, which are to be found either in the Fields, or in the Gardens or Cabinets of the Curious in these Parts." Killyleagh was a center of learning and had a school of philosophy, founded by the Hamilton family; and County Down, with Strangford Lough, presented many opportunities for the study of natural history. Sloane visited Copeland Island and saw "how the sea-mews laid their eggs on the ground, so thick that he had difficulty in passing along without treading on them"; and he was much intrigued with the seaweed on the shore, which the Irish were accustomed to chew in order to cure scurvy. These experiences, involving natural history and medicine, were the basis of his career.

At the age of sixteen, Sloane was taken with spasms of spitting blood and probably suffered from an attack of tuberculosis; but "by temperance, and abstaining from wine, and other fermented liquors, and the prudent management of himself in all other respects, he avoided the consequences of a disorder which must otherwise have proved fatal to him." Three years later, in 1679, he was well enough to go to London to study medicine. He lodged in Water Lane, next to the laboratory of the Worshipful Society of Apothecaries, where he studied chemistry under Nicolaus Staphorst and botany at the Apothecaries' Physick Garden at Chelsea. He attended lectures on anatomy and medicine, but most important at this period of his life were his friendships with two of the greatest English men of science of the day, John Ray and Robert Boyle.

In 1683 Sloane started on his grand tour of Europe. On his way to Paris he met Nicolas Lemery; in Paris he frequented the Charité hospital and heard botany lectures by Tournefort and anatomy lectures by Duverney. It was impossible for a Protestant to take a degree in France, but at that time the town of Orange in Provence was still under the House of Orange. Its university gave examinations and conferred degrees but provided no instruction in medicine. Sloane graduated Doctor of Physick there on 28 July 1683, then went to Montpellier to complete his studies, working under the physicians Charles Barbeyrac, Pierre Chirac, and Pierre Magnol.

The persecution of Protestants in France was starting in 1684, when Sloane returned to London with the intention of practicing medicine. For the contributions that he had already made to botany, he was elected Fellow of the Royal Society on 21 January 1685. Robert Boyle recommended Sloane as a skillful anatomist and good botanist to the surgeon Thomas Sydenham. The latter exclaimed "That is all moghty [sic] fine, but it won't do; . . . no, young man, all this is Stuff: you must go to the bedside, it is there alone that you can learn disease." The secret of Sydenham's fame lay in his systematic approach to the symptoms observed in his patients, and this method fitted perfectly into Sloane's systematic study of botany; diagnosis of disease became a part of natural history. Sloane was admitted a fellow of the Royal College of Physicians of London on 12 April 1687.

Christopher Monck, second duke of Albemarle, was at that time appointed governor of Jamaica; and Sloane accompanied him as physician, sailing on 12 September 1687. The expedition was of great value to Sloane, not only giving him firsthand experience of the flora and fauna of a relatively little-known island but also enabling him to search for new drugs; it was not long since the bark of *Cinchona vera* had been brought to Europe and used as a febrifuge. The description of the voyage and the observations on the inhabitants, diseases, plants, animals (some of which he brought back alive), and meteorology of the West Indies make Sloane's book on the natural history of Jamaica indispensable even today. On his return to England in 1689, Sloane found James II fled and William III on the throne.

On 11 May 1695, Sloane married Elizabeth, daughter of John Langley and widow of Fulk Rose, formerly of Jamaica; they lived in a house that is now 4 Bloomsbury Place. Sloane was now launched not only in the highest and scientifically the most distinguished society—his friends included Ray, Boyle, John Locke, Samuel Pepys, Edmond Halley, and Sir Isaac Newton—but also in his profession of medicine, which became very lucrative. One guinea an hour was the value of his time, although he treated the poor for nothing. His fees, his investments in quinine bark and in sugar, and his wife's fortune—derived from her first husband's estates in Jamaica—made Sloane a rich man.

Sloane had four children, of whom two daughters survived infancy. The elder, Sarah, married George Stanley of Paultons, from whom the family of Sloane Stanley is descended; the younger, Elizabeth, married Colonel Charles Cadogan of Oakley, afterwards second Lord Cadogan, and ancestor of the Earls Cadogan.

Appointed physician to Queen Anne in 1712, Sloane played a small but vital (although unrecognized) part in the history of England. On 27 July 1714, a political battle was fought in the Privy Council between Henry St. John, Viscount Bolingbroke, and Robert Harley, earl of Oxford. The latter lost and was dismissed; and nothing seemed to stand in the way of the succession of the Jacobites upon the queen's death, which appeared imminent because she had fainted at the Privy Council meeting. Rumors circulated constantly that the queen was dead (on which government stocks rose 3 percent) or that she was still alive. Sloane urged that she be bled, which was done; and she recovered sufficiently to preside over another meeting of the Privy Council, at which she had just strength enough to hand the treasurer's staff of office to Charles Talbot, duke of Shrewsbury. The Protestant succession of George, elector of Hanover, in accordance with the Act of Settlement, was then assured. It was her certified death, after so much uncertainty, that gave rise to the expression "as dead as Queen Anne."

On 3 April 1716, George I conferred a baronetcy on Sloane, and on 30 September 1719 he was elected president of the Royal College of Physicians of London. In that post he inspired the petition to Parliament drawing attention to the evils of alcoholism that resulted in the Gin Acts, at a time when dissolute crowds thronged the streets of London shouting "No gin, no King." Another event of his presidency was the publication of the fourth *London Pharmacopoeia*, which reflected Sloane's efforts to rationalize medical prescriptions, get rid of the disgusting ingredients that had hitherto disgraced them, discard the fetishes of superstition, and include a catalog of medicinal herbs with clear definitions of their properties and the methods by which they could be identified. When

he bought his property in Chelsea, including the Physick Garden, he conveyed it to the Society of Apothecaries for £5 a year, on condition that every year for forty years, fifty specimens of plants of different species, grown in the garden, be supplied to the Royal Society.

It may be claimed that Sloane introduced the scientific method into medicine. In a volume of the *Philosophical Transactions of the Royal Society* edited by him, he was at pains to emphasize the difference between "Matters of Fact, Experiment, or Observation, and what is called Hypothesis," in which latter category he included the old notion of "humours." The humoral theory could not explain the fact, made evident by experiment, that quinine reduced fever. "A poor Indian who first taught the cure of an Ague, of which the Lady of the Count of Cinchon was sick, overthrew with one simple medicine, without any Preparation, all the *Hypotheses* and Theories of Agues, which were supported by some Scores not to say Hundreds of Volumes."

A great believer in the importance of diet, Sloane, who became familiar with chocolate in Jamaica, found it to be more digestible when mixed with milk. The resulting product was known as "Sir Hans Sloane's Milk Chocolate," a recipe used by Messrs. Cadbury until 1885. He was consulted by the British government on the preservation of the health of ships' crews in the Royal Navy and on the precautions to be taken against the threat of the plague of Marseilles of 1720. He also played an important part in establishing the practice of inoculation for smallpox, brought to England by Lady Mary Wortley Montagu in 1718.

In 1739 Sloane was associated with Thomas Coram in the foundation of the Foundling Hospital, and in the same year his godson, Sir Richard Manningham, founded the first Lying-in Ward in the parochial hospital of St James's, Westminster. Sloane's secretary, Cromwell Mortimer, started a health insurance scheme, offering to treat patients "for a certain salary, by the year." Since he followed the principle that "Sobriety, temperance and moderation, are the best and most powerful preservatives that Nature has granted to Mankind," it is not surprising that at a time when most remedies were useless if not injurious, Sloane's reputation as a physician was so deservedly high.

Sloane was elected one of the two secretaries of the Royal Society in 1693. In 1727 the president, Sir Isaac Newton, died; Sloane was elected to succeed him, a post he occupied until 1741.

In 1712 Sloane felt it desirable to acquire a country house and bought the manor house at

Chelsea from Lord Cheyne; but he did not move into it until 1742. Throughout his life he amassed collections. The first were botanical specimens collected in France and the West Indies; and they formed the material for his catalog of plants. His herbarium fills 337 folio volumes in the British Museum (Natural History). Sloane soon bought and added other collections of plants, animals, insects, fossils, minerals, precious stones, and ethnographical specimens; he also branched out into Egyptian, Assyrian, Etruscan, Roman, Oriental, American Indian, and Peruvian antiquities. To these were added works of art by Albrecht Dürer, Hans Holbein, and Wenzel Hollar, and a rich collection of coins and medals. Sloane's library contained over 50,000 books and 3,500 bound volumes of manuscripts.

Such treasures demanded careful provisions in a will. Sloane might have left them to his family, but there would then have been no guarantee of their preservation intact. Eventually, in a will made in 1739, to which codicils were added in 1749 and 1751, "desiring very much that these things, tending many ways to the Glory of God, the confutation of Atheism and its consequences, the use and improvement of Physic, and other Arts and Sciences, and Benefit of Mankind, may remain together, and not be separated, and that chiefly in and about the City of London . . . where they may by the great confluence of people be of most use," he offered them to the British nation, provided the sum of £20,000 was paid to his daughters.

After Sloane's death in 1753, the trustees whom he had appointed met; the matter was brought before Parliament, which on 7 June 1753 received the royal assent for the act enabling purchase of the museum or collection of Sir Hans Sloane. To this were added the Harleian collection of manuscripts and the Cotton Library, and the British Museum was founded. It was installed in Montague House, Great Russell Street, and was opened to the public in 1759. The natural history departments, which had been the original kernel of Sloane's collections, were moved to the British Museum (Natural History) in South Kensington, which was opened to the public in 1881.

The names of Sloane and of his sons-in-law are dotted all over his former property in Chelsea: Sloane Street, Hans Crescent, Paultons Square, Cadogan Gardens, Oakley Street; and the Physick Garden, where Sloane worked and its curator Philip Miller established the part played by insects in pollination, still continues to serve botany.

Sloane was a fellow of the Royal Society for all but twenty-one days in sixty-eight years, the longest fellowship. When he was young, Thomas Hobbes, born at the time of the Armada, was alive; when he was old, he knew Thomas Martyn, a botanist who died after the birth of Queen Victoria.

BIBLIOGRAPHY

I. ORIGINAL WORKS. Sloane's writings include *Catalogus plantarum quae in Jamaica sponte proveniunt* (London, 1696); and *Voyage to Madeira, Barbadoes, and Jamaica; With the Natural History of Jamaica*, 2 vols. (London, 1707–1725).

II. SECONDARY LITERATURE. See Gavin de Beer, *Sir Hans Sloane and the British Museum* (London, 1953), which contains bibliographical references to all the chief MSS and printed sources of information on the life and work of Sloane; William Eric St. John Brooks, *Sir Hans Sloane. The Great Collector and His Circle* (London, 1954); and *The Sloane Herbarium. An Annotated List of the Horti Sicci Composing It; With Biographical Accounts of the Principal Contributors*, based on records compiled by James Britten and with an intro. by Spencer Savage, revised and edited by J. E. Dandy (London, 1954).

GAVIN DE BEER

SLUSE, RENÉ-FRANÇOIS DE (*b.* Visé, Principality of Liège [now Belgium], 2 July 1622; *d.* Liège, 19 March 1685), *mathematics.*

Although the family name is variously spelled in the archives and documents, its correct form is de Sluse in French and Slusius in Latin. Sluse was a nephew of Gualthère Waltheri, secretary of papal briefs to Innocent X. Destined by his well-to-do family for an ecclesiastical career, Sluse went to Louvain in the fall of 1638 and remained through the summer of 1642. In 1643 he obtained a doctorate in law from the University of Rome. He lived in Rome for ten years more, becoming proficient in Greek, Hebrew, Arabic, Syriac, and astronomy. But his natural gifts led him to mathematics and a thorough study of the teachings of Cavalieri and Torricelli on the geometry of indivisibles.

On 8 October 1650, Innocent X appointed Sluse canon of the cathedral of Liège. His understanding of law and his great knowledge brought him many high positions. But his success in the administration of a small state severed him from the life he had known in Rome and thrust him into an intellectual vacuum; and his administrative duties left him little leisure for scientific work, particularly after 1659, when he became a member of the Privy Council of Prince-Bishop Maximilian Henry, who was also elector of Cologne. The only way that Sluse could survive as a scientist was, according to the practice of the time, to conduct an extensive correspondence with the leaders of mathematical studies: Blaise Pascal, Huygens, Oldenburg, Wallis, and M. A. Ricci.

In June 1658, Pascal, under the name of A. Dettonville, challenged mathematicians to solve a number of problems related to the cycloid. The evaluation of the area between a cycloid and a line parallel to its base, and the calculation of the volume generated by a rotation of this area around the base or around a line parallel to the base, were among the problems proposed, and already solved, by Pascal. In his work on the cycloid (1658) Pascal paid homage to the elegance of the solutions Sluse had sent to him, and the two remained regular correspondents. In his correspondence with Pascal, Sluse discussed the areas limited by curves corresponding to the equation

$$Y^m = Kx^p (a-x)^n$$

and the cubature of various solids; and as an example he found the volume generated by the rotation of a cissoid around its asymptote. These questions were discussed in his *Miscellanea*, published in 1668 as a section of the second edition of his *Mesolabum*.

One of the questions widely studied by the geometers of Greek antiquity was the duplication of the cube, that is, the construction of a cube of a volume double that of a given cube. This led to the solution of a cubic. More generally, Sluse discussed the solutions of third- and fourth-degree equations. Descartes had shown that their solution corresponds to the intersection of a parabola and a circle, and Sluse demonstrated that any conic section can be substituted for the parabola. He developed his method in *Mesolabum* (1659), particularly in the second edition.

In his *Géométrie*, Descartes had demonstrated the application of geometrical loci to the solution of equations of higher degrees. Sluse was among those who perfected the methods of Descartes and Fermat to draw tangents and determine the maxima and minima. By completing Descartes's construction for the solution of third- and fourth-degree equations and using a circle and any conic section, Sluse generalized the method for the solution of equations through the construction of roots by means of curves. In 1673 he published a digest of the results of his work in the *Philosophical Transactions* and became a member of the Royal Society in the following year.

The discovery of a general method for the construction of tangents to algebraic curves places Sluse among the pioneers in the discovery of the calculus. At Huygens' suggestion, Leibniz learned analytical geometry through the writings of Sluse and Descartes. Sluse deserved the judgment formulated by Huygens in a letter to Oldenburg: "(Slusius) est geometrarum, quos novi, omnium doctissimus candidissimusque."

Sluse was also a historian and wrote a book on the death of St. Lambert, the bishop of Tongres, who was killed on the spot to which St. Hubert, his successor, transferred the seat of his bishopric (which became Liège). Another historical study concerns the famous bishop of Maastricht, St. Servatius. Among his unpublished manuscripts is a history of Cologne.

The breadth of Sluse's interests is attested by the variety of subjects covered in the hundreds of pages of his unpublished manuscripts now preserved at the Bibliothèque Nationale, Paris. Although concerned mainly with mathematics, they also treat astronomy, physics, and natural history.

BIBLIOGRAPHY

Sluse's writings include *Mesolabum seu duae mediae proportionales inter extremas datas per circulum et ellipsim vel hyperbolam infinitis modis exhibitae* (Liège, 1659; 2nd ed., enl., 1668); "An Extract of a Letter From the Excellent Renatus Franciscus Slusius, Canon of Liège and Counsellor of His Electoral Highness of Collen [Cologne], Written to the Publisher in Order to Be Communicated to the R. Society, Concerning His Short and Easier Method of Drawing Tangents to All Geometrical Curves Without Any Labour of Calculation," in *Philosophical Transactions of the Royal Society*, **7** (1672), 5143–5147; "Illustrissimi Slusii modus, quo demonstrat methodum suam ducendi tangentes ad quaslibet curvas . . .," *ibid.*, **8** (1673), 6059; *De tempore et causa martyrii B. Lamberti, Tungrensis episcopi, diatriba chronologica et historica* (Liège, 1679); and *De S. Servatio episcopo Tungrensi, ejus nominis unico: Adversus nuperum de sancto Arvatio vel duobus Servatiis commentum* (Liège, 1684).

M. C. Le Paige published more than 100 letters from Sluse to Pascal, Huygens, Oldenburg, Lambeck, Sorbière, and Pacichelli in "Correspondance de René-François de Sluse publiée pour la première fois," in *Bullettino di bibliografia e di storia delle scienze matematiche e fisiche*, **17** (1884), 494–726, and his introduction is the best available biography of Sluse. Secondary literature also includes C. Le Paige, "Notes pour servir à l'histoire des mathématiques dans l'ancien Pays de

Liège," in *Bulletin de l'Institut archéologique liègeois*, **21** (1890), 457–565; P. Gilbert, *René de Sluse* (Brussels, 1886); F. Van Hulst, *René Sluse* (Liège, 1842); and L. Godeaux, *Esquisse d'une histoire des sciences mathématiques en Belgique* (Brussels, 1943).

MARCEL FLORKIN

SLUTSKY, EVGENY EVGENIEVICH (*b.* Novoe, Yaroslavskaya guberniya, Russia, 19 April 1880; *d.* Moscow, U.S.S.R., 10 March 1948), *mathematics, statistics.*

Slutsky's father was an instructor at a teachers' seminary and, from 1886, director of a school in Zhitomir. After graduating from a classical Gymnasium, Slutsky enrolled in the mathematics department of Kiev University in 1899. He participated in student disturbances there and consequently was inducted into the army in 1901; readmitted to the university shortly thereafter, he was again expelled in 1902. He then studied for three years at the Munich Polytechnikum.

In 1905 Slutsky received permission to continue his studies in Kiev. His interest in political economy led him to enroll at the Faculty of Law, from which he graduated in 1911 with a gold medal. From 1913 he taught at the Kiev Institute of Commerce, and from 1926 he worked in Moscow in the government statistical offices. He began teaching at Moscow University in 1934 and, in 1938, at the Institute of Mathematics of the Academy of Sciences of the U.S.S.R.

Slutsky belonged to the generation of Russian statisticians that developed under the influence of Pearson and his school. His interest in both practical statistical problems (economics and later the natural sciences) and their theoretical background led Slutsky into purely mathematical studies, which although sometimes not fully extended in their generality, nevertheless contained fundamental new ideas.

A pioneer of the theory of random functions, Slutsky generalized or introduced stochastic concepts of limits, derivative, and integral (1925-1928), and obtained the conditions of measurability of functions (1937). In 1927 he discovered that multiple moving averages obtained from a series of independent random variables generate series close to periodic ones; this finding stimulated the creation of the theory of stationary stochastic processes and constituted an important contribution to business cycle theory. An important group of Slutsky's papers is devoted to the classical theory of correlations of related series for a limited number of trials. In 1915

he contributed to economics what is now known as the fundamental equation of value theory, which partitions the effect of a change in the price of a commodity into the income and substitution effects.

Slutsky's applied work included studies of the pricing of grain, the mean density of population, the periodicity of solar activity (using information on aurorae boreales from 500 B.C.), and statistical studies of chromosomes.

BIBLIOGRAPHY

Slutsky's basic writings were collected in *Izbrannye trudy. Teoria veroyatnostey i matematicheskaya statistika* ("Selected Works. Probability Theory and Mathematical Statistics"; Moscow, 1960). Separately published works include *Teoria korrelyatsii i elementy uchenia o krivykh raspredelenia* ("Correlation Theory and Elements of the Theory of Distribution Curves"; Kiev, 1912); *Ser Viliam Petty. Kratky ocherk ego ekonomicheskikh vozzreny* ("Sir William Petty. A Short Essay on His Economic Views"; Kiev, 1914); *Tablitsy dlya vychislenia nepolnoy Γ-funktsii i funktsii veroyatnosti* χ^2 ("Tables for the Calculation of an Incomplete Γ-Function and the Probability Function χ^2"; Moscow–Leningrad, 1950); and "Sulla teoria del bilancio del consumatore," in *Giornale degli economisti*, **51** (1915), 1–26, trans. by American Economic Association, as "On the Theory of the Budget of the Consumer," in *Readings in Price Theory* (Chicago, 1952), 27–56.

On Slutsky and his work, see A. N. Kolmogorov, in *Uspekhi matematicheskikh nauk*, **3**, no. 4 (July–Aug. 1948), 143–151, with bibliography of 47 works by Slutsky (1912–1946); and N. V. Smirnov's obituary in *Izvestiya Akademii nauk SSSR*, Seria mat., **12** (1948), 417–420.

A. A. YOUSCHKEVITCH

SLYKE, DONALD DEXTER VAN. See **Van Slyke, Donald Dexter.**

SMEATON, JOHN (*b.* Austhorpe, England, 8 June 1724; *d.* Austhorpe, 28 October 1792), *civil engineering, applied mechanics.*

One of the foremost British engineers of the eighteenth century, Smeaton also gained a reputation as a man of science and distinguished himself through experimental research on applied hydraulics. He was descended from a family of Scots, one of whom, Thomas Smeton, turned to Protestantism late in the sixteenth century and held important positions in the church and in the University of

Glasgow. By the time of Smeaton's birth, the family resided near Leeds, where his father, William, practiced law. Smeaton was encouraged to follow a legal career, and after a sound elementary education he served in his father's office and was later sent to London for further employment and training in the courts. An early inclination toward the mechanical arts soon prevailed, however; and, with his father's consent, he became a maker of scientific instruments, a pursuit that allowed ample scope for both his scientific interests and his mechanical ingenuity.

Early in the 1750's Smeaton began the experiments that constituted his chief contribution to science; and during this period he also busied himself with several technical innovations, including a novel pyrometer with which he studied the expansive characteristics of various materials. The pace of industrial and commercial progress was quickening in Britain, however, and the attention of technical men was being directed increasingly toward large-scale engineering works. From 1756 to 1759 Smeaton was occupied with his best-known achievement, the rebuilding of the Eddystone lighthouse. By the end of the decade it had become evident that structural engineering and river and harbor works were more profitable than making scientific instruments. Accordingly, Smeaton established himself as a consultant in these fields; indeed, it was he who adopted the term "civil engineer" to distinguish civilian consultants and designers from the increasing number of military engineers who were being graduated from the Royal Military Academy at Woolwich. During the last thirty-five years of his life he was responsible for many engineering projects, including bridges, steam engine facilities, power stations run by wind or water, mill structures and machinery, and river and harbor improvements.

Smeaton became a fellow of the Royal Society, a member of the Royal Society Club, and an occasional guest at meetings of the Lunar Society. He also was a charter member of the first professional engineering society, the Society of Civil Engineers (not to be confused with the later Institution of Civil Engineers), founded in 1771; after his death it became known as the Smeatonian Society. Its founding reflected the growing sense of professionalization among British civilian engineers during the eighteenth century.

In 1759 Smeaton's engineering and scientific careers were crowned with outstanding success. In that year he completed the Eddystone lighthouse, which confirmed his reputation as an engineer, and

published a paper on waterwheels and windmills, for which he received the Copley Medal of the Royal Society.

In his research on waterwheels Smeaton reopened the question of the relative efficiency of undershot wheels (which operate through the impulse of the water against the blades) and overshot wheels (where the water flows from above and moves the wheel by the force of its weight). Through experiments on a model wheel he showed that, contrary to common opinion, overshot wheels are twice as efficient as undershot. Beyond this empirical generalization Smeaton displayed his scientific bent by speculating on the cause of the greater loss of energy ("mechanic power," as he termed it) in the undershot wheel and by concluding that it was consumed in turbulence — "nonelastic bodies [water], when acting by their impulse or collision, communicate only a part of their original power; the other part being spent in changing their figure in consequence of the stroke."

Following this initial success in research on applied mechanics, Smeaton's interests drifted toward natural philosophy and he devoted two further experimental investigations to the *vis viva* dispute and the laws of collision. He maintained that these seemingly abstract studies were of importance in practice, inasmuch as the conclusions of natural philosophers might, if incorrect, mislead practical men to adopt unsound procedures. The results he obtained, however, were more consequential in theory than in practice, for they confirmed not only the belief that mechanical effort could indeed be "lost" but also that mv^2 (*vis viva*) was a measure of "mechanic power." Smeaton recognized that his conclusions were in opposition to those favored by the disciples of Newton, and he diplomatically specified that both mv and mv^2 were useful values when properly interpreted.

Smeaton's career provides an early example of the interaction of engineering and applied science. His technical interests influenced the direction of his scientific research; and he used the results of his research in his own waterwheel designs, consistently favoring breast wheels and overshot wheels and almost never using the undershot system. There is reason to believe that Smeaton's work led other designers to foresake the long-preferred undershot wheel. Moreover, the continued economic importance of waterwheels contributed a sense of urgency to the recurrent controversy over the measure of "force"; and in these discussions Smeaton's research and his support of the *vis viva* school of thought played a prominent role.

Smeaton also performed extensive tests on an experimental Newcomen engine, optimizing its design and significantly increasing its efficiency. These studies, however, never rose above the level of systematic empiricism and, moreover, were soon overshadowed by James Watt's invention of the separate condenser. A few minor contributions to observational astronomy rounded out Smeaton's scientific work.

BIBLIOGRAPHY

I. ORIGINAL WORKS. Many of Smeaton's papers were collected and published posthumously: *Reports of the Late John Smeaton*, 4 vols. (London, 1812–1814). Vol. IV, *The Miscellaneous Papers of John Smeaton* (1814), contains the papers he contributed to the *Philosophical Transactions of the Royal Society*, of which the most important are his Copley Medal paper, "An Experimental Enquiry Concerning the Natural Powers of Water and Wind to Turn Mills and Other Machines Depending on a Circular Motion," **51** (1759–1760), 100–174; "An Experimental Examination of the Quantity and Proportion of Mechanic Power Necessary to Be Employed in Giving Different Degrees of Velocity to Heavy Bodies From a State of Rest," **66** (1776), 450–475; and "New Fundamental Experiments Upon the Collision of Bodies," **72** (1782), 337–354. These three papers were reprinted together as *Experimental Enquiry Concerning the Natural Powers of Wind and Water* (London, 1794) and are also conveniently collected in Thomas Tredgold, ed., *Tracts on Hydraulics* (London, 1826). P. S. Girard translated them into French as *Recherches expérimentales sur l'eau et le vent* (Paris, 1810). For the results of his experiments on the steam engine, see John Farey, *A Treatise on the Steam Engine* (London, 1827), 158 ff.

John Smeaton's Diary of His Journey to the Low Countries 1755, Newcomen Society for the Study of the History of Engineering and Technology, Extra Publication no. 4 (London, 1938); and "Description of the Statical Hydraulic Engine, Invented and Made by the Late Mr. William Westgarth, of Colecleugh in the County of Northumberland," in *Transactions of the Royal Society of Arts*, **5** (1787), 185–210, throw some additional light on the engineering sources of Smeaton's scientific interests.

II. SECONDARY LITERATURE. The fullest biography of Smeaton is still Samuel Smiles, "Life of John Smeaton," in *Lives of the Engineers*, 3 vols. (London, 1861–1862), II, 1–89. John Holmes, who knew Smeaton well, published *A Short Narrative of the Genius, Life and Works of the Late Mr. J. Smeaton, Civil Engineer* (London, 1793). For a recent biographical article, see Gerald Bowman, "John Smeaton—Consulting Engineer," in *Engineering Heritage*, 2 vols. (New York, 1966), II, 8–12. None of these treats Smeaton's scientific work adequately.

D. S. L. Cardwell has interpreted Smeaton's research in the context of the developing relationship between power technology and thermodynamics; see "Some Factors in the Early Development of the Concepts of Power, Work and Energy," in *British Journal for the History of Science*, **3** (1966–1967), 209–224; and *From Watt to Clausius* (Ithaca, N.Y., 1971), see index. The influence of Smeaton's research on the controversy over the measurement of "force" may be seen in Peter Ewart, "On the Measure of Moving Force," in *Memoirs of the Literary and Philosophical Society of Manchester*, 2nd ser., **2** (1813), 105–258. On his water power engineering, see Paul N. Wilson, "The Waterwheels of John Smeaton," in *Transactions. Newcomen Society for the Study of the History of Engineering and Technology*, **30** (1955–1957), 25–48.

The little that is known of the Society of Civil Engineers in the eighteenth century is presented fully in T. E. Allibone, "The Club of the Royal College of Physicians, the Smeatonian Society of Civil Engineers and Their Relationship to the Royal Society Club," in *Notes and Records of the Royal Society of London*, **22** (1967), 186–192; S. B. Donkin, "The Society of Civil Engineers (Smeatonians)," in *Transactions. Newcomen Society for the Study of the History of Engineering and Technology*, **17** (1936–1937), 51–71; and Esther Clark Wright, "The Early Smeatonians," *ibid.*, **18** (1937–1938), 101–110.

HAROLD DORN

SMEKAL, ADOLF GUSTAV STEPHAN (*b.* Vienna, Austria, 12 September 1895; *d.* Graz, Austria, 7 March 1959), *physics.*

Adolf Smekal was the elder child and only son of Gustav Smekal, an artillery officer. Because of repeated shifts of residence, he gained admission to university study by the "back door" of a Realschule diploma (1912) and a year at the Technische Hochschule in Vienna (1912–1913). He then attended the University of Graz for four years, receiving his doctorate (14 June 1917) under Michael Radakovič, to whose family the Smekals were closely related. Extreme nearsightedness had rendered Smekal unfit for military service. Yet neither that handicap nor his short stature prevented the young man from becoming a skilled and tireless alpinist, who however in middle age succumbed to corpulence and heart disease.

From the autumn of 1917 to the spring of 1919 Smekal continued his study of mathematics and physics at the University of Berlin. In June 1919 he took up an assistantship in the physical institutes of Heinrich Mache and Ludwig Flamm at the Technische Hochschule in Vienna, where he immediately joined the circle of young theorists around Hans Thirring at the university. In the autumn of 1920 Smekal accepted an assistantship to Gustav Jaeger at the university, where he qualified simultaneously as *Privatdozent* in both theoretical and experimental physics. The following year this *venia legendi* was extended to the Technische Hochschule, where in 1923 Smekal was appointed *Honorardozent* in the newly established Abteilung für Technische Physik.

In the autumn of 1928 Smekal was appointed professor of theoretical physics at the University of Halle. He was especially pleased by the experimental facilities of the institute attached to that post, in which he continued until deported by the Americans to West Germany in June 1945. After some years of professional and financial uncertainty, Smekal obtained a chair and institute of his own in 1949 at the University of Graz, as professor of experimental physics. Smekal married twice, in 1924 and 1942, and had one child of the latter marriage.

Like virtually all Austrian theoretical physicists of his generation, Smekal was thoroughly trained in statistical mechanics; and his first publications were on the foundations of quantum statistics. (His doctoral dissertation aimed to show "that such radical assumptions as those of the quantum theory are by no means necessary in order to avoid the equipartition of energy.") Smekal also assimilated, and soon exemplified, the Austrian critical-encyclopedic style. In the 1920's he was the principal abstractor of publications on quantum theory for the *Physikalische Berichte* and wrote several extraordinarily learned handbook articles and a veritable fountain of research papers that, although distinguished for their recherché bibliographic citations, generally were conceptually derivative and often lacking in physical as well as personal "tact."

The year and a half in Berlin (1917–1919) was extremely important for Smekal's scientific development and subsequent research. There he took up the quantum, the Bohr theory, and the problem of X-ray spectra. After delivering a crushing blow to the faltering Sommerfeld-Debye theory, which deduced X-ray spectra from hypothetical intraatomic mechanisms, Smekal, in competition with Dirk Coster and Gregor Wentzel, induced from the experimental data the number, arrangement, and allowed transitions between the atomic energy levels resulting in X-ray spectral lines.

Although these papers of 1920–1921 are probably his most original achievement as a theorist, Smekal's name is better known through the effect predicted by him in September 1923 and discovered experimentally by C. V. Raman in 1928. This is the alteration of the frequency of light upon being scattered by an atomic-molecular system—a decrease, or increase, by an amount equal to the frequency of the light that would be absorbed or emitted in transitions between the stationary states of that system. It was the radical light quantum viewpoint that enabled Smekal to foresee this effect; but its necessity was immediately accepted also by theorists who rejected that viewpoint, particularly those around Niels Bohr. Smekal implied that his considerations were independent of A. H. Compton's; he did not, however, preclude influence from William Duane, with whose particulate theory of X-ray diffraction Smekal's considerations appear to bear considerable affinity.

Smekal's interest in the great discrepancy between the mechanical strength of ideal and of real crystals, apparently aroused by contact with the Austro-Hungarians staffing the Kaiser-Wilhelm-Institut für Faserstoffchemie in Berlin, was first expressed in a paper extending A. A. Griffith's theory of fracture (1922). There Smekal first advanced his conception of irregularities in the structure of real crystals arising as a "frozen Brownian molecular motion."[1] After 1925 he advocated it as a kind of "universal remedy"[2] in solid-state physics. In 1925–1927 Smekal turned from fundamental questions in quantum theory and atomic physics to the technical physics of structure-dependent properties of solids. By 1933 he had become, and at his death he remained, a world authority on brittleness and the technology of pulverization.

NOTES

1. A. Smekal, "Kristalleigenschaften und Kristallisationsbedingungen," in *Forschungen und Fortschritte*, **5** (1929), 385–387.
2. A. Joffé, letter to the editor, *Naturwissenschaften*, **16** (1928), 744–745. Cf. Joffé, *Begegnungen mit Physikern* (Leipzig, 1967), 83–84.

BIBLIOGRAPHY

I. ORIGINAL WORKS. The only bibliography of Smekal's publications is in Poggendorff, V, 1176; VI, 2473–2474; and VII, 427–429. Smekal's most important monograph is "Allgemeine Grundlagen der Quantenstatistik und Quantentheorie," in *Encyklopädie der mathematischen Wissenschaften*, V, pt. 3 (Leipzig, 1926), 816–1214. His papers on the X-ray term scheme, "Zur Feinstruktur der Röntgenspektren," in *Zeitschrift für Physik*, **4** (1920), 26–45, and **5** (1921), 91–106, are based upon "le système de Smekal" (adopted by the marginal French but not by better-informed theorists), according to which every degree of freedom of every electron in an atom is entitled to its own quantum number. Smekal's prediction of the Raman effect is "Zur Quantentheorie der Dispersion," in *Naturwissenschaften*, **11** (1923), 873–875; and his initial publication on fracture theory is "Technische Festigkeit und molekulare Festigkeit," *ibid.*, **10** (1922), 799–804.

II. SECONDARY LITERATURE. The informative obituaries are by Ludwig Flamm: *Almanach. Österreichische Akademie der Wissenschaften*, **109** (1959), 421–427; *Acta physica austriaca*, **13** (1960), 140–143; Technische Hochschule, Vienna, *150 Jahre Technische Hochschule in Wien, 1815–1965* (Vienna, 1965), I, 359–361, and II, 166. See also H. Rumpf, "Zur Entwicklungsgeschichte der Physik der Brucherscheinungen; A. Smekal zum Gedächtnis," in *Chemie-Ingenieur-Technik*, **31** (1959), 697–705; and A. Faessler, "Adolf Smekal," in *Glastechnische Berichte*, **32** (1959), 180. For Smekal's criticism of the Debye-Vegard theory of X-ray spectra, see John L. Heilbron, "The Kossel-Sommerfeld Theory and the Ring Atom," in *Isis*, **58** (1967), 451–485.

PAUL FORMAN

SMITH, EDGAR FAHS (*b*. York, Pennsylvania, 23 May 1854; *d*. Philadelphia, Pennsylvania, 3 May 1928), *chemistry*.

After graduating from Gettysburg College, Smith studied under Friedrich Wöhler at the University of Göttingen, where he received the Ph.D. in 1876. He taught chemistry at the University of Pennsylvania (1876–1881), Muhlenberg College (1881–1883), Wittenberg College (1883–1888), and the University of Pennsylvania again (1888–1920). At Pennsylvania he also held the important executive offices of vice-provost (1898–1911) and provost (1911–1920) of the university. Smith was three times president of the American Chemical Society and served as scientific adviser to the federal government.

Smith's most important research was in electrochemistry. In 1901 he developed the rotating anode, which permitted the application of higher cathode current densities and greatly decreased the time required for electroanalysis. In turn this led to broader application of electroanalysis in research and industry.

In studies on atomic weights Smith and his students endeavored to determine more precisely the atomic weights of eighteen elements, using electrolytic and chemical methods. Other research was on complex inorganic acids. He and his collaborators prepared many salts of complex acids (for example, ammonium vanadico-phospho-tungstate) and elucidated their relationships.

A prominent historian of chemistry in the United States, Smith wrote mainly from a biographical viewpoint. He collected and endowed a notable library of books, manuscripts, prints, and other memorabilia, now known as the Edgar Fahs Smith Memorial Collection in the History of Chemistry, at the University of Pennsylvania.

BIBLIOGRAPHY

A complete list of Smith's books, translations, and brochures, and of doctoral theses by his students, is in the biography by Meeker (below) and in *Memorial Service for Edgar Fahs Smith . . . December 4, 1928* (n.p., 1928[?]). His books include *Electrochemical Analysis* (Philadelphia, 1890; 6th ed., 1918); *Elements of Electrochemistry* (Philadelphia, 1913); and *Chemistry in America: Chapters From the History of the Science in the United States* (New York, 1914).

George H. Meeker, "Biographical Memoir of Edgar Fahs Smith, 1854–1928," in *Biographical Memoirs. National Academy of Sciences*, **17** (1936), 103–149, has a portrait and references to seven other biographical accounts of Smith.

WYNDHAM D. MILES

SMITH, EDWARD (*b*. Heanor, Derbyshire, England, 1818[?]; *d*. London, England, 16 November 1874), *physiology, nutrition, public health*.

Strikingly little is known about Edward Smith, a competent and highly influential nineteenth-century British physiologist and public health worker. His birth was not recorded in the parish register at Heanor, and the exact date is unknown. His father was Joseph Smith, apparently a successful businessman in Derbyshire; the surname of his mother, whose Christian name was Martha, is not recorded. Smith obtained his initial medical degree (M.B.) at the Royal Birmingham Medical School in 1841, and the M.D. degree followed in 1843. The medical school shortly afterward became Queen's College, and its certificates were recognized by the University of London. In 1848 Smith received the London B.A. and LL.B. from Queen's College. He is known to have practiced medicine in Birmingham from 1841 to about 1848.

Smith's religion seems to have been Wesleyan, a matter of some significance at the time and for much of the nineteenth century. Candidates for admission to Oxford or Cambridge (although not to the University of London system) were required to subscribe to the tenets of the Church of England and could not be Dissenters. To adhere to doctrines not sanctioned by the official church was a handicap in other ways as well; it made easy access to Britain's highest intellectual and political strata considerably more than a matter of course. Smith's religious convictions, at least as a young man, would seem to have been intense, judging from a prize essay written while he was in medical school. The essay was basically a theological tract that employed the aortic system to support various fundamentalist religious theses. Smith married Matilda Frearson Clarke (an American citizen, according to the census of 1861) at Nottingham on 4 May 1843. Two daughters were born between 1847 and 1850. Smith made a rapid survey of living conditions in Texas in 1849, possibly requested by relatives of his wife, and published the results within a few weeks of his return to Britain.

Sometime in 1851 Smith established practice in London, having become fellow of the Royal College of Surgeons by examination in that year. Late in 1851 he was appointed lecturer in botany at the Charing Cross Medical School, and within a short time he became demonstrator in anatomy there. He also held an appointment as physician-accoucheur at the West London Lying-in Institution. All went well for a short time; but within a year he was involved in acrimonious dispute with the medical committee of Charing Cross Hospital. The incident was the first in a series of controversies with authorities and colleagues that characterized much of his professional career. It resulted in his dismissal from Charing Cross Medical School in 1853.

For several years Smith occupied himself with practice and writing on medical topics for the layman as well as for the professional. On 29 March 1855 he was appointed assistant physician at the Brompton Hospital for Consumption, a post he held for ten years. No later than 1862 Smith became known to public health officials in Britain, and in 1866 he was appointed inspector and medical officer to the Poor Law Board. For a time he was very influential in medical aspects of Britain's welfare system. But he ultimately came into conflict with Sir John Simon, head of the Local Government Board's medical department, and the remaining years of his life were spent somewhat in limbo.

Smith was elected fellow of the Royal Society in 1860. He was also a fellow of the Royal College of Physicians (1863), president of the Physiological Subsection of the British Association, and member of the National Association for the Promotion of Social Science.

Smith's claim to scientific distinction rests on his pioneering work in respiratory physiology, metabolism, and nutrition. It is also clear that scientific curiosity was not his primary motivating force; he was at heart a social reformer and was unique among scientific investigators of his time in that he mobilized his research to support reform movements.

While on the staff at Brompton Hospital, Smith devised ingenious and original methods for measurement of respiratory function and related metabolic phenomena. His interest in measuring the effects of physical exertion on respiration seems to have directed his attention to the punitive treadmill, a device then used in Britain's prisons. Prisoners sentenced to hard labor were required to spend many hours each day on the treadmill; and Smith, perceiving an opportunity, asked prison authorities for permission to use prisoners as experimental subjects. This experience seems to have led him to consider the diet and living conditions of prisoners. His published work dealt with both subjects and ultimately brought him election to the Royal Society.

The work also allied Smith with groups and individuals seeking to reform Britain's prisons and may have been a factor in arousing a more general interest in the plight of Britain's lower classes. In 1863 he testified before a parliamentary commission investigating prison conditions and presented physiological evidence to show that the treadmill, as it was used, was a cruel and inhumane device.

The work on respiration at rest and during exercise led naturally to a consideration of metabolism of foodstuffs and energy sources under the same conditions. Once again Smith's emphasis was on measurement, this time of foods ingested and metabolic products excreted. He made short work, in the process, of Liebig's dogmatic assertion that the energy for muscle exercise comes entirely from protein.

The cotton famine of 1862 (in Lancashire) brought Smith formally into what is now known as public health. In December of that year Sir John Simon, then medical officer of health, asked him to visit six stricken towns in order to determine the general state of nutrition of the unemployed and "the least outlay of money which [will] procure food enough for life." A second and larger survey was carried out in the summer of 1863. The results, although in many respects imperfect, provided very valuable quantitative information concerning diet and economic conditions. Probably more important is that Smith's surveys pointed the way to, and the necessity for, health research on entire populations. They were, in fact, the forerunners of larger and more elaborate field studies done before and after both world wars.

Smith subsequently did additional studies of nutrition and working conditions of specific types of workers, reporting the results in the language of the reformer. Tailors in the London area, he found, had mortality rates that were higher, at all ages, than the rates for farmers. In London's printshops the condition of young boys, working twelve hours a day, six days a week, demanded "instant amelioration." As a Poor Law official, Smith had a great deal of influence on workhouse dietary practice, the provision of medical care, and the design of the workhouses themselves.

The last ten years of Smith's life undoubtedly were anticlimactic. He was involved in almost continuous controversy and, in some instances, did not acquit himself well. One such case was the bitter quarrel with Sir John Simon. An obituary in the *British Medical Journal* said: ". . . medically, Dr. Edward Smith met with little success in practice nor did he contrive to conciliate the affections of his colleagues." This was probably a fair assessment. But neither the *Journal* nor other publications at the time credited him with his great innovations in the quantitative study of respiration and metabolism, and in the nutrition of populations. Smith was the first to devise quantitative methods suitable for studies on the human being during exercise. His monumental data on inspiratory volume, respiratory and pulse rates, and carbon dioxide production at rest and at various levels of exercise served as the basis for much of the work on muscular exercise in the latter part of the nineteenth century. But most British physiologists, in sharp contrast to those on the Continent, seem to have known little of Smith and his work, although both groups built on his concepts and results. His work on nutrition, although innovative and fundamental, fared little better until quite recently.

Smith's reputation may well have suffered during his life and after his death owing to the numerous quarrels in which he was involved. But he was also very much ahead of his time: he was not only scientifically gifted and innovative; he also believed in seeing that his results were applied *pro bono publico*. Partly for this reason and partly because of belated recognition of the excellence of his scientific work, Smith's life and work have recently been rescued from obscurity.

BIBLIOGRAPHY

I. ORIGINAL WORKS. Smith's writings include "The Spirometer: Its Construction, Indications and Fallacies," in *Medical Circular*, **9** (1856), 294, 305, 313–314; **10** (1857), 5, 40, 64–65; "The Influence of the Labour of the Treadwheel Over Respiration and Pulsation, and Its Relation to the Waste of the System, and the Dietary of the Prisoners," in *Medical Times and Gazette*, n.s. **14** (1857), 601–603; "Inquiries Into the Quantity of Air Inspired Throughout the Day and Night and Under the Influence of Exercise, Food, Medicine, Temperature, etc.," in *Proceedings of the Royal Society*, **8** (1857), 451–454; "Inquiries Into the Phenomena of Respiration," *ibid.*, **9** (1858), 611–614; "Experimental Inquiries Into the Chemical and Other Phenomena of Respiration, and Their Modifications by Various Physical Agencies," in *Philosophical Transactions of the Royal Society*, **149** (1859), 681–714; "On the Immediate Source of the Carbon Exhaled by the Lungs," in *Philosophical Magazine*, 4th ser., **18** (1859), 429–436; "Report on the Action of Prison Diet and Discipline on the Bodily Functions of Prisoners. Part I," in *Report of the British Association for the Advancement of Science*, **31** (1861), 44–81, written with W. R. Milner; "Report on the Food of the Poorer Labouring Classes in England," in *Sixth Report of the Medical Officer Privy Council* (London, 1864), 216–329 (app. 5); "Report on the Sanitary Circumstances of Printers in London," *ibid.*, 383–415; "Report on the Sanitary Circumstances of Tailors in London," *ibid.*, 416–430; "Dietaries for the Inmates of Work Houses," in House of Commons, *Parliamentary Papers* (Reports From Commissioners for 1866), XXXV, 321–629; "Metropolitan Workhouse Infirmaries and Sick Wards," *ibid.* (Accounts and Papers) (1866), LXI, 171–388; and "Report on the Sufficiency of the Existing Arrangements for the Care and Treatment of the Sick in Forty-Eight Provincial Workhouses Situated in Various Parts of England and Wales," *ibid.* (1867–1868), LX, 325–483.

II. SECONDARY LITERATURE. See T. C. Barker, D. J. Oddy, and John Yudkin, *The Dietary Surveys of Dr. Edward Smith, 1862–3*, Occasional paper no. 1, Dept. of Nutrition, Queen Elizabeth College (London, 1970); and Carleton B. Chapman, "Edward Smith (?1818–1874). Physiologist, Human Ecologist, Reformer," in *Journal of the History of Medicine and Allied Sciences*, **22** (Jan. 1967), 1–26.

CARLETON B. CHAPMAN

SMITH, ERWIN FRINK (*b*. Gilbert's Mills, New York, 21 January 1854; *d*. Washington, D.C., 6 April 1927), *plant pathology, bacteriology*.

Smith was the son of Rancellor King Smith and Louisa Frink Smith, who left New York to farm near Hubbardston, Michigan. Smith graduated from high school in Ionia, Michigan, then attended the University of Michigan, from which he received the B.Sc. in 1886 and the doctorate in 1889. From his youth, Smith was profoundly interested in botany; he served in the United States Department of Agriculture from 1899 and later was director of the plant pathology laboratory of the Bureau of Plant Industry. His researches made him the most distinguished of the early American plant pathologists. His work fully established that bacteria cause plant disease—a view that was vigorously contested by his European counterparts.

One of Smith's earliest investigations concerned yellows, a perplexing disease of the peach of which the etiology is not yet fully understood. Smith established the infectious nature of the disease and attempted its control by eradication. Although he was personally disappointed by his inability to discover the causative agent, he nevertheless disproved a number of earlier misconceptions about the malady.

Smith was more successful in demonstrating that bacterial pathogens invade plants through wounds and natural openings. He also showed insect transmission in certain diseases and provided a workable classification of genera of bacterial plant pathogens. He demonstrated that certain soil fungi (Fusaria) cause widespread and devastating vascular wilts. Smith's later years were taken up with the study of crown gall disease, which he compared to animal cancer. His studies on tumor formation in plants in its relation to cancer in man and animals won him the certificate of honor of the American Medical Association in 1913. His researches are summarized in a number of papers and, especially, in his three-volume *Bacteria in Relation to Plant Diseases*.

Smith was a member of a number of scholarly societies and received many honors. He had broad interests in biology and was a lover and patron of music, art, and literature (a collection of his sonnets was brought out privately in 1915). He was twice married, first to Charlotte May Buffett, who died in 1906, then to Ruth Warren, who survived him. His home life was simple to the point of austerity, and, although not an active churchman, he was deeply religious.

BIBLIOGRAPHY

I. ORIGINAL WORKS. Smith's most important works are *Bacteria in Relation to Plant Diseases*, 3 vols. (Washington, D.C., 1905–1914); and *An Introduction to Bacterial Diseases of Plants* (Philadelphia–London, 1920). He published widely in journals, and bibliographies of his works may be found in notices by L. R. Jones and R. H. True, cited below.

II. SECONDARY LITERATURE. On Smith and his work, see Florence Hedges, "Dr. Erwin F. Smith, Scientist, is Dead," in *United States Department of Agriculture Official Record*, 6 (1927), 1, 5, 8; L. R. Jones, "Biographical Memoir of Erwin Frink Smith," in *Biographical Memoirs. National Academy of Sciences*, 21 (1939), 1–71, with portrait, synopsis of researches, and bibliography; L. R. Jones, W. H. Welch, and F. V. Rand, "To Erwin Frink Smith," in *Phytopathology*, 18 (1928), 1–5, testimonials to Smith given at a dinner in Philadelphia in December 1926; F. V. Rand, "Erwin F. Smith," in *Mycologia*, 20 (1928), 181–186, with portrait; A. D. Rodgers, III, "Erwin Frink Smith, a Story of North American Plant Pathology," *Memoirs of the American Philosophical Society*, 31 (1952); and R. H. True, "Erwin F. Smith (1854–1927)," in *Phytopathology*, 17 (1927), 675–688, with portrait and bibliography.

ROBERT AYCOCK

SMITH, GRAFTON ELLIOT. See **Elliot Smith, Grafton.**

SMITH, HENRY JOHN STEPHEN (*b.* Dublin, Ireland, 2 November 1826; *d.* Oxford, England, 9 February 1883), *mathematics.*

Smith's contributions to mathematics, although relatively few, were not slight in importance. His best work was done in number theory, but he also wrote on elliptic functions and geometry.

Smith was the youngest of four children of John Smith, an Irish barrister, and the former Mary Murphy. His mother's family were country gentry from near Bantry Bay. After his father's death in 1828, Smith's mother took the family to the Isle of Man in 1829 and to the Isle of Wight in 1831. Smith was taught entirely by his mother until 1838, when he was given instruction by a Mr. R. Wheler. In 1840 the family moved to Oxford, and Henry Highton was engaged as tutor. When Highton went to teach at Rugby School in 1841, Smith accompanied him as a pupil but was soon removed, following his brother's death, and spent some time in France and Switzerland. He won a scholarship to Balliol College, Oxford, in 1844. Benjamin Jowett later described his natural abilities as greater than those of anyone he had ever known at Oxford, and T. H. Huxley made a similar comment. While on a visit to Rome, Smith was obliged by illness to interrupt his studies at Oxford between 1845 and

1847; but during his convalescence in Paris he attended the lectures of Arago and Milne-Edwards. After returning to Oxford he won the Dean Ireland scholarship in classical learning in 1848, and took a first class in the schools of both mathematics and *literae humaniores* in 1849. He was elected a fellow of Balliol and in 1851 was senior mathematical scholar in the university. Smith was long undecided between a career in classics and one in mathematics. He was elected Savilian professor of geometry in 1860, fellow of the Royal Society in 1861, and president of the Mathematical Section of the British Association and fellow of Corpus Christi College, Oxford, in 1873; from 1874 he was keeper of the University Museum, and in 1877 he became first chairman of the Meteorological Council in London. Smith devoted considerable effort to educational administration and reform, and was appointed an Oxford University commissioner in 1877. Smith was an unsuccessful Liberal candidate for Parliament. He died unmarried. The many eulogies to his powers and character are tempered with hints that he was lacking in ambition; and this was, no doubt, the secret of his undoubted popularity.

After graduating, Smith published a few short papers on number theory and geometry but soon turned to an intensive study of Gauss, Dirichlet, Eisenstein, and other writers on number theory. His reports to the British Association between 1859 and 1865, which contain much original work, were the outcome of this study. He presented important papers to the Royal Society on systems of linear indeterminate equations and congruences, and established a general theory of n-ary quadratics permitting the derivation of theorems on expressing any positive integer as the sum of five and seven squares. (Eisenstein had proved the theorem for three squares, and Jacobi for two, four, and six.) Smith's general theory with n indeterminates has been described by J. W. L. Glaisher as possibly the greatest advance made between the publication of Gauss's *Disquisitiones arithmeticae* (1801) and Smith's time.

Smith gave only an abstract of his results in 1864, and in 1868 he provided the general formulas without proofs. In 1882 the French Academy, not knowing of his work, set the problem of five squares for its Grand Prix des Sciences Mathématiques; the last of his published memoirs contains his entry, with proofs of the general theorems so far as they were needed. The prize of 3,000 francs was awarded to Smith posthumously in March 1883. An apology was subsequently made for awarding the prize jointly to a competitor (Minkowski), who seems to have followed Smith's published work.

Smith extended many of Gauss's theorems for real quadratic forms to complex quadratic forms. During the last twenty years of his life he wrote chiefly on elliptic functions; in a field marred by an excessive number of alternative methods and notations, his work is especially elegant. At the time of his death Smith had almost completed his "Memoir on the Theta and Omega Functions," which was written to accompany Glaisher's tables of theta functions. The memoir is a very substantial work running to 208 large quarto pages in the second volume of Smith's collected papers. As an appendix to the same volume there is an introduction written by Smith for the collected papers of W. K. Clifford, and papers written for the South Kensington Science Museum on arithmetical and geometrical instruments and models. Smith was one of the last mathematicians to write an original and significant memoir in Latin, "De fractionibus quibusdam continuis" (1879).

BIBLIOGRAPHY

Smith's mathematical works are assembled in *The Collected Mathematical Papers of Henry John Stephen Smith*, J. W. L. Glaisher, ed., 2 vols. (Oxford, 1894). This collection includes a comprehensive mathematical introduction by the editor (I, lxi–xcv), a portrait, and biographical sketches containing references to nonmathematical writings and to forty mathematical notebooks, more than a dozen of which include unpublished works.

Apart from the introduction to the collected papers, the best biographical notice is the obituary by J. W. L. Glaisher, in *Monthly Notices of the Royal Astronomical Society*, **44** (1884), 138–149. For references to similar notices by P. Mansion, L. Cremona, W. Spottiswoode, and others, see G. Eneström, "Biobibliographie der 1881–1900 verstorbenen Mathematiker," in *Bibliotheca mathematica*, 3rd ser., **2** (1901), 345. For a different collection of references, see A. M. Clerke's article on Smith in *Dictionary of National Biography*. See also A. Macfarlane, *Lectures on Ten British Mathematicians of the Nineteenth Century* (New York, 1916), 92–106.

The introductory material for *Collected Mathematical Papers*, by C. H. Pearson, Benjamin Jowett, Lord Bowen, J. L. Strachan-Davidson, Alfred Robinson, and J. W. L. Glaisher, is reprinted without change in *Biographical Sketches and Recollections (With Early Letters) of Henry John Stephen Smith* (Oxford, 1894). It includes new material in the form of fifteen early letters, one to Smith's mother and the rest to his sister Eleanor.

J. D. NORTH

SMITH, HOMER WILLIAM (*b.* Denver, Colorado, 2 January 1895; *d.* New York, N.Y., 25 March 1962), *physiology, evolutionary biology.*

The youngest of six children, Smith grew up in Cripple Creek, Colorado, where his family encouraged his early fascination with science. He attended high school in Cripple Creek and in Denver, and received his A.B. degree from the University of Denver in 1917. After graduation Smith served in the armed forces, first in a battalion of engineers and ultimately as chemist in the Chemical Warfare Station of the American University in Washington, D.C. Shortly after the end of World War I, he began studies with the physiologist William H. Howell at Johns Hopkins University, where he earned a D.Sc. in 1921. He was a research fellow in the Harvard laboratory of Walter B. Cannon from 1923 to 1925 and subsequently became chairman of the department of physiology at the University of Virginia School of Medicine. In 1928 Smith was appointed professor of physiology and director of the physiological laboratories at the New York University College of Medicine. He retired in 1961 and died of a cerebral hemorrhage a few months later.

Smith's research interests gradually shifted from physical chemistry through chemotherapy to the chemical physiology of the body fluids. By the late 1920's he had focused his energies on problems of renal physiology. Toward the end of his life, however, he wrote: "Superficially, it might be said that the function of the kidneys is to make urine; but in a more considered view one can say that the kidneys make the stuff of philosophy itself." He took the position, originating with Claude Bernard, that an animal's true ambience is its own *milieu intérieur*, not the external environment. The kidneys are the chief regulators of this milieu, upon the constancy of which all other physiological processes depend. Smith thus used questions arising from functional considerations of the kidney to probe phenomena as diverse as paleontology, the biology of consciousness, and the history of religion.

In 1928 Smith began his investigations on the African lungfish *(Protopterus aethiopicus),* summarizing its biological significance in his philosophical novel *Kamongo: The Lungfish and the Padre* (1932, revised 1949) and in his book on the evolutionary history of kidney function, *From Fish to Philosopher* (1953). He also published a number of papers on the comparative renal physiology of the seal, the goosefish *(Lophius piscatorius),* and both fresh- and salt-water elasmobranchs. Smith spent many summers in Maine at the Mount Desert Island Biological Laboratory, and he brought his comparative studies to bear on the problems that became central to his later research: the functions of the mammalian (and especially human) kidney.

Smith played a major part in the development of contemporary understanding of the kidney. In the 1930's he and A. N. Richards independently discovered that inulin, a kind of sugar, is filtered by the human and canine renal glomeruli and is then neither excreted nor absorbed by the tubules and collecting ducts. Inulin thus made possible the accurate measurement of glomerular filtration rate (GFR), a concept introduced in the nineteenth century by Carl Ludwig. Smith did much to make van Slyke's felicitous notion of "renal clearance" fundamental to the study of kidney function; and he and his collaborators elucidated the manner in which the kidney "clears" creatinine, urea, mannitol, sodium, and inulin. He also performed classic experiments on differential blood flow in both normal and diseased kidneys, and investigated the role of the kidneys in the pathogenesis of hypertension.

Smith's New York University laboratory became an international center of renal physiology where he trained and collaborated with more than one hundred clinicians and physiologists. Despite his lack of formal medical training, Smith's work possessed immediate clinical significance; and the ties between his laboratory and clinical departments were close and mutually fruitful.

Smith's preeminence in his specialty was demonstrated by two monographs. *The Physiology of the Kidney* (1937) was the first comprehensive study of renal physiology in English since Cushny's *The Secretion of the Urine* (1917). In his *magnum opus, The Kidney: Structure and Function in Health and Disease* (1951), a massive yet readable tome, Smith judiciously surveyed the entire field of renal physiology and pathology. Its depth and scope made the book definitive.

Smith's last book, *Principles of Renal Physiology* (1956), was an engaging summary written primarily for medical students. He was in the process of revising it when he died.

BIBLIOGRAPHY

In addition to the works mentioned in the text, Smith wrote two historical and philosophical studies in which he spelled out his own naturalistic humanism: *The End of Illusion* (New York, 1935) and *Man and His Gods* (Boston, 1952). which includes an autobiographical account of his Colorado boyhood.

A complete bibliography of his published writings through 1962 is in Herbert Chasis and William Goldring, eds., *Homer William Smith: His Scientific and Literary Achievements* (New York, 1965), 259–268. This volume, edited by two of his colleagues, contains selections from Smith's writings; a list of his awards, honors, and appointments; a partial list of the scientists associated with him at New York University; and a short memoir by Robert F. Pitts that was reprinted (with bibliography) in *Biographical Memoirs. National Academy of Sciences,* **39** (1967), 445–470.

WILLIAM F. BYNUM

SMITH, JAMES EDWARD (*b.* Norwich, England, 2 December 1759; *d.* Norwich, 17 March 1828), *botany.*

Smith was the eldest of the seven children of James Smith, a textile merchant, and Frances Kinderley. During his childhood he showed an interest in botany. Encouraged by several competent botanists who lived in Norwich, Smith wanted to study botany formally, but his father insisted that he should also read medicine. Consequently, in 1781 he went to the University of Edinburgh to study under John Hope, an exponent of the Linnaean method; and in 1783 he moved to London to read anatomy under John Hunter. He came with an introduction from Hope to Joseph Banks, who entertained freely and encouraged young scientists. Smith was with Banks when a letter arrived from Linnaeus' executors offering to sell his library, manuscripts, herbarium, and specimens. Having tried and failed to purchase the collection earlier, Banks was disinclined to take it, but he urged Smith to acquire the collection for himself. Smith negotiated the sale for about £1,000 and deposited the collection in rooms in Chelsea. He later moved to other houses in London, and upon his marriage to Pleasance Reeve in 1796, he took his whole establishment back to Norwich.

The Linnaean collections gave Smith both a purpose to his work and standing in scientific society in London. He studied the material, some of which he rearranged and relabeled, and in 1796 he auctioned off the minerals. His first published works were translations of Linnaeus' *Reflections on the Study of Nature* (1785) and *Dissertation on the Sexes of Plants* (1786). He later published a translation of Linnaeus' *Flora Lapponica* and, in 1821, *Correspondence of Linnaeus and Other Naturalists,* which was based on the manuscripts in his possession. Although a devoted admirer and follower of Linnaeus, Smith—once attacked as

"bigotedly attached to the Linnean system"— was aware of the need for change and latterly acknowledged the importance of Jussieu's system. Probably the most important effect of the purchase was the founding in 1788 of the Linnean Society, with a high proportion of foreign members. Smith was elected the first president and held the office until his death. His inaugural address was a "Discourse on the Rise and Progress of Natural History," and he published many papers in the *Transactions* of the society. After Smith's death there was some resentment that he had not left the Linnaean collection to the Society, but the collection was eventually purchased by the Society for £3,000.

The remainder of Smith's life was shaped and influenced by the collection. He was elected fellow of the Royal Society in 1785; and from 1786 to 1787 he traveled in Europe, where he visited famous sites, libraries, and botanical gardens, and met botanists, including Antoine-Laurent de Jussieu. At Leiden in 1786 he took his M.D. with a thesis "De generatione." He published a very personal account of his tour, including an assessment of the state of science in the countries he visited. Upon his return he did some work on the irritability of vegetables and read a paper on the subject to the Royal Society in 1788, but most of his work was on taxonomy. He instructed the queen and princesses in botany, and was knighted in 1814. A popular teacher, Smith lectured regularly at the Royal Institution and for a time at the University of Cambridge; but he was mortified by the refusal to appoint him professor of botany on the grounds that he was a Dissenter, and wrote two pamphlets protesting against the system. His textbooks *Introduction to Physiological and Systematic Botany* (1807) and *Grammar of Botany* (1821) went through several editions, including some published in the United States.

At his home in Norwich, Smith grew many of the plants that he studied and he was in contact with gardeners who grew specimens from overseas, described in his *Exotic Botany* (1804–1805). He prided himself that whenever possible, he personally checked all descriptions that he issued. It was characteristic of his work that before writing on the genus *Salix,* he spent five years collecting and growing all available kinds of willow. Smith's importance in the history of botany rests on his ability to popularize the subject and on his meticulous accuracy and comprehensiveness in describing the flora of Great Britain and of other countries previously little known.

BIBLIOGRAPHY

I. ORIGINAL WORKS. The most enduring memorial of Smith's work in botany is the series of taxonomic books, which often include fine illustrations. The works are *English Botany*, 36 vols. (London, 1790–1814; 2nd ed., 12 vols., 1832–1846; 3rd ed., 1863), with colored plates (highly regarded for their accuracy) by James Sowerby; *Flora Britannica*, 3 vols. (London, 1800–1804), also condensed into a *Compendium* (1800), both of which were issued in German. The *English Flora*, 4 vols. (London, 1824–1828), which was not merely a translation, but was revised, was the most complete treatise of its kind; it was followed by the *Compendium* (1829). He wrote the botanical part of *Zoology and Botany of New Holland and the Isles Adjacent. . . .* (London, [1793]), the first substantial work on the flora of Australia, and edited vols. I–VII of J. Sibthorp's *Flora Graeca* (London, 1806–1840). Smith wrote most of the articles on botany in Rees's *Cyclopaedia*.

A bibliography of Smith's publications never has been fully worked out. The most accessible comprehensive list is appended to G. S. Boulger's article on Smith, in the *Dictionary of National Biography*, XVIII, 469–472. F. A. Stafleu, *Taxonomic Literature* (Utrecht, 1967), 449–451, gives more bibliographical details and several references, including one to the *Catalogue of Herbarium and Types*, which is available on microfilm (IDC 5074). The Royal Society *Catalogue of Scientific Papers*, V, 725–727, lists 57 papers by Smith; and Stafleu, in an article, "Taxonomic Literature," in *Journal and Proceedings of the Royal Society of New South Wales*, **42** (1928), 80–81, lists all his papers on Australian plants and gives a complete list of the one genus and several species named after Smith. The MS sources are well documented. Lady Smith gave her late husband's library and over 3,000 of his letters to the Linnean Society; the letters have been recorded by W. R. Dawson, in *Catalogue of the Manuscripts in the Library of the Linnean Society of London, Part I: the Smith Papers* (London, 1934).

II. SECONDARY LITERATURE. The official biography of Smith is *Memoir and Correspondence of the Late Sir James Edward Smith . . . Edited by Lady Smith*, 2 vols. (London, 1832). Other accounts shortly after Smith's death are John Nichols, *Illustrations of the Literary History of the Eighteenth Century*, VI (London, 1831), 830–850; and E. B. Ramsay, "Biographical Notice of the Late Sir J. E. Smith . . . With an Estimate of His Character and Influence of His Botanical Labours," in *Edinburgh Journal of Science*, n.s. 1 (1829), 1–16. See also G. S. Boulger, in the *Dictionary of National Biography*, LIII (1898), 61–64. Many more articles are listed in J. Britten and G. S. Boulger, *A Bibliographical Index of Deceased British and Irish Botanists*, 2nd ed., *Revised and Completed by A. B. Rendle* (London, 1831). For a careful analysis of Smith's taxonomic decisions see W. J. Hooker's review of *English Flora*, vols. I–II, in *Edinburgh Journal of Science*, 3 (1825), 159–169;

later vols. also were reviewed extensively. Smith's work in editing Sibthorp was described by W. T. Stearn, "Sibthorp's Smith, The 'Flora Graeca' and the 'Florae Graecae Prodromus,'" in *Taxon*, **16** (1967), 168–178. A comprehensive account of Smith's life, mainly from sources in Norwich, is A. M. Geldart, "Sir James Edward Smith and Some of His Friends," in *Transactions of the Norfolk and Norwich Naturalists' Society*, **9** (1914), 645–692, with bibliography. Smith's relationship to the Linnean Society is covered in A. T. Gage, *A History of the Linnean Society of London* (London, 1938), and in B. D. Jackson, "History of the Linnean Collections, Prepared for the Centenary Anniversary," in *Proceedings of the Linnean Society of London* (1890), 18–34. For probable dates of publication for most of Smith's 3,045 articles in Rees's *Cyclopaedia*, see B. D. Jackson, "Dates of Rees's Cyclopaedia," in *Journal of Botany*, **34** (1896), 307–311.

DIANA M. SIMPKINS

SMITH, PHILIP EDWARD (*b*. DeSmet, South Dakota, 1 January 1884; *d*. Florence, Massachusetts, 8 December 1970), *anatomy, endocrinology.*

Prior to Smith many investigators had attempted to study the effects of the removal of the hypophysis (pituitary gland). Since this gland lies at the base of the brain, most workers used an intracranial approach, which involved some possibility of damage to the brain. For many years there was an intensive controversy as to which of the symptoms of hypophyseal removal (hypophysectomy) were due to brain damage and which to actual removal of the gland. Smith developed a surgical approach through the neck to the base of the skull, where he drilled a small hole that exposed the hypophysis directly without any contact being made with the brain. The results of removal of the gland by this route resulted in a number of symptoms, which were completely reversible by daily implants of the anterior lobe of the hypophysis from donor animals into the operated animal. He further placed lesions in the base of the brain close to the hypophyseal region and demonstrated that these were the cause of the remarkable adiposity that many previous investigators had attributed to hypophyseal insufficiency. He showed that uncomplicated hypophysectomy in mammals resulted in cessation of growth; loss of weight; atrophy of the reproductive system, the thyroid gland, and the cortex of the adrenal gland; and a number of other symptoms.

Philip E. Smith was the youngest of the three children of John E. Smith, a congregationalist minister, and his wife, Lydia Elmina Stratton. Not long after Philip's birth, the family moved to Nio-brara, Nebraska, where his father was both a government agent and a schoolteacher at a Ponca Indian mission.

The Smith family remained in Niobrara until Philip was about six years old, at which time they acquired an eighty-acre farm in Moorpark, California (Ventura County).

Smith attended the local elementary school and spent much of his time helping with the farm work. There was no high school near the farm and all three children had to leave home to attend the Pomona Preparatory School. They subsequently went to Pomona College. Philip and his older brother did plumbing work for the college and helped to run the college heating plant. When he graduated with a B.S. degree in 1908, Philip was president of his class.

After his graduation from Pomona, he worked for a year as an entomologist engaged in the control of certain insect infestations on citrus fruit trees in southern California. During this year his sister Hope died of appendicitis.

Probably influenced by one of his instructors at Pomona, William Atwood Hilton, who had obtained his degree in histology at Cornell, Smith applied for admission to the Cornell Graduate School. He requested a first major in advanced systematic entomology and a first minor in advanced economic entomology. He entered Cornell in the fall of 1909. Shortly afterward, Smith asked to change his minor subject (for his M.S. degree) to histology under Kingsbury in the anatomy department, but it seems that the M.S. (1910) was in entomology. In October 1910 Smith applied for candidacy for a Ph.D. with a major in histology and embryology, a minor in vertebrate zoology, and a minor in systematic entomology. Sometime later he requested a change of his minor in vertebrate zoology to a minor in human anatomy. It appears that this request was granted; Smith obtained his Ph.D. in anatomy under Kingsbury in 1912.

In his doctoral thesis, "Some Features in the Development of the Central Nervous System of *Desmognathus fusca*," Smith discussed in depth the development of the hypophysis cerebri (pituitary gland), but he also devoted an almost equal amount of attention to the pineal body. It is clear from his thesis that even before he left Cornell, Smith faced the scientific crossroads of his career. He was intensely interested in the embryology, morphology, and function of the pituitary gland, but he also appeared to have an equally intense interest in the pineal body. It is also clear

which road he decided to follow. All of his many subsequent published papers were directly or indirectly concerned with the pituitary gland. It is probable that Smith selected this path because he had already foreseen possible approaches to experimental ablation of the hypophysis in the amphibian embryo.

In the summer of 1912 Smith accepted a position as an instructor in the Department of Anatomy at the University of California (Berkeley). Here he shared an office with Irene A. Patchett, an assistant in anatomy. It is significant that she shortly obtained her master's degree with a thesis on the development of the hypophysis of the frog.

Irene Patchett and Philip Smith were married in December 1913.

During this time, Smith was preparing to start his experimental operations aimed at the ablation of the pituitary anlage in the amphibian embryo. Research funds were scarce, and Smith and his wife spent much of their time on field trips collecting frogs, tadpoles, and frog eggs. He set up facilities in the laboratory to raise amphibia and at the same time he devoted many hours to the meticulous task of hand-grinding sewing needles into the microscalpels which he needed for his work. He launched his first classic experiments on the pituitary gland. He surgically removed the anlage of the pars distalis of the hypophysis at a very early stage in the developing frog embryo (early tail bud stage). Although this work progressed rapidly and successfully, it was not published until 1916.

During the same period, B. M. Allen, an investigator at the University of Kansas, was conducting almost identical research. He and Smith chanced to meet socially and thus discovered their common interest and discussed their experiments fully and frankly. It is said that there was an understanding between them in regard to the initial publication of this very important work. Allen's work was published in the same journal as Smith's and in the same year, but several issues later. Throughout his life Smith retained the highest admiration for Allen as a scientist.

Meanwhile an event occurred which significantly influenced Smith's subsequent scientific career. In 1915 Herbert M. Evans came from Johns Hopkins to Berkeley as professor of anatomy (he was only a few years older than Smith). He brought with him from Hopkins another brilliant young anatomist, George W. Corner. During the first few years of this new departmental regime Smith and Evans were the best of friends. But with the passage of time Smith developed rather negative feelings toward Evans. Smith and Corner remained lifelong friends.

This change in Smith's attitude may have been due to Evans' rather close observation of Smith's work. In his memorial biography of Evans (*Anatomical Record*, 1971), Corner stated: "Daily observations of Philip Smith's pioneering study of the effects of ablation of the hypophysis of frog larvae (begun at Berkeley before Evans took over the department of anatomy) turned Evans' attention to the hypophyseal hormones." It seemed that Smith felt that this "daily observation" of his work overstepped the bounds of scientific propriety. Smith remained at Berkeley until 1926, and it appears that he continued to feel that Evans might be taking advantage of his work.

Smith continued his work on hypophysectomy in amphibia, which ultimately led to his classic monograph "The Pigmentary, Growth and Endocrine Disturbances Induced in the Anuran Tadpole by the Early Ablation of the Pars Buccalis of the Hypophysis" (1920).

In 1919 Smith took a six-month sabbatical leave from the University of California. He moved his family to Boston, where he worked with W. B. Cannon. Smith was greatly stimulated to repeat in mammals what he had done in amphibia and he had been considering approaches to this problem. Probably the greatest obstacle blocking the progress of endocrinology was the confusion and controversy about the function of the hypophysis.

Many investigators in Europe had already published reports on hypophysectomy in various mammals. The results were as varied as the investigators. In 1912, in the United States, Harvey Cushing "hypophysectomized" dogs and came to the conclusion that survival for more than a few days was impossible without the hypophysis.

Most of the investigators in this field had used an intracranial approach to the pituitary gland. This procedure led to considerable difference of opinion as to whether the results obtained in various experiments were the consequence of ablation of the pituitary gland or the result of damage done to the brain along the course of the surgical approach employed.

After his return to Berkeley from Cannon's laboratory, Smith continued his work on amphibia, but he now started to concentrate on an approach to the mammalian hypophysis. It is interesting that his first efforts (in the rat) utilized the intracranial approach. But the method Smith used was novel. He designed and constructed a microsyringe which was capable of accurately injecting quantities of

less than .002 milliliters. He used this instrument to inject .010 to .013 milliliters of a chromic acid solution into the anterior lobe of the hypophysis. In 1923, in a little-known paper ("The Production of the Adiposogenital Syndrome in the Rat, With Preliminary Notes Upon the Effects of a Replacement Therapy"), Smith described one single rat in which he had obtained adequate histological evidence of the complete destruction of the anterior lobe and at the same time he could find no evidence of any damage to neural components of the hypophysis or of brain damage. In this single rat, Smith was able to demonstrate that there was ovarian and uterine atrophy with no adiposity. He also showed that the genital atrophy was reversible by replacement therapy with a material derived from anterior pituitary extracts. It may seem strange that Smith should have published a paper which included his findings on a single rat. However, it is typical of the man that his extreme intellectual honesty and his intelligence never permitted any of his intuitive feelings or theories to interfere with the objective realities of an experiment. When one single experimental animal did not fit in with his previous ideas on hypophyseal function, he thought that this was worthy of public comment. Smith later published extensive accounts of this chromic acid injection technique and its results.

Nevertheless, he realized that this intracranial approach was unsatisfactory. He proceeded to develop a parapharyngeal approach to the hypophysis in the rat. This involved a surgical route through the neck to reach the sphenoid bone at the base of the skull; by drilling through the sphenoid at the proper site he exposed the gland. By means of suction applied through a glass cannula he attained a complete hypophysectomy without touching the brain. The very vital hypothalamic region in the floor of the brain is separated from the pituitary body by the very tough double reflection of dura mater (the diaphragma sella) which is pierced by the pituitary stalk. This adequately protected the floor of the brain from any possible damage during the course of the operation, and made it possible to clearly distinguish the effects of hypophysectomy from the symptoms which followed damage to the hypothalamic region. Furthermore he was able to reverse the symptoms which followed this hypophysectomy by daily implants of anterior pituitary tissue into his operated animals.

In 1921 Smith had been promoted to associate professor of anatomy at the University of California. In 1926 he refused an offer of a full professorship in physiology. He felt that he was primarily an anatomist. In the same year he was offered an associate professorship in anatomy at Stanford and he accepted. Both he and his family were quite happy in this new location and he quickly set up his laboratory to continue his work on hypophysectomy in rats. Smith probably did not realize that as a result of his work on amphibia, he was already internationally known in scientific circles. Although at this date he had published little on his mammalian work, the news of his great breakthrough in this field had spread.

He had barely become settled at Stanford when he was offered a position as a full professor of anatomy at the College of Physicians and Surgeons at Columbia. He accepted this position as of July 1927 but proceeded with a planned leave. He visited various laboratories in Europe, spending three months in one laboratory in Vienna. Smith returned to New York in December 1927 to assume his new position.

Smith's years at Columbia were his most productive. His classic paper "Hypophysectomy and a Replacement Therapy in the Rat" (1930) was published there. This work gave a complete account of his great breakthrough in hypophyseal physiology.

Smith continued his studies of the mammalian hypophysis and finally extended this work to the monkey (rhesus). This later work made possible the study of the role of the hypophysis in a primate whose reproductive cycle was very similar to that of the human species.

During this period many young postdoctoral students from the United States and Europe came to spend a year or more in his laboratory. Also a number of graduate students obtained their Ph.D. degrees under Smith's direction.

Smith was also instrumental in bringing two of his former co-workers from the West Coast to Columbia; Earl Theron Engle, an anatomist from Stanford, and Goodwin Lebaron Foster, a biochemist from the University of California. Both these men remained at Columbia for the balance of their careers.

In the late fall and winter of 1939–1940 Smith and his wife spent three months at the School of Tropical Medicine in Puerto Rico. The U.S. Public Health Service maintained a large colony of rhesus monkeys on a small island off the coast of Puerto Rico. Smith thus had access to a supply of monkeys that was far superior to his own colony at Columbia and this gave great stimulus to his work on the rhesus monkey, which he continued at Columbia until his retirement.

Smith's work was interrupted for some months in 1951 when he was run over by a small cultivator tractor while working on a hillside on his property in Westwood, New Jersey. He returned to active work in his laboratory. Smith became professor emeritus of anatomy in 1952; but he continued at Columbia with an appointment as a lecturer in anatomy until 1954, at which time he decided to retire. He and his wife settled in Sunderland, Massachusetts, near their daughter Fredrika, a pediatrician in Northhampton.

Smith was obviously restless without a laboratory, and in 1956 he returned to Stanford as a research associate. His wife received an appointment as his research assistant. Their work at Stanford was supported by the National Science Foundation. The Smiths intended to spend a year at Stanford, but remained in active work there for seven years. Shortly before his final retirement in 1963, the endocrinologists A. S. Parkes and E. C. Amoroso journeyed from London to Stanford to present to Smith the Sir Henry Dale Medal of the Society of Endocrinology of Great Britain.

After the Smiths left Stanford they returned to Florence, Massachusetts. They also had a cabin in Maine, where they spent the summer months. All his life, from the five-year-old boy riding his pony with the Ponca Indians to the eighty-five-year-old man fly-fishing for salmon in Maine, he loved the outdoors. He remained active until a few weeks before his death.

BIBLIOGRAPHY

I. ORIGINAL WORKS. Smith's writings include "A Study of Some Specific Characters of the Genus *Pseudococcus*," in *Journal of Entomology and Zoology*, **5** (1913), 69–84; "Some Features in the Development of the Central Nervous System of *Desmognathus fusca*," in *Journal of Morphology*, **25** (1914), 511–557; "The Development of the Hypophysis of *Amia calva*," in *Anatomical Record*, **8** (1914), 490–508; "Experimental Ablation of the Hypophysis in the Frog Embryo," in *Science*, **44** (1916), 280–282; "On the Effects of the Ablation of the Epithelial Hypophysis on the Other Endocrine Glands," in *Proceedings of the Society for Experimental Biology and Medicine*, **16** (1919), 81; "Studies on the Conditions of Activity in the Endocrine Glands," in *American Journal of Physiology*, **60** (1922), 476–494, written with W. B. Cannon; "The Pigmentary, Growth and Endocrine Disturbances Induced in the Anuran Tadpole by the Early Ablation of the Pars Buccalis of the Hypophysis," in *American Anatomical Memoirs*, **11** (1920), 1–151; "The Production of the Adiposogenital Syndrome in the Rat With Preliminary Notes on the Effects of a Replacement Therapy," in

Proceedings of the Society for Experimental Biology and Medicine, **21** (1923), 204–206, written with A. T. Walker and J. B. Graeser; "The Function of the Lobes of the Hypophysis as Indicated by Replacement Therapy With Different Portions of the Ox Gland," in *Endocrinology*, **7** (1923), 579–591; "The First Occurrence of Secretory Products and of a Specified Structural Differentiation in the Thyroid and Anterior Pituitary During the Development of the Pig Foetus," in *Anatomical Record*, **33** (1926), 289–298; "Hastening Development of the Female Genital System by Daily Homoplastic Pituitary Transplants," in *Proceedings of the Society for Experimental Biology and Medicine*, **24** (1926), 131–132; "The Genital System Responses to Daily Pituitary Transplants," *ibid.*, **24** (1927), 337–338; "Induction of Precocious Sexual Maturity in the Mouse by Daily Pituitary Homeo and Heterotransplants," *ibid.*, **24** (1927), 561–562, written with E. T. Engle; "The Induction of Precocious Sexual Maturity by Pituitary Homotransplants," in *American Journal of Physiology*, **80** (1927), 114–125; "Hypophysectomy and Replacement Therapy," in *Journal of the American Medical Association*, **87** (1926), 2151–2153, written with G. L. Foster; "A Comparison in Normal, Thyroidectomized and Hypophysectomized Rats of the Effects Upon Metabolism and Growth Resulting From Daily Injections of Small Amounts of Thyroid Extract," in *American Journal of Pathology*, **3** (1927), 669–687, written with C. L. Greenwood and G. L. Foster; "Experimental Evidence Regarding the Role of the Anterior Pituitary in the Development and Regulation of the Genital System," in *American Journal of Anatomy*, **40** (1927), 159–217, written with E. T. Engle; "The Disabilities Caused by Hypophysectomy and Their Repair," in *Journal of the American Medical Association*, **88** (1927), 158–161; "The First Appearance in the Anterior Pituitary of the Developing Pig Foetus of Detectible Amounts of the Hormones Stimulating Ovarian Maturity and General Body Growth," in *Anatomical Record*, **43** (1929), 277–297, written with C. Dortzbach; "Hypophysectomy and a Replacement Therapy in the Rat," in *American Journal of Anatomy*, **45** (1930), 205–273; "Disorders Induced by Injury to the Pituitary and the Hypothalamus," in *Journal of Nervous and Mental Diseases*, **74** (1931), 56–61; "The Effect of Hypophysectomy on Ovulation and Corpus Luteum Formation in the Rabbit," in *Journal of the American Medical Association*, **97** (1931), 1861–1863; "The Non-essentiality of the Posterior Hypophysis in Parturition," in *American Journal of Physiology*, **99** (1932), 345–348; "Prevention of Uterine Bleeding in the Macacus Monkey by Corpus Luteum Extract (Progestin)," in *Proceedings of the Society for Experimental Biology and Medicine*, **29** (1932), 12–25, written with E. T. Engle; "Effect of Injecting Pregnancy Urine Extracts in Hypophysectomized Rats. I. The Male," in *Proceedings of the Society for Experimental Biology and Medicine*, **30** (1933), 1246–1250, written with S. L. Leonard; "Effect of Injecting Pregnancy Urine Extracts in Hypophysectomized Rats. II, The Female," *ibid.*, **30** (1933), 1248, written with S. L. Leonard; "Increased Skeletal Effects in Anterior Pituitary Growth Hormone Injections by Administration of Thyroid in Hypophysectomized, Thyroparathyroidectomized Rats," *ibid.*, **30** (1933), 1252–1254; "The Effect of Castration Upon the Sex Stimulating Potency and the Structure of the Anterior Pituitary in Rabbits," in *Anatomical Record*, **57** (1933), 177–195, written with A. E. Severinghaus and S. L. Leonard; "Responses of the Reproductive System of Hypophysectomized and Normal Rats to Injections of Pregnancy Urine Extracts, I. The Male," *ibid.*, **58** (1934), 145–173, written with S. L. Leonard; "Responses of the Reproductive System of Hypophysectomized and Normal Rats to Injections of Pregnancy Urine Extracts, II. The Female," *ibid.*, **58** (1934), 175–203, written with S. L. Leonard; "Differential Ovarian Responses After Injections of Follicle Stimulating and Pregnancy Urine in Very Young Female Rats," in *Proceedings of the Society for Experimental Biology and Medicine*, **31** (1934), 744–746, written with E. T. Engle and H. H. Tyndal; "The Role of Estrin and Progestin in Experimental Menstruation," in *American Journal of Obstetrics and Gynecology*, **29** (1935), 787–798; "Effect of Hypophysectomy on Blood Sugar of Rhesus Monkeys," in *Proceedings of the Society for Experimental Biology and Medicine*, **34** (1936), 247–749, written with L. Dotti, H. H. Tyndal, and E. T. Engle; "The Reproductive System and Its Responses to Ovarian Hormones in Hypophysectomized Rhesus Monkeys," *ibid.*, **34** (1936), 245–247; "Response of Normal and Hypophysectomized Rhesus Monkeys to Insulin," in *Proceedings of the Society for Experimental Biology and Medicine*, **34** (1936), 250–251, written with H. H. Tyndal, L. Dotti, and E. T. Engle; "Is the Blood Calcium Level of Mammals Influenced by Estrogenic Hormones?," in *Endocrinology*, **22** (1938), 315–321, written with L. Levin; "The Endometrium of the Monkey and Estrone-Progesterone Balance," in *American Journal of Anatomy*, **63** (1938), 349–365, written with E. T. Engle; "Responses of Normal and Hypophysectomized Immature Rats to Menopause Urine Injections," in *American Journal of Physiology*, **124** (1938), 174–184, written with L. Levin and H. H. Tyndal; "Certain Actions of Testosterone on the Endometrium of the Monkey and on Uterine Bleeding," in *Endocrinology*, **25** (1939), 1–6, written with E. T. Engle; "Effect of Equine Gonadotropin on Testes of Hypophysectomized Monkeys," in *Endocrinology*, **31** (1942), 1–12; "Continuation of Pregnancy in Rhesus Monkeys (*Macaca mulatta*) Following Hypophysectomy," in *Endocrinology*, **55** (1954), 655–664; "The Endocrine Glands in Hypophysectomized Pregnant Rhesus Monkeys (*Macaca mulatta*) With Special Reference to the Adrenal Glands," in *Endocrinology*, **56** (1955), 271–284; "Postponed Homotransplants of the Hypophysis Into the Region of the Median Eminence in Hypophysectomized Male Rats," in *Endocrinology*, **68** (1961), 130–143; "Postponed Pituitary Homotransplants Into the Region of the Hypophysial Portal Circulation in Hypophysectomized Female Rats," in *Endocrinology*, **73** (1963), 793–806, written with Irene P. Smith; and *Baileys Textbook of Histology*, 7th–13th eds. (1932–1958), co-author.

II. SECONDARY LITERATURE. On Smith and his work, see Frederic J. Agate, Jr., "Philip Edward Smith," in *Anatomical Record*, **4**, no. 1 (1971), 135–138; Nicholas P. Christy, "Philip Edward Smith," in *Endocrinology*, **90**, no. 6 (1972), 1415–1416; and Aura E. Severinghaus, "Philip Edward Smith," in *American Journal of Anatomy*, **135**, no. 2 (1972), 161–163.

FREDERIC J. AGATE, JR.

SMITH, ROBERT (*b*. Lea, near Gainsborough, England, 1689; *d*. Cambridge, England, 2 February 1768), *physics*.

Smith's father, John Smith, was rector of the parish of Lea; his mother, Hannah Smith, was the aunt of Roger Cotes, Plumian professor of astronomy at Cambridge. Smith was educated at the Leicester Grammar School and from 1708 at Trinity College, Cambridge, where he lived with and assisted his cousin Cotes. Smith graduated B.A. in 1711 and M.A. in 1715. He was elected a fellow of his college in 1714, Plumian professor in 1716, and fellow of the Royal Society in 1718. He received the LL.D. in 1723 and the D.D. in 1739. Appointed master of Trinity College in 1742, Smith was vice-chancellor of the University in 1742–1743, and he held the Plumian professorship until 1760. Among his many bequests to the university and to his college, he founded the two Smith's prizes for undergraduate attainment in mathematics and natural philosophy.

Smith wrote on optics and harmonics. In 1738 he published *A Compleat System of Opticks in Four Books, viz. A Popular, a Mathematical, a Mechanical, and a Philosophical Treatise*. Both comprehensive and reliable, the work became probably the most influential optical textbook of the eighteenth century. It was also published in Dutch in 1753, in German in 1755, and in two different French translations in 1767. In 1778 an abridged version was published in English. In turn, its popularity helped to establish the eighteenth-century conviction that light is particulate.

Although Newton had expressed some uncertainty about the nature of light, Smith asserted in the "Popular Treatise" that there was no reason to doubt that light consisted of material particles. He then gave a plausible explanation of most known optical phenomena in terms of particles of light that were acted upon by attractive and repulsive forces. In these explanations Smith never even

suggested that any vibrating medium might exist to produce light or "Newton's rings," nor did he even mention Newton's theory of "fits." Rather he repeated Newton's assertion that the rings were caused by the disposition of varying thicknesses of air or films that reflect or refract different colors of light.

In the "Mathematical Treatise," Smith developed a very comprehensive set of geometric propositions for the computation of the focus, location, magnification, brightness, and aberrations of systems of lenses and mirrors. Apparently he was the first person to construct images by means of an unrefracted central ray and a ray parallel to the axis that is refracted through the focus.[1] He also derived a particular case of the relationship now known as the Smith-Helmholtz formula or the theorem of Lagrange. Using a relationship between the magnification and location of object and image for one lens, Smith showed that the same relationship was invariant within a system of any combination of lenses.[2]

In the "Mechanical Treatise," Smith gave methods for making optical instruments, and in the "Philosophical Treatise," he gave an account of astronomical discoveries.

In 1749 Smith published *Harmonics, or the Philosophy of Musical Sounds*, which had a second edition in 1759 and a postscript in 1762. Although it was partly a textbook, Smith's principal objective was to describe his system of tempering a musical scale by making "all the consonances . . . as equally harmonious as possible. . . ."[3] He derived the "equally harmonic" intervals by a mathematical theory and confirmed his results on an organ and a harpsichord. Smith's temperament was an improvement on existing systems, but its use required impractical mechanical changes in the instruments.

NOTES

1. Ernst Mach, *The Principles of Physical Optics* (New York, 1953), 57.
2. Smith credits Roger Cotes with the discovery of the relationship for one lens. See Smith, *A Compleat System*, bk. II, ch. 5, esp. arts. 247–249, 261–263, 267, 465–474. See also Lord Rayleigh, "Notes, Chiefly Historical. . . ," in *Philosophical Magazine*, 5th ser., **21** (1886), 466–469.
3. Smith, *Harmonics* (1749), p. vi.

BIBLIOGRAPHY

I. ORIGINAL WORKS. Smith's works are *A Compleat System of Opticks in Four Books, viz. A Popular, a Mathematical, a Mechanical, and a Philosophical Trea-*

tise (Cambridge, 1738) and *Harmonics, or the Philosophy of Musical Sounds* (Cambridge, 1749; repr., New York, 1966).

II. SECONDARY LITERATURE. There is no full biography of Smith. Biographical information in this article is from the *Dictionary of National Biography*, XVIII, 517–519. Smith is mentioned in Ernst Mach, *The Principles of Physical Optics* (New York, 1953), 57, 62. The most useful article on the Smith-Helmholtz formula is Lord Rayleigh, "Notes, Chiefly Historical, on Some Fundamental Propositions in Optics," in *Philosophical Magazine*, 5th ser., **21** (1886), 466–476. The best discussion of Smith's historical importance is in an unpublished master's thesis by Henry John Steffens, "The Development of Newtonian Optics in England, 1738–1831" (Ithaca, New York: Cornell University, 1965). On Smith's *Harmonics*, see Lloyd S. Lloyd, "Robert Smith," in *Grove's Dictionary of Music and Musicians*, 5th ed., VII (London, 1954), 857–858; and "Temperaments," *ibid.*, VIII, 377.

EDGAR W. MORSE

SMITH, ROBERT ANGUS (*b.* Glasgow, Scotland, 15 February 1817; *d.* Colwyn Bay, North Wales, 12 May 1884), *chemistry.*

Smith was the son of a manufacturer. Having demonstrated some talent for classics, he prepared at the Glasgow grammar school and at the University of Glasgow for a career in the church. His interest in chemistry, however, was kindled by the public lectures of Thomas Graham at Anderson's College. After leaving the University of Glasgow, Smith spent a few years as tutor to the children of several families, eventually traveling to Germany with the Reverend H. E. Bridgeman.

During his German sojourn, Smith's interests in chemistry were reawakened, and in 1839 he made his way to Giessen, where he worked in Liebig's laboratory. His fellow students at that time included Lyon Playfair and Henry Edward Schunck, both of whom were later his colleagues in Manchester. He took the Ph.D. and returned to Great Britain in 1841. Soon afterward he published a translation of Liebig's paper "On the Azotised Nutritive Principles of Plants." Smith's chemical prospects were dim, and in 1843, when Playfair offered him the post of assistant at the Royal Manchester Institution, he eagerly accepted, ultimately joining Playfair in the "health of towns" investigation. Thus Smith commenced a long and distinguished career as a sanitary chemist.

As early as 1845 Smith began to publish a series of analyses of the air and water of large towns. In 1864 he served as consultant to the Condition of Mines Inquiry and published a report on the analysis of the atmosphere in mines and the methods of analysis that were used. Smith's studies on atmospheric and water pollution are collected in his *Air and Rain* (1872).

Smith pioneered also in the chemistry of disinfection. He joined with Frederick Crace-Calvert and Alexander McDougall in experimenting with sewage deodorants in the River Medlock. With McDougall, Smith took out a patent (1854) on a disinfectant powder (largely carbolic acid), which was later manufactured by McDougall and widely used. At Carlisle, McDougall's powder caught the attention of Lister. By 1869 the *Chemical News* reported that "By common consent Dr. Smith has become the first authority in Europe on the subject of disinfection" (*Chemical News*, **9** [1869], 105). Many of Smith's papers on the subject were integrated into his *Disinfectants and Disinfection* (1869). His national reputation as a sanitary chemist made Smith the logical choice for the position of first inspector under the Alkali Act of 1863. He was popular with the manufacturers, providing both constructive regulation and much-needed technical advice. He served also as an inspector under the Rivers Pollution Act of 1876.

In 1845 Smith was elected a member of the Manchester Literary and Philosophical Society and after 1859 served regularly as its president and vice-president. His honors included election as a fellow of the Royal Society (1857) and honorary degrees from Glasgow (1881) and Edinburgh (1882). His concern with the local scientific community is reflected in his historical works, for example, his sketch of the life of Dalton and the atomic theory (1856) and his history of the Manchester scientific community (1883).

BIBLIOGRAPHY

I. ORIGINAL WORKS. More than 45 of Smith's articles are listed in the Royal Society *Catalogue of Scientific Papers*, V, 731–732; VIII, 974; XI, 440; XVIII, 812. His major works are "Memoir of John Dalton . . . and History of the Atomic Theory Up to His Time," in *Memoirs and Proceedings of the Manchester Literary and Philosophical Society*, 2nd ser., **13** (1856), 1–29; *Disinfectants and Disinfection* (Edinburgh, 1869); *Air and Rain, the Beginnings of Chemical Climatology* (London, 1872); and "A Centenary of Science in Manchester," in *Memoirs and Proceedings of the Manchester Literary and Philosophical Society*, 3rd ser., **9** (1883), 1–475.

II. SECONDARY LITERATURE. The best accounts of Smith's life are P. J. Hartog, in the *Dictionary of Na-*

tional Biography, XVIII, 520–522; H. E. Schunck, "Memoir of Robert Angus Smith," in *Memoirs and Proceedings of the Manchester Literary and Philosophical Society*, 3rd ser., **10** (1887), 90–102; and T. E. Thorpe, in *Nature*, **30** (1884), 104–105. On Smith's career as Alkali Acts inspector, see R. MacLeod, "Alkali Acts Administration, 1863–1884," in *Victorian Studies*, **9** (1965), 85–112. An excellent article, A. Gibson and W. V. Farrar, "Robert Angus Smith, F.R.S. and 'Sanitary Science,'" in *Notes and Records of the Royal Society of London*, **28** (1974), 241–262, appeared too late for inclusion in the preparation of this article.

ROBERT H. KARGON

SMITH, SIDNEY IRVING (*b.* Norway, Maine, 18 February 1843; *d.* New Haven, Connecticut, 6 May 1926), *zoology.*

Smith's parents, Elliot Smith and Lavinia Barton, were of old New England families, and he spent his life studying the fauna of New England. His first work in natural history, a collection of the insects of Maine, was so comprehensive that Agassiz purchased it for Harvard. In 1864 Smith went to the Sheffield Scientific School at Yale to work with Addison E. Verrill. He graduated in 1867 and was then appointed assistant in zoology. As an undergraduate he started his life's work on the little-known subject of marine invertebrates. Spending all his summers on dredging expeditions and afterward identifying the collected specimens, he soon specialized in Crustacea. In 1875 he was appointed first professor of comparative anatomy in the same department of zoology, where he remained until his retirement in 1906. He married Eugenia P. Barber in 1882.

Expeditions in which Smith was involved were first those in which he was working with Verrill between 1864 and 1870 to Long Island Sound and the Bay of Fundy. He was zoologist to the U.S. Lake Survey in 1871, studying the deeper parts of Lake Superior. The following year he joined the U.S. Coast Survey, went to St. George's Bank (Newfoundland), and became a member of the U.S. Fish Commission. He was at Kerguelen Island in 1876, and he dredged off the New England coast with the *Fish Hawk* in 1880 and the *Albatross* (mainly in deep water) in 1883. He did little scientific work thereafter.

The value of Smith's work was the large volume of careful identification and description, with accurate drawings, and some observations of behavior, of many species of aquatic Crustacea, hitherto little studied. He discovered many new species and genera, and worked out their relationships. He also collected extensive data on distribution, including bathymetric distribution, and found that deep-sea samples often contained known species not previously found at great depths, as well as new species.

The large number of available specimens, carefully ordered, allowed Smith to trace the developmental stages of many crustacean larvae formerly thought to be different species from their adult forms. His early papers on the North American lobster are models, and later work on other species threw new light on their relationships. Implications of the details found and generalizations about the group did not come easily to Smith. He never wrote a monograph on the Crustacea, and comments on the size, structure, color, form of eye, and breeding habits of the deep-sea Crustacea collected by the *Albatross* were made by Verrill.

Smith's excellent collections of preserved specimens were given to the Peabody Museum of Natural History at Yale and to the National Museum of Natural History (Washington, D.C.). He was active in the foundation of the Woods Hole Oceanographic Institution.

Smith had a number of minor professional interests. His first published work in 1864 was on the fertilization of orchids, but he did not continue this work. In 1868 he was awarded the Berzelius Prize at Yale for an essay on the geographical distribution of animals, which was concerned largely with fossil forms. He never lost his early interest in entomology. He wrote an occasional paper in the field and was for a time state entomologist of Maine and Connecticut. Smith was also an enthusiastic teacher and started one of the first courses in biology for premedical students.

BIBLIOGRAPHY

I. ORIGINAL WORKS. Smith's most important work, "The Metamorphosis of the Lobster, and Other Crustacea," is in A. E. Verrill, "Report of the Invertebrate Animals of Vineyard Sound and the Adjacent Waters," in *Report of the United States Commission of Fish and Fisheries, 1871/72, Supplementary paper 18*, **2** (1873), 522–537. A fuller account of the development of the lobster is "The Early Stages of the American Lobster (Homarus Americanus Edwards)," in *Transactions of the Connecticut Academy of Arts and Sciences*, **2** (1873), 351–381.

Other papers describing new species, and lists of Crustacea from the various expeditions were published in later *Reports of the Commission of Fish and Fisheries*, the *Bulletin* and the *Proceedings of the United States*

National Museum, Transactions of the Connecticut Academy of Arts and Sciences, and *American Journal of Science*. They can be traced through the Royal Society *Catalogue of Scientific Papers*, VIII, 974–975; XI, 440–441; XII, 690; XVIII, 814; or the bibliography of Coe (see below).

II. SECONDARY LITERATURE. The best account of Smith is Wesley R. Coe, in *Biographical Memoirs. National Academy of Sciences*, **14** (1932), 3–16, with a portrait. There is also an obituary by Smith's colleague A. E. Verrill, in *Science*, **64** (1926), 57–58. For the background to Smith's work at Yale, see R. H. Chittenden, *History of the Sheffield Scientific School, Yale*, 2 vols. (New Haven, 1928).

DIANA M. SIMPKINS

SMITH, THEOBALD (*b*. Albany, New York, 31 July 1859; *d*. New York, N.Y., 10 December 1934), *microbiology, comparative pathology.*

The scope and thoroughness of Smith's researches in bacteriology, immunology, and parasitology produced many discoveries of theoretical import and immediate utility to public health and veterinary medicine. He was the most distinguished early American microbiologist and probably the leading comparative pathologist in the world. His greatest accomplishment—elucidation of the causal agent and mode of transmission of Texas cattle fever—first conclusively proved that an infectious disease could be arthropod-borne. Unlike many contemporary bacteriologists in the United States, he received no training in France or Germany.

Smith's parents were Philipp Schmitt, a tailor and the son of a farmer, and Theresia Kexel, whose recent forebears were village schoolmasters. Both born in Nassau, Germany, they married in 1854 and emigrated to America, settling in Albany, N.Y., where Philipp followed his trade for more than forty years. They took in boarders, worked hard, and were very thrifty. Their second child and only son was baptized in his mother's Roman Catholic faith (the father being Lutheran) and given the surname of his godparent, Jacob Theobald, a friendly immigrant neighbor. He appears in the parish register as Theobald Schmitt but in high school lists as Theobald J. Smith, for he temporarily adopted his godfather's first name. Before entering Cornell University, he discarded the second initial and rejected the Roman church.

At home, German was spoken, and Smith's education began at a German-speaking private academy. As a youth he quoted Goethe, enjoyed Schil-

ler, and subsequently mastered the original reports of Robert Koch and Paul Ehrlich. In 1872, after two years of parish schooling, he entered the recently founded Albany Free Academy, where he excelled in all subjects, became president of the debating society, and was valedictorian in 1876. With a state scholarship supplementing his earnings from piano lessons, organ-playing, and bookkeeping, he entered Cornell in 1877. His industry, versatility, and inclination for scientific studies brought durable friendships with the physiologist and comparative anatomist Burt G. Wilder and with the microscopist Simon H. Gage. In 1881 he received the Ph.B. degree with honors and enrolled at the Albany Medical College, where he headed the 1883 M.D. class. His thesis was entitled "Relations Between Cell-activity in Health and Disease."

Smith felt unready for private practice. With Gage's help he obtained an assistantship with Daniel E. Salmon, chief of the veterinary division of the U.S. Department of Agriculture, commencing December 1883. Six months later he became inspector in the new Bureau of Animal Industry, established by Congress under Salmon's charge to combat bovine pleuropneumonia, glanders, infectious diseases of swine, and Texas cattle fever. Smith taught himself Koch's culture-plate methods and improved on them. Through careful field and laboratory studies of swine epizootics, he differentiated hog cholera from the multiplex swine plague (*Schweineseuche*), implicating distinctive bacillary species for each disease. These findings appeared in the annual reports of the Bureau from 1885 to 1895 and in two monographs, *Hog Cholera: Its History, Nature and Treatment* (1889) and *Special Report on the Cause and Prevention of Swine Plague* (1891). In 1886 pioneer observations involving hog cholera bacilli were made by Smith on bacterial variation—a phenomenon that excited his continuing speculative interest—and on the immunity developed by pigeons inoculated with heat-killed cultures, thus heralding a new approach to bacterial vaccine production. Salmon's assumption of sole or senior authorship of several reports on this bacillus led to the selection of the species in 1900 as the prototype of an eponymous *Salmonella* genus. Some twenty years after Smith discovered *Salmonella cholerae-suis*, however, the actual etiological agent of hog cholera proved to be viral, and the bacillus was accepted thereafter as a secondary invader.

These projects overlapped with another major assignment, Texas cattle fever, on which Smith worked intermittently, restricted by its summer incidence. In November 1892, more than six years after first observing "small round bodies" in red blood corpuscles from stricken cattle, he completed his classic monograph *Investigations Into the Nature, Causation, and Prevention of Texas or Southern Cattle Fever* (1893). He found that the disease resulted from erythrocyte destruction by a protozoan microparasite, *Pyrosoma bigeminum,* carried in the blood of apparently healthy southern cattle and transmitted to susceptible northern cattle by the progeny of blood-sucking ticks (*Boophilus bovis*). The complex, meticulously verified tick-borne mechanism was viewed incredulously by many, but never refuted—a situation that facilitated acceptance within a decade of the mosquito-borne nature of malaria and yellow fever.

Smith's international recognition was hastened by his publications in German journals. In Washington he was promoted in 1891 to chief of the division of animal pathology of the Bureau of Animal Industry, but Salmon sought unduly to divert credit for Smith's work to himself and to other veterinarians. For example, F. L. Kilborne, superintendent of the experimental farm, was overgenerously made coauthor of the Texas fever monograph. Smith chafed under the repeated injustices but delayed resigning because of fresh research opportunities, including a novel protozoal disease of turkeys, and observations on two varieties of tubercle bacilli from mammals, which presaged a lifelong involvement with human and animal tuberculosis. Moreover, he had developed other interests, including the bacteriology of water supplies. Beginning in 1885–1886 with unofficial observations on the total bacterial count of samples from the Potomac River, he systematically examined (1892) microflora in the Hudson River for fecal bacteria. In a report to the New York Department of Health (1893), he advocated quantitative assays of *Bacillus coli communis* as an index of intestinal pollution and introduced the fermentation tube to demonstrate the presence of gas-producing coliforms. His techniques and detailed studies of *B. coli* and related microorganisms were incorporated in recommendations of the committee of American bacteriologists appointed in 1895 (on which Smith served under W. H. Welch's chairmanship) that culminated in the first edition of the *Standard Methods of Water Analysis* (1905) of the American Public Health Association.

Smith recognized that improved sanitation of sewage, milk, and water supplies required an aroused public interest. He disliked the limelight but nevertheless addressed farmers and sanitarians and regularly participated in meetings of the Biological Society of Washington. In 1886 he became lecturer and professor of bacteriology at the National Medical College (the medical department of Columbian University, now known as George Washington University). This appointment, one of the first chairs of bacteriology at an American medical school, was held until 1895. His industry was leavened and his dissatisfactions eased by a happy marriage in 1888 to Lilian Hillyer Egleston, a clergyman's daughter. Her intelligence, high principles, and social graces furthered her husband's work and life aims. They had two daughters (born in Washington) and a son.

In 1895 Smith resigned from the Bureau, becoming director of an antitoxin laboratory for the Massachusetts State Board of Health and professor of zoology at Harvard University. In six months, in improvised quarters at the Bussey Institution in Jamaica Plain, near Boston, he produced potent diphtheria antitoxin. Through a cooperative arrangement devised by H. P. Walcott, chairman of the State Board of Health, and President Charles Eliot of Harvard, both of whom admired Smith, he was appointed in 1896 to the new George F. Fabyan chair of comparative pathology, endowed by a wealthy Bostonian. Smith retained his directorship of the antitoxin laboratory and was privileged to reside in a mansion nearby, commuting daily to Harvard Medical School. Although he took teaching and committee duties seriously, his class lectures were more thorough than inspiring. The dual position intensified his resolute pursuit of new knowledge on the etiology, pathology, and prevention of communicable disease.

During a European trip in 1896, Smith met many leading microbiologists, including Ehrlich and Koch. To the latter he imparted preliminary observations on two varieties of mammalian tubercle bacilli, which he expanded in another classic report, "A Comparative Study of Bovine Tubercle Bacilli and of Human Bacilli From Sputum" (1898). Three years later Koch confirmed these distinctions but failed to acknowledge Smith's priority until 1908. Koch's extreme views on the negligible role of bovine bacilli in human tuberculosis were not endorsed by Smith.

In 1903 Smith inspected several European vaccine lymph manufacturing facilities, prior to designing an enlarged antitoxin and vaccine laboratory, erected in 1904. Smith was the first scientist in North America to adopt Ehrlich's standardized antitoxic unit; and he introduced many improvements in titration methods, which Ehrlich praised on visiting his laboratory in 1904. Irregularities in guinea pig susceptibility to toxin could be reduced through careful breeding and selection of animals (1905), especially by eliminating passively immune progeny of females previously used for titrations (1907). In studying the antigenic properties of toxin-antitoxin mixtures, Smith foresaw their application to the active immunization of humans (1909, 1910). Incidentally, he mentioned to Ehrlich his observation of the sudden death of guinea pigs following second injections of antitoxin. Ehrlich's colleague R. Otto verified this serum-hypersensitivity, designating it the "Theobald Smith phenomenon" (1906).

In 1903 Smith and A. L. Reagh reported agglutination relationships between certain members of the typhoid-paratyphoid-coliform group of bacilli. A second paper, again involving the hog-cholera bacillus, revealed the nonidentity of its flagellar and somatic agglutinogens. These immunologic contributions were fundamental to subsequent development of the Kauffmann-White schema for identifying *Salmonella* organisms serologically.

Research opportunities in parasitology were plentiful. Malaria was endemic in parts of Massachusetts, and as early as 1896 Smith conjectured that the disease was mosquito-borne. The reluctance of the State Board of Health to support his hypothesis, and difficulties in studying malaria in an unfavorable latitude, discouraged him from this field and the palm soon went to Ronald Ross. Other parasitic diseases studied by Smith were murine sarcosporidiosis (1901), coccidiosis of mouse kidney (1902) and rabbit intestine (1910), and amebiasis in the pig (1910). In 1913 he revived investigations into turkey blackhead begun twenty years earlier in Washington.

The output and functions of the pioneer state laboratory multiplied, and administrative duties mounted. Smith was consulted by bacteriologists, veterinarians, and sanitarians; and Eliot encouraged his defense of animal experimentation during an antivivisectionist campaign (1902), his membership on the Charles River Dam committee (1903), and his inquiry into possible damage to animals by smelter smoke from the Anaconda Copper Mining Company in Montana (1906). Before retiring from the presidency of Harvard University in 1909, Eliot persuaded Smith to give eight Lowell lectures (never published) to popularize comparative pathology. In 1912, as Harvard exchange professor at the University of Berlin, Smith took his family to Germany for six months. His inaugural address,

"Parasitismus und Krankheit," abstractly developed the theme of his Harvey lecture (1906) on the parasitism of the tubercle bacillus. His convictions about the importance of comparative pathology were permeated increasingly by the concept of host-parasite interrelationships.

Smith's acknowledged leadership in this field and his unmatched reputation for productive research led early in 1914 to an invitation from Simon Flexner, director of the Rockefeller Institute for Medical Research, to head the newly endowed department of animal pathology to be established at Princeton, N. J. (In 1901, at the inception of the Institute, he had rejected an offer to become its director but had agreed to join the board of scientific directors under Welch's chairmanship.) Although fifty-five years old and not in robust health, Smith resigned from the State Board of Health that summer, but remained at Harvard for a year until preliminaries for the new department had been fulfilled. At a testimonial dinner in June 1915 extraordinary tributes were paid to Smith by distinguished colleagues from all over the world.

For more than two years after this move, Smith was largely responsible, in consultation with Rockefeller Institute representatives, for the general plans of the new division, design of laboratories and animal quarters, and selection of equipment. By the end of 1917 antipneumococcus and antimeningococcus sera were being manufactured in horses, researches into cattle diseases had begun, and a commodious director's house was under construction. In 1920 Smith and H. W. Graybill showed that the protozoon causing turkey blackhead was transmitted in novel fashion, involving ingestion by the healthy host of the embryonated eggs of a small nematode, *Heterakis papillosa*, which was parasitic in the ceca of infected birds. (The infective agent, erroneously designated *Amoeba meleagridis* by Smith many years before, was recognized by E. E. Tyzzer, his successor at Harvard, as a unique flagellate, which he renamed *Histomonas meleagridis*.) Proximity to a large dairy herd infected with Bang's disease reawakened Smith's early interest in *Bacillus abortus* and bovine contagious abortion; and he published, sometimes with R. B. Little, twelve papers and a monograph in this field. In 1926 they described the protection induced by vaccinating heifers with living *B. abortus* culture of low virulence. Other investigations concerned *Vibrio fetus*, a spirillar cause of cattle abortion, hitherto unrecognized in America; a possible new species, *Bacillus actinoides*, producing bronchopneumonia in calves;

the vitally protective antibodies in colostrum for newborn calves; and the pathogenicity of certain bovine strains of *B. coli* in calves deprived of colostrum.

In 1929, at age seventy, Smith relinquished his directorship of the division of animal pathology and was succeeded by Carl Ten Broeck, a long-time associate. He continued working at Princeton as member emeritus of the Rockefeller Institute. Smith had become vice-president of the board of scientific directors in 1924 and succeeded Welch as president in 1933. In November 1934 increasing weakness forced his hospitalization in New York, where he died just before an exploratory operation for intestinal cancer. His ashes and, six years later, those of his wife were buried in the woods at their summer home at Silver Lake, New Hampshire. In 1967 they were reinterred in the nearby Chocorua cemetery.

Honors came to Smith rather late in life. Between 1917 and 1933 he delivered the Herter, Mellon, Pasteur, Gross, De Lamar, Milbank Memorial, Welch, and Thayer lectures. A climactic series of five Vanuxem lectures, given at Princeton in 1933, and published in book form as *Parasitism and Disease* (1934), philosophically embodied his scientific credo. He received a dozen honorary doctorates from renowned American and European universities. He held membership in numerous scientific and medical societies, and was president of the Society of American Bacteriologists (1903), the National Tuberculosis Association (1926), and the Triennial Congress of Physicians and Surgeons (1928). He was elected a trustee of the Carnegie Institution in 1917. Smith was a foreign member of the Royal Society of London and eleven other ancient societies. Further honor awards were the Mary Kingsley, Flattery, Kober, Trudeau, Holland, Sedgwick, Manson, and Copley medals. Recommended several times for the Nobel prize, this ultimate distinction eluded him, although he surely deserved it. His portrait in oils is at the entrance to the Theobald Smith Building of the Rockefeller Institute, now known as the Rockefeller University, which he served so faithfully in various capacities for thirty-three years.

Smith unpretentiously summarized his own life work as "a study of the causes of infectious diseases and a search for their control." His outstanding success in this quest derived from a farsighted, dispassionately critical intelligence, linked to capacities for unsparing industry, punctilious concern for detail, technical inventiveness, and indomitable persistence. These qualities were applied to realis-

tically chosen problems. Between 1883 and 1934 he published at least one scientific research report annually, and in several of these years the annual output was ten or more such publications—a record demanding rare degrees of self-discipline and dedication to the work ethic. As a director he was fair in judgment but sparing of praise, painstakingly conscientious, and abhorrent of waste or extravagance. All who knew Smith and his work respected him. Individuals as diverse as Osler, Welch, Simon Flexner, Prudden, Rous, Richard Shope, and Ten Broeck held him in profound admiration. Smith was too reticent to be popular, and he displayed an element of restraint even with Gage, despite their half-century of close friendship. Hans Zinsser observes, "there was about him an unobtrusive pride, a reserve tinged with austerity which did not invite easy intimacy."

Smith's relaxations were modest and quiet, befitting his nature. When tired, he sought solace in reading, piano-playing, or calculus. During the hot summer months, he found refreshment at his lakeside home, where besides preparing manuscripts, he enjoyed boating, making household repairs, and landscaping. His pattern of life was consistently rational, yet he did not lack emotion. He hated war and wastefulness, for example, but knew that great ideas must be both launched and defended, often at high cost in a hostile environment. That a scientist of such unswerving probity and fine accomplishments should carry so little fame among his countrymen testifies partly to his self-effacing character. Many of his discriminating contemporaries thought him comparable to Pasteur and Koch, and the passage of time has not dimmed the luster of his contributions to the conquest of disease.

BIBLIOGRAPHY

I. ORIGINAL WORKS. Mimeographed bibliographies were prepared by Earl B. McKinley and Ellen G. Acree, and also by the library of the Rockefeller Institute for Medical Research (now Rockefeller University), New York. Published versions accompany the sketch of Smith's life in *Medical Classics*, **1** (1936–1937), 347–371, and the biographical memoir by Hans Zinsser (see below), which respectively cite 224 and 247 items. These are all incomplete and contain inaccuracies. The actual total, excluding unverifiable editorials and multiple publications of the same article, is almost 300 titles.

His more important monographs include *Hog Cholera: Its History, Nature and Treatment* (Washington, 1889), written with D. E. Salmon and F. L. Kilborne; *Special Report on the Cause and Prevention of Swine Plague* (Washington, 1891); *Investigations Into the*

Nature, Causation, and Prevention of Texas or Southern Cattle Fever (Washington, 1893), written with F. L. Kilborne, repr. in *Medical Classics,* **1** (1936–1937), 372–597; *Studies in Vaccinal Immunity Towards Disease of the Bovine Placenta Due to Bacillus Abortus (Infectious Abortion)* (New York, 1923), written with R. B. Little; and *Parasitism and Disease* (Princeton, N.J., 1934), the Vanuxem lectures.

Smith's lasting interest in the bacteriology and immunology of tuberculosis is expressed in "The Diagnostic and Prognostic Value of the Bacillus Tuberculosis in the Sputum of Pulmonary Diseases," in *Albany Medical Annals,* **5** (1884), 193–198; "Some Practical Suggestions for the Suppression and Prevention of Bovine Tuberculosis," in *Yearbook of the United States Department of Agriculture* (1895), 317–330; "Two Varieties of the Tubercle Bacillus From Mammals," in *Transactions of the Association of American Physicians,* **11** (1896), 75–93; "A Comparative Study of Bovine Tubercle Bacilli and of Human Bacilli From Sputum," in *Journal of Experimental Medicine,* **3** (1898), 451–511, repr. in *Medical Classics,* **1** (1936–1937), 599–669; "The Thermal Death-point of Tubercle Bacilli in Milk and Some Other Fluids," in *Journal of Experimental Medicine,* **4** (1899), 217–233; "The Relation Between Bovine and Human Tuberculosis," in *Medical News,* **80** (1902), 343–346; "Studies in Mammalian Tubercle Bacilli. III. Description of a Bovine Bacillus From the Human Body. A Culture Test for Distinguishing the Human From the Bovine Type of Baccilli," in *Transactions of the Association of American Physicians,* **18** (1903), 109–151; "The Parasitism of the Tubercle Bacillus and Its Bearing on Infection and Immunity," in *Journal of the American Medical Association,* **46** (1906), 1247–1254, 1345–1348, the Harvey lecture; "Certain Aspects of Natural and Acquired Resistance to Tuberculosis and Their Bearing on Preventive Measures," *ibid.,* **68** (1917), 669–674, 764–769, the Mellon lecture; and "Focal Cell Reactions in Tuberculosis and Allied Diseases," in *Bulletin of the Johns Hopkins Hospital,* **53** (1933), 197–225, the Thayer lectures.

His main contributions to the bacteriology and immunology of diphtheria are "Antitoxic and Microbicide Powers of the Blood Serum After Immunization, With Special Reference to Diphtheria," in *Albany Medical Annals,* **16** (1895), 175–189; "The Production of Diphtheria Antitoxin," in *Journal of the Association of Engineering Societies,* **16** (1896), 83–92; "The Conditions Which Influence the Appearance of Toxin in Cultures of the Diphtheria Bacillus," in *Transactions of the Association of American Physicians,* **11** (1896), 37–61; "The Relation of Dextrose to the Production of Toxin in Bouillon Cultures of the Diphtheria Bacillus," in *Journal of Experimental Medicine,* **4** (1899), 373–397; "The Antitoxin Unit in Diphtheria," in *Journal of the Boston Society of Medical Sciences,* **5** (1900), 1–11; "The Degree and Duration of Passive Immunity to Diphtheria Toxin Transmitted by Immunized Female Guinea-pigs to Their Immediate Offspring," in *Journal of Medical Research,* n.s. **11** (1907), 359–379; and "Active Immunity Produced by So-called Balanced or Neutral Mixtures of Diphtheria Toxin and Antitoxin," in *Journal of Experimental Medicine,* **11** (1909), 241–256.

Other fundamental contributions to immunology are "On a New Method of Producing Immunity From Contagious Diseases," in *Proceedings of the Biological Society of Washington,* **3** (1886), 29–33, and "Experiments on the Production of Immunity by the Hypodermic Injection of Sterilized Cultures," in *Transactions of the IX International Medical Congress, Washington,* **3** (1887), 403–407, both written with D. E. Salmon; "The Agglutination Affinities of Related Bacteria Parasitic in Different Hosts," in *Journal of Medical Research,* n.s. **4** (1903), 270–300, and "The Non-identity of Agglutinins Acting Upon the Flagella and Upon the Body of Bacteria,"*ibid.,* n.s. **5** (1903), 89–100, both written with A. L. Reagh; "Agglutination Affinities of a Pathogenic Bacillus From Fowls (Fowl Typhoid) Bacterium sanguinarium, Moore) With the Typhoid Bacillus of Man," *ibid.,* n.s. **26** (1915), 503–521, written with C. Ten Broeck; "The Significance of Colostrum to the New-born Calf," in *Journal of Experimental Medicine,* **36** (1922), 181–198, written with R. B. Little; and "The Relation of the Capsular Substance of B. coli to Antibody Production," *ibid.,* **48** (1928), 351–361.

His studies on the properties and differentation of new or unusual bacterial species include "A New Chromogenous Bacillus," in *Proceedings of the American Association for the Advancement of Science,* **34** (1885), 303–309, written with D. E. Salmon; "The Bacterium of Swine Plague," in *American Monthly Microscopical Journal,* **7** (1886), 204–205; "A Contribution to the Study of the Microbe of Rabbit Septicaemia," in *Journal of Comparative Medicine and Surgery,* **8** (1887), 24–37; "Zur Unterscheidung zwischen Typhus- und Kolonbacillen," in *Centralblatt für Bakteriologie und Parasitenkunde* (Original-Mittheilung), **11** (1892), 367–370; "On a Pathogenic Bacillus From the Vagina of a Mare After Abortion," in *Bulletin. Bureau of Animal Industry. United States Department of Agriculture,* no. 3 (Washington, 1893), 53–59; "Spontaneous Pseudo-tuberculosis in a Guinea-pig, and the Bacillus Causing It," in *Journal of the Boston Society of Medical Sciences,* **1** (1897), 5–8; "Ueber die pathogene Wirkung des Bacillus abortus Bang," in *Centralblatt für Bakteriologie, Parasitenkunde und Infektionskrankheiten,* I. Abteilung (Originale), **61** (1912), 549–555; "A Pleomorphic Bacillus From Pneumonic Lungs of Calves Simulating Actinomyces," in *Journal of Experimental Medicine,* **28** (1918), 333–344; "Spirilla Associated With Disease of the Fetal Membranes in Cattle (Infectious Abortion)," *ibid.,* 701–719; "Some Cultural Characters of Bacillus abortus (Bang) With Special Reference to CO_2 Requirements," *ibid.,* **40** (1924), 219–232; "Studies on a Paratyphoid Infection in Guinea Pigs," *ibid.,* **45** (1927), 353–363, 365–377, written with J. B. Nelson; and "Studies on Pathogenic B. coli From Bovine Sources," *ibid.,* **46** (1927), 123–131, written with R. B. Little.

Smith's persistent concern with bacterial variation is apparent from "On the Variability of Pathogenic Organisms, as Illustrated by the Bacterium of Swine-plague," in *American Monthly Microscopical Journal,* **7** (1886), 201–203; "Observations on the Variability of Disease Germs," in *New York Medical Journal,* **52** (1890), 485–487; "Modification, Temporary and Permanent, of the Physiological Characters of Bacteria in Mixed Cultures," in *Transactions of the Association of American Physicians,* **9** (1894), 85–106; "Variations Among Pathogenic Bacteria," in *Journal of the Boston Society of Medical Sciences,* **4** (1900), 95–109; "Animal Reservoirs of Human Disease With Special Reference to Microbic Variability," in *Bulletin of the New York Academy of Medicine,* 2nd ser., **4** (1928), 476–496; and "Koch's Views on the Stability of Species Among Bacteria," in *Annals of Medical History,* n.s. **4** (1932), 524–530.

His pioneering work on the bacteriological analysis of water is expressed in "Some Recent Investigations Concerning Bacteria in Drinking Water," in *Medical News,* **49** (1886), 399–401; "Quantitative Variations in the Germ Life of Potomac Water During the Year 1886," ibid., **50** (1887), 404–405; "The Relation of Drinking Water to Some Infectious Diseases," in *Albany Medical Annals,* **9** (1888), 297–302; "On Pathogenic Bacteria in Drinking Water and the Means Employed for Their Removal," ibid., **13** (1892), 129–150; "A New Method for Determining Quantitatively the Pollution of Water by Fecal Bacteria," in *New York State Board of Health. Thirteenth Annual Report for the Year 1892* (1893), 712–722; "Notes on Bacillus coli communis and Related Forms, Together With Some Suggestions Concerning the Bacteriological Examination of Drinking-water," in *American Journal of the Medical Sciences,* **110** (1895), 283–302; and "Water-borne Diseases," in *Journal of the New England Water Works Association,* **10** (1896), 203–225.

On other aspects of sanitation he wrote "Recent Advances in the Disinfection of Dwellings as Illustrated by the Berlin Rules," in *New York Medical Journal,* **48** (1888), 117–120; "The Sanitary Aspects of Dairying," in *Maine Farmer* (15 Dec. 1898), 1, 4; "The House-fly as an Agent in the Dissemination of Infectious Diseases," in *American Journal of Public Hygiene,* n.s. **4** (1908), 312–317; "What Is Diseased Meat and What Is Its Relation to Meat Inspection?," ibid., n.s. **5** (1909), 397–411; and "Insects as Carriers of Disease," in *Monthly Bulletin of the State Board of Health of Massachusetts,* **5** (1910), 112–119.

The scope and duration of Smith's work in parasitology is exemplified by "Some Observations on Coccidia in the Renal Epithelium of the Mouse," in *Journal of Comparative Medicine and Surgery,* **10** (1889), 211–217; "Preliminary Observations on the Microorganism of Texas Fever," in *Medical News,* **55** (1889), 689–693; "Some Problems in the Etiology and Pathology of Texas Cattle Fever, and Their Bearing on the Comparative Study of Protozoan Diseases," in *Transactions of the Association of American Physicians,* **8** (1893), 117–134; "An Infectious Disease Among Turkeys Caused by Protozoa (Infectious Entero-hepatitis)," in *Bulletin. Bureau of Animal Industry. United States Department of Agriculture,* no. 8 (1895), 7–38; "The Etiology of Texas Cattle Fever, With Special Reference to Recent Hypotheses Concerning the Transmission of Malaria," in *New York Medical Journal,* **70** (1899), 47–51; "The Etiology of Malaria With Special Reference to the Mosquito as an Intermediate Host," in *Journal of the Massachusetts Association of Boards of Health,* **11** (1901), 99–113; "The Production of Sarcosporidiosis in the Mouse by Feeding Infected Muscular Tissue," in *Journal of Experimental Medicine,* **6** (1901), 1–21; "On a Coccidium (Klossiella muris, gen. et spec. nov.) Parasitic in the Renal Epithelium of the Mouse," ibid., **6** (1902), 303–316, written with H. P. Johnson; "The Sources, Favoring Conditions and Prophylaxis of Malaria in Temperate Climates, With Special Reference to Massachusetts," in *Boston Medical and Surgical Journal,* **149** (1903), 57–64, 87–92, 115–118, 139–144, the Shattuck lecture; "Some Field Experiments Bearing on the Transmission of Blackhead in Turkeys," in *Journal of Experimental Medicine,* **25** (1917), 405–414; "Coccidiosis in Young Calves," ibid., **28** (1918), 89–108; and "Encephalitozoon cuniculi as a Kidney Parasite in the Rabbit," ibid., **41** (1925), 25–35.

Technological innovations and elucidations are reported in "A Few Simple Methods of Obtaining Pure Cultures of Bacteria for Microscopical Examination," in *American Monthly Microscopical Journal,* **7** (1886), 124–125; "The Fermentation Tube With Special Reference to Anaërobiosis and Gas Production Among Bacteria," in *Wilder Quarter-Century Book 1868–1893* (Ithaca, 1893), 187–233; "Ueber die Bedeutung des Zuckers in Kulturmedien für Bakterien," in *Centralblatt für Bakteriologie und Parasitenkunde,* I. Abteilung (Originale), **18** (1895), 1–9; "A Modification of the Method for Determining the Production of Indol by Bacteria," in *Journal of Experimental Medicine,* **2** (1897), 543–547; and "One of the Conditions Under Which Discontinuous Sterilization May be Ineffective," ibid., **3** (1898), 647–650.

Smith's advocacy of research in comparative pathology and his emphasis on the host-parasite relationship are expressed in "Comparative Pathology in Its Relation to Human Medicine," in *Bulletin of the Harvard Medical Alumni Association,* **9** (1896), 50–69; "Adaptation of Pathogenic Bacteria to Different Species of Animals," in *Boston Medical and Surgical Journal,* **142** (1900), 473–476; "Some Problems in the Life History of Pathogenic Microorganisms," in *Science,* n.s. **20** (1904), 817–832; "The Relation of Animal Life to Human Diseases," in *Boston Medical and Surgical Journal,* **153** (1905), 485–489; "Animal Diseases Transmissible to Man," in *Monthly Bulletin of the State Board of Health of Massachusetts,* **4** (1909), 264–276; "Parasitismus und Krankheit," in *Deutsche medizinische Wochenschrift,* **38** (1912), 276–279, inaugural address as visiting professor, University of Berlin; "Parasitism as a Factor in Disease," in *Science*, **54** (1921), 99–108; "Some Biological and Economic Aspects of Comparative Pathology," in *Edinburgh Medical Journal*, **31** (1924), 221–240; and "Disease a Biological Problem," in *Bulletin of the Harvard Medical Alumni Association*, **5** (1931), 2–6.

Correspondence and other data relating to Smith are among the Simon Henry Gage Papers at the Olin Research Library, Cornell University, Ithaca, N. Y., and the Simon Flexner Papers in the American Philosophical Society Library, Philadelphia. The Countway Library of Medicine, Harvard University, Boston, Mass., and the Archives of the Rockefeller University, New York, also contain material bearing on his work at Boston and at Princeton. Bronze bas-reliefs of him hang in the School of Public Health at Harvard University and in the library of the Rockefeller University. Small medallions of another such portrait were distributed to all members attending the third International Congress of Microbiology, held in New York in 1939.

II. SECONDARY LITERATURE. Obituaries include J. H. Brown, "Theobald Smith, 1859–1934," in *Journal of Bacteriology*, **30** (1935), 1–3, with photograph of bronze plaque; W. Bulloch, "*In Memoriam* Theobald Smith, 1859–1934," in *Journal of Pathology and Bacteriology*, **40** (1935), 621–635, with portrait; A. E. Cohn, "Obituary: Theobald Smith, 1859–1934," in *Bulletin of the New York Academy of Medicine*, 2nd ser., **11** (1935), 107–116, with portrait; E. G. Conklin, "Theobald Smith," in *Proceedings of the American Philosophical Society*, **75** (1935), 333–335; S. Flexner's minute on Smith, read at the meeting of the board of scientific directors of the Rockefeller Institute for Medical Research, on 20 April 1935; S. H. Gage, "Theobald Smith, 1859–1934," in *Cornell Veterinarian*, **25** (1935), 207–228, with portrait; and "Theobald Smith, Investigator and Man, 1859–1934," in *Science*, **84** (1936), 365–371, with portrait; Preston Kyes, "Theobald Smith, M.D., 1859–1934," in *Archives of Pathology*, **19** (1935), 234–238; E. B. McKinley, "Theobald Smith," in *Science*, **82** (1935), 575–586; C. R. Stockard, "Theobald Smith," *ibid.*, **80** (1934), 579–580; S. B. Wolbach, "Dr. Theobald Smith," in *Bulletin of the Harvard Medical Alumni Association*, **9** (1935), 35–38; and H. Zinsser, "Biographical Memoir of Theobald Smith, 1859–1934," in *Biographical Memoirs. National Academy of Sciences*, **17**, no. 12 (1936), 261–303, with portrait and bibliography. Obituaries in foreign languages include O. Seifried, "Theobald Smith †. 1859–1934," in *Tierärztliche Rundschau*, **41** (1935), 46–48; and V. L. Yakimoff, "Theobald Smith," in *Priroda*, no. 5 (1935), 81–86.

Other references to Smith's life and work are P. F. Clark, "Theobald Smith, Student of Disease (1859–1934)," in *Journal of the History of Medicine and Allied Sciences*, **14** (1959), 490–514, with photograph of bronze medallion; I. S. Cutter, "Theobald Smith and His Contributions to Science," in *Journal of the American Veterinary Medical Association*, **90** (1937), 245–

255; C. E. Dolman, "Texas Cattle Fever. A Commemorative Tribute to Theobald Smith," in *Clio Medica*, **4** (1969), 1–31; and "Theobald Smith, 1859–1934: Life and Work," in *New York State Journal of Medicine*, **69** (1969), 2801–2816; D. Fairchild, *The World Was My Garden. Travels of a Plant Explorer* (New York, 1938), 24–25; M. C. Hall, "Theobald Smith as a Parasitologist," in *Journal of Parasitology*, **21** (1935), 231–243, and "Theobald Smith on Disease," in *Journal of Heredity*, **26** (1935), 419–422, a review of Smith's *Parasitism and Disease;* Paul de Kruif, "Theobald Smith: Ticks and Texas Fever," in *Microbe Hunters* (New York, 1926), 234–251; J. Middleton, "A Great American Scientist," in *World's Work* (July 1914), 299–302; T. M. Prudden, "Professor Theobald Smith and a New Outlook in Animal Pathology," in *Science*, n.s. **39** (1914), 751–754; Anna M. Sexton, "Theobald Smith: First Chairman of the Laboratory Section, 1900," in *American Journal of Public Health*, **41** (1951), 125–131; and P. H. Smith, "Theobald Smith," in *Land*, **8** (1949), 363–368.

CLAUDE E. DOLMAN

SMITH, WILLIAM (*b.* Churchill, Oxfordshire, England, 23 March 1769; *d.* Northampton, England, 28 August 1839), *geology.*

Smith's father, John, was a village blacksmith; but his grandparents and great-grandparents were small farmers in Oxfordshire and Gloucestershire. His mother, Ann, daughter of an unrelated William Smith, was also descended from a farming family. William was the eldest of five children and was only seven when his father died. His first eighteen years were spent in the village of Churchill, with the exception of two years spent in London. He attended the village school until he was eleven; there he learned simple arithmetic and how to write in a good, clear hand. Later, with some older friends and neighbors, he pursued further studies, including mathematics.

The year 1787, when he was eighteen, was a turning point in Smith's life. A local surveyor, Edward Webb of Stow-on-the-Wold, came to Churchill to make a detailed survey of the parish preparatory to the enclosure of the common lands. He needed the assistance of an intelligent lad to hold the chain and to take notes, and Smith got the job. Evidently Webb realized that he had made a good choice, for he took Smith into his business, carried on in the large house (now known as Manor House) on the corner of the market square in Stow. Here Smith lived with Webb and his family for nearly five years. There is no evidence that he was articled to Webb, but he learned all the duties of a land surveyor and valuer and must have become well qualified.

In the autumn of 1791 Webb sent Smith to survey and value an estate in north Somerset. He went there on foot and lodged at Rugborne Farm, near High Littleton, about eight miles southwest of Bath. Smith later designated this farmhouse "the birth-place of English geology," for it was there that he began to think about the succession of the strata. The house is still standing, almost unaltered since Smith lodged there. At that time the district had many active coal mines, and Smith went underground to examine some of them and draw plans. He also prepared a map of High Littleton that still exists.

In 1793 Smith was engaged by a group of local landowners to make a survey for a proposed canal, on which the coal from their mines could be carried to a wider market at a lower cost. In March 1794 he gave evidence before Parliament in connection with the act authorizing the canal construction; and in August he went with two members of the canal committee on a carriage tour to the north of England to see other canals and collieries. The tour provided him with valuable additions to his knowledge of the strata, a subject in which he was increasingly interested. While in London he had visited booksellers in order to find books on geology, but with little success.

Work on the canal, which became known as the Somerset Coal Canal, began in July 1795. Two branches, each extending from the coal-mining areas along nearly parallel valleys and each about six miles long, were to be constructed. From their meeting point a canal two miles long would connect with the Kennet and Avon Canal, also under construction; the latter was intended to link Bath with the towns of Newbury and Reading in the Thames Valley.

Smith was employed by the Canal Company from 1794 to 1799; and during this period, he became familiar with the strata through which the canal passed, from Triassic marls to the Lias and Oolites of the Jurassic. He collected fossils, and his notes show that by January 1796 he had made the great discovery that lithologically similar beds can be distinguished by the assemblage of fossils found in them, a concept virtually unrecognized by the geologists of that period. He also began to color maps to show how the different beds outcropped around the neighboring hills. In June 1799 his engagement with the Canal Company was terminat-

ed; and about this time he dictated to two local clergymen, Joseph Townsend and Benjamin Richardson, both collectors of fossils, a list of the strata found around Bath and the fossils characteristic of each. This list is deservedly famous. A contemporary copy, and also a map by Smith of the country five miles around Bath "colored geologically in 1799," is held by the Geological Society of London.

Smith had already drained some land for local landowners, and this type of work offered him prospects of traveling about the country and seeing more of its geology. In 1800 he was employed by a famous landowner and agriculturist Thomas Coke of Holkham in Norfolk; and in 1801 Coke introduced him to Francis, Duke of Bedford, who then employed Smith on his Woburn estate. Both Coke and the duke held large annual meetings on their estates to coincide with the June sheepshearings, and these were attended by many prominent landowners, including distinguished foreigners. From 1801 Smith went to these meetings, exhibited his maps, and talked about geology and its economic value. Several small maps of England and Wales colored geologically by Smith around 1801–1803 are still extant.

In 1802 Smith first met Sir Joseph Banks, president of the Royal Society, and explained his ideas to him. Banks greeted Smith's proposal for a geological map of England and Wales with enthusiasm and encouraged him to complete it. Smith had already issued a printed prospectus, dated 1 June 1801, of his projected work; and a list of subscribers had been opened. The book was to be called "Accurate Delineations and Descriptions of the Natural Order of the Various Strata That Are Found in Different Parts of England and Wales"; it was to be accompanied by a "correct map of the strata."

Nevertheless, during the next ten years, Smith's only publication was a nongeological book on irrigation and water meadows (1806). From this venture he learned that books are not necessarily profitable to their authors. He continued to make numerous notes and write portions of his proposed work on geology but was constantly employed on different projects and had little spare time for concentrated writing, even though he employed an amanuensis to copy his notes. His work varied from the construction of sea defenses on the east coast of England and in South Wales, to supervising sinkings for coal in Yorkshire and Lancashire, and to reporting on the value of estates. In 1804 he leased a large house in London. There he set up his collection of fossils on sloping shelves to represent the different strata. This collection was inspected in 1808 by members of the newly formed Geological Society of London.

In 1812 a London map engraver and publisher, John Cary, offered to publish Smith's geological map of England and Wales on a scale of five miles to the inch. Plates were specially engraved, and Smith himself decided what place names were to be inserted. During 1813 and 1814 he added the geological lines; and when the coloring was carried out he insisted on the use of a novel feature—each formation was colored a darker shade at its base to make clear how the beds were superimposed. In May 1815 the completed map, *A Delineation of the Strata of England and Wales, With part of Scotland*, was exhibited in London to the Board of Agriculture, to the Royal Institution, and to the Society for the Encouragement of Arts, Manufactures and Commerce. This society had offered annually since 1802 a premium of fifty guineas for a mineralogical map of England and Wales. Smith received this award. By March 1816, 250 copies of the geological map had been colored and issued to subscribers; most of the maps were numbered and signed by Smith and duly noted in his diary. Probably 400 copies in all were issued, of which fewer than a hundred are known to be extant.

The map was sold at five guineas (£5.25) a copy, but the costs of production and coloring must have absorbed most of the proceeds. About this time he found himself in severe financial difficulties. In 1812 he had leased a quarry near the Coal Canal and had set up a sawmill and stoneworks under the management of his brother John; but the stone proved to be of poor quality because the quarry was intersected by unsuspected faults. Smith's debts had rapidly increased. For this reason he decided to sell his vast collection of fossils, arranged stratigraphically, to the British Museum; and he began negotiations with the government in 1815. Unfortunately the sum he eventually received was well below his expectations—£500 in installments with a further £100 in 1818 for some additional fossils. About 2,000 of these fossils, mostly bearing Smith's original reference marks, are still in the collections of the British Museum (Natural History). Also, museum officials demanded a catalog; and Smith had to give much time to its compilation, although he was aided by his nephew John Phillips, then aged fifteen.

Despite his difficulties, during the next few years Smith published several works. *Strata Identified by Organized Fossils*, in four parts (London, 1816–1819), in which fossils from the London Clay (Tertiary) down to the Fuller's Earth Rock (Middle Jurassic) were shown on nineteen colored plates. *Stratigraphical System of Organized Fossils Part I* (London, 1817) described fossils from the London Clay down to the Marlstone of the Lias, with particular reference to those purchased by the British Museum, and contained a "Geological Table of British Organized Fossils Which Identify the Course and Continuity of the Strata in Their Order of Superposition, as Originally Discovered by W. Smith Civil Engineer; With Reference to His Geological Map of England and Wales." This table was also issued separately and later was included in a volume of geological sections (1819), five large folding sheets of hand-colored panoramic horizontal sections, across different parts of southern England. A geological section from London to Snowdon, the highest mountain in Wales, usually included in this work, had also been issued separately in 1817.

Cary, who published these geological sections, now provided Smith with maps of the English counties to color geologically; and in May 1819 geological maps of Kent, Sussex, Norfolk, and Wiltshire were published. This work continued up to 1824. In all, twenty-four maps of twenty-one counties were issued; Yorkshire, the largest English county, required four sheets. Other county maps were in an advanced state of preparation, but never appeared with geological coloring.

From 1820 Smith lived in the north of England. For many years he had no settled home—but lived in lodgings wherever his work or inclination led him. During 1824 and 1825 he and his nephew John Phillips (who later became professor of geology at Oxford) lectured on geology in several Yorkshire towns, but rheumatism and increasing deafness made it difficult for Smith to continue this occupation. In 1828 he was offered a post as land steward to Sir John Johnstone of Hackness (a village near Scarborough in Yorkshire), a great admirer of Smith's work. Smith lived at Hackness for about five years and while there mapped, on a scale of six-and-a-half inches to the mile, the Jurassic rocks of the Hackness Hills. This beautiful and accurate map was published in 1832.

In 1831 Adam Sedgwick, president of the Geological Society of London, announced that the first Wollaston Medal had been awarded to Smith "in consideration of his being a great original discoverer in English Geology; and especially for his having been the first, in this country, to discover and teach the identification of strata, and to determine their succession by means of their imbedded fossils" (*Proceedings of the Geological Society*, **1** [1834], 271). The gold medal (not then ready) was presented to Smith the following year at the British Association meeting in Oxford. This recognition of Smith's fundamental contribution to geology was followed by an award by the government of an annuity of £100. In 1835 whilst at a British Association meeting the LL.D. was bestowed on him at Trinity College, Dublin. In 1834 he left Hackness to live at Scarborough, and in 1835 moved into Newborough Cottage, Bar Street. He regularly attended the annual meetings of the British Association, to which he twice contributed papers; and at Scarborough he spent many hours writing reminiscences, fragments of geology, and notes on many topics.

Smith's last geological task was performed in 1838, when he accompanied Henry de la Beche, director of the newly established Geological Survey, Charles Barry, the architect, and C. H. Smith, a sculptor and mason, on a horse-and-carriage tour of the principal quarries of England and Scotland in order to choose a suitable stone for the new Houses of Parliament. On this tour particular attention was paid to the condition of the stone in old abbeys and churches. In 1839 the official report recommended the use of a magnesian limestone from certain quarries at Bolsover Moor, Derbyshire. As building proceeded, the supply of stone proved inadequate; and further supplies of magnesian limestone were obtained from the Anston quarries eight miles to the north in Yorkshire. Although this stone proved excellent when used for the Museum of Practical Geology, opened in 1851, it failed badly in parts of the Parliament buildings; and as early as 1861 an inquiry was held about its decay. The present view is that the stone was unsuitable for the highly decorated Parliament buildings, although satisfactory for the classic style of the museum. Smith's notes made at the time of the tour indicate that he was well aware of the many factors that can affect the condition of stone buildings. Had he not died suddenly from a chill on his way to a British Association meeting in Birmingham, his specialized knowledge and supervision might have made a marked difference in the selection of stone for the more deeply sculptured portions of the Houses of Parliament.

Smith's contributions to the advancement of geology were chiefly practical and were based on field geology; and to seek in his works, published or unpublished, theoretical considerations of a profound nature is a waste of time. Smith was a surveyor, a working man, not an academic; and he saw his discoveries as tools that could be used to promote the economic development of his country, in agriculture and in industry. Many of his unpublished notes confirm this viewpoint.

It is perhaps not widely realized how the geological succession in England itself contributed to William Smith's rapid progress in interpreting its order. In England it is possible to find sedimentary rocks of every age from Precambrian through Paleozoic to Mesozoic and Tertiary, and only the older Paleozoic rocks are so folded and compressed that interpretation of their succession is difficult. In only a few places in England does the intrusion of granites or other igneous rocks cause some disorder and irregularity; local folding and faulting also occur, but the intense folding and faulting that gave rise to the complicated Alpine structures of Europe reached only the very south of England, as minor ripples. Nor are there vast gaps in the succession, such as occur, for example, in the eastern United States, where the Jurassic beds are entirely absent and Cretaceous sediments directly overlie Triassic ones. In the former kingdom of Saxony, where Werner sought to distinguish "formations," the Jurassic rocks are also absent and the Cretaceous ones rest directly on Paleozoic or even older rocks.

This view is confirmed by T. H. Huxley, in his address to the British Association in 1881, "The Rise and Progress of Palaeontology." He stated that "this modest land-surveyor, whose business took him into many parts of England, profited by the peculiarly favourable conditions offered by the arrangement of our secondary strata . . ." (*Collected Essays*, IV [London, 1895], 37).

Unlike certain naturalists, Smith did not concern himself with the extinction of species or the living analogues of fossils. His knowledge of biology was minimal and he regarded fossils solely as a means of identifying a particular stratum, such as the Cornbrash or the Coral Rag. He did not recognize any age difference in these beds. Hence his approach was quite different from that of the naturalists Buffon and Soulavie, who earlier had concluded that rocks containing fossils of which there were no known living representatives must be older than those containing fossils part or all of which resembled creatures living in modern oceans. Smith did, however, recognize before 1800 that fossils worn by attrition found in alluvial beds indicated that the beds were deposited later than those containing the unworn fossils.

Smith's major achievements were (1) the recognition of a regular succession in the strata of England, first confirmed in the southwest and then established across most of the country; (2) the discovery that many individual beds have a characteristic fossil content that can be used to distinguish them from other beds that are lithologically similar; and (3) the utilization of these two discoveries in the preparation of a large-scale geological map of the whole country. Since this map was Smith's first geological publication, it is necessary to make clear how his discoveries became known long before its publication in 1815.

The earliest extant list of English strata prepared by Smith was written in 1797; more than twenty different strata from Chalk (Cretaceous) down to the Coal Measures and the limestone (Carboniferous) beneath them are briefly described, but no reference is made to fossils. His next list, dictated to Richardson and Townsend in June 1799, enumerated the strata around Bath, their particular characteristics, and the fossils found in certain beds. Twenty-three different beds are named, from Chalk to Coal; and in most cases the thickness of each, as then known to Smith, is given. Under the heading "Fossils, petrifactions, &c., &c." are listed the fossils found in ten of the named beds. Their names were provided mostly by the two clergymen. Besides the general terms ammonite, belemnite, and gryphite, more specific ones—high-waved cockle, prickly cockle, and large Scollop—are also given. These details were written down in tabular form by Richardson, and copies were made. Although the table was not printed until 1815, when it was inserted in a memoir accompanying the map, it is certain that manuscript copies were in circulation within a year or so, at first apparently unknown to Smith himself. To whom copies were distributed is not known, although in 1831 Richardson stated that he "without reserve gave a card of the English strata to Baron Rosencrantz, Dr Muller of Christiana, and many others, in the year 1801" (*Proceedings of the Geological Society*, **1** [1834], 276).

The first printed account of Smith's order of strata and of the fossils contained in different beds was published in the Reverend Richard Warner's *History of Bath* (Bath, 1801). Although brief and incomplete—and clearly not fully understood by Warner himself—on account of the date of publication it is of considerable significance. Warner, a Bath curate and also a well-known author, lived at Widcombe Cottage, situated between Bath and the

Coal Canal. He was acquainted with Smith, who examined Warner's collection of fossils and arranged them stratigraphically. In the *History of Bath*, a short chapter (pp. 394–399) is entitled "Mineralogy and Fossilogy of Bath"; in this Warner stated that he would give a general view of the strata and their "fossilogical contents" and that a "more scientific and particular account" would soon be given in a work written by "the very ingenious Mr. Smith, of Midford, near Bath. . . ." Warner then briefly described the principal strata found near Bath, the "Forest Marble," the Bath freestone (Upper Oolite), the fuller's-earth beds, the lower freestone (Inferior Oolite), the sands and marls beneath it, and the Lias. Each description included the names (in a far more detailed form than in Smith's 1799 list) of the fossils associated with the particular bed. This account seems to be the first printed description of a succession of different strata accompanied by details of their fossil content, and it was certainly derived from Smith. The account could have been read by many visitors to Bath, where there were several subscription libraries.

After 1800 Smith's knowledge of the English strata and its fossils was made known to others by Richardson (whose rectory near Bath was frequented by many persons interested in geology); by Farey, who from 1806 published references to Smith and his discoveries; and by Townsend, who published *The Character of Moses Established for Veracity as an Historian* (Bath, 1813). Although this title appears to have little connection with geology, the book contains a detailed account of the English strata from Chalk to Carboniferous Limestone, with plates illustrating the fossils from different formations. Townsend readily acknowledged his debt to Smith and used a number of Smith's names for different strata, names designated "uncouth" by geologists of the Wernerian school but still familiar to every English geologist.

In 1822, W. D. Conybeare and W. Phillips wrote that Smith "had freely communicated the information he possessed in many quarters, till in fact it became by oral diffusion the common property of a large body of English geologists, and thus contributed to the progress of the science in many quarters where the author was little known" (*Outlines of the Geology of England and Wales* [London, 1822], xlv).

Smith's great work, *Delineation of the Strata of England and Wales With Part of Scotland*, was undoubtedly a major cartographic and scientific achievement. It represented about 65,000 square miles, was the first large-scale geological map of any country, and was based on the scientific principles discovered by Smith himself. Moreover, the coloring was designed to indicate not only the surface area of any one geological formation, but, by using a deeper shade along the base of a formation, an attempt was also made to show how the beds were superimposed; thus a structural factor was introduced.

This map owed remarkably little to the work of others. Smith's manuscript maps of 1801 and 1802 show his early grasp of the general succession across England; and a comparison of his 1815 map with a modern geological map of England, on the same scale, shows the extent of his knowledge. Errors, of course, were made, and the more important were pointed out in 1818 by Fitton (*Edinburgh Review*, **29**, 310–337). But the amount of correct detail that Smith recorded is amazing and still impresses modern geologists. A stratigraphical succession of twenty-one sedimentary beds or groups of beds was shown in different colors, and one more color was used for large masses of granite or other crystalline rocks. Different signs were used to indicate mines of tin, lead, and copper; for collieries; and for salt and alum works. Not content with the map as issued in 1815, during the next few years Smith continued to make small alterations and additions, marked by changes of coloring and engraving. A noteworthy addition, made soon after April 1816, was the insertion of another limestone distinguished by its fossils, the Coral Rag, colored in orange. This outcrop was added first in Berkshire, Oxfordshire, Somerset, and Wiltshire and later, perhaps in 1817, in Yorkshire.

Smith's other cartographic publications—his geological sections across parts of England and his county maps—demonstrate his continued interest in field geology and its economic importance. This interest is particularly well shown by his four-sheet map of Yorkshire (1821), which has many details concerning the coal seams and their accompanying grits and sandstones.

Although Smith's map was superseded in 1820 by the geological map compiled by Greenough (published by the Geological Society), his county maps were used by geologists for many years; and their value was acknowledged by Sedgwick in 1831 (*Proceedings of the Geological Society*, **1**, 278).

Smith's two publications on fossils, *Strata Iden-* *tified* and *Stratigraphical System*, were complementary to his cartographic work. They appeared at a time when some prominent geologists were still unwilling to admit the value of fossils in determining the stratigraphical succession, but within a few years this opposition was overcome. Smith's publications no doubt contributed to the changed outlook. Certainly no one could deny Smith's right to the title "Founder of Stratigraphical Geology."

BIBLIOGRAPHY

I. ORIGINAL WORKS. All known publications by Smith are listed in Joan M. Eyles, "William Smith (1769–1839): A Bibliography of His Published Writings, Maps and Geological Sections, Printed and Lithographed," in *Journal of the Society for the Bibliography of Natural History*, **5** (1969), 87–109. A large collection of his MSS is in the possession of the Department of Mineralogy and Geology, University of Oxford. His portrait in oils by Fourau (1837) is owned by the Geological Society of London.

II. SECONDARY LITERATURE. John Phillips published a biography soon after his uncle's death: *Memoirs of William Smith, LL.D.* (London, 1844). This work was the principal source of information about Smith until the discovery of his papers at Oxford in 1938. These were examined and arranged by L. R. Cox, who gave an account of them in "New Light on William Smith and His Work," in *Proceedings of the Yorkshire Geological Society*, **25** (1942), 1–99. A detailed and well-illustrated, although uncritical, account of Smith's principal publications is in T. Sheppard, "William Smith: His Maps and Memoirs," *ibid*, **19** (1917), 75–253; repr. (Hull, 1920). Both Sheppard and Cox provide extensive bibliographies, that by Cox being supplemental to Sheppard's. An account of Smith's 1797 MS list is given by J. A. Douglas and L. R. Cox, "An Early List of Strata by William Smith," in *Geological Magazine*, **86** (1849), 180–188.

The principal sources of information about Smith available to 1967 are described by J. M. Eyles in "William Smith: Some Aspects of His Life and Work," in C. J. Schneer, ed., *Toward a History of Geology*, (Cambridge, Mass., 1969), 142–158; details of Smith's work as related to the construction of the Somerset Coal Canal are also in this paper. Smith's work in Somerset is also described by John G. C. M. Fuller, "The Industrial Basis of Stratigraphy: John Strachey, 1671–1743 and William Smith, 1769–1839," in *American Association of Petroleum Geologists Bulletin*, **53** (1969), 2256–2273.

A useful collection of quotations by and about Smith is in D. A. Bassett, "William Smith, the Father of English Geology and Stratigraphy: An Anthology," in *Geology: Journal of the Association of Teachers of Geolo-*

gy, **1** (1969), 38–51. One aspect of Smith's economic work is described by J. M. Eyles, "William Smith (1769–1839) and the Search for Coal in Great Britain," in *Geologie*, **20** (1971), 710–714; his interest in technological developments is described in "William Smith, Richard Trevithick and Samuel Homfray: Their Correspondence on Steam Engines, 1804–06," in *Transactions of the Newcomen Society*, **43** (1974), 137–161; and the correct identification of Smith's property near Bath is made in "William Smith's Home Near Bath: the Real Tucking Mill," in *Journal of the Society for the Bibliography of Natural History*, **7** (1974), 29–34. A detailed account of the progressive changes in Smith's 1815 map is in V. A. Eyles and J. M. Eyles, "On the Different Issues of the First Geological Map of England and Wales," in *Annals of Science*, **3** (1938), 190–212.

A journey by Smith in 1813 is described by J. E. Hemingway and J. S. Owen, "William Smith and the Jurassic Coals of Yorkshire," in *Proceedings of the Yorkshire Geological Society*, **40** (1975), 297–308, and an account of his lectures in Yorkshire is given by J. M. Edmonds, "The Geological Lecture-Courses Given in Yorkshire by William Smith and John Phillips, 1824–1825," *ibid.*, **40** (1975), 373–412.

JOAN M. EYLES

SMITH, WILSON (*b.* Great Harwood, Lancashire, England, 21 June 1897; *d.* Newbury, Berkshire, England, 10 July 1965), *microbiology.*

Smith, whose father, John Howard Smith, kept a small retail drapery shop, grew up in a modest and serious-minded environment. This upbringing had a lasting influence on him, and he remained throughout his life a most earnest person with an intense devotion to his work. He served as a private in the Royal Army Medical Corps in World War I, an experience that determined his career in medicine. He qualified at Manchester University in 1923 and afterward practiced medicine for three years. From 1926 to 1927 he took his diploma in bacteriology at Manchester under W. W. C. Topley, who was mainly responsible for Smith's interest in medical research.

In 1927 Smith married Muriel Nutt, one of Topley's demonstrators, and moved to the National Institute for Medical Research in London. Here he associated with many notable scientists, including H. H. Dale, S. R. Douglas, Percival Hartley, Clifford Dobell, P. P. Laidlaw, and C. H. Andrewes, and entered what was then a comparatively new field, the study of viruses. In 1939 he became professor of bacteriology at Sheffield University and in 1946 returned to London as professor of bacteriology at University College Hospital Medical School.

In 1960 he retired but continued his researches at the Microbiological Research Establishment, Porton, and gave up working only a few months before his death.

At the National Institute for Medical Research, Smith first worked with vaccinia and herpes viruses, directing his researches toward elucidating the mechanisms of protection that immunized animals acquire against viral infections (1, 2, 3, 4). But his most important contribution to virology concerned influenza. Intensive efforts were being made at the National Institute to isolate the influenza virus, and many different species of animals were being inoculated with the throat garglings from suspected cases and also with lung material from fatal cases. No success was achieved until Smith decided to inoculate ferrets intranasally. The first isolation of influenza virus was made in February 1933, when two ferrets were inoculated with throat washings from Andrewes, who had suffered a serious attack of influenza. Unfortunately this strain of influenza virus was lost, because of an outbreak of distemper in the ferrets, but a month later Smith himself had influenza. It was suspected he had caught it from a ferret, and the virus was isolated from his throat garglings by inoculating them into ferrets. The WS strain still remains one of the classic strains (5).

The isolation of the influenza virus opened up a wide field of work at the institute, most of it being shared by Laidlaw, Smith, and Andrewes. It was shown that the virus can be serially transmitted in ferrets and that virus material from infected ferrets can infect mice on intranasal inoculation. This finding led to the development of methods of determining the degree of infectivity of viral material and of estimating the neutralizing antibody present in human sera. Further, it was shown that the virus can grow in fertile eggs and in tissue culture and that there are antigenic differences between strains of virus (6–16).

Smith's work in virology was interrupted by his move to Sheffield in 1939 and by the administrative responsibilities he took on because of the war, and also by a heavy load of teaching. On returning to London in 1946, he again worked in virology and led a team of young workers, many of whom now hold important positions in microbiology.

Smith became one of the leading virologists in the United Kingdom; and his advice was sought by many institutions, including the Medical Research Council, the Ministry of Health, the Agricultural Research Council, and the Microbiological Research Establishment. The value of his work was widely recognized, and he was elected a fellow of the Royal Society of London (1949) and of the Royal College of Physicians of London (1959).

BIBLIOGRAPHY

A complete bibliography of Smith's works is given by D. G. Evans in *Biographical Memoirs of Fellows of the Royal Society*, **12** (1966), 479–487. Among those mentioned above are:

(1) "The Distribution of Virus and Neutralizing Antibodies in the Blood and Pathological Exudates of Rabbits Infected With Vaccinia," in *British Journal of Experimental Pathology*, **10** (1929), 93–95.

(2) "Generalized Vaccinia in Rabbits With Especial Reference to Lesions in the Internal Organs," in *Journal of Pathology and Bacteriology*, **32** (1929), 99–100, written with Douglas and Price.

(3) "A Study of Vaccinal Immunity in Rabbits by Means of *in vitro* Methods," in *British Journal of Experimental Pathology*, **11** (1930), 96–99, written with Douglas.

(4) "Specific Antibody Absorption by Viruses of Vaccinia and Herpes," in *Journal of Pathology and Bacteriology*, **33** (1930), 273–276.

(5) "A Virus Obtained from Influenza Patients," in *Lancet* (1933), **2**, 66–68, written with Andrewes and Laidlaw.

(6) "The Susceptibility of Mice to the Viruses of Human and Swine Influenza," *ibid.* (1934), 859–864, written with Andrewes and Laidlaw.

(7) "Cultivation of the Virus of Influenza," in *British Journal of Experimental Pathology*, **16** (1935), 508–511.

(8) "Influenza: The Preparation of Immune Sera in Horses," *ibid.*, 275–282, written with Laidlaw *et al.*

(9) "Influenza: Experiments on the Immunization of Ferrets and Mice," *ibid.*, 291–301, written with Andrewes and Laidlaw.

(10) "Influenza: Observations on the Recovery of Virus From Man and on the Antibody Content of Human Sera," *ibid.*, 566–568, written with Andrewes and Laidlaw.

(11) "Influenza Infection of Man from the Ferret," in *Lancet* (1936), **2**, 121–125, written with Stuart-Harris.

(12) "The Complement Fixation Reaction in Influenza," *ibid.*, 1256–1258.

(13) "Influenza: Further Experiments on the Active Immunization of Mice," in *British Journal of Experimental Pathology*, **18** (1937), 43–46, written with Andrewes.

(14) "Immunological Observation on Experimental Influenza," in *Proceedings of the International Congress of Microbiology* (1937), 107–110.

(15) "A Study of Epidemic Influenza: With Special Reference to the 1936–1937 Epidemic," *Report of the Medical Research Council*, no. 228 (1938), written with Stuart-Harris and Andrewes.

(16) "Serological Races of Influenza Virus," in *British Journal of Experimental Pathology*, **19** (1938), 293–297.

D. G. Evans

SMITHELLS, ARTHUR (*b.* Bury, Lancashire, England, 24 May 1860; *d.* Highgate, London, England, 8 February 1939), *chemistry.*

Smithells was an articulate spokesman for chemistry, chemical education, and the larger cultural dimensions of chemistry. His research contributed to the understanding of combustion and the structure of flames.

Smithells was the third son of James Smithells, a railway manager, and Martha Livesey. From 1875 to 1877 he studied physics and chemistry under Kelvin and Ferguson at the University of Glasgow, where he developed an abiding interest in the latter. Then, for the next five years, he studied chemistry under Roscoe and Schorlemmer at Owens College in Manchester. He received the B.Sc. in 1881 from the University of London and then became an "Associate" of Owens College until 1882, when he went to Munich to pursue his studies in chemistry with Baeyer and then to Heidelberg to study with Bunsen. The following year he returned to Manchester as assistant lecturer in chemistry. In 1885, at the age of only twenty-five, he succeeded T. E. Thorpe as professor of chemistry at the University of Leeds.

Smithells discovered a method for separating the two cones of the flame of a Bunsen burner and found that the inner cone contains residual hydrogen. Since it had been thought previously that hydrogen was burned preferentially, Smithells' discovery led to further investigations in combustion. He also conducted extensive research into the structure of flames and the luminosity of gases, but this work was inhibited by his administrative duties.

In 1901 Smithells was elected a fellow of the Royal Society, and in 1918 he was made a companion of the Order of St. Michael and St. George for his skilled organization of antigas training. He was also involved in educational reform: he strongly opposed specialized universities and advocated the integration of pure and applied science in the university curriculum. He also wished to extend science to the practical problems of daily life.

In 1907 Smithells was elected president of the chemistry section of the British Association for the Advancement of Science. In his presidential address he stressed the importance of atomic research and of the new investigations prompted by the discovery of radioactivity. Because of these scientific developments, the chemist, whose work had previously been "confined to comparatively gross quantities of matter," was called upon to examine and reinterpret earlier theories concerning the ultimate constituents of matter.

In 1923 Smithells resigned his chair at Leeds and moved to London, where, as director of Salters' Institute of Industrial Chemistry, he was influential in admitting students of chemistry. He retired in 1937, two years before his death.

BIBLIOGRAPHY

I. ORIGINAL WORKS. Many of Smithells' scientific works are listed in the Royal Society *Catalogue of Scientific Papers*, **18**, p. 820. His major works are "The Structure and Chemistry of Flames," in *Journal of the Chemical Society*, **61** (1892), 204–216, written with H. Ingle; and "The Electrical Conductivity and Luminosity of Flames Containing Vaporised Salts," in *Philosophical Transactions of the Royal Society*, **193A** (1900), 89–128, written with H. M. Dawson and H. A. Wilson. His address to the British Association appeared in *Report of the British Association for the Advancement of Science* (1907), 469–479; he also published a collection of his addresses in *From a Modern University* (Oxford, 1921). Smithells edited Schorlemmer's *The Rise and Development of Organic Chemistry*, 2nd ed. (London, 1894).

An autobiographical letter dated 2 May 1893 exists in the Krause *Album*, IV, MSS 7766, Sondersammlungen, Bibliothek, Deutsches Museum, Munich. MacLeod, *Archives*, indicates that many of his papers are held by Professor Phillip Smithells, 2 Pollock St., Maori Hill, Dunedin, New Zealand. There are several items in the Royal Institution, Imperial College, and at the University of Leeds.

II. SECONDARY LITERATURE. Smithells' former student J. W. Cobb wrote "Professor A. Smithells, C.M.G., F.R.S.," in *Nature*, **143** (1939), 321–322. He also wrote the article in the *Dictionary of National Biography 1931–1940*, 820–821, and one in *Journal of the Chemical Society* (July 1939), 1234–1236. H. S. Raper, another student, wrote the article in *Obituary Notices of Fellows of the Royal Society*, **8** (1940), 97–107. An anonymous note appeared in *The Annual Register* (1939), 428.

THADDEUS J. TRENN

SMITHSON, JAMES LOUIS MACIE (*b*. Paris, France, 1765; *d*. Genoa, Italy, 27 June 1829), *chemistry.*

Smithson was the illegitimate son of Hugh Smithson Percy, first duke of Northumberland, and Elizabeth Hungerford Keate Macie, who was the widow of James Macie, a country gentleman of Bath, England. While pregnant, his mother had gone to Paris, where Smithson was born in 1765; but no record has been found of the exact date of birth. Until the age of thirty-six he was known as James Louis Macie. He took the surname Smithson on 16 February 1801. Upon the death of his mother in 1800, James acquired a small fortune, which enabled him to support his researches and extensive travels.

At the age of ten, Smithson returned to England with his mother, where he was naturalized a British subject; but a provision was made at the time that he could not hold public office (civil or military) or have any grant of land from the crown. On 7 May 1782 he entered Pembroke College, Oxford, and subsequently took his master of arts on 26 May 1786. He was attracted to the study of chemistry and mineralogy, and having been sponsored by Kirwan and Cavendish, he became a fellow of the Royal Society on 26 April 1787. He also may have worked in the private laboratory of Cavendish at that time.

Smithson read his first paper before the Royal Society on 7 July 1791. The paper concerned a study of the chemical properties of tabasheer, a substance found in bamboo. He also was listed as a charter member of the Royal Institution. His most important work, "A Chemical Analysis of Some Calamines," was read before the Royal Society on 18 November 1802 and published in the *Philosophical Transactions*. He analyzed zinc ores from various European deposits and showed them to be primarily zinc carbonate. Since his analytical techniques were creditable, and as a result of his study, the mineral zinc carbonate was named smithsonite in his honor. Smithson published twenty-seven papers on chemical subjects in the *Philosophical Transactions* and the *Annals of Philosophy*, but his importance as a chemist is minimal.

Smithson is remembered chiefly because he left money for founding the Smithsonian Institution in Washington, D.C. On 23 October 1826 Smithson prepared his will, according to which he left his estate to Henry James Hungerford, his nephew. There was the following stipulation:

> In the case of the death of my said Nephew without leaving a child or children, or the death of the child or children he may have had under the age of twenty-one years or intestate, I then bequeath the

whole of my property subject to the Annuity of One Hundred pounds to John Fitall, & for the security & payment of which I mean Stock to remain in this Country, to the United States of America, to found at Washington, under the name of the Smithsonian Institution, an Establishment for the increase & diffusion of knowledge among men.

On 27 June 1829 James Smithson died in Genoa, Italy, at the age of sixty-four; his estate passed to his nephew. On 5 June 1835 the nephew died without heirs, and the United States Government was notified of its claim to the estate. The claim was prosecuted, and within three years the estate, which amounted to $508,318.46, was shipped to the United States Mint at Philadelphia. It was not until 10 August 1846 that Congress agreed to the disposition of the money and the founding of the Smithsonian Institution. This delay was due to disagreement among Congressional leaders as to the nature of the proposed institution. Among the schemes suggested were an observatory, a library, a university, and a museum. On 3 December 1846 Joseph Henry was elected the first secretary of the Smithsonian Institution, and he guided its development as primarily a scientific institution.

In 1904 the remains of James Smithson were brought to Washington from Genoa and interred in the original Smithsonian building.

BIBLIOGRAPHY

I. ORIGINAL WORKS. Much of the papers, personal library, and mineral collections of James Smithson, which were brought to the United States and housed in the Smithsonian Institution, were destroyed in a fire of 1865. His twenty-seven published scientific papers were edited by William J. Rhees and published in 1879 by the Smithsonian Institution; see *Smithsonian Miscellaneous Collections*, 21 (1881).

II. SECONDARY LITERATURE. Many of the details of Smithson's life have been documented by biographers, but much of his life remains unrecorded. Some of the notable biographical sketches include Leonard Carmichael and J. C. Long, *James Smithson and the Smithsonian Story* (New York, 1965); Samuel P. Langley, "Biographical Sketch of James Smithson," in George Brown Goode, ed., *The Smithsonian Institution 1846–1896, the History of Its First Half-Century* (Washington, D.C., 1897), 1–24; Paul H. Oehser, *Sons of Science* (New York, 1949), 1–25; and W. J. Rhees, *James Smithson and His Bequest* (Washington, D.C., 1880), also in *Smithsonian Miscellaneous Collections*, 21 (1881).

DANIEL P. JONES

SMITS, ANDREAS (*b*. Woerden, Netherlands, 14 June 1870; *d*. Doorn, Netherlands, 13 November 1948), *physical chemistry*.

Smits first studied chemistry at the University of Utrecht and then entered the University of Giessen, from which he received the Ph.D. *magna cum laude* (1896) for his dissertation "Untersuchungen mit dem Mikromanometer." Two years later he was appointed chemist at the Municipal Gasworks in Amsterdam, and in 1901 he became a privatdocent in chemistry at the University of Amsterdam. He subsequently was named professor of general chemistry at the University of Technology at Delft (1906), and in 1907 succeeded Roozeboom as professor of chemistry at Amsterdam. He held this latter post until his retirement in 1940.

During these years, Smits's scientific research covered three fields. Prior to 1905 he studied the relationship in dilute solutions between the decrease in vapor pressure and the elevation of the boiling point. Also, he determined the so-called van't Hoff factor *i* (or ionic coefficient) for different aqueous concentrations of various salts. After 1924 he investigated the possibility of metal interconvertibility—especially of lead into mercury—by various means, using a quartz-lead lamp; sparks or electric arc between lead electrodes; and solar, X, and ultraviolet radiation.

Smits's major research, however, was in phase theory, especially in three-component systems with critical endpoints where two phases are identical; and in so-called pseudobinary systems. Smits thought that every component of a pseudobinary system actually contains two types of molecules. A pseudobinary system is a one-component system the molecules of which are chemically the same, but nevertheless may be divided into two types differing from one another in respect of "physical" properties, and there exists an equilibrium between the two types. He defended this theory throughout his career and tried to apply it to allotropic forms; the phenomena of passivity, polarization, and overvoltage in metals; electromotive equilibriums; ortho- and para-hydrogen systems; intensively dried substances; and continuous transitions.

Many of Smits's investigations (both independent and with his students) are summarized in two books: *Die Theorie der Allotropie* (1921) and *Die Theorie der Komplexität und der Allotropie* (1938).

BIBLIOGRAPHY

I. ORIGINAL WORKS. A complete bibliography of Smits's writing is in *Chemisch weekblad*, **3** (1906), 582–

583; **28** (1931), 561–566; **37** (1940), 435–436; and **45** (1949), 151. His major works include "Untersuchungen mit dem Mikromanometer" (Ph.D. diss., Univ. of Giessen, 1896), which appeared in part in *Verslagen en mededeelingen der Koninklijke Akademie van Wetenschappen*, **5** (1897), 292–295; *Die Theorie der Allotropie* (Leipzig, 1921), with trans. by J. S. Thomas as *The Theory of Allotropy* (London, 1922); and *Die Theorie der Komplexität und der Allotropie* (Berlin, 1938).

II. SECONDARY LITERATURE. On Smits and his work, see J. M. Bijvoet, "Prof. Dr. A. Smits. 4 October 1906– 4 October 1931," in *Chemisch weekblad*, **28** (1931), 555–559; F. E. C. Scheffer, "Het 25-jarig hoogleeraarschap van Prof. Dr. A. Smits. Enkele persoonlijke herinneringen," *ibid.*, **28** (1931), 560–561; the unsigned "Professor Dr. A. Smits. 14 Juni 1870–14 Juni 1940," *ibid.*, **37** (1940), 430–435; and E. H. Buchner, "Andreas Smits 1870–1948," *ibid.*, **45** (1949), 149–151.

H. A. M. SNELDERS

SMOLUCHOWSKI, MARIAN (*b.* Vorderbrühl, near Vienna, Austria, 28 May 1872; *d.* Cracow, Poland, 5 September 1917), *physics.*

Born to a Polish family, Smoluchowski spent his youth in Vienna. His father, Wilhelm, was a senior official in the chancellery of Emperor Franz Josef; his mother was the former Teofila Szczepanowska. Smoluchowski attended the Collegium Theresianum from 1880 to 1885 and was an outstanding student. From 1890 to 1895 he studied at the University of Vienna under the direction of Josef Stefan and F. Exner; his doctoral dissertation was entitled "Akustische Untersuchungen über die Elastizität weicher Körper."

From November 1895 to August 1897 Smoluchowski worked under Lippmann in Paris, with Lord Kelvin in Glasgow, and with Warburg in Berlin. Two papers published during his stay in Paris dealt with thermal radiation (the Kirchhoff-Clausius law). At Glasgow he investigated the influence of Röntgen and Becquerel rays on the conductance of gases. In 1901 Smoluchowski received the LL.D. from Glasgow, where he had been a research fellow. In Berlin he worked on the discontinuity of temperature in gases, a problem suggested to him by Warburg. In 1875 Warburg and A. Kundt, on the basis of the kinetic theory of gases, had predicted that if the temperature of a gas differed from that of the container wall, the former temperature would not pass continuously to the latter: there would be a discontinuity of temperature between the gas and the wall. Their experiments, successful in the case of the analogous phenomenon of the slipping of gases, had not been decisive for temperature discontinuity. Smoluchowski, observing the cooling time of a thermometer in a gas-filled container, demonstrated that such an effect exists and reached significant values with rarefied gas in "Uber Wärmeleitung in verdünnten Gasen" (1898).

This work was of special importance, for by publishing it Smoluchowski joined the dispute on the validity of atomic conceptions. These, represented in physics mainly by the kinetic theory of gases developed by Boltzmann, were far from accepted at the end of the nineteenth century; and their recognition was partly due to Smoluchowski. At that time only a few phenomena were predicted by the kinetic theory or required it for intelligibility. Among them was discontinuity of temperature, for its existence was wholly unexplained from a classical point of view. Moreover, in 1897, after his return to Vienna, Smoluchowski pointed out the quantitative agreement of his experimental results with the kinetic theory. In 1898 the University of Vienna admitted him *veniam legendi*.

From 1899 Smoluchowski worked at the University of Lvov. Appointed professor in 1900, he held the chair of mathematical physics there until 1913. His first works at Lvov concerned atmospheric physics, aerodynamics and hydrodynamics, electrophoresis, and the theory of mountain folding. Recognition for these specialized works was shown by his being asked to write the chapter on endosmosis phenomena in *Handbuch der Elektrizität und des Magnetismus*, edited by J. A. Barth (Leipzig, 1914).

From about 1900 Smoluchowski worked on Brownian movement. He wished to use experimental data to verify the theory he had obtained, a desire complicated by the confused situation of experimental research. In the meantime Einstein, in papers of 1905 and 1906, had presented a solution to the problem. Smoluchowski then decided to publish his results in "Zarys kinetycznej teorii ruchów Browna" ("An Outline of the Kinetic Theory of Brownian Movement," 1906), which presented his different method. Einstein started from general relations of statistical physics, an approach that was universal but did not lend itself to visualization. For example, Einstein said nothing of the collisions between a Brownian particle and the surrounding molecules. Smoluchowski started by examining the effects of successive collisions and obtained a final formula that differed little from Einstein's. Smoluchowski's further works in this field extend through an examination of the Brownian movement of a particle undergoing the influence of a quasi-elastic force to the Brownian movements of macroscopic bodies. At the Conference of Natural Scientists at Münster in 1912, Smoluchowski proposed the observation of the Brownian rotative movement of a small mirror suspended on a thin quartz fiber and the observation of a free end of a similar fiber. The first experiment was performed by W. Gerlach and E. Lehrer in 1927, and later by Eugen Kappler; the second, by A. Houdijk and P. Zeeman, and by E. Einthoven, in 1925. Both experiments confirmed Smoluchowski's calculations.

Another of Smoluchowski's interests concerned fluctuations and was related to the second law of thermodynamics. The kinetic approach to the second law proposed by Boltzmann implied the occurrence of spontaneous deviations from a state of maximum entropy. Because the occurrence of such deviations had not been experimentally confirmed, however, the kinetic theory lay open to attack. Boltzmann, in defense, calculated the time of return for given micro states and showed how rare and difficult to observe these phenomena are. Smoluchowski, on the other hand, laid the foundations of the theory of fluctuations, calculated the times of return for macro states, linked the theory to measurable parameters, and proved the actual existence of fluctuations.

In his first paper dealing with this problem, "Über Unregelmässigkeiten in der Verteilung von Gasmolekülen" (1904), Smoluchowski gave a theoretical approach to the fluctuations of density in a gas. At the same time he indicated the experimental possibility of detecting these fluctuations either by optical methods or when the number of particles is not too great, as in the case of colloidal suspensions. In 1910 Theodor Svedberg based his experiments on Smoluchowski's calculations, observing how many particles of a suspension can be seen in the field of vision of a microscope at a given time and experimentally confirming Smoluchowski's predictions to an astonishingly high degree. Smoluchowski himself had previously proved the existence of fluctuations of density in a pure gas by demonstrating that they are responsible for the known but unexplained phenomenon of the opalescence of a gas at a critical state. His paper "Teoria kinetyczna opalescencji gazów w stanie krytycznym" ("Kinetic Theory of Gas Opalescence at the Critical State," 1907) shows why the critical point plays such an important role and

states that the opalescence of a pure gas also should be observable under normal conditions: "Each of us has observed it innumerable times when admiring the blue of the sky or the glow of the rising sun." Smoluchowski combined the theory of fluctuations with the results of Lord Rayleigh's researches on the blue of the sky; his finding (Einstein also took part in the discussion) was that the blueness of the sky was caused by fluctuations in the density of the air. Smoluchowski's laboratory production of sky blue closed the investigation to a certain extent. In "Experimentelle Bestätigung der Rayleighschen Theorie des Himmelsblaus" (1916) Smoluchowski demonstrated that pure air can opalesce in the laboratory under normal conditions.

Experimental proofs of the existence of fluctuations and the revelation of the causes of Brownian movement have limited the validity of the classical formulation of the second law of thermodynamics. A new, statistical formulation of this principle that had been initiated by Boltzmann was developed by Smoluchowski. He concluded that the deeper one goes into microscopic processes, the more visible the reversibility becomes. Macroscopic processes, although theoretically reversible, are practically irreversible because of the unimaginably long times of return. Statistical interpretation of the second law gave ground, at that time, for hopes of constructing the *perpetuum mobile* of the second kind. Smoluchowski resolved this question in his lectures at Münster (1912) and Göttingen (1914): If we expect great deviations from the state of maximum entropy, the efficiency of the machine will be infinitesimal, since great deviations are extremely rare; and if we hope for microscopic deviations, the valves and other parts needed to eliminate deviations occurring in an adverse direction will have to be so fine that they themselves will undergo Brownian movement and will not be able to perform their task. Thus it is not possible to construct a *perpetuum mobile* of the second kind if we mean a machine with any finite efficiency.

Smoluchowski obtained important results in the physics of colloids. His interest in the methodology of teaching physics can be seen in the chapters he wrote for *Poradnik dla samouków* ("Primer for Home Studies"), edited by A. Heflich and S. Michalski (Warsaw, 1917). The title of the book is a bit misleading, for it is really a report on physical research and aims.

In 1913 Smoluchowski became professor of experimental physics at the Jagiellonian University in Cracow. In 1917 he was elected rector of the university. Later that year he died of dysentery, at the age of forty-five.

BIBLIOGRAPHY

I. ORIGINAL WORKS. Smoluchowski's writings were collected as *Pisma Mariana Smoluchowskiego* ("The Works of Marian Smoluchowski"), W. Natanson, ed., 3 vols. (Cracow, 1924–1928). Besides texts in Polish, this collection includes versions published in other languages.

His most important works are "Über Wärmeleitung in verdünnten Gasen," in *Annalen der Physik und Chemie*, **64** (1898), 101–130; "Uber Unregelmässigkeiten in der Verteilung von Gasmolekülen und deren Einfluss auf Entropie und Zustandsgleichung," in *Boltzmann-Festschrift* (Leipzig, 1904), 626–641; "Zur kinetischen Theorie der Brownschen Molekularbewegung und der Suspensionen," in *Annalen der Physik*, 4th ser., **21** (1906), 756–780; "Molekular-kinetische Theorie der Opaleszenz von Gasen im kritischen Zustande, sowie einiger verwandter Erscheinungen," *ibid.*, **25** (1908), 205–226; "Beitrag zur Theorie der Opaleszenz von Gasen im kritischen Zustande," in *Bulletin international de l'Académie des sciences et des lettres de Cracovie*, ser. A (1911), 493–502; "Experimentell nachweisbare, der üblichen Thermodynamik widersprechende Molekularphänomene," in *Physikalische Zeitschrift*, **13** (1912), 1069–1079; "Gültigkeitsgrenzes des zweiten Hauptsatzes der Wärmetheorie," in *Vorträge über die kinetische Theorie der Materie und der Elektrizität* (Leipzig–Berlin, 1914), 89; "Studien über Molekularstatistik von Emulsionen und deren Zusammenhang mit der Brownschen Bewegung," in *Sitzungsberichte der Akademie der Wissenschaften in Wien*, Math.-nat. Kl., Abt. IIa, **123** (1914), 2381–2405; "Molekulartheoretische Studien über umkehr thermodynamisch irreversibler Vorgänge und über Wiederkehr abnormaler Zustände," *ibid.*, **124** (1915), 339–368; "Über Brownsche Molekularbewegung unter Einwirkung äusserer Kräfte und deren Zusammenhang mit der verallgemeinerten Diffusionsgleichung," in *Annalen der Physik*, **48** (1915), 1103–1112; "Experimentelle Bestätigung der Rayleighschen Theorie des Himmelsblaus," in *Bulletin international de l'Académie des sciences et des lettres de Cracovie*, ser. A (1916), 218–222; "Drei Vorträge über Diffusion, Brownsche Molekularbewegung und Koagulation von Kolloidteilchen," in *Physikalische Zeitschrift*, **17** (1916), 557–571, 585–599; and "Über den Begriff des Zufalls und den Ursprung der Wahrscheinlichkeitsgesetze in der Physik," in *Naturwissenschaften*, **6** (1918), 253–263.

II. SECONDARY LITERATURE. Polish sources are W. Kapuściński, "Poglądy filozoficzne Mariana Smoluchowskiego" ("Marian Smoluchowski's Philosophical Views"), in *Fizyka i chemia*, **6** (1953), 200; W. Krajewski, *Swiatopogląd Mariana Smoluchowskiego* ("Marian Smoluchowski's *Weltanschauung*"; Warsaw, 1956); S. Loria, "Marian Smoluchowski i jego dzieło" ("Marian Smoluchowski and His Work"), in *Postępy fizyki*, **4** (1953), 5; and A. Teske, *Marian Smoluchowski—życie i twórczość* ("Marian Smoluchowski—Life and Work"; Warsaw, 1955).

Other sources are A. Einstein, "Marian von Smoluchowski," in *Naturwissenschaften*, **5** (1917), 737–738; A. Sommerfeld, "Zum Andenken an Marian von Smoluchowski," in *Physikalische Zeitschrift*, **18** (1917), 533–539, with bibliography; A. Teske, "An Outline Account of the Work of Marian Smoluchowski," in *Études consacrées à Maria Skłodowska-Curie et à Marian Smoluchowski* (Wrocław–Warsaw–Cracow, 1970); and S. Ulam, "Marian Smoluchowski and the Theory of Probabilities in Physics," in *American Journal of Physics*, **25** (1957), 475–481.

ANDRZEJ A. TESKE

SMYTH, CHARLES PIAZZI (*b*. Naples, Italy, 3 January 1819; *d*. Clova, near Ripon, Scotland, 21 February 1900), *astronomy, meteorology.*

Smyth was named after Giuseppe Piazzi, the astronomer-friend of Smyth's father, William Henry Smyth (1788–1865), who was stationed in Italy with the Royal Navy when Smyth was born. In 1825 the Smyth family returned to England and settled at Bedford, where Smyth's father established the Bedford observatory, the best-equipped private observatory in England.

Smyth was educated at the Bedford grammar school. From 1835 to 1845 he served as assistant to Maclear at the royal observatory, Cape of Good Hope. There Smyth observed and drew the great comets of 1836 and 1843, and participated in the verification and extension of Lacaille's arc of meridian. In 1845 Smyth was appointed successor to Thomas Henderson as director of the Edinburgh observatory, a position that included the additional titles astronomer royal for Scotland and professor of practical astronomy in the University of Edinburgh. Although under his direction the Edinburgh observatory produced observations of the positions of the sun, moon, planets, and stars, Smyth was primarily interested in more experimental and speculative matters.

In 1856 Smyth led an expedition to the Peak of Tenerife, primarily "to ascertain how much astronomical observation can be benefited, by eliminating the lower third or fourth part of the atmosphere" (C. P. Smyth, *Report on the Teneriffe Astronomical Experiment* [London, 1858]). Financial support for this experiment was given by the British Admiralty, and moral support by the entire British scientific community. In addition to telescopic observations of planets and stars, Smyth measured the radiant heat of the moon; observed the solar spectrum and noted which lines of absorption were of terrestrial origin; and made various other observations of the meteorology, geology, and botany of the island.

With his wife Smyth spent four months in 1865 in Egypt, where he measured the orientation, sizes, and angles of the various parts of the Great Pyramid at Giza, and correlated the results with astronomical phenomena. Smyth's measurements earned him the Keith Prize of the Royal Society of Edinburgh. His speculations on the mysteries hidden within the pyramid sparked an acrimonious debate, which led to his resignation from the Royal Society of London.

Smyth also charted the spectra of the sun, aurora, zodiacal light, the atmosphere under different meteorological conditions, and—in the laboratory—of various luminous gases. In order to resolve difficult solar lines, he obtained Rutherfurd and then Rowland diffraction gratings; and, in search of clearer atmosphere than Edinburgh afforded, Smyth traveled to Palermo, Portugal, Madeira, and Winchester. For his spectroscopic studies Smyth won the Makdougall-Brisbane Prize of the Royal Society of Edinburgh.

Like his father, Smyth was a member of numerous societies. Toward the end of his life he signed his name "C. Piazzi Smyth, F.R.S.E., F.R.A.S., F.R.S.S.A., Corresponding Member of the Academies of Science in Munich and Palermo; honorary member of the Royal Society of Modena, of the Institute of Engineers in Scotland, and of the Edinburgh Photographic Society; Regius Professor of Practical Astronomy in the University of Edinburgh, and Astronomer-Royal for Scotland; also Ex-Member of the Royal Society, London" (*Astronomical Observations Made at the Royal Observatory, Edinburgh*, **15** [1886]).

BIBLIOGRAPHY

I. ORIGINAL WORKS. Smyth's works include *Astronomical Observations Made at the Royal Observatory, Edinburgh*: vols. VI–X (1847–1852) contain observations made by T. Henderson, reduced and edited by C. P. Smyth; vols. XI–XV (1852–1880) contain observations made under Smyth's directorship; *Paris Universal Exposition of 1855 . . . Description of New or Improved Instruments for Navigation and Astronomy* (Edinburgh, 1855); *Teneriffe: An Astronomer's Experiment; or Specialities of a Residence Above the Clouds*

. . . *Illustrated With Photo-Stereographs* (London, 1858); *Report on the Teneriffe Astronomical Experiment of 1856, Addressed to the Lords Commissioners of the Admiralty* (London, 1858), repr. from *Philosophical Transactions of the Royal Society*; and *Three Cities in Russia*, 2 vols. (London, 1862).

See also *Our Inheritance in the Great Pyramid . . .* (London, 1864; 5th enl. ed., 1890); *Life and Work at the Great Pyramid During the Months of January, February, March, and April, A.D. 1865; With a Discussion of the Facts Ascertained*, 3 vols. (Edinburgh, 1867); *On the Antiquity of Intellectual Man, From a Practical and Astronomical Point of View* (Edinburgh, 1868); *A Poor Man's Photography at the Great Pyramid in . . . 1865, Compared With That of the Ordnance Survey Establishment . . .* (London, 1870); *On an Equal-Surface Projection for Maps of the World, and Its Application to Certain Anthropological Questions* (Edinburgh, 1870); *The Great Pyramid and the Royal Society (London)* (London, 1874); *Madeira Spectroscopic, Being a Revision of 21 Places in the Red Half of the Solar Visible Spectrum, With a Rutherfurd Diffraction Grating, at Madeira . . . During the Summer of 1881* (Edinburgh, 1882); and *Madeira Meteorologic* (Edinburgh 1882). In addition to the above, the catalog of the British Museum Library lists several responses to Smyth's ideas about the Great Pyramid. The Royal Society *Catalogue of Scientific Papers*, V, 735–737; VIII, 976; XI, 443–444; XVIII, 823; and Poggendorff, III, 1261–1262, list over 130 articles by Smyth.

II. SECONDARY LITERATURE. On Smyth and his work, see the obituaries by Ralph Copeland, in *Monthly Notices of the Royal Astronomical Society*, **61** (1901), 189–196; in *Popular Astronomy*, **8** (1900), 384–387 (also in *Astronomische Nachrichten*, **152** [1900], 189); and in *Observatory*, **23** (1900), 145–147. See also the article by Agnes Clerk, in *Dictionary of National Biography*, XXII, 1222–1223.

DEBORAH JEAN WARNER

SNEL (Snellius or **Snel van Royen), WILLEBRORD** (*b.* Leiden, Netherlands, 1580; *d.* Leiden, 30 October 1626), *mathematics, optics, astronomy.*

Snel was the son of Rudolph Snellius, or Snel van Royen, professor of mathematics at the new University of Leiden, and of Machteld Cornelisdochter. He studied law at the university but became interested in mathematics at an early age. Through the influence of Van Ceulen, Stevin, and his father, he received permission in 1600 to teach mathematics at the university. Soon afterward he left for Würzburg, where he met Van Roomen. He then went to Prague to conduct observations under Tycho. He also met Kepler, and traveled to Altdorf and Tübingen, where he saw Mästlin, Kepler's teacher. In 1602 Snel studied law in Paris. He returned home in 1604, after having traveled to Switzerland with his father, who was then in Kassel at the court of the learned Prince Maurice of Hesse.

At Leiden, Snel prepared a Latin translation of Stevin's *Wisconstighe Ghedachtenissen*, which was then being published; Snel's translation appeared as *Hypomnemata mathematica* (1608). He also busied himself with the restoration of the two books of Apollonius on plane loci, preserved only in abstract by Pappus. Related tasks on other books of Apollonius also occupied Viète (*Apollonius gallus* [1600]) and Ghetaldi (*Apollonius redivivus* [1607–1613]). Snel's work was in three parts: the first remained in manuscript and is preserved at the library of the University of Leiden; the second appeared under the title Περὶ λόγου ἀποτομῆς καὶ περὶ χωρίου ἀποτομῆς *resuscita geometria* (1607); and the third was published as *Apollonius batavus* (1608).

In 1608 Snel received the M.A. and married Maria De Lange, daughter of a burgomaster of Schoonhoven; only three of their eighteen children survived. After his father's death in March 1613, Snel succeeded him at the university, and two years later he became professor. He taught mathematics, astronomy, and optics, using some instruments in his instruction.

Sharing the admiration of his father and of Maurice of Hesse for Ramus, Snel published Ramus' *Arithmetica*, with commentary, in 1613. He later published *P. Rami Meetkonst* (1622), an annotated Dutch translation by Dirck Houtman of Ramus' *Geometria*. It was the only one of Snel's works to be published in Amsterdam; all the others appeared at Leiden. Snel's *De re numeraria* (dedicated to Grotius), a short work on money in Israel, Greece, and Rome, also dates from 1613.

During this period Snel prepared the Latin translation of two books by Van Ceulen, probably at the request of his widow, Adriana Symons. His rather careless translation of *Van den Circkel* includes some notes by Snel, among them the expression $\sqrt{(s-a)(s-b)(s-c)(s-d)}$ for the area of a cyclic quadrilateral. Although this expression had already appeared in the work of Brahmagupta, it seems to be the first time that it was used in Europe.

Snel's lack of attention to this translation may have been due to preoccupation with geodetic work. In 1615 he became deeply involved in the determination of the length of the meridian, selecting for this work the method of triangulation, first proposed by Gemma Frisius in 1533 and also used by Tycho. Snel developed it to such an extent that he may rightfully be called the father of triangulation. Starting with his house (marked by a memorial plaque in 1960), he used the spires of town churches as points of reference. Thus, through a net of triangles, he computed the distance from Alkmaar to Bergen-op-Zoom (around 130 kilometers). The two towns lie on approximately the same meridian. Snel used the distance from Leiden to Zoeterwoude (about 5 kilometers) as a baseline. His instruments were made by Blaeu; and the huge, 210-centimeter quadrant used for his triangulations is suspended in the hall of the Leiden astronomical observatory. The unit of measure was the Rhineland rod (1 rod = 3.767 meters), recommended by Stevin to the States General in 1604 (Stevin, *Principal Works*, IV [1964], 24); and, following Stevin, the rod was divided into tenths and hundredths. The results were presented in *Eratosthenes batavus* (1617).

In order to locate his house with respect to three towers in Leiden, Snel solved the so-called recession problem for three points. The problem is often named after Snel, as well as after L. Pothenot (1692); and claims have been made for Ptolemy.

Dissatisfied with his geodetic work, Snel began to correct it, aided by his pupils, and extended his measurements to include the distance from Bergen-op-Zoom to Mechelen. Unaided by logarithms, he continued this work throughout his life. His early death in 1626 prevented him from publishing his computations, which are preserved in his own copy of *Eratosthenes batavus* at the Royal Library in Brussels. They were recently checked by N. D. Haasbroek and were found to be conscientious and remarkably accurate. Haasbroek could not say as much for the way in which Musschenbroek handled these notes in his "De magnitudine terrae," in *Physicae experimentales* . . . (1729).

Snel published some observations by Bürgi and Tycho in 1618, and his descriptions of the comets of 1585 and 1618, published in 1619, show Snel to be a follower of the Ptolemaic system. Although he demonstrated from the parallax that the comet was beyond the moon and therefore could not consist of terrestrial vapors, he still believed in the character of comets as omina.

In the *Cyclometricus* (1621) π was found, by Van Ceulen's methods, to thirty-four decimals; and the thirty-fifth decimal, found in Van Ceulen's papers, was added. Snel also explained his own shorter method, following and improving on Van Lansberge's *Cyclometria nova* (1616), establishing the inequality

$$\frac{3\sin\varphi}{2+\cos\varphi} < \varphi < \tan\frac{\varphi}{3} + 2\sin\frac{\varphi}{3},$$

of which the inequality to the left agrees with Cusa's result in *Perfectio mathematica* (1458).

In 1624 Snel published his lessons on navigation in *Tiphys batavus* (Tiphys was the pilot of the *Argo*). The work is mainly a study and tabulation of Pedro Nuñez' so-called rhumb lines (1537), which Snel named "loxodromes." His consideration of a small spherical triangle bounded by a loxodrome, a parallel, and a meridian circle as a plane right triangle foreshadows the differential triangle of Pascal and later mathematicians.

The last works published by Snel himself were *Canon triangulorum* (1626) and *Doctrina triangulorum* (1627), the latter completed by his pupil Hortensius. The *Doctrina*, which comprise a plane and spherical trigonometry, includes the recession problem for two points, often named after P. A. Hansen (1841). It uses the polar triangle for the computation of the sides of a spherical triangle.

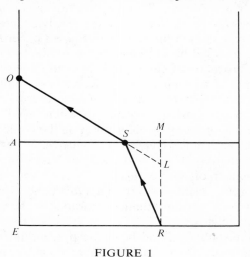

FIGURE 1

Snel's best-known discovery, the law of refraction of light rays, which was named after him, was formulated probably in or after 1621, and was the result of many years of experimentation and of the study of such books as Kepler's *Ad Vitellionem paralipomena* (1604) and Risner's *Optica* (1606), both of which quote Ibn al-Haytham and Witelo. Snel's manuscript, which contained his results, has disappeared, but it was examined by Isaac Vossius (1662) and by Huygens, who commented on it in his *Dioptrica* (1703, 1728). Snel's wording of his law has been preserved in what C. De Waard considered to be an index of the manuscript preserved in Amsterdam, and it checks with the account of

Snel's law given by Vossius: If the eye *O* (in the air) receives a light ray coming from a point *R* in a medium (for example, water) and refracted at *S* on the surface *A* of the medium, then *O* observes the point *R* as if it were at *L* on the line *RM* ⊥ surface *A*. Then *SL*:*SR* is constant for all rays. This agrees with the present formulation of the law, which states that sin *r*:sin *i* is constant, where *i* and *r* are the angles that *OS* and *SR* make with the normal to *A* at *S*.

The priority of the publication of the law remains with Descartes in his *Dioptrique* (1637), stated without experimental verification. Descartes has been accused of plagiarism (for example, by Huygens), a fact made plausible by his visits to Leiden during and after Snel's days, but there seems to be no evidence for it.

Snel was buried in the Pieterskerk in Leiden. The monument erected to him and his wife, who died in 1627, is still there.

BIBLIOGRAPHY

Snel's works are cited in the text. On his life and work see C. De Waard, in *Nieuw nederlandsch biographisch woordenboek*, 7 (1927), 1155–1163; and P. van Geer, "Notice sur la vie et les travaux de Willebrord Snellius," in *Archives néerlandaises des sciences exactes et naturelles*, 18 (1883), 453–468. On his trigonometric work, see A. von Braunmühl, *Geschichte der Trigonometrie*, I (Leipzig, 1900), 239–246. On his geodetic work, see H. Bosmans, "Le degré du méridien terrestre mesuré par la distance de Berg-op-Zoom et de Malines," in *Annales de la Société scientifique de Bruxelles*, 24, pt. 2, 113–134; N. D. Haasbroek, *Gemma Frisius, Tycho Brahe and Snellius, and Their Triangulations* (Delft, 1968), with references to three previous papers by Haasbroek, in *Tydschrift voor kadaster en landmeetkunde* (1965–1967); and J. J. Delambre, *Histoire de l'astronomie moderne*, II (Paris, 1821), 92–110.

On the recession problem, see J. Tropfke, *Geschichte der Elementarmathematik*, 2nd ed., V (Berlin 1923), 97; and J. A. Oudemans, "Het problema van Snellius, opgelost door Ptolemaeus," in *Verslagen en mededeelingen der K. Akademie van wetenschappen*, 2nd ser., 19 (1884), 436–440.

For Snel's formulation of the law of refraction, see E. J. Dyksterhuis, *The Mechanization of the World Picture*, pt. 4, sec. 170 (Oxford, 1961), also in Dutch and German; this follows the text of Isaac Vossius, *De lucis natura et proprietate* (Amsterdam, 1662), 36 (see J. A. Vollgraff's ed. of *Risneri optica cum annotationibus W. Snellii pars I liber I* [Ghent, 1918], 216). Other works are D. J. Korteweg, "Descartes et les manuscrits de Snellius," in *Revue de métaphysique et de morale*, 4 (1896), 489–501; C. de Waard, "Le manuscrit perdu de

Snellius sur la réfraction," in *Janus*, **39** (1935), 51–75); J. A. Vollgraff, "Snellius' Notes on the Reflection and Refraction of Rays," in *Osiris*, **1** (1936), 718–725; and Huygens' remarks in *Oeuvres complètes de Christiaan Huygens, publiées par la Société Hollandaise des Sciences*, XIII, and esp. X (1903), 405–406.

DIRK J. STRUIK

SNOW, JOHN (*b.* York, England, 15 March 1813; *d.* London, England, 16 June 1858), *medicine, anesthesiology, epidemiology.*

Snow was the eldest son of a farmer. Little is known of his early life, but at the age of fourteen he was apprenticed to William Hardcastle, surgeon, of Newcastle upon Tyne. He is said to have been industrious and studious, and at the age of seventeen became imbued with vegetarianism and temperance, beliefs which he held—almost to the extent of obsession—to the end of his life, and which he carried into his medical practice. (Although it is reported that later in life he occasionally and of necessity took a little wine.)

In 1831 the first cholera epidemic struck England, entering through Sunderland, a seaport near Newcastle, which suffered a disastrous visitation. Snow was sent to the nearby Killingworth colliery, where in appalling conditions he worked indefatigably and laid the foundations of his interest in, and knowledge of, cholera, for which no cure was known, and which was frequently fatal.

Moving to London in 1836, Snow studied at Westminster Hospital and became a member of the Royal College of Surgeons and a licentiate of the Society of Apothecaries. In 1843 he graduated M.B. from the University of London and in the following year proceeded M.D. The only course then open to him was to enter general practice, and to use his own words, Snow "nailed up his colours" at 54 Frith Street, Soho. He never married, and his life in practice consisted in assiduous attention to his patients (mostly of the working classes), to posts such as that of visitor to outpatients in Charing Cross Hospital, and to his two great contributions to medicine. Snow's friend and biographer, Benjamin Ward Richardson, described him as reserved and lonely with a dry sense of humor.

In 1841 Snow read to the Westminster Medical Society "Asphyxia and the Resuscitation of Newborn Children." In this paper he described a double air pump and gave some of his ideas on lack of oxygen. Since he had made other similar studies on the physiology of respiration, his knowledge placed him in a favorable position when ether was introduced as an anesthetic in 1846. The first major operation in England in which the new drug was used was an amputation of a leg performed by Robert Liston at University College Hospital, London, on 21 December 1846, with William Squire administering the ether. Snow at once began experimenting with the substance and invented an apparatus for its administration, based on physiological principles. He demonstrated its use at St. George's Hospital with so much smoothness and success that he was invited to work with Liston, and later with most of the well-known surgeons of London. Rapidly Snow became the premier anesthetist of the country. In September 1847 he published his masterly little book *On Ether*, which included a description of his apparatus and of the properties of the drug, together with physiological and practical information regarding its administration. His division of the stages of anesthesia into five degrees was not improved upon until the work of Arthur Guedel in 1917.

When chloroform was introduced into anesthesia by James Young Simpson in November 1847, Snow was quick to appreciate the advantages and disadvantages of the new drug. Snow's expertise in apparatus led him to construct new pieces for the administration of chloroform. He laid emphasis on the use of such apparatus as a means of delivering low and exact percentages of chloroform in air; this was in direct contrast to Simpson's "open method" of dropping chloroform on the corner of a towel or handkerchief. The controversy between protagonists of the two methods lasted for the remainder of the century, but Snow was the pioneer in raising the art and practice of English anesthesia and anesthetic apparatus to its subsequent heights.

Simpson's goal had been the prevention of the pangs of childbirth, and he fought valiantly against religious and medical prejudice. Anesthesia, however, became respectable on 7 April 1853, when Snow administered chloroform to Queen Victoria at the birth of Prince Leopold, the so-called "chloroform à la reine." Snow continued his work in anesthesia for the remainder of his short life. He introduced amylene in 1856, and his great book *On Chloroform*, completed a few days before his death, was edited by B. W. Richardson, who added a definitive biography.

During these years Snow was occupied also with investigations of cholera, which many will consider as giving him an even greater claim to recognition as a benefactor of humanity. Since his interest in the disease had been aroused by his earlier work in the colliery near Newcastle, recurrent outbreaks in London gave him opportunity and experience, and in 1849 he wrote the first of many papers and published his book *On the Mode of Communication of Cholera.* Many physicians still believed in the ancient view that infectious diseases such as cholera and smallpox were carried by "miasmas" or evil humors arising from mud, sewage, or other noxious sources. Snow's theory and proof of transmission by water infected with fecal matter was to provoke controversy between supporters of Snow, such as William Budd, and the "miasmatists." Even workers in the Board of Health were slow to alter their ideas.

In the great London epidemic of 1854, Snow's genius as an epidemiologist and statistician reached fruition. By meticulous survey he established that the areas supplied by water from the Southwark and Vauxhall Water Company, obtained from the fecal-contaminated Thames, were infected nine times more fatally than the areas supplied by the Lambeth Company, which supplied water from an upstream source.

Even more dramatic was the affair of the Broad Street pump, which he showed by careful plotting to be in the center of a cholera outbreak in his own parish of Soho. Within a few hundred yards of this pump, some five hundred fatal cases occurred in ten days. Snow found that a sewer pipe passed within a few feet of the well, and his belief that contaminated water was the source of infection was vindicated when he persuaded the parish councillors to remove the pump handle.

Pasteur and Lister had not yet published their work on microorganisms and infection, and the vibrio of cholera was not to be described by Koch till 1884. Snow's reasoned argument was that cholera was propagated by a specific living, waterborne, self-reproducing cell or germ. He recommended sensible precautions such as decontamination of soiled linen, washing hands, and boiling water. In treatment, he believed in the use of saline fluids, given intravenously, although techniques were hardly sufficiently advanced to make full use of this advice, which is now the basis of modern treatment. Snow's writings and practice were a very considerable influence upon the great sanitary reformers such as Sir John Simon and Sir Edwin Chadwick in the later part of the century. He was a founder of the Epidemiological Society.

Tired and worn by overwork, and perhaps undermined by too ascetic a way of life, Snow died of a cerebral hemorrhage at the age of forty-five. He was buried in Brompton churchyard, where his tombstone, originally erected by Richardson, was reconstructed in 1947 by the Association of Anaesthetists of Great Britain and Ireland.

BIBLIOGRAPHY

I. ORIGINAL WORKS. Snow wrote over 30 papers on cholera and matters of public health, and a similar number on ether, chloroform, other anesthetics, and respiratory physiology. Among the most important works on cholera, are *On the Mode of Communication of Cholera* (London, 1849; 2nd ed., 1855); "On the Pathology and Mode of Communication of Cholera," in *London Medical Gazette,* **44** (1849), 730, 745, 923; "On the Communication of Cholera by Impure Thames Water," in *Medical Times and Gazette,* n.s. **9** (1854), 365–366; "On the Chief Cause of the Recent Sickness and Mortality in the Crimea," *ibid.,* n.s. **10** (1855), 457–458; and "Drainage and Water Supply in Connection With the Public Health," *ibid.,* n.s. **16** (1858), 161, 189.

On anesthesia, see "Asphyxia and the Resuscitation of New-born Children," in *London Medical Gazette,* n.s. **I** (1842), 222–227; *On the Inhalation of the Vapour of Ether in Surgical Operations* (London, 1847); "On the Use of Ether as an Anaesthetic . . .," in *London Medical Gazette,* n.s. **4** (1847), 156–157; "On the Inhalation of Chloroform and Ether," in *Lancet* (1848), **1,** 177–180; "On Narcotism by the Inhalation of Vapours," in *London Medical Gazette,* in 16 parts from n.s. **6** (1848) to n.s. **12** (1851); "Death From Inhalation of Chloroform," in *Association Medical Journal,* **1** (1853), 134; "On the Administration of Chloroform During Parturition," *ibid.,* **1** (1853), 500–502; and *On Chloroform and Other Anaesthetics* (London, 1858).

II. SECONDARY LITERATURE. The chief biographical source on Snow is Benjamin W. Richardson's memoir in *On Chloroform and Other Anaesthetics.* An obituary is in *Medical Times and Gazette,* n.s. **16** (1858), 633–634; and reviews (written almost as obituaries) of Snow's posthumous book are in *Lancet* (1858), **1,** 555–556; and in *British Medical Journal* (1858), **11,** 1047–1049.

See also Lord Cohen of Birkenhead, "John Snow—the Autumn Loiterer?" in *Proceedings of the Royal Society of Medicine,* **62** (1969), 99–106; B. Duncum, *The Development of Inhalation Anaesthesia* (Oxford, 1947), with an extensive discussion of Snow's work on anesthesia; J. Edwards, "John Snow, M.D., 1813–1858," in *Anaesthesia,* **14** (1959), 113–126; and K. B. Thomas, "John Snow, 1813–1858," in *Journal of the Royal College of General Practitioners,* **16** (1968), 85–94.

Snow MSS are in the Clover/Snow Collection of the Woodward Biomedical Library, University of British Columbia (annotated by K. B. Thomas, in *Anaesthesia,* **27** [1972], 436–449).

K. BRYN THOMAS

SODDY, FREDERICK (*b*. Eastbourne, England, 2 September 1877; *d*. Brighton, England, 22 September 1956), *radiochemistry, science and society.*

Soddy developed with Lord Rutherford during 1901–1903 the disintegration theory of radioactivity, confirmed with Sir William Ramsay in 1903 the production of helium from radium, advanced in 1910 the *concept* of isotope, proposed in 1911 the alpha-ray rule leading to the full displacement law of 1913, and was the 1921 Nobel laureate in chemistry, principally for his investigations into the origin and nature of isotopes.

The youngest son of a London merchant, Soddy was raised in the Calvinist tradition by his dominant half-sister. He developed a lifelong sense of extreme social independence, as well as a plague-on-both-your-houses attitude toward religious controversy, later extended to social institutions in general. An aspiring scientist from an early age, Soddy was encouraged by his influential science master, R. E. Hughes, at Eastbourne College to study chemistry at Oxford. After an interim year at co-educational Aberystwyth, Soddy in 1895 received a science scholarship to Merton College, Oxford. In 1898, with Ramsay as external examiner, Soddy received a first-class honors degree; he remained at Oxford for two more years, engaged in independent chemical research.

In May 1900, Soddy adventurously followed up an unsuccessful application to Toronto, with a personal visit to Montreal, accepting a position as demonstrator at McGill University. His childless marriage in 1908 to Winifred Beilby (*d*. 1936) was a source of great happiness and stability in his life. Soddy was "an admirable writer and a clear and interesting lecturer,"[1] noted for originality in demonstrations. A fellow of both the Chemical Society (from 1899) and the Royal Society (from 1910), Soddy was also a foreign member of the Swedish, Italian, and Russian academies of science. In 1913 he was awarded the Cannizzaro Prize for his important contributions to the new chemistry.

Profoundly disturbed by World War I and "enraged"[2] by the death of Moseley, Soddy felt that society was not yet sufficiently mature to handle properly the advances of science. He began to concern himself more with the interaction between science and society. In order to ascertain "why so far the progress of science has proved as much a curse as a blessing to humanity,"[3] Soddy studied economics. He considered the free development of science to be the new wealth of nations and advocated the rise of a "scientific civilization."[4] The "curse" he felt arose from constraints, put upon both the progress of science and the distribution of technological productivity, for the self-maintenance of the existing but decadent economic system.[5]

At McGill, Soddy joined with Rutherford in a series of investigations which produced the theoretical explanation of radioactivity. The constant production of a material "emanation" from thorium was shown to be the combined effect of the production of an intermediate, but chemically separable, substance, thorium X, balanced by its decay. The production of one substance was thus the result of the uncontrollable disintegration of another. The "radiation" proved to be both particulate in nature and a direct accompaniment of the process of disintegration. The rate of the process was found in every case to be as the exponential law of a monomolecular chemical reaction. So complete was their 1903 disintegration theory of radioactivity, that in 1909 only an extension to branching series was required.[6]

In March 1903, Soddy elected to join Ramsay in London to examine more fully the gaseous products of decay. Using Giesel's radium preparations, Ramsay and Soddy experimentally confirmed in July 1903 the prediction of Rutherford and Soddy that radium would continuously produce helium. In 1908 Rutherford "settled for good"[7] the long-suspected identity of the helium, so produced, with expelled alpha particles. During the ten-year period following his 1904 appointment to the University of Glasgow, Soddy helped to clarify the relation between the plethora of radioelements and the periodic table.

McCoy and Ross had reported in 1907 that Hahn's 1905 radiothorium was chemically inseparable from thorium. Boltwood, in turn, indicated a similar difficulty with thorium and ionium. From crystal morphology studies, Strömholm and Svedberg in 1909 confirmed a family resemblance between such radioelements as thorium X and radium. In 1910 the chemical inseparability of mesothorium 1 and radium, reported by Marckwald, as well as Soddy's own experimental evidence, that these two radioelements form an inseparable trio with thorium X, convinced Soddy that such cases of chemical inseparability were actually chemical *identities*. Without the unnecessary continuation of the genetic series of radioelements throughout the entire periodic table, postulated by Strömholm and Svedberg, Soddy declared in 1910 that "the recognition that elements of different atomic weight may possess identical chemical properties seems destined to have its most important application in the region of the inactive elements."[8] "Soddy possessed," as Hahn wrote in admiration,[9] "the courage to declare that these were chemically identical elements."

To be able to refer generically to these active and inactive elements with identical chemical properties, Soddy introduced the technical term "isotope" in 1913.[10] While chemically inseparable, active isotopes were distinguishable by their radioactive properties, and all isotopes differed in atomic weight. Soddy suggested that the 1912 metaneon of J. J. Thomson be considered "a case of isotopic elements outside the radioactive sequences."[11] Following Soddy, Aston announced a partial separation in 1913 on this very basis.[12] The connection between chemical properties and the periodic table became increasingly clarified with concurrent developments in the physics and chemistry of the nuclear atom. From the chemical side, Soddy proposed the alpha-ray rule in 1911, the key to the first of two locks. Applying his general principle that the common elements are mixtures of chemically inseparable elements "differing step-wise by whole units of atomic weight"[13] specifically to the case of the radioelements, Soddy recognized that the expulsion of an alpha particle would result in a lighter element chemically inseparable from those occupying the "next but one"[14] position in the periodic table. The second lock to the displacement law involved the beta transitions.

During 1912 Soddy assigned Lord Fleck the task of sorting out the short-lived beta emitters, especially at the complex branching points in the series. Once these experimental results became available, several partially correct generalizations were published, inducing Soddy, therefore, to publish his own complete and correct form of the law in February 1913. "Fajans," Soddy acknowledged, "worked out the Periodic Law Generalization quite independently of me,"[15] although his conclusions were fundamentally different. On electrochemical considerations, Fajans interpreted the changes among the clusters, "plejade,"[16] of radioelements as evidence against the nuclear origin of radio-changes.[17] Soddy, on the other hand, argued for a crucial distinction[18] between radiochange and chemical change, concluding on chemical evidence, as Bohr had done on physical evidence, that beta decay, like alpha decay, was of nuclear origin. As a result, Soddy considered van den Broek's hypothesis, that successive places in the periodic table correspond to unit differences in the net intra-atomic charge (see Figure 1) "practically proved so far as the . . . end of the sequence,

from thallium to uranium, is concerned."[19]

By early 1914 Moseley,[20] using physical methods, had completed his independent extension of this verification throughout the entire periodic table. During the period 1914–1919, in the chair of chemistry at Aberdeen, Soddy, in addition to his war work, examined two predictions of the displacement law. It was commonly accepted that lead was the end product only of the uranium series, and Soddy had predicted by 1913 that a heavier isotope of lead from thorium must also exist. Separate determinations were undertaken in 1914 on lead from Ceylon thorite by Soddy and on lead from uranium ores by T. W. Richards and O. Hönigschmid, thereby confirming the prediction that common lead was indeed a mixture of isotopes.[21] Soddy suggested that the parent of actinium might be an alpha-decaying member of Mendeleev's missing eka-tantalum. An exclusively beta-emitting homologue of tantalum found by Fajans and O. Göhring in 1913 and called "brevium" [UrX_2], however, caused Soddy to begin to investigate the other alternative. But after proving that the parent of actinium could not be a beta-decay product of radium, he reexamined the first alternative with Cranston. In 1918 they found, isotopic with UrX_2, the direct parent of actinium, produced through the rare UrY branch, which was found in 1911 by G. Antonoff and later linked to uranium 235. Protactinium, element 91, was simultaneously and independently found by Hahn and Meitner.

Soddy was called in 1919 to a chair of chemistry at Oxford. During his seventeen-year tenure, he failed to establish the expected school of radiochemistry, devoting himself rather to the improvement of chemistry teaching and to the modernization of the laboratories.[22] He also continued to treat radioactive minerals for their constituents. After the disturbing death of his wife, Soddy retired early. He went exploring for monazite sand, and patented his 1923 process for thorium extraction in 1940. He then turned his attention to mathematics. Looking beyond to the significance of science, Soddy, who had once confidently spoken of the potential peaceful benefits for society given the key to "unlock this great store of energy bound up in the structure of the element,"[23] and, by controlling it, "virtually provide anyone who wanted it with a private sun of his own,"[24] was profoundly concerned by subsequent developments. He zealously endeavored[25] to awaken the conscience of the scientific community to the social relevance of their own research. Soddy urged that "universities and learned societies should no longer evade their

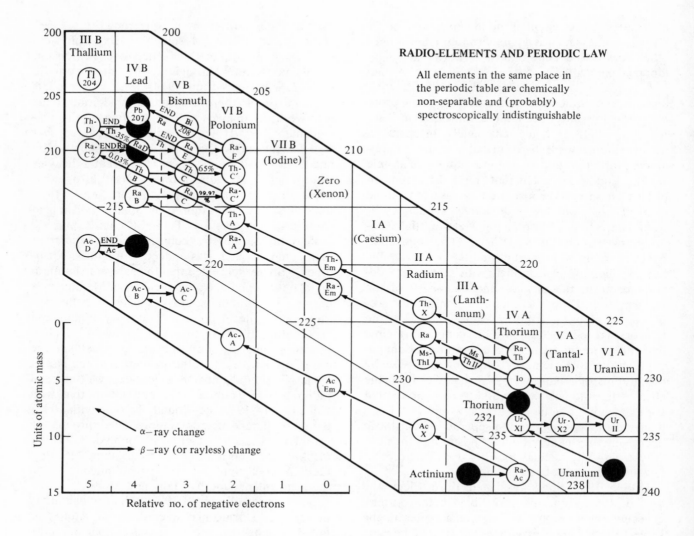

FIGURE 1.　Reproduced with the permission of the British Association for the Advancement of Science.

responsibilities and hide under the guise of false humility as the hired servants of the world their work has made possible, but do that for which they are supported in cultured release from routine occupations, and speak the truth though the heavens fall."[26] He was largely unheeded, however, and he judged at the end that the blame for the plight of civilization "must rest on scientific men, equally with others, for being incapable of accepting the responsibility for the profound social upheavals which their own work primarily has brought about in human relationships."[27]

NOTES

1. Rutherford letter, 15 June 1914. Cf. Howorth, *Pioneer Research*, 192. The original is in Bodleian Library, Soddy-Howorth Collection, 75, 95, courtesy Soddy trustees.

2. Soddy, *Memoirs*, I, 274.

3. Fleck, "Soddy," 210, courtesy the Royal Society.

4. Soddy, "Social Relations of Science," in *Nature*, **141** (1938), 784–785.

5. A comprehensive statement of his general view regarding the monetary system preventing modern Western civilization from distributing its scientific and technological abundance by peaceful means appears in an address, February 1950, partially republished in the 24-page *Commemoration to Professor Frederick Soddy* (London, 1958).

6. Soddy, "Multiple Atomic Disintegration: A Suggestion in Radioactive Theory," in *Philosophical Magazine*, **18** (1909), 739–744; this was developed in "Multiple Disintegration," in *Annual Report*, **9** (1912), 311–316.

7. Referring to his joint paper with Royds, *Philosophical Magazine*, **17** (1909), 281, Rutherford further noted in his letter of 14 Feb. 1909 to Elster and Geitel that "you will have seen that the α particle has at last been proved to be helium." Darmstaedter Collection, G 1, 1896 (26), courtesy Staatsbibliothek, Preussischer Kulturbesitz.

8. Soddy, "Radioactivity," in *Annual Report*, **7** (1910), 286. Strikingly similar views regarding mixtures of similar elements of different atomic weight were expressed by D.

Strömholm and T. Svedberg in *Zeitschrift für Anorganische Chemie*, **63** (1909), 206.

9. Fleck, "Soddy," 208. The rare earths had given ample evidence of chemical "inseparability" without identity.

10. Soddy, "Intra-atomic Charge," in *Nature*, **92** (4 Dec. 1913), 400. "The same algebraic sum of the positive and negative charges in the nucleus, when the arithmetic sum is different, gives what I call 'isotopes' . . . because they occupy the same [*iso*] place [*topos*] in the periodic table" (see diagram). Perhaps the first use of "isotope" for the position of elements was W. Preyer, *Das Genetische System der chemischen Elemente* (Berlin, 1893). The stimulus for Soddy's term arose when he "got tired of writing 'elements chemically identical and non-separable by chemical methods' and coined the name *isotope* . . . ," as he said in "Contribution to a Discussion on Isotopes," in *Proceedings of the Royal Society*, **99** (1921), 98.

11. Soddy, "Radioactivity," in *Annual Report*, **10** (1913), 265.

12. Aston, *Isotopes*, 37, 42.

13. Soddy, "The Chemistry of Mesothorium," in *Transactions of the Chemical Society*, **99** (1911), 82; cf. n. 8.

14. Soddy, *The Chemistry of the Radioelements* (1911), 29. For a remarkable partial anticipation of isotopes and the displacement law, see A. T. Cameron, *Radiochemistry* (London, 1910), 141.

15. Soddy letter to F. O. Giesel, *ca.* 1913/14 in Giesel Archives, courtesy Chininfabrik, Buchler & Co., Brunswick. The generalization of A. S. Russell had not only assumed a *discontinuous* series, *Chemical News*, **107** (31 Jan. 1913), 49, but also questioned the chemical identity notion of Soddy; cf. Russell letter to Rutherford, 14 Sept. 1912, Cambridge Univ. Lib., Add. MSS 7653/R106. Russell "knew of Fleck's results," and "through him they got known to Hevesy and Fajans"; cf. *Report of the British Association for the Advancement of Science* (1913), 446; and the Soddy letter to Howorth, 29 Jan. 1953, Bodleian Lib., Soddy Collection, Alton 29, item no. Trenn S-6.

16. K. Fajans, *Radioaktivität und die neueste Entwickelung der Lehre von den chemischen Elementen* (Brunswick, 1919), 35.

17. Fajans' letter to Rutherford, 10 April 1913, Cambridge Univ. Lib. Add. MSS 7653/F5.

18. Soddy's distinction between chemical change and radio-change was originally based upon the disintegration theory, "Radioactive Change," in *Philosophical Magazine*, **5** (1903), 576. With the development of the nuclear atom, however, it became possible to clarify this distinction by defining the actual locus of the radio-changes. Bohr expressed this clarification in his letter to Hevesy, 7 Feb. 1913, L. Rosenfeld, "Introduction" to *On the Constitution of Atoms and Molecules* (Copenhagen, 1963), xxxii.

19. Soddy, "Intra-atomic Charge," 400.

20. H. G. J. Moseley, in *Nature*, **92** (1914), 554. "My work was undertaken for the express purpose of testing [van den] Broek's hypothesis . . . [and] certainly confirms the hypothesis."

21. "Soddy's prediction concerning the atomic weights of leads from uranium and thorium minerals had been triumphantly vindicated by some of his most severe critics." F. W. Aston, "The Story of Isotopes," in *British Association Report* (1935), Presidential Address to Section A, p. 26. The concurrent investigations comparing uranium lead with ordinary lead could neither confirm nor deny the possibility of thorium lead.

22. Brewer, "Chemistry at Oxford," 185.

23. Soddy, "The Internal Energy of Elements," in *Journal of the Proceedings of the Institution of Electrical Engineers, Glasgow*, **37** (1906), 7. An earlier statement on the latent internal energy of the atom is Soddy, "The Disintegration Theory of Radioactivity," in *Times Literary Supplement* (26 June 1903), 201.

24. Soddy, "Advances in the Study of Radio-active Bodies," two lectures to the Royal Institution on 15 May and 18 May 1915, as recorded in *The Royal Institution Friday Evening Lectures 1907–1918* (privately bound at the Royal Institution, London, n.d.). The original MS is in the Bodleian Library, Soddy-Howorth Collection, 58. The quotation is from this MS, page II, 9. The lectures are apparently unpublished but are reviewed in *Engineering*, **99** (1915), 604.

25. Shortly after Soddy's retirement, Joseph Needham pointed out the importance of such efforts, "Social Relations of Science," in *Nature*, **141** (1938), 734.

26. Soddy, *Frustration in Science*, Foreword.

27. Soddy, Typescript-A, 1953, concluding statement, Bodleian Library, Soddy-Howorth Collection 4.

BIBLIOGRAPHY

A nearly complete list of Soddy's main scientific papers, books, lectures, and other contributions is given by Alexander Fleck, "Frederick Soddy," in *Biographical Memoirs of Fellows of the Royal Society*, **3** (1957), 203–216. For comparisons and additions, including his contributions on economics and on science and society, see Muriel Howorth, *Pioneer Research on the Atom* (London, 1958), 281–286. This unusual account is subtitled *Rutherford and Soddy in a Glorious Chapter of Science*, and further subtitled *The Life Story of Frederick Soddy*. In spite of the author's uncritical attempt to glorify Soddy, this remarkable reference source is the fruit of great effort to preserve the existing documents of Soddy. Soddy gave all his papers to Muriel Howorth of Eastbourne, and his will contained the provision: "I give to Muriel Howorth also the copyright of all my published works," cf. *Pioneer Research*, p. 286. The Soddy-Howorth Collection was deposited in the Bodleian Library and a partial reference key thereto is appended to *Pioneer Research*, pp. 333–339. In 1974 J. Alton of the Contemporary Scientific Archives Centre, Oxford, deposited in the Bodleian a 29-page systematic catalogue of the Soddy Collection incorporating the Howorth portion. This collection must be directly consulted for precision in both quotations and other references. Richard Lucas, *Bibliographie der radioaktiven Stoffe* (Leipzig, 1908), 72–73, provides a useful list of Soddy's early works. Consultation of the British Museum General Catalogue of Printed Books, 1964, amplifies the list of works of Soddy. In addition to the scientific contributions collectively listed in Fleck and Howorth, the following should be noted: "The First Quarter-Century of Radioactivity," in *Isotopy* (Westminster, 1954), 1–25. See the obituaries of "Rutherford," in *Nature* (30 October 1937); "Ramsay," *ibid.* (10 August 1916); and of H. Becquerel, "The Founder of Radioactivity," in *Ion: A Journal of Electronics, Atomistics, Ionology, Radioactivity and Raumchemistry*, **1** (1908), 2–4. Soddy was joint editor of this short-lived serial, *Ion*. In this same issue, Soddy completed his series of investigations concerning whether the alpha particle was charged before, during, or after expulsion. Soddy's abstracts of the papers by Russell, Fajans, and Soddy concerning the displacement law are also of interest; see *Abstracts of Chemical Papers Journal of the Chemical Society London*, pt. 2 (1913), 274–278. Soddy's classic call for scientific responsibility appears as the foreword to *Frustration in Science* (London, 1935).

Soddy's nine joint papers with Rutherford (1902–1903) are reproduced in *Collected Papers of Lord Rutherford of Nelson*, Sir James Chadwick, ed., I (London, 1962). Soddy contributed a series of original reports on "Radioactivity" for the *Annual Reports on the Progress of Chemistry* (London, 1904–1920). These articles contain much otherwise unpublished work on isotopes, as well as a running account of the history of radioactivity. These articles have been published in facsimile and edited with commentary by T. J. Trenn, in *Radioactivity and Atomic Theory* (London, 1975). The diagram "Radio-Elements and Periodic Law" first appeared as a supplement to Soddy's paper "The Radio-elements and the Periodic Law," in *Chemical News*, **107** (28 Feb. 1913), 97–99. Essentially the same diagram appeared in *Jahrbuch der Radioaktivität und Elektronik*, **10**, no. 2 (1913), 193. The actinium series was separated and minor additions were included in the version drafted July 1913 for the *British Association Report* (1913), 446, and here reproduced; it also appeared in the *Annual Report*, **10** (1913), 264.

Soddy's most important books are *Radio-Activity: an Elementary Treatise From the Standpoint of the Disintegration Theory* (London–Leipzig, 1904), based upon a series of lectures at the University of London from Oct. 1903 to Feb. 1904, carried concurrently in *The Electrician*, **52** (1903), 7 et. seq.; *The Interpretation of Radium* (London, 1909; 4th ed., 1920), translated into several languages). In the series edited by Alexander Findlay, *Monographs on Inorganic and Physical Chemistry*, Soddy contributed *The Chemistry of the Radio-Elements*, pt. I (London, 1911; Leipzig, 1912); pt. II (1914), containing "Radioelements and the Periodic Law"; and pt. I, 2nd. ed. (1915). See also *The Interpretation of the Atom* (London, 1932).

Soddy's most important lectures were The Wilde Lecture VIII, "The Evolution of Matter as Revealed by the Radioactive Elements," 16 March 1904, in *Memoirs and Proceedings of the Manchester Literary and Philosophical Society*, **48** (1904; Leipzig, 1904); The Nobel Lecture, 12 Dec. 1922, "The Origin of the Conception of Isotopes," in *Les Prix Nobel en 1921–1922* (Stockholm, 1923).

Information concerning the life and work of Soddy can be obtained from *Pioneer Research*. Howorth also edited the *Memoirs* of Soddy, as *Atomic Transmutation, Memoirs of Professor Frederick Soddy*, vol. I (London, 1953), subtitled *The Greatest Discovery Ever Made*. Volume one deals with the period until 1904. There were no further volumes. There are numerous sketches of Soddy's life and work. Alexander Fleck, in *Nature*, **178** (1956), 893, is an interesting personal account. Fleck also contributed the note for the *Dictionary of National*

Biography (1951–1960), 904. Alexander S. Russell, "F. Soddy, Interpreter of Atomic Structure," in *Science*, **124** (1956), provides insights into Soddy the man. Russell published further on Soddy, in *Chemistry and Industry*, no. 47 (1956), 1420–1421, and in Eduard Farber, ed., *Great Chemists* (New York, 1961), 1463–1468. Perhaps the best account is F. Paneth, "A Tribute to Frederick Soddy," in *Nature*, **180** (1957), 1085–1087; repr. in the Paneth Collection, H. Dingle, ed., *Chemistry and Beyond* (London, 1964), 85–89. A more recent sympathetic account is that of A. Kent, "Frederick Soddy," in *Proceedings of the Chemical Society* (November 1963), 327–330. Besides his brief editorial "Frederick Soddy and the Concept of Isotopes," in *Endeavour*, **23** (1964), 54, T. I. Williams wrote the article on Soddy for his *Biographical Dictionary of Scientists* (London, 1969). The account of I. Asimov, *Biographical Encyclopedia of Science and Technology* (New York, 1964), no. 398, is subject to the limitations imposed by this effort. An extremely concise and accurate summary is included in W. A. Tilden and S. Glasstone, *Chemical Discovery and Invention in the Twentieth Century* (London, 1936), 140. There is a supplementary account in Eduard Farber, *Nobel Prize Winners in Chemistry 1901–1961* (London, 1963), 81–85. It is of interest to compare the biographical account in *Nobel Lectures in Chemistry* (Amsterdam, 1966), 400–401, with the original in *Les Prix Nobel en 1921–1922* (Stockholm, 1923), 128–129. See also the account in H. H. Stephenson, *Who's Who in Science* (London, 1914), 535, and *Journal of Chemical Education*, **8** (1931), 1245–1246.

Relevant sketches of Soddy's work are to be found in F. W. Aston, *Isotopes*, 2nd ed., 1924. His work on lead isotopes, pp. 17–19, is particularly valuable. See A. Kent and J. A. Cranston, "The Soddy Box," in *Chemistry and Industry* (1960), 1206, 1411, which describes Soddy's original 1910 preparation, a deliberate mixture of radium and mesothorium, which led him to the concept of the isotope. In Gleditsch, "Contribution to the Study of Isotopes," *Norske Videnskaps-Akademi I. Mat.-Natur. Klasse* no. 3 (Oslo, 1925), E. Gleditsch notes, p. 7, that "The theory of isotopes put forward . . . by Soddy in the years 1911–1914 has proved to be fully in accord with our present views on atomic structure." See also Fleck, "Early Work in the Radioactive Elements," in *Proceedings of the Chemical Society* (1963), 330. In this same issue, J. A. Cranston contributed "The Group Displacement Law," pp. 330–331, and an even more detailed documentation in the following issue (1964), 104–107. Soddy's work with Rutherford is considered by A. S. Eve, *Rutherford* (Cambridge, 1939); N. Feather, *Lord Rutherford* (London, 1940); A. Romer, *The Restless Atom* (New York, 1960); Howorth, *Pioneer Research*; and T. J. Trenn, "Rutherford and Soddy: From a Search for Radioactive Constituents to the Disintegration Theory of Radioactivity," in *Rete*, **1** (1971), 51–70. M. W. Travers, *A Life of Sir William Ramsay* (London, 1956), ch. 14, pp. 210–221, deals with his work with Ramsay. At the request of Travers, Soddy contributed a portion of this account. The original transcript is in Soddy-Howorth 4.

In addition to F. M. Brewer, "The Place of Chemistry at Oxford," in *Proceedings of the Chemical Society* (July 1957), 185, Soddy's work at Oxford is considered by Sir Harold Hartley, "The Old Chemical Department," in *Journal of the Royal Institute of Chemistry* (1955), 126. J. A. Cranston's "The Discovery of Isotopes by Soddy and his School in Glasgow," in *Isotopy* (1954), 26–36, and "Concept of Isotope," in *Journal of the Royal Institute of Chemistry*, **18** (1964), 38, provide important historical and scientific distinctions in the use of the term "isotope."

A. Romer, ed., *The Discovery of Radioactivity and Transmutation*, Classics of Science, II (New York, 1964), provides not only some of the papers of Soddy in collaboration both with Rutherford and with Ramsay but also valuable comments on this pre-1904 work. Soddy's hypothesis concerning an isotope of lead as the final product of the thorium series is dealt with in S. I. Levy, *The Rare Earths* (London, 1915), 107–108.

For a partial account of Soddy's work on isotopes, emphasizing the contributions of Fajans and Richards, see O. U. Anders, "The Place of Isotopes in the Periodic Table: the 50th Anniversary of the Fajans-Soddy Displacement Laws," in *Journal of Chemical Education*, **41** (1964), 522–525. Additional information about Soddy as others saw him is in L. Badash, ed., *Rutherford and Boltwood: Letters on Radioactivity* (New Haven, 1969), which exposes Soddy's research on the parent of radium. Soddy as a public figure and social rebel, who ushered in the atomic age, is epitomized in C. Beaton and K. Tynan, *Persona Grata* (London, 1953), 87. Besides the Soddy-Howorth Collection, extensive correspondence exists also at the Cambridge Univ. Library, Add. MSS 7653/S. There is also correspondence with W. H. Bragg at the Royal Institution, with J. Larmor courtesy the Royal Society, and with O. Lodge at University College London. The Soddy Memorial at Glasgow was reported in "Unveiling of the Soddy Memorial," in *Chemistry and Industry* (8 Nov. 1958), 1462–1464. There is one collection of Soddy's apparatus and equipment at the Chemistry Department of the University of Glasgow and another at the Inorganic Chemistry Laboratory of the University of Oxford.

THADDEUS J. TRENN

SOEMMERRING, SAMUEL THOMAS (*b*. Torun, Poland, 18 January 1755; *d*. Frankfurt, Germany, 2 March 1830), *comparative anatomy, human anatomy, anthropology, physiology.*

Soemmerring's maturity coincided with the French Revolution and the subsequent political disorders in Germany. Yet despite the instability of his career, his writings made him the most famous German anatomist of the early nineteenth century. His works were characterized by a fully developed presentation of the text and by scientifically accurate illustrations of considerable artistic merit.

After attending the Gymnasium in Torun (1769–1774), Soemmerring studied medicine from 1774 to 1778 at Göttingen, where he was inspired by the zeal for research of two of his teachers, Heinrich August Wrisberg and Johann Friedrich Blumenbach. While still a student Soemmerring had decided to become an anatomist; the prerequisites for this career were a gift for observation and skill in drawing. Soemmerring's choice of profession did not meet with the approval of his father, Johann Thomas Soemmerring, the municipal physician of Torun. (Soemmerring's mother was the former Regina Geret, a pastor's daughter.) The family was upper middle class; old-fashioned thrift and Lutheran convictions were the foundations of its way of life. By accepting various privations, Soemmerring was able to pursue his plans without his father's assistance and earned the M.D. on 7 April 1778 with a dissertation on the base of the brain and the origin of the cranial nerves. With the aid of his own illustrations he criticized earlier accounts and proposed the order of the twelve cranial nerves that is still taught. Although not the first to adopt this order, Soemmerring provided such solid grounds for it that ultimately it was generally accepted. To complete his training, he traveled in Holland, England, and Scotland. Among the physicians and scientists he met, those who most impressed him were Peter Camper, John Hunter, and Alexander Monro (Secundus).

In April 1779, shortly after his return to Göttingen, Soemmerring became professor of anatomy and surgery at the Collegium Carolinum in Kassel. He remained there until the fall of 1784, when he assumed the professorship of anatomy and physiology at the University of Mainz (until 1797). Both Kassel and Mainz were important cultural centers where the arts and sciences were encouraged by discerning rulers. At Kassel, Soemmerring was permitted to dissect animals that had died in the menagerie and to examine the corpses of members of the city's Negro colony. These investigations resulted in a study on the bodily characteristics of Negroes and Europeans (1784); Soemmerring concluded that despite several differences, both belonged to the same species.

One of Soemmerring's chief fields of research was neuroanatomy. His demonstration of the crossing of the optical nerve fibers (1786) was followed by a publication on the brain and spinal cord (1788), the annotations to which contained a wealth of findings in comparative anatomy. Further evidence of Soemmerring's extensive knowledge can be seen in the footnotes to his translation of Haller's *Primae lineae physiologiae* (1788). Soemmerring no longer considered the spinal cord to be a "great nerve" but, rather, a part of the central nervous system. Further, he gave the hypophysis its current name, replacing the outmoded term *glandula pituitaria*. Soemmerring's interest in the nervous system and the sense organs also resulted in the publication of illustrations and descriptions of several deformities (1791); the frontispiece to the work showed a series of progressive duplications of the face and head.

In 1796 Soemmerring published a work on "the organ of the soul." The anatomical part was well received, especially the assertion that the cranial nerves originate (or, as the case may be, terminate) in the ventricle wall. But the speculative claim—based on the ideas of *Naturphilosophie*—that the intraventricular cerebrospinal fluid is the seat of a *sensorium commune*—was rejected by more perceptive readers. In contrast, the illustrations of the base of the brain (1799) were widely admired. In Mainz, Soemmerring had met Christian Koeck, whom he trained as a scientific draftsman. Koeck enriched the literature with many excellent illustrations, including those of the brain of a three-year-old boy.

Soemmerring presented further anatomical findings in a book on the effect of corsets. He showed that the deformation of the thorax that they produced led to the displacement of the stomach and liver, and consequently he opposed the tight lacing of the waist and the crinoline. His handbook of human anatomy, *Vom Baue des menschlichen Körpers* (1791–1796), was based as far as possible on his own observations and was conceived as a supplement to Haller's *Primae lineae physiologiae*; the work was still in use in expanded form a half century later. The foreword to the first edition reveals Soemmerring's preoccupation with the use of clear, unmistakable terminology. It also contains an impressive list of his anatomical discoveries, although not all of them have proved to be valid. Among those that are still accepted, two of the most remarkable are the observation that arterial trunks always lie on the bent side of the joints and the discovery that the small part of the trigeminal nerve always lies against the third branch. Soemmerring also sought to illustrate the ideal form of a female skeleton (1797) and published a collection of illustrations of the human embryo (1799).

During his last years in Mainz, Soemmerring's life changed in important ways. In March 1792 he married Margaretha Elisabetha Grunelius, who came from a prominent Frankfurt family. Wartime conditions precluded his establishing a permanent residence in Mainz, and he supervised his office from Frankfurt. Soemmerring was offered various posts but accepted none of them because of the low salaries. He sold a portion of his anatomical and anthropological collections, dedicated himself to medical practice in Frankfurt, and completed scientific studies that he had begun earlier. After his wife died in January 1802, he was sought by many universities and in March 1805 finally accepted nomination as member of the Bavarian Academy of Sciences in Munich.

Soemmerring's abiding interest in the anatomy of the sense organs was especially stimulated after his discovery in 1791 of the fovea centralis in the macula lutea. Between 1801 and 1810 he published four groups of illustrations of the human sense organs, with descriptions in German and Latin. The work was greatly enriched by copperplate engravings based on Koeck's drawings. In preparing the drawings Soemmerring was less concerned with correct perspective than with the architectonically correct representation of the material.

Soemmerring translated and commented upon the works of others and did extensive reviewing for the scholarly journals published in Göttingen. In addition he often participated in prize competitions, but the material he submitted was mostly of a clinical nature—except for a paper on the structure of the lung (1808). Of Soemmerring's communications published in the *Denkschriften* of the Bavarian Academy of Sciences the most important by far was one on electric telegraphs (1809–1810).

Soemmerring's rheumatic fever and chest ailments were aggravated by the severe climate of Munich, and in 1820 he returned to Frankfurt, where he practiced medicine for ten years and studied sunspots. Soemmerring was awarded many titles and honors. He became a privy councillor and was named to several orders. As a knight of the Order of the Civil Service of the Bavarian Crown (1808) he was granted personal nobility.

BIBLIOGRAPHY

I. ORIGINAL WORKS. A bibliography of Soemmerring's writings is in Adolph Callisen, *Medicinisches Schriftsteller-Lexicon der jetzt lebenden Verfasser*, supp., XXXII (Altona, 1844), 348–359.

His works include *De basi encephali et originibus nervorum cranio egredientium libri quinque* (Göttingen, 1778); *Ueber die körperliche Verschiedenheit des Mohren vom Europäer* (Mainz, 1784; 2nd ed., slightly different title, Frankfurt–Mainz, 1785); *Vom Hirn und Rückenmark* (Mainz, 1788; 2nd ed., Leipzig, 1792); *Ueber die Schädlichkeit der Schnürbrüste* (Leipzig, 1788; 2nd ed., slightly different title, Berlin, 1793); *Abbildungen und Beschreibungen einiger Missgeburten* (Mainz, 1791); *Vom Baue des menschlichen Körpers*, 5 vols. (Frankfurt, 1791–1796; 2nd ed., 1796–1801), also in Latin, 6 vols. (Leipzig, 1794–1801); *Ueber das Organ der Seele* (Königsberg, 1796); *Tabula sceleti feminini juncta descriptione* (Frankfurt, 1797); "De foramine centrali limbo luteo cincto retinae humanae," in *Commentationes Societatis regiae scientiarum Göttingensis*, **13** (1799), 3–13; *Icones embryonum humanorum* (Frankfurt, 1799); *Tabula baseos encephali* (Frankfurt, 1799); *Abbildungen des menschlichen Auges* (Frankfurt, 1801), also in Latin (Frankfurt, 1804); *Abbildungen des menschlichen Hörorganes* (Frankfurt, 1806), also in Latin (Frankfurt, 1806); *Abbildungen der menschlichen Organe des Geschmackes und der Stimme* (Frankfurt, 1806), also in Latin (Frankfurt, 1808); *Ueber die Structur, die Verrichtung und den Bau der Lungen* (Berlin, 1808), 57–126, written with F. O. Reisseisen; *Abbildungen der menschlichen Organe des Geruches* (Frankfurt, 1809), also in Latin (Frankfurt, 1810); and "Ueber einen elektrischen Telegraphen," in *Denkschriften der Bayerischen Akademie der Wissenschaften zu München* for 1809–1810 (1811), 401–414.

II. SECONDARY LITERATURE. See Gerhard Aumüller, "Zur Geschichte der Anatomischen Institute von Kassel und Mainz (I–III)," in *Medizinhistorisches Journal*, **5** (1970), 59–80, 145–160, 268–288; Ignaz Döllinger, *Gedächtnisrede auf Samuel Thomas Soemmerring* (Munich, 1830); W. Riese, "The 150th Anniversary of S. T. Soemmerring's Organ of the Soul," in *Bulletin of the History of Medicine*, **20** (1946), 310–321; Wilhelm Stricker, *Samuel Thomas Soemmerring nach seinem Leben und Wirken geschildert* (Frankfurt, 1862); and Rudolph Wagner, *Samuel Thomas Soemmerrings Leben und Verkehr mit seinen Zeitgenossen*, 2 vols. (Leipzig, 1844).

ERICH HINTZSCHE

SOHNCKE, LEONHARD (*b*. Halle, Germany, 22 February 1842; *d*. Munich, Germany, 1 November 1897), *crystallography, physics, meteorology.*

Sohncke's chief scientific contribution was the extension of the lattice theory of Bravais to arrive at sixty-five of the 230 possible space groups.

Sohncke's father was professor of mathematics at the University of Halle and was known for his translation into German of Chasles's *Aperçu historique sur l'origine et la développement des méthodes en géométrie*. Sohncke was thus stimulated to study mathematics. He received his early education at the Gymnasium in Halle and then pursued mathematics and physics at the University of Halle and then at the University of Königsberg. There, Franz Neumann, professor of mineralogy and physics, influenced Sohncke greatly and urged him to direct his attention toward theoretical physics. Sohncke received his doctorate in 1866 and taught for a brief period in the Gymnasium in Königsberg before becoming *Privatdozent* at the university.

In 1871, on the recommendation of Kirchhoff, Sohncke was called as professor of physics to the Technische Hochschule in Karlsruhe. In 1883 he moved to the University of Jena in the same capacity, and in 1886 he became professor of physics at the Technische Hochschule in Munich, where he remained until his death. Sohncke acquired a reputation as a fine teacher and instructed many young physicists who later attained success. He also won prominence in educational circles because of his campaign, while in Munich, to break the monopoly held in secondary education by the Gymnasiums and to aid the Realschulen in reaching parity with them.

Early in his career Sohncke was concerned primarily with pure mathematics, particularly series; and he published several short papers in mathematical journals. In the mid-1870's, however, he turned his attention to the internal symmetry of crystals. Working under Neumann at Königsberg, Sohncke had become aware of this field and knew of the previous work of Bravais in determining the fourteen types of space lattices and the thirty-two symmetry classes. He had also followed the investigations of Camille Jordan in group theory. During his research, Sohncke discovered that Hessel, in 1830, had anticipated the work of Bravais; and Sohncke saw to it in his later publications that Hessel's contribution was recognized.

The fourteen Bravais lattices accounted for only seven of the thirty-two classes of external symmetry. In studying internal symmetry, Sohncke realized that previous investigators had looked upon internal symmetry from a completely external orientation. They had imposed, as a condition of symmetry, translational equivalence; and Sohncke saw that this restriction was not justifiable. Inasmuch as symmetry is defined as the equivalence of internal configurations, it is of no consequence whether the direction of one's view has been altered in being transported from one point to another within an object. Thus Sohncke insisted that the view of the system of points is the same from every point and that it need not be a parallel view.

Sohncke eventually arrived at sixty-five different spatial arrangements of points, introducing two new symmetry elements: the screw axis, in which a rotation around an axis is combined with a translation of the system along the axis; and the glide plane, in which the reflection in a mirror plane is combined with a similar translation without rotation along the axis. (Using only simple translation, Bravais had arrived at fourteen lattices.) Sohncke, however, failed to consider two additional symmetry elements of the thirty-two classes of external symmetry: rotation-reflection and rotation-inversion axes. Their inclusion by Fyodorov and almost simultaneously by Arthur Schoenflies and William Barlow in the late 1880's added the additional 165 space groups. Sohncke published his results in his major work, *Die Entwicklung einer Theorie der Krystallstruktur* (1879). He also made models from cigar boxes to demonstrate his derived space groups; and while in Munich, where he became a close friend of the crystallographer Groth, he extended his study of crystal physics.

When Sohncke accepted his post at Karlsruhe he simultaneously took over the direction and administration of the network of meteorological stations in the province of Baden. He also became editor of the *Jahresbericht über die Beobachtungsergebnisse der Badischen meteorologischen Stationen*, and in order to popularize meteorology he published *Über Stürme und Sturmwarnungen* (1875). He also wrote several articles on temperature changes in humid, rising streams of air; the derivation of the formula for barometric height; and the green rays of the sunset. Further, he proposed a theory of the presence of electricity in thunderstorms. Sohncke further conducted research in optics. He determined the thickness of a drop of oil when placed on water after having diffused on its surface.

While in Munich, Sohncke became particularly interested in the flight of aerial balloons, and he directed the activities of a society that encouraged this activity. He organized a number of ascents, and as a reward for his efforts the group named one of its balloons "Sohncke."

BIBLIOGRAPHY

I. ORIGINAL WORKS. Sohncke's chief publications are *Über Stürme und Sturmwarnungen* (Berlin, 1875); *Die Entwicklung einer Theorie der Krystallstruktur* (Leipzig, 1879); *Über Wellenbewegung* (Berlin, 1881); and *Ge-mein Verständliche aus dem Gebiete der Physik* (Jena, 1892), his last publication. He also published more than forty articles in scientific journals.

II. SECONDARY LITERATURE. See S. Günther, "Leonhard Sohncke," in *Allgemeine Deutsche Biographie*, LIV, 377–379; F. Erk, "Leonhard Sohncke," in *Meteorologische Zeitschrift*, **15** (1898), 81–84; and C. Voit, "Leonhard Sohncke," in *Akademie der Wissenschaften (München), Sitzungsberichten*, **28** (1898), 440–449.

JOHN G. BURKE

SOKHOTSKY, YULIAN-KARL VASILIEVICH (*b.* Warsaw, Poland, 5 February 1842; *d.* Leningrad, U.S.S.R., 14 December 1927), *mathematics*.

Sokhotsky was the son of Vasili Sokhotsky, a clerk, and Iozefa Levandovska. After graduating from the Gymnasium in Warsaw, he joined the department of physics and mathematics of St. Petersburg University in 1861 but returned to Poland the following year to study mathematics independently. In 1865 he passed the examinations at the mathematics department of the University of St. Petersburg and received the bachelor of mathematics degree in 1866. After defending his master's thesis in 1868, Sokhotsky began teaching at the university as assistant professor and in 1869–1870 delivered the first course taught there on the theory of functions of a complex variable. He defended his doctoral thesis in 1873 and was elected extraordinary professor, becoming professor in 1883; from 1875 he also taught at the Institute of Civil Engineers. His lectures, especially on higher algebra, the theory of numbers, and the theory of definite integrals, were extremely successful. Sokhotsky was elected vice-president of the Mathematical Society at its founding in St. Petersburg in 1890 and succeeded V. G. Imshenetsky as president in 1892. He taught at the university until 1923.

Sokhotsky belonged to the school of P. L. Chebyshev; and the latter's influence, while not exceptional, is strong throughout his work. Thus, in his master thesis, which was devoted to the theory of special functions, Sokhotsky, besides employing expansions into an infinite series and continued fractions, made wide use of the theory of residues. (Chebyshev avoided the use of functions of a complex variable.) Elaborating the foundations of the theory of residues, Sokhotsky discovered and demonstrated one of the principal theorems of the theory of analytical functions. According to this theorem, a single-valued analytical function assumes in every vicinity of its essential singular point all complex values. This result was simultaneously published by Felice Casorati, but the theorem attracted attention only after its independent formulation and strict demonstration by Weierstrass in 1876. In his doctoral thesis Sokhotsky continued his studies on special functions, particularly on Jacobi polynomials and Lamé functions. One of the first to approach problems of the theory of singular integral equations, Sokhotsky in this work considered important boundary properties of the integrals of the type of Cauchy and, essentially, arrived at the so-called formulas of I. Plemel (1908).

Sokhotsky also gave a brilliant description of E. I. Zolotarev's theory of divisibility of algebraic numbers and wrote several articles on the theory of elliptic functions and theta functions.

BIBLIOGRAPHY

I. ORIGINAL WORKS. There is no complete bibliography of Sokhotsky's writings. His principal works include *Teoria integralnykh vychetov s nekotorymi prilozheniami* ("The Theory of Integral Residues With Some Applications"; St. Petersburg, 1868), his master's thesis; *Ob opredelennykh integralakh i funktsiakh upotreblyaemykh pri razlozheniakh v ryady* ("On Definite Integrals and Functions Used for Serial Expansion"; St. Petersburg, 1873), his doctoral dissertation; and *Nachalo naibolshego delitelia v primenenii k teorii delimosti algebraicheskikh chisel* ("The Application of the Principle of the Greatest Divisor to the Theory of Divisibility of Algebraic Numbers"; St. Petersburg, 1898).

II. SECONDARY LITERATURE. See the following (listed chronologically): S. Dickstein, "Wspomnienie pośmiertne o prof. J. Sochoskim," in *Wiadomośći matematyczne*, 30 (1927–1928), 101–108; A. I. Markushevich, "Vklad Y. V. Sokhotskogo v obshchuyu teoriyu analiticheskikh funktsy" ("Y. V. Sokhotsky and the Development of the General Theory of Analytic Functions"), in *Istoriko-matematicheskie issledovaniya*, 3 (1950), 399–406; and *Skizzen zur Geschichte der analytischen Funktionen* (Berlin, 1955)—see index; I. Y. Depman, "S.-Peterburgskoe matematicheskoe obshchestvo" ("The St. Petersburg Mathematical Society"), in *Istoriko-matematicheskie issledovaniya*, 13 (1960), 11–106, esp. 33–38; and I. Z. Shtokalo, ed., *Istoria otechestvennoy matematiki* ("A History of Native Mathematics"), II (Kiev, 1967)—see index.

A. P. YOUSCHKEVITCH

SOKOLOV, DMITRY IVANOVICH (*b.* St. Petersburg, Russia [now Leningrad, U.S.S.R.], 1788; *d.* St. Petersburg, 1 December 1852), *geology.*

Sokolov's father, a locksmith, invented a machine for turning screws that was used in the fountains of St. Petersburg. After his death in 1796 Sokolov was sent to the preparatory class of the St. Petersburg Mining School, which was reorganized in 1804 as the Mining Cadet Corps and, in 1834, as the Institute of the Corps of Mining Engineers (now the Leningrad Mining Institute). He remained associated with this institution throughout his life.

After graduating in 1805, Sokolov worked in the laboratory of the Mining Cadet Corps as assayer and then began lecturing, first in metallurgy and assaying and later in geognosy and mining. In 1813 he was appointed supervisor of the mineralogical laboratory and from 1818 was also in charge of the collection of models. He continued his chemical research following his appointment as director of the joint laboratory of the Mining Cadet Corps and the Department of Mining and Salt Works. In 1817 he participated in the creation of the Mineralogical Society in St. Petersburg.

In 1822 Sokolov became professor of geognosy and mineralogy at St. Petersburg University, where he taught for more than twenty years while serving at the Mining Cadet Corps as class inspector, from 1826, and as assistant director in charge of teaching (1834–1847). He was also active in the creation in 1825 of *Gorny zhurnal* ("Mining Magazine"), the first specialized periodical in Russian, of which he became editor. In 1839 Sokolov was elected member of the Russian Academy of Sciences and, in 1841, honorary member of the division of language and philology of the St. Petersburg Academy of Sciences for the compilation of a dictionary of Old Church Slavonic and Russian.

Sokolov's marriage to Ekaterina Nikolaevna Prytkova was childless; their adopted daughter took care of him during his old age and illness.

Sokolov's early training as an experimental chemist enabled him to approach the problem of the classification of minerals in a new way. An advocate of the atomistic concept, which was then only beginning to win recognition, he believed that the physical properties of minerals depend upon vectorial atomic forces. Sokolov attached primary importance to chemical composition and classified minerals according to their cations. On the basis of similar properties, he outlined their natural groupings long before Mendeleev's periodic system. Thus his first category included lithophile elements of the first four groups of Mendeleev's periodic table, while his second category of metals comprised only the chalcophile elements.

Sokolov correctly attributed the physical properties of minerals—morphology, shape, color, luster, hardness, cleavage—to their chemical composition and regarded physical features as secondary. Although he did not take them into consideration in his classification, he did stress their importance for determination. Following the traditions of the Russian mineralogical school of Lomonosov and Severgin, Sokolov was especially concerned with paragenesis and, in his textbook of mineralogy, discussed the association of minerals, indicating that some, like cinnabar, have no characteristic paragenesis.

By the late 1820's Sokolov's early neptunist views had been superseded by the plutonist approach; and he came to regard most minerals, rocks, and ores as the result of crystallization from a magmatic melt. In his petrographic descriptions he consistently attempted to give not only a chemical and mineralogical description but also an account of their genesis. He distinguished magmatic from sedimentary rocks, adding in the 1830's the category of metamorphic. His interest in problems of petrology led him to conclude that the comparatively limited variety in rock composition could be explained by affinities of chemical elements. He also emphasized the importance of geological conditions in the physicochemical environment (pressure, temperature, presence of water vapor), assuming that with a variation in these conditions a change in petrology occurs.

In associating the conditions of ore formation with the specific features of geological environment, Sokolov attached decisive importance to processes occurring deep within the earth, and tried to distinguish periods of ore formation and provinces. He considered pneumatolytic and contact deposits the most important genetic types and distinguished the hydrothermal concentration of ore.

Sokolov's papers were also devoted to the problems of metalliferous placers. Contrary to prevalent concepts, he demonstrated that placers are located near the outcrops from the weathering of which they originated; this fact substantially altered methods of prospecting for gold. His distinction of metalliferous zones and belts of varying composition laid the foundation of modern concepts of metallogenetic provinces.

On Sokolov's initiative, work was begun on the geological mapping of Russia, and his suggestions resulted in the undertaking, in 1834, of geological surveys of individual mining regions. These surveys were carried out on his instructions and proceeded under his direct supervision. By 1841 the

first geological maps of European Russia had been published. The development of geological surveying had necessitated detailed stratigraphic subdivision. Of great importance for mapping was the summarized stratigraphic table compiled by Sokolov in 1831 and substantially modified in 1839.

Sokolov's three-volume textbook Kurs geognosii ("Course of Geognosy," 1839) contains a large section that would now be described as a history of geology; it gave a detailed description of all the geological systems then known. In this work Sokolov first distinguished an independent system at the top of the Paleozoic deposits, which two years later was named the Permian system and was included in international stratigraphic tables by Murchison. Sokolov also pointed out the essential differences between Lower and Upper Silurian deposits that warranted his describing each as an independent system.

In sedimentology Sokolov asserted that sedimentary masses are originally deposited horizontally and are only later compressed into folds, whereas lamination is an indication of episodic changes in the environment.

As early as 1820 Sokolov began to advocate that current geological phenomena can serve as a prototype for past events reflected in the geological sequence and that geological processes proceeded in the same way throughout history. He stressed this approach even more insistently, especially after the appearance of Lyell's Principles of Geology (1830–1833), which contributed to the development of the actualistic method and the worldwide dissemination of uniformitarian concepts, certain elements of which were contained in Sokolov's views.

Here, however, he did not take the position of formal uniformitarianism but systematically stressed that physicochemical conditions in the earth's interior, as well as paleogeographical conditions on its surface, are consistently changing and that, consequently, mineral and petrogenesis are not uniformitarian. In this respect Sokolov developed Lomonosov's views on the continual transformation of the world environment and believed that there is a continuous development process in both organic and inorganic nature. Advocating the existence of transformistic phenomena, which alter the organic world, Sokolov pointed out that this process is directed toward the origin of increasingly more highly organized forms—ideas that contained the rudiments of evolutionary concepts.

Sokolov actively disseminated advanced geological ideas through his textbooks and public lectures, which attracted a broad audience. His highly praised books were awarded three prizes by the St. Petersburg Academy of Sciences. Despite the rapid development of geology, Sokolov's textbooks remained valid; *Kurs mineralogii* ("Course of Mineralogy"), for example, was still in use after his death.

BIBLIOGRAPHY

I. ORIGINAL WORKS. Sokolov's writings include "Kratkoe nachertanie gornykh formatsy po noveyshemu sostoyaniyu geognozii" ("A Brief Outline of Rock Formations According to the Latest State of Geognosy"), in *Gorny zhurnal*, nos. 4–5 (1831); *Rukovodstvo k mineralogii s prisovokupleniem statisticheskikh svedeny o vazhneyshikh solyakh i metallakh* ("Textbook on Mineralogy With Additional Statistical Data on the Most Important Salts and Metals"), 2 vols. (St. Petersburg, 1832) and supplement (St. Petersburg, 1838); *Kurs geognosii* ("Course of Geognosy"), 3 vols. (St. Petersburg, 1839); and *Rukovodstvo k geognozii* ("Handbook of Geognosy"), 2 pts. (St. Petersburg, 1842).

II. SECONDARY LITERATURE. On Sokolov and his work, see E. A. Radkevich, *Dmitry Ivanovich Sokolov 1788–1852* (Moscow, 1959); and V. V. Tikhomirov, "Dmitry Ivanovich Sokolov (k 100-letiyu so dnya smerti" (". . . on the Centenary of His Death"), in *Byulleten Moskovskogo obshchestva ispytatelei prirody*, Ser. geolog., **27**, no. 6 (1952).

V. V. TIKHOMIROV

SOLANDER, DANIEL CARL (*b*. Piteå, Sweden, 19 February 1733; *d*. London, England, 13 May 1782), *natural history*.

Solander's father, Carl Solander, was a delegate to the Riksdag as well as a Lutheran rector and rural dean for Piteå; his mother, Magdalena Bostadia, was the daughter of a district judge. After spending his formative years in Swedish Lapland, he entered the University of Uppsala on 1 July 1750, apparently to prepare for a legal or clerical career; but his study of natural history with Linnaeus and of chemistry with Wallerius turned his interest toward science.

In 1752 Solander helped Linnaeus classify and index the royal natural history collections at Ulriksdal and Drottningholm, as well as the collection of Count Carl Tessin; the results were published in *Museum Adolphi Frederica, Museum Ludovicae Ulricae*, and *Museum Tessinianum*. In 1756 Solander published *Caroli Linnaei Elementa botanica*, an epitome of Linnaeus' general botany.

Two years later he examined a supposedly parasitic worm and reported his findings in "Furia infernalis, vermis" (1772). In 1758 Solander assisted Linnaeus in examining Patrick Browne's herbarium; his efforts, coupled with his frequent visits to Linnaeus' home, aroused in the latter such esteem for his student that he wanted Solander to be both his successor and son-in-law.

During the 1750's Solander made two botanical expeditions to Lapland. In 1753 he traveled up the Piteå River, crossed the Kjölen Mountains into Norway, botanized in the vicinity of Rörstad, and returned to Uppsala. In 1755 he studied the natural history of the Tornio basin.

Chosen by Linnaeus to help popularize the Linnaean system in England, Solander left Uppsala on 6 April 1759; but a severe attack of epidemic influenza detained him in southern Sweden until 30 May 1760, and he did not arrive in England until 30 June. Solander was readily accepted into English society. Boswell once said of him, "Throw him where you will, he swims." Frances Burney found him "very sociable, full of talk, information, and entertainment, . . . a philosophical gossip." Richard Pulteney wrote that "the urbanity of his manners, and his readiness to afford every assistance in his power, joined to that clearness and energy with which he affected it," made Solander and the Linnaean system popular with naturalists.

A few weeks after his arrival, Solander had firmly established himself as the link between Linnaeus and English naturalists. In this capacity he collected plants for Linnaeus, obtained favors for English naturalists from him, and toured southern and southeastern England. He proved so useful an authority in decisions on Linnaean taxonomy that John Ellis and Peter Collinson worked to secure him a post, which he received in 1763, with the newly established British Museum.

Solander immediately began organizing the natural history collection, and by 1768 he had completed the first-draft descriptions. In the meantime he had described the collections of Gustavus Brander, the duchess of Portland, John Ellis, John Bartram, and Alexander Garden.

In June 1764 Solander became an active member of the Royal Society, and his friendship with Joseph Banks dates from that year. He advised Banks on how to prepare for his voyage in 1766 to Labrador and Newfoundland, and two years later, consulted with him on preparations for the voyage of the *Endeavour*, headed by Captain James Cook (1768–1771). After several days of planning, the Swedish naturalist asked to join the voyage and was accepted. Although Solander and Banks spent their time routinely collecting plant and animal specimens and conducting observations of the natural history and inhabitants, the voyage was not without its dangers: a fierce snowstorm on Tierra del Fuego, the near wreck of the *Endeavour* on the Great Barrier Reef, and an outbreak of malaria followed by dysentery in Batavia, Java.

The published results of their joint efforts are disappointing. Solander and Banks collected an estimated 100 new families and 1,000 new species of plants, in addition to hundreds of new species of animals. Of these new specimens, only 100 plant descriptions, printed by the British Museum (Natural History) in the early 1900's, and 200 insect descriptions by Johann Christian Fabricius, have been published. A lazy streak in Solander has often been blamed for the neglect; but the descriptions were, in fact, finished before Solander's death, and in 1785 Banks wrote that the project was nearing completion. Thus it was Banks who, for his own reasons, did not complete the project.

Solander Island (off the southern coast of South Island, New Zealand) and Cape Solander (the south side of Botany Bay, New South Wales, Australia) were named during the voyage, and Solander devised the Solander case, a book-shaped box that is still used, to guard the manuscript records of the voyage. On 21 November 1771 he was awarded the D.C.L. by Oxford for his part in the voyage.

Solander's manuscript descriptions were used by John Latham in his ornithological studies, by Johann Reinhold Forster for his reports on the second Cook voyage, and by Joseph Gaertner for his studies on plants. Furthermore, Solander's efforts helped establish a precedent: naturalists were subsequently included in government-sponsored voyages of exploration. Charles Darwin held the post aboard H.M.S. *Beagle*. Solander's and Banks's praise of the breadfruit contributed to attempts to introduce it to the West Indies and on the voyage of H.M.S. *Bounty*. Their influence on Cook stimulated his awareness of the importance of human and natural history, which was put to good use on his subsequent voyages.

Solander was part of a team of scientists recruited by Banks for Cook's second voyage. Banks refused to go because of inadequate quarters, and in the summer and fall of 1772 the group went instead on a four-month journey to the western coast of Britain, Iceland, and the Orkneys. Again, the results were never published, but the data were made available through Banks's collection and, later, through the collections of the British Museum.

After returning to London, Solander resumed his busy schedule, regaining his post as assistant keeper at the British Museum and becoming keeper in June 1773. He increased its collection, testified at Parliamentary hearings on the museum, organized and described its collection, and conducted tours. As Banks's private secretary, he was in charge of one of the largest natural history collections; as Banks's librarian, he named all of the new plants received by the Royal Botanical Gardens at Kew and assisted William Aiton in the early planning of *Hortus Kewensis*.

Solander helped several others as well. The duchess of Portland continued to employ him to help with her collection, and he aided Thomas Pennant in his studies in zoology, John Fothergill in his Upton garden, and John Lightfoot with his *Flora scotica*. Descriptions for Ellis' works were also written by Solander, and he participated in experiments by Charles Blagden and Benjamin Franklin. According to Thomas Krok, Solander contributed to sixty-six publications.

An active member of several scientific societies, Solander regularly attended meetings of the Royal Society and dined with the Royal Society Club, of which he was treasurer from 1774 until his death. He also regularly attended the meetings of a nameless society of scientists that met at Jack's Coffee House, an affiliation that led him to visit the Lunar Society of Birmingham. He met with Fothergill's medical society and was a corresponding member of the Académie des Sciences, the Royal Swedish Academy of Sciences, the Gesellschaft Naturforschender Freunde, the Royal Academy of Arts and Sciences of Göteborg, and the Academy of Sciences of Naples.

BIBLIOGRAPHY

I. ORIGINAL WORKS. Solander's MSS and letters are dispersed. The British Museum (Natural History) has a large collection of his biological notes and MSS, as well as his "Memoranda Connected With the Visit to Iceland. . . ." The Linnean Society of London has a collection of Solander's letters, particularly those written to Linnaeus. British Museum MSS Add. 45,874 and 45,875 are Solander's diaries of his work at the museum; MS Add. 29,533 contains some of John Ellis' correspondence with Solander. Many of Solander's letters were included in James Edward Smith, *A Selection of the Correspondence of Linnaeus and Other Naturalists* (London, 1821).

Solander's own publications include *Caroli Linnaei Elementa botanica* (Uppsala, 1756); "An Account of the Gardenia. . . ," in *Philosophical Transactions of the Royal Society*, **52**, pt. 2 (1762), 654–661; and "Furia infernalis, vermis, et ab eo concitari solitus morbus," in *Nova acta Regiae Societatis scientiarum upsaliensis*, **1** (1795), 44–58.

II. SECONDARY LITERATURE. On Solander and his scientific work, see Roy A. Rauschenberg, "A Letter of Sir Joseph Banks Describing the Life of Daniel Solander," in *Isis*, **55** (1964), 62–67; "Daniel Carl Solander, the Naturalist on the *Endeavour* Voyage," *ibid.*, **58** (1967), 367–374; and "Daniel Carl Solander. Naturalist on the 'Endeavour,'" in *Transactions of the American Philosophical Society*, n.s. **58**, pt. 8 (1968), 1–58, with extensive bibliography of primary and secondary sources. Earlier accounts and reminiscences include Frances Burney d'Arblay, *Diary and Letters of Madame d'Arblay*, C. F. Barrett, ed., I (London–New York, 1904), 318; James Boswell, *The Journal of James Boswell*, G. Scott and F. A. Pottle, eds., XIV (Mt. Vernon, N.Y., 1930), 182; and Richard Pulteney, *Historical and Biographical Sketches of the Progress of Botany in Britain*, II (London, 1790), 350–351. See also Thomas Krok, *Bibliotheca botanica suecana* (Uppsala–Stockholm, 1925), 655–660.

ROY A. RAUSCHENBERG

SOLDANI, AMBROGIO (or **Baldo Maria**) (*b.* Pratovecchio, Arezzo, Italy, 15 June 1736; *d.* Florence, Italy, 14 July 1808), *geology.*

The son of Dr. Soldano Soldani and Benedetta Nesterini, Soldani dropped the name Baldo Maria and assumed that of Ambrogio in 1752, when he entered the Camaldolese Congregation. Although an exemplary monk, he was active in the cultural life of Siena, where he spent most of his life. With the economist Sallustio Bandini, he reorganized the celebrated Accademia dei Fisiocritici, and in 1781 he was appointed professor of mathematics at the University of Siena.

Soldani was not only a mathematician, however, but also an ardent naturalist. In his studies of Pliocene marine formations of Tuscany and of preexistent ones bordering the Pliocene sea he proved to be an accomplished geologist, describing with great accuracy the lithological, stratigraphic, and paleontological characteristics of the deposits. Although his emphasis on the study of microscopic fossils (he described and drew hundreds of them, from mollusks to foraminifers) entitles him to be considered a paleontologist, Soldani never approached paleontological research as an end in itself. His desire to study the microfauna of the Mediterranean, which was almost unknown in his time, derived from his conviction that knowledge of present zoological conditions would have decisive consequences for the correct interpretation of the deposits left by the ancient seas.

In deposits of the Pliocene Tuscan sea, especially in those found near Siena and Volterra, Soldani identified three layers: an abyssal, formed by material of marine origin; a littoral, formed by material of terrestrial origin; and an intermediate, formed by both marine and terrestrial material. In addition he showed the presence of lacustrine microfauna in the sediments of the early Pleistocene Tuscan lakes that have since disappeared (such as Valdarno Superiore). Soldani's studies distinguished him as a leader in establishing the interrelation of zoology and paleontology; and he deserves considerable credit for his efforts to derive, from present conditions, material for the study of the geological past. Unfortunately he made no effort to classify systematically the many species that he described and distinguished. The harsh criticism, especially in this respect, of his most important work, the *Testaceographia*, so embittered him that he burned all copies of the work in his possession.

Soldani also studied the celebrated meteorites that fell in the region around Siena in 1794. Reasoning on the basis of eyewitness reports and on lithological study of the meteorites, he excluded any possibility of their terrestrial origin, proposing instead that they originated from a condensation of atmospheric vapor. This fallacious hypothesis, as well as that of their extraterrestrial origin, appeared absurd at the time to scientists; and Soldani was derided in a lively polemic.

An impartial evaluation of Soldani's entire work was not possible until after his death. Charles Lyell assigned him a prominent place among eighteenth-century naturalists, and not merely that of a founder of micropaleontology.

BIBLIOGRAPHY

I. ORIGINAL WORKS. Three basic works by Soldani are *Saggio orittografico, ovvero osservazioni sopra le terre nautiliche ed ammonitiche della Toscana* (Siena, 1780); *Testaceographia ac Zoophitographia parva et microscopica*, 4 vols. (Siena, 1789–1798), very rare, partly because Soldani destroyed copies; and *Sopra una pioggetta di sassi accaduta nella sera de' 16 giugno del MDCCXCIV in Lucignan d'Asso nel Sanese* (Siena, 1794).

II. SECONDARY LITERATURE. See E. Manasse, "Commemorazione di Ambrogio Soldani," in *Atti dell' Accademia dei fisiocritici di Siena*, 4th ser., **20** (1908),

365–376; O. Silvestri, "Ambrogio Soldani e le sue opere," in *Atti della Società italiana di scienze naturali*, **15** (1872), 273–289; and F. Rodolico, "Ambrogio Soldani e Ottaviano Targioni Tozzetti; carteggio sulla 'pioggetta di sassi' del 1794," in *Physis*, **12** (1970), 197–210.

FRANCESCO RODOLICO

SOLDNER, JOHANN GEORG VON (*b.* Georgenhof, near Feuchtwangen, Germany, 16 July 1776; *d.* Bogenhausen, Munich, Germany, 18 May 1833), *geodesy, astronomy.*

Soldner's father, Johann Andreas Soldner, was a farmer in Georgenhof. The boy's schooling began in Banzenweiler and continued in Feuchtwangen and at the Gymnasium in Ansbach, where he was taught by the physicist Julius Yelin. His education was interrupted by periods of work on his father's farm. Soldner's diary indicates that his earliest interest in land surveying was aroused by neighboring farmers and by some geometric notes in an early Ansbach calendar. He was largely self-taught and he devised his own instruments for measuring the altitude of the sun.

Soldner was a pupil of the astronomer Bode and first became known through his contributions to Bode's *Astronomisches Jahrbuch*. In 1805 Frederick William of Prussia made him director of the survey of Ansbach, but the battles of Jena and Auerstedt made this work impossible.

Soldner was patronized by the astronomer Ulrich Schiegg, who in 1808 was appointed technical member of the Tax Rectification Commission. Correct land-tax assessments required an accurate survey of the province, and in 1811 Soldner was appointed to the Bavarian Land Survey. In May 1810 he had published his famous memorandum on the calculation of a triangle network, in recognition of which he was made an ordinary member of the Munich Academy of Sciences. This work was lost until 1873, when Karl Orff published an account of the Bavarian survey.

Soldner's method for calculating the spherical triangles of the main network was an improvement on Delambre's method, which was adequate for degree measurement but was not suitable for land surveying, that is, when the lengths of arcs of spherical triangles must be known. Soldner used a system of coordinates that introduced three points into every mesh of the network plan, accurately aligning it with the land to be surveyed. Thus he was able to measure arcs to an accuracy of one centimeter. Also, his solution of spherical triangles was more convenient than that of Legendre; Soldner kept two angles of the spherical triangle the same and altered the length of the sides when comparing it to a plane triangle.

In 1813 Soldner prepared a paper on a new method of reducing astronomical azimuths. His method depended on the observation of the maximum east and west displacements from the North Star. This discovery aroused the jealousy of the astronomer and councillor Karl Felix Seyffer, who, in 1815, was removed from office and was succeeded by Soldner as director of the Bogenhausen observatory. Soldner was responsible for supervising the construction and equipping of the new observatory. He also retained his post as consultant to the land-tax commission. In 1820 Delambre defended the originality of Soldner's work in the *Connaissance des temps*.

The observatory was completed in 1818, and instruments from the workshops in Utschneider, Reichenbach, and Fraunhofer were installed. Soldner had previously tested many of these instruments for their accuracy and suitability. By 1820 he had begun to observe the positions of the stars and planets on the meridian circle. With Nicolai, he worked on the measurement of a degree; and, independently, he studied lunar methods for determining longitude.

From 1823 on Soldner confined himself to the administration of the observatory; his assistant, Lamont, who succeeded him as director, undertook the observational work. During these years, Soldner's health deteriorated because of a liver ailment.

Soldner was simple and reserved in manner, and he valued real scholarship for its own sake. His painstaking observational work on the detection of motion among the fixed stars could be of value only to future generations of astronomers and illustrates the unselfish spirit of his work. His writings are clear and concise, and he avoided repetition of what was already common knowledge.

BIBLIOGRAPHY

I. ORIGINAL WORKS. Soldner's papers appeared in many leading astronomical journals. His books include *Théorie et tables d'une nouvelle fonction transcendente* (Munich, 1809); *Bestimmung des Azimuths von Altomunster* (Munich, 1813); *Neue Methode, Beobachtete Azimuthe zu Reduzieren* (Munich, 1813); and probably *Astronomische Beobachtungen von 1819 bis 1827 auf der Sternwarte zu Bogenhausen,* 3 vols. (Munich, 1833–1835). Hostile colleagues prevented the publication of later volumes of this last work.

II. SECONDARY LITERATURE. Biographical information is given by C. M. Bauernfeind, in *Allgemeine Deutsche Biographie,* XXXIV, 557–563; F. J. Müller, *Lebenslaufe aus Franken,* II (Wurzburg, 1922), 417–427; and *Poggendorff,* II, 955–956.

Studies of Soldner's geodetical work include F. J. Müller, *Johann Georg von Soldner der Geodät* (Munich, 1914); and Gunther Rutz, *Die Alte Bayerische Triangulation von Johann Georg Soldner* (Munich, 1971). Müller and Rutz draw extensively on Bauernfeind. See also R. Sigl, "Johann Georg von Soldner zum Gedächtnis," in *Mitteilungsblatt des Deutsches Landesvermessungswerk, Landesverein Bayern,* no. 2 (1966). Archives relating to Soldner exist at the Bayerisches Landesvermessungsarnt, Munich. Letters between Soldner and Gauss exist at Niedersächsische Staats- und Universitätsbibliothek, Göttingen.

SISTER MAUREEN FARRELL, F.C.J.

SOLEIL, JEAN-BAPTISTE-FRANÇOIS (*b.* Paris, France, 1798; *d.* Paris, 17 November 1878), *optical instruments.*

Soleil learned his craft while working under the engineers Hareing and Palmer. From 1823 to 1827 he was intimately associated with the work of Fresnel. Soleil directed the construction of the annular lenses and the mechanism to rotate them that Fresnel had designed for use in lighthouses. Soleil's first lens was made under the direction of Fresnel and was based on his theory. In 1841 it was presented to the Académie des Sciences by his son, Henri. Soleil constructed most of the apparatus used by Fresnel in his optical research based on the experimental demonstration of the wave theory by Thomas Young. This work brought Soleil into contact with those scientists who, following Fresnel, developed the new optics: François Arago, Jacques Babinet, Charles Delezenne, Fredrik Rudberg, and Johann Nörrenberg. A notable piece of apparatus constructed by Soleil—the diffraction bench—was intended for use in public demonstrations of interference and diffraction phenomena with either sunlight or lamplight.

Soleil produced and sold a wide variety of optical instruments at 35 rue de l'Odéon. His inventions included an apparatus for measuring the interaxial angle in biaxial crystals and an improved model of Biot's saccharimeter. In 1849 he retired from business and was succeeded by his son-in-law and former apprentice Jules Duboscq. In November 1850 Soleil was named chevalier of the Légion d'Honneur.

Soleil received a number of exhibition awards, including a gold medal in 1849; and the physical optics section of the Great Exhibition of 1851 in London was devoted solely to the products and inventions of Duboscq and Soleil. This exhibition won the highest award—a council medal.

BIBLIOGRAPHY

I. ORIGINAL WORKS. Soleil's works include "Appareils pour la production des anneaux colorés à centre noir ou blanc," in *Comptes rendus hebdomadaires des séances de l'Académie des sciences,* **18** (1844), 417–419; "Note sur la structure et la propriété rotatoire du quartz cristallisé," *ibid.,* **20** (1845), 435–438; "Note sur moyen de faciliter les expériences de polarisation rotatoire," *ibid.,* 1805–1807; "Note sur perfectionnement apporté au pointage du saccharimètre," *ibid.,* **24** (1847), 973–975; "Notice sur l'horloge polaire de M. Wheatstone, construite et perfectionnée par M. Soleil," *ibid.,* **28** (1849), 511–513; and "Note sur un nouveau caractère distinctif entre les cristaux à un axe, positifs et négatifs," *ibid.,* **30** (1850), 361–362. Brief notices of apparatus shown to the Academy are also printed in *Comptes rendus hebdomadaires des séances de l'Académie des sciences.*

II. SECONDARY LITERATURE. See *Exhibition of the Works of Industry of All Nations 1851. Reports by the Juries on the Subjects in the Thirty Classes Into Which the Exhibition Was Divided* (London, 1852), 272; and G. Vapereau, *Dictionnaire universel des contemporains* (Paris, 1880).

G. L'E. TURNER

SOLLAS, WILLIAM JOHNSON (*b.* Birmingham, England, 30 May 1849; *d.* Oxford, England, 20 October 1936), *geology, paleontology, anthropology.*

Sollas was educated at the Royal School of Mines, London, where he studied under A. C. Ramsay and T. H. Huxley, and then at St. John's College, Cambridge, where T. G. Bonney, deputy to Adam Sedgwick, was his tutor. For six years he gave university extension courses in geology and biology to adult students in various parts of the country. He was afterward successively professor of geology at Bristol (1880–1883), Trinity College, Dublin (1883–1897), and Oxford (1897–1936). Sollas was a versatile investigator and experimentalist, and in the course of a long and active life he made significant contributions to many branches of the geological sciences and to biological research. He received numerous British and foreign academic awards, including the Royal Medal of the Royal Society of London.

Sollas' early work was on fossil and modern sponges and culminated in a paper on Tetractinellida in the *Report of the Scientific Results of the Voyage of H.M.S. Challenger* (1888). While at Dublin, Sollas carried out petrological and mineralogical investigations of the granites of Ireland. In this work he concluded that the "pleochroic haloes" surrounding the zircon crystals—which are contained in the dark micas of Wicklow granite—were probably caused by an unknown element. Joly, his colleague at Dublin, later proved that this phenomenon resulted from the presence of radioactive elements of the uranium decay series in the zircons.

Sollas published papers on riebeckite and zinnwaldite, and he also studied the intimate architecture of crystals. To form crystalline solids he adapted the methods and ideas of Haüy and approached the subject from the standpoint of the structural arrangement of the atomic units. The structures later revealed by more sophisticated X rays fully supported his contention that the closest packing of atoms does not provide the only style of crystal architecture: in many cases the packing is as open as he had postulated. To obtain a preliminary separation of minerals for chemical analysis, Sollas devised the "diffusion column," in which heavy liquids are superposed in layers of graded density that are indicated by floating markers of known specific gravity. He adapted the method of serial sections, which he had used for his sponge research, to the examination of fossils, thus obtaining data on their internal structure and then constructing models of the skeleton. Using this method, he made important investigations on fossils ranging from Silurian echinoderms to Jurassic reptiles. In some of this work his daughter Igerna was his collaborator. His elder daughter, Hertha, was responsible, under his direction, for the English translation of Eduard Suess's *Das Antlitz der Erde*.

In 1911 Sollas published *Ancient Hunters and Their Modern Representatives*. Thereafter he engaged increasingly in research on various aspects of anthropology and prehistoric archaeology. He excavated the Paviland Cave in South Wales, which had first been explored by a predecessor at Oxford, William Buckland. With Breuil, Sollas showed that the flint implements associated with the skeleton of the "Red Lady" belonged to the Aurignacian age,

and he suggested that the skeleton itself belonged to Cro-Magnon man. This conjecture has been confirmed by isotope dating.

BIBLIOGRAPHY

I. ORIGINAL WORKS. Few of Sollas' MSS survive. They consist mainly of notebooks and lecture notes at the department of geology, University of Oxford. He published some 180 works from 1872 to 1933. His major writings include *The Age of the Earth and Other Geological Studies* (London, 1905); *The Rocks of Cape Colville Peninsula, Auckland, New Zealand*, 2 vols. (Wellington, New Zealand, 1905–1906); and *Ancient Hunters and Their Modern Representatives* (London, 1911; 2nd ed., 1915; 3rd ed., 1924).

II. SECONDARY LITERATURE. Accounts of Sollas' work and life are in *Obituary Notices of Fellows of the Royal Society of London*, **2** (1938), 265–281; and *Proceedings of the Geological Society of America for 1937* (1938), 203–220. Both of these contain a complete list of his publications and a portrait.

J. M. EDMONDS

SOLVAY, ERNEST (*b*. Rebecq-Rognon, near Brussels, Belgium, 16 April 1838; *d*. Brussels, 26 May 1922), *industrial chemistry*.

After a modest education at local schools, Solvay entered the salt-making business owned by his father, Alexandre Solvay. Then, at the age of twenty-one, he joined his uncle Florimond Semet in managing a gasworks in Brussels—his particular concern being the discovery of better methods of concentrating ammoniacal liquors. In 1861 he noted the ease with which ammonia, salt solution, and carbon dioxide react to form sodium bicarbonate, which can be converted easily to the soda ash of commerce. The glass and soap trades had created a demand for this chemical that was met only by the economically unstable and chemically complex Leblanc process. At this time Solvay was unaware that the ammonia-soda reaction had been known for fifty years; that several industrial chemists, including William Gossage and Henry Deacon, had been unsuccessful in employing it for large-scale production; and that James Muspratt had lost £8,000 on unsuccessful experiments.

Solvay's knowledge of the industrial preparation of salt and ammonia was useful when he turned to the problems of the ammonia-soda process: particularly those involving the loss of expensive ammonia and the practical difficulties of mixing liquids and gases on a large scale. With his brother

Alfred, Solvay established (1861) a small works in the Schaerbeek district of Brussels. Following some small success, and supported financially by the family, the Solvay brothers built, in 1863, a factory at Couillet, near Charleroi; production started in 1865. Solvay patented every stage of the process but granted licenses to soda manufacturers in other countries. In 1872 a license was acquired by Mond, who introduced the Solvay process in England and later achieved great success with it. Solvay's key contribution to the soda trade was his invention of a carbonating tower in which ammoniacal brine could be mixed thoroughly with carbon dioxide. By 1890 Solvay had established plants in most European countries, in Russia, and in the United States.

Solvay was a member of the Belgian senate and a minister of state. He founded the Solvay International Institutes of Chemistry, of Physics, and of Sociology. By the terms of Solvay's gift, the institutes held periodical international conferences at which such broad areas of science as electrons and photons (1928), the solid state (1951), and the origin and structure of the universe were discussed. The names of the participants testify to the quality of the contributions. The 1928 physics congress, for example, was addressed by Bragg, de Broglie, Bohr, Born, Heisenberg, and Schrödinger.

BIBLIOGRAPHY

I. ORIGINAL WORKS. Solvay's only descriptions of his process are contained in his patents—the most important are the British patents 3131 of 1863, 1525 of 1872, 2143 of 1876, and 999 of 1904. The Royal Society *Catalogue of Scientific Papers*, XI, 450, XVIII, 845, lists nine papers of only minor importance.

II. SECONDARY LITERATURE. A meticulously detailed description of Solvay's plant and methods appears in G. Lunge, *The Manufacture of Sulphuric Acid and Alkali*, 2nd ed., III (London, 1896), 9–100. Obituaries, with portraits, appear in *Industrial and Engineering Chemistry*, **14** (1922), 1156; *Journal of the Society of Chemical Industry*, **41** (1922), 231–R; and *Revue de métallurgie*, **19** (1922), 696, which contains several errors. See also Jacques Bolle, *Solvay: L'homme, la découverte, l'entreprise industrielle* (Brussels, 1968); E. Farber, *Great Chemists* (New York, 1961), 773–782; and P. Heger and C. Lefebure, *La vie d'Ernest Solvay* (Brussels, 1919).

W. A. CAMPBELL

SOMERVILLE, MARY FAIRFAX GREIG (*b*. Jedburgh, Roxburghshire, Scotland, 26 December 1780; *d*. Naples, Italy, 29 November 1872), *scientific and mathematical exposition, experimentation on the effects of solar radiation.*

One of the foremost women of science of the nineteenth century, Mrs. Somerville was through her writings and example influential in gaining wider acceptance among a literate public for various nineteenth-century scientific ideas and practices and in opening new opportunities to women. Her notable career, spanning more than half a century, brought her in contact with many of the foremost scientific, literary, and political personages of Europe and America. Public recognition accorded her had profound and beneficial effects in advancing the cause of science and of women's education and emancipation.

Through her father, Vice-Admiral Sir William George Fairfax, R.N., a hero of the Battle of Camperdown, she was connected with the distinguished Fairfax family of England that produced the great Cromwellian general Sir Thomas Fairfax and the Fairfaxes of the Virginia Colony. Through her mother, Margaret Charters, his second wife, daughter of Samuel Charters, solicitor of customs for Scotland, she was related to several ancient Scottish houses, among them the Murrays of Philiphaugh, the Douglases of Friarshaw, the Douglases of Springwood Park, the Charterises of Wemyss, and John Knox.

Fifth of their seven children (only three of whom survived to majority), Mary Fairfax was born in the manse at Jedburgh, the home of an aunt, Martha Charters Somerville, who later became her mother-in-law. Her childhood was spent in Burntisland, a small seaport on the Firth of Forth opposite Edinburgh. In a house sold to the Fairfaxes by Samuel Charters and still standing, her easygoing, indulgent mother thriftily reared four children—the eldest surviving son Samuel, Mary, and two younger ones, Margaret and Henry—on slim navy pay. Customarily in the Charters, as in many well-connected Scottish families, sons received excellent educations, attending university and entering the kirk, the legal profession, or service in the East India Company. For daughters, mastery of social and domestic arts and a minimum of formal book learning was considered sufficient. Mary's father, returning from a long period of sea duty, was "shocked to find . . . [his daughter] such a savage," hardly able to read, unable to write, and with no knowledge of language or numbers. He dispatched her, at the age of ten, to a fashionable, expensive boarding school at Musselburgh—a drastic step for a man of such strong Tory convictions. There for twelve months she had the only full-time instruction of her long life, emerging from the experience, she recounts in her autobiography, "like a wild animal escaped out of a cage" but with a taste for reading, some notion of simple arithmetic, a smattering of grammar and French, poor handwriting, and abominable spelling.

Over the next years she had occasional lessons in ballroom dancing, pianoforte playing, fine cookery, drawing and painting (under Alexander Nasmyth), penmanship, needlework, and the use of the globes. A lively and persistent mind, immense curiosity and eagerness to learn, supported by a robust constitution and quiet, unswerving determination, enabled her to take advantage of every opportunity for enlightenment. At Burntisland she had freedom to roam the Scottish countryside and seashore, observing nature at first hand. She read through the small family library, teaching herself enough Latin for Caesar's *Commentaries*. In Edinburgh during the winter months, family position brought her in contact with intellectual and professional circles and the rich artistic life of the Scottish capital. A charmingly shy, petite, and beautiful young woman—Edinburgh society dubbed her the "Rose of Jedburgh"—she delighted in the parties, visits, balls, theaters, concerts, and innocent flirtations that, with domestic and daughterly duties, filled the days of popular Edinburgh belles at the turn of the century.

Another and less conventional interest absorbed her during these years. Between the ages of thirteen and fifteen, the chance glimpse in a ladies' fashion magazine of some strange symbols, said to be "algebra," aroused her curiosity. None of her close relatives or acquaintances could have told her anything of the subject, even had she the courage to ask. Mrs. Somerville, in contrast to other scientific women of the nineteenth century, had no family incentive to investigate science or mathematics and no household exposure to these subjects. Her unguided efforts to learn something of this mysterious but strangely attractive "algebra" were fruitless until, overhearing a casual remark by Nasmyth, she was led to persuade her younger brother's tutor to buy for her copies of Bonnycastle's *Algebra* and Euclid's *Elements*, which she then began to study on her own. Discovering her reading mathematics, her father instantly forbade it, fearing that the strain of abstract thought would injure the tender female frame. This view was widely and long held and shared to a degree by Mrs. Somerville herself: she believed her injudicious encouragement of her oldest daughter's intel-

lectual precocity had been a factor in the child's death at age ten. In the late 1790's Captain Fairfax's strictures against arduous mental effort, combined with outspoken criticism of "unwomanly behavior" by aunts and female cousins, drove Mary Fairfax to secret, intermittent application to mathematics but sharpened her resolve to learn the subject.

In May 1804 she married a cousin, Samuel Greig, commissioner of the Russian navy and Russian consul general in Great Britain. Greig's father, Admiral Sir Samuel Greig, a nephew of her grandfather Charters, had been one of five young British naval officers who, at the request of Empress Catherine II, went to Russia in 1763 to reorganize her navy and had been chiefly responsible for the success of that undertaking. He and his English wife reared their children in Russia: their sons made careers in Russian service, and their daughters married into Russian families, but ties with Britain were never broken. Young Samuel Greig, a captain in the Russian navy, trained aboard Admiral Fairfax's ship and, on his marriage to the admiral's daughter, was given an appointment in London, where he and his bride lived until his death in September 1807, at the age of twenty-nine. For Mary Greig, this period in London—away for the first time from family and Scotland—was a difficult one. She was much alone, and although she could read and study more freely than ever before, her husband had, in her words, "a low opinion of the capacity of . . . [the female] sex, and had neither knowledge of nor interest in science of any kind." After his death she returned with their two young sons to her parents' home in Scotland.

With the newly acquired independence of widowhood and a modestly comfortable fortune, she set out openly to educate herself in mathematics, ignoring the ridicule of relations and acquaintances. The greater part of each day was occupied with her children, and her evenings with social and filial obligations; yet she read Newton's *Principia* and began the study of higher mathematics and physical astronomy. Moreover, she found help and encouragement among Edinburgh intellectuals. John Playfair gave her useful hints on study. A group of young Whigs in her social circle, among them Henry Brougham, Francis Jeffrey, the Horner brothers, and Sydney Smith—who had some years earlier launched the successful *Edinburgh Review* and who urged, as social reforms, widened educational opportunities—became and remained her champions, finding in this pretty, quiet, and liberal-minded young widow the capacity and zest for learning that they asserted for her sex. Mrs. Somerville's most helpful mentor in these days was William Wallace, later professor of mathematics at Edinburgh. Wallace advised her by correspondence and rewarded her efforts with a silver medal for solving, in his *Mathematical Repository*, a prize problem, the first of many awards she would receive over the next sixty years.

In May 1812 Mary Greig married, as his second wife, her first cousin, William Somerville, son of the historian and minister at Jedburgh, Thomas Somerville. A cosmopolitan army doctor who had served in Canada, Sicily, and South Africa, and an affable, generous, and intelligent man of liberal convictions, William Somerville, from the first, staunchly supported his wife's aspirations. Throughout their half century of marriage, until his death in 1860 in his ninetieth year, he was her invaluable aide, taking great pride and satisfaction in her fame. Soon after their marriage she began, at age thirty-three, to continue on her own a rigorous course of readings (laid out by Wallace) in French higher mathematics and astronomical science. At her husband's urging she also devoted an hour each morning to Greek and, when the young naturalist George Finlayson became tutor to her son Woronzow Greig, she commenced the systematic study of botany. Together she and Somerville interested themselves in geology and mineralogy under the casual tutelage of their friends John Playfair, Robert Jameson, and James Hall. Among their other Edinburgh intimates were Sir James Mackintosh, Sir Walter Scott, John Leslie, James Gregory, and David Brewster.

When in 1816 William Somerville was named to the Army Medical Board, the family moved from Edinburgh to London, their chief residence for the next twenty years. Through Scottish friends and connections they were immediately introduced into the best intellectual society of the British capital, where they were soon popular figures. Dr. Somerville became a Fellow of the Royal Society. Mrs. Somerville's mathematical and scientific pursuits made her a minor lioness. In 1817 on their first tour abroad, Biot and Arago, who had been charmed by her in London, introduced them to Laplace, Gay-Lussac, Bouvard, Poisson, Cuvier, Haüy, and other French savants, who received them both as colleagues, entertained them during their stay in Paris, and afterward maintained friendly correspondences with the couple. In Switzerland the Somervilles were welcomed by Candolle, de La Rive, Prevost, and Sismondi; and in

Italy by the English colony and various celebrated Italians. Many of Mrs. Somerville's friendships with scientific, literary, and political personages date from this time.

In London their familiar circle included the Henry Katers, the Thomas Youngs, the Alexander Marcets (Mary Somerville and Jane Marcet always held each other in affectionate esteem), Sir Humphry and Lady Davy (whom Mrs. Somerville had known in Edinburgh as Mrs. Apreece), the poets Thomas Campbell, Samuel Rogers, and Thomas Moore, Maria Edgeworth, John Allen, the Misses Berry, Harriet Martineau, Joanna Baillie and her family, Francis Chantrey, John Saunders Sebright, Henry Warburton, and, above all, William Hyde Wollaston. From such natural philosophers she learned science directly, as they discussed their latest findings, described and demonstrated their newest apparatus, and shared their enthusiasms and ideas with fellow guests at the small, convivial gatherings typical of the day. In an age of gentleman amateurs, Mrs. Somerville's informal apprenticeship to these scientific masters was in many respects identical with the nurture of male scientists.

Her first paper, a report in *Philosophical Transactions* of some experiments she designed and carried out on magnetizing effects of sunlight, appeared in 1826; it was communicated to the Royal Society by her husband. This work, widely praised and accepted for some years, although its conclusions were later disputed and disproved, had a vitalizing effect on investigations of the alleged phenomenon. Ten years later Arago presented to the Académie des Sciences an extract from one of her letters as a paper entitled "Experiments on the Transmission of Chemical Rays of the Solar Spectrum Across Different Media," which appeared in the *Comptes rendus*. Her third and final experimental paper, "On the Action of Rays of the Spectrum on Vegetable Juices" (1845), came out in *Abstracts of the Philosophical Transactions*. All her experimental work is characterized by rationality in approach, delicacy and simplicity in execution, and clarity in presentation. Her only essay for a popular journal was a long one on comets in the December 1835 issue of *Quarterly Review*, soon after Halley's comet had been seen.

Research, writing, and study were always unobtrusively carried on in the midst of a full social life and numerous maternal and domestic responsibilities. In 1824 William Somerville was appointed physician to the Royal Hospital, Chelsea; and, with three children, they moved to Chelsea College, on the outskirts of London. Margaret, their oldest daughter, had died the previous year. In 1814 they had lost two children: the younger Greig boy at the age of nine, and their own only son, an infant. Of Mrs. Somerville's remaining offspring, her son Woronzow Greig, a successful and esteemed barrister, graduate of Trinity College, Cambridge, died in 1865 at the age of sixty, while her two daughters, Martha and Mary Somerville, both unmarried, survived their mother. She herself supervised the education of the girls, determined that they should not lack, as she did, for systematic learning. At the behest of Lady Byron, widow of the poet, Mrs. Somerville directed also the early mathematical studies of Ada Byron, later Lady Lovelace; the Byron and Somerville families were, from the mid-1820's onward, intimate friends.

In 1827 Henry Brougham wrote William Somerville to ask him to persuade his wife to put Laplace's *Mécanique céleste*, which she had studied in Edinburgh, into English for the Library of Brougham's Society for the Diffusion of Useful Knowledge. Unsure of her ability, she finally gave in to their urgings, provided the manuscript, if unsatisfactory, would be destroyed. A rendition rather than mere translation—since full mathematical explanations and diagrams were added to make Laplace's work comprehensible to most of its English readers—her treatise when completed in 1830 was too long for the Library series. Somerville, however, submitted it to their great friend J. F. W. Herschel, who urged the publisher John Murray to bring it out. Dubious of the success of a book on such a subject, Murray printed 750 copies in 1831. To his and Mrs. Somerville's amazement, *The Mechanism of the Heavens* sold well and won praise for her. It was put to use in advanced courses at Cambridge. Its preface—the necessary mathematical background—was reprinted in 1832 as *A Preliminary Dissertation on the Mechanism of the Heavens*; both it and the previous volume were immediately pirated in the United States and were used in Britain as textbooks for almost a century. The Royal Society hailed the work by voting to place a portrait bust of Mrs. Somerville in their Great Hall. Acclaim came also from the younger generation of British scientists, including Babbage, Brewster, Buckland, Faraday, Herschel, Lyell, Murchison, Sedgwick, and Whewell, who gave her the same unstinting admiration, respect, and assistance their elders had bestowed, regarding her as the spokeswoman for science and offering her honors and opportunities unique for a woman.

Thus at age fifty-one Mary Somerville embarked on a professional career as a scientific expositor.

Her gift for clear and cogent explanation, a quick and lively mind, access to the best scientific thought of the times, patience and perseverance, together with a sweet simplicity and charm of manner and a "womanliness," which demonstrated that learning and comfortable domesticity could successfully be combined, sustained this career for the next four decades. Mrs. Somerville was fortunate too in her times: industrialization had popularized notions of self-help, expanding opportunities, changes, and new freedoms. Her second book, *On the Connexion of the Physical Sciences*, a synthetical consideration of the mutual dependence of the physical sciences, came out in 1834 to even greater acclaim. The Royal Astronomical Society (1835) elected her and Caroline Herschel their first female honorary members. Sir Robert Peel awarded her a civil pension of £200 annually (later increased to £300 by Lord Melbourne) in recognition of her work. The Royal Academy of Dublin (1834), the Société de Physique et d'Histoire Naturelle (Geneva, 1834), and the Bristol Philosophical Institution (1835?) voted her honorary memberships. In the ten editions of this work Mrs. Somerville put forward the newest, most penetrating, and authoritative ideas and practices, avoiding fads and gimcrackery. Clerk Maxwell, forty years afterward, classed the work as one important in advancing scientific thought through its insistence on viewing physical science as a whole. J. C. Adams attributed his first notions about the existence of Neptune to a passage he read in its sixth edition.

In the late 1830's her husband's health failed, and the family migrated to Italy, where Mrs. Somerville spent the remaining thirty-six years of her long life, a valued guest and brilliant part of the Italian scene during the Risorgimento. Not only was she offered access to Italian libraries and scientific facilities, but she was given membership in six of the leading Italian scientific societies (1840–1845). Although, as the years passed, it became more and more difficult to stay abreast of British science, she managed through letters, visits, journals, and books to keep in touch with major developments. In 1848, at age sixty-eight, she published her third and most successful book, *Physical Geography*, a subject which had always interested her deeply. Its seven editions brought her numerous honors: the Victoria Gold Medal of the Royal Geographical Society (1870); election to the American Geographical and Statistical Society (1857), to the Italian Geographical Society (1870), and to five additional provincial Italian societies (1853–1857); several medals; and praise from Humboldt. In this book, as in the *Connexion of the Physical Sciences*, Mrs. Somerville strongly endorsed the new geology of Lyell, Murchison, Buckland, and their school—a stand that brought her some public criticism.

Twenty-one years later, when she was eighty-nine, her final work, *On Molecular and Microscopic Science*, appeared in two volumes. It deals with the constitution of matter and the structure of microscopic plants. At this date its science was considered old-fashioned, but young John Murray published it out of loyalty to and affection for its author, on the recommendation of Sir John Herschel, who had also been instrumental in persuading Mrs. Somerville to bring out her *Physical Geography*. The public received it with kindly interest and deference to its venerable creator. In the same year she was made a member of the American Philosophical Society (she had warm regard for Americans) and completed her autobiography—a vivid and spritely account of her life in Scotland, England, and Italy; of her visits to Switzerland, France, and Germany; and of the many interesting personages she had known. After her death her elder surviving daughter, Martha, published parts of this manuscript as *Personal Recollections From Early Life to Old Age of Mary Somerville* (1873).

In her later years Mrs. Somerville gave powerful but always temperate support to the cause of the education and emancipation of women. Hers was the first signature on John Stuart Mill's great petition to Parliament for women's suffrage, solicited by Mill himself. An early advocate of higher education for women, many of her books were given after her death to the new Ladies College at Hitchin (now Girton College, Cambridge). Somerville College (1879), one of the first two colleges for women at Oxford, is named after her. Although frail and deaf in her last years, Mary Somerville's spirit and intelligence, her interest in friends, in the cause of women, and in science never faltered. At the time of her death, at ninety-two, she was revising a paper on quaternions.

BIBLIOGRAPHY

I. ORIGINAL WORKS. The Somerville Collection (MSS, papers, letters, documents, diplomas, and memorabilia owned by Mrs. Somerville's heir) is deposited in the Bodleian Library, Oxford.

Mary Somerville's works are "On the Magnetizing Power of the More Refrangible Solar Rays," in *Philosophical Transactions of the Royal Society*, **116** (1826), 132; *The Mechanism of the Heavens* (London, 1831); *A Preliminary Dissertation on the Mechanism of the Heav-*

ens (London, 1832); *On the Connexion of the Physical Sciences* (London, 1834; 2nd ed., 1835; 3rd ed., 1836; 4th ed., 1840; 5th ed., 1842; 6th ed., 1846; 7th ed., 1848; 9th ed., 1858; 10th ed., A. B. Buckley, ed., 1877); "Art. VII.-1. Ueber den Halleyschen Cometen . . .," in *Quarterly Review*, **105** (1835), 195–233; and "Experiments on the Transmission of Chemical Rays of the Solar Spectrum Across Different Media," in *Comptes rendus hebdomadaires des séances de l'Académie des sciences*, **3** (1836), 473–476.

See also the extract from a letter by Mrs. Somerville to Sir John Herschel, Bart., F.R.S., dated Rome, 20 September 1845, entitled "On the Action of the Rays of the Spectrum on Vegetable Juices," in *Philosophical Transactions of the Royal Society*, **5** (1845), 569; *Physical Geography* (London, 1848; 2nd ed. 1849; 3rd ed., 1851; 4th ed., 1858; 5th ed., 1862; 6th [1870] and 7th [1877] eds. revised by H. W. Bates); and *On Molecular and Microscopic Science* (London, 1869).

II. SECONDARY LITERATURE. On Mary Somerville and her work, see the following works by Elizabeth C. Patterson: "Mary Somerville," in *British Journal for the History of Science*, **4** (1969), 311–339; "A Washington Letter," in *Bodleian Library Record*, **8** (1970), 201–205; and "The Case of Mary Somerville: An Aspect of Nineteenth-Century Science," in *Proceedings of the American Philosophical Society*, **118** (1974), 269–275. See also Martha Somerville, ed., *Personal Recollections From Early Life to Old Age of Mary Somerville, With Selections From Her Correspondence* (London, 1873).

ELIZABETH C. PATTERSON

SOMMERFELD, ARNOLD (JOHANNES WILHELM) (*b.* Königsberg, Prussia [now Kaliningrad, U.S.S.R.], 5 December 1868; *d.* Munich, Germany, 26 April 1951), *theoretical physics.*

Sommerfeld's father, Franz Sommerfeld (1820–1906), had been married to Cäcile Matthias (1839–1902) six years when his son Arnold Johannes Wilhelm was born. Franz Sommerfeld had himself been born and raised in Königsberg, where his father, Friedrich Wilhelm Sommerfeld (1782–1862), had been Hof-Post-Sekretär. The family was Protestant; and although Sommerfeld was not religious, he never renounced his faith. "My father, the practicing physician . . . , was a passionate collector of natural objects (amber, shells, minerals, beetles, etc.) and a great friend of the natural sciences"; he was also a member of the semipopular Physikalisch-Ökonomische Gesellschaft in Königsberg. "To my energetic and intellectually vigorous mother I owe an infinite debt," Sommerfeld also acknowledged in 1917 in his autobiographical sketch.[1] At the humanistic Altstädtisches Gymnasium (Collegium Fridericianum) in 1875–

1886, where Hermann Minkowski and Max and Willy Wien were a few years ahead of Sommerfeld, "I was almost more interested in literature and history than in the exact sciences; I was equally good in all subjects including the classical languages."

Passing the *Abitur* at the end of September 1886, Sommerfeld matriculated immediately at the University of Königsberg. "After some irresolution" he opted for mathematics but heard lectures on philosophy and political economy, as well as natural sciences. Active participation in fraternity life (Burschenschaft Germania), with its compulsory drinking bouts and fencing duels, prevented systematic and concentrated study in his first few years at the university. His instructors in mathematics were David Hilbert, *Privatdozent*; Adolf Hurwitz, *extraordinarius*, and Ferdinand Lindemann, *ordinarius*. To the latter Sommerfeld expressed particular and continuing thanks in his doctoral dissertation, "Die willkürlichen Funktionen in der mathematischen Physik,"[2] which "I conceived and wrote out in a few weeks" during the summer of 1891.

The dissertation was indeed but an exposition of the general mathematical foundation for a harmonic analyzer that Sommerfeld and Emil Wiechert, who in those years served Sommerfeld "as the highest model of a deep mathematical-physical thinker," had conceived and constructed in 1890 at the institute of Paul Volkmann, professor of theoretical physics. Their mechanical instrument was, moreover, only part of a comprehensive attack on the problem of interpreting the earth-thermometer observations at the meteorological station in Königsberg, which had been set as a prize question by the Physikalisch-Ökonomische Gesellschaft. The analyzer would reduce the observed temperature curve (arbitrary function) to a trigonometric series; this same series and these same numerical coefficients must then be shown to result from a solution of the heat conduction equation with the appropriate boundary conditions. Sommerfeld tackled this latter problem alone; and although he was not entirely successful, he developed the methods that were to underlie his most important scientific work in the following decade—the application of the theory of functions of a complex variable to boundary-value problems, especially diffraction phenomena.

Although he hoped for a university career, Sommerfeld, as was customary, spent the following academic year preparing for examinations to qualify as a Gymnasium teacher of mathematics and physics. Then, in the autumn of 1892, not yet twenty-four, small in stature, and still very youthful in appearance—but with virility attested by a long fencing scar on the forehead—he entered upon his year of obligatory military service, choosing his reserve regiment in Königsberg. Discharged in September 1893, Sommerfeld, at his own option, participated in eight-week military exercises in 1894, 1896, 1898, 1901, and 1903, in the latter three as lieutenant. Despite his squat build, by middle age, with the aid of a turned-up waxed moustache, he managed to give the impression of a colonel of the hussars.

Drawn to Göttingen as "the seat of mathematical high culture," Sommerfeld first obtained, through personal connections, an assistantship in the Mineralogical Institute from October 1893 to September 1894. During the following two years he was Felix Klein's assistant, managing the mathematical reading room and writing out Klein's lectures for the use of the students. "Consciously and systematically Klein sought to enthrall me with the problems of mathematical physics, and to win me over to his conception of these problems as he had developed it in lecture courses in previous years. I have always regarded Klein as my real teacher, not only in things mathematical, but also in mathematical physics and in my conception of mechanics." In particular, although continuing the line of research in mathematical physics that he had begun at Königsberg, Sommerfeld recognized that Klein's program for applying analytical mechanics, and higher mathematics generally, to engineering problems promised manifold mutual advantages.

In March 1895, Sommerfeld became a *Privatdozent* in mathematics at Göttingen, presenting as his *Habilitationsschrift* the first exact solution of a diffraction problem, which he gave as a complex integral in closed form suitable for numerical evaluation. Henri Poincaré immediately adopted "my 'méthode extrêmement ingénieuse,'" and in the following decades the reduction of a problem in mathematical physics to the evaluation of a complex integral became Sommerfeld's hallmark. Sommerfeld gave an account of this work in September at the Lübeck Naturforscher-Versammlung, where he had a ringside seat at the bullfight in which the agile Wilhelm Ostwald was charged by Ludwig Boltzmann; he sided with the Bull.

Sommerfeld lectured on advanced topics at Göttingen for five terms before accepting a full professorship in mathematics at the Bergakademie in Clausthal in October 1897. Although the teaching was elementary, the salary allowed him to marry. His bride, Johanna Höpfner, was the daughter of the new *Kurator* of the University of Göttingen, Ernst Höpfner, a close associate of Friedrich Althoff and an enthusiast for Klein's schemes.[3] The distance between Clausthal and Göttingen was short enough for Sommerfeld to remain in close contact with the university and Klein.

At Clausthal, Sommerfeld applied his extraordinary ingenuity in boundary-value problems to the propagation of electromagnetic waves along wires of finite diameter (obtaining the first rigorous solution) and to the diffraction of X rays by a wedge-shaped slit. Both calculations were of considerable interest to experimental physicists at that time. The collaboration with Klein on *Theorie des Kreisels* (1897–1910), which grew out of Klein's lectures in 1895–1896 and became a thousand-page treatise, continued at Clausthal. Sommerfeld also undertook the editorship of the physics volume of the *Encyklopädie der mathematischen Wissenschaften*, initiated and directed by Klein. The last part of this multivolume "volume" was not issued until 1926.

In April 1900, as the result of Klein's energetic wire-pulling, Sommerfeld became full professor at the Technische Hochschule of Aachen—significantly, however, not in mathematics but in technical mechanics. Sommerfeld was expected to show in his courses, as well as in his research, that even this classical engineering discipline could be developed on a consistent mathematical foundation: "Although my Aachen colleagues and students at first regarded the 'pure mathematician' with suspicion, I soon had the satisfaction of being accepted as a useful member not merely in teaching but also in engineering practice; thus I was requested to render expert opinions and to participate in the Ingenieurverein." There resulted fruitful collaborations with several theoretical engineers—with August Föppl on problems of resonance phenomena in the vibration of bridges, with Otto Schlick on the analogous phenomena in ships, and with August von Borries on problems of locomotive construction. Of fundamental importance, however, were Sommerfeld's investigations of the hydrodynamics of viscous fluids, aiming at an explanation of the onset of turbulence and a theory of the lubrication of machines.

Recognition was not withheld by the engineers. Sommerfeld declined a highly complimentary offer of the chair of mathematics and technical mechanics at the Berlin mining academy. In 1903 the council of the Gesellschaft Deutscher Naturforscher und Ärzte invited him to deliver one of the plenary addresses at the Kassel congress. There he pointed to the felicitous collaboration between engineering and mathematics that he had done so much to initiate and that appeared—characteristically, but erroneously—destined to absorb not merely most of Sommerfeld's future efforts but also much of the attention of physicists and mathematicians in general.

At the same time Sommerfeld, with his tremendous capacity for work, continued in mathematical physics and joined the advancing front of fundamental physical research with a series of extensive papers intended to provide a general dynamics of electrons, with special attention to motion faster than the speed of light. In this area his urgent need of discussion to clarify his thoughts could not be satisfied by technical colleagues; instead two bright engineering students, Peter Debye and Walter Rogowski, were invited to dinner two or three times a week and afterward were talked at for two or three hours in Sommerfeld's study.[4]

Although the electron theory papers of 1904–1905 were soon rendered utterly passé by relativity (regaining some interest and currency only after the discovery of Čerenkov radiation), they made Sommerfeld a name among the most advanced theoretical physicists—Boltzmann, Lorentz, Wilhelm Wien. In the summer of 1906 this growing reputation brought a call to the chair of theoretical physics at the University of Munich. It was only under pressure from Roentgen, then professor of experimental physics at the university, that this chair, one of the very few in the field, had recently been funded after having been defunct for several years. Curiously, Sommerfeld's appointment was opposed by Ferdinand Lindemann, now professor of mathematics at Munich, who was hostile to the electron theory in all its various forms and was disturbed by the want of mathematical rigor in its development.

At Munich an institute was established for Sommerfeld—a dozen rooms were fitted up for collections, seminars, assistants, and experimental work. Determined to check his own theories, Sommerfeld directed a considerable program of experimental research—even experimental doctoral dissertations. In the spring of 1912 his experimental assistant, Walter Friedrich, using covertly the facilities of the institute, discovered the diffraction of X rays by crystals.[5]

Sommerfeld always had a very ambitious conception of what he had to offer in his courses: the most recent results of research. Now, as a professor of theoretical physics, he felt obliged to work his way intensively into all the important problems of modern physics. At the September 1907 *Naturforscherversammlung* he defended Einstein's relativity theory—thus placing himself, after Planck, among the earliest converts. In subsequent publications he cast the theory into vector form (1910) and applied it to various problems. One of the most striking applications was the prediction of a forward shift and narrowing of the direction in which an electron decelerated from relativistic velocities emits the greatest amount of energy (distribution of *Bremsstrahlung*).

Sommerfeld met Einstein for the first time at the September 1909 *Naturforscherversammlung* in Salzburg. Despite the great difference in background and talents of the two men, they felt an immediate attraction—"a magnificent fellow" was Einstein's reaction.[6] At Salzburg the subject of the liveliest and most urgent interest was not, however, relativity but the quantum theory. Einstein pressed his radical view of a radiation field containing discrete atoms of light, while Planck and virtually all his colleagues resisted this revolutionary break with Maxwell's electrodynamics. Sommerfeld, accepting Planck's view that one must proceed as conservatively as possible, had in fact been led to his discovery of the forward shift of the *Bremsstrahlung* maximum while seeking an alternative explanation for a group of phenomena in which Johannes Stark, one of the very few advocates of light quanta, had seen strong evidence for the radical view.

During the year following the Salzburg *Naturforscherversammlung*, however, Sommerfeld gradually became convinced of the fundamental importance of the quantum and spent a full week with Einstein at Zurich "in order to parley over the problem of light and a few questions in the relativity theory. His presence was a real festival for me," Einstein reported, especially pleased at the extensive concessions that Sommerfeld made to his views on quantum statistics.[7] Influential in this reorientation, as in so many other shifts of Sommerfeld's scientific opinion, were the work and enthusiasms of his students and assistants—in this case especially Peter Debye and Ludwig Hopf.

It was, however, only after the announcement of the Compton effect in 1922–1923 that Sommerfeld, or his colleagues, accepted Einstein's literally particulate structure of light even tentatively. Thus in 1910–1912, before the introduction of Bohr's theory, Sommerfeld sought to add to the classical Maxwell-Lorentz theory a formal postulate regulating the interaction of atoms and electromagnetic radiation—a postulate that, although in no way demanded or suggested by the classical theory, was also, in contrast with Einstein's and Bohr's postulates, not inconsistent with it. This postulate, that the "action" (the integral of the energy with respect to the time over the duration of the interaction) is always equal to Planck's constant $h/2\pi$, Sommerfeld applied to the production of *Bremsstrahlung*, and to the inverse phenomenon, the photoelectric effect.

Sommerfeld placed great importance upon this work, and his presentations at the first Solvay Congress (October 1911) and elsewhere attracted considerable attention. Although it led nowhere and had been abandoned by the end of 1913, it had nonetheless most effectively emphasized two points of view that were adopted in the more fruitful efforts of J. W. Nicholson and Bohr: it is the action primarily, and the energy only secondarily, that is quantized; the ubiquity of an *h* in the interactions of atoms and radiation is not to be regarded as a secondary expression of the size, structure, and internal energy of atoms, "but rather the existence of molecules [atoms] is to be regarded as a function and consequence of the existence of an elementary quantum of action."[8]

The breakthrough came then in the summer of 1913, with the appearance of Niels Bohr's first paper on the "Constitution of Atoms and Molecules." Sommerfeld studied the paper immediately and closely, for, as he wrote Bohr early in September, "the problem of expressing the Rydberg-Ritz constant [in the exceedingly precise yet empirical formulas for the frequencies of the spectral lines] by means of Planck's *h* has been with me for a long time. I discussed it with Debye a few years ago. Even though I remain for the present in principle somewhat skeptical toward atomic models still your calculation of that constant is undoubtedly a great contribution."[9] And he closed by courteously announcing that he would like to try applying Bohr's model to the Zeeman effect (the splitting of spectral lines emitted in a magnetic field).

That application, as Bohr himself discovered, was not as simple as it appeared. Sommerfeld found himself obliged first to find a generalization of the various quantization prescriptions that could be applied to mechanical systems with more than one degree of freedom. In the winter semester of 1914–1915 he was already lecturing to his students on the astonishing initial results of this investigation: a quantitative theory of the fine structure of the spectral lines of hydrogen and of the X-ray spectra of the heavy elements, regarded as arising from the relativistic increase in mass of an electron by an amount depending upon the eccentricity of its orbit.[10] It was only in the spring of 1916, however, that Sommerfeld found the definitive formulation of his quantization rules yielding a quantum theory of the normal Zeeman effect and, in the hands of his student Paul S. Epstein, of the Stark effect (the splitting of spectral lines emitted in an electric field).[11] In the course of this work Sommerfeld entered into (and afterward maintained) very close contact with experimental spectroscopists, especially Friedrich Paschen, with whom he exchanged some fifty letters within six months in 1916.

This extraordinary extension, enrichment, and precision of Bohr's theory by Sommerfeld contributed decisively to its rapid and widespread acceptance. Only five years after Bohr's first publication Sommerfeld, recognizing that the mathematical development of this quantum-theoretical atomic model had reached a conclusion of sorts, undertook a comprehensive exposition of the field. His *Atombau und Spektrallinien*, of which the first edition appeared late in 1919, immediately became the bible of atomic physics and its successive editions, appearing almost annually in the early 1920's, chronicled the progress of this field up to the eve of the introduction of quantum mechanics.

In these years, 1919–1926, Sommerfeld remained in the forefront of theoretical atomic physics; but he did so by largely reorienting his method and approach. Persuaded that the detailed structure of the spectra of atoms with more than one electron—and the close contact with current experimental work that he valued so highly—could not be obtained deductively by calculations from first principles, Sommerfeld pioneered a new style of theoretical spectroscopy. In this a posteriori approach, in contrast with the older a priori, the theorist began by immersing himself in the spectroscopic data, and worked back, by means of the combination principle, to the atomic energy levels. These levels he then tried to characterize by quantum numbers and selection rules—on the basis of established mechanical and quantal laws if possible, or, if not, ad hoc. Thus where Sommerfeld had previously spoken of "numerical harmonies" in the quantum theory, he now began to speak of "number mysteries" (1919)—in the first instance, and most particularly, in the Zeeman effect. An adequate understanding of this phenomenon, and of the complex structure of spectral lines which was so intimately connected with it, was then widely regarded as the specific content or contribution of

a satisfactory atomic mechanics. Consequently the success of Sommerfeld and his students in the ordering of X-ray, atomic, and molecular spectra was followed with excitement and widely imitated. This approach did not, however, prove to be what it was then widely supposed to be, namely the highroad to quantum mechanics. Still, the results obtained were taken over with but slight alteration into the post-1925 quantum-mechanical theory of atomic structure.

Although not among the inventors of quantum mechanics, of quantum statistics, or of electron spin, Sommerfeld immediately became one of the most adept in the exploitation of these new concepts and prescriptions for the calculation of energies and rates of atomic processes, and the macroscopic properties of matter resulting from them. It was Schrödinger's form, the wave mechanics, the partial differential equation, that Sommerfeld found most congenial. In 1929 he published one of the first textbooks of wave mechanics, the *Wellenmechanischer Ergänzungsband* to *Atombau und Spektrallinien*. That favorite phenomenon, the relativistic forward shift of the *Bremsstrahlung* maximum, was recalculated with wave mechanics; and the reciprocal phenomenon, the distribution of photoelectrons, was given considerable attention. But in these years, and into the early 1930's, the problem that drew most of Sommerfeld's interest was the joint application of wave mechanics and Fermi statistics to the behavior of electrons in metals. With the aid of his students—especially Hans Bethe—Sommerfeld rehabilitated the electron theory of metals, which, after a promising beginning at the turn of the century, had languished under classical statistics and mechanics.

"What I especially admire about you," Einstein wrote to Sommerfeld in January 1922, "is the way, at a stamp of your foot, a great number of talented young theorists spring up out of the ground."[12] In the twenty-five years following his arrival in Munich—the period in which theoretical physics became a recognized, indeed glamorous, subdiscipline—Sommerfeld had more advanced students and turned out more doctorates than any other theorist. The near-monopoly that he held for the first fifteen of these years was seriously challenged only after Max Born arrived at Göttingen in 1921. The first, prewar, generation of doctorates included (in order of seniority) Peter Debye, Ludwig Hopf, Wilhelm Lenz, P. P. Ewald, Paul S. Epstein, Alfred Landé; the second, early postwar, generation included Erwin Fues, Gregor Wentzel, Wolfgang Pauli, Werner Heisenberg, Helmut Hönl, Otto

Laporte; the third, postquantum-mechanical, generation included Hans Grimm, Albrecht Unsöld, Walter Heitler, Hans Bethe, Herbert Fröhlich. To this latter group must be added the American postdoctoral students then flocking to Germany. Partly in consequence of Sommerfeld's visits to the United States (September 1922–April 1923; January–May 1929, as part of a trip around the world begun in October 1928; June–August 1931), the Americans made a point of spending some time in Munich; and several of them (Carl Eckart, William Houston, N. H. Frank) collaborated in Sommerfeld's work on the electron theory of metals.

Sommerfeld took real pleasure in the company of his students, at least of those who had shown the requisite talent and *Sitzfleisch*. With a disregard of social distance almost unheard of before the war, Sommerfeld took his students on strenuous outings in the Bavarian Alps. These occasions too were used for vigorous discussions of the physics that filled Sommerfeld's life and that he insisted be the exclusive intellectual occupation of his students as well. With them he discussed not merely his own and their own work, but also the news that his extensive correspondence and travels brought him. His liberality and enthusiasm for new results were not always welcomed by Sommerfeld's colleagues, who often saw their brain children nostrified and propagated in his conversations, lectures, and papers, or exploited by his protégés.

Although Sommerfeld never received a Nobel Prize, from 1917 on, a steady stream of honors — prizes, memberships in foreign academies, honorary doctorates — flowed to him. The most valuable of these marks of recognition were the offers of the chairs of theoretical physics at the University of Vienna in 1917 (as successor to Hasenöhrl) and at the University of Berlin in the spring of 1927 (to succeed Planck). The first brought the title *Geheimrat* and a substantial increase in salary; the second brought a great deal of publicity, a doubling of his institute budget, and a far larger increase in personal income.

In return for the compliment Sommerfeld had paid his university by refusing the call to Berlin, it was anticipated that his colleagues would elect him rector of the university for 1927–1928. But grotesque as it may seem, this native of East Prussia, for whom the "Prussian virtues" — devotion to duty and love of the fatherland — had always been the norms of thought and action, was regarded by his colleagues as insufficiently "national" for this post. Properly patriotic as a Burschenschaftler, a

member of the National Liberal Party while at Aachen, a (moderate) annexationist during the war, disgusted and despairing at the revolution, Sommerfeld nonetheless compromised himself irreparably in the Weimar period by his manifest distaste for anti-Semitism, by openly siding with Einstein, by want of intransigence in international scientific relations, by accepting a visiting professorship at the University of Wisconsin in 1922–1923, and by favoring a political party (Deutsche Demokratische Partei) committed to parliamentary democracy. Right-wing groups put forward an opposition candidate with "reliable, national convictions"; and in the election, 16 July 1927, Sommerfeld was defeated 68–50.

Sommerfeld's progress away from the antidemocratic chauvinism in which the great majority of German academics were mired had begun at age fifty; at sixty-five, after fifteen months of Hitler's regime, he noted in the draft of a letter to Einstein: "Moreover I can assure you that the misuse of the word 'national' by our rulers has thoroughly broken me of the habit of national feeling that was so pronounced in my case. I would now be willing to see Germany disappear as a power and merge into a pacified Europe."[13]

As early as 1915 Johannes Stark had labeled Sommerfeld the "energetic executive secretary" of the "Jewish and philo-Semitic circle" of mathematicians and theoretical physicists,[14] and Stark's enmity grew more intense and more open in the Weimar period as Sommerfeld continued to protect the interests of these circles and to frustrate Stark's ambitions. At the Nazi take-over, Stark immediately attained positions of power and influence; and he sought to use them to extirpate, root and branch, the "Jewish" spirit in German physics. A tug-of-war now developed over Sommerfeld's chair, for in the spring of 1935 he passed the obligatory retirement age and continued to function as professor only provisionally, from semester to semester, pending appointment of a successor. The faculty took the position that only a theorist of first rank, if possible from Sommerfeld's own school, could maintain the tradition and placed Werner Heisenberg at the top of its list. This choice was resisted strenuously by the advocates of a "German" physics. In July 1937 an article in the magazine of the SS labeled Sommerfeld and Heisenberg, among others, "white Jews of science" and "agents of Judaism in German intellectual life" who will have to "disappear just like the Jews themselves."[15]

Stark himself added an unreserved endorsement

of the article, although he was more discreet in his rhetoric and charges. Sommerfeld, and other physicists as well, entered official protests; and as Stark went into eclipse, Heisenberg's appointment seemed assured. But the "German physics" faction won the final round; in 1940 Sommerfeld received, as he himself said, "the worst conceivable successor," Wilhelm Müller, one of the stalwarts of the movement.

Through the war Sommerfeld occupied himself with the preparation for publication of his six-semester cycle of lectures on theoretical physics. At its end, now approaching eighty, he resumed the directorship of the Institute of Theoretical Physics—but not his lectures—for several years. Early in April 1951, while strolling with his grandchildren, he was struck by an automobile and died a few weeks later.

NOTES

1. "Autobiographische Skizze," in *Gesammelte Schriften*, IV, 673–682; unless otherwise indicated, all first-person quotations are from this sketch.
2. *Gesammelte Schriften*, I, 1–76; henceforth, where a bibliographic citation is not given for a piece of scientific work, the publication in question has been reprinted in Sommerfeld's *Gesammelte Schriften*.
3. Karl-Heinz Manegold, *Universität, technische Hochschule und Industrie . . .*, Schriften zur Wirtschafts- und Sozialgeschichte, XVI (Berlin, 1970), 164–165.
4. P. Debye, interviews, 3 and 4 May 1962, at Archive for History of Quantum Physics.
5. Paul Forman, "The Discovery of the Diffraction of X-Rays by Crystals; a Critique of the Myths," in *Archive for History of Exact Sciences*, 6 (1969), 38–71.
6. Einstein to J. J. Laub, 31 Dec. 1909, as quoted by Carl Seelig in *Albert Einstein, Leben und Werk eines Genies unserer Zeit* (Zurich, 1960), 145.
7. Einstein to Laub, Sept. [?] 1910, as quoted *ibid.*, 197.
8. "Das Plancksche Wirkungsquantum und seine allgemeine Bedeutung für die Molekularphysik," address at the Karlsruhe *Naturforscherversammlung*, Sept. 1911 in *Gesammelte Schriften*, III, 19.
9. German original published by L. Rosenfeld in his ed. of Niels Bohr, *On the Constitution of Atoms and Molecules* (Copenhagen–New York, 1963), lii.
10. John L. Heilbron, "The Kossel-Sommerfeld Theory and the Ring Atom," in *Isis*, 58 (1967), 451–485.
11. A. Hermann, ed., *Der Stark-Effekt*, Dokumente der Naturwissenschaft, Abt. Physik, VI (Stuttgart, 1965).
12. Albert Einstein and Arnold Sommerfeld, *Briefwechsel . . .* (1968), 98.
13. *Ibid.*, 114–115.
14. Quoted from the draft of a letter to the Prussian Education Ministry by A. Hermann, in *Sudhoffs Archiv . . .*, 50 (1966), 280.
15. *Das Schwarze Korps* (15 July 1937), 6.

BIBLIOGRAPHY

I. ORIGINAL WORKS. An excellent bibliography of Sommerfeld's publications and also of articles about or honoring him is included in Sommerfeld's *Gesammelte* [in fact, selected] *Schriften*, F. Sauter, ed. (Brunswick, 1968), IV, 683–728. The following significant omissions have been noted: "Die Überwindung der Erdkrümmung durch die Wellen der drahtlosen Telegraphie," in *Jahrbuch der drahtlosen Telegraphie . . .*, 12 (1917), 2–15; "En Ensartet Opfattelse af Balmers og Deslandres Serieled," in *Fysisk Tidsskrift*, 18 (1920), 33–40, and in the original German in *Arkiv für matematik, astronomi och fysik*, 15 (1921), 1–5; and "Spectroscopic Interpretation of the Magneton Numbers in the Iron Group," in *Physical Review*, 29 (1927), 208. Sommerfeld's printed remarks in discussion at scientific meetings are cited only irregularly, as are his newspaper articles.

Sommerfeld's literary remains—some 2,000 pages of lecture notes and some 1,000 letters, almost all to Sommerfeld—are at the Bibliothek des Deutschen Museums, Munich. Microfilms are available at the Archive for History of Quantum Physics, for which see T. S. Kuhn *et al., Sources for History of Quantum Physics* (Philadelphia, 1967), 87–89, where about 200 letters by Sommerfeld in various other collections are also listed. Sommerfeld's correspondence with Albert Einstein and with Johannes Stark, not listed by Kuhn *et al.*, has now been published: A. Einstein and A. Sommerfeld, *Briefwechsel. Sechzig Briefe aus dem goldenen Zeitalter der modernen Physik*, edited and annotated by A. Hermann (Basel–Stuttgart, 1968); and A. Hermann, "Die frühe Diskussion zwischen Stark und Sommerfeld über die Quantenhypothese," in *Centaurus*, 12 (1967), 38–59. Letters from Sommerfeld to Léon Brillouin and W. F. Meggers are included in the collections of the Niels Bohr Library, American Institute of Physics, New York; letters to Felix Klein are in the Klein-Nachlass, Niedersächsische Staats- und Universitätsbibliothek, Göttingen.

Sommerfeld's "Autobiographische Skizze," prepared in 1917 and supplemented in 1950, has been edited and amplified by Fritz Bopp: "Arnold Sommerfeld," in *Geist und Gestalt. Biographische Beiträge zur Geschichte der Bayerischen Akademie . . .*, II (Munich, 1959), 100–109, repr. in Sommerfeld's *Gesammelte Schriften*, IV 673–682. Sommerfeld also supplied the biographical data for his entry in the *Reichshandbuch der Deutschen Gesellschaft* (Berlin, 1931), 1802. Much additional biographical information has been drawn from the Universitätsarchiv, Munich (Akten des Rektorats, Pers. Akt. EII-N, "Sommerfeld, 1905 bis–   ") and from the Bayerisches Hauptstaatsarchiv, Munich (Abt. I, MK 35736, "Sommerfeld, Dr. Arnold").

II. SECONDARY LITERATURE. Numerous commemorative and obituary notices are cited in *Gesammelte Schriften*, IV, 723–727, of which the most detailed and reliable is Max Born, "Arnold Johannes Wilhelm Sommerfeld," in *Obituary Notices of Fellows of the Royal Society of London*, 8 (1952), 275–296, repr. in Born's *Ausgewahlte Abhandlungen*, II (Göttingen, 1963), 647–659. Publications omitted from or issued subsequent to the listing in the *Gesammelte Schriften* are "Arnold

Sommerfeld, Recipient of the 1948 Oersted Medal," in *American Journal of Physics*, **17** (1949), 312–314; Linus Pauling, "Arnold Sommerfeld: 1868–1951," in *Science*, **114** (1951), 383–384; Helmut Hönl, "Memoirs of Research on Zeeman Effect in Munich in Early 1920s," unpublished typescript at the Archive for History of Quantum Physics; P. P. Ewald, "Erinnerungen an die Anfänge des Münchener physikalischen Kolloquiums," in *Physikalische Blätter*, **24** (1968), 538–542; and A. Hermann, "Sommerfeld und die Technik," in *Technikgeschichte*, **34** (1967), 311–322; *Frühgeschichte der Quantentheorie* (Mosbach, 1969), esp. ch. 6; and "Arnold Sommerfeld," in *Die Grossen der Weltgeschichte*, IX (Munich–Zurich, 1970), 702–715.

The addresses at the Sommerfeld Centennial Memorial Meeting, Munich, 10–14 Sept. 1968, have been published in F. Bopp and H. Kleinpoppen, eds., *Physics of One- and Two-Electron Atoms* (Amsterdam–New York, 1969): F. A. Bopp, "Opening Address," 1–7; P. P. Ewald, "Arnold Sommerfeld als Mensch, Lehrer und Freund," 8–16; A. Hermann, "Sommerfeld's Role in the Development of Early Quantum Theory," 17–20; B. L. van der Waerden, "The History of Quantum Theory in the Light of the Successive Editions of Sommerfeld's *Atombau und Spektrallinien*," 21–31; H. Welker, "Impact of Sommerfeld's Work on Solid State Research and Technology," 32–43; and W. Heisenberg, "Significance of Sommerfeld's Work Today," 44–52.

P. Forman, "Alfred Landé and the Anomalous Zeeman Effect, 1919–1921," in *Historical Studies in the Physical Sciences*, **2** (1970), 153–261, discusses Sommerfeld's work on the Zeeman effect and includes several letters; and Roger H. Stuewer, "William H. Bragg's Corpuscular Theory of X-Rays and γ-Rays," in *British Journal for the History of Science*, **5** (1971), 258–281, publishes correspondence between Bragg and Sommerfeld.

PAUL FORMAN
ARMIN HERMANN

SOMMERING. See **Soemmerring, Samuel Thomas.**

SOMMERVILLE, DUNCAN MCLAREN YOUNG (*b.* Beawar, Rajasthan, India, 24 November 1879; *d.* Wellington, New Zealand, 31 January 1934), *mathematics.*

Sommerville, the son of Rev. James Sommerville of Jodhpur, India, was educated in Scotland, first at the Perth Academy, then at the University of St. Andrews, where he was awarded Ramsay and Bruce scholarships and in the mathematics department of which he served as lecturer from 1902 to 1914. During that time he met, and in 1912 married, Louisa Agnes Beveridge, originally

of Belfast, Ireland. From 1915 on Sommerville was professor of pure and applied mathematics at Victoria University College, Wellington, New Zealand. He was active in the Edinburgh Mathematical Society, to whose presidency he was elected in 1911. He helped to found the Royal Astronomical Society of New Zealand and became its first executive secretary. Sommerville presided over the mathematics section at the Adelaide meeting (1924) of the Australasian Association for the Advancement of Science. In 1928 the Institute (Royal Society) of New Zealand awarded him its Hector Medal.

Although primarily a mathematician, Sommerville was interested in other sciences, particularly astronomy, anatomy, and chemistry. Crystallography held special appeal for him, and crystal forms doubtless motivated his investigation of repetitive space-filling geometric patterns. Also, his abstract conceptions called for the construction of clarifying models, which revealed an artistic skill that was even more evident in his many watercolors of New Zealand scenes.

Sommerville contributed to mathematics both as a teacher and as an original researcher. His biographer, H. W. Turnbull, who considered him (in 1935) Scotland's leading geometer of the twentieth century, stated that his pedagogic style was scholarly, unobtrusive, and much appreciated at St. Andrews. One of his most distinguished pupils, A. C. Aitken, revealed that when the New Zealand University of Otago was without a mathematics professor, Sommerville willingly provided a sort of "correspondence course" in higher mathematics. Further evidence of his teaching ability is reflected in his four textbooks, which are models of deep, lucid exposition. Among them are *The Elements of Non-Euclidean Geometry* and *An Introduction to the Geometry of n Dimensions*, books whose titles indicate his two major research specialties and whose contents develop geometric concepts that Sommerville himself created. In addition to his texts, his *Bibliography of Non-Euclidean Geometry* is also a bibliography of *n*-dimensional geometry.

Sommerville wrote over thirty original papers, almost all on geometric topics. Notable exceptions were his 1928 "Analysis of Preferential Voting" (geometrized, however, in his 1928 "Certain Hyperspatial Partitionings Connected With Preferential Voting") and two 1906 papers that gave pure mathematical treatment to statistical questions arising from notions in Karl Pearson's biometric research.

In his texts Sommerville explained how non-

Euclidean geometries arose from the use of alternatives to Euclid's parallel postulate. Thus, in the Lobachevskian or hyperbolic geometry, it is assumed that there exist two parallels to a given line through an outside point. In Riemannian or elliptic geometry, the assumption of no parallels is made. By suitable interpretation Klein, Cayley, and then Sommerville showed that Euclidean and non-Euclidean geometries can all be considered as subgeometries of projective geometry. For Klein any geometry was the study of invariants under a particular transformation group. From his point of view, projective geometry is the invariant theory associated with the group of linear fractional transformations. Those special plane projective transformations leaving invariant a specified conic section, Cayley's "absolute," constitute a subgroup of the plane projective group; and the corresponding geometry is hyperbolic, elliptic, or Euclidean according to whether the conic is real (an ellipse, for example), imaginary, or degenerate. This conception makes it possible in all three geometries to express distance and angle measure in terms of a cross ratio, the fundamental invariant under projective transformation.

Even in two of his earliest investigations, namely, "Networks of the Plane in Absolute Geometry" (1905) and "Semi-Regular Networks of the Plane in Absolute Geometry" (1906), Sommerville used the Cayley-Klein notion of non-Euclidean geometries, in particular the projective measurement of lengths and angles. These two papers indicated a trend that he was to follow in much of his research, namely, the study of tesselations of Euclidean and non-Euclidean spaces, a theme suggested by the repetitive designs on wallpaper or textiles and by the arrangement of atoms in crystals. Sommerville showed that whereas there are only three regular tesselations in the Euclidean plane (its covering by congruent equilateral triangles, squares, or regular hexagons), there are five mosaics of congruent regular polygons of the same kind in the elliptic plane, and an infinite number of such patterns in the hyperbolic plane. In all cases the variety is greater if "semi-regular" networks of regular polygons of different kinds are permitted. Moreover, as Sommerville pointed out, still further variations are attainable because the regular patterns are topologically equivalent, if not aesthetically so, to nonregular designs. In several papers and in his text on n-dimensional geometry, he generalized his earlier results and methods to include honeycombs of polyhedrons in three-dimensional spaces and "honeycombs" of polytopes in spaces (Euclidean and non-Euclidean) of 4, 5, $\cdots$, n dimensions.

Many of Sommerville's geometric concepts have algebraic counterparts in the theory of groups. Thus, since his repetitive patterns can be considered as the result of moving a single basic design to different positions, it is possible to asssociate with each tesselation or honeycomb one or more "crystallographic groups," each a set of motions that displace a fundamental region so that it will cover an entire plane, space, or hyperspace. Thus, if a square (with sides horizontal and vertical) is the fundamental region in a Euclidean plane, one can cover that plane with duplications of the square by two basic motions or their inverses, namely translation of the square one side-length to the right, and a similar translation upward. Those two motions are said to "generate" a crystallographic group corresponding to the network of squares. For that same network a different crystallographic group is generated by three basic motions—the two reflections of the square in its vertical sides, and the translation of the square one side-length upward.

There are also associations with group theory in Sommerville's "On Certain Projective Configurations in Space of n Dimensions and a Related Problem in Arrangements" (1906), in which he showed interrelationships between certain finite groups and the finite projective geometries of Veblen and Bussey. Such groups also played a role in his "On the Relation Between the Rotation-Groups of the Regular Polytopes and Permutation Groups" (1933).

BIBLIOGRAPHY

I. ORIGINAL WORKS. Among Sommerville's many research papers are "Networks of the Plane in Absolute Geometry," in *Proceedings of the Royal Society of Edinburgh*, **25** (1905), 392–394; "Semi-Regular Networks of the Plane in Absolute Geometry" in *Transactions of the Royal Society of Edinburgh*, **41** (1906), 725–747; "On the Distribution of the Proper Fractions," in *Proceedings of the Royal Society of Edinburgh*, **26** (1906), 116–129; "On the Classification of Frequency Ratios," in *Biometrika*, **5** (1906), 179–181; "On Links and Knots in Euclidean Space of *n* Dimensions," in *Messenger of Mathematics*, 2nd ser., **36** (1906), 139–144; "On Certain Projective Configurations in Space of *n* Dimensions and a Related Problem in Arrangements," in *Proceedings of the Edinburgh Mathematical Society*, **25** (1906), 80–90; "The Division of Space by Congruent Triangles and Tetrahedra," in *Proceedings of the Royal Society of Edinburgh*, **43** (1923), 85–116; "The Regular Divisions of Space of *n* Dimensions and Their Metrical Constants," in *Rendiconti del Circolo matematico di Palermo*, **48** (1924), 9–22; "The Relations Connecting the Angle-Sums and Volume of a Polytope in Space of *n* Dimensions," in *Proceedings of the Royal Society of London*, **A115** (1927), 103–119; "An Analysis of Preferential Voting," in *Proceedings of the Royal Society of Edinburgh*, **48** (1928), 140–160; "Certain Hyperspatial Partitionings Connected With Preferential Voting," in *Proceedings of the London Mathematical Society*, 2nd ser., **28** (1928), 368–382; "Isohedral and Isogonal Generalizations of the Regular Polyhedra," in *Proceedings of the Royal Society of Edinburgh*, **52** (1932), 251–263; and "On the Relations Between the Rotation-Groups of the Regular Polytopes and Permutation-Groups," in *Proceedings of the London Mathematical Society*, 2nd ser., **35** (1933), 101–115.

Sommerville's books are *Bibliography of Non-Euclidean Geometry* (London, 1911); *The Elements of Non-Euclidean Geometry* (London, 1914, 1919); *Analytical Conics* (London, 1924); *An Introduction to the Geometry of n Dimensions* (London, 1929); and *Analytical Geometry of Three Dimensions* (Cambridge, 1934).

II. SECONDARY LITERATURE. On Sommerville and his work, see H. W. Turnbull, "Professor D. M. Y. Sommerville," in *Proceedings of the Edinburgh Mathematical Society*, 2nd ser., **4** (1935), 57–60.

EDNA E. KRAMER

SOMOV, OSIP IVANOVICH (*b*. Otrada, Moscow gubernia [now Moscow oblast], Russia, 1 June 1815; *d*. St. Petersburg, Russia [now Leningrad, U.S.S.R.], 26 April 1876), *mathematics, mechanics*.

Somov graduated from the Gymnasium in Moscow and enrolled at the Faculty of Physics and Mathematics of Moscow University. After graduating in 1835, he published a work on the theory of determinate algebraic equations of higher degree (1838), in which he manifested not only deep knowledge but also extraordinary skill in presenting the newest achievements of algebraic analysis.

Somov's pedagogic career began in 1839 at the Moscow Commercial College. After defending his master's dissertation in Moscow, he was invited to St. Petersburg University in 1841 and taught various courses in mathematics and mechanics there for the next twenty-five years. Somov defended his doctoral dissertation at St. Petersburg and was awarded the title of professor of applied mathematics.

In 1857 Somov was elected an associate member of the St. Petersburg Academy of Sciences, and in 1862 he succeeded Ostrogradsky as academician.

Turning his attention to problems of theoretical mechanics, Somov applied results obtained in analytical mechanics to specifically geometric problems. He is rightfully considered the originator of the geometrical trend in theoretical mechanics in Russia during the second half of the nineteenth century. In the theory of elliptical functions and their application to mechanics, he completed the solution of the problem concerning the rotation of a solid body around an immobile point in the Euler-Poinsot and Lagrange-Poisson examples.

The first in Russia to deal with the solution of kinematic problems, Somov included a chapter on this topic in his textbook on theoretical mechanics. His other kinematic works include studies of a point in curvilinear coordinates. Somov's theory of higher-order accelerations of a point, and of an unchanging system of points, was a significant contribution. His works were the first special studies in Russia of *n*th-order accelerations of both absolute and relative motions of points. His studies of small oscillations of a system around the position of equilibrium are also important.

BIBLIOGRAPHY

I. ORIGINAL WORKS. In addition to more than fifty papers on mechanics and mathematics, Somov published *Teoria opredelennykh algebraicheskikh uravneny vysshikh stepeny* ("Theory of Determinate Algebraic Equations of Higher Degree"; Moscow, 1838); *Analiticheskaya teoria volnoobraznogo dvizhenia efira* ("Analytic Theory of the Undulatory Motion of the Ether"; St. Petersburg, 1847); *Osnovania teorii ellipticheskikh funktsy* ("Foundations of the Theory of Elliptical Functions"; St. Petersburg, 1850); *Kurs differentsialnogo ischislenia* ("Course in Differential Calculus"; St. Petersburg, 1852); *Analiticheskaya geometria* ("Analytic Geometry"; St. Petersburg, 1857); *Nachalnaya algebra* ("Elementary Algebra"; St. Petersburg, 1860); *Nachertatelnaya geometria* ("Descriptive Geometry"; St. Petersburg, 1862); and *Ratsionalnaya mekhanika* ("Rational Mechanics"), 2 pts. (St. Petersburg, 1872–1874), translated into German by A. Ziwet as *Theoretische Mechanik* (Leipzig, 1878).

II. SECONDARY LITERATURE. Bibliographies of Somov's works are included in Y. L. Geronimus, *Ocherki o rabotakh korifeev russkoy mekhaniki* ("Essays on the Works of Leading Russian Mechanists"; Moscow, 1952), 58–96; T. R. Nikiforova, *Osip Ivanovich Somov* (Moscow–Leningrad, 1964); and E. I. Zolotarev, "Ob uchenykh trudakh akademika O. I. Somova," in *Zapiski Imperatorskoi akademii nauk*, **31** (1878), 248–266.

A. T. GRIGORIAN

SONIN, NIKOLAY YAKOVLEVICH (*b.* Tula, Russia, 22 February 1849; *d.* Petrograd [now Leningrad], Russia, 27 February 1915), *mathematics.*

The son of a state official who later became a lawyer, Sonin received his higher education at the Faculty of Physics and Mathematics of Moscow University (1865–1869). His first scientific work was a report on differentiation with arbitrary complex exponent (1869). After defending his master's thesis in 1871, Sonin was appointed *Dozent* in mathematics at Warsaw University in 1872 and, after defending his doctoral dissertation in 1874, was promoted to professor in 1877. He taught at Warsaw for more than twenty years, was twice elected dean of the Faculty of Physics and Mathematics, and was an organizer of the Society of Natural Scientists. In 1891 Sonin was elected corresponding member of the Russian Academy of Sciences and, in 1893, academician in pure mathematics. In connection with the latter rank he moved to St. Petersburg, where from 1894 to 1899 he was professor at the University for Women, from 1899 to 1901 superintendent of the Petersburg Educational District, and from 1901 to 1915 president of the Scientific Committee of the Ministry of National Education. With A. A. Markov, Sonin prepared a two-volume edition of the works of Chebyshev in Russian and French (1899–1907).

Sonin made a substantial contribution to the theory of special functions; the unifying idea of his researches was to establish a few convenient definitions of initial notions and operations leading to broad and fruitful generalizations of these functions. Especially important were his discoveries in the theory of cylindrical functions, which he enriched both with general principles and with many particular theorems and formulas that he introduced into the contemporary literature. He also wrote on Bernoullian polynomials, and his works on the general theory of orthogonal polynomials were closely interwoven with his research on the approximate computation of definite integrals and on the various integral inequalities; in the latter area he continued Chebyshev's research. Also noteworthy are Sonin's works on the Euler-Maclaurin sum formula and adjacent problems.

BIBLIOGRAPHY

I. ORIGINAL WORKS. Sonin's writings include "O razlozhenii funktsy v beskonechnye ryady" ("On the Expansion of Functions in Infinite Series"), in *Matematicheskii sbornik*, **5** (1871), 271–302, his master's thesis; "Ob integrirovanii uravneny s chastnymi proizvodnymi vtorogo poryadka" ("On the Integration of Partial Differential Equations of the Second Order"), *ibid.*, **7** (1874), 285–318, translated into German in *Mathematische Annalen*, **49** (1897), 417–447, his doctoral dissertation; "Recherches sur les fonctions cylindriques et le développement des fonctions continues en séries," *ibid.*, **16** (1880), 1–80; "Sur les termes complémentaires de la formule sommatoire d'Euler et de celle de Stirling," in *Annales scientifiques de l'École normale supérieure*, **6** (1889), 257–262; "Sur les polynômes de Bernoulli," in *Journal für die reine und angewandte Mathematik*, **116** (1896), 133–156; "Sur les fonctions cylindriques," in *Mathematische Annalen*, **59** (1904), 529–552; see also *Issledovania o tsilindricheskikh funktsiakh i o spetsialnykh polinomakh* ("Research on Cylindrical Functions and on Special Polynomials"), N. I. Akhiezer, ed. (Moscow, 1954).

II. SECONDARY LITERATURE. See N. I. Akhiezer, "Raboty N. Y. Sonina po priblizhennomu vychisleniyu opredelennykh integralov" ("The Works of N. Y. Sonin on the Approximate Computation of Definite Integrals"), in Sonin's *Issledovania o tsilindricheskikh funktsiakh*, 220–243; A. I. Kropotov, *Nikolay Yakovlevich Sonin* (Leningrad, 1967), with complete bibliography of Sonin's works, pp. 126–130; and G. N. Watson, *A Treatise on the Theory of Bessel Functions* (Cambridge, 1922).

A. P. YOUSCHKEVITCH

SONNERAT, PIERRE (*b.* Lyons, France, 18 August 1748; *d.* Paris, France, 31 March 1814), *natural history.*

Sonnerat started his career as secretary to his godfather and relative, Pierre Poivre, enlightened intendant of Île de France (now Mauritius). Through influential patrons he rose from clerk in the overseas service of the Ministry of Naval Affairs to commissioner of the colonies, ending his career as commandant of the French settlement at Yanam, India, then a center for the manufacture of salt and cotton goods. He would have ended his life in affluence if the French Revolutionary Wars had not broken out; the English invaded the French settlement in 1793 and took Sonnerat to Pondicherry. He remained in captivity there until 1813, when he was repatriated on account of the joint intervention of Joseph Banks and Antoine-Laurent de Jussieu. Soon afterward Sonnerat died in Paris.

Sonnerat's fame rests on his determination to adhere, despite the lack of sympathy of his traditionally oriented bureaucratic superiors, to the enlightened policy initiated by the last naval ministers under the royal government: that of collecting essential scientific information on the overseas territories they administered. Indeed, he insisted on his title of "naturaliste pensionnaire du roi et correspondant de son cabinet."

The right opportunities had been given to Sonnerat by Poivre, who had sent him at the beginning of his career on an expedition to Poelau Gebe, in the Moluccas, in search of the spice plants that he sought to acclimatize in the Mascarenes. This was an auspicious beginning to extensive travels in Asia. The botanical and zoological collections that Sonnerat brought back, mainly from the Philippines and the Moluccas, formed the basis of his first major publication, *Voyage à la Nouvelle Guinée*, and no doubt promoted his admission to the Académie des Sciences, Belles-Lettres et Arts of Lyons as associate member, and his election on 19 January 1774 to the Académie Royale des Sciences as correspondent of the botanist Adanson (communication of December 1773: "Description du coco de mer [*Lodoicea maldivica* Pers.] de l'Isle Praslin"). His accomplishments were well summarized by Adanson: "sachant parfaitement bien le dessin, la peinture et la miniature."

Sonnerat's success with the academies may have been due in great measure to the legend that he created and sedulously fostered: that he had been a student and "disciple" of Philibert Commerson, who was the naturalist on Bougainville's expedition round the world. Sonnerat also claimed to have accompanied Commerson on his explorations of Île de France, Bourbon (Réunion), and Madagascar. Chronological and contemporary evidence disprove this assertion, however; and official documents invariably cite Paul Philippe Sauguin de Jossigny as Commerson's constant companion and draftsman. Sonnerat's awareness of the prestige that science commanded in the Enlightenment is exemplified in his use of his membership in academies to open doors for him. Relying on a chance acquaintance with Joseph Banks in 1771 at the Cape of Good Hope, where Cook's expedition had called on the last lap of the journey round the world, Sonnerat sought to secure election in 1783 as foreign associate of the Royal Society. He confided to Banks that membership would be of great help to him in the travels he proposed to make in Tibet and Central Asia, and would serve as an introduction to British governors in India. Disappointed in London, he was later honored by Revolutionary France by election as correspondent in the Botanical Section (First Class) of the Institut National on 28 November 1803. In 1806 the Société d'Émulation de l'Île de France elected him corresponding associate.

The success of the *Voyage à la Nouvelle Guinée* (1776) was doubtless on account of the very powerful, but anonymous, patronage. In the field of natural history, the work appears to have been a supplement to Brisson's *Ornithologie* (1760) and a link with *Histoire naturelle des oiseaux* of Buffon and Guéneau de Montbéliard.

Heartened by this first success, Sonnerat confidently launched his second publication, the *Voyage aux Indes orientales et à la Chine* (1782), dedicated to his lifelong patron, the Comte d'Angiviller, intendant of the Jardin Royal des Plantes. Severe censure of the frivolity of his observations on the countries he had visited came from many sources: the missionaries of Peking criticized what he had written on China; J. A. B. Law de Lauriston, his account of India; and J. F. Charpentier de Cossigny, his strictures on Île de France. Nevertheless, the success outlasted the criticism, for a second, less lavish edition (that does not seem to have been authorized), with critical notes by C. N. Sonnini de Manoncourt, was published in 1806. The work was probably a prey to literary piracy, for as late as 1816 Jean-Amable Pannelier published anonymously a work entitled *L'Hindoustan, ou religion, moeurs, usages, arts et métiers des Hindous*, with descriptions of crafts literally transcribed from Sonnerat's text of 1782.

Sonnini's edition of the *Voyage* must not be confused with a proposed publication that Sonnerat planned to entitle *Nouveau voyage aux Indes orientales*. He had worked on the manuscript during the latter part of his stay in India; and he intended the publication to be in three, later extended to four volumes. After his death the completed manuscript, brought to France on his last journey, was entrusted by his daughter to Antoine-Laurent de Jussieu, who was requested to edit it. Despite exhaustive searches in the Paris archives, the manuscript has not been traced; and the work is known only from a prospectus that was distributed shortly after 1803.

Sonnerat was an avid, if admittedly indiscriminate, collector. Botanical specimens were sent to Adanson, A.-L. de Jussieu, Linnaeus the younger, and Lamarck; collections of reptiles from India and of tropical fishes were sent to Lacépède; and his notes and drawings were used by Cuvier. Sonnerat had a great interest in tropical fishes—attested by the handsome collection of seventeen undated plates in the collection of *Vélins du roi*; and he seems to have been among the first to study, in a scientific spirit, those fishes from the lagoons of Île de France that were reported to cause poisoning.

Sonnerat was the first to give an account of the indris (*I. brevicaudatus*) and of the aye-aye (*Daubentonia madagascariensis*) from Madagascar. His elegant drawings of exotic birds, if not free from error in attribution or habitat, are fundamental for the study of ornithology. His name is commemorated in the genus *Sonneratia* (mollusk) and in six or eight species of mangrove swamp plants (*Sonneratia* L. *f*) of the eastern tropics.

Unfortunately Sonnerat's fame rests on his achievements as a young man. Little is mentioned of his accomplishments as a skillful administrator, or of his understanding of contrasting cultures and civilizations that made him a forerunner of modern social anthropologists. In his two major publications it is evident that his insight into other civilizations gave a strong impetus in Europe to the spread of a fashionable interest in the religion, arts, and customs of India and the Indian Archipelago; this marked the second part of the eighteenth century in Europe, in contrast to the interest in the arts and civilization of China that had prevailed earlier. Sonnerat was responsible for nurturing in France a taste for the exotic style of painting known in England as "company painting." This style is evident in the collections of prints and drawings preserved in the Bibliothèque Nationale, Paris, dating from the end of the eighteenth century to almost the 1840's, and representing French artistic interest in the racial types and crafts of India and Southeast Asia.

BIBLIOGRAPHY

I. ORIGINAL WORKS. Sonnerat's published writings are *Voyage à la Nouvelle Guinée* . . . (Paris, 1776), also in English ed. (Bury St. Edmunds, 1781); and *Voyage aux Indes orientales et à la Chine fait par ordre du roi, depuis 1774 jusqu'en 1781* . . ., 2 vols. (Paris, 1782); also rev. ed. by C. N. Sonnini de Manoncourt, 4 vols. (Paris, 1806); other eds. are in German (Zurich, 1783), Swedish (Uppsala, 1786), and English (Calcutta, 1788).

There is also abundant archival material preserved in the following institutions: Académie des Sciences, Paris; Muséum National d'Histoire Naturelle, Paris; Archives Nationales, Paris and Depôt des Archives d'Outre-Mer at Les Fenouilères, Aix-en-Provence; Bibliothèque Nationale, Paris; Archives Municipales, Lyons; British Museum, London; Hunt Botanical Library, Pittsburgh; and Archives of the Royal Society of Arts and Sciences of Mauritius.

II. SECONDARY LITERATURE. A short note on Sonnerat appears in *Index biographique des membres et correspondants de l'Académie des sciences, du 22 décembre 1666 au 15 décembre 1967* (Paris, 1968), 509. The basic biographical article is Alfred Lacroix, "Notice historique sur les membres et correspondants de l'Académie des sciences ayant travaillé dans les colonies françaises des Mascareignes et de Madagascar au XVIIIe siècle et au début du XIXe . . .," in *Mémoires de l'Académie des sciences de l'Institut de France*, 2nd ser., **62** (1936), 70–75, also in *Figures de savants*, IV: *L'Académie des sciences et l'etude de la France d'outre-mer de la fin du XVIIe siècle au début du XIXe* (Paris, 1938), 25–31. M. J. van Steenis-Kruseman has given a factual account in *Cyclopaedia of Collectors* ser. 1 I: *Flora Malesiana: The Botany of Malaya, Indonesia, the Philippines, and New Guinea*, C. G. G. J. van Steenis, ed. (Groningen, 1950). A short, balanced notice is given by Jean Vinson in *Dictionary of Mauritian Biography*, no. 18 (Port Louis, 1945), 561–562.

It should be noted that none of these writers gives any account of the latter part of Sonnerat's life, spent mainly in India. These biographies are based on Sonnerat's testimony or on his published work. Berthe Labernadie, who has done pioneer work in the Pondicherry archives, has written a spirited account of the French Revolution in Yanam in *La révolution et les établissements français dans l'Inde* (Pondicherry, 1929).

See also Madeleine Ly-Tio-Fane, "The Career of Pierre Sonnerat (1748–1814): A Reassessment of His Contribution to the Arts and to the Natural Sciences" (Ph.D. diss., University of London, 1973), which is based on a critical study of all the extant archival material preserved in the above-mentioned institutions.

More general works are *Adanson. The Bicentennial of Michel Adanson's "Familles des Plantes,"* 2 vols. (Pittsburgh, 1963–1964); Mildred Archer, *Company Drawings in the India Office Library* (London, 1972); L. H. Bailey, "Palms of the Seychelles," in *Gentes herbarum*, **6**, fasc. 1 (1942), 9–29; Joseph Banks, *The Banks Letters, a Calendar of the Manuscript Correspondence of Sir Joseph Banks Preserved in the British Museum, the British Museum (Natural History) and Other Collections in Great Britain*, Warren R. Dawson, ed. (London, 1958), 774; Bibliothèque Nationale, Paris, *Trésors d'Orient (Exposition organisée sous l'égide du Comité national des commémorations orientalistes, . . . réalisée avec le concours du 29e Congrès international des Orientalistes)* (Paris, 1973); J. F. Charpentier de Cossigny, *Lettre à M Sonnerat* (Port Louis, 1784); R. Decary, *La faune malgache: Son rôle dans les croyances et les usages indigènes* (Paris, 1950), 19, 26–30; A.-A. Fauvel, "Le cocotier de mer des Îles Seychelles (*Lodoicea Sechellarum*)," in *Annales du Musée colonial de Marseille*, 3rd ser., 1 (1915), 169–307; *The Journals of Captain James Cook*, J. C. Beaglehole, ed., 5 vols. (London, 1961–); Berthe Labernadie, *Le vieux Pondichéry (1673–1815)* (Pondicherry, 1936), 175–187; Madeleine Ly-Tio-Fane, *Mauritius and the Spice Trade*: I, *The Odyssey of Pierre Poivre* (Port Louis, 1958), 13, 34, 96–97, II, *The Triumph of Jean Nicolas Céré and His Isle Bourbon Collaborators* (Paris–The

Hague, 1970), 26–28, 178–179, 186; Madeleine Ly-Tio-Fane, "Pierre Poivre et l'expansion française dans l'Indo-Pacifique," in *Bulletin de l'Ecole française d'Extrême-Orient*, **53** (1967), 453–511, with specific reference to Sonnerat on pp. 473–478; S. P. Sen, *The French in India, 1763–1816* (Calcutta, 1958), 486–490; and C. P. Thunberg, *Voyages de C. P. Thunberg au Japon, par le Cap de Bonne Espérance, les Îles de la Sonde, etc.*, I (Paris, 1796), 275–278.

MADELEINE LY-TIO-FANE

SORANUS OF EPHESUS (*fl.* Rome, second century), *medicine.*

Soranus of Ephesus can be considered one of the major Greek physicians in the Roman Empire at the beginning of the second century. According to Suidas he was the son of Meandros and Phoibe, but a second article of the lexicographic *collectaneum*, "Sōranos Ephesios, iatros neōteros," does not justify a belief that two physicians with the same name had historical importance.[1] Scheele's careful research has confirmed this opinion; and even the fact that many other physicians were named Soranus and that in the families of physicians a name often was given to the son or the grandson, is not sufficient reason for new doubts.[2] Some of the statements about his life and sites of activity from late antiquity and from Byzantine literature are legends.[3] But it is certainly correct that he was a member of the methodist sect, that he practiced at Rome during the reigns of Trajan and Hadrian, and that he had studied in Alexandria.[4]

Although there is no direct evidence, priority must be given to Ephesus as the site of Soranus' medical training and scientific development; for in the first two centuries of the Christian era it gradually became necessary to offer the professional studies, as taught at Alexandria, in schools and academies throughout the Roman Empire.[5] And in this respect Ephesus was of special importance for the whole of Asia Minor.[6] Soranus thus was trained primarily at the medical school of Ephesus; and if his references to medical experience in Egypt and Rome are taken as proof for his training and activity there, serious consideration must likewise be given to his remarks concerning Caria.[7] It must be left undecided if he only took his training there—perhaps under Magnus Ephesius, whom he repeatedly cites[8]—or was a lecturer. His numerous textbooks, obviously meant for practical instruction, and the fact that he had pupils are evidence for such a supposition. In any case, he ranks among the important physicians of the Ephesian school.[9]

These statements are not meant to diminish Soranus' merits of the methodist school. The methodist doctrine rejected the theory of humors and, influenced by Epicurus' philosophy and its skepticism, had developed ideas stating that the human body consists of movable and immovable atoms, interlaced by fine pores, the tension of which is responsible for health and sickness. This cellular-pathological structure allowed certain vaguely defined communities of the human organism — τερατώδεις ἐκεῖναι κοινότητες ("communities miraculous," as Galen caustically called them).[10] This type of structure provided the opportunity to classify diseases into three conditions according to the state of the pores: *status laxus* (grossly relaxed), *status strictus* (grossly contracted), and *status mixtus* (mixed).[11] Thus the method basically renounced any etiology and pathology, as well as basic anatomical and physiological knowledge, and was guided in its practice by observing "certain communities of sicknesses."[12] This kind of thinking made it possible for Thessalus of Tralles, Soranus' predecessor, to develop the distinction between acute and chronic illnesses, which proved successful where the old theory of the crasis had failed.[13]

Such a simplified method impressed the Romans—and thus imperial physicians were predominantly representatives of methodism—but this "method" could not satisfy the advanced, highly developed, and occasionally contradictory standard of knowledge attained after the early Alexandrian epoch. It was therefore Soranus' main contribution "to have reestablished the 'method' by ordering its principles," and Caelius Aurelianus called him "methodicorum princeps."[14] From the existing theoretical suppositions he had to direct his attention to consolidating diagnostics, and thus in his work differential diagnostics gained importance for the first time. Soranus also sought to place the vague and extremely hypothetical "communities" on a firm basis and to give them a distinct definition; and the strict separation of acute and chronic diseases was made with remarkable clarity and excellent power of clinical observation in his practical instruction on diseases.[15] In his time the "method" became a genuine alternative to the older theories, especially for those who did not cling slavishly to the details and had a solid medical training.

Soranus retained his own views, which sometimes diverged from those of the methodist school. Even if he considered the science of the healthy body, including anatomical and physiological knowledge, to be useless, as a scholar at Ephesus and Alexandria he frequently used it and declared the former to be necessary.[16] His gynecological works demonstrate to what extent he valued Herophilus' teachings on obstetrics, so that it is incorrect to call him—as did Diepgen—merely a *Vertreter methodischer Gynäkologie* ("representative of methodist gynecology")[17] who added nothing to the development of this specialty. His knowledge comprised the whole of medicine and even extended to philosophy and grammar, fields in which he also was outstanding. Therefore Galen, who expressed contempt for the masters of "method," never attacked Soranus; on the contrary, he recommended some of his prescriptions. Even Tertullian, a theologian not at all on friendly terms with the physicians, characterized him as "methodicae medicinae instructissimus autor."[18]

Soranus' works deal with many fields of medical science and are noted for their clarity and the rigorous treatment of the stated problems; they also give the reader a more comprehensive biological view by using vivid comparisons from zoology and agriculture.[19] Both his manner of citing the sources and his exact observance of their chronological sequence in mentioning the doctrines and theories of older physicians, to whom he gave considerable attention, were remarkable.

Soranus' major extant work, *Gynaecia*, comprised four books.[20] Book I records the necessary qualities of a prospective midwife (integrity, zest for work and strong constitution, smooth hands, good theoretical knowledge and practical experience, refusal to perform a criminal abortion) and her work (gynecological physiology with exact representation of the anatomy; feminine hygiene, including comments on menstruation and conception; how to have healthy children; hygiene during pregnancy and abortion). Book II deals with obstetrics (symptoms of and preparations for delivery, parturition, complications, nursing by women in childbed, the nursing of the baby and the choice of a wet nurse, confinement and infant hygiene, and childhood diseases). Books III and IV deal with women's diseases. In Book III, Soranus concedes that women have diseases (πάθη) that men cannot, a controversial thesis in antiquity, and comments on diseases to be treated dietetically; and Book IV deals with diseases that can be treated surgically and pharmaceutically. Although the *Gynaecia* was a comparatively complete work in the original text, it is necessary to warn against the prevalent view that Soranus was "the" gynecologist of antiquity. His work would have been impossible

without the preliminary studies of the Herophileans, however independent and superior his mastery and exposition of the subject; in addition, his knowledge far surpassed this specialty.

A shorter compendium, a sort of catechism for midwives, has been lost in its original edition; but it may be preserved in Muscio's sixth-century translation, as well as in a Greek retranslation that was formerly considered the original edition, by the Greek physician Moschion.[21] Περὶ σπέρματος καὶ ζῳογονίας, on sperm and the genesis of creatures, now lost, counts in the same interrelation.[22] Parts of the work were translated into Latin in a treatise by Vindicianus.[23]

Soranus' magnum opus, Περὶ ὀξέων καὶ χρονίων παθῶν, on acute and chronic diseases, also was lost; but there is a sufficient substitute in Caelius Aurelianus' *Celerum sive acutarum passionum*, Books I–III, and *Tardarum sive chronicarum passionum*, Books I–V, because Caelius made a faithful translation into Latin and introduced very few of his own ideas.[24] This work is solidly grounded in the methodist doctrine; and when treating each of the major "internal" diseases, it quite distinctly shows, even in the Latin, the disposition, systematic manner, and wording of Soranus. This work also regularly cites the doctrines of earlier authors, although Soranus nearly always agrees with the views of his own school; and when there are divergencies, he takes a conciliatory standpoint. He is as critical of every sort of medical superstition as he was in the *Gynaecia*, a practice that was no longer a matter of course in science; but when searching for the natural causes of diseases, he reaches beyond both the therapeutic frame of the work and the intentions of his school.

The lost work Αἰτιολογούμενα, on causes of diseases, seems to treat that subject exclusively; and Περὶ κοινοτήτων, on the "communities," apparently seeks a more distinct definition of that vague concept.[25] Soranus also wrote Περὶ πυρετῶν, containing instructions for treating fever, and Περὶ βοηθημάτων, on medical resources, both of which are probably supplements to the work on acute and chronic diseases that have been lost.[26] Caelius Aurelianus often quoted the latter in such a way that one is inclined to consider it as a systematic description of nursing, bloodletting, purgations, and physical therapy.[27]

When prescribing remedies Soranus used only medicaments approved by his teachers and friends—or so Galen said—and recorded them in Περὶ φαρμακείας ("Instruction on Medicaments") and in a pharmaceutical booklet, Μονοβιβλίον φαρμακευτικόν.[28] He largely agreed with the theories of his school in pharmaceutical practice but disagreed in matters of surgery and the closely related techniques of bandaging. Here the methodist doctrine was unable to support him because it was opposed to anatomy and consequently to surgery as well.[29] Although his great work on surgery, Χειρουργούμενα, is lost, an apparently extant fragment, Περὶ σημείων καταγμάτων ("On the Symptoms of Fractures"), reveals not only an exact knowledge of the normal skeletal anatomy but also a precise conception of the anatomicopathologic misposition of the fragments of bones.[30] It is characteristic that Demetrius, a Herophillean, is the only physician quoted in this fragment.[31] Soranus' completely extant Περὶ ἐπιδεσμάτον ("On Bandages") gives numerous examples of conformity with the pseudo-Galenic instruction on bandages and with that of Heliodorus, as it is presented in Oribasius' work.[32] Two other lost works on medical practice are Ὀφθαλμικόν, on ophthalmology, and Ὑγιεινόν, a general work on hygiene. The latter was also translated by Caelius Aurelianus, but the Latin edition has not survived.[33] In addition to hygiene Soranus was deeply concerned with the human psyche and wrote the four-book Περὶ ψυχῆς βιβλία δ΄, on the human soul.[34] Although this work has not survived, it is possible to obtain an idea of its contents because Tertullian used it as the main source of his *De anima*.[35] There will always be uncertainty, however, whether Soranus really composed commentaries on Hippocratic writings.

The work on the soul extends into philosophical as well as allied fields of medicine, as do his last two works. The first is Βίοι ἰατρῶν καὶ αἰρέσεις καὶ συντάγματα δέκα, ten books containing biographies of physicians and information on their schools and writings.[36] This biographical work, together with the doxographic description of the existing medical groups and their writings on theoretical and practical problems, is Soranus' main contribution to medical history. An extant fragment is Ἱπποκράτους γένος καὶ βίος κατὰ Σωρανόν ("The Noble Origin and the Life of Hippocrates According to Soranus' Statements").[37] Written many centuries after the death of Hippocrates, Soranus' statement must necessarily contain some traces of legend.[38] Nevertheless, apart from occasional remarks in Plato and later authors, it is the oldest extant complete biography of him. The second work, Περὶ ἐτυμολογιῶν τοῦ σώματος τοῦ ἀνθρώπου, on the origin of bodily terms, concerns the nomenclature for the parts of the human body and its linguistic origin.[39] Its loss is less serious, for later etymologists made extensive use of it; thus a judgment is still possible concerning its conception, range, and quality.[40]

The extent of Soranus' work demonstrates that, with Galen, he was the greatest medical author of late antiquity. That almost all his works were lost, whereas Galen's were widely preserved, is a result of the fact that Galen and his theory of crasis dominated medical thought during the following 1,500 years and deprived the atomistic and cellular-pathological approach of any chance of acceptance. And yet these latter theories exerted a decisive effect. In addition to the works of translators and physicians of the Western Empire, of etymologists, of lexicographers, and of theologians, important chapters of Soranus' works appeared in the compilations of Byzantine medical science;[41] even Galen used parts of them. Soranus' surviving works reveal a physician with an unprejudiced view of the substance of a medical science that was threatened with being swamped by its own abundance of knowledge. Essential parts of his work show that he was a master of "method"—but by no means indoctrinated in a way that might prevent him from looking beyond the limits of his school. His liberal views permitted Soranus to use the Herophileans as a base in obstetrics and osteology and to accept principles of other schools—for instance, in matters of bandaging. This cultivation of a liberated mind was characteristic of the school of Ephesus, and Soranus proves that he belonged to it through his use of terminology: only this school produced such personages as Rufus and, enlarging the circle a bit, Charmides, who wrote the *Onomasticon*.[42] Galen, on the other hand, deliberately neglected distinct diction in the nomenclature.[43]

Soranus' moral and intellectual freedom also enabled him to write about Hippocrates and his followers in a way that presented an analysis of their doctrines without advocating a return to them. If we knew more about him, we would undoubtedly conclude that Soranus was Galen's only great intellectual antagonist, intellectually his peer—if we are allowed to use Tertullian's *De anima*—and in character his superior. His pupil Statilius Attalus, unjustly defamed by Galen, held an eminent place at the court of the emperor at Ephesus.[44]

NOTES

1. Suidas, *Lexicon*, T. Gaisford, ed., rev. by G. Bernhardy, II (Halle–Brunswick, 1853), 850.
2. L. Scheele, *De Sorano Ephesio medico etymologo* (Strasbourg, 1886), 3 ff. (Ph.D. diss.); also see R. Fuchs, "Ge-

schichte der Heilkunde bei den Griechen," in T. Puschmann, M. Neuburger, and J. Pagel, eds., *Handbuch der Geschichte der Medizin*, I (Jena, 1902), 340.

3. See F. E. Kind, "Soran," in Pauly-Wissowa, 2nd ser., III A, pt. 1, 1114.

4. For Rome, Suidas, *loc. cit.*; Soranus, *Gynaeciorum libri quattuor*, II, 44, edited by J. Ilberg, in *Corpus medicorum Graecorum*, IV (Leipzig–Berlin, 1927), 85; Caelius Aurelianus, *De morbis acutis*, II, 130, edited and translated by J. E. Drabkin (Chicago, 1950), 218; and M. Albert, "Les médecins grecs à Rome," in *Les grecs à Rome* (Paris, 1894), 197 ff. For Alexandria, see Suidas, *loc. cit.*; Soranus, *op. cit.*, I, 6, Ilberg, ed., p. 55; Caelius Aurelianus, *De morbis chronicis*, V, 30, J. E. Drabkin, ed., 924.

5. See U. Kahrstedt, *Kulturgeschichte der römischen Kaiserzeit*, 2nd ed. (Bern, 1958), 276.

6. Concerning Ephesus at this time see Kahrstedt, *op. cit.*, 169 f. For the medical school and the association of the physicians, see J. Keil, "Ärzteinschriften aus Ephesos," *Jahreshefte des Österreichischen archäologischen Instituts Wien*, **8** (1905) and **23** (1926); and *Forschungen in Ephesos*, IV, pt. 1 (Vienna, 1932).

7. See Kind, *loc. cit.*; and Fuchs, *loc. cit.* Cf. Caelius Aurelianus, *De morbis acutis*, III, 124; and *De morbis chronicis*, V, 30, Drabkin, ed., 378, 924.

8. M. Wellmann, following H. Haeser, placed Magnus, among others, in the Pneumatic school because he was a pupil of Athenaeus; and he has remained uncontradicted. M. Wellmann, "Die pneumatische Schule," in *Philologische Untersuchungen*, A. Kiessling and U. v. Wilamowitz-Moellendorff, eds., XIV (Berlin, 1895), 178 ff., 187; H. Haeser, *Lehrbuch der Geschichte der Medizin und der epidemischen Krankheiten* (Jena, 1875; repr. Hildesheim–New York, 1971), I, 334 ff. At first this view gives the impression that a connection between Magnus and Soranus would be incompatible. There is no doubt that the investigation and intellectual definition performed by the medical schools of late antiquity were important and instructive; but the classic science of antiquity and the historians of medicine of the past century, influenced by Galenic polemics, have attached too much importance to the differences among these schools; and in doing so they have failed entirely to notice the importance of the schools' belonging to the same academy and its physicians' association. Haeser, referring to Athenaeus, the teacher of Magnus, said, "The methodists had made him make so many concessions that they could call him one of theirs." Thus it is not necessary to state that belonging to a school separates more strongly than belonging to an academy can bind. It seems that this statement becomes valid with Magnus, for Galen sees Magnus' view concerning the cause, origin, and importance of the pulse in total contrast with that of Archigenes, who was a faithful follower of the Pneumatic school. Only a few pages later in Galen's work there is this statement: ". . . καὶ αὐτὸς ἀπὸ τῆς πνευματικῆς αἱρέσεως εἶναι προσποιούμενος . . ." (". . . and he himself makes us believe that he is"—or, to put it more distinctly, ". . . and he himself claims to be"—"a member of the Pneumatic school"). Galen, *De pulsuum differentiis* III.1, in Galen's *Opera omnia*, C. G. Kühn, ed., VIII (Leipzig, 1824), 640, 646. Therefore, even if Magnus styled himself a follower of the Pneumatic school, he must have remained much more of a methodist than his master. This view is also proved by the title Περὶ τῶν ἐρευρηνέvων μετὰ τοὺς Θεμίσωνος χρόνοµς ("[Medical] Discoveries After the Time of the Methodist Themison"), *ibid.*, 640. It seems impossible that in such a work a qualified follower of the Pneumatic school would base his chronology on such a confirmed methodist. Thus it is not astonishing that we find Magnus thoroughly incorporated into the school of the methodists by Caelius Aurelianus in *De morbis acutis*, II, 58, Drabkin, ed., 160.

9. To the list of Rufus of Ephesus, Titus Statilius Kriton, and Statilius Attalus given by J. Benedum in his archaeologically oriented essay "Statilios Attalos," in *Medizinhistorisches Journal*, **6** (1971), 274, we can add from the literature—besides Soranus and Magnus—Heraclides of Ephesus as a traumatologist. See M. Michler, *Die Hellenistische Chirurgie*, I, *Die Alexandrinischen Chirurgen* (Wiesbaden, 1968), 89, 132 ff., 148 f. A recommendation of the school can be seen in the fact that the author Athenaeus of Naucratis makes an Ephesian physician join the discussions in his *Deipnosophistae* ("The Learned Banquet"). From this, G. Kaibel, in his pref. to the Teubner ed. (Stuttgart, 1965), iv, expressed the idea that Athenaeus might have derived the names of the two physicians who had been the interlocutors—Daphnus Ephesius and Rufinus Nicaeensis—from Rufus of Ephesus. In any case, during the discussions about medical problems the two physicians are referred to as "the Ephesians and the like-minded persons": Οἱ μὲν Ἐφέσιοι καὶ οἱ τούτοις ὅμοιοι. Lib. III, sec. 33 (87 c), Kaibel, ed., I, 202.

10. Galen, *De methodo medendi*, I, 4, in his *Opera omnia*, C. G. Kühn, ed., X, 35.

11. See T. Meyer-Steineg, "Das medizinische System der Methodiker," in *Jenaer medizin-historische Beiträge*, nos. 7–8 (1916), 23.

12. See Celsus, *De medicina*, "Prooemium" 54, edited and translated by W. G. Spencer, I, 30.

13. Until then only the acute diseases were distinguished from the rest. See Meyer-Steineg, *op. cit.*, 33.

14. Caelius Aurelianus, *De morbis acutis*, II, 46, Drabkin, ed., 150 f.

15. See Meyer-Steineg, *op. cit.*, 38 ff.; and Fuchs, *op. cit.*, 341 f.

16. Soranus, *Gynaeciorum libri quattuor*, I, 2, 3, p. 4, Ilberg, ed., and I, 5, p. 6.

17. See Michler, *op. cit.*, 142 f.; and Wellmann, *op. cit.*, 118. See also P. Diepgen, "Geschichte der Frauenheilkunde. I: Die Frauenheilkunde der Alten Welt," in W. Stoeckel, *Handbuch der Gynäkologie*, XII, pt. 1 (Munich, 1937), 107.

18. Tertullian, *De anima* 6.

19. See J. Ilberg, "Die Überlieferung der Gynäkologie des Soranus von Ephesos," in *Abhandlungen der Königlich-Sächsischen Gesellschaft der Wissenschaften*, Phil.-hist. Kl., **28**, no. 2 (1910), 36, 76 ff.

20. The ed. by J. Ilberg, *Corpus medicorum Graecorum*, IV, 3–152, is still the authoritative one; an English trans. is O. Temkin, *Soranus' Gynecology* (Baltimore, 1956). For a systematic order other than Ilberg's, see Kind, *op. cit.*, 1118 ff.

21. *Gynaecia ex Muscionis ex Graecis Sorani in Latinum translatum sermonem*, Valentin Rose, ed. (Leipzig, 1882); the Greek retrans. is Μοσχίωνος, Περὶ γυναικείων παθῶν. See Ilberg, *op. cit.*, 102 ff.; and Diepgen, *op. cit.*, 108.

22. See Ilberg, *op. cit.*, 38, and n. 1.

23. Bruxellensis, 1342–1350 (12th century).

24. The ed. and trans. by Drabkin is the authoritative publication today, but for reliability of text one should also consult G. Bendz, "Caeliana, Textkritische und sprachliche Studien zu Caelius Aurelianus," in *Acta Universitatis lundensis*, n.s. **38**, no. 4 (1943); and "Emendationen zu Caelius Aurelianus," in *Publications of the New Society of Letters at Lund*, **44** (1954). For Caelius Aurelianus, see M. Wellmann, in Pauly-Wissowa, III, 1257 ff.; and Meyer-Steineg, *op. cit.*, 42 ff. On the treatment of paralysis, see M. Michler, "Die physikalische Behandlung der Paralysis bei Caelius Aurelianus," in *Sudhoffs Archiv . . .*, **48** (1964), 123.

25. On Αἰτιολογούμενα, see Caelius Aurelianus, *De morbis chronicis*, I, 55, Drabkin, ed., 474. To the overcoming doxographic reports from this work, see Kind, *op. cit.*, 1127. Referred to as Περὶ κοινότητων, see Soranus, *Gynaeciorum libri quattuor*, I, 29, 3, Ilberg, ed., 19.

26. See Caelius Aurelianus, *De morbis acutis*, II, 177, Drab-

kin, ed., 254; and Soranus, *Gynaeciorum libri quattuor*, III, 28, 6, Ilberg, ed., 112.

27. See Kind, *op. cit.*, 1128.

28. Galen, *De compositione medicamentorum secundum locos*, in his *Opera omnia*, I, 7, C. G. Kuhn, ed., XII, 493 f.

29. See M. Michler, *Das Spezialisierungsproblem und die antike Chirurgie* (Bern–Stuttgart–Vienna, 1969), 37.

30. Cited from Soranus, *Gynaeciorum libri quattuor*, I, 7, 4 [76], Ilberg, ed., 56. See also Soranus, *De signis fracturarum*, J. Ilberg, ed., in *Corpus medicorum Graecorum*, IV, 155–158.

31. *Ibid.*, §9, p. 156.

32. Soranus, *De fasciis*, J. Ilberg, ed., in *Corpus medicorum Graecorum*, IV, 159–171. See also Pseudo-Galen, *De fasciis*, in Galen's *Opera omnia*, C. G. Kühn, ed., XVIII A (Leipzig, 1829), 768–827; for Heliodorus, see Oribasius, *Collectiones*, XLVIII, J. Raeder, ed., in *Corpus medicorum Graecorum*, VI, 2, 1 (repr. Amsterdam, 1964).

33. Cited from Soranus, *Gynaeciorum libri quattuor*, I, 32, 1, and 40, 4, Ilberg, ed., 21, 28. For the Latin trans. by Caelius Aurelianus, *Salutaria praecepta*, see Wellmann, in Pauly-Wissowa, *loc. cit.*

34. Tertullian, *loc. cit.*

35. See H. Diels, *Doxographi Graeci* (repr. Berlin, 1958), 206 ff.

36. See Suidas, *loc. cit.*

37. *Vita Hippocratis secundum Soranum*, J. Ilberg, ed., in *Corpus medicorum Graecorum*, IV, 175–178; on the origin of this, see the pref. to IV, xiv f.

38. See also H. E. Sigerist, *Anfänge der Medizin* (Zurich, 1963), 697 ff. This is a trans. of *A History of Medicine* (New York, 1955).

39. See Orion, *Etymologicon*, F. W. Sturz, ed. (Leipzig, 1820), 34, ll. 9 f.; also 131, ll. 4 f., and 159, l. 18.

40. Such a judgment is possible from Orion, who cites him some twenty times, occasionally in long and detailed passages; there are also citations from him in *Etymologicum magnum*, Gudianum, and other Greek etymological dictionaries. Fragments are in Pollux, *Onomasticon* II. See also Kind, *op. cit.*, 1117.

41. For instance, in works of Philumenus of Alexandria, Aëtius of Amida, and Paul of Aegina; see Fuchs, *op. cit.*, 341.

42. See Rufus of Ephesus, Περὶ ὀνομασίας τῶν τοῦ σώματος μορίων, Daremberg and Ruelle, eds., (repr. Amsterdam, 1963), 237 ff. For Charmenides and his *Onomasticon*, see J. Benedum, "Charmenides," in Pauly-Wissowa, Supp. XIV (1974), 96.

43. See E. Marchel, *Galens anatomische Nomenklatur* (Bonn, 1951), 117 (M.D. diss.).

44. See J. Benedum, "Statilios Attalos," 264 ff.

MARKWART MICHLER

SORBY, HENRY CLIFTON (*b*. Woodbourne, near Sheffield, England, 10 May 1826; *d*. Sheffield, 10 March 1908), *microscopy, geology, biology, metallurgy.*

Most of Sorby's ancestors since the seventeenth century had been middle-class cutlers in Sheffield. His father, Henry Sorby, owned a small cutlery factory; his mother was the daughter of a London merchant. Sorby attended local schools and at age fifteen he won, as a prize for mathematics, a book entitled *Readings in Science*, published by the Society for Promoting Christian Knowledge (first edition 1833), which set the direction of his life. During the next four years he completed his education with a full-time private tutor, the Rev. Walter Mitchell, a competent scientist who later wrote on crystallography and mechanical philosophy in the popular compendium *Orr's Circle of the Sciences*. Sorby attended no university—he later said that he planned his education "not to pass an examination but to qualify myself for a career of original investigation." Closely tied to his mother, he never married. He inherited a modest fortune after his father's death in 1847 and thereafter devoted himself entirely to science while continuing to live in Sheffield, a flourishing steel manufacturing town with somewhat limited intellectual resources. Sorby became very active in the local Literary and Philosophical Society, which had fortnightly discussions on a wide range of subjects and provided him with diverse intellectual stimulation and the opportunity for both leadership and service that he could not have obtained as a young man in a metropolis.

Isolated from the most active scientific circles, Sorby worked quietly on unfashionable topics in a laboratory in his own house. Cast in much the same mold as many other English country gentlemen whose education, isolation, and leisure enabled them to make original observations, he initiated two major areas of science—and carried neither to the point of maturity. Often called an amateur, he was one only in the sense that he was not working in an institutional environment or at the expense of anyone else. He was, in fact, a full-time independent research scientist at a time when there were few such.

Sorby's most influential scientific work was done between 1849 and 1864 on the application of the microscope to geology and metallurgy. In both fields his work had a certain elegance derived from a mixture, in about equal proportions, of simple quantitative observation, meticulous new experimental technique, and novel interpretation based on the application of elementary physicochemical principles to complex natural phenomena. "My object," he said in his last paper, "is to apply experimental physics to the study of rocks." His most famous achievement is the development of the basic techniques of petrography, using the polarizing microscope to study the structure of thin rock sections. The geological conclusions that Sorby drew from such studies were of utmost importance. He started this work in 1849 with studies of sedimentary rocks. In 1851 he became involved in a widely noticed debate on the origin of slaty cleavage, and in an 1853 paper he showed conclusively that cleavage was a result of the reorientation of particles of mica accompanying the deformation (flow) of the deposit under anisotropic pressure. Sorby later studied organisms in limestone and discovered the presence and significance of microorganisms in chalk. In a paper published just before his death (1908) he returned to sedimentation and summarized his whole approach.

Sorby first studied the rate of settling and angle of repose of sand and silt particles in still and turbulent water, and the transport of grains along the bottom by currents of various velocities; then, observing bedding angles, ripple marks, and the variation of particle size with depth (with porosity measurements to allow correction for compaction, solution, or compression of strata) in actual sandstones, he deduced rather precisely the conditions under which the sediments had been deposited.

From slate Sorby moved to schists and metamorphic rocks in general. Of great importance was his 1858 paper on liquid inclusions in crystals, both natural and artificial. Inclusions in large crystals had been observed by David Brewster and Humphry Davy in the 1820's, but Sorby used the microscope to find abundant smaller ones within the microcrystals in many metamorphic rocks. He measured the size of the bubbles that resulted from liquid shrinkage after the cavity had been sealed, and he performed laboratory experiments to measure the expansion of liquids in sealed tubes under pressure that enabled him to deduce the temperature and pressure at which the rocks had been formed. This information revealed large differences in the temperature of formation of granites from various localities and led Sorby to realize the great role played in rock formation by water-bearing magma at high temperature and pressure. (In 1863, after further experiments, he wrote: "Pressure weakens or strengthens chemical affinity according as it acts against or in favour of the change in volume"—a clear anticipation of Le Chatelier's principle.) The 1858 paper was illustrated with 120 drawings made under the microscope at magnifications between 60 and 1,600, transferred to the lithographer's stones by Sorby himself. He concluded: "There is no necessary connection between the size of an object and the value of a fact, and . . . though the objects I have described are minute the conclusions to be derived from the facts are great."

There were still eminent geologists who saw little good to come from studying mountains with microscopes, and Sorby's work was rather slow to be widely appreciated. He was not one to wring the last shred of meaning from a topic, however; and it was fortunate for geology that while touring the Rhine valley with his mother before a conference in the summer of 1861, he met the young geology student Ferdinand Zirkel (1838–1912), whom he inspired to take up the new methods. Zirkel did so with Germanic thoroughness, and his two-volume *Lehrbuch der Petrographie* (1866) established petrography as a broad and systematic science.

In 1863–1864 Sorby turned briefly from rocks to metals. Although he began with a general interest in the structure of meteorites (the only metallic bodies to have an easily visible crystalline structure), the principal stimuli seem to have been two evenings at the Literary and Philosophical Society during which ornamental etching and the manufacture of iron and steel were discussed—combined, of course, with the omnipresence of these metals in his native city. On 28 July 1863 he recorded in his diary, "Discover the Widmannstättischm structure in Ⓛ iron." Circle-L was the brand mark of the Swedish wrought iron preferred over all others by Sheffield steelmakers for conversion to blister steel. (Sorby was probably using a piece that had already been converted to steel and had large grains containing easily visible plates of iron carbide. The iron itself, being free from carbon, could not have had a true Widmannstätten structure.) Always somewhat of a showman, Sorby prepared six different samples to display under the microscope at a soirée of the meeting of the British Association at Newcastle in August, by which time he had already identified in steel three separate crystalline compounds that differed in their reaction to nitric acid.

Early in 1864 Sorby recorded his structures by nature printing, an old and simple process in which a relief-etched surface was inked and pressed to paper. A superb print showing the structure in the Elbogen meteorite made by Aloys von Widmannstätten and Karl von Schreibers in 1813 had inspired later ones of which Sorby knew; but before 1863 the etching of terrestrial irons was done only decoratively or to reveal gross texture, not microstructure.

Sorby worked with a local photographer to make several photomicrographs of steel, which he showed and discussed at the British Association meeting in September 1864. This paper was the true foundation of metallography, although it was published only in abstract. Sorby mentions "various mixtures of iron, two or three well-defined compounds of iron and carbon, of graphite and of slag; and these, being present in different proportions, and arranged in various manners give rise to a large number of varieties of iron and steel." Despite considerable interest at the time, no one followed this start. There was no Zirkel of metallurgy, and Sorby himself moved on to other fields, not returning to steel until 1882 and not publishing anything in detail until 1885. By that time interest in metal structure had been aroused by papers by Chernoff (1868, 1879), Martens (beginning in 1878), and Osmond (1885), none of whom knew Sorby's earlier work. Sorby's 1885 paper was circulated in preprint form, but final publication was delayed for two years by a search for suitable photogravure methods.

In the meantime Sorby had shown, by the use of higher magnifications, that the feature that he had earlier called the "pearly constituent" because of its iridescence was an extremely fine duplex lamellar mixture of iron and iron carbide resulting from the decomposition on slow cooling of a constituent that was stable at high temperatures. Earlier he had identified graphite and iron oxide in iron samples and had described the true nature of recrystallization and transformation: "Iron and steel are not analogous to simple minerals, but to complex rocks." The structural origin of many age-old differences between various kinds of iron and steel was now clear. After 1885 people in many countries took up the new field. By 1900 a range of structures had been observed in many alloys and cataloged in relation to composition and heat treatment, and a beginning was being made in the application of thermodynamics to the study of alloys and of the effects of mechanical deformation.

The interest that supplanted metals and meteorites in Sorby's mind in 1864 was spectrum analysis. Four new elements had been discovered by emission spectroscopy since Bunsen and Kirchhoff's announcement of 1860. G. G. Stokes described the use of absorption spectra for identifying organic substances in March 1864, and Sorby at once saw a new application for his favorite instrument. Quickly developing the necessary combination of microscope and spectroscope, he first examined minerals in rock sections, then moved on to study the coloring matter in animal and plant tissues. Carotene was one of his discoveries, and his work on chlorophyll, autumn colors, and blood identification aroused popular interest, the last involving him in a famous murder trial in 1871. Sorby's observation of unrecorded lines in the absorption spectrum of the mineral jargon (jargoon) led him to announce, in March 1869, the discovery of a new element that he named jargonium. He became involved in an unpleasant priority dispute and more embarrassment when, six months later,

he had to retract, for he had found that the lines were due to uranium. Also in 1869 he used his microspectroscope to study the color of borax beads, thus refining an old, and at the time very important, method of mineral analysis.

After his mother's death in 1874 Sorby widened his activities. He frequently traveled to London for scientific meetings, became a member of the Council of the Royal Society, and was elected president of several societies: the Royal Microscopical Society in 1874, the Mineralogical Society (of which he was the first president) in 1876, and, in 1878–1880, the Geological Society of London. Although he continued to conduct his own research on a purely personal basis, he became increasingly concerned with public policy in support of science. Sorby advocated separation of research and teaching, and in the contribution "On Unencumbered Research—A Personal Experience," to *Essays on the Endowment of Research* (1876), edited anonymously by Charles Appleton for a group of scientists at Oxford, he used his own work as an example of the value of unencumbered and undirected—but not isolated—research. In 1871 he had discussed the possibility of endowing a Royal Society professorship in experimental physical research that would be free from teaching duties. In 1874 he planned a marine biological research station that he proposed to endow and direct; but when one was established ten years later by the Royal Society, he was not asked to participate either financially or scientifically. This was apparently the result of a quarrel with Cambridge biologists including Alan Sedgwick (great-nephew of the geologist of the same name), whose intolerant antireligious attitude disgusted him. Though a revolutionary in science, Sorby was a pillar of the Church of England and very conservative in general outlook.

Sorby's public activities thereafter assumed a more local focus. Beginning in 1880, he worked to promote the formation of Firth College in Sheffield and served as its president from 1882 to 1897. In 1897 the University College was formed in Sheffield by the amalgamation of Firth College with the local technical and medical schools. Despite his international reputation, Sorby was not a great enough local figure to be chosen as its president, and he noted in his diary that he was "a trifle disappointed at being thus superseded after so many years of work." He was appointed vice-president of the college, however, and his research continued. He had bought a thirty-five-ton yawl, carrying a crew of five, in 1878 and had equipped it as a floating laboratory. Thereafter he spent five summer months of almost every year until 1903 cruising off the east coast, studying marine biology and geology but also developing new interests, especially architectural history based on a close study of brick dimensions and construction details in Roman, Saxon, and Norman buildings in East Anglia. History and archaeology remained at the level of serious hobbies, however, and Sorby did not carry them to the point of professional publication. He undertook important studies on temperatures and on silt and sewage movements in the Thames estuary for the Royal Commission on the Thames. For some years, marine biology was his dominant interest. This work has not been critically evaluated by historians, but it seems that his only lasting contribution was the technique he developed in 1889 for differentially staining biological tissues and mounting soft-bodied animals as permanent lantern slides for demonstration and study.

Sorby became lame in 1902 and suffered partial paralysis after an accident in 1903 that confined him virtually to his room. For the next five years he worked over the notes that he had accumulated throughout his life, returning to the geology that had begun his career and that resulted in a last major, although retrospective, paper on sedimentary rock formation. It was read at the meeting of the Geological Society of London on 8 January 1908, two months before his death.

In his will Sorby left some journals and £500 to the Literary and Philosophical Society, but his main bequests were to the University of Sheffield for the establishment of a professorship and a research fellowship, the latter under the control of the Royal Society. The bulk of his library also went to the university.

BIBLIOGRAPHY

I. ORIGINAL WORKS. The library of the University of Sheffield has Sorby's diary, containing terse daily entries for 1859–1908 (except for most of 1871–1882, 1894–1895, and 1903–1905) and a 2-vol. bound assembly of his printed papers and notices. A collection of letters to Sorby from many correspondents is in the Sheffield Central Library, Cat. no. SLPS 51. The metallurgy and geology departments of the University of Sheffield have preserved many of Sorby's original microsamples of rock and steel, the earliest bearing the date 1849. Many of his magnificent preparations of marine animals are in the zoology department.

Sorby wrote no book. G. H. Humphries, "A Bibliography of Publications—H. C. Sorby," in C. S. Smith, ed., *The Sorby Centennial Symposium on the History of Metallurgy* (New York, 1965), 43–58, contains 233 entries. The most important of these are "On the Origin of Slaty-Cleavage," in *Edinburgh New Philosophical Journal*, 55 (1853), 137–150; "On the Microscopical Structure of Crystals Indicating the Origin of Minerals and Rocks," in *Journal of the Geological Society*, 14 (1858), 453–500; "Bakerian Lecture. On the Direct Correlation of Mechanical and Chemical Forces," in *Proceedings of the Royal Society*, 12 (1863), 538–550; "On a Definite Method of Qualitative Analysis of Animal and Vegetable Colouring Matters by Means of the Spectrum Microscope," *ibid.*, 15 (1867), 433–456; "On Unencumbered Research—A Personal Experience," in *Essays on the Endowment of Research* (London, 1876), 149–175; "The Application of the Microscope to Geology, etc. Anniversary Address of the President," in *Monthly Microscopical Journal*, 17 (1877), 113–136; "On the Structure and Origin of Limestone," in *Quarterly Journal of the Geological Society of London*, 35 (1879), 56–95; "On the Application of Very High Powers to the Study of the Microscopical Structure of Steel," in *Journal of the Iron and Steel Institute*, 31 (1886), 140–144; "The Microscopical Structure of Iron and Steel," *ibid.*, 33 (1887), 255–288; "On the Preparation of Marine Animals as Lantern Slides to Show the Form and Anatomy," in *Transactions of the Liverpool Biological Society*, 5 (1891), 269–271; "Fifty Years of Scientific Research," in *Annual Report of the Sheffield Literary and Philosophical Society* (1898), 13–21; and "On the Application of Quantitative Methods to the Study of the Structure and History of Rocks," in *Quarterly Journal of the Geological Society of London*, 64 (1908), 171–233.

II. SECONDARY LITERATURE. The only complete biographical study is Norman Higham, *A Very Scientific Gentleman. The Major Achievements of Henry Clifton Sorby* (Oxford, 1963). Obituary notices with more than usual perception are J. W. Judd, "Henry Clifton Sorby, and the Birth of Microscopical Petrology," in *Geological Magazine*, 5th ser., 5 (1908), 193–204; and Archibald Geikie, "Henry Clifton Sorby, 1826–1908," in *Proceedings of the Royal Society*, B80 (1908), lvi–lxvi; and W. J. Sollas, "Anniversary Address of the President," in *Proceedings of the Geological Society* (London), 65 (1909), 1–lvii. Shortly before Sorby's death Geikie had discussed nineteenth-century achievements in petrology as part of his anniversary address as president of the Geological Society—*Transactions of the Geological Society of London*, 64 (1908), 104–111—which presents Sorby's great impact on geology from a contemporary viewpoint. See also George P. Merrill, "The Development of Micro-Petrology," in *The First One Hundred Years of American Geology* (New Haven, 1924), 643–647.

For later analyses, see W. H. Wilcockson, "The Geological Work of Henry Clifton Sorby," in *Proceedings of the Yorkshire Geological Society*, 27 (1947), 1–22; and G. H. Humphries, "Sorby: The Father of Microscopical

Petrography," in C. S. Smith, ed., *The Sorby Centennial Symposium on the History of Metallurgy* (New York, 1963), 17–41. This centennial volume depicts the changes in metallurgy following Sorby and contains other comments on Sorby himself by the editor (ix–xix) and N. Higham (1–15). Sorby's metallurgical contributions also were analyzed by C. H. Desch in his 20-page pamphlet *The Services of Henry Clifton Sorby to Metallurgy* (Sheffield, 1921); and by C. S. Smith in "Metallography in Sheffield," ch. 13 of his *A History of Metallography* (Chicago, 1960). Records and memorabilia of Sorby are described by A. R. Entwisle, "An Account of the Exhibits Relating to Henry Clifton Sorby . . . ," in *Metallography 1963*, Special Report no. 80, Iron and Steel Institute (London, 1964), 313–326. A rather personal view of the role of microscopic petrography in the broader science of rocks is given by F. Y. Levinson-Lessing, *Vvedenie v istoriyu petrografii* (Leningrad, 1936), English trans. by S. I. Tomkeieff as *A Historical Survey of Petrology* (Edinburgh–London, 1954).

CYRIL STANLEY SMITH

SØRENSEN, SØREN PETER LAURITZ (*b*. Havrebjerg, Slagelse, Denmark, 9 January 1868; *d*. Copenhagen, Denmark, 12 February 1939), *chemistry.*

The son of a farmer, Sørensen was educated at the high school at Sorø and entered the University of Copenhagen at the age of eighteen. He planned to study medicine; but under the influence of S. M. Jorgensen, an important investigator of inorganic complex compounds, he chose chemistry for his career. While at the university Sørensen received two gold medals, the first for a paper on the concept of the chemical radical and the second for a study of strontium compounds. While working for the doctorate he assisted in a geological survey of Denmark, acted as assistant in chemistry at the laboratory of the Danish Polytechnic Institute, and served as a consultant at the royal naval dockyard. His doctoral dissertation (1899) concerned the chemistry of cobaltic oxides. Thus most of his training was in inorganic chemistry.

All this was changed when, in 1901, Sørensen succeeded Johann Kjeldahl as director of the chemical department of the Carlsberg Laboratory in Copenhagen, where he remained for the rest of his life. Kjeldahl had worked on biochemical problems, and Sørensen continued this line of inquiry. His investigations can be divided into four classes: synthesis of amino acids, analytical studies, work on hydrogen ion concentration, and studies on proteins. The first, beginning in 1902, was concerned with synthesis of such amino acids as ornithine,

proline, and arginine. The following year he demonstrated that the Kjeldahl method for determination of amino nitrogen was of much greater generality than its discoverer had claimed. After working out the Formol titration method for analysis of proteins, he turned to a study of the effects of such buffers as borates, citrates, phosphates, and glycine on the behavior of proteins, with especial attention to enzymes.

This work led Sørensen to study the action of quinhydrone electrodes and the effect of ion concentration in the analysis of proteins. His most notable suggestion came from this work. In 1909 he investigated the EMF method for determining hydrogen ion concentration and introduced the concept of pH as an easy and convenient method for expressing this value. He was particularly interested in the effects of changes in pH on precipitation of proteins. After 1910 Sørensen made many studies on the application of thermodynamics to proteins and the quantitative characterization of these substances in terms of laws and constants. In much of this work he was assisted by his wife, Margrethe Høyrup Sørensen. They studied lipoproteins and the complexes of carbon monoxide with hemoglobin and in 1917 succeeded in crystallizing egg albumin for the first time.

Sørensen always encouraged visiting scientists at the Carlsberg Laboratory to work on medical problems. He also was active in chemical technology, contributing to the Danish spirits, yeast, and explosive industries. He received many honors from both scientific and technological societies. Sørensen retired in 1938 after a period of poor health and died the following year.

BIBLIOGRAPHY

I. ORIGINAL WORKS. There is a complete bibliography in *Kolloidzeitschrift*, **88** (1939), 136–139. The pH concept is presented in "Enzymstudien. II. Über die Messung und die Bedeutung der Wasserstoffionkonzentration bei enzymatischen prozessen," in *Biochemische Zeitschrift*, **21** (1909), 131–200. The isolation of crystalline egg albumin is described in "On the Composition and Properties of Egg-Albumin Separated in Crystalline Form by Means of Ammonium Sulphate," in *Comptes rendus du Laboratoire de Carlsberg*, **12** (1917), 164–212.

II. SECONDARY LITERATURE. Biographical sources are the Sørensen memorial lecture by E. K. Rideal, in *Journal of the Chemical Society* (1940), 554–561; K. Linderstrøm-Lang, "S. P. L. Sørensen," in *Kolloidzeitschrift*, **88** (1939), 129–136; and Edwin J. Cohn, "Søren

Peter Lauritz Sørensen," in *Journal of the American Chemical Society*, **61** (1939), 2573–2574.

HENRY M. LEICESTER

SOSIGENES (*fl*. Rome, middle of first century B.C.), *astronomy*.

Sosigenes helped Julius Caesar with his reform of the calendar. Caesar is said to have made use of Egyptian astronomy, but this may mean only that he discussed astronomy with Greeks from Alexandria. It is, in any case, not certain that Sosigenes was an Alexandrian, and he is not the only person whom Caesar consulted. Plutarch (*Caesar*, 59) simply states, without mentioning any names, that Caesar consulted the best philosophers and mathematicians before producing an improved calendar of his own. Caesar's adoption of the 365-1/4-day solar year may have been one result of Sosigenes' advice, and the statesman's seasonal calendar another. The 365-1/4-day year could even have been borrowed directly from Callippus at the suggestion of Sosigenes. All that Pliny says in this connection, however, is that during Caesar's dictatorship Sosigenes helped him to bring the years back into conformity with the sun (*Naturalis historia* 18.211). He adds (*Naturalis historia* 18.212) that Sosigenes wrote three treatises, including corrections of his own statements.

Sosigenes agreed with Cidenas in giving the greatest elongation of Mercury from the sun as 22° (Pliny, *Naturalis historia* 2.39). It is therefore possible, but far from certain, that he made use of Babylonian astronomical knowledge. Lucan (*Pharsalia* 10.187) implies that Caesar tried to improve upon the seasonal calendar of Eudoxus— "nec meus Eudoxi vincetur fastibus annus" ("and my year shall not be found inferior to the calendar of Eudoxus"). Theodor Mommsen maintains that Caesar ". . . with the help of the Greek mathematician Sosigenes introduced the Italian farmer's year regulated according to the Egyptian calendar of Eudoxus, as well as a rational system of intercalation, into religious and official use." Mommsen here alludes to the calendar in the papyrus *Ars Eudoxi*, but there is no proof of any close connection between the ideas of Sosigenes and the doctrines in the *Ars*.

BIBLIOGRAPHY

On Caesar's alleged use of "Egyptian" sources, see Appian, *Bella civilia* 2.154; Dio Cassius, *Hist. Rom.* 43.26; and Macrobius, *Saturnalia* 1.16.39 and 1.14.3. There are useful discussions regarding Caesar and Eudoxus' seasonal calendar in A. Böckh, *Ueber die vierjährige Sonnenkreise der Alten* (Berlin, 1863), 340–342; F. K. Ginzel, *Handbuch der mathematischen und technischen Chronologie*, II (Leipzig, 1911), 274–277; and Pauly-Wissowa, *Real-Encyclopädie*, 2nd ser., III (Stuttgart, 1927), s.v. Sosigenes (b) 1153–1157—compare Theodor Mommsen, *The History of Rome*, IV (London, 1887), 555. The calendar in the *Ars Eudoxi* is discussed in C. Wachsmuth, *Ioannis Laurentii Lydi Liber de ostentis et calendaria Graeca omnia* (Leipzig, 1897), lxviii–lxix, 299–301.

G. L. HUXLEY

SOTO, DOMINGO DE (*b*. Segovia, Spain, 1494 or 1495; *d*. Salamanca, Spain, 15 November 1560), *logic, natural philosophy*.

Born to parents of modest means who gave him the baptismal name of Francisco, Soto received his Latin training at Segovia under Juan de Oteo and Sancho de Villaveses. He continued his education in arts at the newly founded University of Alcalá, where he studied logic and natural philosophy under Thomas of Villanova and earned the baccalaureate in 1516. Shortly thereafter he transferred to the College of Santa Barbara at the University of Paris; his preceptors included Juan de Celaya, under whose tutelage he became acquainted with the terminist physics then current in Paris, where he completed the master's degree in arts. He then began the study of theology, while teaching the arts, and came under the influence of the Scottish nominalist John Major, who was then teaching at the Collège de Montaigu (along with two of Soto's fellow Segovians, Luis and Antonio Coronel), and the Spanish Thomist Francisco de Vitoria, who was lecturing at the Dominican priory of Saint-Jacques.

In 1519, however, Soto's longing for Spain and for his close friend Pedro Fernández de Saavedra prompted his return to Alcalá, where he completed the course in theology under Pedro Ciruelo and immediately (October 1520) occupied the chair of philosophy at the College of San Ildefonso. Here he taught logic, physics, and metaphysics until early in 1524, when internal difficulties in the college led him to resign his post. By this time he had received the licentiate in theology at San Ildefonso. He withdrew temporarily to the Benedictine abbey of Montserrat and was advised there to enter the Dominican order. In the summer of 1524 he became a Dominican novice at the priory of San Pa-

547

blo in Burgos, changing his name to Domingo and being professed on 23 July 1525.

Assigned to the priory of San Esteban in Salamanca, Soto taught theology until 1532, a period of service interrupted only by a stay in Burgos during 1528–1529 while supervising the publication of his first work, the *Summulae*. During the academic year 1531–1532 he substituted for his former mentor, Francisco de Vitoria, who held the "prime chair" of theology at the University of Salamanca. The next year Soto was elected to the "vesper chair" of theology at the same university, a post he held for sixteen years. During this period he prepared a second edition of the *Summulae* (1539), a *Dialectica* (1543), and a commentary and questions on the *Physics* of Aristotle (1545). Immediately adopted at both Salamanca and Alcalá, these works went through many editions in Spain and elsewhere.

The works on the *Physics* are particularly important for the history of science, since in his questions on Book VII Soto was the first to apply the expression "uniformly difform" to the motion of falling bodies, thereby indicating that they accelerate uniformly when they fall and thus adumbrating Galileo's law of falling bodies. Soto accounted for the velocity increase in terms of an accidental impetus built up in the body. He assimilated the "calculatory" techniques developed at Merton College, Oxford, in the fourteenth century and the terminist physics perfected at Paris during the early sixteenth century within a Thomistic framework, and thus dealt with most of the physical problems that interested the nominalists and realists of his day. On this account he is sometimes charged with eclecticism, although he tried to work out a position intermediate between those of Duns Scotus and Ockham and more consistent with Aquinas' teaching. Soto had distinctive views on the nature of motion, time and space, infinity, movement through a vacuum, maxima and minima, and the ratios of velocities. He subscribed to the Ptolemaic theory of the universe and generally defended the Scholastic Aristotelian theses of natural philosophy.

Soto was called to the Council of Trent early in 1545, having just completed his questions and commentary on Book VII of the *Physics*; the incomplete texts were printed immediately but did not include the passages of interest to present-day historians of science. He returned from Trent in 1550 and finished both texts, which were published at Salamanca in 1551. (In all, these works went through nine editions, the penultimate appearing at Venice in 1582, when Galileo was beginning his studies at Pisa. Soto's questions on the *Physics* are cited by Galileo in his *Juvenilia*, although not in the context of discussions of falling bodies.) While at Trent, Soto was closely associated with the Spanish ambassador to Venice, Diego Hurtado de Mendoza, who had studied the science of weights under Niccolò Tartaglia; Mendoza's correspondence shows him critical of Soto's physics, probably more because of Mendoza's Averroist and classical leanings than because of any particular attachment, on his part, to Archimedean statics.

Soto held various professorial and administrative positions at Salamanca until his death. He achieved renown in this university city for his extensive knowledge of both philosophy and theology, and is best known for his work in political philosophy, *De iure et iustitia* (1553–1554), in which he developed concepts of natural law and a "translation theory" of the origin of political authority. His competence is attested by a saying current in sixteenth-century Spain: "Qui scit Sotum, scit totum" ("Whoever knows Soto, knows everything").

BIBLIOGRAPHY

I. Original Works. For a complete listing of Soto's writings, see Vicente Beltrán de Heredia, O.P., *Domingo de Soto: Estudio biográfico documentado* (Salamanca, 1960), 515–588. Brief Latin and English texts from Soto's works on the *Physics* are in Marshall Clagett, *The Science of Mechanics in the Middle Ages* (Madison, Wis., 1959), 257, 555–556, 658. Pierre Duhem, *Études sur Léonard de Vinci*, III (Paris, 1913), gives excerpts from the same in French translation.

II. Secondary Literature. See William A. Wallace, O.P., "The Concept of Motion in the Sixteenth Century," in *Proceedings of the American Catholic Philosophical Association*, 41 (1967), 184–195: "The Enigma of Domingo de Soto: *Uniformiter difformis* and Falling Bodies in Late Medieval Physics," in *Isis*, 59 (1968), 384–401; and "The 'Calculatores' in Early Sixteenth-Century Physics," in *British Journal for the History of Science*, 4 (1968–1969), 221–232. See also Erika Spivakovsky, "Diego Hurtado de Mendoza and Averroism," in *Journal of the History of Ideas*, 26 (1965), 307–326; Vicente Muñoz Delgado, *La logica nominalista en la Universidad de Salamanca* (1510–1530), Publicaciones del Monasterio de Poyo, XI (Madrid, 1964); *Logica formal y filosofia en Domingo de Soto*, Publicaciones del Monasterio de Poyo, XVI (Madrid, 1964); and W. A. Wallace, "Galileo and the Thomists," in Armand Maurer *et al.*, eds., *St. Thomas Aquinas Commemorative Studies 1274–1974*, II (Toronto, 1974), 293–330.

William A. Wallace, O.P.

SOULAVIE, JEAN-LOUIS GIRAUD (*b.* Largentière, Ardèche, France, 8 July 1752; *d.* Paris, France, 11 March 1813), *geology.*

Ordained in 1776, Soulavie was one of the many philosophical *abbés* and pamphleteers active before the Revolution. He became an early member of the Jacobin Club, supported the Civil Constitution of the Clergy (1790), and served as the diplomatic resident of the First Republic in Geneva (1793–1794). He married in 1792 and was later permitted to return to secular life by Pope Pius VII. After Thermidor (27 July 1794) he devoted himself to the writing and editing of memoirs concerned with the history of France; perhaps best known is his *Mémoires historiques et politiques du règne de Louis XVI* (1801), which, like his other works, remains difficult to evaluate for its accuracy and historical significance.

Soulavie's scientific activities occupied a relatively short period of his life. A self-taught amateur, he was widely read and spent some time during the 1770's exploring the volcanic regions of Vivarais and Velay. On his arrival at Paris in 1778, his geological views had already been formulated; after additional field trips, he returned to Paris in 1780, established permanent residence there, and became a familiar figure at several salons.

Soulavie's major geological publication was the eight-volume *Histoire naturelle de la France méridionale* (1780–1784). Beginning with the then common idea that most sedimentary formations had been deposited by a universal, gradually diminishing ocean, he went on to stress and develop the principle of superposition. Soulavie, however, used superposition not only to determine the relative ages of strata, but he also attempted to correlate age with fossil remains. He argued that the oldest strata also contain the largest proportion of extinct species, while the youngest show a predominance of forms with living analogues.[1] He then attempted to work out a local geochronology for Vivarais by taking note of those sedimentary formations in which volcanic debris could be found.[2] These observations and ideas were expressed on geological maps of his own design, using a combination of symbols, hachures, and color.[3]

Soulavie's geological ideas were actually less clear and consistent than is suggested by any one of his publications. Although he always insisted that volcanic activity was more important and widespread than some contemporaries believed, he was vague and contradictory about the source of volcanic heat. On occasion, he seems to have held Neptunist views of the nature of the earth's core and oldest formations; elsewhere, however, he discussed the probable existence of a central heat within the earth.[4] In different portions of the *Histoire naturelle*, he emphasized both the extinction of species and the likelihood that seemingly extinct species had merely migrated to warmer climates.[5]

Although extravagant claims have been made for Soulavie's originality, his place in the history of geology cannot yet be assessed. Certain of his contemporaries admired his boldness and imagination, while others condemned the very same traits. He himself condemned system-building and was complimented on his "method of philosophizing" by Benjamin Franklin.[6] However, Buffon (to whom Soulavie was indebted for many of his ideas) roundly condemned Soulavie as an observer and thinker. Such negative views were not shared by some members of the Académie Royale des Sciences, which awarded its *privilège* to the first two volumes of Soulavie's *Histoire naturelle*. The conflicting evaluations can be attributed in part to factionalism within the French scientific community, but also, in part, to the fact that naturalists were not wholly in agreement about two of Soulavie's major ideas: the extinction of species and the importance of volcanic activity.[7]

NOTES

1. *Histoire naturelle*, esp. I, 161–163, 317–332. References to this work are to the copy at the Bibliothèque Nationale, *cotes* S. 21194–21200 and Rés. S. 1158. See Bibliography.
2. *Ibid.*, esp. II, 362–377; IV, 16, 42–44.
3. *Ibid.*, I, 143–149, for his method of constructing maps. A good example of the result is in vol. II, and the same map is in *Géographie* (in color in the copy at the Bibliothèque Nationale).
4. For Neptunism and the recent origins of volcanoes, see *Classes*, 101, 140–141, 149 (table), 157. The role of central heat is treated in *Oeuvres*, 290–297, and *Histoire naturelle*, I, 167.
5. *Ibid.*, V, 217–221, and above, n. 1.
6. Carl Van Doren, *Benjamin Franklin* (1938; reissue New York: Viking Press, 1964), 659–660. For Soulavie on system-building, *Oeuvres*, 280–281, 288.
7. *Correspondance inédite de Buffon*, H. Nadault de Buffon, ed., II (Paris, 1860), 109. Also, letter of Faujas de St.-Fond, in Bibliothèque Municipale de Nîmes, MS 94, fols. 59–60. Specific scientific issues separating Buffon and Soulavie are mentioned by Aufrère, p. 38. The copy of *Classes* at the Bibliothèque Nationale includes a prefatory statement by the Imperial Academy that it admires some of Soulavie's ideas and information, but "ne prétend pas autoriser par son suffrage [ses] hypothèses . . . hazardés." The work received a second *accessit* (third prize).

BIBLIOGRAPHY

I. ORIGINAL WORKS. *Géographie de la nature* (Paris, 1780) is a brochure presenting the principal ideas of

Soulavie's *Histoire naturelle de la France Méridionale*, 8 vols. (Paris, 1780–1784). Some sections of the latter work bear different titles and have their own publication history; the order in which parts are bound may vary in different copies. The first 7 vols. deal with "minerals," and another vol. on this subject was apparently planned (see Bibliothèque Nationale, *Catalogue des livres imprimés, s.v.* Soulavie, entry S. 21206). Vol. VIII deals with the plant kingdom. *Prospectus de l' "Histoire naturelle de la France méridionale"* (Nîmes, 1780) was reprinted in the *Histoire naturelle*, I, 3–51. *Oeuvres complettes de M. le Chevalier Hamilton* (Paris, 1781) has extensive notes and commentary by Soulavie; part of the volume consists of Sir William Hamilton's *Campi phlegraei* (Naples, 1776). *Les Classes naturelles des minéraux et les époques de la nature correspondantes à chaque classe* (St. Petersburg, 1786) was written in response to a prize question posed in 1785 by the Imperial Academy of St. Petersburg. He published articles in *Observations sur la physique, sur l'histoire naturelle et sur les arts*, as well as many works on nonscientific subjects. Soulavie's library and other possessions were sold in several lots after his death; of the extant sales catalogs, the one listing his science library is *Notice des principaux articles composant le cabinet de livres, tableaux, gravures, et collection d'estampes* . . . (Paris, 1813), B.N., Δ13478.

II. SECONDARY LITERATURE. See Albin Mazon, *Histoire de Soulavie (naturaliste, diplomate, historien)*, 2 vols. (Paris, 1893), and *Appendice à l' "Histoire de Soulavie"* (Privas, 1901); E.-J.-A. d'Archiac de St.-Simon, *Introduction à l'étude de la paléontologie stratigraphique*, 2 vols. (Paris, 1864), I, 348–354, and *Géologie et paléontologie* (Paris, 1866), 142–145. Mazon relies heavily on Archiac in discussing Soulavie's geology. Léon Aufrère. *De Thalès à Davis. Le relief et la sculpture de la terre. Tome IV. La fin du XVIIIᵉ siècle. I. Soulavie et son secret* (Paris, 1952), 71–83, discusses the "unexpurgated" and "expurgated" versions of vol. I of the *Histoire naturelle*; the final version omitted Soulavie's evidence for the great age of the earth.

RHODA RAPPAPORT

SOULEYET, LOUIS-FRANÇOIS-AUGUSTE (*b.* Besse, Var, France, 8 January 1811; *d.* Martinique, 7 October 1852), *zoology.*

Souleyet is one of several health officers in the French navy who won renown for zoological work connected with a voyage of circumnavigation. Nothing is known about his family and childhood. His entry into the health service of the navy was fairly late; and after some difficulty at the outset of his career, he obtained in 1835 the opportunity to sail on the voyage of the *Bonite* around the globe under the command of August-

Nicholas Vaillant. The main purpose of the voyage was to transport French consular agents to various parts of the world. In conformance with the practice begun on Louis-Claude de Freycinet's voyage of the *Uranie* (1817–1820), all scientific research was to be handled by members of the navy. Fortuné Eydoux, surgeon major of the expedition, was charged with zoology, and Souleyet, as second surgeon, became associated with his research. Souleyet also benefited from the friendship and guidance of Charles Gaudichaud-Beaupré, pharmacist and adjoint to the expedition for research on natural history. Gaudichaud introduced Souleyet to the study of pelagic mollusks, which became his field of specialization. The *Bonite* left Toulon in February 1836, stopped in South America, Hawaii, the Philippines, and various parts of the Indian and Chinese seas, and returned to Brest in November 1837. Because the brevity of the stopovers precluded an investigation of the ecology or anthropology of the areas visited, the naturalists concentrated on collecting new species, particularly microscopic mollusks. The collections were deposited at the Museum of Natural History and cataloged by the professors there.

Eydoux, who after the voyage became physician in chief at Martinique, died there in 1841, leaving Souleyet with most of the work of publishing the zoology of the voyage. Another tour of duty during 1846–1849 interrupted publication. Souleyet finally returned to Paris in January 1850 and completed the work in 1851. Free to pursue his career, he was preparing for the competitive examination to become a professor when he received orders to embark for the Antilles. Leaving France reluctantly, he reached Martinique in July 1852, during an epidemic of yellow fever, and soon fell victim to the disease.

The results of Souleyet's anatomical and physiological investigations of pteropod and gastropod mollusks are contained in the second volume of the zoology of the *Bonite* voyage and in various memoirs. Zoologists had been divided over the composition of the Pteropoda, a group created by Cuvier, and the position of this group among the mollusks. Souleyet reworked its classification, removing the Heteropoda and dividing it into four natural families. Accepting Blainville's ranging of the Pteropoda among the Gastropoda, Souleyet pointed out the analogies between these two groups of mollusks. He demonstrated that the alary expansion of the pteropods is merely the foot of the gastropod in disguise. In 1852 Souleyet completed a monograph on the pteropods begun in 1830 by P.-C.-A.-L. Rang. Besides writing the entire text, which appears to be an abbreviated version of the zoology of the *Bonite*, Souleyet added several plates depicting newly discovered species.

Also important is Souleyet's description of the nervous collar of mollusks, which he believed to correspond to both the brain and the spinal cord of vertebrates. He argued that the apparently anomalous nervous collar of pteropods was merely a modification of the general form of that structure in mollusks.

From 1844 Souleyet was involved in a controversy with Armand de Quatrefages on the subject of "phlebenterism." Quatrefages had established a new order of gastropod mollusks, the Phlebenterata, which he defined as degraded gastropods with no proper respiratory organs and an imperfect or absent circulatory system. He claimed that a system of intestinal canals, a "gastro-vascular apparatus," took over part of the functions of respiration and circulation. Elaborating on the general ideas of his master, Henri Milne-Edwards, Quatrefages believed that the animal kingdom was composed of several series in which the type was effaced at the lower limits. Phlebenterism was a general phenomenon of degeneration among animals.

Souleyet, in a series of memoirs and notes beginning with "Observations sur les mollusques gastéropodes designées sous le nom de *Phlébentérés* par M. de Quatrefages" (*Comptes rendus . . . de l'Académie des sciences*, **19** [1844], 355–362), attacked Quatrefages's position. He argued that the Phlebenterata were not essentially different from other Gastropoda, that they did have a complete circulatory system including veins, and that the "gastro-vascular" canals were in reality hepatic canals not without analogues in other mollusks. The controversy touched on the large questions of whether types degraded, whether exterior and interior conformation can be independent, and whether organs can degenerate and be replaced by other organs developed especially for the purpose. The issues were important enough to merit two commission reports, one by the Academy of Sciences and the other by the Society of Biology. The former report was noncommittal, and the latter declared in favor of Souleyet.

BIBLIOGRAPHY

I. ORIGINAL WORKS. With Fortuné Eydoux, Souleyet published *Voyage autour du monde exécuté pendant les* années 1836 et 1837 sur la corvette la Bonite commandée par M. Vaillant, Zoologie, 2 vols. plus atlas (Paris, 1841–1852). This work contains Blainville's instructions to the zoologists of the voyage in the name of the Academy of Sciences and his report on the zoological results of the voyage. The instructions and report had previously been published in *Comptes rendus . . . de l'Académie des sciences*, **1** (1835), 373–377, and **6** (1838), 445–460, respectively. With P.-C.-A.-L. Rang, Souleyet published *Histoire naturelle des mollusques ptéropodes* (Paris, 1852). A list of Souleyet's memoirs written alone or in conjunction with Eydoux is in the Royal Society *Catalogue of Scientific Papers*, V, 760.

II. SECONDARY LITERATURE. Apparently the only biographical notice on Souleyet is S. Petit, "Louis-François-Auguste Souleyet," in *Journal de conchyliologie*, **4** (1853), 107–111. For a helpful review of Souleyet's second volume of the *Bonite* voyage, see Pierre Gratiolet, "Zoologie du voyage de la *Bonite*, par MM. Eydoux et Souleyet," in *Journal de conchyliologie*, **4** (1853), 93–107. On the phlebenterism controversy, see the commission report of the Academy of Sciences, presented by Isidore Geoffroy Saint-Hilaire in *Comptes rendus . . . de l'Académie des sciences*, **32** (1851), 33–46, and in *Mémoires de l'Académie des sciences . . .*, **23** (1853), 83–104; and the commission report of the Society of Biology, presented by Charles Robin in *Comptes rendus des séances de la Société de biologie*, **3** (1851), 5–132.

TOBY A. APPEL

SOUTH, JAMES (*b.* Southwark, London, England, October 1785; *d.* Campden Hill, Kensington, London, 19 October 1867), *astronomy.*

South's great disappointments in, and severe criticisms of, contemporary scientific institutions often overwhelmed his actual scientific accomplishments. The son of a pharmaceutical chemist, he had studied surgery, become a member of the Royal College of Surgeons, and acquired an extensive practice when, through marriage in 1816, he became sufficiently wealthy to forgo medicine and devote himself to astronomy. He established several observatories, in the environs of London and Paris, where he observed with some of the finest telescopes available. From its inception in 1820 South held various offices in the Astronomical Society of London; barred by a technicality from serving as first president of the chartered Royal Astronomical Society in 1831, he thereupon left the organization. He was knighted in 1831 and awarded an honorary LL.D by Cambridge in 1863; in addition he was a fellow of the Royal Society of London (1821), the Linnean Society, and the Royal Society of Edinburgh, and a member of the Royal Irish Academy, the Académie Royale des Sciences, des Lettres, et des Beaux-Arts Belgique, and the Academia Scientiarum Imperialis Petropolitana.

Double stars, essentially discovered by William Herschel, were of great interest throughout the nineteenth century—new ones being found and position measurements made more precise with each improvement in telescope construction. South, working with John Herschel during the years 1821–1823, reobserved the double stars charted originally by William Herschel, mainly for the purpose of detecting position changes. Their observations helped verify the newly recognized orbital motion of these neighboring stars. Their resulting catalog of 380 double stars, presented to the Royal Society in 1824, earned them the gold medal of the Astronomical Society and the grand prize of the Institut de France. For his second catalog of double stars, two years later, South was awarded the Copley Medal of the Royal Society.

South was concerned, perhaps rightly so, by the decline of science in Britain. In 1822 he published a criticism of the *Nautical Almanac,* alleging its inferiority to Continental ones; and in 1829 he presided over an Astronomical Society committee charged with suggesting improvements in this institution. In 1830 South publicly criticized the Royal Society, but to no avail. His major disappointment, however, came from his quarrel with Edward Troughton. In 1829 South bought a French achromatic objective of 11.7 inches aperture, one of the largest in the world, and contracted with Troughton—who had made many of his other instruments—for an equatorial mount. South's dissatisfaction with Troughton's work led to a court suit that Troughton eventually won, extended and acrimonious debates, and South's public destruction of the mount in 1836.

BIBLIOGRAPHY

South's articles are listed in the Royal Society *Catalogue of Scientific Papers,* V, 761–762; his books, in the British Museum *General Catalogue of Printed Books.*

Secondary literature includes J. C., "Sir James South," in *Monthly Notices of the Royal Astronomical Society,* **28** (1867–1868), 69–72; A. M. C[lerk], "Sir James South," in *Dictionary of National Biography,* LIII, 272–274; and T. R. R., "James South," in *Proceedings of the Royal Society,* **16** (1867–1868), xliv–xlvii.

DEBORAH J. WARNER

SOWERBY, JAMES (*b.* London, England, 21 March 1757; *d.* London, 25 October 1822), *natural history, geology.*

Sowerby, son of John and Arabella Sowerby, was trained as an artist and studied at the Royal Academy of Arts. He married Anne de Carle, of Norwich; and their sons, particularly the eldest, James de Carle Sowerby (1787–1871), and the second, George Brettingham Sowerby (1788–1854), from an early age assisted him with his work. Their children, too, were artists and naturalists, so that throughout the nineteenth century there were Sowerbys illustrating works of natural history.

Sowerby is best known for his illustrations to *English Botany; or Coloured Figures of British Plants, With Their Essential Characters, Synonyms, and Places of Growth* (1790–1814). The text was supplied by James Edward Smith, whose name was at first withheld, at his own request; and the work became widely known as Sowerby's *Botany.* Sowerby's skillful drawings, beautifully colored, and Smith's accurate descriptions made it a highly esteemed work that was frequently reissued, later with supplements.

In 1802 Sowerby began to issue *British Mineralogy,* also with colored plates, in parts and followed it with *Exotic Mineralogy.* More important was his *Mineral Conchology of Great Britain,* illustrating "remains of Testaceous Animals or Shells," issued in parts from 1812 and continued after his death by his son James de Carle. Although lacking any systematic arrangement, it was a valuable aid to collectors and is still important as a reference work. Sowerby also prepared illustrations for many natural history works, including William Smith's *Strata Identified by Organized Fossils.*

Not the least of Sowerby's contributions to natural history was his vast correspondence with naturalists in Britain and abroad, in which he encouraged and advised collectors of plants, birds, insects, fossils, and minerals. Specimens were sent to him for identification, and he sent in return other specimens as well as parts of his publications, thus stimulating further research. His own museum, at 2 Mead Place, Lambeth, was regularly visited by naturalists.

BIBLIOGRAPHY

I. ORIGINAL WORKS. There is a useful, although incomplete, list of Sowerby's publications, with many bibliographical details, in *Catalogue of the Books, Manuscripts, Maps and Drawings in the British Museum (Natural History),* V (London, 1915), 1981–1983. See also R. J. Cleevely, "A Provisional Bibliography of Natural History Works by the Sowerby Family" in *Journal of the Society for the Bibliography of Natural History*, **6** (1974), 482–559. Sowerby's herbarium and more than 2,500 original watercolor drawings for *English Botany* were purchased in 1859 by the British Museum (Natural History), which also bought his collection of about 5,000 fossils in 1861. A large collection of his correspondence is also in the museum. See Jessie Bell MacDonald, "The Sowerby Collection in the British Museum (Natural History): A Brief Description of Its Holdings . . .," *ibid.*, **6** (1974), 380–401.

II. SECONDARY LITERATURE. There is no definitive biography, but A. de C. Sowerby *et al.*, *The Sowerby Saga* (Washington, 1952), has much information about the family. An obituary notice appeared in *Gentleman's Magazine*, **92**, pt. 2 (Dec. 1822), 568. See also R. J. Cleevely, "The Sowerbys, the *Mineral Conchology*, and Their Fossil Collection," in *Journal of the Society for the Bibliography of Natural History*, **6** (1974), 418–481.

JOAN M. EYLES

SPALLANZANI, LAZZARO (*b*. Scandiano, Italy, 12 January 1729; *d*. Pavia, Italy, 11 February 1799), *natural history, experimental biology, physiology.*

Among the many dedicated natural philosophers of the eighteenth century, Spallanzani stands preeminent for applying bold and imaginative experimental methods to an extraordinary range of hypotheses and phenomena. His main scientific interests were biological and he acquired a mastery of microscopy; but he probed also into problems of physics, chemistry, geology, and meteorology, and pioneered in volcanology. Acute powers of observation and a broadly trained and logical mind helped him to clarify mysteries as diverse as stone skipping on water; the resuscitation of Rotifera and the regeneration of decapitated snail heads; the migrations of swallows and eels and the flight of bats; the electric discharge of the torpedo fish; and the genesis of thunderclouds or a waterspout. His ingenious and painstaking researches illuminated the physiology of blood circulation and of digestion in man and animals, and also of reproduction and respiration in animals and plants. The relentless thoroughness of his work on the animalcules of infusions discredited the doctrine of spontaneous generation and pointed the way to preservation of foodstuffs by heat.

Spallanzani's father, Gianniccolò, was a successful lawyer. He was of locally established stock; and his wife, Lucia Zigliani, came of good family from Colorno, in the duchy of Parma. The natal house is still preserved in Scandiano, a small town in the province of Emilia, northeast of the Apennines. They had a large, closely knit family, but of Lazzaro's siblings only two sisters and a brother feature in his letters. His younger brother Niccolò, who acquired a doctorate in law and wide knowledge of agronomy, and his sister Marianna, who became a naturalist, shared many of his scientific interests; both survived him.

After attending the local school, Lazzaro went at age fifteen to a Jesuit seminary in Reggio Emilia, seven miles away, where he excelled in rhetoric, philosophy, and languages. The Dominicans wanted him to join their order. Instead, he left Reggio Emilia in 1749 to study jurisprudence at the ancient University of Bologna, where Laura Bassi, a cousin on the paternal side, was professor of physics and mathematics. Under this remarkable woman's influence, Spallanzani liberalized his education. New subjects included mathematics, which impressed him with the significance of quantitative exactitude, while physics, chemistry, and natural history aroused his curiosity and revealed his bent. His classical talents, stimulated and polished, brought lasting advantages in historical awareness and aptness of self-expression; and he acquired an invaluable knowledge of French. For some three years he also worked toward his doctorate in law, a project that familiarized him with logic but otherwise grew distasteful. With Laura Bassi's support, Antonio Vallisneri the younger, professor of natural history at Padua and a fellow Scandianese, secured paternal consent for Lazzaro to abandon jurisprudence and follow his predilections.

In 1753 or 1754 Spallanzani became a doctor of philosophy. Then, having received instruction in metaphysics and theology, he took minor orders. Within a few years he was ordained priest and attached to two congregations in Modena. By 1760 he was designated "l'Abate Spallanzani" and was generally known as such thereafter. His priestly offices were performed irregularly; nevertheless, even in later life he still officiated at mass. Since he had no private income, the financial assistance (and moral protection) of the church facilitated his investigations of natural phenomena.

Apart from casual religious commitments, and despite an insatiable enthusiasm for travel, his career was wholly academic and centered in Lombardy. The main features of his last thirty years (1769–1799) as professor of natural history at Pavia are documented adequately, and his period of tenure in the chair of philosophy at Modena now seems settled as 1763 to 1769; but some aspects of his initial appointments at Reggio Emilia remain unclear. Loss of earlier letters and lack of other records misled biographers into discrepant conjectures. Fortunately, fresh and dependable data are available in his *Bibliografia* (1951) and *Epistolaria* (1958–1964).

Reggio Emilia. Early in 1755 Spallanzani began teaching logic, metaphysics, and Greek at the ancient College of Reggio Emilia. Two years later he was appointed lecturer in applied mathematics at the small, recently founded University of Reggio Emilia. In 1758 he was concurrently professor of both Greek and French at Nuovo Collegio, which, presumably, replaced the old seminary. The university lectureship remained unaltered in title, but in 1760 his chair at the college was designated languages and by 1762 had become Greek.

Spallanzani assisted in the public oral examinations of graduating students, and his first publication was possibly the anonymously compiled booklet of astronomical questions headed *Ex coelestibus corporibus*, and variously titled *Theses philosophicae . . .* and *Propositiones physico-mathematicae . . .* (1757–1759), which served as the basis for interrogating candidates. During this period also, certain literary and philosophic papers read by him to the Accademia degli Ipocondriaci at Reggio Emilia may have appeared anonymously in print. His first acknowledged publication is a critique of A. Salvini's Italian translation of the *Iliad*. This work, *Riflessioni intorno alla traduzione dell'Iliade del Salvini . . .* (1760), comprising three letters addressed to Count Algarotti, chamberlain to Frederick the Great of Prussia, displayed intimate understanding of Greek style and metaphor. In numerous examples Spallanzani showed that the translator's prolixity enfeebled the vigor of the original and that the Italian language, when chosen felicitously, conserved the beauty and pith of Homer.

In the summer of 1761 Spallanzani set out for the Reggian Apennines and Lake Ventasso, on the first of many scientific excursions to various parts of Italy and elsewhere, in the multiple capacities of natural historian in the broadest sense, field investigator of unexplained occurrences, aggressive collector of museum specimens, and observer of humanity. An indefatigable walker and daring climber, his main concern on this journey was the origin of springs and fountains gushing from the mountain slopes. Descartes's contention that the source was seawater, purged of salinity by subterranean fires after reaching the mountain sides through devious channels, was superseded after

1715 by an unconfirmed hypothesis of Antonio Vallisneri the elder, who stated that water precipitated near the summit, whether as snow, rain, or mist, insinuated itself between sloping strata of the mountain and descended by gravity until arrested by an impervious stratum, whereupon it emerged from some hidden reservoir. Spallanzani verified the latter concept by observing such factors as the relationship between the number and size of springs and total precipitation in the area; the water-condensing characteristics of the mountain involved; and the disposition, nature, and water affinity of the constituent strata. Further, he disproved a local belief that a great whirlpool existed in the middle of Lake Ventasso. Embarking on a raft improvised of beech stumps, he sounded the depth of the lake at various points, including its center, afterward tracing its origin to two fountains. His report (1762) appeared in the form of two letters to Vallisneri the younger.

Spallanzani was introduced by Vallisneri in 1761 to works by Buffon and to those by his occasional collaborator, the English priest and microscopist John Turberville Needham. For some twenty years thereafter Spallanzani recurrently focused his attentions upon the fundamental phenomena of vitality and reproduction and on the doctrines of Buffon and Needham concerning them. Buffon had claimed in the second volume (1749) of his *Histoire naturelle* that all plant and animal matter (including seminal fluid) decomposed ultimately into minute motile particles, termed "organic molecules," which served as elementary building blocks for the reconstitution of every form of life. Buffon contended he could identify these particles in Needham's microscopic preparations. Needham, an enthusiastic but erratic experimenter, hurriedly published further surmises in *Nouvelles observations microscopiques . . .* (1750), an expanded French version of a letter to the Royal Society of London on the generation, composition, and decomposition of animal and vegetable substances. Needham described animalcules that developed in many kinds of infusions, despite precautions to exclude external air. These animalcules, and likewise those in spermatic fluid, eventually languished, died, and disintegrated. Their debris, with that of decomposing plants, resolved into filaments yielding "animals of an inferior species." He traced various other fancied modes and sequences of renascence. The agent provoking such spontaneous generation was designated a productive "vegetative force," present in the most minute component of organic matter.

Equipped in 1762 with an adequate microscope, Spallanzani began to repeat Needham's experiments. The work was interrupted by his departure from Reggio Emilia. Supposedly, chairs had been offered him from as far afield as Coimbra and St. Petersburg and also from neighboring Modena and Cesena. In 1763 he went to Modena as professor of philosophy at the university and at the College of Nobles. Here he was still only fourteen miles from the family home at Scandiano.

Modena. Francesco Redi's experiments on fly maggots in 1668 had dispelled the myth of spontaneous generation for complex animals. But the notion was reapplied to lesser forms of life after the pioneer microscopist Leeuwenhoek described the little animals teeming and cavorting in his infusions (1674); he and several of his successors supposed that these animals were of atmospheric origin. Spallanzani verified this surmise experimentally and proved the animalcules did not arise spontaneously. His infusions of vegetables or cereal seeds, whether boiled or unboiled, in plugged or open vessels, yielded various microorganisms, possessing such attributes of animality as definite shape, orderly motion, and ability to withstand certain degrees of heat or cold. Whereas Needham had abandoned as unavailing and superfluous all precautions designed to control these infusoria, Spallanzani redoubled efforts to prevent their appearance. In hundreds of experiments he tested various rituals for rendering infusions permanently barren and finally found that they remained free of microorganisms when put into flasks that were hermetically sealed and the contents boiled for one hour. The entrance of air into the flask through a slight crack in its neck was followed by proliferating infusoria. His masterful essay, dedicated to the Bologna Academy of Sciences, *Saggio di osservazioni microscopiche . . .* (1765), reported no spontaneous generation in strongly heated infusions protected from aerial contamination. Further, the causes of Needham's misinterpretations were analyzed and Buffon's assertions about organic molecules refuted. The work first appeared jointly, under the title *Dissertazioni due . . .* , with a short thesis in Latin about the mechanism of stone skipping on water, *De lapidibus ab aqua resilientibus.* This latter tract, dedicated to Laura Bassi, explained "ducks and drakes" physicomathematically.

Charles Bonnet at Geneva had predicted (unknown to Spallanzani) in his *Considérations sur les corps organisés* (1762) that Needham's claims would prove fallacious. A uniquely constructive and durable friendship developed between Bonnet and Spallanzani following the former's receipt of a copy of the *Saggio*. In 1765, after cutting up thousands of earthworms and exploiting the ability of the aquatic salamander to regrow its tail, Spallanzani resolved to investigate reproductive phenomena in animals and plants. He received encouragement from Bonnet (coupled with a warning against spreading his energy among too many problems) and began to study methodically the regeneration of lost parts in lower animals. This phenomenon, brought to attention twenty years earlier by Trembley's work on regrowth in polyps, had been extended to earthworms by Réaumur and Bonnet himself. Spallanzani found the precise location of those cuts in earthworms that affected the segmental regenerative response. Other species of worms displayed different reactions after being divided. Amputation of the tail of the freshwater boat worm, the young aquatic salamander, and the tadpole was followed by vascularization of the transparent growing stump, observable microscopically. Regenerative capacities of remarkable complexity and repetitiveness were noted in the horns of the slug; in the foot, horns, and head of the land snail; the limbs and jaw of the salamander; and the limbs of the toad and frog. Besides adding to the knowledge of the potentialities of the mechanism, Spallanzani established the general law that in susceptible species an inverse ratio obtains between the regenerative capacity and age of the animal.

Early in 1768 he reported these findings in *Prodromo di un opera da imprimersi sopra le riproduzioni animali,* which he intended as a prelude to a major work on animal reproduction. Reactions ranged from surprised interest to disbelief, particularly as regards the ability of decapitated snails to produce completely new heads. Some of his peers, on attempting to repeat the work, reported deaths or only partial regenerations in such snails; others confirmed his findings, including Bonnet and Senebier at Geneva, Laura Bassi, Lavoisier, and the Danish naturalist Müller. Spallanzani promptly sent a copy of the *Prodromo* to the secretary of the Royal Society of London (who translated it into English), and in that same year he was elected a fellow of the society. He detailed his experiments involving more than 700 decapitated snails in *Resultati di esperienze sopra la riproduzione della testa nelle lumache terrestri* (1782, 1784).

Two other publications appeared in 1768. The first, *Memorie sopra i muli . . .* , was a collection of communications about hybrids by Bonnet and various authors, edited by Spallanzani, who urged that experiments on insect hybridization were a possible means of disentangling the problem of generation. He did not pursue this particular path but later attempted to cross batrachian species and even such diverse animals as cats and dogs. A second booklet, *Dell'azione del cuore ne' vasi sanguini*, outlined his findings on the action of the heart upon the blood vessels and was addressed to the great physiologist Albrecht von Haller.

Haller's microscopic observations of blood movements in his *Deux mémoires sur le mouvement du sang* (1756) had been made by refracted light on medium-sized vessels in the isolated mesentery of the frog. Spallanzani, using P. Lyonet's novel dissecting apparatus, conducted his observations mostly in a darkened room with reflected light from sunbeams impinging upon exposed parts of the aquatic salamander. He systematically noted how the cardiac systolic force motivated the blood circulation. The rhythmic inequality of blood flow in the aorta and large vessels disappeared in medium and small arteries, becoming regular and uniform. The velocity diminished in the smaller vessels, but sinuosities did not retard the flow. In the smallest vessels, individual red corpuscles negotiated acute angles and folds by elastically changing shape. The blood velocity in the venous system increased as the caliber of the vessels enlarged. Haller responded to the many amplifications and corrections of his work by securing Spallanzani's election to the Royal Society of Sciences of Göttingen.

Pavia. His scientific accomplishments and growing renown as an eloquent, informative lecturer brought offers of chairs at Parma and Pavia. The latter city had been in Austrian hands for more than fifty years, and Maria Teresa's government sought to restore some of its ancient dignity by appointing new professors to a reconstituted university. The prospect of higher emoluments and greater distinction proved irresistible, and Spallanzani became professor of natural history at Pavia in November 1769. He had just completed a painstaking Italian translation of Bonnet's philosophic and eloquent *Contemplation de la nature*; and he looked forward to his official duties being confined to natural history—"my dominating passion for several years." A recent French version of the *Saggio*, elaborately annotated by Needham, contended that excessive heat enfeebled or destroyed the vegetative force of infusions and impaired the essential elasticity of air within sealed flasks. Spallanzani had lost patience with his opponent: "Quelle confusion, quelle obscurité règne-t-il dans ses notes à mes observations microscopiques!

Quelles monstruosités dans ses pensées!" he complained to Bonnet. His inaugural address, delivered in Latin and published as *Prolusio . . .* (1770), made clear his intention to settle the dispute with Needham and to rebut the peculiar views of his supporter Buffon. Meanwhile, educated laymen aligned themselves with scientists on each side of the spontaneous generation controversy, as happened a century later over Darwinian theories.

Spallanzani found his daily lectures taxing, and other duties interrupted his researches. He took charge of the public Museum of Natural History of the university, the development of which the court at Vienna supported through its minister plenipotentiary, the governor of Lombardy, Count Carlo di Firmian. The acquisition of exhibits proved congenial to Spallanzani's aggressive instincts and broad vision, so that within a decade the collections of the museum were among the most magnificent in Italy. In the summer of 1772, when the government sent him to visit the mines and collect fossils in the Alps north of Milan, his itinerary included lakes Como and Maggiore, and the towns and villages of Ticino. Less agreeable distractions ranged from the procurement and disposition of new specimens to the preparation of a complete catalogue. Nevertheless, because of his energy, versatility, and enterprise, Spallanzani secured monumental collections for posterity and also made lasting contributions to science in Pavia.

Spallanzani launched countless experiments relating to infusion animalcules and "spermatic worms," with results that soon made a chimera of the vegetative force and undermined the doctrine of organic molecules; but unforeseen complications or fresh ideas demanded further investigation, and publication was postponed. His previous observations on the physiology of circulation were expanded to include species of frogs and lizards. Through a chance discovery in 1771 that the vascular network in the umbilical cord of an embryonated hen's egg could be seen clearly with the Lyonet apparatus, he first established the existence of arteriovenous anastomoses in a warm-blooded animal. He also studied the effects of growth (in the chick embryo and tadpole) upon circulatory mechanisms; the influence of gravity and the consequences of wounds on different parts of the vascular system; and changes in the languid or failing circulation in dying animals. Finally, Spallanzani demonstrated that the arterial pulse is due not to mere cardiac displacements but to lateral pressure upon an expansile wall from cardiac impulsions conveyed by the blood column. A total of 337 experiments were outlined and expounded in four dissertations, forming a treatise on the dynamics of circulation that appeared as *De' fenomeni della circolazione . . .* (1773).

Spallanzani's next outstanding publication, *Opuscoli di fisica animale e vegetabile . . .* (1776), contained five reports that displayed unexcelled experimental skill, remarkable powers of observation, and lucid literary talent. The first volume included the long-deferred treatise on infusoria, "Osservazioni e sperienze intorno agli animalculi delle infusioni . . ." This work challenged Needham's concept of a heat-labile vegetative force by comparing the growth-promoting qualities of various infusions preheated to different extents and left in loosely stoppered flasks. After several days infusions that had been boiled for two hours generally showed better growth than corresponding preparations boiled for shorter periods. Profuse growth appeared in infusions made from vegetable seeds reduced to powder in a coffee roaster or burned to a cinder by a blowpipe. Spallanzani concluded that the vegetative force was imaginary. He disposed of Needham's other objection by instantaneously sealing the capillary end of the drawn-out neck of each infusion flask. Thus rarefaction of the enclosed air was avoided, and the sterility of boiled infusions could be attributed no longer to diminished elasticity of air in the sealed flasks.

Spallanzani also found that complex infusoria are more susceptible to heat and cold than the "infinitely minute" germs of lower class, whose relative resistance he ascribed to their eggs. He sought comparisons between such effects and the influence of temperature upon seeds and their respective plants; upon frog spawn, tadpoles, and adult frogs; and upon species of insects and their ova. Finally, he developed a technique for isolating single animalcules in water drops; he then observed their modes of reproduction, whether by transverse division, longitudinal fission, budding, or the peculiar daughter-colony system of *Volvox*, which he followed through thirteen generations.

The main treatise in the second volume of the *Opuscoli* confirmed and extended Leeuwenhoek's observations on spermatozoa (which began in 1677) and refuted Buffon's erroneous concepts of their nature and origin. The latter claimed that both male and female gonads contained a fluid teeming with nonspecific, incorruptible organic molecules, of which all living matter was fundamentally constituted. Tailed spermatic vermiculi, if present, developed in stale semen from mucilaginous and filamentous components and eventually disintegrated into smaller animalcules and organic molecules. Spallanzani always used fresh semen in studying spermatozoal form, size, motion, and reactions to heat, cold, and drying. He even demonstrated their unaltered appearance in the epididymis of live dogs and a ram. Initially he had considered the spermatic vermiculi analogous to infusion animalcules; but he later concluded that the former are distinctive components of the living animal body and that they neither shed their tails nor divide. Buffon's perversity stemmed from poor microscopy, metaphysical confusion, and overconfident eloquence; but since he had a considerable following, Spallanzani (who sincerely admired some of Buffon's accomplishments) at first hesitated to refute him. Five years earlier, however, Bonnet had urged him to overcome such diffidence: "You have cherished no theory, but are satisfied with interrogating nature, and giving the public a faithful account of her responses." Spallanzani therefore minced no words. Describing Buffon's theory as "completely destroyed," he urged him to repeat experiments "with better microscopes, forgetting his beloved organic molecules, and imposing the rule on himself to receive as truth only the images transmitted by the senses, without adding the corrections of his imagination."

The remaining three tracts were of lesser significance. The first concerned the effects of stagnant air upon animals and vegetables, which Spallanzani was led to investigate after observing the proliferation of infusoria and germination of vegetable seeds in sealed vessels. Impaired vitality occurred among his specimens after widely varying exposure periods. Death was accelerated when the volume of the container was diminished or the temperature increased. Since the animals did not die from lack of air, he cautiously postulated a toxic exhalation acting upon the nervous system. Another tract concerned animalcules that "enjoy the advantages of real resurrection after death"—unlike his infusoria, which were nonrevivable after death. This resurgence from the torpid state had been earlier observed in Rotifera found in a roof gutter by Leeuwenhoek (1702) and in the Anguillae of blighted wheat described by Needham (1745). In the latter species, Spallanzani induced eleven revival cycles by alternate humidification and desiccation, without significant casualties, but found its immortality limited. He discovered in roof-tile sand two novel animalcules that had this property, a sloth (Tardigrada) and another Anguilla species. The final communication reported that the black dust on the ripened heads of a mold engendered new moldiness when implanted on moistened bread. Powdered roots, stalks, or unripened heads of the mold were ineffective. The unusually heat-resistant "seeds" would not develop on naturally mold-resistant substrates but their proven germinal power eliminated spontaneous generation as a factor to be considered in mold production.

Late in 1780 another two-volume work appeared, of similar title but altogether different content, *Dissertazioni di fisica animale e vegetabile*. Volume one, a treatise on digestion, comprised six dissertations arranged in 264 sections. The second volume, on the generation of certain animals and plants, partly fulfilled the intention expressed in his *Prodromo*. As in the *Opuscoli*, two letters of analytic comment and constructive suggestions from Bonnet were incorporated. The first work, completed in not more than two years, shows Spallanzani at his best as a thorough, resourceful, and courageous physiologist, dedicated to the scientific investigation and understanding of what he called "a subject of so much beauty and utility as the function of Digestion."

In 1777 he publicly demonstrated the great force exerted by the gizzards of fowls and ducks in pulverizing hollow glass globules, thus confirming Redi's century-old account (1675). The French physicist Réaumur had opened a different line of investigation by persuading a kite to swallow openended tubes containing foodstuffs, which when regurgitated showed partial digestion, for which gastric fluid was apparently responsible (1752). Spallanzani greatly extended Réaumur's experiments. He administered food samples, generally in perforated metallic tubes or spherules, to an astonishing variety of animals. The containers were recovered by regurgitation, by passage in the feces, or by sacrificing the animal, and the contents examined for weight loss and other changes. Sometimes pieces of meat were fed, to which string was attached, permitting withdrawal at will. The test animals included many bird species, from turkeys and pigeons to herons, owls, and an eagle, grouped according to stomach wall structure and feeding habits. A miscellaneous category was formed of frogs and newts, water snakes and vipers, fish, and ruminants (sheep, ox, and horse). The final group comprised cats, dogs, and one man. Spallanzani experimented on himself to the limit of endurance. The fate of foodstuffs swallowed in linen bags or wooden tubes was noted, remnants being sought in the voided containers. One piece of resistant membrane was returned twice for further digestion in his alimentary tract before it finally dissolved.

Samples of gastric fluid were procured by inducing himself to vomit on an empty stomach. Crows yielded more liberal samples through sponges placed in the tubes, as described by Réaumur.

The solvent action of this fluid on foodstuffs was determined *in vitro* at different temperatures. Comminuted meats were most readily dissolved; but tendon, cartilage, and soft bone disappeared slowly. Bread, broken grains, and vegetable products were also susceptible. The speed of dissolution of a given food was related to the quantity of available juice, but more particularly to the prevailing temperature. Since body heat gave optimal reactions, Spallanzani sometimes kept the glass tubes containing such mixtures in his axillae. He concluded that the basic factor in digestion is the solvent property of the "gastric juice"—a term introduced by him. Trituration only makes food particles more accessible to this juice. In gizzardless animals, mastication substitutes for trituration. Nor is the digestive process associated with putrefaction; indeed, the gastric juice is strongly antiputrefactive. The acidity that mainly accounts for this property was overlooked by G. A. Scopoli, professor of chemistry and botany at Pavia, on analyzing a specimen of gastric juice from the crow. His nugatory report was capped by Spallanzani's own assertion that the juice "is neither acid nor alkaline, but neutral." Nevertheless, Spallanzani suggested some "latent acid" might account for its milk-curdling properties. These experiments concerned only digestion in the stomach, and he realized secretions of the small intestine might "complete the process." Despite errors and gaps, the work successfully illuminated many phenomena of gastric digestion. Biochemical techniques were not applied to such studies during Spallanzani's lifetime; but he cleared the way for John Richardson Young, whose M.D. thesis for the University of Pennsylvania, "An Experimental Inquiry Into the Principles of Nutrition and the Digestive Process" (1803), emphasized the acidity of gastric juice. Only in 1824 was this reaction identified as due to hydrochloric acid.

The first dissertation in the second volume related detailed observations and original experiments (some carried out many years earlier) on natural generation in four species of frogs and toads and in the water newt. The prolonged amours of the mating season for the frogs and toads culminated in fertilization, shortly after the eggs were extruded from the female cloaca, through semen bedewed upon them by the tightly clasping male. Spallanzani showed that this clasp reflex persisted after severe mutilation, including amputation of limbs or even decapitation. The nuptials of the newt followed a different pattern. Without firm contact, the couple remained in close proximity until the male discharged semen into the water near the female's cloaca, whence fertilized ova soon emerged without intromission. He adduced abundant evidence that, notwithstanding the absence of true copulation in these amphibians, actual contact between eggs and seminal fluid is essential to fecundation. When the hindquarters of the green frog were covered with waxed taffeta breeches—a device used earlier by Réaumur and Nollet—the male's amatory clasp was undiminished, but impregnation was prevented.

The next dissertation reported Spallanzani's recent findings on artificial fecundation. Only slight contact between mature ova and homologous seminal fluid was necessary to achieve fertilization. This fluid was sufficiently prolific to fecundate after being diluted 1:8,000 in water. Admixture with amphibian blood, bile, urine, and various tissue juices, or with human urine, was not inhibitory. But there were no cross-reactions: attempts at artificial hybridization between toads, frogs, and newts were fruitless. Bonnet, who had discovered parthenogenesis in the aphid more than three decades earlier, suggested that an electric current might serve as a nonspecific fertilizing agent. But parthenogenesis could not be induced in frog eggs either by electricity or by various body fluids; and "stimulating agents"—for example, vinegar, and lemon or lime juice—failed to replace the appropriate seminal fluid. Last-minute findings allowed Spallanzani to end this section on a positive note. By impregnating silkworm eggs with seed from male silkworms, he succeeded where Malpighi had failed. Further, Spallanzani recorded the first artificial insemination of a viviparous animal. A spaniel bitch in heat, carefully isolated throughout the experiment, received by vaginal syringe some fresh semen from a dog of the same breed. Two months later, three healthy whelps resembling both parents were born—an event that provoked Spallanzani to aver, "I never received greater pleasure upon any occasion since I cultivated experimental philosophy."

These two dissertations illustrated the indispensability of seminal fluid to the generative process, but Spallanzani obscured rather than elucidated how it functioned. He demonstrated that the *aura seminalis* could not fertilize but left the role of spermatozoa undefined, despite his previous studies of them. Indeed, he apparently welcomed the fallacious results of a single experiment that yielded tadpoles from eggs touched with droplets of semidried sperm "quite free from worms"; for this permitted him to deride the "vermiculists," followers of Leeuwenhoek's concept that the spermatozoon solely embodies the preformed future individual. Spallanzani thought that this mode of impregnation equally demonstrated "the falsehood of epigenesis, or of that system which has been raised from the dead, protected and caressed by Buffon. . . ." He himself was, in fact, a convinced preformationist, but of the "ovist" persuasion, like Malpighi, Haller, and Bonnet before him. Thus he held that the embryo was already within the ovum: a small, coiled-up tadpole awaiting only vitalization by seminal fluid in order to uncoil and grow. This belief can be traced back to 1767, when he informed Bonnet that by "rigorous comparison" (including microscopic examination) of their external and internal structure, unimpregnated and freshly impregnated eggs of frogs were identical. Since tadpoles visibly unfolded in the fertile spawn, he felt entitled to assert that "the tadpole that becomes a frog preexists fecundation." In the following year his *Prodromo* contained a hint to that effect. Now, with unsubstantial evidence buttressing a fruitless concept, he publicly exemplified those very faults that he condemned in others, particularly his old adversary the archepigenesist Needham. Spallanzani's espousal of this doctrine encouraged much futile disputation and perhaps helped to delay until the mid-nineteenth century the discovery that the spermatozoon fertilizes the ovum by actually penetrating it.

Publication of the whole work was delayed pending completion of a final section on generation in diverse plants. His observations were made principally on common vegetables and flowering plants, during summer and autumn visits to Scandiano. He thought that the striking analogies noted between animal and plant life might include their reproductive arrangements. After removing the anthers from flowering hermaphrodite plants, and safeguarding female from male blossoms in other selected species, he studied their ovaria for seed and embryo development. Again influenced by preformationist leanings, he contended that embryos appeared in all seeds prior to and irrespective of fecundation. In hermaphrodites such as sweet basil and Syrian mallow, and in the female plants of annual mercury, want of pollen rendered their seeds sterile. Here, he compared the role of pollen to the effect of seminal fluid upon dormant embryos in amphibian ova. Productive seeds were borne, without benefit of pollination, by gourds, spinach, and hemp—a claim that Spallanzani anticipated would receive disfavor from "all modern naturalists and botanists." He admitted accumulating experimental data as foundation for a speculative disquisition on generation in plants. Unfortunately, his industrious and novel contribution did little to resolve the prevailing confusion and rancor over this complex problem.

The *Dissertazioni* brought Spallanzani additional recognition at home and abroad. A French version of the *Opuscoli*, translated with unmatched promptitude and accuracy by Senebier, the distinguished naturalist-librarian of Geneva, had expanded the circle of his readers. Followers, competitors, and opponents again increased when the same translator duly produced French editions of the dissertations on digestion and reproduction. Spallanzani's latest publication climaxed a period of such unbounded experimentation that he may have sensed the dangers of overextension. For although during the early 1780's he intermittently decollated snails, observed the breeding habits of his amphibia, and planned a monograph on artificial fecundation, he did not turn to new researches in experimental biology and physiology until the last five years of his life. Other possible reasons for a change of direction included the risk, largely unrecognized or ignored, of antagonizing his Pavian colleagues and even foreign specialists by overconfident pronouncements or trespassings. Tardy and unjustified evidence that he had given this kind of offense came in 1786, when the choleric John Hunter insulted Spallanzani for his work on digestion. Hunter was angered by some mild and gentlemanly criticism in "Digestione" of his vitalistic explanation of digestion of the stomach wall observed in some cases of sudden death, as reported to the Royal Society of London in 1772. A still more delayed reaction was that of a spiteful colleague, G. S. Volta, who alleged in 1795 that experiments on plant generation described in the *Dissertazioni* were never performed. Spallanzani replied effectively, with dignity to Hunter and bitterness to Volta.

More powerful influences were the demands made upon him for teaching, at which he excelled: he was elected rector for the scholastic year 1777–1778. Enrollment in his natural history course had increased each year and in 1780 exceeded 115. Finally, as there were large gaps in the museum collections and he was starved for travel, his curiosity and talents could be exercised on specimen-gathering excursions. His appetite was whetted in 1779, when, after several annual postponements, he enjoyed a month-long summer tour

of Switzerland; during this time, he stayed for several days at Bonnet's villa outside Geneva. There he met Senebier and other naturalists, Abraham Trembley and his nephew Jean, and H. B. de Saussure, all of whom he deeply impressed. On the return journey he called on Haller's widow and son at Bern and visited other Swiss cities. In his letter of thanks to Bonnet, Spallanzani stated that of all the natural history museums he visited in Switzerland, only Zurich possessed one where the collections and curator were not amateurish. A few months later, with self-assurance fully harnessed to new objectives, he wrote that his sole remaining pleasure was to see the Royal Museum enlarge daily under his direction.

During the next five years, beginning in the spring of 1780, Spallanzani made several marine and overland excursions, mostly during summer vacations. He thereby corrected the deficiencies of the museum in marine biology (while he himself developed broad scientific interests in that field) on expeditions to Marseilles and the Genoese gulf in 1781; to Istrian and other northern Adriatic ports in the autumn of 1782; to Portovenere on the Gulf of Spezia in 1783; and to Chioggia, near Venice, in 1784. He also traveled to Genoa and vicinity during Easter vacations in 1780 and 1785; on the latter occasion he was equipped with meteorological instruments. In October 1783 he returned from Portovenere on foot to Scandiano, through the Carrara marble quarries and over the Apuan Alps, collecting many geological and fossil specimens on the way. He wrote happily to Bonnet of this sea and land expedition: "During the whole time I was occupied always in observing and interrogating Nature; I have assembled an astonishing collection of observations and facts, of which several appear to me very interesting and until now unknown."

At Marseilles, Spallanzani collected 150 fish species, many of them large and rare, and in improvised quarters studied and dissected marine fauna. During the visit to Portovenere, he instituted the first marine zoological laboratory and while there described new species of fireflies and conducted studies on deep-sea phosphorescence. He refuted the claim that the torpedo fish was attracted by magnets, intrepidly showing that its greatest shock was delivered when the fish was laid on a glass plate. Excising the heart did not lessen the shock until the circulation began to fail. He showed the animal nature of corals and many other minute marine organisms and assigned several sponges and sea moss to the vegetable kingdom. He also studied marine infusoria, testaceans, and crustaceans. The Adriatic waters proved more plentiful in many species than did the Mediterranean. His intention to write a major work on the natural history of the sea did not materialize; but in open letters to Bonnet he recorded observations on "diverse produzioni marine," from sponges, corals, and sea-mussels, to a freshwater fountain gurgling through the salt water of the Gulf of Spezia. (For further details and an amplified bibliography of his voyages and scientific excursions, G. Pighini's account should be consulted.)

In October 1784 a prominent Venetian patron of science, Girolamo Zulian, conveyed the offer of the chair of natural history at Padua, vacant through the recent death of Vallisneri the younger. Spallanzani informed Count Giuseppe Wilzeck (successor to Firmian) that the humidity in Pavia so aggravated his gout that he must relinquish his teaching post. Although lacking the requisite length of service for a pension, he solicited special consideration because of the arduous expeditions for the museum that were made without recompense. Wilzeck offered financial concessions but overcame Spallanzani's obduracy only by granting a leave of absence for a prolonged visit to Constantinople, in addition to a substantial salary increase. Moreover, Joseph II refused to permit his resignation and granted an ecclesiastical benefice. In August 1785 he sailed from Venice in a gunboat with a flotilla escort, as a guest of Zulian, now the Venetian envoy to the Porte. Two months later they reached Constantinople, where Spallanzani was given quarters in Zulian's palace. Their ship had nearly foundered in a gale off Kíthira, where they refitted; then, having threaded the Cyclades, they reached Tenedos, whence the sultan's emissaries escorted them to the locality that excavators a century later identified as the site of Homer's Troy. Spallanzani never forgot that he was a natural historian, even while exploring territory suffused with classical and Homeric reminders. He recorded a waterspout in the Adriatic; collected medusae in the Sea of Marmara; and, during the last part of his visit, spent on the Bosporus, made elaborate geological studies, described local semi-precious stones, and studied marine fauna. Many of the social and political customs of the country, the apathy, polygamy, and excessive wealth and poverty distressed him. He secured information from a friendly seraglio physician that developed into one of his most popular lectures; collected eudiometer samples at dances to determine the effect of overcrowding on atmospheric vitiation; negotiated for crocodiles and skins of lions and ti-

gers with the British and French ambassadors; and was received in audience and regaled with a sixty-course repast by the sultan.

In August 1786, having dispatched the valuable museum collections by ship, Spallanzani set out with a single attendant on the unimaginably difficult return overland. Despite hazardous mountain passes, floods and torrents, brigands and cutthroats, detours were made to inspect mines and geological structures, and more specimens were collected. Reaching Bucharest through the eastern Balkans, he crossed the Transylvanian Alps to the Hungarian plain and also Buda and Pest. In December, although welcomed in high circles in Vienna and bemedaled by Joseph II, he encountered rumors that he had enriched his personal museum at Scandiano by transferring exhibits from Pavia. In Milan he learned that certain university colleagues had circulated a defamatory letter after a subordinate, Canon Serafino Volta, curator of the Royal Museum, whom Spallanzani had recommended as a suitable temporary substitute during his absence, had visited incognito the Scandiano collection and there discovered specimens missing from the Royal Museum. Spallanzani importuned Wilzeck to establish a judicial enquiry into the calumny.

Notwithstanding the spreading scandal, Spallanzani was welcomed at the gates of Pavia by enthusiastic students, more than 400 of whom attended his first lecture. The conspirators, besides Volta, were identified as Gregorio Fontana; Antonio Scarpa, the distinguished anatomist and surgeon; and Scopoli. In common they envied his fame and resented his authoritarianism: also, each had some personal grudge against "the pasha." Spallanzani submitted his evidence to the Royal Imperial Council, and in August an imperial decree exonerated him completely; Volta was dismissed from the university and banished from Pavia; the other parties were reprimanded and ordered to desist from troublemaking. While Volta nursed his grievances in Mantua, Fontana acknowledged his transgression and Scarpa sought to make amends; Scopoli bore the brunt of a malicious reprisal. Spallanzani considered him a plagiarist and knew that he had signed the libelous letter sent to Bonnet, Senebier, and other well-known scientists. A curious specimen, purportedly excreted by a patient, was sent to Scopoli, who designated it *Physis intestinalis*, a novel species of intestinal worm, and illustrated it in a text, dedicated to Sir Joseph Banks, president of the Royal Society of London. The worm was actually a cunningly teased-out portion of chicken gullet. In 1788 the hoax was revealed in a pseudonymous publication addressed to Scopoli. This work was followed by another volume, of anonymous authorship, disparaging Scopoli's earlier text on natural history.

In 1788 Spallanzani journeyed to the Two Sicilies, mainly in order to correct deficiencies in the volcanic collections of the museum. Southern Italy had suffered for five years from intense eruptive and seismic activities. Messina was still in ruins and the countryside devastated. Vesuvius, near Naples, Stromboli and Vulcano in the Eolian Isles, and Etna on the island of Sicily, remained active. Spallanzani visited them all, undauntedly making several perilous ascents that involved great physical endurance. Vesuvius was tranquil on his first visit, but a later attempt to reach the summit was frustrated by a violent eruption. He went to within five feet of the lava pouring from the rent mountainside and accurately measured its flow rate. Just short of the crater of Etna, toxic gases rendered him unconscious; but later he peered from the rim at the boiling lava. From a cavern near the summit of Stromboli he noted that bellowing gas explosions forced up the red-hot lava and ejected massive rocks—an observation fundamental to the science of volcanology. He descended alone into the crater of Vulcano and retired with burned feet and his staff afire when sulfurous fumes prevented further identification of mineral structures. Field observations were correlated with laboratory analyses and thermal tests on volcanic specimens. A glass furnace and the Wedgwood pyrometer made it possible to determine the composition and fusion temperatures of lava and the identity of gases liberated from the melted igneous rocks.

Throughout these volcanic travels, Spallanzani's attention focused upon innumerable phenomena, from Scylla and Charybdis to the annual passage of swordfish through the Strait of Messina; from the punctuality of migrant birds to the kindness of the stricken peasantry. Becalmed off Laguna di Orbetello, noted for its eels, he went inland to investigate by mass dissection their mysterious mode of propagation. In 1789 and 1790 he climbed the Modenese Apennines carrying chemical apparatus for examining the natural gas fires of Barigazzo and the salses. Two years later he made further studies on eels at Lake Comacchio, south of Venice. The new wonders of the Pavia museum attracted many distinguished visitors, including Joseph II himself in 1791. A fascinating five-volume account of these journeys, *Viaggi alle due Sicilie e in alcune parti dell'Appennino*, appeared in 1792

and was dedicated to Count Wilzeck. A sixth volume was added in 1797.

Although now more than sixty years old, Spallanzani undertook several new researches. In 1794 he reported that blinded bats could fly without striking artificial obstacles. After apparently eliminating other explanations, he reluctantly postulated a sixth sense. Two Italian scientists, and also Senebier, were invited to repeat the work. All confirmed Spallanzani's findings; but a French scientist, L. Jurine, demonstrated that blinded bats blundered helplessly into obstacles after their ears were effectively plugged. Spallanzani promptly accepted the ear hypothesis, but the notion of a chiropteran sixth sense prevailed. The extraordinary sensitivity of the ear of the bat to self-emitted supersonic notes as the basic mechanism in the directional sense of this animal was first clearly demonstrated by D. R. Griffin and R. Galambos in 1941.

Spallanzani had adopted the new chemical doctrines that developed following the discoveries, mainly by British chemists, of carbon dioxide, hydrogen, nitrogen, and oxygen during the period 1755–1774. Contrary to the claim of Johann Göttling of Jena that phosphorus would burn in nitrogen, Spallanzani denied in 1796 that this element would burn in either nitrogen, hydrogen, or carbon dioxide. When plunged into oxygen, however, phosphorus ignited with a luminosity proportional to the amount of that gas present in the eudiometer. In that year, Napoleon's armies were overrunning Lombardy. One of Spallanzani's biographers, the French military surgeon J. Tourdes, found his laboratory full of vessels containing different gases, the effects of which were being tested upon various substances. He was engaged already in researches that, although incomplete, formed a major contribution to the understanding of animal and plant respiration. His last personal report, appearing in 1798, contained the novel observation that whereas plants kept in water and in sunlight furnish oxygen and absorb carbon dioxide, they reverse this exchange in deep shade.

Spallanzani suffered from an enlarged prostate, complicated by a chronic bladder infection. Early in February 1799, shortly after his seventieth birthday, he became anuric and after a restless night fell unconscious. Among the medical attendants were Tourdes and Scarpa. In the ensuing week, during lucid intervals between bouts of uremic coma, he discussed experiments with colleagues, reviewed personal affairs with relatives, and recited passages from the classics. He died peacefully at night, after receiving religious offices.

He was buried in the cemetery at Pavia. The heart was placed by his brother Niccolò in the church at Scandiano, while by his own wish the bladder became an exhibit in the historical museum of the university.

Three manuscript memoirs, translated and assembled by Senebier, were published posthumously in 1803 as *Mémoires sur la respiration*. Lavoisier's suggestion that respiration was a form of slow combustion, with direct oxidation of carbon and hydrogen occurring in the lungs, was disputed by the French mathematician Lagrange. Spallanzani's experimental data resolved this controversy and laid the groundwork for modern conceptions of respiratory physiology. Snails kept in an atmosphere of nitrogen or hydrogen exhaled almost as much carbon dioxide as when breathing air. Even after lung removal, snails absorbed oxygen and gave up carbon dioxide. Excised individual organs, including the stomach, liver, and heart, respired similarly. In concluding that the blood transported carbon dioxide as a product of tissue oxidation, Spallanzani discovered parenchymatous respiration—usually accredited to the biochemist Liebig half a century later. Spallanzani left additional notes on many thousands of experiments concerning the respiratory processes of animals and plants; Senebier again loyally edited these. They appeared as *Rapports de l'air avec les êtres organisés . . .* (1807). The first two volumes comprised fourteen additional memoirs on respiration, which established the basic uniformity of the respiratory process throughout the animal kingdom. The third volume concerned respiration in plants.

Spallanzani received many honors, including membership in the ten most distinguished Italian academies, and foreign associateship in a dozen famous European scientific societies. Frederick the Great personally arranged his election to the Berlin Academy of Sciences in 1776. Spallanzani made fortunate friendships with generous-minded scientists, especially Haller and Bonnet. The latter once assured him, "You have discovered more truths in five years than entire academies in half a century." His fame was commemorated at Scandiano, Reggio Emilia, Modena, Pavia, even at Portovenere—wherever he had lived and labored—by statues, busts, tablets, museums, manuscripts, or other memorabilia.

Of middle stature, with dark eyes and complexion, domed head, aquiline nose, and pensive countenance, Spallanzani had a resonant voice and firm gait. Masterful in personality, his character was complex. He conversed eagerly and forthrightly about scientific problems but avoided political or personal topics. He did not underrate his accomplishments, often resented criticism, and was not above canvasing friends and influential acquaintances to obtain election to learned societies. Although disliking formal restraints, he flattered and cajoled authorities from whom he sought favors. His life-style was frugal, but he enjoyed good food and wine and the company of high-minded women. If his religious vows ever vexed his robust temperament, they spared him many distracting and time-consuming obligations, and perhaps secured him from persecution by church, state, or invading armies. Considerate of relatives and friends, he could become ruthlessly angry when wronged. In unraveling the secrets of nature, every aspect of which intrigued or inspired him, he maimed and slaughtered countless animals. Among his few relaxations were fishing and hunting, and he was expert at chess. Athletic in his youth, he remained vigorous to the end.

About a decade after his death, Spallanzani was portrayed as a genius-wizard in one of E. T. A. Hoffmann's fantastic *Tales*. In more recent times the overspecialized have ignored the rare scope and stature of his accomplishments and have disparaged his prodigious output as dilettantism. Allegations that nothing practical came of his splendid studies on infusorial microorganisms overlook the importance of food canning. Nicolas Appert's . . . *l'Art de conserver pendant plusieurs années toutes les substances animales et végétales* (1810) was made possible through Spallanzani's work on heat sterilization. To suggest that he should have gone further and discovered the germ theory of disease is to forget that 100 years later Pasteur had to repeat Spallanzani's work before he finally laid to rest the specter of spontaneous generation; and only then could he convince a reluctant medical profession and skeptical fellow scientists that man might be brought low and killed by parasites of almost incredible minuteness. Pasteur paid his tribute daily; he commissioned a full-length portrait of Spallanzani, which hung in the dining room of his apartment.

The particular indifference of the English-speaking world to Spallanzani's significance stems partly from linguistic difficulties. Certain translations were made into English, but some were of poor quality and none was widely circulated. There is still no version of his complete works in English. Besides, in Spallanzani's day Britain had a number of brilliant investigators who, collectively, covered his many fields. Stephen Hales, John Hunter, Erasmus Darwin, Joseph Black, Henry Cavendish, and Joseph Priestley were among those whose combined luster outshone the multifaceted achievements of any foreign priest-polymath. Spallanzani's countrymen view him in different perspective. In his birthplace a bust of Spallanzani stands on a marble mantel. Above it is a plaque inscribed: *Natus Scandiani Clarus Ubique.* Even after allowing for local pride, this assertion is surely close to truth.

BIBLIOGRAPHY

I. ORIGINAL WORKS. The only two eds. of Spallanzani's collected writings are in Italian: *Opere di Lazzaro Spallanzani* (Milan, 1825–1826) and *Le opere di Lazzaro Spallanzani* (Milan, 1932–1936), each 6 vols. The latter, compiled by Filippo Bottazzi and ten collaborators under the auspices of the Royal Academy of Italy, was to include Spallanzani's letters, edited by Benedetto Biagi, but World War II intervened and Biagi died. Dino Prandi became coeditor, adding to the collection until 1,475 letters, written to 173 individuals or institutions (and twelve anonymous addressees), appeared as *Lazzaro Spallanzani. Epistolario* (Florence, 1958–1964). Prandi's *Bibliografia delle opere di Lazzaro Spallanzani* (Florence, 1951) is a detailed and generally dependable bibliography. It also cites writings about Spallanzani in various languages.

Several lengthy monographs on natural history and physiology, published in Italian during the thirty-year period 1773–1803, reached wider circles through trans. into French, English, or German. The first of these, *De' fenomeni della circolazione . . .* (Modena, 1773), was translated by J. Tourdes as *Expériences sur la circulation . . .* (Paris, 1800), and by R. Hall as *Experiments Upon the Circulation of the Blood . . .* (London, 1801). The next great treatise, *Opuscoli di fisica, animale e vegetabile . . . ,* 2 vols. (Modena, 1776), finalizes his famous work on the animalcules of infusions, "Osservazioni e sperienze intorno agli animalculi delle infusioni . . . ," and records observations and experiments on human and animal spermatozoa, "Osservazioni e sperienze intorno ai vermicelli spermatici . . . ." Also included are two letters to the author from Bonnet about the animalcules and reports on the effects of stagnant air on animal and plant life, the killing and resuscitation of Rotifera, and the origin of moldiness. This work was translated into French by Jean Senebier as *Opuscules de physique, animale, et végétale,* 2 vols. (Geneva, 1777), and into English by T. Beddoes as *Tracts on Animals and Vegetables,* 2 vols. (London, 1784, 1786). Another trans., by J. G. Dalyell, appeared as *Tracts on the Nature of Animals and Vegetables,* 2 vols. (Edinburgh, 1799), the second edition of which, entitled *Tracts on the Natural History of Animals and Vegetables,* 2 vols. (Edinburgh, 1803), was augmented by Dalyell's intro-

ductory observations and by "Tracts on Animal Reproduction," which included accounts by Spallanzani and by Bonnet of experimental reproduction of the head of the garden snail.

Another important treatise, *Dissertazioni di fisica animale e vegetabile . . . ,* 2 vols. (Modena, 1780), of which a second ed. was entitled *Fisica animale e vegetabile,* 3 vols. (Venice, 1782), contains his experimental enquiries into digestion in various animal species ("Digestione"), reproduction in animals and plants ("Della generazione di alcuni animali . . . di diverse piante"), and artificial fecundation ("Sopra la fecondazione artificiale in alcuni animali"). An English version of the *Dissertazioni,* translated and prefaced by T. Beddoes, is *Dissertations Relative to the Natural History of Animals and Vegetables,* 2 vols. (London, 1784, 1789). The sections on digestion and on reproduction, translated by Senebier, appeared separately as *Expériences sur la digestion de l'homme et de différentes espèces d'animaux . . .* (Geneva, 1783, 1784; facs. ed., Paris, 1956) and as *Expériences pour servir à l'histoire de la génération des animaux et des plantes . . .* (Geneva, 1785). These two dissertations in French, added to the *Opuscules,* were republished as *Oeuvres de M. l'Abbé Spallanzani,* 3 vols. (Pavia–Paris, 1787).

The long account of his travels, *Viaggi alle due Sicilie e in alcune parti dell'Appennino . . .* (Pavia, 1792–1797), became available in French, German, and English under the respective titles *Voyages dans les deux Siciles, et dans quelques parties des Apennins,* 6 vols. (Bern, 1795–1797); *Des Abtes Spallanzani Reisen in beyde Sicilien und in Gegenden der Appenninen,* 5 vols. (Leipzig, 1795–1798); and *Travels in the Two Sicilies, and Some Parts of the Apennines,* 4 vols. (London, 1798).

Spallanzani translated and annotated Bonnet's *Contemplation de la nature,* 2 vols. (Amsterdam, 1764–1765), under the title *Contemplazione della natura del Signor Carlo Bonnet,* 2 vols. (Modena, 1769–1770). Several eds. of this work appeared. Published posthumously and translated by Senebier, were *Mémoires sur la respiration* (Geneva, 1803) and *Rapports de l'air avec les êtres organisés . . . ,* 3 vols. (Geneva, 1807). The former appeared in Italian as *Memorie su la respirazione,* 2 vols. (Milan, 1803), and in English as *Memoirs on Respiration* (London, 1804).

Spallanzani's earliest publication, variously titled *Theses philosophicae . . .* (Parma, 1757) and *Propositiones physico-mathematicae . . .* (Reggio Emilia, 1759), probably served as basis for public disputations by university degree candidates. A better-known early work is the essay of classical criticism, *Riflessioni intorno alla traduzione dell'Iliade del Salvini . . .* (Parma, 1760). A short monograph, *Prodromo di un opera da imprimersi sopra le riproduzioni animali* (Modena, 1768), translated by M. Maty as *An Essay on Animal Reproduction* (London, 1769), which included an account of regeneration of the decapitated head of the snail, was the first work of Spallanzani to appear in English.

Many reports in the foregoing vols. were published initially as tracts or booklets; others first appeared as articles in scholarly periodicals, often in the form of letters to well-known personages. Among Spallanzani's characteristic shorter communications are "Lettere due . . ." [to Antonio Vallisneri], in *Nuova raccolta di opuscoli scientifici e filologici* . . . , **9** (1762), 271–298, on the circulation of subterranean waters and the sources of fountains observed during his travels in the Reggian Apennines; and *Dissertazioni due* . . . (Modena, 1765), comprising "Saggio di osservazioni microscopiche concernenti il sistema della generazione dei Signori di Needham e Buffon," and "De lapidibus ab aqua resilientibus" (stone skipping on water). The former appeared in a French trans. by Abbé Regley, with added critical commentary by J. Needham, as *Nouvelles recherches sur les découvertes microscopiques, et la génération des corps organisés* (London–Paris, 1769). In *Prolusio* (Modena, 1770), his University of Pavia inaugural address, given in Latin, Spallanzani again disputed Needham's support of spontaneous generation; and six years later the doctrine received further rebuttals in the *Opuscoli*, I, pp. 3–221.

Meanwhile, Spallanzani reviewed sterility in hybrids, *Memorie sopre i muli* . . . (Modena, 1768), and reported studies on the circulation of the blood, *Dell'azione del cuore* . . . (Modena, 1768). These two short monographs, along with *Dissertazioni due* (1765) and *Prodromo* (1768), were republished in a German trans. as *Herrn Abt Spallanzanis physikalische und mathematische Abhandlungen* (Leipzig, 1769). A pioneering interest in artificial fecundation, revealed by the article "Fecondazione artificiale," in *Prodromo della nuova enciclopedia Italiana* (Siena, 1779), 129–134, culminated in an account of the artificial insemination of a bitch, "Fecondaziona artificiale di una cagna," in *Opuscoli scelti sulle scienze e sulle arti* . . . , **4** (1781), 279–282. Continued investigations of regeneration phenomena, especially of the decapitated head of the snail, are summarized in "Resultati di esperienze sopra la riproduzione della testa nelle lumache terrestri," in *Memorie di matematica e fisica della società Italiana*, **1** (1782), 581–612, and **2** (1784), 506–602.

Spallanzani's range of interests continued undiminished in later life, as witness "Osservazioni sopra alcune trombe di mare formatesi sull' Adriatico," *ibid.*, **4** (1788), 473–479, which describes waterspouts in the Adriatic; "Memoria sopra le meduse fosforiche," *ibid.*, **7** (1794), 271–290, on a phosphorescent jellyfish; *Lettere sopra il sospetto di un nuovo senso nei pipistrelli* . . . (Turin, 1794), and "Lettere sul volo dei pipistrelli acciecati," in *Giornale de letterati*, **13** (1794), 120–186, which record correspondence about his experiments on the sense of direction in bats; *Chimico esame degli esperimenti del Sig. Gottling, professor a Jena, sopra la luce del fosforo di Kunkel* . . . (Modena, 1796), which criticizes Göttling's chemical explanation of the luminosity of phosphorus; and "Lettera . . . sopra le piante chiuse ne' vasi dentro l'acqua e l'aria, ed esposte all'immediato lume solare, e all'ombra," in *Opuscoli scelti sulle scienze e sulle arti* . . . , **20** (1798), 134–146, on exposure to sunlight or shade of plants kept in water or air.

Spallanzani's reply to John Hunter is "Lettera apologetica in risposta alle osservazioni sulla digestione del Sig. Giovanni Hunter . . . ," *ibid.*, **11** (1788), 45–95. Many of his writings contained polemic passages; and vengefulness marked the anonymous letters to G. A. Scopoli, one of four colleagues who had accused him of stealing museum specimens: "Lettere due . . . al Sig. Dottore Gio. Antonio Scopoli . . . ," and "Lettere tre . . . al chiarissimo Signore Gio. Antonio Scopoli, professore di chimica e di botanica . . ." (Modena, 1788–ostensibly "In Zoopolis").

Details of Spallanzani's journey to the Near East, selected from his letters and diaries, were edited by N. Campanini, *Viaggio in Oriente* (Turin, 1888). At Reggio Emilia, the municipal library has custody of about 200 Spallanzani MSS, many still unpublished, and also correspondence with contemporary scientists. The Natural History Museum contains a unique collection of animal, plant, fossil, and mineral specimens catalogued by A. Jona, *La collezione monumentale di Lazzaro Spallanzani* . . . (Reggio Emilia, 1888), and historically documented by N. Campanini, *Storia documentale del Museo di Lazzaro Spallanzani a Reggio Emilia* (Bologna, 1888). His other great zoological collection, now at the Institute of Zoology, University of Pavia, is described by C. Jucci in *L'Istituto di Zoologia "Lazzaro Spallanzani* . . ." (Pavia, 1939). The Historical Museum of that university exhibits relics of Spallanzani, besides a small MS collection. Additional MSS and memorabilia are in the state archives at Milan.

II. SECONDARY LITERATURE. Short biographical accounts in English, often unevenly selective and containing minor inaccuracies, include A. E. Adams, "Lazzaro Spallanzani (1729–1799)," in *Scientific Monthly*, **29** (1929), 529–537; T. Beddoes, "Translator's Preface," in *Dissertations Relative to the Natural History of Animals and Vegetables* (London, 1784), vii–xl; W. Bulloch, "L'Abbate Spallanzani. 1729–1799," in *Parasitology*, **14** (1922), 409–411; G. E. Burget, "Lazzaro Spallanzani (1729–1799)," in *Annals of Medical History*, **6** (1924), 177–184; B. Cummings, "Spallanzani," in *Science Progress in the Twentieth Century*, **11** (1916), 236–245; G. Franchini, "Lazzaro Spallanzani (1729–1799)," in *Annals of Medical History*, n.s. **2** (1920), 56–62; J. B. Hamilton, "The Shadowed Side of Spallanzani," in *Yale Journal of Biology and Medicine*, **7** (1934–1935), 151–170; P. de Kruif, "Spallanzani," in *Microbe Hunters* (New York, 1926), 25–56; A. Massaglia, "Lazzaro Spallanzani," in *Medical Life*, **32** (1925), 149–169; J. G. M'Kendrick, "Spallanzani: A Physiologist of the Last Century," in *British Medical Journal* (1891), **2**, 888–892; F. Prescott, "Spallanzani on Spontaneous Generation and Digestion," in *Proceedings of the Royal Society of Medicine*, **23** (1930), 495–510; J. G. Rushton, "Lazzaro Spallanzani (1729–1799)," in *Proceedings of Staff Meetings of the Mayo Clinic*, **13**

(1938), 411–415; and W. Stirling, *Some Apostles of Physiology* (London, 1902), 60–64.

Important biographic writings in French range from J. L. Alibert, "Éloge historique de Lazare Spallanzani," in *Mémoires de la société médicale d'émulation*, **3** (1800), i–ccii, to J. Rostand, *Les origines de la biologie expérimentale et l'Abbé Spallanzani* (Paris, 1957). Jean Senebier prefaced his trans. of three major works with lengthy essays, "Des considérations sur sa méthode de faire des expériences et les conséquences pratiques qu'on peut tirer en médecine de ses découvertes," in *Expériences sur la digestion de l'homme et de différentes espèces d'animaux par l'Abbé Spallanzani* (Geneva, 1783), i–cxlix; "Réflexions générales sur les volcans pour servir d'introduction aux voyages volcaniques de M. l'Abbé Spallanzani," in *Voyages dans les deux Siciles et dans quelques parties des Apennins*, I (Bern, 1795), 1–74; and "Notice historique sur la vie et les écrits de Lazare Spallanzani," in *Mémoires sur la respiration par Lazare Spallanzani* (Geneva, 1803), 1–58. Another translator, J. Tourdes, wrote "Notices sur la vie littéraire de Spallanzani," as a preface to *Expériences sur la circulation . . .* (Paris, 1800), 5–112.

Among many biographic contributions in Italian are B. G. De'Brignoli, "Dell'Abate Lazzaro Spallanzani scandianese," in *Notizie biographiche . . .*, IV (Reggio Emilia, 1833–1841), 247–387; P. Capparoni, *Spallanzani* (Turin, 1941, 1948); A. Fabroni, "Elogio di Lazzaro Spallanzani," in *Memorie di matematica e fisica della società Italiana delle scienze*, **9** (1802), xxi–xlviii, which reappeared as "Vita di Lazzaro Spallanzani," in the *Opere*, I (Milan, 1825–1826), vii–xxvi; P. Pavesi, "L'Abate Spallanzani a Pavia," in *Società Italiana di scienze naturali di Milano. Memorie*, **6** (1901), fasc. III; P. Pozzetti, *Elogio di Lazzaro Spallanzani* (Parma, 1800); and L. Salimbeni, *L'Abbate Lazzaro Spallanzani . . .* (Modena, 1879). A short eulogy pronounced two days after Spallanzani's death by the leader of the 1786 "conspiracy," Gregorio Fontana, "Mozione . . . in proposito della morte di Lazzaro Spallanzani," is reproduced in *Memorie e documenti per la storia dell'università di Pavia . . .*, I (Pavia, 1878), 421–422. An account of the final illness and autopsy is given by V. L. Brera, in *Storia della malattia e della morte del Prof. Spallanzani* (Pavia, 1801).

Scopoli's libelous letter of 2 February 1787, conveying the accusations about Spallanzani's museum curatorship, appears in P. Leonardi, *Centenario del Prof. Giovanni Antonio Scopoli . . .* (Venice, 1888). Lazzaro's nephew, G. B. Spallanzani, vigorously defended his late uncle's reputation in *L'ombra di Spallanzani vendicata . . .* (Reggio Emilia, *ca.* 1802), but was rebutted in *Lettera di Giovanni Martinenghi . . .* (Pavia, 1803). Nearly a century later the alleged calumny was reviewed by P. Pavesi, "Il crimine scientifico Spallanzani giudicato," in *Rendiconti del Reale Istituto Lombardo di scienze e lettere*, 2nd ser., **32** (1899), 564–568. A fuller modern account appears in Capparoni's *Spallanzani*, ch. 4, 113–127.

Special appraisals are by E. Franco, "Lazzaro Spallanzani precursore dell' industria delle conserve," in *Atti della reale stazione sperimentale per l'industria delle conserve alimentari* (Parma, 1943); C. Massa, in *Modena a Lazzaro Spallanzani* (Modena, 1888), celebrating the dedication of the monument at Scandiano, 21 October 1888; A. Stefani, "In omaggio a Lazzaro Spallanzani nel centenario della sua morte," in *Atti e memorie della reale accademia di scienze, lettere ed arti in Padova*, n.s. **15** (1899), 209–220; and T. Taramelli, "Ricordo dello Spallanzani come vulcanologo," in *Rendiconti del Reale Istituto Lombardo di scienze e lettere*, **46** (1913), 937–951. Commemorative papers by Italian and foreign scientists honoring the centenary of his death are collected in *Nel primo centenario dalla morte di Lazzaro Spallanzani . . .*, 2 vols. (Reggio Emilia, 1899). A booklet, *Nelle feste centenarie di Lazzaro Spallanzani 1799–1899* (Reggio Emilia, 1899), contains portraits and Italian tributes. The second centenary of his birth also was celebrated by addresses published in *Onoranze a Lazzaro Spallanzani nel II centenario dalla nascità* (Reggio Emilia, 1929), the most notable being a detailed review of his travels by G. Pighini, "Lazzaro Spallanzani viaggiatore," pp. 1–441. In 1939 the University of Pavia was host to a meeting of the Italian Society of Experimental Biology and other organizations honoring Spallanzani. The memorabilia exhibited are listed by A. Lo Vasco, *Catalogo della mostra in onore di Lazzaro Spallanzani, 11 Aprile-18 Maggio 1939* (Pavia, 1939). The scientific communications appear in *Commemorazioni Spallanzaniane. 11–14 Aprile 1939*, 4 vols. (Pavia–Milan, 1939–1940).

Writings that relate Spallanzani to the scientific setting of his century include L. Belloni, "Antonio Vallisneri ed il contagio vivo," in *Il metodo sperimentale in biologia da Vallisneri ad oggi* (Padua, 1962); N. Campanini, "Lazzaro Spallanzani, Voltaire e Federico il grande," in *Rassegna Emiliana di storia, letteratura ed arte*, **1**, fasc. VII (1888), 389–406, reprinted in *Nelle feste centenarie . . .* (Reggio Emilia, 1899); A. Castiglioni, "Eighteenth Century Physiology. Haller, Spallanzani, English School," in *A History of Medicine* (New York, 1946), E. B. Krumbhaar, trans. and ed., 609–614; A. Clark-Kennedy, *Stephen Hales, D.D., F.R.S. An Eighteenth Century Biography* (Cambridge, 1929); C. Dobell, *Antony van Leeuwenhoek and His "Little Animals" . . .* (New York, 1932; 2nd ed., 1958; paperback ed., 1960); M. Foster, *Lectures on the History of Physiology During the Sixteenth, Seventeenth and Eighteenth Centuries* (Cambridge, 1924), 200–254; A. von Haller, *Elementa physiologiae corporis humani*, 8 vols. (Lausanne–Bern, 1757–1764); A. von Muralt, "Lazzaro Spallanzani e Albrecht von Haller," in *Commemorazioni Spallanzaniane . . . 1939*, III (Pavia–Milan, 1939–1940), 116–118; R. Savioz, *Mémoires autobiographiques de Charles Bonnet, de Genève* (Paris, 1948); J. Senebier, *Éloge historique d'Albert de Haller* (Geneva, 1778); P. Vaccari, *Storia della università di Pavia* (2nd ed., Pavia, 1957), ch. 8, pp. 177–218; and G. S. Volta, "Nuove ricerche ed osservazioni sopra il sessualismo di alcune piante," in *Memorie della reale accademia di scienze, belle lettere ed arti, Mantova*, **1** (1795), 225–267.

The following works illustrate the significance of Spallanzani's main scientific contributions. SPONTANEOUS GENERATION: C. Bastian, *The Beginnings of Life*, 2 vols. (London, 1872); L. Belloni, *Le "contagium vivum" avant Pasteur* (Paris, 1961); W. Bulloch, "Spontaneous Generation and Heterogenesis," in *The History of Bacteriology* (London, 1938, repr., 1960), ch. 4, pp. 67–125; H. Dale, *Viruses and Heterogenesis. An Old Problem in a New Form* (London, 1935), the Huxley lecture; J. T. Needham, *Nouvelles observations microscopiques, avec des découvertes intéressantes sur la composition et la décomposition des corps organisés* (Paris, 1750); and G. Pennetier, *Un débat scientifique. Pouchet et Pasteur 1858–1868* (Rouen, 1907). REGENERATION, EMBRYOLOGY AND FECUNDATION: C. Bonnet, "Expériences sur la régéneration de la tête du limaçon terrestre," in *Journal de physique*, **10** (1777), 165–179; Comte de Buffon (G. L. Leclerc), "Histoire des animaux," in *Histoire naturelle*, **2**, pt. 1 (Paris, 1749); A. W. Meyer, *The Rise of Embryology* (Stanford, Calif., 1939), chs. 5, 9–11, pp. 62–85, 132–211; and J. Needham, *A History of Embryology* (Cambridge, 1934), 179–229. DIGESTION: D. G. Bates, "The Background to John Young's Thesis on Digestion," in *Bulletin of the History of Medicine*, **36** (1962), 341–362; J. Hunter, "Some Observations on Digestion," in *Observations on Certain Parts of the Animal Oeconomy* (London, 1786), 147–188; R.-A. F. de Réaumur, "Sur la digestion des oiseaux," in *Académie des sciences* (Paris, 1752), 266–307, 461–495; and J. R. Young, *An Experimental Inquiry into the Principles of Nutrition and the Digestive Process* (Philadelphia, 1803), repr. with intro. essay by W. C. Rose (Urbana, Illinois, 1959). FLIGHT OF BATS: R. Galambos, "The Avoidance of Obstacles by Flying Bats: Spallanzani's Ideas (1794) and Later Theories," in *Isis*, **34** (1942), 132–140; D. R. Griffin, *Echoes of Bats and Men* (New York, 1955), 27–33, 87–88; and L. Jurine, "Experiments on Bats Deprived of Sight," in *Philosophical Magazine*, **1** (1798), 136–140, trans. from *Journal de physique*, **46** (1798), 145–148. Spallanzani's chief monographs were reviewed anonymously at some length in *Giornale de' letterati di Pisa* (1774–1795).

CLAUDE E. DOLMAN

SPEMANN, HANS (*b.* Stuttgart, Germany, 27 June 1869; *d.* Freiburg im Breisgau, Germany, 12 September 1941), *embryology.*

Spemann was the eldest of four children of a well-known book publisher, Johann Wilhelm Spemann, and the former Lisinka Hoffmann. His father's family, of Westphalian peasant stock, had a number of members in the legal profession, and his mother's family contained several doctors. There were three other children in the family, and he grew up in a fairly large house well provided with books, with parents who led an active social and cultural life. Spemann attended the Eberhard-Ludwigs-Gymnasium, where he was particularly attracted by classics. He first decided, however, to study medicine; and for that purpose he attended the University of Heidelberg as soon as he had completed his year of military service in the Kassel hussars.

At Heidelberg, Spemann formed a friendship that without doubt greatly influenced the direction of his life. Gustaf Wolff, a few years older than he, had begun experiments on the embryological development of newts and had shown that if the lens of the eye is removed, a new lens may be formed— not from the tissue that gives rise to the lens in normal development, but from the edge of the retina. This "Wolffian lens regeneration" intrigued Spemann throughout most of his life, and it still retains some of the air of mystery that originally surrounded it. At that time (1892) Wolff interpreted it as strong evidence against Darwin's "selection theory" and as proof of "organic purposiveness." Thus Spemann was introduced, at the beginning of his academic career, to the animal that was to remain his favorite experimental material; acquired an insight into the character of a well-planned and clean experiment; and developed an inclination toward what might now be considered a somewhat mystical conception of the nature of biological processes. He retained strong traces of these influences throughout his life.

As a young biologist Spemann began work in 1894 at Würzburg as a doctoral student and teacher, and was the favorite pupil of Theodor Boveri. It was there, just after taking his doctorate, that he married Clara Binder. After fourteen years at Würzburg, Spemann became professor at Rostock (1908–1914). He spent the years of World War I as director of the Kaiser Wilhelm Institute of Biology in Berlin-Dahlem; and in 1919 he succeeded Weismann as professor at Freiburg im Breisgau. He remained at Freiburg for the rest of his life, retiring in 1938 and dying in his country house nearby in 1941. He was awarded the Nobel Prize for physiology and medicine in 1935.

Spemann combined great persistence, foresight, and careful planning with beautifully precise manipulative skill and an insistence on *Sauberkeit* (cleanliness, in all aspects). His first two major works in biology were fully thought out, before he started writing them, to answer quite clearly defined questions. The time for this creative thinking had been forced on him by a lung illness that necessitated a rest cure in Switzerland.

For one line of work Spemann chose the object about which he had first learned from his friend Wolff—the lens of the amphibian eye—but asked himself a question more basic than any raised by Wolff's work on regeneration: how the lens came to develop in the first place. It is formed from the outer layer of cells in the embryo (the ectoderm, which also gives rise to the skin and the nervous system), and it appears at the point where an outgrowth from the brain reaches the surface. At an early stage of development, the region from which this outgrowth arises is exposed on the surface of the egg, and Spemann was able to kill this group of cells by burning with a minute hot needle. He found that the remainder of the embryo could develop normally without the retina that should have developed from the brain outgrowth and, most important, also without the lens. This discovery strongly suggested that the brain outgrowth, when it reaches the ectoderm, exerts an influence that causes the cells to develop into the lens.

In order to study further the reality of this postulated "induction" of the lens by the retina, Spemann had to perfect his experimental methods. He invented a number of very simple but elegant and refined instruments, mostly made from glass, which made it possible to carry out complicated surgical operations on eggs and embryos only a millimeter or two in diameter. In this way he became almost solely responsible for founding the techniques of microsurgery, certainly one of his greatest contributions to biology. Using such instruments, Spemann could remove the region of ectoderm from which the lens would be expected to form, and substitute some other piece for it before the development of the retina; he found that this foreign ectoderm was induced by the retina to develop into a lens.

Spemann's other early problem did not demand as much technical originality, but led even deeper into the major questions of development. The newt's egg, when laid, is enclosed in an oval capsule of jelly. A thin hair—Spemann maintained that it should be from the head of a blond infant less than nine months old—can be tied around it and pulled tight enough to cut the egg in half or compress it to a dumbbell shape. Spemann found that if this constriction is carried out soon after laying, each separate half of the egg may develop into a complete larva; or, alternatively, one may develop into a whole larva and the other only into a more or less formless mass of cells. If the constriction is not complete, and produces only a dumbbell, one may obtain an embryo with a single tail and two complete heads.

The important point is that a half egg (or half region) never produces a half embryo, but always either a complete embryo (or organ such as the head) or nothing at all. The production of a complete embryo from half the egg shows clearly that at this early stage the various parts of the egg are not fixed in their "developmental fate" ("Determined"). On the other hand, if the same constriction experiment is carried out considerably later, after gastrulation but still before the first embryonic organs can be recognized, the halves form only half embryos, each part developing exactly as if there had been no constriction. Some process of "fixing the developmental fate of the parts" must have occurred between the early stage and the later; and Spemann called this process "determination."

Spemann was thus led to take two further steps that, in combination, opened a new era in the understanding of biological development. From the constriction experiments it seemed to follow that, long before any particular organs can be recognized in the embryo, some process of "determination" decides, more or less irrevocably, the nature of the end product into which any given region will develop. It might seem obvious to ask what the nature of this process is. But Spemann was too wary to get involved in such philosophical traps and too good an experimentalist. He posed the more restricted but more manageable problem of whether we can discover any causal antecedent that brings about this determination.

Calling on the microsurgical techniques elaborated in his study of the lens, Spemann devised experimental procedures that did indeed reveal a causal sequence of events leading up to the determination of the main organ that appears in early stages of development: the central nervous system. By transferring small fragments of tissue from one location in the embryo to another, he (and some of his student collaborators, particularly Hilde and Otto Mangold) showed that any part of the ectoderm of the embryo, if brought into contact with the mesoderm before or at the time of gastrulation, would be induced to become neural tissue; whereas if it were not allowed to contact the mesoderm, it would not become neural tissue, even if its original location in the embryo would have led one to expect it to develop in that way. By this achievement Spemann had discovered the first known example of a causal mechanism that makes it possible to control precisely the direction in which a part of the embryo will develop; by surgical manipulation of its neighboring cells, it can be determined whether this embryonic part will develop into nerves or into skin.

When a piece of ectoderm is placed in contact with the mesoderm, it is induced to form not a mere mass of neural cells but a part of a neural organ, such as the brain, with a greater or lesser degree of organization. This finding led Spemann to approach the problem of the mechanism of "the induction of determination" with considerable caution. His biological philosophy, while not explicitly vitalist, tended to fall within the "organicist" framework characteristic of German biology at the turn of the century. He seems at first to have felt that the process of induction occurs in cells that are so biologically complex that an attempt to analyze it would necessarily entail an oversimplification. He therefore used, as a name for the region that develops into mesoderm and that induces the neural plate, the word "Organisator", and he stated, "It creates [schafft] an organization field out of the indifferent material in which it lies."

In later experiments, devoted to the study of induction in other regions of the embryo, Spemann again found that what is induced usually is an organ, with its own characteristic shape. But some of these experiments, in which fragments from frogs' eggs were transplanted to newts' eggs, or vice versa, led to what should probably be considered Spemann's second major contribution: the discovery that the character of the induced organ depends much more on its own intrinsic (presumably genetic) constitution than on that of the inducer. Thus a frog inducer, acting on newt tissues, produces a newt organ. The reacting material is by no means indifferent, as he had earlier thought. Further, under the influence of younger, more analytically oriented students, Spemann gradually accepted the importance of experiments designed to discover the extent to which the effects of his "Organisator" can be produced when the cells of it have been killed or chemically extracted. He never seems, however, to have considered induction from the point of view that now seems so natural: as involving the genetic potentialities of the cells. Perhaps only T. H. Morgan, among his contemporaries, would have been tempted to approach the subject from that angle during the period when Spemann was most active. The communication between German experimental embryologists and American geneticists was so slight that the connection was made only toward the end of Spemann's

life. It was, however, the precision and rigor of Spemann's experiments that led him to formulate clear questions concerning the causal sequences of particular and well-defined developmental performances by identifiable groups of cells, and thus to provide the foundations on which the more recent advances have been based.

BIBLIOGRAPHY

I. Original Works. All but the very last of Spemann's publications were summarized by the author himself in his *Experimentelle Beiträge zu einer Theorie der Entwicklung* (Berlin, 1936), translated into English as *Embryonic Development and Induction* (New Haven, 1938). An autobiography is *Forschung und Leben, Errinerungen* (Stuttgart, 1943).

II. Secondary Literature. An extensive discussion by a long-time pupil and collaborator is O. Mangold, *Hans Spemann, ein Meister der Entwicklungsphysiologie, sein Leben und sein Werk* (Stuttgart, 1953). Less extensive surveys of Spemann's work, and discussion of it by authors not so closely associated with him, are in J. Needham, *Biochemistry and Morphogenesis*, 2nd ed. (London, 1969); L. Saxén and S. Toivonen, *Primary Embryonic Induction* (London, 1962); and C. H. Waddington, *Principles of Embryology* (London, 1956).

Biographical writings are F. Baltzer, "Zum Gedächtnis Hans Spemann," in *Naturwissenschaften*, **30** (1942), 229–239; and O. Mangold, "Hans Spemann als Mensch und Wissenschaftler," in *Wilhelm Roux Archiv für Entwicklungs-mechanik der Organismen*, **141** (1942). 385–425; and "Hans Spemann," in *Freiburger Professoren des' 19. und 20. Jahrhunderts* (Freiburg, 1957), 159–182.

C. H. Waddington

SPENCER, HERBERT (*b.* Derby, England, 27 April 1820; *d.* Brighton, England, 8 December 1903), *philosophy, biology, psychology, sociology.*

Spencer was the only surviving child of William George and Catherine Spencer; his father, a private school teacher of very modest means, was inclined to a deist rationalism and frequented Quaker meetings. Spencer was educated privately, first by his father (author of an original system of teaching geometry, by "discovery" methods) and then by his uncle, the Rev. Thomas Spencer, a radical and scientifically inclined parson. He also participated in the Derby Philosophical Society, a coterie of amateur "natural philosophers" founded by Erasmus Darwin in 1783 along the lines of the Birmingham Lunar Society, and thus became an heir to that provincial tradition of political radicalism, religious free thought, and scientific endeavor of which the key figure had been Joseph Priestley. Above fairly elementary levels he was a virtual autodidact, learning his science from casual reading, attending lectures, and, later, associating with working scientists.

In 1837 Spencer took up railway engineering, during the boom period of railway construction in England, and was active in radical, middle-class, dissenting politics. Dissatisfied with engineering, he hovered long over other choices, finally taking a job in London in 1848 as subeditor of the *Economist*. There he moved among leaders of literary and scientific opinion and gradually shaped his career as an independent writer and reviewer. Spencer never married (despite his celebrated affair with Marian Evans [George Eliot]) and from 1855 suffered, despite good physical health, from a neurotic condition that intermittently prevented him from sleeping, working, or being in company. Despite his friendships (notably with T. H. Huxley and with the other scientists who composed the X-Club) and his membership in the Athenaeum, Spencer was socially an isolate and took pride in declining all the many honors that were offered him. His considerable reputation as a proponent of extreme laissez-faire liberalism, of the claims of science against traditional religion, and of evolutionary philosophy was at its height in the late 1870's (most especially in the United States) but had diminished dramatically by the time of his death.

Spencer is important less for specific discoveries or for his contribution to particular sciences (except sociology and psychology) than for his synthesis of so much of the accepted science of his day in the integrating framework of evolution. In an age when natural science was becoming institutionalized and differentiated, both internally and externally, Spencer was the last of the *Naturphilosophen*. This accounts for the vagaries of his reputation among his scientific contemporaries: low among specialist working scientists and high among many of the most original boundary-crossing innovators (Darwin, Galton, A. R. Wallace). Spencer's unified vision of science, expressed in methodological writings as well as in the synthesis itself, contributed greatly to the acceptance of science as a major component in the intellectual culture of industrial society. If the evolutionary totality had many of the attributes of a theology, it was not merely because of its place in the last decisive battle between science and religion, but because its

original and enduring motive had been to establish "the secularization of ethics," now that the unanimous hold of religion had been weakened. It was necessary to integrate physical science with social science and ethics in order to invest the latter with the authority that only science could truly claim. The unity of the whole, and the fertile but misleading cross references between the parts, were essential to both Spencer's scientific and his social-ethical interests.

The precise origin of Spencer's evolutionary views is impossible to date, but it is likely that they were imbibed in some form during his youth from the "Darwinians" of Derby. He had become a Lamarckian through his reading of Lyell in 1840, and although critical of the *Vestiges of Creation* (1843), he accepted its basic tenet, "the development hypothesis." *Social Statics* (1850), his first book, is evolutionary; and his long essay "A Theory of Population" (1852), in attempting to show that progress is necessitated by population pressure, comes within an ace of anticipating the main elements of Darwinian natural selection. In the latter Spencer applied Malthusian principles to animal populations, deduced a struggle for survival, and coined the phrase "survival of the fittest"; but the perspective remained Lamarckian. It is clear from his subsequent essay "Progress: Its Law and Cause" (1857) that the goal of his theory was quite distinct from Darwin's: to show how progress or development in all areas of the universe—the solar system, the totality of organic species, the maturation of each organism, the psychic development and socialization of the individual, the evolution of society and culture—consists of one fundamental, determinate motion from an incoherent homogeneity to a complex and interdependent heterogeneity.

The path to this vision had been cleared by Spencer's reading of K. E. von Baer's work on embryology and H. Milne-Edwards' theme of "the physiological division of labor"—a notion that, introduced into biology from political economy, was now to be reapplied by Spencer to the social world, in an extensive use of the organic analogy. *First Principles* (1862), with its doctrine of an ultimate unknowable force, sought to reconcile science and religion, and to lay the metaphysical underpinnings of all evolution. The necessity of differentiation was derived from "the Persistence of Force," the instability of all homogeneous physical conditions, and the tendency of all changes to produce multiple effects, leading to ever more heterogeneous and complex results. He resisted Clerk Maxwell's suggestion that the second law of thermodynamics implied increasing entropy, not increasing heterogeneity, as the cosmic trend.

Starting from a definition of life as "a definite combination of heterogeneous changes, both simultaneous and successive . . . in correspondence with external coexistences and sequences," Spencer saw higher forms emerging from a gradual process of adaptation to the environment. *The Principles of Biology* (1864–1867) analyzes the principal mechanisms by which this occurs and relates them to the specialized structures and functions of plants and animals. Although Darwinian natural selection was easily incorporated into Spencer's system (as "indirect equilibration"), Spencer was always concerned to insist on the inheritance of acquired characteristics as a major mechanism of evolution. Long after most professional biologists had abandoned it, Spencer, in *The Factors of Organic Evolution* (1886), his last important scientific essay, argued that "use-inheritance" was necessary to explain most organized systems of behavior or physiological structure. The fatal weakness in his case was his inability to explain how modifications of organs derived from use and the direct effects of the environment could become embodied in the genetic stock. There was one powerful reason—quite apart from the unresolved difficulties that the neo-Darwinism of Weismann and others had left—why Spencer was unwilling to abandon his Lamarckism. It would have undermined what he most wanted to maintain: the fundamental identity of biological evolution and of psychic and social evolution. This tenet of the unity of evolution also led him to blur differences between processes in which the outcome is in some sense "programmed" at the outset (such as the maturation of the embryo) and those in which it is not (such as the evolution of species or the socialization of children).

The Principles of Psychology (1855) was an important and original work, a real milestone in the history of the subject, marking its transition from a heavily epistemological phase to one in which it was closely dependent on physiology. In it Spencer paved the way for Wundt, William James, and Pavlov. Spencer had in his youth accepted phrenology, which, although abandoned, provided him with a critique of the associationist psychology of Hartley, Jeremy Bentham, and J. S. Mill, on the grounds of its not embracing the fact of species or racial character. Spencer presented this character not as an innate essence, but as the "organized" residue of the past experience of the species, a factor that interacted with present experience. But he could give no firm account of just how fixed or fluid this "character" was; and he never distinguished properly between the (in our terms) racial and cultural components of "character" in man. The end of psychological evolution was the emergence of ever more complex powers of "representation" in response to environmental stimuli. The same basic processes operated at all levels, so that abstract thought and developed moral sympathy differed only in degree from the automatic contractions of microorganisms.

Spencer's social theory rested on his psychology, just as his psychology presupposed his biology, because of his basic principle that the character of any aggregate, whether society or physical substance, is fixed by that of its constituent units. The mental development of man, Spencer argued, lay from egotism to altruism: thus society developed from a "militant" phase, in which rigid coercion was needed to hold men together, to an "industrial" phase, in which altruism and a marked individualism permitted the decline of external control and the complex interdependence of an advanced division of labor. The "social state," or end product of evolution, was the ideal of Spencer's youth: a society with the minimum of state control over its members' activities and associations, in which altruism permitted the harmonious free play of each person's individual interest. His lifework was to try to show that this ideal was uniquely in accord with natural principles.

Spencer failed, yet it was a grand failure. Apart from his major contributions to the nascent fields of sociology and psychology, he performed a major function for the science of his day by drawing out and integrating its principal themes with the general culture of his age. It is a function that no one has performed since.

BIBLIOGRAPHY

I. ORIGINAL WORKS. There is a complete bibliography of Spencer's writings in J. Rumney, *Herbert Spencer's Sociology* (London, 1937), 311–323. His principal works, all published in London, are *Social Statics: Or the Conditions Essential to Human Happiness Specified, and the First of Them Developed* (1850); *The Principles of Psychology* (1855; 2nd ed., enl. and rev., 2 vols., 1870–1872); *Education: Intellectual, Moral and Physical* (1861); *First Principles* (1862); *The Principles of Biology*, 2 vols. (1864–1867); *The Study of Sociology* (1873); *The Principles of Sociology,* 3 vols. (1876–1897); *The Principles of Ethics,* 2 vols. (1879–1893); *The Man Versus the State* (1884); and *An Autobiography* (1904).

Many of his essays were reprinted in his *Essays, Scientific, Political, and Speculative*, 3 vols. (various eds., rev. London, 1890). The principal essays of scientific interest are "A Theory of Population, Deduced From the General Law of Animal Fertility," in *Westminster Review* (1852); "The Development Hypothesis," in *Leader* (1852); "The Genesis of Science," in *British Quarterly Review* (1854); "Progress: Its Law and Cause," in *Westminster Review* (1857); "The Social Organism," *ibid.* (1860); "The Factors of Organic Evolution," in *Nineteenth Century* (1886); and "The Inadequacy of Natural Selection" and further rejoinders, in *Contemporary Review* (1893–1894).

Several selections from Spencer's writings, mostly on sociology, have recently been published: *The Man Versus the State, With Four Essays on Politics and Society*, D. G. MacRae, ed. (Harmondsworth, 1969); *Herbert Spencer: Structure, Function and Evolution*, S. L. Andreski, ed. (London, 1971); and *Herbert Spencer on Social Evolution*, J. D. Y. Peel, ed. (Chicago, 1972).

Original MSS of most of Spencer's books are at the British Museum. The remains of his personal papers (seemingly only a small part) are at the Athenaeum (London). Otherwise his letters are widespread in the collected papers of his correspondents, especially T. H. Huxley (at Imperial College, London) and Beatrice Webb (Passfield Papers, British Library of Political and Economic Science).

II. Secondary Literature. The indispensable work is D. Duncan, *Life and Letters of Herbert Spencer* (London, 1908). Among the great volume of contemporary or near-contemporary discussion, criticism, and paraphrase of Spencer's work are F. H. Collins, *An Epitome of the Synthetic Philosophy* (London, 1889); J. Arthur Thompson, *Herbert Spencer* (London, 1906); and William James, *Memories and Studies* (New York, 1911), which contains a judicious obituary assessment. J. Rumney, *Herbert Spencer's Sociology* (London, 1937), stands almost alone in the period when Spencer was all but forgotten. Most present interest is in Spencer as social philosopher and forerunner of sociology. J. D. Y. Peel, *Herbert Spencer: The Evolution of a Sociologist* (London, 1971), gives the fullest account of the social and intellectual background; see also J. W. Burrow, *Evolution and Society* (London, 1966); S. Eisen, "Herbert Spencer and the Spectre of Comte," in *Journal of British Studies*, **7** (1967); and "Frederic Harrison and Herbert Spencer: Embattled Unbelievers," in *Victorian Studies*, **12** (1968); D. Freeman, "The Evolutionary Theories of Charles Darwin and Herbert Spencer," in *Current Anthropology*, **15** (1974); M. Harris, *The Rise of Anthropological Theory* (New York, 1968); J. D. Y. Peel, "Spencer and the Neo-evolutionists," in *Sociology*, **3** (1969); W. H. Simon, "Herbert Spencer and the Social Organism," in *Journal of the History of Ideas*, **21** (1960); and G. W. Stocking, *Race, Culture and Evolution* (New York, 1968). For a greater emphasis on Spencer as natural scientist, see P. B. Medawar, "Herbert Spencer and the General Law of Evolution," in *The Art of the Soluble*

(London, 1967); and R. M. Young, "Malthus and the Evolutionists," in *Past and Present*, **43** (1969); and *Mind, Brain and Adaptation in the Nineteenth Century* (Oxford, 1970).

J. D. Y. Peel

SPENCER, LEONARD JAMES (*b.* Worcester, England, 7 July 1870; *d.* London, England, 4 April 1959), *mineralogy*.

Spencer was the eldest of the eight children of the former Elizabeth Bonser and James Spencer, for many years headmaster of the school attached to Bradford Technical College, from which the boy won a Royal Exhibition to the Royal College of Science for Ireland, Dublin, in 1886. He graduated in chemistry in 1889 and immediately entered Sidney Sussex College, Cambridge, to read geology, mineralogy, and chemistry. In 1893 Spencer won the coveted Harkness scholarship in geology. From September to December of that year he studied at Munich under Groth, Ernst Weinschenk, and Wilhelm Muthmann. Earlier in 1893 he had been appointed to the staff of the mineral department of the British Museum. When he took up this post on New Year's Day 1894, and the following month joined the Mineralogical Society of Great Britain and Ireland, the course was set for the remainder of Spencer's long life: about these two institutions his professional career, and indeed his whole life, were to revolve. In 1899 he married Edith Mary Close of Mortimer, Berkshire; they had one son and two daughters. Almost all Spencer's time was devoted to mineralogy; he allowed himself the occasional relaxation of gardening in London and at his country cottage.

From the beginning Spencer's curatorial duties involved him in widely ranging descriptive mineralogy, and this is reflected in his original publications. His establishment of the relationship between the three lead antimony sulfides plagionite, heteromorphite, and semseyite is notable, and his description of enargite was a model of its kind. The eight new minerals he named—miersite, tarbuttite, parahopeite, chloroxiphite, diaboleite, schultenite, aramayoite and bismutotantalite—are all nonsilicates. Spencer's interest later turned to meteorites and especially to the origin of tektites, which he thought to be impact products; he was responsible for important additions to the already notable meteorite collection at the British Museum and, in 1934, made an expedition to the silica-glass occurrences in the Great Sand Sea of Egypt but failed to find associated meteorite craters. His meticulous curatorial work was largely responsible for making the British Museum mineral collection the best-documented and best-indexed in the world at the time of his retirement in 1935 as keeper, a post to which he had been appointed in 1927.

While still an undergraduate at Cambridge, Spencer began his long career as an abstractor, probably through financial necessity, by preparing abstracts of patents for H. M. Patent Office. In his first years at the British Museum, he abstracted mineral chemistry for *Journal of the Chemical Society*, reviewed the same field for the annual *Report on the Progress of Chemistry* of the Chemical Society, and compiled and edited the mineralogy volumes of the *International Catalogue of Scientific Literature* from 1901 to 1914. In 1900 he was appointed editor of the *Mineralogical Magazine*, an office he held until 1955, and immediately began to publish a few pages annually of abstracts of significant papers. In 1920 Spencer persuaded the Mineralogical Society to start publication of *Mineralogical Abstracts* with coverage from the expiry of the *International Catalogue* in 1915. He edited twelve volumes of *Mineralogical Abstracts* (1920–1955), contributing two-thirds of the text himself. His triennial lists of new mineral names and obituary notices were features of *Mineralogical Magazine* throughout his long editorship.

Spencer wrote two books, *The World's Minerals* (1911) and *A Key to Precious Stones* (1936); he translated, with his wife's assistance, two important German works, Max Bauer's *Edelsteinkunde* (1904) and R. Braun's *Das Mineralreich* (1908–1912).

Spencer's eminence in mineralogy was not unrecognized in his lifetime. He was elected fellow of the Royal Society in 1925; correspondent in 1926, later honorary fellow, and in 1940 Roebling medalist of the Mineralogical Society of America; Murchison medalist of the Geological Society of London in 1937; and honorary member of the German Mineralogical Society in 1927. At the Mineralogical Society he was president (1936–1939) and foreign secretary (1949–1959).

Spencer's service to mineralogy at the British Museum and in the Mineralogical Society, especially through its publications, was long and outstanding. His deep knowledge of his subject and his untiring energy were remarkable. His brusque manner and his single-minded devotion to his science were allied with a sense of humor and an essential kindness that encouraged others to emulate his high standards of scientific scholarship.

BIBLIOGRAPHY

I. ORIGINAL WORKS. Spencer wrote more than 150 original papers. His two books are *The World's Minerals* (London–Edinburgh, 1911; rev. American ed., New York, 1916); and *A Key to Precious Stones* (London–Glasgow, 1936; 2nd ed., 1946). *Precious Stones* (London, 1904) is his translation, with additions, of M. Bauer, *Edelsteinkunde* (1896); and *The Mineral Kingdom*, 2 vols. (London, 1908–1912), is the translation, with additions, of R. Brauns, *Das Mineralreich* (1903–1904). A useful selected bibliography is in C. E. Tilley, *Biographical Memoirs of Fellows of the Royal Society*, **7** (1961), 243–248.

II. SECONDARY LITERATURE. Detailed accounts of Spencer's curatorial and bibliographical work are W. Campbell Smith, in *Mineralogical Magazine and Journal of the Mineralogical Society*, **29** (1950), 256–270; and J. Phemister, *ibid.*, **31** (1956), 1–4. A critical account of Spencer's scientific contribution is given by Tilley, *loc. cit.*

DUNCAN MCKIE

SPENCER JONES, HAROLD (*b*. Kensington, London, England, 29 March 1890; *d*. Greenwich, England, 3 November 1960), *astronomy.*

Spencer Jones was the third child of Henry Charles Jones, an accountant. His early interest in mathematics was fostered at Hammersmith Grammar School, from which he won a scholarship to Jesus College, Cambridge. Thereafter his career followed, with minor modifications, the course usual for men of his abilities and background.

Spencer Jones's scholarly career at Cambridge culminated in his election to a research fellowship of his college in 1913. That year the astronomer royal, F. W. Dyson—following the established pattern for recruiting chief assistants at the Royal Observatory—appointed Spencer Jones to Greenwich as replacement for A. S. Eddington, who returned to Cambridge as Plumian professor; he remained there until 1923, when he was appointed H.M. Astronomer at the Cape of Good Hope. Spencer Jones spent the next decade in South Africa, from which he returned early in 1933 to become astronomer royal—an office he held until his retirement at the end of 1955.

During most of his life Spencer Jones held high positions in the scientific civil service of his country and a number of honorary posts in international professional bodies, including the presidency of the International Astronomical Union in 1944–1948 (to which he succeeded following Eddington's death) and the secretary-generalship of the International Council of Scientific Unions in 1955–

1958. In Britain he received most of the national honors due a man of his station (including knighthood in 1943) and for most of his life was active in the Royal Astronomical Society, of which he was president from 1937 to 1939.

Spencer Jones's scientific interests were connected mainly with the tasks of the observatories with which he was associated: primarily the problems of positional and fundamental astronomy. His outstanding personal contributions were a study of the speed of rotation of the earth, and one of the solar parallax. In his epochal paper "The Rotation of the Earth and the Secular Acceleration of the Sun, Moon, and Planets" (1939) he proved—qualitatively, but quite definitely— that the fluctuations in the observed longitudes of these celestial bodies can be attributed not to any peculiarities of their motions but, rather, to fluctuations in the angular velocity of rotation of the earth.

Spencer Jones devoted several years to a new determination of the mean distance of the earth from the sun through measurements of the parallactic displacement of the asteroid Eros during its favorable opposition in 1930–1931. Still at the Cape of Good Hope, he contributed more than 1,200 photographic observations to this program and later, by international agreement, was entrusted with the reductions of all observations of Eros made in 1930–1931.

The principal result of this work (1941) disclosed that the value of the solar parallax was equal to $8.7904'' \pm 0.0010''$—a considerable improvement over previous results. This figure did not, however, remain unchallenged for long; in 1950 the German astronomer E. K. Rabe, then working in the United States, found from a reduction of all observations made between 1926 and 1945—a feat facilitated by the advent of automatic computers—that a more accurate value was $8.7984'' \pm 0.0004''$, significantly larger than the value deduced by Spencer Jones.

A postscript to the age-long quest for determination of the solar parallax was added after 1961 by a completely different technique, based on the direct measurements by radar of the distance to Venus. The most recent (1967) value of the parallax—$8.79410'' \pm 0.0001''$—rendering the semimajor axis of the terrestrial orbit close to 149,597,890 kilometers (with an uncertainty of a few units of the penultimate digit), is more than ten times as accurate as Spencer Jones's result; and its remaining error hinges, in fact, on limitations of the present knowledge of the velocity of light.

All considered, Spencer Jones's principal original contributions suffered—as did those of many others—from the fact that they were concerned with problems eminently suitable for treatment by automatic computing machinery, but were carried out ten years or so before its advent.

BIBLIOGRAPHY

A complete bibliography of papers published (alone, or jointly) by H. Spencer Jones since 1913 includes 59 separate entries; their sequence terminated rather abruptly in 1945. Many of these papers dealt with subjects of more routine nature (reporting on work carried out at observatories under Spencer Jones's direction). Some of those which should remain of permanent interest for the historian of science are "The Rotation of the Earth and Secular Accelerations of the Sun, Moon, and Planets," in *Monthly Notices of the Royal Astronomical Society*, **99** (1939), 541; and "The Solar Parallax and the Mass of the Moon From Observations of Eros at the Opposition of 1931," in *Memoirs of the Royal Astronomical Society*, **76**, pt. 2 (1941), 11–66. See also his earlier papers on the motion and figure of the moon, in *Monthly Notices of the Royal Astronomical Society*, **97** (1937), 406; and his redetermination of the constant of nutation, *ibid.*, **98** (1938), 440; and **99** (1939), 211.

ZDENĚK KOPAL

SPERRY, ELMER AMBROSE (*b.* Cortland County, New York, 21 October 1860; *d.* Brooklyn, New York, 16 June 1930), *technology, engineering.*

The son of Mary Burst, a schoolteacher, and Stephen Sperry, a farmer, Sperry was raised by his Baptist grandparents after his mother died in childbirth. He later moved from their farm to the village of Cortland and attended the normal school, where the professors interested him in applied science. A visit to the Philadelphia Centennial Exposition, regular reading of *Scientific American* and the patent abstracts in the *Official Gazette*, and the publicity then given to such inventors as Thomas Edison persuaded Sperry to embark upon a career as an inventor and an engineer.

Cortland capitalists with Chicago affiliations helped Sperry to develop a generator and an arc light and, in 1882, to found his own company in Chicago to market them. Among the prominent Chicagoans who backed him was the president of the first University of Chicago. When the company demanded too much routine engineering, Sperry withdrew in 1888 to found his own invention and development company. For the next two decades, in Chicago, Cleveland, Ohio, and Brooklyn, New York, he was successful as an independent inventor concentrating successively upon the electric streetcar, mining machinery, the automobile, and industrial chemistry. Sperry was particularly adept at identifying critical problems, especially those of automatic control, in rapidly expanding areas of technology and in defining with clarity and force his inventive responses—and his patents.

Sperry investigated gyro applications in 1907, stimulated by reports from Germany of Ernst Otto Schlick's gyrostabilizer and Hermann Anschütz's gyrocompass for ships. He committed himself fully to the field after the United States navy adopted his improved gyrocompass, used his gyrostabilizer, and tested his airplane stabilizer. Before World War I, Sperry founded the Sperry Gyroscope Company, which was to become a world-renowned small research and development firm, staffed by resourceful young development engineers and specializing in complex technology and precision manufacture.

During World War I, Sperry, after Edison, was the most active member of the Naval Consulting Board, an early effort to organize science and technology for the military-industrial needs of wartime. Later, Secretary of the Navy Charles Francis Adams said of Sperry, "No one American has contributed so much to our naval technical progress."

During the postwar decade Sperry emerged as a leader in the engineering profession and was elected president of the American Society of Mechanical Engineers, chairman of the Division of Engineering and Industrial Research of the National Research Council, and member of the National Academy of Sciences. He was a major advocate of industrial research, symbolizing for many the transition of America from the era of heroic invention to that of industrial science. Only after his death, and after guidance and automatic control became a major field of science and technology popularized by the concepts of automation and cybernetics, was Sperry's role as a pioneer recognized. The widespread adoption of the Sperry automatic ship pilot, and later the Sperry automatic airplane pilot, further enhanced his reputation. An analysis of his 350 patents reveals his consistent focus upon automatic controls, even in diverse fields of endeavor.

Sperry's sons, Edward, Lawrence, and Elmer, Jr., joined their father at the Sperry Gyroscope Company.

BIBLIOGRAPHY

I. ORIGINAL WORKS. On Sperry's gyrostabilizer, gyrocompass, and automatic ship pilot, see especially his articles in *Transactions of the Society of Naval Architects and Marine Engineers*, **18** (1910), 143–154; **20** (1912), 201–215; **21** (1913), 181–187; **23** (1915), 43–48; **24** (1916), 207–214; **25** (1917), 293–299; **27** (1919), 99–108; and **30** (1922), 53–57. His other articles and patents are listed in the first two references below.

II. SECONDARY LITERATURE. See Thomas Parke Hughes, *Elmer Ambrose Sperry: Inventor and Engineer* (Baltimore, 1971); and J. C. Hunsaker, "Biographical Memoir of Elmer Ambrose Sperry, 1860–1930," in *Biographical Memoirs. National Academy of Sciences*, **28** (1954), 223–260. See also Preston R. Bassett, "Elmer A. Sperry," in *Nassau County Historical Journal*, **21** (Fall 1960).

THOMAS PARKE HUGHES

SPEUSIPPUS (*b.* Athens, *ca.* 408 B.C.; *d.* Athens, 339 B.C.), *philosophy.*

Speusippus' father was Eurymedon, and his mother was Plato's sister Potone. A member of the Academy, he became its head after Plato's death. He was a friend of Dion and supported his political plans. Diogenes Laërtius lists the titles of thirty writings by Speusippus, but his catalog is incomplete. Only scattered fragments have survived. The thirtieth of the so-called letters of the Socratics is presumably from Speusippus to King Philip of Macedonia, in which the writer boasts that his devotion to the king excels that of Isocrates.

Speusippus distinguished several levels of existence, none of which was assigned to the "ideas." The highest level is that of the mathematics, specifically that of the numbers. For the interpreter the main problem is the cohesion of these levels. Aristotle charged that there is no more cohesion among them than between the episodes of a bad tragedy. In spite of this criticism, it has been maintained in recent times that the Neoplatonic conception of a gradual "procession" of being from the absolute One can be traced back, if not to Plato himself, at least to the younger members of the Academy, Speusippus and Xenocrates. That Speusippus was one and possibly the first of the thinkers who conceived of an ontological step-by-step descent without any traces of a dualistic idealism was most impressively argued by Merlan, who supported his claim by information on Speusippus' doctrine, which he had discovered in Iamblichus' *De communi mathematica scientia.*

Regarding Aristotle's criticism of Speusippus as a disjointer, Merlan suggests that it is not aimed at the absence but at the weakness of the vinculum that connects the levels. Not only must any temporal connotation be kept away from this "procession," but the bond that holds together the parts amounts to no more than an analogy (Merlan, p. 118).

Speusippus recognizes two principles of being, which are of unequal rank. The absolutely first principle is the One, which transcends existence like the One of Plotinus, from which it differs, however, in that it is not identical with the Good. Its inferior counterpart is called the Multitude. On each level of existence being is generated by principles that are analogous to the absolute One and its counterpart. In arithmetic, for example, the first principle is the One, in geometry it is the point.

A large proportion of Speusippus fragments comes from his work Ὅμοια ("Similar Things") and deals with zoological classifications. They contain a wealth of detailed information, which has been shown to be very similar to accounts of genera and species of animals in Aristotle's *Historia animalium* (Lang, pp. 9–15). The principle of Speusippus' classifications is a modified form of the Platonic diaeresis. The importance of such classifications can be inferred from Speusippus' assumption that every being is fully determined by the totality of its logical relations to all other beings. This seems to correspond to the doctrine that the concatenation of the levels of existence is constituted by logical relations (analogies).

BIBLIOGRAPHY

I. ORIGINAL WORKS. A collection of fragments is in Paulus Lang, *De Speusippi Academici scriptis. Accedunt fragmenta* (Bonn, 1911; repr., Frankfurt, 1964). A new fragment was discovered by R. Klibansky; see *Parmenides . . . nec non Procli Commentarium in Parmenidem interprete Guillelmo de Moerbeka*, R. Klibansky and C. Labowsky, eds. (London 1953), 38. For the letter to King Philip of Macedonia, see E. Bickermann and J. Sykutris, "Speusipps Brief an König Philipp . . .," in *Berichte. Sächsische Akademie der Wissenschaften*, **80**, fasc. 3 (Leipzig, 1928).

II. SECONDARY LITERATURE. See also the following, listed chronologically: J. Ravaisson, *Speusippi de primis rerum principiis placita* (Paris, 1838); Ernst Hambruch, *Logische Regeln der platonischen Schule in der aristotelischen Topik* (Berlin, 1904); E. Zeller, *Die Philosophie der Griechen*, 5th ed., II, pt. 1 (1922), 986, n. 3, and 996–1010; E. Frank, *Plato und die sogenannten Pythagoreer* (Halle, 1923), 239–261; J. Stenzel, "Speu-

sippos," in Pauly-Wissowa, *Real-Encyclopädie der classischen Altertumswissenschaft*, 2nd ser., III (1929), 1636–1669; H. Cherniss, *Aristotle's Criticism of Plato and the Academy* (Baltimore, 1944), and *The Riddle of the Early Academy* (Berkeley, 1945; repr., New York, 1962), 31–43 and *passim*; H. J. Krämer, *Der Ursprung der Geistesmetaphysik* (Amsterdam, 1964), 207–223; and P. Merlan, in *The Cambridge History of Later Greek and Early Medieval Philosophy* (Cambridge, 1967), 30–32, and esp. *From Platonism to Neoplatonism*, 3rd rev. ed. (The Hague, 1968).

ERNST M. MANASSE

SPHUJIDHVAJA (*fl.* western India, A.D. 269), *astronomy, astrology.*

Sphujidhvaja, who was a Yavanarāja or "official in charge of foreigners," apparently in the kingdom of the Mahākṣatrapas of Ujjayinī in western India, wrote a *Yavanajātaka* in 269, when Rudrasena II (*ca.* 255–277) was reigning. His work was a versification (in *upendravajrā* meter) of a prose translation into Sanskrit of a Greek astrological textbook made by Yavaneśvara in 149. This poem became the foundation of genethlialogy and of interrogational astrology in India, adapting the foreign Greco-Egyptian material for an Indian context; with a lost translation of another Greek text available to Satya (*ca.* 300) it formed the basis of the *Vṛddhayavanajātaka* of Mīnarāja (*ca.* 325–350) and of the *Bṛhajjātaka* of Varāhamihira (*ca.* 550). But besides this Indianized Greek material from Yavaneśvara, Sphujidhvaja drew upon traditional Indian *āyurveda* for his materia medica, and upon the Indian adaptations of Mesopotamian astronomy presented in the *Jyotiṣavedāṅga* of Lagadha (fifth or fourth century B.C.?) and of Greco-Babylonian linear planetary theory in his chapter on astronomical computations (see essay in Supplement). His curious mixture of various traditions indifferently comprehended is characteristic of the exact sciences in India.

BIBLIOGRAPHY

Several passages from the *Yavanajātaka* are discussed by D. Pingree in the following articles: "A Greek Linear Planetary Text in India," in *Journal of the American Oriental Society*, **79** (1959), 282–284; "The Yavanajātaka of Sphujidhvaja," in *Journal of Oriental Research*, **31** (1961–1962), 16–31; "The Indian Iconography of the Decans and Horās," in *Journal of the Warburg and Courtauld Institutes*, **26** (1963), 223–254; and "Representation of the Planets in Indian Astrology," in *Indo-*

Iranian Journal, **8** (1965), 249–267. The text is edited, translated, and furnished with an elaborate commentary by D. Pingree, *The Yavanajātaka of Sphujidhvaja, Harvard Oriental Series* (Cambridge, Mass., in press).

DAVID PINGREE

SPIEGEL, ADRIAAN VAN DEN (also **Spieghel, Spigelius, Spiegelius, Adriano Spigeli**) (*b.* Brussels, Belgium, 1578; *d.* Padua, Italy, 7 April 1625), *botany, anatomy, medicine.*

Spiegel was named for his father and his grandfather, both of whom were surgeons; his mother was Barbara Geens. In 1588 his father was appointed inspector general of the military and naval surgeons of the Dutch Republic; he died in 1600, leaving two sons. Adriaan's (probably younger) brother Gijsbertus became a surgeon at the ducal hospital in Florence. Adriaan studied at the universities of Louvain and Leiden, and later at Padua, where he inscribed his name in the register of the Natio Germanica on 28 March 1601. At Padua he studied under Fabrici and Casserio. It is generally believed that he graduated before 1604, but his name has not been found in the registers of the Sacrum Collegium that granted the degrees to Catholic students. (It is possible that he had not yet become a Catholic and therefore graduated privately.)

In 1606 Spiegel was appointed ordinary physician to the students of the Natio Germanica. Probably he assisted Fabrici in his private practice; certainly he accompanied the old man on a trip to Florence and on another to Venice, where Fabrici gave a consultation. During these years Spiegel studied botany and wrote an introduction to the science, *Isagoge in rem herbariam libri duo* (1606), which he dedicated to the students of the Natio Germanica. In 1607 he competed unsuccessfully for the chair of practical medicine at Padua, left vacant by the death of Ercole Sassonia. In 1612 he left Italy for Belgium. He remained there briefly, however, then traveled through Germany and finally settled in Moravia. Soon afterward he became *medicus primarius* of Bohemia.

On 22 December 1616 the Venetian Senate appointed Spiegel professor of anatomy and surgery. He succeeded Casserio, who had replaced Fabrici after the latter's retirement in 1608. On 17 January 1617 Spiegel performed a public anatomy demonstration in the famous theater at Padua, where in the following years he attracted many foreign students to his public performances. On 25 January 1623 he was elected knight of St. Mark.

He died two years later after an illness of some six weeks—according to one version, as a result of an infection resulting from an injury caused by the breaking of a glass at the wedding of his only daughter Anzoletta (7 February 1625); according to another version because, weakened by his studies, he had no resistance to a feverish disease that ended in a liver abscess.

During Spiegel's lifetime only the *Isagoge*, a work on the tapeworm, and one on malaria (*febris semitertiana*) were published. He did, however, leave some important manuscripts. His son-in-law, Liberalis Crema, published a book on embryology (*De formatu foetu*); and Daniel Rindfleisch, better known as Bucretius, edited his great anatomical work, *De humani corporis fabrica*. It is said that Spiegel entrusted the editing of this book to Bucretius on his deathbed or in his will. Since the manuscript lacked illustrations, Bucretius obtained the beautiful plates that Casserio had had made by a German draftsman and engraver named Josias Murerus (Joseph Maurer). Bucretius added ninety-eight of these fine copperplates to Spiegel's work, separately paginated and under the name of Casserio. These splendid engravings contributed much to the success and fame of the work. Some faults in the text have been indicated, however. J. Riolan the younger blamed Bucretius for them, accusing him of having altered the original text. Nevertheless, the work established Spiegel's renown as an anatomist.

Spiegel's name appears in two anatomical terms: the *linea Spigelii* (the semilunar line between the muscle and the aponeurosis of the *transversus abdominis*) and the *lobus caudatus hepatis* (Spigelii), which, however, had already been described by Eustachi and others.

BIBLIOGRAPHY

I. ORIGINAL WORKS. Spiegel's writings include *Isagoge in rem herbariam libri duo* (Padua, 1606, 1608; Leiden, 1633, 1673; Helmstedt, 1667), 1633 ed. with *Catalogus plantarum* of Leiden and the surrounding area; *De lumbrico lato liber, cum notis et ejusdem lumbrici icone* (Padua, 1618), with a letter, *De incerto tempore partus*; *De semitertiana libri quatuor* (Frankfurt, 1624); *De formatu foetu liber singularis, aeneis figuris ornatus. Epistolae duae anatomicae. Tractatus de arthritide, opera posthuma* (Padua, 1626; Frankfurt, 1631), see also Meyer (below); *De humani corporis fabrica libri X, cum tabulis 98 aeri incisis,* Daniel Bucretius, ed. (Venice, 1627; Frankfurt, 1632), also with other works (Venice, 1654); "Consultatio de lithotomia, *sive calculi vesicae sectione,*" a letter included in Johan van Beverwijck, *De calculo renum et vesicae* (Leiden, 1638), in all eds. and translations of this book but not in vander Linden; and *Adriani Spigelii Bruxellensis . . . opera quae extant omnia*, edited, with a preface, by J. A. vander Linden, 2 vols. in 3 pts. (Amsterdam, 1645), which includes works by others—such as Harvey's *De motu cordis*—a short biography, and a portrait.

II. SECONDARY LITERATURE. Spiegel's accomplishments as a botanist are evaluated in M. Morren, "Adrien Spiegel," in *Revue de Bruxelles*, **1** (Feb. 1838), 51–79.

The following articles deal with Spiegel's life and his contributions to medicine: C. van Bambeke, in Académie royale . . . de Belgique, *Biographie nationale*, XXIII (1921–1924), 330–334; C. Broeckx, *Essai sur l'histoire de la médecine belge* (Brussels, 1838), 311–312; Pietro Capparoni, "Cinque lettre inedite di Adriaan van den Spiegel (Adriano Spigeli)," in *Bollettino dell'Istituto storico italiano dell'arte sanitaria*, **10** (1930), 248–253; Giuseppe Favaro, "Contributo alla biografia di A. Spigeli (Adriaan van den Spiegel) nel terzo centenario della sua morte (1625–1925)," in *Atti del Istituto veneto di scienze, lettere ed arti*, **85**, pt. 2 (1925–1926), 213–252; J. B. Marinus, "Éloge de van den Spiegel (Adrien)," in *Bulletin de l'Académie royale de médecine*, **5** (1846), 842–860, also issued separately (Brussels, 1846); A. W. Meyer, "The Elusive Human Allantois in Older Literature," in E. Ashworth Underwood, ed., *Science, Medicine and History* (London–New York–Toronto, 1953), 510–520, with an English trans. of Spiegel's work on the allantois in ch. 5 of his *De formatu foetu*, 512–513; and A. Portal, *Histoire de l'anatomie et de la chirurgie*, II (Paris, 1770), 449–455.

G. A. LINDEBOOM

SPIEGELIUS. See Spiegel, Adriaan van den.

SPIX, JOHANN BAPTIST VON (*b.* Höchstadt an der Aisch, Germany, 9 February 1781; *d.* Munich, Germany, 15 May 1826), *zoology.*

Spix was the son of a surgeon and *Bürgerrath*. He studied theology at Bamberg and Würzburg, where he decided in 1804 to pursue medicine instead. He graduated M.D. and in 1811 was made an *Adjunkt* of the Munich Academy; he later became a full member and curator of the zoological collections. In 1815 he and K. F. P. Martius were selected by the Bavarian government to take part in an expedition to South America; in April 1817 they left Trieste in the retinue of the Austrian Archduchess Leopoldina, who had just married the crown prince of Brazil (later Emperor Dom Pedro I). Their party, which also included a number of Austrian scientists, reached Rio de Janeiro in July,

and by December of the same year Spix and Martius had set off into the interior to work independently.

Spix and Martius visited the provinces of São Paolo and Minas Gerais, then continued through Minas Novas to Salvador, where they arrived in November 1818, having suffered heat and drought. They sailed to the Amazon estuary by way of Pernambuco, Piauí, and São Luís, at the mouth of the Itapecuru. Spix then left Martius and proceeded upstream as far as the Peruvian frontier, reaching Tabatinga in January 1820. He explored the Rio Negro, then returned to Manaus, where he met Martius. Together they returned in April 1820 to Pará, from which they embarked for Europe two months later. They were back in Munich in December 1820, having accomplished one of the most important scientific expeditions of the nineteenth century.

Spix and Martius were the first European scientists to visit the Amazon after La Condamine. Their collections—including specimens of eighty-five species of mammals, 350 species of birds, nearly 2,700 species of insects, and fifty-seven living animals—provided material for a vast number of works by other scientists. Spix himself was occupied entirely with publishing his findings after his return to Munich. He had planned to expand his study of the skull, published in 1815, and he also had projected a study, for which he had collected a considerable amount of material, of the subterranean zoography and phytography of Bavaria, but he was unable to realize either of these works. Weakened by the fevers that he had suffered on his voyage, he died, leaving other zoologists, including Louis Agassiz, to complete the publication of his works.

BIBLIOGRAPHY

I. ORIGINAL WORKS. *Geschichte und Beurtheilung aller Systeme in der Zoologie nach ihrer Entwicklungsfolge von Aristoteles bis auf die gegenwärtige Zeit* (Nuremberg, 1811) includes a discussion of the difference between "natural" and "artificial" systems. In this book, which follows the German *Naturphilosophen*, Spix differs with Lamarck, who stated that the species was the only natural group. *Cephalogenesis s. capitis ossei structura, formatio et significatio . . .* (Munich, 1815), on the structure of the skull throughout the animal kingdom, lacks importance for zoology. *Reise in Brasilien . . . in den Jahren 1817–20 gemacht und beschrieben von J. B. von Spix und C. Fr. Ph. von Martius.* is in

3 vols. (Munich, 1823–1831), vols. II and III written by Martius. Vol. I was translated into English by H. E. Lloyd, 2 vols. (London, 1824). See also Royal Society, *Catalogue of Scientific Papers*, V, 779.

II. SECONDARY LITERATURE. See *Allgemeine deutsche Biographie*, XXXV (1893), 231–232; and C. F. P. von Martius, *Akademische Denkreden* (Leipzig, 1866), 599–601. On the expedition see Hermann Ross, "Dem Andenken der Forschungsreise von Spix und Martius in Brasilien 1817–1820," in *Berichte der Deutschen botanischen Gesellschaft*, **35** (1917), 119–128.

A. P. M. SANDERS

SPOERER, GUSTAV FRIEDRICH WILHELM (*b.* Berlin, Germany, 23 October 1822; *d.* Giessen, Germany, 7 July 1895), *astronomy.*

Spoerer attended the Friedrich Wilhelm Gymnasium in his native city and in 1840–1843 studied mathematics and astronomy at Berlin University, where he attended the lectures of Encke and Dove. He concluded his studies with the dissertation "De cometa qui a. 1723 apparuit," defended on 14 December 1843. Thereafter Spoerer worked with Encke at the Berlin observatory, performing astronomical computations, and passed the examination for teaching mathematics and sciences in 1846. He taught subsequently at the secondary schools of Bromberg (now Bydgoszcz, Poland) and Prenzlau, and in 1849 moved to Anklam, where he was awarded the title of professor.

In this era such a position afforded an opportunity for scientific work, which was expected by the Ministry of Education. Spoerer gave much more than was customary, especially in his astronomical observations, most of which concerned sunspots. For these efforts in 1868 the crown prince of Prussia presented Spoerer with a parallactic mounted refractor with a five-inch aperture, which made possible more and better observations. In the same year Spoerer participated in an expedition with Friedrich Tietjen and F. W. R. Engelmann to observe a total solar eclipse in India.

In 1874 Spoerer was appointed an observer at the planned astrophysical observatory in Potsdam and moved to that city. Until the building was completed, he continued to make observations of sunspots with his own instrument, mounted on a tower in the city.

Spoerer's main accomplishments were his very careful observations of the sun. He determined the elements of the solar rotation and improved the law for the decrease of the rotation and improved the law for the decrease of the rotation of the sun from the equator to the poles, already derived by Carrington. His statistics for 1879–1893 contain much material on proper motions and on the evolution and distribution of sunspots during a sunspot period.

BIBLIOGRAPHY

Nearly each vol. of *Astronomische Nachrichten* from **55** (1861) to **125** (1890) contains contributions by Spoerer concerning observations of sunspots. Compilations are *Beobachtungen der Sonnenflecken zu Anclam*, 2 vols. (Leipzig, 1874–1876); and "Beobachtungen von Sonnenflecken in den Jahren 1871 bis 1873," *Publikationen des Astrophysikalischen Observatoriums zu Potsdam*, **1**, no. 1 (1878); ". . . in den Jahren 1874 bis 1879," *ibid.*, **2**, no. 1 (1880); ". . . in den Jahren 1880 bis 1884," *ibid.*, **4**, no. 4 (1886); and ". . . in den Jahren 1885 bis 1893," *ibid.*, **10**, no. 1 (1894).

There is an obituary by O. Lohse, in *Vierteljahrsschrift der Astronomischen Gesellschaft*, **30** (1895), 208–210.

H.-CHRIST. FREIESLEBEN

SPORUS OF NICAEA (*fl.* second half of third century), *mathematics.*

Little is known of Sporus. The juxtaposition of available historical data makes it likely that he came from Nicaea, was a pupil of Philo of Gadara, and was either the teacher or a slightly older fellow student of Pappus of Alexandria. Our knowledge of Sporus' activities stems only from such secondary sources as the works of Pappus and the writings of various commentators, among them Eutocius and Leontius, a seventh-century engineer.[1] Most historians, with the notable exception of J. L. Heiberg, agree that Sporus was the author of a work entitled Κηρία, noted by Eutocius.[2] They interpret a second reference by Eutocius to an anonymous Κηρία ’Αριστοτελικά[3] as a subsection of Sporus' work, but Heiberg believes this to be a reference to Aristotle's *De sophisticis elenchis*.

From the above sources it appears that Sporus concerned himself intensively with two mathematical problems: that of squaring the circle and that of doubling the cube.[4] Like many Greek mathematicians who attempted to solve them, he was aware that neither has a solution by means of ruler and compass alone. The close relationship of both problems to limiting processes[5] suggests that Sporus was also interested in questions dealing with approximation, since he reportedly criticized Archimedes for having failed to approximate the value of π more accurately.[6]

The value of the ancient Greeks' preoccupation with special mathematical problems of this type clearly lies in the by-products that this study produced. The squaring problem led to the development of special curves, the quadratrix of Hippias, for example; and the doubling-of-the-cube problem resulted in Menaechmus' discovery of the theory of conic sections and produced a refinement of the theory of proportions. Sporus seems to have contributed to the study of these problems chiefly through his constructive criticism of existing solutions. Indeed, his own solution of the doubling-of-the-cube problem essentially coincides with that of Pappus.

Sporus' writings seem to have been a fruitful source of information for Pappus and later scholars. Pappus, in particular, appears to have valued Sporus' reputation and judgment, since he quoted Sporus in support of his own criticism of the use of the quadratrix in the solution of the squaring problem.

Sporus' nonmathematical writings are known essentially only by topics, through references to them in Maass's *Analecta Eratosthenica* and *Commentariorum in Aratum reliquiae*. They consist of scientific essays on subjects such as the polar circle, the size of the sun, and comets. His literary achievements are reported to include a critical edition of the Φαινομενά of Aratus of Soli.

NOTES

1. E. Maass, *Analecta Eratosthenica*, pp. 45, 47–49, 1939, and *Commentariorum in Aratum reliquiae*, p. lxxi.
2. J. L. Heiberg, *Archimedis opera omnia*, 2nd ed., III, p. 258.
3. *Ibid.*, p. 228.
4. T. L. Heath, *A History of Greek Mathematics*, I, pp. 226, 229–230, 234, 266–268.
5. *Ibid.*, pp. 230, 269.
6. J. L. Heiberg, *Archimedis opera omnia*, 2nd ed., III, p. 258.

BIBLIOGRAPHY

See T. L. Heath, *A History of Greek Mathematics*, I (Oxford, 1921), 226, 229–230, 234, 266–268; J. L. Heiberg, ed., *Archimedis opera omnia*, 2nd ed., III (Leipzig, 1915), 228, 258; F. Hultsch, ed., *Pappus, Collectionis quae supersunt*, I (Berlin, 1878), 252; *Lexikon der alten Welt* (Zurich, 1965), 2863; E. Maass, *Analecta Eratosthenica* (Berlin, 1883), 45–49, 139; and *Commentariorum in Aratum reliquiae* (Berlin, 1898), lxxi; Pauly-Wissowa, *Real-Encyclopädie der classischen Altertumswissenschaft*, 2nd ser., III, 1879–1883; G. Sarton, *Introduction to the History of Science*, I (Baltimore, 1927), 331, 338; P. Tannery, "Sur Sporos de Nicée," in *Annales de la Faculté des lettres de Bordeaux*, 4 (1882), 257–261.

MANFRED E. SZABO

SPRAT, THOMAS (*b.* Beaminster, Dorset, England, 1635; *d.* Bromley, Kent, England, 20 May 1713), *history of science.*

Sprat was one of several children born to Thomas Sprat, a poor parish curate who held B.A. and M.A. degrees from Oxford, and his wife, who was the daughter of a Mr. Strode of Parnham, Dorset. From this "obscure birth and education in a far distant country," as he later described it, he entered Wadham College, Oxford, in November 1651, receiving the B.A. in June 1654 and an M.A. three years later. At Wadham, Sprat became a member of the active and soon influential circle that launched him on his surprising and varied career as the historian and defender of the Royal Society and as a man of the church. He became the favorite and protégé of John Wilkins and formed close associations with other members of the scientific group that gathered around Wilkins during those years, especially with Christopher Wren, Seth Ward, and Ralph Bathurst. Although Sprat may possibly have attended their meetings, there is no record of his having done so and no indication that he ever engaged in the sort of scientific work that was their interest.

In 1659 Sprat's first publications appeared. One was a poem, "To the Happy Memory of the Late Lord Protector," dedicated to Wilkins for "having been as it were moulded by your own hands, and formed under your government," and charged with devoted admiration for Cromwell as the great savior who had led his people into the promised land. Sprat's loyalties were always pliable, a fact often noted by his contemporaries, for he later served Charles II, James II, and William and Mary with the same devotion he had expressed for Cromwell. In politics he became a staunch Tory, a defender of the divine rights of kings, and a strong exponent of high church doctrines. He has, not without reason, been called a time-server. In 1659 Sprat also published a poem in praise of the poet Abraham Cowley, written "in imitation of his own Pindaric odes," thus gaining the nickname "Pindaric Sprat." Cowley, who was also known as a promoter of natural philosophy, returned the favor in his "Ode to the Royal Society," prefixed to Sprat's *History of the Royal Society*, when he said that "ne'er did Fortune better yet / Th' Historian to the Story fit."

Sprat's close association with Cowley had important consequences. In accordance with the poet's will, Sprat was charged with the publication of his *English Works*, published in 1668 and often reprinted, for which he wrote "Account of the Life and Writings of Abraham Cowley." Cowley may also have brought Sprat to the attention of George Villiers, second duke of Buckingham, who by the late 1660's was Wilkins' patron. Having been ordained early in 1660, Sprat later that year gained his first ecclesiastical office through the influence of Cowley and the duke, who also helped him to some of his later preferments. During most of the 1660's and perhaps longer, Sprat was the duke's chaplain, and in 1675 he was appointed one of the three trustees for part of Buckingham's estate, a position that may have helped to pay for his well-known love of good living. In August 1676 he became one of the king's chaplains, soon rising steadily to canon of the Chapel Royal, Windsor, near the end of 1680; dean of Westminster in September 1683; and bishop of Rochester in November 1684, holding the last two offices until his death. In 1676 Sprat married Helen, Lady Wolseley, of Ravenstone, Staffordshire, an event that later in that year led Robert Hooke to note in his diary that he "saw fat Tom Sprat joyd him of marriage." Sprat was survived by his wife, who died in February 1726, and by a son, Thomas, archdeacon of Rochester. They were all buried in Westminster Abbey.

Although he was a prominent figure in his time, Sprat's fame today rests entirely on his *History of the Royal Society*, first published in 1667. Its 438 pages constitute a large and puzzling work on an institution barely seven years old when the book appeared. Since its concerns and their implications touched all major aspects of contemporary affairs, not least religion, the infant Royal Society quickly became involved in controversy and detraction, against which even the good fortune of royal patronage proved insufficient. Neither its present position nor its controversial origins during the past twenty years, open to many unwelcome interpretations, was strong enough to allow it to ignore this opposition without risking serious damage to its reputation and success, which depended on wide cooperation and not least on considerable financial support.

It was the first design of the *History* to explain the nature, organization, work, and aims of the Royal Society to the public, thus showing that the promotion of its affairs was a national, even a patriotic, enterprise that promised both a healing of the wounds left by the recent turbulent events and great material benefits. The *History* was a piece of public relations, even of propaganda. The material that went into it was carefully supervised and selected, and its omissions and suppressions are as significant as its contents. It is not an impartial document; and it gave such strong impetus to renewed controversy that it may be doubted whether the Royal Society would not, at least in England, have been better off without this premature piece of justification. The formidable Henry Stubbe said a few years later that Sprat's work was "a nonsensical and illiterate history." It is a curious irony that the early Royal Society has been the center of similar debate in the extensive recent literature on its history.

The *History* is divided into three parts without separate titles. Part one (pp. 1–51) presents a survey of ancient, medieval, and Renaissance philosophy that is meant to show "what is to be expected from these new undertakers, and what moved them to enter upon a way of inquiry different from that on which the former have proceeded." With exaggeration that mars the conciliatory tone, the Royal Society "most unanimously" follows in the footsteps of antiquity except in "matters of fact: for in them we follow the most ancient author of all others, even Nature itself." It proposes to honor the ancients by being their children rather than their pictures. Here, and often in the rest of the work, Sprat's strong words are reserved for "downright enthusiasts" and the "modern dogmatists," whom he compares to the recent "pretenders to public liberty," who became the greatest tyrants themselves. This political theme recurs forcefully throughout the work—for instance, in a later passage eulogizing Charles I as the royal martyr who followed the "divine example of our Saviour." Sprat finds agreement between the growth of learning and of civil government. Already in this first part Bacon is, as it were, the Royal Society's patron saint, "who had the true imagination of the whole extent of this enterprise, as it is now set on foot." Even members and friends of the Society must have known that this respect for Bacon as the sole intellectual ancestor was exaggerated; but it was necessary in order to rule out the thought of any foreign influence or indebtedness, which an impartial judge would readily have admitted. Both Gassendi and Descartes had been read and admired in England.

Part two (pp. 52–319) contains the history

proper and is chronologically divided into three sections. The first (pp. 52–60) relates the prehistory of the Royal Society up to the first regular meeting on 28 November 1660, tracing its origin exclusively to the meetings that were held "some space after the end of the Civil Wars at Oxford, in Dr. Wilkins his lodgings, in Wadham College." Contradicted repeatedly in the seventeenth century, this brief account has dominated most discussions of Sprat's work and has recently formed the center of much fruitless argument. It will be considered more closely below.

The second section (pp. 60–122) covers the period between the first meeting and the granting of the second royal charter in the spring of 1663. At this time Sprat was proposed for membership by Wilkins and was duly elected. When ninety-four original fellows were elected on 20 May, in accordance with the provisions of the new charter, the Royal Society was firmly established with a large and varied membership. This section is not historical. It explains the nature and aspirations of the Society; its organization, membership, meetings, subject matter, and method of inquiry; its careful interpretation of evidence; and "their manner of discourse," a subject that has received more than its fair share of comment in the secondary literature. These pages contain a panegyric on "the general constitution of the minds of the English" and the special prerogative of England, "whereby it may justly lay claim to be the head of a philosophical league above all other nations in Europe," owing to the "unaffected sincerity," "sound simplicity" of speech, and "universal modesty" that characterize the English as a nation.

This section contrasts the need for cooperative labors and shared verification with Descartes's contemplative method, explains that a division between teachers and scholars is not "consistent with a free philosophical consultation," and suggests that the Royal Society seeks to satisfy the same ambition as the one which at Babel was punished by a "universal confusion" because it "was managed with impiety and insolence." But true knowledge cannot be separated from "humility and innocence": since the Society's ambition "is not to brave the Creator of all things, but to admire him the more, it must needs be the utmost perfection of humane nature."

At this point Sprat observed that the preparation of the *History* had been interrupted for more than a year by the plague (which caused the Royal Society to discontinue its meetings from June 1665 to February 1666) and the great fire of London during the first week of September 1666. Thus, although parts of the rest of the work may have been written before, it was not printed until after this date. The third section (pp. 122–319) of part two tells the story of the Royal Society's work since the spring of 1663. Its first division (pp. 122–157) deals with its reputation and correspondence abroad, and with the encouragement it has received at home from professional and social groups and from the royal family. It concludes with epitomes of the charter of 1663 and of the statutes that had been prepared between June and the end of that year. As late as April 1667 Wilkins was, by order of the Council, directed to prepare these epitomes for the *History*.

The second division (pp. 158–319) of the third section of part two presents fourteen instances "of this their way of inquiring and giving rules for direction . . . from whose exactness it may be guessed, how all the rest are performed," interspersed with Sprat's comments. These papers, not in chronological order, were read before the Royal Society between February 1661 and November 1664. Most of them were also printed in other contemporary publications; and they were drawn from the Society's records under the careful supervision chiefly of Wilkins, who received orders regarding their selection between the end of 1664 and April 1667. The choice was clearly designed to be representative and to have wide appeal both to scientific and, not least, to practical and even lucrative interests. At the end of this part, Sprat confidently observes, "If any shall yet think [the Society] have not usefully employed their time, I shall be apt to suspect, that they understand not what is meant by a diligent and profitable laboring about Nature."

Finally, part three (pp. 321–438) is an apology for the Royal Society that tries to meet all conceivable objections to its enterprise, thus giving a telling picture of the Society's conception of itself in relation to contemporary society, thought, and opinion. Among the many points raised, the following are the most important. The Royal Society poses no threat to learning, education, or the universities. This matter obviously caused some concern, for in the brief account of the Wadham meetings Sprat claimed that they had not only armed many young men against "the enchantments of enthusiasm" but also had helped to save the university itself from ruin. The Society is also a great ally of religion, leading man "to admire the wonderful contrivance of the Creation" so that his praises "will be more suitable to the Divine Nature than the blind applauses of the ignorant," unlike the "enthusiast that pollutes his religion with his own passions." Indeed, experiments are necessary to separate true miracles from falsehoods, and they especially support the Church of England by the agreement that exists "between the present design of the Royal Society and that of our Church in its beginning: They both may lay equal claim to the word reformation." It is Sprat's conviction that "The universal disposition of this age is bent upon a rational religion." Finally, the Society offers great benefits to all manual arts, to trade, to "wits and writers," and to "the interests of our nation." The *History* concludes with a list of all present fellows of the Society up to June 1667.

By early summer the *History* was in the press; in mid-August Pepys saw a copy at the booksellers; at the end of September several persons had read it; and on 10 October 1667 it was presented to the Society by Wilkins, hearty thanks being "ordered to the author for his singular respect to the Society shewed in that book." That it did not, although issued by its printers, bear the Royal Society's imprimatur may indicate some hesitancy to grant it, since this procedure was normal for books encouraged by the Society or written by its fellows, in accordance with the provisions of the royal charter. The *History* sold well, for some six months later Oldenburg reported that the first printing—presumably 1,000 copies—was nearly gone. The work was greatly praised in England and immediately gained the somewhat exaggerated reputation for eloquence and style that has been conventional ever since.

For nearly three years Oldenburg had been announcing the work's imminent publication to his correspondents on the Continent, and at last he could satisfy the inquiries that had kept streaming in. He sent out copies with elaborate covering letters inviting cooperation, an effort that soon proved successful in the form of further inquiries about details of experiments and other matters described in the *History*, although several correspondents complained that their poor English would deny full benefit until they had Latin or French translations. The Royal Society immediately tried to supply them, but only an unsanctioned French translation was published, in identical versions at Geneva and Paris in 1669 and 1670. Thus, as a careful exercise in public relations, Sprat's work confirmed his hope that "this learned and inquisitive age will . . . think [the Society's] endeavours worthy of its assistance."

From the seventeenth century to the present day, the main problem raised by Sprat's work has always been its historical reliability. As early as 1756, Thomas Birch explained in the preface to his own *History of the Royal Society* that part two of Sprat's account was less admired than the others; and he could cite well-informed contemporary opinion for his wish that the history of the Royal Society's "institution and progress" had omitted less and given more facts, and that "the order of time in which they occurred had been more exactly marked." At the very least, Birch was certainly thinking of Sprat's silence on the London meetings in 1645 attended by John Wallis, who in the meantime had given two detailed accounts of them. Their relevance to the prehistory of the Society has been denied—at the cost of creating an unconvincing, ad hoc image of the early post-1660 Society, built on interpretations and arguments so bizarre and ill-informed that they disprove themselves.

In addition to Wallis' two accounts, both of which include Wilkins, it is well-known that since the early 1640's Wilkins had taken a strong interest in natural philosophy, was present in London in the mid-1640's, and in other ways was associated with the people he met at those early gatherings. A further piece of information must be accorded high authority, although it seems not to have been previously cited in this context. It occurs in the Royal Society's official memorial on Wilkins' death, read on 27 November 1672, and it plainly says: "He had been one of that assembly of learned men, who met as early as 1645, and continued their meetings at London and Oxford, until they were formed into the Royal Society" (Birch, *History*, III, 68).

Clearly, the *History* is not reliable on matters of fact; it cannot, as has been claimed, be considered an impartial account written under the supervision of those who had all the information. They may have had it but not wished to use it all. And if less than dependable on this point, the *History* may be so on others, where similar interests were at stake. Well aware of French competition that might challenge its priority, and understandably concerned about some prominent members' actions and allegiances during the 1640's, the Society's interests demanded that its official history omit information that cast doubt on its pure Englishness, on its agreement with the Church of England, and on its loyalty to the restored monarchy. The early London meetings were embarrassing on all counts. There is good reason to accept Wallis' statement that they were suggested by Theodore Haak, a foreigner who had received the suggestion from his French connections, especially from Marin Mersenne, with whom he had then for some years been in correspondence on matters of this sort. At the time both Haak and Wallis were active in the Westminster Assembly; and Haak was associated not only with Comenian circles but also with Comenius, who by 1660, if not earlier, had become anathema owing to his strong millenarianism and defense of the apocalyptic prophets—no doubt Sprat's strong words against enthusiasm are also aimed at Comenius. Tracing the Royal Society's origins only to Wilkins' Wadham group and Oxford ensured respectability. But there would seem to have been more involved than this.

Fortunately we have a great deal of information about the composition, supervision, and uncertain progress of Sprat's work. Referring to the two secretaries, Oldenburg and Wilkins, and their authority "to publish whatever shall be agreed upon by the Society," Sprat said that he was not usurping their prerogative, "for it is only my hand that goes, the substance and direction came from one of them." That man undoubtedly was Wilkins, and Oldenburg seems to have had little to do with the project. The records abundantly demonstrate that the historical part was closely supervised, not only by Wilkins but also by several small groups of fellows, from 21 December 1664 until it went to press. That the choice fell on Sprat is perhaps not surprising: young, energetic (although, before it was finished, Oldenburg complained that the *History* was in "lazy hands"), available, an intimate of Wilkins and perhaps also recommended by Cowley, he had the time that others, especially Wilkins, could hardly have spared. But it might also—for the sake of distance—have been thought useful that the writer not have lived through the entire history of the last decades. Given the sort of image the Royal Society needed, Wilkins was safer in the background, unknown to the public, than as the official historian. Capable but busy, Oldenburg was no doubt ruled out by his German origin and perhaps by other considerations as well. Sprat was a useful and willing tool.

The first mention of a history dates from May 1663, immediately after the second charter and the election of the ninety-four additional original fellows, when Robert Moray wrote to Huygens that the Royal Society would soon publish a small treatise about itself. At the end of the year, again in a letter to Huygens, this work was for the first time called the "history of the Society," intended to accompany the statutes when they were printed, which was believed to be soon. Nothing was heard of the project until November 1664, when Olden-burg wrote to two correspondents that the history was nearly finished and would "we hope, be published soon," and informed Boyle that Sprat intended to give it to the printer in early December. Brouncker, Moray, Wilkins, Evelyn, and others had read it; "but we are troubled," Oldenburg added, "that you cannot have a sight of it, before the publication," for he was worried "whether there be enough said of particulars, or . . . whether there are performances enough for a Royal Society, that has been at work so considerable a time." So far there was no indication that the Society had supervised the work; but within a month, and then repeatedly, well before the plague caused the meetings to be discontinued, it began active supervision and selection of suitable materials for Sprat. There is no doubt that this change was caused by the publication, in May 1664, of Samuel Sorbière's *Relation d'un voyage en Angleterre*, addressed to the French king in the form of a letter dated 25 October 1663 (with a dedication dated 12 December). Sorbière had spent three months in England, beginning in early June 1663, seeing several prominent members of the Society, attending a number of its meetings, and becoming a member on the same day as Christiaan Huygens.

The *Relation* was soon answered by Sprat in the form of a long letter addressed to Christopher Wren, dated 1 August 1664 and published in 1665 under the title *Observations on Monsieur de Sorbière's Voyage Into England*. It was an unfair and defamatory pamphlet, in which Evelyn may have had a share, commensurate with the provocation Sprat felt, "for having now under my hands the History of the Royal Society, it will be in vain for me to try to represent its design to be advantageous to the glory of England, if my countrymen shall know that one who calls himself a member of that assembly has escaped unanswered in the public disgraces, which he has cast on our whole nation." A brief view of the reasons for this violent reaction will explain the aim and reliability of Sprat's *History*.

Sorbière was a somewhat unsteady and superficial character with considerable talent and flair. Some unwise political implications of the *Relation* had caused such strong displeasure in both England and Denmark that Louis XIV banished Sorbière to Nantes; but before the end of the year he had been pardoned, partly owing to the intercession of Charles II through diplomatic channels. Having also heard that some members of the Royal Society were preparing an answer, the king ordered them to desist. Thus the issue in fact concerned the Royal Society alone. As was usual in contemporary travel accounts, Sorbière had made some critical observations on individuals and on English history and institutions, but in general the *Relation* gave a very favorable picture of England and especially of the Royal Society. Sprat, however, dealt only with the criticism, often with obvious misrepresentation of his source. He chided Sorbière for reducing the Society to triviality in his account of its meetings, although clearly no such effect was either intended or expressed. He rejected Sorbière's statement that Hobbes was Bacon's follower in natural philosophy, "between whom there is no more likeness than there is between St. Gregory and the Waggoner." Sorbière's intimacy with Hobbes was a strong irritant: they met several times during Sorbière's English visit—in fact, one of his reasons for going to England was to see Hobbes, whose early work he had translated into French in the 1640's. Worst of all, Sprat claimed— again incorrectly—Sorbière had said that the Royal Society relied upon books for its knowledge of nature and that it divided into sects and parties, the mathematicians holding to Descartes and the men of general learning to Gassendi, "whereas neither of these two men bear any sway amongst them." With the exception of Hobbes, wisely not mentioned in that work, these matters were all made prominent in the *History*, which also shared with the *Observations* Sprat's patriotic defense of English politics and religion, about which Sorbière had said much that was now better forgotten.

Sprat's suppressions are equally telling. He does not refer to Sorbière's statements that he went to England to see his friends and to inform himself about the state of science in England; that the Royal Society's history was being prepared (the first public mention); that as secretary of the Montmor Academy in Paris he knew Oldenburg, who while in Paris as tutor to Boyle's nephew Richard Jones had "constantly" attended its meetings from the spring of 1659 to the spring of 1660, a matter easily attested by other sources and well known to the Society, which in fact had very cordial relations with that Academy during the early 1660's; and that the establishment of the Royal Society had been preceded by the establishment of the Montmor Academy. The official beginning of the latter is placed in 1657, but it was known to have its ancestry in Mersenne's meetings during the 1640's, which through Haak connect with the early London meetings in 1645. Sprat's silence on the Montmor Academy is notable also in the *History*, which cites only a single institution akin to the

Royal Society, although only as a "modern academy for language"—the French Academy, well-known for its hostility to natural philosophy in those very years. This undoubtedly was the crux of the matter: only by suppressing all mention of the London meetings and of the Montmor Academy in favor of Wilkins' Wadham circle was it possible to preserve priority and originality for the Royal Society, for Bacon, and for England.

There is finally one aspect of Sorbière's *Relation* that could not escape any informed reader. Addressed to Louis XIV, the work clearly had as its primary aim, very cleverly pursued by a judicious balance between praise and criticism, to goad the king into official support and patronage for a French academy of science, an effort Sorbière is known to have begun before he went to England, just as the French king is also known to have sought secret intelligence of the state of learning in England at the same time, the eventful spring of 1663. The publication of Sorbière's *Relation* had created a crisis. Sprat put aside his *History* to write his *Observations*, and the Royal Society intervened with its supervision late in 1664 because it felt that it was now openly in a race with Paris and shared Oldenburg's fears that the *History*, in its late 1664 version, did not say enough about details and accomplishments for a society that had "been at work so considerable a time." Thus Sprat's *History* is thoroughly unreliable as history in our sense of the word. Far from ensuring impartiality and truth, the supervision was designed to suppress known but discomfiting facts. That the work also, in this respect, was transformed into a piece of propaganda shows the Society's sense of its vulnerability.

For these reasons, whatever the truth (which may not now be ascertainable and may not matter much), Sprat's *History* cannot be used to refute such accounts as Gian Domenico Cassini's, in *Recueil d'observations faites en plusieurs voyages* (1693), to the effect that on his return to England in 1660, Oldenburg "gave the occasion for the formation of the Royal Society." The general attitude of the Society to the whole matter may be reflected in its reaction to the suggestion that Sorbière be omitted from the lists of the Society, made and favored on 13 November 1666 in a meeting of the Council—which, however, did not have the power to do so. The following day, a vote taken at a meeting of the Society showed fourteen in favor of continuance, eight against.

Sprat was only thirty-two when the *History* was published but never again took any part in the Society's affairs, although he remained a member until his death. Owing to his increasingly conservative politics and his services to changing monarchs, he soon assumed many high offices in the church, although not so high as he had hoped and others expected; he did not become archbishop of York when that see fell vacant in 1686. During the reign of James II, he was an active member of the "infamous" ecclesiastical commission but ultimately terminated its effectiveness when he resigned in August 1688, refusing to prosecute the clergy who had not read the king's Declaration for Liberty of Conscience, although three months earlier he had himself caused much displeasure in London by insisting that it be read in Westminster Abbey. In May 1685 Sprat brought out a tendentious account of the Rye House Plot, written at the request of the king; but he later evaded James's command to write an account of the Monmouth Rebellion. Only a few years later he assisted at the coronation of William and Mary.

Sprat often preached in London, where Evelyn heard him no fewer than seventeen times between 1676 and 1694, always with the greatest praise for "that great wit Dr. Sprat." The sermons extol the monarchy and reason with as much spirit as they denounce "the Romish tyranny" and "the Anabaptistical Madness and Enthusiastical Phrensies of these last ages."

In May 1692, Sprat was the victim of a fantastic blackmail attempt, complete with a forged incriminating document secretly placed in a vase in his palace, purporting to show that he was involved in a conspiracy to restore James II. It caused him great embarrassment, with house arrest and close examination by his peers, before the forgery was found out. For the rest of his life Sprat celebrated the day of his deliverance. He wrote a vastly entertaining account of the plot and the intriguing characters who perpetrated it. It may be argued that this is his best piece of writing.

Estimates of Sprat's character have not been unanimous, either by his contemporaries or by posterity. Gilbert Burnet was not one of his friends, but on Sprat's death he wrote a sketch for which there is support in other contemporary sources: "His parts were bright in his youth, and gave great hopes: but these were blasted by a lazy, libertine course of life, to which his temper and good nature carried him without considering the duties, or even the decencies of his profession. He was justly esteemed a great master of our language, and one of our correctest writers." Swift said that Burnet's estimate was false. Still, both the

Observations on Sorbière's Voyage and the *History of the Royal Society* show qualities that would seem to have belonged also to the man.

BIBLIOGRAPHY

I. ORIGINAL WORKS. There is a mimeographed bibliography by Harold Whitmore Jones and Adrian Whitworth, "Thomas Sprat 1635–1713, Check List of His Works and Those of Allied Writers" (Queen Mary College, Univ. of London, 1952). The *Observations on Monsieur de Sorbière's Voyage* was reissued in 1668, and again in 1709 in a volume that also contained the first English translation of Sorbière's *Relation* and of François Graverol's "Memoirs for the Life of M. Samuel Sorbière." The *Relation* was published in German and Italian in 1667 and 1670. Sprat's *History* was reissued at London in 1702, 1722, and 1734. It has recently been made available in a facsimile reprint "edited with critical apparatus by Jackson I. Cope and Harold Whitmore Jones" in the series Washington University Studies (St. Louis, Mo., 1958). The introduction and notes are useful also for bibliography, but are weak on the actual *History* itself, paying more attention to contemporary controversy, especially to Joseph Glanvill and Henry Stubbe. Unfortunately, this edition does not supply an adequate table of contents and, astonishingly, has no index. Some of Sprat's sermons were printed during his lifetime, but these are all among the ten printed in *Sermons Preached on Several Occasions* (London, 1722). None of the sermons heard by Evelyn is among them. *A True Account and Declaration of the Horrid Conspiracy to Assassinate the Late King Charles II at the Rye-House* was reissued in 3 vols., Edmund Goldsmid, ed., as Collectanea Adamantaea, XIV (Edinburgh, 1886). *A Relation of the Late Wicked Contrivance of Stephen Blackhead and Robert Young, Against the Lives of Several Persons, by Forging an Association Under Their Hands*, 2 pts. (London, 1692), is in *Harleian Miscellany*, VI (London, 1745), 178–254. Sprat's few poems were often reprinted in various collections during the eighteenth century.

II. SECONDARY LITERATURE. There is no full life of Sprat, and the materials for one hardly exist. There is a very brief life in E. Curll, *Some Account of the Life and Writings of the Right Reverend Father in God, Thomas Sprat, D.D.* (London, 1715); it also contains Sprat's will. Much detail is in H. W. Jones, "Thomas Sprat (1635–1713)," in *Notes and Queries*, **197** (5 Jan. 1952), 10–14 and (15 Mar. 1952), 118–123; this is meant to supplement the entry in the *Dictionary of National Biography*. There is much information about Sprat and his *History* in most of the well-known seventeenth-century sources. The most important are the following. Thomas Birch, *The History of the Royal Society*, 4 vols. (London, 1756–1757), has been reissued in facsimile reprint by A. Rupert and Marie Boas Hall as Sources of Science, no. 44 (New York–London, 1968): there is a very incomplete index of names and subjects at the beginning of vol. I. Since Birch's order is strictly chronological, the information drawn from that work can be readily identified. *The Correspondence of Henry Oldenburg*, A. R. and M. B. Hall, eds., II–VII, covers 1663–1672 (Madison, Wis., 1966–1970). Of comparable importance is the correspondence of Christiaan Huygens, in *Oeuvres complètes de Christiaan Huygens*, 22 vols. (The Hague, 1888–1950), with relevant material in II–VII. Sprat's name occurs often in *The Diary of John Evelyn*, 6 vols. (Oxford, 1955), with much information in the excellent notes by the editor, E. S. de Beer.

Balthasar de Monconys, *Journal des voyages de Monsieur de Monconys*, 3rd ed., 3 vols. in 4 pts. (Paris, 1695), III, 1–170, deals with the six weeks he spent in England, where he often saw Sorbière and attended meetings of the Royal Society; where the two accounts cover the same matters, Monconys agrees with Sorbière. *Parentalia, or Memoirs of the Family of the Wrens*, Christopher and Stephen Wren, eds. (London, 1750), has some information about Sprat. Sprat has an entry in Anthony à Wood, *Athenae Oxonienses*, Philip Bliss, ed., IV (London, 1820), cols. 727–730.

Indispensable for reference is *The Record of the Royal Society*, 4th ed. (London, 1940), which reprints the charters and the statutes. General bibliography for the Royal Society can be found in *Isis Cumulative Bibliography 1913–1965*, Magda Whitrow, ed., 2 vols. (London, 1971), II, 749–751; and in Marie Boas Hall, "Sources for the History of the Royal Society," in *History of Science*, **5** (1966), 62–76. Two relevant studies that have appeared since the Hall work are Charles Webster, "The Origins of the Royal Society," *ibid.*, **6** (1967), 106–128 (a review and critique of Margery Purver, *The Royal Society: Concept and Creation* [London, 1967], which is informed by a doctrinal faith in the historical integrity of Sprat's *History*, but the arguments that support this faith are unbelievable); and Quentin Skinner, "Thomas Hobbes and the Nature of the Early Royal Society," in *Historical Journal*, **12** (1969), 217–239.

Valuable for information and bibliography about French academies is Harcourt Brown, *Scientific Organizations in Seventeenth-Century France (1620–1680)* (Baltimore, 1934), to be supplemented by Albert J. George, "The Genesis of the Académie des Sciences," in *Annals of Science*, **3** (1938), 372–401. To Vincent Guilloton goes the credit for first showing that the cause of the Sorbière-Sprat controversy lay in the Royal Society, although he uses only a few of the available sources: "Autour de la *Relation du voyage de Samuel Sorbière en Angleterre*," in *Smith College Studies in Modern Languages*, **11**, no. 4 (July 1930), 1–29. Important further information is in three studies by André Morize: "Samuel Sorbière et son *Voyage en Angleterre*," in *Revue d'histoire littéraire de la France*, **14** (1907), 231–275; "Samuel Sorbière," in *Zeitschrift für französische Sprache und Literatur*, **33** (1908), 214–265, with bibliography of Sorbière's MSS and printed works on 257–265; and "Thomas Hobbes et Samuel Sorbière. Notes

sur l'introduction de Hobbes en France," in *Revue germanique* (Paris), **4** (1908), 193–204. There is a useful essay on Sorbière's philosophical orientation in A. G. A. Balz, *Cartesian Studies* (New York, 1951), 64–79.

Sprat's *History* and the problem of English prose style have been treated in a number of not very fruitful literary studies. The most significant is Francis Christensen, "John Wilkins and the Royal Society Reform of Prose Style," in *Modern Language Quarterly*, **7** (June 1946), 179–187 and (Sept. 1946) 279–290. For general background there are the relevant chapters in R. F. Jones, *Ancients and Moderns*, 2nd ed. (Berkeley–Los Angeles, 1961; paperback, 1965).

HANS AARSLEFF

SPRENGEL, CHRISTIAN KONRAD (*b.* Brandenburg, Germany, 22 September 1750; *d.* Berlin, Germany, 7 April 1816), *botany.*

Sprengel was the last child of Ernst Victor Sprengel (1686–1759), Archdeacon of St. Gotthardt-Gemeinde, and his second wife, Dorothea Gnadenreich Schaeffer (*d.* 1778); and he was the fifteenth child of two marriages.[1] He entered Halle University in 1770 to study theology and philology. Four years later he began teaching in Berlin at the Friedrichs-Hospital School and at the royal military academy. In 1780 he was appointed rector of the Great Lutheran Town School at Spandau, where he taught languages and natural science. His friend Ernst Ludwig Heim, a physician and mycologist, gave instruction in botany to both Sprengel and Alexander von Humboldt. Sprengel in turn shared his knowledge of the Spandau flora with Carl Willdenow. Sprengel's two brothers, Johann Christian and Joachim Friedrich, also studied theology at Halle.[2] The latter taught for a time at the Realschule in Berlin, specializing in history, botany, and mineralogy, and was the father of the botanist Kurt P. Sprengel.

In the summer of 1787 Sprengel began observing the pollination of *Geranium* flowers. These relationships of flower structure, insect visitors, and pollination mechanisms occupied him for the next six years and culminated in the publication of his great work, *Das entdeckte Geheimniss der Natur im Bau und in der Befruchtung der Blumen*, in 1793.[3] Printed in double-column format, it had twenty-five copperplates crowded with 1,117 drawings of floral parts representing 461 species. The striking title page also served as a plate, since the wide border comprised twenty-eight insect and flower drawings. Although it became a milestone on the road to understanding the biology of flowers, Sprengel was greatly disappointed at the book's reception.

Although J. G. Koelreuter had already noted some of the relationships of floral parts, nectar, and insects to pollination, Sprengel went much further in stating that the structure of the flower can be interpreted only by considering the role of each part in relation to insect visits. He noted that color and scent are attractions; that the corolla markings are guides to the hidden nectar; and that grasses have light pollen and are wind-pollinated. His rediscovery of dichogamy (the maturation of anthers and stigmas at different times in the same flower, such that self-pollination cannot occur) led him to one of his major conclusions: "Nature appears not to have intended that any flower should be fertilized by its own pollen."[4] This doctrine, together with the even more important view of the close integration of floral structures with insect visitation, was the first attempt to explain the origin of organic forms from definite relations to the environment. "Since Darwin breathed new life into these ideas by the theory of selection, Sprengel has been recognized as one of its chief supports."[5]

There are a few early comments on Sprengel's book—including a book review, a 1794 letter by Goethe,[6] and a later commentary by his nephew Kurt P. Sprengel. In England, Robert Brown published an article on pollination in 1833, citing two observations from Sprengel.[7] Furthermore, Charles Darwin noted that it was on Brown's advice in November 1841 that he obtained and read "C. K. Sprengel's wonderful book."[8] Perhaps through the work of Brown (and later of Sprengel), Darwin became interested in pollination by insects with observations that began in the summer of 1838. In the chapter "Natural Selection" of the *Origin of Species* (1859) he refers to these observations, and confirms Sprengel's similar ones, on dichogamy: "These plants have in fact separated sexes, and must habitually be crossed. . . . How simply are these facts explained on the view of an occasional cross with a distinct individual being advantageous or indispensable."[9]

Further comments on Sprengel's work are in Darwin's two botanical works, *Orchids* (1862) and *Cross and Self Fertilization* (1876). In the former he refers to Sprengel's "curious and valuable work," to "Sprengel's Doctrine," and again to Sprengel's work that ". . . until lately was often spoken lightly of. No doubt he was an enthusiast, and perhaps carried some of his ideas to extreme length. But I feel sure, from my own observations, that his work contains an immense body of

truth."[10] In the latter book Darwin cites Sprengel's notes on the essential role of insects in the pollination of many plants, and says: "He was in advance of his age, and his discoveries were for a long time neglected." Further, he states that Sprengel, while noting that cross-pollination between flowers of the same species occurred, was not ". . . aware that there was any difference in power between pollen from the same plant and from a distinct plant."[11] It remained for Darwin to assess the importance of this for his theory of natural selection.

Sprengel's difficulties at Spandau with school superintendent D. F. Schulze (noted by R. Mittmann [1893] and by O. Recke [1913]) resulted in his being pensioned, leaving Spandau at age forty-four, and moving to Berlin, where he became a private tutor. A proposed second part of *Das entdeckte Geheimniss* was never published, but a small work on bees appeared in 1811 and a work on philology in 1815.[12] Heinrich Biltz,[13] who studied with Sprengel from 1809 to 1813, published a closely drawn character study (1819) that refers to his botanical excursions, to his work in philology, and to his criticism of Linnaeus and Willdenow for their ignorance of Greek. Sprengel did not marry and died in relative obscurity—but not in poverty.[14] Some archival items are known, including seven letters and notes by Sprengel.[15] He is commemorated by the plant genus *Sprengelia* (Epacridaceae).[16]

Were it not for his remarkable book Sprengel would be forgotten today. This work reached Darwin, and the insect-plant mutualism so elegantly and minutely described there profoundly influenced him. Although the two were poles apart in religious beliefs, the elemental natural processes revealed by their studies provided Darwin with evidence for his theory of evolution.

NOTES

1. D. E. Meyer ("Goethes . . .") has clarified the Sprengel genealogy: the first wife of E. V. Sprengel (1686–1759) was Katharina Elisabeth Krause (1692–1732), whose fourth child was Joachim Friedrich, and his son was Kurt P. Sprengel. Thus, the half brothers Joachim Friedrich and Christian Konrad had different mothers. His second wife was "Dorothea, geb. Hopf . . ." (see *Evangelischen Pfarrerbuch für die Mark Brandenburg*, II, pt. 2 [Berlin, 1941], a letter dated 2 October 1972). The account of Christian's birth has been found in the church record of St. Gotthardtgemeinde in Brandenburg as recorded by his father, the archdeacon, with a later note that this was his fifteenth child (occurring in his sixty-fourth year). A copy of this item is in the Botanical Museum in Berlin (D. E. Meyer, "Biographisches . . .").

A. Krause states that Sprengel's ancestors on his mother's side included the Grunow and Goedicke families; his grandfather was Peter Schaeffer; that in 1799 he was elected an honorary member of the Königliche Bayerische Botanische Gesellschaft, Regensburg; and that a street in Berlin was named for him.

The large E. V. Sprengel family resulted from two marriages, for Peter Nathanael Sprengel (1737–1814) was a "stepbrother of both J. F. and J. C. G. Sprengel." *Das gelehrte Teutschland*, VII (Lemgo, 1798), 588–589. In publications some confusion exists between C. K. Sprengel and Kurt Polykarp Sprengel when first initials are used (the latter often used the Latin form, Curtius Sprengel); and signing letters just "Sprengel" provides further problems.

2. Joachim Friedrich Sprengel wrote *Vorstellung der Kräuterkunde in Gedächtnisstafeln* (Greifswald, 1754); see G. A. Pritzel, *Thesaurus literaturae botanicae* (Leipzig, 1872), no. 8858, p. 303; he also wrote two other works (1751, 1753); see C. G. Kayser, *Vollständiges Bücher-Lexicon, 1750–1832*, V (Leipzig, 1835), 295; and J. G. Meusel, *Das gelehrte Teutschland*, VII (1798), 582–583.

3. "The Newly Revealed Mystery of Nature in the Structure and Fertilization of Flowers" (Berlin, 1793). The title page at lower left reads "Gezeichnet v. C. K. Sprengel," and at lower right W. Arndt is credited as the engraver. The preface is dated "18 Dec., 1792—C. K. Sprengel, Rektor." The plates were engraved mainly by Johann S. Capieux and bear dates 1791 and 1792; some were by J. Wohlgemuth.

4. Sprengel, *Das entdeckte Geheimniss*, 43.

5. J. von Sachs, *History of Botany*, 415.

6. D. E. Meyer ("Biographisches . . .") noted a letter from Goethe to August Batch (professor of botany, Jena), 26 Feb. 1794. J. W. Goethe, *Goethe's Werke*, edited by order of Grand Duchess Sophie of Saxony, pt. 4, *Goethes Briefe*, X (Weimar, 1892), letter 3044, pp. 143–144. An unsigned book review also has been found in *Göttingische Anzeigen von gelehrten Sachen*, **20** (1793), 1105–1114.

7. R. Brown, "On the Organs and Mode of Fecundation . . .," 687, 717.

8. F. Darwin, ed., *C. Darwin, Life and Letters*, 47. The copy that Brown utilized may be the one now in the library of the Linnean Society of London. Although the copy is not dated or identified (except for a heraldic bookplate transferred from the former binding), the library stamp would indicate the first half of the nineteenth century (letter, 5 Jan. 1973, G. Bridson, Linnean Society of London). Darwin's copy (with his signature) is on permanent deposit in the Cambridge University Library; it contains Darwin's numerous marginal annotations with evidence that he acquired (presumably by purchase) this work in Aug. 1841 and read it on 10 Nov. 1841 (letter, 5 Jan. 1973, P. J. Gautrey, Cambridge University Library).

9. Pp. 98–99.

10. *Orchids* 2, 27, 275.

11. *Cross and Self Fertilization*, 5, 6

12. *Neue Kritik der classischen römischen Dichter* (1815). E. Strasburger (1893) located a copy in the Leipzig University Library and noted that it comprised 142 pp.; Strasburger comments at some length (pp. 116–117); J. D. Fuss (1824) published a refutation of some of Sprengel's emendations of Ovid's works.

13. See biography of Friedrich Heinrich Biltz (d. 1835) in *Archive der pharmazie*, **54** (1835), 1–25. See also Royal Society, *Catalogue of Scientific Papers*, XII, 82.

14. No notice at his death could be found; Landesarchiv, Berlin, searched the newspaper *Berlinische Nachrichten von staats- und gelehrten Sachen* from 7 Apr. 1816 to the end of May 1816, with no results.

Previous accounts indicate the place of burial was unknown; Hoffmann ("Urkundliches . . .") cites the church register of the Werderschen Kirche for burial "auf dem Kirchhof vor dem Oranienburger Thor"; Landesarchiv Berlin notes also that the Sprengel burial ground was "Friedhof der Dorotheenstädtischen und Friedrich-Wer-

derschen Gemeinden, Chausseestrasse 126" (W. von Wohlberedt), destroyed when roads to Hannover were constructed. G. Hintze, "Rundgang," 84–85.

15. One MS source is D. F. Schulze's "Chronicle" (to 1804), which in 1893 existed as a folio vol. of 1,071 handwritten pp. in St. Nicholas Church at Spandau, later edited by O. Recke (1913); the diaries of E. L. Heim are additional sources (G. L. Kessler, *Der alte Heim*; see also D. Meyer).

P. Hoffmann ("Urkundliches . . .") has provided data from a large number of sources and original documents: birth record; matriculation at Halle; first teaching position; note on H. Biltz (and his article of 28–29 Dec. 1819 in *Morgenblatt für gebildete Stände*); the full text of the original draft of the will (30 Jan. 1816) and the will (5 Feb. 1816); the death certificate of 9 Apr. 1816, signed by Dr. Kohlrausch. Hoffmann notes that Kerner von Marilaun (*Pflanzenleben*, 3rd ed., II, 310) states that although Sprengel's pension was only 150 taler, he left the Berlin orphanage 5,000 taler. In addition he furnishes the death entry in the register of the Werderschen Kirche, Sprengel's correct street address, and his burial record. He provides a picture of the memorial tablet in the Berlin-Dahlem Botanical Garden. See an anonymous article, "J. G. Kölreuter et C. K. Sprengel: Souscription pour leur élever des monuments," in *Isis*, **1** (1913), 243–244.

D. Meyer ("Biographisches . . .") examined some sheets from Willdenow's herbarium (of more than 20,000 specimens) in the Berlin-Dahlem Herbarium and noted three sheets (*Carex, Juncus*, and *Stipa*) that bore the full Latin names in Sprengel's writing (reproduced in Fig. 1, p. 121).

Following an inquiry to Friedrich Vieweg and Son, Brunswick (the firm publishing Sprengel's book of 1793), five Sprengel letters dating from 11 Sept. 1794 to 23 July 1803 have been discovered in their archives (letter, V. Schlecht, 11 Oct. 1972); see P. Forman, "The Archive of Friedr. Vieweg et Sohn, Braunschweig . . .," in *Isis*, **60**, pt. 3 (1969), 384–385, which notes that the archives contain possibly between 150,000 and 300,000 items representing roughly 5,000 correspondents. Hoffmann ("Urkundliches . . .") quotes in full a Sprengel letter of 18 Nov. 1781 to the magistrate in Spandau (original in the Darmstadt collection, Deutsche Staatsbibliothek, Berlin). A Sprengel letter of 1793 exists in the Archiv der Akademie der Wissenschaften in Göttingen—card catalog indicates C. C. Sprengel, but the letter is signed "Sprengel." Letter (Jäykkä), Niedersächsische Staats- und Universitätsbibliothek, Göttingen, 27 Sept. 1972. A photocopy of this letter reveals that it is not in the handwriting of C. K. Sprengel. It was sent from Halle and is a note of thanks for election to the Academy (at the age of twenty-seven). This letter was probably from Kurt P. Sprengel. C. K. Sprengel was not elected to this Academy.

16. J. E. Smith, "Description of a New Genus . . ."; see also *Index Kewensis* (Oxford, 1905), IV, 970, which cites a Swedish publication of 1794 in which this new genus was first described. A "Biographische Tafel" in honor of Sprengel has been established at the Heimatsmuseums, Bezirksamt Spandau von Berlin, which owns a Sprengel handwritten note of 1784 (letter, E. Blume, 5 July 1973).

BIBLIOGRAPHY

I. ORIGINAL WORKS. Sprengel's books are *Das entdeckte Geheimniss der Natur im Bau und in der Befruchtung der Blumen* (Berlin, 1793), also facs. ed. as Wissenschaftliche Classiker in Facsimile-Drucken, no. 7 (Berlin, 1893) and edited by P. Knuth as Ostwalds Klassiker der Exakten Wissenschaften, nos. 48–51 (Leipzig, 1894), with biographical notes in I, 180–181; *Die*

Nützlichkeit der Bienen und die Nothwendigkeit der Bienenzucht, von einer neuen Seite dargestellt (Berlin, 1811), also edited, with epilogue, by August Krause (Berlin, 1918); *Neue Kritik der classischen römischen Dichter, in Anmerkungen zu Ovid, Virgil und Tibull* (Berlin, 1815). On the last, see W. Engelmann, *Bibliotheca scriptorum classicorum* (Leipzig, 1882), 467. D. E. Meyer ("Goethes . . .") has revealed two journal articles of Sprengel: "Über die Nectarien; Ankündigung," in *Botanisches Magazin*, **4** (1788), 186; and "Versuch die Konstrucktion der Blumen zu erklärung; Ankündigung," in *Botanisches Magazin*, **8** (1790), 160–164.

II. Secondary Literature. See the anonymous "Hermann Müller's 'Fertilisation of Flowers,'" in *Nature*, **27** (1883), 513–514, a book review; P. Ascherson, "Christian Konrad Sprengel als Florist und als Frucht-Biolog," in *Naturwissenschaftliche Wochenschrift*, **8** (1893), 140–141; and "Zur Erinnerung an Chr. K. Sprengel und sein vor 100 Jahren erschienenes Werk: 'Das entdeckte Geheimniss der Natur im Bau und in der Befruchtung der Blumen,'" in *Verhandlungen des Botanischen Vereins der Provinz Brandenburg,* **35** (1894), viii–xiii; H. B., "Erinnerung an Christian Conrad Sprengel, nebst einigen Bemerkungen aus seinem Leben," in *Flora oder Botanische Zeitung*, **2** (1819), 541–552, repr. in *Mitteilungen des Thüringischen botanischen Vereins* (Weimar), **15** (1900), 24–29 (a footnote here states the author is H. Biltz of Erfurt, father of Dr. E. Biltz); J. B. Barnhart, *Biographical Notes Upon Botanists*, III (Boston, 1965), 312; W. Bastine, "Christian Konrad Sprengel, ein vergessener märkischer Botaniker," in *Jahrbuch für brandenburgische Landesgeschichte*, **12** (1961), 121–131; and R. Brown, "On the Organs and Mode of Fecundation in Orchideae and Asclepiadae," in *Transactions of the Linnean Society of London*, **16** (1833), 685–745.

Also of value are Charles Darwin, *On the Origin of Species . . .* (London, 1859); *On The Various Contrivances by Which . . . Orchids Are Fertilized by Insects . . .* (London, 1862; 2nd ed., London, 1877; New York, 1895); and *The Effects of Cross and Self Fertilization in the Vegetable Kingdom* (London, 1876; 2nd ed., New York, 1895); Francis Darwin, *Charles Darwin, His Life Told in an Autobiographical Chapter and in a Selected Series of His Published Letters* (London, 1887; New York, 1893); Francis Darwin and A. C. Seward, eds., *More Letters of Charles Darwin. A Record of His Work in a Series of Hitherto Unpublished Letters*, 2 vols. (New York, 1903), which mentions Sprengel's work in letters to Asa Gray (II, 254) and to Sir Joseph Hooker (I, 446); A. Engler, "Bericht über die Enthüllung des Denksteins für Christian Konrad Sprengel," in *Notizblatt des Botanischen Gartens und Museums zu Berlin*, **7**, no. 62 (1917), 417–420; K. Faegri and L. van der Pijl, *The Principles of Pollination Ecology* (London, 1966), 2, 3, 220; K. von Frisch, "Christian Konrad Sprengels Blumentheorie vor 150 Jahren und heute," in *Naturwissenschaften*, **31** (1943), 223–229, repr. in *Chronica botanica*, **12** (1951), 242–245; and J. D. Fuss. *Ad J. B.*

Lycocriticum . . . epistola, in qua loci Metamorphoseon et Fastorum Ovidii, nec non alii nonnulli sive defunduntur . . ., C. C. Sprengel emendationes . . . refutantur (Cologne, 1824)—*British Museum General Catalog of Printed Books to 1955*, compact ed., IX (London, 1967), 1215; and W. Engelmann, *Bibliotheca scriptorum classicorum* (Leipzig, 1882), 462.

Further works that may be consulted are K. Goebel, "The Biology of Flowers," ch. 20 of A. C. Seward, *Darwin and Modern Science* (Cambridge, 1909); H. A. Hagen, "Christian Conrad Sprengel," in *Nature*, **29** (1883), 29, 573; R. J. Harvey-Gibson, *Outlines of the History of Botany* (London, 1919), 60–61, 132, 160; G. Hintze, "Rundgang über die Berliner Friedhöfe," in *Brandenburgia*, **24** (1933), 84–85; P. Hoffmann, "Einiges über Christian Konrad Sprengel," in *Mitteilungen des Vereins für die Geschichte Berlins*, **36** (1919), 37–39; and "Urkundliches von und über Christian Conrad Sprengel," in *Naturwissenschaftliche Wochenschrift*, **19** (1920), 692–695; G. L. Kessler, ed., *Der alte Heim, Leben und Wirken Ernst Ludwig Heims' . . . aus Hinterlassenen Briefen und Tagebuchern . . .* (Leipzig, 1846), 194; O. Kirchner, "Christian Konrad Sprengel, der Begründer der modernen Blumentheorie," in *Naturwissenschaftliche Wochenschrift*, **8** (1893), 101–105, 111–112, which cites (p. 111) "Herrn Forstmeisters Sprengel in Bonn" as a source of oral and written biographical data; O. Kirchner and H. Potonié, *Die Geheimnisse der Blumen. Eine populäre Jubiläumsschrift zum Andenken an Christian Konrad Sprengel* (Berlin, 1893); P. Knuth, "Christian Konrad Sprengel, Das Entdeckte Geheimniss der Natur. Ein kritisches Jubiläums-Referat," in *Botanisch jaarboek* (Ghent), **5** (1893), 42–107; and *Handbook of Flower Pollination Based Upon Hermann Müller's Work, "The Fertilization of Flowers by Insects,"* translated by J. R. Ainsworth Davis, 3 vols. (Oxford, 1906–1909); G. Kraus, *Der botanische Garten der Universität Halle* (Leipzig, 1894), no. 2, 57, 59, 60; and A. Krause, "Christian Konrad Sprengel," in *Mitteilungen des Vereins für die Geschichte Berlins*, **36** (1919), 32.

Additional secondary works are R. Lamprecht, "Der Rektor Sprengel (1780–1794)," in *Die grosse Stadtschule von Spandau von ca. 1300 bis 1853* (Spandau, 1903); D. E. Meyer, "Biographisches und Bibliographisches über Christian Conrad Sprengel," in *Willdenowia*, **1** (1953), 118–125; D. E. Meyer, "Goethes botanische Arbeit in Beziehung zu Christian Konrad Sprengel (1750–1816) und Kurt Sprengel (1766–1833) auf Grund neuer Nachforschungen in Briefen und Tagebüchern," in *Berichten Deutschen botanischen Gesellschaft*, **80** (1967), 209–217; R. Mittmann, "Material zu einer Biographie Christian Konrad Sprengel's," in *Naturwissenschaftliche Wochenschrift*, **8** (1893), 124–128, 138–140, 147–149; C. Nissen, *Die botanische Buchillustration, ihre Geschichte und Bibliographie*, II (Stuttgart, 1951), 174, which notes that Sprengel prepared the drawings, and Johann Stephan Capieux (1748–1813) the plates, for *Das entdeckte Geheimniss*;

O. Recke, ed., *Zur Beschreibung und Geschichte von Spandow. Gesammelte Materialien von D. F. Schulze*, 2 vols. (Spandau, 1913), I, 237, 251–252, 256–260, 274–279, 315, 424; J. von Sachs, "Further Developments of the Sexual Theory by Joseph Gottlieb Koelreuter, and Konrad Sprengel, 1761–1793," in his *History of Botany (1530–1860)*, translated by H. E. F. Garnsey, revised by A. C. Bayley Balfour (Oxford, 1890), 406–422; J. E. Smith, "Description of a New Genus of Plants Called *Sprengelia*," in *Tracts Relating to Natural History* (London, 1798), 269–274; Kurt P. Sprengel, "Geheimnis der Natur in Befruchtung der Blumen," in *Repertorium des Neuesten und Wissenwürdigsten aus der gesammten Naturkunde*, **5** (1813), 356–364; and *Geschichte der Botanik* (Altenburg–Leipzig, 1817), 266–267; and E. Strasburger, "Zum hundertjährigen Gedächtnis an 'Das entdeckte Geheimnis der Natur,'" in *Deutsche Rundschau*, **20** (1893), 113–130.

Also see G. Wichler, "Kölreuter, Sprengel, Darwin und die moderne Blütenbiologie," in *Sitzungsberichte der Gesellschaft naturforschender Freunde zu Berlin* for 1935 (1936), 305–341; M. Wieser, "Der märkische Darwin Konrad Sprengel," in *Brandenburgische Jahrbücher*, **3** (1938), 48–57; W. von Wohlberedt, *Verzeichnis der Grabstätten bekannter und berühmter Personlichkeiten in Gross-Berlin und Potsdam*, pt. 2 (Berlin, 1934), 125–126; and G. Wunschmann, "Christian Konrad Sprengel," in *Allgemeine deutsche Biographie*, LIII (1893), 293–296.

LAWRENCE J. KING

SPRENGEL, KURT POLYCARP JOACHIM (*b.* Boldekow, Germany, 3 August 1766; *d.* Halle, Germany, 15 March 1833), *botany, medicine.*

In Hermann F. Kilian's survey of German universities of 1828, Sprengel was viewed as the most prestigious professor in Germany. His reputation was principally the result of his erudite and detailed publication in medical history and some botanical contributions, especially in phytotomy. At the time of his death, Sprengel was a member of almost fifty German and foreign academies and learned societies, a shining star in the otherwise bleak sky of contemporary medicine in Germany.

Sprengel was born in a small Pomeranian village, the son of a local preacher and nephew of the distinguished botanist Christian K. Sprengel. Under the direction of his father, a former teacher at the Berlin Realschule, Sprengel learned Greek, Latin, and Hebrew, and also received a solid background in the natural sciences. Later he taught himself Arabic and began the study of five modern European languages, which he soon mastered.

Short of funds and barely seventeen years old, Sprengel found employment as a private tutor near Greifswald, studying theology and philology in his spare moments. In 1784 he successfully passed his religious examinations and was allowed to preach. In 1785 Sprengel matriculated at the University of Halle, determined to study medicine (and not theology, as adduced in some accounts). At the end of five semesters under the direction of Phillip F. T. Meckel and Johann F. G. Goldhagen, he graduated in 1787 with a dissertation on nosology.

Two years later Sprengel began to teach legal and historical subjects at the university as an unsalaried instructor. During the same years he successfully established a medical practice in Halle. In 1795, however, he courageously accepted an invitation to become a full-time academician at the University of Halle, thereby terminating his more lucrative private practice and his higher status.

Versatile and talented, Sprengel taught pathology, legal medicine, semeiology, medical history, and botany at the university. He was popular with the students and well-known for his charity to the needy. After 1800 Sprengel devoted more attention to botany than to medicine. This shift possibly reflected his growing dissatisfaction with the prominence of philosophical German medicine. As professor of botany, he was also director of the university's botanical gardens, where he resided with his family. He established an extensive herbarium and conducted research tours in the nearby countryside.

His contemporaries considered Sprengel to be a keen classical scholar and historian. His most important publication was a medical history, *Versuch einer pragmatischen Geschichte der Arzneikunde*. Although it became the standard work on the subject for nearly a century, Sprengel modestly labeled it an "attempt" to portray medicine chronologically in the various historical periods. He deemphasized the strictly biographical aspects, stressing instead the connections between medicine and contemporary cultural and philosophical forces.

Sprengel called his work a "pragmatic" history of medicine written with a definite utilitarian purpose. In this approach he followed the historical conceptions prevalent during the Enlightenment, which raised the hope of a perfected future, if only the shackles of superstition could be unfastened and the path of reason followed. Therefore Sprengel's goal was to present the medical past with all its errors and pitfalls, in the hope that these aberrations would provide valuable lessons and reveal the basic truths on which a more rational medicine could be developed.

Sprengel's fame was further enhanced by his numerous translations—many of them from English authors—and his editorship of five journals dealing with medical and botanical subjects. When defending his beliefs or attacking those trends in medicine that he profoundly disliked, he wrote clearly, incisively, and to the point, without allowing petty personal arguments to vitiate his criticism.

Sprengel was a vigorous critic of the emerging speculative currents in German medicine. He opposed Brownianism and its modified *Erregungstheorie*, and wrote a monograph against animal magnetism. Moreover, he disproved Hahnemann's claims of classical roots for homeopathy, thus incurring the wrath of its founder, who sought vindication in the courts. Sprengel's analysis of German medicine during the last decade of the eighteenth century provided an invaluable document, written by a strict adherent of Kant's critical philosophy, who found himself averse to the new *Naturphilosophie*.

Although hampered by inadequate optics and preparation techniques, Sprengel strongly promoted the microscopic examination of plants and studied their structure, developing his own theory of plant-cell formation. Although soon superseded, his ideas about the nature of cells and fibers provided an essential stimulus for further investigations by other notable botanists, such as Heinrich F. Link, Johann J. Bernhardi, and Ludolf C. Treviranus.

BIBLIOGRAPHY

I. ORIGINAL WORKS. A complete list of Sprengel's publications is in Rohlfs (see below), 212–218; and Adolph C. P. Callisen, *Medicinisches Schriftsteller-Lexicon*, XXXII (Altona, 1844), 389–399. His most famous work, *Versuch einer pragmatischen Geschichte der Arzneikunde*, 5 vols. (Halle, 1792–1799), was reprinted with corrections (1800–1803, 1821–1828) and was translated into French (1810) and Italian (1812), running to several eds. Other historical publications include *Geschichte der Medicin im Auszuge* (Halle, 1804) and *Geschichte der Chirurgie*, 2 vols. (Halle, 1805–1819).

Among Sprengel's numerous translations of classical and modern medical authors are *Galen's Fieberlehre* (Breslau–Leipzig, 1788); *Apologie des Hippocrates und seiner Grundsätze* (Leipzig, 1789); and *William Buchan's Hausarzneikunde* (Altenburg, 1792).

Sprengel summarized the contemporary medical knowledge in two textbooks: *Handbuch der Pathologie*, 3 vols. (Leipzig, 1795–1797), and *Handbuch der Semi-*

otik (Halle, 1801). In addition he broadly criticized the medical developments of 1790–1800 in *Kritische Uebersicht des Zustandes der Arzneikunde in dem letzten Jahrzehend* (Halle, 1801).

His principal botanical works are *Anleitung zur Kenntniss der Gewächse*, 3 vols. (Halle, 1802–1804); *Vom Baue und der Natur der Gewächse* (Halle, 1812); and *Geschichte der Botanik* (Altenburg, 1817). One section of the first was translated into English and published as *An Introduction to the Study of Cryptogamous Plants* (London, 1807).

II. SECONDARY LITERATURE. The most extensive treatment of Sprengel's life and writings is in Heinrich Rohlfs, *Geschichte der deutschen Medicin*, II (Stuttgart, 1880), 212–279, under the heading "Kurt Sprengel, der Pragmatiker." An early biography and list of his writings are in Julius Rosenbaum, *Curtii Sprengelii opuscula academica* (Leipzig–Vienna, 1844), xii–xx. Shorter biographical sketches appeared in *Allgemeine deutsche Biographie*, XXXV, 296–299; August Hirsch, *Biographisches Lexikon*, 2nd. ed., V (Munich, 1932), 374–375; and *Neuer Nekrolog der Deutschen*, XI (1833), 200–208.

A discussion of Sprengel's medical historiography is in E. Heischkel, "Die Medizinhistoriographie im XVIII. Jahrhundert," in *Janus*, 25 (1931), 67–151. Goethe's minor relationship with Sprengel is mentioned in D. E. Meyer, "Goethes botanische Arbeit in Beziehung zu Christian Konrad Sprengel (1750–1816) und Kurt Sprengel (1766–1833) auf Grund neuer Nachforschungen in Briefen und Tagebüchern," in *Berichte der Deutschen botanischen Gesellschaft*, 80 (1967), 209–217. A more recent article stressing Sprengel's opposition to the prevailing medical systems is S. Alleori, "Il sistema dottrinario medico di Curzio Sprengel avversario dei sistemi," in *Pagine di storia della medicina. Collana miscellanea*, 19 (1968), 119–131.

Some of Sprengel's botanical contributions are mentioned in Julius von Sachs, *History of Botany (1530–1860)*, translated by H. E. F. Garnsey, 2nd imp. (Oxford, 1906). A more extensive account can be found in Gregor Kraus, *Der Botanische Garten der Universität Halle* (Leipzig, 1894), no. 2: "Kurt Sprengel." See also Hermann F. Kilian, *Die Universitaeten Deutschlands in medicinisch-naturwissenschaftlicher Hinsicht betrachtet* (Heidelberg–Leipzig, 1828), 114, 120.

GUENTER B. RISSE

SPRING, WALTHÈRE VICTOR (*b*. Liège, Belgium, 6 March 1848; *d*. Tilff, Belgium, 17 July 1911), *chemistry, physics.*

Spring was the son of Antoine Spring, professor of physiology at the medical school of the University of Liège, a competent man of science, and author of a body of published work in medicine, botany, and anthropology. The scholarly physician was disappointed by his son's slow progress in school and by the boy's dislike of classical languages, in which the father was proficient. Spring failed his university entrance examination; and rather than endure his father's reproaches, he left home and found employment as a gunsmith. In the workshop his manual dexterity was well-paid and further developed. He repeated his examinations, this time successfully, and in 1867 enrolled at the school of mines of the University of Liège, from which he graduated with a diploma in mining engineering in 1872. This, however, was merely preparation for a career in experimental chemistry, toward which he had been strongly influenced by Jean Stas, an eminent Belgian chemist and friend of the family. Guided by Stas's advice, Spring went to study under Kekulé at the University of Bonn. Here he also worked in physics with Clausius, who impressed on him the need for disciplined patience and the ability to sustain drudgery while in quest of a long-range objective. Kekulé's work in organic chemistry showed Spring the value of intuition and imagination.

In 1875, after two years at Bonn, Spring returned to Liège to teach theoretical physics at the university. He was appointed assistant professor of organic chemistry in 1876 and full professor in 1880, a post that he retained for the rest of his life. Early in his career Spring was concurrently an engineer with the Belgian Bureau of Mines. The Belgian Academy of Sciences elected him a corresponding member in 1877, titular member in 1884, and president in 1899. He was permanent examiner for the Military School of Belgium from 1884 to 1906.

Spring's earliest researches dealt with the molecular structures of the polythionic acids. He followed Kekulé in denying the possibility of more than one valence to an atom. This principle was an erroneous extension to all atoms of Kekulé's productive theory of the linking of carbon atoms—which are almost the only ones to have that property. Kekulé and his followers were therefore required to write formulas for complex radicals in the form of chains—for example, H-O-O-S-O-O-H for sulfuric acid. Spring's early papers on the inorganic chemistry of sulfur are flawed by too slavish an adherence to this spurious principle. He nevertheless produced a valuable series of papers on the oxyacids of sulfur and on the polythionates, in which he synthesized new compounds and found new chemical reactions.

Spring's most important work, however, was in physical chemistry. He was prompted to investigate the effect of high pressure on the compaction of powdered solids by the lively controversy on the flow of glaciers that was then arousing great interest, fanned by such masters of the polemic arts as Tyndall, Tait, and Ruskin. Spring found that sodium nitrate, potassium nitrate, and even sawdust, when subjected to great compression in a screw press, become hard, solid masses of unusually high density. These observations were the beginning of a series of researches in which he cleverly used the same experimental technique to investigate the effects of pressure on phase equilibria, on chemical equilibria, on the chemical reactions of solids, and on the ability or inability of one metal to diffuse into another. In this way he was able to explain the formation of solid solutions in certain alloys. Geologists also were interested by his discoveries that the application of high pressure could transform peat into lignite and that layers of clay between which organic humus is introduced can, by the same means, produce schist rocks.

In 1870 Tyndall created much public interest by his partial explanation of the blue color of the sky. It stimulated Spring to ask a cognate question about the color of water. After much labor he succeeded in observing the actual color of natural and of chemically pure waters, as well as of aqueous solutions and alcohols. These investigations required the exercise of his utmost skill as an experimentalist. By ingenious techniques he produced optically empty water, free from all traces of suspended particles. The water was to be contained in glass tubes fifteen millimeters in diameter and up to twenty-six meters long. The difficulty lay in making a tube of this length coaxial with the beam of light that is required to pass through it. Almost six weeks of work was required for the alignment of the apparatus. Spring succeeded in completing these exacting experiments, and reported that the natural color of water is "a pure cerulean blue similar to that of the sky at its zenith when seen from a high mountain." He discovered that convection currents, caused by differences of temperature as small as 0.6°C., were enough to render a twenty-six-meter column of water opaque to transmitted light. In extensive discussions of the use of a Tyndall beam to detect the presence of colloidal particles in water, Spring supplied ideas and emphasis that contributed significantly to the development of the ultramicroscope of Siedentopf and Zsigmondy.

In another series of researches Spring found that soap solutions perform their detergent action by preferential adsorption of the soap on the particles of dirt, which are thereby detached and suspended in water.

Spring's work was characterized by the selection of problems dealing with entire natural phenomena that had not yet received adequate explanation, by his originality of viewpoint combined with experimental ingenuity and manipulative skills, and by the clarity and force of his writing. The versatility of his interests was also remarkable.

BIBLIOGRAPHY

Most of Spring's papers were published in the *Bulletin de l'Académie royale de Belgique. Classe des sciences.* His *Oeuvres complètes*, comprising more than 100 papers, was published by the Société Chimique de Belgique in 2 vols. (Brussels, 1914–1923).

The memoir by L. Crismer, prefixed to vol. I of the *Oeuvres complètes*, is the principal source of biographical information. Briefer sketches are F. Lionetti and M. Mager, in *Journal of Chemical Education*, **28** (1951), 604–605; and F. Swarts, in *Chemikerzeitung*, **35** (1911), 949–950.

SYDNEY ROSS

SPRUCE, RICHARD (*b.* Ganthorpe, near Malton, England, 10 September 1817; *d.* Coneysthorpe, Castle Howard, near Malton, 28 December 1893), *botany.*

Spruce, the only child of Richard and Etty Spruce, emulated his father by becoming a schoolmaster, first at Haxby and then at the Collegiate School of York. His principal recreation was the study of the local flora, particularly Bryophyta, on which he published several papers. Upon the closing of the school at York in 1844, he resolved to make botany his career.

From April 1845 to April 1846 Spruce collected plants in the Pyrenees, where he discovered bryophytes previously unrecorded in the region. The results of this expedition were published in 1849–1850. In June 1849 he sailed to South America, where he spent the next fifteen years in botanical exploration.

Undeterred by constant ill health and incredible hardships, Spruce studied the rich vegetation of the Amazon valley with characteristic thoroughness, dispatching to England specimens of more than 7,000 species, many of them previously unknown. A commission from the British government sent him to Andean Ecuador in 1860 to collect cinchona plants suitable for cultivation in India. He procured 100,000 seeds and many young plants, which were sent to India for the production of quinine to alleviate malaria. He spent his remaining years in South America exploring the coastal regions of Ecuador and Peru.

On his return to England in 1864, Spruce acquired a modest cottage in Coneysthorpe, in his native Yorkshire. Despite comparative poverty and constant ill health brought about by his years in South America, he worked hard on his immense plant collections. "Palmae Amazonicae" (1869) is a scholarly elucidation of the geographical distribution of the palms of the Amazon, with a new classification of the genera. "Hepaticae Amazonicae et Andinae" (1884) convinced Sir Joseph Hooker that this would be Spruce's enduring monument.

Spruce's sound botanical judgment, his accuracy, and his meticulous detail were widely recognized. The Royal Geographical Society acknowledged his skill as a cartographer by electing him an honorary fellow in 1866, and in the year of his death the Linnean Society of London made him an associate.

BIBLIOGRAPHY

I. ORIGINAL WORKS. Spruce's writings include "The Musci and Hepaticae of the Pyrenees," in *Transactions and Proceedings of the Botanical Society of Edinburgh*, **3** (1850), 103–216; "Palmae Amazonicae . . .," in *Journal of the Linnean Society. Botany,* **11** (1869), 65–183; "Hepaticae Amazonicae et Andinae . . .," in *Transactions and Proceedings of the Botanical Society of Edinburgh*, **15** (1884), 1–588; and *Notes of a Botanist in the Amazon*, A. R. Wallace, ed., 2 vols. (London, 1908).

II. SECONDARY LITERATURE. See V. W. von Hagen, *South America Called Them* (London, 1949), 291–374; C. Sandeman, "Richard Spruce, Portrait of a Great Englishman," in *Journal of the Royal Horticultural Society*, **74** (1949), 531–544; and R. E. Schultes, "Richard Spruce Still Lives," in *Northern Gardener*, **7** (1953), 20–27, 55–61, 87–93, 121–125.

R. G. C. DESMOND

SPRUNG, ADOLF FRIEDRICH WICHARD (*b.* Kleinow, near Perleberg, Germany, 5 June 1848; *d.* Potsdam, Germany, 16 January 1909), *meteorology.*

The son of a schoolteacher, Sprung demonstrated an early inclination for the natural sciences and especially for chemistry, which led him to study pharmacy. He gave up a career as a pharmacist, however, because of a serious illness. Instead, he studied mathematics, physics, and chemistry at Leipzig from 1872 to 1876, in which year he received the doctorate for an experimental investigation on the hydraulic friction of salt solutions. Sprung turned his attention to meteorology when his teacher, the physicist G. Wiedemann, recommended him to the newly established naval observatory in Hamburg, the director of which, G. Neumayer, was seeking qualified young workers. There, in the department of synoptic meteorology, Sprung collaborated closely with Wladimir Köppen, Wilhelm van Bebber, and Louis Grossman from August 1876 until the spring of 1886. In his daily concern with atmospheric conditions, Sprung became the first to apply the theorems of mathematical physics to the interpretation of meteorological processes. He thereby laid the foundations for the theory of the dynamics of the atmosphere, with which meteorology became an exact science.

Sprung's field of study expanded when, on 1 April 1886, he was appointed director of the instrument division of the Prussian Meteorological Institute in Berlin, which had been reorganized by J. F. W. von Bezold. Six years later he became director of the meteorological-magnetic observatory in Potsdam, constructed according to his proposals, which he made into an institute of worldwide importance. Most of Sprung's works on instruments occurred during this period. With the collaboration of R. Fuess, a master maker of fine instruments, he enriched the field with remarkable new designs. In his last years a nervous ailment increasingly crippled his creative powers. He died suddenly in 1909.

Sprung was one of the first to expand meteorology into a physics of the atmosphere. While in Hamburg he applied the laws of statics and of dynamics to atmospheric problems, which express themselves both in aperiodic phenomena and in occurrences subject to daily cycles. He investigated the relationship of wind strength to barometric gradient (1876, 1879) and the influence of frictional resistance of the ground (1880). From the curvature of the inertial path of a particle with respect to rotating surfaces, he derived the influence of the deflecting force of the earth's rotation on atmospheric circulation (1881). Sprung's theorem for the effect of the sun on the direction of the wind provided an explanation for the daily period of the wind (1881, 1884). His law of areas in meteorology (1881), which stems from Kepler's second law of planetary orbits, is still applied. In 1885 Sprung gathered all this data; combined it with ideas of

Buys-Ballot, Ferrel, Guldberg and Mohn, Hadley, and others; and presented the whole in his *Lehrbuch der Meteorologie*, the first complete work on dynamic meteorology.

Sprung's work on instruments was equally progressive. As early as 1877 he had developed his steelyard for the precise recording of air pressure; and in 1908 he applied its ingenious principle to the measurement of rain, snow, and humidity. He also employed the conversion of the measuring process into electrical impulses to achieve remote recording of wind and precipitation. Other measuring devices determined snowfall and precipitation intensity. Sprung's formula for ascertaining vapor pressure from observations with the aspiration psychrometer became the basis for the calculation of the psychrometer table, which is indispensable to the networks of meteorological stations. It is still considered sufficient for practical operations, even at subfreezing temperatures.

For the International Cloud Year 1896–1897, Sprung built for the Potsdam observatory an automatic cloud instrument capable of the simultaneous photogrammetric surveying of several points and a reflector for measuring cloud motion. Only with the introduction of electronic structural elements into meteorological measuring techniques were Sprung's designs superseded.

BIBLIOGRAPHY

I. ORIGINAL WORKS. Many of Sprung's writings on dynamic meteorology were brought together in his *Lehrbuch der Meteorologie* (Hamburg, 1885). Such works include "Studien über den Wind und seine Beziehungen zum Luftdruck," in *Aus dem Archiv der Deutschen Seewarte*, 2, nos. 1–2 (1879)—articles on the same topic in *Zeitschrift der Österreichischen Gesellschaft für Meteorologie*, 17 (1882), 161–175, 276–282; and *Meteorologische Zeitschrift*, 11 (1894), 197–200, 384–387; "Zur Theorie der oberen Luftströmungen," in *Zeitschrift der Österreichischen Gesellschaft für Meteorologie*, 15 (1880), 17–21; "Die Anwendung des Prinzips der Flächen in der Meteorologie," *ibid.*, 16 (1881), 57–63; "Über die Bahnlinie eines freien Teilchens auf der rotierenden Erdoberfläche und deren Bedeutung für die Meteorologie," in *Wiedemanns Annalen*, 14 (1881), 128–149—articles on the same subject in *Zeitschrift der Österreichischen Gesellschaft für Meteorologie*, 15 (1880), 1–21; 17 (1882), 75; and *Meteorologische Zeitschrift*, 1 (1884), 250–252; "Eine periodische Erscheinung im täglichen Gang der Windrichtung," in *Zeitschrift der Österreichischen Gesellschaft für Meteorologie*, 16 (1881), 419–424—articles on the same subject in *Meteorologische Zeitschrift*, 1 (1884), 15–22, 65–70; 3

(1886), 223–225; **11** (1894), 252–262; and "Die vertikale Komponente der ablenkenden Kraft der Erdrotation in ihrer Bedeutung für die Dynamik der Atmosphäre," in *Meteorologische Zeitschrift,* **12** (1895), 449–455.

Sprung's articles on instrumental techniques include "Waagebarograph mit Laufgewicht nach Sprung," in *Zeitschrift der Österreichischen Gesellschaft für Meteorologie*, **12** (1877), 305–308; **16** (1881), 1–4; **17** (1882), 44–48; "Bestimmung der Luftfeuchtigkeit durch Assmanns Aspirationspsychrometer," in *Wetter*, **5** (1888), 105–108; "Über den photogrammetrischen Wolkenautomaten und seine Justierung," in *Zeitschrift für Instrumentenkunde*, **19** (1889), 111–118; **24** (1904), 206–213; "Registrierapparate für Regenfall und Wind mit elektrischer Übertragung," in *Meteorologische Zeitschrift*, **24** (1889), 344–348—also in *Zeitschrift für Instrumentenkunde*, **9** (1889), 90–98; and *Meteorologische Zeitschrift*, **32** (1897), 385–388, written with R. Fuess; "Über die automatische Aufzeichnung der Regenintensität," in *Wetter*, **22** (1905), 56–58, also in *Zeitschrift für Instrumentenkunde*, **27** (1907), 340–343; and "Die registrierende Laufgewichtswaage im Dienste der Schnee-, Regen- und Verdunstungsmessung," in *Meteorologische Zeitschrift*, **25** (1908), 145–154.

Additional publications on climatology and maritime meteorology appeared between 1880 and 1908 in *Meteorologische Zeitschrift, Annalen der Hydrographie und maritimen Meteorologie*, and *Wetter*.

II. Secondary Literature. See Richard Assmann, "Professor Dr. Adolf Sprung†," in *Wetter*, **26** (1909), 25–27; and Wladimir Koeppen, "Dr. Adolf Sprung. Nachruf," in *Meteorologische Zeitschrift*, **26** (1909), 215–216.

J. Grunow

SPURZHEIM, JOHANN CHRISTOPH (*b.* Longuich, near Trier, Germany, 31 December 1776; *d.* Boston, Massachusetts, 10 November 1832), *psychology, psychiatry, neuroscience.*

Spurzheim's family, who were Lutheran, farmed the land of an abbey in a small town on the Moselle about sixty miles from Koblenz. His early education was intended to prepare him for a clerical career. He studied Greek and Latin in his native village and at the age of fifteen entered the University of Trèves (now Trier), where he studied Hebrew, divinity, and philosophy. Around 1799 he moved to Vienna and was engaged as a private tutor. The following year he met Franz Joseph Gall, with whom he collaborated on neuroanatomical research for the next thirteen years. From 1800 to 1804 Spurzheim completed his medical studies at Vienna, where he was formally awarded his medical degree in 1813, after he and Gall had ceased working together. He received licensure in London from the Royal College of Physicians; was awarded a second degree, possibly a medical one, at Paris around 1821; and received recognition from many learned societies, including honorary membership in the Royal Irish Academy. He remained a theorist all his life, however, for his skepticism regarding medicine as it was then understood led him to avoid private medical practice.

Spurzheim's unique contributions to the behavioral sciences have traditionally been intertwined in those of his mentor Franz Gall, who, although considered the founder of what later came to be known as phrenology, never used that term to describe his own system. Furthermore, Spurzheim was often accused of being a popularizer of Gall's views on cerebral localization of mental functions because he was responsible for making them into a complete system of phrenology and teaching it widely. Spurzheim accepted the basic assumptions of this theory of mind, brain, and behavior—(1) that the brain is the organ of the mind, (2) that the moral and intellectual faculties are innate, (3) that their exercise or manifestation depends on organization, (4) that the brain is composed of a congeries of as many particular organs as there are propensities, sentiments, and faculties that differ from each other, and (5) that the shape and size of the skull faithfully reflect the shape and size of the underlying cerebral mass. Nevertheless, he extended Gall's basic views in a singular way and made them in many respects more utilitarian and also more acceptable to a wider audience. In contrast with the more conservative view that Gall held in regard to his own doctrines, Spurzheim took the position that phrenology was capable of ameliorating most of the social ills of his day.

Shortly after his professional and personal break with Gall in 1813, Spurzheim formalized his views and presented them in his first major publication, *The Physiognomical System of Drs. Gall and Spurzheim: Founded on an Anatomical and Physiological Examination of the Nervous System in General, and of the Brain in Particular; and Indicating the Dispositions and Manifestations of the Mind* (London, 1815). It was from this major effort that Spurzheim later extracted sections, extended them, and published them as separate works. For example, his *Essai philosophique sur la nature morale et intellectuelle de l'homme* (Paris, 1820) elaborated upon his philosophical position. *A View of the Elementary Principles of Education Found-*

ed on the Study of the Nature of Man (Edinburgh, 1821) applied phrenology to education, and *Observations on the Deranged Manifestations of the Mind, or Insanity* (London, 1817) applied phrenology to psychiatry. Like many of Spurzheim's works, the latter was published in other languages, including German (Hamburg, 1818) and French (Paris, 1818). The work on insanity influenced the development of early American psychiatry; and when it was first published in America (Boston, 1832), it was edited by the well-known alienist Amariah Brigham.

Spurzheim believed that he had been denied the recognition due him as Gall's collaborator. Moreover, he felt that he was personally responsible for many of the neuroanatomical discoveries traditionally credited to Gall alone—especially those made between 1805 and 1813. He was constantly placed on the defensive, furthermore, because he was seen by his critics as only parroting what Gall had taught. But he did not simply take Gall's position unaltered and present it as his own. He contributed to their joint efforts almost from the beginning, and he placed his unique stamp on the history of that very special nineteenth-century doctrine of brain and mind.

Spurzheim taught that there were no fewer than thirty-five innate faculties of the mind—Gall had claimed to have discovered twenty-seven. He placed great emphasis on individual differences in cerebral organization and held that education had to be individualized. He discarded the view that mental faculties have determinate functions, offering instead a more "dignified" view of man that allowed for a greater emphasis on his more positive traits. Spurzheim separated what he believed to be the combined actions of faculties from what individual faculties were held to do, and added a more theological and philosophical perspective.

As a result of Spurzheim's conviction of phrenology's truthfulness, he convinced a great number of auditors to support it. In this regard he was able, unintentionally, to play a unique historical role. By inducing a wide audience to investigate for itself the truthfulness of his enthusiastically espoused belief, Spurzheim inspired inquiries that in some cases led to the establishment of phrenology's inherent incorrectness. Thus, although much of phrenology and Spurzheim's assumptions were essentially wrong—and were shown to be so as neuroanatomical science advanced in the nineteenth century—they were just right enough to further scientific thought.

BIBLIOGRAPHY

I. ORIGINAL WORKS. The small collection of Spurzheim MSS at Harvard Medical School, Boston, includes fragments of Spurzheim's American journal and his correspondence with his wife around the time they were married. A complete bibliography of the many eds. of his works is not available. A listing of his major works is in A. A. Walsh's intro. to *Observations on the Deranged Manifestations of the Mind, or Insanity,* by J. G. Spurzheim (Gainesville, Fla., 1970).

II. SECONDARY LITERATURE. Two biographies of Spurzheim were prepared by his American publisher, Nahum Capen of Boston: *Reminiscences of Dr. Spurzheim and George Combe: And a Review of the Science of Phrenology, From the Period of Its Discovery by Dr. Gall to the Time of the Visit of George Combe to the United States, 1838, 1840* (Boston, 1881); and "Biography of the Author [Spurzheim]," in J. G. Spurzheim, *Phrenology in Connexion With the Study of Physiognomy* (Boston, 1833), 9–168. Andrew Carmichael has published the only complete book on Spurzheim's life, which, despite its errors, is generally well done: *A Memoir of the Life and Philosophy of Spurzheim* (Boston, 1833). W. M. Williams, *A Vindication of Phrenology* (London, 1894), discusses Spurzheim in England (328–340); and A. A. Walsh discusses Spurzheim in America in "The American Tour of Dr. Spurzheim," in *Journal of the History of Medicine and Allied Sciences,* **27** (1972), 187–205, and in "Johann Christoph Spurzheim and the Rise and Fall of Scientific Phrenology in Boston: 1832–1842" (doctoral diss., University of New Hampshire, 1974). Finally, the reader should consult the many references appended to the article on Franz Joseph Gall in the DSB.

ANTHONY A. WALSH

ŚRĪDHARA (*fl.* India, ninth century), *mathematics.*

Śrīdhara, of whose life nothing is known save that he was a devotee of Śiva, wrote two works on arithmetic, the *Pāṭīgaṇita* and the *Pāṭīgaṇitasāra* or *Triśatikā,* and one work, now lost, on algebra. Since he seems to refer to the views of Mahāvīra (*fl.* ninth century), and was used by Āryabhaṭa II (*fl.* between *ca.* 950 and 1100) and cited by Abhayadeva Sūri (*fl.* 1050), it can be concluded that he flourished in the ninth century.

The *Pāṭīgaṇita* is divided into two sections. The first, after metrological definitions, covers the mathematical operations of addition, subtraction, multiplication, and division; finding squares and square roots; finding cubes and cube roots; fractions; and proportions; the second gives solutions for problems involving mixtures, series, plane figures, volumes, shadows, and zero. The text, preserved in a unique manuscript in Kashmir, breaks off in the middle of the rules for determining the areas of plane figures in the second section. The *Triśatikā* summarizes much of the material in the *Pāṭīgaṇita*, including the parts no longer available to us. In the Kashmir manuscript there is an anonymous commentary on the *Pāṭīgaṇita*, and the *Triśatikā* was commented on by Śrīdhara himself and in Kannaḍa (Kanarese), Telugu (by Vallabha), and Gujarātī; the commentaries on the *Triśatikā* ascribed to Śambhūnātha or Śambhūdāsa (*fl.* 1428; *Gaṇitapañcaviṃśatikā* or *Gaṇitasāra* and to Vṛndāvana Śukla (*Pāṭīsāraṭīkā*) are still uncertain, pending an investigation of the manuscripts.

BIBLIOGRAPHY

The best work on Śrīdhara is the introduction to K. S. Shukla's valuable ed. and trans., *The Patiganita of Sridharacarya* (Lucknow, 1959). There is also a Russian trans. and study of the *Pāṭīgaṇita* by A. I. Volodarsky and O. F. Volkovoy in *Fiziko-matematicheskie nauki v stranakh vostoka* (Moscow, 1966), 141–246. The *Triśatikā* was edited by Sudhākara Dvivedin (Benares, 1899) and was largely translated into English by N. Ramanujacharia and G. R. Kaye, "The *Triśatikā* of Śrīdharācarya," in *Bibliotheca mathematica*, 3rd ser., **13** (1912–1913), 203–217.

DAVID PINGREE

ŚRĪPATI (*fl.* Rohiṇīkhaṇḍa, Mahārāṣṭra, India, 1039–1056), *astronomy, astrology, mathematics.*

Śrīpati, who was the son of Nāgadeva (or Nāmadeva) and the grandson of Keśava of the Kāśyapagotra, is one of the most renowned authorities on astrology in India, although his works on astronomy and mathematics are not negligible; in many he follows the opinions of Lalla (*fl.* eighth century; see essay in Supplement). His numerous works include not only Sanskrit texts but also one of the earliest examples of Marāṭhī prose extant. They include the following:

1. The *Dhīkoṭidakaraṇa*, written in 1039, a work in twenty verses on solar and lunar eclipses. There are commentaries by Harikṛṣṇa (*fl.* 1708–1714 at Delhi) and Dinakara. The *Dhīkoṭidakaraṇa* was edited by N. K. Majumdar in *Calcutta Oriental Journal*, **1** (1934), 286–299 — see also his "Dhikoti-Karanam of Śrīpati," in *Journal of the Asiatic Society of Bengal*, n.s. **17** (1921), 273–278 — and by K. S. Shukla, in *Ṛtam*, **1** (1969), supp.

2. The *Dhruvamānasa*, written in 1056, is a short treatise in 105 verses on calculating planetary longitudes, on gnomon problems, on eclipses, on the horns of the moon, and on planetary transits. It is very rare and has not been published.

3. The *Siddhāntaśekhara*, a major work on astronomy in nineteen chapters, follows, in general, the *Brāhmapakṣa*. The chapters are on the following subjects:

    1. Fundamentals.
    2. Mean motions of the planets.
    3. True longitudes of the planets.
    4. On the three questions relating to the diurnal rotation.
    5. Lunar eclipses.
    6. Solar eclipses.
    7. On the syzygies.
    8. On the *pātas* of the sun and moon.
    9. On first and last appearances.
    10. On the moon.
    11. On transits of the planets.
    12. On conjunctions of the planets with the constellations.
    13. Arithmetic.
    14. Algebra.
    15. On the sphere.
    16. On the planetary spheres.
    17. On the cause of eclipses.
    18. On the projection of eclipses.
    19. On astronomical instruments.

A commentary on this work, the *Gaṇitabhūṣaṇa*, was composed by Makkibhaṭṭa (*fl.* 1377); unfortunately, only the portion on the first four chapters survives. The *Siddhāntaśekhara*, with Makkibhaṭṭa's commentary on chapters 1–4 and the editor's on chapters 5–19, was edited by Babuāji Miśra, 2 vols. (Calcutta, 1932–1947).

4. The *Gaṇitatilaka* is a mathematical treatise apparently based on the *Pāṭīgaṇita* or *Triśatikā* of Śrīdhara; there is a commentary by Siṃhatilaka Sūri (*fl.* 1269 at Bijāpura, Mysore). Both text and commentary were published by H. R. Kapadia (Baroda, 1937).

5 and 6. The *Jyotiṣaratnamālā*, in twenty chapters, is the most influential work in Sanskrit on *muhūrta* or catarchic astrology, in which the success or failure of an undertaking is determined from the time of its inception. It is based largely on the *Jyotiṣaratnakośa* of Lalla. Śrīpati himself wrote a Marāṭhī commentary on this (edited and studied for its linguistic content by M. G. Panse [Poona, 1957]); but of much greater historical importance is the commentary *Gautamī* composed by Mahādeva in 1263, for it contains numerous citations from lost or little-known astronomical

and astrological texts. There are also commentaries by Dāmodara (*Bālāvabodha*), Paramakāraṇa (*Bālabodhinī* in Prākṛt), Śrīdhara (*Śrīdharīya*), and Vaijā Paṇḍita (*Bālāvabodhinī*). The *Jyotiṣaratnamālā* was published twice with Mahādeva's *Gautamī*: at Bombay in 1884 and by Rasikamohana Caṭṭopādhyāya (2nd ed., Calcutta, 1915). The first six chapters were edited by P. Poucha, "La Jyotiṣaratnamālā ou Guirlande des joyaux d'astrologie de Śrīpatibhaṭṭa," in *Archiv orientální*, **16** (1949), 277–309.

7. The *Jātakapaddhati* or *Śrīpatipaddhati*, in eight chapters, is one of the fundamental textbooks for later Indian genethlialogy, contributing an impressive elaboration to the computation of the strengths of the planets and astrological places. It was enormously popular, as the large number of manuscripts, commentaries, and imitations attests. The more important of these commentaries are Sūryadeva Yajvan (*b*. 1191), *Jātakālaṅkāra*; Parameśvara (*ca*. 1380–1460); Acyuta (*fl*. 1505–1534), *Bhāvārthamañjarī*—see D. Pingree, *Census of the Exact Sciences in Sanskrit*, ser. A, I (Philadelphia, 1970), 36a–36b; Kṛṣṇa (*fl*. 1600–1625), whose *udāharana* was edited by J. B. Chaudhuri (Calcutta, 1955)— see also D. Pingree, *Census*, II (Philadelphia, 1971), 53a–55b; Sumatiharṣa Gaṇi (*fl*. 1615); Mādhava; and Raghunātha. Acyuta Piṣāraṭi (*ca*. 1550–1621; see D. Pingree, *Census*, I, 36b–38b) wrote an imitation, the *Horāsāroccaya*. The *Jātakapaddhati* was edited with an English translation by V. Subrahmanya Sastri (Bombay, 1903; 4th ed., Bangalore, 1957).

8. A *Daivajñavallabha* on astrology, in fifteen chapters, sometimes is attributed to Śrīpati and sometimes to Varāhamihira (*fl. ca*. 550); its real author remains unknown. It was published with the Hindī translation, *Subodhinī*, of Nārāyaṇa (*fl*. 1894) at Bombay in 1905, in 1915–1916, and in 1937.

There is no reliable discussion of Śrīpati or study of his works.

DAVID PINGREE

STÄCKEL, PAUL GUSTAV (*b*. Berlin, Germany, 20 August 1862; *d*. Heidelberg, Germany, 12 December 1919), *mathematics*, *history of science*.

Stäckel studied at Berlin and defended his dissertation in 1885. He wrote his *Habilitationsschrift* at Halle in 1891 and then held chairs at various German universities, teaching finally at Heidelberg. His interests were varied, for he worked with equal ease in both mathematics and its history. The chief influence was the work of Weierstrass. He specialized in analytical mechanics (particularly in the use of Lagrangians in problems concerning the motion of points in the presence of given fields of force), related questions in geometry, and properties of analytical functions. A linking problem for these fields was the solution of linear differential equations; Stäckel also explored the existence theorems for such solutions. His other interests in mathematics included set theory and, in his later years, problems concerning prime numbers. He was renowned among his students for delivering new sets of lectures every academic year, and he wrote on problems in mathematical education.

In the history of mathematics Stäckel's interests centered on the eighteenth and early nineteenth centuries. He was especially noted for his role in instituting the publication of Euler's *Opera omnia*; and he also published editions of works, manuscripts, and correspondence of J. H. Lambert, F. and J. Bolyai, Gauss, and Jacobi. In addition, he edited several volumes in Ostwald's Klassiker der Exacten Wissenschaften. His interpretive articles dealt largely with the history of the theory of functions and of non-Euclidean geometry. From indications in his and others' writings, it seems clear that locating his *Nachlass* is highly desirable.

BIBLIOGRAPHY

The most comprehensive list of Stäckel's works is in Poggendorff, IV, 1427–1428, and V, 1194–1195.

For a sympathetic obituary, see O. Perron, "Paul Stäckel," *Sitzungsberichte der Heidelberger Akademie der Wissenschaften*, Math.-naturwiss. Kl., Abt. A (1920), no. 7.

I. GRATTAN-GUINNESS

STAHL, GEORG ERNST (*b*. Ansbach, Germany, 21 October 1660 [1659?]; *d*. Berlin, Germany, 4 May 1734), *medicine*, *chemistry*.

Stahl has aroused much controversy. As a physician he was outstanding; he held the highest academic positions, enjoyed a very active practice, and through his writings became vastly influential. As a philosopher he supported the viewpoint known as vitalism and wove that concept into the fabric of his medical system. As a chemist he elaborated and maintained the doctrine of phlogiston, which, until outgrown later in the eighteenth century, provided a reasoned explanation for many chemical phenomena. But his teachings, particularly his stand on vitalism, in large part ran counter to the trend of the times; his chemical theories were overthrown; his vitalist doctrine, in the form that he elaborated it, could not stand up against the onrushing tide of research and experimentation; while his system of medical practice faded away before numerous competitors. Furthermore, his personality was often antagonistic, his style of writing obscure and hard to understand. Yet, even though he seemed to be discredited, Stahl influenced the whole of eighteenth-century medicine; and his imprint is being increasingly appreciated as historians trace his role in the drama of eighteenth-century medical thought.

Although the generally accepted date of Stahl's birth is 21 October 1660, Gottlieb disputes this and claims that the baptismal register, in the parish of St. John in Ansbach, shows 1659. There is little information about Stahl's early life. Even as a youth he had considerable interest in chemistry. He studied medicine at Jena and received his degree in January 1684. He then devoted himself to scientific work and lectured in chemistry at the university, attaining considerable reputation. In 1687 Stahl was invited to become court physician at Weimar, where he remained for seven years. He subsequently joined the medical faculty of the new University of Halle.

Elector Frederick III of Brandenburg (Frederick I of Prussia), eager to surpass his neighbors, decided to establish a new university at Halle. The great liberal jurist Christian Thomasius, who had been expelled from Leipzig, settled at Halle at Frederick's invitation. The elector built his university around Thomasius and attracted such men as August Francke in oriental languages and Friedrich Hoffmann in medicine. In 1693 the university received the imperial privilege and was officially inaugurated in 1694.

Hoffmann, needing help, was instrumental in securing Stahl's appointment as the second professor of medicine. In 1694 Stahl went to Halle, where he remained until 1715, lecturing particularly on the theory of medicine and on chemistry. Hoffmann and Stahl, although different in many respects, formed a very strong faculty, and Halle became a leading medical school. In 1715, at the request of Frederick William I of Prussia, Stahl left Halle and went to Berlin to be court physician. He remained there until his death.

Stahl's personality has received much unfavorable comment. He has been condemned as misan-

thropic and harsh, narrow-minded and intolerant. These qualities also have been contrasted unfavorably with the sunnier and more open disposition of Hoffmann. Much of the evaluation rests on rather slender evidence and stems particularly from the statements of Haller, which many historians have repeated. On the other hand, Stahl had many defenders. The truth is virtually impossible to establish, unless further primary source material should be discovered.

Stahl was a devout Pietist—Halle, in the early days, was the center for Pietism as well as rationalism—and this background undoubtedly colored his doctrines. Many misfortunes attended his personal life. His first wife died in 1696 of puerperal fever and his second wife in 1706, of the same disease. A daughter died in 1708. These were the years of his greatest productivity, and it is only reasonable to see in his outward attitudes some reflection of his personal life. He was a prodigious worker, and pride and self-confidence apparently were notable qualities. Gottlieb quotes Stahl's personal motto as *E rebus quantumcumque dubiis quicquid maxima sententium turba defendit, error est*, for which I suggest the translation, "Where there is doubt, whatever the greatest mass of opinion maintains . . . is wrong."

Stahl's style of writing is prolix and convoluted, and difficult to understand. Perhaps the style is the man himself.

Stahl, who lived well into the eighteenth century, was nevertheless part of the seventeenth-century rebellion against tradition; and his doctrines reflect, in a way, the intense turmoil of that period. The rebellion, of course, involved all phases of intellectual life. In medicine the Galenic theories, which rested largely on Aristotle, had come under severe attack. In astronomy and physics, especially mechanics, new experimental methods had shown how untenable were the older views. In physiology and biology Harvey was the greatest among many investigators who introduced new concepts that were firmly grounded on empirical demonstration. In philosophy Descartes offered new vistas and new methodology, while Gassendi helped to reintroduce atomism. In chemistry the arch-rebel Paracelsus had given great impetus to a movement of which the leading representatives were van Helmont, in the early seventeenth century, and Sylvius and Willis in the later part. At the same time traditional religion remained powerful and entrenched. Piety and orthodoxy were strong values, and atheism and materialism were epithets dreaded by most scientific workers. The new philosophy and the new science, which threatened orthodox religions, had to come to terms with religious tenets.

In the medical world of the later seventeenth century, many different theories competed actively; but no clear victory was in sight for any one. The traditional Galenic theories were in retreat but by no means annihilated. Iatrophysicists tried to explain all medical phenomena on the basis of matter, motion, and the simple laws of mechanics; iatrochemists relied on the chemical "principles" as their explanatory terms. The dichotomy between mind and body had taken deep root and was influencing medical theory, while the close relation between the "mind" of medical doctrine and the "soul" of religious orthodoxy was troublesome indeed.

Although Stahl did not provide any straightforward or systematic exposition of his doctrines, his prolix writing contains certain recurrent themes that serve as a foundation for his more specific discussions. Foremost among his basic concepts is the irreducible difference between the living and the nonliving. Mind and matter are distinct and ultimate. Matter, particulate in its nature, exists as a real entity in its own right and comprises the material aspect of the universe. But equally real and equally deserving of the designation *ens* are the immaterial aspects, of which the *anima* is the key manifestation. While both the living and the nonliving are composed of matter, only living creatures have an *anima*. The immaterial vital principle serves as the ultimate differential feature that distinguishes the living from the nonliving.

A second major principle involves the concept of goal or purpose. The philosopher (or scientist) who tries to describe and explain the phenomena of life must take account of the goal activity. Behavior is not blind or mechanical. Living creatures can be understood only if we pay attention to their striving toward particular ends or purposes. This striving, in turn, implies a directive agency controlling the goal-seeking effort. The agent is the *anima*.

A third principle concerns the place of mechanism in the scheme of things. All nonliving creatures—the inorganic or "mixed"—are entirely mechanical. Living creatures, up to a point, also are mechanical; but the mechanism involved represents only the instrument of the directing agent or *anima*. The agent exerts itself, manifests itself, through mechanical principles.

These major doctrines give rise to certain corollaries: the *anima* that directs the purposive activities of the body acts in an intelligent fashion. It is

rational and exhibits foresight to bring about the desired ends. Furthermore, the directing force can be understood only as a process involving a time span. It implies wholes rather than parts, and only a false philosophy will focus exclusively on the parts and neglect the whole.

On this framework Stahl elaborated a rather detailed and intricate, if rambling and untidy, superstructure. The entire doctrine of animism rests on the ultimate distinction between the living and the nonliving. Stahl pointed out certain significant differences—the nonliving, which may be either homogeneous or heterogeneous, is relatively inert, remains stable over an indefinite time span, and is not readily changed or decomposed. Living creatures, however, are always heterogeneous and always have a great tendency to decomposition and putrefaction. Yet the components of the living body, despite this tendency to putrefaction, remain stable over the limited time that life persists. The tendency to decomposition is held in check by a conserving agent. This agent, the essence of life, is the *anima*, which thus preserves the body from corruption.

The living body depends on motion—most obviously the motion of the heart and the circulation of the blood. The *anima* exerts its control over the body through this very property of motion. By using this concept Stahl engages in some remarkable semantic juggling that enables him to construct a formal and orderly system.

He explicitly denies that motion is in any sense a function of matter. Although material objects do exhibit motion, this property is not intrinsic to them. On the contrary, matter—consisting of material particles—is inert; motion is something added, superimposed from outside. For Stahl motion derives from the *anima*. It is in no sense material (this, indeed, follows clearly from the concept that motion is not a property of matter). On the contrary, he considers it to be immaterial. Motion, then, is somehow reified into an immaterial entity representative of the *anima*, which is also immaterial. The immaterial *anima* acts through motion—also immaterial—and in turn affects material particles.

This sequence is Stahl's answer to the problem, how can an immaterial entity act on something material. This difficulty, the crux of Cartesian dualism, had remained a stumbling block. One popular "solution," widely accepted, involved an intermediary—the animal spirits, which were extremely subtle matter. The soul acted on the animal spirits, which, being material, could then act on the coarser material elements. This formulation employs a sort of Neoplatonic maneuver whereby extremely subtle matter seems to mediate the transition between the conceptual realm and the material world.

This "solution," however, simply begs the question. Stahl, by calling motion an immaterial entity subject to control through the *anima*, believed that he had solved the problem. The concept of motion served as a link whereby an immaterial *anima* could act on the body.

The most important motions of the body are primarily the circulation of the blood, and then those motions that activate the processes of excretion and secretion. Without these, life could not exist. Motion, operative on the humors and the solids, maintains life and health. If the motions are impaired, disease occurs. Motion, Stahl emphasized, is not life but merely its instrument. This concept of instrumentality has extreme importance for his system, which centered on the *anima*.

We do not perceive the *anima*, nor can we study it directly. Instead, we perceive and study the bodily activities in health and the changes in disease, that is, physiology and pathology, and from these data we infer the nature of the *anima*. Stahl was not in any sense an obscurantist or mystic but, rather, a hard-headed clinician. He taught that the proper study of medicine involves the functions of the body, and his voluminous writings concerned themselves with physiology, pathology, and clinical medicine. But while emphasizing the importance of these aspects, he placed them in a suitable perspective: that the motions of the body, in health and in disease, are subordinate to a certain directing and integrating force—the *anima*.

The mind acts on the body in various ways. So-called voluntary actions, depending on a deliberate exercise of will, are quite obvious. But more important for Stahl's system are those bodily effects that result from other psychic causes. Stahl repeatedly referred to the effects on the body produced by psychic disturbance, and he offered examples in two major areas. What we today call emotions—anger, fear, disgust, hatred, love—produce certain significant changes in the bodily functions. The alterations in pulse, respiration, or various digestive activities that result from emotional stress were well-known and obviously were quite different from "voluntary" motions that involve skeletal muscle. Stahl's reasoned explanation presupposed an immaterial *ens*, the *anima*, that felt the emotion and reacted on the bodily organs by inducing changes in their motion.

In a second major illustration, used repeatedly, Stahl fell back on the belief that in a pregnant woman the emotions of the mother exert a material effect on the body of the fetus. It was a firm article of belief that a mother could "mark" the baby *in utero*. In emotional states such as fright or desire, the psychic state could have a physical effect upon the baby. Stahl could not explain in detail how this came about; but he did use these examples to bolster his claims that the *anima*, of immaterial psychic character, had power over the body.

The physical effects of emotion provide merely a striking example of a general situation: that primary changes in the mind are, through motion, transferred to the solids and fluids of the body. But this is merely a special case. Actually, according to Stahl, the *anima* affects the body at all times. It is continually exerting regulatory and directing functions over all bodily activities—and these functions are all purposive.

This is the crux of animism. The *anima* regulates all bodily actions in accordance with certain goals. Life is purposeful. The *anima* is the source not merely of motion but also of directive, purposeful motion. Purpose thus involves the deliberate activity of mind, whereas chance concerns the activity of matter alone, without the intervention of mind. Purpose, implying a goal toward which activity is directed, has what we may call a forward reference, comparable with the final cause of Aristotle. Whatever happens by chance depends solely on backward reference, the *vis a tergo*.

Around these ideas Stahl constructed his important distinction between mechanism and organism. Mechanical properties depend only on the configuration, size, position, and movement (or disposition to movement) of the component parts. The movements have purely mechanical causes. In an organism, however, movements are combined toward a specific end, with a responsible agent that regulates and integrates them. To be sure, in any organism the activities involve mechanism, but merely in an instrumental fashion, subordinate to the purpose of the organism. The purpose, goal, or intent constitutes the reason for existence of the organism. A mechanism does not have such a reason for existence.

Stahl provides numerous examples. In a watch, for instance, a skillfully constructed mechanism, the parts act on each other in mechanical fashion. But the watch also has properties of an organism, insofar as it has the goal of keeping time. If, because of defective parts, it fails to keep correct time, it no longer has the properties of an organism but remains a mere mechanism.

In entities that are not man-made, the question

of possible goal or purpose lies beyond human knowledge. Thus, no one can penetrate the ends served by celestial bodies or the existence of so many species of insects. The answers can lie only in the will of God.

The *anima*, Stahl made explicitly clear, exists only in the body, is inseparable from it, and cannot be thought of apart from the body. It is not a religious concept, nor is it an obscurantist or mystical doctrine. It is understood by rational analysis, which discloses that the *anima* is entirely dependent upon the body for perception and ideas. The body is the instrument of the *anima*, which must have sensory organs to aid the intellect and locomotive organs to aid the will.

Stahl rejected the view that bodily action is carried out solely by the motion of particles. The mechanical philosophy, depending on the *vis a tergo*, simply did not explain observed phenomena. For example, a certain noise—a stimulus falling on the ear—may induce a turning of the eyes toward the noise. Mechanists explained this by the activity of certain particles of the body, incited by the sense organs and reaching the motor organs to act upon them directly and to induce movement. In contrast, Stahl gave the example of a miser who hears a noise like a falling coin. He not only turns his eyes (what we would call a simple reflex) but also searches the entire room, with all the complicated associated movements. He does not stop until he finds the object of his search. That such complex behavior should have a simple mechanical explanation in the motion of particles seemed utterly absurd to Stahl.

Stahl propounded some views that may also seem absurd unless we relate them to the Aristotelian background and see the relationship between the *anima* of Stahl and the form of Aristotle (especially as the Aristotelian doctrine is manifest in the sixteenth- and seventeenth-century Galenists). Stahl declared that the body exists only because of the *anima*, and its form and structure are determined through the energy of the mind. More important, there must reside in the *anima* some special knowledge of the organs, knowledge that regulates growth, shape, and function, and keeps the organs in proper proportion and relationship— proper, that is, for undertaking their functions and achieving appropriate goals. This concept of the immaterial entity controlling both growth and other activity ties in with the Galenists' notion of "substantial form." The form of the oak inheres in the acorn and determines its development into an oak rather than into a pine or a dahlia. Because of

an appropriate form, the seed of a radish develops into a radish and not into a chrysanthemum. Stahl's *anima* includes among its many other functions the directive activities that earlier writers had attributed to form.

Stahl met one obvious objection head on. The *anima* is a conscious agent, but many of the activities attributed to it do not appear in consciousness. We are, for example, consciously aware of sensations but not of directing the growth of bodily parts. Stahl disposed of this objection through a verbal distinction. He distinguished *logos* from *logismos*, *ratio* from *ratiocinatio*. *Logos* is simple "intelligence" or "perception," which cannot be the subject of reasoning or memory and can inhere in the *anima* without being perceived. This is quite different from ideas, which, derived from the external senses, serve as the subject of reasoning and memory (*logismos*). Stahl was saying that the soul has ideas and activities of which it is not fully aware.

All this, of course, merely begs the question; but it does maintain the empirical analysis. At work are forces the nature of which can be identified through observation. Stahl drew into a single concept all the forces operative in living organisms and bestowed on this aggregate the name *anima*.

In carrying out a particular activity, the *anima* must regulate all actions necessary to achieve that goal, even those performed unconsciously. Stahl gave the example that in jumping over a ditch, one controls various muscular movements without any awareness of doing so. Although conscious only of the final goal, the *anima* regulates all the activities needed to achieve that end. Similar considerations apply to the efforts of the *anima* in combating disease. For carrying out the proper end—the restoration of health—it performs all functions necessary for the task and for this goal calls into action various mechanical activities. The *anima*, in brief, is an intelligent agent that wills certain ends and therefore must have organs suitable for achieving them. The human body is the organ of the rational *anima* and is formed for its needs.

The goals of medicine are intensely practical: to maintain health, to keep the body free from threatening ailments, and to combat disease. To achieve these goals, practical medicine must rely on established experience, assisted by sound reason. Speculations that deflect medicine from its goal, even if buttressed by skillful arguments and experiments, have no use. Stahl emphasized that many aspects of knowledge are of little or no positive help. For example, he denied the value of detailed anatomical or chemical studies because attention to precise anatomical findings draws attention away from the body as a whole. And, as Stahl repeatedly indicated, medicine has to do with the whole living organism, presided over by the *anima*, rather than with specific actions of specific parts, which are only instruments. He also rejected from the confines of medicine the specific study of chemistry. Although one of the leading chemists of his era, he expressly denied that chemistry was advantageous in the theory or practice of medicine.

In his strictures Stahl was referring specifically to the teaching of the iatrochemists, who explained bodily activity by the use of a small number of concepts—acidity and alkalinity, coagulation and liquefaction, fermentation, volatility, acrimony. He cogently pointed out that a wide range of ailments was being attributed to essentially the same causes, and that the explanations had neither a priori support from assured theory nor confirmation in solid a posteriori experience, such as a concrete demonstration of fermentation or acidity. This type of doctrine he condemned as useless, unscientific, and sterile. The humors in health or disease simply did not show the various chemical changes that were being invoked as explanatory principles.

We need not concern ourselves with specific details of Stahl's physiology or pathology. He stated that free and orderly circulation, secretion, and excretion are necessary in maintaining health. Harmful material must be eliminated, and if refractory to direct elimination, it must be converted into a state suitable for elimination. The basic reactions of the body exemplify the healing power of nature, through which the *anima*, as an intelligent active force, conserves life and restores health. If nature falters, the physician must use appropriate means to aid the natural processes.

In chemistry as in biology, Stahl strongly disavowed the mechanical viewpoint, which in turn opposed the qualitative philosophy of Aristotle. Aristotelian elements, four in number, embodied qualitative properties that served as the "principles" for material objects. In the mechanical philosophy, on the other hand, the particulate elements had quantitative attributes—size and shape, position and motion—while qualities and properties depended on the interaction of atoms. All atoms obeyed the laws of mechanics; but those that were round, for example, would react differently from those that were pointed or angular.

In addition to the Aristotelian and corpuscular philosophies, a third and more specifically chemi-

cal tradition with three principles (salt, sulfur, and mercury) had developed. These principles had a rather ambiguous status, for they represented not only certain qualities or predictable modes of behavior—such as hardness, inflammability, or volatility—but also concrete material substances. This ambiguity—being characteristic objects and at the same time representing properties that, by inhering in many discrete objects, explained the phenomena of change—went unresolved. Nevertheless, the spagyric elements helped to provide an essentially chemical mode of explanation.

Stahl realized that the simple mechanical atomist viewpoint could not adequately explain the phenomena in chemical operations. He was also aware that the atomist viewpoint must be not totally rejected but merely regarded as inadequate. Matter is not infinitely divisible but exists in elementary particles that are indivisible and impenetrable. Simple elementary bodies, however, are not found in isolation. Everything that we observe is composite, and these composites exist in a hierarchy of increasing complexity. Atoms, never existing by themselves, join to form simple molecules. These in turn unite into more complex molecules, to produce visible objects.

Correlative to this aspect of Stahl's philosophy was the need to explain properties or qualities. In order to account for observed phenomena, the atoms must have specific reactive and qualitative characteristics. Stahl's atoms do indeed bestow particular qualities on the various compounds. However, precisely how a particular quality or functional property relates to the size and shape of atoms he could not satisfactorily explain.

In his attempts to find adequate explanations, Stahl relied heavily on the doctrines of Becher, who believed in three elementary principles—air, water, and earth. Air, however, did not enter into combinations, so that water and earth formed the material bases of objects. "Earth," however, was not a unitary principle. It comprehended three different types: the first, having to do with substantiality, rendered bodies solid and vitrifiable; the second, of moist oily character, provided color, odor, and combustibility; the third supplied weight, ductility, and volatility. These three kinds of earth, despite denials, have rather obvious relationship to the spagyric salt, sulfur, and mercury.

Stahl adopted these views from Becher and used the name "phlogiston" to designate the second earth—the principle of combustibility. It differed from the other two earths, for the first and third could not be separated from the bodies in which they existed, whereas phlogiston did not form any such stable compound.

While the concept of three kinds of earth applied to the vegetable and animal kingdoms as well as to the mineral, the experimental bases for phlogiston rested principally on the behavior of minerals and metals. Striking evidence arose from the reversible relationship of metals and their calxes. A metal contained all three kinds of earth. When the metal was heated intensely, the phlogiston was driven off and a calx appeared. However, when phlogiston was reintroduced into the calx, the metallic form reappeared. Here we have a reversible process: calx plus phlogiston yields the metal; the metal minus phlogiston yields the calx. When phlogiston was driven from a substance, the properties of that substance changed very markedly. And, similarly, if a substance lacking phlogiston received this element through appropriate chemical manipulation, the properties would change. This process also applied to nonmetallic minerals. Sulfur, for example, which had lent its name to the spagyric terminology, was deemed a compound consisting of vitriolic acid plus phlogiston. Far from being synonymous with sulfur, phlogiston was merely a constituent of it.

For minerals the expulsion or reception of phlogiston was a reversible process, but this was not the case in the animal or vegetable kingdoms. Plants were particularly rich in phlogiston; but once it was driven out, the original compound could not be reestablished.

Phlogiston was an element or substance and not an abstract quality or property. As a substance it combined with other chemical substances to form compounds. But, unlike other elements, which could not exist in isolated form, phlogiston could exist in relatively pure form. Finely divided carbon or lampblack, obtained, for example, by holding a cold object close to burning turpentine, was relatively pure phlogiston, visible and palpable. But ordinarily the phlogiston was not directly perceptible. From its combined form it was set free and passed into the air. Flame was considered the whirling motion produced by the escape of phlogiston, and air was necessary for the production of this motion. The air did not enter into the compound but was the receptacle for the phlogiston. In the absence of air, phlogiston could not escape and consequently combustion could not occur. This provided a reasonable explanation for the observation that calcination could not take place in a closed vessel.

Stahl recognized the close relationship between phlogiston and air. Where the quantity of air was limited, the amount of combustion was correlatively limited. He explained this phenomenon through the postulate that air could absorb only a limited amount of phlogiston; and when the limit had been reached, no more combustion—no more liberation of phlogiston—could take place. What happened to the phlogiston that was poured into the air? The Stahlian chemistry held a theory of recycling—the phlogiston in the air passed into plants and thence could pass into animal bodies through the ingestion of plant material.

A retrospective analysis can point out innumerable flaws in the phlogiston theory; and the way in which later eighteenth-century chemists quite demolished the theory forms an important chapter in the history of science and of thought. But this should not blind us to its important role as a bridge between the older concepts and the new. It tried to modify an existing intellectual framework in order to explain experimental observations. It succeeded, but only at the expense of ignoring certain other observations. And it proved unable to encompass new observations as scientific ingenuity devised new experiments.

BIBLIOGRAPHY

Stahl was a very prolific writer, and his doctrines form the subject of a vast secondary literature. Only a few of the more important writings can be mentioned here.

I. ORIGINAL WORKS. Three relatively short essays provide a background for Stahl's general medical philosophy. *Disquisitio de mechanismi et organismi diversitate* (Halle, 1706); *Paraenesis, ad aliena a medica doctrina arcendum* (Halle, 1706); and *De vera diversitate corporis mixti et vivi* (Halle, 1707). His greatest single medical work is *Theoria medica vera, physiologiam et pathologiam . . . sistens* (Halle, 1708), which provides in quite massive detail his doctrines of physiology and pathology, and presents his animistic philosophy as incidental to the exposition. The work includes, as intrinsic introductory material, the three essays mentioned above, plus a fourth, *Vindiciae & indicia de scriptis suis*. A further ed. of the *Theoria medica vera* was published at Halle in 1737. A more recent, 3-vol. ed. edited by Ludovicus Choulant was published at Leipzig in 1831–1833. The work has never been translated into English.

A French rendition of Stahl's writings, *Oeuvres médico-philosophiques et pratiques*, translated with commentaries by T. Blondin, II–VI (Paris, 1859–1864), apparently was intended as 6 vols., although vol. I was not published. This ed. is, unfortunately, rare. The translation is fairly good, although sometimes quite verbose and excessively interpretive.

A German text expounding Stahl's doctrines is Karl Wilhelm Ideler, *Georg Ernst Stahl's Theorie der Heilkunde*, 3 vols. (Berlin, 1831–1832). Although often referred to as a translation, it is only an abbreviated paraphrase. At best it can serve as a "finder"—a rapid way of getting an overview and of locating significant passages that must then be studied in the original Latin.

Stahl and his pupils published extensive clinical studies, describing particular problems, that are significant for the light they throw on the medical practice of the times—for instance, *Collegium casuale magnum* (Leipzig, 1733)—but these are not especially relevant here. He also wrote or was coauthor of a considerable number of dissertations, many of which are available at the National Library of Medicine, Bethesda, Md., and are important for any definitive study of Stahl.

A further work that deserves special mention is *Negotium otiosum seu Σκιαμαχια* (Halle, 1720), an attempt to answer some objections that Leibniz had made to Stahl's animistic doctrines.

Stahl's more specific chemical writings are very numerous. Partington, II, 659–662, has devoted almost four pages to the bibliographic listing, taken from an eighteenth-century bibliography prepared by J. C. Goetz.

One of the earliest is *Zymotechnia fundamentalis* (Halle, 1697). Other important works are *Specimen Beccherianum*, appended to Stahl's ed. of Becher's *Physica subterranea* (Leipzig, 1703); and *Zufällige Gedancken . . . über den Streit von den sogenannten Sulphure* (Halle, 1718), which was translated into French by Holbach as *Traité du soufre* (Paris, 1766). *Fundamenta chymiae dogmaticae et experimentalis* (Nuremberg, 1723), prepared by Stahl's pupils, was translated by Peter Shaw as *Philosophical Principles of Universal Chemistry* (London, 1730). *Fundamenta chymiae dogmatico-rationalis & experimentalis* (Nuremberg, 1732) had a 2nd ed. in 1746. Another important work that presents Stahl's doctrines was written by his pupil Johann Juncker: *Conspectus chemiae theoretico-practicae* (Halle, 1730).

II. SECONDARY LITERATURE. Only a few of the more significant secondary sources can be listed. A relatively recent work is Bernward Josef Gottlieb, "Bedeutung und Auswirkungen des Hallischen Professors . . . Georg Ernst Stahl auf den Vitalismus des XVIII Jahrhunderts, insbesondere auf die Schule von Montpellier," in *Nova acta Leopoldina*, n.s. **12**, no. 89 (1943), 423–502, which covers a great amount of literature but does not exhibit any specially penetrating insight and is far too concerned with Teutonic chauvinism. Walter Pagel, "Helmont, Leibniz, Stahl," in *Archiv für Geschichte der Medizin*, **24** (1931), 19–59, is an important contribution. Albert Lemoine, *Le vitalisme et l'animisme de Stahl* (Paris, 1864) is an important nineteenth-century analysis; an older but very helpful discussion is under the heading "Stahlianisme," *Dictionnaire des sciences*

médicales, 60 vols. (Paris, 1812–1822), LII, 401–449. Every general history of medicine devotes space to Stahl. Of especial value, and a source for many subsequent but more shallow discussions, is Kurt Sprengel, *Versuch einer pragmatischen Geschichte der Arzneikunde*, 2nd ed., V (Halle, 1803), 9–47. Stahl's animism, especially in relation to Friedrich Hoffmann's mechanistic views, is discussed in two recent articles by Lester S. King: "Stahl and Hoffmann: A Study in Eighteenth Century Animism," in *Journal of the History of Medicine*, **19** (1964), 118–130; and "Basic Concepts of Early Eighteenth Century Animism," in *American Journal of Psychiatry*, **124** (1967), 797–807.

A very important secondary source dealing with Stahl's chemistry is J. R. Partington, *History of Chemistry*, II (London, 1961), 637–690. An indispensable analysis of the chemical doctrines is Hélène Metzger, *Newton, Stahl, Boerhaave et la doctrine chimique* (Paris, 1930).

LESTER S. KING

STALLO, JOHANN BERNHARD (*b*. Sierhausen, Oldenburg, Germany, 16 March 1823; *d*. Florence, Italy, 6 January 1900), *philosophy of science.*

Stallo's father was a teacher, as were his ancestors on both sides for many generations. After a private education, at the age of thirteen he entered the normal school at Vechta and then the Gymnasium. In 1839, when he was ready to enroll at a university, lack of money led him to emigrate to the United States, where he joined the German colony in Cincinnati. At first Stallo taught mainly German in a Catholic parish school; at that time (1840) he published a textbook, *ABC, Buchstabier und Lesebuch, für die deutschen Schulen Amerikas*, which went through several editions and was widely used. While teaching at St. Xavier College from 1841 to 1844, he continued to study Greek and mathematics. From 1844 to 1848 he was a professor of physics, chemistry, and mathematics at St. John's College (now Fordham University), New York City.

In 1847 Stallo returned to Cincinnati and began to study law, a program he pursued until 1849, when he was admitted to the bar. From 1852 to 1855 he was judge of common pleas, before returning to private practice. In 1870, before the superior court at Cincinnati, he defended the Cincinnati School Board against the Protestant clergy who tried to enforce the retention of Bible reading and hymn singing as a part of the curriculum. From 1884 to 1889 he was American ambassador in

Florence, where he remained even after his official period of service had ended.

The development of Stallo's thought can be characterized as a gradual transition from post-Kantian idealism in his youth to his own version of positivistically oriented phenomenalism in his mature years. He studied Leibniz, Kant, Herbart, Schiller, and Goethe while still in Germany. This influence is visible in the poem "Gott in der Natur," published in *Wahrheitsfreund* (Cincinnati, 1841) which enthusiastically proclaims, in Goethe's fashion, the unity of God and nature. In 1848 he published *The General Principles of the Philosophy of Nature, With an Outline of Its Recent Developments Among the Germans*; *Embracing the Philosophical Systems of Schelling and Hegel and Oken's System of Nature*, the title of which indicates Stallo's exclusive commitment to *Naturphilosophie*. The book is now of value only as a document illuminating the early stage of Stallo's philosophical development, and Stallo himself later conceded its worthlessness: "That book was written while I was under the spell of Hegel's ontological reveries—at a time when I was barely of age and still seriously affected with the metaphysical malady which seems to be one of the unavoidable disorders of intellectual infancy" (preface to *The Concepts and Theories of Modern Physics* [New York, 1881]). Realizing that these words may have been too harsh, he added: "The labor expended in writing it was not, perhaps, wholly wasted, and there are things in it of which I am not ashamed, even at this day . . ." (*ibid*).

It is not difficult now to see what those things were in Stallo's first book that he was not ashamed of in his mature years. One idea was common to all phases of his thought: that things are not "insular existences" but complexes of relations. Also, the influence of the book was not as negligible as Stallo believed. There is definite evidence that his "evolutionary idealism" greatly appealed to Ralph Waldo Emerson, who copied passages from Stallo's book into his journals beginning in November 1849; in 1873 he even credited Stallo with having anticipated Darwin's theory of evolution by regarding animals as "foetal forms of man." Even later Thomas Sterry Hunt, the American chemist and geologist, conceded Stallo's influence and dedicated *A New Basis for Chemistry* (Boston, 1887) to him.

In 1855 Stallo wrote a critical essay on materialism, published at Cincinnati in *Atlantis* (pp. 369–386) and then as a separate offprint entitled *Naturphilosophische Untersuchungen. Der Materialismus*. It represents an intermediate stage between Stallo's early Hegelianism and his mature work in the philosophy and epistemology of science. Its main target was materialism, which had been revived in Germany and was becoming popular in the United States, especially among German immigrants. Stallo explicitly mentioned Karl Vogt and Ludwig Feuerbach. His criticism was written in a semipopular style and clearly was directed to a wide audience. It consists of three distinct parts: an analysis of the epistemological assumptions of materialism; a questioning of its scientific adequacy; and a sharp rejection of the materialistic reduction of thoughts to the brain processes.

In the first part Stallo showed the intrinsic inconsistency of the materialistic epistemology that claimed that all knowledge comes from sensory perception and at the same time postulated the existence of atoms that can never be perceived; furthermore, in basing all knowledge on the sensations—that is, on conscious data—it unwittingly conceded their primary character (contrary to its professed reductionism). In the third part Stallo criticized the then famous statement of Karl Vogt, "The brain secretes thought as the kidney does urine . . . ." He concluded by recommending that materialists read Kant's *Critique of Pure Reason* and Hegel's *Phenomenology of Mind*.

Stallo's original idealism is even more apparent in the middle part. His main argument is the alleged inadequacy of the atomistic explanation of chemical phenomena. At that time the distinction between the atom and the molecule was still not clarified, and thus Stallo's doubts about the indivisibility of the atoms appeared plausible. Similarly, chemical formulas were still far from being established (there was, for instance, still a dispute over whether the formula of water was HO or H_2O) and structural formulas were unknown; thus it was easy for him to speak of "the chaos in organic chemistry." Stallo obviously mistook the temporary incompleteness of the atomic theory for its basic inadequacy. That he was still far short of positivism is indicated by the polemical note against Comte's veto of any metaphysics. On the other hand, it contains the idea found in his first book: that every material thing is a network of forces and relations (*das Gewebe von Kraften und Beziehungen*).

The preface to Stallo's *Concepts and Theories of Modern Physics* is dated 1 September 1881. Three years later, in the extensive introduction to the second edition, he stressed that the main purpose of the book was an epistemological criticism of the corpuscular-kinetic theory of nature. In the first eight chapters he restated the basic theses of the "atomo-mechanical theory" (as he called it) and showed its empirical difficulties. These basic theses were the following: (a) The primary elements of all natural phenomena are mass and motion. (b) Mass and motion are disparate. "Mass is indifferent to motion, which may be imparted to it, and of which it may be divested, by a transference of motion from one mass to another. Mass remains the same, whether at rest or in motion." (c) Both mass and motion are independently conserved. From these theses two general corollaries followed: (a) All phenomenal diversity is reducible to quantitative differences in configuration and motion. (b) All apparently qualitative changes are reducible to the quantitative changes of position and configuration. These two corollaries thus imply the absolute homogeneity, hardness, rigidity, and passivity of the basic elements of matter.

In subsequent chapters Stallo showed how all previous propositions followed from the basic principles of the "atomo-mechanical theory" and how they simultaneously contradicted the empirical findings of his own time. Thus the claim that the basic units of matter are homogeneous followed from the qualitative unity of matter assumed by classical atomism, but it was at variance with the irreducible diversity between the atoms of different elements. Stallo believed that the persistent hope of reducing these differences to different configurations of the more basic and truly homogeneous units had been definitively buried by the failure of Prout's hypothesis. Similarly, the classical assumption of absolute hardness and rigidity of the basic material units was contradicted in the area where the triumph of mechanistic explanations seemed to be the most spectacular — in the kinetic theory of gases, based on the very opposite assumption of the elasticity of the bouncing particles. Yet the concept of an elastic (compressible) atom was a contradiction. The passivity of the basic material units implied the denial of action at a distance, and Stallo again emphasized how persistent the tendency is to reduce all dynamic interaction to the direct pressure and impact of bodies. This led him to point out the difficulties of the kinetic explanations of gravitation, of the mechanical models of ether, and of the theory of atoms as "vortex rings" proposed by William Thomson (later Lord Kelvin).

After showing what he regarded as the empirical inadequacies of the corpuscular-kinetic models, Stallo proceeded to trace their shortcomings to "the structural fallacies of the intellect," which he attributed to the following four assumptions: (a) that to every concept there corresponds a distinct objective entity; (b) that the alleged entities corresponding to more general concepts exist prior to the entities corresponding to less general concepts; (c) that the order of the genesis of concepts is identical with the order of the genesis of things; (d) that things exist independently of and antecedently to their relations. Thus had arisen "the four radical errors of metaphysics," characterizing "the atomo-mechanical theory." The concept of homogeneous matter, devoid of concrete sensory qualities and existing separately from motion and force, was merely a reified abstraction, comparable with the Hegelian "Being," even though Stallo conceded that it was "somehow less hollow." With equal severity he censured Bošković and the dynamists for their reification of the concept of force. The assumption of the absolute solidity of the basic material units was a mere prejudice due to man's psychobiological conditioning: the fact that "the most obtrusive form of matter is the solid," which was first recognized and manipulated by "the infant intellect of mankind," was the basis of man's tendency to interpret every physical phenomenon in corpuscular terms and every physical interaction by direct contacts and impacts of such solid particles. But there was no absolutely solid body, nor was there any absolutely gapless contact between bodies in nature.

The most valuable and most prophetic insights are expounded in this part of the book. Stallo's firm opposition to the fiction of "insular existence," originally motivated by the metaphysical idea of "the relatedness of reality," was placed on a more convincing basis of concrete physical considerations. He pointed out that the concept of the isolated material body, whether on the atomic or the macrophysical scale, as well as the concept of the single isolated force, was physically meaningless. All physical properties were relational and owed their existence to the physical interaction between various parts of the world. Even inertia, which, according to Newton, represented the very core of matter (*vis insita*) was no exception. Stallo demonstrated it in analyzing Carl Neumann's argument (1870) in favor of the Newtonian absolute space. His thesis was a modification of Newton's classical argument: the absolute rotations (the rotations with respect to absolute space) are physically distinguishable from mere relative rotations, since in the first case centrifugal forces arise while in the second case they do not.

Neumann inferred from this that centrifugal

force would appear on an absolutely rotating body even if it were completely alone in space. Against this Stallo claimed that not only the rotation of a solitary body, but even its very existence, would be utterly meaningless if no other bodies existed: "All properties of a body which constitute the elements of its distinguishable presence in space are in their nature relations and imply terms beyond the body itself" (*Concepts and Theories*, p. 215). He clearly anticipated a similar criticism of Newton by Ernst Mach, who concluded that in the principle of inertia there is "an abbreviated reference to the entire universe" and that "the neglecting of the rest of the world is impossible." This passage from Stallo's book had appeared in *Popular Science Monthly* in 1874 and shows that Mach's criticism of Newton was anticipated a full decade before *Mechanik in ihrer Entwickelung* appeared. Stallo arrived at the same conclusion in showing that Newton's third law requires the existence of at least two bodies in the universe. He quoted Maxwell's comment on this law: Action and reaction are two complementary aspects of the same phenomenon—stress; thus the concept of a single, isolated force is as meaningless as that of a solitary body. Such a body would be not only without weight but also without inertia.

The second anticipatory insight in Stallo's book was his epistemological criticism of mechanical models in general. Its basic idea was concisely expressed in his statement that "a phenomenon is not explained by being dwarfed." In other words, to explain the properties of macroscopic matter by postulating the very same properties on the microphysical scale was no explanation at all. Stallo thus raised doubt about the adequacy of the mechanical and, more generally, intuitive models of the microcosmos. It is hardly necessary to stress how prophetic his view proved to be and how bold it was in the era when William Thomson equated the understanding of any physical phenomenon with the possibility of making a mechanical model of it.

On the other hand, Stallo dogmatically excluded non-Euclidean geometries from physics. His position in this respect was not consistent; and his arguments against "transcendental geometries," as he called them, were inconclusive. First, he claimed that the curvature of space implies the absolute finitude of the material universe, which is "a necessary complement of the assumption of its absolute minimum, the atom." This clearly is not true of Lobachevsky's space, which is infinite. Second, he claimed, as Poincaré did later, that no empirical proof can ever be given for a non-Eu-

clidean character of space; even the discovery of a nonzero parallax of very distant stars would not be conclusive, since it would be more natural to explain it by some deviation of the light rays *in* space rather than by some intrinsic curvature *of* space. He dogmatically insisted that by its own nature space must be homogeneous, confusing homogeneity with Euclidean character; he did not realize that Riemannian and Lobachevskian spaces of a constant curvature are also homogeneous. Finally, he claimed that the concept of a straight line, purportedly eliminated by Riemann and others, was surreptitiously reintroduced in the form of "the radius of curvature." Stallo clearly misunderstood the metaphorical nature of this term, which can be taken literally only for two-dimensional illustrative models of non-Euclidean spaces and not for those spaces themselves, for which the term "space constant" is less misleading than "radius of curvature." Thus today we see that it was Helmholtz and Clifford who were on strong ground, rather than Stallo, who criticized them.

In *Essay on the Foundations of Geometry* (1897) Bertrand Russell pointed out the irrelevance of Stallo's criticism of non-Euclidean geometry (p. 88), and his demonstration caught Ernst Mach's attention. While disagreeing with Stallo about non-Euclidean geometry, Mach welcomed his criticism of what he called "mechanistic mythology" and referred to him in the fourth edition of his *Science of Mechanics*. He regretted not having known Stallo's work earlier, and began a correspondence with him. Mach wrote the preface to H. Kleinpeter's German translation of Stallo's *Concepts and Theories of Modern Physics* (Leipzig, 1901), and dedicated his *Principien der Wärmelehre* (1896) to him.

This was not the only recognition that Stallo received, although his influence was greater on philosophers, particularly philosophers of science, than on scientists. Josiah Royce, in his introduction to G. B. Halsted's translation of Poincaré's *Foundation of Space*, recalled how scientific orthodoxy was shocked by the appearance of Stallo's book. Its early translation into French (1884) also attracted considerable attention: it probably inspired Arthur Hannequin's *Essai critique sur l'hypothèse des atomes* (1895); and Bergson, Meyerson, and L. Brunschvicg referred to it in their books. With the mounting crisis of the mechanical models in physics, Stallo's work began to receive increasing attention: Rudolf Carnap referred to him in *Der Raum* (1922) and *Physikalische Begriffsbildung* (1926), and P. W. Bridgman reedited

Concepts and Theories with his own introduction (1960).

In a broader historical perspective Stallo appears as one of the prophets of twentieth-century physics. Like Mach, he correctly recognized the inadequacy of the classical corpuscular-kinetic models on the microphysical scale and questioned the absolutistic assumptions of Newton. But, also like Mach, he failed to appreciate the fruitfulness of the same models on the macrophysical and molecular level. This failure was revealed in Stallo's rejection of the kinetic theory of gases; his comments on this theory, especially his claim that it is simpler to assume "gaseousness" as a primary attribute of matter, are mere historical curiosities today. On the other hand, his lucid analysis of the logical structure of classical atomism is on a par with similar analyses by Kurd Lasswitz and Émile Meyerson, and will remain a lasting contribution to the history of ideas.

BIBLIOGRAPHY

On Stallo and his work, see M. Čapek, "Two Critics of Newton Prior to Mach: Boscovich and Stallo," in *Actes du XII Congrès international des sciences* (Paris, 1968), IV, 35–37; Stillman Drake, "J. B. Stallo and the Critique of Classical Physics," in *Men and Moments in the History of Science* (Seattle, 1959); Lloyd D. Eaton, *Hegel's First American Followers* (Athens, Ohio, 1966); H. Kleinpeter, "Stallo als Erkenntniskritiker," in *Vierteljahrsschrift für wissenschaftliche Philosophie*, **25** (1901), 401–440; and *Die Erkenntnistheorie der Naturforschung der Gegenwart. Unter Zugrundelegung der Anschauungen von Mach, Stallo, Clifford, Kirchhoff, Hertz, Pearson und Ostwald* (Leipzig, 1905); J. H. Rattermann, *Johann Bernard Stallo, deutsch-amerikanischer Philosoph, Jurist und Staatsmann*: . . . (Cincinnati, 1902); and Lancelot Whyte, "Stallo Versus Matter," in *Anglo-German Review*, **1** (1961).

M. ČAPEK

STAMPIOEN, JAN JANSZ, DE JONGE (*b.* Rotterdam, Netherlands, 1610; *d.* The Hague, Netherlands [?], after 1689), *mathematics.*

Stampioen's father (of the same name, whence the cognomen *de Jonge*) made astronomical instruments and was an official surveyor and gauger until his removal from office in 1660 for breach of trust.[1] The son began his own career in 1632 with an edition of Frans van Schooten the Elder's sine tables, to which Stampioen appended his own fully algebraic treatment of spherical trigonometry.

In 1633, while a mathematics teacher in Rotterdam, Stampioen took part in a public competition, during which he challenged Descartes to resolve a quartic problem involving a triangle with inscribed figures. Descartes derived the correct equation but did not solve it explicitly, and Stampioen rejected the solution as incomplete.[2] The issue was dropped for the moment, but Stampioen had made an enemy of Descartes and would soon feel the effects.

After being named tutor to Prince William (II) in 1638, Stampioen moved to The Hague, where he opened up a printing shop and in 1639 published his *Algebra ofte nieuwe stelregel* (*Algebra, or the New Method*), which he had completed in 1634. Despite the general title, the work focused on a new method of determining the cube root of expressions of the form $a + \sqrt{b}$ and on the method's application to the solution of cubic equations. In order to attract attention to his forthcoming book, he assumed in 1638 the alias Johan Baptista of Antwerp and posed two public challenges, the more difficult of which demanded the calculation of a traversing position for a siege gun (as stated, a cubic problem). Then he immediately published a solution under his own name. Soon thereafter, he presented another challenge requiring the determination of the position of the sun from the condition that three poles of given heights placed vertically in the ground each cast shadows reaching to the feet of the other two.[3]

A young surveyor in Utrecht, Jacob van Waessenaer, also published a solution to the first, or Antwerp, challenge, employing the methods of Descartes's *Geometry*. Stampioen's rejection of this solution prompted Waessenaer to publish a broad-scale critique of Stampioen's mathematics, emphasizing the inadequacies of the "new methods."[4] The exchange of pamphlets lasted two years and soon involved Descartes, who may in fact have been behind Waessenaer from the beginning.[5] With a wager of 600 gulden riding on the outcome, the issue was adjudicated in 1640 by Van Schooten and Jacob Gool, who found in Waessenaer's favor.[6]

Judging from their correspondence, Constantijn Huygens agreed with Descartes that Stampioen behaved badly during the dispute. Nonetheless, in 1644 Huygens engaged Stampioen for a year as mathematics tutor for his two elder sons, Constantijn, Jr., and Christiaan.[7] Thereafter, Stampioen faded from public notice. A brief reprise of the Waessenaer dispute in 1648, a topographical map published in 1650, and a mention of his having served in 1689 as an expert in a test of a method for determining longitude at sea are the only traces left of Stampioen's later life.

NOTES

1. The British Museum Catalogue (Ten-year supplement, XXI, col. 148) lists a copy of the *Sententien, by den Hove van Hollant gearresteert, jegens Ian Ianssz Stampioen en Quirijn Verblas. Gepronuncieert den acht en twintichsten Iulij Anno 1660*. For further details, see Bierens de Haan's "Bouwstoffen," cited in the bibliography.
2. For details, see the letter from Descartes to Stampioen in Charles Adam and Paul Tannery, eds., *Oeuvres de Descartes*, I (Paris, 1897), 275–280, and the editorial note in *ibid*. 573–578.
3. Newton, in his deposited lectures on algebra, published in *Universal Arithmetick* (London, 1707), states the problem as follows: "When, somewhere on Earth, three staves are erected perpendicular to the horizontal plane at points *A*, *B*, and *C*—that at *A* being 6 feet, that at *B* 18 feet and that at *C* 8 feet, with the line *AB* 33 feet in length—, it happens on a certain day that the tip of stave *A*'s shadow passes through the points *B* and *C*, that of stave *B*, however, through *A* and *C*, and that of stave *C* through the point *A*. What is the sun's declination and the polar elevation? in other words, on what day and at what place do these events occur?" (Derek T. Whiteside, ed., *The Mathematical Papers of Isaac Newton*, V [Cambridge, 1972], 267.) Newton's complete solution of this problem, which involves conic sections, occupies pp. 266–278 of this edition.
4. Waessenaer's two major tracts are *Aanmerckingen op den nieuwen Stel-Regel van J. Stampioen, d'Jonge* (Leiden, 1639); and *Den On-wissen Wis-konstenaer I. I. Stampioen ontdeckt Door sijne ongegronde Weddinge ende mis-lucte Solutien van sijne eygene questien. Midtsgaders Eenen generalen Regel om de Cubic-wortelen ende alle andere te trecken uyt twee-namighe ghetallen: dewelcke voor desen niet bekent en is geweest. Noch De Solutien van twee sware Geometrische Questien door de Algebra: dienstich om alle te leeren ontbinden* (Leiden, 1640).
5. That is the conclusion of Bierens de Haan in his "Bouwstoffen," cited in the bibliography below, 79ff.
6. Stampioen published the judgment, entitling it so as to make himself appear the winner: *Verclaringe over het gevoelen by de E. H. professoren matheseos der Universiteit tot Leyden uyt-ghesproken, nopende den Regel fol. 25 van J. Stampioen, Welcke dese Verclaeringhe soodanigh ghetstelt is, dat yeder een daer uyt can oordeelen dat den Regel fol. 25 beschreven van Johan Stampioen de Jonge in sijnen Nieuwen Stel-Regel, seer licht, generael, ende der waerheydt conform is, om daer door den Teerling-wortel te trecken uyt twee-naemighe ghetallen* (The Hague, 1640). Judging by Descartes's complaints to Henricus Regius (Bierens de Haan, "Bouwstoffen," 99), the decision was formulated in a manner vague enough to permit such a contrary reading.
7. See Christiaan Huygens, *Oeuvres complètes*, XXII (The Hague, 1950), 399ff., and *ibid*., I (The Hague, 1888), 15, for the list of mathematical works suggested by Stampioen as a syllabus for Huygens' sons.

BIBLIOGRAPHY

I. ORIGINAL WORKS. Stampioen's major works include his *Kort byvoeghsel der sphaerische triangulen* appended to his edition of Frans van Schooten, *Tabula sinuum* (Rotterdam, 1632); and his *Algebra, ofte Nieuwe stel-regel, waer door alles ghevonden wordt*

inde wisconst, wat vindtbaer is . . . (The Hague, 1639).

The challenges and disputes of 1633 and 1638–1640 gave rise to several polemic pamphlets, a complete listing of which can be found in either of Bierens de Haan's articles listed below. The more important items are "Solutie op alle de questien openbaer angeslagen ende voorgestelt door Ez. de Decker" (Rotterdam, 1634); "Questie aen de Batavische Ingenieurs, Voor-gestelt door Johan Baptista Antwerpiensis. Volghens het spreeckwordt: Laet const blijken, Met goet bewys" (1638; for a more easily accessible statement of the problem, see *Oeuvres de Descartes*, C. Adam and P. Tannery, eds., II [Paris, 1898], 601ff.); "Wiskonstige Ontbinding. Over het Antwerpsch Vraegh-stuck toe-ge-eyghent alle Lief-Hebbers der Wis-Const" (The Hague, 1638); and "Wis-Konstigh Ende Reden-Maetigh Bewijs. Op den Reghel Fol. 25, 26 en 27. Van sijn Boeck ghenaemt den Nieuwen Stel-Regel" (The Hague, 1640).

Bibliographical details of Stampioen's map are given by Bierens de Haan in his "Bouwstoffen" (see below), 114.

II. SECONDARY LITERATURE. The most complete biography is that of Cornelis de Waard in *Nieuw Nederlandsch Biografisch Woordenboek*, II (Leiden, 1912), cols. 1358–1360. See also David Bierens de Haan, "Bouwstoffen voor de geschiedenis der wis- en natuurkundigen weteschappen en de Nederlanden, XXX: Jan Jansz. Stampioen de Jonge en Jacob à Waessenaer," in *Verslagen en Mededeelingen der koninklijke Akademie van Wetenschappen, Afdeeling Natuurkunde*, 3rd ser., III (Amsterdam, 1887), 69–119; and "Quelques lettres inédites de René Descartes et de Constantyn Huygens," in *Zeitschrift für Mathematik und Physik*, 32 (1887), 161–173. Further bibliography can be found in these three articles.

MICHAEL S. MAHONEY

STANNIUS, HERMANN FRIEDRICH (*b*. Hamburg, Germany, 15 March 1808; *d*. Sachsenburg, Germany, 15 January 1883), *comparative anatomy, physiology*.

Stannius was the son of a merchant, Johann Wilhelm Julius Stannius, and the former Johanna Flügge. After attending the Johanneum in Hamburg, he began his medical studies at the Akademisches Gymnasium there (1825). To complete his studies, Stannius went to Berlin in 1828 and then to Breslau, where he finished a doctoral dissertation in comparative anatomy on 26 November 1831. He returned to Berlin, where he became an assistant at the Friedrichstädter-Krankenhaus (1831–1837) while working as a general practitioner. Simultaneously he investigated a great number of questions in entomology and pathological anatomy.

On 3 October 1837, Stannius, then aged twenty-nine, was offered an appointment as full professor of comparative anatomy, physiology, and general pathology at Rostock University and as director of the institute for the same fields. He lectured on these subjects and also taught histology from 1840 to 1862. Although Stannius had been in poor health since 1843, he succeeded to the rectorship of the university in 1850 and carried out much fruitful scientific research until 1854. Beginning in 1855 his illness, a serious nervous disease connected with mental disturbances, grew worse, and in 1862 it obliged him to abandon his work. The last twenty years of his life were spent in a mental hospital at Sachsenburg.

Although his health and his position at the university allowed Stannius to undertake scientific work for only seventeen years, he nevertheless gained a reputation in various fields of research. He first worked in entomology, dealing with the structure of the diptera and with deformities of insects (1835).

In Berlin he dealt with general pathology (1837). His outstanding monograph was the second volume of *Lehrbuch der vergleichenden Anatomie der Wirbeltiere* (1846). He also investigated the nervous systems and the brains of sturgeons and dolphins (1846, 1849), and he conducted pharmacological studies on the effects of strychnine and digitalis (1837, 1851). In a noted work (1852), he ligatured a frog heart and established the location of the stimulus-building center within the *sinus venosus*. This experiment has since been known as "Stannius' experiment." He was a long-time friend of Rudolph Wagner, a physiologist at Göttingen University. Of the contributions he undertook to write for Wagner's *Dictionary of Physiology*, he was able to finish only the article on fever (1842), which he said resulted from a "changed mood" of the nervous system.

BIBLIOGRAPHY

I. ORIGINAL WORKS. A bibliography of forty of Stannius' papers is in Royal Society, *Catalogue of Scientific Papers*, V, 797–798. His works include *De speciebus nonnullis Mycethophila vel novis vel minus cognitis* (Bratislava, 1831), his inaugural dissertation; *Beiträge zur Entomologie, besondere in Bezug auf Schlesien, gemeinschaftlich mit Schummel* (Breslau, 1832); "Über den Einfluss der Nerven auf den Blutumlauf," in *Notizen aus dem Gebiet der Natur- und Heilkunde*, **36** (1833), 246–248; "Ueber einige Missbildungen an Insekten," in *Archiv für Anatomie, Physiologie und wis-*

senschaftliche Medizin (1835), 295–310; *Allgemeine Pathologie*, I (Berlin, 1837); "Ueber die Einwirkung des Strychnins auf das Nervensystem," in *Archiv für Anatomie . . .* (1837), 223–236; "Ueber Nebennieren bei Knorpelfischen," *ibid.* (1839), 97–101; "Ueber Lymphherzen der Vögel," *ibid.* (1843), 449–452; "Ueber den Bau des Delphingehirns," in *Abhandlungen aus dem Gebiet der Naturwissenschaften* (Hamburg), **1** (1846), 1–16; *Lehrbuch der vergleichenden Anatomie der Wirbelthiere*, 2 vols. (Berlin, 1846; 2nd ed., 1852), of which Carl von Siebold wrote the first, and Stannius the second volume; "Untersuchungen ueber Muskelreizbarkeit," in *Archiv für Anatomie . . .* (1847), 443–462, and (1849), 588–592; *Das peripherische Nervensystem der Fische, anatomisch und physiologisch untersucht* (Rostock, 1849); "Ueber die Wirkung der Digitalis und des Digitalin," in *Archiv für physiologische Heilkunde*, **10** (1851), 177–209; and "Zwei Reihen physiologischer Versuche," in *Archiv für Anatomie . . .* (1852), 85–100.

II. SECONDARY LITERATURE. See J. Pagel, *Biographisches Lexicon der hervorragenden Aerzte*, 2nd ed., V (Berlin–Vienna, 1934), 390; Wilhelm Stieda, "Hermann Stannius und die Universität Rostock 1837–1854," in *Jahrbuch des Vereins für mecklenburgische Geschichte*, **93** (1929), 1–36; Richard N. Wegner, *Zur Geschichte der anatomischen Forschung an der Universität Rostock* (Wiesbaden, 1917), 2, 127–128; and Axel Wilhelmi, *Die mecklenburgischen Ärzte von den ältesten Zeiten bis zur Gegenwart* (Schwerin, 1901), 107–108, with bibliography.

K. E. ROTHSCHUH

STANTON, THOMAS ERNEST (*b.* Atherstone, Warwickshire, England, 12 December 1865; *d.* Pevensey Bay, Sussex, England, 30 August 1931), *engineering.*

Stanton was educated at Atherstone Grammar School, Owens College, Manchester (1884–1891), and University College, Liverpool (D.Sc. 1898). While a student at Owens, he was an articled pupil at Gimson & Co. Engineers, Leicester (1884–1887). He began his academic career as demonstrator in the Whiteworth Engineering Laboratory at Owens College (1891–1896), then advanced steadily; senior lecturer in engineering at University College, Liverpool (1896–1899); professor of civil and mechanical engineering, University College, Bristol (1899–1901); and superintendent of the engineering department at the National Physical Laboratory, Teddington (1901–1930). At Owens College, Stanton was assistant to Osborne Reynolds and later experimentally established the latter's theoretical laws of fluid flow in pipes. He was associated with the National Physical Laboratory from its inception in 1901. Now administered by the Department of Trade and Industry it was originally established under the overall control of the Royal Society, to pursue scientific research, particularly for application in industry.

Throughout his career Stanton was occupied with hydrodynamics, strength of materials, heat transmission, and lubrication. His papers on the flow of water in channels of varying cross section (1902), the resistance of thin plates and models in a current of water (1909), the mechanical viscosity of fluids (1911), and comparisons of surface friction and eddy-making resistance in fluids (1912) led to his definitive text *Friction* (1923), in which he postulated the boundary theory. Experiments in alternating stress and impact testing machines (1905–1906) and on hardness testing (1916–1917) resulted in awards by the Institution of Civil Engineers (1899, 1906, 1921) culminating in the Howard Quinquennial Prize for his research and writing on the properties of iron and steel.

Stanton was involved with aerodynamics from the early years of airplane development, the engineering department of the National Physical Laboratory having charge of this work until it was assigned to a separate department. He built a vertical wind tunnel and other equipment for wind velocity investigations; and as late as 1928, although no longer engaged in administration, Stanton made tests for the British Airscrew Panel on airfoils at speeds near the velocity of sound in air. He was elected a Fellow of the Royal Society in 1914 and also served on its Council in 1927–1929. He was a member of the committee for the restoration of St. Paul's Cathedral, London, in the early 1930's. Stanton was awarded the C.B.E. for his service in World War I and knighted in 1928.

BIBLIOGRAPHY

I. ORIGINAL WORKS. Stanton's writings include "Flow of Water in Channels of Varying Cross Section" in *Engineering*, **74** (1902), 664; "Resistance of Thin Plates and Models in a Current of Water," in *Transactions of the Institution of Naval Architects* (1909), 164–169; "Mechanical Viscosity of Fluids as Affected by the Speed and by the Dimensions of the Channel . . .," in *Proceedings of the Royal Society of London,* **A85** (1910), 366–376; "Similarity of Motion in Relation to the Surface Friction of Fluids," in *Philosophical Transactions of the Royal Society of London*, **A214** (1914), 199–224. written with J. R. Pannell; *Friction* (London, 1923); "Tests Under Conditions of Infinite Aspect Ratio of Four Airfoils in High-Speed Wind Channel" and "Distribution of Pressure Over Symmetrical Joukowski Section at High Speeds," both in *Technical Report of*

Aeronautical Research Committee of National Physical Laboratory, 1929–1930, I, Aerodynamics (London, 1931), 290–296; and "Engineering Research," in *Engineer*, **151** (1931), 506; and *Nature*, **127** (1931), 748–749.

II. SECONDARY LITERATURE. See S. H. Hooker, "Compressibility Effects in High Speed Air Flow," in *Journal of the Royal Aeronautical Society*, **35** (1931), 665–674, with illustration of high-speed wind tunnel at 669; and the unsigned "The Late Sir Thomas Stanton," in *Engineering*, **132** (4 Sept. 1931), 280.

P. W. BISHOP

STARK, JOHANNES (*b.* Schickenhof, Upper Palatinate, Germany, 15 April 1874; *d.* Traunstein, Upper Bavaria, Germany, 21 June 1957), *experimental physics.*

Stark, the son of a farmer, entered the University of Munich in 1894 and received the doctorate in May 1897 for a dissertation entitled "Untersuchungen über Russ" ("Investigations on Lampblack"). After passing the two state examinations required for the teaching of mathematics at an advanced level, he began work on 1 October 1897 at the University of Munich, spending most of his time as private assistant to Eugen Lommel. On 1 April 1900 he became assistant to Eduard Riecke at the University of Göttingen, where the following autumn he qualified as lecturer and was appointed *Privatdozent* on 24 October 1900. While noting Stark's highly developed technical and experimental skill and his gift for conceptualization, the Faculty observed in its confidential report to the university curator that "it is accompanied by a definite deficiency in the area of exact mathematical formulation of the problems under consideration."

Stark's main field of interest was electrical conduction in gases, and his first book dealt with this subject. In 1904 he founded *Jahrbuch der Radioaktivität und Elektronik* to publish studies in the newly developing field of particle physics. He rapidly acquired a broad understanding of this field through intensive reading of the literature and his own experiments. This knowledge prompted him to test for the existence of the optical phenomenon known as the Doppler effect in canal rays, which had been recognized as fast-moving particles by Wilhelm Wien. The effect had never been detected in terrestrial light sources; but, as Stark later said, he found it "almost effortlessly" in the hydrogen lines. He made this discovery immediately after moving into the new physics institute at Göttingen.

Seeking to determine the scientific significance

of his discovery, Stark attempted to make the optical Doppler effect a proof of Einstein's theory of special relativity and, a year later (1907), with the quantum hypothesis as well. Stark was thus one of the earliest defenders of the hypothesis, and he remained in the forefront of research until 1913. Curiously, after that year he turned vehemently against both the quantum theory and the general theory of relativity. He therefore is still remembered by older physicists as a conservative, indeed as a reactionary.

In his initial campaign for the quantum hypothesis, Stark, following Einstein, sought new natural phenomena that could be understood in terms of it—and only in those terms. Along with many correct deductions, such as the interpretation of the ultraviolet boundary of X radiation due to electron deceleration, he arrived at a considerable number of erroneous interpretations. At the end of 1909 he initiated a vigorous debate with Arnold Sommerfeld on quantum theory that was conducted in the pages of *Physikalische Zeitschrift* and in their correspondence. Stark sought to understand the distinctive features of the radiation—especially the directional dependence—with the aid of the quantum hypothesis of light; and he tried, virtually by brute force, to wrest agreement from Sommerfeld. "I am so bold as to hope," he wrote, "that just as you changed your position regarding the theory of relativity, so you will modify your stand on the quantum hypothesis and . . . that the differences between us will have been the reason for it." But it was not difficult for Sommerfeld to point out that Stark had made serious errors in physics. The classical electromagnetic theory was already leading to experimentally discovered directional dependence. Thus Sommerfeld wrote: "Nothing is further from my intention than to begin a feud with you. That would be a very unfair match. For you far surpass me in experimental ideas, just as I excel you in theoretical clarity." The discussion eventually veered off into a polemic. Only a few months before, Sommerfeld had supported Stark's appointment to Aachen; the resulting enmity later overshadowed the careers of both men.

On 1 April 1906 Stark had become *Dozent* in applied physics and photography at the Technical College in Hannover, where he also received the associated post of assistant and was named professor. He soon came into conflict with his superior Julius Precht, who repeatedly requested the Ministry of Education to dismiss Stark and finally succeeded in having him temporarily transferred to Greifswald on 1 October 1907. Following Stark's return

the tensions between the two increased until Stark, with the energetic support of Sommerfeld, was appointed full professor at the Technical College in Aachen on 1 April 1909.

With the relatively elaborate equipment available to him in his own laboratory at Aachen, Stark completed his research in progress, chiefly on the dissymmetry of *Bremsstrahlung*, and undertook a new series of experiments on the splitting of spectral lines in an electric field. While at Göttingen he had been encouraged by Woldemar Voigt to investigate this electrical analogy to the magnetic Zeeman effect. Stark's first preparatory experiment, at the beginning of 1906, had been a failure; but he was successful in October 1913. He described the experiment in a short autobiographical account. Having procured all the necessary equipment—"a high-intensity spectrograph of rather large dispersion, high-tension sources, and Gaede pumps"—he looked for the effect "simultaneously in the hydrogen and helium lines." An electric field of between 10,000 and 31,000 volts/cm. was established in the canal-ray tube.

One afternoon soon after courses resumed in October, I began recording the canal rays in a mixture of hydrogen and helium. About six o'clock I interrupted the exposure and . . . went to the darkroom to start the developing process. I was naturally very excited, and since the plate was still in the fixing bath, I took it out for a short time to look at the spectrum in the faint yellow light of the darkroom. I observed several lines at the position of the blue hydrogen line, whereas the neighboring helium lines appeared to be simple ["Die Entdeckung des Stark-Effekte"].

At the beginning of July 1913, several months before Stark's discovery, Niels Bohr published his concept of a quantum-mechanical model of the atom. This provided, in principle, the possibility of understanding the reason for the Stark effect, which the classical theory was powerless to explain. Stark therefore had an opportunity to be doubly gratified, having also been one of the first, after Max Planck and Einstein, to stress the "fundamental significance" of Planck's elementary law (since 1907), which he had championed in many polemical discussions. Yet, almost incomprehensibly, Stark denied himself the satisfaction of seeing his own experiments confirm a theory for which he had helped prepare the way conceptually, even if he had not directly participated in its creation. Apparently he always had to oppose the accepted point of view. Thus, as Bohr's theory continued to gain adherents in 1914–1916, Stark set out to reverse this inexorable development. From the moment that Eduard Riecke's retirement from his chair at Göttingen was announced in June 1914, Stark hoped to be his successor. The process of selection, however, dragged on for years and ended with a bitter dispute between Stark and the Göttingen faculty. In 1917, therefore, Stark went to Greifswald. While there he received the Nobel Prize for physics in 1919, for the "completeness and reliability" of his measurements.

More than ever Stark claimed to be Germany's leading physicist; and in 1920 he founded, in opposition to the Deutsche Physikalische Gesellschaft, the Fachgemeinschaft der deutschen Hochschullehrer der Physik. The latter remained an essentially insignificant organization, but with its help Stark managed to join the board of physics (Fachausschuss) of the Notgemeinschaft der Deutschen Wissenschaft and the Helmholtz-Gesellschaft, the two most important science foundations of German scientists. Since the members of the physics boards had to be confirmed by vote of the entire membership, Stark was obliged to resign from the Notgemeinschaft in 1922 and from the Helmholtz-Gesellschaft in 1924.

In 1920 Stark was appointed successor to Wilhelm Wien at Würzburg. He soon quarreled with almost all his colleagues, and tensions reached a peak in a fight over the habilitation of a student. Unhappy with his situation at Würzburg, Stark was able, by virtue of the large sum of money he received with the Nobel Prize, to resign and move to his native Upper Palatinate, where he devoted himself to the development of a porcelain industry. In the difficult postwar years, however, he was unable to make his project a financial success. He decided to return to science but could not obtain another teaching post: he had made too many enemies. The only physicist whom he had not alienated was Philipp Lenard, who had become an outsider and whose battle against "Jewish, dogmatic physics" Stark unreservedly supported.

After the Nazi seizure of power, Stark, as a partisan of Hitler, was appointed president of the Physikalisch-technische Reichsanstalt (PTR), effective 1 April 1933. He conceived ambitious—not to say grandiose—plans for the expansion of the institution, which was to be part of a total reorganization of physics in Germany. Further, in accord with the "Führerprinzip" and in his capacity as president of the PTR, he hoped to become permanent president of the Deutsche Physikalische Gesellschaft. At the physics congress at Würzburg, however, he encountered such strong resistance,

crystallized above all by a courageous speech by Max von Laue, that he had to abandon this ambition. Similarly, Laue again intervened successfully against Stark when he was about to be forced upon the Prussian Academy—with Friedrich Paschen, Max Planck, and Karl Willy Wagner proposing his nomination on the ground that the academy had no real alternative to accepting him.

In June 1934 the government appointed Stark president of the Notgemeinschaft der Deutschen Wissenschaft (which was renamed "Deutsche Forschungsgemeinschaft"), replacing Friedrich Schmidt-Ott, who had been dismissed. In his new, powerful position as head of both this organization and the PTR, Stark intensified his fight against modern theoretical physics, designating its proponents, led by Laue, Arnold Sommerfeld, and Werner Heisenberg, as "white Jews in science" and "viceroys of the Einsteinian spirit" in Germany. Stark emphasized that the Nazi seizure of power and the Nuremberg laws had brought only a partial victory in the fight against the Jews. The "viceroys of Judaism in German intellectual life," he insisted, must disappear. The violence of his attacks made Stark increasingly appear pathologically quarrelsome.

Struggles within the hierarchy of the Third Reich compelled Stark to retire from the presidency of the Forschungsgemeinschaft at the end of 1936, and he was subsequently pensioned by the PTR. Shortly before the outbreak of war Stark, on bad terms with virtually everyone, retired to his estate of Eppenstatt, near Traunstein in Upper Bavaria. Even more than Philipp Lenard, Stark has been condemned, even despised, by contemporaries and posterity. Nevertheless, he contributed significantly to the development of physics, especially in his earlier years; and he later exerted a strong, if disastrous, influence.

BIBLIOGRAPHY

I. Original Works. A complete list of Stark's writings is in Poggendorff, IV, 1430–1431; V, 1196–1198; VI, 2524; VIIa, S–Z, 1, 485–486.

Papers on the Doppler effect include "Der Doppler-Effekt bei den Kanalstrahlen und die Spektra der positiven Atomionen," in Physikalische Zeitschrift, 6 (1905), 892–897; and "Über die Lichtemission der Kanalstrahlen in Wasserstoff," in Annalen der Physik, 4th ser., 21 (1906), 401–456.

On the quantum hypothesis, see "Elementarquantum der Energie, Modell der negativen und positiven Elektrizität," in Physikalische Zeitschrift, 8 (1907), 881–884;

"Beziehung des Doppler-Effekts bei Kanalstrahlen zur Planckschen Strahlungstheorie," *ibid.*, 913–919; "Zur Energetik und Chemie der Bandenspektra," *ibid.*, **9** (1908), 85–94, 356–358; "Neue Beobachtungen zu Kanalstrahlen in Beziehung zur Lichtquantenhypothese," *ibid.*, 889–900; and "Über Röntgenstrahlen und die atomistische Konstitution der Strahlung," *ibid.*, **10** (1909), 826.

His major work on the Stark effect is "Beobachtungen über den Effekt des elektrischen Feldes auf Spektrallinien. I–VI," in *Annalen der Physik*, 4th ser., **43** (1914), 965–1047, and **48** (1915), 193–235, repr. as *Johannes Stark. Paul S. Epstein, Der Stark-Effekt*, which is vol. 6 of Dokumente der Naturwissenschaft (Stuttgart, 1965).

Among his books are *Die Elektrizität in Gasen* (Leipzig, 1902); *Die Prinzipien der Atomdynamik*, 3 vols. (Leipzig, 1910–1915); *Die gegenwärtige Krisis in der deutschen Physik* (Leipzig, 1922); *Die Axialität der Lichtemission und Atomstruktur* (Berlin, 1927); *Atomstruktur und Atombindung* (Berlin, 1928); *Adolf Hitler und die deutsche Forschung* (Berlin, 1935); and *Jüdische und deutsche Physik* (Leipzig, 1941), written with Wilhelm Müller.

Letters from the Stark estate, including 90 percent of his scientific correspondence up to 1920, are at the Staatsbibliothek der Stiftung Preussischer Kulturbesitz, Handschriften Abteilung, Berlin-Dahlem.

II. SECONDARY LITERATURE. See Armin Hermann, "Die Entdeckung des Stark-Effekte," in *Johannes Stark. Paul S. Epstein, Der Stark-Effekt* (Stuttgart, 1965), 7–16; "Albert Einstein und Johannes Stark. Briefwechsel und Verhältnis der beiden Nobelpreisträger," in *Sudhoffs Archiv für Geschichte der Medizin und Naturwissenschaften*, **50** (1966), 267–285; "H. A. Lorentz–Praeceptor physicae. Sein Briefwechsel mit dem deutschen Nobelpreisträger Johannes Stark," in *Janus*, **53** (1966), 99–114; and "Die frühe Diskussion zwischen Stark und Sommerfield über die Quantenhypothese," in *Centaurus*, **12** (1967), 38–59; and Kurt Zierold, *Forschungsförderung in drei Epochen. Deutsche Forschungsgemeinschaft . . .* (Wiesbaden, 1968), 173–212.

ARMIN HERMANN

STARKEY, GEORGE (*b.* Bermuda, 1628; *d.* London, England, 1665), *medicine, alchemy.*

Starkey, who used his family name Stirk until adopting the more familiar cognomen found on his title pages, was the son of George Stirk, a Puritan minister in Bermuda, and Elizabeth Painter. During his youth he developed an interest in natural history and devised experiments to test the theory of the spontaneous generation of insects. While a student at Harvard College, he began the study of medicine and alchemy and was soon attracted to Helmont's doctrines. He graduated in 1646, took the master's degree, and practiced medicine in the Boston area, where he married a daughter of Israel Stoughton. Sometimes relying on his friend John Winthrop, Jr., for books, chemicals, and apparatus, he pursued research along Helmontian lines.

In 1650 Starkey emigrated to England. Associating with Samuel Hartlib's circle of investigators, he engaged in a wide range of experiments, including the production of alchemical metals and the preparation of chemical medicines. Starkey told Hartlib's friends of his contact with an "adept" in New England from whom he supposedly had obtained the secret of transmutation and a number of unpublished alchemical treatises. The account was expanded and clothed in more mystification as part of Starkey's contribution to alchemical verse, *The Marrow of Alchemy* (1654–1655).

With the publication of *Natures Explication and Helmont's Vindication* (1657), Starkey entered the dispute between those physicians who adhered to the "Paracelsian compromise" and those who advocated more frequent use of chemical remedies. Starkey's book, an outspoken defense of Helmontian doctrines, was followed by *Pyrotechny Asserted and Illustrated* (1658), in which he continued the style of rhetoric that won him the friendship of only a small fraternity. Starkey claimed that his books were based on the experimental method and a desire to reform the state of medicine, but only vague processes were given for his medicaments.

After a brief excursion into the polemical exchange accompanying the Restoration, Starkey returned to iatrochemical controversies. He frequently resorted to print to combat fellow chemical practitioners, both to secure priority for his own medicines and to condemn those whose preparations invited censure of the chemical cause. A last exchange with the so-called "Galenists" was cut short by the Plague in 1665. Starkey remained in London to treat victims of the disease, of which he himself died at an unknown date.

During his lifetime, only one of his "American" alchemical manuscripts was published, apparently without his permission, but a number of treatises were printed after his death under the pseudonym Eirenaeus Philalethes. Several of these works on the theory and practice of transmutation and related matters became classics of alchemical literature, especially the *Introitus apertus* (1667), which appeared in English as *Secrets Reveal'd* (1669). Evidence points to Starkey as the probable author of the tracts, although John Winthrop, Jr., may have served as the inspiration for the story of the New England adept.

BIBLIOGRAPHY

I. ORIGINAL WORKS. Starkey's *The Marrow of Alchemy* (London, 1654–1655) was published in 2 pts. under the pseudonym "Eirenaeus Philoponos Philalethes." *Natures Explication and Helmont's Vindication* (London, 1657) and *Pyrotechny Asserted and Illustrated* (London, 1658) were followed by political publications concerning the Restoration. Other polemical tracts were *The Admirable Efficacy, and Almost Incredible Virtue of True Oyl, Which is Made of Sulphur-Vive* (London, 1660); *George Starkeys Pill Vindicated From the Unlearned Alchymist* [London, 1663 or 1664]; *A Brief Examination and Censure of Several Medicines* (London, 1664); and *A Smart Scourge for a Silly, Sawcy Fool* ([London], 1664). Starkey's *An Epistolar Discourse to the Learned and Deserving Author of Galeno-Pale* was printed with George Thomson's ΠΛΑΝΟ-ΠΝΙΓΜΟΣ *or, a Gag for Johnson* (London, 1665). *Liquor Alchahest* (London, 1675), J. Astell, ed., was a posthumous work.

The most influential of the "Philalethes" essays was *Introitus apertus* (Amsterdam, 1667), printed in English as *Secrets Reveal'd* (London, 1669); also important are those collected in W. Cooper, ed., *Ripley Reviv'd* (London, 1678). Others are in M. Birrius, ed., *Tres tractatus de metallorum transmutatione* (Amsterdam, 1668), and in W. Cooper, ed., *Opus tripartitum de philosophorum arcanis* (London, 1678). Philalethes' *The Secret of the Immortal Liquor Called Alkahest* (London, 1683) was advertised as a separate imprint by bookseller William Cooper and was included in his *Collectanea chymica* (London, 1684).

II. SECONDARY LITERATURE. On Starkey, see the following, listed chronologically: John L. Sibley, *Biographical Sketches of Graduates of Harvard University*, I (Cambridge, Mass., 1873), 131–137; John Ferguson, "The Marrow of Alchemy," in *Journal of the Alchemical Society*, 3 (1915), 106–129; George H. Turnbull, "George Stirk, Philosopher by Fire," in *Publications of the Colonial Society of Massachusetts*, 38 (1949), 219–251; Ronald S. Wilkinson, "George Starkey, Physician and Alchemist," in *Ambix*, 11 (1963), 121–152; and "The Problem of the Identity of Eirenaeus Philalethes," *ibid.*, 12 (1964), 24–43; J. W. Hamilton-Jones, "The Identity of Eirenaeus Philalethes," *ibid.*, 13 (1965), 52–53; Ronald S. Wilkinson, "The Hartlib Papers and Seventeenth-Century Chemistry, Part II: George Starkey," *ibid.*, 17 (1970), 85–110; "Further Thoughts on the Identity of Eirenaeus Philalethes," *ibid.*, 19 (1972), 204–208; "Some Bibliographical Puzzles Concerning George Starkey," *ibid.*, 20 (1973), 235–244; and "George Starkey, an Early Seventeenth-Century American Entomologist," in *Great Lakes Entomologist*, 6 (1973), 59–64. The most recent discussion of the Philalethes matter is in Wilkinson, *The Younger John Winthrop and Seventeenth-Century Science* (London, 1975), ch. 6.

RONALD S. WILKINSON

STARLING, ERNEST HENRY (*b*. London, England, 17 April 1866; *d*. Kingston, Jamaica, 2 May 1927), *physiology, education.*

In circulatory physiology the Starling legacy is conceptually one of the most influential in the twentieth century. The "Starling sequence," embracing both central circulatory function and fluid exchange at the capillary level, was and remains the unifying theme of contemporary circulatory theory.

Starling was born into a family of limited financial means and fundamentalist religious belief. His father, Matthew Henry Starling, was a barrister and served for many years as clerk of the crown at Bombay, returning to England once every three years. Starling's mother, the former Ellen Watkins, remained in Britain and had the responsibility of rearing their children, of whom Ernest was the eldest. He received his early education at Islington (1872–1879) and at King's College School (1880–1882). In 1882 he entered Guy's Hospital Medical School (London), where he set a record for scholarship and received his qualifying degree (M.B., Lond.) in 1889.

One of the most influential periods in Starling's formative years was the summer of 1885 spent in Kühne's laboratory at Heidelberg. It probably marked the beginning of his strong rejection of empiricism as the basis for clinical practice, and it played a role in directing him toward physiology as a means of bringing basic science to the bedside. In 1887 he became demonstrator in physiology at Guy's and in 1890 began part-time work in Schäfer's laboratory at University College, where he began a lifelong association with William Maddock Bayliss.

It was a highly productive and complementary union. Bayliss was the learned, methodical, and cautious partner; Starling was the aggressive, impatient, and sometimes incautious visionary. The first of their joint papers appeared in 1891, and in January 1902 they presented a preliminary communication that opened the door to the vast field of hormonal function. Published in full in September 1902, the paper established the existence and role of secretin, a product of the duodenum; and in 1905 Starling coined the word "hormone" to designate the body's "chemical messengers" produced by the endocrine glands.

In 1892 Starling again went to Germany, this time to work with Rudolf Heidenhain at Breslau; and on his return he attacked the problems of lymph production, capillary permeability, and the physiological effects of osmotic forces. On the basis of his findings, he began the synthesis of what came to be called the "Starling equilibrium," referring to the balance between intravascular pressure and osmotic forces at the capillary level.

With his acceptance in 1899 of the Jodrell professorship at University College, Starling finally joined Bayliss full time. It was at University College that he shifted his interest from the peripheral circulation to the heart itself. His career as a scientific investigator reached a peak in the years immediately preceding World War I with the publication of two papers on control of the heart, written with S. W. Patterson, who later became his son-in-law.

Starling's wartime service was turbulent, largely because of his outspoken impatience with the obtuseness, where scientific matters were concerned, of his military superiors. He was ultimately sent to Thessaloniki, Greece, with no specific assignment and little opportunity to apply his extraordinary talents in the service of his country. Paradoxically, the only recognition he ever received from his government (the comparatively minor Companion of the Order of St. Michael and St. George) came for his "services at Salonika."

In 1919 Starling delivered the most significant lecture of his career, correcting some of his earlier oversimplified statements on circulatory control and anticipating many present-day workers. Unfortunately, the lecture received little attention at the time and was published in a journal of very limited circulation. His remaining years were vigorous but somewhat anticlimactic. In 1922 Starling accepted the Royal Society's Foulerton research professorship. Despite deteriorating health, he continued his research work with fellows and students from all over the world. He died aboard ship, while on a Caribbean cruise, and was buried at Kingston, Jamaica.

Starling was elected fellow of the Royal Society in 1899 and was a prominent member of the Physiological Society. He was honorary member of many foreign scientific organizations and delivered the Harvey lecture for 1908 in New York.

Starling influenced several areas of physiology and medicine. His work with Bayliss on capillary function and on hormones would, in itself, guarantee him great prominence. But he is best-known for his work on the heart. Focusing primarily on the intrinsic response of the isolated heart to increased filling, Starling formulated a widely quoted law of the heart and summarized the concept in his Linacre lecture (1915). Unfortunately, in the lecture he attempted rather uncritically to extend his findings

on the isolated heart to the intact organism at rest and under stress. Within the next few years he recognized the inadequacy of his earlier concepts, and in 1919 he extended and refined them, incorporating intrinsic myocardial response as one feature of a highly complex control system. Subsequent work in the field has, in large measure, consisted of extensions and elaborations of the views set out by Starling in 1919.

Starling was astonishingly gifted in synthesizing disparate views and information to produce meaningful and effective generalizations. His experimental work was often of simple design and yielded data that have sometimes been thought inadequate to support his conclusions. But using those data and building on the work of several German physiologists, of whom Otto Frank was the most significant, Starling arrived at an expansive and surprisingly accurate understanding of circulatory function as a whole.

Starling's eloquent, often anguished, and biting comments on education still carry great conviction. He called for ". . . educational reform, or even revolution, for the maintenance of our place in the world," and added that ". . . in matters of urgent necessity [such as education] it is unprofitable to count the cost." Writing soon after the Armistice, he reserved his most forceful words for the great peril into which, in his view, the British educational system had brought the nation. "The astounding and disastrous ignorance [of science] . . . displayed by members of the government in the early days of the war raised some doubts . . . as to the efficiency of the education imparted to . . . the upper classes." Noting that Germany had, since Napoleon's day, laid great emphasis on education as a means of enhancing national power, he said that Britain, against such a force, could oppose only " . . . a kindly, gentlemanly stupidity." He attributed Britain's salvation to the self-sacrifice and bravery of her young people: "The great rally of the nation occurred in spite of an education which taught the ruling classes that their first duty was to their clan, their party, or their service. . ." ("Science in Education," p. 474).

Views like these fill in the gaps left when one limits his attention to Starling's scientific publications. Physiology was his passion, but it was not in itself sufficient. He emerges not only as a great scientist but also as a responsible and pragmatic activist in education. He and his colleagues, notably Bayliss, changed the face of classical physiology more, probably, than any other group since Harvey's time. And in the course of his work on the circulation, Starling succeeded to a remarkable degree in replacing empiricism with scientific understanding as the basis for medical practice at its best. But if his pungent and highly critical attacks on British education affected the system, the results were, at least in his lifetime, very difficult to discern.

BIBLIOGRAPHY

Starling's publications include "The Arris and Gale Lectures on Some Points in the Pathology of Heart Disease. Lecture I. On the Compensatory Mechanisms of the Heart. Lecture II. The Effects of Heart Failure on the Circulation. Lecture III. On the Causation of Dropsy in Heart Disease," in *Lancet* (1897), **1**, 569–572, 652–655, 723–726; "The Mechanism of Pancreatic Secretion," in *Journal of Physiology*, **28** (12 Sept. 1902), 325–353, written with W. M. Bayliss; "On the Chemical Correlation of the Functions of the Body," in *Lancet* (1905), **2**, 339–341, 423–425, 501–503, 579–583; "On the Mechanical Factors Which Determine the Output of the Ventricles," in *Journal of Physiology*, **48** (8 Sept. 1914), 357–379, written with S. W. Patterson; "The Regulation of the Heart Beat," *ibid.* (23 Oct. 1914), 465–513, written with S. W. Patterson and Hans Piper; *The Linacre Lecture on the Law of the Heart* (London, 1918), delivered at St. John's College, Cambridge, in 1915; "Natural Science in Education: Notes on the Position of Natural Sciences in the Educational System of Great Britain," in *Lancet* (1918), **2**, 365–368; "Science in Education," in *Science Progress*, **13** (1918–1919), 466–475; and "On the Circulatory Changes Associated With Exercise," in *Journal of the Royal Army Medical Corps*, **34** (1920), 258–272.

Carleton B. Chapman

STAS, JEAN-SERVAIS (*b.* Louvain, Belgium, 21 August 1813; *d.* Brussels, Belgium, 13 December 1891), *chemistry*.

Stas was the son of a stovemaker of Louvain. As a boy he was taken to Paris by his elder brother Guillaume, a pupil of the sculptor François Rude, to pose for the statue *Neapolitan Fisherboy Playing With a Tortoise*, now at the Louvre. In 1832 Stas entered the medical faculty of the University of Louvain. After receiving the M.D. in 1835, he became assistant to his former professor of chemistry, Jean-Baptiste Van Mons, an expert in pomology who owned an apple nursery and experimented on fruit trees. When the nursery was displanted, Stas and Van Mons's other assistant,

L. G. De Koninck, were provided with a large supply of fresh roots of apple trees from which they isolated a crystalline glucoside that they named phlorizin. Stas conducted research on this substance in a small laboratory that he had equipped in the attic of his father's house.

In 1837 Stas moved to Paris, to profit from the scientific milieu. As a collaborator of Dumas he performed a complete study of phlorizin, splitting it into phloretin and glucose. Stas published papers with Dumas on the composition of carbonic acid and on chemical types, and worked with him on the composition of water and on the action of potassium hydroxide on alcohols.

Stas's career was markedly influenced by contemporary research on the composition of carbonic acid. Following Dalton's hypothesis that the constant ratio combining weights of elementary substance is in the same ratio of their relative atomic weights, the determination of atomic weight had become an objective of prime importance to chemists.

In September 1840 Stas was appointed to the chair of chemistry at the Military School in Brussels. He began teaching in February 1841 under unfavorable conditions, for he lacked adequate facilities for research. During the next four years he contributed to the determination of the atomic weight of carbon. Proceeding by way of the combustion of carbon monoxide, he deduced the atomic weight as between 75 and 75.06 through comparing the weight of carbon dioxide formed by the reduction of a known weight of copper oxide under the action of carbon monoxide.

Following the publication in 1860 of his "Recherches sur les poids atomiques," Liebig arranged financial help for Stas from the king of Bavaria, a gesture that inspired the Belgian government to grant him a subsidy of 6,000 francs, to cover three years' expenses.

Like Dumas, Stas had been inclined to accept Prout's hypothesis that atomic weights are whole-number multiples of the atomic weight of hydrogen. In 1860 he published his main work, "Recherches sur les rapports réciproques des poids atomiques," a study devoted to a number of elements (nitrogen, chlorine, sulfur, potassium, sodium, lead, and silver) that were considered by Dumas to support Prout's hypothesis. In this work Stas demonstrated that the values of the atomic weights he had determined were neither multiples of unity, nor of one half, as Marignac believed, nor of one quarter, as Dumas maintained. This publication led Marignac to doubt the universality of the law of definite proportions. In three papers collectively entitled "Nouvelles recherches sur les lois des proportions chimiques, sur les poids atomiques et leurs rapports mutuels" (1865), Stas presented the results of an extensive series of experiments devoted to the new demonstration. By painstaking and accurate measurements he established that atomic weights were incommensurable, thereby disproving the facile conclusion that discrepancies with whole-number values were due merely to experimental errors. Prout's hypothesis was thus discredited.

Stas determined atomic weights using indirect methods. To determine the atomic weight of nitrogen, for example, he measured (1) the amount of ammonium chloride required to precipitate the chloride of silver from a silver nitrate solution. The result was $Ag : NH_4Cl = 100 : 49.6$; (2) the weight of potassium nitrate obtained by repeated evaporation of potassium chloride with nitric acid, and the weight of potassium chloride obtained from potassium nitrate by evaporation with hydrochloric acid; the result gave an atomic weight of 14.03; (3) the weight of silver nitrate obtained by the dissolution of silver in nitric acid and by evaporation; the result was that 100 Ag yielded 157.4952 $AgNO_3$ not fused, and 157.484 fused, which gives $AgNO_3 = 169.99$ (Ag = 107.94) $\therefore$ N = 14.05. The mean result of these experiments was N = 14.09, which differs little from the currently accepted value, 14.008, achieved through the refinements introduced by recourse to physical methods.

Stas was also active in toxicology and published several papers on this subject in the early 1850's. For a criminal case in which nicotine had been used he developed a method for detecting the poison in the victim's body. It was later generalized to detect alkaloids in cases of poisoning.

Suffering from a herpetic disorder of the respiratory tract, Stas became professor emeritus in 1868. His pension covered only his living expenses. His modest inheritance was exhausted by the research expenses of the laboratory that he had equipped in his own house, where he accomplished the first part of his work on atomic weights.

In 1872 Stas resigned his post as commissioner of currency at the mint, to which he had been appointed in 1865. His simple way of life, his dedication to his work, his undaunted independence, exemplary tolerance, and commitment to the progress of higher education and research were made manifest in the deep influence he exerted in Belgium, which occasionally earned him disfavor in official circles.

BIBLIOGRAPHY

I. ORIGINAL WORKS. Stas's writings were collected in *Oeuvres complètes*, W. Spring and M. Depaire, eds., 3 vols. (Brussels, 1894).

II. SECONDARY LITERATURE. On Stas and his work, see R. Delhez, "Jean-Servais Stas, 1813–1891," in *Florilège des sciences en Belgique pendant le XIXe siècle et le début du XXe* (Brussels, 1968), 285–321; L. Errera, "Jean-Servais Stas," in *Revue de Belgique*, 2nd ser., 4 (1892), 192–210; L. Henry, "Une page de l'histoire de la chimie générale en Belgique: Stas et la loi des poids," in *Bulletin de l'Académie royale de Belgique*. Classe des sciences (1899), 815–848; J. W. Mallet, "Stas Memorial Lecture. Jean-Servais Stas and the Measurement of the Relative Masses of the Atom of the Chemical Elements," in *Journal of the Chemical Society*, 63 (1893), 1–56; and J. R. Partington, *A History of Chemistry*, IV (London–New York, 1964), 876–878.

See also the following works by W. Spring: "Lecture sur la vie et les travaux de Stas," in *Bulletin de l'Académie royale de Belgique*, Classe des sciences, 3rd ser., 21 (1878), 736–761; "Notice sur la vie et les travaux de Jean-Servais Stas," in *Annuaire de l'Académie royale de Belgique*, 59 (1893), 217–376; and Académie Royale de Belgique, *Biographie nationale*, XXIII (Brussels, 1921–1924), cols. 654–684.

Also useful is *Prout's Hypothesis* (Edinburgh, 1932), which contains papers by Prout, Stas, and Marignac.

MARCEL FLORKIN